True Enlightenment:
From Natural Chance to Personal Creator

(History of Modern Science from 1500 A.D. to Present)
Volume 1

Carl W. Wilson

Andragathia Books

But the path of the righteous is like the light of dawn,
which shines brighter and brighter until full day (Proverbs 4:18).

Andragathia Books
P.O. Box 621421
Oviedo, FL. 32762-1421
Email address: andragathia@bellsouth.net
ISBN 0-9668181-1-3

CONTENTS

Acknowledgements iii
Introduction .... v-x
Illustrations .... xi-xii

SECTION ONE: REBIRTH OF TRUE ENLIGHTENMENT AND BIASED PERVERSION

Part I: The Historical Perversion of Modern Science

1. Man's Wisdom Biased by Sin versus God's Wisdom 5-18
2. Birth of Modern Science by Christians 19-34

Part II: The Biased Perversion of Natural Revelation

3. Doubts of Christian Religion and Growth of Faith in Human Wisdom (Seventeenth Century) 37-46
4. Satan's Perversion of Modern Science: Philosophical Claim of Evolution of All Things (Eighteenth Century) 47-58
5. Finding Uniformity in Nature to Support Determinism 59-74
6. Determined Knowledge of Evolution of Life 75-94

SECTION TWO: THE UNIFORMED NATURAL DETERMINED UNIVERSE FOUND TO BE INDETERMINATE

Part III: Enlightenment about the Physical Universe

7. Indeterminacy in the Physical Cosmos 99-112
8. Search for Evidence of Nebula Hypotheses in Heavens 113-124
9. Mystery in the Starry Universe 125-142
10. Solar System Found Mysterious 143-156

Part IV: Enlightenment about Life's Origin and Development

11. Molecular Neo-Darwin Mechanism Search 159-176
12. Biological History Reversed in Fossils and Cell Functions 177-192
13. Evolutionary Theories of Non-Evidence 193-206
14. The Impossibility of Chance Origin of Life 207-218
15. Molecular Biology, Genetics and Ancestry 219-240

SECTION THREE: MAN AND HIS HISTORY ON EARTH

Part V: Enlightenment about Uniqueness of Man and Civilization

16. Fossils of Human Evolutionary Descent 245-266
17. Human Uniqueness from Animals 267-286
18. Civilization Reflects Recent Creation 287-304

Part VI: Enlightenment about World Geology and Climate

19. Enlightenment About the Worldwide Judgment by Flood 307-322
20. Enlightenment About Glacial Ice Sheets and Climate 323-350
21. Enlightenment About Tectonic Continental Separation 351-382

SECTION FOUR: EARTH HISTORY AND AGE OF THE EARTH

Part VII: Enlightenment about the Finding Time in Nature

22. Enlightenment about Age of Earth and Geological Column 387-400
23. Enlightenment about Dating Methods and Ages of Evolution 401-430

SECTION FIVE: USE OF SCIENCE IN PERVERSION OF GOD'S REVELATION

Part VIII: Bias to Distort Biblical Truth Special Revelation

24. Man's Evolutionary Creation of the God of History 435-444
25. Efforts to Conform Jesus to Scientific Determinism 445-460
26. Claim of the Death of God and Intellectuals as Priests 461-470
27. Denial of the Incarnation for Religious Syncretism 471-486

Part IX: Enlightenment about God in Biblical History and Creation

28. Intellectual Bias Led to Rejection of Christ 489-500
29. Credibility of Documentary Sources for Christianity 501-508
30. Enlightenment about Origin of God Idea 509-522
31. Enlightenment to Scientific History and Divine Judgment in Conflict 523-540

Part X: Review and Conclusion

32. Summary of the History of Modern Science 543-558
33. Present and Future Secular Science 559-572
34. Christian and Return of Modern Science To Christ as Creator Today 573-590

Addendum 591-598
Index 599-604
More About the Author 605-606

Achowledgements

While many of us as Christians claim divine leading, often it is superficial. But in a true sense the two volumes of this book have been motivated and guided by the Spirit of God, which is not a claim of inerrancy. His providential leading allowed me to study under some prominent secular and Christian scientists, and also led when I did not really want to continue this research and writing, I encountered new information from new teachers or reading that motivated me to continue on. This has enabled me to do 40 years of research to know a little about science, and as a reporter write this book.

My wife Sara Jo, with a scientific background as well as Christian commitment, and one who knows books as few individuals do, has been an intellectual stimulus and critic. She has helped me get the resources I needed and helped guide my thinking. Her patience and that of my children has also allowed me to pursue the subjects for many hours.

While many have assisted me through the years, Linda Koval has typed, retyped, edited, done art work and correspondence for not only this book but all my work for a number of years. Only God could have kept her in pursuit of excellence and patience with such a poor writer and servant. Primarily her commitment to the Lord motivated her and made our work together interesting and fun. I am grateful for David, her husband who has supported both her and me.

I am very grateful to Judy Hagar for her editing, and for Dr. James Cook, a retired history professor who has also read this book to especially make sure historical data is correct. I am also grateful to the students at the conferences and graduate schools that have responded to this teaching, especially those at Reformed Theological Seminary, Orlando.

Any errors in this book are attributed to me. I and all those named as helping have tried to make this book correct and understandable in content, when it is often technical and difficult. I encourage you as a reader to plow through the areas that are difficult to see the whole picture that fits together amazingly well. I believe that this book on the hard sciences is more truthful, accurate and up to date at its time of publication than most science histories now available. I believe it is likely to be valuable for many years, and pray that God will make it so.

Carl W. Wilson

Introduction

The Story of this Book is about the History of the Modern Hard Sciences

True Enlightenment is a two-volume, highly documented study of modern science from 1500 A. D. to the present time. This book on the hard sciences is the first volume that recounts the slow changes in Western cultural thought from faith in the biblical God of creation to demonic gods that exalt the wisdom of man and pervert mankind's worship to materialism by evolutionary naturalism. Jacques Barzun, dean and historian of Columbia University, has described this general trail of Western history as *From Dawn to Decadence.*[1] The dawn was the time of the reformation of science and theology from the perversions of the Roman Catholic Church which in the fourteenth century had shifted both science and theology from the biblical faith in Christ of the Apostles Paul and Peter to Aristotelian philosophy. This new dawn of Western civilization in the Reformation (early 1500 A. D. ff.) produced the movement toward a new value of humans and ultimately toward democracy. The Magna Carta of England in the thirteenth century primarily profited barons of the feudalistic system, but by the seventeenth and eighteenth centuries those principles and other ideas about freedom gathered meaning for the common man. In America, where the persecuted English and other European Protestants had gone, the colonies produced a democracy ***with restraints*** against sin by a common faith in the biblical Christ.[2]

In Europe the seeds of decadence began in the sixteenth and seventeenth centuries, with men rebelling against kings and abusive clergy using unscientific perversions of human wisdom to what became known as the Enlightenment. That movement sowed the seeds of individual rebellion against God and moral corruption that is producing twenty-first century decadence. The decadence had reached a peak of the "death of God theology" in Europe by the early twentieth century, shifting the leadership of the false enlightenment to America, which reached a peak of opposition to God and the Bible in the "death of God theology" about 1960. This is now nearing its final decadence because of man's natural sin bias inclined toward materialism, which destroys democracy and changes sexual roles that destroy the family. All the events of perversion by biased men that lead to current problems in America and the West, while demonically motivated, are under God's sovereign wise control that we can trust (Isaiah 40:21-26; John 19:10-11).

Recent Recognition of a Near Turing Point in History

The validity of my claim of the approaching end of Barzun's cycle of history, and my belief that this book reveals that we are nearing a final phase, is providentially revealed in articles from authoritative periodicals that are just published as I write this introduction. They reveal the conflict between the devil and the Son of God, born of woman (Genesis 3:15), who is the Creator and Recreator, who is the only Savior for mankind and true Son of Man. The conflict is between true disciples of God in the church and the false religions conceived by natural sinful men to solve man's problems by one world political government led by one humanist. Democracy that once worked under a faith in a common biblical God in America and the West is now subject to humanism based on pride, greed and lust. It seems postmodern anti-church leaders are ready to ***sacrifice America and the West for a chance at a man-made U. N. world government.***

Simultaneous recent magazine articles show our crisis. The lead article in *The Economist* is "Welcome to the Anthropogenic, Geology's new age" (superseding the last 10,000 years of the

---

[1] Jacques Barzun, *From Dawn to Decadence: 500 years of Western Cultural Life, 1500 to Present, (New York: Harper Collins Publishers*), 2000.

[2] Cf. Carl W. Wilson, *Liberty in an Evil Age*,(Longwood, FL: Xulon Press, 2008), pages 13-18, 161-171 review liberal Alexis de Tocqueville's evaluation of why the French Revolution failed and common Christianity in America made democracy work. Charles Darwin hesitated to publish, fearing moral effects, and Sigmund Freud said religion was needed to restrain evil until man could be pragmatically educated to do right.

Holocene). Anthropos in Greek means 'man.' This is a claim of the editors of a new age when man has reached a point of population crisis that is converging with an environmental, climate and food crisis so that ***man must,*** by political unity under one government, solve the looming worldwide problems of mankind. The writers see science as now weak and impotent to influence, and the wisdom of one government must take control. The editors say, "It is one of those moments where a scientific realization, like Copernicus grasping that the Earth goes round the sun, could fundamentally change people's view of things far beyond science. It means more than rewriting some text books. It means thinking afresh about the relationship between people and their world and acting accordingly." By this, the editor means intelligent world politicians of action must go beyond the natural scientists who somehow have led us to the crisis.[3] In *True Enlightenment* we will show that the false science begun in the eighteenth century is weak and dark for the West and the world. Western leaders hope to try unification of man as at the tower of Babel (Genesis 10).

At the same time of *The Economist*'s article, the lead article in an authoritative Christian magazine, *Christianity Today* was, "The search for the Historical Adam (anthropos): some scholars believe genome science casts doubt on the existence of the first man and woman (Adam and Eve). Others say the integrity of the faith requires it. The state of the debate." The argument in the article is that molecular science now seems to indicate there was an emerging population from the primates of 10,000 humans from various places, and not from one Adam—***removing the idea man began as a creation of God***. Ted Olsen, the managing editor, introduces the article on page 3 and refers to their editorial on page 11, 'Where we Stand: No Adam, No Eve, No Gospel." [4] He suggests that ***there must be some resolution*** of compromise of this view of modern science and theologians and says, "Hebrew thought offers one clue to resolving the tension: the corporate nature of humanity." On page 61 he appeals, "We ***don't need*** another fundamentalist ***reaction against science***. We need instead a positive interdisciplinary engagement that ***recognizes the good will of all*** involved and the creative thinking takes time." (Emphasis added). Olsen is suggesting this ***science of unregenerate biased men accept Adam as the symbol of many emerging humans as the truth.*** Thus biblical Christians must in this way find a way to comply with the logic of false science. In *True Enlightenment* the evidence shows that it is time to end the efforts of compromise, to reject perverted science, and renew proclamation of the biblical Creator as truth.

The Fundamentalist Movement recognized a crisis of truth in the first half of the twentieth century, but became divisive and judgmental in ecclesiastical conflicts. But it is important that Jesus clearly taught there is a division of humanity, with the whole world following a demonic seduction of all mankind that would persecute his elect disciples, who are those that had died to self with him to live unto God alone (Genesis 3; John 15:18-16:4; Matthew 5:10-16; Revelation 12:9a). The crises of Jesus' ministry usually centered around him as ***Lord of the Sabbath,*** who had created and was doing the work of redemption in new creation. If we are in a final phase of modern history, the dividing line between men who claim they are wiser than God and those who have uncompromising humility to be obedient unto death for God's glory ***must be in a reaction,*** which is contrary to Olsen. The great need for Christians is to stand by the truth of Scripture.

The true modern science of the scientists in the Reformation era has been, and still is, the basis of modern productive good. The sciences, based on Reformation faith, two millenniums later began to be perverted due to ecclesiastical conflict. By the 18th century a ***false science*** began that was based on the ***evolutionary theory of all things beginning with a nebular hypothesis.*** That theory is now leading to efforts to present a final rejection of man as created by and accountable to God. Carl F. H. Henry, considered by many as the dean of American theologians, was the

---

[3] Editor's presentation of the issue is on 11, "Welcome to the Anthropocene," *The Economist*, May 28-June 3, 2011. Main article is on 81-83, and also linked with a putdown of Christians on page 3 in article of false prediction of the coming of the Lord by Harold Camping on May 21, "Paradise postponed, the World still awaits God's judgment," 38.

[4] *Christianity Today,* June, 2011.

founding editor of *Christianity Today* and leader in a more loving neo-evangelical defense of the Gospel. He stated at the crisis point of 1960, "The basic tension is still between the concept of a personal Creator-God and that of an impersonal chance process. To resolve that debate ***delineates in truth, the very destiny of our century.***"[5] (Emphasis added). Marvin Olasky in his *World* article, "2011 Books of the Year," makes it clear that the theory of evolution is now the focal point causing Christians to reject God as personal Creator. *World* magazine chooses books of the year according to what is the main issue— "we tend to see what's under assault."[6]

The present leaders of *Christianity Today* and many Christians do not seem to know the history, and the perversion of the truth from the personal Creator to a natural process that now is reaching an end. They believe they can reason with unregenerate biased men, whose power rests in their false views that they created and will not give up. Is the neo-evangelical movement that was to be friendlier, now willing to give up to the false views for compromise and worldly credibility? This seems to be happening at the very time when many scientific evolutionists, who recognized that theory was false about 1960, are ***now admitting there is no evidence*** of evolutionary creation and development of life at all.[7]

The Importance of Science History for You and All Christians

This first volume (612 pages) is on the hard sciences and the second is on the soft sciences, which are influenced by the hard sciences, as the above-mentioned articles reveal. You may yawn at such a lengthy, seemingly boring subject. The fact is that this book may be the most important one you could read, not because this author is brilliant, but because of the hard facts it presents which he has been providentially privileged to learn. The increasing number of problems, and the seriousness of the problems you face in life, will probably not be understood or solved by your investment banker or economist, your congressmen, your doctors, your educators or psychologists, except in part, and only temporarily, and certainly not by the world politicians.

A clear understanding and basis for solutions will best come by seeing the long progressive course of Western thought, which is most profoundly influenced by modern science that history shows has made a radical, slow shift ***that totally changed our values*** from being centered in God in Christ, to the wisdom of sinful man. Current science that today is led by most intellectual elite, believed and followed by many common men, is highly faulty and based on ***a demonic delusion influenced by a bias in human nature***. These statement are shown and documented by the facts in this book and will be seen to influence the soft sciences. Blind men leading the people will all fall into a ditch of destruction. These two volumes will show that the facts that are so important to you and to me have been ***repeatedly exposed as error before at least in part,*** by some leading scientists, but these facts have been conveniently in ignorance or malice, hidden, and passed over. The repeated refusal to see the facts as perversions ***reveals the essential cause of our problem as human bias.*** Chapter One of this book will emphasized that Christian believers and the Western world must face these unsavory facts, repent for renewal, or suffer and die, as have all previous civilizations.

Little Known Truth about the History of Modern Science, Even by Professionals

The actual history of modern science is little known and the faulty science, even with repeated exposure is ignored. This is admitted by the most famous of our scientists. Most scientists learn the difficulties of their specific science but ***know little about the history of science*** and the changes in science philosophy. Thomas S. Kuhn (1922-1996), professor at Massachusetts Institute

---

[5] In the symposium, Russell L. Mixter, ed. *Evolution and Christian Thought Today*, (Grand Rapids: Wm. B. Eerdmans Publishing Company 1959), 190. Francis Schaeffer and others later concurred.

[6] Marvin Olasky, "Books of the Year: two new books are important responses to the rapidly growing promotion of theistic-or more properly, deistic-evolution," *World*, July 2, 2011, 36-41

[7] See *True Enlightenment*, chapter 13.

of Technology, is still considered the world's leading philosopher of science. He has shown that he and many other scientists did not have an accurate idea of the history of science. He states that in textbooks on science history, "We have been misled by them in fundamental ways."[8] When Kuhn was completing his graduate study in theoretical physics in mid-century he became exposed to the history of science and he discovered how out-of-date it was. This led to his interest and to his changing the whole concept of the history of science, which views will be discussed later.[9] The teaching of false science was acknowledged by Stephen Jay Gould, professor of zoology at Harvard University, and a recognized world evolutionary science leader. Gould learned of errors of science and he and others wrote of these. This led to his theory of Punctuated Equilibrium. He especially spoke about illustrations used to communicate and spoke of "the evocative power of a well-chosen picture," many which were wrong, saying, "These are the most potent sources of conformity, since ideas passing as descriptions led us to equate the tentative with the unambiguously factual."[10] Jonathan Wells, a senior fellow at the Discovery Institute, with Ph.D.s from Yale University and U. C., Berkeley, has shown that the illustrations which Gould admitted were fraudulent have continued to be used in modern textbooks.[11] The recent long book *On The Shoulders of Giants* produced by renowned scientist Stephen Hawking, centers almost entirely on these early reformers but with no overt admission of their faith in God.[12] Hawking's book is an admission these believing scientists formed foundations for modern times that are contrary to evolutionary views.

This book, *True Enlightenment,* will reveal that almost all the founders of different areas of the hard sciences from 1500 A.D. continuing for the following quarter of a millennium, worked with a philosophy of science that had the perspective of faith in God as personal Creator. Moreover, most were not casual Christians, but serious students of the Bible and servants in various churches. The problem of the Western world today is this persistent desire to reject not only the Christian philosophy of a Creator in science, but the cultural structures that emerged.

How Demonic Delusion of America and the West Came through Perverted Science

In the mid-eighteenth century a turn toward rejecting knowledge of the biblical God was initiated by a small German of only five feet in stature with a deformed shoulder. He will later be identified. He had been in search for a sense of self-worth in Lutheran legalistic pietism, with plans for ministry. However, he found escape from moral bondage in the writings of a Roman classical hedonist, Lucretius, who wrote during fits of insanity and was a follower of the Greek philosopher Epicurus who believed in a chance world. This little German formed a theory to deny any real knowledge of God, and taught that ***everything progressively developed by an evolutionary process,*** with designing power only ***in and by nature***. His theory was created from thoughts in writings of other discontented intellectual philosophers from the previous fifty years in other parts of Europe. They too were in rebellion against God and institutional authorities and had placed ***authority of knowledge in themselves as individuals***, and often were influenced by what many believed to be demonic visions and voices. The little German's biased view spread throughout Europe and England, invading all areas of science and producing a series of supporting natural theories to eliminate faith in God and amazingly rewriting the whole of established history. This theory spread by growing literary media, incredibly capturing world intellectual centers.

That evolutionary theory of everything was ***driven by man's sin bias*** against God and ***hope for natural material progress by man's wisdom.*** Later, secular scientists with anti God bias went

---

[8] Thomas S. Kuhn, *The Structure of Scientific Revolutions*, (Chicago: University of Chicago Press, second edition 1970 and third edition 1996), in his preface and in The Introduction on page 1.

[9] Ibid. Preface, V.

[10] Stephen Jay Gould, *Wonderful Life*, (New York: W. W. Norton, 1989), 28.

[11] Jonathan Wells, *Icons of Evolution*, (Washington, D. C.: Regnery Publishing, Inc., 2002).

[12] Stephen Hawking, *On the Shoulders of Giants: The Great Works on Physics and Astronomy*, (Philadelphia: Running Press Book Publishers, 2002), 1160 of 1266 pages on early Christians.

beyond the pantheistic views of its German author who had abandoned Lucretius' view of Epicurus' creation by chance, for pantheistic evolution. He hoped a pantheism might minimize church accusations against him. But the anti God desire to reject all ideas of intelligence in the late nineteenth century, produced a retuned to a view of evolution ***without any intelligently designing power,*** driven ***by the chance of natural selection.*** This chance selection claimed to have created a progressively higher family tree of life that explained the origin of everything, including man.

Today it is commonly believed that Charles Darwin discovered the philosophical evolutionary view in the facts of nature and convinced the world of its truth. But in fact, various forms of the evolutionary theory had long grown to acceptance for over 100 years by intellectuals in Europe, and were the subject of some of the most popular books before Charles Darwin, especially in England. These books were without credible facts, and were rejected by leading Christian scientists. Many evolutionists before Darwin had claimed a substitution of evolutionary theory as an explanation for the biblical creation story.

The book *True Enlightenment* reports the history and shows that the evolutionary theory that Charles Darwin contributed to 100 years after it began, has never been supported by facts and now, in 2011 this is being admitted. This book, *True Enlightenment, volume I,* shows that the main evolutionary and supporting deterministic theories of every area of naturalistic science, sought to transfer the creative power from Christ, the Son of God, to a determined nature. Many supporting theories claimed a regularity of natural laws that required long ages for them to gradually work. This book shows that all these theories were perverted from the facts, unsupported, and ***refuted by subsequent findings*** that other secular scientists discovered. This ***documented record disclosing the refutation*** of these theories to support evolution is primarily given in this book ***from footnoted*** sources on each page.

True Enlightenment is for Believers Because Sin Bias of Unbelievers is Source of Seduction

A primary purpose of this two-volume history of science is to inform Christians, since the perversions were developed and maintained as a ***result of the natural sin bias of men*** who think they can be "wise as God." Only Christians who are taught by the Spirit of God can discern the wisdom of God in creation that is visible and invisible. Demonic ***perversions*** as shown in this book are the ***basis of authority of intellectuals*** and ***they must, by bias,*** reject this book as garbage.

America and the Western nations, through their material progress, have ***influenced the developing nations*** to want our materialistic, godless philosophy, education, technology and resulting products. But this materialism is leading the world toward a point of confusion and major crisis. Not only is biblical reformed Christianity, which was the foundation of this modern age (1500 A. D. ff.), being intellectually rejected, but ***the cultural structures*** based on Christianity are often blamed as a major cause of the modern economic and social dilemma by postmodern and neo-Marxist intellectuals, the Muslim jihadists, and Jesuit Aristotelians. The western intellectual elite have gained dominant control of educational and governing institutions around the world.

The tragedy is that many Christian leaders have accepted these false teachings to find peace with accepted evolutionary science and gain approval with intellectuals of prominence, and favor with Christians. The false science is a ***denial that the simple creation stories are true*** as written in the Bible, promoting ***doubts about Christ as personal Creator***. The intelligent design (ID) movement of Christians in academic circles has correctly appealed to the complex mysteries now known in astronomy and in microbiological systems as evidence that there is purpose in the universe. But their methodology of trying to use this knowledge to persuade intellectuals who are naturally biased is unscriptural. Unregenerate people will willingly accept a design ***purpose in nature*** from a deistic and pantheistic point of view, because these views leave nature to operate on its own and ***exempt man from accountability to God,*** with men still in control. Intellectual humanists do not want to yield to a personal God as sovereign Creator, and these disbelievers first need repentance and faith in God's love. Christ is the personal Creator of all the cosmos, visible and invisible and the only Savor. Natural laws are Christ's ***ordinary ways of upholding everything*** but he also has and can do miracles to reveal himself as Lord: he created wine from water, bread for 5,000, stilling storms, healed the sick, and cast out demons. But his forgiveness can enlighten.

Understanding Science as Revealing Christ as Creator Can Strengthen Our Witness

Many professing Christians have great concerns about the national debt, retirement, and family problems. But the root cause of these anxieties grows out of real doubts about God our Father as Creator who loves us, and our little faith that he can really provide what we can eat or cloth us. We are little motivated about treasure in heaven and promoting the kingdom of Christ. In America there are many thousands of organizations with Christian traditions that are in conflict and change, reflecting a lack of headship of Christ in the churches, and the failure to find a proper understanding of the problem of the spiritual quandary. The main focus of this history of modern science is a revelation of ***man's sin bias as the main problem in the church and the world.*** Therefore, understanding man's sin in rejecting the Creator, offers ***an opportunity to revive the church for spreading the good news of Christ and his kingdom***. Evidence of the demonic world perversion is a call for church renewal, for Christian unity to fulfill Christ's Great Commission.

This book on modern science is obviously not antiscientific, but is rather an effort to see correct ***unbiased science***. Being made in the image of God to have dominion for him means we have the ability to see his mind in creation and use scientific knowledge for his glory. Paul taught that after repentance, faith in Christ's cross, and the new birth, the Holy Spirit in the believer will then be a teacher of the wisdom of God that centers in Jesus Christ (1 Corinthians1:30-2:6). Proverbs makes it clear that ***the wisdom of God importantly involves his created work*** (Proverbs 8:22-31). This was what ended the debate of Job and his friends and restored Job's credibility and fortune (Job 38-42). True faith in the risen glorified Christ will call for real faith in the unseen hope of the resurrection and involves laying up treasure in heaven ***more than*** *material success now* (Matthew 6:19-34; 16:24-27; Luke 12:15-21; 1 Corinthians 15:17-19; Colossians 3:1-6; et al.). Our work and success now should be ***as stewards for him*** and eternity (Matthew 25:14-30). If the good news we proclaim is without God as the real Creator and mainly about this present world, it is a false gospel (1 John 2:15-17). Accepting the fact that the world has come under a slow demonic working, and trusting the simple obvious teachings of the Scripture as true, ***requires repentance,*** or changing of our mind, to put first the more important things of the kingdom.

Correct Expectations of Limits of this Book

No presentation of views by any finite fallible human is completely accurate, which includes the work of this author. This book and its contents are a product of years of research, revision and reorganization. As its author, I make no claim to being a practicing scientist, but simply a science reporter who has had considerable privilege of study under a number of well-known scientists in various fields and to have faithfully read scientific books and journals for over 60 years.[13] Those assistants who helped produce *True Enlightenment* have worked hard to make it accurate.

The objective of these two volumes on modern science is not primarily to change the world, to save the United States, or even prove the disbelieving world is wrong, but to glorify Christ and motivate Christians to unity and sacrificial service. Many sinful men, all who have gifts from God, ought to be respected as rational creatures of God. By our faith and witness to the Creator and Savor, and by God's grace, some of them will be given hearts to be called as God's elect people, and to hear the voice of the Holy Spirit about God's creation as in this book. But the one main objective of these two volumes is to enlighten and unite Christians in a greater faith in the Christ of the Bible and to motivate us to be obedient to Christ, the King of kings, before he returns in glory with our rewards. May we also, as we lose our lives for Christ, be an influence for righteousness in America and the present world.

Carl W. Wilson, a servant of Jesus Christ, 2011 A. D.

---

[13] A resume of my scientific and theological background is given with some biography on pages 605-606.

List of Illustrations

Waves of Civilization....5
Four Progressive World Views....11
Diagram of the Universe by Copernicus....23
Diagram of Fundamental Particles of the Standard Model....105
Diagram of Three Basic Models of Inflation....120
WMAP 5 year temperature....121
WMAP 5 year Microwave Map....122
Sloan Telescope Picture of Heavens....123
Percentages of Dark Energy and Matter in Universe....139
Pictorial Presentation in Strata Periods....178
Geological Timescale....179
Moore's Vision of the Developing Cone of Tree of Life....181
Syke's Female World Clans....232
Well's Map of Male Y-Chromosomes....233
Human Genetic Diversity....236
Huxley's Ape to Man Graduations....247
Icon of Man's Descent....248
Pictures of fossil finds of hominoids and humans....251
Speculative Family Tree....256
Ardi's Bones....260
Hominid Family....261
Miocene Ape Fossil Localities....262
Chart of Human and Ape Differences....268
Map of Glamar Challenger Drilling of Mediterranean 1970....317
Map of La Marmotta Area....317
Map of Lake Agassiz....317
Map of Colorado Plateau....318
Paleozoic Glaciation in Gondwana....329
Northern Ice Sheet....330
North and South Pole Shifts....335
Plimer's Ice Accumulation of Two Glaciations....338
Temperature Warming Trends....348
Computer Projection of CO_2 Emissions....348
Northern Hemisphere Sea Ice Anomaly....349
Southern Hemisphere Sea Ice Anomaly....350
North Polar Magnetic Wanderings....357
Magnetic Drift of India....358
Early Model of Mantle Runaways....360
Later Model of Mantle Runaways....361
Map of Tectonic Plates....363
Indian and S. W. Pacific Oceans....366
Canadian Shield Rock Collisions....371
Age of Rocks of Canadian Shield....371
Ring of Fire....372
Breakup and Dispersal of Pangaea into Continents....373
GPS Map of Recent Plate Movements....374
Example of Overthrusting....393

Overturned Fold or Anticline........ 393
Branson's Lava Emergence........ 410
Discord of Grand Canyon Rocks........ 412
Setterfield's Recording and C curve of Light Speed........ 425
Contrast of Wellhausen's View of History........ 441
Christian Linear View of History........ 498
Paul's Steps of Sin That Destroy........ 526
Outline History of Dominant Modern Scientific View........ 544
Cone of Life's Development........ 551
Failure of Scientific Determinism........ 554
Seven Days of Creation........ 578-580

SECTION ONE

REBIRTH OF TRUE ENLIGHTENMENT AND BIASED PERVERSION

PART I

THE HISTORICAL PERVERSION OF MODERN SCIENCE

For my thoughts are not your thoughts, neither are your ways my ways, declares the Lord. For as the heavens are higher than the earth, so are my ways higher than your ways and my thoughts than your thoughts. “For as the rain and the snow come down from heaven and do not return there but water the earth, making it bring forth and sprout, giving seed to the sower and bread to the eater, so shall my word be that goes out from my mouth; it shall not return to me empty, but it shall accomplish that which I purpose, and shall succeed in the thing for which I sent it (Isaiah 55:8-11).

Chapter 1
Man's Wisdom Biased by Sin versus God's Wisdom

Introduction

Today is a time of confusion and there is need for light from the past to guide the future. This book on the history of modern science will inform the average Christian of what is true in order to understand what is happening in the world. It is viewed from the perspective of the Bible as the word of God. Hopefully, it will also serve in the future as a reliable text book for our mature youth. The civilization and culture of America, the West and the world increasingly has the symptoms of many past civilizations that have risen to brilliance and power and then collapsed into confusion, conflict, subjection, and chaos. This has usually occurred when the civilization was at a peak and its people were confident in man's wisdom and they could meet every emergency.

I have illustrated the pattern of past civilizations as waves that grow by faith in God, and then turn from God to human wisdom, and in time break and crash on the sands of judgment, leaving monuments as tombstones of greatness such as the pyramids of Egypt, the Parthenon of Athens, and the Coliseum of Rome. Each civilization has a distinctive nature—Egypt with its monumental buildings, Greek democracy with its science, philosophy and art, Rome with its organization and military might, and the West and America with its scientific technology, and so on. This was previously illustrated as below and furnished with detailed data to support this view of history in my book *Man Is Not Enough*.[1] The waves of civilization usually are expended in about 500 years.

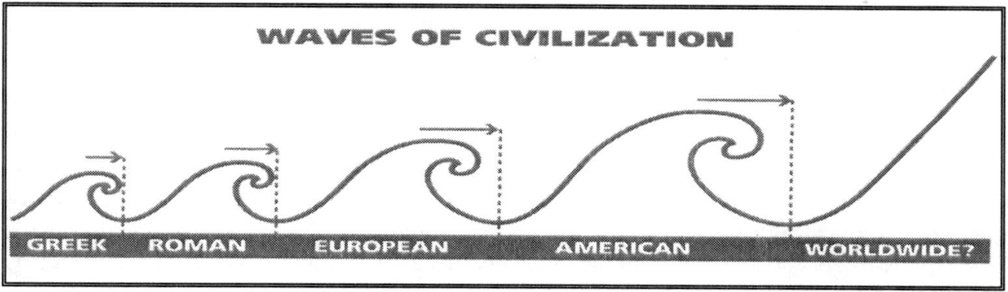

The second volume on ***the soft sciences of economics, psychology, sociology and politics*** show how our wave of civilization is crashing because of the demonic deceptive delusion from the influence of the hard sciences that is presented in this first volume. To understand the growing crisis in America, the West and the world, the Christian must understand the historical development of the wave in the hard sciences as follows: (1.) the origin and progressive growth of the wave of science of civilization under faith in God (chapters 1, 2) (2.) the demonic perversion through the bias of man's sin nature to trust only in the reality of a nebula hypothesis of the naturalistic existence and evolutionary development of all things of material creation (chapters 3-6), (3.) the reemerging truth of mystery and invisible power of God in the world and the denial of this by the intellectually deceived to try to maintain a faith only in natural materialistic laws (7-23) and (4.) the revelation for God in creation and biblical and scientific history, and man's folly and destructive wisdom seen in scientific history (24-33). The wave of American civilization has the usual evidences of near breaking soon. Exorbitant debt, growing gaps between classes, large, confusing, and unmanageable government, world military involvement in many places, et al.

---

[1] Carl W. Wilson, *Man Is Not Enough*, (Oviedo, FL: Andragathia Books, 1998; distributed by New Life Resources, Peachtree City, GA), illustration on 45 of 483 pages.

God's Wisdom and Man's Bias

The biblical picture of a wave of civilization is always one beginning with faith in the grace of God that has a prolonged period of growth of blessing and prosperity over 200 years until man in sin turns away from God. Man began in the blessings of Eden and then by sin was expelled by God into a world of struggle and pain. The nations or Gentiles all have gone though brilliant growth and blessing before turning away to bias and trouble. Israel, God's people, was given a promise land, but through sin bias turned away to trouble. The law given to guide for their blessing was placed in the Ark of the Covenant under the mercy seat that offered a way of forgiveness and restoration by the sacrificial offering of a high priest for those who turned to faith. So every civilization is marked with a long period of blessing and creative growth. The Bible repeatedly talks about God's "loving kindness endures forever."

God's Revelation of His Undeviating Righteousness and Loving Mercy in History

But strangely, most of the Bible is filled with evidence of man's failure, his sin, and the inevitable justice and judgment of God. Late in the book of Romans the apostle Paul sums up God's wisdom in allowing ***all civilizations to fail***. He reviewed the olive tree symbol of the nation of Israel as having its branches broken off by God because of unbelief. Then God grafted in the branches of the other nations (gentiles) by faith, until they too would all lapse into unbelief. Paul said that at that point God would consummate history by grafting a believing Israel back into his olive tree. Thus all men would have sinned and failed, showing that ***only God is worthy*** to rule and, by his saving mercy, use repentant men. ***"For God has consigned all to disobedience, that he may have mercy on all.*** Oh the dept of the riches and wisdom and knowledge of God! How unsearchable are his judgments and how inscrutable his ways!" (Romans 11:32, 33). Without seeing the rebellion and sin of man, there could be no reason or revelation of God's suffering love.

This was the hidden mystery of God's eternal will according to his purpose "which he set forth in Christ as a plan for the fullness of time, ***to unite all things in him/Christ***, things in heaven and things on earth" (Ephesians 1:9, 10). Elsewhere, having discussed the inclusion of the nations/gentiles with his people Israel (Ephesians 2:11-22), Paul said this was ***God's eternal purpose in Christ Jesus*** our Lord which the church would make known as the manifold wisdom of God "to the rulers and authorities in the heavenly places" (cf. Ephesians 3:8-11). The purpose is to show that man cannot save himself, and only Jesus Christ, God's Son, can give the life of eternal blessing. As you as a Christian read and understand, you will be shocked at the story of demonic seduction and human ***perversion documented in this book***. It is importance for Christians to know ***Christ is Creator*** to believe the Scriptures and hope in him as ***the Re-creator***.

The crucial issue in this history presented in this chapter is ***the cause of the wave to begin to be hindered by under-tow and finally break. That hindering cause*** is man's individual self-wisdom that ***is biased against his Creator's*** wisdom to rule and save. In great prosperity man thinks he does not need God and therefore rejects him. Man's sin nature will be demonstrated in most of the book. Briefly conditions in the broken Holy Roman Empire will be discussed to show how the Reformation faith of science and theology gave birth to the new wave of faith (1500ff.)

The Effects of Original Sin on Civilization in the History of the Nations of Mankind

Man's Limitations by Sin: No Knowledge of God Except as the Spirit Teaches

Man became spiritually dead to God in his sin and could only learn from God at God's initiative. When Nicodemas, who was considered "***the*** master teacher of Israel," asked how he could enter the kingdom of God, Jesus replied, "You must be born of the Spirit." He explained that except a man be born of water (cleansing from sin) and of the Spirit, he cannot see/enter the kingdom of God.

> Nicodemus said to him, "How can a man be born when he is old? Can he enter a second time into his mother's womb and be born?" Jesus answered, "Truly, truly, I say to you, unless one is born of water and the Spirit, he cannot enter the kingdom of God. That which is born of the flesh is flesh, and that which is born of the Spirit is spirit. Do not marvel that I said to you, 'You must be born again.' The wind blows where it wishes, and you hear its sound, but you do not know where it comes from or where it goes. So it is with everyone who is born of the Spirit. ...If I have told you of earthy things and you do not believe, how can you believe it I tell you of heavenly things?" (John 3:4-8, 12).

The apostle Paul also spoke of the inability of men to understand God and the invisible:

> What no eye has seen, nor ear heard, nor the heart of man imagined what God has prepared for those who love him"—these things God has revealed to us through the Spirit. For the Spirit searches everything, even the depths of God. For who knows a person's thoughts except the spirit of that person, which is in him? So also no one comprehends the thoughts of God except the Spirit of God. Now we have received not the spirit of the world, but the Spirit who is from God, that we might understand the things freely given us by God. And we impart this in words not taught by human wisdom but taught by the Spirit, interpreting spiritual truths to those who are spiritual. The natural person does not accept the things of the Spirit of God, for they are folly to him, and he is not able to understand them because they are spiritually discerned. The spiritual person judges all things, but is himself to be judged by no one. "For who has understood the mind of the Lord so as to instruct him?" But we have the mind of Christ (1 Corinthians 2:8-14).

This was manifested after the original sin. Man (Genesis 3) became a mortal individual and selfish, thinking he was wise as God to know good and evil and was offended or fearful of God. When Adam and Eve recognized their nakedness, they became aware of their differences. Their awareness of these (gender/physical) differences and later other differences among humans became a threat to their individual autonomy. This self-righteous knowledge is the basis of all offenses of differences in gender, race, class, etc. Feminism, racism and class hate are born from the individual deceived self-righteous wisdom of mankind. The pursuit of the survival of the fittest of individual men was born and social adhesion was removed. All subsequent offspring of Adam and Eve were born with their sin nature and their spiritual death, and exclusion from Eden's tree of life introduced mortality, leading to physical and eternal death (Romans 5:12). Man's self-centeredness tends to isolate him, not only from God, but also from others. This is minimized by trying to cover over or deny the differences. Confined by guilt and knowledge of coming death, man pursues happiness by exalting himself, accumulating wealth for his own security and freedom, and enjoying bodily pleasures. He is restrained only by God's law as evidenced of God in creation and written on his heart, and by human government.

Demonic forces Deceptively Drive Man's Selfish Desires to Gain Control in the World

The devil continually deceives by demonic suggestions, moving people toward sin and death (James 1:13-15), to obsession and addiction (demonic possession). The devil progressively deceives leaders, thereby moving man's ability to reason, to discern good from evil, further from the law of God. He used the fear of death to keep men in lifelong bondage to fleshly pursuit (Hebrews 2:14, 15). This is the way the devil seeks to rule the world. The drive toward becoming a superman of the world has moved man towards the devil's objective of ruling by antichrist. This began with Nimrod at Babel to all other dictators (Genesis 10, 11; Isaiah 14:12-15; Ezekiel 27:11-18, Daniel 7:19-25; 11:21-44; Revelation 13). The devil's ambition to rule the world is seen in his efforts to get Christ to worship him, and thus making Christ the antichrist (Matthew 4:8-10). Satan is "the great dragon, the ancient serpent who is called the devil and Satan, *the deceiver of the whole world"* (Revelation 12:9). He has been given opportunity to become the invisible "prince of this world" (John 12:31; 14:30; 16:11). Paul said every man was born following the devil in deceptive bondage to sin.

> And you were dead in the trespasses and sins in which you once walked, following the course of this world, following the prince of the power of the air, the spirit that is now at work in the sons of disobedience—among whom we all once lived in the passions of our flesh, carrying out the desires of the body and the mind, and were by nature children of wrath, like the rest of mankind (Ephesians 2:1-3).

Christians are commanded to, "love not the world, the desires of the flesh/body, the desires of eyes to possess, and the pride of life, which are not of the Father, but of the world, and the world is passing away. But he who does the will of God, abides for ever" [paraphrase] (1 John 2:15-17).

Satan uses man's worldly desires to reject evidence of God's control, leading men to hate God and his will. Natural man thus becomes progressively deceived to reject knowledge of God and his will. Anti-Christian efforts and anti-Semitism are a natural desire to seek to remove God's restraints and are the main cause for perversion and the inextricable product of the secular mind. Satan uses those claiming the most intellectual abilities to deceive other men. They present themselves as apostles of righteousness, for the devil himself is revealed as an angel of light (2 Corinthians 11:13-15). Belief in the light-reflecting brazen serpent, not the Son of God himself, is man's destruction. Paul says the devil uses the world to blind man to the truth of God.

> And even if our gospel is veiled, it is veiled only to those who are perishing. In their case the god of this world has blinded the minds of the unbelievers, to keep them from seeing the light of the gospel of the glory of Christ, who is the image of God (2 Corinthians 4:3, 4).

Evidence of One Evil Spiritual Power who Communicates through the Wisest Serpent

From the earliest documents of ancient civilizations we find evidence of one devil or superior invisible spirit or god who leads man into evil. These written accounts developed after the tower of Babel and the confusion of tongues. Every early language and culture records one demonic power.

The first men lived for nearly 1000 years (Adam et al.) until the time of the flood. From the time of the flood until the division of the nations and confusion of tongues human lifespan was about 450 years. During this time the human story was maintained by oral tradition. Stephen Olson and scholars of various sciences, using computerized actuarial data, shockingly concluded that the known cultures were only about 6000 years old. John Hein of Oxford, reporting on this data in *Nature* Magazine, said that if one entered any village on earth in 3000 B. C. he would meet some of his ancestors.[2] Therefore, at about 3000 B. C., when mankind was losing accurate touch with the past, writing began in Babylon, Egypt, and also was recently found in cultures in Latin America. More will be said about this later in the book.

---

[2] Stephen Olson, in *Mapping Human History* first reported this and after later work with others, John Hein of Oxford discussed these shocking findings in *Nature* and they were discussed by Matt Crenson, Associated Press science writer (cf. "Human's common link is found not so far back," *Orlando Sentinel*, July ,2006 A4.

Significantly, all the major ancient cultures record an early demonic force of the underworld. He was considered the god of nature who blessed agriculture to give prosperity. He was often identified with the cult of death. This god was known by different names in different cultures/records: Tammuz in the Cuneiform Texts from Babylonian tablets; Alein in the North Syrian Ras Shamra tablets, and elsewhere Aleyn Baal; Baal Zebul (highest god) in Phoenician and Canaanite writing; Osirus in Egypt; Adonis in Greece; Barcus in India; Baccus among Romans.

Bar Cush means the son of Cush, Nimrod or the first antichrist leader that demonically by astronomical knowledge wanted to be exalted to the throne of God.[3] These gods were later humanized and various stories and images proliferated, identifying them with fertility goddesses and sexual love. But the earliest god was clearly one evil spirit. This evil power was worshiped for prosperity, fertility, and pleasure. Also, the worship of the serpent and dragon is found worldwide in all early cultures. He was associated with the evil god in most of these cultures. The word "serpent" is the same Hebrew word for "bronze" used for polished mirrors that reflected bright light. Serpent worship was associated with metallurgy. The serpent was also associated with wisdom or cunning.[4]

How could these ideas of one evil invisible power that was connected to the serpent be in the earliest written documents all over the world if the story of Genesis 3 were not true? Modern psychiatry, psychology, and neuroscience cannot control or understand why humans are driven by voices and visions to commit inhuman evil acts. Every human being, man or woman, thinks thoughts that they know are wrong and that keep coming back. These repeated temptations demonstrate that man desires to be free from God, for his own happiness, and he is thus driven by spiritual deception of equal wisdom for self-esteem.

A Historic Process of Demonic Perversion Recurs in Society through Man's Sin

The Recurring Process of Demonic Motivated Sin Destroys Society and Civilization

In Romans the apostle Paul describes how, despite God's clear revelation of himself in creation, man's sinful nature progressively rejects God with devastating consequences. I have given details to historically document the Greek-Roman process which Paul is describing here.[5] Therefore, only the steps of the process are given here. After a civilization grows under the knowledge of one God for about 250 years, intellectually proud men see the brilliance in creating the culture and the power and prosperity as from their own wisdom. Then the intellectual elite begin to slowly reject God and morals collapse. In about 200 years the civilization is destroyed. The slow process of perversion is deceptive.

The slow steps of degeneration are as follows.

1. The ***intellectual elite slowly suppress and pervert the truth*** replacing it with lies.
2. Their trust is shifted to ***worshiping the creation or material prosperity*** and to considering men as just higher animals of a natural world.
3. The ***different sexual role of women is debased*** because they cannot equally produce materially when they have and raise children and this causes a women's movement to change the natural use/role of women and then practice sexual lusts also with men.
4. Strong ***individual desires and competition create anarchy*** breaking down the authority of the home and community.

---

[3] George A. Barton, *Archaeology and the Bible* (American Sunday School Union, 1949), chapter XXIII, et al. G. Earnest Wright, Biblical Archaeology (Philadelphia: Westminster Press, 1960) cf. 6 ff.; et al.

[4] Cf. "Serpent Cults," *Encyclopedia Britannica, Volume 20*, (Chicago, 1953), 369, 370.

[5] Carl W. Wilson, *Man Is Not Enough,* cf. chapters 2-6, pages 19-112.

5 The ***people do not want God,*** so God's Spirit releases them to do all kinds of evil.
6. J***udgment and conflict over injustice*** arises and God sends his judgment to end the pain of the breakdown of the civilization. The following is Paul's description of how each change slowly produces suppression of truth.

Slow Intellectual Trust in Human Wisdom and Depreciation of God

For the wrath of God is revealed from heaven against all ungodliness and unrighteousness of men, who by their unrighteousness ***suppress the truth***. For what can be known about God is plain to them, because God has shown it to them. For his invisible attributes, namely, his eternal power and divine nature, have been clearly perceived, ever since the creation of the world, in the things that have been made. So they are without excuse. For although they knew God, they did not honor him as God or give thanks to him, but they became futile in their thinking, and their foolish hearts were darkened. ***Claiming to be wise, they became fools*** (Romans 1:18-22). (Emphasis added)

Worship or Trust Creation and Man as Part of Nature

....and exchanged the glory of the immortal God for images resembling mortal man and birds and animals and creeping things. Therefore God gave them up in the lusts of their hearts to impurity, to the dishonoring of their bodies among themselves, because they exchanged the truth about God for a lie and ***worshiped and served the creature rather than the Creator***, who is blessed forever! Amen (Romans 1:23-25). (Emphasis added)

Individual Sexual Roles Equalized Allowing Lust

For this reason God gave them up to dishonorable passions. For their women exchanged natural relations for those that are contrary to nature; and the men likewise gave up natural relations with women and were consumed with passion for one another, men committing shameless acts with men and receiving in themselves the due penalty for their error (Romans 1:26-27).

Knowledge of God Rejected, Producing Anarchy

And since they did not see fit to acknowledge God, God gave them up to a debased mind to do what ought not to be done. They were filled with all manner of unrighteousness, evil, covetousness, malice. They are full of envy, murder, strife, deceit, maliciousness. They are gossips, slanderers, haters of God, insolent, haughty, boastful, inventors of evil, disobedient to parents, foolish, faithless, heartless, ruthless. Though they know God's decree that those who practice such things deserve to die, they not only do them but give approval to those who practice them (Romans 1:28-32).

Judgmental Conflict over Justice and God's Righteous Judgment

Therefore you have no excuse, O man, every one of you who judges. For in passing judgment on another you condemn yourself, because you, the judge, practice the very same things. We know that the judgment of God rightly falls on those who practice such things. Do you suppose, O man—you who judge those who practice such things and yet do them yourself—that you will escape the judgment of God? (Romans 2:1-3).

God's Wisdom and Man's Bias

The shift in the culture from faith in God to intellectual trust in man's wisdom or claim of self-enlightenment about nature is the folly that perverts all truth. It progressively, over a couple of hundred years, produces the rejection of God and the worship of man as the superior creature in nature. The bias of natural man's sin to desire to be free to possess and control the world and to fulfill his lusts leads to man suppressing the evidence of God as it has been revealed.

Bias of Intellectually Trusting Self-wisdom over God Progresses through Three Steps

Man begins with belief in a sovereign personal God, defined as ***theism.*** God is sovereign over every area of all things, and knows and is wise in all things. But man's thinking, however, is demonically perverted and directed away from God in three ways. First, natural man may say God created the world and left it to run mechanically for itself, but he no longer is present or controls it. This is like a watchmaker who makes the clock and leaves it to run on its own. Thus man must understand natural laws and help control them for his own good. This mechanical view is ***deism.*** A second way to deny God as Creator is to say that nature or ***creation has an intelligent designing power within.*** Nature is God. This innate designing power is unchanged or unmoved and gradually actualizes a potential within it, producing higher, more complicated things by evolution. Since man is obviously the superior being in nature, man believes he should be in control. Aristotle's pagan philosophy, which believes nature has a potential to progressively evolve the highest form, ends up believing one man is superior and should be the leader—Superman. Each individual man seeks to control, to be superman in his own place in the world. This is ***pantheism.*** Finally, in ***atheism*** man may deny there is any God or any intelligent power in control and that all has and is working by chance and is mysteriously changing as ***naturalism***. A graph of these follow and the progression is from left to right.

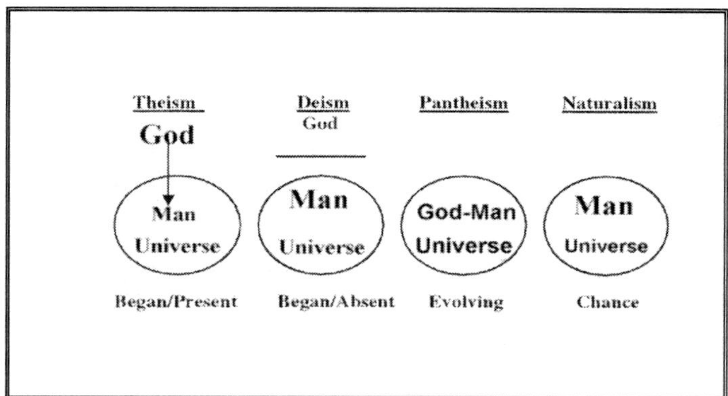

First, ***Deism*** recognizes an intelligent Creator but says he no longer controls his creation. Man determines the control of the world. Second, ***Pantheism*** says nature is god and has within it intelligent purpose to evolve or develop the world and produce higher beings up to man. Man is therefore the one to be worshiped; man should direct the world, and man can progress to superman. ***Naturalism*** also sees man as the main creature but without purposeful direction so there is no intelligent control and man is free from any meaningful restraint. Only ***theism*** accepts God as Creator presented as personal Lord in control and holding man accountable for his actions.

All these three philosophical and religious views—deism, pantheism, and naturalism—***suppress the truth of God,*** allowing man to think he is equal with God. Man then exalts himself to be God and worships that which is created rather than the Creator. He reinterprets sexual roles as equal, changing the natural use of roles to that which is contrary to nature to freely express his passion without accountability. Material greed and sexual perversion are promoted by demon-driven idolatry (Ephesians 5:5; Galatians 3:5; 1 Corinthians 10:19, 20; Deuteronomy 32:17; Psalm: 37, 38). Using an idolatrous created image, men told stories about the god and how that god blesses materially and gives great pleasures. In modern times this is done by created images projected on a screen to stimulate desires of greed, lust and pride. This book will later present evidence that man's inclination to suppress God's truth and reject God, is the demonic pattern for destroying all known civilizations.

No matter how much sinful man perverts the truth about God, the evidences for God reemerge because they are embedded in creation and have been there all along. But proud intellectuals foolishly continue to push naturalism and antichrist ideas by appealing to human desires until these ideas become dominant. Invisible demonic thoughts can easily sway man's sinful nature (Ephesians 6:10 ff.; 2 Corinthians 4:4). The devil will progressively pursue deception against evidence for God in nature until man's sinful acts reach a point at which God lets them remove moral constraints until that society collapses into conflict and anarchy and God sends judgment to end it all.

Sinful Men Demonically Pervert Truth by Claiming to Deny God or Speak for Him

Man's bias to be equal with God and his sinful, anti-God desire is expressed in many ***deceptive ways through religious and secular efforts*** to seek to usurp God's power over the world. One demonic way of deception is ***through religion*** so that man falsely claims to be a leader ***speaking for God*** but is a false prophet for self-gain. Man thereby changes values to please himself, claiming this is God's will and authority. Perverted religion always exalts man, focusing not on God, but on what is good for mankind, and claiming God's authority to act. The false prophets of the Old Testament and the Jewish Sadducees or Pharisees who condemned Jesus are models of this perversion. The Sadducees embraced Greek Aristotelian pantheism, while the Pharisees hypocritically used their knowledge of the law to appear righteous and control people. Jesus identified them as serpents or vipers (Matthew 12:34). These ways of using religion to selfishly control others are still used today by other names. Others try to appropriate God's authority through visions or claiming written tablets to add to the Bible. Secular attempts to deny God's power are often subtly mixed with religious claims, making them especially deceptive.

Biblical History Reveals Repeated Deceptive Sinful Bias that is Judged as Destructive

Man's natural bias is seen throughout biblical history and confirms the justice of God in pronouncing all men of all nations under judgment. Sin was manifested in Cain, the first antichrist man and his progeny (Genesis 4:16-24, cf. 1 John 2:18, 3:10-12). As mankind multiplied, so did sin, until it reached the point of rejection of God's general and prophetic revelation. By the time of Enoch, man had rejected God and God translated this righteous prophet up to heaven. Soon after this event, God brought the shock of the physical death of Adam who had fathered the race, reminding them of their mortality for sin. The same progression toward the rejection of God was reached by the time of Noah. At that time "God saw that every thought and imagination of men's hearts was only evil continually and God was sorry He had made man" (Genesis 6:5). So God destroyed the world by a great flood (Genesis 6-8). About 300-400 years later, at the time of Nimrod, the whole world joined in rejecting God and began to build a tower that would unite them in their rebellion. This tower was to be an instrument for obtaining astronomical knowledge for unity in the knowledge of God. So God divided the earth and confused the languages, dispersing man into separate nations throughout the world (Genesis 10:8-12, 25; 11:1-9; 18, 19; Acts 17:24-

26). Pyramids or similar towers are found all over the world as evidence that they carried this unique cultural desire to be united in nature worship rather than God.

History repeatedly traces the process described in Romans 1:18-2:3 in which civilizations rejected God after God prospered them. The Chaldean kingdom at Ur was destroyed after God brought Abram out. The Egyptians fell into great sin and completely rejected God at the time of the Exodus. Forty years after the Exodus, the sin of the Amorites in Canaan had reached its fullness and Israel was sent to destroy them and occupy their land—the fulfillment of God's promise to Abraham 400 years earlier. Even the nation of Israel, God's chosen people, went though repeated cycles of such sin, rejection, and judgment. All the Major and Minor Prophets declare God's judgment on all the surrounding nations for their sin.

Nearly all the Old Testament prophets state that there was "*not a man*" in Israel who would represent God (Isaiah 41:28, 29; 51:18; 59:16; 63:4, 5; Jeremiah 5:1-3; 6:28-30; 12:7-13). Perhaps Ezekiel 22:30 is the saddest and most graphic, God said, "And I sought for a man among them who should build up the wall and stand in the breach before me for the land, that I should not destroy it, but I found ***none***" (emphasis added). Jesus accused the leaders of Israel of always rejecting God's prophetic message and by rejecting him they bear the guilt of all God's revelation.

> That the blood of all the prophets, which was shed from the foundation of the world, may be required of this generation; From the blood of Abel unto the blood of Zacharias, which perished between the altar and the temple: verily I say unto you, It shall be required of this generation (Luke 11:50, 51).

The deacon Stephen, before being stoned to death, reviewed Israel's repeated sin (Acts 7). Even God's elect people of Israel became completely corrupt and rejected God. He had to vindicate his holiness by judging and destroying them. But the prophets saw all the surrounding nations as sinful and predicted their judgment as well. The Scriptures proclaim this universal tendency of all men's bias against God. "The fool has said in his heart, 'There is no God.' They are corrupt; they have ***all gone*** out of the way. There is none that does good, ***no not one***" (Psalm 14:1-3; 50:1-3) (Emphasis added). Paul clearly states that because of man's spiritual death to God, ***all*** have walked after "the prince of the power of the air, according to the desires of the flesh and the world" (Ephesians 2:1-3). After reviewing the sins of Israel as well as the Gentiles, Paul said, "For there is no distinction, for all have sinned and fall short of the glory of God" (Romans 3:23). More emphatically he said, "The carnal mind is ***enmity against God***, and is not subject to the law of God, ***neither can be***. So then they that are after the flesh ***cannot*** please God" (Romans 8:7, 8).

The sinful depravity of all men in all nations is a cardinal doctrine of Scripture, but not one that most intellectual Christians count important. Many intellectual Christians assume that all unbelieving people need is more *information* which will logically persuade them to become Christians. Many believe knowledge alone will make Christians better Christians. Unfortunately, even most leaders, who say they believe the Bible, talk little about man's sin bias and tendency toward demonic deception. While "total depravity of man" is counted as one of the five cardinal doctrines of the Bible, when have you heard a message about it? This is why the message of the Gospels and the book of Acts is 'repent and turn to God in faith.'

Previous Broken Wave of Darkness in Europe from which Modern Light Arose

Bias Using Religion Produced Dark Ages and God's Grace Initiated Rebirth of Faith

The degradation in the Greco-Roman civilization, which the apostle Paul describes, repeated itself in Europe a millennium later. A renewal of faith occurred which led to an even worse cycle of degradation around 1400 A.D.—a later Dark Ages. From that time of troubles the Reformation produced modern scientific enlightenment. How did human bias produce the conditions of the Dark Ages from which modern culture emerged? Primitive Christianity had gained prominence, producing a strong witness for Christ that was centered in Rome. The Roman church was a powerful witness for Christ. But after several hundred years of imperial rule that gave power to the church, the eastern Roman civilization had been overrun by the barbarians. Out of this, by God's grace, there was a reemergence of faith that led to the power of the Roman Church influencing all of Europe. During the eleventh century the Roman Catholic Church experienced significant spiritual renewal. It exercised growing power as the Holy Roman Empire in Europe.

That renewal in the second millennium occurred because new groups of believers arose in most European nations such as England, Germany, France, and Italy. These people had gained biblical knowledge that revealed corruption in the church and society. People demanded change in both the church and civilization. These enlightened believers pressured the organized church for reform. A Benedictine monk, born Hildebrand of Sovana, Italy, arose to the papacy as Gregory VII in 1073. He had a bent to moral reform and order due to the influence of these faith groups. But his ability to influence turned to ambition.

> His life-work was based on his conviction that the Church was founded by God and entrusted with the task of embracing all mankind in a single society in which divine will is the only law; that, in her capacity as a divine institution, she is supreme over all human structures, especially the secular state; and that the pope, in his role as head of the Church, is the vice-regent of God on earth, so that disobedience to him implies disobedience to God.[6]

After a long conflict, the church, as the Holy Roman Empire in the fifteenth century, controlled the state. This followed many years, in which the church grew in wealth and power, a pattern of intellectual pride again emerged, asserting the pope's right to speak for God. Once again things deteriorated, beginning in the thirteenth century, reaching a crisis point by the fifteenth century. Roman Catholicism's union with politics resulted in oppression of the common man, leading to a hunger and struggle for a rebirth of human dignity.

Philip Schaff, the great church historian, said of Hildebrand's time, "The papacy was a necessity and a blessing in a barbarous age, as a check upon brute force, and as a school of moral discipline."[7] The influence of the reforming Latin Church spread until it reached all of Europe. But after a couple hundred years of prosperity, having gained great power, Europe deteriorated into the Dark Ages. Philip Schaff also says about the five hundred years before the Reformation and Renaissance, "But the papal theocracy carried in it the temptation to secularization. By the ***abuse of opportunity*** it became a hindrance to pure religion and morals."[8] (Emphasis added)

Religious Intellect Usurpation of Christ's Authority became Linked to Greek Thought

The Roman clergy in the Holy Roman Empire claimed intellectual power and became interested in ancient Greek thought, especially that of Aristotle. The Aristotelian philosophy that gained ascent in the end of the fourth century is not easy for contemporary Americans to understand.[9] The relevant ideas are as follows: The real world of particular things is based on a

---

[6] Pope Gregory VII, *Wikipedia.*
[7] Philip Schaff, *The Middle Ages,* Vol. V, (Grand Rapids: Wm. B. Erdmann Publishing Co., 1950), 5.
[8] Ibid.
[9] B. A. G. Fuller, *A History of Philosophy,* (Revised edition), Volume One, (New York: Henry Holt and Company), see "Aristotle" pages 176-194 for better understanding of these views. For the Ptolemaic system

complete reason that is eternal and unchanging in nature that is called the Unmoved Mover. This natural power is in the process of producing or actualizing the potential of everything. It is an eternal rationalistic pantheism (the world is god) that says things are in progress but man's *only knowledge is from the individual things he sees and knows. Man is the highest creature in the* world so he is now the highest actualization of potential of the rational universe. This view of nature was the preamble of a developmental or modern evolutionary theory. Aristotle's philosophy was a pantheism that led man to think he could become as God. It produced a loss of democratic control leading to a dictator. Alexander the Great, Aristotle's student, became the dictator and declared himself Zeus-Ammon or God of Greece and the Orient bringing demise of Greek culture.

In the mid thirteenth century Albertus Magnus of Cologne of the Dominican order became a bishop and sought to harmonize science and religion by Aristotle's philosophy. Thomas Aquinas, also a Dominican monk and scholar, produced an organized theology for the church centered in Aristotle's thought. Colin Brown has said,

> Taking his cue from Aristotle's pre-Christian idea of the Unmoved Mover..., [Thomas] believed that one could argue back from the things that we observe in the world to a prime mover, a first cause of a great designer behind it. ...Every event must have a cause.[10]

Aquinas gave his famous rational arguments for the proof of God and tied Aristotle's philosophy to the Christian God and the Roman Catholic Church. His arguments are weak (or fail) because he attempts to prove the infinite transcendent biblical God from finite creation. In this step God was a conception of man's rational thought. The Jesuit order later defined this.

In contrast, the apostle Paul taught that men could see God's power and deity in creation because being made in the image of God, he had "shown it to them," or given them an internal revelation in their nature (Romans 1:19, 20). Through Aquinas, the whole Roman Catholic "Christian" system was tied to Aristotle's pagan philosophy, which influenced the church's thinking about nature and science. This marriage of Aristotle's pantheism to the Creator God and his biblical revelation was intellectually appealing. By the fifteenth century, however, it consistently and progressively weakened European accountability to God and increased man's power ecclesiastically. Roman Catholic theism or true faith in God moved to a deism and toward pantheism.

Roman Catholic clerics held authority over all institutions in the fourteenth and fifteenth centuries. The right of the common man to know and decide religious matters was denied him. Rather the educational institutions taught what the church theological intellectuals decided was true and right. Laymen were denied the opportunity to read the Scriptures. Church theologians extended their authority to every area of life by debate or ***scholasticism,*** and the clergy's teaching and tradition replaced interest in and submission to Scripture. The church maintained control of the clergy through a pattern of intolerant education or indoctrination. But much scholastic religious discussion of detail was beyond the common man's thinking and ***irrelevant*** to life.[11] By the fourteenth century, this form of study was oppressive, tedious, boring, and had even reached

---

built on Aristotle's views see Herbert Butterfield's, *The Origins of Science 1300-1800*, (New York: Collier Books, 1962), chapters 1, 2; pages 13-48.

[10] Colin Brown, *Philosophy of the Christian Faith: A Historical Sketch from the Middle Ages to Today,* (Downers Grove, IL: InterVarsity Press, 1976), 26, 27.

[11] The following is scholasticism. "To give one illustration: the species 'cow' does not exactly reflect what was in the mind of God; and what man perceived in the species 'cow' is not altogether what is in that species."

some major contradictions. Denied by intellectual clergy the freedom to study for himself, common man was belittled, and he hungered for a new meaningful humanism.

Roman Catholics had Accepted Aristotle's False Science the Reformers Removed

Around 127 B.C. post-Aristotelian Greek scholars, Hipparchus and Ptolemy, had developed a mathematical astronomical planetary system. Aristotle's system of philosophy was tied to this Ptolemaic view of the planetary and heavenly movements as ***centered around the earth*** and everything was moved toward the earth by natural innate reason. The Roman Catholic Church made this a part of its teaching. Herbert Butterfield, Professor of History at Cambridge, has said,

> On the Aristotelian theory all heavy terrestrial bodies had a natural motion towards the centre of the universe, which for medieval thinkers was at or near the ***centre of the earth***; but motion in any other direction was violent motion, because it contradicted the ordinary tendency of a body to move to what was regarded as its natural place. ...The essential feature of this view was the assertion or the assumption that a body would keep in movement only so long as a mover (Unmoved Mover) was actually in contact with it, imparting motion to it all the time. Once the Mover ceased to operate, the movement stopped—the body fell straight to earth or dropped suddenly to rest.—It was motion, not rest that had to be explained.[12] (Emphasis added)

The Roman Catholic Church based its scientific teaching on this idea of motion. Thomas Aquinas made Aristotelian thought the center of Roman Catholic teaching of its whole doctrinal system. As the primary intellectual group, however, the Jesuit monastic order was responsible for promoting and defending Ptolemaic astronomy as the center of orthodoxy in the growing Roman Catholic schools. This bias to the pagan view of God became the main conflict in reforming science.

Corruption of Church Conduct by Paganism Required Change

The loss of accountability to the Creator led to corruption of the Roman Catholic Church that damaged its credibility. Under the influence of Aristotle's philosophy that exalted individual knowledge by the mid-fourteenth century, clergy were greedy, immoral, cruel, unjust, and exercising unwarranted oppressive political power. Increased commerce and industry produced a myriad of new worldly pleasures to be desired, which easily prompted greed. The clergy controlled a large portion of the land and wealth of Europe (e.g. one-fifth of Germany). Philip Schaff has described the conditions.

> The scandalous lives of the popes whose names fill the last paragraph of the history of the Middle Ages would have excluded them from decent modern (20th century) circles and exposed them to sentence as criminals. They were perjurers, adulterers. Avarice, self indulgence ruled their life. They had no mercy. The charges of murder and vicious disaster were laid to their door. They were willing to set the states of Italy one over against the other to allow them to lacerate each other to extend their own territory or to secure power and titles for their own children and nephews. ...In all history, it would be difficult to discover a more glaring inconsistency between (religious) profession and practice that is furnished by the careers of the last popes of the Middle Ages. ...Upon freedom of thought, the papacy continued to lay the mortmain of alleged divine appointment.[13]

---

[12] Herbert Butterfield, *The Origins of Modern Science*, (New York: Collier Books, 1962), 15, 16. Cf. whole chapter I, "The Historical Importance of a Theory of Impetus."

[13] Schaff, *The Middle Ages; A.D. 1294-1517, Volume VI,* 772, 773.

In fighting over papal power the papacy moved from Rome, Italy to Avignon, France, and back to Rome. At one time, there were four popes and two centers of government. Church councils denounced papal authority and vice versa. Such luxury, lust, abuse of power, and corruption—all of this was a scandal to the European intellectuals and brought resentment to many people.[14] The schism and corruption in the Dark Ages weakened papal authority so much that a further blow to papal claims came later in the sixteenth century when Henry VIII forced the English clergy to back his illegal divorce. Upon their refusal, as the head of the church, he separated the English church from Rome. These schismatic conditions undercut the credibility of the Roman Church's claim to be God's authority in much of Europe. Consequently, its strength declined in the last of the fourteenth and early fifteenth centuries. Its pagan philosophy and lifestyle called out for rejection and a new understanding of man and God. This was a repeat of the conditions prior to the renewal that begun in the eleventh century. The intolerance of others, linked with the loss of moral and institutional credibility, opened the door to new thought—rather demanded it. This produced groups seeking reforms and more intellectual light.

Emergence of Forces of Truth in the Degraded Holy Roman Culture

A renewed interest in the Scriptures and an evidence of a creator reemerged to revive Christian faith in the Dark Ages. Remnants of the eleventh century reform groups remained, such as the Waldenseans, the poor laymen's movement outside the Catholic Church. Also, within the Catholic Church there were movements such as the Cistercians. Their most prominent leader was Bernard of Clairvaux who believed the Bible and salvation by grace, and Francis of Assisi and his followers who exhibited trust in God and love for the poor. Through the centuries, these groups had spawned others that believed the Bible. All these emerged from remnants and challenged the oppressive Roman Catholic authority. But the Roman Catholic Church either assimilated or actively persecuted those groups that protested its religious authority and abuses.

The question of what was good and evil, according to the authority of the clerical leaders, was not to be tolerated. During the thirteenth century the Catholic Church as a power of the Empire was threatened by the Muslim expansion. During that time the Catholic Church attacked and slew many Muslims by the sword in senseless crusades that ended in the thirteenth century. Jews also were slaughtered in the Inquisition. Godly Christians and religious movements such as Wycliffe's, the Lollards, the Hussites, and the Albagensians et al. who dared to speak against the church for her perversions were persecuted. In England, Wycliffe translated the Bible; the Latin Scriptures, contrary to Rome, and reintroduced Christ and the primitive church for the benefit of common man. This act produced persecution of these groups. John Huss was burned at the stake. Wycliffe also was put to death as a result of Catholicism's intolerant claims of being the only source of truth.

Slow Emergence of Independent Thinking about the World in the Dark Ages

It was not only independent religious groups that challenged Rome, but also emerging, primitive universities with a Christian humanistic emphasis. A few such preliminary educational schools had begun in the eleventh century. These emerging universities differed from Roman theological scholasticism in that some were more influenced by Platonic systems of thought passed down through Augustine, a Christian thinker. Greek science, along with the Scriptures,

---

[14] See Kenneth Scott Latourette, *A History of Christianity,* Vol. I, 624-659, 675-676, and Phillip Schaff, *The History of the Christian Church*, Vol. VI, 5-185.

influenced man's examination of the world to see God's work in the natural order. Hence, the university centers of learning were open to these new and more exciting ways of non Roman Catholic thinking, which had already secured entrance into their classes.

But the true beginning of universities at Paris, Oxford, and elsewhere did not occur until the fourteenth century. In time the universities became more independent in their teaching. Commercial expansion and contact with Islamic conquests reintroduced Greek and Latin with religions, science, and inventions of the past. This especially involved Archimedes' science from Syracuse which the Romans had accepted and developed after the first Punic war.[15] Archimedes had studied in Alexandria, learning especially from the Greeks, but also from the Egyptians. Commerce from Italy brought Europe into contact with ideas and learning from as far as Asia.

Islam had risen to power and dominated the eastern or orthodox church. Baghdad had become a renowned center of learning. The study of Greek and Latin writings, including the biblical Scriptures and early Greek and Roman science became a part of Islamic thought by a circuitous route. The Christian emperor of Constantinople conveyed a large collection of Greek manuscripts to the Islamic caliph Mammon at Baghdad. These were translated into Arabic by Syrian Christians and then rendered into Latin. But these Muslim mathematicians, philosophers, astronomers, and linguists advanced the research as from one god and established great centers of learning. The Mediterranean world "looked to the Bait al–Hikmah (the House of Wisdom) during the Golden Age in Baghdad.[16] In the tenth century, Islamic scholar Ibn al-Haytham described the scientific method of observation, measurement, experiment, and conclusion in a rational "Search for Truth." Likewise, in the thirteenth century, Ibn al-Nafis followed and encouraged the scientific method under a view of Allah as Creator. Studies of the science of creation and a reexamination of Hebrew Christian Scriptures found their way into emerging European universities.

These Islamic scholars later expanded and extended their studies to Islamic Spain and transmitted their discoveries, concepts, and theories to Europe. In time, these became a basis for learning in Italy, for medicine taught at Salerno, Italy and a school of legal studies at Bologna. The powerful influence of Islam on the Roman Catholic Christian West through trade, and by conquest, brought interaction with many ancient manuscripts. New universities grew because students found exciting independent thought from Roman control. The Roman Catholic Church often ignored this growing independence because it was caught up in the pursuit of wealth and power and controversies between various factions. The primitive Christian knowledge and science thus became the source of Reformation ideas that were refined and made powerful by the emerging new birth of faith in the biblical Creator in sixteenth century. Therefore, the remnant of God's people of faith brought forth a knowledge of a good God who Created and who in mercy came in Christ to forgive and renew his church by mercy. When in the fifteenth century, God's people were oppressed and cried to God and searched for him, the Reformation of science and revelation in Scripture was reborn.

In the twenty first century, man's sin bias again is rejecting God and the cycle of 500 years is approaching the time of God's judgment. With it will come a renewal of a remnant to witness to the truth of God and lead to a new cycle begun by grace or to the return of Christ and his eternal kingdom. This book explains the destruction of demonic deception and sin in modern times.

---

[15] Wilson, *Man Is Not Enough*, 76, 77.

[16] *Aramco World Magazine,* January-February, 1986, 2; see Kenneth Scott Latourette, *A History of Christianity* Volume I, (New York: Harper & Row Publishers, 1975 Revised Edition), 497-498.

Chapter 2
Birth of Modern Science by Christians

Birth of New Faith in Man's Intellect was Rekindled from the Bible about 1500 A. D.

Roman Catholic intolerance of religious Reformation efforts caused the struggle and conflict with new scientific thought. This struggle produced the new humanism. This was enhanced by merchants that brought new information from their voyages, the discovery of the New World by Columbus, and by new inventions, like improved printing and paper, that spread knowledge. Old science from Egypt, Greece and Rome and Christian writings of the New Testament and the ideas about these from the Muslims were recovered as previously mentioned.

This book cannot begin to trace the religious Reformation; many well-written church histories do that. But by the encounter with Greek and Latin writings, especially of the New Testament, the ***Protestant Reformation was the most significant event*** in breaking open Catholic control over thought, both theological and scientific. The Reformers were also instrumental in releasing other ideas of art, science, and invention which constituted the true Renaissance humanism. Catholic Church science, which the educated elite of the church had adopted from the pagan Greeks and Romans, was slowly contested as new Christian science emerged to challenge it. Twenty-first century Western scientists in general do not understand the pantheism which controlled Roman church thought.

The struggle for truth was against Catholic Church paganism. Twentieth century intellectuals misrepresented the conflict. The early scientific conflict in the sixteenth and seventeenth centuries was ***not Christians against new science*** as the modern intellectual elite say. It was between the science of the new biblical believers and the pagan Aristotelian science of the Catholic Church. The Catholic Church, in an effort to protect their clerical authority—derived from pagan philosophy—persecuted those who subscribed to true biblical faith. Their pagan Greek thinking allowed their exorbitant claims of papal and hierarchical authority. It was this pantheistic view that allowed the emergence of antichristian authority over the church. The Reformers recognized the antichristian nature of the papacy and called it that. It was this perverted pagan thought that hindered scientific development.

Herbert Butterfield, professor of modern history at Cambridge, commented on the hindrance this dogmatic Catholic theory caused.

> It was as though science or human thought had been held up by a barrier (the Aristotelian view of movement toward rest) until this moment – the waters damned because of an initial defect in one's attitude in everything the universe that had any sort of motion. ...This made the issue momentous.[1]

The emergence of the biblical view of God in the sixteenth century Reformation led to the clash with Aristotle's Roman Catholic philosophy. As early as the middle of the fourteenth century, ***based on biblical revelation of God's working***, Jean Buridan and Nicolas of Oresme of Oxford put forth the first ideas of the alternate view of ***impetus.*** Their ideas, which were in contrast with those who subscribed to the concept of the Unmoved Mover, spread to schools in Italy, as well as Paris where they continued till the sixteenth century. Others, in reaction to

---

[1] Herbert Butterfield, *The Origins of Modern Science*," chapter 1, The Historical Importance of a Theory of Impetus," (New York: Collier Books, 1962) 19.

Romanism and for independent thought, improved on and clarified this alternate view of ***impetus*** and its influence grew.

Butterfield points out that at times Christians, such as Jean Buridan, referring to pagan Roman church views "even noted that ***the Bible provided no authority for these*** spiritual agencies."[2] (Emphasis added) The Catholic Church was opposing science from creation that confirmed the Bible and God's authority, which was against their pagan philosophy that was the basis of their claim for ecclesiastic authority. The Reformers were discovering and seeking to explain a more accurate view of how God's creation works in the world to replace the Catholic pagan views. They believed this was a better understanding of creation by the God of the Bible.

A true biblical faith was emerging and found expression by the time of Wycliffe who preached reform in England and translated the Latin Vulgate Bible for the laymen in the fourteenth century. Martin Luther's experience of acceptance by justification by grace through faith alone was the main doctrine that made it possible to challenge the Catholic claims to religious authority. The effect of ***Luther's 95 theses (1517)*** was felt around the world and Luther stood firm under threat of death. God saved Martin Luther from being killed by Catholic Church authorities by King Frederick's protection in Germany, where Luther's teachings had widely spread. Luther's followers influenced Scandinavian countries and Iceland and his influence spread all the way to Italy and Spain. Remnants of the Bohemian Brethren, the Waldenseans and others who fed into the Anabaptists movement, along with Zwingli, were an initiating and continuing influence. The Huguenots were initiated in 1512 by Jacobus Faber's publication of Reformed principles and later by his French translation of the New Testament. Faber's followers spread in France and, while later identified as Lutherans, they were precursors of Calvinism. They armed themselves early and gained recognition in the Edict of Nantes but were later slaughtered by Catholic armies. Calvin came from Faber's French followers and Calvin's school in Geneva spread in Switzerland and resulted in sending out over 600 missionaries and forming thousands of churches in Europe. Notable was John Knox's influence in Scotland. The new freedom and sense of human independence encouraged by the Reformation promoted many varied ideas based on Scripture. Different political leaders took sides and slowly the once unified Europe was divided with some giving protection to Protestants and separating itself from the Catholic solidarity.

The Protestant movement, through its rediscovery of the Scriptures, brought about a renewed interest in the created order of the world and man's ability to understand it. Scientific groups began to gather in major centers, such as the fledgling schools at Oxford in London, the University of Paris, and especially the schools in Italy. Renewal within and freedom from Catholicism were slow and diverse. The movement ***began*** with people ***within*** the Roman Catholic Church who differed with the church's pagan interpretation. In time these differences evolved into full-scale conflict from which a 'protesting' or reformed Christian approach to science emerged. The story of reforming science from Aristotelian paganism of Roman Catholicism is best viewed by looking at key individuals and what they thought and discovered.

Early Reformation to Modern Science

Leonardo da Vinci (1452-1519)

Leonardo da Vinci was the illegitimate child of a peasant girl. He was raised by his father, Ser Piero da Vinci. As a youth he was attracted to science and culture by the athletic and gifted

[2] Ibid., 20.

Leon Battista Alberti. The brilliant Alberti served as a role-model, especially challenging Leonardo's interested in studying the secrets of science. Alberti was also born illegitimate and overcame this stigma. Leonardo was arrested for youthful deviancy, an event that seems to have turned him back toward the church. He is said to have cared for his mother in later life. While Protestant teachings had not yet emerged, it is said, "He ***accepted*** the essential ***Christian doctrine*** as an outward form for his inward spiritual life."[3]

He applied math to practical science and even painting. Most of his famous paintings were religious. He collaborated on *The Baptism of Christ*, painted the *Adoration of the Three Kings*, *The Virgin and Child with Saint Anne*, *The Lord's Supper* and more. His devotion to Christ is seen in the way that he broke with the usual portrayal of Christ. For example in *The Lord's Supper*, Jesus is seated separately *in the center,* with the disciples in groups. In *The Adoration of the Three Kings* the baby Jesus is *in the center* instead of to one side as in other paintings of the day. He spent tedious years of work doing these paintings.

He was one of the first Christian scientists who worked within the Catholic Church. We know how the Greek science was recovered by Islam from Christians and how this knowledge, "passed into Italy, how it was promulgated in the universities of the Renaissance, and how Leonardo da Vinci picked it up." (da Vinci recorded this in his notebooks.) These "were in reality transcriptions from fourteenth-century Parisian scholastic writers."[4] Leonardo was profoundly influenced in mechanics by Archimedes of Syracuse and by his contemporary, Aristarchus of Somas, in astronomy in concepts that preceded Copernicus. Leonardo encountered these scientists while studying in French universities. He was one of the first to question the Roman Catholic Aristotelian idea of force of objects to rest rather than impetus. He grasped the ***principle of inertia,*** or that every body has weight in the direction of its movement and held that the sun stood still, rather than moved around the earth. Leonardo broke with the Aristotelian philosophy of movement in nature and believed in inertia as did Copernicus and Galileo.

Leonardo da Vinci deduced the law of the lever as the primary machine. His amazing scientific ventures included designs for military and industrial machines, including the crane, study of human anatomy, exploration of flight, and much more. In general, he held to regular uniform operation of the forces of nature but seemed never to have challenged the Catholic Church's view of God or its more pantheistic theories. He worked under papal and church leaders but had differences with monastic leaders. Leonardo da Vince's ideas mark the beginning of the rebirth of new thinking. His approach to science was similar to that of later Christian scientists who viewed nature as mystical but created so as to expect reasonable regularity. He said that, first he would do some experiments before proceeding farther, because his intuition is to cite experience first and then with reasoning show why such experience is bound to operate in such a way. Thus he followed the inductive method as did others who followed him. Leonardo's thinking was from the context of God as creator, but prior to the Protestant breakthrough which also opened God's "book of creation."

His choices in art do not exhibit the lust of his time exhibited even in many clergy. He chose the Lord's Supper as the center of the New Covenant of Grace, and would abhor the idea of moving the apostle John and identifying him as Mary Magdalene as some modern speculations suggest. The whole Hebrew Christian tradition held to twelve male leaders. On the other hand he painted many pictures of women such as the Mona Lisa. But this famous painting focused on the

---

[3] Sir William Cecil Dampier, *A Shorter History of Science*, 48.
[4] Butterfield, *The Origins of Modern Science*, 20.

character of her face, did not portray her sensual feminine aspects (her breasts and curves). All other pictures of women he painted had similar discretion.

Nicolaus Copernicus (1473-1543)

Copernicus was educated at the Universities of Craccow, Bologna, Padura and Farrara of Italy. His early work was on sundials. At Bologna he studied under Dominico Maria de Navara, professor of astronomy, who questioned the Ptolemaic system and Aristotle's view of movement by the Unmoved Mover. At Farrara in 1503 he read the Arab astronomer al-Farghani's ***Elements of Astronomy*** that influenced the revival of science in Europe. Nicolaus Copernicus probably gained the idea of the earth rotating on an axis around the sun from two theories of Arabic Islamic scholars, Nasir al-Din al-Tusi and Muáyyd al-Din al-úrdi, who adopted the concept of the solar system from Greek and Roman sources before them (Cicero, Hicetas, Plutarch, and Aristarchus). The perspective of Islamic monotheism was changed by Christian Biblical views. He graphically described his theory first in 1530 A.D. and it was published in a book, *De Revolutionibus orbium Caelestium* in 1543. What Copernicus discovered about the sun and the questions he raised about movement all came together in such a way that he had to reject the Catholic, Aristotelian system. While Copernicus knew that his views were provocative, he also knew that they were a more accurate understanding of God's working. He said his views were "difficult, almost inconceivable and quite contrary to the opinion of the multitude, nevertheless in what follows we will with God's help make them clearer than day–at least for those who are not ignorant of the art of mathematics."[5] He sought to dedicate his work to the Church which was accepted by some leaders and rejected by others.

Copernicus pursued his science with a sense of "loving duty to seek the truth in all things in so far as God has granted that to human reason."[6] As an adult he was made a canon in the church, was the assistant to the Bishop, served as an interim administrator-general of a diocese, and was supported by the church for years. But he commented that the world system "has been built for us by the Best and Most Orderly Workman of all." He also said, "The universe has been wrought for us by a supremely good and orderly Creator."[7] His tombstone notes his dependence on ***the grace of God as His Savior***.

Copernicus' great book, *On the Revolutions,* was so radically different from the accepted system and highly mathematical that it was not widely read. He withheld its publication for a while due to unsolved scientific and theological problems which made it hard for others to believe his theories. While some astronomers acquired it for the tables, Copernicus had no more than ten known followers until the last part of the seventeenth century, when his book became very popular. It was not until another half century that other brilliant men accepted his views. Copernicus profited by the revival of the older science and was more deliberate than Leonardo in attributing it to the work of the Creator. The following is a diagram of the universe as mathematically constructed by Copernicus, showing the earth as rotating around the sun.

---

[5] Charles E. Hummel, *The Galileo Connection*, (Downers Grove, Illinois: Intervarsity Press, 1986), 55.

[6] Ibid.

[7] Ibid.

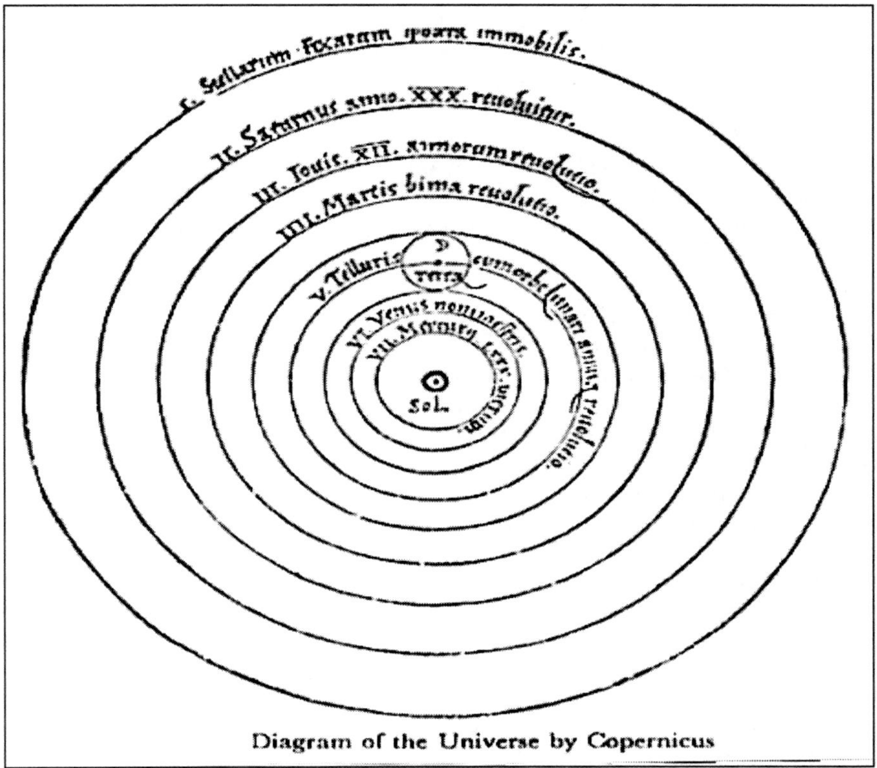

Diagram of the Universe by Copernicus

William Gilbert (1540-1603)

William Gilbert was the President of College of Physicians, and personal doctor to England's Queen Elizabeth and her successor. Elizabeth had restored the Episcopal Church and was its head. Gilbert was a strict follower of scientific methodology. He discovered and named electricity (Greek, electron, "amber") and demonstrated magnetism and its relationship to electricity. He discovered the earth was a ball rotating with magnetism. He also studied the influence of the earth and the moon as the cause of the tides. He first identified things having ***mass*** and objects influenced each other. Gilbert was the first Englishman to accept and support the Copernican theory. Electricity and magnetism were invisible unknown controlling forces; he called these forces *soles*, and said the sun and solar objects had *soles* or operated as secondary powers of God. He had a profound influence on seeing that science could measure quantities, but that forces beyond the senses, as God's working, limit man's full understanding of the universe. His thinking influenced Kepler more directly, but also Galileo and even Newton.

Blaise Pascal (1623-1662) and Pierre De Fermat (1601-1665), Air, Math Numbers and Probabilities

These two Frenchmen made tremendous advancement in mathematics. Pascal's influence is also seen in some areas of science. Fermat read the classics of antiquity but had no formal math training. He developed a form of calculus and analytic geometry which he applied to three dimensions. He claimed to solve the "Last theorem" known by his name without recording it so that remained as a puzzle. He inaugurated the modern theory of numbers. With Pascal he formulated the Theory of Probabilities which says that all empirical knowledge is a matter of probabilities and therefore not certain. Newton gave credit to Fermat for calculus. By age 16 Pascal developed the foundation for conic sections in math, advanced infinitesimal calculus, and progressed toward differential calculus. He solved the problem of general quadrate of the cycloid, and much more. As an experimental physicist he invented a barometer and showed the pressure of the atmosphere, beginning the study of air. He had a profound Christian experience and wrote *Pense'es,* a significant Christian apologetic that was published posthumously and still is widely read.

John (Johanus) Kepler (1571-1630) and Tyco Brahe (1546-1601), Astronomical Data

In God's providence John Kepler provided a transition to a more accurate historical analysis and scientific interpretation through a more objective study of God's revelation in creation as well as Scripture. Kepler's grandfather, a staunch Lutheran believer and mayor in a Catholic community, was from a formerly noble family. His father and mother were nonbelievers. While a youth, Kepler was profligate. He became interested in astronomy when as a child his mother showed him a comet and his father an eclipse of the moon. The desertion by his father after his tavern business failed motivated him toward Christ. He studied under one of the great mathematicians and astronomers at the Christian Protestant University of Tubengin, Michael Maestlin, who often referred to Copernicus.

Kepler was heir not only to the work of Copernicus but also to the work of Tycho Brahe, the leading astronomer of Europe in Denmark. He took both of their work to a much more complete and accurate level. Tycho Brahe had developed a more accurate and portable sextant for mapping the stars and planets. He also built the most accurate observatories of the time, having acquired the patronage and financial backing of the king of Denmark. The king furnished him the island Hven as residence and place for research. Brahe was committed to inductive study and compiled a huge amount of data about the stars and especially the sun and planetary system. While Brahe read Copernicus, he held to the Ptolemaic system partly because of the churches' stationary view of the earth. But his observations led him to a sort of a partial view toward acceptance of Copernicus' view of the planets with circular orbits. He worked under a world view of acceptance of God as Creator but was somewhat influenced by Aristotelian pantheism and became involved in astrology for a time. Brahe chose Kepler as his assistant and successor, bequeathing to him all his data to compile and his instruments to use under the continued endorsement of the king. While Kepler was required as the king's astronomer to make predictions, he did so more from political and economic trends, having given up belief in astrology. He believed he could find logical reasons in nature since he was created in the image of the Creator.

Kepler's tenacious commitment to observational accuracy, his divinely guided belief in interpreting the mind of God in creation and his mathematical proficiency enabled him to make amazing contributions to science and Christian faith. Kepler was in search of the mathematical harmonies in the mind of the Creator. He established three laws: (1) the planets travel in ellipses with the sun in one focus; (2) the straight line from the sun to a planet (radius vector) passes over equal areas in equal times: (3) the squares of the periodic times which the different planets take to describe their orbits are proportional to the cubes of their mean distances from the sun. While Copernicus had described planetary revolutions around the sun from an ***earth perspective***, Kepler described the revolutions from the ***view of the sun***. He also discovered the ***elliptical orbits*** of the planets and worked out their very difficult mathematical formulas. Some of Kepler's ideas asserting the mystical harmony of geometry, music, astronomy, etc. were published in a book,

Harmonies of the World, which he called "a sacred sermon, a veritable hymn to God the Creator." He gave a certainty to the idea of the planets in orbit around the sun, confirming Copernicus, and published the most accurate tables for them and the stars. His planetary positions were considered thirty times better than those before. He clearly defined light as rays and laid the foundations for modern geometrical optics in his *The Optical Part of Astronomy,* defining parallax refraction and eclipse instruments. He applied these to the telescope in 1611.

In astrophysics Kepler formed a transition to those who followed. He accepted, perfected, and gave wide acceptance to Copernicus' work, and later, according to Galileo who solved the problem of dynamics of the motion of heavenly bodies, Kepler was the first great astronomer to accept and encourage him. Without Kepler's work the synthesis achieved by Newton would never have been possible. Newton introduced his great *Principia Mathematica* to the Royal Society as a "mathematical demonstration of the Copernican hypothesis ***as proposed by Kepler***."[8]

Kepler spent hours studying the Scriptures and while he considered the Bible the guide for everything, accurate science instead of Greek philosophy was to be used in properly interpreting God's work in creation. He said, "Nothing about nature can be known completely except quantities or by quantities. And it so happens that the conclusions of mathematics are most certain and indubitable."[9] He thus confessed that the power behind the nature he described was God's. By searching the Roman and Greek records along with Hebrew and Babylonian calendars, he discovered a mistake in the Latin calendar by which he corrected chronological errors Roman Catholics had given Christian events. He was the first to calculate that Jesus was actually born in 4 B. C. He wrote a thesis defending the accuracy of Luke's gospel. After much study he concluded that some Lutheran teachings were wrong and refuted the idea of Christ's physical presence in the communion elements, becoming Calvinistic in believing in Christ's spiritual presence. For this he was excluded from his beloved Lutheran churches and they refused to accept him when he repeatedly sought admission, even when he was famous in his last years.

Kepler was born prematurely into a dysfunctional family. He suffered from ill health and grief—losing to death two wives consecutively and nearly all his children, all of whom he deeply loved. He defended and gained acquittal from his mother's charges of witchcraft. Despite these difficulties, he had a jovial spirit and accepted suffering joyfully understanding Christ's suffering for him, believing in God's providential working in all things. He was a Christian humanist, appreciating all of creation.

Galileo Galilei (1564-1642), Astrophysics

Galileo was a man God endowed with a mathematical bent for physical astronomy and a gift for inventive inductive study and the ability to combat inaccurate ideas. His work became a turning point for objective scientific study. His God given gifts of persistent argumentation, which Copernicus lacked, enabled him to move astronomical and physical science away from the Catholic pagan Aristotelian view of motion established by university intellectuals. He moved physical science to the astronomical regular working of nature of the Copernican theory. Kepler was the bridge to observation mathematical application but Galileo completed the Copernican theory of motion by math and physics, and was able to communicate the reality of it to the public by his advancement in telescopic observation. His ability to offer a persuasive literary argument helped him gain public acceptance of his theories.

[8] Hummel, *The Galileo Connection*, 77.
[9] Ibid., 76.

He was born into a noble family in Florence, Italy. Galileo's father was a talented musician and knowledgeable in music theory. His father held strong views and was involved with a professor in Venice in a conflict over publishing a *Dialogue on Ancient and Modern Music*. His father set an example of argumentative tenacity when he thought he was right. The son, Galileo, was a heavy-set passionate man who like his father was easily angered but quick to return to good humor. In time he gained the nickname, The Wrangler. His early education was in a Benedictine school and his ***spiritual aspirations*** were expressed in an early desire to join the religious order, which his father opposed. Galileo was musical but early in life became interested in Archimedes's science and the problems of motion. This strongly influenced him toward the inductive method and study of mechanics. In his college days he rejected and became ***antagonistic toward Aristotelian philosophy and his dogmatic professors.*** Conflict with that philosophy characterized his struggle for the rest of his life.

Italy was a conglomerate of states with varying independence, under the dominance of the Roman Catholic Church. The pope had virtually dictatorial authority by using religious threats. In states with a more independent rule, papal control was not complete. Italy was the most advanced country in Renaissance studies having thirteen universities compared to three in England and Scotland. (Bologna alone had 18,000 students.) These institutions were influence by the new humanism but were brought under the control of the Catholic Church and were committed to teaching the Aristotelian philosophy. The University at Pisa where Galileo attended the school of medicine was in a province that was more politically independent. It had a rather free spirit, but was under a closed authoritarianism of Catholic persuasion. The morals at Pisa were relaxed and later as a professor Galileo had a mistress who bore his children, although he professed faith in the Christian world view.

In science Galileo aimed at the primary characteristics of quantity and not those of quality, and focused on the primary ideas. He admitted to not understanding the qualitative powers concerning gravity or the powers of God in Creation. He would posit a mechanical theory and then perform experiments to prove it. He did not seek just data. He insisted on objective scientific experimentation and mathematical demonstration. He introduced "imaginary rotations of massive spheres and defined 'neutral' motions, an idea that eventually led to the concept of inertia in terrestrial physics."[10] "He was not a philosopher but a scientist; he did not propose a new theory of science but a new science as he laid the foundations for modern mathematical physics. Yet in doing so he pioneered a new path that ultimately led to a new conception of the scientific enterprise."[11] When it came to the stars and planetary heavens he admitted direct experiments were not possible but based his conclusions on analogy.

Upon learning of the new telescope, he invented one for use in spying ships. His had the capacity to sight an enemy ship much sooner than an enemy could see his. He finally made a twenty power telescope that offered amazing discoveries of the heavens. He found that the moon was not a perfect sphere and did not shine with its own light. He discovered there were mountains and valleys on the moon; he measured their height by their shadows. When sun spots were discovered he showed that they were immense clouds on the solar surface that changed. He discovered the planet Venus and showed that it had phases like the moon. He discovered Jupiter had moons that revolved around the planet and used this as an analogy of the earth revolving around the sun. He allowed prominent people who were not scientists to see these things through his telescope, and published his findings for the public to read. All of these findings disproved the

---

[10] Ibid., 85.
[11] Ibid., 102.

assumptions of Aristotle and the Ptolemaic model which with divine implications assumed perfection. He therefore popularized the scientific method and refuted the theory of movement and astronomy that the religious Jesuit astronomers had led the church to adopt.

Opposition to Galileo and Christian Science

At first the leaders of the Jesuit order accepted Galileo's findings. But in time they organized a movement against him and the Copernican theory. Astronomers of the Jesuit order, who were the primary teachers in the universities, opposed Galileo. They formed a group called the *Liga* and nicknamed it the "pigeons" (*columbi)* after the founder, Ludovico delle Columbi. These opponents first claimed Galileo's teachings were ***against logic*** and the argument of the earth as stationary. Then it was said Galileo's views were ***against the church's biblical authority***; after that they attacked ***his physics***; then they began to ***spin word definitions*** to refute him. Failing in these attacks the Liga took their efforts ***to the courts and accused him*** of being anti-religious. In the public mind Galileo's teachings became the watershed of the new scientific method and he became a symbol of intellectual freedom. He was accused of contradicting the Bible in saying that the earth moved, while the Aristotelian interpretation was that the earth was still. At the Council of Trent (1545-1563) a counter-reformation was launched against the growing Protestant churches' challenge and the *Index* of prohibited books was begun.

Galileo never rejected God and the church, but continued to work under that Biblical world view. He seemed to grow in his experience with God. After termination of teaching at Pisa, he gained an appointment in 1592 to the University of Padua where he had a mistress but for some reason would not marry her. She bore him two daughters and a son. But when he went to Florence in 1610 to become the Philosopher and Mathematician to the Grand Duke where he cared for his mother and family, his common law wife did not go with him and eventually married someone else. His two daughters entered a convent to serve God. The oldest, Virginia, was called Maria Celeste, had a close relationship with her father and his son helped care for him in his later years. Galileo's work was favorably accepted by Pope Paul V and Cardinal Maffeo Barberini who later became Pope. Galileo was encouraged by their openness to his ideas. In contrast was the opposition from his university professors who held to Aristotle's views. In his battles he increasingly prayed and asked his friends to pray for him. The encouragement of Johan Kepler, the brilliant biblical Protestant also must have impressed him. After publishing his slim volume, The *Starry Messenger*, he sent it to Kepler for comments. Kepler, in turn, published a pamphlet, *A Discussion with the Starry Messenger* extolling the work of Galileo which was clearly an acceptance of the Copernican system. These two works made Galileo famous and spurred interest in the telescope and did much to advance the Copernican views.

Galileo understood the conflict. Even though he had sought and gained the approval of many in the hierarchy, as well as the papacy, he knew the Copernican theory hit at the heart of the intellectual elite's teachings which gave them power with the church. Maintaining the ***political power of the educated elite was the real issue***. Galileo tried to make a distinction between the phenomenal language of the Bible which was understood by the common man, and the efforts of science. He argued there were two books: the Bible that revealed God, His relationship to man and (the means of) salvation, and God's work in nature. He argued that the interpretations of creation should be left to the findings of science which required experimentation and demonstration, and the Bible should be the source for theologians.

"Galileo was pressing less for ecclesiastical approval of Copernicanism than for the right of science to develop in its own domain free from the authority of theology. Yet he saw no need for a breach between science and theology since ***God is the Author of both*** books—of nature and of

Scripture."[12] He insisted that a scientist must provide "conclusive demonstration" of his hypothesis before the theologian interprets. But Galileo wrote, "Yet even in those propositions which are not a matter of faith, this ***authority for the Bible ought to be preferred*** over that of all human writings which are supported only by bare assertions of probable arguments, and set forth in a demonstrative way."[13] Elsewhere he said, "The Holy Bible can never speak untruth—whenever its true meaning is understood."[14] When there seems conclusive demonstration about creation, theologians must judge how to interpret it and reconcile it with the Bible. Galileo knew that the ***real problem was that of struggle for political power*** "and not a simple conflict between science and religion so commonly pictured." He indicated this in the prologue of his book *Dialogue*.[15] The Dominicans, to whom Thomas Aquinas belonged, backed the Jesuit intellectual elite of the universities. After a protracted series of debates and trials Pope Urban VIII, a friend of Galileo's, finally gave into the anger of the academic elite leaders, like Pilot to the Sanhedrin. He began to oppose Galileo's teachings.

Santillana said in his assessment of the controversies in the church courts,

> In reality it was a confused free-for-all in which prejudice, inveterate rancor, and all sorts of special and corporate interests were prime movers... It has been known for a long time that a major part of the ***church (Christian) intellectuals*** were on the side of Galileo, while the clearest opposition to him ***came from secular ideas***... The tragedy was the result of a plot of which hierarchies themselves turn out to be the victims no less than Galileo–and intrigue engineered by a group of obscure and disparate characters in strange collusion.[16] (Emphasis added)

After reviewing the facts, Hummel himself also concluded,

> The real authoritarianism that engineered Galileo's downfall was that of the Aristotelian scientific outlook in the universities. Only after Galileo had attacked that establishment for decades did his enemies turn their controversy into a theological issue. Even then it was ***the natural philosopher*** who worked behind the scenes with pliable church authorities to torment Galileo's trial, and finally to rob him of the reasonable solution worked out by the Inquisition.[17] (Emphasis added)

The power of the intellectuals to control the church and society with their pagan views was the real issue. This is the bias of original sin. They could not let Galileo win, for they had staked their whole power on Aristotle's false philosophy of motion.

Robert Boyle (1637-1691), Chemistry, Air, et al.

Robert Boyle, born of Irish nobility, was brilliant already as a child. At the age of 14, while traveling with a tutor, stayed for a period in Italy and studied the science of Galileo. Early on he was involved with others of like interest in a group that called themselves "the invisible college." He resided at Oxford and while working with Robert Hooke developed the pneumatical engine or air pump. He became involved in experiments with air, published writings on properties of air, and stated Boyle's Law that the volume of a gas varies inversely as the pressure. He discovered the

[12] Ibid., 119.
[13] Ibid., 107.
[14] Ibid., 105.
[15] Ibid., 116.
[16] Ibid., 123 from Santillana, *Crime of Galileo*, xii, xiii.
[17] Ibid., 122, 123.

effect of pressure on the boiling point of water, explained heat as a "brisk" agitation of particles. He investigated the relationship of the speed of sound to air and studied the expansive force of freezing water, studied specific gravities, refractive powers, crystals, electricity, color, hydrostatics, combustion, respiration and even conducted experiments in physiology. Most importantly he studied chemistry. He prepared phosphorus and collected hydrogen in a vessel over water. He distinguished elements, mixtures and compounds. He began to understand that matter was ultimately composed of "corpuscles" of various sorts and sizes capable of arranging themselves into groups, and that each group constituted a chemical substance. He thus prepared the way for modern chemistry. After Boyle's death, others placed on his epitaph, "The Father of Chemistry" and he is so considered by many today. He did much to discredit alchemy toward which Newton was inclined. Boyle sought a union between chemical practitioners and natural philosophy and rejected the old Greek foundations. Joseph Freind of Oxford (1712) gives Boyle credit with much concentration on the processes of combustion, calcinations, and respiration. Boyle and contemporaries came near to discovering gases and especially the suggested existence of oxygen and the discovery of nitro-aerial particles like gunpowder.

Based on his own findings and the ideas generated by 'the invisible college,' Boyle wrote the proposal for what eventually became the Royal Society of London for Science. That also became the model for the French Academy etc. He was on the council and was once offered the presidency of the Royal Society by the king, but turned it down.

Boyle was a highly committed Christian believer. He studied Hebrew and Greek in order to study the Bible, spent large sums of money on Bible translations, founded the Boyle lecture for proving the Christian religion against "notorious infidels"... He was so knowledgeable in theology he was once offered the office of provost of the theology school of Eton, but turned it down because he would not agree to subscribe to orders or take an oath, the same reason he turned down the presidency of the Royal Society.

John Ray (1627-1705), Biology

Ray was a professor at Trinity College, Cambridge but gave his office up for the freedom to study nature. Ray studied both plants and animals extensively. He distinguished monocotyledons and dicotyledons as two groups of plants and evaluated their fruit, flower, leaf and other characteristics and began organizing classifications. His observation of animals led him to conclude that fossils were shells or bones of formerly living things, giving direction to later studies of Cuvier and others. His published books include: *A History of Plants, a History of Insects* and *Synopses of Animals, Birds, Reptiles and Fishes.* He was the founder of biological sciences. While he believed that regular natural causes ought to be studied by observation and sought to eliminate superstitions in biology, he also remained a churchman and wrote *The Wisdom of God in the Works of Creation.*

Sir Isaac Newton (1642-1727), Principles in Astronomy—Solar System

Newton is called "the most illustrious name in the long role of science."[18] Using Kepler's research, Newton conceived the solar system as movements that could be mathematically calculated by gravity to agree with the observations. Newton, in communication with Haley in 1687, wrote the *Principia, Mathematical Principles of Natural Philosophy*, a book presented in geometric form that by most is considered "the greatest book in the history of science." As William Cecil Dampier has said,

[18] William C. Dampier, A Shorter History of Science, 68.

> The heavenly bodies...were brought by Newton within range of man's inquiry, and shown to move in accordance with the dynamic principles established by terrestrial experiments. Their behavior could be deduced from the one assumption that every particle of matter attracts every other particle with a force proportionate to the product of the two masses and inversely proportional to the square of the distance between them. The movements so calculated were found to agree accurately with the observations of more than two centuries.[19]

Hailey found that comets also moved and could be tracked according to gravity and its principles. Whereas Galileo had never found the answer to the tides, as a proof against Aristotle's view of motion, Newton produced a sound tidal theory. It seems he devised some new mathematical methods, chief of which was his method of fluxions—finding the rate of one variable x with another y.

The concept of mass as inertia, distinct from weight, is implied in the work of Galileo. "Newton defined mass as 'the quality of matter in a body as measured by the product of its density and bulk', and force as 'any action on a body which changes, or tends to change, its state of rest, or of uniform motion in a straight line'."[20] He showed that mass and weight must be proportional. Newton seems to have had an amazing intuition which enabled him to see the solution of a problem; sometimes his proofs may have been merely helps to weaker minds.[21] He also made advances concerning light. He showed that white light is made up of light of different colors, differently refracted by passing through the prism, the most refracted being violet and the least red. Much of his work prepared the way for future use of light in astronomy. Newton said he dealt with ***how*** things appear to work and not ***why.*** He said that science deals with appearances and not necessarily with reality. Newton clearly distinguished between a successful mathematical description of natural phenomena and a philosophical explanation of their causes. "Newton like Galileo accepted the atomic theory which could not then be put in numerical form. Newton and his immediate followers regarded the new discoveries as a revelation of the power and wisdom of God. He said,

> The most beautiful System of the Sun, Planets, and Comets could only proceed from the counsel and dominion of an intelligent and powerful Being.... God...endures forever and is everywhere present, and by existing always and everywhere, he constitutes duration and space. All these things being considered, it seems probable to me that God in the Beginning formed matter in solid, massy, hard, impenetrable, moveable particles, of such Sizes and Figures, and with such other Properties, and in such proportion to space, as most conduced to the End for which he formed them.[22]

From these statements Dampier concluded that, "Thus to Newton, God is not only a First Cause, but is also Imminent in nature." As Hummel pointed out, Newton did not hold to a deistic view of God as a principle or "watchmaker" but as the God of the Bible and a personal God to be worshiped.[23]

According to Hummel, Newton published little about religion.

---

[19] Ibid., 70.

[20] Ibid., 71.

[21] Ibid.

[22] Ibid., 76.

[23] Hummel, *The Galileo Connection*, 144.

> Yet during his life Newton spent more time on theology than on science. He wrote about 1,300,000 words on biblical subjects. This vast manuscript legacy lay hidden from public view for two centuries until the Southeby sale of the Portsmouth Collection in 1936. ...Newton's religion was no mere appendage to his science; he would have been a theist no matter what his profession. His understanding of God came primarily from the Bible over which he pored days and weeks at a time.[24]

Hummel goes on to explain how Newton, like Kepler, did extensive checking on the historical accuracy of events and teachings of Scripture. While Newton did not want biblical revelation brought into the scientific discussions, for he like others of the period saw God's book of Revelation in the Bible and God's working in nature to be found in science. But his Christian faith influenced his science. Newton accepted the miracles, was a faithful member of the Anglican Church all his life. His faith was never questioned. Like Boyle he gave money to distribute the Bible and to help the poor. However, Newton had a leaning toward Arianism, without feeling he was unbiblical. "He recognized Christ as a divine mediator between God and humankind, but subordinate to the Father who created him. Christ had earned the right to be worshiped (though not on a par with the Father) through his obedience and death."[25] The mystery of the Trinity was an unresolved struggle for him. Others extended, but did not alter Newton' mechanics. Joseph L Lagrange (1736-1813) of Italy, Germany and France transformed Newton's mechanics into a branch of analysis, and William R. Hamilton (1805-1865) of Ireland applied Newton's second law to optics and astronomy.

Development of Atomic Physics, Chemistry, and Electro Magnetism

Greek Atomic Theory and Aristotle's Error

The Greeks held to the atomic nature of the universe. Scholars in the Ionian school (ca 460 B.C.) such as Leucippus of Abder in Thrace, Empedocles and Democritus believed in atoms—Greek for invisible. Aristotle later rejected these theories, but Epicurus, in his extreme naturalism, embellished the idea and it was passed on centuries later to his follower, the Roman Lucretius from whom Kant got his inspiration. In the Arab science that preceded the Reformation science, Arabs had held that Allah created atoms from moment to moment.

The science of chemistry and physics began in the seventeenth century but fully developed in the transition period from Reformation scientist to the beginning of uniform natural determinism in the mid eighteenth century. Many of the founders were genuine Christians. Important searches were made of minerals and significant efforts were made for chemistry. This science was developed significantly in the transition period when modern Christian science was used, but under attack by anti-God attitudes and views that were arising. Boyle and Newton had accepted the idea of atoms in their physical speculations.

The breakthrough in understanding chemistry was delayed because of the ***phlogiston theory***. This theory was derived from Aristotle who believed that fire was the element released during combustion, implying ***loss of weight or substance***. Nothing that could be burned could be an element. G. E. Stahl (1731) gave the theory the name, "philogiston," regarding it as an actual physical substance. Around 1750 the theory occupied prominence. It was associated with pharmacology and later as physics. Air was not thought of as a mixture although having various fumes. But when burning occurred or metals calcined there was an action of increased weight.

---

[24] Ibid., 142-143.

[25] Ibid., 144.

This was known by Arabic scientists, observed by Boyle, and known by the Royal Society in 1660, which was the inversion of the phlogiston theory. Boyle sought to explain the weight gain because of the acceptance of the idea of phlogiston.

Breakthrough from Aristotle to Modern Atomic Theory

The Greeks had speculated about atomic theory and years later the Roman poet Lucretius had a somewhat philosophical, not scientific idea, of such a basis to the material world. The phlogiston theory was ancient Greek tradition, and the accepted theory barring the discoveries of chemistry until it was disproved. This view began to be disproved after reigning about a quarter of a century.[26] Joseph Priestley and Swedish chemist Scheele, both isolated oxygen in the early 1770s. But it was ***Antoine L. Lavoisier (1743-1794)*** who surveyed the study of air and gases and in 1775 published a paper, *The Nature of the Principle that combines with Metals in Calcination and that increases their Weigh.* In 1783 he openly attacked and broke through the false theory of phlogiston. He launched the identification of chemical elements by proving that a chemical compound, however obtained, is always made up of the same amount of constituent parts. Lavoisier was from a prominent family, gained a good education, and associated with brilliant scientists, including Pierre Laplace. He was honored by the king and held leadership positions and was not supportive of the French intellectuals in the revolutionary efforts. He was condemned to death and guillotined in 1794. But it was not until the atomic theory was developed that chemistry, light and energy came together in what is now considered physics.

But it was ***John Dalton (1766-1844)*** who explained the properties of gases by atoms and found combined weights in chemical combinations. He gave relative weights of the atoms and drew up a list of 20 such atomic weights. Being from a poor background, Dalton struggled for his knowledge. He was a strong Christian Quaker who attended church twice a week and was a leader in Quaker councils. His work was carried to fruition in the last of the nineteenth century when a brilliant Russian scientist, Mendeléeff (1834-1907) arranged the elements in the ascending order of atomic weights with a periodicity table of the stable ones. Since then many unstable ones have been found. With the findings of Mendeléeff several researchers including Prout suggested that all elements were composed of hydrogen, but proof was for a later time. New chemical elements artificially created in a laboratory which are unstable have continued to be found. Russians recently created ununseptium with 117 proteins in the nucleus and is the 26th artificial element heavier then uranium that has been added to the periodic table.[27]

These findings in inorganic chemistry of the elements soon led to the discovery of the importance of carbon, and to organic chemistry with carbon, and attachments of hydrogen, oxygen and other elements in different amounts and ways. Some discoveries follow. Berzelius (1779-1848) discovered the differences in connections between atoms in the molecules. Discoveries by Kekulė (1863)—benine, W. H. Perkin, Sr. (1856) and E. and O. Fisher (1878) found the basis of dyes was triphenylmethane, Mitscherlich (1844) the structural formula of isomers of tartaric acid, Pasteur found two varieties of crystal related to each other as a right hand to a left hand. Wislicenus (1863) and LeBel and Van't Hoff showed atoms in molecules of the varieties were arranged differently. As many organic compounds were isolated and better understood, Lieberg grouped them as member or derivatives in three great classes. All this led into the understanding of the differences in Crystalloids (e.g. sugar or salt) and Colloids (protoplasm of living cells) that

---

[26] Herbert Butterfields, *The Origins of Modern Science*, (New York: Collier Books, 1962) cf. Chapter 11, "The Postponed Scientific Revolution in Chemistry," 203 ff.

[27] *The Economist*, April 10th-16th 2010, 9.

are longer and have two phases synthesizing of organic chemicals. All this led much later to the creative work of today of pharmaceutical discoveries, and to molecular understanding.

Michael Faraday, (1791-1861) made discoveries of the physics of electricity, magnetism, and radiation forces, especially of light, that united physics from the micro studies in the laboratories to the macro stage of astronomy. Sir Humphry Davy (1778-1829) discovered other elements by electric currents. Michael Faraday arose out of poverty and little educational opportunity, to break through the mysteries of electricity, magnetism, and radiational waves of light. Faraday's mentor was Davey and through him he drew on the work of Andrė Marie Ampėre who showed that parallel currents moving in the same direction attract each other and oppose if going opposite. Faraday profited from Alessandro Volta's discovery of the decomposition of water by electric current and of forming a storage battery and measure of volts. Faraday assimilated and composed previous knowledge and showed all forms of electricity were the same-static, chemical, thermodynamic and showed the relationship to magnetism. He isolated chlorine by electrolysis, measured the quantity of electricity, invented the first transformer, discovered inductance, and created the first primitive dynamo. He gave terms commonly used in electricity – e.g. electrolysis, electrode, anode, and cathode. He gave credit to others—volt as measure of force, ampere as measure of current per second. Faraday established laws of electricity and His work paved the way for other great scientists such as James Joule, James Clark Maxwell, and Lord Kelvin, that led to the laws of heat and energy and the idea of conservation of energy in work and the second law of thermodynamics. Without Faraday's work, all the efforts with cyclotrons and other aspects of modern physics would be impossible. He became the head of the Royal Institution and was offered the presidency of the Royal Society, which he refused. In this time of growing skepticism he was a lay preacher of a Presbyterian offshoot called Sandemanians who believed the Bible as the basis of all things.

Galileo and especially Newton had studied light and Newton suggested light was a wave that was near the modern view. ***Francesco Maria Grimaldi (1618-1663)*** was a devote Jesuit priest, and like Galileo he believed the interpretation of Scripture was at stake.[28] He showed that light was a wave by shining it through a pinhole. He showed that it bended around an obstacle, which particles could not do. But it remained for ***Thomas Young (1773-1829)*** to give a break through. He was brilliant in math and learned about 12 languages – Semitic, Egyptian and Greek and Latin et al. He was prominent in deciphering and publishing the text of the famous Rosetta stone. He studied, practiced, and became a professor of medicine. He made discoveries about the eye which he knew was the source to understand light, which was his principle interest. He presented a paper to the Royal Society showing that the lens of the eye changes to focus on things at various distances. He also showed that the retina contains nerve endings that respond to the primary colors and explained that astigmatism was a result of a curvature of the cornea. He showed that light that passed through two slits formed wave patterns on a screen, which vindicated the earlier work of Grimaldi. ***James Clark Maxwell (1831-1871)*** was a Christian, an associate with Michael Faraday. He serviced as a Christian lecturing in F. D. Maurice's Christian schools for the poor and serving in other ways. He gave Faraday's work mathematical bases. Maxwell postulated the theory of electromagnetic wave, and discovered radio waves bringing much of physics together. He showed that electromagnetic waves moved at the constant speed of light, expanded on the spectrum of light colors (earlier begun by Melville and W. Herschel), and many other scientific findings. Many consider him the father of modern physics.

---

[28] Dan Graves, *Scientists of Faith*, 54.

By the last forty years of nineteenth century many Christians had turned to Deism and Pantheism and there was a hostility to Christianity. Maxwell was hesitant to use his work to promote Christianity, but he was recognized as a devote Christian, who prayed much and expressed in a letter to his wife that a scientist in union with Christ should do his work to benefit the body of Christ.[29] But most atomic theory progressed from Maxwell's work even though these scientists were not known Christians and while mentioned here will be discussed later. His successor Ernest Rutherford along with others investigated chemical radiation of radium, uranium, and other chemicals which led to the theory of radioactive decay being discovered. Rutherford identified the alpha and beta forces, and developed the theory of protons and the nuclear theory, seeing the atom similar to a solar system with a nucleus and electrons revolving around. In 1911 Rutherford's work led to the breakthrough of Niels Bohn and the indeterminacy of the Quantum Theory, and was used by Einstein in his theory of relativity. Einstein considered Maxwell next to Newton in physics.

These great scientist pioneers also experimented with gases and light that pointed to later developments. In the eighteenth and nineteenth centuries, when the European continent was experiencing the rejection of God, England went though hard times and a spiritual renewal. In the last of the century were the revivals of Dwight L. Moody and the birth of the Student Volunteer Movement. Many of the scientists working in atomic and electro-magnetic theory held a Christian world view. Dan Graves said,

> It has been often remarked that the nineteenth century British physics was completely dominated by Christians. The recognized giants of scientific advancement ... were also men of Christian faith....[30]

Conclusions from First Period of 275 Years of Founding Modern Science

The rebirth of faith in God and a new Christian humanism grew out of resistance to the oppressive power of the Catholic Church. With this came the rejection of Aristotelian pantheistic philosophy that removed the transcendent God and endowed nature with design and power of movement. The recovery of the Scriptures, of salvation by grace through faith, and the rediscovery of inertia in nature from the Creator by science, refuted Catholic Aristotelianism.

Today many of the most knowledgeable scientists acknowledge that the foundations of modern science were laid in the first three centuries after 1500. In 2002 Stephen Hawking, who is considered the world's leading astrophysicist, published a book, *On The Shoulders of Giants: The Great Works of Physics and Astronomy*.[31] This book is 1266 pages, which Hawking considers the basis for modern science. All but 98 pages (that are devoted to Albert Einstein and two at the end of the book to the editor) Hawking refers to scientists who did their work before 1727 A.D. ***All of these*** men ***were devoted believers in God*** and felt they were discovering the work of God in nature. They all ***accepted the Bible*** as the word of God to guide them and believed in the authority of the ***transcendent God who created the world and revealed Himself.***

[29] Ibid., 153.

[30] Dan Graves, *Scientists of Faith*, 106.

[31] Stephen Hawking, *On The Shoulders of Giants: The Great Works of Physics and Astronomy,* (Philadelphia: The Running Press, 2002).

PART II

THE BIASED PERVERSION OF NATURAL REVELATION

The Gospel "is the power of God for salvation to everyone who believes."… For the wrath of God is revealed from heaven against all ungodliness and unrighteousness of men, who by their unrighteousness suppress the truth. For what can be known about God is plain to them, because God has shown it to them. For his invisible attributes, namely, his eternal power and divine nature, have been clearly perceived, ever since the creation of the world, in the things that have been made. So they are without excuse. For although they knew God, they did not honor him as God or give thanks to him, but they became futile in their thinking, and their foolish hearts were darkened. Claiming to be wise, they became fools, and exchanged the glory of the immortal God for images resembling mortal man and birds and animals and creeping things. Therefore God gave them up in the lusts of their hearts to impurity, to the dishonoring of their bodies among themselves, because they exchanged the truth about God for a lie and worshiped and served the creature [creation] rather than the Creator, who is blessed forever! Amen (Romans 1:16, 18-25)

Chapter 3
Doubts of Christian Religion and Growth of Faith in Human Wisdom (Seventeenth Century)

Secular Humanism in Exalting Man Moved to Ascendancy

As previously mentioned, in the seventeenth century a true Christian humanism emerged in Europe. The Roman Catholic Religion, a major influencer of thought and culture, was helpless in the face of the fourteenth century's Bubonic Plague (Black Death) which had eradicated about one-fourth of Europe's population. That, along with the corruption, immorality, and division within the Roman clergy, diminished the common man's confidence in the church. The advent of the printing press in the fifteenth century promoted freedom of thought, as did the availability of the original Greek Scriptures that had been regained from Muslims. These factors, along with the Reformation's renewed interest in God's book of creation led to a growing number of scientists holding to a Christian world view—first within the Roman Catholic Church such as Leonardo da Vinci, Copernicus, and Galileo. Later, Protestants Kepler, Newton, and Boyle, rejected the Aristotelian views that dominated the Roman Catholic Church.

At the same time other intellectual thinkers were leaning toward a ***secular humanism***. Men like Frances Bacon and Rene Descartes, two prominent secular humanists, professed faith in God, but they really believed in man and his power of reason more than the Creator. The scientific methodology revealed and extensive regularity in the operation of nature. This provided an opportunity for the bias mind to see this only as a machine that worked without God. This bias was encouraged by disappointment in religion. In the first half of the seventeenth century a new outbreak of plague in London, in which more than 150,000 died with no solution, combined with religious wars and their accompanying deaths and devastation to create a desire for something more. Confidence in man's secular knowledge found opportunity to assert dominance in the seventeenth century.

Butterfield described the urgency of the times.

> The early seventeenth century was more conscious than we ourselves (...as historians) of the revolutionary character of the moment that had now been reached. While everything was in the melting pot—the older order undermined but the new scientific system unachieved—the conflict was bitterly exasperated. Men were actually calling for a revolution–not merely for an explanation of the existing anomalies but for a new science and new method. ...They seemed to be curiously lacking in discernment ...for they tended to believe that the scientific revolution could be carried out entirely in a single lifetime.[1]

Two Viewpoints of Man Claiming Capability of All Knowledge: Francis Bacon (1561-1636 by induction) and Rene Descartes (1576-1650 deduction)

Francis Bacon and Rene Descartes are primarily responsible for the view of knowledge that relies on man's individual observations and thinking, rather than on God's revelation in creation and special revelation. While both preferred to accept the prevailing world view that there is a God, both began with a ***confidence in man***. It is curious that working individually they started at different approaches of the human mind to find a method for solution. In the first quarter of the century the Englishman, Francis Bacon, glorified the ***inductive*** method of hypotheses from observation. A hypothesis could be made and proven, or made and other scientists, following the

[1] Butterfield, *The Origins of Modern Science*, chapter 6, Bacon and Descartes, 108.

inductive method, could improve it and arrive at the big picture. Accumulating data was secondary to intuitive observation. Bacon's inductive method of connecting mind with nature was almost a law in itself. If man places his mind on the right road, it "would proceed with unerring and mechanical certainty to the invention of new arts and sciences."[2] By inductive arrival at a hypothesis, another hypothesis would be found and eventually all nature could be understood. Deduction followed ***induction***.

Descartes, in Holland in the second quarter of the century, focused on the ***deductive*** and philosophical mode of reasoning. Beginning with self and God as a logical certainty, Descartes reasoned from God downward with a strong emphasis on mathematics for solutions to a theoretical idea. While Descartes saw the value of induction, he preferred to apply it on the later end of knowledge. Descartes' was a ***method of explanation*** of primary or ordinary phenomena to lead to the discovery of later phenomena.

Both Bacon and Descartes wanted to start over with a clean slate seeing past knowledge and method as all wrong, and by complete critical ***doubt*** arrive at correct and comprehensive knowledge: Bacon started from the bottom up by induction and Descartes from the top down by deduction which coincided with inductive proofs. Descartes thought that in his lifetime he could see the whole picture or arrive at all basic principles. Today Stephen Hawking shares that same hope for finding the "mind of God." Bacon thought that there were a very limited number of principles in the universe and these all could be discovered in a couple of decades by using the right methodology. Both believed they could arrive at the mind of God to control the world. Bacon actually saw his quest as a restoration of man's ability to control the world that was lost at the fall of Adam. He saw that man's past history of this quest to control the world had made only a little progress a couple of times in human history. Bacon saw his method as unique and it would accomplish this easily and quickly.

Both Bacon and Descartes displayed an arrogance that led them to think they were created to help man save himself. They felt called to ***rewrite history for man***, so that mankind who had been so slow at making progress could be set on the path to understand and control all nature. Bacon's error was in seeing the cosmos as static instead of dynamic and in process. He rejected the Copernican theory. For Bacon, only data in use, not accumulated, was considered valuable. He placed little value on the geometric and mathematical. Descartes relied on philosophical reason as being accurate and correct while at the same time he dismissed all the Greek philosophers as failures. He relied heavily on math. Both began by doubting all knowledge but their own individual mind, believing man could come to or eventually understand it all.

Biographical Background Leading to Their Two Views

Francis Bacon was wealthy, highly educated, and knowledgeable about the history of science of the time. A powerful man, he became the Lord Chancellor of England. He described himself as having a "mind nimble and versatile enough to catch the resemblance of things."[3] His underlying error was his arrogant optimism about his mental ability. As Butterfield said, "The root of the error in him lay in his assumption that the number of phenomena, the number even of possible experiments, was limited, so that the scientific revolution could be expected to take place in a decade or two." Bacon once said, "The particular phenomena of the arts and sciences are in reality but as a handful."[4] He was under the demonic delusion that the educated elite could save mankind.

Rene Descartes, commonly called "the father of modern philosophy," articulated this new

---

[2] Encyclopedia Britannica, vol. 2, (Chicago Encyclopedia Inc., 1953), 885.
[3] Ibid, 882.
[4] Butterfield, *Origins of Modern Science*, 116.

way of thinking from a point of ***doubt about everything*** except self. He had rebelled against his Jesuit teachers in the Roman Catholic Church, who in typical scholastic methodology quoted many conflicting authorities. In disgust he entered the military. Feeling that he was only increasingly discovering his own ignorance, he became skeptical of all knowledge. He said, "I had learned what others had learned ...and I did not feel that I was esteemed inferior to my fellow-students. ...And this made me take the liberty of judging all others ***by myself....***"[5]

On November 10, 1619 while near Ulm on the Danube, Descartes fell into a deep depression, filled with darkness and doubt about human knowledge. At this time he had a burst of insight. The method of analytical geometry, starting with the simple and obvious and reasoning to the complex, was the way to learn, he posited. This was followed by a series of dreams that encouraged him. Bypassing experience from inductive reasoning, he came to the basic fact, "I think, therefore ***I am***." He held this to be true even if his thinking was faulty or he was dreaming. He went on to construct his dualistic view of the mind and the world *by* ***deductive reasoning***. He reserved a place for God, saying "I could think of God beyond which I could think of none greater." Therefore God must exist. God existed in Descartes' reason. He thus established the ***authority of the individual*** in deductive reasoning, as scientific thinkers had by inductive reason.

G. Voetius of the University of Utrecht suggested this view of Descartes was practically atheistic, although Descartes denied this. Voetius was right in that the acceptance of the idea of God depended on Descartes' individual judgment at the time and was not based on man's innate sense of objective truth in nature or revelation given to God's people in the Bible. Even though Bacon and Descartes included the idea of God in their thought processes, they established themselves as men with the standard of good and evil to control the world. This is the original sin nature of Adamic thinking. The Scriptures say that even the Devil believes in his mind. But neither Bacon nor Descartes displayed a humble trust in God.

These men established truth as dependent on their ***individual self*** and what they reasoned was right, one primarily by induction, and the other by deduction or reflection. In essence, they had assumed they could decide good and evil equally with God to determine knowledge. Many educated men tended to believe ***individual self knowledge to be deterministic!*** Many, including John Locke, were looking for a way to exclude God from the rational world.

John Locke (1632-1704), Limiting Individual Man's Knowledge

John Locke was born to a lawyer who was also a small landowner. He studied and taught at Oxford and practiced medicine for a time. He lived during the influence of Francis Bacon, Thomas Hobbes, an atheistic mechanical materialist, and strong Christian scientists like Robert Boyle. He also was negatively influenced by the Puritans, especially John Owens. For fifteen years he lived with, was political advisor to, and physician and tutor for Lord Ashley of Shaftsbury. He was implicated with Ashley in a conspiracy to assassinate the king and fled with him to Holland where he gained intellectual stimulation from Descartes and Spinoza. Following a rebellion in England in which King James II was removed, Locke returned to England where he held a position on the Board of Trade. These experiences caused him to react to oppressive leaders.

Locke's Philosophy

John Locke, who wrote near the end of the seventeenth century, reacted to all authority. He doubted only what he himself learned. He loved independent and critical thinking. He wanted freedom in the state and in religion. Locke studied and was repelled by scholasticism. He reacted

---

[5] This and the following quotes are taken from the biographical note of Descartes in the book by Alburey Castell, *An Introduction to Modern Philosophy,* (New York, Macmillan Co., 1950), 95-109.

to and refuted the claims of papal infallibility. While in Holland Locke was influenced by the popular Descartes and Spinoza. Spinoza rejected the Mosaic Pentateuch, the first five books of Scripture, and its God. Spinoza taught an idealistic pantheism and was expelled from his Jewish synagogue. Locke was influenced by this criticism. He was offended by extreme Calvinism in Holland and later by the extreme efforts of control by English Presbyterians. As a result Locke's teaching embraced the individualism of Bacon and especially Descartes. He sought to move the pendulum away from the sovereignty of God to the sovereignty of the individual. Locke's essays on toleration and government later influenced and became the basis for Thomas Jefferson's efforts in the *Declaration of Independence.*

He brought together the deductive reasoning of Descartes with the inductive science advocated by Bacon. He described man's mind at birth ***as a blank slate*** and real knowledge as coming only through sense perception. ***No ideas are innate***. Sensation and reflection are the operation of minds. He too, like Descartes advocated starting from ***simple primary ideas and reasoning to the complex.*** Locke believed that man began with a blank slate of mind and developed it by sense perceptions and reflection. Locke proposed that a child at the moment of its birth had no knowledge and step by step the child progressed in all its understanding. He saw that the understanding of beasts was limited and therefore man's continuous learning set him above them.

Locke, whose views on education profoundly affection Western education, devised four classifications of knowledge. To Locke only the first category is coextensive with our human perception. The fourth is *independent of our perception* and has to do with God, heavens, and any ontological ideas. Bacon and Descartes gave him the idea of the independent self as the ultimate authority. Descartes influenced his idea of basing religion on reason from nature (naturalism); Spinoza influenced his criticism of the Bible.

While Bacon, Descartes, and Spinoza centered all knowledge in the independent self, Locke was the first to take the bold step to arbitrarily draw a line between man's knowledge and God, putting a wall, or separation between God and the world. Locke denied man's innate ability to know God in creation and denied man had any ability to communicate with God or demonic forces. He recognized no innate sense of morals. Morality depended on man's ability to reason. Although he professed acceptance of Biblical revelation and Jesus Christ as a result of his own sound logic, Locke based his belief in God on "natural religion." At the core, Locke believed, each man should base his religion ***on his own individual judgment by reason,*** as had Descartes. He believed men would logically come to Christianity as the reasonably superior religion, but there should be ***tolerance because there is no certain revelation***. Thus in the name of tolerance, he established intolerance for any claim of uniqueness which Christianity makes. Locke did not profess deism, the idea the world is like a big clock or machine which God made and let run, but he was a primary influence and laid the foundation for deism and naturalism in his book, *Reasonableness of Christianity*. Locke's influence continued in developing a concept of knowledge that exists today.

According to Locke's view of knowledge man was cut off from responsibility to God. Since he held there were ***no innate moral principles***, ethics come from experience based on ***self-preservation and the laws of happiness***. Hence these ethics are not universally applicable except as individuals agree. This view set the stage for moral relativism and opened the door to anarchy, a charge Locke would deny. He made the statement that "the taking away of God dissolves all,"[6]

---

[6] Encyclopedia Britannica, Vol. 14, 272.

deceptively holding that this would remove morality. But his idea of individual freedom led him to say, "A man cannot determine arbitrarily what his neighbor must believe."[7] Individual reason would be the only valid authority. Revelation about God and the Bible were acceptable only if the individual judged that they agreed with the standards in the laws of nature as the individual interpreted them. While denying absolutes, he professed the Scriptures were infallible ***because they were the most reasonable to him.***

In 1695 Locke wrote *In Reasonableness of Christianity as Developed in Scripture.* He "sought to separate the essence of the teaching of Jesus from later assertions." He also sought to remove mystery or supernatural in *A Paraphrase on the Notes on the Epistles of Paul* and in his *Miracles.*[8] Locke sought to conform Christianity to empirical reason and to remove the supernatural. He redefined miracles so he never actually concedes anything occurred beyond the laws of nature. John Locke sought to establish ***natural religion***. He was deceptively defined miracles so there was no supernatural or violation of natural laws. In writing about *The Reasonableness of Christianity* and on *A Discourse of Miracles* he gave the appearance of believing in supernatural workings in nature, but in fact never acknowledges any work beyond nature.

L. T. Ramsey of Oxford evaluated his discourse on miracles and said,

> Here Locke emphasizes that the definition of miracle must inevitably have reference to the observer. ...On the other hand the reference to the observer means that a miracle must needs be considered in relation to a wider situation which in the end not only includes the miraculous event as such and the observer, but someone claiming to have a mission from God with the miracle as evidence of that claim. ...The miracle is therefore the power of God it exhibits in the totality of a certain situation, and the criteria for a miracle will thus be criteria of such power.[9]

In other words, the miracle is the influential power of the person who believes he has a miraculous (or supernatural?) call from God. Such a sense of call, therefore, motivated the leader who motivated other men. The power of the person is the miracle. Thus Locke never clearly admitted any event involved God's power beyond that in nature. In this sense, Locke suggests there were only two great miracles – the call and mission of Moses and Jesus. It was the dynamic and influences of these two events that were miraculous. Therefore, reason from sense experience validates the reasonableness of the morals in the Mosaic Law and in Jesus. But Locke places miracle in his fourth category of knowledge—***outside the limits of real perception***. The power that gives miracles credibility goes beyond reason and is as mysterious "as is knowledge of myself to others."[10] In essence Locke avoids saying that anything contrary to commonly understood natural law occurred. He seems to profess acceptance of Christ and the teaching of the Bible as a revelation of God, but acceptance of God and any part of the Christian message must fit in with his reason and under his criteria of what is perceived.

In 1696 Bishop of Worchester, Stillingfleet in *Vindication of the Doctrine of the Trinity* "charged Locke with ***disallowing mystery*** in human knowledge, especially in his accountability of

---

[7] Ibid., 276.
[8] Ibid., 272.
[9] L. t. Ramsey, an introduction to, John Locke's The *Reasonableness of Christianity, With a Discourse of Miracles*, (Stanford, California: Stanford University Press; 1958), 71.
[10] Ibid., 79.

the metaphysical idea of substances."[11] John Sunjent in *Solid Philosophy Asserted Against the Fancies of the Idealists* (1697) also identified the anti-mystical or anti-supernatural bias, biases which also appear in the writings of Thomas Burnet, Dean Sherlock, et al. Although Locke was a champion for toleration, he meant freedom only for those who differed with the state or religious authority and with accepting or rejecting revelation from God to man. He thus freed the individual from all control except his own individual fleshly judgment. But he rejected the idea of spiritual forces, either demonic or divine, influencing or communicating with man. In other words he pled for tolerance for all views except those which opposed his ideas of man's knowledge. Locke, like Adam, was the standard for knowledge of good and evil.

Locke's Faulty Assumption about Man's Knowledge

Locke reportedly formed his views about human understanding and knowledge in troubled times when he was confined for health problems. He presented his views to about five or six friends in impromptu discussions in his home. When he published his ideas in *An Essay Concerning Human Understanding* in 1690,[12] these thoughts were not some profound result of his study, but were off the top of the head sporadic discussions when resting, during a time of illness. One cannot but recall that Descartes' ideas came out of a time of deep depression and at other times from subjective vision.

At that time very little was known about the human brain or mind, but Locke admitted that he was ***not even interested in how man perceives in his mind***. He said,

> I shall not at present meddle with the physical consideration of the mind, or trouble myself to examine wherein its essence consists, or by what motions of our spirits or alterations of our bodies we come to have any sensation by our organs, or any ideas in our understandings; and whether those ideas do, in their formation, any or all of them depend on matter or not. ... I shall decline [these], as *lying out of my way in* ***the design*** I am now upon....[13] (Emphasis added)

What was Locke's ***designed aim***? Isaiah Berlin informs us that he was not, as he thought, a Newtonian reformer,

> but in fact he far more resembles the physician (that he was) who seeks to cure the diseases (in this case ***delusions about the eternal world*** and the mental faculties of men). It (the essay on human understanding) is written in plain and lucid language which does a good deal to conceal the ***vagueness and obscurity*** of much of the thought itself.[14] (Emphasis added)

There is no mistake that while Locke professed faith, his aim was to dismiss God and eternal things. It is important to note that Locke himself was biased and all his writing was opposed to the control of the state and religious authority. Locke's efforts were to persuade men that there was no place for God. His definition of knowledge determined to ***exclude any ideas except those of this world***. Man, in his design, was to be king who could understand and control the world. This limitation of knowledge must ***remove mystery***, and as we shall see, also any catastrophe God might send. His design to free man to do what he desires in the flesh has influenced almost all modern secular teaching. While professing God, this view of knowledge of God was beyond perception. Locke tried to identify his ideas with science, but they were his ideas of what he

---

[11] Britannica, vol. 14, 272.

[12] John E. Bentley, *Philosophy- An Outline History*, (Ames, Iowa: Littlefield, Adams and Company, 1956).

[13] Isaiah Berlin, *The Age of Enlightenment*, (New York: A Mentor Book, 1956), 34, 35.

[14] Ibid., 31.

imagined from superficial observation. However, due to strong anti-religious biases at the time, his teachings were readily received.

This removal of God from perception was Locke's ***great failure*** (and the failure of other "determinists" in epistemology). In the late twentieth and early twenty-first century modern science discovered a spirit or mind of man separate from the biological brain which receives thoughts and at times words from spiritual forces influence man which are equally as real to the human mind as sense perception. Moreover, Locke's faulty conception of how man knows was soon manifested in the skepticism about man's knowledge in the philosophers who studied his views. The weakness of Locke's views was revealed by the philosophers of the eighteenth century.

Leading Scientists Rejected Individual Deterministic Views as Wrong

For about a quarter of a millennium the leading scientists not only believed in God, they ***saw His regular working as natural laws*** of creation. The leading Christian scientists renounced the tendency toward deism. Locke's views of knowledge were rejected. Only the philosopher Hobbs boldly interpreted regularity of nature as a materialistic machine. The leading Christian scientists insisted that the force behind creation was the mysterious power of God, and that only the consequences of His working could be observed and tested as His regular will for sustaining the world.

The historical scientist William Cecil Dampier stated,

> The world of Galileo and Newton, undoubtedly suggested, when translated into philosophy, a mechanical or materialist creed, and, as we shall see later, this was its outcome in the eighteenth century, especially in France, where it helped forward, first deism and then atheism. But to Newton and his contemporaries that conclusion would have been quite foreign. Firstly, Newton himself, as shown above, clearly distinguished between a successful mathematical description of natural phenomena and a philosophical explanation of their causes. Secondly, Newton and his immediate followers regarded the new discoveries as a revelation of the power and wisdom of God. ...To Newton God is not only a First Cause, but is also immanent in Nature. [15]

Boyle and other prominent scientists opposed this step toward mechanical naturalism. Even Gilbert, Kepler, and Galileo limited science to quantity and saw supernatural involvement behind science. These scientists understood that natural laws were God's ordinary way of using his power and providence, while special miraculous revelations, as recorded in the Bible, were his extraordinary way of using his power.

While Christian scientists scorned rationalism's rejection of revelation and the Bible they had a deep appreciation for man's ability to study nature using the inductive scientific method. They maintained this was possible only because God was the Creator who had made man in his own image with the intellect to be able to see his working and use it for man's good. Foremost among the Christian biologists was John Ray, who wrote to applaud the triumph of the new approach to learning and the information it produced. He said,

> Those who scorn and decry (inductive scientific) knowledge should remember that it is knowledge that makes us men, superior to the animals and lower than the angels, that makes us capable of virtue and of happiness such as animals and the irrational cannot attain. [16]

---

[15] Sir William Cecil Dampier, *A Shorter History of Science*, 76.

[16] *Synopsis of British Plants,* in the preface.

About thirty years later (1690 A. D.) he wrote *The Wisdom of God Manifested in the Works of Creation*, clearly affirming that these discoveries coalesce with Christian orthodoxy. At that time, the whole culture—including both educational and political institutions, were dominated by this theistic point of view. Many of the thinkers like Newton, Ray, Boyle, and others were genuine, devout, personal believers in and worshipers of God. They saw Locke's efforts as an error in understanding man's nature and the nature of creation. But man's bias moved him beyond this to see himself as wise enough to understand nature.

Christian's Followed Demonic Temptations for Rejecting God

Individualism that Could Lead to Pride

The Christian humanists of the Protestant Reformation allowed confidence in their thinking to give in to two demonic deceptions. Ironically these were the same two fallacies that centuries earlier had led to perversions within the Catholic Church. Protestant Christians had experienced great success in interpreting the Bible as the word of God, allowing them to correct the Roman view of salvation by works, and refuting the Catholic view of Aristotle. They were now deceived in their successes.

This confidence in man's individual ability to reason expressed itself in science and in biblical interpretation. Individual ability and reason, promoted by Descartes and those who followed him, was also basic to the Protestants. Instead of seeking Christ's wisdom in interpreting Scripture and the book of creation, they began to trust their own individual thinking.

The judgment of ***the individual was central to Protestantism for interpreting Scripture and nature, went too far with this view.*** As Kenneth Scott Latourette points out:

> However, common to all its forms, but not always equally stressed, was what had been emphasized by Luther, salvation by faith. As a corollary, even when not acknowledged in practice, was a principle, also formulated by Luther, the priesthood of all believers. Since salvation by faith meant the faith of the ***individual,*** by implication Protestantism entailed the right and the duty of the ***individual to judge for himself*** on religious issues.[17]

While the mind of the individual is the final place of faith, respect for God and others require humility to be open to criticism and government of our views. Over time the faith of Protestants abandoned the importance of love for all brothers in an effort to hold the views of their sect as the only right one. Protestant Christians let the pendulum swing too far and began to see their intellectual abilities as competent to speak for God in regard to the book of creation and Scriptures. They began to see everything as definite and fixed in both theology and nature. In addition, they turned to political power for defense and control.

Protestant Individualistic Pride Led to Two Demonic Errors of Destruction

Claims for the Mind of God Was Extended Toward Everything

The Reformers, in recognizing how God works in the book of creation or nature and rediscovering God's revelation in Scripture, began to go too far. Having found regularity in many areas of science and theology, they succumbed to demonic deception and human depravity and claimed too much. In the mid eighteenth century Carolus Linnaeus went beyond John Ray's

---

[17] Latourette, *A History of Christianity,* Vol. II, 873.

discovery of orderliness in the biological world. Linnaeus' effort set nearly every individual creature apart; he even went so far as to list man with the primates. He did recognize humans as distinct from primates, but nevertheless included them among the animal kingdom. Cuvier identified 60 species of men. Dalton established a fixity for finality in chemistry as well as other areas of science. Light and matter were said to be distinct along with the forces of gravity, electrical, and magnetic. While these were valid observations, some went to extremes. Isaac de la Payrene, for example, identified the black race with animals.

Protestant theologians had rightly tried to establish correct doctrinal confessions of faith. But they went too far. Luther's successor was Philippe Melanchthon who drew strong doctrinal lines for Lutheranism. Beza, Gomar and others extended Calvinism. This process of close definition was done by Augustinian Confessions, the Anabaptist views, et al. This resulted in new heresies such as the Socinians. Having a childlike faith gave way to doctrinal self righteousness and inflexibility. Brotherly love and cooperation, which would have given great power, gave way to insistence on conformity which led to many controversies and divisions. The Synod of Dort insisted on God's sovereignty to such an extent that man's responsibility was minimalized leading to the Arminian movement in reaction. Malancthon Lutheranism produced the Pietistic movement in reaction. Warring factions within Catholicism and opposition to Protestantism led to the Council of Trent. Other contentions arose over similar details, creating numerous divisions. In one extreme example, Johan Gottlob Carpzov, professor of Oriental languages at Leipzig, in claiming the total inerrancy of Scripture, went so far as to say that even the vowel points of the Hebrew Old Testament were inspired. These, however, were a much later addition by scribal copiers. Because Protestant Christians believed in individual interpretation of Scripture, they extended their knowledge of God's will beyond the clear meaning of Scriptures to philosophical systems. This logical extension was similar to the way Roman Catholics clergy claimed they could speak for God from Scripture and tradition. So each Protestant group claimed their individualistic interpretation was God's voice.

Protestant Protection by Kings/Rulers Produced Devastating Wars

Princes of political power often took sides for their own interests. On occasion they would do so out of faith convictions. In Germany Frederick protected Luther and his movement survived. French Protestants in town after town armed themselves. The various kings chose sides in the religious battles. Each religion in a state was supported by their king. Because the Roman Catholic political power was being threatened, this could not be tolerated. The Catholic Church had become a kingdom of this world in the name of Christ. In the name of freedom of religion, the Protestant groups forsook the word of God as their sword and fought with the sword of the flesh, also mingling the power of the kingdom of Christ with the world. Today we cannot understand what their dilemma was, or what side we might have chosen.

The religious wars to destroy the new biblical faiths that were perpetrated by Catholic powers were devastating for all Europe. For example, at the start of the Thirty Years War, Bohemia had a population of two million people, about 80 percent Protestant. By the end of the war 800,000 Roman Catholics remained and no Protestants. Historian John H. Newman said, "Taking Germany and Austria together, we may safely say that the population was reduced by one half, if not by two-thirds." Newman concluded his review of this tragedy of slaying, pillaging, and hunger by saying,

> We have grave reason for doubting whether the destroyer of old evangelical Christianity and the father of the great politico-ecclesiastical Protestant movement, which called for the counter-

> Reformation and the Jesuits, and which directly and indirectly led to the Thirty Years' War, was after all so great a benefactor of the human race and promoter of the kingdom of Christ as has commonly been supposed.[18]

While every state has a responsibility to protect its citizens and there is a time for the sword, Christians do well to remember that Jesus said, "My kingdom is not of this world. If my kingdom was of this world, then would my servants fight" (John 18:36). Unfortunately, Christians forget the kingdom is Christ's both in this age and the age to come, and begin to build their own little kingdoms. The political contest into which the Christians of this era had been drawn confused that issue. Biblical Christians began to contest for souls and territory. In so doing they became more dogmatic, making the mistake of setting themselves up as speaking for God in the same way the Roman Catholics had. Love for one another might have allowed them to hear each other.

Desire for Unity, Prepared the Way for Unity without God

The establishment of detailed fixed order in every area of science and theology began to go far beyond what was observed in nature or Scripture. As R.E.D. Clark said, "The very extremism of the advocates of the fixity of species set the stage for reaction."[19] This led to the deep desire to find unity from the Creator that seemed to be lost in absurdities. Some answers were correct such as Faraday's discovery that energy had many forms such as heat, light, electricity, kinetic and potential, chemical energy.

But most efforts went beyond science. For example William Whiston, who succeeded Newton as professor at Cambridge, extended the idea of regularity to a theory of all things using Scripture and nature. He wrote *A New Theory of the Earth, from Its Origin to the Consummation of All Things (1696).* Whiston went so far as to deny the Trinity in order to rationalize union in nature and Scripture. Edward Tyson claimed there were no unbridgeable gaps and proposed, without evidence, transitions "from minerals to plants, from plants to animals and from animals to men,"[20] thus paving the way for early views of biological evolution. Such was the strong desire for a rational explanation of unity.

Spiritual and Political Religious Conflict Left Pain and a Vacuum

Europe had been shattered by confusion and conflict and was looking for relief. Will and Ariel Durant commented on the sad state of human affairs at the time when the desire to remove God gained dominance.

> The story of the eighteenth century in Western Europe had a double theme: the collapse of the old feudal regime, and the near-collapse of the Christian religion that had given it spiritual and social support. State and faith were bound together in mutual aid, and the fall of one seemed to involve the other in a common tragedy.[21]

Man's natural bias—to avoid God—was motivated by the conflict, pain, and confusion of this time thus opening the door to a human solution devoid of God.

---

[18] Alfred Henry Newman, A Manual of Church History, volume II, (Philadelphia: The American Baptist Publication Society: 1931), 419, 411.

[19] Robert E. D. Clark, *Darwin: Before and After* (Grand Rapids: Grand Rapids Publications, 1958), 42.

[20] E. Tyson, *Orang-Otang Sive Homo Sylvestris*, 1699.

[21] Will and Ariel Durant, *The Age of Voltaire,* (New York: Simon and Schuster, 1965), 116.

Chapter 4

Satan's Perversion of Modern Science: Philosophical Claim of Evolution of All Things (Eighteenth Century)

Introduction Reviewing the Conditions Ripe for another Theory

As conflicts grew among the various Protestant groups and disagreements with Roman Catholics continued, many became increasingly skeptical of the Christian faith and its God. The growing confidence in the human mind promoted intensive efforts to understand and control the world for good. Scientific efforts had gone too far in trying to explain all the complexities of nature. This resulted in a fragmented understanding of biblical revelation. Until this time men were unified in accepting that God had created all things with order and beauty. But in order to claim to speak for God religious men began to claim to know more than was revealed in creation and in the Bible.

Man's natural sinful bias now drove him to look for a philosophical solution to unify knowledge apart from God so that man himself could control the working of nature. The turbulent seventeenth century was characterized by deep skepticism about everything except what man could logically observe and reason. Through induction and deduction the individual mind was to be considered the only real source of knowledge.

Frances Bacon and Rene Descartes had argued that the human mind could understand all things. The brilliant Jewish philosopher Spinoza raised rational questions about the Mosaic authorship of the Pentateuch, resulting in his excommunication from his synagogue. Due to his stays in Holland and England, John Locke was aware of these teachings. As a result he was hostile toward Roman Catholicism and all Protestants. During a period of illness, lacking any knowledge or interest in scientific study of the human mind, he created a philosophical theory of knowledge which excluded any real objective revelation about God. His ideas promoted deistic thinking. Others were returning to a god in nature working as the Unmoved Mover of Aristotle. They claimed that all things were governed by providence without any intervention. So confident were intellectuals about the human mind, that they often considered their theories as scientific truth.

Isaiah Berlin, a Fellow of All Souls College at Oxford, who was selected to write on *The Age of Enlightenment* in the book series, *The Mentor Philosophers* has described the thinking of this time:

> The philosophers of the eighteenth century tried to prove that everything–or almost everything–in the world moved according to unchangeable and predictable physical laws. ...It seemed that philosophy might almost have been converted into a natural science.[1]

The methodology of the great scientists of the sixteenth and seventeenth centuries, in which they observed objective data, formed and tested a theory, was abandoned. However, some men in the eighteenth century had so much confidence in the human mind that they assumed their theories were scientifically correct by their logic. It was believed that everything could be explained by empirical knowledge by the senses. Voltaire was neither a scientist nor philosopher of note.

> [But he] became the most famous individual in the eighteenth century... His claim to immortality rests on his polemical genius and his power to ridicule in which, he knows no equal. The friend

---

[1] Isaiah Berlin, *The Age of Enlightenment* (New York: Mentor Books, 1956), 1.

> of kings and the implacable enemy of the Roman Church and indeed, of all institutional Christianity, he was the most admired and dreaded writer of the century, and by his unforgettable and deadly mockery did more to undermine the foundations of the established order than any of its other opponents. He died in 1778, the intellectual and aesthetic dictator of the civilized world of his day.[2]

Voltaire owed much to the Roman Catholic Jesuit order who educated him and the Epicurean philosophy he gained from a group of deviant monks. These encounters gave him freedom for his immoral lifestyle of lying and sexual promiscuity. He held to the basic Aristotelian pantheism, believing in an intelligent designer in nature and in some form of immortality. But he was a chief spokesman against revealed religion and Jesus Christ. Voltaire's endorsement of Locke's and others' anti-Christian views sheds light on the attitudes of the eighteenth century, which paved the way for Kant's views.

Forming a Theory for Determinism in Nature: Immanuel Kant (1724-1804)

Kant's Background Experience in Religion and Science

Immanuel Kant is highly important because he was born at the time when natural human bias against God was motivated by painful religious conflicts, enthusiasm about human wisdom in discovering regularity in nature, and a skepticism about any knowledge beyond the senses such as biblical revelation.

Kant's background prepared him to become a turning point in dividing science and religion. Kant's grandfather had emigrated from Scotland to Königsberg, Germany (now Chojna, Poland), near Berlin, where Immanuel was born, spent his life, and taught. His father, a tradesman, raised Kant under a strict Lutheran Pietism.

In the early sixteenth century Lutheranism had accepted a firm doctrinal statement from Philippe Melanchthon. This rigid intellectual requirement eventually produced a reactionary movement emphasizing personal piety—Bible study, prayer and home fellowship. Initiated by Philipp Jakob Spener (1635-1705), the movement initially brought a new sense of freedom and devotion. But by the early eighteenth century, in reaction to moral laxity, pietism produced a new legalism, calling for escaping the world and imposing strict rules regarding worldly amusements such as dancing, card playing, and public games.

As a youth Kant entered a Lutheran school with the intent of studying theology and entering the ministry. As an adult he stood only five feet tall. He also had a shoulder deformity. These physical shortcomings may have given him a feeling of inferiority and a desire to seek and serve God. Kant took courses on theology and even preached a couple of times.

Freedom from Religious Legalism

Kant's studies led him away from God and to conflict with the church. He studied the classic Latin scholars. His favorite, according to a classmate David Runkhen, was Titus Lucretius. Others have also said the views of Lucretius continued to attract him and influence his thinking, furnishing many of Kant's basic ideas for his studies, and teaching.[3]

Titus Lucretius (98-55 B.C.) was a disciple of the Greek philosopher, Epicurus (341-270 B.C.), writing some 250 years later. Epicurus, a contemporary of Aristotle, wrote after him. Aristotle had established the ethical and ultimate aim in life as ***individual happiness***. For

---

[2] Ibid., 113.

[3] Immanuel Kant, Encyclopedia Britannica, Vol. 13, 266.

Aristotle, knowledge was based on an individual's sense perception and proceeds by cause and effect from the particular to the universal similar to scientific logic. Nature had an Unmoved Mover that actualizes the potential of all things by intelligent design. This is a form of pantheism.

Both Epicurus and Lucretius agreed with Aristotle that the ethical aim was individual happiness and individual knowledge based on sense perception and logical progression. But they broke with Aristotle by saying ***chance*** rather than an intelligent designer produced the universe and things in it. This was a rejection of any intelligent control outside of the individual man. Lucretius taught the world is without design or rule of gods and the human soul is made of natural atoms that end at death, which is extinction.

Lucretius fulfilled Kant's desire to find a scientific basis that would free individuals from Divine legalism. Lucretius' purpose was to free men from fear of God and death, and from moral rules for freedom and to find happiness by enjoying the pleasures one desires. He sought to separate the natural universe from the knowledge or influence of the gods to give man freedom. The bodies and the world are made of atoms and formed by swerve of perpetual motion known only by sense perception. The gods do not communicate and are known only by intuition or feeling. The dilemma in the Epicurean world view was how to have moral control while every man seeks his own selfish pleasure. At issue was how to achieve ethical accountability and restrain desires that might destroy both individual and societal happiness, while rejecting accountability to God and His law.

Skepticism of Scriptural Inspiration at that Time Influenced Kant

In Kant's time the advances in manuscript evidence of Scriptures encouraged his skepticism about the extreme claims of supernatural inspiration of Scripture held by some reformers such as the Pietistic movement in which he was involved. Biblical inspiration had been pushed by Carpzovius and others beyond the claims of Christ and the Scriptures themselves. The search for early Christian writings had been promoted by scholars.

Johann Albrecht Bengel, prelate and professor at Alpirspach School near Stuttgart, Germany laid some of the foundations for the later science of lower criticism of documents that actually established the reliability of the biblical documents. But Bengel had at that time found over 20 manuscripts and other fragments and discovered over 30,000 variant readings, causing him and many to give up the extreme statement of the doctrine of biblical inspiration as it was held, although Bengel still believed the Scriptures. In time he devised the idea of families of manuscripts that later enabled biblical scholars to trace back to the earliest manuscripts, which actually gave confidence as to what was nearest the original writers. But at the time these many variations seemed to discredit the views of biblical reliability. This liberated Kant from Scriptural confidence held by Lutheran Pietism.

Early Interest in Science Led to Idea of Development of All

Kant repeatedly sought confirmation to a professorship. Due to his interest and ability in math and physics he was finally successful in securing a position at his alma mater, the University of Königsberg. Kant's early focus was on English science, especially Newton's work. Newton was still the premier scientist and Kant either had to use his work in some way or ignore it, which no scientist could do. The movement toward a more comprehensive theory of a fixed nature by Whiston's astronomy, Linnaeus' biology, and Edward Tyson's theory, attracted him to a naturalistic explanation that fit with a theory of determinist knowledge proposed by Locke and others.

Growing confidence in human wisdom was producing a search for a rational explanation for uniting everything. Newton's successor, Whiston, had already suggested a complete, unified development of the universe involving God, but included a rejection of the Trinity resulting in him being condemned as a heretic. Emanuel Swedenborg had described the development of the universe and earth's solar system by a nebular hypothesis which became the basis for complete mechanical origin of everything.

Swedenborg furnished the key idea for Kant that offered a naturalistic explanation of the development of all things by an evolutionary process of the entire universe from a beginning nebula. Emanuel's father, Jesper Swedberg who was from a mining family became professor of theology at Uppsala and a bishop of Lutheran Pietism. Jasper, his father also rebelled against the legalism, which would have attracted Kant. Swedberg's views were toward virtues of faith and love and came under the accusation of heterodoxy. The son Emanuel later changed his name to Swedenborg. He was greatly hindered with a problem of stuttering, and felt inferior to preach.

Emanuel became a brilliant scientist traveling in England, Holland, France and Germany studying natural philosophy and Latin verses. He published a collection of Latin poetic verses but it is not known whether these were a source for Kant. Swedenborg, brilliant in Paleontology, anatomy, physiology crystallography, mineralogy, physics, astronomy, other sciences and mathematics, was ahead of his time. Appointed by Charles XII to the Swedish Board of Mines, he not only wrote about inventions, but also made some significant ones of his own.

As early as 1721 Swedenborg was seeking a scientific explanation for the universe. In 1734 he published a three volume work, *Principia,* in which he presented his nebular hypothesis for the evolutionary formation of stars and planets. Kant's reading of this provided his theory of the development of the whole universe by natural laws. However, by the time Kant wrote his theory in 1755, Swedenborg had been discredited as a scientist by claims of visions and supernatural voices. Kant no longer referred to him, deferring instead to Thomas Wright, whose scientific thought was more accepted at that time.

Thomas Wright (1711-1786) was born in Durham and moved for a brief time to London. A designed of gardens, he also taught mathematics and navigation. He became an English astronomer, mathematician, engineer, and inventor. In 1750 he wrote *An original theory or new hypothesis of the Universe* in which he took the nebular hypothesis further, explaining "an optical effect due to our immersion in what locally approximates to a flat layer of stars" or the appearance of the milky way. While these were new observations, they offered no evidence of origins except for those who believed the philosophy of the nebular hypothesis of development. He also explained the Grand Orrery to include the planet Saturn. These were seemingly additionally information added to Newtonian findings that could be used for a theory of development of all things which Kant had gained from Lucretius and Swedenborg. Kant was also influenced by Locke's view of knowledge that excluded revelation from gods as had Lucretius. Hume's skepticism also involved him.

Kant's Theory of the Evolution of the Whole Universe

While pursuing a position at the University in Königsberg at age 31, Kant's studies in English and European science resulted in writing his *Universal Natural History and Theory of the Heavens: an Essay on the Construction and Mechanical Origin of the Whole Universe Treated According to Newton's Principles* (1755).[4] Since Kant wanted a professorship and knew the state

[4] Immanuel Kant, *Universal Natural History and Theory of The Heavens: or an Essay on the Constitution and Mechanical Origin of the Whole Universe Treated According to Newton's Principals, 1755 A.D.* (Ann Arbor, MI: The University of Michigan Press, 1969, a translation and republication) Notes concerning introduction by Milton K. Munitz.

Lutheran church would not approve of his atheism, Kant presented a view of development of nature based on pantheism. Although not acknowledged, this was essentially a return to an Aristotelian view of the universe. Kant's was not a biblical theist who believed in a supernatural creation of and control of the world or of God's revelation in the Bible. Rather his view presented ***a god within nature*** and the development of everything entirely by natural laws. But to hold to belief in God and not offend the religious leaders, he broke with Epicurean creation of nature by chance. Because Kant's theories influenced science from the mid eighteenth century until now, we will give considerable attention to his views as stated in his *Universal Natural History.*

Kant admitted he used Greek and Roman pagan philosophy in forming his view but claimed there was one major difference. He said,

> I will therefore *not deny* that the theory of Lucretius or his predecessors, Epicurus, Leucippus and Democritus has ***much resemblance with mine***. I assume, like these philosophers, that the first state of nature consisted in universal diffusion of the primitive matter of all the bodies in space, or of the atoms of matter, as these philosophers have called them. ...Notwithstanding the similarity indicated, there yet remains ***an essential difference*** between the ancient cosmogony and that which I present, so that the very opposite consequences are to be drawn from mine ... [They] derive all the order which may be perceived in it from **mere *chance***... In my system, on the contrary, I find matter *bound to certain necessary laws.* Out of its universal dissolution and dissipation I *see a beautiful and orderly whole quite* ***naturally developing itself.*** This does *not take place by accident, or of chance*; but it is perceived that ***natural qualities necessarily*** bring it about. (24, 25)

> [Why] matter must just have had laws which aim at order and conformity. Their origin at first, must have been a universal Supreme Intelligence, in which the natures of things were devised for common combined purposes? (26)

> I have tried to remove the objections which seemed to threaten my positions from the side of religion. ...Thus it will be said, that although it is true that God has put a secret art ***into the forces of nature*** so as to enable it ***to fashion itself*** *out of chaos into a perfect world system.* (27)

> Isn't it vain to try to discover the immeasurable, and what took place in nature before there was yet a world? (Kant's answer) Just as among all the problems of natural science none can be solved with more correctness and certainty than that of the true constitution of the universe as a whole, the laws of its movements, and the inner mechanism of the revolutions of all the planets – that department of science in which the Newtonian philosophy can furnish .... (28)

> It seems to me that we can here say with intelligent *certainty* and *without audacity*: "*Give me* ***matter,*** *and I will construct a world out of it!*" i.e. give me matter and I will show you how ***a world shall arise*** out of it. For if we have ***matter existing endowed*** with an essential force of attraction, it is not difficult to determine those causes which may have contributed the arrangement of the system of the world as a whole. (29)[5] (Emphasis added)

Epicurus and his follower Lucretius (98-55 B.C.) sought to move philosophy beyond any intelligent controlling design and accepted chance as the cause instead of a pantheistic god. While not denying gods may exist, their view of knowledge excluded any real knowledge of them as did Kant's view. Kant essentially returned to Aristotelian pantheism. For Kant a self existing god did not create, but natural powers within nature created.

---

[5] Kant, *Universal Natural History*, 24-29.

Kant deceptively sought to establish a naturalism by referring to a "*Supreme Intelligence,*" *and "to providence"* which would be the unfolding of intelligence. To say Kant believed in a 'transcendent Creator' as some did is not true. Nowhere does he present his views to be even deistic. Moreover, Kant and those from whom he took his ideas, e.g. Swedenborg and Wright, were not biblical theists. Kant was aware that his views were contrary to the Christian religion, a fact that the church later officially decreed.

Kant relied on his limited observations of natural things, not scientific observations or any knowledge of origins or natural development. His was a philosophical theory that seemed posited how all things ***might*** be united by evolutionary development. His theory required millions of years when no such measurement of time was recognized. Contrary to all known history and previous beliefs and even to Newton's claims, Kant's theory required long ages. Without any explanation for the origin of the first nebula, Kant presented his ***philosophy*** claiming to be science.

Demonic Influences Motivating this Developmental View

The earlier influences on Kant by Descartes and Locke involved conditions indicating demonic influence. Descartes reported his insights came at a time of deep depression from dreams and instructions at Ulm on the Danube, November 16, 1619. Locke also reported his understandings came when he was depressed and ill.

The men that furnished the thinking for Kant's views of the development of everything in the universe and Kant himself had mysterious spiritual illuminations as callings. Lucretius' life is not extensively recorded in history. But Jerome's *Chronica Eusebii* has under the year 94 B. C. the entry: "Titus Lucretius the poet is born. Afterwards he was rendered insane by a love potion and, after writing in the intervals of insanity, some books, which Cicero afterwards emended, he killed himself by his own hand in the 43rd year of his age."[6] He was driven by irrational thinking.

Swedenborg likewise had strange influences that drove him. In 1745 he left the domains of physical research for psychical and spiritual inquiry. He claimed the Lord commissioned him by His spirit to teach the doctrines of the New Church within the churches. However, later his followers formed the Theosophic New Church denomination. What is important is that, "Late in life Swedenborg wrote to Oetinger that 'he was introduced by the Lord first into the natural sciences, and thus prepared, and, indeed, from the year 1710 to 1746, when ***heaven was opened to him***" (that illumination was his call to form the New Church of the heavenly Jerusalem). (Emphasis added) Thus in forming his theories of the origin of the universe, Swedenborg "had been instructed by dreams and enjoyed extraordinary visions, and heard mysterious conversations."[7] This was during the time he was driven to find a theory in nature to explain everything and before his great illumination to change the church. Regarded as weird by many scientists, and rendered a heretic by the church, Swedenborg left his office in 1747.

But Kant himself also had a mystical experience in formulating his views. Fearful of the opposition of biblical Christians to his views, he described how he supernaturally received approval of, and a call to present his views. He said,

> On the other hand, religion threatens to bring a solemn accusation against the audacity which would presume to ***ascribe to nature by itself*** results in which the immediate hand of the Supreme Being is rightly recognized… I feel all the strength of the obstacles which rise before me, and yet I do not despair. I have ventured, on the basis of a slight conjecture, to undertake a dangerous expedition….

---

[6] "Lucretius," *Encyclopedia Britannica*, Volume 14, 466, 467.

[7] "Emanuel Swedenborg", *Encyclopedia Britannica*, Volume 21, 654.

> I saw the clouds dispense… and ***saw the glory of the Supreme Being break forth with the brightest splendor.***[8] (Emphasis added)

From this vision Kant claimed, as "an advocate of the faith" the right to make his case of the forming of the universe. He then argued that "the beauty and order of the universe" required a natural designer ***in nature*** and repudiated the idea of the world being formed by pure chance as held by Epicurus and Lucretius. Kant accused them of "***a profane philosophy***" ***of chance beginning and development*** that "tramples underfoot the faith which furnished the clear light needed to illuminate it."[9] (Emphasis added) Thus Kant was convinced he was divinely ordained as God's defender by returning to a pantheistic view of nature. He thereby appeared to be a champion for God when in reality he rejected the biblical idea of a Creator held by the early modern scientists. The founders of modern science opposed the pantheism endorsed by the Roman Catholic Church for 275 years. They held to the notion of the God of the Bible as the transcendent Creator who revealed himself both in Scripture and in the book of nature. It was not long before those who accepted Kant's theory that the later supporters of the theory again rejected any intelligent designer in nature for ***chance creation***, as had Epicurus and Lucretius. Natural man wants to accept no intelligence beyond his own. The Bible warns that mysterious illuminating visions, such as those described by Descartes, Lucretius, Swedenborg, and Kant, are demonic. The apostle Paul said, "I am afraid that as the serpent deceived Eve by his cunning, your thoughts will be led astray from a sincere and pure devotion to Christ." He warned against "receiving a different spirit from the one (the Holy Spirit) you received" from "false apostles, deceitful workmen, disguising themselves as apostles of Christ. And no wonder, for even Satan disguises himself ***as an angel of light***" (2 Corinthians 11:3, 4, 13-15). (Emphasis added)

It was Kant's sense of divine calling that drove him to formulate a definition of knowledge that would exclude God's revelation "by His Son, whom he appointed the heir of all things, through whom also He created the world. He is the radiance of the glory of God and the exact imprint of his nature, and he uphold the universe by the world of his power" (Hebrews 1:2, 3). This demonic vision of nature has destroyed all known civilizations. It now threatens the United States and all the Western countries.

Kant's Struggle to Formulate a View of Knowledge to Exclude God

Most of Kant's later life was devoted to defining knowledge in a way that would defend his rejection of any authoritative revelation of God in nature or Scripture, and to place moral conduct in another category because man esteemed this was pragmatically practical. Paradoxically, his pantheistic view of nature was derived from what he reported was a divine influence. Milton K. Munitz, the authority on Kantian philosophy has said,

> We cannot understand that career (of Kant's) without giving due place and importance to the roles that cosmological speculation had for him. He did not abandon the interest in cosmology when he wrote *The Critique of Pure Reason.* It is rather that the perspective from which it is viewed had undergone a major change in orientation. *The Critique of Pure Reason* cannot be fully understood apart from its continuity with the *Theory of the Heavens.*[10]

---

[8] Kant, *The History of the Universe*, 18.

[9] Ibid.

[10] Immanuel Kant, *Universal Natural History and Theory of the Heavens*, Milton Munitz, Introduction, page v.

Kant's epistemology was driven by his desire to exclude a personal creator from his view of the universe.

Lucretius had liberated Kant from Pietistic legalism and offered a naturalistic interpretation of cosmology. Lucretius view of knowledge denied any communication with the gods. John Locke also presented an epistemology which denied any real knowledge from God, placing this in a separate fourth category. He held to knowledge being real only through the ideas of sensations on a blank slate of the mind that was used in reflection. Lock's views of knowledge were deficient and faulty, but fed the rising confidence of man's knowledge through science.

Philosophers George Berkeley and David Hume revealed the weaknesses of Locke's views. Seeking to establish a basis for philosophic Christians, Bishop Berkeley said visual ideas are not located in independent space but reside in the mind. He said there is no proof of any outside material substance; all that we know are sensations and ideas. Hence he argued, since only ideas are real, our ideas of the perfect archetype of God are inadequate.

David Hume was Kant's contemporary and made a powerful impact on him. "Hume carries the destruction of Lockeanism a step further (from Berkeley), and Locke and Berkeley to their logical conclusion."[11] Hume, with Locke, taught that knowledge consists of impressions (sensations, inner and outer feelings) and ideas which are copies or images which appear in thinking and reasoning, with impressions always first and basic. Therefore Hume argued, since all knowledge comes from experience about which we have no absolute or certain knowledge, this invalidated universal truths. Hume was therefore a supreme skeptic. Hume denied Locke's fourth level of knowledge. But Hume's view of knowledge was also incomplete. Sensations such as touch and sight have in themselves no sense of space or time, a fact Hume did take into account. Nor did his views have any basis for mathematical knowledge.

Kant was seeking to eliminate skepticism from knowledge of sensations and at the same time find a way of separating the eternal and any real knowledge of absolutes. He wrote a solution from the two-fold perspective of Hume's skepticism and Neo-Platonism. This produced a dualism of the intelligible and phenomenal worlds. Kant brilliantly ***created a philosophical solution*** to divide science from religion. In *Critique of Pure Reason* and *Critique of Practical Reason* he divided knowledge. Pure reason is that which is derived from sense perception or from empirical knowledge. It is mathematical and physical ***in the mind***, or an idealism from sense perceptions. Practical reason is intuitional or ***by feeling***, which he applied to ethics and religion. The distinction was between the desired and the desirable, between what is believed by perception and what is ***worthy of belief***. In essence, he better defined Locke's view of knowledge, especially the first category of knowledge by perception and the fourth of gods which were not by perception but desirable. By insisting that the individual mind was ***not completely blank at birth*** (as with Locke) ***but the blank slate had innate intuitions or categories***, which made sense perceptions possible, Kant removed skepticism from pure knowledge. He reasoned there were about twelve categories of thought; space and time were just such innate categories among others. He thereby denied the spiritual realm of Berkeley by having these intuitive categories a part of natural man. While greatly influenced by Hume, he sought to put an end to skepticism by saying the mind was not just a blank slate but the mind had categories involved in sensations. This gave a naturalistic answer to Berkeley who had placed ideas in spirit and God. For Kant all real knowledge came from sense experience producing ideas according to his intuitive categories. These categories were contrived and ***defined logically by Kant alone*** so as to exclude revelation. At this time, little was known

---

[11] Bentley, *Philosophy: and Outline-History*, 65, 66.

about the amazing human mind; neither Locke, Hume nor Kant gave any consideration to this aspect of knowledge.

Kant's critique of practical reason gave a ***subjective*** place for morals that are not based on empirical fact, as in Locke's fourth category. He distinguished "hypothetical imperative" as what a person ought to do to achieve his desires (e.g. work, so he can eat; treat his wife right, so he can make love), from a "categorical imperative." This fundamental principle is, "Act always so that you can will the maxim or the determining principle of our action to become universal law."[12] These are ethical needs, without preference or taste, but are not derived from God and revelation. Kant said that man has no free will (responding like Pavlov's dog), but man must act as free to have laws (cf. *Critique of Judgment*). But these are not based on objective fact within history but are subjective feeling of ought and not perceived real or pure knowledge.

Saying all was determined by natural laws not only confronted Kant with the problem of free will, but of moral obligation toward others. Kant ***assumed*** there is an *Ideal Kingdom,* that is a systematic union of different rational beings under a phantom self-imposed and universal law of reason in man's natural brain. He assumed this was to man real and could be tested by the maxim–could you ***will this*** to be universal law. Thus, this was a ***'categorical imperative.'***[13] Therefore the autonomy of the will tells us that the fact of pure practical reason legitimatize these as universal laws valid for every rational being. So the autonomy of the will is the supreme principle of morality and moral law is made by each man himself as a pragmatic decision. Thus he says that the categorical imperative is transcendental or passes all limits of our experience as pure-practical reason. But Kant ignores the fact that the individual feels he must surpass others for self-esteem and to survive as a moral being. Therefore in the end each autonomous individual will act for selfish ends transgressing what he has imagined in his brain is right. This inevitable selfishness will become evident in the second volume on modern science on the soft sciences. These definitions by Kant were readily received as the answer to the problem of knowledge and adopted many educators. They were Kant's ***philosophical definitions,*** not scientific discoveries.

Result of Kant's Determined View of Knowledge

Carl F. H. Henry, the honored modern American theologian under whom I studied, repeatedly said that Kant's epistemology "drew a line between time and eternity." Some people mistakenly believed Kant was giving a new "transcendental method." Henry Aiken, Professor of Harvard, Columbia University, and other schools, who was chosen as best to write the series on *The Age of Ideology*, which included Kant, also agreed with Carl Henry when he said,

> They are mistaken. Kant had, in fact, very stringent views about the possibility of our knowing any transcendent reality whatever. ...***A very large part of his intellectual effort*** was spent in discrediting metaphysical application beyond the bounds of possible experience.[14] (Emphasis added)

Kant's view of knowledge excluded revelations from God or gods, similar to Lucretius' and Locke's, and Hume's views. This upheld his earlier view of development of nature from an innate pantheism and not from a transcendent Creator who revealed Himself to man. While excluding

---

[12] Bentley, *Philosophy, An Outline History*, 78.

[13] T. N. Pelegrininis, *Kant's Conceptions of the Categorical Imperative And The Will*, (New Jersey: Humanities Press, 1980), 221 pages, especially chapter 3 ff.

[14] Henry D. Aiken, *The Age of Ideology,* (New York: Mentor Books, 1956), 30, 31.

knowledge of God as real, he inconsistently was motivated by his beatific vision as the real basis of his call. By separating God and ethics into what felt right and was desirable from real knowledge of the senses, this led to liberal theology where God and what was right is a matter of feeling and emotion. His view of a rational universal mind also led to Hegel's philosophical views. So in the highly enlightened Germany based on Kant's epistemology, under Hitler they could eradicate over 6 million Jews to build a super race.

Results of Natural Mechanism of Laws that Exclude Divine Revelation

Over 80 civilizations of nations have accepted such a view of mechanical natural laws, leading to a rejection of God and the collapse of morals. A similar process occurred in Greece, leading to its decline. In Greece, Anaxagoras, having adopted Pythagoras' idea that there was a line between time and eternity, brought this to Athens, introducing a determinism of knowledge that opened the gates to the Sophist's rejection of the one god, Zeus, and all other supernatural beliefs, establishing a fatal mechanical materialism.[15]

This change of epistemology is the beginning of the death of all civilizations. Kant likewise gave the final philosophical line for determinism in the West. In essence he drew a firm circle around man's real determined knowledge, with a more acceptable definition than Locke's. Again, Kant's view was an idealism based on empiricism. God was only a subjective feeling in each individual. Morals were based on what was reasonably, universally needed for man's good life, but not absolute. There was no place for accountability and judgment by God.

Descartes and Locke deceptively had left room to choose the existence of God by individual logic, which Kant also seemed to do. But John E. Bentley said, "The real God of Kant is freedom in the service of the ideal."[16] Kant was at war with the church and in practice left no room for objective knowledge of God or God's will, but he knew man's sinfulness and need for some morals. In summary, Kant's teaching ***eliminated*** any objective knowledge for ***accountability*** to God since ethical conduct was based on a subjective categorical imperative of universal need. But all history shows that while this is demanded by conscience, man will not obey due to his selfish desire for individual happiness.

Kant's View of Knowledge Produced Liberal Theology

Kant's epistemology became the standard for most educators in the West. He closed the door for many intellectuals on the certainty that man could have any objective revealed knowledge of God or morals. It also changed social and governmental concepts. His teaching *was the **turning point*** away from accepting revealed theology in the Bible, toward a subjective epistemology for religion and ethics. On the other hand, knowledge of nature was perceived by the senses from an eternal process.

Kant's teaching was a turning point for all philosophy and ideas of God excluding them from reality. His thinking led to Hegelian idealism and to ***man's progressive development*** through a process of movement and contradiction. It produced liberal theology which was based only on subjective feeling. Liberals such as Schleiermacher, taught religion was based on a feeling of dependence and Harnack taught the feeling of God as Father of all and the brotherhood of man. Descartes, Spinoza, Locke and Kant were men with a deceptive bias to revolt against church and God and inclined toward individual authority to be free to do what the flesh and mind desires. The

---

[15] See Carl W. Wilson, *Man Is Not Enough*, 54-56 where historians assert this. Cf. chapters 2-5 for this pattern in history.

[16] Bentley, *Philosophy: An Outline—History*, 76.

definition of how man knows was ***not derived from inductive science*** but was a ***philosophical definition*** that gave individual freedom from God.

Kant's brief scientific efforts to develop a nebular hypothesis of development, along with his tendency to limit knowledge to sense perceptions, laid the foundation for scientists who followed to claim ***a uniform determined view*** of the cosmos. Kant formulated his philosophy at the same time the body of evidence for regularity in nature was growing. Kant's theory was based on his philosophical bias to eliminate God.

Spread to Intellectuals of Scientific Philosophy as New Authority

This new conviction that the Bible and religion were not based on real knowledge spread quickly among the intellectuals. Two developments facilitated its spread. Printing, which had been invented in Europe in the fifteenth century, became more practical and widely available. New inventions, such as iron presses, replaced the hand block presses and made wide distribution of writing possible and quicker. Along with a growing interest in knowledge, information was gathered in a more encyclopedic form. The publication of Kant's teachings and Albrecht Von Haller's *Elements of Physiology, Volume I*, which presented all the studies of the physical laws as evidence for regularity in creation, marked a turning point in science history. W. C. Dampier has observed, "It has been said that the year 1757 marks *the* ***dividing line*** *between modern physiology and that that went before.*"[17] (Emphasis added)

In addition to the spread through encyclopedic writing, a second reason these new naturalistic philosophies spread quickly and gained influence can be attributed to the availability of the scientific academies. In the seventeenth century objective Christian scientists established scientific institutes that created distinctive disciplines and published their findings. The writing of scientists at the Royal Society of London and the Academe des Sciences in France were generally accepted as authoritative. While the academies held to objective science for more than 100 years, beginning in the eighteenth century and continuing into the nineteenth, the academies subscribed to more speculative views, such as Kant's.

Many intellectuals attempted to explain the uniform development of all things by natural laws from the evolutionary nebula hypothesis. This theory requiring long ages was imposed on all aspects of science. The nebular hypothesis was transferred from Kant, to others in a spreading effort to prove man's determined understanding of all of nature. It went first from Kant to William Herschel the English astronomer, to Hutton in geology, to Lyell from John Herschel (William's son), and from Lyell and his colleagues to biological evolution of Charles Darwin, and to the scientific elite of today. The spread and development of this evolutionary concept initiated by Kant was expanded in meaning to all areas of science through long ages of efforts to explain all working of nature. This theory changed the definition of science from observation and verification of the natural ways of God's working in creation, to a philosophy of science that is completely relativistic. It changed the concept of science from the foundations that had been laid in the first 275 years of the modern era. Kant's bias to give a rational rejection of God was readily received because of the natural sin bias of others and led the thinking of modern science to this day. The next chapter will trace how this occurred.

---

[17] W. C. Dampier, *A Shorter History of Science,* 85 (emphasis added).

Chapter 5
Finding Uniformity in Nature to Support Determinism

The Need for Uniformity for Determined Knowledge that Excludes God

Philosophical ideas about the nature of man's mind were developed to limit human knowledge. In order for this concept of determined knowledge to be meaningful and influence human worth, these thinkers had to show that everything in the world operated in a uniform manner by rational cause and effect, that man could know, understand, control and give hope. This determined view would have little credibility or usefulness unless determined knowledge actually could demonstrate evidence of evolutionary development in the world.

Christian scientists had already shown that man could understand the regular way things work in nature, especially in the way the solar system worked. Moreover, the Roman Catholic Church had been shown to be in error for endorsing the pagan Aristotelian and Ptolemaic views of the Unmoved Mover in the solar system. Copernicus, Galileo, Kepler, and Newton had already demonstrated using both mathematical calculations and the telescope how the solar system works. While all the Christian scientists had clearly argued that a transcendent Creator produced the forces that caused the regularity, this mystery was exchanged for an unseen power in nature or by unknown forces of chance that the secularists believed they would understand. Thus the men arguing for a determined view of knowledge began by arguing for a progressive development of the starry universe and planetary system. The philosophy of uniform understandable operation, based on astronomy would soon be extended to geology, and from geology to biology. We will follow these steps of deception.

Proposal of Uniformity in Astronomy

As new scientific facts were uncovered, it gave impetus to the desire to know more about the origins of the universe. While the ancient Greeks and others knew of Uranus and held myths about the planets, only the planets through Jupiter were visible. Galileo discovered the moons of Jupiter and showed their revolutions around the planet by the principle of inertia. However, the mathematical realities that were known led to predicting other planets and gave credibility to the mechanical view of how the solar system seemed to work. As Fred L Whipple said in the Encyclopedia Britannica,

> An empirical rule expressing very closely the relative distances from the sun of the primary planets known to the ancients was discovered about the middle of the century (in 1772) by Titius of Wittenberg... and is generally know as Bode's Law. If a series of 4's be written down, and 3 be added to the second, 6 to the third, 12 to the fourth and so on, each of the added numbers being double its predecessor, the sums of the first 7 numbers are 4m 7m 10m 16m 28m 52m 100. These numbers divided by 10 were found to give with a surprising closeness the relative distances of the planets as far as Saturn, except that no planet was known to revolve in the orbit corresponding to 2.8 between the orbits of Mars and Jupiter.[1]

This discovery, just before Kant wrote his developmental views, encouraged the search for what turned out to be asteroids and other planets, following this mathematical law.

---

[1] "Fred L. Whipple", *Encyclopedia Britannica*, 998.

Sir (Frederick) William Herschel (1738-1822), Observation of Heavens

Born in Germany William Herschel was about 19 years old when Kant, also a German, wrote his mechanical view of the development of the heavens. Shortly thereafter, as a member of the band of the Hanover Guard, he was ordered to England where he became a famous musician. He returned home to Germany in 1773 when Kant was becoming popular. Returning to England, he gained fame as an astronomer with a passion for Kant's nebula hypothesis. An avid user of Newton's six- foot telescope, Herschel, with the help of his sister, Caroline, his brother and his son, John, built better telescopes—up to 40 feet—that enabled him to make some astonishing progress in astronomy. Caroline and his brother discovered comets and studied their behavior. William became one of the most brilliant and famous astronomers of that time. Some of his astonishing findings are: 1781, determined the rotation of planets and moons; 1789, studied Saturn and found 6 satellites/moons: 1815, discovered the new planet Uranus. He is credited for founding the present day system of star astronomy and surveyed and mapped the northern skies, while his son, John Frederick surveyed the southern. He began to calculate star distances and study double stars. Late in life he believed one could discover the metals by the colors of light as in a prism which later led to the spectrum lines and spectrometers. He found that the sun is situated not far from the bifurcation of the Milky Way and that most stars are in clusters that are scattered comparatively thin across the universe. Most importantly for this discussion, he discovered a canopy of nebulous masses from what was "supposed" to be condensation and the formation of a stellar system.

William Hershel thought his discoveries fit with Immanuel Kant's nebular hypothesis and he applied it to the solar system and the whole universe. He had no evidence that this evolution happened over millions of years, but thought the analogy of the theory fit his findings. According to his theory the galaxies, stars, and planets were formed by condensing gases that operated with mathematical precision based on gravity.[2] While his son, (John) Frederick also became a very noted astronomer, John later became a Christian and turned to biblical faith. However, William's theory of progressive development based on things he saw dogmatized the ***idea of uniform development back from an unknown beginning.*** His findings did not scientifically prove a long ranged development by the nebular hypothesis but gave data that he thought suggested it. William shared his evolutionary theory with friend James Hutton who extended the view to geology. In 1801 Giuseppe Piazzi discovered Ceres (plus other objects), thought to be a small planet between Mars and Jupiter.

Pierre-Simon Laplace (1749-1827), Development of Solar System

Pierre-Simon Laplace was born to a small farmer. His education was paid for by rich neighbors. He became a math teacher in a military school and moved to Paris where he got an appointment at the Ecole Militaire where he continued to excel in mathematics. He tried to avoid making political commitments, which saved his neck several times and allowed him to gain favorable positions. He was recognized as a brilliant mathematician. After several attempts to gain an appointment to the Academy of Science, he finally gained entrance and later in life became vice president, then president of the Academy. Stephen Hawking said: "Laplace never hesitated to let anyone know the high opinion of himself. Even in his early twenties he considered himself the

---

[2] William Herschel, "On the Nature and Construction of the Sun and Fixed Stars," *Scientific Papers, I,* 484; originally in *Philosophical Transaction*, 1795.

best mathematician in France."[3] Laplace was a contemporary and admirer of William Herschel and wrote soon after Herschel's discovery of Saturn and his description of the moons and their rotations.

By the end of the eighteenth century Pierre-Simon Laplace wrote his book, *Exposition of the System of the World* (1796). John C. Greene described it:

> Laplace used Herschel's idea of a nebular condensation to solve the problem of the origin of the solar system. Beginning by supposing that the sun's atmosphere had once extended far beyond the orbits of the present planets, he tried to show how known laws would operate to form this atmosphere into concentric rings and these rings into planets accompanied by satellites.[4]

Laplace felt so confident of his hypothesis drawn from the growing knowledge of certainty about the planets that he optimistically applied this to man's ability to ***eventually know everything***. With the development of the solar system, he said,

> If we trace the history of the progress of the human mind, and of its errors, we shall *observe final causes perpetually receding*, according as the boundaries of our knowledge are extended. ... [These final causes] in the view of the philosopher, they are therefore only an expression of our ignorance of the true causes.[5]

Laplace's belief in the growth of human knowledge led him to ***assume*** there was no ultimate supernatural cause and predicted the rejection of God. It is said that when Laplace presented his theory to the Emperor Napoleon of France who once employed him, Napoleon asked, "Where is the final cause or God?" Laplace replied. "There is no need for one."

In regard to the planets at the time Hawking says, "All the known planets and moons revolved in the same direction and almost in the sun's equatorial plane. The sun, planets, and the moons all rotated in the same direction as the earth. If the commonality in all twenty-nine motions was merely the result of chance, then the probability of the chance would be exceedingly small, about 1 in 100 billion."[6] Later discoveries proved this wrong, however.

In 1846 Neptune was discovered and a tremendous amount has been learned about it since. A later chapter on the solar system will show that Laplace's knowledge was very limited and that conditions were actually much different than he thought and therefore he was wrong. But his presentation was extremely persuasive for claiming nature by regular laws had formed the solar system.

However, Laplace was well aware of Blaise Pascal's and Pierre De Fermat's mathematical demonstrations for probabilities in the universe. These seemed to threaten such claims of certainty for the solar system, much less the universe. In the last years of his life he worked extensively on the question of probabilities, writing, and rewriting on it several times. He tried to popularize his conclusions with examples of coin tossing and lotteries. He argued that probabilities are controlled by natural law as well and that our doubts of certainty lie in our ignorance. He said,

> We look upon a thing as the effect of chance when we see nothing regular in it, nothing that manifests design, and when furthermore, we are ignorant of the causes that brought it about. Thus chance has no reality in itself. It is nothing but a form of expressing our ignorance of the way in

---

[3] Stephen Hawking, *God Created the Integers,* (New York: Running Press, 2005), 284.
[4] John C. Greene, *The Death of Adam,* (New York: A Mentor book, 1959), 45.
[5] Pierre-Simon Laplace, *Exposition of the System of the World*, 1796, 30.
[6] Hawking, *God Created the Integers*, 386.

which the various aspects of the phenomenon are interconnected and related to the rest of nature.[7]

However, a later chapter in this book will show that this brilliant scientist and mathematician was himself ignorant. The probabilities throughout creation are so exorbitant that the law of chance becomes more like miracle. Laplace's arguments for the regularity of nebula hypothesis garnered acceptance of a general regularity in all nature so that God is not needed. He also established in the mind of the scientific community the idea that ***chance can be a cause itself.***

Man's ability to comprehend the planetary system's rotation around the sun was a formidable and amazing intellectual effort, confirming for many man's brilliance and eventually promoting a *transition to faith in* ***man*** that rejected God. A similar transition occurred in the sixth century B.C. in Greece, after Thales shockingly predicted an accurate eclipse of the sun. This, too, led to an unqualified confidence in man's intellectual capacity promoted by the educated elite in Greece.

Uniform Regularity Claimed in Geology

Introduction: Uniformity Extends to Geology

Western liberal thought turned on the diamond pivot point of gradualistic geology. In rejecting the geology that called for a cataclysmic flood, men eventually rejected the Bible itself in favor of natural gradualism. Kant initiated the theory; William Herschel brought it to England. Its public acceptance represented a significant shift in public thought. Biological evolution was built on this foundation and would never have existed without it. Because the mathematical concepts underlying astronomy are complex, common man's interest in astronomy was more as a curiosity which only secondarily affected daily life. But the public interest in geology was intense in the middle of the eighteenth century because of the importance of minerals and its role in the industrial revolution. Moreover, geology was a science that the common man could examine without special instruments like telescopes, etc. The claims of uniformity and regularity in nature related to the earth, the environment in which man lives. Interest in the idea of uniform determined knowledge became pertinent to the common person in regards to the development of the earth. Geological knowledge gradually emerged and the formation of the strata were traced and named. Those who were biased toward a gradual evolution of all things turned the evidence to support their philosophical theory of development over millions of years in a slow uniform process. The development of evolutionary geology was, as in astronomy, a conflict of evolutionary gradualism with that of a shorter period of a young earth as held by Christian scientists over a quarter a millennium.

Until the middle of the eighteenth century, the creationist's arguments for a young earth involving cataclysms and God's supernatural acts were centered primarily on the idea of the worldwide flood sent by God in judgment for man's sin. This was based on the fact that the earth is primarily filled with sedimentary rocks and this was the initial point of scientific focus. The biblical record had a much more important and extensive emphasis on geology than that one event, however. For example the fifth day of creation involved the processes of separating the seas and forming the dry land, which was one-sixth of God's creative work (Genesis 1:9-10). Such formation of the land involving sedimentary and possibly other rocks could involve extensive geological activity. Numerous other geological events mentioned in Scriptures were not considered. This primary focus on the worldwide flood opened the door for valid criticisms.

---

[7] Ibid., 388.

History of the Science of Geological Knowledge

Preliminary Geological Interests

The Greek scientific movement under Thales of Miletus and Anaximander in sixth century B. C. recognized fossil fish as extinct life forms and the influence of water for sedimentary rocks. Lucius Seneca (60 A.D.) recorded information on earthquakes, volcanoes and surface underground waters. Pliny the Elder and Younger made geological observations. Around 1020 A.D. an Arabian physician, Avicenna, wrote on earth erosion, meteorites, and the origin of rocks, and mountains, as did one or two other Arabs at that time. The resurgence of some of this information in the Renaissance stimulated modern thought.

Reformation Geological Knowledge

In the early Reformation period a Saxon physician, Georgius Agricola, published books on fossils (*Der Natura Fossilium*, 1546), minerals and mining (*De Re Metallica*, 1550) which furnished basic ideas for modern books on metallurgy. In 1669 a Danish physician, Nicolaus Steno, taught that layers of rock or strata are deposited with the oldest on the bottom and the youngest on top, which is *the law of superposition* that later became key to geological study. This fit with God's work in nature and a young earth. Many geological observations were accumulated by Peter S. Pallus, Horace Benedict de Saussure, S. C. de Dolomien, and John Mitchel in the early eighteenth century before the shift toward determinism. However, in the last half of the eighteenth and early nineteenth centuries geological conflict developed with the spread of the uniform deterministic views.

Emergence of Geology and Conflict with Evolutionary Gradualism of Uniform Determinism

Abraham Gottlob Werner (1750-1817): Water, the Principle Force

Abraham G. Werner's father was an inspector for an iron works in Saxony. However, Abraham chose not to follow his father's profession, choosing instead to study mining in Freiberg and later mineralogy and law at Leipzig. He returned to head the mining school at Freiberg University. He developed the science curriculum and was a principle teacher there for 40 years. His primary interest became lithology, or rock layers, which he had gained some ideas about from other teachers. He found that the sedimentary rocks followed each other in a definite chronological order, which he believed had been laid down by primeval oceans in a succession. He was one of the first to *assume* these layers *existed worldwide* as he saw them in his small district. His book, *Kurze Klassifikation und Beschreibung der verschiedenen Gebirgsarten,* (1787), was important in that it insisted on *confining study in geology to observation of present earth dynamics.* Also he was one of the first who clearly demonstrated and popularized the *idea of studying the chronology of rock strata.* He was called the "father of German geology." However, he mistakenly attributed *all rock to successive floods* giving *little prominence to volcanic action.* He treated volcanic rocks as late and of minor importance. More importantly he assumed material solidifying from fusion never assumed a crystalline structure and that *volcanic layers originally precipitated from solution* and subsequently melted and were ejected by volcanic action. But there was no explanation of how this happened and how basalt occurred in defined formations far from volcanoes. More importantly, he limited his study to the Hartz district of Germany, having little worldwide evidence. His teaching came to be known as Neptunism, for the god of the sea, Neptune. But Werner's claims were extended beyond his evidence; his work was a tremendous advance in geology. His views in earth study were influenced by the extensiveness of sedimentary rocks. Modern geologist E. B. Branson said,

> The sedimentary rocks are the most common ones at the surface of the earth. They are estimated to cover 75 per cent of the land surface which leaves only 25% for the igneous and metamorphic rocks together. ...The sedimentary rocks on the earth's surface range in thickness from a thin film to 40,000 or 50,000 feet. We have good reason for believing that at one time sedimentary rocks were much more extensive than they are at present.[8]

James Hutton (1726-1797): Cycles by Pressure and Heat Gradually in Endless Time

James Hutton is the key figure who promoted evolutionary gradualism for geology. The son of a successful merchant in Scotland, he came from a well-to-do family. He received a doctorate in medicine and practiced in Edinburgh and ***also in Paris***, a center of unbelief. He turned to agriculture where he made many scientific improvements. Following his retirement his interests turned to geology. He knew of Werner's Neptunism. His interest in mineralogy and study of geology led some to call him the "father of modern geology." In 1784 after conducting research on moisture and rain around the globe he presented a *Theory of Rain* to the Royal Society. He said that rainfall is regulated everywhere by the humidity of the air on the land, and the causes which promote mixtures of different aerial currents in the higher atmosphere, pointing the way to some aspects of meteorology. The cycle of rain caused great erosion and deposited sediments in the oceans.

Through his involvement in the Royal Society, he knew William Herschel with whom he corresponded. He was influenced by Herschel's insistent claim of uniform development of the heavens involving millions of years. In Paris he learned of Laplace's uniform determinism about planet formation and was fully convinced evolution had occurred over millions of years. In 1752 J. E. Guettard showed from the volcanoes of Auvergne that basalt lava had been poured out extensively and repeatedly in a molten state. He presented this as a refutation of Werner's view of chemical precipitation. Hutton used this discovery of error to zealously launch his cause and determined that many rocks were of volcanic nature as a result of heat. This theory had received little attention because the London Society was dominated by scientists who accepted the Neptune theory of water as the main geological cause. Hutton started a major angry controversy over the importance of heat and volcanic rocks in rock formation rather than primarily water. In 1788, after 20 years of research and soon after Werner's book on geology, Hutton published his preliminary writing, *Theory of the Earth: an Investigation of the Laws Observable in the Composition, Dissolution, and Restoration of Land upon the Globe.* Hutton revived and expanded his material and in 1795 he republished *Theory of the Earth* in two volumes. Hutton's books drew considerable criticism from the secular scientific community for asserting the major view of heat over against the Neptunist's views as well as from Christian scientists who saw it for what it was, an effort to repudiate the biblical flood.[9] But the biblical defenders were seeing almost all geology as a result of the catastrophic flood of Noah, which failed to take into account other biblical emphases. Hutton held a comprehensive uniform view of the formation of geological strata involving heat and water. He asserted,

> First, regular strata were formed by gradual deposition in the ocean. Second, these strata were consolidated, elevated, and altered by ***subterranean pressure causing great heat causing volcanic activity and change***. Thirdly, the face of the earth was worn away by wind, water, and

[8] E. B. Branson and W. A. Tarr and revised by others, *Introduction to Geology*, third edition, (New York: McGraw-Hill Book Company, Inc. 1952), 181.

[9] For a discussion of this controversy see Charles Coulston Gillespie, *Genesis and Geology*, (New York: Harper & Row, Publishers, 1959), Chapter II, Neptune and the Flood, 41-72. Gillespie had such prominence in geological history that he became the editor of The Encyclopedia of Geology.

> organic decay. The three types of processes, viewed in relation to each other, formed a self-regulating and self-preserving system of mater in motion. From the present state of the system, previous states could be inferred and its future states conjectured. ...The surface of ***the globe [is] a law-bound system*** of matter in motion.... From it he drew a revolutionary conclusion: that the operation of the terrestrial system had produced a long succession of worlds on the face of the globe and must continue to do so in the ages to come.[10]

Hutton's assumptions are the result of Herschel and Laplace's influence - ***not any geological facts***. His bias is revealed in Hutton's own statements:

> For having, in the natural history of this earth, seen a succession of worlds, we may from this conclude that there is a system in nature; in like manner *as from seeing* ***revolutions of the planets, it is concluded that there is a system by which there are intended to continue these revolutions. But if the succession of worlds is established in the system of nature, it is in vain to look for anything higher in the origin of the earth.*** The result, therefore, of the physical inquiry is that we find ***no vestige of a beginning– no prospect of an end.***[11]

For Hutton, this uniformity of the earth was rigidly law bound, undeviating, and had occurred over millions of years. He had no evidence for his views, only his theory. He was so committed to the uniform deterministic view of the astronomers that he said, "We are not to make nature act in violation of that order which we observe, and in subversion of that end which is perceived to be in the system of created things."[12]

There is a curious paradox in Hutton's Vulcanist views in opposition to Werner's Neptune theory. Gillispie points out the following:

> In the Vulcanist-Neptune debate, ***the antiquity of the earth was the issue***... even though the ostensible difference between the two schools centered around the primacy of heat as opposed to water in the formation of the crust of the earth.
>
> ...Hutton insisted upon confining the attention of geologists to earth dynamism, and he deplored any attempts to account for the origin of the processes which could not be observed in current operation.[13]

He insisted that all earth development occurred gradually from the distant past. He is often quoted as saying there is "no vestige of beginning—no prospect of an end"[14]. However, his defining of the process has based only on the current observation.

John C. Greene observed, "It was no valid rejection to this method of explaining natural phenomena that ***it required the assumption of millions of years on earth history***." Hutton went on to assert uniformity in ***the development of life and animals***, another assertion that he had never observed. This may have impacted Buffon and especially Lamarck and Robert Chambers who claimed a progressive development of evolutionary biology.

In time Hutton conceded to ***a late appearance of man*** as in the Mosaic record. Greene

---

[10] James Hutton, *Theory of the Earth*, as quoted by John C. Greene, *The Death of Adam*, (A Mentor Book, 1959, pp 85, 96), 44.

[11] Ibid

[12] Greene, *The Death of Adam*, 45.

[13] Gillespie, *Genesis and Geology,* 43.

[14] Ibid., 46.

comments that this concession by Hutton was a costly mistake.[15] The history of man and the obvious differences between man and animals made it difficult for even Charles Darwin to believe in human biological evolution. There was no evidence of development between lower life and man.

Hutton's dogmatic law of nature involved the ***assumptions,*** "no vestige of beginning – no prospect of an end," and that there had not been nor could there be ***any cataclysms*** in climate or the events of the earth from what he perceived. His statement on the assumed progressive uniform change in geology ***was admittedly based on the analogy of the progressive nebular hypothesis of the starry universe and solar system***. The development of the heavens was not based on scientific data but on ***a priori assumptions.*** And Hutton had not seen the indefinite past and did not know the indefinite future.

Hutton's theory of geology was based on heat caused by pressure on sediments in the oceans. This ***major assumption*** was basic to his theory of how uniformity worked. He had observed volcanic rock in between the sedimentary layers in many places, but he had not seen nor did he then see this complete process at work. So he had no definitive evidence of how these formed. One hundred years later his successor, the famous geologist Charles Lyell, would reject this assumed theory as unsound and replace it with another. But because Hutton's assumption required millions of years to produce the heat and lava, as had the assumption of the nebula progressive formation of the heavens over millions of years, the idea of extending the formation of the heavens and earth to ***millions of years*** to allow these slow uniform processes to occur was introduced as correct science. This was in known conflict with the biblical history and marked the beginning of rewriting earth history.

John Playfair, a young pupil of Hutton, had been introduced to a circle of skeptics that included Adam Smith, David Hume, and others. Playfair became the bull dog promoting and insisting on Hutton's geology of evolutionary gradualism in *Illustrations of Huttonian Theory of the Earth,* (1802). He, more than Hutton himself, gave this theory worldwide prominence and launched the conflict between evolutionary gradualism and creation geology.

Conflict for Reality between Evolutionary Naturalism and Biblical Creation and Catastrophism

Introduction

During the eighteenth century a human philosophy of evolutionary creation, fueled by the pain of religious wars and driven by an anti-God bias, took root. During the first half of the nineteenth century ***science contested whether the evidence supported biblical*** or man's wisdom. Biased men sought evidence for their theories that all was by nature and that man could be king of the world. We highlight here only a few of the many who were involved in this conflict and their most significant contributions and strategic findings.

William ("Strata") Smith (1768 - 1839): Select Fossils Identify Strata

William Smith was the firstborn son of a blacksmith at Churchill of Oxfordshire in England. He learned surveying as a young man and worked in irrigation canals. Fortunately for him it was a time of intense interest in the collection of fossils in Britain for their beauty and curiosity. Though uneducated, Smith was fortunate to know some of 177 prominent men and women "like Townsend and Richardson, Cunnington and Warner" listed in the *Dictionary of National*

[15] Ibid., 89.

Biography who were collecting these fossils. [16] A contemporary of Hutton, "he formed his theories and multiplied his observations between 1791 and 1799."[17]

Winchester described Smith's method of solving the sorting out the rock layers.

> He might find a succession of sandy beds in one valley, and another succession of sandy beds in another valley, that looked to all intents and purposes the same, but that his knowledge of their dip and strike and distance apart persuaded him were not the same at all – that the bed lying on top was younger (...been deposited more recently than) the bed that lay below. ...How then to tell the rocks apart? The answer lay in Smith's sudden realization that there was one aspect of the two types of rock, and only one that differed. The blocks of stone found in the cuttings may have all had the same color, and acid bottle would show them all to have the same chemistry, a magnifying lens would show the sandstones as all having the same grain size. But the fossils that were to be found in the two rocks – the bivalves, the ammonites, the gastropods, the corals – they were all different. Every single one of the specimens of one kind of fossil might be the same throughout one bed, but would be subtly different from those of the same kind of fossil found in another bed.[18]

He discovered that certain fossils existed in the same sedimentary rock layers, helping him identify and map these. He published his important findings in his book, *Strata Identified by Organized Fossils*. Just prior to this, he had published his large *Geological Map of England and Wales, with part of Scotland* in fifteen large sheets, (1815). At the time of this publication he was fifty years old, had just gotten out of Fleet debtor's prison, and was virtually unknown.

Before Smith published his book, his colleague, Joseph Townsend, with Smith's support, published in 1813 the book, *The Character of Moses established for veracity as an historian, recording events from the Creation to the Deluge,* which was based on Smith's findings of the English strata which supported the biblical data. At this time the controversy was raging in Edinburgh over Neptunism or water geology with Volcanism or heat. It was discussed in Oxford. Gillispie said that Smith's faith in the flood and young earth was naïve, but Smith's corroboration and statements reveal otherwise. Gillispie said,

> Since speculative controversy concerned him so little, it is scarcely surprising to find Smith in 1798, like most of his contemporaries, simply assuming the deluge as a historical fact. Its effects, he once remarked casually, "are very visible upon the surface of the earth, and to the great extend beneath." He was referring to the evidence of unpetrified animal remains often found in alluvial deposits, the same evidence which Buckland later exploited as the basis for an entire system of (flood) geology.[19]

Before Smith's death he was given the highest honors by the Geological Society. In the 1831 presidential address to the Geological Society Adam Sedgwick, a Christian, said, "Having succeeded in identifying groups of strata by means of their fossils, he saw the whole importance of the inference—gave it its utmost extension–seized upon it as the master principles of our science..."[20]

---

[16] Simon Winchester, *The map that Changed the World*: *William Smith and the Birth of Modern Geology*, (New York: Harper and Collins Publishers, 2001), 115.
[17] Gillispie, *Genesis and Geology: A Study in the Relations of Scientific Thought, Natural Theology, and Social Opinion in Great Britain, 1790-1850,*(New York: Harper & Row, Publishers), 1959, 84.
[18] Winchester, *The map that Changed the World*, 117, 118.
[19] Gillispie, *Genesis and Geology*, 92.
[20] Ibid., 84.

Townsend's book, exceedingly popular, was reprinted and followed by another book. Gillispie says that Townsend, while giving full credit to Smith, actually *explained the geological process and importance better than Smith* did. Smith did comment on the regularity of the operation of nature but attributed it to the "regularity, order and utility of *the Creator's arrangements*."[21] Smith and Townsend launched an objective way of tracing strata that spurred greater interest in fossils and geological strata. It is important to note that the prevailing scientific method of using fossils to discern strata did not yet lend itself to ceding a young earth theory in favor of gradual development. The bias of the determinists would accomplish that later.

William Buckland (1784-1856) became the most famous geologist and paleontologist at Oxford in England, attaining many leadership positions and honors. He taught and was associated with most of the prominent geologists of the period. His brilliance and knowledge enabled him to be objectively honest with all data and to weave them together in a synthesis to support with modifications Christian revelation. Hutton's views were modified by Cuvier's comparative anatomy and belief in successive diluvial creations. Gillispie said of Buckland's efforts,

> By adopting Huttonian dynamics and by classifying their observations according to organized fossil remains, they had (they thought) put their science on a firm, unassailable basis. He exploited and extended Cuvier's methods very ably, and he returned natural history to the explicit service of religious truth.[22]

Buckland later met Agassiz and, as we shall discuss, incorporated his glacial views. Buckland retained authority for a biblical view but began to extend it to what is known as an age day and gap concept with a modifying view of the worldwide flood. An excellent teacher and leader, he was an authority on fossils, describing the first dinosaur. His students became leaders in the conflict over theories of creation and evolution. In the end he yielded his diluvial view that encouraged evolutionary ideas. Some of his important students are as follows.

Adam Sedgwick (1785-1873) was born in Dent, Yorkshire, the third child of an Anglican vicar. He was educated at Sedbergh School and Trinity College, Cambridge. He studied mathematics and theology, and obtained his BA (5th Wrangler) from the University of Cambridge in 1808 and his MA in 1811. He became a Fellow of Trinity College, Cambridge, and Woodwardian Professor of Geology at Cambridge, a position he held for 55 years until his death in 1873. Sedgwick studied the geology of the British Isles and Europe. Charles Lyell and Charles Darwin studied under him as well as under Buckland. He collaborated with and debated the data with other important scientists who were trying to establish geology theory.

Roderick I. Murchison (1792-1871) worked with Sedgwick in studying and defining the geology of the strata of the Alps. Murchison was highly interested in Livingston's missionary work in Africa at this time. Murchison and Sedgwick did important work together, investigating the rocks of Wales and the western Midlands. In general Sedgwick worked upwards in the succession of the stratified rocks and Murchison downwards but their work overlapped. Murchison's work on the Selurian strata was most important. With the help of Charles Lapworth (1842-1920) who was an outstanding paleontologist, the strata were settled with Sedgwick in the Cambiran, with the lower Silurian designated as the Ordovician, and to the higher Silurian. Lapworth gave the name Ordovician.

Sedgwick and Murchison, both Christians who held modified views of creation, worked together in the Rhineland. Together they named and devised a brilliant way to deal with most of

---

[21] Ibid., 92.

[22] Ibid., 98.

the oldest strata. Murchison led an expedition to Russia and produced a geological study, *Russia and the Ural Mountains*. Sedgwick investigated the phenomena of metamorphism and concretion, and was the first to distinguish clearly between stratification, jointing, and slaty cleavage. With Murchison, he also worked out the order of the Carboniferous and underlying Devonian strata. Lapworth's insights helped pave the way for the proper understanding for the geology of other areas. He used fossils called graphtolites to follow strata in agreement with William Smith's insights. These Christians with modified views of creation worked out and named most of the oldest strata. Murchison did important work on the Perian strata. He also worked with Lyell in the volcanic region of Auvergne in France where Werner's views of precipitation had been refuted. Charles Lyell mainly worked on the Tertiary or Cenozoic younger period and divided it into three parts: the Eocene, Miocene, and Pliocene, with the last as most recent.

At times it was difficult to trace the geological strata because metamorphic rocks were formed and comprised of the various igneous rocks that were mixed with sedimentary rocks in relationships of unconformity, disconformity and paraconformity. Some of the difficulties of interpretation, such as changes in rock strata, will be discussed later.

The limited locations of the study of strata are seen in their names. Cambrian is the Latin word for Wales, Ordovician, and Silurian are names for Welsh tribes. Jurassic is the name of the Jura Mountains in Europe and Permian is the Perm in Russia. Other names were based on certain minerals being studied, such as Cretaceous for chalky rocks and Carboniferous from coal strata. Often there were no marked visible transitions in stratification. The Paleozoic ends with the Permian and is followed by the three Mesozoic divisions of Triassic, Jurassic, and Cetaceous. But as Martin Ice said, "The Mesozoic, starts with the Trias (or the Triassic), which is so continuous with the Permian that they are sometimes lumped together as the Permo-Trias." The order of the strata was based on studies of limited areas and there were little marked divisions among them. Gillispie and others have commended ***these early distinctions of strata as a major step in modern geology,*** and while the various strata were based on only a small percentage of the world, they were applied to world geology.

Charles Lyell, (1797-1875): Dogmatic Uniformity in Strata Identification—Transition to New Views of Geology

It was Charles Lyell more than any geologist who was a ***major influence in promoting gradualistic evolution*** and deliberately rejecting biblical ideas. Nevertheless, Lyell presented himself as a Christian. For that reason much attention must be devoted to his views. Born in Scotland he studied geology under William Buckland, professor of mineralogy and geology at Oxford, and the dominant flood geologist at the time. Lyell became a lawyer for two years but continued his interest in geology and finally through his involvement with the Geological Society devoted full time to it. He wrote articles on English rock formations, and became secretary of the Society. He traveled with Buckland and others working in France, Scotland, Italy and Sicily. He drew up tables for extinct shells in the Tertiary classifications. Like Charles Darwin later, Lyell organized and presented a significant amount of data in promoting his views.

Lyell was influenced by John (Fredrick) Herschel, (1792-1871), the son of William Herschel who had developed the nebula uniform developmental theory of the heavens. John had attended the same law school but turned toward astronomy and became one of the most prominent scientists. John Herschel insisted that "the only acceptable causal agents were to be those that could be observed producing the kinds of effects that needed explanations."[23] John's influence on

---

[23] Martin J. Rudwick, Introduction to Charles Lyell, *Principles of Geology,* (Chicago: University of Chicago Press, reprint 1990), ix.

Lyell turned him away from catastrophes, especially the flood, and began his insistence on science involving strict rational cause and effects. Millions of years of heat, water and wind persistently working produced the present geology. In later years John Herschel was converted to Christianity by his wife, Margaret. Under her influence John tried to retain biblical revelation by compromising it with geology. Lyell was greatly influenced by the French Enlightenment and the views of G. D. Deshayes of the French Museum of Historic Naturalism. Just before Lyell wrote his influential Principles of Geology, the two used all the work on stratification and brilliantly produced a table of the strata with the fossils that identified each giving long ages, even for those determined by Christians under a biblical time scale of a young earth.

Lyell's Persuasive Presentation for Gradualism and Long Ages of Natural Causes

In presenting uniformity, Lyell adopted the term "actualism" emphasizing present events and causes. He was little interested in mineralogy and focused more on biological life in tracing geological history. At first he, like many uniformitarians, adopted a directional or even "progressive" history of life on the earth. But in reading Lamarck, Lyell began to see this directional view was leading to a progressive evolutionary view requiring purpose and design, which he wished to eliminate. Moreover, Robert Chambers' adaptation of Lamarck's views in his book, *Vestiges of the Natural History of Creation*[24], was popular in England but was rejected by English scientists for its errors. This involved development from one common original form of imperfection to more complex and looked forward to the future of perfectibility of man in his physical, intellectual, and moral attributes. Lamarck saw progression from lower forms in even ability to reason. Lamarck saw entirely new organs and bodily aspects in response to the "felt needs" of organisms. In promoting uniformity Lyell avoided Aristotelian progressive pantheism by holding to a cyclic view. Lyell saw no physical improvement of man, believing he was just a mammal, but inconsistently he saw the "superiority is due to his power of reason and particularly his power of improvable reason."[25]

Lamarck's apparent acceptance of occult and mystical intelligent powers appeared out of agreement with the uniform determined knowledge that was then basic to the scientific movement. This was alarming to Lyell. Lyell wanted to avoid the problem of origins and therefore modified his thinking to a uniform cyclic one. His interest in Hutton's volcanic theory was greatly influenced by George Poulett Scrope's attack on the diluvial theory in favor of heat and volcanoes. Scrope's convincing arguments led Lyell to Hutton's cyclic view of earth history, but Lyell rejected Hutton's basic thesis that volcanoes were mainly caused by sedimentary build up. But he strongly adopted his idea of perpetual endless cycles without beginning or ending. Lyell's philosophy attacked the idea that the world is degenerating as Scripture and some geologists held, and proposed cycles of slow destruction ***and then rebuilding.*** The following sums up Rudwick's theory of cycles in the first volume:

> Lyell… interpreted the whole range of inorganic geological processes in such a way as to demonstrate that the physical features of the earth are in a state of perpetual flux…–of change consists of endless fluctuation around a stable mean, not of overall directional alteration.[26]

Lyell attacked Lamarck early pointing out "limits of organic vulnerability, the ***absence of***

[24] Robert Chambers, *Vestiges of the Natural History of Creation,* (London: John Churchill, Princes Street, Soho), 1844.
[25] Ibid., xxiv.
[26] Lyell, *Principles of Geology*, xxix.

intermediate forms. And the purely hypothetical character of the causes invoked to explain the supposed development from monad to man."[27] Initially he rejected Charles Darwin's evolution for these and other reasons. While seeing a great gap between man and other animals, in *The Antiquity of Man* (1863), written after Charles Darwin's *On The Origins of Species,* he argued for long ages in man and his early appearance on earth. After Darwin's book gained popularity and became accepted as science, Lyell gave in to Darwin's progressivism by survival of the fittest.

Lyell's Principles of Geology

While working with G. P. Deshayes in France Lyell formulated his ideas for the *Principles* and began to write the first volume of his *Principles of Geology* (1833), in which he brilliantly set forth the uniformitarian view as it applied to stratification and fossils. He wrote in a letter to Murchison with whom he had earlier worked in France saying,

> It (*The Principles…*) will endeavor to establish the principles of reasoning in the science; and all my geology will come in as illustration of my views of those principles, and as evidence strengthening the system necessarily arising out of the admission of such principles, which as you know, are neither more nor less than that *no causes* whatever have from the earliest time to which we can look back, to the present, ever acted, *but those now acting*; and that they never acted with different degrees of energy from that which they now exert (Lyell, 1831, vol. I, 234).[28]

He insisted that all change was uniform, ***there were no cataclysms.*** This encouraged him to oppose the use of the Bible and miracles. Anticlerical feelings pervaded European thinking at the time, especially in France, providing an accepting climate for Lyell's views. While Lyell avoided offending by making direct attacks against God (a mistake Hutton had made), he knew his chief opponents were physic-theologians.

Charles C Gillispie has said, "The ***real purpose*** of his (Lyell's) book was 'to ***sink the diluvialists***, and in short, ***all the theological sophists'*** (emphasis added)."[29] In Rudwick's introduction to the three volumes he also shows that Lyell was attacking the biblical views of the Pentateuch in subtle and knowingly deceptive ways as well as geologists who differed with him, resorting at times to dogmatic intimidation, and ad hominem arguments against classes of geologists from whom he had learned and worked. He avoided saying these ideas outright, being inspired to equivocation "by social expediency." He wanted to avoid the antagonism of the Royal Society. Writing to a friend who had access to expose these ideas through the Quarterly Review, Lyell encouraged ***seducing the whole public*** by persuading the religious intellectuals without a strong reaction. He said,

> If we don't irritate, which I fear that we may…we shall carry all with us. If you don't triumph over them, but compliment the liberality and candor of the present age, the bishops and enlightened saints will join us in despising both the ancient and modern physicotheologians. It is just the time to strike, so rejoice that… the Q. R. is open to you.[30]

Lyell's weaknesses were apparent even to those who agreed with him. He used history in general as an analogy and ***gave an inaccurate review of geological history*** to impress but also to

---

[27] Greene, *Death of Adam*, 249.

[28] Ibid., xii.

[29] C. C. Gillispie, *Genesis and Geology,* 133 and quoted from K. M. Lyell, *Life of Charles Lyell*, I, 309-311.

[30] Gertrude Himmelfarb, *Darwin and the Darwinian Revolution,* (Chicago: Elephant Paperbacks, Ivan R. Dee, Publisher, copyright 1959 and republished 1996), 189, 190.

convince of the need for his rewriting it. Rudwick said,

> Like many a would-be reformer, Lyell rewrites history in order to demonstrate that those whose present views he is attacking are inheritors of a historic tradition that has had a retarding influence on the progress of the science. He contrasts this with the ***more enlightened*** tradition in which he finds his own antecedents….[31]

Lyell critically spoke of Mosaic young earth believers' "scientific imagination that needed transforming" to accept millions of years of age, for which no data existed. Despite his ***many inconsistencies***, Lyell enjoyed success. At that time geologists knew little of the extreme powers of earth's forces and the brief time that sediments or volcanic rocks could form. Rudwick commented that his contemporaries at that time questioned his principles.

> "Lyell consistently conflates these two assertions (observable cause or processes, and the assertion that ***only these*** in kind and degree existed in the past), although this to contemporaries was not clear that the first could be true without entailing the second." They suggested there is no reason "within recorded human history might have acted more rarely with paroxysmal or catastrophic intensity.[32]

Indeed, Lyell's insistence on science being based only on present sensual observation seems to contradict Hutton's dogmatic assertions. Moreover, ***most data he used was not from his experience***. He did minimal research. He primarily took data of others, which he had not observed and organized it to support the interpretation of his philosophical theories. While Lyell insisted on basing science on what is observed, he claimed to deny previous forces which he has not seen.

His principles contradicted his assertions not only in extending only observed data to the past and future of time, but he extended what he had observed ***to be applicable to the whole world.*** His observations are based only on his work in Eastern Europe and Scandinavia, along with some data from places such as Senegal, and the Indian Ocean. He **assumed** that what he had observed, even when he used selective data to illustrate his principles, ***applied to geology over all other parts of the world***. His observations involved only a small fraction of the world, ***for little was known at that time of the geology of the vast majority of the earth.*** Lyell dated the Pleistocene (most recent) era now dated 1.8 million years from a distinctive set of fossil mollusks ***that still exist***.

Lyell's frequent flip flops demonstrate his reliance on philosophy rather than data. Along with Buckland and others in the Society in London he first accepted Werner, the great Neptune geologist's principle of strata and "Strata" Smith's brilliant key for identifying strata by certain fossils. Both of whom believed in the worldwide flood. He then shifted his views to John Herschel's and Hutton's theories of uniform determined directional science. Lyell at first rejected Lamarck's occult directional evolution progression in favor of Hutton's cyclic view of nature, but he rejected the forces Hutton identified as causing the cycles. He later reverted to a directional progression of Charles Darwin that was inconsistent with his cyclic theory.

However, his *Principles of Geology* was an immediate success and was widely accepted because Europe's intellectuals were biased to reject religion and accept a uniformed determined view in science, and because of the large impressive accumulated data of geology that Lyell claimed to be from determined uniform science. This appealed to the common man. He convinced the world that the age of the layers went from the younger on top down to the older, which was often true, but to which there are also many exceptions. Lyell's geology, that he claimed was

---

[31] Rudwick, Introduction to Lyell, *Principles of Geology*, xvi.
[32] Ibid., xiii.

completely based on uniform forces daily at work all over the world, was a powerful motivation for giving in to the origin of all things based on the analogy of the nebula hypothesis.

As Charles Coulston Gillispie, one of the most renowned scientists of the twentieth century[33] had reported, Lyell's motive was to establish the nebular hypothesis in order to remove confidence in the biblical story, especially of faith in God and the flood. He had succeeded in building on this bias by influencing the acceptance of the idea of the long ages of geology. It was not so much that Lyell had a conscious hate for God, as he wanted to establish the wisdom of man, and especially his own. This is the same temptation that confronted Adam and Eve in the Garden which carried a hidden rejection of the Creator.

Conclusions of Efforts to Find Evidence for Evolutionary Uniform Determinism

William Herschel and others applied Immanuel Kant's nebula hypothesis to their planetary studies, searching for evidence that the stars and planets formed from clouds of matter into great galaxies. Laplace attempted to demonstrate mathematically how all the planets formed and operated without supernatural creation and upholding. Herschel had persuaded Hutton that all things evolved from an unending past, and continued by the same regular laws into the unknown future without any evidence of supernatural or cataclysmic events. Hutton found that the importance of heat and igneous rocks had been given minimal attention and were more important, raising questions about earlier geological findings.

Most of the great contributions to early basic geology were made by men who believed in biblical revelation and God's imminent work by regular laws of nature but also in cataclysmic events like the worldwide flood. Some of these contributors and their views were: Nicolas Steno's "the law of superposition;" Werner's belief in worldwide strata established by floods; William Smith's use of key fossils to follow strata; Cuvier's application of the finding of the same fossils in multiple superposition of strata; Sedgwick, Murchison and others distinguishing and naming of the first and oldest strata; Agassiz's discovery of the great influence of ice movements. While these Christian scientists found some of their discoveries difficult to explain, their faith in God who controls the world helped, not hindered their science and gave them insight.

The work of early Christian scientists who emphasized the flood was challenged, first by Hutton who rightly emphasized heat, but based this operation on a false premise. About one hundred years later Lyell dogmatically, yet brilliantly claimed principles of geology that supported a gradual development of the earth. He and others extended the age of the geological strata to millions of years to fit their efforts to remove theological biblical truth. Charles Darwin was a contemporary of Lyell and his theory of biological evolution was built on this foundation of slow gradual natural development in geology. Those who followed this biased philosophy of evolution imposed long ages, denying God's creation and judgment in the flood. An anti-religion, anti-church spirit led educators and the public to accept these age changes with little question or evidence. In later chapters, subsequent scientific studies and evidence will be examined to see if later astronomers and geologists were truthful and if there was any truth to the theory of evolutionary development.

---

[33] Cf. *Dictionary of Scientific Biography*, Vol. 1 Ed. Charles Coulston Gillispie (New York: Charles Scribner's Sons, 1970).

Chapter 6

Determined Knowledge of Evolution of Life

Introduction

The effort to remove the evidence of God revealed in nature was incomplete. The seventeenth century efforts toward naturalism (Descartes, Locke, et al.) had led to Kant's nebular hypothesis of evolution in mid eighteenth century. Kant derived his naturalistic view from Lucretius, who was an Epicurean philosopher who rejected Aristotle's pantheism of rational design. The Epicureans based all on chance, removing all evidence of intelligent working. To avoid the attacks of the church, Kant included a pantheism involving design. In the eighteenth century unbelieving scientists' demands for a complete naturalism were paralleled by a claim of fixity in all things by those who sought religious dominance.

Mystical Evolutionary Development: Reaction to a Rigid Fixity of Life

These conflicting views of God, man, and nature led to an intense effort by theists in the eighteenth century to remove any intellectual power to make all the rational power of nature fixed. This effort to fix and categorize everything produced errors that discredited theism. The claim of divine fixity led to a reaction that produced a theory of organized natural development, the basis of Kant's nebular hypothesis. In the eighteenth and nineteenth centuries this progressed toward pure naturalism.

The claim that all was fixed produced the ideas of developmental progress in astronomy and the solar system, as well as biology. Kant's theory, accepted by Herschel in astronomy, and from Herschel to Hutton in geology, pushed human wisdom toward a pure naturalism without the intelligent design of pantheism. As Robert E. D. Clark said, "The very extremism of the advocates of the fixity of (biological) species set the stage for reaction."[1] The evolutionary development of life was not explained and left man still facing a Creator. This led to theories of organic evolution, the first involving a pantheism of innate force and then moving to naturalism by chance.

Transitional Findings of Unexplained Forms of Life (1750 ff.)

The biologists of the middle to the late eighteenth century were strongly influenced by the claims of uniformity in astronomy and by the emerging geological discoveries. The argument that the strata of rock required long years of weathering and other forces gained ground. At the same time they were also strongly influenced by the Christian claims of God and revelation in Scripture that were still dominant, so they were at a mid-point of shifting worldviews of faith in God toward secular natural uniform forces. Men who believed in Creation and sincere naturalists were contemporaries Carl von Linnaeus, a Swedish naturalist, (1707-1778) and Conte de (George L.) Buffon, of France (1707-1788). Frenchman Georges Cuvier (1769-1832) came later. They found life forms that were inexplicable by previous science saw one time successive creative acts with no significant change. Their new findings opened the door to transitional views.

Buffon saw nature as a system of controls by natural law, an organism with balance and equilibrium, destroying and renewing the organisms. His first publication, a history of animals, was written in search of a hidden method of operation, rather than as a classification of animals. John C. Green said of Buffon,

> Although he stressed the variability of nature and recognized that some variants were more likely to endure than others, he never pushed the idea of natural selection to its logical conclusion. He

---

[1] Robert E. D. Clark, *Darwin: Before and After,* (Grand Rapids: International Publications, 1958), 42.

> used it to explain how species disappeared but not how they were modified. He saw the elimination of the weak and maladapted as an occasional extraordinary event rather than as a process operating relentlessly within each species as well as among species. He viewed change more in terms of degeneration than of improvement.[2]

Buffon recognized surviving changes occurring over time, but he found ***only degenerative forms.*** He gradually moved from his early writing to a more thorough naturalistic view. He still advocated the creation by God but did not see the habits and mechanism of change as purposeful from creation. These were to be discovered, not in Scripture, but by observations of science as temporal processes.[3] While advocating a balance of death and birth, it is significant that he only observed their demise. He never saw evidence of new creatures generated either in degeneration or for improvement to a new species. He opposed that idea.[4] Man was the highest but unique mammal. But he drifted more away from the unchangeableness of nature to variability. He saw some species as more likely to survive than others. His strong emphasis on the effects of environment for limited change was a contribution for, but not toward evolution. But his appeal was increasingly toward observation and experimentation.[5]

Linnaeus saw all creatures as God given to be named, understood, and used for the good of mankind. Man's responsibility was to understand and classify them. Nature was therefore constructed, at least in part, in such a way as to be discoverable.[6] He saw the pattern of nature and absolute stability appropriate to its divine origin. He emphasized the harmony and interdependence of nature, concepts that are held by environmentalists today. He recognized fossils as previously living and some as extinct. While Buffon described animals, Linnaeus observed them for defining, arranging them in relationships and cataloging them in such a way as to give the first extensive effort at taxonomy. He saw species beginning in primitive genera and then these impregnated and ***changing "accidentally" as "daughters of time," being influenced by climate and geography***. In "looking for a historical explanation of the origin of genera and species, he conceived the outcome as determined to a considerable degree ***by chance***."[7] "Later he questioned whether the genera might not be the ***separately created forms***," which was an opening for evolutionary ideas.[8] It is important to note that in cataloging, Linnaeus listed ***man with the primates***. Linnaeus was not an evolutionist but offered ideas used by them.

Cuvier was a very respected and effective follower of the scientific methodology. He was the perpetual secretary of The Institute of the Museum of National History, was made the President of the Imperial University, prominent in the Academy of Science, and at one time the president of the Council of the French State. His family had been persecuted as strong Lutheran believers and he himself superintended the teaching of theology in the university. He focused on studies of mollusks, fishes, and mammoths but studied other animals. He was probably the leading paleontologist of the time. The discovery of mammoths in America and Siberia which were very different from modern elephants and lions two or three times the size of known lions, led Cuvier to postulate about the possibility of ***successive creations***. Subsequent studies showed that while they were indeed much larger and different in other ways, as well, the Indian and African elephants were descendents of the same species. He held to changes caused by catastrophes and

---

[2] Greene, *The Death of Adam*, 157.
[3] Ibid., 160.
[4] Ibid., 150.
[5] Ibid., 157.
[6] Ibid., 137.
[7] Ibid., 139.
[8] Encyclopedia Britannica, vol. 8, 916.

rejected the idea of millions of years of uniformitarianism that was accepted by those holding to evolutionary views.

It is significant that all these biologists, Buffon, Linnaeus, and Cuvier were dedicated to the scientific methodology with extensive and authoritative research and ***none found intermediaries*** either in modern species or in the record of paleontology. Linnaeus and Cuvier had uncovered many deviant species that needed explanation. R. E. D. Clark reviewed the efforts to show development in biology that had begun at the turn of the eighteenth century and produced evolutionary developmental views. All biologists recognized a certain adaptability to environment and change which we since have learned is genetic variation and limited. None of these expert scientists saw or espoused directional progressive evolution in the biological record. Lyell who followed in the nineteenth century surveyed ***almost all the limited but known strata*** with their fossil remains arguing for uniform determined knowledge over millions of years. He reported there were ***no known intermediate fossils*** and none known among the living species.

However, for those who held to uniform determinism and rejected catastrophes and miracles, the fact of finding different and strange species in the ancient fossil record offered opportunity for another explanation. Human sin bias in uniform deterministic science had not yet removed God. There was still no explanation of origins of living things and no explanation of how they all seemed to point to man and his amazing intellect which was the greatest mystery. These transitionalists paved the way for evolutionists.

Evolutionary Progress by Pantheism to Remove a Creator in Nature

The next step to remove God had to explain development of life for the picture to be complete. Some of these theories were influenced by previous Greek thought.[9] The idea of evolution was not original with the West but rather was resurrected from the Greeks. Heraclitus taught that everything is involved in change and the essence of being is becoming. Empedocles even mentions that change may be linked to "the survival of the fittest." A new human solution other than religious dogma was desired. The mystery of life and especially of man seems to require an unknown power and the first theories accepted that idea.

Lamarck (1744-1829) was the Most Prominent of Early Evolutionists

He briefly studied for the priesthood and also medicine. A French biologist, he specialized in invertebrate fossils and their classification. He rose to the position of professor of zoology at the Museum of Natural History in Paris. He was very interested in the weather, named some of the types of clouds, and was one of the first to try to forecast the weather. By his classifications, and those of Linnaeus before him, a biologist was able to ***see similarities*** in an emerging order in the different genera and species. The idea of natural selection that improved species by survival was already being considered. The emphasis of other biologists on environmental change saw probable successive creations, accidental change, and intermediaries. W. C. Wells (1813) suggested that races emerged by natural selection. Diderot in France thought that in the combat of animals, those best adapted survived; and Erasmus Darwin (1731-1802), Charles' grandfather espoused survival of the fittest as a basis of evolution in his *The Laws of Organic Life.* He was known to say at the dinner table the first law of nature is "Eat or be eaten." Robert Malthus' (1766-1834) *Essay on Population* (six editions between 1798 and 1826) laid out his ideas on poverty and population. He believed that government aid to the poor only served to increase the number of people who lived in squalor. He presented the law of the survival of the fittest; nothing could be done, so most die, while the fittest survived. Others offered birth control, euthanasia, etc. for artificially averting the

---

[9] Carl W. Wilson, *Man Is Not Enough*, 55 when quotes are given.

law of Malthus. Erasmus Darwin certainly knew this book and doubtless Charles Darwin's father spoke of this to his son.

Lamarck presented a view which eliminated the need for biblical creation and a way that nature itself could produce a progressive evolution that would explain origins and development. He espoused a Catholic pantheistic view that nature had designing power within it and that as emerging creatures "felt the need" for more complex functions, there was purposeful improvement. His defined nature as laws and forces governing the motions of matter, "an order of things separate from matter, determinable by the observation of bodies, the ***whole of which constitutes a power unalterable*** in its essence, determined in all its acts and ***constantly acting*** on every part of the universe."[10] Lamarck held a progressive directional evolution which denied the influence of a transcendent and imminent Creator. Lamarck saw these occult changes or developments (e.g. giraffes grew longer necks as they needed to reach higher foliage) as inherited by the next and successive generations producing the directional progress. ***Nature was an organism.***

Strangely, this was a reversion to the idea of pagan Aristotelian impetus which modern science had rejected (1500 ff.). It was a reversal of objective scientific methodology which the Christian scientists developed. Lamarck was trained for the priesthood by the Roman Catholic Jesuits in Aristotle's pantheism of the unmoved mover. So this was the source of Lamarck's thinking. Lamarck ignored the distinctive boundaries between species that other scientists recognized. Lamarck's work, which was widely read, inserted a ***hope for progress from nature*** which was becoming basic to French thinking. Lamarck's evolution was on the one hand a denial of proof from human reason of objective science of observation and on the other hand a denial of a transcendent but imminent God working in the world. It gave nature the power of God and the educated man as king to understand and control this. Lamarck was known and read by the scientists of England, including Charles Robert Darwin, and in all of Europe.

Preliminary Evolutionary Views and Events in England

Lamarck's idea of straight line progress in evolutionary development was met with skepticism, especially by Lyell who accepted Hutton's pure naturalism. However, Herbert Spencer adopted the idea of evolution and applied it to society as an organism. This had marked impact in social thought.

In England Robert Chambers responded to the powerful naturalistic attack on Christianity with his book, *Vestiges of the Natural History of Creation* (1844).[11] Professedly Christian, Chambers book went through twelve editions. It sought to identify the natural forces of evolution as secondary powers of God. Chambers took the pantheism of Lamarck and identified the intelligent force as the Christian God of the Bible as an apologetic while discrediting the Genesis actual account and the biblical idea of a transcendent Creator and personal imminent Sustainer. Published anonymously the author of *Vestiges* was not known for many years. Robert Clark describes its primary thesis:

> According to the author, a universal law of ***nature was forever creating*** order out of chaos. Chaotic matter in the heavens condensed to give orderly worlds, where highly complex

---

[10] Greene, *The Death of Adam*, 160. Found in Lamarck's, Histoire naturelle des animaux vertebras, 1815 ff. Charles Gillispie had written two articles on "The Formation of Lamarck's Evolutionary Theory," in which he argued this philosophy was what governed all Lamarck's thinking.

[11] Robert Chambers anonymously, *Vestiges of the Natural History of Creation* (Chicago: The University of Chicago Press, originally 1844, this edition 1994).

> organisms developed. 'We see a gradual evolution of high from low, of complicated from simple, of special from general, all in unvarying order, and therefore all natural, all of divine ordination.'[12]

Most of the arguments used by evolutionists argued for gradual directed development, such as recapitulation, survival, et al. Chambers even promised that in time a species greatly superior to modern man might arise—superman.

Vestiges... sought to reconcile the secular scientific movement with Christianity, to harmonize evolution with Moses' writing. His book seeks to show progressive creation from science, not comparing it with Scripture but claiming it can be reconciled. [13] But nature replaced the work of the transcendent sovereign Creator and cancelled the sense of direct personal accountability inherent in God's personal immanence. Moreover, since Chamber's response was not what Scripture actually taught, it turned interpretation away from faith in Scriptural authority and Christ. This is the error of most efforts to reconcile Christianity with the secular educated elite even today.

Vestiges spread the idea of evolution throughout England and made it acceptable to the many in the church. It was the best seller for over a decade with eleven editions. But the intellectuals who were seeking to remove the biblical God did not accept it or Lamarck's views as science. *Vestiges* was also read by Charles Darwin and motivated a hunt for change by chance. The century following, Hutton had moved toward a stricter view of naturalism without intelligent design. The pantheistic intelligent design of Lamarck and Chambers left a controlling intelligence in the picture. These views were opposed by Lyell and many others. Many newspapers opposed *Vestiges* and the Cambridge geologist, Sedgwick spoke for the scientific community calling it 'unfounded and nonsense.' Disbelieving science continued to search for a view that completely eliminated the God of the Bible and any intelligent design that would imply morality.

Concurrent with these evolutionary debates, Queen Victoria reigned in England, overseeing a period of great internal conflict and hostile political division. The Irish potato famine and the failure of England's wheat harvest left many poor and dying. The Industrial Revolution was causing women and children to work long hours with little pay. While it is was a time of growing prosperity and expansion of the Empire, a growing power struggle between the common people and the upper class led, at times to rioting. The church gave no answers. The conditions generated Marxist thinking and soon led to the tragic French revolution.

Determined Knowledge Excluding Intelligent Design

Charles Robert Darwin's Background (1809-1882)

Charles Darwin sought and seemed to find an answer for pure naturalism in biology. Charles' overpowering father likely contributed to his hesitancy, as well as his perfectionist tendencies. He had given up medical training, turning to theology. Some say he had also been asked to leave theological training, but the best evidence is that he passed with moderate grades and could have become a reluctant moderate pastor. His theological experience produced a tension toward the church.

He developed an interest in collecting and evaluating beetles and in natural forces, that commended him as the third choice as a naturalist for the voyage of the ship Beagle in 1836 that changed his life. He began to lean toward the mutability of the species. He was probably

---

[12] Robert E. D. Clark, *Darwin: Before and After*, 47, 48.

[13] Robert Chambers, *Vestiges...*, 388, 389.

influenced by the evolutionary theory of his grandfather, Erasmus Darwin, and the other evolutionists, especially Lamarck, whom he once said was to biology what Hutton was to geology.[14] Lamarck gave Darwin an extensive view of the evolutionary process and taught "that modifications due to the environment, if constant and lasting, would be inherited and produce a new type."[15] This idea of conflict in nature and inheritance of characteristics were basic for Darwin's theory. Darwin had studied Lyell who in his *Principles of Geology* in 1832 had dealt with species in some length, having examined many known fossils in the available strata. But Lyell saw these as in "stasis" or stable, and with no progressive development. Darwin's pursuit of naturalism following the Beagle trip led to eight years of study of barnacles or mollusks. These studies gained him a place of acceptance as a scientist. Darwin sought to gain credibility by setting forth a geological theory that claimed the roads in Glen Roy, Scotland were ancient beaches formed by the ocean sinking and draining away as the earth was raised. This was similar to what he had seen in Chile on his voyage. His effort for geological recognition was completely rejected as unacceptable.[16] Darwin was not seen as having great scientific ability.

Charles Darwin owned and read most of Lamarck's works, and Lamarck taught development was guided progressively by natural intelligent purpose. Darwin also adopted the idea of inheritance of acquired characteristics that Buffon suggested; but Buffon, looking at evidence, saw only degeneration, never improvement by the inherited characteristics. All directional evolutionists at the time sought to explain deviant individual characteristics as intermediaries for ***progressive improvement*** rather than progressive degeneration. The pessimistic idea of degeneration was contrary to hope for improvement.

These writers of evolutionary ideas had already established the idea of evolutionary progress in the biological world before Charles Darwin. Lamarck contributed the Aristotelian idea of progress from a pantheistic nature, and Robert Chambers in *The Vestiges*...claimed a pantheistic power of successive creations as the way of biblical creation. Evolutionary theory had spread all over England in the fifteen years before Darwin wrote. So he did not have to convince people of the evolutionary theory which was done by the Roman Catholic pantheism of Lamarck or the identification of the biblical God as pantheistic of the *Vestiges.* What he had to do was to remove all intelligent design and purpose in the theory.

Thus, ***the optimistic belief in man's progress was parent to the idea of biological progress***. Progressive improvement was a presupposition, not a conclusion from observational data. *Vestiges* and similar writings promoted an optimism that man would continually improve, eventually producing a more perfect human. Charles Darwin believed that the mechanism that contributes to the survival of the fittest would not only produce higher-functioning species, but would also influence other characteristics of a species. Unknown to most modern-day evolutionists, Darwin argued in his book, *The Descent of Man,* that this mechanism was responsible for the superiority of the male over the female. To Darwin men were biologically and intellectually superior to women.[17] His notebook recordings in 1837 and 1838 resulted in the publication of his ideas of biology in his Journal in 1839. He wrote his first 35-page abstract on evolution in 1842 and expanded this to 230 pages, copies of which he circulated to friends. Darwin formed his theories of evolution ***before he had an understandable mechanism***. He, with others of his time, was

---

[14] Gertrude Himmelfarb, *Darwin and the Darwinian Revolution,* 179.

[15] Dampier, *A Shorter History of Science*, 120.

[16] Bruno Maddox, "Deconstructing Darwin," *Discover Magazine,* November, 2009, 39-41.

[17] For a discussion of this, see Glenna Matthews, *Just a Housewife: The Rise and Fall of Democracy in America*, (New York: Oxford University Press, 1987), 117 ff.

driven by the desire to find a complete naturalism in biology. The missing piece of the puzzle was a mechanism for change in nature – a piece Darwin spent twenty years looking for.

Darwin's Discovery of the Key Mechanism

Darwin found the cause of mutability by reading Malthus in 1843. Drawing on Malthus' law of "survival of the fittest," Darwin posited that natural selection of the species was the key to biological improvement. He says of this flash of insight in reading, "It at once struck me that under these circumstances favorable variations would tend to be preserved, and unfavorable ones to be destroyed."[18] Significantly Darwin ***did not claim he found this in the data of biology***. He had delayed publishing his views for years until he got this insight for a mechanism.

Soon after this insight in 1856, when writing his opus, Darwin was given an abrupt motivation to publish his views. He had an acquaintance with Alfred R. Wallace, who was a fourteen year younger man who was also a beetle collector and was involved in studying nature in Malay and New Guinea. Wallace sent him a manuscript containing the same insights on survival of the fittest, requesting his help in publishing it. Wallace had also read Malthus and arrived at the same ideas about natural selection. Darwin, however, was not ready to publish and had his hopes crushed. Darwin consulted with his friends Hooker and Lyell, who did not fully agree with his theory but due to the uniqueness of the mechanism had become interested in Darwin's ideas. Both of them had held to the idea of fixed species. They suggested that Darwin ***announce that both he and Wallace held these views***. They then went to the prestigious Linnean Society on July 1, 1858 and presented extracts from Darwin's 1844 sketch along with a recent letter to Asa Gray explaining his mechanism and Wallace's paper. They explained the circumstances for giving both credit. Darwin had accumulated data for the theory for twenty years while Wallace had spent considerably less time developing the theory. The public announcement was not given much attention. It was not until Darwin's publication of *The Origin of Species* in 1859 that the theory gained prominence.

While Lyell still had private reservations he saw the potential in the idea and sought to get John Murray, a publisher who had published Darwin's *Voyage of the Beagle* to publish *The Origins...* Lyell had published his geological views opposing the Genesis creation account in a deceptive way so as to try to avoid the kind of conflict that engulfed Hutton. Darwin's work was equally biased and was seeking advice in how to present his theory. In Darwin's letters to Lyell about Murray's persuasion he makes it clear that he knows his ideas are a rejection of biblical faith. His goal was to ***stealthily get his views into circulation*** without incurring the criticism of Christian scientists and the public. In order to do so Darwin deliberately avoided the areas in biology that opposed his views. He said,

> Would you advise me to tell Murray that my book is not more *un*-orthodox than the subject makes inevitable. That I do not discuss the origin of man, etc., etc., and only give facts, and such conclusions from them as seem to me fair.
>
> Or had I better say nothing to Murray, and assume that he cannot object to this much unorthodoxy, which in fact is not more than any Geological Treatise which runs slap counter to Genesis.[19]

---

[18] Robert E. D. Clark, *Darwin Before and After*, 51.

[19] Quoted from Gertrude Himmelfarb, *Darwin and the Darwinian Revolution*, (Chicago: Elephant Paperbacks, 1996), 251 from Darwin's *Life and Letters, II*, March 28, 1859, 152.

Darwin therefore wrote from an uncertain bias, aware of the weaknesses and tremendous obstacles to his theory, which he deliberately hid. Strangely enough, *The Origin...* did not deal with origins, but with what he regarded as the development of new species from old. He omitted the idea of origins and the problem of man and deferred trying to explain that for over a decade.

Darwin's Presentation of *The Origin of Species*

Darwin began his book with natural micro-variations of domestic species to establish the idea of change occurring. He then presented the struggle for existence followed by the argument of survival of the fittest as the key to evolution. This was presented as speculation on how variation took place in lowly organized structures and then how it would lead to their development. Hoping to establish these ideas, he then discussed the absence or the rare evidence for transitional variation. He tried to deal with the lack of evidence and imperfection of the geological record by showing succession and distribution. He predicted confidently that the little known history of fossils would prove his theory. He then argued from mutual affinities in the various classifications which he had researched, such as morphology and, embryology to claim recapitulation. He did this with comparison of insects and mollusks by anatomy. He argued that rudimentary organs were remnants from the past. His widespread accumulation of data for comparison and the use of unfamiliar life forms from the Galapagos Islands were impressive. He ended dealing with the objections of the theory of "natural selection" and arguments in favor of his views.

Darwin's Main Points of His Different Theory of Evolution

Darwin's theory was different from all others before him and offered what anti-god forces wanted, removal of any intelligent design. The crucial points about his "means of the change" are:

1. Change occurred by natural selection in individuals enabling them to survive adversity when others died out ("survival of the fittest" from Malthus).
2. The gradual minute multitude of changes occurred over long periods of time (many thousands of years).
3. These minute changes or acquired characteristics were inherited by their offspring and thus perpetrated to successive generations.
4. Many acquired characteristic changes produced transmutations from one species to another.
5. These transmutations gradually produced improvements that progressed to higher species in a family tree.

Darwin had interacted with three friends, Joseph Hooker, Julian Huxley, and Charles Lyell, and tried without success to persuade them of his idea of natural selection. ***After*** the publication of *The Origin...*, Darwin's ideas gained wide acceptance, including from prestigious scientists. Joseph Hooker was a prestigious biologist and authority on species who spent hours furnishing information and interacting with Darwin, but ***did not see the data allowing for transforming mutability.*** He claimed he had seen only minor changes in species. Lyell had long argued there was ***no evidence of transmutation*** of species rejecting Lamarck's ideas, and said that ***the origin of man was completely unexplainable.***

The Reason for Scientists Accepting Darwin's Key Idea

Charles Darwin made a ***definitive statement*** about biological development that ***appealed to the spirit of scientific determinism*** prevalent at the time. He seemingly identified a natural explanation for mutability. The theory avoided intelligent design that uniform determinists like

Lyell, Herschel, and the other evolutionists wanted to claim. He made the very important distinction to ***limit development to a cause and effect mechanism,*** and ***not to a mysterious designing force.*** While several others before him taught evolution by ***natural selection***, he alone made it the pivotal point and ***the result of chance working of nature***.

Improvement by chance in the solar system had been identified by Laplace who mathematically sought to demonstrate that there was a ***natural law in chance*** and removed probabilities as a problem. Darwin saw variations of species from various causes such as climate change, needs for getting food and the preservation of these acquired characteristics which allowed them to survive in the difficult conflict in nature. ***Others before*** had ***not believed*** that this mechanism of ***natural selection was sufficient*** to produce development without ***some occult or innate natural control to guide*** things. In this way Darwin escaped the idea of the unmoved mover of Aristotelianism pantheism which modern science had rejected for over two millenniums and which Kant had resorted to in order to avoid church criticism.

To Darwin's credit he brilliantly understood the thinking of his time. Again, it is important to remember that while there was wide spread acceptance of evolution by natural intellectual design, he knew the intellectuals wanted to remove any kind of design. At first he strongly rejected the occult and argued for this cause-and-effect mechanism as a purely natural force. He therefore called his view of change "***natural selection***," implying that only natural physiological processes were involved. When he posited his theory ***he had no real evidence*** that such actually occurred.

Questions about Darwin's Survival of the Fittest

Darwin's Own Struggle to Exclude Internal Purpose or Design

Darwin deeply disliked design and miracle. When Lyell asked, "Must you not assume a primeval creative power which does not act with uniformity...?" Darwin quickly dismissed such a "miraculous intervention," saying, "I would ***give absolutely nothing for the theory*** of natural selection, ***if it requires miraculous additions at any stage of descent***." He said if such additions to natural selection were necessary, he "would reject it as rubbish."[20] When Asa Grey skillfully adopted Darwin's view of evolution as an act of God's design, Darwin expressed a clear difference with him.[21] When many knowledgeable and accepted scientists, such as Lyell and Wallace, accepted evolution but insisted that at certain distinct points, especially in man's origin, something more than natural selection was needed, Darwin would have nothing of it. He thus ***divested man of any special act of creation,*** which was seen as putting man on the level with other animals. This had grave implications to many. But it was ***this more than anything else that made Darwin's view accepted*** as a thoroughgoing scientific theory. Man was an extension of chance selection and nothing more.

While at times Darwin said there ***were other natural elements involved*** in the progressive transmutation from one species to another, and from the simplest to the complex animals, he never said what the others were. He insisted he was a thoroughgoing naturalist, yet at times he weakened his views and referred to God in a deistic way. In chapter VI of the first edition of *The Origin...* he tried to deal with the difficulties that had emerged.

Darwin intuitively knew that natural selection by chance was not adequate to explain the biological world. For example, in dealing with the complicated development of the eye, he said,

---

[20] Charles Darwin, *Life and Letters,* II, 210-211.

[21] Ibid., 311-312.

> We must suppose that there is ***a power,*** represented by natural selection or the survival of the fittest, *always intently watching each slight alteration* in the transparent layers (scil of the eye) and *carefully preserving each* which under varied circumstances, in any way or in any degree, each new state of the instrument to be multiplied by the million; *each to be preserved* until a better one is produced, and then the old ones to be all destroyed. In living bodies, variation will cause the slight alternations, generation will multiply them almost infinitely, and natural selection *will pick out with unerring skill each improvement.* Let this process go on for millions of years and during each year on millions of individuals of many kinds; may we not *believe that a living optical instrument thus be formed as superior* to one of glass, as the ***works of the creator*** are to those of man.[22]

In a letter to Darwin dated March 11, 1865, Lyell rightly accused Darwin and Huxley of "deifying secondary causes" and of penetrating "into the realm of the unknowable." [23] In Darwin's later editions of *The Origin...* he attempted to explain eye development.

Darwin's Views Questioned about Lack of Evidence or Errors

Darwin was challenged with the fact that in modern times, the various animals and plants appear distinct and there are not millions of transitional forms. To this he simply said, "We have no right to expect (excepting in rare cases) to discover directly connecting links between them, but only between each and some extinct and supplanted forms." He argued, "We have no just right to expect often to find intermediate varieties in the intermediate zones... Only a few species of a genus ever undergo change; the other species becoming extinct and leaving no modified progeny."[24] But he never answered the important question as to why the millions of intermediate forms he insisted were needed were ***not apparent at all.*** When challenged that there were also no immediacies in the fossil record where he said they should be, he argued the fossil record was only now being uncovered and would uphold his view.[25] These views ***based on similarity*** made them persuasive but could be equally attributed to the same Creator. The only evidence he gave was for micro-change, such as in the beaks of finches. He made other arguments that had skewed evidence. ***Those from embryology were fraudulent.***

A few years before Darwin published *The Origin...* the abbot, J.G. Mendel, published his important theory on genetic inheritance in one of the prominent scientific journals. It revealed the methods and limitations of deviant characteristics of interbreeding. Darwin and his early followers ignored this significant work. W. Bateson, one of the great early genetic scholars, accused him of deliberate neglect and of hindering this science.[26] Mendel had already shown there were controlled predictable micro changes. When Darwin wrote *The Origin...* the science of paleontology was in infancy but it was extensive. However, a majority of the record across the whole earth was untouched. While his knowledgeable friends, who knew the record to that point, argued against him Darwin believed that when the data was in it would support his view of origins as the evidence for his theory. Also, as Loren Eiseley points out, he later extended his theory "to man ***without having available as evidence a single subhuman fossil*** by which he could have

---

[22] Charles Darwin, *The Origin of Species,* 1st ed., 162; 3rd ed., 208, last 146.

[23] See Darwin, *Life and Letters II*, 363 and 384 and Lyell, *The Antiquity of Man*, 469.

[24] Darwin, *The Origin of Species,* 455.

[25] Ibid., 456.

[26] W. Batson on *Mendel's Principles of Heredity,* 1913; Robert E.D. Clark refers to this accusation and R.A. Fishers, *Annals of Science,* 1963, Vol. I, 115, which discounts this. Clark, *Darwin Before and After*, 125-126.

satisfactorily demonstrated the likelihood of man's relationship to the world of subhuman primates."[27] In Darwin's day, none had been found.

Darwin had said in *The Origin…*, "It seems to me the leading facts in ***embryology***, which are *second to none* ***in importance***, are explained on the principle of variations in the many descendants from some one ancient progenitor." In *The Descent of Man*, Darwin argued for extending this argument to the human embryo. He based his arguments on Ernst Haeckel's work and quoted the leading embryologist of the day, Karl Ernst von Baer for support.[28] In the early 1900s embryology was more highly developed, and the arguments for evolution that the embryo recapitulated the species (e.g., embryonic arches as tentative gill slits in fish as being in humans)[29] and others were discounted. However, this idea continued to be taught as fact after it was know to be fraudulent. It was subsequently shown that Haeckel completely misrepresented the facts and that von Baer had in fact rejected the parallelism that was claimed. This was later exposed by Jane M. Oppenheimer.[30] Jonathan Wells, with PhDs from Yale and UC Berkeley, has thoroughly evaluated this argument and shown how the modern embryologists rejected these arguments from embryology as distortion and fraud.

Vestigial organs, such as appendage bones in some snakes, the appendix in man, etc. were said to be useless, slowly-disappearing vestiges from the past. While many of these have been found to have present usages, the real problem is that this claim of slow change is not accompanied by evidence of nascent organs being born. Comparative anatomy revealed no living intermediaries being formed, and the increased development of paleontology based on the evolutionary hypothesis shows *few fossil specimens* that could be considered as intermediary. Homology of vertebrate structures was dismissed as circular reasoning based on similarity.[31] Gradual progressive development began to be questioned due to *lack of a clear mechanism* for evolving change.

Authoritative Rejection of Inherited Characteristics Admitted by Darwin

Soon after Darwin published *The Origin...* Richard Owen, a paleontologist and the greatest anatomist of that time, wrote an article in the *Edinburgh Review* pointing out that since ***variations are not normally transmitted at all***, it was difficult to see how natural selection could work. He insisted this was an error in the working of biology. Owen did not accept Scripture so he was not defending biblical Christianity but was arrogantly defending his findings.[32] Darwin's friend, Joseph Hooker, who had became a famous botanist and followed his father as director of the botanical garden at Kew, spent many hours with Darwin. Hooker was his source of authority in botany; Darwin admitted learning much from him.[33] Hooker repeatedly warned Darwin that he could not see any evidence of transmutation. This also was the opinion of Huxley and Lyell. It was

---

[27] Ibid., 256.
[28] Jonathan Wells, *Icons of Evolution: Science or Myth?,* (Washington, D. C.: Regnery Publishing , Inc., 2000), 81-109.
[29] The fraud and error in embryology is adequately presented in Jonathan Wells, *Icons of Evolution: Science or Myth,* (Washington, D.C.: Regnery Publishing, Inc., 2000), chapter 5, 81-109.
[30] Jane M. Oppenheimer, "An Embryological Enigma in *The Origin of Specie,*" *Essays in the History of Embryology and Biology,* (Cambridge, MA: The M.I. T. Press, 1967), 221-255.
[31] Jonathan Wells, *Icons of Evolution: Science or Myth?* 59-80.
[32] Robert E. D. Clark, *Darwin: Before and After*, (London: Paternoster Press, 1959), 63-65.
[33] Himmelfarb, *Darwin and the Darwinian Revolution,* 207-211.

not until five years later that Lyell endorsed Darwin's view, even though they helped Darwin publish.

More importantly within ten years Fleming Jenkin's article in *The North British Review* demonstrated that in crossbreeding, rare mutants would be swamped out of existence. He maintained that only changes in whole populations could survive. Darwin admitted in a letter to A. R. Wallace in February 1869, that Jenkin's argument "against single variations ever being perpetrated has convinced me.... I did not appreciate how rarely single variations, whether slight or strongly marked, could be perpetrated. The justice of these remarks cannot, I think, be disputed."[34]

Many years of scientific research show that the scientists of that day rightly saw that Darwin's mechanism for evolutionary change and development was false. Carrol L. Fenton, Associate Editor of *American Midland Naturalist*, said,

> At the time Darwin wrote, no one knew much about the nature of variation or heredity.... He did not distinguish these variations from acquired characters mentioned by his grandfather and Buffon. *According to Charles Darwin, all variations belonged to one general class and might be inherited. Later studies showed that* ***this was not true****....* The difference lies in heredity itself, which is a precise and often complex process.... Changes produced by genes, therefore, are passed on, but those caused *only by outside conditions are* ***not****.*[35]

Darwin's Shift in Cause and Final Brief Admission of Need of Intelligent Purpose

Darwin acknowledged his earlier errors regarding acquired characteristics and later held to pangenesis—the view that the body gathered material particles from all over the body, thus permitting somatic modifications that were passed on. This too was a Lamarckian idea revised from the early Greeks. It also was proven wrong as scientists learned more about genetics and molecular biology.

Subsequent evidence found no support for, but more evidence against Darwin's main thesis. Darwin's argument for a theory that removed God and had gained him fame was now losing strength. His long efforts to eliminate purpose from his theory seemed to lose credibility. In Darwin's final edition of *The Origin of the Species* he made an almost hidden astonishing admission when he said,

> There *must be* ***some efficient cause*** for each slight individual difference, as well as for more strongly marked variations which occasionally arise; and if ***the unknown cause*** were to act persistently, ***it is almost certain that all the individuals of the species would be similarly modified.***[36] (Emphasis added)

Concerning this, Loren Eiseley, the learned evolutionary scientific historian said, "Darwin with his gift for compromise has here accepted a point of view which, if pursued, would be ***metaphysically***

---

[34] Loren Eiseley, *Darwin's Century: Evolution and the Men Who Discovered It,* (Garden City, NY, Anchor Books of Doubleday & Company, Inc. 1961), 210.

[35] Carroll L. Fenton, "Evolution," *The World Book Encyclopedia*, published by Field Enterprises Inc., 1959, 2426-2427.

[36] Loren Eiseley, *Darwin's Century,* (Garden City, New York: Anchor Books, 1961), 216.

fatal to his system*....”[37] (Emphasis added) Thus Charles Darwin ***never escaped the need for an intelligently ***designing cause in accounting for the species***.

Dominance of Desire for Material Progress was the Climate

When Charles Darwin conceived and wrote his view of biological evolution the intellectual world was excited about new scientific methods and theories. The prevailing philosophies held man in high esteem and endorsed a pantheistic ideal world (Hegel, et al.) in which man hoped and conceived of material progress though natural forces (Adam Smith, Karl Marx). Human pride was embracing a determined view of the world as working by uniform and gradual natural forces that could be understood and controlled to produce progress. Herschel and Kant deemed this to be true of the stars, Laplace the planets, and (earlier) Hutton and contemporary, Lyell, the earth. Their views were all false assumptions. But the ***spirit of the times*** was like a team of horses pulling all thought in a direction ***to promote human progress.*** This idea had already promoted a number of biological evolutionary concepts such as those of Lamarck, Erasmus Darwin (Charles' grandfather), and Robert Chambers. While Lyell had removed God and catastrophe from geology, he at first denied any linear progress, such as with Lamarck, to avoid designing power.

Only after Darwin's theory made progress popular, did they concede to it. The reason it became popular was that it offered biological evolutionary progress without including any natural innate intelligent force, but relied instead on chance in nature. Darwin protested Lamarck's view of progress because it was caused by an occult intelligent process. But he once admitted, "the conclusions I am led to are not widely different from his (Lamarck's), though the ***means of the change*** are wholly so."[38] The previous views of evolutionary progress helped set the stage for Darwin to offer his hope of the whole organic world tending inevitably to "progress toward perfection."[39] Those preliminary evolutionary views laid the foundation for the acceptance of Charles Darwin's ideas toward progress by chance in the modern mind.

Worship of Man's Wisdom and Hostility to God Propelled Acceptance

The real reason Charles Darwin's theory was accepted was that it gave a reason to replace God with naturalism and give great hope in man alone. Very important was the optimism of human ability for ***progress*** through science. In the early twentieth century naturalistic science was the dominant view in public education and became a dominant force in politics. While Darwin did not explicitly give a complete description of ***inevitable progress***, it was explicit in the method which he had borrowed from Lamarck. He, however, shifted from innate natural power (as in pantheism) to chance by natural selection. The descent of all forms from lower to higher in a tree of life made it a method of progress and Darwin expressly said so at times. In the late nineteenth century, the secular universities were having a powerful influence on business and politics, and Darwin's theory of natural selection appealed to the strong anti-god and anti-intelligent design view of the intellectual elite.

Darwin clearly understood the implications for religion and the desire of many intellectuals, including himself, to exclude God. He waited a long time to publish *The Origin...* in order to avoid the controversial question of the origin of life and man's chance ascent to dominance. By

---

[37] Ibid.

[38] Darwin to Hooker, *Life and Letters*, II, 23, cf. Himmelfarb, *Darwin and the Darwinian Revolution*, 179-180.

[39] Loren Eiseley, *Darwin's Century*, 283; *The Origin of Species*, 93.

the time he published this book he professed his ***disbelief in God***. It was his great supporter, Thomas Huxley, who added the idea of the origin of life by chance. Darwin added the naturalistic development of man in his *Descent of Man* over a decade later (1871). Man was merely the highest animal. But it was Huxley who played on the anti-religious theme and feelings of the time, and publicly intimidated those with religious beliefs. Darwin, whose character would never allow him to do so publicly, expressed similar sentiments in private. When Darwin died in 1882 his theory was widely accepted but ***not yet dominant***.

Darwin's untested theory quickly became ***the new religion*** of man and worship of man in the Western world, profoundly influencing human thought. In the introduction of the re-publication of *The Origin*... at the Darwinian Centennial Celebration the noted evolutionist W. R. Thompson said, "The concept of organic evolution was an object of genuinely religious devotion...." The keynote speaker at this event, Sir Julian Huxley, said,

> It is only through possessing a mind that man has become the dominant portion of the planet and the agent responsible for its future evolution.... And he must face the job unaided by outside help. In the evolutionary pattern of thought there is no longer either need or room for the supernatural. The earth was not created; it evolved. So did all the animals and plants that inhabited it, including our human selves, man and soul as well as brain and body. So did religion... Such supernaturally centered religions are early organization of human thought... They are destined to disappear in competition with other truer and more embracing thought organizations, .new religions which are surely destined to emerge on this world scene.[40]

Great Gap of History Was between Man and Animals Yet Unexplained by Determinism

Emergence of Man by Natural Evolution- the Major Issue with No Evidence

Arguing that man was the result of natural selection processes so that he emerged from lower animals was ***difficult to explain***. Man is so different from any primate or any part of nature. Human intellect and consciousness, language ability and erect mobility, hand dexterity with mental coordination, and the vast differences in human sexual practices offered unexplainable problems for determinism. At the time leading Christian scientists such as Adam Sedgwick spoke of the advent of man "as breaking in upon any supposition of zoological continuity–and utterly unaccounted for by what we have any right to call the laws of nature."[41] Lyell, as the leading geologist, and other determinists emphasized there was no evidence for bridging the gap from highest animals to man. Darwin fully realized this great difficulty. He deliberately omitted this issue from *The Origins*... in 1859, while admitting to Wallace in a letter that it was "the highest and most interesting problem."[42] He did not make such a claim of transgressing this gap until eleven years later when his theory of evolution had gained acceptance by many scientists.

Loren Eiseley, the pro-evolutionary historian, pointed out Darwin had not one single fossil to base his views on when he wrote *The Descent of Man*.[43] So Darwin's claim of human evolution was not based on historical evidence, but entirely on his theory of chance development of survival qualities and of inheritance of these characteristics, both which were rejected by leading biologists

[40] Julian Huxley, "The Evolutionary Vision," *Centennial Review of Arts and Science*, vol. III, (Davney-Marx, 1859-1959), 252-253, 1958.

[41] Adam Sedgwick, *A Discourse on the Studies of the University of Cambridge*, 5th ed., London 1850, p. xlv. *Also Proceedings of the Geological Society of London*, 1831, Vol. 1, 305.

[42] LLD, Vol. 2, p. 109.

[43] Eiseley, *Darwin's Century*, 256.

and his friends in the scientific community. Darwin denied progressionist theories based on design. As Eiseley said, "His work had destroyed the man-centered romantic evolutionism of the progessionists. It had, in fact left man one of innumerable creatures evolving through the play of secondary forces and it had divested him of his mythological and supernatural trappings."[44] But Darwin truly believed in a progressive direction in his chance and survival theory. In the final edition of *The Origins...* he wrote, "All corporeal and mental endowments will tend to ***progress toward perfection***."[45] The idea of natural perfectibility of man without God lay at the root of the deterministic view. Their ***main objective*** of the intellectual Enlightenment movement was to free man from God. The whole enlightenment movement was the idea that man could become wise as God, better and more right knowing good and evil, becoming superman. Darwin knew quite well the implications for man. It linked man to the animals but made him the superior in nature to make his own way without any accountability to God. This was the final step of arrogance in the progressive evolutionary argument.

Diligent Data Research Yielded No Evidence of Gap Transition

However, as all his naturalistic friends had warned Darwin, the knowledge gap between men and animals was extremely wide and without any evidential data. For this reason, Charles Darwin who was sensitive to criticism and rejection waited a dozen years to argue for bridging that gap. In the *Descent of Man* (1871) Darwin speculated that man descended from what he considered his closest allies—extinct apes such as the gorilla and chimpanzee in Africa. ***Similarities do not prove*** extension from animals but may be evidence of the same Creator. In Darwin's and modern times the early location of primates was thought to be in Africa.

Feverish efforts were made by scientists to find such evidence because it was the central factor of rebellion of human bias. After *The Origins...*, but especially after *The Descent of Man,* tremendous efforts in paleontology, anthropology, and biology were made to find evidence of missing intermediaries and trends that supported the idea. The great pursuit in paleontology that followed found different types of men such as Cro-magnum, Neanderthal and others in Europe, and Java man in Asia. But these, while different, were not seen as the missing links. The greatest pursuit for intermediary forms was in Africa, which Darwin suggested was the place of evolutionary transition.

Philosophy of Man's Origins Determined by Biased Interpretation of Data

During the last half of the nineteenth century the battles raged over man's origins. Referring to this time of the early fossil finds, pro-evolutionary Eiseley said, "That the activities of confirmed evolutionists, as the Darwinian enthusiasm began to mount, are sadly revelatory of a state of mind in its way as *dogmatically fervid* as that of those opposed to the evolutionary point of view."[46] The nature of various fossils was perverted to meet the demands of each side. For example, in referring to fragmentary Neanderthal remains of La Naulette, one science book described the specimen as extremely apelike "with huge projecting canines," when in fact *the teeth of the specimen were* ***missing***. Eiseley described the unscientific speculation saying, "No known form of fossil man possesses gorilloid canines. These descriptions are *the product of*

---

[44] Ibid., 195, 196.

[45] Charles Darwin, *The Origins of Species,* Modern Library Edition, 155-156, cf. Loren Eiseley, *Darwin's Century,* 216.

[46] Eiseley, *Darwin's Century*, 277.

imagination...."[47]

After fifty years of searching the nineteenth century ended with ***no evidence*** that bridged the gap between animals and men. In fact the few fossil finds showed that the skulls of these earlier men revealed much larger brains than modern Homo sapiens which went counter to the views of Darwin that man's brain increased gradually. Eiseley has commented,

> Neither Cro-Magnons nor Neanderthals showed the rapid mental regress which had been assumed, in the underestimated time scale of that day, to characteristics the skulls of genuine primates, particularly in the light of the assumptions which had been made about various of the living races of man. ... Later as the forests were cleared and the apes were seen in the sunlight, the gap *loomed a* ***little larger*** between man and his beasts.[48]

A major jolt to this absence of information leaving the gap between animals and men was the findings and report of Alfred R. Wallace who was the published cofounder with Darwin of the theory of "natural selection by survival of the fittest" derived from the Malthusian theory. Wallace had recognized the gap of evidence between animals and men in the claim that primitive man has progressed slowly from animal intellect to human. More than any man, Wallace sought evidence for this and later visited many of the known primitive tribes especially in New Guinea, Indonesia, and elsewhere. Wallace concluded that all the primitive people had superior intelligence as do western men, and later publicly rejected the idea of man evolving from primates and questioned the evolutionary process.

The findings of ancient history in Egypt and Babylon began about the mid-nineteenth century but the science of modern archeology began to emerge only toward the end of the nineteenth century and more fully in the early twentieth century. That science began to produce much of the earliest history of man. Much of this history produced stories that corresponded to biblical history in Mesopotamia, Palestine, and Egypt. But these facts of archeology that revealed early history of man were ignored by the evolutionists or treated as myths.

Genetical studies also seemed contrary to change by inheritance. Mendel's great genetic thesis was resurrected and became prominent after 1900. Soon the modern science of genetics was understood as controlled variations that were transmitted. In the twentieth century August Weismann "actually diverted the study of evolution into the pathway which has led on to the great modern advances in the field of genetics."[49] He said,

> The transmission of acquired characters *is an impossibility*, for if the germ plasma is not formed anew in each individual but is derived from that which preceded it, its structure and above all its molecular constitution *cannot depend upon the individual* in which it happens to occur....[50].

In the first quarter of the twentieth century, evidence was against Charles Darwin's theory. Darwinism declined as the certainty of the genetic mechanisms were explored and became general

---

47 Ibid., 277.

48 Ibid., 284, 285.

49 Loren Eiseley, *Darwin's Century*, 218.

50 August Weismann, *Essays on Heredity,* (Oxford, 1892), Vol., 80-81. See Loren Eiseley, *Darwin's Century*, quote on 219.

knowledge. Reviewing the rise of genetic knowledge, Eiseley said, "For a time there was an understandable feeling that Darwinism was moribund."[51]

Burning Search to Confirm Man's Wisdom to Solve His Problems without God

By the time Darwin's *Descent of Man* was published twelve years after *The Origin...* society had accepted the new skepticism about God of determinism, and was becoming highly materialistic, worshiping "that which is created rather than the Creator" (Romans 1:25). Europe was in the throes of industrial revolution and major class divisions with the rich getting richer and the poor poorer. Karl Marx had published his Communistic beliefs, *Das Kapital,* in 1867. He wanted to dedicate his work to Darwin, but Darwin declined the honor. Anti-god biblical criticism especially of the Old Testament also reached its peak about this time. Efforts to improve the understanding of Adam Smith's view of capitalism to eliminate the problem of inequality and class resentment were made and will be discussed under the soft sciences. The turmoil in society searched for ***human*** wisdom to solve their problems and support this radical blatant change in interpreting human history and the explanation of human nature.

As the twentieth century dawned, scientific determinists arrogantly plunged ahead to change the description of man to make man's history fit their evolutionary views of progress by human wisdom. Scientific anthropologists have long made unsubstantiated claims to knowing how man evolved from higher primates. Many charts of how man descended from apes or other primates have been drawn and taught. Most significant was Henry Fairfield Osborne's exhibition in the American Museum of Natural History, prominently displayed and shown to thousands of children and adults every year as the evidence of man's descent from apes.

A Fraudulent Public Media Drama was the Turning Point for Intellectuals in America

Triumph for Darwin's Evolution through the Scopes Monkey Trial

When Charles Darwin published, Robert Chambers' *Vestiges* had flooded England with the idea of evolutionary creation by the Christian God. The passion for evolution as a theory of progress without God was so great that even without evidence Charles Darwin's theory was accepted. Carl Zimmer, in his nationally recognized book, *Evolution: The Triumph of an Idea*[52] introduced by primer evolutionist Stephen Jay Gould, comparative zoologist of Harvard, claims the victory of Darwin's evolutionary theory came at the crisis called the Scope's Monkey Trial in Dayton, Tennessee in 1925. Zimmer also offers what he believes is subsequent evidence for what he believes is the truth of Darwin's theory.

Denial of Freedom of Speech to a Threatening Theory

Darwin's theory (published 65 years earlier) had subsided with little substantiating evidence in the nineteenth century, but was reemerging and being taught in many public schools. The country was growing in prosperity and the educated elite were influencing politics and sought to prove everything could be understood by natural laws. In 1922 the Kentucky Baptist State Board of Missions had passed a resolution calling for a state law to prohibit the teaching of evolution in the public schools, which would have prohibited freedom of speech on the subject of origins. It lost by one vote in Kentucky but resulted in a crusade by William Jennings Bryan, a prominent national politician, and many others to pass laws to prohibit the teaching in other states. Tennessee

---

[51] Eiseley, *Darwin's Century*, 229.

[52] Carl Zimmer, *Evolution: The Triumph of an Idea,* (New York: Harper Collins, 2006).

was the first to actually pass such a law in what was known as The Butler Act. Darwin's theory of "the survival of the fittest" was presented as the mechanism for its operation. The theory taught that individuals in combat with the environment and with each other would result in the best chance of survival. Social Darwinism and Monism, as well as Karl Marx's Communism produced inhumane and cruel and oppressive efforts. Bryan and many people saw Darwin's views as a great threat arising from the loss of faith and moral accountability. Bryan and others wrongly sought to curtail the freedom of speech on evolution as the answer, which was a violation of the first amendment and unconstitutional. What those in opposition should have done was to launch an effort to express their views in the classrooms to help people evaluate the issue. Instead, they resorted to efforts to control by force. This had ignited the anger of the strong believers in evolution.

Promotion of a Massive Deception

The Scopes Monkey Trial was a planned promotion of evolution. Very few people understand what happened. It was planned to deceive the public and be propaganda rather than a true science report. The little city of Dayton, Tennessee wanted a show trial for an attraction for business that was lost from a diversion of the highway. The American Civil Liberties Union had offered to defend anyone who would break the law against evolution, so they could take it to the Supreme Court to overturn the law as violating teacher's freedom of speech. The Dayton town leaders along with the ACLU talked a physics teacher, John Scopes, who had as a substitute taught human evolution from the textbook, *Civic Biology,* into being the test. Watson Davis saw this as an opportunity for his informational organization, Science Service to work with the ACLU. The trial was boosted into national prominence by those involved. Amy Maxmen recently reported in a recent book disclosing the recovery of old history of planned perversion in the reporting of the trial. Her statements show the bias.

> The Science Service Executive Committee agreed to give its reporters $1,000 (more than enough to buy a new car then) to cover the trial. The committee also decided to reject neutrality, supporting the defense on the side of evolution. Davis and Frank Thone, the senior biology editor of Science Service's newsletters, acted as journalists as well as informal assistants to the defense. They sought out top scientists to comment on the trial and lived among the scientists and biological teachers (in a big house rented for the defense). Magazines and newspapers across the country ran columns published by Science Service, now Society for Science & and the Public, publisher of Science News.[53]

Bryan, a well known national political figure who was in Tennessee offered to prosecute the case. An attorney, Clarence Darrow who had just gained national prominence by acquitted two young brutal murderers on the grounds of insanity, volunteered to the ACLU to defend Scopes. Neither Bryan nor Darrow were fully informed or representative of the issues they argued. Bryan had actually accepted a form of evolution and was not arguing the Genesis creation story (as most people today think) but the social consequences of Darwin's views. Darrow appealed on evidence not yet certain or proven. He won mainly by skillfully causing Bryan to be publicly contradictory and look foolish. While Bryan won the trial, the sentiment was turned in favor of evolution across the nation. The Tennessee law was repealed.

---

[53] Marcel Chotkowski LaFollette, *Reframing Scopes: Journalists, Scientists, and Lost Photographs from the Trial of the Century* (Kansas City: University Press of Kansas, 2008), 172 p.; as reported by Amy Maxmen, *Science News*, August 16, 2008, 31.

Evidence of Scopes Trial Reevaluated as Fraud

The arguments of Clarence Darrow to promote the truth of the evolutionary point of view went beyond Darwin's old arguments to include dramatic illustrative data that Henry Fairfield Osborn, curator of the American Museum of Natural History in New York furnished, the results of which Osborn highly promoted (Osborn, *The Earth Speaks to Bryan*, 1925). Osborn had studied under Thomas Huxley, Darwin's attack dog, who had taught Osborn to put together fossil illustrations to present the descent of animals (e.g. Huxley, composed what was claimed to be the evolution of the horse). Osborn composed the "Hall of Men" showing man's evolution from the ape. He used fossils (some found thousands of miles apart and some drawn only from a tooth) beginning with the ape and then naming other intermediaries to man. In this line up of progression was included the Piltdown Man, the Nebraska Man (Hesperopithecus haroldcookti), a hunched over Neanderthal, Cro-magnum and a modern homo sapien. Osborn's progression was strongly argued by Darrow as evidence of evolution.

In subsequent years honest scientists proved this picture of man's evolution used in the trial was a complete hoax. Osborn's "Hall of Men" was later embarrassingly removed to the attic of the Museum. Incidentally Stephen Jay Gould, the spokesman for evolutionary theory in the midst of his 1432 page book, casually admits on page 760 that all the icons or classic examples of evolution such as that of the evolution of the small to large horse, of man from apes et al. used in text books have been disproved.[54] But they continue to be used. After 140 years of searching for ape–human intermediaries (hominoids), there is still no certain line of descent. Moreover, William Jennings Bryan's fears of degrading of morals by the rejection of a Creator for evolution of man have since been proven to be true.

Results of the Scopes Trial for Worldviews Involving Evolution and Creation

As a result of the Scopes trial, the supreme court of Tennessee later overturned the law of the Butler Act. But the perverted picture of the facts communicated to the American public changed the belief of many of the intellectual elite of the educators and their attitude toward creation. Zimmer correctly said that the trial changed the sentiment of many in the nation and is now dominant in science. The lie of the Scopes trial was a turning point of scientific thought for America.

Rewriting Human History to Present Man as Produced by Evolution

The whole culture of man as evolutionary theory required that man's ***art** must have **emerged slowly*** from stick figures to detail and color. Language was first thought to be only musical communication; man was said to know nothing of agriculture at first and for that reason was a hunter–gatherer, who gradually learned to build structures, and to gather into towns and communities. By evolution, man gradually emerged from an animal to greater intelligence and creativity. This was said to occur by Darwin's chance emergence and not intelligent design in nature or otherwise.

By the first half of the twentieth century archeologists and experts in almost every field were scouring the world and compiling much scattered scientific data about man's early tools for hunting, agriculture, art, literacy, industry, medicine and various aspects of life in an effort to show that man slowly emerged. Evolutionists assumed at first that man worshiped nature and then higher gods as aspects of nature. The whole change in the philosophy of how religion and the biblical God was known changed to embrace a view that man by evolution created God. This view

---

[54] Stephen J. Gould, *The Structure of Evolution*, (Cambridge, Massachusetts: The Belknap Press of Harvard University, 2002).

is presented in Section Five. Using scattered limited data to fit their philosophy, ***the whole of human history was rewritten showing a gradual progression from apes to humans.*** All known ancient history was perverted, virtually ignored, or never studied in order to rewrite this history. The brilliance of all the ancient civilizations as uncovered by archaeology was overlooked and in many ways perverted because such evidence of early man's intelligence did not fit scientists' pre-conceived ideas of man's development.

With the extension of paleontology to human bones as fossils and the birth of archaeology, the whole science of anthropology became prominent. The so-called science of "comparative religion" along with higher criticism of the Bible had cause a powerful Enlightenment movement in the churches toward unbelief. But by the 1940s and especially the 1950s, a complete effort to review the history of man's development from a naturalistic and evolutionary perspective emerged. Scientific determinism required that ***everything known about man be revised to fit the evolutionary theory.*** Some examples of the scholarly imaginative authors and their books are: V. Gordon Childe, Professor of Prehistoric Archaeology in the University of Edinburgh, who wrote *Man Makes Himself (1951)*, and *What Happened in History;* and Ashley Montague, Chairman of the Department of Anthropology at Rutgers University, who wrote *Man: His First Million Years* (1957) These claimed to trace human development from the Paleolithic (savage), Mesolithic, and Neolithic or progressive development of the Stone Age into the times of civilization during the Copper, Bronze, and Iron Ages, especially in Eurasia and North Africa and later other parts of the world. Along with human development, these authors had to explain how the system of natural laws worked to progressively improve man's conduct.

It is incredible that contrary to scientific facts, without evidence of a mechanism of transmutation, and contrary to emerging known human history, the intellectual educational community that had taken over the schools of Europe and America managed to present as truth a gigantic lie that progressively removed God from the public arena. How this was legally done in America will be presented in later chapters.

SECTION TWO

THE UNIFORMED NATURAL DETERMINED UNIVERSE FOUND TO BE INDETERMINATE

PART III

ENLIGHTENMENT ABOUT THE PHYSICAL UNIVERSE

The heavens declare the glory of God, and the sky above proclaims his handiwork. Day to day pours out speech, and night to night reveals knowledge. There is no speech, nor are there words, whose voice is not heard. Their measuring line goes out through all the earth, and their words to the end of the world. (Psalm 19:1-4a)

Chapter 7
Indeterminacy in the Physical Cosmos

Review and Orientation to Modern Century of Error and Mystery

Chapter 1 outlined man's anti-God bias. Over a period of 275 years the Reformation brought about a return to biblical Christianity and the re-discovery of God's work in creation. During this time—a period of material prosperity—man's pride reasserted itself. In the mid to late eighteenth century a deterministic theory of knowledge and claims of uniform natural law emerged. These became the grounds for rejecting biblical teachings about God as the Creator and the continuing ruler and judge of the world responsible for past cataclysmic judgments. Determinism was not based on scientific data, but on the nebular hypothesis, a theory about the origins of the solar system that required millions of years for the system to form. Scientists then applied the same theory to geology, assuming the entire universe was much older than originally believed. Adoption of this a mechanical materialistic view of nature which claimed to eliminate evidence for God eventually led to an attack on biblical history, first questioning the historical reality of the Old Testament and later the crux of Christianity—the incarnation, life and ministry of Jesus Christ. By the beginning of the twentieth century the uniform deterministic view of knowledge was a major force in the intellectual circles and educational systems of Europe and increasingly in the United States. The educated elite, operating on the belief that religion is based on man-made myths and that all mystery could be removed from scientific discovery, propagated their theories of deterministic science. By 1865 they had achieved control of most of the oldest religious colleges in the eastern United States.

But just as the biblical world view of a sovereign God who created all things was rejected in the nineteenth and twentieth centuries, more objective and sophisticated modern science was providentially uncovering many new discoveries that began to show the error of determinism. As our vacation experience of moving from a canoe to a motor boat to see the landscape illustrated that what we found was much different than what it first appeared, so too scientists, with new scientific instruments began to gradually extend their vision and reveal that the universe is filled with mystery. Now less than a century later what was thought to be uniform and determined is shown to be beyond man's understanding and increasingly mystical.

In the eighteenth century scientists' information about the universe was very limited. Their *a priori* assumptions regarding determinism seemed revolutionary. They assumed greater prosperity and held out an optimistic view of man's goodness. As Stephen Hawking said, "In Newton's time it was possible for an educated person to have a grasp of the whole of human knowledge, at least in outline."[1] The 1758 publication and circulation of an encyclopedia of knowledge marked a major turning point, creating a tremendous sense of pride in some of the educated people. But even Hawking admits that while ***man felt he knew about everything***, many scientists realized that much was still unknown. Some felt that honesty required a place for God to operate, at least deistically. Hawking said that for these, "God was confined to the areas that nineteenth-century science did not understand."[2] But for Laplace and others in the early nineteenth century, they *a priori* adopted a philosophy in which there was already no place for God, and the faith of Christians and Jews was seen as unenlightened and unprogressive.

---

[1] Hawking, *A Brief History of Time*, (New York: Bantam Books, 1988), 167.

[2] Ibid., 172.

New Twentieth Science Began to Show Uniform Determinism Too Small

As the twentieth century dawned determinisms' limitations and errors became evident as Newton, Boyle and other leaders had warned earlier. Whole new areas of science were further explored; earlier theories were found to be incorrect? New areas of science developed: archaeology, paleontology, genetics, micro-molecular biology, and others; all revealed new findings and perspectives. By the middle of the twentieth century new knowledge began to exceed man's comprehension; man could not understand much of the new data about the physical universe. This confirmed what the early Christian scientists had argued in their opposition to determinism: excluding God was wrong. But faith in man's mind continued like a tidal wave, denying any supernatural acts in nature. Intellectuals' blind intolerance of God solidified along with a greater determination to eradicate belief.

New Indeterminism in the Basic Atom of the Universe

A Complete Overturning of Certain Understanding in Physics

Early Greek studies by Democritus, and the Epicureans, and later Gassendi had explored atomic theory. Newton and Boyle used atomic ideas in their physical speculations. Although the individual atomic elements had not yet been discerned, John Dalton found that the properties of gases were explained by atoms and that combining weights in chemical combinations gave the relative weight of the gases. With the developments of Mendeleev and his periodic table, there was a conviction there was centrality about the knowledge of the atomic theory. In the closing years of the eighteenth century J. J. Thomson and Rutherford shattered the chemical atom into elements. The discovery of radioactivity allowed the discernment of individual atomic composition. The hydrogen nucleus was identified as the proton with various electrons added for the various elements. This seemed to determine the atomic theory.

These last discoveries opened the way to radical change in physical thought. In study that led beyond the established atomic ideas, they found a new physics. From 1900 to 1930 scientists' conception of the universe underwent radical change. The popular Russian scientific writer, George Gamow was involved in many of the studies of that time. He was the first to suggest the chemical development idea of the Big Bang. He describes the shift in thinking that occurred:

> The opening of the twentieth century heralded **an unprecedented era of turnover** and reevaluation of the classical theory that had governed physics since Newtonian times.[3]

Again—an ***unprecedented turnover occurred in classical theory of the world*** since the Christian world view had been abandoned.

Change in Definition of Science

The radical shift from the unknown of the quantum theory indeterminacy in atomic theory expanded the definition of science. Scientists began to assert that a theory could be proven to be completely true. Early Christian scientists such as Newton, Boyle, et al. argued that theories based on observations could be shown to be the natural working but ***may have unknown probabilities***. In the end the probability or "chaos" factor had always been present, but pride and confidence in

---

[3] George Gamow, *Thirty Years that Shook Physics,* (Garden City, New York: Anchor Books, Doubleday & Company, Inc., 1966), Chapter 3, The Revolution in Physics, 1.

human knowledge was blind to this. As probabilities mounted, a new view of proof was introduced. The philosopher Karl Popper changed the idea of proof to "***falsification***" as an element of scientific determinism. He taught a theory is scientific only if it is not proved to be false. For example, he said that the fact that sun has risen every day is not a demonstration of proof that it will rise. It is true because it has not failed to rise.

At the end of the book, I will suggest limits to understand God's normal way of working or "natural laws" and limits to God's special way of working or special providence and miracles. Also, there will be an effort to show the freedom of true biblical faith. But these ideas will be limited.

Atoms Act in Quantum and Indeterminately

George Gamow described this major turnover within the field of physics:

> Two great revolutionary theories ***changed the face of physics*** in the early decades of the twentieth century: The Theory of Relativity and the Quantum Theory. The former was essentially the creation of one man Albert Einstein, and came in two installments: The Special Theory of Relativity, published in 1905 and The General Theory of Relativity, published in 1915. Einstein's theory of relativity called for radical changes in the classical Newtonian concept of space and time as two independent entities in the description of the physical world, and led to a unified four-dimensional world in which time is regarded as the fourth coordinate, though not quite equivalent to the three space coordinates. The Theory of Relativity introduced important changes in the treatment of the motion of electrons in an atom, the motion of stellar galaxies in the universe.
>
> The Quantum Theory, on the other hand, is the result of the creative work of several great scientists. Max Planck (1900) stated that radiant energy can exist only in the form of discreet packages called light quanta. Einstein (1905) used this to explain the emission of electrons from metallic surfaces irradiated by violet and ultraviolet light. Arthur Compton then showed that the scattering of X-rays by free electrons followed the same law as the collision between two spheres. These established the novel idea of quantumization of radiant energy. This was expanded to mechanical energy of electrons in an atom by Niles Bohr (1913), and he established 'quantumization rules' for the mechanical systems of atomic sizes confirming Ernest Rutherford's planetary model of an atom. This 'stood in sharp contradiction of all the fundamental concepts of classical physics.'... Bohr's first paper on the quantum theory of the atom led to cataclysmic developments. Within a decade, due to the joint efforts of theoretical as well as experimental physicists of many lands, the optical, magnetic, and chemical properties of various atoms were understood in great detail.'[4]

While further refinements and developments are detailed and complicated, it is important to recognize that Werner Heisenberg, with his colleague Max Born wrestling with these problems, demonstrated the "principle of ***indeterminacy,***" or uncertainty, which says that when an electron is in motion, both its position and velocity cannot be measured with high precision at the same time. As Gamow said:

---

[4] See Gamow, *Thirty Years That Shook Physics*, 2.

> Einstein was probably the first to realize the important factor that the basic notions and laws of nature, however well established, were valid only within the limits of observation and did not necessarily hold beyond them. For people of the ancient cultures the Earth was flat, but it certainly was not for Magellan, nor is it for modern astronauts. [Similarly] the basic physical notions of space, time, and motion were well established and subject to common sense until science *advanced beyond the limits* of the confirmed scientists of the past. Then *arose a drastic contradiction*...which forced Einstein to abandon the old 'common sense' ideas of the reckoning of time, the measurement of distance and mechanics and led to the formulation of the non-common-sensical Theory of Relativity. It turned out...things were not as they should have been. Heisenberg argued that the same situation existed in the field of Quantum Theory....' Under classical physics in studying the trajectory of a moving material body we could decrease the errors of the coordinates and velocity of the moving particles. But he pointed out, 'the existence of quantum phenomena reverses the situation.'[5]

The fact that negative electrons jumped from orbit to orbit and radiated in quanta according to the size of the jump indicated that the electrons had an innate knowledge of frequency to vibrate and where to stop which defied common logic. This troubled Einstein. Dietrick E. Thomsen observed that quantum theory also violated the normal Aristotelian logic of cause and effect Martin Klein of Yale University said, this view is an "ingenious mixture of Platonism and old physics, difficult to understand."[6]

Logical Inconsistency of Quantum Physics Challenged

The quantum theory challenged the whole uniform deterministic philosophy—a challenge the unbelieving mind could not endure. Einstein and his students, along with others, felt the quantum theory was incomplete. They performed the EPR experiment to show that the paradox did not exist. Moreover, he felt that if the *observer moved* ***from the microscopic*** *atomic level* ***to the macroscopic*** *or astronomical one*, a logical unity could be found that fit classical physics. But Einstein and his researchers repeatedly failed to find discrepancies in the quantum theory.

Moreover, John S. Bell of CERN Laboratory in Geneva and other scientists using electron microscopes consistently showed results that supported the mystery of the quantum theory. While extensive atomic experimentation has been carried on using highly sensitive instruments, including cyclotrons, three families of leptons and quarks have been isolated. Yet there is still no evidence to change the quantum theory. This mystery, which Einstein called "spooky action at a distance," remained. Niles Bohr believed one could never explain or understand this quanta action and insisted on this as a complementary principle, namely, contrary modes being somehow united in a single entity.

Practical Regularity in Atoms with Irregular Probability

William Cecil Dampier, an outstanding science historian and honored lecturer in physics at Trinity College, Cambridge has said,

---

[5] Ibid., 106-107.

[6] Ibid.

> Thus, twenty-five years after the atom was resolved into electrons, electrons themselves were resolved into an unknown type of radiation or into a disembodied wave-system. The last trace of the old, hard, messy, atom has disappeared, mechanical models of the atom have failed, and the ultimate concepts of physics have, it seems, to be left in the decent obscurity of mathematical equations.

He also said, "Thus the quantum theory, like many physical and chemical problems is an *exercise **in probability***."[7] Once again scientists had to confront the possibility of probability in understanding the physical universe, just as the godly Pascal had suggested centuries earlier.

With the development of the quantum theory the old ideas of time and space changed. The categories of thought that Kant held so dogmatically were no longer seen in the same way. A new perspective formed in which man could no longer feel he was God because the center of the modern physical world was beyond his comprehension. He could *describe* how it worked, not why nor understand the power behind it, but developed uses for atomic energy.

Most of the recent advances of modern science including the atomic bomb, nuclear energy, the laser, the electron microscope, the semiconductor, the transistor, the superconductor, etc. has developed through the observations of the quantum theory and the theory of relativity. Through these theories scientists found that the very essence of matter did not work completely according to logical cause and effect. But even in the face of this mystery the idea of a determined scientific world was still dogmatically proclaimed by many scientific leaders. Therefore, scientists diligently sought to discover a ***standard model*** of the atomic world that they could understand and control as uniform and determined.

Quest for the Standard Atomic Model

New Instruments to Probe the Atom for New Theory

These modern theories of physics and the practical developments resulting from them continue to expand our knowledge of matter as well as ***increased its mystery***. Determinists tried even more diligently to remove mystery from atomic discoveries. The opening of the door to mystery went contrary to the natural deterministic theory and left an opening for God which they wanted to eliminate. Cyclotrons, instruments that speed up atomic particles to near the speed of light, create high speed collisions and fragment the atom so that atomic energy can be studied and described further. Some major results are summarized here only to emphasize the further uncertainty that resulted.

In the late 1960s and 1970s physicists developed a "standard model" of atomic particles with three families of leptons and quarks, the first family being lightest in mass and the last family being the heaviest. Each proton or neutron in an atom's nucleus theoretically is composed of three quarks bound together. The first family has up and down quarks, an electron and electron neutrino leptons. The second has strange and charm quarks with muon and muon neutrino leptons, and the third has bottom and top quarks with tau and tau neutrino leptons. This model was conceived and tested in cyclotrons in the United States and Europe. Soon after the model was proposed by a group of scientists experiments confirmed the existence of most, not all of the particles.

Researchers poured millions of dollars into searching for the ***top quark.*** It was not until

---

[7] See William Cecil Dampier, *A Shorter History of Science,* (New York: Meridian Books, 1957), 151, 154.

March 1995 when two teams of about 450 physicists using a particle detector at Fermilab demonstrated conclusively that this elusive subatomic particle does exist. It is not found in ordinary matter as other quarks are. It is as ***heavy as an entire atom of gold***, having a mass that has been measured at 176 billion electron volts (GeV). This top quark is 40 times larger than the next heaviest known particle. The tau neutrino was found in 2000.[8]

Standard Atomic Model Becomes Mystery

Gravity has *not* been included in the standard theory and remains a mystery. Other than gravity there are three interactive forces in matter: (1) the strong nuclear force that holds atomic nuclei together, (2) the weak nuclear force that causes radioactive decay, and (3) electromagnetic force in magnetism and radio waves. A simulation of the standard model with the quarks and leptons is shown below.

Diagram of Fundamental Particles of the Standard Model

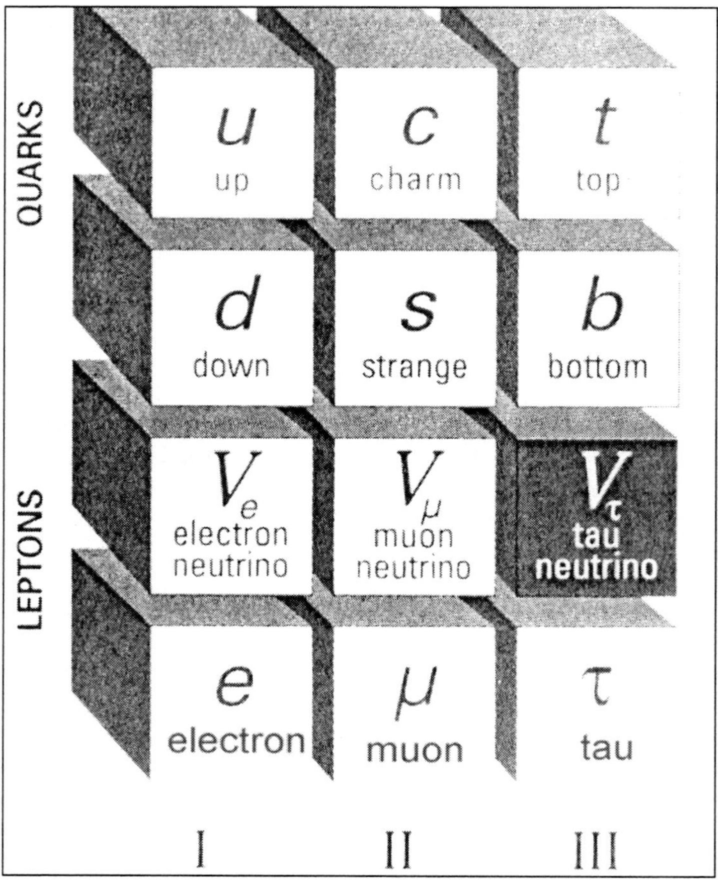

[8] P. Weiss, "Matter's Missing Piece Shows Up," *Science News*, Vol. 158, July 29, 2000.

Rows contain similar particles. Columns show family groupings (I, II, III) that grow more massive from left to right. Each family has two quarks and two leptons. This is taken from *Science News*.[9] While this atomic model was said to be "standard" and proven, many consider that an overstatement

Problems Causing Questions about this Model as Standard or True

Oscillations or morphs, behavior not found in conventional or standard particle physics, has been found in neutrinos and antineutrinos emitted from the sun and from Japanese nuclear plants.[10] By the standard theory there should have been equal amounts of antimatter as matter existing, whereas there is little antimatter in the universe. The Belle team at the High Energy Accelerator Research Organization (KEK) in Tsukuba, Japan calculated just how different the decay rates of b and anti-b quarks should be and the so-called charge-parity violation is 0.73 +- 0.06. The Belle team reported to a conference a finding of –0.9 +- 0.50, which seems to indicate the presence of ***unknown*** subatomic matter. A Stanford team has found rates nearer what a standard model would show, but not exact. Helen R. Quinn of the Stanford team said that there could yet be undiscovered heavy particles that interact differently with quarks than with antiquarks.[11]

While the particles for the standard model have been found, many questions and disagreements persist. Experiments at a particle accelerator called HERA in Hamburg, Germany have revealed another mysterious object at almost the speed of light in about ten percent of the collisions. When an electron smashes a proton at this speed, it usually knocks loose a quark inside the proton, and because an individual quark cannot exist in isolation, it quickly transforms into a motley spray of subatomic particles that jet away. The scientists observed the jet of particles away from the struck quark, but didn't see the jet of particles in the direction of the proton. In ten percent of the cases, the proton did not break up, raising a question about the working of the standard model.

It is possible that the researcher detected an entirely new type of particle inside the proton or hit an aggregate of gluons that hold quarks together in a proton. There seems to be a peculiar sort of lumpiness that apparently doesn't correspond to individual quarks or gluons. Some suggest that an electron sometimes penetrates deeply enough to encounter a new kind of object buried within the proton. Scientists simply do not yet understand what is happening.

New experiments with muons, at Brookhaven National Laboratory in Upton, NY, have raised further questions about the standard model. A muon is a cousin to the electron, which is 200 times as heavy but very unstable. Muons have a magnetic trait like tiny bar magnets, which makes them into minute gyroscopes that swing horizontally. In measuring this trait of muons it was found that while they originally pointed in the direction of the particles' overall motion, they then veered away and swept through 360 degrees and traveled in a circle. According to members of the research team, John P. Miller of Boston University and Vernon W. Hughes of Yale University, this was contrary to what should occur according to the standard model. This indicates something is wrong with the standard model and may require some new kind of physics.[12]

To complicate matters more, a five-quark particle, a pentaquark has been found. It has five

---

[9] Ibid.

[10] *Science News*, Dec. 14, 2002, Vol. 162, 371.

[11] "Particle decays hint of new matter," *Science News*, September 20, 2003, Vol. 164, 189.

[12] P. Weiss, "Muon orbits may defy main physics theory," *Science News*, Vol. 159, February 17, 2001, 102.

quarks and one anti-quark.[13] This pentaquark, which is a subatomic oddity, has been confirmed by other prominent physicists. Gunther M. Roland of M.I.T. has said, "This is really the ***beginning of a new era***."[14] Harris J. Lipkin, Weizmann Institute in Rehovot, Israel has said, "Do not believe ***any theoretical model*** at this stage."[15]

The theory of the standard model has become highly complicated. It involves ten dimensions, seven in addition to the three families already mentioned. It is hoped that when this is tested, it will prove the standard model and give explanation for why there are now eighteen crucial numbers, including the masses of thirteen fundamental particles and give reason why they developed as they have. A number of other particles have continued to be found. Those working on this theory are Lawrence Hall, at University of California, Berkeley; Savas Dimopoulos of Stanford University; Stuart Raby of Ohio State University; and others. In recent times, there is no general consensus of physicists backing a particular theory, because there are great irregularities; for example, the top quark being 40 times larger than the next particle and much heavier and more powerful, and a Muon is 200 times as heavy and unstable as electrons.

In commenting on atomic particles, Michio Kaku, physicist at the City University of New York, expressed the feeling of many physicists when he said,

> The so-called standard model... contains some 60 different particles with a broad variety of properties. Though a tremendous achievement of modern science, this theoretical model is simply too complicated to be the ultimate description of reality. It's like using Scotch tape to pull together a mule, a whale, a tiger and a giraffe.[16]

He suggests a number of other theories that other physicists are examining. Too much mystery remains.

In 2003 additional discoveries only added more mystery to the understanding of any standard model. At the BaBar Detector of the Stanford Linear Accelerator Center they "found a previously unknown particle they called Ds (2317)–thought to be short-lived union between a charm quark and a strange anti-quark, peculiar with a mass 9 percent lower than expected–like seeing an elephant do a disappearing act. Some think it rather is a charm-anti-strange composite but a 4 quark 'molecule'." Svitil commented that this is the discovery of a new sub-atomic particle with properties that defy conventional theory.[17] Also there are particles that have from four to six quarks. Five quarks were first found and then evidence of others. Peter D. Barnes of Los Alamos New Mexico National Laboratory said, "Exactly what form of the theory makes it work now becomes very interesting."[18]

Moreover, recent studies reveal that neutron quarks do not behave as expected. According to Gerald Miller of the University of Washington, "Long awaited neutron spin-data indicate that,

[13] P. Weiss, "Wild Bunch: First five-quark particle turns up," *Science News,* July 5, 2003, Vol. 164, 3.

[14] P. Weiss, "New Quarktet: Subatomic oddity hints at pentaparticle family," *Science News*, October 18, 2003, Vol. 164, 245.

[15] Ibid.

[16] Michio Kaku, *Hyperspace: A Scientific Odyssey Through Parallel Universes, Time Warps, and the 10th dimension,* (New York: Oxford University Press, 1994), as quoted by William Allman, "Alternative Realities," *US News & World Report*, May 9, 1994, 59.

[17] Kathy A. Svitil, "Particles and Theory Collide," Discover January 2004.

[18] Peter Weiss, "Wild Bunch: First five-quark particle turns Up," *Science News*, July 5, 2003, Vol. 164, 3; Kathy A. Svitil, "New Matter Detected at Japanese Accelerator," *Discover*, January 2003; et al.

under some conditions, the previously overlooked orbital motion of valence quarks make a major contribution to nucleon spin." Experiments done at the Thomas Jefferson National Accelerator Facility in Newport News, Virginia were reported by Xiangdong Ji of the University of Virginia. "They (the team) have found that a single quark may hog most of the energy in the neutron, yet spin in the opposite direction of the neutron itself." Since neutrons and protons are sister particles this probably applied to both. Ji has said, "The finding suggests that ***scientists have erred in calculations using fundamental theory*** to predict quark behavior with neutrons." Weiss comments, "Physicists peering inside the neutron are seeing glimmers of what ***appears to be an impossible situation***. The vexing findings pertain to quarks, which are the main components of neutrons and protons."[19]

Additional findings add further questions to our fundamental understandings of matter. "A New particle ... calls into question what we know about the composition of Matter. David MacFarlane and team using the BaBar detector at the Stanford Linea Accelerator center announced over 100 appearances of Y (4260) with properties that imply it contains one particular combination of two quarks, a charm and anticharm quark that 'decays in ***a manner counter to theoretical expectations***.' ...It suggests there may be a whole new spectrum of matter due to more exotic combination of quarks."[20]

Some findings about neutrinos challenge the standard model of 12 particles which involved the belief neutrinos have no mass. Ray Davis and John BaCall began experimenting on neutrinos in 1965; M. Hoshiba of Japan began his work on neutrinos in 1983. BaCall and Hoshiba, who earned Nobel peace prizes in physics, believe their findings may call for rethinking the fundamental theory of the universe. ***Neutrinos may have been the original matter from the Big Bang and then formed into everything else. Scientists identify three kinds of neutrinos:*** electron, muon, tau, labeled ***Ve Vu, Vt.*** These were found to have mass and travel less than the speed of light and oscillate in kind. There may be more than three as well as other unknown factors about neutrinos. Experiments in Tevatran at Batavia, Illinois found that about half of muon neutrinos changed, ***showing mass***. They originally have no mass but change in travel, gaining mass by interaction with gravity and electromagnetic forces.[21]

Mystery of Mass and the Higgs Field

The origin of mass presents another significant atomic mystery. A mass theory has been proposed which if proven will help explain the development of matter since the "Big Bang" and show the relationships or symmetries of elements. Scientists hope this in turn will lead to explaining puzzles in chemistry. Current thinking is that the quarks that were created from the "Big Bang" flew through the Higgs field of energy and created subatomic particles. These acquired weight and sometimes broke up, thus forming the various particles and ultimately the chemicals of the universe that formed into galaxies, stars and planets. The theory is that "when subatomic particles flew through the Higgs field, they grew massive, like geese winging through a magic, magnifying hoop."[22] The Higgs energy field works like a mud puddle. The big particles like big shoes pick up the most particles and little shoes the least. Using the large electron proton

---

[19] J. Weiss, "Topsy Turvy: In neutrons and protons, quarks take wrong turns," *Science News*, January 2004, Vol. 164, 3.

[20] Susan Kruglinski, "Mystery Particle Shakes Up Physics," *Discover*, January 2006, 39.

[21] Mark Alpert, "the Neutrino Frontier," *Scientific American*, August 2006, 20, also cf. internet of Nova, Feb.21, 2006.

[22] Sharon Begley, *Newsweek*, April 19, 1993, 53.

collider at the European Laboratory for Particle Physics in Geneva, scientists thought they were on the verge of evidence for the Higgs. That research is currently on hold since the search requires an even larger collider. Until that is built the origin of mass is still a mystery. The new findings that neutrinos probably have mass may question this.

Instruments for Atomic Studies are Particle Accelerators or Colliders

The particle accelerator or "atom smasher" was invented in the early twentieth century. It is a device that uses electric fields to propel ions or charged subatomic particles to high speeds and to contain them in well-defined beams. An ordinary CRT television set is a simple form of accelerator. There are two basic types: linear accelerators and circular accelerators. The largest and most powerful particle accelerators, such as the RHIC, the Large Hadron Collider (LHC) at CERN (which came on-line in mid-November 2009) and the Tevatron, are used for experimental particle physics. The LHC involved 25 years of planning and ten billion dollars on construction intended to smash protons together at 99 percent the speed of light in a 17 mile long tunnel creating a subatomic fireball mimicking the first trillionth a second of what might have been the universe's existence. These have been used to try to discover the basic particles of a standard model and to experiment to discover the Higgs field forming the basic elements. The Large Hadron Collider is intended to test various predictions of high energy physics, including the existence of the hypothesized Higgs boson and the large family of new particles predicted by super symmetry. The Large Hadron Collider had problems when first used in September 19, 2008 and had to be shut down for repairs, but in early 2011 it was in full operation. The number of particles produced is 25 % higher than expected and there are 1.25 gigabits of data to go through to try to find what is happening. In early June the scientists trapped an elusive form of hydrogen antimatter for 17 minutes. They are studying this data. [23]

Because of the confusion surrounding the standard model, physicists are eager to find Higgs particles in order to complete the picture of the fundamental units. Government deficits caused the U.S. government to cancel construction of a giant super conducting Supercollider (SSC) that was to look for the Higgs. The much bigger one, the Large Hadron Collider, is scheduled to be in use with the hope the Higgs will be found.

But many feel that discovering the source of mass is a very long shot. Roger Penrose, the noted University of Oxford scholar and leading physicist previously cited, has co-authored a science fiction work about the Higgs.[24] The story says that when the LHC could find nothing when it was to go online in 2005, it was boosted one hundred percent by 2009, still with no results. Then another collider, the SHC, built in 2024 across the two states of Utah and Nevada as a cooperative effort of nations found only what they called Alpha Smudge or Higgs Smudge. The discovery of another smudge led to the view these smudges were not particles, but a sequence of higher and higher energies, with the belief the finding of a Gamma Smudge would reveal the Higgs perimeters. So another super-duper called the Luna Collider was built on the moon by 2030, but with no results. Then as a part of a Mars colony, a science project shifted to building a ring-shaped tube filled with superfluid around Mars to detect a HIGMO, which is speculated to be a hidden-symmetry gravitation monopole revealing density. But when complete in 2065 there were only unexpected oscillating signals not like the speculated HIGMO. So a hundred years later scientists built the Cheeth-Rosewall detector or a ring the size of the diameter of one of Saturn's outer rings

---

[23] Tim Folger, "Laragae Hadron Collider Gets Going with a Bang", *Discover,* January 2, 2011.

[24] Roger Penrose with Brian W. Aldiss, *White Mars: or, The Mind Set Free, A 21st-Century Utopia,* (New York: St. Martin's Press, 1999), 323, cf. 127-129, 155 ff.

filled with superfluid to detect not only HIGMOs but to discover the nature of universal consciousness as well as the riddle of mass. In the story the hope is this instrument will enable man to project the human mind across the universe. Penrose's fictional speculations show the exorbitant miracles physicists now hope for.

William F. Allman said, "Many physicists today assume that there must be a deeper, more comprehensive theory that unites all the phenomena in the universe from light to gravity to quarks."[25] The findings about neutrinos add great questions. There is now no agreed-upon testable theory that can explain how matter came into being. Recent discoveries have only added to the mystery. It's a mystery that also relates to the question of gravity.

Discovery of Chemical Elements Continues to Expand as Indeterminate

With increasing knowledge of atoms, the sense of uncertainty has also expanded into the field of chemistry. At one time scientists thought that all elements were known. Believing there would be no more than 92 elements they established the uniform periodic table of chemical elements. Yet in the last half-century scientists have artificially created up to 112 elements. At this writing they believe they will soon find 113 and 114. We do not know how many there may be. Modern equipment has isolated additional isotopes of many of the elements. Although scientists have discovered the means to produce these elements, there is growing uncertainty as to how many there can be and what value they have. Whereas there was once a sense of understanding and being able to control all chemical matter, there is now a sense of mystery concerning chemical matter composed of atoms. Much is not known about the alpha force or the four main forces – both of which we will look at later.

Conclusions about Mysteries of Atomic and Chemical Physics

While some scientists believe science is coming to the end of new discoveries, others point to the March 1995 discovery of a top quark as one more important finding. Yet much about atomic and chemical physics remains mysterious and confusing. Fran Wiczek comments,

> The inability of current experiments to crack the standard model means that experimenters must create *almost inconceivable extreme conditions* in order to produce any phenomena that might give new insights: they must recreate, in effect, temperatures and pressures that existed back at the time of the Big Bang.[26]

The humorous story of the famous Roger Penrose about the inability to solve the mystery of particle physics by larger and larger cyclotrons also accentuates the seeming impossibility of solving the mystery of matter. Modern physics has difficulty giving a determined view of the universe and is faced with great mystery. Einstein was disturbed when indeterminacy could not be refuted. In 1930 he wrote a paper in which he adopted the metaphysical philosophy of Benedict Spinoza that ***assumed*** the view of an idealistic pantheism or intelligent design in nature. In 1940 he wrote a paper, "Science and Religion" for a Symposium on Science, Philosophy, and Religion for the Jewish Theological Seminary where he expanded on this. Rejecting the facts of probabilities, this theory ***assumed*** the ***absolute determinism of everything***, including human thoughts and acts. In the effort to declare an absolute determinism, Einstein with Spinoza

---

[25] William Allman, "Alternative Realities," *U.S. News & World Report*, May 9, 1994, 59-60.

[26] Frank Wiczek, "The End of Physics?" *Discover*, March 1993, Vol. 14, No. 3, 30.

continued to reject a personal God and human freedom. But Einstein argued inconsistently that man must act as if he is free and responsible ***to man***, not God. Einstein saw nature as the only God and man as its supreme creation. This is a return to much the same view as Aristotelian paganism that da Vinci, Galileo, Kepler and reformed Christians rejected.

Deterministic Theory Yields to Chaos, or Probability

The problem or randomness and probabilities had been mathematically explored from the 16th century, and in the 17th century had been advanced by Pierre de Fermat with Blasé Pascal. After Maxwell's great work Henry Poincaré in 1889 A.D. moved ideas toward special relativity and the phenomenon of some chaos was set forth. But because the strong commitment to determinism was based on a bias against God, these arguments were largely ignored until after 1970 A.D. Since that time, scientists in many fields have encountered unexplainable irregularities in the working of nature that could not be solved within the deterministic framework. Edwin Lorenz of MIT encountered it in regard to weather forecasting; David Ruello, Belgium Mathematician, found it in regard to turbulence; Robert May at The Institute for Advance Study, at Princeton, encountered it in biological problems; Benoit Mandebrant of MIT encountered the matter of chaos in regard to puzzling fractals; Jack Wisdom of University of California encountered it in regard to a gap in the asteroid belt; Gerald Sussmar encountered it in regard to the outer planets; and it is so in many other areas. Mitchell Feigenbaum of The Los Alamos National Laboratory discovered "the route to chaos was universal, and had a universal constant associated with it." This means that while one may be able to generally predict things, there is never certainty.

Indeterminacy in most of nature was long known by many, but science had such a reductionist mentality that few would discuss the matter. When James Gleick wrote *Chaos: Making a New Science* in 1987, the recognition of the phenomena about nature began to be the subject of many books, and the impression of many was that *nothing* is predictable and the natural order is really chaotic. The idea of chaos was associated with complexity, but the two are not the same, although both are beyond reason and chaotic happenings are related to complexity. The theory has been received with two extremes. The determinists seek to dismiss it. Feigenbaum said that to call the theory "Chaos" is a fraud. Others now see it as a rule of everything. The fact is that quantum physics and *most areas of nature have regularity with possibility that the unexpected may occur.* Barry Parker, who has reviewed recent studies on the subject, has said,

> At one time Scientists were convinced that everything in the universe could be predicted if enough computing power was available. In other words, all of nature is deterministic. We now know, however, that *this isn't true.* ...Even things that we might think of as deterministic can become chaotic under certain circumstances.
>
> ...At one time the universe was considered to be deterministic. With enough machinery, we could follow every particle in the universe throughout its entire history. ...But this assumed that nature was linear, satisfying linear equations that were easy to solve, and scientists gradually began to realize this wasn't the case. Much, it not most, of nature is nonlinear, and with this nonlinearity comes unpredictability and chaos.[27]

By now science had, with the computer and other instruments, found that the physics of the

---

[27] Barry Parker, *Chaos in the Cosmos: The Stunning Complexity of the Universe*, (New York: Plenum Press, 1996), 2, 257. For a fairly simple but more complete presentation see, Garnett P. Williams, *Chaos Theory Tamed*, (Washington, D. C. Joseph Henry Press, 1997).

universe was much more mysterious and beyond determinism that early scientists of the European Enlightenment ever anticipated. Basic physics of the universe operated by a higher knowledge.

Uniform Determinism was now Clearly not True, a New Direction was Needed

Science motivated by bias toward rejection of God must have a new direction. The turnover of traditional science was so clearly evident in all areas and mystery was so prevalent that Dr. Vannevar Bush wrote to alert the scientific community to the open door this development presented for faith. Bush, the former dean of the prestigious Massachusetts Institute of Technology, is referred to as the "Grandfather of Modern Computers." In 1965 when he wrote his warning he was the Director of the Office of Scientific Research and Development of the United States. There was no more prestigious scientist in America. In his article, *Science Pauses*, after referring to new scientific breakthroughs that revealed great mystery beyond the circle of uniform determinism, Bush said,

> No longer can science prove, or even bear evidence. Those who base their personal philosophies or their religion upon science[28] are left beyond that point, without support. They end where they began; except that the framework…is far more probable….Science proves nothing absolutely. On the most vital questions, it does not even produce evidence….And what is the conclusion? He who follows science blindly and follows it alone, comes to a barrier beyond which he cannot see…. And with a pause, he will admit a faith.[29]

But the whole educational system in the United States was now committed against faith in the Christian God. James B. Conant was a brilliant scientist who had played a key role, along with his friend Vannevar Bush, in ramping up the Manhattan Project which developed the first nuclear weapons. Conant earlier had respect for German science and had brought in Nazi speakers, and was anti-Semitic and racially discriminate. Conant later become the president of Harvard, with its theological seminary renouncing all Christian doctrines, and he, like leaders at Columbia University who brought in neo-Marxist thinkers, moved toward postmodernism that saw all things changing. Conant conceived of and introduced Thomas Kuhn to the idea of history of science as changing by revolutions of temporary paradigms, having no continuing structures, and in effect having no real history, since all was changing. Kuhn used the suggestions of Conant and developed the philosophy of science that replaced the previous anti-god determined views.[30]

Bush had suggested that beyond science there were mysteries that the Creator allow man to see as his work of upholding all things by the word of his power. Indeterminacy in science was suggested that it took man to the door of faith. Instead of making the choice of faith in Christ from the revelations that were beyond man's ability and indeterminate, Kuhn chose the false view that there is mystery because all nature is in change, and going through deconstruction. Kuhn's views became the leading philosophy of modern science to explain away indeterminacy. This has led the way to chaos, not only in nature, but of morality. It has led to the darkness in all the soft sciences of economics, psychology, sociology, and politics..

---

[28] Referring to modern liberals.

[29] Dr. Vannevar Bush, "Science Pauses," *Fortune,* May 1965, Vol. LXXI, No. 5, 116-172.

[30] Thomas S. Kuhn, *The Structure of Scientific Revolutions*, (Chicago: University of Chicago Press, 2nd edition 1970) preface xi.

Chapter 8

Search for Evidence of Nebular Hypothesis in the Heavens

Introduction

Immanuel Kant proposed the natural evolutionary development of everything in the universe from a cloud of heavenly dust or gas, or the nebula hypothesis. About that time Abbe Francoise MaMartre in Belgium proposed this nebular hypothesis of Kant may have begun with a Big Bang and expanded outward. The focus and insistence by Kant was that everything developed by natural laws and not by the biblical self existent Creator and that man could understand these. The beginning point and expansion was not specifically described.

William Herschel, who became interested in astronomy, after he picked up on Kant's theory and took this with him from Germany to England. There he sought to advance the theory by enlarging Newton's telescopes. He discovered what he thought were dust clouds that might be forming stars and galaxies. With the advanced telescopes that he had constructed, he and his son John sought to map the heavens and learn more about how they developed. Laplace tried to define exactly how the solar system had formed as Swedenborg and Kant had attempted previously. In England, the family of William Herschel spread the theory of gradual evolution to geology, and in a successive generation the theory went to biology. For many this implied an eternal machine and eliminated a personal God. The deists and pantheists however, still referred to an Author of nature. The nebular hypothesis—that everything worked by natural laws—formed the underlying assumption in all these efforts.

Uniform gradual geology and then biology in the nineteenth century took the spotlight from astronomy until the slump in uniform Darwinism in the late nineteenth century, at which time an interest in astronomy was revised and soon took a leading place, especially in America.[1] The significant advance in astronomy, as in other sciences, began in the first quarter of the twentieth century. Accepting the mystery of the quantum theory opened the way to consider the mystery of a beginning of the universe, promoted by Einstein's theory of general relativity and a measurement of space-time.

Efforts to unfold the mysteries of the universe continue today employing thousands of scientists, huge sums of money, and ingenious inventions uncovering great amounts of significant data. The accumulation of this information is highly important to the philosophy of origin by chance and evolution of all things. If the Big Bang and inflation of the universe as proposed is true, there is data to set aside the creation by God. Charles Lineweaver has said,

> [This] has been the unifying theme of cosmology, much as Darwinian evolution is the unifying theme of biology. Like Darwinian evolution, cosmic expansion provides the context within which simple structures form and develop over time into complex structures. Without evolution and expansion, modern biology and cosmology ***make little sense***. The expansion of the universe is like Darwinian evolution in another curious way: Most scientists think they understand it, but ***few agree on what it really means***. A century and half after *On the Origin of the Species*, biologists still debate the mechanisms and implications (though not the reality) of Darwinism. ...Similarly, 75 years after its initial discovery, the expansion of the universe is still widely misunderstood ...even among those making some of the most stimulating contributions to the flow of ideas. ...The Big Bang model is based on observations of expansion, the cosmic microwave

[1] Rudolf Thiel, see chapter five, "Rejuvenation," *And There Was Light,* (New York: Mentor Books of Alfred A. Knopf, Inc., 1957), 286-299.

> background, the chemical composition of the universe and the clumping of matter. Like all scientific ideas, ***the model may one day be superseded***. But it fits the current data better than any other model we have. [2]

The hypothesis of the Big Bang and the inflation of the universe is the foundation of even Darwinism. The nebular hypothesis involving the Big Bang is the lynchpin, the fulcrum of the theory that gives authority to modernism and validity to postmodernism. All secular institutions and culture rest on it. Because of the large amount of data required, and its importance, the mysterious findings of the heavens will be given separately in Chapter 9 and new knowledge about the solar system will be discussed in Chapter 10. Factors of long time and space distance are quoted in these chapters as the deterministic scientists estimated them. Their accuracy will be reevaluated later.

New Technical Instruments Reveal New "Lakes of Knowledge" Leading to the Theory

New instruments opened the doors for extensive discovery. Astronomy had several divisions of new amazing technology: telescopes for visible light, infrared astronomy, radio telescopes, gamma ray astronomy, neutron astronomy, and cosmic ray astronomy. Different instruments and methods of research may be used because of differences in lengths and frequencies in the electromagnetic spectrum. From short to long and high frequency to low, the order is: gamma rays, x rays, ultra violent rays, visible light rays, infrared rays, and radio rays. Some unusual technological advances were made. H. A. Rowland of Johns Hopkins invented diffraction gratings; Samuel Langley invented a thermometer to measure heat radiations. G.E. Hale, working in Mount Wilson, California, where the skies were clear most of the time, began to use photography in his mountain observatories. A real breakthrough came when Ejnar Hertzsprung began to use spectrometry with photography to analyze the chemical composition of the stars. Physicists had learned four hydrogen elements form helium, then there is a gap of unstable elements and then three helium form carbon, and helium and carbon form oxygen. These and other elements burn with various colors and have gaps in the light spectrum. An analysis of speed and distance were made by study of the redshift as proposed by Einstein. By 1965 radio telescopes were invented to reveal some of what had previously been invisible.

The distance, size, and temperatures of stars began to be measured in various ways. The closer ones, under 3000 light-years away, can be measured with *the parallax,* taking two readings of the star's exact position in the sky from each tangent of the earth, and measuring the star's slight shift as a result of the earth's motion around the sun. This very limited way of measuring was technically expanded with Einstein's concept of light years. The distant, size and temperatures also use Cepheids and lately supernovas as stars whose brightness are known, and then the study of the redshift of starlight. Einstein's theory was a major factor in space time projections.

With the naked eye one may see about 3000 stars on a clear night, but with the newest telescopic instruments it is estimated that the number is ten with twenty-two zeros after it (a trillion has 12 zeros). Some estimate there are a great many more than these. Every star is believed to be different. Scientists developed a modern classification, identifying stars according to size, brightness, temperature, speed of travel and the like. Names such as red giants, Cepheids, type 1a and 1b supernovas, white dwarfs, binary stars, nebulas, neutron stars helped to bring some clarity and organization to the study of the stars. Scientists also discovered that the stars are in galaxies of

---

[2] Charles H. Lineweaver and Tamara M. Davis, "Misconceptions about the Big Bang," *Scientific American*, March 2005, on page 38, 45 of 36-45).

various sizes and kinds such as spiral, and nebula.[3] The number of these galaxies is estimated from a hundred to two hundred and fifty billion. Galaxies are extremely large and seem relatively close together (e.g. like dinner plates about ten feet apart) while stars in the galaxies are much smaller and further apart like peas 100 million miles away.

Astrophysics Seeking the Mystery of Physics Led to a Finite Universe in Movement

The data on astrophysics is immense and growing. Only a small amount can be examined to show the radical transformation that has occurred.[4] But it is important to understand the basics. As seen in the "mass theory" of atomic physics, the hope of some leading physicists was to link physics with astrophysics to resolve the mysteries. Albert Einstein advanced the uniform deterministic view involving the nebular hypothesis. Einstein believed that if the macrocosm of the heavens could be understood, there would be no remaining mystery as seen in quantum physics.

About 1917 Einstein published ***a memoir*** on the cosmological consequences of his *theory of general relativity*. He showed that the gravitational interaction of all material bodies could be formulated free from the gravitational paradox of the Newtonian universe. He said the values of the average density of matter and of the gravitational constant could infer the value of the total mass of the universe and its overall radius of curvature produced by that mass. This memoir included *three verifiable predictions*, each independent of the other: (1) the gravitational redshift, (2) the gravitational bending of light, and (3) the advance of the perihelion of planetary orbits. In time these were demonstrated to be true with only a slight margin of error. Other predictions have been proven, furnishing further confirmation. All this was based on believing in a uniform speed of light.

Only later were the implications of this fully understood. *Einstein had clearly set forth* ***a finite world***. This view, therefore, required a beginning point for the universe. Einstein theorized there had to be an antigravity in the universe or it would collapse. He called this *antigravity the* ***cosmological constant***. Later Einstein called this his greatest blunder. But later we will see that Einstein's prediction of another force will be justified. We will also see that theory of relativity came to be questioned in other ways. This accelerated the search for the beginning nebula that produced the finite universe. While some prominent astronomers proposed an alternate eternal view of the universe called the "steady state" universe, this was abandoned by most as the evidence built from Einstein's proposals for a finite universe seemed to be confirmed.[5] Other theories of the universe have been made. Einstein's predictions were soon demonstrated to be true for the universe in his limited time frame.

Conception of an Expanding Finite Universe Suggested a Big Bang Beginning

Edwin Hubble, for whom the Hubble Space Telescope was later named, showed that there were many galaxies. He found a way to measure their distances of stars and their brightness.

---

[3] For a good illustrative presentation, cf. Mark A Garlick, *Astronomy: a Visual Guide,* (Westport, CN: Firefly Books, Ltd. 2004).

[4] Roger Penrose, *The Road to Reality: a complete guide to the laws of the universe*, (New York: Alford A. Knopp, 2005). This book furnishes all of the complicated math and in the last part discusses many of the problems and possible answers that have arisen in the technical terms if one would like that technical data.

[5] Bondi, Gold, and Hoyle's view of "steady state" in Hermann Bondi, *The Universe at Large,* (New York: Anchor Books, 1960).

Based on Alexander Freedman's 1929 demonstration that the universe was ***not static,*** as Newton and Einstein thought, Hubble showed that galaxies are moving away from the earth and the universe is expanding. As Stephen Hawking said, "The discovery that the universe is expanding was one of the ***great intellectual revolutions of the twentieth century.***"[6] Georges Lemaitre was early involved with Einstein and Hubble and contributed to astronomical theory and proposed the Big Bang idea. While some astronomers have questioned the expansion of the universe, the idea gained general acceptance within the scientific community, though atheists remained hostile to the idea.

In 1933 George Gamow, in pondering the relative abundance of chemical elements in universe and the fact that hydrogen was basic, speculated that this might be the source of the heavier elements that appear at the beginning of the universe. This 'Yelm,' or ***hot ball of gas*** produced an expansion of gases based on an assumed "Hubble constant."[7] Other findings added to the idea of a nebular beginning and expansion. The idea of a Big Bang beginning and an expanding universe assumed an intelligent cause. The atheistic astronomer Fred Hoyle named the theory of beginning "the Big Bang" but could not accept this. In 1948, with Thomas Gold and Hermann Bondi, he proposed the "steady state theory." This theory postulated that the amount of matter in the universe was constant, without explaining the creation and disappearance of matter. By mid-century the theory of a Big Bang and an expanding universe was under speculation while atheists continued to look for evidence to support their claims of an eternal universe. Ernest J. Opik suggested that the universe was continuously oscillating with an explosive beginning and coming together again in a yoyo fashion, resulting in an eternal universe.[8] Joe Silk, Rainer K. Sachs, and Arthur M. Wolfe argued there must be residual perturbations that gave rise to the stars, galaxies, and galactic clusters from the Big Bang. Soon after Herman and Ralph Alpher at John's Hopkins predicted a microwave background if this was true. Scientists continued to search diligently for a way to determine the rate at which the universe expanded—a "Hubble constant"—and appeared to be close to their goal. [9]

Apparent Discovery and Demonstration of Evidence for Big Bang

In 1965 Bob Dick and Jim Peebles developed a computer model of how they thought the cosmic background might occur, thus launching the search for the origins of the Big Bang. According to the Big Bang theory the universe originated from nothing, became a super dense state, and then expanded in a giant explosion that produced the various elements. The first elements, hydrogen and helium, evolved into heavier elements, and later the stars and galaxies formed.

Working together at Bell Labs Arno Penzias and Robert Wilson devised an ultra-sensitive cryogenic microwave receiver for radio astronomy observations. They encountered anisotropic radio noise with different values in different directions. This was believed to be a serendipitous discovery of residual microwave radiation from the Big Bang. They believed that they had discovered the smoking gun that could be used to trace back to the initial explosion. In 1978 they received the Nobel Prize for discovering the remnant of what was thought to be the primordial

---

[6] Stephen Hawking, *A Brief History of Time,* (New York: Bantam Books, 1988), 39.

[7] I had the privilege of meeting and studying briefly under Gamow, formerly of Ukraine in the late 1950s.

[8] Ernest j. Opik, *The Oscillating Universe*, (New York: The Mentor book of the New York Library, 1960), 120-122.

[9] *Science News*, Vol. 142, July 4, 1992, 4.

fireball of the universe's beginning.

Discovery Mystery and Singularities: Black Holes and then the Big Bang

Emerging Evidences of Mystery

Around 1965 scientists began to uncover evidence of singularities in the universe. A *singularity* is a point or boundary in space-time at which the space-time curvature becomes infinite. This is an inexplicable event that stands apart from any succession of events. *A singularity is therefore a boundary situation beyond human comprehension*–a mystery and possible miracle. In 1782 John Michell of Cambridge University suggested that some stars were so massive and compact that their light would be dragged back by gravity. Subrahmanyan Chandrasekhar worked out the size limit for such stars in the 1920's. A Russian, Lev Davidovich Landau, determined some stars would contract to a great density of about ten miles; he named these neutron stars. He also saw other stars above the Chandrasekhar limit, called white dwarfs, which would only contract to a density of a few thousand miles. In 1930 Fritz Zwicky determined that there had to be dark matter to account for the universe. In 1939 Robert Oppenheimer figured out what would happen to neutron stars and black holes. He also worked on the theory of quantum mechanics, quantum field theory, and the interactions of cosmic rays.

Numerous astronomical observations called for some great force like black holes. In 1965 the British mathematician Roger Penrose, using the way light behaves in general relativity, showed mathematically that a large star which is collapsing under its own weight would compress the matter in the star into a region of zero volume. The density of matter and curvature of space-time become infinite forming ***a black hole***, or a singularity. In 1970 Vera Rubin discovered galaxies rotating too fast for their own axis. Even our Milky Way galaxy was rotating so fast it should fly apart unless there was something at the center that held it together. These and other mysteries motivated Penrose's research.

A black hole is defined as intense gravity that pulls matter such as gas and dust into it with such force that even light cannot escape. When anything reaches the edge of the black hole, *the event horizon,* it cannot escape and is sucked into it like water going down a drain. It develops greater and greater mass with high velocity and creates temperatures of millions of degrees. At a certain point a singularity occurs and time vanishes.

Big Bang Singularity, the Reverse of a Black Hole

In 1965 Stephen Hawking, a graduate student looking for a Ph.D. thesis subject, saw promise in Penrose's theory of black holes. Working with Penrose, Hawking envisioned the reversing Penrose's black hole theorem. Penrose theorized that a star collapsing into a black hole might produce a modified finite inflationary universe expanding from the condensed state of the Big Bang. This idea had already been proposed from the discovery of the expanding universe and expanding residue. Such a Big Bang would begin with a singularity just as a black hole ends with one. Hawking and Penrose produced a joint paper on the Big Bang needed for the inflationary theory in 1970. Hawking commented on the implications and reception of this work. He said,

> [This] at last *proved* that there must have been a big bang singularity, provided only that general relativity is correct and the universe contains as much matter as we observe. There was ***a lot of opposition*** to our work, partly from the Russians because of their Marxist [atheistic] belief in

> scientific determinism, and partly from people who felt that the whole idea of singularities was repugnant and spoiled the beauty of Einstein's theory. However, *one cannot really argue with a mathematical theorem.* So in the end our work became generally accepted and nowadays nearly everyone assumes that the universe started with a big bang singularity.[10]

Hawking's mathematical demonstration of the Big Bang earned him great notoriety. But he was rebuffed by committed naturalists and atheists, including the Russian scientists, over the implication of a creation point. Hawking felt the offense from these brilliant scientists. But his work inspired further exciting and fervent research.

Determination of Physical Specifics of the Universe and World

Beginning with the radiation source, physicists determined the specificity of the various chemical elements and their numbers in the universe. The concept of the specificity of the elements prompted physicists such as Hawking and Martin Rees to study all the earth's forces and elements. Rees, Britain's Astronomer Royal and an astronomer at King's College in Cambridge, England, has written a book, *Six Numbers.* In it he lists six phenomena identified by one Greek letter that sets our universe apart as unique for maintaining life. He claims these six encompass everything from quantum weirdness, to biological imperatives to galactic clumping. If any of the six were different, life would not exist on the earth. They are as follows: (1) Strength of force binding nuclei of hydrogen for energy from the sun; the force of hydrogen that produces helium. (2) The strength of force holding atoms together divided by the force of gravity. (3) The density of material of universe relative to gravity in expansion of universe. (4) Strength of anti-gravity forces controlling expansion of universe. (5) The amplitude of complex irregularities or ripples in expanding universe. (6) The number of spatial dimensions in our universe, which is three. Rees argued, "These six numbers constitutes the recipe for the universe. If any one of the numbers were different even to the tiniest degree, there would be no stars, no complex elements, no life."[11] Hawking and many other astrophysicists agree that the definiteness of the universe makes it adaptable for man, and the composition and forces on earth make it the only known place suitable for human habitation.

Robert Jastrow, founder-director of NASA's Goddard Institute of Space Studies, explained the implications of proving the exactness of these forces. He pointed out that the specificity of such a beginning was so precise that even a slight change in the nuclear force would *not have allowed life on the earth.* He also said, "It seems to me astronomy has proven that forces are at work in the world that are ***beyond the present power of scientific description****: these are literally* ***supernatural forces***, because they are outside the body of natural science."[12] Jastrow was not a believer, but his comments confirm uniform development of nebular condensation, a concept that followed from Gamow's idea of the specificity of the elements.

The idea of nuclear condensation of mater and the universe led to a theory of how this was achieved mechanically. Later we shall see how this was applied to the research for the microwave background.

---

[10] Hawking, *A Brief History of Time*, 49-50.

[11] Brad Lemely, "Why is There Life?" *Discover*, November 2000, 66.

[12] Robert Jastrow, "The Astronomer and God," *The Intellectuals Speak Out About God,* (Chicago, IL: Regnery Gateway, Inc., 1984), 19, 21.

Search for Data of Expansion of Universe: Mapping of Starry Heavens and of Microwave Background

Models for the Universe's Expansion after the Big Bang

Scientists then proposed various theories for how the universe developed. ***Three basic models of inflation describe what different scientists believe happened to the universe following the Big Bang***. The first, the new inflationary theory, is similar to a balloon with spots on it, in which the spots became further and further away as the balloon is blown up, representing the development of stars and galaxies. The illustration was used of tracing the expanding smoke back to the nozzle of a gun. The search was then made to try to follow the development of the expansion back to its source. The astronomers believed that from very mature late galaxies they could trace the story back to younger and younger galaxies and then follow the microwave background to its source.

The ***three basic models describe what different scientists believe happened to the universe following the Big Bang*** with graphic illustration on the next page. The first, the new inflationary theory, is similar to a balloon with spots on it, in which the spots became further and further away as the balloon is blown up, representing the development of stars and galaxies. The illustration was used of tracing the expanding smoke back to the nozzle of a gun. The search was then made to try to follow the development of the expansion back to its source. The astronomers believed that from very mature late galaxies they could trace the story back to younger and younger galaxies and then follow the microwave background to its source.

The second theory is a view of inflation of the universe from the Big Bang to a point in which matter continues to move through space in a parallel way but *flattens* out, so that after a time it neither gets further apart nor closer together. This flat universe appears as *a continuing universe* after creation. The expansion at this point would be the Hubble constant.

The third view holds that after the initial explosion, or Big Bang, the dots of matter continue to expand until they reach a certain point where the explosive energy is spent and gravity causes matter to collapse back together, deflating and colliding again into super density or a big crunch (the reverse of the Big Bang). After this, it explodes again. This is a cyclic type universe where the expansion goes on until matter reaches the point where the momentum decreases until it is exceeded by gravity, then proceeds to condense and crunch at the opposite pole from which it exploded. According to this model the universe is a perpetual yo-yo with singularities on the opposing poles. In this view there is no final point for a beginning by God or an ending, but a repeated yo-yo affect.

The particular pattern that the inflation took after the Big Bang is thought to be determined by the amount of matter that exists and the resultant forces exerted by gravity and the other forces. The three are (1) the open, (2) the flat, or (3) the closed or yo-yo which may also be called a global model. Each view now holds to a Big Bang and corresponding expansion by inflation. Until recently most astronomers held to the closed or flat universe. In 1978 Arno Penzias and Robert Woodrow picked up unusual radiation with their equipment at Bell Labs and Robert Dicke later identified this as cosmic wave background of the results of the Big Bang. Later, a computer simulation was made that traced the explosion of the Virgo supercluster which seemed to demonstrate the new inflationary view of the universe.

Diagram of three Basic Models of Inflation

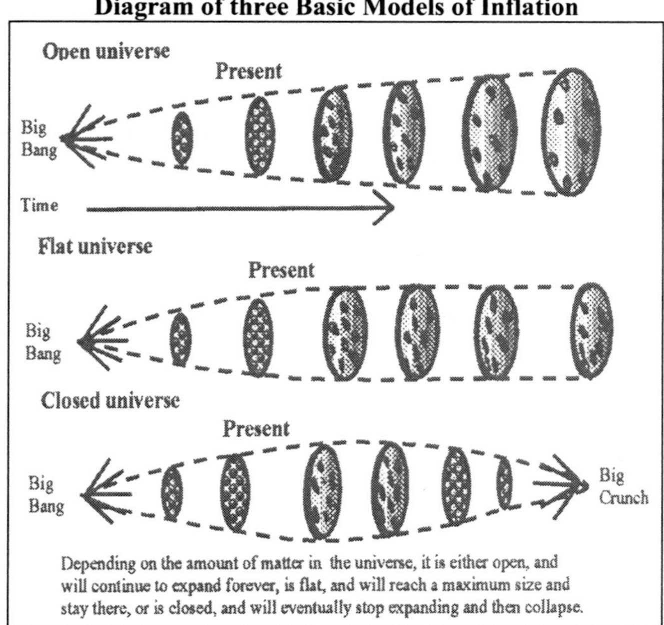

Depending on the amount of matter in the universe, it is either open, and will continue to expand forever, is flat, and will reach a maximum size and stay there, or is closed, and will eventually stop expanding and then collapse.

Wide arrays of theories about the development of the universe have been offered. Alan Guth suggested a "phase transition" that may have been needed for a beginning with a uniform temperature. Andrei Linde suggested a single big bubble joining many bubbles. La Moss rejected this for another explanation to allow space bubbles to join together. Later Linde suggested a spin-0-field without a phase transition and called it the chaotic inflationary model. In time many of these ideas were rejected. Most astrophysicists continued to believe in the Big Bang, and in the five years after Hawking wrote his mathematical proof, astronomical research sought to give experimental evidence for this. The tracing of this development required new instruments not then known. New telescopes overcame the limitations of the past. The Hubble space telescope overcame the earth's atmospheric limitations; Keck I and II, using a conglomerate of mirrors, bypassed the limitations of one mirror. Now radiation telescopes in space and on the ground and other magnetic instruments transport the astronomer to new unexpected horizons of knowledge or lakes of mystery. New instruments are often invented.

Search of the Microwave Background

The study of microwave background has been a highly technical and complicated project involving many able scientists. Many aspects of physical forces were brilliantly thought out. Early efforts involved the U2 plane and balloon experiments. The effort to study the microwave background has centered around two major instruments sponsored by universities and NASA. They are the Cosmic Background Explorer (COBE) and Wilkinson Microwave Anisotropy Probe (WMAP, formerly known as the deep space explorer). George F. Smoot III led one microwave study and received a Nobel Prize. NASA launched the first satellite in 1989 and in 1992 announced that the COBE, orbiting at 560 miles outside the earth's atmosphere, began to give results that are the most important discovery for astrophysics in twenty years. It reported on the oldest area of the universe (now said to be 13.7 billion years old) and analyzed the cosmic microwave background, first believed in 1965 to be the residue of the Big Bang. In the early 1990s COBE detected ancient radiation from the universe in a picture of plasma present from about 3 minutes to 300,000 years after the Big Bang. Strangely, COBE found that the universe's radiation is uniform to one part in 25,000 with a temperature of 2,735 degrees above absolute zero, hardly varying from one side of the sky to the other. At first it seemed the beginning of the inflation of the universe was unexpectedly smooth, but in time evidence suggested formations of materials.

Evidence from the COBE indicated miniscule differences in depression and in temperature (ca 30 millionths degree variation). In 1990 John Mather and colleagues in the Firas Experiment presented a perfect backbody curve to the America Astronomical Society in Washington, D. C. In 2002 a team from the California Institute of Technology, using radio telescopes in Chilean Andes, sought to produce more definite pictures of temperature differences showing very slight structures in cosmic radiation from the Big Bang which were previously unknowable, according to Alan Guth of MIT. David Wilkinson led a second WMAP study using more sensitive instruments. Placed in a satellite one million miles from the earth or four times the distance of the moon, they measured the cosmic microwave background radiation. The theory is that eventually the universe cooled enough so that protons and electrons could combine to form neutral hydrogen. This is estimated to have occurred sometime around 380,000 years after the Big Bang. At that time a new era occurred and radiation was allowed to travel more or less directly. The result is that very cold CMB now fills the universe. Some of Smoot's significant observations of the microwave studies are as follows: the CMB comes from far beyond remote galactic clusters, is not a local phenomenon but a general one and the polarization seems to be in early phases. Smoot observed many cold spots with no microwave evidence. Some were very large and one at least one billion light years across. This is a phenomenon unaccounted for and can be seen in the composite results pictured. The almost uniform temperature of 2.7 K was a discovery of surprising proportions. The results of the survey is not like following a smoking gun back to its source as expected, but of finding an underlying microwave blanket with holes. More observations of underlying technical questions have since emerged. A better and more definitive picture of microwave background is given by WMAP.

WMAP 5 year temperature

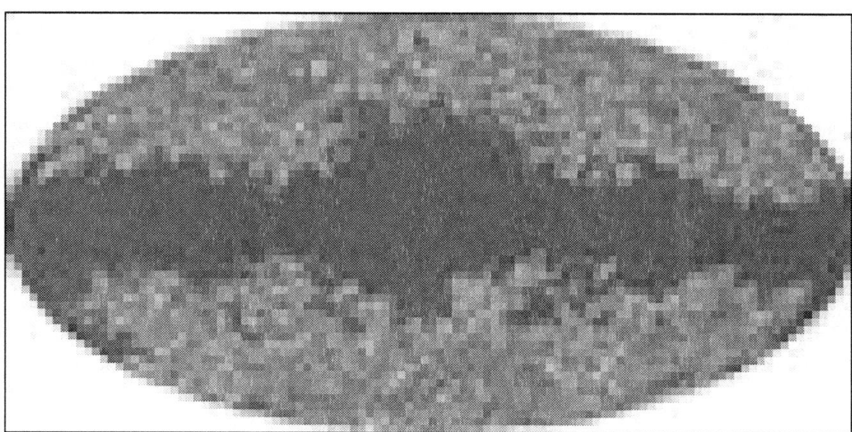

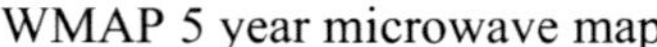

WMAP 5 year microwave map

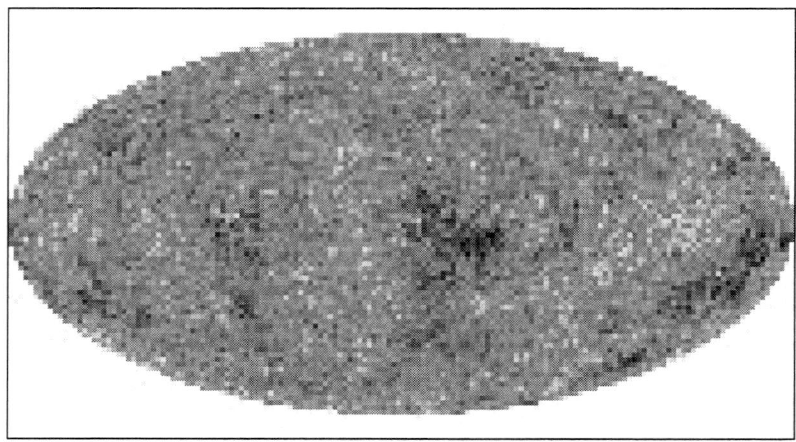

There have been various pictures drawn from the beginning to 330,000 years by NASA, various magazines, including the recent article in *Smithsonian*, April 2010. The fact is that expansion has not really been observed from a local beginning, but only shows there is a microwave background to the whole universe with no proof there is a progression from a Big Bang.

Mapping the Stars and Galaxies

Herschel's minimal but heroic efforts at the turn of the nineteenth century to map the sky and discern the nebular beginnings and workings have become much clearer and vastly more extensive in recent years. The 1950s Palomar Sky Survey was one big step forward in sky mapping. In 2000 the Apache Point Observatory in Sunspot, New Mexico, released what was the first detailed sky map, using its Sloan digital sky telescope, which exposes the universe's deepest secrets. While the Hubble space telescope and the Keck telescopes in Hawaii see the dimmer stars and galaxies far away, the Sloan digital telescope has a field of view that is 360 times that of Hubble and can survey and record its pictures digitally. The Sloan digital has already recorded 350 million celestial objects and in 2009 released the most detailed sky map of the visible universe that has ever been made (which is about 30% of what is known). It has revealed much about the clusters and superclusters of galaxies, found areas or blobs of almost pure dark matter, found mini galaxies, new kinds of stars, and much more. *Discover* gives the picture of the Sloan work as seen below. Michael Lemonic reviews the Sloan findings and said,

> Taken together, the Sloan results lay out one of the most astonishing stories in science. The visible universe is merely the foam atop of a much grander cosmic sea. The vast majority of what is out there is more dynamic and complicated and just plain weirder than the tiny fraction we know. Only now are we starting to see the universe as it truly is.[13] .

---

[13] Michael Lemonick, "The Great Cosmic Census," *Discover*, March 2009, 61-65.

The Sloan telescope picture of the heavens follows.
(Line one fourth from left is page blur)

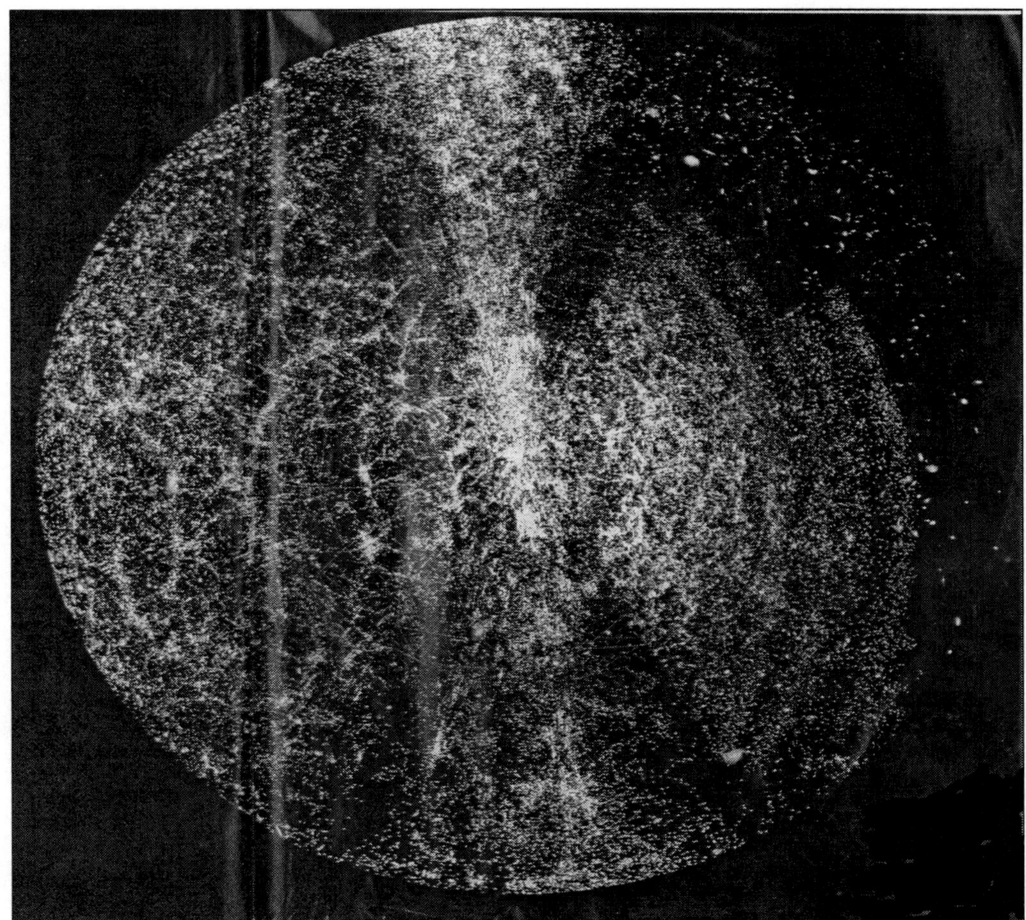

Conclusions

From about 1960 the astronomers had the idea of a beginning of the nebular hypothesis with a Big Bang that produced an inflation expansion of the universe. They devoted research to discover the point of singularity from which it began and found and mapped a microwave source and sought to map the entire heavens to see how the inflation occurred. The next chapter evaluates if the data supports heaven's creation by natural law.

Chapter 9
Mystery in the Starry Universe

Introduction

Since about 1960 a growing number of astronomical scientists, using increasingly accurate instruments, have accumulated more data producing more certain findings, while, at the same time, uncovering new mysteries of the universe. This chapter will look at some of these important, but discoveries of mysteries beyond determined understanding. While it is a very technical chapter, it presents critical information that Christian leaders need in order to truly understand creation and a creation world view.

The critical issue for the modern world is whether the evolutionary theory of all things is supported by the evidence of the big bang inflationary theory. The prevailing popular opinion is that the Big Bang provides the data for the nebula condensation theory. The question of biblical revelation hinges on whether this is fact or fiction. Is this based on truth? Or is it an existential leap in the dark with no foundation or basis in fact? Does the evidence better fit biblical evidence of God's creation?

Nebular Theory of Astrophysics of Neucleosynthesis and Star and Galactic Formation

Some of the data given by astrophysics include the following: the nebular hypothesis had a beginning and what was called a Big Bang with a singularity that expanded in an inflationary way with a constant rate of unfolding. A mathematical presentation was demonstrated. The theory was presented that beginning gases expanded in radiation and while inflating formed various forms of matter accumulating mass as it went through a theoretical Higgs field. It then proceeded to form stellar dust that condensed into stars and baby galaxies which then grew and became larger as more infant galaxies were formed. A search has been made for the beginning gellum of gas by collision and giant colliders. Research was made for the radioactive beginnings and for the point from which the inflation expanded and maps of the starry universe have been drawn from huge telescopes to find the picture of this expansion. The research on the microwave background went back to around 300,000 years. Speculative estimates of phases were made on how the microwave progression occurred at 10^{12} seconds, 100 seconds, one month, 10,000 years and 380,000 years. Old maps of the universe were replaced by more comprehensive sensitive pictures viewing 30 % of the heavens.

What has ***actually been observed*** in these phases? We have seen earlier that attempts to establish a standard atomic model of quarks and leptons were inconclusive. While physicists and chemists have tested the temperature of the elements and have found new forms of matter in cyclotrons, they have not observed these elements actually forming from such a state as the earliest phases indicate. There is no explanation for the transition from one step to the other. Moreover, there is no experimental observation until the last two steps. In addition, serious questions remain about both the Big Bang and the findings of CMB.

Question about Primordial Beginning from the Big Bang

Scientists at the Brookhaven National Laboratory in New York attempted to simulate the Big Bang. Using a colossal atom smasher, the Relativistic Heavy Ion Collider (RHIC), they charged and collided gold atoms at near the speed of light, producing a fireball 150,000 times as hot as the center of the sun. Some of the following are statements about their experiment. "Their experiment implies that ***the cosmos started out, not as a hot, dense cloud of gas but as a strangely sublime friction free liquid***... The findings mostly ***have physicists puzzled.***" Sam Aronson, associate director of the lab said, "Many of the people working at RHIC were surprised by this striking finding." [1] This new finding ***requires a complete change in the theory of how the universe began.*** The surprising beginning of a brief liquid, which they consider quarks and gluons, was one trillionth of an inch wide with temperatures exceeding a trillion degrees. It lasted fifty trillionths of a trillionth of a second. Scientists are presented with "a labyrinth of ***strange, heady, hitherto seemingly unrelated theories of physics***"[2] to form a theory of what happened. In addition, it is said to be, "challenging the conventional model of how the universe behaved in its earliest instance." (Emphasis added)

Some believe this may turn out to support a string theory. This is, in fact, a rewriting of the whole idea of the Big Bang. At a meeting of the American Advancement of Science in early 2009 Barbara Jacak of RHIC, John E. Thomas of Duke University, Peter Steinberg of Brookhaven National Laboratory, and Clifford Johnson of USC offered ideas to save the big bang theory with a liquid beginning. [3] They resorted to an idea related to the string theory using "***shadow*** physics" with "***invisible*** dimensions or dimensions ***unknown***" which are "***inaccessible*** to experimental study" "in the physics version of *The Twilight Zone*, ***disconnected*** from the ordinary world of sight and sound." (Emphasis added). These efforts with the RHIC using a quark-gluon soup and the Duke University experiment using cold lithium are aiming at a perfection ratio of viscosity to entropy. This is an escape to non evidence. This experiment by the Brookhaven National Laboratory in February 2010 measured the temperature of formation of the plasma at 4 trillion degrees and confirmed that this liquid plasma was composed of quarks and gluons.[4]

Steps of Formation of Elements with Mass

The issue had been raised back in 1964 of how nuclear plasma developed into mass when the inflation occurred. In the discussion under physics (Chapter 7), the idea of a Higgs mechanism, or quantum field theory, was discussed about how particles gained mass in a spontaneous symmetry with various elements having their specific mass. Larger and more powerful particle accelerators have been built to try to find this. The CERN Hadden Collider, began working correctly in 2010. Cal Tech physicist Barry Barish, who is the designer of the future International Linear Collider projected to be finished by 2020, has commented to finding the Higgs boson,

> One of the biggest surprises may come, in my mind, in what's publicized as the Higgs boson.... We have a theory of elementary particles, which works pretty, but [its particles are] massless and we put this [the Higgs boson] in by hand. It's one guy's idea and life is probably different than that.[5]

The theoretical problem is so large that the brilliant scientist Penrose joked about having to build a particle accelerator as big as a ring around Saturn to try to discover the operation of the Higgs. Field. [6] The mechanical theory is there, but there is no knowledge how this occurs.

---

[1] Alex Stone, "Big Bang Reenacted in the Laboratory," *Discover*, January 2006, 54.
[2] Ibid.
[3] Tim Folger, "The Big Bang Machine," *Discover,* February, 2007, 32-38; Tom Siegfried, "**Strings** link the ultracold with the superhot," *Science News*, April 25,2009, 26.
[4] Laura Sanders, "Physicists cook cosmic soup to 4 trillion degrees," *Science News,* March 13, 2010, 8.
[5] Barry Barish, "Contemplating future plans for particle colliders," *Science News,* March 27, 2010, 32.
[6] Roger Penrose, *The Emperor's New Mind: Concerning Computers, Minds, and the Laws of Physics,* (Oxford: Oxford University Press, 1999), 263.

The CMB in the WMAP Itself does not Produce what was Expected

The expectation was a location of the place where the smoking gun went off and the trail could be followed. Instead they have discovered a blanket of microwave that reveals a constant temperature with a range of 2.7251 K to 2.7249 K. Some notable astrophysicists believe that this revelation of an almost uniform temperature for the whole universe is not what would be expected from the Big Bang and the inflationary model.[7] Even Smoot and other leaders in the efforts admit that finding hot and cold spots presents a difficulty. Very large cold spots were found with no evidence to explain their existence. Some of these are huge. One is over one billion light years across; the whole universe is said to be only 13.7 billion light years. Ron Cowen has said, "The remnant radiation is riddled with hot and cold spots, most of which reflect the lumpiness of the infant universe, from which galaxies grew. But some of the energy in the hot spots may have been acquired later as light traveled for billions of years to reach Earth." [8] The unexpected findings have had to be explained since there are marked differences between what scientists have thought and what they have actually found.

More recently Roger Penrose of Oxford and Vahe Gurzadyan of Yerevan State University in Armenia have given up the standard inflationary theory of the Big Bang because of circular patterns discovered in the cosmic microwave background. These are smaller than average temperature variations that they say cannot be explained from an instant growth spurt. They believe inflation could not easily generate or erase them. They suggest a new model of a cyclic universe.[9]

In addition, the map shows an ***unexpected twist*** which does not fit. Astrophysicists from the Max Planck Institute of Astrophysics in Germany, the University of Oslo in Norway, and the Jet Propulsion Laboratory in the United States have uncovered evidence that the microwave background looks slightly ***different in different directions*, *as if the whole sky has been twisted*.** The implication is that the ***universe did not expand uniformly in sharp contradiction to expectations*** from the big bang theory. Ned Wright of UCLA and others think this may be a distorted perspective.[10]

Scientists have discovered other unexpected things that don't fit with earlier theories or findings, including ***gravitational lensing***. Physics professor Dr. Richard Lieu and his associate Dr. Jonathan Mittaz of the University of Alabama at Huntsville reported in the *Astrophysical Journal*[11], "evidence of gravitation lensing is missing" in the map of the cosmic microwave background. Lieu said, Einstein's theory says that gravity attracts light, and the beginnings of the microwaves traveling though uneven distributed matter in the near universe should show a spread of sizes of the microwave background cold spots. Lieu says, "The problem is that cold spots in the microwave background are too uniform in size to have traveled across almost 14 billion light years from the edges of the universe to Earth.... Not only is the average about right but far too many of the spots themselves are 'just right' with too little variation in sizes. Given the uneven distribution of matter in an expanding universe," he says, "we should see a broader size distribution among the cool spots by the time that the radiation reaches Earth. An expanding universe would tend to 'stretch' space, causing radiation to disperse as it flies through.... That dispersion would make

---

[7] Adam Frank, "Cosmic Abodes of Life," *Discover*, May 2009, 46.

[8] R. Cowen, "Repulsive Astronomy," *Science News*, August 2, 2003, Vol. 164, 67.

[9] Ron Cowan, "Cosmic radiation analysis hints at series of universal reincarnation," *Science News,* December 18, 2010.

[10] Corey S. Powell, "Field Guide to the ENTIRE UNIVERSE," *Discover*, December 2005 35.

[11] UPI, "New Look at Microwave Background May Cast Doubts On Big Bang Theory," *Space Daily*, August 10, 2005.

objects appear to an observer to be smaller than they really are, as if the light went through a concave lens." Lieu mentions several possible sources of error such as those mentioned and others.

Scientists have also discovered that the elements of the microwave background ***seem out of tune.*** [12] Glenn D. Starkman and Dominik J. Schwarz, experts in cosmological physics, are working at the CERN lab near Geneva. They explain what many scientists believe to be the problem with data of the cosmic microwave background soon after the Big Bang occurred and the inflation of the universe was beginning. The standard model is the inflationary lambda dark matter model of how the universe developed. "The puzzling data comes from studies of the cosmic microwave background radiation. Astronomers divide the CMB's fluctuations into 'modes,' similar to splitting an orchestra into individual instruments. By the analogy, the bass and tuba are out of step, playing the wrong tune at an unusually low volume. The data may be contaminated, such as by gas in the outer reaches of the solar system, but even so, the otherwise highly successful model of inflation is seriously challenged." They conclude, "Like the discord of key instruments in a skillful orchestra quietly playing the wrong piece, mysterious discrepancies have arisen between theory and observations of the 'music' of the cosmic microwave background. Either the measurements are wrong or the universe is stranger than we thought."[13] So far scientists have not uncovered any error that would lead to these discrepancies.

Complete Mapping of the Universe of Galaxies and Stars Reveal Unexpected Mysteries

Extensive Black Hole Singularities with Mystery Were Studied

Until about 1980 black holes were still a theory. Many scientists, after examining the data made available through the orbiting Chandra X-ray Observatory avail, now think there are three main types of black holes that have been formed in different ways.

1. ***Minor*** or petite black holes such as Penrose located that may have formed by the collapse of a massive star about a dozen times greater than our sun. These are stars that probably collapse under their own weight. A second theory is that a supernova explodes, hurling outer layers into space, leaving a burned out remnant called a neutron star, into which the debris from the explosion falls back into it turning into a black hole.[14]

2. ***Major*** or giant black holes formed at the center of galaxies. The galaxies, which are millions to billions of times greater than the sun, probably formed by eating up entire groups of stars. Because of the rotation effect outside the event horizon these emit massive cosmic rays. One-third of this comes from bright shining galactic nuclei where there are massive black holes. These caused the surrounding gas to produce x-rays and visible light outside the event horizon because of great spin. These became big electro magnetic generators.[15] Fuvio Melia reports that super massive black holes may have been involved in building the early universe, spewing bursts of stars and planets.

Our Milky Way seems to have a black hole equal to 2.6 million suns. The acceleration rate of three suns rotating around this have been measured as similar to that of the rotation of Earth around the sun.[16] In 1997, Douglas Richstone using the Hubble telescope presented evidence that

---

[12] Glenn D. Starkman and Dominik J. Schwarz, "Is the Universe Out of Tune?," *Scientific American*, August 2005, 49, cf. 48-55.
[13] Ibid., 50.
[14] "Is there a super way to make black holes?," *Science News*, Vol. 156, September 11, 1999.
[15] Andrew Fabian of Cambridge and Jörn Wilms of University of Tübingen have produced a model of how these long/major black holes work. Cf. Robert Kunzig, "Black Holes Spin?," *Discover,* July 2002, 30-37.
[16] "Stellar motions provide hole-y data," *Science News*, Vol. 158, October 7, 2000.

suggests that ***every galaxy*** has a black hole center. Most astronomers now believe that all galaxies with a bulge or about half [17] have giant black holes, which may start from 0.2 percent of the mass of the first galaxy fragments. This grows when galaxies merge. Others surmise a black hole starts small and grows by feeding on the same gas as the galaxy as it forms. It is now estimated that about half of all galaxies have black holes, which is twice as many as had been counted in visible light.[18]

3. ***Midsize*** black holes of 500 to 800,000 times the mass of our sun, each having a region the size of the moon, probably formed by the merger of many stellar clusters or the implosion of a large star cluster. Most black holes traditionally have a temperature of about a billion K, but a super massive black hole at the heart of the Andromeda galaxy has a temperature of only a few million K and emits much lower rations of x-rays and radio waves, which is baffling.[19]

Scientists estimate that there may be 250 million galaxies, which means that about 125 million black holes may exist in the universe. While Hawking and others have tried to explain away singularities, scientists have discovered an unimaginable number of mysteries throughout the universe.

New Mysteries about Black Holes

One mystery surrounding black holes concerns the limit of the influence of the bulge. Scientists such Karl Gephardt from the University of Texas at Austin, and Laura Ferrarese and veteran astronomer, David Merritt who are at the Dominion Astrophysical Observatory in Victoria, Canada have documented some of this. They have stated some of the following observations. [20] In comparing the mass of each black hole with the average velocities of the billion or so stars that surround it, out to a distance of several thousand light years, scientists have discovered a mysterious ratio. This close swarm of stars is called a bulge. The mystery is that, "regardless of their size ***the bulges always turn out to be 500 times as massive as the giant black hole*** at the hub of their galaxies."

The puzzling mystery is that a super massive black hole can only suck in matter that resides less than a light year from its own location at the center of the galaxy. But *most of the stars in the bulge lie as far as twenty thousand light years away* ***and can't be affected by the black hole's gravity***. Ferrarese said, "The black hole and the bulge should really not know about each other because they are on completely different scales. Somehow, something at the very center of the galaxy knows about the overall structure of the galaxy." This mystery and other factors have spurned over ***thirty theoretical models of black holes*** and how they develop, in addition to the three prevailing theories, on which there is consensus.

A mysterious ***magnetic thrust seems to be needed for an accretion disc of gas*** to be pulled into a black hole. Jon Miller from the University of Michigan at Ann Arbor, using X-ray wind data from star system J1655-40, found from computer simulations that the gas of stars forms an accretion disc that revolves around the black hole until it seems a magnetic pressure pushes the X-ray wind out like a spring and results in the material going into the black hole.[21] ***Gamma ray bursts*** are associated with forming black holes. Much is unknown and being discovered.

---

[17] "Black holes and galaxies may grow up together," *Science News*, Vol. 157, June 17, 2000.
[18] "X-ray Data Reveal Black Holes Galore," *Science News*, Vol. 157, January 15, 2000.
[19] "Chandra eyes low-temperature black hole," *Science News*, Vol. 157, January 29, 2000.
[20] Ron Cowen, "The Hole Story," *Science News,* Januarys 22, 2005, Vol. 167.
[21] E. Jaffe, "Magnetic Thrust: Fields force matter into black holes," *Science News*, June 24, 2006, Vol. 169, 387.

Giant Gamma Ray Bursts Discovered

Scientists believe that on average once a day there is a flash of gamma rays stronger than the sun which results from the death of an aging heavyweight star whose core collapses into a black hole. The overlying material forms *a cone-shaped jet* of these gamma rays. They occur in all types of galaxies in all directions.[22]

Using NASA's High Energy Transient Explorer Satellite, Krzysztof Stanek discovered a gamma ray burst in the Leo Constellation that emitted 100 million trillion times more radiation than a solar flare. He said, "In just a few seconds it produced as much energy as our sun would produce in 10 billion years." This was spit out by a hyper nova jet of material that interacts with stars' outer layers.[23]

Huge gamma ray bursts with extremely high energy are occurring randomly all across the sky at nearly one per day. Gerald Fishman of the Compton Gamma-Ray Observatory has said that these bursts are "one of the biggest mysteries in astronomy today."[24] The center of black holes may be emitting great cosmic radiation in an unknown way.

David De Young of Kitt Peak National Observatory has discovered ***massive jet beams*** of charged particles traveling at more than 1,000 miles per second. These are ***the largest single objects in the universe*** and so old it is ***hard to account for them***. Researchers think they may be driven from massive spinning black holes that eat their surrounding discs of rotating gas. The best speculation accredits these to massive collapsing hyper-nova that produce awesome streams of cosmic radiation.

Mysteries of Giant Cosmic Currents

Alan Dressler and his associates at the Carnegie Institution's Observatory in Pasadena discovered another mystery in 1986, the ***Great Attractor***. They found that the Milky Way and its neighbors are rushing headlong toward the Virgo constellation, apparently pulled by ***a powerful gravitational siren,*** 50,000 quadrillion times the mass of the sun, a cosmic structure *larger than is* ***plausible.***

Sam Flamsteed has recently reported on some of the mysteries of the cosmos and quotes scientists.[25] Marc Postman of the Space Telescope Science Institute in Baltimore, and Todd Lauer of the National Optical Astronomy Observatories in Tucson, recently discovered cosmic convergence far beyond the Great Attractor in a ***great cosmic current***. They measured the brightest stars, of around 600 million light-years in 119 distant galaxies, against the fixed cosmic expansion of CMB of the Big Bang, which is essentially uniform in all directions. What they found, said Lauer, "was that the Earth is indeed moving with respect to these distant galaxies. But it's ***moving in a different direction*** with respect to the microwave background–the two are off by 75 degrees." The only plausible conclusion, reports Sam Flamsteed, "was that some ***cosmic current, far larger*** even than the one caused by the hypothetical *Great Attractor*, is sweeping along the entire collection of hapless galaxies—including, of course, the Milky Way and its neighbors—at a velocity of 435 miles per second with respect to the CMB. From Earth out to a distance of 600 million light-years on all sides, everything is being carried toward some distant point, somewhere beyond Orion." Only a very large mass could cause enough gravitational force

---

[22] *Discover*, December 2005, 23.

[23] Kathy A. Svitil, "Gamma Ray Burst Source Located," *Discover*, January 2004, 3.

[24] William J. Cook, "Knocking on Heaven's Door: Powerful Orbiting Observatories Reveal a Violent, Puzzling Universe," *U.S. News & World Report*, January 27, 1992, 62.

[25] Sam Flamsteed, "Crisis in the Cosmos," *Discover*, March 1995, 64-77, quotes on 73-74.

to pull in this wrong direction, and such a ***force is not known***. Either this is a very great mystery or there is some large error in the data, which does not seem likely. Such a cosmic flow has been found in the microwave background causing the galaxies to be racing in the direction of the constellation Centauries.[26]

Mystery Stars and Galaxies Either Dark or Too Much Dark Matter

Scientists are just now discovering and investigating ***huge galaxy clusters.*** With the aid of the Hubble Space Telescope scientists believe the visible universe contains 80 billion bright galaxies with several hundred thousand galaxy clusters, ***only a few which have been mapped***. One such cluster, the Abell 2029, has about 1000 galaxies in a shape vaguely like an egg in a frying pan with a humongous elliptical galaxy at the center. The temperature that indicates the density is "a torrid 200 million degrees Fahrenheit. That implies a huge, impossible amount of ***dark matter*** which is ***more than the visible matter***.[27] (Emphasis mine). Scientists have also found mysterious ***dark galaxies*** alone. Robert Minchin of Arecibo laboratory in Puerto Rico, using the Lovell radio telescope of the Jodrell Bank Observatory at the University of Manchester in England, recently found a new galaxy, VirgoHi 21, that is ***dark or without visible stars***. It is thought to be a cloud of hydrogen rich gas 100 million times the mass of the sun. Based on computer simulations, scientists speculate that there should be ***more such galaxies*** and there may be many such dark galaxies that are in perforation. The Kepler telescope is revealing multitudes of new kinds of stars.

Powerful Mysterious Magnestars and Radio Transient Stars

Scientists have reported finding magnestars with magnetism and radiation. [28] Neutron stars known as magnestars "turn magnetism into radiation, unleashing as much energy in the blink of an eye as the sun does in 250,000 years." They do not function as we think most stars do by nuclear fusion energy, but from erupting magnetic fields. These neutron stars are thought to be formed by the explosion of supernovas blowing off their outer layers while its core collapses. An implosion of the core causes it to rotate up to hundreds of times per second and the super spin causes the remnant star's magnetic field to wrap around itself, ***reaching a thousand trillion times the strength of the earth's magnetic field which could reshape all atoms***. This "intense magnetism bends and deforms its crust to produce a series of star quakes and the shape of the star's magnetic field changes catastrophically." This new configuration releases a tremendous amount of energy in the form of electrons, positrons (antimatter electrons) and gamma rays. There may be 10 million to 100 million of these in the universe. These stars were first found on March 5, 1979. The biggest ever was seen on December 27, 2004. "A flood of X-rays and gamma rays swept through our neighborhood. The radiation was so intense that it ionized atoms in Earth's upper atmosphere for five minutes disrupting some radio communications. It turned out to be a neutron star, just 12 miles wide, located about 50,000 light years away in the direction of the constellation Sagittarius." Because of the short life of about 10,000 years it is estimated that one of these stars must form every 1000 years in our galaxy. In the last six thousand years we have no idea what kind of forces like these have influenced all the radioactive forces on earth.

A new class of stars from the ***radio pulsars,*** possibly cousins of pulsars, are known as ***Radio Transient*** (RRATs). They have a staccato pulse of strong radio waves for 2 to 30 milliseconds

---

[26] Andrew Moseman, "What Lies Beyond the Edge of the Universe," *Discovery,* 01.02 2011, 71.

[27] *Discover*, December, 2005, 34.

[28] Bob Berman, "Stellar Blowhards," *Discover*, February, 2006.

before going silent for minutes to hours. Astronomers have found 11 of these and think the Milky Way may have several hundred thousand of these elusive stars.[29]

The Mother Who is Younger than Her Babies

One of astronomers' biggest surprises came in 1995 when Wendy Freedman and her leading team of astronomers at the Carnegie Observatory in Pasadena finished their efforts to date the universe and announced the results of experiments with the repaired Hubble telescope. It was especially designed to find Cepheid variable stars, at great distances, to use in calculating the universe's expansion and to calculate the time since creation at the Big Bang. They measured the distance from M100 to Virgo as a yardstick to Coma. They measured the distance from the earth to the distant galaxy as being 56 million light-years and concluded that it has been ***8 to 12 billion*** *years since the Big Bang.* Many astronomers, using other means such as studying the chemistry and other features of stars in globular clusters, got an estimate for the *age of the stars as 15 to 20 billion years old.* Thus, Freedman's evidence using the Hubble telescope indicates that the babies, the stars, are older than their mother, the universe that begot the stars. The Freedman team calculated the Hubble constant at 80 k. This poses a gigantic paradox or mystery. A later more accurate time was figured to be 13.7 billion years old for the universe.

Inflationary Growth from Primitive and Midlife of Stars and Galaxies Not Found

Using the Cosmic Background Explorer, large new telescopes, and a group of Very Large Array (VLA) or 27 large coordinated radio telescopes, along with other tools, efforts have been made to map the sky. As a result scientists are discerning a ***different structure of the heavens than they first theorized***, along with some big surprises that have introduced more unexpected mysteries. Radio galaxies, quasars, and black holes ***were found earlier than should occur in the inflationary process.*** The process of astrophysics would not allow them at that location. The VLA radio telescope records outpourings of energy from objects five billion or more light-years away, usually radio galaxies and quasars at this stage, suggested a very homogeneous universe rather than one which is large scale and lumpy as was expected. Astronomers observed, "A crucial part of the puzzle remains unsolved: how did the universe form monster quasars so early in its history?"[30] Others are asking the same question about black holes: "The recent finding of super massive black holes already in place soon after the Big Bang raises the question: how could these heavy weights have time to form?"[31] The Chandra and XMM\Newton X-ray Observatories have found "quite unexpectedly: many of the active super massive black holes in relatively nearby, luminous galaxies."[32] While reports seem to indicate some difference in the nature of the early stars and black holes, these things are not certain. Most of the quasars are thought to have spawned the earlier ones.

But more and more studies reveal that the universe did not unfold like the big bang projected. ***Scientists found stars earlier than expected***. They thought the universe was a microwave state for 380,000 years after the Big Bang, and then it began to form into the stars, then into the galaxies that we see today and finally into the most mature stars and galaxies. But they found that stars

---

[29] Andrew G. Lyne of U. Of Manchester in England, R. Cowan, Radio Daze, *Science News*, Feb. 18, 2006, Vol. 169, 99, 100.

[30] Amy J. Barger, "The Midlife Crisis of the Cosmos," *Scientific American,* January 2005, 53.

[31] Ron Cowen, "In the Beginning: How to Grow a Black Hole," *Science News,* January 22 2005, Vol. 167, 57.

[32] Barger, *Midlife Crisis…*, pg 52.

appeared at 200 million years, or extremely early and contrary to expectations.[33] Ron Cowen commended that it was like peering in a nursery and seeing grown men. After the Big Bang, when the galaxies should be forming as babies, scientists found large mature galaxies. Using the scenario astronomers have held for two decades to explain the features of the universe, this should not be that way. [34] In this regard, Cowen quotes others. Andrew Bunker and his colleagues at the University of Exeter in England found two galaxies where the universe was suppose to be only five percent of its age or less than a billion years old. Bahram Mobasher of the Space Telescope Science Institute in Baltimore found one "outlandish result [of a galaxy that is] perhaps the most massive ever detected from the early universe.... [It is] likely to be one of the earliest galaxies ever found when the universe was only .7 of a billion years old or 13 light years from the earth. It is six times as massive as our Milky Way."[35] Mobasher's team "is currently studying several other galaxies that are within the Ultra Deep Field data that might be nearly as massive and as distant .... There are others that are 'surprisingly massive....' Bunker's team found a similarly old, red population of stars in the young galaxies they have detected."[36] Alex Hutchinson also reports, "Two recent sky surveys show the mature galaxies existed in the early universe and oddly infantile galaxies are hanging around in the present."[37] These findings are in complete contradiction of the Big Bang inflationary theory – adults should not be lying down in the nursery and babies should not be born in the living room.

The assumption of a development of radiation into atomic materials and then into stars and galaxies presented an evolutionary view of the universe coming into being and gradually growing. The facts do not support this interpretation. Newer telescopes of various kinds are powerful enough to uncover earlier events. Unexpectedly, many scientists have discovered that galaxies and stars already early existed that shouldn't be and are quoted as follows by D. Shiga. [38] One of the scientists said, "If, 20 years ago, you said you were going to detect [galaxies at this distance], they wouldn't have believed you." Two teams using different instruments reported on primordial clusters. "One of the teams, led by Massimo Stiavelli of the Space Telescope Science Institute in Baltimore, Maryland, used the Hubble Space Telescope for its observations. The other, led by Masami Ouchi of the same institute, uses the Subaru Telescope on Mauna Kea, Hawaii." Ouchi's team found two clusters that had to be formed within about one-fourteenth of the time since the Big Bang (12.7 billion years ago) and reported that these clusters of stars have been formed "at a prodigious rate." Another team led by Charles C. Stendel of Cal Tech used the infrared Spitzer Space Telescope. They found "a considerable number of mature-looking galaxies that had already formed large populations of stars which were thought to be only a billion years later.... They observed that the proto cluster galaxies possessed bigger star populations than galaxies outside the proto cluster and seemed to have formed their stars earlier." Peter van Dokkum of Yale University said, "[It] suggests that the differences between field and cluster galaxies that we see today ***were already imprinted at an early time***, when the clusters were still in the process of formation." Another team, led by Christopher R. Mullins of the University of Michigan, Ann Arbor, found a **mature** cluster from 4.6 billion years which led to the conclusion major formation was over.

---

[33] Kathy A Svitil, "Probe Reveals Age, Composition, and Shape of the Cosmos," *Discover*, January 2004, 37.
[34] Ron Cowen, "Crisis in the Cosmos? Galaxy-formation theory is in peril," *Science News*, October 8, 2005, Vol. 168, 235, 236.
[35] Ibid.
[36] Ibid.
[37] Alex Hutchinson, "New and Old Galaxies Show up in all the Wrong Places," *Discover*, January 2006, 60.
[38] D. Shiga, "Nursery Pictures: Astronomers glimpse primordial clustering," *Science News*, March 5, 2005, Vol. 167, 148-149. The quotes are all found in this reference.

These researchers also were surprised at finding the ***universe so transparent*** so early and speculated that the cleaning out of the universe was a result of the "ionizing the opaque gas that previously permeated the universe." (These quotes are all in Shiga's article).

Views of Maturing Universe Also Had Error

The developmental expansion model led scientists to believe that after six billion years, the early fireworks of forming stars and galaxies was over. Amy Barger has made the following quotes and comments, [39] "In the following eight billion years, in contrast, galactic mergers became much less common, the gargantuan black holes were dormant, and star formation slowed to a flicker. In the past few years, however, new observations have made it clear that the reports of the universe's demise have been ***greatly exaggerated.*** " "By examining the x-rays emitted by the cores of these relatively close galaxies, researchers have discovered ***many tremendously massive black holes still*** devouring the surrounding gas and dust. Furthermore, a more thorough study of the light emitted by galaxies of different ages has shown that ***the star formation rate has not declined*** as steeply as once believed." The Wilkinson Microwave Anisotropy Probe, which is a cosmic background radiation study "indicates that star formation began just 200 million years after the big bang." The Sloan Digital Sky Survey "has discovered quasars that existed when the universe was only one sixteenth of its present age…." Collapsed quasars are believed to have produced earlier massive black holes. Fabian Walter of the National Radio Astronomy Observatory detected the presence of carbon monoxide in emissions from one of these quasars "which suggests that a ***significant amount of star formation occurred in the universe's first several hundred million years***." (Emphasis added)

To maintain the idea of inflationary development it is now presumed that in the ***earlier universe*** a small number of ***large black holes dominated and greater bursts of star formation occurred***, whereas today these things are more dispersed and the black holes are more medium in size. However, the discovery of close galaxies minimizes the difference, and does not exactly support the theory. The question is, why have prolific stars and galaxies been found at such a supposedly early age of development and ***would these be considered primordial if the developmental theory of expansion was not used to interpret*** the data? Star surveys from DEEP2, Deep Extragalactic Evolution Probe, conducted by Alison Coil of U. C. Berkley, first detailed pictures of galaxies when the universe was about one-half its present age. Using a spectroscope, the young stars are blue and the old stars appear red. Rees of London comments, "Like older galaxies the young stars don't clump as much as the mid life ones. The mid life ones show more clumping than in young galaxies, thought to support the *idea that gravity increases the clustering* over time."[40]

Structures of the Universe Different and Unexplainable

In the 1980s, some scientists speculated that the structure of the galaxies and stars was like meatballs, then like bubbles in a sink, and then like a lattice work. More accurate instruments present the structure like sheets with large voids between them.

There was a surprise in the finding of a ***Great Wall*** or a thin sheet of galaxies half a billion light-years long with 1,700 galaxies, which lie 200 to 300 light-years from the earth. Also in 1981,

---

[39] Amy J. Barger, "The Midlife Crisis of the Cosmos," 47-53.

[40] S. McDonagh, "Sky Prospecting: surveying the Universe's middle aged galaxies," *Science News*, July 26, 2003, Vol. 164, 52. Three quadrants of 50,000 in this period.

Robert Kirshner of the Harvard-Smithsonian Center for Astrophysics discovered in the constellation Bootes the biggest empty region in space that is now known as the ***Great Void.***

Will Saunders of Oxford University, with other astronomers, report that 2,163 galaxies, covering 74 percent of the sky, probably have formed ***superclusters*** hundreds of millions of light-years across. A. S. Szalay of Johns Hopkins and colleagues found *that galaxies clumped together every 400 million light-years.*[41]

Recent mappings reveal even clearer pictures of the visible universe.[42] Studies such as the Sloan Digital Sky Survey indicate that galaxies cluster ***much closer than expected*** and indicate that unexplained dark matter is present as a gravitation force. Recently an even ***greater cluster of galaxies,*** the **Sloan Great Wall**, has been found. It ***stretches 1.37 billion light years*** across the sky. The idea of a great wall and great void alternating, with perhaps others like them is very surprising and seems ***to violate "the accepted model of galaxy evolution.***"[43] But some computer simulations give a 15% chance of this happening, although that is ***not the expected model.*** The galactic maps suggest the age of the universe is 14.1 billon years. Cowan commented these recent celestial maps suggest the blueprint for the present distribution of galaxies was set at the time of the Big Bang by random subatomic fluctuations.

There are other mystical findings. Two surveys, one by Shaun Cole and a team at the University Durham, England, and another by Daniel Eisenstein and a team at the University of Arizona, Tucson, found acoustical circles as a possible remnant of a pattern when analyzing the maps. David N. Spergel of Princeton suggested these may have been a remnant of the Big Bang.[44] Several studies suggest sound waves may be the source of these acoustical ripples. The ripples, separated by about 35,000 light years, produce a B-flat octave below middle C, which is the lowest note ever detected. Chandra X-ray Observatory found jets from a black hole in the Perseus cluster of galaxies that plow into the cluster's intergalactic gas, creating heat of 50 million K, heating the gas within 150,000 light years of the cluster. The ***sound waves heat the core and drive material*** outward.[45] Similar huge radio blasts have been found coming from the black hole at the center of the Milky Way, and in other galaxies.

On September 30, 2004, other strange huge radio blasts came from a source 600 light years from the Milky Way galactic core. They came at regular intervals for about ten minutes, spaced 76 minutes apart. Magnetic stars of exotic magnetic fields seemed to be the source, boosting energy and slowing rotation.[46]

Visible Appearance of Map by Sloan Telescope Is Weird, Not of Simple Inflation

A review of the Sloan telescope map on page 120 reveals a very strange scattering of the stars. In the middle, from top to bottom there is the appearance of a bow tie effect with a very bright belt in the center. To the right is a bright area that progresses into darker and fewer stars into darkness. Up beside both of the narrow bows, above and below, and on both sides are what appear to be dark areas reminding one of dragon fly wings of darkness. One can also see the sheets of stars and blank areas. All these represent amazing space. In no way does this suggest a Big Bang and a simple inflationary model.

---

[41] Ron Cowen, "Big, Bigger, Biggest?" *Science News*, Vol. 158, August 12, 2000, 104, 105.
[42] *Science News,* November 25, 1989, Vol. 136, 340.
[43] R. Cowen, "Cosmic Survey," *Science News*, November 1, 2003, Vol. 164, 275, 276.
[44] R. Cowen, "Ultimate Retro: Modern echoes of the early universe," *Science News*, January 15, 2005, Vol. 167, 35.
[45] Cf. Andrew Fabian of University of Cambridge, in *Science News,* September 13, 2003, Vol. 164, 163.
[46] R. Cowan, "Puzzling radio blasts," *Science News*, March 19, 2005, Vol. 167, 188.

Failure of the Big Bang Theory to Support the Nebular Hypothesis

Scientists have failed to prove that the universe developed as an inflating universe from a Big Bang. Other, non-scientific theories are being proposed in an effort to sustain the nebular hypothesis. Big Bang history is covered in the book by Harvard astrophysicist, Michael D. Lemonick, *Echo of the Big Bang*. Early in the book he says,

> Among all the sciences, cosmology is especially prone to the danger of premature conclusions. One reason is that it's not an experimental science.... It's hard, moreover, to gather information about the cosmos; the photons of electromagnetic overlap with each other in a tangle of data that must be untangled. As a result, astronomers have always been forced to build their models of the universe, initially at least, on meager information.[47]
> But the feeling is that CMB is sort of a genome ***that would exist*** if the Big Bang occurred, supporting it."[48]

The evidence does not support the Big Bang and inflationary model of the nebular hypothesis as the genome of the universe. All the early great Christian scientists such as Kepler, Boyle, and Newton warned that the uniform deterministic theory was wrong, and now they have been shown to be right!

Avi Loeb of the Harvard-Smithsonian Center for Astrophysics in Cambridge, Massachusetts, has said, "Astronomers still have only the vaguest of notions about how and when the universe emerged from darkness."[49] The beginning view of scientific determinism, that the history of the universe formed by a nebular hypothesis of gas condensations over millions of years, has not been sustained. By analogy, this became the basis for other uniform sciences. Christians may now therefore ask, "If the secular scientist now admits he doesn't know, why is it so outlandish to claim that the beginning was from God and that science may support the history of Genesis?"

New theories are therefore emerging, suggesting that time and space didn't begin with a single, cataclysmic explosion. One new theory, proposed by Paul J. Steinhard and Neil Turok, addresses some of the flaws in the inflationary model. "In recent decades, the theory has been revised on the basis of information that astronomers have gleaned from their ongoing observations of the skies as well as of new findings relating to apparent acceleration of the universes expansion.... They suggest that their theory (of parallel universes) addresses ***some of the flaws*** in the inflationary model of the universe...."[50] Adam Frank comments on an alternate cyclic view to that of the Big Bang. "In the competing theory, our universe generates and regenerates itself in and endless cycle of creation."[51] Since the Big Bang failed to be as expected, this and other multi-universe theories have been discussed, which are all speculations outside scientific discovery. Cliff Burgess and Fernando Quevedo, Peter Byrne, and John Rennie and others write on this with variations. [52] To have millions of possible universes is one way for them to claim it all happened by an exorbitant chance.

---

[47] Michael D. Lemonick, *Echo of the Big Bang*, (Princeton, New Jersey: Princeton University Press, 2003), 6.
[48] Ibid., 11, 12.
[49] Ron Cowen, "Blasts from the Past," *Science News*, Feb. 11, 2006, Vol. 169, 88.
[50] Paul J. Steinhardt and Neil Turok, *Endless Universe: Beyond the Big Bang,* (New York: Doubleday), 2007.
[51] Adam Frank, "The Day Before Genesis: Three radical theories revise everything we thought we knew about the history of the universe," *Discover*, April 2008, 54-60.
[52] Paul J. Steinhardt and Neil Turok, *Endless Universe: Beyond the Big Bang*; Cliff Burgess and Fernando Quevedo, "The Great Cosmic Roller-Coaster Ride." Scientific American, November 2007, 53-59; Peter Byrne, "The Many Worlds of Hugh Everett," Scientific American, December 2007, 98-104; John Rennie, "Worlds Apart," Scientific American, December 2007, 12.

The Greatest Astronomical Mystery Is Dark Matter and Dark Energy

Discovery of Dark Matter

About 1935 Swiss astronomer Fritz Zwicky theorized dark matter was an unseen substance which he thought bound clusters of galaxies together. It seemed that small galaxies were trapped in the larger ones by an invisible gravity resulting in cannibalization. Only a small amount of the mass of the universe is visible as light. Recently it has been discovered at the Chandra X-Ray Observatory that gas clouds between galaxies contain twice as much visible matter as the galaxies in intergalactic clouds.[53] Before the recent discovery of large areas of dark energy, it was estimated that 90-99 percent of the mass of the universe was invisible. Scientists dubbed this cold dark matter.

Various attempts have been made to explain dark matter, but scientists are in general agreement that such a force is necessary to explain modern cosmology. The most recent speculation is that cold dark matter is a liquid, not gas. Ron Cowen has given various views about this subject.[54]

Shocking Discovery of Dark Energy

A progressive search was underway to account for these unknown forces. Karl Fredric Gauss, Directory of the Astronomical Observatory at Gottingen, presented the idea of curved space in 1820. But in 1917, Einstein, in his general theory of relativity, had predicted expansion or contracting of the universe. Later he speculated there was a comparable invisible power that existed, but eventually he concluded that idea was wrong. E. P. Hubble of Mount Wilson Observatory discovered some time earlier that the universe was expanding and suggested that it was expanding at a constant rate.

Hubble derived a law for the expansion time using the distance of expansion and the speed increase. He calculated a constant for the expansion of the universe using certain Cepheids, but his figures were in error. However, Hubble himself was not dogmatic about his figures as is reflected in some of his statements. He was not positive that "the redshift was a genuine measure of expansion."[55] But his work marked the way for those who succeeded him to continue studying the expansion of the universe. In 1977 Bob Wagner found that observing supernovae was a better method for discovering the constant. He classified the supernova as 1A and later 1B, with the 1A as preferable. The formation of the big bang theory and its mathematical proof, and the subsequent discovery of the radiation residue and the acceptance of the Big Bang theory a couple of decades later, increased the interest in the Hubble constant. The various theories about the inflationary pattern of the universe heightened the interest in the search for the unknown forces in the universe.

In 1998 two rival teams, one put together by Robert P. Kirshner of Harvard with the observation research led by Brian Schmidt, and another team led by Saul Perlmutter, were studying the titanic explosion of distant, elderly stars—type 1A supernova. These seem to be cosmic speedometer markers for the effects of gravity. In his book, *The Extravagant Universe,* Kirshner tells how the two teams, working separately and almost simultaneously, uncovered data that overturned the prevailing belief that the cosmos has, because of gravity, been slowing down

---

[53] Ron Cowen, "Visible Matter: Once Lost But Now Found," *Science News,* August 10, 2002, Vol. 162, 83.

[54] Ron Cowen, "Have Milky Way MACHOs Been Found?" *Science News,* Vol. 156, September 18, 1999; Cf. Ron Cowen, "A Dark View of the Universe," *Science News,* Vol. 157, January 8, 2000.

[55] Robert P. Kirshner, *The Extravagant Universe,* (Princeton: Princeton University Press, 2002), analyzes the Hubble constant and law, 89-91, and its continued pursuit by others elsewhere.

its rate of expansion since the Big Bang. In disbelief these teams revealed that the universe is actually flying apart at a faster rate than thought possible—it is a "Runaway Universe."[56] Stephen Hawking checked the measurements mathematically and confirmed the acceleration of the expansion as correct. (He did not mention that this contradicts his view of a global closed universe.) At the American Physical Society's 100th annual meeting in Atlanta in March 1999, this radical new view of the universe's acceleration, implying an open universe, was established as correct. Other experiments have since confirmed this.

Robert Kirshner underscored the importance of these discoveries when he said,

> The observations of distant supernovae show that we live in a universe that is not static as Einstein thought, and not just expanding as Hubble showed, but accelerating! We attribute this increasing in expansion over time to a dark energy with an outward pushing pressure. In its simplest form this might be Einstein's cosmological constant, which for 60 years was theoretical poison ivy. Dark energy makes up the missing component of mass-energy that theorists have sought, reconciles the ages of objects with the present expansion rate of the universe, and complements new measurements of the lingering glow of the Big Bang itself to make a neat and surprising picture for the contents of the universe.[57]

> The expansion will literally be exponential: the bigger it gets, the more it speeds up. The universe will run away in headlong expansion.[58]

This was further confirmed by two experiments, MAXIMA (Millimeter Anisotropy Experiment Imaging Array), and BOOMERanG (Balloon Observations of Millimetric Extragalactic Radiation and Geophysics). Using telescopes lifted 25 miles high to take pictures of cosmic rays, scientists confirmed the density of matter and energy to expand forever.[59] Experiments in other areas confirm these findings.

Alan Guth proposed that the universe had two phases of expansion. He speculated that the universe expanded very slowly for some six billion years until the dark energy overcame the gravitational pull as the stars and galaxies were forming. After that, he surmised the dark energy broke free and began to accelerate. During this period the Hubble constant would apply, lengthening the time period of existence of the universe. For many years astronomers had sought, unsuccessfully, to arrive at a Hubble constant. Even Hubble himself had not been dogmatic about his findings.

Astronomers began to focus on finding data that would substantiate a much slower acceleration during the beginning phase of the universe. Studying ten or more supernovas, Adam G. Riess of Space Telescope Science Institute, Baltimore, showed brighter supernovas indicated less or slower acceleration than later supernovas had shown. But a later, comprehensive study of 1A supernovas using the Hubble Space Telescope, including six of the oldest supernovas doubled scientists' knowledge and revealed that ***this was not correct.*** The evidence shows that the dark energy was accelerating at the beginning.[60]

---

[56] M. N. Jensen, "Red Glimmer Reveals Most Distant Galaxy," *Science News*, March 21, 1998, 182; R. Cowen, "Studies Support An Accelerating Universe," *Science News*, November 31, 1998, 277.

[57] Kirshner, *The Extraordinary Universe*, preface xi.

[58] Ibid., 258.

[59] Ron Cowen, "More evidence of a flat universe," *Science News*, Vol. 157, June 3, 2000.

[60] Ron Cowen, "Wrenching Findings," *Science News*, February 28, 2004, Vol. 165. 132; R. Cowen, "Super Data: Hail the cosmic revolution," *Science News*, October 11, 2003, Vol. 164, 227.

Dark energy is a very disturbing concept for astronomers. Joseph Lykken of Fermi National Accelerator Laboratory in Batavia, Illinois has said, "Cosmic acceleration is not just another mystery. It is something fundamental in our understanding of gravity, energy, and quantum theory." Cowen has said, "It has been described as, 'an unchanging property of empty space that imbues the universe with a constant acceleration.' "[61]

Conclusion about the Mystery of Darkness

Ron Cowen has called the search for dark energy, "Dark doings: searching for signs of a force that may be anywhere or nowhere."[62] Kathy A. Svitil said, "There is little consensus about what these things [dark matter and dark energy] are. The idea that our universe is dominated by two separate mystery components, utterly unlike ordinary matter or energy, strikes many researchers as absurd."[63] Up to now the evidence has shown that because of dark energy, the universe is expanding at ***an ever-faster rate,*** as Kirshner and others say.[64]

The implications for matter and energy in the universe have been radically altered.[65] While the APS has authorized scientific approval of these views, other studies continue. The percentages of matter and energy are shown on the graph below.

These new findings have thrust astronomy into theories about which it knows little. ***Most of the universe is mystery***. This in itself destroys the deterministic view of the universe. To overcome this, Mordehai Milgrom and several others have proposed a ***departure from the standard laws of physics*** to do away with dark matter. They propose that when acceleration is small, Newton's second law is altered: force becomes proportional to the square of the acceleration. In galaxies where this might apply, the MOND (Modified Newtonian Dynamics) theory eliminates the need for dark matter.[66]

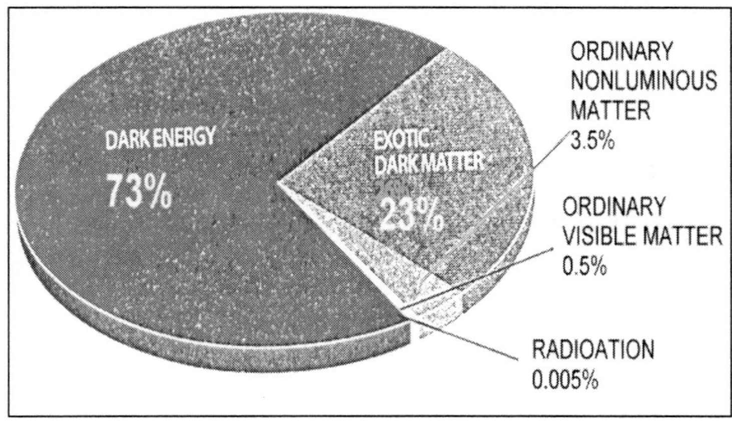

[61] Ron Cowen, "Dark Doings," *Science News*, May 22, 2004, Vol. 165, 330.
[62] Ibid.
[63] Kathy A. Svitil, "Darkness Demystified," *Discover,* Oct 11, 2004, 11.
[64] Ron Cowen, "Lonely Universe: Runaway Expansion Will Lead to Dark Times Ahead," *Science News,* August 31, 2002, Vol. 162, 139, 140.
[65] Robert P. Kirshner, *The Extravagant Universe*, 89-91, analyzes the Hubble constant and law and its continued pursuit by others elsewhere.
[66] Mordehai Milgrom, "Does Dark Matter Really Exist," A *Scientific American* Special Report, 2003, 1-9, cf. 8, 9 for reference to others who advocate this.

But evidence continues to refute theories of dark matter and dark energy. A map by Sloan Digital Sky Survey, covering 200,000 galaxies over six percent of the sky, "Supports previous evidence that most of the universe's matter is invisible. Without the gravity provided by the unseen material, dubbed dark matter, galaxies wouldn't have clustered as tightly as the Sloan map indicates they do."[67] Therefore, many scientific determinists are searching for a way to explain this. In reviewing the new theories it is said to be, "The Wild West of theoretical cosmology right now."[68]

The discovery of dark energy, causing the accelerated expansion, must change all the calculations of the speed of light. The expansion of the universe has to be counted in distance of light speed. The illustration of one airplane flying in the jet stream and one outside the jet stream establishes a constant for the rate of expansion (or steady rate of a jet stream.) But with an increasing rate of expansion, the time of the speed of light must be considered as rapidly shortening. Even a small **increase of *rate*** from the beginning can radically change all figures of the astronomical formations. For example, if one accelerates on a freeway to 100 M.P.H. and drives 13.7 hours, he will have driven 1370 miles. (The universe is said to have expanded for 13.7 light years to measure distance). But if the person gets in a rocket ship that has a constantly ***accelerating rate*** of speed, he will pass the mark of 1370 miles in only a matter of minutes. Now that it is known that that ***rate*** of the universe's ***expansion*** has always been increasing ***at an accelerating rate***, is it possible that the age of the universe is only several thousands of years old? This fact means that all the figures about the universe in modern times are no longer accurate. To even admit this presents chaos for astronomers. In addition to the rate of expansion of the universe, the reported tests for the speed of light through several centuries show the ***rate of light speed*** seems to be slowing down. If that is true it would also diminish the age of the universe and the distances between galaxies and stars. The distances and size of the universe and the position of the stars and galaxies have been calculated on Einstein's determination of the speed of light as a constant. If this constant is also in question, it presents even more confusion. Astronomers will not even allow consideration of such possibilities, much less admit to them. We will look at the physical standards of the forces more completely in Chapter 23 on age dating.

Conclusions: Uniformed Determinism Not Demonstrated by Astronomy

Questions about Dark Matter and About New Inflationary Model

Ninety six to ninety seven percent ***of the universe is either dark energy or dark matter,*** about which science knows nothing. The amount of dark matter is exorbitant and the dark energy that dominates and changes the rate of expansion throws previous astronomical calculations into chaos. In other words, almost all the universe is a mystery and outside the realm of human understanding.

Moreover, the ***whole big bang and new inflationary theoretical model did not prove*** to be that which is found in the natural universe. Laboratory experiments were not found to be condensed gas that expanded, but hot liquid. Scientists thought these gases were extremely hot (3000 K) and gradually cooled, but unexpectedly, the whole universe was found to be only 2.7 Celsius above absolute zero. Leading astronomers have asked, "How could different regions of the universe, separated by such enormous distances all have the same temperature? In the standard

---

[67] Ron Cowen, "Cosmic Survey: Galaxy map reveals dark business as usual," *Science News*, November 1, 2003, Vol. 164, 275, 276.

[68] Kathy A. Svitil, *"Darkness Demystified,"* Discover, Oct. 2004, 11.

version of the Big Bang, they couldn't.'"[69] Also, the radiation is not progressive, but uniform to one part in 25,000. Scientists have also found the universe to be uniformly luminous.

The ***mapped universe does not fit the theory***. The mapped dispersion of the ***background cosmic radiation was not as would be expected:*** it deviated from a naturally unfolding pattern; it seems twisted, out of tune, or not harmonious as expected, and gravitational lensing is missing. In addition, scientists found a repetitive pattern of formation in the universe that was based on ***mysterious acoustical sounds*** and not on progressive condensation. The primordial beginning of the universe has ***mature stars, galaxies, and black holes that do not fit*** the expected primordial view. These are adults that are sleeping in the nursery of space. The maturing galaxies do not reflect the pattern that was expected from the Big Bang expansion. There are varied kinds of galaxies and stars of all descriptions that would not be expected, or for which development cannot be explained. Some are entirely of dark matter. Great walls or sheets of stars and great voids which do not fit the pattern of uniform developmental theory are quite surprising and mysterious. Great clumps of galaxies, some with more dark matter than visible matter, are unexplainable. While black holes are now accepted as a fact, major ones have a mysterious ratio of 1 to 500 to the bulge of stars far outside the influence of the black hole. Black holes emit acoustical waves that create extremely powerful ripples. Unknown sources outside the black holes have been known to emit unexplainable acoustical waves in a regular cycle of emission and rest. There are millions of singularities of the three major black holes, but perhaps as many as 30 of one, all of which are mysteries.

Many other large, strange, and unaccountable phenomena in the universe can be affecting the solar system and the earth. Those mentioned here are only a few. Huge mysterious cosmic currents are totally unexplainable, such as the Great Attractor, a large gravitational siren. One huge cosmic current, similar to the Great Attractor, seems to be moving the earth at 75 degrees different than the cosmic background. Powerful gamma ray bursts, jet beams, the largest objects in the universe, moving at tremendous speeds, huge radio transients that pulse at regular times, and a large cluster of galaxies, 1.37 billion light years, contrary to any other galaxy formation, are just some of the mysteries that continue to confound scientists.

Questionable Cosmological Theory Is from Wish of Bias, Not Facts

The early theories regarding the origin of the universe were the basis for uniform determinism. They paved the way for uniform determinism to be accepted and adapted in geology and biology. But it is a failing model.

Even toward the end of the twentieth century, before the most recent discovery of major discordant, dark matter and energy, Sam Flamsteed reviewed the emerging data. Observing that it didn't fit the pattern, he said, "One could almost have imagined a time when we would have figured it all out. These days, that looks distant indeed: these days ***cosmology seems to be collapsing in upon itself.***"[70] Robert Kirshner has said, "It's a strange picture we have painted. The universe has dark energy and dark matter, neither of which is familiar to us from our everyday experience, nor detected from any experiment on Earth."[71] Astronomers present impressive mathematical data and brilliant ways of figuring out things based on certainties that no longer exist.[72]

---

[69] Tim Folger, "A Universe Built For Us," *Discover,* December 2008, 55, about Andrei Linde, Stanford University et al., 52-58.

[70] Sam Flamsteed, "Crisis in the Cosmos," *Discover*, March 1995, 68.

[71] Kirshner, *The Extraordinary Universe,* preface xi.

[72] Cf. Neil Degrasse Tyson and Donald Goldsmith, *Origins: Fourteen Billon Years of Cosmic Evolution*,

The bias to prove a nature without God has led to claims to know and understand a nature that evidence does not support and is not understood. Astrophysicists are opposed to a God who works by unseen forces. They believe in invisible powers of nature to support the evolutionary theory that they have said must be determined to observational knowledge of a uniform world. God still reveals "His eternal power and divine nature" (Roman's 1:20). It is foolish to turn from Him to worship that which is created. True enlightenment is seen in Christ who created and upholds all things, visible and invisible (Colossians 1:16, Hebrews 1:2-3).

(New York: Norton, 2004, 345 pages).

Chapter 10
Solar System Found Mysterious

Importance of Accretion Theory of Solar System

Laplace's claim to mathematically demonstrate the incremental formation of the solar system marked a major shift in rejecting God's creative power and presence. It propelled Kant's theory of evolution to acceptance in astronomy and geology. But as Hawking notes, Laplace projected his theories of the solar system on what he then ***thought*** to be neatly and similarly formed planets, all created in the same way, with similar orbits, each rotating on a similar axis in the same direction. His calculations led him to conclude that natural laws were sufficient; there was no God. His views influenced many others who exchanged belief in a supernatural God, who created and allowed biblical catastrophes, for the popular view of uniform progressive evolution. This provided the platform on which the evolution of biological life was later built.

This theory of natural development of the solar system and of the earth changed the popular world view and man's values. Man himself became the supreme value, and with his education to understand and use the world of uniform nature, he replaced God and His will.

Laplace's Determined Science Early Seemed to Indicate Uniform Principles

By the turn of the nineteenth century, under the influence of Bode's Law, the planets as far out as Neptune had been discovered. Throughout the nineteenth century, the scientific community, using Newtonian physics, and by virtue of better telescopes and better exchange of information, arrived at greater mathematical certainty about the solar system, the size of the planets, their mass, their rotation, and the path of their orbits. Many great universities displayed models of the planets with their relative size and their moons, extended and suspended on metal guides, showing their orbits in the relative distances from the sun. These ***orrerys*** were displayed for students and the public to see at Oxford, Harvard and elsewhere, impressing many with claims of great understanding of the solar system's operations. Man's brilliance, as demonstrated in many inventions at this time, made this demonstration more persuasive.

Birth of a New View of the Solar System by New Instruments and Knowledge

In the twentieth century, especially in the last quarter, a tremendous number of new instruments had been invented for studying the solar system. These new instruments enabled men to discover whole new greater unperceived areas of knowledge about our planet and the others in the solar system. Giant reflective telescopes, some with new kinds of lenses, were placed atop high mountains; space and radio telescope, and telescopes measuring temperatures and magnetism were built.

Modern science led man into new information about the most obvious planets. By the end of the nineteenth century all the planets through Uranus were known. Pluto was found in 1930 and designated the ninth planet. Scientists thought they had uncovered all they could about the solar system. But they continued to learn, through various calculations, many new things about the size, temperature, elements, and moons within the solar system. Some appearances led to imaginary theories (e.g. irrigation canals on Mars) that were not correct.

Inadequacy of the Old Laplace Theory

As the first phase of modern planetary study ended, and even before the revolutionary space exploration of the twentieth century, Laplace's theories had become untenable. Robert Fox reviews how new findings showed the inadequacy of Laplace's theory.[1] Speaking in 1950 E. B. Branson, one of America's renowned geologists, said,

> The hypothesis [of Laplace's theory] is simple and its implications are clear and satisfying, *if not carefully scrutinized.* So many things are now known about the earth and the heavens that were unknown in the days of Laplace that *the task of bolstering and modifying the nebular hypothesis has become burdensome.* When modified to fit present knowledge it has *lost nearly all of its distinctive features* and really deserves an altogether new name.[2]

Efforts, therefore, were made to adapt the theory to new data. Victor Safronov (1969) built on Laplace's idea that the stellar dust stuck together particle by particle to form first grains, then pebbles, building until planets were formed. Safronov's theory, ***the standard theory*** of planet formation by accretion, which is the most popular, was a revision and extension of Laplace's theory. George Wetherill developed a computer model for the theory in 1970. This model has phases with shifts that are hard to explain; some don't fit. The first 50,000 years of accretion by gravity of the sun's gases are said to have formed balls with metallic molten centers and smooth, rocky exteriors. But at a certain point, this had to shift so that these were violently bombarded for 100 million years with fragments that left craters. This was then followed by the formation of the other four large planets, which have a small core and cold gases, presumable because it is much colder at their much farther distance from the sun. Wetherill's computer model says it took 100 million years for these cold planets to form. But as McNab and Younger have said,

> This was a problem because it was common knowledge among planetary scientists that Jupiter and Saturn had to have formed much faster than these theoreticians were predicting. The history of the Sun itself imposed a severe time limit.[3]

The inner two of the colder four planets seem different than Uranus and Neptune. Jupiter and Saturn probably have water vapor and gases such as carbon dioxide frozen into ices, while Uranus and Neptune have frozen gases such as methane and ammonia. But mysteriously, these have moons, some of which are large with craters like the inner planets and our moon. Then beyond the so called "snow line" and beyond the big four, was little Pluto which is very different still—more of a rocky planet with a very large moon. Scientist speculated that icy planets existed beyond this.

Even the near planets are very different than Laplace calculated and described. The first two planets, Mercury and Venus, which are close to the sun, are hot; Earth and Mars are more moderate in temperature; the temperatures of the large middle planets drop quickly and immensely after Mars. The size in diameter, mass, or volume of the middle planets that are past the "snowline" is much greater. Jupiter and Saturn are much larger, especially Jupiter. Uranus and Neptune are only about a third of Jupiter, while still four times the size of Earth and Venus. Jupiter is so large it could contain all the planets combined. Jupiter's gravitational pull is thought to affect the comets and possible asteroids and meteorites. The time of orbital revolution jumps from just days for the inner planets to many years the farther away from the sun they are. The inclination of the orbit or ellipsis around the sun of most of the planets is from nearly one to three degrees,

[1] Cf. Robert Fox, "The Rise and Fall of Laplacian Physics," (*Historical Studies in the Physical Sciences,* 1972), 89-136.

[2] E.B. Branson, *Introduction to Geology*, third edition, 325.

[3] David McNab and James Younger, *The Planets,* (New Haven: Yale University Press, 1999), 32.

except Mercury, the closest, is seven degrees and Pluto the farthest away is 17.2 degrees. Pluto's orbit starts higher than the others and ends up very much below them, mysteriously cutting across the plains of the rest. After the mid twentieth century the best efforts to patch up the accretion theory of Laplace could not be given a clear uniform explanation.

Discovery of the Kuiper Belt Area

But the solar system has since become much more complicated. After the discovery of Pluto in 1930, Fredrick C. Leonard speculated that there were a series of ultra-Neptunian bodies. In 1943 Kenneth Edgeworth speculated that beyond Neptune, "the outer region of the solar system, beyond the orbits of the planets, is occupied by a very large number of comparatively small bodies," [4] some of which wander as comets to visit the solar system. Gerard Kuiper's studies and speculations in the 1950s led to the discovery of the Kuiper Belt or Kuiper-Edgeworth Belt.[5] But the direct evidence of these objects did not come into view until the last decade of the century. This led to further searching and mapping the path of comets and finding other Kuiper Belt objects (KBOs).

With advanced telescopes and space probes, the area around Neptune and the cold planets has been researched. The Kuiper belt has an inner orbiting group called ***plutinos*** that describe objects orbiting two times while Neptune orbits three times. These include the larger one-fourth of the KBO's such as Pluto, its moons etc. This belt is considered dynamically stable. Beyond this is a further ***scattered disc.*** This is a region that extends as far as 100AU which is the place of the comets. Comets were discovered much earlier by Carolyn Herschel, William's sister. Astronomers have found a huge reservoir of comets of amazing variation in sizes. At least 70 very large Apollo-size comets have been identified and named, and many with orbits to the sun have been tracked. It is thought that Jupiter may kick comets out of place, but this is not known for certain. These comets are very cold and probably composed of dust/ice. The scattered disc also includes centaurs. Some of these, like Eris, are extremely large objects with highly elliptical orbits.

At least 1,000 KBOs have been identified. Astronomers think there are more than 70,000 that are as big as 60 miles in diameter. Among these is a new classification of ***dwarf planets*** into which astronomers have classified Pluto, with Haumea and Makemake and others. Pluto has a tilt of 58 degrees. Its equinox comes at the innermost part of its orbit to the sun when the frozen surface is temporarily warmed. Each season lasts 62 years. One day is equal to 6.4 earth days. Pluto's binary dwarf, Chiron, is twenty times closer to it than Earth's moon. They both always have the same face toward each other.

Beyond this Kuiper belt astronomers speculate there are Hills cloud objects and much, much farther out, the Oort cloud objects, from which scientists think long-period comets, like Hal-Bopp, originate. Short-period comets seem to cluster near the plane of the solar system in the scattered disc. Some of the largest trans-Neptune objects such as Eris, Pluto, Makemake, Haumea, Sedna, Orcus, Quadar, and Varuna are from 300 miles to 744 miles in diameter with satellites. There had been speculation of the existence of a Planet X, which had generated a cult that has been shown not to exist. The search for a Planet X that generated a cult is known not to exist and is a phantom. These KBOs are a great mystery.

The Asteroid Belt Is Now Better Known

The asteroid belt between Mars and Jupiter, dividing the inner planets from the outer and

---

[4] John Davies, *Beyond Pluto: Exploring the outer limits of the solar system*, (Cambridge University Press, 2001) xii, 2; et al.

[5] Ibid.

cold planets are still a mystery. Where another planet should have been, there is the asteroid belt. More than half the mass of the main belt is contained in the four largest objects: Ceres, 4 Vesta, 2 Pallas, and 10 Hygiea. All of these have mean diameters of more than 400 km, (ca. 250 miles) while Ceres, the main belt's only dwarf planet, is about 950 km (ca. 500 miles) in diameter. Scientists have discovered thousands of asteroids of varying size; many smaller ones hit the earth every year as meteorites. "Nevertheless, collisions between large asteroids do occur, and these can form an asteroid family whose members have similar orbital characteristics and compositions. Collisions also produce a fine dust that forms a major component of the zodiacal light. Individual asteroids within the main belt are categorized by their spectra, with most falling into three basic groups: carbonaceous (C-type), silicate (S-type), and metal-rich (M-type)."[6] Space probes have investigated, and photographed these objects and have learned more about their composition and shape. Still, we have no explanation why comets and asteroids exist and how they were formed. These are a great mystery to our solar system, and increasingly a threat to the earth. These solar objects do not fit any natural laws we know.

Much is known now to be very complicated about the solar system. Laplace, and those who accepted his view, ignorant of these things when he formed his simple accretion theory.

New Era of Seeing the Solar System by Space Probes

Space exploration has done much to change our understanding of the solar system. Sophisticated new telescopes, precise cameras and recording devices offer a completely different view of the universe than was previously thought. America, Russia, and Japan collectively have sponsored several dozen space probes; computer analysis provides more accurate information quickly.

In the last half of the twentieth century, the first superficial observations gave way to a new phase with space probes. These dispelled many false ideas by producing new, more accurate data. By the end of the twentieth century, David McNab and James Younger described the transition from the old views to more objective information. They said,

> Astronomers had trained their telescopes on them (earliest known planets in orbit), measured their size and their speed, and even added three more planets unknown to the Ancients to the family of the Solar System. But the planets remained faceless spots of light for the most part, fuzzy discs at best. Just as nature deplores a vacuum, so, the human mind detests lack of detail, and imaginations ran riot. ...The sky was filled with alien, alternative earths. Three decades after Mechto (the first satellite probe), America's unmanned probe, Voyager 2, floated past Neptune and brought the first stage of planetary exploration to a glorious end. In those 30 short years, human beings have sent robotic emissaries to every planet apart from Pluto, discovered dozens of new moons in orbit around other planets, and put to rest the myths and fantasies that had been accepted for centuries. The space age replaced fantasy with even more fantastic fact.[7] (Emphasis added).

Our study of the sun has begun to give vital information about its workings and its solar winds, flares, and Corneal Mass Ejections. The 1973 Skylab was occupied by our scientists for six months with constant observations. Other probes from other nations have assisted. Additionally, the jointly sponsored, Solar Heliospheric Observatory (SOHO), unmanned space lab, equipped with highly sophisticated instruments, has been placed at a point of equal gravitational pull of the earth and sun. It sends thousands of images to many scientists each day. The Ulysses probe that

---

6 "Asteroids," Wikipedia.

7 McNab and Younger, *The Planets*, 7.

swung around the poles of the sun revealed the spiraling nature of fast and slow winds. New information from new probes occurs almost daily, and the twenty first century showed a new arrangement as shown below. These explorations provide information about the potential benefits of the sun as well as its threats.

New Millennium Revised Nomenclature and Description of the Solar System

Four Inner Planets (metal and rock)	Mercury Venus Earth-one moon Mars-two moons
Asteroid Belt	Ceres and other large asteroids-many named and tracked. Thousands of small asteroids Meteorites and blank spaces
Four Middle Planets (core with mostly outer Gases)	Jupiter-8 regular and 55 irregular moons, Comets Saturn-21 regular and 26 irregular moons, comets-ca. 70 plus large, Uranus-6 regular and 7 irregular moons, Apollos type comets, Neptune, 1 large (Triton, retrograde) and 12 small moons
Kuiper Belt	Millions of small ***plutinos*** (more stable) one fourth of KBOs such as Pluto and moons etc. some are binary planets (e.g. Pluto and Charion) (more than 70,000 KBOs) ***scattered disc*** (unstable, SDOs) centauries such as Eris, short-peroid comets.

Hills Cloud

Oort Cloud

Observations Showing Much Difference between Solar Planets

We can make some general observations about the planets in the solar system. Of the inner planets, the first two are hot; the second two are moderate. The planets beyond the asteroid belt are very cold. The revolution of the inner planets ranges from 88 to 687 days, but the revolution of the middle, cold planets ranges from 119 years to 285 years. There is no gradual progression in either of these groups in their distance from the sun; it varies greatly. The inner planets have very few moons and the middle to outer have many moons of various sizes. ***Proportionately***, the earth's moon is much larger than those of other planets. The planets with many moons have all kinds of sizes. The diameter of the inner planets is about one-tenth of the first two cold, middle planets and one-fifth of the other two beyond them, so size goes from small to large to medium large. The inner planets are four or five times more dense than the outer ones. The inner ones have an atmosphere (except Mercury) of carbon dioxide, nitrogen, oxygen and similar gases, but the atmosphere on the middle, cold planets consists of hydrogen, helium, and methane. The inner two have virtually no tilt to their axis; the earth and Mars tilts about 24 degrees; Jupiter, Saturn and Neptune tilt 26 to 30 degrees, but Uranus tilts more than 90 degrees. While these are generalizations, the planets within the groups have marked differences, some of which are inexplicable.

Many specific differences are also evident. Mercury has virtually no atmosphere. Venus has dense clouds, mostly of carbon dioxide, while Earth, which is about the same size, has atmosphere mostly of nitrogen and oxygen. But Earth's large moon has none. Yet Mars, like Earth, has no oxygen, but Venus, on the other side of Earth, has mostly carbon dioxide and some argon. There are uncertain signs of water on Mars. While there is the appearance of polar ice, the composition is not yet known, and it appears there may have been water on Mars in the past, in the form of lakes. Jupiter and Saturn have frozen water and carbon dioxide, and Uranus and Neptune probably frozen methane and ammonia.

Mercury is small, about one-third the size of Venus and Earth, and about one-half the size that would be expected based on computer models. A catastrophic collision may explain the difference in their sizes. Mercury has one huge crater, indicating an impact so hard it caused the mountainous surface on the opposite side. It has a metallic core, much larger proportionally than any other planet. Venus, which is about the size of Earth, was expected to be much cooler under its creamy clouds. But satellites show it has over 167 large volcanoes, 100 kilometers across and over 50,000 small volcanoes. Its surface temperature is over 400 degrees centigrade. Yet, there do not seem to be any present eruptions and strangely, very few impact craters. There are pancake domes, which are over 50 kilometers across, and plains that appear like formerly flowing lava. It is about four-fifths plains and one-fifth highlands. It is "a searing lava-filled hell-hole beneath a cloud cover."[8] Venus, surprisingly, is considered a ***much younger planet*** than the other three rocky planets closest to the sun. Venus has no evidence of plates or plate tectonics like Earth.

On the other side of Earth is Mars, which is about one-half the size of Earth. Instead of a multitude of volcanoes, it is pocked on one side with craters, and on the other side with four of the largest known volcanoes. They originate from a hot spot that has caused them to grow and rise higher than any such counterpart on Earth. On that side of Mars there is a giant gash of a canyon, 180 kilometers wide. Violent dust storms rage on Mars most of the year.

On Saturn there is a mysterious eruption which occurs every thirty years, causing an oval spot over about five percent of the circumference, about 20 degrees below Saturn's equator. Saturn seems to have two rings around it, "but Pioneer 11 revealed a ring system of baffling complexity that seemed to defy the laws of physics."[9]

Jupiter's size is awesome. It has a great red spot on it that is thought to be "the greatest storm in the Solar system."[10] Mysteriously, Jupiter has a giant magnetic radiation field that is 20,000 times stronger than Earth's, engulfing all its moons. If it could be viewed from Earth, it would appear as large in the sky as our sun. The energy radiated from it is hotter than the sun's surface and generates 500 times the lethal dose of radiation for a human being. In spite of this hot energy, the planet is minus 150 degrees centigrade[11]. Jupiter has a core of molten rock and ice fifteen times the mass of Earth.[12] In 1994, scientists discovered a giant comet, Shoemaker-Levy SL 9 that had broken up into twenty-one pieces. They were able to observe and record the explosive collision with Jupiter as a giant catastrophe.

The tilt of the axis and speed of rotation of the planets are equally unique and unexplainable. Some tilt very low: Mercury is 0.0 degrees; Venus is 2 degrees rotating East to West; Jupiter is 3.05; Pluto tilts 3.05 and rotates east to west. Most of the others are tilted nearly a quarter, forming seasons: Earth 23.27, Mars 24:46, Saturn 26:44, and Neptune 28:48. Uranus is completely on its side, 97.53, with its moons going round like a Ferris wheel. The time or speed of a complete rotation of the planets varies widely, but there are some that seem to pair together: Mercury is 59

---

[8] McNab, *The Planets,* 16.
[9] Ibid., 115.
[10] Ibid., 110.
[11] Ibid., 112.
[12] Ibid., 113.

days and Venus 243 days. Earth and Mars are hours and minutes 23:56 and 24:37. Jupiter and Saturn—9:55 and 10:39, and Uranus and Neptune—17:14 and 16:7. Pluto's rotation is 6 days and 9 hours.

The tilts affect the equinoxes of the various planets, which as Bob Berman notes, are very different.[13] Mercury's is permanent with poles vertical at all times. Earth's is in March with 23.5 tilt. Mar's is nearly the same tilt as Earth but it has seasons twice as long because of its slower orbit. Venus, with only 2½ degrees tilt, has virtually no seasons and with a heavy atmospheric cover is very hot all the time. Jupiter's tilt is like Venus's so there is little seasonal variation. While Saturn's equinoxes are similar to Earth's, it has 1/100 of the sunlight and its atmosphere is driven by the planet's internal heat. Its rings make it independent of seasons. Neptune is much like Saturn. Uranus, with a tilt of 98 degrees, has a 21-year-long summer of continuous light with the high northern latitudes remaining dark. It has an 84-year orbit to a brief vernal equinox (which last occurred in May, 2007). Some of the planets are extremely unique. Jupiter is the ruling planet.[14] Jupiter's mass is greater than all planets combined and doubled or equal to 1,400 Earths. It has a Great Red Spot, which is a hurricane twice the size of Earth, varying in color from salmon pink to ruddy. Three smaller white storms have been discovered, which, though smaller, are still massive. The planet is covered with bumblebee stripes or dark clouds called belts, and lighter clouds known as zones, running parallel to its equator. Its 89,000- mile bulk rotates at breakneck speed in less than 10 hours. It crackles with lightning like a thunderstorm.

Saturn is also unique. The moons mysteriously herd together and are in resonance with the seven bands of rings. The rings are a blizzard of particles, some as fine as smoke, and others as big as a barn, are mostly water and ice mixed with dust. They span a region equal to the distance between the earth and moon. There is a DCB gap A (major) and FGE (faint) called the Cassini division.[15] No other planet has these strange phenomena. Neptune also has rings that are less pronounced rings.

Satellites or Moons of the Planets

Space probes and new telescopes have revealed a great deal about the satellites/moons of solar planets, much of it difficult to explain. Recent discoveries reveal the giant or cold planets have 150 moons of two kinds, regular and irregular. Regular moons are like the earth's, "being large and having comparatively tight circular and nearly equatorial orbits with prograde rotation or turning in the same direction as the planet." [16] The irregular moons have large and highly elliptical orbits and are tilted with respect to the equators of their host planets and they turn retrograde to the planet. Titan of Jupiter is thought to have large methane seas. Saturn's tiny moon, Enceladus, has geysers spewing water or vapor. Neptune's moons circle in the reverse of its sister moons. The number of known moons shown by Jewitt is given in the chart (above). It is hard to distinguish the color of these satellites and Jupiter's are all dark. "***Neither*** the source region nor the mechanism of capture ***is well understood***."[17]

The moons defy easy explanation. Mercury and Venus have no moons; Earth has one moon which is proportionately large, a full quarter of the size of Earth's width. Mars has two smaller

---

[13] Bob Berman, "When Night Equals Day," *Discover*, March 2006, 28.

[14] Bob Berman, "Jupiter Rules," *Discover*, June 6, 2006, 34, (data excerpted).

[15] Ron Cowen, "Groovy Science: Cassini gets the Skinny of Saturn's Rings," *Science News*, Nov. 19, 2005, Vol. 168, 128, 328, 329, 332.

[16] David Jewitt, Scott S. Sheppard and Jan Kleyna, "The Strangest Satellites in the Solar System," *Scientific American*, August 2006, 40-47.

[17] Ibid.

ones. Earth's moon consistently causes tides on the earth, changing Earth's shape slightly. No other planet has oceans of water or this phenomenon. Earth's moon is crater-pocked, and has a large deep crater 2,500 kilometers across at the South Pole. Mars has two moons, which are not round but like big asteroid rocks. Then there is the area of the asteroids beyond Mars, where another planet should be, according to Bode's Law. Then farther out are the huge gaseous planets with many moons of varying size, kind and groupings. Saturn's moon, Enceladus, is cratered on one side but smooth on the other, which is similar to the planet Mars.[18] Earth is 80 times as massive as its moon, but the moon is *proportionately larger* than any other planetary satellite other than Pluto's. Moon rocks revealed that Earth's moon has many of the same elements as its crust; it also has a small metallic core. Scientists continue to explore the heavens, adding new information.

These and other factors led a young scientist, Bill Hartmann, to postulate in 1974 that Earth's moon was created by a collision with another planet or body of similar size. This theory was endorsed by Alistair Cameron, a top Harvard planetary scientist, and since by others.[19] Such a large planetary collision would have been an amazing catastrophe. George Darwin conjectured that the moon has gradually drifted away from the earth. A British scientist, Harold Jeffreys, studied records of lunar and solar eclipses back to Babylonian days and agreed that the moon was getting further away by about three centimeters a year. In 1969 scientists on Apollo 11, who landed on the moon, confirmed the rate of recession to be 3.8 centimeters a year. Satellites have measured it even more precisely. It is speculated that when the moon first formed, Earth's day was only six hours long and as the moon receded, it has gradually increased to 24 hours.

To reiterate, a number of findings about the planets and their moons do not follow uniform activity. Gravity holds the heavy side of Earth's moon toward the earth so the same face is always showing toward the earth. At least four of the moons on Jupiter, one on Saturn, and one on Neptune are rotating in the opposite direction from the normal direction for planets and moons, or retrograde. Venus has such a slow rotation around the sun it has the same face toward the earth as does our moon. Mercury always has the same face toward the sun. Pluto strangely rotates from east to west and has the largest moon proportionately. Scientists hold to three speculative theories of how moons were formed: (1) the gas drag theory, (2) the pull down theory, and (3) the three body interactions theory.

Amazing Unique Things about the Earth

New Findings about Earth Dynamics of Core and Mantle

Using new geomagnetic instruments and gravitational drilling, scientists continue their extensive research of Earth, raising even more questions. Why does the earth's spin axis trace counterclockwise? Why has the earth's magnetic field turned upside down every so often? Scientists think that has occurred 300 times in the last 170 million years. Or does the magnetic field shut down and reverse repeatedly? Is this caused by the earth's inner solid core, 1,500 miles in diameter that spins on a 10-degree axis to the Earth and moves 20 km per year related to the liquid core?[20] Some authorities speculate that the formerly dominant view, that the earth's core was all iron, is questioned by some. Others now believe it may be made of radioactive potassium, thorium or uranium with iron and that may be causing the continuous heat.[21]

---

[18] See McNab, *The Planets,* for much of this discussion.

[19] Ibid., 71, 72.

[20] "Putting a New Spin on Earth's Core," *Science News*, Vol. 150, July 20, 1996, 36.

[21] Kathy A. Svitil, "A Strange Brew in Middle Earth," *Discover,* August 2003, 16 where she quotes V. Rama

Moreover, researchers at Harvard and the University of California-Berkeley have discovered that the whole liquid iron, outer core of the earth is turning about three degrees a year faster than the mantle; this may be caused by thermal wind. The mantle rotates once in 24 hours, while the inner core turns once in 23 hours and 48 minutes. Scientists think real ocean currents may change the center of the yoke from time to time. These tensions within the earth and the record number of magnetic flip-flops, which are just being researched, seem to indicate a past buildup of enough energy to cause great Earth cataclysms and tremendous forces that may explain previously unexplainable plate shifts and mountain building. Does that also explain sudden glacier changes and radical changes in climates? These are just some of the questions that on-going research raises.

The Mystery of Earth's Suitableness to Life and Man

As more details about the universe have come to light, they all seem to show a preciseness that if slightly changed would ***not*** be suitable for man to live on Earth. One of the great problems that scientists confront is the fact that the earth seems designed to accommodate life as necessary to sustain man. According to this anthropocentric principle the universe and everything in it is designed for a purpose that ends in and is centered in man. The biblical account of creation also seems to support this principle.

This was one of the major problems that Hawking wrestled with in his book, *A Brief History of Time.*[22] Indeed, I shall argue that this is the major problem that unbelievers have with the biblical view. They cannot accept the way the world is so fit for man because it means that there is a God who created and sustains the universe with man as supreme and accountable to God.

This is really why Hawking rejected the big bang theory, which he himself had proved mathematically. He defines two possible views of the anthropic principle: the strong one which holds all things from the beginning of the universe were designed with purpose, and a weak one which holds there is limited purpose from change. Of the strong anthropic view he said,

> This means that the initial state of the universe must have been very carefully chosen indeed, if the hot big bang model was correct right back to the beginning of time. It would be very difficult to explain why the universe should have begun in just this way, *except as the act of God who intended to create beings like us*."[23] (Emphasis added)

Einstein was intrigued by the idea of purpose in the universe. Hawking could not escape some purposeful meaning in the universe, so he opted for the limited one. He admits,

> The remarkable fact is that the values of these numbers (of our universe) seem to have been very finely adjusted to make possible the development of life. For example, if the electric charge of the electron had been only slightly different, stars either would have been unable to burn hydrogen and helium, or else they would not have exploded.[24]

But Hawking then argues from *non-existent data*, saying there ***may be other universes*** in which

---

Murthy at the University of Minnesota.

[22] Hawking, *A Brief History of Time*, 124 ff.

[23] Ibid., 127.

[24] Ibid.

the numbers do not hold and other forms of life could exist in different worlds. But Hawking cannot define how far his limits must go or even why a limited part of the universe that is created by chance should be meaningful. Chance is the opposite of logical cause and effect.

Buffon's biology saw a harmony in nature and with man that he based on his observations. Astrophysicists who study earth's unique chemical makeup also suggest a harmony between man and nature. Jastrow, of the U.S. Space Center, pointed out that the universe has developed very specifically in regard to the elements and the proportion of elements and their formation, and that leads down to our world, which is strangely adapted to life and to man. Robert Gange, of the David Sarnoff Research Center, has effectively argued that physical reality all is purposefully designed to terminate in man.[25]

These observations create problems for environmentalists they realize that nature's delicate balance is important to man. They make nature the object of worship, but dare not consistently extend this to a religious motive of accountability to God. To confront clear evidence that in many ways the world is adapted for man indicates there is purposeful meaning and the mind of God, which human bias rejects.

Search for Extra Solar Planets that Are Friendly to Man/life

The fact that the earth is the only solar planet that is anthropic or uniquely designed to sustain life implies divine accountability. Scientists spend billions of dollars trying to discover other stars that have planets similar to Earth's conditions, which might sustain life. Such a discovery would make it easier to argue for a limited anthropic world and increase the minute probability of chance beginnings. Many astronomers believe there are planets around other stars, which might be like the earth. Scientists infer that some are planets because of light influence, which indicates a gravitational pull, causing a star to waltz or wobble. Discovery of three stars near a pulsar in Virgo, some stars around Beta Pictories, a Jupiter-size object circling Star 51 in the constellation Pegasus, another star at 47 Ursae Majoris, and 70 Virginis lead scientists to these inferences. These are usually very large, like Jupiter. Other discoveries are claimed almost daily. This search has not revealed anything that meets the desired end. Jack Kelley has said, "After a decade of searching for planets orbiting stars like our own, astronomers had ***found nothing*** but giant planets, most of them gas balls like Jupiter, around other stars."[26] Recently a smaller one was found, but was not suited for life as on Earth. As of April 2009, scientists think there are several hundred extra solar planets, only one or two actually seen but ***none fitted for human life***.[27]

After all this search, Paul Butler from Berkeley has said, "We find a planet-great! But where does it fit in the great scheme of things?"[28] Recently, information from the Galileo space probe indicates dark stripes on ice-covered Europa, one of the inner moons of Jupiter, suggesting ice flow plates on water. This and some conditions on Mars (possibly polar ice and evidence that looks like lake beds) furnish assumptions there might have been life on the only two remotely possible locations in our planetary system. Most have given up on Mars as a source of life. Scientists' final attempt to explain the possibility of life on other planets is the phantom of multi

---

[25] Robert Gange, chapter 15 on "Why Does Physical Reality Terminate in Man?," *Origins and Destiny,* (Waco, TX: Word Book Publishers, 1986, 110-122).

[26] Jack Kelley, "Hunt for another Earth Broadens," *Discover,* January 2006, 26.

[27] Adam Frank, "Cosmic Abodes of Life," *Discover*, May 2009, 48. Cf. the other articles of many pages in this issue that show many desperate efforts but without success.

[28] Jeffrey Winters, "Planets, Planets, Everywhere," *Discover*, April 1996, 32.

universes which is an escape to non-evidence.

Through all our expensive exploration of the solar system, we now know that none of the other planets, not even the moon, Titan, or Saturn, can sustain life. Finding other stars with planets conducive to life is the last hope of secular scientists who hold to a limited anthropy. Millions of dollars are spent for listening devices, hoping to pick up signals from other space creatures, which also might confirm this view. But even if there should be other planets and other intelligent creatures (though there is *not one shred of evidence* for this), this would not prove a limited anthropy or remove the need to explain how chance can produce a limited anthropic principle.[29]

Because of the growing threats to man and our world, many scientists are saying that we need to find another planet that is friendly to life for man to inhabit in order to preserve humanity in the days ahead. ***But the time to travel to any other planet and the very short life span of man allows no hope***.

Return to Recognition of Cataclysms among Planets

Major Asteroid and Comet Collisions with the Earth

Cataclysmic threats now seem much more likely than previously thought. Early in the twentieth century, when Samuel Shoemaker began to claim the earth had been hit by gigantic meteors or comets, because geologists were all uniformitarians, only a few would listen to what he claimed. In time, he proved these occurred and he and his wife discovered the one that collided with Jupiter and was observed by astronomers and geologists all over the world. Comets are now thought to invade our solar system; areas of Earth have been mysteriously destroyed by heavenly fires. The United States has focused on trying to discover all the major asteroids of a mile in diameter that might strike our world and destroy it, and has already discovered a number which paths cross the earth. Such was the asteroid that is said to have destroyed two-thirds of the life on Earth and all the dinosaurs. Scientists think there may be hundreds of big asteroids. The British are searching for the thousands that are large enough to destroy a city. The Shoemakers have said, "It is no longer a matter of if an asteroid or large meteor will strike the earth, but when."

An errant asteroid 1,000 feet in diameter (ten times larger than the devastating one in Siberia in 1908) came within less than 500,000 miles of the earth in 1989. Scientists believe an asteroid one mile in diameter strikes the earth once in 300,000 years, which means the chance of dying from this is six times greater than dying in an airplane crash. The drama of a planetary catastrophe was demonstrated by the pictures of the January 1994 collision of comet fragments into Jupiter. Hurtling through space at 134,000 miles per hour these fragments collided with tremendous explosions, scattering debris and gas from Jupiter for five or six days.

Discovery of Plate Tectonics and Drifting Continents

Kant's uniform deterministic view led geologists, including Lyell, to rule out any cataclysmic activity. Uniform naturalism did not account for the breakup of the continents or the

---

[29] See *Science News,* "Is There Anybody Out There?," Vol. 142, November 7, 1992, 420; "Hubble Scopes Possible Planet-Forming Disks," Vol. 142, December 7, 1992, 316; "Low Mass Stars: Born to Make Planets?," Vol. 143, January 16, 1993, 36; "Life at Other Stars: A Matter of Climate," Vol. 143, January 30, 1993, 74; "Little Discs, New Planets Grow," *Newsweek,* December 1992, 53; Sam Flamsteed, "Planets by the Carload," Gap in Stellar Disk Hints at Planet," *Science News*, June 25, 1994, Vol. 145, 404; Vol. 140:53, et al.

powerful plate tectonics that scientists now recognize as some of the most significant geological forces which shaped the Earth. It is apparent that some planets have no plate tectonics, which suggests there was a major cataclysm dividing the Earth.

Earthquake and Volcanic Consequences Much Greater

The study of plate tectonics leads scientists to believe that the effects of earthquakes and volcanoes are much greater than uniformitarian scientists once thought. The eruption of Mount St. Helens and of volcanoes in the Philippines, Japan, and the Hawaiian Islands have shown that if plate tectonics produced a number of these at the same time, the climate of the earth would be significantly altered. Scientists now believe that many giant earthquakes are possible, yet geologists have not yet found a way to predict accurately when an earthquake will occur. In fact, Richter's measuring device may not even be the most valid measure of intensity. The effects on climate from volcanic dust are greater than originally thought. Islands have been formed and become inhabited by life in a very brief period of time, events which were once thought to take much longer. Recovery from volcanic eruptions has been much faster than thought, as was seen after the devastation of Mount St. Helens. There are other cataclysmic mysteries that are now acknowledged such as "rogue waves" as tall as buildings in the North Atlantic and elsewhere, by giant avalanches of rocks that "mysteriously flow," the flow of earth in giant earthquakes.

Growing Confusion concerning planets

Charles W. Petit describes the shift from Laplace's neat picture of the origin of the universe to the more confusing picture today. He said,

> In the old days, ...astronomers knew a planet when they saw one. The word means 'wanderer.' Planets were the things that move around among the stars. Today, forget it. An avalanche of discovery, with 55 planets identified around other stars in just the last five years, has experts gasping for words. Many are too hot, or have orbits too weird, or are too big, to fit categories found in our own orderly solar system. 'We thought we had our labels straight, but Mother nature seems to have other ideas,' said Alan Boss, a Carnegie Institution astronomer.
>
> Head scratching hit a frenzy ...at an American Astronomical Society meeting in San Diego. A team led by Geoffrey Marey of the University of California-Berkeley, and by Paul Butler of the Carnegie Institution of Washington, said five years of analysis of the wobbles in the star HD168443 some 123 light years away, reveals that it's orbited by one planet with at least 7.7 times the mass of Jupiter...and by a second with a jaw-dropping heft at least 17 and perhaps as much as 42 times Jupiter's. The greater bulk is not only far more than any 'planet' detected before but puts it in territory reserved for stars and similar things.
>
> The latest system is helping to blow a lot of old theories on planets and star formation to smithereens. Alan Stern of the Southwest Research Institute in Boulder, Colo., says 'it's like we'd been living in this small town, and we thought we knew what people are like. Now we're visiting other lands and are asking, what are those guys?'
>
> Solar failures. Gas balls with a mass beyond 13 and 70 times that of Jupiter are called brown dwarfs, or failed stars. They don't have enough mass to sustain the internal fusion reactions that keep stars bright. But as with true stars, the going theory was that brown dwarfs are born from collapsing nebulae, or clouds of gas, while planets form in disks of debris that subsequently form around nascent stars.
>
> That's not the only muddle. Last year British astronomers reported 'free-floating planets' in the constellation Orion, smaller than brown dwarfs but circling no parent stars. Can something that doesn't even go around a star be a planet? If Pluto, much closer to home at the small end of the scales, is a planet, some say the definition should include a dozen or so larger asteroids. The

> biggest asteroid is Ceres, some 578 miles across. Pluto and its moon, Charon, are 1,470 and 730 miles in diameter, respectively. Good science needs consistent nomenclature.[30]

Ron Cowen reported that because of new planetary research, the International Astronomical Union adopted a new definition of planets at their August 24, 2006 meeting.

> A planet is any body that orbits a star, is neither a star nor a satellite of a planet and has gravity strong enough to pull it into a rounded shape. A planet must be heavy enough to clear other objects from its path.

On this basis, only the eight major planets retained that designation. Pluto was demoted to a dwarf planet with Charon, Xena or UB313 and Ceres. Some 41 other objects lack any clear planetary definition. Neil deGrasse Tyson, director of Hayden Planetarium of New York City said, "This is a scientifically informed cultural decision." R. Cowen commenting on it said, "Scientists don't have any new understanding of these bodies or how they are grouped in nature."[31]

Scientists now have no idea how the planetary system or an Earth, *mysteriously designed **for man,*** were formed. On the one hand, mankind has a strong bias against believing in a possible catastrophe because he does not want to recognize he cannot control the future. On the other hand, he would like to remove strong evidence of purpose in the world. To admit these things makes him confront his problem of sin and a future accountability to God.

The vast amount of information that scientists have uncovered about the solar system creates more questions than answers. The mysteries of the universe are immense. While Laplace offered a brilliant hypothesis, his theory and all the others since come woefully short of explaining so much that is unknown and outside the sphere of a deterministic science.

Conclusions: Old Uniformitarianism Yields to New Solar System

Radically New Information Makes Laplace's View Obsolete

Increased information about the universe has introduced a new view of cosmology. A number of cosmologists have accumulated the data and computerized it, producing new orreries that allow a rapid evaluation of orbits. An orrery is a device that plots a model of the revolutions of the planets around the Sun. Jacques Laskar at the French Bureau of Longitudes, David Freedman with Jack Wisdom and Gerald Sussman at MIT, along with others, have created new computer models of the current data. These men have all arrived at the conclusion *one **cannot predict the origin and future*** of the solar system.

Based on their more accurate orreries, these scientists have concluded:

> If the solar system is sculpted by chaos, the possibility that it might be unstable begins to seem a bit more real. The beautiful eighteenth century orreries are museum pieces, not just because they are old and we have learned how to make a better one, the way we have supplanted mechanical timepieces, quartz watches and atomic clocks- but because ***the cosmology** that underlay them, and which they epitomized **is dead**.*[32] (Emphasis added)

---

[30] Charles W. Petit, "A celestial conundrum: Planets are planets, Stars are stars. Or are they?" *U.S. News & World Report*, January 22, 2001), 56.

[31] R. Cowen, "New Solar System" *Science News*, August 19, 2006, Vol. 170, pg. 119 and see "Doggone! Pluto gets a planetary demotion," *Science News,* September 2, 2006, Vol. 170, 149.

[32] David H. Hunter, *Discover*, May 1990, 54-60.

Dead! Dead! Our new understanding of the earth and the other planets is radically different from the old dogmatic uniformitarian deterministic views. The facts reveal a creation that is beyond the mind of man. This is true enlightenment about the solar system and all the heavens and earth.

PART IV

ENLIGHTENMENT ABOUT LIFE'S ORIGIN AND DEVELOPMENT

O Lord, how manifold are your works!In wisdom have you made them all; the earth is full of your creatures (Psalm 104:24).

Chapter 11
Molecular Neo-Darwin Mechanism Search

Introduction

By the last quarter of the twentieth century it became increasingly obvious that there was no known mechanism of evolution causing transmutation of the species. Darwin had predicted that the historical record of life in the then meager fossil records would give historic validity to his theory. Darwin's friends, Hooker, Huxley, and Lyell, had warned that the evidence did not seem to support his optimism about fossils. Darwin's theory had been widely accepted because of enthusiasm for natural progress by chance and without intelligent design. The Scopes trial, with the encouragement of the scientific community, did much to promote public acceptance of Darwin's theory. Despite prolific fossil study, no evidence appeared to support evolutionary theories. Known authorities did not support Darwin's theory. Richard Owens, the leading biologist in Britain pointed out that acquired characteristics were never inherited. F. Jenkin demonstrated that the survival of individual fluctuations was impossible, and "only if the same trait emerged *simultaneously* throughout the majority of the species could it be expected to survive."[1] This involved a large group changing in a similar direction, contemporaneously, which would not be "chance natural selection." Jenkin's study showed the population of a species swamps and wiped out the acquired characteristics. Darwin conceded the evidence supported Jenkin's observation.

These facts, plus Darwin's occasional concessions to a guiding purposeful influence, produced an overwhelming motivation for the intellectual elite to find a new mechanism. Unwilling to cede their rejection of a creator God, scientists looked for alternative explanations for evolution—without or contrary to the evidence.

Development of Neo-Darwinism to Give New Credible Methodology

Search for a New Mechanism

For Darwinian evolution to survive, it had to find a new mechanism. The definition of evolution changed from transmutation and the descent of new species which had always been accepted by evolutionary naturalists. However, finding slight mutations that reproduced and survived became the focus of neo-Darwinism. In the twentieth century, molecular biology, a science unknown to Darwin, emerged as an important new area of study. It seemed to be the answer to explain evolution. Scientists hotly pursued this as a mechanistic path to account for evolution.

New Science of Molecular Biology Showed Hope

Cell theory began about the middle of the nineteenth century as Darwin was forming his views. It was not until nearly the mid twentieth century that molecular biology became a definitive science. Matthias Schleiden (1838), Theodor Schwann (1839), and Rudolf Virchow (1858), were some of the first to contribute to cell theory, which is essential to modern biology. They found nucleic acids, long-chain polymers of nucleotides of sugar, phosphoric acid and nitrogen. Only later did they find that the sugar in the nucleic acid would be ribose or deoxyribose in form of RNA and DNA. Molecular theory had three parts: (1) all living things are made of cells, (2)

---

[1] Jane M. Oppenheimer, "An Embryological Enigma in *The Origin of the Species*," *Essays in the History of Embryology and Biology*, (Cambridge, MA: The M.I.T. Press, 1967), 210.

metabolism occurs with cells, and (3) all cells come from pre-existing cells.

Key developments in molecular biology are: In 1944 American Oswald Avery showed DNA carries genetic information and might be the gene. In 1948 Linus Pauling found the alpha helix of proteins as a spiraling spring coil. In 1950 Erwin Chagaff found the arrangement of nitrogen bases in DNA varied, but always occurred in a one-to-one ratio. In 1953 Rosalind Franklin, using X-ray diffractions, showed that all DNA was helical, Franklin's associate, Maurice Wilkins, working with Francis Crick and James Watson theorized there were two chains of nucleotides, arranged as a helix. They added Chagaff's findings of base pairs and showed each strand was a template for the other. Their findings earned them the Nobel Prize in 1962 for deciphering DNA,[2] launching exciting research in molecular biology. But when Eiseley wrote in 1961, evolution by mutations was not yet proven as a mechanism. He could honestly say, "There is still much that is unknown: the cellular location and nature of the *great mechanisms that control* the structure of phyla and classes ***escape us still***."[3] (Emphasis added)

New Tools Revealed Molecular Functions and Operations

The mid twentieth century ushered in revolutionary new microscopes (illuminating immunofluorescent, electron, scanning electron, transmission electron) with their many magnifying improvements and delicate diversities. Add to this the photographic, X-ray technology, and computers with their powerful enhancements, and scientists now have the ability to understand more about cells and how they work, the structure of DNA and RNA, to map genes, and even count the various cells, proteins, and their amino acid sequences. These were expensive and it took several years to develop and get these inventions into the hands of biologists.

Molecular cells are different in size (10-30 micrometers), shape, and function; they have complicated membranes and at least four or five different types of proteins. The cells store, communicate, nurture, and reproduce. A later chapter will look at the complicated nature and work of cells. The important fact to note here is that ***Charles Darwin had no idea about what was involved in life or reproduction when he sought to describe how the species developed.***

Molecular geneticists study two forms of genetic materials, the DNA, which has long been seen as the message-giving center for development, and RNA, which contains the information to replicate and carry out the development. More recent studies also have shown that RNA may have some limited message replicating powers, as well. The whole science of gene construction and reproduction has proven reliably repetitive and unchanging, so that today we talk about genetic fingerprints (language accepted in courts), and footprints. We know which genes ***control*** reproduction. Scientists use chemical and radiation exposure to produce mutations of all kinds, and enzymes to perform genetic surgery for improving life. Objective biological data has literally exploded in the last half of the twentieth century. This new science became the evolutionary scientists' hope to support their theory—a theory they have dogmatically proclaimed as true, but without facts to support it.

Neo-Darwinism Claim Molecular Mutations as Evolutionary Mechanism

When it became obvious that Darwin's original theory was not true, those scientists committed to Darwinian evolution met to find a mechanism for it to work, especially in regard to

---

[2] *People and Discoveries*, "Watson and Crick describe structure of DNA, 1953," Internet.

[3] Jane M. Oppenheimer, "An Embryological Enigma in *The Origin of the Species*," *Essays in the History of Embryology and Biology*, 231.

understanding and tracing descent. Modern molecular science was only beginning and therefore did not have a major voice in the development of that new view. Molecular biology came into being within the next few decades.

As researchers began to understand the DNA double helix and found that copying errors were made, or mutants, they thought this would be the mechanism whereby new species would evolve and survive as Darwin envisioned. The neo-Darwinian view held that with many variant mutations occurring, natural selection would produce individual beneficial changes that would survive. After a half century of experimentation and observation, scientists have not shown that mutations can produce developmental improvement by replication.

In 1982 when Thomas Cech at the University of Colorado discovered that RNA also had some self-replicating abilities, scientists took new hope that RNA might be an agent of change. DNA and the errors produced by reproduction did not yield new creatures. But Cech's discoveries seemed to offer selective changes. For example, Gerald Joyce, of the Scripps Research Institute in La Jolla, California, has found that certain RNA can be developed that have the ability of slicing DNA into shorter pieces. By *selecting* those pieces that have this property, out of millions of RNA, he has developed RNA mutants he calls ribozymes. After two years and 27 cycles, he has produced what he considers improved ribozymes.

There are several reasons Joyce's findings cannot be evolution ***by chance***. Joyce himself is a designer. He has to supply *two primers* that prepare the RNA for copying and *two enzymes* that expedite the replicating process. He has had no success without his supplying the enzymes. Moreover, Joyce must *observe and select* those that show DNA residue on them. He is thus carrying out the job of ***intelligently creating toward a specific purpose***, which he has designed. Without him adding this superior information, it would not occur. Moreover, even by his creative work he has ***not produced*** any evidence of producing ***an improved, functioning creature***.

In fact, Joyce speculates that there is *an internal selective choice*, like giraffes purposely growing a longer neck to reach higher vegetation. This is the concept of Lamarck that Darwin first rejected as occult and unscientific, but later returned to. Thus, RNA has shown it has limits of control and it was not self-sustaining. Moreover, the RNA can only cut DNA into ***smaller*** parts, which is not promising for producing something more complicated and larger with more information. Thus, ***RNA has also been a dead-end*** for hope of improving or evolving upward.[4] Moreover, this is not Darwin's idea of evolution by chance through natural selection.

Conclusion about Evolution from Mutations

Pierre P. Grassé was the editor of the 28 volumes of *Traite de Zoologie*, author of numerous original biological investigations, ex-president of the Academie des Sciences, and is considered the most distinguished of French zoologists. In his book on evolution, he denied that evolution could occur from mutations. Neo-Darwinian evolution *presupposes* ***adding*** *by chance* to the enormous amount of ***information*** that is condensed on a molecular scale in the chromosomes of a life form that gives it the limits of what it is. But gene *mutations* ***never add*** *more information*. The ***real problem*** *of evolution is accounting for* ***the origin of new genetic information***, which cannot happen by chance, but must come from somewhere beyond the gene. When genes break apart and reunite to make new and diverse individuals, errors occur. The wrong nucleotide is inserted,

[4] Cf. brief report by Peter Radetsky, "Speeding Through Evolution," *Discover*, May 1994, 83-87.

forming a mutation. Grassé only recognized mutations as fluctuations.[5]

Philip Johnson of the University of California, Berkeley, has observed,

> Grassé argued that, due to their uncompromising commitment to materialism, the Darwinists who dominate evolutionary biology have failed to define properly the problem they are trying to solve. The real problem of evolution is to account for the origin of new genetic information, and it is not solved by providing illustrations of the acknowledged capacity of an existing genotype to vary within limits.[6]

Grassé accused the Darwinists of following their assumptions, arriving at presumed conclusions, and overlooking the reality of the facts.[7]

While Grassé continued to be an evolutionist, he ended his book with the statement, "It is possible that in this domain biology, impotent, yields the floor to metaphysics." Theodosium Dobzhansky and other leading evolutionists clearly acknowledged this as a frontal assault on all kinds of Darwinism, which was not answered.

But what Grassé said has been confirmed by many other biologists. Coline Tudge, in his massive book on biodiversity has said,

> These small changes are mutations; and although it has always been assumed that most mutations are harmful, while a few prove to be beneficial, the surprise has been that most mutations simply do not matter. *They **do not seem to affect the survival of the organism one bit***.... (Emphasis added). The point is, however, that these changes tend to affect all the organisms in any one lineage in much the same way, because sexual congress leads to swapping of genetic material.[8] (Emphasis added)

While mutations do not contribute to Darwin's theory of survival of the fittest, as hoped by neo-Darwinists, some think they offer some evidence as a molecular clock, or how much time occurs between changes.

Certain things became increasingly clear. There was *no known mechanism for evolution by individual "mutations."* Many diversities that were thought to be evolutionary changes are now known not to be so. The peppered moth, which turned dark, but in time has turned white again, was a famous example. The differences in Darwin's finches have turned out to be only variations within genetic controls. Biologists studied these finches for years and found differentiation of beaks because of the seeds available by varying weather, but there have been no great changes in evolution. As a result, the evolutionists claim every minor or micro-change as an evidence of evolution. These micro-changes have always been evident and the limitations and explanation of regulating these were set forth by the Austrian monk, Gregor Johann Mendel, (1822-1884) before Darwin.

Human genetic *engineering* involves the ability of human ***minds*** to use their accumulated

---

[5] Pierre P. Grassé, English trans. from French, *The Evolution of Living Organisms*, 2.

[6] From a symposium at Southern Methodist University, Jon Buell and Virginia Hearn, editors, *Darwinism: Science or Philosophy?,* (Richardson, TX: Foundation for Thought and Ethics, 1992), 7.

[7] Grassé, *The Evolution of Living Organisms*, 7-8.

[8] Colin Tudge*, The Variety of Life,* (Oxford University Press, 2000), 76.

knowledge—a knowledge that does not exist in lower species. The geneticist can use ***his*** greater information as *an act of superior creation* and can correct, and to a degree, change the genetic structure, producing many so-called long-range small improvements. But even ***as human creators, nothing has ever been done that shows significant transmutation*** from one major "kind" (as the Bible describes the groupings) of plants or animals. As of now at least, there are limits even to human acts of creation in genetics.

Again, ***no natural mechanism has been found*** that allows going from one major "kind" to another. Study of mutant changes, whether on individuals or populations, has not shown how this could occur. The statistical probabilities of this occurring by natural influences, as we now know them, are so infinitesimally small that it would be more than a miracle for even one such transmutation in one major organ to occur. Genetics predominantly reveals natural control.

Lynn Margulis, professor of biology at the University of Massachusetts and a member of the National Academy of Science, frequently asks her audiences of molecular scientists "to name a single, unambiguous example of the formation of a new species by the accumulation of mutations." Her challenge goes unmet. She has said, "Neo-Darwinism, which insists on the slow accrual of mutations, is in a complete funk."[9] This absence of a mechanism for evolution has led to several new theories of how evolution *might* have occurred other than by Darwinian gradualism.

Status of Sharply Defined Groups Becomes Clear

Cladism in Taxonomy Defines Distinct Species and Relationships

It is important to note that the science of taxonomy demonstrates the distinctions of species—all the way down to the microscopic molecular level. Darwin's theory that the distinctive species originated by slow intermediary individual changes, eventually producing the world family or tree of life, involved a progressive repeated transmutation of thousands of individuals organisms. This left scientists with the task of defining species. Tudge refers to Ernst Mayr's definition as a species being "a group of creatures that can breed together sexually to produce 'full viable' offspring." Some, which have once been considered species, have later been considered in the same species with another one within controlled limits of genetics. Classifications run in an ascending order as numbered:

1. Species	5. Class
2. Genus	6. Subphylum
3. Family	7. Phylum
4. Order	8. Kingdom

Scientists have differing opinions on classifications. Linnaeus was the first modern scientist to extensively classify things. He did so by structure or morphology as only plant and animal kingdoms. As biology developed, the system required greater distinctions. The explosion of scientific information in the twentieth century led to new insights in classification, as well. Willi Henning, a German entomologist, developed the cladistic method which he details in his book, *Phylogentic Systematics* (1966). Others have since taken Henning's work a step further by including genetics and other factors. A multitude of creatures have not been classified; it is estimated that there are from eight million to 100 million species. Some speculate that since the beginning of life there have been many more. Most of these are the smaller creatures such as bacteria, shellfish, and insects. A realistic estimate may be about 30 million species.

---

[9] C. Mann, "Lynn Margulis: Science's Unruly Earth Mother," *Science, 1991*, 252, 378-381.

As science disproved supposed mechanisms for Darwinian evolution, so it has disproved the idea of many intermediaries showing progress from one species to the other. Darwin believed that gradual evolution continues to occur. He could not answer the comparative anatomists and taxonomists of his day when asked why we did not see any intermediary forms. Darwin answered, "We do not have the right to expect them"... because they are not prevalent enough to be observed, and the weak deviant forms die out. But in time, it has been shown that ***intermediates are universally not visible.***

Cladism is a new technical scientific approach to evaluating and classifying various individual groups of forms. It goes beyond superficial observation, rejecting any preconceived ideas of relationships, including presupposed ancestral lines of evolution. Using sophisticated computers, relationships are diagrammed in a sphenogram, providing a very accurate way of grouping and relating types. While many cladists believe in evolution, they view this, and any other preconceived idea of family or other relationships, as a hindrance to the science.

As a result of this new taxonomy, distinct types of all living creatures have emerged, with ***no sign of gradualism or individuals in between them***. But the cladists insist that the classifications are so specific that the ***intermediaries do not exist*** and, and cladist say to search for ancestors is a fool's errand; that all we can do is determine sister group relationships based on the analysis of derived characters. Beverly Halstead, not a follower of cladism, has said, "No species can be considered ancestral to any other." Cladism is seen as an antithesis to Darwinism.[10] Darwinists argue this systemic approach is individualistic and does not look at the population approach, but the cladists insist *there is* ***no visible population change***, No intermediaries can occur because of population swamping. Some scientists want to reject the old taxonomy for entirely new genetic classifications. Some who reject the evidence have tried to devise a new taxonomy to allow evolutionary claims.

Accusations of Over Simplistic Record in Genesis Doesn't Fit the Facts

Some scientists have made jest at the limitations of the number of species that are mentioned in Genesis 1 and taken by Noah into the ark in Genesis 7. Christians argue that the Bible is not intended as a science textbook, but a record that may be accepted as historically true. Theologians assert that God revealed creation to Moses as a vision, placing him as an observer of a sequence of events. The picture would not necessarily include Moses seeing every bacterium or insect that God made. But the picture God gave Moses was adequate to convince him and God's people of the fact that the transcendent God was the creator. Adam did not have to name every individual species to show there was none adequate to be his equal as a mate, but only those God placed before him in the garden. The account of the flood mentions that Noah took two of everything that had *the breath of life*, excluding many species and families. The obvious intent was to preserve those creatures who God wanted to preserve in the new world. Noah took *those* ***God caused*** *to enter the ark.*

The arguments of this book are not primarily to prove creationism, but to show that the theory of evolutionary determinism as Charles Darwin and others taught is not in agreement with the facts, and that creation by a transcendent God better fits the facts. Taxonomy became a

---

[10] For discussions of this see K. Thompson, "A Radical Look at Fish Tetrapod Relationships," *Paleobiology,* 7:153, cf. 153; B. Halstead, "Halstead's Defense Against Irrelevancy", *Nature,* 292:403-404; E. Myar, *Populations, Species and Evolution,* (Cambridge: Harvard University Press, 1970), 4. See Michael Denton, *Evolution: A Theory in Crisis,* (Bethesda, MD: Adler & Adler, 1986) 128-130, 138-140.

twentieth century challenge for Darwinian evolutionists to demonstrate gradual intermediary forms for change. Tudge and others found great distinctiveness of species, but as far down in the molecular studies as one goes, even to the number and arrangement of proteins involved, there proved to be no change—only distinctiveness from one to another in the elements of life.

Greater Distinctiveness Presents More Problem for Gradualism

The new taxonomy, with its extensive knowledge of the groupings of living organisms, reveals much clearer and more rigid distinctions among the species than imagined. Evidence at every level is contrary to Charles Darwin's theory of gradual inheritance of many intermediary changes. The microscopic distinctiveness magnifies the challenge of maintaining a theory of gradual evolutionary changes to higher forms.

The clear distinction of species ***multiplies exponentially the improbability*** of gradual evolutionary change. Assuming Darwin's theory of gradual natural selection, it is estimated that not **one** eye must be accidentally produced, but the eye must have been accurately, gradually produced, accidentally ***at least forty times.*** Moreover, it must occur in large populations at once. When this is applied to all the organs and organisms, the mathematical problem even to imagine it happening is exorbitant.

Sequencing of Amino Acids in Proteins Shows Distinctiveness

The sequence of amino acids in the proteins of organisms has been used to try to trace ancestry. There is a comparison of the positions at which they are identical or similar, and the places where they are not in amino-acid-by-amino-acid of two different proteins, or of a nucleotide in comparison with two different pieces of DNA. Scientists assumed that by this comparison, the differences would suggest the lines of descent. For example, in comparing human hemoglobin proteins with other kinds, frogs had 46 differences, chickens had 26, horses had 17, and monkeys had 5.

Of this effort to trace ancestry, Michael Denton, a molecular biologist at the Prince of Wales Hospital in Australia and a naturalist, has made several quotes in this regard.

> The molecular biological revolution has dramatically changed this situation by providing an entirely new way of comparing organisms at a biochemical level. ...The amino acid sequence of a protein from two different organisms can be readily compared by aligning the two sequences and counting the number where the chains differ.... As more protein sequences began to accumulate during the 1960s, it became increasingly apparent that the molecules were *not going to provide any evidence* of sequential arrangements in nature. Moreover, *the divisions turned out to be more mathematically perfect* than even the most die-hard taxonomist (one who puts in order) would have predicted.[11]

Denton then provides a graph of protein cytochromes from Dayhoff, showing the percent sequence in different matrices and adds,

> The most striking feature of the matrix is that each identifiable subclass of *sequences is isolated and* ***distinct***. Every sequence can be unambiguously assigned to a particular subclass. ***No sequence*** *or group of sequences can be designated as intermediate* with respect to other groups. All the sequences of each subclass are equally isolated from the members of another group.

[11] Michael Denton, *Evolution: A Theory in Crisis*, cf. 275, 277, 278.

> *Transitional or intermediate classes are* ***completely absent*** from the matrix. (Emphasis added).[12]

He points out that this holds true *of all the classes* of proteins. The most important thing this shows is that while there are similarities in *classes*, all *are distinct, with* ***no intermediaries***. Moreover, this is ***a mystery.***

Molecular Clocks and Sequencing to Discover Evolution

The effort to sequence evolution by protein differences is not the same as mutations. But not only was protein sequencing used to try to show ancestry, it has also been argued that the differences indicate relative time in evolving. Emile Zuckerkandl and Linus Pauling proposed a molecular clock theory. Thomas H. Jukes, professor of biophysics at the University of California-Berkeley (speaking in the Phillip Johnson debate and elsewhere), argued that a broad sequence in molecules between species indicates evidence of an earlier point in evolution. He then claims this as a molecular clock for tracing descent. The idea that *distance in sequence shows time of development* is a questionable way to measure.

Scientists using the molecular protein sequence as a clock have applied this to proteins in different parts of the ***same organism*** and gotten different readings, indicating it cannot be used to show time of progression of separate species. And most in this field now agree that much more must be understood to support the protein molecular clock idea. In fact, some feel this effort *accents the differences* between various species rather than their common origin.[13] It reveals the gaps between sequences and because of the differences that have been shown to exist, sometimes in the same organism, the gaps may, in fact, fit the idea of special creation better than an evolutionary clock sequence.

Grouping of Lower and Higher Organisms Shows Distinctiveness

It was once believed that there were only two divisions of living forms, according to the fundamental divisions in the living world. The *eukaryotes* are composed of all plants and animals, and the *prokaryotes* of bacteria. But Carl Woese, a microbiologist at the University of Illinois, and others have discovered the prokaryote kingdom includes a new-found group with two divisions: the *archaebacteria* which live in the normal or commonly enjoyed conditions for life, and the *anaerobic* group which seem to thrive in unusual environments where every other form of life dies. (Anaerobic means living in the absence of free oxygen.) Several extremes are found. One thrives in high temperatures (212 degrees and above, with metabolism that is often fueled by tungsten instead of iron). Another thrives in salt-saturated environments, and another manufactures methane gas. The latter is the most recent discovery and is *assumed* to be the oldest form of life. Woese's efforts have recently been challenged. But these forms of life are again separated from each other in protein sequences and *show no evidence of having come from intermediaries or of producing any intermediaries today*. While one may detect ***similarities*** between these, there is ***no evidence of origin*** *by progressive development of one group of life forms to the other.*[14]

---

[12] Ibid., 278, 280.

[13] Ivan Amato, "Ticks in the Tocks of Molecular Clocks," *Science News*, Vol. 131, 75.

[14] See Michael Denton as above. Karl Stetter discovered the anaerobacteria in 1982 and is now being experimented with by Mike Adams at the University of Georgia and others. See Will Hively, "Life Beyond Boiling," *Discover,* May 1993, 86-91.

New Tools Show Distinctive Divisions Everywhere, Never Gradualism

As evidence accumulates and increasingly sophisticated tools are available for the study of the species, it becomes clear that *the more the evidence,* ***the more specific the organisms*** *appear*—both living and fossil. Scientists have now documented a level of specificity far beyond what Darwin or other nineteenth century scientists could have imagined. Those who argue to the contrary are either ignorant of the facts or deliberately refuse to accept them.

Some have also hoped that understanding sexual development would give evidence of evolution. By original Mendelian laws, a child with two chromosomes, 15 segments from male or female would be a normal set of genes, but it is now known that a child must have a contribution from mother and father to be normal.[15] The sexual development of an individual is now seen as much ***more complicated*** than originally thought. It is not simply a male Y chromosome joining with a female X chromosome. There may be portions of these, but the more scientists know, the more complicated they realize reproduction is. One of the greatest mysteries in modern science is how two cells merge to form an embryo that develops all the details of an individual that were endemic to the original cell.

Some have assumed that various environmental factors cause evolutionary development. All the arguments and experiments for evolutionary change caused by such things as predators, parasites, or climate, only demonstrate changes of *activity* in a species and no evidence of transmutation of a species as Darwin argued.

Similarity became the only factor remaining to claim evolution in the face of distinctiveness. Almost all of the remaining arguments for Darwinism are based on structural similarities, from the largest to the most microscopic. The most recent challenge to Darwinism focuses on Phillip Johnson's presentation at Berkeley; published in his book, *Darwin on Trial.*

All Darwin's supporters now defend his theory from instances of the similarity of species. For instance, William B. Provine, teaching in the history of science and division of biological sciences at Cornell University, recently responded to Phillip Johnson's book. He refers to these arguments of Darwin. (1) The *similarities* of fossil species to living species in one district suggest descent. (2) The *similar* but different species in one ecological niche in South America. (3) The *similarities* of features in species in the Galapagos and Cape Verde Island, the Galapagos and the West Coast of South America and between Cape Verde Island and the West Coast of Africa.

Thomas H Jukes, responding to the same debate, likewise pleaded *similarities* between bacteria, algae, and animals which have similar sequences of amino acids, proteins, nucleotides, and ribosomal RNA molecules, as a reason for accepting a common descent. But studies all show distinctiveness, separation, and not intermediaries that would suggest development from one to the other. Denton and other thorough-going naturalists who hold to determinism, have argued this distinctiveness. Their arguments, therefore, do not indicate bias against evolution, but are based in facts.

One might argue because two beings have hemoglobin they are descendants, but researchers in Australia now report hemoglobin in plants. Adams found hemoglobin may even exist in the low anaerobic group. This certainly does not indicate only one line of descent between these.[16]

Commonality of ***structure*** cannot demonstrate origins but only the same information. The same argument from similarities can be applied equally to suggest something coming from a

---

[15] *Science News*, Vol. 139, April 6, 1993, 213.

[16] See *Nature*, January 14, 1988.

common designer and creator. Artists, architects, and automobile designers all become known for certain distinctive characteristics of their *similar* works. This shows the ***same designer***, not that one work comes from another.

If one claims that similar species were produced from each other, in a sequence of thousands of mutations, one has the *right to expect* that least some of the time there would be evidence of *intermediaries*. On the other hand, the origin from a common creator does not require or even involve an expectation of such intermediaries. Moreover, for one to believe that the production of higher and more complex forms *requires a rational purpose of design for such a direction across the gap* toward that higher level. If evolution were to occur by pure chance, *why would it proceed in any specific* ***direction,*** much less toward the more complex forms? For higher species to move in a designed direction, requires a designer. But that would completely repudiate Darwin's theory. While he originally refused to have anything to do with a designed purpose, he invoked this discussion himself in his last edition of *Origin* when he included a designing force in the development of the eye. Moreover, his overall theory, using the tree of life illustrations, requires moving from the simple to the complex.

Molecular Complexity Renders Chance Selection beyond Reason

Studies Evaluating the Simple to Complex Developmental Model

A primary tenet of evolutionary theory is that development progresses from the simple to the complex. Today scientists are very interested in the complexity of species. The basic premise behind complexity is that a natural power enables the earth to order itself. Many fascinating aspects of nature have been studied, such as the power of sand dunes to maintain their shape. James Lovelock, in *The Ages of Gaia: a Biography of Our Living Earth*, argues that life has evolved not just by adapting to its surroundings, but by a living force remaking them. Lynn Margulis of the National Academy of Science also believed in an organizing force in nature. Living things take control and change things. This involves a mystical pantheistic idea that the earth has innate powers to correct itself.

There is evidence that animals' physical bodies have an innate ability to heal, and natural forms can spread and re-grow quickly. However, these are closed organic systems, which have innate programmed information systems. But scientists have not been able to produce evidence for an overall power of nature to reconstruct itself. Such a view is a departure from science into magic or miracle.

Darwinists deny a pantheistic power of purpose like those who hold the self-organizing theories. But Darwin and all evolutionists believe that when life evolved, it did so from the simple to the complex. Evolutionary scientists have focused on this matter, assuming that this is the way life had to evolve. But it has been shown that there is no evidence that the simple evolved to the complex. In spite of the lack of evidence, some scientists still assert that this must have been true because evolution could not have gone any other way.

Dan McShea, an evolutionary biologist at the University of Michigan, recently published his findings on vertebral columns, "Complexity and Evolution: What Everybody Knows." Organisms may become more, or less complex, but after studying many vertebral columns, he came to the conclusion that successive generations of vertebrae have nothing to do with survival. The data, therefore, disproves the theory that the process of development leads to greater complexity. Scientists agree the McShea's simple premise was sound. McShea said, "In most of the comparisons, there was no significant change in complexity in either direction." And the few cases in which complexity seemed to increase from ancestor to descendant were offset by complexity decreases in other parts. He said, "The bottom line is that this showed no preferred tendency for complexity to increase. Increases and decreases tend to happen about as often."[17]

Other significant studies indicate that development of organisms, through so-called millions of years, shows no evidence of growth toward complexity. George Boyajian, paleontologist at the University of Pennsylvania, collaborating with Tom Lutz of West Chester University, compared ammonoids, free-swimming spiral-shelled mollusks from 330 million years ago. They focused on mathematical fractal of septa. They "determined the fractal dimension of 615 ammonoids sutures, one from each of 615 different ammonoid genera." When they looked at ancestor-descendant pairs, they concluded, "In those pairs there's an equal chance of the ancestor being more complex or less complex than the descendant. In other words, we *don't see any direction to the change in complexity*."[18] Moreover, they found no influence of complexity on survival. Boyajian said, "Complex or simple, very few ammonoid genera lasted more than 15 million years, and ***those that did weren't necessarily the most complex*** ones. So complexity didn't help them, but it didn't hurt them either."[19] (Emphasis added).

So the assumption of evolution progressing from the simple to the complex is a logical one only if you accept ***the premise*** of evolution producing progress. But there is no demonstration of this fact. Scientists believe it because ***it supports their bias of*** man as the highest creature, and man's desire to be free from moral right and wrong before God.

Complexity in Parts and Timing Require Simultaneous Creation

Michael J. Behe, professor of biochemistry at Lehigh University, said,

> Darwin was ignorant of the reason for variation within a species [one of the requirements of his theory], but biochemistry has identified the molecular basis for it. Nineteenth-century science [Darwin's world] could not even guess at the mechanism of vision, immunity, or movement, but modern biochemistry has identified the molecules that allow those and other functions.
>
> It was once expected that the basis of life would be exceedingly simple. That expectation has been smashed.... Science has made enormous progress in understanding how the chemistry of life works, but the elegance and complexity of biological systems at the molecular level have paralyzed science's attempt to explain their origins. There has been virtually *no attempt to account for the origin of specific, complex bimolecular systems*, much less any progress. Many scientists have gamely asserted that explanations are already in hand, or will be sooner or later, but no support for such assertions can be found in the professional science literature. More importantly, there are compelling reasons–based on the structure of the systems themselves–to think that a Darwinian explanation for the mechanism of life will ***forever prove elusive***.[20]

Behe believes in evolution at certain micro levels, but denies evolution can explain molecular life or that macroevolution changes species. Behe continues:

---

[17] Lori Oliwenstein, "Evolutionary Watch: Onward and Upward?" *Discover*, June 1993, 23.

[18] Ibid., 22-23.

[19] Ibid.

[20] Michael J. Behe, *Darwin's Black Box: The Biochemical Challenge to Evolution,* (New York: The Free Press, 1996), x cf., 5 ff.

Biochemistry has pushed Darwin's theory to the limit. It has done so by opening the ultimate black box, the cell, thereby making possible our understanding of how life works. It is the astonishing complexity of sub-cellular organic structures that has forced the question, 'How could all this have evolved?' The understanding has gone beyond the anatomical structures and superficial observations about their workings to the molecular level where cellular operations can be seen in details. The scientific disciplines that were part of the evolutionary synthesis are all non-molecular. Yet, for the Darwinian theory of evolution to be true, it has to account for the molecular structure of life.[21]

Complexities of Single Molecules are too Great for Chance

Michael Denton, an Australian who is also a specialist in molecular biology and a naturalist, has explained the complexity of one cell. He said,

> To grasp the reality of life as it has been revealed by molecular biology, we must magnify a cell a thousand million times until it is twenty kilometers in diameter and resembles a giant airship large enough to cover a great city like London or New York. What we would then see would be an object of unparalleled complexity and adaptive design. On the surface of the cell, we would see millions of openings, like the portholes of a vast space ship, opening and closing to allow a continual stream of materials to flow in and out. If we were to enter one of these openings we would find ourselves in a world of supreme technology and bewildering complexity. We would see endless highly organized corridors and conduits branching in every direction away from the perimeter of the cell, some leading to the central memory bank in the nucleus and others to assembly plants and processing units. The nucleus itself would be a vast spherical chamber more than a kilometer in diameter, resembling a geodesic dome inside of which we would see, all neatly stacked together in ordered arrays, the miles of coiled chains of the DNA molecules. A huge range of products and raw materials would shuttle along all the manifold conduits in a highly ordered fashion to and from all the various assembly plants in the outer regions of the cell. We would wonder at the level of control implicit in the movement of so many objects down so many seemingly endless conduits, all in perfect unison. We would see all around us, in every direction we looked, all sorts of robot-like machines. We would notice that the simplest of the functional components of the cell, the protein molecules, were astonishingly, complex pieces of molecular machinery, each one consisting of about three thousand atoms arranged in highly organized 3-D spatial conformation. We would wonder even more as we watched the strangely purposeful activities of these weird molecular machines, particularly when we realized that, despite all our accumulated knowledge of physics and chemistry, the task of designing one such molecular machine---that is one single functional protein molecule–would be completely beyond our capacity at present and will probably not be achieved until at least the beginning of the next century. Yet the life of the cell depends on the integrated activities of thousands, certainly tens, and probably hundreds of thousands of different protein molecules.
>
> We would see that nearly every feature of our own advanced machines had its analogue in the cell: artificial languages and their decoding systems, memory banks for information storage and retrieval, elegant control systems regulating the automated assembly of parts for quality control, assembly processes involving the principle of prefabrication in modular construction. In fact, so deep would be the feeling déjà vu, so persuasive the analogy, that much of the terminology we would use to describe this fascinating molecular reality would be borrowed from the world of the late twentieth-century technology.
>
> What we would be witnessing would be an object resembling an immense automated factory, a factory larger than a city and carrying out almost as many unique functions as all the manufacturing activities of man on earth. However, it would be a factory which would have one

---

[21] Behe, *Darwin's Black Box*, 22, 25.

> capacity not equaled in any of our own most advanced machines, for it would be capable of replicating its entire structure within a matter of a few hours. To witness such an act at magnification of one thousand million times would be an awe-inspiring spectacle.[22]

Denton said that each such cell is composed of ten million atoms. So the complexity of one simple cell is *beyond the imagination* of most people.

Biological Systems of Many Cells Multiplies Complexity Exponentially

Michael J. Behe has expanded the complexity of the molecular world to that of whole systems, of which he only gives examples of three main ones. He not only argues about the impossibility of gradual evolution because of the complexity of the cell's operation, but he shows conclusively, for any who think logically, that gradual evolution of the systems would have been totally impossible. The operation of the biological system, which is essential for life to continue, had to ***have all the complex parts to exist at once*** before it would have worked to keep the animal alive. He argues this in regard to the eye, the clotting of blood, and the immune system, as well as much simpler systems.

The blood system is an extremely complicated system. Millions of cells form veins, arteries, and miles of capillaries that carry blood to and from the heart. These must repeatedly, with intricate timing, pump the blood to the lungs for oxygen, through the digestive system for food, and through the many other organs such as the liver for storage and to the kidneys for cleansing. Within the blood, are many platelets, white and red blood cells. Each red blood cell has hemoglobin, which is made up of 574 amino acids of different kinds, each precisely arranged in order to perform the functions of the cell. This is only part of the intricate blood system.

Behe presents one of the most complicated and mysterious aspects of the blood—its ability to coagulate and thus promote healing.[23] This occurs in at least two cascades of reaction, intrinsic and extrinsic, and involves many parts working together harmoniously to insure the right amount of clotting, at the right place, at the right time, when a wound occurs. There are at least thirty-two (32) different complicated proteins involved in this one function. Each involves enzymes that help cut and catalysts that act at the right time and place, and in proper proportions, in order that the wound is sealed so it can heal. But its accuracy also involves regulations so that too much clotting does not occur that would stop the circulation and destroy the whole body. He showed convincingly that all aspects of blood clotting had to exist from the beginning and development one by one would be useless or even dangerous if only a few of these different functioning proteins came into being separately. Moreover, each function has to switch on and off at the right times for the system to work. Behe reviews in detail how the system works and that such a complex machine could not possibly have evolved gradually from the simple to the complex.

Behe presents the same arguments for the eye, the immune system, the brain, the body's transportation system, and for other less significant, but essential operations for life. He shows that the complexity of each molecular system could not have come about gradually as Darwin projected. Moreover, the many innate controls that turn on and off, that cause alterations at the right time in the right way, ***are mysterious***. No one understands this power of molecular control, nor can they explain how the complexities of the system could gradually evolve. Scientists made formal, but not specific descriptions of molecular working.

---

[22] Denton, *Evolution, a Theory in Crisis*, 328-329.

[23] Behe, *Darwin's Black Box*, 77-97.

Thus, just at the time Darwin launched his theory, the concept of the cell was born, and the scientists were beginning to understand genetic processes. Only after neo-Darwinism was formulated did the full extent of molecular science come into being. Molecular science shows conclusively that gradualistic evolution could not possibly account for the complex organisms of life. Molecular structures present an impassable barrier against any logical scientific explanation for evolution. Only willfully blind scientists can continue to proclaim gradualistic formation of species as Darwin did.

Edge of Microevolution Proves Fixed Limits for Species

Many years of fossil observation fail to provide evidence of transmutation, confirming stasis as an undeviating rule. Molecular studies reviewed by Behe present millions of replications of malaria, sickle cells, and other molecular studies which have been observed. The rapidity and numbers of these replications are far greater than most other species and show that no transmutation ever occurs from mutations. These all confirm that mutations produce only limited changes, but never produce new species. Neo-Darwinists hoped that these studies would reveal survival by inherited characteristics transferred by DNA mutations. But these studies show clearly that never happens. Behe reports,

> Now that the molecular changes underlying malaria resistance have been laid bare, they tell a much different tale than Darwinists expected – a tale that highlights the incoherent flailing involved in a blind search. Malaria offers some of the best examples of Darwinian evolution, but that evidence points both to what it can, and more important, what it cannot, do. Similarly, changes in the human genome, in response to malaria, also point to the radical limits on the efficacy of random mutation.
>
> Because it has been studied so extensively, and because of the astronomical number of organisms involved, the evolutionary struggle between humans and our ancient nemesis malaria is the best, most reliable basis we have for forming judgments about the power of random mutation and natural selection. Few other sources of information even come close. And the few that do tell similar tales.
>
> Straightforward extrapolations from malaria data allow us to set tentative, reasonable limits on what to expect from random mutation, even for *all of life on earth in the past several billion years*. Not only that but studies of the bacterium E. *coli* and HIV, the virus that causes AIDS, offer to Darwin hoped otherwise, random variation doesn't explain the most basic features of biology. It doesn't explain the elegant sophisticated molecular machinery that under girds life. ...We can begin to find the edge of evolution with some precision. ...Most mutations that built the great structures of life must have been nonrandom.[24]

It is now clearly shown from the history of the fossil records that there is stasis and absence of any molecular transmutation. Therefore no evolutionary development occurred. It is even questionable that any minor changes, such as sickle cell and malaria resistance, are an improvement. Even if sickle cell may help prevent malaria in babies, it is recognized as a sickness and greatly reduces the life span of all who have it. How is that survival of the fittest? Behe has demonstrated there is an edge of evolution on the microscopic level showing gaps and denying evidence for transmutations.

Progression to Complexity is denied for Darwinism

The lack of evidence for development from the simple to the complex is another formidable

---

[24] Ibid., 83.

blow to the evolutionary theory. Stephen Jay Gould, who was the most astute scientist in following the developments of science sought to make adjustments to claims for refuting evolution. First, he strongly denies that Darwin ever taught progress or development from the simple to the complex, or that Darwin approved the idea of "evolution" with a view of the development from the simple to the complex. Gould flatly says, "The basic theory of natural selection *offers no statement about general progress*, and supplies no mechanism whereby overall advance might be expected."[25] Even Gould gives some of the clearest quotations from Darwin, which shows development toward perfection. He refers to the last page of *The Origin...*, which says, "As natural selection works solely by and for the good of each being, all corporeal and mental endowments *will **tend to progress** towards perfection*."[26] (Emphasis added). Darwin widely uses the word in his later works. Gould's effort to deny Darwin's progressive evolution is the height of hypocrisy because this concept permeates and motivated Darwin's works and was endemic to his illustration of "the tree of life."

In the midst of his effort to describe the evolutionary process, Darwin said,

> Natural Selection acts exclusively by the preservation and accumulation of variations, which are beneficial under the organic and inorganic conditions to which each creature is exposed at all periods of life. The ultimate result is that each creature tends to become *more and **more** improved* in relation to its conditions. This improvement inevitably *leads to the gradual advancement* of the organization of the greater number of living beings throughout the world.[27] (Emphasis added)

In this vital description of how natural selection works, Darwin links the adjustment to the environment and improvement (biotic) to the "advancement of the organization of the greater number of living beings throughout the world" (abiotic). His concluding statements in *The Descent of Man* make clear that demonstrating the development from simple forms to complex man is his main objective: He said, "The main conclusion here arrived at, and now held by many naturalists who are well competent to form a sound judgment is that man is descended from some less highly organized form."[28] Darwin's writings contain so many claims for natural progression that Gould's attempt to deny them is embarrassing for evolutionists. Darwin did have inner conflicts over the idea of progress. He criticized Lamarck for espousing progress caused by occult power in nature, but he admitted that, in general, he agreed with him. He also had doubts about progress because he himself could not see how the idea of development of conditions to fit the environment could necessarily produce a descent of progressive improvement. A variation in one situation that might enable survival might later cause the demise of the organism in new conditions. Moreover, after Jenkin's arguments about the swamping by populations and Owen's denial of inheritance of developed characteristics, Darwin could not see how improvement could occur. As a result, in his last edition of *The Origins...*he had to revert to the idea of the occult and a power in nature causing development.

Gould gives what *seems* to be an ingenious argument. He argues that there is a wall on one

---

[25] Stephen Jay Gould, *Full House: The Spread of Excellence from Plato to Darwin,* (New York: Harmony Books, 1996), 136.

[26] Ibid., 141.

[27] Darwin, *The Origins...*, (Garden City, New York: A Dolphin Book, Doubleday), 127.

[28] Darwin, *Descent of Man*, (Chicago: Great Books of the Western World, *Encyclopedia Britannica, Inc.*), in chapter xxi, General Summary and Conclusion, 590.

side where the simple begins, and that competition tends to eliminate the less able, and produce the better qualified because there is *no other direction development can go*. That is, bacteria as the simplest form cannot proceed any way except to the more complex, so that progress ***must*** result. One must ask, "By ***what compulsion*** does life develop at all ***in any direction***?"

Gould resorts to the argument for the general improvement. But again, ***intelligent design for improvement is added*** in his illustrations of baseball and all other sports. The fatal flaw in Gould's argument is that he uses the area of human endeavor within baseball and sports in which man uses his creative intelligence to cause improvement, while Darwin insistently denied there was any intelligent design in the process of natural development. Moreover, Gould begins with the ***assumption*** that a simple spontaneous form of bacteria was the beginning form; essentially he attempts to prove an unproved assumption. The lengthy studies referred to above have, in fact, not shown any advancement toward complexity, as Gould's arbitrarily chosen wall would produce. Gould's desperate effort to contradict scientific findings was an attempt to prop up Darwin's ideas of evolutionary progress from simple to complex. This argument is in fact an admission that Darwin's theory has failed.

New Partial Non-Darwinistic Theories Indicate Failure

Lynn Margulis, who exposes the failure of genetic mutations, has put forth a theory of symbiosis about cell development. Her view may explain the development of a number of molecular organisms. Her theory goes something like this. On the ancient earth, one larger cell in some way "swallowed" a bacterial cell, but did not digest it. It continued to live inside the host cell. Gradually, they adapted to helping each other. They both reproduced at the same time, but the symbiotic cell lost many of the systems that the free-living cells need and specialized more and more in providing energy for its host. Eventually it became a mitochondrion, of which one cell may have as many as 2,000 of them.

Behe argues that this process cannot explain the origin of complex biochemical systems. He said,

> The essence of symbiosis is the joining of two separate cells, or two separate systems, *both of which are already functioning*. In the mitochondrion scenario, one preexisting viable cell entered a symbiotic relationship with another such cell. Neither Margulis nor anyone else has offered a detailed explanation of how the preexisting cells originated. Proponents of the symbiotic theory of mitochondria explicitly assume that the invading cells could already produce energy from foodstuffs; they explicitly assume that the host cell already was able to maintain a stable internal environment that would benefit the symbiotic. Because symbiosis starts with complex, already-functioning system, it cannot account for the fundamental biochemical systems we have discussed in this book. ...It cannot explain the ultimate origins of complex systems.[29]

Summary of Failure in Evidence of Biological Transitions to New Species

The summary of these technical efforts in biology to prove Darwin's proposition for chance development by the survival of the fittest from simple to complex all failed. The claim that acquired characteristics could be transmitted to successive generations was refuted by contemporaries of Darwin, and continues to be refuted. With the acceptance of this fact, and the emergence of the science of genetics, a neo-Darwinian position claimed that small mutations would be replicated to improve the species and create new developments and species. Studies

---

[29] Behe, *Darwin's Black Box*, 188-189.

show, however, that changes are either genetically controlled by laws or are micro-mutations, which have no effect on survival. Only by adding new creative information can genetic changes cause improvements. Scientists hoped that certain elements of RNA would add new information, but even in the experiments in which scientists added new information to create changes, no new species were produced. While the genetic code of individuals allowed for minor change, due to environment or cross breeding, these could be reversed. Again, no new species were created.

Darwin's claim that the simple developed into more and more complex biological forms has been refuted by many experiments, showing there is no development toward complexity or simplicity in genetic reproduction. The study of complex organisms shows conclusively that the extreme complexity of organisms coming by chance from survival of the fittest would itself be unbelievably miraculous.

The new techniques of Cladism accurately categorize distinctions, but fail to show intermediaries, as Darwin predicted. Moreover, the use of new instruments to study microbiological entities shows that sequencing of proteins and molecular formations are equally distinct on the micro-level. Molecular biology shows that the molecule is extremely complex and beyond the possibility of chance evolution.

Even in the face of the contrary evidence, the evolutionists are trying to rebuild the idea of "the tree of life." With 560 networked computers using taxonomic criterion of many kinds, such as genetic sequences, shapes of ankle bones, seed pods, and the like, a group of scientists seek to reconstruct a rough outline of the tree. They are contradicting the views of many others and giving absurd conclusions such as humans are kin to fish, mushrooms are closer to humans than potatoes, and the like.[30] This is only another desperate effort to prop up a theory that appeals to man's hope of progress from nature, without God. This is not enlightenment, but seeking to impose darkness.

---

[30] Thomas Hayden, "All in the Family: Putting every creature with its kin on the tree of life," *U. S. News & World Report*, June 3, 2002, 56.

Chapter 12
Biological History Reversed in Fossils and Cell Functions

Introduction: Fossils – Real History of Biological Origins and Development

To support his theory, Darwin relied on the new science of paleontology, which was beginning to study fossils preserved in sedimentary rocks, and bones of extinct plant and animal life. These, of course, are *the **preserved history** of what had occurred* in the past and were *crucial to demonstrate his theory as science.* It is important to remember that Darwin knew little of the fossil record and nothing of primitive men. The first modern evolutionists (e.g. Lamarck, et al.) *incorporated an **intelligent design** force,* working in the various species to produce gradual descent to higher life forms. While Darwin's theory eliminated this occult force for chance survival of the fittest, in the final revision of *The Origin,* he incorporated a designing force in his theory of natural selection by a casual obscure statement. He thus, in the end, *gave in to **intelligent design***, which he had incorporated for development of the eye and other organs. This sly admission was an "ace in the hole," though he never openly conceded this. Eiseley said Darwin's admission, if understood and admitted, would have been "fatal to his theory." By the mid-nineteenth century, when natural selection by survival of the fittest had been shown not to work, the Neo-Darwinists moved to the idea of accomplishing transmutations by genetic mutations in DNA and RNA. But, this too failed.

Twentieth Century: Revolution in Paleontology and History of Life

In Darwin's time, Lyell et al. began a vigorous study of the earth's strata. The Cambrian, Silurian, and Ordovician regions had begun to be established. By the middle of the twentieth century, all major and minor divisions of stratification had been worked out and dated in millions of years, even though the times of sedimentary rock formations were mere guesses, based on the time needed for uniform forces to work. "Strata" Smith, who held to a young earth, conceived of special fossils as markers to put a chronological picture together. It was primarily individual scientists, not a worldwide group, who established these strata. For example, Sir Roderick Murchison identified the Silurian from strata in south Wales (1830s), joining with Adam Sedgwick in England, who had already identified the Cambrian. Together they established the germ of the modern time scale. Charles Lapworth differed with them, arguing for Ordovician strata. Joachim Barrande, a Frenchman examining Bohemian strata, contended for eight stages in the Silurian rocks. Later, Edward Forbes (1854) raised questions about these and others; he developed the Devonian. In these early stages of paleontology, as scientists were just beginning to understand and map geological forces, they frequently disagreed over some designations. Glaciation was just being recognized as a geological force. Oceanography and meteorology were in infancy. Scientists were a century away from recognizing the importance of continental break up and tectonic plate movement, including the ring of fire of volcanoes.

In the middle of the nineteenth century Lyell's geology extended all these periods to long ages and they were extended even further when Charles Darwin introduced his theory of evolution that required slow, gradual changes in development. During the first half of the twentieth century geology became a serious study that sought to organize around the use of the strata and long time periods that were required. The following is a pictorial presentation of what was envisioned to have occurred. [1] This is followed by a geological timescale.

PICTORIAL PRESENTATION IN STRATA PERIODS

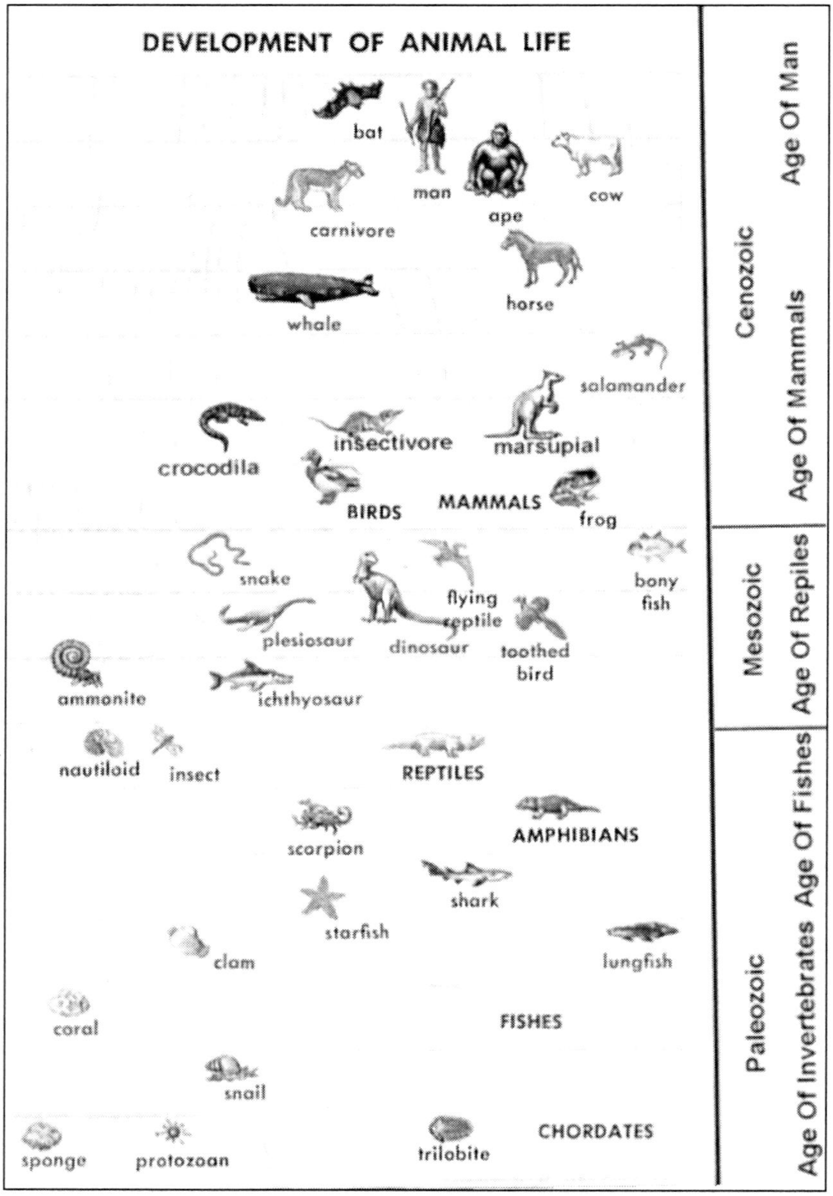

DEVELOPMENT OF ANIMAL LIFE

bat
man
ape
cow
carnivore
horse
whale
salamander
insectivore
marsupial
crocodila
BIRDS
MAMMALS
frog
snake
bony fish
flying reptile
plesiosaur
dinosaur
toothed bird
ammonite
ichthyosaur
nautiloid
insect
REPTILES
scorpion
AMPHIBIANS
shark
starfish
clam
lungfish
coral
FISHES
snail
sponge
protozoan
trilobite
CHORDATES

Cenozoic
Age Of Man
Age Of Mammals
Mesozoic
Age Of Repiles
Paleozoic
Age Of Fishes
Age Of Invertebrates

[1] R. Will Burnett, Harvey I. Fisher, Herbert S. Zim, *Zoology: An Introduction to the Animal Kingdom,* (New York: Golden Press, 1958), 16. The illustration is by James Gordon Irving.

GEOLOGICAL TIMESCALE

Era	Period	Epoch	Duration in Millions of years	Millions of years ago Tudge 2000	Moore 1958
Cenozoic	Quaternary	Holocene		00.1	
		Pleistocene	1.8	1.8	
	Neogene	Pliocene	3.5	5.3	28
		Miocene	18.5	23.8	
	Palaeogene	Oligocene	9.9	33.7	60
		Eocene	21.1	54.8	
		Palaeocene	10.2	65	
Mesozoic	Cretaceous		77	142	100
	Jurassic		63.7	205.7	155
	Triassic		42.5	248.2	185
Palaeozoic	Permian		41.8	290	210
	Carboniferous				
	Pennsylvanian		33	323	240
	Mississippian		31	354	265
	Devonian		63	417	320
	Silurian		26	443	360
	Ordavician		52	495	440
	Cambrian		50	545	520
Precambrian				4055	4600

The timescale on the previous page shows the time of the strata by Moore on the far right, which was mid-ninetieth century estimates. On the left of Moore's figures is given the timescale of Trudge at the turn of the twenty first century after the fossil record had been more completely explored. When Moore wrote, radioactive dating was in its infancy, but by the time of Trudge's figures much was know about half-life of element disintegration. However, except for work in the late Holocene period when radiocarbon dating was useful, radioactive dating was so long that it was difficult to use from the Cambrian period when life and fossils were in the strata. To the left of Trudge I have given the duration of millions of years of each geological period that Trudge has given.

In this graph, the dates differ by many millions of years. The Cambrian by 25 m, the Ordovician by 55 m., Silurian 83 m., Devonian 97 m., Carboniferous 172 m., Triassic 63.2 m., Jurassic 50.7 m., and so on. Such a great difference, by experts, certainly implies lack of firm data, and a high probability of guesswork. A small percentage of difference might be expected, but these are sizable deviations. Since Tudge wrote, the differences have widened even more Scientists had already added longer periods of time to their original ideas because of the need for the fossil record to support Darwin's theories of evolution.

During this period of explosive research, the beginning of evolutionary development was thought to be in the Precambrian Era, which is said to have lasted over 4.05 billion years, beginning about 4.6 billion years ago or most of earth's existence. Life is said to have begun with single-celled organisms that ***left no record*** in the rocks during that time. The rocks of Precambrian era, Proterozoic (before life) and Archeozoic (early life), are often treated together because "their history is difficult to determine accurately" and "fossils are rare." Interestingly, one-fifth of Precambrian rocks, which are the oldest, are on the surface of the earth and many are on the top of some of the highest mountains.[2]

The Cone of Data Has Turned Out Differently Than Expected by Evolution

Preconceived Cone of Data Went From Beginning Point to Wide Base

At the mid-nineteenth century ***the cone of data*** *of the fossil record* was beginning to emerge and was expected to follow Darwin's tree of development. Scientists expected it would begin with few, small, simple forms of life and grow to a widening cone with multiple large and increasingly complex life forms, similar to the cone effect of a growing tree. This preconceived philosophy of the survival of the fittest required the widening cone effect. By the time Moore's standard textbook was published, the fossil record was in its mid-stage of discovery. The science of radioactive dating was in primitive infant form. Moore admitted that dating by sedimentary rock, in which most fossils are found, was at best a guess, since sedimentation depends on volume and movement of the force of water, which was greatly underestimated at the time, and on the kind of rock or earth that is being eroded. At that time, the uniform deterministic theory formulated in geological principals completely excluded cataclysmic events so the great eradication of species that have occurred were not even envisioned. It wasn't until the later, mid- twentieth century discoveries that the true shape of the history of life was revealed. The graph on the next page depicts the way evolutionary geologists viewed the cone of data at mid century. The beginning of the cone, with the long center ridges, was mostly speculation at the beginning, with some reality to the ridges

[2] E.B. and Carl C. Branson, W.A. Tarr, and W.D. Keller, *Introduction to Geology,* Third edition, 333.

representing later life forms. Moore's markers of time are given on the right in signs alongside the names of the strata as he and others had estimated then.[3] Markers on the left signify certain areas that were involved in the cone formation.

MOORE'S VISION OF THE DEVELOPING CONE OF TREE OF LIFE

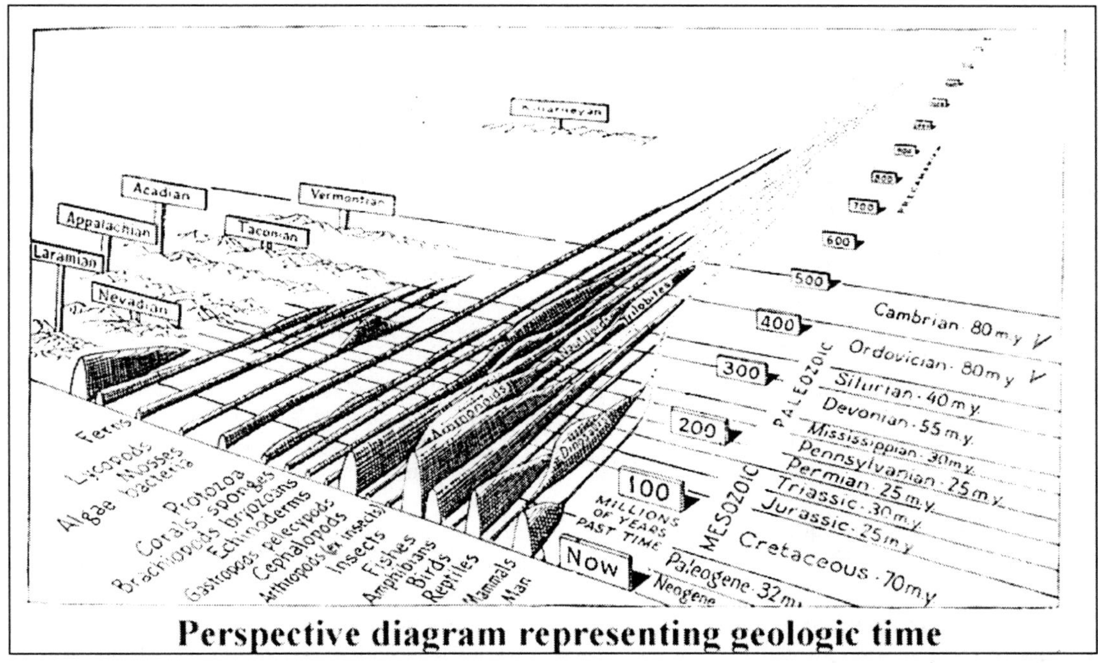

Perspective diagram representing geologic time

The Actual Cone Begins Wide and Narrows with Points of Mass Extensions

However, this picture by Moore was not true to the facts that had emerged even then. Well before the mid-twentieth century, it was known that most forms of life went back to the Cambrian period, which is the earliest rock strata having prominent fossils. In their *Introduction to Geology*, Carl C. Branson, W.A. Tarr, and W.D. Keller, in discussing the early Cambrian life, confess, "*Probably **nine-tenths of all basic organic structures** in animals had appeared by the beginning of the Paleozoic* period"[4] (Emphasis added). Evolutionary paleontologists sought to explain away this fact by saying these forms had already evolved in Precambrian times and were covered over by rock thrusts, leaving no record, not by scientific observation. But where did all these structures come from, and why is it only one-tenth of the structures had to evolve in the rest of time shown in the rocks? Shouldn't there be ***some evidence*** in the rocks for the evolution of some of the nine-tenths of the structures?

Today, many years later, geologists who have embraced the idea of ***bursts of evolution,*** are rediscovering the prevalence of most of the life forms, as reported by Branson years ago. They

[3] Raymond C. Moore, *Introduction to Historical Geology,* Second addition, (New York: McGraw-Hill Book Co. Inc., 1958), 31.

[4] Branson, *Introduction to Geology,* 349.

claim an unexplained explosion of life forms ***happened at once, by nature, mysteriously***. In the last decade of the twentieth century, United States geologists were granted access to remote northeast Siberia. They "succeeded in dating the evolutionary explosion at the beginning of Earth's Cambrian period, a biologic burst that ***produced almost all major groups*** of modern animals in an astonishingly short span of time"[5] (Emphasis added). Similar finds have recently been made in Canada and others more recently in China. It is now asserted footprints of four legged land vertebrate have been found as early as 395 million years ago.[6] It is now clear that at least nine-tenths of the forms, and probably more that were thought nonexistent, already existed in most of the world by the time the first fossils were found in the rocks. Some refer to this as the ***"zoological big bang"*** that began the forms of life. This has never been the picture that evolutionists have presented. They have insisted that all life forms gradually evolved from the minutest forms of life to man by millions of individuals, by chance survival of the species, in an expanding cone of life (2nd graph).

The cone of evidence did not turn out as Moore illustrated for Darwin's theory, where a few living things emerge and the cone becomes a broader tree with more and more species coming into the picture. Dr. Kurt P. Wise, PhD in paleontology from Harvard, and professor at Bryan College, has said,

> In the fossil record, however, such a "cone of increasing diversity" is not observed. Instead, the number of major groups we have today was achieved early in earth's history, when species diversity was low. In fact, the number of classes of arthropods and echinoderms at the time of the first appearance of each of these groups was actually higher than it is at present. [7]

Repeatedly, in the new millennium, paleontologists are announcing data that supports this. Complex animals of bilateral symmetry, admittedly, existed in Precambrian times, as documented by studies in Guizhou, China and elsewhere.[8] Zhe-Xi-Luo of Carnegie Museum of Natural History recently announced that extremely large mammals, Castrorocauda lutrasimilis, "which was ***ten times as big as any previously known mammal***" (emphasis added) existed 100 million years in Mongolian inner lakes and rivers before swimming mammals were thought to exist." "The discovery supports the emerging view that mammals were already a part of earth's menagerie tens of millions of years before dinosaurs left the scene."[9] It is now well known that most animal species existed in the Cambrian period and expanded their variations until the P T catastrophe in the Permian Triassic period and then this expansion occurred again until fewer, later reductions, especially at the K T catastrophe in The Cretaceous Tertiary period. This will be explained in later chapters.

The *facts have turned the predicted* **cone upside down** and showed significant decreases as the inversed cone developed. Where it was predicted there were no living things, there really were almost all kinds of animals and plants in the earliest paleontological record. These diversified and

---

5 R. Monastersky, "Siberian Rocks Clock Biological Big Bang," *Science News,* Vol. 144, September 4, 1993, 149; cf. *Science News,* Vol. 145, June 11, 1994, 145.

6 Amy Barth, "Fossil Prints Rewrite History," *Discover,* 01.02 2011, 67.

7 Kurt P. Wise, chapter 6, J. P. Moreland, editor, *The Creation Hypothesis,* (Downers grove, IL: InterVarsity Press, 1994), 220. He refers to Gould's, *Wonderful Life.*

8 David J. Bouujer, "The Early Evolution of Animals," *Scientific American*, August 2005, 42-47.

9 Jessica Ruvlinsky, "Mammals Stake their Place in Jurassic Park," *Discover*, June 6, 2005, 17.

greatly diminished through time. The historical record turns out to be the exact opposite of scientists' predictions, which they formed to uphold Darwin's theory of evolution (cf. pg. 550).

Only Limited Microchange in Specific Species

Darwinian evolutionists also expected visible transitional forms in the record, but even by mid century the record did not show this, and now the distinction is quite clear that there are no intermediaries. Only a microchange or development within the major groups is evident. This was partly revealed in early Mendelian genetics and the control mechanisms that limit change by the genes are now more fully known. Darwin had envisioned a tree of life, progressively developing from one species to another, but the lack of intermediary forms contradicts this prediction.[10] In other words, there are no transmutations, as Darwin argued. Hooker, his more learned botanist friend, warned Darwin earlier these did not exist according to the data; Lyell said the same.

Colin Tudge points this out in a number of cases. In regards to the plant *composite* he said,

> In traditional classification these apparent close relatives of the composites have generally been scattered among several orders (Asteridae, Campanulocae, Dipsacaceae, Goodeniaceae and Colycersceae). Clearly the Asteridae as a whole needs more sorting out. But the family compositae itself *forms a clearly defined and coherent group.* As the great English botanist George Bentham (1800-84) commented in 1875 (in Darwin's time), 'I cannot recall a single ambiguous species as to which there can be any hesitation in pronouncing whether it does or does not belong! It is indeed monophylectic. But because the family is so extensive and so various, it has long been subdivided.' (split sometimes into 19 or 14 tribes).[11]

Tudge gives a number of such illustrations. Evolutionists have sought for over a century[12] to find more species than really exists in order to broaden the evidence and find intermediates for the Darwinian tree.

Current genetic discovery shows that mobile genetic elements are ***restrained by protein information mechanisms*** within the cells.

> Left to its own devices the transcription machinery of the cell would express every gene in the genome at once: unwinding the DNA double helix, transcribing each gene into single-stranded messenger RNA and, finally, translating, the RNA messages into their protein forms. No cell could function amid the resulting cacophony. So cells muzzle most genes, allowing an appropriate subset to be heard. In most cases, a gene's DNA code is transcribed into messenger RNA only if a particular protein assemblage has docked onto a special regulatory region in the gene.[13]

Built-in controls determine how far genetic mechanisms can change. These limit change and protect the health of the individual. Individuals do not inherit acquired characteristics. Research shows that information within the genes does allow divergence in species, but only to a controlled limit. Inherited protein information and operation do censor to what distance change can normally happen and be reproduced.

---

[10] Wise, *The Creation Hypothesis*, 219, 220.

[11] Colin Tudge, *The Variety of Life*, 598, 599.

[12] Gould, *The Structure of Evolutionary Theory,* (Cambridge: The Belknap Press of Harvard University Press, 2002), 779.

[13] Nelson C. Lau and David P. Bartel, "Censors of the Genome," *Scientific American*, August 2003, 34-41.

Not only had Darwin's mechanism of survival of the fittest by inheriting acquired characteristics failed, the Neo-Darwinists genetic hopes failed to produce intermediates by genetic mutations, which would lead to improvements and transitions. Stephen Jay Gould discusses the various unsuccessful efforts to expand and rephrase, or reform the statements of mechanisms for gradualism.[14] This led to certain attempts to insert ***hierarchical views,*** in which the individual organism changed. While claiming to include various views, Gould's evidence does not show transition. He forcefully states that the paleontological facts are against such change.[15] Gould says that because Darwinian evolution had wide acceptance in the nineteenth century and with Darwin's insistence on gradual change in individuals in transmutation, Darwin's theory was an "intellectual novelty" and professionals went along with it sheepishly, even when they knew it was wrong. He says Darwin repeatedly appealed to "imperfect" paleontological record. Gould makes a strong statement that there is virtually no evidence for gradualism. He said,

> If most fossil species changed gradually during their geological lifetimes, biostratigraphers would have codified "stage of evolution" as the primary criterion for dating by fossils. But, in fact, biostratigraphers treat species as stable entities throughout the documented ranges–because the vast majority so appear in the empirical record.[16]

Gould offers repeated statements by other prominent paleontologists that gradual transitions do not exist.[17] Therefore, it is clear that the data in the fossils do not support Darwin's views, or the modified views of Darwinism. However, efforts to maintain the idea of gradualism have produced many perversions, which scientists continue to offer up in an attempt to persuade the public.

Illustrations of Perversion from the Fossil Records

The Descent of Mammals and Whales

It is not possible to show all the perverse claims about the lines of descent. Only samples of some distortions in paleontology will be given here. While new and different ideas are offered regularly, we will examine a sampling of the more frequent claims and their lack of support.

Dr. Phillip Johnson, U. C. Berkeley law professor, published his book, *Darwin on Trial,* in 1991. In the debate over Johnson's views, Dr. William B. Provine argued that functional legs on a recently discovered whale ancestor indicate a bridging of the gap to land mammals. Yet, the scientists reporting this in the July 13, 1990, *Science* magazine said the *legs* were inadequate either for walking or swimming and did not meet the definition of legs, therefore, they must have been an aid in copulating. So, the argument is, again, one of similarity.[18] The truth of the matter is that how the different kinds of whales developed is still in dispute.

Scientists offer various speculations as to how land animals descended from whales. One such claim holds that tetrapods—salamandar-like creatures with mitten-like fins are intermediates between fish and amphibians. While the purpose of the fin is not known, scientists admit that these fins "were virtually useless for helping Acanthostega walk on land" and that it was "an animal that

[14] Stephen Jay Gould, *The Structure of Evolutionary Theory*, chapters 7, 8.
[15] Ibid., chapter 9, 745 ff.
[16] Ibid., 751.
[17] Ibid., 746-755.
[18] See "Whale of a Change for Cetacean History," *Science News*, Vol. 143, February 20, 1993, 127.

couldn't walk on land."[19] Farish Jenkins and a team from Harvard University have found a large fish with a crocodile-shaped head, with similar finlike appendages, which they ***assume were primitive legs***. Dubbing their finding Tiktaalik roseae, Jenkins, et al. ***speculated*** the fish adapted fins to escape predators in the water.[20] Are the scientists saying that these creatures had ***a desire to change design to move toward land mobility*** to survive? Or are they creatures that have a distinct design of their own? No matter how scientists phrase their speculation, it is a mysterious or supernatural explanation, much like Lamarck's contention that giraffes that grew longer necks to get higher tree vegetation, an explanation which Darwin soundly rejected. Scientists repeatedly make claims that animals and plants make adaptations to help them, such as their conjecture that birds developed UV vision to see fruits better. While evolutionary scientists repeatedly theorize in this way, they don't realize they are declaring purposeful change that involves intelligence.

Descent of Birds

As biological scientists have accumulated more data, the lines of evolutionary descent have become more muddled and less convincing. Several species with reptile features that seem to have feather-like aspects have been unearthed in different places and are interpreted to be support for birds emerging from reptiles. The six or more specimens of *Archaeopteryx*, a peculiar crow, or pheasant-sized species found in Germany and estimated to have lived 150 million years ago, were once hailed as a missing link (one of thousands needed) from reptiles to birds. Ten *Archaeopteryx* were found intact with skeleton and feathers encased in limestone. They had teeth and a skimming membrane fastened to a large fourth finger that was considered a wing. An apparent small dinosaur found in Lioning Province in China is considered related to the coelurosaur Compsagnathus. Larry D. Martin of the University of Kansas says it has a remnant of dermal scales, or possibly feather-like scales. But it comes *after Archaeopteryx*, so does not fit in for descent.[21] Terry D. Jones of Oregon State University, with American and Russian teammates, unearthed a ten-inch long lizard in Kyrgyzstan and dated it at least 75 million years, before *Archaeopteryx. Longisquama insignis,* with its feather staff and a calamus or feather-like end, appears to be related to the dinosaurs. A Soviet paleontologist declared, "that the long structures on its back are featherlike scales." Martin made the same observation about a similar China specimen, dating millions of years later. Alan Feduccia of the University of North Carolina "points out that the most birdlike theropods date from 70 million years after Archaeopteryx, 'you can't be your own grandmother,' he teases." Bird tracks have been found that are 212 million years old.[22]

At least two claims to bird fossils confuse the so-called line of descent.[23] Scientists claim a new find, Protavis, discovered in the mud flats of West Texas, to be a bird fossil, 225 million

[19] Carl Zimmer, "Coming on to the Land," *Discover*, June 1995, 116-127.

[20] Alex Stone, "How Life Got a Leg Up," *Discover*, July 2006, 14, 15.

[21] R. Monastersky, "Hints of Downy Dinosaur in China," *Science News*, Vol. 150, October 26, 1996, 260; cf. *Science News,* March 3, 1997, 271.

[22] S. Milius, "Overlooked fossil spread first feathers," *Science News,* Vol. 157, June 14, 2000, 405; "Unknown Creatures Made Birdlike Tracks," *Science News*, July 27, 2002, Vol. 162, 62.

[23] See Peter Wellnhofer, "Archaeopteryx," *Scientific American*, May 1990, 70-78; Pat Shipman, "Sixth Find is a Feathered Friend," *Discover,* January 1989, 63; Paul C. Sereno, "China Bird Fossil: Mix of Old and New," *Science News,* Vol. 138, October 1990, 138; Richard Monastersky, "The Lonely Bird," *Science News*, Vol. 140, August 17, 1991, et al.

years old. At least four specimens of similar shaped birds, with a claw appendage at the wing spot, have been found in Mongolia. *Mononychus* is dated at 75 million years. It seems to be a small bird that cannot fly. Another flock of dinosaurs, found in China, is a form of predatory theropod dinosaurs with down-like coats or fibers.[24] Other isolated specimens, even from different continents, while similar, do not show a line of descent. Paul Sereno's finding of aerosteon riocoloradensis, near the Rio Colorado in Argentina, confirms the connection between dinosaurs and birds, rather than reptiles. The specimen that Sereno, a paleontologist from the University of Chicago, uncovered was ten meters long, weighed as much as an Indian elephant, and contained a bird-like system of breathing, similar to a chicken's. [25] If birds and reptiles or dinosaurs are involved in a line of descent, the whole story of how is very confusing at present and getting more so. Are scientists saying reptiles developed wings because they wanted/needed to fly? This is, again, occult design that Darwin labeled as unscientific.

Radical Discoveries Change Original Claims for Dinosaurs

In the 1820s, scientists began discovering fossil bones of previously unknown, strange animals. The British anatomist, Richard Owen, called the creature Dinosauria, meaning "terrible lizards." Owen's name seems to agree with the view of the great paleontologist, Georges Cuvier, that these creatures were *probably* reptiles. Dinosaurs quickly became the major exhibit of intermediaries between reptiles and mammals and humans. They have become *the symbol of evolution* in museums, books, toys, and every way imaginable.

But ***modern research has discredited much*** that was first taught and exhibited ***about dinosaurs***. Two prominent paleontologists, Robert Bakker, of the University of Colorado Museum, and Jack Horner, of the Montana State University Museum of the Rockies, along with many others, in the last few years have shown that their habitat, the manner of their breeding, the speed and manner of their movement, the birth and care of their young, the way they looked when they stood, the construction of their bones, the fact that most usually traveled in herds, and the texture of their skin are ***all different than thought***. Many now believe they were not reptiles at all but mammals.

Stephen Jay Gould, Harvard paleontologist and the leading spokesman for American evolutionists candidly admitted that the former ideas were misleading.[26] Gould said,

> Bakker and Horner provide an elegant and coherent solution to this paradox of dinosaurs as stupid, slow, inefficient, and torpid-overgrown cold-blooded reptiles of little brain who yet lived with mammals. ***We were wrong***. Dinosaurs were sleek, anatomically efficient, probably warm-blooded creatures with complex social behaviors and average-sized brains for *reptiles* of their bulk.... The full implications have yet to be assimilated, and they're both disturbing and wonderfully enlightening. Dumb and torpid dinosaurs fit well with our most cherished notion of ***evolution as progress*** leading inevitably to us. But in the new view, dinosaurs are as worthy as mammals (only different) and their success (apparently beyond the power of mammals to challenge) ***implies that life doesn't proceed in lockstep toward increasing efficiency and***

---

[24] R. Monastersky, "China yields a flock of downy dinosaurs," *Science News,* Vol. 156, 183.

[25] Laura Sanders, "Forget bird-brained", *Science News,* October 25, 2008, p 14.

[26] Virginia Morell, "Announcing the Birth of a Heresy," *Discover,* March 1987, 26-50; Robert Bakker, *The Dinosaur Heresies: New Theories Unlocking the Mystery of Dinosaurs and Their Extinction*, ed. Maria Grarnoscilli, (Great Neck, NY: Morrow Publisher, 1986), 488 pages.

mentality, eventually (and inevitably) to culminate in us.[27] (Emphasis added)

Much additional evidence has been found since then. The study of dinosaur bone structure and even some residual genetic material preserved in a mine, shows they do not fit with reptiles or birds, but remain in a distinctive class of their own.[28] Studies of dinosaurs' teeth indicate they were warm-blooded. Paleontologist Michael Hammer of Oregon believes they found a dinosaur heart in South Dakota that, under x-ray scan, looked more like that of a bird or mammal.[29] New evidence suggests that dinosaurs are distinct from other groups and do not necessarily lead to higher animals. Much controversy yet rages about this. There is some reason to believe that dinosaurs are neither reptiles nor birds, but are a distinct kind of animal that fits between birds and mammals rather than reptiles. They may be nearer ostriches and flightless birds.

Human bias perverted the image of dinosaurs as striking and unusual creatures. It was this image more than any other that deceived people, especially children. Museum scientists wrongly constructed dinosaurs and placed them in the wrong environment. For example, the exhibit of Triceratops in the Smithsonian Institute in Washington was put together in 1905. It has been viewed in Dinosaur Hall by over 100 million museum-goers. It has been in children's books, in the movies of *The Lost World* (1925s), and *Jurassic Park* (1993) and is now recognized as one of the 20 best-known animals.

But this construct was not a creature at all, and not a true representation of the species. Bones from twenty different dinosaurs were used. The head was not that of a full-grown, but of a young specimen. Only about a dozen of the 25 foot-long spines are real bones. The rest were made of plaster and added. One of the humerus or upper forelimbs was much longer than the other, since they came from different animals. Since there were no Triceratops feet scientists doing the construction added feet from the duck-billed dinosaur instead. The creature's posture and gait were incorrect. We must applaud the scientists who discovered its misrepresentation and are boldly trying to correct these inaccuracies, which have never been made public.

For years it was established knowledge that dinosaurs that had been extinct for millions of years could not possibly have flexible tissues. Leading dinosaur specialists have found shocking evidence to the contrary. A team of dinosaur experts led by Mary H. Schweitzer, under John Horner, has found several dinosaurs in the Hell Creek Formation above the Fox Hills Sandstone in Montana. The first was a Tyrannosaurus Rex, of which some bones were slightly deformed and crushed, but excellently preserved, estimated to be 18 million, plus or minus 2 million years old. In the laboratory Schweitzer found unusual bone tissue on the endospeal surface: "fibrous matrix, what appeared to be supple bone cells, their three-dimensional shapes in tact; and translucent blood vessels that looked as if they could have come straight from an ostrich at the zoo."[30] The bone tissue was de-mineralized. After seven days, Schweitzer found fragments of living tissue with great elasticity and resilience. In addition, there were small oval features on the surface,

---

[27] Stephen Jay Gould, "The Lesson of Dinosaurs: Evolution Didn't Inevitably Lead to Us," *Discover*, March 1987, 51.

[28] Kathy A. Svitil, "Evolution Watch: Dinosaur Mine," *Discover*, May, 1995, Vol. 16, No. 5, 36-37 telling of DNA by Jack Horner and Mary Schweitzer and especially about Scott Woodward and DNA from a mine in Utah.

[29] S. Perkins, "Teeth tell tale of warm -blooded dinosaurs," *Science News,* and cf. *Geology*, September 1999. "Telltale Dino Heart Hints at Warm Blood," *Science News,* Vol. 157, April 22, 2000.

[30] Barry Yeoman, "Schweitzer's Dangerous Discovery," *Discover*, April 2006, 37-41, 77.

consistent with endothelial cell nuclei. Subsequently, other dinosaur fossils were recovered that also yielded similar structures. When compared under the microscope with extant ostrich bone both were flexible, pliable and translucent. More important than the similarity to the ostrich, is the discovery of tissues, previously unimagined for dinosaurs that were assumed to be millions of years old. Such conditions, existing after millions of years, are unexplainable and elicit many suggestions and questions, "Whether preservation is strictly morphological and the result of some kind of unknown geo-chemical replacement process or whether it extends to the sub-cellular and molecular levels is uncertain."[31]

Scientists have long accepted that no soft tissue of animals can last long and that no fossils have had soft tissue, certainly not after millions of years. Yeoman commented in his article, "When this shy paleontologist (Schweitzer) found soft, fresh looking tissue inside a T. Rex femur, she ***erased a line between past and present***. Then all hell broke loose."[32]

Numerous paleontologists expressed their dismay at finding such tissue. Cameron J. Tsujita, at University of Western Ontario, London said [the new study] "Is improbable, but obviously not impossible." Derek E. G. Briggs at Yale University described this as a, "totally novel discovery."[33] Previously, the development of dinosaurs from reptiles and how the dinosaurs died was questioned. It was thought they came from a reptile named dinosauromorphs, but now it is known they coexisted at the same time. Again, recent discoveries seem point to a relationship between dinosaurs and birds, rather than reptiles. Chris Organ, Harvard biologist, and his team have concluded "the genomic structure of birds ends up being—like feathers and parental care—not avian but dinosaurian in origin'[34]

The nature, structure, and almost everything about dinosaurs has been misrepresented to the public. Yet the theory of evolution is linked more to this icon than any other thing. Their place in the so-called evolutionary descent is now in question, as is their age. Their quick death, which has been widely acclaimed, is not agreed upon.

The mysterious death of the dinosaurs has been attributed to the impact of an asteroid. Many dinosaurs are now known to have died from a possible great flood that caused neural damage, or possibly from oxygen deprivation known as hypoxia from a bacterial infection. A study by Cynthia Marshall Faux of the Museum of the Rockies in Bozeman, Montana, Matthew Carrano, curator of dinosaurian at the Smithsonian Institution's Natural Museum of Natural History, Kevin Padian, curator of the Museum of Paleontology of the University of California, Berkeley, and others recently reviewed the fact that many dinosaurs died in the "dead bird pose." C. Barry said, "Now, scientists report that the posture probably came about because dinosaurs or other animals died of central nervous system damage. Fossils of nearly all bird-like Archaeopteryx, as well as some Tyrannosaurus Rex and other ancient creatures, exhibit the curious pose." Barry said that "paleontologists traditionally attribute to rigor mortis, desiccation of the carcass, or the shifting of

---

[31] "Soft-Tissue Vessels and Cellular Preservation in Tyrannosaurus Rex," *Science Magazine,* March 25, 2005, Vol. 307, 1952-1955; cf. 1955.

[32] Yeoman, "Schweitzer's Dangerous Discovery," *Discover*, April 2006, 37-41, 77.

[33] S. Perkins, "Old Softy: Tyrannosaurus fossil yields flexible tissue," *Science News,* March 26, 2005, Vol. 167, 195 (cf. 39).

[34] Jennifer Barone, "Did T. Rex Taste Like Chicken? Protein and DNA analyses cement the link between dinosaurs and birds," *Discover* August 1, 2007, 16. Cf. S. Perkins, "Big and Birdlike: Chinese dinosaur was 3.5 meters tall," Science News, June 16, 2007, vol.171, 371 (a creature11+ feet tall and 24 feet long.)

bones by water currents."[35] In visiting Dinosaur National Monument in Colorado and Utah, I noticed that many seemed to have died by crushing force of water, and not by asphyxiation.

Plants not slowly Evolving

Scientists have long held that grasses slowly evolved to what they are today. Caroline A. E. Stromberg, of the Swedish Museum of Natural History, Stockholm, found grass phytoliths in fossil dinosaur feces that showed modern grasses existed much earlier than thought—56 to 70 million years ago. She identified remains of five types of grasses. Perkins commented on her data, "The finding not only provides the first evidence of grass eating dinosaurs but also shows that grasses evolved diverse forms ***much earlier*** than scientists had previous recognized." Elizabeth A. Kellogg, University of Missouri, St. Louis, said, "If this research holds up it would completely revise what we've thought about the origin of grasses. ... To find such diversity at that time is surprising."[36]

Admission by Highest Authorities of No Gradualism in Evolution

When all the forms of the plants and animals are evaluated, there is no evidence of any intermediaries. In 1958, this author took a brief course under Harland Banks, paleobotanist of Cornell University. In our course, he stated that much of the geological record had been studied and that most paleontologists now admitted that there were distinct gaps between the species. Moreover, he also said that Russian scientists had found fossils of iris flowers and other late life forms in the Precambrian layers. He still held to evolution, but he said Darwin's view of natural selection by gradualism must give way to change by quick jumps from one life form to another, without leaving any record. Yet, Darwin had rejected this idea.

In 1940, Richard Goldschmidt, professor at the University of California, Berkeley, and one of the world's greatest geneticists, published *The Material Basis of Evolution,* in which he insisted that the *neo-Darwinist theory of micromutations was no longer tenable* as a general theory of evolution. Instead, he said macroevolution accounted for larger steps in evolution. Although attacked and reviled at first, *about forty years later,* the book was reissued and the prominent evolutionist, Stephen Jay Gould of Harvard, wrote the introduction. In that introduction Gould said, "I do...believe that its (Goldschmidt's book) general vision is uncannily correct (or at least highly fruitful at the moment) in several important areas where conventional Darwinian Theory has become both hidebound and unproductive."[37] Elsewhere, Gould (with Niles Eldredge) said,

> The history of most fossil species includes two features particularly inconsistent with gradualism.
> (l) *Stasis.* Most species exhibit no directional change during their tenure on earth.
> (2) *Sudden appearance.* In any local area, a species does not arise gradually by the steady transformation of its ancestors; it appears all at once and 'fully formed'.[38]

About that time, Niles Eldredge and Ian Tattersall, both of the prestigious American Museum of Natural History, wrote in their book, *The Myths of Human Evolution*, that the idea of

[35] C. Barry, "Jurassic CSI, Fossils indicate central nervous system damage," *Science News*, June 23, 2007, vol. 171, 390.

[36] S. Perkins, "Ancient Grazers," *Science News*, Nov. 19, 2005, Vol. 168, 323.

[37] Richard Goldschmidt, *The Material Basis of Evolution*, Silliman Milestones in Science, 1940.

[38] Stephen Gould and Niles Eldredge in their essay on the theory of "Punctuated Equilibrium."

progressive change in the fossil record from primates to men was a myth. To continue the idea that discovering new fossils would reveal intermediates was also a myth.[39]

Colin Patterson, paleontologist for the British Museum of Natural History, addressed the Systematic Discussion Group of the American Museum of Natural History in 1981. He confessed his blind devotion to the theory of evolution for over twenty years and, yet, admitted he knew nothing about it. He said, "It's quite a shock to learn that one can be so misled for so long. Either there was something wrong with me or there was something wrong with evolutionary theory." Responding to a question about transitions in the record, Patterson said,

> There is not one such fossil for which one could make a watertight argument.... Is Archaeopteryx the ancestor of all birds? Perhaps yes, perhaps not: There is no way of answering the question. It is easy enough to make up stories of how one form gave rise to another, and to find reasons why the stages should be favored by natural selection. But such stories *are not part of science*, for there is no way of putting them to the test.[40]

He became noncommittal on evolution.

There are no higher authorities than Niles Eldredge and Ian Tattersall of the American Museum of Natural History, Colin Patterson, senior paleontologist for the British Museum of Natural History, Stephen Jay Gould, professor of geology at Harvard, Richard Goldschmidt of the University of California, Berkeley, and Pierre Grassé of France. All of these authorities agree there is no present known mechanism for Darwinian gradual evolution that can be supported by evidence from genetics or paleontology, while they all still claim that in some way evolution occurred across gaps in a macro way. Their statements show *the death of neo-Darwinism.*

Still, most scientists refuse to dismiss Darwin's theory of evolution, despite lack of evidence. To give up evolution would be to abandon the faith of the educated elite minority that there is a natural basis for human *progress,* and against belief in God. And to give up Darwin would be a tragic *admission of mistaken confidence in man* and in his modern basis for leftist social and political efforts. That would be the supreme act of intellectual suicide.

In spite of the fact that these leading experts in many fields say there is *no **evidence** of descent* at any level from one species to another, the media and some scientists, including Tattersall, continue to publish charts of how forms grew from one to the other. The April 26, 1993 issue of *Time* magazine contains an illustration showing how dinosaurs descended one from the other. In spite of all the graphic lines, there are no intermediary forms that show this happened. These and all such diagrams are ***purely imaginary and a fraud.*** There is also a marked difference in the order of the fossils in the record from what evolutionists claim. Selecting some fossils of major groups as significant for dating, the oldest being at the bottom and the youngest at the top strata and using these all over the world to date sedimentary strata seems a valid procedure. But as Kurt Wise has reported,

> The correspondence between phylogeny and the fossil record is not as strong as it might first seem. When *the order of all* kingdoms, phyla and classes is compared with the most reasonable

---

[39] Niles Eldredge and Ian Tatters all, *The Myths of Human Evolution,* (New York: Columbia University Press, 1982), 127.

[40] Tom Bethell, "Agnostic Evolutionists: The Taxonomic Case Against Darwin," *Harper's*, Vol. 270, No. 1671, February 1985, 49 ff.

> phylogenies, ***over 95 percent*** *of the lines are not consistent with the* ***order in the fossil record.***"[41] (Emphasis added)

Classic Icons of Chance and Intermediate Transitions Now Lied About

Certain classical icons are regularly used to persuade the public that life began spontaneously, and developed, with intermediates, from simple to higher life forms. These include Darwin's tree of life, a homology in vertebrate limbs, recapitulation of different embryos, *Archaeopteryx* as link from reptiles to birds, peppered moths, Darwin's finches, pictures of the evolution of the horse and apes to men, etc. In 2002 Gould has said,

> Most nonpaleontologists never learned about the predominance of stasis, and simply assumed that gradualism must prevail, as illustrated by the exceedingly few cases that became textbook "classics": the coiling of Gryphaea, the increasing body size of horses, etc. (Interestingly, nearly ***all these*** *"classics" have been* ***disproved,*** thus providing another testimony for the temporary triumph of hope and ***expectation over evidence.***[42] (Emphases added).

Jonathan Wells, with Ph.D.s from Yale University and University of California at Berkeley, in his book, *Icons of evolution: Science or Myth?*, has presented evidence showing the icons used to persuade the public about evolution were all false. Moreover, he has shown that in ten of the current major biological textbooks used in American schools, seven of these still include all of the classic icons, and the other three have all but one. Millions of people, including high judges, government leaders, and people in the media had been taught and our children are still taught these lies.[43] Man's bias to reject God and his moral standards of good and evil are promoted by these.

Paleontologists and teachers who try to use the evidence of fossils to uphold evolutionary gradualism or any other kind are writing with what Gould labeled intellectual "publishing bias."[44] Evolutionary theory is now based on non-evidence when it comes to fossils. Marvin L. Lubenow of Christian Heritage College of El Cajon, California who wrote an exhaustive and comprehensive review of all the fossil record has shown this.

> The human fossil record, like the fossil record in general, has failed to furnish evidence for evolution. Evolutionists, understandably, are reluctant to admit it. One way they now handle the problem is to claim that the fossils really are not important. Indicative of this new trend is the incredible quotation by Mark Ridley of Oxford University: "... ***no real evolutionist***, whether gradualist or punctuationist, ***uses the fossil record as evidence*** in favor of the theory of evolution as opposed to special creation. This does not mean that the theory of evolution is unproven."
> The heading of Ridley's article reads: "The evidence for evolution simply does not depend upon the fossil record."[45]

Because of its importance and the amount of data involved the discussion of the claims that fossils show descent of men from apes must be reserved for the chapter on anthropology and man's uniqueness.

---

[41] Kurt Wise, *The Creation Hypothesis*, 225. This is substantiated in unpublished papers on the subject.
[42] Gould, *The Structure of Evolutionary Theory,* 760.
[43] Jonathan Wells, *Icons of Evolution: Science or Myth?* 338 pages.
[44] Gould, *The Structure*, 764.
[45] Marvin Lubenow, *Bones of Contention*, (Grand Rapids: Baker Book House, 1992), 181, cf. Mark Ridley, "Who Doubts Evolution?" *New Scientist 25* (June 1981): 830, 831.

Conclusions about the Fossil Records

The fossil record does not support evolution. Neither the cone of data, selective arrangement of fossils, nor the classic icons support the evolutionists' theories of simple life forms developing into higher, more complex forms, and eventually, humans. Rather, the more scientists discover, the more the evidence points to a transcendent, intelligent Designer who created and implanted information that controls the development of the various plants and animals.

Chapter 13
Evolutionary Theories of Non-Evidence

Unproven Theories of Evolution by Jumps between Species

The best evidence that there are no lines of descent (apart from the admission of the leading evolutionists) is the fact that every current view of evolution assumes this lack of evidence to be true. All theories are conjectures of how evolution happened, ***without any evidence*** *of lines of descent.* They all require jumps from one form to another, either rapidly or invisibly, which does not leave any evidence of how one species got from one to the other. The ***absence of gradualism*** was in itself a demonstration that Darwin was wrong, as were the biologists that followed him.

Wolfgang Smith, formerly a teacher at M.I.T. and later at Oregon State University, has said concerning these conclusions,

> As the matter stands, the ***only avenue of escape from the negative evidence of paleontology*** seems to lie in some feasible concept of cryptogenesis or ***'hidden evolution,'*** of which a number of variants have been proposed. One possible approach (and this applies especially to the higher stages of evolution, corresponding to the fossil ferrous strata) is to postulate special phases of development during which the transformation of species *takes place with such rapidity as to* ***elude detection*** via the fossil record. In line with this general idea, one encounters such concepts as Severtzoll's 'aromorphosis', Schindwolf's 'explosive evolution', Zeuner's 'episodes of intense evolution' and Simpson's 'tachytely.' Somewhat different types of cryptogenesis have also been considered, such as de Beer's 'clandestine evolution.' Yet, all these theories suffer apparently from the same fundamental drawback, which is simply **the *lack of positive evidence.*** The best that can be hoped for in this domain, it would seem, is to avoid obvious conflict with known facts.[1] (Emphasis added)

Claims of Saltation in Theories of NonEvidence

Neither Darwin nor the neo-Darwinians were able to show lines of descent in the paleontological record. Molecular biologists demonstrated that the chance, gradual development of complex organisms was both irrational and impossible. Faced with these realities, evolutionists were forced to rethink their ideas of natural progress.

F. Jenkin's devastating observation in Darwin's day, that populations would eradicate slight differences in time, after further study, was adopted in 1908 as the Hardy-Weinburg Law. As a part of the law, the authors ***supposed*** that a few individuals could have developed diverse characteristics, and could be ***cut off*** from the main population so that they would not be swamped. They assumed this small splinter group would likely have a gene frequency different from the main group, where some would die, but others would retain differences from the main population.

Because of Jenkin's observations and the Hardy-Weinburg law, the neo-Darwinists, who were trying to find answers in population studies, observed that the ***whole population*** *would have to evolve in gradual increments.* Darwin acceded to the accuracy of this. The paleontological record shows no evidence of whole population progress. Lacking evidence, the answer seemed to be saltation, as Harlan Banks, Cornell paleontologist suggested when I studied under him in 1958. Macroevolution by saltation or jumps soon ran its course for lack of a known mechanism, as well as R. A. Fisher and others' arguments noting the great improbability of such a development. Similar theories, lacking evidence, include Stuart Kaufman's computer-simulated theory of complexity and Marguila's assumptions that symbiosis did this in a mysterious and even occult

---

[1] Wolfgang Smith, *Cosmos & Transcendence*, (La Salle, IL: Sherwood Sugden and Co. Publisher, 1984), 70-71.

way.

Evolution by Punctuated Equilibrium?

The recognition of the rule of population swamping as a major obstacle forced new creative theories. Niles Eldredge and Stephen Jay Gould, two of the world's leading evolutionists, finally brought the focus on rejection of gradualism to the surface, since no viable theory had taken its place. They were well aware of Jenkin's arguments, the Hardy-Weinberg Law, and the history of evolutionary theory. They had to find a way that saved face for evolution and allowed the fossil record to stand. They seized upon the Hardy-Weinberg speculation of splintering off, even though there was no evidence for this theory. Eldredge and Gould adopted a ***hierarchical view*** that allowed the possibility of other views of evolutionary change in order to keep traditional Darwinians happy. This was a tactic to attract all the followers of Darwin, including those who espoused gradualism.

However, Gould and Eldredge really created one ***macroevolutionary view with new non-Darwin definitions***. They abandoned Darwinian evolutionary gradualism as such. Instead of the individual organism, they redefined the word ***individual*** to mean "whatever is the focus," either the gene, the organism, the species and all each constitutes. They chose the ***species*** as representing the ***basic units*** in theories and mechanisms of macroevolutionary change because only species are seen in the record.[2] They said,

> We ***define selection*** as occurring when plurification results from a causal interaction between traits of an evolutionary individual (a unit of selection) and the environment in a manner that enhances the differential reproductive success of the individual. Thus, and finally, units of selection must above all, be interactors. Selection is a causal process....[3] (Emphasis mine).

> Selection is only as an engine for generating diversity, not an agent for changing average form within a clad.[4]

> Once we agree to ***define*** higher-level selection by differential proliferation of relevant unites based on interaction between their traits and the environment.[5] (Emphasis mine).

Although the individual as a single organism could not inherit any acquired changes, they said the species that developed within genetic controls did rapidly reproduce the changes as a micro development. They said,

> The species-individual does not maintain integrity (as the organism does) by suppressing differential proliferation at low-level units.... Therefore they maintain that directional speciation suffers no constraint or suppression–and may represent one of the most common modes of macroevolution. [6]

In other words, new species may be created some way and genetically perpetrated. The finding or discovery of new species might indicate that they were created anew. Thousands of new species are being discovered that may have been there all along, but they thought that these may have been

---

[2] Gould, *The Structure of Evolutionary Theory*, 674, 703.
[3] Ibid., 623.
[4] Ibid., 777.
[5] Ibid., 656.
[6] Ibid., 725.

created, improved, and continued to evolve.

Working of Punctuated Equilibrium

Allowing other evolutionary views, along with theirs, allowed Eldredge and Gould to avoid Darwin's errors of population swamping etc., without being an affront to those who hold Darwinist views. At the same time they supplanted and avoided gradual change by survival of the fittest individuals by focusing on another level of hierarchy of macroevolution by species. These two leaders of the evolutionary school first formulated this view in the early 1970s (Eldredge, 1971; Gould and Eldredge, 1971; Eldredge and Gould, 1972; Gould and Eldredge, 1977). The following are Gould's sayings:

> Punctuated equilibrium does not signify macromutational transformation *in situ* but an origin of the later species from an ancestral population living elsewhere (as a small population peripherally isolated at the edges of parental range) followed by migration into the local region.[7]
>
> It is now 'promoted' by reproductive isolation to full separation.[8]
>
> It has extensive consequences of its key implication that conventional mechanisms of speciation scale into geological time as the observed punctuations and stasis of most species, and ***not as elusive gradualism*** ...[9] (Emphasis added).

The change is from those forces that characterize other phenomena at other scales, and catastrophic mass extinction triggered by bolides (meteorite type) impacts, for example. These take in different mechanisms responsible for similarities in the general features of stability and change across nature's domain. Gould mentions that while he and Niles Eldredge speculated and developed the theory, he worked with Elizabeth Vrba on speciations. The authors never show how forces cause the punctuated change into the new species, avoiding proof of mechanism. Gould names three circumstances involved:

> New species arise rapidly, usually instantaneously which amounts to a blink of the eye of geological time, and is therefore imperceptible.
> It occurs in a small geographical region, which is peripheral and isolated (not seen).
> It would at origin be elsewhere in an area paleontologists were not collecting fossils. Hence there would be no record observed.

Gould has said,

> We dimly recognized, in short, that if species act as stable units of geological scales, the evolutionary trends—the fundamental phenomenon of macroevolution—could be conceptualized as results of a "higher order" selection upon a pool of speciational events that ***might occur*** at random with respect to the direction of a trend. In such a case, the role of species in a trend would become directly comparable with the classical status of organisms as units of change within a population under natural selection. We wrote (1972, p.112)
>
> A reconciliation of allopathic (going in many unintended directions of) speciation with long-term trends can be formulated... We ***envision*** multiple... invasions, or a stochastic (random chance)

---

[7] Ibid., 748 with 779.
[8] Ibid., 780.
[9] Ibid., 779.

> basis, of new environments by peripheral isolates. There is nothing inherently directional about these invasions. However, a subset of these new environments ***might…lead*** to new improved efficiency…. The overall effect ***would*** then be one of not apparently directional change: but, as with the case of selection upon mutations, the initial variations [species] ***would be*** stochastic with respect to the change [trend].[10] (Emphasis added)

This was admittedly an entirely imaginative speculative theory with words like "might," "could," and "would be." These changes supposedly happened, not with minute acquired characteristics, but with a group species on an unknown edge of a population. But at the time, Gould and Eldredge envisioned no reason why these fringe individuals would be isolated or splintered off, what unknown force could be the instant cause, and why it might result in such significant biological change. On the other hand, they claim no directional or designing intelligent force for guiding the change so that a new and higher species would occur, but claim that ***direction by chance did occur***. They also give no explanation of why these forms would arise and become immediately stable and identifiable, and why in small numbers they would be superior to the population from which they emerged and, therefore, survive. Moreover, these imaginary new group improvements are not identified.

Gould and Eldredge have been able to offer no observable proof that this ever occurred as Douglas Futuyma and other scientists commented. Moreover, this theory involves a powerful ***mysterious or unknown force***, which motivated this isolated, quick evolution, and has left no trace of how it produced emergence meaningful to higher forms. While good science allows the use of imagination, it only becomes science when it is supported by observable data as proof. This was not Darwinian gradualism as other prominent scientists recognized. E. M. Stanley was one who recognized the departure from Darwin and approved.

Gould makes it plain that E. M. Stanley was correct in that this macroevolutionary theory "***decoupled***" from the forms of Darwin's inherited characteristics, neo-Darwinistic mutations and all other micro evolutionary forms.[11] These critics rightly saw this theory as ***an entirely new theory and a break from Darwinian thought***. The authors dogmatically denied such a break, while approving Stanley's insights. It was really a forsaking of the evolutionary theory as it had been known and taught.

Gould's and Eldredge's "punctuated equilibrium" might be added to Smith's list of the theories claimed to be true ***from non-evidence***. The punctuations occur in a small, isolated, unknown place that would be unrecorded; they occurred so quick in geological time that there would be no evidence; the theory cannot be tested in a laboratory, and the only evidence is that new/previously unknown, distinct, enduring stasis of species are said to occur. But punctuated equilibrium never argues for creation of new kinds of plants or animals. Only new forms of shells, new species of similar plants, and new species in the elephant family are presented as evidence. However, most of these "new" species were, in time, shown as only genetic deviations of the species that had been known. Their theory deviated from known biological laws to an imaginary theory to escape proven laws. This is similar to efforts in other areas of science (Hawking's black holes, e.g.) which departed from scientific fact to imaginary ideas to claim improvement.

Vrba's Addendum to Punctuation by Pulse Theory

The current, popular version of the non-evidence theories is Elizabeth Vrba's "pulse theory." Vrba, from Yale University, had assisted Gould and Eldredge in formulating their "punctuated

---

[10] Stephen Jay Gould, *The Structure of Evolutionary Theory,* 714, 715.

[11] Ibid., 715.

equilibrium" theory. Her view, an addition to punctuated equilibrium, supplies the missing force. [12] This hypothesis holds that there have been curious evolutionary bursts (which leave no fossil record) that are driven by a radical shift in *global climate changes.* These changes are caused ***by cataclysms***, especially by radical shifts in tectonic plates in the earth's outer shell, which cause a change in world temperature. Vrba sees three such geological catastrophes, which caused species to be destroyed or created. One occurred 5 million years ago, another 2.5 million, and a third 900,000 years ago. This view is a ***departure from geological gradualism as a part of uniform determinism that rejected catastrophe.***

Vrba says that as a result of these catastrophes, life forms had three ways of confronting them. She explains this in terms of *Hindus' pantheism*. She said, "The Hindus believe in a triad with three deities—Brahma the creator, Vishnu the preserver, and Siva the destroyer." She asserts that species follow one of these three paths, some taking the creative path of Brahma, others splitting off into new, better-adapted species, and others going into extinction.[13] The creative path is that of macroevolution. Vrba published her 24-page thesis in 1980, and it has become increasingly popular.[14]

While this has the semblance of science, in that it studies geologic evidence of great catastrophes, it ***has no basis in fact as the mechanism*** which causes evolution of new forms. Why would sudden heat or cold cause jumps from one species to another? This never happens in a laboratory or genetic censors that control species development and prevent macroevolution. Moreover, the ***average recorded change in temperature*** in successive periods of the earth has only ranged ***from about 50 to 73 degrees***, which is not that great. Also, there are no intermediary forms to show it happened, and there is no reason for *a purposeful change to higher forms,* except to attribute this to *miraculous power of design in nature, or* ***pantheism***. This is occult or New Age religion. It is paganism. It is, again, an argument from no evidence. It is catastrophic, and it is contrary to Darwin's scientific principles as well as Lyell's geological principles.

The pulse theory is attractive to Gould and others who hold that evolution is motivated *from without for changes within* and on the fringes of populations. It also appeals to Vrba who sees the cause *from without*, since their theories and all other current theories depend on quick periodic jumps that ***leave no record*** of life in fossils. In addition, creation by the pulse theory goes against the paleontological evidence that natural ***catastrophes are known to extinguish species***; not create them.

Disagreement on Evidence of Catastrophes Creating/Changing Life

Geologists argue over the source of the pulses. Some, such as Alfred Fisher and Michael Arthur, propose a pattern of mass extinctions at intervals of about 32 million years. More recently, David Raup and John Sepkoski of the University of Chicago, based on compilations of the stratigraphical occurrences of families of marine organisms, have suggested periodicity of 26 million years. But there is no evidence that pulses occur at those intervals, because at points where there ought to be a pulse, there is no evidence. Moreover, at which periods did the pulses occur? At 32 or 26 million-year intervals? A six-million-year difference is quite a lot. The whole idea of periodicity is still in question. Those who hold to periodicity link the pulses to an astronomical

[12] Gould dedicated his 1443 page work, *The Structure of Evolutionary Theory*, to the three as compatriots in working together.

[13] Ellen Ruppel Shell," Waves of Creation," *Discover*, Vol. 14, No. 5, May 1993, 54-61.

[14] William F. Altman with Betsy Wagner, "Climate and the Rise of Man," *U.S. News & World Report,* June 8, 1992, 60-67.

event, such as a comet, which *might be* triggered when the solar system ***might*** go through the spiral arms of the Milky Way. All of this is ***speculation***, and the data on which it is based is unclear and unscientific. Moreover, it is ***entirely catastrophic in nature***, which is not known or understood, and ***against uniformitarian modern geology***.

Many paleontologists now seem to lean toward the idea that earth plate tectonics caused climatic changes as the continents broke up and drifted. This, of course, is also speculative, even though some data may seem to point this way. The reason for leaning toward this view, instead of the astronomical periodicity one, is that an astronomical one does not allow enough time for the species eradication, while glaciation by plate tectonics seems to allow for this. Also, no regular periodicity can be tracked in the fossil and geological record.

The fact is that there is no conclusive evidence that new forms originated through pulses or anything else. This is seen in scientists' diverse views of how higher species would have emerged. If there was clear evidence, there would be a consensus. There is not. But Vrba links the ***creation of new species*** to events that also lead to the ***destruction of species***. About this aspect of the pulse theory, other evidence will be presented later that will make it even clearer that the pulse theory does not produce evolution.

Richard Dawkins' Theory of Evolution by Cumulative Chance Selection

Crisis Point of Man's Theory of Gradual Development

The integrity of the western intellectual elite was based on the false enlightenment worldview of a uniform determined universe. The data supporting that view was collapsing in the last of the twentieth century. Richard Dawkins, professor of zoology at Oxford, in his book, *The Blind Watchmaker, (1987 ed.)* sought to rephrase Darwin's theory as a basis for the whole universe, not just for biology. All of the knowledge of the universe had turned out to be a great mystery, beyond understanding. Especially in the field of biology, the complexity of molecular biology showed consistent genetic reproduction, revealing distinct separation of all species of life from the micro to the macro level. Michael Denton, a committed naturalist and evolutionist, critiqued Darwin's theories in his 1985 book *Evolution: A Theory in Crisis.* Among intellectuals the Intelligent Design Movement was in beginning discussions as a result of biological findings.

Dawkins' purpose was to argue, "Why the evidence of evolution reveals ***a universe without design***." He tied the whole developmental theory of the universe to his own interpretation of the theory. He aims at scientifically demonstrating the factual operation of Charles Darwin's theory, and that the method of the theory is free from intelligent designing purpose. He does not deny there is amazing design in nature, but he is arguing that there was not ***purposeful*** design by any divine power, pantheistic in nature, or by a transcendent God. Design, as he sees it, is completely by accident and can only be ***seen after the fact***. Thus the title is *The* ***Blind*** *Watchmaker*. Even Kant, who started the evolutionary theory, had a beatific vision of nature doing the designing.

Dawkins took his cue from a fellow Oxford physical chemist, Peter Atkins, who is a thoroughgoing uniform scientific determinist, in spite of all the world's mystery. Atkins claimed to present a journey of the mind about which he said, "On it I shall argue that there is ***nothing that cannot be understood***, that there is nothing that cannot be explained, and that everything is extraordinarily simple."[15] (Emphasis added). Dawkins said that Atkins, as a physicist, is faced with the explanation of the problem of "of ultimate origins and ultimate natural laws." He then

[15] P. W. Atkins, *The Creation*, (Oxford: W. H. Freeman, 1981).

says, "The biologist's problem ***is of complexity***."[16] He claims there is a distinct difference in simple things like pebbles, rocks, dirt, etc., and ***living things*** that the biologist deals with. Dawkins is committed to scientific determinism and upholding it in the realm of biology in such a way as to give credibility to Charles Darwin's theory. He is aware of the conclusions and theories that threaten, and he seeks to restate Darwin's theory.

Brilliantly, Dawkins seems to understand that the theory of evolution in every modified form had failed and that uniform scientific determinism had failed to erase the possibility to interpret evidence for a transcendent God who created and sustains the world. The probability of accountability to God should be removed. Moreover, Dawkins saw that to ***give up Darwin's theory is a terrible admission*** that unbelieving scientists were wrong in accepting evolution from the beginning, and in dogmatically rejecting creation. So Dawkins created a new theory which appears to do what Darwin's claimed to do. He presented this as the theory of ***Darwin gradualism***. Dawkins was familiar with all the views and restraints. In his theory he seeks to give some validity to each of the various ideas mentioned above in a way that will save face for all evolutionists, especially Darwinian evolution.

He realizes the concept of biological evolution is logically, highly improbably in the light of modern complex mystery, which he knows is extreme, beyond human imagination. The statistical improbability of biological life originating and evolving into millions, possibly hundreds of millions of distinct species, with such amazing complexity, by chance, without innate design or directional force is infinitesimally small. Dawkins ***insists*** *no design or mystical force can be involved* and said,

> What I do care about is improbable in a direction specified without hindsight, it as an important quality that needs a special effort to explanation. It is the quality that characterizes biological objects as opposed to the objects of physics. The kind of explanation we come up with must not contradict the laws of physics. Indeed it will make use of the laws of physics, and physics in a special way that is not ordinarily discussed in physics textbooks. That special way is Darwin's way...***cumulative selection.*** [17] (Emphasis added).

Cumulative selection is his key idea. Dawkins insists that he is truthful, which we interpret to mean he is not deceiving, or distorting, or hiding facts, not contradicting what he says, and is presenting things as they may be observed to actually occur. Since he claims this to be modern science, this involves (1) ***Observation of what occurs***. (2) A ***logical theory*** of cause and effect based on the observation. (3) ***Testing for repeated regularity*** to see if the observed data actually supports the logical theory. While there are new philosophies of science, these elements are still recognized. Historically, these are the scientific criteria for evaluating all data for truth, which will be applied to Dawkins' claims about his theory. If his purpose is to exclude ***intelligent design and power***, to be truthful, Dawkins ***can't include this in any way***. No intelligent designed effort can be added, or his theory is false. He is arguing for an ***accidental progressive addition of informational complexity***.

The Criteria of Normal Human Logic

Is Dawkins' claiming a ***logical*** theory? In the preface and first chapter Dawkins states that his explanation of Darwin's theory requires a different way of thinking, not normal logical thinking. Here are some of Dawkins statements.

[16] Dawkins, *The Blind Watchmaker,* (New York: W. W. Norton & Company, Inc. 1987), 14, 15.
[17] Ibid., 15.

> Our brains are built to ***deal with events radically different***….
>
> All of ***our intuitive judgments of what is probable turn out to be wrong*** by many orders of magnitudes.
>
> I want to ***inspire the reader with a vision*** of our own existence as, on the face of it, is a spine chilling ***mystery*** and simultaneously to convey the full excitement of the fact that it is a mystery with an elegant solution.
>
> …a very ***large leap of imagination***.
>
> See that ***contrary to all intuition***, there is another way.
>
> Complicated things deserve a very ***special kind of explanation***.
>
> We find it convenient to ***use words in different senses***.
>
> [There are] answers that question in general terms, even ***without being able to comprehend*** the details. [18] (Emphasis added).

These statements indicate Dawkins is saying I am going to ***let the reader in on a special kind of knowledge*** that only brilliant people see. Is Dawkins really trying to present science that fits normal logic?

This is a deceptive technique, often used to brainwash people into thinking they are joining an elite group. Throughout the centuries, cults, secret societies, even the ancient Gnostics promised to give special insights—beyond logic. If the teaching is true, why must there be different ways of logic? This can hardly be science that has a logical explanation about what is observed. In other areas of science, when observed facts of science do not fit a policy, there has been a tendency to violate the laws of science with new unscientific theories to deceptively promote an idea.

Dawkins repeatedly insists there is no designing purpose in Darwin's theory of evolution. But is Dawkins honest about this? He states that he is basing his theory on Darwin's ***original edition*** of *The Origins…*, and not the sixth and final one. This is highly important because Darwin admits that the final edition was responding to his critics who pointed out the weaknesses in his theory. A final edition of an author ought to be the most definitive. Loren Eiseley pointed out that in the final edition of *The Origins...* Darwin makes the statement that concedes to necessary progression with a designing purpose. Darwin said, "All corporal and mental endowments ***will tend to progress toward perfection***." (Emphasis added). Eiseley, an evolutionist, points out this was fatal to Darwin's theory, resulting in support for the occult power of design in nature that he had rejected in Lamarck and others.

As a prominent zoologist, Dawkins had to openly confront the extreme magnitude of complexity that modern biology had revealed. Dawkins refers to W. Paley's argument about the complications of the eye, and even goes beyond this to concede that modern science has shown that the eye is exceedingly complicated far beyond what Paley knew.[19] But Dawkins does not mention Darwin's concession to designing purpose in speaking of the eye. Perhaps because of Darwin's concession, Dawkins chooses the eye to illustrate he is not ignorant of that magnitude of

---

[18] Ibid., x, xi, xii, 1-3, et, al.

[19] Ibid., 15-18.

complexity when he claims non-directive, accidental evolution is extremely amazing and a mystery. But a mystery that he claims has an "easy solution."

Dawkins' "Easy Solution" of Cumulative Chance Selection

The key to Dawkins' book and theory is chapter 3, "Accumulating Small Change." To demonstrate Darwinian evolution, Dawkins suggests "***cumulative chance,***" rather than Darwin's "simple chance." His answer to complexity is to demonstrate how "cumulative chance" operates with the intent of minimizing the great magnitude of probabilities. He presents his theory of cumulative chance in slow steps, without purpose or design. He presents the theory first by illustrating the idea of accumulated chance with limited probabilities, design, and purpose; then with disguised design and more limited probabilities; and then he addresses the high probabilities with a theory that he claims has similar elements in nature, as he has demonstrated with only cumulative chance. He identifies this with Darwin's view. Later in the book he changes this to address the more radical, improbable evidence in the data that contradicts his theory. The following evaluation is to see if Dawkins' theory conforms to consistent logic, based on data observed (not suppositions or assertions to fit the theory), and if it fits the facts of science in observed testing.

Cumulative Chance Selection as a Way of Reducing Probabilities

To impress the reader with his awareness of the tremendous complexity that has been discovered about living beings, Dawkins addresses the example of hemoglobin, with four chains of 20 different amino acids and 146 links that must be perfectly aligned to function. This offers unimaginable probabilities of simple chance arrangements. Dawkins reports there is one chance in a number with 190 zeros after it to form hemoglobin. The number trillion has only twelve zeros/nougats (compared to 190), and this much larger, mind-boggling number is hard for the average person to conceive. This is only one of the many other complicated molecules in nature. The difficulty of probability of hemoglobin is hardly even suggested as to its extreme magnitude, He then submits, "If a monkey bangs on a typewriter long enough, he would eventually produce the entire works of Shakespeare by accident." This too is dismissed as very improbable, but a lesser problem. Dawkins jokingly proposes that one consider the reduced probability of the monkey producing the one line of ME THINKS IT IS LIKE A WEASEL, by using 27 possible letters, including the space bar, to put together the sentence of 28 letters. He points out the simple, chance possibility of even this would be one in ten thousand million, million, million, million, million, million. That too is mind-boggling, but less of a challenge. With this the reader is persuaded that ***simple chance*** could not produce a world or ***life*** by accident.

With this mental preparation, Dawkins then offers "cumulative chance selection" as a viable alternative. He reports that he designed a program for his computer (which is a highly ***complex, designed*** machine) to do cumulative chance selection—that is, when the monkey hits any letters that go together ***it will save those for the monkey's next bang***, thus accumulating more progressively toward the designed purpose. Dawkins reports that with 30, 40, 43, and 64 runs of the computer, in a short time the sentence is quickly arrived at. Thus, he persuades you that cumulative selection by chance is a feasible way to overcome improbabilities, which he says is "quintessentially non random."

This is a flawed illustration. Why would the typewriter begin with the few correct letters the monkey hit the first time? That would not happen. It only happened because an ***intelligent being*** wrote a program for a machine. If the tiger, by mutation, inherited sharp teeth, why would his offspring inherit another progressive step for survival? For such accumulation to repeatedly continue to work would be extremely illogical.

However, Dawkins then confesses that this cannot be the way "nature works," because he used the objective purpose or end of the program—the sentence ME THINKS IT IS LIKE A WEASEL. He has dogmatically said nature has no objective, purpose, or target in its work. The progress, from nothing to small and simple life, up to man, was an accident of nature by "cumulative selection of chance." But Dawkins has, with this example, sought only to persuade the reader that cumulative selection by chance is a very viable way. What has Dawkins really done? He has illustrated the power of very small ***purposeful design*** in action. He has demonstrated how one small mind of a man (Dawkins), with his intelligence, (which he does not know how or from where it was derived) has created by his own design, a program to make a computer that man has creatively designed, and produced by accumulation one silly sentence of only twenty eight letters. He presents this as if a dumb animal or monkey, from whom he claims we descended, accidentally banged out on a typewriter. He has only ***really shown*** a small ***example of how a creator (himself) can logically achieve a purposeful end***—the exact thing he is trying to deny. He has not, in any way, shown how the universe and life works with all its extreme complexity, which he admits is beyond imagination.

Progressive Accumulation Identified with Embryology and Genetics

Continuing his argument, Dawkins says he creatively designed another program (for his sophisticated, intelligently-designed computer) based on simple numbers from one to nine. He designates these numbers as 'genes'. It is important to note that the new program significantly reduced the probability from the monkey program (from twenty-eight to nine) involving millions less nougats or zeros. He designs this new program because he says, "We must have something equivalent to embryonic development, and something equivalent to genes that can mutate."[20] Genes are living and complex, numbers are not, but the reader is led to think of them as genes, working as embryos, and therefore the way life works. However, Dawkins concedes that this is not what is known about nature, but is an "allegoric" or "metaphorical" way embryology works. He admits the real way these living entities work is "too elaborate to simulate."[21]

Dawkins' design is to "have a simple picture drawing rule that the computer can easily obey and which then can be made to vary under the influence of the 'genes.' "[22] One number ('gene') controls the angle of drawing a branch, another the length of a particular branch, another depth of recursion or succession of the drawing, and so on. Nine positives and nine negatives are said to "***represent all possible*** single step mutants" derived in the tree of life. Dawkins says, "You will notice that all the shapes are symmetrical about a left/right axis. This is ***a constraint that I imposed*** on the development procedure"[23] He then instructs the computer to follow the program, and draws figures until some have a minimal similarity to insects, imaginative scorpions, etc. To many figures that emerge he gives silly names like "Swan tail, man in hat, lunar lander, precision balance," etc., which have nothing to do with living beings or have never been seen in nature.[24] Dawkins later stops calling the figures "twigs" of a tree and starts referring to them as "biomoths," which he says will help us think of them as animals. About the drawings, he says, "these are arbitrary conventions, they could have been different and still remain biologically realistic." In other words, these are his design choices, made by computer, while, as the creator he could have

[20] Ibid., 51.
[21] Ibid., 51.
[22] Ibid., 51.
[23] Ibid., 55.
[24] Ibid., 61.

chosen otherwise. Dawkins designates the branches or the 'genes' to progressively draw new changes. The 'genes' therefore ***do not reproduce faithfully as genes*** really do. Thus, his program aims at repeated cumulative change and he labels these repeated changes called "mutations." But Dawkins admits "The very high mutation rate is a distinctly unbiological feature of the computer model. In real life the probability that a gene will mutate is often less than one in a million."[25] He further says, "In the computer model...the selection criterion is ***not survival***, but the ability to appeal to human whim... This too is not unlike certain kinds of natural selection. The human tells the computer which one of the current litter to breed from."[26] He, therefore, admits the unbiological nature of his program, and that it does not really indicate anything about survival of the fittest. This ***seeming honesty disguises his intent to deceptively convince you*** that he is claiming this is really the way nature works.

Certain factors need to be pointed out. This second program is not a chance, devised program to illustrate cumulative ***chance selection***. It is a program ***designed by human intelligence*** (his). The design is for the one purpose of showing cumulative effect, and it is ***produced by a complex machine, designed by human intelligence, and admittedly not by any kind of chance.*** While the impression is given that there is no target or purpose, Dawkins knew, planned, administered, and admits the program was controlled to proceed according to 'his constraints,' produce figures that would progressively differ, and at times suggest more and more complex living things to the viewer. If Dawkins had been smart enough to know the end of the program from the beginning he would not have had any surprises as to what was produced, but because of his intelligent limitations he did not foresee ***the targets that would have and had to emerge because of the preceding design***. This is to deliberately give the impression that the program is really not designed and not controlled except by the multiplication shown.

Dawkins' program is designed with very low probabilities and complexity (only nine 'genes') when compared to the working of real life information systems, such as hemoglobin. Moreover, in the program the 'genes' are not repeatedly doing what genes do—reproducing themselves, but he has these so-called 'genes' always producing new things, called 'mutations.' The supposed 'genes' cause drawings until they produce new figures that are supposedly a new individual. There are no corrections for any errors, though he admits the corrective power of real genes is near perfection. Later, Dawkins admits that real life genes are faithful, and almost undeviating in reproducing the previous species. That is not the case here. Dawkins' apparent objective was to persuade one to believe that slight, random accumulations will produce new species repeatedly and quickly. He insists elsewhere and repeatedly, a very long time is needed for evolution to occur, even by cumulative selection. Many readers come away with the idea that he has demonstrated that cumulative chance selection happens in real life.

What has Dawkins really done? He has, as an intelligent ***creator, designed*** a program to draw lines to produce imaginative, silly images that must, by way the program is designed, become increasingly complex. But the complexity is not by cumulative ***chance,*** and nowhere in the magnitude of real, living biological creatures. And by his own admission, his analogy does not function like real life. He has demonstrated ***nothing*** that has to do with what living genes in embryology do, and, in fact, is contrary to what happens. Has Dawkins given "a simple solution" to how evolution works, or is this a false "analogy" driven by a bias against God?

Dawkins then identifies his drawing analogy with his ideas of evolution. The genes continue to reproduce; with each reproduction the bodies that are produced differ, in random directions.

---

[25] Ibid., 55, 56.
[26] Ibid., 57.

Thus, the genes develop mutations or errors which they consistently pass on to the next generation. The survivability of the individual is in the errors in the body that enables it to survive. The "better" body, therefore, preserves the genes that have accumulated mutations, which then reproduce other better bodies. But it was shown that in real, living body mutations do not pass on changes of strength as Darwin thought, (reproduce acquired characteristics). Yet, Dawkins claimed the mutated ***genes are preserved,*** and thereby pass on better bodies, which reproduce the mutated genes. This was intended to be what was illustrated in nine numbers (called 'genes') in the computer drawing program. This is a dishonest claim drawn from his drawing illustration, because the intelligent creator (Dawkins) is designing, and directing. That designing power is exactly what Dawkins denies to nature.

Dawkins' computer program distorts not only the frequency of reproduction, but also the probability. The greater problem with his theory is that his claims for mutations ***not what scientists have observed in nature***. The consensus among biologists is that mutations have ***no effect on producing survivability,*** as has been shown in the data. The evidence is that neither acquired characteristics, as Darwin thought, nor mutations, as the neo-Darwinist thought, produce intermediate, gradual species change. So the effort to link reproduction and development as "cumulative selection by chance" has not in any way been demonstrated to be true.
Dawkins concludes his drawing analogy with the following comments:

> What is all this telling us about real evolution? Once again, it is ramming home the importance of gradual, step-by-step change. There have been evolutionists who have denied that gradualism of this kind is necessary in evolution. ...If genuinely random jumps really occurred, then a jump from insect to scorpion would be perfectly possible. Indeed it would be just as probable as a jump from insect to one of its immediate neighbors. But it would be just as probable as a jump from to any other....[27]

This is a very confusing statement. On the one hand he is insisting on ***gradualism like Darwin***, and on the other he says the probabilities are equal, if it is not gradual or slight and even extreme jumps. This is mathematically illogical, but he says this because he knows the facts do not support gradualism, even with his contrived "accumulated chance selection." He plans to resort to jumps later. He insists that gradualism can be traced in the history of real living beings,[28] a statement he tries to tie to facts, while knowing gradualism is not in the fossil history.

Transfer His Cumulative Chance Selection to Darwinian Gradualism

In chapter 5, "The Power and the Archives," Dawkins seeks to transfer this idea of cumulative chance to the real life idea of Darwinian natural selection. He claims that real life has the same properties of cumulative change as his designed program. Jenkin's argument of population swamping is dispelled by misrepresenting his argument. Dawkins says Jenkin's idea is like mixing black and white paint and getting grey, which can't be corrected by adding more grey. He then turns to the unanswered problem of life's origin, and suggests that RNA may be the answer. He tries to tie his "accumulated chance" to Darwin, using speculative descriptions. He illustrates how, if tigers inherited sharper teeth, then the next generation of tigers would inherit sharper teeth, which continue to let them survive. Dawkins gives no demonstration from observation of what has happened in real biology, living or fossil, but gives only speculations about what "must" have, or "probably" happened if accumulative selection by chance were to

---

[27] Ibid., 72.
[28] Ibid., 73.

work. For what he claims to occur, he argues that nature "***has to" or "must"*** do certain things. The argument is not that we observe nature doing these things. Rather, Dawkins gives the wish for what he wants to happen as proof of the facts for his theory. For example,

> At some point in either our ancestry of chimps ***there must have been*** a change in chromosome number.... There ***must have been*** at least one individual who had a different number of chromosomes from his parents.[29] ...The same ingredients... ***must have*** arisen spontaneously...otherwise cumulative selection, and therefore life, would never have started in the first place. ...There ***must somehow,*** as a consequence of the ordinary laws of physics, come into being self-copying entities... replicators. ...We ***may suspect*** that the first replicators of the primitive earth were not DNA molecules. It is ***unlikely that*** a fully fledged DNA molecule ***could*** spring into existence without the aid of other molecules that normally exist only in living cells. ...It **seems likely** that the first replicators on Earth were much more erratic and at least some of the replicators ***should*** exert power over their own future. ...Each new copy ***must be*** made from now materials, smaller building blocks knocking around.[30] (Emphasis added).

Thus the first replicators and early building blocks are ***all speculative,*** as well as how they came into being. Dawkins claims that "accumulative selection" is constantly happening, as in his drawing program, and in such a way as to avoid population swamping.

> The nonrandom survival and reproductive success of individuals within the species effectively 'writes' improved 'instructions' to survival into the collective genetic memory as species as the generations go by. ...The statistical profile of location contents (in the population) changes as the centuries go by. [31]

His writing program analogy is based on the random "mutation" rate. But he says "evolution by natural selection could not be faster than the mutation rate." He then concedes "a conservative estimate is that, in the observed natural selection, DNA replicates so accurately that it takes five million replication generations to miscopy 1 percent of the characters."[32] He also adds "the DNA copying mechanism does...error correcting automatically. ...About 5,000 DNA letters degenerate per day in every human cell and are immediately replaced by a repair mechanism."[33]

If mutations were half-good and inherited, would the one billion years of fossil history which Dawkins refers to, allow time for the origin of life and all the species to gradually evolve, even if there was such a mechanism as "accumulative selection"? He therefore adds ***another imaginary speculation***, that the good mutation may have a "***sticky***" quality to allow more survival.[34] There is an admission that real life survival stories are hypothetical.[35] Moreover, Dawkins never clearly addresses the definition of "better" or "fittest" that allows a species to survive. Survival in nature demands different qualities. In some cases it would be better to be a big strong animal, but there would be the alternate that a small animal might survive because it was small and could escape in a hole in the rocks. There are many ways this idea can be exploited.

Dawkins' speculations are not restricted to the possibilities on earth. In his chapter, "Origins

---

[29] Ibid., 119.
[30] Ibid., 128, 129.
[31] Ibid., 119.
[32] Ibid., 125.
[33] Ibid., 126.
[34] Ibid., 130.
[35] Ibid., 136.

and Miracles," he extends it to involve the speculation of life as having begun in one of possibly 100 billion billion planets. He forsakes his early theory of origins in his book, *The Selfish Gene,* for that of Cairns-Smith, that life began by crystals from which RNA develops. All of this is to extend the opportunity for chance to work. This is only wild, deceptive speculation.

The discussion on numerical probabilities is postponed to a later chapter in which origins and the new information theory are joined to the new, second law of thermodynamics. The review of these things makes it clear that Dawkins' theory is speculative, and not really science.

In complete contradiction of Dawkins' main arguments for Darwinian gradualism, he concedes to the actual findings of the fossil record as showing an initial biological explosion, followed by at least two major ***catastrophes.*** He also concedes to ***stasis*** and ***sudden appearance,*** as do most biologists, and, therefore, to the lack of evidence of gradual transmutation of the species. But he does not fully accept the theory of punctuated equilibrium. Instead he offers an undocumented theory of genetic laws being different, and allowing early interbreeding to cross from one species to another. For this there is no fossil evidence. Gould and Tattersall are more honest with the data. Dawkins' answers, "of all conceivable evolutionary pathways, only a minority actually ever happened. We can think of most of this tree of all possible animals as ***hidden in the darkness of non existence."***[36] (Emphasis added).

The conclusions of Dawkins' learned book is to save Darwin's evolutionary theory. He said, "If there are versions of the evolutionary theory that deny ***slow gradualism***, and deny the central role of natural selection (of Darwin), they may be true in particular cases. But they cannot be the whole truth for *they deny* ***the very heart*** *of the evolution theory."*[37] (Emphasis added). But for all of his efforts to present a view of "cumulative chance selection" in contrast to the simple chance view of Darwin, he still must conclude non-evidence and a mystery of great improbabilities. Dawkins thus ended up claiming there ***is no evidence for evolution***. By admission, he joins all other desperate views to uphold Darwin's theory that never had evidence to begin with. He says, "Let us hear the conclusion of the whole matter. The ***essence*** of life is ***statistical improbability on a colossal scale.***"[38] (Emphasis added).

Conclusions of Efforts to Demonstrate Darwinian Evolution

After over 150 years of efforts to find evidence of how Darwin's concept and all other efforts to know how evolution of life occurred, there is no scientific support. The latest efforts seem to be by so called 'wet' biologists with environmental efforts to change the definitions involved. Two evolutionary atheists, Jerry Fodor, a philosopher and cognitive scientist of Rutgers University and Massimo Piattelli-Palmarini, a biophysicist, molecular biologist, and cognitive scientist at the University of Arizona have recently reviewed all efforts of study and published in 2010-2011 *What Darwin Got Wrong.* In their last chapter of summary they said,

> 'Ok; so if Darwin got it wrong, what do you guys think is the mechanism of evolution?' Short answer: we don't know what the mechanism of evolution is. As far as we can make out, ***nobody knows*** exactly how phenotypes evolve. We think that, quite ***possibly***, they evolve in lots of different kinds of causal routes to the fixation of phenotypes as there are different kinds of natural histories of the creatures whose phenotypes they are (see previous chapter).[39] (Emphasis added)

---

[36] Ibid., 315.
[37] Ibid., 318.
[38] Ibid., 317.
[39] Jerry Fodor, Massimo Piattellie-Palmarini, *What Darwin Got Wrong* , (New York: Picador of Farrar, Straus and Giroux, 2011), 153.

Chapter 14
The Impossibility of Chance Origin of Life

Origin of Life Not in Original Evolutionary View

Charles Darwin did not himself extend his theory of evolution to the origin of life. Darwin later wished that he had not used the word *origin* in the title of his book. His friend, T.H. Huxley, who added the idea that life began by chance, admitted there was no evidence for chance origin of life. But, referring to what he as an observer might have seen if present at the beginning, Huxley said, "I should expect to be a witness of the evolution of living protoplasm from not-living matter."[1] So the idea was a philosophical speculation made by Huxley and accepted by evolutionary scientists ***without any scientific data*** because it was essential to a view of evolution by chance. Molecular science did not begin until after 1940 so this was seventy five or eighty years before any real understanding of how life functions.

In 1864 Louis Pasteur decisively ended speculation on the spontaneous generation of life and exposed it as a myth. Haeckel and German critics had held to spontaneous generation of life and did not want to accept Pasteur's findings simply because they said the only other alternative was special creation. But at the time, most of the scientific world did accept Pasteur's proof. However, about a decade later, uniformitarians resurrected spontaneous generation by adding the ingredient of long periods of time, which were the speculations of Herschel, Hutton and Lyell. By evolutionists calling it by another name they got many to accept it. No one ever explained why time could change Pasteur's conclusive results. Even after proof was presented, man continued to hold to his bias believing what is essentially the same idea as spontaneous generation, namely creation of life from elements by chance.[2] Bias triumphed over science.

Theories and Efforts to Discover the Origin of Life by Chance

In the early twentieth century, A.I. Oparin of Russia and J.B.S. Haldone of England offered theories that life could be formed accidentally in a molecular soup in the universe and especially on the earth. Under the right conditions, matter could come together into the necessary compounds to form living structures. These, supposedly, formed into replicating cells, and after eons of time, various species emerged by natural selection.

This theory resulted in considerable research to prove that life could really originate in such an accidental manner. After many unsuccessful efforts by a number of scientists, Stanley Miller produced some simple amino acids from a molecular soup in 1953. From this, some efforts have been made to show that amino acids, under certain conditions, *might* organize themselves with electrical properties similar to cells. But nothing more happened, and it seemed the molecular soup concept had not proven correct.[3] Moreover, these were efforts by intelligent beings who were trying to facilitate creation.

Desperate Efforts Turned to New Ideas, especially RNA

As scientific determinism reached a crisis, unbelieving scientists began a vigorous effort to find a theory that would explain the beginning of life. A. G. Cairns-Smith, a chemist at the University of Glasgow, proposed life came from "low-tech" clay crystals. Gustaf Armenious suggested that minerals such as iron hydroxides (rust-like compounds) might be likely candidates

---

[1] T.H. Huxley, "Biogenesis and Abiogenesis," Collected Essays, 1894.
[2] Robert E.D. Clark, Darwin Before and After, 15.
[3] Science News, June 30, 1984, 408.

for life's origins.[4] None of the new theories showed promising results.

The discovery in 1980 of anaerobic bacteria, which thrive at extremely high temperatures or hyperthermophilic prompted some to suggest these locations might be the source of life. Scientists began investigating the oceans where molten rock enters fractures in the sea floor, some 3,000 geysers and hot springs in Yellowstone National Park, and in simulations in the laboratories of many universities. But these bacteria have also proven stable and reproductive of their kind. Others have argued that life began in ice.

In the following chapter we will take an in-depth look at genetic developments, but we will touch upon genetics briefly here. By the mid 1960s scientists began to understand the operation of DNA and RNA. By the late 1980s this became the focus of finding life's beginning. Currently, efforts have been made to link the origin with the action of RNA to possibly produce higher DNA, and then to develop higher bacteria. The key components of RNA and DNA are four bases that make up the genetic alphabet. In the 1960s, two of these, *adenine* and *guanine,* were synthesized in the laboratory. Recently Stanley Miller and associates synthesized *cytosine* and *uracil* in urea. Each of these bases is attached to sugar molecules of ribose or deoxyribose, in alternation with phosphate molecules surrounded by four oxygen atoms. But even Miller admits this phosphate ribose backbone is too unstable and won't produce life. Peter Nielsen, of Copenhagen, tried an artificial backbone of carbon and nitrogen to form a peptide bond (PNA). It is not believed that this could form spontaneously, and has shown no signs of the catalytic ability of RNA, so is admittedly a blind alley. Gene Levinson at Genetics & IVF Institute, Fairfax, Virginia, thinks that recombination of DNA sequences has a better chance of survival and evolution than simple mutations for producing evolution. But it is admitted that the survival ratios are small. Moreover, all these are *acts of human creative intelligence*, which contributes much information from outside, and not from chance.[5]

Richard Dawkins latched on to the experiments with RNA to try to link his theory of replicating natural selection to this. He drew on several experiments. Sol Spiegelman and his colleagues isolated two forms, RNA V2 and RNA Q-beta, which is a virus, and stored them. When these were united, they began to replicate. Dawkins hijacks the word replication from the process of repeating the molecule, and implies this is the same as "cumulative," of what he claims are modifications to perform natural selection. But he had nowhere shown that remembering, saving, and adding modifications or mutations (accumulation) has occurred in nature. That was his idea in his computer models. He proposes that the two forms of RNA "might" replicate at different speeds as different kinds. He speculates that his imaginary theory of "stickiness" would apply to one, so this would happen. He says,

> ***If for any reason*** the new variety (of ***accidental possible*** slightly different mutant of DNA) is competitively superior to the old one, superior in the sense that, ***perhaps*** because of its low 'stickiness,' it gets itself replicated faster or otherwise more effectively, the new variety will obviously spread through the test-tube in which it arose, outnumbering the type outnumbering the parental type which gave rise to it. Then when a drop of solution (of the superior RNA) is

---

[4] See "The First Organism," *Scientific American,* June 1985; Charles Thaxton, "Theoretical Clay Feet," *Eternity,* September 1985, 16; "Rusty Path of Life's Origin," *Science News,* March 17, 1987, 152.

[5] Carl Zimmer, "Life Takes Backbone," *Discover*, December 1995, 38-39; "Could Recombination Drive Evolution?" *Science News*, Vol. 146, July 30, 1994, 77.

removed from the test tube, it will be the new variety that does the seeding.[6] (Emphasis added).

Thus Dawkins radically proposes two *speculations that might happen to produce a* new superior RNA.

To this he adds an experiment by Leslie Orgel and his team in California, who added a poison, ethidium bromide to inhibit the replication of RNA and repeatedly increased its strength ten times. They found the RNA overcame each increase and was able to replicate. They called the new resistant V40 RNA. This did slow down the process. But this did not produce a new or different type RNA, as such. Dawkins also referred to Manfed Elgen in Germany whose team provided RNA building blocks in a test tube and produced a larger RNA molecule that reproduced itself. These are all speculations that "superiors" would be produced and survive if created by these experiments. But each superior survivor *is accumulated* according to Dawkins speculative theory. In addition, this is *not beginning life,* but using RNA which is a very complicated information-bearing molecule that already exists.

Subsequently, David Deamer, a biophysicist at the University of California at Santa Cruz, has chosen as his field the study of the origin of life. Concluding that for life to begin it *had to have* ***a cell wall***, which could take in nutrients, and craft them into genetic material and then expel excretions. He and his assistant, Chakrabarti, decided that lipids 16 to 18 carbons long, appearing in some microbes, might have formed the first cell membranes. By placing together selected RNA in a bath of loose nucleotides with two enzymes and two primers, it was hoped that all five molecules would get trapped in some of the liposomes and begin to produce a simple cell. His theory is that after "impacts of meteorites in the oceans billions of years ago, such substances could have washed up onshore in a tide pool, died, and then been rehydrated." Thus life from another planet might have come to earth and reformed.

There are several problems with life originating this way. One is that the meteorites have so far furnished nonionic acid with a chain of only *nine carbons* that have been found in membranes. A *cell membrane needs 16 atoms* to form bilayer lipids as in living creatures. Another big problem is that the lipids form a cell wall that has *too rigid* a separation from the environment. "A cell needs to pull in ions and toss them out all the time, so it overcomes its membrane's impermeability with intricate channels, pumps and shuttles. Swallowed by a liposome, a primitive genetic molecule would have been ***unequipped to manufacture channels*** through the membrane. The liposome would not be a shelter but a prison."[7] (Emphasis mine). Also, Deamer has observed about his possible living cell, "It'll be technically alive, but if we put it out to compete in any natural environment, something will eat it long before it has a chance to make its way up the evolutionary ladder."[8] The last fallacy is that Deamer has *selected* the kinds of things which must be put together in the way he conceived to produce life, using his observations of how life functions. His experiments are not the workings of chance, but the *efforts of a creator of higher intelligence*, namely his own.

In his previous book, Dawkins had postulated the beginning of life from a primordial soup[9]. But in *The Blind Watchmaker* he adopts Graham Carns-Smith's more current theory that life began from self-replicating inorganic crystals. Dawkins adds that "organic replicators, and

---

[6] Dawkins, *The Blind Watchmaker*, 132.

[7] Carl Zimmer, "First Cell," *Discover*, November 1995, Vol. 16, Number 11, 78, cf. 71-78.

[8] Ibid.

[9] Richard Dawkins, *The Selfish Gene*, (Oxford: Oxford University Press, 1976).

eventually DNA, must have taken over or usurped the role." He admits,

> The process still lacks a vital ingredient in order to give rise to evolutionary change. That ingredient is hereditary variation or something equivalent to it. ...Nearly all naturally occurring crystals have flaws. And once a flaw has appeared, it tends to be copied as subsequent layers of crystal encrust themselves on top of it. ...Once infected with the right sort of dust, a new stream starts to grow crystals of dam-making clay, and the whole depositing, damming, drying eroding cycle begins again. To call this a 'like' cycle would be to bet an important question, but it is a cycle of sort, and it shares with true life cycles the ability to initiate cumulative selection."[10]

Later he says, "Cairns-Smith discussed in more detail than I can accommodate here, early uses that his clay-crystal replicators might have had for proteins, sugars and most important of all, nucleic acids like RNA."[11]

Dawkins goes on to make a radical assertion regarding his superior opinion. He said,

> My personal feeling is that, once cumulative selection has got itself properly started, we need to postulate only a relatively ***small amount of luck*** in the subsequent evolution of life and intelligence. Cumulative selection once it has begun seems to me powerful enough to make the evolution of intelligence probable, if not inevitable.[12] (Emphasis added).

But Dawkins has found no facts that it has or can be started! He only speculates how it might, by what he admits, is the remote possibility it *might have gotten started.*

In the chapter, "Origins and Miracles," Dawkins argues we are "allowed a certain amount of luck...but ***not too much***." He says the amount requires more than logic can conceive. It must go ***"beyond what our earthbound"*** minds can conceive. He tries to argue that this requires probabilities beyond our imagination, based on his "cumulative selection" theory. [13] How much? He says the odds "depends whether there is life anywhere else in the universe. ...*But we have **no idea at all** whether there is life anywhere* else in the universe" [14] But he also adds, it depends also on "billions and billions of years" for chance to operate by his undemonstrated, "cumulative selection." He admits "it is entirely possible that our backwater of a planet is literally the only one that has ever born life"[15] Years of research have not found one other place in the universe that has the conditions for life. He then throws in another unknown—an estimate that it took a billion years for life to form on earth. He never mentions a similar time period would be required for life to form anywhere else. He estimates there is "an upper limit of 1 in 100 billion, billion planets and it would take a "billion, billion, billion planet years" for life to form by chance.[16] For Dawkins ***"this is not too much luck."*** But he admits it is way ***beyond our imagination***. But as we will show, even this exorbitant imaginative theory is not anywhere near enough possibilities because of the new information theory.

Dawkins ends his speculation about how life began with the following:

---

[10] Dawkins, *The Blind Watchmaker*, 153-155.
[11] Ibid., 157.
[12] Ibid., 146.
[13] Ibid., 139, 140.
[14] Ibid., 142.
[15] Ibid., 143.
[16] Ibid., 144, 145.

> So we have arrived at the following paradox. If a theory of the origin of life is sufficiently 'plausible' to satisfy our subjective judgment of plausibility, it is then too plausible to account for the paucity of life in the universe as we observe it. According to this argument, the theory we are looking for has got to be the kind of theory that seems implausible to our limited, Earthbound decade bound imaginations....We still don't know exactly how natural selection began on Earth. This chapter has had the modest aim of explaining only the kind ***of way in which it must have happened.***[17] (Emphasis added).

He admits the theory is speculative, beyond our logic as humans, based on no idea at all, yet it is *inevitable* and *must* have happened this way.

Most scientists see replicator-first theory an unproductive speculation. Chemical and molecular biology have produced new tools and much research since the idea of the spontaneous origin of life first started in the early twentieth century. They reveal obstacles that make it unbelievably difficult for life to begin in this way. Dean H. Kenyon, professor of biology at San Francisco State University has summarized some of these concerns. The experimental conditions are "so artificially simplified as to have virtually no bearing on any actual process that might have taken place on the primitive earth."[18] For example, if prebiotic mechanisms of condensation of free amino acids to polypeptides occurred, and sugars or aldehydes were present, polypetides would not form because of interfering cross-reaction. Second, there is an "enormous gap between the most complex 'protocol' model systems produced in the laboratory and the simplest living cells."[19] Third, there is an intractable problem concerning "the spontaneous origin of the optical isomer preferences found universally in living matter,"[20] which scientists "really are ***no nearer to a solution today*** than we were thirty years ago."[21] Fourth, the non-randomness has no biological relevancy of genetic subunits for providing a clue in prebiotic polynucleotides or polypeptides. Fifth, there has been unacceptable interference by investigators, as creators, to give any meaningful suggestions as to how the forming of life might have occurred. This relates to the highly important new understanding of thermodynamics, which we will discuss next.

Computer specialists have tried to devise accidental creation of life with a computer program. Computer programs have been designed at Harvard and elsewhere which have produced some interesting configurations that suggest life. But these are only on the screen. And most importantly, these are *the work of human intelligence.* The oldest rocks in various places of the world, some estimated at about four billion years, have been thoroughly examined. There is *no trace of rudimentary organic compounds* as building blocks for life. So there is no evidence that this actually happened. After millions of dollars in research, in every conceivable way, with efforts of a multitude of scientists over a century and a quarter, no plausible theory for the origin of life has been found. None of these efforts are observation of the facts, or formation of a normal logical theory, tested and retested to be confirmed as true. They are not science, but human philosophical speculation. Why does this effort to generate life by spontaneous chance continue, except for *the bias of man* to eliminate God and his responsibility to Him?

---

[17] Ibid., 165, 166.
[18] Charles B. Thaxton, Walter L. Bradley, and Roger L. Olsen, *The Mystery of Life's Origin: Reassessing Current Theories,* (New York: Philosophical Library, 1984) vi-vii.
[19] Ibid.
[20] Ibid.
[21] Ibid.

New Second Law of Entropy Resulting From New Information Theory

In the last half of the twentieth century it became clear that life did not, and could not begin accidentally. All the efforts to claim chance creation of life, and to show the evolution of all the species upward into more complex forms is directly in contradiction to the laws of nature.

In 1946 C. Shannon and W. Weaver of the University of Illinois introduced the new information theory, which for the first time made it possible for scientists to qualify the degree to which they could specify a physical system, or assign numbers to the amount of information it contains which causes it to function toward a useful end. This number describes the information location and motions that produce its designed end or intended functions.[22] This law was developed and tested by scientists in the secular universities, and has since led to *a radical modification of the second law of thermodynamics*, often called the law of entropy, and now called the new generalized second law. Charles Thaxton, who did postdoctoral work in molecular biology at Harvard and Brandeis University, said,

> There is a general relationship between information and entropy. This is fortunate because it allows an analysis to be developed in the formalism of classical thermodynamics, giving us a powerful tool for calculating the work to be done by energy flow through the system to synthesize protein and DNA (if indeed energy flow is capable of producing information)[23]

Scientists like Jeffrey S. Wicken, of the Division of Science and Engineering at Pennsylvania State, object to joining the new information theory to the law of entropy. He argues that the information theory should not be joined to thermodynamics because it is too complex and involves probabilities, which should exclude it from mathematical demonstration. He feels this changes the meaning of math. Wicken's argument, however, is not consistent with the facts. The laws of quantum physics involve probabilities and mathematics and were used to produce nuclear energy, lasers, and much modern science. When the second law is applied to the macro level, or astronomy, again, there is a wide range of probabilities. Moreover, since Wicken wrote, new instruments have been devised which allow a tremendous amount of accuracy in counting biological systems. The new information theory is not speculation, but is demonstrated by facts.

There is no scientific reason for not joining the new information theory and the second law of thermodynamics, because the law of entropy also involved probabilities and uses mathematics for applications. The main reason Wicken and other scientists reject this new generalized second law seems to boil down to the devastating results this has to the evolutionary philosophy, which claims to begin by chance and produce higher and more complex forms of life, each of which involves adding more information.[24] This law is so important that it is worth the effort to try to understand it and its implications for the concept of evolution.

The new information theory of the new generalized second law differentiates substances into *three classes* that may be illustrated by the following letter arrangements:

---

[22] C. Shannon and W. Weaver, *The Mathematical Theory of Communication,* (Urbana, IL: University Press, 1949).

[23] Charles Thaxton, et al., *The Mystery of Life's Origin*, 131.

[24] Jeffery S. Wicken, "Entropy and Information: Suggestions for Common Language," *Philosophy of Science: Official Journal of the Philosophy of Science Association*, Vol. 54, 1987, 176-193.

(1) An *ordered* (periodic) and therefore *specified* arrangement:
e.g., THE END THE END THE END
Examples: nylon, or a crystal

(2) A *complex* (aperiodic) *arrangement*:
e.g., AGDCBFE GBLCAFED ACEDFBG
Example: random polymers (polypeptides)

(3) A *complex* (aperiodic) *specific* arrangement:
e.g., THIS SEQUENCE OF LETTERS CONTAINS A MESSAGE
Examples: DNA, protein, organic life

The meaning or bits of information in the third group is exceedingly high, and the formation by chance of the third group, in which all letters have a meaningful sequence, is highly unlikely to occur by chance. There is a distinct difference in these categories of substances, and it is a mistake to reason that one category substance can lead to another.

For example, some believe that life may have come from various forms of crystals. H. Yockey and Wicken have shown that this is not a logical or viable approach. After evaluating it, they said, "In short, the redundant order of crystals cannot give rise to specified complexity of the kind of magnitude found in biological organization: Attempts to relate the two have little future."[25]

New Generalized Second Law and No Chance Beginning or Evolution

Explanation of the Generalized Second Law

Robert Gange, who earned his Ph.D. for research on the application of cryophysics to information systems, has spent 25 years in research with the David Sarnoff Research Center at Princeton. He has said, "In the words of the New Generalized Second Law, an observer functioning with a closed system will lose, but ***never gain***, information." (Emphasis added). In a sense, this gives reason to Dollo's Law, and says that systems degenerate and cannot become more complex without an outsider creating or contributing new information. In simple language, on average, things mix, and with the passage of time *natural processes will destroy* patterns. *To create* or unmix will conversely *require the activity of the intellect*, and does not occur by chance. For more explanation, see Gange's discussion in *Origins and Destiny*.[26]

In other words, the new generalized second law predicts that complexity in organisms will decrease if left to the function of nature, and will not be changed to a higher level except by some outside intellect adding information. Therefore, life could not have organized itself by chance, or produced increasingly complex beings.

Moreover an array of sophisticated microscopes, including the new stimulated emission depletion microscope has revealed an almost unimaginable complexity in the microbiological world.[27]

---

[25] J. Yockey and J. Wicken, *Journal of Theoretical Biology,* 191, 579; cf., Charles B. Thaxton, *The Mystery of Life's Origin*, 130-131 and all of chapters 7 and 8.

[26] Gange, *Origins and Destiny*, (Waco, TX: Word Book, 1986), 46-47.

[27] James Trefil, "Seeing Atoms," *Discover,* June 1990, 55-60; "Opening New Frontiers in Molecular

The complexity may be illustrated with the hemoglobin protein molecule that is so essential to our human life, and is now found even in some plants and bacteria. In every protein some twenty or more different kinds of amino acids are present, and must be selected and lined up in just the right order to make up each particular protein. This is like lining up twenty different kinds of railroad cars in an exact sequence. In the case of *hemoglobin,* its configuration is like lining up two trains *composed of 574 cars* all of which must be in the exact place.

In spite of the opposition to the revision of the second law of thermodynamics, it has now been well established. In 2006 Charles Seife, an associate professor of journalism at New York University, who has an MS in probability theory from Yale, reported,

> [The new laws of information], they describe the behavior of the subatomic world, all life on Earth, and even the universe as a whole. ...And information seems to be at the heart of the deepest paradoxes in science—the mysteries of relativity and quantum mechanics, the origin and fate of life in the universe, the nature of the ultimate destructive power of the black hole, and the hidden order in the seemingly random cosmos.[28]

Moreover, Seife makes it clear that the new second law shows the degrading of the universe. He began his book by saying,

> No matter how advanced our civilization becomes, no matter if we develop the technology to hop from star to star or live for six hundred years, there is only a finite time left before the last living creature in the visible universe will be snuffed out. The laws of information have sealed our fate, just as they have sealed the fate of the universe itself.[29]

The fact of the matter is that there is always a loss in transfer of information and never a gain. This is absolutely contrary to the developmental and evolutionary theory, in spite of Seife's efforts to gloss over this fact by stating the information theory is a witness to the fact that life began. He refuses to be consistent and face what he has already stated about the entropy of all energy and information.[30] The new information theory is a devastating contradiction to the philosophical theory of evolution. It demands an instant beginning.

The Improbabilities of Life Originating or Evolving

Thus, *living organisms are of* ***the third kind*** *of substance*, with information as listed in the letter sequence noted previously. *Cytochrome C* is one of the simplest proteins involved in living forms and has been studied extensively. It is much simpler than hemoglobin. It has *only 101 amino acids or box cars* to line up in the right order, or is less than one-fifth as complex as hemoglobin. The probability of accidentally, or by chance, lining up each amino acid in Cytochrome C in the exact right place in the sequence is unbelievable.

Robert Gange figured this out and illustrates as follows:

> Picture an 8 1/2 x 11 inch sheet of paper with letters printed on both sides. Let's allow eighty columns by sixty-six lines of letters, giving us just under 5,300 letters on each side of the paper, or 10,600 letters per sheet. Putting the sheets into piles, we can stack about 320 sheets per inch,

---

Biology," *Discover,* March 1987, 14.

28 Charles Seife, *Decoding the Universe,* (New York: The Penguin Group, 2006), 2. 3.

29 Ibid., 1.

30 Ibid., 110, 111.

> giving us just over thirty-six thousand letters in a cube one inch on a side. Now what volume of space do we need to store enough sheets whose total number of letters equals the certainty that chance did not produce Cytochrome C protein? When I first did the calculation the answer astounded me. We need the space of almost forty thousand universes, each 30 billion light-years wide!
>
> Light travels at a speed of just over 186,000 miles per second. This means that in one second light can travel seven and a half times around the world. If light can travel that far in one second, imagine how far it will go in a year. Scientists call the distance light travels in one year a 'light-year.' It is roughly six thousand billion miles. We believe that the universe out to the visible horizon–is about 30 billion light-years wide. Yet, the certainty that an accident did not create Cytochrome C is the number of letters filling both sides of 8 1/2 x 11-inch sheets packed into the space of forty thousand universes![31]

Moreover, the *most primitive organism* is made up of some *2000 enzymes, which are very complex proteins*. Gange estimated the size of our universe is 30 billion light years, which is now measured at 13.7 billion. Since the size of the earth is less then half of what he estimated, it would take twice as many sheets as he calculated.

Other outstanding scientists, including H. Yockey, the theoretical biologist working with Wicken, Fred Hoyle, the famous astronomer, and Francis Crick, discoverer of the structure of DNA, have made and recorded similar calculations.

Fred Hoyle and his colleague, Chandra Wickramasinghe, found that the mathematical probability of life evolving from non-life is one chance out of 10 to the 40,000 power. They likened this to the equivalent of a tornado going through a junk yard, producing a 747 Boeing jet airplane with all its complicated instruments and parts in the right places, and emerging ready to fly. It is obviously illogical to believe this.

Using computers, Hoyle and Wickramasinghe calculated the possibility of the evolution of life from one form changing to the other as the evolutionary biologists claim. Wickramasinghe describes their experiment as follows:

> If you start with a simple micro-organism, no matter how it arose on the earth. ...If you just have the single organizational, informational unit and you said that you copied this sequentially time and time again, the question is, does that accumulate enough copying errors, enough mistakes in copying, and the accumulations of copying errors to lead to diversity of living forms that one sees on the earth.
>
> ...We looked at this quite systematically, quite carefully, in numerical terms. Checking all the numbers, rates of mutation, and so on, we decided that *there is no way* in which that could even marginally approach the truth. On the contrary, any organized living system that developed or emerged, say in the form of a microbe, 4 billion years ago, if it was allowed to copy itself time and time again, it would have destroyed itself essentially.
>
> The pile-up of copying errors could have two effects: (1) it could improve the genotype for survival, or (2) it could decrease the survival characteristics of this particular living form. If you consider the balance between the two effects, it turns out that *it's always the destructive component that wins.* For every favorable mutation, there will be hundreds of unfavorable mutations. So it has to be a steady downward procedure... it has to ultimately decline and

---

[31] Gange, *Origins and Destiny*, 74-75.

> degenerate.
>
> The way out of this is to suppose that natural Selection operates, not in relation to the copying errors, but in relation to continual *inputs of new information* that organisms could imbibe from the outside world.[32] (Emphasis added).

Evolutionists have argued that evolution *occurred by chance* multiplication. They present no mathematical calculations to show this happened. But when these calculations have been made with accurate modern computers, they want to dismiss them because they show chance cannot produce the desired effect. These prove that the degeneration of the species is exactly as observed in the geological record, that species eventually die out. They also confirm the new generalized second law and the loss of information.

Metabolism First—Last Effort for Chance Beginning

Since Huxley, without any evidence, speculated that life began by chance, many efforts have followed speculative trails to find some way it might reasonably have happened. The utter failure of them all, including the molecule replicator theory espoused in recent years have proven to be impossible to the extreme. Nobel Laureate Christian de Duve speaking of the molecular replicator theory called it "a rejection of improbabilities so incommensurably high that they can only be called miracles, phenomena that fall outside the scope of scientific inquiry." [33] The frustrations of secular scientists are further complicated in not being able to find any place in the universe like the earth from which life may have come.

Robert Shapiro, emeritus professor of chemistry, and senior research scientist at New York University, agrees with Christian de Duve. He and others have envisioned another speculative avenue for the origin of life, the metabolism-first theory. This theory posits that small molecules, driven by energy to interact with minerals chemically, and through a network of reactions, draw material into it faster than it loses it, forming compartments with information that increase in complexity. This idea assumes five requirements: a boundary to separate life from non-life, an energy source, a coupling mechanism to link release of energy to organization process, formation of a chemical network for adaptation in evolution, and growth of the network to reproduce. Such a process has never been observed. Shapiro knows this goes against the second law of thermodynamics as a whole, and that the fossil record lacks evidence of any development of complexity. Lack of evidence is reflected in his use of the words "probably," "might," "assumption," "if," "must" and "assumption." This theory also goes against the revealed evidence and is driven against the rules of biology and the unbelievably large possibility of it occurring by chance. This continued drive to explain life against what is known to happen in nature can only be explained by human bias to remove evidence of a Creator.

Final Options in Search for Life's Beginning

Cell research has progressed to the point that humans are now attempting to create life. Patrick Barry reports the latest in this effort by John Glass of the J. Craig Venter Institute in Rockville, Maryland. "Scientists are on the verge of ***creating*** living cells by piecing together small

---

[32] Chandra Wickramasinghe, "Science and the Origin of Life," *The Intellectuals Speak Out About God,* Roy Abraham Varghese, ed. (Chicago: Regnery Gateway, 1984), 28-29.

[33] Robert Shapiro, "A Simpler Origin of Life," *Scientific American,* June 2007, 46-53.cf. 50 His web description gives more suggested details.

molecules that are themselves not alive. The result would be the world's first human-made life forms, synthetic cells made more or less from scratch." [34] They are removing single genes from the parasite M. Genitalium to determine which are dispensable and which are not. By finding the essential genes, they hope to systematically string together the right sequence of nucleotides for life. Other scientists have other approaches to try to recreate life. The problem with what they are doing is that it is ***"creating"*** and shows that the intelligence given to man is a mystery that ***requires creative intelligence***.[35] (Emphasis added).

In the light of modern knowledge about the complexity of biological life and man, knowledgeable scientists resulted to one option for the origin of life and its development. Fred Hoyle, Francis Crick, and others who have honestly faced the statistics, support the pansperma theory. Developed by the first atheists in ancient Greece, pansperma theory posits that life must have originated on another planet and come to Earth somehow. Hoyle and Crick have speculated that billions of years ago the universe and the solar system were peppered with biological "seeds," which took root wherever conditions were right. This is why there is such an extensive effort to find other habitable planets and why 100 million dollars is being spent by SETI (The Search for Extraterrestrial Intelligence) on listening devises to receive some communication from life elsewhere in the universe.[36] Finding life elsewhere would seem to open the door, and increase probabilities that require new creative information. But this is the only way to avert the complex problem and is an *escape* ***to non-evidence,*** *which is totally* ***unscientific***.

There are tremendous hazards to life on all known planets. The planet Mars, with its great dust storms, presents great hazards to life. Costs for researchers to travel there seem exorbitant. NASA's estimate given to President George H. W. Bush for traveling to Mars was $400 billion. Even so, scientists may find evidence of life beyond the earth tomorrow. "*Such a discovery would* ***transform all of science*** *and philosophy and human thought*."[37] (Emphasis added). Some breakthrough in how life began has been desperately needed if a Creator is to be avoided or even placed at a distance. So the costly exploration of Mars, and plans to go there continue, while very great needs in the nation go unmet.

Claims of ***wishful evidence*** of extraterrestrial life were soon produced to give reason to this theory. A rock labeled ALH84001 was found in the Antarctica ten years before plans to go to Mars. The rock which been studied for about three years, was suddenly announced in an unusual press release to be a Martian rock with indications of life. But William Schopf, a UCLA paleobiologist, chosen by NASA as an expert in this field said, "I happen to regard the claim of life on Mars, present or past, as an extraordinary claim. And I think it is right for us to require extraordinary evidence in support of the claim." [38] He said of PAH's (organic compounds such a

---

[34] Patrick Barry, "Life From Scratch: Learning to make synthetic cells," *Science News*, January 12, 2008, Vol. 173, 27-29.

[35] Ibid.

[36] "Low-Mass Stars: Born to Make Planets?," *Science News,* Vol. 143, January 16, 1993, 36; "From Little Disks, New Planets Grow," *Newsweek*, December 28, 1992; D. Pendick, "Hubble Scopes Possible Planet-Forming Disks," *Science News,* Vol. 142, December 19 & 26, 1992, 421; "Life at Other Stars: A Matter of Climate," *Science News,* Vol. 143, January 30, 1993, 74; "Is There Anybody Out There?," *Science News,* Vol. 142, November 7, 1992; 317.

[37] John Horgan, *The End of Science: Facing the Limits of Knowledge in the Twilight of the Scientific Age* (Reading, MS: Addison-Wesley Publishing Company, 1996), 14.

[38] David Warmflesh and Benjamin Weiss, "Did Life come from ANOTHER WORLD?," *Scientific American,* Nov. 2005, 64-71.

benzopyrene) found on this and many rocks, "In none of these cases have they ever been interpreted as being biological." Also, he warned that the images, which are inferred by NASA to be fossils by electron-microscope images "are a hundred times smaller than any found on Earth, too minuscule to be analyzed chemically or probed internally.... There was no evidence of a cavity within them, a cell.... The biological explanation is unlikely."[39] An August 16, 1996 *Time* magazine article concluded, "This led him (Schoff) to believe that the structures NASA was touting as fossilized life-form were probably made of a "mineralic material like dried mud." All other so-called Martian characteristics such as carbonate globules and magnetite are found on Earth as well. Andrew Ingersoll of Caltech, also interviewed by *Time* said, "This could be another cold fusion flash in the pan." Other so-called Mars rocks are coming forth, such as the British rock designated 79001. Some are sold as Mars rocks.

Some speculate that two tons of Martian material and an equal amount from the Earth land on each other every year. Others suppose that giant meteors or some other *cataclysm* dislodge rocks that find their way from one to the other in 10 million years. But chances of more such rock remaining in the earth's gravitational field and falling elsewhere on the earth are greater, even in Antarctica, than that they would make their way to Mars. There is no certain knowledge that ***any rock*** found on the earth is from Mars, or that life could have survived on a rock. David Warmflesh and Benjamin Weiss speculate how this might have occurred, if life ever existed on Mars.

Other scientists propose other options to explain the origins of life. Vrba suggests creation by a mystical earth force (with an external impetus), Gould (with an internal impetus), and all saltationist evolutionists call in an unknown force as a cause. Again, all this is unscientific. All these unknown factors are contrary to modern science and to Darwin's dogmatic rejection of occult forces. The only realistic option is that God created life!

Conclusions about Origin of Life

Nature seems to open the door to a transcendent intelligence that created life. Dean H. Kenyon has told why the origin of life has eluded the science.

> Perhaps these scientists fear that acceptance of this conclusion would open the door to the possibility (or necessity) of a supernatural origin of life. Faced with this prospect many investigators would prefer to continue in their search for a materialistic explanation of the origin of life in spite of the many serious difficulties of which we are now aware.[40]

---

[39] Ibid.

[40] Thaxton, *The Mystery of Life's Origin*, viii.

Chapter 15
Molecular Biology, Genetics and Ancestry

History of Molecular and Genetic Biological Sciences

While a number of ancient civilizations had some understanding of genetic rules of inheritance, genetics is another exciting science that has slowly emerged. Now in the twenty-first century, it is front and center in the quest for the origin of life. Those who began to remove the idea of God from modern geological and biological thinking, including Lyell and Charles Darwin, knew virtually nothing about the amazing science of molecular biology and genetics. Indeed, molecular biology was born about 1940 and genetics has exploded into prominence with monumental strides in the last decade of this past century and will likely occupy the center of scientific thinking for the years just ahead. Some see the year 2001, the year the human genome was deciphered, as one of the most significant of all scientific events.

For seventy-five years, from the time Mendel devised his genetic principals in the mid-nineteenth century, until the second quarter of the twentieth century (ca. 1925 ff.) molecular biology was a relatively unimportant science. Darwin ignored it. The cell theory for plant life had been proposed by the mid-nineteenth century by Matthias Schleiden in connection with animal tissue by Theodore Schwann. In 1902 Walter Sutton and Theodore Boven discovered genes on the chromosomes. In this second stage, genetic science and mathematical principals of genetics were slowly defined and introduced great knowledge and mystery.

But the greatest breakthroughs in molecular biology occurred from about 1940 to the end of the century. The whole science involves the processes of metabolism and digestion, delivery, and absorption of nutritive elements, as well as elimination. Significant breakthroughs are as follows.[1] Progress was enhanced by the development of the electron microscope and production of mass spectrograms along with computers for recording and analyzing revolutionized biology and the work on cells. In 1940 George Beadle and Edward Tatum demonstrated the existence of a precise relationship between genes and proteins. New directions in molecular biology began in 1944 when O. T. Avery and two colleagues studied deoxyribonucleic acid (now known as DNA) and found it was capable of changing one strain of bacteria into another, while remaining bacteria.

The breakthrough in genetic science came in the mid-twentieth century and has rapidly expanded, reaching a peak in 1980s, when molecular biology became an observable science. Exploration of the genetic world also led to the discovery of variations by minor mutations of the genes. This produced the neo-Darwinist theory of evolution that gave hope of gradual change by the means of mutations. Thus, with molecular biology, science entered a whole unlimited source of knowledge that man never knew existed.

Around 1953 J. D. Watson and F. H. C. Crick, using the suggestion from Rosalind Franklin's brilliant x-ray work, discovered the double helix by working backwards from x-ray distraction data. This allowed concentrated efforts in the area of genetic sequencing. In 1961 Francois Jacob and Jacques Monod hypothesized the existence of an intermediary between DNA and its protein products, which they called messenger RNA. Between 1961 and 1965 scientists discovered that the relationship between the information contained in DNA and the structure of proteins determined that the genetic code creates a correspondence between the succession of nucleotides in the DNA sequence and a series of amino acids in proteins. At the beginning of the 1960s, Monod and Jacob also demonstrated how certain specific proteins; regulative proteins latch onto DNA at the edges of the genes and control the transcription of these genes into messenger RNA.

---

[1] "History of molecular biology," *Wikipedia, the free encyclopedia*, 1- 3, cf. Internet.

These direct the "expression" of the genes. In 1961 Marshall Nuremberg and Heinrich Mathaei broke part of the DNA genetic code.

They found four base pairs of phosphate and sugar: A (adenine), C (cytosine), G (guanine), and T (thymine). Linked by hydrogen bonds, in various orders, these base pairs copy determining information in the genes. Each human cell has 46 chromosomes. Each chromosome has thousands of genes, composed of 100–100,000 nucleotides. There are triple-folded proteins, 28,000 nucleotides long, which are composed of many different amino acids.

Development of Genetic Engineering

The most important phase of genetics—the idea of molecular engineering—began in the 1980s when restriction enzymes allowed scientists to snip DNA into tiny, easy-to-read bits. In 1983 Kary Mullis of UCLA (Nobel prize in 1993) discovered that PCR or polymerize chain reaction allowed for unlimited copies of DNA strands in the test tube. This allowed laboratory work with previously unknown genes. The present phase seems to be a continuation of that which began in the 1980s, and recently produced the mapping of the genomes of all forms of life, including the human.

Genetics is now opening the door to exciting and, perhaps, dangerous developments. Scientists now know that the two strands of chromosomes from the male and female contribute the controlling development of the successive individuals, and the scientists have a better genetic mathematic understanding. Armed with this information, scientists are manipulating genetic materials, creating new kinds of crops and animals for agricultural and medicinal purposes. They have already cloned sheep, mice, pigs, and other animals, and human cloning might soon be a reality. At the turn of the century, only fifteen percent of the animal clones have been successful, which portends serious results when working with human genes. The use of genetics in curing diseases is now considered one of the most promising aspects of medicine, and is becoming highly commercialized. This is just the beginning. The outcome is unsure.

Mapping of Human and Other Genomes

The Process of Mapping

In 1990 the breakthrough in genetic science opened the door to the race to decipher the complete human genome and to study the genomes of all life beginning.[2] In February 2001, two science research organizations, one government, Human Genome Project Consortium and a private biotechnology company, Celera Genomics, of Rockville Maryland, after a decade of effort, announced the whole human genome had been read. This constitutes a tremendous amount of work with ingenious methods and a monumental achievement. Thus, the genetic controls and deviations of men are a definite science for identification and new developments.

Francis Collins, of the National Human Genome Research Institute, announced this as "The first draft of the human book of life."[3] Scientists have essentially read three billion or so letters, coiled up in every human cell, that spell out the human genome, which is encoded within the six feet of DNA. This is said to represent "a working draft in which scientists have put in order DNA fragments covering 97 percent of the genome and sequenced more than 85 percent of them to

---

[2] The whole process is described by Bryan Sykes, professor of genetics at the Institute of Molecular Medicine at Oxford University, *The Seven Daughters of Eve,* (New York, W. W. Norton & Company, 2001), 12-17.

[3] "Human Genome Work reaches Milestone," *Science News*, Vol. 158, July 1, 2000, 4. Actually, the mapping was not completed until 2003.

varying degrees of accuracy."[4] The mapping of genomes is considered to be a major breakthrough in science, not only because of the potential for humans to change the information that may improve plants and animals, but also for the potential to diagnose and cure diseases.

Unfortunately, scientific research was not the only motivation for this project. Greed and the desire for world-wide recognition also fueled "the gold rush." According to *Science News,* "In the race for genes, a desire for publicity seems to be winning out over the peer review process. Instead of scientific publication, genome researchers are increasingly turning to press releases and news conferences to present their findings."[5]

Revelations from Reading the Genome

A number of surprises have come to light thus far. The number of meaningful genes in humans is much fewer than expected, and 90 percent of the protein motifs are seen in flies, worms, and in many bacteria. The number of human genes is now known to be about one-third of what was expected. The great majority of genetic materials are only slightly understood and their meaning unknown. Much is like mixed up letters instead of communications with meaning. Prior to 2000 the function of fewer than 5,000 genes was understood. Knowing the location in the genome accelerates that understanding.

People need to understand that humans can use their creative intelligence to change and mutate genes, but genes do not improve apart from human creativity. Moreover, 99.9 percent of human genes are the same in all human beings. Only about one-tenth of one percent account for the differences in human beings. This is extremely important because it shows that genetic material has, throughout history, constantly been changing to produce millions of different people with family markers. These changes are so controlled it has not allowed change that shows any links between animals and men. Humans remain distinct.

False Finality Believed in the Mapped Genome

Mapping Was Inconclusively Done

This is not the end, but the beginning of an understanding of human genetics. In an article entitled, "The Emperor's New Genome?" Josie Glausiusz refers to David Schwartz of the University of Wisconsin, Madison, as saying,

> You wouldn't know it from the press conferences, but the scientists are still far from deciphering the human genome. About 20 to 30 percent of our genetic code – containing enigmatic chunks of repetitive DNA – is difficult to read using current sequencing methods. The rest has been sorted out only in bits and pieces Schwartz likens the situation to reconstructing a book whose pages have been torn, ripped, and scattered. "You're missing some pages, and some are out of order, but you tape it together and say 'Aha! It's finished! I've got the book.' No, you don't. It's far from complete."[6]

Many Mysteries beyond the Genome

The public was given the impression that when the scientists had mapped the genome they would know all about biological life and could discover how it worked. Moreover, great similarity in chromosomal genes, especially the fact that only a few of the genes of the higher apes were

[4] Ibid.

[5] J. T., "Genes, genes, and more genes," *Science News*, Vol. 157, May 6, 2000, 298; cf. Sykes, *The Seven Daughters of Eve,* 19 ff.

[6] Josie Glausiusz, "The Emperor's New Genome?" , *Discover*, October 2000, 18.

much like human genetics, led some to claim that this showed that evolution was a simple process. But scientists have discovered that protein-coding genes account for less than two percent of the total DNA of each human cell. As far back as 1960 some realized there were many more.

Two other levels of complicated control, beyond the protein-coding genes of the mapped genome, have been uncovered. The second layer consists of myriad "RNA only" genes sequestered within stretches of noncoding DNA that had been considered junk DNA. This profoundly affects the behavior of the genes.[7] It is now becoming clear that these 20,000 or more so-called pseudogenes may have an important part to play in the function of various individuals. For example, Bertha Madras and Gregory Miller of Harvard pinpointed hyperactive disorder (ADHD) in junk DNA that is in the mass of unmapped genetic material. "Shinji Hirotsune of the Saitama Medical School in Japan and his colleagues found that 'real' genes might not be able to function properly without the pseudo ones."[8] This was discovered when substituting fruit fly genes into mice without the proper pseudogene led to deformity and death. A third important and extensive part is "the epigenetic" layer of information stored in the proteins and chemicals ***that surround*** and stick to DNA. W. Wayt Gibbs described epigeneses:

> The DNA sequence is not the only code stored in the chromosomes. So-called epigenetic phenomenon of several kinds can act like volume knobs to amplify or mute the effect of genes. Epigenetic information is encoded as chemical attachments to the DNA or to the histone proteins that control its shape within the chromosomes.[9]

W. Wayt Gibbs reported on these three levels and quotes Carmen Sapienza of Temple University who said, "The Human Genome Project was just the beginning of the job." Timothy H. Bestor of Columbia University says, "There is a whole new universe out there that we have been blind to. It is very exciting."[10] Having now mapped the genome, another two levels of mystery are to be explored.

Scientists are finding tremendous complexity beyond these levels in the proteins, how they assemble amino acids, and how the proteins operate. The genes were very complex, but in comparison to the protein problem, they were relatively simple. Genes are two-dimensional, but proteins are three-dimensional, and have an exceedingly complex structure with multi-folds. In a protein, each of the hundreds of amino acids must be created in the exact right place with right folds for the protein to do its function. So there is a very high improbability for getting them in the right order by chance. Moreover, proteins are much more complicated than long thought. "Proteins are traditionally viewed as rigid or semi-rigid 'blocks' whose specificity and catalytic power are determined by the unique three-dimensional structure. ...But the discovery of proteins that are wholly disordered or contain lengthy disordered segments, yet are functional, has wreaked havoc on the lock-and-key world view...."[11] Research to understand proteins has proven so lengthy and expensive that most of these efforts have been discontinued.

But the problem is even more extensive. Scientists have discovered millions of unknown forms of DNA in the ocean,[12] as well as on land. Many viruses serve as genetic sensors. While

---

[7] "The Unseen Genome: Gems among the Junk," *Scientific American*, November 2003.
[8] "More Gene Than Junk," Discover, September 2003, 16.
[9] W. Wayt Gibbs, "The Unseen Genome: Beyond DNA," *Scientific American*, December 2003, 112.
[10] Ibid., 106-113. He refers to several articles and books on these whole areas of important mysteries that are beginning to be explored, 113.
[11] Vladimir N. Uversky and A. Keith Dunker, "Controlled Chaos," *Science*, vol. 322, 28 November 2008, 1340-1341.
[12] "Gene Saavant sifts life from seas," *Discover,* December 2004, 18-19.

many phages inhabit and eat bacteria, only 28 percent of phage sequences are similar to documented genomes, and may represent the major form of life in the biosphere. Graham Hatfull and Marisa Pedulla at the University of Pittsburgh are authorities and admit that about phages, "We are thoroughly ignorant."[13] Life is a highly complicated mystery.

Basic Genetic Operation Replaced by Mystery

Early research led to early understanding that DNA transferred information to RNA which carried out the message for work and duplication. The RNA machine edited out "junk" portions of the gene, called introns, and used other portions, called exons, to form a new version of the gene with the same, but diversified, information. By 1980 Randolph Wall of UCLA formed a basic standard view that pre mRNA discarded all introns, and all exons were included in the new individual. Scientists believed that new individuals were formed and functioned differently by this simple, cutting machine-like process. Certain copying errors, or mutations, that were not eliminated were transmitted from one generation to another, marking a line of generation. Thus, operating by this mechanism, all genes were seen as pretty much the same in the various life forms. At first it seemed there was not much difference between genes in worms, primates, or men, but it was expected that men had developed more different genes.

The neo-Darwinists thought that accidental good mutations would lead to improvement, by the survival of the fittest; bad mutations, not eliminated, would be descendent markers. By chance changes in genes, the individual would develop good qualities to survive, producing higher forms of life. Man was expected to have the most genes, about 150,000, that produced about 90,000 proteins. This basic view, influenced by Darwinian deterministic thought, seemed to be a neat understanding of a machine that progressively produced better individuals. A history of mutations occurred, but it turned out to be all wrong, much more complicated, and not transmutational. Moreover, no one knows what power makes the genetic system work so amazingly well with an innate ability to correct errors and continue control.

Recent discoveries reveal a very different picture. The human genome "has been revised downward still further (from 30,000 in first mapping), to fewer than 25,000. Geneticists have come to understand that our low count might actually be viewed as a mark of our sophistication because humans make such incredibly versatile use of these few genes."[14] While the basic idea of DNA having a four letter string (A, G, C and T) of an order of nucleotides (three billion strung together in humans), the workings of the genes of different forms of life have been found to be very distinct, complicated, and mysterious in exercising controls.

The higher the organism, the more complex and versatile the genetic working is. In humans, each gene can produce two or more proteins (90,000+) and do other amazing functions, such as detecting where to splice DNA, place and affix the exons in the new individual in a correct complex order, determine life and death of cells (apoptosis), and do some housekeeping and maintenance work. "The average human protein-coding gene...is 28,000 nucleotides long, with 8.8 exons (of only 120 nucleotides each) separated by 7.8 longer introns (of 100 to 100,000 nucleotides)."[15]

Contrary to the early basic machinery idea, there is an amazing, very complex, and diverse

---

[13] John Travis, "All The World's a Phage," *Science News,* July 12, 2003 Vol. 164, 24-28; cf. Nelson C. Lou, David P. Bartel, "Censors of the Genome," *Scientific American*, August 2003, 34-41.

[14] Gil Ast, "The Alternative Genome," *Scientific American*, April 2005, 58-65. This article gives an excellent presentation of what is now known about these mysterious workings in genetics and is the source of the unidentified quotes. I have given a limited idea of the functions.

[15] Ibid.

way of creating new individuals. "In fact, the cellular machinery can 'decide' to splice out an exon or to leave an intron, or pieces of it," in the final mRNS transcript. This gives "the splicing mechanism tremendous power to determine how much of one type of protein a cell will produce over the other possible types encoded by the same gene."[16] The genes have built-in information that limits the organism, yet at the same time creates diversity within predetermined limits. In the process, there may occasionally be seeming errors causing cancer and other diseases. This is such a great mystery that over 3,000 biologists are thought to be investigating.

Several important facts are manifested in these new genetic discoveries. First DNA does not function mechanically, but the way it works is a complete mystery. Second, the higher the organisms, especially in humans, the more complex the genetic ability is to do diverse things, and the greater information is involved in the operation of the genes. Their working seems to be specified abilities of information in the organism. This means that lower organisms cannot create or evolve more complex genes of the higher organisms. To do so would be contrary to the new information theory incorporated in the new second law of thermo dynamics, which will soon be discussed. Information is always lost and never gained or increased. So this offers no mechanics for evolution to higher organisms.

Third, the changes in the kind of individuals of a species, such as various mammals, are not caused primarily by mutations, but mostly by shifting of tandem segments of DNA sequences. John W. Fondon, University of Texas Southwestern, in Dallas, has traced the development of different kinds of dogs and other animals in which this was the case. He found that such changes happen 100,000 times as often as single point mutations in affecting successive generation. This is highly important as it discredits the use of mutational genetic time clocks to determine age.

Fourth, geneticists are surprised to now find that there are duplicated or missing blocks of DNA, known as copy number variations (CNVs), that are frequently occur. Some are large, with millions of DNA bases, and some are as little as 500 base pairs in length. Studies in Africa and Europe uncovered 10,000 CNVs and identified 1500 common ones. But these are spot checks and are likely missing many. Some are more common than others. Jan Friedman of the University of British Columbia in Vancouver, Canada, at the American Society of Human Genetics in Philadelphia said, "Many of us in the field were just blown away when we realized how often all of us have regions of the genome that are missing or present in extra copies. We just had no idea [the genome] was so plastic."[17] This new discovery raises many questions. The CNVs are associated with some diseases such as autism, diabetes, colorectal cancer, and body mass. But there are false positives and they are found in both healthy and sick people. It is not known if they may be the cause or result of disease. The key may be where they are, rather than how many. It has been said that "Like a kaleidoscope, the human genome keeps offering up new views."[18]

Other influences in biological developments are surfacing that also are not understood. In addition to genes and segments of genes, scientists have learned that cell division and differentiation are influenced by various hormones. But they don't know how. They recognized this activity in humans, but now see its effects in other forms of life, as well.[19] Bonnie Bassler, leading microbiologist with a team at Princeton University discovered that bacteria can communicate in multi-languages with each other and form a collusion of an army to cause a

---

[16] Ibid.

[17] Jennifer Couzin, "HUMAN GENETICS: Interest Rises in DNA Copy Number Variations—Along With Questions," *Science,* vol. 322, 28 November 2008, 1314. Cf. Science, 7 September 2007, 1315.

[18] Ibid.

[19] Raffaele Dello Ioio et al., "The Genetic Framework for the Control of Cell Division and Differentiation in the Root Meristem," Science, vol. 322, 28 November 2008, 1380-1384.

disease. Antibiotics have been devised such as penicillin to penetrate and destroy the bacteria. But by innate abilities to mutate they can resist the disease. Bassler and others are now working on how to interrupt this communication so the hundreds of forms of bacteria cannot communicate to form a collusion and therefore cannot cause the disease and cannot build up the immunity or resist an antibiotic. This is a whole new important discovery of an amazing ability of molecular life forms.[20]

In the study of mutations in species, studies of shifting tandem sequences in a species, and the DNA duplications and omissions are now believed to have more influence than changing a single gene. Duplications are not at random, but controlled by chunks of genes that have no known function as yet. The primate differences in millions of duplication bases are as follows: macaque, 45.8; orangutan, 20.3; and chimpanzee, 6.2. For humans it was 13.6. The rate of duplications accelerates to double for the last two. Humans have about 20 percent of duplications which are unique to humans and set them apart with a higher rate of change.[21] The importance will be seen in the dating of ancestry.

Genetic Distinctiveness, Ancestry, and Dating

Distinctiveness of Individual DNA and Tracing of Identity

Scientists hoped that tracing the origins of men and using genes to date fossils would help them uncover man's age and route of descent. But genetics is just beginning to inform the paleontological claims about prehistoric men. Three different forms of human genetic evidence in DNA used for tracing family ancestry and dating it, have been developed: mitochondrial DNA in women, male Y markers of reproductive gene diversity, and nucleic DNA that has material from male and female. The DNA of individual humans is so distinctive that today it is used in courts to convict criminals, to definitely link children and parents, and to understand how diseases progress. DNA identifies each unique individual. It is also useful for tracing relationships of ancestral lines. Each individual cell contains distinctive DNA from both mother and father.

The mother's DNA, mitochondria, is outside the cell nucleus in the cytoplasm. It is enclosed in a membrane in a molecule, called ATP, containing enzymes to use oxygen and release energy. One-quarter of a million mitochondria are found in a mother's egg. The nucleus chromosomes have 3000 million bases in the length, but in the middle of the mitochondria there are mini-chromosomes sixteen and a half thousand bases long. Mitochondria DNA is circular and has a control region of one short section of this DNA of 500 bases that has no specific function. Areas of specific function tend to correct mutations, which this does not. Thus, it has more tendencies to retain the mutations that occur. The nuclear DNA has extremely low mutations of about one in one thousand million, while the mitochondria DNA allow only twenty to one thousand million. Mitochondria DNA, therefore, records mutations at a much more manageable rate than can be counted by a sequencing machine.

In contrast, the male Y chromosomes are full of junk DNA and few genes. Having no function, it should have more mutations. The mutational changes are of two types. There is a simple change from one base to another (as in mitochondria), but in the Y, these are spaced out at irregular intervals along the chromosome, which make it hard to test. In addition, the DNA double up or duplicate when reproducing. This is common in nuclear and Y chromosomes. Some of the mutations or errors tend to be corrected in the process in the Y, but not in the mitochondria.

[20] Natalie Angier, "Listening to Bacteria," *Smithsonian*, Special Issue, July-August 2010, 76-82.

[21] Tina Hesman Saey, "Genome duplications may separate humans from other great apes," *Science News*, March 14, 2009, 14.

Nuclear tracing is more difficult than following mitochondria and Y chromosomes, but it is thought the study of the nucleus of the cells will reveal more family evidence. Individual differences and male ancestry can be traced in these repetitions. However, it is much more difficult to trace in the male than in the mitochondria of the mother.

The mutations are random changes that occur in the DNA as it is copied and cells divide. Over time mitochondria DNA shows greater variance in individuals. Those with the same or close to the same DNA are grouped together in tribes. The assumption is that the greater the variance, and the more mutations in the DNA, the longer the time it took to change. The greater the genetic distance, the longer they had been apart. The grouping of individuals allows seeing the diagram of development of genetic trees for groups of populations. The more alike, the closer the ancestry is. Bryan Sykes of Oxford University has said, "By using the mutation process... we can estimate the rate at which mitochondria DNA changes with time."[22]

Genetic Tracing of Ancestry

In 1987 Allan Wilson, Rebecca Cann, and Mark Stoneking of the University of California-Berkeley announced that, using mitochondrial evidence, they traced human female ancestors back some 200,000 years to Africa. The oldest woman's DNA was dubbed "Eve's." Later discovery of mathematical flaws in the research pushed the date to about 150,000. Some think humans came out of Africa as early as 100,000 B.C. However, it seemed this work had opened a new way of tracing ancestral relationships!

The male Y chromosome, with fewer markings, took longer to trace its ancestry. Michael Hammer of the University of Arizona traced the Jewish Cohanim priestly line in 1995. This identified black Jewish people in Africa as having the lineage of priests. The lineage of "Adam" was also claimed to be traced back to Africa. Peter Underhill, Stanford University, has published 87 Y markers, which he has used to trace all the world's men into ten branches. Spencer Wells, who studied with Underhill, compiled all the evidence for the male Y chromosomes, as they had been traced throughout most of the world. His work became a documentary for PBS, *The Journey of Man,* as well as a book by the same title. [23] Bryan Sykes of the University of Oxford, along with a team, traced the mitochondrial DNA for most of Europe for a book, *The Seven Daughters of Eve*. Because this was incomplete, Sykes added a chapter at the end, presenting the picture of mitochondrial DNA that geneticists had developed for most of the world. Sykes did a brilliant job of tracing descendents from China to Polynesia and Australia. Sykes and others admit the present picture of DNA ancestors is far from complete. "This picture will no doubt change in the years to come as people from previously unsampled regions volunteer their DNA."[24]

DNA has been used to trace ancestry in most of the world. Douglas Wallace of Emory University Medical School has traced genetic markers from Africa to Europe to Asia to America all the way to the tip of South America. His discoveries join with new paleoanthropologic evidence that gives new views about man coming to America. Wallace believes there probably was more than one migration into the Americas by coming across a land bridge during the Ice Age and coming down the west coast at later times as the glaciers melted. Sykes and Wells accept the idea of two migrations from northeast Asia across the Bering Straits to Alaska and North and South America.

Various scientists attempt to support these theories using the fossil skeletons that have been

---

[22] Sykes, *Seven Daughters of Eve*, 49.

[23] Spencer Wells, *The Journey of Man: A Genetic Odyssey,* (Princeton, New Jersey: Princeton University Press, 2002).

[24] Sykes, *Seven Daughters of Eve*, 274.

discovered through the years. Amazingly, in addition to the Native American skulls with high cheekbones and other Indian characteristics, there are Caucasian skeleton remains in Western China, in Japan among the Imu people, along the coast of Alaska and in Canada, Arizona, and elsewhere. Jim Chatters found a "beach man" along the Colombian River and named him Kennewick Man for the place of the find in Washington State. There is also the Western Nevada Spirit Cave Man, and a jawbone found by Tim Heaton along the Alaskan Coast with a similar skull. The oldest skull, Peñon II Woman, found in Mexico's Teotihuacan Valley is considered to be the oldest American at 12,700 years, and may be from Pericu bones.[25] Silvia Gonzalez, John Moores, and a team from London University found footprints in lava in Mexican rock quarry that were dated by CO 14 in sedimentary layer as 40,000 years old. This meant that humans were in Mexico before the claims for earliest humans as being 11,500 years. Paul Rene at University of California, Berkley has challenged whether these really are footprints, much as Mary Leakey's find in Africa was challenged.

The Clovis story of the first Americans being the tribal Indians, who came across a bridge of land as the Ice Age receded, is now in serious question by some genetic and paleontological data. The speculation has been that these Americans all arose possibly from Neandertals (sometimes called Neanderthals) in Europe and Homo erectus in Asia and came out of North and East Africa. There is a strange absence of such fossils from West Africa.

Genetic Relationship of Neandertals and Homo sapiens

At First a Puzzle

First Neandertals were thought to be strange humans. French paleontologist, Boule said they were bent over and were a missing link to modern man. His views misguided W. F. Osborn in his Hall of Men, and the entire world. Desperate to locate intermediaries, even Tattersall tried to include Neandertals to close the gap between apes and men. He said they were primitive in thought and creative abilities, and were a small first step toward cognition and social skills. However, he thought that about 100,000 years ago they encountered another strain of evolving men in the Levant and Homo sapiens.[26] There was no clear evidence this occurred, so he said this capacity was born "as an ability that lay fallow until it was activated by a cultural stimulus of some kind." This was speculation of occult power at work. Besides, brain size was considered the key to intelligence, and the average Neandertal's was 1625 cc. and Homo sapiens only 1450 cc. It became common to teach this idea that superior Homo sapiens outdid or led to the demise of Neandertal.[27] The genetic differences at first added more speculations.

Sykes found that the idea that the Neandertals were transformed into Cro-Magnon men by replacement, as Tattersall thought, was not reasonable. They coexisted in some places for thousands of years. But Sykes thought the period was too short for Neandertals to be related to Cro-Magnon. Neandertal DNA showed twenty five mutations, while Cro-Magnon had an average of only three to eight mutations, or less. Sykes commented on such a transformation, saying,

> It is inconceivable that a mutation of such magnitude could have occurred as to transform the heavily built Neandertals into the thoroughly modern-looking Cro-Magnon man more or less (in evolutionary time) overnight.[28]

---

[25] "A new Look For The First Americans," *Discover Magazine,* March 2003, 11.
[26] Tattersall, "The Roots of our Solitude," *Scientific American*, January 2000, 61.
[27] Lubenow, *Bones of Contention,* 73.
[28] Ibid., 114.

Sykes, in an experiment conducted in Wales, concluded that people were either 100 percent Neandertal or 100 percent Cro-Magnon. He could not explain the relationship of Neandertal to Europeans. Later we will consider more recent research on the higher mutation rate for Neandertals.

Sykes also claimed to disprove the multi-regionalist view. But many Neandertals have been found to live all over Eurasia from Gibraltar, South Spain, France, Croatia, Israel, Iraq, and Uzbekistan. No solution was found to the Neandertal puzzle of far greater mutations and distribution. Wells and others agree with Sykes. This genetic difference in mutation showing longer life is important, and will be shown to be the key to understanding.

Neandertals now Considered to be Fully Human

Chris Stringer, paleontologist at the National History Museum in London, says that genetics is moving ahead of the fossil historical record. The earlier limited genetic evidence seemed to see the Neandertals as distinct from modern men.[29] However, about 3.7 billion bases or 63 percent of a Neandertal female thigh bone fossil from a Vindija cave in Croatia has been decoded. Three other specimens have also been researched, but less extensively. Stringer's team makes various speculations about evolution, indicated by the use of the word "may have," but the following are some facts from the Croatia specimen.

> Neandertals are similar to humans across most of the genome–any given Neandertal was about as similar to a human as two humans are to each other. ...But humans and Neandertals did share a version of the FOXP2 gene associated with speech in humans. …Neandertals appear to have been less genetically diverse than humans are.[30]

While admitting there is no more difference than in Homo sapiens, they give an evolutionary developmental spin by comparing them to "humans." Recognition of the greater genetic stability in Neandertals is important.

Other discoveries confirm the similarities. ***An infant*** Neandertal of Mezmais Kay cave in the Caucuses near the Black Sea was found to be *comparable to modern H. sapiens,* based on the genetic evidence.[31] The fact that this was an infant is important and will be considered later. *Nature* magazine said, "variation between the Neandertal sequences equals that seen in random pairs of either modern Europeans, modern Asians, and exceeds that of modern Africans.[32] Only a few months later B. Bower pointed out,

> The mitochondrial DNA extracted from early H. sapiens fossils in Australia *may usher Neandertals back into the modern fold*, according to a team led by Gregory J. Adcock of Pierre and Marie Curie University of the University of Paris.... Genetic material from the oldest specimen, which was found at Lake Mungo in southwestern Australia differs more from that of living people than do the previously isolated Neandertal sequences.[33]

Evidence continues to mount that, while there are genetic differences, Neandertals were fully

---

[29] *Science News*, July 19, 1997, 37.
[30] "Team decodes Neandertal DNA," *Science News*, March 14, 2009, 5.
[31] "Salvaged DNA adds to Neanderthal's Mystique," *Science News*, Vol. 157.
[32] *Nature*, March 30, 2000.
[33] Bower, "Gene, Fossil data back divers human roots," *Science News*, January 13, 2001, 21.

human and contemporaneous, and often found together with Homo sapiens.[34] Neandertal and H. sapiens coexisted together. B. Bower has said, "The Neandertals were the intellectual equals of H. Sapiens at least in tool making."[35] "Excavations of Neanderthal artifacts at two caves in northern Spain have yielded an unexpected discovery—a trove of thin doubled–edged stone blades that researchers usually regard as the work of Stone Age people who lived much later (47,000 – 42,000 years)."[36] The finding of both large and small sharp tools in the cave Grotte des Fées Chatel Perma, which characterize Neandertals, indicated interaction of Neandertals and modern humans. Paul Mellars of Cambridge University says this indicates modern humans spread into Western Europe around 40,000 years ago and coexisted with Neandertals for 10,000 years until the later species died out.[37] They are now found together in most of Eurasia. While there are differences of Neandertals from Homo sapiens, these actually represent a superiority by evolution.

The long argument as to whether Neandertals intermarried with Homo sapiens was settled from research published in *Science* journal and other publications in May 2010. Slvant Paabo, a German genome researcher of Max-Plant Institute for Evolutionary Anthropology announced that after fossil analysis they found that the Europeans and Asians inherited a small amount, an average of 1% to 4% of their genes from Neandertals. The conclusion was that Neandertals were not very distinct from modern men. This indicated a small amount of interbreeding occurred. This finding left the question on why this occurred.

Explanation of Neandertals as Fully Human and Why They Differ

Previous paleontological data revealed a number of structural differences in Neandertals: a larger cranial capacity, low, broad and elongated skulls with a pointed rear with a bun, large heavy brow ridges, low forehead, a large long face with the center of the face jutting forward, a weak rounded chin, and a rugged posterior skeleton with very thick bones which are stronger than most Homo sapiens. A number of unconvincing speculations were offered to explain the differences: unique stress when jaws and teeth were subjected to use as tools or weapons; slow brain growth due to slow postnatal brain growth relative to cranial vault growth; rickets from vitamin D deficiency, and others.[38] These and other explanations have not seemed adequate.

In 1998 Jack Cuozzo, a specialist in orthodontics who is trained in oral biology and has done extensive research in dental and cranial problems, studied many Neandertal skulls, using x-ray technology and dental measuring devices. He was aware of studies which have revealed that after normal human growth has ended in adulthood, growth still continues in the face and head, albeit at much slower rates. Cuozzo refers to more than 16 studies by credible scientists, which he uses as documentation. Working with a computer expert, Brian Garner, who was doing Ph.D. work in biomedical engineering, Cuozzo used these studies and applied the computer program design for a man the size of Neandertal. The program showed that if Neandertal had lived extra long (200 to 500 years) it accounted for these differences. This coincides with the age of men in the Bible just

---

34 "Revolutionary anthropologists from Germany's Max Planck Institute, EXTRACT NUBLEAR DNA FROM A NEANDERTHAL, *Discover*, August 20, 2006, 20.

35 B. Bower, "Neandertals take out their small blades," *Science News*, May 13, 2006 Vol. 169, 302.

36 B. Bower, "Digging up debate in a French cave," *Science News*, May 13, 2006 Vol. 169, 302.

37 B. Bower, "Mysterious Migrations," *Science News*, Vol. 171, March 24, 2007, 184-185.

38 Richard G. Klein, *The Career: Human Biological and Cultural Origins,* (Chicago: University of Chicago Press, 1989), 279-282; Valerius Geist, "Neanderthal the Hunter," *Natural History,* 90:1, January, 1981, 34; Lawrence Angel, "History and Development of Paleonpathology," *American Journal of Physical Anthropology*, 56:4, December, 1981, 512 et. al.

following the worldwide flood. These men would have grown to look as the Neandertal skeleton does.[39] This explains the growth of the brow ridges, the shifting forward of the chin and the lower teeth, and other such features. This idea of aging, accentuating these trends, has been acknowledged and approved by Rolf G. Behrents, professor and chairman of the department of orthodontics, University of Tennessee College of Dentistry and others.[40]

The DNA evidence also reveals that Neandertals have characteristics that are best explained by living much longer (e.g. continued growth of brow ridges). The mtDNA markers of normal Homo sapiens are reported at about eight and the Neanderthal markers are about twenty-four, which is about three times as many. The biblical answer to this discrepancy is that after the flood men lived three times longer (450 years) than the children who succeeded them. Also, after the scattering of mankind and the confusion of the languages in the time of Peleg, the medium longevity dropped to about 140 years or about one-third. The heavy brow ridges and other features of Neanderthal continue to grow longer. These findings support the premise that Neandertal characteristics are the result of a longer lifespan. It supports the fact that the Neandertal child was most like Homo sapiens. This also explains why they seemingly seldom intermarried with the later Homo sapiens, since they were their offspring who moved out into different communities. Thus, the differences between these so-called "prehistoric men" can be attributed to the fact that they are from different generations, and a radically changed world following the catastrophe of the flood. This also explains why young Neandertal DNA fit with Homo sapiens, and older, mature DNA was seen as different, with more mutations. It also fits with the findings that show the growth of their teeth was nearly the same.[41] The greater stability in DNA indicates changes that would have reflected diminishing body life.

For years it has been argued that Neandertals and Cro-Magnon men and women did not intermarry as would be expected. This view was established on superficial evidence. Now that DNA for these groups has been recovered, that argument has been contested.[42] The evidence suggests intermarriage of Neandertal men with Cro-Magnon women. This could be expected if Neandertals were the older generation, from Noah after the flood, and the Cro-Magnon were their offspring, but Cro-Magnon did not live before the flood.

This agrees with the biblical claims of the descendents of Noah, Shem, Ham, and Japheth. "These three were the sons of Noah and from these *the* ***whole earth*** *was populated*" (Genesis 9:19) (Emphasis added). "Out of these, the nations were separated on the earth after the flood" (Genesis 10:32). The population of man expanded from Eurasia. This agrees with evidence that all plants and animals, capable of domestication, support this expansion.

One Line of Origin or Multi-regional Lines?

The debate over multiple regions for Homo sapiens or one line of replacement out of Africa is still much in doubt. Milford Wolpoff of the University of Michigan and a leading critic of the

---

[39] Jack Cuozzo, *Buried Alive: The Startling Truth About Neanderthal Man,* (Green Forest, AR 72638: Master Books, Inc., 1998), see Chapter 27, "Age Changes in Our Head and Face," and following.

[40] Ibid., 161.

[41] *Nature*, December 7, 2006. CF. B. Bowers, "Living Long in the Tooth," *Science News*, Vol. 16, July 10, 2004, 20.

[42] Jeff Nicoll and Joan Carter, "Stone Age Genetics: ancient DNA enters humanities heritage," *Science News,* May 17, 2003, 309; cf. also Science News, Vol. 163:307, 371; cf. also Vol. 165:181, 328 & Vol. 166:183 and Vol. 166;12/18 & 25/04, 405.

"Out of Africa" theory has said, "The molecular stuff has been very important, but in the end it has the same problem fossils have—the sample size is very small." In a *US News & World Report* article, they refer to *Science* magazine about Wolpoff.

> A Wolpoff study of early human skulls... suggests that Africans may have mixed with earlier hominids rather than supplanting them. The small number of living humans sampled by geneticists and the effects of natural selection over the millennia makes it foolhardy to say with assurance that Out of Africa is right. The geneticists, for their part readily admit they need more samples, more markers, and more precise calculations.[43]

The replacement or multi-regional controversy has not been resolved. Until the new century, the early DNA tracing claimed to support the Out of Africa scenario. Recently the Neandertals were tested for DNA from what is thought were earlier skeletons. Now diversity or multicultural origins are arising from DNA. B. Bower reported the following about two studies.

> One, slated to appear in the *Proceedings of the National Academy of Sciences,* finds that ancient mitochondrial DNA retrieved from a 62,000 year old Australian H. sapiens fossil differs greatly from the DNA of living people.... The second study published in the Jan. 12, *Science* compares the anatomy of H. sapiens skulls from different regions and times. ...The study's authors conclude that modern humans evolved separately in Central Europe and Australia, each lineage arising as a regional variation on an anatomical theme that originated in Africa nearly 2 million years ago.[44]

"Different mitochondrial DNA lineages could have flourished and then disappeared as modern H. sapiens evolved simultaneously in two or more points of the world over the past 1 million or 2 million years." [45] Milford Wolpoff, of the University of Michigan in Ann Arbor said, "Modern humans evolved separately in Central Europe and Australia, each lineage rising as a region's variation on an anatomical theme that originated in Africa nearly 2 million years ago."[46]

The problem of DNA in Australia that keeps this controversy alive is that Wells and others hold to a hypothetical migration of early humans out of Africa to Australia near the time of the first migration into Eurasia. This requires a very early, unimaginably long migration around the coast and across oceans, or a long sea migration which does not seem reasonable. Why? The world was tropical and plush.

The Small Amount of DNA Evidence for Out of Africa, or for Origin

Sykes' diagram of the mitochondrial DNA is found on the following page. I have added the zigzag line to show DNA in Africa on the left, from Europe on the right. The indication of Neandertal on the left as Out of Africa is incorrect, since Neandertals have been found in Eurasia. Sykes indicates that the evidence for clans emerging out of Africa was small. The mtDNA evidence indicates an unusual phenomenon in that there were thirteen clans of mtDNA, with several of these relating to each other before reaching the clan from which the African mtDNA exited. If one also counts Eve on the diagram, there would be fourteen. Sykes makes this remark:

> Of the thirty-three clans we recognize worldwide, thirteen are from Africa. Many people have left Africa ***over the last thousand years***, a lot of them forcibly taken as slaves to the Americas or to

[43] Nancy Shute, "Where We Come From," *U.S. News & World Report*, January 29, 2001, 36-41.
[44] B. Bower, "Gene, Fossil data back divers human roots," *Science News*, Vol., 159, January 13, 2001, 21.
[45] Ibid.
[46] Ibid.

Europe. Although Africa has only 13 percent of the world's population, it lays claim to 40 % of the maternal clans. The reason for this is that Homo sapiens have been in Africa for a lot longer than anywhere else.

Estimates of the size of the human population one hundred and fifty thousand years ago (when man came out of Africa) are bound to be not more than guesswork, but it may have been in the order of one or two thousand individuals.

Incredible as it may seem, we can tell from the genetic reconstructions that this settlement of the rest of the world involved only one of the thirteen African clans. It could not have been a massive movement of people. Had hundreds or thousands of people moved out, then it would follow that several African clans would have been found in the gene pool of the rest of the world. But that is not the case. Only one clan, which I have called the clan of Lara, was involved. It is theoretically possible from the mitochondrial DNA evidence that only one modern human female, one woman, left Africa, and that from this one woman, all of us in the rest of the world can claim direct maternal descent...But ***the numbers must have been very small.*** This was no mass exodus. Lara herself was not in the party.[47] (Emphasis added).

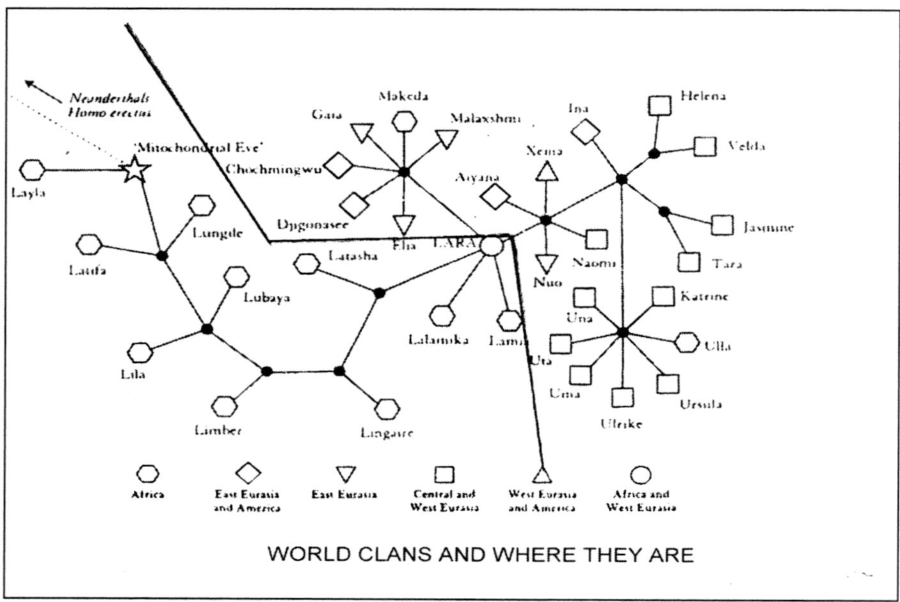

WORLD CLANS AND WHERE THEY ARE

Eleven of these clans never exited, nor are they found elsewhere, with the exception of markers from Lara in other clans. I have drawn the line where African mtDNA is indicated and only the hexagon symbol is African in origin. This is indeed, "incredible." Yet there are two African clans, Markeda and Utla, associated with other families *outside* Africa. Would it not be more logical to say that this indicates one of the Eurasian clans migrated to Africa where the African tribes proliferate?

But Sykes and Wells each admit that Neandertals and Homo erectus both exited Africa long before this and were intermediary. They say that Neandertals were in Eurasia long before and contemporary. Sykes has said, "The dashed line on Figure 7 indicates this even deeper genealogy through which our species, Homo sapiens, is connected to the other, extinct humans, the

[47] Sykes, *Seven Daughters of Eve*, 278, 279.

Neandertals and Homo erectus, and eventually back to the common ancestor of humans and other primates."[48] Connecting Neandertals with Africa is contradictory and very confusing. Also, neither Sykes nor Wells are being honest in using the names for Adam and Eve as the beginning point, when they therefore admit there are other clans that probably preceded them.

The map of the locations of the male Y chromosomes by Wells follows.

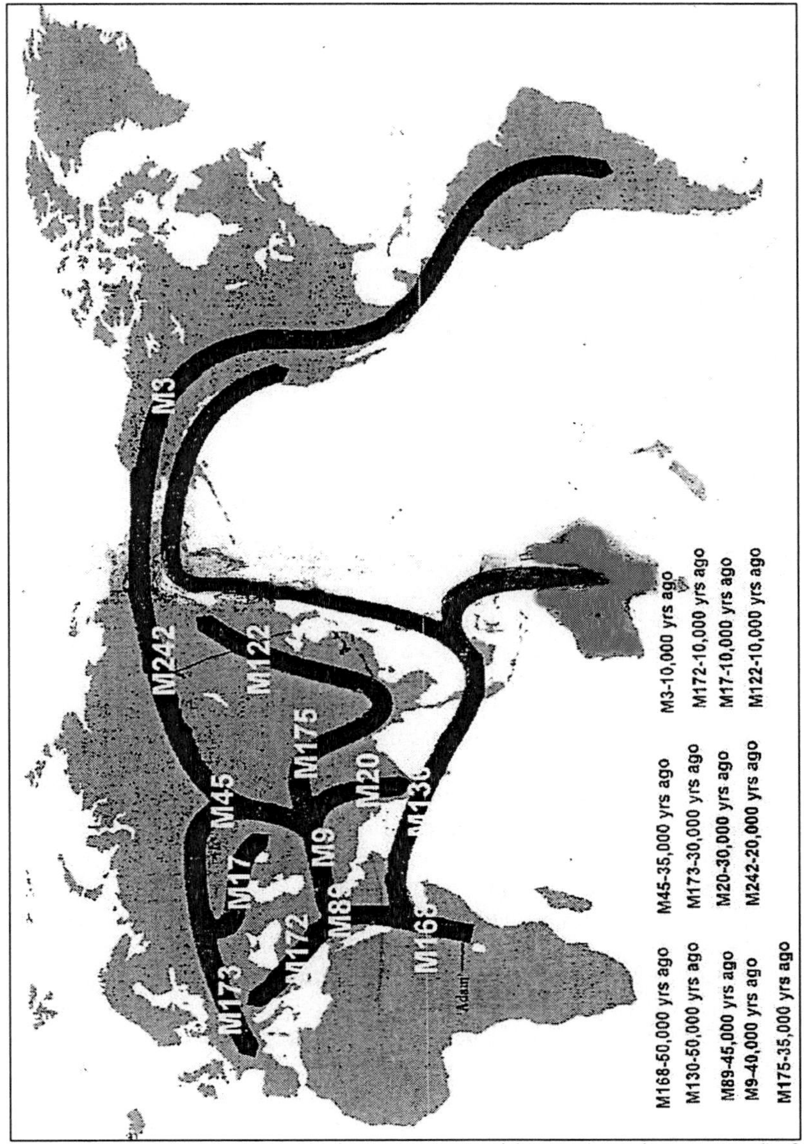

Moreover, Wells gives slight Y chromosome evidence (no clans named) from Africa to match the African mtDNA for Sykes. He names the male clan, which exited Africa, M168 as

[48] Ibid., 277.

Adam, or the first clan, and from this he traces all Y chromosome clans out of Africa. The principle evidence he refers to is the San Bushmen found in Ethiopia and Sudan. He admits the San people have a "'non-African' physical appearance" being lighter skin, smaller, curled hair, and an epicanthic fold over the eyes, which he says "characterize people from east Asia." Moreover their language is unique and not like any other African tribe.[49] How can this be? Wells said, "If M168 defines a marker common to all non-African populations, then in our genetic recipe is the equivalent to impala—the marker that unites everyone outside Africa."[50] He then points out that M89 is also found in Africa. Why could not M89 have been the first clan equally as well to leave Africa and then relate to M168 some other way? He then says there are genetic Negroid indications found in Thailand, which link them to the bushman and the pigmies. This is also true in Malaysia, Philippines, and Mongolia. Thus, M168 is not clearly the one source from Africa. He also, as did Sykes, admits the genetic link is slim, involving few, maybe less than one hundred of M168 who exited Africa to the Levant. Why could not Lara and M168 or M86 both migrate from the Middle East into Africa?

Improbable Guesses as to Routes of Passage

Wells' assertions about genetic development are quite improbable. He admits that M130 is equally as old as M168 (50,000 years for each) and it is not found in Africa at all.[51] It is located in the Levant and costal India. He assumes that these men (M130) walked around the coast of Arabia, the Persian Gulf, to and around India, and around southeastern Asia, and then went across the waters hundreds of miles to Australia. They supposedly did this as hunter-gatherers seeking new resources of shellfish, etc. The highest percentage of M130 (60 percent) is found in Australia and then decreases further away: 15 percent in New Guinea, 10 percent in Malaysia, and only 5 percent or less in India. A rare mitochondrial cluster, named M, is 100 percent in Australia and 20 percent in India. It is virtually absent from the Middle East and Europe, and M130 is likewise absent west of the Caspian Sea. These diminishing progressions would suggest the clan of M130 and M went out of Australia and not toward it. Also, M130 is found in Mongolia, Eastern Siberia, and in America.[52] Thus, they diverted from their coastal culture and went inland. In discussing the coastal route, Wells is clearly speculating. He uses words such as "A good estimate," "suggests," "no doubt," "almost certainly," "almost certainly," "suggesting that."[53] Paleontologists have evidence of stone tools in Sri Lanka, India and Australia, but this could be used in migration either direction.

Wells and Sykes also posed questions about migrations to America. They argue for two migrations from Siberia to Alaska via the Bering Straits. This is a widely accepted anthropological view, first suggested genetically by Allan Wilson. I believe there is a better view that I will offer later. There are four clans found in North and South America. Three are considered to have migrated across the Bering Straits. The clan of Ina seems to be an exception. There are no known inhabitants from these Americans in Siberia and Alaska. However, there is a Y chromosome marker 92R7 in Siberia that is associated with M3 in America. But this would appear to show the development in America with M3 and migration of 92R7 to Siberia on its own. Otherwise, why is M3 missing in Siberia? This M3 clan is found at 90 percent in South and Central American and 50 percent in native North America, as far north as Vancouver Island, making it a dominant

---

[49] Wells, *The Journey of Man*, cf. 56, 57.
[50] Ibid., 104.
[51] Ibid., 104, 109.
[52] Ibid., 74, 75.
[53] Ibid., 95.

American male chromosome. M3 is in most Native American men and on no other continent.[54] All men with M3 also have M242, which is in part of the men in 24 Eurasian populations who don't have M3. This indicates a strong Eurasian origin. Intriguingly, the Ina clan of America is also one that is closely associated with colonization of the Polynesian Islands from Southeast Asia. Sykes speculates this is done by seaborne colonization. It is thought that three clans from Siberia and Mongolia appearing in America apparently were already established from each other before reaching America. A fifth clan, Xenia, is found in one percent of Americans, and its origin is found in the borders of Europe and Asia. How did it get to America?[55] Does the Out of Africa migration answer these questions well?

Most of the mtDNA and Y Chromosome Evidence Originates in Eurasia

From Wells' map, it is evident that the main lines of Y chromosome markers proceed out of the Eurasian area, M89, M9, M45, M120, M175, M130, and even M168, which are the basis of the other male markers. He notes that M173, which branches from M45 in Western Asia, is the marker for 90 percent of western Europeans, for which all other lineages are lost.[56] Wells speaks of how M168, as the Eurasian Adam, is paired with L3, "Eurasian Eve." Wells says, "More evocatively, he could be seen as the Eurasian Adam—the great-great-grandfather of every non-African man alive today." He comments that without these, the chromosome evidence in the world would indeed be sparse.[57] He said, "It seems that the central Asian clan had made it to the new world as well, picking up the defining M3 marker in the process."

In Sykes' diagram of the world clans, he gives six family symbols: East Eurasia and America, East Eurasia, Central and West Eurasian, West Eurasia and African and West Eurasia, and Africa. Five of the six have to do with Eurasia. Only the African, a hexagon symbol, is separate. Sykes makes several statements agreeing that all emanates from Eurasia. Having found the seven daughters of Eve, Sykes comments, "With the possible exception of the Helena/Velda ancestor, these common ancestors lived way before modern humans ever reached Europe, most probably in the Middle East." Towards the top center of the figure (of his Seven Daughters diagram) is the common ancestor of all Europeans, where the Xenia branch leads off from the rest. Through this woman, the whole of Europe is joined to the rest of the world."[58] Again he said, "All the evidence points to the near East as the jumping-off point for the colonization of the rest of the world by modern humans."[59] The predominance of Eurasia as the location fits exactly with what we will see in Chapter 21, and was found to be the case in the origin of apes and men as David Begun indicates. How is this accounted for? More recent studies tracing worldwide human relationships that "constructed a phylogenetic tree by the maximum likelihood method using orthologous chimpanzee alleles as the outgroup" claimed Out of Africa as the source. But again, they show only one link from there to the rest of mankind. They admit that among "those in the Middle East and South/Central Asia, there are multiple sources of ancestry." In their chart, these families are in the center and all other branches come from there. The more logical interpretation, omitting African chimp connections, is that mankind branched from the genetic mainstream as they interbred. All the migrations presented by Sykes and Wells ignore the possibility of

---

[54] B. Harder, "New World Newcomers: Men's DNA supports recent settlement of the Americas," *Science News*, August 9, 2003, 85.
[55] Sykes, *Seven Daughters of Eve*, 281.
[56] Wells, *The Journey of Man*, 127, 128.
[57] Ibid., 71.
[58] Sykes, *Seven Daughters of Eve*, 274.
[59] Ibid., 278.

dispersion of humans in the break up of the continents from Pangaea (see Chapter 21).

Amazing is the dating of the male Y chromosomes for Adam as 59,000 years or 80,000 years after the mtDNA for Eve in Africa. This is an astonishing discrepancy, which Wells dismisses saying, "The dates have very little significance" because the data shows a convergence of male and females in Africa. *Converge* in Webster's dictionary means to "tend to a common point" which is not the case for Adam and Eve in Africa if they are apart by 80,000 years. If Adam was 80,000 years late for a wedding with Eve, we wouldn't be here.

Tracing DNA Racial Ancestry

Scientists have used polymorphisms or tiny genetic variations that recur in DNA reproduction and indicate previous ancestry, to distinguish large, related population groups. There are millions of such polymorphisms, but certain distinctive ones have been identified and used to trace the history of groups according to continents. The Alu sequence enables scientists to identify groups by geographic origin. Michael J. Bamshad and Steve E. Olson have found four major groups. Two in sub-Sahara are related and one of these is entirely of the Mbuti Pygmies. This leaves three main geographic origins of race. Bamshad and Olson report the two main Alu sequences as distinguished in the following quantities: (1) Asians, including the Mesopotamian area eastward: 60 percent Chromosome 1 Alu, 50 percent Chromosome 7 Alu (with 10 percent both); (2) Europeans and Northern Africans 75 percent Chromosome 1 Alu, 50 percent Chromosome 7 Alu (with 25 percent both); (3) Sub-Saharan Africans 95 percent Chromosome 1 Alu, 5 percent Chromosome 7 Alu. Only the Sub-Saharans show no mixture of both and have far less of 7 Alu. While they profess to hold to an Out of Africa theory, the evidence seems to indicate otherwise. The maximum interrelatedness would seem to be in Europe and North Africa; the second seems to be from Europe across Asia, and the least in Sub-Sahara. The map with the figures is given below.[60]

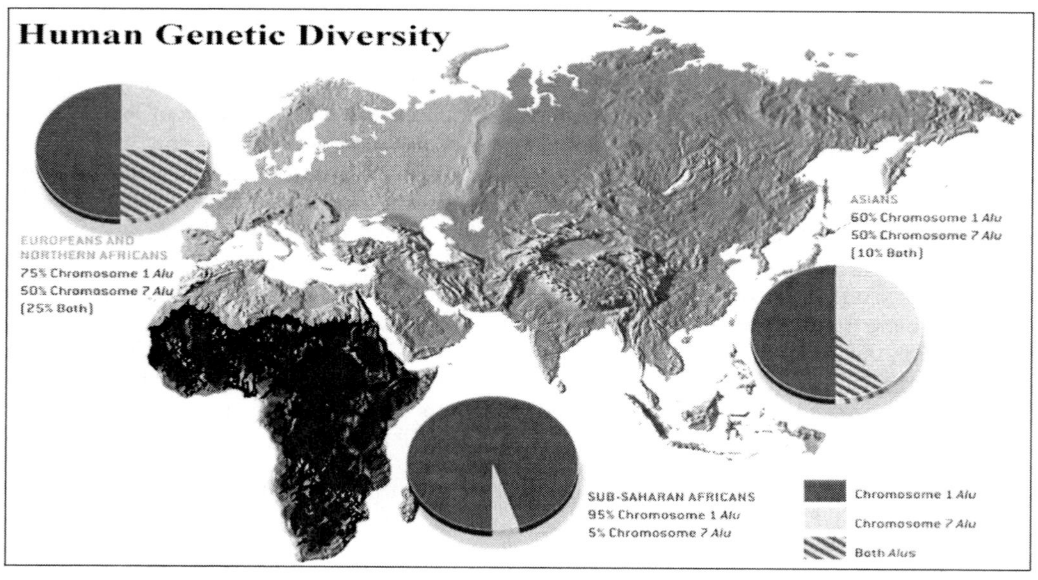

[60] "Worldwide Human Relationships Inferred from Genome-Wide Patterns of Variation," *Science*, 22. February 2008, vol. 319, 1100-1104 (cf. chart on 1101).

This map seems to confirm the biblical data concerning the three sons of Noah: that Japheth and Shem settled together while Ham (meaning black) and his son, Canaan, migrated alone to the south (Genesis 9:18, 19, 27).

Absolute Date Setting by Chromosome Markers?

Instead of the millions of years previously proposed from the fossil record for the transition from apes to men, the emergence of man as human was reduced to around 200,000 years ago. Genetic dating is based on the time it takes for mutations to be developed in the genes of the chromosomes and established as polymers. The mutation mistakes that occur are repeated in successive generations. Sykes says that the guess, one mutation in mtDNA, occurs in 20,000 years in a maternal lineage. Only those that occur in female eggs are passed on to the next generation. The long time needed is because re-combinations occur which scramble the data. By finding the sequence of mutations and comparing them, we can fit the genetic dating into clans and trace the relationship from each clan to the other.[61]

Human DNA is compartmentalized in four flavors of three billion nucleotides in the chromosomes. In copying, these chromosomes break and reattach. In the copying process, there are duplicates of duplicates of duplicates which introduce random errors. The recombination of polymorphisms allows independent behavior after hundreds of thousands of generations, and the patterns of the parent are changed with the remaining mutations. The deck is reshuffled. The molecular clock is based on polymorphisms that appear through mutations in the Y chromosomes. MtDNA do not recombine like Y chromosomes so that there is one polymorphism per 100, but there are 10 times more polymorphisms in the mtDNA than there are in the rest of the genome. The Y chromosomes more often tend to recombine and remove mutations. MtDNA genes have only one copy so they don't often recombine, making maternal dating easier. The difficulties of tracing Y chromosomes have been reasonably solved.

This method of dating is based on assumptions made by Luca Covalli-Sforza of Stanford that may or may not be true. Two landmark assumptions apply when analyzing human genetic diversity: (1) the genetic drift/mutations ***occur by chance***, according to Ockham's rule, which means that the simple will have precedence over the absurd. (2) This minimizes evolution so that the ***ones with the most genes frequency are closest together***. In addition, there is the assumption that the chance ***formation is at a regular rate*** and that they are added as permanent markers, so in time they become more and more complicated. Those with the greatest number of gene differences are assumed to be farther apart in time.

There are difficulties with these assumptions. The mtDNA markers may even be lost. Also, chance can't be trusted in information systems. Geneticists differ in how many years in a generation and how many generations it takes for a mutation to show up in using them as markers. Wells has admitted, "The ***possibility of error is high*** for absolute genetic dating methods, because there are quite a few assumptions involved in the way the dates are calculated."[62] (Emphasis added)

Moreover, using space and number of DNA mutation markers as a clock are based on these early assumptions that will be shown to now be faulty. This is in part because the later finding that much greater change is made more often by tandem repeat sequences of DNA. Also, as mentioned, it was shown that human DNA changes are much faster than other primates. This implies a much faster time clock with great uncertainties.

---

[61] Sykes, *Seven Daughters of Eve*, 155-159.

[62] Wells, *The Journey of Man*, 105.

What is the basis for *absolute* genetic dating?

Ann Gibbons, in *Science* "Research News," said,

> The most widely used mutation rate for noncoding human mtDNA relies on estimates of the date when humans and chimpanzees shared a common ancestor, taken to be 5 million years ago. That date is based on counting the mtDNA and protein differences between all the great apes and timing their divergence using dates from fossils of one great ape's ancestor. In humans this yields a rate of about one mutation every 300 to 600 generations, or one every 6000 to 12,000 years (assuming a generation is 20 years), says molecular anthropologist Mark Stoneking of Pennsylvania State University in University Park. Those estimates are also calibrated with other archaeological dates, but nonetheless, yield wide margins of error in published dates. But a few studies have begun to suggest that **the actual rates** are much faster, prompting researchers to think twice about the mtDNA clock they depend upon.[63] (Emphasis added).

So the first guesses about the dates of change in humans were based on the assumption of man coming from apes, as estimated by Allan Wilson who assumed the same rate of change of primates as in humans. How absolute were these?

Objective Evaluations to Standardize Absolute Genetic Dating of Humans

Until recently the efforts to calibrate this standard for absolute genetic dating have been based on little objective genetic data. Present efforts to find an actual basis for calibration of time were motivated by new data discovered when trying to identify the bodies of the Russian czar's family, who were shot by a firing squad in 1918. Because two bodies were thought to be missing, they were exhumed in 1992 and tested for mtDNA. What was thought to be Czar Nicholas II, had two different sequences of mtDNA, known as heteroplasmy, which was different from other members. This was not confirmed as the czar until it was found that his brother, the grand duke of Russia Georgij Romanov, whose body was later exhumed, had inherited the same condition from their mother. This was published in *Nature Genetics* in 1996. More conclusive evidence has further confirmed the czar's family.[64]

Thomas J. Parsons of the Armed Forces DNA Identification laboratory in Rockville, Maryland, helped identify this and has continued the study. Michael Holland, director of the lab says that heteroplasmy may be a frequent event, occurring in 10 to 20 percent of humans. This has led to other studies that show that ***human genetic markers occur much faster*** than thought. Paradoxically, in Sykes' research into his own genetic ancestry, he found he was related to Nicholas II. He therefore certainly knows of this study of actual calibrations.

Many systematic and extensive studies have been conducted to better establish the rate and time of human descent. Parsons, along with colleagues in the United States and England who helped identify the czar, has studied several groups, such as the Amish, soldier's families and old-line British families. Ann Gibbons said,

> The researches sequenced 610 base pairs of the mtDNA control region in 357 individuals from 134 different families, representing 327 generational events, or times that mothers passed on mtDNA to their offspring. Evolutionary studies led them to expect about one mutation in 600 generations (one for every 12,000 years). So they were "stunned" to find 10 base-pair changes, which gave them a rate of one mutation every 40 generations, or one every 800 years. The data were published last year in Nature Genetics, and ***the rate has held up*** as the number of families

[63] Ann Gibbons, "Calibrating the Mitochondrial Clock," *Science*, 2 January 1998 (volume 279, 5347), 28.
[64] "Czar's Missing Children ID'd," *Science*, vol. 323, 20 March 2009, 1543.

(studied) has doubled.[65] (Emphasis added).

Other studies have upheld a much-accelerated rate. Neil Howell, a geneticist at the University of Texas Medical Branch in Galveston, has studied 40 members of an Australian family with a disease. He uncovered three different mtDNA sequences and traced them back to 1861, showing *accelerated divergence rates*. These have been challenged by other teams of scientists who have found higher rates than Parsons, Howell, and others.

Since most of the calibrations of the mitochondrial DNA clock were done in 1995, the rate of the clock studied in these more recent studies has been found to be much faster than Sykes and fellow geneticists thought. Ann Gibbons published these new findings in *Science* magazine. She reported as follows.

> Regardless of the cause, evolutionists are most concerned about the ***effect of a faster mutation*** rate. For example, researchers have calculated their "mitochondrial Eve"—the woman whose mtDNA was ancestral to that in all living people- lived 100,000 to 200,000 years ago in Africa. Using the new clock, she would be ***a mere 6,000 years old.*** (Emphasis added)
> No one thinks that's the case [that Eve is 6000 years old], but at what point should models switch from one mtDNA time zone to the other?[66]

The rates are so much changed that the prevailing time is 6,000 years for the oldest genetic dating, with some extending it to possibly 12,000. The facts demonstrated by all these studies go contrary to evolutionary science and ***are rejected for that reason only***. The media and scientists have not made this significant fact of accelerated rate for humans known. In spite of early denials, there are no longer prehistoric men – Neandertal and Homo erectus are equally as modern humans as Homo sapiens have been counted to be. Equally important is the fact that dating by existing mutations is not reliable, since shifting whole tandem segments of sequences of DNA has been found to happen 100,000 times more often than single point mutations. The actual time of change is much shorter. The scientific truth of the age of man has diminished to near that which biblical creationists believe. And the map of human genetic diversity (p. 232) shows three different groups which concord with the three sons of Noah that survived the worldwide biblical flood.

The molecular and genetic search reveals extreme complexity and amazing, mysterious controls that show there is a superior intelligence involved. There is more mystery in microbiology than has been discovered and understood by microbiological science.

---

[65] Ann Gibbons, "Calibrating the Mitochondrial Clock," *Science,* 28, 29.
[66] Ann Gibbons, "Calibrating the Mitochondrial Clock," 28, 29.

SECTION THREE

MAN AND HIS HISTORY ON EARTH

PART V

ENLIGHTENMENT ABOUT UNIQUENESS OF MAN AND CIVILIZATION

…what is man that you are mindful of him, and the son of man that you care for him? Yet you have made him a little lower than the heavenly beings and crowned him with glory and honor. You have given him dominion over the works of your hands; you have put all things under his feet, all sheep and oxen, and also the beasts of the field, the birds of the heavens, and the fish of the sea, whatever passes along the paths of the seas (Psalm 8:4-8).

Chapter 16
Fossils of Human Evolutionary Descent

Early Search for Human Descent Began to Reveal Errors

The concept of evolution that came to maturity with Lamarck in France in 1809 and soon flooded England with the ideas in Chambers' *Vestiges,* of successive evolutionary creations turned the fossil search to the relationship between animals and humans. *Vestiges* had turned many, even in the churches, and the general population of England to befriend the evolutionary theory which prepared for Charles Darwin's view in *The Origins....* However, this became much more serious fifty years later, after Charles Darwin published his theory of evolution. Then a purely naturalistic, secular and anti supernatural view gained great popularity in Europe and America.

In *Descent of Man* (1871), Darwin speculated that man descended from what he considered man's closest allies—extinct apes, such as the gorilla and chimpanzee in Africa. In Darwin's and modern times, primates had not been found in Europe, but many were located in Africa. Loren Eiseley, the pro-evolutionary historian, pointed out that Darwin had not one single fossil of prehistoric men to base his views on when he wrote this book.

But beginning in the last decade of the nineteenth century and especially in the first quarter of the twentieth century, the search for the missing links for humans was intense. A variety of prehistoric men, thought to be hominoid, were found with different features.

During that time, the battles raged. Referring to this time of the early fossil finds, pro-evolutionary Eiseley said, "That the activities of confirmed evolutionists, as the Darwinian enthusiasm began to mount, are sadly revelatory of a state of mind in its way as *dogmatically fervid* as that of those opposed to the evolutionary point of view."[1] The nature of various fossils was perverted to meet the demands of what each side wanted. For example, in referring to fragmentary Neanderthal remains of La Naulette, one science book described the specimen as extremely apelike, "with huge projecting canines," when *the teeth of the specimen were missing.* Eiseley described the unscientific speculation that entered the effort saying, "No known form of fossil man possesses gorilloid canines. These descriptions are *the product of imagination....*"[2]

Multitude of Claims of Intermediaries

The search, with resultant claims, to find intermediaries between apes and men emerged all over the world and are far too numerous to mention. Some important discoveries *in Europe* include a Neanderthal man found in Germany in 1865 who was considered fully man until later Marcell Boule, a Frenchman, declared these fossils humped over and intermediates to apes and men. The Heidelberg jaw, uncovered in 1909, was considered related to Neanderthal and too human to be a link. Charles Dawson found a skull and jaw in England, near a gravel pit at Piltdown, which Andrew Smith Woodward of the British Museum presented to the Geological Society of London as an intermediary between man and ape. He was named, Eoanthropus dawsoni, or Dawn Man in 1912. Other discoveries include the Swanscombe skull found with tools and weapons in Barfield Pit near the Thames River in 1935, between Danform and Gravesend; Gally Hill man near London; Fonechevade man in France; and Mentone skeletons on the Italian Rivera. Lartet found a different, but fully human, Cro-Magnon in rock shelter/caves near the

[1] Eiseley, *Darwin's Century*, 277.
[2] Ibid.

village of Les Eyzies in France. This appeared Negroid or American Indian with proportions of 5′6 ½″ (long forearms cf. upper, and long shins cf. to thighs).

When the focus shifted to Asia for a time, significant earlier finds were thought to be intermediates. *Java* produced some fossils. In 1890 Dubois, a Dutch army physician, found the Trinil man, along with a skull cap (914 cc.), two molar teeth, a pre-molar, and a femur 50 yards away. Dubois said the ratio of skull to length of femur would correspond to a gibbon. This, plus successive finds in Java, were called Pithecanthropus. About nine different finds were given this name, including some juvenile skulls. Von Koenigswald, who labored in the Java area until about 1940, named two very large fossils with huge teeth, Magganthropus. Von Koenigiswald's 1930 find was along Solo or Bengaway River in central Java. Eleven fossil skulls, with two tibias, were found with capacity ranging from 1160 cc. to 1316 cc. These were human, but different, and were claimed as intermediates. Near Peking, China between 1922 and 1938, a find consisting of 38 fossils of teen jaw fragments and odd bits of skulls, six skulls with vaults preserved, was unearthed. Dubbed the Peking Man (Sinanthropus), forty percent were children under the age of 14. Peking brains were 850-1300 cc. or average 1075 cc. The skull had a prominent ridge like Eskimos and Australians and broad nasal aperture like Negroes. In Hong Kong in 1935 von Koenigswald purchased a gorilla molar in a drugstore in Hong Kong and named it Gigantopithecus. Weidenrich considered it the largest and most primitive member of the human race. In *Australia* the Talgai skull was publicized as the Cohuna specimen, and Keilor skulls from Keilor village north of Melbourne (1593 cc.) were all within the range of modern aboriginal skulls.

Gradually *Africa,* the source first suggested by Darwin, became the focus, with a search from apes toward Homo sapien forms. Earlier in 1924, Raymond A. Dart had discovered the Taung Baby, a six year old primitive hominoid in South Africa. He considered this find, named Australopithecus africanus, meaning "southern ape African," to be one million years old. Later this A. africanus was re-dated about 2.8 to 2.4 million years old. Dr. Robert Broom, of the Tansvaal Museum, Pretoria, South Africa, made similar finds, which he called Plesianthropus, Paranthropus, and Australopithecus Prometheus. Dart and Broom's findings were small apelike skulls (450-650 cc) and reduced to one genus and two species. They were Australopithecus africanus – Taung, and another with Sterkfontein and Makapansgar and Australopithecus Robustus from Slwarkrans and Kromdraai and East Turkana. To these later ones from East and North Africa were: Rhodesian man, an apelike skull with other bones that are human. Boskop man, 1913 found in the Transvaal, a very large skull (1,630 cc.), with teeth and a structure like small Bushman 5′ 6″, and were definite Homo sapien; and the Asselar man, in Sahara Desert east of Timbuktu 1927. In Bougie, Algeria there were 48 finds, some full skeletons, like the Galley Hill type of Cro-Magnon men.

Graphic Icons Made for Speculative Claims

During this period, graphic drawings were made of Darwin's idea of the evolution or descent of men from higher primates. After some time, Thomas Huxley gave qualified acceptance to Darwin's *Origins....* Seven years before Darwin wrote his *Descent of Man,* Huxley wrote *Man's Place in Nature* (1863), comparing skeletons of apes to man's, showing gradations as depicted on the next page.[3] He compared, looking for similarities, without any fossil support for how any species of apes produced another or how man might have developed from others. Earlier in his career Huxley had argued to Charles Darwin that there were no known intermediaries. However,

[3] This figure was taken from J. Wells, *Icons of Evolution*, 215.

this illustration proved to be a strong motivation to build such a lineup of apes to men.

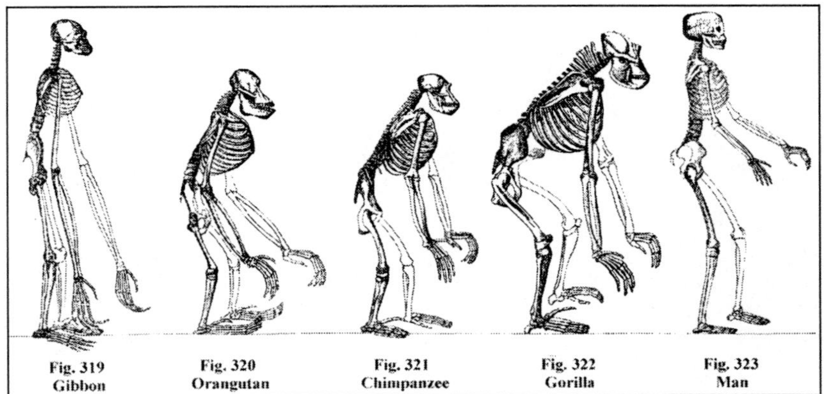

Based on similarities, in his treatise, *Man's Place in Nature,* Huxley said,

> But if Man be separated by no greater structural barrier from the brutes than they are from one another, then, there would be ***no rational ground*** for doubting that man ***might have*** originated… by the gradual modification of a man-like ape [or] as a ramification of the same primitive stock as those apes. (Emphasis added)

His arguments were philosophical, based only on similarities, and not from any claim of known evidence of transmutation.

However, Henry Fairfield Osborn, paleontologist from Princeton, studied under Thomas Huxley, who had assembled skeletons of different animals. Osborn, the curator of the American Museum of Natural History in New York, became a leading authority on vertebrate zoology. At the museum he outdid Huxley and developed halls to display various reconstructed animals. Osborn used much from Huxley, but Osborn, not a fan of Huxley, was fully dedicated to Darwin's claim of natural selection and *transmutation* of the species. In his biography of many naturalists, Osborn said of his famous teacher and his idea of natural selection, "He [Huxley] never contributed a single original or novel idea to it."[4] Yet, Osborn adopted the ideas of chance beginning of life, of the evolution of the horse, his reconstruction of vertebrates, and Huxley's illustration of the progression from primates to man. These became Osborn's basis of his "Hall of Men."

Almost all the modern textbooks that include a graphic design of man's evolution do so based on Osborn's Hall of Men. Osborn began with a full-fledged ape that was hunched over and supplied skeletons showing a progression to upright modern man. In his arguments he included the Nebraska man, the Piltdown man, the humped over Neanderthal, and other so-called links. Following is a more correct illustration, similar to Osborn's, which became the icon in all biology textbooks.[5]

---

[4] H. F. Osborn, *Impression of Great Naturalists*, 1924, 91.
[5] J. Wells, *Icons of Evolution*, 210.

Osborn's writings were the chief source of information for the evolutionary theory. His works include *Huxley and Education* (1910*), Men of the Old Stone Age* (1915), *Origin and Evolution of Life* (1917), and *The Earth Speaks to Bryan* (1925), written after the Scopes trial. He and his Hall of Men have been the strongest influence for excluding the teaching of Creation in American public schools. During the second quarter of the twentieth century Osborne's exhibition in the American Museum of Natural History in New York City was prominently displayed and shown to thousands of children and adults every year as the evidence of man's descent. The argument for man's evolution involved associating fossil specimens of male and females, young and old from all over the world, oftentimes sequencing fossils found some *thousands of miles apart*. Some of the most important reconstructed fossils presented in that line of descent from apes to man have since been proven fraudulent.

Errors and Fraud in Fossil Claims by Osborn Used in the Dayton Scope Trial

Henry F. Osborn's Hall of Men was a main resource in the Dayton trial. The Piltdown man, with its human skull and apelike jaw, came under suspicion because it did not correspond to other fossils found at the same time and in the same location. In 1953 Joseph Weiner, Kenneth Oakley and Wilfred Le Gross Clark, discerning that the fossil did not seem to fit, proved the skull was that of an old, modern human, but the jaw came from an ape. The jaw, which was actually from a more recent modern orangutan, had been chemically treated to look older; the teeth were filed and colored. Authorities acknowledged the evidence was fraudulent.[6] Jane Maienschein, a biological historian, referring to the Piltdown fraud said, "How easily susceptible researchers can be manipulated into believing that they have actually found just what it was they had been looking for."[7]

[6] J. S. Weiner, E. P. Oakley, and W. E. Le Gros Clark, "The Solution of the Piltdown Problem," *Bulletin of the British Museum (Natural History), Geology 2,* 1953, 139-146; Stephen J. Gould, " The Piltdown Conspiracy," *Natural History 89 (August 1980),* et al.

[7] Jane Maienschein, "The One and the Many; Epistemological Reflections on the Modern Human Origins Debate," in G. A. Clark and C. M. Willermet, ed. *Conceptual Issues In Modern Human Origins Research,* (New York: Aldine de Gruyter, 1997), 413-422.

The Nebraska man was another fraud. In 1922 experienced geologist Harold Cook submitted a single tooth to Henry Fairfield Osborn, who stated it was a new genus of an Anthropoid ape. Osborn named it Hesperopithecus haroldcookti. William K. Gregory and Mile Hellman of the American Museum of Natural History, specialists in fossil teeth, wrote several publications to designate its place in the missing links. Hellman sided with Osborn that it was more apelike, and Gregory that it was nearer human. G. Elliot Smith called it the oldest of all human remains known to science and probably a migrant from Asia or Africa. H. H. Wilder, professor of zoology at Smith College, had said it was halfway between then-Java man, Pithecanthropus (Homo erectus) and Neanderthal. As a new fossil, it was used as evidence in the Scopes trial. Further exploration near the site in Nebraska uncovered more extensive and complete fossils and the tooth was definitely identified as that of *an extinct pig* called, Prosthennops.[8]

Osborne accepted Boule's idea of Neanderthal as humped over like an ape. He misrepresented the Neanderthal man because the particular specimen was later found by Neanderthal paleontologists to be an arthritic one. Many more well-preserved fossil finds have shown Neanderthal walked erect. The old exhibits of human descent, such as Osborn's, were quietly abandoned. The Hall of Men was transferred to the attic, but the public, influenced by the media to believe evolution, were not told these were false. Moreover, almost every textbook still used in the public schools contains this false icon as a scientific fact.[9] The Scopes trial thus made monkey of the truth and propelled evolution into favor. Even the stalwarts of Presbyterian theology, such as B. B. Warfield and J. Gresham Machen, who had studied in Europe, accepted much of the evolutionary theory.

Conclusions of a Century of Search

Nearly a century after Darwin, most of the mistakes and frauds had been found. All of these, with a few exceptions, were either apelike with small cranial capacity and long jaws and teeth, or larger skulls with more fully human features. Many evolutionary scientists began to feel that the evolution of man and primates began and separated much earlier than any fossils yet found (an argument from non evidence), and that their relationships were not obvious. It was hoped that in time older finds would tell the story. Commenting on this fact, the evolutionary historian Loren Eiseley said, "Later, as the forests were cleared and the apes were seen in the sunlight, ***the gap loomed a little larger*** between man and his beasts." Man made his efforts to know the story of how apes became humans but had not found the evidence. Eiseley continued:

> In the end it could no longer be done. The tale will tell itself and man will listen. ***He is quite alone now.*** In spite of the claims that persisted into the beginning of this century, his brothers in the forest do not speak. Utterly alone, man senses the great division between his mind and theirs. He has completed a fearful passage, but of its nature and causation even *the modern biologist is still profoundly ignorant.*[10] (Emphasis added)

Fervent Search Back to Africa for the Next Generation

The efforts to find a clear line of intermediary descent from primates to man had not only

[8] William K. Gregory, "Hesperopithicus Apparently Not an Ape nor a Man," *Science,* December 16, 1927 Vol. 66:1720:579.

[9] Jonathan Wells, *Icons of Evolution,* 209 ff. for more discussion of this.

[10] Loren Eiseley, *Darwin's Century*, 284-285.

failed, but spread to other parts of the world and added confusion. Therefore, the serious search shifted back to Africa, as Darwin first suggested. In the late 1960s and through the 70s, scientists made important new discoveries using brilliant methodologies for finding and assembling fossils, sometimes from dozens of pieces. Notable explorations were those of Louis and Mary Leakey in Olduvai Gorge; Clark Howel and Richard Leakey near the Omo River; Richard Leakey in East Turkana; Donald Johanson and Maurice Taieb in Hadar and the Afar region of Ethiopia; and Louis and Mary Leakey at Laetoli. A multitude of cranial and other bones, along with Stone Age implements and weapons, were found. Moreover, Mary Leakey at Laetoli found *human footprints with ape footprints* in lava and also Homo sapien fossils which were earlier than many apelike ancestors. Also disturbingly, Richard Leakey's ER 1470 was human and came before the significant find of Lucy and her clan that came near the end of this period and were hailed as intermediaries. East African Australopithecus Afarensis were categorized with those of South Africa as much older. These were found about 2,500 miles apart. Using this for dating is straining facts, but allows an earlier date as an intermediary for all of them.

Until the 1980s none of these finds offered any conclusions about the gap between man and apes. Efforts continued to clear up Mary Leakey's old human footprints. Others found since have caused problems for evolutionary theory. Therefore, efforts have been made to trace a development of footprints from apes to men.[11] However the comparisons of adjacent prints dated 1.5 million years old from Lleret, Kenya, indicated that the Homo erectus was distinct from apes.[12] Other efforts have been made to show a progression of birthing, age of weaning, tooth development, and lifespan from childbirth. Where the data was available, chimpanzees and Australopithecus afarensis (i.e. Lucy) were shown to be much different from Homo erectus and consistently so for Homo sapiens. Fossil pelvises of women show that Homo erectus like Homo sapiens could birth babies with much larger heads than apes. Also, it is said, "Key events show that modern humans live slower and die later…"[13]

The 1980s ended with *much new data but with no clear relationships*. Niles Eldredge and Ian Tattersall, both curators of the American Museum of Natural History, reviewed the evidence and presented a summary with conclusions about human evolution. As the rest of the natural sciences had reached a point of mystery instead of uniform determinism, so the effort to trace the tree of life and give evidence of human evolution had ended in greater mystery, creating a crisis for paleontology.

In Tattersall's book that labels finding intermediaries a mystery, he presents a graph as of about 1980 of the finds of the pre-human or hominoid fossils, (see following page). I have drawn a perpendicular line in the graph showing a divide between the ape or chimpanzee-like finds which are on the left and those on the right which are the homo or humans. The far right shows the stone implements or weapons. The vertical lines below the skulls show when that type began and ended in the geographical record. The bottom lines show the beginning in the strata and extending upward. The dotted lines show these in time relationship.

The graph indicates that the homo or *human fossils often began contemporaneously or before*

---

[11] Robin How Crompton and Todd C. Patsky, "Stepping Out," and Matthew R. Bennett et al., "Early Hominin Foot Morphology…." both in *Science,* vol. 323 , 27 February 2009, 1174, and 1197-2001.

[12] Bruce Bower, "Footprints suggest H. erectus walked the modern human walk," *Science News*, March 28, 2009, 14

[13] Ann Gibbons, 'The Birth of Childhood," *Science*, 14 November 2008, vol. 322 1040-1043, from sciencemag.org, cf. Bruce Bower, "Stone age gal had wide hips," *Science News*, 1 December, 2008 14.

the apelike fossils, and the Stone Age *instruments begin even earlier* and indicate the existence of human intelligence. There is a clear overlapping of evidence of humans with the apes they are supposed to have evolved from.

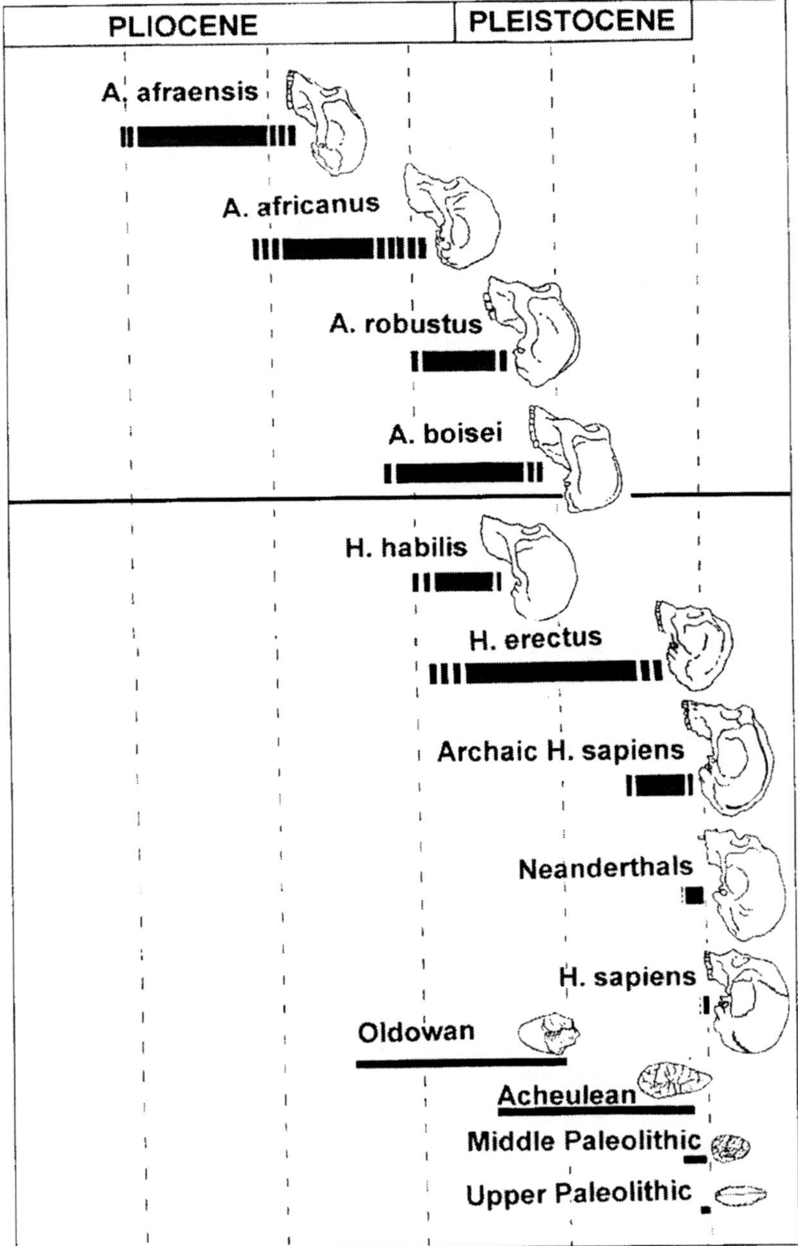

The Quandary over Human Descent with Many Theories of Descent that Arose

Finding Australopithecus Afraensis in East Africa seemed to furnish a breakthrough in locating an intermediary. A. Afraensis, such as found in South Africa earlier (dated 2.5 - 3 million years), added to the specimens and other evidence of early humans. The most famous of these is Lucy, a three and a half-foot female, found by Don Johanson, assistant to Tim White of Berkeley. The lower jaw, some skull fragments, most of both arms and a leg, parts of the spine and rib cage were found, but no "hands" or "feet." *Nova* made a video, widely shown on public television, in which the artist has presented her as a graceful upright lady with apelike head and hair (though no hair was found) gliding along through the jungle paths as the precursor to modern humans. Since finding Lucy, there have been 52 or more males found who are *larger than the modern ape*. Because of the difference in size, some feel these should be two species, and often some are designated differently, but male apes are always larger. These were dated about three million years old, speculated to walk upright, and, some hoped, the first real intermediary to man. But these and other finds in different parts of East Africa *disturbed all previous charts* about the descent of man.

Finding a *route of evolution became mystery or myth*. By the last quarter of the twentieth century the lack of intermediaries for most species of the fossil record was now clear. Niles Eldredge and Ian Tattersall collaborated and argued that because stasis, the idea of progressive change in the fossil record, was attended by another myth that if one finds enough new fossils he can trace the descent of man. They argued that there was no clear evidence of the descent of man, and that the clear lack of intermediary progression in the fossil record for other species was likely to continue showing a lack of evidence for man, even as more fossils are found.[14] Moreover, they argued, "In no sense can we explain human cultural evolution (e.g. stone tools) in terms of biological evolution."[15] They said similarity of patterns of change "is only a similar pattern of episodic change."[16] The importance of the disjunction of culture and biological evolution will be seen in Chapter 17 on human uniqueness and chapter 18 on evolution and culture.

Three years later Pat Shipman, a paleontologist at John Hopkins University, who had participated in some of the East African digs said, "An extraordinary 2.5 million year-old skull (of Lucy) found in Kenya [summer 1985] has overturned all previous notions of the course of early humanoid evolution. We no longer know who gave rise to whom–perhaps not even how, or when, we came into being." She continued:

> It's a new era in paleoanthropology. The things we thought we understood reasonably well, we don't.... Bill Kimball, one of the scientists most knowledgeable about A. Afarensis and one of the first to see a cast of the new skull, acted for all of us the other day, at the end of a lecture on australo-pithecine evolution; he erased all the tidy, alternative diagrams and stared at the blackboard for a moment. Then he turned to the class and threw up his hands.[17]

Soon new revisions of man's evolution were forthcoming, but articles reveal these arguments were unconvincing to many. Reviewing the situation in 1989, William F. Allman said, "Paleontologists are coming to realize that all these scenarios are ***unsupported by*** *a closer scrutiny*

---

[14] Niles Eldredge and Ian Tattersall, *The Myths of Human Evolution,* (New York: Columbia University, Press, 1982).

[15] Ibid., 5.

[16] Ibid., 4.

[17] *Discover*, Vol. 7, No. 9, September 1986, 91-93.

of the evidence in the fossil records."[18] (Emphasis added). By 1992, other efforts were made. Donald Johanson, who discovered Lucy, published three optional graphs and Richard Leakey gave five hypotheses in optional graphs.[19] While these two lists and data of fossils are very much the same, there are a few differences in similar species and some slight differences in dates. While both admitted at one time that men/homo fossils existed *with or before the Lucy family* of Australopithecus A., they still cited *these as intermediates*. How could this be? The focus in dating was mostly in trying to reconcile finds in Africa, ignoring those found elsewhere.

Problems with Australopithecus Afarensis as an Intermediary

Initial data about the first A. Afarensis fossil, Lucy, was not clear. While the skull was incomplete, it was believed she walked upright. There was a question about the shape of the leg and hip for walking, but it was conceded, at first, that she walked, even though her phalanges were missing. Of the 52 other fossil fragments found, most are male and larger than modern apes. In fact, the *teeth are larger than any found before* and they have a long muzzle, like a chimpanzee. Looking at Laletoli-Hadar fossils, the upper jaw is longer, but not quite as long as a chimpanzee; slightly more rounded like the human, but *much larger than a human upper jaw*. The canines are not as large as that of the chimpanzee, but there is a gap in front of the canine, like the chimp's, and a little smaller. There is no gap in human teeth. Comparing the fossils, most anthropologists admit one would not confuse the teeth with the average human's.

The early claims and the *Nova* video of Lucy's graceful walking and running have been shown to be incorrect. The feet of Australopithecus Afarensis specimens found later are proportionately thirty percent longer than human feet and have long curved phalanges, so walking would be awkward. The hands have long curved phalanges. So, the feet and hands are like apes', adapted for climbing. The short hind limbs and long arms would indicate the walking or running would be like the chimpanzee, not like humans.

Peter Schmidt thoroughly examined the skeleton of Lucy in Zurich and said,

> What you see in Australopithecus is ***not*** what you'd want in inefficient *bipedal* running animal. The shoulders were high, and combined with the funnel shaped chest, would have made arm swinging improbable in the human sense. It wouldn't have been able to lift its thorax for the kind of deep breathing that we do when we run. The abdomen was potbellied, and there was no waist, so that would have restricted the flexibility that's essential to human running.[20]

In spite of Schmidt's analysis, Leakey concluded, "In other words, Lucy and other australopithecines *were bipeds*, but they weren't humans, at least in their ability to run. ...The australopithecine species almost certainly were not adapted to a striding gait and running as humans are. Here the adaptations to tree climbing can be seen in Australopithecus Afarensis."[21]

---

[18] B. Bower, "Humanoid Evolution, A Tale of Two Trees," *Science News*, July 4, 1987, Vol. 132, 7; B. Bower, "Retooled Ancestors," *Science News*, May 28, 1988, Vol. 133, 344-345; William F. Allman, "The First Humans," *U. S. News and World Report*, February 27, 1989, 50-59.

[19] Donald C. Johanson and Maitland A. Edey, *Lucy: The Beginnings of Humankind* (New York: Simon and Schuster, 1981) and Richard Leakey and Roger Lewin, *Origins: Reconsidered,* (New York: Doubleday, 1992), Richard Leakey, 131.

[20] Richard Leakey and Roger Lewin, *Origins: Reconsidered* 194-195.

[21] Ibid.

In comparing the location of the larynx, the human's is very low, allowing a wide range of sound productions, while in the ape/chimpanzee and also in Australopithecus, it is very high, allowing a restricted range of sound production. This shows clearly that expression of language for speech, which is a primary human characteristic, was not practiced.

Many feel that this species is more a *deviation in chimpanzees.* The complete ulna, which was found, curves like a chimpanzee's. C.F. Spoor of the University of Liverpool, England, and Frans W. Sooneveld, a radiologist at the University of Utrecht, Netherlands, did computerized tomography scans of the area around the ear in skulls of humans, chimps, gorillas, and orangutans. After doing the same with Australopithecus Afarensis, Spoor announced to the American Association of Physical Anthropologists, "Australopithecines are *more similar to chimpanzees than to modern humans* in their inner ear anatomy. This supports the view that Australopithecines combined arboreal [tree climbing] and terrestrial movement." Sir Solly Zuckerman of Edinburgh assembled all the arguments above and especially emphasized that the foramen magnum (the place the spinal column enters the head) is positioned to the rear of the head in apes/chimpanzees while man's is almost in the center. He argues that *it **is myth** to make them walk* in any other position than that of apes because erect biped walking would be awkward and uncomfortable.[22] (Emphasis added).

A later significant event indicated that full men already existed when A. Afarensis existed, making them brothers in history, *not ancestors* of one another. After Donald Johanson's review of Richard Leakey's find of the almost complete skull of the hominoid fossil, 1470, which Leakey dated as 2.9 to 3 million years old, Johanson said this:

> To summarize: when the australopithecine collections were hastily reexamined after Richard Leakey's astonishing announcement of 1470, it became disturbingly clear that here was a remarkably advanced *Homo* that was about as old as any of them. Or, to put it in the other way round, *one of the best-preserved and **oldest bona fide hominid fossils** in the world was a **human** and not an australopithecine.* And yet, Australopithecines were clearly more primitive in a number *of characteristics*. That just *did not make sense.*[23] (Emphasis added)

Previously Johanson had said, "To begin with, there could no longer be any doubt that robust Australopithecines were ***not ancestral*** to Homo."[24] (Emphasis added). In spite of the admissions of both Donald Johanson and Richard Leakey, and these facts which contradict *their diagrams* of man's descent, they both *list Australopithecus as **ancestor** to the Homo* fossils, simply because their characteristics were more primitive; not because they come first in the paleontology record. Is this scientific? Morphology cannot prove descent in fossils any more than it proves descent today.

A large body of evidence is accumulating that indicates Australopithecines were not ancestral. Recently, two fossil hunters from the Afar tribe in Ethiopia's Hadar region, where Lucy was located, found two pieces forming an upper jaw, with ten complete teeth and with many fragments. *Human tools were also located in this find.* The team, led by William Kimbel from the Institute of Human Origins in Berkeley, announced in the December 1996 edition of *The Journal*

---

[22] Sir Solly Zuckerman, "Myths and Methods in Anatomy," Journal of Royal College of Surgeons, (Edinburgh, Vol. 11, No. 2, 1966), 87-114.

[23] Johanson, *The Beginnings of Human Kind,* 140-141.

[24] Ibid.

of Human Evolution that the pieces were 2.33 million years old. This skull, with tools, is between the two branches of Homo habilis and Homo rudalfensis and *contemporaneous with A. afarensis*. This finding, along with other evidences, shows A. afarensis can't be intermediary. Mary Leakey and others identified footprints at Lawetoli, Tanzania, dated 3.6 million years old as indistinguishable from modern humans. Vigorous efforts to deny these footprints as fully human have been refuted.[25] Also, a modern humerus found at Kanapoi, Kenya is 4 million years old; and ER 1481 femur from Lake Turkana, Kenya is dated 2 million years old. These are only a few of the inconsistencies uncovered.[26]

Admission of Simultaneous Diversity while Holding to Developmental Tree

In January 2000 Ian Tattersall, the leading voice in the field, along with Jay H. Matternes, an artist, sought to present a speculative partial solution to the origins of man.[27] He presented a central thesis with a graph in which there is no evidence for one line to the human species, but for several sudden appearances. He said,

> Sharing a single Landscape, four kinds of hominids lived about 1.8 million years ago in what is now part of northern Kenya. Although paleoanthropologists have no idea how–or if–these different species interacted, they do know that *Paranthropus boisie, Homo rudolfensis, H. Habilis and H. ergaster* foraged in the same area around Lake Turkana.
>
> Their (the paleoanthropologists) tendency was to downplay the number of species and to group together distinctively different fossils under single, uninformative epithets such as "archaic Homo sapiens." As a result, they tended to lose sight of the fact that many kinds of hominids have regularly contrived to coexist. Although the minimalist tendency persists, recent discoveries and fossil reappraisals make clear that the biological history of hominids resembles that of most other successful animal families. It is ***marked by diversity rather than by linear progression***. Despite this rich history–during which hominid species developed and lived together and competed and rose and fell–H. sapiens ultimately emerged as the sole hominoid. *The reasons for this are generally* ***unknowable***, but different interactions between the last coexisting hominids–H. sapien and H. neanderthalensis–in two distinct geographical regions offer some intriguing insights.[28] (Emphasis added)

---

[25] William F. H. Harcourt Smith of American Museum of Natural History, Chares Hilton of W. Michigan Univ. say they are as human's as Leakey's.

[26] See also Michael A. Cremo and Richard L. Thompson, *Forbidden Archeology: The Hidden History of the Human Race,* (Los Angeles: Bhaktivedanta Book Publishing, Inc., 1996), xxiii-xxxvii.

[27] See Ian Tattersall and Jay H. Matternes, "Once We Were Not Alone," *Scientific American*, January 2000, 56-62 for his (Tattersall's) popular summary and more technical book, *Becoming Human: Evolution and Human Uniqueness,* (Harcourt Brace, 1998).

[28] Ibid., 57, 58.

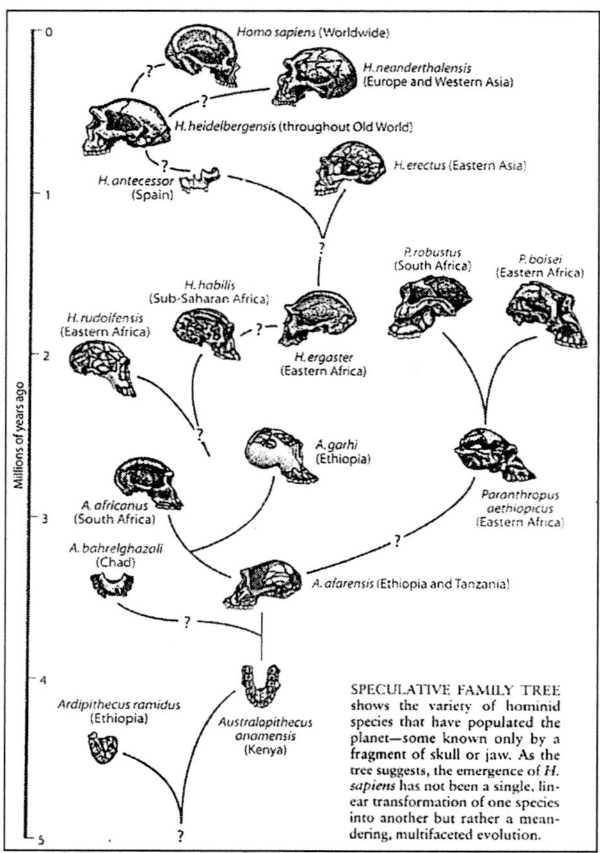

SPECULATIVE FAMILY TREE shows the variety of hominid species that have populated the planet—some known only by a fragment of skull or jaw. As the tree suggests, the emergence of *H. sapiens* has not been a single, linear transformation of one species into another but rather a meandering, multifaceted evolution.

In the graph, fossil specimens (read from bottom to top) are *not joined by lines* between a grouping of hominids or men. Some are known only by a fragment of skull or jaw. As the graph suggests, the emergence of H. sapiens could not have been a single, linear transformation of one species into another, but rather a meandering, multifaceted evolution. The gaps are present. He does not connect all the lines together as one would in what is normally called *a family tree,* because that is contrary to his correct claim of existence in diversity, together in mysterious and unknown relationships. But by *calling it a family tree and by the apparent arrangement in an ascending order*, the impression is given that these someway are all related, which he admits no one knows if they are. He admits there is very scarce evidence for many of these.

Is not this a distortion of the data? He also includes Australopithecus as the *beginning point of the first skull evidence*. Tattersall's article is, on the one hand, intended to be a deliberate denial that one can draw a line of descent (which is what a family tree is), but is on the other, veiled to imply there was a line of descent. The border designation of his graph claims "millions of years," to imply *gradualism in progressive development* which is absent from the fossil record.

Last Quarter of the Century Data Produced Changes in Concepts

The absence of an intermediary route to man from apes and the frustration as expressed in Eldredge and Tattersall's book, *The Myths of Human Evolution,* began to produce changes in the methods and locations of the search for an intermediary. Beginning in 1967 and standardized in 1980 everything in a site was considered. Paleontologists also expanded the African finds from the "savannah" terrain to the desert and more recently into forest territory. New fossil evidence of man continues to emerge and add to the confusion.

In Oma in 1967 the team learned that the search in paleoanthropology had been driven by the passion to find hominids, to the neglect of the flora and fauna of the environment that would shed light on the time and circumstances of the fossils. The lack of clarity of proposed human descent in 1980 increased the fervor and expanded the search, but was still unclear in 2000.

Ann Gibbons, primary evolutionary writer for *Science* Magazine, reviews many of the significant later discoveries in her book, *The First Human.* A review of these is important to see the trend.

In 1992 Gen Suwa and Tim White found a primitive baby molar, a lower right jaw of a child with milk molar, another ten teeth together, and an early skeleton at Aramis, Ethiopia. Claiming this was an older intermediary; they named it Ardipithicus ramidis, called it Root Ape, and dated it 4.4 million years old. The teeth were like a chimpanzee and not like human teeth. In 1994 Meave Leakey and Hominid Gang found a shinbone in Kanpoi, Kenya, which they claimed showed upright walking of a chimpanzee/hominid named Australopithicus andamensis, dated at 3.9 to 4:1 million years old. The finds included half an upper jaw, later a complete lower jaw, some teeth together, later baby molars like White's, an ear region of a skull, and two parts of a shin bone similar to Lucy's. The teeth were like a chimpanzee's. A year later Michel Brunet and Mission Paleoanthropologique Franco Tchadienne (French: Franco-Chadian Palaeontology Mission; est. 1995 (MPFT) found fossil remains in Koro Toro, Chad, similar to the Lucy clan, named Austraopithicus barelghazali and called Abel, dated at 3.5 million years. This supposedly showed that hominids had existed in north central, as well as eastern Africa. Some felt this was the same as the Lucy clan. A scapula of a young girl found in Dikika, Ethiopia, was like Lucy and considered 3.3 million years old. Daniel E. Lieberman of Harvard University commented that her "scapula indeed resembles scapulas of gorillas" who were tree climbers. She had "a flat nose and projecting face that looked like chimpanzees."[29]

In 1996 Yohannes Haile-Salassie found W. Margin in Middle Awash, Ethiopia. This finding included a toe bone that might fit upright walking, and teeth more like humans than apes. The specimen, named Ardipithicus kaddaba, was dated at 5.8 million years. In 2000 Martin Pickford and Brigitte Senut found a lower leg/tibia, broken thighbones, ribs, vertebra, collarbones, pelvis, shoulder blades, upper arm bone and a canine tooth at Togen Hills, Kenya. It was *claimed* that these indicate a hominid that walked upright, but these lacked complete skeletons and phalanges. Lacking these, the evidence certainly is not clear. Dated six million years, these were named Orrorin tugenensis, and called the Millennium man. This was not scientifically reported, not properly photographed, and restricted from others who wanted to view these.

Frenchman Michel Brunet and a team made an important discovery at Toros-Menalla, Chad, in 2002. They unearthed an apelike, crushed skull with teeth, but missing the lower jaw. Two pieces of one jaw, perhaps related, were later found nearby. These were dated 6 to 7 million years. The crushed skull was reassembled. (Brunet was accused of misplaced gluing of a tooth). Sahelanthropus tchadensis was Burnet's name for this new intermediary. It is nearly a complete

[29] B. Bower, "Evolution's Child," *Science News*, Sept. 23, 2004, Vol. 199, 195.

cranium, fractured jawbones and teeth about which Harvard University anthropologist Daniel Lieberman said, "it is the oldest skull by far of a human ancestor. This will have the scientific impact of a small nuclear bomb."[30] It was *speculated* to have the brain stem enter the skull more on the vertical like a human. After trying to fit many of the pieces together, the measurements of the angle of entry were not conclusive, nor were computer speculations. Moreover, the brain size was only 360-380 cc. That is smaller than an average chimpanzee and is much smaller than modern humans of 1,400 cc. Neandertals have even larger brains. The shape of Sahelanthropus tchadensis is longer, like a chimpanzee and not round like humans, and would require the head to point down like a chimpanzee, and therefore require the spinal column to enter at the rear for comfort. The Chad fossil needs additional evidence and is not considered a good intermediary. But with this as evidence, White, Meave Leakey and others saw a line from their fossils leading to Lucy.

However, many questions were raised about the chimpanzee Lucy line as the descent of men. Specialists regarding primate ***sexual practices*** show a disjunction between the Lucy type A. Afraensis and man. Studies show that chimps were highly promiscuous and therefore have many mutations that have developed in their Y chromosomes. Human Y chromosomes are much more intact because they tend to be monogamous. This seems to indicate a disconnect rather than descent.[31] The male chimpanzee is much larger than the female which gives credence to the theory of sexual promiscuity in Australopithecus. afarensis.[32]

Considering all the fossil finds, Tim White took the lead and said that there was one single root for producing the Homo species. He lists them: (1) Ardipiphus ramidus, 4.4 million years old that has a partial jaw with a ramus like a modern chimp (2) Australopithius anamensis, 4 million. (3) Australopithius afranensis, 3.6 million. Other anthropologists believe there were many roots, as far back as six to seven million years for branching. So there was no agreement as to what occurred. Alan C. Walker of Pennsylvania State University shows that there were different sized jaws and teeth from earlier fossils. Jeffrey Schwartz of the Buffalo Museum of Science says that the picture of Australopithius Afraensis or Lucy is wrong in drawings and should be more like an orangutan than a chimp.[33] The presentation of how primitive man may have evolved is built on speculations and is obviously confusing.

The so-called primitive finds, such as Homo erectus and others, were thought to have originated in Africa. The vertebrae of apes and humans are very different and only human vertebrae are now thought to be able to support upright posture for walking. In some finds the chimpanzee and human vertebrae are found together. Discovery of "the *oldest known vertebrae for the genus homo*" have been found in a 1.8 million-year-site in central Asia at Dmanisi, Georgia. Marc R. Meyer of the University of Pennsylvania and two colleagues from Georgia State Museum in Tbilisi found five vertebrae among 2200 bones of people, chimpanzees, and gorillas. The finding of these Homo erectus in Georgia of Asia, with other bones of other primates confuses the picture. Homo erectus was considered one of the earliest men who came out of Africa and that story of immigration is now called into question. David Prayer of the University of Kansas, Lawrence, agrees with Meyer that these men "possessed a speech ready vocal tract" and all the

---

[30] Michael Lemonic and Andrea Dorfman "Chimplike Creature Roamed The Woods of Central Africa 7 Million Years Ago," *Time*, July 22, 2002, 40-47.
[31] Kathy A. Svitil, "Chimps' Promiscuity Could Damage DNA," *Discover*, Jan. 2006, 70.
[32] Jocelyn Selln, "How Loyal Was Lucy?" *Discover,* 2005, 10.
[33] Mary Carmichael, "Evolution, Lucy Who?" *News Week*, April 5, 2004, 10.

Australio A forms showed short necks.[34] Kate Wong, another anthropologist, commenting on the fact that the Dmanisi finds with hand axes, had skulls of about 600-770 cc. which were much smaller than the Homo erectus finds in Africa.[35] Most of the Lucy-type skeletons from chimps are about half that size.

Some want to designate small humans that have been found in *Indonesia* as a specific homo species. The idea that man emerged from primates, one step at a time and gradually became a superior human race, Homo sapiens, is brought into question by other forms of humans that existed long after the development of Homo sapiens. A skeleton of an adult female that was the height of a three-year-old average Homo sapien was found on West Flores, an Indonesian island, in a large limestone cave at Lianne Bua. Earlier, imperfect fossils found in 1940 were deemed Australopithecus that somehow came from Africa hundreds of years earlier. This led to the assumption this was a development from them. Milford H. Wolpoff of the University of Michigan has said that this scenario of descendents from African parents, transported over 3000 miles across the Indian Ocean, needs now to be reevaluated. He also points out that the very small brain, even for her diminutive size, raises the evolutionary question of, "Why would selection downsize intelligence?" Advanced Stone Age tools show good intelligence. Dean Falk, a Florida State University anthropologist, found this pygmy brain to have advanced features and higher cognitive powers. Falk said, "Science is full of surprises." Its small teeth and narrow nose, the overall shape of the braincase and the thickness of the cranial bones all evoked Homo." Other related human fossils, of the same family, have been found. Some researchers have concluded that these 13,000–18,000-year-old skeletons are "miniature homo erectus."[36]

Thus, in many ways the idea of descent from chimpanzees/A. Afraensis seemed wrong. While thighbone measurements offer some validity for upright walking, none of the findings are conclusive. All of these fossils comprise a small number in a world that is six million years old. All the interpretations are biased to show a link to man.

Changes in Thinking about Finding Human Descent

Tim White, the recognized paleontologist from the University of California, Berkeley, is one of the leading authorities promoting the idea of A. Afarensis as ancestor to man. He has been one of the most knowledgeable spokesmen for Africa, helping establish Tattersall's 1980 graph. In a significant interview with Carle Zimmer, White was asked why paleontologists were able to find so many specimens from different historical ages in one small area of Africa. He replied, "The reason we are able to do it is that ***it's just a geological nightmare***. You have ***a patchwork quilt of different aged sediments*** on the surface. You can step across a fault and step back 2 million years. It allows you to look through many different windows in one small area." [37] (Emphasis added) White also said that former paleontologists were dominated by a "savannah mentality," but "now we're finding an association of early hominid fossils with faunas more characteristic of woodland, or more closed habitats. We've also learned that the idea that ***the last common ancestor*** *of hominids was* ***like a chimpanzee is just wrong.***"[38] This is a clear startling admission that the designation of A. Afraensis and related fossils as intermediaries was wrong. White also now

---

[34] B. Bower, "Evolutionary Back Story," *Science News,* May 9, 2004, Vol. 169, 375.

[35] Kate Wong, "Strangers in a New Land," *Scientific America,* Nov. 23, 2003, 74083.

[36] Kate Wong, "The Littlest Human," *Scientific American*, February 2005, 56-65.

[37] Carle Zimmer, "Human Origins," in an interview with Tim White "Dialogue: Beyond the Savannah Mentality," and "Digital Ancestors Walk Again," on use of CT scans on skulls etc, *Discover*, Oct.2005, 40, 42.

[38] Ibid.

claims to have found 5.5 million year old hominids and early humans of 160,000 years close together. White claimed that Ardipithecus ramidus is oldest intermediary that is a mix of apelike and monkeylike traits. White, the primary influence in naming A. Afarensis or the chimpanzee as the ancestor of man, now confesses this *was wrong*. A jaw with a few other partial jaws of A. Afraensis (Lucy clan) and a study by Yoel Rak of Tel Aviv University rendered the decision that supports the "long-standing minority viewpoint that Lucy's kind occupied only a ***side branch*** of human evolution" and "didn't contribute to the evolution of modern people."[39](Emphasis added)

Summary of Recent Fossil Considerations

Since 1994 four new steps or genera from primates to man were recognized. These fossils went back to 7 or 6 million years ago and included new ones from 5.2 million years. They are Sahelanthropus (only a skull), Orrorin (from fossil teeth and a leg bone pieces), Kenyanthropas, and Ardipithecus rumidus. The last called Ardi was found in Ethiopia and is the most upsetting to the other of Africa fossils. Ardi – is known from most of a skull and teeth as well as extremely rare bones of the pelvis, hands, arms, and feet—these few bones are the claim to human evolution. [40]

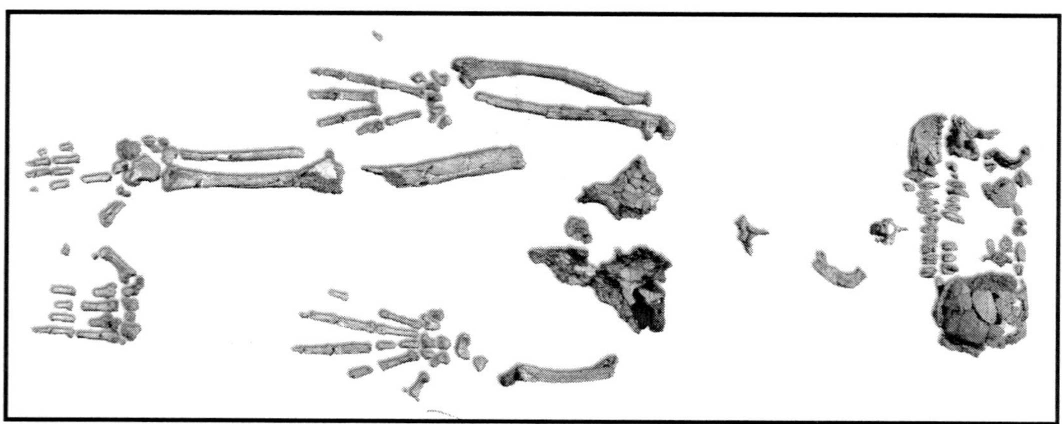

There are parts of 36 others from her group. She is not kin to the chimpanzee and found in a different wooded environment. Therefore, she disrupts the whole line of hominoids being more ape and monkey like.[41] On the next page is a graph of the 2010 list and dates of fossils thought to lead up to man in Africa. [42] These fossils come from many other places many thousands of miles apart—South Africa, Tanzania, Kenya, Ethiopia, and Chad. It is estimated that the evolution took place beginning perhaps seven million years ago until various homo specimens recently appeared. They come from strata of very different kinds.

[39] B. Bower, "Disinherited Ancestor: Lucy's kind may occupy evolutionary side branch," *Science News*, Vol. 171, April 14, 2007, 230.

[40] Ann Gibbons, "Our Earliest Ancestors," *Smithsonian.com*, March 2010 page 34.

[41] Bruce Bower, "Evolution's Bad Girl: Ardi shakes us the fossil record," *Science News,* January 16, 2010, 22-25. Cf. Ann Gibbons, "Our Earliest Ancestors," *Smithsonian.com,* March 2010, 34-41.

[42] Ibid., 25.

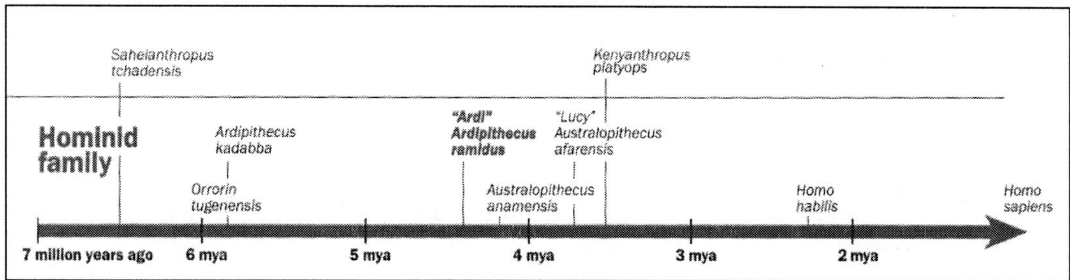

Moreover, some are supported by only partial and few bone parts and no one can prove they belong to the same fossil being. The claim is that some walked upright but they have long phalanges that would make it very difficult to walk upright and certainly are made to climb and walk on all fours. The deceptions that are presented from these few fossils that are spread out over millions of years over thousands of miles with incomplete, scanty evidence are given artistic drawings, presenting them with features and hair that are mostly fabrication. There have been several suggested routes, none positive of development from hominoids to men.

Passion and Concentration Devoted to Africa Was Bias to Darwinism

Search in Africa to Find Primates Was Misguided

The focus on Africa, after earlier embarrassments, was mainly to uphold Darwin's suggestions and because most known living primates were there. As a result, the discovery of primate fossils elsewhere has been ignored. David R. Begun, professor of anthropology at the University of Toronto has, in this new millennium, been trying to reconstruct the landscapes and mammalian dispersal patterns that characterize the Old World, before two million years ago. He identifies as apes the fossil apes that are primates that more closely resemble living apes. He includes the proconsuls and great apes. Both lack tails and have apelike jaws and teeth, but do not have highly flexible limbs.[43] Begun said,

> Today's apes are few in number and in kind. But between 22 million and 5.5 million years ago, a time known as the Miocene epoch, apes ruled the primate world. Up to 100 ape species ranged throughout the Old World, from France to China in Eurasia and from Kenya to Namibia in Africa. Out of this dazzling diversity, the comparatively limited number of apes and humans arose. Yet fossils of great apes–the large bodied group represented today by chimpanzees, gorillas and orangutans (gibbons and simians make up the so-called lesser apes)–have turned up ***only in western and central Europe, Greece, Turkey, South Asia and China.*** It is thus becoming clear that, by Darwin's logic, ***Eurasia is more likely than Africa*** to have been ***the birthplace*** of the family that encompasses great apes and humans, the hominids.[44] (Emphasis added)

He has illustrated Miocene ape localities on his graphic map, showing Dryopithecus, Sivapithecus, and other apes. Begun's distribution of greater apes, when hominids might have been found, may be seen on the map on the following page with dots for location and names.

---

[43] David R. Begun, "Planet of the Apes," *Scientific American*, August 2003, 74-83.
[44] Ibid., 76.

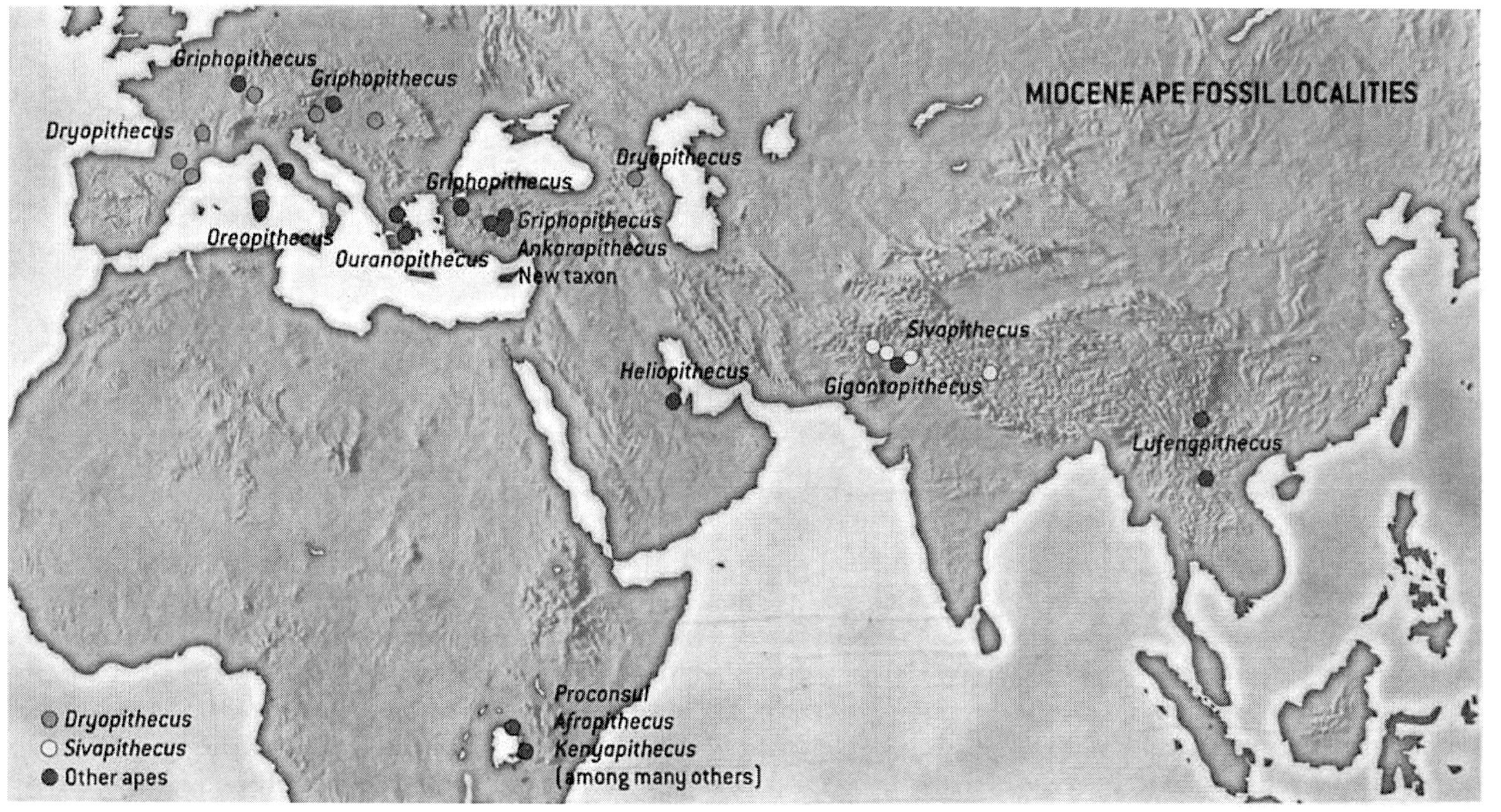
MIOCENE APE FOSSIL LOCALITIES
Griphopithecus
Griphopithecus
Dryopithecus
Dryopithecus
Griphopithecus
Oreopithecus
Ouranopithecus
Griphopithecus
Ankarapithecus
New taxon
Sivapithecus
Heliopithecus
Gigantopithecus
Lufengpithecus
Proconsul
Afropithecus
Kenyapithecus
(among many others)
Dryopithecus
Sivapithecus
Other apes

In an "Overview/Ape Revolution" he makes three important notes:

- Only five ape genera exist today, and they are restricted to a few pockets of Africa and Southeast Asia. Between 5.5 million and 22 million years ago, in contrast, dozens of ape genera lived throughout the Old World.
- Scientists have long assumed that the ancestors of modern African apes and humans evolved solely in Africa. But a growing body of evidence indicates that although Africa spawned the first apes, Eurasia *was the birthplace of the great ape and human clad.*
- The fossil record suggests that living great *apes and humans are descended from two ancient Eurasian ape lineages*: one represented by Sivapithecus from Asia [the probable forebear of the orangutan] and the other by Dryopithecus from Europe [the likely ancestor of African apes and humans].[45] (emphasis added)

Therefore, the *evidence has increasingly turned from Africa and toward Eurasia*, and against the preponderance of paleontologists who still argue for migration out of Africa. If they had not been biased to support Darwin's direction to Africa and followed the data, they would have been searching in Eurasia. Moreover, data from all over the world adds evidence and more confusion, because the finds in other areas of the world that emerged were ignored.

Location of Most Human or Homo Species Now Acknowledged

The seven or more proposed species of human fossil forms are now generally agreed to be mainly three with suggested sub groupings: Homo erectus, Homo Neandertal, and Homo sapiens. Seemingly the oldest, Homo erectus (meaning "upright man') fossils are found in the following places. Indonesia (Java man): Trinil 2 (holotype), Sangiran collection, Sambungmachan collection, Ngandong collection. China: Lantian (Gongwangling and chenjiawo), Yunxian, Zhoukoudian (Peking Man, Nanjing, Hexian. Many fossils in caves and elsewhere with many thousands of stone artifacts have been found that are now dated 800,000 years. India: Narmada (some debate about these). Vietnam: Norther, Tham Khuyen, Hoa Binh. Turkey: Kocabas fossil. Kenya WT 15,000 (Nariokotome), ER 3883, ER 3733. Tanzania: OH 9.[46] The Asian site in Georgia called Dmanisi found conclusively that Homo erectus and their tools were found before any evidence of their existence in Africa.[47] The preponderance of fossils of Homo erectus is in Eurasia and Indonesia and only a few in the one area of Africa of the Rift Valley. We will come back to the significance of that one area later.

Homo Neanderthal were intelligent, with the ability to perform sophisticated tasks. They started and controlled fire, constructed complex shelters, skinned animals, and built traps. Some of their characteristics, along with differences from other species will be considered later. The important thing here is that their fossils and artifacts are found plentifully and almost entirely in Eurasia. This includes Portugal, France, Spain, Britain, Germany, Czech Republic, Slovakia, Croatia, Greece, Iraq, Iran, Israel, Romania, Russia, Uzbekistan and even 1250 miles east into southern Siberia's Altay Mountains.[48]

The conclusion is that if Charles Darwin and the early evolutionary scientists had known that

[45] Ibid.

[46] Cf. Homo erectus, *Wikipedia.*

[47] Bruce Bower, "Homo may have originated in Asia: Find suggests non-African origin for human genus.", *Science News,* July 2, 2011, p 8.

[48] Cf. Neanderthals, "Habitat and Range," *Wikipedia.*

the fossils of large primates and of all of the known early humans are in Eurasia, they would not logically have picked Africa as their source. Such logic does not fit the facts, and only fits what seemed to be the center of primates at Darwin's time.

Multi-regions as Well as Multi-lines of Descent, Not Just Africa

Primate hominid finds have emerged in many places other than Africa. In the last half of the 20th century when there was confusion about African origins, many other claims of the geographical location of the evolution from primates to man came to the forefront. There is now not only the idea that there may have been diverse lines of descent, but *multi-regions in which man arose*. The areas where fossils of hominids and humans have been found are sometimes thousands of miles apart. Others have recently tried to redraw the hominoid family.[49] Within the last decade, Russell Ciochou, of the State University of New York, said a jawbone found in Burma, estimated to be 40 million years old "may be the remains of the earliest known ancestor of monkeys, apes, and humans." Robert B. Eckhardt, of Penn State University, claims that a fossil found in Greece that is nine to ten million years old, called Ouranopithecut, may be the earliest known link to hominoids, or the earliest hominoid. Yuri Mochanov, a Soviet expert, found human artifacts in Siberian stones that are 1.8 million years old and suggested that man may have originated near Diring Yuriakh and migrated from there.[50]

First humans are found in multi-regions with very confusing appearances. Most anthropologists continued to insist on the one origin from Africa because this was Darwin's theory. Some scholars believe there are hints that *Homo ergaster* was the first to emerge from Africa with simple tools. Similar skulls have been found in the *Republic of Georgia.* But anthropologist G. Philip Rightmire sees *ergaster* as *a regional variation of H. erectus* with a smaller brain. Stone tools and shellfish remains *in Eritrea's Red Sea coast* area leading into central Asia suggest the route that H. sapiens *may have* taken out of Africa.[51] In recent years, strong evidence surfaced that at approximately *the same time* similar implements were discovered in Africa and Asia where early forms of hominoids were discovered.

The arguments surrounding the first humans, where they arose or how they came from Africa is in confusion. Some Homo sapiens and Neandertals go back 125,000 to 150,000 years in Morocco, Israel, etc. A partial Homo sapien skull found in Hofmeyr, South Africa, mysteriously looks like a European specimen. Fred E. Grine, Stony Brook, New York University, sees a common ancestry of Europeans and sub Sahara Africans. But Homo sapiens have been found many places. There is a Romanian skull 40,000 years old; Southern Russian specimens from Kostenki, 250 miles south of Moscow, 45,000 years old with unusual cultural advances and some modern teeth; Chinese specimens in Tianyuan Cave, southwest of Beijing, dated 60,000 years old; Borneo specimens 45,000-39,000 years old; Portugal specimens 25,000 years old, and so on. In many of these places there are Neandertal finds, as well, and indications that they were living together and using the same tools (e.g. southern Russia, Portugal, Romania, etc.). The battle rages as to whether Homo sapiens and homo Neandertals intermarried. In addition, Richard Fullagar of Wollagong University has recently documented thousands of circles on sandstone monoliths, with

[49] Cf. Tudge, *The Variety of Life*, 500.

[50] "Evolutionary Cusp", *Science 85*, Vol. 6, No 9, November 1985; "Fossil Skull Goes Ape," *Science News,* Vol. 139, April 20, 1991, 254; *U.S. News & World Report,* Vol. 101, No. 9, 62.

[51] B. Bower, "Fossils Hint at Who Left Africa First," *Science News*, Vol. 157, May 13, 2000, 308. B. Bower, "Africa's east coast netted ancient humans," *Science News*, Vol. 157, May 6, 2000, 292.

human tools, that show evidence of man in outback *Australia* over 60,000 years ago, shattering the certain claim that man came from Africa earlier. Eric Trinkaus of the University of Washington, Saint Louis, has said, "Far more complex population interactions occurred as modern humans dispersed into Southern Africa and Europe than many researchers have recognized."[52]

Most fossil records seem to point to origins in the area of Eurasia in which the Ark of Noah is said to have landed. There are far more complete clear fossil men in the Eurasian area, with Neandertals, Cro-Magnon and others now recognized as fully humans, than from Africa. This is consistent with the genetic evidence, as detailed in Chapter 15.

All of the different dates for origins in different locations have confused the story of "so-called" human descent. Robert Martin, provost and curator of biological anthropology at the Field Museum in Chicago, has written a book, *Primate Origins and Evolution*, which he claims is history. He speculates that primates arose 85 million years ago and people 20 million. Martin uses phrases like, "I'm not sure," "It's possible," "Our suggestion" and others. He has devoted his life to this study and his projection differs in date from the conclusions of primate anthropologists by 15 or 20 million years.[53] There is no agreement on date of origins, or clear place of origin, and certainly no clear line of descent.

In another approach to argue for migration out of Africa, Andrea Manico, of the University of Chicago, presented evidence from measuring 37 structural characteristics of 4,600 male and 1,500 female skulls less than 2,000 years old and claimed confirmation from DNA. She found very high variations in Africa with few in Eurasia. She claimed the fewer variations in Eurasia, along with the few remaining in Africa, indicated Africa as the source. But usually the argument for plants and all other creatures is that fewer variations indicate the place of the original source. Her data would show the opposite.[54]

Conclusions about Human Descent from Apes or Any Animal

When Darwin wrote *The Descent of Man*, there was not one fossil hominid or any certain early fossil man. Before the publication of *The Origin...,* his scientific friends, Hooker, Huxley, and Lyell, all warned him there was a gap between the species, especially between apes and men. Fifty years after Darwin's *Origins...* the gap between men and apes was still very definite. But soon afterwards a deceptive picture was presented to the public that later scientists proved wrong. It is scarcely known that Darwin did little research about primitive man. His co-publisher, Alfred R. Wallace, later spent years visiting primitive human societies all over the world. He found no gradualism in human intelligence and eventually gave up the idea of man's development from apes. He never adopted atheism.

In 1958, at the hundred-year of anniversary of *The Origin...* the scientific consensus was that the gap between animals and men still existed. An army of highly skilled paleontologists made desperate efforts to find some evidence of transition, by focusing primarily on Africa, upholding Darwin's idea. Their brilliant and heroic searches found many fossils and produced the view that Austrialinopithicus afarensis, a chimpanzee-type creature was the intermediary. But after the dust of that generation had settled, the existence of fossils, footprints and tools of humans had been

---

[52] B. Bower, Asia Trek: "Fossil puts Ancient humans in Far East," *Science News*, Vol. 171, April 7, 2007, 211. For more definite listings and discussion of these see also, B. Bower, "Mysterious Migrations," *Science News*, vol. 171, March 24, 2007, 184-188.

[53] "Man of Apes Resets Clock of Evolution," *Discover,* July 2004, 20 ff.

[54] Jennifer Barone, "Out of Africa, All of Us," *Discover,* 2007, 15.

found to be contemporaneous and even prior to Austrialinopithicus afarensis. By 1980 Tattersall and Eldredge presented the data, with no proof, of how man arose, along with warnings about the myths involved.

All the finds really left the theories about human descent in a mixed quandary. Richard Leakey gave two models from which the races of modern men arose—one a candelabra, which shows an early branch into three prongs, interrelated and producing three races of men in Europe, Africa, and Asia. The other, Noah's Ark, represents two branches terminating before modern men and one progressing a long time until it branches into modern men of Europe, Africa, and Asia. Richard Leakey agreed with Stephen Gould that the Noah's Ark model was correct, seeing all men as related by recent migration from Africa.

But objective study showed the majority of great apes at the supposed time of man's evolution were not in Africa, but in Eurasia. During this period there had been claims of hominids from all over the world as possible sources of man's evolution. But these were ignored because of the fixation on Africa. Moreover, vertebrae structure, sexual habits, etc. were incongruent with A. afarencis as an intermediary. Also, leading authority Tim White, who had endorsed the line from the chimpanzee, admitted this was wrong, and other studies agreed.

Tattersall's graph of human descent, made in 2000, contained question marks and he admitted development of man was not known and involved myths. Later, in 2006, Ann Gibbons also drew a new graph with the latest finds, showing lines of descent. She also included question marks as to how man's descent occurred and gives two routes of possibility.[55] Both suggested two or more possibilities, with no certainties.

Uncertainties continue regarding the environment, geographic area, number of lines of descent, appearance of the primary descendent and how long ago man emerged, with speculation ranging from 7 to 65 million years. There is no clear scientific data, only philosophical speculations regarding man's origin.

At a scientific meeting in Phoenix in 2003 the scientists said, "The ancient members of the human evolutionary family branched into as many as 20 different *now–extinct species.*" Efforts at the Phoenix meeting were made to somehow prune "this species–laden scenario." Thus the bush or tree of human descent seems even more confusing.[56] Ann Gibbons has reviewed the situation in her book and said, "The human story was beginning to look as complicated as a Tolstoy novel, with new characters appearing unexpectedly as the book of life unfolded."[57]

Carle Zimmer has summarized the situation of eight of the most important questions about how humans differ from other primates and said, "Still what we ***don't know*** about our evolution vastly outweighs what we do know. Age-old questions defy a full accounting, and new discoveries introduce new questions."[58] (Emphasis added)

Almost all scientific thinking has been perverted by this evolutionary mythical theory, in spite of repeated claims to be true. It has been a major hindrance to science. It seems clear that this whole effort to trace the descent of man is *driven by bias* of the evolutionary biologists themselves against evidence of natural creation and biblical revelation *and not by scientific evidence*!

---

[55] Ann Gibbons, *The First Human: The Race to Discover Our Earliest Ancestors*, xviii.

[56] B. Bower, "Ancestral Bushwhack," *Science News,* May 3, 2003, Vol. 163, 275.

[57] Ann Gibbons, *The First Human*, 6.

[58] Carle Zimmer, "Great Mysteries of Human Evolution: New Discoveries Rewrite the Book on Who We Are and Where we Came From," *Discover*, September 2003), 34; cf. 34-43.

Chapter 17
Human Uniqueness from Animals

Introduction:

Past Views of Man and Animals

Previous chapters reviewed various ways the relationship between humans and animals has been understood. In the mid-eighteenth to the mid-nineteenth centuries, some Christian scientists spelled out "The Chain of Beings" and located man at the top of all the world's creatures, next to angels who were next to God. During this period Christians were using the explosion of new knowledge to claim great specificity for everything, to the extent that they held that they almost had the mind of God.

Early in the nineteenth century the theory of scientific determinism began to be applied to man by extending uniformitarianism to the area of biology. Lyell had prepared the way in geology and, therefore, anticipated it. But the lack of evidence of intermediaries in any species, living or fossil, and the great distinctions between plant and animal life, and especially between animals and humans were obstacles. Darwin admitted that human evolution was of major importance, but in *The Origin...,* he avoided extending the tree of life to man. After the popular acceptance of his views on evolution, about twelve years later, he had the courage to do so in *The Descent of Man,* though he had no fossil remains to show such a development.

Speculation about Progress across Gaps between Men and Animals

Darwin acknowledged the difficulties Lyell and other friends had pointed out: the hand dexterity, the ability to walk upright, to speak, to coordinate thought, the intellectual ability of logic, and the ability to create, etc. His speculations regarding how these gaps might have been crossed are still referred to today. For example, he suggested speech grew out of birds' abilities to sing, and other forms of animal communication. This was pure speculation without any real data.[1]

By the twentieth century, the search for intermediaries between primates and man was in full force and graphic displays of man emerging from apes were presented in the textbooks and museums, e.g. Osborn's "Hall of Men." The history of man, or anthropology, and civilization was dominated by evolutionary philosophy. Man was considered only a chance extension of animal life, first to primitive and then slowly to full human abilities. This was applied to sociology, economics, psychology, and government by Herbert Spenser, Karl Mark, Sigmund Freud, John Dewey, and many others. This exciting concept of man's ability to make progress dominated thought. Men saw themselves as superior to previous generations. The science of anthropology itself generally began about the second and third decade of the twentieth century. From about mid-nineteenth to mid-twentieth centuries, the pendulum swung from evolution to theological fixity. Because of the gap that was even more evident, the evolutionists positioned man as just an animal, but *the highest* in a natural process. Since the mid-twentieth century, the trend has been to say man is at the top of the evolutionary chain, but *set apart* with an unusual *animal* uniqueness.

Speculative History and Search for Evidence

With the extension of paleontology to human bones as fossils and the birth of the science of

---

[1] Jaron Lanier, "Sing a Song of Evolution: Is language descended from musical mating calls?" *Discover,* August 2006. "Darwin's speculation," 24-25.

archaeology that explored early civilizations, the whole science of anthropology became prominent. By the 1940s and especially the 1950s, a complete effort to review man's history of development from a naturalistic and evolutionary perspective emerged. In spite of the lack of evidence, the scientific determinists required that everything known about man be revised to fit the theory.

Some examples of the many authors and books of the story of man's emergence from apes to modern men were: V. Gordon Childe, professor of prehistoric archaeology at the University of Edinburgh, who wrote *Man Makes Himself* and *What Happened in History,* and Ashley Montague, chairman of the department of anthropology at Rutgers University, who wrote *Man: His First Million Years*. The story traced human development from the Paleolithic (savage), Mesolithic, and Neolithic, through the progressive development of the Stone Age, into the eras of copper, bronze, and iron, especially in Eurasia and North Africa and later across land bridges to other parts of the world.

In the latter twentieth century, archeologists and experts in almost every field have scoured the world and gathered much objective scientific data about man's tools, hunting, agriculture, literacy, industry, medicine, and other details of their civilizations. Even though the perspective was scientific determinism, the gap was still there. Moreover, it did not fit with the brilliance of man as evidenced in the oldest cultures. Moreover, since then medical, psychological, neurological, and biological science has greatly extended and defined the gap between man and animals, revealing that man is much more than a superior ape. The emerging picture of man is now more in agreement with the Bible, that man was uniquely created in God's image to rule over all creatures and creation. Unbelieving elite still struggle to twist the facts to free man from God.

The Physiological Uniqueness of Man to Other Animals

The intent here is not to give all the arguments for the physiologically unique characteristics of man. The list that follows offers a brief comparison of some of the physical differences between men and apes that will be shown to be relevant to the ability to form civilizations. Space does not allow an extensive comparison of the organic and other physical differences.[2]

MAN	**APES (Gorilla, Chimp, Orangutan)**
Legs much longer than arms	Arms longer than legs
Spinal column enters head in center	Spinal column enters from rear and slightly from front
Comparatively vertical face	Slightly inclined face
Low projection of jaws	Great projection of jaws
Reduced canine teeth	Large canine teeth
Nose elongated beyond nasal bone	Nose ends at nasal bone
A well marked chin	Chin not prominent
A forward lumbar convexity	Receding lumbar
Non-opposable toes in line	Opposable big toe not in line
Foot arched transversely	Foot not arched transversely
Hairless body and no tactile hairs	Very hairy body
Usually small neck	Massive neck

[2] For more information, see Gange, *Origins and Destin,* 116-145.

The brain size is much larger	Brain size averages much less
Year round fertility	Seasonal heat
Young are long time dependent	Young are soon independent
Usually expansive of offspring	Usually regulate offspring
Wide temperature regulation	Narrow temperature regulation
May be herbivorous or carnivorous	Usually herbivorous?
Expansive territorially	Narrow limit territorially

All these differences can easily be rationalized by the evolutionary scientists as a mere speculative extension of the physiology of the primates by adaptations. But no evidence is given of how these developed. Therefore, they fit within the trend since 1960 to give man a higher status as an animal, first as a noble savage and then as one with an evolved ghost in the machine, and only in that sense unique. However, these differences must be included in man's uniqueness for a complete understanding.

Molecular biology has shown that the human body is far more complex than that of any other animal. While there is a claim of a high degree of similarity in the genes of the chimpanzee (95 to 97 percent) this is a superficial judgment. The differences in many other areas show that man is distinct and different. Considering the three levels of genetic workings and everything in the human DNA, biologists have seen much greater distance and many differences than gene figures indicate. One speculated there were as many as 47,000. Because of the differences heart transplants and other efforts to use chimpanzee's organs in humans have not worked long-range. The comparative anatomy in living, as well as fossil remains of humans and higher primates, are very distinct.

Physical Brain of Animals and Men Are Very Different

Determinists Claim Intelligent Closeness of Primates to man

Human intelligence is distinctly different from that of other animals. For years evolutionists have tried to prove that animals can reason and communicate in ways that are similar to man's abilities. That effort has been extended widely with much experimentation in recent years. Much of this is an obvious effort of the liberal educated elite, with the media support, to prove a close connection between animals and humans in order to persuade the public that evolution occurred. A review of some of the more recent training and experiments with animals may be summarized in the words of Eugene Linden, "Something less than true creativity... but ***something more than*** simple mimicry seems to be at work."[3] (Emphasis added) A few of the experiments will be mentioned to show similarity, but that line between the two is highly significant.

Animals have done some impressive things in the various research centers in the United States and around the world. Kanzi, a chimp at the Georgia State University Language Research Center, can punch special symbols on a board that *man has designed* for him, and in this way ask for certain things. He can open a closed box to get food and other things. His accomplishments have been compared to a two-year-old child's, and they were considered equal in many attainments.

Several significant factors need to be acknowledged. Animals have built-in intuitive

---

[3] Eugene Linden, "Can Animals Think," *Time,* March 22, 1993.

knowledge, and this can be sharpened to a high degree by long hours of training by human trainers with the stimulus of food and other low-level pleasures. Thus, their trainer may expand the animal's intuitive abilities beyond what is normal and to their full innate abilities. The trainer-developed symbols, taught to Kanzi, are an example of this occurring because of *human creative contribution by a skilled trainer*. Human children, in their early stages, respond to similar motivations and can equally be enhanced and motivated with work and mimicry. This also helps them develop. But, there are significant differences.

There is now evidence that unborn human babies understand language, and have amazing triggers for change already pre-programmed in their development. These are present from birth and kick in at certain points. Human children develop certain abilities at predictable points i.e. crawling, walking, and talking. This never occurs in animals. Animals do not have these developmental abilities, thus they cannot go beyond a certain level.

No animal has ever been able to talk except as biblical miracles (Balaam's ass). Although animals can communicate with one another, the number of distinct signals available to them is intuitive and is severely limited. Gange has pointed out that the maximum number of distinct animal signals registered in 1986 were only thirty-seven. Current research of other animals such as whales extends the number. In contrast, humans' vocabulary is *continuingly increasing*. All animal communication is genetically preprogrammed, while humans, with language syntax, have the ability to learn and expand their communication skill to a high level of sophistication. Language is the expression of the ability to reason, in the true sense of the word.

Most importantly, only humans can be taught *information and morals which they can transmit to their offspring*. Only humans can transmit knowledge they have learned to their children, and grandchildren, and therefore they are the only animals that can build a civilization. By training in rituals to gain food and other animal needs some profitable memories have been transferred to the young. Kanzi cannot, on his own initiative, train his offspring to respond to the symbols, open the box, and do other tricks he is trained to do. Chimps have been trained to do amazing things, even to chip stones to sharpen them, but they do not transfer that information. Such acquired characteristics of knowledge are not hereditary. Much that has been claimed as animal intelligence has been the result of intense hours of training, and to claim comparison with human learning has been deceptive.[4]

This is why the idea of animals developing tools and through subsequent generations developing higher culture is not possible. There is *no real record or history of any sub-human animal civilization*. All such books with these ideas are fiction or myth. Studies show that even though the abilities of animals are often surprising, the animal mind is limited to sub-human abilities.[5]

The Human Brain Compared to that of Animals

Structurally, the *human brain is created significantly different*. The human brain is much larger than animals. Most ape brains are around 400 cc; the australopithecines about 800 cc, while the human brain averages 1350 cc, and may range from about 875 cc in small races to 1660 cc in

---

[4] Ibid., 54-61; see chapters 17-19 in Gange, *Origins and Destiny*.

[5] A Schrier and F. Stolinitz, ed., *Behavior of Non-Human Primates*, Vol. 4, Academic Press; J. Lilly, *The Mind of the Dolphin-A Non-Human Intelligence*, Doubleday; and Robert Gange, *Origins and Destiny*, 129-145.

larger races. Human beings have been known to function as a man with half their brain removed, and some reasonably intelligent people have been on the low end of brain size. Although brain size is not directly related to intelligence, the fact that man has the largest brain places him at the top of mental abilities among all animals. While size is important, the mechanical operative ability is very different.

Maya Pines describes some significant differences between animal and human brains. He said,

> Frontal lobes scarcely exist in laboratory animals. They are fully developed only in human beings.... Prefrontal cortex barely exists in mice and rats, is 3.5% [of the brain] in cats, 7% in dogs, is 17% in chimps, and in the much larger human brain is 29% or almost one third. So, humans have much larger frontal lobe function. The hemispheres balloon out and are convoluted with vastly increased layers of gray cells. Moreover, humans have far more well-defined columns or bands of neurons that operate in an organized and modular way. This seems to give humans some very distinctive abilities that other animals lack.[6]

Physiologically, there is a gap between the brains of all animals, but a very distinct gap between higher primates and humans, with no evidence of transitional forms between. There are similarities, as well as significant differences.

The major difference in man and animals relates to mental abilities. Man is able to see reasonable logical relationships as seen in the new information theory. Man can discern God's thoughts that are revealed in creation, and man is able to create, and to talk. Man *intuitively knows the world is rationally connected* by some superior intelligence that he can study and discover. Man's uniqueness is his *ability to talk and to form language* of unlimited symbols. This ability lies in the realm of his mental abilities, but it also involves his spiritual nature, indicated in the idea that mankind, both male and female, is made in the image of God. Several areas of modern study have revealed clear distinctions of man and animals.

Man's Uniqueness of Mind Revealed in Current Scientific Studies

Idea of Deterministic Knowledge on a Blank Slate Was Unscientific

The modern Enlightenment philosophy of Descartes, R. Bacon, John Locke, and Emmanuel Kant that individual man's only real knowledge was based on sense perception and reflection, with certain innate categories of thought, and that man therefore began with a blank slate was a perversion. At first, these philosophers allowed for a God of their creation from reason. This meant their god was a product of reason and not experience, and that they could limit or dismiss God at any time. This led to philosophical atheism.

In essence these men decided from the outset that knowledge could not include God revealing to a human mind His nature in creation; nor could God have communicated to individuals His message directly in history, as in the Bible. With those rules of knowing, all the study of nature and history automatically ruled out God by philosophical presupposition. This *philosophy of the mind*, and it was a philosophy, not scientific observation, was really *a perversion of human nature*. It was the basic and only reason for disallowing knowledge about a Creator at work. This false philosophy, not the constitutional separation of church and state, is the real reason educators in America hostilely reject God from involvement in science in the public schools.

---

[6] Maya Pines, "The Human Difference," *Psychology Today,* September 1983, 62-68.

The man who brought this view into focus was John Locke, a rebel against most authority, in state and religion. More importantly, John Locke *knew little about the human brain or mind, admitted that he didn't know,* and plainly said that such considerations were not in *his designed purpose*. As previously quoted, he said, "I shall declare [these] as laying outside of my way in design I am now upon." This theory immediately manifested flaws, which Kant tried to solve. But Locke and Kant had launched the ship of modern education in a false direction by ignoring the study of man's mental workings.

This view of how man knows, based supremely on *faith in man's mind,* soon replaced science based on faith in God. Instead of studying God's book of nature with the history of His working with His people as recorded in the Bible, as the great scientists had done, humanistic scientists sought to fit the data about man into the idea of the mind as a machine with a blank slate. But it was soon evident that man was more than an animal beginning with a blank slate. This gave way to the idea of man as having become, by evolution, a "noble savage," a view that asserted that man was animal, but had mysteriously risen to a unique level in a noble way. Scientists soon realized that some aspects of the human mind are mysterious and transcendent even beyond a superior machine of the noble savage. In time the "noble savage" was seen as insufficient to explain man, and yielded to the idea that man had evolved a "ghost in the machine." These various views were used to interpret human actions, social relationships, and development of culture without God.

Four Areas of Modern Scientific Research: Man's Mind Found Different

New Research Instruments and Experiments Give New Perspectives

Psychological testing and experimentation, along with new instrumentation for examining the brain, opened new vistas of knowledge that were unknown even to Locke and Kant. In the same way new instruments have affected atomic physics and astronomy, new and different instruments in biology are a great help in learning more about the mind/brain.

Four Areas of Study of the Mind Are Examined and Defined

Steven Pinker, former professor of psychology at MIT and now at Harvard, has recently written a book, *the blank slate: The Modern Denial of Human Nature.* The title makes clear that the basic modern theory of the human mind by Locke, Kant, et. al was a denial of what is true about humanity. This leading psychologist is admitting that the view which launched uniform deterministic science on the dogma to deny God in science was wrong—an error. Locke's and Kant's views were an effort to adjust modern scientific findings to a new naturalism, denying God, thus dismissing the foundational theory that perverts human nature. Pinker's book is a comprehensive exploration of all the ramifications of what is known about the working of the mind/brain from current science (since about 1980). He shows that the philosophical theory of a blank slate was wrong, but he never rises above many of the errors it produced, and he continues to reject the supernatural. Pinker's views will be examined as a prime example of representatives of those of the secular elites.

Pinker sees four bridges to a better understanding of how the human brain works and affects thought and conduct. This is important because this was not even an interest for Locke, whose dedication was to end religious authority. Moreover, the excellent instruments available today were not available in the eighteenth century. The four bridges Pinker refers to are: (1) *Cognitive science* which sees the physical world as linked by information. (2) *Neuroscience* which examines how emotions, thoughts and actions are influenced by brain operations and conditions. (3)

Behavior genetics looks at how genes affect standard, but unique and nonstandard brain function and personality. (4) *Evolutionary psychology* or how man's conduct is possible as a result of adaptations to change that resulted in improvements. These four are not as distinct as Pinker implies, but overlap.

Pinker reveals his bias when he links psychology to evolution in his fourth category. This bias causes perverted interpretations of all the amazing scientific data as seen in the first three categories. He attributes the amazing mystical human mind to an unobserved mystical power of chance, and argues the mind/brain is nothing more than *a sophisticated animal machine.* Indeed, it is *an exceedingly more sophisticated* one. In fact, it is this that causes Pinker to omit important data and be blind to a correct understanding of other data.

Cognitive Science of Information

Cognitive Science of Computers Have Gap to Human Intelligence

In regard to cognitive science, Pinker makes a distinction between a computational theory of mind and a computer metaphor of the mind.[7] Pinker is not saying the mind works like a computer. He is arguing that information processing is a form of "reasoning, intelligence, imagination and creativity." He points out that programmed computers can do many things that human intelligence does—such as beat Garry Kasparov, the world's chess champion, write short stories, and the like. But he denies that computers will replace human intelligence or have first person subjective experiences. By linking the mind to the idea of a computer, he falsely argues that this somehow removes the "ghost" further from the machine. It really does not do that. A creative human mind wrote the chess program with the options for decisions built into the game, and also created choices for writing a short story that could be made a computer program. This only reveals the mystery of the human mind at work. The mystical mind has only been veiled, or placed at a distance because the creative ghost has transferred its brilliance by writing the programs into the magnificently built computer. But by this, *the mystery* of the ghost in the mind has only *been made greater* because the human person has been able to build such an amazing machine with such an excellent program.

While much work has been done and conferences have been held on artificial intelligence, as of yet, nothing like the human brain has been produced. All that has been done in making and programming computer intelligence has been the result of intelligent human input, and is therefore only a testimony of a superior creative human mind. In fact, the human physical brain is more than an electronic computer. It is a complex electronic machine, mixed with a complex of chemical molecules and hormones, perhaps some 200 chemical substances, which affect mood, the immune system, and the ability to act.[8] Moreover, this is controlled by a conscious personal being. The human brain is so complex and operates in such inexplicable ways that one researcher who spent ten years trying to discover how it functioned, abandoned the search saying it is an impossible mystery.

Recognizing the difference in computer memory and human thought, scientists have

---

[7] Stephen Pinker, *the blank slate: The Modern Denial of Human Nature,* (New York: Penguin Books, 2002), 33, 34.

[8] Richard Bergland, *The Fabric of Mind: A Radical New Understanding of the Brain and How it Works,* (Washington, DC: Viking), 202; August Pert, "The Body's Own Tranquilizers," *Psychology Today,* September 1981, 11; Joseph Carey, "The Brain Yields its Secrets to Research," *U.S. News & World Report,* June 3, 1985, 64-65.

developed a silicon neuron to operate in an analog mode much like brain cells. With this new model, scientists hope to create a new computer simulation more like humans that may move toward artificial intelligence, within certain limitations. But scientists like Gunter Stent of the University of California, Berkeley, Roger Penrose, and others believe the power of the human brain is awesome and unlikely to be copied except in some limited ways. The mystery of the brain is only part of the greater mystery of human thought and acts.

New Information Theory Extends Cognition to Whole Universe

Cognitive science has also discovered and is *built on the new information theory* which reveals *information behind everything* in the cosmos. How could there be meaningful information in the whole universe without a personal cognitive mind behind it so that the information is logical or has meaning? Cognitive science really shows the physical world is designed by one Great Mind that transcends the human mind and matter containing it.

It also points to the fact that the human mind has been given unique, amazing abilities to think, so as to evaluate the information in creation, and to create as God created. Cognitive science really points to God. The Bible teaches that by wisdom God created the world (Proverbs 8:22-31). He brought order into existence by speaking (cf. Genesis 1:3, 6, 9, 14, 20, 24, 26, 29) and that his Son, the *Word* (*logos* in Greek, from which we get logic) was the creator of everything (Jn 1:1-3).

Cognitive science has revealed the fallacy of "the blank slate" idea. Pinker refers to new theories to bolster the blank slate idea, saying, "Nothing comes out of nothing, and the complexity of the brain has come from somewhere. It cannot come from the environment alone…."[9] It also shows that the idea that man developed upward to "the noble savage" is wrong. The law of entropy of information (evolution cannot add information), and the "ghost in the human machine" is inadequate because, not only man, but the whole universe exists by information, or one unifying logical personal mind behind it all. The idea of cognition of information inevitably links man's mind/brain to the concept of a comprehensive God. While God's understanding is "inscrutable" (Isaiah 40:28), or beyond understanding, man can partially trace his thoughts in creation within its finite limits because of his uniqueness.

Behavior Genetics

Genetics Shows Different Writing for Mental Control

The bridge of *behavior genetics* is thought to also discredit the blank slate idea of human nature. Pinker gives several arguments. "Different genes produce people of different behavior." "Small genetic differences may produce large behavior differences." [10] This is true for abnormal behaviors, but different genes also produce different variations in ability, temperament etc., that are considered standard or normal. Neurology, with genetics, has revealed that every brain of every man and woman is different, but all have the same essential operational areas or parts. This is true for almost all brains, unless there is some defect. Genetic mutations in copying produce differences and defects. In some rare cases, genetic toxins interfere with normal development. One nucleotide or protein may cause different character traits, or several or many genetic differences may produce different behavior. "But genes do not determine every detail of the mind."[11] Environment and conditions influence behavior which is added in specific different ways because

---

[9] Pinker, *the blank slate: The Modern Denial of Human Nature*, 75.
[10] Ibid., 45-51.
[11] Ibid.

of genetic differences.

This behavior may come from not only one gene, but many. John W. Fordon, University of Texas of Southwest Medical division in Dallas, has said that there are not only slow mutation changes or mistakes, but tandem segments of DNA repeat sequences that control animal and human development. Fred Gage, neuroscientist at the Salk Institute in La Jolla, California, reported in *Nature* that he found, while investigating L I retrotransposon DNA, that human virus-like genes jump from spot to spot in the genome and help shape the nerves in our human brains, which explains why *brains differ even in identical twins.* This is a mysterious and deliberate development.

Pinker claims that about one-half of the differences in behavior have some genetic influence, while the rest are from environmental and other causes. Pinker's arguments, of course, invalidate the blank slate theory of human nature, because the slate is written on by genetic inheritance; if genes affect conduct, the slate cannot be blank. But this does not invalidate the idea of the mystery of every individual as genetically designed. The genes are control mechanisms that are filled with mystery not understood by human psychology or science. Genetic activity is beyond understanding. Pinker appeals to a *hypothetical* story of twins making the same decisions at the same time, unknowingly, to claim uniform natural law is invisibly in control. But this unexamined and undocumented story is hardly evidence against the mystery of individual personal intelligence and responsibility. While there is much commonality to identical twins, there is not complete sameness, even genetically, as mentioned by Gage.

Genetics Extends beyond Information but Informational Abilities

Pinker omits other important studies. One of the great problems of education in America is to limit mankind's cognitive abilities primarily to information on a blank slate theory. In the past, deterministic science has judged individuals on intellectual informational performance and has, therefore, been discriminatory. Modern educators have viewed man's intelligence more in terms of empirical information that may be put into language. The I.Q. tests and most education have been based on this presupposition. But Ulrich Neisser of Cornell University, Howard Gardner of Harvard, and others see much more to human intelligence than this, including emotion, plans, motivations, and ways of expressing one's thoughts and personality.[12] Gardner argues for five or more different kinds of human thinking, which extends beyond stored information to expression of innate abilities and skills in art and other areas. These differences in humans are inherited and continue to influence who the person is and what he and she can do.

Neuroscientific Revelations

New Research Instruments and Experiments

Psychological testing and experimentation in classes, along with new instrumentation for examining the brain, are providing objective data. Neuroscience, with genetics and psychological testing, has overlapping influence, as would be expected. As a consequence, the next decade will enable the mapping of the functions of the brain in detailed ways. Research is already discovering how to interpret thinking before it is expressed. Perhaps magnetic resource imaging (MRI) is most

---

[12] Dan Goeman, "A Conversation with Ulric Neisser," *Psychology Today,* May 1983, 54-62; also Neisser's book, *Cognition and Reality, Memory Observed;* Alvin Sanoff, "Human Intelligence Isn't What We Think It Is," *U.S. News & World Report,* March 19, 1984, and Howard Garner's book, *Frames of Mind: the Theory of Multiple Intelligences.*

important, but positron emission tomography (PET), superconducting quantum interference device (SQUID), single-photon emission computerized tomography (SPECT), and the electroencephalogram (EEG) will all be used. There is an effort to develop an MRI that can see inside the atom.

These studies will improve neurosurgery, and may cure brain problems, but they will also, assuredly, make clearer all the functions of the brain. Whether or not there will be major changes in understanding is yet to be seen. Mapping has its limitations, for the human brain has amazing abilities to relocate memory and controls, as has been discovered by surgical excisions of portions of the brain. The amazing brain abilities to compensate are important for these bridges in thought for mapping the functions of the various parts of the brain. This is true in the differences in men and women, the differences in every individual, the way the brain works in humans in contrast to animals because of differences in brain structure, and the differences in how humans think in contrast to animals.

Neuroscience tends to extreme interpretations. Pinker argues against mystery in *brain plasticity,* restricting too much to natural law. He says,

> The brain's ability to reweight (rewrite) its inputs is ***indeed remarkable***, but the kind of information processing done by ***the taken-over cortex*** has not fundamentally changed; the cortex is still processing information about the surface of the skin and angles of the joints. And the representation of a digit or part of the visual field cannot grow indefinitely, no matter how much it is stimulated; the intrinsic wiring of the brain would prevent it.[13] (Emphasis added)

This conclusion passes over the mysterious, or what is *remarkable* in the brain's ability "*to take over,*" and focus on the obvious. The brain's ability to compensate for loss of eyesight, or the loss of a finger, even as it continues to process information, is a mysterious power. Deliberate human activity also affects changes in the physical neurons and functions of the brain. Braille readers develop a physical brain change that enhances this ability. These are beneficial compensations that arise out of need. But wrong human attitudes, practices, and desires also cause changes in the brain.

The old idea was that when the brain was fully formed in adolescence there were no changes, but it is now known that even in old age the brain still makes changes. Claims for brain differences in those with homosexual tendencies are now known to have been wrong. Conflict in society or especially in the home between husband and wife often causes a hormonal change and may influence the fetal areas of the brain during pregnancy involving sexual inclinations and produce some feminine traits. But those same areas of the brain in men, such as the hypothalamus develop more like the female's also *by the practice* of homosexuality, increasing the desire. Homosexuality is not produced without these and other factors.[14]

Normal Sexual Brain Differences in Humans Are Unique

Already much is known about male and female differences and differences between individuals within their normal sexual inclinations. Much is also known about the functions of brain areas in men and women. Whole books have been written on these differences. The female brain is less lateralized and less tightly organized than the male brain. In male right-handers, for example, language seems to be rather rigorously segregated to the left hemisphere, while their

---

[13] Pinker, *The blank slate*, 94, referring to *Recanzone*, 2000, 245.

[14] Carl Wilson, *Man Is Not Enough*, chapter 18. I have not repeated all the evidence given there. See also Susan Andersen of Harvard Medical School on hippocampus et al.

visual-spatial skills are as rigorously segregated to the right. This does not seem to be true in right-handed females. Their hemispheres seem to be less functionally distinct from each other and more diffused in organization. And switching between them seems easier.[15]

Louann Brizendine, neurophysicist at the University of California, in her book on the female brain, points out that of the one percent of the genes in the brain of the male and female that differ, those in women make the perceptions of the world *profoundly different*. A woman's hormones and other biological chemicals affect both the structure and function of her brain over her lifetime. The male's brain is relatively stable from day to day; female brains fluctuate according to the concentrations of estrogen, progesterone, and other hormones, and subject women to a range of emotional experiences. These may help the woman adapt to finding a mate, raising children, developing relationships, and even avoiding danger. Women use twice the number of words that men do, remember details of emotion-laden events which men forget. Women are twice as likely to suffer from depression as men.[16]

Based on genetic and neurological studies, the differences in men's and women's are very important in influencing what is normal practice for both sexes and are key to culture and civilization. Feminism argues men and women are the same. Pinker, on the one hand, says, "There is, in fact, no incompatibility between the principles of feminism and the possibility that ***men and women are not psychologically identical.*** No one has to spin myths about the indistinguishability of the sexes to justify equality." (Emphasis added) On the other hand, he *refers to the many differences* and even says that these are evolved adaptations (or writing on the blank slate over millions of years?). He approves of the saying, "It is better to have the male adaptations to deal with male problems and the female adaptations to deal with the female problems."[17] Pinker and all determinists have a blind bias to these differences because they endorse a system of values of equality that is based on their scientific materialism which overrides their proper recognition and interpretation of these differences. These neurological differences have profound effects on emotions, needs, desires, and conduct.

The reason for these developmental differences is not and cannot be explained logically as adaptations. The liberal elite dogmatically, without data, claim male and femaleness occurred by chance for survival. Jared Diamond listed distinct human sexual aspects as logically *unexplainable*; a mystery.[18] The human diversity from other animal primates are:

1. Constant female receptivity and male desire.
2. Unnecessary and exceptionally large male penis compared to other primates.
3. Repeated enjoyable recreational sex.
4. Desire generally for one sexual partner.
5. Private practice of sex.
6. Multiple reproductions of offspring.
7. Offspring have larger heads and long-term dependency requiring two parents.

---

15 Jo Durden-Smith and Diane Desimone, *Sex and the Brain,* (New York: Warner Books, Inc., 1984). They have reviewed the development and conformation of male and female differences, and the understanding in this has been more confirmed and extended. See chapter 4, "The Tale of Two Hemispheres," 47-61.

16 Louann Brizendine, *The Female Brain,* (New York: Broadway, 2006), 279pp.

17 Pinker, *The blank slate*, 337-348, cf. 340.

18 Jared Diamond, *Why Sex is Fun: The Evolution of Human Sexuality*, (New York: Basic Books, 1997).

The difference in brain construction and function, when united with these, are highly important to the way men and women act and in forming family and culture. The scientific determinists are blind to these implications and give them a different and damaging spin.

Human Uniqueness in Intellectual Ability to Use Memory/Information

The Human Brain's Power to Deliberate and Visualize

Michael E. Goldberg, a Harvard neurologist, points out that humans seem to have the mystical power to pause, deliberate, and the will to act without sensory stimulation, which animals do not. Goldberg says, "But we find that most human behavior is organized in large bits. Given a stimulus that may or may not evoke a movement, we look at it and ask, 'Should we respond now? Later? Or, should we respond at all? Does it fit into our plans?' "[19] Goldberg found that brain cells seem to be able to direct the eye in a remembered direction in the dark without benefit of sensory stimulus because of previous experience. These experiments may only be confirming what has already been discovered through neurosurgery and experimentation. Animals operate by instinct for physiological satisfactions, while humans can *pause and choose by reason from information.* This power to choose and remember choices enables them to transfer the learned knowledge that produces civilization. If man follows his biological and physiological instincts as persistently as animals do, he will eventually destroy civilization.

Humans also have a superior capacity for memory and memory organizing for speech. We also have a mental *ability to visualize*, as on a screen. These abilities are unknown in animals. Humans seem to have the capacity either to logically reason something out or to visualize it, whichever seems best. Some use visualization more than others, and this vision capacity may be used in dreams. Visualization has been the key to many great breakthroughs in human thinking. For example, Einstein visualized himself riding a sunbeam traveling at the speed of light and thereby conceived his seminal idea of relativity.[20]

Mystery of Man's Conscious Transcendent Thought and Control

Physiological or Reductionist's Attempts to Explain Consciousness

Several attempts have been made to understand human consciousness and thinking from a purely materialistic viewpoint. After Francis Crick discovered the nature of DNA, he turned toward the complex problems of neurology and consciousness. He and his assistant, Christof Koch, tried to explain consciousness by an oscillation theory, believing that the way the visual cortex responds to stimulation shows the way attention is changed. Certain oscillating neurons start joining together to promote attention. But Crick admits there is no mechanical solution yet.

Gerald Edelman developed a reentrant *loop hypothesis* whereby the environment influences select groups of neurons, strengthening connections as nature selects the fittest species. But many feel that his views are a rehash of an old feedback theory, and his views do not offer any definitive answer.

---

[19] Pines, "The Human Difference," 64, 68.

[20] Michael Guillen, "Mindview," *Psychology Today,* May, 1984, 73-74; Roger N. Shepard, "The Kaleidoscopic Brain: Spontaneous Geometric Images May Be the Key to Creativity," *Psychology Today,* June 1983, 61-68; Edward Ziegler, "Dreams: the Genie Within," *Readers Digest,* September 1985, 72-82; et al.

Consciousness is similar to the mystery of quantum physics. David H. Freedman has commented, "Nothing in science is as mysterious as quantum mechanic–except, perhaps, the mechanics of the mind."[21] The great atomic physicist, Roger Penrose, collaborating with Stuart Hameroff, a University of Arizona anesthesiologist, is seeking to find a link between human consciousness and quantum mechanical-like structures in the brain. Penrose hopes this will shed light on the matter of human consciousness and on the quantum theory.[22]

The mystery in quantum theory is that when an electron is moving one way or the other and encounters matter or energy, the electron "chooses" a single state. This process is not logical in the ordinary sense of being computable. Hameroff and other biophysicists found that microtubules in the brain have some unique properties, like quantum mechanics. The microtubules seem to insulate a pulse in the brain, and, as long as the pulse wasn't forced to choose a single state, it would be free to explore, simultaneously, any number of possible patterns within and among the bundles of tubules. But, when the brain seeks to solve a problem, the microtubules become an information switch, helping it to decide by some sort of sympathetic vibration, as in a tuning fork.

This is also an amazing feat of the brain. Freedman says,

> According to Hameroff, the brain's conventional neural network alone is simply far too under-powered even to account for such tasks as a person's ability to walk into a room and instantly recognize every object in it. Though this may seem entirely unremarkable to us, it is in fact a near-miraculous feat of information processing. Even a dozen of the world's largest super computers couldn't come close to replicating it.... The brain seemed to require more computing power than neurons could provide, and microtubules signaling each other through sympathetic vibrations seem to offer a plausible additional mechanism.[23]

Penrose and Hameroff have not solved the problem of consciousness, but have only shown there is a resemblance between quantum activity and brain activity. They have only identified a similar *area of great mystery*, with the hope that they may yet offer insights into how consciousness works. The problem with the Penrose-Hameroff Theory is that microtubules are found in many cells and not just in the brain. While the work of the tubules doesn't explain consciousness or even locate it, these microtubules may relate to innate created information systems.

The vast number of neurons and their actions are only partly known. Michael Denton, a prominent Australian microbiologist, illustrates their vastness, saying,

> The human brain consists of about ten thousand million nerve cells. Each nerve cell puts out somewhere in the region of between ten thousand and one hundred thousand connecting fibers by which it makes contact with other nerve cells in the brain. Altogether, the total number of connections in the human brain approaches 10^{15} or a thousand million million. Numbers in the order of 10^{15} are of course completely beyond comprehension. Imagine an area about half the size of the U.S.A. (one million square miles) covered in a forest of trees containing ten thousand trees per square mile. If each tree contained one hundred thousand leaves, the total number of

---

[21] David H. Freedman, "Quantum Consciousness," *Discover,* June 1994, 89-98.

[22] Roger Penrose, *The Emperor's New Mind,* (New York: Oxford University Press, 1989) 480 pages.

[23] Freedman, "Quantum Consciousness," 97.

leaves in the forest would be 10^{15}, equivalent to the number of connections in the human brain![24] Such a density of trees and leaves would be impossible. Simply trying to conceive of something as complex as Denton illustrates confirms what a complex organ the brain is. We can quickly see that the brain is able to contain an exorbitant amount of information. It is important to remember that all animals have limited abilities of consciousness, but there is a *tremendous difference* in the highest animal consciousness and human consciousness. Moreover, human consciousness has an ability to think in terms of historic sequence, as well as the future, and therefore can be aware of sin, guilt, and coming accountability at judgment. This also allows man to plan, and to impart ideas to one another. Animals lack a sense of right and wrong, except for basic physiological instincts. Their consciousness is limited mainly to rewards, and solely for their preservation.

Continued Support of Naturalistic Determinism In Spite of Data Against

In spite of the biological complexity and exorbitant improbability, the bias of unbelief has continued to argue against the mystery of man created by God. Pinker appeals to neurological malfunctions of the brain as an argument against a spiritual, individual consciousness, and accountability to God. He refers to the brain accident of Phineas Gage, a railroad man who in the late nineteenth century had a railroad spike driven through his ventromedial prefrontal cortex of his brain and lived. His body miraculously preserved his memory and motor functions. But he "turned into a different person, from courteous, responsible, and ambitious to rude, unreliable and shiftless." Pinker also refers to the brain surgeries on Michael Gazzaniga and Roger Sperry in which doctors cut the corpus callosum of the cerebral hemisphere, which seemed to inhibit the person making right decisions using both sides of the brain. The relationship between the knowledge of the left and right does not properly correlate. When the left hemisphere is shown a chicken and the right a snowfall, and the patient is asked to make a decision, the left selects a chicken claw and the right a shovel. But when asked why, the person says, "The chicken claw goes with the chicken, and you need a shovel to clean out the chicken shed."[25]

Pinker argues this discredits a conscious mind making decisions, as if there is a "control panel with gauges and levers operated by a user—the self, the soul, the ghost, the person, the 'Me'. But cognitive neuroscience is showing that the self, too, is *just* another network of brain systems."[26] Is that the right analysis and conclusion? Is not the correct view that the operator or self is using a broken machine? In the case of Gage, his misfortune caused him to choose a victim mentality and make wrong decisions. In the case of the divided brain, the person made an effort as a transcendent self trying to use both sides, but the broken machine won't allow it to function correctly. Is there still not a self working? Pinker ignores this fact. But he also apparently deliberately omits some highly significant neurosurgical findings that completely contradict determinism.

Surgical Experiments Show Conscious Transcendent Mind to Brain

Of great significance is the work of Sir Charles Sherrington and his student successors, Drs. William Penfield and John Eccles. Following strict scientific methods and experimentation, these outstanding neurosurgeons, discovered a remarkable "double consciousness." In open brain

---

24 Denton, *Evolution: A Theory In Crisis,* 330.

25 Pinker, *the blank slate*, 42, 43.

26 Ibid.

surgery, often with conscious patients, using brain electronic probes, Penfield said, "The mind of the patient was as independent of the reflex action of the brain as was the mind of the surgeon, who listened and strove to understand." This was repeated many times with many patients. The person seemed to have a brain-like computer but not *be* the computer. Much as a person uses the TV remote control and directs and uses it as *he* pleases, so the mind transcends the brain.

It was not possible to explain these actions from a neural point of view. Eccles was forced to the conclusion that *the brain and the conscious mind of a person are two different things that interact* and normally work together, but that the brain is the instrument of the conscious mind.

Upon reviewing the evidence, Karl Popper, the great philosopher of science, collaborated with Eccles, writing about the amazing phenomenon of the brain and an independent mind. They agreed that the mind is a spiritual entity. Eccles, therefore, assumed it was created, capable of spiritual influences and of continuing after death. Popper agreed that interaction and dualism existed, but argued that this amazing human power was somehow physiologically generated by a mysterious evolutionary process, but no longer subject to physiological evaluation.[27]

Eccles' and Penfield's findings agree with the findings of others in other fields. V. S. Ramachandran, M.D., PhD, Director of the Center for Brain and Cognition and Professor of Psychology and Neuroscience, University of California and adjunct professor of biology at the Salk Institute has reviewed numerous findings about the brain. He concluded, "A crucial ***yet exclusive*** aspect of self is its self-referential quality, the fact that it is aware of itself." [28] (Emphasis added).

Mapping Cognitive Functions Shows Mind/Brain Spiritual Interaction

Two prominent neurobiologists, Andrew Newberg, M.D., Eugene D'Aquill, M.D., PhD, and Vince Rause, doing research in mapping of the higher cognitive functions in the cerebral cortex, have located the area of the interaction of the transcendent mind and the brain in religious experiences. They are both naturalists and believe this ability in man evolved. They have written the book, *Why God Won't Go Away.* Herbert Benson, M.D., President of Mind/Body Medical Institute, has called this, "A wonderful assessment of the brain and its activity when God is experienced." These researchers were surprised at this discovery while doing SPECT scans on live humans.

They confirmed the dual nature discovered by Sir Charles Sherrington and extended our understanding of the way the mind/brain works. They say,

> [Somehow] the brain, with its great perceptual powers, began to perceive its own existence, and human beings gained the ability to reflect, as if from a distance, upon the perceptions produced by their own brains. There seems to be, within the human head, an inner, personal awareness, a free-standing, observant self. We have come to think of this self… as the phenomenon of mind.[29]

---

[27] Sir Charles Sherrington, *Integrative Action of the Nervous System;* Wilder Penfield, *The Mystery of the Mind,* (Princeton: Princeton University Press, 1975); John C. Eccles and Karl Popper, *The Self and the Brain,* (Springer Verlag International, 1977); cf. Arthur C. Custance, *The Mysterious Matter of the Mind,* (Grand Rapids, MI: Zondervan, 1980), 64-66, 70-71, 75-86.

[28] V. S. Ramachandran, *A Brief Tour of Human Consciousness: From Impostor Poodles to Purple Numbers,* (New York: P I Press & Pearson Education, Inc., 2004) III.

[29] Andrew Newberg, Eugene D'Aquill, and Vince Rause, *Why God Won't Go Away: Brain Science & The Biology of Belief,* (New York: Ballantine Books, 2001, epilogue 2002) 32.

As they identified the locations of various functions in the cortex, they found an area where spiritual or religious experiences occurred. From this study they said,

> Gradually, we shaped a hypothesis that suggests that spiritual experience, at its very root, is intricately interwoven with human biology, that biology, in some way, compels the spiritual urge.
>
> We believe that we were seeing colorful evidence on the SPECTs computer screen of the brain's capacity to make spiritual experiences real.... [We] further believe that we saw evidence of a neurological process that has evolved to all us humans to transcend material existence and acknowledge and connect with a deeper, more spiritual part of ourselves perceived as an absolute universal reality that connects us to all that is. [30]

Skeptics answer that this is a delusion caused by the chemical misfiring of a bundle of nerve cells. But because of the regular and objective evidence, the authors interpreting the SPECT scans suggest the other possibility, of an innate connection of the self with a power in nature, or a pantheistic experience.

Yet the authors have already identified the double nature of man as having a transcendent self or mind that was nonmaterial. Is there not then the equally valid possibility that these persons were communicating with transcendent spiritual powers instead of just forces from nature? Popper speculated that this transcendent self was naturally evolved from forces unidentifiable, but Eccles identified the mind as able to communicate with a nonnatural transcendent spiritual force.

Moreover, Newberg and D'Aquill believed this evolved, but say,

> As we've seen, mystics are not necessarily the victims of delusion; rather, their experiences are based in ***observable functions*** of the brain. The neurological roots of these experiences would render them as ***convincingly real*** as any other of the brain's perceptions. In this sense, the mystics are not asking nonsense; they are reporting ***genuine, neurobiological events***. This is the conclusion to which our research draws us; it forces us to ask a provocative question about the ultimate nature of human spirituality. (Emphasis added)
>
> Are we saying, then, that God is just an idea, with no more absolute substance than a fantasy or a dream? Based upon our best understanding of how the mind interprets the perceptions of the brain, the simple answer is no.
>
> Our own brain science can neither prove nor disprove the existence of God, at least not with simple answers. The neurobiological aspects of spiritual experience support the sense of the realness of God. Yet we interpret and funnel that which our brain tells us is real through our subjective self-awareness.[31]

Thus, these science authors show that there is real objective information received from a spiritual experience, different from bodily sense perception, and is from a transcendent creator God or from spiritual forces such as demons. But they have preferred to interpret these as some spiritual power within nature that the mind discerns. They show the brain of the body also can convey messages from spiritual forces to the mind or human spirit.

---

[30] Ibid., 6, 9.
[31] Ibid., 143.

Uniqueness of Human Language

The uniform deterministic scientists, who held to the idea that all is learned by the senses and that the mind is a blank slate with certain categories that excluded God, held that all language in every aspect was based on learning. B. F. Skinner was a chief proponent of learning theory. But in 1957 MIT linguist Noam Chomsky, after much research, wrote *Syntactical Structure,* in which he showed that certain aspects of linguistic knowledge and ability are the product of "universal innate ability," or what he called "language acquisition device." He said, LAD

> enables each *normal* child to construct a systematic grammar and generate phrases. This theory claims to account for the fact that children acquire language skills more rapidly than other abilities, usually mastering most of the basic rules by the age of four. As evidence that an inherent ability exists to recognize underlying syntactical relationships within a sentence children readily understand transformations of a given sentence into different forms–such as declarative and interrogative–and can easily transform sentences of their own.[32]

While all languages have many and varied differences, and their own unique words and lexicon, all language has basic universal syntactical principles.

The secular linguists say that this intuitive ability, which man did not learn, came about by evolution. But there is no observed data that can illustrate gradualism in learning. This innate ability is present at birth. Animals have limited sounds for relating, but this intuitive syntactical linguistic ability is missing.

Intuitive Moral Code in the Human Mind is Unique

Uniform deterministic scientists have also predominantly believed that morals were learned from society and culture. This was the basis by which Freudian psychology dismissed guilt, maintaining guilt was induced to the ego by social influence. Now, in the twenty-first century, that is challenged by objective scientific data. Much as Chomsky demonstrated the innate syntactical ability among all people, Marc Houser, who is professor of psychology and human evolutionary biology at Harvard University, has recently reported from his extensive studies that there are innate moral principles that are intuitive to all people. Elliot Turiel, cognitive scientist at the University of California at Berkley, had previously determined that there is a very important *distinction between a social convention and a moral rule.* Children, by at least the age of three, are able to understand that distinction.

Marc Hauser had done extensive experiments with animals and children.[33] While he presents much evidence of common cognitive abilities, learning and deception in animals, he found no evidence anywhere of any sense of thought, or is any sense of transgression in animals. His new book shows how humans have—

> a universal moral grammar, a set of principles that every human is born with. It's a tool kit in some sense for building possible moral systems. ...a suite of universal principles that dictates how we think about the harming and helping others. But each culture has some freedom – not unlimited—to dictate who is harmed and who is helped.[34]

---

[32] Noam Chomsky, *Wikipedia Encyclopedia.*

[33] Marc Hauser, *Wild Minds: What Animals Really Think.*

[34] Josie Blausiusz, the Discover Interview, Discover, May 2007, 62-66, cf. Marc Hauser, *Moral Minds: How Nature Designed Our Universal Sense of Right and Wrong,* (New York: Ecco of Harper Collins, 2006), 489 pp.

The way of administering justice and punishment may vary greatly in any society. But right and wrong are intuitive.

Some moral principles hold true across all societies. The Golden Rule applies in every culture. There are almost universal interpretations of acceptable ways to treat others. When people were asked, "would you switch a trolley car to kill one person, but save five others?" The answer was always, "Yes." But to the question, "Would you kill a person for his organs, to save five who need them?" The answer was always, "No." Hauser also found understanding of moral questions differed from cultural and social ones. One group held it was right for children to raise their hands if they wanted to ask a question. When these kids were asked if it was all right in France that they never raise their hands, the students agreed it was ok for the French. But when asked if someone annoyed you, was it OK to punch him, the answer was, "No." But if told the French said it was OK to punch him, the response was, "The French are weird." Hauser says that on moral tests the answers are 100 percent the same for all people, but he points out that there is a difference in application of reciprocity of justice which is followed by "an in group, or an out group."

Hauser, however, admits that there is a wide distinction between what people judge is right or wrong *and what they do*. For selfish reasons they often do not do what they know is right. There is also evidence that there are differences in response to wrong when it is experienced. Hauser admits this "moral grammar" is universal and intuitive in all people, and it is not known in animals. Therefore, there is no known evidence for gradualism by learning or acquisition. But he insists it is the result of an evolutionary process. Since discovering this "moral grammar" others have tried to explain it as evolution.[35] Hauser's findings contradict Freud's super-ego.

Conclusions: Man's Unique Transcendent Consciousness and Body

Scientists Confirm Transcendent Mind over Brain and Body

Science has confirmed the gap between the brain of men and animals. Cognitive science shows human brains have the capacity for more information than animals' brains. Human brains are different from computers, which are linked to information that pervades the world. Genetics control the development of the brain, but mystically and uniquely impart more than animal abilities to man. Neuroscience reveals a brain working through, and processing information, and communicating not only sense perceptions, but also spiritual impressions to the mind or spirit. This involves mystery.

In reviewing the modern understanding of the human mind, Gordon Rattray Taylor, an outstanding science reporter on the human mind, said,

> Man's mind may show capacities outside the scope of current physics.... A whole new wing of science is perhaps waiting to be opened up. LeShane puts it more forcibly: 'To explain damned facts like this one, you need a new concept, a ***new definition of man*** and his relationship to the cosmos.'[36] (Emphasis added)

In reviewing his book, *The New Story of Science*, in *The Brain/Mind Bulletin*, George Stanciu said, "Physics, neuroscience and now psychology are throwing off 19th century materialism.... But the new worldview is just in the throes of being born, so people feel lost, caught up in old

---

[35] Richard Joyce, *The Evolution of Morality*, (Cambridge, MA: M.I.T. Press, 2006), 288 pp.

[36] Gordon R. Taylor, *The Natural History of the Mind,* (New York: S.P. Dutton, 1979), 308-309.

thinking that no longer works."[37] The author of the book quotes Roger Sperry, a Nobel Laureate, as saying, "Current concepts of the mind-brain relation involve a *direct break with the long established materialist* and behavior doctrines that have dominated neuroscience for many decades."[38] Thus, another barrier beyond physical science has been revealed beyond which man does not seem to be able to penetrate. It exceeds scientific deterministic theory.

Man's Body, Mind, and Familial Sexual Uniqueness Produce Culture

Research shows that men and women are biologically different in every cell and their brains operate differently. They are equally intelligent above animals, but don't desire the same objectives. Except in rare perverted cases, man and women don't enjoy, and therefore don't do the same things equally well. Their differences allow them to complement each other for oneness in producing offspring and produce successive offspring of equal ability. Biblically, they reveal two different aspects of the image of God. The sexual differences cannot be explained from evolution, but are denied by the scientific determinists. By oneness in marriage they can have offspring to whom they transmit information that is accumulated to build culture.

Only man has the deliberative mental process to evaluate information, to choose and decide. Man is *different in kind from all other animals*. Only he can envision how things are or might be. Only he has the long-term memory capacity to remember his plans and carry them out. Only he can truly talk and communicate faith, morals and objectives to his young. Only man, not apes, can walk and sit upright in order to do creative work for hours on end. His mind and body structure allows him to do repetitive tasks for business, artistic and musical purposes.

Apes have only limited intuitive information. None of their sexual traits are familial. Apes cannot do more than simple tasks. Because their spinal column enters their head from the rear and not from the center, they cannot sit or stand erect. Apes, with a receding lumbar, must bend over with their long arms and phalanges near the ground and their face inclined upward, making it uncomfortable to sit and work. Apes cannot continually look ahead and down conveniently; their arms and legs are not constructed in a way to make walking long distances convenient. Their arms and legs are constructed primarily to climb and swing. In summary, these are some of the most prominent ways that man is unique, and in which he is the only known animal that can use the world to create civilization.

The Scientific Findings Support the Biblical View of Man

God created them male and female in His image (reflecting two aspects of His image) and gave them dominion over all creation. The woman was created from Adam to be his equal because no animal was his equal and he was alone. The man had a physical body with a brain similar to animals (from the dust of the ground) but man was inbreathed by God to have a spirit and he became a living soul or person. His ability to rule over creation for God reflected his ability to think God's thoughts after Him about creation (Genesis 1:25-28; 2:7, 18-24, et al.). Adam and Eve together were to multiply and fill the earth, thereby extending culture or civilization of God's kingdom. The conception, therefore, involved the contribution of both man and woman. The woman with "the help of the Lord" was the giver of new life (Eve, "the life giver"), but God

---

[37] Robert Augors and George Stanciu, *The New Story of Science,* (Lake Bluff, IL: Regnery Gateway, 1984), "The New Story of Science: Including Mind in the World," Brain/Mind Bulletin, Vol. 10, No. 2, December 10, 1984, 1.

[38] Ibid.

controlled the formation/information of each new individual for His purposes (Genesis 3:20; 4:1, 2, 25. Psalm 139:13). The ability to recognize God as Creator and His law are *within* man (Romans 1:19, 20; 2:15). But man's sin nature now gives him a bias against God and His will, so that man oppresses or perverts the truth (Romans 1:18 ff.; 7:14 ff). When man repents and trusts Christ, He again breathes on him, imparting the Holy Spirit within his consciousness to restore communication with God's Spirit (Genesis 2:7; John 20:22; John 3:5-8).

Man alone can do science and understand moral relationships. The human mind is a miracle that can observe physical created phenomena. Einstein once remarked that the great miracle is that man can observe and understand nature. C. S. Lewis, the great English scholar, pointed out in his book, *Miracles* that reason cannot come from non-reason or chance. He argued that man's ability to reason and to understand the world is a great testimony to the fact that man had an origin beyond nature.[39] Eccles' belief that man has a transcendent mind or spirit and Newberg's and D'Aquill's understanding that there is a location in the brain that can be influenced by either demonic spirits or the spirit of God seems true in experience.

This all seems to conform to Augustine's biblical interpretation that the light of man is from Christ. The mind or spirit of man seems to be implanted with innate knowledge and thought forms, which include morals that come from God and which are used with the sense images and interpreted according to their choice. Moreover, the mind may be enlightened or influenced by other spiritual forces. Man's light thus includes the spiritual consciousness with these imprinted thought forms and may be influenced by God's Spirit or demonic forces as man also receives and considers natural physical stimulations from the brain. Man's transcendent mind is created to control his body through his brain. He intuitively knows God, what is morally right, and how to communicate using language. Man's bias against God has rejected his uniqueness to enable him to be free from his responsibility to God. His imagination has let him commit idolatry.

[39] C.S. Lewis, *Miracles,* (New York: Macmillan Co., 1977), 35-36, cf. 23-66.

Chapter 18
Civilization Reflects Recent Creation

Introduction and Orientation

Only Man's Unique Abilities Produce Culture and Civilization

Humans, with their God-given nature and unique characteristics, make man the only creature suited to produce, promote, and perpetuate civilization. Their physical design, along with their mind-brain capabilities, equips them to use and adapt their environment to shape a culture. Most important of all is the fact that man forms a continuing family. This enables him to develop civilization. When the family functions as the most basic unit, creative abilities and products are extended and improved, from one generation to the next, thereby, developing culture. Each successive civilization learns from and builds on the previous ones. As the early scholars said, "We stand on the shoulders of giants."

No animal can build a civilization, only man can. Stories about ape civilizations, or even about apes raising men, are all myths. While research shows that chimps and orangutans can mimic certain different patterns, such as making a rasping-like noise before retiring at night, such limited thought and communication cannot produce a civilization. The anthropologists that make these claims are stretching the fact of animals' innate information to support their own biases.[1]

Civilization is possible because of the differences in men and women for complementation and for being companions for having, nurturing, and raising children who can be trained to follow their teachings and training. Civilizations decline when the complementary sexual relations for marriage oneness is changed to competition for material egalitarianism. The intellectual change of values causing denial of differences of male and female ends constructive culture building and brings decline of civilization.

The unique abilities of man from animals, male and female enable them to form families and produce as a group which is not true of any animals. The human male is usually committed to the female and the offspring they produce together. The female's reproductive and accompanying emotional cycle furnish humans with abiding sexual desire and pleasure. The woman's sexual attractiveness, along with the ability to have a child that is dependent on her nurture over a long period of time, gives the woman a sense of great importance. Her sense of value may also come from being a meaningful partner with her husband. Human offspring-remain dependent far longer than animals, creating a greater incentive for mutual long-term faithfulness and support. This motivates the female to respect and defer to her mate, and gives the male a sense of importance and motivation to provide. The wife's respect for her husband is his value source, since he cannot have children. His aggressive nature and greater power from his muscular build and hormones allows him to protect and provide for his family. At least two people are needed to care for children, and when either husband or wife is absent, surrogate help must be obtained.

Male primates, such as apes, have no continuous responsibilities or desire toward the female or the offspring. At times, the male may even become hostile. *The family is a human phenomenon* that demands integrity and sexual fidelity to continue society. The higher institutions that emerge, protect and provide for it. The uniform determinists have claimed that sexual differences are taught, and all these social traits are learned and emerged gradually, and are not innate to human nature.

Evolutionary philosophers were not able to show any missing links from apes to men, nor have they any unified, reasonable explanation or evidence for how men acquired these unique

---

[1] Meredith F. Small, "Orangutans Show Signs of Cultured Behavior," *Discover*, January 2004, 52.

characteristics. They speculate that through intermittent links from apes to men, humans gradually learned and acquired all these characteristics involved in culture. The intellectual speculation is that through evolution, men gradually acquired tool making and other cultural skills, first as hunters and gathers, and then as they learned how to live together in villages and towns, domesticate crops and animals, and devise and build the structures and infrastructure to sustain civilizations. The evolutionists have insisted that ***these abilities originated and emerged in Africa*** and migrated from there around the world. They make such claims, even though similar stone tools and other signs of early culture are also found elsewhere in the world.

History Shows Man's Early Brilliance, Not Gradual from Ape to Men

Mature Science of Anthropology Reveals Early Creation of Culture

In the last half of the twentieth century anthropology, involving thousands of scientists, sought to rewrite human history, based on the philosophy of evolution. In the twenty-first century, anthropology has matured to the point that it offers a clearer worldwide picture of early human culture. The anthropologists and historians have sought to present a picture of gradual development from apes to men.

While many sources are available, this work focuses on two efforts to accumulate evidence of man's history: Jared Diamond (1999), *Guns, Germs, and Steel: The Fates of Human Societies* and Steve Olson (2002) *Mapping Human History*. Both men offer a well-researched, assimilated, and comprehensive picture of the development of civilizations.[2] Diamond focused on the development of crops and animals, conquest and disease, while Olson incorporated genetic sources along with these. Many dates are based on radio carbon dating. (The reliability and limitations of radio carbon dating will be discussed in later chapters.) Both of their books won national acclaim in the scientific community and were widely read. Both espouse Darwinian evolutionary views, and Darwin's idea of Africa as the point of man's origin. Their foremost premise is that humankind emerged as primitive hunter-gathers and developed culture later. As demonstrated in previous chapters, this view of evolution held by the educated elite is a perversion of the facts and while Jared and Olson interpret data to try to support this, the data really supports the biblical history.

Evidence of Man's Cultural Development out of Eurasia, not Africa

Diamond compiled a vast amount of observable, complex data, dated mainly from 9000 BC, and simplifies it in many charts and maps. Some of his early evidence is from about l2000 BC. He shows that civilization *first began in the Fertile Crescent or center of Eurasia* about 8500 BC, spread westward, existed in China about 7500 BC, and spread westward across Europe and into England by 3500 BC. Diamond says that some aspects of civilizations were known in the Andes

[2] Jared Diamond, professor of physiology at the UCLA School of Medicine, *Guns, Germs, and Steel: The Fates of Human Societies,* (New York: W. W. Norton and Company, 1999) 480 pages He was awarded the Pulitzer Prize, Steve Olson is a journalist writing for the National Academy of Sciences, White House Office of Science and Technology and the Institute for Genomic Research. *Mapping Human History* (New York: Mariner Books of Houghton Mifflin Company, 2002), 292 pages. Award in Science and the book was a best seller. The book was endorsed by Ian Tattersall, who has been quoted, and also selected by *Discover* as one of the best science books of the year. The major insights from these research efforts will be used. In the first part of both books they present their evolutionary story over a number of pages.

of South America, Amazonia, Mesoamerica around 3000 BC, and in the eastern U. S. by about 2500 BC. Since Diamond wrote, more recent data reveals much earlier development in the Americas that will be furnished from other sources.

Diamond argues that in this progressive geographic expansion there was also a progression in the kinds of culture. For example, in the Fertile Crescent, he says, there were villages of hunter-gatherers in 9000, domestication of plants by 8500, domestication of animals by 8000, government by chiefdoms by 5500, widespread use of metal tools of copper and bronze by 4000, formation of states by 3700, writing by 3200, and iron tools by 900. He believes a similar progressive development occurred in China and England, or across the Eurasian continent, but not as clearly or regularly in the Americas.

Many aspects of civilization were spread from one group of people to another, but in some cases, there was independent development. For example, he believes there may have been about nine locations, but possibly as few as five places, throughout the world, where plant domestication occurred. These are the Fertile Crescent, Eastern China, Mesoamerica, Amazonia and the Eastern U.S. Often a "founder package" of farming was imported from somewhere else; it may also have included some local contributions. He says cattle were domesticated independently in India and Western Eurasia within the last 10,000 years.

Diamond speculates that where there were a few plants that would have been useful for domestication in some locations, the people were not willing, or really able to give up their life as hunter-gatherers, because that would not have sustained them if they had settled down in towns. It was only when a larger package of domesticated plants and animals were brought into the area from elsewhere that they were willing and able to give up the hunter-gatherer lifestyle and domesticate their own local plants or animals. Accepting the "founder package" enabled them to give up hunter-gathering and attend to local plants or animals. Therefore, the spread of domesticated plants or animals was often the main catalyst for domestication and growth of civilization.

He points out, for example, that it was only after receiving the arrival of founder crops from elsewhere that Western Europe developed poppy and oats plants (6000-3500 BC). The same was the case in the Indus Valley where they domesticated sesame and eggplant, and humped cattle (7000 B.C). Egypt did not develop the sycamore fig and chufa, nor the donkey and cat until after getting founder crops (6000 BC). The reception of founder crops was, in many cases, the means for spreading civilization.[3] It is worthy of note that all these areas were accessible *from the Fertile Crescent.*

Eurasia the Recognized Center of Domestication of Food and Animals

Jared Diamond acknowledges the great significance of the *Fertile Crescent*, the leader in domestication, as *the key to spreading civilization.* While there were a few other areas from which founder crops seemed to come, such as Southwest Asia, *the Fertile Crescent is the primary migration point for most of the world.* He, therefore, spends many pages discussing and explaining the development of this central geographic area of Eurasia, why it developed faster and better than anywhere else in the world, and why these crops, after development, spread rapidly.

In speaking of centers of domestication, Diamond said,

---

[3] Ibid., 100

> As we had seen, ***one*** of them, ***the Fertile Crescent***, was perhaps ***the earliest*** center of food production in the world, and the ***site of origin*** of several of the modern world's major crops and almost all of its major domesticated animals. The other two regions, New Guinea and the eastern United States, had domesticated local crops, but these crops were very few in variety, only one of them gained worldwide importance, and the resulting food package ***failed to support extensive development*** of human technology and political organization as in the Fertile Crescent.
>
> That area appears to have been the ***earliest site*** for a whole string of developments including cities, writing, empires, and what we term (for better or worse) ***civilization.*** All those developments sprang, in turn, from the dense human populations, stored food surpluses, and feeding of non-farming specialists made possible by the rise of food production, the form of crop cultivation and animal husbandry. Food production was the first of those major innovations to appear in the Fertile Crescent. Hence, any attempt to understand the ***origins of the modern world*** must come to grips with the question why ***the Fertile Crescent's domesticated plants and animals gave it such a potent head start.***[4] (Emphasis added)

The choicest plants for domestication are found in Mediterranean areas of the Near East. Diamond points out that Mark Blumler's study listed *the 56 heaviest–seeded wild grass species* (excluding bamboo). He said that *virtually all of these are native to Mediterranean zones* or other seasonally dry environments, with 32 occurring in the Mediterranean zone or other seasonally dry environments. Furthermore, they *are overwhelmingly concentrated in the Fertile Crescent or other parts of western Eurasia,* both Mediterranean zones, which offered a huge selection to incipient farmers—about 32 of the world's 56 prize wild grasses, specifically, barley and emmer wheat, the two earliest important crops of the Fertile Crescent, ranking respectively 3rd and 13th in seed size among those top 56. Diamond concludes,

> The ***Fertile Crescent's biological diversity*** over small distances contributed to its wealth in ancestors not only of valuable ***crops*** but also of ***domesticated big mammals***. As we shall see, there were few or no wild mammal species suitable for domestication in the other Mediterranean zones of California, Chile, southwestern Australia, and South Africa. In contrast, four species of the big mammals–the goat, sheep, pig, and cow–were domesticated very early in the Fertile Crescent, possibly earlier than any other animal except the dog anywhere else in the world. Those species remain today four of the world's five most important domesticated mammals. All four lived in sufficiently close proximity that they were readily transferred after domestication from one part of the Fertile Crescent to another, and the whole region ended up with all four species.
>
> *Agriculture was launched in* ***the Fertile Crescent*** by the early domestication of eight crops termed "Founder crops" (because they founded agriculture in the region and possibly **in the world**). Those eight founders were the cereals emmer wheat, einkorn wheat, and barley; the pulses lentil, pea, chickpea, and bitter vetch; and the fiber crop flax. Of these eight, only two, flax and barley range in the wild at all widely outside the Fertile Crescent and Anatolia.
>
> Eventually, thousands of years after the beginnings of the animal domestication and food production, the animals also began to be used for milk, wool, plowing, and transport. Thus, the crops and animals of the Fertile Crescent's first farmers came to meet humanity's basic economic needs: carbohydrate, protein, fat, clothing, traction, and transport.
>
> In the Fertile Crescent the transition from hunting-gathering to food production ***took place relatively fast***: as late as 9000 BC people still had no crops and domestic animals and were entirely dependent on wild foods, but by 6000 BC some societies were almost completely dependent on crops and domestic animals.[5] (Emphasis added)

---

[4] Ibid., 134, 135.

[5] Ibid., 139-142.

Diamond explained that not only were these important crops, all *conveniently located* for development and accumulation, but the area of the Fertile Crescent just happened "to provide a ***wide range of altitudes and topographies*** within a short distance—allowing the staggering of harvest seasons."[6] Thus, produce would be available continually.

Moreover, Diamond points out the *genetic stability* of the crops found and selected by humans in the Fertile Crescent. He said,

> However, for most of the Fertile Crescent foundry crops, *all cultivated varieties in the world today share only one arrangement of chromosomes* out of the multiple arrangements found in the wild ancestor; or else they share only a single mutation (out of many possible mutations) by which the cultivated varieties differ from the wild ancestor *in characteristics desirable for humans.*[7]

Diamond also makes two other important observations. There is the *mystery*, which he attributes (with no evidence) to evolution. He says, "Each plant is *genetically programmed*, through natural selection, to respond appropriately to signals of the seasonal regime under which it has evolved."[8] Also, "the plants seemed to be *adapted to the diseases* prevalent at their latitude."[9] We ask, "Did this happen by chance, or creation and providence?" Genetic improvement as Diamond speculates does not occur without adding new information by creation. The new information theory excludes gaining more information. He affirms that genetic tracing of these crops leads back to the center of Eurasia, and that they were exactly what man needed to begin domestication.

The fact that civilization arose in the Fertile Crescent is seen by Diamond as due almost entirely to the fact that this was *the one place in the world in which nature provided the wild species of plants and animals that could be domesticated.* He comments on the fact that all humans, wherever in the world, seem to have the ability to study and evaluate thousands of plants and animals, but were not able to domesticate the products.[10] He later says, "All these facts indicate that the explanation for the lack of *native mammal domestication* outside *Eurasia* lay with the *locally available wild mammals* themselves, not with the local people," an observation he already made for plants.

This *mystical way* the species were concentrated, their adaptability to the environment and disease, their genetic stability, and ability to be grown at strategic altitudes in most seasons, all strangely, *mysteriously, prepared the region* of the Fertile Crescent for beginning and expanding human civilization to the rest of the world. This correlates with the biblical account of Noah and his family, along with the animals, exiting the ark in that area, and re-establishing civilization.

Expansion of Agri*culture* from the Fertile Crescent

Diamond presented this evidence, along with a map, showing that the agricultural culture of the Fertile Crescent spread across the whole Eurasian continent much sooner, and progressed more rapidly and further than in America and Africa. He said,

> Soon after food production arose there, somewhat before 8000 BC, a centrifugal wave of it

---

[6] Ibid., 140.
[7] Ibid., 183.
[8] Ibid., 140.
[9] Ibid., 184.
[10] Ibid., 142.

> appeared in other parts of western Eurasia and North Africa farther and farther removed from the Fertile Crescent, to the west and east. ....The wave had reached Greece and Cyprus and the Indian subcontinent by 6500 BC, Egypt soon after 6000 BC, central Europe by 5400 BC, southern Spain by 5200 B. C., and Britain around 3500 BC. In each of those areas, food production was initiated by some of the same suite of domestic plants and animals that launched it in the Fertile Crescent. In addition, ***the Fertile Crescent package penetrated Africa Southward to Ethiopia*** at some still-uncertain date. However, Ethiopia also developed many indigenous crops, and we do not yet know whether it was these or the arriving Fertile Crescent crops that launched Ethiopian food production.[11] (Emphasis added)

Diamond, however, admits that all sub Saharan domesticated animals came from Eurasia, as well as all domesticated agricultural crops.[12] This is the exact opposite of Darwinian views that civilizations emerged from Africa.

Great Expansion Because of East-West Continental Axis

Diamond shows that because Eurasia is the biggest continent, and extends east and west on its axis, domesticated products of the Fertile Crescent would have a similar climate as civilization expanded in those directions. On the other hand, Africa and the American continents extend on a north and south axis, which required expanding over a broad range of climates. They could not have hosted such domestication. Diamond began his study by showing the expansion of production from southwest Asia into Polynesia, New Guinea, Australia and New Zealand, and the Philippines, which he dates in the second century BC. He suggests that the beginning of expansion from the Fertile Crescent extended later, by boat, and over the Russian land bridge to Alaska and the Americas; however, he has no documentation for this theory. He also covers late domestication of America's distinct plants and animals, which presumably began around the fourth century BC and in West Africa.

S. Christopher Caran, a research geologist, and James A Neely, professor emeritus of anthropology of the University of Texas, Austin, uncovered ruins near San Marcos Necoxtla, in the Tehuacan Valley of Mexico that seem to place that earlier. They found an excavated well and terraced hillsides, with canals many kilometers long, extending over irregular terrain, with a continuous downward gradient of less than two degrees. This was an amazing feat of hydraulic engineering, even for today. This excavated well, five meters deep and ten meters in diameter, is dated 10,000 years ago. It was used for over 2000 years. Was this extensive effort not done for agriculture and human survival, as their illustrated art shows? Or was it only for landscaping? This new evidence does not correspond with dates of primitive man and animals coming across the Alaska Bridge, as Diamond and evolutionists believe.[13] It is important to note that Diamond *does not consider the breakup of the continents*, or the possible presence of man and animals in that catastrophic division. If man were present at the division of the continents, as the Bible indicates, that would make a significant difference.

---

[11] Ibid., 180, 181. The map was assembled by geneticists Daniel Zohary and botanist Maria Hopf.

[12] Ibid., 398, 399.

[13] S. Christopher Caran and James A Neely, "Hydraulic Engineering in Prehistoric Mexico," *Scientific American,* 2006.

Method of Expansion both Voluntarily and Later by Empire Building

Diamond maintains that in the initial stages, the products of agriculture were spread mainly by willing acceptance, and only partly by conquest, as the farmers extended their territory and expanded trade. Diamond saw the emergence of agriculture and food surpluses and storage as the basis and prelude to the gathering of large, dense sedentary, stratified societies. Out of these developments grows technology for weapons, such as steel swords and guns, and transportation, along with political organization, writing and education for business and records. Hence, the title of the discussion is "Farmer Power." He also emphasized the importance of the domestication of the horse in southern Russia about 4000 BC, and the expansion of its use for transportation and in war.

Diamond said city development and later empire development began with an egalitarian vision in villages and cities, but ended up in what he calls "kleptocracy" in the empire. This process of intelligent farming, and development and expansion provided the resources for political conquest and empire building. Along with economic prosperity and technology to produce superior weapons, an empire spreads diseases to which the conqueror is immune, but the people in the new location are not. These diseases, it is thought, were first contracted from the domestication of animals. While he expands greatly on describing this process, since he argues the foundation of superiority and civilization is based on the domestication and storage of food and to transportation, this process is not the center of our focus and is mentioned only in passing here. [14] The emphasis on economic prosperity and empire building, along with war, will be important in a future chapter.

The Explosion of Culture by Human Brilliance, Not by Slow Gradualism

Unexplained Sudden Transition from Hunter-gathering to Farming

The appearance of human intellect with conscious logical thinking, along with the ability to communicate, enables man to build culture. This is a major phenomenon that evolutionists have not been able to explain. Diamond speculated that cultures began to develop when the first man began to speak. Tattersall believed that early Neandertal went though some crisis that produced this advance. Others think it might have been produced by some climate crisis. Other ideas abound, but none with any evidence of how speech began.

A major question confronting Diamond was, "Why did even people in the Fertile Crescent wait until 8500 BC, instead of becoming food producers already around 18500 or 28500 BC?"[15] Again he said, "Hence, we must ask: what were the factors that tipped the competitive advantage away from the former (hunter-gathering) and toward the latter (farming)?"[16] Not only did food domestication *appear suddenly*, but it was accomplished rapidly. Diamond said,

> In the Fertile Crescent the transition from hunter-gathering to food production took place relatively fast: as late as 9000 B.C. people still had no crops and domestic animals and were entirely dependent on wild foods, but by 6000 B. C. some societies were almost completely dependent on crops and domestic animals.[17]

---

[14] Diamond, *Guns and Steel*... 87. He has diagramed it in the chapter on Farmer Power in a chapter entitled "Factors Underlying the Broadest Pattern of History."

[15] Ibid., 104.

[16] Ibid., 109.

[17] Ibid., 142.

Recent studies have shown that farming began even earlier than Diamond said. Natalie D. Munro of the University of Connecticut, Storrs, CT, claims that the earliest farmers in the Middle East were from the Natufian culture, which lasted from 12800 to10200. They posit that the transition from hunting to farming took place about 11000, either because of climate or political changes. Prehistoric Mexican culture shows it was earlier, even in America. However, the Harvard University archeologist Ofer Bar-Yosep has said, "The idea that foragers made a seamless, ***gradual transition to farming is unrealistic and has no sound evidence*** to support it."[18] (Emphasis added) Sue Colledge of the University College, London, studied 166 sites in the Middle East and Europe. She found that farming was quickly adopted in Europe about 10,000 years ago. Referring to these studies, Bruce Bowers says, "Over the next 3,000 years, local variations on this basic crop repertoire appeared in central Turkey and then in Cyprus, Crete and Greece."[19] R. Cedric Leonard also spoke clearly,

> The Natufian culture was considerably advanced for a Mesolithic culture, and most anthropologists believe it was among them that the domestication of animals and agriculture began – although I have shown elsewhere that both were practiced earlier by the Cro-Magnon people.[20]

Since Diamond wrote, new findings by Ken–ichi Tanno of the Research for Humanity and Nature in Kyoto, Japan have pushed the domestication of wheat back to 10,200 years ago. George Wilcox of the National Center for Scientific Research in Berrias, France, found hundreds of samples from ancient villages in southeastern Turkey and northern Syria (Eurasia) that furnished evidence from early dates through the centuries to 6,500 years in numerous villages.[21] Wheat domestication and farming appeared suddenly, which does not agree with evolutionary theory. According to over a thousand samples reported by Linda Perry of the Smithsonian Institution Archaeological Laboratory in Suitland, Maryland, maize, and probably potatoes were grown in the South American Andes in Waynuna, Peru, 4,000 years ago.[22] So, where is gradualism found?

Community Living, Engineering, and Building Was Sudden and Early

The deterministic view presented man as slowly developing over millions of years, from a primitive savage to a skilled worker. However, the discovery of a massive early Neolithic community at Goekli Tepe in southeastern Turkey raises doubts about this view. Goekli Tepe reveals a well-developed culture, based around ritual and monumental sculptures. Ian Hodder, professor of culture and social anthropology at Stanford, thinks this finding *contradicts the traditional view of man slowly developing from hunting, farming* and then into cultural

[18] B. Bower, "Cultivating Revolutions; early farmers may have sown social upheavals from the Middle East to Europe," *Science News*, February 5, 2005, Vol. 167, 88.

[19] Ibid., pg. 89.

[20] R. Cedric Leonard, "An Atlanian Outpost" And anomolias cro Magnon Site in Palestine," Wikipedia: the Free Encyclopedia.

[21] B. Bowers, "Early farmers took time to tame wheat," *Science News*, April 15, 2006, vol 169, 237; cf. *Science*, March 30, 2006.

[22] B. Bower, Ancient Andean Maize Makers, *Science News*, March 4, 2006, Vol. 169, 132; cf. *Nature,* March 2, 2006.

communities. According to evolutionary views, the developments uncovered at Goekli Tepe represent much later developments.[23]

Findings all over the world contradict the idea that engineering skills developed gradually. Diggings at Hebrew University, near Jerusalem, found modest stone houses from over 5,000 years ago.[24] Decorative pottery has been unearthed at Hassuna, an early site of Nineveh, across the Tigris from Mosul, and in Samara, north of Baghdad. Archaeological digs provide evidence of copper usage as early as 4,500 BC. The early Semitic and Chaldean culture contained many amazing buildings, irrigation and plumbing. The construction of the pyramids, found in various locations around the world, reveals advanced knowledge of astronomy. Early Babylonians were apparently skilled in mathematics. A tablet from Susa indicates 1 40 should be interpreted as 1 + 40/60 or 1.666—a calculation for the area of a pentagon. Many think they understood the golden ratio of pi.[25]

Recent discoveries by Rita Wright, New York University, Louis Flam, New York City University, and Gregory Posschl, University of Pennsylvania, reveal that the Indus civilization was as large and complex as the early Babylonian and Egyptian civilizations, and existed at about the same time. Located between Pakistan and India, this civilization extended from the Arabia Sea to the Himalayas. With at least three major cities, Harappa, Mohenjo Daro, and Mehrgarh, and many smaller towns, well-laid streets of burned brick, copper tools, standardized measurements, water tanks, irrigation and sewage, it was a structured society. What was previously thought to be a Buddhist stupa is now recognized as an "original structure ... a series of platforms, perhaps similar to the Ur ziggurat in Mesopotamia."[26]

Confirming early knowledge of astronomy, observatories have been found throughout Europe. One of the oldest, dating back to at least 7000 BC was found near Goseck, Germany. Known as the German Stonehenge, it is considered the *oldest representation of the cosmos yet found.* This observatory appears to have been *linked with agriculture* and religious ceremonies. It has solstice gate circles; an angle on a bronze disk was unearthed on a hilltop 25 kilometers away. More than two hundred similar circles have been found across Europe.[27]

Over a period of thirty years Stefan Kropelin and Rudolph Kuper of University of Cologne, Germany, explored over one hundred sites in Africa. They found the Sahara to be lavish with lakes, supporting a culture at least 10,000 to 5,500 years old. Kropelin says, "For five millennia humans thrived in the Sahara, fishing, herding cattle, and making pottery and art."[28] These were settled, industrious people, and not hunter-gatherers. Due to significant climate change, these people are thought to have later migrated and contributed to primitive Egypt, which had been quite cold and had warmed. We are just now realizing how brilliant the Egyptians were. Discoveries at deir Tasa, opposite Abutig in Middle Egypt, in the Fayun, reveal knowledge of copper, slate palettes, fine pottery, eye-paint, etc. In the same latitude Amaratian villages were uncovered and

---

[23] Ibid. "The Most Significant Developments in Archaeology."

[24] "Highlights," Science 85, November, 13.

[25] Michael Schneider, *A Beginners Guide to Constructing the Universe*, Helene Hedians discussion in The Fibonacci Quarterly, in late 2005 issue of *Proceedings of the National Academy of Sciences,* Mario Livio, *The Golden Ratio,* 44-47. These sources should be read for more explanation and documentation.

[26] Andrew Lawler, "Boring No More, a Trade-Savvy Indus Emerges." *Science*, 6 June, 2008, Vol. 320, 1280, cf. 1276-1285.

[27] Madhusree Mukerjee, "Circles for Space," *Scientific American*, December 2003, 32-34.

[28] Alex Stone, "Swimming in the Sahara," *Discover*, October 2006, 20.

vessels resembling animals which they created were found. These findings are dated 5000 BC. The Egyptian pyramids, obelisks, and temples are recognized as amazing feats of planning and construction. With their perfectly cut stones, precisely placed and raised to great heights. These too, reflect the mathematical genius of this early civilization. Some believe these people also had knowledge of pi.[29]

In addition to the discovery of irrigation techniques in Mexico, William Altman and Joannie M Schrof reported other evidence of advanced cultures in South America.

> A recent series of stunning archaeological finds in South America has revealed that the Incas were merely the final act in an Andean civilization that was far older and far more sophisticated than ever imagined. New observations have turned up huge stone pyramids and other monuments that date back nearly 5,000 years to about the time when the great pyramids were being constructed in Egypt.[30]

Archaeologist Tom D. Dillehay and his team from Vanderbilt University, in Nashville, Tennessee, found four irrigation canals in the Zana Valley in the Andean foothills of Peru. Located at five sites, beside a river bed near the Pacific Coast, at elevations of 6,000 to 9,000 feet, the majority are from 4,500 to 5,000 years old. Irrigation sites at three villages were inhabited between 6,500 and 4,700 years ago. They were filled with silt and buried deep under sediments; some were lined with pebbles, others with burnt clay, etc. These show major cultural advance arising independently in the ancient world.[31] Such irrigation ditches were not for just drinking water and washing, but must also have been used for farming and livestock. Men have had high-tech, even their version of computers, in early civilizations.[32]

The neat stages of evolutionary development presented by Jared Diamond and Steve Olson may have seemed right earlier. However, the data that has emerged shows man's brilliance was evident already in the early days of the world's great civilizations.

Human Industry and Art Reevaluated as Early and Not Evolutionary

Though it has long been assumed that the Chinese first made silk, around 2400 BC, recent discoveries point to Eurasia as the origin of food and textile development. Irene Good at Harvard University's Peabody Museum says, "Now it looks like some of the early silk industry outside China was earlier than thought and more widespread. Textiles, including silks, found in cool, well-sealed tombs indicate that weaving techniques, using dyes and pigments to color them *began early in the Eurasian area*."[33] (Emphasis added) Based on Sumerian texts, it appears that beer brewing began in Eurasia. Early evidence of alcohol production is even found as far as Peru in South

---

[29] Livio *The Golden Ratio*, 47- 61.

[30] William Altman and Joannie M Schrof, "Lost Empires of the Americas," *U. S. News & World Report*, April 2, 1990, 40.

[31] B Bower, "Gone with the Flow: Ancient Andes canal irrigated farmland," *Science News*, Nov. 12, 2005.

[32] Cf. Stephen Ornes, "The First Computer," *Discover*, October 2006, 17, Charles Hapgood, *Maps of the Ancient Sea Kings: Evidence of Advanced Civilization in the Ice Age*, (Philadelphia: Clinton Books, 1966), and Donald E. Chittick, *The Puzzle of Ancient Man: Advanced Technology in Past Civilization*, (NewBerg, Oregon: Creation Compass, 1998).

[33] Diana Parsell, "Remnants of the Past: High-tech Analysis of Ancient Textiles Yield Clues to Cultures," *Science News*, December 11, 2004, Vol. 166, 376-377.

America.[34] Studies in Egypt show white and other refined wines were produced there very early. Seedless, domesticated grapes were found at Jericho 10,000 years ago. A glass manufacturing center, believed to be in operation around the time of the exodus, has been found in Egypt.[35]

In central Turkey where the old civilization known as Catalhöyük has been referred to, there was a very large Stone Age town having art, farming and cooking. They did trading and produced tools in a way that revealed apparent equality between men and women. Begun over 9,000 years ago, it is still being explored.[36] Ian Hodder had referred to the fact that Goekli Tepe is extremely advanced and contradictory to evolution.[37]

Art spread all over the world with the migration. But the earliest art is found in European and Asian caves. The Chauvet cave in southeast France contains paintings and exhibits that have been dated 30,000 years ago. The art is exquisite, apparently the work of people who had artistic talent and who may have been trained. Ivory figurines found at Hohle Fels cave in the Swabian Mountains of Ulm, Germany, appear to be produced by what researchers called, "astonishing, precocious artists." Contemporary with the Chauvet cave art, these works have been dated 32-34,000 years ago.[38] Over 1,000 paintings by the Tassili people have been found in the previously lush Sahara Desert, showing mysterious figures and exquisite drawings of people and domestic animals. These have been dated at least 9,000 years ago.[39] Art is found on rocks in the eastern desert in Egypt, dated 6,150 years ago,[40] but also as far as the Guasalel cave in southwest Borneo, dated 9,900 years ago. All of this is contrary to the idea of the gradual development of art. Jean Clottes, former general inspector of archeology and scientific advisor for prehistoric rock art at the French Ministry of Culture, reflecting on the discovery in the Chauvet cave and elsewhere, has commented how *the evidence contradicts the slow evolutionary development of art.* She said, "The age-old paradigm of art having crude beginnings and evolving in Europe to more and more sophisticated form is now a thing of the past. ***Art did not evolve in an ascending line***."[41] (Emphasis added)

The cave art that is readily found across modern Europe, Asia, and North America depicts adolescent pornography and manly prowess, typical of the Paleolithic times. Experts, who have re-evaluated this art, dated 50,000 to 10,000 years ago, have determined that it is ***not based on primitive shaman worship, as evolutionary views of religion claim.*** Rather the subjects who are 4 to 1 male are adolescents are shown hunting big animals (mammoth, elk, bison, horses copulating, bellowing, biting- fighting, etc), and sex (pictures of voluptuous women), and human figures. [42]

An advanced culture has also been discovered on a plateau at about 7,000 ft in the Grand Canyon. Identified as the Anastacy people, they are considered to be 10,000 years old, similar in

---

[34] Carrie Lock, "Original Microbrews," *Science News*, October 2, 2004, Vol. 166, 216-217.

[35]Science, "Excavations unearth secrets of Egyptian glassmaking," *Orlando Sentinel*, June 12, 2005 A24.

[36] Ian Hodder, "Women and Men at Catalhöyük," *Scientific American*, January 2004, 77-83.

[37] Ian Hodder, Cf. "Archaeology: Great scientist discuss the breakthroughs…," 76.

[38] Thomas H. Maugh II. "Ivory Figurines Reveal Prehistoric Artistry," *Orlando Sentinel*, December 21, 2003, A-24.

[39] Robert Kunzig, "Memories of a Lush Sahara," *U.S. News & World Report*, October 13, 2003, 52-53.

[40] Toby Wilkinson, *Genesis of The Pharaohs: Dramatic New Discoveries Rewrite the Origin of Egypt,* (London: Thames & Hudson LTD. 2003), 54 ff.

[41] "Archaeology: Great scientist discuss the breakthroughs….," *Discover*, March 2005, 73.

[42] J. R. Minkel, Paleolithic Juvenilia: "were cave artists sex, and hunting obsessed teenage boys*?" Scientific American,* August 2006, 27; John Broadman, *The World of Ancient Art,* (Thames and Hudson, 2006), 404 pp. For other considerations of ancient art see these sources.

age and culture to some of the oldest people in Mexico. They had animals, such as goats, cats, and mammoth, as evidenced from the dung among the grasses and art work.

Conclusions: Culture Was Early, Sudden and Brilliant from Eurasia

These findings reveal advanced cultures, which developed over several centuries, a relatively short period of time compared to Darwin's evolutionary theory, which insists on millions of year of gradualism. Most of these facts were unknown, ignored, or misinterpreted when Darwinian evolutionary gradualism was proposed as a theory. Recent discoveries and scientific data contradict Darwin's theories.

Linguistic Evidence: Gradual Change from Africa or Eurasia Man

Challenging the Tradition of Origin from Eurasia

In presenting man as evolving from primitive apelike intermediaries from Africa, Diamond and Olson need evidence of a trail of acquired intelligence and language. The evidence that civilization began in Eurasia is already overwhelming. It was imperative, then, to find linguistic data to support their theory that human culture evolved from Africa. Olson makes an effort to do just that.

Aware of the strong evidence for linguistic development from Eurasia, Olson begins by presenting the view of Eurasian origins, and then claims that there was language from Africa even before this. Recognizing that genetic ancestry and linguistic origins and routes must be associated, he argues for genetic linguistic emergence from Africa. He begins by admitting the data for *one originating language*, but his evaluations and conclusions are based on those of linguists, who argue that language originated in Eurasia.

> The classic example of genetic correspondences comes from western Asia and Europe. For more than two centuries, scholars have known that most of the languages spoken in a broad swath from Britain to India are derived from a single original language. The languages belong to what is called the Indo-European family, and they are spoken by more people than are the languages of any other family in the world.... Today many of these languages seem to bear only the slightest resemblance to each other. Yet most linguists believe they all are descended ***from a single ancestor language spoken by a small group*** of people – probably a few thousand – living in a relatively small area, perhaps just a few thousand square miles. (Emphasis added)
>
> A remarkable amount is known about the original speakers of what is called Proto-Indo-European. They raised cattle, kept dogs as pets, used bows and arrows in battle, and worshiped a male god associated with the sky. All this information comes from linguistic reconstruction of the language. If a number of Indo-European languages have words that derive from the same root word, linguists can conclude that Proto-Indo European contained that word. What's more, by studying how individual words have changed in different Indo-European languages, linguists can work out an earlier form from which all the derived forms descend. For example, the word "father" in English, "padre" in Spanish, "pater" in Latin, and "pita" in Sanskrit all come from the Proto-Indo-European for *pater (the asterisk means that the word is a reconstruction).

Just prior to claims for Eurasian origins Olson argued there is a relationship for English and Australian language.

> The word for "heaven" in eleventh century English, for example was (in Australian) "heofonum," different but not unrelated. Similarly, grammatical aspects of the two languages – for example, whether the object of a sentence follows or precedes the verb – would reveal commonalities[43].

In these statements Olson admits there was a common ancestry from a small group in Eurasia that spread from Britain to India, and even to Australia. He had previously said, "If languages diversify the way groups do, then the worldwide patterns of language should match the worldwide patterns of human DNA."[44]

Diamond links the expansion of language in the Old World to the expansion of food production and culture. According to him, language families can be traced back to about 6000 BC. In the chapter, "Hemispheres Colliding," Diamond devises a table of inferred dates including the language, where it expanded to, and the driving force of food production. He notes the expansion of the *Indo-European languages* began in the Ukraine or Anatolia, and extended into Europe, central Asia, and India.[45] He also agrees that almost half of the population of the world speaks some form of Indo-European language as a first language.[46] However, what is so significant is that the earliest and most far-spreading languages also began in the *same area, between the Black and Caspian Sea in the Fertile Crescent.*

The Bible presents all men as coming from Noah and his three sons who settled in different areas. It is believed that there was one language that diversified among the three descending groups. Olson's linguistic evidence for one original language would fit this account. In this book, *True Enlightenment* in the chapter on genetics is a map of genetic ancestry, showing three major groups.[47]

Claim a Proto-Afro-Asiatic Source Preceded the Prot-Indo-European

Olson argues that there was a Proto-Afro-Asiatic language that exited Africa, which precedes Proto-Indo European languages. To give some credibility to this theory, he links the claim to the small group Diamond referred to as an example for the early transition from hunter-gatherers. These people, found at Wadi Al Natuf, a spring north of Jerusalem, probably numbered a few thousand. There is some indication they camped briefly at Jericho, which was thought to be a transit point to or from the Fertile Crescent. Many scholars have studied the Natufian culture; none have found evidence they brought a Proto-Afro-Semitic language into the country from Africa. Olson claims this isolated group is the basis of the first, and all language in Eurasia. His whole theory is identified with *an unknown cause* for migration, and the transition from hunter-gatherers to farmers out of Africa. But Olson's arguments are based on suppositions, or speculations, such as "climate change, population fluctuation, or even the appearance of a solitary genius." His lack of data is reflected in phrases such as: "*must have been* subject to a climate whiplash," "They *undoubtedly* began," ... "*might be* called," ..."*appear to have been.*" In addition, he speculates about the cause of migration from Africa, "Among the Natufians, for instance, a sudden drying of the climate about 11,000 years ago *must have* caused food supplies to dwindle."[48] He then *admits*,

---

[43] Olson, *Mapping Human History*, 141.
[44] Ibid., 139, Cf. Chapter 15 on genetic ancestry.
[45] Diamond, *Guns, Germs and Steel*, 369.
[46] Thomas V. Gamkrelidze and V. V. Ivaov, "The Early History of Indo-European Languages," *Scientific American*, March 1990, 110.
[47] Genetic map of 3 groups, p 232.
[48] Olson, *Mapping Human History,* 98-100, Olson for some reason links this to Kathleen Kenyon's

"Proto-Afro-Asiatic *appears to* have predated the invention of agriculture. *No one knows for sure*, but *perhaps* it was spoken by the first people to establish permanent settlements, the Natufians, and their way of life spread, *so did their language*."[49] (Emphasis added) Again, he admits, "No one knows for sure." This is a slim theory, with no objective data on which to hang such a complete radical worldwide cultural change and to claim that all language originated in Africa. Was the culture and language of the entire world changed by this small group of people, without any known evidence?

Reference to Joseph Greenberg's African influence on Language

One aspect of language presented by Diamond is important for future information which is a complete mysterious phenomenon. A previous chapter on biblical history furnished logical insights for this and with these, further evidence of Indo-European as the source for all languages. Diamond spends much time examining why Africa has remnants of Austronesian language, musical influence and foods which he considers one of the great mysteries of the world. After tracing language influence, He said,

> Other traces of Austronesians are very thin on the ground in East Africa: mainly just Africa's possible legacy of Indonesians musical instruments (xylophones and zithers) and, of course the Austronesian crops that became so important in African agriculture. Hence, ***one wonders*** whether Austronesians, instead of taking the easier route to Madagascar via Indian and East Africa, somehow (incredibly) sailed straight across the Indian Ocean, discovered Madagascar, and only later got plugged into East African trade routes. Thus, some ***mystery remains*** about Africa's most surprising fact of human geography. [50] (Emphasis added)

The mystery of Austronesian influence in E. Africa and Madagascar will be discussed in the chapter on the breakup of the continents which gives a more logical explanation for the diversification of language and other elements.

Eurasia is the More Certain Source of Language and People Migration

The linguistic data seems to run contrary to the out of Africa theory, and toward the idea of mankind moving *from the Middle East* and the Levant into other parts of the world *and into Africa*. At least two-fifths of North Africa has had nineteen Semitic languages, with twelve still living and spoken in Ethiopia, which is located across the Red Sea from Arabia. According to the Cambridge Encyclopedia of Language,

> The Semitic languages have the longest history and the largest number of speakers. They are found throughout southwest Asia, including the whole of the Saudi Arabian peninsula, and across the whole of the North Africa, the Atlantic to the Red Sea. The oldest languages of the group, now extinct, date from the 3rd millennium BC; they include Akkadian, Amoritic, Moabite, and Phoenician, all once spoken in and around the Middle East. There was a vast literature in Akkadian, written in cuneiform script (p. 200). Hebrew dates from the 2nd millennium BC.[51]

---

explorations of Jericho that were later discredited.

[49] Ibid., 146.

[50] Diamond, *Guns, Germs and Steel,* 393, His source of discussions on this are from Joseph H. Greenberg, Stanford linguist who had unorthodox presuppositions rejected by many but with these insights about Austronesians in Africa were accepted by others, see *The Languages of Africa,* (Bloomington: Indiana University Press, 1966).

[51] *Cambridge Encyclopedia of Language,* second edition (New York: Cambridge University Press, 1997),

These particular languages are now identified as Afro-Asiatic, with five divisions, "which are thought to have ***derived from a parent language*** that existed around the 7th millennium BC."[52] (Emphasis added) All of the oldest Semitic languages originated near the Fertile Crescent, or Eurasia, and spread southward. In arguing for language origination in Africa, Olson chose the idea of the Natufians and Semitic language in Palestine because there is no evidence any languages left Africa anywhere near the origin of Proto-Indo-European language. The Niger-Congo languages, with the exception of Bantu, go across most of the western center of Africa, except for Nilo-Saharan. The southern third of Africa is occupied by the Bantu, and the southwestern by the Khoisans. Hence, instead of the traditional African languages being in the north, where they would go out with a migration, they are in the south. Olson and other Darwinists are suggesting that Semitic languages began in Africa and migrated to the Middle East. This is contrary to the out of Africa view and to the generally accepted understanding that the Semites were originally in the Mesopotamian area. The Encyclopedia Britannica, in a lengthy article on Semitic languages, traces and *graphs the spread* of all the Semitic languages, noting that, "The 'Semitic' or 'Shemitic' languages ...were spoken in Arabia, Mesopotamia, Syria and Palestine, ***whence they spread into*** Abyssinia, Egypt, ***northern Africa*** and elsewhere."[53] (Emphasis added)

One Group with One God and Not Evolving Hunter-Gathers

According to Olson, linguists found the original Proto-Indo-European group worshiped one supreme sky God (as J. Schmidt, the great comparative religionist showed). They had invented bows and arrows, domesticated animals, and raised cattle. Therefore, they were not hunter-gatherers. Linguistic evidence, therefore, shows these are not primitive people who rose from apes. Rather, the evidence supports those who assert that the Indo-European languages, of which Semitic languages were one main group, began in the Euro-Asia region, and the speakers of these languages *were capable of culture from the earliest time.*

The theory of evolution was used to contest the idea of languages spreading from the Eurasian area. Russell Gray and Quentin Atkinson, from the University of Auckland, New Zealand, borrowing some of the tools that evolutionary biologists use for data crunching, did an extensive analysis between basic words and structures in language. The results concurred with the traditional idea of languages spreading from the Eurasian area, as reported and reviewed by Andrew Curry in *Nature* magazine (December, 2003). Curry says, "They concluded that the language family began branching out nearly 10,000 years ago, supporting the idea that ***it was farming*** and not fighting that spread indo-European languages far and wide."[54] (Emphasis added)

Writing Also began and spread from Eurasia

Writing began in the Fertile Crescent before 3000 BC. Hieroglyphic writing appeared first in Sumer, then in Egypt, Iran, Crete and Turkey. Diamond commented, "It would be a remarkable coincidence if, after millions of years of human existence without writing, all those Mediterranean and Near Eastern societies had *just happened* to hit independently on the idea of writing within a few centuries of each other."[55] They seem to have learned from each other and put together their

---

318.

[52] Ibid.

[53] *Encyclopedia Britannica*, 1953, vol. 20, pg 314.

[54] Andrew Curry, "Language: Before Babel," *U. S. News & World Report*, December 8, 2003, 78.

[55] Diamond, *Guns, Germs and Steel*, 232.

own writing from what was first started at Sumer. The early writing consisted of no phonetic logograms (sounds). Diamond said, "Perhaps the most important single step in the whole history of writing was the Sumerian's introduction of phonetic representation, initially by writing an abstract noun (which could not be readily drawn as a picture) by means of the sign for a depictable noun that had the same phonetic pronunciation."[56] They later developed the phonetic alphabet. With some exception, he said, "All other writing systems devised *anywhere in the world*, at any time, appear to have been descendants of systems modified from or at least inspired by Sumerian or early Mesoamerican writing."[57] (Emphasis added) Sumer is, of course, *in the heart of the Fertile Crescent or Eurasia* The Phoenicians, who are related to the Amoritic people in Ugarit, developed these early scripts into the phonetic alphabet, which was used by the Greeks and others, and later came down to the Western nations.

Carmen Rodriguez Martinez of the National Institute of Anthropology and History in Vera Cruz reports finding hieroglyphic writing on a stone in a quarry near Vera Cruz in Southern Mexico. (The finding is confirmed by Stephen D. Houston of Brown University of Providence, Rhode Island.) The writing is over 3,000 years old, the oldest in the Americas. One side of this Cascajal block has 62 signs of Olmec writing running horizontally. Earlier, samples of Olmec and Maya writing dated nearly 2,600 years later, had been found.[58]

Age of Humans and Culture

Diamond estimated the beginning of human domestication of crops and animals to be about 8,600 years ago. But, others had-discovered similar advances among the Cro-magnum men earlier. Olson's extensive research, using genetic and archaeological data resulted in the search for *a scientifically accurate date for the beginning of human life*. Steve Olson, working with a statistician and a scientist, using a supercomputer and the data available to him, *calculated actuarially how long* it had taken to reach current population levels. Assuming that everybody living today began with the same set of ancestors, they determined human history to span *about 5,000-7,000 years*. These facts are based on exponential growth. John Hein of England's Oxford University published a commentary on the results in the journal, *Nature*, saying that this is *"especially startling."* (Emphasis added) Hein marvels at the finding that if you entered any village on earth *in 3000 BC, the first person you would have met would probably be your ancestor.* Hein speculated that the origin of the primary ancestor was likely East Asia, Taiwan, Malaysia, or Siberia. B. [59] The age of humans, as proposed by paleontologists (three to five million years), and geneticists (200 to 150,000 years) is actuarially impossible.

Moreover, the extended family would still have been intact until 3000, as Hein mentions. Evidence is that the ancient peoples had incredible memories and relied on oral transmission. By about 3000 BC oral transmission would begin to be inadequate and writing would become a human necessity. All of the *early written accounts of ancient people record God's creation of man and of a worldwide flood* as seen in the next chapter.

---

[56] Ibid., 220.

[57] Ibid., 224.

[58] B. Bower, "Scripted Stone: Ancient block may bear Americas' oldest writing," *Science News*, September 16, 2006, Vol. 170, 170.

[59] Matt Crenson, (AP) "Human's common link is found not so far back," *Orlando Sentinel*, Sunday July 2, 2006, A4.

Conclusions: Cultural Beginnings, Instant- not Gradual, in Eurasia

Only humans are uniquely designed—physically, mentally, emotionally, and spiritually—to develop culture. Humans everywhere manifest intelligence for art, engineering, and production and not a long gradual development. They show ability to communicate and interact in culture. Nearly all the scientific evidence currently available points to Eurasia as the geographic area where civilization originated, flourished, and spread to other parts of the world. The same evidence makes it clear that well-advanced civilizations appeared suddenly, not after years of evolution, as Darwin and others asserted.

Graham Hancock makes the following observation, "What is remarkable is that there are *no traces of evolution from simple to sophisticated*, and the same is true of mathematics, medicine, astronomy and architecture and of Egypt's amazingly rich and convoluted religion-mythological system…."[60] Kurt Mendelssohn has commented that these "emerged ***all at once and fully formed***. Indeed, the period of transition from primitive to advanced society appears to have been so short that it makes no kind of historical sense."[61] (Emphasis added)

Evidence of Man's Spread from Eurasia Agrees with Biblical Record

Such admission by secular intellectuals about culture and civilization uphold the biblical theory of history and not the evolutionary theory. The area from which all this spreading occurred is precisely below or near the area where biblical/historical scholars believe the ark of Noah is landed after the world wide flood. *Therefore, the preponderance of evidence is that human origins, development and civilization occurred suddenly in Eurasia* and not in Africa.

All the facts point to the work of God and biblical truth. After the worldwide flood, the Bible and tradition says the Ark of Noah landed on Mount Ararat, which has long been thought to be in the Caucuses, or in the area northeast of Tehran, and near Turkey. The family of Noah would have had to descend and begin to repopulate the earth in the area of the Fertile Crescent. According to Genesis 9:20, this is the area where Noah began farming. Some 300 years after the flood, the people were disbursed by languages into various geographical locations, according to God's plan.

This is much more reasonable and scientific than some mythical ape-man from some unknown place in Africa suddenly getting mysteriously zapped so his genes jump him up to all the higher abilities of a human—thinking, speaking, writing, designing, farming, and building, and then creating a brilliant culture. The idea of man being created in the image of God, so he can logically interpret the world, relate to, and control it is "true enlightenment" to the facts of science.

---

[60] Graham Hancock, *Fingerprints of the Gods*, (New York: Crown Trade Paperbacks, 1995), 135, 136.

[61] Kurt Mendelssohn, "A Scientist Looks at the Pyramids," *American Scientist*, 1971, March-April, 210.

PART VI

ENLIGHTENMENT ABOUT WORLD GEOLOGY AND CLIMATE

…knowing this first of all, that scoffers will come in the last days with scoffing, following their own sinful desires. They will say, “Where is the promise of his coming? For ever since the fathers fell asleep, all things are continuing as they were from the beginning of creation.” For they deliberately overlook this fact, that the heavens existed long ago, and the earth was formed out of water and through water by the word of God, and that by means of these the world that then existed was deluged with water and perished. But by the same word the heavens and earth that now exist are stored up for fire, being kept until the day of judgment and destruction of the ungodly (2 Peter 3:3-7).

Chapter 19
Enlightenment about the Worldwide Judgment by Flood

Introduction: Modern Data Shows Rejection of Noah's Flood a Mistake

In chapter five we noted that the intellectual elite began to accept the deterministic worldview, based on Kant's claim of the mechanical working of the universe and solar system as demonstrated by Newton. While Herschel first proposed the uniform working of the heavens, with the nebular hypothesis, it wasn't until Hutton and Lyell transferred uniform gradualism to their geological findings, that the public was persuaded of its validity. The *presumption* for a mechanical uniformitarian view required a selective interpretation of the facts, which led to the claim that there was *no great worldwide flood* that influenced geological history. It is this rejection of catastrophic events that marked a major turning point in accepting the lie that there are no catastrophes or unexpected events and so, no special acts of grace or judgment by God.

Recall that Abraham Gottlob Werner, the leader of the Neptune geology, first emphasized geological stratification and held to a worldwide flood. The man who more than any other developed the idea of using shell and other fossils to distinguish geological layers, and who was honored by the Royal Society for discovering this as the key to modern geological stratification, was William (Strata) Smith. Smith continued to advocate the reality of the worldwide flood of Noah's time. Moreover, William Buckland, under whom Lyell learned geology, was the leading advocate of the worldwide flood at the time.

Lyell borrowed the idea of long ages of time from Herschel's work in astronomy, and from Hutton in geology who was influenced by Herschel. Hutton needed long ages for his theory for formation of layers in the oceans to support his volcanism. But, Lyell had no scientific data of long ages when he constructed his geological principles. Gillespie presented evidence that Lyell's principle objective in his geology was to remove the evidence of catastrophe in geology by rejecting the worldwide flood. It is, therefore, important to devote space to presenting the data scientists have found since that very significant shift to accepting uniform gradualistic geology.

Emergence of Facts to Support the World Wide Flood

About the same time that unbelieving uniformitarian scientists attacked and abandoned the biblical report of a worldwide flood, events of divine providence came to light, providing new evidence that a worldwide flood reshaped the world. Note two aspects of this data. (1) Stories about the flood are found in the oral and ancient written histories *from all over the world.* (2) We have striking new geological evidence and understanding about the power of water. In recent years, professional geologists, using modern methods and equipment, have produced a more accurate history of the Earth. Their evidence reveals unusually large bodies of floodwater in locations, and of such proportions, that they can only be accounted for by a worldwide flood. Geologists have also discovered evidence, not understood at the time that great bodies of water, moving at great speeds, make unexpected changes in nature very quickly. These findings reveal that rejecting catastrophes, such as the flood, which were acts of God, were distortions of major proportions.

I. Discovery of Stories of Great Worldwide Flood Everywhere in the World

Early Discoveries of Extra Biblical Stories at a Providential Timing

At almost the same time (ca. 1840) that William Buckland of the British Royal Society was influenced by Agassiz to abandon the biblical flood as a chief cause of geologic change (cf. page 42), written and oral records of such a flood began to be found in very ancient sources. *Historical documents* about the worldwide flood were providentially discovered. Henry Cheswick Rawlinson had been stationed in Bombay, where he heard a High Court judge, Sir William Jones, recount to the Asiatic Society of Bengal that a mother language linked the many languages dispersed into Europe, Asia, and India. Jones told of an epic legend of a great flood in the Rig Veda in India, similar to the Noahic flood. Rawlinson was motivated to study many of these languages, hoping to follow the language of Noah's son, Japheth, and his descendants. Because he was fluent in the Persian language, Rawlinson was sent as a military advisor to the Shah of Persia. Here, while exploring ancient ruins, he found the Behistun Stone with writing in several languages, some of which were familiar to him. In a major historical linguistic breakthrough, he was able to decipher the Akkadian language.

But it wasn't until 1872 that Rawlinson's brilliant student, George Smith, studying under him in the British Museum in London, pieced together fragments of a shattered clay document he had discovered. When translated, it was discovered to be the ancient Akkadian flood story. Rawlinson's flood story is drawn from documents by Berossos, a Babylonian priest of about the third century BC. These documents were ***said to be transmitted from the first Mesopotamian kingdom about 5150 BC.***[1] These were perhaps remnants of the museum of King Nabonidus and his daughter who by archeological interests had preserved ancient documents in sixth century BC.[2]

In a few weeks, three separate and slightly different versions of the deluge story coalesced from this ancient literary collection and were translated. In response, the museum reopened the excavations, sending Smith to Mesopotamia to the tel of Kuyunijk, which contained the royal library of King Assurbanipal. At this time, he discovered the long Gilgamesh Epic which contained the most famous of the flood stories, similar to that of Noah's, along with a creation story similar to that found in Genesis.[3] Findings of Mesopotamian people go back to c5000 B. C.[4]

When these significant flood documents were found, Lyell's uniformitarian view of geology was already widely accepted in academic circles. These revelations made news around the world, but did not change the trend of rejecting a universal, cataclysmic judgment by a worldwide flood. Some biblical scholars had already begun to shift the interpretation of the Noahic flood to a small local flood, for which evidence had been found at Ur, Kish, and elsewhere. This shift in interpretation was a capitulation to scientific determinism, and an attempt to make the worldwide flood an exaggeration about a major local flood. This removed the concept of God's worldwide judgment against the wickedness of men.

The Extensive Reports of a Worldwide Flood

In modern times, anthropologists have accumulated about 150 stories of a worldwide flood, from every area of the world. Arthur C. Custance says there is "a fairly complete annotated bibliography of works which deal with these traditions, along with a list of some 140 accounts (according to the tribes or nations which carry them) with fairly accessible source references."[5] Dr. Johannes Riem of Germany, in his extensive study in his book *Die Sintflut in Sage und Wissenschaft,* mapped many of these, showing stories of the flood are most common in Asia, islands south of Asia, in Europe, and on the North American continent. Though found in Africa,

---

[1] Roberts, *Cuneiform, Parallels to the Old Testament*, (New York, 1912), 114 ff., cf. George A. Barton, *Archaeology and the Bible,* (American Sunday School Union, Philadelphia, second revised edition, 1937), 331.

[2] Finegan, *Light From The Ancient Past: The Archeological Background of Judaism and Christianity,* (Princeton, NJ: Princeton University Press, 1959), 13, 227-229.

[3] Barton, *Archaeology and the Bible*, 327 ff.

[4] Finegan, *Light From The Ancient Past*, 12ff.

[5] Arthur C. Custance, *The Flood, Local or Global,* (Grand Rapids, MI: Zondervan, 1979), 68.

they are not nearly as common as on other continents.[6] It is important to note that there was virtually no knowledge of flood stories before the uniformitarianists rejected the biblical flood of Noah's time Had this much evidence surfaced earlier, it is highly unlikely that faith in the worldwide flood would have been rejected for uniformitarianism that excluded catastrophe. Many of the flood stories have been preserved only in oral traditions. Some were written down later. Liberal scholars have argued for oral tradition in many other fields, so they cannot, without being hypocritical, reject the value of these when it comes to narratives of the flood. But many stories in *very old written form* are of great interest, especially those found in the written records of the oldest cultures in the area of Eurasia, near where it is believed Noah's ark landed, and are similar.

Review of Some Earliest Written Stories

The earliest flood narratives are found in the Lower Mesopotamia area and were told by those who called themselves "the black-headed people." They were "neither Semites nor Aryans and their language in which many texts are now available, was neither Semitic nor Indo-European"[7] from the fourth millennium. Statues of these people have also been preserved.[8] These flood stories are presented in connection with two king lists which provide a chronology of antediluvian history. One was found on an ancient Sumerian cuneiform tablet near Ur (probably from early fourth century BC) and the other, mentioned previously, by a Babylonian Priest, Berossos (d. 260 BC). His map was based on earlier sources, near the same time in the fourth century. Berossos also presented a list of 14 dynasties of some of the cities after the flood.

Both Barton and Finegan go into great detail discussing whether these pre-flood lists are of the same people or whether there were two different lists of two different groups leading up to the flood.[9] Barton takes into account the various linguistic possibilities and harmonizes these pre-flood men with a list from the Hebrew Bible.[10] They resolve some of the differences in the lists. Our interest here is not to argue the discrepancies between the lists, but rather to note their relevance in substantiating the validity of the worldwide flood. In dealing with what was undoubtedly oral transmission in different communities, for hundreds of years prior to written record, differences are bound to develop. The point here is the striking agreement between the biblical story and stories from the lists.

The lists confirm the biblical narrative that mighty and powerful men lived on the earth (Genesis 6:4). They also show that for some time, men of power gained fame and were in conflict with each other, asserting their power as one dictator in one city replaced another. It is difficult to explain the length of time they ruled. The king's lists are likely exaggerated. Assuming their *year* meant a shorter unit of time, such as a moon time (e.g. a month), it would correspond to the biblical record. It is interesting that the king list after the flood shows a significant decline in longevity, similar to the biblical record.

Barton, Finegan, and others also point out that some of the names on the lists bear a close resemblance to the pre-flood biblical names. For example, in Akkadian, Amelon became amelu, "man." In Hebrew 'adham is "man." Obviously, there is a linguistic similarity between the two. But Amelon is third in the list, while Adam is first. Likewise, there are also ten pre-flood patriarchs from Adam to Noah as in Berossos' list. Almost all scholars recognize agreement

---

[6] Alfred M Rehwinkel, *The Flood in the Light of the Bible, Geology, and Archaeology*, (Saint Louis, MO: Concordia, 1951), 129.

[7] Jack Finegan, Light From the Ancient Past, 29.

[8] Ibid, figures 10, 11.

[9] Barton, Ibid., 327-336, Finegan, Light From the Ancient Past, 27-42.

[10] Barton, Ibid., 321.

among many elements in the flood stories and the events of the Genesis narrative. Many admit that this clearly attests to actual happenings.

Quotations from Written Documents Much Like Those in the Bible

In the cuneiform list, the final king is Ubar-Tutul. His son is not listed as one of the eight, but is elsewhere known as Ziusudra, who is the hero of faith (Noah) who provided salvation through the flood. Thus, he would have become the ninth king. In an early Sumerian fragment of six columns, found at Nippur, in the third column, Ziusudra received his revelation of the coming flood as follows:

> Zusudra, standing at its (the wall's) side, listened. 'Stand by the wall at my left side....By the wall I will say a word to thee, take my word, give ear to my instruction: by our [command] a flood will sweep over the cult-centers, to destroy the seed of mankind is the decision, the word of the assembly of the gods. [11]

Column five continues the narrative when the terrific deluge has begun:

> All the windstorms, exceedingly powerful, attacked as one; at the same time, the flood sweeps over the cult-centers. After, for seven days and seven nights, the flood had swept over the land, and the huge boat had been tossed about by the windstorms on the great waters.
> Utu (the sun) came forth, who sheds light on heaven and Earth.
> Ziussudra opened a window of the huge boat, the hero Utu brought his rays into the giant boat. Ziusudra, the king, prostrated himself before Utu, the king kills and ox, slaughters a sheep.[12]

The flood and storm having ended, column six tells how the gift of eternal life is given:

> Ziusudra, the king, prostrated himself before Anu (the Father in Heaven) and Enlil (His son). Anu and Enlil cherished Ziusudra. Life like that of a god they give him, breath eternal like that of a god they bring down for him. Then Ziusudra the king, the preserver of the name of vegetation and of the seed of mankind, in the land of crossing, the land of Kilmun, the place where the sun rises, they caused to dwell.[13]

In Berossos' story, the hero of the flood is the last king, who is the son of King Otiartes (who is listed ninth) and is the tenth king named Xisouthoros, who was his son. The sounds of Ziusudra are linguistically similar according to the placing of the tongue on the teeth (Xis=Zius, outhoros=udra). Such changes occur when going from one language pronunciation to another. The record names Xisouthorous as the hero of the flood. The king list states: "After the flood swept thereover, when the kingship was lowered from heaven, the kingship was in Kish."

Then follows the king list, giving fourteen dynasties of kings. Agga and his family of twenty-three kings in the first Dynasty of Kish are revealed to overlap with the second dynasty of Uruk. Gilgamish is the fifth dynasty of Uruk. Another source, a Sumerian poem, reveals that he was dissatisfied with a political decision of the upper house and appealed to the lower house of congress in Kish, and led a revolt against Agga. This gives additional validity to the story.

Jack Finegan dates the early dynastic period c.2800–c.2360 BC and the revolt in the early part of the third century. The first Dynasty of Kish, of 23 kings, is said to have lasted 24,510

---

[11] Finegan, *Light From the Ancient Past*, 31-32.
[12] Ibid.
[13] Ibid.

years; the first dynasty of Uruk, of 12 kings, lasted 2,300 years. The First Dynasty of Ur had four kings and lasted 177 years. The Dynasty of Awan had three kings and lasted 356 years. The original cuneiform texts also show that after the flood there was a significant drop in the exaggerated length of life and reigns. King Gaur at Kish reigned 12,000 years and King Khulla-Nidaba for 960 years. Successive columns show the reigns continued to decrease until times equivalent to those of more historical times.

The first dynasty could have reigned before the flood, which may explain the longevity of the First Dynasty of Kish. Interestingly, the story is told in this dynasty of one, Etana, "a shepherd, the one who *to heaven ascended.*" This is also "attested by representations of early Sumerian seals and by fragments of Babylonian and Assyrian literature." Because he received "the plant/tree of life," Etana rode high into the heavens on the back of an eagle…."[14] In early Egyptian tradition, the one spirit sky God, Horus was symbolized by the outspread wings of a hawk or eagle, supported by the invisible power of God. Spirit means wind. This unusual story of Etana is like that of Enoch in Genesis 5:21-24, "who walked with God and was not, because God took him." Historically, Enoch's translation preceded the flood about 670 years. Such a strange story in both documents is not likely to be an accident of chance, but of an actual, unusual happening.

The famous Gilgamesh Epic of Babylon bears a great deal of similarity to the Genesis account and other similarities to Berossos' account about the area of Ur and the Sumerians.[15] In it, God of wisdom of the Earth, Ea, spoke to the hero of the flood, Utnapishtim, through a wall of a hut, similar to the Nippur account, and instructed him to build a ship, giving instructions for the way to build it (lines 19-84). God, who is known as Elille, then discloses that he is going to send a great rainstorm and a flood on the Earth because of his displeasure for man's sin (36-48). The story then tells how Utnapishtim built the ship with his family and kinsmen, and taking cattle, beasts, etc., he embarked (56-88). The mighty rainstorm comes and he closes the door. The waters then covered the mountains, and continued six days and seven nights, as a mighty deluge and overpowering flood (89-135). When it subsided, he opened the window and the sunlight fell on his cheek and he wept. The ship landed on mount Nizir and held fast. On the seventh day, he sent out a dove, and it returned having no resting place. He then sent a raven, and it did not come back. He then disembarked and made a sacrifice, pouring out a libation in four directions (136-158). The story, which is close to the biblical narrative to that point, then departs from the biblical story into idolatries and to relating of how Elille was still not pleased.

Observations of Same Emphases about Many Flood Stories

Four points are consistent in nearly all the flood accounts. (1) The God [or gods] was angry with mankind for his sin. (2) There was a worldwide flood to destroy all men as punishment. (3) One man and his family found favor and survived, usually in a boat the man constructed. He then sacrificed to God and praised Him. (4) The boat almost always lands on a mountain, and a seed of mankind survives and perpetuates the race.

In many stories, animals and eight souls are saved. Numerous other details, some graphic, are the same as the biblical narrative. The slight discrepancies between the accounts do not invalidate the testimony of the witnesses. Despite the differences, the main points are corroborated, establishing credibility. How could this story be re-told and recorded all over the world if the event had not actually occurred?

Arthur Custance also points out radical differences from the biblical account—fantasy events, mythical embellishment of details, and gods other than the one true God. In many

[14] Finegan, *Light From the Ancient Past*, 37.

[15] The whole is given by Barton, Archeology of the Bible, 323-331.

accounts, there is a feeling that God is still angry, which is not the case in the Bible.[16]

II. Geological Evidence for a Worldwide Flood

When the nineteenth century geologists rejected the biblical flood, much of the world had not been extensively studied and many of the present instruments were either primitive or nonexistent. Today many new instruments are available, bringing to light startling facts and geological evidence that is unexplainable without a worldwide food. A great deal of evidence reveals exorbitant bodies of water that once existed and has since drained. Other evidence reveals the unbelievable power of water moving at great speeds, and creating a powerful force. These facts have recently been discovered with the aid of modern laboratory equipment and experiments.

Mediterranean and European Area

Sudden Flooding of Whole Mediterranean Desert into Black Sea Area

Geological evidence that refutes uniform gradualism and sustains the view of a massive sudden flood has been found in the area that Noah's ark landed in Eurasia and elsewhere. Leading educational institutions, the government of the United States, along with cooperative ventures with Bulgaria, Russia, and Turkey contributed to the discovery of new geological evidence, including a number of geophysical climate and oceanographic studies reported by William B. F. Ryan and Walter C. Pitman of the Lamont-Doherty Earth Observatory of Columbia University.

Ryan was first involved in an expedition which discovered mysterious countercurrents in the Zoospores, located between the Mediterranean and the Black Sea. He called this "the hidden river." Subsequently, he cooperated with other researchers in a succession of studies in the Mediterranean, and later in the Black Sea. Numerous studies corroborated his discoveries and understanding. While Ryan identified his findings with the story of the flood, as a deterministic scientist, he *sought to spin the data* to maintain a completely naturalistic view of the flood events. He ignored and refused to point out the very significant fact that the massive flood he reports completely refutes the uniform gradualistic view that Lyell and others adopted to undermine faith in a biblical flood by a sovereign God. That evidence follows, and involves the mass of water in the Mediterranean Sea and the Black Sea flood.

W. Pitman traced the drifting of continents after they had broken up, and showed that Africa and Europe collided from five to seven million years ago. Ryan and Pitman *assumed this happened at the time* that the Mediterranean Sea was closed off from the Atlantic Ocean. But Ryan and Pitman uncovered evidence that the Mediterranean Sea had once been a desert area. On a coring expedition on the Glomor Challenger, Ryan was surprised he *did **not** find* "sedimentary rock containing the particles of even older mountain chains," as expected. Rather he found species with dwarfs and forms that are normally found in lagoons, beaches, and coastal environments, some of which dates from radioactivity to three to five million years.[17]

Ryan and Pitman describe the Mediterranean as "an inferno of heat twenty to thirty times farther below sea level than Death Valley in California."[18] They demonstrated this by coring the bottom from Gibraltar eastward across the sea, even in the center. By coring, they found the entire Mediterranean area had been dry. According to Ryan and Pitman,

> The cores cut from beneath the abyssal plain ... were the most breath taking of all. ... Clearly

[16] Custance, *The Flood, Local or Global*, 67, 68.

[17] William Ryan and Walter Pitman, *Noah's Flood: The New Scientific Discoveries About the Event That Changed History,* (New York, Simon & Schuster, 1998), 80-81.

[18] Ibid., 84.

> visible on the fresh smooth face of rock [in one last core] was the cross-section of a desiccation crack, filled with salt crystals. Indeed, for one brief moment even the briny lake in the center of the Mediterranean had withered to an empty puddle![19]

They found anhydrite of a variety of selenite that mineralizes under temperatures exceeding 110 degrees Fahrenheit. Their findings reveal that the Mediterranean Sea was formerly a vast desert that was suddenly flooded by millions of cubic miles of water.

While Ryan implied from this information that the continental collisions caused the drying out of the Mediterranean by cutting it off from the Atlantic Ocean, he and Pitman offered no evidence this occurred over millions of years. In fact, Ryan admitted this was not the real cause.

> For example, the Mediterranean would never have dried out when its gate to the Atlantic was shut by the collision of Morocco with Spain were it not for climatic aridity. An inland sea in which evaporation did not exceed the supply of incoming water from rivers and rain would not dry up.[20]

Ryan's research reveals that streams from Europe and Egypt had been flowing in to the Mediterranean basin. When Russians contracted to build the Aswan Dam, they did coring to find foundations. Russian scientist I. S. Chumakov was in charge of drilling a series of bore holes into the Nubian bedrock from bank to bank, to locate a secure foundation for the dam. Ryan tells of the Russians' surprise discovery,

> When it came time to drill at the [Nile] River's center line, the hole had penetrated the usual twenty or thirty feet of riverbed silt, and sand but then continued another *nine hundred feet* in these sediments before the bit struck the granite substrate. The engineers had discovered an extraordinarily deep and narrow gorge belonging to an ancient hidden river. More astonishing was the recovery of deep-sea ooze in the bottom of the gorge, sandwiched between Nile mud and granite bedrock. The ooze was exactly the same age as the sediment cored by the *Glomor Challenger* [in the bottom of the Mediterranean by Ryan shortly before] right above the anhydrite.... What was perplexing was the location of the gorge more than six hundred miles inland from the present [Mediterranean] coast. However, there had been no doubt about this deep canyon connecting with the Mediterranean. In the ooze Ryan had seen not only the tiny shells and marine plankton but shark's teeth. In order to reconcile how salt water could have invaded a stream so far inland, Chumakov constructed an explanation just as outlandish as the evidence in his cores. He concluded on his own that the surface of the Mediterranean had once dropped more than five thousand feet below its present level. While the Mediterranean was drying up, the Nile was incising a deep valley to continually adjust its stream gradient to the depressed coastline.[21]

This protruding arm of the Nile into the increasingly desert Mediterranean was found to be "in fact a spaghetti-thin arm"—a gorge more than six hundred miles inland from the present coast. However, there had been no doubt about this deep canyon connected."[22]

A Mediterranean Desert Changed Suddenly into a Sea

What is even more amazing in the evidence Ryan and others found is that the Mediterranean desert was abruptly changed into a mighty sea by flooding. The team at first rejected the idea that this large area was immediately flooded. They sought to hold to a theory of gradual change over

---

[19] Ibid., 86.
[20] Ibid. 114.
[21] Ibid., 88.
[22] Ibid., 88.

many years. They continued coring, however, until the entire scientific group was persuaded. The evidence they found was too compelling, and they concluded that the Mediterranean Sea was the result of a rapid flooding. Some of their statements on this follow.

> According to Cita [a scientist of *Glomor Challenger*], who used the bottom-dwelling fauna as a measure of water depth, the site of the dried-out salt lake [of the Mediterranean area] had *suddenly* been transformed into a new marine sea thousands of feet deep and far removed from land. *No transition was observed.* The seabed creatures living directly on the anhydrite once the sea water returned were indicative of a "bathyal realm"–a term the paleoecologists generally reserved for the cold dark internal ocean three thousand or more feet below the warm sunlit sea surface. This observation supported the idea that the deserts had formed in a depression that had dried out and the deserts had drowned suddenly by flooding of the depression under thousands of feet of newly supplied seawater.[23]

The expedition performed a final drilling as they entered the Tyrrhenian Sea, about seventy miles east of Sarinia, at a water depth of ten thousand feet. Ryan said, "The cores brought on deck and then into the lab left no room for ambiguity." The typical ooze, which accumulates at about an inch in a thousand years, was razor thin, *showing a sudden "**abrupt** change*." Ryan described this core as "the most breathtaking of them all."[24] (Emphasis added)

This abrupt filling of the entire Mediterranean basin with thousands of feet of water is *assumed* to have come from water that flooded over a wall or dam at Gibraltar, but the source and explanation for this sudden appearance of a vast amount of water is unknown. The indisputable evidence shows instantaneous flooding.

Ryan and the other scientists assumed the desert conditions of the Mediterranean occurred about five million years ago, when the plates of Africa collided with Europe, leading progressively to the Mediterranean desert conditions. Then, as the glaciers melted many years ago, the ocean water level slowly increased until it poured over the Gibraltar dam. They believed this must have occurred in an earlier glacier period. This does not agree with the instant change they found from coring from the bottom of the whole Mediterranean Sea. Glacial melting would result in a slow increase in water levels, not instant flooding. The question of glacial involvement will be considered in more depth in the next chapter.

The Black Sea's Strange Desiccation and a Later Great Flood

Ryan argues that the Black Sea discoveries and flood were different from those in the Mediterranean. The Mediterranean shrank from a salt sea to a desert, but the Black Sea had been a large, fresh water lake and shrank to a smaller, fresh water lake, until it suddenly flooded and became a sea both of salt water and fresh. Moreover, Ryan attributed the drying out of the Mediterranean to continental collision millions of years ago. He attributed the shrinkage of the freshwater Black Sea to the weight of the Euro-Russian Glacier causing a depression of the Earth's crust that deflected the fresh water away from the Black Sea, creating in turn, a depression and drying it out. This hypothesis is supported by some facts, but is only a plausible theory, lacking the facts to substantiate it.

Ryan's speculative interpretation of the earlier history of the Black Sea area is as follows. From about 20,000 years ago, the Black Lake built up as a freshwater lake from a growing glacier, which may have covered about two-sevenths of the globe. In time, the weight of the glacier deflected the water in other directions. During this time of increasingly dry climate, desiccation

[23] Ibid., 84.
[24] Ibid., 85.

dropped the water level over 350 feet, to two-thirds of its former size. The present area of the sea of Azov was dry land, with a river and chasm going out with *spaghetti*-like tendrils (similar to the Nile into the drying Mediterranean area), into the lake for distances of ten to 100 miles. Beach deposits formed 350 to 450 feet lower, forming three submarine canyons of the sunken river. A number of streams descended the mountains, forming deep gorges and carrying increasing sediments into the lake. The lowest shoreline reached about 550 to 520 feet below today's sea, with the average level being about 330 feet. A ribbon of sand dunes spread across the lake. It seems there were blowing and shifting seasonal storms up to about 7,500 years ago. These desert beach-like conditions continued for about 2,000 years. In some cores, at about 225 feet, there were fossil roots of plants, like those in the cores of the Mediterranean when it dried to a desert.[25]

Ryan speculated there were corridors through the glacial ice between Eastern and Western Europe through which men and animals migrated. Many inhabited the area of the Black Lake. There, the famous, amply-endowed Paleolithic "Venus figurines" were found, revealing an obsession with sex. This reflects the biblical description in the days of Noah when the men took whomever they chose for wives and engaged in perverted sex and violence.

Then suddenly, *an enormous flood came into the lake at 7,500 years ago* or in 5500 BC. An avalanche of water poured into the lake, instantly affecting all life. Ryan said,

> First arrival of marine species occurred simultaneously with suffocation of the Black Sea by poisonous hydrogen sulfide at all depths.... This meant that the increase in salinity had been more rapid than previously thought.... And with massive amounts deficient in oxygen.[26]

The specimens were all the same age, pointing to a single event. In other words, the evidence indicates such a rapid flooding of the freshwater Black Lake, with sea water, that in one fell swoop all life forms were obliterated. In its place, salt water flowed in and salt water marine life began to grow. Ryan assumed that the water came through the Bosporus at an intense rate, and continues to flow into the Black Sea from there. The Black Sea expanded so that it occupied the Azov and other areas as it does today.

Speculation attributes the cause of the flood to the melting of the glaciers which had been going on for years, previously flowing over the Gibraltar dam into the Mediterranean, and later reaching the place where it crested a dam at the Bosporus, and flooded the Black Sea. At the time of the Black Sea flood, Fairbanks estimated that the oceans were full, to within 50 feet of what they are today. Fairbanks believes the melting of the glaciers occurred in two rapid occasions, separated by about 1,000 years.

Consider! Is it really feasible that the gradual rising ocean, from slow-melting glaciers could have flooded the whole Mediterranean basin suddenly? Even Ryan considered that an inadequate explanation. There was little evidence to show that the Mediterranean filled millions of years earlier. In fact, the pictures of the Mediterranean and the Black Sea are very much the same before the flooding, with extended coastal waters, rivers, and canyons *to deeper levels.* Was there a dam at Gibraltar and at the Bosporus that burst, resulting in flooding that caused many people to flee, and abruptly killing all life forms almost at once? Why would people flee in alarm if the Mediterranean Sea filled gradually?

This sudden infusion of millions of gallons of water occurred in such a way that it abruptly changed the marine life and killed off all freshwater forms almost at once. This, in itself, is *a giant catastrophe* that would invalidate the acceptance of a uniformitarian view of nature. If these two

---

[25] Ibid., all these details are taken from pages 101-161.

[26] Ibid., 146.

massive extremes of flooding occurred at once, by a flood of such great proportions that it inundated the whole world, it would be feasible. But if the Mediterranean filled slowly, why would the people in the Black Sea area be suddenly surprised? Moreover, a limited giant flood that caused the people living in the Black Sea area to flee does not fit the data. Most of the stories of the flood that are told worldwide say that the flood came as judgment against all mankind for his sin and that *only one man* survived with his family, usually in a large boat. This is not explainable by even a massive flood, where many families escape to tell their story. Ryan's interpretation cannot account for these stories worldwide. There are other evidences in the area.

Italy and Europe Probably Covered with Water at Same Time?

Italian archeologist Maria Antonietta Fugazzola Delpino has discovered a town, La Marmotta, dated 5700 BC. Her discovery occurred about the same time as Ryan's discovery of this massive flood of the Black Sea. La Marmotta was located 20 miles northwest of Rome, a few hundred yards outside the village of Anguillara Sabazia. It had been submerged and preserved under water all these years. La Marmotta is considered to have been in existence about 400 years before it was submerged, thus it goes back 7,800 years. The findings reveal an advanced culture: animals had been domesticated; a large collection of plants, as well as painted pottery was found; large posts and roof timbers were used in building. The archeologists found a ten-inch ceramic model boat. This site is in the process of being excavated, but may be one of the few known towns that existed before the worldwide flood.[27]

At the time of the flood there was one continent, Pangaea that soon after the flood broke into the continents as we now know them. It is now known that the present continents contained massive reservoirs of water that drained after the division into their particular plates. This will be discussed more fully in chapter 21 on the division of the continents. These massive reservoirs of water are evidence of the world wide flood and the following are such in North America.

The map of Ryan and Pitman's work with the Glomer Challenger, and other evidences for the inundation of the Mediterranean area plus the massive reservoirs from the flood in the United States are given on the next page followed by a discussion of these and other evidences of a massive flood in the past. These graphs were used from the following sources. [28]

---

[27] Robert Kunzig, "La Marmotta," *Discover*, November 2002, 33-40.

[28] The Glomar Challenger drilling, William Ryan, *Noah's Flood*, 76; the pre-flood town, "La Marmotta," Italy, Kunzig, *Discover*, November 2002, 33-40; Lake Agassiz, in Canada, Perkins, "Once Upon a Lake," *Quarterly Science Review*, April, 2002, and The Colorado Plateau, Steven A. Austin, *Grand Canyon : Monument to Catastrophe,* (Santee, CA: Institute for Creation Research, 1994), 93.

Map of Glomar Challenger Drilling of Mediterranean 1970

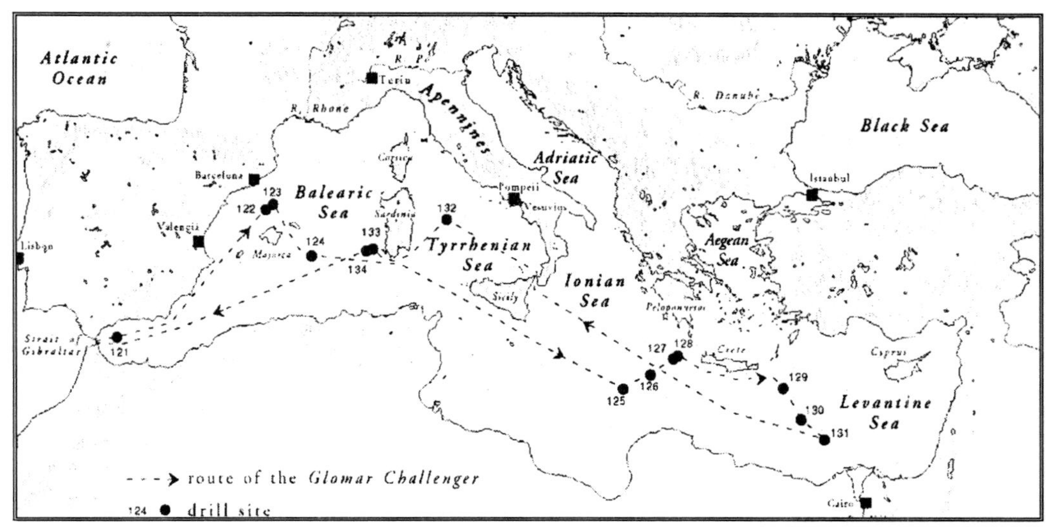

Submerged La Marmotta

Map of Lake Agassiz

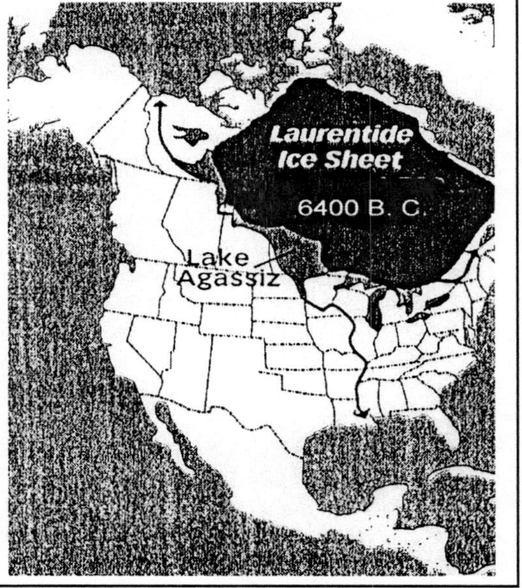

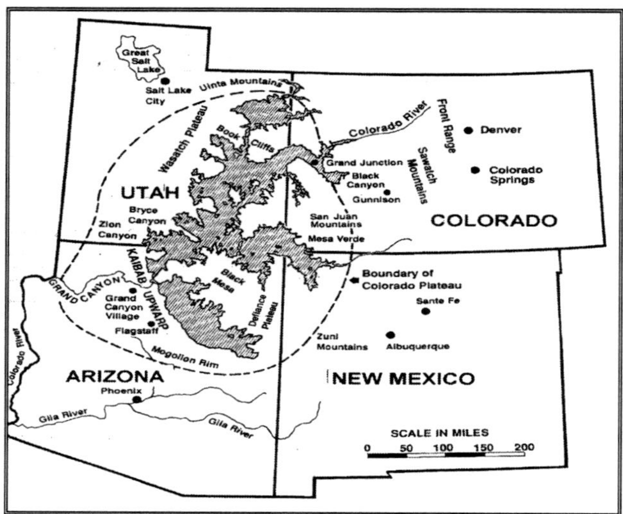

North America Plateaus Covered by Flood Leaving Vast Reservoirs?

Laurentide Ice and Lake Agassiz Area in Canadian—Northern U. S.

There are two areas of North America that resulted from the continental break ups, the first in Canada, and the other in the southwestern Colorado Plateau. The Canadian reservoir was found by recent studies accumulated by geologists Timothy Fisher, Indiana University, NW, David W. Leverington, of the Smithsonian Institute, Peter U. Clark, Oregon State University, and James T. Teller, University of Manitoba indicate that *the largest freshwater lake in the world,* Lake Agassiz which is associated with the Laurentide Ice Sheet, once existed over the Hudson Bay region of North America and contributed to the estuary of *three major rivers,* the Mississippi, the Saint Lawrence, and Mackenzie River in northwestern Canada. It stretched 2,000 km (1,247 miles) with an area of more than 841,000 km^2 (52,257 square miles), but is believed to have drained about 8,400 years ago. Sid Perkins reported,

> Before the collapse, the lake drained southeastward through an outlet that led to the St. Lawrence River. At spots along the northern edge of the ice sheet, which defined the lake's northern edge of the ice sheet, the water was more than 500 m (1,640 feet) deep. When the sheet's mass of ice suddenly gave way, at least 163,000 km^3 (191,283 cubic miles) of water- *about 30 percent more water than is contained in all of the world's lakes today*–spilled northward through Hudson Bay into the North Atlantic.[29]

The estimated date of 6400 BC for this flow is within 500 to 1,000 years of the other floods. The 500-year time range is incidental when dealing with events of the distant past. Perkins said that Teller has suggested the flow of Agassiz Lake at that time "could have been the ***source of the stories of massive floods*** recorded in Babylonian history and the Bible." (Emphasis added) Where did all this water come from? Was it not a residue from the worldwide flood? Teller associates the large amount of water at the time of the biblical flood with the Noahic flood.

---

[29] Sid Perkins, "Once Upon A Lake: The life, times, and demise of the world's largest lake," *Science News*, Vol. 162, November 2, 2002, 283, 284. This was reported in the *Quaternary Science Review*, April 2002.

Western United States Areas that Also Indicate Massive Flood

Recently, new evidence of a worldwide flood has been uncovered in the Colorado Plateau, through which the Grand Canyon cuts. The plateau, ranging from 5,000 to 11,000 feet in elevation, is 250,000 square miles and includes large areas of Utah, Arizona, Colorado, and New Mexico. As more and better data has been gathered, the theories about the Grand Canyon have changed. Estimates on the age of the canyon range from 340,000,000 years for some places to 1,000,000 years for other areas. Evidence now seems to require much less. Three different theories have been formulated about how the canyon was formed.

The antecedent river theory was first offered by J. W. Powell, who first explored the canyon. He posited that the drainage that caused the canyon was antecedent to the faulting and folding and erosion.[30] This view logically fit the uniformitarian gradualistic geology of deterministic science. A few years later, geologist C. E. Dutton confirmed the conviction that the river was older than the structural features of the country.[31] This became the accepted theory for all the textbooks. It required a slow uplift of the Kaibab Upwarp in the Cretaceous period, and an equally slow erosion by the river.

While preparing to build the Glen Canyon Dam, the engineers measured the load sediment of the Colorado River, extending the load over 25 years. The average load was 168 million tons per year. In times of big floods, it carried 55 times the average. But using the average of 168 million tons, the average density of 9.3 billion tons per cubic mile for 70 million years would equal 1.3 million cubic miles of sediment. This amount of sediment *would be 1,500 times that of the Grand Canyon.* The time it took to carve the canyon was therefore much shorter because there was nowhere near this amount of sediment from the river. Moreover, the exit of the river, at Pierce Ferry, does not have the kinds of deposits nor a river flowing that long. Geologist C. R. Longwell studied the Muddy Creek Formation intensely to determine its age. He concluded, "There is no possibility that the river was in its present position west of the Plateau in Muddy Creek time."[32] The evidence gradually discredited the antecedent river theory. A symposium of geologists unanimously rejected the theory in 1964.

A second proposal, the stream capture theory, allows for a much younger Colorado River. In this theory the Little Colorado River flowed southeast of the Kaibad and Coconino Plateaus and another Hualapai stream. The enormous energetic erosion of the Little Colorado cut down the plateau, forming an enormous gully, and capturing and diverting the flow of the Colorado. This theory was derided as the "precocious gully theory." However, there is no east-west trending fault or zone of rock weakness, no trough-like sag in the plateau to guide it eastward to explain why a hundred-mile-long drainage would be where the Colorado is today. C. B. Hunt commented, "It would indeed have been a unique and precocious gully that cut headward more than 100 miles across the Grand Canyon section to capture streams east of the Kaibab Upwarp"[33]. Trying to date such events became highly confusing, controversial, and contradictory. The theory could not

---

[30] J. W. Powell, "Exploration of the Colorado River of the West and Its Tributaries," *Smithsonian Institution Annual Report*, 1875, 198.

[31] C. E. Dutton, Report on the Geology of the High Plateaus of Utah, (Washington*: U. S. Geological Survey*, 1880), 16.

[32] C. R. Longwell, "How Old Is the Colorado River?" *American Journal of Science* 244 (1946), 831, 832, 823.

[33] C. B. Hunt, "Cenozoic Geology of the Colorado Plateau," *U. S. Geological Survey Professional Paper,* 279 (1956), 1-99.

answer how such a gully could erode the Grand Canyon in so brief a time; nor was there any evidence of an abandoned channel where the ancestral Colorado River would have flowed. A number of geologists questioned the theory because of the absence of evidence in the various formations. M. Collier has said, "No one has ever found the ancestral river bed of the Colorado where it was supposed to flow east and south across Arizona, New Mexico and Texas."[34] The precocious gully theory collapsed for lack of evidence.

A more credible theory, supported with evidence, was proposed about fifty years ago by Harlan Bretz, one of the foremost geologic authorities. His breached dam theory suggests that the Grand Canyon formed from lakes, following *a giant sudden vast flood.* This theory, however, required acceptance of a catastrophe, thus, it was bypassed. Further studies by Bretz and others confirmed the idea of a diluvial origin of the scablands.[35]

The most recent theory, the major flood theory, is better supported by scientific data, and is more widely accepted. A June 2000 Grand Canyon conference supported a scenario in which the overflow of a lake or lakes caused the Grand Canyon to form. The lakes would either overtop or develop piping. By overtopping, a spillway would develop, and with increasing speed, increase the volume of flow. Piping occurs when the water pressure behind the dam builds up to such a level that tunnels of water begin flowing through the dam. A great storm, or great agitation by the earth could initiate this and an enormous flow follows. Caviting occurs when a great flow causes vacuum cavities that implode with great force, hammering the rock surfaces at tremendous rates. High-speed water flow also causes great rocks to be "plucked" and moved with great force. Such forces have been exhibited at Glen Canyon Dam and elsewhere, and could erode the canyon much more rapidly than what occurs normally, exponentially reducing the time required to erode the area.

This theory achieved prominence at the 2000 Grand Canyon conference. Billingsley said, "Circumstantial evidence is mounting that erosion of the gorge could have been started by the floodwaters of a small lake that stood near where the eastern Grand Canyon sits today." Rocks from the bouse formation, with strontium in sediments in the lower Colorado River suggest, "that a lake fed by the ancestral upper Colorado River began to overflow about 5.5 million years ago. After the water broke through the edge of the basin, it spilled across the Colorado Plateau and began to gorge the Grand Canyon," says Norman Meek, of California State University in San Bernardino. He believes the resulting flow carved much of the Grand Canyon and other gorges in eastern Arizona. Huge amounts of sediments resulted, some as thick as 600 feet, possibly thicker in the Muddy Creek formation. Those who first proposed the idea suggested that the lake was small. But later estimates have increased the amount of water needed. Scott Lundstrom, of the U. S. Geological Survey in Denver has estimated that 1,000 cubic miles of water could have been involved in this flood. That certainly would have been a major catastrophe.

It now appears that three lakes may have been involved—Hopi Lake in Arizona, Canyonlands Lake in Utah, northwest New Mexico, and eastern Colorado, and a smaller lake, Lake Vernal, mostly in Utah. These three lakes occupy about one fourth of the drainage basin above the Grand Canyon and would have three times the volume of Lake Michigan, or over 3000

---

[34] M. Collier, An Introduction to Grand Canyon Geology, (Grand Canyon, AZ: *Grand Canyon Natural History Association*, 1980), 36.

[35] J. H. Bretz, H. T. V. Smith, and G. E. Neff, Channeled Scabland of Washington: New Data and Interpretations, (*Bulletin of the Geological Society of America*, Vol. 67, August 1956) 957-1049.

cubic miles of water, which is three times the amount suggested by Scott Lundstrom.[36] The Colorado Plateau forms a huge saucer in this region. This entire saucer area could have been filled with water. If the area were dammed at Kaibar Upwarp on the southwest, and broke through, tremendous forces would be unleashed, causing the initial formation of the canyon as it flowed through the many layers of rock, present in the plateau the beginning of the world. Many scientists now acknowledge this as a logical explanation for the Grand Canyon, occurring about 5000-6000 BC. How did all this residue or water get up there if it were not for a worldwide flood?

In regards to fossils, it is said, "Of the land vertebrates, no bone or actual physical portion of the animals has been discovered in any of the strata within Grand Canyon."[37] There are disputed tracks, which may have been amphibians or reptiles. None of the animals destroyed by the flood are found in the sediments, so most of the water in the lake/lakes could have been deposited on the plateau by the great flood and drained afterwards.

More recent investigations in caves and surroundings of the canyon have produced more data. In a TV documentary by NOVA, paleontologists tell how all kinds of advanced higher animal life, including humans as far back as the first ice age or 11,000 or 10,000 years ago included large goats, advanced cats, and even mammoths. There was a very different climate that changes almost immediately. These are found in the thousands of caves off various levels of the canyons. Mammoth dung that is petrified shows various grasses. These cave people who lived in the side of the Grand Canyon were escaping the cold that set in from the reduction of heat from darkness from volcanic eruptions that occurred during the break up of Pangaea.

The ancient people called the Anasazi who are like some of the oldest in Mexico are said to have lived there and had a very high advanced culture of pottery, knives, and weapons. The people and culture disappeared almost immediately. There are very old hieroglyphic drawings in the rocks. The later Walapaie (Hualapai) tribe says there has long been a story that the Grand Canyon resulted from a worldwide flood that drained and cut this great canyon. The Havasupai Indians also still tell the story that an immense flood covered the world and caused the canyon.

Washington State Scablands Illustrates Mysterious Flooding Water

A large area in the farmland of southeast Washington reveals a powerful flow of water that caused strange and enormous changes. A mysterious mega flood in the Channel Scablands area formed 1000-feet-deep gorges, evidence of previous waterfalls five times wider than Niagara, great holes in the valley floors, and left huge boulders, sitting as if dropped there. The enormous pot holes are ten times larger than a river could make. Erratics, 100-ton granite rocks which are not native to the scablands, cover the area. Thick sedimentary layers are repeated. From the air, it appears as if a sea left giant ripples. The Columbia River is 50 miles west. There is no evidence a river created this, nor evidences that Canadian ice sheets ever reached here.

On January 12, 1927, J. Harland Bretz proposed to a meeting of the Geological Society that the only cause for this phenomenon was a single giant body of water, 900 feet deep, with 500 cubic miles of water, instantly flooding the area at great speed. An outrageous concept to uniform geologists, it was rejected, yet no one offered any other explanations for these phenomena. J. T. Pardee speculated that the cause was a giant glacial ice dam, 250 miles away near Missoula,

[36] I am very much in debt to the book edited by Steven A. Austin, *Grand Canyon: Monument to Catastrophe* (Santee, California: Institute for Creation Research, 1994). The writers furnished these estimates of the lake area, 91-107 and I have been led by them to many of the quotations they gave. The geologists involved contributed much.

[37] Steven A. Austin, *Grand Canyon*, 146, see chapter 7.

Montana, that formed a lake 522 miles long and 1000 feet deep. This would have been larger than Lake Erie and Lake Ontario combined. If such a glacial dam had broken it would have released a fast, powerful flow. The whole glacial dam would have exploded, and released two-and-one-half million tons of water, one mile wide, roaring from Montana, across Idaho over mountains, into the scablands of Wisconsin.

Hydrologists at the University of Minnesota have run mini tests and found that the deep channels, vast pot holes, and erratics found in this area of Washington could have been caused by a major surge of water. Surging water, going around rocks, forming great bubbles, can cause swirls that carve out giant holes in the ground, move big rocks, and make great gorges. The scientists accepted Pardee's giant flood theory. In 1980, Bretz was awarded the Penrose Metal, the highest award of the Geological Society, for proposing this theory earlier.

Giant Flood Created English Channel and British Island

Sanjeev Gupta, geologist of Imperial College, London, discovered evidence that the English Channel may have been formed by a similar catastrophic event. Writing in the July 19, 2007 issue of *Nature,* Gupta says it appears a broad chalk ridge, connecting England and France was violently eroded by water. The lowest spot on the ridge, now submerged, was once 30 meters above sea level. Grooves have been found in the bedrock that are100 meters wide, 15 kilometers long, and 20 meters deep. Violent, colossal torrents of water covered, carved and carried one million cubic meters every second for months, leaving chaotic remains and flat-topped islands in the middle of the channel, along with other amazing features. The event that isolated the main British island or islands from Europe has been described as "quickly created by a colossal deluge." Timothy J. Walsh, a geologist with the Washington State Department of Natural Resources in Olympia, says these findings resemble those of the Washington scablands. Philip Gibbard, geologist at the University of Cambridge, compared this to the Washington Scabland catastrophe, which is speculated to have been caused by a glacial ice blockage that suddenly released. Similarly, it is speculated that the English Channel formed at the conclusion of an Ice Age about 450,000 years ago. Northern flowing rivers collected in big lakes along the southern boundary of a kilometer-thick ice sheet that once smothered Scandinavia and much of Britain. The ice sheet gave way suddenly, forming the English Channel. The theories of such glacial and interglacial ages are far from certain, and slow melting ice causing such a chaotic torrent is questionable.[38]

Conclusions about a Worldwide Flood

The data related to flooding needs a complete restudy. Hutton and Lyell operated under Kant's and William and John Herschel's biased philosophical views that the universe developed by natural processes, over long ages. They rejected the possibility of a catastrophic event, such as a biblical worldwide flood. With the geologic data now available, continuing to deny such a possibility stands in utter contradiction to the facts and reason.

The Bible declares that the vast amount of water came from two sources. "The fountains of the great deep burst open, and the floodgates of the sky were opened" (Genesis 7:11). It appears that the flood occurred because subterranean water was forced out, and that the waters above the Earth fell from some catastrophe. Years ago Dr. Roger Rusk, physicist at the University of Tennessee, using astronomical research, set forth the theory that there was once a watery canopy over the Earth was destroyed and released the water. I have suggested a comet was the source.

---

[38] S. Perkins, "Birth of an Island: Megaflood Severed Europe from Britain," *Science News,* July 21, 2007, Vol. 172, 35, 36.

Chapter 20
Enlightenment about Glacial Ice Sheets and Climate

Introduction

Modern science began in the early sixteenth century and it was not until the early nineteenth century that the importance of glaciers was recognized. By that time natural deterministic science was becoming a dogma in conflict with the concept of science as determining the working of God as creator and sustainer. Very soon an effort began to explain glaciers and ice sheets as a cyclic repetition of natural epochs that governed these phenomena by a theory of nature alone. The ice sheets and local glaciers that formed in great valleys are the focus of the ice epoch(s). Glaciers are classified as *cirques*, which are step-walled basins on a mountain, *valleys*, in cold mountains; and then the *ice sheets,* or *polar* glaciers, that cover large areas.

Ice sheets and glaciers—their magnitude and influence on the world—became important when the uniformitarian deterministic scientists were developing their geological theories. Their discovery profoundly affected the philosophy of geological science. Hutton promoted gradualism in geology at the time the glacial theory was being formulated. His ideas prompted questions about the influence of the worldwide flood. Glaciers, as a geological force, were accepted in the last two-thirds of the nineteenth century. Scientific inquiry into how they formed and melted continued in the twentieth century.

In the twenty-first century, the glaciers' and ice sheets' affect on the world's climate have become a major item of interest. The third millennium also brought politics into the "global warming" discussion. For the purposes of this chapter, we will focus on the science of glaciers and climate.

Climate science, ocean coring, submarine research, space photography, and global positioning satellites shed new light on glaciology. It is only in the last quarter of the twentieth century and the beginning of the new millennium that some of the factors are better understood Most theories about glaciers ***were formed*** by the scientific determinists ***before scientists were aware of the more significant forces of geology*** and had access to modern geological instruments Research focuses on two areas: glaciers as a force for geological change, and the search for a theory of a cyclic, determined natural uniform law to explain ***glacial formation and movement, without a catastrophe***. Recent discoveries about glaciers have changed modern thinking, showing that glaciers are compatible with biblical teaching and flood geology. In this chapter we will look at their history, the theories of uniform determinism, a reasonable theory to explain the mystery of glacial ice, and the trends in global temperature. The statement of times given by scientist are the estimates of geologists that later will be shown to be relative and not absolute.

History of Discovery and Acceptance of Work of Glaciers and Ice Sheets

Bernhard Kuhn first wrote (1787) about glaciers. A. Bernhardi first discussed ice sheets at the time Lyell was establishing the geological view of uniform gradualism over long periods (in 1830). Jean-Pierre Perrauding, a mountaineer and amateur geologist, called attention to big rocks (*erratics*) dumped by glaciers, and large rock piles, *moraines*. His friend, Jean de Charpentier, a salt mine geologist, persuaded Ignace Venetz to publish on this subject. Venetz first suggested that glaciers were once spread in a vast ice sheet across Switzerland, the Jura Mountains, and other parts of Europe. Louis Agassiz, a Swiss biologist and geologist, first rejected this idea. But after discovering and studying the evidence for glacial effects in his country, as well as Scotland, he was convinced that ice sheets had once existed. As the president of the Swiss Society of Natural Science, Agassiz advocated for the great ice age. In 1840 he published *Vtves sur los Glaciers,* and in 1846 he came to America to teach.

Agassiz persuaded Lyell and Buckland of his idea. All three presented papers on the subject, garnering wide acceptance for their views. It took about thirty years for most of the geologists to give up their belief in floodwater as the major cause for the erratics and moraines, and to attribute these to glaciers. Agassiz's influence on Buckland, the leading defender of flood geology, tipped the scales toward glaciers as a major influence. But, Buckland's acceptance of glacial influence, rather than floodwater, occurred prior to modern understanding of the great effects of water power and more recent flood evidence. Much that was attributed to glaciers may be from water. Agassiz and Buckland were strong believers in catastrophic forces and opposed evolutionary geology. But Buckland's shift away from the flood, to the power of glaciers, diminished acceptance of the flood catastrophe, and encouraged doubts about biblical accuracy, since there was no mention of ice sheets in Scripture.

Agassiz, Buckland, and other glaciologists discovered massive glacial effects. Glaciers and huge ice sheets of one to two miles thick were thought to exist over about 28 to 30 percent of the land at both the poles, causing massive *abrasion* scouring bedrock and distributing rock debris and *plucking,* or removal of large amounts of fractured, joined or layered rock. *Tills* of earth and boulders were spread over large areas and *bosses* or outcrops of bedrock smoothed, scratched, and sculptured were attributed as being peculiar to heavy glacial work.

Quest for a Deterministic Uniform Cause of Glaciations and Ice Sheeting

Theory of Astronomical Causes of Natural Predictable Cycles

As scientists became more accepting of uniform gradualism and a deterministic view, they began to ***search for a regular and uniform gradual theory of glaciers,*** even though there was no clear evidence for a cyclic regular theory. Just two years after the publication of Agassiz's book, *Estudes sur les Glaciers*, Joseph Adhemar, a French mathematician and tutor, published *Revolutions de la mer,* in which he presented a theory of glacial formation and melting. He used three astronomical factors: Kepler's brilliant discovery of the earth's elliptical orbit and a time of ***perihelion*** or distance or closeness of earth to the sun; the ***tilt*** of the earth in its rotation; and the ***precession*** of the equinox, or the earth's wobble or tilt.[1] Adjemar presumed that when ***these three things were in concordance,*** placing the Northern Hemisphere its greatest distance away from the sun, it would be colder and glaciers would form. As the earth tilted toward the sun, the glaciers would begin to melt, forming interglacials. While his theory was not based on precise data, Adhemar pointed toward a theory that would ***seemingly*** give regular logical cause for a succession of glaciations, which would support a uniform gradualism over long periods of millions of years. He said no great catastrophes or miracles were needed for glacial formations and melting. From this theory, which fit uniform deterministic natural law, the ***idea of a succession of gradual glacial ice ages*** developed. He had no actual earth data showing such extreme ages. This was only uniform *theory*.

Urbain Le Verrier, a French astronomer, calculated the changes in the orbital eccentricity of Earth, Mercury, Venus, and Mars. He showed the eccentricity of the Earth's orbit is about one percent and reaches its most extreme at six percent. This gives a high eccentricity of 10 to 20 thousand years, and a low eccentricity of 100,000. In 1864, James Croll, writing in *Philosophical Magazine,* proposed that seasonal heat distribution would cause an ice age every 11,000 years, first in one hemisphere, and then the other. But he calculated that the cold and heat from these

---

[1] For these historical details see many of them in John and Mary Gribbin, *Ice Age,* (New York: Barnes & Noble Books, 2001), 7 ff.

three factors (perihelion, tilt and precession) and the seasons was ***not enough to account for such changes in climate.*** He proposed the ***added influence of ocean currents*** with trade winds as another major factor. However, he assumed currents as they are today, which did not take into account that Pangaea was still one continent and the change that would make in ocean currents. For instance, the Gulf Stream or the powerful jet streams, influenced by the current configuration of the continents, would not have existed. Ocean currents were not known or accepted at the time of Pangaea when the first formation of ice occurred. So, Le Verrier and Croll were simply speculating about what might have occurred thousands and millions of years earlier.

Milankovitch's Efforts to Establish Astronomical Determinism—Resultant Problems

In 1920, Milutin Milankovitch was attracted to the three major astronomical causes of climate change. He concluded that the earth's spin, tilt, and orbit must definitely be the cause of the rhythm of glaciers. His deterministic thinking ***ruled out any catastrophic event***. These astronomic factors formed the philosophical bias for the events that seemed to change temperatures in the glaciers by two or three degrees. Milankovitch ***was obsessed*** with working out the calculations for the heat arriving on each square meter of earth, for certain latitudes, for one hundred days, for the past 650,000 years using astronomical factors. He arrived at an empirical rule for the snow cover to solar heating and the advance or retreat of glaciers. Adhemar's theory became known as the Milankovitch theory.

However, in 1920, Vladimir Köppen perceptively discerned it was an ***error*** to focus on the spread of ice by snow fall in the Northern Hemisphere, because in the Arctic ***it is always cold enough for snowfall,*** and therefore cold enough to build glacial ice. He said, "What matters is that ***interglacials only occur when the astronomical influences*** conspire to produce unnecessarily ***warm summers*** encouraging the ice to retreat."[2] (Emphasis added) He understood that this was the only way the earth could come out of an ice age. The ***key to regular cycles was the summer warmth from the astronomical factors.*** Milankovitch identified the geological regions of greatest influence, and Köppen sought to match known geological dates with Milankovitch's graphs. However, ***this did not account for formation of the huge ice sheets to begin with*** – only for ***rhythmic melting***. Where did two or three miles of ice over nearly 30 percent of the earth come from originally? This was the major issue left unexplained!

No such huge ice formation is now observed, as deterministic science required in the theory. Still, lacking this data, Milankovitch and Köppen proceeded to estimate the math of the rhythm of interglaciation. They determined it was 100,000 years tied to the elliptical orbit, with 41,000 years associated with the tilt (varying from 22 to 25 degrees) and 22,000 years associated with precession (wobble of the tilt cause by irregular spin of lopsided equator). The ***conclusion of the theory*** was that "***Given the present day geography*** of our planet – the distributions of the continents and oceans – the *natural state of* ***the earth is in full Ice Age***"[3] (Emphasis added) This theory put the world in the midst of the 100,000-year cold cycle now—a theory that continued to gain support. But the Milankovitch ***theory*** had ***no geologic evidence*** of natural glacial cycles. This was pure philosophy to tie glaciation to a regular natural process. In that way they could explain glaciers in a way that ruled out God's acts and cataclysms.

---

[2] Ibid., 54.

[3] Ibid., 56.

Quest for Evidences to Support Milankovitch Theory of Cycles of Natural Glaciation

Attempts to prove glacial cycles focused on geological, chemical, and paleontological (fossils) evidence. Again, any evidence of a cataclysm as a cause was excluded *a-priori*. What evidence was found to uphold the theory?

Most geological evidence is based on marks and remains from valley glaciers. Geologists admit, "Successive glaciations tend to distort and erase the geological evidence, making it difficult to interpret." [4] Chemical evidence is found in isotopes, sedimentary rocks, and ocean sediment cores and ice cores. "This evidence is also difficult to interpret since other factors can change isotope ratios." Fossil dating is based on the fact that glaciers cause cold-adaptive organisms to spread into lower latitudes and other organisms to become extinct. "This evidence is also difficult to interpret…" for several reasons. In order to determine more exact dates of glaciations, oceanographers have taken thousands of cores out of the oceans and from layers of deposits. These show plants and animals that were washed by rivers and deposited in the sea, or that grew in the sea and flourished at different temperatures. As a result of all these efforts, Wallace Smith Broecker and J. van Donk thought they found six ice age cycles of 100,000 years.

Certain tests were used to check indications of when it was cold and warm to discover cycles. Scientists have used Thorum 230, with a half-life of 80,000 years, to test raised beaches in the Pacific and Atlantic to confirm the Milankovitch model. Their findings suggest the ice ages correspond to 12,000 and 80,000 years ago—not the same as Broecker's and van Donk's 100,000. They checked for layers of *loess*, or fine silt, that is thought to be deposited during ice ages. This is found covering large areas in Kansas, the Ukraine, Eastern Europe, and China, Jiri Kukla near Brno, Czechoslovakia and from this the suggestion was 10 cycles with 100,000-year rhythms using this idea. These different approaches somewhat ***resembled*** Milankovitch's estimates, but in spite of much manipulating of the data, were ***not what was wanted nor expected*** to affirm natural theory. Ocean coring was first tried in the *Pacific,* but didn't produce the desired results. Next it was tried in the *Atlantic*, which has a higher deposit rate, again with disappointing results. Finally, scientists attempted coring in the *Indian Ocean,* where the sedimentation rate is *three times higher than other oceans.* These results came the closest to matching Milankovitch's theory. Within the cores, *specific species* in different layers did not fit the theory. John Imbrie then compiled as many as 25 *species together,* and was able to show temperature fluctuations of two percent instead of six percent to find the figures that fit Milankovitch's model. Imbrie and Nicolas Shackleton agreed that more O^{18} in the plankton in the oceans indicated more O^{16} had evaporated and had fallen as snow and was in the ice sheets as an ice age.[5] These intense efforts to prove the Milankovitch theory had ***many variations,*** thus, they did not supply firm evidence of a natural cycle of ice formation and climate variation.

Official Endorsement of Astronomical Cycles and Determination of Ice Now

In 1976, John Mason, director-general of the U.K. Meteorological Office, who ***had opposed the Milankovitch theory***, made a shocking about-face, endorsing Milankovitch's model. Many considered this a major scientific decision—equal to Agassiz's endorsement of the glacial theory in the nineteenth century. ***On Mason's authority,*** the scientific community accepted the repetitive ice age theory, based on astronomical factors. Relying on this model, consensus developed that *the* ***world is in a slight warming phase of a continuing, on-going ice age.*** In 2001, Gribbins, reporting on glacial deterministic views said,

---

[4] Ice Age-Wikipedia, the free encyclopedia (Internet), (all quotes in paragraph above are here).

[5] John and Mary Gribbin, *Ice Age*, 71.

> We are living in an Ice age. ...we are living in an Ice Age now.... Since 6,000 years ago, all these factors have turned around, and conditions for Northern Hemisphere summer warmth are becoming less favorable. The ***prospect*** is for a ***return*** *of Northern Hemisphere* ***glaciations***, on a timescale of thousands of years."[6] (Emphasis added)

This became the dogma of uniform determinate science: ***glaciation cold is in process now***. Using astronomical variants, determinists developed a model for the timing and length of glaciations and interglacials, which they referred to as The Pacemaker of the Ice Ages. Writing in *Science* in 1976, Jim Hays, John Imbrie and Nick Shackleton said,

> It is concluded that changes in the Earth's orbital geometry are the fundamental cause of the Quaternary ice ages. ...Orbital–climate relationships... predict that ***the long-term trend*** *over the next several thousand years is toward extensive Northern Hemisphere* ***glaciation***.[7] (Emphasis added)

The authoritative scientific view was settled on this idea of a mechanical rhythm, which placed the ***world in the midst of an ice age*** with minor periodic vacillations.

Continuing Questions and Complications about the Astronomical Model

But not everyone accepted Milankovitch's model view. Using astronomical factors to explain warmer summers only helps to understand interglacials or possibly warming periods but not the cold and massive build up of ice in a cycle. ***The reason for the huge mile-high ice sheets, over thousands of square miles remained unexplained, while scientists tried to find and plot the ages!*** Many still thought the ice buildup had to occur slowly, over a period of 100 million years or more. But there was no certain evidence for this explanation. Modern variations of weather at the poles have been recorded over several centuries, but these are only short variations not giving evidence of long ages as geologists claim. There was no evidence of how these developed except by a slow buildup of ice and snow as a cause for the long glacial formation. Moreover, there was no explanation of how the valley glaciers join together to help form ice sheets.

Scientists used a variety of techniques and technology in assorted attempts to reconcile Milankovitch's theory with the remaining, unresolved questions. At times, data was altered to support the theory. Gribbin notes, "Milankovitch's Ice Age graphs were used by many geologists to place dates on Ice Age debris found in different parts of Europe...," "but it didn't work." Independent dating techniques, including radio carbon dating (C^{14}), also raised questions. Gribbin said,

> It soon showed that the pattern of glacial advance and retreat *was more complicated than had been thought*, with what had been thought to be one set of glacial debris really being a mixture of material too old to be dated with radio carbon techniques and younger material dated around 18,000 years ago suggesting a late advance of the ice following a *rapid retreat* around 10,000 years ago.[8] (Emphasis added)

Forces in Glaciers Were Not Adequately Considered in Milankovitch Theory

Many inconclusive factors raise questions about Milankovitch's uniform gradual theory. While there are clearly some ***short term weather variations*** related to snow and ice, how can the

---

[6] Ibid., 2, 5, 89.
[7] Ibid., 85, 86.
[8] Ibid., 60.

astronomical theory account for these? S***ome land areas***, which were said to be covered by glaciers, had little or ***no indication of significant ice erosion***. It was *assumed* that the ice was not as thick in these areas, and the glaciers were slow-moving. Glacial ice sheets were interpreted on the basis of the theory of a slow, long ice age. The ***South Pole*** has limited ***snowfall***, and is built up primarily by accumulation of ice buildup from the cold. This seemed inadequate for the theory. The origin of the ice over a large portion of the earth is unknown.

Milankovitch predicted that cyclic changes occurred with the Earth's orbital parameters, but his claims could not be proved. Scientists admitted that a ***multitude of factors***—atmospheric composition, tectonic plate variations, dynamics of the Earth-Moon system, and impacts of large meteorites, super-volcanic eruptions, and possible changes in heat and magnetism from sun and earth—may be involved in forming and melting ice. They admitted, "Some of these factors are casually related to each other." Scientists also acknowledge, "While [some factors] can be expressed in the glaciation record, additional explanations are necessary to explain which cycles are observed to be most important in the timing of glacial/interglacial periods."[9]

Milankovitch Theory Established Before Consideration of Continental Breakup

The Milankovitch theory was developed before scientists accepted the theory of one continent (Pangaea) breaking up into the present continents. Rocks, animals, and plants reveal that the ***impact of polar ice was different before the continents broke up,*** thus we must consider that the ice sheet was formed first, and covered water and land before the land areas of Pangaea were separated. If Asia was still a part of the mega continent until the Paleozoic era, or later times of continental movement, the ice sheet would have been smaller. It would not have covered the ocean areas around Greenland, and especially not the areas of Europe or Siberia. Thus, all the ice sheets of the northern continents would have been one.

Croll and others had figured that the model included the ocean currents (e.g. Gulf Stream) which would not have been operative when Pangaea existed. The partial landmass of Gondwana that was yet undivided during the Paleozoic Age included the continents of South America, Africa, Australia and Islands, and Antarctica.

Since there is evidence that Pangaea and other parts such as Gondwana, as shown in the diagram of the Antarctica on the follow page, were a united continent that was originally all tropical, the evidence of plants, animals, minerals and other factors as discovered and drawn indicates ***the ice sheets formed before the breakup***. The illustration indicates that lower South America, South Africa, India, the Antarctic, Mozambique, and lower Australia were covered with ice ***before they were divided***. These are not the same as the ice sheets as they exist today. The ocean currents, winds, magnetic forces were not the same as they are today. The dotted lines on the diagram indicate the continental breaks, which occurred later.[10]

---

[9] Ice Age, *Wikipedia, the free encyclopedia on the* Internet is the source for most of the quotes in this paragraph.

[10] Jon Erickson, *Plate Tectonics: Unraveling the Mysteries of the Earth, revised,* (New York: Facts On File, Inc.) 2011, 39.

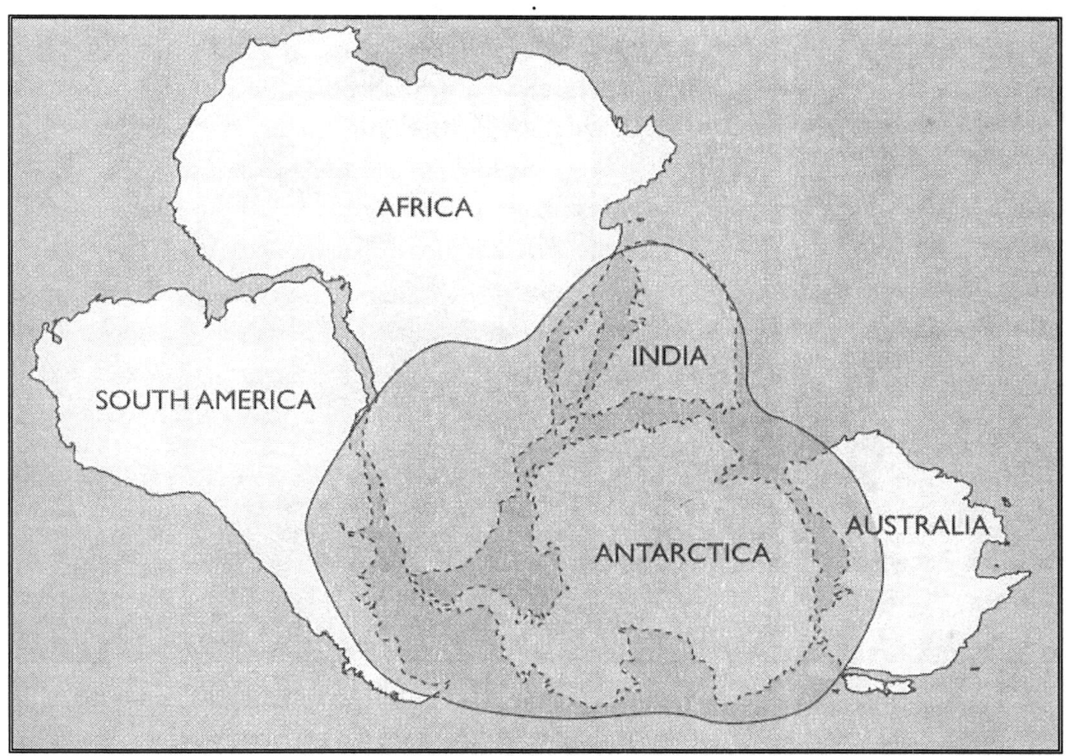

The extent of late Paleozoic glaciation in Gondwana

One reason the effort of study for the Milankovitch theory was inadequate, was because it did not include the study of the breakup of the one continent called Pangaea into the present continents and their tectonic plates. The inclusion of geological factors such as currents, winds, and animal and plant life would be different with Pangaea and the calculations he used based on these as they now exist were therefore not correct. The following drawing of Arctic is thought to indicate previous ice sheets and the direction of forces suggesting these may have influenced the ice sheets.[11]

[11] Encyclopedia Britannica, (1952) "Glacial Epoch of Pleistocene Ice Age," base from Hydrographic and Map Service, Canada, 1944. This was before acceptance of separation of Pangaea.

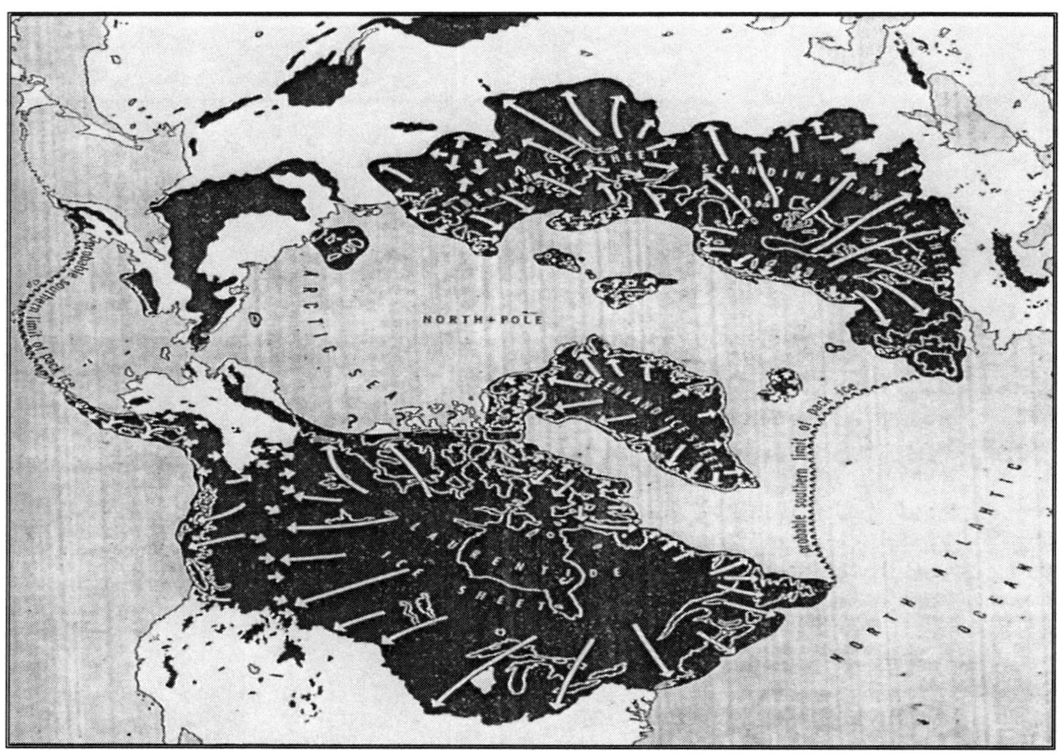

Questions about the Arctic Ice Sheets

The locations and age of present glacial ice sheets in the Arctic raise questions. Study has revealed the long existent primitive ice sheet, which is now near a shifting North Pole. This covers a very large area, much of which was over water. The northern glacial, believed to have formed during the Pleistocene period, covers not only the land area, but extends to the borders of the Atlantic and Pacific Oceans. In North America it is estimated to have piled up from 8,000 to 10,000 feet thick, and covered 5,200,000 square miles. In Europe, the Scandinavian Peninsula was the center of glaciation. Ice accumulated about 10,000 feet thick, and flowed southeast about 800 miles, almost to Moscow.[12] The Greenland Ice sheet is still nearly this thick. Most glaciations studies were based on the configuration and location of continents after the ***breakup of the continents and continental drift***. Therefore, the studies to prove the ***Milankovitch's* model *are not valid***. The discovery of tectonic movements and the volcanic forces that accompanied the continental breakup made Lyell's ideas obsolete. Other major discoveries changed scientists' understanding about glaciers and their influence. Some of the following information was not available when scientists rejected the possibility of biblical catastrophes.

Water Power, Not Fully Understood before Theory of Glaciers Was Formed

When the study of glaciers came into significance, hydrolysis was not well understood. As an agent of geological cause, the power of water was losing favor with scientists. Past ideas of the

---

[12] William R. Farrand, "Ice Age," *World Book Encyclopedia,* 7.

biblical flood as only regional meant calculating water only at the rate of forces ordinarily seen as daily erosion, and not at cataclysmic rates. The effects of cataclysmic water forces are much different than those estimated in uniform gradualism.

Since then, the science of hydrolysis, has become much more developed, both by studies in laboratory experiments and by experience with natural, worldwide cataclysms. For example, early on when the fossils of sedimentary rocks were studied, little was known about how the ***mass of the plant or animal affected the rate of settlement*** and accumulation in the rock layer. That mass is now known to affect the level of the fossil in the sediment. But more importantly, in regard to glaciers, the great power of massive amounts of water to cut, break, dissolve, and move rock and heavy objects was not well understood. Scientists now know that what was attributed to the great force of heavy glacial ice, might in many cases, also be caused by powerful water forces from cataclysms. This is a fact not recognized by uniformitarians.

A 1983 incident at the Glen Canyon Dam on the Colorado River provides a good example. Following a large seasonal snow that created a massive runoff, resulting in great volumes of water behind the dam, engineers released tremendous amounts of water to keep the dam from breaking. That release caused damage which scientists had not believed possible—moving large blocks of steel reinforced concrete, huge rocks, and massive amounts of dirt in a matter of hours.[13] Scientists now understand that in a very large dynamic flood, water alone may account for some things that were formerly attributed to glaciers. The massive tsunami of 2004 wreaked devastation from India and throughout Indonesia, destroying whole towns and cities, and killing hundreds of thousands of people—a catastrophic water event—not unlike the Channel Scablands in Washington State. The phenomena in the Scablands were finally attributed to glacial ice and forces of water power.

Differences of Ice Sheets and Other Forms of Glaciers

Since the last half of the twentieth century glaciologists have learned a great deal more about glaciers themselves.[14] While there are some ***differences between ice sheets and*** valley and cirque glaciers, there is little question that ice sheets are greatly influenced by excessive snowfall, and are impacted seasonally and by major climate changes. But glaciologists have found differences between ice sheets and local glaciers, some of which are significant in our considerations. The ***equilibrium line*** (elevation where snow accumulation and melting are equal) of the large ice sheets is near the margin, while that of the valley glaciers is near the middle. Scientists have also discovered that ice formed ***in very cold temperatures has more of the normal isotope of oxygen*** (O^{16}), while that formed in warmer periods has a much higher ratio of the isotope (O^{18}). These ratios can be found by coring a glacier; this can also establish climate shifts. ***Ice sheets*** *in general are* ***very cold*** *and have a much* ***higher ratio of normal oxygen*** *than most valley glaciers* where it is easier to measure seasonal alterations. This indicates that contrary to what many expected ice sheets form during periods of extreme cold.

---

[13] U. S. Department of Inferior, Bureau of Reclamation, *Challenge at Glen Canyon Dam,* (Salt Lake City, Film/Video, 1983).

[14] Roger LeB. Hooke, "Glaciology," *McGraw-Hill Encyclopedia of Science & Technology*, eighth edition, 127-133. This offers these and other facts about modern glaciology from a professional perspective.

Review of Other Great Earth Forces Affecting Glaciers Barely Now Understood

Feedback Systems Affecting Glaciation

Systems of natural forces, such as ocean currents, heat absorption, et al. are called feedback systems which affect the amount of ice in the glaciers. In the new locations and shapes after break up of Pangaea, the continents cause tremendous ocean currents that affect climate and polar melting. This is important because the astronomical changes that the Milankovitch model relied on were not adequate to change temperatures without these ocean currents.

The current configuration of the continents shields the poles from melting. North America and Greenland shield the North Pole from the warm Gulf Stream that could reduce the temperature there by six degrees. The Pacific currents also flow toward the north, and the Arctic is protected by the landmasses of Russia and Alaska. The water flows from the warm tropics north toward Greenland and Europe. The surface current is warmed and therefore flows over colder deep water, 30 million cubic meters of water per second carrying 1 million watts of heat to Europe. As it nears Europe, it loses heat to trade winds that carry the heat to Europe. The melting of the Arctic ice cap and fresh water from rivers dilutes the saltiness and it cools the water so that it sinks, carrying it back under the warm water. The South Pole has had little currents of warmth because the huge continent of Antarctica became stationed there after the continental breakup. Moreover, the currents in the south circulate in the opposite direction. A number of scientists agree that the continental arrangement causing the ocean currents occurred in more recent geological history invalidating Milankovitch's model. John and Mary Gribbin describe current thinking about ocean currents:

> As far as climate is concerned great ocean currents are much more important[15]
>
> The whole system forms a kind of conveyor belt, driven by upside-down convection, pushed by the descending dense salty water of the North Atlantic. The flow of this 'river' in the ocean is twenty times greater than the flow of all the rivers on all the continents of the Earth put together.
>
> If this (increasing of temperature) stopped the water from sinking, the push that drives the conveyor belt would be turned off, and the whole flow would stop. ...The whole flow could even reverse.[16]

The more the ice cap melts, the more warmth from the Gulf Stream goes further north, further accentuating the rise in temperature.

Effect or Percent of Reflected Light from Ice and Snow, the Albedo

The albedo effect, or the amount of light that is reflected back from the earth instead of absorbed as heat, is another important consideration in glaciology. Strum, Perovich, and Serreze explain the albedo effect this way:

> The ice albedo feedback is the granddad of all these systems. It works this way: land, ocean and ice reflect a fraction of the incoming sunlight, which consequently escapes into space and does not contribute to heating the climate. This fraction is called the albedo [high albedo is more reflective]. In the late spring the ice pack is snow-covered-bright and white. Some of the ice melts, causing the ice edge to retreat and replacing the bright, highly reflecting snow-covered ice

---

[15] John Gribbin, *Ice Age*, 2.

[16] Ibid., 4, 5.

> with the dark, absorbing ocean water. Moreover, away from the ocean's edge, melting snow produces ponds of water that also have a low albedo. Melting in both of these areas decreases the albedo (reflection ratio), which leads to even greater melting, and so on and on.[17]

The albedo effect is complicated because more heat and evaporation cause clouds that reflect and reduce rays, but the clouds also hold heat in. Climatologists are still sorting out the full effect.

Gribbin thinks it probably causes ***a cumulative rate*** and at some point will become irreversible without other influences. He said, "You can't have half the north polar icecap; the feedback makes it an all or nothing choice."[18] Matthew Sturm has said of the feedbacks involved, "These are the ones—such as the ice-albedo feedback—that can amplify changes already under way, speeding them up and magnifying them. They are the ones that *can push the system over the edge*."[19] Others see counteracting forces to the albedo effect that re-thicken the ice. The claim of pushing the system over the edge so that it cannot recover is still a matter of contention.

Vegetation and Peat Exposure Releasing Greenhouse Gases affecting Albedo Effect

Many argue that using fossil fuels creates a greenhouse effect, and that we ought to reduce use of fossil fuels. But reducing ice coverage decreases the amount of reflection, and exposes more arctic ground and peat moss. As a result, the shrubs grow, and the tree line moves northward. This further reduces the albedo. Peat underlies much of the tundra in Alaska and Russia – "about 600 cubic miles of peat are currently in cold storage." "Warming has produced a shift: the Arctic now appears to be a net source of carbon dioxide." "Carbon dioxide and methane constitute the primary greenhouse gases in the atmosphere, returning heat to the earth instead of allowing it to escape into space." [20] Aerial photos comparing the tundra today with the same area fifty years earlier, are dark with growth, a reflection of extreme peat moss change and vegetation growth. A comparison of 1986 and 2001 studies show a temperature rise of four to seven degrees and a change in precipitation. As the arctic summers are changing, the freshwater lakes in the northern locations dry up and further lower the reflection of light.[21]

Influx of Freshwater from Rivers

The influx of freshwater and its thawing effect is another significant feedback factor. Sturm writes,

> Another recently detected change: over the past 60 years, the discharge of freshwater from Russian rivers into the Arctic Basin has increased by 7 percent–an amount equivalent to roughly one-quarter the volume of Lake Erie for three months or three months of the outflow of the Mississippi River. Scientists attribute the change partly to greater winter precipitation and partly to a warming of the permafrost and active layer, which they believe is now transporting more groundwater. This influx of fresh-water could have important implications for global climate: the paleo-record suggests that when the outflow of water from the Arctic Basin hits a critical level of freshness, the global ocean circulation changes dramatically. When ocean circulation changes,

---

[17] Matthew Sturm, Donald K. Perovich and Mark C. Serreze, "Meltdown in the North," *Scientific American*, October, 2003, 66.
[18] John Gribbin, *Ice Age*, 3.
[19] Matthew Sturm, "Meltdown in the North," *Scientific American*, October, 2003, 66.
[20] Ibid.
[21] Michel W. Robbins, "Arctic waterfowl sanctuary dries up," *Discover*, January 2005, 43.

climate does as well, because the circulation system... is one of the prime conveyors of heat to the North Pole.[22]

Polar and Electromagnetic Changes Associated with Ice Melting

Any glaciation theory needs to take polar shifts and migration into account. Scientists have long understood that regularly. Iron cores reveal that the *oxygen isotopes* (O^{18}, O^{16}) *change with the magnetic switches*. So these assumptions of polar shifts were not just affected by evaporation and snow fall. The ***ice sheets were tied to magnetic change***. How is magnetism related to the astronomical model? Earth's magnetism seems to be tied more ***directly to the working of the sun***, and the activities of the sun are known to be highly important in earth warming and cooling.

There is a direct relationship between the strength of electromagnetism and the reversal of magnetic poles, the fluctuation in the strength of magnetism, and diminishing of ice. Isotope formation may be related to magnetic force. There is a direct connection with the amount of ice and the amount of O^{16} formed in cold weather and isotope O^{18} formed in warmer weather.

Mario Acuna, a leading authority in electromagnetic fields, says that the electromagnetic field in our solar system fluctuates, and that our solar system has even lost some magnetism. Mars once had a magnetism 20 or 30 times that of the earth, but today has little magnetic field. The sun goes through a loss and flip-flop of its magnetic field every eleven years and is now in that process with ***the lowest magnetism yet recorded***. The Earth has experienced changes in its magnetic fields and its present rate of change is accelerating. The Earth's magnetic field protects it from the radiation belts that surround it, from the sun's radiation, and from the great bursts of radiation that occur in the universe.[23]

Some estimate that about ***11,000 years ago*** the earth gradually began losing the strength of its electromagnetism. About ***5000 years ago it seemed to accelerate again.*** The rate of loss is now rapidly accelerating. John Shaw at the University of Liverpool has studied early pottery. When the pottery is fired the iron particles in it set as it cools, showing the direction of magnetism at the time. One theory posits that the loss of magnetism occurs because an iron core in the earth, which is liquid when hot, diminishes when it is cool. But Peter Olson's study shows that the magnetism in pottery is changing rapidly and too rapidly to fit this theory. Mario Acuna says, "We are experiencing a decrease much faster than expected."

The Hawaiian Islands are believed to be formed by the buildup of volcanic lava, revealing spectacular magnetic upheavals. For as long as we know, the volcano on the large island of Hawaii has been flowing into the sea and cooling. Fifty years ago all the iron particles in the lava pointed north. Today the iron in the cooled lava points south—a radical shift.

The logbooks of the British Navy reveal another clear record of rapid change. The 300-year logs record the direction of the ships' compasses, which showed the direction of magnetic north, as well as north as determined through celestial navigation. These records show that changes in magnetism began in the South Atlantic Ocean in patches, and spread to join other patches, gradually changing the whole area of magnetism. Jeremy Glassmer built a computer model from all known data about the earth's magnetic field. For four years he ran this model on some of the world's fastest computers to see how the magnetic field works. After persistent effort, Glassmer found that the ***strength of the magnetic field is diminishing***. As the model anticipated, patches of magnetism joined together, and then the poles flipped direction. The important thing to note is that

---

[22] Matthew Sturm, "Meltdown in the North," *Scientific American*, 64.

[23] Some of the data on this comes from the Nova video presentation of Magnetic Storm presented in 2003. Other references are given. See also *The Space Show,* Dr. David Livingstone, November 10, 2006, Dr. Mario Acuma as guest discussed magnetic fields.

these patches of ***magnetic loss are beginning*** to occur in the South Atlantic and are spreading, while the ***whole magnetic field seems to be diminishing***. The magnetic field has gotten 30 percent weaker in the last 100 years and the rate is accelerating. This corresponds to the rapid diminishing of the polar ice. In 2009 there was a shift in the solar cycle and may be the cause of the temperature shift which is now projected to extend for several years.

Solar radiation is increasing. As the sun loses magnetism, it flings out solar electronic and radiation clouds. How this is occurring has been reported by the U. S. Air Force satellite, P78-1 for the years 1979 to 1985 and by the Solar and Heliospheric Observatory from 1996 until today. [24] The greatest effect on earth is coronal mass ejections in space sending ionized gas on a collision course with the Earth. This involves a sudden rearrangement of magnetic field lines called "Reconnection." The magnetic fields break down and then reconnect.[25]

As the sun flings these radiation and magnetic clouds toward the earth they break through the earth's weakening magnetic field, in a sort of "handshake" of force, transferring the radiation to the earth. These events have been observed to occur over four hours at a time. In some way this seems to be adding heat to the earth, increasingly affecting the loss of magnetism and its effect on the earth. The largest solar flares ever recorded have occurred since 2003, after which there was a sharp change from warming to cooling. The radiation has apparently also changed the isotopes of oxygen and other radioactive elements.[26] At an American Geography Union meeting, Joseph Stoner of Oregon State University reported, that the earth's magnetic shield has decreased 10 % in the past 150 years. The following are pictures of the movement of the poles.

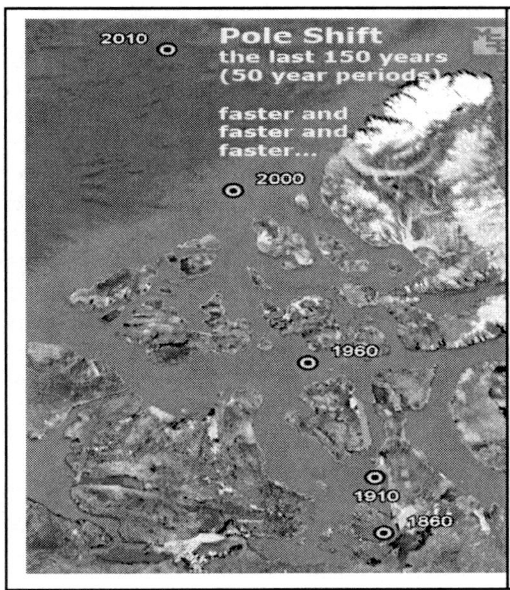

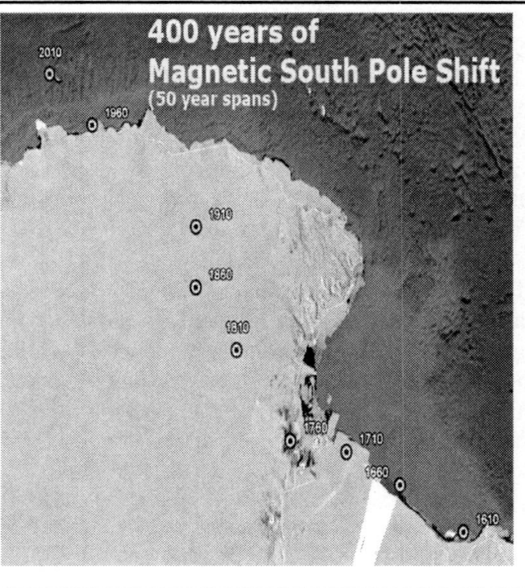

---

[24] Cf. R. Cowan, "Solar Flip-flops: Sun storms spawn magnetic reversal," *Science News*, December 6, 2003, vol. 164, 355, 356.

[25] Gordon D. Holman, "The Mysterious Origins of Solar Flares," *Scientific American*, April 2006, 38-45; Cf. R. Cowen, "Explosive Aftermath: Sluggish neutron star puzzles astronomers," *Science News*, July 15, 2006, Vol. 170, 35, 36.

[26] See chapter 35 on radio active elements for more.

It is important to note there is a ***correlation of time the solar flares and activity of earth's magnetism*** given above with this evidence of when the weather turned cold. The ***rate of the two magnetic pole's movements has increased*** in the past century with fairly steady movement in the previous four centuries. At the present rate the north magnetic pole is swinging out of northern Canada into Siberia wandering about 685 miles out into the Arctic. This seems to be connected at least in part to sun cycles. Pictorial maps of these changes as recorded by government agencies are given below for both the North and South Poles. It appears the world may be moving toward a major magnetic shift as seems to have occurred before. Influence by world political forces to consider carbon emissions without taking this into account would be a major mistake.

Conclusions about Milankovitch's Long Ice Ages that Make the Theory Unacceptable

Admission that Knowledge of Forming of Glacial Ices Sheets was not yet Discovered

As scientific knowledge continues to grow, Milankovitch's model appears less viable. The lack of convincing data prompted many others to offer ***other explanations to add or modify the officially approved Milankovitch theory***, to give it a more scientific basis. In 2006 some scientists still believed, "The 'traditional' Milankovitch explanation struggles to explain the dominance of the 100,000 year cycle over the last 8 cycles." Richard A Miller and Gordon J. MacDonald said Milankovitch's calculations were based on a two-dimensional orbit of Earth and there is a 100-thousand-year, three-dimensional orbit. Ruddiman suggested modifications to this. Peter Huybers suggests that the 41,000-year cycle has been modified by Earth entering a new mode of climate behavior, so that only the 2nd and 3rd cycle triggers an ice age. Charles D. Keeling and Timothy P. Whorf of Scripps Institution of Oceanography suggest a new theory about warming and cooling. Their theory suggests that the fact that about every 1,800 years the sun, moon, and Earth come into alignment and cause unusually high tides, may drastically affect the climate.[27] Physicist William Hyde of Texas A & M and Harvard geologist Paul Hoffman, studying old rocks at the equator, believe they reveal the entire surface of the earth may have been frozen over at least once. Others see a "slushball" Earth, with a possible belt at the equator that was open for life to survive. This is speculation based on data prior to the Pangaea breakup.

As William R. F. Farrand has pointed out, "Scientists ***do not know exactly*** when the glaciations occurred."[28] According to the astronomical factors in Milankovitch's theory, the earliest the glaciers might have formed would be during the Precambrian time, 600 million years ago. However, little verifiable evidence is available for that period of time. Other possibilities are thought to be the early Cambrian, Carboniferous, and Permian periods. The latter was thought to be between 350 and 230 million years ago. It is now clear that we know ***more about glaciers in the later, more recent period*** of Pleistocene. Farrand said,

> When the glaciations were named, scientists thought there had been only four glaciations and three interglacials. However, research in the 1960's and 1970's indicated that the earth ***may have*** undergone from 6 to 20 glaciations during the Pleistocene epoch.[29] (Emphasis added)

This definitive dating of glaciers mainly in the last geological era is highly important.

---

[27] T. Hesman, "Its High Tide for Ice Age Climate Change," Science News, Vol. 157, April 15, 2000.
[28] William R. Farrand, "Ice Age," The World Book Encyclopedia, Vol. 10, (Chicago: Field Enterprises Educations Corporation 1977), 6.
[29] Ibid., 6, 7.

By some current ***estimates*** there may be four ice ages. The first is estimated to be 2.7 to 2.8 billion (10^9) years old, the second, 800 to 600 million years old. Some believe this ice age may have reached to near the equator. The third was 460 to 430 million years ago. And the fourth covers the last 40 million years. Others believe there were ten ice ages and others three. It is obvious from the millions of years of variation that these are ***guesses,*** based on an uncertain theory. They are all linked to the ages of the geological strata, which were estimated or guessed from sedimentary rock, according to Lyell's principles, which have now been superseded.

The Milankovitch theory is therefore to be rejected as objective science, along with the claims of data for millions of years of glaciation, which are dogmatically defended by natural determinists. Only data of recent glacial influence is accepted on the basis of objective science, and not all of that can be pinpointed to exact times. This data ***only extends back about 15,000 years*** and these years have some clear history. Moreover, since there is no proven scientific cyclic natural deterministic theory, an understanding of ice ages and glaciers ***must include biblical historic information and the possibility of cataclysmic forces.***

Science Records Two Major Times of Ice Formations Similar to Two Biblical Catastrophes

Much reliable, objective data has been accumulated from thousands of studies. It yields some information that sheds light on glacial history. But we recognize that the speculative claims from Milankovitch's theory cannot be trusted as proof of natural determinism linked to millions of years. We should use the major facts revealed by these studies to try to evaluate the many details found in geology, oceanography, climatology, archeology, core studies, and fossils. Deterministic scientists do not know the causes of the amazing changes in the earth. Ian Plimer, Australia's leading geologist and world-recognized authority on earth science, repeatedly uses phrases such as: "There ***must be*** other causes.... There ***must be*** other global-scale processes at work...."[30] Some suggest glaciations caused by the flood, by dust clouds, by eruptions from volcanoes, et al.

The following graph by Plimer[31] reveals records of estimated temperatures and ice accumulations during the past 20 thousand years. This graph showing time and indicating two major dips is to be read from the right to the left, and his descriptions, which refer to hundred of thousands of years prior to the times of the dips, are speculative estimates with slight evidence.

Plimer's analysis begins with what he labels and interprets the last Pleistocene Ice Age or glaciation, which he says had lasted for 110,000 years, obviously based on guesses of long ages of strata that Plimer and other geologists accepted from Lyell's column have made, not actually knowing the cause or length of that era.

---

[30] Ian Plimer, *Heaven and Earth: global warming and missing science*, (New York: Taylor Trade Publishing, 2009) 24.

[31] Ibid., 43.

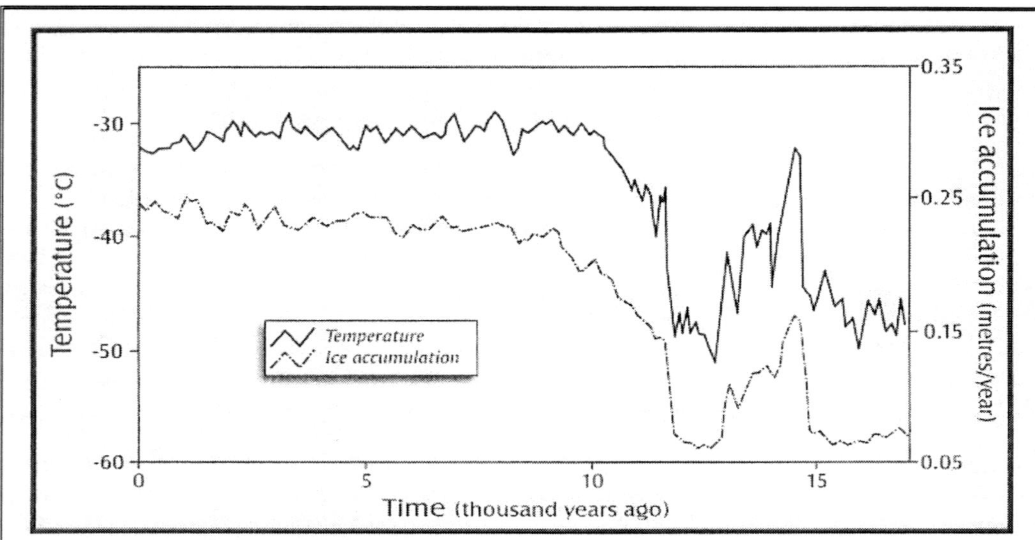

Figure 10: Extreme variations in temperature towards the end of the last glaciation with the sudden cooling in the Younger Dryas, even more rapid warming after the Younger Dryas and temperature stabilisation during the current interglacial. The warmer more humid times resulted in higher precipitation and greater ice accumulation.

Ian Plimer says the last major glaciation began 116,000 years ago, and reached its peak at 20,000 years, which is where the graph begins on the right.[32] He says the ice age ***ended at about 14,000 years ago***. Temperature had a sharp cooling in 13,900 to 13,600 and then ***warmed quickly for a short time of about 400 years*** as shown on the right side of the graph. This was followed by another major zigzag cooling called the Younger Dryas until about 10,000 years ago, slowly beginning warming more than a millennium before. The temperature and ice accumulation then restored to present. For about 6,000 years there seems to have begun a warming with variations, and a little ice age[33]. Other data will be included to show that these two times of intense cooling are associated with the mass P-T extinction of the flood when tropical temperatures ended, and the second one relates to a milder freeze of the K-T extinction which resulted from the break up of the continents. The major glaciation or the first big freeze is the focus of this chapter as the formation of the huge ice sheets and glaciers. The shorter period glacial cooling or Younger Dryers (following the graph from right to left) will be discussed in the next chapter.

First Earth's Major Freeze Identified with Greenhouse Loss at Permian

Despite the determinists' dogma of millions of years of ice ages, no convincing, satisfying explanation has been proffered for how or when the thousands of square miles of ice sheets, one-to-two miles thick were formed. Some maintain they formed during and beyond the Devonian Age. Yet, at that time the whole known world was tropical, with temperatures some places up to 70 degrees centigrade. If the ice sheets had existed before, they were certainly gone by then. The ice sheets had to be formed after that time to coincide with the known facts. Even the frozen

---

[32] Ibid., 38.

[33] Ibid., 35-53, especially 37-39.

animals in Siberia have been found with tropical grasses in their stomachs. The evidence now emerging seems to confirm the association of the ice ages with the great Permian destruction, which admittedly was the greatest catastrophe the world has seen.

It is generally agreed that about 90 to 95 percent of the life forms known were destroyed at that P-T event. Before that catastrophic event, most known organisms existed from the earlier Cambrian period (or early Ordovician) and continued into the Ordovician, when there was an explosion and proliferation of all life forms. This Ordovician period is acknowledged to be a long period of equatorial tropical temperatures ***having what is described as a greenhouse effect.*** It seems the ice ages had to follow this worldwide tropical temperature time. Scientific findings are now confirming this.

Oxygen isotopes are useful in determining previous climate and temperature changes. Oxygen isotopes range in mass numbers from 12 to 28. Four are significant, but two are used to decipher climate temperatures for the ages. The isotope ^{16}O is lighter and composes 99.759 of Earth's oxygen, and ^{18}O is heavier, and amounts to only 0.204percent. Normal water molecules with lighter ^{16}O are more likely to evaporate and fall as snow than ^{18}O, so cold fresh water and polar ice contain less of ^{18}O. This can be used to tell when ice was formed. Several efforts have been made, using this principle, to determine when the ice sheets formed. Previously, scientists thought that oxygen isotopes evolved in the sea water slowly and gradually over a long period of time, indicating a gradual change in temperature.

Recent evidence shows that at the P-T time, ***there was a sudden change*** indicating significant cooling from the tropical temperatures. Other new studies seem to back this theory. The study by Julie A. Trotter, et al. on thermometry, based on conodont microfossils (small eel-like animals) and other factors, states the following:

> The proposition that $\delta^{18}O$ seawater changed significantly ***over time*** is, however inconsistent with altered seafloor basalt compositions and models of the seawater isotopic budget which imply that $\delta^{18}O$ seawater has remained essentially consistent since the Achaean. (Emphasis added)
>
> The temperature record of Ordovician oceans is central to understanding links between seawater chemistry, climate change, major bio-events and thus fundamental Earth processes. The marine biosphere Great Ordovician Biodiversification Event (GOBE), recognized as the longest period of sustained biodiversifications, increasing family and genus numbers three to fourfold. This unparalleled event is characterized by the replacement of the Cambrian Evolutionary Fauna with considerably more complex Paleozoic and Modern Evolutionary Faunas. ***The GOBE was terminated with a sudden and catastrophic extinctions during the latest Ordovician (Hirnantian), probably associated with rapid ice sheet growth at the south Paleopole.*** (Emphasis added)
>
> The causal mechanism(s) that drove these radiations (during the diversification of the Ordovician) has been elusive and enigmatic given the long-standing ***belief that supergreenhouse conditions prevailed.***[34] (Emphasis added)

The study observed a slight cooling of temperature from previous Cambrian and earlier Ordovician time, but this is not known to be the cause for diversification. Also, that was not a

[34] Julie A Trotter, Ian S. Williams, Christopher R. Barnes, Christophe Lecuyer, Robert S. Nicoll, "Did Cooling Oceans Trigger Ordovician Biodiversification? Evidence from Conodont Thermometry," 25 July 2008, Vol. 321, *Science*, 550-557, cf. 551 ff.

period of glaciation, a side issue that is not relevant to the issue being discussed here. The main issue is the ***instant ending of the greenhouse tropical temperatures*** with a sudden ice age.

The long greenhouse period ended with a dramatic glaciation. Of great significance is that the ***dramatic change in temperature was worldwide***. The details of the study concerning temperature change are given and introduced by stating the extent of these findings. It said,

> The measured conodont $\delta^{18}O$ apatite compositions are internally consistent across geographically disparate sites from two cratons, with no systematic patterns related to facies, and are thereby interpreted as ***a global temporal trend***. Futhermore, the earliest Ordovician conodonts from Australia have compositions equivalent to those from Texas reported by Bassett et al. and the Wenlock data from Cornwallis Island are consistent with coeval samples from Gotland. [35] (Emphasis added)
>
> The conodont $\delta^{18}O$ apatite record indicates that the sea-surface temperatures for much of Ordovician to Early Silurian were well within present-day ranges. Even our highest temperature estimate (42 C, when considering analytical (3 C range) as well as $\delta^{18}O$ seawater uncertainties, is essentially in the upper limit of present–day surface waters (e.g. Red Sea, Persian Gulf, Sunda Sea).
>
> The late Ordovician temperature fall in $\delta^{18}O$ apatite record coincides with the Himantian glaciation.... The actual temperature range, however, is difficult to ascertain the ice volume component, which was clearly substantial during this interval, cannot be discriminated.[36]

The study indicates a three degree C in Baltic ice samples than samples in other places for which the study could not discriminate. But ***these all indicate an extreme drop in temperature and the beginning of the ice sheets.***

In 1955, Italian Cesare Emiliani, in an effort to show glaciation by oxygen isotopes, and to prove the Milankovitch theory, used shells of forams of calcium carbonate to test ocean sediment forms with no success. John Imbrie and Nick Shackleton decided to try samples from the Indian Ocean, in which there is a much higher sediment rate. The sediment from the Indian Ocean gave false readings. While they registered an oscillation of cooling over the geological ages where the forams were found, they showed no major ice age as these other worldwide studies do, and were therefore of limited worth. In the next chapter we will look in more detail at the reason for this.

Further evidence of the extreme drop in temperature at the time of the Permian crisis is that helium and argon gases have been found trapped in large amounts at that time. ***Helium has been found at fifty times the amount in normal background helium***. Argon and helium are two of a few natural gases that require much lower temperature to liquefy than other gases. ***Helium is most exceptional because it requires -268.6 C with pressure to liquefy it.*** Luann Becker, whose studies will be mentioned in the next chapter in regard to impact craters, discovered this in regard to study of fullerenes or bucky balls. A radical temperature drop would remove other gases as liquids and leave a higher content of helium in contrast. This discovery at the P-T boundary is a further confirmation of the study of oxygen isotopes.

Based on early rock strata, the oxygen isotope studies indicate a drop in temperature to a slightly cooler, but tropical temperate, during which time there was a great proliferation of biodiversity, under a greenhouse effect. This period ended with a sudden drop in temperature, ushering in an age of extensive ice sheets. The helium studies and others also show a quick,

---

[35] Ibid.

[36] Ibid., 552.

extreme drop, resulting in the formation of great ice sheets. The drop in temperature was connected to mass extinction. This was later followed by a moderate climate as exists today.

Confusion of Timing of Geological Column, Worldwide Extinction, and Glaciation

Trotter's oxygen isotope study is decisive about the timing of the ending of the greenhouse temperatures, the massive extinctions, and glaciations. Some suggest the greenhouse temperatures existed at the end of Ordovician or early Silurian periods. Recall that these geological ages and strata were set by limited studies in small areas; they were often the determination of one man, and arbitrarily applied worldwide. The lack of real scientific geological understanding creates many problems and great confusion in establishing a reliable, accurate geological timeframe. The anti-God bias that dominated the whole period, led by Lyell, excluded objectivity because of his utter denial of catastrophes such as is evidenced in the P-T and K-T, and especially the continental breakup that is now accepted. Even the great world-changing event of one continent dividing into many, producing tectonic plate collisions is minimized, being explained by slow gradualism. This a-priori rejection of catastrophe has confused the data.

The establishing of the age boundaries is based on a scarcity of evidence in what is considered the oldest rocks, compared to that since the Mesozoic Age and subsequently. Many of what were thought to be oldest rocks are now known to have been raised from tectonic collisions. Moreover, since the establishment of the oldest ages, most organisms have been found to already exist in the Cambrian and some even in Precambrian data. The inconclusiveness of these strata and ages is seen in that there were tremendous arguments over the early periods. This was so even when the data was very limited to Europe and there was ignorance of most of the world. Sir Roderick Murchison identified the Silurian from strata in south of Wales (1830s) joining with Adam Sedgwick in England who had already himself identified the Cambrian, and together they established the modern strata sequence that became the basis of the time scale. Charles Lapworth differed with them arguing for Ordovician strata. Joachim Barrande, a Frenchman examining Bohemian strata, contended for eight stages in the Silurian rocks. Later, Edward Forbes (1854) raised questions about these and others and created the Devonian.

While there are certain chosen fossils to indicate the periods, these fossil layers are often mixed in with others of other ages. For example when Raymond C. Moore in his significant text book discusses the history of the Permian, he says,

> The lowermost Permian strata, in places containing thick basal conglomerates, are found resting on the upturned truncated edges of Pennsylvanian and older rocks, including both lower Paleozoic and Precambrian formation.[37]

The point is that the greatest extinction, which is labeled P-T, occurred at a time when the greenhouse tropical climate ended and the ice sheets were created and the duration of the eras leading up to it may not end in the time periods named in the study. Permian was the time all this occurred.

Biblical Data Supports the Relationship of Worldwide Extinction by Flood and Glaciation

Evidence of Widespread Instant Deep Freeze, Not a Slow Glacial Buildup

The evidence as presented is therefore that a major catastrophe involving much water and ***an immediate very cold drop in temperatures*** at near the same time on the earth which seems

---

[37] Raymond C. Moore, *Introduction to Historical Geology*, 259.

associated with the late formation of ice sheets (14,000 years or less). There was a tropical temperature with tropical plants worldwide—in Northern Siberia, Northern Europe and northern North America, Africa, South America, et al. These areas were suddenly covered by ice sheets and there were significance previously polar shifts indicated by magnetism in various ways. For centuries Chinese, Siberians, Eskimos, and others observed and recorded thousands of animals, of all sorts, ***instantly frozen*** in deep muck. These reports include bear, wolf, fox, badger, wolverine, saber toothed tiger, jaguar, lynx, cats, horses, camels, saiga antelope, bison, caribou, moose, ground sloth, rodent, rhinoceros, wooly mammoth, and mastodon. Some believe these have been frozen for about ten thousand years. The stomach contents, still visible in some of these animals, were frozen, undigested ***tropical vegetation***. This shows they were instantly taken while in a warm climate, the vegetation—palm trees, flowers, and grasses—is normally found in tropical areas. Some of the animals had literally been ***violently torn apart before they were frozen***. Modern hunters have found as many as 50,000 mammoth tusks in thawing ice. When they removed the ivory, it was still fresh, with pulp. The flesh of these recently exposed animals was edible for both animals and humans.

A number of scientists with expertise in DNA, holding to various points of view, have examined this evidence. Stephan Schuste, and a team from Penn State University, recently decoded 80 percent of the genetic code of 3 billion basic DNA from hairballs.[38] Over 39 mammoths from frozen muck have been examined, and dissected by scientists studying their condition. Douglas Kennett of the University of Oregon claims to have found nanodiamonds in numerous North American locations. Based on these findings, he proposes that mammals, such as the mammoths, large saber tooth cats, and others, (amounting to about 30 species in North America and 50 others later in South America) were instantly killed by comets that ***began an ice age*** about 12,900 years ago. These scientists believe this also ended the Clovis human culture.[39] But instant death of the mammals has not been seriously discussed or considered in connection with the extensive forming of the great ice sheets and initiation of worldwide glaciation.[40]

The fact that the mammals were instantly covered in mud and water, and frozen while in a warm climate is supported by other evidence. Geologist Edgar B. Heylmun observes,

> For example, there is little evidence that climatic belts existed in the earlier history of the earth; yet climatic zonation, both latitudinal and vertical, is clearly apparent in all parts of the earth today. This anomalous situation is difficult to explain. ...Any rotating planet, orbiting the sun on an inclined axis of rotation, must have climatic zonation. It is obvious therefore, that climatic conditions in the past were significantly different from those in evidence today.[41]

He is saying that there had to be a change in the tilt of the earth's rotation axis from one that was formerly vertical to explain change from all the earth having a consistently temporal climate.

This correlates with the data of the history of the biblical worldwide flood. The greenhouse cover which maintained the tropical climate until the flood was followed by seasonal climate

---

[38] Seth Borenstein, "Woolly mammoths: Back to the future?" The Associated Press, *Orlando Sentinel*, November 20, 2008, A8.
[39] Dan Vergano, "Soil points to comet storm that was fatal to mammoth," report from *Science* in *USA Today*, January 2, 2009, 2A.
[40] Joseph C. Dillow, *The Waters Above*, (Chicago: Moody Press, 1982) 311-420 has covered this evidence from all points of view and need not be repeated here. He has given a length bibliography 427-459 listing many scientists and engineers, zoologists, and historians reporting and studying these.
[41] Edgar B. Heylmun, "Should We Teach Uniformitarianism?" *Journal of Geological Education,* 19 January 1971, 36.

changes. We can attribute these anomalies to the climatic changes that followed the Noahic flood. After the flood, God told Noah that there would then be, "Seedtime and harvest, and cold and heat, and summer and winter, and day and night." God gave Noah a new sign of His covenant—a rainbow that appeared in the sky. With this, God promised there would never again be a worldwide flood (Genesis 8:22; 9:14, 15). Joseph Dillow asserts that this indicates there was no longer a canopy covering the Earth, screening sunlight, and that the axis of the earth had been tilted by that cataclysm.[42] As Trotter's study shows, the Earth's greenhouse effect ended at the same time as the worldwide extinction. This instant change from a warm condition to extreme freezing requires a very powerful event. What major event could cause the shift of the earth's axis about 25 degrees and cause so much ice to form?

Dillow relates experiments conducted by Birds Eye and other frozen food and meat companies in 1961 to check the temperatures required to quick freeze the animals and their stomach contents. [43] Temperatures as low as –250 F. were considered probable, but a temperature no higher than –150F was agreed as absolutely essential to produce the results seen in the frozen mammoths. Such instant cold temperatures that encased animals in ice for years, across all the northern continental areas of the world must have required some unusually cold catastrophe.

Significant efforts were made to explain catastrophic extinction by the impact of an asteroid or comet. Twenty six scientists participating in the October 2007 proceedings of the National Academy of Sciences concluded different opinions. David Kring of the Lunar Planetary Institute of Houston reported, "Obviously something really interesting happened 13,000 years ago. ... A world ... suddenly plunged back into a millennium of near-glacial climate, before emerging into the current warmth. It was about then—emphasis on the uncertainties summed up 'about' —that the mammoths and other great beasts disappeared from North America. And the Paleo-Indian Clovis culture vanished from the archeological record around then, too." [44] However, it is of interest that Kennett's more recent, extensive evidence connects this to comets.

Science of Twenty-First Century Calls for New Science Theory of Catastrophic Glaciation

Current climate trends seem to demand another look at historical understanding in light of more recent geologic and glacial discoveries. Only unbelieving scientists who want to hold to a naturalistic uniformitarianism will object to looking for a ***cataclysmic explanation*** that may help us understand what is going on. There is no decisive reason to believe there were a repeated number of long ice epochs, unless we tie them to a uniform long continuing naturalism. Even after a century of research, "Some authorities, however, believe that any real distinction between glacial conditions and present conditions is artificial, and regard the glacial epoch as continuing uninterruptedly through to the present day."[45] If the earth is young, or the changed tilt has occurred since the flood or the wobble since the breakup of one continent around the equator, these astronomical factors may have no or perhaps little influence. A growing body of evidence points to one cataclysmic event that brought on great glaciations, followed by a shorter one only a few hundred years latter followed by continuous, irregular warming trend that may now be accelerating to approaching a crisis in climate and nature.

---

[42] Dillow's discussion of the rainbow and seasons, *The Waters Above*, 95-99.
[43] Ibid., 387.
[44] "Experts Find No Evidence for a Mammoth-Killer Impact," *Science* vol. 319, 7 March 2008, 1331, 1332.
[45] "Glacial Epoch," *Encyclopedia Britannica*, 374a.

Time of Ice Sheets Began with an Event Related to Noah's Flood

The actual length of the ice ages is, at best, the last fourteen to ten thousand years. Some authorities still speak of millions of years, and acknowledge, "More colloquially, when speaking of the last few million years, ice age is used to refer to colder periods with extensive ice sheets over the North American and Eurasian continents: in this sense, the last ice age ended about 10,000 years ago."[46] That places the ice ages since the P-T extinction of the Flood and division of the continents with the K-T as they now are known. The facts of the global ice sheets and glaciers seem to point to one great cataclysm for the major glaciation followed by a brief freeze shortly afterwards, both in more recent history as the source for understanding of what is happening. The different time figures of millions or of hundreds of thousand years are not accurate.

It appears that the ice sheets in various locations around the world began to melt around the same time: the Laurentide Ice Sheet and Lake Agassiz from about 8400 to 6400 BC and the Mount Kilimanjaro melt around 8300 and 4000 BC. South America scientists found that a plant that was recently exposed in the Quelccaya ice cap in the Peruvian Andes was frozen about 5,200 years ago, according to radio carbon dating. Pittman and Ryan estimated the glacial melt occurred about 7400 BC in connection with the Noah flood. Jiri Kuklas of Brno Czechoslovakia traced the desert sands or loess as continuing into the eighth century, when they ended.[47] David Ross and Egon Degens on the Atlantic II expedition traced the high water in the Atlantic to after 9000 BC. The Russians thought the flooding into the Black Sea occurred about the same time.[48] Martin Claussen, Caudia Kubatzki and Andri Ganopolski have written about a rapid and radical transformation of North African Sahara area. They note that nine to six thousand years ago the area went through "comparatively abrupt" changes, going from cooler to warmer and dry between 6,700 and 5,500 thousand years ago.[49] While all of these may not fully agree, their speculative estimates show shifts to within a few thousands of years of each other in a young earth.

While the Gribbins endorse the astronomical cycles, they too observed that forces begin to "drag the Earth into a ***peak of warmth about 6000 years ago***."[50] The evidence does not show glaciers advancing, but ***glaciers retreating*** leaving moraine, scratching and the like. Could it be there was a sudden glaciation, withdrawing much of the world's water resources into glaciers and ice sheets and great cold, followed by a shift toward an irregular warming trend and then the shorter Younger Dryas glaciation followed by a slow warming that now is accelerating by albedo and other effects?

Optional Theories of One Major Glaciation and Meltdown

Various speculative theories have been advanced to explain the sudden drop in the earth's temperature and the formation of great ice sheets. Dr. Dillow speculated that the collapse of the water canopy over the earth caused a deep drop in the temperature. A second option suggests that great volcanic eruptions blotted out the sun, causing a great drop in temperature which fits the second Younger Dryas. (This second, less severe drop in temperature related to this event will be addressed in the next chapter.) The volcano option would likely produce massive fires and heat. The "ring of fire," produced by multiple volcanoes appears to lineup with tectonic plates and the

---

[46] "Ice Age," *Wikipedia, the free encyclopedia,* cf. Gibbon, *Ice Age*, 89.
[47] Pitman and Ryan, *Noah's Flood*, 103.
[48] Ibid., 115, 116.
[49] Martin Claussen, Caudia Kubatzki and Andri Ganopolski, "Stimulation of abrupt change in Saharan Vegetation in the mid-Holocene," *Geophysical Research Letters*, Vol. 26. No 14, 2037-2040, July 15, 1999.
[50] Gribbins, *Ice Age*, 89.

new division of continents. The evidence of instant cooling to form world extensive ice sheets requires a major catastrophe. I suggest another cause for the instant cooling could be a comet.

A Comet or Comets Possible Atmospheric Close Encounter

The chapter on the new view of planets indicated that the average temperature at Jupiter and planets beyond is -150 degrees centigrade (-302 degrees Fahrenheit). Beyond Neptune, in the Oort Cloud, circling in the Kuiper belt are over a trillion comets. Many are small; some are as large as a few miles or more in diameter, with temperatures as low as -400 degrees Fahrenheit. Many of these are nudged out of the Kuiper belt by gravity. One of the middle or cold planets, especially Jupiter, will push a small comet into orbit around the sun, traveling at over 100,000 miles per hour. Louis Frank at the University of Iowa says about 10 million house-size comets, weighing 40 tons each, hit earth's atmosphere every year and vaporize, leaving their water. It is estimated there are enough of these between Earth and Jupiter to fill all the world's oceans. There are some 500 visible large size comets known in solar orbits now and possibly many more to come. One comet, Shoemaker-Levy 9, was estimated to be as long as two or three football fields. It fragmented into twenty pieces and crashed into Jupiter in July 1994, leaving a mark in the planet bigger than the earth.[51] When Shoemaker and his wife announced this was about to happen, many scientists refused to believe this because it went contrary to uniformitarian science. One scientist commented on the TV report, "there goes the uniform world." Many comets are known to venture near the earth in the past.

Possible Cause of Flood and Recession

The worldwide flood, even by today's scientific standards, was a catastrophe of biblical proportions. The biblical account says that great reservoirs of underground water ("fountains of the great deep") released, and the floodgates of the sky ("waters above the earth," or possible canopy and comets) opened. Torrential rains fell for 40 days and 40 nights, causing the flood. Rains of one-half to two inches an hour over prolonged periods are known, but in monsoons the figures have been recorded much higher.[52] After forty days of rain, flood waters reached 22.5 feet (15 cubic), above the highest mountains, and remained over the whole earth for 150 days. Then at the end of the 150 days God sent a wind that restrained the waters, and the water abated (Genesis 8:1-3). What was this wind? How did it stop fountains of the deep, close the windows of heaven, and cause the water to decrease? Was the great wind associated with a major catastrophe which changed the earth's axis 23.27 degrees at that time?

How might this worldwide historical event throw light on what happened? God is sovereign. He could have caused the flood and its recession without any secondary catastrophic causes. In fact, He could have killed all life except those in the ark without using the secondary cause of the flood. Therefore, another secondary cause for the flood and the deep freeze may be providential. What seems likely to account for the large amount of water, and then of its receding, that allowed life to continue afterwards? Might glaciers be involved?

The maps of areas of the world unaffected by tectonic plates indicate a general mountain height of about 1,000 feet. The highest mountains were built later by the collision of tectonic plates. Because the water covered the mountains, the biblical estimate would then be that the water rose to above 1022.5 feet. Some scientists believe the water on the continental shelves is a recent occurrence, and the shelf averages about 500 feet deep. It is estimated that the amount of water in

---

[51] "The Age of Comets," *National Geographic*, Vol. 192, No. 6, December 1997, 94-109.

[52] In 1953 New Orleans had 4.4" per hour; Cherapunji, Indian had 446" in 31 days.

the glacial ice sheets, if melted, would raise it 525 feet.[53] The estimated general total of water in the continental shelves and in the glacial ice sheets together would raise the level to 1025 feet. ***These are generalized speculations*** but it is one that includes the deep freeze temperatures and the subsequent changes in climate.

If a large comet or comets entered or passed near or through Earth's atmosphere and went on to orbit close to the sun, what would happen? The gravity of the large comet could cause the disruption of the earth, releasing underground water and causing volcanic eruptions. It would probably leave huge amounts of water. The comet's intense cold (–400 degrees Fahrenheit) would cause lower temperatures and precipitation from the water canopy surrounding the earth.

The same comet or comets (or another one) could proceed to orbit around the sun and return, then actually breaking up and colliding into the ocean. That could have changed the earth's axis and caused the cold winds and ice that would have frozen the animals. The extreme cold could initiate the buildup of the glacial ice sheets, causing the water to progressively decrease out of the oceans, until it left only about half, or the 500 feet that is now on the continental shelf. A circle around the sun like other comets would probably equal the 150 days when the water prevailed on the earth, after which God caused the winds that made the flood recede. (At 100,000 plus miles per hour to travel directly to the sun it would take 77.5 days. An orbit would be about twice this.) Such an impact by a comet in the ocean would not leave evidence of such a major event. Others have associated the impact of an asteroid or comet in the Indian Ocean with the worldwide flood. (An impact in that location fits a later catastrophe, to be discussed in the next chapter.) "Scientists are looking at the possibility that super massive quantities of water ice were rudely delivered by a comet which passed close to the Earth and the Moon at the end of the Pleistocene era…."[54]

Glacial ice sheets accompanied the flood, freezing the animals and leaving mankind in a much colder climate in most of the world. We may never know, with any certainty, how the glaciers formed. But a cataclysm, such as a comet is the most likely source of such cold in the solar system. Pitman and Ryan linked glaciers to the flood in Eurasia as did scientists giving the evidence regarding Lake Agassiz, both about seven to eight thousand years ago. The removal of the water canopy over the earth would expose man to health hazards such as the increase of C^{14}, as would colder weather.[55] Following the flood, human life expectancy dropped from more than nine hundred years to about half, or about 440 years. This great drop in life expectancy corresponds to other ancient flood stories. Glacial ice could have been a result of the flood event and part of the judgment of God. The next chapter on the continental division that occurred 300 - 400 years later will show that there was a short glaciation period which also resulted in a decrease of life expectancy of about half.

Conclusion of True Enlightenment about Ice and Global Temperatures

Some evidence points to a very large comet or comets passing close to Earth, removing the gaseous canopy and gravitationally disrupting the subterranean waters and causing the flood. Then in a return these may have showered the earth with extreme cold, causing the water to be drawn into ice sheets and glaciers so that the waters receded. The account of the ending of the flood says,

---

[53] *McGraw-Hill Encyclopedia of Science and Technology*, eighth edition, (New York: McGraw Hill, 1997) 115b.

[54] *Morien Institute*, "Underwater discoveries news archive," 2002.

[55] Cf. Dillow, *The Waters Above*, 165-182.

"And God made a ***wind*** blow over the earth, and the waters subsided. The fountains of the deep and the windows of the heavens were closed, the rain from the heavens was restrained, and the waters receded from the earth continually. At the end of 150 days the waters had abated...." (Genesis 8:1-3). While the Hebrew word translated ***wind*** may mean just breath or wind, in Genesis 3:8 wind is translated ***cool*** of the day in Genesis 3:8. Cold atmospheric high pressure involves high winds. These are speculations based on the evidences given, but the time of the flood may have been consummated by God providentially using the cold of a comet to produce this wind and to end the day of the flood and thereby progressively over 150 days end the sources of rain and draw up half of the water into ice sheets.

The scientific proposal of glaciations and interglaciation had no real evidence of working in regular repeated natural ways over millions of years as Milankovich and others sought to prove. Climate change may be explained in three main providential workings. The once formed ice sheets after the flood have had a very slow progressive warming. This has reached an accelerating rate toward a crisis by the feedback systems. There has been a zigzag fashion of warming and cooling caused by the rhythms of the sun and the effects on the gravitation and polar activity of the earth. That warming may now be accelerating, not primarily from man's use of fossil fuels, but by the feed-back mechanisms as presented above, especially by the albedo effect. The effect of God's direct one time great glaciation forming the ice sheets working as in Genesis 8 better fits with the data than theories to claim by repetitive natural forces. This theory of the formation of the ice sheets after the flood, and after the separation of the continents, which will be discussed in the next chapter, seems to fit with all the data that is recently known

Recent Ice and Temperature Data Indicate Oscillations but Warming

The initial freeze after the flood and the one after the division of the continents were followed by a warming to normal temperatures as previously shown by Plimer's graph. Since the last half of the twentieth century more recordings from satellites, submarines, and other studies have provided objective and helpful information about weather since glaciers became known. History and geology indicate that since the fourteenth century there have been periods of cooling and warming oscillations, of varying lengths and temperatures. The most recent, from 1300 AD to 1850 AD, is known as "the little ice age" did not form ice sheets. The Little Ice Age (LIA) was a period of cooling that occurred after the Medieval Warm Period (Medieval Climate Optimum). While not a true ice age, the term was introduced into scientific literature by François E. Matthes in 1939. While the world was generally colder, certain areas like Alaska, New Zealand Patagonia did increase their glaciers. This was followed with a renewed warming trend that continued. When the earliest records (from 1880) are added to the recent, more objective findings, they present a picture of a warming trend, with minor fluctuations until the year 2000, as indicated in the following graph.[56] Since then until 2011 there's a cooling trend.

[56] Ibid., 25.

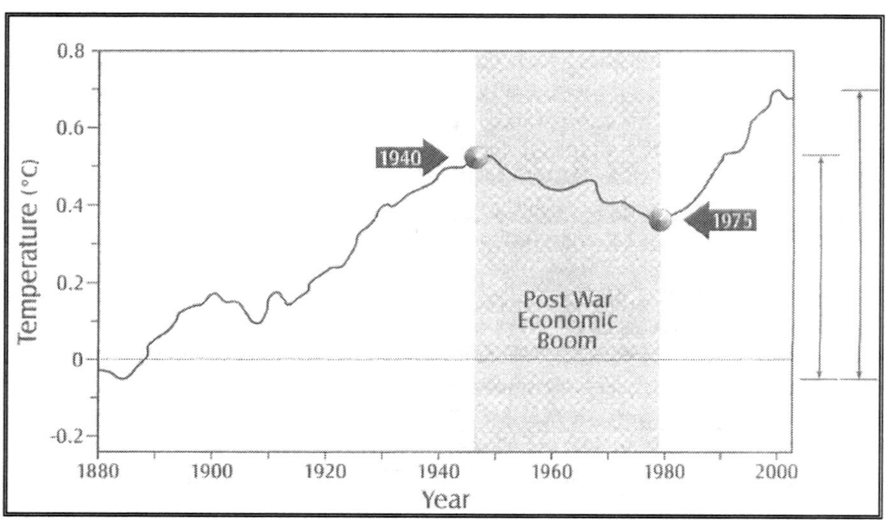

There was warming until 1940, a slight decrease until 1975, and then a return to a sharp increase after 1975 until a dip after 1999. The warming trend seemed to accelerate up to about 2003 or 2004. From that data, scientists developed ***computer models projecting*** continued warming, in what seemed to be alarming and damaging increases. This is shown as follows.[57]

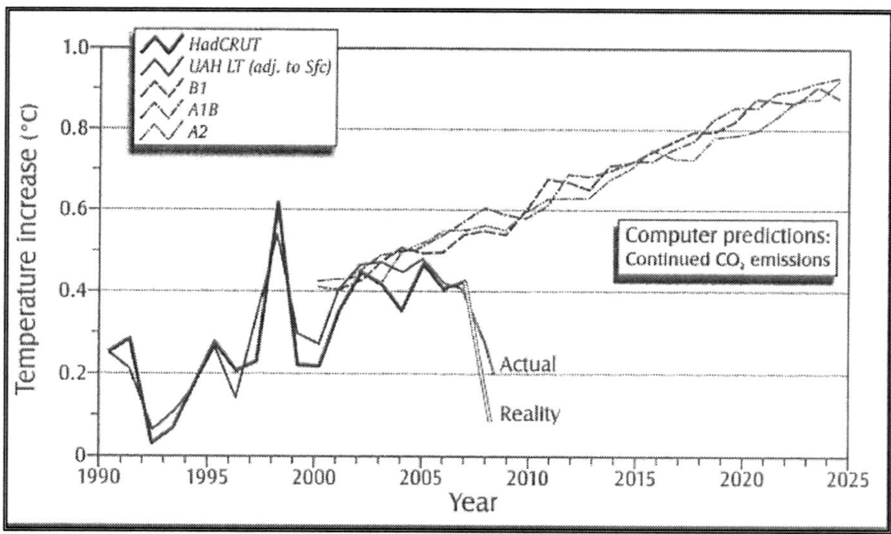

Because these computer trends seemed alarming, the Intergovernmental Panel on Climate Change (IPCC) predictions for the United Nations made a bad mistake and the postmodern/neo-Marxist politicians led by Al Gore used this as a political agenda calling for changes significantly

[57] Ibid., 11.

affecting nations and peoples. The postmodern politicians jumped to the conclusion the warming temperatures were caused by use of fossil fuels used by the industrialized nations and conceived the scheme of forcing the rich nations by charging them for CO2 emissions which should be paid to the United Nations and given to developing nations. They present themselves as saviors to equalize all developing nations. However, from about 2003 to 2004, there was a drastic unexpected and unexplained drop in temperature all over the world, revealing the failure of the computer models. [58]

The mistake from the computer models became the political tool of the United Nations from the IPCC which is given in the addendum. The anomaly of melting sea ice has been recorded. The National Snow and Ice Data Center (NSIDC) has kept records from 1979-2011 and the graphs are given below and on the following page. The graph of the Arctic sea ice shows that there has been a consistent decline of sea ice in a zigzag fashion and is apparently rapidly disappearing at an accelerating rate. In the northern hemisphere the slow loss up until about the year 2000 accelerated in rate. This could be explained by slow melting from the time of the flood, and the break up of the continents until the albedo and other effects began to accelerate the rate of melting.

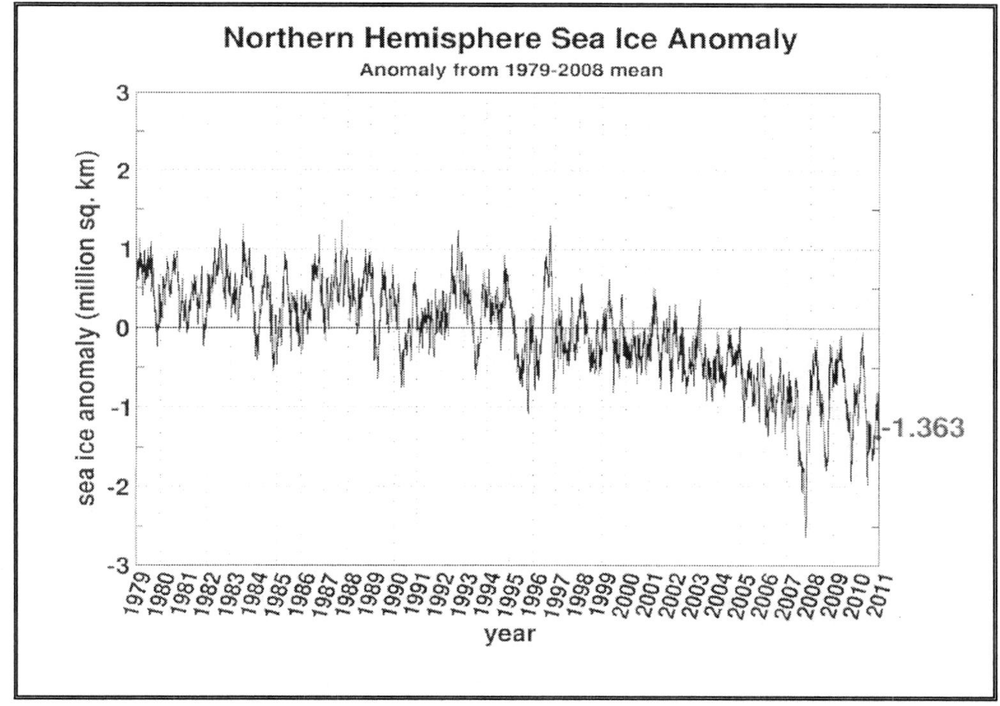

The graph on the Antarctic seems to be reasonably stable in a zigzag fashion and the November 2010 sea ice was 16.90 million square kilometers, which exceeded the prior record of 16.76 million square kilometers in November 2005. The Antarctic, including the sea ice and the continent contains 90 % of the world's ice.

[58] To retain the focus on glacial theory, the various political comments and arguments generated by these findings have been placed in an addendum at end of this book.

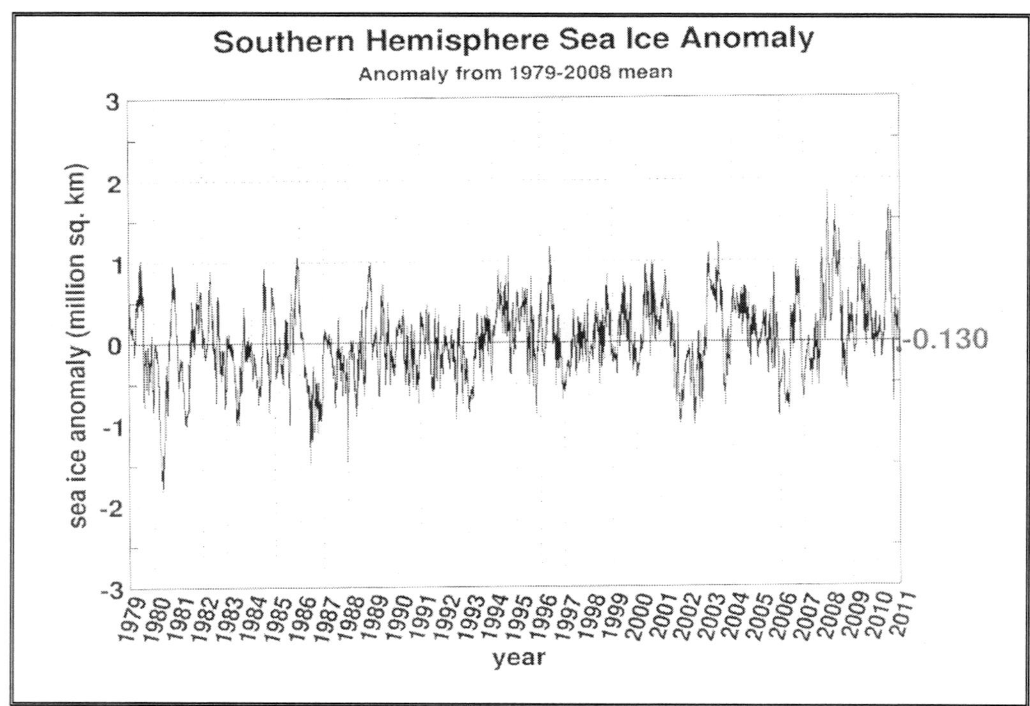

One cannot claim to know all that is occurring in regard to ice loss. The alarm of the world environmentalists is based primarily on the Arctic, but this is not to be attributed primarily to the use of fossil fuels. The melting of the sea ice which is over polar water in the Arctic is explainable as slow melting since the time of the break up of the continents and last cooling over about 6,000 years.

The Antarctica consistency of sea ice is more complicated. Antarctica is the fifth largest continent covering 5.4 million square miles and 98 % is covered by ice about one to three miles thick. The magnetic south pole is there that has been rapidly shifting toward New Zealand and Australia. Sea ice there in austral winter (April to October) is 7.3 million square miles around the continent and melts to about half that in the summer in February. The recent spell of world cooling since after 2003 has affected the Australian side of the continent and is probably affected by the shift of the pole and cold Antarctic Circumpolar Current. The calving of large ice sheets has been mainly on the South American side of the continent with pieces of ice sheets very large, one reported about the size of the state of Connecticut. The size and location of the ice over such a large continent divided by mountains makes it harder to understand how slow warning may be involved since the cold periods after the flood and the division of the continents.

However, the preponderance of evidence fits the data of the Bible and does not fit natural determinism.

Chapter 21
Enlightenment About Tectonic Continental Separation

Introduction

Since the mid-twentieth century, the principles of geology have been greatly altered. In the same way that quantum theory has transformed physics, and the complexity of molecular biology has transformed biological science, new information about the division of the continents reveals the bias in the principles of geology, these being much different than Lyell ever conceived.. Advancements in meteorology, the discovery of plate tectonics and the possibility that cataclysmic forces led to the separation of the continents, have created a major shift in scientists' understanding of geologic history. Not only do these later discoveries challenge Lyell's theories, they give more credibility and plausibility to the biblical record.

History of the Continental Division Theory

Maps outlining the shape and locations of the continents had been available for centuries, but after Columbus opened the way across the Atlantic in 1492, expeditions around South America and into the Pacific followed. General maps were developed and 100 years later the general shapes of the continents were known. Frances Bacon, Comte de Buffon, and E. C. Pickering recognized the way continents might fit together. Two hundred years later, with the help of advanced astronomical instruments, the Ordnance Department of England began trigonometric surveys, which enabled d'Anville and many others that followed to make accurate maps. But there was no serious consideration that they had previously composed one continent that broke into many.

More than any one person, a German, Alfred Wegener, presented the scientific theory and evidence that the continents had formerly been one. Wegener had a Ph.D. in astronomy from the University of Berlin. He was interested in all sciences, especially meteorology to which he made significant contributions. He pioneered the use of balloons to trace wind currents, and was perhaps the first man to suggest the concept of a jet stream. While a tutor at Marburg University, he read a scientific paper that identified identical fossils and organisms on opposing sides of the Atlantic at the point that the continents would have been connected. About the same time, in 1910, Frank B. Taylor proposed a radical new idea that the drifting of the continents caused mountain building. The idea of continental separation was born. Wegener was sent to Greenland to study polar air circulation. In figuring the exact location of an island, he discovered the island had moved from where it had been located by a survey years earlier. He discovered that tropical plants once grew in Greenland and that glaciers once covered equatorial regions of Africa and Brazil.

As a soldier recovering in a hospital during the First World War, Wegner conceived the theory of continental drift. In 1915 he wrote *The Origins of Continents and Oceans* and later became a professor at the University of Grog in Austria. He named the one continent Pangaea, saying it had broken up and drifted apart some 300 million years ago. He realized this theory was contrary to existing geological and physical science. He wrote, "We have to be prepared always for the possibility that each new discovery, no matter which science furnishes it, it may modify the conclusions we draw." Wegener's theory was contrary to previous uniformitarian geological work. Even though he embraced gradualism, his ideas appeared to require cataclysmic events. The theory of one continent breaking up into the present continents, drifting apart and colliding, was so revolutionary that it was stillborn with geologists. Uniform determined scientific bias was closed to it, as Jon Erickson's statement reveals,

> The major problem with early theories dealing with the separation of the continents was that the phenomenon was thought to have commenced *early in the Earth history*. Therefore, scientists were forced to devise complex theories to account for the similarities among plants and animal fossils that have been separated by oceans for eons. It seemed highly unlikely that so large a variety of species could have evolved along parallel lines in such diverse environments over such a lengthy period.[1]

Wegener's theory did not have a satisfactory explanation for the breakup and movement of the continents. As a uniform determinist, he simply attributed it to gravitation and centrifugal force—the rotation of the earth as it affected the continents. Also, Wegener was so certain his theory was right, and he was so zealous to prove it, that he sometimes used uncertified data and surveys to support his arguments.

Resistance and Conflict from Established Uniform Gradualism and Such Major Change

Wegener's view of mountain building and of the spread of plant and animal life ran contrary to the old theory of evolutionary views. It was especially contrary to the prevailing idea of land bridges across the continents, by which the slow movement of animal and human life was explained. Scientists that believed in divided continents were now confronted with the idea of one land mass, which negated major theories that many scientists had developed about mountain building, animal migration, and more.

The uniform determined theory required major changes to make Wegner's theory credible. Jon Erickson said,

> Once a theory is accepted as a fact by the scientific community, and in this case, land bridges, it becomes a solid doctrine... Geologists were caught in the middle of a scientific revolution in which hard ***evidence was simply ignored*** for fears painstaking research would have to be thrown out.[2] (Emphasis added)

Because Wegener's theory suggested cataclysm, along with matching animal and plant life on various continents, it threatened unbelieving scientists' entire gradualistic uniform developmental theory of the universe. So, Erickson said, "When Wegener died...in 1930 the continental drift theory largely died with him."[3] But as more evidence surfaced a generation and a half later, scientists sought diligently for an explanation for continental breakup and drift that included slow uniform gradualism. New evidence emerged supporting Wegener's theory, but the resistance was so strong that secular scientists resorted to redefining the data to make it fit their preconceived theories

The new definitions limited scientists to observing only forces that could be found in connection to tectonic plates. Geologists now say, "*Continental drift was one of many ideas about tectonics proposed* in the late 19th and early 20th centuries. The ***theory has been superseded*** and the concepts and data have been incorporated within plate tectonics."[4] (Emphasis added) Since Wegner, other geologists' and physicists' research endorse a gradualism that might have contributed to the division of the continents.

---

[1] Jon Erickson, *Plate Tectonics, Unraveling the Mysteries of the Earth,* (New York: Facts on File, Republica Books, 1999), 7.
[2] Ibid., 13.
[3] Ibid.
[4] *Wikipedia, Plate Tectonics*, under "Historical Context, Continental drift."

Time and space do not permit an exhaustive analysis of all the theories surrounding the current location of the continents. We will limit this discussion to the main ideas, which rely, primarily, on the use of modern instruments for measuring sound, studying radioactive elements, studying the earth, measuring magnetisms and directions, space probes, satellite observations and pictures with computer evaluations and others. A great deal of science in the geophysical year (ca. 1960) focused on the theory of the continental breakup to test it beyond the early small evidence and claims. Amazing data emerged motivating further study and openness. All of this has thrown tremendous light on the idea of the continental drift theory and of plate tectonics, part of what is real science and other that is speculation to support gradualism.

Confirmation of Continental Division and Drift Was Produced Out of Scientific Research

New Efforts to Understand Earth's Contents and Operation of Mantle and Crust

About mid-twentieth century, Yugoslavian Andrija Mohorovicic produced some evidence of a division between the mantle and the lithosphere or crust, known as the Mohorovicic (Moho) discontinuity. The idea of the Moho encouraged more study of the continental drift. Major efforts were made to dig through the crust to the mantle. But such a major task was abandoned in 1966. In 1970, Russian geologists began a venture to dig a cantaloupe size tunnel to Earth's center, but after 22 years they had to stop at 7.6 miles, only 0.2 percent of the way to the core. They were "hoping to learn more about Earth's enigmatic insides… The rest of earth's interior remains as frustratingly out of reach as it was three centuries ago."[5] Additional facts are emerging from the NASA's Gravity Recovery and Climate Experiment (GRACE) experiment. Using satellites, it measures mass by gravity, enabling us to see under water, water movement, ice, and its density and height. Much about the earth remains a mystery, and exactly how continental division occurred may never be known.

British geophysicists Dan McKenzie and Robert Parker unified many concepts into one theory of continental drift. In 1967, they first publicized their theory that the currents of the mantle interior were joined to the concept of the sea floor spreading to produce new crust. They presented a comprehensive model to explain the process. They marked the boundaries of plates as earthquake zones and began plotting the movement of the earth's crust. They associated volcanic activity and much radical mountain building with continental breakup and plate collisions. Jason Morgan at Princeton University supported this view, seeing the continents as movable plates 60 miles or more thick.

A general description of the earth is now thought to be is as follows. The earth seems to contain an inner and outer core, and a larger mantle. The mantle is composed of the lower mantle (1,200 miles thick), the asthenosphere (100-300 miles) with low density rock and seduction zones. Above is the lithosphere (30 to 200 miles thick, or others say 62 to 124 miles thick) with the ocean plates about half as thick as the continental tectonic plates, including the crust (3 to 40 miles thick) which is like the peel on an orange. The upper lithosphere rides on the asthenosphere, which has relatively low viscosity and sheer strength. The ocean crusts usually have a crust of basalt rock about one-third as thick as the continents that are composed of lighter granite rock. These are estimated by various means of measurement and lack absolute certainty. The plates have three types of boundaries: (1) *transform*, which grind past each other; (2) *divergent*, where two plates slide apart; and (3) *convergent*, when two plates come together and either are seduction plates, one

[5] Susan Kruglinski, "Journey to the Center of the Earth," *Discover*, June, 2007, 54-56.

going under the other, or collision plates. Geologists continue to learn how these plates operate, hoping that gradualism offers an explanation.[6]

One theory suggested a ***preliminary building*** of the continents, including a propensity for plate units in Earth formation and the later breakup. This held out the possibility of a more cyclic uniform theory. Ward and Brownlee, from the University of Washington, speculating about the early earth, believe "The Earth underwent a 'continental explosion' that resulted in a rapid formation of land area." But they admit, "Many lines of evidence suggest that by far the greatest growth ***took place rather rapidly...***," which for them is two to three billion years.[7] There are several theories of what preceded Pangaea. It is presumed that earlier plates united with a preliminary Rodena that had three parts—Laurasia, Gondwana, and perhaps, a Congo craton. This, of course, corresponds to the biblical act of creation. Some believe the continental stress lines were providentially predetermined in creation, as suggested in Genesis 1:9, Psalm 104:5, 6, and Job 38:4, 8. A theory of initiating cause for the division of the continents with evidence is to be presented for the process. But neither science nor Scripture offers a definitive explanation.

Search for How Plates Could Break Up and Drift by Slow Gradual Forces

Although they do not fully understand the entire process, for many years scientists have theorized that the continents are composed of a low-density mass of granite that is lighter and floats on a thin high-density, heavier bed of basalt in the oceans' crust. Radioactive elements generate heat and liquid, low-viscosity convection cells in the mantle caused rising currents that move the brittle outer layer or plates. This is a modification of Arthur Holmes' theory of boiling water. The magna then enters a crack in the surface of two plates, creating a conveyer belt-type action. The rocks change: basalt moves, absorbs gabbros, and expands; granite absorbs water; some sinks and melts again. There are also many mid-plate hot spots (e.g. Hawaii-Emperor).

Secular geologists suggested the following forces are at work: (1) ***Dissipation*** of heat from the mantle; (2) ***Excess density*** of ocean lithograph sinking in subduction zones; (3) ***Friction*** from basal drag and slab suction; (4) ***Gravitational sliding*** (called by a misnomer "ridge push'); (5) ***Slab pull*** (cold plates sinking into mantle trenches); and (6) external forces such as ***tidal friction*** from the moon caused by the earth's rotation. Since little is known about how these forces work, much is assumed.

New technologies have unveiled more specific data about how these forces work. As a result, the theory has shifted from an earlier assumption that ***Earth's gravity*** was the main force, to a belief that internal ***convection currents of the mantle*** (about which little is known), along with the force of ***ocean spreading*** are responsible for plate movements. Alvin Silverstein comments on these new findings on seafloor spreading, saying, "Some scientists now believe that this, rather than convection in the mantle, is the main driving force in plate tectonics."[8] Japanese geologist Seiya Uyeda thinks that ***gravity and subduction of the colder denser ocean plate*** has dragged the rest of the plate.[9] The Atlantic ridge is forming much of the new crust and the Atlantic Ocean is slowly growing. On the other hand, the Pacific Ocean is slowly getting smaller, with Asia pushing eastward and Australia pushing northeast toward California. Some think that the Atlantic ridge may have been the source of the breakup of Pangaea. The expansion of the ocean floor by heat from the edges of plates is poorly understood; some scientists attribute it to the Atlantic ridge and

---

[6] Refer to Chapter 10 on the solar system for more mysteries of the deeper parts of the earth.
[7] Peter D. Ward, Donald Brownlee, *Rare Earth: Why complex Life is Uncommon in the Universe,* (New York: Copernicus of Springer-Verlag, 2000), 216.
[8] Alvin Silverstein, *Plate Tectonics*, (Brookfield, Conn. Twenty first Century Books, 1998), 26.
[9] Ibid.

other similar sites. That does not explain the lack of continuity of the ridge in North and South hemispheres, much less the vast differences in the Indian and Pacific Oceans. But these are some of the efforts to naturally explain the continental breakup and tectonic actions, using a gradual naturalistic theory.

New Mysteries Were Found and Forces Were Discovered that Are Inadequate

All these efforts to find out what happened and how—apart from a major catastrophe—have yielded much mysterious information. They have also unveiled much that cannot be explained by uniform law. These new findings raise additional questions. Scientists are learning that the crust is thick in some places; thin in others. Some of the ocean floors are remarkably 390 feet lower than average, and others are 300 feet higher. Sections of ocean water mysteriously move about as gigantic hockey pucks. Some sections of ocean have very little crust. The basalt rock on the ocean floor is different from continental rock, granite, and andesite, and there are mountainous chains of rock. These strange, diverse findings are not easily explained by natural forces. Also, it is certain that higher forms of animal life existed when the breakup occurred, which limits the time for these slow forces to work.

Jon Erickson comments about the ***inadequacies of all the long-acting uniform forces*** to explain a sudden breakup. He said,

> Despite the mounting evidence supporting continental drift, skeptics continued to doubt the breakup and drift of the continents and ***questioned*** *whether the currents in the Earth's interior* ***were powerful enough*** to propel the continents around. Even some supporters of the continental drift theory thought this energy source might ***not be sufficient*** and suggested that ***additional mechanism***, such as gravity, were needed to move the continents. Furthermore, if Earth processes were essentially uniform throughout geologic time, why did the breakup of the continents happen ***so late in the Earth's History?*** And were there earlier episodes of continental collision and break up? Regardless of these objections, the overwhelming majority of Earth's scientists accepted continental drift as a scientific certainty.[10] (Emphasis added)

Others also admit the findings are still not sufficient, saying, "The sources of plate motion are a matter of intense research and discussion among earth scientists."[11] The relatively small forces, some operating locally and sporadically, are totally inadequate to move huge continents of rock 60 to 120, or even 300 or more miles thick; nor can they explain the melting of rock of the plates in breaking up ***without any known large initiating force***. Scientists' limited explanations for continental drift are no more compelling than their theories of glaciations, using inadequate astronomic data or their explanations for the complete absence of giant ice sheets over much of the earth. The forces these scientists use to explain plate tectonics are like expecting six blind mice to pull a mature elephant with string. Erickson and others honestly described what was obviously a ***struggle to find a regular repetitive law to fit in with uniform gradualism.*** They struggle to avoid the possibility of a catastrophe. This question of sufficiency of the mechanisms for causing the break up and spreading as conceived was obviously being questioned as adequate, even with the addition of millions or billions of years to increase chances of it happening.

---

[10] Jon Erickson, *Plate Tectonics,* 17.

[11] *Plate Tectonics,* Wikipedia.

Proof from Unexplainable Location of Animal and Plant Life, Rocks and Minerals

The new discoveries gave credibility to old theories that scientists now took more seriously. For instance, *Lystrosaurus*, which is an unusual mammal-like reptile with a tusk that points downward, has been found in the mountains of Antarctica, China, India and South Africa. This animal is a unique-looking ***freshwater*** reptile. It is impossible to conceive of how this reptile could have crossed the saltwater oceans, and even the land bridges, to make it to all of these continents. Evidences of opossums were discovered in Central Siberia, North and South America, and Australia. Mesosaurus, a three-foot long reptile that lived in ***shallow freshwater lakes***, also has been found in Brazil and South Africa. It's unlikely it could have crossed the oceans. Fossils of the late Paleozoic fern *Glossopteris* have been found in Australia and India, but are missing in the northern continents, which suggest there were two landmasses: one in the Southern Hemisphere and another in the Northern. Though there are differences between them, monkeys are found in Asia, Africa, and South America. All kinds of plant and animal life seem to have had common ancestors in various continents and dispersed from a common area. In November and December of 1960, after studying the evidence from the Geophysical Year, Dr. Edwin H. Colbert, curator of vertebrate paleontology at New York museum said the 450 reptile and amphibian fossils found proved these could not have migrated and were divided and distributed by a break up of Pangaea.

There were similarities in rocks as well as mountain ranges. The Cape Mountains in South Africa are very similar to the Sierra Mountains south of Buenos Aries in Argentina. The mountains of Canada, Scotland, and Norway are also similar. Rocks of the same type and age are laid down in the exact same order in these various continents. We have already seen glaciation tills and grooves, as well as ancient rocks excavated by boulders, correspond in the continents, as though they were joined. Moreover, the lines of ice flow were strangely ***away from the equator*** and toward the poles—contrary to naturalistic theory. If the continents were in their current configuration, this would be reversed. Also, the ice moved across a single landmass, radiating from a glacial center near the South Pole. Erickson thinks ten composite rock types, called erratics, also match up with rocks on the opposing continent. Glacial deposits, which were overlaid by basalt lava flows, also tend to match. Coal beds in disparate locations, contain similar fossilized plants. The veins of various minerals in different mountain ranges match as they would if the continents had been joined. Facts are clear—the continents were once joined—***after glaciation***.[12]

Surprise Evidence of Paleomagnetism Traces of Continental Movement

As Earth's core rotates, it is like a generator, creating a magnetic field, which can then be measured in the iron-rich deposits from lava flows on the various continents. As the magnetic, iron-rich lava cools, it locks the existing magnetism in place, revealing the direction of the magnetic field at the time it settled. With the introduction of magnetometers in the 1950s, scientists expected magnetic indications to be entirely toward what we know now as the North Pole. The readings of Earth's directions, according to the magnetic iron in lava rocks in England, India and Eurasia, were shocking. Erickson reports,

> The North Pole had wandered some 13,000 miles over the last billions of years from Western North America, across the northern Pacific Ocean and northern Asia, finally coming to rest at its present location in the Arctic Ocean.[13]

---

[12] Erickson, *Plate Tectonics*, 7-10.

[13] Ibid., 15.

> [When the] similar magnetometer measurements were taken in North America, however, the results came as a complete surprise. Although the polar paths derived from data on Eurasia and North America both were much the *same shape* and had a *common point of origin* at the North Pole, the curves gradually veered away from each other. ***Only by hypothetically joining the continents together*** *did the two curves overlap*. Thus, in their efforts to disprove continental drift, scientists inadvertently provided the strongest evidence in its favor."[14] (Emphasis added)

The evidence for the break up of the theory of the continental break up presented by Wegner had grown to scientific certainty. The magnetic evidence added evidence of the paths of movement. Erickson's graphs of magnetism and continental drift are given on this and the next page. The first diagram shows the lines of magnetism, indicating the drift of the magnetic North Pole. The second diagram indicates the magnetism found in India. The rotations and final location demonstrate movement that had to be involved in the continental breakup and movement from East Africa. This cannot be explained by ocean floor expansion or gradualism. Therefore, these magnetic evidences, the life forms, and other evidences, and every effort to find a naturalistic theory of the division and drifting of the continents point, instead, to a relatively recent occurrence, when higher forms of life existed.

This was a shattering discovery for modern scientists, presenting facts contrary to Lyell's geology and threatening biological findings. The time of the break up of Pangaea had to be in recent history because of the mammalian life forms, the evidence of migration of animals and man, the indication of glacial overlay and other like data that was astonishing. Much of speculation and imaginative ideas of scientists about how evolution had worked over millions of years must now be redone to conform to the evidence that was produced and is accumulating.

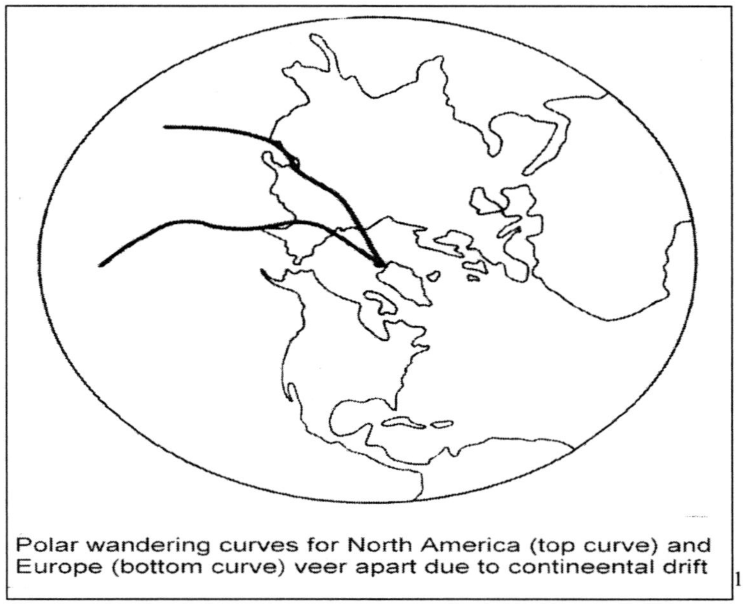
Polar wandering curves for North America (top curve) and Europe (bottom curve) veer apart due to contineental drift [15]

[14] Ibid., 15, 16.

[15] Jon Erickson, *Plate Tectonics,* 15.

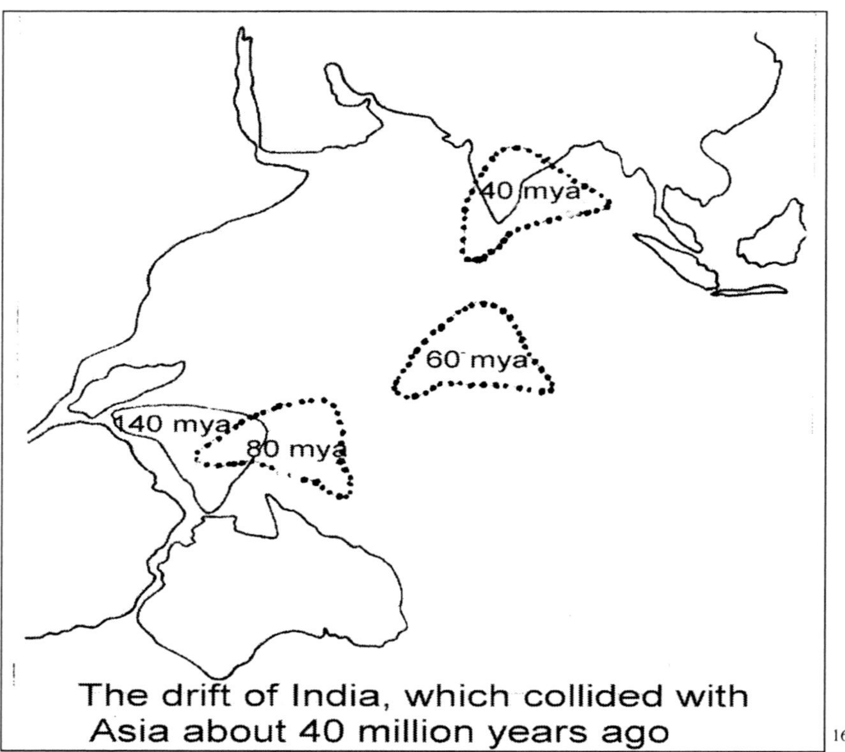

The drift of India, which collided with Asia about 40 million years ago [16]

Objective science was radically changing not only what had happened and how it happened, but when it happened. There were formerly unseen and not understood forces at work that deterministic science had not considered. These dynamic forces causing the division of Pangaea, and formation into specific giant plates of great movement were not found or understood. The search for the answers after the Geophysical Year became a major objective for new understanding.

Discovery of Unknown Dynamics That Allow Continental Edges to Break and Move

Finding Silica's Extreme Melting and Runaway under Stress

The idea of one huge land mass breaking up, drifting apart, and forming the continents as we now know them is such a major event that it is hard to believe that it occurred. But modern studies of rock functions suggest some possibilities. Scientists discovered that at preexistent stress lines in the lithospheric slabs ***a thermal runaway can occur.*** This is such an amazing phenomenon that the rather complex process is described for those who need evidence. The ground or crust of the earth, as we know it, seems inflexibly hard and impossible to break up and become moveable like seemingly floating rafts on a lake.

Since the 1960s scientists have discovered that the viscosity of mantle materials changes at varying stress points and temperatures. Not until the last part of the twentieth century did scientists

---

[16] Ibid., 14.

develop a mathematical solution for this situation. John R. Baumgardner of the Los Alamos Laboratory said,

> The crucial final piece of the puzzle has come from laboratory experiments that carefully measured the way in which silicate minerals deform under conditions of high temperature and high stress. These experiments reveal silicate material can ***weaken dramatically***, by factors of a billion or more, at mantle temperatures and for stress conditions that can exist in the mantels of planets the size of the Earth.[17]

Baumgardner referred to the fact that in 1963, I. J. Gruntfest showed a layer's temperature and deformation can increase without limit, or runaway, when subject to constant, applied sheer stress, and a viscosity with Arrhenius temperature dependence. In the early 1970s, O. L. Anderson and P. C. Perkins investigated the possibilities of thermal runaway of lithospheric slabs in the mantle. They found an average temperature at least 1,000K lower than the upper mantle, but with a bulk chemical composition several percent denser. Thus, the surrounding upper mantle rock had a natural ability to sink. Baumgardner said, "As a slab sinks, most of its gravitational potential energy is released in the form of heat in these regions of high deformation. ...The weakening arising from heating can lead to an increased sinking rate, and increased heating rate, to greater weakening."[18] ***Once started, this can result in runaway.*** In recent years, scientists have discovered that the strength of materials depends not only on temperature, but the deformation rates that differ by mechanisms such as dislocation creep, plastic yield where there are higher levels of shear stress, or a combination of deformation-rate-weakening. Until recently it was not understood that "these weakening mechanisms can reduce the silicate strength by ***ten or more orders of magnitude without the material ever reaching its melting temperature***."[19] (Emphasis added)

Y. S. Yang at the University of Illinois developed numerical ***methods for modeling*** and investigating this runaway mechanism. He showed that a matrix-dependent transfer multi-grid approach allows one to treat such problems with a high degree of success. He applied this to 3D spherical shell geometry and to a simpler 2D Cartesian version, capable of much higher spatial resolution. This multi-grid solver gave a breakthrough in treating large local variation in rock strength, and allowed the complete modeling of the runaway process. Baumgardner used this for a plastic flow regime that "had not been included in previous efforts to model the runaway process."[20] These have been used in laboratory experiments to plot primary deformation regimes for the common mantle mineral olivine. Baumgardner presented three snapshots (the following are the first two) from a 2D calculation to show the progressive temperature range and log viscosity, using arrows to denote flow velocity scaled to the peak velocity "unmax." Using the 2D result, he presented a horizontal resolution at the earth's surface of about 120 km, in an approach to solve equations of mass and energy conservation and a balance of forces for each cell in the computational grid. Concerning this process Baumgardner said,

---

[17] John R. Baumgardner, *Catastrophic Plate Tectonics*, (Pittsburgh, PA: Publish in Proceedings of the Fifth International Conference on Creationism of the Creation science Fellowship), 113-126, 2003. It contains a review of developments from 27 footnotes by leading authorities in scientific papers and journals.

[18] Ibid., cf. Baumgardner's footnote [2] by Anderson and Perkins, *Runaway Temperatures in the Asthenosphere Resulting from Viscous Heating.*

[19] Ibid., cf. Baumgardner's footnote [5] for his study, *Numerical Simulation of the Large-Scale Tectonic Changes Accompanying the Flood.*

[20] Ibid., 2.

> A special method [is used] for treating tectonic plates at the top boundary of the spherical shell domain ...a set of rules for the particles governs the interactions of the plates at their boundaries. Where plates diverge, new particles are added in a manner that represents symmetric cooling on earth side of the existing plate boundary. Where plates converge, particles are removed to represent subduction if ocean plate lied on at least one side of the common boundary. Where one side is continent and the other side is ocean, it is the ocean plate that disappears. When both sides are ocean, symmetric removal of plate is enforced. If both sides are continent, equal and opposite normal forces are applied to both plates to model continent-continent collision. [21]

Baumgardner's relevant pictorial presentation of the events of Pangaea breakup follows.

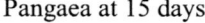

Pangaea at 15 days

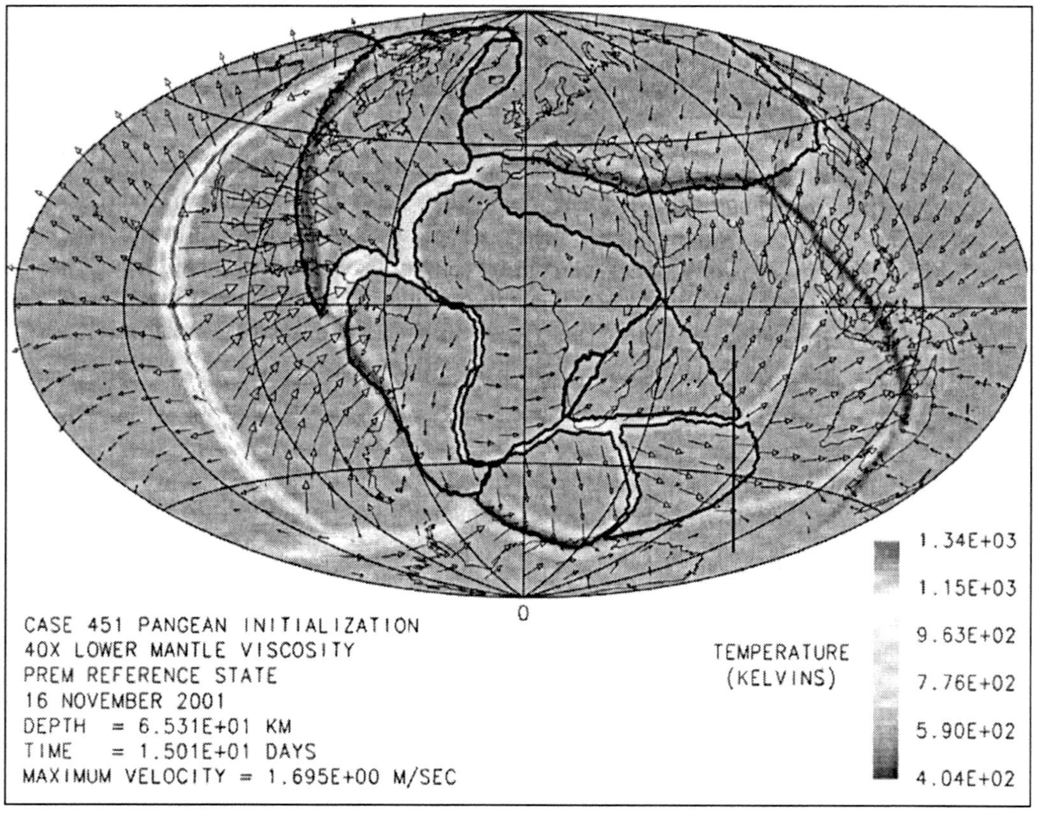

[21] Ibid., 3-6.

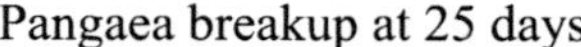

Pangaea breakup at 25 days

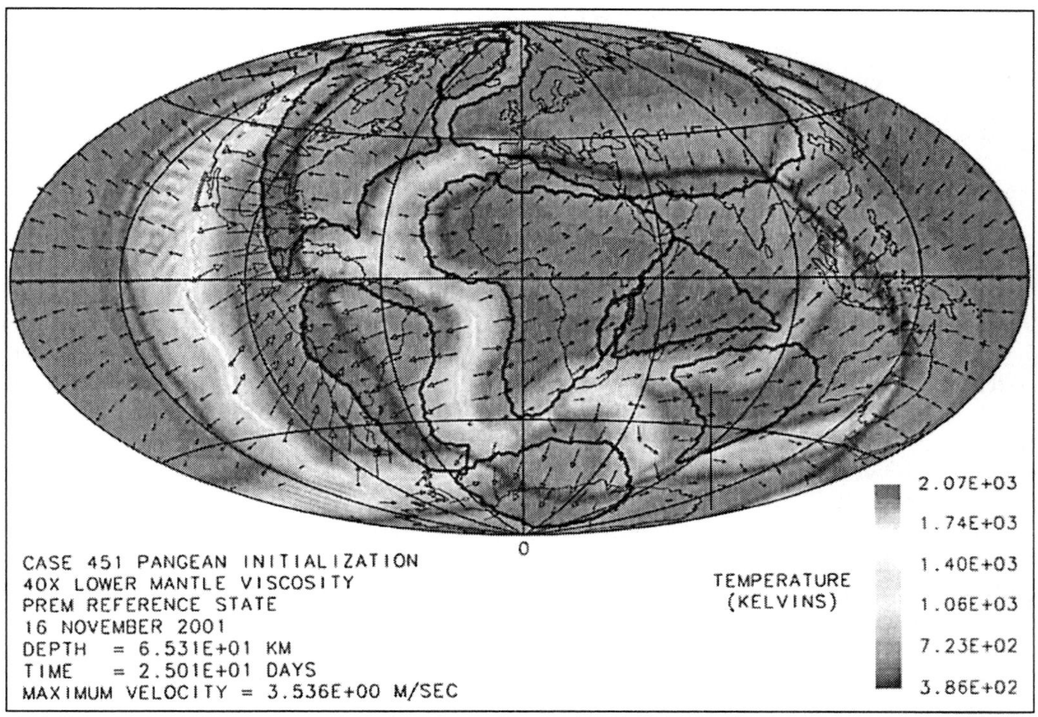

The picture of the mantle viscosity has been omitted as not relevant to illustrate the process. The illustrations indicate temperature, lower mantle viscosity, depth, time, and maximum velocity. He includes an equal area projection of a spherical surface 65 km below the top surface in which grayscale denotes absolute temperature, and arrows denote velocities in the plane of the cross section.[22] Baumgardner created a mathematical pictorial presentation of mantle runaway continental land meltdown in the Pangaea breakup. He attributes ***the shapes*** of the continents involved in the breakup as ***an initial condition***, which he derived from the present continental configurations. Baumgardner's representations are presented on the following page to illustrate how the science was applied to the breakup of the continents, not as a complete technical explanation. The reader ought to realize that some scientists now understand how the earth's mantle did undergo such a meltdown at plate boundaries.

The Huge Initiating Force Producing the Stress for Runaway Was Not Named

No one ***uniform force*** or all the forces mentioned in uniform natural theories seem sufficient to cause such a momentous event as the earth's continental breakup. Neither gravity as we know it, currents of the mantle nor ocean spreading offer adequate explanations. The theory that the Atlantic ridge and seafloor spreading were the main force raised questions for naturalistic scientists. If that is the primary force driving the continent apart, why was the division at the

---

[22] Ibid. these figures of 3 (a) (b) are on pages 7, 8.

center of the land mass? Why did it take so many millions of years for the continents to begin to divide? Is there evidence that other forces were involved? Did there not have to be some cataclysmic force to initiate the break up and land meltdown, beyond the forces that are slowly moving things today? Sylvia A. Earle, National Geographic Explorer-in Residence and an eminent authority, has commented on the need for such a force to initiate this. She said,

> Not all the processes still shaping the planet are as obvious as the demolition of Krakatau volcano, but others may be far more spectacular in their ultimate results. Consider the *forces necessary to split the huge landmass* called Gondwana into the components now recognized as Africa, India, Australia, South America and Antarctica. ***How can the uniform gradual forces seen today cause such a division of the earth that drove them into each other?***[23] (Emphasis added)

To repeat: *how can the normal uniform gradual forces cause such a division of the earth and move the continents?* Common logic suggests that a crust, perhaps miles thick, was not so thoroughly broken up by regular currents that had been circulating in the mantle for thousands of years. Some powerful impact or impacts had to occur to demolish the one continent of Pangaea into the many tectonic plates that currently exist. Fairly recently, some ***great force had to initiate this momentous*** movement. The theory of uniform gradualism forces scientists to propose inadequate theories. None includes a powerful initiating force. Baumgardner suggests the worldwide flood as a starting point. The force needed to heat up, break up and move the continents must exceed their thermal division and location. The force initiating this event was so great that it drove the plates apart and into each other with extreme collisions. Scientists understand the thermal process, but fail to adequately explain its initiation by uniform natural forces.

Recognition of Need for Giant Initiating Force

Under a uniform gradualistic philosophy, the previous explanations are as good as one can concoct, but even some scientists acknowledge their inadequacy. Earle, a leading geographic authority, says the obvious. ***"Something much more than uniform processes are needed"*** (Emphasis added) to account for what occurred. What could that force be? Uniform gradual theories have thus far been wrong in every area of science. The long epochs required to move the tectonic plates are speculations, based on reasonable sequences that carry little weight. While there is virtually no scientific evidence for long periods of time, there is much evidence for a young earth with unusual and extraordinary forces at work at times. If the reader will assume that the breakup occurred ***by a recent cataclysmic event***, a good scientific case can be made for what happened.

Baumgardner and his team explained the meltdown, but offered no initiating force. He attributes the originating motion to "an initial temperature perturbation within the spherical shell domain." Therefore ***he does not give the beginning of the process for the perturbation*** and admittedly does not describe it. In his word, "The initial condition used for the calculation… does not capture the earliest portion of the cataclysm…"[24] While he believed a catastrophe, such as the worldwide flood is someway involved, the source is not given. However, he later reports the spacecraft Magellan photographed Venus showing similar seduction and runaway events that

---

[23] Sylvia A. Earle, *National Geographic Atlas of the Ocean: The Deep Frontier,* (Washington, D.C.: National Geographic, 2001), 115.

[24] Baumgardner, *Catastrophic Plate Tectonics*, 6.

revealed over 1,000 meteorite craters were involved. This showed he suspected large meteorite involvement as the possible initiator.

Earth's Continental Plates as Understood from Research Today

It is important to see the continental plates as most geologists see them. The illustration below will help the reader understand the various impacts and direction of the plates as they are discussed. The plates, as interpreted at the end of the twentieth century by Silverstein, are given with one addition.[25]

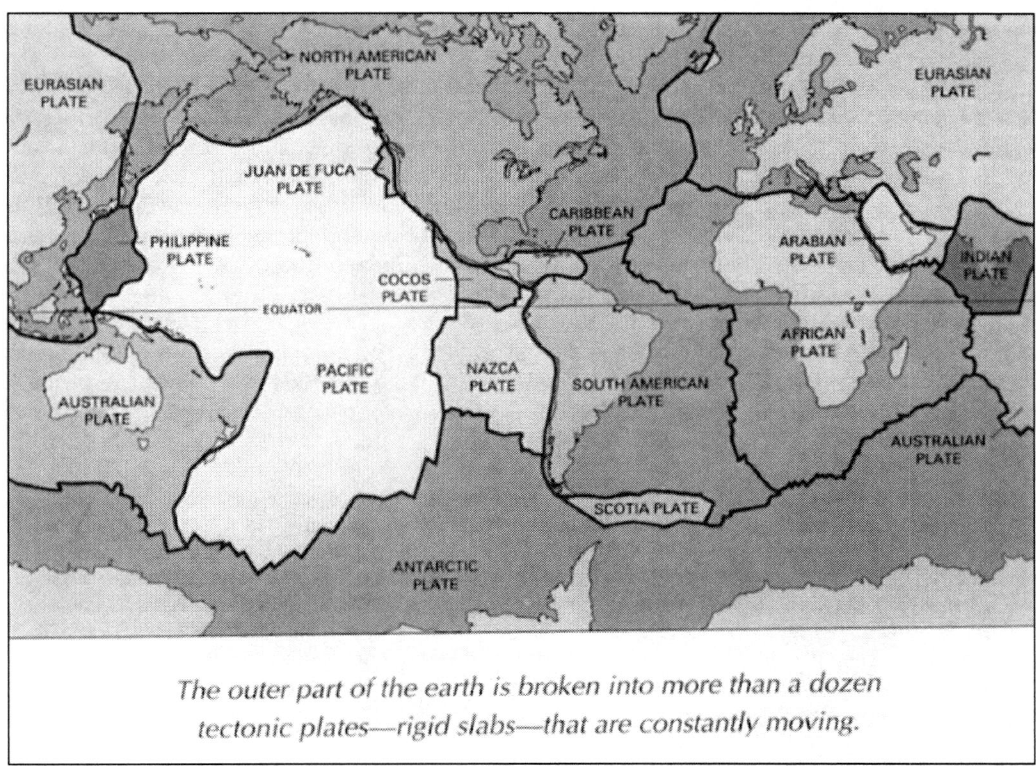

The outer part of the earth is broken into more than a dozen tectonic plates—rigid slabs—that are constantly moving.

Scientists hold different views, but from all the studies, Silverstein's seems most definitive. Satellite positioning and other studies will lead to further refining of these views. The chief difference seems to be related to the area of the Indian Ocean as it affects the Antarctica, and Australian-Polynesian areas. Silverstein has a number of separate smaller plates such as Arabian Plate, Indian Plate, Philippine Plate, Juan De Fuca Plate of North America, Cocos Plate off Mexico, Nazca Plate off South America, Caribbean Plate, and Scotia Plate off tip of South America and Antarctic. Some show a Madagascan Plate. A later diagram suggests how the process of breakup occurred, forming these plates.

---

[25] Silverstein, *Plate Tectonics*, 29.

Celestial Evidence of Irregular Astronoid Bodies as Possible Impact Sources

Until recently, uniformitarian geologists have avoided the cataclysmic view of asteroid impacts. Baumgardner suggests that a growing body of evidence points to asteroid or meteor portions showering the earth at certain points. How reasonable is it to believe asteroids were the force? Is such a force available?

The Titus-Bode law would lead one to expect another planet between Mars and Jupiter, a space now occupied by asteroids and meteorites. In the last three decades of the twentieth century, teams of scientific specialists, using new instruments, have discovered the orbits, as well as the rotation rate, size, shape, origin and composition of asteroids, elements that are more indicative of their composition, and how they heat up. On March 22, 1989, an asteroid passed less than twice the distance to the moon, indicating how close and possible a major impact might be.[26]

More recent evidence shows other possible impacts in the past or future. Scientists have long thought that the Kuiper Belt harbors millions of comets, and that the great gravity of the cold planets, especially Jupiter, can cause disturbances that can come earthward. Scientists have recently discovered ***irregular or Wild Moons*** or satellites in the Kuiper Belt area. These have a mysterious instability with "strange features" that appear "downright unruly… swoop in and out of the plane in which the planets orbit the sun… highly elongated rather than circular paths. Some orbit in the direction opposite to the rotation of their host." Some of these "have colors and orbital characteristics of the Hilda group of asteroids in the outer part of asteroid group—and may be from a collision and fragmentation." [27] Matija Cuk and J. Burns of Cornell believe there must have been a "much larger initial satellite population or a much larger initial population of comets and asteroids than scientists now observe."[28] David Jewitt of the University of Hawaii referring to these irregularities said, "In the last 10 years or so… we have made observations that fire a new and different view of our solar system."[29] Elizabeth Roemer said,

> There are thousands of known asteroids, and new ones are being discovered constantly. About 30,000 asteroids are bright enough at one time or another to be photographed with medium-sized telescopes. Astronomers have studied about 1,700 asteroids, and know the orbits of these small planets well. .... A total of about 20 asteroids have diameters that are *larger than 100 miles* (160 kilometer). (Emphasis added) A few small asteroids move in oval-shaped orbits that at certain times bring them near the earth. …The orbits of the asteroids change because of the attraction of Jupiter and other planets. Over a long period of time, these changes lead to collisions. …The total mass of all the asteroids cannot be reliably estimated because so many asteroids are still undiscovered.[30]

These large Apollo asteroids are known to break up into large meteors. Many of them have hit the earth.[31] Some of the meteors weigh several hundred tons and are known to create craters or basins as large as 400 miles in diameter. But those meteors are small compared to other larger, known asteroids. There are 200 million meteorites visible in earth's atmosphere each day. They add 1,000

---

[26] Curtis Peebles, *Asteroids: A History*, (Washington: Smithsonian Institute Press, 2001), 181, 185, 237.

[27] Ron Cowen, "Moonopolies: the solar system's outer planets host a multitude of irregular satellites, " *Science News*, November 22, 2003, Vol. 164, 328-330.

[28] Ibid.

[29] Ibid.

[30] Elizabeth Roemer, "Asteroids," *Worldbook Encyclopedia*, Vol. 1, (Chicago, Ill. Field Enterprises Incorporation, 1977), 788.

[31] See a listing of the largest, Frank D. Drake, "Meteor, Meteorites," *The World Book Encyclopedia*, Vol. 13, (Chicago: Field Enterprises Education Incorporation, 1977), 365, 366.

short tons of weight, daily, indicating the accessibility of the earth to these objects. Asteroids and meteors can collide with the earth at combined speeds of 44 miles per second and bring temperatures around 4,000 degrees Fahrenheit. Basins, called craters often formed as a result of large meteors hitting the earth. On land, craters do not contain the asteroid because impact blows them all around and away. Evidence is seen in the crater formed by the asteroid impact, and surrounding rocks formed by the heat? There is little evidence if these occurred in the oceans.

Evidence of Meteorite Impacts Dividing the One Continent, Pangaea

Observers note that the geological phenomena on the floor of the Indian Ocean and the nearby southwestern Pacific Ocean are different than those found in other oceans. Many of the differences between these two connecting oceans and other oceans can best be explained by assuming this is the area at which impact and separation began. A picture of this area is given on the next page.[32] They are *divided by the Ninety East Ridge*, which is "the ***longest linear feature in the world.***" It is near this highly unstable area that the earthquake and great tsunami of 2004 occurred. Earthquakes and volcanic eruptions are frequent there. The Indian Ocean and the southwestern Pacific are both marked by islands, ***basins***, "drag marks" caused by plates. See map on following page, dark on map shows continents, the light shows the oceans.

The Indian Ocean has ***seven of the world's largest basins***, all but one of which is about a mile deep. The adjacent ***Western Pacific area has eleven basins***, all but two of which are about a mile deep. Antarctica, not shown on this map, is in the South Indian Ocean, which is a part of this area. The large mid-Indian Basin (# 7) is about 900-1,000 miles long and 400 miles wide with other large basins in the area. Most of these basins slope from the north to the south from about one or two thousand feet deep and generally progressively reaching four to six thousand feet deep. This could be explained by asteroids hitting at an angle. Land craters, such as the Rio Cuarto in Argentina, also show this grading characteristic. If these are basins made by huge craters, most ocean basins would not be visible because of silt coverage and other reasons. Craters formed from asteroids of this size would be so catastrophic that uniform deterministic scientists would not consider it. There is no clear evidence of what caused these basins—impacts or dragging continents, or something else. But evidence continues to grow that major asteroids may have created these craters.

It is possible, and quite likely that the impact of great asteroids initiated the breakup of the continents and other impacts continued the effect. Many believe Pangaea was composed of two major parts—the northern conglomerate, Laurasia, and the southern portion, Gondwanaland. Other divisions and names have been suggested: Northern Rodina and Southern Rodina. The two main divisions are thought to have begun separating on the east side of the one continent, at an opening between them called the Tetheys Sea (now the Indian and southwestern Pacific Oceans). The division sent Laurasia in a north and west counterclockwise direction, and pushed Gondwana south, separating the two main parts across northern Africa and Eurasia about across at the Mediterranean Sea.

---

[32] Sylvia A. Earle, *National Geographic Atlas of the Ocean*, (Washington, D. C.: National Geographic, 2001), 106, 107.

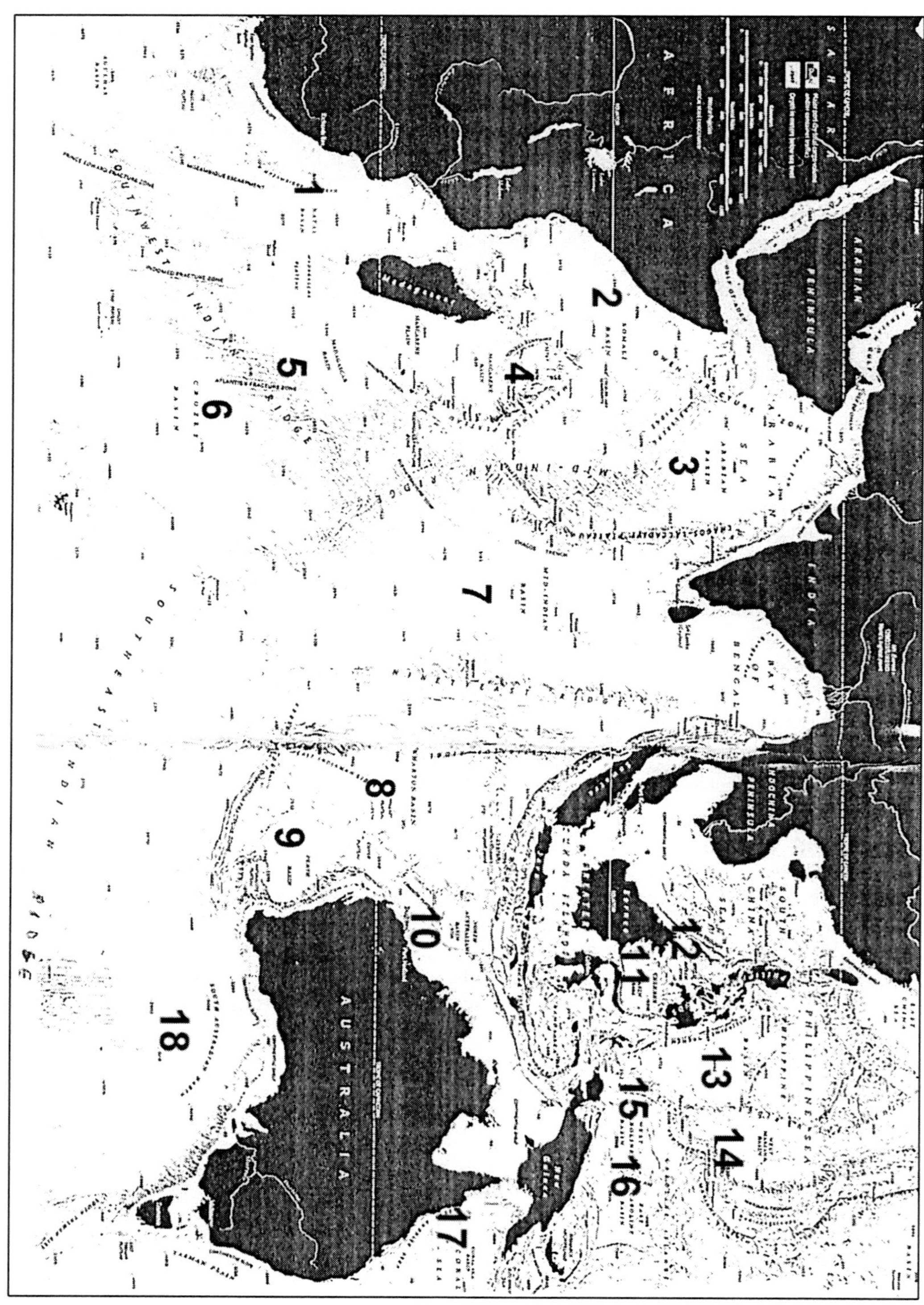
A F R I C A
ARABIAN PENINSULA
I N D I A
A U S T R A L I A
ARABIAN SEA
BAY OF BENGAL
MID-INDIAN RIDGE
SOUTHWEST INDIAN RIDGE
SOUTHEAST INDIAN RIDGE
PHILIPPINE SEA
SOUTH CHINA SEA
INDOCHINA PENINSULA
1
2
3
4
5
6
7
8
9
10
11
12
13
14
15
16
17
18

Evidence of Points of Impacts and Continental Division Beginning in Tetheys Ocean

Ten years ago, little research supported the view that an asteroid impact initiated the continental breakup. The evidence in the Indian Ocean was minimal. The Glomer Challenger had done about fifty cores in the Indian Ocean, which covers about 20 percent of the earth; the Navy had done study, yet, *the National Geographics of the Oceans* exploration as "mostly too sparse."[33]

A growing body of evidence suggests significant geological forces at work in that area, leaving unique qualities. Researchers from the Woods Hole Oceanographic Institute found rich deposits of iron, manganese, zinc, and copper in the Indian Ocean, which includes the Red Sea. Some deposits were as much as 300 feet thick for more than 6,000 feet. The area is also rich in hydrogen sulfate and low in oxygen. Large deposits of titanium have been found on the east coast of India. Meteorites have twice the percentage of manganese as most of the earth. Manganese nodules, which form under high heat, have been found there. Meteorites could well be the source of that heat. The 2004 tsunami left miles of titanium, a high-heat element found in asteroids and meteors, on beaches. We've already noted the Indian Ocean has ***three times the sedimentation rate*** of other oceans. These findings support the theory of a meteor impact in the area.

In addition, Earth studies indicate ***strange factors about the mass in these regions***. GRACE studied and mapped the places on earth with the strongest gravity pull. The strongest gravitational force is found at the equator, due to the earth's rotation. But strangely, "...Gravity's pull is strongest near the southwestern Pacific and weakest just off the southern tip of India."[34] Could this be caused from impacts of asteroids blowing away material and compressing the adjacent ocean area?

In May 2004, scientists reported that Bedout, northwest of Australia (#10 on the previous map) is a 125-mile crater, formed by a "Mount Everest-size meteor." The impact is indicated by melting that produced glass inside crystal and quartz fracturing in multiple directions. These impacted areas support the geologists' claim that the continental breakup and mass extinctions began in the Tetheys Sea.[35]

As Laurasia was progressively dislodged from Gondwana, and slowly the division occurred across North Africa and Europe. Later, the Indian plate was removed by an impact and migrated as shown in the magnetic readings on page 353. The large, later impact dislodged the continental plate of India from Africa, at what is now Tanzania and Kenya, and it drifted north. The impact involved tremendous heat and volcanic flow, followed by runaway melting. In the present area of Bombay, which was the break off point from Tanzania, there are huge Deccan Traps of multiple layers of solidified flood basalt. They are over a mile thick, with lava flows possibly as large as 930,000 square miles, leaving a volume of basalt, estimated to be around 317,440 square miles. Initially, this area may have covered half of India. The Shiva crater, recently discovered off the west cost of India, is thought to have been involved. The discovery of dinosaurs here indicates their survival from the earlier mass extinction. The Ngorongoro Crater in Tanzania, where the break off occurred, is the world's largest unbroken, unflooded volcanic caldera. It involved a giant volcano—2,001 feet deep, its floor covering 102 square miles, estimated to have been fifteen to nineteen thousand feet high—that collapsed on itself. In nearby Tanzania, near Kenya, is Mount Kilimanjaro, the highest peak in Africa at 19,330 feet, with three volcanic cones.

The plate spreading, brought on by the initial impacts, and that broke off the Indian plate, was so unbelievably large, that it shattered and pulled at the African plate, and above the

---

[33] Ibid., 202.

[34] Bjorn Carey, "Map Earth's Fourth Dimension: A gravitational rainbow points to our planet's invisible topography," *Discover*, August 2006, 22.

[35] Charles Choi, Permian Percussion, *Scientific American,* July 2004, 36 cf. *Science,* May 6, 2004.

continental plate. It caused the Rift Valley that begins at the Olduvai Gorge in Tanzania and likely extends to the Dead Sea. Mysteriously, at the point of the disconnect of India is ***a Hot Zone that runs across mid Africa,*** beginning near Mombasa, Kenya, up across north of Tanzania and south of Uganda, down through Zaire and across north Angola to Pointe Noire on the Atlantic Ocean. From this line comes many of the known most deadly viruses such as Ebola, HIV etc. indicating something unique occurred at this location. The mighty Congo River empties into the Atlantic at about this point, and across the Atlantic in South America the great Amazon River empties into the Atlantic. All across the Atlantic in this area is a fracture zone, and the Romanche Gap where the Mid-Atlantic Ridge shifts 2,500 miles from west to east.

The giant impact that broke the Indian plate off from Africa impacted the Asian plate, with the force causing the formation of the Himalayan Mountains and a huge plateau. In the process, this formed the Ninety East Ridge and may have caused the Indonesian and other islands on the east side of the Ninety East Ridge. While the Arabian plate may have broken off from India, it is possible that a northern part of the Cimmerian may have been the Arabia plate that was the last to migrate and hit Asia.

Madagascar is a unique mystery. An asteroid must have divided it from part of Antarctica and pushed the island back toward Africa. Madagascar is extremely different from Africa. The flora and fauna are more similar to plant life in Australia and the far Eastern and Galapagos Islands, rather than Africa. Diamond commented that the common use of some African seeds in Madagascar may be a result of later borrowing. The impact area of Antarctica that formed these islands may have been the now hot spot of Reunion Island. This may have created the Coco and Nazca plates, with their similar life forms, driving them in front of the Pacific plate far to the east.

Other major meteorites are believed to have hit areas of Antarctica, which were then warm and in the south Indian Ocean area (not shown on the previous map). According to an updated article on the Weddell Sea, "It is believed that the breakup of Gondwanaland started in the Weddell Sea." The Weddell Plain is a basin 4,000 to 5,000 feet deep that runs 2,000 miles into the south Indian Ocean area.[36] This impact on west Antarctica would have initiated the division of South America and Africa, as well as the separation and spreading of the South Atlantic Ocean.

Other evidences of a giant ***meteor impact have been found recently in Antarctica*** at Graphite Peak in the Wilkes Land area that fronts the southern Indian Ocean near Queen Mary. Other evidence has been found as far away as New Zealand. Ralph von Frese, professor of geological sciences at Ohio State University, and Laramie Potts, a post doctoral researcher, in collaboration with scientists from other nations and NASA, have identified a giant impact crater in Antarctica. Using gravity fluctuations measured by NASA's GRACE satellites, they were able to peer beneath Antarctica's icy surface. When they overlaid their gravity image with airborne radar images of the ground beneath the ice, they found the mascon (plug of mantle material in the crater) perfectly centered inside a circular ridge 300 miles wide—a crater easily large enough to hold the state of Ohio. It has been suggested that the low- mass density of the Indian Ocean area is related to the Southwest Pacific, as are Antarctica's high density impact areas.

The Wilkes Land crater in Antarctica is more than twice the size of the Chicxulub crater in the Yucatan peninsula in the Mexican Gulf. The Chicxulub meteor was thought to be six miles wide. The Wilkes Land meteor was near 30 miles wide, or four or five times wider initiating a powerful push for Australia and the Pacific plate.

---

[36] Weddell Sea, Wikipedia free encyclopedia article upgraded December 31, 2009.

The Vredefort crater in South Africa is about 25 miles in diameter; it was formed by a smaller meteor which has lost its mascon. Von Frese believes the Wilkes Land Antarctica crater is identified with these continental divisions.

> Approximately 100 million years ago Australia split from the ancient Gondwana subcontinent and began drifting north, pushed away by the expansion of a rift valley into the eastern Indian Ocean. This rift cuts directly through the crater, so the impact may have helped the rift to form.[37]

The asteroid that hit Antarctica at Wilkes Land separated Australia and New Zealand and drove them apart, and moved Australia northeastward toward the Pacific plate. The Pacific plate would have, thereby, been moved northward, meeting eastern Asia and North America and pushing the smaller Cocos and Nazca Plates eastward against Central and South America. They had been divided earlier from Madagascar, going toward Africa for the impact. The impacts in the Tetheys Sea—Indian, southwest Pacific and Antarctic Ocean area—are all related. In order to support their claims of gradualism, scientists place the times of some of these impacts at millions of years ago.

Possible Later Companion Asteroids Continuing the World Wide Continental Drift

After the initial division of the north Laurasia and Gondwanaland, other rifts appear to have continued the division. Photographs generated by space shuttle flights provide evidence.[38] A massive asteroid may have landed in ***Central Ontario*** between the Great Lakes and the Hudson Bay in Canada, moving North America, and creating a crater 900 miles wide. This movement caused such heat that it melted vast amounts of basalt, and formed the granite crust and stable basement rocks driving the Canadian Shield.[39] The ***Chicxulub crater*** could have initiated the Gulf of Mexico, causing the Central American chain linking South America and North America. This crater is believed to have created a basin 180 kilometers (or about 112 miles) wide. It has likely been filled with sediment since that time. Many craters and volcanoes in Central America are associated with this event.

The meteorite craters are unquestionably ***connected to a worldwide catastrophe, such as the continental breakup***. In their article, "The Day the World Burned," David A. King and Daniel D. Durda report a worldwide calamity at the time of the Chicxulub impact.[40] Asish R. Baso and his team from the University of Rochester, and Michael R. Rampino of New York University have reported on the findings in Antarctica. Speaking of "the Antarctic fragments," Rampino said this "was part of ***a worldwide phenomenon***..." (Emphasis added.)[41] While many scientists exaggerate, it is obvious from the data supplied by different teams that various scientists felt they were reporting very large world-affecting cataclysms, which are quite likely related to continental breakup.

---

[37] "Ancient Killer crater found under Antarctic ice," *Ohio State University News Release*, June 4, 2006, cf. Pam Frost Gorder, "Big Bang in Antarctica-Killer Crater Found Under Ice," *Research News*, June 1, 2006.

[38] "Killer Crater; Shuttle-borne radar detects remnant of dino-killing impact," *Science News,* Vol. 163, No. 11, March 15, 2003, 163.

[39] Erickson, *Plate Tectonics*, 21.

[40] David A King and Daniel D. Durda, "The Day the World Burned*," Scientific American*, December 2003, 98-105.

[41] S. Perkins, "Pieces of a Pulverizer?," *Science News*, November 22, 2003, vol. 164, 323, 324; for other evidence see "sediment fragments may be from Killer Space Rock," *Science*, November 21, 2003, et. al.

King and Durda report finding a centimeter of clay laced with exotic elements of soot ***around the world*** —spherical particles of carbon clustered like grapes, which are like residue from forest fires. They refer to the work of Walter Alvarez and Frank Asaro to support their claim that the contents are from an ***extraterrestrial object***.[42] Globally, they report 70 billion tons of residue, which they attribute to a fiery plume coming from a crater and rocketing through the atmosphere, carrying crystals of quartz from as deep as 10 kilometers. They think it swelled to 100 to 200 kilometers of material which fell back to the earth and which may have struck areas of all other continents. They believe these areas were located differently and this occurred ***before the breakup and drift of continents***.[43]

Formation of North America, North Atlantic, and Siberia

The major initiating impacts around the Tetheys Sea caused Laurasia and Gondwanaland to break apart; separating across all Pangaea at what is now the Mediterranean Sea and pushing Laurasia counterclockwise. The Laurasia group was propelled north and counterclockwise causing the western Canadian and Siberian plates to eventually collide with the American plate. The Canadian plate was driven over and down into what is now North America. The pictures of the North American Plate on the following page reveal a complex geography. Part of the Siberian plate was driven across the middle of the North American plate, thrusting the Canadian Shield into it, and pushing other rocks down the middle. This left a deep tongue of differing rocks and orogenic belts, bordered by the Appalachian Mountains on the east and Rocky Mountain chain on the west. This, with the Hudson Bay area meteor, caused the Colorado plateau, and the hot area of Yellowstone near the Cordillera and Wyoming. The picture on the left shows the Paleolgeological origins of "basement" rock. The image on the right shows the age of the underlying bedrock. The boundaries of the Canadian thrust are Wyoming, Yavapai Mazatlan and Grenville. The hot volcanic zone of Yellowstone and the earthquake fault of the Superior rocks generally follow the Mississippi. The northern expansion of the Mid-Atlantic Ridge ends where the impact occurred. This was below Iceland, forming Iceland and leaving Greenland, and North America to the west with clear remnants from the collision and overthrust. Europe was pushed eastward by this separation.

The ***area of Iceland*** is one of the hottest and most volcanic areas of the world. The north Mid-Atlantic Ocean Ridge leads up to and is interrupted with Iceland, which separates Scandinavia, Greenland, and Canadian North America. Iceland has smaller ridges, straits, and sea channels paralleling it. South and north of Iceland indicate meteorite impacts. Something cataclysmic caused the breakup of the two joined continents of North America and Europe after the northern part of the Mid-Atlantic Ridge was formed. These impact meteors, dividing North America and Europe, pushed North America counter clockwise to the west with the intruded Canadian Shield in the area of the United States, and it pushed Europe clockwise to the east.

---

[42] Walter Alvarez and Frank Asaro, "An Extraterrestrial Impact," *Scientific American*, October 1990.
[43] King and Durda, "The Day the World Burned*,"* 102.

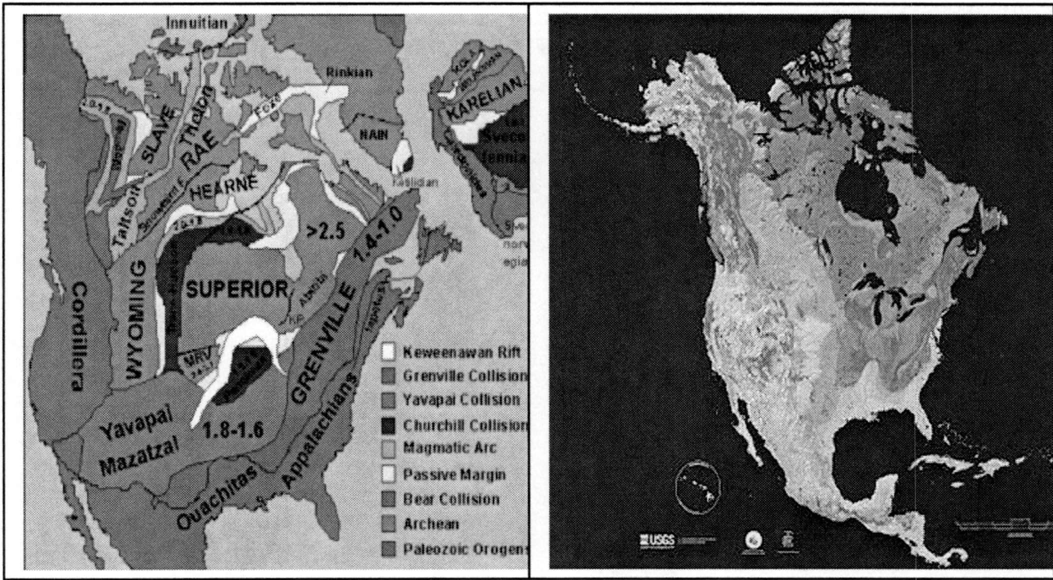

Erickson said, "Most of the continent, comprising the entire area of North America, Greenland and Northern Europe evolved ***in a relatively brief period****....* A major point of the continental crust underlying the United States from Arizona to the Great Lakes to Alabama formed in one great surge of crustal generation ... that has no equal."[44] This can be seen in the graphs above.

The Unique Action of the Pacific Plate Reveals Departure from Tetheys Sea Area

The Pacific Plate is in different condition than all the others. The action of the Pacific Plate indicates it received the mighty force of initiation at Tetheys and the subsequent impacts, causing the Australia-New Zealand push northeast. The northern push against the continents caused the ring of fire of volcanoes that is different from all other oceans. The powerful movement east drove the creation of the Coast Mountains, the Cascade Rang and the Sierra Nevada Mountains in North America and the and Andes in South America. Studies of the volcanoes involved in the "ring of fire," as illustrated on the next page, confirm this. ***Only the Pacific Plate has great seduction troughs or canyons all the way around it***, except in minor isolated areas. Scientists discovered CO^{12} organic forms that are produced on the Pacific Ocean floor were found in the volcanic lava from the ring of fire, indicating this was carried by seduction, down under the heavier continental plates as a result of the collision with the Pacific plate. The seduction of the cold floor reached the hot interior under the crust and promoted volcanic eruptions. These seduction troughs occur all around the Pacific Ocean, producing the ring of fire. Seduction troughs are rare anywhere else.

The Atlantic Ocean, with the pronounced Mid-Atlantic Ridge, indicates its separation was caused by the forces from the Weddell Antarctic impact area in the south and the northern Iceland impacts. These impacts may also be the source of ocean spreading. The Atlantic also is unique in that North and South America both have wide continental shelves along their long borders showing the spreading eastward of the Atlantic. While slight continental shelves occur in other

---

[44] Erickson, *Plate Tectonics, Revised*, 2001, 34.

areas, the wide shelves are clearly unique to the forced separations eastward forming the Mid-Atlantic ridge.[45]

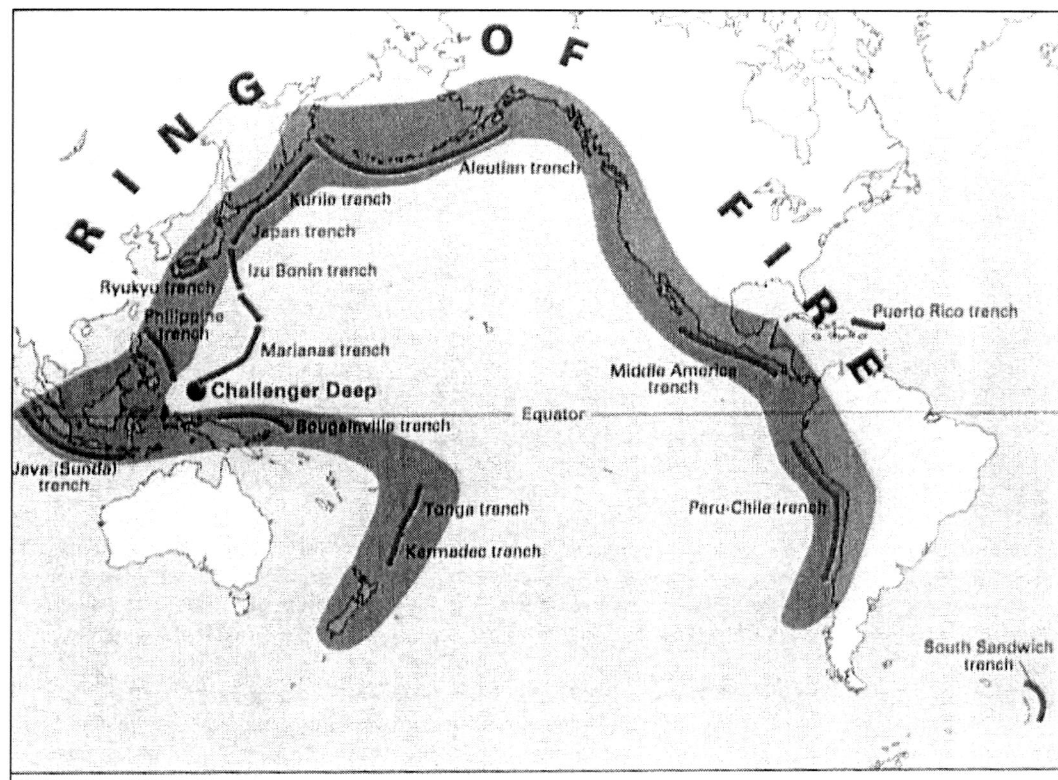

There's nothing pacific (peaceful) about the Ring of Fire, a huge belt of volcano and earthquake activity that nearly surrounds the Pacific Ocean.

A summary description with Graphs of the Process of Division of Pangaea into Continents

The breakup of Pangaea appears to be a large and complicated process that began with an asteroid shower. The asteroids then broke into meteors and were attended by other smaller meteors. More than 42 major land impact craters have been mapped.[46] The progression is illustrated below. On each step I have numbered the area of the meteor or meteorites impacts. Many geologists have suggested illustrations like this, but because they deny catastrophic causes, I

---

[45] The features may be observed as presented by Sylvia A. Earle in the *National Geographic Atlas of the Ocean* in pictures on pages 48 and 49.

[46] Cf. Geology.Com, "Meteor Crater Map."

believe their illustrations to be incomplete, at best. These illustrations do not convey accurate sizes and shapes, but are simply meant to help the reader understand the progression in division.[47]

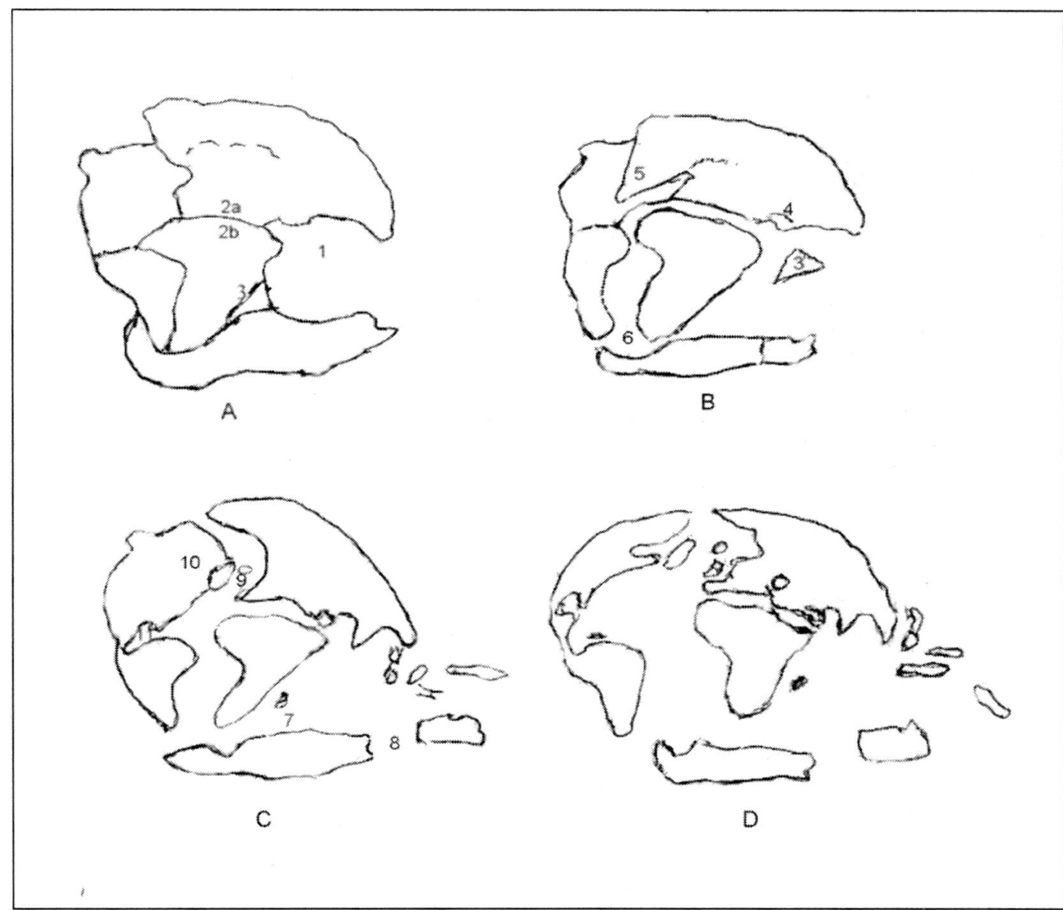

The major event seems certain to have begun (1) in the Tetheys Sea or the region of the Indian and South Pacific Oceans. (2) This initiated the major division of Laurasia, pushing it north in a counter clockwise direction, dividing it from Gondwanaland, and splitting the two across current Mesopotamia and Eurasian areas toward the Atlantic. (3) The Indian Plate was then broken off by another meteor at East Africa, sending it spinning toward Asia as magnetic records show. (4) The Arabian plate was separated and attached to Eurasia. (5) The eastern or Euro and Siberian part of Laurasia was pushed into the North American section, inserting the Canadian Shield, and forcing the separation that formed the Atlantic Ocean. (6) A meteor hit the Weddell, the northern part of Antarctica, dividing and pushing it south from, and dividing off the united South American plate and African plate, breaking them apart and forming the southern Atlantic, pushing Africa in a northeasterly direction and shifting Europe and Africa in a more clockwise direction. (7) This

[47] I have adapted from Erickson, *Plate Tectonics...,* 16.

was attended by a meteor hitting Antarctica at what is now Resume Island, breaking off a part which is now Madagascar, sending it northwest toward Africa, other islands into the South Pacific, and the Nazca and probably the Cocos plates eastward. They were then pushed by the expanding Pacific plate. (8) Then the eastern Wilkes Area of Antarctica was hit by a meteor, splitting off Australia and sending it north eastward, pressing the Pacific plate northward and slightly east. (9) The Laurasia continent was then hit somewhere near Iceland, at what is now the northern site of the Mid-Atlantic Ridge, breaking off and leaving Greenland and Canada to the west, and dividing and pushing Europe even more strongly to the east in a clockwise direction. (10) This was followed by the Canadian meteor near the Hudson Bay. (11) The Chicxulub impact, which formed the Gulf of Mexico, and the Mexican and Central American string of nations with South America. These impacts produced the modern continents as in diagram D. There may have been smaller meteorite impacts then or later which may or may not have been involved in the continental separation. There have been over 24 on land and no one knows how many in water. The events precipitating continental division were highly complex and may never be accurately traced. This progression seems to give a reasonable explanation for what occurred.

Satellite Pictures of Plate Movements of Today Seem to Fit this Sequence

Based on ideas of other geologists, I have diagrammed the general process as follows, giving the steps by numbers and the major meteor impacts in letters. Other evidence points to the accuracy of this interpretation. Satellite studies also uphold these general directions of plate movements. Below is a picture of the present plate motion ***based on Global Positioning System (GPS) satellite*** data from NASA JPL. Vectors show direction and magnitude of motion.

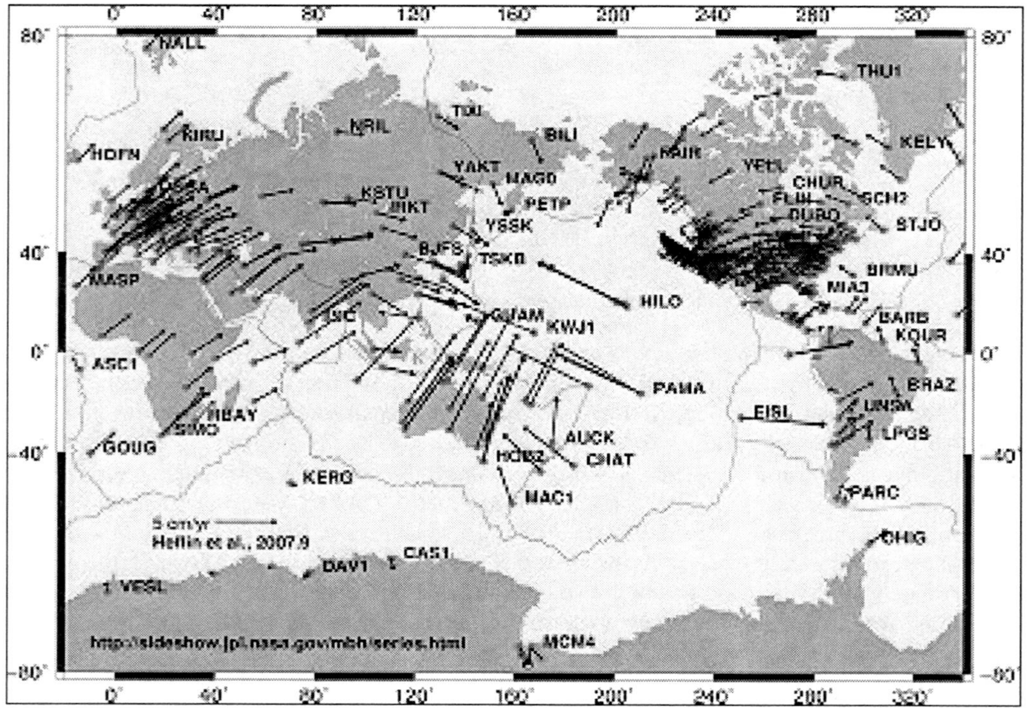

Central Plateaus and Rivers on Many Continents Make Them Habitable

After continental separation some of the existing continents have high plateaus that form rivers draining from them to the oceans that made them habitable. This is especially true of areas of Africa, South America, and Asia which would be exceedingly hot without the raised plateaus. Studies by Thomas Jordan of cratons and rock density, and how these were possibly created suggested that the hot lava in boundary tectonic divisions caused the areas with lighter rocks to rise forming plateaus.[48] These continents had large reservoirs of water that were massive water residues from the worldwide flood that drained to form these rivers as discussed in that chapter. For example, most of eastern and central Africa, South America et al. are high with good temperature for their latitudes. In the north there is the Nile flowing from Lake Victoria to the north and Mediterranean; in the northwest there is the Niger River; on the west the Congo, and on the south the Zambezi. In North America Lake Agassiz formed the Mississippi going south, the Saint Lawrence going southeast, and the McKenzie going northwest. Before the Canadian thrust on the west the great lake formation making the Channel Scabland's probably drained east in the Columbia before the Pacific plate caused the Cascade Range. The reservoir on the Colorado Plateau made the Colorado and Rio Grand draining to the Gulf of Mexico. In South America there is the plateau area of Bolivia and Paraguay with the Amazon that drained not only to the east, but to the west before the Pacific plate formed the Andes. Much research is needed to better understand why these things occurred with the formation of the continents from Pangaea.

Other Confirmation by Second Ice Age, Life Destruction and Preservation, and Migration

Geological Timing of the Division of the Continents Coincides with Second Ice Age

Scientists generally agree that a catastrophe that resulted in the division of the continents occurred at the time of the K-T boundary (K stands for the Cretaceous period) and that this event caused the second worse destruction of life forms. Chris Turney discusses the geological timing in depth. He gives evidence of a worldwide event involving much volcanic activity, darkening the skies around the world. The whole world has a layer of elements that are rare at the K-T boundary, especially a large amount of iridium. Many life forms managed to survive during these events. Dinosaurs have even been found between the various layers of lava flows in the Deccan Traps of India, which was not the friendliest area and occurred later in the drifting of plates. This showed ***that much life continued during the time.*** Scientists have had a difficult time explaining this. Turney says the scientists were asking, "Was it possible that instead of one large meteorite making a big impact crater, there could have been a storm of them striking the Earth at the same time? The result would have been several smaller impact craters on the surface."[49] But, being persuaded by others, he finally concedes that the whole event was caused by the Chicxulub crater off Mexico. This conclusion does not consider the other significant events occurring around the world at this time, which are also placed at the K-T boundary.

Since much plant and animal life survived, scientists have determined that this event was most likely to happen in particular areas. Some have speculated that the skies, darkened by debris, would cause extreme cold and ice formation, resulting in the plant and animal extinctions that are associated with this event. As indicated in the previous chapter, this caused a minor second Ice Age. It explains why Cro-Magnon men dwelt in caves in Europe (e.g. Mousterian caves sites) and

---

[48]Charles Petit, "Continental Hearts: Ancient expanses called cratons pose a geological puzzle, *Science News,* December 18th, 2010, Vol. 137, 22-26.

[49]Chris Turney, *Bones, Rocks, and Stars*, 144.

as far as China and North America, even in the Grand Canyon caves. Cave pictures illustrate many animals frozen in tundra at the time of the flood. The people who first lived in what is now earliest Egypt also indicate a cold climate, as illustrated by the wearing of fur coats, etc.[50] These events would have occurred after the great flood and the loss of most life in the P-T, followed by the great glaciation, and, therefore, during the K-T time after the division of the continents.

Where Life Existed in Eurasia and North Africa was Most Protected to Survive

The scientists making these reports admit, ironically, that the *other* ***unlikely location for such a catastrophe*** *was in the area of* ***India*** (or Indian Ocean vicinity), which would have been four to ***five thousand miles away from the center of most living things***.[51] Most primates and other large animals were formerly in Eurasia. According to the Bible, people and animals from the time of Noah would be in Eurasia and outside the area of worst events. Paleobotanists have confirmed that northeastern North America, and ***especially Europe, escaped the worst*** *of the devastation*. Many admit the early Eurasian and African continental place of separation was in the European latitude where living plants and animals (possibly humans) would have escaped. They refer to evidence from studies by Arthur Sweet of the Geological Survey of Canada, by Kirk R. Johnson of the Denver Museum of Nature and Science, Conrad C. Labandier of Smithsonian Institute and others to support this.[52]

Asish R. Baso and his team attribute their findings to an earlier period. Rampino believes the particles were little pieces of meteorites that were scattered elsewhere in the world. Using magnets, ultrasound, and other instruments, scientists found these magnetic particles from powdery materials contain iron, nickel-rich oxide, sulfide materials, and silicates manganese. They submit that the low concentrations of iron oxide and the very high ratio of manganese to iron "strongly suggests that the minerals didn't crystallize on earth. Such a chemical mix is *found only in a type of meteorite* that formed 4.5 billion years ago…"[53] Moreover, these do not show up in the coal bed below the clay they were in. Again, it should be emphasized that King and Durda identified very high concentrations of these materials *in India.*

Second Short and Less Severe Ice Age Coincides as Evidence of Catastrophic Division

Many believe major species extinctions, caused by catastrophe, have occurred twice in Earth's history.[54] The first (older) occurred at the Permian and Tertiary boundary (called P-T). The second at the later or Tertiary and Cretaceous boundary (called K-T). Walter Alvarez of U. C. Berkley found evidence of iridium in Italian clay in 1970, and later in Denmark in 1980 that point to asteroid impacts. After much controversy, his findings were accepted. The cause of the older, P-T boundary catastrophe has been hotly debated. Karen Wright reports,

> The superlative [death to life species] belongs to a more severe crisis at the P-T boundary…. Fossil records show that about 250 million years ago, *90 % of the species on Earth were snuffed out* in ***an abrupt event that spanned the globe***. Its causes, like *its sediments*, are buried more deeply. ***No one has even come close to proving what happened***. The Permian extinction

[50] George A. Barton, *Archaeology of the Bible,* (American Sunday School Union, 1949 edition), 8-10. This is based on the findings of W. M. Flinders Petrie, *Prehistoric Egypt*, London, 1920.
[51] Chris Turney, *Bones, Rocks, and Stars* 102, 103.
[52] Ibid., 104.
[53] King and Durda, "The Day the World Burned, " 102.
[54] Others question this. Atul Gawander, ed. *The Best American Science Writing 2006*, (New York: Harper Perennial, 2006), 223-236.

> obliterated ecosystems as complex as any on Earth today. On land, 10-foot long saber-tooth reptiles succumbed, and grazing root-grubbing, and insect-eating lizards vanished, along with the plants and bugs they ate. In the ocean, reefs teeming with life were reduced to bare skeletons. The Permian even finished off the lowly trilobite....[55] (Emphasis added)

Wright also said,

> Doug Erwin of the National Museum of Natural History in Washington showed that the extinction happened abruptly ***in the oceans***. The same year Peter Ward documented ***a sudden die-off*** of vegetation on land in the present-day South Africa. Lines of evidence were converging, and figures kept ratcheting downward (*from millions of years*) Although they couldn't resolve time in terms of days, weeks, or months, many experts came to believe that the whole doomed Permian assemblage-flora, fauna, and for-minifera–might have bought it ***overnight***.[56] (Emphasis and insertion added)

Luann Becker's theory about asteroid off Australia seemed a possibility. However, Wright says that for many, asteroid material does not fit. "The dates are not unequivocally 250 million years'' or Permian. Wright continues, "They say a gravity model of the site, a kind of topographical map of buried geologic structure, looks much like the gravity model of Chicxulub, the K-T impact structure." While there is an increase in iridium at the P-T boundary, it is nowhere as big as the later K-T. "If an impactor landed in the deep ocean, it wouldn't create much shocked quartz either, because the ocean floor has less quartz in it than continental crust. If a king size comet landed in the deep ocean, it would be like stabbing a man with an icicle: a murder with a weapon that vanishes."[57] The K-T extinction, while extensive, did not destroy nearly so many species as the earlier P-T one. We know that the division of the continents occurred while animal and plant life survived all over the world, especially in Eurasia.

Some have tried to find repetitive catastrophes, but there is a general consensus there were two main events. Dinosaurs are now known to have existed all over the world, even in New Zealand, North Africa, Russia, China, in North America, and in the Dyars of India. It is not likely that all of them were destroyed in the Yucatan Chicxulub impact. Later studies indicate the death of ***most*** dinosaurs occurred over 300,000 years earlier, perhaps at the P-T (cf. Wright's arguments). The findings of iridium all over the world support the whole event of continental division. Thus the two catastrophes fit the biblical catastrophes of the flood and the parting of the continents at the tower of Babel. The first was the ***worst, most likely by water, accompanied by major cold and enormous glaciations.*** The second, not long afterwards, was the ***continental division by asteroid impacts*** which darkened the skies with meteorite material and volcanic eruptions, placing the earth in great cold for a brief time. It is commonly agreed that the time between the recent ice ages, shown in the graph in the previous chapter, was short.[58]

Migrations of Animals and Humans More Reasonable by Continental Division

Animal migration is more easily explained by the division of the continents. Jon Erickson notes that scientists were reluctant to accept the theory of drifting continents because they developed immaculate schemes to explain the migration of animals and the existence of similar

---

[55] Karen Wright, "The Day Everything Died," *Discover*, April 2005, 66, 67.

[56] Ibid., 67, 68; cf. also Chares Choi, "Permian Percussion", Scientific American July 2004, 36 for two major catastrophes.

[57] Ibid., 69.

[58] See graph in Plimer, *Heaven and Earth*, 51 and in previous chapter 21.

species on different continents. The breakup and drift of continents seemed to conflict with their evolutionary views. Erickson revealed data showing that certain ***rare life forms on each continent*** could ***not be explained in any other way than breakup of the continents***.

Many higher forms of common animals were also found at points where the continents divided that neither Erickson nor others mention. Science and geologists have never fully come to grips with the data that reveals that Homo sapiens also may have lived in these areas at the time of the continental divide. If this many other mammals survived, and if the safest area at the time of the cataclysm was Eurasia, and divided where some men were living, why could they not also have survived? Denying the possibility of cataclysm, uniformitarians cling, rather, to impossible ideas of migration over inhospitable terrain and across miles of water. According to them, these migrations, occurred while there were warm temperatures and plentiful food, so why would they migrate? Some of the DNA views and the language histories we have previously given also gives support to this argument. The distribution of man at the continental division fits the evidence better than Wells' and Sykes' views. It also fits the biblical account of the dispersion of language and culture.

Though the migration patterns of some rare species are difficult to explain, Douglas S. Fox report that Tom Rich, paleontologist at Monash University in Melbourne, offers evidence that mammals, even primates, did not migrate across Siberia to North and South America and around to Antarctica and Australia. Rather, he argues, "In the Garden of Eden Theory, placental mammals migrated from the south as the continents were breaking up." Rich drew diagrams showing many animals' probable migration resulting from continental breakup. This is a ***more feasible concept than long migrations previously*** proposed.[59] Uniform determinists, however, bypassed such ideas because they are contrary to their theories and time scales.

Diamond also documented that most of the domestication of plants and animals, and other cultural factors spread from Eurasia.[60] The cave drawings found across Eurasia are similar to the 260 sites with rock art, such as those found at Boqueirao da Pedra Furad in Piaui, Brazil, which are said to correspond to the time of early men in Eurasia. It also explains the belief that Clovis stone weapons of earliest Americans in the eastern United States and elsewhere are linked to the Salurians of France, since there is no other reasonable explanation of how they got there.

The scientific data about the division and distribution of the continents fits more with a biblical cataclysm than with long slow ages. The facts are also a powerful witness to the biblical account of a common point of man's dispersion by the continental break up in Eurasia. The biblical history is still reliable. It confirms the statement in Hebrews 2:2, "He upholds all things by the word of His power."

As mentioned, the break up of Pangaea appeared to be along the lines of a divine composite in Geneses 1:9, 10. After the division, probably from plate collisions of separate continents, most of the continents had central raised plateaus and remaining water from the floods caused drainage and the formation of major rivers. In South America it seems that the Amazon also ran in the opposite direction until the Andes was formed by the late motivation of the Pacific Plate. [61]

---

[59] Douglas S. Fox, "Pouch or No Pouch," *Discover,* July 2004, 69-74, cf. maps 71.

[60] Ibid., 375 et al.

[61] "Flow West, Young River: Ancient Amazon ran opposite today's route", *Science News,* Vol. 170, November 4, 2006, 293.

Biblical History of Continental Separation is Similar and Supported by Geological Findings

Since the geological data has reinstated the evidence pointing to the biblical flood, the breakup and drifting of the continents are also a confirmation of the historical biblical account (Genesis 10:25) and other historical records. Other interpretations have been suggested for the biblical statement, "the Earth divided," but they do not involve geographical divisions. None of these views have been widely accepted.[62] God could have, by fiat, confused the languages in the mind of men. But God usually uses secondary causes, such as a flood, drought, plagues, or earthquakes, to demonstrate that he controls nature, even in rendering judgment. The cataclysmic separation of the continents, with the formation of mountain and water barriers to confuse the languages, is, therefore, not contrary to Divine patterns. Many critics say the brief statement about the divisions of the earth has some weight. If men survived the division of the continents and the eruptions caused another cold period, the division would not have been recognized as a major worldwide event.

The biblical record says, "In his (Peleg's) days the earth was divided" (Genesis 10:25; 1 Chronicles 1:19). Peleg means "division." Scholars believe Peleg lived as long as 308 years after the flood, tracing the earth's division to some time prior to 308 years after the flood. This agrees with the short periods found between the greater and lesser glaciations. Ian Plimer said, "Evidence suggests that the shift from interglacial to glacial conditions occurred in only 400 years. Snowlines throughout the world were 900 meters lower than today. Air temperature at the glacier was some 5 degrees C colder than today and the tropical sea surface was 3 degrees C cooler."[63]

For five generations following the flood of Noah, people lived an average of 450 years, propagating children for over three hundred years. If each person had a child every few years after reaching fertility, there could have been, perhaps, as many as ten million people by the time of the Tower of Babel. Larger and more fertile animals would have spread widely through Pangaea. The growth and expansion of the population of the world would be a cause to fear that mankind would lose contact. The concern for lose of unity of the people of the world in the days of Nimrod is a concern even today in our search for world unity.

The cataclysmic breakup of the continents involved the formation of mountains, oceans, seas, rivers, valleys and other features that form national boundaries. Paul addressed the intellectuals of Greece telling them the God who created the world also determined their national boundaries.

> The God who made the world and everything in it, being Lord of heaven and earth, does not live in temples made by man, nor is he served by human hands, as though he needed anything, since he himself gives to all mankind life and breath and everything. And he made from one man every nation of mankind to live on all the ***face of the earth***, having determined allotted periods and ***the boundaries*** of their dwelling place (Act. 17:24-26). (Emphasis added)

The cataclysmic division of the continents is associated with Nimrod's reign as recorded in Genesis 10:8-12. The Jewish Targums the Greek translation of the Hebrew Old Testament present Nimrod as a hunter of men *against* (literally, "before/to the face of) the Lord."[64] Tradition says he was a dictator who founded a kingdom and built a number of cities in what was known as Babylon

---

[62] John W. Davis, *The Westminster Dictionary of The Bible,* (Philadelphia: The Westminster Press, 1994), see Peleg, 465.

[63] Plimer, *Heaven and Earth*, 37.

[64] C. F. Keil, *Biblical Commentary of the Old Testament,* (Grand Rapids, Michigan: Wm. B. Erdmann Publishing Company, 1951), 165-168.

and Assyria in the plain of Shinar. He built what the Hebrews called Babilu (Bab-ilu in Akkadian), which means "Gate of God."[65] After these early times, the city disappeared from prominence. Many years later (1830-1550 BC) Babylon again emerged as a power. Babylon came to power again in the sixth century BC, under Nebuchadnezzar, as the nation that conquered and deported Judah. The biblical story about the confusion of tongues follows below. Notice the phrase, ***"face of the earth,"*** is the same phrase Paul used in Acts 17:24-26.

> Now the whole earth used the same language and the same words. And it came about as they journeyed east, that they found a plain in the land of Shinar and settled there. And they said to one another, "Come, let us make bricks and burn them thoroughly." And they used brick for stone, and they used tar for mortar. And they said, "Come, let us build for ourselves a city, and a tower whose top will reach into the heaven and let us make for ourselves a name; lest we be scattered abroad over *the face of the whole earth*." And the Lord came down to see the city and the tower, which the sons of men had built. And the Lord said, "Behold they are one people, and they all have the same language. And this is what they began to do, and now nothing which they purpose to do will be impossible for them. "Come let Us go down and there confuse their language, that they may not understand one another's speech." So the Lord scattered them abroad from there over ***the face of the whole earth***; and they stopped building the city. Therefore its name was called Babel, because there the Lord confused the language of *the whole earth* and from there the Lord scattered them abroad over *the face of the whole earth* (Genesis 11:1-9). (Emphasis added)

Most historians agree that building the tower had religious implications. Such towers were typically a type of pyramid (or in the shape of a pyramid), and often associated with the study of astronomy. Knowledge of the heavens, no matter how crude, is common among ancient cultures, leading to human pride. The name initially given the city, Bab-ilu, implies that the people saw themselves involved in a project of divine worship. This was pure humanism, which is the failure of all civilizations that exalt their knowledge and themselves.

At the uppermost level of such a tower, one can see and study stars and planets more clearly. The worshipers' intent was ***to keep a unity of mankind and to gain understanding*** that would exalt man and give them security. God, however, viewed their actions as a rebellious worship of the creatures rather than the Creator. God knew that *by united knowledge man could achieve any purpose of rebellion.* "Nothing, which they propose to do will be impossible for them" (Genesis 11:6). This may be a similar circumstance we are approaching in worldwide computer knowledge today.

The use of the Hebrew word *saphah,* while denoting speech/language ***refers to a lip*** of a jar, or the lip of a river into the ocean. More than speech, it describes what comes out. In this case, what is from the nature of the heart that is speaking from within, for example, "uncircumcised *lips*," "*lips* of praise," etc. Scripture, therefore, reflects a unity of language that proceeded from a heart of humanistic culture and thought. The issue was not just unity in language, but a self-centered language. God's purpose in scattering the people, by dividing the earth and confusing their tongues, was to hinder or end this process of man's united heart rebellion against Him.

Geology Confirms the Biblical Area of Separation for Language Separation

Biblical scholars generally agreed that the land of Shinar was in the Mesopotamian region. Assuming it was Pangaea before it divided, the area of worship is generally known. Noah's son ***Japheth*** settled "in the tents of ***Shem***," or they cohabited together. Japheth's descendents

[65] Finegan, *Light From the Ancient Past*, 10.

(including Shem's) all settled along the lands now known as Eurasia, which was the north shore of the Mediterranean area. Many geologists believe an early major split/division of Pangaea occurred there. The Scriptures agree saying, "*From these (places of Japheth's descendants)* the coastlands of ***the nations were separated*** into their lands, every one according to his language, according to their families, into their nations" (Genesis 10:5). Africa and the Arabian plate joined this area after the division, so their relationship to Shinar is not certain. It is likely that the Eurasian area led into the plain that extended into all of Pangaea. Some people were already residing far away on areas of other plates, composing the one landmass along the equator where Pangaea was located. Many geologists' maps show the breakup of Pangaea occurred at the same place as identified by the Bible as the place of the tower of Babel. This is also the ***area that was* least *impacted*** by great debris from meteorite collisions and volcanic eruptions. This area was near where Noah and his family landed after the flood, and the origin of the early human outmigration was most protected at the time of the divisions.

The disappearance of the city of Bab-ilu or Babel, for succeeding centuries, probably indicates the judgment of dividing the continents occurred near there and caused the people to cease their building and forsake the city. The people were thus scattered to form their own nations and languages. Human nature did not change at that point, but soon after God called Abraham to be the father of His elect people, through whom He would bless the world (Genesis 12:1-3). Nimrod may have continued to propagate his heresy. The average age of beginning fertility after the flood was 35 years and the average life span was 450 years, so from Ham to Cush and then to Nimrod would be about 70 years. It's possible that Nimrod was still living 520 years after the flood, or over two hundred years after the judgment of the division of the earth. The perverted religion of Nimrod could have been continued after the Pangaea's division and pyramid like towers are found all over the world in ancient civilizations. The towers exhibit their involvement in the mechanics of astronomy and astrology, which is at the root of unbelief in one Sovereign Sky God.[66] Some anthropologists believe that Nimrod is the same as Egyptian Osiaris and in other lands is Tammuz, Adonis, Bacchus, and Baal identified with the devil worship. His wife is thought to be the source of all nations' sex goddesses.

Independent Historical Testimonies to the Biblical Account Substantiated Elsewhere

If such a large cataclysm occurred, why isn't more said in the Bible, and why are there not more historic records? Dr. Robert Jamieson points to ***a number of historical recordings*** referring to the cataclysmic dispersion of people and confusion of language:

> Beside, the Mosaic record of this memorable occurrence is confirmed by a variety of independent testimonies. The account of Berosus, the Chaldean historian, is substantially the same as that of Moses, as also is the Hindu tradition, according to Sir William Jones. The Egyptian monuments attest the fact of the dispersion at Shinar (Osburn's "Egypt and her Testimony"), and the cuneiform inscriptions speak of Chaldea or Babylonia as 'the land of tongues' (Fox Talbot). The most eminent ethnologists also have come to this conclusion. 'There is the greatest probability that the human race, no less than their language go back to one common stock-to a first man- and not to several, dispersed in different parts of the world. And it is asserted, with the greatest confidence, that from an extensive examination of languages, the separation among mankind is shown to have been violent; not, indeed, that they voluntarily changed their language, but that

[66] Cf. the ziggurat of ur-Nammu, dynasty II of Ur, *The Westminster Dictionary of the Bible, Revised,* (Philadelphia, The Westminster Press, 1944), 55.

they were rudely and suddenly (brusquement) divided from one another' (Wiseman's 'Lectures').[67]

Australian astronomer George Dodwell became interested in the historic astronomic records of ancient Egypt, Babylon, and elsewhere. He did reverse reconstruction calculations from the planets and stars as they are today, especially in connection to the earth's perihelion, tilt, and rotation as would affect their locations. Among others, he used the records of the Egyptian Solar Temple of Amen-Ra, the records found and reported by the Greek mathematician and astronomer, Eudoxus, who lived from 400 to 347 BC, Stonehenge, etc. He discovered ***two major points of change***—one from 2000 to 1900 BC and the other about 1500 BC. By devising a graphed curve, he discovered a pattern similar to a spinning top that had been hit by an object and then recovered. He also concluded that the cause of the resultant wobble ***was an asteroid impact*** in the area of the Pacific. The western Pacific would, be the area of the Tetheys Sea, which is where the continents would have begun breaking up, as geologists and biblical records suggest. The time would have been about 2000 BC. Later, however, he associated it with the flood.[68]

Important Conclusion

The scientific evidence regarding the division of the continents agrees with the biblical record, an indication that the event was God's judgment for mankind's prideful rebellion. ***Scientific darkness about the division of the continents is because it was such a huge worldwide catastrophe beyond natural determinism*** that man's bias against God and his power rejected the facts. The history of these two great catastrophes shows God's judgment of the world for sin, and points to the future judgment in fire mentioned by Peter in 2 Peter and often in Old Testament. But God acted in grace to start again after the Flood, and by the choice of his people through Abraham after the tower of Babel and division of continents. Paul indicated the boundaries of the nations and periods were formed by God (Acts 17:26, 27). A similar bias exists today as expressed at Babel that is evidenced in the efforts at world unity in the United Nations, and natural catastrophic events as Christ warned, for example, "The roaring of the seas and the billows" (Luke 21:25-26) and earthquakes (Matthew 24:7). His elect need to be witnessing now and looking for the final judgment of the world and his eternal kingdom. Christians need true enlightenment about God's judgment of all nations and the amazing grace now offered before Christ returns.

---

[67] Robert Jamieson, *A Commentary of the Old and New Testaments, Volume I*, (Grand Rapids, Michigan: Wm. B. Erdmann Publishing Co. 1945), 125.

[68] Paul D. Ackerman, Chapter 10, "The Top That Reeled," *It's a Young World After All*, (Grand Rapids, MI.: Baker Book House, 2004), 67-99; Cf. Carl Wieland, "An Asteroid Tilts the Earth," *Ex Nihilo*, January 1983 and Barry Satterfield, "An Asteroid Tilts the Earth," *Ex Nihilo*, April 1983.

SECTION FOUR

EARTH HISTORY AND AGE OF THE EARTH

PART VII

ENLIGHTENMENT ABOUT THE FINDING OF TIME IN NATURE

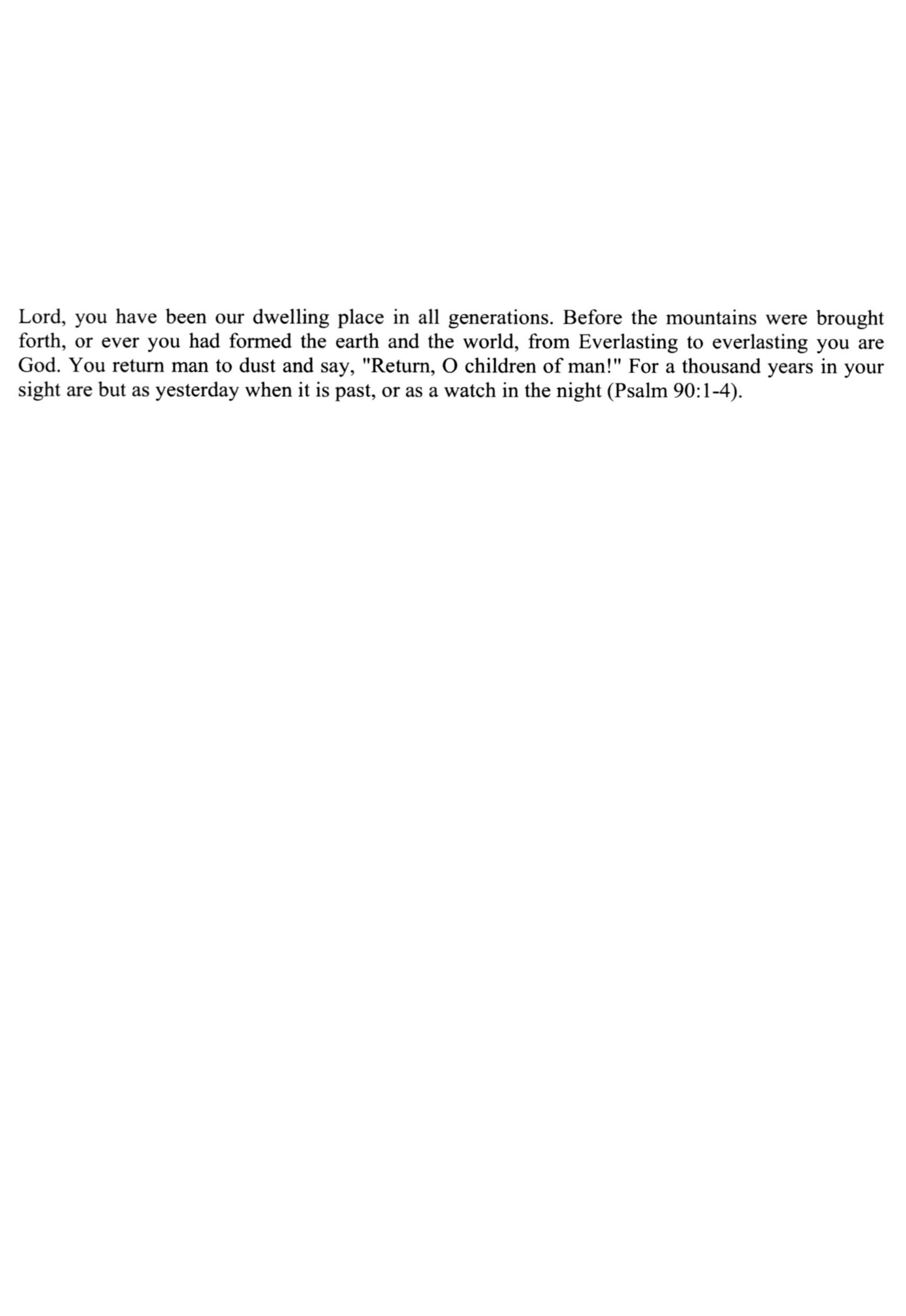

Lord, you have been our dwelling place in all generations. Before the mountains were brought forth, or ever you had formed the earth and the world, from Everlasting to everlasting you are God. You return man to dust and say, "Return, O children of man!" For a thousand years in your sight are but as yesterday when it is past, or as a watch in the night (Psalm 90:1-4).

Chapter 22
Enlightenment about Age of Earth And Geological `Column

Introduction: Endless Time Adopted as a Philosophy without God

Immanuel Kant wrote his dissertation, *Universal Natural History and Theory of the Heavens: an Essay on the Constitution and Mechanical Origin of the Whole Universe Treated According to Newton's Principles,* describing the developmental concept of the universe by a nebular hypothesis. This evolutionary theory was transferred from Kant, to Herschel, to Laplace, to Hutton, to Lyell. Lyell then wrote a philosophy of the earth's development and principles of geological operation using the idea of long ages which the development required. The scientists who were trying to support Kant's evolutionary theory had no scientific data to give their ideas any credibility, nor any real estimate as to how long such development would take.

Lyell used the past works of Christian young earth flood geologists with his Triassic ones in defining the strata, and with G. P. Deshayes of France worked out the progressive table of the main strata with marker fossils. He published the *Principles of Geology* in 1833, claiming long ages for the buildup of the geological strata of the earth, based on the nebular hypothesis. Lyell's main objective was to establish a deterministic naturalistic view of geology, to present his views of geology as definitive, and to ***discredit all cataclysms, especially the biblical flood and theology***. Lyell's overriding purpose was shown from excerpts from his private letters and later by C. C. Gillispie, perhaps the leading historian on geology. Bias against God was the main reason for the way the geological column was developed.

Lyell, however, ignored the opposition of the great creation scientists and like Hutton he promoted long circular ages of natural development. His theories went contrary to the data of the recognized geologists of the day, including his teacher Buckner. Moreover, in his *Principles of Geology,* Lyell claimed that the long ages of strata development ***applied worldwide***, but his claims were based on strata only in limited parts of the world. He was influenced by the culture of his day bent on materialism, and highly influenced by Locke and Kant's deterministic view of knowledge. He was also influenced by Lamarck's and Chambers' biological evolutionary concepts that included pantheistic or deistic naturalistic intelligent design. Lyell (who rejected the linear biological views) wrote taking advantage of the growing intellectual prevailing anti-religious and anti clerical sentiment.

Lyell's uniform deterministic view of geology had a profound motivating effect on Charles Darwin's (1809-1882) idea of survival of the fittest. Above all, Darwin's philosophy of evolution required age after age to produce millions of intermediaries to progress up to man, even though Lyell and other more knowledgeable scientific friends told Darwin there was no evidence of development of new species in the geological record and none of man's development. Later when Darwin's theory became popular, Lyell endorsed Darwin's views which ***required even more time.*** Again, there was no scientific data about how life began or how long it took the animals to develop or fossilize in sedimentary rocks.

Man's belief in progress and his sense knowledge of the world led him to follow his bias to claim long ages of natural development of all things without God or without any semblance of objectivity or standard measurements. Not until over one hundred years later in the twentieth century was there any claim for a standard measurement of the heavens or in geology. These philosophical theories were proposed, accepted, and popularized to reject the biblical and other history of the world. In contrast, the leading scientific authorities held to God's creation that required no long ages. They felt the scientific data supported a young earth. The column was also developed before any objective measurements of the age of strata.

Equally important is the fact that these theories and philosophies were adopted ***before significant geological discoveries of the twentieth century***, such as geological evidence of worldwide flood waters or of continental breakup and drift and the force and impact of tectonic plates. Many scientists acknowledge that these discoveries change their understanding of how the earth was formed and developed. The geological bias by Lyell and others was also influenced by the idea that the heavens also took long ages to develop, which also had little objective data for support until a hundred years later.

Further Consideration of Bias as Motivation for Requiring Long Ages

So, the appeal to *long periods of time in constructing the geological column grew out of a bias against God*, not out of any compelling data. Gillispie observed, "The question had become very much more than a geological one, and *the root of Lyell's ideas* ***lay outside the bounds of that science***, wide though they then were.... *Uniformitarian presuppositions*, then, were simply ***those of optimistic materialism***."[1] (Emphasis added) To sustain the idea that all geologic change occurred only from regularly observed forces and without major catastrophes, long periods of time were needed for water, glaciation, fire, wind, and other regular natural forces to produce the earth changes.

It seems *strange* that the very objective of uniformitarian ideology was *to remove the idea of a major flood*, while at the same time the basis of arriving at a chronological order of the fossils in the rock was *dependent mainly on the fast burial* ***in water*** of the living organisms and the strata formed and dated by them. R. C. Moore, professor of geology at the University of Kansas, points out that fossils are formed by being frozen in glaciers, mummified by dry air in a desert or cave, and by entombment in volcanic ash. But by far the *great majority is formed by being covered by water* in fine sediments. He said,

> Preservation of organic remains as a fossil commonly depends on two chief requisites: (1) The organism be buried quickly in some protecting medium, and (2) that it possesses some sort of hard parts such as a skeleton or shell. The death of an animal or plant normally is followed quickly by decay that destroys tissues and eventually obliterates all traces of the organism.[2]

A quick burial and decay were key elements in fossilization, and ***catastrophes*** were the surest way to obtain these—flooding, major torrents of a river, etc. Flooding, which is the most important cause of index fossils and identifying geologic periods, was treated as incidental.

When Darwin accepted uniformitarian geology, he also felt he needed to extend the time periods to gain *even more time* for the gradual development of plants and animals. The theory of natural selection required an extended time to allow for these developments. When Mendelian genetics were rediscovered, revealing great controlled restrictions on change, Lord Kelvin (Sir W. Thomson) the outstanding physicist, argued that the earth's heat would be lost after a few million

---

[1] Gillispie, *Genesis and Geology*, 133, 135; see K. M. Lyell, *Life of Charles Lyell,* I, 271; Lyell to Scrope, June 14, 1830.

[2] Raymond C. Moore, *Introduction to Historical Geology*, second edition (New York: McGraw-Hill Book Company, Inc., 1958), 10-13.

years. Darwin responded, "I am greatly troubled at the short duration of the world according to Sir. W. Thomson, for I require for my theoretical views *a very long period* before the Cambrian formation."[3] (Emphasis added) Darwin was not compelled by any evidence that there were long periods of time, but he accepted this assumption because he had to have it for his unproven theory of gradualism to be credible.

In Darwin's first edition of *The Origin* he wrote, "In all probability a far longer period than 300,000,000 years has elapsed since the latter part of the secondary period."[4] After Thomson's arguments, late in the century, geologist W. J. Sollas said there is less than 17 million years since the beginning of the Cambrian system, and early in the twentieth century, Hugo DeVries placed the most probable estimate at 20 to 40 million years.[5] These were obviously guesses, based on their whims, but these speculative guesses influenced those who actually fixed the dates. The best authorities on ***sedimentary rocks*** say there is ***no way to date them accurately*** based on the amount of sediment. This sediment forms rapidly from a fast moving large volume of water while the same amount of sediment may, by slow small volume, take very much longer.

Later, after the discovery of the radioactive disintegration of atomic elements became known as a new source to heat up the earth, the extension of time was no problem. So, ***before*** *radioactive dating was developed* with any degree of accuracy, scientists began to extend the estimated dates for the various geologic periods. Since the motivation was to establish a developmental theory to discredit the flood and biblical geological ideas involving catastrophes, the estimated length of time was often extended to support uniformitarianism and evolution. In a previous chapter it was shown that even after radiometric dating, the times were repeatedly significantly changed.

Geological Age Dating and Construction of the Geological Column

How the Ruler for Measuring Geological Time Was Put Together

How did the idea of a column with different rock strata originate and how did scientists determine the time from the rocks? Patrick Hurley of the Massachusetts Institute of Technology (MIT) explains "age dating" in this way,

> The term "geological age" refers to a relative time scale *not based on years* ago but *based on the* ***established sequence*** *of the evolution of plants* and animals. Therefore, in order to establish the geological age of the section of sedimentary strata in any area, it was necessary to find fossils in these rocks and to recognize their position in the sequences.[6] (Emphasis added)

The approximate date was based on the *slow build-up by uniform ordinary day-by-day forces* of the "geological column" of rocks, especially sedimentary rocks with fossils. But as Hurley admits, the system ***establishes sequence***, not dates. The dates are based on presuppositions.

William "Strata" Smith suggested the construction of the column as a basis for dating. The lowest level of sedimentary rock would be considered the oldest and, therefore, the oldest fossils and each layer nearer the surface would have the newer or more recent fossils. Using certain shells

---

[3] Charles Darwin to J. Croll, January 31, 1869.

[4] Eiseley, *Darwin's Century*, 237, 241.

[5] Hugo DeVries, "The Evidence of Evolution," *Science*, 1904, Vol. 20, 398.

[6] Patrick Hurley, *How Old Is the Earth?* , (Garden City, NY: Anchor Books, 1959), 116.

and fossil in England, Smith sorted out the strata in different places in the rock. The same fossils, in other sedimentary rocks located elsewhere, were assumed to be the same age. Using the thickness of the sediments and the supposed period of time for forming the various fossils, the age could be estimated. This seems a reasonable and meaningful way to at least arrive ***at a chronology*** of relative ages. But the sequence ***does not establish length of time***.

In his geological textbook, Branson writes,

> The eras are *delineated by great breaks in physical history* and great changes in animal and plant life. Each is terminated by broad uplifts of the land and by mountain making.... The names of the eras are descriptive of the stages of development of the fossils in the rocks of each.[7] (Emphasis added)

To help in dating the sediments, the geologists determined two kinds of rock units, "local rock units" and "time-rock units." The latter are widely recognized and defined as having paleontological value for a time basis. In other words, *some rocks and their fossils were not considered* significant in age dating, while *others* were. The names of the eras are descriptive of the stage of development of the fossils in the rocks of each unit. Hutton inserted the importance of the volcanic rocks, which might be interspersed, but the column was initially based on certain selected fossils in the sedimentary rocks. Choosing which rocks were valuable for timing was a very subjective task and varied among geologists.

Raymond Moore has said,

> The geologic column of a large area such as a continent or the whole world consists of time-rock units solely. It is made up by *bringing together* in proper order the time-rock divisions recognized in *different places* so as to represent as completely as possible all known geologic time. It is *a composite yard stick*.[8] (Emphasis added)

In other words, certain sections of geologic columns, which were chosen to have time-rocks, were chosen as significant for piecing together as part of the total time column, while others were rejected. The subjectively selected columns were chosen as valuable because of the subjectively chosen *time-rocks*. The column was thus formed even from different locations of the world.

However, it is highly important that the time estimate was much dependent on large land uplifts and mountain building. All of these should have been completely restudied and reconstituted after the understanding of the breakup of the continents and discovery of plate tectonics. Indeed, those events were so cataclysmic that one must question how they can be measurements of time.

The Geological Column—A Product of Imagination

Using the geologic column for dating lacks evidence, thus discrediting this approach. There is no place on the earth where a complete geological column exists with all the different fossils in their respective, chronological sequence. Geologists and the media talk about this as though it is an objective fact which can be visited and viewed, while in fact, it is a very subjective exercise.

---

[7] E.B. Branson, W.A. Tarr, Carl C. Branson, W.D. Keller, *Introduction to Geology,* (New York: McGraw Hill Book Co., Inc. 1952), 331-332.

[8] Raymond C. Moore, *Introduction to Historical Geology*, 26.

This ideal "geologic column" has been fabricated by the use of select fossils in a number of different columns. Thus, while a certain segment of a column might be seen in Arizona, one may have to pick up a part of the column, which has some of the same fossils, elsewhere, and continue the imaginary column in another geographic area. So, the "geological column" is pieced together from various places in the geologist's imagination. Hence, in a true sense, it is ***not a real geological column, but an imaginary one.*** Though this angers some geologists, they admit there is no place in nature one can actually view the column.

Improbable Assumptions Are Used to Construct the Column

But scientists have made assumptions, without a clear demonstration of their validity, that are now blindly accepted, and affirm this imaginary column of rocks. Some of these *assumptions* follow:

(1) It ***assumes that the evolutionary hypothesis is true*** and, therefore, that the smallest and simplest forms of life are the oldest, and the larger, more complex forms come later. But this is a presupposition of Darwin's evolutionary hypothesis that has not been proven.

(2) It assumes development ***from the simple to the complex***. On the contrary, most organisms are found very early, by a "biological big bang" and there is no scientific data to document the tendency to develop from the simple to the complex.

(3) Geologists date the rock strata, on the basis of the theory of evolution, and then turn around and say the sequence of the strata proves evolution. This is ***invalid circular reasoning***.

(4) Geologists ***assumed the same fossils found in two different sets of rocks would be from the same age.*** But must this necessarily be true if the evolutionary hypothesis is not true? Or even if it is true? Many fossils are found in a number of eras. While this is a helpful idea for establishing strata, as Smith showed, it is not always so. Why couldn't the same fossils be from another period of time?

(5) Those who formed the column must assume most of the living creatures alive at the time the index fossils were formed, ***died at that same time in water all over the world.*** If this were not so, they would not represent the same date. Yet, evolutionary uniformitarians often assume that the fossils were in restricted areas of water. The land is thought to have been slowly submerged, or high lakes spilled out and ran to lowlands, or rivers formed fossils in isolated ponds, or limited floods caused the animals to be caught. But unless it was a worldwide flood that trapped all the animals, why would this fossil indexing be true across the world? Would not agile or higher animals and other forms of life be outside the water, and if a limited flood threatened, why would they not migrate or escape to high ground?

(6.) It assumes that ***hydrolytic principles apply similarly for both large and small animals***, and plants, and that the same organisms, of different densities, would be buried at the same time. Yet, that has not been shown to be true. Hydrolytic principles require the more dense forms to be buried first.[9] Density, size, and shape affect the way moving water deposits organisms and the sediment on top of the forms when the movement slows down.

Therefore, it must be argued that *the very opposite of these assumptions is the most probable*. But is it possible that the logical implications about these principles are ignored because the

[9] See Morris' discussion on the hydraulic principles, *The Genesis Flood*, 273.

uniformitarians have a bias toward constructing a timetable that will support a worldwide sequence, which is required by evolution? Wish, rather than fact, gave birth to acceptance of the geologic column.

The Geological Column is Inaccurate in Many Ways

Column Formed on Few Limited Areas Long Before World Studied

Many questions about the arrangement and dating of the geological column remain. Raymond Moore points out, "Sedimentary rates are *much **too variable and little known***, however, *to put much reliance in such a method* of computing geologic time."[10] (Emphasis added)

Contrary to what Branson said about how clear events and fossil changes mark divisions in the geological ages, E. M. Spieker, professor of geology at Ohio State University who was committed to uniformity, argued that *there have been few major worldwide geological events* and therefore *few real identifiable boundaries* between the ages. He also points out that these geological ages, which were *worked out for a **very limited section** of Europe*, have been applied, but never tested, elsewhere. In addition, they were established long *before most of the modern geological research was done*. Spieker says,

> Does our time scale, then, partake of natural law? No... I wonder how many [geologists] realize that the time scale was frozen in essentially its present form by 1840... How much world geology was known in 1840? A bit of Western Europe, not too well, and a lesser fringe of eastern North America. All of Asia, Africa, South America, and most of North America were virtually unknown. How dared the pioneers ***assume*** that *their scale would fit* the rocks in these vast areas, by far most of the world? Only in ***dogmatic assumption*** – a mere extension of the kind of reasoning developed by Werner from the facts in his little district of Saxony. And in many parts of the world, notably India and South America, ***it does not fit***. But even there, it is applied! The followers of the founding fathers went forth across the earth and in Procrustean fashion *made it fit* the sections they found, even in places where *the **actual evidence** literally proclaimed **denial***. So, flexible and accommodating are the 'facts' of geology (emphasis added).[11]

Thus, the geological column is contrary to ***globally*** *applied* scientific methodology.

Raymond Moore pointed out that the "geological column" is a composite, based on a limited area. He goes on to refer to areas in the United States where it does not fit. He said, "The geologic columns of Pennsylvania and Illinois, unlike this general column, *entirely lack some divisions because rocks representing these are **not found** there*."[12] (Emphasis added)

A Number of Places Have Reverse Ages, Older on Top

Moreover, in many places worldwide there are very large, extensive sections of rocks which have the fossils upside down, so that so-called *older fossils* are on the top and *younger fossils* are on the bottom. This is attributed to several different kinds of geological activity, such as *overthrusting*, and *folding*. *Overthrusting* occurs when the earth's tectonic plates, or possibly glacial activity, push large chunks of rocks up over other rocks (exhibit A, illustrated on following

---

[10] Raymond C. Moore, *Introduction to Historical Geology,* (New York: McGraw-Hill Book Co.) 1958, 30.

[11] Edmund M. Spieker, "Mountain-Building Chronology and Nature of Geologic Time-Scale," *Bulletin of American Association of Petroleum Geologists*, Vol. 40, August 1956, 1803.

[12] R.C. Moore, *Introduction to Historical Geology*, 27.

page). *Folding* takes place when forces from one or both sides of a segment of rock push against it and it bulges up and folds over (exhibit B).

However, there are problems in using these as explanations for many of the contradictions in the geological column. When a huge and extremely heavy layer of rock is pushed up over or under another segment of rocks, the sliding of one over the other has some very necessary and significant effects. This sliding, or *slickensides*, causes grooves in the rock, showing there has been movement of these heavy rocks. Also, *fault gouge,* with pulverized rock, occurs between the layers of the thrust rock. Fault *breccias*, or broken and ground-up rock, occurs and builds up between the layers. When the rock's layers are exceedingly large and of tremendous weight, and long overthrusting occurs, breccias will be quite thick, while smaller and shorter and lighter rocks accumulate less. Such tremendous forces also produce *metamorphism* from the intense heat. The ***absence*** *of these evidences must* ***rule out any overthrust*** *occurring.*

Example of Overthrusting with Reversal of Age
Exhibit A

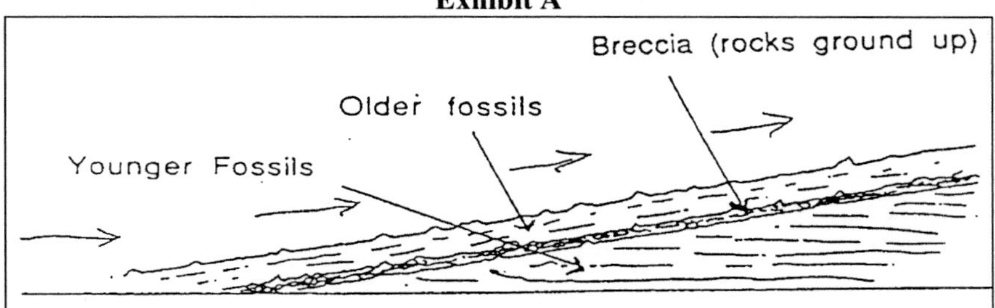

Overturned Fold or Anticline
Exhibit B

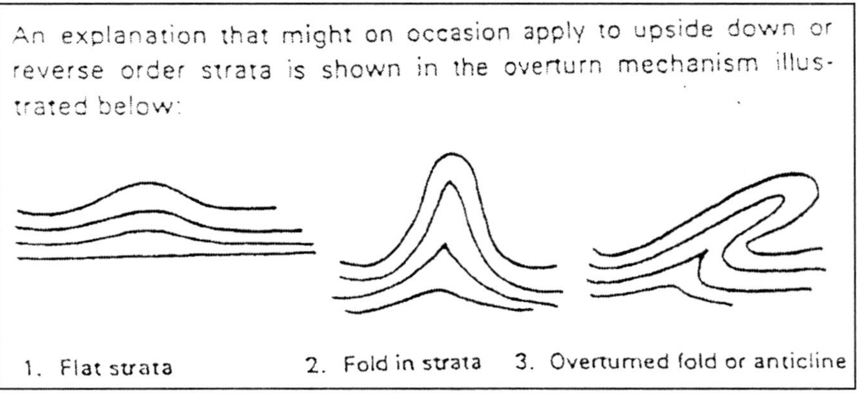

We know that these and similar phenomena have occurred. But there are many places throughout the world where the fossil record is upside down and cannot be explained by these phenomena. The Lewis Overthrust on Chief Mountain in Glacier Park is one such place. It has produced more than 12,000 square miles of upside-down strata. Some 10,000 feet of shale, plus other strata, are considered to have overridden the lower rock surface at the thrust plane, involving 800,000 billion tons of rock overriding. From such a situation, one would expect 75 to 500 feet of rock rubble as a result of the grinding and crushing. But the rock rubble *does not exist.* Similar contradictions can be found at Franklin Mountain near El Paso, Texas, and at nearby Empire Mountain. These contradictions can only be explained by a cataclysmic event.

Some try to explain these in another way, calling them overturned folds or anticlines (as illustrated exhibit B). If this were the case with the Lewis Overthrust, which is 350 miles long and 40 miles wide, it is estimated that the fold would have extended eight times the height of Mount Everest's isostasy. The earth's crust could not have supported such weight. Moreover, if such a monstrous rock column and overturn had occurred, it would have certainly resulted in tremendous breakup. Breakup, however, is not evident. Such problems leave geologists with calling it an "overthrust," *without overthrust evidence.* These contradictions are usually omitted from the data presented to the public.

Geographical Phenomena Unaccounted for by Uniformitarian Column

As noted in the previous chapter, modern science now recognizes the role of catastrophic events in the formation of the earth, thus eliminating the uniformitarians' claims for long periods of development. However, the age-times established earlier, have not been adjusted to accommodate the current thinking. In addition, much of the geological and paleontological data cannot be explained apart from a cataclysmic event, leaving geologists with mysteries and unsolved problems that continue to puzzle them.

Exceedingly Large High Plateaus

George C. Kennedy, professor of geology at U.C.L.A., has said, "The *problem* of the uplift of *large plateau areas* is one that has *puzzled* students of the Earth's crust for a very long time"[13] (Emphasis added) While plate tectonics are now somehow considered the source of these wonders, they certainly cannot be explained on the basis of the uniformitarian day-by-day forces we see today, and no one can explain the amazing forces involved here.

Henry M. Morris, while professor of hydraulic engineering and chairman of the department of civil engineering at Virginia Polytechnic Institute, and a member of many national and international organizations on irrigation and drainage, accumulated data on these matters. He points out that the Colorado plateaus, which cover a region of 250,000 square miles, including Utah, Arizona, and large segments of Colorado and New Mexico, are covered with flat-lying sedimentary rock that was formed by uplifting from below sea level several times with several disconformities in the stratigraphic sequences. How could this be if there have been no cataclysms?

Describing this plateau province, N. M. Fenneman has said,

> The first distinguishing feature is approximate horizontality of its rocks... The second distinguishing feature of the province is great elevation. Aside from canyon bottoms, no considerable portion of it is lower than 5,000 feet. Between this and 11,000 ft., there are plateaus of all altitudes, some of them being higher than the nearby mountain ranges.[14]

---

[13] George C. Kennedy, "The Origin of Continents, Mountain Ranges, and Ocean Basins," *American Scientist*, Vol. 47, December 1959, 495.

[14] N.M. Fenneman, *Physiography of the Western United States,* (New York: McGraw-Hill, 1931), 274.

The Grand Canyon was actually cut mysteriously through this plateau area, which shows little disturbance.

George C. Kennedy continues the discussion of this awesome phenomenon, saying,

> The Tibetan plateaus present a similar problem, but on a vastly larger scale. There, an area of 750,000 square miles has been uplifted from approximately sea level to a mean elevation of roughly three miles, and the Himalayan mountain chain bordering this region has floated upward some five miles, and rather late in geologic time, probably within the last 20,000,000 years.[15]

Most major continents have very high plateaus in their central areas.

Another amazing phenomenon is the fact that many of the highest mountains in the world have the oldest kind of rocks, with no exposed fossils. The top of Pike's Peak is Precambrian igneous rock. Branson has referred to this and said, "Many of the great mountains of the other continents have Precambrian rock exposed at their summits."[16] In America, these are west of Albany. The absence of fossils in these old rocks is attributed to the claim that all life at the Precambrian was like the amoeba, which had no hard body parts to form fossils. But in the West, where vast areas of sedimentary rock lie, the igneous rocks are of a recent date, such as those of Mount Rainier. One would expect the opposite of Precambrian rock is exposed by long, slow erosion.

Many Vast Geosynclines are Evidence against Uniformitarian Forces

Another extraordinary phenomenon is the many great geosynclines found throughout the world. These are vast trough-like depressions. George Kennedy says,

> These deep troughs filled with sediments may contain 50,000 to 100,000 feet of sediments and may be 1000 or more miles long and 100 miles in width. ...*The mystery*, then, of the downsinking of the sedimentary troughs, in which low density sediments apparently displace higher density rocks, is heightened when we note that these narrow elongated zones in the Earth's crust, downwarped the most, with the greatest accumulation of rock debris, shed by the higher portions of the continents, become in turn the mountain ranges and highest portions of the continents.[17]

L. H. Adams, of the Carnegie Institute and president of the American Geophysical Union, says these geosynclines are *one of the great* ***unsolved problems*** *of geophysics.*[18]

Unexplainable Fossil Beds Including Many Species Together

Morris refers to the varied large fossil beds that are quite contrary to anything that is observed today. Dr. Heribert-Nilsson, director of the Swedish Botanical Institute, says,

> In the pieces of amber (of the Baltic Amber deposits), which may reach a size of 5 kilos or more, especially insects and parts of flowers are preserved, even the most fragile structures. The insects

---

[15] Kennedy, "The Origin of Continents, Mountain Ranges, and Ocean Basins," 494.

[16] Branson, *Introduction to Geology*, 333.

[17] Kennedy, "The Origin of Continents, Mountain Ranges, and Ocean Basins," 495.

[18] L.H. Adams, "Some Unsolved Problems of Geophysics," *Transactions, American Geophysical Union,* Vol. 28, October 1947, 676.

> are of modern types and their geographical distribution can be ascertained. It is then *quite astounding* to find that they *belong to all regions of the earth*, not only to the Paleoarctic region, as was to be expected.... The geological and paleobiological facts concerning the layers of amber are *impossible to understand* unless the explanation is accepted that they are the final result of an allochthonous process, *including the whole earth.*[19]

Other similar phenomena are found in Geisettal, Germany; the Cumberland Boar Cave, Agate Springs, Nebraska; the La Brea Pits in Los Angeles; the Korroo formation of South Africa, and other areas.[20] These great graveyards of different fossils, found throughout the world, are unexplainable apart from some great sudden cataclysm. Robert Broom, a South African paleontologist, said there are eight hundred thousand million skeletons (800,000,000,000) of vertebrate animals in the Korroo formation. In Agate Springs, 9,000 complete animals, like rhinoceroses, camels, and giant wild boars, are all buried in one hill.

Grand Canyon Doesn't Display Record of the Column and is Young

Grand Canyon, Place Expected to Display Much of the Column

The Grand Canyon, perhaps the most famous canyon in the world, is 275 miles long, up to 18 miles wide, and in places more than a mile deep. Located in the Colorado Plateau, which is 150,000 square miles, the region is shaped like a kidney and covers large portions of Arizona, New Mexico, Utah and Colorado. The Grand Canyon is the major spot in the United States and one of the places in the world that has mostly layers of sedimentary rock a mile high, and large expanses of walls that run for miles. Geologists thought the geological column would be visible here. The many layers of mostly sedimentary rock are not only visible in the canyon walls but in temples or large protruding rocks carved by the river, some over one mile high, that have remnants of the bed that forms the plateau. "It is of immense scientific interest because it discloses a great perspective of the geologic past."[21]

Until about 1860, only a few people had seen the canyon except for several tribes of the American Indians. In 1869, Major John W. Powell took a team all the way through the canyon and made it the study of his life, becoming a foremost authority. His work drew much attention to this wonder of nature. He was the first to set forth a theory about the development of the canyon. He submitted it was millions of years old.

Downward Adjustment in Age of the Canyon

In the early twentieth century, scientists thought the granite layers in the deepest part of the canyon were 1.7 billion years old, and some of the Carboniferous rocks were over 300 million years old. A consensus emerged that the canyon was about 70 million years old. In the 1930s, data began to emerge that indicated a relatively young area. In 1960, other data led to further revisions, and at a conference of geologists in 2000 there was further revision. Much of the data used by geologists represent improvements in geological method. Geologists are to be praised for their willingness to adjust. Richard A. Young, who is a geologist at the State University of New York in Geneseo, has participated in these studies since the early 1960s and attended the recent

---

[19] N. Heribert-Nilsson, *Synthetische Artbildung,* (Lund, Sweden: Verlag CWE Gleerup, 1953), 1194-1195.

[20] See Morris, *The Genesis Flood*, 157-161.

[21] Grand Canyon, *Encyclopedia Britannica*, 1953.

conference. He said, "Fifty years ago, geologists didn't realize how fast erosion could occur. When there's a depression in the rock and the river flows through, it can erode incredibly rapidly."[22] Revisions on the age came down at first to a minimum of 40 million years, and now, with the use of radioactive dating, and discovery of additional fossils, these figures have been reduced to from four to twelve million years, depending on the site.

In reviewing the studies on the canyon after a 1987conference, Richard Monastersky said,

> Moved by such an immense monument to erosion, the early geologists felt that the canyon must be an old structure, representing 50 million years or more of cutting.... In the 1930s and 1940s, scientists who studied the western part of the canyon and the lower Colorado River dated it "back no farther than 6 million years.... Scientists who studied the river's upper section in the 1960s found evidence that it was at least 20 million to 30 million years old. Researchers *puzzled over* this river that *seemed to be both old and young.*[23]

Questions on Direction of Early Flow and Different Ages of Parts

Scientists now even disagree as to the early direction of the Colorado River's flow. Some hold that the river flowed northwest at first and that the plate tectonics that formed the Gulf of California caused a great cleavage in the Colorado plateau, and that the river then flowed southwest. Thus, a great earth divide formed the canyon, which grew as the river continued to erode the rock. But there is no evidence such a fault was involved. So, the date and direction of the flow of the Colorado River are in doubt.

More importantly, the Grand Canyon has significant age gaps. Richard Monastersky says, "While some sections of time are totally absent, appearing nowhere in the canyon, other gaps in the sedimentary record may appear in selected parts of the canyon." In Surprise Canyon, there are Carboniferous rocks from a period of 340 million years ago, or before the Grand Canyon was formed. George Billingsley of United States Geological Survey (U.S.G.S.) said, "The rocks that later filled these valleys (in Surprise Canyon) represent a 20 million year period missing in most other parts of the canyon.... And it has yielded a greater variety of fossils *than any other formation* in Grand Canyon."[24] Also, the rocks at the bottom of Grand Canyon are said to be Precambrian, or about 2 billion years old.

At the June 2000 conference, the first extensive meeting on the Grand Canyon in about 35 years, geologists suggested, "that substantial portions of the eastern Grand Canyon are geological youngsters, having been eroded only within the past million years."[25] Thus, the geological column is not the basis of the dating within the canyon. In fact, the column is contradicted in much of this dating. Obviously, the dating and formation of the Grand Canyon are in disarray. Some scientists calculate that even by today's sedimentary rates the deposits would be nearer 10,000 to 6,000 years. And the original rate of sedimentation would have been much greater with the massive flow from the original plateau reservoir.

---

[22] Sid Perkins, "Carving this beloved hole in the ground may not have been such a long-term project," (*Science News*, Vol. 158, September 30, 2000), 219.

[23] Richard Monastersky, "What's New in the Ol' Grand?, Geology's Great Monument Continues to Baffle and Amaze," *Science News*, Vol. 132, December 19 & 26, 1987, 392-395.

[24] Ibid., 394.

[25] Perkins, "Carving this beloved hole in the ground may not have been such a long-term project," 218.

Many Mysteries about the Canyon

The Grand Canyon holds other unexplainable mysteries. It contains a great many breccias, ground up rock, and pipes, tunnels through the ground, filled with the breccia. More than one thousand breccia pipes, some less than 300 feet across, extend several thousand feet vertically. About eight percent contain valuable minerals, while others don't. How these pipes were formed and the reason some have minerals and others don't is a mystery. Monastersky says, "For geologists, the breccia pipes have been *stimulating questions* that are *yet unanswered.... No one can explain* this selectivity or how they were caused."[26] There are many other mysteries that are associated with such geosynclines such as vast and intricate dry canyons or coulees, hanging valleys, dry waterfalls, rock-rimmed basins.

A number of similar plateaus, marked with tremendous geosynclines akin to the Grand Canyon, exist throughout the world. While the sedimentary beds have been classified, the process of their formation has not been adequately addressed and remains mysterious. A number of prominent geologists have commented on the absence of any defining cause for these plateaus, and the massive erosion that formed the canyons.[27] Of the Grand Canyon, George H. Billingsley, of the U.S.G.S. in Flagstaff, expresses doubt that anyone may ever be able to verify the early history of the canyon. He said, "Most of the evidence is gone, because the canyon swallowed the clues to its early history as it grew wider and deeper. It's a puzzle with too many pieces missing."[28]

Summary of Major Thinking of Grand Canyon Ignored the Column

The estimates of age of the Grand Canyon have been reduced from hundreds of millions of years to fifty million, and less, so that now some areas have been reduced to less than a million years. All of these dates have been in deference to the uniformitarian view to exclude catastrophe. Isn't it possible that a theory, which was based on a bias against God, as demonstrated earlier, has really caused a blindness that has hindered geological thinking? Is an even earlier date for all of these major mysterious geological phenomena possible?

Geologists have confirmed that about seven thousand years ago an amazing amount of water flooded into the Mediterranean and the Black basins. This may also be the source of the Grand Canyon and similar areas. For those so long committed to the geological column for dating to support the evolutionary theory the theory was a blinding influence against a massive flood at such a recent date.

In other words, a larger, instant flood created different phenomena than a continuous one. Is this not the possible answer to the unusual phenomena that exists in these wide plateaus and the geosynclines, as J. H. Bentz and others have mentioned? The cataclysmic breakup of the continents may also have contributed.

---

[26] Monastersky, "What's New in the Ol' Grand?, Geology's Great Monument Continues to Baffle and Amaze," 395.

[27] Paul D. Krynine, "A Critique of Geotectonic Elements," (*Transactions, American Geophysical Union*, Vol. 32, October 1951), 743-44; L. H. Adams, "Some Unsolved Problems of Geophysics," (*Transactions, American Geophysical Union*, Vol. 28, October 1947), 673.

[28] Perkins, "Carving this beloved hole in the ground may not have been such a long-term project," 220.

Conclusions about the Geological Column and Dating

Philosophical Bias Drove the Column's Creation, not Scientific Evidence

As many of its supporters admit, the creation of the geological column was motivated by bias. The concept of the column was designed to bolster a scientific determinism founded on deism, and later naturalistic philosophy. Those who backed the column looked at the data from a distorted perspective. While Werner was deistic and used the flood catastrophe too rigidly in explaining data, Hutton, Lyell, and their successors sought to remove any vestige of catastrophe in constructing the column, in order to remove the possibility of miracles and support a deterministic materialistic view of science.

A few men, in largely one restricted area of the world that had been studied, constructed the idea of the column, before the vast majority of geological data was known, at a time when the dating for each successive layer of sedimentary rock worldwide was speculative. But because it was an effective offense against God, religion, and church authorities and in support of deterministic science, it was held tenaciously and *applied worldwide,* even when other geologists pointed out that much of the data did not fit. Great gaps in time existed and other locations were used to help fill these gaps, sometimes with no index fossils visible anywhere, and reversals or contradictions in the fossil record exist that could not adequately be explained. A great deal of data does not agree with or cannot be explained by the uniformitarian view. Gigantic cataclysms are required to explain many phenomena and are a contradiction to the basic premise of the slow formation of the column only by normal daily observed forces over millions of years.

The formation of the geological column required slow sedimentation of huge areas which may have contained 50,000 to 100,000 feet of sediment possibly 1,000 or more miles long and 100 miles wide. Those advocating the formation of the column then said these areas were suddenly uplifted from one to three or more miles. Moreover, these areas have all contained multitudes of marine fossils, which were supposedly slowly formed. This is true of the Rocky Mountains and even of the higher Himalayas. Common sense and good science discredit the old geological column and open the possibility of catastrophe for some of these unexplainable occurrences.

The basic premise that the older rocks are on the bottom and the younger ones on top allowing for a sequence of chronology is basically, logically correct. But the dates assigned to the various rock ages have no certainty to them. And there are huge overthrusts and other formations that contradict the chronology of the column.

Can uniformitarianism explain the attendant phenomena that have been denied to catastrophes? When I studied briefly under Harland Bank, paleontologist at Cornell, he mentioned much discordant data that has been discarded because it did not fit, including iris found by Russian scientists in Precambrian rocks. Footprints of different animals have been found in various places in the Connecticut River Valley and in the Cretaceous period in Texas. This includes footprints of dinosaurs with human footprints together and, in one case, the dinosaur's prints were superimposed on the man's. Human footprints next to three-toe dinosaur footprints were reported in Russia.[29]

The scientists holding to the evolutionary theory and the long ages represented in the column reflect their bias to this unproven view by rejecting any evidence which might contradict it. For example, one publication was actually devoted to discrediting the finding of the Texas dinosaur's

---

[29] See *The Moscow News,* (1983, no. 24) 10.

print. They argued against the findings on the basis that the finders were not trained geologists. But how trained was Darwin and the original pioneer paleontologists? Or the early geologists? The photographs and evidence of the dinosaurs and men being contemporaneous should be given recognition as viable evidence against the long ages. Another illustration of bias is the discovery of ***living tissue in dinosaur*** bones by a team of the eminent paleontologist John Horner which is strong evidence that the dinosaurs cannot be dated millions of years old.

The use of certain fossils to form the geological column, in order to promote evolution, must be questioned by the fact that nine-tenths of all animal organisms already exist in the early Cambrian layers. The claim of a sudden explosion of new forms of life by many scientists, caused by some unproved event, certainly reflects further deliberate blindness to fact in an effort to protect evolution, which is calling on catastrophes to promote another form of extended uniformitarianism. No effort to patch up that now discredited premise can be continued and be credible to men who know the scientific data. The science of geology is of extreme value to mankind. It is needed now more than ever. It has been limited and distorted by its anti-God bias.

It is highly important that the geological principles of Lyell on which the geological column was founded are no longer acceptable. The discovery of great cataclysmic forces which were denied are now known to have certainly occurred. The evidences of a worldwide flood have emerged such as that of the study of the Grand Canyon. This started out with the uniformitarians' gradualist theory of scientific determinism, but God has led geological scientists to slowly produce the truth: the most likely source of the canyon is a catastrophic lake overflow that had forces that could have reduced the time from millions to thousands of years to conform to the worldwide flood. Others have been presented. The break up of the one continent of Pangaea into the great tectonic plates and these tectonic forces were undiscovered, yet these are some of the most important geological forces. This much is true: the use of the geological column in estimating the age of the canyon reveals the scientific evidence as false and confirms the possibilities of the biblical data. The wisdom of God revealed through science again discredited the perversions of intellectual human bias.

Moreover, the use of absolute dating by radioactive elements, which are supposed to prove the geological column, usually require ***first*** having *a preliminary beginning point in the column* for the arrived-upon radioactive dates. In other words, the dates they have guessed at in the geological column are used to help determine the radioactive answer. What math teacher would allow that *method*? True enlightenment about geology should produce some Christian geologists who can rewrite geological history with greater understanding to help mankind.

Chapter 23
Enlightenment about Dating Methods and Ages of Evolution

Importance of New Dating Methods

The major credibility problem for evolutionary science for years was the inability to convincingly measure time in the development of astronomy and also of geology. This was critical for the theory and the discovery of ways to measure were not formed until the first half of the twentieth century.

As shown in the previous chapter, the long ages of the earth required for Kant's evolutionary theory came to a focus with Charles Lyell. Using the brilliant work of previous Christian geologists in establishing geological strata, Lyell, along with G. P. Deshayes had developed a basis for the geological column. This had to be expanded after popular acceptance of Charles Darwin's estimate of at least 300,000,000 years needed in biological evolution. But as stated in the last chapter, the lengths of time for the various strata were estimates with no real objective standards of measurement.

Scientists have assumed that the earth has an inner core of crystallized iron/nickel that is probably five miles in diameter; that it is surrounded by a fluid intermediary core of iron/nickel and other elements in successive layers, then a lower mantle of silicate perovskite with oxidized iron, and an upper mantle of olivine peridotite. This mantle has a crust, like the peeling around an orange or grapefruit. Geologists have also thought that most of the radioactive elements are in the upper 22 or more miles of the mantle, and that radioactive materials were formed in the near surface areas of the earth.[1] While no one knows exactly when or how, scientists believe the radioactivity began quite early, based on isotopes of neodymium that have decayed to samarium in rocks dated back to about 3.8 billion years ago.[2] Discovery of radioactivity as a way to measure earth time enabled geologist to claim some objective long age dates to the strata. From radiometric dating, long ages became a basis of conviction of the certainty of the evolutionary theory, and began to be called "Absolute Dating." Some qualified Christians in this field have worked at evaluating this important technical mater.[3] This chapter also seeks to evaluate this dating.

Explaining Absolute Age Dating in Geology, Especially Radioactivity

History of Development of Radiometric Dating

At the turn of the twentieth century Ernest Rutherford's work opened the possibility that radioactive decay could more accurately date the age of the earth. But the real breakthrough occurred in 1931 when Knopf, Schuchert, Kovaric, Holmes, and Brown published *The Age of the Earth,* demonstrating how the half-life of radioactive uranium, decomposing into lead, could be used to determine the age of old igneous rocks. From 1935-1940 new methods using mass spectrometry were developed. From 1938-1941 A. O. Nier of Harvard led in making corrections of daughter elements at crystallization. Most important was the work by W. F. Libby in the use of radio carbon dating. Since that time, an explosion in information, methods, efforts to standardize,

---

[1] Recent speculation is that even the inner core is radioactive. Cf. Brad Lemley, "Nuclear Planet: Is there a five-mile-wide ball of hellaciously hot uranium seething at the center of the Earth? " *Discover*, August 2002, Vol. 23, number 8, 34-42. More about this new theory later.

[2] R. Cowen, "Taking a Chemical Look at the Early Earth," *Science News*, Vol. 141, April 4, 1992, 214.

[3] Larry Vardiman, Andrew A. Snelling and Eugene F. Chaffin, editors, *Rdioisotopes And The Age Of The Earth*, (El Cajon, California: Institute For Creation Research 2000), 676 pages.

and new instruments to evaluate samples more precisely and easily have contributed great advances to the process. Chemical dating methods complement the more traditional methods. The huge amount of data, variation in methods, possibilities for error, and other factors make this a very complicated process. It is even difficult for professional scientists to evaluate the results. It is partly because of the complexity of radioactive dating that many do not venture to try to understand or critically evaluate it, and why scientists tend to claim this as a holy place, where no one but they should enter. This chapter is aimed at discovering the reliability of their claims.

Simplistic Explanation of how Radioactive Elements Work

Atoms are made up of a nucleus of protons and neutrons. Those with more than 82 protons are radioactive, as well as some naturally lighter ones. The elements with the same number of protons in the nucleus, but with different neutrons, are called isotopes of the same element. The heavy elements are uranium, thorium, and the actinium series. These elements undergo radioactive decay or breakdown of the nucleus and lose various kinds of particles so that they are changed into new isotopes, and in so doing lose heat. By discovering the rate of decay and measuring the amount of the parents and the daughter isotopes in certain minerals, a time can be determined. "It is ***the principal source of information about the absolute age*** of rocks and other geological features…." [4] (Emphasis added). It has been assumed that at the time of the Earth's origin the radioactive elements were homogeneously distributed and moved that way to the location that under normal circumstances, certain elements change only into certain other elements or isotopes.[5] Although there is no way of knowing this and some geophysicists don't agree,

There are three ways of radio-active decay. The first and most prominent one is alpha decay. An alpha (first letter in Greek, a) particle consists of two protons and two neutrons bound together into a particle identical to a helium nucleus in decay. Alpha particles, like helium nuclei, have a net spin of zero, and classically a total energy of about 5 MeV. They are a highly ionizing form of particle radiation, and have low penetration depth. A nucleus emits an alpha particle in decay and changes it into an isotope. Alpha particles may be loosely used as a term when referring to stellar helium nuclei reactions and also when they occur as ***components of cosmic rays***. Second is Beta decay which occurs with the transfer motion of a neutron into a proton or proton into a neutron, thus increasing or decreasing the atomic number. Thirdly, gamma ray decay is a change in energy level of nucleus to lower state emitting electric magnetic radiation. Dating usually focuses on the alpha decay.

Uranium has several isotopes, such as uranium 238, which decays into an atom of lead isotope 206. In so doing, it gives off eight alpha particles, numerous gamma rays and beta particles, which in turn give off 47.4 million electron volts of energy. Lead has four sister isotopes, two of which are also formed from uranium. Uranium forms lead 206, which is 26 percent of the earth's lead, and lead 207, which is 21 percent of the earth's lead. One isotope, lead 204, is stable. It is not formed from disintegration, and is only 1.4 percent of earth's lead.

Some elements, such as uranium 238, are parent elements, while others, such as lead 206 are daughter elements. If one has a sample of both parent and daughter in a closed mineral and knows how fast the radioactive transformation occurs, by measuring the amount of the parent and the daughter elements, one may measure the time the reaction has been going on in the closed mineral that has cooled and locked it in. If the mineral has been disturbed at some time, and there is an error, there are various ways for correcting. For example, the proportionate amount of stable lead 204 in the sample of mineral may tell what proportionate amount of sample has been lost.

---

[4] *Rdiometric dating* , Wikipedia the free encyclopedia.

[5] Hurkey, *How Old Is The Earth*, 68

The Way Radioactivity is Used to Date Fossils and Evolution

The assumption is that radioactive elements were ***homogenously distributed*** and began to decay soon after their formation, adding a great deal of heat to the initial heat from the earth's formation. As the convection heat from these radioactive elements occurred, melting began and the molten lava then migrates to the surface with the radioactive materials. They had already disintegrated some to produce the heat. Since they continued to release heat, they were pushed upward in the liquid while the solid igneous material settled downward, leaving a stable mantle. As the radioactive materials near the surface they cool, and as minerals form and cool, these radioactive elements are locked into the rock mineral. They then continue to decay in the mineral at the set rate for the particular type of atomic isotope. If heat or other means do not disturb this mineral, one can measure the amount of parent (e.g. uranium) and daughter (e.g. lead) and thereby arrive at an age for the mineral formation associated with sedimentary rocks and fossils in them. Geophysicists call this an atomic clock.

These radioactive elements for dating are found only in the igneous rocks (those heated and melted) that contain no fossils. Some radioactive material may be found in sedimentary rock. However, one cannot date sedimentary rock since the clocks were not "reset" at formation. As radioactive elements come up in molten lava, they cool and form into minerals on top of sedimentary rocks where the fossils are already formed. Or they may fill in spaces between sedimentary layers. The radioactive minerals are thought to be useful in measuring the age of the fossils in the sedimentary rocks above and below, if the minerals have not been affected and the mineral sample is kept pure until tested. By using radioactive decay to date the igneous rocks above and below the sedimentary rock, the adjacent sediments with fossils are given an "absolute age date." Thus, by using the rate of decay of the parent element to the daughter element, the amount of the change of the parent to the daughter is thought to reveal the age of the fossils in the sedimentary rocks to which they are related. Since the elements decay exponentially, the age is more easily measured by calculating the half-life of the element. This seemed to be an ingenious idea for measuring time dates in geology.

Difficulties of Using This Radiometric "Absolute Age Dating"

Problem of Complexity of Method

The credibility of the geological column came into question and the passion for a more accurate method of dating grew more intense. Since there have been wide differences in radioactive dating, even in the same layers of rock, other dating methods have been devised to assist. New dating methods are complex and may be calculated in multiple ways. To give some idea of the complexity of these new dating methods, I have counted sixty-one (61) radiometric dating methods, using a number of different elements that decay, six (6) chronostratigraphic methods evaluating magnetism, measuring oxygen isotopes, radionuclides, geochemical and other time markers, and ten (10) chemical dating methods. This comes to a total of seventy-seven (77) procedures requiring different methods and instruments. I made this estimate a few years ago and there are probably more today. These are the most common, but are by no means all the methods; new ones are continually being devised. The radiometric methods are considered by far more reliable than the chronostratigraphic and chemical methods; these others are mainly assisting methods. Mebus Geyh and Helmut Schleicher have said, "Radioactive decay is the most important of these processes, on the other hand, methods that are based on example, or stratigraphic anomalies or utilize chemical reaction rates as a measure of time can generally yield only relative ages."[6] Because of the basic importance of radiometric dating, this is the main focus here.

Understanding radioactive decay has practical purposes other than dating the strata and following evolution, including the possibility of understanding climate jumps, insights into tectonic activity of earth structures, and even claims to predict the future by comparing long-term records. Amazing new instruments, such as laser and ion microprobes, accelerator mass spectrometers, and scintillation instruments linked to on-line computers make the determinations and interpretations faster and more accurate.

While these new methods contribute more information, they also introduce more uncertainty. M.A. Geyh and H. Schleicher, in their highly authoritative review of these methods, have said,

> The spectrum of physical and chemical dating methods now covers the entire range of Earth history. But there are so many methods that it is becoming *more and more difficult to select* those that are appropriate for solving a specific problem. It is also becoming *increasingly difficult to assess the meaning* of the data obtained; for example, the question may arise whether the determined age is the age of formation, early or late digenesis, or some stage of metamorphism. Moreover, different components of a sample *may yield different kinds of ages*, depending on the method applied. In *some cases*, even *no date at all* may be obtained. Hence, the selection of the most appropriate method for a particular task remains a problem.[7]

Problem of Time of Igneous Rock Being In Sediment with Fossil

Earlier dating had many errors, some of which have been addressed over time. In the beginning, the various isotopes were unknown, and the crude method of weighing the uranium and lead (or parent and daughter) produced a crude answer. Hurley emphasized a key problem. He said, "One of the most difficult problems in the development of these methods has been the accurate determination of the half-life of the parent element."[8] Since he wrote in 1960, the new instruments are more accurate, and the isotope abundances better known. Still, it was necessary for the Sub Commission for Geochronology of The International Union of Geological Sciences, (I.U.G.S.) to establish values at the 24th Geological Congress in 1976.[9] Today values are accepted, with exceptions.

A main problem is that radioactive dating cannot be done in the sedimentary rocks where the fossils are, and the age of the fossils and the igneous rock intrusion may not be the same. As Moore said, "Sedimentation rates are much too variable and little known, however, to put much reliance on such a method of computing geologic time." He further says,

> Here lies the main difficulty in calibrating the early parts of geologic time. The uranium, thorium, and other radioactive elements do not occur in fossiliferous stratified rocks, but in igneous rocks that intrude the sedimentary strata. *How long a time lapsed* after the sediments were deposited before the igneous rocks invaded them generally ***is unknown***."[10] (Emphasis added)

But this unknown is generally ignored and the time is assumed to be the same.

---

[6] Mebus A. Geyh and Helmut Schleicher, *Absolute Age Determination: Physical and Chemical Dating Methods and Their Application,* R. Clark Newcomb, translator (New York: Springer-Verlag, 1990), 3.

[7] Ibid. vi.

[8] Hurley, *How Old Is the Earth?* , 104.

[9] Mebus A. Geyh and Helmut Schleicher, *Absolute Age Determination....* 378.

[10] Moore, *Introduction to Historical Geology* , 30, 33.

Problem of the Condition of the Sample Igneous Rock with Radioactive Elements

Many factors are involved in determining the value of the sample of igneous rock to be measured. The *conditions for the formation of the mineral may determine if it is a valid sample to read.* Did it cool quickly or did it take some time to form and lock in the radioactive element?[11] Hurley said,

> So, what we really are measuring is the time that has elapsed since the materials in granites and other rocks crystallized. Our measurements will reflect one or two events: ***the time*** when molten materials ***came up*** into the crust, or ***the time*** when crystal materials were so deformed and heated that they ***recrystallized*** into their existing forms. Also, we are able to determine the age of the earth itself (within fairly narrow limits), and the *age of the elements* that make up our part of the universe, *by methods that differ* from those used on crystal materials." (Italics added; I will refer back to this qualification that in radioactive dating one must determine the age beforehand by a different method.)

Then Hurley said, "When a crystal is formed, it is a tight system that sometimes lasts for billions of years without disturbance. On *other occasions* the system is not tight and contains *elements that may diffuse through it or exhange with other elements* in the environment."[12] Geyh especially emphasizes avoiding samples that have even come in close association with certain other conditions—

> It must be certain that no younger igneous rock or hydrothermal veins occur near the sampling site that could have caused allochemical overprinting, thus putting the presence of a closed system in doubt. *Younger thermal overprinting can reset* or partially reset the radiometric clock, resulting in ages that are too young.[13]

Concerning some of these difficulties, Hurley said, "The drawback with the method is the ***lack of general distribution*** *of radioactive minerals* and the fact that these radioactive minerals have suffered such a large amount of radiation damage that *they freqently* ***show ages that are not concordant.***"[14] Geyh also says, the methods "assume that at the time of the formation of the rock, mineral or meteorite, they were either completely separated from one another, or the amount of daughter isotope originally present can be estimated. *This condition is fulfilled in nature* ***only in exceptional cases.***" He adds,

> After its formation, the object being studied must have formed a ***closed system***, i.e., there has been no transport of isotopes between the system and its surroundings (i.e., by diffusion, chemical exchange, weathering, evaporation, condensation, and contamination). *This condition is also* ***not always fulfilled***; many rocks have been proven to represent an open system.[15]

Hurley also said, "The relative abundance of radio-genic isotopes can vary considerably, depending whether or not these isotopes have been in close association with the parent element

---

[11] Geyh, Absolute Age Determination, 11-12.

[12] Hurley, How Old Is the Earth ?, 76-78.

[13] Geyh, Absolute Age Determination, 13.

[14] Hurley, How Old Is the Earth?, 105.

[15] Geyh, Absolute Age Determination, 26.

over a long period of time." He further says, "In very *young* materials it is ***almost impossible to eliminate all contamination***; to obtain accurate age measurements the correction is essential."[16] Since ***most*** *fossils are young*, this is significant. Geyh says, "***Errors*** arising from ***incorrect sample collection*** are often considerably *larger than the analytical error* of the age determination."[17] So, it is clear that it is *very* ***difficult to find a reliable sample***, that certain assumptions are made about it, and that a *different method for an approximate age* is needed before this "most reliable" method is used.

Frequent Errors in Packaging, Delivery, and Analysis

In addition to these difficulties, *wrong packaging and unsafe transporting of the samples* can cause error. Geyh warns about mixing packaging material and the effect of moisture on the sample. Losing labels is also a problem. Samples that decompose must be dried or frozen for storage. "An important but *unfortunately not always fulfilled condition for satisfactory dating* is that the samples arrive undamaged at the laboratory."[18] Once the sample is delivered to the laboratory in good condition, there are still many errors that can occur in its analysis. In evaluating several different radioactive elements and their daughter isotopes in the same sample, sometimes there is concordance, and sometimes there is not. This leaves one with great uncertainty. This early method of estimating with single samples, "the model-age" method, was not authoritative, and, therefore, not credible. These difficulties questioned "absolute dating" and led to efforts to solve them, but geologists do not always follow them.

Variations in Model-age Samples Led to Isochron-Age Method to Resolve Differences

Because the model-age method was not reliable, scientists conceived of another method they felt gave greater certainty—the isochron-age method. This is based on multiple isotopic analyses of various rocks or minerals from one entire geologic unit or area. (Isos is Greek for "equal," chronas for "time.") By locating the amount of parent and daughter of each sample, a linear graph is formed on an inclined line, ascending to the different amount in each sample. The linear array is interpreted as an effect produced by the radioactive decay. Lines can be projected to the horizontal line, showing the rocks as they originally formed. ***This assumes the daughter elements were homologous at the beginning***, so the amount of daughter elements was the same, but the amount of the parent's was different. The isochron method attempts to account for initial daughter atoms in a rock at its formation. Scientists devised a formula for figuring the isochron and wrote a computer program to automatically arrive at a date. While this is not a complete description of the method, it gives the reader the idea of the approach. But it is important to note that the very ***need for the method shows the variation of the individual samples in a specified area.***

Questions about Corrections of Sample Errors

With many questionable samples, ways of correction may be attempted. For some daughter elements—lead, argon, and others—there are, besides daughter isotopes, stable isotopes that occur in the sample. For example, lead 204 is stable in nature, while 206 and 208 are daughters and are unstable. If the daughter isotope does not appear correct, an adjusted assumption can be made by the amount of known stable isotope (i.e., 204) in the sample. The stability of these isotopes and the

---

[16] Hurley, *How Old Is the Earth,* 80-81, 85.

[17] Geyh, *Absolute Age Determination*, 11.

[18] Ibid., 8.

measures for correction seem a logical way for assumed corrections. But, if there is an error in the daughter amount, *how can we be sure* there is not escape or diffusion *of the stable isotope*? Also, can we be sure there has always been that proportion of stable isotopes? Moreover, in any sample, how can we be sure there had not been a *loss or intrusion of the daughter or parent?* While the analysis may be correct most or some of the time, how can we be sure? To ask these questions upsets those doing this tedious work, but are they not valid?

Contradiction: Using Geological Column of Strata to Find Right Radiometric Samples

In addition to all these problems, scientists' *aim* is usually to fit the dating of the fossils found in the sedimentary rocks *within stratigraphic boundaries* (those in the geological column). But, we have seen that the ages in the column are an imaginary composite, often from different locations. Sometimes one has to correlate these boundaries between continents. It is very difficult to find rocks associated with boundaries which can be reliably dated. Geyh commented on this difficulty, saying, "This, however, is *happenstance*, and is *just barely fulfilled if large regions are taken* into consideration. *Extrapolations* to locally defined *chronostratigraphic boundaries in other regions are often necessary.*" But in doing this, one confronts the problems of dissimilar time scales based on dates of different dating methods of different rocks and minerals, and that many dates set before the standardizing of decay rates done in 1976, and these all have not been readjusted.[19] These problems require more examination. Geyh said,

> However, it illustrated that each radiometric 'age' is ***never more than*** *an analytically determined geochemical* ***parameter*** [date], which can provide information about the time of a specific geological event *only when all known geological, petrographic, and geochemical aspects are included in the interpretation.*[20] (Emphasis added)

It is no wonder that Geyh recommends that those gathering samples for analysis include someone who has previous experience with the established dating. He said, "It is highly recommended that a geochronologist participate in the initial sample collection and the setting up of the pilot study.... If at all possible the samples should be taken *from stratigraphically known positions*."[21] Hurley also recommended that the sample come from a position *where the age had* ***previously been determined*** *by a method that differs."* So, there must be an appeal first to the uncertain geological column or to some other source to assist with the radioactive dating. Hurley said, "Strangely enough, it has been *qite* ***difficult to find out in actual years*** how long each part of the geological age scale is."[22] (Emphasis added)

In other words, to use radioactive dating, the sample must come from a site from which the date has already been determined on the basis of the fossils in the sedimentary rocks. But we have already seen the fallacy of this dating method. Thus, the dating of radioactivity to prove the column is also dependent on the column. This is the paradox, and both involve a certain amount of guesswork.

---

[19] Ibid., 378.

[20] Ibid., 12.

[21] Ibid. 7.

[22] Hurley, *How Old Is the Earth?*, 116.

Problem of Suitability of Radioactive Element for Dating Age of Particular Rocks

Carbon Fourteen Dating, Reliable for Short, Recent Periods

The most credible method of radioactive dating has been to determine the amount of carbon fourteen in organic substances. As Moore said,

> In quite a separate class is the carbon–14 (^{14}C) age-measurement method because of the relatively rapid rate of its radioactive decay. This adapts it for pinpointing determinations of age within the last 25,000 years and for making somewhat less precise age computations to 40,000 years ago or a little more.

He describes this method,

> The isotope ^{14}C is produced from nitrogen atoms (N^{14}) in the atmosphere by bombardment of neutrons (cosmic rays) coming from outer space. Each N^{14} atom yields one ^{14}C proton. The ^{14}C combines with oxygen (O) to make radioactive carbon dioxide ($^{14}C\ O_2$), which eventually reaches the earth's surface and is absorbed by living matter.... When incorporated in organic tissue, such as a piece of wood or fragment of bone, the quantity of ^{14}C can be measured accurately. Knowing the rate of decay, or half-life, one can figure the age of the organic sample.[23] (The half-life of ^{14}C is 5730 years)

Radiocarbon dating has gained credibility for the more recent years because its samples have been verified by tree rings and other collaborating evidence, and found reasonably in agreement. Before the most recent ten thousand years, accuracy declines exponentially. One problem with radiocarbon dating is that ^{14}C and ^{12}C ratios in the atmosphere have not always been constant; plants discriminate against carbon dioxide containing ^{14}C. Also, of great significance is the fact that ***exposure of the elements to cosmic radiation*** boosts the age determination. In recent years huge cosmic radiations were found that could have influenced the age.

Some significant corrections in radiocarbon dating have been made as recently as the 1980s. Many of the difficulties of sample collections of other methods also apply to organic objects. The possibilities of error are significant. But ^{14}C has been very useful in archaeology and in studying bones and other human organic materials. However, radio carbon dating has been a significant factor in dating many human activities, such as food domestication, to within a few thousand years of the Noahic flood. But science dogmatically asserts that radio carbon dating for short-term dating is reliable.

Radioactive Dating with Many Elements Are Not Applicable

While many of the early uncertainties and difficulties with radioactive dating have been significantly reduced, all of the recorded deviations have not, and probably can never be corrected. Holms and his team were the first to use the half-life of the elements with uranium. As methods improved, others expanded this to other elements (U/Pb, Pb/Pb, K/Ar, and Rb/Sr and later with Re/Os, K/Ca and others). The primitive methods of dating were replaced by improved accelerated mass spectrometers. Many dates set by different half-lives were replaced by agreed-upon standards. Early collection methods were crude and allowed for errors. Today they have been refined and eliminate most errors. The earliest dates set by Holms in 1937 have been stabilized through the years by others: Kulp (1961), Harland (1964), Lombert (1971), Fitch (1974), Palmer

---

[23] Moore, *Introduction to Historical Geology,* 31-32.

(1983), and Salvador (1985). However, those using radioactive dating can argue that the method has yielded some ***reasonably similar long age dates***.[24] They have impressively produced a method that seems to give a progressively lower time from the Precambrian times or the oldest rocks where fossils are scarce. Only certain elements have usefulness and one must choose the half-life time they want to apply: potassium 40 to argon 40 has a half life of 1.3 billion years; uranium 235 to lead 207 is 700 million years; Uranium 234 to thorium 230 has a half-life of 80,000 years, Uranium 235 to protactinium has half-life of 34,000 and so on.

The half-life of such radioactive elements is not reliable for dating most fossils, which are too young to measure by slow decaying element. Potassium-argon is the preferred element for studying some of the older strata. Potassium is the eighth most abundant element and has three natural isotopes: masses 39, 40, 41. The decay of ^{40}K, a liquid, continually produces radiogenic ^{40}Ar, a gas. Argon is a noble gas, not combining with any other elements chemically, but normally incorporated into minerals, and is not bound in the minerals formed, but is retained and accumulated by crystal lattices at low temperatures due to the large atomic radius. But Geyh pointed out the real problem with potassium to argon dating,

> But the system is very sensitive to thermal influences, loss by diffusion occurs at elevated temperatures; *in melts, accumulation of argon is* ***no longer possible***. ...Below a critical temperature ...all the argon produced is retained by the crystal; at high temperatures, none. In between there is a transition range in which partial accumulation occurs.[25] (Emphasis added)

Heat loss is a critical issue with this method.

There are other elusive problems. For example, one strong advocate of the use of the system has been Brent Dalrymple, a geochronologist. In his work with historic lava flows he found "non-zero concentrations of ^{40}Ar in violation of the key assumption of the K-Ar dating method. He went on to state that some cases of initial ^{40}Ar remaining in rocks have been documented...." While claiming this to be uncommon, nearly 20 percent of his samples contained excess.[26]

All these efforts to document the geological column by uniform determinists deny cataclysms, such as the worldwide flood, formation of ice sheets, and the breakup of the continents, and exaggerated cosmic rays and magnetism connected, despite the strong evidence we have given. These cataclysms would therefore render K/Ar measurements very questionable. Yet, this is the main way of measuring radioactive time for ages longer than when ^{14}C is useful. And the other influences of radiation and magnetism apply to this method as well.

While it is logical that the top layers of sedimentary rock are the youngest and the bottom layers are oldest, the use of radioactive dating was developed primarily to show that the chronological record supported evolution. Uniform geologists expected that as they found fossils, they would experience a cone of growth from small and simple to greater and complex. However, what they discovered was not what they expected. Also, beginning with the early organisms, Darwin's theory of macroevolution turned out not to be true, having no demonstrable mechanism that worked. The whole paleontological record failed to show transmutation from one major species to another in the Darwinian tree of development. And the vast majority of fossils do not fit into the evolutionary tree as diagrammed. Moreover, the geological column is often uncertain because of over thrusting et al. Hence, even if the dating by radioactive elements is correct, the

---

[24] See Geyh, *Absolute Age Determination*, 374.

[25] Ibid., 57.

[26] Larry Vardiman, Andrew A. Snelling, Eugene F Chaffin, ed. *Rdioisotopes and the Age of the Earth,* 126.127. Dalrymple's work, 1969, 1991.

long ages in the imaginary column is not substantiated because of contradictions. Using radioactive dating to assign verifiable dates to a speculative column is a decidedly unscientific approach.

The imprecise, sometimes confusing, and even contradictory dating of the geological strata of the rocks seem to give a reason for upholding an older earth. But has the rate of decay now seen by the heavier elements always been constant? If there is evidence of a younger earth, the long rate of decay could be defective and even a perversion in trying to prove an old earth in order to give credibility to a uniform gradualism. There are several evidences linked to the field of formation of samples for dating radioactivity that require the earth be young. The last two given below are very persuasive.

Reasons to Question Old Age Dates

Beginning Mixing of Radioactive Elements before Entering Minerals

There are a number of reasons to question older dates. In the past, scientists have generally agreed that the radioactive elements were created in the mantle of the earth and that they began to decay and rise toward the surface soon afterwards. The radioactive decay caused the heat that made them rise as lava, allowing the liquid come to the surface and the solid to sink and form a solid section of the mantle. The diagram below illustrates this process.

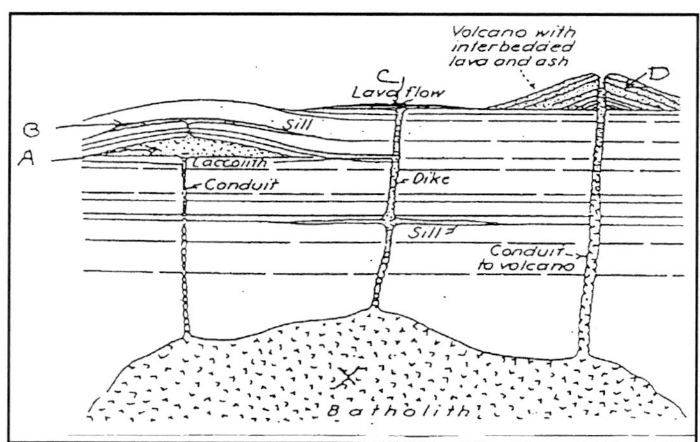

From Branson's illustration above[27] it is clear that the decay of the parent to the daughter element begins deep in the mantle. The radioactive decay forms molten rock that turns into the batholith, or a mass of deep-seated igneous rock. When this is a molten element, the fusion continues to occur and it flows in the conduits and dikes to the surface, where the surface igneous rocks are formed on top of or in between the sedimentary layers.

What is therefore being measured at positions A, B, C, and D is not the time the parent elements have been decaying to the daughter elements within the mineral, but how long it has been decaying since the elements began after creation somewhere in the mantle below position X. Therefore, the rock in the sills or laccoliths does not reveal the time the sedimentary layers have been above and below it, since this rock has intruded there between the layers. If the rock is

[27] E. B. Branson, *Introduction to Geology*, 17.

eroded away, it may appear it was laid on top, when it wasn't.

Radioactive measurements by parent and daughter elements may indicate the amount of time since its creation and the beginning of its decay. This can be only approximate, since as the molten lava emerges, the parents and daughters most probably are mixed and redistributed in various ways. A surface sample therefore indicates the antiquity and depth in the mantle at some X point. Unless the radioactive elements were ***created and distributed completely homogenously and they stayed homogenous*** throughout the turbulent path to the surface, it might not even indicate anything definite about the antiquity of the place it is found in the sedimentary rocks with fossils.

In discussing this problem, Hurley has said, "It is reasonable *to infer* because the earth's heat is coming from radioactive breakdown, that *some process* has moved the radioactive elements to this [surface] location from ***presumably*** *homogenous distribution* at the time of the earth's origin." (Emphasis added). This presumption is evident that later he refers to a disagreement about a homogeneous beginning among geologists,

> There has been much discussion and *difference in opinion about the possibility* of major convection overturns in the mantle down to the core boundary as a result of *inhomogeneous distribution* of heat sources. This process of convection *could* give rise to surface activity also, and *could* be the cause of mountain-building events.[28] (Emphasis added)

In fact, it is not known if there was homogeneity at the beginning. Many believe there would have been a mixing in rising to the surface. Those using the isochron method have had to assume that there are equal amounts of the daughter elements in the samples. But the very need for the isochron method suggests this was not so.

Referring to the model-age method, Hurley noted that the minerals are ***not evenly distributed in the samples*** of an area, but a concordance had to be found to get a date, showing that the condition on the surface shows a considerable disparity, which is evidence of the beginnings of the radioactive materials.[29] Can this dating process really be called *absolute* dating? The age the decay of parent to daughter began in the heart of the mantle, not on the site in the sedimentary rock when the minerals are formed in the strata. Moreover, they may have been changed by cosmic radiation.

Isochrones using Different Elements Yield Different Ages for Rocks

Isochron dating is often not concordant, and sometimes yields conflicting facts as graphed by results in part of the Grand Canyon seen on the following page.[30] Of notice is the results in the basaltic rocks of the Uinkaret Plateau differ from 117 million years to 2600 million years.[31] The differences of the Cardenas Basalt (Precambrian) and Diabase Sills (Precambrian) also have significant differences. Why would the different radioactive elements on the same rocks yield such different ages? At times, some of the ages are so unreasonable that they are rejected outright and go unreported.

---

[28] Hurley, *How Old is the Earth?* , 67, 69-70.

[29] Ibid., 105 as mentioned before.

[30] Steven A. Austin, "Grand Canyon, Monument to Catastrophe," (Santee, California: Institute for Creation Research), 126.

[31] Ibid., 126.

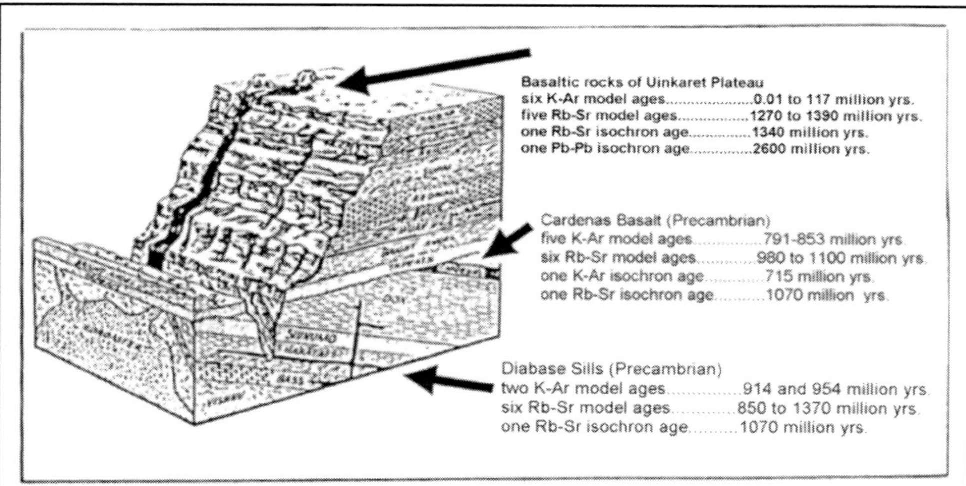

Are Grand Canyon rocks one billion years old? Radioisotope dating methods give strongly discordant "ages." The lava flows at the rim of Grand Canyon, impose an incredible contradiction on radioisotope dating. The flows at the rim appear to be older than deeply buried lava flows. Data sources shown in test.

Isochron Methods Sometimes Yield Great Age for Younger Rocks

Some of the findings produce extremely old ages for what are obviously younger rocks. Dr. Stephen Austin said,

> The 1.34 ± 0.04-billion-years Rb-Sr-(rubidium-strontium)-isochron "age" cannot be the true age for the Quaternary flows on the north rim of Grand Canyon. It is the oldest isochron "age" yet published for Grand Canyon rocks, but it represents the *youngest* lava flow in the Canyon. The Rb-Sr isochron must be grossly in *error!* These lava flows are among the most recent geological formations in Arizona, some of which even fill modern stream valleys. [32]

Radio Active Dates, Too Old for Important Evolutionary Fossils

The radioactive dating of most elements is not suitable for dating the sedimentary rocks that contain fossils. Because decay occurs slowly, radioactive measurements seem to indicate *accurate* dating of 500 million years ago or earlier. Some methods, such as radio potassium, produce dates to 500,000 years. Geyh shows that argon, potassium, and uranium radioactive dates are most useful in the estimated Precambrian times, but run out of usefulness before the Pleistocene Age, when primates and man appear. Thus, the fossils in the sedimentary rock were mostly formed since the elements of more accurate radioactive dating times. Since the unit amount of the radioactive element breaks down exponentially with time, it is less and less accurate.[33] Both Hurley and Geyh indicated that young samples are "*almost impossible*" to date by radioactivity.

New Model of Earth's Core May be Radioactive

An alternative model to the earth's central core is under discussion but not widely accepted.

---

[32] Ibid., 125.

[33] Hurley, *How Old is the Earth*, 100-101.

It uses *uranium in reaction*, rather than the standard theory of iron/nickel. It is supported by the finding that the core of the earth is much hotter than predicted. Daniel Hollenbach of the Oak Ridge Laboratory thinks this a better explanation for the power of the earth's magnetic field that protects from solar radiation. The earth's magnetic field is decreasing, as it would if the element was reacting, having steadily decreased by a factor of 2.7 over the past 1,000 years. This would point to a length of time that is no more than 10,000 years. This is contrary to the premise of radioactive dating—to support long ages.

In its lead article in the August 20, 2002 issue, *Discover* discussed others who hold to the possibility of a uranium reactor in the center of the earth by saying,

> J. Marvin Herndon, an independent geophysicist...believes the preponderance of evidence reveals that Earth is a natural fission reactor, with a massive ball of hot uranium at its center. The idea starkly contrasts with the traditional view that the inner core is a solid ball of iron and nickel, but Herndon feels his theory is in good company. "Geophysics is a conservative science," he says. "Plate tectonics was opposed for 60 years."

Not only does Earth have a reactor at its core, says Herndon, but so do many of the other planets in our solar system. And what about those extra-solar bodies? Could be, says Herndon. Natural nuclear reactors, he says, "are probably a general feature of planets," no matter how far away.[34]

V. Rama Murthy, a geochemist at the University of Minnesota, now says he and his colleagues have evidence that the center of the Earth is heated by large amounts of radioactive potassium, mixed with radioactive uranium and thorium and iron and iron sulfide at the core. If there are radioactive elements at the core of the earth, this raises serious questions about what is going on.[35] If such a theory should gain acceptance, then action of radioactive elements in other areas of the earth will have to be reevaluated. The whole question of radioactive dating is still filled with unproven assumptions that are denied to the public.

Thermo Luminescence Dating of Rocks a Half of a Million to 40,000 years

Because of the distance between these elements with slow radioactive dates to ^{14}C, there has been a strong desire to find some dating method applicable to fossils. In the seventeenth century Robert Boyle, a Christian, discovered the phenomena of thermo luminescence. Recently, Martin Aitken of Oxford developed this as a way to estimate time. Helene Vallada, from the Center for Low-Level Radioactivity of the French Atomic Energy Commission, is using this revised method to date human fossils. Thermo luminescence attempts to fill the gap between the dates from radioactive elements of a half a million years. It seems to be most useful for Precambrian times, and radiocarbon dates of 30,000 years ago or less. Because most of the evolutionary history of the higher animal forms is earlier, supposedly from four million years to the carbon-14 period, the method does not provide definitive answers to dating.

The technique works as follows: mineral rocks have a certain number of electrons, normally captured with crystal impurities or electronic aberrations in the mineral structure itself. These are held in place until the rock is heated and they are released in the form of light. At about 900 degrees, all of it will be released in a luminescent puff, which can be measured by a device called a photomultiplier. The more time that has elapsed since the rock was exposed to heat, the brighter the luminescence will be. Hence, the age of human flint instruments and other stone implements of

[34] Brad Lemely, "Nuclear Planet," *Discover*, August 20, 2002, 35.

[35] Kathy A. Svitil, "A Strange Brew in Middle Earth," *Discover*, August 2003, 16.

prehistoric men can be measured by the photomultiplier, ***if there is no significant environmental disturbance***.

Many factors affect the luminescence of a rock sample, such as radioactive elements in the rock (which may be measured), radioactivity in the soil around it, cosmic rays from the atmosphere, exposure to strong sunlight, an external electromagnetic field, and absorption of external radioactive elements in the ground (such as uranium). Also, the amount of moisture around it affects its absorption rate. Hence, only under rare circumstances can one find a sample that can be securely tested.

Like all radioactive tests for age, a number of the conditions named (such as cataclysms) may make the whole method erroneous, or the individual sample flawed. Hence, this method is very tenuous. Scientists still hope to use other methods to determine the age of the higher animals at strategic points in the column. The thermo luminescence method is upsetting previous dates set for certain prehistoric men, such as Neanderthals and others. James Shreeve said, "To some, this simply does not make sense." He then quotes Anthony Marks, of Southern Methodist University, "If these dates are correct, what does this do to what else we know, to the stratigraphy, to fossil man, to archaeology?"[36] These dates do not come close to agreeing.

Some Radio Evidence Points to a Young Earth

Radiohalos Raise Questions about Rates of Radio Active Decay

Radiohalos are a phenomenon that has been found in 40 minerals. Although they are distributed inconsistently, they are prolific in biotite, a major mineral. Radio halos represent circular areas in minerals as minute circular zones of color or darkening. They represent a historical record of radioactive decay over a period of time that is detailed enough to estimate decay energies. These depict decay steps due to α-emissions. The decay chains in the U-(uranium) and Th- (thallium) series are frequent and reflect that those rates probably have been constant. Only minerals transparent in 30 mm sections are suitable for detailed ring analysis. However, these small fossils of radioactivity show rates of decay. Radio halos of 'squashed' polonium 210 are found throughout the Jurassic, Triassic, and Eocene formations in the Colorado plateau. These were deposited, not in the hundreds of millions of years as required by conventional time scales, but ***within months of each other***. Gentry says certain polonium 218 radio halos are said to show no evidence of mother elements, and thus imply either miraculously short decay rates or instant creation.[37] Can science just ignore this data?

Helium Retention in Minerals is Way too High

In the last twenty-five years more exact methods have produced two other major factors that clearly ***prove that radioactive dating of fossils as millions of years old were wrong.*** One instance occurs when volcanic lavas erupt to the surface and form igneous rocks. The second occurrence is when minerals that contain radioactive materials between the sedimentary rocks trap helium that forms radioactive alpha particles. The alpha force is identified with the emission of the basic particle of helium. Zircon, which is a semiprecious stone, loses its helium, which is an inert light gas, over time at a measurable rate. Rather than being old, studies have shown that zircons in supposedly very old rocks ***have high retention levels of helium***, which suggests that they formed

---

[36] James Shreeve, "The Dating Game," *Discover,* Vol. 13, No. 9, September 1983, 75-83, see 81.

[37] R. V. Gentry, "Radiohalos in coalified Wood: new evidence relating to time of uranium introduction and coalification," *Science*, 194, (15 October 1976), 315-318.

much more recently than thought. The findings are as follows for the oldest or Precambrian rocks.

> Two decades ago, Robert Gentry and his colleagues at Oak Ridge National Laboratory reported surprisingly high amounts of nuclear-decay-generated helium in tiny radioactive zircons recovered from Precambrian crystalline rock, the Jemez Granodiorite on the west flank of the volcanic Valles Caldera near Los Alamos, New Mexico. Up to 58% of the helium (that radioactivity would have generated during the alleged 1.5 billion year age of the granodiorite) was ***still in the zircons***. Yet the zircons were so small that they should not have retained the helium for even a tiny fraction of the time. The high helium retention levels suggested to us and many other creationists that the helium simply had not had enough time to diffuse out of the zircons, and the recent accelerated nuclear decay had over a billion years worth of helium within only the last few thousand years during Creation and/or the Flood. Such acceleration would reduce the radioisotopic time scale from mega years down to months.
>
> The RATE project did the experimental and theoretical studies necessary to confirm this conclusion quantitatively. In 2000 the RATE project began experiments to measure the diffusion rates of helium in zircon and biotite specifically from the Jemez Granodiorite. The data, reported here, are consistent with data for a mica related to biotite, with recently reported data for zircon and with a reasonable interpretation of the earlier zircon data. We show that these data limit the age of these rocks to between 4,000 and 14,000 years. These results support our hypothesis of accelerated nuclear decay and represent strong scientific evidence for the young world of Scripture.[38]

According to the studies, most of the helium is still in the zircons. This is indicated also by the following study.

> In the 1970s, geoscientists from Los Alamos National Laboratory began drilling core samples at Fenton Hill, a potential geothermal energy site just west of the volcanic Valles Caldera in the Jemez Mountains near Los Alamos, New Mexico. There, in borehole GT-2, they sampled the granite Precambrian basement rock, which we will refer to as the Jemez Granodiorite. It has a radioisotopic age of about 1.5 billion years, as measured by various methods using the uranium, thorium, and lead isotopes in the zircons themselves. [39]

This Los Alamos study took into account the greater diffusion for the increased temperature at greater depths. While the radioactive decay of lead present indicated an older age of about 1.5 billion years, the large retention of helium indicated that its—

> ...diffusion took place over thousands of years not billion. ...The zircon data (from the Jemez Granodiorite) are also quite consistent with all published zircon data. ...Considering the estimates of error the data indicates an age between 4000 and 14000 years. ...The data offer no hope for the uniformitarian model. It is unlikely that the zircon data will continue down on the intrinsic line for 5 more orders of magnitude. ...The data in our analysis show that over a billion years worth of nuclear decay have occurred very recently between 4000 and 14000 years ago. This strongly supports our hypothesis of recent episodes of highly accelerated nuclear decay. These diffusion data are not precise enough to reveal detail about the acceleration episodes. ...Our most important result is this: ***helium diffusion casts doubt on the uniformitarian – long age***

---

[38] D. Russell Humphreys, PhD. Institute for Creation Research, Steven A. Austin, PhD. Institute for Creation Research, John R. Baumgardner, PhD, Los Alamos National Laboratory, Andrew A. Snelling, PhD. Institute for Creation Research, *Helium Diffusion Rtes Support Accelerated Nuclear Decay,* (RATE papers presented in June 2003) pg. 1.
[39] Ibid., 2.

interpretations of nuclear data and strongly supports the young world of Scripture.[40]

Helium Lost from Radioactivity is Missing in Atmosphere

Radioactive decay produces helium. If the earth were five billion years old the rate of helium loss from diffusion would result in an accumulation in the atmosphere. In reality, the helium in the atmosphere is only 0.05 percent of what would have accumulated in five billion years. The rocks of the earth are supposed to be over one billion years old, but their helium retention suggests they are only thousands of years old.

Older Rocks ought to be Carbon Fourteen Dead, but are Often Still High in it

In addition to the high retention of helium in zircon, which indicates an exaggerated accelerated rate of radioactive decay, another very significant factor has emerged through many and repeated measurements of ^{14}C. An unexpected occurrence of ***^{14}C has occurred in old rocks that one would expect to be ^{14}C dead.*** "Given the short ^{14}C half-life of 5730 years, organic materials purportedly older than 250,000 years, corresponding to 43.6 half-lives, should contain absolutely no detectable ^{14}C." [41] The radioactivity should have been long over. But even more importantly, the fossil remains, which these older rocks are supposed to validate as older, have too much radioactive carbon. Phanerozoic samples are those from the three periods of earliest living creatures.

> An astonishing discovery made over the past twenty years is that, almost without exception, when tested by highly sensitive accelerator mass spectrometer (AMS) methods, organic samples from every portion of the Phanerozoic record show detectable amounts of ^{14}C! ^{14}C/C ratios from all but the youngest Phanerozoic samples appear to be clustered in the range 0.1-0.5 pmc (percent modern carbon), regardless of geological 'age.' A straightforward conclusion that can be drawn from these observations is that all but the very youngest Phanerozoic organic material was buried contemporaneously much less than 250,000 years ago.[42]

Since the fossil evidence shows almost all organisms existed in the earliest Cambrian era, ***there can be no radio carbon earlier than 100,000 years*** (since its life is then expended). The radioactive dates using the other elements that have very long life cannot be correct. This supports the fact that there had to be accelerated rates in radioactive decay of heavy elements. P. Giem has reviewed over 70 reported AMS (Accelerated Mass Spectrometer) measurements of ^{14}C in organic materials from the geological record. In older rocks of the geological timescale, which should be ^{14}C dead, "The surprising result is that organic ***samples from every portion of the Phanerozoic record show detectable amounts of ^{14}C*** with the ratios that appear to fall in the range of the percent of the modern ^{14}C/C ratio."[43] John A. Baumgardner et al. has evaluated a number of the most important studies of these uniformitarian deterministic geologists. Every effort to remove supposed contamination, even very extreme ones, produced potentially the same results. For example, the Leibnitz Christian-Albrechts Laboratory in Kiel, Germany concluded,

---

[40] Ibid., pg. 4-15.

[41] John R. Baumgardner, PhD. Los Alamos National Laboratory, Andrew A. Snelling, PH.D, Institute For Creation Research, D. Russell Humphreys, PhD, Institute For Creation Research, Steven A. Austin, PH.D, Institute For Creation Research, *Measurable* ^{14}C *in Fossilized Organic Materials: Confirming The Young Earth Creation-Flood Model,* (RATE reports, June 2003).

[42] Ibid.

[43] Ibid., 1.

> It was not possible to reach lower ^{14}C levels through cleaning indicating the contamination to be **intrinsic to the sample**. So far, no theory explaining the results has survived all the tests. No correction between surface structure and apparent ages could be established.[44]

Many other experiments produced similar results.

The fact is that this anomaly has been reported from the earliest days of ^{14}C dating and has been ignored by those trying to give an early date to the strata. Baumgardner refers to R. L. Whitelaw's study of ***15,000 radiocarbon dates,*** that covered specimens of all kinds of alleged ancient material including coal, oil, natural gas, etc.[45] These facts led to the extended study of the Radioisotopes and the Age of the Earth (RATE) team. Their results, using the best and more recent AMS tests, with much greater sensibility, confirmed all of these findings. They said,

> We conclude that the possibility this ^{14}C is primordial is a reasonable one. Finding ^{14}C in diamond formed in the earth's mantle would provide support for such a conclusion. Establishing the non-organic carbon from the mantle and from Precambrian crustal settings consistently contains inherent ^{14}C well above the AMS detection threshold would, of course, argue the earth itself is less than 100,000 years old, which is orders of magnitude younger than the 4.56 GA currently believed by the uniformitarian community.[46]

Baumgardner et al. have concluded that this ^{14}C evidence, Performed largely by uniformitarians who hold no bias in favor of this outcome places, "Very strong constraints on the age of the earth itself. …If this conclusion proves robust, these reported ^{14}C levels then place a hard limit on the age of the earth of less than 100,000 years, even when viewed from a uniformitarian perspective."[47] It is Baumgartner's conviction that this indicates catastrophic influence, which changed radioactive rate acceleration, both for the heavier elements and ^{14}C. From a creationist perspective, Baumgardner attributes this primarily to the flood and possibly earlier geology involved in creation.

Isotopes and New Elements Created in Other Ways than Decay Rates Change Isotope Ratios

Forces That Cause Creation and Change in Radioactive Elements

After the closure temperature of a mineral sample it is thought that there is no interference with the sample. However, changes may occur before the closure that produces daughter isotopes that appear in the sample and give false readings in the sample. Mebus A. Geyh and Helmut Schleicher in their comprehensive review entitled, *Absolute Age Determination* report on the forces of creation and change and review a number of these studies. They say, "Factors affecting cosmogenic production are (a) the intensity of the primary ***cosmic radiation***, (b) ***solar activity***, and (c) ***the geomagnetic field***.[48] (Emphasis added) They refer to a number of studies that were done in the 1980s that show these forces. They present a graph of changed dating with ***Cosmogenic Radionuclide***. They refer to other studies that demonstrated modifications by the earth's magnetic dipole movement and some other references to possible effect to "changes" in oceanic abyssal water circulation."[49] They also refer to atmosphere and ocean influences concerning southern and

---

[44] Ibid, pg. 6.
[45] Ibid, pg. 2.
[46] Ibid, pg. 10.
[47] Ibid, pg. 12.
[48] Geyh, *Absolute Age Determination*, 161.
[49] Ibid., 167.

northern hemispheres and to the effect of volcanic active areas which lower ^{14}C concentration of atmospheric CO_2.[50] Geyh also refers to several studies of time scales from theoretical models of glaciers, and accumulation and ablation rates of ice sheets.[51]

Outside influences supporting this were previously referred to in this book was the fact that ^{16}O was formed in colder weather (sheet ice) and ^{18}O in warmer weather. The production of these isotopes is related to the change in the strength of the electromagnetic field and the reversal of the magnetic poles in the past. I pointed out that currently it appears that there is the beginning of the change of the magnetic fields. In the year 2003 the earth had the largest solar flares ever recorded and that magnetic clouds from the sun have broken through and weakened the earth's magnetic field in what they have called sort of a "hand shake" of force transferring electrical force to the earth as previously mentioned. The breakdown of the earth's magnetic field intrudes new electrical forces but exposes the earth to much more intense cosmic radiation. New findings suggest that the whole solar system is entering a cosmic cloud of radiation that is warming all the planets and may affect magnetism.

This agrees with Geyh's research on forces that effect radioactive elements. If there were other mysterious cosmic forces that hit the earth at such a time, it could easily influence the radioactivity of the elements. While it must be admitted that *when* such influences may have caused a change in rates and amounts of isotopes of radioactivity is unclear, the fact that they do influence them is clear. Since the worldwide flood, formation of the ice sheets, and the breakup of the continents were enormous cataclysmic events, these would likely be a time to expect such changes to have occurred.

In discussing radioactivity and ***laboratory procedures*** in regard to producing isotopes, Dr. Darlene C. Hoffman, PhD, professor of graduate school, at the University of California, Berkeley, and nuclear chemist at Lawrence Berkeley National Laboratory, said,

> These isotopes are *produced* by still another method, which can be adapted for use with many elements. An electrical discharge *ionizes* a vapor of the element or of a compound containing the element in ionization; one of the electrons that orbit around the atoms nucleus is knocked off the atom then has a positive charge.
>
> An electric field accelerates the charged atoms, called *ions*, to a definite energy. This process produces a beam of ions; in which all the ions have the same energy. When *a magnetic field* bends the ion beam, the ions with different masses separate into circles of different radii. Each circle consists of a different isotope of the element. (Emphasis added)

Thus, what Geyh found as means of producing radioactive elements in nature is also demonstrated in the laboratory.

All this would tend to substantiate that catastrophes, such as the flood that shifted the earth's axis, and the breakup of Pangaea and the shift of the glacial ice in the days afterwards, could involve radical readjustments of the indications of radioactive decay. Baumgardner estimated that there was a change in ^{14}C of about 20 percent that would have occurred, and would make the pre-flood data misleading. This would produce a date of the flood of about 5920, which is close to the biblical date of the flood. The shift of other radioactive materials may have involved both of these catastrophes or possibly some others.

---

[50] Ibid., 175.

[51] Ibid., 176.

Considerations Electro-Magnetic Force Changes Radioactive Rates

The earlier view that radioactive decay involves a different force than electromagnetic energy is no longer true. The view that the weak nuclear force was entirely different from the electromagnetic force and gravitational force has been modified. The weak nuclear force of radioactivity discovered by Henri Becquerel in 1896 was described and accepted by Enrico Fermi in 1933. But there were unexplained aspects involving the relative orientation of the spans of the participating particles. However, Steven Weinberg, a Nobel Prize winning physicist et al.[52] moved this to a theory in 1967 that brought together the weak force of Fermi and the electromagnetic forces, which gained acceptance in 1980. Of this electroweak, now weak force, Weinberg has said,

> It turned out to be a theory not only of the weak forces (as by Fermi), based on an analogy with electromagnetism; it turned out to be a unified theory of the weak and electromagnetic forces that showed that they were both just different aspects of what subsequently became called an electroweak force.[53]

Are Radioactive Elements also Boosted by Magnetic Change and Cosmic Radiation?

In the years 2003 and 2004 the earth had the largest solar flares that were ever recorded and magnetic clouds from the sun broke through and weakened the earth's magnetic field in what they have called sort of a "hand shake" of force transferring electrical force to the earth. Evidence has been mentioned that the earth now seems to be entering a time of the weakening and changing of the magnetic poles. The breakdown of the earth's magnetic field intrudes new electrical forces but exposes the earth to much more intense cosmic radiation. It is entirely possible that the radioactivity of the elements has been changed by cosmic radiation at one or more times in the earth's past. When one accepts cataclysms, then the various higher radiation levels that existed in many places in the universe may have influenced these elements. Also, we know that 450 miles above the earth, in the Van Allen radiation belt, the cosmic radiation is 1,000 to 3,000 times more than expected. A new radioactive field that accumulated and built up from interstellar heavy nuclei, including nitrogen, oxygen, and neon, has also been recently discovered by NASA's SAMPEX satellite. This new radiation belt shares space with the inner Van Allen belt 200 miles or more out. These are held in place by the earth's magnetic fields.

But we also know that the earth's magnetic fields have turned over many times, possibly causing the earth to be influenced by these radiation belts. Moreover, we know that certain radioactive clouds in the Milky Way may encompass the earth for brief times. And we have discovered unexplained powerful radioactive bursts that are constantly occurring and bombarding the earth. Hypo nova or colliding black holes shoot out powerful streams of radiation across the universe, and the earth may have received such a bombardment. And most importantly, the discovery in 2000 AD that the universe is two-thirds composed of previously unknown or dark energy surely presents questions about the radioactive elements being reset at times. To summarize, there are various ways and times when cosmic radiation is influencing and has influenced the earth and *these surely could have an effect on all these radioactive elements.* These forces are beyond human knowledge at this time and who can dogmatically say changes in radioactive elements have not occurred? One other extremely important fact about the alpha force involved in radiation change will also be considered.

---

[52] Abdus Salam, John Ward, Gerard T. Hooft, and Ben Lee were also involved.

[53] Steven Weinberg, *Dreams of a Final Theory: A Scientists Search for the Ultimate Laws of Nature*, (New York: Vintage Books of Random House, 1992), 119.

An Example of Young, When Radioactivity Indicates Old

An amazing geological event demonstrates that radioactive dating really occurs at the time they begin to decay in the mantle. Iceland, which is atop the Atlantic ridge, is one of the most volcanic countries in the world, with 200 volcanoes that have furnished about one-third of the lava flow of the world since 1500 AD. For three and one-half years, beginning November 1963, in the ocean off Iceland, an underwater volcano erupted and built up a brand new island. Surtsey was one square mile in area, with elevations more than 560 feet above sea level and 950 feet above the ocean floor. After the island had cooled and minerals formed, samples were taken from the surface and sent to various laboratories for potassium/argon radioactive tests, without indicating where they had come from. The samples were dated many millions of years old on an island that was little more than a decade old.[54]

Development of Astronomical Time and Distance

Astronomical calculations of time and distance developed slowly and became more definite in the twentieth century. Kepler's meticulous mapping of the planets and stars was a continuation of the work begun by Tyco Brahe, a Danish astronomer recognized for his accurate astronomic and planetary measurements. Recovering higher math from the Greeks, Kepler incorporated other modern advances, which led to his discovering the planets' elliptical orbits and applying Newton's laws of gravity for a more precise measurement of the planets. He found that by looking at a planet or a very close star from two distant places, he could measure the change ***(the parallax),*** and determine the distance between them. Kepler's and Newton's contributions are still fundamental to the whole exploration of the solar system, and NASA's work today.

William Herschel used "his superior telescope to measure the distance to hundreds of stars, using the ***rough and ready assumption*** that all stars emit the same amount of light and the fact that brightness falls away with the square of the distance"[55] (Emphasis added). Since this is a false assumption, this method was inexact. In 1810, young Frederick Bessel, using the concept of the parallax, demonstrated the first real scientific measurement. He measured the distance from the earth of the very close star 61 Cygne, and after 6 months of earth's changed position found it was 720,000 times as far away as the sun. Today that distance is short (11.4 compound to most distant of 1.4 billion light years).[56] In 1892, John Goodricke conceived of using variable stars. Building on Henrietta Leavitt's meticulous work with Cepheid comparisons, he found the variables 12 times as distant when fainter. But these were only relative distances, not measurements.[57]

While light speed was measured very early, it was not until about 1915 that Einstein used the concept of space-time to actually measure distances of stars. Prior to much of Newton's work, and using Kepler's brilliant discoveries, in 1675, Danish astronomer O. C. Roemer successfully first measured the speed of light at 192,000 miles per second, by noting the disappearance (eclipses) and reappearance of the moons of Jupiter.[58] Roemer and Jean Picard observed over 140 eclipses of Jupiter's moon. In comparison, and considering the difference in longitude, Io and Giovanni D. Cassini observed the same in Paris. Roemer and Cassini observed the time between eclipses got shorter as the earth approached Jupiter and longer as the earth moved away. All these efforts produced varying readings of the speed of light—from 135,000-140,000 miles per second. As

---

[54] *Encyclopedia Britannica*, Vol. 11, Micropaedia, 414; vol. 20, Micropaedia, 760 (15th edition, Chicago, 1992). See *New Scientist*, July 3, 1975, 20.

[55] Simon Singh, *Big Bang: The Origin of the Universe,* (New York: Harper Perennial, 2004), 173.

[56] Ibid., 174-177.

[57] Ibid., 212, 213.

[58] William C Dampier, *A Shorter History of Science*, 159.

Roemer had no accurate value for the astronomical unit, he gave no value for the speed in his paper beyond the aforementioned lower bound. However, he was recognized by the Royal French Academy for determining that if light travels 3,000 leagues, or the approximate diameter of the earth, it would take one second of time, which was somewhat greater than today's accepted value for light speed. James Bradley confirmed and accepted Roemer's work in 1727. In 1809, Delambre used the same technique and established the speed of light at 300,000 kilometers per second. Some have tried to discredit the amazing achievement of Roemer and his associates.

The nature of light and how it performed was gradually discovered. Newton had intuitively seen it as particles with wavelike action, which is close to the modern theory.[59] Basic to Newton's calculations for the solar system and its planets was the importance of mass and gravity. Newton had found that waves travel through a medium—water waves through water, sound waves through air. For this reason, in the nineteenth century, scientists who accepted the wave theory postulated that light waves also traveled through a medium that they named ether/aether or called luminiferus ether. James Clerk Maxwell, founder and director of Cavendish Laboratory and teacher of experimental physics at Cambridge University, did not accept the ether hypothesis. In his *Treatise on Electricity and Magnetism* (1875), he predicted wavelike disturbances in the combined electromagnetic field ***would travel at a fixed speed***.

Experiments Showing Light Travels at a Fixed Speed for a Standard

No serious efforts after Roemer's were made to measure light speed until the last quarter of the nineteenth century or two hundred years later when Cornu-Helmert, Michelson, S. Newcomb and others continued to build on Roemer's discoveries. In 1887 Albert Michelson and Edward Morley of Case School of Applied Science in Cleveland discovered that the speed of light in the direction of the earth's motion was the same as that at right angles to the earth. This led to the rejection of the ether theory. It is thought that Einstein knew of Michelson's findings. Basing his calculations on light as having a ***fixed speed that does not change,*** Einstein proposed his special theory of relativity in 1905 and the general theory in 1915. He also held that ***nothing can travel faster than the fixed speed of light***. Several of Einstein's theoretical predictions confirmed his theory and led to its acceptance. The proofs were over a relatively short period of time, which means Einstein's proofs would be possible even with a very slight change in light speed if it were not fixed.

Einstein's special theory did not fit with Newton's physics which require gravitational effects traveling at an infinite velocity. Einstein had to discard Newton's gravity as a force in his special theory, and he derived a new view of gravity for the general theory which required a curved universe. According to the theory, the energy which an object has due to its motion will add to its mass. Therefore, because of this increasing mass, the faster it goes it would be harder to increase its speed. So, according to Einstein, nothing can reach the speed of light because the mass would become infinite. A new unit of length was adopted by Einstein, ***a light-second,*** to define the distance light travels in a second. Long distances were measured in ***light years.*** The theory required a fundamental change in the idea of space and time, since time is not separate from independent space, and forms an object called ***space-time***. Therefore, all measurements were founded on the fixed speed of light that these scientists accepted.

Light Speed Used to Measure Astronomical Distance and Time

In the early twentieth century, the speed of light became vital to measuring objects in the universe after Einstein's theory of general relativity. The theory of relativity and the idea of the

---

[59] Ibid., 74.

redshift, based on the constant speed of light, gave some standard for measurement with other ideas. The ***unchangeableness of light speed in a vacuum became the basis*** of a preponderance of astronomical work in measuring distance, time and size of galaxies and stars. The use of light speed has led to other ways of estimating distance and time. By the 1980s the age of the universe was estimated at 15 billion years. With refined instruments, various new radiation measurements were made, giving a more exact time and distance. Around the turn of the century, using light measurements along with other techniques, such as angles of other known astronomical objects, scientists declared that the universe was 13.7 billion years old. However, ***the speed of light in a vacuum*** as a constant was ***the basic standard*** for measuring distance in the universe and became the basis for the nebular hypothesis given prominence years before by William Herschel that claimed the universe developed gradually over a long period of time. According to Hawking, the theory of a cone of light assumes extension to infinity that fits original assumptions. At the same time the concept of light speed was determined, quantum theory was born, introducing indeterminism. This shift—from a determined/creation view of nature, to indeterminism, not only changed the perspective, but it also led to confusion in the scientific world.

New Evidence Indicates Decline of Natural Forces and Cannot Be Uses for Time

Evidence of Irregular Change in Radioactive Elements by Turn of Century

Nuclear and all forces seem to be diminishing. Since 2001 Flambaum and his colleagues have presented evidence for the change of the alpha force.[60] This was challenged by others.[61] But clear evidence now exists that this is so and that nature can create new elements and isotopes before they are locked in to the minerals to undergo normal decay. Reactors in nature are producing fission that produced less-massive atoms. In old uranium mines near Oklo in the Republic of Gabon in W. Africa, scientists explain the formation, "without invoking exotic physics," of what has turned out to be 18 reactors.[62] Alexander P. Meshik, a Russian scientist now at Washington University in St Louis, attained from a scrap of mineral from a depleted uranium mine in France, and found a deficit in the concentration of U235 that led to the discovery of what had been happening.

He and other scientists found that spontaneous fission was occurring in a little understood cycle. It happens in an active and inactive cycle: thirty minutes fission and then off two and a half hours. How this occurred is a mystery, but scientists think it is influenced by water. The main fission of uranium apparently shattered into less-massive atoms as precursors of xenon gas that migrated into crystallizing aluminum phosphate. The precursors are presumed to have later decayed into volatile xenon atoms held by the aluminum phosphate mineral. Xenon results when fission creates isotopes of other elements such as tin, antimony, and iodine that undergo a further series of radiation decay.

In checking the results, Steve K. Lamoreaux of the Los Alamos National Laboratory and Justin B. Torgerson found a fissioning uranium atom releases two or three neutrons. If another uranium atom takes in one of these neutrons, that atom itself fissions, and releases others. This was discovered by precisely pinpointing particular isotope concentrations. The action is strange because ***alpha (fine structure constant) of the electromagnetic force was considered a constant***. Lamoreaux has said these ancient reactors are "one of the greatest natural phenomena that ever

---

[60] *Science News*, 10/6/01, 222.

[61] Science News 5/14/ 5, 318.

[62] Peter Weiss, "Primordial Nukes: the 2 billion-year-old tale of Earth's natural nuclear reactors," *Science News*, March 12, 2005, Vol. 167, 170-173, cf. 170).

occurred."[63] These natural phenomena are important **because *the amount of isotopes in minerals cannot be said to only indicate the normal decay measured by the half-life of the material*** such as uranium.

These natural reactions seem to be associated with the fluctuation of the electromagnetic field that was seen to be one with the weak force and probably associated with the breakdown of the earth's magnetic field. This was shown in the studies of the heavens that the decline of alpha was also discovered in studying supernovas. Michael T. Murphy of Cambridge who worked with Flambaum has studied 143 quasar systems and finds evidence of change in alpha force there also. There is a mystery of naturally produced new elements or isotopes in minerals. Lennox L. Cowie of University of Hawaii wants more proof, saying the implications are so profound.[64]

In regard to the uncertainty of scientists about natural forces Tim Folger has said recently,

> According to quantum mechanics space… is filled with fields and particles that constantly pop in and out of existence. The problem is that when physicists estimate how much energy is contained within these fields and particles, they come up with a number – called the cosmological constant – is insanely large, 10^{120} times larger than what we observe. A cosmological constant of that magnitude would rapidly tear the universe apart. It is an embarrassing error…. [65]

New Millennium Evidence Confirms Sun Influences the Decline of Radioactive Forces

Evidence from many places is confirming that radioactive decay is not a constant, and date setting by the long term declining elements and even C^{14} is unreliable. In 2006 Jere Jenkins, nuclear engineer at Perdue University, noticed that there was a decay in manganese 54 just before a solar flare. Changes in decay rate have been found in other elements. Peter Sturrock, professor emeritus of Stanford has suggested this might be due to neutrinos, but this is not known. It was found that decay rates of elements were affected repeatedly every 33 days, which matches the rotational period of the core of the sun. It was also found that the decay rates varied in summer and winter so there seems to be some predictable pattern of change. The change in decay rates from the measurable half life has been recognized by many scientists around the globe. The recent evidence of changes has many scientists alarmed. When the techniques for "absolute dating" by radioactive elements seemed to have been given some resolution, all those dates are now known to be unreliable and even with C^{14} the evidence is at best an estimate. All the dogma about long ages and evolutionary theory is now not to be trusted and the dates used are virtually useless.[66]

Gravity Theories Still Inconclusive but Newest Linked to Second Law and Decline

Numerous efforts have been made to resolve many of the conflicting views of gravity. In 2004, NASA launched Gravity Probe B to test Einstein's theory of gravity in general relativity. This contains four very sensitive gyroscopes that will check (1) warping of nearby space-time by Earth's gravity, and (2) the Earth's dragging of space-time as the planet rotates. The probe is gathering a year of data.[67] This expensive quest, and ongoing uncertainty and disparity among other theories demonstrates how little about gravity is understood.

Two professors at the Weizmann Institute, Rahovot, Israel, Mordehai Milgrom and Jacob

---

[63] Ibid.

[64] P. Weiss, "Universe in Flux: Constant of nature might have changed," *Science News,* 159.

[65] Folger, "Gravity, Tabletop Physics" *Discover,* 56, 57.

[66]Ian O'Neill, "Is the Sun Emitting a Mystery Particle?, Discovery News, August 25, 2010; Terrence Aym, "Strange emissions by sun are mutating matter," Project World Awareness, October 5, 2010 by rockingiude, et al.

[67] B. B., "The matter of gravity," *Science News*, Nov.5, 2005, Vol. 168. 302.

Bakenstein, propose that Newton's theory that works with the solar system should be modified when gravitational forces become very small, as with stars at a great distance. They propose such a modification only if acceleration falls below one ten-billionth of a meter per second, every second. They believe that this will do away with the need for dark matter. The result is so radical it is said "to tear up our whole picture of how the universe is put together." This is yet to be demonstrated and new evidence for dark matter continues to emerge.[68]

This theory is similar that of Raman Sundrum, professor at John's Hopkins University, who suggests that Newton's gravitational laws must work differently at great distances. He submits that gravity is transmitted by a particle called ***a fat graviton*** that may be as large as 1/200 of an inch wide, which is large compared to atoms. These gravitons barely interact with particles and energy in space. A lab at Washington University in Seattle is experimenting with a "torsion pendulum to measure gravity's strength across small distances with unprecedented accuracy, but with inconclusive evidence thus far." [69]

> Gia Dvali, from New York University, believes gravity is weak compared with other forces, because ***most of it escapes into higher dimensions outside the three we experience.*** He is testing his views with lasers and reflectors.[70] This appeal to another dimension is found in another view that also seeks to find an answer to gravity with laws that do not fit present laws of nature. It views ***gravity as an illusion***. This theory postulates that the three dimensional world may be viewed as two dimensions without gravity and these may be completely equivalent. The two-dimensional world would be like an event on the boundary of the three dimensional world. "The mathematics of the theory has not yet been rigorously proved…."[71]

Under present laws of physics, calculations don't fit. The problem is that, while we have described relationships in various ways, the forces of nature are not understood and known. All the early great scientists through Newton and Boyle believed the invisible forces were the work of the Creator.

From all these speculations and experiments it is clear that the whole measure of mass and gravity is still a mystery. David Darling, in reviewing the findings about gravity from Aristotle until today, concludes, "Scientists don't really know what gravity is or how it is created. Despite this wealth of knowledge, much about gravity remains unexplained."[72] In commenting on the views of physics about gravity, Lawrence Krause recently said, "Generations of physicists have remained stumped by the utter strangeness of gravity. Not only is it the weakest of the four natural forces, but it is also the ***only one*** that appears to be directly related to the nature of space and time. …As always, nature is mystifying."[73]

But new thinking shows the theory or gravity may be joining the forces of slowing corruption. As mentioned, some physicists have suspected that all the forces, including gravity, were in some way united. Eric Verlinde of the University of Amsterdam has presented a New Theory of Gravity linking the New Information theory and entropy as in the New Revised Second Law of Thermodynamics extended to black holes. He says that the link of energy, entropy, and the

---

[68] Adam Frank, "Gravity's Gadfly: Mordehai Milgrom's new physics could overthrow Newton and Einstein – and" *Discover*, August 2006. 31-37 His new formula is ($F=ma$ to $F=ma^2/a_o$).

[69] Tim Folger, "Gravity, Tabletop Physics," *Discover*, October 2005, 56, 57.

[70] Ibid.

[71] Juan Maldacena, "The Illusion of Gravity," *Scientific American*, Nov. 2005, 57-63.

[72] David Darling, *Gravity's Arc: The Story of Gravity, from Aristotle to Einstein and Beyond,* (Wilely, John and Sons), 2006.

[73] Lawrence Krause, "Nice Going Einstein," *Discover*, August 2006, 38-39.

disordering of the universe in the New Information theory replicates Newton's laws of gravitational attraction. He extends this to Einstein's general theory of relativity based on the work of Jacob Bekenstein and Stephen Hawking that showed a parallel between ordinary thermodynamics and the physics of black holes done in 1970s. He included the work of Gerard't Hooft on holograms done in 1993. Physicists Damien Easson, Paul Frampton and George Smoot, a Nobel Prize winner, assume from this theory a holographic principle encompassing the whole visible universe and explaining dark energy. The result is gravity is linked to entropy and disorder of information and a corrupting universe. This effort of "extending Varlinde's ideas to encompass the history of the universe. …And beneath it all may lurk ***a new worldview emphasizing the primacy of information*** over matter and energy"[74] suggests the Christian belief of the Creator being the Word of God. Verlinde admits his theory is tentative and yet incomplete.

New Evidence Light Speed Is Not an Unchanging Standard of Natural Forces

Many of the current theories about light are based on data and assumptions that go beyond the original findings. The Michelson-Morley experiment proved that the speed of light did not change with the direction of motion. Maxwell demonstrated, mathematically, that the various electromagnetic forces would have definite speeds that could be measured. But he ***did not prove*** that ***these had never changed,*** or were not slowly changing, or could not change. The results of Michelson and associates reported changes in light speed which were thought to be through better instrumental measurements. The speeds are as follows: 1879- 299,910 +/-50 km/s; 1883- 299,853 +/-60 km/s; 1906-299,781 +/-10 km/s; 1924- 299,769 +/-4km/s; 1930- 299,744 +/- 11 km/s. These results of Michelson also do indicate a progressive slowing down of light speed.

The speed of light based on the assumption its speed was fixed has become the dominant standard and basic to measurements for stars, galaxies, and outer space. The constancy of the speed of light is ***highly important*** to support and explain the ***nebular hypothesis*** and the evolution of the universe. Is the speed of light constant? Since the question has varied meanings, it is answered in a variety of ways.

> Yes. Light is slowed down in transparent media such as air, water and glass. The ratio by which it is slowed is called the refractive index of the medium and is always greater than one. Einstein claimed light constant at 186,282 m.p. second in empty space/vacuum and the story of the stars is based on this. Photons may be absorbed by an atom in material. Light speed in water 140,000 mps, glass at 125,000 mps, diamonds slow it to 38 mps.[75]

Jean Foucault had discovered the slowing of light in various substances in 1850.

Various theories and experiments about light speed slowing have been developed recently by Michael Fleischhauer of Kaiserslautern University in Germany, by Lene V. Hau, of Rowland Institute of Science of Harvard University, by Stephen E. Harris of Stanford University, Ronald L. Walsworth, Mikhail D. Lukin, Harvard-Smithsonian Center of Astrophysics in Cambridge, Philip R. Hemmer of Hanscom Air Force Base, Boston and some physicists at Perdue. Hau and her associates at Harvard inserted lasers at normal light speed into a vacuum chamber, where the beams collided with a sodium cloud that slowed and stopped light. The first response was that this violated Einstein's theory of special relativity, but "Einstein's theory places an upper, ***not a lower***, limit on the speed of light."[76] At least, until recently, the upper limits were thought to be fixed.

---

[74]Tom Siegfried, A New View of Gravity: Entropy and information may be crucial concepts for explaining roots of familiar force, *Science News*, September 25, 2010, 26

[75] Bob Berman, "Astronomy at the Speed of Light," *Discover,* December 2006, 24.

[76] Chris Kirkman, quoting *Sciences*: Rowland Institute for Science: Harvard University, *Harvard Magazine*

This change of light speed is such a threat to astronomical time that the determinists have been quick to say that "circumstances which slow light in space ***are not likely to occur.***"[77] But space is not empty, neutrinos have mass, and no one knows what else affects light speed and, therefore, history. If the ***rate*** of light is being slowed by any of the unknown forces in space, or is itself a declining force, then this also opens the door for a younger universe. This would be devastating to long ages of astrophysicists.

However, light speed has also exceeded that which was considered fixed. Similar to sound breaking the sound barrier: in Cerenkov radiation, charged particles break the speed of light. In 2000, NBC Research institute in Princeton, New Jersey, ***pushed a pulse of light*** through a gas-filled chamber ***at 310 times the regular speed of light.*** Swiss scientist Nicolas Gisin, in 1997, found that when two elementary particles are created and bonded together, the property of one twin changes the other instantaneously in a spooky communication, even across distances, making it swifter than light. It is thought that at the speed of light, time stands still and one would experience being everywhere in the universe at the same time.[78]

Diminishing Rate of Light Speed and Young Earth Rejected

In 1983, an Australian student of astrophysics, Barry Setterfield, set forth a thesis that the speed of light was diminishing and the rate allowed a curve on light (C), which he said demonstrated that light began in the universe about 4040 BC or about 6,000 years ago based on the history of past recorded measurements. Mr. Setterfield's work was bitterly attacked by the scientific community as simply a source to defend the Bible. Among other criticisms, he was accused of not giving a quotation exactly, so the whole thesis was rejected. At that time the change in the speed of light was not understood as it is now. Despite the fact that Mr. Setterfield's work was rejected, **his *major observation,*** that from 1675-1965 the measurement of light *declined* in a zigzag or erratic fashion has been affirmed by others. The listing of Setterfield's recordings of light speeds he used and graph of C are given on the following page.[79]

---

in the Orlando, Sentinel, January 7, 2001, G8.

[77] Cf. Peter Weiss, "Light Stands Still In Atom Clouds," *Science News,* January 27, 2001, Vol. 159 #4, 52; see also *Science News*: 3/27/99, 207; 8/26/00, 132; 6/1/91, 340.

[78] Ibid.

[79] Barry Setterfield, "The Velocity of Light and the Age of the Universe," *Science at the Crossroads: Observation or Speculation?*(papers of the 1983 National Creation Conference)122-127.

EXPERIMENTER	DATE	OBSERVED VALUE Km/sec.
Roemer[2]	1675.0	301,300 ± 200
Bradley[3]	1728.0	301,000
Cornu[8]	1871.0	300,400 ± 200
Cornu-Helmert[4]	1874.8	299,990
Michelson[5]	1879.5	299,910 ± 50
Newcombe[6]	1882.7	299,860 ± 30
Michelson[7]	1882.8	299,853 ± 60
Anonymous[8]	1885.0	299,940
Perrotin[9]	1902.4	299,901 ± 84
Perrotin[9]	1902.8	299,895
Perrotin[9]	1906.0	299,880
Michelson[6,9]	1924.0	299,802 ± 30
Michelson[9]	1926.5	299,796 ± 4
Mittelstaedt[10]	1928.0	299,778 ± 10
Pease-Pearson[11]	1932.5	299,774 ± 11
Anderson[13]	1939.0	299,771 ± 12
Huttel[14]	1940.0	299,768 ± 10
Essen[15]	1947.0	299,797 ± 3
Aslakson[12]	1949.0	299,792.4 ± 5.5
Bergstrand[15]	1949.0	299,796 ± 2
Essen[15]	1950.0	299,792.5 ± 1
Bergstrand[15]	1950.0	299,793.1 ± 1
Bergstrand[16]	1951.0	299,793.1 ± 2.5
Aslakson[15]	1951.0	299,794.2 ± 1.4
Froome[15]	1951.0	299,792.6 ± 1.3
Kraus[16]	1953.0	299,800
Froome[15]	1954.0	299,792.75 ± 0.35
Florman[15]	1954.0	299,795.1 ± 3.1
Scholdstrom[15]	1955.0	299,792.4 ± 0.4
Plyler, Blaine & Cannon[16]	1955.0	299,792.0 ± 6
Plyler, et. al.[21]	1955.0	299,793.0
Cohen, et al.[22]	1955.0	299,793.0 ± 0.3
Bergstrand[21]	1956.0	299,793.0 ± 0.3
Wadley[16]	1956.0	299,792.9 ± 2
Rank, Bennett & Bennett[15]	1956.0	299,791.9 ± 2
Edge[15]	1956.0	299,792.4 ± 0.4
Wadley[15]	1957.0	299,792.6 ± 1.2
Bergstrand[21]	1957.0	299,792.9 ± 0.2
Rank et. al.[21]	1957.0	299,793.7 ± 0.7
Rank et. al.	1957.0	299,793.2
Mulligan & McDonald[21]	1957.0	299,792.8 ± 0.6
Froome[15]	1958.0	299,792.5 ± 0.1
Corson & Lorraine[17]	1962.0	299,790.0
Karolus[Y15]	1966.0	299,792.1 ± 1
Helmberger[23]	1966.0	299,792.44 ± 0.2
Simkin et. al.[15]	1967.0	299,792.56 ± 0.11
I.T.T. Staff[18]	1970.0	299,793
Bay, Luther & White[19]	1972.0	299,792.462 ± 0.018
Evenson[24]	1973.0	299,792.4574 ± 0.0011
Blaney[25,26]	1974.0	299,792.4590 ± 0.0008
C.C.D.M.(France)[26]	1975.0	299,792.458 ± 0.004
Annonymous[20]	1976.0	299,792.456

(Continued in next column)

C = speed of light

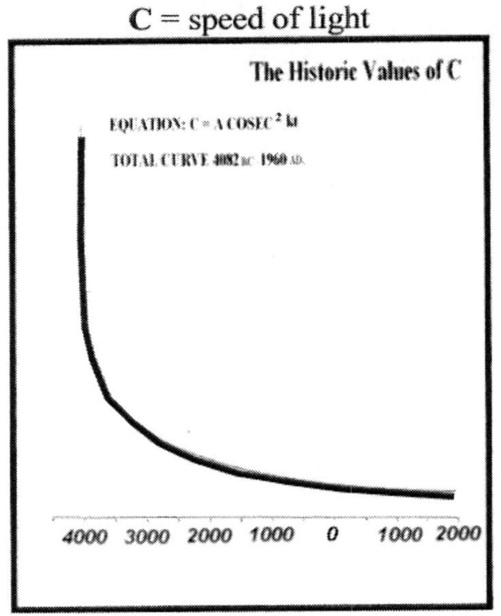

Other Modern Studies Confirm Diminishing Speed

A growing body of evidence indicates that Einstein was wrong about the speed of light being fixed. Even Michelson's later dates after Einstein showed a slowing and agree with Satterfield's demonstration of change, and a minimal change in rate would not affect the proof of Einstein's theories. The discovery of dark energy and the rate of the universe's expansion are involved. Joao Magueijo of Cambridge University theorizes that light traveled much faster in a first stage after the big bang, and then stabilized at the speed it now travels. This would suggest change to uniformity of the universe and its inflation. According to Magueijo, "A varying speed of light could actually explain where the cosmic unity of the universe comes from."[80]

New evidence shows that the basic constants of protons and electrons are diminishing, as the following quote substantiates. Standard ratio of mu of mass of proton and that of the electron is changing. A team of physicists and astronomers in Netherlands, Russia, and France in April 21, 2006 *Physical Review Letters* report indications that one of the constants seems to have undergone a subtle shift. They report,

> The new findings indicate that the ratio between the mass of the proton and that of the electron– a number known as mu– might have decreased to about two thousandth of a percent in the past 12 billion years, say Elmer Reinhold, now of the European Space Agency in Noordwijk, the Netherlands, and his colleagues. The evidence for the change in the constant, which has a current value of the 1,836.15, emerged from light absorption patterns of hydrogen molecules.
>
> To arrive at the new findings for mu, Alexandre V. Ivanchik of the Ioffe Institute in St Petersburg, Russia and Patrick Petitjean of the Astrophysics Institute of Paris made extraordinarily precise telescope measurements of radiation coming from two quasars. The researchers focused on wavelengths absorbed by frigid clouds of hydrogen molecules in space.
>
> Whatever the speed of light and the time that involves, it shows that there has been a change. Using the idea of constancy of light they estimate the change occurred in 12 million years, but this could be ***shorter if light speed is decreasing***. ...If correct, it is ***a revolutionary result***. It doesn't matter that the variation is small. "If mu varies, we need ***new theoretical physics and cosmology***." comments Victor V. Flambaum of the University of New South Whales in Sydney, Australia.[81] (Emphasis added)

Other studies and evaluations of the speed of light (symbolized by C) have supported the idea that the ***speed of light has not been constant*** in the past. An evaluation is as follows.

> According to SI (System International) definition the metre is the length of the path traveled by light in vacuum during a time interval of 1/299 792 458 of a second. This defines the speed of light in vacuum to be exactly 299,792,458 m/s. This provides a very short answer to the question "Is c constant": Yes, c is constant by definition ... The SI definition makes certain assumptions about the laws of physics. For example, they assume that the particle of light, the photon, is massless.
>
> If the photon had a small rest mass, the SI definition of the metre would become meaningless because the speed of light would change as a function of its wavelength. Unlike the previous definitions, these depend on absolute physical quantities which apply everywhere and at any time. Can we tell if the speed of light is constant in those units?... [For various reasons] it is nonsense to say that the speed of light is now constant just because the SI definitions of units

[80] Tim Folger, "At the Speed of Light: What if Einstein was wrong?," *Discover*, April 2003, 34-41.

[81] P. Weiss, "Universe in Flux: Constant of nature might have changed," *Science News*, April 29, 2006, Vol. 169, 259.

define its numerical value to be constant. By eliminating the dimensions of units from the parameters we can derive a few dimensionless quantities, such as the fine structure constant and the electron to proton mass ratio. These values are independent of the definition of the units, so it makes much more sense to ask whether these values change.

If they did change, it would not just be the speed of light which was affected. The whole of chemistry is dependent on their values, and significant changes would alter the chemical and mechanical properties of all substances. Furthermore, the speed of light itself would change by different amounts according to which definition of units you used. In that case, it would make more sense to attribute the changes to variations in the charge on the electron or the particle masses than to changes in the speed of light.[82] (Emphasis added)

Conclusions: Declining Unknown Forces Fit Reformation Science of Faith

More modern experimental data demonstrates that all the major forces known in the universe are now slowly changing. When the measuring tool changes, everything based on fixed measurements must be considered wrong, except for the moment, and the alteration of the rate of change alters the whole picture. If the store owner changes the weights on the scales, the amount of chicken you buy may seem the same, but is actually less. If the surveyor uses fake measurements, the results are confusing, harmful and chaotic. Kant's lie of evolution of all things to exclude real evidence of God has disrupted all relationships

All the early modern astronomers, Copernicus, Galileo, Kepler and Newton held to a created universe and solar system that required only the time limits of biblical history. As a believer in God, Newton held to fixed time and a determined creation (not determined knowledge) with unknown space. He believed the world began ***at creation*** and will culminate in ***judgment*** that leads to eternal life or eternal death. But beginning in the late 1700s, astronomy of unbelief was changed to the ***nebular hypothesis*** that required lengthy, unknown ages of time, all in an effort to find evidence to support that philosophy of determined forces known by sense perception. But then there was no accurate, reliable way of dating and measuring great distances out in the universe. Fixed light speed became the answer, but now cannot be trusted.

The universe continues to confound scientists. Even after years of study and research, they acknowledge there is more ***dark (invisible) matter than*** visible, known matter. The universe is expanding at an accelerated pace, a phenomenon scientists dub dark energy, but cannot explain. ***Black holes*** have influence ***far beyond what their sphere*** and confront science with many singularities beyond which is mystery. New studies with x-ray instruments reveal new structures, but no explanation for this mystery.[83] ***Adult stars and galaxies*** are calculated to be lying ***in the nursery of the universe,*** yet according to the big bang theory they should not be near the beginning. Also with cosmic background evidence many aspects of the big bang theory don't fit natural expectations. All of this leaves only a small percentage of the universe that fits explanation by the developmental nebular hypothesis philosophy of deterministic uniform scientists.

The true enlightened position is a young earth. The evidence for the worldwide flood and the ice ages are less that 14,000 years and probably should be shortened because of cosmic radiation and electromagnetic influences. The division of Pangaea has been within the life of advanced mammals and also man. The study with human families rather than guesses from genetic mutations from chimps was 6,000 years. The actuarial figure for the length of time man had been on the earth studied by scientists with Steve Olson turned out shockingly to be 6,000 years. The

---

[82] Excerpts from an article, "Is the Speed of Light Constant?," by Philip Gibbs and updated by Steve Carlip,1997. Carlip is at UC Davis.

[83] Ron Cowen, "A New Spin: x rays shed light on black Holes," *Science News*, Vol. 171, January 6, 2007.

age of the universe based on a declining speed of light turns out to be 6,000 years. Unbelieving scientists do not accept these things because they have a bias against God. Many factors reported above about radioactive dating require a young earth.

While the theories of quantum activity and indeterminacy, of relativity and gravity, and of chaos have helped solve many problems, these all have left science with many ***other major problems***. Many scientists, including Stephen Hawking, have questioned these problems, prompting the search for another unified theory that solves the unanswered questions. Hawking and others want a theory of quantum gravity that can be extended to a theory of all things, such as the string theory. Hawking called such, "the mind of God." The Reformation scientists believe it was God working. In chapter 33 efforts to find a new determinism by Hawking will be considered.

The Apostle Paul refers to the degeneration of all nature which is in accord with the new information theory with the second law of thermodynamics. He said,

> For the creation was subjected to futility, not willingly, but because of him who subjected it, in hope that the creation itself will be set free from its ***bondage to corruption*** and obtain the freedom of the glory of the children of God. For we know that the whole creation has been groaning together in the pains of childbirth until now. And not only the creation, but we ourselves, who have the firstfruits of the Spirit, groan inwardly as we wait eagerly for adoption as sons, the redemption of our bodies. For in this hope we were saved (Romans 8:20-24a).

The bias of man to reject God in evolutionary science has progressively perverted religion and brought the soft sciences or human relationship and government to a crisis. The study of this theological perversion will be considered in the next section and then it will be shown how this process of rejecting God has affected human behavior as seen in the soft sciences.

SECTION FIVE

USE OF SCIENCE IN PERVERSION OF GOD'S REVELATION

PART VIII

BIAS TO DISTORT BIBLICAL TRUTH SPECIAL REVELATION

To whom then will you compare me that I should be like him? says the Holy One. Lift up your eyes on high and see: who created these? He who brings out their host by number, calling them all by name, by the greatness of his might, and because he is strong in power not one is missing (Isaiah 40:25, 26).

Chapter 24
Man's Evolutionary Creation of the God of History

Scientific Determinism Applied to Biblical History

Introduction

The civil wars that involved religion and class in Europe led to destruction and death (1618-1648) and prepared the way for doubt and religious skepticism. The religious conflict gave man's bias against God a corporate reason to reject the clerical corruption and control and paved the way for the secular and humanistic movement, which originated with philosophers Descartes, Locke, Hume and others. Even Presbyterians' commitment to orthodoxy waned in light of their declining political power. During Voltaire's time in England he discovered that religion among the upper class was itself becoming a joke. Encouraged in his skepticism, he returned to France where he launched an all-out humorous attack on Christianity. Fortunately, the revival under John Wesley and his colleagues in England brought about spiritual renewal, especially among the common people.

The deistic and materialistic philosophy that accompanied these conditions had already denied God's participation in history. These philosophies held that the Creator had left nature as a great machine. Deism left no room for God's special revelation in the Bible; it had to devise an explanation to dispose of the biblical evidence for God. Biblical criticism had already started in Holland with Descartes and Locke, and continued with Spinoza's rejection of the Old Testament. He had endorsed an idealistic pantheism, which was the basis for the Hegelian evolutionary philosophy of idealistic pantheism. After the development of the uniformed deterministic view of nature, philosophers became more insistent on explaining away God. If there was no evidence for a Creator in nature, there could no longer be any place for Him to speak and reveal Himself to man. This chapter traces the use of the uniformed deterministic view to remove God from the scene.

W. Schmidt, the great German comparative religionist, said, "The eighteenth-century rationalism… tried all religions alike by its touchstone of 'good sense'."[1] The uniformitarian deterministic view left no place for supernatural creation or for an apocalyptic end to creation, and a future judgment. Religion had to be explained as a natural phenomenon and fit within the circle of empirical knowledge as a natural machine, as these men speculated. By excluding God the unbelieving scientists removed a sense of moral obligation before God and toward others.

New Proposals on the Origin of Religion

The intellectual elite reasoned that if determined knowledge excludes God, then all claims about God must be imagined by men to meet certain of their needs. Thus, evolutionary scientists and rational philosophers had cleared the way for the "science" of religion that ***presupposed*** there is no God. Therefore, the ideas of God had to be developed by men. The first efforts at studying primitive peoples and "nature myths" began in the last third of the eighteenth century. These studies expanded slowly, until a decade after Charles Darwin's *Origins*, Max Muller founded "the science of religion" or comparative religion.[2] The theory of uniformitarian deterministic science

---

[1] W. Schmidt, *The Origin and Growth of Religion,* 34, H.J. Rose trans., (New York: The Dial Press, 1931).

[2] Ibid., 39.

turned more specifically toward discrediting the biblical God.

Historical Relevancy and Reliability Is Essential to Christian History

The Jewish and Christian religions are based on the belief that the Scriptures recorded and preserved what God ***really did*** and revealed about Himself ***within history***. What God revealed about the beginning of the world, man's sin, His chosen people and His purposes through Israel, His fulfillment in sending His Son Jesus Christ, His church, its mission to call all men and to teach Christ's truth, and the final consummation in judgment are presented as *historical events.* Specific dates, rulers, and circumstances place biblical events within history. Because it is God's message, given to and through real men in historical events, ***Christianity is a historically-based revelation.*** No other religion is historical in that sense, although Islam, over half a millennium later, makes some claim to this. The history of God's revelation stands as God's standard for judgment and redemption of all men everywhere. To undermine it destroys its relevancy. To falsify its history ***removes it from the realm of real revelation*** and makes it fairy tales or ***myth,*** and thereby removes all sense of man's responsibility to God. By denying its historicity, man ***removes his sense of guilt.***

When faith in God is removed, the motivation to live morally toward others is also removed. There is an innate sense of danger to man in removing moral accountability because we know our own selfishness. Even Immanuel Kant wanted to claim a ***divine imperative*** for ethics, because he recognized and said, man was "like a crooked stick that can't be straightened." But Kant's feeble effort at moral accountability by a logical, universally applicable standard did not hold back the tide of evil in Germany that in time produced Hitler. Man does not sit down to determine whether his actions are universally applicable, nor does he agree with other men as to what is universally right or wrong. He does what is best for his own interest at the time. But if God has revealed His character and will in history, then man must be accountable to God.

Part I: Deterministic Efforts to Remake Israel's History Evolutionary

Questions by Skeptics about Historicity of Documents

It is beyond the scope of this book and contrary to its purpose to try to trace out the whole of higher criticism of the Bible; but it is necessary to see the general trend of thought, and examine and document some of the highlights of the emergence of higher critical studies that resulted from trying to fit information about God and His people into the circle of scientific determinism. Larger résumés of critical studies are available elsewhere for those who want more.[3]

In all the centuries of writings of the rabbis in the Talmud and in commentaries, even though there were many points of debate, no one had called the basic story of the history of Israel in the Old Testament into question, except for the obvious distortions to support Samaritan claims. Historical inconsistencies, duplications, and in some points apparent contradictions were recognized. Even Luther, Calvin, and other reformers acknowledged duplications and inconsistent reporting in the Gospels and other biblical events. But they, along with others, never believed this threatened the acceptance of these as real historical events. Questioning Israel's history was new and innovative until deistic science developed.

---

[3] Dr. A. Noordtzy, Rector Magnificus of University of Utrecht, *The Old Testament Problem,* Miner B. Stearns, Trans. (3909 Swiss Ave., Dallas, TX: Seminary Book Room), and a less academic, but excellent review, by Herman Wouk in *This Is My God,* an addendum on Wellhausen, (New York: Doubleday & Company, Inc., 1973).

Earliest Critical Conjectures

But once Descartes and others (ca. 1650) made doubt and individual rational criticism prominent, the hostility and distrust against the church and religion furnished the right atmosphere to find fault with Christianity's historical foundations. Benedict Spinoza, an admirer of Descartes and Thomas Hobbes, rejected the idea of miracles, and attacked the traditional origin of books of the Bible, thus launching hostile criticism. As a Jewish Talmudic student who had been excommunicated from his synagogue, Baruch Spinoza knew all the rabbinical observations against various styles, chronological discrepancies, oddities in grammar and vocabulary, and repetitious passages. He seems to have been the first to assume another story behind the Old Testament. He split up the Old Testament, with the first section including Genesis through Kings. He ***speculated*** this was probably compiled and written by Ezra after the captivity in Babylon, and these united writings were viewed as just a common book like others, and not inspired. In many ways, his suggestion pointed the way for the criticism that followed.

Conjecture of Source Documents Using the Names of God

Soon after, in 1753, Jean Astruc, a French physician sought to divide Genesis into documents based on the two names for God, Jahweh and Elohim, suggesting two major documents and six smaller ones. Other scholars quickly followed his documentary approach. Wilhelm M. L. de Wette took a major step and placed the whole Pentateuch as a composition of documents much later in the time of David to support the worship in Jerusalem. Critics questioned other books in the Old Testament as to reliability and sources, and attributed them to different authors than those credited by church tradition. By 1780, Johann Gottfried Eichorn had claimed the documents of J and E from Astruc, and suggested others. In the early nineteenth century, Wilhelm M. L. DeWette submitted that all the Pentateuch was put together after the return from captivity and an effort to establish Jerusalem as the center of worship and strengthen the city for kingdom control. In the decade after Darwin republished his *Descent of Man,* when evolutionary concepts were the intellectual rage, three men, Herman Hupfeld, K. H. Graf, and Abraham Kuenen made contributions that arrived at four main source documents. Named the J.E.D. (Deuteronimist writer) and P. (priestly work), they were said to have been assembled by Ezra and read to unite the people at the time of their return from Babylonian captivity (Nehemiah 8). This was conjecture and required a complete rewrite of Israel's history. The next chapter will show how attacks on the deity of Jesus accompanied the attacks on the Old Testament.

Part II: Evolutionary Basis for all Comparative Religion: Pluralism Renders Religion Irrelevant

Rationalistic Assumed a Naturalistic Beginning

From the time of John Locke at the end of the seventeenth century, under the developing deistic philosophy, scholars tended to deny special revelation and tried to discover religion by reason from nature. ***Before Darwin***, suggestions were made from various studies of primitive peoples about how religion began. The nature myth theory was proposed by comparative religionists in the eighteenth and revised again in the nineteenth century. They especially examined and gave symbolic explanation to the star myths about gods. Also, De Brosses (1760) proposed a fetish-worship theory. The focus was first on the Indo-Germanic group of people and then later on Africa, Oceanic Islands, and America. This early method of speculation under deism sought historical sources, not necessarily evolutionary.

Evolutionary Methodology Applied to All Religions

After Darwin's *The Origin...* (1859), religious scholars with vastly divergent views sought to interpret world religions using the evolutionary method, searching for early simple origins. These attempts ranged from the study of horticultural people and the worship of souls, to Herbert Spencer's similar ghost theory, to E. B. Tylor's comprehensive animistic view. They included new astral nature myths of Babylon, Assyria, and Egypt, and totemism, based on descent from certain animals and magic as the source. Other evolutionary scholars postulated a type of preanimism, based on an undifferentiated mixture of religion and magic.

Such post-Darwinian thought could reach no final consensus. W. Schmidt said of this period,

> All the theories... were under the sway of progressivist Evolutionism, this is to say they assumed that religion began with lower forms, and explained all its higher manifestation, especially monotheism, as the latest in time, the products of a long process of development.[4]

With the 1873 publication of Max Müller's *Introduction to the Science of Religion,* his "science of comparative religion" became very popular. Wellhausen's efforts to base Israel's religious origins on animism made the study of comparative religions popular, and at the same time, drew credibility from the ideas of Müller and others. But no one theory was final in the field, one holding sway for a time and then another. From the time of the biological evolutionary theory of Darwin, who published in 1859 ff., the unbelievers had to explain the source of the idea of God. They said that the idea of God began with mystery in nature and they gradually evolved a higher view to a monotheistic God.

Resistance to Theories Which Undermined Evolutionary Presuppositions

At the end of the nineteenth century, Andrew Lang, a former pupil of Tylor, in his book, *The Making of Religion,* called attention to an interesting new fact. ***Among primitive peoples a high spirit God***, who was both kind and good, was recognized as creator and the founder of the moral code. He was adored as the 'father' who eternally existed, and could not be the product of a long process of development. Schmidt says, "He had indeed, by his recognition of the ancient Supreme Being, existing even among very low savages, destroyed one of the fundamental propositions of Evolutionism." Andrew Lang's work on the subject says Schmidt, "was received at first with general opposition, and then with still more general silence."[5] Obviously this finding ran head-on into the concept of evolutionary progress, and was unacceptable to the educated elite.

All data was expected to fit evolutionary patterns. This theory did not. As Schmidt said,

> All the other theories, however, came into being after the outbreak of materialism and Darwinism, and their work was all done on the lines of Evolutionist natural science. This puts all that is low and simple at the beginning, all that is higher and worth being regarded only as the product of a longer or shorter process of development. They found this easier to do, because the principal objects of their study were savages who had not yet made the acquaintance of any sort of writing by which to date the monuments of their culture. Hence the question of earlier and later, of the chronological sequence of religions and forms of religion, which must be settled before the causal interaction of the facts can be exactly determined, could be answered only by help of the Evolutionist method, which is really no method at all....[6]

---

[4] Schmidt, *The Origin and Growth of Religion*, 12.

[5] Ibid., 13.

[6] Ibid., 13-14.

Wellhausen's evolutionary interpretation of the history of Israel, and this application of evolution to other religions resulted in a growing consensus among the educated elite that man created all ***ideas of God out of psychological or other needs***. Thus, no religion was unique. This faulty conclusion led to the next step of concluding that each religion was equally acceptable, and any religion claiming uniqueness was a bigoted and disturbing force in the world. Since religion existed, society should aim for a syncretism of the best religious ideas.

A New Evolutionary View of Israel's History: A Man-Centered Approach

Julius Wellhausen used these critical, speculative steps to take major strides and develop and rewrite Israel's history with an evolutionary basis for the Old Testament. He gathered up many of the previous critical ideas, changed and modified some, and molded them into a final form that was called in English, *Introduction to the History of Israel,* but usually is referred to by the first German word, *Prolegomena.*[7] There are many books that give more of the details of the development to the point of Wellhausen's work.[8]

The average person does not appreciate the influence this book had on undermining faith in biblical history. This faith has never been fully restored. As A. Noordtzy has said, "We have ***no other book*** on the subject of the Old Testament which has exerted so much influence or dominated thought in so broad a sphere."[9] Herman Wouk, the literary historian and novelist, points out, "For a generation and more the *Prolegomena* took the field of Bible criticism and held it. Most Bible critics went down before it like ninepins."[10] This book became central for teaching the Old Testament in graduate schools and theological seminaries in Europe and America, and many other places in the world. It is still taught, in part, in many places. It influenced hundreds of pastors and others who went into church leadership. What was so radical about what Wellhausen taught?

The Object of Wellhausen's Attack

Wellhausen "really only handles one single question in this work, namely: "Is ***the Mosaic Law*** *the starting point* of Israel's history (at the Exodus), or from the history of Judaism which grew out of the nation destroyed by the Assyrians and the Chaldeans?"[11] Wellhausen rejected the traditional view that the Law and the tabernacle were given at Mount Sinai in a covenant after the Exodus, and argues that the law became central to Jewish worship only through priestly leadership during the captivity in Babylon, nearly a thousand years later. Noordtzy commented on Wellhausen's supposition about Mosaic Law and worship. He said,

> [Mosaic law] was not yet in existence, being in flagrant apposition to it. And if someone should be inclined to try to prove the contrary by an appeal to the so-called historical books of the Old Testament, then Wellhausen answers that these same books cannot be taken as they lie before us,

---

[7] Julius Wellhausen, *Prolegomena zur Geschichte Israels,* (Introduction to the History of Israel, 1878).

[8] For a liberal perspective see Robert H. Pfeiffer, *Introduction to the Old Testament,* (New York: Harper & Brothers, 1948), 41-49; and for a conservative view see, Gleason L. Archer, Jr., *A Survey of Old Testament Introduction,* (Chicago: Moody Press, 1964), 73-82. See also chapter 6.

[9] A. Noordtzy, *The Old Testament Problem*, 468.

[10] Herman Wouk, *This Is My God,* (New York: Simon & Schuster's Pocket Book Edition, 1973), 266-267; cf. Donald Grey Barnhouse's, book review of addendum of Herman Wouk's, *This Is My God,* (Doubleday & Co., Inc.) in *Eternity* magazine, January 1960, 18.

[11] A. Noordtzy, *The Old Testament Problem*, 467.

> that Israel's tradition has had a very complicated history, and that one must first determine by 'critical' methods what in this tradition is trustworthy, what is less trustworthy, and what is untrustworthy.[12]

One of Wellhausen's main premises was that in Mosaic times the art of *writing was unknown.*

Wellhausen's Methodology

In other words, Wellhausen saw himself as a literary critic who was able to take the historic documents and discover other ***preliminary documents*** behind them that told an entirely different story. He saw a ***story of deception*** by a priestly group to promote their interests, creating a centralized worship that did not even exist under Solomon.

His major arguments are: evolution was working in Israel, preliminary oral and written sources that previously existed were put together in documents; a group of deceptive power-hungry priests composed final documents, and an interpolator[13] repeatedly intervened in the documents throughout the centuries.

According to Wellhausen, Israel was a small primitive group of people in Palestine who lived without contact with other people, and whose religion began as a "crude anthropomorphic polytheism." Being isolated, like Darwin's finches that evolved on an island, the Israelites evolved a higher and more sophisticated religion until priestly canonizers under Ezra, in the time of the second temple, produced Israel's history as it now appears. Using earlier documents under the pretense of fixing the canon of Israel's sacred books, these priests "performed a work of massive manufacture with one end in view: to shore up their own claims to power and money." They invented one God and His law, which they had authority to enforce. Everything in the book of Leviticus and in the entire Pentateuch about Hebrew worship around the Law and "the account of central worship of Solomon's temple in the history book was also a mass of priests' inventions."[14]

The idea of a proposed forgery, plus the many literary arguments he offered, were borrowed from previous critical speculations and added some new speculative theories of his own. Yet, Wellhausen gave ***no real historical evidence*** that such a priestly group ever even existed or did this.

In his imaginative mind, the historical books had been worked over three times—first by a prophetic narrator, second by the Deuteronimist, and finally by the men of the Priestly Code. He presented the ***four major documents*** listed here from earliest (about 870 BC) to most recent (between 500 and 280 BC): J (for using the name, Jahweh), E (for the name, Elohim), D (for the Deuteronimist), and P (for the Priestly code.).

Herman Wouk, a German scholar, noted literary author, and a capable Hebrew linguist, carefully worked though the *Prolegomena,* and described Wellhausen's grandiose method as follows,

> Whatever passages of Scripture *support his thesis*, or at least do not oppose it, are *authentic.* Wherever the text *contradicts him*, the verses *are spurious*. His attack on each verse that does not support him is violent. He shows bad grammar, or internal inconsistency, or corrupt vocabulary, or jerkiness of continuity, every time. There is no passage he cannot explain away or annihilate. If he has to change the plain meaning of Hebrew words he does that too. He calls this 'conjectural

---

[12] Ibid., 467-468.

[13] Wellhausen never identified the interpolator, remaining his imaginary person.

[14] Wouk, *This Is My God*, addendum, 266-267, Barnhouse on Herman Wouk, *Eternity* Magazine, 17.

emendation'.

> Early in the game he seems to realize that he will not quite be able to shout down one *haunting question:* how is it after all that hundreds and *hundreds of Bible verses refute his theory* in plain words? Wellhausen answers this challenge by unveiling an extraordinary hypothetical figure, *the Interpolator, a sort of master forger.* Seeing across a span of 23 centuries, this man (or men) obviously anticipated the Wellhausen theory, and went through all of Holy Scripture carefully inserting passages that refuted it!... Wellhausen, of course, does not name the Interpolator. He does not even personify him as a single figure. He merely summons an interpolator, perhaps once on every other page, to do his duty. When all else fails Wellhausen—grammar, continuity, divine names, or outright falsifying of the plain sense of the Hebrew—he works an interpolator.[15]

With the help of this interpolator, Wellhausen produced a new history of Israel that captured the minds of the educated elite in the West because he eliminated any real acts or revelation of God. His work was almost 500 pages, and included about 5,000 textual references to the Old Testament.

Wellhausen's View of History Contrasted with the Traditional View

The contrast between Wellhausen's new history of Israel and the traditional view of Hebrew history is as follows:

Wellhausen's View of History	**Traditional View of Hebrew History**
Anthropocentric—man-centered	God-centered
Evolutionary – developing over a long period of time	God created, acts, and reveal himself in history
Deceptive fabrication of power-hungry priests	Revealed through God's prophets calling men to integrity, honesty and fairness
Long development, early men inept	One God known in high civilizations in Ur, the Amorites, in Egypt, et al.
Man develops higher and higher and better	Man as sinner, from first man throughout Hebrew history
Religion develops as a result of man's felt needs for God	Religion based on objective revelation initiated by and from God
View of God, Jahweh, was "a 'horrible and man-hating god', or a 'powerful, blustering Wotan, or a 'fetish in the form of a serpent', or a 'god of storm and rain', or a 'fire-god,' perhaps of Egyptian or even Aryan origin,"	God of Israel is one who is just to judge sin, but repeatedly is willing to be gracious and forgive if man will forsake evil and turn to him.

Inverting the Story of the Bible: *Man* Created God

In traditional history, ***God created man, called one nation*** of people to whom he gave his law, and through whom He would redeem the world. In Wellhausen's history, ***men created God*** to meet their needs and put together a fraudulent law to establish their power to control the people of Israel.

In the traditional biblical view, God miraculously delivered the children of Israel out of Egypt. He then met with them at Mount Sinai in a cloud of glory, and entered into a covenant with

[15] Wouk, *This Is My God*, addendum, 267; Barnhouse referring to Herman Wouk, *Eternity*, 18.

them. The Old Testament covenant focused on the law (the Ten Commandments) and a system of sacrifices carried out and supervised by the priest. These were the means whereby forgiveness could be offered when the law was broken and the people repented. God's throne was symbolized in the Holy of Holies over the Ark of the Covenant in which was contained the Ten Commandments, written on the stone tablets He had given to Moses (cf. Exodus 19, 20, 24, et al.). The temple of Solomon was also based on worship of God and His law.

God's Rule through His Law, the Focus of Critical Rejection

Higher criticism attacked precisely ***the point of God's rule*** over His people and all men through the law in the tabernacle. Wellhausen aimed directly at the center of the idea of ***man's accountability to God*** through the Mosaic Law, the focal point of Israel's worship. Nothing could more dramatically demonstrate man's bias against God and against any accountability to Him. The net result was to reject God and to enthrone man as king.

A Pluralistic Equality Suggested a Human Syncretism

Yet, synchronized religion's prominence in the early twentieth century emerged from an unexpected direction. Christian leaders recognized that Hinduism, Islam, and other religions in countries being affected by Western culture were seizing Christian ideas and trying to show similarities to their religions. The competition, caused by the encroachment of Christianity, sparked a missionary spirit in these religions and an effort to claim affinity to Christianity. John R. Mott referred to this observation made by outstanding missionary leaders in the early part of the twentieth century, at the first International Missionary Conference in Edinburgh.[16] At that time, a member of the Chinese delegation questioned the uniqueness of Christianity to other religions. They saw it as intolerance to other religions.

In 1928, the International Missionary Council met in Jerusalem with what has been called "the first representative global assembly of Christians in the long history of the church."[17] Christianity's uniqueness, in the face of religious evolution, was the primary concern of the meeting. And it would have divided the assembly except for William Temple's skillful wording of the report.[18] But this did not satisfy the growing number of people who pushed for doing away with the claims of Christian uniqueness. One such person was William Earnest Hocking of Harvard.

All this reached a turning point in 1930 when lay members of seven American Protestant denominations, acting independently, engaged the Institute of Social and Religious Research to survey foreign missionary activity in India, Burma, China, and Japan. The resulting data was submitted to a Commission of Appraisal, chaired by Dr. William E. Hocking, professor of philosophy at Harvard University. This commission published its report under the title *ReThinking Missions: A Layman's Inquiry after One Hundred Years*. Concerning this it was said, "There ran throughout the implication that Christian truth is relative rather than absolute and that Christianity should cooperate with, rather than try to supplant, the non-Christian religions."[19]

---

[16] John R. Mott, *The Decisive Hour for Christian Missions,* (New York: The Student Volunteer Movement, 1910), 56-61.

[17] William Richey Hogg, *Ecumenical Foundations,* (New York: Harper and Brothers Publishers, 1952), 244.

[18] Ibid., 248.

[19] Leferts A. Loetscher, *The Broadening Church,* (Philadelphia: University of Pennsylvania Press, 1954), 149-150.

The denominational leaders immediately rejected these views. But the educated elite, in their certainty that Christianity must give up its uniqueness, would not let this issue die, in spite of the fact that evidence continued to emerge to the contrary. However, by the last part of the nineteenth century, the biblical historicity of Israel was rejected. It was supplanted by a contrived history, based on the evolution and development of nature worship into the idea of one God.

Using the theory of scientific determinism to excuse a fraudulent history of Israel and evolutionary view of all religious ideas, the educated elite then turned to discredit the supernatural in the person of Jesus Christ.

Chapter 25
Efforts to Conform Jesus to Scientific Determinism

Introduction

The Historical Reality of Jesus the Christ

The Gospel writers present Jesus Christ *in specific historical situations* that relate Him to humanity and its redemption from sin—a redemption God had promised throughout the Old Testament. Christ was born in the "days when Caesar Augustus issued a decree that a census should be taken" and at the time Herod, the son of Antipater, was procurator of Judah. The genealogies in Matthew link his coming to the lineage of Abraham through whom God promised to bless all nations and to David through whom would come a king who would rule the world for God eternally. Joseph and Mary, the mother of Jesus, had to go to Bethlehem, the city of David, to register his birth. The Jewish Sanhedrin accused Jesus when Caiaphas was high priest and he was sentenced to death under the Roman ruler, Pontius Pilate. Jesus was a man identified by specific events in history.

Samuel J. Andrews, in his outstanding account of Jesus' life, said of the gospel writers,

> It is true that they do not enter into any great minuteness of detail. Yet they do not neglect those relations of time and place which are necessary to convince us of the reality of his earthly existence, and to give us a distinct picture of his labors.[1]

The gospel writers saw their faith based on real facts that occurred in history.

Indeed, for Matthew, Mark, Luke and John, all history before this came into focus with the events associated with Jesus Christ. Jesus is recorded as saying, "The *time is fulfilled,* and the kingdom of God is at hand, repent and believe the good news" (Mark 1:15). The Apostle Paul wrote, "When the *fullness of time was come*, God sent forth his Son" (Galatians 4:4, emphasis added). These and many more such statements show that these writers saw these events as central to all God had revealed about himself to men. John makes it clear that the eternal Word of Son of God in the Old Covenant came as a man incarnate (John 1:14) and Matthew and Luke tell of the virgin birth (Matthew 1:18-25; Luke 1:26-38). It was this coming of the eternal Son of the Father into the determined world of knowledge that was central to the act of God to save man. John warns that it is the denial of this coming of God's Son in the flesh that divides those with Christ's Spirit from those of the world who are of the antichrist (1 John 4:2, 3). The coming of Christ was the fulfillment of God's revelation of himself. And it should be remembered that since the third century, for about two thousand years, the whole world has dated all events BC, (or before Christ), and AD, (or anno Domini, meaning since the year of our Lord). In that sense, the whole world has acknowledged the events associated with Jesus Christ to be central to the meaning of real history.

How Determinism Incredibly Made Possible Rejecting Christ

In the previous chapters, we noted that the Christian view of history involves a linear view of time—a beginning point followed by a sequence of events leading to a consummation point where

---

[1] Samuel J. Andrews, *The Life of Our Lord Upon the Earth,* (New York: Charles Scribner's Sons, 1900), vii. Andrews was professor at Hartford Theological Seminary, whose work in the gospels was a standard of excellence for over half a century.

every person will be called into account for what they have done. The central events concerning Jesus Christ give meaning to the whole historical process. We have also seen that the main objective of demonic strategy is to discredit Jesus Christ, who is the fulfillment of God's message to man. In the last chapter we saw that all of the Old Testament historical events that prepared for the meaning of Christ had been attacked as untrue. The next logical point of demonic attack was the person of Jesus Christ.

By mid-eighteenth century, satanic strategy had slowly excluded God from history and substituted only natural forces operating through sense perception of uniform determined laws. This was so well accepted (though not demonstrated) for long enough that most elite intellectuals had forgotten Christians began modern science. By claiming the eternity of natural processes that excluded miracles, intellectuals had attacked the credibility of creation and any possibility of a final judgment.

This also required denying the possibility of God's working and revealing himself in the Old Testament. Therefore, the Old Testament history must exclude God's acting or speaking to his people supernaturally. Higher criticism changed the story of the Bible, beginning in Genesis, to make the Hebrew religion a work of man and a deception about God acting or speaking. Wellhausen's rewrite of the history of a sovereign God calling his people and revealing Himself to them by miracles had been accepted.

Having destroyed the evidence for God in creation and the preparatory message in the Old Testament, the educated elite undermined the entire context and purpose for Christ's coming, which was to fulfill the revelation of the person of God and to achieve his promises to redeem man from his guilt and the power of sin. The critics especially had attacked the credibility of God's law, the sacrifices for sin, and his covenant with his people as revealed in the tabernacle and temple. By perverting this preliminary revelation, they had made it easy to try to impose new meaning on the event of the coming of Jesus. To deny that Jesus was God with a divine mission, everything supernatural or miraculous about Jesus Christ had to be denied and cast aside.

Just as the philosophical trend through the previous centuries had removed the reality of God's existence, now human bias sought to discredit the divine and miraculous Jesus and to reduce Him to a mere man. That would remove any divine imperative for all men to receive him as God. It made him appear as just another human religious leader like others.

The secular humanists had focused on questioning miracles and using logic as the means of evaluating all religion. Not until the deistic period of the mid-eighteenth century and following, were there consistent outright attacks against Jesus such as those of Voltaire. Then men began to systematically separate [that in his life, which was confined to the regular laws of nature from the miraculous] the natural from the miraculous in Jesus' life.

Perverting the Meaning of Jesus Christ, the Center of Biblical Revelation

Successive Stages of Rejection of the Miraculous in Jesus

Religious scholars' efforts to remove the supernatural from the life of Jesus Christ to make him fit into deterministic science went through successive stages, each more boldly denying the supernatural in a quest for a purely historical person. The stages were: (1) **hesitant** devious beginnings, (2) increasingly bold **rationalizations,** (3) making the supernatural into **myth**, (4) denying and supplanting **documentary sources**, (5) creating an **imaginative Jesus**, (6) seeking sources from **other religions**. Finally, scholars admitted all these were failures. (7) Claiming

Jesus' disciples had deluded hopes of God bringing in **an eschatological kingdom**. (8) The failure to find any meaning to worshiping a historical Jesus, the critics turned to searching for the context or literary units of worship stories in Form Criticism which all were created by the critic. (9) When that failed they searched for an earthly purely historical meaning to worship Jesus, from the preaching of the early disciples, or Kerigma. (10) When that failed to have credibility, they claimed that disciples had a subjective earthly experience at the edge of history with a transcendent God beyond history. They never allowed for the coming of God across the line into the circle of determined knowledge that Kant had drawn. The following is a review of these speculative ideas.

Hesitant Deviant Beginnings of a Jesus Who Was Not Supernatural

About 1750, H. S. Reimarus sought to divide the "real" historical Jesus from the traditional one in the Gospel records. He taught that Jesus was merely a Jewish man who proclaimed the end of the world. It was his disciples who deliberately imposed miraculous powers upon him, e.g., stole his body and claimed he was raised from the dead. But Reimarus knew what he wrote was radical for his time, and he never published his 4,000-page work. It wasn't until after his death in 1778 that Leesing published this controversial material. During the last third of the eighteenth century, the battle over the miraculous had not emerged as the important issue, but was skirted by rational evaluation of other issues tangentially related to the nature of Jesus.

Others who skirted the supernatural include: J. J. Wettstein who said the Gospels should be interpreted like other literature. J. P. Gabler made a distinction between the history of religion and the rise of religious theological ideas. J. J. Hess made an effort to look at the divine and the human separately, while Franz V. Reinhard tried to discover the plan of Christ for mankind. J. A. Jakobi argued that the miraculous events were later additions. J. G. Herder sought to question the historicity of the traditional Gospels, seeing John as a protest against the other Gospels, and saying that Mark was an archetype of the other Synoptic Gospels on which they rested. The original gospel of John was questioned as history. These were followed by writers, such as K. F. Mahrdt, who sought to discover causal relationships between the events, plans, and aims of Jesus in a fictitious life. At the turn of the century (1800), in one such life of Jesus, K. H. Venturini sought to apply consistent logic to interpret the miraculous. But a bold outright attack to explain away the miraculous aspects of Jesus as untrue was likely withheld because scholars intuitively knew the tremendous implications for Christianity and the world, and for fear of opposition by the church.

Increasingly Bold Rational Rejection of the Miraculous

In the second and third quarters of the nineteenth century (ca 1825-1875) after deism became the dominant intellectual view and as naturalism began to grow, deliberate efforts were made to reject the miraculous and find a Jesus who was purely human, but reasonable as an object of worship. This was the age of skepticism. Kant taught that man's knowledge did not extend to the infinite or eternal and locked out accepting anything beyond natural laws. So any supernatural evidence was automatically excluded. Hegel's philosophy of evolving idealism with the claim that man's understanding of the world was the apex of his idealistic pantheism, followed. This philosophy introduced complete rationalism and prepared for Darwin's evolution. These views made it pure folly to accept miracle. Full rationalism was soon to come.

In 1828, H. E. G. Paulus reacted to a spiritualist craze of his father and sought to give a rational explanation for the miracles of Jesus. Like most critics, he felt he was saving Christianity by making it acceptable to the modern mind. He claimed the eyewitnesses did not know secondary causes, which were really natural. The only true miracle was the human Jesus, and he was placed into a position so that he had to force himself to declare he was the Messiah, thereby becoming a traitor to his holy human character. Many writers began to give logical explanations for everything supernatural in the life of Christ and to assume that the original story about Jesus did not have any miraculous aspects. This produced the clear search for the historical, totally human, Jesus, as well as a search for the oldest and original documents. With this *presupposition* that the miraculous was false, the traditional gospel records as they were known could not represent the original information about Jesus. Thus, the documentary hypothesis of the New Testament emerged.

Liberal Rationally Rewriting of "Lives of a human Jesus"

There ensued a number of writings that are considered the liberal "Lives of Christ." The miraculous virgin birth stories of Jesus were abandoned as late efforts to deify the historical person; there was a tendency to establish two periods of Jesus' ministry, and other ingenious efforts to remove miracles. John's gospel was held to by some in order to retain theological meaning for the person of Christ. Some of these writers were Schenkel, Weizsacker, H. J. Holtzmann, Kein, and Hase. E. D. Schleiermacher, who sought to reconstruct a view of Jesus using intuitive subjective knowledge and give a rationalization of miracles in an ascending order, also had a significant voice. This effort to give a rational explanation of everything in the gospels gradually lost historical validity. Scholars needed some way to remove all miracles in one sweep.

Bold Rejection of Miracles as Myth

One of the early and most radical writings was by D.F. Strauss (1835), who relegated miracle to *myth*, or religious ideas in historic form. As Albert Schweitzer said, "He made an end of miracle as a matter of historical belief, and gave the mythological explanation its due."[2] He rejected personal immortality in every form and considered it a present quality of the spirit. Myths were present spiritual realities which appear as "moments" in the eternal being and in humans becoming absolute spirit. By myths, the human spirit is emancipated "in the consciousness that Jesus is the creator of the religion of humanity" and is beyond criticism. Myth comes from religious feeling. DeWitt had suggested this idea in the Old Testament offering a view of religion which was merely a feeling of dependence, self-surrender, and inner freedom. He obtained such a view by merging the dialectic of Hegel and Schleiermacher. But Strauss was more radical. He saw miracle as a gate to enter, yet escape and exit gospel history. Public reaction to Strauss did not end the rationalization of miracle, but produced a more subdued way of stating things.

Shift to Find Documents that Gave Historical Development

Many efforts were made to uncover the sources. While Mark's gospel had been seen as an archetype for the other Synoptic Gospels (Herder), C. H. Weisse suggested Mark was prior to the other gospels on the basis of simplicity. He gave five reasons to support this view, and C. G. Wilke offered further confirmation. Mark does not mention the virgin birth or Davidic kingly claims. And some of the oldest copies showed no report of seeing Jesus alive after the crucifixion, but told only of finding an empty tomb and a young man there who said Jesus had gone to meet

---

[2] Albert Schweitzer, *The Quest of the Historical Jesus: A Critical Study or Its Progress from Reimarus to Wrede,* (New York: MacMillan Publishing Co., ninth printing in 1975, first in 1906), 95.

them in Galilee (verses 16:9-20 were missing). Scholars argued that there was less of the supernatural in Mark. Hence, Jesus' followers had not yet added the embellished and evolving ideas of Jesus as God. Mark was the shortest gospel, and its chronological outline might suggest Matthew and Luke used it as their historical base.[3]

Julius H. Holtzmann spearheaded the conflict between the Synoptic Gospels and the Johannine Gospel. The Gospel of John stood apart, in that it was more theological and centered more in Jerusalem than Galilee. John didn't seem to fit as neatly with the Synoptic Gospels. He referred often to light and darkness and to other ideas that seemed more Greek or even Gnostic. Generally, it was agreed that someone other than the apostle John had written this gospel much later.

But even the Synoptic Gospels seem to require earlier sources, before the idea of deity was attributed to the human historic Christ. Claims of authorship, place of writing, historical data, and the like were examined by literary criticism using means such as style, language, etc., as had been done with the Old Testament books. While Luke claimed to know of other efforts to write about Jesus, the critics claimed earlier documents to enable them to remove the supernatural elements. Over time, the critics decided there must be several sources behind Mark. There was a document called proto-Luke, a hypothetical Q document (for Quelle, meaning "source") which furnished other data that was common to Matthew and Luke, a hypothetical M document for the source of the other data of Matthew, and a hypothetical L for the other data in Luke, which also had an F document. No evidence existed for any of these.

There were other sources suggested for the various Gospels. For example, Albrecht B. Ritchel refers to an apocryphal gospel of Marcion. Some believed there was a fifth gospel from which the writers drew their ideas. For about two decades, critics focused on Mark's Gospel and then little by little turned their gaze toward John's. Hellenistic thought, Gnosticism, and Mysticism were suggested as sources of influence in John.[4]

Gradually, Mark, Q, M (for other material behind Matthew), and L (for other material in Luke) became the most accepted basic documents. John's data continued to be considered late. All the liberals who sought to construct a "Life of Jesus" combined philosophy of the Enlightenment with a critical approach to the documents. Practically everyone had a different idea for the supernatural in the life of Christ. Some believed the miraculous came from the preaching of an unknown evangelist, others believed the disciples exaggerated Jesus' work and imposed the miraculous, while others believed these were myths, or from the preaching of his community of followers through "reflection." Some authors, such as Strauss and Schleiermacher, wrote a second life of Christ and changed some of their views. After a long string of these hypotheses, it was obvious none knew how to explain the miraculous in Jesus.

Albert Schweitzer, the liberal critic who has read and studied them all, commented about the liberal "Lives of Christ."

> The only real advance in the meantime was the general recognition that the Life of Jesus was not to be interpreted on rationalistic, but on historical lines. All other, more definite, historical results

---

[3] For a complete discussion on the arguments, see Albert Schweitzer, *The Quest of the Historical Jesus,* 121-160, or George Eldon Ladd, *The New Testament and Criticism,* (Grand Rapids, MI: William B. Erdmann Publishing Co., 1967), 109-140. For a detailed study on the whole issue, see B.H. Streeter, *The Four Gospels: A Study of Origins,* (London: MacMillan and Company, 1936), fifth printing.

[4] Donald Guthrie, *Biblical Criticism: Historical, Literary, and Textual,* (Grand Rapids, MI: Zondervan), 95.

> had proved more or less illusory; there is no vitality in them. ...You feel that you have read exactly the same thing in the others, almost in identical phrases.[5]

Thus, Schweitzer, the leading German critical scholar, admitted no one had actually found a historical basis to explain Jesus, and all had resorted to speculation.

"Imaginative" Jesus Was Created for Psychological Purposes

As the rationalistic efforts to explain away the miraculous and the evolutionary and the developmental trend of Hegel's philosophy became more accepted, Jesus followers began to imagine "Lives of Jesus" with psychological purposes for meeting the needs of the Christian community. There also was more of an effort to tie these psychological approaches to some other historical data. These trends were seen even as early as Bruno Bauer (1840). He said there never was a historical Jesus, but that the "body of Christ," or church, represented a new religion from the West based on a new Roman Judaism related to Philo, who had fused Jewish and Greek ideas. He received little recognition. F. C. Baur sought to date the New Testament documents on an evolutionary theory. He said this evolution arose out of a conflict between the legal Jewish point of view and Paul's more liberal views, which came later.

An enormous range of hypotheses were offered. Ritchl proposed a *pure spiritual-ethical religion.* C. C. Hennel traced the development of the supernatural Jesus to the Essenes, a Jewish sect whose legends about the cross and the resurrection took on a significant role. A. F. Groren created a Jesus that emerged from the factious conflict between Palestinian theology and Alexandrian theology. E. Renan proposed that while there was no miraculous person, certain ideas of spiritual values jumped out of the text and supplied a subjective revelation that presents a Jesus who meets needs. Adolph Harnack based his view of liberal Christianity on man's subjective dependence on God as Father and the brotherhood of men.

The critical scholars rejected the traditional authors of most of the New Testament books, and the books were dated later to allow evolutionary embellishment of ideas. For example, they considered the possibility that the pastoral epistles of I and II Timothy, Titus, and also Ephesians, and Colossians may not have been written by Paul. Thus, literary criticism questioned the historicity of almost all the New Testament books, not just the gospels. By the turn of the century (1900), Schweitzer as one of the critics, could say, "Criticism of the Gospels ***has 'run to seed.'"***[6] He thus confessed a lack of any historical fruit in all these efforts.

Comparative Religion to Explain the Historical Jesus

Those pursuing comparative religion sought to find sources for Jesus and his message from other religions, but with little success. Max Muller confessed that all his life he had looked for origins from Eastern religion, especially Buddhism, but that in spite of many "startling coincidences between Buddhism and Christianity," no one could point out "historical channels through which Buddhism influenced Christianity."[7] Rudolf Seydel expressed the view that it was possible that some of the Buddhist legends might have influenced Christianity. But Schweitzer

---

[5] Schweitzer, *The Quest of the Historical Jesus*, 200.

[6] Ibid., 194.

[7] Max Muller, *India, What Can It Teach Us?,* (London: Longmans Green, & Co., 1883), 279; see A. Schweitzer, *The Quest of the Historical Jesus,* 291.

comments, "We are justified in waiting until new discoveries are made in that quarter before asserting the necessity of a Buddhist primitive Gospel."[8] And Eduard von Hartmann renounced any relation of Jesus to Buddha, only parallel information, as did Muller. No serious claim of Buddhist influence has been made.[9]

It was suggested that some of Jesus' eschatological ideas had Persian origin, but with little evidence to support this. Some writers assert that influence during the exile may have contributed to development of covenant ideas. Since comparative religion was also fruitless, other directions were therefore inevitable.

Conclusive Failure to Find a Benevolent Purely Human Jesus

After almost one hundred and fifty years of suspicion and rationalization to try to find a benevolent or psychologically meaningful human Jesus, no concept of Jesus had received widespread acceptance. New Testament scholars had run the gamut of critical creativity to find a purely human Jesus. This led to the one remaining naturalistic explanation for why so many worshipped Jesus of Nazareth, namely, Jesus had a mistaken belief or delusion about himself and taught that he was the Jewish Messiah. Criticism now turned in this direction.

All previous "Lives of Jesus" and efforts to find a psychological meaning for his ministry had struggled with the idea that Jesus had to do with ***a present kingdom***. But much data in the gospels speaks of and looks to a future kingdom and finding no other solution, the last critics said that Jesus mistakenly believed his kingdom would come eschatologically.

Jesus' Quest for Future Eschatological Kingdom

Mistaken Messiah Looking To the Future

In his book, *The Quest for the Historical Jesus*, Albert Schweitzer reviewed the situation at the close of the nineteenth century. It was believed that the Marcan hypothesis had ended and that John was late and there was no document in which to find the human Jesus. Schweitzer referred to "a certain historical skepticism." He went on to show that the critics at that point were turning toward Jesus' own teaching about his glorious coming at the end time as recorded in the Synoptic Gospels, especially to Mark's view. The word eschatological came from the Greek "escatos" meaning "ending," or here in Schweitzer, "after history."

Schweitzer argued that many critics did not see this issue clearly and struggled to find a present kingdom. Even many of the first people who saw the importance of Jesus' eschatological teaching, such as Timothee Colani, Gustav Volkmar, Wilhelm Weiffenbach, and W. Baldensperger, tried to hold a place for a present kingdom. According to Schweitzer, these men were in error. For him, the only remaining evidence for the historical Jesus is in the eschatological, or the idea of a future kingdom. There were those who saw both, but "struggled against" the future apocalyptic idea. Wilhelm Bousset held to the liberal fatherhood of God idea. He admitted the apocalyptical was a part of early Christian teaching, but viewed it as not being Jewish, but Persian, and as uncharacteristic of Jesus, who emphasized the joy of life.[10]

---

[8] Rudolf Seydel, *The Gospel Of Jesus in Its Relation to the Buddha Legend and the Teaching of Buddha,* (Leipzig, 1882), 337; Schweitzer, *The Quest of the Historical Jesus,* 293.

[9] Eduard von Hartmann, *The Christianity of the New Testament,* (1905).

[10] Wilhelm Bousset, The Antithesis Between Jesus; Preaching and Judaism: A Religious-Historical

Schweitzer, reporting on the search, said that some who saw the eschatological earlier failed because "they allowed a false conception of the Kingdom of God to keep its place among the data." He said there is *no place for a present kingdom,* but "we arrive at a kingdom of God which is wholly future.... Being still to come, it is at present purely supra-mundane."[11]

Only Eschatological Human Jesus Is Historical and Not Conjecture

Johannes Weiss was the first to focus on the importance of the general conception of the kingdom. He said the kingdom is not something Jesus establishes; He exercises no Messianic functions; He only proclaims it and waits "for God to bring about the coming of the Kingdom by supernatural means." Weiss said,

> The sole object of this argument is to prove that the Messianic self-consciousness of Jesus, as expressed in the title 'Son of Man,' shares in the transcendental apocalyptic character of Jesus' idea of the Kingdom of God, and cannot be separated from that idea.[12]

Schweitzer said,

> He (Weiss) lays down the third great alternative, which the study of the life of Jesus had to meet. The first was laid down by Strauss: *either* purely historical *or* purely supernatural. The second had been worked out by the Tubingen school and Holtzmann: *either* Synoptic *or* Johannine. Now came the third: *either* eschatological *or* non-eschatological[13] (Emphasis added)

Schweitzer dedicated a whole chapter on "Thoroughgoing Skepticism and Eschatology." He based this study on his and W. Wrede's books that were published on the same day, with almost the same name, similar views and often even similar phraseology. One focused on literary criticism and the other on historical recognition of eschatology.[14] Both aimed at ending the idea of a *present* historical kingdom, as found in the liberal views of a "Life of Christ." Schweitzer believed Mark had something about the historical Jesus, but it was of no value because Jesus was an apocalyptic man without relevance to modern times. Wrede saw Mark only as apologetic literature.

The earlier critical view is simply, as Wrede says, "each critic retains whatever portion of the traditional sayings can be fitted into *his construction* of the facts and *his conception* of historical possibility and rejects the rest."[15] Then Schweitzer comments,

> The psychological explanation of motive, and the psychological connection of the events and actions which such critics have proposed to find in Mark, *simply do not exist.* That being so, nothing is to be made out of his account by the application of a priori psychology.[16]

Comparison, (Gottengen, 1892).

[11] Schweitzer, *The Quest of the Historical Jesus.* 239.

[11] Ibid.

[12] Schweitzer, *The Quest of the Historical Jesus,* 240, referring to Johannes Weiss, *Preaching of Jesus Concerning the Kingdom of God, 1892.*

[13] Ibid., 238.

[14] Ibid., 330-397.

[15] Ibid., 333.

[16] Ibid.

Later he says, "Either the Marcan text as it stands is historical, and therefore to be retained, or it is not, and then it should be given up. What is really unhistorical is any softening down of the wording, and the meaning which it naturally bears."[17] Wrede and Schweitzer are saying that the many *efforts to find ideas behind the historical books* as they stand, especially Mark, *are pure conjecture that failed to show any truth.* The eschatological writers, Weiss, Wrede, and Schweitzer, swept all other critical views of the "Life of Christ" out as fraud, based on the critic's own *a priori* assumptions.

The Presentation of the Eschatological Views of the Human Jesus

Wrede said that Jesus did not really profess to be the Messiah at all. Mark is not historical at all, but is of a dogmatic character. Jesus saw himself as "the bearer of a special office to which He was appointed by God...." Jesus did not go to Jerusalem to die, but to work. The community of followers and Mark *secretly believed that Jesus was the Messiah*, and, after what Jesus' disciples mistakenly thought was the resurrection, this secret was made known publicly. Tradition passed on the secret much later, and when John's gospel was written, this conception of Jesus had crystallized so the writer of John's gospel was able to more openly present Jesus as the Messiah. But Wrede's claim that Jesus had gone to Jerusalem simply to work dramatically altered the clear motivation presented in Mark's gospel.

Schweitzer criticized Wrede for what is "so destructive of the real historical connection," and comments that where to draw the line "between what is traditional and what is individual cannot always be determined even by a careful examination directed to this end." Moreover, if Jesus did not profess to be the one designated to be the Messiah, "on what grounds was He condemned?"[18] So he shows Wrede is guilty of the same sin as previous critics, of making up data. Schweitzer demolishes and rejects his views.

Schweitzer also embraced the idea of the messianic secret. But he saw this as a notion conceived by one lone evangelist, not the community of disciples. He does not make it clear who this is. As Wrede had also said, Jesus had the sense of divine call, but Schweitzer goes one step further. He presents Jesus as having a bold consciousness of himself as Messiah-designate. The source of the messianic concept for Schweitzer was the intertestamental apocalyptic vision. Many Jewish writers during intertestamental times of suffering foresaw God cataclysmically coming in power and glory to save his people.[19] For Schweitzer, Jesus, the gospel writers, and Paul were influenced by these intertestamental writings.

According to Schweitzer, Jesus had two phases of ministry, marked by a clear turning point. That line of demarcation occurred when Jesus sent out the twelve (Matthew 10:23). He sent them out to proclaim the coming of the Son of Man in judgment and to establish the kingdom, not to teach or witness to Him. Jesus fully thought this eschatological event would occur before the disciples finished going through the cities of Israel. The second phase began when this did not occur. Jesus was disappointed and foresaw the coming of suffering. The concept of suffering led

---

[17] Ibid., 333, 336.

[18] Ibid., 338-339.

[19] See R.H. Charles, ed., *The Apocrypha and Pseudapigrapha of the Old Testament,* Volumes I, II (Oxford: Clarendon Press l, 1913); (I have written a 75 page unpublished review of the eschatological thinking of all of these with the intent of publishing this after I could also add a review of eschatology in intertestamental Dead Sea Scrolls).

to the idea of resurrection and his glorious coming.

In Schweitzer's view, resurrection is a metamorphic, and not a real resurrection of his body from the dead. The secret of Jesus' hidden Messiahship is a concrete definite event, which is a sudden revelation to his disciples by the pouring out of the Holy Spirit. To Schweitzer, these ideas were doctrinal, not historical, events. "The expectation of sufferings *is therefore doctrinal and unhistorical*, as is, precisely in the same way, the expectation of the pouring forth of the Spirit uttered at the same time."[20] (Emphasis added) Jesus did no real miracles. Schweitzer applies his conceived idea of the secret Messiah to many events in the Synoptic Gospels, apparently giving no priority to Mark, but using anything in any gospel he wishes.

Acceptance of suffering with a misguided hope of victory became the disciples' central belief. According to Schweitzer, the Jesus the church worships today is not a Jesus who really lived. He was created by the Jesus of history who transmitted these views as dogma through the church. He deliberately went to the cross in Jerusalem and was crucified on the basis of his own confession that he was the Son of Man. This was like the suffering of God's people at the end of the age. Jesus thereby introduced the religion of the Christian church which is a spiritual force and not historical. What Christians worship, therefore, is dogma and not historical fact. Schweitzer said,

> It is in truth surprising that he (Jesus) succeeded in transforming into history this resolve which *had its roots in dogma*, and really dying alone. Is it now almost unintelligible that His disciples were not involved in His fate?[21] (Emphasis added)

In evaluating Schweitzer's view of Jesus, he presents nothing that Jesus did that would have caused his followers to remember Him or to be committed to following Him sacrificially. There was no real purpose in his death on the cross. It was not out of commitment to God or for the atonement for man's sin. Moreover, Schweitzer denies Jesus' ministry had any psychological benefit for man. It is, therefore, not clear why Jesus died. Moreover, Jesus did not really rise from the dead, nor does he promise real bodily resurrection for mankind. Schweitzer offers nothing of real meaning to Christianity that would cause people to love or follow Christ.

Schweitzer, for the most part, bypasses all the major eschatological discourses of Jesus. Instead, he focuses on the one verse connected with the sending out of the Twelve as his key. Yet the traditional meaning of the verse about the Son of Man "coming" to the Twelve is not an eschatological idea at all, but seems to mean simply that Jesus would shortly come after his disciples to follow up and support their work. This is clearly stated when He sent forth the seventy later. "He sent them two by two ahead of him to every town and place *where he was about to go*" (Luke 10:1, emphasis added). Schweitzer proceeds to try to force his invented eschatological idea into many places where it does not fit.

Wrede and Schweitzer do the same thing they condemn all the writers of the 'Lives of Jesus" of doing. They *make up their own story* and try to impose it on the historical documents. Thus, neither Wrede nor Schweitzer allows the texts to say what they really mean. They both do with eschatology what those with "Lives of Jesus" and the psychological approaches of a present kingdom did. They superimpose their own ideas of what happened, rather than what Mark or any

---

20 Schweitzer, *The Quest of the Historical Jesus,* 362.

21 Ibid., 392.

of the gospel writers said. But Wrede and Schweitzer are worse than their predecessors, because they did what they said others should not have done. Have they thus become the worst deceivers and hypocrites of them all?

Conclusions on Rationalistic, Psychological and Eschatological Quests

Schweitzer and other critics saw the quest as a failure. As Schweitzer concluded his long review of the efforts of men to find a Christ of history who is purely human, he summarized this quest for the historical Jesus.

> What surprised and dismayed the theology of the last forty years was that, despite all forced and arbitrary interpretations, it could not keep Him in our time, but had to let Him go. He returned to His own time, not owing to the application of any historical ingenuity, but by the same inevitable necessity by which the liberated pendulum returns to its original position. The historical foundation of Christianity as built up by rationalistic, by liberal, and by modern theology *no longer exists*; but this does not mean that Christianity has lost its historical foundation. ...Jesus means something to our world because a mighty spiritual force streams forth from Him and flows through our time also.[22] (Emphasis added)

By this, he means that the eschatological Christ, supposedly as he conceived him, is a memory to influence us. In the end, Schweitzer still held to Albrecht Ritchl's belief in Christianity as a spiritual force. Schweitzer became a complete humanitarian and went to Africa as a missionary.

Searching for Meaning in the Early Church Apart From Jesus

Introduction: Question–Why the Dynamic of Christianity?

While most of the critical scholars generally agreed that the quest to find the real human behind the Jesus in the New Testament documents had ended in failure, there had to be some explanation of the Jesus that has been worshipped by the churches through the centuries. How could such a movement as Christianity have originated and become such a growing and persuasive force? Following Schweitzer and the eschatological debate, the attention was turned away from the historical Jesus and toward early Christian worship.

With the failure of the quest for the historical Jesus, it would seem that the scholarly community might conclude that the problem was perhaps with their worldview and the misguided beliefs in deterministic science. But the quest was driven by man's bias and not by logic, and the quest therefore turned in a new direction. While many conservative scholars continued to adhere to the supernatural Jesus, the educated elite moved in a new direction to give some natural explanations for the phenomenon of the dynamic Christian religion that was still spreading worldwide.

The search for meaning in Christ and Christianity focused first on the forms of worship and then shifted to the preaching of the churches.

Form Criticism to Find Essential Elements of Church Worship

There was long-standing general agreement that oral traditions about Jesus had been passed

---

[22] Ibid., 399.

down and were the basis for the gospels and for what was preached in the early church. A methodology called form criticism was used to discover some of the literary units by which historical teaching was transmitted. Each Bible critic had his own idea of the forms that were preserved. M. Dibelius looked for paradigms, tales, legends, exhortations, and myths, while R. Bultmann looked for wisdom sayings, apocalyptic or prophetic sayings, parables, miracle stories, and the like. The meaning of each was to be interpreted on the basis of the conditions in which it was found ("Sitz in leben," German). Were the people under persecution? Were they worshipping? C. H. Dodd transferred the apocalyptic idea into realized eschatology—that these events were to be realized now. And so on. Each scholar saw one main controlling motive of Christian worship behind these units, but each one was different. For Bultmann, it was the alleged debates between the early community and Judaism. For Dibelius, the "missionary goal" was the motive, and "preaching" the means. These subjective ideas of the critic were used as the test for whether or not the *unit of form* was genuinely primitive.

As New Testament scholar I.J. Peritz has said, "The great fault of Form Criticism is its *imaginative subjectivity* in evaluating tradition."[23] (Emphasis added) So, the results were as varied and unsure as the search for sources. James M Robinson saw the failure in that the form critics were guided by personal subjectivism of their own desires. He said, "...their method was at best only indirectly relevant, they tended to arrive at the conclusion which their general orientation suggested, rather than a conclusion which form criticism as such required."[24] Robinson concluded that Form Criticism was inconclusive and not producing decisive results in finding remnants about the historic Jesus in the church. [25]

By the early twentieth century, Christian scholars had conceded to the claims of naturalistic science that the world was run by understandable laws, and miracles were unacceptable. Very few intellectuals, other than the religious fundamentalists, held to a transcendent God who revealed Himself by miraculous disclosure. The Wellhausen view of the Old Testament and the evolutionary concepts of comparative religion had won control of the major denominations and centers of higher learning. Immanuel Kant's view of the limitation of man's knowledge to time and space and man's inability to know eternal things ruled in higher education. If there was a God, no objective knowledge of him was available. The quest for the historical Jesus was fruitless, and the Christian community had produced nothing generally acceptable through form criticism. This forced the critics to turn to philosophy to try to find the early relevant meaning to Christianity in its existential expressions. To fit determinism, these had to have meaning apart from anything supernatural.

Reaction for Existentialism

By the middle of the nineteenth century, the insistence on all knowledge as rational, as the only basis of knowledge for real truth, produced both a loss of faith and hypocrisy in religion. In response to these conditions, a Danish Christian philosopher, Soren Kirkegaard (d.1855), introduced what came to be known as existential thinking. In a protest against rationalism he said that truth, good, and the beautiful are all essential to *the existence of the individual.* He said there is a *discontinuity between faith and reason.* By the early twentieth century, a number of

---

[23] Ismar J. Peritz, "Form Criticism as an Experiment," *Religion in Life,* Vol. 10, No. 2, Spring, 205.

[24] James M. Robinson, *The New Quest For The Historical Jesus,* (London, England, S. C. M. Press, 1959), 36, 37.

[25] For an extensive consideration of Form Criticism see, George Eldon Ladd, *The New Testament and Criticism,* (Grand Rapids MI: William B. Eerdmans Publish Company, 1967), 141-169.

philosophers joined to emphasize this existential theme from numerous perspectives. Jean-Paul Sartre, a first cousin to Albert Schweitzer was influenced by Kant, Hegel, Kierkegaard, Husserl, Heidegger and Henri Bergson; he embraced communism and wrote existentially in politics making it a popular philosophy. There was Nicolas Berdyaev of the Orthodox Church, Jacques Maritain of the Roman Catholics, Martin Buber of Judaism, and numerous Protestants. Several sought to give credibility to the Christian message by appealing to existential experiences.[26]

Bypassing Scientific Rationalism by a Dialectical Existential Message

The new focus of critical scholars was to look at the preaching (called "kerigma" in Greek) in the early church to discover how the Christians defined their existence and to find those ideas that are relevant for the individual today. Having given up on a historical Jesus, this was another way of finding something to explain the origin and purpose of Christianity in history. Otherwise, why did it ever exist? The Form Critics had moved the focus to the experience of worship in the church, which led away from the historical Jesus himself, to what the church thought about and created through its preaching.

First came the *dialectical theologians* who used the idea of thesis (God), antithesis (man and his need), and synthesis (man's encounter with God based on his divine initiative) to make sense of the development of Christianity. (The dialectic was a method used by Hegel, Marx, and Kierkegaard earlier.) Karl Barth led the new dialectical theological movement. He emphasized *God's utter transcendence,* which *man encountered* at the edge of history, resulting in *change in the individual.* This view was dominant 1930-1950. Next, the *existential theologians* emerged, basing their ideas on myth, reality, and experience. Rudolph Bultmann led this trend that held dominance between 1950 and 1960. The dialectical scholars talked in terms of biblical doctrine and practical application.

Existentialists focused on worship experiences from the primitive church's preaching (in Greek, kerigma). This led the critics to try to trace history backwards and discover how the church looked back on Jesus from what the churches preached about him. These existentialists were of all views, first Abertz as a conservative to the radical positivism of Bultmann. None of these writers claimed any historical certainties, and *each successive one progressively rejected more biblical content* and identified with the secular world. They chose biblical themes as existential for preaching to the church today. They adhered to Kant's line between knowledge of time and eternity, while sometimes this was hypocritically transgressed for the sake of content.

The Dialectical Theologians

Karl Barth asserted that God was "wholly other," or completely transcendent. Man could only encounter God and understand his Word at the edge of time and eternity. The knowledge of God was subjective, with no interruption of the line of the circle of determined knowledge. Thus scientific determinists could not criticize what Barth affirmed because this subjective experience did not have anything to do with the physical world. While Barth did not say the word of God was subjective and he claimed one could put the word in propositions, yet this content was *never truth that could be certified in history* by historians or scientists. He also said that the word was only authoritative as God made it so *subjectively* to the individual. He avoided biblical criticism. While Barth preached most Scripture and traditional doctrines, he never ever admitted that there was a

[26] For more on existential thinkers cf. H. J. Blackhow, *Six Existential Thinkers,* (New York: Harper Row, Publishers), 1952; Will Herberg, *Four Existentialist,* (New York: Doubleday Anchor Books), 1958.

real resurrection of Christ from the dead and there was never any mediation of time and eternity for any Christian fact.[27]

Moreover, he taught that justice, as an aspect of the God's love and grace, is an eternal possession ***of all***. He thus held to universal salvation. Therefore, the death of Christ was not relevant to the atonement for sin, though it was considered an evidence of the love of God. Barth admittedly took a philosophical position, which he claimed was exempt from historical and scientific knowledge, but he spoke inconsistently to these areas of life with biblical truth.

Some of those who followed his dialectical approach, with differing emphases, were Emile Brunner and Reinhold Niebuhr. Paul Tillich was an existentialist, but his emphasis on Christ was as the ground of being. He seemed to imply pantheism, but he never admitted this was so.

Church Existentialism Reawakened Need to Replace the Focus on the Person of Jesus

As liberal evolutionary philosophy produced relativity in morals and religious thinking, and as modern science produced rapid change, and pragmatic psychological thinking released people from accountability for acting out their desires, world conditions became more and more uncertain. The meaning and security of the individual's existence became uncertain. After two world wars and the emergence of brutal fascism in Germany and Italy, and callous Communism in Russia, man's hope for the future in science and human goodness was deeply shaken, but not shattered. *The emphasis and authority for religion* in these insecure and violent human conditions became totally *individualistic* and according to the feelings of existential needs. All relevant meaning was to be found in man's struggles for existence. All that remained was subjective truth for the individual—a truth that was only relevant for him. And there was no objective word of God that could hold any or all people accountable.

It was Rudolph Bultmann who consistently shifted theology to existential needs. But he too, like the dialectical theologians, argued to hold the line between time and eternity. Geovannie Miegge said,

> Bultmann argues that *all* conceptions of the (religious) world have to be called 'mythological' insofar as they are 'different from the conception of the world which has been formed and developed by science.' ...Bultmann calls for us in our existence to be open to the mighty deeds of God which are mysteriously hidden within the events of this time and this world, but are revealed to the eyes of faith here and now.[28]

Thus, according to Bultmann, as man studies the early church and discovers what God said through their experience and worship, these may become relevant to his existential need today. This encounter with the 'acts of God' (which are nebulous and non-historical) is based on the early preaching of the church. Bultmann thus saw himself as demythologizing Christianity by removing it from the myths of the supernatural, and yet finding a basis of faith in God's mighty deeds as embellished *in the memory* of the churches' worship and preaching.

Bultmann was also, by this subjective methodology, escaping from the world's rational

---

[27] Carl F. H. Henry tells of his question about the resurrection to Barth very late in life and his non-commitment to this. See *Jesus of Nazareth, Savior and Lord,* (Grand Rapids, MI: William B. Erdmann's Publishing Company, 1966), 247-249. Barth never crossed the line of natural determinism objectively.

[28] Giovannie Miegge, *Gospel and Myth: The Thought of Rudolf Bultmann,* (Richmond, VA: John Knox Press, 1960).

criticism, as had Barth, by claiming knowledge of the supernatural, without acknowledging any objective truth which could be examined. Bultmann mostly ***ignored the historic personality of Jesus*** in his early writings, and based his message on the kerigma (preaching) of the church.[29] Thus, he was not dealing with what the church actually experienced in the world but what was subjectively remembered. One must question Bultmann as to how there could be a memory of something, which never occurred. This lack of anything historically objective left his most brilliant scholars uncertain.

The New Quest for the Historical Jesus with an Existential Basis

In time, Bultmann's followers were discontented with having no connection to the historic person of Jesus. The continuity from the worship community to the existential message of today was not adequate. There had to be some explanation for a connection to Jesus Christ. This led to a new quest for the historical Jesus by Bultmann's followers. E. Kasemann denied that one could construct a "Life of Christ" as the liberals earlier tried to do, nevertheless he insisted on the need for a connection of the historic Christ to the primitive church. He feared what a disengagement from the person of Jesus might mean. He said,

> If this [having no connection to the historical Jesus] were to happen, we should either be failing to grasp the nature of the primitive Christian concern with the identity between the exalted and the humiliated Lord; or else we should be emptying that concern of the real content, as did the docetists (an early Christian heresy). We would also be overlooking the fact that there are still pieces (German, stucke) of the Synoptic tradition, which the historian has to acknowledge as authentic *if he wishes to remain an historian at all.* (Emphasis added)[30]

Toward the end of his arguments he says, "Our investigation has led to the conclusion that we must ***look for the distinctive element*** in the earthly Jesus in his preaching and interpret both his other activities and his destiny in the light of this preaching."[31] Other Bultmann followers such as Gunther Bornkamm, Ernst Fucks, and James M. Robinson[32] also felt some connection to the person of Jesus was needed.

In their search for the essence of the churches' preaching, they came face to face with the importance of the belief of the disciples that one must accept Jesus' death as his own and then have a new resurrection power to be existentially transformed in life. The cross and resurrection had been brought into center stage by the brilliant work of Richard R. Niebuhr which is to be considered next. But the Form Critics refused to accept Christ's resurrection as real fact. To escape the crossing of the line of the determined natural circle of knowledge, they viewed this preaching backwards as an existential church creation, so the New Quest by-passed the essential link to Jesus out of bias

Conclusions: Can Christianity Have Meaning Apart From Jesus?

The fact that the liberal critics themselves could not feel credible without some known

---

29 Bultmann, in his book, *Jesus*, 212, ff. may have recognized some knowledge of Jesus.

30 E. Kasemann, *Essays on New Testament Themes,* trans. W. J. Montague (London: SCM Press, 1964), in his famous lecture "The Problem of the Historical Jesus", 15-47.

31 E. Kasemann, *Essays on New Testament Themes,* 44-46.

32 James M. Robinson, *The New Quest For The Historical Jesus.*

connection to the meaning of Jesus Himself reveals their failure to find any meaning that was generally acceptable. The plain fact is that the church cannot be explained and cannot continue with any meaning without a clear connection to the historical Jesus. The birth, growth, and spread of the Christian church as a phenomenon of history cannot be explained without this. So, the Barthian movement and the Bultmannian movement lapsed back into the old liberal effort to find or claim some essential link to a historical Jesus Christ. But being unwilling to go against the modern naturalistic deterministic view, the historical method did accept ideas beyond nature.

Reference to a Third Quest in the 1980s to find the essential teaching of Jesus will be mentioned later. The futility of the three was in the biased method to exclude the supernatural.

Chapter 26
Claim of the Death of God and Intellectuals as Priests

Introduction to the Death of God Movement

By the last half of the nineteenth century, the German Enlightenment, with the use of scientific determinism, had exalted man and swept away faith in the supernatural. On the one hand, it created existentialism, and on the other it left religion as meaningless. The German Friedrich Nietzsche saw man's "will to power" as supreme and pronounced the death of God. He struggled with the moral and ethical results and predicted "barbaric brotherhood" under dictatorial leadership. At the time, Nietzsche's teaching was not well received in England and America, but by the mid-twentieth century the "death of God" idea found full fruition in America. The result of unbelief in Germany should have been a warning to America, but it was not.

By 1955, the concept of God had been removed from general education in the Western world, and by the early 1960s, the concepts behind the death of God theology were generally discussed in the American media and public. Theology had progressed to a place where God as a supernatural being who was involved in the world was no longer acceptable. The existential theology of Europe was a last effort to circumnavigate scientific determinism. Barth and Bultmann declared discontinuity.

Three Steps in the Death of God Thinking

The last phase or final steps of removing God from public knowledge came in three ways. The first was the *death of God teachers*, who believed God is no longer relevant in our secular world. The second was the *secularizing of the church* for man's use to save it from obscurity. The third was *church leader's' outright denial of the incarnation of Christ as God* in history. Moreover, this third step identified the transcendent God, active in the world, as the enemy of man, or satanic. Thus, these three final steps completed the rejection of the Christian message of the Bible and sought to turn the church into an institution under the control of intellectual man rather than God. Worship was now focused on man and helping him. All of these developments contributed to America's moral breakdown.

The Development of Death of God Theology

Initial Suggestions of the Idea of the Death of God Theology

While the whole anti-God movement had its beginnings in the European Enlightenment, especially Germany, its final phase moved to the center of the English-speaking world—Britain and America.

Herbert Braun planted the seeds of the death of God movement when he recommended a theology without God in an essay proposing God is *found in the world* in relation with one's fellowman. (Such a concept is really an oxymoron, since theology means study of God.) Dorothee Solle, in a short statement, based this idea more on the sense of God's absence from the world. While Dietrich Bonhoeffer should not be considered a death of God theologian, he developed one of the main ideas, which was used by the death of God theologians. He used the illustration that when a child grows up, he no longer needs his parent to take care of him and meet his needs. Since the world is governed by natural laws, and man through science has been able to understand the world and use it for solving his problems, he is like an adult who no longer needs God. He has outgrown the need for his parent, God.

The rejection of God idea was a powerful influence in 1960 and following. At that time I went to a conference for all ministers working with college students in the Presbyterian Church (US). At this church conference, both speakers asserted that they were thorough-going empiricists and their message was, "To save the church from irrelevance, it must be awakened to the fact that there is no supernatural person or power and the word 'God' is now meaningless." Their argument was that in this scientific age the church must make its message relevant to a secular world or die. They said, "We must take the cross of Christ seriously and die to self to communicate this to the church." These men were popular speakers on the campuses. One was from an institution in Dallas, Texas, and the other was from a graduate school in Atlanta, Georgia. The striking similarity of their messages was even more remarkable in that they had not spoken together or even known each other before.

Three Leading Death of God Theologians Compared

Although there were others, three death of God theologians emerged to prominence: Paul Van Buren, William Hamilton, and Thomas J. J. Altizer. Van Buren sought to build his case for the death of God on the meaning *of language to communicate.* Hamilton went beyond Bonhoeffer's idea of man's outgrowing the need for God and asserted that *God has disappeared from meeting human need.* Altizer sought to merge the idea of oriental mysticism with the idea of Christ as the logos in calling man to a *bold trust in the process of nature,* and not to return to the past genesis, but rather move on boldly to the emerging future. All of them employed eschatological and existential emphases, although Altizer less in regard to existential thought. Hamilton and Van Buren drew heavily on ideas of the goodness of the Jesus of history. Van Buren wanted to speak to and for the church. Hamilton was indifferent to the church, while Altizer was hostile to her.

William Hamilton's Core Belief of Caring for Our Neighbor

Hamilton did not seek a comprehensive thought system, but was satisfied with fragmentary thinking and sought to reduce theology to an essential core of belief. He saw God as meaningless and irrelevant in this secularized world, more because of the problem of suffering and the feeling that He has disappeared as a need-fulfiller. He believed there was no longer any meaning in speaking about God.

Thomas W. Ogletree who reported on these thinkers said,

> In summary, Hamilton candidly confesses the loss of God in his experience, especially the God who is a need-fulfiller and problem-solver. Man must solve his problems and meet his needs without God as he lives and works in the world. At the same time the primary meaning of our lives as Christians is to be in the world in the place defined by Jesus, that is, alongside our neighbor, participating in his struggles and sufferings.[1]

But Hamilton failed to give a reason in this individualistic world why one should care for others.

Hamilton wanted to translate "God-talk" into talk of the kinds and quality of human experience. He speaks of "unmasking" the "hidden Jesus" by "becoming Jesus" in and to the world by accepting our own place and identifying with our neighbor in his/her struggles and

---

[1] Thomas W. Ogletree, *The Death of God Controversy,* (Nashville: Abingdon Press, 1966), 39. I am greatly in debt to Ogletree since the primary sources for some of this evaluation of the death of God theologians were not available to me.

suffering. One could do this by engaging in the Civil Rights movement, the war on poverty, or other egalitarian struggles in politics and elsewhere. Hamilton called "Christians" to worldly responsibility in promoting a better, more just, social order. The core of Hamilton's *faith is in man carrying out, on his own, Jesus' command to love our neighbor.*
While he abandoned any theology of God, he confessed,

> Jesus is the one to whom I repair, the one before whom I stand, the one whose way with others is also to be my way because there is something there, in his world, his life, his way with others, his death, that I do not find elsewhere. I am drawn, and I have given my allegiance.[2]

He thus chooses the *Jesus of the gospel* records as *the ideal man* and places the hope of the future in men seeking to act as Jesus would.

Hamilton gave the Jesus of the gospels a central, even unique, position in his thought and worship, and called all men to act like him. At the same time he confessed that humans are sinners in need of forgiveness.

Ogletree insightfully comments on Hamilton's weakness.

> How does this self-interested man come to find himself alongside his neighbor in love? Clearly some sort of redemptive process is involved which frees man from the drive to grasp life for himself and for the possibility of giving himself to his neighbor. Yet it is precisely such a redemptive process that has been traditionally understood as the work of God in and through the announcement and personal appropriation in faith of Jesus Christ.[3]

I would add to these criticisms the question, how does Hamilton *know what sin is or what is the right way to act* in the place of Jesus without some objective knowledge of Jesus as an absolute? While denying any objective knowledge of God, he has confessed to something he objectively believes himself and feels he has the right to call others to do as a Christian. Is this not inconsistent and hypocritical?

Paul Van Buren Dismissed God on the Basis of Linguistic Meaning

Paul M. Van Buren was convinced that men today are secular in a secular world, which renders God meaningless. Therefore, the Christian message must be a secular one, and any language used in theology must be precisely understood. Using the highly acclaimed writing of Ludwig Wittgenstein as a point of departure for definition, he proceeded to evaluate what could be meaningful to hold to as Christian faith in our world.

He held to a "verification principle," which says there are two kinds of statements: (l) those that refer to empirical data (that which is knowable by the senses) which can be tested, and therefore are relevant to truth or falsity, and (2) statements of formal definitions of concepts and their relationship to one another, which don't therefore tell about reality, but are empty and useless. To Van Buren, these are only tools of thought. For example, the word *games* is only a convenient grouping of a number of different kinds of activities and speaks of nothing specific. While Van Buren does not follow Wittgenstein consistently, he proceeds to claim that the word "God" is meaningless or misleading and useless in our world. He therefore says the word "God"

---

[2] William Hamilton, *Christian Century,* "The Shape of a Radical Theology," Vol. LXXXII, 1221.

[3] Ogletree, *The Death of God Controversy*, 45.

may only be understood in relationship *to man*. He divides words into cognitive and non-cognitive categories and puts God into the latter one.

Ogletree comments on Van Buren's method, saying, "He dispenses with the word 'God' by means of a procedure which can only be called arbitrary."[4] He dismisses the reality of God because the word does not fit the way he gains finite knowledge, and thus rejects any knowledge that is beyond himself. The only god he could accept as real is a god limited to his size of knowledge, or man. This is precisely what he does to remain acceptable in a deterministic scientific world.

Van Buren said,

> I am trying to raise a more important issue: whether or not Christianity is fundamentally about God or about man.... I am trying to argue that [Christianity] *is fundamentally about man*, that is language about God is one way—a dated way, among a number of ways—of saying what it is, Christianity wants to say about man and human life and human history... I want to argue that what Christianity is basically about is a certain form of life—patterns of human existence, norms of human attitudes, and disposition and moral behavior.[5]

Van Buren thus proceeds to draw a picture of Jesus which he claims is witnessed and reported on in the Christian community, and he sees him as a remarkably free man—free from religious traditions and ceremonies, free from anxiety and fear, and free to give himself for his neighbor. Van Buren said the "secular meaning" of the resurrection is the idea that Jesus' followers "found themselves caught up in something *like the freedom of Jesus himself*, having become men who were free to face even death without fear."[6] The Christian believer is said to be "grasped" and "held" by this view. Van Buren appeals to the ecumenical councils of the fourth and fifth centuries for interpreting Jesus. He sees the Christian as having a freedom that is imparted by some external force to the individual man. But Van Buren is inconsistent, because it is certain the ecumenical councils believed in an objective knowledge of God.

Van Buren seeks to deny any objective uniqueness to Jesus other than that *he frees other men*. Hence, he really appeals to the Jesus in the historic Christian community without admitting the historical gospels as the source, and he claims the working of a special force of "grace," while denying there is any supernatural redemption occurring. Moreover, like Hamilton, he sees the *Christian life as unselfishly loving our neighbor*, but fails to identify the redemptive power that allows man to do this. One must, therefore, genuinely question whether the gospel of Van Buren is truly "secular."

Altizer's Mystical Process Exalts Men to Future Transcend Gods

Thomas J. Altizer has been called the Christian atheist. He tried to use much of the eschatological and existential thinking to dismiss God while avoiding being called pagan. Much that he says seems contradictory and radical. He calls Christians to be historical in the sense of facing up to what is happening in the present world and not to regress to the traditional supernatural ideas of the past as archetypes, but rather move into the future. This means accepting

---

[4] Ogletree, *The Death of God Controversy*, 59.

[5] See Ogletree's quote "Profiles: The New Theologian" from *New Yorker*, November 13, 1965, *The Death of God Controversy*, 63.

[6] Paul M. Van Buren, *The Secular Meaning of the Gospel,* (New York: The Macmillan and Company, 1963), 128.

the idea that we live in a world in which God is more than just absent. Rather, God is unreal and the God who is dead is precisely the Christian God. Ethics and moral obligations imposed by the traditional God have died with him. Yet, Altizer insists a Christian message (his ideas, not the traditional one) is relevant to the situation of contemporary man.

Like Van Buren, he says relevance is *centered in man himself.* Altizer says,

> If there is one clear portal to the twentieth century, it is a passage through the death of God, the collapse of any meaning or reality lying beyond the newly discovered radical immanence of modern man, an immanence dissolving even the memory of the shadow of transcendence.[7]

The gospel he seeks to find is a new form of faith, which he claims, is true to the underlying meaning of the Christian message.

Altizer does not see this gospel as uniquely found in the Christian canon of the Old and New Testament books, but in a canon including the world's religions and human thinkers. For him, the truth is to be found in the idea of the "logos" of God Incarnate and also in Oriental mysticism, rather than in the historical Jesus. He sees a dialectical tension between these. Oriental mysticism seeks to negate the profane and disclose the sacred by drawing man backward *toward the reality of primordial unfallen totality*. He is thus seeking to avoid classification with the pagan cyclic and typical pantheistic views with this idea (as Mircea Eliade pointed out). For Altizer, the Incarnate Word calls man to participate eschatologically in the forward movement of history toward a final end in which the profane is affirmed and redeemed. He sees a dual movement in which the sacred descended to become flesh, while the flesh must ascend and become Spirit. The death of God was a fundamental precondition for this to happen. Man is involved in the spiritual *process* in nature *toward godness as man.* The *God* who is worshipped in the churches *is in fact the enemy, Satan, the Antichrist,* according to Altizer.

Altizer draws his mysticism from an archaic religion set forth by Mircea Eliade and his sensitivity to the profanity of our civilization from the vision of men such as Melville, Sigmund Freud, Marcus, Sartre, Hegel, and Nietzsche. From this synthesis of ideas he puts together a new form of faith as a means of life affirmation in the midst of that profane civilization.

Altizer's view really seems to be a form of pantheism in which man is moving toward being god. But Altizer denies this, since he sees the god-man as not trapped by the process of nature, implying *man* can move to transcendence. He thus advocated that modern man assert himself as God, which is biblically, Antichrist.

Altizer admits that his Christian atheism involves cutting off all previous forms of faith, existing outside the church, the loss of Christian values, and separation from solidity of biblical authority. He admits the danger that *faith* may be lost altogether, and means submitting to the darker currents of our history, which embrace a life-destroying nihilism and a choice of the way of madness, dehumanization, and the most totalitarian society known to man. But for him, a faith that is not open to risk the loss of faith is one that does not believe Christ is an ever-present reality, even at the time of the death of God. He rejected the idea of the church as the body of Christ, saying that teaching about the church isolates one from the world with which one should be united.

Altizer was not seeking to promote existential ideas. The announcement of God's death was

---

[7] Thomas J. J. Altizer, *The Gospel of Christian Atheism,* (Philadelphia: Westminster Press, 1966), 22, 115, 136, 138.

not a description of personal experience or of a cultural fact, but a theological assertion about Christ and the meaning of his presence in the ongoing process of the world. For Altizer, the Christ is the logos or concept of the ideal man toward which men should move in the future. Yet, his radical teachings see Christ's presence as the Eternal Recurrence taught by Nietzsche, which will perpetually recur in an endless cycle of existence with no direction, meaning, or purpose. Altizer thus ends with a cycle whereby man hopes to move back toward his original innocence, but there are no standards or absolutes for that man. He is open to evil spirits in the process of the world of nature. Altizer brilliantly understands the pitfalls of past paganism and while denying these, he actually deceptively embraces and teaches them.

Conclusions about Death of God Theologians

These three death of God theologians move Western thought one step further in its bias against God. The Bartians and the Bultmannians placed their message outside the realm of the rational and physical to escape the criticism of modern intellectuals who hold that the world is closed to anything but natural laws. They thereby exempted their message from the world's area of knowledge. Certainly, Barth put the origin of his theology completely in the transcendent, making the knowledge of God beyond the rational. These men had little influence on the world because they lacked an objective message about God. But the death of God theologians escaped the deterministic natural worldview by dispensing with God and the transcendent altogether and trying to make the Christian message immanent.

Gabriel Vahanian, also writing about the death of God, candidly warns that their teachings are nothing less than capitulation to the immanent form of thought characteristic of our times.[8] In essence, the educated elite have arrived at pagan nature worship that will make man into god.

In summary, and most importantly, these theologians not only dispense with the God of the Bible, but they put man in the place of God. Hamilton talks about man "becoming Jesus." Van Buren denies any uniqueness to Jesus and sees the Christian as free to be a man like Jesus. Altizer sees the believer as involved in the process of the future where he can reach a point of transcendence or become the God who has died in the world. In a veiled or deceptive way, they all make secular man the center of worship. These views are an expression of the Antichrist.

The Secularization of the Church Because of God's Death

Church's Call to Serve the World under the Intellectual University

Theologians and the press had barely announced the death of God when Christian leaders began to call for the church to join the secular world in promoting the building of the temporal heaven on earth. Two of the most prominent leaders were the Bishop of Woolwich, John A. T. Robinson of England, and Harvey Cox, professor of religion at Harvard Divinity School, the first school built by the Puritans in New England to train ministers.

A. T. Robinson: Abandoning God for Serving Man in the World

Bishop Robinson wanted to reduce Christianity to serve man. His book, *Honest to God,* created a swarm of news coverage because of his prominence in the official Anglican Church of

---

[8] Gabriel Vahanian, "Swallowed Up by Godlessness," *Christian Century*, December 8, 1965, 1505-1507, as referred to by Ogletree, *The Death of God Controversy*, 23.

England.[9] He began by saying he was reluctantly one of the leaders who believed the church's teaching does not need to be "reformatted," but needs to have a "radical recasting," with most fundamentals of the church needing to be remade, including God and religion itself. Then he claimed he had never "really doubted the fundamental truth of the Christian faith—though I have constantly found myself questioning its expression."[10]

In his chapter on "The End of Theism?" Robinson argued as the other critics, that the modern scientific secularized world no longer accepts the biblical teaching and that it has become "essential to the defense of Christian truth to recognize and assert that these [supernatural] stories were *not* history… " (Emphasis added) He agrees with Bultmann that these are myth. He proceeds to argue that biblical faith needs to be expressed in terms of the modern secular world. Like others before, he argues along with Bonhoeffer that man has matured and no longer needs God. The thought of God as a personal Being, wholly other than man, is, according to him, collapsing into meaninglessness. He also taught that belief in a personal God is "the enemy of man's coming of age." He says that the doctrine of transcendence is an outmoded view in the world and considers "that sort of 'religion' as *an unmitigated evil*, far, far more anti-Christian than atheism…. If Christianity is to survive, let alone to recapture 'secular' man, there is no time to lose in detaching it from this scheme of thought …."[11] He essentially endorses the death of God theology.

Robinson accepted Paul Tillich's view of God as the "ground of our being," or of our ultimate concern, or what we take seriously without reservation. Robinson then endorsed the idea that we know *love in relation to other men* as the deepest structure of *reality*. Along with Hamilton and Van Buren, he says God can only be known through *men:* "For the eternal Thou is met *only* in, and with and under the finite Thou, whether in the encounter with other persons or in the response to the natural order."[12] (Emphasis added) While these statements are clearly pantheistic, by using double talk he claims the transcendent is brought into the finite, but is not identical to it. This is plainly a transgression of the most basic principle of communication, the law of non-contradiction.

For Robinson, Jesus was merely a human example. He said that Jesus never claimed he was God, and argues that in the title "son of man," Jesus set the example of suffering as "the man for others."[13] Jesus exemplifies the ideal man in suffering to come. Thus, the Christian participates in the suffering of God in the life of the world. He then turns the meaning of holiness (which is 'separateness') on its head by calling for "worldly holiness." He calls the Christian to go into the secular world and find his communion with God out there in the world. He argues that the Ten Commandments must be forsaken for a relativistic view of love. These views are the opposite of both Jesus' and Paul's teachings in the New Testament. They, in contrast to Robinson, interpreted love in terms of obedience to the Ten Commandments of God (Matthew 5; 22:37, 38; Romans 13:8-10, etc.).

Robinson wants to recast the church for community service in the state. In his final chapter, Robinson defines his ideal of recasting the mold of the church, saying,

> I would see much more hope for the Church if it was organized not to defend the interests of

---

[9] John A.T. Robinson, *Honest to God,* (London: SCM Press, 1963).

[10] Ibid., 27.

[11] Ibid., 33, 42-43 (see the whole chapter).

[12] Ibid., 53.

[13] Ibid., a chapter by that title, 64-83.

> religion against the inroads of the state... but to equip Christians, by the quality and power of its community life, to enter with their 'secret discipline' into all the exhilarating, and dangerous secular strivings of our day, there to follow and to find the workings of God.[14]

But Robinson does not tell how one knows what is right or what will motivate him to perceive it. He ends *Honest to God* by saying, "I have not attempted in this book to propound a new model of the Church or of anything else.... I have tried simply to be honest.... "[15] But is Robinson really being honest? In his opening statements he declares that everything about Christianity needs melting down and recasting. In the three books that follow, Robinson directly aims at restructuring the church according to the ideas he has laid down.[16]

In his book, *The New Reformation?*, he argues that the whole world has turned secular in its thinking and that the church needs to send its laymen into the world to participate and to reinterpret Christianity in terms of the world's thought forms. Presumably he means the ideas he has set forth in *Honest to God.* He is arguing for the people of the church to leave the church and be a part of the world. He claims that there must be a transition, or double life of "living in the overlap" until the "new reformation" occurs. He describes what this means (1) in doctrine, (2) in liturgy, and (3) in the structure of the church. While Robinson expresses some good and genuine concerns, his primary concern is that he wants *the world to set the agenda* for the church, and he sees the most genuine people in the church as thinking *not of serving or worshipping God* in the local congregation, but of being involved in the world itself. While he wants to maintain the congregation in the local parishes, this is an interim to reorient them to worldly involvement.[17]

Educated Elite Want the Church to be Controlled by the State for Man

The Universe is Now Man-Centered and God is Meaningless

It remained for Harvey Cox in his book, *The Secular City,* to propose the manner of the complete secularization of the church.[18] He talked about "the end of religion," and made the statement that metaphysical *systems* will never again integrate whole societies nor still men's persistent questions as once they did.

> In our day the secular metropolis stands as both the pattern of our life together and the symbol of our view of the world.... We experience the universe as the city of *man.* It is a field of *human* exploration and endeavor from which the gods have fled. The world has become *man's* task and *man's* responsibility.

This has come about by the process of secularization, or as he defines it, the *deliverance of man from religious and metaphysical control* over his reason and language. Cox sees this as a "historical process, almost certainly irreversible, in which society and culture are delivered from

---

[14] Ibid., 139.

[15] Ibid., 140.

[16] John A.T. Robinson, *On Being the Church in the World,* (Philadelphia: Westminster Press, 1960), *Christian Morals Today,* (Philadelphia: Westminster Press, 1964), and *The New Reformation?,* (London: SCM Press, 1965).

[17] Robinson, *The New Reformation?* 90, 92.

[18] Harvey Cox, *The Secular City,* (New York: The Macmillan Company, 1965), 244.

tutelage to religious control…."[19]

Man Has a Sense of Deity in Political Involvement

Cox used the ideas of most of the death of God theologians. By secularization, Cox means man has outgrown God, and he is no longer useful or meaningful. Language has changed, and the word "God" is no longer relevant to the secular culture. While the word God was "the conceptual linchpin" for a metaphysically oriented society, it is now "empty." Therefore, "Christendom is disappearing," but it is important to learn to speak about the God of the Bible in a secular and non-metaphysical way. We must not go back to tribal mystical views, but go on to the future.

Cox therefore advocates a post-metaphysical rather than a pre-metaphysical theology, similar to Altizer. Like Hamilton and Van Buren, man can only know God in relation to man. But Cox takes a step further in that, for him, *politics* replaces metaphysics or God. "In the epoch of the secular city, politics replaces metaphysics as the language of theology."[20] Politics involves using all the sciences for the good of the city. To speak of God to secular man means to speak about man. God is "his partner, as the one charged with the task of bestowing meaning and order in human history."

The University Elite Should Replace Church Clergy

For Cox, all the myths of the traditional Christian worldview are to be replaced with secular ones, which means basically with *his, Cox's own views*. The implication is that *the university replaces the church*; the morals of the Bible are replaced by his in regard to work, sex, community, etc. Thus, the educated elite should serve as the new priests controlling, through the university, thought and worship–a worship centered on man.

God Is Hidden and Jesus Is Not a Revelation of God

Cox uses the idea of the hiddenness of God, who makes himself known as he chooses. He is "at once different ***from*** man, unconditionally ***for*** man, and entirely unavailable for coercion and manipulation ***by*** man." But this must always include a relationship to man.[21] The transcendent God is in the secular and manifested in events of social and political change, especially as one works with a team for the social good. The team of equals becomes the new authority. There is no clear recognition of how Jesus of Nazareth fits.

> God does not 'appear' in Jesus; He hides himself in the stable of human history. ...In Jesus, God does not stop being hidden; rather he meets man as the unavailable 'other.' He does not 'appear' but shows man that He acts, in His hiddenness, in human history.[22]

Cox goes on to suggest that the time has come to stop talking about "God" for a while, until some other designation appears.[23] This will only be when he manifests himself in secular events. Thus, all of Cox's talk about the Christian God has only been about man working together to build a better secular city. He clearly rejects the God and church of the Bible for a mystical meaning in

---

[19] Ibid., xii, 1, 18, cf., 26. The italics are mine to emphasize the kingdom is now man's and not God's.

[20] Ibid., 223.

[21] Ibid., 225-229.

[22] Ibid., 226.

[23] Ibid., 232-233.

the secular world where the educated elite control politics.

Conclusions about Secular "Theologians"

The scholars who followed the death of God theologians sought to lead the church to abandon any separateness as the people of God and to make the church purely a tool of service to man and the state. God is irrelevant, and for the sake of man, people in the church should treat the state as God and live for the future hope of building a better secular society. Thus, God is no longer significant, and the church has no theological purpose in itself. Any sense of sacredness or holiness is to be lost in secularization. God is dead, and there is no place for his people as distinct from the world. The university should be the center for the church, and the educated elite who control it and the state are the new priesthood to lead the people.

Conclusions and Consequences of Dismissing God from History

In limiting all knowledge to uniformed determined natural forces there was no meaning left for a purely human Jesus or for God's revelation in history. Fragmentation was the net result of the trends that culminated in the death of God theologies. These trends transformed Christianity into a tool for humanism and for the secularization of the church—a secularization which ended in advocating the replacement of the church with the university and state. Ben Witherington III has gathered these ideas together as a third quest.[24] He continued the eschatological ideas and the concept of subjective experience with Jesus, and more importantly, a greater hostility to the person of Jesus and the effort to make Jesus the tool of the egalitarian human social reform movement. In Witherington's view, man alone matters.

The next, climactic step after the death of God theology was for leading churchmen and women to marshal all the arguments, both rational and psychological, against the man Jesus and to seek to deny his incarnation and uniqueness. After about 1980 critical scholars began a third quest for the historical Jesus and sought to take aspects of Jesus in the gospels and in later second and third century spurious writings, such as the Gospels of Thomas, and of Peter, and the Secret Gospel of Mark, and dress him in the garb of an itinerant preacher and social reformer promoting egalitarian causes. Gerd Theissen pointed the way and John Dominic Crossan added to this. Elisabeth Schusseler Fiorenza created a feminist Jesus as the radical prophet for the goddess Sophia. The Jesus' Seminar, led by Crossan and others went so far as to cast votes by lot on sayings as to whether they were made by Jesus and thus democratically decide the truth.

The critical scholars could not arrive at a consensus view by any objective standards. The conclusions of the efforts to find a completely human Jesus without any supernatural powers turned to a hostile rejection of the biblical Jesus, and the scholars tried to demonstrate that Christianity should be synchronized as only one the world's religions.

---

[24] Ben Witherington III, *The Jesus Quest: The Third Search for the Jew of Nazareth* (Downers Grove, IL: InterVarsity Press, 1995). This gives excellent details.

Chapter 27
Denial of the Incarnation for Religious Syncretism

Introduction: God's Incarnation Should be Removed if God is Dead

The death of God theologians went to extremes to declare that God was not relevant to modern times, and therefore, nonexistent and hidden to a modern scientific world. Needless to say, they were not greatly concerned for the church. But those who followed them proclaimed that *the church* must give up claiming to be the people of God and be an agent for the world. For Harvey Cox, this meant *political* involvement and the loss of the church in the "secular city." Cox believed that only intellectuals of the universities were qualified to achieve major goals in development for humanity. Since those who searched to find a Jesus who could be explained as the historical figure beyond Christianity, these who followed sought to deny the importance of Jesus and by denial engage the church to lose itself in the world. This movement having also failed, the next step in rendering Christianity as of no special value is to deny its uniqueness and advocate the church as only one of many religions. This would end the evidence for the transcendent God who had revealed Himself in history.

In the midst of denying God, a group of clergy in the Church of England, under the leadership of John Hick aimed specifically at getting Christians to reject what Christians have always considered cardinal, the teaching *that Jesus Christ was God incarnate,* or that He came in the flesh.[1] This denies God's unfolding revelation of His redemptive plan that reached its fulfillment in Christ. The incarnation is the very basis on which traditional Christianity has always claimed its uniqueness. Moreover, the incarnate deity of Christ is central to the meaning to his life, ministry, death, and resurrection. The previous studies seeking a natural historical Jesus had presumed against incarnation, but never had ***leaders within the church*** so directly targeted this doctrine.

This attack on the distinctive meaning of Jesus Christ actually presents little that differs from earlier critics. Hick's book is highly significant, however, in that it seeks to gather many critical ideas from previous failed writings into one final effort to completely remove the supernatural aspect of the person of Jesus Christ. Indeed, in Hick's work important theological scholars of the Church of England accuse traditional Christian believers of ignorance and hindering the church of the future.

This position is presented by an organized series of arguments. The seven authors met five times over three years to design and compile the book. It was published in 1977, at the height of English and American rebellion against Christianity on university campuses.

Incarnational Theology to Be Rejected for a Syncretistic Human Religion

Plea to Adjust Christianity to Scientific Determinism

Generally, in the preface of a book an editor or writer will present the views which motivated him or her to write the book. In *The Myth...* the preface clearly shows the presuppositions of at least most of the contributors, as the following excerpt reveals:

> In the 19th century, Western Christianity made two major new adjustments in response to

[1] John Hick, editor, *The Myth of God Incarnate,* (Philadelphia: The Westminster Press, 1977), 211.

> important enlargements of human knowledge: it accepted that *man is part of nature and has emerged within the evolution* of the forms of life on this earth, and it accepted that the books of *the Bible* were written by a variety of human beings in a variety of circumstances, and *cannot be accorded a verbal divine authority.*[2]

If Christians don't accept these two concepts, the authors say that they are:

> '...kicking against the pricks' of the facts. ...The pressure upon Christianity is as strong as ever to go on adapting itself into something which can be believed—believed by *honest and thoughtful people* who are deeply *attracted by the figure of Jesus* and by the light which his teaching throws upon the meaning of human life.[3]

In the introduction, Hick asserts that new light has emerged on the origins of the meaning of Christianity (which he and his coauthors are about to present). Thus, modern intellectual understanding must produce "another major theological development...in this last part of the twentieth century" (which Christianity must accept with the previous two). This theological movement must be vigorously mounted in the "interest of truth" and because of the "practical implications for our relationship to the peoples of the other great world religions." Hick then proceeds to deny that there has ever been a fixed set of beliefs for Christians, but rather Christianity has always been diverse and changing. He dogmatically proclaims, "Orthodoxy is a myth." He claims his intent is to free people to talk about and serve God "with greater integrity." [4]

Christianity Must be Dynamic Adapting to Change

Maurice Wiles, professor of divinity at Oxford and chairman of the Church of England's Doctrine Commission, asserts that the doctrine of the church has continuously changed. He says, doctrine is "an evolving living tradition within Christianity," and he gives examples such as interpreting the Eucharist as "transubstantiation," views on the "inerrancy of scriptures," and modern differences over "the virgin birth."[5]

He goes on to argue that the "gospels stop short of the assertions that have come to characterize the *later* doctrine of the incarnation." For Wiles, the meaning of language changes as the environment changes. While there is evidence that the ideas at the time of Christ lend themselves to this incarnational thinking, new knowledge has come to light about how the teaching of the incarnation originated.

He is willing to admit that according to "the narrower definition of Christianity the central tenet affirms the Incarnation of God in the particular individual Jesus of Nazareth" and that in the minds of most incarnation and Christianity are "virtually interchangeable."[6] But he argues that it is *not essential* to Christianity, and we should surrender this view to more recent evidence that incarnational theology was not held by the first Christians, but rather was developed later by the church.

Wiles believes that in the modern world we should give up the idea of Jesus, the perfect human ideal, as an absolute authoritative model. Since this model isn't available to us anyway, we should adjust our language to empirical evidence that denies this for "a looser sense of the phrase

---

[2] Ibid., ix.
[3] Ibid., ix.
[4] Ibid., ix, x.
[5] Ibid., 2.
[6] Ibid., 1-10.

'incarnational faith'." He says the "physical world can be the carrier of spiritual values," which is a "much broader sense in which those words (of Incarnation) are so often used."[7]

Later Orthodoxy said to turn from Diversity to Intolerance

Frances Young, New Testament lecturer at Birmingham University, argued that original Christianity embraced diversity, but later Christianity embraced an orthodoxy with incarnational belief that was intolerant and bigoted. She said that atonement and salvation ended up with one view that became a test of orthodoxy which appears "as improbable scientific fact, and that has encouraged *arrogant and intolerant attitudes* among the faithful" when there are "other equally valid responses in a pluralistic world."[8] Jesus, as the "hidden Messiah," proclaimed the kingdom of God was now coming and stirred up all kinds of expectations. Christians searched for how to respond to Jesus. Young concluded the early Christians found ideas in the pre-Christian literature, which they applied to Jesus and filled these categories with new content. Originally, this produced many different pictures of Jesus. John saw Jesus in one way and Paul another, but not as incarnate Messiah. But as the early Christians searched for an explanation of their experience with Jesus, their 'environment' led the [church] fathers to the dogmatic position from which the New Testament has traditionally been interpreted.

Suggested Sources for the Myth of Incarnation

Galilean Eschatology and Samaritan Gnosticism as Sources

Michael Goulder, tutor in theology, Birmingham University, says an unidentified Galilean eschatology and Samaritan Gnosticism were the source of the incarnation ideas. Goulder said that Jesus saw himself as a significant person of destiny, like many leaders such as Churchill, Mao Tse-tung, and others. Jesus identified his significance with the Son of Man in Daniel's prophecy. He expressed this sonship in his fervent prayers to God as Father. This contributed to the disciples' expectations. Goulder claims that in response to these expressions, Christians drew from two sources in the environment for the incarnation dogma. John, he asserted, arrives at the doctrine by inspiration, and the circumstances of his "doctrinal development are misty; and mist, notoriously, tends to foster mystery." Thus, he admits no evidence from where John derived his view, other than it was a cloudy or "misty" source.[9]

Goulder's second source was *Samaritan Gnostical myth.* He said the Christian community arrived at the incarnation from this Samaritan Gnosticism, while other scholars (presumably Francis Young) have found the early Christians coming to this idea from other sources. Together these produced the dogma of incarnation. He argues that Paul appropriated the idea of Jesus' incarnation in the course of dialectic with Samaritan missionaries in Corinth and Ephesus between AD 50 and 55. However, none of these missionaries are named or known in any literature.

He admits that a *major difficulty* with this view is that there is no evidence until "documents from *the fourth century AD,* notably the Memar [teacher of] Marqah, and later."[10] (Emphasis added).

From this fourth century writing, Goulder claims a source for the dualism of Gnosticism. In

[7] Ibid., 7.

[8] Ibid., 13, 14.

[9] Ibid., 64.

[10] Ibid., 68.

Memar VI, 2, the Marque has two titles for the Godhead, "the Divine One" and "Glory." He claims these two are spoken of as having two crowns. He then asserts that when Simon Magus was called "the great *power of God,*" it is the same as what Jesus said about Himself as "the Son of Man...seated at the right hand of the *power of God*" (Acts 8:9-10; Luke 22:69). (Emphasis added)

Goulder then refers to Justin Martyr, a prominent early church father from Samaria, who identified Simon Magus with a Simon who so influenced the Roman people with his magic that they made a statue to him, Simoni Deo Sancto, acclaiming his divinity and worshipping him. Justin's reference (Apol. I.C. 26, 56) was passed on through other patristic writers—an error pointed out pointed out by Philip Schaff. The error was first demonstrated by Gott many years ago and was investigated and confirmed as an error by a number of others.[11] The statue was actually identified in Rome in 1574. Goulder then points out that Origen, an early Christian scholar, and Eusebius, the early historian, refer to this identification with the Simoni, whose statue was in Rome. Thus, Goulder believes the early Christians claimed Christ was incarnate God to top the claims of Simon Magus who also claimed to be God. To him, this was the motive for the incarnation myth.

Pagan Sources for Incarnational Myth

With this important claim to deification from a Samaritan influence, Frances Young then presents claims of possible sources from the Gentile, mostly pagan, environment. First, phrases like 'son of god' were commonly used in that society. Second, it was often claimed that men were gods or were born to a virgin by divine intervention, and ascended to heaven. In addition, many believed in heavenly intermediaries, some of whom descended to succor men or to serve as God's vice-regent in judgment.

Young also refers to Lucian of Samosata's work, *The Passing of Peregrinus.*[12] She includes bizarre stories of mystery cult leaders, such as Alexander of Abonoteichos, Peregrinus, and Proteus, who affixed a false human head to a tame serpent, birthed a serpent from an ostrich egg, along with other magic, and claimed he was a god from heaven. Proteus immolated himself at the Olympic Games in 165 A. D., at which time onlookers said they had seen him flying to heaven. Other strange death accounts and visible disappearances, supposedly to heaven, are referred to in regard to Heracles, Asclepius, Dionysus, and Empedocles. Young uses these to show the people's gullibility, at that time, to miracles and claims of deity. This would thus indicate they lived in an "environment" in which the people were open to the naive idea of the incarnation.

She then refers to cases where men were said to be begotten of a virgin by the gods and were called "god" or "son of god." Among these were Plato, Apollonius, a Pythagorean philosopher (who denies these), Egyptian leaders, Alexander the Great, and Romulus of Rome. She points out many Homeric myths refer to gods coming to men to have children. This environment, Young claims, was conducive to myths like the virgin birth and incarnation.

Denial That Incarnation Was a Consistent Essential Doctrine

Don Cupitt, university lecturer in divinity and Dean of Emmanuel College, Cambridge,

---

[11] Philip Schaff, *History of the Christian Church,* Vol. I. (Grand Rapids, MI: William B. Eerdmans, 1950), 257 and footnote 4, and 258 and footnote 1, which give lengthy discussion.

[12] Frances Young, Hick, ed., *Myth of God Incarnate*, chapter 5, 88-97; P. Lucian, The Passing of Peregrinus, 2, 4, 11-16, 29, 39-40.

argues that throughout its history the church has had no historical consistency on the doctrine of the incarnation. Cupitt maintains the doctrine was clearly formulated in the fourth and fifth centuries, especially at the Chalcedon Council. It prevailed until the late nineteenth century, shortly after Bishop H.P. Liddon's book, *The Divinity of our Lord and Savior Jesus Christ,* in 1865, and during the next generation under Bishop Gore.

Cupitt said, "Somewhere between Liddon and Gore a view of Christ which took shape in the fourth and fifth centuries began to collapse, not just in the minds of rationalist critics, but in the minds of the leading churchmen of the day."[13] But he concedes that Gore also held to the incarnation. He asserts "that the doctrine of the Incarnation has had some *harmful effects* upon the understanding of Jesus' message, on the understanding of Jesus' relation to God, and even upon faith in God."[14]

Wiles points out that the idea of myth came into prominence with Keightley's 1831 publishing of *Mythology of Ancient Greece and Italy*, in which he recorded popular traditions or legendary tales about the Greek gods and goddesses. Four years later, Strauss published his famous *Life of Jesus,* in which he claimed that the supernatural in the New Testament must be attributed to myths about Jesus of Nazareth. Bultmann, in the twentieth century, emphasized the same in seeking to find an existential faith from the primitive church. Wiles then quotes Chadwick, "Myth provided a way out for those who were also unhappy to have to choose 'between unmiraculous miracles and lying evangelists'."[15] The claim that Jesus' power was a myth of being God is actually an admission of the recognition of his incarnate power.

Incarnation Must be Yielded Because it's Not Unique to Other Religions

The World is Growing out of Incarnational Myth

This critical book ends with Hick's chapter on "Jesus and the World Religions." *The Myth...* is really an effort to draw together the rationalistic attacks for removing the supernatural and exalting man as god and king of the world. Hick argues that the authors have shown the claim of the incarnation was only *one way of responding* to the person of Jesus of Nazareth. It was a logical way of explaining him in the Greek and Roman world when men believed in magic and the heroic interaction between gods and men. But, Hick argues man's knowledge has evolved to greater and clearer understanding of the world, and the incarnation is inappropriate for modern times. Moreover, Hick believes it is an arrogant position to hold in the midst of a world where, from comparative religions, we now see that the founders of most religious movements were given deity. Christianity should continue to be spread, but *only as one* of the most important religions.

Argument of Similar Claims of Mahayana Buddhism

Hick chooses Mahayana Buddhism as his example of another great religion and lists a number of the parallels between Jesus and Buddha. He names some of the parallels but not all. I furnish here a more complete list as given by R.E. Hume in his book:[16]

---

[13] Hick, *The Myth of God Incarnate* 137.

[14] Ibid., 145.

[15] Ibid., 165.

[16] Ibid., Hick lists them on 169 ff., R.E. Hume, *The World's Living Religions,* (New York: Charles Scribners and Sons, 1946), 67.

Pre-existent, planfully incarnate, supernaturally conceived.
Miraculously born.
Sinless, yet suffering inexplicably.
Entered the world with a redemptive purpose.
All-knowing and all-seeing.
Savior of gods and men.
He is everlasting.

Hick concedes there is one thing *distinctive* that was claimed about Jesus—his resurrection *from the dead.* But, he points out that there were other resurrections in the New Testament, as Irenaeus also admitted. So he concludes: "Thus the claim that Jesus had been raised from the dead did not automatically put him in a quite unique category."[17]

Moreover, Hick says the claims were not based on Jesus' innate power, but on God. He dismisses the resurrection as contrary to modern science, and argues against it because the accounts in the gospels are conflicting in details.[18] For Hick, Jesus is only "a man approved by God" (Acts 2:22).

Jesus: A Mere Man with a Superior Awareness of God

Though Hick considered Jesus to be only a man, he acknowledged him as a man with an extreme sense of eschatological call and consciousness of God in His life, which led Him to appropriate Messianic titles. Thus, Jesus was a man of extreme spiritual power. This influenced the primitive church community in three ways. He met many of their existential needs; he exerted a feeling of absolute authority, which demanded total discipleship; and as a result, he gave them a contagious, existential joy that empowered them to spread the message of who He was. This produced a sense of acceptance and reconciliation to God which led the Christian community to try out additional titles for describing Jesus. In turn, it led the church fathers to accept the idea of Jesus' deity, virgin birth, and incarnation. Over time, these ideas about Jesus began to be literally believed and made into orthodox dogma.

Incarnational Theology Denounced as Divisive For the Modern World

But, Hick asserts, this myth of the incarnation created by the church fathers, and which lasted for a long period of the church's history, is *no longer acceptable.* Indeed, it is "deeply damaging" and "harmful" in that it hinders man's advance in scientific knowledge and puts the church in a position of rejection in this modern scientific and pluralistic age. Thus, this remnant of the age of Greek and Roman myths must give way to a new concept of Jesus that fits the language of modern times and advances human well-being. Christianity is, after all, a religion *for **man's** good.* Hick then argues that claims of Jesus Christ's uniqueness and as the only way of salvation are arrogant and must go.

I quote extensively some of Hick's final paragraphs, because they represent the thinking and arguments of most of the world's educated elite at the present time. Remember, he is speaking for the religious leaders of the official English church who are teachers in universities and colleges at Oxford, Cambridge, Birmingham, etc., six of whom wrote this book with him. For them, this represents what is *politically and scientifically correct* according to modern human knowledge,

---

[17] Ibid., 171.

[18] Ibid., 170-171.

which is pursuing *a one-world new order*.

He says,

> Transposed into theological terms, the problem which has come to the surface in the encounter of Christianity with the other world religions is this: If Jesus was literally God incarnate, and if it is by his death alone that men can be saved, and by their response to him alone that they can appropriate that salvation, then the only doorway to eternal life is Christian faith. It would follow from this that the large majority of the human race so far has not been saved. But is that credible, that the loving God and Father of all men has decreed that only those born within one particular thread of human history shall be saved? Is not such an idea excessively parochial, presenting God in effect as the tribal deity of the predominantly Christian West?[19]

Hick then chides modern theologians who try to find a way to convert and include men of other religions under Christian belief. He calls these efforts "an anachronistic clinging to the husk of the old doctrine after its substance has crumbled."

Insistence on Necessity of a One-World Global Religion

Hick continues,

> It seems clear that we are *being called today* to attain *a global religious vision* which is aware of the unity of all mankind before God and which at the same time makes sense of the diversity of God's ways within the various streams of human life. On the one hand, we must affirm God's equal love for all men and not only for Christians and their Old Testament spiritual ancestors. And on the other hand we must acknowledge that a single revelation to the whole earth has never in the past been possible, given the facts of geography and technology, and that the self-disclosure of the divine, working through human freedom within the actual conditions of world history, was bound to take varying forms. We must thus be willing to see God at work within the total religious life of mankind, challenging men in their state of 'natural religion', with all its crudities and cruelties, by the tremendous revelatory moments which lie at the basis of the great world faiths, and we must come to see Christianity within this pluralistic setting.[20]

Hick then suggests the following as a basis for developing a theology.

> All salvation, that is, all creating of human animals into children of God, is the work of God. The different religions have their different names for God acting savingly towards mankind. Christianity has several overlapping names for this—the eternal Logos, the cosmic Christ, the Second Person of the Trinity, God the Son, the Spirit. If selecting from our Christian language, we call God-acting-towards-mankind the Logos, then we must say that *all* salvation, within all religions, is the work of the Logos and that under their various images and symbols men in different cultures and faiths may encounter the Logos and find salvation. But *what we cannot say is that all who are saved are saved by Jesus of Nazareth.* The life of Jesus was one point at which the Logos—that is, God-in-relation-to-man—has acted; and it is the only point that savingly concerns the Christian; but we are not called upon nor are we entitled to make the negative assertion that the Logos has not acted and is not acting anywhere else in human life. On the contrary, we should gladly acknowledge that Ultimate Reality has affected human consciousness for its liberation or 'salvation' in various ways within the Indian, the Semitic, the Chinese, the African . . . forms of life.[21]

---

[19] Ibid., 180.
[20] Ibid., 180.
[21] Ibid., 181.

Hick suggests that Christians share the insights of Jesus with the other people as a gift and gain from the insight of other religions. He says we must not think of religions as "monolithic entities each with its own unchanging character. They are complex streams of human life, continuously changing." He said Christianity has emerged from its stagnate view of the incarnation from the Middle Ages, and that now all the other religions are also awakening to "enter the rapids of scientific, technological and cultural revolution. Further, *the religions* are now meeting one another in a new way as *parts of the one world of our common humanity."* He predicts, "They will inevitably exert a growing influence upon one another, both by the attraction of elements which each finds to be good in the others and by a centripetal tendency to draw together in face of increasing secularization throughout the world."

Hick insists the old Christian view (the traditional view of incarnation) should give way to this evolving new religion. He says,

> In the case of Christianity the older missionary policy of the conversion of the world, proceeding largely along the highways opened up by Western arms and commerce, can now be seen to have failed; and any hope of renewing it has been ruled out by the ending of the era of Western political and religious imperialism.[22]

He ends with a warning to the Christians who wish to continue with incarnational Christianity:

> The outgrowing of biblical fundamentalism was a slow and painful process which has unhappily left the church scarred and divided, and we are still living amidst the tension between a liberal and a continuing and today resurgent fundamentalist Christianity. The church has not found a way to unite the indispensable intellectual and moral insights of the one with the emotional fervor and commitment of the other. Will the outgrowing of theological fundamentalism be any easier and less divisive? If not, the future influence of Jesus may well lie more outside the church than within it, as a 'man of universal destiny' whose teaching and example will become the common property of the world....[23]

Evaluation of Discrediting Claims of Mythical Sources for Incarnation

Hick's words are either magnificent, or deceptive to the long-standing meaning of the gospel of Jesus Christ and to the church. Apart from the claim that the church must concede to a deterministic view of the world, are there other reasons the church should change its long-standing views? Are these arguments for a change in the church's message compelling? What are the facts?

Was there a Samaritan Source for an Incarnation Myth?

Goulder's argument was given priority because he claimed to have more current evidence in the Samaritan's connection. Graham Stanton, another of *The Myth...* authors, in later writings, questions this as a source for the idea of Jesus' deity on the grounds that in his studies, the first century Samaritans showed remarkable monotheistic conservatism, which would not seem to be open to dualism.[24] He questions if Marqah VVI.3, an early writing referred to by Goulder, has any

[22] Ibid., 182.

[23] Ibid., 184.

[24] Graham Stanton in *Incarnation and Myth: The Debate Continued,* Michael Goulder, ed., (Grand Rapids: William B. Eerdmans Publishing Company, 1979), 143.

more biatarian (meaning two persons in the Godhead) or incarnational emphasis than first century Samaritans, or even Philo, who boldly referred to the Logos as a 'second God,' even though he himself was firmly monotheistic. In fact, the reference to two crowns referred to in the Marqah statement seems, in the context, to refer to the fact that God gave Adam a crown like his. Stanton says that in the Marqah, the two titles of Glory and Divine One are but alternate titles for God.

Stanton argues that it is unlikely that Simon Magus was being called the incarnate God in Acts 8:10, when it says, "This man is what is called the Great Power of God" (NASV). It is one thing to be identified with "the great power of God" and quite another to be called "God incarnate." Moreover, the passage in Acts 8:17-19 shows that Simon himself was awe-struck by the greater power of the Holy Spirit demonstrated by Philip, Peter and John, which manifestly surpassed anything he himself demonstrated. Simon Magus was cursed by Peter for his wrong intentions.

It is also important to point to Goulder's erroneous claims regarding the statue, supposedly built for Simon Magus. Both Gott and Schaff submitted evidence proving the error of that claim. Such a claim of widespread acceptance and deification of Simon Magus is even more untenable in that coauthor Frances Young points out that Simon's followers seemed to be limited to 30 people, according to his successor Dositheus.[25] Simon Magus' condemnation by the apostle Peter was known to the whole church. The consideration of Simon Magus and his followers seems to be *one of the most **unlikely** sources* for the idea of Jesus' deity and incarnation. Thus, some of the authors making these claims actually discount them.

Were Earlier People Willing to Accept Pagan Source for Incarnational Myth?

Young cited sources that referred to shyster miracle workers as possible influences for the idea of Jesus' incarnation. She referred to the fraud Alexander of Abonuteichos and to other eccentrics. She also pointed out that Celsus mentions those who said, "I am God" or "a son of God" or "a divine spirit." But none of these persons were the kind of persons Christians would have wanted to identify with Jesus. She appealed to the Greek legends of divine births from the gods and the claims to deity made by Plato, Alexander the Great, and Romulus. But all of these claims for the various origins of the virgin birth and incarnation were answered, first, in part by James Orr in his *The Virgin Birth of Christ* in 1907, and extensively by J. Gresham Machen in his book by the same name in 1930.[26] If these writers in the *Myth*... knew of these great books, they purposely ignored them.

Machen concluded his examination of pagan legends with the following,

> Finally, it should be observed that these early Christian writers are not really unaware of the profound difference between the pagan stories of the births of gods and demigods and the Christian story of the birth of Jesus. They regard the pagan stories with horror... Justin Martyr (a church father of ca. 150 A.D.) displays a true moral indignation against the view that the gods, who supposedly are worthy of imitation, should have become slaves of pleasure and should have entered into adulterous unions. Here Justin seizes upon the really important point: the analogies between the pagan stories on the one hand and the Christian story on the other, however useful

---

[25] Hicks, *Myth of God Incarnate,* 88.

[26] James Orr, *The Virgin Birth of Christ*, especially Chapters V and VI, and J. Gresham Machen, *The Virgin Birth of Christ,* (New York: Harper and Brothers, 1930). Machen's book was highly acclaimed and even the liberal commentator, Walter Lippmen, commended its provocative thinking. One should read this if he wished to examine each argument in detail.

> they may be in an argumentum ad hominem, are superficial merely; and at bottom the two representations are different in kind. Still more significant is the following passage: Justin Martyr said, 'But lest certain men, not understanding the aforementioned prophecy (of Isaiah 7:14), should bring the same objection against us that we brought against the poets when we said that Zeus approached women for the sake of carnal pleasure, we shall try to make the words clear. The words, 'Behold the virgin shall conceive,' signify that the virgin conceived without intercourse; for if she had had intercourse with anyone whatever, she would no longer be a virgin. But the power of God coming upon the virgin overshadowed her and caused her to conceive though she was a virgin.... Accordingly the Spirit and the power which was from God must be understood as nothing else than the Logos, who also is the firstborn of God, as Moses the aforementioned prophet declared; and this, coming upon the virgin and overshadowing her, caused her to be with child not by intercourse but by power'.[27]

In addition, it should be noted that the stories of the Greek gods were accepted tongue-in-check by many Greek writers. The incidents they refer to were not given real historical settings, as is clearly the case concerning the birth of Jesus in Bethlehem.

One of the most important arguments against the pagan sources being the origin of the incarnation is *the diametrical difference in the motivation* evident in the narratives. Dennis Nineham, who was asked to be a collaborating author of *The Myth*... but was unable to participate, writes in the epilogue that the authors fail to acknowledge the unique model presented by Jesus in contrast to pagan myths. He said,

> Descriptions of the unique facts in question vary a good deal but, whatever precise wording may be used, the view I have in mind usually boils down to something like this: whereas before the time of Christ all men had, in varying ways and degrees, put *themselves rather than God*, in the center of their lives and so turn out *'self-centered'* in the usual sense of the term. Jesus' life was at every stage and at every level centered entirely upon the being, grace, and demands of God, and so introduced into history a new humanity, a new way and possibility of being human.[28]

Nineham's observation hits at the core of the reason these pagan sources are not only unacceptable, but also antithetical to the concept of the incarnation. The idea in Scripture is always that the man Jesus was the eternal Logos, the eternal Son with the Father who humbled Himself to enter the world to be rejected by men He had created in order to redeem them. (See John's account in John 1:1-18; Paul's in Philippines 2:5-11; Matthew's in 1:18- chapter 2; and Luke's in chapter 2.) The Son gave up everything to obey and serve; he was born in a stable, raised by a carpenter and his wife, and rejected by the nation's leaders. Quite in contrast are the stories about the magicians, philosophers, and rulers like Alexander or Romulus. These were all men who were seeking to control the crowds and gain power for themselves. The same is true of the leaders and followers who passed on the stories. Their carnality and lust for power is in stark contrast to the servant life of Jesus Christ.

Were Primitive People More Gullible Than Today?

Hick and those who follow him say that the thinking at the time of the birth of Jesus and Christianity was different from our scientific educated world, in that people today are no longer gullible as they were then. But is this really so? Many today believe Elvis Presley is still alive and

---

[27] Machen, *The Virgin Birth of Christ*, 334. See Justin Martyr, *The First Apology of Justin*, chapter xxxiii.

[28] Nineham, Hick, *Myth of God Incarnate*, 186-187.

worship him; we have our legends of Superman and heroes who major in sexual escapades; Hitler demanded, in educated Germany where much biblical criticism was born, to be called "Heil Hitler." The German word 'heil' can connote savior. The emperor of Japan is said to be divine, and the former founder-dictator of North Korea was worshipped. Millions in the modern world believe pagan astrology, witchcraft, and other mystical occult ideas. Most newspapers recognize this and publish astrological data daily. But Jesus Christ and what he stands for is in strong contrast to these pagan ideas.

Admissions by Authors There is No Clear Source for Incarnation

Young herself makes startling statements about the uncertainty of pagan or any pre-Christian origin as a source. She said,

> ***Whether or not*** *we can unearth the precise origins* of incarnational belief, it is *surely clear* that it belongs naturally enough to a world in which supernatural ways of speaking seemed the highest and best expression of the significance and finality of the one they identified as God's awaited Messiah and envoy. (Emphasis added) [29]

But we must ask of Young, Is it "surely clear"? And where has she or anyone else looking for the origins of Christianity found one that is "the highest and best expression of the significance and finality of the One"? No pagan origins exhibit such! Young herself says,

> ***There does not seem to be a single, exact analogy*** to the total Christian claim about Jesus in material which is definitely pre-Christian; full scale redeemer-myths are *unquestionably found A.D. but* ***not B.C.*** (emphasis added)[30]

This is a significant admission by one of the book's primary authors.

Leslie Houlden, principal of Ripon College and another contributor to *The Myth...,* supports Young's views of many early pagan pictures that she sees as contributing to incarnational claims. But she also asserted the titles were not labels, but 'oblique statements about God' which carry no validity in our attempts to speak of Jesus today.

When Goulder comes to the conclusion of his argument for a Samaritan source, he says,

> It has *not* been my intention to suggest that ***any of the evidence*** presented in this chapter, or indeed ***any of the theories outlined***, makes it possible to reconstruct a *definitive account* of the rise of incarnational belief in the early church. (Emphasis added)[31]

Thus, Goulder, Young and Houlden admit they don't have any clear evidence for what they argue. If so, what is the motive for their arguments? The answer is they are unwilling to accept anything about a supernatural God because there is no place for it in their belief in deterministic science.

Did Authors Intend Deception To Cause Rejection of Christ?

In a sequel book, *The Myth of God Incarnate: The Debate Continued*, Goulder argued in the foreword that people reacted to the first book because "many people missed the double entendre

---

[29] Hick, *Myth of God Incarnate*, 119.

[30] Ibid., 118.

[31] Ibid., 117.

intended: between myth as a story of profound meaning by which men guide their lives, and myth as a fairy-tale, not true."[32] The amazing thing about this statement is that the whole purpose of the book, *The Myth...* was aimed at saying the belief in the incarnation was not based on fact, but generated as stories by the followers of Jesus from other non-factual stories and was, therefore, not true. Just because Goulder and his co-authors want to induce the rejection of the incarnation as a lie with a high meaning for guidance to our lives does not make it more true.

Cupitt expresses an argument that most of the authors share—that the teaching of the incarnation of Christ is *not of the essence of Christianity* and has been taught "*only to a certain period* of church history."[33] I would like to believe that these authors were saying this in ignorance. But, they are professed enlightened Christian scholars; it is obviously their intent to deceive.

Cupitt, I'm sure, would respond that he means only "the classical doctrine of the Incarnation" as asserted in the church councils of Nicaea and especially Chalcedon. However, his chapter is not aimed at rewording those statements, but at getting Christians to give up the incarnation altogether as a believable fact. Moreover, Cupitt surely knows there are very early statements about the incarnation that precede the statements of the councils. Almost every Christian is familiar with the Apostles' Creed that goes back to early Christianity, probably the second century as a baptismal confession.[34] Justin Martyr clearly defends the incarnation (ca. 150 AD). Origen, who was a highly educated Christian scholar in the third century, clearly states faith in the incarnation. Eusebius, the church historian quoted for other matters by the authors of *The Myth...* argues extensively for the incarnation in his *The Proof of the Gospel,* especially in chapter 3, entitled, "That we rightly teach that there are not many sons of the Supreme God, but One only, God of God."[35]

Not only did this belief exist in the patristic writings from the time of the New Testament onward, but the number of people and churches in modern times that believe the New Testament documents and that doctrine have constantly been expanding. According to the Lausanne Statistics Task Force, only 1 of 49 people were Bible-believing Christians in 1790 who professed the incarnation, but today 1 out of 9 people believe. This growing acceptance of this truth in the modern world does not support the argument that the doctrine of the incarnation is an obstacle for the church. Is it not really only an obstacle to the pride of these scholars who wrote *The Myth...* who believe science is deterministic?

Are Other Religious Founders Like or Equal to Christ?

In his final argument, Hick insists that Christianity is *not unique* and should be accepted syncretistically with other religions to help fashion the future of a new world order. According to Hick, because the idea of God's incarnation and saving work in Christ is not really true, but based on natural origins, it must be seen as only one way to God among many. Christianity should be presented as an important religion, but alongside of other equally good religions, he argues.

Hick appeals to Mahayana Buddhism to exemplify this syncretism. But his argument lacks

[32] Michael Goulder, *The Myth of God Incarnate: The Debate Continued,* Grand Rapids, MI, William B. Eerdmans Publishing Co. 1979) vii.

[33] Hick, *Myth of God Incarnate*, chapter 7, 134, et al.

[34] Phillip Schaff, *The Creeds of Christendom,* (Grand Rapids: MI: Baker Book House, reprinted in 1990), Vol. 1, 19.

[35] Eusebius, *The Proof of the Gospel,* Vol. I (Grand Rapids: MI: Baker Book House, W.J. Ferrar, editor and translator, reprinted 1981), 166 ff.

credibility. Mahayana Buddhism is a perversion of original Buddhism. R.E. Hume, one of the most respected authorities on world religions, draws upon most of the Buddhist authorities of his day in *The Living World Religions*. He points out that while Buddhism likely developed about 500 years before Christ, and was one of the first beliefs to spread worldwide, it *originally was not a religion at all* and made no reference to God. Hume quotes from other sources: "Though for historical purposes we may class it as a religion ... it comes short of the notion of a religion, and is not properly entitled to that name" (Menzies, "History of Religion", pp. 353, 380, 424).[36] "Buddhism, at least in its earliest and truest form, is no religion at all, but a mere system of morality and philosophy founded on a pessimistic theory of life" (Monier-Williams, "Buddhism," pp. 537, 539).[37]

Hume goes on, "Doubtless the founder did not set out to found a new religion. His main emphasis was on saving oneself from a world which is thoroughly infected with misery."[38] Gautama Buddha reacted to the Hindu caste system and belief in reincarnation and wanted to escape from the suffering of the world into nirvana or nothingness. While sitting under a bo tree, he had a vision of how to do this and of the eightfold path for living, which involves meditation. He then *left his wife and son,* gave away his wealth, and started going around preaching his doctrines. It did not include worship of God. It is thought that this was about 500 years before Christ. Gautama Buddha died in 483 BC. Before his death, four councils were held with his followers. They were organized into monastic orders and their regulations and instructions recorded in the *Vinaya.* The original form of Buddhism was later perverted into different sects, while the original one seemed to die out.

The form of Buddhism that Hick presents did not exist when Christ was born. According to the *Encyclopedia Britannica*, Mahayana Buddhism was first known when Fa Hien and other pilgrims returned from India to China in the fourth century, and then others with Hiuen Tsiang in the seventh.[39] It is thought that this form of Buddhism began in the first or second century AD. This form of Buddhism is *post Christian* and the parallels were thought to be *adopted by his followers to compete with Christianity*. Buddhism has many forms and is highly polytheistic in its later forms. Hick does not give a genuine presentation of what Buddhism was like originally.

But why did Hick not refer to other religions? What about Hinduism with its reincarnations and terrible caste system, with its burning of wives, with its sale of child brides? This would not appeal to the modern world of political correctness in regard to ethnicity and equality. The highest form of Hinduism involving Brahma is a nebulous pantheism.

The greatest threat to Christianity is Islam—a militant, post-Christian religion. Muhammad was born of the Koreish tribe in Mecca, which was a center of animism and idolatry. Hume said,

> He learned something of monotheism from Jews and Christians. ...In a period of mental depression Muhammad suddenly felt himself appointed to go forth and preach a religion of one absolute God, Creator, Potentate, and Judge of the world. For twelve years he continued to have visions.[40]

---

[36] R.E. Hume, *The World's Living Religions: An Historical Sketch,* (New York: Charles Scribner's Sons, 1949), 59.

[37] Ibid.

[38] Ibid.

[39] *Encyclopedia Britannica*, Vol. 4, 325 ff., and Vol. 14, 675 ff., 1953 edition.

[40] Hume, *The World's Living Religions: An Historical Sketch,* 214.

While hiding in a cave with one other man, Muhammad had a vision that God's hosts would support him "to set up the rule of Allah, with himself as the immediate dictator upon the basis of the six-fold Pledge of Akaba," including the pledge to worship only one God, not to steal, not to commit adultery, not to kill our children, not to slander, "nor will we disobey the Prophet in anything that is right." This event, known as a Jihad, took place in 622 AD, when Muhammad was 52. It was the turning point in his life. He went forth with the sword, killing any who did not worship his God Allah, especially those with idols, but also Jews and Christians. He even slaughtered those of his own Koreish tribe that opposed him. Islam expanded rapidly by use of the sword. While he limited the number of wives to four, he himself had eleven. He even married Zainab, the divorced wife of an adopted son, Zaid.[41] When Mohammad died, his followers divided over who was Mohammed's rightful heir. That political controversy continues to this day between the Shiites and Sunnis. They violate the six-fold pledge of Akaba in their controversy with each other. One of the pledges included not to kill children. They indiscriminately destroy each other and even use children in suicide efforts. Much later, during the Ottoman rule, Islam became more moderate and intellectual and there are other moderate Islamic groups today, which would be more acceptable to Americans.

What about Shintoism, or Confucianism, which is not a religion? None of these appeals to educated people in the West who are being told all religions are alike. Moreover, virtually all of these religions are hostile to Christianity and have no desire for a syncretism with it. Is Hick ignorant of these facts?

Conclusions about Rejection of God Incarnate

Rejection of the Incarnate Christ for a One-World Religion

The Myth... implies that the intellectual elite of the world are moving toward a pantheistic view of religion that comes from the universities instead of the churches, and that *this new religion, in link with science,* will replace the purely naturalistic or secular view. Man and his needs will be dominant, and man will be the king, directing their progressive development. God will be reduced to the mysterious forces of nature that find their highest expression in man. This is the same pagan, syncretistic, pantheistic view of god that prevailed when most previous civilizations died. The implications of this book are profound because they are made ***by the leaders of the organized church***, the body of Christ. The leaders of the church of Jesus Christ have become his crucifiers. *The Myth...* is not an epitome of brilliance or even of truth and would have had little or no recognition except that the voice of supposed leading church scholars saying such a shocking message appealed to the biased media who are ignorant of theological facts.

The radical terrorism that is threatening the world in the twenty-first century is putting pressure on the believing Christian community who capitulate to the idea of one religion as a unifying factor. Pope John Paul II began efforts to bridge the gap of the Roman Church with other religions. And the new pope, Benedict XVI, has already indicated that unifying religion is one of his objectives. It remains to be seen if he will hold to the uniqueness of Jesus Christ and the fact that Jesus came to live a perfect life of love for God and man and to shed His blood and be raised to deal with the problem of forgiveness and victory over sin. This alone is the answer to man's sinfulness.

The Apostle John said to the church about testing the spirits,

---

[41] Ibid., 214, 215. He gives references in Koran. See 212-233.

> This is how you can recognize the Spirit of God: every spirit that acknowledges that Jesus Christ has come in the flesh is from God, but every spirit that does not acknowledge Jesus is not from God. This is the spirit of the antichrist, which you have heard is coming and even now is already in the world" (1 John 4:2-3; cf. John 8:44-45).

Intellectual Elite Accept Fraudulent Jesus in Twenty-First Century

Having failed to effectively dismiss the Jesus of the historic gospel records, the secular world has reverted to emphasizing false ideas about Jesus that were created by late second and early third-century intellectuals. As Christianity impacted the Greek and Roman world that emphasized human knowledge, the heresy of Gnosticism was born—a heresy that made Jesus entirely human. The Gnostics wrote mythical stories about Jesus, claiming they were written by Thomas, Phillip, Mary Magdalene et.al. These were attacked by Christian leaders such as Irenaeus, Bishop of Lyons, a third generation Christian, in his book, *Against Heresies,* which he wrote in the late second century These false gospels were written at least five generations after Christ and had no real historical credibility.

Copies of these Agnostic writings were found in 1945 at Nag Hammadi. More recently manuscripts by Egyptians Coptic's were found in Egypt and have been given prominence by critical scholars. One about Judas was recently uncovered. These have been popularized by fictional books and liberal scholars that failed to point out that there is no credibility to what they say. They were written to deliberately discredit Christianity. Thus, the final effort to delude the public about Jesus the Christ is to disseminate lies.

PART IX

ENLIGHTENMENT ABOUT GOD IN BIBLICAL HISTORY AND CREATION

Then he said to them, "These are my words that I spoke to you while I was still with you, that everything written about me in the Law of Moses and the Prophets and the Psalms must be fulfilled." Then he opened their minds to understand the Scriptures, and said to them, "Thus it is written, that the Christ should suffer and on the third day rise from the dead, and that repentance and forgiveness of sins should be proclaimed in his name to all nations, beginning from Jerusalem. You are witnesses of these things (Luke 24:44-48).

Chapter 28
Intellectual Bias Led to Rejection of Christ

Introduction: Rejection of God Driven by Human Bias?

In this book we reviewed the progressive efforts to limit all knowledge to scientific methodology and an epistemology which resulted in scientific determinism and excluded any evidence of God and revelation in creation or history. The sin nature of man to believe he is wise as God formulates all his thinking based on his own limited intellect and the belief this all developed by evolution and not by creation. This process began with denial of the beginning point of human history, ***creation***, then moved to rejection of subsequent biblical history, and ended with the denial of the supernatural Jesus Christ. Jesus Christ has been recognized as the center of biblical meaning about God, and also the most important person giving meaning to world history and God's purposes for the world and man. Christ's incarnate life led to his death and resurrection. God's work in revelation centered in Jesus

This chapter seeks to answer the question, were the repeated and various efforts to deny the miraculous in the life of Jesus and finally to deny the incarnation of Jesus, driven *by **man's bias** to destroy faith in a God* who created the world and revealed himself in biblical history? These protracted efforts of the intellectual elite culminated in existentialism, the death of God theologies, and the denial of Jesus Christ's supernatural incarnation. These intellectuals ended up saying that all religious efforts should be controlled by the intellectuals of the universities, through the government to direct the church ministry toward social betterment ***of man,*** rather than to worship God. The current effort is to unite all religion under government of the United Nations.

Were these evaluations really unbiased objective scholarly research of the facts, or did they *begin* with the presupposition there is no God that created and revealed Himself? This issue of unbiased research is critical, because if the predominant reason for rejecting evidence of a supreme sovereign God is that they have dogmatically limited knowledge to scientific determinism, and if scientific determinism is later shown by other liberal scholars to be a result of perversion, then the deterministic theory and the conclusions from its application will fall as mass deception.

Evidence that Bias Drove the Search for a Jesus who was Not Supernatural

Bias in the *Myth of God Incarnate*

Hick's book, *The Myth...,* brought to light a major departure from traditional biblical understanding. It denied the uniqueness of Jesus Christ, thus generating considerable controversy. This book and others related to it were circulated throughout American higher education in the 1970s. As the death of God theologies were not widely read or understood outside the intellectual elite, neither was the book. However, because of its shocking claims, the mass media, many of whom are uninformed in theology, made these claims significant in the public mind. The express objective of the Anglican university scholars in *The Myth...* was to deny, *within the churches,* the uniqueness of Jesus and to promote one religion that a world government might control. This denial of the incarnation of God in Jesus Christ was pivotal in turning all religious worship in the world away from one God to man himself.

As noted, *the authors **did not even accept each other's arguments*** for mythical sources that attributed supernatural powers to a purely human Jesus. *They confessed that they had not found **any** certain source to explain the claims* for His deity. Yet, the authors were dogmatic in denying the incarnation and made many statements that imply dishonesty and even harm from scholars who hold to the long-standing traditions of the incarnation.

The authors of *The Myth...* made many ***derogatory statements*** that were unnecessary for simply presenting facts. They contrasted their views in terms such as "for honest and thoughtful that people," "in the interest of truth," "to free people to talk," and for "greater integrity."[1] They imply incarnationists are not honest, thoughtful or truthful. They clearly described those who hold to the incarnation as having "a narrower definition of Christianity," of holding "improbable scientific fact" and as those who "encouraged arrogant and intolerant attitudes"[2] as causing "harmful effects" and as "deeply damaging."[3] Other, similar derogatory statements are found throughout the book.

Since they presented no significant evidence for their claims of mythological sources for the church's view of the incarnation, and indeed, denied the substantial evidence for the belief in the incarnation of Christ, their bias against the church's belief in a supernatural Christ is obvious. Lacking truth, they sought to win by deprecation. *Ad homonym* arguments against the opponent reflect weakness. The authors are driven by their blind faith in the theory of scientific determinism.

Repeated Failed Efforts to Find a Purely Human Jesus Shows Bias

Bias is also evident in the long quest for a historical Jesus. This can be seen in the fact that, despite their efforts, not *even a large minority of scholars could reach a* ***lasting consensus*** to deny the incarnation. Most *importantly, the liberal critics fully familiar with this long quest* ***admit that bias*** dominated the research. The noted scholar and ardent critic, F.C. Burkitt, who wrote the preface to Albert Schweitzer's *The Quest of the Historical Jesus* said, "Among the many bold paradoxes enunciated in this history of the Quest, there is one that meets us at the outset... the paradox that the greatest attempts to write a Life of Jesus have written with ***hate***."[4] Schweitzer himself commented on this, saying, "It was not so much ***hate of*** the Person of Jesus as of ***the supernatural*** nimbus with which it was so easy to surround Him and with which He had in fact been surrounded. They were eager to picture Him as truly and purely human..."[5] Thus, at the outset, these two men who fully understood and supported the quest for a purely human Jesus pinpointed the critical writers' glaring bias against God.

Bias Caused Omission of the Central Message of the Gospels

Biased Oversight of Final Climactic Events of Jesus' History

It is important to recognize that this long, unconscious process of exalting man, at the expense of God, has moved progressively toward rejecting the New Testament message of Jesus Christ. In the process, the critics have virtually ***ignored*** the *important part* of that message—***Christ's death and resurrection***.

A total of ninety chapters in the four gospels and Acts 1 tell of Jesus Christ's earthly ministry. Five of these tell of Christ coming into the world. The genealogies and the birth narratives clearly present Jesus Christ as an historical figure who continued and fulfilled God's promises to the world through Abraham and David. Luke 3 traces his lineage back to Adam. These chapters make it clear that at his Son's coming, God entered history by a supernatural working. John links God's coming in the flesh to the eternal ***Logos*** revealed to the Jewish prophets of old. But the events are

---

[1] Hick, *The Myth,* 138.

[2] Ibid., 139.

[3] Ibid., 142, 144.

[4] Schweitzer, *The Quest of the Historical Jesus*, Preface by F.C. Burkitt, Preface, vii.

[5] Ibid., 4.

presented as clearly historical, naming the contemporary rulers, the conditions of history, the people, and the places. These historical gospel events are detailed in extra-biblical documentation, including governmental decrees, requirements to go to the place of origin for registration, and other details that show their accuracy.[6]

Only fifty-two chapters focus on the life and ministry of Jesus Christ, while the last 33 focus on his death and resurrection. The critical writers focused almost exclusively on the fifty two telling about his ministry and teaching. From the early writing of Paulus (1825) the effort to make the life of Jesus fit within human reason, and those books that were written subsequently about the psychological life of Jesus, and those about his ethical emphasis that followed, dealt mainly with these chapters of the historical record. The end of the critics work was of those who emphasized that Jesus mistakenly saw his kingdom in the future (Weiss, Schweitzer ca 1900), and this emphasis was not mainly focused on Jesus' great eschatological discourses presented with the purpose of his death and second coming. From the 52 central chapters of the gospels, each scholar made his own specific and different effort to find some aspect of Jesus's ministry as key. Yet, not one of these gained lasting credibility as an explanation for the importance of why Jesus was followed and worshipped. We also have shown that some of first critics demonstrated uncertainty in that they were hesitant and delayed in publishing and writing against the supernatural events in Jesus' life, although clearly seeking a purely human Jesus.

However, the last 33 chapters (of the ninety) covering the story of Jesus's death and resurrection were largely ignored as the key for understanding Jesus. Even the eschatology involved in the preliminary teaching before Jesus's trial was only looked at briefly in the eschatological interpretations. The critics virtually ignored the data in the remaining chapters, which make up ***more than one-third*** *of the gospel records.* Why? Surely, the sheer space that the followers of Jesus gave to these historical records would indicate that they considered these accounts very important. But, the critics almost always found the key to interpreting Jesus of Nazareth elsewhere.

Bias of the Critics Led Them to Overlook Death and Resurrection

By mid-twentieth century, the search to find a meaning for Christianity in something other than the supernatural working of God and the fulfillment of his revelation in Jesus Christ had run its course. The next step was the death of God theologies, the rejection of the incarnation, and the secularization of the church. The quest for a historical Jesus, the search in philosophical theology and in forms of the early church, and the new quest ***produced nothing considered of major significance.***

It remained for the brilliant Harvard scholar, Richard R. Niebuhr, to focus on the fact that the ***bias*** *of the critics had led them* ***to ignore the only key*** to understanding Jesus Christ, namely His ***death and resurrection***. Niebuhr did this in his book, *Resurrection and Historic Reason: A Study in Theological Method.* He attacked the critics by saying that the laws of nature cannot give meaning to specific events, such as the important events of Jesus's death and resurrection, which the critics missed ***because of their naturalistic inclination***.

Ignoring of Final Events Removes the Meaning of the Whole

Most of the critics did not seek the gospel meaning of Christ's death. It is true that some like

---

[6] E.M. Blaiklock, Professor of Classics at Auckland University College, accumulated these in an excellent article, "Archaeology and the Birth of Christ," *Eternity,* December 1954, 10-11.

Strauss, Herrmann, and Schweitzer spoke of historical revelation in the death of Jesus, but they **only *made him a victim*** of his own inner secret messianic conviction, or a hero who died. Richard R. Niebuhr said,

> The minimizing of Jesus' death forces the interpreter of Jesus to look for a clue to the real Jesus of history in the Galilean ministry. But the narratives belonging to this ministry period are so fragmentary that they require reorganization on a new basis; and this new basis has to be provided by the critic himself, since *the evangelists offer only one key, the passion-death-resurrection* complex, which has been rejected.[7]

The critics who dealt with the death of Jesus usually trivialized the resurrection narratives, making them into *a psychological phenomena and a "spiritual force" of a nebulous ethical nature.* Richard R. Niebuhr said,

> This understanding of the causality operative in historical events has dictated the assessment of the resurrection narratives as nothing more than expressions of psychological certitude, though the relation between psychological certitude and the kind of certainty that any historical science can claim has been left unexplored. At the same time, it is clear the single fact of which the theologians of this period felt they could be certain, namely Jesus' crucifixion, has no interest for them, unless they can surreptitiously invest it with the significance of the resurrection.[8]

By this, Niebuhr means they did not see significance in Christ's death, but use it simply to give it a psychological significance as a continuing spiritual force. And they imply a spiritual resurrection and ***not a real bodily transformation***.

Niebuhr continues, "The paradox is that the excision of the resurrection tradition from the fabric of the gospel history is followed by the disintegration of the entire historical sequence of the New Testament." He then gives the example of Renan's *Life of Jesus* in which Renan "pushes the cross to the periphery of the historically important picture by making Jesus' death trivial." He adds, "The New Testament itself puts up a stiff resistance to the subordination of the resurrection texts. For one thing, *all of its analogies* for the meaning of the death of Jesus ***collapse***, when deprived of this support."[9] (Emphasis added)

Niebuhr then argues what is clearly true, that having ignored the key of the death and the resurrection together which unify the meaning of Christ's mission, they find some romantic reason for a victory for Christianity in a nebulous immortality of the soul. He says,

> The difficulty of successfully performing this revisionary operation is shown in the inability of the redactors to be consistent. It is hardly possible to conceal the fact that in the gospel history the resolution to go to Jerusalem is set against a messianic, not a heroic, consciousness. No amount of patching with the concepts of hero and of immortality can make a unity of the history again, once the passion and death are surrendered through the dissolution of the resurrection as the key to the meaning of the New Testament. Considerations of this nature indicate that insofar as the New Testament history is indispensable to the Christian consciousness, recognition of the priority of the resurrection tradition is required by an adequate theological ratio cognoscenti. Our

---

[7] Richard R. Niebuhr, *Resurrection and Historical Reason,* (New York: Charles Scribner's Sons, 1957), 15.

[8] Ibid., 14.

[9] Ibid., 14.

ability to catch sight of the historical Jesus depends on it.[10]

Determinism of Critics led Elsewhere for the Key to Jesus

Niebuhr assigns their failure to their biased theory.

> When the resurrection texts are set aside, as distorting rather than revealing the true dimensions of the historical Jesus, a basic error has been committed, for historical thinking cannot go behind the New Testament; it can only penetrate into it. The history of biblical criticism has given an irrefragable demonstration of that fact. ...An a priori conception of historical causality led them to discount the meaning of the resurrection texts too easily.[11]

In essence, every critic held the same bias that led them to miss the key, namely that Jesus's purpose in coming was to redeem sinful man. He did this by living a holy earthly life that ended in his death and resurrection.

Niebuhr has put his finger on the central reason that all the writings in the search for a historical Jesus failed to be convincing, and for the failure of those who sought the remnant of the Jesus tradition in the Christian community, or existentially. The reason was that they missed the key of Jesus death and resurrection which gives the New Testament meaning. Niebuhr is right in saying, "As long as theologians hope to take seriously the historical origins of the church and of its proclamation, understanding the resurrection narratives constitute the first step in the study of the New Testament."[12]

But the hostile trend of man's bias against God continues. In 1985, Robert W. Funk, owner of independent Polebridge Press, founded what he calls the "Jesus Seminar" and its sponsoring organization, the Westar Institute, a liberal think tank. Liberals have sought to exclude believing Christians from education, claiming separation of church and state. But the liberals have established religious teaching in most universities, and excluded any who believe the historicity of the New Testament. Therefore, while most religion teachers in universities are liberal, little was done to get the attention of Americans who were gaining an increased interest in religion. So, Funk launched this effort to revive liberal criticism. While some 200 liberal scholars have been involved in these efforts, only about 50 to 100 attend and participate. Three of these have had books that sold well, namely, Dominic Crossan, John Shelby Spong, and Marcus Borg.

This is not a gathering of scholars that truly evaluate all the evidence. They look at various data and then vote with beads (red for true of Jesus, pink—probably true, gray—possibly true, and black—not true) to gain a consensus. Anything that might involve the supernatural is out. Birger A. Pearson, professor emeritus of religious studies at the University of California, Santa Barbara, and interim director of religious studies at UC Berkeley, disputes the methodology and believes its hidden agenda is really to debunk Christianity. Such a bias is in line with all past criticism of Jesus by scientific determinists. It is hard to believe that intellectual scholarship and research has fallen to such a low level. The Jesus Seminar is again a shocker for the liberal media and has gotten much attention, despite the fact that the participation of the members seems to be sagging. Peter Jennings of ABC did a special on it that received attention, but avoided data from evangelical scholars. Recently, Funk has announced that the focus will be on turning their attention to forming a one world religion and deciding how Jesus might fit. Thus, the failed quest for the historical

[10] Ibid., 14-16.
[11] bid., 18-19.
[12] Ibid., 18.

Jesus is bypassing the issue and using the scientific deterministic bias to persuade the world to accept their worldview.

Since Richard Niebuhr's work showing the importance of the cross and resurrection, there has been more study emphasizing the last part of the gospels. Most notable are the works of Raymond E. Brown and John P. Meier.[13] These and others show a swing back in the right direction in studies about Jesus.

Biblical Meaning Centers on Sin against God and Atoning Sacrifice

But, why did all the educated elite scholars in the quest, or the new quest, miss the centrality and key of the death and resurrection which is so obvious in the gospel writing and in the sheer volume of space devoted to it by the gospel writers? The answer is certainly that man in the flesh has an unconscious bias against acknowledging his sin against God and his need for redemption. The whole message of the Bible, Old and New Testament, came to focus in the historical events of man's redemption in the death and resurrection of Christ.

It is these events—and these alone—which give consistent meaning to the whole of the Bible and to all of God's acts and revelation in history. It ties together the *character* of God and *his purposes* from the creation of Adam and his sin to the gigantic consequences of reward and punishment of final judgment. The cross and resurrection reveal the depths of man's sin and rebellion against God, the certainty of God's judgment against sin, and the love of God for those who respond to His call of repentance and commitment through His grace. Apart from the death and resurrection and His gracious forgiving love, God is seen as stern, harsh, and unrelentingly judgmental. Apart from the cross, where sin is first punished before it can be forgiven, sin against God is made trivial and God's character, which is abused by sin, is made unimportant.

Human Bias Confirmed in Biblical Idea of Antichrist

Biblical Teaching about Covenant, Sin and Sacrifice

The law of God and sacrifice are central to God's covenant; these are also the targets of the Antichrist. The whole story of the Bible centers around the covenant of God's will as revealed in his law, the Ten Commandments, and God's grace in forgiving sin through the shed blood of a substitutionary sacrifice. These concepts develop progressively through Scripture. The elementary beginning was the slaying of animals to cover the shame of Adam and Eve after their sin. Next was Abel's simple belief that God would accept him through his sacrifice of a lamb (Genesis 4:3-4; Hebrews 11:4). Cain's anger and slaying of Abel was the first manifestation of antichrist (Genesis 4:5; cf. 1 John 2:18 ff.; 3:11 ff. Following the flood, God established a new covenant with Noah, designating that clean animals be offered on an altar. In return, God vowed never again to judge the world with a flood (Genesis 8:20-9:16; Hebrews 11:7).

After God's judgment of the world following the tower of Babel, God made a covenant with Abram to bless the whole world through his seed (Genesis 12:2-3). Abram built an altar and offered a sacrifice to the Lord. Later, God made a covenant sacrifice with Abraham. Abraham believed God's promise that his descendants would inherit Palestine, which was a type of an eternal promised land (Genesis 15; 17; Hebrews 11:8 ff.). Still later, Abraham's faith was challenged when he was directed to offer his son Isaac. When it was clear that Abraham trusted

---

[13] Raymond E. Brown, *The Death of the Messiah*, 2 Vols. (New York: Doubleday, 1994); and John P. Meier, *A Marginal Jew*, 2 Vols. (New York: Doubleday, 1991 and 1994), et al.

God, God revealed a substitute for Isaac—a ram caught in a thicket. The idea that a miraculously born son must be sacrificed on Mount Moriah was initiated (Genesis 22; Hebrews 11:17-19). Progressively different aspects of sacrifice were added: burnt offerings, sin and trespass offerings, and peace offerings were revealed.

The center of the worship of God's people in the tabernacle and the temple was in the covenant at Mount Sinai. He reigned in the holy of holies over the Ten Commandments in the ark with a mercy seat covering them. The high priest gained forgiveness by the blood of the sacrifices on the brazen alter that had been burned by the fire from God and the coals from the sacrifice that allowed access to God above the ark. All the sacrifices led to the cross and resurrection of Christ to give forgiveness of sins that were broken. Jesus himself taught this after his resurrection (Luke 24:44-47).

Prophetic Fulfillment of Atonement in Jesus Christ

The prophets, especially Isaiah 53:6 ff., had prepared the people to see that, as Israel the servant of God had suffered, there had to be a person sent of God who would suffer as his lamb to atone for the people's sins. This was fulfilled in Jesus's baptism, when God announced that Jesus was his Son who pleased the Father. John the Baptist also points to Jesus as the "lamb of God that takes away the sin of the world" (John 1:27). Jesus's death and resurrection proclaim man's justification as central to the new covenant and a fulfillment of the old covenant.

Antichrist Efforts Aim at Removing Sin and Sacrifice

The concept of antichrist is manifest throughout biblical history at key points, beginning with Cain's hostility toward Abel's sacrifice. It is expressly stated in Daniel and referred to by Jesus. For both, the Antichrist exalts himself as God, and his focal point of attack is distorting and making desolate the sacrifice. Daniel prophesied about the Antichrist and the sacrifice beginning in chapter 8. There, the interpretation is clear that the events apply to the time of Media and Persia, Greece, and the successors to the Greek kingdom. Daniel 10 and 11 again refer to the prince of Persia and Greece, but with greater specifics.

Daniel tells us,

> The king will do as he pleases. He will exalt and magnify himself above every god and will say unheard-of things against the God of gods. He will be successful until the time of wrath is completed, for what has been determined must take place. He will show no regard for the gods of his fathers or for the one desired by women, nor will he regard any, but will exalt himself above them all. Instead of them, he will honor a god of fortresses; a god unknown to his fathers he will honor with gold and silver, with precious stones and costly gifts (Daniel 11:36-38).
>
> Great wealth and military might, which he achieves by wealth, are his god (cf. 11:23-24, 28).

Daniel shows clearly that the Antichrist's objective is to remove the covenant of God with his people and the sacrifice for sins, which are the diamond pivot point of the relationship of God, with his people.

> His heart will be set against the *holy covenant*. He will take action against it and then return to his own country. He will turn back and vent his fury against the holy covenant. He will return and show favor to those who forsake the *holy covenant*. His armed forces will rise up to desecrate the temple fortress and will ***abolish the daily sacrifice***. Then they will set up **the**

> **abomination that causes desolation**. With flattery he will corrupt those who have violated the covenant, but the people who know their God will firmly resist him (Daniel 11:28-32). (Emphasis added)

Just as there is a progressive revelation to fulfillment of the sacrifice to that in Jesus' death, so there is a progressive unfolding of the idea of antichrist until a final world human ruler who opposed God and His Son. The Greek ruler Antiochus Epiphanes in 165 B. C. offered the sacrifice of swine for bacchanalia orgiastic feasts in the place of the designated sacrifices as required by Mosaic law (lamb, et al.) for the forgiveness of sins. In this Daniel foresaw the abomination that causes desolation by the antichrist by an offering for pagan pleasure. However, Daniel later saw a final Antichrist during one final tribulation (Daniel 12:1-3). The sacrifice and the abomination that makes desolate is again referred to in concluding this vision (12:11). This Antichrist and removal of the sacrificial system for forgiveness in Daniel 12 is the final antichrist with the last and greatest tribulation, which is to precede the coming of the Messiah and the resurrection of the elect saints.[14]

Jesus foresaw the world's climax of evil and final antichrist as a future event that just ***precedes his glorious dynamic eschatological return*** in his resurrection body when he will raise his followers from the dead. He warned that his return will be preceded by effective mass worldwide deception by false teachings and terrible persecution of his elect (Matthew 24:4-10). He said,

> This gospel of the kingdom will be preached in the whole world as a testimony to all nations, and then the end will come. So when you see standing in the holy place *the abomination that causes desolation* spoken of through the prophet Daniel—let the reader understand—then let those who are in Judea flee to the mountains. ...For then there will *be **great distress**, unequaled from the beginning of the world until now–and never to be equaled again"* (Matthew 24:14, 15, 21 ff.). (Emphasis added)

The Temple Sacrifice Fulfilled in Jesus' Death and Resurrection

In Jesus's first visit to Jerusalem after John publicly proclaimed him the Messiah, he cleansed the temple and announced that if they destroyed the temple (of His body), he would rebuild it in three days. He came, he said, not to destroy the law and the prophets, but to fulfill them.

The aim of Jesus's mission, to go to Jerusalem to die, was manifested in his teachings throughout his life (e.g., John 6:53, 56; Matthew 16:21; Luke 24:44-48). When Jesus died on the cross, the veil in the temple was rent in two by an earthquake, symbolizing that his death opened the way into the Holy Place. The author of Hebrews explains clearly that the animal sacrifices were not sufficient since they had no moral merit, but that Jesus who obeyed the law and pleased the Father was the only and final sacrifice, which was sufficient for all. Moreover, Jesus offered himself as the ultimate and final high priest, providing a new way into God's presence in the true tabernacle of heaven itself. The earthly tabernacles and temples were only types or shadows of the real dwelling of God in heaven.

The seriousness of man's sin is made clear in the cross. Man's sin against his holy Creator dishonors God. The degradation and misery of the world must be punished to its full dessert.

---

[14] Carl W. Wilson, *The Quest to Understand the End of History*, (unpublished manuscript, cf. bravegoodmen.org).

When man met God in the flesh in Jesus Christ, he rejected, mocked, spit upon, and crucified him. But God's eternal plan was to let his Son die as the atonement for the sins of those who would respond and believe in his love. And through that grace, those who respond would be changed to follow and be remade into Christ's likeness. In this, the sacrifice on the altar was fulfilled.

Just before Jesus predicted the final Antichrist and the rejection of sacrifice for sin, he predicted the fall of Jerusalem and the destruction of Herod's temple. Jesus, throughout his ministry, foresaw that by rejecting him as the God of the temple where people worshiped, his resurrected body would become the center of worship and the temple of God. From his resurrection, all those who trusted him would then become a human temple of living stones. (1 Corinthians 3:16; 1 Peter 2:4-8). Some biblical scholars see a rebuilding of the temple and resumption of sacrifices. But since the sacrifice is finished in Christ's death, the abomination of desolation must be the forbidding of the preaching of the cross and of the celebration of the Lord's Supper in remembrance of His death. Since the temple is now fulfilled in the people of God the Antichrist will establish himself as the head of the church and demand worship. The sacrifice of celebration of Christ's death will be replaced by a feast **to a Man** as the world's ruler (cf. 2 Thessalonians 2:1-12). Paul warns that this will be done by "***powerful delusion*** to believe a lie" (2:9-12). (Emphasis added)

Demonic Influence Appealed to Human Bias

Demonic Application of the Theory of Scientific Determinism

There is a more ominous aspect of the whole modern development and use of the theory of uniformitarian deterministic science. The way the progressive movement occurred over centuries, through many successive generations and many scientists who did not know each other but knew how to use the theory of scientific determinism, reflects a logical strategy that transcends the mind of man. No one man, no group of men could have set forth the path to discredit the Bible that many men, in many nations followed for over two hundred years. Yet, that path was generally very logical. In retrospect, one who is familiar with the Bible and its philosophy of history can see that the path used to develop and apply deterministic science is the only one that would likely have been successful to destroy the powerful religion of Christianity. The path of attack against God and the Bible reveals that some trans-human logical spirit was guiding it. Christianity is a religion anchored in history, and that is what was attacked. Voltaire argued to King Frederick in his letter that Christianity can never be destroyed by the sword, but ***only by logical reasoning***.

Description of the Christian's Linear View of History

Christians, scholars, and others have long recognized that the Bible presents ***a linear view** of history*. Oscar Cullmann, in his book, *Christ and Time*, explains the Christian view.[15] It is important to have that linear view clearly in mind to see that the spiritual influence was demonic. See the following diagram of the biblical, linear view of history.

[15] Oscar Cullmann, published in German as *Christus und die Zeit* was translated by Floyd V. Filson as *Christ and Time: The Primitive Christian Conception of Time and History,* (London: SCM Press LTD, 1952).

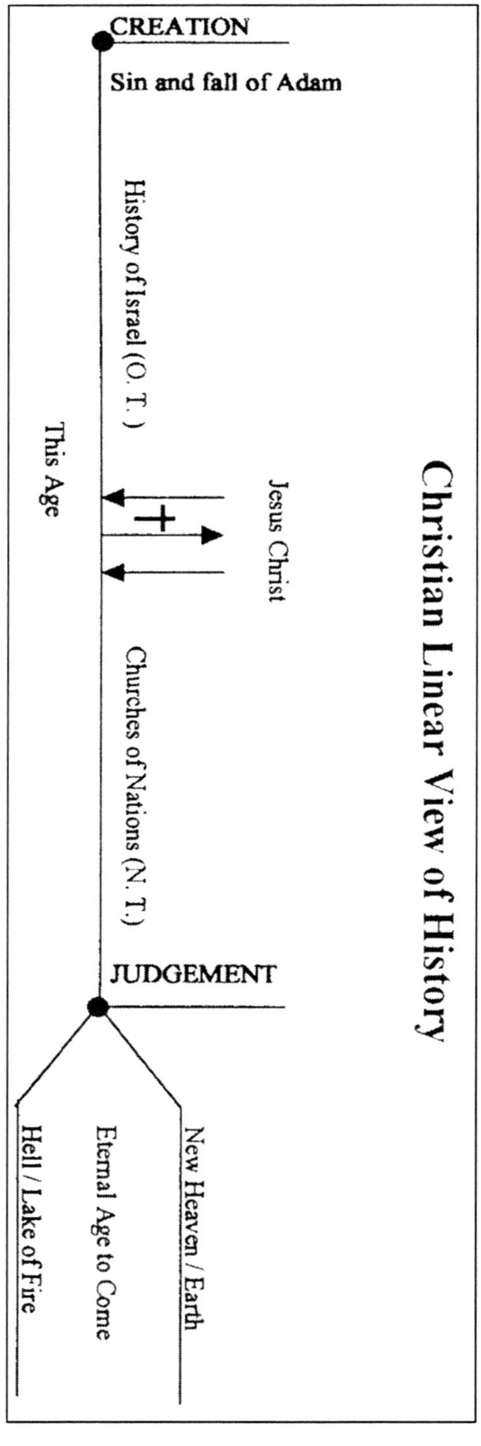
Christian Linear View of History
CREATION
Sin and fall of Adam
History of Israel (O. T.)
This Age
Jesus Christ
Churches of Nations (N. T.)
JUDGEMENT
New Heaven / Earth
Eternal Age to Come
Hell / Lake of Fire

Any line must have a *beginning point* and an *ending point,* noted by the • on the diagram. The Bible begins with God's creation, the first dot. He made the world and man to rule over his creation under his sovereign control. This marks the beginning of time, as well as history. As man's Creator, God holds man accountable to Him. God, who is transcendent, is also imminent or *present sovereignly guiding the line of history* to its consummation at the other dot, which is his final judgment of all men and angels. Time and history end with the judgment and eternity, and eternal time begins. Thus, all along the line of unfolding history, each individual man and nation is held accountable. Biblical history centers on God's elect—beginning with Abraham, Isaac, and Jacob—through whom he reveals his plan to redeem a people for himself. God is in control and moves the line of history. The center point is the coming of Christ and his death and resurrection (upward arrow) to save those who believe in him. The meaning of the center furnishes the message for the church, which is empowered by the Holy Spirit (downward arrow) to preach until the final judgment. *These Christ events are the diamond pivot point of meaning of biblical historical revelation.* Since the Christ event in the center has been accepted by the whole world for its importance, BC and AD, the facts about Jesus Christ would be the hardest to get people to give up. I believe that the way scientific theory has been used to discredit belief in the God of the Bible is the only way that would have been effective.

Sequence of Application of Scientific Determinism Was Demonic.

Had the rationalistic attack first been on the center of the line, representing the Christ event, or on any point of the line, it would have had little acceptance. But the first attack was at the two points at the ends of the line where super historical events are involved in the creation beginning and the future proclaimed ending of judgment.

For many years the scientists who believed in God had demonstrated the regularity of a heleo-centric solar system, insisting that this was the way God worked by little-understood forces. But Herschell, Kant, and LaPlace presumed an eternal universe which involved the solar system, thus, by their presupposition arguing there was no point of beginning or ending. They separated history from God's working of beginning and ending. Then, geologists Hutton, Lyell and others claimed there were no evidences of God's control in the world, but only natural uniform geological forces. Thus, they eliminated God from the time line of history. Having severed the points of beginning and the line itself from God, the attack was then aimed at saying the history of Israel was not one created by God's supernatural working (e.g. the exodus, meeting with and giving of the law at Sinai), but was really a product of evolutionary thinking of men—beginning with nature worship and power hungry priests who concocted the story of one God to give them authority over the people. Spinoza, Astruc, et al. formulated one story by Julius Wellhausen on a new view of evolution that was diametrically opposed to the traditional history of Israel. Moreover, now the scholars of comparative religion speculated with different theories to show that all ideas of God evolved from nature worship to a monotheistic god. They posited these views on their evolutionary speculations in spite of the fact that very early Lange produced evidence to the contrary.

There is no way that this very logical effective strategy of beginning at the end points, then with the line of history, and finally attacking the center could have been from any source but demonic. The mass deception by the educated elite has now captured the minds of many as being true and has caused tremendous doubts in the minds of millions of Christians themselves. Man has followed a similar path to trouble and destruction in many other cultures. If the strategy is demonstrated to be filled with a perversion of facts, then the devil is guiding the educated elite, and they would appear to be leading America and the entire world into what may be the greatest tribulation it has known. The perversions based on the presuppositions of uniformitarian scientific deterministic theory seem to constitute the greatest massive antichristian deception the world has ever seen.

Just as men, in their wisdom, think they have discredited God in any area, God has providentially raised up other facts to discredit that area. The rejection of the cross and the resurrection of Christ reveal the bias of man's sin nature as no other act has done.

Chapter 29
Credibility of Documentary Sources for Christianity

False Assumptions by Critics

Although the evangelical doctrine of inspiration of scripture is vital to Christian faith, this chapter focuses on the historical documents concerning Jesus Christ and what the primitive church believed about him. From the very beginning of a rationalistic approach to the New Testament, when the educated elite was certain all things must conform to natural laws of science, they began to look for evidence for a real, purely human, historical Jesus. They *assumed* there must be *other earlier written documents, preceding the gospels* and other New Testament books, *that did not contain the supernatural ideas* from which Jesus' image evolved. They thought that if and when these were discovered they would reveal only a man with, perhaps, unusual talents. Finding these earliest or oldest documents was central to the search for a historic Jesus.

As C. H. Weisse indicated (cf. Chapter 25) the critics believed Mark was the earliest Gospel and this was agreed on by others, both critics and evangelical scholars. Therefore, the synoptic Gospels were deemed later and John much later.

Churches Search for Reliable Documents

From its early days, the church had been interested in the historical record about Jesus. The church fathers collected documents as they could, and often referred to and quoted from them in their writings. These writings also became part of the historic record. Apparently, after Constantine made Christianity the state religion in the early fourth century, he ordered the collection and recording of scrolls of the Scriptures. The earliest Greek manuscripts written were on fragile papyrus, but in the fourth century, more durable leather was used. In the late fourth century, Jerome made an authorized version for the Pope from Latin copies and a few Greek manuscripts. The emperor Constantine gave a large collection of Greek manuscripts to the Mammon at Baghdad which had become a center of learning. Here Syrian Christians translated them into Arabic. As Islam made its way into Europe, these ancient manuscripts emerged. Later, during the sixteenth century, Martin Luther made a German translation.

The Dutch scholar Desiderus Erasmus (1466-1536) suggested that a serious effort be made to evaluate the historical Greek records. Evidently, he had four Greek texts of late origins and from them composed one manuscript that he published in 1517. Erasmus argued there were errors in Jerome's long-accepted Latin version. Using the *Vulgate of Jerome* and two other texts, he made later revisions. Erasmus spent time in England and his text was the basis for the authorized King James Version in English in 1611.

The gathering of Latin and Greek manuscripts led to the production of another Latin version by Pope Clement VIII in 1592. By 1633, these evolved into the *Textus Receptus*. In the eighteenth century, John Will made an edition of the Greek New Testament, which prompted J.A. Bengel to study the variants in the texts and to gather the many editions, manuscripts and early translations of the New Testament. He began to assemble them into families of variant readings. In 1751, J. J. Wettstein initiated a logical system of categorizing manuscripts.

A Russian, Constatiaus Tischendorf, recovered a leather copy of the Old Testament Septuagint in Greek and the New Testament from a monastery on Mount Sinai. He traced these back to the fourth century. Around 1870, he published a two-volume Greek New Testament with a monumental apparatus indicating all variant readings known to scholars at the time.

In the late nineteenth century, two English scholars, B. F. Westcott and F. J. A. Hort, did the real science of textual evaluation. They gathered many manuscripts, and using the family system, began to show which were the oldest. If a group had the same errors, they were copied from the same manuscript. If other variants were found, they were put in a later family. The earliest families were traced back to the fourth century. While they identified other small families, three main families of manuscripts were established:

1. Western texts based on a fifth century Greek Codex Beza found in the Old Latin and Old Syrian translation of the gospels and in Old Latin fathers such as Marcian, Irenaeus, Tertullian, and Cyprian.
2. Revised Alexandrian texts supported by Egyptian Coptic translations.
3. "Neutral text" based on the two oldest fourth century manuscripts, called *Siniaticus* and *Vaticanus* that did not have corruption found in the other two families.

Credibility of Earlier Apostolic Documents

When the quest for the historical Jesus began in the eighteenth century, scholars had accumulated only 330 New Testament manuscripts. In 1924, when Burnett Hillman Streeter wrote his treatment of the manuscript tradition, sources, authorship, and dates, there were 1,400. When F. F. Bruce produced his examination of the reliability of the New Testament documents in 1943, there were over 4,500. Since then, a number of others with significance have been found, resulting in a total of about 5,400 today. The critics had little of the textual evidence that now exists.

Thus, *when the debate* over the critical examination of the manuscript evidence of the gospels *was hottest, less than one-fifteenth of the manuscript data was available* (1/15th) that has now been found. In addition, the science of textual criticism was little understood. Hence, the critics seeking the historical Jesus had some good manuscripts, but little evidence compared to what has emerged since then. The whole science of lower criticism, with a more modern dating system, has been developed. This enables one to establish clearly which were the earliest manuscripts. Lower criticism developed reasonable principles to decide which are the oldest and more reliable texts.[1] A number of critical texts have been published since them.[2] In addition, there are many earlier Syriac, Latin, and other versions, with families, that developed from them.

None of the Hypothetical Documents Have Been Found

Of great significance is the fact that while scholars have discovered thousands of ancient manuscripts, there has ***not been a shred of evidence*** *for the supposed preliminary documents*. There is no proto-Luke, or Q, or other sources for Matthew and Luke. Moreover, none of the early church fathers mention other such documents in their writing. So, the speculation about other written origins has no evidence to support them. Quite to the contrary, the New Testament documents have themselves been established as reliable and written close to the time of the apostolic gospel writers.

Disagreements and Distortions over the Priority of Mark

From the beginning, there was difference of opinion concerning the documents. Strauss rejected the priority of Mark, saying that for every five arguments for it, there were at least five against it; and he ignored John's Gospel. Many years later, Streeter, a critic himself, said, "I have refrained from discussing recent attempts to reach by critical analysis the sources used by Mark;

---

[1] For information about the principles of lower criticism, read George E. Ladd, *The New Testament and Criticism,* (Grand Rapids, MI: William B. Eerdmans, Publishing Co., 1967).

[2] For details of this cf. George Eldon Ladd, *The New Testament and Criticism,* see chapter III, Textual Criticism.

brilliant as some of these are, they leave me unconvinced."[3]

While some critics determined that Mark's gospel was one of the earliest manuscripts, (for reasons named previously) some modern—even many very competent translators agree that the early manuscripts do not contain the twelve verses (16:9-20, e.g., New International Version). They imply that these verses did not exist on the original copy and were added in some of the later Christian copies—as late as the *fourth century*. They are missing on the highly important Codex Beta and Codex Aleph, dated in that century.

But further study reveals evidence that contradicts this, and there is evidence that explains the omission in the fourth century manuscripts. I believe the arguments are clearly in favor of the resurrection account being in the earliest Mark manuscripts.

Evidence for Inclusion of the Resurrection in Mark

Based on the evidence, it is impossible that these verses were not in the original manuscript of Mark. Five out of ten of the *early ante-Nicene Church Fathers* actually quote or refer to these verses, including Irenaeus (in AD 185) and Hyppolytus (AD 190-227), who both quote six of them. The longer version of Mark, containing the resurrection, is in harmony with the second century Syriac gospels by Tatian. These verses occur in *every other manuscript* of Mark (other than Cod. B and A) that are in existence, uncial and cursive, and in *every version*, and in *every lectionary* read at Easter and Ascension Day. Many of these *existed 100 years before the discovery of the two manuscripts with the missing passage (Cod. B and Cod. A).*[4]

Moreover, Cyprian relates that at the *eighth council of Carthage (AD256)*, one of the 87 African Bishops assembled,[5] *Vicentius a Thiberi, quoted* the seventeenth verse of chapter sixteen in the presence of the council. There is, therefore, no way that the last twelve verses could have been omitted on the author's original copy, even though the two good fourth century codices omit them.

One can only speculate on their omission in Codices B and A. There are indications that these two manuscript copiers knew of the twelve verses. Cod. B never left any space anywhere else in the copying, but he left a space exactly adequate to include the twelve verses at that very point. Moreover, the copies of Aleph enlarged the letters at this place in order to take up the space that would be required for the twelve verses; something done nowhere else. What other explanation can be given for this other than that these writers were aware of the twelve verses?

There is one reasonable explanation. The omissions from fourth century documents occur after the popular writings of Eusebius, the early church historian. In a discussion in Eusebius' writing, Marinus quoted the twelve verses of Mark on the resurrection and then asked Eusebius for an explanation for the contradiction in Matthew 28:1 and Mark. Eusebius then proceeded with an explanation, *assuming their genuineness,* and makes the suggestion that one might dismiss the problem by saying the last section of Mark is not found in all copies. Eusebius quotes one of the twelve verses repeatedly as fact. So, he confirmed their existence, but might have sowed the seed of doubt, causing their omission shortly after that in the fourth century.

None of the church fathers ever mention the abrupt ending of Mark in Cod. B and Cod. A in any of their writings. The space in these manuscripts actually suggests the copyists knew of the existence of these verses. All of the evidence points to the existence of the resurrection account at the end of Mark, as in the other gospels.

---

[3] B.H. Streeter, *The Four Gospels,* xxx.

[4] For more details on this see Ladd, *The New Testament and Criticism*, 71-74, et, al.

[5] These include the two oldest Syriac Version, the Old Latin translation, the Gothic Version, the Egyptian Version plus later church Fathers, Ambrose, Cyril of Alexandria, and Augustine, and Codices A and C.

Moreover, after years of scholars trying to find a Jesus in Mark who is less supernatural than in the other Synoptics, H. Hoskyns and Noel Davey demonstrated that Mark is impregnated with the supernatural element from beginning to end. "The consistent witness of the entire New Testament (is) that God acted to reveal Himself for man's salvation in the historical Jesus." The critical method, they insisted, "has clearly revealed the living unity of the New Testament documents; and the historian is compelled to state that both the unity and the uniqueness of this claim are historical facts."[6]

Reversal of Claim of Lateness of Gospel of John

Most critics considered the gospel of John, which is very theological and clearly presents Christ's deity, as inferior and written much later. But, some liberal scholars accepted it. Schleiermacher looked on it with favor; A.F. Groner, exalted it; Bruno Bair saw it as a work of art, and most important; and Weizsacker saw in it genuine reminiscences of the early apostles.

Later, Albert Schweitzer claimed that the liberal scholars had dismissed John as a valid source in order to push for his (Schweitzer's) view of the eschatological as the only remaining viable view to interpret the historical Jesus. John's gospel has little to say about the eschatological. Schweitzer took two pages to claim scholars' dismissal of John, but he himself named many of the prominent scholars who did not dismiss it. Referring to the claim that John was no longer a viable source with the Synoptics, Schweitzer says, "That seems surprising, since the chief representatives of this school, Holtzmann, Schenkel, Weizsacker, and Hase, took up a *mediating position* on this question, not to speak of Beyschlage and Weiss." Schweitzer goes on to say, "Individual attempts to combine John and the Synoptics which appeared after this decisive point (of liberal lives of Jesus) *are in some cases deserving of special attention,*" (Emphasis added) and he refers to Wendt and his common link with Weiss, Alexander Schweitzer, and Renan.[7] Thus, it appears that even in Schweitzer's eyes, the issue of John's gospel as a source was not settled.

Some of the critics dismissed John's gospel as being written from Ephesus by someone else, much later, around the fourth century. The critics asserted that the frequent references to "light and darkness," "logos," etc., showed it was not Palestinian but Hellenistic and probably influenced by Gnosticism.

The facts, however, soundly refute a late date for John's gospel. The discovery of the Dead Sea Scrolls completely exploded the idea of excluding Palestine as the setting for John's writing. The writings that were found in the Qumran caves were almost certainly done between the Roman invasions of 70 and 135 AD and contained writings prior to that, probably by the Essenes or whatever other groups of devout Jewish people fled from the Romans. One document, *The War of the Sons of Light and the Sons of Darkness,* as well as others, is filled with terminology used in John's gospel. In an interview with William F. Albright at Fuller Theological Seminary, I heard this outstanding archaeologist make the statement that because of the Dead Sea Scrolls he felt John may be the first and earliest gospel written.

The discovery of some of the earliest gospel manuscripts confirms John's early date. A papyrus codex found in Egypt was later discovered in the John Ryland's Library in Manchester, England. It contains a fragment of John 18:31-33, 37 ff. This has now been dated, on paleographical grounds, around AD 130. Christian tradition says that John the Apostle wrote the

---

[6] George Eldon Ladd, *The New Testament and Criticism*, 49, referring to H. Hoskyns and Noel Davey, *The Riddle of the New Testament,* 1931.

[7] Schweitzer, *The Quest of the Historical Jesus*, 218-219.

gospel while in Ephesus between AD 90 and 100, which means it was circulated in Egypt within forty years of its writing. Moreover, another papyrus manuscript found earlier, Bodmer II, was rediscovered in the Bodmer Library of Geneva in 1956. It is well preserved and contains the first fourteen chapters of the gospel of John with but one lacuna (of 22 verses) and with considerable portions of the last seven chapters. It is dated at AD 200. So, the evidence from these two papyri pushes the date of the writing of John's gospel exceedingly close to the apostolic author, who died about 100 AD.

Also, of great significance is the fact that in this early manuscript, John 1:18 reads, "No one has ever seen God, but the only begotten *God*, who is in the bosom of the Father, he has explained Him." This statement of the incarnation, "God only begotten," is supported by most other early data, (cf. Leon Morris).[8]

Moreover, in every reference to John 1:13 written before the third century, it reads in the singular thus, "Even to those who believe in His name," (Christ's name) and then is followed by "Who *was* [not "were," pl.] *born* not of blood, nor of the will of the flesh, nor of the will of man, but *of God*." These references, therefore, said that Christ was virgin born from God. The early church Fathers who quote it this way are Justin Martyr (150 AD), Irenaeus (185 AD), Tertullian (190-220 AD), and Augustine (late fourth century) and is found thus also in the old Syrian version, the Old Latin version, and others.

Tertullian accused the Valentinian Gnostics of changing this verse (1:13) to the plural to say believers are those born of God to discredit Christ's incarnation. The apocryphal *Ephistul Apostolorum* (dated 150 AD) seems to refer to this verse, saying, "He is the Word made flesh, born in the sacred virgin's womb, conceived by the Holy Ghost, *not by the will of the flesh*, but by the will of God." Considering John's statement about the eternal Word who was God (1:1) who became flesh and tabernacled with men (1:14) along with both verses (1:13, 18), there is no question that John's gospel early testified to the incarnation and suggested Christ's virgin birth. Thus, three of the four Gospels—Matthew, Luke and John—begin with the incarnation idea.

Multiplicity of Gospel and Other N. T. Manuscripts and Other Evidence

Thousands of manuscripts, including the Chester Beatty Biblical Papyri consisting of eleven codices, were written in the early third century (200 AD and following). Three contain most of the New Testament writings—one with the four gospels with Acts. Another contains Paul's letters to the churches and the Epistle to the Hebrews. F. F. Bruce has pointed out that there were quotations and references from most of the New Testament books in the earliest catholic Apostolic Fathers who wrote around 100 AD.

Even groups who were hostile to Christians witness by the quotes in their writings to the early existence of the New Testament books. Bruce said that in, "the writings of the Gnostic school of Valentinus before the middle of the second century, most of the New Testament books were as well known and as fully venerated in that heretical circle as they were in the Catholic Church.[9] There are many spurious claims that texts found at Nag Hammadi, Egypt written by hedonistic Gnostics were prior to biblical documents, but these are considered no earlier than the mid-second century by competent textual critics. They have no valid historical relevancy. By 350

[8] Leon Morris, *The Gospel According to John,* (Grand Rapids, MI: William B. Eerdmans Publishing Company, 1971), 113, found that way in 66, 75, XBC, L33 and in much patristic evidence.

[9] F. F. Bruce, *The New Testament Documents: Are They Reliable?,* (Downers Grove, IL: InterVarsity Press, 1970), Fifth edition, 19.

AD, there are major codices with the most of the whole New Testament (Vaticanus and Sinaiticus) and many more soon afterwards.[10]

Conclusion

It is important to recognize how formidable the early written historical evidence is for Jesus and his sayings. *No other historical event has so many thousands of manuscripts* that go back, some within the times of those who heard the apostles or knew those who heard the apostles preach. In great contrast, even when B.F. Streeter wrote in 1924, he pointed out "the paucity and late date of MSS" in the field of classical literature. He said,

> No portion of Tacitus, for example, survived the Dark Ages in more than one [manuscript]; and the number of famous works of which, apart from Renaissance reproductions, there are less than half with a dozen MSS, which is very large. Again, apart from fragments, there are no MSS. of the Greek classics earlier than the ninth century [AD], and very few older than the twelfth.[11]

He points out the opposite is true of the New Testament—too much very early data is the problem with gospel study.

F. F. Bruce makes a similar statement about the thousands of manuscripts,

> Perhaps we can appreciate how wealthy the New Testament is in manuscript attestation if we compare the textual material for other ancient historical works. For Caesar's *Gallic War* (composed between 58 and 50 BC) there are several extant MSS, but only nine or ten are good, and the oldest is some 900 years later than Caesar's day. Of the 142 books of the *Roman History of Livy* (59 BC-AD 17) only thirty-five (35) survive; these are known to us from not more than twenty MSS of any consequence, only one of which, and that containing fragments of Books iii-iv, is as old as the fourth century.[12]

Bruce goes on to say,

> To sum up, we may quote the verdict of the late Sir Frederic Kenyon, a scholar whose authority to make pronouncements on ancient MSS was second to none: 'The interval then between the dates of the original composition and the earliest extant evidence becomes so small as to be in fact negligible, and the last foundation for any doubt that the Scriptures have come down to us substantially as they were written has now been removed. Both the *authenticity* and the *general integrity* of the books of the New Testament may be regarded as finally established'.[13]

But this is only the beginning of the certainty of the early historical documents of the New Testaments as valid witnesses. While many manuscripts were discovered since the late nineteenth century, the science of papyrology, which studies ancient texts written on all sorts of materials like papyrus, leather, etc., (except stone) has become much more precise in the last half century. The Congress of International Papyrologists' Association held only its twenty-first meeting in 1995. It has developed many standards for examining and dating texts and has new, more exact instruments. One is the epifluorescent confocal laser scanner microscope which can differentiate

---

[10] Ibid., 120.

[11] B.H. Streeter, *The Four Gospels*, 33.

[12] F. F. Bruce, *The New Testament Documents*, 16 on other historical works and 20 on F. Kenyon.

[13] Ibid.

between twenty separate micrometer layers of a papyrus manuscript, select individual layers, measure height and depth of ink, and put the results on video printout. These scientists have done away with many myths that caused late dating of New Testament texts (such as the codex did not begin until the third or fourth century) and have reconstructed and restored damaged texts with great certainty.

They have demonstrated that a fragment from Qumran Cave 7, called 7Q5, contains Mark 6:52-53 and must be dated before AD 68 and possibly by AD 50. Moreover, fragments of Matthew's gospel found in Luxor, Egypt and preserved at Magdalene College, Oxford, designated $P^{64,}$ (containing Matthew 26:32-33 and another from Matthew at the Fundacion San Lucas Evangelista in Barcelona), number $P^{67,}$ (containing two leaves with part of chapters iii and v), are considered from the same uncial codex which is dated approximately AD 66. The Paris papyrus, designated P^4, containing Luke chapters 1-6, has been dated early second century or earlier. Papyrologists have re-dated one papyrus codex each of John and of the Pauline Epistles to the first and second centuries, and are rapidly now beginning to give earlier dates to many other manuscripts.[14] Thus, the dates of the gospels and other New Testament books correspond to the lives of those who were eyewitnesses to Jesus.

Unbelieving Critics Admit Overwhelming Evidence

B. H. Streeter, who wrote when less than one-third of the manuscript evidence was known, said, "The mass of material to be considered is *crushing*."[15] Even Bishop John A.T. Robinson, who rejected all the evidence for the supernatural in his book, *Honest to God*, later conceded, "the wealth of manuscripts, and above all the narrow interval of time between the writing and the earliest extant copies, make it by far *the best attested text of any ancient writing in the world.*"[16] (Emphasis added) Robinson then tells how this changed him to an openness to Christ. Robinson's final work, *The Priority of John*, was published posthumously, and in it, he dated the writing of John as AD 66.

He also observed that the scholars involved in the quest for the historical Jesus displayed an alarming inconsistency and skepticism toward the documents they considered. In restudying the quest for the historical Jesus, Robinson concluded that the critics were faithless to the very scientific and historical method which they used in every other field of study. He indicted them for their dishonesty in dealing with the New Testament documents. This rejection by the elite shows the deep bias against the God who is revealed in Jesus.

Tradition says Matthew wrote his gospel first; Mark was second (which was probably used to chronologically revise Matthew's). Luke's was next and John was later written from Ephesus. All were members of the early witnesses of Jesus.

---

[14] Read the story of these rapid scientific changes in Carsten Peter Thied and Matthew D'Ancona, *Eyewitness to Jesus: Amazing New Manuscript Evidence about the Origin of the Gospels,* (New York: Doubleday, 1996).

[15] B.H. Streeter, *The Four Gospels*, 33.

[16] John A.T. Robinson, *Can We Trust the New Testament?,* (Grand Rapids, MI: Wm. B. Eerdmans Publishing Company, 1977), 36; *The Priority of John* (Oxford University Press, 1985).

Chapter 30
Enlightenment about Origin of God Idea

Comparative Religion Changed by Scientific Development

Evolutionary Speculation and Scientific Ethnology

As scientific determinism progressed, efforts grew, not only to remove God, but also to develop a theory to explain God as a man-made myth. While various religions were compared, the main focus was on the Hebrew-Christian God. By the beginning of the twentieth century, scholars had accumulated a tremendous amount of material on comparative religions and had written much about primitive and post literate societies and their religions. The early focus on nature myths was replaced by the evolutionary hypothesis. The idea of progress became widely accepted and greatly influenced thinking. Gradually, it became obvious to some scholars that there were many speculative ideas presented in the name of science that were only guesses based on evolutionary philosophy. Some scholars began to establish principles by which to judge various kinds of cultural evidence more objectively.

Without considering the faltering beginnings, it is important to look at studies of cultural history that produced a truly reasonable and scientific way of sorting out the facts of a culture. Ethnologists established some brilliant principles for determining the age of a culture, even where there are few or no written documents or monuments. Using these principles, scientists could then trace the origin of the particular element of interest in that culture, whether domestic, military, economic, or religious. The following elementary explanation, while incomplete, is sufficient to demonstrate that this is an objective, scientific approach to determining the age of a culture.

Development of the Ethnological Methodology

Drawing upon key insights by earlier scholars, E. Graebner and B. Ankermann developed the science of ethnological methodology at the Berlin Ethnological Museum. W. Foy and W. Schmidt joined them, and similar ideas emerged in America. They did not develop this approach to study religion specifically, but rather to study all aspects of culture. In time, W. Schmidt applied this scientific approach to the development of religion and the idea of God.

Graebner produced the main points in his *Method of Ethnology*, but he gives credit to some others.[1] In this method, he looked first at criticism of real (not supposed) sources, then at interpreting the true meaning of each fact, and then at a combination of these to show the more extensive course of development.

Graebner explained how to examine for the two main criteria within a culture—form and quantity. Quantity strengthens the data on the form. Other factors, such as agreement in language and frequent repetition indicated associated cultures, organic closeness and historical relationships between cultures. Hence, the large number of individual characteristic agreements or quantities proves the cultural relation. Form refers to agreement in type of weapons, worship, etc.

One can assume connections even when the distribution of items is discontinuous, even when

---

[1] W. Schmidt, *The Origin and Growth of Religion: Facts and Theories*, Chapter XIV, "The Historical Method and It's Result for Ethnology," 219-250 is the main source of these quotes. For a complete study of this method see F. Graebner, *Method der Ethnologie*, Heidelberg, 1911. He drew upon the work of others such as that of F. Ratzel, who developed a theory of migration through the history of the bow in Africa. Leo Frobenius developed the theory into the doctrine of 'culture-circles' or "spheres" and suggested a criteria of quantity of a material. There were others.

and where they are separated by great distances. But, to do so, form and quantity must be proportionately stronger. In the case of a wide separation, the ethnologist must look for the possibility of continuous connection at an earlier date by appealing to the continuity within the culture. The possibility of a former connection is strengthened if there are intervening groups of peoples with the same or similar cultural elements. These can only be the remains of the connection which once existed.

A third criterion—degree of relationship—may also be considered. In this criterion, resemblances in question increase in strength and in number in proportion to the two chief areas separated from each other in the site. Thus, one would know for certain that the resemblance does not arise independently, but owes its origin and existence solely to a historical connection between the two principal areas.

Thus, by these and similar criteria, one is able to objectively determine, with a high degree of scientific certainty, the spatial connection of cultural areas. This method makes it possible to determine the temporal connection or which elements in a culture appeared or occurred earlier. Schmidt says,

> If two cultural areas of different character touch one another, then either they will overlap at the borders and so produce crossings, or else will merely touch at the edges and so produce the phenomena of contact. It is obvious that forms arising either out of a crossing or a contact must be more recent than the parent forms. This is the first objective ethnological criterion of time, and it may lead to still further chronological results; for those forms due to mixture or contact in which the two components are still clearly to be recognized as such are thereby shown to be less ancient than those in which the component elements have blended into a unity; since this process must needs demand a longer time. The forms arising from crossing and contact must always be later than the forms in the two cultures from whose combination they arose. This is naturally true likewise of the similar forms, which arise from the contact of two entire circles of culture. The areas where such forms occur and the forms themselves must be later than the cultures themselves and the simple forms belonging to them.2

Thus, it is possible to lay down two useful rules to help solve the matter of absolute sequence from cultural history.

> Firstly, a culture which appears as the oldest in every part of the world in which it occurs, must be regarded as absolutely the oldest. Secondly, a culture-area which divides another in two, or overlays it, can in no case have arisen on the spot where the dividing or overlaying occurs.[3]

The Berlin school then proceeded, using these standards, to decide the origin of the individual cultural forms. First, they established to what particular culture the element belongs. Second, they followed the individual phases of the development of the element through the whole of that culture, thereby, making sure which is the oldest form of the element. That element is then put in the foreground of origin.

From this procedure, they had two general rules: (1) Every cultural element can be explained, as regards its origin, only from the ideas and association belonging to that culture to which it belongs and not from any general guesses as to what may have been; still less, of course, from the

---

[2] Ibid., 235-237.

[3] Ibid., 235.

ideas and association of a foreign culture. (2) Within a given culture, the oldest forms of any element are especially significant for explaining its origin, for they come nearest to reflecting the influences, physical and mental, to which the first appearance of that element was due. These rules proved very reliable. This is an oversimplification of the methods, but can give the reader some idea of how they work. For a more comprehensive explanation of this method, I recommend reading W. Schmidt's work on the process.[4]

Ethnological History Reveals Supreme High Spirit God as Oldest

Increasingly, ethnologists adopted and applied this approach to the origin of religions, including Lang's findings of a priority of one great high Spirit God. The application of these ethnological principles showed that the oldest form of God that exists in every culture was the one most high God. This completely refuted evolutionary claims and was made entirely on a scientific and not a religious basis.

More than any other, W. Schmidt studied the field of comparative religion and applied this scientific approach. He sought to cover the whole of the studies of religious origins and completed five of the seven volumes he intended to write. In his book, *The Origin and Growth of Religion: Facts and Theories*, he summarized the research up to that point. He applied the ethnological method to Lang's work and further surveyed the many efforts to study much of the other data. Based on these principles, he determined that the most High God is the earliest god referred to in every culture. While he acknowledged that Lang was correct about the one Supreme Being, he felt that Lang's treatment needed a better psychological explanation of how it meets the simple needs of man.

Schmidt then said,

> The other point has been left almost as completely unnoticed. It is *the universal distribution of **the worship of a Supreme Being over the whole earth***. ...The equal value for the proof of the materials from all over the world, increases immensely the cogency of that proof as a whole, by excluding *ab initio (from the beginning)* all purely external attempts at explanation, especially the theory of borrowing, and shows convincingly that we are dealing with ideas which go back to the earliest days of humanity.[5] (Emphasis added)

He discerned three primary attributes of God from the data. This Supreme Being *furnished primitive man with the ability and power to live* in the world. Secondly, Schmidt goes on to say,

> From his eternal heaven he invades nothingness, and *begins the time-series with his creative activity.* Throughout all the periods of creation and all successive periods he is *lord of man's history,* although he does not always actively interpose. He stands, moreover, at the beginning of each human life, accompanies it through all its length of days, *awaits its appearance before his tribunal and determines the nature of its eternity.* For such a unity as this primitive man has no model and no corroboration in what he sees and experiences in his own time. (Emphasis added)

Thirdly,

> He *dominates all* space. ...One power extending over these distances and joining them one to

[4] Ibid., 220-250.

[5] Ibid., 184.

> another. For the God of these primitive men is not the god merely of one tribe and its environment; if only because he is the creator of the whole world and of all men, he has no neighbor whose realm could limit his. The thoughts and feelings of these men have no room for more than one Supreme Being.[6] (Emphasis added)

Schmidt struggled with the idea of why man can conceive of, and begin with, the belief in such a God who has these attributes which he has not experienced (since according to Schmidt, there is no God).

In spite of the fact that the ethnological method was a scientific approach to discovering history in a culture, those who held to the evolutionary concept of scientific determinism fought this effort and rejected its findings. This scientific approach refuted the idea of evolutionary rational progress.

Precisely at the time Schmidt and his colleagues were working out this new method of scientifically dating religious origins, Albright's archeological evidence and other areas of scientific discovery that refuted Wellhausen's theory were being compiled. But the evolutionary view of progress in religion fit the whole view of the liberal educated elite, and it persisted in spite of all evidence to the contrary.

Archaeology Supports Evidence Ethnological Findings

Based on archaeological evidence and deciphering Sumerian, Amorite, and Egyptian languages, these cultures began with a belief in one high sky God. I noted the same in my review of Greece and Rome in, *Man is Not Enough*. All these cultures have substantiated the findings of the ethnologists. Yet, most of the archaeological discoveries, as well as scientific theology, *came* ***after*** *the comparative religionists dogmatically proclaimed the evolution of religion.*

In my unpublished book, *The Call*, I present archeologists' data about God found in the Sumerian cities, and in the prolific writings at Ebla and elsewhere.[7] All of this shows that there was one supreme sky God, An, the creator. His Son Spirit, Enlil ruled for him. Enlil was the earliest God among the Sumerians, and he is shown to rule over all other gods that are said to come into being later. More and more gods of nature were formed in these cultures, and later this supreme spirit God was supplanted. I also point out in *The Call* that the most primitive God of Egypt was Horus, who likewise was the sky God, an omnipresent Spirit who created and holds men accountable. Schmidt's studies, as well as my own, show that all the views of God of all the early civilizations ***degenerated*** from one sky God into base sensual families of gods, or a pluralism. The evidence does not uphold evolution.

In his book, *The Canaanites,* Ulf Odlenburg, an outstanding Semite scholar, reviewed the studies of the Bible lands of the Semitic cultures, including early Semitic, Phoenician, Ugaritic, Hebrew and other peoples, and concluded in all of them that the God, El, and the Hebrew Yahweh were the original monotheistic sovereign God of those peoples. Their views later degenerated to a pluralism of gods, which ultimately supplanted El. He refers to studies of Sanchuniathon, I. J. Geib, Maria Hofner, Eissfeld, and others. He points out that the God, El, is stated to be the "builder" or "maker" of all things in Hebrew, Ugaritic, and Phoenician writings. This high God has existed before all things and appears before any other gods and is over the later pantheon of

---

[6] Ibid., 283-284.

[7] Carl W. Wilson, *The Call,* (as yet an unpublished mss.) Cf. Finegan, *Light From The Ancient East,* for each culture.

gods. Moreover, he is the Father of all mankind.

What happened in the Greek and the Roman civilization, beginning with a belief in the sky God, Uranus, and his son—Zeus in Greece and Jupiter in Rome, respectively, is exactly the trend Paul describes in the steps of degeneration of a culture in Romans 1 and 2. (For further discussion, see *Man Is Not Enough,* pp. 19-89) The degeneration of the view of God followed as man's sin carried him along.

Man's Bias Exposed in the Study of Comparative Religion

The evolutionist opposed and fought scientific facts. Schmidt was well aware that Darwin's evolutionary theories dominated in Europe. Even after he and his ethnological colleagues applied this new scientific approach to culture, a universal bias against the idea of the sky-God or one Supreme Being persisted. He said,

> If they agreed in nothing else, [all framers of theories] were at one in this, that the figure of the sky-god *must be got rid of from the earliest stages of religion,* as being too high and incomprehensible.... The strength of this universal current of thought was so great, and the resulting discredit into which it brought the notion of the great age of the sky-god so complete, that hardly anyone found courage to oppose it and to draw attention to the quite frequent examples of this exalted sky-god appearing among decidedly primitive peoples, where not the least trace of Christian influence was to be found.[8]

Moreover, Schmidt points out the irony of the comparative religion movement's earlier theories. He says, "The sequence of the theories and of the religions they are founded upon, as they come one after another into prominence, is ***exactly the reverse*** *of the order in which those religions were actually and historically developed.*"[9] (Emphasis added) In other words, the comparative religion teachers saw man as progressing in his thoughts about God, while actually in all primitive cultures man was progressively debasing the original high idea of God.

Tragic Unbelief Even When Etymological Science Proves Otherwise

In explaining why a Supreme Being is found at man's origins, Schmidt said,

> If we apply that criterion to the abundant mass of data which we can now produce regarding the primitive Supreme Being, the first thing to notice is that the total sum of the facts is of a nature to satisfy the total sum of human needs.[10]

He lists these needs for which man wants a God as follows: a rational cause, social needs or Father of mankind, moral needs, a lawgiver, overseer and judge of the good and the bad, and one who is free from moral taint, meets emotional wants, and a protector. Having said all this, Schmidt concludes, "Hence we are *not yet in a position to answer,* positively and with scientific accuracy and certainty,[11] the question ***how*** *the primitive high god and the religion of which he is the center originated."*

---

[8] Schmidt, *The Origin and Growth of Religion*, 171.

[9] Ibid., 14, 15.

[10] Ibid., 283.

[11] Ibid., 286.

Significantly, Schmidt did not rise to the level of faith in God. He seems to have gone along with Lang and the spirit of unbelief in his time by believing that all *men everywhere needed that kind of God* to meet his needs *and thus man conceived of or created Him*. Schmidt does this even though the idea of one Supreme Being is evident as the oldest and the universal view of God. Moreover, he has no answer to the question he himself asked, "Where was man to get the conception of pure spirit, which he does not meet with as such in nature?" Moreover, he admits man has "no model" of such a God.[12]

Schmidt has no evidence that such a God was created by man. To the contrary, he admitted this Supreme God just existed, or was from the beginning in man's consciousness and that he was recognized to have existed throughout the entire world. Schmidt used the ethnological method honestly and arrived at one Supreme Being that was recognized by all men as the only original God. But, he himself refused to accept that this God really existed, maintaining that man created this God to meet his needs. Thus, all men are born with a consciousness of one God and recognize evidence of him in creation.

But, as we have seen from historical studies, man has a bias to dispose of God to whom he is accountable and who will bring him into judgment at the end of history. Thus, Schmidt's assertion that man wanted and created this kind of God is in contradiction to the facts. This belief that one Supreme Being existed was not a need of men, but rather a teaching man did *not* feel he needed or wanted. Yet, Schmidt found this idea about God existed among all people from the earliest times.

Thus, while Schmidt followed his scientific objective approach to its honest conclusions, he did not escape the bias of the evolutionary speculators because *he too makes man the creator of God*, and he *does it without any evidence*. Indeed, the evidence is that one spirit sky God has been revealed in the consciences of all men, even in their most primitive stage from nature. This makes the evidence for man's bias against accepting God even stronger. And this points strongly to the fact that no man, apart from God's grace, is willing and desirous to believe in the God who created and holds him accountable until judgment.

Wellhausen's Evolutionary History of Israel was Erroneous

The Literary Critical Approach Failed To Find Documents

As previously shown, the critics tried to discredit belief in the biblical view of God in the history of the Old Testament. That Old Testament history is vital to understanding the fulfillment in Christ. Julius Wellhausen's rewriting of the history of the religion of Israel in his book, *The Prolegomena of the History of Israel*, published in 1870, was one of the most important efforts to discredit the God of the Bible. There Wellhausen used the documentary hypothesis developed before him. His hypothesis proposed that the books of the Old Testament were composed by a group of power hungry priests during the captivity period. They pieced together earlier documents into a fraudulent story of a monotheistic God. They place this about seven to nine hundred years after the traditional meeting of Moses with God at Sinai. Wellhausen said the history of Israel was one whereby the idea of God evolved from a primitive pluralism of the gods of nature to the one God, Yahweh.

The impact of Wellhausen's work spread through the schools and churches of the Western world and throughout the world. In conjunction with the prevailing skepticism motivated by deterministic science, Wellhausen's view gradually changed a great many of the most prominent

---

[12] Ibid., 181, 284.

religious educational centers from faith in the supernatural God to faith in secularized man and his ability to solve men's problems through social action. But as religious educators and scientists embraced and promulgated Wellhausen's views, they also began to devote themselves to intense study of every aspect of these teachings.

The Various Critical Techniques Fell into Disrepute.

In time, many of the critics themselves came to see the weakness of his arguments. The literary critical approach was seen for the absurdity that it really was. The basis for discovering background *documents* was shown to be invalid. What's more, the number of "documents" critics named as the primary "documents" proliferated until there was a complex jigsaw puzzle that was impossible to piece together. The mysterious figure of the interpolator was invoked to such an extent that it came to be seen as ridiculous and embarrassing. Even the higher critics' *dating of* the various materials in *the documents* proved to be false. In the end, the science of literary criticism was shown to be mere whimsical imagination and not a science at all.

Critical scholars discovered that the names for God, which were the main and earliest basis for the discovery of the documents, did not consistently appear in the Greek translation of the Old Testament (called the Septuagint) in the same way as in the text that Wellhausen used (called the Hebrew Masoretic text). Moreover, other Hebrew texts of some of the books were found that preceded the Masoretic Hebrew text. The names did not appear exactly the same there either. Hence, the use of names was seen as an unreliable way of discovering sources.

Eminent scholars, using the same critical approach, kept finding more source documents, and documents within the documents. There were more than thirty background documents and subdivisions (e.g., D1, D2, D3, and S1, S2, S3). The idea that a group of priests put all this together in a consistent story from all the sources that were still evident was ridiculous. The literary analysis approach was revealed for what it was. As Herman Wouk said,

> What scholars had found out at long last, of course, was that literary analysis is not a scientific method. Literary style is a fluid, shifting thing, at best a palimpsest or a potpourri. The hand of Shakespeare is in the pages of Dickens; Scott wrote chapters of Mark Twain; Spinoza is full of Hobbs and Descartes. Shakespeare was the greatest echoer of all, and the greatest stylist of all. Literary analysis has been used for generations by obsessive men to prove that everybody but Shakespeare wrote Shakespeare.[13]

The foolishness of the ghost interpolator who was the key to *Prolegomena* became evident. Herman Wouk said,

> But when a historian finds in a long stable text dozens and scores and hundreds of verses that directly contradict his pet theory about the text, and reaches the conclusion that this state of things proves a clairvoyant interpolator's hand, his work seems to cross the red line into the curious literature of systemized delusion.[14]

---

[13] Herman Wouk, *This Is My God*, addendum, 268, cf. Barnhouse referring to Wouk, 33.

[14] Ibid., 18b.

Archaeology Disproved Presuppositions of Evolution of Israel's Religion

New Science of Archeology Emerged After Wellhausen

It wasn't until after Wellhausen wrote *The Prolegomena* that the science of archaeology was fully developed. Most of the archaeologists who went to Palestine had a firm conviction that their findings would uphold Wellhausen's developmental document theory. Once, while I was sitting under the teaching of William Foxwell Albright, of John's Hopkins University, a member of the National Academy of Science and perhaps the world's greatest archeologist, he said that when he went to Palestine he held to the liberal critical views and was gradually forced by evidence to leave them.

Invalid Dating of "Documents"

The archaeologists gradually began to dig up and piece together a much clearer and more accurate picture of the facts about the Near East. They found that writing was clearly as old as Moses (Wellhausen denied Moses could write), tabernacles were used by Bedouin tribes as centers of worship quite early, and that the central worship of Israel emerged far sooner than Wellhausen thought. Moreover, archaeologists placed data in the documents at different dates than Wellhausen had placed them. For example, aspects of J document were found to be three or four hundred years earlier. Also, by mid-twentieth century, they discovered that aspects of P document, which was the last, according to Wellhausen, were really the earliest. It was crucial to Wellhausen's view that P was last, so this completely upset his argument.

Documentary critics considered the book of Ezekiel to be an argument for the lateness of the P document. Carl Gordon Howie, a noted Semitic scholar, points out in his book on Ezekiel that the P document material was early and he comments on the significance of this fact. He said,

> The problem of this book has indeed become the problem of the Old Testament studies since this prophecy was the pole-star on which much of our present dating of other books depends. A dislodgment of the traditional date would bring utter confusion of biblical literary chronology as far as many scholars are concerned.[15]

Howie upheld the captivity date for Ezekiel's works.

The Law Was Early, Not Late

Wellhausen's central thesis was that the Mosaic Law was late. Archaeological studies refute this theory. William F. Albright's writing shows the law was early. In a chapter on "The Antiquity of the Mosaic Law," he demonstrates this and later gives "extrinsic arguments for an original date of the Decalogue no later than the thirteenth century B. C."[16] Also, many archeological finds verify the location of the worship of religion in Jerusalem by the time of Solomon. They have even unearthed a silver chalice in the Jerusalem Temple area with the Aaronic benediction engraved on it that goes back to that time. Thus, Wellhausen's documentary hypothesis, which depends on Israel's central religion, based on the law and the tabernacle, being composed during

---

[15] Carl Gordon Howie, *The Date and Composition of Ezekiel,* (Philadelphia: Society of Biblical Literature, 1950), cf. 2, also 1, 29, 43, 90, 104.

[16] John Maier and Vincent Tollers, eds., *The Bible in It's Literary Milieu,* (Grand Rapids, MI: William B. Eerdmans Publishing Company, 1979), 149, see also Albright's, *The History of Religion of Israel.*

the exile in Babylon, is completely contrary to the facts.

Israel Not Isolated to Evolve Uniquely

Wellhausen's teaching that Israel was an isolated little group in Palestine, not interacting with other views of God so that it evolved its own distinct monotheistic view, should have never had credibility. After years of struggling with the Old Testament problem, Noordzy observed:

> If there is one thing, which has become clear through the study of Israel's religion, it is certainly this: that two influences were active. The one tended constantly to eliminate all distinction between Israel and the other nations: the other exerted pressure toward sharply accentuating Israel's unique character.[17]

Archeological studies of Palestine clearly reveal that Israel was at the crossroads of the world, with the various civilizations coming back and forth and interacting with them from the time of Abraham to the captivity. The Fertile Crescent was the center of transportation of tradesmen, armies, and travelers of all sorts. Israel was by no means an isolated group.

Thus, Wellhausen's and others' rewriting of Israel's history along evolutionary lines, claiming the history was a fraud by power hungry priests, has itself been shown to be a perversion to discredit the history of God's revealing himself to Israel.

Invalidation of Evolutionary View of Israel's History

Leading Religious Scholars Abandon Wellhausen's View

After two generations (about 1930), the literary critical approach fell out of favor. At the leading of Ivan Engnell and the entire Scandinavian school, Wellhausen's theory was completely renounced. Engnell called Old Testament scholars back to a traditional-historical view, which conformed more closely to the Old Testament texts. From that point on, Wellhausen lost his place of leadership. New approaches were adopted, although the dogma of the J, E, D, P theory is still maintained by some liberal institutions in modified forms.[18]

Finally, after years of archaeologists uncovering the historical data that relates to Israel and the nations around it, it became clear that there was no historical evidence to contradict the traditional view of the Bible. In 1959, Nelson Glueck, the outstanding Jewish archaeologist, wrote in his book, *Rivers in the Desert*, "It may be stated categorically that no archaeological discovery has ever controverted a biblical reference" and repeatedly made the assertion of "the almost incredibly accurate historical memory of the Bible, and particularly so when it is fortified by archaeological fact."[19]

Archaeological Results from Secular Schools Uphold Biblical History

Almost immediately, J. J. Finkelstein of the University of California gave a lengthy review of Glueck's book in which he accused him of taking a blind fundamentalist stance of proving the Bible's historicity rather than writing as a professional archaeologist. He accused Glueck of

---

[17] Noordtzy, *The Old Testament Problem*, 223.

[18] See H. H. Rowley, *Old Testament and Modern Study,* (Oxford: Clarendon Press, 1951), 63 ff.; cf. Barnhouse ref. Wouk, *This Is My God*, addendum, 33-34.

[19] Nelson Glueck, *Rivers in the Desert,* (New York:, Farrar, Straus and Cudoby 1959), 31.

distorting the picture by bias.

G. Ernest Wright, archaeologist of Harvard University, not only took up his pen to defend Glueck, but to charge that this attack by Finkelstein brought forward a more important issue that needed to be faced. That issue was that the great efforts in archaeological research had been sponsored by the liberal universities and the data found can in no way be attributed to fundamentalist efforts. He pointed out that Glueck was educated in an atmosphere skeptical toward biblical traditions in America and in Germany, and therefore had not approached the data from the traditional view. Moreover, Wright pointed out that William Foxwell Albright, the greatest Palestinian archaeologist "led the attack in the English-speaking world on the unexamined presuppositions of Wellhausen from the standpoint of ancient history and particularly archaeology."[20] Albright stated in 1932, "The theory of Wellhausen will not bear the test of archaeological examination." He subsequently demonstrated this in three areas: the patriarchal period of Genesis, biblical law, and the Babylonian exile. Albright did not hold to inspiration of Scripture or inerrancy of the Bible and disclaimed fundamentalist beliefs, yet he held firmly to the *invalidity* of Wellhausen's views.[21]

While Wright and Glueck went on to take theological positions for Christianity and Judaism, Wright backed away from this after much criticism and after failing to find the support for his theology by using a new approach to digging used by Kathleen Kenyon. Others, led by William G. Dever, launched an attack on the findings of Albright in favor of an "anthropological interpretation," which may be seen as a reaction in the extreme to Albright's assertions. Subsequently, Kathleen Kenyon's findings that refuted the biblical story of Jericho have now been questioned, calling her method into question.[22] Most scholars now favor Garstang's traditional understanding of the Jericho account.

Unfortunately, the battle over methodology and paradigms has clouded the important fact that *archeology has supported the general history of Israel, showing Wellhausen's history to have been a fabricated lie.* This does not prove all biblical history nor prove any theological position. But it shows the bias of men to reject the facts about God and a willingness to distort them.[23]

Conclusions: Many Unbelieving Scholars Rejected Wellhausen's History

Herman Wouk, the Hebrew German writer, asked the burning question,

> The puzzle today is how such a work [as *The Prolegomena*] ever captured, even for a few decades, a serious scholastic field. But the history of science shows that any vigorously asserted hypothesis can have a good run, in the absence of solid facts. The main thing, probably, was that in 1875 evolution was in the air. The battles over Darwin were still being fought, but it was obvious who was going to win. A theory that imposed evolution on Old Testament religion radiated chic and excitement, even though it stood the Bible on its head. Wellhausen's job of documentation, shrill and twisted though it was, lacking any scientific precision, nevertheless was

---

[20] G. Ernest Wright, "Is Glueck's Aim to Prove the Bible True?," *Biblical Archaeological Reader,* Vol. I (Cambridge, MA: American School of Oriental Research, 1978), 14-21.

[21] William F. Albright, *Archaeology of Palestine and the Bible,* (New York: Fleming H. Revel, 1935), 129.

[22] See Bryant G. Wood, *Biblical Archaeology Review,* "The Battle Over Jericho," March-April and September-October, 1990.

[23] Thomas W. Davis, "Faith and Archeology: A Brief History to the Present," *Biblical Archaeology Review,* March-April, 1993, Vol. 19, No. 2 gives a short inadequate survey.

overpowering in its sheer mass of minute scholarly detail.[24]

The bias of human sin was Wouk's answer as evidence revealed. By the 1960s and 70s many Old Testament scholars had abandoned Wellhausen's documentary theory. Umberto Cassuto, professor at the University of Florence, University of Rome, and Hebrew University, examined the five main presuppositions of the theory and concluded:

> I did not prove that the pillars were weak or that each one failed to give decisive support, but I established that they were not pillars at all, that they did not exist, that they were purely imaginary. In view of this, my final conclusion that the documentary hypothesis is null and void is justified.[25]

The strong questioning or outright rejection of the critical theory in favor of traditional history is also supported by scholars such as M.H. Segal, Yehezkel Kaufmann, K.A. Kitchen, Orlansky, et al.[26] While many scholars admit the weaknesses of Wellhausen's documentary hypothesis, they refuse to give it up because the only alternative is to go back to traditional biblical history, and it would be a disgrace to thousands of scholars who have taught it for years. Moreover, the dominant remaining alternative is the traditional Old Testament view of Israel's history, which holds men accountable to a Creator.

Biblical Christians should be cautious about claiming that there are no inconsistencies in the Bible, that there are no interpolations, that there were no oral traditions or source documents used by the authors. The Jewish Talmudic writers evaluated these matters and admitted them, as did Luther and Calvin and other Reformers. Vitringa, a godly and able biblical teacher, and others point out the evidence for the use of sources in the existing texts. If some fundamentalists have gone too far in pushing an absolute revelation that allows almost no human involvement, this does not invalidate holding to historical validity and the inspiration and revelation of the Bible. This question of biblical inspiration will be discussed more in the last part of this book. A view of human involvement based on the data discussed here strengthens the fact of the Bible as real history that can be accepted, rather than weakening it. It upholds divine inspiration.

The tragedy is that the liberal elite in the church and in educational circles of America and Europe have embraced the lies of Old Testament criticism that have removed a sense of accountability before God. These scholars have never admitted, and in some cases, never realized the lies and bias against God that they have taught. Yet, it is this "lie that there is no God who has revealed himself that damns a nation and civilization" because, if God has not revealed himself and his will, man is freed from any accountability at the future judgment. Remember, Wellhausen's attack was on the *law of God* as the center of the religion of Israel. Removing a sense of accountability to God removes the moral motivation that controls the sensuous and violent nature of man. Distorting and discrediting biblical history still confirms millions of people in their unbelief. This bias removes the sin of man and the purpose of God's grace in Christ.

---

[24] Barnhouse ref. to Wouk, *This Is My God*, "addendum," 186.

[25] Umberto Cassuto, *The Documentary Hypothesis,* (Jerusalem: Magnes Press, Hebrew University, first English edition, 1961), 100-101.

[26] Josh McDowell, *More Evidence that Demands a Verdict,* (San Bernardino, CA: Campus Crusade for Christ, 1975), Vol. II, 49-50, 167-169.

Subsequent Modern Views in Old Testament and Religion

In recent decades modern Old Testament and religious studies became much more restrained and scholarly. Speculation is less likely now to stray from broad tradition. H. H. Rowley said, "It will be seen that the movement towards more conservative views has been along more than one line...."[27] Historical accuracy of Old Testament documents was settled in the Dead Sea Scrolls.[28]

Since discrediting Wellhausen, archaeology began to consume more attention, along with epigraphy (the study of poems and sayings), and textual criticism. As archaeology produced new finds, these were compared with the Hebrew text, such as the Ugaritic psalms, etc. With the finding of new and older texts, such as the Dead Sea Scrolls, the study of the texts became more exact. The Dead Sea Isaiah Scroll, which is thought to be from the second century BC, has shown that there are fewer changes in the text of Old Testament Scriptures than critics thought. Historical information from the other Palestinian cultures led to comparisons of forms of religion. The former approach of setting the prophets against the priests of Israel gave way to seeing them as co-agents in worship. Many of the biblical texts have been interpreted by scholars as ritual texts for worship. The efforts to find sources in the Old Testament books gave way to searching for the oral tradition behind them. Form-criticism was applied to the Psalter and elsewhere. With a return to belief in central worship, critical scholars desire to discover the "psychology of inspiration."

Many scholars have not returned to a belief in the supernatural God of the Bible, and the old J, E, D, P documentary hypothesis has been maintained in altered but less dogmatic ways. Critical scholars are less confident in their attacks on the idea of God. But, educational institutions have established a worldview that all is controlled by natural laws which man can know and conquer as dogma. The bias against God's revelation in the Bible is entrenched in most modern schools, even though Wellhausen is greatly discredited. Some men, who claim to be scholars, continue to fabricate new stories to replace the Bible. Gary Greenberg's *101 Myths of the Bible* is but one example.[29]

Conclusions about Man's Creation of God

Comparative religions studies on the origin of religion that were actually scientific show that the early efforts to fit the data to natural law and then to the evolutionary theory were false and imaginary. J. Schmidt, using the scientific standards of the oldest ideas in a culture that were developed by the ethnologists and confirmed by archeologist, shows the story of man's evolutionary creation to be false. Wellhausen's efforts to link evolutionary development to a monotheistic God to the history of Israel were also shown to be completely fallacious. The concept of a completely human historic Jesus to fit the evolutionary development of the Jesus in the New Testament ended in failure as did the wildly ridiculous efforts to put together ideas claiming the church gathered men's claims to be God from sources wide in time and geography. This ended the dogmatic claim that Christianity was one of many religions to be put together in a syncretistic religion for men who needed religion. The death of God was widely proclaimed. This was a desperate effort to fulfill man's bias to remove God.

But as true scholarship and scientific research showed, these *evolutionary views* ***were make-believe myths***, and the real truth about the history of God in the world is again confirmed as that which is in the Bible (cf. Romans 1:19-23). Paul said that all men began by knowing God, and

---

[27] H.H. Rowley, *Old Testament and Modern Study*, XXX, see the whole introduction, XIV-XXXI.
[28] John A. T. Robinson, Can we trust the New Testament? 35, et al.
[29] Gary Greenberg, *101 Myths of The Bible,* (Naperville, IL: Sourcebooks, 2002).

being prosperous they were not thankful and turned to worship the material creation. When man no longer trusts God, his dependence is on materialism. The Bible gives detailed historical data throughout with names, genealogies, historical events, and places about God's working. Even in the Koran, Islam has obviously borrowed from biblical history because it does not have detailed historical data and accuracy of the biblical events. These perversions in religious studies of sinful men are like Adam and his descendents who have wanted to be wise as God. Their efforts are ending like all other histories of human cultures that started with a knowledge of God and rejected him as discovered in scientific history. The approaching self-inflicted end by man's sinful bias is apparently to be in modern times like all other civilizations as will be seen in the trends of the soft sciences of economics, psychology, sociology and political history leading to today.

Chapter 31
Enlightenment to Scientific History And Divine Judgment in Conflict

Introduction

Review of the Evolutionary Deterministic Concept of History

For the intellectual elite, the purpose of nature and the history of man is material progress. This is based on the belief that life has, by chance, evolved from a single cell generated by nature, by chance, and produced man. Man has evolved from primate, to cave man, to civilized man. Enlightened intellectuals believe that mankind will naturally continue to produce a more brilliant, more comfortable, freer, and more pleasurable culture with more humane attitudes and actions toward each other. This vision of a new and wonderful world in man's improved understanding is the hope of the educated and enlightened. The idea of God (which is said to be myth) caused man to hope in the supernatural rather than in his own efforts. Therefore, the idea of God must be removed as an ignorant deterrent. This book shows that their destructive vision is based, not on scientific fact, but on bias. Yet, it drives modern intellectuals. Having become committed to evolutionary theory in the nebula hypothesis by the early twentieth century, they have rewritten the whole history of man using this hypothesis. A scientific study of history was made using their view of scientific natural determinism.

Scientific History to Help Guide Humanity

Scholars inevitably applied the *scientific approach to history*. R. G. Collingwood, in his book, *The Idea of History*,[1] deals mainly with the idea of scientific history. In the pursuit of deterministic science, the methodology narrowed to pure positivistic methodology. Scientific historians tried to expand knowledge to include historic knowledge, but "under the shadow of positivism."[2] In the first half of the twentieth century, the new, more empirical approach to history was supposed to furnish better and more definitive answers to problems from a much longer, more realistic view of man's history than ever before, and produce better progress. But it aimed at looking more at the specific objective facts left by man as a scientist would, and then seeking to see what they say.

Some of the most significant scientific historians were Oswald Spengler of Germany, the Russian-born historical sociologist, Pitirim Sorokin who taught at Harvard, and Arnold Toynbee, the great English comprehensive historian from Cambridge. I will refer to some of the other scientific efforts, but I have had to select some of the greatest to see their main discovery and emphasis.

Various Scientific Views of Scientific History

Oswald Spengler

Spengler saw history as a ***succession of self-contained individual*** units, each with its own character and each civilization as going through the same pattern: birth, youthful strength, full maturity, old age and death. This pattern repeated itself in a cyclic manner like earlier views of Greece, the Orient, etc. Spengler's view was very ***deterministic*** in its outcome, because, after all,

---

[1] R.G. Collingwood, *The Idea of History,* (Oxford University Press: A Galaxy Book, 1956).

[2] Ibid. 134.

science shows that man is a part of a strictly regulated naturalistic order. Spengler was not thorough and complete, and at times dealt superficially with data to force it into his interpretation.[3] The finding of a uniform repetitive pattern was in agreement with a determined destiny (not determinism in knowledge as the theory).

Arnold Toynbee

But Arnold Toynbee was ***extensive, detailed, and comprehensive*** while still being ***analytic***. More than all historians before him, Toynbee reviewed all the data more with a view to evaluating it for repetitive patterns. At the time, the early history of certain parts of the world in Asia, Polynesia, Africa and South America, was not understood as it is today, but he reviewed most of the known civilizations. He saw civilizations repeatedly arising from a committed spiritual minority that influenced a broader growing covenant-type group to accept their way of life, which in turn influenced the majority of the populace. The power of this minority grew and generated a creative and productive influence from which the civilization emerged.

Then, after achieving dominance, the committed elite, who had produced the moral power, lost it through the growth of pride and greed. When their views dominated, the originating minority became arrogant and rich and their morals declined. In place of moral example they imposed a legal moralism against which men rebelled. Feeling superior and without need, they lost a sense of accountability to anyone but themselves. Still professing to follow God, they became hypocritical, hiding behind their legalism. Their position, though still dominant, began to give way to what Toynbee called, "original sin." This led to a time of trouble involving dissent and war, within and without. The pax shifted from a moral intrinsic motivation to self-privilege and extrinsic police enforcement, and brought the civilization down. He traced this pattern in over twenty.[4] Gordon H. Clark and others did not agree with Toynbee's interpretation and rejected the idea of validity to any "scientific history."

Pitirim Sorokin

If Toynbee was extensive, then Pitirim Sorokin of Harvard, the great historical sociologist, was ***intensive***. He sent teams of students all over the world, into the museums, libraries, art galleries, etc., to count the various motifs in the art, literature, and music of each historic period. He evaluated this data, charting the course of history for each civilization. Sorokin saw civilizations going through three major phases. The civilization began with an ***ideational*** phase of spiritual vision and commitment, moved to an ***idealistic*** (or *rationalistic)* phase of great productivity and growth, and finally moved into a ***sensate*** phase where spirit and reason gave way to lust, and lost its creativity, succumbing to disintegration and violence.[5]

Comparison of the Scientific Historian

Spengler's pessimism and determinism, although less accurate than later scientific historians, hung like a shadow over others' scientific approaches. Although these historians differed greatly in how they stated their views, they were extremely similar in their evaluation of the various stages of history, and in their pessimism. Spengler, with definite repeated phases or cycles, tried to escape the accusation of determinism and pessimism, but with little success. Toynbee denied

---

[3] Oswald Spengler, *The Decline of the West,* 2 Vols. trans. Charles Francis Atkinson, (New York: Alfred A. Knopf, 1922).

[4] Arnold Toynbee, *Civilization on Trial,* 1948 and *The World and the West,* (Oxford: University Press, 1953).

[5] P.A. Sorokin, *The Crisis of Our Age: The Social and Cultural Outlook* (New York: E. P. Dutton, 1941).

determinism and held that history could be linear, or going somewhere, like earlier Christian historians had proposed (e.g. Augustine). But he saw this linear view as also having a cyclic nature—like one riding a bicycle in a straight line. Toynbee believed that the death of Western civilization would lead to a new form of more empirical faith.

Sorokin also denied determinism. He felt the sensate culture had not yet merged with the idealistic one and that a new form of ideational society could emerge before the culture was completely destroyed. He optimistically hoped it would be "a creative altruism." Harold O. J. Brown has tried to apply Sorokin's theories to produce an optimism in American renewal. But he recognized that "our culture and civilization ... inevitably will (end) if certain present trends are prolonged to their logical conclusion and more promising ones are choked off or suppressed before they have a chance to bear fruit."[6] Since he wrote a decade ago, the process has accelerated to a near crisis, based on the predictions of J. D. Unwin.

Conclusions of These Prominent Scientific Historians

These major voices of scientific history (and there are others), all found a similar pattern: an optimistic beginning, with human greed and lust resulting in ***a time of trouble***. While there were efforts to find an escape from such a pessimistic view of history, even those hoped-for solutions were seen to involve terrible times of trouble.

One Indian historian, Paul Sarkar, has tried to escape this pessimism by a unique approach. Instead of starting the cycle or pattern with the spiritual emergence and growth of society, he *begins with the last step–the disintegration and warring of society*, and thereby is able to end with man in a glorious intellectual phase. This would be similar to starting the story of an accused convict with the execution of the prisoner, then moving to his more innocent childhood and development, and showing his good points in his adulthood before he became fully immersed in crime. The story is more pleasant to read and has some merit for evaluation at points, but not realistic as to cause and effects for the emergence of a productive civilization. Sarkar sees an era of Warriors, an era of Intellectuals, and an era of Acquisitors with each going through five steps of a cycle culminating in turmoil similar to Spengler. For Sarkar, the era of Warriors is really the end of the last stage of the civilization's decline, which produces a strong patristic spiritual leader, along with the beginning of the first step of warriors. His dividing lines are thus different. When seen as a whole, Sarkar does not escape the pessimism. He really confirms it.

Apostle Paul Outlined Same Path to Decadence as Presented by Science

In the first six chapters of *Man is Not Enough* I have presented the biblical teaching given by the Apostle Paul in Romans 1:18-2:3 on the pattern of the decline of civilizations with the Greek Democracy and the Roman Republic in view. The last chapters trace how this has been the pattern of America since the founding of the colonies in the 16th century to today. The scientific historians demonstrated the fact that civilizations rise to greatness and then decline into conflict and defeat. It is important to see that the steps given by Paul, as I diagramed them, have been the same for all civilizations, as far as we know. Paul traces the growth of civilization through a series of steps of degradation that suppress knowledge of God and foolishly follow man's bias to the flesh under demonic deception. That graph from *Man Is Not Enough* (page 14) along with discussion on these steps of sin that destroy a nation is given again below.

---

[6] Harold O. J. Brown, *The Sensate Culture,* (Dallas, TX: Word Publishing, 1996). This book came out too recently for me to incorporate many of his valuable insights.

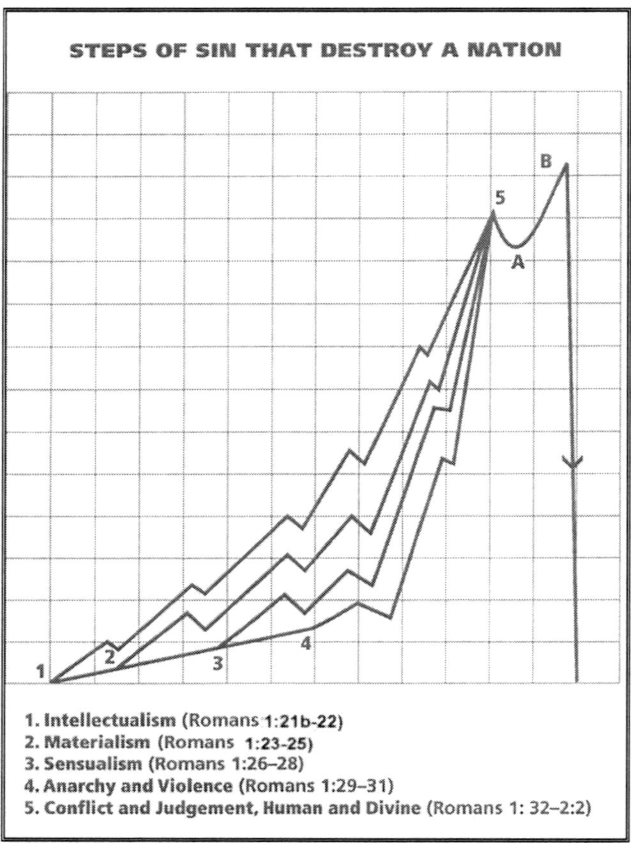

Man Knows God from Creation

God imparted to the mind/spirit the ability to see evidence of God in the order of created nature, revealing his power and sovereignty/deity and to know his moral law in the conscience. The scientific study of comparative religions in the last chapter confirms Paul's original view of God.

The German ethnologists developed scientific canons for determining the oldest elements in a culture. Later W. Schmidt, the authority on comparative religions, applied these to religious ideas in the culture. He found "the universal distribution of the worship of a Supreme Being over the whole earth." "He is the lord of man's history… at the beginning of each human life… awaits its appearance before his tribunal and determines the nature of its eternity." [7] These findings confirm the statements of Paul, and Schmidt said that the claim for the evolution of the idea of God was wrong.

But after all his study, Schmidt was baffled because he did his research from a deterministic view of science, believing man began with a blank slate on which he accumulated all knowledge from his experience. For Schmidt, this idea of one God had to be derived from experience. He

[7] W. Schmidt, *The Origin and Growth of Religion: Facts and Theories*, 184.

said, "For such a unity as this primitive man has no model and no corroboration in what he sees and experiences in his own time."[8] Schmidt never came to faith and trust in this God, and struggled to explain how man begins with a monotheistic God. In the previous chapter we saw that Schmidt's findings were confirmed by archeological studies of many nations and that the faith in one Supreme Sky God degenerated into pluralism and nature worship.

Schmidt's observations showed that such degeneration is in accordance with the steps in the graph from Paul in Romans 1and 2. At this point, it is important to trace the steps of some of these civilizations into sin, degeneration, and destruction as submitted by Paul and confirmed by archeology.

Review of Pattern of Rise and then Steps of Decline in Some of Many Significant Cultures

Early Sumerian Civilization

The earliest Sumerian period in the Babylonian area northeast of present Baghdad leaves enough information to see the pattern. The earliest Sumerian texts, written around 3000 BC, reveal the history of creation and the flood very much like the biblical account. The events and writing also disclose the confusion of tongues prior to this time of writing. The early dynasties reflected in the Kings lists, that are said to be "not much more than 500 years," tell of very long lives and reigns. The city states were united in the worship of the one supreme sky God. At first, city states had their own names for Him (Nanna for Ur, Inanna for Uruk and Enlil for Nippur). An or Annu was the Father sky God in two of the names, but focused on worship of the son, Enlil (meaning "ruler god"). An Archaic Temple was built to Enlil and was the center of worship for the cities and was the dominant institution in the early days. The cities united together and there ended up with a primitive congress of two houses. Evidence of writing, engineering, domestication of plants and animals appear to make this a prosperous civilization. This union under God appears to have lasted through the first dynasty or nearly 200 years.

Prosperity soon led to human intellectual confidence. As humans came to understand astronomy and regularity in nature, they also developed a religious belief in a naturalistic pantheistic god. Materialism became the main focus. A trade network became the single supervisory body for the cities. Material prosperity made for further cooperation, resulting in the construction of a storage facility containing materials to which the various cities had access. With this material pursuit, a women's movement emerged. In Lagash, the best-documented state, a fundamental change in the central administration. The *e-mi*, the household of the wife, was renamed *e-Bau*, the household of the goddess *Bau*, wife of Nirgirsu. The Archaic Temple was redesigned and rebuilt into a square with idolatrous pluralism to Abu and his goddess wife. A royal class developed. Wealth and chariots were plentiful. Amazing jewelry was found buried with the corpses. In the later dynasties, there was the temple for the goddess Ninhursag built by Mes-Anne-pado, a dictator who declared himself god. There was a big bureaucracy and high taxes. After about 500 years the civilization had internal problems and there was an attack with weapons of war.

A similar cycle followed with the rise of a culture that shared leadership with Semites and Sumerians. A leader, Uukaina overthrew the government at Lagash, established a return to one god, Enlil and a covenant for justice and freedom. The leading city was changed and finally leadership was located at Ur. Pantheism emerged and the worship of Nidaba; the goddess of writing and wisdom was exalted, focusing on education and materialism as the pursuits. In time

---

[8] Ibid.

Narim-Sinm, grandson of Sargon continued the women's movement, making his daughters priestesses. He proclaimed himself god. Under the wealth and amazing culture, thousands of foreigners immigrated. Finally, under Shulgi a pyramid tower of great proportions was built; another tower was built to women. The dictator "proclaimed himself 'the divine Shulgi, god of his land'." Ur fell in the same manner.

The Mesopotamian area then fell to the leadership of the Amorites. The Amorites probably had four major centers, Mori, Ugarit, Ebla, and finally Canaan. They began with a culture of worship of God and moral integrity giving rise of to great civilization which declined through the steps of degradation. Abram/Abraham was called out Ur just before its fall, sojourned among the Amorites and other Canaanites and was told by God that the promise to inherit the land would not occur for 400 years until "the iniquity of the Amorites was full" (Genesis 15:16).

Early Egyptian Civilizations

The civilizations in Egypt also began about 3000 BC and went through similar cycles. After a radical change in climate in Egypt, the Thinites gained a central leadership and established a government at the border of upper and lower Egypt below Thebes. At that time the Egyptians worshipped the great sovereign sky god, Horus. The symbol of the Hawk with outstretched wings supported by invisible force or wind represented the Father God, and He ruled through Re, or light represented by the sun. Commerce, ship building and other aspects of trade emerged, leading to great prosperity. After about 200 years, astronomy and engineering developed, and the culture shifted to human wisdom, pantheism and control of man. Imhotep, a man of brilliance, emerged and directed education and progress. The worship of Hathor, the women's fertility goddess, entered the worship along with other gods, including the cult of the snake or Cobra. Foreigners began to immigrate and even later gained leadership in cities. The leaders began to build the great buildings and pyramids. The great pharaohs, Khufu, Khafre, and Menkaure built the great pyramid monuments to themselves and claimed they were gods. Finally there was a rebellion by the nobles and the kingdom declined from the fifth dynasty.

A second cycle began after unification in the 11th dynasty with an agreed support of the city-area nobles to the Theban king, Nebhepetre Mentuhotpe. This is called the Middle Kingdom when rule was mainly at Memphis. There was a return to the worship of the god of light, Re. The needs of the poor and rich people were attended to, new acres were converted to irrigation, canals were built, trade revived, art and study began. They conquered Nubida and consolidated it and connected the Nile to the Red Sea for commerce. After two hundred years naturalism took over, a pluralism of gods emerged, and the worship of Osiris, the lord of the underworld became popular. Hathor, the sexual goddess, conducted the dead to the underworld. Sexual lust was centered in this cult. Women's jewelry became fashionable and entertainment and literature flourished. Asiatic people brought their entertainment, and eye cosmetics. Semitic immigrants began to occupy cities and even gain control of some. By the fifteenth dynasty or after 400 years the Hyksos, who were Semitic/Asiatic like the children of Israel, invaded and took over with little opposition.

In 1540, Ahmose led the Egyptian nobles to rebel and reestablish Egyptian rule, expelling the Hyksos and subjecting all Semitic people to slavery, including the Israelites and. He tried to reestablish a monotheism, calling God Amon/Amen-Re. Amen is Semitic for *truth* or unchangeable. But the pluralism was too pervasive and the degradation continued. The women's movement reached a peak with Hatshepsut who was of royal linage and achieved power as Pharaoh. She probably adopted Moses to prevent Thutmose III, the son of her husband's concubine, from gaining the throne. It is thought that Thutmose assassinated her and sought to remove her name from the inscriptions from the monuments. He oppressed the Hebrews and Moses fled from him. The exodus pharaoh was Amenhotep II, who worshiped the sexual cow goddess, Hathor, was a pervert and a hard hearted dictator. During the reign of Thutmose IV, the Egyptian army was rebuilding and he made no invasions into Palestine. At that time, the Israelites were in the wilderness and at the end of their wilderness wanderings began conquering the Canaanite cities (reflected in the Tell El-Amarna Tablets). In fear, the new Pharaoh, Akhenaton of Egypt restored monotheistic worship of Re the early God of light and removed all other pagan deities. God, as King of Israel, commanded the Israelites under Joshua to obliterate the terrible idolatrous Amorites whose sin had reached the full.

These early civilizations in Mesopotamia and Egypt and other civilizations moved from belief in a sovereign God to belief in the sovereignty of the individual through the steps outlined by Paul in Romans. In chapter four of *Man Is Not Enough*, I give in detail the way the Greek democracy arose under faith in the God of Heaven, Uranus and his son Zeus, begun with a feast to the one God on Mount Olympus. It then went through these steps of decline and ended with the dictator Alexander the Great who practiced sex with women and men and declared himself God. In chapter five I detail the rise and decline of the Roman Republic. It began with a dedication to Uranus and his son, Jupiter, and ended about five hundred years later with the dictator Julius Caesar who also practiced sex with women and men and declared himself God. Most of the book, *Man is Not Enough,* traces the pattern that has been followed in America.

In a shorter book, *Liberty in an Evil Age,* Alexis de Tocqueville's map of democracy showed that France followed the same pattern. He observed that the thing that caused American democracy to continue to work was the consensus of faith in a Christian civil religion. The following were his observations about why French democracy was forced by unrestrained desires to despotism. [9]

Map of Decline of France Compared to Conditions in United States

Educated Elite Claim Just and Equal Changes

According to Alexis de Tocqueville, the educated elite in France lead the efforts for freedom of equality. De Tocqueville shows the promotion of revolution for freedom for equality was sponsored by the "most civilized," educated elite of the nation. They were the Ivy League of that day; they treated all religion as antiquated and unenlightened. A growing skepticism about God and a very high view of man, especially themselves, preceded their efforts. The graph of Paul's description of the steps of decline, and an *intellectual revolt is the initiating point* or driving growing concept as the first step.

De Tocqueville repeatedly mentions the growing and vicious attack on Christianity and the churches in general. By the eighteenth century "Christianity has lost most of its hold on men's minds," "Skepticism was in fashion in the royal courts," but "the middle class and the masses were impervious to it." The leader's hostility to religion "had a preponderant influence..." and "was in fact, its most salient characteristic, and nothing did more to shock the contemporary observers."[10] All history of nations reveal the effort to remove God and to remove man's moral failures.

---

[9] Carl Wilson, *Liberty in an Evil Age,* (Longwood, FL: Xulon Press, 2008) for quotes on Tocqueville in chapter 6, 161-170.

[10] Carl W. Wilson, *Liberty in an Evil Age,* (Longwood, Florida, Xulon Press, 2008), see my discussion of de Tocqueville, pages 162-169.

Elite Motivated Poorer People by Uncontrollable Natural Passions

To gain power, the elite attacked the church and the aristocratic class of which they were a part and in which they participated. But the elite, with all the privileges they enjoyed, appealed to and carried out the revolution by motivating the most uneducated and unruly elements.[4] The elite promised by their knowledge to satisfy the desires of the oppressed for freedom of equality. Ironically, the poorer are motivated to want the things that they hate others for having. Once started, the desires created were uncontrolled. "Our contemporaries are ***driven on by a force,*** *that we may hope to regulate or curb, but cannot overcome*, and it is ***a force impelling*** them, sometimes gently, sometimes out of headlong speed, to the destruction of the aristocracy." "More than any other kind of regime, it fosters the growth of all the vices to which they are congenitally prone, and indeed, incites them to go still farther on the way to which their natural bent inclines them." Money has become the "sole criteria of a man's social status," there is "love of gain, fondness for business concerns, the desire to get rich at all costs, a craving for material comforts." He points out that these ruling passions affect all classes and "tend to lower the moral standards of the nation as a whole…."[11] (emphasis added) De Tocqueville sees revolution for freedom of equality in democracy driven by inclinations of human nature that, once incited, could not be controlled.

Elite's Strong Self-Confidence of Attaining Good by Their Will and Insights

The elite also displayed a *confident expression of superior wisdom* in the individual human abilities of the leaders. De Tocqueville was a product of the secular humanistic movement of the liberal Enlightenment. He saw the individualistic humanism that marked the leaders who headed the movement as one of its strengths. They were cocky about the certainty of their cause for the poor. He recognized a *contrast of their character to the Christian man.* He said,

> For though the men who made the Revolution were more skeptical than our contemporaries as regards the Christian verities, they had anyhow *one belief,* and an admirable one, that we today have not: they *believed in themselves*. Firmly convinced of the perfectibility of man, they had faith in his innate virtue, placed him on a pedestal, and set no bounds to their devotion to his cause. They had that arrogant self-confidence which often points the way to disaster yet, lacking which, a nation can but relapse into a servile state. In short, they had a fanatical faith in their vocation—that of transforming the social system, root and branch, and regenerating the whole human race. Of this passionate idealism was born what was *in fact a new religion*, giving rise to some of those vast changes in human conduct that religion has produced in other ages.[12] [Emphasis added]

In France, the elite were sure man would act right if just released from oppression. In the Greek decline of democracy, Aristotle commented on this characteristic of youth leaders, and this was certainly true of the youth involved from the Gracchi revolts in Rome that reached a peak in about 44 BC. Sallust, Julius Caesar and others also mention the arrogance of Roman youth. In America, the bold confidence was the attraction to the youth of the SDS, the Black Panthers and educational leaders and even others today. They are committed to social ***change*** in a sort of bold dedication of self to faith and hope in man's wisdom. In an age in which confusion is growing in the economy, sexual psychology, sociology and politics, they claim to have a direction. But it has

[11] Alexis de Tocqueville, *The Old Regime and the French Revolution,* translator, Stuart Gilbert (New York: A Doubleday Anchor Book, 1955) XII.

[12] Ibid., 156.

been their liberal insights that have caused this growing confusion.

As representative governments have, by economic prosperity, yielded to socialism and then to chaos of anarchy, in past civilizations the youth have always stepped forward to speak and lead when the older generation has no sense of direction. Moreover, the elderly were increasingly blamed. These confident youth made bold general promises, but de Tocqueville mentioned that they contained little specifics. Youth had moved more into prominence as degeneration of the societies progressed. After attacking God and pursuing greed, lust, and anarchy, this led (as previously mentioned) to dictators such as Alexander the Great at the climax of Greece, Julius Caesar and Augustus at the end of the Roman Republic, and Napoleon of France. These were examples of relatively young men of confidence that took despotic control. But in their efforts to bring order, liberty in democracy was lost and the lives of millions of young men and the people who followed them were sacrificed. Their despotic rule *ended people's abilities to govern* for themselves.

This self-assurance of right and good had major flaws in leading. Rush to equality was in conflict with liberty. Liberty was lost in efforts to change, change, change! In the ideas and aspirations for revolution "the concept and desire of political liberty, in the full sense of the term, were the last to emerge, as they were also the first to pass away." They could not reconcile freedom with equality when each was seeking more for self.[13]

Today, as in these former civilizations, the older generation has the wealth. The corporations, educators, and government are led mostly by white men. For the brilliant elite, their control must be broken and transferred to those who want equality. The capitalist system must be destroyed by taxes for the benefits of the poor to gain freedom, the male leadership of the family must be ended for equality of single mothers, the conservative government and military must be weakened. The elite feel they are the only ones qualified and will seek to end old conservative areas of communications. These claims would be like reading the newspapers of Paris and the anger vented against the aristocrats. Of course, there is truth of the greed of the rich and powerful, but the destruction of the system by the young and inexperienced has never produced better. Only the knowledge of true justice by faith in God and His will can do that.

Isolation and Loss of Cohesion and Community to Individualism

One of the reasons liberty failed was that selfish individualism produced increasing isolation. Intellectual leaders exchanged the sovereignty of a good God for the sovereignty of self-righteous greedy individuals. The concept was freedom for equality based on ***self***-esteem to gain wealth and pleasure—this was the passionate driving force motivating the movement toward socialism that had no power of cohesion, but only the power to isolate. De Tocqueville observed,

> For in a community in which the ties of family, of cast of class, and craft fraternities (labor unions) no longer exist, people are far more disposed to think exclusively of their own interests, to become self seekers practicing a narrow individualism and caring nothing for the public good.[14]

He observed that no ten or even two people could stand together without appealing to the government. Self interest destroyed mutual trust. Self confidence to achieve yielded to confidence in the state to help ~~as~~ victims. Karl Marx said about France, "One also had to convert the

---

[13] Carl. W. Wilson, *Liberty...*, see my discussion on p. 167, 168.

[14] Alexis de Tocqueville, *The Old Regime and the French Revolution*. XII.

discussion of material progress into a discussion of social cohesion. Socialists seem to have argued that *liberal democracy does not permit* ***social cohesion***."[15] (Emphasis added) In accentuating self interest with self worth based on material gain, this intensification of individual selfishness alienated toward individual aloneness.

De Tocqueville emphasized that this individualism lowered and destroyed moral standards. The pursuit of liberal democracy, with the emphasis on enhancing individual material self worth, not only did not permit social cohesion, it was the very thing that isolated and destroyed all relationships and democratic control. Thus, in pursuing equality for every individual, they destroyed that which allowed people to be a meaningful society with freedom toward accepting each other and for allegiance to the nation. All morals disintegrated and marriage became almost meaningless. The same loss of morality occurred in Germany with the collapse of the democracy before Hitler.[16]

De Tocqueville was a liberal of the Enlightenment and even then when little was known about comparative religion, while not being anti-religious he believed that God and the various religions were the creations of men. In spite of the destruction of cohesion, he believed that ***somehow*** *there could be a rational solution found* to reconcile freedom and equality. He saw ***the answer*** as found in *human reason based on simple natural principles.* But De Tocqueville, like Christopher Lasch, in seeking rules to control the economy, did not say what these were. All history showed that human individualistic desires for gain had caused passions that destroyed democracy and equality. De Tocqueville even admitted these forces had not been controllable. He believed that the idea of democracy is ultimately threatened with either *tyranny of the majority*, or *despotism.* The philosophy of democratic government needed some power to give cohesion with equal concern for others by the ruling majority or they would lose liberty to despotism. The intellectual forces and a weakened and divided Christianity have excluded the cohesive force of a civil religion of a common Christianity.

Conclusions about the Pattern of Historical Decline

About 150 civilizations have been found; about 20 to 25 are considered major. Unwin studied the feminist movements in 86 civilizations and found that after sexual anarchy through feminism gained dominance, they all declined. Toynbee observed that the time of troubles was accompanied by major class conflicts and then often by an enemy gaining conquest. Great materialism and sexual pleasure attracted many immigrants that then joined the other ethnic poor in revolt. Selfish individualism seeking the survival of the fittest produced anarchy and God gave the civilization up to judgment.

With all values being material and in the hands of males, women's value as wives and especially as mothers was degraded and produced a sexual revolt. Without the acceptance of equal value of men and women for oneness in marriage, individual competition is the only way for women to be equal and free. The revolt of women and disparagement and rejection of children always transformed patriarchy from kindness to harshness, producing conflict and sexual immorality. Self-centered individualism becomes unrestrained, destroying society. Pitirim Sorokin of Harvard described this:

---

[15] Wm. B. Allen, *Radical Challenges to Liberal Democracy,* Wikipedia free encyclopedia on Marx's observations.

[16] Otto Friedrich, *Before the Deluge*, (New York: Avon Books, 1972.) in Chapter Vii, "A Kind of Madness."

Unwin finds that the Babylonians, the Egyptians, the Athenians, the Romans, the early Arabians, and the Anglo-Saxons had a strict monogamy during the early period of their social expansion and cultural and intellectual growth. The authority of the pater families over the members of his family, and the husband over his wife was unlimited. Sexual life was confined within marriage, and the morals were chaste and temperate. Violations of the prescribed rule of conduct did occur now and then, of course, but they were few, and were unanimously disapproved and severely punished. These limitations of sexual activity permitted such societies to accumulate an enormous reserve of social energy, which found its outlet in creative growth–intellectual, aesthetic, religious and social. Hence there occurred a vigorous expansion of these societies, accompanied by an astounding ability to defend themselves against their enemies. With the expansion and growth of these societies, however, the stern regulations of sex relationships were progressively replaced by weaker ones. Sexual freedom widened until it encompassed the whole society, and eventually turned into anarchy. Wives and children were emancipated from the absolute power of the *pater families*, and their newly won equality brought with it sexual freedom. Within three generations from the moment of significant expansion of sexual freedom, the culture and creativity of these societies began to decline.[17] (Emphasis added)

My observation is that all of these changes occur after man accepts a mechanical material view of nature and the changes progress slowly over two to three hundred years. The only way to plain this is a demonic deception that promotes a fleshly bias. The rejection of faith in God and moral accountability is slow and unperceived, but desired by the people. The selfish individualism then extends that freedom to demand that government devote itself to supplying the desires of the individual instead of control, and thus anarchy slowly occurs. This selfish individualism results in living together outside of marriage, children left as individuals, and finally sexually seeking to make homosexuality normal. Homosexuality, by definition, involves rejection of the opposite sex for the same sex. But history shows that in only a small number of cases, is there any long-term accountability between such people who by definition cannot produce new individual children.

The critical issue—the diamond pivot point for civilization, is faith in one sovereign God. Exchange God for the demonic deception that nature runs as uniform natural laws that have produced man as the highest reigning creature without any power beyond himself, and this world. This view ushers man down the other steps to greed, lust, conflict, misery and judgment. Man is continually motivated by his guilt and resists returning to God and his divine control. The selfish individualism that produces this anarchy almost always results in a rejection of linear history and the past as an extension of its rejection of God. Past ethical standards and heroic actions showed greater moral courage and produce guilt. The result is to transform real history into an imaginary cycle.

Rejection of Scientific Historiography Crucifies Human Wisdom and Lust

The modern educated elite have never embraced scientific history. It was much too hard to objectively analyze all the facts of history since it lacked clearly repeatable, extensive data. Many other factors were unknown and somewhat subjective. But the real reason was that the modern educated mind was so *fully convinced of the validity of human wisdom and* ***scientific progress*** and evolution toward greater prosperity and security, that it *could not accept the united* ***pessimism of***

[17] Pitirim Sorokin, *The American Sex Revolution,* (Boston: P. Sargent, 1956), 111, 112. Cf. also my book, *Our Dance Has Turned to Death,* (first Renewal Publishing, 1976; then Tyndale House, still available from Christian Growth Books, P. O. Box 621412, Oviedo, Fl., 32762). This book has been quoted by *Parade Magazine* and a number of newspapers as an effective appeal against feminism and for a return to kind patriarchy that recognizes equality of the sexes before God.

scientific history. The *liberal optimism of human nature* was that man would make errors, but would improve as he understood his world more fully. This optimism *runs contrary to the conclusions of all scientific history*. It was certainly contrary to the idea of the survival of the fittest of evolutionary biology. Scientific history had to be rejected because it ran contrary to all modern education, all political aspirations, and all business ambitions. The modern view of evolutionary man does not fit with the findings of the scientific historians. In the scientific histories, man, the king, is dethroned because he could not rule himself. In short, scientific, good, socialistic *progress, and the pessimistic outcome of "scientific history" were* ***diametrically opposed.*** But the very fact that most of the scientific historians found data contrary to the evolutionary optimism of the thinking of the time is clear evidence of their general objectivity and efforts to present the truth. It discloses the myth of evolutionary progress in history, as the earlier review of cultural development revealed and the review of the rise and fall of the early civilizations in Mesopotamia, Egypt, Greece, and Rome.

The Myth of Historical Progress in the Long Run

The fact is that *the modern view of man as improving* ***is the real myth***. Modern archaeological findings have shown the view that "only modern men are intellectually astute" is *now known to be the biggest lie of all.* Early civilizations have been found to be exceedingly brilliant in the things they invented and created; some of their feats of engineering and science are not yet understood by modern man. Any knowledgeable archaeologist now knows that the earliest known civilizations were all exceedingly brilliant: Sumer, Assyria, Babylon, Egypt, Ebla, and others. Fragments of supposed cranial and other bones and the skimpy findings of so-called prehistoric man can't judge man's past intelligence. Man's history is the only place his previous accomplishments can be judged scientifically, and the repeated story of ancient historical sites shows amazing intellectual and inventive genius of mankind everywhere. But today, few archaeologists and anthropologists are really willing to fight the accepted view of progress held by the educated elite. The peer pressure of intellectual establishments is too strong.

Modern man in his pride has been able to delude himself into thinking he is smarter than those who came before him, because each successive civilization has always used the intellectual ideas of the previous civilizations to build on. Thus, in comparison, the successive civilizations seem much more brilliant. By standing on the shoulders of previous civilizations, the one on the top—now modern man—seems by far bigger than any before. By denying the existence of previous brilliant human civilizations, modern man deludes himself and propagates a lie to his children who in turn carry on the delusion.[18]

Human Bias Must Forsake History Even as a Story Process

But modern popular historians have gone beyond replacing the lie of evolutionary progress for the truth about previous human brilliance. Having propounded the dogma that man has progressively gotten more brilliant and more humane; the historian has forsaken history as such and is *trying instead to find aspects of history to be a major contributor to progress.*

In the mid-twentieth century, thousands entered the study of history and the number of Ph.D.s in the field swelled rapidly because many young men and women wanted to contribute by helping design and chart future progress. Most did not want to teach or recount history, but desired to do research so they could contribute to making progress happen. But this surge in historical investigation produced evidence that did not show repeated progress, but rather produced evidence

[18] Wilson, *Man Is Not Enough,* chapters 2 and 3 ff.

to the contrary.

Hence, thousands of these historians turned historical research into merely looking for *bits and pieces of sociological clues* to help produce progress. They *claimed* this was an effort at *practical application of history*. This diversion only produced greater fragmentation in historical research and has led to the abandonment of the pursuit of real history. So, in addition to replacing the pessimistic outcome of true history with a lie claiming progress, many historians ***have forsaken the search for history as story***, presenting it only in bits and pieces for pragmatic use.

Leading historians warn of the dangers of the loss of real history. Bernard A Weisberger, one of the leading members of the Society of American Historians who has been a professor at the Universities of Chicago, Rochester, and Vassar College, has written in *American Heritage* magazine about this chaotic condition in history.[19] While he spoke specifically about the demise in teaching American history, he applies what has happened to history in general.

Until the mid-twentieth century, the great historians all believed there was a whole story to tell, even though they quarreled over it. Weisberger said,

> They were united in believing [history] had direction. There was a story that made sense.... Parts added up to a whole adventure in which everyone shared. For the past forty years academically training historians have been picking apart this fabric. But nothing has been left in its place. That one learns virtue and character from the lives of the ancients is also gone.[20]

Weisberger warns, "The sense of the wholeness of history has to be restored and passed on, whatever that may take... It is that sense of wholeness that makes for good judgment, for good ground for a nation to stand upon."[21] Children are being denied any sense of values as ground to stand on to make judgments.

He then concludes,

> Surely, those who are involved in history in any way can do better. If they accept this state of affairs, they lose touch with the unifying sense of a shared past. And then they have abandoned themselves and their fellow citizens to a world with neither joy, nor love, nor light, nor certitude, nor peace, nor help for pain.... It may be intellectually fashionable to talk of living in a post historic age, but it is a bit like Noah's family commenting that they seem to be in for an extended period of wet weather. Unless we can remember how to build an ark, we're going under.[22]

Modern man desperately needs to return to history and take it seriously. While it threatens his arrogant view of progress, it alone could save his corporate life. Community, patriotism and civility are dying.

Imaginary History Founded on Bias is Substituted for Truth

So in substance, the Western intellectual elite set scientific history aside, and substituted new

---

[19] Bernard A. Weisberger, "American History is Falling Down," *American Heritage, Feb/Mar 1987, 26-32.* For more about "historic skepticism" see Hans Meyerhoff, editor, *The Philosophy of History in Our Time,* (New York: Doubleday Anchor Books, 1959), 28-86 on Dilthey, Croce, Collingwood and others. Cf. also Patrick Gardiner, editor, Theories of History (The Free Press of Glencoe, fifth printing, 1964), "Can History Be Objective?" by Christopher Blake, 329ff.

[20] Ibid.

[21] Ibid.

[22] Ibid.

philosophies of history based on their ambitions toward pride and power. The intellectual elite passed judgment on scientific history, accusing it of being entirely unscientific and based on one's subjective reference.

Wilhelm Dilthey expressed the view well, claiming that every historian imposed his own philosophy on his interpretation of history. To him, every historian views history like a man floating down the river of time. He sees it from his position in the boat as it drifts down. Another man in a boat at a different location and in a different time would describe it differently.

This "historic skepticism" was really driven by the fact that scientific history did not produce the results that agreed with the philosophical history of the prevailing view of progress essential to deterministic science and the secular faith on which it was built. As Hans Meyerhoff, in his introduction to such philosophy, said,

> The failure of this secular faith, therefore, has left a deep mark upon modern culture in general. Man seems to have lost both the rational key to, and the practical mastery over, his own history; and this twofold loss has contributed to the prevailing mood of our age that history is full of sound and fury signifying nothing.[23]

There is an element of truth to subjectivity in historical science, as the differences in the various efforts to report history scientifically indicate. While there is perhaps more subjectivity in history, there is subjectivity in selection in all of science. And there is, of necessity, a particular view in science and history, as there is in this book. But *there is a great deal of commonality* that most relativistic historians refuse to acknowledge. Two different men in two different boats on the same river looking at the same shore would still see similarities in most of what passes, but with differing details. Those differences do not render that which is common invalid, but rather gives it more credibility. They show the common points were not from collusion, but from an individual perspective, and therefore are true witnesses to that which is common to mankind.

There must be some standard criteria for finding data and most important, there must be honesty in interpreting the data. Raymond Aaron commenting on how today, for the first time in history, historians have piece together "a picture of the majority of former civilizations." He writes,

> It is the inevitable weakness (of dealing with fragments) but at the same time the triumph of science and in this, history does not differ from the natural sciences. The age of the encyclopedist has passed and everyone now has to reconcile himself to limitations. What has been achieved is that the past, forever gone yet still surviving in its monuments and documents, has been bit by bit reconstructed to its precise dimensions and infinitely varied perspectives by generations of patient inquiry.[24]

This core of shared historical truth has *showed the **sinfulness** of modern man*, revealing his pride and folly, which destroys both him and the civilization that he creates. The various scientific historians saw a similar process of development in most civilizations, as well as a pattern of judgment when sinful man violates moral law. This scientific view of man and *history went against the grain of the philosophy of man's progress of the educated elite,* and had to be rejected, or they would need to reject their whole philosophy of life that exalted man and not God.

---

[23] Meyerhoff, *The Philosophy of History in Our Time,* 9.

[24] Ibid., Raymond Aaron, "Relativism in History," 155.

Denial of God and History is Escape from Accountability and Guilt to God

Motivation of Escape from History

The failure (or disintegration) of true history, as Weisberger described, occurs precisely because *man does not want to be accountable to God*; he *wants ultimately to be accountable to none but himself.* As Paul put it in Romans, "He holds down or suppresses the truth in unrighteousness" (Romans 1:18). If man removes his historical past, he has no record of his failures or evil. The more he removes the idea of God or good, the more he can avoid facing his past. And the less history is available and credible, the freer he feels to engage in more evil and deny it.

Man cannot live without meaning, and by rejecting history as a story with meaning, which reveals his sin; *he must inevitably turn to a cyclic view of history* to shut out the point of creation and the concluding point of history—judgment. Proceeding from a point of creation lineally to a final point of judgment must involve a succession of events for which each man is accountable to God.

Thus, fleshly mortal *man inclines away from a linear view of history and toward a cyclic one* for which there is no creator and no accountability in judgment. Looking back, God and early man present a higher standard. If history is really only a cycle linked to nature, then man is a part of the highest expression of a mystical power of nature, which he sees as god. He is thus a part of god himself. This, of course, is New Age thinking that is nothing but *old* paganism. For such a man history does not have an ending and man is only a part of the whole of nature as a great mechanism of physical forces. But he also is able to imagine himself as the highest expression of nature and as god. Exalting man as god is the inevitable end of pagan history. It was so in Greece with Alexander the Great who proclaimed himself Zeus-Ammon and of the emperors ending the Roman Republic who announced they were gods.

Studies Show Escape from History Has Been a Quest to Escape Guilt

One major study, and some lesser ones, show that this anti-historical and cyclic inclination has been true in most societies. Mercea Eliade, professor in the department of history of religions at the University of Chicago, did a comprehensive study of archaic societies. Some of his conclusions are given here: In his statements, he used the term "archetype," which he later defined as "exemplary model" in order to avoid another meaning, now the popular one used by Carl Jung in psychology, so I have used this other phrase to avoid that conclusion.[25] By archaic societies, Eliade also considers ancient ones such as Greece, Hindu societies and others.

> Archaic societies, ...although they are conscious of a certain form of 'history,' *make every effort to disregard it....*One characteristic has especially struck us: it is their *revolt against concrete, historical time*, their nostalgia for a periodical return to the mythical time of the beginning of things, to the 'Great Time.' ...These societies ***will*** *to refuse concrete time*, their *hostility toward every attempt at autonomous 'history',* that is, at history not regulated by 'exemplary models.' *This dismissal, this opposition*, is not merely the effect of the conservative tendencies of primitive societies, as this book proves. In our opinion, it is justifiable to read in *this depreciation of history* (that is, events without transhistorical models), and in *this rejection* of profane, continuous time,

[25] Mercea Eliade, *Cosmos and History: The Myth of the Eternal Return,* (New York: Harper Torchbooks, 1952), ix.

> a certain metaphysical 'valorization' of human existence.[26] (Emphasis added)

This rejection of true historical men for primitive mythical men as heroes or gods is an escape from human failure and guilt.

Eliade makes it clear here and throughout the book that there is a definite opposition to the concept of time as a continuous line from creation to consummation in judgment. Mankind's history has been to turn toward nature and a cyclic interpretation of history that goes back toward innocence and to the tendency to exalt man and his achievements, or as he puts it, to the "valorization" of human existence. He also says, "This conception of a periodic creation is, of the cyclical regeneration of time, poses the problem of the abolition of 'history,' the problem which is our prime concern in this essay."[27]

Eliade put his finger on the source of the opposition to history. He includes also an authoritative study by R. Pettazzonie on confession of sins in his book, *La confessions dei peccati,* saying,

> This shows us that, even in the simplest human societies, 'historical' memory, that is, the recollection of events that derive from no 'symbolic model' the recollection of personal events ('sins' in the majority of cases) is intolerable..." But what interests us...is primitive man's need to free himself from *the recollection of* ***sin***, i.e., of a succession of personal events that, taken together, constitute history. (Emphasis added)

Eliade speaking of forsaking real history for a mythical beginning later says,

> For the cosmos and man are regenerated ceaselessly and by all kinds of means, the past is destroyed, evils and sins are eliminated, etc. Differing in their formulas, all these instruments of regeneration tend toward the same end; to annul past time, to abolish history by a continuous return in ill temper, by the repetition of the cosmogonic acts. ...What is of chief importance... in these archaic systems is the abolition of concrete time, and hence their anti-historical intent.[28]

Eliade shows that the tendency to tie life to the cycles of nature often involves astrology and the solar system and various expressions of what is at root pantheism, involving reincarnation and the like. Many religions are built around these cyclic concepts. Almost all Eastern religions have this tendency, including the so-called New Age teachings.

But the alarming thing is that seeking identity and meaning with the mystical power of nature puts man into the direct influence of the demonic. Having rejected God, he *must have meaning and help* as his heaven on earth crumbles, so he pursues the psychic, the occult, and the demonic. Nature is devoid of person and man's passion for significance and his utter dependence on the material opens him up to demonic dependence and control. As meaning is lost and selfish individualism grows, he turns to destructive forces beyond his control; he turns to these satanic forces for help.

When Eliade wrote (1958), he warned of the tendency of Western thinking to move in that direction. Since that time, people have rushed into the New Age Movement. Many cyclic religions and philosophies have come into the USA and the West, in general. Even some Christian theology

---

[26] Ibid., xi.
[27] Ibid., 52.
[28] Ibid., 75, 81.

has turned from objective historical revelation and toward experience and "process theology."

As Weisberger pointed out above, when the concept of linear history is removed, the people lose any concept of standards of conduct. Everything becomes more and more relative, and only the standard of "what pleases me as an *individual*" remains. As morals degenerate and the sense of history is lost, individual greed destroys *progress,* the very thing that drives secular society. Thus, the bias of sinful man is suicidal to his culture and his life.

Modern psychiatry and psychology have not solved the problem of guilt, but made it worse. ***No religion other than Christianity has an answer to the problem of guilt.*** God cannot deny himself and must punish with wrath all unrighteousness by the oppression of truth about him. The good news that God sent his Son, who in obedience bore the punishment of our guilt for us, is the only real meaningful message of forgiveness. Jesus bore the penalty of death to deliver us from hell and the devil who keeps mankind in bondage by deception and the fear of death. Jesus was raised for our justification.

But the Western nations, especially the United States, are now also discovering their educational systems are faltering, and there seems to be no solution. Money, higher standards, and better management only give temporary relief. One of the most pressing political issues today is how to revive education in society. The social democracies of the West are now obviously failing in skills and character values, especially the United States, which has been the world's example of progress. Ironically, it is in the area of math and science that is so vital to the intellectual elite that America's youth are falling behind.

Crane Brinton of Harvard long ago pointed out that while Marxism and liberal democracy have many differences, they have much in common.

> [They] can fairly claim common origin in the Enlightenment and that both are in important ways opposed to traditional Christianity. *Both reject the doctrine of original sin* in favor of a basically optimistic view of human nature, *both exclude the supernatural*, both focus on the ideal of a happy life on this earth for everyone, both reject the ideal of a stratified society with permanent inequalities of status and great inequalities of income.[29] (Emphasis added)

Thus, the social democracies in the West are moving to the same conclusive failure as has communism in the Soviet Union. The educated elite have changed the textbooks on American history to support their politically correct ideas. Christopher Lasch, former professor of history at the University of Rochester, in *The True and Only Heaven,* showed that unless there can be a restoration of ethical and moral force, progress will soon cease. Progress is made to satisfy consumer greed and lust, but that greed and lust are now getting out of control and will lead to anarchy and mob violence. This has ended the progress in most known civilizations. Progress is highly desirable, but unsustainable when God is denied and moral control is lost. Had there not been the collapse of the previous great civilizations from the denial of one sovereign God the progress of humanity would have been slower but continuous, long-term, and would have produced a far greater progress for humanity and a much more amazing civilization.

Western Nations Approaching Time of Troubles

The western nations and especially the United States are now approaching a very serious time of troubles. Arnold Toynbee, the comprehensive historian, who had the perspective of

[29] Crane Brinton, *The Shaping of the Modern Mind*, (New American Library, 1953), 201.

centuries instead of decades or years, has warned of the terrible conditions that we seem to be approaching. He warned,

> If the analogy between our western civilization's modern history and other civilization's 'times of troubles' does extend to the points of chronology, then a Western 'time of troubles' which appears to have begun sometime in the sixteenth century may be expected to find its end sometime in the twentieth century, and this prospect may well make us tremble; for in other cases the grand finale that has wound up a 'time of troubles' and ushered in a universal state has been a self-inflicted knock-out blow from which the self-stricken society never has been able to recover.[30]

P. A. Sorokin, the Harvard historical sociologist, writing earlier about the moral decline and foreseeing these days, warned saying,

> The oblique rays of the sun still illuminate the glory of the passing epoch. But the light is fading, and in the deepening shadows it becomes more and more difficult to see clearly and to orient ourselves safely in the confusions of the twilight. The night of the transitory period begins to loom before us, with its night mares, frightening shadows, and heartrending horrors."[31]

Jacques Barzun, history professor, dean of the faculties and provost of Columbia College of New York University, and certainly one of few prominent historians today, entitled his recent (2000) monumental book on the five hundred years of Western and American history, 1500 to the present, *From Dawn to Decadence.*[32] He shows that the pride and unlimited desires of the selfish individual now possess our soon terrible time of troubles. The problems of ethnicity and race have not been solved.[33] All American politicians admit that the coming crisis of government welfare provisions will soon bankrupt the government. Neither party can really solve the 14 million illegal immigrants in America that come because business greed wants their cheap labor. The failure of the government to lead and solve problems soon will fall to women, and when they fail, to youth who are most anarchistic.

Most prominent civilizations throughout human history have gone down *the same path to their own destruction*, just when their culture seemed so brilliant and prosperous. I believe that the turning point has been man in his pride accepting a mechanical materialism based on the perversion of a gradual uniformity of nature in order to be free from God.

It is exceedingly important that we understand this tragic fact, especially since modern technology is so powerful. A future demise could spell the end not only of our civilization, the emerging world order but also of the whole human race. We have heard that warning so often that our ears are deaf to it. England's Royal Astronomer, Malcolm Rees has predicted the doom of mankind. Stephen Hawking says we must find another planet soon to which we can escape or mankind is doomed, and there are many more such predictions. There is an answer, which is ***brave good men*** turning the nation back to faith in the sovereign loving God revealed in his Son Jesus Christ. Paul proclaimed "the gospel of Christ is the power of God unto salvation to everyone that believes" (Romans 1:16). The solution might mean our emerging world civilization may revive and prosper to grant amazing good for all nations.

---

[30] Arnold Toynbee, *Civilization on Trial*), 9, 10.
[31] P. A. Sorokin, *The Crisis of Our Age,* 13.
[32] Jacques Barzun, *From Dawn to Decadence,* 877 pages.
[33] Wilson, *Man Is Not Enough*, chapters 1 and 16.

PART X

Review and Conclusion

He lavished upon us [his grace]in all wisdom and insight making known to us the mystery of his will, according to his purpose, which he set forth in Christ as a plan for the fullness of time, to unite all things in him, things in heaven and things on earth (Ephesians 1:8-10).

Chapter 32
Summary of the History of Modern Science

Introduction

This chapter is a summary outline and evaluation of the history of modern science and the perversions of human bias. While the foundations of modern science from the Reformation faith are the basis of productive science today, the perversions begun in the mid eighteenth century have gained dominant control of western education and the media and are destroying social relationships in the soft sciences. It is important to see that the bias of man's sin nature is now threatening to produce deliberate rejection of the Trinitarian God now known through Jesus Christ. The common faith in Jesus Christ by the early colonies that proclaimed the Declaration of Independence and formed the constitutional republic of the United States now leads the free world. That faith was formed by the Reformers in Europe and the various peoples of the Reformation came to America to escape the persecution in Europe.

Trust in Christ as Creator of all things, visible and invisible, is extremely important for the Christian to believe Christ is now Lord in heaven and earth. The weakening of the churches today is directly related to the loss of the biblical certainty of Christ as Creator and Re-Creator. Without this certainty there can be no faith to believe he has all power to give eternal life and to live as risen with him now for his eternal kingdom (Colossians 3:1-4). The result is the church becomes chained by the blind eye of trusting in material creation (Proverbs 28:22; Matthew 6:22-24) and the pride in material prosperity will be shown to drive the greed, and lust that destroys civilization. Hope must rest on Christ as Creator who is now Lord and Re-Creator. Faith in God in Christ is what motivates honesty, belief in man's creative ability, hard work, and commitment in marriage and business and prevents conflict (James 3:13-4:4). When a people no longer want to trust the only true God, he no longer restrains their conscience, and their sin leads to his judgment to uphold righteousness.

Summary of the History of Modern Science and Civilization

The evidence presented in *Man Is Not Enough* and given in *True Enlightenment* in the second chapter traces the beginning of civilization with a high view of a sovereign God, or Theism. This was demonstrated by J. Schmidt to be true of every known civilization using ethnological science, and has been confirmed by archeologists. I have shown that the waves of civilization grew for many years, usually for a quarter of a millennium under faith in God, and after many years, through hard work and creative reason, man developed a prosperous culture. At a certain point the intellectual elite were demonically motivated to believe human wisdom is all there is, and life's blessings consist of success, wealth, pleasure and health. They then led a rational revolt and began a turn to deism, pantheism and chance creation by nature, thereby eliminating God. This process in the western world which was the same as other civilizations has been traced in much detail through the philosophy and theories of modern science with documentation by the words of the scientists. Because this has been a long story, this is a summary to help the reader review and see the perversions of human bias in the nebula hypothesis of evolution of all things, with many theories to try to support that theory, and how scientists themselves have shown this perversion to exclude God has been a complete failure. For the reader who is historically savvy, and has been able to keep the picture in mind, this chapter may not be needed. But even for those, this chapter will bring home the crisis of our times. A graph of modern scientific history is on the next page followed with an explanation of each phase of development in the rest of this chapter.

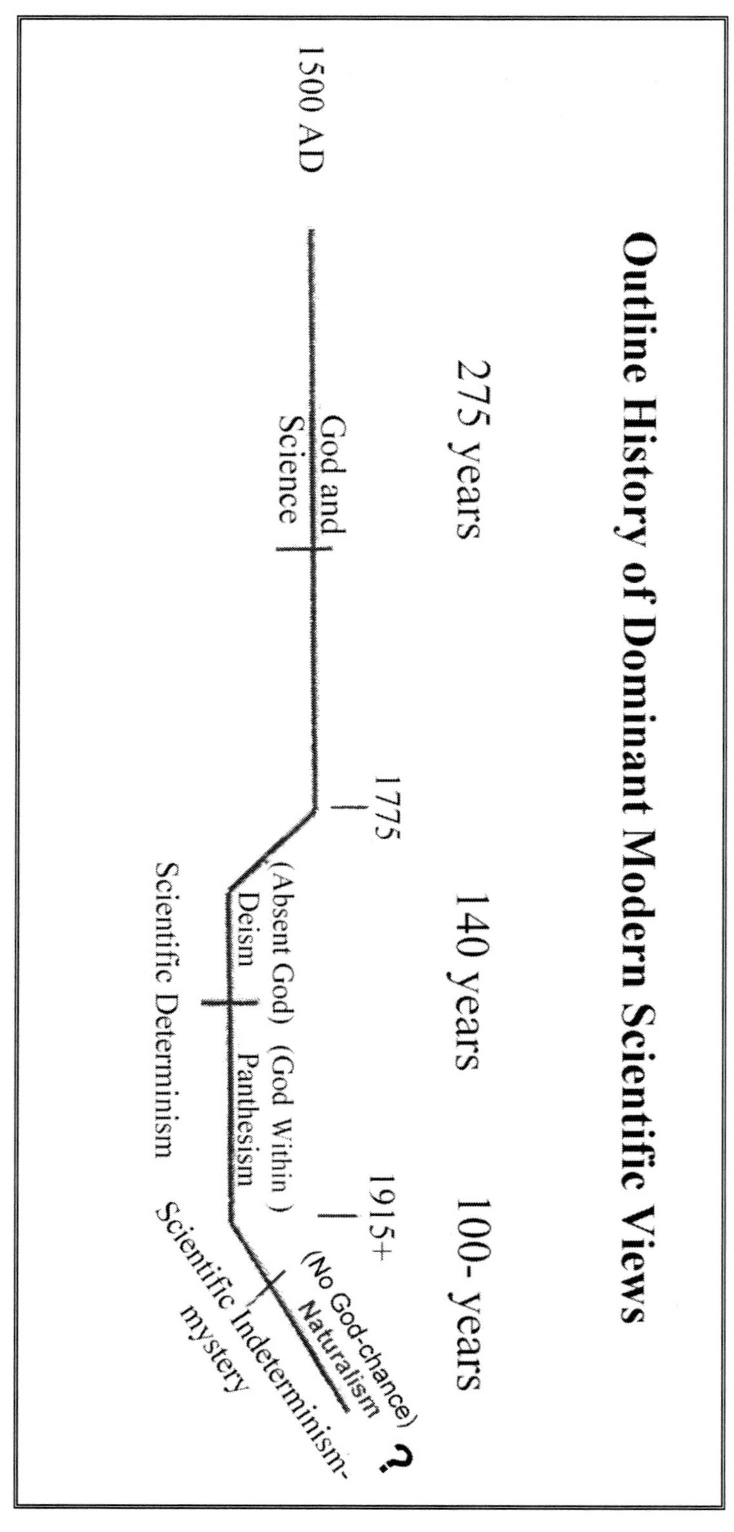

Outline History of Dominant Modern Scientific Views
275 years
140 years
100- years
1500 AD
God and Science
1775
(Absent God) Deism
(God Within) Panthesism
Scientific Determinism
1915+
(No God-chance) Naturalism
Scientific Indeterminism- mystery
?

Conditions of Reformation Science

Modern science and Reformation theology took root in Europe, prompted by the oppressive control and corruption of the Roman Catholic Church, and motivated by the recovery and access to many original early documents of the Bible and of Egyptian, Greek, Archimedes and Islamic science. This occurred at first especially in the primitive universities throughout Europe. The reformation in modern scientific understanding in regard to general revelation in nature was paralleled by the theological reformation though the study of the Bible and formed a union in the Reformation by refuting the pagan philosophy that had captured the Roman Catholic hierarchy and control in the Holy Roman Empire. One aspect after another of pagan science introduced through the Holy Roman Empire and the controlling Roman Catholic clergy was gradually point by point removed for a return to God in Christ as Creator.

Reforming Pagan Church Science and Theology

The foundations of modern science with faith in God as the Creator and Sustainer of all the universe were discovered from 1500 A. D. to about the early nineteenth century. The first two hundred and seventy five years were dominated by science based on faith in God as Creator (theism) and a search for how he works in nature or his "book of creation" before man's sin bias reacted. While working for the church Leonardo de Vinci used the scientific process of observation, theory and testing for proof, and recognized the errors of the Aristotelian view of movement, but without challenging the false views of the Roman Catholic clergy. Copernicus formulated a clear presentation of the planetary solar system, which was in conflict with the Ptolemaic system of the Aristotelians taught by the Catholics, but his views were not widely understood or studied. But Galileo fully accepted Copernicus theory and understood the principals of motion and using his telescope, he strongly, publically conflicted with the Roman Catholic Jesuit teachers who attacked him and his science. Pascal, De Fermat, and others advanced mathematics for use in science and Pascal studied fluids and pressure. Kepler, a Lutheran, was a brilliant mathematician and scientist with access to latest astronomical data. He accepted Copernicus' theory, encouraged Galileo, and accurately figured the path of the planets in elliptical orbits around the sun. He also greatly advanced the calculus that was then used by Newton. Newton, more than any other, as an Anglican biblical believer developed the principles of mass and gravity in physics and astronomy, and developed an advanced telescope for more exploration to define the solar system. Newton and Boyle, a brilliant biblical Christian, explored gases, light and other areas of matter. John Ray explored development and relationships of plants and animals. Later Linnaeus sought to expand and make this taxonomy more specific.

There was a transition from a firm Reformed science to one of emerging bias to remove God in the late 1600s, and strongly in the mid 1700s. But Christians continued to lead in the develop of science in the late eighteenth and early nineteenth centuries when human wisdom came to prominence and man's anti-God bias was strong. One of the last of Aristotle's errors was the phlogiston theory about fire which hindered understanding, but was finally refuted. Then the Christian John Dalton advanced the science of chemistry, and atomic physics began to be understood and developed. Later other Christians such as Michael Faraday discovered the physics of electricity, magnetism and radiation forces. Grimaldi, and later Young, led to an understanding of light and the human eye. These discoveries opened the way for Maxwell to understand electromagnetic and radio waves, and led to a new understanding of the atom. On the basis of his work Earnest Rutherford, in 1911, identified the alpha and beta forces, and developed the protons and nuclear theory that led to the quantum theory and the theory of relativity of new physics.

The Christian science of geology also emerged in the transition time before a focused opposition to God. A. G. Werner described strata of the earth formed mostly by water, giving little notice to volcanic rocks, and speculated these strata were worldwide. William Smith learned to use shell fossils as the key to determine sedimentary layers of the strata, A. Sedgwick and R. Murchison further developed the science of stratification and defined older layers. Agassiz discovered and researched glacial activity and ice sheets, which brought new geological data but raised questions about the worldwide flood and the timing of geological events.

All these major areas of science were developed by Christian believers with faith in the biblical concept of God as sovereign Creator and Sustainer of the universe. The Christian view of science thus lasted and dominated more than a quarter of a millennium.

Demonic Influence Motivated Human Sin Bias to Uniform Determined Natural Science

By 1775 the bias against God and religion found reasons to slowly emerge into a force. After the Thirty Years Wars (1618-1648), great religious conflict and destruction had influenced anti-religious feelings, and the clear emergence of man's worth in the newborn Christian humanism of the renaissance and Reformation led to a pride in man's ability. With the new manifestation of man's ability, skepticism about religion emerged among intellectuals, giving them a cause for criticism because of corruption and oppression by royalty and clergy. The Reformers had long broken the back of Roman control and new improvements were clear from human wisdom. But man's pride in his achievements led unbelievers by sin bias to see themselves as wise as God. Francis Bacon emphasizing induction, and Rene Descartes by deduction both ***declared the individual's reason as the source of auth*****ority**. Spinoza challenged Mosaic revelation of the Bible and launched ***biblical criticism***. John Locke, against all authority, said all real knowledge in the human mind came by sense perception on a blank slate with morals and religion in a second tier of knowledge that was ideally derived by individual reason.

Prominent Christians, such as Newton and Boyle who dominated science at this time of growing skepticism, rejected these views. But some Christians reacted, and went too far for religious views, and began to claim ***all knowledge was fixed by God***, e.g. by Linnaeus in biology, by Whiston in astronomy and all of the universe, and others held a view of inspiration of the Bible that claimed perfection to the letter and limited human participating. This extreme claim of control by religion set the stage for unbelieving man's sin bias to put ***change in nature*** and to minimize a personal God and promote a deism or a new pantheism.

The Enlightenment thinking of the late seventeenth century was brought into focus by Kant in the mid-eighteenth century. He expressed the idea of two tiers of knowledge and a universe of matter formed by chance, by natural forces from Lucretius, an early Roman Epicurean; the nebular hypothesis from Swedenborg that had been enhanced by T. Wright; and Kant's view from these was confirmed by a beautify vision of a pantheistic development of all things by nature's powers alone. Kant improved his view of two aspects of knowledge removing God by a revision of Locke's views with a strong influence from Hume and by adding categories of thought. He thus redefined knowledge so that it excluded revelation, or any supernatural, making this desirable, but not real. Kant thus launched the theory of the evolution of all things from a nebula beginning and limited real knowledge as ideal, drawn entirely from sense perception. He adopted Newton's regularity of the universe for claim of a purely natural working of things. His whole philosophy of knowledge was driven by his naturalistic theory of the development of everything from the nebula hyposthesis.

Kant, *a priori,* ***redefined science as only ideal knowledge determined from nature by the senses through natural categories on a blank slate,*** By him all faith in God and ethics were thus only ***what is desired knowledge without objective reality***. He drew a determined circle around ***real knowledge*** as only from the senses excluding knowledge of God. After Kant, liberal theology developed that was based on subjective feeling with mainly humanistic objectives. Twenty years later in 1775 Diderot's encyclopedia of Enlightenment thinking was published and circulated. It included naturalistic philosophical thinking from the last of the eighteenth and the nineteenth centuries as science. This launched the concept of man's Enlightenment wisdom as adequate and supreme. The spreading of this view of confidence in man was accelerated by improvement and proliferation in printing at this time, and this denigrated the revelation of God.

Enlightenment Definition of Knowledge Supporting Evolution Spread to All Science

The views of Enlightenment spread and were dominated by Kant's nebular hypothesis of the theory of evolution. A young German, William Hershel, took it to England, and with members of his family developed much larger telescopes, and made many astronomical findings which he sought to fit as an analogy under a deistic philosophy within the nebular hypothesis of Kant. Laplace, a Frenchman, influenced by Herschel, presented a mathematical model of the development by accretion of the solar system based on the nebular hypothesis, and claimed things happened by nature alone and God was no longer needed. W. Herschel passed this nebular hypothesis along to his friend Hutton, who was also a friend of the skeptic Hume. Hutton applied the evolutionary theory to geology claiming a theory of earth's development giving a prominence to volcanic rocks, and challenged Werner's view of development of worldwide strata mostly by water. He emphasized long ages of gradual development by regular forces without cataclysms.

In the early 1800s Hutton's views of geology were seen as incorrect and were superseded by Lyell whose principles of geology were aimed at removing faith in biblical creation and cataclysms, especially the flood. He presented the strata as developed by long ages of cyclical natural geology based only on what is commonly daily experienced. Lamarck, a Frenchman, had developed the theory of progressive evolutionary of life under the Jesuit Aristotelian pantheism. Robert Chambers in England made the idea of evolution of life popular for most intellectuals as a series of evolutionary progressive creations which he identified as God's way of creation.

The presence of intelligent design in these biological theories left man's bias to remove evidence of God incomplete and Darwin's theory of development by chance of nature fit man's strong desire of human progress without God. Darwin made his voyage on the Beagle in 1831 accumulating data. It was not until 1837 that he arrived at what was virtually the theory expressed in *The Origin....,* at which time he *doubted* God. Darwin knew about Lamarck's theory and Lamarck's admission there were no evidences of intermediaries. He also knew that Cuvier, the greatest paleontologist, also a Frenchman, said there were no intermediaries in the fossil records. Darwin wrote his final sketch of this theory in 1844. His friends Lyell, who was by then the famous geologist, Hooker who was an authority in botany, and Huxley all told him there was no evidence of intermediaries or transmutation. By that time his disbelief in God was complete. He hesitated even after concluding the mechanistic clue was the survival of the fittest until Alfred Wallace forced him into publishing in 1859. Even in publishing and thereafter, Darwin did not express his unbelief, following the advice of Lyell to avoid examination and conflict by believers in the church to thereby gain a seductive following. Also, Emma, his believing wife, opposed.[1] .

After publication Darwin's views were rejected by Owens, the leading English biologist who argued the evidence was against any inheritance of acquired characteristics, and by Jenkins who proved the theory was unworkable because of population swamping. Darwin's theory had no evidence of truth. But the idea of progress in nature was already widely accepted in the evolutionary theory and along with the removal of intelligent purpose, Darwin's theory fit the bias of man at the time, and led to Darwin's view being widely and popularly accepted. Darwin's view was that a family tree of life developed progressively from small to large and more complex, from bacteria into intelligent man and skills for civilization. He admitted the truth of Jenkin's evidence of population swamping and even in the last edition of *The Origin* added the admission of a driving power of intelligent purpose. After Darwin wrote *The Descent of Man* the search began as to how life began, believing that the right concoction of elements instigated by lightning or electric energy would initiate things. The atom was more clearly understood and the nineteen century ended with a certainty that man was beginning to understanding everything in a determined natural universe.

---

[1] For a review of this see Gertrude Himmelfarb, *Darwin and the Darwinian Revolution*, chapter 7, Genesis of the Theory

The nineteenth century ended with amazing advances in transportation and new communication that reflected the brilliance of man. The educational elite had control of the National Education Association, and by the first quarter of the twentieth century many of the Northeastern colleges were promoting secular Enlightenment education through the progressive thinkers in politics and the churches. They were promoting Darwin's evolutionary theory and threatening the biblical concept of creation and were replacing supernatural teachings of Christianity with biblical criticism and the humanistic social gospel. The Scopes Trial in Dayton, Tennessee was planned and effectively publicized to promote Darwin's theory of the evolution of man. From 1924 there was a strong push to get the evolutionary theory into the public schools and into all colleges. How this was done will be briefly described after reviewing the surprising shift in scientific findings. Reliance on money with trust in mammon was the growing motivation and surpassing faith in God.

Thus the Kantian evolutionary idea of the uniform natural deterministic view of development of everything without the personal God of the Bible expanded in theories for support. It produced a progressive materialistic philosophy based on human wisdom without God, increasing in strength in about 140 years. It swept the European nations and found strong roots in America. It invaded and changed the soft sciences that will be presented in the second volume. During this time the evolutionary philosophy after Darwin produced efforts in comparative religion to say the idea of a monotheistic sovereign God evolved by nature worship, and from a pluralism of gods over many centuries. The whole known history of man was rejected for an effort to rewrite a gradual development of man from primates and his gradual development of intelligence. But no clear evidence of evolution of man from primates was found to bridge the gap. Also, Alfred Wallace, the coauthor of the evolutionary theory with Darwin did a study of many so-called "primitive tribes" and concluded that there was no evidence of gradual development of human intelligence. He then rejected the idea of gradual evolution of humans, which was a main objective in removing accountability to God.

Indeterminate and New Science Introduced Mystery in Early Twentieth Century

In that first quarter of the twentieth century the graph of history shows the discovery of the quantum theory that changed the concept of theoretical science to one of indeterminacy, and Einstein's view of relativity challenged the traditional Newtonian view of the universe. While evolutionary science based on chance naturalism continued to gain ground by dogma and authority in many schools, change to include mystery, and the unknown powers of nature, now gained prominence in scientific study. Also, the foundational Christian scientific theories were still used by business for producing practical products and ideas.

With quantum indeterminacy, secular Enlightenment scientists tried all the harder to seek and teach new naturalistic theories based entirely on regularity and repetition of nature with no cataclysms or supernatural events. The flood, rejected by Lyell's geology, was said to be only man's exaggerated remembrance of ordinary limited floods that archeologists found in the Mesopotamian area. The glacial ice sheets and local glaciers were said to be a result of a theory of a collusion of three astronomical factors associated with Milankovich who tried to mathematically demonstrate cycles of forming glaciers and then interglaciation of melting. A new concept was proposed of the construction of the standard theory of an atom. Chemicals were theorized as formed by passing through a speculated Higgs field. Experiments with great cyclotrons were made larger and larger to find the secrets in matter, at increasingly great costs. Black holes were theorized and evidences for their presence were found. A theory arose of the nebular formation of the whole astronomical heavens based on a big bang of a gaseous explosion followed by its inflation and the progressive formation of the elements that then condensed into stars, galaxies, and solar systems.

A continued, more intense search for fossils was made to show Darwin's idea of the history of transition of man from apes in Africa. Many claims were made of missing links, especially among chimpanzees. Lucy, found by Johansen in the Rift Valley, was highly acclaimed. Later, with the birth of molecular biology and genetics in the twentieth century, the claim of tracing the DNA ancestry of the first woman and the first man back to Africa was made. Animals and man and woman were believed to have migrated all over the world over long ages by land bridges from continents. There were new theories about how the first life was formed: from liquid mixtures of elements transformed by electricity, or from clay, and later, possibly from RNA. The idea of the geological column of geological strata developed to show long ages of gradual development were efforts of speculation. As new understanding of the atom developed the idea of dating fossils by the use of radioactive decay produced what was called absolute dating of fossils by lava rocks nearby. And Einstein's concept of measuring the universe by light years also gave a new way of astronomical determinations.

Failure of Evolutionary Development from Nebular Beginning by Natural Determinism

However, as the twentieth century progressed, all of the theories of formation by cycles of nature observed by regularly experienced natural forces did not work out as anticipated. The Quantum theory that introduced indeterminacy was followed by the acknowledgement of the Chaos theory in every area of science introducing indeterminacy everywhere by about 1960. Characteristics of survival of the fittest causing transitions from one species to another were not found leaving gaps from the highest to lowest species. So new evolutionary theories by saltation (jumping the gaps) from one species to another leaving no evidence were developed. A neo-Darwinist theory of transitions by mutations in genes was adopted, but evidence revealed no transmutations and primarily degenerations. Single molecules were found to be so complicated in design and functions, like complex microscopic factories, that chance development seemed impossible. The way of forming any one molecule seemed impossible. One simple protein with the smallest number of amino acids was too complex for formation by chance. New imaginative theories were created for explaining how transitions from one species to another happened, such as punctuated equilibrium. This admittedly had no evidence of transitions with only a speculative occult cause. Use of the most sophisticated computers to figure the probabilities of the formation of a living cell and for the evolution of life produced exorbitant figures like a miracle. An information theory of everything was proven and added to the second law of thermodynamics showing the loss of information, not evolution to higher. Therefore, the idea of life being formed elsewhere and coming to earth was considered the option, even though there was no evidence of life coming from Mars or elsewhere. Planets that might be hospitable to life were sought, especially ones possibly containing water, but no certain one that was friendly to life has been found anywhere close to earth. Listening devices were created costing millions of dollar to try to see if there were messages from other alien creatures in the universe. But to no avail.

The evidence of the descent of man from primates became even more unclear. Evidence of hominids between humans was conceded to exist prior to Lucy. Questions by authorities about her clan as walking and erect as an intermediary were raised. So the development through her and other chimpanzees was abandoned for a few fragments of another much earlier line called Ardie from a million years earlier and not from the same environment. Moreover, fossils from other parts of the world led to the speculation of origin from places other than Africa, and possible of several different lines of developments of men. Three different lines were suggested. Moreover, speculations of the development of men from chimpanzees showed the first Adam to be 120,000 years and from Eve to be about 150,000 years, and the genetic evidence of emerging from Africa were very small. Moreover, the tracing of the primates showed that most of the large apes were earlier in Eurasia, with very few in Africa. The tracing of racial descent by DNA revealed three major divisions, which fit the number of sons of Noah in the Bible. When actual studies of human families (not from chimps) was done to discover genetic beginnings it was found that they showed an origin of man from only about 6,000 years ago.

The light from this archeological historical research agreed with the fact that all three branches of men (erectus, Neanderthal, and Sapiens) were recently related, even though the process was nowhere evident in history. All larger apes or primates from which man was said to come were found, not in Africa but in Eurasia. And all the three branches of man of the homo fossils were found first in Eurasia and not Africa. Humans were found to be different from primates and other animals physiologically, mentally, emotionally, sexually, and in their ability to communicate. Only man is able to build culture and a civilization.

Man's erect body, spinal entry to the head, opposing thumbs, and ability to sit for extended periods, enable him to understand and create and empower him with a dexterity and brain coordination which no other animal has. He can play a musical instrument, sit at a desk, draw and design, use tools—all skills which no primate has ever exhibited. Man's sexual habits normally motivate him to want a permanent mate, to want and enjoy sexual pleasures repeatedly, and stay together to train and raise children. The human baby is more completely and longer dependent, requiring extended parental care. Men think and find satisfaction in action and dominating; women communicate twice as much and think differently. Their hormones are different and motivate them differently.

Man has a much larger frontal lobe than animals. He has an ability to visualize. He has cognitive abilities, allowing him to deliberate, choose, and repeat. Animals have only instinctive memories for communicating to obtain sustenance, for protection and for reproduction. The greatest difference is man's spiritual ability. Surgically, it has been demonstrated he has a mind that is transcendent to his brain and an area has been mapped that the brain not only receives motor impulses but spiritual messages through the mind/soul to the brain. He has been found to have a mind capable of understanding morals and guilt. Locke's and Kant's views were incorrect demonic ideas.

The tracing of civilization revealed it spread primarily from Eurasia and not from Africa. Moreover, the research in archeology revealed that there was no long period of gradual transition from primitive hunter-gathering societies to high culture, but domestication, city building, governing organization was early in the Eurasian area, and spread to the rest of the world as was all intelligent cultural elements. Computer assimilation of data showing human beginnings from all human societies surprisingly revealed only 6,000 years since beginning. These studies showed that up until 3,000 years B.C. most humans were related and knew each other so that oral transmission was likely up until the beginning of written records at that time.

The process of creation and development, and measurement of time of development of life turned out to different than evolutionists had taught. Science discovered that the oldest strata revealed all biological forms of fossils and that the cone of development was much different than the evolutionists proposed. Instead of a cone or triangle developing from a small beginning and spreading to larger and complex forms it was from a wide cone of a sudden biological big bang of all forms of life, then an extinction of most forms followed by a rapid expansion to multiple forms again, then followed by another partial extinction. The speculated geological column of long ages of development build on the speculated graphed concept at the left (which did not exist) was later supported supposedly by the decay of half life of radio active elements. But this was shown to be inaccurate because the formation of isotopes was shown to be affected by radiation exposure and by magnetism done in nature and even in the laboratory. Moreover, the formation of isotopes was found to occur in nature in uranium mines, so that the amounts in minerals were not trustworthy as measurements. These studies revealed, as did study of certain stars, that the alpha particle in the universe and matter is diminishing. Graphs of cones of development on the next page show that all the evolutionary claims had been wrong.

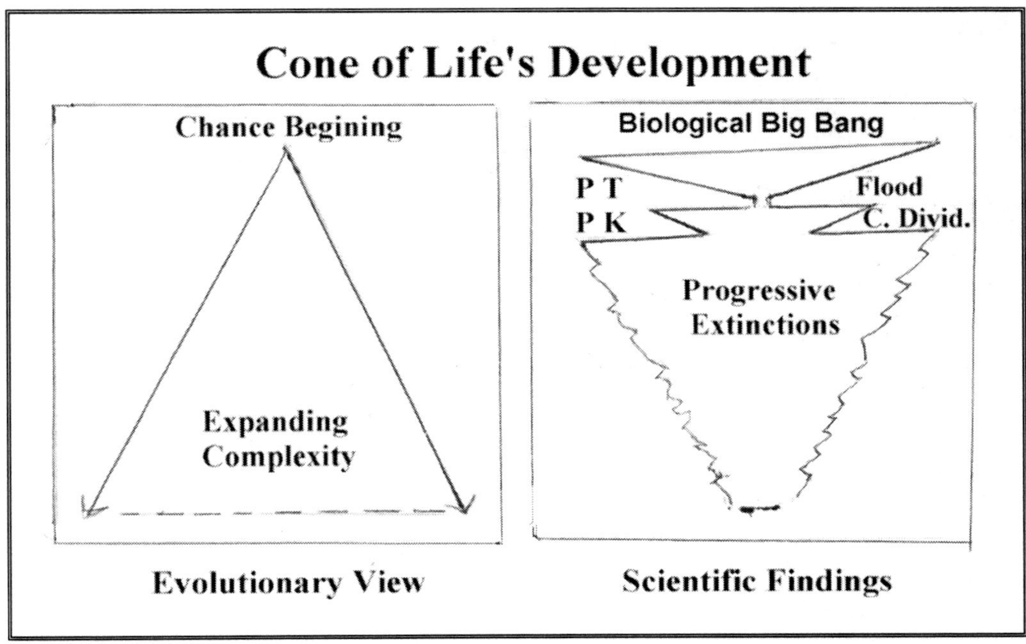

The discovery of the division of the continents was made about 1960. Shortly after the P-T extinctions and loss of the canopy of tropical life conditions of several hundred years, the P-K extinction occurred with another short freeze. This revolutionized the whole evidence of geology in the events involving plate tectonics. How the runaway lithospheres occurred was discovered and the break up into continents after mammalian life was proven. This showed Lyell's principles of geology were out-dated and incorrect and further invalidated the evolutionary view of the expanding cone as projected in the illustration on the left above. Lyell's concept of mountain building, volcanoes et al. had to be revised. The evidences of large reservoirs of water on the continents and other studies reconfirmed the concept of a worldwide flood as the explanation of catastrophic events. Moreover, the Milankovitch natural theory of glaciation and interglaciation has been abandoned for other speculations about formation of glaciers and ice sheets.

The studies of the solar system turned out to be much different than the speculated accumulation development by Laplace with no one theory rationally explaining the great diversity and extent of formations of the planets and other objects in the solar system. Also, the evolutionary Big Bang and evolutionary inflation of the universe as projected was not supported by the microwave evidence or the sky mapping. There were spaces, radioactive waves, flow of parts of the universe with many black hole singularities, and different kinds of galaxies that did not function as expected. Instead of young galaxies and then more mature ones there were adults lying in the nursery of the so called, beginning heavens. The mapping of the heavens did not reveal a progressive inflation as expected and the radioactive background had many unexpected aspects such as hot and cold spots, disharmony, circles, and uniform temperature that did not fit a gradual inflation. Tests by cyclotrons unexpectedly revealed a beginning with a liquid and not a gas expanding.

The universe was found to have large amounts of dark matter in order to explain the way the stars and galaxies worked. Moreover, the universe was found not to be diminishing in energy at a Hubble constant rate but to be rapidly accelerating in expansion rate from unknown or dark energy. Therefore, science revealed that the universe was made up of 96 % or more of dark matter and energy about which nothing was known. Thus the universe did not reveal the nebular hypothesis of gradual inflation after a Big Bang.

Thus the whole universe and everything in it did not fit the nebular hypothesis of evolutionary development. Moreover, studies have revealed that light speed is and has been diminishing gradually in rate. This was not sufficient to invalidate Einstein's theories in the short period they were tested. But the record of light speeds shows a consistent slowing of the speed and that a curve of this showed the earth only 6,000 years old, instead of the 13.7 billion light years. All of these facts as documented by science given above reveal the evolutionary creation by nature was contrary to facts. Thus the uniform natural determined definition about knowledge was completely wrong. The Reformers' science based on a Creator was more correct.

"God Is Dead!" (1960), Rejection of Biblical God by Seductive False Evolutionary Science

The spread of Kant's nebular hypothesis of evolution and his redefinition of science by what is real knowledge was aimed at excluding objective evidence of the Christian God as Creator in general or natural revelation began about 1750 and continued until the 20th century. At the same time other Enlightenment critics were slowly developing their case for rejecting the knowledge of God in the Bible. Both of these biased attacks on the biblical God had begun because of corrupt religion and conflict about1650 ff. that furnished a popular reason to attack clergy. The efforts at scientific literary criticism directed against the Bible itself gained momentum among intellectuals, especially in the second quarter of the nineteenth century, and were directed toward the supernatural Jesus in the Gospels in an effort to make him purely human.

The quests for the historical Jesus failed because through bias the critics ignored the historical data in the Gospel records which made the cross and resurrection central to the fulfillment of the whole Bible. Instead, the intellectual critics sought to build theories around various speculative themes (psychological, ethical, mythical, eschatological, and others) to give a natural reason for why the church worshiped Jesus. By the twentieth century the first quest for a historical Jesus based on bias had run its course. In an effort to find meaning for Jesus, religious intellectuals began seeking this in existential reasons and church preaching, putting the meaning of Jesus in complete transcendence or in myth. These existentialists thereby evaded the circle of determined natural knowledge while often even using biblical themes but excluding them from real objective truth.

By the last of the nineteenth century the evolutionary theory for biological evolution, especially by Charles Darwin, produced claims of and faith in the gradualism in the evolution of man from primates. The claim of gradual evolutionary emergence of man in body, mind, communication and intelligent civilization was the main objective of bias. So critics argued that religion, or ***the idea of God, also gradually was created by man*** as a part of his evolution. In the last of the nineteenth century the efforts to explain God resulted not only in biblical criticism, but in a study of comparative religion in the world. The theory of ***comparative religion based on evolution*** was said to have begun by believing the idea of God evolved from primitive man worshiping many aspects of nature as gods, and as his intelligence increased. They taught man that gradually narrowed the idea of many nature gods to that of one Supreme Being with moral and ethical rules. This was the opposite of their view of the cone of biological evolution given above. This was said to be done by priests or religious leaders to gain control by claiming God's authority over the people and society. In this theory the critics were claiming there was no true existing God, but *that the idea of God was developed by man for social purposes.* Later Freud used this idea to form his psychology.

The most significant effort at rejection of the biblical God was J. Wellhausen's rewriting of the history of the nation of Israel. He radically changed Israel's history, and built it around Charles Darwin's theory. He used several methods: incorporation of previously speculative suggested divisions into documents and the creating of others; a mythical person labeled a "redactor" who operated over centuries to change things and weave the documents together; the claim that Israel was an isolated people who for their survival created Jehovah; and the idea that in the Babylonian exile selfish priests weaved this all together to gain control of the people. Because the bias against God was so strong, and man wanted to reject biblical morals, Wellhausen's theory of the evolution of God was rapidly accepted and taught by religious liberal theological schools, by liberal universities, and by the public media.

During the early efforts of comparative religion, Andrew Lang had noticed that in early cultures there was one primary sovereign God, but Lang's work was passed over, not fitting with the evolutionary hypothesis. By the early part of the twentieth century, the German ethnologists developed objective standards for historically determining which element of a culture was oldest– which weapons, form of government etc. and their findings proved accurate. One of the ethnologists, W. Schmidt found that in every civilization, the oldest idea of God was of one sovereign ruling Father, with a Son. Schmidt could not explain how this occurred and never became a believer in the one God he found as oldest. Archaeologists confirmed Schmidt's findings that the idea of God began with one sovereign God and then ***degenerated to the worship of many gods*** to suit their desires.

By the second quarter of the twentieth century, the claim of the evolutionary creation of God was dominant in educational circles, even though most people had never read comparative religion or Wellhausen's revised history of Israel. During that time the past history, not only of man and God, but of all written recorded ancient history of man was rejected and man's history was rewritten as a history of gradual development over millions of years from an ape to modern man. The critics' claim and hope was that man would continue to develop toward becoming superman. In time archaeology and literary criticism found Wellhausen's theory was contrary to the facts, ridiculous, and that W. Schmidt's findings were correct, diminishing the acceptance of Wellhausen's theory.

But in spite of the evidence of evolutionary lies and the discovery of a completely new era of mystery in science (quantum, chaos and microbiological theories etc.), in that second quarter of the century, by about 1950 to 1960, the strongest proclamation of bias against God and Christianity emerged. At this time scientists, the elite educators, and the media made extreme dogmatic insistent claims that religion was not to be considered real knowledge for science. By 1960 the death of God theologies dominated the media. These "theologians" had nothing profound to say, except to claim that the culture reflected that man was so smart *he had outgrown and no longer needed God.* At this time in America leading educators, the media, and the laws of the Supreme Court made religion unacceptable in the public arena. In the 1970's the educated elite also returned to strongly promoting a syncretism of all religions (which had been suggested about 1925). They said such religion could be a support and a unifying force for mankind. To achieve this syncretism, it was advocated that the church should yield its teaching to the educated elite as high priests in the universities for government control. These writings by A. T. Robinson, Harvey Cox, John Hick, and others to advocate this new religious humanism under the state were not profound or brilliant but fit the driving bias of Western education and were therefore popular.

All through the centuries that these biased efforts were made to remove God, the intellectual engineers of the theories sensed, and even expressed, the need for men to have ethics that control man's moral actions. Kant had his "categorical imperative" because he admitted man is "crooked as a stick." Darwin hesitated to write *The Descent of Man* for a dozen years because he feared releasing moral restraints from religion. Sigmund Freud said religion was needed until men could be fully educated to accept the pragmatic view of unselfishness. The teachings of the soft sciences of economics, psychology, sociology, and politics developed, and were influenced by these false theories (as will be shown in Volume II).

Now a century later, after the full fruit of the rejection of God and his revealed moral law in 1960, the selfish bias of man seems to have been released and is exploding, as past civilizations did when they rejected God. The sin of the West and America will certainly lead to great punishment from God in the near future if there is no genuine repentance and faith soon. It is unreasonable that man plunges ahead, following these theories to support human bias against God, even though the elite's claims have been shown to be completely false. Reformation scientists had warned that man could only quantitatively measure the regular working of the sustaining power of God, and could not find a uniform natural determined science. How has man become so deluded?

Illustrations of Failure of the False View of Knowledge by Uniform Scientific Determinism

The following circle graphs show the failing epistemology of evolutionary scientific natural determinism. The graph on the left illustrates Kant's circle of determined knowledge from the senses. He theorized that all nature operates within this circle of knowledge so that knowledge of God and ethical standards were outside of real knowledge and are only to be accepted if desirable to live; they were not real, but imaginary. However, science, since the first third of the twentieth century, shows nature involves mysterious powers outside the circle of sense perception as the arrows on the left circle show. The bent arrows in the graph on the right show that the theories to support natural determinism failed. The arrows penetrating the circle show that the source and meaning were from beyond man and his intellectual ability from sense perception. Deterministic scientists found the powers of nature to be mysterious and beyond understanding—and better fit the biblical claim of miracles of providence and special acts from God. Moreover, the arrows penetrating the circle show that revelation of God in created nature, known in man's heart and special revelation from Moses, the prophets and the Psalms/poetic literature, prepared the way for the fulfillment of meaning of God and his kingdom in Christ.

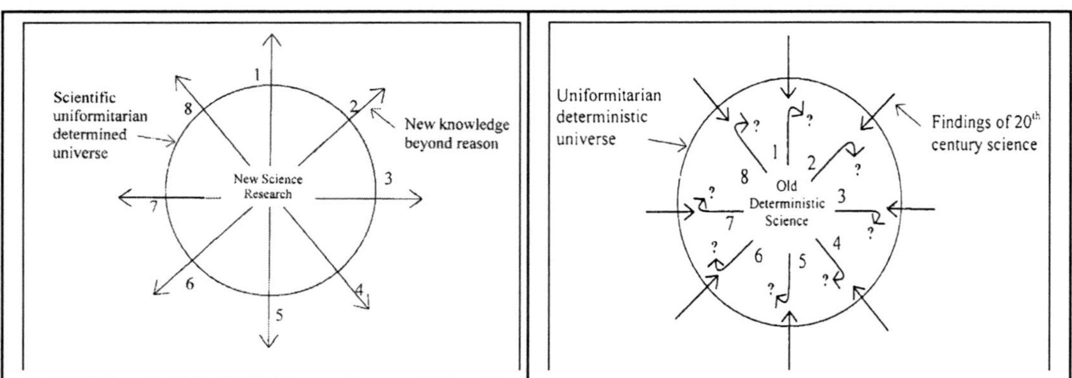

The graph of the history of modern science previously given indicating phases of change in understanding also suggests why the philosophy of science went through transitions to adjust to the emerging facts and to retain credibility without God.

Changes of Man's Understanding of Nature by Science Required Adjustment in Philosophy

The philosophy of science had to make changes to try to adjust to what was claimed to be true. Reformers' philosophy was a sustaining Creator allowing continuing paradigms and catastrophes. Man's biased speculation about uniform determined nature required another philosophy. Then mystery and power beyond determined knowledge and efforts to adjust to mystery required other changes. The graph outlining the history of changes in science helps understand these changes. The philosophies of science as they developed are as follows:

1. **Inductive Creation science.** Reformers phenomenal observation of created nature for inductive facts about what occurs; theorized why and how it works; record repeated tests and prove its regularity and reliability because it is sustained and upheld by the Creator. This assumes improved observation and minor modifications in the paradigm. Truth is

defined in the consistent success and usefulness of the theory. All the reforming scientists from da Vinci, Galileo, Kepler, Newton, et al.

2. **Naturalistic Uniform science.** Presumption that all objective data can fit and progress toward evolutionary presuppositions meant all data interpreted as catastrophic or beyond daily observation was to be rejected for only regular daily gradual working over millions of years. Truth exists only if the theoretical teaching fits individual evolutionary development in some way. Kant, Herschel, Hutton, Lyell, C. Darwin, et al.
3. **Falsification science.** Since evolution means all nature is changing without known purpose or by mystery, the accumulation of experienced knowledge does not produce truth, so there is no truth. There can be no anticipations to be dogmatically upheld. Science cannot be a system of established statements or experiences to be accumulated and organized. So truth is no longer based on repeated reliability. If anticipations are proven ***not to be false*** they are acceptable. Karl Popper proposed and was known for this falsification theory in the late 1930s.[2] The old science of proof by testing of a theory was ruled out by mystery. By the falsification theory, only a paradigm of science is acceptable if it cannot be proven to be false. In this way, the evolutionary theory, the big bang and the inflationary theory, determined geological principles, the evolution of man from primates, can be acceptable even when progressive knowledge is mystery or unfounded.
4. ***Structural Revolutions of change*** was Thomas S. Kuhn's theory in the late 1960s after the Chaos theory of everything. In pursuing science history, Kuhn proposed that science is always changing and making significant structural shifts, as revolutions in thinking occur when scientists face new problems. In this way the scientists develop new useful temporary paradigms.[3] Kuhn believes that all observations, theories, and paradigms are in flux, and hence there are no firm fixed structures. There are only revolutions or breakthroughs in major paradigm structures which in themselves are not structures that abide. All is relative and changing so that there is no real history of science. He goes beyond and rejects Popper's falsification theory and really applies postmodern deconstruction to the whole of scientific theory. This view acknowledges that all determinism in science was false and another philosophical view of science was needed. Kuhn's theory excuses the scientific community from having final solutions but mystery in science. It only has a process of discovery. His views gained dominance in universities about 1970. The Chaos theory led to scientific uncertainty of everything. Kuhn's theory was an effort to adjust to deconstruction of Neo-Marxist postmodernism. But this complete relativism tends to violate the great principles of science that all modern developments in science used to give stability. Only by believing the fact that Christ upholds all things, visible and invisible by the word of his power can a person trust natural laws and do science.

Since the Structural Revolutions theory has gained dominance in science education, enlightenment liberal science has lost its scientific base and is struggling to offer new meaning and find new authority for their teachings. This theory was needed as they lost credibility in science. They now believe that ninety six to ninety seven percent of the universe is dark matter or energy; that they are not certain about how the universe or man developed. The distribution of life across continental bridges or across oceans is more reasonably interpreted by the break of Pangaea into continents but is not certain. The origin of life is unknown and its development was immediate and extinctions fit the biblical history of the flood and the continental break up. Long ice glaciation

---

[2] Karl Popper, *The Logic of Scientific Discovery* (New York: Rutledge Press, 1957), Chapters IV and X, cf. 275-280. He built on David Hume's skepticism.

[3] Thomas S. Kuhn, *The Structures of Scientific Revolutions* (Chicago: University of Chicago Press, 1962, 1970, ff.).

and ice sheets that formed at the P-T catastrophe have gone through slow melting that is now accelerating because of feedback systems. Old Enlightenment natural deterministic science is destroyed by most recent modern scientific findings, and the authority of liberalism that rested on the deterministic theory has been discredited. Unbelieving mankind is struggling to find a new basis for authority. They have declared God is dead and all religion is man-made and needs to be synchronized for world unity and peace. The basis of uniform deterministic science has collapsed, resulting in a return to mystery and catastrophe, but without God. A history of relativism in science with perpetual uncertain change is the result of not having a God who sustains the universe as his regular way of working. The theory of evolution by nature of all things was a dogma without any proof with an optimism about man to free him from accountability to God.

Anti-God Educational Propaganda Spread Unsubstantiated False View of Evolution,

These amazing perversions based on human optimism were perpetrated by self-righteous educational leaders who were not scientists with the intent of promoting material prosperity by putting revolution over faith in God. Almost all the early American colleges and town schools were created by the church or church and state to support faith among the American people. Of the first eighteen colleges and universities, all but one was formed by church or church and state and that one was saved from bankruptcy by an offering from Whitfield. Even the Northwest Ordinance of 1789 in the Constitutional Convention said "the purpose of these schools is religion, morality and knowledge…"[4] The First Amendment rights had in mind protecting the rights of Christian teaching in schools and in the public square more than anything else. The movement for public education from the time of Horace Mann was to take over public education for Enlightenment education[5] and the Virginia Dynasty of the Jeffersonian Republicans had that in mind until the Jackson Democratic revolt. The First Amendment has been turned on its head by the use of the new definition of science that had been started by Kant, gained prominence in America in the nineteenth century. By the end of the nineteenth century, John Dewey's pragmatic functional instrumentalism was widely promoted and controlled by his and Nicolas Murray Butler's concept of teacher training schools and requirement that teachers be certified by liberal standards.

The progressive political intellectuals at the turn of the twentieth century in government, beginning with Theodore Roosevelt and Woodrow Wilson, was achieving the purpose to replace Christian schools with state run schools of enlightenment. But Wilson, in order to carry the South, maintained segregation, which set the stage for a racial natural rights rebellion later in mid-twentieth century. However, as the twentieth century began, through the money of tycoons like Andrew Carnegie, who was an avid evolutionist, teacher organizations (such as the NEA—the largest) promoted and controlled secular education by a leading secular educated elite. By 1920 all states had public schools that required elementary school attendance. By the first quarter of the twentieth century the centers of the enlightened educated elite were gaining control and seeking to remove religion, thus stealing public education to remove religion.

In higher education, the supremacy of uniform deterministic natural science replaced God and his revealed will, especially in prestigious schools in the Northeast and West. Since this was the accepted view of the elite, many professional anthropologists completely rewrote history based on gradual evolution of man. After the fake Scopes trial many leading educators were determined to replace creation by evolution. While in Europe, hope in man and science to build supermen was then dimmed by war and lack of evidence, in America the concept of man's wisdom and progress was like a hurricane sweeping away religious influence. In the interest of promoting progress, material prosperity and social action, government intruded into religious ideas, especially from Theodore Roosevelt forward, and promoted the social gospel for the churches to help replaced worship and glorifying God by man and mammon. By 1960 liberal education had gained

---

[4] Ibid, 136.

[5] Wilson, *Man Is Not Enough*, 128-136, especially 134 ff.

dominance in the Supreme Court to eliminate Christian teaching in the public schools. The NEA had unionized by 1960 and gained the rights to deduct dues from the salary of teachers, giving them huge funds to promote politics to remove Christianity.

The American Council on Education (ACE) was formed in 1918 and now involves most of the higher education organizations. It is committed to the politically correct ideas and natural equality in contrast to equality by creation by God. ACE leaders are adamant that Christianity must be excluded from education. The vice president of curriculum studies of the American Education Research Association is William Ayers of the radical neo-Marxist SDS and Weathermen. Also in recent years there is a smaller group of about 700 college and university presidents that are specifically anti-Christian. These organizations are making a maximum push to eliminate Christianity while allowing all other kinds of religious studies.

The hindrance to the liberal elite achieving this goal of secularization has been the fact that most boards of education are local. Also, since 1960 there has been a resurgence of private schools, many which are Christian, and there has been a strong home schooling movement, both of which have produced superior education. There has been an effort to make all teachers comply with the standards of public secular teacher training for certification and to therefore eliminate any education that does not comply with this by court action. The new educational director under the Obama administration of the Federal government is against Christianity.

America is at a very dangerous point. The apostle Paul tells of the crisis point of evil in Greece and Rome as nations. "And since they did not see fit to acknowledge God, God gave them up to a debased mind to do what ought not to be done… Though they know God's decree that those who practice such things deserve to die, they not only do them but give approval to those who practice them. …We know that the judgment of God rightly falls on those who do such things" (Romans 1:28, 32; 2:2). Can America and the West expect any other outcome? It is important the churches return to a firm faith in Christ as Creator and Re-Creator to fulfill Christ's great commission before he comes.

Chapter 33

Present and Future in Secular Science

The Status of Disbelieving Science Today Shows Closing Opportunities in Science

Is modern perverted science reaching a terminus? The feeling has been growing that humans will soon reach the limits of their knowledge. As early as 1932, J. B. Burry expressed this in his book, *The Idea of Progress*, as did Gunther Stent in 1978 in *The Paradox of Progress*. In an essay in *Science* written in 1971 Bentley Glass, while president of the American Scientific Association for the Advancement of Science, questioned whether there would come a time when human understanding would reach its limits. John Horgan, a reliable staff writer for *Scientific American,* published a revealing and controversial book in 1996 entitled, *The End of Science.*[1] Horgan did not seriously mean scientific work would end; he was pointing out that science had probed all major areas of science to the extremes that current instruments could reasonably reach. New instruments are so costly and they would not likely produce anything radically new.

Neither Horgan nor other prominent writers see science as useless or science as a dying profession. Practical science will still probably offer jobs for thousands and be profitable for people around the world for years to come. American greed for new products and the pride of presenting something new, as well as the market in the developing world, will motivate secular science. But science and public media are increasingly uncovering dishonesty and deceit in scientific work. Many scientists feel that science is reaching a point where no more major theories can be devised, found, and understood. Therefore, the opportunity for heroes who break through to new major principles or revolution of structure is more limited than before. However, the human mind is very creative and who knows what it may yet find and prove? But Horgan's arguments are more manifestly true.

Practical science has not benefited, but been hindered by the theories that scientists concocted to rule out God for evolution. Practical science will go on using the foundational paradigms mostly found by the Reformers until there is an economic or moral breakdown and research can produce few benefits. As recently as 2000 the large book stores that I frequent, had five book shelves on science. Now they have two, and most of the books are about the human mind, psychology, and the astronomical universe, all which are still mostly a mystery. Science magazines feature many major articles that are pure speculation about multi universes, planets that may be inhabited, possible creation of life, or life from somewhere unknown in the universe, or unknown aliens that may communicate.

The urgency to discover beyond the known rational limits is calling for enormous government and corporate expenditures of money on ever more exaggerated projects that are producing a minimum of scientific evidence and western debts are very prohibitive. Even many notable scientists are expressing skepticism about the extreme efforts of science. For example, the search for Higgs bosons has been desperately pursued by use of existing cyclotrons with only negative success of where it isn't, and newer, much bigger and much more expensive ones are now being built. As Peter Weiss has pointed out, "While theorists have had a heyday with Higgs physics ever since (1964), experimentalists have run into brick walls for nearly 30 years." Scientists have estimated that Higgs can't have a mass smaller than 113 GeV. He comments, "It is easier to rule out particular masses for Higgs boson than it is to establish the particle's reality."[2] To find this legendary particle, scientists must bash together a billion trillion or more familiar

---

[1] John Horgan, *The End of Science: Facing the Limits of Knowledge in the Twilight of the Scientific Age*).

[2] Peter Weiss, "Jiggling the Cosmic Ooze," *Science News*, Vol. 159, 153, 154.

subatomic particles at energies higher than any previously achieved, and a single accelerator being built to do this job will cost $4 billion. The largest cyclotrons have, in 2010, drawn some limits of where the Higgs cannot be, but that proves nothing as real. This search is such a long shot that some leading scientists have made it a joke.[3]

Thus, the nature of matter and its mass are a giant mystery. All areas of science have reached vast lakes of data that are beyond human comprehension. Secular science has more knowledge, but less understanding. The probability theory leaves openness to the probability of God and indeed points to the mind of God. While most scientists reject God because they have accepted the secular definition begun by Kant, they are not legitimately doing so on the basis of man's determined knowledge but because of dogma. They do not know the history of science and that the major paradigms were developed under faith.

The complexity of all fields, especially molecular biology has produced a move toward intelligent design. But many scientists reject it because of Darwin's claim, in his foundational theory, to reject any unseen purpose or powers of the occult in evolution. But, as we've noted before, Darwin was not consistent. In his final edition of *The Origin*...he admitted a driving purpose for evolutionary progress. Modern science had rejected Aristotelian pantheism of the Unmoved Mover who was actualizing the potential of all things with incorrect views of many things that was adopted by the perverse Roman Catholic Church. A similar pantheism emerged in different ways. Lamarck, in studying for the priesthood was influenced by Aristotle and used pantheism in presenting biological evolution driven by a natural directive force in the form of progressive creationism. Chambers in *Vestiges*... presented evolution as a series of natural creations from nature as a substitute for acts of a personal God. Modern knowledge confronts modern science with intelligent reason behind the universe that was accepted by the Reformers as the personal Creator.

Present Alternatives to God with Intelligent Design

Return to Occult or Quasi Pantheism as an Option to Claim Determinism

While the majority of scientists are struggling with new views to which they give a pantheistic interpretation, most of them do not yet openly accept a total pantheistic view. They interpret pantheistically, without acknowledging pantheism, e.g. a whale developed finlike legs because it wanted to walk on land, or the dinosaur that grew feathers in order to fly. But there are many scientists, in the most prominent universities, who have already been advocating a pantheistic philosophy for the entire universe. I've mentioned some previously. Verba, whom Stephen Gould claimed as his close helper, appealed to Hinduism to account for pulses of weather she claimed could create new species.

But some scientists, who embrace pantheism completely, occupy prominent positions: Ilya Prigogine (a Nobel prize winner who died in 2003) of the Free University of Brussels, who also lectured at the University of California, Berkeley, and the University of Texas in Austin, is the chief "catalyst" for promoting the occult power of nature. One of his main followers, co-conceptionists, and promoters is Erich Jantsch (1929-1980) at the Center for Research in Management, U.S. Berkeley.[4] Fritjof Capra, the author of *The Tao of Physics* and *The Turning*

---

[3] Roger Penrose and Brian W. Aldiss, *White Mars: the Mind Set Free,* 2000. Roger Penrose, who pioneered black holes and mentored Hawking, has jokingly indicated skepticism that Higgs particles can be found even with experimental colliders as large as around the rings of Saturn.

[4] Erich Jantsch, *The Self-Organizing Universe: Scientific and Human Implications of the Emerging Paradigm of Evolution* (New York: Pergamum Press, 1980), 343 pages.

Point seeks to give pantheistic religious meaning to all areas of science and medicine.[5] The concept of the self-organizing universe is becoming accepted by many who argue for organized chaos or complexity, such as some at the Sante Fe Institute in New Mexico. James Lovelock of England is a leader of the self-organizing theory.[6] Lynn Margula, of the National Academy of Science favors powers in nature to organize in biology. Marvin Minsky believed that at the end of evolution, forces would cooperate and merge into a single metamind. Freeman Dyson speculates that intelligence might spread through the entire universe, transforming it into one great mind or world soul as god. Frank Tipler's cyclic omega point, based on an eternally oscillating universe, is a speculation about a dualism in nature that is pantheistic and occult.[7] Carl Rogers, who popularized psychology in the universities and a leading authority in this field for years, ended up in having séances with the dead. Carl Jung adopted Freud's belief in Gnosticism and his use of this has become a prominent part of Jung's popular psychology. William Straus and Neil Howe are leading historians advocating an innate principle in successive generations that denies a linear concept of history and advocates a cyclic view that promises a better future. This is essentially a pantheistic approach.[8] The list could be expanded to scientists in all fields. But most scientists have been influenced by Darwin to reject intelligent design, and not because they have studied the issue.

Pseudo Christian and other religious pantheistic views are gaining wide acceptance in America. The resurgence of witchcraft from Africa and in South America, and the influence of modern Buddhism, Taoism, Shintoism, and Hinduism are growing. In the last twenty-five years, temples of worship to all these religions have sprung up across America. Liberal Christian theology, based on deterministic science, is moving toward pantheism or a limited God within nature. Thus, the trend in science is moving toward a meeting with liberal theology. Alfred North Whitehead long ago suggested a process based on the worship of life and a pantheistic philosophy. Paul Tillich was accused of holding to a pantheism with his faith in "the ground of being." Wolfhart Pannenberg and other process theologians deny present miracles and the resurrection in Christ, but speak of a theology of hope in a future resurrection. Roman Catholic theologian Tielhard de Chardin has become a central figure in the modern New Age movement and his view is basically pantheistic. The death of God theologian T. J. Altizer drew on the occult in pointing to the man of the future. Even some evangelicals who are now calling for "an open theism" are advocating a limited God who is not sovereign and is subject to the acts of man and events of nature. While these men are not now pantheists, their views are a big step toward limiting God in nature and exalting man. But none of these views gain prestige as modern science.

The "Mind of God Theory" or Comprehensive Natural Determinism

Without consciously admitting it, the scientific world has been pushed to admit that there is one logical mind behind the universe that is transcendent to their own and in whose image they are made which allows them intuitively to search to understand it.[9] Their human mind is the greatest

---

[5] Fritjof Capra, *The Turning Point* (New York: Bantam Books, 1983), 464 pages.
[6] James Lovelock, *The Ages of Gaia: A Biography of Our Living Earth* (New York: W. W. Norton, 1988), 252 pages.
[7] Cf. Michael Demonick, "Life, the Universe, and Everything," and Laura Marlowe, "Deadly Science," *Time*, February 22, 1993, 62-64.
[8] William Straus and Neil Howe, *Generations: The History of America's Future*, 1584-2064 (New York: Quill of William Morrow, 1991).
[9] Paul Davies, *The Mind of God: A Scientific Basis for a Rational World*, 1992.

mystery and they use it to reject God. Some of them even speak of this final theory as "the mind of God."[10] While Steven Weinberg denies there is a God, he has even said,

> If there were anything we could discover in nature that would give us some special insight into ***the handiwork of God***, it would have to be the final laws of nature. Knowing these laws we would have in our possession the book of rules that governs stars and stones and everything else.[11] (Emphasis added)

Brian Greene has explained the complications of the various string theories which goes beyond what is needed in this discussion. [12]

While there are numerous final theories, the most popular one is the string theory, which has many diverse definitions. The string theories began about 1970, after the chaos theory reveal indeterminacy in all science. But string theories made a number of predictions that were in direct conflict with observations, and the success of the point-particle quantum field theory was so successful in describing the strong force that it led to the dismissal of the string theory at that time. However, an earlier suggestion had been made that the string theory could go beyond explaining the strong force to include all four forces. But some comprehensive determinism was needed to support unbelieving science and the search could not die.

In 1984-1986, the first revolution for a superstring theory occurred, producing over a thousand research papers. But the venture died again because the approximate equations to solutions were inadequate or too difficult. A second revolution occurred at a string conference in 1995. Edwin Witten suggested a method for uniting the various string theories by postulating space with many, six or seven dimensions, all unobserved, and not just the three we know, some being reflections of others. He added an element M, saying a string could expand to one or many membranes as big as universes that could collide causing a big bang.[13]

Difficulty of Experimenting to Prove Such a String Theory is True

Greene illustrated that string theories involve diminishing levels from an apple, atoms, electrons, proton, neutrons, quarks, and the idea of strings. His illustration can be very misleading because his graph conveys nothing about the size of these diminishing particles. Greene points out that a string is ten to the twentieth power (10^{20}) times smaller than an atom. He also said "...The huge string tension causes the loops of string theory to contract to a minuscule size. Detailed calculation reveal that being under Planck tension translates into a typical string having Planck length—10^{33} (or ten to the thirty third power) centimeters."[14] The size is so small our imagination fails to picture it.

In discussing "The Road to Experiment," Greene is honest at the seeming impossibility of direct experimentation to prove the string theory or even at developing technology for it. In essence, he says that the only hope is by physical implications it will appear more likely true, which is not proof. He describes the difficulties as follows.

---

[10] Stephen Hawking, *The Universe in a Nutshell*, (New York: Bantam Books, 2001), 175.
[11] Steven Weinberg, *Dreams of a Final Theory: A Scientist's Search for the Ultimate Laws of Nature* 241, 242.
[12] Brian Greene, *The Elegant Universe*, (New York: A Vintage Book of Random House, Inc., 2000).
[13] Michael J. Duff, "The Theory Formerly Known as Strings," *Scientific American*, February 1998.
[14] Brian Greene, *The Elegant Universe,* 148.

> Without monumental technological breakthroughs, we will never be able to focus on the tiny length scales necessary to see a string directly. As the Planck length [of a string] is some 17 orders of magnitude smaller (10^{17}) than what we can currently access, using today's technology we would need an accelerator the size of the galaxy to see individual strings. In fact, Shumuel Nussinov of Tel Aviv University has shown that this rough estimate based on straightforward scaling is likely to be ***overly optimistic***; his more careful study indicates that we would require an ***accelerator the size of the whole universe***.[15] (Emphasis added)

Greene added, "Don't hold your breath while waiting for the money for a Planck-prodding accelerator."[16] While strings are normally projected to be extremely short of Planck length, since what happened, if at all, at the big bang is a complete mystery, scientists fantasize the possibility that mysteriously some may have been produced under such huge energies that there would be strings that could be visible. Whitten has said, "Although somewhat fanciful, this is my favorite scenario for confirming string theory as nothing would settle the issue quite as dramatically as seeing a string in a telescope."[17] Greene's new book[18] is projecting a grand design like Hawking, which we will discuss.

Thus, the authorities of the string theories admit that for them to know the "mind of God," which is involved in such a theory, it would require a mysterious miracle, because the possibilities of proving the theory are more improbable than a miracle. Yet, most of these scientists dogmatically assert all past theories of deterministic science have been proven true, and they passionately assert the truth of superstrings, while in the same breath admit it is unproven and probably cannot be proven. Hawking admits that a final theory probably can't ever be proved and that,

> We could never be quite sure that we had indeed found the correct theory.... But if the theory was mathematically consistent and always gave predictions that agreed with observations, we could be reasonably confident that it was the right one.[19]

However, mathematical consistency or computer simulations are not always a correct demonstration of what happens. Even Hawking admitted this fact when he was willing to renounce his mathematical demonstration of the big bang for another model. Barry Parker has said, "Many things appear beyond our mathematical models, and this means a theory of everything may not exist. Chaos may forbid us from knowing everything."[20] What Hawking and others were seeking in the ultimate theory is to again be able to have some sense of determinism of the universe, which would seem to give man the mind of God and rule out the need for God. They think this would restore waning scientific authority.

Speculation of Need for a Grand Design of New Deterministic Rational Universes

In the twenty-first century, evidence of a precise, rational intelligent design emerged, requiring an explanation for our universe's anthropic friendliness. Several efforts have been made

---

[15] Ibid., 215.
[16] Ibid.
[17] Ibid., 224, as quoted from and interview with Edward Whitten, March 4, 1998.
[18] Brian Greene, *The Hidden Reality: Parallel Universes and the Deep Laws of the Cosmos,* (New York: Alfred A. Knopf, Division of Random House, 2011).
[19] Hawking, *A Brief History of Time,* 167.
[20] Barry Parker, *Chaos in the Cosmos: The Stunning Complexity of the Universe* (New York, NY: Plenum Press, 1996); 2, 4, 285.

to construct a unifying theory that claims to include everything in the universe that fits the facts. The universe is known to be specific and the only known place specific for life is the earth. The new science, since the discovery of the quantum theory, the theory of relativity, and the revolutionary explorations with new technology in the twentieth century, has revealed an unprecedented uniqueness of the earth and its universe. No other planet has the capacity to sustain life; the universe is designed in such a way that even if there were slight differences in its makeup and location, life would not be possible. Scientists such as Martin Rees, and earlier contributors Hoyle and Gamow have examined this and marvel at the world's precise friendliness to man.

Martin Rees, named Britain's Astronomer Royal by the queen, and astronomer of King's College in Cambridge, England, has written a new book, *Six Numbers*. In it he lists six phenomena identified by one Greek letter that sets our universe apart as unique for maintaining life. He claims these six encompass everything from quantum weirdness, to biological imperatives to galactic clumping. If any of the six were different, life would not exist on the earth. Brad Lemely presented a condensation of Rees' views from an interview for *Discover* magazine.[21] Rees' six figures are as follows:

ε The strength of the force that binds atomic nuclei of .007 of the mass of hydrogen, which becomes energy when the sun produces helium. It could not be .006 or a proton could not bond to a neuron. If it were larger than .008 fusion would be so rapid, no hydrogen would have survived the big bang.

N is equal to the number measured for the strength of the force that holds atoms together, divided by the force of gravity between them. (10^{32}) If this is weaker than the force holding atoms together and if it was any smaller there could be only a miniature universe that would not have lasted.

Ω (Omega) is the measure of the density of the material in the universe and shows the relative importance of gravity in an expanding universe. If gravity were stronger the universe would have collapsed, but had it been weaker there would have been no galaxies or stars.

λ represents the strength of cosmic antigravity forces that controls the expansion of the universe. If it were larger, it would stop galaxies and stars and life from forming.

Q represents the amplitude of complex irregularities or ripples in the expanding universe that seeds the growth of galaxies and planets and is equal to 1/100,000. A smaller ratio would produce a cloud of dust and a larger one would produce globs of matter that would contract to black holes.

Δ is the number of spatial dimensions in our universe, which are three. Life could not exist with two or four.

Rees argues that "These six numbers constitutes the recipe for the universe. If any one of the numbers were different even to the tiniest degree, there would be no stars, no complex elements, no life."[22] One would think that Rees and many others who agree with him would assume that if the world is so ordered for man and could be no different, there must be a wise Creator who formed it. With all the mystery introduced by modern science, and the evidence there is intelligent

[21] Brad Lemely, "Why is There Life?" *Discover*, November 2000, 66.
[22] Ibid.

reason behind the world, there would be room for God to work in the world. The intent of the string theorists and of Rees and his followers is to try to create another but more complete fixed view of the universe. However, this view also leads to a determinism that could breed fatalism that arises from pantheism. Unbelieving scientists such as Rees cannot rest with the facts that point to a wise Creator.

Martin Rees has boldly accepted the uniqueness of the universe for life, and especially for mankind. But like most of the scientific community, he is unwilling to attribute the precise conditions to a transcendent God. He therefore fled to the theory of Andre Linde at Stanford called "the self reproducing inflationary universe." He holds that as the universe expanded or inflated from the big bang, there were quantum, wavelike fluctuations in the universe. Linde theorizes that these waves could "freeze" one on the top of another and thereby magnify and cause very intense disruptions in scalar fields and exceed a cosmic critical mass that would start birthing other universes in other domains. Each of these domains would then spread and cool into a new universe. Thus, there could be a multiple of universes created which are unavailable and unknown to us. This is sheer speculation with no known evidence that this has or could occur.

Bruce Lemely, who was sympathetic to Rees's views of multiple universes, said, "The problem then, as now, is that most theories say the universes must remain forever inaccessible to one another even in principle, which makes the multiverse seem little more compelling than the conjured-by God hypothesis." Rees admits that, "at present the premise upon which many multiverse calculations rest are highly arbitrary."[23] Brian Greene, Weinberg, and many string theorists adopted Linde's speculations about a multiverse or many universes.[24] These men speculate that there are perhaps thousands or millions of universes and that in them there are other life forms with intelligent beings. The search for communication with living beings in our universe is extended as a possibility of life in other universes. This is all speculation with no evidence.

The purpose of this arbitrary speculation is that, if there were a huge number of universes, it would be ***more likely that one or more by chance*** could turn out with conditions for life. In their efforts to escape the idea of a God who creates, they are forced to speculate on other universes with other intelligent beings. Hawking also argues for multiple universes in order to hold to a very limited anthropic view.

There were a number of efforts near the turn of the twenty-first century to find a complete view of the universe, such as Weinberg's, Greene's, Rees' and Linde's. Timothy Ferris, emeritus professor of the University of California, Berkeley wrote *The Whole Shebang*. He points out that the greatest obstacle is that all things originated miraculously from a single source. Ferris points to the paradox, "You can't get something from-or for-nothing. The 'origin' of the universe, if that concept is to have meaning, must create the universe out of nothing." At the same time he warns, "The doctrine of causation erodes considerable when applied to the subatomic realm of quantum physics, and there seems a dubious tool for understanding the early universe, when virtually all structures were subatomic."[25] No viable explanation for the universe had been offered using the natural deterministic view. But, now an effort has been made to describe a grand final understanding and because Stephen Hawking, whom many consider the world's greatest physicist, gives such a view, it may be the last effort to persuade the unbelieving world that God is not needed. For that reason an extensive look at Hawking's book is important.

---

[23] Ibid.
[24] Greene, *The Elegant Universe*, 366-370.
[25] Timothy Ferris, *The Whole Shebang: A state of the universe(s) report* (New York: Simon and Schuster; 1997), 247.

Grand Design: A Naturalistic Beginning and a Deterministic Universe without God

Stephen Hawking has been deeply interested in a final naturalistic theory which he called "the mind of God." Together, with Leonard Mlodinow, a physicist at Caltech, they produced what they claim is probably an answer to a final theory which they say eliminates the need for God or miracle. They name questions of interest: "How can we understand the world in which we find ourselves? How does the universe behave? What is the nature of reality? Where did all this come from? Did the universe need a Creator?" In their book, *The Grand Design* they said,

> The purpose of this book is to give the answers that are suggested by recent discoveries and theoretical advances. They lead us to a ***new picture of the universe and our place in it that is very different*** from the traditional one, and different even from the picture we might have painted just a decade or two ago.[26]

The authors declare that ***the universe is ruled by natural law***, but they define this differently ***with a new definition of natural law*** than in traditional science with three questions: What is the origin of the laws? Are there any exceptions to the laws, i.e. miracles? Is there only one set of possible laws? The first and third on origin and the possibility of alternative laws occupy much of the book, and obviously these preparatory presentations relate to the answer to number two about miracle. They end chapter 2 on natural law by saying, "This book **is *rooted in the concept of scientific determinism***, which implies that the answer to question two is that ***there are no miracles***, or exception to the laws of nature."[27] But they are obviously dealing with mystery that they can not allow miracle. In the final chapter, "The Grand Design" [of the universe] they use Richard Feynman's functional integral formulation of quantum mechanics, and also claim gravity as being the creative power which fulfills the M-Theory of strings. In the four chapters the authors present ideas to prepare the reader to reject the idea of God and to accept their closing arguments for how creation and the universe exist.

In "What Is Reality?" (Chapter 3) they argue for "model dependent realism" which is Thomas Kuhn's basic postmodern idea that there are revolutionary structural changes in models that are sufficient only for that time the scientist is making a paradigm. For their argument Hawking/Mlodinow, used real models such as Ptolemy's Aristotelian view of the solar system that changes to the model of Copernicus, and Matrix fiction, etc. It is said that in classical science the external world is independent of these observers. This prepares the reader to accept their new ***speculative models as* real**.

In chapter 4, "Alternative Histories," they argue that two alternative theories may be true, such as particles and waves, and exemplify this with two different scientific experiments with minuscule buckyballs. ***Imaginary players*** kick balls through two different slits in a wall; behind the slits are screens that open and close and catch that balls that go through. By opening a second gap, fewer molecules (balls) landed at certain points. This was to illustrate that quantum mechanics in nature operates differently than had been seen in science up until the quantum theory in early twentieth century. The Werner Heisenberg's theory of ***indeterminacy***, which they and others call, the theory of ***uncertainty*** showed we could not know the velocity and position of a particle at the same time. These ideas are to prepare for Richard Feynman's brilliant theory that somewhat altered the picture of uncertainty.

---

[26] Stephen Hawking and Leonard Mlodinow, *The Grand Design*, (New York: Bantam Books, 2010), 5

[27] Ibid., 34.

> In the double-slit experiment Feynman's ideas mean the particles take paths that go through only one slit or only the other; paths that thread through the first slit, back out through the second slit, and through the first again.... This, in Feynman's view, explains how the particles acquire the information about which slits are open—if a slit is open, the particle takes paths through it. When both slits are open, the paths in which the particle travels through one slit can interfere with the paths in which it travels through the other, causing the interference.[28]

Thus, in Feynman's theory there are an infinite number of paths for the particle to travel and therefore there ***may be*** multiple histories. In Newton's theory there is only one path with a definite series of logical events, and therefore only one history. Moreover, Feynman explained the interference pattern—paths that that go through one slit interfere with paths that go through the other.

> Thus Quantum physics tells us that no matter how thorough our observation of the present, the (unobserved) past, like the future, is indefinite and exists only as a spectrum of possibilities. The universe, according to quantum physics, has no single past, or history.[29]

One must respond to Hawking's conclusions by pointing out that the sum or the conglomerate of the various particles' history has occurred ***within*** our universe's history. In reality there is one history of multiple histories that are unknown. To not know past events does not exclude them from one real history.

In chapter 5, "The Theory of Everything," after tracing the discovery of the four forces in the universe, the authors attempt to see the relationship and the understanding of the unity of the forces, e.g. Weinberg's resolution of the electromagnetic and weak forces united into the electroweak force. They proceed to imply the probability for a theory of everything. They considered Feynman's Quantum Electrodynamic force (QED) and the renormalization of infinities to give a sum of all histories and other considerations of the forces. They *admit the difficulty of merging all the forces*, especially in a quantum gravity theory.[30] After mentioning other technicalities of dealing with infinities, Hawking/Mlodinow arrive at a point where the only possible solution to a unified theory would be in a string theory.

Early considerations of string theories envisioned four or six dimensions. Hawking and Mlodinow briefly describe their view of stings.

> According to string theory, particles are not points, but patterns of vibration that have length but have no height or width—like infinitely thin pieces of string. String theories also led to infinites, but it is believed that in the right version they will all cancel out. They have another unusual feature. They are consistent only if space time has ***ten dimensions***, instead of the usual four. ... If they are present, why don't we notice these extra dimensions? [Because] they are curved up into a space of very small size.
>
> The most fundamental theory is called M-theory. ...Whether M-theory exists as a single formulation or only as a network, we do know some of its properties. First, M-theory ***has eleven space-time dimensions, not ten***.... one dimension had indeed been overlooked. ... [The M-theory] determines the ***apparent*** *laws of nature.* We say 'apparent' because we mean the laws that we observe in our universe—the laws of the four forces, and the parameters such as mass and charge that characterize the elementary particles. But the more fundamental laws are those of M-

[28] Ibid., 75, 76.
[29] Ibid., 82.
[30] Ibid., 112.

> theory. The ***laws of M-theory therefore allow for different universes with different apparent laws***, depending on how the internal space is curled.[31] (Emphasis added)

The authors insist that this view of M-theory is testable for probabilities. In "Choosing Our Universe," Hawking discusses the problem of a singularity in regard to the big bang theory which leads to a beginning, which seems to require a miracle. There is lengthy discussion of how this beginning might have occurred, and evaluation of the microwave background as evidence that it did happen. Then they discuss how Feynman's summary of histories in quantum theory might be involved in a beginning and in creating many universes. The authors end up proposing a new view of the universe that requires ***a new view of physics***.[32] For them, classic physics must be discarded in order to find a final theory.

Chapter 7 evaluates Rees's "The Apparent Miracle" and why the universe is so specific for life and man. After evaluation and discussion, the authors admit the following.

> Our universe and its laws appear to have a design that both is tailor-made to support us and, if we are to exist, leaves room for alteration. This is ***not easily explained***, and raises the natural question of why it is that way.[33] (Emphasis added)

Their answer is that by M-Theory with many universes beginning possibly as with multi-results in Feynman's quantum theory, physical laws produced this kind of ***luck*** that produced this world.[34]

The final chapter in *The Grand Design* is an argument that a universe designed for life and man, is based on arguments, not from science but from Conway's Game of Life. They present this as an example "that even a very simple set of laws can produce complex features similar to those of intelligent life."[35] They argue that life can occur by natural chance and be meaningful and even more complex than Conway's simple game. In the end, Hawking's and Mlodinow's premise for a final theory comes down to the last three pages in their book. They argue that empty space must have equal positive and negative energy or be stable, otherwise there would be motion. Notice their use of the word ***must,*** and ***if*** rather than give facts. They say,

> If the total energy of the universe must always remain zero, and it costs energy to create a body, how can a whole universe be created from nothing? That is why ***there must be*** a law like gravity. Because gravity is attractive, gravitational energy is negative.
>
> Because gravity shapes space and time, it allows space-time to be locally stable but globally unstable. On the scale of the entire universe, the positive energy of the matter can be balanced by the negative gravitation energy, so there is no restriction on the creation of whole universes. Because there is a law like gravity, the universe can and will ***create itself from nothing***.... Spontaneous creation is the reason there is something rather than nothing, why the universe exists, why we exist. It is ***not necessary to invoke God*** to light the blue torch paper and set the universe going.
>
> We have seen that there ***must be*** a law like gravity, and ...for a theory of gravity to predict finite quantities, the theory ***must have*** what is called supersymmetry between the forces of nature, and the matter on which they act. M-theory is the most general supersymmetric theory of gravity. For

---

[31] Ibid., cf. 115-118.
[32] Ibid., 139,143.
[33] Ibid, 162.
[34] Ibid., 162.
[35] Ibid., 179.

> these reasons ***M-theory is the only candidate for a complete theory of the universe***. It is finite—and this has yet to be proven—it will be a model of a universe that creates itself. We must be a part of the universe, because there is no other consistent model.
>
> ***If*** the theory is confirmed by observation, it will be the successful conclusion of a search going back more than 3,000 years. ***We will have found the grand design***.[36] (Emphasis added)

Critique of the Grand Design as Demonic Delusion

Hawking's and Mlodinow's theory of a grand design presents many difficulties. The most fundamental is that it rejects God in favor of the two main creative events—gravity and Feynman's multiple directions of particles in quantum physics with the M-theory for promoting multiple universes. First, the theory of gravity as a force is not yet scientifically understood, which is also true with God. However, gravity is a force ***involving mass*** that exists as understood by Newton, Einstein, et al known ***in our universe*** and nowhere else. Therefore, gravity is a force involved ***after creation and now in creation?*** It is not known to exist until after our creation has occurred. Secondly, quantum physics deals with the atomic activity of particles, already in the universe, on the micro level and not in the macro universe, as Ferris pointed out and Hawking admitted. More importantly, the description of atomic particles that are multidirectional with multiple histories ***are already created particles*** and therefore cannot be involved in the act of creation, but only as a result of it. Lastly, it is highly important to know that ***Feynman rejected the proponents of the string or M-theory*** before his death. James Gleick, who in 1992 published *Genesis: The Life and Science of Richard Feynman* reported Feynman as saying about those seeking string theories,

> I don't like that they're not calculating anything… I don't like that they don't check their ideas. I don't like that for anything that disagrees with an experiment, they cook up an explanation—a fix-up to say "Well it still ***might*** be true."[37] (Emphasis added)

Also, in the idea of a progressive improvement of "model-dependent realism," it is important to distinguish those improvements on any accepted model, such as the path of planetary circulation in the Ptolemaic system and in Copernicus' model, should be from actual observed data leading to improvement. This is required in order to have a reasonable belief that improvements may be seen in future theoretical models. Their past models described real observed events that were later improved upon and had a basis of observed reality, but models of multiple universes or string theories are not observed models of reality. This is a deceptive argument. While they claim their ***new*** physics is based in reality, it's not; it's a myth. Hawking has had a tendency in the past to speculate outside of known physics such as rejecting his mathematical proof of the big bang and questioning Einstein's theory to escape a singularity to find a quantum gravity theory, or again when he proposed a leakage around black holes.

Also the ***claim*** that "this book is rooted in scientific determinism" appears that their claim to determinism is in agreement with the view of "uniform natural determinism" of past science, and that is false. The deterministic science of the eighteenth and nineteenth centuries claimed to base its views only on what could be known by sense perception or what is observed, and can be tested in everyday experience. This was a rejection of catastrophes and anything not uniform that they could not observe. The theory of indeterminacy from the quantum theory and the chaos theory as

---

[36] Ibid., 180, 181.

[37] James Gleick on Feynman in an interview by Robert Crease, February 1985, cf. also Feynman on Wikipedia free encyclopedia, and Gleick's biography.

destructive of deterministic science are not fully acknowledged. What Hawking and Mlodinow mean by determinism is that they believe there are unchanging natural laws that persist, and not events that have been and are observed and tested. The repeated use of words that show what is speculated or hoped for such as "must be," "could be," etc., show they are ***not presenting objectively observed science***. And on the one hand, they admit there is no proven theory of quantum gravity or of the M-theory, yet they use these ideas as if they were certain and real.

The authors speak of "recently discovered and theoretical advances," and it is not clear what they are referring to involving objective science. They never mention recent findings that are extremely important modern discoveries, e.g. that 96 percent of our universe is made of dark matter and dark energy or refer to the findings of much of past science that has been disproven, e.g. Lyell's principles of geology. Their grand design is nothing but ***a new form of pantheism***. Immanuel Kant got his ideas on the development of all things in the nebula hypothesis from Lucretius and Swedenborg who were inspired by visions and voices that many would say were demonic. Kant had a beatific vision of how all began by natural forces and produced the grand universe which was extra sensory and not scientific. This was a pantheistic idea that all came from forces in nature. Hawking and Mlodinow see forces ***in nature creating*** and producing our universe with specific conditions for life and man, and millions of other ***imaginative and unknown*** universes from gravity and quantum atomic forces.

The year after Hawking and Mlodinow published *The Grand Design*, Brian Greene, who is known as one of the most knowledgeable scientists on string theory and multiple universes, published his book, *The Hidden Reality: Parallel Universes and the Deep Laws of the Cosmos* that reviews most of the parallel universe theories. In his book he says he believes that the parallel universe and string theories should be considered science, but by this he does not mean they fit within observable proof of natural science, but that it is valid to theorize about the possiblilities. Studies of quantum properties of black holes with stunning results from string theory can be used to give a feeling of reality. Here are his statements,

> It suggests remarkably, that all we experience is nothing but a holographic projection of processes taking place on ***some distant surface*** that surrounds us. You can pinch yourself and what you feel will be real, but it mirrors a parallel process taking place in a ***different distant reality.***
>
> ***No experiment or observation (of parallel universes) has established that any version of the idea is realized in nature***. I'm not convinced—and speaking generally, ***no one should be convinced- of anything not supported by hard data.***
>
> Rather all of the parallel-universe proposals that we will take seriously emerge unbidden from the mathematics of theories developed to explain conventional data and observation. [38] (Emphasis added)

These theories he reports are computer models and are theories and no more, visions of possible reality with no hard observational data in nature. They are not proven science—only speculative theories. Greene says the parallel universe theories "***suggest*** this astonishing" He speaks of them "as yet, thus ***explanatory approach*** from being fully realized" and "what ***migh***t be…" This to him is a search from "a lonely vantage point in the inky black stillness in a cold and forbidding cosmos…."[39] Such are the false hopes from searching unbelievers.

---

[38] Brian Greene, *The Hidden Reality: Parallel Universes and the Deep Laws of the Cosmos, (New York: Alfred A. Knopf' of Random House, 2011)*, 8, 9.

[39] Ibid., 9.

Unknown Intelligent Designing Forces in a Mythical Nature are Twenty-First Century Hypocrisy

The new pantheism of Hawking is a myth—completely undemonstrated and unproven speculation. Going beyond traditional science is not new to Hawking. Previously he said that man ***must*** find another unknown planet that is inhabitable, that ***unseen living aliens exist*** having evolved elsewhere, that ***life came from some unknown place*** in space, and geneticists are creating life (others show they only use existing life's building blocks to rearrange them), and that there are millions of other unobserved speculated universes.[40] Hawking and Mlodinow claim these unseen worlds and beings ***are real***, but these unbelievers say that Christians who believe in a God in heaven with angels and unseen powers are trusting myths. These authors and others who reject God for their own created ideas have escaped to demonic myths that destroy any accountability to God. Are we to assume they do not know they are lies and they are speaking for the devil who is the father of lies? Many scientists use these myths, claiming they are based on scientific mechanism, while accusing Christians of using magic.[41]

Some scientists who disbelieve in God are presenting devised events that occur in our universe that are similar to Stephen Hawking's claim of events that go beyond singularities or mysteriously happen beyond the edge of physics of our universe in black holes or the reverse, such as the Big Bang in his *Grand Design*. Hawking suggested his idea furnishes a belief in Multiuniverses and a reason why one such as ours just happened to be friendly to man. These other scientists say the idea of event horizons ***within*** our universe, because they are similar, support Hawking's claims of events ***beyond*** our known universe. William Unruh of University of British Columbia in Vancouver passed water over a piece of wood shaped like an aircraft wing making it go faster and when water going the opposite way encountered it, it slowed to a stop. He claimed this was an event horizon giving analogical proof of the idea of a black hole. Daniele Faccio of University of Insubria in Italy did a similar event horizon by pulsing laser light through glass, which caused a small increase in the density of the glass that propagates like a wave at the speed of light. Because light passing from a less dense to a denser medium, a light jogging to catch up won't make it past the increased density. Photons therefore popped into view showing an event horizon. This was published November 12, 2010 in *Physical Review* letters as will be Unruh's experiment soon. Such analogical experiments are simply that—***analogies*** occurring according to the known laws of physics and are no proof of events beyond the edge of our universe in an unknown place and contrary to known physics.[42] Roger Penfield, as an atheist who laughed at the Higg's field idea has now joined those who use this unproven idea to support multiuniverses in physics.[43]

Contemporary scientists who hold to such views are pantheists who believe in intelligent designing forces ***in nature*** that create and guide its development, with man as supreme over all other animals and nature. This is no different than Aristotle's Unmoved Mover who actualizes the rational potential of all things and which is now working in man who they believe can become supermen or God. As with Adam, demonic forces appeal to the bias of men who want to claim they are wise as God and selfishly and self-righteously claim they know what is good and evil. These pantheists, like Aristotle, always associate themselves with animals over which they are

[40] Cf. Janie B. Cheaney, "Hawking's Wager," *World,* October 9, 2010 et al.
[41] Cf. Robert Pennock, *Tower of Babel: The Evidence Against the New Creationism,* (Cambridge, MA: MITPress, 1999).
[42] Marissa Cevallos, "Black Holes in the Bathtub: Science observe Hawking' radiation in unexpected materials" *Science News,* December 18, 2010, 28, 29.
[43]Roger Penfield,

superior (Romans 1:21-23). Francis S. Collins of genome notoriety later claimed he believed in a designing evolutionary pantheistic god and gained John Templeton's multimillion dollar financial backing. But Collins' efforts to promote theistic evolution by genetics were done before the genome was discovered much more complicate than he had discovered. That science became mysterious and his views are now inadequate. Collins claims now overlap the admission of scientist now admitting one knows how evolution occurred.[44]

Believing that man is the highest evolved creature who is becoming God, these intellectuals end up like Nimrod at Babel believing they can be superman who can reach to heaven. Aristotle taught Alexander the Great that he could become a god and in the end he declared himself Zeus-Amon, the one God known in Greece and Egypt. The Romans who followed such science, proclaimed their emperors to be God. It is surprising that when the Roman Catholic Church departed from biblical Christianity to Aristotelian philosophy that they ended up claiming papal infallibility? Jesus taught that the world would end in an Antichrist who claimed he was the supreme god.

The pantheists, from Kant to Hawking, believe there is an unknown creative intelligent designing purpose and power in nature. Both the deists who believe God created but left the world to run on its own, and the pantheists who believe intelligent nature itself is designing the world, are without accountability or any sense of repentance toward God. They live as they please, believing they are wise as God. Many in the I.D. movement base their arguments on those of two molecular biologists, Denton and Behe, who questioned the evolutionary theory. But in the end, both these scientists said they believe evolutionary ***forces in nature*** caused the jumping of the gaps for a progressive creation. While saying there is a God, this is not the Creator of the Bible. Is this different than Robert Chambers' successive natural creations, Lamarck's designing force, or is it really better than Hawking and Mlodinow, or the old pagan Roman Catholicism from which the Reformers broke away?

For over two hundred years the early modern scientists rejected these views for the sovereign Creator who sustains his creation and their views have been trustworthy. There is a true event horizon between the seen and the unseen throne of Christ in heaven where he has all power of the seen and unseen and where there are all the heavenly hosts. The resurrection indicates Christians now have an access into heaven by Jesus Christ, and to the wisdom of God by the Holy Spirit. The right interpretation of creation and recreation will be seen in the next chapter.

---

[44] Francis S. Collins, The Language of God, and Beliefs, which is readings

Chapter 34
The Christian and Modern Science Today

Introduction

In the first chapter the evidence was presented that, after most civilizations grew to prosperity for over two hundred years, then some intellectuals began to assume their wisdom was adequate and God was not needed. The educated elite began to transfer faith in God as sovereign Creator to faith in man. When this movement of criticism grew to dominance with a controlling influence, the wave of each civilization began to break. Today the governments in America and the West are approaching breakdown and, while there may be temporary solutions, they are seemingly in irreversible debt and controversy. There is a growing class division, a strong competition and confusion sexually from strong individualism, with an abandonment of marriage in the lower and middle class youth, with millions of women raising children without fathers, and efforts to compel respect for homosexuality. The United States president and the United Nations officially approve. The values of the Creator are replaced by those of what every man thinks is right in his eyes. The churches are increasingly divided and without a united voice on how to influence the problems. What should the Christian do about modern science and perversion against the Creator today, if the church is to understand and be Christ's witness?

This concluding chapter will seek to furnish and show the following.

1. Understanding and trusting what Jesus Christ teaches about creation.
2. Reasonable biblical principles for interpreting creation and God's purpose for man.
3. The Genesis creation accounts, their purpose and proper interpretations.
4. What is possible about the teaching of the length of time of the days of creation?
5. Christian use of modern science today, with warnings and limitations.
6. Breaking waves of all civilizations are in God's sovereign plan for elect mankind.

The second volume, on the soft sciences of economics, psychology, sociology and politics, will reveal how the demonic delusion of perverted hard sciences in denying God in Christ as Creator has affected all human values and relationships, and is now causing the wave of modern civilization in America, the West, and the world to possibly break into chaos. These may lead to the final Great Tribulation before the Lord returns. Such chaos should awaken Christians to see the importance of reflecting God's image in our witness in these times.

Understanding and Trusting What Jesus Christ Teaches about Creation

The trust of Christians rests on the works and teachings of Christ our Lord, which were given to the common man rather than the intellectual elite. The Gospel writers show that Jesus referred to and trusted the Mosaic Genesis accounts of creation at least six times in his teaching, showing his belief these events really occurred and are true (Matthew 19:4, 8; Mark 13:6, 19; John 8:25, 44). John begins his Gospel with the idea of Jesus as the Creator who performed all the works of creation and "gave life that was the light to men." Jesus repeatedly said that he was "the Lord of the Sabbath," which was a claim that he was the Creator who rested after his six days of work. While the work of creation was ended after the initial six days, God in Christ continued to ***work in sustaining his creation*** and carrying out his sovereign plan for what he had created (John 5:17; Hebrews 1:1-4). There are many other parts of the Bible, and other Scriptures in the Gospels that teach God is Creator. The Gospel records about Christ show that he accepted these teachings as true. If he is our Lord, we also should believe them. There are over eighty passages using words for "create" that teach direct creation by the only living personal God, with hundreds that indicate

God and Christ as Creator. Jesus and the apostles taught the whole biblical record of God's relationship to man and creation to be inspired by God (2 Timothy 3:16). Jesus and his apostles also taught the ***sovereignty of Christ, and that the Son of God's work in creation is essential*** to his work of redemption (John 5:18-24). His obedience to the Father was to redeem the whole creation and man, and his power as Creator will lead to his return tin glory to rule with all authority, with power to recreate (John 17:4, 5; Ephesians 1:20-23; Philippians 2:15-3:9, et al.). If we take the lordship of Christ seriously, and his redemption which is based on his authority as Creator, we must understand and not tamper with the meaning of the creation accounts with a motive to gain acceptance and approval by compliance with the world's views. Adam and Eve's original sin, that established the natural bias of all men to sin, was based on eating the one fruit of the knowledge of good and evil that certified the ownership of God over his creation. Any claim that Jesus, Moses and the other spokesmen for God were compromising to their culture is an excuse to change Scripture.

Reasonable Biblical Principles for Interpreting Creation and God's Purpose for Man

To correctly understand the teaching on creation, the Genesis records must be understood by principles of interpretation taught by the Bible itself. A complete course in biblical interpretation would be diversionary in this book, but a few important principles to interpret Genesis will help Christians to see science from God's perspective and how to apply them in using science in our witness for Christ. The following principles are essential.

1. The Bible is ***a real historical record, of a vision written from the perspective of man on the earth to reveal his responsibility to God.*** God has revealed this through men of his choosing. Throughout, Scripture refers to the dates and times, people and places of actual historical events, e.g., "In the day Uzziah died…" (Isaiah 6:1); "In the days of Herod, king of Judea" (Luke 1:5); etc. The events and messages are given in a context of exact time and place, making them a reliable record of what was said and done. God used chosen prophets to give his heavenly perspective, which is beyond our understanding, to his people in real situations on earth. The prophets each spoke from their own limited understanding, experience, and historical context. No other so-called "sacred books," such as the Koran, give specific historical details in precise historical contexts unless borrowed from the Bible. Proper understanding of God's word requires understanding the historical situation and each writer's understanding. This is the perspective the common man would normally expect as a creature of God.

2. God gives ***his message or word in phenomenal human language*** (not scientific or technical language) for the purpose of showing man God's purpose for him in his historical context. The word "phenom" is from the Greek word "phainein," meaning "to show" or "to appear." The vision or message conveyed is ***how it would have appeared*** or have occurred to a person at that time; not as a complex scientific description for a modern scientist. God conveyed the eternal word through a representative of his choosing, who speaks the ordinary language of God's people (Hebrew, Greek, etc.) in a vision or in simple recorded terms of the event that all can understand. Phenomenal language from God who is a Spirit may therefore be ***spiritually given directly to the human mind*** and not just by sense perception. If it were given in scientific and technical language, only a few people could understand. Elect men, such as Moses, the prophets, the priests by the Urim and Thummin, the apostles, et al., received God's message by common language or by visions describable in common words. God's Spirit spoke to the prophet's mind within the person's knowledge: "the word of the Lord" was given to Moses; the word of the Lord came to Ezekiel; David was inspired by the Spirit to write his emotions of faith in a situation that typified other experiences to be revealed according to God's principles.

The words of the message should be accepted as from God's chosen authority communicating it. Critics argue that Moses or David were not the authors of everything given under their name. But even today, writings are ascribed to an author even though others assist him or her and use other documents in support. Our president of the United States tells speech writers what to write for him and we consider this message his. The work is still considered the primary author's message which he chooses to give. In God's revelation in the Bible, the Word of God was the person of God's eternal Son who gave the meaningful information to the chosen prophet. The Word made known the words by his presence as a person, or by some other means, that is God's eternal information. The Greek in these cases was "logikē" as logos was "logic", or meaning that was spoken. In a sense, all of creation is God's message to man to help him know God the Creator (Romans 1:20). The scientific New Information Theory says that everything is created with measurable information. This fits with the biblical claim that Christ is the Word of God who created by his eternal wisdom and upholds all things by his power.

3. God's recorded word to finite man was not comprehensive but ***limited, with the specific intent to show man his worth and responsibility toward God, toward others and the world***. The vision of creation that God gave to Moses, for God's elect, was only to communicate the simple purpose of ***man's place in the world***, which would have been obscured if God had given every detail about everything in creation. All the different animals later described in names by Adam, and all the fruit trees, the insects are not mentioned in Genesis 1. Also, in the New Testament every teaching of Jesus' ministry was not recorded, but only the ones that help us believe that Jesus was the Christ and Lord and how to have eternal life through him (John 21:24, 25; 1 John 5:13). Therefore the vision as recorded in Genesis 1 is described from the perspective of man as if he was on the earth. In describing this I will indicate this by a stick man on the world seeing the vision of the first days, when he was not actually yet created, until the day of man's actual creation. The vision is of what would appear to him if he had been there so that he sees the meaning of those events for him in the world that he is to have dominion over.

4. God revealed His purposes in ***an ever-unfolding progressive meaning with principles that become clearer until they reach fullness in His Son Jesus Christ***. Each time a prophecy revealing a principle of God's working is given in a certain limited context, it helps lead to an expanded understanding of that principle of God's working. Each successive event involving that principle reveals more understanding of the principle and then leads to a fulfillment in the final Word in Jesus the Christ. It is a mistake to say prophecy is a prediction of one event of the future, since each predicted event gives understanding of a principle leading to the final event. That principle is repeated multiple times with expanded meaning. Authoritative biblical scholars recognized this was the Hebrew idea of prophetic fulfillment.[1]

The following are examples. Each sacrificial death, such as Abel' sacrificial lamb, Abraham's ram to replace Isaac as the sacrifice, the lambs on the tabernacle and temple altar or at the Passover, all point to Jesus, the eternal final Lamb of God that takes away the sin of the world (1 Peter 1:10). Each miraculous birth of God's elect person points to the Seed of Eve who defeats the devil and evil (Genesis 3:15), such as the birth of Isaac to Sarah, Jacob to Rebecca, Samson to the wife of Manoah, John the Baptist to Elizabeth, and finally the virgin birth of Jesus to Mary. Another principle is God's required judgment of death for unjust acts of idolatrous worship seen in the flood, at the Tower of Babel, of the people of Sodom and Gomorrah, of the Canaanites, and his own people Israel when they also turned to idol worship- all of these point to his future final judgment for injustice by Christ at the "day of the Lord."

To fail to interpret in the light of the ***principle of progressive revelation*** leading to Christ, makes some prophetic claims ludicrous (e.g. Matthew 2:15) and things taught to guide in one cultural context without understanding in another. All Scripture progresses to show the sovereign

---

[1] cf. Alfred Edersheim, *Prophecy and History in Relation to the Messiah*, reprinted (Grand Rapids: Baker book House, 1955), chapter iv; Brevard S. Childs, "Prophecy and Fulfillment", *Interpretation*, July 1958. 1-13, et al.

love of God in Christ revealed in his life, death, and resurrection, in the sending of the Holy Spirit and in the eternal hope at Christ's future coming (cf. Luke 24:27,44).

5. The Bible is a ***verbally inspired and a preserved correct historical human record of all God's revelation in Scriptures***. Verbal inspiration and truth must be understood in the light of the Bible and especially ***Jesus' teachings of truth***. This means that the Bible says what God intended it to say, and that it is all fulfilled in Jesus Christ's redeeming work. Jesus claimed that every detail of Scripture will be fulfilled to uphold God's words of what is right (Matthew 5:17-20, ff.) and Paul taught that all the Hebrew and Greek canon of national writings were given by inspiration of God (Romans 3:2) and are profitable for righteousness (1 Timothy 3:16). Paul meant this to be understood as from a logical human perception of what was the meaning of what was said or occurred. Many people make false claims about inspiration, such as that God dictated all of them to the prophet. Some go as far as Carpzovius in trying to defend every aspect of written detail, with no scribal error, which is a view of inspiration that is not the way or necessary for common people to interpret real history.

6. God's intent is that ***Scripture is reliable from two or three witnesses***, which is a historical standard for truth in all cultures. Many aspects of history that are generally accepted in a culture do not require two or three witnesses. But the historical validity of an event or trend is confirmed when two or three truthful observers agree on the primary events (Exodus 20:16; Deuteronomy 17:6, 7; Matthew 18:6; 2 Corinthians 13:1; 1 Timothy 5:19). They may differ in the details, but the ***diversity of details and perspective confirms the truth*** and does not invalidate what was said or done as God's intended message. The Bible records a number of ***apparent*** contradictions by the witnesses (for example, see Matthew 8:5-13; Luke 7:1-10; and the different Gospel's crucifixion and resurrection accounts). But, if one *reads all the Scriptures* ***with a heart of faith*** and love for God and neighbor, and takes into account the traditions of the times that are given from different perspectives, he or she may ***by the Holy Spirit's guidance*** discern what exactly was said or occurred (cf. John 16:13, 14; 1 John 2:27). This principle shows why the Bible gives several books that record the same history (Samuel, Kings, and Chronicles, four Gospels, and why Jesus took two or three apostles with him at significant times, and why two or three were involved in ascertaining a person's sin (Matthew 18:16, 19).

The Talmudic rabbis and reformers such as Luther and Calvin acknowledged recorded differences that present minor contradictions, but they accepted the claims of plenary inspiration of all Scripture. When I worked with college students who did not understand this doctrine of inspiration, I found they lost faith in the Bible when unbelieving professors pointed out minor contradictions that did not discredit but actually validated the witness of what happened. The natural bias to discredit God and his revelation will not recognize or take into account these principles of interpretation of the Bible.

The above principles are necessary to properly interpret the creative narratives of Genesis 1 and 2. The passages must be interpreted ***as phenomenal language in visions*** of the earth viewed by man on the earth to help him understand his purpose and relationships. The first chapter is to help us understand ***the world and all that is in it*** for us to reflect the image of God as we have dominion over it for him. Many other Scriptures confirm the importance of God's wisdom in creation, encouraging men in worship and warn against perverting his creative works (cf. Psalm 104, Proverbs 8:22-31; Job 38-42, and many more). Genesis 1 is a beautiful presentation to man of God's work in the *order and harmony of* creation. From this, man was to be able to understand and relate properly to the world. Genesis 2 shows *man's relationships* to plants, other living animals, and his wife, involving her with him to serve God. The two different Genesis documents therefore are given with different purposes and perspectives, but with a unity.[2]

---

[2] P. J. Wiseman, D. J. Wiseman editor with Preface by R. K. Harrison, *Ancient Records and the Structure of Genesis: A Case for Literary Unity*, (Nashville: Thomas Nelson Publishers, 1985).

The Genesis Creation Accounts, Their Purpose, and Proper Interpretations

Genesis 1 is an amazing vision of the world's order and harmony for man. Moses' records of creation (Genesis 1, 2) are visions of what occurred, given in brief scenarios as would appear to man, like a motion picture replay for man to understand his role in it. One summer in a course, a Christian geologist explained how Genesis 1 presents the wisdom of God's work of creation, if you view it as a vision with man seeing it as if he were present on the earth. The passage starts by saying, "In the beginning God created the heavens and the earth, and the earth was without form and void" [As if seen to a person sitting on the earth]. This indicates that the act of making the universe out of nothing was done before the six days of designing the earth for man, which is given in most of the rest of the first chapter.

The vision of the six days had the important purpose of God's design of the world for man. The galaxies, stars, sun and planets, including the earth, as material elements, are said to have been made before the six days. "By faith we understand that the universe was created by the word of God, so that what is seen was not made out of things that are visible" (Hebrews 11:3). The Apostle John also tells us that Jesus Christ, as the Word of God, was the Creator of everything (John 1:1-5), and Paul said, "For by him [Christ] all things were created in heaven and on earth, visible and invisible, whether thrones or dominions or rulers or authorities—all things were created through him and for him. And he is before all things, and in him all things hold together" (Colossians 1:16, 17). God's Son was the one "through whom He created the world...and he upholds the universe by the word of his power" (Hebrews 1:2, 3). The making of the universe of stars and galaxies and planets was done before the description of the forming of the earth for man. "In the beginning, God created the heavens and the earth. The earth was without form and void, and darkness was over the face of the deep. And the Spirit of God was hovering over the face of the waters" (Genesis 1:1, 2).

Genesis 1:2 ff. describes the process of how God's word by the Holy Spirit formed the earth in the six days that followed, in logical progressive preparation for man. The Spirit was the person of imminent power present with God the Father and God the Son. The six-day account is a vision of the history of how God designed all about the things in earth up to man, with light, plant life, animals, etc. Men and women were last as given power to rule over it as sons in his image and for his glory. Seeing the vision from the earth where man would live, and as if God was ***showing him the important interdependence of all created things***, it makes perfect sense and agrees with the sequence of how nature had to be made. Man who was created to have dominion needed to understand this.

The series of graphs on the following pages depicts the vision of each day, from a human perspective. Although man was not created until the sixth day, to show this from man's seeing how it appeared, as suggested, I include a stick figure on the earth to remind the reader that the graphs demonstrate the six days of creation from a human perspective.

Observations about the Six Days Work of Creation

The six days begin with no form, only void in darkness. The Holy Spirit was present and involved, with the earth already formed and covered with water (Genesis 1:1, 2). The Hebrew word for "void" used here only occurs elsewhere in Scripture in Jeremiah 4:23-26, where the prophet described the judged land as empty or without organization, "I looked on the earth, and behold, it was without form and void."

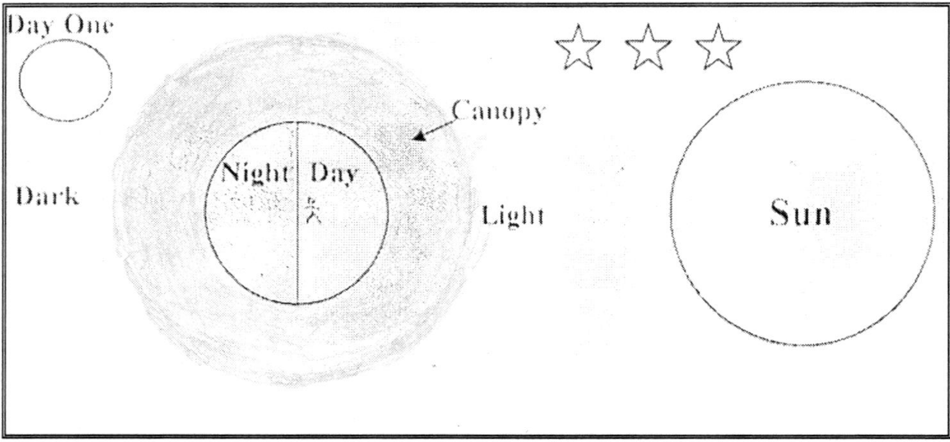

The first day began in the evening when the dark was initiated, with the earth covered with water, apparently included forming a gaseous water canopy of darkness. Some planets such a Venus have such a canopy today. At the word of God, the Spirit instigated light from the sun which penetrated through the canopy of gases. What appeared to the man was the phenomenon of ***indirect diffused light and not the source*** of the sun. It may be speculated that the moon was either close within the canopy not receiving light, or its reflected light could not penetrate the canopy, so at the turning of the earth the night was ***dark***. This refracted sunlight was the basic factor for energy for subsequent natural activity. Light and darkness continued on the earth at that time, apparently from man's vision of the turning earth. While moon and stars are shown here, they are not named, so not seen. The first day was composed of the evening and the morning.

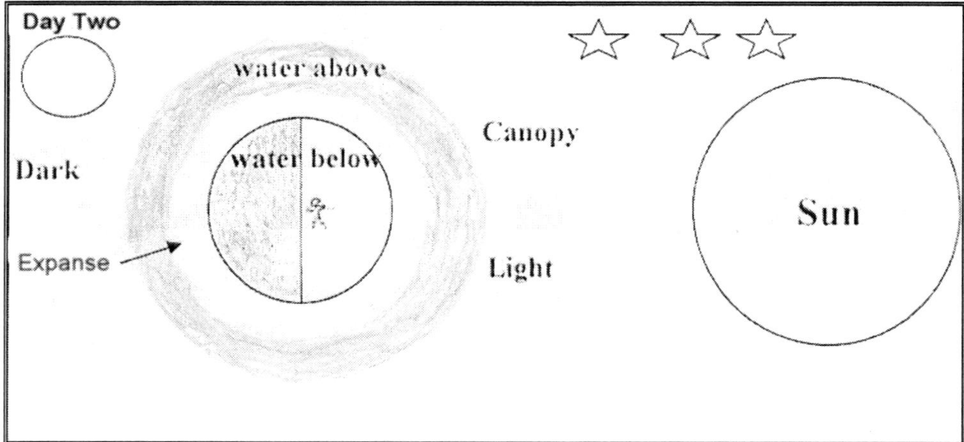

On the second day the water covering the earth was by the light energy apparently separated from the water in the canopy. A vapor arose from the earth and separated into two bodies of water and caused the growth of an expanse between the canopy above and the water below. The indirect light was day, and dark was night, as illustrated above.

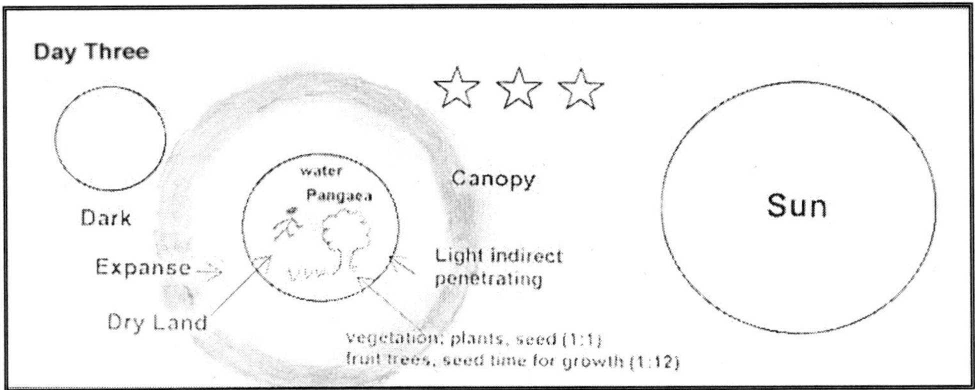

The third day, either by the loss of water to the canopy above or by direct formation by the Spirit, one continent of dry land known as Pangaea emerged from the waters. Most scientific evidence points to the earth having a tropical or semi-tropical climate under the canopy until what is called the PT catastrophe. Also on this third day God commanded that the one unified land (called Pangaea) sprout plants with seeds and fruit trees with fruit seeds. This is a picture of vegetation beginning and growing on the earth while there was indirect light and water. From Genesis 2, this occurred in the limited Garden of Eden and not throughout the whole world's land mass. This sprouting and growing implies a period of time, but could have occurred rapidly or instantly by divine command. This could not have naturally occurred without the previous events of day one and two. Small insects to pollinate the flowers and fruit trees would not have been seen, but would have been a necessary part of creation at that time.

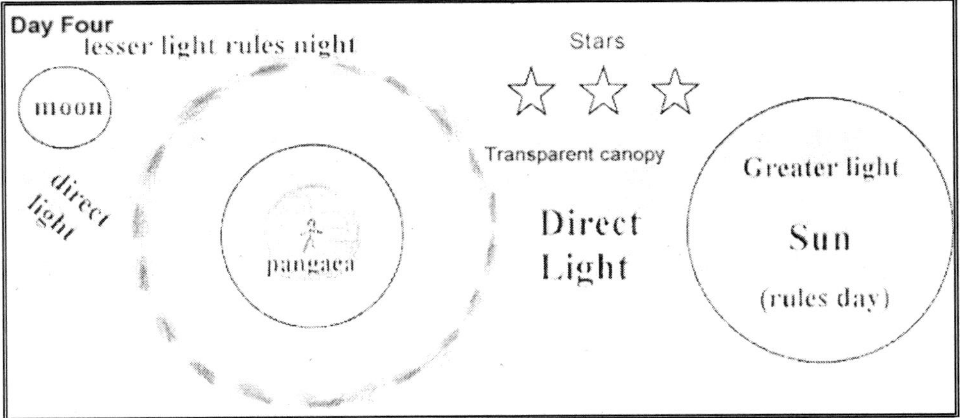

On the fourth day "God said, 'Let there be lights in the expanse of the heavens to separate the day and the night' " (Genesis 1:16). It **appeared** to the man seeing the vision from the earth's perspective that ***God made*** the sun, moon, and the stars at this time. That was the phenomenon as it appeared. But the vision previously revealed these were already created (1.1). The direct visibility of these was said to be useful for man in determining the time of seasons, days and years which he could not have done by indirect light.

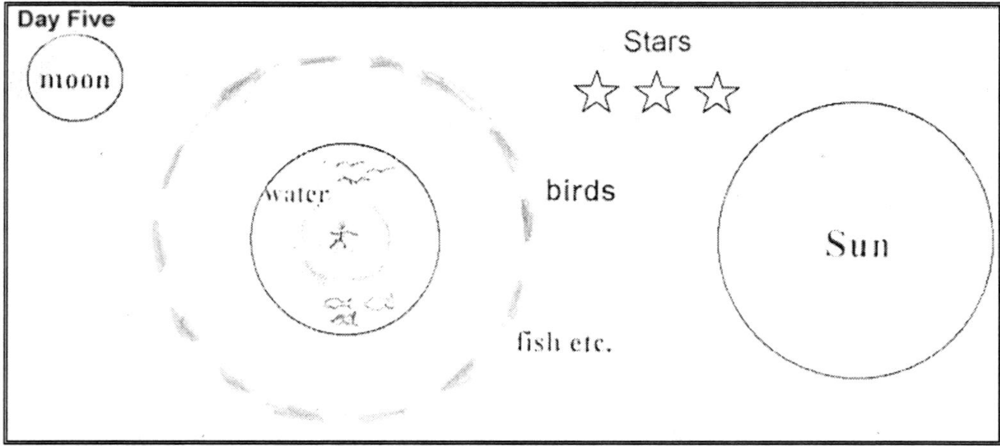

On the fifth day God said to let the waters swarm with fish and great sea creatures, such as whales and possibly amphibians. At the same time he made birds to fly in the heavenly expanse. The birds and fish could not exist without the grasses and seed, fruit of the trees and insects from the third day before.

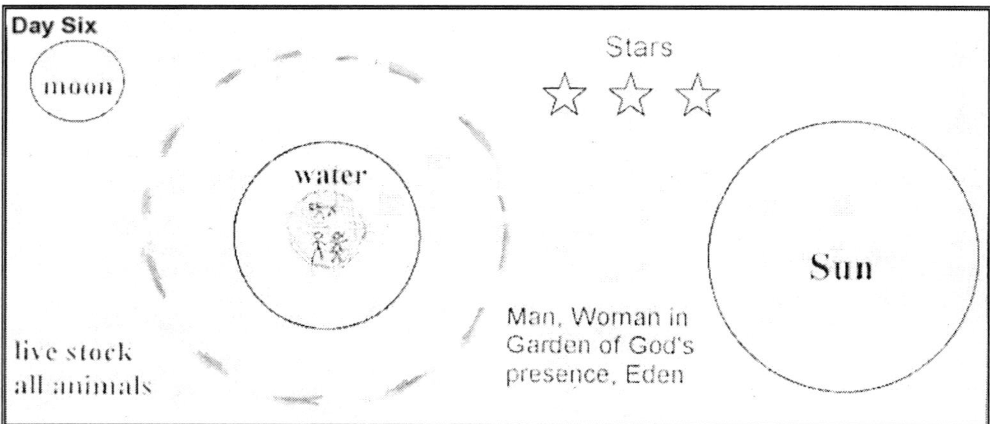

On day six God's word created the various kinds of mammals and livestock and creeping things, which implies all kinds of animals that the man might see in the vision. These were all interdependent for food, with the birds and fish of previous days, supported by the grasses and fruit. After all these things were done, God made a deliberate decision in the persons of the Trinity, to create man in his likeness or image to have dominion over all the creatures. He made them male and female in his image, both reflecting two different aspects of his image, in unity together. God "blessed them" and told them to be fruitful and multiply and fill the earth under their dominion. Filling the earth seems to be involved in having dominion over it. God then told them that they and the birds and animals were to have all green plants and fruit trees for food. The fact that the man and woman were created on the same day as the higher animals indicates their physical relationship to them, but the distinct decision and way of man's creation in God's image makes a gap and sets man and woman apart over all things.

Day seven (Sabbath). On the seventh day God rested from his work of creation and continued to sustain or uphold what he had made, seeing it as good and making it a day of significance to be celebrated for his work (Genesis 2:1-3). This clearly indicates a juncture in creative work that was once and for all completed and his following work of upholding or sustaining the natural order (cf. Hebrews 1:2, 3; John 5:16-19).

Observations, Evaluations, Interpretation of Genesis 1, 2 of a Personal Wise Creator

The accounts of the visions clearly say that ***the personal transcendent sovereign God deliberately spoke,*** so the events of creation were his act and ***not a process of nature.*** Each day it is recorded, "God said, 'Let there be…' " Moreover, when God ***created man,*** the trinitarian Godhead made ***a deliberate decision:*** "Then God said, 'Let us make man in our image, after our likeness. And let them have dominion over the fish of the sea and over the birds of the heavens and over the livestock and over all the earth and over every creeping thing that creeps on the earth'" (Genesis 1:26). The decision involved imparting his own personal image with the ability to rule. Moreover, ***God gave the man, male and female, instructions*** for procreation, food and dominion and explained how the animals would be sustained, saying,

> Be fruitful and multiply and fill the earth and subdue it and have dominion over the fish of the sea and over the birds of the heavens and over every living thing that moves on the earth …Behold, I have given you every plant yielding seed that is on the face of all the earth, and every tree with seed in its fruit. You shall have them for food. And to every beast of the earth and to every bird of the heavens and to everything that creeps on the earth, everything that has the breath of life, I have given every green plant for food (Genesis 1:22, 29-30).

And it was so. When have the natural powers of the elements and energy in this earth given specific instructions? Moreover, the personal God then recorded his opinion about what he had done: "And God saw everything that he had made, and behold, it was very good" (Genesis 1:31). This was a personal opinion of the transcendent Creator which he expressed about his creation.

In Genesis chapter 2 the way in which man was created is given in detail. The personal God formed man from the elements of the ground (dust) and God with a face to face confrontation inbreathed his image into the man so that he "was made a living soul" (2:7; cf. John 1:4). After man's sin, God's people's desire was for God's face to shine on them (Numbers 6:25). God then led Adam to evaluate the animals and to name each of them, showing man's intelligence to discern each animal's abilities, and none was able to be his equal companion. As man was created, he was ***alone*** in the universe, not gradually emerging from a group of similar primates. The personal ***Creator deliberately acted*** to make the woman to be an equal companion able to be his helper. She was made from his genetic essence, from his side, and his equal of understanding, to partner in having dominion. But being equal in value, she was ***different to reveal a different aspect of God's image.*** The differences of male and female could thereby reflect God's full image by union of love in marriage and procreation. The differences of male and female and their sexual practices that are different from other primates, admittedly could not be explained by scientists from a natural process of evolution.

The accounts of the visions clearly say that ***the personal transcendent sovereign God deliberately spoke,*** so the events of creation were his act and ***not a process of nature.*** Each day it is recorded, "God said, 'Let there be…' " Moreover, when God ***created man,*** the trinitarian Godhead made ***a deliberate decision:*** "Then God said, 'Let us make man in our image, after our likeness. And let them have dominion over the fish of the sea and over the birds of the heavens and over the livestock and over all the earth and over every creeping thing that creeps on the earth'" (Genesis 1:26). That decision imparted his own personal image with the ability to rule, defining a gap between man and all other creatures. Moreover, ***God gave the man, male and***

female, instructions for procreation, food and dominion and explained how the animals would be sustained, saying,

> Be fruitful and multiply and fill the earth and subdue it and have dominion over the fish of the sea and over the birds of the heavens and over every living thing that moves on the earth ...Behold, I have given you every plant yielding seed that is on the face of all the earth, and every tree with seed in its fruit. You shall have them for food. And to every beast of the earth and to every bird of the heavens and to everything that creeps on the earth, everything that has the breath of life, I have given every green plant for food (Genesis 1:22, 29-30).

And it was so. When have the natural powers of the elements and energy in this earth given specific instructions? Moreover, the personal God then recorded his opinion about what he had done: "And God saw everything that he had made, and behold, it was very good" (Genesis 1:31). This was a personal opinion of the transcendent Creator which he expressed about his creation.

Creation made by the personal God initiated the ***reasonable or logical processes*** which were his power being expressed so that ***the creation is linked by meaningful information*** that Adam and Eve could examine and evaluate. Meaningful reasonable unity characterized the whole. There were no independent natural laws that work by chance. The six days of creation events show one united mind at work. Unity of reason in nature is recognized even by unbelieving scientists. Paul Davis argued this must be the basis of science, Hawking spoke of "the mind of God," and so with many others. The New Information Theory clearly asserts this reasonable continuity, with everything having measurable information showing the whole was done by the ***Logos*** or Word of God.

Also, this meaningful connected creation is ***completely interdependent***. The succession of the six days, each later ones depend on those of the days before. The forces and elements of the universe were shown by Martin Rees to be so precise they could not be one bit different for life to be sustained and for man to exist. All chemical elements are interrelated so they could be seen to fit in a scale. Everything is so interrelated and interdependent that the environmentalists are adamant about saving every species, even though they lack the common sense to see the ascending value up to man, who has the greater value over the whole. The days of creation *could not have come in a different order* or way. Light energy is basic for all processes; water is the most essential element and the water cycle from above must perform as a result of light energy; land must exist for grasses and fruit trees to grow and all must have light and water. Each day of creation shows a higher form of life dependent on what went before.

The creation also had an ***expansive process*** from the limited Garden of Eden where forms of life originated. These were to expand and fill the earth. The basic genetic forms of plants, of animals, and of man could then reproduce, diversify and expand throughout the whole earth by mutual interdependence and human cultivation and control. This was a comprehensive process, meaningfully created by the mind of God for a limited like-mind of man to progressively understand and rule over. This description of the created order was certainly *not a chance survival of the fittest in a wild runaway threatening world* described by Malthus or Charles Darwin, which could not be proclaimed "good."

This logically interdependent and expansive process also involved a ***limited end for a delegated control***. Man's mind was like God's, but was highly limited, so that God's upholding sustaining work had to be experienced and complied with. Adam and Eve, with a limited likeness of God's intelligence, had to learn from, and depend on, the greater mind of God for them to have dominion. In the book of Job, the evidence of God's creation was in the power and complexity of nature's creatures and forces that were beyond Job's comprehension (chapter 38 ff.), and required his humility.

God's Creative Work Transcended the Limited Intelligence God Gave Man

In modern science man discovered how much of the world exceeded his knowledge and was a mystery. The discovery of the indeterminacy of the atomic world, the chaos theory of everything, and a genetic theory with genes and "junk," of proteins and amino acids, and how they work are much beyond even the most educated man's ability. Man has discovered atomic forces, cybernetic technology, biological diseases that he now knows can be used to destroy himself and all of life in a brief moment of time, and he does not know how to establish a control to stop these if they are unleashed. The forces of magnetism, climate and weather, the tsunamic power of the oceans, the volcanic and quaking power of the earth are still beyond his control.

At Babel man thought he could reach the heavens in knowledge of everything. He has grown physical giants in size, and seemingly extended life a few years, and in our time thinks he can achieve eternal life and make supermen. In eating of the tree of knowledge of good and evil man's ambition was to be wise as God. Nietzsche's aim was to be superman (uberman). But strangely, Nietzsche predicted that when the world completely excludes God it might go mad, and after writing *The Antichrist*, he did go insane. Our intellectual elite today are reaching for and through technology and are near Babel again. Through the serpent the devil deceived Adam and Eve, so they failed to see their limits in the creation and sinned with a biased desire to know like God does. However, as of yet man cannot even understand the limited intelligent mind that he has, much less the universe, which is now considered 96% dark or unknown of matter and energy.

The creation stories present things beyond human understanding. Thus, everything in Genesis 1 and 2 indicates ***the spoken acts of a personal God who is present, and not some impersonal providential known power built into nature to evolve***—not an unmoved mover or a universe with an impersonal built-in intelligent designing power in nature. Moreover, the conditions and processes that were initiated are not nature working by survival of the fittest, described by Charles Darwin, or a series of providential natural workings as in *Vestiges*, or by occult powers in nature by Lamarck, or any other evolutionary theory. God himself worked, not an impersonal nature.

The Sabbath day of rest after creation indicates the creative work was ended and ***was not a process that gradually continued on and on until today***. God sustained all things by the word of his power by which he created (Hebrews 1:1-4). The Scriptures seem to indicate the day of rest ended with the sin of man. Man was excluded from the Garden of Eden and the woman was to have ***"labor"*** in childbirth and the man was "***to work*** by the sweat of his brow." Before sin, it was an easy task to keep the garden, enjoying the pleasures, and having peace with the animals. After man's sin, the whole world was thrust into a process of degeneration with "groaning and travail" (Romans 8:20-25) until Christ returns and the resurrection occurs with a new heaven and a new earth. Jesus' claim to be the Lord of the Sabbath was demonstrated by healing various people, showing the groaning and pains of nature could be ended by his redemptive work. He created food indicating he could end the struggle of man to gain a living.

The ending of rest by man's disobedience is also taught by Hebrews 4:1-11. Rest may begin by Christians who respond by submitting to the kingly authority of Christ now. Based on his conscious sovereign authority with the Father, Jesus said, "Come unto me, all you who labor and are heavy laden, and ***I will give you rest***. Take my yoke upon you and learn of me and you shall have rest for your souls" (Matthew 11:25-29). The rest from Christ is begun now by obedience and is consummated in an eternal condition of complete rest in him at his return.

What is possible About the Teaching of the Length of Time in the Days of Creation?

This discussion is not about the philosophical question of time which arises for those who reject creation. For a Christian, time is the unfolding of the conscious historical process that only began with God's act of creation and moves toward judgment and eternity. The issue questioned in this section is, Could these days described in Genesis be millions of years long, making possible a gradual development of life so that all species up to man occur by evolution as a creative force? What do the visions of God to man in Genesis imply about the length of time?

In Genesis 1 and 2 the word “day,” (Hebrew, “yom”) is used to have several meanings—as being one of a sequence of the seven days of creation events in: day one, day two, et al.; of a division of the time of evening and morning, of light and darkness, day (usually 12 hours) and night (12 hours). Elsewhere it also denotes a specific unit of time of one event, “the day that you eat, you will surely die”, or also of a period of a series of extended actions—“in the day God created the heavens and the earth." about which we do not know. Today we use the word “day” in different ways—day as the time it is light, and day as 24 hours on a calendar. We usually think of a day as the period including a day and night, which is 24 hours, and much of Scripture uses the word “yom” that way.

However, there are at least two events recorded in the Bible when it was ***not 24 hours***—One was the long day of Joshua (Joshua 10:14), which was unique when God answered Joshua’s prayer prolonging daylight and helped Israel in battle for the promised land. The second was an extended day as a sign to King Hezekiah to indicate he would be healed (Isaiah 38:8). It seems that there may have been changes in time, especially after the flood, when the implications of the change of earth’s axis is given to affect constant seasons for man to work, although seasons had existed before (Genesis 2:14; 8:22). Earth time might have varied, though we do not know this. But that is not the issue here. The important issue here is whether the word yom can define the times of the six successive acts of creation in Genesis as long days of millions of years to allow nature ***to gradually produce everything on its own,*** so that man is not made by God but comes from apes. Do the six days allow that?

Each day of Genesis 1 clearly involved an act in a series of God’s works followed by a break. The day describing the time of light and then darkness involved the short process of the turning of the world for exposure to the hidden sun, moon, and stars, which probably would be about a 24-hour day as we know it. There seems to be a period of time involved in the separating of the waters below and the firmament of waters above that implies an extended process of our normal days. There also would seem to be a longer time process involved in the gathering of the waters so that the dry land of the continent of Pangaea was formed (perhaps in united plates) and exposed. A period of time seems implied in the “sprouting” of vegetation and the yielding of seed and of the bearing of fruit by trees—“they brought forth fruit.” The appearance of the sun and the moon seems to have been involved in the dissipation of the gases in the canopy, which gave the phenomenon that they were “made.” The creation of the fish which “swarmed,” implies a time process of the reproduction and multiplication of fish in large groups and numbers with a break before the next work of God. Evolution as a continuous natural gradual process of the development of all things is contrary to six successive works with a break between.

In other words, the days of creation seemed to have involved ***an act from God’s command that he personally at a point initiated*** as a sustaining law that occurred in the period designated a day which he successively upheld. The beginning of light and designation of day and night would involve only one turning of the earth. The separating of the expanse of waters above from those below would seem to require a period of weeks, as would the gathering of waters for the dry land of Pangaea to appear. Of course, God could have done this instantly if he wished, but the description seems a process involving some limited time. The creation of grasses and fruit trees and the description of them sprouting and bringing forth fruit seem to involve a process. But the description of these days ***could not allow millions of years for gradualism***? The creation of grasses, fruits, and procreation of animals was in the limited Garden of Eden, and even a man can plant a garden or fruit in days or at most a week or two. Also, the events of the days as described would certainly not be millions of years, because ***the interdependence of these aspects of nature requires that they occur in short periods.*** Water and light are required for plants, plants for fish and birds and animals etc.

The claims of current evolutionists is that the process began with a big bang of multiple universes; that in our universe over millions of years, there was a nebula beginning that produced a more complex nature that added more information to produce extremely diverse stars, galaxies and planets, and life by chance. This led to higher and higher life forms up to intelligent man from primates in one continuous gradual process without obvious breaks. This cannot be traced or demonstrated from causes in nature. They say this gradual process is still continuing, and never has stopped or will stop. This is a foolish theory without evidence, and has no resemblance to the accounts of creation by acts of a personal God in Genesis. What evolutionists describe could not possibly have occurred over millions of years because of the interdependence required by God's upholding power in nature. Those professing Christians who interpret the Genesis accounts as age-days of millions of years overlook these requirements in God's sustained natural laws.

Efforts to find support from Church fathers for a process of long`age days are not valid. It would appear Christians who seek to do so are accommodating naturalistic theories simply to gain approval of the world, rather than to glorify the Creator. And the evolutionary theory of all things is a theory of bias against God and of deceptive delusion of America and the West by the devil. Augustine, the church leader at Hippo in North Africa in the fourth century, recognized and spoke of these processes of creation. But it must be remembered that Augustine was influenced by and taught Greek natural philosophy before conversion, and he once taught Aristotle's pantheism of the Unmoved Mover actualizing the potential of things as an extended process. But even Augustine taught a variety of ideas over a long ministry of 30 or so years and also declared that the creation days were instant acts of God.[3] Christians with science knowledge who seek a reconciliation with chance evolution are well-intended, but weaken faith in Christ as personal Creator and Lord of the Sabbath.

Christian Use of Modern Science Today and Warnings and Limitations

Evaluating and Fully Accepting Man's Natural Bias as Main problem of Perversion

This book on *True Enlightenment* began by asserting that Christian reformers began modern science by faith in the biblical God and by the guidance and discernment of the Holy Spirit. Both previous studies in scientific history, and the great ethnologist in comparative religion, J. Schmidt, revealed that in the dawn or beginning of every major civilization, the people believed in God's sovereign creative power in nature and trusted him. After the civilization grew strong and prosperous through several hundred years, because man asserted his wisdom and rejected God, it began to break like a wave. This reveals that man's sin bias since Adam is to remove knowledge of God, resulting in his own hurt.

By *true enlightenment*, Christians can avoid the destruction of Satan's delusion. God's elect, who were created to learn and dominate the world for God, can use true science to help God's people be informed and protected. "God resists the proud but gives grace to the humble" (James 4:7-10; 1 Peter 5:5). The present delusion about the world was created by the devil to bring death to every civilization since Eden. As the devil in Eden in the form of a serpent deluded mankind, so in modern times his demonic delusion has misled some educated intellectuals to pervert the sciences.

Repeatedly, in modern Western civilization supernatural lying spirits imparted voices and visions that led the authors of modern delusions. Descartes in depression had a vision of the importance of man's individual spirit, the mad Lucretius and Swedenborg had visions of the nebular hypothesis of evolution; Locke, when recovering from depression and illness, envisioned man's knowledge as only by sense perception, Kant had a beatific vision of nature for his nebular

---

[3] I have eight volumes of Augustine's teaching's of 400 to 700 pages, probably being 4,000 pages, so as most of us, he often had changes of thought and it is doubtful he ever conceived of millions of years.

hypothesis producing everything. All who followed Kant, including Darwin, were men depressed by worldly failures and who had abandonment the biblical God for delusions produced in men's minds by the devil. Kant, Darwin and others rejected serving as ministers of God and the church to free man for a life of finding happiness for one's self.

These deceptions have thrown the world off course into darkness for the possible breaking of the final wave of civilization before the Lord returns in his glory. The whole world is moving into chaotic conditions by worshipping mammon and rejecting God in Christ. God at the time of the tower of Babel divided the world with its boundaries for the nations and their times (Acts 17:26, 27). Today God may be allowing man, by human wisdom, to again seek to try to find unity against him. This may be the final breaking of the wave of world civilization, when their hatred will be against his people in the Great Tribulation before Christ returns. It has been demonstrated how in the West, and especially in America, the idea of God as Creator was gradually replaced by the theory of the origin of all things by nature in a nebular hypothesis of evolution over millions of years. The delusion by demonic influence on the natural bias of man is the main, basic, and continuing problem, and is the source of all temptation, controlling the unbeliever and influencing the believer. The crest of the wave of civilization seems to be breaking because of this gradual process of substituting human wisdom and control for that of the one sovereign good Creator God, now known through his Son Jesus Christ. Man's bias of his sin nature is the main issue for Christians in understanding and using science today.

Humans cannot understand or accept and obey the truth unless they repent and humble themselves. Paul makes it clear that the wisdom of God will not be understood and accepted by unbelievers, but rather we are to offer them the cross and forgiveness of Christ and his resurrection power over evil. Only after a man repents and is born of and taught by the Spirit can he know God and understand his wisdom in the world. The natural man cannot discern the things of God, for they are foolishness to him, for the Spirit discerns them (1 Corinthians 2:6-12). The fear of the Lord is the beginning of wisdom, and that applies to the wisdom of God's work in creation by Christ (cf. Proverbs 8:22-31; Luke 11:31; 1 Corinthians 1:24, 30). Job's argument with his friends ended only after God showed him his creative power—a power that Job could not fully understand. Understanding the sovereignty of God in creation enabled Job to endure his suffering (Job 38-41). Job then acknowledged God's sovereignty and his constant need of repentant humility and a willingness to care for and pray for his friends (Job 42:1-9).

A Disciple in Christ Ends His Bias by Death with Christ to be His Witness

Only a person who has arrived at true understanding of his sin nature, and dies with Christ to love his enemies, has arrived at a point of true discipleship. Death with Christ alone ends the bias of the sin nature. The essence of the New Covenant given by Christ as he instituted the sacrament of the Lord's supper (the Eucharist) was made clear by Jesus after washing the disciples' feet, exemplifying they must give up wanting to be first and be the least and die with him. He said, "A new commandment I give unto you, that you ***love one another as I have loved*** you (in my broken body and shed blood). By this shall all men know that you are my disciples" (John 13:34, 35). This is what he repeatedly had taught.

> And he said to all, 'If anyone would come after me, let him deny himself and take up his cross daily and follow me. For whoever would save his life will lose it, but whoever loses his life for my sake will save it. For what does it profit a man if he gains the whole world and loses or forfeits himself?' (Luke 9:23-25; cf. Matthew 16:24-28).

Only he who is willing to die and show Christ's love is a disciple and ambassador for the kingdom. Paul taught that when a man accepts the love of God and dies with Christ he becomes a new creation and has a different worldview (2 Corinthians 5:14-6:1; cf. Colossians 2:13, 14; 3:1-5;

1 John 3:15-17). As partakers of Christ's life and love, we witness to Christ as Creator and Re-creator in control of the invisible and the eternal life for which we live. Our unselfish life manifested by the Spirit in our words of witness and life, makes disciples known to the world.

To try to reason with biased men and show them true enlightenment is a misguided idea of Christians who think more logical information alone can by arguments change the unbeliever's mind. While we must treat unbelievers with respect and answer their questions, we must logically point them to Christ's cross and resurrection, which alone will change them. Some Christians who are some of our greatest heroes of the faith have misguided views of science which they teach, contributing to the breaking of the wave of faith of our civilization. They apparently do not understand the biblical teaching on creation, nor do they know the history of modern science with perversion by bias against God. If they knew the history, they apparently have not carefully compared that with the teachings of Genesis and all of Scripture about creation.

The church must make clear the Creator is Jesus Christ and he alone can save men, especially in this crisis time. When a person does not trust God as Creator to supply his daily needs, his only alternative is to trust in mammon, and all his values will be distorted so that he cannot work for the kingdom of God putting treasure in heaven or leading his children to seek eternal life (Matthew 6:19-33). The problem of many American Christians is that they have their values in mammon that promotes greed and lust and they are ashamed to witness to the unseen power of God and the hope of eternity. All surveys of modern Christians show many have put Baal, the lord of the material, and Ashteroth, the feminist goddess that means "to be rich", as central in their worship, as the Jews did in the temple of God before its destruction.

As repeatedly emphasized, the basic issue is the natural man's bias and unless regenerated by the Spirit, he cannot be changed by logical arguments. This is why some in the current Intelligent Design (I.D.) movement are convinced of the Creator but are misguided and are failing. Their first and *primary efforts have been to try to present to intellectuals the complex and specifically designed mysterious world* that current science has unveiled. In this way they hope to get a foot in the door and present the idea of a Creator God behind the intelligent design. For example, many I.D. Christians have boasted that a leading world atheist, Antony Flew, after debates over I.D. issues, announced that he had "been led to accept the existence of a divine mind."[4] Flew is to be commended for accepting his error of atheism, but he did not repent and become a Christian, but a deist.

A deist believes God designed and created and left the world to run on its own, and therefore the universe has an intelligent designer who is absent. Deists never worship the personal triune God as sovereign Creator or become accountable to him. He still trusts self and lives in sinful unbelief of the Creator/Sustainer/Redeemer. Deists may talk about "God" that fits ***his own ideas of design*** and want his help, but does not experience it. Many pantheists, like Aristotle, also talk about a designing reason in nature, but they worship man as the one to be in control and live as man thinks is good. The pantheists, from Kant to Hawking, believe there is an unknown creative intelligent designing purpose and power ***in nature***. Both the deists who believe God created but left the world to run on its own, and the pantheists who believe intelligent nature is actualizing its design in the world, are also without accountability or with any sense of repentance toward the living personal God revealed in Christ. There have been many man-conceived ideas of God such as those of Mohammed, Roman Catholic Aristotelians, modern Syncrestists, and New Agers who seek to politically control men for themselves.

Genuine Christians forget that the world is demonically controlled and that we Christians still have a fleshly nature through which the devil tempts us. Many of the finest Christian organizations and institutions are legitimately trying to get Christians to abandon the secular

---

[4] Antony Flew, *There is a God*, (New York: Harper One: 2007), 185.

worldviews behind Enlightenment views for a true Christian worldview without emphasizing repentance. But the basic issue of man's sin bias must first be addressed by asking people to repent and accept Christ's death as theirs to live unto him. And unless Christians continue to repent to trust Christ, all the best information about the ***Christian worldview will not work. Even Christians*** will go on trying to save their lives and living for the world's values and in the world's wisdom. Many of the good Christian schools are seeking to escape the secular public school teachings, but then paradoxically the schools and parents want to send their children to the prestigious Ivy League schools, all of which have rejected their Christian foundations, or send them to other prominent secular colleges that reject Christ as Creator God. Unless there is a good foundation to understand the bias of unbelievers, the truth about Creation, and the school has a strong Christian fellowship, the students can have their faith sabotaged.

Christians must be serious about accepting true science for their witness now. It has never been more important than today for Christians to believe in the scientific method as begun by the sixteenth century Reformers. We believe man was made in the image of God to understand and rule his creation, and we as Christians who are led by the Spirit are the only men who can truly understand the wisdom of God in creation, and that only partially and imperfectly. In Christ are hidden all the treasures of wisdom and knowledge, and that is beyond the wisdom of Solomon. By our humility we can receive the grace from God to guide us in understanding science, and using it to glorify him and help people. Our challenge is to search out and use corrected science. It is hoped this book, *True Enlightenment,* will motivate young men and women to do that. In the years just ahead, mankind needs better understanding for provision of food which is becoming scarcer and poorly distributed; more and better water is needed in urban areas; and there is need to harness the forces of nature for use. These aspects of life are seemingly more disrupted as we approach the end times of tribulation.

Breaking Waves of All Civilizations are in God's Sovereign Plan for Elect Mankind

God's plans in Christ are not haphazard, but move with certainty to his tellos, or end. The unbelieving world uses the evil of the world to persuade that God cannot be both sovereign and good. In their bias they do not see that the evil is man's choosing and the whole is planned by God to reveal both his justice and his mercy. Without man's acts of evil, God's eternal abiding character of his truth, honor, justice, and power to judge and punish now and in hell would not be revealed. But the evil of all men makes him utterly dependent on the Creator, and reveals that God's love is eternally merciful and gracious. God in Christ alone is able to save and enable repentant elect men to rule with him as eternal sons through Jesus Christ. While Adam failed to rule for him in Eden, God the Creator will fulfill his plan for man to have dominion through the perfect Son of man, Jesus Christ and his elect. This record of evil and redemption was all to reveal his glory in the wisdom of his eternal decrees, and to glorify himself through his Son Jesus Christ.

The Bible tells us the failure of all people was his eternal plan; to show that every nation ended without a single man who would obey him, even Israel. Therefore all deserved eternal judgment, but are offered his mercy. "God has consigned all to disobedience that he may have mercy on all," revealing his awesome wisdom (Romans 11:32, 33). The times and seasons of all nations has been set, "having determined allotted periods and the boundaries of their dwelling place" (Acts 17:26). The job of his elect people today is to call all men to repent and believe in Christ as Creator and Redeemer. But the bias of the demonic deception by false science will certainly be unrelenting until the Antichrist temporarily rules, is destroyed and death ended (I Corinthians 15:24-26).

Christians Will Be Persecuted as Witnesses but Hope in the Recreation

The lapse of the whole gentile world civilization into unbelief will, in God's time unknown to man, turn all under Antichrist against the true Israel of God to destroy God's people. God will then restore the natural branches of Israel when the believing remnant will cry out to him for salvation. Then all Israel will be saved. The Christ will come to save those he elected before the foundation of the world, those who have repented (Romans 11). Then in Christ all his elect will be gathered and united for his heavenly blessings (Ephesians 1:3-10). These true disciples who have died to self in the cross of Christ, and who have shown his love and mercy to the world will be transformed, male and female, as equal sons to reign with his Son, the second Adam. We will be sons of a new human race with the true Son of Man (1 Corinthians 15:42-58; 2 Corinthians 5:14-6:1) in a new heaven and new earth at his coming. With Christ we will judge the world and in eternity in the new Eden, the New Jerusalem, we will judge angels (1 Corinthians 6:2, 3). As his elect believers, we now are to proclaim the mystery of God's plan hidden for ages "in God ***who created all things*** ...through the church [that] the manifold wisdom of God might now be made known to the rulers and authorities in the heavenly places. This was according to the eternal purpose that he has realized in Christ Jesus our Lord" (Ephesians 3:9, 10).

Only God knows if and when the wave of American civilization will crash, but it appears imminent. However, God is a God of grace and in spite of the confusion and divisions, evangelical Christianity is still professed by nearly a third of the nation and many may be humbled to lead repentance and fulfillment of the Great Commission. God was about to judge Nineveh, and they repented at the preaching of Jonah and were spared for a time. In the divided kingdom of God's people, the southern kingdom of Judah had several brief revivals, while the northern one never did, and was judged nearly 150 years earlier.

Today we Christians most of all need to repent and cry to God that he may give us, his elect, grace to experience renewal of his churches to witness to his love. The darkness coming in today's world is a time for light from the Creator. If God could save Israel from the Midianites by Gideon's handful of men, a united witness by believing Americans could, by God's grace, renew and extend faith through repentant leadership. Our Creator and Redeemer Jesus Christ has all power. "He is able to do far more abundantly than all we ask or think, according to the power [of his Spirit] that works in us, to him be glory in the church in Christ Jesus throughout all generations, forever and ever. Amen" (Ephesians 3:20, 21).

Addendum
Climate and Political Change

Global Warning to Distribute Wealth from Rich to Poor Nations

From about 1850 to 2000 AD, following the little ice age, glacial ice began to slowly melt and the feedback systems slowly increased the rate of melting. The rate of melting is now fast. Some scientists interpreted this with alarm. The neo-Marxists or postmodern movement incorrectly or deliberately misinterpreted this loss to the industrial nations' use of fossil fuels to try to promote redistribution of wealth from prosperous nations to those developing. The postmodern anti-capitalists said capitalist Western nations have "unjustly" attained wealth by oppressing poorer nations. This is an extension of the efforts to make the poor see themselves as victims so postmodernists can gain and use political power, and extend their worldwide political aims which they have previously sought through the United Nations. The claim of redistribution because of warming from fossil fuels has proven to be a hidden means to gain power. Every communist socialist national dictatorship has assumed dictatorial power with privileges for the communist elite. The claim that the rich industrial nations are causing global warming is a ploy to get richer nations to give up wealth for the United Nations in order to distribute and move toward a one-world government.

The neo-Marxists charge that fossil fuels emit so much CO^2 that it affects the ozone layer and causes climate temperatures to rise which has evil effects, especially on undeveloped nations. This effort began at the Kyoto conference in Japan and was to have been achieved at Copenhagen in 2009. At Copenhagen, they proposed that the rich nations give $10 billion a year for the next three years to the United Nations for this purpose. But, George Soros said this was not enough and proposed an increase to $100 billion as a just effort of the industrial nations for causing the climate abuse to the poor. These efforts to maintain what some thought to be a righteous cause have proven to be based on false science.

Is CO^2 The Cause of Warming?

The accusations against CO^2 are questionable and even disproved. There is probably truth that the destruction of the forests that absorb this gas, and the burning of fossil fuels in the growing industrialization of the world have increased the levels to some slight degree, but these were not a major factor of warming. Many other, more important factors caused the trend. In the past, about half of all carbon dioxide emissions have been absorbed by natural 'sinks' in ecosystems and oceans. However, rising temperatures might cause some reversal of this process. But there is still much to learn about this gas. Recently, Japanese scientists were shocked to find a large lake of liquid CO^2 at 4,600 feet on the ocean floor east of Taiwan in the Yonaguni Knoll IV Hydrothermal Field of the East China Sea.[1]

Roy Spencer, senior scientist for climate studies at NASA's Marshall Space Flight Center, believes efforts to stop the meltdown are futile because there are countless unknown variables and it would be painful to millions to make the changes. The use of fuels is a small part of the problem. Farm CO^2 emissions (18%) are far larger than fuel emissions in the U.S. Moreover, growing nations are increasingly using more fossil fuels, and those such as China and India with larger populations than any Western nation, will not abide by restrictions. The Evangelical Climate Initiative should be viewed with caution.[2] Most importantly, just when a trend seemed certain, there has been a radical shift in global temperatures. The effects of man-made CO^2 and other factors are seemingly a small contribution to the problem. Turning it into a political issue may cause serious economic problems throughout the world. Ian Plimer provides strong scientific

---

[1] Anne Casselman, "Global Warming sinks: Greenhouse gases could go undersea," *Discover*, December 2006, 13.

[2] Mark Bergin "Greener Than Thou," *World*, April 22, 2006.

data to refute this argument.[3] In Chapter 22, I provide evidence that actions of the sun and effects on earth magnetism and long waves increasingly appear to be a major factor in ice breakup.

Computer Models of Climate and Ice on Past Warming Made Wrong Predictions

The computerized global warming theory ***was faulty.*** The amount of Earth's surface warming seemed to be less than should have been created by greenhouse gases that have been measured in the atmosphere. Scientists think the unaccounted for gases disappeared into the ocean. More importantly, ***radical change occurred about 2002 AD,*** from warming to cooling. The computer models did not anticipate this.

The History of What Really Has Happened

Recent Arctic Trends Were Formerly Warming and Then Cooling. Why?

Obvious changes have occurred in the Arctic. The north and central portion of Greenland, where great annual snowfalls and buildup remained much the same for many years, even as temperatures rose. But, the margins and southeastern portion of the ice sheet had been thinning by almost 300 mm. annually.[4] The North Pole was becoming ice-free, making possible the Northwest Passage of ships. Sea ice in the Northern Hemisphere had shrunk six percent in the last twenty years and the thickness decreased forty-two percent in the last three decades. In the last twenty years, the temperature was up seven degrees Fahrenheit over a large part of the Arctic. Fresh water was freezing later and thawing earlier than any time in the last 150 years. Science writer Tim Appenzeller said,

> A great thaw has set in at the top of the world. The crust of floating ice that covers the Arctic Ocean almost from edge to edge is shrinking, and it has thinned by almost half since the late 1950s. No one knows exactly what's causing the melting, though global warming is an obvious suspect. And no one knows how far it will go, although a few scientists are predicting that within decades the Arctic Ocean could be completely ice free in summer....[5]

In 1999, satellite measurements showed ice has shrunk by fourteen percent or an area larger than France in the last twenty years; and studies by submarines over the past forty years, from the late 1950s, showed it had thinned from an average of ten feet to less than six feet thick. At these rates, the ice would be gone in another 50 years—a total polar meltdown. Scientists drilled into Dasuopu Glacier at 23,500 feet and found "that the last 50 years were warmer than any other equivalent period in the last 1000 years."[6] The total area covered by sea ice has shrunk by three percent during each of the past three decades. Thickness has decreased even more over the same period—as much as 40 percent in some places.[7] Normal gradual melt down since the post flood formation of the ice sheets have reach an acceleration point. The following are some of the reasons for the recent acceleration of changes:

The albedo effect seems to be building. The forces that accelerate meltdown were already in progress. Sturm, et al. have described what was occurring in the ice albedo effect as follows:

> Of all the changes we have catalogued, the most alarming by far has been the reduction in the Arctic sea-ice cover. Researchers tracking this alteration have discovered that the area covered by the ice has been decreasing by about 3 percent each decade since the advent of satellite records in 1972. ...With the sea ice covering an area approximately the size of the U. S. the reduction per

---

[3] Ian Plimer, *Heaven and Earth*, see whole first chapter, Introduction.

[4] S. Perkins, "Greenland ice is thinner at the margins," *Science News*, vol. 158, July 22, 2000, 54. Cf. "Grace in Space," *Discover*, March 2007, 43, 44 shows loss in S. Greenland.

[5] Tim Appenzeller, "Playing a fabled waterway," *U.S. News & World Report*, August 28, 2000, 49.

[6] Deborah Hudson, *Discover*, Special Issue, January 1, 2001, 60.

[7] Ibid., 62.

> decade is equivalent to an area the size of Colorado and New Hampshire combined. ... The change in the thickness of the ice (determined from submarines) is even more striking: as much as 40 percent lost in the past few decades.[8]

Glaciers have been Shrinking. While there may be data discrepancy, the evidence is clear that not only the ice sheets, but also the valley and cirque glaciers are shrinking. An analysis of the Columbia Glacier, extending from the Chugach Mountains to Prince William Sound in Alaska, indicates it retreated seven miles since 1982. It has thinned significantly, breaking off new icebergs.[9] Climatologists drove poles deep into the ice of a glacier on Ellesmere Island in 1882, but in 2002 the poles were lying on the ice. At that time it was said, "In Alaska, they have been shrinking for five decades, and more startlingly, the rate of shrinkage has increased threefold in the past 10 years. The melting glaciers translate into a rise in sea level of about two millimeters a decade or 10 percent of the total annual rise of 20 millimeters."[10]

Roger G. Barry, Director of the National Snow and Ice Data Center at the University of Colorado at Boulder, says around 100,000 km^3 of ice is locked in glaciers. He says that glacial melting has skyrocketed since the mid1900s. Between 1961 and 1976, the world's glaciers posted an average ice loss of 56 km^3. Since then, the melt rate has almost tripled. Water entering the oceans from melting glaciers is now boosting sea levels by about 0.4 millimeter per year to provide 15 to 20 percent of the current annual rise. Asia and the Coast Range of Alaska make up only 30 percent of the world's glaciers, but contribute 70 percent of the total glacial melt-water. The Central Asian mountain range in Kyrgyzstan has 170 glaciers of about 436 square kilometers. Aerial photos taken in 1943 and 1977 and by satellite in 2001 show accelerated rates of melting in recent decades. The Davydov Glacier lost only 0.5 km^2 between 1943 and 1977, but melted back another 4.8 km^2 by 2001.Similarly, the Sary-Tory Glacier lost just 0.1 km^2 between 1943 and 1977, but then dropped 0.9 km^2 by 2001. The British Antarctic Survey confirmed this from an analysis of 100 satellite images and 2,000 aerial photos and found that since 1953 about 212 of 244 glaciers on the west side of the peninsula had retreated an average of about 2,000 feet. Thus, about 90 percent of them have shrunk in the last half century. The factors involved in warming and melting continued to affect the area even after the shift to climate cooling in 2000. In the Beaufort Sea, a remote part of the Arctic Ocean north of Alaska, during a period of six weeks in July and August 2006, "A huge 'lake' bigger than the state of Indiana has melted out of the sea ice ...forming a kind of bay in the planet's northern ice cap." Dr. Mark C. Serreze of the National Snow and Ice Data Center in Boulder, Colorado said, "We have never seen anything like that before."[11]

The same accelerating trend was clear in Africa and South America. Alexa Stanard of the Associated Press has reported that Eric Rignot of the National Aeronautics and Space Administration, and Andres Ribera and Gino Casassa of the University of Chile have said that the Patagonian glaciers melted at twice the rise between 1995 and 2000 as in previous decades. They are losing ten cubic miles per year of ice and account for nine percent of the annual global sea-level change from mountain glaciers. Raymond Bradley, director of the Climate System Research Center of the University of Massachusetts says that radar imaging shows this is a world-wide phenomenon. Some glaciers, such as Chacaltaya Glacier near La Paz, Bolivia are nearing disappearance. In North America, similar rapid changes have been recorded in Washington State

---

[8] Ibid., 66.

[9] Tad Pfeffer of University of Colorado, *EOS Journal*, November 28, 2000.

[10] Ibid.

[11] Frank D. Roylance, "Huge lake melts out of Arctic ice cap," *Orlando Sentinel* from *The Baltimore Sun*, August 24, 2006, A27.

Cascade Mountains and in Glacier National Park in Montana and Canada. In Africa, the glaciers on Mount Kilimanjaro in Kenya were 80 percent gone.[12]

Matthew Sturm, Donald K. Perovich and Mark C. Serreze in their article, "Meltdown in the North" began by saying,

> Sea ice and glaciers are melting, permafrost is thawing, tundra is yielding to shrubs and scientists are struggling to understand how these changes will affect not just the Arctic but also the entire planet. The list is impressively long: the warmest air temperatures in four centuries, a shrinking sea-ice cover, and a record amount of melting on the Greenland Ice Sheet, Alaskan glaciers retreating at unprecedented rates. Add to this the increasing discharge from Russian rivers, and Arctic growing season that has lengthened by several days per decade, and permafrost that has started to thaw. Taken together, these observations announce in a way no single measurement could that the Arctic is undergoing a profound transformation. Its full extent has come to light only in the past decade, after scientists in different disciplines began comparing their findings.[13]
>
> Scientists are collaborating and trying to predict the future "because the Arctic exerts an outsize degree of control on the climate. Much as a spillway in a dam controls the level of a reservoir, the polar regions control the earth's heat balance.[14]

NASA scientist James Hansen says the Earth's temperature will still climb at least another one-half degree. Biodiversity in nature is already being affected and an additional three-degree increase could probably have "catastrophic consequences," says Stephen H. Schneider of Stanford University.[15].

In all of these studies a trend toward warming, ***and then an abrupt change*** occurred again toward cold and ice buildup. What is happening is not related to the old uniformitarian Milankovich theory and is a mystery.

Antarctic Ice trends in East and West had Other Conditions of Melting and Cooling

Antarctica is centered asymmetrically on the South Pole. It "covers more than 14 million km² (5.4 million sq mi), making it the fifth-largest continent, about 1.3 times as large as Europe. The coastline measures 17,968 kilometers (11,160 mi) and is mostly characterized by ice formations."[16] There is only a small part that is ice-free. The ***continent is divided north to south by very high mountains***, some 16,000 feet. It contains volcanoes, a few of which have been active, and many lakes and ice streams. The west is exposed to Pacific Ocean currents, the southeastern part borders the Indian Ocean, and the northeast is on the Atlantic Ocean. Since the time of the worldwide flood until the last part of the 20th century, it appears the ice has been decreasing.

Strangely, in East Antarctica, which is the largest region of the continent, altitudes of surface ice ***steadily rose as much as 6 centimeters per year between 1992 and 2003***. In West Antarctica there was a wide fluctuation, from a loss of 10 cm per year to a gain of 19 cm. This is attributed to large snow falls that could cause a slight sea level drop of 0.12 mm per year.[17] In the winter of

---

[12] These reports of glacier melting came from Sid Perkins, "On Thinning Ice: Are the world's Glacier in moral danger", *Science News*, October 4, 2003, 215, Vol. 164.and Alexa Stanard, Global Warming melts Patagonian glaciers: Scientists worry about possible worldwide effects, Science and Medicine, *Orlando Sentinel*, G6, and elsewhere as indicated.

[13] Sturm, Perovich, Serreze, "Meltdown in the North," 60-62.

[14] Ibid., 62.

[15] These reports of glacier melting came from Sid Perkins, "On Thinning Ice: Are the world's Glacier in moral danger", *Science News*, October 4, 2003, 215, Vol. 164.and Alexa Stanard, Global Warming melts Patagonian glaciers: Scientists worry about possible worldwide effects, Science and Medicine, *Orlando Sentinel*, G6, and elsewhere as indicated.

[16] *Wikipedia*, "Antarctica."

[17] SP, Antarctica's gaining ice in some spots," *Science News*, July 2, 2006, vol. 168, 13.

2007, cold and snow records were set in Australia, South America, and South Africa. According to NASA GISS data for the South Pole winter (June-August), the temperature cooled about one degree Fahrenheit since 1957. The interior of Antarctica has been colder and ice elsewhere has been more extensive and longer lasting. The 2007-2008 winters had record levels of Antarctic sea ice. But in March of 2009 there may have been a shift again toward loss of ice.

A different occurrence is found in the northwest. West Antarctica extends in a peninsula toward South America. It contains the Larsen ice shelf on its east point. "The immense and supposedly permanent Larsen Ice Shelf began to disintegrate in 1995. Nearly 1,000 square miles of shelf have collapsed just in two years before 2006, with thousands of square miles more appearing ready to go." These climate changes began to show up on the radar screen thirty years ago, but went unheeded.[18] The Grace in Space Mission showed the Antarctic ice sheet has shrunk "by an average of 36 cubic miles per year."[19] In 2003, the second part of the Larsen Shelf crumbled. This cannot be blamed on human impact or toxic pollution because there is very little human contact or urbanization. In the Antarctic, this was having a significant impact on the seal population, and other plants and animals.

At Palmer Station, the glacial front was fifty yards away thirty years ago; during warming it has moved within a quarter mile of the station. Since the mid 1940s, the year-round temperature increased three to four degrees Fahrenheit and in the winter, it is up seven to nine degrees. The rate of change there was ten times the global average. Charles Petit refers to scientists that said, "The Earth is unstoppably entering a heat wave that could last centuries.... The computerized climate models used to forecast global warming reveal no reason for this place to be warming more rapidly than the rest of the planet."[20] These computer models have not proven reliable. But this trend of warming took a turn.

The Wilkins ice shelf is on the western side of the peninsula. It has been stable for most of the last century, but began retreating in the 1990s. Toward the end of the first decade of twenty-first century, there has been a fracture and loss in the Wilkins Sea shelf area as large as the state of Connecticut. This and other fractures are not just melting from the edges. The reasons for these large breakoffs need more study. Satellite images suggest that part of the ice shelf is disintegrating, and will soon crumble away. So the ice trends at the South Pole also offer mystery. The Scripps Institution of Oceanography thinks this may be caused by long Pacific waves.

Scientific Confusion about Glaciation and Loss of Ice when the Trend Transitioned to Cold

It is necessary to reconsider glacial theories because computer data does not fit the reality. Scientists once thought the Arctic summer ice extent is largely determined by variable oceanic and atmospheric currents such as the Arctic oscillation. In the summer of 2008, NASA claimed that "***not all the large changes*** seen in Arctic climate in recent years ***are a result of long-term trends*** associated with global warming."

A review of the polar meltdown, based on data from the National Snow and Ice Data Center, National Oceanographic and Atmospheric Administration, shows a map and graph in September 1979 when the Arctic ice covered 3.8 million square miles. In 2005, it was 3.2 million square miles.[21] This does not show the loss of thickness. Mark Serreze of the U. S. Ice Data Center in Boulder, CO said, "At this rate the Arctic Ocean could be nearly ice free at the end of summer by 2012 much faster than previous predictions."[22] Again, "The European space agency said nearly 200 satellite photos, taken in September 2007, showed an ice-free passage along northern Canada,

---

[18] Ibid.

[19] Ibid., 48.

[20] Charles W. Petit, "Polar Meltdown," U.S. News & World Report, February 28, 2000, 66.

[21] "Polar meltdown? Floating arctic ice is cited as evidence of global warming, but scientist differ on the significance of the 25 year shrinkage," *World Magazine*, April 22, 2006, 21.

[22] Seth Borenstein, "The Arctic is Screaming." Accelerated Ice Melt Stuns, Scares Experts, *Orlando Sentinel,* Dec. 12, 2007, A 1.

Alaska and Greenland, and ice retreating to the lowest level since such images were taken in 1978."[23]

The Radical Unexpected Shift toward Cold and Ice Build Up Unexpectedly Occurred with Confusion

But surprisingly, a rapid cooling change in climate occurred, and uniform theory clouded previous evidence of interglaciations. David A. Peel, of the British Antarctic Survey in Cambridge has said, "We have quite *a lot of rethinking to do* about our basic concepts of climate change." This is a result of deep boring through Greenland's 3,000-meter icecap, which records the previous interglacial period. This gives a "hint that the planet's thermostat has a habit of shifting back and forth with ease, making the climate much less stable than previously presumed."[24] Mike Toner, reporting on evidence from Richard Alley of Pennsylvania State University and William Tanner of Florida State University at the meeting of the American Association for the Advancement of Science said, "New evidence from the Greenland icecap and from the floor of the North Atlantic Ocean shows that large changes in climate, and perhaps in sea level, are sometimes ***stunningly abrupt***."[25] (Emphasis added) A new finding from a study of the North Atlantic sea floor sediments, posits,

> ...Earth repeatedly swayed from extremely frigid conditions to warmth and back again with startling speed. As part of these shifts, the great North American ice sheet vomited huge numbers of icebergs that filled the North Atlantic. The new revelations have left scientists reeling, ***because the steady orbital cycles thought to control the ice ages cannot easily account for the evidence*** of quick climatic jitters. ***It is clear that the climate theory is not complete.***[26] (Emphasis added)

Gerard Bond of Columbia University's Lamont-Doherty Earth Observation, at a 1994 meeting of the American Geophysical Union said, "It's getting so complicated, I sometimes think we're going backwards."[27] Understanding the glacial ice is critical to the future of the Earth.

But the shift from warm to cold had begun and the temperature ***began to drop in 2003***. The warming trend reversed to cold and the ice began to rebuild. However, the pessimism about warming continued because many were committed to it. The following data is taken from a Steve Goddard report, with minor editing.[28]

> The National Snow and Ice Data Center (NSIDC) in Boulder, Colorado released an alarming graph on August 11, 2008 showing that Arctic ice was rapidly disappearing, back towards last year's record minimum. Their data shows Arctic sea ice extent only 10 per cent greater than this date in 2007, and the second lowest on record. Data sources show Arctic ice having made a nice recovery this summer. NASA Marshall Space Flight Center data shows 2008 ice nearly identical to 2002, 2005 and 2006. Maps of Arctic ice extent are readily available from several sources,

---

[23] Jamey Keaten, The Associated Press, "Arctic ice has hit record low, satellites show," Orlando Sentinel, September 16, 2007, A15.

[24] R. Monastersky, "Ancient Ice Reveals Wild Climate Shifts," *Science News,* Vol. 144, July 17, 1993, 36; cf. *Science News,* July 3, 1993, 7.

[25] Mike Toner, "Past Rapid Shifts in Climate Cast New Light on 'Greenhouse Effect,' " *Atlanta Journal,* February 18, 1995, A6.

[26] Ibid.

[27] Richard Monastersky, "Staggering Through the Ice Ages: What Made the Planet Green Between Climate Extremes?" *Science News*, Vol. 146, July 30, 1994, 74-76.

[28] The reports given in the paragraphs below are from Steven Goddard, "Arctic ice refuses to melt as ordered: There's something rotten north of Denmark," *The Register Science Environment*, 15th August 2008 10:02 GMT.

> including the University of Illinois, which keeps a daily archive for the last 30 years. A comparison of data derived from NSIDC shows that Arctic ice extent was 30% greater on August 11, 2008 than it was on the August 12, 2007. (2008 is a leap year, so the dates are offset by one.). Ice has grown in nearly every direction since last summer—with a large increase in the area north of Siberia. Also note that the area around the Northwest Passage (west of Greenland) has seen a significant increase in ice. Some of the islands in the Canadian Archipelago are surrounded by more ice than they were during the summer of 1980. The 30% increase was calculated by counting pixels which contain colors representing ice. This is a conservative calculation, because of the map projection used. As the ice expands away from the pole, each new pixel represents a larger area—so the net effect is that the calculated 30% increase is actually on the low side. The Arctic did not experience the meltdowns forecast by NSIDC and the Norwegian Polar Year Secretariat. It didn't even come close. Additionally, some current graphs and press releases from NSIDC seem less than conservative. There appears to be a consistent pattern of overstatement related to Arctic ice loss.[29]

Also, the Antarctic ice extent is well ahead of the recent average. Goddard asked,

> Why isn't NSIDC making similarly high-profile press releases about the increase in Antarctic ice over the last 30 years? So how did NSIDC calculate a 10% increase over 2007? Their graph appears to disagree with the maps by a factor of three (10% vs. 30%)—hardly a trivial discrepancy.[30]

In 2007-2008 there also was a radical shift to cold in the Arctic. According to R.D. Walker,

> Over the past year, anecdotal evidence for a cooling planet has exploded. China has its coldest winter in 100 years. Baghdad sees *its first snow* in all recorded history. North America has *the most snow cover in 50 years*, with places like Wisconsin the highest since record-keeping began. There is record cold in Minnesota, Texas, Florida, Mexico, Australia, Iran, Greece, South Africa, Greenland, Argentina, Chile—the list goes on and on. All four major global temperature tracking outlets (Hadley, NASA's GISS, UAH, and RSS) have released updated data. All show that over the past year, global temperatures have dropped precipitously.[31] (Emphasis added)

Meteorologist Anthony Watts compiled the results of all the sources. The total amount of cooling ranges from 0.65C up to 0.75C—a value large enough to erase nearly all the global warming recorded over the past 100 years, all in one year time.

Controversy over Trends through Error but also Perversion Promoted by Postmoderns

Controversy on climate trends has had confusion which could be expected. There have been errors by technical reports that are highly relied on. The hardware of DMSP satellite NSIDC sensor and *Cryospace Today* reports seemed to have a HD error. While this was repaired and was not a major factor since there are many other ways of gauging what has happened, it led to some conflict because of the high interest politically. Our technology is not without error, and requires caution. But it is now known that there has been also legitimate use of perversion.

When the climate data shifted toward warming those with political motives used perverted facts. In mid-August 2009 the University of East Anglia's Climate Research Unit (CRU) disclosed that it had destroyed the raw data for its global surface temperature data set because of an alleged lack of storage space. The CRU data has been the basis for several of the major international studies that claim we face a global warming crisis. CRU's destruction of data, however, severely

[29] Steven Goddard, "Arctic ice refuses to melt as ordered: There's something rotten north of Denmark," *The Register Science Environment,* 15th August, 2008, 10:02 GMT.
[30] Ibid.
[31] R.D. Walker Conspiracies, Environmentalism, Global Warming, Media, Politics, Science, from the internet and reported by Associated Press.

undercuts the credibility of those studies. ***Even before the East Anglia e-mail explosion in 2009,*** NASA was caught red-hot handed in falsifying temperatures by blatantly assigning September 2008's temperatures to October 2008, just to make October of that year look warmer. It has been discovered that internal discussions relating to the e-mail sent to James Hansen and/or Reto A. Ruedy from a Stephen (Steve) McIntyre admitted errors in NASA/GISS online temperature data (August 2007). On Dec. 17, 2009 the Moscow-based Institute of Economic Analysis (IEA) issued a report claiming that the Hadley Center for Climate Change based at the headquarters of the British Meteorological Office in Exeter (Devon, England) ***had probably tampered with Russian-climate data and thereby distorted the facts.*** Also, there was a report by the IPCC warning of climate change so great that the Himalayan Glacier would be gone by 2035 A.D. This has been admitted to be a claim made without any objective measurements, only reports written by mountain climbers. The exposure of these and other false data has discredited the claims of the IPCC on climate change, and has confused the glacial and climate data. Moreover, the U.N. panel of climate experts overstated how much of the Netherlands is below sea level.

The British Broadcasting Co. reported significant omissions from the climate data.

> Colleagues say that the reason Professor Phil Jones has refused Freedom of Information requests is that he may have actually lost the relevant papers. Professor Jones told the BBC yesterday [February 13, 2010] there was truth in the observations of colleagues that he lacked organizational skills, that his office was swamped with piles of paper and that his record keeping is 'not as good as it should be'. The data is crucial to the famous 'hockey stick graph' used by climate change advocates to support the theory. Professor Jones also conceded the possibility that the world was warmer in medieval times than now—suggesting global warming may not be a man-made phenomenon. And he said that for the past 15 years there has been no 'statistically significant' warming.[32]

The admission of fraud or error is fresh evidence that there are serious flaws at the heart of the science of climate change and the orthodoxy that recent rises in temperature are largely man-made.

The evidence is now clearer. In April 2009 the Heartland Institute held a conference at which more than 700 scientists from all over the world came together to testify that man-made ***Global Warming does not exist***, which is more than 12 times the number of the UN IPCC group. Top that with the fact that more than 31,000 American scientists have signed a petition saying there is no convincing scientific evidence that human release of carbon dioxide, methane, or other greenhouse gases is causing disruption of the Earth's climate. Wolfgang Knorr of the Department of Earth Sciences at the University of Bristol reanalyzed available atmospheric carbon dioxide and emissions data and reported in *Geophysical Research Letter* that the airborne fraction of ***carbon dioxide has not increased*** either during the past 150 years or during the most recent five decades when the carbon fuel was most used.

Conclusions about Global Warming Controversy

The long slow meltdown from the original glacial formation and the growing albedo effects are causing loss of ice. But, the sun's activity and changes in the earth's magnetic field also yield surprising effects. These less understood causes of temperature changes and computer models gave postmodern politicians a false pretence to try for a power grab.

Scientists now believe the sun may be slipping into "several decades of hibernations that could exert a cooling effect on earth's climate…"[33]

---

[32] Jonathan Petre, *Mail online,* February 14, 2010.

[33] Ron Cowan, "Next Solar Cycle may be a No-Show: Studies suggest sun may be headed for decades of dormancy," *Science News,* July 16, 2011, 12.

Selective Index

This index does not include hundreds of scientists and reporters who are referred to. Most are to be found by using the table of contents for related chapter subjects and footnotes in the chapter. Only major scientists, subjects, or books that are significant are given here for help.

Adam, mankind, vi, 7, 8, 12, 38, 39, 42, 73, 164, 226, 233-236, 285, 309, 479, 490, 494, 521, 549, 571, 574, 575, 581-583, 585, 588, 589
Albedo, 332, 333, 344, 347, 592, 598
Absolute Age Determination, see chapter 23
Absolute Genetic Dating, 237, 238
The Age of the Earth, 401
American Civil Liberties Union, ACLU, 92
Africa, out of Africa, 226, 230-236, 239, 258, 263-265, 299-301
Agassiz, Louis, 68, 73, 307, 308, 323, 324, 326
Aiken, Henry D., 551
Albertus Magnus of Cologue, 15
Albright, William Foxwell, 504, 512, 516, 518, 599
Alpha particle, 402, 414, 550
Altizer, Thomas J. J., 462, 464-466, 469, 561
America, v, viii, ix, 5, 8, 76, 91, 94, 100, 113, 146, 226, 234, 235, 245, 275, 289, 292, 294, 296, 302, 323, 395, 437, 439, 461, 499, 509, 518, 519, 525, 529, 530, 539, 540, 543, 548, 553, 556, 557, 561, 573, 585-587
Amino acid, amino acids, 160, 165, 171, 201, 207, 211, 214, 219, 220, 222, 549, 583
Anaerobic, 166, 167, 208
Asthenosphere, 353
Anthropic, 151-153, 563, 565
Anthropology, 89, 94, 191, 229, 261, 265, 267, 268, 288, 292, 294, 302, 599
Antichrist, 7-9, 12, 445, 465, 466, 485, 494-497, 572, 583, 588
Aquinas, Thomas, 15, 16, 28
Archaebacteria, 166
Archeology, 90, 297, 337, 516, 518, 527, 550
Ardipithecus ramidus, 260
Aristotle, Aristotelian, v, ix, 11, 14-16, 19-22, 24-28, 30-32, 34, 37, 44, 47-49, 51, 59, 70, 75, 78, 80, 83, 103, 110, 424, 530, 545, 547, 560, 566, 571, 572, 585, 587
Art, 5, 19, 21, 38, 51, 93, 275, 292, 295-298, 303, 378, 524, 528
Asteroids, Apollo asteroids, meteorites, 59, 63, 111, 144-147, 150, 153-155, 188, 209, 328, 343, 346, 364-369, 372, 376, 377, 382
Astronomy, ix, 16, 21, 22, 24, 25, 27, 29-31, 33, 34, 49, 50, 60, 62, 69, 75, 100, 113-116, 118, 130, 131, 134, 139, 140, 143, 212, 272, 295, 303, 307, 351, 380, 381, 401, 428, 527, 528, 545, 546
Astruc, Jean, 437
Atheism, 11, 43, 51, 265, 271, 465, 467, 572, 587
Augustine, 18, 286, 505, 525, 585
Australopithecus afarencis, A afarencis, 246, 250, 252-254, 256, 258, 259, 266
Babylon, Babylonian, 8, 9, 25, 90, 150, 295, 308, 309, 311, 318, 379-382, 437-439, 517, 518, 527, 532, 534, 553
Bacon, Francis, 37-39, 546
Bakker, Robert, 186
Balloon Observations of Millimetric Extragalxitic, 138
Babel, vi, 8, 377, 379-382, 494, 572, 575, 583, 586
Barth, Karl, 457
Barzun, Jacques, v, 540
Basins, 313-315, 320, 323, 333, 364, 365, 369, 398,
Baumgardner, John H., 359-362, 364, 416-419
behavior genetics, 273, 274
Behe, Michael J., 169, 171, 172, 174
Bentz, J. H., 398
Berkeley, George, 54
Bersossos, 308-311
Bible, biblical, Scriptures, v, viii-x, 5, 6, 12, 13, 15, 17-20, 24, 25, 27-31, 33, 34, 37, 39-41, 43-48, 51, 53, 56, 57, 62, 70, 71, 75, 76, 79, 85, 94, 163, 164, 229, 268, 271, 272, 274, 291, 299, 303, 309, 310, 312, 308, 322, 324, 350, 354, 376, 380, 381, 415, 416, 426, 435-437, 439-441, 446, 456, 457, 461, 466, 469, 472, 480, 482, 494, 497, 499, 501, 506, 512, 514, 517-521, 545, 546, 548, 548, 552, 572-577, 583, 584, 587, 588
Big Bang theory, 116, 126, 127, 136, 137, 151, 429, 468
Black hole, black holes, 117, 128-130, 134, 135, 141, 146, 419, 424, 425, 429, 548, 551, 564, 569-571
Black Sea, 228, 312, 314-316, 344
The Blank Slate, 272
Bode's Law, 59, 143, 150
Bohr, Niles, 102,103
Bondi, Hermann, 116,
Bosporus, 315,
Boyle, Robert, 28, 39, 413,
Brahe, Tyco, 24, 420
Branson, E. B. and Carl C., 63, 144, 181, 390, 392, 395, 410
Breccia, breccia pipes, 393, 398
Bruce, F. F., 502, 505, 506
Bryan, William Jennings, 91-93
Buckland, William, 67-69, 72, 307, 324
Buffon, George L., 65, 75-77, 80, 86, 152, 351
Bultmann, Rudolph, 456-461, 466, 467, 475
Burkitt, F. C., 490
Bursts of evolution, 181
Bush, Vannevar, 100
Butterfield, Herbert, 16, 19, 20, 37, 38
Cairns-Smith, A. G., 206, 207, 210
Calvin, John, 20, 436, 519, 576
The Canaanites, 512
Carpzovius, 49, 576
Plate Tectonics, 148, 153, 154, 198, 351-355, 390, 394, 397, 413, 551
Catastrophes, 70, 76, 77, 143, 197, 206, 307, 324, 330, 337, 341, 377, 382, 388, 389, 399, 400, 418, 554, 569
Cell, 32, 159, 160, 197, 170-172, 174, 177, 183, 187, 188, 205, 207, 209, 211, 216-220, 222-226, 271, 274, 278, 279, 282, 285, 354, 359, 523, 549
Cell wall, 209
Central Plateaus, 375

Chambers, Robert, 65, 70, 78-80, 87, 91, 245, 387, 547, 560, 572
Chaos theory, 111, 549, 555, 562, 569, 583
Chicxulub, 368, 369, 374, 375, 377
Childe, Gordon, 94, 268
Christian uniqueness, 442,
Christian worldview, 469, 588
Cladism, 163, 164, 175
Clovis people, 227, 342, 378
cognitive science, 273-275, 280, 281, 283, 284, 464, 550
Coliseum of Rome, 5
Collingwood, R. G., 523
Collins, Francis, 220, 572
Comparative Religion, 94, 301, 435, 437, 438, 450, 451, 456, 475, 499, 509, 511-513, 520, 526, 532, 548, 552, 553, 555
Cone of development, 550
Congress of International Papyrologists' Association, 506
Continental division, drift, 153, 198, 237, 312, 330, 346, 351-355, 357, 358, 367, 369, 370, 374, 375, 377-379, 388
Copernicus, Nicolaus, vi, 22, 24, 25, 37, 59, 429, 545, 566, 569
Cosmic Background Explorer, COBE, 120, 121, 132
Cosmic radiation, 121, 130, 141, 408, 411, 417-419, 429
Cox, Harvey, 466, 468, 469, 471
Creator, creation, , v-x, 5-7, 9-12, 15-18, 20-22, 24-29, 34, 37, 40, 43, 44, 46, 47, 51-55, 57, 59, 62, 66-70, 73, 75-81, 83, 84, 88, 89, 91-93, 99, 102, 110, 113, 116, 118, 119, 123, 125, 132, 136, 151,,156, 159, 161, 163, 164, 166, 168, 169, 191, 194, 197-199, 202-204, 207-209, 211, 212, 216-218, 248, 266, 268, 271, 274, 282, 285, 286, 288, 291, 302, 308, 323, 354, 380, 388, 399, 410, 411, 415, 417, 422, 424, 425, 429, 430, 435, 438, 446, 448, 484, 489, 494, 496, 499, 512, 514, 519, 520, 526, 527, 537, 538, 544-548, 550
Crick, Francis H. C. 160, 215, 217, 219, 278
Cro-Magnon, 90, 227, 228, 230, 245, 246, 265, 294, 375
Cryogenic microwave receiver, 116
Cupitt, Don, 474, 475, 482
Cumulative selection, 199, 201-205, 210
Cuneiform list, 310
Custance, Arthur C, 308, 311
Cyclotrons, accelerator, 33, 101, 103, 104, 110, 126, 548, 551, 559
Cytoplasm, 225
D'Aquill, Eugene, 281, 282, 286
Dampier, Sir William Cecil, 29, 30, 43, 57, 103
Dark Age, later dark ages, 13, 14, 17, 506
Dark energy, 137-141, 419, 425, 428, 429, 551, 570
Dark matter, 117, 122, 128, 131, 135, 137, 139-141, 424, 551
Darrow, Clarence, 92, 93
Darwin, Charles Robert, ix, 57, 66, 68, 69, 71-73, 75, 78-93, 113, 114, 150, 159-175, 177, 180, 182-186, 189-191, 193-201, 204, 206, 207, 219, 245-247, 249, 250, 261, 263-267,288, 298, 303, 387-389, 400,401, 409, 435, 437, 438, 440, 447, 513, 518, 547, 548, 552, 555, 560, 561, 582, 583, 586, 599
Darwin, Erasmus, 77, 78, 80, 87
Dawkins, Richard, 198-206, 208-210
De Fermat, Pierre, 23, 61, 545
Death of God, 461, 462, 465-467, 469, 471, 489, 491, 520, 553, 561
Deism, 11, 15, 34, 40, 43, 399, 435, 437, 447, 543, 546
Denton, Michael, 165, 167, 170, 171, 198, 279, 280, 572
Descartes, Rene, 37-40, 42, 44, 47, 52, 53, 56, 75, 271, 435, 437, 515, 546, 585
Deshayes, G. P., 70, 71, 387, 401
Determinism, 31, 48, 56, 60-63, 64, 73, 77, 82, 88, 91, 94-99, 100, 101, 110
Devil, demonic, Satan, 111, 118, 136, 140, 141, 164, 167, 197, 199, 207, 250, 267, 268, 280, 308, 323, 325, 337, 350, 382, 399, 400, 430, 435, 436, 443, 445, 456, 461, 471, 489, 490, 493, 497, 499, 509, 512, 523-525, 549, 554, 555, 560-563, 565, 566, 569, 570
DeWette, Wilhelm M. L., 448
Diamond, Jared, 277, 288-294, 296, 298-300, 302, 343, 368, 378
Dick, Bob, 116
Diderot, 77, 546
Dillow, Joseph C., 343, 344
Dinosaur, 68, 153, 182, 185, 186-190, 367, 399, 400, 560
Dissipation, 51, 354, 584
DNA, 159-161, 165, 170, 172, 177, 183, 205, 208, 210, 212, 213, 215, 219-228, 230-233, 235
Dollo's Law, 213
Durant, Will and Ariel, 46
Earle, Sylvia A., 362, 365
Earth's core, 150
Eccles, John C., 280-282, 286
Edge of evolution, 172
Einstein, Alfred, 34, 102, 103, 110, 113-116, 118, 127, 137, 138, 548, 549, 552, 569,
Eiseley, Loren, 84, 86, 88-91, 160, 177, 200, 245, 249
Eldredge, Niles, 189, 190, 194-196, 250. 252, 257, 266
electroencephalogram, EEG, 276
electromagnetic, electromagnetism, 33, 105, 108, 114, 136, 334, 414, 418, 419, 421-423, 425, 429, 545, 567
Element M, 562
Eliade, Mercea, 465, 537, 538
Encyclopedia, encyclopedic, 57, 59, 99, 300, 301, 483, 536, 546
The End of Science, 560
Engineering, 162, 212, 220, 229, 292, 294, 296, 303, 394, 527, 528, 534
English Channel, 322
Epicurus, viii, ix, 31, 48, 49, 51, 53
Epifluorescent confocal laser scanner, 506
Epigenetic, 222
Erasmus, Desiderus, 77, 78, 80, 87, 501
Erickson, Jon, 351, 355-35, 369, 371, 377, 378
Eschatological kingdom, Eschatology, 447, 451-456, 462, 464, 465, 470, 473, 476, 491, 496, 504, 552
Ethnology, ethnologist, 381, 509-515, 520, 526, 543, 553, 585
Eurasia, 94, 228-231, 232, 235,261, 263-266, 268, 288, 292, 294, 296, 298-302,309,312, 344, 346, 356, 357, 365, 373, 376-378, 381, 549, 550
Eve, vi, 7, 53, 73, 226, 231, 233-236, 239, 285, 494, 549, 574, 575, 582, 583
evolutionary psychology, 273

Excess density, 354
Existentialism, 456
Falsification, falsification theory, 101, 555
Families of manuscripts, 49
Faraday, Michael, 33, 46, 545
Fertile Crescent, 288-293, 299, 301-303, 517
Feynman, Richard, 566-569
Finegan, Jack, 30-,310
Flew, Anthony, 572, 587
folding, 319, 392, 393
Fossil, fossils, 29, 63, 66-73, 76, 77, 80-82, 84, 85, 89, 90, 93, 94, 159, 160, 167, 172, 177, 180-182, 184-186, 188-195, 197, 198, 204-206, 216, 218, 225-229, 231, 236, 238, 245, 246, 248-250, 252, 261, 263-265, 267, 269, 307, 315, 321, 326, 331, 333, 337, 341, 347, 349, 352, 356, 357, 387-393, 395-397, 399, 400, 403, 404, 406, 407, 409, 411-414, 416, 545, 547-550
Franklin, Rosalind, 160, 219
Friction, 125, 354
From Dawn to Decadence, v, 540
Funk, Robert W., 493
Galaxy, galaxies, 60, 73, 102, 108, 113-117, 119, 122, 125, 127-135, 137-142, 425, 429, 523, 548, 551, 563, 564, 577-585
Galilei, Galileo, 21, 23, 25-28, 30, 33, 37, 43, 59, 110, 152, 429, 449, 473, 492, 545, 555
Gamma ray, gamma rays, 114, 129-131, 141, 402
Gamow, George, 101, 102, 116, 118, 564, 599
Gange, Robert, 152, 213-215, 270
Gene, genotype, 160-162, 167-183, 193, 194, 202-204, 206, 215, 217, 219-229, 232, 237
General theory of relativity,
Genesis, 78, 81, 92, 136, 164, 308, 310, 311, 347, 437, 446, 462, 518, 569, 573-577, 579-581, 583-585, 587
Gentiles, 6, 13
Geological age, 340, 341, 389, 392, 407
Geology, v, 57, 59, 60, 62-64, 66-73, 75, 80, 87, 99, 100, 113, 141, 143, 177, 181, 190, 198, 267, 307, 308, 319, 324, 337, 347, 351, 357, 380, 387, 388, 392, 394, 400, 401, 403, 417, 545, 547, 548, 551, 570
Geosynclines, 395, 398
Gravitational sliding, 354, 393
Guns, Germs and Steel, 288
Geyh, Mebus A., 403-407, 409, 412, 417, 418
Gibbons, Ann, 237-239, 257, 266
Gilbert, William, 23, 43
Gilgamish, 310
Gillespie, Charles Coulston, 307
Glacier, ice sheets, valley and cirque, 26, 26, 67, 151, 226, 227, 221, 268, 289, 292, 296, 312-314, 315, 321, 323-332, 334, 336-342, 344-347, 349, 350, 355, 368, 369, 379, 388, 393, 397-399, 409, 412, 418, 546, 548, 549, 551, 556, 592-594, 598
Gleick, James, 111, 569
Glomor Challenger, 312-314
Glueck, Nelson, 517, 518
Gondwana, 328, 329, 354-362, 365, 367, 370, 373
Gould, Stephen J., viii, 91. 93, 173, 174, 182-184, 186, 189-191, 194-197, 206, 218, 266, 560
Goulder, Michael, 473, 474, 478, 479, 481, 482
Graebner, E., 509
Grains, wheat, oats, 79, 144, 289-291
Grand Canyon, 297, 319-321, 376, 395, 400, 411, 412
The Grand Design, 566, 568, 570
Great apes, 237, 261, 263, 266
Great Attractor, 130, 141
Great cosmic current, 130
Great Wall, 134, 135
Greek, vi, viii, 5, 9, 12, 14-19,21-23,25, 29, 31-33, 37, 38, 48, 51, 59, 63, 77, 86, 101, 118, 274, 302, 379, 382, 402, 406, 420, 449-451, 457, 475, 476, 479, 480, 485, 496, 501, 502, 506, 513, 516, 525, 529, 530, 545, 564, 574-576, 585, 599
Greene, Brian, 562, 563, 565
Guth, Alan, 120, 121, 138
Haldone, J. B. S., 207
Half-life, 180, 326, 401, 403, 404, 408, 409, 416, 423, 550
Hall of men, 227, 247-249, 267
Hameroff, David H., 279
Hardy-Weinberg Law, 194
Hawking, Stephen, viii, 34, 38, 60, 61, 99, 116-118, 120, 129, 138, 143, 151, 152, 196, 422, 425, 430, 540, 563, 565-572, 582, 587
Heisenberg, Werner, 102, 103, 566
Helium retention, 414-416
Henry, Carl F. H., vi, 55, 599
Herschel, John vi, 55, 599
Herschel, William, 57, 69, 70, 72, 322
Higgs, Higgs field, 108, 109, 125, 126, 548, 559, 560
High Spirit God, Supreme Being, 52, 53, 438, 511, 512-514, 526, 552
Hills Cloud, 145, 147
Holy Roman Empire, 145, 147
Holy Spirit, Spirit , iii, ix, x, 6-10, 25, 26, 42, 43, 52-54, 73, 82, 87, 282, 284-286, 311, 430, 435, 438, 442, 445, 448, 454, 465, 477, 479, 480, 485, 497, 499, 511, 512, 514, 524, 526, 572, 574, 576-579,585-589
Hominids, 231, 255-257, 259-261, 264, 266, 549
Homo erectus, 227, 232, 233, 239, 249, 250, 258, 259, 263
Homo sapiens, 90, 227-230, 232, 239, 250, 255, 259, 263, 264, 378
Hooker, Joseph, 81, 82, 85, 159, 183, 265, 547
Horner, Jack, 186, 187
Houser, Marc, 283
Hoyle, Fred, 116, 215, 217, 564
Hubble constant, 116, 119, 132, 137, 138, 551
Hubble space telescope, 115, 120, 122, 128, 131-133, 138
Hubble, Edwin, 115, 116, 137, 138
Human genetic dating, 237-239
Human genetic diversity, 237, 239
Human Genome Project Consortium, 220
Hume, David
Hume, R. E., 50, 54, 55, 66, 435, 475, 483, 546, 547
Hutton, James, 57, 60, 64-68, 70-73, 75, 78-81, 87, 207, 307, 322, 323, 387, 390, 399, 499, 547, 555
Huxley, Julian, 82, 84, 85
Hyperthermophilic, 208
Icons, *Icons of Evolution,* 93, 191, 192, 246
The Idea of History, 523
Inflationary Theory, 117, 119, 125, 127, 133, 555, 599
Intellectual elite, intellectuals, iii, vii, viii, ix, 8-17, 19, 25, 27, 28, 32, 37, 39, 44, 47, 48, 55-57, 62, 70-72, 75, 79, 83, 87, 89, 91, 93, 94, 99, 100, 116, 159, 184, 190, 191, 198, 229, 267, 275, 278, 287, 288, 303, 307, 379, 387, 400, 435, 437, 446, 447, 456, 461, 466, 471, 472,

478, 484, 485, 489, 493, 523, 525, 527, 529, 531-535, 539, 543, 546, 547, 552-554, 556, 572, 573, 583, 585, 587
Intelligent design, ID, ix, 11, 48, 49, 53, 75, 79, 80, 82, 87, 93, 110, 159, 174, 177, 192, 198, 199, 387, 547, 560, 561, 563, 571, 572, 583, 587
Interpolator, 440, 441, 515
Interpretation (principal of), 574, 576
Islam, 18, 21, 22, 436, 442, 483, 484, 501, 521, 545
Jacob, Francois, 219
Jenkin, Fleming, 86
Jesuit order, 15, 27, 48
Jesus Christ, Son of God v, ix, x, 6-40, 48, 99, 436, 443, 445, 446, 459, 460, 463, 471, 475, 476, 478, 480, 481, 484, 485, 489-491, 495, 497, 499, 501, 540, 543, 572-577, 586-589
"Jesus Seminar", 470, 493
Johanson, Donald, 250, 252-254
John, Apostle John, 21, 445, 446, 448, 449, 451, 453, 473, 479, 484
Kant, Immanuel, 31, 48-57, 59, 60, 62, 73, 75, 83, 87, 104, 113, 143, 153, 198, 271, 272, 307, 322, 387, 401, 429, 436, 447, 456, 457, 499, 546548, 550, 552-557, 560, 572, 585-587
Kanzi, 269, 270
Keck telescope, 122
Kepler telescope, 122
Kepler, John (Johanus) , 23, 27, 29, 31, 36, 43, 59, 110, 131, 136, 324, 420, 429, 545, 555
Kerigma, 447
Kirshner, Robert, 35, 137-139, 141
Kuhn, Thomas S., vii, viii, 323, 555, 566,
Kuiper Belt, 145, 147, 345, 364
Lake Agassiz, 317, 318, 344, 346, 375
Lamarck, John-Baptiste, 8, 70, 72, 77, 80, 82, 86, 87, 161, 173, 177, 185, 200, 245, 387, 547, 560, 572, 583
Language, 8, 12, 27, 33, 42, 45, 88, 93, 160, 170, 213, 224, 230, 234, 254, 267, 269-271, 275, 276, 283, 286, 298-301, 303, 308-310, 378-381, 449, 462-464, 468, 469, 472, 476, 477, 509, 512, 574, 576, 599
Laplace, Pierre-Simon, 32, 60-62, 64, 65, 73, 83,
Large Hadron Collider, 109
Leakey, Mary, 227, 250, 255, 258
Leakey, Richard, 250, 253, 254, 266
Leonardo da Vinci, 20, 21, 37, 110, 555
Lepton, 103, 105, 125
Light Speed, 140, 420-422, 425, 426, 428, 429, 552
Light years, 114, 121, 127, 129-135, 140, 141, 154, 215, 420, 421, 549, 552
Linnaeus, Carl von, 44, 45, 49, 75-77, 163, 545, 546
Locke, John, 44, 47-50, 52, 54-56, 75, 271, 273, 387, 402, 403, 422, 435-437, 447, 546, 550, 585
Lucretius, Titus, viii, ix, 31, 32, 48-55, 75, 546, 570, 585
Lucy, 250, 252-254, 257-260, 549
Luther, Martin, 20, 44, 45, 436, 501, 519, 576
Lyell, Charles, 57, 66-73, 77-88, 153, 159, 177, 183, 197, 207, 219, 265, 267, 307, 308, 312-324, 330, 337, 341, 351, 357, 387, 388, 399-401, 499, 547, 548, 451, 455, 470
Lystrosaurus, 356
Macroevolution, 169, 189, 193-1 97, 409
Magnestars, 131
magnetic resonance imaging, MRI , 275, 276
Malthus, Robert, 77, 78, 81, 82, 90, 582
Man is Not Enough, 5, 512, 513, 525, 529, 543
Mantle, 150, 151, 353, 354, 358, 359, 361, 362, 368, 401, 403, 410, 411, 417, 420
Mapping cognitive functions, 281
Mapping Human History, 288
Margulis, Lynn, 163, 168, 174, 561
Maxwell, James Clark, 33, 34, 421, 425, 545
Mendel, Mendilian, 84, 90, 162, 167, 219, 389
Microevolution, 172
microscopes, microscopic, 103, 104, 160, 163, 165, 167, 172, 188, 213, 218, 219, 506, 549, 599
Microwave background, 116, 118-122, 125, 127, 128, 130, 131, 568
Miller, Stanley, 207, 208
Millimeter Anisotropy Experiment Imaging Array, MAXIMA, 138
Mitochondria, mtDNA, 174, 225, 226, 228, 230-239
Modern Science, v, vi, vii, x, 5, 20, 34, 53, 55, 57, 78, 83, 91, 99, 100, 104, 107, 143, 167, 199, 200, 212, 218, 323, 394, 446, 458, 476, 543, 545, 554, 560, 561, 564, 573, 583, 585, 587
Mohorovicic, Andrija, 353
Molecular clock, 162, 166, 237
Molecular mutations, mutations , 82, 159-163, 166, 168, 172, 174, 177, 184, 196, 203-205, 208, 215, 220, 223-228, 230, 237, 239, 258, 274, 291, 429, 549
Monod, Jacques, 219
Montague, Ashley, 94, 268
Moore, Raymond C., 180-182, 341, 388, 390, 392, 404, 408
Morphology, 82, 163, 254
Mullis, Kary, 220
Muon, 104, 106-108
Murchison, Roderick I., 68, 69, 71, 73, 77, 341, 545
The Myth of God Incarnate, 471, 475, 478-482, 484, 489
nation, nations, ix, 6, 8, 12-14, 56, 92. 93, 109, 146, 217, 230, 3002, 349, 368, 374, 379-382, 439, 441, 445, 480, 489, 496, 497, 499, 517, 519, 525, 527, 529, 530, 532, 535, 539, 540, 548, 552, 557, 573, 586, 588, 589
National Human Genome Research Institute, 220
natural selection, ix, 75, 77, 81-85, 87-90, 159, 161, 165, 172, 173, 176, 189, 190, 195, 203-208, 211, 216, 231,247, 291, 388
nature, naturalism, v, vi, vii, ix, 5-12, 15, 19, 21, 24, 25, 27, 29-32, 34, 37-41, 43, 44, 46-49, 51-53, 55-57, 59, 61-68, 70, 72, 73, 75-81, 83, 86-89, 91, 93, 99, 100, 102, 109-111, 113, 132, 142, 143, 146, 147, 152, 154, 155, 160, 165, 168, 173, 175, 182, 188, 197, 198, 200-205, 208, 212-214, 216, 218, 228, 238, 245, 247, 249, 271,272, 274, 275, 278, 281, 282, 286, 287, 291, 294, 301, 302, 315, 322, 323, 343, 376, 379-381, 391, 396, 405, 406, 418, 421, 422, 424, 430, 435, 437, 438, 443, 446, 447, 460, 462, 465, 466, 472, 484, 490-492, 499, 500, 509, 511-514, 519, 524-528, 533, 537-540, 543, 545-550, 552, 554-556, 560-562, 566-568, 570-572, 577, 581-588, 594
Neanderthals, Neandertals, 89, 90, 93, 227-233, 239, 245, 247, 249, 255, 258, 263-265,293, 550
Nebula Hypothesis, 5, 57, 60, 62, 73, 113, 543, 570
Neo-Darwinism, 159, 160, 163, 172, 190
Neuroscience, 9, 272, 275, 276, 280, 281, 284, 285
New Generalized Second Law, 212, 213, 216

New Information Theory, 206, 210, 212, 214, 224, 271, 274, 291, 424, 425, 430, 575, 582
Newberg, Andrew, 281, 282, 286
New Covenant, 21, 494, 495, 586
New Testament, 19, 20, 448, 450, 451, 555, 456, 465, 467, 473, 475, 476, 482, 490, 492-494, 501, 502, 504-507, 520, 575, 599
Newton, Sir Isaac, 23-25, 29-31, 33, 34, 43, 44, 46, 49, 50, 52, 59, 60, 99-101, 113, 116, 132, 136, 139, 307, 387, 420, 421, 424, 425, 429, 545-555, 567, 569
Nicodemas, 6
Nimrod, 8, 9, 12, 379, 381, 572
12, 64, 164, 230, 236, 239, 265, 266, 291, 299, 303, 307-310, 312, 315, 318, 343, 344, 376, 379-381, 494, 535, 549
Nonevidence 193, see all chapter 13
North Pole, 330, 332, 334, 356, 357, 390
Ocean currents, 328, 332, 394
Ocean spreading, 354, 361, 371
Odlenburg, Ulf, 512
Oklo unranium mines, 422
Old Testament, 12, 13, 45, 91, 99, 379, 382, 435, 437, 439, 441, 442, 445, 446, 448, 449, 456, 477, 502, 514-520, 599
Olson, Steve, 236, 288, 296, 298, 299-302, 429
One world religion, world syncretism, 484, 493
Oparin, A. I, 207
Orreries, 50, 143, 155
Oort Cloud, 145, 147, 345
The Origin and Growth of Religion, 511
The Origin of Species, 71, 81-89, 91, 173, 177, 200, 245, 246, 265, 267, 389, 435, 438, 547, 560
Osborn, Henry Fairfield, 91, 93, 247,-249,267
overthrusting, 292, 293
Owen, Richard, 85, 159, 173, 186, 547
Oxygen isotopes, $O^{16,}$ O^{18} , 334, 339, 340, 403
Paganism, 16, 19, 20, 110, 197, 466, 537
Palomar Sky survey, 122
Pansperma, 217
Pantheism, ix, 11, 12, 15, 19, 24, 34, 40, 48, 49, 51, 53, 55, 70, 75, 77, 78
Parallel universes, Multi-universes, 136, 559, 564, 569-571
Parthenon of Athens, 5
Pascal, Blaisé, 23, 24, 61, 104, 111, 545
Pasteur, Louis, 32, 207
Patterson, Colin, 190
Paul, Apostle Paul, v, x, 6-10, 13, 15, 41, 53, 379, 380, 382, 430, 445, 450, 507, 513, 520, 525-527, 529, 537, 540, 557, 576, 577, 586
Penfield, William, 280, 281
Penzias, Arno, 116, 119
Peter, Apostle Peter, v, 382, 470, 479, 497, 575, 585
Phages, 223
Physics, vii, 25-27, 31-34, 49, 50, 92, 100-106, 108-111, 115, 126-128, 139, 143, 148, 170, 199,205, 212, 272, 279, 284, 421, 422, 424, 425, 428, 545, 559, 560, 565, 567-569, 571, 599
Pinker, Stephen, 272-277, 280
Planck, Max, 102-127, 562, 563
Plutinos, 145, 147
Popper, Karl, 101, 281, 282, 555
positron emission tomography, PET, 276
Prokaryote, 166
Prophet, prophets, 12, 13, 440, 441, 456, 470, 480, 484, 490, 493, 496, 520, 554, 574-577
Protein, proteins, 32, 160, 165-167, 170, 171, 175, 183, 210, 212-215, 219, 221-224, 237, 274, 290, 549, 583
Punctuated Equilibrium, viii, 194197, 206, 549
Pyramids of Egypt, 5
Quantum theory, 34, 101-104, 113, 139, 279, 351, 422, 545, 48, 49, 64, 65, 68
Quark, 103-108, 110,125, 126, 562
The Quest of the Historical Jesus, 451, 490
Radio carbon dating, 344, 401, 407
Radio transient stars, 131
Rawlinson, Henry Cheswick, 307, 308
Ray, John, 29, 43, 44, 545
Reformation, reformers, Protestant, v, vi, viii, 6, 14, 16, 18-20, 27, 31, 37, 42, 44, 46, 49, 63, 72, 99, 429, 430, 468, 470, 519, 543, 545, 546, 552, 554, 559, 560, 572, 576, 585, 588
Reimarus, H. S., 447
Relativistic Heavy Ion Collider, RHIC, 109, 125, 126
Religion, v, 12-15, 18, 28-31, 37, 39-41, 45, 46, 48, 51, 52, 54, 56, 57, 72, 73, 87, 88, 93, 94, 99, 100, 110, 197, 272, 297, 303, 381, 399, 430, 435-442, 446-451, 454-456, 458, 461, 465-472, 475-478, 482-484, 489, 493, 497, 499, 501, 511-514, 516-520, 526, 529, 530, 532, 537-539, 546, 548, 552, 553, 556, 561, 585
Rethinking Missions, 442
Rivers in the Desert, 517
RNA, 159-161, 167, 175, 177, 183, 204-206, 207-210, 219, 222, 223, 229, 549
Robinson, John A. T., 456, 459, 66-68, 507, 553
Roman, viii, 6, 9, 13, 14, 17, 18, 22, 25, 31, 32, 44, 51, 53, 447, 450, 474-476, 485, 504, 506, 513, 525, 529-532, 537, 545
Roman Catholic Church, Catholic Church, v, 14-18, 20, 21, 26, 34, 37, 39, 44, 45, 46, 48, 49, 61, 78, 80, 452, 484, 505, 545
Rutherford, Ernest, 34, 101, 103, 545
Ryan, William B. F., 312-316, 344
Saltation, 143, 218, 549, 599
Schaff, Philip, 14, 16, 474, 479
Schleicher, Helmut, 403, 404, 417
Schmidt, W., 435, 438. 509-514, 520, 526, 527, 553
Schweitzer, Albert, 448-455, 457, 490-492, 504
Schweitzer, Mary H., 187, 188
Scope, John; Scope's Monkey Trial, Dayton, TN, 248
The Secular City, 468
Sedgwick, Adam, 67-69, 73, 79, 88, 179, 341, 545
Sedimentary rock, sediments, 177, 180, 190, 227, 259, 296, 312, 313, 315, 319-321,326, 331, 337, 340, 369, 376, 388-392, 394-399, 403, 404, 407. 409-412, 414, 545, 596
Seduction troughs, trenches, 354, 371
Setterfield, Barry, 426
Seven Daughters of Eve, 226, 235
Shannon, C., 212
Sherrington, Sir Charles, 280, 281
Silica, silica's, 358, 359, 376, 401
Silverstein, Alvin, 354, 363
single-proton emissions computerized tomography, SPECT, 254, 276, 281, 282
Singularity, singularities, 117-119, 123, 125, 128, 129, 141, 429, 551, 568, 569
Skylab, 146

Slab pull, 354
Sloan telescope, 123, 135
Smith, William ("Strata"), 66-69, 72, 73, 177, 308, 389-391, 545
Smith, Wolfgang, 193, 196
Smoot III, George F., 120, 121, 127, 425
Solar Heliospheric Observatory (SOHO), 146
Solar system, 22, 29, 34, 50, 59-61, 66, 75, 83, 99, 100, 114, 128, 141, 143, 145-148, 153-156, 198, 217, 307, 344, 346, 364, 413, 418, 420, 421, 424, 429, 499, 538, 545, 547, 548, 551, 566
Solar winds, solar flares, Corneal Mass Ejections, 146, 335, 336, 418, 419
Sorokin, Pitirim, 523-525, 532, 540
South Pole, 150, 328, 332, 336, 350, 356, 394, 395
Space probes, 145, 146, 149,152,353
Special theory of relativity, 102, 421
Species, 45, 46,75-88, 113,159-161, 163-168,172,174, 175, 179,180,182-185,187,189, 190, 193-199, 203-206,212, 216, 224, 225, 229,233,246, 247, 252-256, 258, 259, 261,263, 265-267, 278, 290, 291, 312, 315, 326, 342, 352, 376-378, 387,395,409,549, 560,582, 584,600
Spencer, Herbert, 78, 438
Spengler, Oswald, 523,525
Spinoza, Baruch, 39, 40, 47, 56, 110, 435-437, 499, 515, 546
Spiritual experiences, 282
Standard model, 104-109, 128
Stanley, E. M., 196
Strauss, D. F., 448, 449, 452, 475, 492, 502
Steady state theory, 116
Stone Age, 94, 229, 248, 250, 251, 259, 268, 297
Streeter, Burnett Hillman, 502, 506, 507
String theories, 562, 563, 567, 569, 570
Superconducting quantum interference device, SQUID, 276
Supernova, 114, 128, 131, 137, 138, 423
Swedenborg, Emanuel, 50, 52, 53, 113, 546, 570, 585
Sykes, Bryan, 226-228, 231-235, 237-239,378
Symbiosis, 174, 193
Talmud, 436, 437
Tammuz, 9, 381
Tattersall, Ian, 189, 190, 206, 227, 250, 252, 254, 256, 257, 259, 266, 293
Taxonomy, 76, 163-165, 545
Tectonic plates, plates, 177, 197, 328, 330, 341, 344, 345, 352, 353, 360, 362, 388, 392, 400
Tetheys Sea
Textus Receptus, 501
Theism, 11, 15, 75, 467, 543, 545, 561
Thermal runaway, runaway, 358, 359, 361, 362, 367, 551
Thermo Luminescence, 413, 414
Tidal friction, 354
Tillich, Paul, 458, 467, 561
Time-rocks, 590
Tischendorf, Constatiaus, 501
Tocqueville, Alexis de, 529-532
Townsend, Joseph, 66-68
Toynbee, Arnold, 523-525, 532, 539
The True and Only Heaven, 539
Tudge, Colin, 162, 163, 165, 180, 183
Two or three witnesses, 576
Unmoved Mover, 15, 16, 19, 22, 47, 49, 59, 78, 83, 560, 571, 583, 585
Van Buren, Paul, 462-467, 469
Verlinde, Eric, 424, 425
Vestiges of the Natural History of Creation, Vestiges, 70, 78-80, 85, 91, 245, 560, 583
Volcano, vocanic, 63, 64, 66, 69, 70, 72, 148, 154, 177, 321, 328, 330, 334, 337, 344, 346, 353, 362, 365, 367, 369, 370, 371, 375, 377, 381, 388, 390, 414, 415, 418, 420, 545, 547, 551, 583, 594
Von Haller, Albrech, 57
Vrba, Elizabeth, 195-198, 218
Wallace, Alfred R., 81, 83, 86, 88, 90, 265, 547
Washington State Scablands, 321, 331
Watson, J. D., 160, 219
Waves of civilization, 5, 543
Weaver, W., 212
Wegener, Alfred, 351, 352
Weisberger, Bernard A., 535, 537, 538
Weisse, C. H., 449, 501
Wellhausen, Julius, 438-442, 446, 456, 499, 512, 514-520, 553
Wells, Jonathan, viii, 77, 85, 191
Wells, Spencer, 226, 228, 231-236, 378
Werner, Abraham Gottlob, 63-65, 69, 72, 73, 307, 392, 399, 545, 547
Westcott, B. F., 501
Whiston, William, 46, 49, 546
White, Tim, 252, 257, 258
Whitten, Edwin, 563
Wickramasinghe, Chandra, 215
Wild moons, 364
Wiles, Maurice, 472, 475
Wilkes Land, 368, 369, 374
Wilkinson Microwave Anisotropy Probe, WMAP, 120-122, 134, 137
Wilkinson, David, 121
Wisdom of God, ix, x, 6, 29, 30, 43, 44, 400, 522, 577, 586, 588, 589
Wise, Kurt, 182, 190
Woods Hole Oceanographic Institute, 367
Worldwide catastrophe, worldwide phenomenon, 369, 382
Wouk, Herman, 439, 440, 515, 518, 519
Wright, G. Ernest, 518
Wright, Thomas, 50, 52, 546
Writing, Hieroglyphics, Phoenician, 9, 300-302, 321, 512
Y chromosomes, 225, 226, 233, 236, 237, 258
Yang, Y. S., 359,
Yockey, H., 213, 215
Young, Frances, 473, 474, 479, 481
Young, Thomas, 33
Zoological big bang, 182

More About the Author

Carl W. Wilson was born and grew up in Montgomery, Alabama. He was a newspaper reporter before enlisting in the naval air corps. In the military he accepted the glorified Christ as his Savor. After leaving military service he gained degrees in science and theology from accredited colleges and graduate schools, Maryville College and Fuller Theological Seminary. He was a college laboratory assistant in inorganic and organic chemistry and in Theological Seminary was an assistant in Old Testament Hebrew. He graduated with honors in all his studies. He worked briefly in medical technology for a hospital. He has planted and been pastor of several churches. As a leader in two major Evangelical organizations he worked extensively with students and pastors and traveled and taught with national scholars and pastors in twenty five countries. He writes here as a science history reporter and not as a practicing technical scientist.

Through the years Wilson has briefly studied under various outstanding scientists: Dr. George Gamow, former Russian scientist who became Professor of Theoretical Physics at George Washington University who participated in the early development of the quantum theory and was later the first man to suggest the chemical development of the Big Bang inflationary theory of the universe. Dr. Roger Rusk, Physics Professor at the University of Tennessee was an affiliated scientist at Oak Ridge Laboratory for the development of the atomic bomb; Dr. Harlan Banks, Professor of Paleontology of Cornell University, who was one of the first to abandoned a dogmatic Darwinism and embraced a saltation evolutionary theory; Dr William Foxwell Albright, Professor of Archaeology at Johns Hopkins University, considered the leading archeologist of his time and the authority on Palestinian archaeology; Dr. Kenneth Pike, Professor of Linguistics at University of Michigan, Spring Arbor, and one of the greatest pioneers and authorities in primitive anthropology; Dr. Gleeson Archer, a Semitic, Egyptian and Greek scholar at Fuller and Trinity Theological Seminaries, and a leading authority on Old Testament Criticism; Dr. William S. LaSore, was a leading Semitic language scholar; Dr. Carl F. H. Henry, was considered by many the dean of American Theologians, an authority on the issue of epistemology for understanding the shift in modern thought, and the founding editor of the magazine, *Christianity Today*.

After studying under the above scientists and biblical scholars he also read many of their books and for forty years has read three monthly science news magazines recording new findings in related areas of science. He was a contributing scholar to *Baker's Dictionary of Theology* and has written books for several major Christian publishers. He writes from a position of biblical and a moderate Reformed theological view. He taught seven years as an adjunct professor for Reformed Theological Seminary, Orlando, FL, and has been a guest lecturer at many Evangelical Theological Graduate schools in America and abroad. He is the author of 16 books and materials for group study. *True Enlightenment* is a work of forty years research.

He now works with Andragathia Inc. Ministries doing training for churches, research, writing and publishing. The web page is bravegoodmen.org. He is married to Sara Jo Wilson and they have five children, seventeen grandchildren and some great grandchildren. Carl and Sara Jo live in Winter Springs, Florida.

Other Useful Books by Carl W. Wilson
Andragathia@bellsouth.net

Three Books as a Set or Separately For Knowing Christ, and Leaders Training Guide
Unique Harmony of the Gospels: a new parallel outline for our time of the documents written by Christ's early disciples, revised 2011, 192 pages, paperback, See what Jesus actually said and did and how these progressed to his ascension as glorified Lord.

The Fulfillment: An Objective Look at the Person and Ministry of Jesus Christ in a Harmony of the Gospels. revised 2011, 272 pages, paperback, a nine month personal or small group study with guide and questions. Use any harmony but best studied by use of the *Unique Harmony….*

With Christ in the School of Disciple Building, revised 2011, 398 pages, paperback. Includes the study of Christ's method of building disciples, important principles of discipleship, and the need for relationships and for disciple building, and how gifts of leadership should be used in the church.

Biblical Essentials for Training Discipleship Leaders, published by Andragathia Books, 2002, 145 pages. A textbook for senior pastors or associates to use *With Christ in the School of Disciple Building* in training pastors of home churches to help people to relate, grow and be accountable within a larger community of the church.

Other Books for Understanding Our World
Man Is Not Enough, Published by Andragathia, Inc., Books, 1998, 496 pages, hardback. Understand why America's morals have declined to a dangerous point and what is likely to happen soon in our country. This predicted the debt, immigration, class and family crises now occurring.

Liberty in an Evil Age: Winning the Spiritual Battle Over Perversion by Intellectuals, 219 pages, a preliminary brief presentation of how the soft sciences have led to corrupting of our civilization. www.xulonpress.com.

Our Dance Has Turned to Death, published by Renewal Publishing Company, 1979, and republished by Tyndall House Publishers, 1981, 267 pages, paperback. An analysis of the causes of the decline and the problems in marriage, the family and nation, andragathia@bellsouth.net.

Books for Study in Living the Christian Life
From Uncertainty to Fulfillment: A comprehensive guide to God's guidance. Published in 1993 by Worldwide Discipleship Association, and now available through New Life Resources, 393 pages, paperback. For study personally or in small groups.

Pursuing Christ Daily, Leader's Discussion Guide, 115 pages. Includes instructions for teaching disciples how to establish a daily quiet time, a commentary on the readings in the Gospel of John that the disciples are covering each week, and guidelines for leading group discussions to hold disciples accountable, (spiral bound for leader to use in small groups).

Pursuing Christ Daily: *Meditations and Discussions in John*, Disciple's Notebook, 54 pages. Provides a tool for helping disciples establish a quiet time. John's Gospel has been divided into six readings per week for nine weeks (spiral bound for disciples to use daily and in their small group).

Publisher Provided Course Homepage

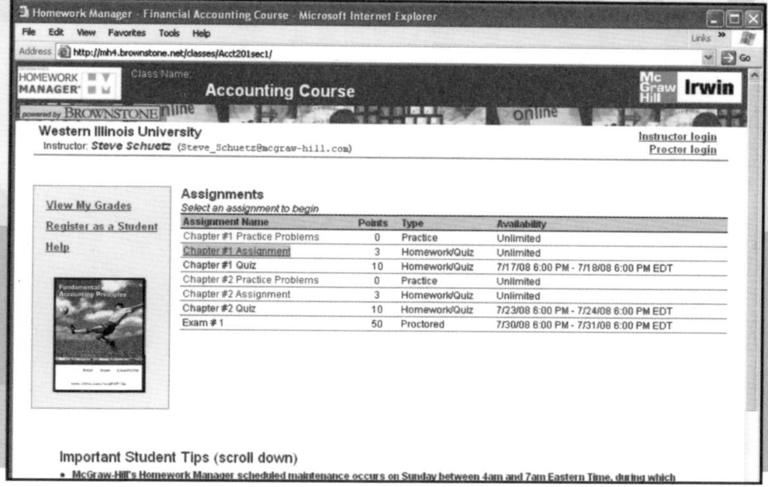

Easily Assign Online Homework

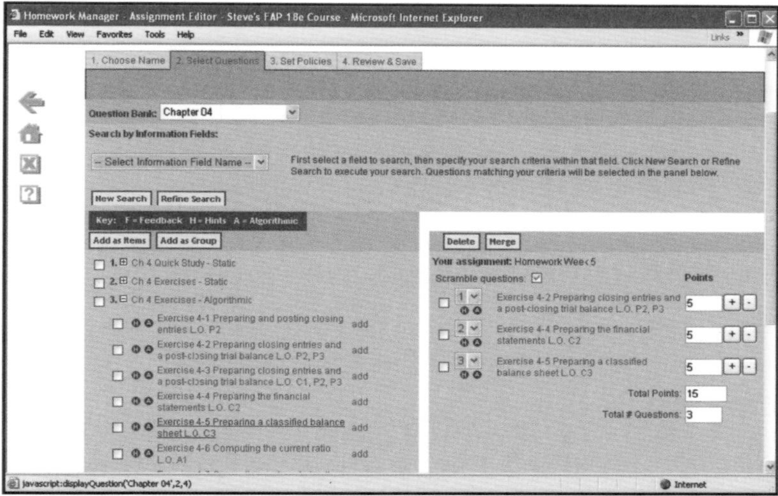

Track Student Results

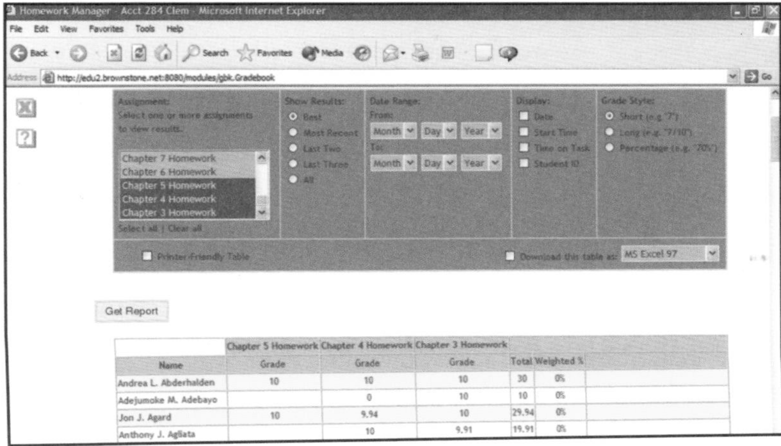

It's that easy.

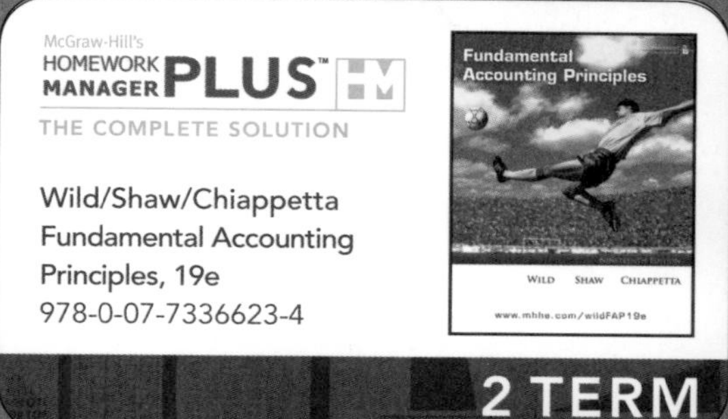

Continuously Improving!

McGraw-Hill is committed to listening to our customers and making changes based on our customer desires; hence we are updating our system to bring you the following new features:

- Single Entry Point, Single Registration, Single Sign-on
- Local Time Zone Support
- Enhanced Grade Book Reporting Capabilities
  - Run reports on multiple sections at once
  - Export reports to WebCT and BlackBoard
- Enhanced Question Selection
  - Select by learning objective, AACSB Accreditation Criteria
- Enhanced Assignment Policies
- Enhanced eBook*
- Integrated eBook – Simple Toggle (no secondary log-in)
  - Topic Search
  - Adjustable text size
  - Jump to Page #
  - Print by Section
- Student Assignment Preview for Instructors

Volume 2 Chapters 12–25

19 edition

Fundamental Accounting Principles

John J. Wild
University of Wisconsin at Madison

Ken W. Shaw
University of Missouri at Columbia

Barbara Chiappetta
Nassau Community College

Boston Burr Ridge, IL Dubuque, IA New York
San Francisco St. Louis Bangkok Bogotá Caracas Kuala Lumpur
Lisbon London Madrid Mexico City Milan Montreal New Delhi
Santiago Seoul Singapore Sydney Taipei Toronto

To my wife **Gail** and children, **Kimberly, Jonathan, Stephanie,** and **Trevor.**
To my wife **Linda** and children, **Erin, Emily,** and **Jacob.**
To my husband **Bob,** my sons **Michael** and **David,** and my **mother.**

McGraw-Hill Irwin

FUNDAMENTAL ACCOUNTING PRINCIPLES
Published by McGraw-Hill/Irwin, a business unit of The McGraw-Hill Companies, Inc., 1221 Avenue of the Americas, New York, NY, 10020.

This book is printed on acid-free paper.

3 4 5 6 7 8 9 0 DOW/DOW 0

ISBN-13: 978-0-07-337954-8 (combined edition)
ISBN-10: 0-07-337954-9 (combined edition)
ISBN-13: 978-0-07-336629-6 (volume 1, chapters 1–12)
ISBN-10: 0-07-336629-3 (volume 1, chapters 1–12)
ISBN-13: 978-0-07-336628-9 (volume 2, chapters 12–25)
ISBN-10: 0-07-336628-5 (volume 2, chapters 12–25)
ISBN-13: 978-0-07-336630-2 (with working papers volume 1, chapters 1–12)
ISBN-10: 0-07-336630-7 (with working papers volume 1, chapters 1–12)
ISBN-13: 978-0-07-336631-9 (with working papers volume 2, chapters 12–25)
ISBN-10: 0-07-336631-5 (with working papers volume 2, chapters 12–25)
ISBN-13: 978-0-07-336627-2 (principles, chapters 1–17)
ISBN-10: 0-07-336627-7 (principles, chapters 1–17)

Vice president and editor-in-chief: *Brent Gordon*
Editorial director: *Stewart Mattson*
Publisher: *Tim Vertovec*
Executive editor: *Steve Schuetz*
Senior developmental editor: *Christina A. Sanders*
Executive marketing manager: *Sankha Basu*
Managing editor: *Lori Koetters*
Full service project manager: *Sharon Monday®, Aptara, Inc.*
Lead production supervisor: *Carol A. Bielski*
Lead designer: *Matthew Baldwin*
Senior photo research coordinator: *Lori Kramer*
Photo researcher: *Sarah Evertson*
Lead media project manager: *Brian Nacik*
Cover and interior design: *Matthew Baldwin*
Cover photo: *© Photolibrary*
Typeface: *10.5/12 Times Roman*
Compositor: *Aptara®, Inc.*
Printer: *R. R. Donnelley*

The Library of Congress has cataloged the single volume edition of this work as follows

Wild, John J.
Fundamental accounting principles / John J. Wild, Ken W. Shaw, Barbara Chiappetta.—19 ed.
p. cm.
Includes index.
ISBN-13: 978-0-07-337954-8 (combined edition : alk. paper)
ISBN-10: 0-07-337954-9 (combined edition : alk. paper)
ISBN-13: 978-0-07-336629-6 (volume 1 ch. 1–12 : alk. paper)
ISBN-10: 0-07-336629-3 (volume 1 ch. 1–12 : alk. paper)
[etc.]
1. Accounting. I. Shaw, Ken W. II. Chiappetta, Barbara. III. Title.
HF5635.P975 2009
657—dc22

2008035921

www.mhhe.com

Dear Colleagues/Friends,

As we roll out the new edition of *Fundamental Accounting Principles*, we thank each of you who provided suggestions to enrich this textbook. As teachers, we know how important it is to select the right book for our course. This new edition reflects the advice and wisdom of many dedicated reviewers, focus group participants, students, and instructors. Our book consistently rates number one in customer loyalty because of you. Together, we have created the most readable, concise, current, accurate, and innovative accounting book available today.

We are thrilled to welcome Ken Shaw to the *Fundamental Accounting Principles* team with this edition. Ken's teaching and work experience, along with his enthusiasm and dedication to students, fit nicely with our continuing commitment to develop cutting–edge classroom materials for instructors and students.

Throughout the writing process, we steered this book in the manner you directed. This path of development enhanced this book's technology and content, and guided its clear and concise writing.

Reviewers, instructors, and students say this book's enhanced technology caters to different learning styles and helps students better understand accounting. *Homework Manager Plus* offers new features to improve student learning and to assist instructor grading. Our *iPod* content lets students study on the go, while our *Algorithmic Test Bank* provides an infinite variety of exam problems. You and your students will find all these tools easy to apply.

We owe the success of this book to our colleagues who graciously took time to help us focus on the changing needs of today's instructors and students. We feel fortunate to have witnessed our profession's extraordinary devotion to teaching. Your feedback and suggestions are reflected in everything we write. Please accept our heartfelt thanks for your dedication in helping today's students understand and appreciate accounting.

With kindest regards,

John J. Wild   Ken W. Shaw   Barbara Chiappetta

#1 with Customers

Fundamental Accounting Principles rates #1 in Instructor and Student satisfaction over the prior three editions, and we are now proud to report that both independent research and development reviews show that *Fundamental Accounting Principles* is now #1 in customer loyalty!

#1 CUSTOMER LOYALTY

#1 in Accuracy

#1 in Assignments

#1 in Topic Coverage

#1 in Supplements

#1 in Readability

#1 in Organization

#1 in Overall Textbook Satisfaction

With ratings such as these, it is no surprise that *Fundamental Accounting Principles* is the fastest growing textbook in the accounting principles market. Take a look at **what instructors are saying** about *Fundamental Accounting Principles.*

Ken Coffey, Johnson County Community College
"There is nothing about this text that I do not like. We have been using it since I started teaching the course in 1971. We switched texts three times, but always went back to this one and I do not think we will try anything else."

Linda Rose, Westwood College
"Very thorough, readable, and graphically pleasant! Engages students in a variety of activities that create learning opportunities and [...link] materials to real-work scenarios and situations. End-of-chapter materials are engaging learning tools and broaden the learner's opportunity to practice and enhance retention of the chapter content."

Shafi Ullah, Broward CC
"The book contains detailed information with alternative exercises, problems, and cases with 'Beyond the Numbers'."

Marilyn Cilolino, Delgado CC
"I have always been a big fan of this book and praise it every chance I get. The authors are great people and are really concerned with their product. They listened to the instructors and take our comments and suggestions seriously. And they are always ready to help or answer any questions one might have."

Gloria Worthy, Southwest Tennessee CC
"FAP is a very good text and it is also good for our students because it addresses a variety of learning styles"

Phillip Lee, Nashville State Tech CC
"The Wild text packs more useful material into fewer pages than other principles of accounting textbooks."

John J. Wild is a distinguished professor of accounting at the University of Wisconsin at Madison. He previously held appointments at Michigan State University and the University of Manchester in England. He received his BBA, MS, and PhD from the University of Wisconsin.

Professor Wild teaches accounting courses at both the undergraduate and graduate levels. He has received numerous teaching honors, including the Mabel W. Chipman Excellence-in-Teaching Award, the departmental Excellence-in-Teaching Award, and the Teaching Excellence Award from the 2003 and 2005 business graduates at the University of Wisconsin. He also received the Beta Alpha Psi and Roland F. Salmonson Excellence-in-Teaching Award from Michigan State University. Professor Wild has received several research honors and is a past KPMG Peat Marwick National Fellow and is a recipient of fellowships from the American Accounting Association and the Ernst and Young Foundation.

Professor Wild is an active member of the American Accounting Association and its sections. He has served on several committees of these organizations, including the Outstanding Accounting Educator Award, Wildman Award, National Program Advisory, Publications, and Research Committees. Professor Wild is author of *Financial Accounting, Managerial Accounting*, and *College Accounting*, each published by McGraw-Hill/Irwin. His research articles on accounting and analysis appear in The Accounting Review, Journal of Accounting Research, Journal of Accounting and Economics, Contemporary Accounting Research, Journal of Accounting, Auditing and Finance, Journal of Accounting and Public Policy, and other journals. He is past associate editor of Contemporary Accounting Research and has served on several editorial boards including The Accounting Review.

Professor Wild, his wife, and four children enjoy travel, music, sports, and community activities.

Ken W. Shaw is an associate professor of accounting and the CBIZ/MHM Scholar at the University of Missouri. He previously was on the faculty at the University of Maryland at College Park. He received an accounting degree from Bradley University and an MBA and PhD from the University of Wisconsin. He is a Certified Public Accountant with work experience in public accounting.

Professor Shaw teaches financial accounting at the undergraduate and graduate levels. He received the Wiliams Keepers LLC Teaching Excellence award in 2007, was voted the "Most Influential Professor" by the 2005 and 2006 School of Accountancy graduating classes, and is a two-time recipient of the O'Brien Excellence in Teaching Award. He is the advisor to his School's chapter of Beta Alpha Psi, a national accounting fraternity.

Professor Shaw is an active member of the American Accounting Association and its sections. He has served on many committees of these organizations and presented his research papers at national and regional meetings. Professor Shaw's research appears in the Journal of Accounting Research; Contemporary Accounting Research; Journal of Financial and Quantitative Analysis; Journal of the American Taxation Association; Journal of Accounting, Auditing, and Finance; Journal of Financial Research; Research in Accounting Regulation; and other journals. He has served on the editorial boards of Issues in Accounting Education and the Journal of Business Research, and is treasurer of the American Accounting Association's FARS. Professor Shaw is co-author of *Financial and Managerial Accounting* and *College Accounting*, both published by McGraw-Hill.

In his leisure time, Professor Shaw enjoys tennis, cycling, music, and coaching his children's sports teams.

Barbara Chiappetta received her BBA in Accountancy and MS in Education from Hofstra University and is a tenured full professor at Nassau Community College. For the past two decades, she has been an active executive board member of the Teachers of Accounting at Two-Year Colleges (TACTYC), serving 10 years as vice president and as president from 1993 through 1999. As an active member of the American Accounting Association, she has served on the Northeast Regional Steering Committee, chaired the Curriculum Revision Committee of the Two-Year Section, and participated in numerous national committees.

Professor Chiappetta has been inducted into the American Accounting Association Hall of Fame for the Northeast Region. She had also received the Nassau Community College dean of instruction's Faculty Distinguished Achievement Award. Professor Chiappetta was honored with the State University of New York Chancellor's Award for Teaching Excellence in 1997. As a confirmed believer in the benefits of the active learning pedagogy, Professor Chiappetta has authored *Student Learning Tools*, an active learning workbook for a first-year accounting course, published by McGraw-Hill/Irwin.

In her leisure time, Professor Chiappetta enjoys tennis and participates on a U.S.T.A. team. She also enjoys the challenge of bridge. Her husband, Robert, is an entrepreneur in the leisure sport industry. She has two sons—Michael, a lawyer, specializing in intellectual property law in New York, and David, a composer, pursuing a career in music for film in Los Angeles.

Helping students achieve their goal!

Fundamental Accounting Principles 19e

Assist your students in achieving their goals by giving them what they need to succeed in today's accounting principles course.

Whether the goal is to become an accountant or a businessperson, or even just gain an understanding of the principles of accounting, *Fundamental Accounting Principles* (FAP) has helped generations of students succeed by giving them the support in the form of leading-edge accounting content that engages students, with state-of-the-art technology.

With FAP on your team, you'll be passed **engaging content** and a **motivating style** to help students see the relevance of accounting. Students are motivated when reading materials that are clear and relevant. FAP runs ahead of the field in engaging students. Its chapter-opening vignettes showcase dynamic, successful entrepreneurial individuals and companies guaranteed **to interest and excite students**. This edition's featured companies–Best Buy, Circuit City, RadioShack, and Apple–engage students with their annual reports, which are a pathway for learning financial statements. Further, this book's coverage of the accounting cycle fundamentals is widely praised for its clarity and effectiveness.

FAP also delivers innovative technology to help students succeed. **Homework Manager** provides students with instant grading and feedback for assignments that are completed online. **Homework Manager Plus** integrates an online version of the textbook with Homework Manager. Our algorithmic test bank in Homework Manager offers infinite variations of numerical test bank questions. FAP also offers students portable **iPod-ready content** to help students study and raise their scores.

We're confident you'll agree that **FAP will help your students achieve their goal**.

Engaging Content

FAP content continues to set the standard in the principles course. Take a look at Chapters 1, 2 and 3 and you'll see how *FAP* leads with the best coverage of the accounting cycle. We are the first book to cover equity transactions the way most instructors teach it and students learn it–by introducing the separate equity accounts upfront and not waiting until a chapter or two later. Chapter 2 has the time-tested 4-step approach to analyzing transactions: [1] Identify, [2] Analyze, [3] Record, and [4] Post. And Chapter 3 offers a new 3-step process to simplify adjusting accounts. *FAP* also motivates students with engaging chapter openers. Students identify with them and can even picture themselves as future entrepreneurs. Each book includes the financial statements of Best Buy, Circuit City, RadioShack, and Apple to further engage students.

State-of-the-Art Technology

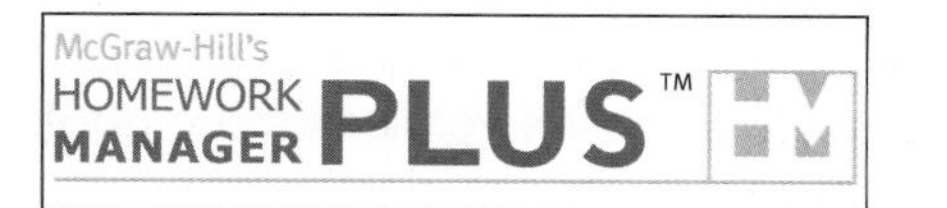

FAP offers the most advanced and comprehensive technology on the market in a seamless, easy-to-use platform. As students learn in different ways, *FAP* provides a technology smorgasbord that helps students learn more effectively and efficiently. Homework Manager, eBook options, and iPod content are some of the options. Homework Manager Plus takes learning to another level by integrating an online version of the book with all the power of Homework Manager. Technology offerings follow:

- Homework Manager
- Homework Manager Plus
- iPod content
- Algorithmic Test Bank
- Online Learning Center
- Carol Yacht's General Ledger
- ALEKS for the Accounting Cycle
- ALEKS for Financial Accounting

Premier Support

McGraw-Hill/Irwin has mobilized a new force of specialists committed to training and supporting the technology we offer. Our commitment to instructor service and support is top notch and leads the industry. Our new "McGraw-Hill Cares" program provides you with the fastest answers to your questions or solutions to your training needs. Ask your McGraw-Hill sales rep about our Key Media Support Plan and the McGraw-Hill Cares Program.

Through contemporary and engaging content, state-of-the-art technology, and committed service and support, *FAP* provides you and your students everything you need to achieve your goals!

How Technology helps

What Can McGraw-Hill Technology Offer You?

Whether you are just getting started with technology in your course, or you are ready to embrace the latest advances in electronic content delivery and course management, McGraw-Hill/Irwin has the technology you need, and provides training and support that will help you every step of the way.

Our most popular technologies, Homework Manager and Homework Manager Plus, are optional online Homework Management systems that will allow you to assign problems and exercises from the text for your students to work out in an online format. Student results are automatically graded, and the students receive instant feedback on their work. Homework Manager Plus adds an online version of the book.

Students can also use the Online Learning Center associated with this book to enhance their knowledge of accounting. Plus we now offer iPod content for students who want to study on the go.

For instructors, we provide all of the crucial instructor supplements on one easy to use Instructor CD-ROM; we can help build a custom class Website for your course using PageOut; we can deliver an online course cartridge for you to use in Blackboard, WebCT, or eCollege; and we have a technical support team that will provide training and support for our key technology products.

How Can Students Study on the Go Using Their iPod?

iPod Content

Harness the power of one of the most popular technology tools students use today–the Apple iPod. Our innovative approach allows students to download audio and video presentations right into their iPod and take learning materials with them wherever they go. Students just need to visit the Online Learning Center at **www.mhhe.com/wildFAP19e** to download our iPod content. For each chapter of the book they will be able to download audio narrated lecture presentations and financial accounting videos for use on various versions of iPods. iPod Touch users can even access self-quizzes.

It makes review and study time as easy as putting in headphones.

How Can My Students Use the Web to Complete Their Homework?

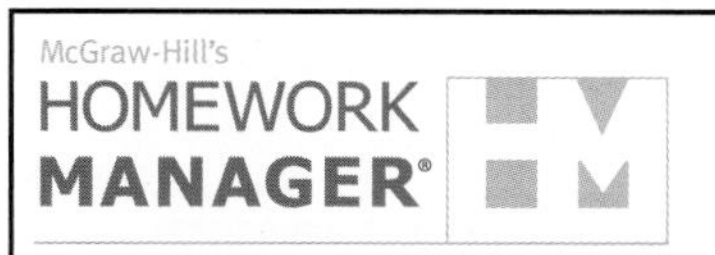

McGraw-Hill's Homework Manager®

is a Web-based supplement that duplicates problem structures directly from the end-of-chapter material in your textbook, using algorithms to provide a limitless supply of online self-graded assignments that can be used for student practice, homework, or testing. Each assignment has a unique solution. Say goodbye to cheating in your classroom; say hello to the power and flexibility you've been waiting for in creating assignments. Most Quick Studies, Exercises, and Problems are available with Homework Manager.

McGraw-Hill's Homework Manager is also a useful grading tool. All assignments can be delivered over the Web and are graded automatically, with the results stored in your private grade book. Detailed results let you see at a glance how each student does on an assignment or an individual problem—you can even see how many tries it took them to solve it.

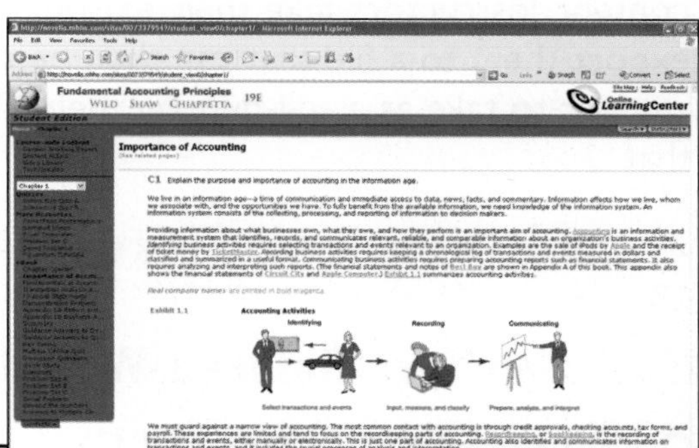

Barbara Gershowitz, Nashville State Technical Community College

"Very thorough . . . and there is a wide variety of supplemental materials that can be used by students. Homework Manager is a real time saver for instructors . . . as it provides immediate feedback to the students and gives them the opportunity to rework problems so that if they did not understand the concept the first time, they can repeat several times until they get it. Homework Manager Plus has worked very well for us. The instructors like it and I have lots of positive feedback from students."

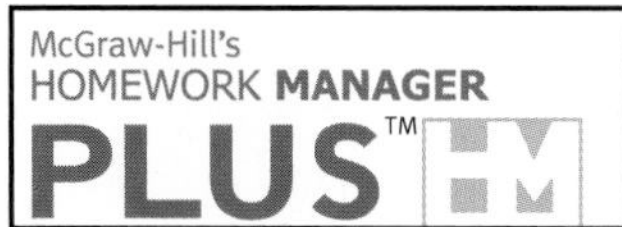

Homework Manager Plus

is an extension of McGraw-Hill's popular Homework Manager System. With Homework Manager Plus you get all of the power of Homework Manager plus an integrated online version of the book. Students simply receive one single access code which provides access to all of the resources available through Homework Manager Plus.

Paula Ratliff, Arkansas State University

"I like the idea that there are online assignments that change algorithmically so that students can practice with them."

When students find themselves needing to reference the textbook to complete their homework, they can now simply click on hints and link directly to the most relevant materials associated with the problem or exercise they are working on.

How Technology helps

Use EZ Test Online with Apple iPod® iQuiz to help students succeed.

Using our EZ Test Online you can make test and quiz content available for a student's Apple iPod®.

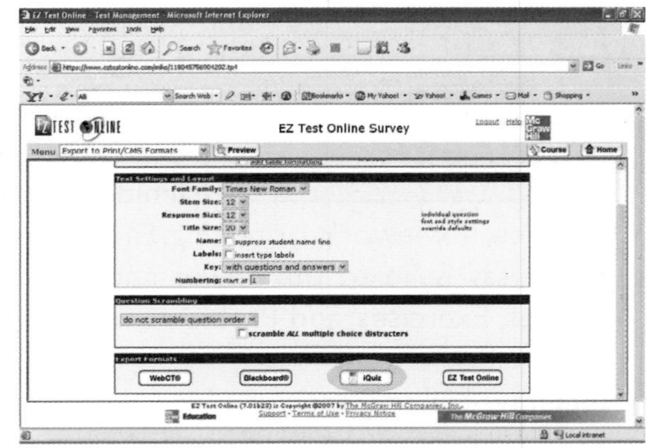

Students must purchase the iQuiz game application from Apple for 99¢ to use the iQuiz content. It works on the iPod fifth generation iPods and better.

Instructors only need EZ Test Online to produce iQuiz-ready content. Instructors take their existing tests and quizzes and export them to a file that can then be made available to the student to take as a self-quiz on their iPods. It's as simple as that.

How Can Text-Related Web Resources Enhance My Course?

Online Learning Center (OLC)

We offer an Online Learning Center (OLC) that follows *Fundamental Accounting Principles* chapter by chapter. It doesn't require any building or maintenance on your part. It's ready to go the moment you and your students type in the URL: **www.mhhe.com/wildFAP19e**.
As students study and learn from *Fundamental Accounting Principles*, they can visit the Student Edition of the OLC Website to work with a multitude of helpful tools:

- Generic Template Working Papers
- Chapter Learning Objectives
- Interactive Chapter Quizzes
- PowerPoint® Presentations
- Problem Set C
- Narrated PowerPoint® Presentations
- Video Library
- Excel Template Assignments
- Animated Demonstration Problems
- iPod Content
- Peachtree Templates

A secured Instructor Edition stores essential course materials to save you prep time before class. Everything you need to run a lively classroom and an efficient course is included. All resources available to students, plus . . .

- General Ledger and Peachtree Solution Files
- Problem Set C Solutions
- Instructor's Manual
- Solutions Manual
- Solutions to Excel Template Assignments
- Test Bank

The OLC Website also serves as a doorway to other technology solutions, like course management systems.

Lillian Grose, Delgado CC
"Logical, concise, comprehensive with excellent publisher support material."

Save money. Go green. McGraw-Hill eBooks.

Green...it's on everybody's mind these days. It's not only about saving trees, it's also about saving money. At 55% of the bookstore price, McGraw-Hill eBooks are an eco-friendly and cost-saving alternative to the traditional printed textbook. So, do some good for the environment...and do some good for your wallet.

CourseSmart

CourseSmart is a new way to find and buy eTextbooks. CourseSmart has the largest selection of eTextbooks available anywhere, offering thousands of the most commonly adopted textbooks from a wide variety of higher education publishers. CourseSmart eTextbooks are available in one standard online reader with full text search, notes, and highlighting, and email tools for sharing between classmates. Visit **www.CourseSmart.com** for more information on ordering.

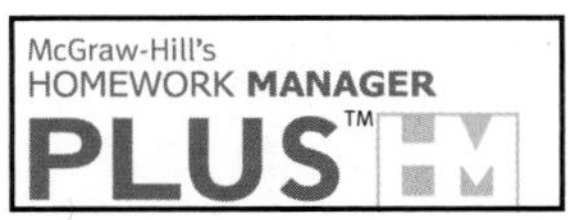

McGraw-Hill's Homework Manager Plus

If you use Homework Manager in your course, your students can purchase McGraw-Hill's Homework Manager Plus for *FAP* 19e. Homework Manager Plus gives students direct access to an online edition of the text while working assignments within Homework Manager. If you get stuck working a problem, simply click the "Hint" link and jump directly to relevant content in the online edition of the text.

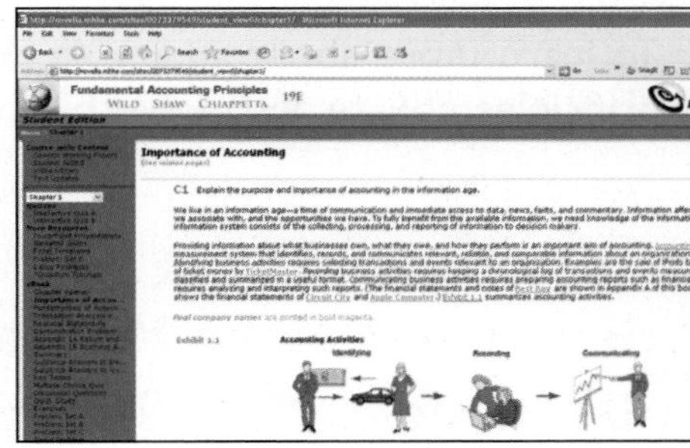

Visit the Online Learning Center at www.mhhe.com/wildFAP19e to purchase McGraw-Hill's Homework Manager Plus.

McGraw-Hill/Irwin CARES

At McGraw-Hill/Irwin, we understand that getting the most from new technology can be challenging. That's why our services don't stop after you purchase our book. You can e-mail our Product Specialists 24 hours a day, get product training online, or search our knowledge bank of Frequently Asked Questions on our support Website.

McGraw-Hill/Irwin Customer Care Contact Information

For all Customer Support call **(800) 331-5094**
Email **be_support@mcgraw-hill.com**
Or visit **www.mhhe.com/support**
One of our Technical Support Analysts will be able to assist you in a timely fashion.

How Can McGraw-Hill Help Me Teach My Course Online?

ALEKS®

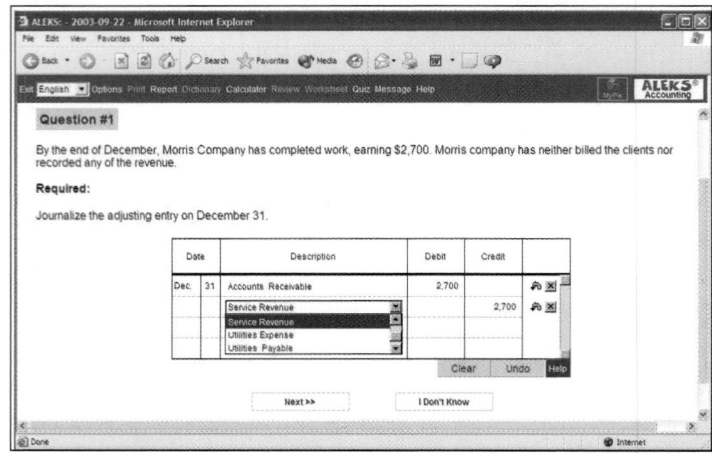

ALEKS® for the Accounting Cycle and ALEKS® for Financial Accounting

Available from McGraw-Hill over the World Wide Web, ALEKS (Assessment and LEarning in Knowledge Spaces) provides precise assessment and individualized instruction in the fundamental skills your students need to succeed in accounting.

ALEKS motivates your students because ALEKS can tell what a student knows, doesn't know, and is most ready to learn next. ALEKS does this using the ALEKS Assessment and Knowledge Space Theory as an artificial intelligence engine to exactly identify a student's knowledge of accounting. When students focus on precisely what they are ready to learn, they build the confidence and learning momentum that fuel success.

To learn more about adding ALEKS to your principles course, visit www.business.aleks.com.

Janice Stoudemire, Midlands Technical College
"The supplemental material that this principles accounting text provides is impressive: homework manager, the extensive online learning center, general ledger application software, as well as ALEKS."

How Can I Make My Classroom Discussions More Interactive?

CPS Classroom Performance System

This is a revolutionary system that brings ultimate interactivity to the classroom. CPS is a wireless response system that gives you immediate feedback from every student in the class. CPS units include easy-to-use software for creating and delivering questions and assessments to your class. With CPS you can ask subjective and objective questions. Then every student simply responds with their individual, wireless response pad, providing instant results. CPS is the perfect tool for engaging students while gathering important assessment data.

eInstruction®

Liz Ott, Casper College
"I originally adopted the book because of the tools that accompanied it: Homework Manager, ALEKS, CPS."

Online Course Management

No matter what online course management system you use (WebCT, BlackBoard, or eCollege), we have a course content ePack available for *FAP* 19e. Our new ePacks are specifically designed to make it easy for students to navigate and access content online. They are easier than ever to install on the latest version of the course management system available today.

Don't forget that you can count on the highest level of service from McGraw-Hill. Our online course management specialists are ready to assist you with your online course needs. They provide training and will answer any questions you have throughout the life of your adoption. So try our new ePack for *FAP* 19e and make online course content delivery easy and fun.

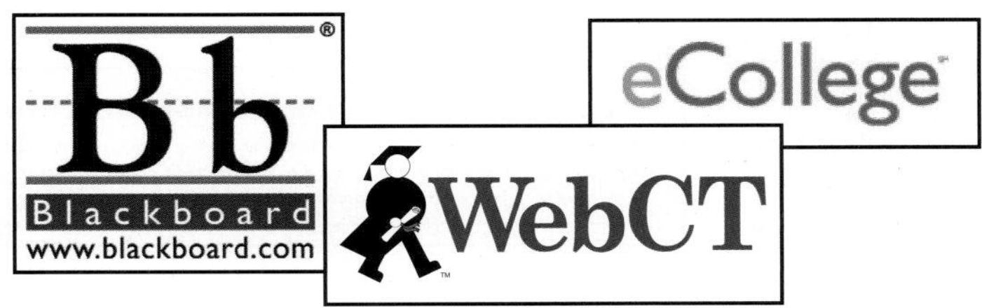

PageOut: McGraw-Hill's Course Management System

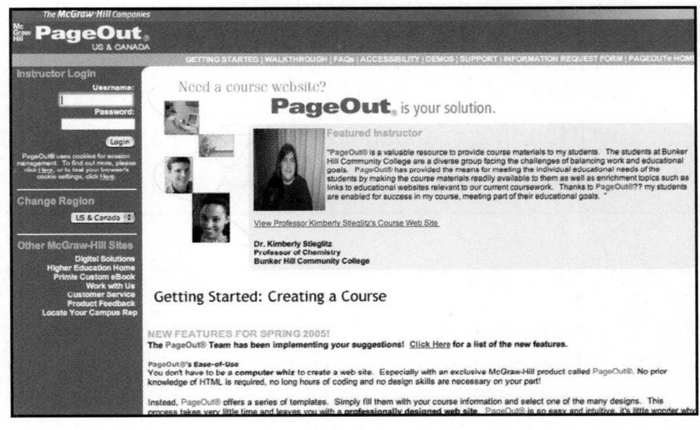

PageOut is the easiest way to create a Website for your accounting course. There is no need for HTML coding, graphic design, or a thick how-to book. Just fill in a series of boxes with simple English and click on one of our professional designs. In no time, your course is online with a Website that contains your syllabus!

Should you need assistance in preparing your Website, we can help. Our team of product specialists is ready to take your course materials and build a custom Website to your specifications. You simply need to call a McGraw-Hill/Irwin PageOut specialist to start the process. To learn more, please visit www.pageout.net and see "PageOut & Service" below.

Best of all, PageOut is free when you adopt *Fundamental Accounting Principles*!

PageOut Service

Our team of product specialists is happy to help you design your own course Website. Just call 1-800-634-3963, press 0, and ask to speak with a PageOut specialist. You will be asked to send in your course materials and then participate in a brief telephone consultation. Once we have your information, we build your Website for you, from scratch.

Decision Center

Whether we prepare, analyze, or apply accounting information, one skill remains essential: decision-making. To help develop good decision-making habits and to illustrate the relevance of accounting, *Fundamental Accounting Principles* 19e uses a unique pedagogical framework called the Decision Center. This framework is comprised of a variety of approaches and subject areas, giving students insight into every aspect of business decision-making. Answers to Decision Maker and Ethics boxes are at the end of each chapter.

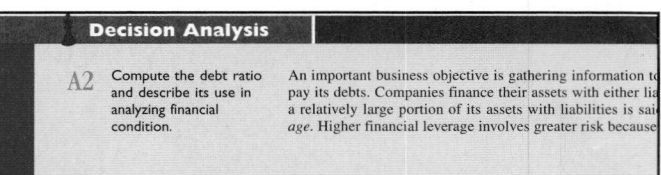

Decision Insight

IFRSs IFRSs require that companies report the following four financial statements with explanatory notes:
—Balance sheet —Statement of changes in equity (or statement of recognized revenue and expense)
—Income statement —Statement of cash flows
IFRSs do not prescribe specific formats; and comparative information is required for the preceding period only.

Decision Analysis

A2 Compute the debt ratio and describe its use in analyzing financial condition.

An important business objective is gathering information t pay its debts. Companies finance their assets with either li a relatively large portion of its assets with liabilities is sai *age*. Higher financial leverage involves greater risk because

Decision Ethics

Credit Manager As a new credit manager, you are being trained by the outgoing manager. She explains that the system prepares checks for amounts net of favorable cash discounts, and the checks are dated the last day of the discount period. She also tells you that checks are not mailed until five days later, adding that "the company gets free use of cash for an extra five days, and our department looks better. When a supplier complains, we blame the computer system and the mailroom." Do you continue this payment policy? [Answer—p. 203]

Decision Maker

Entrepreneur You open a wholesale business selling entertainment equipment to retail outlets. You find that most of your customers demand to buy on credit. How can you use the balance sheets of these customers to decide which ones to extend credit to? [Answer—p. 71]

"This text has the best introductions of any text that I have reviewed or used. Some texts simply summarize the chapter, which is boring to students. Research indicates that material needs to be written in an 'engaging manner.' That's what these vignettes do—they get the students interested."
Clarice McCoy, Brookhaven College

CAP Model

The Conceptual/Analytical/Procedural (CAP) Model allows courses to be specially designed to meet your teaching needs or those of a diverse faculty. This model identifies learning objectives, textual materials, assignments, and test items by C, A, or P, allowing different instructors to teach from the same materials, yet easily customize their courses toward a conceptual, analytical, or procedural approach (or a combination thereof) based on personal preferences.

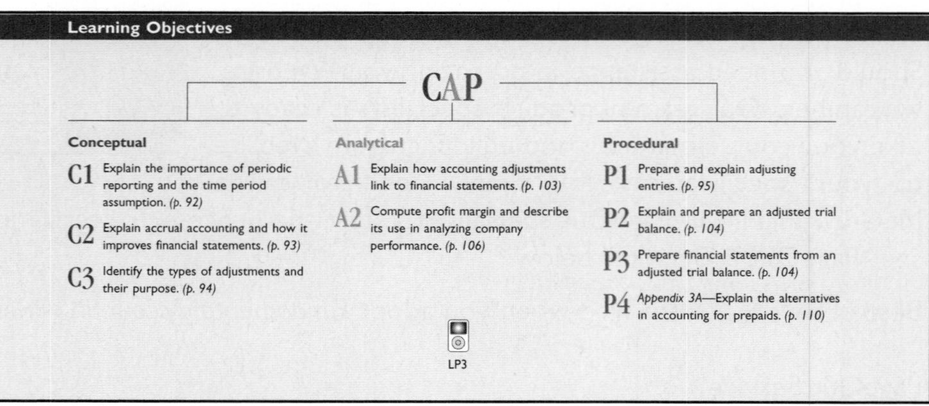

Learning Objectives

CAP

Conceptual

C1 Explain the importance of periodic reporting and the time period assumption. *(p. 92)*

C2 Explain accrual accounting and how it improves financial statements. *(p. 93)*

C3 Identify the types of adjustments and their purpose. *(p. 94)*

Analytical

A1 Explain how accounting adjustments link to financial statements. *(p. 103)*

A2 Compute profit margin and describe its use in analyzing company performance. *(p. 106)*

Procedural

P1 Prepare and explain adjusting entries. *(p. 95)*

P2 Explain and prepare an adjusted trial balance. *(p. 104)*

P3 Prepare financial statements from an adjusted trial balance. *(p. 104)*

P4 *Appendix 3A*—Explain the alternatives in accounting for prepaids. *(p. 110)*

LP3

Chapter Preview with Flow Chart

This feature provides a handy textual/visual guide at the start of every chapter. Students can now begin their reading with a clear understanding of what they will learn and when, allowing them to stay more focused and organized along the way.

Buyers of merchandise expect many products, discount prices, inventory on demand, and high quality. This chapter introduces the accounting practices used by companies engaged in merchandising. We show how financial statements reflect merchandising activities and explain the new financial statement items created by merchandising activities. We also analyze and record merchandise purchases and sales, and explain the adjustments and the closing process for these companies.
Accounting for Merchandising Operations
Merchandising Activities
• Reporting income
• Reporting inventory
• Operating cycles
• Inventory systems
Merchandising Purchases
• Purchase discounts
• Purchase returns and allowances
• Transportation costs
Merchandising Sales
• Sales of merchandise
• Sales discounts
• Sales returns and allowances
Accounting Cycle
• Adjusting entries
• Preparing financial statements
• Closing entries
Financial Statement Formats
• Multiple-step income statement
• Single-step income statement
• Classified balance sheet

Quick Check

These short question/answer features reinforce the material immediately preceding them. They allow the reader to pause and reflect on the topics described, then receive immediate feedback before going on to new topics. Answers are provided at the end of each chapter.

Quick Check Answers—p. 204

4. How long are the credit and discount periods when credit terms are 2/10, n/60?
5. Identify which items are subtracted from the *list* amount and not recorded when computing purchase price: (*a*) freight-in; (*b*) trade discount; (*c*) purchase discount; (*d*) purchase return.
6. What does *FOB* mean? What does *FOB destination* mean?

"(This book is) visually friendly with many illustrations. Good balance sheet presentation in margin."
Joan Cook, Milwaukee Area Technical College

Marginal Student Annotations

These annotations provide students with additional hints, tips, and examples to help them more fully understand the concepts and retain what they have learned. The annotations also include notes on global implications of accounting and further examples.

ransactions is to post journal entries to
dger is up-to-date, entries are posted as
ime permits. All entries must be posted
o ensure that account balances are up-
ts in journal entries are transferred into

Point: Computerized systems often provide a code beside a balance such as *dr.* or *cr.* to identify its balance. Posting is automatic and immediate with accounting software.

FastForward

FastForward is a case that takes students through the Accounting Cycle, chapters 1-4. The FastForward icon is placed in the margin when this case is discussed.

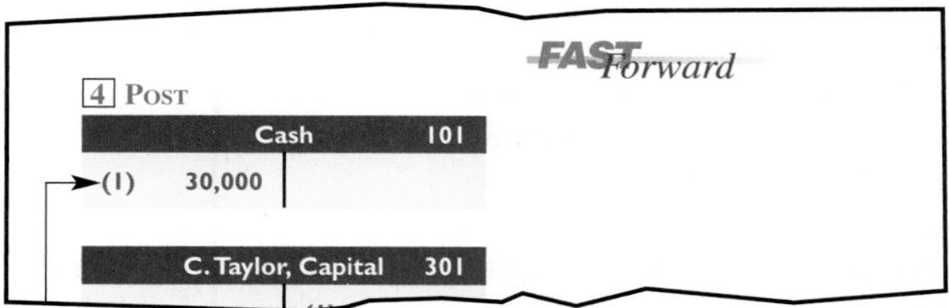

How are chapter concepts

Once a student has finished reading the chapter, how well he or she retains the material can depend greatly on the questions, exercises, and problems that reinforce it. This book leads the way in comprehensive, accurate assignments.

Demonstration Problems present both a problem and a complete solution, allowing students to review the entire problem-solving process and achieve success.

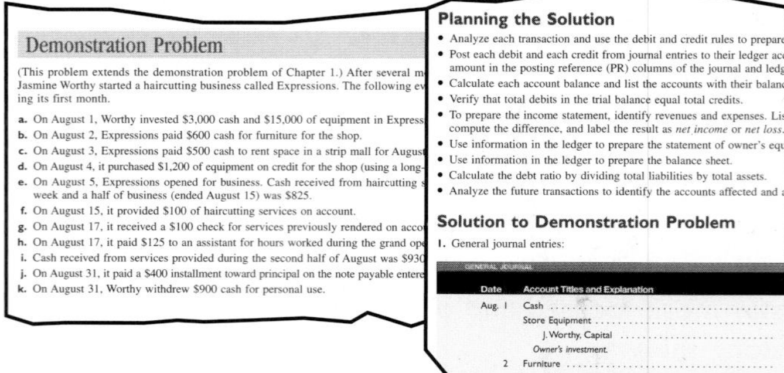

Demonstration Problem

(This problem extends the demonstration problem of Chapter 1.) After several m… Jasmine Worthy started a haircutting business called Expressions. The following ev… ing its first month.

a. On August 1, Worthy invested $3,000 cash and $15,000 of equipment in Express…
b. On August 2, Expressions paid $600 cash for furniture for the shop.
c. On August 3, Expressions paid $500 cash to rent space in a strip mall for August…
d. On August 4, it purchased $1,200 of equipment on credit for the shop (using a long-…
e. On August 5, Expressions opened for business. Cash received from haircutting s… week and a half of business (ended August 15) was $825.
f. On August 15, it provided $100 of haircutting services on account.
g. On August 17, it received a $100 check for services previously rendered on acco…
h. On August 17, it paid $125 to an assistant for hours worked during the grand op…
i. Cash received from services provided during the second half of August was $930…
j. On August 31, it paid a $400 installment toward principal on the note payable entere…
k. On August 31, Worthy withdrew $900 cash for personal use.

Planning the Solution

- Analyze each transaction and use the debit and credit rules to prepare…
- Post each debit and each credit from journal entries to their ledger ac… amount in the posting reference (PR) columns of the journal and ledg…
- Calculate each account balance and list the accounts with their balanc…
- Verify that total debits in the trial balance equal total credits.
- To prepare the income statement, identify revenues and expenses. Lis… compute the difference, and label the result as *net income or net loss.*
- Use information in the ledger to prepare the statement of owner's equ…
- Use information in the ledger to prepare the balance sheet.
- Calculate the debt ratio by dividing total liabilities by total assets.
- Analyze the future transactions to identify the accounts affected and a…

Solution to Demonstration Problem

1. General journal entries:

GENERAL JOURNAL

Date	Account Titles and Explanation
Aug. 1	Cash
	Store Equipment
	J. Worthy, Capital
	Owner's investment.
2	Furniture
	Cash

Chapter Summaries provide students with a review organized by learning objectives. Chapter Summaries are a component of the CAP model (see page xiv), which recaps each conceptual, analytical, and procedural objective.

Key Terms are bolded in the text and repeated at the end of the chapter with page numbers indicating their location. The book also includes a complete Glossary of Key Terms.

Key Terms

Key Terms are available at the book's Website for learning and testi…

Account (p. 49)	**Debtors** (p. 49)
Account balance (p. 53)	**Debt ratio** (p. 67)
Balance column account (p. 56)	**Double-entry accounting** (p. 53)
Chart of accounts (p. 52)	**General journal** (p. 54)
Compound journal entry (p. 59)	**General ledger** (p. 49)
Credit (p. 53)	**Journal** (p. 54)
Creditors (p. 50)	**Journalizing** (p. 54)
Debit (p. 53)	**Ledger** (p. 49)

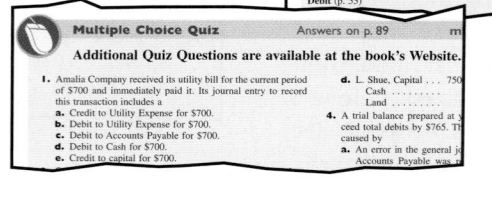

Multiple Choice Questions Multiple Choice Questions quickly test chapter knowledge before a student moves on to complete Quick Studies, Exercises, and Problems.

Multiple Choice Quiz   Answers on p. 89

Additional Quiz Questions are available at the book's Website.

1. Amalia Company received its utility bill for the current period of $700 and immediately paid it. Its journal entry to record this transaction includes a
   a. Credit to Utility Expense for $700.
   b. Debit to Utility Expense for $700.
   c. Debit to Accounts Payable for $700.
   d. Debit to Cash for $700.
   e. Credit to capital for $700.

d. L. Shue, Capital . . . 750… Cash . . . Land . . .

4. A trial balance prepared at y… ceed total debits by $765. Th… caused by
   a. An error in the general jo… Accounts Payable was p…

Quick Study assignments are short exercises that often focus on one learning objective. All are included in Homework Manager. There are usually 8-10 Quick Study assignments per chapter.

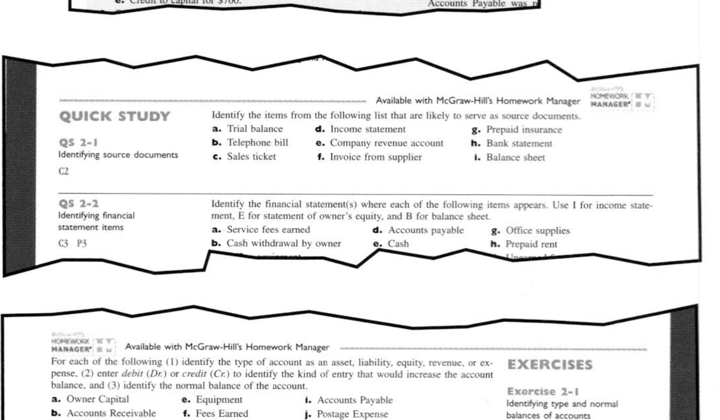

QUICK STUDY — Available with McGraw-Hill's Homework Manager

QS 2-1 Identifying source documents C2

Identify the items from the following list that are likely to serve as source documents.
a. Trial balance b. Telephone bill c. Sales ticket d. Income statement e. Company revenue account f. Invoice from supplier g. Prepaid insurance h. Bank statement i. Balance sheet

QS 2-2 Identifying financial statement items C3 P3

Identify the financial statement(s) where each of the following items appears. Use I for income statement, E for statement of owner's equity, and B for balance sheet.
a. Service fees earned b. Cash withdrawal by owner d. Accounts payable e. Cash g. Office supplies h. Prepaid rent

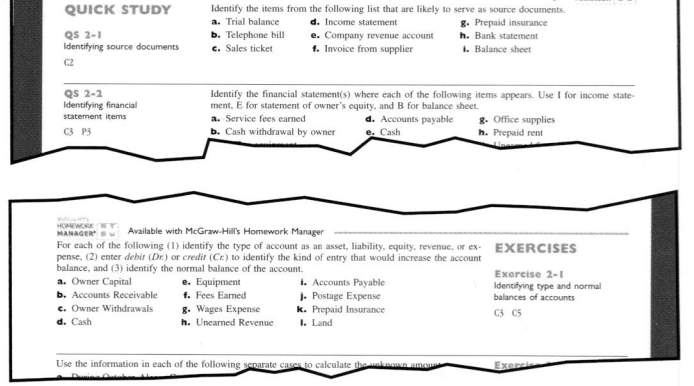

Exercises are one of this book's many strengths and a competitive advantage. There are about 10-15 per chapter and all are included in Homework Manager.

Available with McGraw-Hill's Homework Manager

EXERCISES

Exercise 2-1 Identifying type and normal balances of accounts C3 C5

For each of the following (1) identify the type of account as an asset, liability, equity, revenue, or expense, (2) enter *debit (Dr.)* or *credit (Cr.)* to identify the kind of entry that would increase the account balance, and (3) identify the normal balance of the account.
a. Owner Capital b. Accounts Receivable c. Owner Withdrawals d. Cash e. Equipment f. Fees Earned g. Wages Expense h. Unearned Revenue i. Accounts Payable j. Postage Expense k. Prepaid Insurance l. Land

Use the information in each of the following separate cases to calculate the unknown amount…

Problem Sets A & B are proven problems that can be assigned as homework or for in-class projects. Problem Set C is available on the book's Website. All problems are coded according to the CAP model (see page xiv), and Set A is included in Homework Manager.

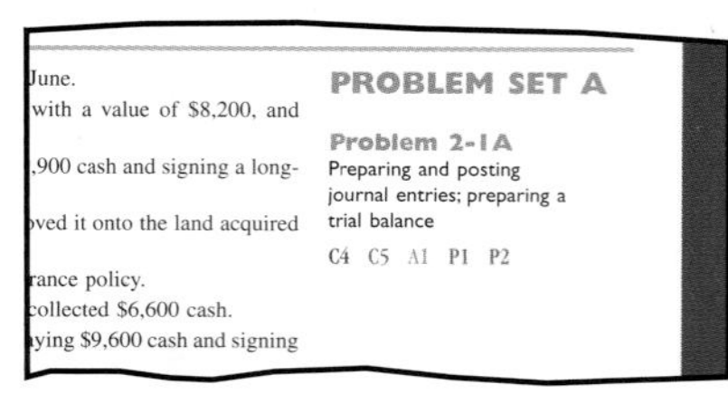

PROBLEM SET A

Problem 2-1A Preparing and posting journal entries; preparing a trial balance C4 C5 A1 P1 P2

…June. …with a value of $8,200, and …,900 cash and signing a long- …ved it onto the land acquired …rance policy. …collected $6,600 cash. …ying $9,600 cash and signing

PROBLEM SET B

Problem 2-1B Preparing and posting journal entries; preparing a trial balance C4 C5 A1 P1 P2

At the beginning of April… Softworks. The company h…
a. Vanessa Wende investe… computer equipment in…
b. The company purchase… term note payable for $…
c. The company purchase… in *b.*
d. The company paid $3,…
e. The company provided…
f. The company purchase… signing…

"One of the best features in FAP is the Serial Problem. I find the continuation of a company from a service to a merchandiser, to a manufacturing, and from a sole proprietorship form or business to a corporation, provides the student a real picture of a company's development. It also provides a consistency from one lesson to another."

Barbara Marotta, Northern Virginia Community College, Woodbridge

Beyond the Numbers exercises ask students to use accounting figures and understand their meaning. Students also learn how accounting applies to a variety of business situations. These creative and fun exercises are all new or updated, and are divided into sections:

- Reporting in Action
- Comparative Analysis
- Ethics Challenge
- Communicating in Practice
- Taking It To The Net
- Teamwork in Action
- Hitting the Road
- Entrepreneurial Decision
- Global Decision

BEYOND THE NUMBERS

REPORTING IN ACTION

A1 A2

BEST BUY

BTN 2-1 Refer to Best Buy's financial statements in Appendix

Required

1. What amount of total liabilities does it report for each of the fisca and March 3, 2007?
2. What amount of total assets does it report for each of the fiscal ye March 3, 2007?
3. Compute its debt ratio for each of the fiscal years ended February
4. In which fiscal year did it employ more financial leverage (Febru Explain.

Serial Problem uses a continuous running case study to illustrate chapter concepts in a familiar context. The Serial Problem can be followed continuously from the first chapter or picked up at any later point in the book; enough information is provided to ensure students can get right to work.

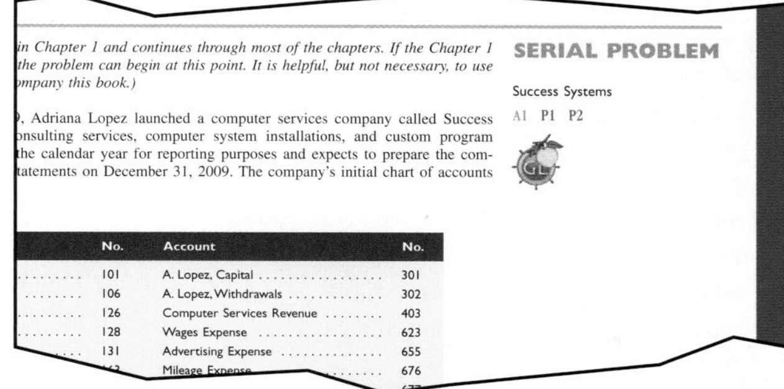

SERIAL PROBLEM

Success Systems

A1 P1 P2

in Chapter 1 and continues through most of the chapters. If the Chapter 1 the problem can begin at this point. It is helpful, but not necessary, to use ompany this book.)

), Adriana Lopez launched a computer services company called Success onsulting services, computer system installations, and custom program he calendar year for reporting purposes and expects to prepare the com- tatements on December 31, 2009. The company's initial chart of accounts

	No.	Account	No.
....	101	A. Lopez, Capital	301
....	106	A. Lopez, Withdrawals	302
....	126	Computer Services Revenue	403
....	128	Wages Expense	623
....	131	Advertising Expense	655
		Mileage Expense	676

"The best feature of this book is the use of real (financial) information in the Beyond the Numbers section. This is something that I do on my own, which can be very time consuming. I also like the Entrepreneurial questions, which are not even addressed in most textbooks."

Cindy Navaroli, Chaffey Community College

The End of the Chapter Is Only the Beginning

Our valuable and proven assignments aren't just confined to the book. From problems that require technological solutions to materials found exclusively online, this book's end-of-chapter material is fully integrated with its technology package.

- Quick Studies, Exercises, and Problems available on Homework Manager (see page ix) are marked with an icon.

- Problems supported by the General Ledger Application Software or Peachtree are marked with an icon.

- Online Learning Center (OLC) includes Interactive Quizzes, Excel template assignments, and more.

- Problems supported with Microsoft Excel template assignments are marked with an icon.

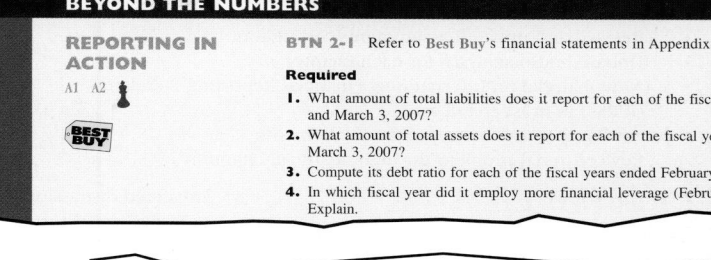

- Material that receives additional coverage (slide shows, videos, audio, etc.) available in iPod ready format are marked with an icon.

Put Away Your Red Pen

We pride ourselves on the accuracy of this book's assignment materials. Independent research reports that instructors and reviewers point to the accuracy of this book's assignment materials as one of its key competitive advantages.

The authors extend a special thanks to accuracy checkers Barbara Schnathorst, The Write Solution, Inc.; Helen Roybark, Radford University; Beth Woods, CPA, Accuracy Counts; David Krug, Johnson Community College; ANSR Source; David Burba, Bowling Green State University; and Marilyn Sagrillo, University of Wisconsin - Green Bay.

Enhancements for FAP 19e

This edition's revisions are driven by feedback from instructors and students. Many of the revisions are summarized here. Feedback suggests that this is the book instructors want to teach from and students want to learn from. General revisions include:

- Revised and updated assignments throughout
- Updated ratio analyses for each chapter
- New material on International Financial Reporting Standards (IFRSs) in most chapters
- New and revised entrepreneurial elements
- Revised serial problem through nearly all chapters
- New art program, visual graphics, and text layout
- New Best Buy Annual Report with comparisons to Circuit City, RadioShack, Apple, and DSG (UK) with new assignments
- New graphics added to each chapter's analysis section
- New iPod content integrated and referenced in book

Chapter 1

Life is Good NEW opener with new entrepreneurial assignment

Revised section on accounting principles and assumptions

New visual layout on building blocks for GAAP

New graphic listing recent publicized accounting scandals

Transaction analysis with expanded accounting equation

Updated compensation data in exhibit

New discussion of conceptual framework linked to IFRSs

Chapter 2

SPANX NEW opener with new entrepreneurial assignment

New 4-step process to learn accounting for transactions: Identify, Analyze, Record, and Post

New sequential art layout for visualizing transaction accounting

New arrow lines link journal entries to the ledger

New material on financial statements related to IFRSs

Chapter 3

PopCap Games NEW opener with new entrepreneurial assignment

New 3-step process for accounting adjustments: (1) Determine what is, (2) Determine what should be, (3) Record adjustment

New summary tables show the accounting adjustment AND account balances before and after adjustment

New adjustment tables for prepaid expenses, including insurance, supplies, and depreciation

New adjustment tables for unearned revenue, accrued expenses, and accrued revenues

Chapter 4

Kathryn Kerrigan NEW opener with new entrepreneurial assignment

New graphic for learning the 4-step closing process

New graphic to aid learning of accounting cycle

Streamlined presentation of classified balance sheet

Chapter 5

BigBadToyStore NEW opener with new entrepreneurial assignment

Enhanced discussions on purchase allowances and purchase returns

Revised discussions on sales returns and sales allowances

Streamlined 4-step process for merchandisers' closing entries

New material on income statement formats for U.S. GAAP vis-à-vis IFRSs

Chapter 6

Beauty Encounter NEW opener with new entrepreneurial assignment

Specific Identification explanation revised for additional clarity

New explanation added to First-in, First-Out

Enhanced explanation to Last-in, First-Out

New description of weighted average

Enhanced explanations for LCM

New material on inventory methods for U.S. GAAP vis-à-vis IFRSs

Illustration of inventory errors is revised and enhanced

Revised exhibit on inventory errors for balance sheet

Appendix 6A, enhanced discussion of periodic inventory

Appendix 6A, expanded explanations of periodic FIFO, LIFO, and weighted average

Chapter 7

TerraCycle NEW opener with new entrepreneurial assignment

Enhanced exhibits for special journals and ledgers

New explanations for controlling and subsidiary ledgers

Updated discussion of enterprise resource planning

Chapter 8

Dylan's Candy Bar NEW opener with new entrepreneurial assignment

New material added on Sarbanes-Oxley and internal controls required

New section on cash management

Additional description of electronic funds transfer

Enhanced explanation of bank reconciliation with new assignments

New discussion of internal controls for IFRSs conversion

Chapter 9

Under Armor NEW opener with new entrepreneurial assignment

Reorganized sections on valuing accounts receivable

Enhanced explanation of recovering a bad debt

New material on assessing the direct write-off method

Reorganized section on estimating bad debts using percent of receivables

New enhanced summary Exhibit 9.13 on methods to estimate bad debts

Added new exercises on bad debts estimation

Chapter 10

Sambazon NEW opener with new entrepreneurial assignment

Updated all real world examples and graphics

Enhanced explanations on declining-balance and impairments

New description of ordinary repairs and its journal entry

Reorganized discussion of betterments and extraordinary repairs

New material on depreciation estimates for U.S. GAAP vis-à-vis IFRSs

Enhanced illustration for depletion including journal entry

Reorganized section on types of intangibles

Enhanced material related to franchises, licenses, trademarks and other intangibles

Enhanced explanation of exchanging plant assets

Chapter 11

Feed Granola Company NEW opener with new entrepreneurial assignment

Updated real world examples including those for Univision, Six Flags, AMF, and K2

New material on contingent liabilities for U.S. GAAP vis-à-vis IFRSs

Updated tax illustrations and assignments using the most recent rates

New section on "Who Pays What Payroll Taxes and Benefits"

Chapter 12

Samanta Shoes NEW opener with new entrepreneurial assignment

Updated real world examples including those for Macadamia Orchards and Big River

Productions

New example of partnership accounting using Trump Entertainment Resorts

Streamlined explanation of partnership liquidation

Chapter 13

Inogen **NEW opener** with new entrepreneurial assignment

New info graphic on subcategories of authorized stock

Updated many real world examples

New info graphic on stock splits

New material on preferred stock classification for U.S. GAAP vis-à-vis IFRSs

Additional explanation on closing process for corporations

Updated statement of stockholders' equity using Apple

Chapter 14

Rap Snacks **NEW opener** with new entrepreneurial assignment

Enhanced graphic and explanation of determining bond discount and premium

Revised bond illustration to use a 4-period bond

Streamlined bond illustration to show entire amortization process from issuance to maturity

Streamlined bond illustration for effective interest amortization

New material on bond interest computation for U.S. GAAP vis-à-vis IFRSs

Added new assignments that require accounting for bonds given an amortization table

Chapter 15

Tibi **NEW opener** with new entrepreneurial assignment

Updated real world examples and graphics including those for Gap, Pfizer, Starbucks, and Brunswick

Enhanced summary graphic on accounting for securities

Updated discussion related to FAS 157 and 159

New material on consolidation per U.S. GAAP vis-à-vis IFRSs

Chapter 16

Jungle Jim's **NEW opener** with new entrepreneurial assignment

Updated graphics for operating, investing and financing cash flows

Enhanced steps 1 through 5 for preparing the statement of cash flows

Simplified summary Exhibit 16.12 for indirect adjustments

Updated real world examples and graphics including that for Harley, Starbucks, and Nike

New info on indirect vs direct method for U.S. GAAP vis-à-vis IFRSs

Chapter 17

The Motley Fool **REVISED opener** with new entrepreneurial assignment

New Best Buy, Circuit City and RadioShack data throughout chapter, exhibits, and illustrations with comparative analysis

Enhanced presentation on comparative financial statements

Chapter 18

Kernel Season's **NEW opener** with new entrepreneurial assignment

New section on fraud and the role of ethics in managerial accounting

Added discussion on Institute of Management Accountants and its road-map for resolving ethical dilemmas

Updated real-world examples including that for Apple

Added balance sheet to exhibits that show cost flows across accounting reports

New discussion on role of nonfinancial information

Chapter 19

Sprinturf **NEW opener** with new entrepreneurial assignment

New info on custom design involving Nike

Enhanced exhibit on job order production activities

Added discussion linking accurate overhead application for jobs to both product pricing and performance evaluation

Streamlined explanation of closing over- and underapplied overhead

New discussion of employee payroll fraud schemes

Chapter 20

Hood River Juice Company **NEW opener** with new entrepreneurial assignment

New discussion on impact of automation for quality control and overhead application

Added explanation for use of a process cost summary in product pricing

Chapter 21

RockBottomGolf.com **NEW opener** with new entrepreneurial assignment

Enhanced explanation of evaluating investment center performance with financial measures

New discussion of residual income

Added explanation of economic value added

New section on evaluating investment center performance with nonfinancial measures including balanced scorecard

New Appendix 21A on transfer pricing

Decision Analysis: new explanation of investment center profit margin and investment turnover with new assignments

Chapter 22

Moe's Southwest Grill **NEW opener** with new entrepreneurial assignment

New section on working with changes in estimates for CVP analysis

New graphics illustrating how changes in estimates impact break-even analysis

New discussion on weighted average contribution margin in multiple product CVP analysis

New Appendix 22A on using Excel to estimate least squares regression

New assignments on break-even and changes in estimates

Chapter 23

Jibbitz **NEW opener** with new entrepreneurial assignment

Enhanced discussion of master budgets

New assignments on preparing budgets and budgeted financial statements

Chapter 24

Martin Guitar **NEW opener** with new entrepreneurial assignment

Reorganized discussion of overhead variance analysis

Revised graphics on framework for understanding overhead variances

New discussion of increased automation for overhead application

Revised explanation of journal entries for standard costing

Simplified discussion of closing variance accounts

New assignments on variance analysis

Added journal entries to chapter demonstration problem

Chapter 25

Prairie Sticks Bat Company **NEW opener** with new entrepreneurial assignment

New info graphic on cost of capital estimates across industries

Added discussion and example on use of profitability index to compare projects

New discussion on incorporating inflation in net present value calculations

Added explanation on conflicts between meeting analysts' forecasts and choosing profitable long-term projects

New Appendix 25A on using Excel to compute net present value and internal rate of return

New assignments on profitability index

Instructor Supplements

Assurance of Learning Ready

Assurance of learning is an important element of many accreditation standards. *Fundamental Accounting Principles* 19e is designed specifically to support your assurance of learning initiatives.

Each chapter in the book begins with a list of numbered learning objectives which appear throughout the chapter, as well as in the end-of-chapter problems and exercises. Every test bank question is also linked to one of these objectives, in addition to level of difficulty, AICPA skill area, and AACSB skill area. EZ Test, McGraw-Hill's easy-to-use test bank software, can search the test bank by these and other categories, providing an engine for targeted Assurance of Learning analysis and assessment.

AACSB Statement

The McGraw-Hill Companies is a proud corporate member of AACSB International. Understanding the importance and value of AACSB accreditation, *Fundamental Accounting Principles* 19e has sought to recognize the curricula guidelines detailed in the AACSB standards for business accreditation by connecting selected questions in the test bank to the general knowledge and skill guidelines found in the AACSB standards.

The statements contained in *Fundamental Accounting Principles* 19e are provided only as a guide for the users of this text. The AACSB leaves content coverage and assessment within the purview of individual schools, the mission of the school, and the faculty. While *Fundamental Accounting Principles* 19e and the teaching package make no claim of any specific AACSB qualification or evaluation, we have, within the *Fundamental Accounting Principles* 19e test bank labeled questions according to the six general knowledge and skills areas.

Instructor's Resource CD-ROM

FAP 19e, Chapters 1-25
ISBN13: 9780073366456
ISBN10: 0073366455

This is your all-in-one resource. It allows you to create custom presentations from your own materials or from the following text-specific materials provided in the CD's asset library:

- Instructor's Resource Manual
  Written by Barbara Chiappetta, Nassau Community College, and Patricia Walczak, Lansing Community College.
  This manual contains (for each chapter) a Lecture Outline, a chart linking all assignment materials to Learning Objectives, a list of relevant active learning activities, and additional visuals with transparency masters.
- Solutions Manual
- Test Bank, Computerized Test Bank
- PowerPoint® Presentations
  Prepared by Jon Booker, Charles Caldwell, and Susan Galbreth.
  Presentations allow for revision of lecture slides, and includes a viewer, allowing screens to be shown with or without the software.
- Link to PageOut

Test Bank

Vol. 1, Chapters 1-12
ISBN13: 9780073366524
ISBN10: 0073366528

Vol. 2, Chapters 13-25
ISBN13: 9780073366487
ISBN10: 007336648X

Revised by Barbara Gershowitz and Laurie Swanson of Nashville State Technical Community College.

Algorithmic Test Bank

ISBN13: 9780073366395
ISBN10: 0073366390

Solutions Manual

Vol. 1, Chapters 1-12
ISBN13: 9780073366517
ISBN10: 007336651X

Vol. 2, Chapters 13-25
ISBN13: 9780073366470
ISBN10: 0073366471

Written by John J. Wild, Ken W. Shaw, and Marilyn Sagrillo.

Geoffrey Heriot, Greenville Technical College
"The text is well presented and has excellent materials for both students and instructors. It is certainly one of the top texts in an entry level principles of accounting marketplace."

Excel Working Papers CD *(Chapters 1-25)*

ISBN13: 9780073366371
ISBN10: 0073366374

Written by John J. Wild.

Working Papers delivered in Excel spreadsheets. These Excel Working Papers are available on CD-ROM and can be bundled with the printed Working Papers; see your representative for information.

Working Papers

Vol. 1, Chapters 1-12
ISBN13: 9780077289515
ISBN10: 007728951X

Vol. 2, Chapters 12-25
ISBN13: 9780073366364
ISBN10: 0073366366

PFA, Chapters 1-17
ISBN13: 9780073366340
ISBN10: 007336634X

Written by John J. Wild.

Study Guide

Vol. 1, Chapters 1-12
ISBN13: 9780073366388
ISBN10: 0073366382

Vol. 2, Chapters 12-25
ISBN13: 9780073366357
ISBN10: 0073366358

Written by Barbara Chiappetta, Nassau Community College, and Patricia Walczak, Lansing Community College.

Covers each chapter and appendix with reviews of the learning objectives, outlines of the chapters, summaries of chapter materials, and additional problems with solutions.

Carol Yacht's General Ledger CD-ROM

ISBN13: 9780073366401
ISBN10: 0073366404

The CD-ROM includes fully functioning versions of McGraw-Hill's own General Ledger Application software. Problem templates prepared by Carol Yacht and student user guides are included that allow you to assign text problems for working in Yacht's General Ledger or Peachtree.

Contributing Author

The authors and book team wish to thank Marilyn Sagrillo for her excellent contributions.

Marilyn Sagrillo is an associate professor at the University of Wisconsin at Green Bay. She received her BA and MS from Northern Illinois University and her PhD from the University of Wisconsin at Madison. Her scholarly articles are published in *Accounting Enquiries*, *Journal of Accounting Case Research*, and the *Missouri Society of CPAs Casebook*. She is a member of the American Accounting Association and the Institute of Management Accountants. She previously received the UWGB Founder's Association Faculty Award for Excellence in Teaching. Professor Sagrillo is an active volunteer for the Midwest Renewable Energy Association. She also enjoys reading, traveling, and hiking.

Acknowledgments

John J. Wild, Ken W. Shaw, Barbara Chiappetta, and McGraw-Hill/Irwin would like to recognize the following instructors for their valuable feedback and involvement in the development of *Fundamental Accounting Principles* 19e. We are thankful for their suggestions, counsel, and encouragement.

Janet Adeyiga, Hampton University

Audrey Agnello, Niagara County Community College

John Ahmad, Northern Virginia Community College

Sylvia Allen, Los Angeles Valley College

Donna Altepeter, University of North Dakota

Juanita Ardavany, Los Angeles Valley College

Charles Baird, University of Wisconsin-Stout

Scott Barhight, Northampton Community College

Richard Barnhart, Grand Rapids Community College

Mary Barnum, Grand Rapids Community College

Allen Bealle, Delgado Community College

Susan Beasley, Community College of Baltimore County-Catonsville

Beverly Beatty, Anne Arundel Community College

Irene Bembenista, Davenport University

Laurel Berry, Bryant and Stratton College

Jaswinder Bhangal, Chabot College

Rick Bowden, Oakland Community College-Auburn Hills

Deborah Boyce, Mohawk Valley Community College

Nancy Boyd, Middle Tennessee State University

Debbie Branch, Surry Community College

Karen Brayden, Front Range Community College-Fort Collins

Nat Briscoe, Northwestern State University

Linda Bruff, Strayer University

Scott Butterfield, University of Colorado-Colorado Springs

James Capone, Kean University

Roy Carson, Anne Arundel Community College

Trudy Chiaravelli, Lansing Community College

Siu Chung, Los Angeles Valley College

Marilyn Ciolino, Delgado Community College

Ken Coffey, Johnson County Community College

Edwin Cohen, DePaul University

Kerry Colton, Aims Community College-Main Campus

James Cosby, John Tyler Community College

Kenneth Couvillion, San Joaquin Delta College

Ana Cruz, Miami-Dade College-Wolfson

Walter DeAguero, Saddleback College

Suryekaent Desai, Cedar Valley College-Lancaster

Mike Deschamps, Chaffey College

Susan Dickey, Motlow State Community College

Beth Dietz, South Texas College

Andy Dressler, Walla Walla College

Betty Driver, Murray State University

Michael Farina, Cerritos College

Jeannie Folk, College of DuPage

Jim Formosa, Nashville State Technical College

Ron Dustin, Fresno City College

Anthony Espisito, Community College of Baltimore County-Essex

Steve Fabian, Hudson County Community College

Patricia Feller, Nashville State Technical Community College

Anna Fitzpatrick, Gloucester City College

John Gabelman, Columbus State Community College

Dan Galvin, Diablo Valley College

Barbara Gershowitz, Nashville State Technical Community College

Lillian Grose, Delgado Community College

Betty Habershon, Prince George's Community College

Patricia Halliday, Santa Monica College

Jeff Hamm, University of Arkansas-Little Rock

John Hancock, University of California Davis

Jeannie Harrington, Middle Tennessee State University

Sara Harris, Arapahoe Community College

Laurie Hays, Western Michigan University-Kalamazoo

Kathy Hill, Leeward Community College

Patricia Holmes, Des Moines Area Community College

Constance Hylton, George Mason University

Vincent Huygen, North Greenville University

Bill Johnstone, Montgomery College

Jeffery Jones, Community College of Southern Nevada

Richard Irvine, Pensacola Jr. College

Irv Jason, Southwest Tennessee Community College-Macon Campus

Frank Jordan, Erie Community College South Campus-Orchard Park

Thomas Kam, Hawaii Pacific University

John Karbens, Hawaii Pacific University

George Katzt, Phillips College

Charles Kee, Kingsborough Community College

Howard Keller, Indiana University/Purdue University Indiana

Monique Kelly (Byrd), Fresno City College

Chula King, University of West Florida

Debra Kiss, Davenport University

Shirley Kleiner, Johnson County Community College

Robert Koch, St. Peter's College

Jerry Kreuze, Western Michigan University

David Krug, Johnson County Community College

Tara Laken, Joliet Junior College

Michael Landers, Middlesex County Community College

Sherman Layell, Surry Community College

Philip Lee, Nashville State Technical Community College

Paul Lee, Cleveland State University

Natasha Librizzi, Milwaukee Area Technical College

Danny Litt, University of California-Los Angeles

Steve Ludwig, Northwest Missouri State University

Maria Mari, Miami Dade College

Diane Marker, University of Toledo-Scott Park

Barbara Marotta, Northern Virginia Community College-Woodbridge

Brenda Mattison, Tri-County Technical College

Lynn Mazzola, Nassau County Community College

Cynthia McCall, Des Moines Area Community College

Patricia McClure, Springfield Technical Community College

Robert Mcwhorter, Northwest Vista College

Audrey Morrison, Pensacola Junior College

Andrea Murowski, Brookdale Community College

Tim Murphy, Diablo Valley College

Joe Nicassio, Westmoreland County Community College

Jamie O'Brien, South Dakota State University

Kathleen O'Donnell, Onondaga Community College

Ralph Ostrowski, Illinois Central College

Shelley Ota, Lee Ward Community College

Susan Pallas, Southeast Community College

Thomas Parker, Surry Community College

Ash Patel, Normandale Community College

Anahid Petrosian, South Texas College

Yvonne Phang, Boro of Manhattan Community College

Mary Phillips, Middle Tennessee State University

Gary Pieroni, Diablo Valley College

Greg Prescott, University of Southern Alabama

Paula Ratliff, Arkansas State University-State University

David Ravetch, University of California-Los Angeles

Jenny Resnick, Santa Monica College

Jim Riley, Palo Alto College

Richard Roding, Red Rocks Community College

Linda Rose, Westwood College

Al Ruggiero, Suffolk County Community College-Selden

Roger Sands, Milwaukee Area Tech-Milwaukee

Marilyn Sagrillo, University of Wisconsin-Green Bay

Richard Sarkisian, Camden County College

Wallace Satchell, St. Phillips College

Christine Schalow, University of Wisconsin-Stevens Point

William Schmalz, Sanford Brown College

Brad Smith, Des Moines Area Community College

Nancy Snow, University of Toledo

Laura Solano, Pueblo Community College

Charles Spector, State University of New York-Oswego

Gene Sullivan, Liberty University

Laurie Swanson, Nashville State Technical Community College

Larry Swisher, Muskegon County Community College

Margaret Tanner, University of Arkansas-Fort Smith

Domenico Tavella, Pittsburgh Technical Institute

Thomas G. Thompson, Madison Area Technical College

Debra Tyson, Bryant & Stratton College

Shafi Ullah, Broward Community College

Bob Urell, Irvine Valley College

Patricia Walczak, Lansing Community College

Valerie Walmsley, South Texas College

Keith Weidkamp, Sierra College

Dale Westfall, Midland College

Gloria Worthy, Southwest Tennessee Community College

Ray Wurzburger, New River Community College

Lynnette Yerbury, St. Louis Community College

Judith Zander, Grossmont College

In addition to the helpful and generous colleagues listed above, we thank the entire McGraw-Hill/Irwin *Fundamental Accounting Principles* 19e team, including Stewart Mattson, Tim Vertovec, Steve Schuetz, Christina Sanders, Sharon Monday of Aptara, Lori Koetters, Matthew Baldwin, Carol Bielski, and Brian Nacik. We also thank the great marketing and sales support staff, including Krista Bettino, Sankha Basu, and Randy Sealy. Many talented educators and professionals worked hard to create the supplements for this book, and for their efforts we're grateful. Finally, many more people we either did not meet or whose efforts we did not personally witness nevertheless helped to make this book everything that it is, and we thank them all.

John J. Wild *Ken W. Shaw* *Barbara Chiappetta*

Brief Contents

12 Accounting for Partnerships 476
13 Accounting for Corporations 504
14 Long-Term Liabilities 548
15 Investments and International Operations 590
16 Reporting the Statement of Cash Flows 626
17 Analysis of Financial Statements 678
18 Managerial Accounting Concepts and Principles 724
19 Job Order Cost Accounting 768
20 Process Cost Accounting 806
21 Cost Allocation and Performance Measurement 850
22 Cost-Volume-Profit Analysis 900
23 Master Budgets and Planning 938
24 Flexible Budgets and Standard Costs 978
25 Capital Budgeting and Managerial Decisions 1022
Appendix A Financial Statement Information A-1
Appendix B Time Value of Money B

Contents

Preface iii

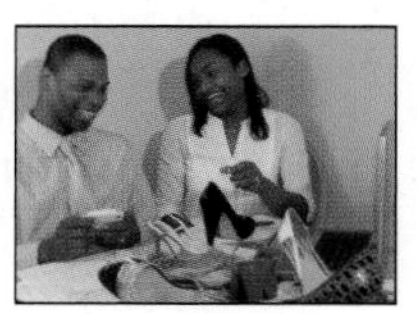

12 Accounting for Partnerships 476

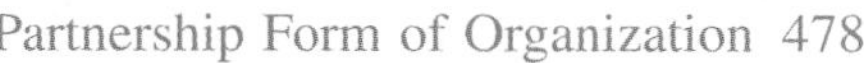

Partnership Form of Organization 478
Characteristics of Partnerships 478
Organizations with Partnership Characteristics 479
Choosing a Business Form 480
Basic Partnership Accounting 481
Organizing a Partnership 481
Dividing Income or Loss 481
Partnership Financial Statements 483
Admission and Withdrawal of Partners 484
Admission of a Partner 484
Withdrawal of a Partner 486
Death of a Partner 487
Liquidation of a Partnership 487
No Capital Deficiency 488
Capital Deficiency 489
Decision Analysis—Partner Return on Equity 490

13 Accounting for Corporations 504

Corporate Form of Organization 506
Characteristics of Corporations 506
Corporate Organization and Management 507
Stockholders of Corporations 508
Basics of Capital Stock 509
Common Stock 510
Issuing Par Value Stock 510
Issuing No-Par Value Stock 511
Issuing Stated Value Stock 512
Issuing Stock for Noncash Assets 512
Dividends 513
Cash Dividends 513
Stock Dividends 514
Stock Splits 516
Preferred Stock 516
Issuance of Preferred Stock 517
Dividend Preference of Preferred Stock 517
Convertible Preferred Stock 518

Callable Preferred Stock 519
Reasons for Issuing Preferred Stock 519
Treasury Stock 520
Purchasing Treasury Stock 520
Reissuing Treasury Stock 521
Retiring Stock 522
Reporting of Equity 522
Statement of Retained Earnings 522
Statement of Stockholders' Equity 523
Reporting Stock Options 523
Decision Analysis—Earnings per Share, Price-Earnings Ratio, Dividend Yield, and Book Value per Share 524

14 Long-Term Liabilities 548
Basics of Bonds 550
Bond Financing 550
Bond Trading 551
Bond-Issuing Procedures 552
Bond Issuances 552
Issuing Bonds at Par 552
Bond Discount or Premium 553
Issuing Bonds at a Discount 553
Issuing Bonds at a Premium 556
Bond Pricing 558
Bond Retirement 559
Bond Retirement at Maturity 559
Bond Retirement before Maturity 559
Bond Retirement by Conversion 560
Long-Term Notes Payable 560
Installment Notes 560
Mortgage Notes and Bonds 562
Decision Analysis—Debt Features and the Debt-to-Equity Ratio 563
Appendix 14A Present Values of Bonds and Notes 567
Appendix 14B Effective Interest Amortization 569
Appendix 14C Issuing Bonds between Interest Dates 571
Appendix 14D Leases and Pensions 573

15 Investments and International Operations 590

Basics of Investments 592

Motivation for Investments 592

Short-Term versus Long-Term 592

Classification and Reporting 593

Accounting Basics for Debt Securities *593*

Accounting Basics for Equity Securities *594*

Reporting of *Non*influential Investments 595

Trading Securities 595

Held-to-Maturity Securities 596

Available-for-Sale Securities 596

Reporting of Influential Investments 598

Investment in Securities with Significant Influence 598

Investment in Securities with Controlling Influence 599

Accounting Summary for Investments in Securities 599

Decision Analysis—Components of Return on Total Assets 600

Appendix 15A Investments in International Operations 605

16 Reporting the Statement of Cash Flows 626

Basics of Cash Flow Reporting 628

Purpose of the Statement of Cash Flows 628

Importance of Cash Flows 628

Measurement of Cash Flows 629

Classification of Cash Flows 629

Noncash Investing and Financing 631

Format of the Statement of Cash Flows 631

Preparing the Statement of Cash Flows 632

Cash Flows from Operating 634

Indirect and Direct Methods of Reporting 634

Application of the Indirect Method of Reporting 635

Summary of Adjustments for Indirect Method 640

Cash Flows from Investing 641

Three-Stage Process of Analysis 641

Analysis of Noncurrent Assets 641

Analysis of Other Assets 642

Cash Flows from Financing 643

Three-Stage Process of Analysis 643

Analysis of Noncurrent Liabilities 643

Analysis of Equity 644

Proving Cash Balances 645

Decision Analysis—Cash Flow Analysis 645

Appendix 16A Spreadsheet Preparation of the Statement of Cash Flows 649

Appendix 16B Direct Method of Reporting Operating Cash Flows 652

17 Analysis of Financial Statements 678

Basics of Analysis 680
Purpose of Analysis 680
Building Blocks of Analysis 681
Information for Analysis 681
Standards for Comparisons 682
Tools of Analysis 682
Horizontal Analysis 682
Comparative Statements 682
Trend Analysis 685
Vertical Analysis 687
Common-Size Statements 687
Common-Size Graphics 689
Ratio Analysis 690
Liquidity and Efficiency 691
Solvency 695
Profitability 696
Market Prospects 697
Summary of Ratios 698
Decision Analysis—Analysis Reporting 700
Appendix 17A Sustainable Income 703

18 Managerial Accounting Concepts and Principles 724

Managerial Accounting Basics 726
Purpose of Managerial Accounting 726
Nature of Managerial Accounting 727
Managerial Decision Making 729
Managerial Accounting in Business 729
Fraud and Ethics in Managerial Accounting 731
Managerial Cost Concepts 732
Types of Cost Classifications 732
Identification of Cost Classifications 734
Cost Concepts for Service Companies 734
Reporting Manufacturing Activities 735
Manufacturer's Balance Sheet 735
Manufacturer's Income Statement 737
Flow of Manufacturing Activities 739
Manufacturing Statement 740
Decision Analysis—Cycle Time and Cycle Efficiency 742

19 Job Order Cost Accounting 768

Job Order Cost Accounting 770
Cost Accounting System 770
Job Order Production 770
Events in Job Order Costing 771
Job Cost Sheet 772
Job Order Cost Flows and Reports 773
Materials Cost Flows and Documents 773
Labor Cost Flows and Documents 775
Overhead Cost Flows and Documents 776
Summary of Cost Flows 778
Adjustment of Overapplied or Underapplied Overhead 781
Underapplied Overhead 781
Overapplied Overhead 782
Decision Analysis—Pricing for Services 782

20 Process Cost Accounting 806

Process Operations 808

Comparing Job Order and Process Operations 809

Organization of Process Operations 809

GenX Company—An Illustration 809

Process Cost Accounting 811

Direct and Indirect Costs 811

Accounting for Materials Costs 812

Accounting for Labor Costs 813

Accounting for Factory Overhead 813

Equivalent Units of Production 815

Accounting for Goods in Process 815

Differences in Equivalent Units for Materials, Labor, and Overhead 815

Process Costing Illustration 816

Step 1: Determine Physical Flow of Units 817

Step 2: Compute Equivalent Units of Production 817

Step 3: Compute Cost per Equivalent Unit 818

Step 4: Assign and Reconcile Costs 818

Transfers to Finished Goods Inventory and Cost of Goods Sold 821

Effect of Lean Business Model on Process Operations 823

Decision Analysis—Hybrid Costing System 823

Appendix 20A FIFO Method of Process Costing 827

21 Cost Allocation and Performance Measurement 850

SECTION 1—ALLOCATING COSTS FOR PRODUCT COSTING 852

Overhead Cost Allocation Methods 852

Two-Stage Cost Allocation 852

Activity-Based Cost Allocation 854

Comparison of Two-Stage and Activity-Based Cost Allocation 856

SECTION 2—ALLOCATING COSTS FOR PERFORMANCE EVALUATION 858

Departmental Accounting 858

Motivation for Departmentalization 858

Departmental Evaluation 858

Departmental Reporting and Analysis 859

Departmental Expense Allocation 859

Direct and Indirect Expenses 859

Allocation of Indirect Expenses 860

Departmental Income Statements 861

Departmental Contribution to Overhead 866

Evaluating Investment Center Performance 867

Financial Performance Evaluation Measures 867

Nonfinancial Performance Evaluation Measures 868

Responsibility Accounting 869

Controllable versus Direct Costs 870

Responsibility Accounting System 870

Decision Analysis—Investment Center Profit Margin and Investment Turnover 872

Appendix 21A Transfer Pricing 875

Appendix 21B Joint Costs and Their Allocation 877

22 Cost-Volume-Profit Analysis 900

Identifying Cost Behavior 902
Fixed Costs 902
Variable Costs 903
Mixed Costs 903
Step-Wise Costs 904
Curvilinear Costs 904
Measuring Cost Behavior 905
Scatter Diagrams 905
High-Low Method 906
Least-Squares Regression 907
Comparison of Cost Estimation Methods 907
Using Break-Even Analysis 908
Contribution Margin and Its Measures 908
Computing the Break-Even Point 909
Preparing a Cost-Volume-Profit Chart 910
Making Assumptions in Cost-Volume-Profit Analysis 911
Applying Cost-Volume-Profit Analysis 912
Computing Income from Sales and Costs 913
Computing Sales for a Target Income 913
Computing the Margin of Safety 914
Using Sensitivity Analysis 915
Computing Multiproduct Break-Even Point 915
Decision Analysis—Degree of Operating Leverage 918
Appendix 22A Using Excel to Estimate Least-Squares Regression 920

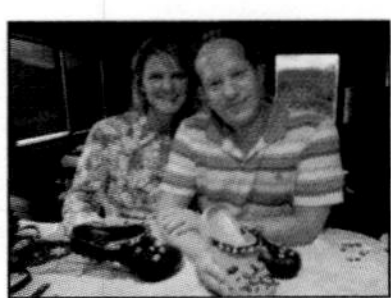

23 Master Budgets and Planning 938

Budget Process 940
Strategic Budgeting 940
Benchmarking Budgets 940
Budgeting and Human Behavior 941
Budgeting as a Management Tool 941
Budgeting Communication 941
Budget Administration 942
Budget Committee 942
Budget Reporting 942
Budget Timing 942
Master Budget 944
Master Budget Components 944
Operating Budgets 946
Capital Expenditures Budget 949
Financial Budgets 949
Decision Analysis—Activity-Based Budgeting 953
Appendix 23A Production and Manufacturing Budgets 959

24 Flexible Budgets and Standard Costs 978

SECTION 1—FLEXIBLE BUDGETS 980

Budgetary Process 980

Budgetary Control and Reporting 980

Fixed Budget Performance Report 981

Budget Reports for Evaluation 982

Flexible Budget Reports 982

Purpose of Flexible Budgets 982

Preparation of Flexible Budgets 982

Flexible Budget Performance Report 984

SECTION 2—STANDARD COSTS 985

Materials and Labor Standards 985

Identifying Standard Costs 985

Setting Standard Costs 986

Cost Variances 986

Cost Variance Analysis 987

Cost Variance Computation 987

Materials and Labor Variances 988

Overhead Standards and Variances 990

Setting Overhead Standards 990

Computing Overhead Cost Variances 992

Computing Variable and Fixed Overhead Cost Variances 992

Extensions of Standard Costs 996

Standard Costs for Control 996

Standard Costs for Services 996

Standard Cost Accounting System 996

Decision Analysis—Sales Variances 998

25 Capital Budgeting and Managerial Decisions 1022

SECTION 1—CAPITAL BUDGETING 1024

Methods Not Using Time Value of Money 1025

Payback Period 1025

Accounting Rate of Return 1027

Methods Using Time Value of Money 1028

Net Present Value 1029

Internal Rate of Return 1031

Comparison of Capital Budgeting Methods 1033

SECTION 2—MANAGERIAL DECISIONS 1034

Decisions and Information 1034

Decision Making 1034

Relevant Costs 1035

Managerial Decision Scenarios 1035

Additional Business 1036

Make or Buy 1037

Scrap or Rework 1038

Sell or Process 1039

Sales Mix Selection 1040

Segment Elimination 1041

Qualitative Decision Factors 1042

Decision Analysis—Break-Even Time 1043

Appendix 25A Using Excel to Compute Net Present Value and Internal Rate of Return 1047

Appendix A Financial Statement Information A-1
- Best Buy A-2
- Circuit City A-19
- RadioShack A-24
- Apple Computer A-29

Appendix B Time Value of Money B

Glossary G

Credits CR

Index IND-1

Chart of Accounts CA

A Look Back

Chapter 11 focused on how current liabilities are identified, computed, recorded, and reported. Attention was directed at notes, payroll, sales taxes, warranties, employee benefits, and contingencies.

A Look at This Chapter

This chapter explains the partnership form of organization. Important partnership characteristics are described along with the accounting concepts and procedures for its most fundamental transactions.

A Look Ahead

Chapter 13 extends our discussion to the corporate form of organization. We describe the accounting and reporting for stock issuances, dividends, and other equity transactions.

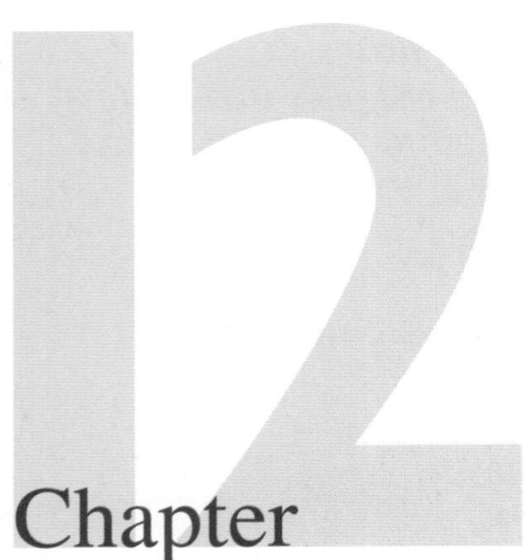

Chapter 12

Accounting for Partnerships

Learning Objectives

CAP

Conceptual

C1 Identify characteristics of partnerships and similar organizations. *(p. 478)*

Analytical

A1 Compute partner return on equity and use it to evaluate partnership performance. *(p. 490)*

Procedural

P1 Prepare entries for partnership formation. *(p. 481)*

P2 Allocate and record income and loss among partners. *(p. 481)*

P3 Account for the admission and withdrawal of partners. *(p. 484)*

P4 Prepare entries for partnership liquidation. *(p. 488)*

LP12

Decision Feature

A Fitting Pair

"Be your best, make a difference, and live with passion"—Samanta and Kelvin Joseph

NEW YORK—Business recipe: Take one information systems graduate, mix a bit of international flair, add an accountant, and stir. The result is **Samanta Shoes (SamantaShoes.com),** a start-up shoe manufacturer. Founded by Samanta and Kelvin Joseph, their partnership is aimed at providing "stylish, comfortable, and affordable" shoes, explains Samanta. "I design every shoe, and nothing less than the best material is used."

Kelvin's focus is on the accounting and financial side of Samanta Shoes. "The knowledge gained from my years at Ernst & Young LLP [a major accounting firm]," explains Kelvin, "has enabled me to be more helpful." Kelvin's knowledge of partnerships and their financial implications are important to Samanta's success. Both partners stress the importance of attending to partnership formation, partnership agreements, and financial reports to stay afloat. They refer to the partners' return on equity and the organizational form as key inputs to partnership success.

Success is causing their partnership to evolve, but the partners adhere to a quality first mentality. "It's all in the design," insists Samanta. "[A quality design] allows for more comfort and support." But quality also extends to style and uniqueness. "Women don't like other women having their shoe," explains Samanta. "We don't want to dilute our brand by being too mass market." Kelvin explains that the smallest manufacturing run they can have is 18 pairs of a special line. In a world of 6 billion people, that is unique.

The partners also continue to apply strict accounting fundamentals. "The partnership cannot survive," says Kelvin, "unless our business is profitable." They regularly review the accounting results and assess the partnership's costs and revenues. Although he adds, "money does not equal happiness." Samanta explains, "Live your dreams . . . make a difference . . . give back to your community"—advice that we can all live by. "If you're not enjoying it," continues Samanta, "there's no point to doing it."

[Sources: *Samanta Shoes Website,* January 2009; *New York Resident,* August 2004; *Caribbean Vibe,* August 2004; *Black Enterprise,* June 2006; *Regine Magazine,* Spring 2004; *Inc.com,* July 2007]

Chapter Preview

The three basic types of business organizations are proprietorships, partnerships, and corporations. Partnerships are similar to proprietorships, except they have more than one owner. This chapter explains partnerships and looks at several variations of them such as limited partnerships, limited liability partnerships, S corporations, and limited liability companies. Understanding the advantages and disadvantages of the partnership form of business organization is important for making informed business decisions.

Partnership Form of Organization

C1 Identify characteristics of partnerships and similar organizations.

A **partnership** is an unincorporated association of two or more people to pursue a business for profit as co-owners. Many businesses are organized as partnerships. They are especially common in small retail and service businesses. Many professional practitioners, including physicians, lawyers, investors, and accountants, also organize their practices as partnerships.

Characteristics of Partnerships

Partnerships are an important type of organization because they offer certain advantages with their unique characteristics. We describe these characteristics in this section.

Voluntary Association A partnership is a voluntary association between partners. Joining a partnership increases the risk to one's personal financial position. Some courts have ruled that partnerships are created by the actions of individuals even when there is no *express agreement* to form one.

Partnership Agreement Forming a partnership requires that two or more legally competent people (who are of age and of sound mental capacity) agree to be partners. Their agreement becomes a **partnership contract,** also called *articles of copartnership.* Although it should be in writing, the contract is binding even if it is only expressed verbally. Partnership agreements normally include details of the partners' (1) names and contributions, (2) rights and duties, (3) sharing of income and losses, (4) withdrawal arrangement, (5) dispute procedures, (6) admission and withdrawal of partners, and (7) rights and duties in the event a partner dies.

Point: When a new partner is admitted, all parties usually must agree to the admission.

Limited Life The life of a partnership is limited. Death, bankruptcy, or any event taking away the ability of a partner to enter into or fulfill a contract ends a partnership. Any one of the partners can also terminate a partnership at will.

Point: The end of a partnership is referred to as its *dissolution.*

Taxation A partnership is not subject to taxes on its income. The income or loss of a partnership is allocated to the partners according to the partnership agreement, and it is included in determining the taxable income for each partner's tax return. Partnership income or loss is allocated each year whether or not cash is distributed to partners.

Point: Partners are taxed on their share of partnership income, not on their withdrawals.

Mutual Agency **Mutual agency** implies that each partner is a fully authorized agent of the partnership. As its agent, a partner can commit or bind the partnership to any contract within the scope of the partnership business. For instance, a partner in a merchandising business can sign contracts binding the partnership to buy merchandise, lease a store building, borrow money, or hire employees. These activities are all within the scope of a merchandising firm. A partner in a law firm, acting alone, however, cannot bind the other partners to a contract to buy snowboards for resale or rent an apartment for parties. These actions are outside the normal scope of a law firm's business. Partners also can agree to limit the power of any one or more of the partners to negotiate contracts for the partnership. This agreement is binding on the partners and on outsiders who know it exists. It is not binding on outsiders who do not know it exists. Outsiders unaware of the agreement have the right to assume each partner has normal agency powers for the partnership. Mutual agency exposes partners to the risk of unwise actions by any one partner.

Point: The majority of states adhere to the Uniform Partnership Act for the basic rules of partnership formation, operation, and dissolution.

Unlimited Liability **Unlimited liability** implies that each partner can be called on to pay a partnership's debts. When a partnership cannot pay its debts, creditors usually can apply their claims to partners' *personal* assets. If a partner does not have enough assets to meet his or her share of the partnership debt, the creditors can apply their claims to the assets of the other partners. A partnership in which all partners have *mutual agency* and *unlimited liability* is called a **general partnership.** Mutual agency and unlimited liability are two main reasons that most general partnerships have only a few members.

Point: Limited life, mutual agency, and unlimited liability are disadvantages of a partnership.

Co-Ownership of Property Partnership assets are owned jointly by all partners. Any investment by a partner becomes the joint property of all partners. Partners have a claim on partnership assets based on their capital account and the partnership contract.

Organizations with Partnership Characteristics

Organizations exist that combine certain characteristics of partnerships with other forms of organizations. We discuss several of these forms in this section.

Limited Partnerships Some individuals who want to invest in a partnership are unwilling to accept the risk of unlimited liability. Their needs can be met with a **limited partnership.** This type of organization is identified in its name with the words "Limited Partnership" or "Ltd." or "LP." A limited partnership has two classes of partners, general and limited. At least one partner must be a **general partner,** who assumes management duties and unlimited liability for the debts of the partnership. The **limited partners** have no personal liability beyond the amounts they invest in the partnership. Limited partners have no active role except as specified in the partnership agreement. A limited partnership agreement often specifies unique procedures for allocating income and losses between general and limited partners. The accounting procedures are similar for both limited and general partnerships.

Decision Insight

Nutty Partners The Hawaii-based **ML Macadamia Orchards LP** is one of the world's largest growers of macadamia nuts. It reported the following partners' capital balances ($ 000s) in its balance sheet:

General Partner	$ 81
Limited Partners	$43,297

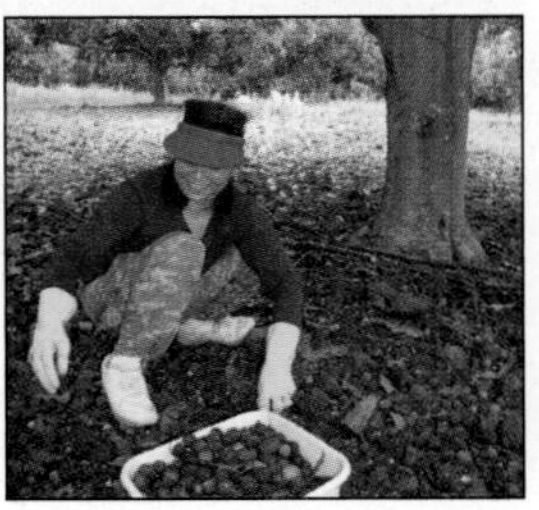

Limited Liability Partnerships Most states allow individuals to form a **limited liability partnership.** This is identified in its name with the words "Limited Liability Partnership" or by "LLP." This type of partnership is designed to protect innocent partners from malpractice or negligence claims resulting from the acts of another partner. When a partner provides service resulting in a malpractice claim, that partner has personal liability for the claim. The remaining partners who were not responsible for the actions resulting in the claim are not personally liable for it. However, most states hold all partners personally liable for other partnership debts. Accounting for a limited liability partnership is the same as for a general partnership.

Point: Many accounting services firms are set up as LLPs.

S Corporations Certain corporations with 75 or fewer stockholders can elect to be treated as a partnership for income tax purposes. These corporations are called *Sub-Chapter S* or simply **S corporations.** This distinguishes them from other corporations, called *Sub-Chapter C* or simply **C corporations.** S corporations provide stockholders the same limited liability feature that C corporations do. The advantage of an S corporation is that it does not pay income taxes. If stockholders work for an S corporation, their salaries are treated as expenses of the corporation. The remaining income or loss of the corporation is allocated to stockholders for inclusion on their personal tax returns. Except for C corporations having to account for income tax expenses and liabilities, the accounting procedures are the same for both S and C corporations.

Point: The majority of proprietorships and partnerships that are organized today are set up as LLCs.

Limited Liability Companies A relatively new form of business organization is the **limited liability company.** The names of these businesses usually include the words "Limited Liability Company" or an abbreviation such as "LLC" or "LC." This form of business has certain features similar to a corporation and others similar to a limited partnership. The owners, who are called *members,* are protected with the same limited liability feature as owners of corporations. While limited partners cannot actively participate in the management of a limited partnership, the members of a limited liability company can assume an active management role. A limited liability company usually has a limited life. For income tax purposes, a limited liability company is typically treated as a partnership. This treatment depends on factors such as whether the members' equity interests are freely transferable and whether the company has continuity of life. A limited liability company's accounting system is designed to help management comply with the dictates of the articles of organization and company regulations adopted by its members. The accounting system also must provide information to support the company's compliance with state and federal laws, including taxation.

Point: Accounting for LLCs is similar to that for partnerships (and proprietorships). One difference is that Owner (Partner), Capital is usually called *Members, Capital* for LLCs.

Choosing a Business Form

Choosing the proper business form is crucial. Many factors should be considered, including taxes, liability risk, tax and fiscal year-end, ownership structure, estate planning, business risks, and earnings and property distributions. The following table summarizes several important characteristics of business organizations:

	Proprietorship	Partnership	LLP	LLC	S Corp.	Corporation
Business entity	Yes	Yes	Yes	Yes	Yes	Yes
Legal entity	No	No	No	Yes	Yes	Yes
Limited liability	No	No	Limited*	Yes	Yes	Yes
Business taxed	No	No	No	No	No	Yes
One owner allowed	Yes	No	No	Yes	Yes	Yes

* A partner's personal liability for LLP debts is limited. Most LLPs carry insurance to protect against malpractice.

Point: The Small Business Administration provides suggestions and information on setting up the proper form for your organization—see SBA.gov.

We must remember that this table is a summary, not a detailed list. Many details underlie each of these business forms, and several details differ across states. Also, state and federal laws change, and a body of law is still developing around LLCs. Business owners should look at these details and consider unique business arrangements such as organizing various parts of their businesses in different forms.

Quick Check

Answers—p. 493

1. A partnership is terminated in the event (*a*) a partnership agreement is not in writing, (*b*) a partner dies, (*c*) a partner exercises mutual agency.
2. What does the term *unlimited liability* mean when applied to a general partnership?
3. Which of the following forms of organization do not provide limited liability to *all* of its owners? (*a*) S corporation, (*b*) limited liability company, (*c*) limited partnership.

Basic Partnership Accounting

Since ownership rights in a partnership are divided among partners, partnership accounting

- Uses a capital account for each partner.
- Uses a withdrawals account for each partner.
- Allocates net income or loss to partners according to the partnership agreement.

This section describes partnership accounting for organizing a partnership, distributing income and loss, and preparing financial statements.

Organizing a Partnership

P1 Prepare entries for partnership formation.

When partners invest in a partnership, their capital accounts are credited for the invested amounts. Partners can invest both assets and liabilities. Each partner's investment is recorded at an agreed-on value, normally the market values of the contributed assets and liabilities at the date of contribution. To illustrate, Kayla Zayn and Hector Perez organize a partnership on January 11 called BOARDS that offers year-round facilities for skateboarding and snowboarding. Zayn's initial net investment in BOARDS is $30,000, made up of cash ($7,000), boarding facilities ($33,000), and a note payable reflecting a bank loan for the new business ($10,000). Perez's initial investment is cash of $10,000. These amounts are the values agreed on by both partners. The entries to record these investments follow.

Zayn's Investment

Jan. 11	Cash	7,000	
	Boarding facilities	33,000	
	Note payable		10,000
	K. Zayn, Capital		30,000
	To record the investment of Zayn.		

Assets	= Liabilities +	Equity
+7,000	+10,000	+30,000
+33,000		

Perez's Investment

Jan. 11	Cash	10,000	
	H. Perez, Capital		10,000
	To record the investment of Perez.		

Assets	= Liabilities +	Equity
+10,000		+10,000

In accounting for a partnership, the following additional relations hold true: (1) Partners' withdrawals are debited to their own separate withdrawals accounts. (2) Partners' capital accounts are credited (or debited) for their shares of net income (or net loss) when closing the accounts at the end of a period. (3) Each partner's withdrawals account is closed to that partner's capital account. Separate capital and withdrawals accounts are kept for each partner.

Point: Both equity and cash are reduced when a partner withdraws cash from a partnership.

Decision Insight

Broadway Partners **Big River Productions** is a partnership that owns the rights to the play *Big River.* The play is performed on tour and periodically on Broadway. For 2006, its Partners' Capital was $288,640, and it was distributed in its entirety to the partners.

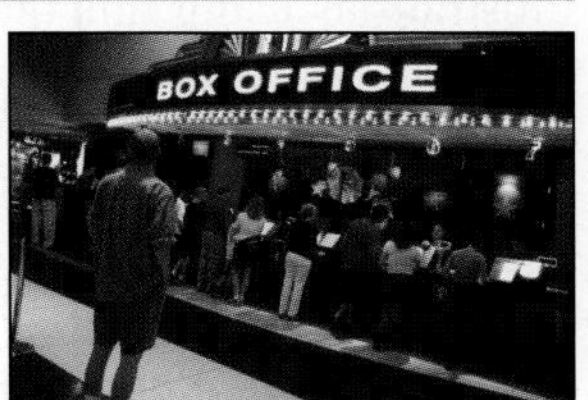

Dividing Income or Loss

P2 Allocate and record income and loss among partners.

Partners are not employees of the partnership but are its owners. If partners devote their time and services to their partnership, they are understood to do so for profit, not for salary. This means there are no salaries to partners that are reported as expenses on the partnership income statement. However, when net income or loss of a partnership is allocated among partners, the partners can agree to allocate "salary allowances" reflecting the relative value of services provided. Partners also can agree to allocate "interest allowances" based on the amount invested. For instance, since Zayn contributes three times the investment of Perez, it is only fair that this be considered when allocating income between them. Like salary allowances, these interest allowances are not expenses on the income statement.

Partners can agree to any method of dividing income or loss. In the absence of an agreement, the law says that the partners share income or loss of a partnership equally. If partners agree on how to share income but say nothing about losses, they share losses the same way they share income. Three common methods to divide income or loss use (1) a stated ratio basis, (2) the ratio of capital balances, or (3) salary and interest allowances and any remainder according to a fixed ratio. We explain each of these methods in this section.

Point: Partners can agree on a ratio to divide income and another ratio to divide a loss.

Allocation on Stated Ratios The *stated ratio* (also called the *income-and-loss-sharing ratio,* the *profit and loss ratio,* or the *P&L ratio*) method of allocating partnership income or loss gives each partner a fraction of the total. Partners must agree on the fractional share each receives. To illustrate, assume the partnership agreement of K. Zayn and H. Perez says Zayn receives two-thirds and Perez one-third of partnership income and loss. If their partnership's net income is $60,000, it is allocated to the partners when the Income Summary account is closed as follows.

Point: The fractional basis can be stated as a proportion, ratio, or percent. For example, a 3:2 basis is the same as 3/5 and 2/5, or 60% and 40%.

Assets = Liabilities + Equity
−60,000
+40,000
+20,000

Dec. 31	Income Summary	60,000	
	K. Zayn, Capital		40,000
	H. Perez, Capital		20,000
	To allocate income and close Income Summary.		

Allocation on Capital Balances The *capital balances* method of allocating partnership income or loss assigns an amount based on the ratio of each partner's relative capital balance. If Zayn and Perez agree to share income and loss on the ratio of their beginning capital balances—Zayn's $30,000 and Perez's $10,000—Zayn receives three-fourths of any income or loss ($30,000/$40,000) and Perez receives one-fourth ($10,000/$40,000). The journal entry follows the same format as that using stated ratios (see the preceding entries).

Point: To determine the percent of income received by each partner, divide an individual partner's share by total net income.

Allocation on Services, Capital, and Stated Ratios The *services, capital, and stated ratio* method of allocating partnership income or loss recognizes that service and capital contributions of partners often are not equal. Salary allowances can make up for differences in service contributions. Interest allowances can make up for unequal capital contributions. Also, the allocation of income and loss can include *both* salary and interest allowances. To illustrate, assume that the partnership agreement of K. Zayn and H. Perez reflects differences in service and capital contributions as follows: (1) annual salary allowances of $36,000 to Zayn and $24,000 to Perez, (2) annual interest allowances of 10% of a partner's beginning-year capital balance, and (3) equal share of any remaining balance of income or loss. These salaries and interest allowances are *not* reported as expenses on the income statement. They are simply a means of dividing partnership income or loss. The remainder of this section provides two illustrations using this three-point allocation agreement.

Illustration when income exceeds allowance. If BOARDS has first-year net income of $70,000, and Zayn and Perez apply the three-point partnership agreement described in the prior paragraph, income is allocated as shown in Exhibit 12.1. Zayn gets $42,000 and Perez gets $28,000 of the $70,000 total.

Illustration when allowances exceed income. The sharing agreement between Zayn and Perez must be followed even if net income is less than the total of the allowances. For example, if BOARDS' first-year net income is $50,000 instead of $70,000, it is allocated to the partners as shown in Exhibit 12.2. Computations for salaries and interest are identical to those in Exhibit 12.1. However, when we apply the total allowances against income, the balance of income is negative. This $(14,000) negative balance is allocated equally to the partners per their sharing agreement. This means that a negative $(7,000) is allocated to each partner. In this case, Zayn ends up with $32,000 and Perez with $18,000. If BOARDS had experienced a net loss, Zayn and Perez would share it in the same manner as the $50,000 income. The only difference is that they would have begun with a negative amount because of the loss. Specifically, the partners

Point: When allowances exceed income, the amount of this negative balance often is referred to as a *sharing agreement loss or deficit.*

Point: Check to make sure the sum of the dollar amounts allocated to each partner equals net income or loss.

EXHIBIT 12.1

Dividing Income When Income Exceeds Allowances

	Zayn	Perez	Total
Net income			$70,000
Salary allowances			
Zayn	$ 36,000		
Perez		$ 24,000	
Interest allowances			
Zayn (10% × $30,000)	3,000		
Perez (10% × $10,000)		1,000	
Total salaries and interest	39,000	25,000	64,000
Balance of income			**6,000**
Balance allocated equally			
Zayn	3,000		
Perez		3,000	
Total allocated			6,000
Balance of income			**$ 0**
Income of each partner	$42,000	$28,000	

EXHIBIT 12.2

Dividing Income When Allowances Exceed Income

	Zayn	Perez	Total
Net income			$50,000
Salary allowances			
Zayn	$ 36,000		
Perez		$ 24,000	
Interest allowances			
Zayn (10% × $30,000)	3,000		
Perez (10% × $10,000)		1,000	
Total salaries and interest	39,000	25,000	64,000
Balance of income			**(14,000)**
Balance allocated equally			
Zayn	(7,000)		
Perez		(7,000)	
Total allocated			(14,000)
Balance of income			**$ 0**
Income of each partner	$32,000	$18,000	

would still have been allocated their salary and interest allowances, further adding to the negative balance of the loss. This *total* negative balance *after* salary and interest allowances would have been allocated equally between the partners. These allocations would have been applied against the positive numbers from any allowances to determine each partner's share of the loss.

Point: When a loss occurs, it is possible for a specific partner's capital to increase (when closing income summary) if that partner's allowance is in excess of his or her share of the negative balance. This implies that decreases to the capital balances of other partners exceed the partnership's loss amount.

Quick Check Answers—p. 493

4. Denzel and Shantell form a partnership by contributing $70,000 and $35,000, respectively. They agree to an interest allowance equal to 10% of each partner's capital balance at the beginning of the year, with the remaining income shared equally. Allocate first-year income of $40,000 to each partner.

Partnership Financial Statements

Partnership financial statements are similar to those of other organizations. The **statement of partners' equity,** also called *statement of partners' capital,* is one exception. It shows *each* partner's beginning capital balance, additional investments, allocated income or loss, withdrawals, and ending capital balance. To illustrate, Exhibit 12.3 shows the statement of partners' equity for BOARDS prepared using the sharing agreement of Exhibit 12.1. Recall that BOARDS' income was $70,000; also, assume that Zayn withdrew $20,000 and Perez $12,000 at year-end.

EXHIBIT 12.3
Statement of Partners' Equity

BOARDS Statement of Partners' Equity For Year Ended December 31, 2009		Zayn		Perez	Total
Beginning capital balances		$ 0		$ 0	$ 0
Plus					
Investments by owners		30,000		10,000	40,000
Net income					
Salary allowances	$36,000		$24,000		
Interest allowances	3,000		1,000		
Balance allocated	3,000		3,000		
Total net income		42,000		28,000	70,000
		72,000		38,000	110,000
Less partners' withdrawals		(20,000)		(12,000)	(32,000)
Ending capital balances		**$52,000**		**$26,000**	**$78,000**

The equity section of the balance sheet of a partnership usually shows the separate capital account balance of each partner. In the case of BOARDS, both K. Zayn, Capital, and H. Perez, Capital, are listed in the equity section along with their balances of $52,000 and $26,000, respectively.

Decision Insight

Gambling Partners Trump Entertainment Resorts LP and subsidiaries operate three casino hotel properties in Atlantic City: Trump Taj Mahal Casino Resort ("Trump Taj Mahal"), Trump Plaza Hotel and Casino ("Trump Plaza"), and Trump Marina Hotel Casino ("Trump Marina"). Its recent statement of partners' equity reports $979,000 in partners' withdrawals, leaving $594,230,000 in partners' capital balances.

Admission and Withdrawal of Partners

P3 Account for the admission and withdrawal of partners.

A partnership is based on a contract between individuals. When a partner is admitted or withdraws, the present partnership ends. Still, the business can continue to operate as a new partnership consisting of the remaining partners. This section considers how to account for the admission and withdrawal of partners.

Admission of a Partner

A new partner is admitted in one of two ways: by purchasing an interest from one or more current partners or by investing cash or other assets in the partnership.

Purchase of Partnership Interest The purchase of partnership interest is a *personal transaction between one or more current partners and the new partner.* To become a partner, the current partners must accept the purchaser. Accounting for the purchase of partnership interest involves reallocating current partners' capital to reflect the transaction. To illustrate, at the end of BOARDS' first year, H. Perez sells one-half of his partnership interest to Tyrell Rasheed for $18,000. This means that Perez gives up a $13,000 recorded interest ($26,000 × 1/2) in the partnership (see the ending capital balance in Exhibit 12.3). The partnership records this January 4 transaction as follows.

Assets = Liabilities + Equity
−13,000
+13,000

Jan. 4	H. Perez, Capital	13,000	
	T. Rasheed, Capital		13,000
	To record admission of Rasheed by purchase.		

After this entry is posted, BOARDS' equity shows K. Zayn, Capital; H. Perez, Capital; and T. Rasheed, Capital, and their respective balances of $52,000, $13,000, and $13,000.

Two aspects of this transaction are important. First, the partnership does *not* record the $18,000 Rasheed paid Perez. The partnership's assets, liabilities, and *total equity* are unaffected by this transaction among partners. Second, Zayn and Perez must agree that Rasheed is to become a partner. If they agree to accept Rasheed, a new partnership is formed and a new contract with a new income-and-loss-sharing agreement is prepared. If Zayn or Perez refuses to accept Rasheed as a partner, then (under the Uniform Partnership Act) Rasheed gets Perez's sold share of partnership income and loss. If the partnership is liquidated, Rasheed gets Perez's sold share of partnership assets. Rasheed gets no voice in managing the company unless Rasheed is admitted as a partner.

Point: Partners' withdrawals are not constrained by the partnership's annual income or loss.

Investing Assets in a Partnership Admitting a partner by accepting assets is a *transaction between the new partner and the partnership.* The invested assets become partnership property. To illustrate, if Zayn (with a $52,000 interest) and Perez (with a $26,000 interest) agree to accept Rasheed as a partner in BOARDS after an investment of $22,000 cash, this is recorded as follows.

Jan. 4	Cash	22,000	
	T. Rasheed, Capital.		22,000
	To record admission of Rasheed by investment.		

Assets	= Liabilities +	Equity
+22,000		+22,000

After this entry is posted, both assets (cash) and equity (T. Rasheed, Capital) increase by $22,000. Rasheed now has a 22% equity in the assets of the business, computed as $22,000 divided by the entire partnership equity ($52,000 + $26,000 + $22,000). Rasheed does not necessarily have a right to 22% of income. Dividing income and loss is a separate matter on which partners must agree.

Bonus to old partners. When the current value of a partnership is greater than the recorded amounts of equity, the partners usually require a new partner to pay a bonus for the privilege of joining. To illustrate, assume that Zayn and Perez agree to accept Rasheed as a partner with a 25% interest in BOARDS if Rasheed invests $42,000. Recall that the partnership's accounting records show that Zayn's recorded equity in the business is $52,000 and Perez's recorded equity is $26,000 (see Exhibit 12.3). Rasheed's equity is determined as follows.

Equities of existing partners ($52,000 + $26,000)	$ 78,000
Investment of new partner	42,000
Total partnership equity	$120,000
Equity of Rasheed (25% × $120,000)	$ 30,000

Although Rasheed invests $42,000, the equity attributed to Rasheed in the new partnership is only $30,000. The $12,000 difference is called a *bonus* and is allocated to existing partners (Zayn and Perez) according to their income-and-loss-sharing agreement. A bonus is shared in this way because it is viewed as reflecting a higher value of the partnership that is not yet reflected in income. The entry to record this transaction follows.

Jan. 4	Cash	42,000	
	T. Rasheed, Capital.		30,000
	K. Zayn, Capital ($12,000 × ½).		6,000
	H. Perez, Capital ($12,000 × ½)		6,000
	To record admission of Rasheed and bonus.		

Assets	= Liabilities +	Equity
+42,000		+30,000
		+6,000
		+6,000

Bonus to new partner. Alternatively, existing partners can grant a bonus to a new partner. This usually occurs when they need additional cash or the new partner has exceptional talents. The bonus to the new partner is in the form of a larger share of equity than the amount invested. To illustrate, assume that Zayn and Perez agree to accept Rasheed as a partner with a 25%

interest in the partnership, but they require Rasheed to invest only $18,000. Rasheed's equity is determined as follows.

Equities of existing partners ($52,000 + $26,000)	$78,000
Investment of new partner	18,000
Total partnership equity	$96,000
Equity of Rasheed (25% × $96,000)	$24,000

The old partners contribute the $6,000 bonus (computed as $24,000 minus $18,000) to Rasheed according to their income-and-loss-sharing ratio. Moreover, Rasheed's 25% equity does not necessarily entitle Rasheed to 25% of future income or loss. This is a separate matter for agreement by the partners. The entry to record the admission and investment of Rasheed is

Assets	= Liabilities +	Equity
+18,000		−3,000
		−3,000
		+24,000

Jan. 4	Cash	18,000	
	K. Zayn, Capital ($6,000 × ½)	3,000	
	H. Perez, Capital ($6,000 × ½)	3,000	
	T. Rasheed, Capital		24,000
	To record Rasheed's admission and bonus.		

Withdrawal of a Partner

A partner generally withdraws from a partnership in one of two ways. (1) First, the withdrawing partner can sell his or her interest to another person who pays for it in cash or other assets. For this, we need only debit the withdrawing partner's capital account and credit the new partner's capital account. (2) The second case is when cash or other assets of the partnership are distributed to the withdrawing partner in settlement of his or her interest. To illustrate these cases, assume that Perez withdraws from the partnership of BOARDS in some future period. The partnership shows the following capital balances at the date of Perez's withdrawal: K. Zayn, $84,000; H. Perez, $38,000; and T. Rasheed, $38,000. The partners (Zayn, Perez, and Rasheed) share income and loss equally. Accounting for Perez's withdrawal depends on whether a bonus is paid. We describe three possibilities.

No Bonus If Perez withdraws and takes cash equal to Perez's capital balance, the entry is

Assets	= Liabilities +	Equity
−38,000		−38,000

Oct. 31	H. Perez, Capital	38,000	
	Cash		38,000
	To record withdrawal of Perez from partnership with no bonus.		

Perez can take any combination of assets to which the partners agree to settle Perez's equity. Perez's withdrawal creates a new partnership between the remaining partners. A new partnership contract and a new income-and-loss-sharing agreement are required.

Bonus to Remaining Partners A withdrawing partner is sometimes willing to take less than the recorded value of his or her equity to get out of the partnership or because the recorded value is overstated. Whatever the reason, when this occurs, the withdrawing partner in effect gives the remaining partners a bonus equal to the equity left behind. The remaining partners share this unwithdrawn equity according to their income-and-loss-sharing ratio. To illustrate, if Perez withdraws and agrees to take $34,000 cash in settlement of Perez's capital balance, the entry is

Assets	= Liabilities +	Equity
−34,000		−38,000
		+2,000
		+2,000

Oct. 31	H. Perez, Capital	38,000	
	Cash		34,000
	K. Zayn, Capital		2,000
	T. Rasheed, Capital		2,000
	To record withdrawal of Perez and bonus to remaining partners.		

Perez withdrew $4,000 less than Perez's recorded equity of $38,000. This $4,000 is divided between Zayn and Rasheed according to their income-and-loss-sharing ratio.

Bonus to Withdrawing Partner A withdrawing partner may be able to receive more than his or her recorded equity for at least two reasons. First, the recorded equity may be understated. Second, the remaining partners may agree to remove this partner by giving assets of greater value than this partner's recorded equity. In either case, the withdrawing partner receives a bonus. The remaining partners reduce their equity by the amount of this bonus according to their income-and-loss-sharing ratio. To illustrate, if Perez withdraws and receives $40,000 cash in settlement of Perez's capital balance, the entry is

Oct. 31	H. Perez, Capital	38,000	
	K. Zayn, Capital	1,000	
	T. Rasheed, Capital	1,000	
	Cash		40,000
	To record Perez's withdrawal from partnership with a bonus to Perez.		

Assets	= Liabilities +	Equity
−40,000		−38,000
		−1,000
		−1,000

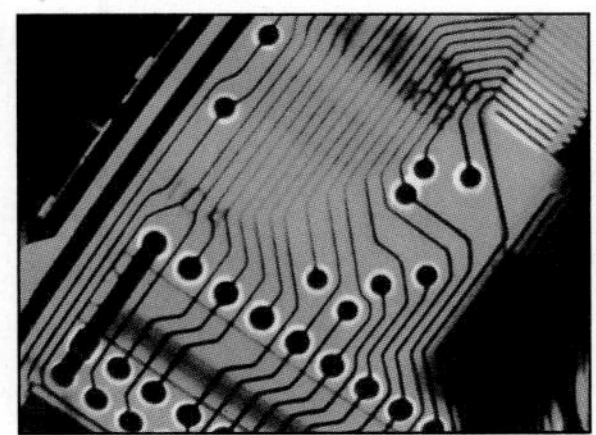

Falcon Cable Communications LLC set up a partnership withdrawal agreement. Falcon owns and operates cable television systems and had two managing general partners. The partnership agreement stated that either partner "can offer to sell to the other partner the offering partner's entire partnership interest . . . for a negotiated price. If the partner receiving such an offer rejects it, the offering partner may elect to cause [the partnership] . . . to be liquidated and dissolved."

Death of a Partner

A partner's death dissolves a partnership. A deceased partner's estate is entitled to receive his or her equity. The partnership contract should contain provisions for settlement in this case. These provisions usually require (1) closing the books to determine income or loss since the end of the previous period and (2) determining and recording current market values for both assets and liabilities. The remaining partners and the deceased partner's estate then must agree to a settlement of the deceased partner's equity. This can involve selling the equity to remaining partners or to an outsider, or it can involve withdrawing assets.

Decision Ethics

Financial Planner You are hired by the two remaining partners of a three-member partnership after the third partner's death. The partnership agreement states that a deceased partner's estate is entitled to a "share of partnership assets equal to the partner's relative equity balance" (partners' equity balances are equal). The estate argues that it is entitled to one-third of the current value of partnership assets. The remaining partners say the distribution should use asset book values, which are 75% of current value. They also point to partnership liabilities, which equal 40% of total asset book value and 30% of current value. How would you resolve this situation? [Answer—p. 493]

Liquidation of a Partnership

When a partnership is liquidated, its business ends and four concluding steps are required.

1. Record the sale of noncash assets for cash and any gain or loss from their liquidation.
2. Allocate any gain or loss from liquidation of the assets in step 1 to the partners using their income-and-loss-sharing ratio.

3. Pay or settle all partner liabilities.
4. Distribute any remaining cash to partners based on their capital balances.

Partnership liquidation usually falls into one of two cases, as described in this section.

No Capital Deficiency

P4 Prepare entries for partnership liquidation.

No capital deficiency means that all partners have a zero or credit balance in their capital accounts for final distribution of cash. To illustrate, assume that Zayn, Perez, and Rasheed operate their partnership in BOARDS for several years, sharing income and loss equally. The partners then decide to liquidate. On the liquidation date, the current period's income or loss is transferred to the partners' capital accounts according to the sharing agreement. After that transfer, assume the partners' recorded equity balances (immediately prior to liquidation) are Zayn, $70,000; Perez, $66,000; and Rasheed, $62,000.

Next, assume that BOARDS sells its noncash assets for a net gain of $6,000. In a liquidation, gains or losses usually result from the sale of noncash assets, which are called *losses and gains from liquidation.* Partners share losses and gains from liquidation according to their income-and-loss-sharing agreement (equal for these partners) yielding the partners' revised equity balances of Zayn, $72,000; Perez, $68,000; and Rasheed, $64,000.[1] Then, after partnership assets are sold and any gain or loss is allocated, the liabilities must be paid. After creditors are paid, any remaining cash is divided among the partners according to their capital account balances. BOARDS' only liability at liquidation is $20,000 in accounts payable. The entries to record the payment to creditors and the final distribution of cash to partners follow.

Assets	= Liabilities	+ Equity
−20,000	−20,000	

Assets	= Liabilities +	Equity
−204,000		−72,000
		−68,000
		−64,000

		Debit	Credit
Jan. 15	Accounts Payable	20,000	
	Cash		20,000
	To pay claims of creditors.		
Jan. 15	K. Zayn, Capital	72,000	
	H. Perez, Capital	68,000	
	T. Rasheed, Capital	64,000	
	Cash		204,000
	To distribute remaining cash to partners.		

It is important to remember that the final cash payment is distributed to partners according to their capital account balances, whereas gains and losses from liquidation are allocated according to the income-and-loss-sharing ratio.

[1] The concepts behind these entries are not new. For example, assume that BOARDS has two noncash assets recorded as boarding facilities, $15,000, and land, $25,000. The entry to sell these assets for $46,000 is

		Debit	Credit
Jan. 15	Cash	46,000	
	Boarding facilities		15,000
	Land		25,000
	Gain from liquidation		6,000
	Sold noncash assets at a gain.		

We then record the allocation of any loss or gain (a gain in this case) from liquidation according to the partners' income-and-loss-sharing agreement as follows.

		Debit	Credit
Jan. 15	Gain from liquidation	6,000	
	K. Zayn, Capital		2,000
	H. Perez, Capital		2,000
	T. Rasheed, Capital		2,000
	To allocate liquidation gain to partners.		

Capital Deficiency

Capital deficiency means that at least one partner has a debit balance in his or her capital account at the point of final cash distribution. This can arise from liquidation losses, excessive withdrawals before liquidation, or recurring losses in prior periods. A partner with a capital deficiency must, if possible, cover the deficit by paying cash into the partnership.

To illustrate, assume that Zayn, Perez, and Rasheed operate their partnership in BOARDS for several years, sharing income and losses equally. The partners then decide to liquidate. Immediately prior to the final distribution of cash, the partners' recorded capital balances are Zayn, $19,000; Perez, $8,000; and Rasheed, $(3,000). Rasheed's capital deficiency means that Rasheed owes the partnership $3,000. Both Zayn and Perez have a legal claim against Rasheed's personal assets. The final distribution of cash in this case depends on how this capital deficiency is handled. Two possibilities exist.

Partner Pays Deficiency Rasheed is obligated to pay $3,000 into the partnership to cover the deficiency. If Rasheed is willing and able to pay, the entry to record receipt of payment from Rasheed follows.

Jan. 15	Cash	3,000	
	T. Rasheed, Capital		3,000
	To record payment of deficiency by Rasheed.		

Assets = Liabilities + Equity
+3,000 +3,000

After the $3,000 payment, the partners' capital balances are Zayn, $19,000; Perez, $8,000; and Rasheed, $0. The entry to record the final cash distributions to partners is

Jan. 15	K. Zayn, Capital	19,000	
	H. Perez, Capital	8,000	
	Cash		27,000
	To distribute remaining cash to partners.		

Assets = Liabilities + Equity
−27,000 −19,000
−8,000

Partner Cannot Pay Deficiency The remaining partners with credit balances absorb any partner's unpaid deficiency according to their income-and-loss-sharing ratio. To illustrate, if Rasheed is unable to pay the $3,000 deficiency, Zayn and Perez absorb it. Since they share income and loss equally, Zayn and Perez each absorb $1,500 of the deficiency. This is recorded as follows.

Jan. 15	K. Zayn, Capital	1,500	
	H. Perez, Capital	1,500	
	T. Rasheed, Capital		3,000
	To transfer Rasheed deficiency to Zayn and Perez.		

Assets = Liabilities + Equity
−1,500
−1,500
+3,000

After Zayn and Perez absorb Rasheed's deficiency, the capital accounts of the partners are Zayn, $17,500; Perez, $6,500; and Rasheed, $0. The entry to record the final cash distribution to the partners is

Jan. 15	K. Zayn, Capital	17,500	
	H. Perez, Capital	6,500	
	Cash		24,000
	To distribute remaining cash to partners.		

Assets = Liabilities + Equity
−24,000 −17,500
−6,500

Rasheed's inability to cover this deficiency does not relieve Rasheed of the liability. If Rasheed becomes able to pay at a future date, Zayn and Perez can each collect $1,500 from Rasheed.

Decision Analysis — Partner Return on Equity

A1 Compute partner return on equity and use it to evaluate partnership performance.

An important role of partnership financial statements is to aid current and potential partners in evaluating partnership success compared with other opportunities. One measure of this success is the **partner return on equity** ratio:

$$\textbf{Partner return on equity} = \frac{\textbf{Partner net income}}{\textbf{Average partner equity}}$$

This measure is separately computed for each partner. To illustrate, Exhibit 12.4 reports selected data from the Boston Celtics LP. The return on equity for the *total* partnership is computed as \$216/[(\$84 + \$252)/2] = 128.6%. However, return on equity is quite different across the partners. For example, the Boston Celtics LP I partner return on equity is computed as \$44/[(\$122 + \$166)/2] = 30.6%, whereas the Celtics LP partner return on equity is computed as \$111/[(\$270 + \$333)/2] = 36.8%. Partner return on equity provides *each* partner an assessment of its return on its equity invested in the partnership. A specific partner often uses this return to decide whether additional investment or withdrawal of resources is best for that partner. Exhibit 12.4 reveals that the year shown produced good returns for all partners (the Boston Celtics LP II return is not computed because its average equity is negative due to an unusual and large distribution in the prior year).

EXHIBIT 12.4

Selected Data from Boston Celtics LP

(\$ thousands)	Total*	Boston Celtics LP I	Boston Celtics LP II	Celtics LP
Beginning-year balance	\$ 84	\$122	\$(307)	\$270
Net income (loss) for year	216	44	61	111
Cash distribution	(48)	—	—	(48)
Ending-year balance	\$252	\$166	\$(246)	\$333
Partner return on equity	128.6%	30.6%	n.a.	36.8%

* Totals may not add up due to rounding.

Demonstration Problem

DP12

The following transactions and events affect the partners' capital accounts in several successive partnerships. Prepare a table with six columns, one for each of the five partners along with a total column to show the effects of the following events on the five partners' capital accounts.

Part 1

4/13/2007 Ries and Bax create R&B Company. Each invests \$10,000, and they agree to share income and losses equally.

12/31/2007 R&B Co. earns \$15,000 in income for its first year. Ries withdraws \$4,000 from the partnership, and Bax withdraws \$7,000.

1/1/2008 Royce is made a partner in RB&R Company after contributing \$12,000 cash. The partners agree that a 10% interest allowance will be given on each partner's beginning-year capital balance. In addition, Bax and Royce are to receive \$5,000 salary allowances. The remainder of the income or loss is to be divided evenly.

12/31/2008 The partnership's income for the year is \$40,000, and withdrawals at year-end are Ries, \$5,000; Bax, \$12,500; and Royce, \$11,000.

1/1/2009 Ries sells her interest for \$20,000 to Murdock, whom Bax and Royce accept as a partner in the new BR&M Co. Income or loss is to be shared equally after Bax and Royce receive \$25,000 salary allowances.

12/31/2009 The partnership's income for the year is \$35,000, and year-end withdrawals are Bax, \$2,500, and Royce, \$2,000.

1/1/2010 Elway is admitted as a partner after investing \$60,000 cash in the new Elway & Associates partnership. He is given a 50% interest in capital after the other partners transfer \$3,000 to his account from each of theirs. A 20% interest allowance (on the beginning-year capital balances) will be used in sharing any income or loss, there will be no salary allowances, and Elway will receive 40% of the remaining balance—the other three partners will each get 20%.

12/31/2010 Elway & Associates earns $127,600 in income for the year, and year-end withdrawals are Bax, $25,000; Royce, $27,000; Murdock, $15,000; and Elway, $40,000.

1/1/2011 Elway buys out Bax and Royce for the balances of their capital accounts after a revaluation of the partnership assets. The revaluation gain is $50,000, which is divided in using a 1:1:1:2 ratio (Bax:Royce:Murdock:Elway). Elway pays the others from personal funds. Murdock and Elway will share income on a 1:9 ratio.

2/28/2011 The partnership earns $10,000 of income since the beginning of the year. Murdock retires and receives partnership cash equal to her capital balance. Elway takes possession of the partnership assets in his own name, and the partnership is dissolved.

Part 2

Journalize the events affecting the partnership for the year ended December 31, 2008.

Planning the Solution

- Evaluate each transaction's effects on the capital accounts of the partners.
- Each time a new partner is admitted or a partner withdraws, allocate any bonus based on the income-or-loss-sharing agreement.
- Each time a new partner is admitted or a partner withdraws, allocate subsequent net income or loss in accordance with the new partnership agreement.
- Prepare entries to (1) record Royce's initial investment; (2) record the allocation of interest, salaries, and remainder; (3) show the cash withdrawals from the partnership; and (4) close the withdrawal accounts on December 31, 2008.

Solution to Demonstration Problem

Part 1

Event	Ries	Bax	Royce	Murdock	Elway	Total
4/13/2007						
Initial investment	$10,000	$10,000				$ 20,000
12/31/2007						
Income (equal)	7,500	7,500				15,000
Withdrawals	(4,000)	(7,000)				(11,000)
Ending balance	$13,500	$10,500				$ 24,000
1/1/2008						
New investment			$12,000			$ 12,000
12/31/2008						
10% interest	1,350	1,050	1,200			3,600
Salaries		5,000	5,000			10,000
Remainder (equal)	8,800	8,800	8,800			26,400
Withdrawals	(5,000)	(12,500)	(11,000)			(28,500)
Ending balance	$18,650	$12,850	$16,000			$ 47,500
1/1/2009						
Transfer interest	(18,650)			$18,650		$ 0
12/31/2009						
Salaries		25,000	25,000			50,000
Remainder (equal)		(5,000)	(5,000)	(5,000)		(15,000)
Withdrawals		(2,500)	(2,000)			(4,500)
Ending balance	$ 0	$30,350	$34,000	$13,650		$ 78,000
1/1/2010						
New investment					$ 60,000	60,000
Bonuses to Elway		(3,000)	(3,000)	(3,000)	9,000	0
Adjusted balance		$27,350	$31,000	$10,650	$ 69,000	$138,000

[continued on next page]

[continued from previous page]

Event	Ries	Bax	Royce	Murdock	Elway	Total
12/31/2010						
20% interest		5,470	6,200	2,130	13,800	27,600
Remainder (1:1:1:2)		20,000	20,000	20,000	40,000	100,000
Withdrawals		(25,000)	(27,000)	(15,000)	(40,000)	(107,000)
Ending balance		$27,820	$30,200	$17,780	$ 82,800	$158,600
1/1/2011						
Gain (1:1:1:2)		10,000	10,000	10,000	20,000	50,000
Adjusted balance		$37,820	$40,200	$27,780	$102,800	$208,600
Transfer interests		(37,820)	(40,200)		78,020	0
Adjusted balance		$ 0	$ 0	$27,780	$180,820	$208,600
2/28/2011						
Income (1:9)				1,000	9,000	10,000
Adjusted balance				$28,780	$189,820	$218,600
Settlements				(28,780)	(189,820)	(218,600)
Final balance				$ 0	$ 0	$ 0

Part 2

Date	Accounts	Debit	Credit
2008			
Jan. 1	Cash	12,000	
	Royce, Capital		12,000
	To record investment of Royce.		
Dec. 31	Income Summary	40,000	
	Ries, Capital		10,150
	Bax, Capital		14,850
	Royce, Capital		15,000
	To allocate interest, salaries, and remainders.		
Dec. 31	Ries, Withdrawals	5,000	
	Bax, Withdrawals	12,500	
	Royce, Withdrawals	11,000	
	Cash		28,500
	To record cash withdrawals by partners.		
Dec. 31	Ries, Capital	5,000	
	Bax, Capital	12,500	
	Royce, Capital	11,000	
	Ries, Withdrawals		5,000
	Bax, Withdrawals		12,500
	Royce, Withdrawals		11,000
	To close withdrawal accounts.		

Summary

C1 **Identify characteristics of partnerships and similar organizations.** Partnerships are voluntary associations, involve partnership agreements, have limited life, are not subject to income tax, include mutual agency, and have unlimited liability. Organizations that combine selected characteristics of partnerships and corporations include limited partnerships, limited liability partnerships, S corporations, and limited liability companies.

A1 **Compute partner return on equity and use it to evaluate partnership performance.** Partner return on equity provides each partner an assessment of his or her return on equity invested in the partnership.

P1 **Prepare entries for partnership formation.** A partner's initial investment is recorded at the market value of the assets contributed to the partnership.

P2 **Allocate and record income and loss among partners.** A partnership agreement should specify how to allocate partnership income or loss among partners. Allocation can be based on a stated ratio, capital balances, or salary and interest allowances to compensate partners for differences in their service and capital contributions.

P3 **Account for the admission and withdrawal of partners.** When a new partner buys a partnership interest directly from one or more existing partners, the amount of cash paid from one partner to another does not affect the partnership total recorded equity. When a new partner purchases equity by investing additional assets in the partnership, the new partner's investment can yield a bonus either to existing partners or to the new partner. The entry to record a withdrawal can involve payment from either (1) the existing partners' personal assets or (2) partnership assets. The latter can yield a bonus to either the withdrawing or remaining partners.

P4 **Prepare entries for partnership liquidation.** When a partnership is liquidated, losses and gains from selling partnership assets are allocated to the partners according to their income-and-loss-sharing ratio. If a partner's capital account has a deficiency that the partner cannot pay, the other partners share the deficit according to their relative income-and-loss-sharing ratio.

Guidance Answers to **Decision Ethics**

Financial Planner The partnership agreement apparently fails to mention liabilities or use the term *net assets.* To give the estate one-third of total assets is not fair to the remaining partners because if the partner had lived and the partners had decided to liquidate, the liabilities would need to be paid out of assets before any liquidation. Also, a settlement based on the deceased partner's recorded equity would fail to recognize excess of current value over book value. This value increase would be realized if the partnership were liquidated. A fair settlement would seem to be a payment to the estate for the balance of the deceased partner's equity based on the *current value of net assets.*

Guidance Answers to **Quick Checks**

1. (*b*)

2. *Unlimited liability* means that the creditors of a partnership require each partner to be personally responsible for all partnership debts.

3. (*c*)

4.

	Denzel	Shantell	Total
Net income			$40,000
Interest allowance (10%)	$ 7,000	$ 3,500	10,500
Balance of income			$29,500
Balance allocated equally	14,750	14,750	29,500
Balance of income			$ 0
Income of partners	$21,750	$18,250	

Key Terms

mhhe.com/wildFAP19e

Key Terms are available at the book's Website for learning and testing in an online Flashcard Format.

C corporation (p. 480)
General partner (p. 479)
General partnership (p. 479)
Limited liability company (LLC) (p. 480)
Limited liability partnership (p. 479)
Limited partners (p. 479)
Limited partnership (p. 479)
Mutual agency (p. 478)
Partner return on equity (p. 490)
Partnership (p. 478)
Partnership contract (p. 478)
Partnership liquidation (p. 488)
S corporation (p. 480)
Statement of partners' equity (p. 483)
Unlimited liability (p. 479)

Multiple Choice Quiz

Answers on p. 503

mhhe.com/wildFAP19e

Additional Quiz Questions are available at the book's Website.

1. Stokely and Leder are forming a partnership. Stokely invests a building that has a market value of $250,000; and the partnership assumes responsibility for a $50,000 note secured by a mortgage on that building. Leder invests $100,000 cash. For the partnership, the amounts recorded for the building and for Stokely's Capital account are these:

a. Building, $250,000; Stokely, Capital, $250,000.
b. Building, $200,000; Stokely, Capital, $200,000.
c. Building, $200,000; Stokely, Capital, $100,000.
d. Building, $200,000; Stokely, Capital, $250,000.
e. Building, $250,000; Stokely, Capital, $200,000.

2. Katherine, Alliah, and Paulina form a partnership. Katherine contributes $150,000, Alliah contributes $150,000, and Paulina contributes $100,000. Their partnership agreement calls for the income or loss division to be based on the ratio of capital invested. If the partnership reports income of $90,000 for its first year of operations, what amount of income is credited to Paulina's capital account?
   a. $22,500
   b. $25,000
   c. $45,000
   d. $30,000
   e. $90,000
3. Jamison and Blue form a partnership with capital contributions of $600,000 and $800,000, respectively. Their partnership agreement calls for Jamison to receive $120,000 per year in salary. Also, each partner is to receive an interest allowance equal to 10% of the partner's beginning capital contributions, with any remaining income or loss divided equally. If net income for its initial year is $270,000, then Jamison's and Blue's respective shares are
   a. $135,000; $135,000.
   b. $154,286; $115,714.
   c. $120,000; $150,000.
   d. $185,000; $85,000.
   e. $85,000; $185,000.
4. Hansen and Fleming are partners and share equally in income or loss. Hansen's current capital balance in the partnership is $125,000 and Fleming's is $124,000. Hansen and Fleming agree to accept Black with a 20% interest. Black invests $75,000 in the partnership. The bonus granted to Hansen and Fleming equals
   a. $13,000 each.
   b. $5,100 each.
   c. $4,000 each.
   d. $5,285 to Hansen; $4,915 to Fleming.
   e. $0; Hansen and Fleming grant a bonus to Black.
5. Mee Su is a partner in Hartford Partners, LLC. Her partnership capital balance at the beginning of the current year was $110,000, and her ending balance was $124,000. Her share of the partnership income is $10,500. What is her partner return on equity?
   a. 8.97%
   b. 1060.00%
   c. 9.54%
   d. 1047.00%
   e. 8.47%

Discussion Questions

1. ♝ If a partnership contract does not state the period of time the partnership is to exist, when does the partnership end?
2. What does the term *mutual agency* mean when applied to a partnership?
3. ♝ Can partners limit the right of a partner to commit their partnership to contracts? Would such an agreement be binding (*a*) on the partners and (*b*) on outsiders?
4. ♝ Assume that Amey and Lacey are partners. Lacey dies, and her son claims the right to take his mother's place in the partnership. Does he have this right? Why or why not?
5. ♝ Assume that the Barnes and Ardmore partnership agreement provides for a two-third/one-third sharing of income but says nothing about losses. The first year of partnership operation resulted in a loss, and Barnes argues that the loss should be shared equally because the partnership agreement said nothing about sharing losses. Is Barnes correct? Explain.
6. Allocation of partnership income among the partners appears on what financial statement?
7. What does the term *unlimited liability* mean when it is applied to partnership members?
8. How does a general partnership differ from a limited partnership?
9. ♝ George, Burton, and Dillman have been partners for three years. The partnership is being dissolved. George is leaving the firm, but Burton and Dillman plan to carry on the business. In the final settlement, George places a $75,000 salary claim against the partnership. He contends that he has a claim for a salary of $25,000 for each year because he devoted all of his time for three years to the affairs of the partnership. Is his claim valid? Why or why not?
10. ♝ Kay, Kat, and Kim are partners. In a liquidation, Kay's share of partnership losses exceeds her capital account balance. Moreover, she is unable to meet the deficit from her personal assets, and her partners shared the excess losses. Does this relieve Kay of liability?
11. After all partnership assets have been converted to cash and all liabilities paid, the remaining cash should equal the sum of the balances of the partners' capital accounts. Why?
12. Assume a partner withdraws from a partnership and receives assets of greater value than the book value of his equity. Should the remaining partners share the resulting reduction in their equities in the ratio of their relative capital balances or according to their income-and-loss-sharing ratio?

♝ ***Denotes Discussion Questions that involve decision making.***

McGraw-Hill's HOMEWORK MANAGER® Available with McGraw-Hill's Homework Manager

QUICK STUDY

QS 12-1
Partnership liability
C1 

Otis and Hunan are partners in operating a store. Without consulting Otis, Hunan enters into a contract to purchase merchandise for the store. Otis contends that he did not authorize the order and refuses to pay for it. The vendor sues the partners for the contract price of the merchandise. (*a*) Must the partnership pay for the merchandise? Why? (*b*) Does your answer differ if Otis and Hunan are partners in a public accounting firm? Explain.

QS 12-2
Partnership income allocation
P2

Ramona Stolton and Jerry Bright are partners in a business they started two years ago. The partnership agreement states that Stolton should receive a salary allowance of $30,000 and that Bright should receive a $40,000 salary allowance. Any remaining income or loss is to be shared equally. Determine each partner's share of the current year's net income of $104,000.

QS 12-3
Partnership income allocation
P2 

Hiram and Tyrone are partners who agree that Hiram will receive a $50,000 salary allowance and that any remaining income or loss will be shared equally. If Tyrone's capital account is credited for $1,000 as her share of the net income in a given period, how much net income did the partnership earn in that period?

QS 12-4
Liability in limited partnerships
P1

Vernon organized a limited partnership and is the only general partner. Xavier invested $40,000 in the partnership and was admitted as a limited partner with the understanding that he would receive 10% of the profits. After two unprofitable years, the partnership ceased doing business. At that point, partnership liabilities were $170,000 larger than partnership assets. How much money can the partnership's creditors obtain from Xavier's personal assets to satisfy the unpaid partnership debts?

QS 12-5
Partner admission through purchase of interest P3

Dresden agrees to pay Choi and Amal $20,000 each for a one-third (33⅓%) interest in the Choi and Amal partnership. Immediately prior to Dresden's admission, each partner had a $60,000 capital balance. Make the journal entry to record Dresden's purchase of the partners' interest.

QS 12-6
Admission of a partner P3

Neal and Vanier are partners, each with $80,000 in their partnership capital accounts. Brantford is admitted to the partnership by investing $80,000 cash. Make the entry to show Brantford's admission to the partnership.

QS 12-7
Partner return on equity
A1 

Paulson and Fleming's company is organized as a partnership. At the prior year-end, partnership equity totaled $300,000 ($200,000 from Paulson and $100,000 from Fleming). For the current year, partnership net income is $49,000 ($38,400 allocated to Paulson and $10,600 allocated to Fleming), and year-end total partnership equity is $400,000 ($280,000 from Paulson and $120,000 from Fleming). Compute the total partnership return on equity *and* the individual partner return on equity ratios.

McGraw-Hill's HOMEWORK MANAGER® Available with McGraw-Hill's Homework Manager

EXERCISES

Exercise 12-1
Characteristics of partnerships
C1

Next to the following list of eight characteristics of business organizations, enter a brief description of how each characteristic applies to general partnerships.

Characteristic	Application to General Partnerships
1. Life	
2. Owners' liability	
3. Legal status	
4. Tax status of income	
5. Owners' authority	
6. Ease of formation	
7. Transferability of ownership	
8. Ability to raise large amounts of capital	

Exercise 12-2
Forms of organization
C1 

For each of the following separate cases, recommend a form of business organization. With each recommendation, explain how business income would be taxed if the owners adopt the form of organization recommended. Also list several advantages that the owners will enjoy from the form of business organization that you recommend.

a. Sharif, Henry, and Saanen are recent college graduates in computer science. They want to start a Website development company. They all have college debts and currently do not own any substantial computer equipment needed to get the company started.

b. Dr. LeBlanc and Dr. Liu are recent graduates from medical residency programs. Both are family practice physicians and would like to open a clinic in an underserved rural area. Although neither has any funds to bring to the new venture, a banker has expressed interest in making a loan to provide start-up funds for their practice.

c. Novato has been out of school for about five years and has become quite knowledgeable about the commercial real estate market. He would like to organize a company that buys and sells real estate. Novato believes he has the expertise to manage the company but needs funds to invest in commercial property.

Exercise 12-3
Journalizing partnership transactions
P2

On March 1, 2009, Eckert and Kelley formed a partnership. Eckert contributed $83,000 cash and Kelley contributed land valued at $66,400 and a building valued at $96,400. The partnership also assumed responsibility for Kelley's $77,800 long-term note payable associated with the land and building. The partners agreed to share income as follows: Eckert is to receive an annual salary allowance of $30,500, both are to receive an annual interest allowance of 11% of their beginning-year capital investment, and any remaining income or loss is to be shared equally. On October 20, 2009, Eckert withdrew $32,000 cash and Kelley withdrew $25,000 cash. After the adjusting and closing entries are made to the revenue and expense accounts at December 31, 2009, the Income Summary account had a credit balance of $86,000.

1. Prepare journal entries to record (*a*) the partners' initial capital investments, (*b*) their cash withdrawals, and (*c*) the December 31 closing of both the Withdrawals and Income Summary accounts.

2. Determine the balances of the partners' capital accounts as of December 31, 2009.

Check (2) Kelley, $87,860

Exercise 12-4
Income allocation in a partnership
P2

Daria and Farrah began a partnership by investing $64,000 and $58,000, respectively. During its first year, the partnership earned $175,000. Prepare calculations showing how the $175,000 income should be allocated to the partners under each of the following three separate plans for sharing income and loss: (1) the partners failed to agree on a method to share income; (2) the partners agreed to share income and loss in proportion to their initial investments (round amounts to the nearest dollar); and (3) the partners agreed to share income by granting a $52,000 per year salary allowance to Daria, a $42,000 per year salary allowance to Farrah, 8% interest on their initial capital investments, and the remaining balance shared equally.

Check Plan 3, Daria, $92,740

Exercise 12-5
Income allocation in a partnership
P2

Assume that the partners of Exercise 12-4 agreed to share net income and loss by granting annual salary allowances of $52,000 to Daria and $42,000 to Farrah, 8% interest allowances on their investments, and any remaining balance shared equally.

1. Determine the partners' shares of Daria and Farrah given a first-year net income of $98,800.

2. Determine the partners' shares of Daria and Farrah given a first-year net loss of $16,800.

Check (2) Daria, $(3,160)

Exercise 12-6
Sale of partnership interest P3

The partners in the Biz Partnership have agreed that partner Estella may sell her $111,000 equity in the partnership to Sean, for which Sean will pay Estella $88,800. Present the partnership's journal entry to record the sale of Estella's interest to Sean on September 30.

Exercise 12-7
Admission of new partner
P3

The Josetti Partnership has total partners' equity of $558,000, which is made up of Dopke, Capital, $392,000, and Hughes, Capital, $166,000. The partners share net income and loss in a ratio of 85% to Dopke and 15% to Hughes. On November 1, Nillsen is admitted to the partnership and given a 10% interest in equity and a 10% share in any income and loss. Prepare the journal entry to record the admission of Nillsen under each of the following separate assumptions: Nillsen invests cash of (1) $62,000; (2) $97,000; and (3) $32,000.

Exercise 12-8
Retirement of partner
P3

Edison, Delray, and West have been partners while sharing net income and loss in a 5:4:1 ratio. On January 31, the date West retires from the partnership, the equities of the partners are Edison, $330,000; Delray, $231,000; and West, $165,000. Present journal entries to record West's retirement under each of the following separate assumptions: West is paid for her equity using partnership cash of (1) $165,000; (2) $192,000; and (3) $129,000.

Exercise 12-9
Liquidation of partnership
P4

The Red, White & Blue partnership was begun with investments by the partners as follows: Red, $153,000; White, $183,000; and Blue, $180,000. The operations did not go well, and the partners eventually decided to liquidate the partnership, sharing all losses equally. On August 31, after all assets were converted to cash and all creditors were paid, only $39,000 in partnership cash remained.

1. Compute the capital account balance of each partner after the liquidation of assets and the payment of creditors.

Check (1) Red, $(6,000)

2. Assume that any partner with a deficit agrees to pay cash to the partnership to cover the deficit. Present the journal entries on August 31 to record (*a*) the cash receipt from the deficient partner(s) and (*b*) the final disbursement of cash to the partners.

3. Assume that any partner with a deficit is not able to reimburse the partnership. Present journal entries (*a*) to transfer the deficit of any deficient partners to the other partners and (*b*) to record the final disbursement of cash to the partners.

Exercise 12-10
Liquidation of partnership P4

Brewster, Conway, and Ogden are partners who share income and loss in a 1:5:4 ratio. After lengthy disagreements among the partners and several unprofitable periods, the partners decide to liquidate the partnership. Immediately before liquidation, the partnership balance sheet shows total assets, $117,000; total liabilities, $87,750; Brewster, Capital, $1,600; Conway, Capital, $11,600; and Ogden, Capital, $16,050. The cash proceeds from selling the assets were sufficient to repay all but $20,500 to the creditors. (*a*) Calculate the loss from selling the assets. (*b*) Allocate the loss to the partners. (*c*) Determine how much of the remaining liability should be paid by each partner.

Check (b) Ogden, Capital after allocation, $(3,850)

Exercise 12-11
Liquidation of limited partnership
P4

Assume that the Brewster, Conway, and Ogden partnership of Exercise 12-10 is a limited partnership. Brewster and Conway are general partners and Ogden is a limited partner. How much of the remaining $20,500 liability should be paid by each partner? (Round amounts to the nearest dollar.)

Exercise 12-12
Partner return on equity
A1

Hunt Sports Enterprises LP is organized as a limited partnership consisting of two individual partners: Soccer LP and Football LP. Both partners separately operate a minor league soccer team and a semipro football team. Compute partner return on equity for each limited partnership (and the total) for the year ended June 30, 2009, using the following selected data on partner capital balances from Hunt Sports Enterprises LP.

	Soccer LP	Football LP	Total
Balance at 6/30/2008	$331,000	$1,357,100	$1,688,100
Annual net income	35,405	725,803	761,208
Cash distribution	—	(65,000)	(65,000)
Balance at 6/30/2009	$366,405	$2,017,903	$2,384,308

McGraw-Hill's HOMEWORK MANAGER® Available with McGraw-Hill's Homework Manager

PROBLEM SET A

Problem 12-1A
Allocating partnership income
P2

Alex Jeffers, Jo Ford, and Rose Verne invested $32,500, $45,500, and $52,000, respectively, in a partnership. During its first calendar year, the firm earned $391,200.

Required

Prepare the entry to close the firm's Income Summary account as of its December 31 year-end and to allocate the $391,200 net income to the partners under each of the following separate assumptions: The partners (1) have no agreement on the method of sharing income and loss; (2) agreed to share income and loss in the ratio of their beginning capital investments; and (3) agreed to share income and loss by providing annual salary allowances of $40,000 to Jeffers, $35,000 to Ford, and $46,000 to Verne; granting 8% interest on the partners' beginning capital investments; and sharing the remainder equally.

Check (3) Verne, Capital, $136,760

Problem 12-2A
Allocating partnership income and loss; sequential years
P2

mhhe.com/wildFAP19e

Jasmine Watts and Liz Thomas are forming a partnership to which Watts will devote one-half time and Thomas will devote full time. They have discussed the following alternative plans for sharing income and loss: (*a*) in the ratio of their initial capital investments, which they have agreed will be $30,000 for Watts and $45,000 for Thomas; (*b*) in proportion to the time devoted to the business; (*c*) a salary allowance of $3,000 per month to Thomas and the balance in accordance with the ratio of their initial capital investments; or (*d*) a salary allowance of $3,000 per month to Thomas, 10% interest on their initial capital investments, and the balance shared equally. The partners expect the business to perform as follows: year 1, $18,000 net loss; year 2, $45,000 net income; and year 3, $75,000 net income.

Required

Prepare three tables with the following column headings.

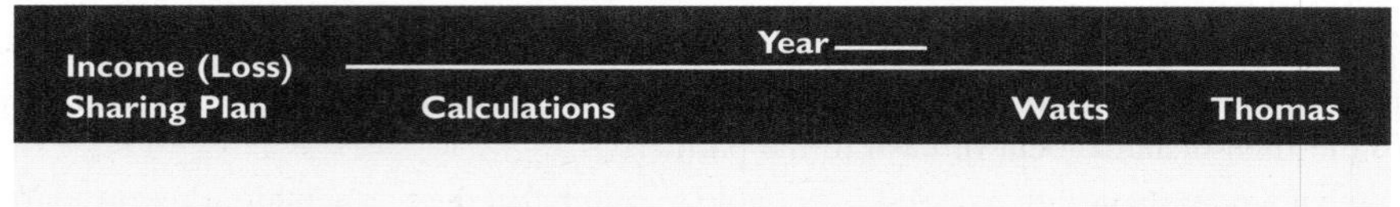

Income (Loss) Sharing Plan	Calculations	Year ____	
		Watts	Thomas

Complete the tables, one for each of the first three years, by showing how to allocate partnership income or loss to the partners under each of the four plans being considered. (Round answers to the nearest whole dollar.)

Check Plan d, year 1, Thomas's share, $9,750

Problem 12-3A
Partnership income allocation, statement of partners' equity, and closing entries
P2

mhhe.com/wildFAP19e

Will Beck, Trevor Beck, and Barb Beck formed the BBB Partnership by making capital contributions of $142,500, $118,750, and $213,750, respectively. They predict annual partnership net income of $210,000 and are considering the following alternative plans of sharing income and loss: (*a*) equally; (*b*) in the ratio of their initial capital investments; or (*c*) salary allowances of $38,000 to Will, $28,000 to Trevor, and $43,000 to Barb; interest allowances of 10% on their initial capital investments; and the balance shared equally.

Required

1. Prepare a table with the following column headings.

Income (Loss) Sharing Plan	Calculations	Will	Trevor	Barb	Total

Use the table to show how to distribute net income of $210,000 for the calendar year under each of the alternative plans being considered. (Round answers to the nearest whole dollar.)

2. Prepare a statement of partners' equity showing the allocation of income to the partners assuming they agree to use plan (*c*), that income earned is $87,500, and that Will, Trevor, and Barb withdraw $18,000, $25,000, and $34,000, respectively, at year-end.

3. Prepare the December 31 journal entry to close Income Summary assuming they agree to use plan (*c*) and that net income is $87,500. Also close the withdrawals accounts.

Check (2) Barb, Ending Capital, $221,125

Problem 12-4A
Partner withdrawal and admission
P3

Part 1. Meir, Zarcus, and Ross are partners and share income and loss in a 1:4:5 ratio. The partnership's capital balances are as follows: Meir, $43,000; Zarcus, $179,000; and Ross, $228,000. Zarcus decides to withdraw from the partnership, and the partners agree to not have the assets revalued upon Zarcus's retirement. Prepare journal entries to record Zarcus's February 1 withdrawal from the partnership under each of the following separate assumptions: Zarcus (*a*) sells her interest to Garcia for $160,000 after Meir and Ross approve the entry of Garcia as a partner; (*b*) gives her interest to a son-in-law, Fields, and thereafter Meir and Ross accept Fields as a partner; (*c*) is paid $179,000 in partnership cash for her equity; (*d*) is paid $215,000 in partnership cash for her equity; and (*e*) is paid $20,000 in partnership cash plus equipment recorded on the partnership books at $70,000 less its accumulated depreciation of $23,200.

Check (1e) Cr. Ross, Capital, $93,500

Part 2. Assume that Zarcus does not retire from the partnership described in Part 1. Instead, Potter is admitted to the partnership on February 1 with a 25% equity. Prepare journal entries to record Potter's entry into the partnership under each of the following separate assumptions: Potter invests (*a*) $150,000; (*b*) $110,000; and (*c*) $196,000.

(2c) Cr. Zarcus, Capital, $13,800

Problem 12-5A
Liquidation of a partnership
P4

Kendra, Cogley, and Mei share income and loss in a 3:2:1 ratio. The partners have decided to liquidate their partnership. On the day of liquidation their balance sheet appears as follows.

KENDRA, COGLEY, AND MEI
Balance Sheet
May 31

Assets		**Liabilities and Equity**	
Cash	$199,100	Accounts payable	$258,000
Inventory	548,400	Kendra, Capital	97,900
		Cogley, Capital	220,275
		Mei, Capital	171,325
Total assets	$747,500	Total liabilities and equity	$747,500

Required

Prepare journal entries for (*a*) the sale of inventory, (*b*) the allocation of its gain or loss, (*c*) the payment of liabilities at book value, and (*d*) the distribution of cash in each of the following separate cases: Inventory is sold for (1) $625,200; (2) $452,400; (3) $321,000 and any partners with capital deficits pay in the amount of their deficits; and (4) $249,000 and the partners have no assets other than those invested in the partnership. (Round to the nearest dollar.)

Check (4) Cash distribution: Mei, $104,158

PROBLEM SET B

Problem 12-1B
Allocating partnership income
P2

Erin Rock, Sal Arthur, and Chloe Binder invested $47,840, $64,400, and $71,760, respectively, in a partnership. During its first calendar year, the firm earned $361,800.

Required

Prepare the entry to close the firm's Income Summary account as of its December 31 year-end and to allocate the $361,800 net income to the partners under each of the following separate assumptions. (Round answers to whole dollars.) The partners (1) have no agreement on the method of sharing income and loss; (2) agreed to share income and loss in the ratio of their beginning capital investments; and (3) agreed to share income and loss by providing annual salary allowances of $33,000 to Rock, $28,000 to Arthur, and $40,000 to Binder; granting 10% interest on the partners' beginning capital investments; and sharing the remainder equally.

Check (3) Binder, Capital, $127,976

Problem 12-2B
Allocating partnership income and loss; sequential years
P2

Maria Selk and David Green are forming a partnership to which Selk will devote one-third time and Green will devote full time. They have discussed the following alternative plans for sharing income and loss: (*a*) in the ratio of their initial capital investments, which they have agreed will be $208,000 for Selk and $312,000 for Green; (*b*) in proportion to the time devoted to the business; (*c*) a salary allowance of $8,000 per month to Green and the balance in accordance with the ratio of their initial capital investments; or (*d*) a salary allowance of $8,000 per month to Green, 10% interest on their initial capital investments, and the balance shared equally. The partners expect the business to perform as follows: year 1, $72,000 net loss; year 2, $152,000 net income; and year 3, $376,000 net income.

Required

Prepare three tables with the following column headings.

Income (Loss) Sharing Plan	Calculations	Year____ Selk	Green

Check Plan d, year 1, Green's share, $17,200

Complete the tables, one for each of the first three years, by showing how to allocate partnership income or loss to the partners under each of the four plans being considered. (Round answers to the nearest whole dollar.)

Problem 12-3B
Partnership income allocation, statement of partners' equity, and closing entries
P2

Sally Cook, Lin Xi, and Sami Bruce formed the CXB Partnership by making capital contributions of $169,750, $121,250, and $194,000, respectively. They predict annual partnership net income of $270,000 and are considering the following alternative plans of sharing income and loss: (*a*) equally; (*b*) in the ratio of their initial capital investments; or (*c*) salary allowances of $40,000 to Cook, $29,000 to Xi, and $42,000 to Bruce; interest allowances of 12% on their initial capital investments; and the balance shared equally.

Required

1. Prepare a table with the following column headings.

Income (Loss) Sharing Plan	Calculations	Cook	Xi	Bruce	Total

Use the table to show how to distribute net income of $270,000 for the calendar year under each of the alternative plans being considered. (Round answers to the nearest whole dollar.)

Check (2) Bruce, Ending Capital, $208,380

2. Prepare a statement of partners' equity showing the allocation of income to the partners assuming they agree to use plan (*c*), that income earned is $124,500, and that Cook, Xi, and Bruce withdraw $19,000, $26,000, and $36,000, respectively, at year-end.

3. Prepare the December 31 journal entry to close Income Summary assuming they agree to use plan (*c*) and that net income is $124,500. Also close the withdrawals accounts.

Problem 12-4B
Partner withdrawal and admission
P3

Part 1. Davis, Brown, and Nell are partners and share income and loss in a 5:1:4 ratio. The partnership's capital balances are as follows: Davis, $303,000; Brown, $74,000; and Nell, $223,000. Davis decides to withdraw from the partnership, and the partners agree not to have the assets revalued upon Davis's retirement. Prepare journal entries to record Davis's April 30 withdrawal from the partnership under each of the following separate assumptions: Davis (*a*) sells her interest to Leer for $125,000 after Brown and Nell approve the entry of Leer as a partner; (*b*) gives her interest to a daughter-in-law, Gibson, and thereafter Brown and Nell accept Gibson as a partner; (*c*) is paid $303,000 in partnership cash for her equity; (*d*) is paid $175,000 in partnership cash for her equity; and (*e*) is paid $100,000 in partnership cash plus manufacturing equipment recorded on the partnership books at $269,000 less its accumulated depreciation of $168,000.

Check (1e) Cr. Nell, Capital, $81,600

Part 2. Assume that Davis does not retire from the partnership described in Part 1. Instead, McCann is admitted to the partnership on April 30 with a 20% equity. Prepare journal entries to record the entry of McCann under each of the following separate assumptions: McCann invests (*a*) $150,000; (*b*) $98,000; and (*c*) $213,000.

Check (2c) Cr. Brown, Capital, $5,040

Problem 12-5B
Liquidation of a partnership
P4

London, Ramirez, and Toney, who share income and loss in a 3:2:1 ratio, plan to liquidate their partnership. At liquidation, their balance sheet appears as follows.

LONDON, RAMIREZ, AND TONEY Balance Sheet January 18			
Assets		**Liabilities and Equity**	
Cash	$179,500	Accounts payable	$241,500
Equipment	552,000	London, Capital	98,000
		Ramirez, Capital	220,500
		Toney, Capital	171,500
Total assets	$731,500	Total liabilities and equity	$731,500

Required

Prepare journal entries for (*a*) the sale of equipment, (*b*) the allocation of its gain or loss, (*c*) the payment of liabilities at book value, and (*d*) the distribution of cash in each of the following separate cases: Equipment is sold for (1) $605,400; (2) $474,000; (3) $301,200 and any partners with capital deficits pay in the amount of their deficits; and (4) $271,200 and the partners have no assets other than those invested in the partnership. (Round amounts to the nearest dollar.)

Check (4) Cash distribution: Ramirez, $98,633

SERIAL PROBLEM

Success Systems

(This serial problem began in Chapter 1 and continues through most of the book. If previous chapter segments were not completed, the serial problem can begin at this point. It is helpful, but not necessary, to use the Working Papers that accompany the book.)

SP 12 At the start of 2010, Adriana Lopez is considering adding a partner to her business. She envisions the new partner taking the lead in generating sales of both services and merchandise for Success Systems. Lopez's equity in Success Systems as of January 1, 2010, is reflected in the following capital balance.

A. Lopez, Capital	$90,148

Required

1. Lopez is evaluating whether the prospective partner should be an equal partner with respect to capital investment and profit sharing (1:1) or whether the agreement should be 4:1 with Lopez retaining four-fifths interest with rights to four-fifths of the net income or loss. What factors should she consider in deciding which partnership agreement to offer?
2. Prepare the January 1, 2010, journal entry(ies) necessary to admit a new partner to Success Systems through the purchase of a partnership interest for each of the following two separate cases: (*a*) 1:1 sharing agreement and (*b*) 4:1 sharing agreement.
3. Prepare the January 1, 2010, journal entry(ies) required to admit a new partner if the new partner invests cash of $22,537.
4. After posting the entry in part 3, what would be the new partner's equity percentage?

BEYOND THE NUMBERS

REPORTING IN ACTION

C1 

BTN 12-1 Take a step back in time and imagine Best Buy in its infancy as a company. The year is 1966.

Required

1. Read the history of Best Buy at BestBuymedia.tekgroup.com. Can you determine from the history whether Best Buy was originally organized as a sole proprietorship, partnership, or corporation?
2. Assume that Best Buy was originally organized as a partnership. Best Buy's income statement in Appendix A varies in several key ways from what it would look like for a partnership. Identify at least two ways in which a corporate income statement differs from a partnership income statement.
3. Compare the Best Buy balance sheet in Appendix A to what a partnership balance sheet would have shown. Identify and explain any account differences you would anticipate.

COMPARATIVE ANALYSIS

C1 

BTN 12-2 Over the years Best Buy and Circuit City have evolved into large corporations. Today it is difficult to imagine them as fledgling start-ups. Research each company's history online.

Required

1. Which company is older?
2. Which company started as a partnership?
3. In what years did each company first achieve $1,000,000 in sales?
4. In what years did each company have its first public offering of stock?

ETHICS CHALLENGE

P2 

BTN 12-3 Doctors Maben, Orlando, and Clark have been in a group practice for several years. Maben and Orlando are family practice physicians, and Clark is a general surgeon. Clark receives many referrals for surgery from his family practice partners. Upon the partnership's original formation, the three doctors agreed to a two-part formula to share income. Every month each doctor receives a salary allowance of $3,000. Additional income is divided according to a percent of patient charges the doctors generate for the month. In the current month, Maben generated 10% of the billings, Orlando 30%, and Clark 60%. The group's income for this month is $50,000. Clark has expressed dissatisfaction with the income-sharing formula and asks that income be split entirely on patient charge percents.

Required

1. Compute the income allocation for the current month using the original agreement.
2. Compute the income allocation for the current month using Clark's proposed agreement.
3. Identify the ethical components of this partnership decision for the doctors.

COMMUNICATING IN PRACTICE

C1

BTN 12-4 Assume that you are studying for an upcoming accounting exam with a good friend. Your friend says that she has a solid understanding of general partnerships but is less sure that she understands organizations that combine certain characteristics of partnerships with other forms of business organization. You offer to make some study notes for your friend to help her learn about limited partnerships, limited liability partnerships, S corporations, and limited liability companies. Prepare a one-page set of well-organized, complete study notes on these four forms of business organization.

TAKING IT TO THE NET

P1 P2 

BTN 12-5 Access the March 6, 2007, filing of the December 31, 2006, 10-K of **America First Tax Exempt Investors LP**. This company deals with tax-exempt mortgage revenue bonds that, among other things, finance student housing properties.

1. Locate its December 31, 2006, balance sheet and list the account titles reported in the equity section of the balance sheet.
2. Locate its statement of partners' capital and comprehensive income (loss). How many units of limited partnership (known as "beneficial unit certificate holders") are outstanding at December 31, 2006?
3. What is the partnership's largest asset and its amount at December 31, 2006?

TEAMWORK IN ACTION

P2

BTN 12-6 This activity requires teamwork to reinforce understanding of accounting for partnerships.

Required

1. Assume that Baker, Warner, and Rice form the BWR Partnership by making capital contributions of $200,000, $300,000, and $500,000, respectively. BWR predicts annual partnership net income of $600,000. The partners are considering various plans for sharing income and loss. Assign a different team member to compute how the projected $600,000 income would be shared under each of the following separate plans:
   a. Shared equally.
   b. In the ratio of the partners' initial capital investments.
   c. Salary allowances of $50,000 to Baker, $60,000 to Warner, and $70,000 to Rice, with the remaining balance shared equally.
   d. Interest allowances of 10% on the partners' initial capital investments, with the remaining balance shared equally.
2. In sequence, each member is to present his or her income-sharing calculations with the team.
3. As a team, identify and discuss at least one other possible way that income could be shared.

ENTREPRENEURIAL DECISION

C1 

BTN 12-7 Recall the chapter's opening feature involving Samanta and Kelvin Joseph, and their company, **Samanta Shoes**. Assume that Samanta and Kelvin, partners in Samanta Shoes, decide to expand their business with the help of general partners.

Required

1. What details should Samanta, Kelvin, and their future partners specify in the general partnership agreements?
2. What advantages should Samanta, Kelvin, and their future partners be aware of with respect to organizing as a general partnership?
3. What disadvantages should Samanta, Kelvin, and their future partners be aware of with respect to organizing as a general partnership?

GLOBAL DECISION

C1

BTN 12-8 Access **DSG international**'s Website (www.DSGiplc.com) and research the company's history.

1. When was the company founded, and what was its original form of ownership?
2. Why did the company change its name from Dixons to DSG international?
3. What are some of the companies that are part of DSG international?

ANSWERS TO MULTIPLE CHOICE QUIZ

1. e; Capital = $250,000 − $50,000
2. a; $90,000 × [$100,000/($150,000 + $150,000 + $100,000)] = $22,500
3. d;

	Jamison	Blue	Total
Net income			$ 270,000
Salary allowance	$120,000		(120,000)
Interest allowance	60,000	$80,000	(140,000)
Balance of income			10,000
Balance divided equally	5,000	5,000	(10,000)
Totals	$185,000	$85,000	$ 0

4. b; Total partnership equity = $125,000 + $124,000 + $75,000 = $324,000
   Equity of Black = $324,000 × 20% = $64,800
   Bonus to old partners = $75,000 − $64,800 = $10,200, split equally
5. a; $10,500/[($110,000 + $124,000)/2] = 8.97%

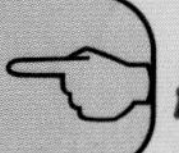

A Look Back

Chapter 12 focused on the partnership form of organization. We described crucial characteristics of partnerships and the accounting and reporting of their important transactions.

A Look at This Chapter

This chapter emphasizes details of the corporate form of organization. The accounting concepts and procedures for equity transactions are explained. We also describe how to report and analyze income, earnings per share, and retained earnings.

A Look Ahead

Chapter 14 focuses on long-term liabilities. We explain how to value, record, amortize, and report these liabilities in financial statements.

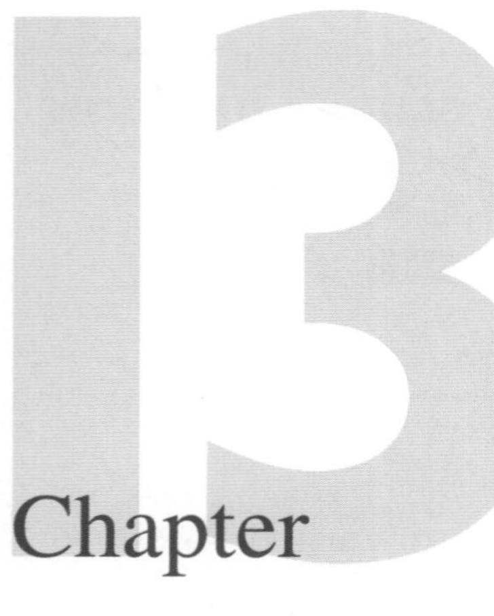

Chapter 13 Accounting for Corporations

Learning Objectives

CAP

Conceptual

C1 Identify characteristics of corporations and their organization. *(p. 506)*

C2 Describe the components of stockholders' equity. *(p. 509)*

C3 Explain characteristics of common and preferred stock. *(p. 517)*

C4 Explain the items reported in retained earnings. *(p. 522)*

Analytical

A1 Compute earnings per share and describe its use. *(p. 524)*

A2 Compute price-earnings ratio and describe its use in analysis. *(p. 525)*

A3 Compute dividend yield and explain its use in analysis. *(p. 525)*

A4 Compute book value and explain its use in analysis. *(p. 526)*

Procedural

P1 Record the issuance of corporate stock. *(p. 510)*

P2 Record transactions involving cash dividends. *(p. 513)*

P3 Account for stock dividends and stock splits. *(p. 514)*

P4 Distribute dividends between common stock and preferred stock. *(p. 517)*

P5 Record purchases and sales of treasury stock and the retirement of stock. *(p. 520)*

LP13

Decision Feature

Breathing New Life

"We weren't planning on starting a company"
—Ali Perry

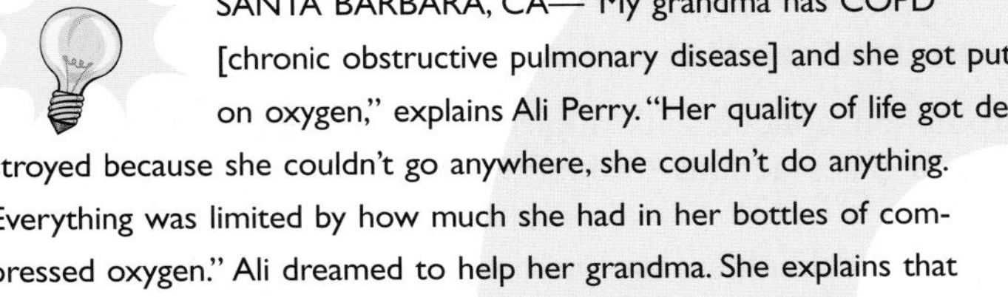

SANTA BARBARA, CA—"My grandma has COPD [chronic obstructive pulmonary disease] and she got put on oxygen," explains Ali Perry. "Her quality of life got destroyed because she couldn't go anywhere, she couldn't do anything. Everything was limited by how much she had in her bottles of compressed oxygen." Ali dreamed to help her grandma. She explains that COPD is the fourth leading cause of death, and is predicted to soon be third on that list. COPD is a disease where the airways of the lungs narrow, which limits the flow of air and causes shortness of breath.

To make her dream a reality, Ali enlisted the aid of two college classmates, Byron Myers and Brenton Taylor. The three of them designed a portable oxygen system, wrote a business plan, and set off to secure financing. Their company, named Inogen, which is a combination of the words innovation and oxygen **(Inogen.net),** soon had a portable oxygen supply unit whose sales exceed 10,000 units to date. "We worked really hard at the technology," explains Ali. "We saw the value of the company was in creating technology that was thought impossible in the marketplace."

The three founders insist that proper financing was a key to their success. To make it happen, says Ali, they needed equity (stock) financing. With their business plan and prototype, the three raised a whopping $4 million from a venture capital firm. Still, explains Ali, the focus is on helping folks, including her grandma. Adds Byron, "Oxygen users can now take off on a moment's notice, without having to watch the clock or guess at how long their oxygen will last."

Their equity financing "brings both opportunities and challenges," explains Byron. "New patients do not know anything about oxygen therapy. All they know is that their life has changed and they now need to have a supply of oxygen with them whenever and wherever." Inogen answers that call. Their focus on people continues to reap rewards as they recently secured another $22 million in equity financing. As Byron put it: "[Inogen] makes old ways of thinking and operating inadequate."

[Sources: *Inogen Website,* January 2009; *HME Business,* January 2006; *Daily Nexus,* January 2004; *Goleta Valley Voice,* December 2003; *Inc.com,* July 2007]

Chapter Preview

This chapter focuses on equity transactions. The first part of the chapter describes the basics of the corporate form of organization and explains the accounting for common and preferred stock. We then focus on several special financing transactions, including cash and stock dividends, stock splits, and treasury stock. The final section considers accounting for retained earnings, including prior period adjustments, retained earnings restrictions, and reporting guidelines.

Accounting for Corporations

- **Corporations**
  - Characteristics
  - Organization and management
  - Stockholders
  - Stock basics
- **Common Stock**
  - Par value
  - No-par value
  - Stated value
  - Stock for noncash assets
- **Dividends**
  - Cash dividends
  - Stock dividends
  - Stock splits
- **Preferred Stock**
  - Issuance
  - Dividend preferences
  - Convertible preferred
  - Callable preferred
- **Treasury Stock**
  - Purchasing treasury stock
  - Reissuing treasury stock
  - Retiring stock
- **Reporting on Equity**
  - Statement of retained earnings
  - Statement of stockholders' equity
  - Stock options

Corporate Form of Organization

Video13.1

A **corporation** is an entity created by law that is separate from its owners. It has most of the rights and privileges granted to individuals. Owners of corporations are called *stockholders* or *shareholders.* Corporations can be separated into two types. A *privately held* (or *closely held*) corporation does not offer its stock for public sale and usually has few stockholders. A *publicly held* corporation offers its stock for public sale and can have thousands of stockholders. *Public sale* usually refers to issuance and trading on an organized stock market.

Characteristics of Corporations

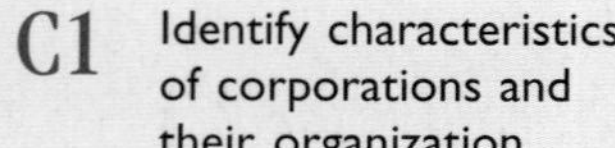

C1 Identify characteristics of corporations and their organization.

Corporations represent an important type of organization. Their unique characteristics offer advantages and disadvantages.

Advantages of Corporate Characteristics

- **Separate legal entity:** A corporation conducts its affairs with the same rights, duties, and responsibilities of a person. It takes actions through its agents, who are its officers and managers.
- **Limited liability of stockholders:** Stockholders are neither liable for corporate acts nor corporate debt.
- **Transferable ownership rights:** The transfer of shares from one stockholder to another usually has no effect on the corporation or its operations except when this causes a change in the directors who control or manage the corporation.
- **Continuous life:** A corporation's life continues indefinitely because it is not tied to the physical lives of its owners.
- **Lack of mutual agency for stockholders:** A corporation acts through its agents, who are its officers and managers. Stockholders, who are not its officers and managers, do not have the power to bind the corporation to contracts—referred to as *lack of mutual agency.*
- **Ease of capital accumulation:** Buying stock is attractive to investors because (1) stockholders are not liable for the corporation's acts and debts, (2) stocks usually are transferred easily, (3) the life of the corporation is unlimited, and (4) stockholders are not corporate agents. These advantages enable corporations to accumulate large amounts of capital from the combined investments of many stockholders.

Point: The *business entity assumption* requires a corporation to be accounted for separately from its owners (shareholders).

Global: U.S., U.K., and Canadian corporations finance much of their operations with stock issuances, but companies in countries such as France, Germany, and Japan finance mainly with note and bond issuances.

Disadvantages of Corporate Characteristics

- **Government regulation:** A corporation must meet requirements of a state's incorporation laws, which subject the corporation to state regulation and control. Proprietorships and partnerships avoid many of these regulations and governmental reports.
- **Corporate taxation:** Corporations are subject to the same property and payroll taxes as proprietorships and partnerships plus *additional* taxes. The most burdensome of these are federal and state income taxes that together can take 40% or more of corporate pretax income. Moreover, corporate income is usually taxed a second time as part of stockholders' personal income when they receive cash distributed as dividends. This is called *double taxation.* (The usual dividend tax is 15%; however, it is less than 15% for lower income taxpayers, and in some cases zero.)

Point: Proprietorships and partnerships are not subject to income taxes. Their income is taxed as the personal income of their owners.

Point: Double taxation is less severe when a corporation's owner-manager collects a salary that is taxed only once as part of his or her personal income.

Decision Insight

Stock Financing Marc Andreessen cofounded **Netscape** at age 22, only four months after earning his degree. One year later, he and friends issued Netscape shares to the public. The stock soared, making Andreessen a multimillionaire.

Corporate Organization and Management

This section describes the incorporation, costs, and management of corporate organizations.

Point: A corporation is not required to have an office in its state of incorporation.

Incorporation A corporation is created by obtaining a charter from a state government. A charter application usually must be signed by the prospective stockholders called *incorporators* or *promoters* and then filed with the proper state official. When the application process is complete and fees paid, the charter is issued and the corporation is formed. Investors then purchase the corporation's stock, meet as stockholders, and elect a board of directors. Directors oversee a corporation's affairs.

Organization Expenses **Organization expenses** (also called *organization costs*) are the costs to organize a corporation; they include legal fees, promoters' fees, and amounts paid to obtain a charter. The corporation records (debits) these costs to an expense account called *Organization Expenses*. Organization costs are expensed as incurred because it is difficult to determine the amount and timing of their future benefits.

Management of a Corporation The ultimate control of a corporation rests with stockholders who control a corporation by electing its *board of directors,* or simply, *directors.* Each stockholder usually has one vote for each share of stock owned. This control relation is shown in Exhibit 13.1. Directors are responsible for and have final authority for managing corporate activities. A board can act only as a collective body and usually limits its actions to setting general policy.

EXHIBIT 13.1

Corporate Structure

Stockholders
↓
Board of Directors
↓
President, Vice President, and Other Officers
↓
Employees of the Corporation

A corporation usually holds a stockholder meeting at least once a year to elect directors and transact business as its bylaws require. A group of stockholders owning or controlling votes of more than a 50% share of a corporation's stock can elect the board and control the corporation. Stockholders who do not attend stockholders' meetings must have an opportunity to delegate their voting rights to an agent by signing a **proxy,** a document that gives a designated agent the right to vote the stock.

Point: *Bylaws* are guidelines that govern the behavior of individuals employed by and managing the corporation.

Day-to-day direction of corporate business is delegated to executive officers appointed by the board. A corporation's chief executive officer (CEO) is often its president. Several vice

Global: Some corporate labels are:

Country	Label
United States	Inc.
France	SA
United Kingdom	
Public	PLC
Private	LTD
Germany	
Public	AG
Private	GmbH
Sweden	AB
Italy	SpA

presidents, who report to the president, are commonly assigned specific areas of management responsibility such as finance, production, and marketing. One person often has the dual role of chairperson of the board of directors and CEO. In this case, the president is usually designated the chief operating officer (COO).

Decision Insight

Seed Money Sources for start-up money include (1) "angel" investors such as family, friends, or anyone who believes in a company, (2) employees, investors, and even suppliers who can be paid with stock, and (3) venture capitalists (investors) who have a record of entrepreneurial success. See the National Venture Capital Association (NVCA.org) for information.

Stockholders of Corporations

This section explains stockholder rights, stock purchases and sales, and the role of registrar and transfer agents.

Rights of Stockholders When investors buy stock, they acquire all *specific* rights the corporation's charter grants to stockholders. They also acquire *general* rights granted stockholders by the laws of the state in which the company is incorporated. When a corporation has only one class of stock, it is identified as **common stock.** State laws vary, but common stockholders usually have the general right to

1. Vote at stockholders' meetings.
2. Sell or otherwise dispose of their stock.
3. Purchase their proportional share of any common stock later issued by the corporation. This **preemptive right** protects stockholders' proportionate interest in the corporation. For example, a stockholder who owns 25% of a corporation's common stock has the first opportunity to buy 25% of any new common stock issued.
4. Receive the same dividend, if any, on each common share of the corporation.
5. Share in any assets remaining after creditors and preferred stockholders are paid when, and if, the corporation is liquidated. Each common share receives the same amount.

Stockholders also have the right to receive timely financial reports.

Stock Certificates and Transfer Investors who buy a corporation's stock, sometimes receive a *stock certificate* as proof of share ownership. Many corporations issue only one certificate for each block of stock purchased. A certificate can be for any number of shares. Exhibit 13.2 shows a stock certificate of the **Green Bay Packers.** A certificate shows the company name, stockholder name, number of shares, and other crucial information. Issuance of certificates is becoming less common. Instead, many stockholders maintain accounts with the corporation or their stockbrokers and never receive actual certificates.

EXHIBIT 13.2

Stock Certificate

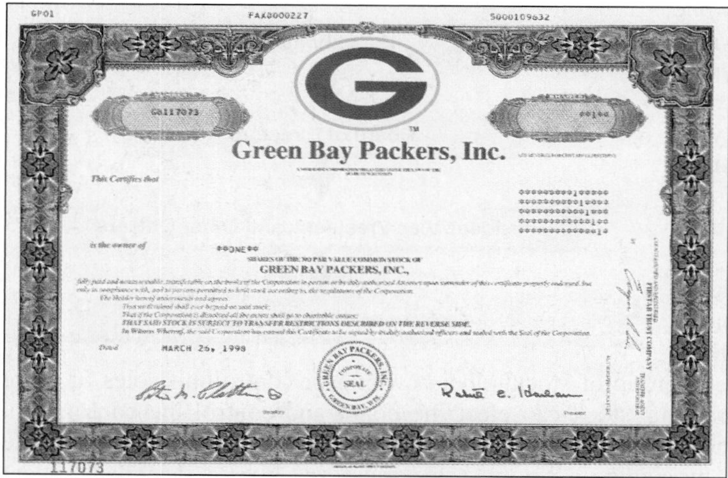

Registrar and Transfer Agents If a corporation's stock is traded on a major stock exchange, the corporation must have a registrar and a transfer agent. A *registrar* keeps stockholder records and prepares official lists of stockholders for stockholder meetings and dividend payments. A *transfer agent* assists with purchases and sales of shares by receiving and issuing certificates as necessary. Registrars and transfer agents are usually large banks or trust companies with computer facilities and staff to do this work.

Decision Insight

Pricing Stock A prospectus accompanies a stock's initial public offering (IPO), giving financial information about the company issuing the stock. A prospectus should help answer these questions to price an IPO: (1) Is the underwriter reliable? (2) Is there growth in revenues, profits, and cash flows? (3) What is management's view of operations? (4) Are current owners selling? (5) What are the risks?

Basics of Capital Stock

C2 Describe the components of stockholders' equity.

Capital stock is a general term that refers to any shares issued to obtain capital (owner financing). This section introduces terminology and accounting for capital stock.

Authorized Stock **Authorized stock** is the number of shares that a corporation's charter allows it to sell. The number of authorized shares usually exceeds the number of shares issued (and outstanding), often by a large amount. (*Outstanding stock* refers to issued stock held by stockholders.) No formal journal entry is required for stock authorization. A corporation must apply to the state for a change in its charter if it wishes to issue more shares than previously authorized. A corporation discloses the number of shares authorized in the equity section of its balance sheet or notes. Best Buy's balance sheet in Appendix A reports 1 billion shares authorized as of the start of its 2008 fiscal year.

Subcategories of Authorized Stock

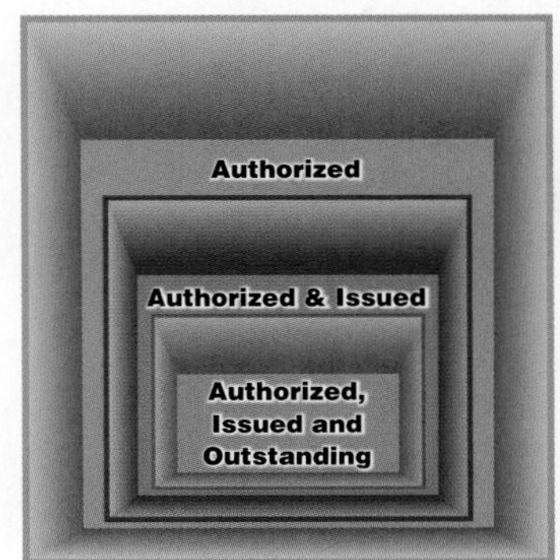

Selling (Issuing) Stock A corporation can sell stock directly or indirectly. To *sell directly,* it advertises its stock issuance to potential buyers. This type of issuance is most common with privately held corporations. To *sell indirectly,* a corporation pays a brokerage house (investment banker) to issue its stock. Some brokerage houses *underwrite* an indirect issuance of stock; that is, they buy the stock from the corporation and take all gains or losses from its resale.

Market Value of Stock **Market value per share** is the price at which a stock is bought and sold. Expected future earnings, dividends, growth, and other company and economic factors influence market value. Traded stocks' market values are available daily in newspapers such as *The Wall Street Journal* and online. The current market value of previously issued shares (for example, the price of stock in trades between investors) does not impact the issuing corporation's stockholders' equity.

Classes of Stock When all authorized shares have the same rights and characteristics, the stock is called *common stock.* A corporation is sometimes authorized to issue more than one class of stock, including preferred stock and different classes of common stock. American Greetings, for instance, has two types of common stock: Class A stock has 1 vote per share and Class B stock has 10 votes per share.

Par Value Stock **Par value stock** is stock that is assigned a **par value,** which is an amount assigned per share by the corporation in its charter. For example, Best Buy's common stock has a par value of $0.10. Other commonly assigned par values are $10, $5, $1 and $0.01. There is no restriction on the assigned par value. In many states, the par value of a stock establishes **minimum legal capital,** which refers to the least amount that the buyers of stock must contribute to the corporation or be subject to paying at a future date. For example, if a corporation issues 1,000 shares of $10 par value stock, the corporation's minimum legal capital in these states would be $10,000. Minimum legal capital is intended to protect a corporation's creditors. Since creditors cannot demand payment from stockholders' personal assets, their claims are limited to the corporation's assets and any minimum legal capital. At liquidation, creditor claims are paid before any amounts are distributed to stockholders.

Point: Managers are motivated to set a low par value when minimum legal capital or state issuance taxes are based on par value.

Point: Minimum legal capital was intended to protect creditors by requiring a minimum level of net assets.

No-Par Value Stock **No-par value stock,** or simply *no-par stock,* is stock *not* assigned a value per share by the corporate charter. Its advantage is that it can be issued at any price without the possibility of a minimum legal capital deficiency.

Point: Par, no-par, and stated value do *not* set the stock's market value.

EXHIBIT 13.3
Equity Composition

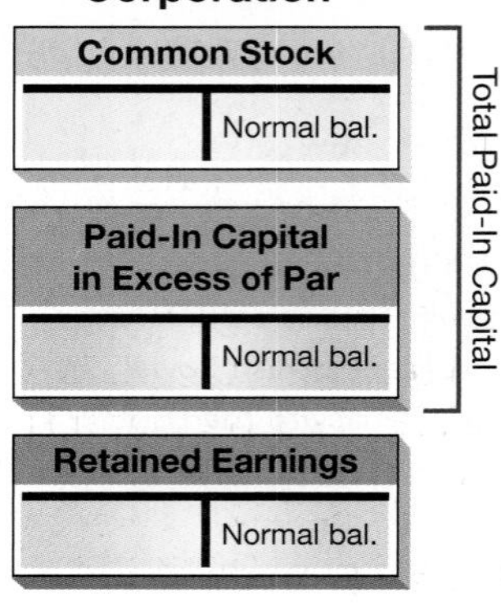

Point: Paid-in capital comes from stock-related transactions, whereas retained earnings comes from operations.

Stated Value Stock **Stated value stock** is no-par stock to which the directors assign a "stated" value per share. Stated value per share becomes the minimum legal capital per share in this case.

Stockholders' Equity A corporation's equity is known as **stockholders' equity,** also called *shareholders' equity* or *corporate capital.* Stockholders' equity consists of (1) paid-in (or contributed) capital and (2) retained earnings; see Exhibit 13.3. **Paid-in capital** is the total amount of cash and other assets the corporation receives from its stockholders in exchange for its stock. **Retained earnings** is the cumulative net income (and loss) not distributed as dividends to its stockholders.

Decision Insight

Stock Quote The Best Buy stock quote is interpreted as (left to right): **Hi,** highest price in past 52 weeks; **Lo,** lowest price in past 52 weeks; **Sym,** company exchange symbol; **Div,** dividends paid per share in past year; **Yld %,** dividend divided by closing price; **PE,** stock price per share divided by earnings per share; **Vol mil.,** number (in millions) of shares traded; **Hi,** highest price for the day; **Lo,** lowest price for the day; **Close,** closing price for the day; **Net Chg,** change in closing price from prior day.

52 Weeks				Yld		Vol				Net
Hi	Lo	Sym	Div	%	PE	mil.	Hi	Lo	Close	Chg
54.15	41.85	BBY	0.13	0.98	19	7.2	53.14	52.36	52.91	+0.20

Quick Check

Answers—p. 531

1. Which of the following is *not* a characteristic of the corporate form of business? (*a*) Ease of capital accumulation, (*b*) Stockholder responsibility for corporate debts, (*c*) Ease in transferability of ownership rights, or (*d*) Double taxation.
2. Why is a corporation's income said to be taxed twice?
3. What is a proxy?

Common Stock

P1 Record the issuance of corporate stock.

Accounting for the issuance of common stock affects only paid-in (contributed) capital accounts; no retained earnings accounts are affected.

Issuing Par Value Stock

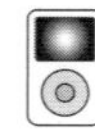

Video13.2

Par value stock can be issued at par, at a premium (above par), or at a discount (below par). In each case, stock can be exchanged for either cash or noncash assets.

Issuing Par Value Stock at Par When common stock is issued at par value, we record amounts for both the asset(s) received and the par value stock issued. To illustrate, the entry to record Dillon Snowboards' issuance of 30,000 shares of $10 par value stock for $300,000 cash on June 5, 2009, follows.

Assets	= Liabilities +	Equity
+300,000		+300,000

June 5	Cash	300,000	
	Common Stock, $10 Par Value		300,000
	Issued 30,000 shares of $10 par value common stock at par.		

Exhibit 13.4 shows the stockholders' equity of Dillon Snowboards at year-end 2009 (its first year of operations) after income of $65,000 and no dividend payments.

EXHIBIT 13.4

Stockholders' Equity for Stock Issued at Par

Stockholders' Equity	
Common Stock—$10 par value; 50,000 shares authorized; 30,000 shares issued and outstanding	$300,000
Retained earnings	65,000
Total stockholders' equity	$365,000

Issuing Par Value Stock at a Premium A **premium on stock** occurs when a corporation sells its stock for more than par (or stated) value. To illustrate, if Dillon Snowboards issues its $10 par value common stock at $12 per share, its stock is sold at a $2 per share premium. The premium, known as **paid-in capital in excess of par value,** is reported as part of equity; it is not revenue and is not listed on the income statement. The entry to record Dillon Snowboards' issuance of 30,000 shares of $10 par value stock for $12 per share on June 5, 2009, follows

Point: A *premium* is the amount by which issue price exceeds par (or stated) value. It is recorded in the "Paid-In Capital in Excess of Par Value, Common Stock" account; also called "Additional Paid-In Capital, Common Stock."

June 5	Cash	360,000	
	Common Stock, $10 Par Value		300,000
	Paid-In Capital in Excess of Par Value, Common Stock		**60,000**
	Sold and issued 30,000 shares of $10 par value common stock at $12 per share.		

Assets	= Liabilities +	Equity
+360,000		+300,000
		+60,000

The Paid-In Capital in Excess of Par Value account is added to the par value of the stock in the equity section of the balance sheet as shown in Exhibit 13.5.

Point: The *Paid-In Capital* terminology is interchangeable with *Contributed Capital.*

EXHIBIT 13.5

Stockholders' Equity for Stock Issued at a Premium

Stockholders' Equity	
Common Stock—$10 par value; 50,000 shares authorized; 30,000 shares issued and outstanding	$300,000
Paid-in capital in excess of par value, common stock	**60,000**
Retained earnings	65,000
Total stockholders' equity	$425,000

Issuing Par Value Stock at a Discount A **discount on stock** occurs when a corporation sells its stock for less than par (or stated) value. Most states prohibit the issuance of stock at a discount. In states that allow stock to be issued at a discount, its buyers usually become contingently liable to creditors for the discount. If stock is issued at a discount, the amount by which issue price is less than par is debited to a *Discount on Common Stock* account, a contra to the common stock account, and its balance is subtracted from the par value of stock in the equity section of the balance sheet. This discount is not an expense and does not appear on the income statement.

Point: Retained earnings can be negative, reflecting accumulated losses. Amazon.com had an accumulated deficit of $1.8 billion at the start of 2007.

Issuing No-Par Value Stock

When no-par stock is issued and is not assigned a stated value, the amount the corporation receives becomes legal capital and is recorded as Common Stock. This means that the entire proceeds are credited to a no-par stock account. To illustrate, a corporation records its October 20 issuance of 1,000 shares of no-par stock for $40 cash per share as follows.

Oct. 20	Cash	40,000	
	Common Stock, No-Par Value		40,000
	Issued 1,000 shares of no-par value common stock at $40 per share.		

Assets	= Liabilities +	Equity
+40,000		+40,000

Frequency of Stock Types

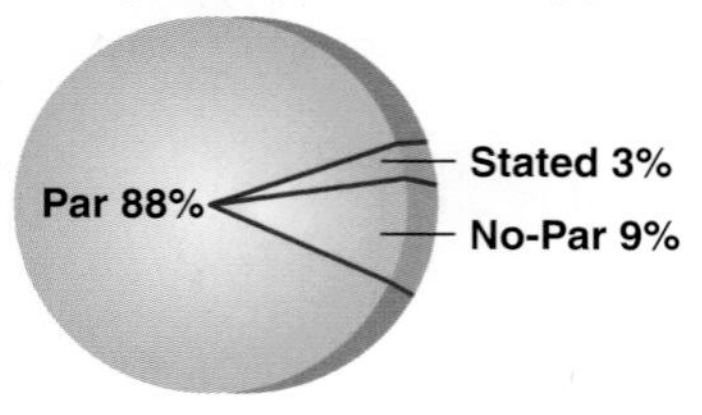

Issuing Stated Value Stock

When no-par stock is issued and assigned a stated value, its stated value becomes legal capital and is credited to a stated value stock account. Assuming that stated value stock is issued at an amount in excess of stated value (the usual case), the excess is credited to Paid-In Capital in Excess of Stated Value, Common Stock, which is reported in the stockholders' equity section. To illustrate, a corporation that issues 1,000 shares of no-par common stock having a stated value of $40 per share in return for $50 cash per share records this as follows.

Assets	= Liabilities	+ Equity
+50,000		+40,000
		+10,000

Oct. 20	Cash	50,000	
	Common Stock, $40 Stated Value		40,000
	Paid-In Capital in Excess of Stated Value, Common Stock		10,000
	Issued 1,000 shares of $40 per share stated value stock at $50 per share.		

Issuing Stock for Noncash Assets

Point: Stock issued for noncash assets should be recorded at the market value of either the stock or the noncash asset, whichever is more clearly determinable.

A corporation can receive assets other than cash in exchange for its stock. (It can also assume liabilities on the assets received such as a mortgage on property received.) The corporation records the assets received at their market values as of the date of the transaction. The stock given in exchange is recorded at its par (or stated) value with any excess recorded in the Paid-In Capital in Excess of Par (or Stated) Value account. (If no-par stock is issued, the stock is recorded at the assets' market value.) To illustrate, the entry to record receipt of land valued at $105,000 in return for issuance of 4,000 shares of $20 par value common stock on June 10 is

Assets	= Liabilities	+ Equity
+105,000		+80,000
		+25,000

June 10	Land	105,000	
	Common Stock, $20 Par Value		80,000
	Paid-In Capital in Excess of Par Value, Common Stock		25,000
	Exchanged 4,000 shares of $20 par value common stock for land.		

Point: Any type of stock can be issued for noncash assets.

A corporation sometimes gives shares of its stock to promoters in exchange for their services in organizing the corporation, which the corporation records as **Organization Expenses.** The entry to record receipt of services valued at $12,000 in organizing the corporation in return for 600 shares of $15 par value common stock on June 5 is

Assets	= Liabilities	+ Equity
		−12,000
		+9,000
		+3,000

June 5	Organization Expenses	12,000	
	Common Stock, $15 Par Value		9,000
	Paid-In Capital in Excess of Par Value, Common Stock		3,000
	Gave promoters 600 shares of $15 par value common stock in exchange for their services.		

Quick Check

Answers—p. 531

4. A company issues 7,000 shares of its $10 par value common stock in exchange for equipment valued at $105,000. The entry to record this transaction includes a credit to (*a*) Paid-In Capital in Excess of Par Value, Common Stock, for $35,000. (*b*) Retained Earnings for $35,000. (*c*) Common Stock, $10 Par Value, for $105,000.
5. What is a premium on stock issuance?
6. Who is intended to be protected by minimum legal capital?

Dividends

This section describes both cash and stock dividend transactions.

P2 Record transactions involving cash dividends.

Cash Dividends

The decision to pay cash dividends rests with the board of directors and involves more than evaluating the amounts of retained earnings and cash. The directors, for instance, may decide to keep the cash to invest in the corporation's growth, to meet emergencies, to take advantage of unexpected opportunities, or to pay off debt. Alternatively, many corporations pay cash dividends to their stockholders at regular dates. These cash flows provide a return to investors and almost always affect the stock's market value.

Video13.2

Accounting for Cash Dividends Dividend payment involves three important dates: declaration, record, and payment. **Date of declaration** is the date the directors vote to declare and pay a dividend. This creates a legal liability of the corporation to its stockholders. **Date of record** is the future date specified by the directors for identifying those stockholders listed in the corporation's records to receive dividends. The date of record usually follows the date of declaration by at least two weeks. Persons who own stock on the date of record receive dividends. **Date of payment** is the date when the corporation makes payment; it follows the date of record by enough time to allow the corporation to arrange checks, money transfers, or other means to pay dividends.

Percent of Corporations Paying Dividends

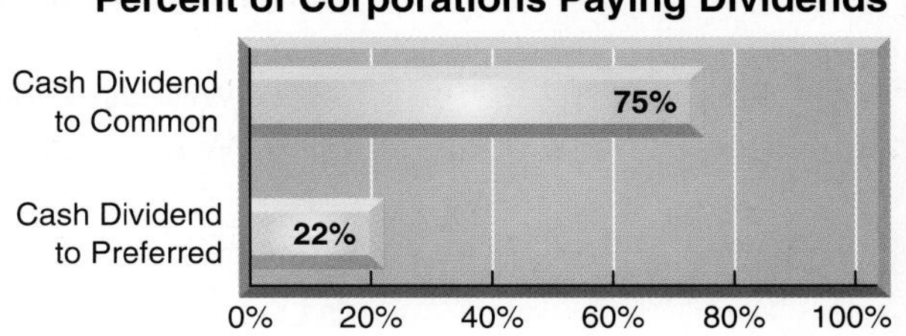

To illustrate, the entry to record a January 9 declaration of a $1 per share cash dividend by the directors of Z-Tech, Inc., with 5,000 outstanding shares is

Date of Declaration

Jan. 9	Retained Earnings	5,000	
	Common Dividend Payable		5,000
	Declared $1 per common share cash dividend.[1]		

Assets = Liabilities + Equity
+5,000 −5,000

Common Dividend Payable is a current liability. The date of record for the Z-Tech dividend is January 22. *No formal journal entry is needed on the date of record.* The February 1 date of payment requires an entry to record both the settlement of the liability and the reduction of the cash balance, as follows:

Date of Payment

Feb. 1	Common Dividend Payable	5,000	
	Cash		5,000
	Paid $1 per common share cash dividend.		

Assets = Liabilities + Equity
−5,000 −5,000

Deficits and Cash Dividends A corporation with a debit (abnormal) balance for retained earnings is said to have a **retained earnings deficit,** which arises when a company incurs cumulative losses and/or pays more dividends than total earnings from current and prior years. A deficit is reported as a deduction on the balance sheet, as shown in Exhibit 13.6. Most states prohibit a corporation with a deficit from paying a cash dividend to its stockholders. This legal restriction is designed to protect creditors by preventing distribution of assets to stockholders when the company may be in financial difficulty.

Point: It is often said a dividend is a distribution of retained earnings, but it is more precise to describe a dividend as a distribution of assets to satisfy stockholder claims.

Point: The Retained Earnings Deficit account is also called *Accumulated Deficit.*

[1] An alternative entry is to debit Dividends instead of Retained Earnings. The balance in Dividends is then closed to Retained Earnings at the end of the reporting period. The effect is the same: Retained Earnings is decreased and a Dividend Payable is increased. For simplicity, all assignments in this chapter use the Retained Earnings account to record dividend declarations.

EXHIBIT 13.6

Stockholders' Equity with a Deficit

Common stock—$10 par value, 5,000 shares authorized, issued, and outstanding	$50,000
Retained earnings deficit	**(6,000)**
Total stockholders' equity	$44,000

Some state laws allow cash dividends to be paid by returning a portion of the capital contributed by stockholders. This type of dividend is called a **liquidating cash dividend,** or simply *liquidating dividend,* because it returns a part of the original investment back to the stockholders. This requires a debit entry to one of the contributed capital accounts instead of Retained Earnings at the declaration date.

Point: **Amazon.com** has never declared a cash dividend.

Quick Check Answers—p. 531

7. What type of an account is the Common Dividend Payable account?
8. What three crucial dates are involved in the process of paying a cash dividend?
9. When does a dividend become a company's legal obligation?

Stock Dividends

P3 Account for stock dividends and stock splits.

A **stock dividend,** declared by a corporation's directors, is a distribution of additional shares of the corporation's own stock to its stockholders without the receipt of any payment in return. Stock dividends and cash dividends are different. A stock dividend does not reduce assets and equity but instead transfers a portion of equity from retained earnings to contributed capital.

Reasons for Stock Dividends Stock dividends exist for at least two reasons. First, directors are said to use stock dividends to keep the market price of the stock affordable. For example, if a corporation continues to earn income but does not issue cash dividends, the price of its common stock likely increases. The price of such a stock may become so high that it discourages some investors from buying the stock (especially in lots of 100 and 1,000). When a corporation has a stock dividend, it increases the number of outstanding shares and lowers the per share stock price. Another reason for a stock dividend is to provide evidence of management's confidence that the company is doing well and will continue to do well.

Accounting for Stock Dividends A stock dividend affects the components of equity by transferring part of retained earnings to contributed capital accounts, sometimes described as *capitalizing* retained earnings. Accounting for a stock dividend depends on whether it is a small or large stock dividend. A **small stock dividend** is a distribution of 25% or less of previously outstanding shares. It is recorded by capitalizing retained earnings for an amount equal to the market value of the shares to be distributed. A **large stock dividend** is a distribution of more than 25% of previously outstanding shares. A large stock dividend is recorded by capitalizing retained earnings for the minimum amount required by state law governing the corporation. Most states require capitalizing retained earnings equal to the par or stated value of the stock.

To illustrate stock dividends, we use the equity section of Quest's balance sheet shown in Exhibit 13.7 just *before* its declaration of a stock dividend on December 31.

EXHIBIT 13.7

Stockholders' Equity *before* Declaring a Stock Dividend

Stockholders' Equity (before dividend)	
Common stock—$10 par value, 15,000 shares authorized, 10,000 shares issued and outstanding	$100,000
Paid-in capital in excess of par value, common stock	8,000
Retained earnings	35,000
Total stockholders' equity	$143,000

Recording a small stock dividend. Assume that Quest's directors declare a 10% stock dividend on December 31. This stock dividend of 1,000 shares, computed as 10% of its 10,000 issued and outstanding shares, is to be distributed on January 20 to the stockholders of record on January 15. Since the market price of Quest's stock on December 31 is $15 per share, this small stock dividend declaration is recorded as follows:

Point: Small stock dividends are recorded at market value.

Date of Declaration

		Debit	Credit
Dec. 31	Retained Earnings	15,000	
	Common Stock Dividend Distributable.		10,000
	Paid-In Capital in Excess of Par Value, Common Stock		5,000
	Declared a 1,000-share (10%) stock dividend.		

Assets = Liabilities + Equity
−15,000
+10,000
+5,000

The $10,000 credit in the declaration entry equals the par value of the shares and is recorded in *Common Stock Dividend Distributable,* an equity account. Its balance exists only until the shares are issued. The $5,000 credit equals the amount by which market value exceeds par value. This amount increases the Paid-In Capital in Excess of Par Value account in anticipation of the issuance of shares. In general, the balance sheet changes in three ways when a stock dividend is declared. First, the amount of equity attributed to common stock increases; for Quest, from $100,000 to $110,000 for 1,000 additional declared shares. Second, paid-in capital in excess of par increases by the excess of market value over par value for the declared shares. Third, retained earnings decreases, reflecting the transfer of amounts to both common stock and paid-in capital in excess of par. The stockholders' equity of Quest is shown in Exhibit 13.8 *after* its 10% stock dividend is declared on December 31—the items impacted are in bold.

Point: The term *Distributable* (not *Payable*) is used for stock dividends. A stock dividend is never a liability because it never reduces assets.

Point: The credit to Paid-In Capital in Excess of Par Value is recorded when the stock dividend is declared. This account is not affected when stock is later distributed.

EXHIBIT 13.8

Stockholders' Equity *after* Declaring a Stock Dividend

Stockholders' Equity (after dividend)	
Common stock—$10 par value, 15,000 shares authorized, 10,000 shares issued and outstanding	$100,000
Common stock dividend distributable—1,000 shares	**10,000**
Paid-in capital in excess of par value, common stock	**13,000**
Retained earnings	**20,000**
Total stockholders' equity	$143,000

No entry is made on the date of record for a stock dividend. On January 20, the date of payment, Quest distributes the new shares to stockholders and records this entry:

Date of Payment

		Debit	Credit
Jan. 20	Common Stock Dividend Distributable	10,000	
	Common Stock, $10 Par Value.		10,000
	To record issuance of common stock dividend.		

Assets = Liabilities + Equity
−10,000
+10,000

The combined effect of these stock dividend entries is to transfer (or capitalize) $15,000 of retained earnings to paid-in capital accounts. The amount of capitalized retained earnings equals the market value of the 1,000 issued shares ($15 × 1,000 shares). A stock dividend has no effect on the ownership percent of individual stockholders.

Point: A stock dividend does not affect assets.

Recording a large stock dividend. A corporation capitalizes retained earnings equal to the minimum amount required by state law for a large stock dividend. For most states, this amount is the par or stated value of the newly issued shares. To illustrate, suppose Quest's board declares a stock dividend of 30% instead of 10% on December 31. Since this dividend is more

Point: Large stock dividends are recorded at par or stated value.

than 25%, it is treated as a large stock dividend. Thus, the par value of the 3,000 dividend shares is capitalized at the date of declaration with this entry:

Assets = Liabilities + Equity
−30,000
+30,000

Date of Declaration

Dec. 31	Retained Earnings	30,000	
	Common Stock Dividend Distributable.		30,000
	Declared a 3,000-share (30%) stock dividend.		

This transaction decreases retained earnings and increases contributed capital by $30,000. On the date of payment the company debits Common Stock Dividend Distributable and credits Common Stock for $30,000. The effects from a large stock dividend on balance sheet accounts are similar to those for a small stock dividend except for the absence of any effect on paid-in capital in excess of par.

Before 5:1 Split: 1 share, $50 par

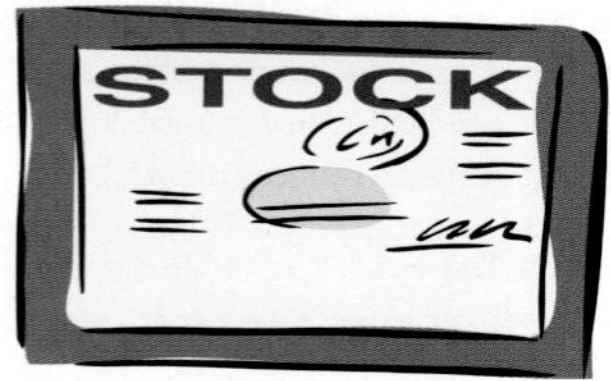

After 5:1 Split: 5 shares, $10 par

Point: Berkshire Hathaway has resisted a stock split. Its recent stock price was $150,000 per share.

Point: A **reverse stock split** is the opposite of a stock split. It increases both the market value per share and the par or stated value per share with a split ratio less than 1-for-1, such as 1-for-2. A reverse split results in fewer shares.

Stock Splits

A **stock split** is the distribution of additional shares to stockholders according to their percent ownership. When a stock split occurs, the corporation "calls in" its outstanding shares and issues more than one new share in exchange for each old share. Splits can be done in any ratio, including 2-for-1, 3-for-1, or higher. Stock splits reduce the par or stated value per share. The reasons for stock splits are similar to those for stock dividends.

To illustrate, CompTec has 100,000 outstanding shares of $20 par value common stock with a current market value of $88 per share. A 2-for-1 stock split cuts par value in half as it replaces 100,000 shares of $20 par value stock with 200,000 shares of $10 par value stock. Market value is reduced from $88 per share to about $44 per share. The split does not affect any equity amounts reported on the balance sheet or any individual stockholder's percent ownership. Both the Paid-In Capital and Retained Earnings accounts are unchanged by a split, and *no journal entry is made*. The only effect on the accounts is a change in the stock account description. CompTec's 2-for-1 split on its $20 par value stock means that after the split, it changes its stock account title to Common Stock, $10 Par Value. This stock's description on the balance sheet also changes to reflect the additional authorized, issued, and outstanding shares and the new par value.

The difference between stock splits and large stock dividends is often blurred. Many companies report stock splits in their financial statements without calling in the original shares by simply changing their par value. This type of "split" is really a large stock dividend and results in additional shares issued to stockholders by capitalizing retained earnings or transferring other paid-in capital to Common Stock. This approach avoids administrative costs of splitting the stock. **Harley-Davidson** recently declared a 2-for-1 stock split executed in the form of a 100% stock dividend.

Decision Maker

Entrepreneur A company you cofounded and own stock in announces a 50% stock dividend. Has the value of your stock investment increased, decreased, or remained the same? Would it make a difference if it was a 3-for-2 stock split executed in the form of a dividend? [Answer—p. 531]

Quick Check

Answers—p. 531

10. How does a stock dividend impact assets and retained earnings?

11. What distinguishes a large stock dividend from a small stock dividend?

12. What amount of retained earnings is capitalized for a small stock dividend?

Preferred Stock

A corporation can issue two basic kinds of stock, common and preferred. **Preferred stock** has special rights that give it priority (or senior status) over common stock in one or more areas. Special rights typically include a preference for receiving dividends and for the distribution of assets if the corporation is liquidated. Preferred stock carries all rights of common stock unless the corporate charter nullifies them. Most preferred stock, for instance, does not confer the right to vote. Exhibit 13.9 shows that preferred stock is issued by about one-fourth of large corporations. All corporations issue common stock.

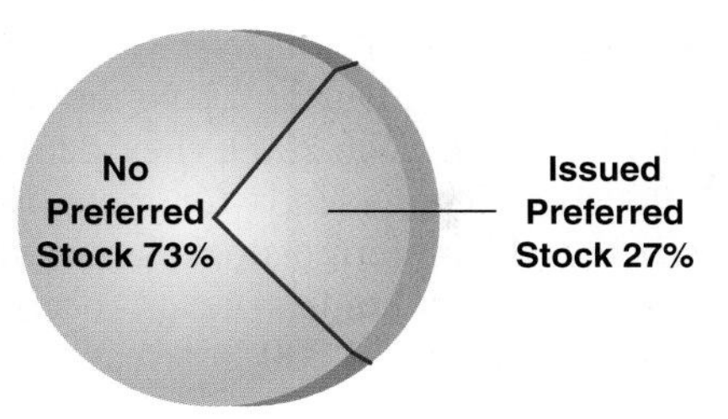

EXHIBIT 13.9

Corporations and Preferred Stock

Issuance of Preferred Stock

C3 Explain characteristics of common and preferred stock.

Preferred stock usually has a par value. Like common stock, it can be sold at a price different from par. Preferred stock is recorded in its own separate capital accounts. To illustrate, if Dillon Snowboards issues 50 shares of $100 par value preferred stock for $6,000 cash on July 1, 2009, the entry is

July 1	Cash	6,000	
	Preferred Stock, $100 Par Value		5,000
	Paid-In Capital in Excess of Par Value, Preferred Stock		1,000
	Issued preferred stock for cash.		

Assets	=	Liabilities	+	Equity
+6,000				+5,000
				+1,000

The equity section of the year-end balance sheet for Dillon Snowboards, including preferred stock, is shown in Exhibit 13.10. (This exhibit assumes that common stock was issued at par.) Issuing no-par preferred stock is similar to issuing no-par common stock. Also, the entries for issuing preferred stock for noncash assets are similar to those for common stock.

EXHIBIT 13.10

Stockholders' Equity with Common and Preferred Stock

Stockholders' Equity	
Common stock—$10 par value; 50,000 shares authorized; 30,000 shares issued and outstanding	$300,000
Preferred stock—$100 par value; 1,000 shares authorized; 50 shares issued and outstanding	**5,000**
Paid-in capital in excess of par value, preferred stock	**1,000**
Retained earnings	65,000
Total stockholders' equity	$371,000

Dividend Preference of Preferred Stock

P4 Distribute dividends between common stock and preferred stock.

Preferred stock usually carries a preference for dividends, meaning that preferred stockholders are allocated their dividends before any dividends are allocated to common stockholders. The dividends allocated to preferred stockholders are usually expressed as a dollar amount per share or a percent applied to par value. A preference for dividends does *not* ensure dividends. If the directors do not declare a dividend, neither the preferred nor the common stockholders receive one.

Point: Dividend preference does not imply that preferred stockholders receive more dividends than common stockholders, nor does it guarantee a dividend.

Cumulative or Noncumulative Dividend Most preferred stocks carry a cumulative dividend right. **Cumulative preferred stock** has a right to be paid both the current and all prior periods' unpaid dividends before any dividend is paid to common stockholders. When preferred stock is cumulative and the directors either do not declare a dividend to preferred stockholders or declare one that does not cover the total amount of cumulative dividend, the unpaid dividend amount is called **dividend in arrears.** Accumulation of dividends in arrears on cumulative preferred stock does not guarantee they will be paid. **Noncumulative preferred stock** confers no right to prior periods' unpaid dividends if they were not declared in those prior periods.

To illustrate the difference between cumulative and noncumulative preferred stock, assume that a corporation's outstanding stock includes (1) 1,000 shares of $100 par, 9% preferred

Video13.2

stock—yielding $9,000 per year in potential dividends, and (2) 4,000 shares of $50 par value common stock. During 2008, the first year of operations, the directors declare cash dividends of $5,000. In year 2009, they declare cash dividends of $42,000. See Exhibit 13.11 for the allocation of dividends for these two years. Allocation of year 2009 dividends depends on whether the preferred stock is noncumulative or cumulative. With noncumulative preferred, the preferred stockholders never receive the $4,000 skipped in 2008. If the preferred stock is cumulative, the $4,000 in arrears is paid in 2009 before any other dividends are paid.

EXHIBIT 13.11

Allocation of Dividends (noncumulative vs. cumulative preferred stock)

	Preferred	Common
Preferred Stock Is Noncumulative		
Year 2008	$ 5,000	$ 0
Year 2009		
Step 1: Current year's preferred dividend	$ 9,000	
Step 2: Remainder to common		$33,000
Preferred Stock Is Cumulative		
Year 2008	$ 5,000	$ 0
Year 2009		
Step 1: Dividend in arrears	$ 4,000	
Step 2: Current year's preferred dividend	9,000	
Step 3: Remainder to common		$29,000
Totals for year 2009	$13,000	$29,000

Example: What dividends do cumulative preferred stockholders receive in 2009 if the corporation paid only $2,000 of dividends in 2008? How does this affect dividends to common stockholders in 2009? *Answers:* $16,000 ($7,000 dividends in arrears, plus $9,000 current preferred dividends). Dividends to common stockholders decrease to $26,000.

A liability for a dividend does not exist until the directors declare a dividend. If a preferred dividend date passes and the corporation's board fails to declare the dividend on its cumulative preferred stock, the dividend in arrears is not a liability. The *full-disclosure principle* requires a corporation to report (usually in a note) the amount of preferred dividends in arrears as of the balance sheet date.

Participating or Nonparticipating Dividend **Nonparticipating preferred stock** has a feature that limits dividends to a maximum amount each year. This maximum is often stated as a percent of the stock's par value or as a specific dollar amount per share. Once preferred stockholders receive this amount, the common stockholders receive any and all additional dividends. **Participating preferred stock** has a feature allowing preferred stockholders to share with common stockholders in any dividends paid in excess of the percent or dollar amount stated on the preferred stock. This participation feature does not apply until common stockholders receive dividends equal to the preferred stock's dividend percent. Many corporations are authorized to issue participating preferred stock but rarely do, and most managers never expect to issue it.[2]

Convertible Preferred Stock

Preferred stock is more attractive to investors if it carries a right to exchange preferred shares for a fixed number of common shares. **Convertible preferred stock** gives holders the option to

[2] Participating preferred stock is usually authorized as a defense against a possible corporate *takeover* by an "unfriendly" investor (or a group of investors) who intends to buy enough voting common stock to gain control. Taking a term from spy novels, the financial world refers to this type of plan as a *poison pill* that a company swallows if enemy investors threaten its capture. A poison pill usually works as follows: A corporation's common stockholders on a given date are granted the right to purchase a large amount of participating preferred stock at a very low price. This right to purchase preferred shares is *not* transferable. If an unfriendly investor buys a large block of common shares (whose right to purchase participating preferred shares does *not* transfer to this buyer), the board can issue preferred shares at a low price to the remaining common shareholders who retained the right to purchase. Future dividends are then divided between the newly issued participating preferred shares and the common shares. This usually transfers value from common shares to preferred shares, causing the unfriendly investor's common stock to lose much of its value and reduces the potential benefit of a hostile takeover.

exchange their preferred shares for common shares at a specified rate. When a company prospers and its common stock increases in value, convertible preferred stockholders can share in this success by converting their preferred stock into more valuable common stock.

Callable Preferred Stock

Callable preferred stock gives the issuing corporation the right to purchase (retire) this stock from its holders at specified future prices and dates. The amount paid to call and retire a preferred share is its **call price,** or *redemption value,* and is set when the stock is issued. The call price normally includes the stock's par value plus a premium giving holders additional return on their investment. When the issuing corporation calls and retires a preferred stock, the terms of the agreement often require it to pay the call price *and* any dividends in arrears.

Point: The issuing corporation has the right, or option, to retire its callable preferred stock.

Decision Insight

IFRSs Like U.S. GAAP, IFRSs require that preferred stock be classified as debt or equity based on analysis of the stock's contractual terms. However, IFRSs use different criteria for such classification.

Reasons for Issuing Preferred Stock

Corporations issue preferred stock for several reasons. One is to raise capital without sacrificing control. For example, suppose a company's organizers have $100,000 cash to invest and organize a corporation that needs $200,000 of capital to start. If they sell $200,000 worth of common stock (with $100,000 to the organizers), they would have only 50% control and would need to negotiate extensively with other stockholders in making policy. However, if they issue $100,000 worth of common stock to themselves and sell outsiders $100,000 of 8%, cumulative preferred stock with no voting rights, they retain control.

A second reason to issue preferred stock is to boost the return earned by common stockholders. To illustrate, suppose a corporation's organizers expect to earn an annual after-tax income of $24,000 on an investment of $200,000. If they sell and issue $200,000 worth of common stock, the $24,000 income produces a 12% return on the $200,000 of common stockholders' equity. However, if they issue $100,000 of 8% preferred stock to outsiders and $100,000 of common stock to themselves, their own return increases to 16% per year, as shown in Exhibit 13.12.

EXHIBIT 13.12

Return to Common Stockholders When Preferred Stock Is Issued

Net (after-tax) income	$24,000
Less preferred dividends at 8%	(8,000)
Balance to common stockholders	$16,000
Return to common stockholders ($16,000/$100,000)	16%

Common stockholders earn 16% instead of 12% because assets contributed by preferred stockholders are invested to earn $12,000 while the preferred dividend is only $8,000. Use of preferred stock to increase return to common stockholders is an example of **financial leverage** (also called *trading on the equity*). As a general rule, when the dividend rate on preferred stock is less than the rate the corporation earns on its assets, the effect of issuing preferred stock is to increase (or *lever*) the rate earned by common stockholders.

Point: Financial leverage also occurs when debt is issued and the interest rate paid on it is less than the rate earned from using the assets the creditors lend the company.

Other reasons for issuing preferred stock include its appeal to some investors who believe that the corporation's common stock is too risky or that the expected return on common stock is too low.

Decision Maker

Concert Organizer Assume that you alter your business strategy from organizing concerts targeted at under 1,000 people to those targeted at between 5,000 to 20,000 people. You also incorporate because of increased risk of lawsuits and a desire to issue stock for financing. It is important that you control the company for decisions on whom to schedule. What types of stock do you offer? [Answer—p. 531]

Quick Check

Answers—p. 531

13. In what ways does preferred stock often have priority over common stock?
14. Increasing the return to common stockholders by issuing preferred stock is an example of (*a*) Financial leverage. (*b*) Cumulative earnings. (*c*) Dividend in arrears.
15. A corporation has issued and outstanding (i) 9,000 shares of $50 par value, 10% cumulative, nonparticipating preferred stock and (ii) 27,000 shares of $10 par value common stock. No dividends have been declared for the two prior years. During the current year, the corporation declares $288,000 in dividends. The amount paid to common shareholders is (*a*) $243,000. (*b*) $153,000. (*c*) $135,000.

Treasury Stock

P5 Record purchases and sales of treasury stock and the retirement of stock.

Corporations acquire shares of their own stock for several reasons: (1) to use their shares to acquire another corporation, (2) to purchase shares to avoid a hostile takeover of the company, (3) to reissue them to employees as compensation, and (4) to maintain a strong market for their stock or to show management confidence in the current price.

Corporations and Treasury Stock

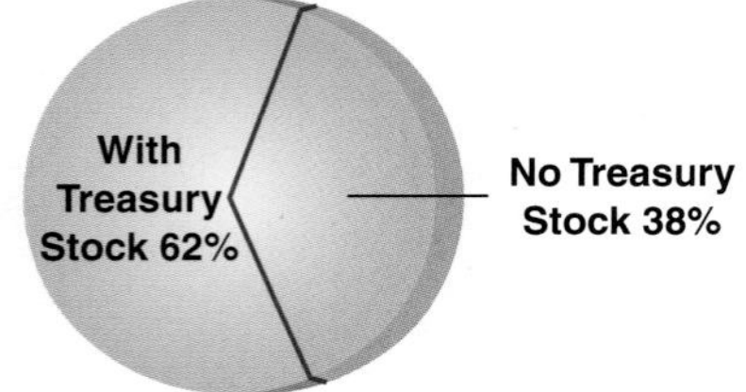

A corporation's reacquired shares are called **treasury stock,** which is similar to unissued stock in several ways: (1) neither treasury stock nor unissued stock is an asset, (2) neither receives cash dividends or stock dividends, and (3) neither allows the exercise of voting rights. However, treasury stock does differ from unissued stock in one major way: The corporation can resell treasury stock at less than par without having the buyers incur a liability, provided it was originally issued at par value or higher. Treasury stock purchases also require management to exercise ethical sensitivity because funds are being paid to specific stockholders instead of all stockholders. Managers must be sure the purchase is in the best interest of all stockholders. These concerns cause companies to fully disclose treasury stock transactions.

Purchasing Treasury Stock

Purchasing treasury stock reduces the corporation's assets and equity by equal amounts. (We describe the *cost method* of accounting for treasury stock, which is the most widely used method. The *par value* method is another method explained in advanced courses.) To illustrate, Exhibit 13.13 shows Cyber Corporation's account balances *before* any treasury stock purchase (Cyber has no liabilities).

EXHIBIT 13.13

Account Balances *before* Purchasing Treasury Stock

Assets		Stockholders' Equity	
Cash	$ 30,000	Common stock—$10 par; 10,000 shares authorized, issued, and outstanding	$100,000
Other assets	95,000	Retained earnings	25,000
Total assets	$125,000	Total stockholders' equity	$125,000

Cyber then purchases 1,000 of its own shares for $11,500 on May 1, which is recorded as follows.

Assets = Liabilities + Equity
−11,500 −11,500

May 1	Treasury Stock, Common	11,500	
	Cash		11,500
	Purchased 1,000 treasury shares at $11.50 per share.		

This entry reduces equity through the debit to the Treasury Stock account, which is a contra equity account. Exhibit 13.14 shows account balances *after* this transaction.

EXHIBIT 13.14

Account Balances *after* Purchasing Treasury Stock

Assets		Stockholders' Equity	
Cash	$ 18,500	Common stock—$10 par; 10,000 shares authorized and issued; 1,000 shares in treasury	$100,000
Other assets	95,000	Retained earnings, $11,500 restricted by treasury stock purchase	25,000
		Less cost of treasury stock	**(11,500)**
Total assets	$113,500	Total stockholders' equity	$113,500

The treasury stock purchase reduces Cyber's cash, total assets, and total equity by $11,500 but does not reduce the balance of either the Common Stock or the Retained Earnings account. The equity reduction is reported by deducting the cost of treasury stock in the equity section. Also, two disclosures are evident. First, the stock description reveals that 1,000 issued shares are in treasury, leaving only 9,000 shares still outstanding. Second, the description for retained earnings reveals that it is partly restricted.

Point: The Treasury Stock account is *not* an asset. Treasury stock does not carry voting or dividend rights.

Point: A treasury stock purchase is also called a *stock buyback.*

Reissuing Treasury Stock

Treasury stock can be reissued by selling it at cost, above cost, or below cost.

Selling Treasury Stock at Cost If treasury stock is reissued at cost, the entry is the reverse of the one made to record the purchase. For instance, if on May 21 Cyber reissues 100 of the treasury shares purchased on May 1 at the same $11.50 per share cost, the entry is

Date	Account	Debit	Credit
May 21	Cash	1,150	
	Treasury Stock, Common		1,150
	Received $11.50 per share for 100 treasury shares costing $11.50 per share.		

Assets	=	Liabilities	+	Equity
+1,150				+1,150

Selling Treasury Stock *above* Cost If treasury stock is sold for more than cost, the amount received in excess of cost is credited to the Paid-In Capital, Treasury Stock account. This account is reported as a separate item in the stockholders' equity section. No gain is ever reported from the sale of treasury stock. To illustrate, if Cyber receives $12 cash per share for 400 treasury shares costing $11.50 per share on June 3, the entry is

Point: Treasury stock does not represent ownership. A company cannot own a part of itself.

Date	Account	Debit	Credit
June 3	Cash	4,800	
	Treasury Stock, Common		4,600
	Paid-In Capital, Treasury Stock		**200**
	Received $12 per share for 400 treasury shares costing $11.50 per share.		

Assets	=	Liabilities	+	Equity
+4,800				+4,600
				+200

Selling Treasury Stock *below* Cost When treasury stock is sold below cost, the entry to record the sale depends on whether the Paid-In Capital, Treasury Stock account has a credit balance. If it has a zero balance, the excess of cost over the sales price is debited to Retained Earnings. If the Paid-In Capital, Treasury Stock account has a credit balance, it is debited for the excess of the cost over the selling price but not to exceed the balance in this account. When the credit balance in this paid-in capital account is eliminated, any remaining difference between the cost and selling price is debited to Retained Earnings. To illustrate, if Cyber sells its remaining 500 shares of treasury stock at $10 per share on July 10,

Point: The phrase *treasury stock* is believed to arise from the fact that reacquired stock is held in a corporation's treasury.

Point: The Paid-In Capital, Treasury Stock account can have a zero or credit balance but never a debit balance.

equity is reduced by $750 (500 shares × $1.50 per share excess of cost over selling price), as shown in this entry:

Assets	= Liabilities	+ Equity
+5,000		−200
		−550
		+5,750

July 10	Cash	5,000	
	Paid-In Capital, Treasury Stock	**200**	
	Retained Earnings	**550**	
	Treasury Stock, Common		5,750
	Received $10 per share for 500 treasury shares costing $11.50 per share.		

This entry eliminates the $200 credit balance in the paid-in capital account created on June 3 and then reduces the Retained Earnings balance by the remaining $550 excess of cost over selling price. A company never reports a loss (or gain) from the sale of treasury stock.

Retiring Stock

Point: Recording stock retirement results in canceling the equity from the original issuance of the shares.

A corporation can purchase its own stock and retire it. Retiring stock reduces the number of issued shares. Retired stock is the same as authorized and unissued shares. Purchases and retirements of stock are permissible under state law only if they do not jeopardize the interests of creditors and stockholders. When stock is purchased for retirement, we remove all capital amounts related to the retired shares. If the purchase price exceeds the net amount removed, this excess is debited to Retained Earnings. If the net amount removed from all capital accounts exceeds the purchase price, this excess is credited to the Paid-In Capital from Retirement of Stock account. A company's assets and equity are always reduced by the amount paid for the retiring stock.

Quick Check

Answers—p. 531

16. Purchase of treasury stock (*a*) has no effect on assets; (*b*) reduces total assets and total equity by equal amounts; or (*c*) is recorded with a debit to Retained Earnings.

17. Southern Co. purchases shares of Northern Corp. Should either company classify these shares as treasury stock?

18. How does treasury stock affect the authorized, issued, and outstanding shares?

19. When a company purchases treasury stock, (*a*) retained earnings are restricted by the amount paid; (*b*) Retained Earnings is credited; or (*c*) it is retired.

Reporting of Equity

Statement of Retained Earnings

C4 Explain the items reported in retained earnings.

Retained earnings generally consist of a company's cumulative net income less any net losses and dividends declared since its inception. Retained earnings are part of stockholders' claims on the company's net assets, but this does *not* imply that a certain amount of cash or other assets is available to pay stockholders. For example, **Circuit City** has $1,336 million in retained earnings, but only $141 million in cash. This section describes events and transactions affecting retained earnings and how retained earnings are reported.

Video13.1

Restrictions and Appropriations The term **restricted retained earnings** refers to both statutory and contractual restrictions. A common *statutory* (or *legal*) *restriction* is to limit treasury stock purchases to the amount of retained earnings. The balance sheet in Exhibit 13.14 provides an example. A common *contractual restriction* involves loan agreements that restrict paying dividends beyond a specified amount or percent of retained earnings. Restrictions are usually described in the notes. The term **appropriated retained earnings** refers to a voluntary transfer of amounts from the Retained Earnings account to the Appropriated Retained Earnings account to inform users of special activities that require funds.

Prior Period Adjustments **Prior period adjustments** are corrections of material errors in prior period financial statements. These errors include arithmetic mistakes, unacceptable accounting, and missed facts. Prior period adjustments are reported in the *statement of retained earnings* (or the statement of stockholders' equity), net of any income tax effects. Prior period adjustments result in changing the beginning balance of retained earnings for events occurring prior to the earliest period reported in the current set of financial statements. To illustrate, assume that ComUS makes an error in a 2007 journal entry for the purchase of land by incorrectly debiting an expense account. When this is discovered in 2009, the statement of retained earnings includes a prior period adjustment, as shown in Exhibit 13.15. This exhibit also shows the usual format of the statement of retained earnings.

Point: If a year 2007 error is discovered in 2008, the company records the adjustment in 2008. But if the financial statements include 2007 and 2008 figures, the statements report the correct amounts for 2007, and a note describes the correction.

EXHIBIT 13.15

Statement of Retained Earnings with a Prior Period Adjustment

ComUS Statement of Retained Earnings For Year Ended December 31, 2009	
Retained earnings, Dec. 31, 2008, as previously reported	$4,745,000
Prior period adjustment	
Cost of land incorrectly expensed (net of $63,000 income taxes)	**147,000**
Retained earnings, Dec. 31, 2008, as adjusted	4,892,000
Plus net income	1,224,300
Less cash dividends declared	(301,800)
Retained earnings, Dec. 31, 2009	$5,814,500

Many items reported in financial statements are based on estimates. Future events are certain to reveal that some of these estimates were inaccurate even when based on the best data available at the time. These inaccuracies are *not* considered errors and are *not* reported as prior period adjustments. Instead, they are identified as **changes in accounting estimates** and are accounted for in current and future periods. To illustrate, we know that depreciation is based on estimated useful lives and salvage values. As time passes and new information becomes available, managers may need to change these estimates and the resulting depreciation expense for current and future periods.

Point: Accounting for changes in estimates is sometimes criticized as two wrongs to make a right. Consider a change in an asset's life. Depreciation neither before nor after the change is the amount computed if the revised estimate were originally selected. Regulators chose this approach to avoid restating prior period numbers.

Closing Process The closing process was explained earlier in the book as: (1) Close credit balances in revenue accounts to Income Summary, (2) Close debit balances in expense accounts to Income Summary, and (3) Close Income Summary to Retained Earnings. If dividends are recorded in a Dividends account, and not as an immediate reduction to Retained Earnings (as shown in this chapter), a fourth step is necessary to close the Dividends account to Retained Earnings.

Statement of Stockholders' Equity

Instead of a separate statement of retained earnings, companies commonly report a statement of stockholders' equity that includes changes in retained earnings. A **statement of stockholders' equity** lists the beginning and ending balances of key equity accounts and describes the changes that occur during the period. The companies in Appendix A report such a statement. The usual format is to provide a column for each component of equity and use the rows to describe events occurring in the period. Exhibit 13.16 shows a condensed statement for Apple.

Reporting Stock Options

The majority of corporations whose shares are publicly traded issue **stock options,** which are rights to purchase common stock at a fixed price over a specified period. As the stock's price rises, the option's value increases. Starbucks and Home Depot offer stock options to both full- and part-time employees. Stock options are said to motivate managers and employees to (1) focus on company performance, (2) take a long-run perspective, and (3) remain with

EXHIBIT 13.16

Statement of Stockholders' Equity

APPLE
Statement of Stockholders' Equity

($ millions, shares in thousands)	Common Stock Shares	Common Stock Amount	Retained Earnings	Other	Total Equity
Balance, Sept. 30, 2006	**855,263**	**$4,355**	**$5,607**	**$22**	**$9,984**
Net income	—	—	3,496	—	3,496
Issuance of Common Stock	17,066	364	(2)	—	362
Other	—	649	—	41	690
Cash Dividends ($0.00 per share)	—	—	—	—	—
Balance, Sept. 29, 2007	**872,329**	**$5,368**	**$9,101**	**$63**	**$14,532**

the company. A stock option is like having an investment with no risk ("a carrot with no stick").

To illustrate, Quantum grants each of its employees the option to purchase 100 shares of its $1 par value common stock at its current market price of $50 per share anytime within the next 10 years. If the stock price rises to $70 per share, an employee can exercise the option at a gain of $20 per share (acquire a $70 stock at the $50 option price). With 100 shares, a single employee would have a total gain of $2,000, computed as $20 × 100 shares. Companies report the cost of stock options in the income statement. Measurement of this cost is explained in advanced courses.

Video13.1

Decision Analysis

Earnings per Share, Price-Earnings Ratio, Dividend Yield, and Book Value per Share

Earnings per Share

A1 Compute earnings per share and describe its use.

The income statement reports **earnings per share,** also called *EPS* or *net income per share,* which is the amount of income earned per each share of a company's outstanding common stock. The **basic earnings per share** formula is shown in Exhibit 13.17. When a company has no preferred stock, then preferred dividends are zero. The weighted-average common shares outstanding is measured over the income reporting period; its computation is explained in advanced courses.

EXHIBIT 13.17

Basic Earnings per Share

$$\textbf{Basic earnings per share} = \frac{\textbf{Net income} - \textbf{Preferred dividends}}{\textbf{Weighted-average common shares outstanding}}$$

To illustrate, assume that Quantum Co. earns $40,000 net income in 2009 and declares dividends of $7,500 on its noncumulative preferred stock. (If preferred stock is *non*cumulative, the income available [numerator] is the current period net income less any preferred dividends *declared* in that same period. If preferred stock is cumulative, the income available [numerator] is the current period net income less the preferred dividends whether declared or not.) Quantum has 5,000 weighted-average common shares outstanding during 2009. Its basic EPS[3] is

$$\text{Basic earnings per share} = \frac{\$40{,}000 - \$7{,}500}{5{,}000 \text{ shares}} = \$6.50$$

[3] A corporation can be classified as having either a simple or complex capital structure. The term **simple capital structure** refers to a company with only common stock and nonconvertible preferred stock outstanding. The term **complex capital structure** refers to companies with dilutive securities. **Dilutive securities** include options, rights to purchase common stock, and any bonds or preferred stock that are convertible into common stock. A company with a complex capital structure must often report two EPS figures: basic and diluted. **Diluted earnings per share** is computed by adding all dilutive securities to the denominator of the basic EPS computation. It reflects the decrease in basic EPS *assuming* that all dilutive securities are converted into common shares.

Price-Earnings Ratio

A2 Compute price-earnings ratio and describe its use in analysis.

A stock's market value is determined by its *expected* future cash flows. A comparison of a company's EPS and its market value per share reveals information about market expectations. This comparison is traditionally made using a **price-earnings (or PE) ratio,** expressed also as *price earnings, price to earnings,* or *PE*. Some analysts interpret this ratio as what price the market is willing to pay for a company's current earnings stream. Price-earnings ratios can differ across companies that have similar earnings because of either higher or lower expectations of future earnings. The price-earnings ratio is defined in Exhibit 13.18.

Point: The average PE ratio of stocks in the 1950–2008 period is about 14.

EXHIBIT 13.18

Price-Earnings Ratio

$$\textbf{Price-earnings ratio} = \frac{\textbf{Market value (price) per share}}{\textbf{Earnings per share}}$$

This ratio is often computed using EPS from the most recent period (for Amazon, its PE is 31; for Altria, its PE is 13). However, many users compute this ratio using *expected* EPS for the next period.

Some analysts view stocks with high PE ratios (higher than 20 to 25) as more likely to be overpriced and stocks with low PE ratios (less than 5 to 8) as more likely to be underpriced. These investors prefer to sell or avoid buying stocks with high PE ratios and to buy or hold stocks with low PE ratios. However, investment decision making is rarely so simple as to rely on a single ratio. For instance, a stock with a high PE ratio can prove to be a good investment if its earnings continue to increase beyond current expectations. Similarly, a stock with a low PE ratio can prove to be a poor investment if its earnings decline below expectations.

Point: Average PE ratios for U.S. stocks increased over the past two decades. Some analysts interpret this as a signal the market is overpriced. But higher ratios can at least partly reflect accounting changes that have reduced reported earnings.

Decision Maker

Money Manager You plan to invest in one of two companies identified as having identical future prospects. One has a PE of 19 and the other a PE of 25. Which do you invest in? Does it matter if your *estimate* of PE for these two companies is 29 as opposed to 22? [Answer—p. 531]

Dividend Yield

A3 Compute dividend yield and explain its use in analysis.

Investors buy shares of a company's stock in anticipation of receiving a return from either or both cash dividends and stock price increases. Stocks that pay large dividends on a regular basis, called *income stocks,* are attractive to investors who want recurring cash flows from their investments. In contrast, some stocks pay little or no dividends but are still attractive to investors because of their expected stock price increases. The stocks of companies that distribute little or no cash but use their cash to finance expansion are called *growth stocks*. One way to help identify whether a stock is an income stock or a growth stock is to analyze its dividend yield. **Dividend yield,** defined in Exhibit 13.19, shows the annual amount of cash dividends distributed to common shares relative to their market value.

EXHIBIT 13.19

Dividend Yield

$$\textbf{Dividend yield} = \frac{\textbf{Annual cash dividends per share}}{\textbf{Market value per share}}$$

Dividend yield can be computed for current and prior periods using actual dividends and stock prices and for future periods using expected values. Exhibit 13.20 shows recent dividend and stock price data for **Amazon** and **Altria Group** to compute dividend yield.

EXHIBIT 13.20

Dividend and Stock Price Information

Company	Cash Dividends per Share	Market Value per Share	Dividend Yield
Amazon	$0.00	$90	0.0%
Altria Group	3.32	70	4.7

Point: The *payout ratio* equals cash dividends declared on common stock divided by net income. A low payout ratio suggests that a company is retaining earnings for future growth.

Dividend yield is zero for Amazon, implying it is a growth stock. An investor in Amazon would look for increases in stock prices (and eventual cash from the sale of stock). Altria has a dividend yield of 4.7%, implying it is an income stock for which dividends are important in assessing its value.

Book Value per Share

A4 Compute book value and explain its use in analysis.

Case 1: Common Stock (Only) Outstanding. **Book value per common share,** defined in Exhibit 13.21, reflects the amount of equity applicable to *common* shares on a per share basis. To illustrate, we use Dillon Snowboards' data from Exhibit 13.4. Dillon has 30,000 outstanding common shares, and the stockholders' equity applicable to common shares is $365,000. Dillon's book value per common share is $12.17, computed as $365,000 divided by 30,000 shares.

EXHIBIT 13.21

Book Value per Common Share

$$\textbf{Book value per common share} = \frac{\textbf{Stockholders' equity applicable to common shares}}{\textbf{Number of common shares outstanding}}$$

Point: Book value per share is also referred to as *stockholders' claim to assets on a per share basis.*

Case 2: Common and Preferred Stock Outstanding. To compute book value when both common and preferred shares are outstanding, we allocate total equity between the two types of shares. The **book value per preferred share** is computed first; its computation is shown in Exhibit 13.22.

EXHIBIT 13.22

Book Value per Preferred Share

$$\textbf{Book value per preferred share} = \frac{\textbf{Stockholders' equity applicable to preferred shares}}{\textbf{Number of preferred shares outstanding}}$$

The equity applicable to preferred shares equals the preferred share's call price (or par value if the preferred is not callable) plus any cumulative dividends in arrears. The remaining equity is the portion applicable to common shares. To illustrate, consider LTD's equity in Exhibit 13.23. Its preferred stock is callable at $108 per share, and two years of cumulative preferred dividends are in arrears.

EXHIBIT 13.23

Stockholders' Equity with Preferred and Common Stock

Stockholders' Equity	
Preferred stock—$100 par value, 7% cumulative, 2,000 shares authorized, 1,000 shares issued and outstanding	$100,000
Common stock—$25 par value, 12,000 shares authorized, 10,000 shares issued and outstanding	250,000
Paid-in capital in excess of par value, common stock	15,000
Retained earnings	82,000
Total stockholders' equity	$447,000

The book value computations are in Exhibit 13.24. Equity is first allocated to preferred shares before the book value of common shares is computed.

EXHIBIT 13.24

Computing Book Value per Preferred and Common Share

Total stockholders' equity		$447,000
Less equity applicable to preferred shares		
Call price (1,000 shares × $108)	$108,000	
Dividends in arrears ($100,000 × 7% × 2 years)	14,000	(122,000)
Equity applicable to common shares		$325,000
Book value per preferred share ($122,000/1,000 shares)		**$ 122.00**
Book value per common share ($325,000/10,000 shares)		**$ 32.50**

Book value per share reflects the value per share if a company is liquidated at balance sheet amounts. Book value is also the starting point in many stock valuation models, merger negotiations, price setting for public utilities, and loan contracts. The main limitation in using book value is the potential difference between recorded value and market value for assets and liabilities. Investors often adjust their analysis for estimates of these differences.

Decision Maker

Investor You are considering investing in BMX, whose book value per common share is $4 and price per common share on the stock exchange is $7. From this information, are BMX's net assets priced higher or lower than its recorded values? [Answer—p. 531]

Demonstration Problem 1

DP13

Barton Corporation began operations on January 1, 2008. The following transactions relating to stockholders' equity occurred in the first two years of the company's operations.

2008

Jan. 1	Authorized the issuance of 2 million shares of $5 par value common stock and 100,000 shares of $100 par value, 10% cumulative, preferred stock.
Jan. 2	Issued 200,000 shares of common stock for $12 cash per share.
Jan. 3	Issued 100,000 shares of common stock in exchange for a building valued at $820,000 and merchandise inventory valued at $380,000.
Jan. 4	Paid $10,000 cash to the company's founders for organization activities.
Jan. 5	Issued 12,000 shares of preferred stock for $110 cash per share.

2009

June 4	Issued 100,000 shares of common stock for $15 cash per share.

Required

1. Prepare journal entries to record these transactions.
2. Prepare the stockholders' equity section of the balance sheet as of December 31, 2008, and December 31, 2009, based on these transactions.
3. Prepare a table showing dividend allocations and dividends per share for 2008 and 2009 assuming Barton declares the following cash dividends: 2008, $50,000, and 2009, $300,000.
4. Prepare the January 2, 2008, journal entry for Barton's issuance of 200,000 shares of common stock for $12 cash per share assuming
   a. Common stock is no-par stock without a stated value.
   b. Common stock is no-par stock with a stated value of $10 per share.

Planning the Solution

- Record journal entries for the transactions for 2008 and 2009.
- Determine the balances for the 2008 and 2009 equity accounts for the balance sheet.
- Prepare the contributed capital portion of the 2008 and 2009 balance sheets.
- Prepare a table similar to Exhibit 13.11 showing dividend allocations for 2008 and 2009.
- Record the issuance of common stock under both specifications of no-par stock.

Solution to Demonstration Problem 1

1. Journal entries.

2008			
Jan. 2	Cash	2,400,000	
	Common Stock, $5 Par Value		1,000,000
	Paid-In Capital in Excess of Par Value, Common Stock		1,400,000
	Issued 200,000 shares of common stock.		
Jan. 3	Building	820,000	
	Merchandise Inventory	380,000	
	Common Stock, $5 Par Value		500,000
	Paid-In Capital in Excess of Par Value, Common Stock		700,000
	Issued 100,000 shares of common stock.		
Jan. 4	Organization Expenses	10,000	
	Cash		10,000
	Paid founders for organization costs.		

[continued on next page]

[continued from previous page]

Jan. 5	Cash	1,320,000	
	Preferred Stock, $100 Par Value		1,200,000
	Paid-In Capital in Excess of Par Value, Preferred Stock		120,000
	Issued 12,000 shares of preferred stock.		
2009			
June 4	Cash	1,500,000	
	Common Stock, $5 Par Value		500,000
	Paid-In Capital in Excess of Par Value, Common Stock		1,000,000
	Issued 100,000 shares of common stock.		

2. Balance sheet presentations (at December 31 year-end).

	2009	2008
Stockholders' Equity		
Preferred stock—$100 par value, 10% cumulative, 100,000 shares authorized, 12,000 shares issued and outstanding	$1,200,000	$1,200,000
Paid-in capital in excess of par value, preferred stock	120,000	120,000
Total paid-in capital by preferred stockholders	1,320,000	1,320,000
Common stock—$5 par value, 2,000,000 shares authorized, 300,000 shares issued and outstanding in 2008, and 400,000 shares issued and outstanding in 2009	2,000,000	1,500,000
Paid-in capital in excess of par value, common stock	3,100,000	2,100,000
Total paid-in capital by common stockholders	5,100,000	3,600,000
Total paid-in capital	$6,420,000	$4,920,000

3. Dividend allocation table.

	Common	Preferred
2008 ($50,000)		
Preferred—current year (12,000 shares × $10 = $120,000)	$ 0	$ 50,000
Common—remainder (300,000 shares outstanding)	0	0
Total for the year	$ 0	$ 50,000
2009 ($300,000)		
Preferred—dividend in arrears from 2008 ($120,000 − $50,000)	$ 0	$ 70,000
Preferred—current year	0	120,000
Common—remainder (400,000 shares outstanding)	110,000	0
Total for the year	$110,000	$190,000
Dividends per share		
2008	$ 0.00	$ 4.17
2009	$ 0.28	$ 15.83

4. Journal entries.

a. For 2008 (no-par stock without a stated value):

Jan. 2	Cash	2,400,000	
	Common Stock, No-Par Value		2,400,000
	Issued 200,000 shares of no-par common stock at $12 per share.		

b. For 2008 (no-par stock with a stated value):

Jan. 2	Cash	2,400,000	
	Common Stock, $10 Stated Value		2,000,000
	Paid-In Capital in Excess of Stated Value, Common Stock		400,000
	Issued 200,000 shares of $10 stated value common stock at $12 per share.		

Demonstration Problem 2

Precision Company began year 2008 with the following balances in its stockholders' equity accounts.

Common stock—$10 par, 500,000 shares authorized, 200,000 shares issued and outstanding	$2,000,000
Paid-in capital in excess of par, common stock	1,000,000
Retained earnings	5,000,000
Total	$8,000,000

All outstanding common stock was issued for $15 per share when the company was created. Prepare journal entries to account for the following transactions during year 2008.

Jan. 10 The board declared a $0.10 cash dividend per share to shareholders of record Jan. 28.
Feb. 15 Paid the cash dividend declared on January 10.
Mar. 31 Declared a 20% stock dividend. The market value of the stock is $18 per share.
May 1 Distributed the stock dividend declared on March 31.
July 1 Purchased 30,000 shares of treasury stock at $20 per share.
Sept. 1 Sold 20,000 treasury shares at $26 cash per share.
Dec. 1 Sold the remaining 10,000 shares of treasury stock at $7 cash per share.

Planning the Solution

- Calculate the total cash dividend to record by multiplying the cash dividend declared by the number of shares as of the date of record.
- Decide whether the stock dividend is a small or large dividend. Then analyze each event to determine the accounts affected and the appropriate amounts to be recorded.

Solution to Demonstration Problem 2

Jan. 10	Retained Earnings	20,000	
	Common Dividend Payable		20,000
	Declared a $0.10 per share cash dividend.		
Feb. 15	Common Dividend Payable	20,000	
	Cash		20,000
	Paid $0.10 per share cash dividend.		
Mar. 31	Retained Earnings	720,000	
	Common Stock Dividend Distributable		400,000
	Paid-In Capital in Excess of Par Value, Common Stock		320,000
	Declared a small stock dividend of 20% or 40,000 shares; market value is $18 per share.		

[continued on next page]

[continued from previous page]

May 1	Common Stock Dividend Distributable	400,000	
	Common Stock		400,000
	Distributed 40,000 shares of common stock.		
July 1	Treasury Stock, Common	600,000	
	Cash		600,000
	Purchased 30,000 common shares at $20 per share.		
Sept. 1	Cash	520,000	
	Treasury Stock, Common		400,000
	Paid-In Capital, Treasury Stock		120,000
	Sold 20,000 treasury shares at $26 per share.		
Dec. 1	Cash	70,000	
	Paid-In Capital, Treasury Stock	120,000	
	Retained Earnings	10,000	
	Treasury Stock, Common		200,000
	Sold 10,000 treasury shares at $7 per share.		

Summary

C1 Identify characteristics of corporations and their organization. Corporations are legal entities whose stockholders are not liable for its debts. Stock is easily transferred, and the life of a corporation does not end with the incapacity of a stock holder. A corporation acts through its agents, who are its officers and managers. Corporations are regulated and subject to income taxes.

C2 Describe the components of stockholders' equity. Authorized stock is the stock that a corporation's charter authorizes it to sell. Issued stock is the portion of authorized shares sold. Par value stock is a value per share assigned by the charter. No-par value stock is stock *not* assigned a value per share by the charter. Stated value stock is no-par stock to which the directors assign a value per share. Stockholders' equity is made up of (1) paid-in capital and (2) retained earnings. Paid-in capital consists of funds raised by stock issuances. Retained earnings consists of cumulative net income (losses) not distributed.

C3 Explain characteristics of common and preferred stock. Preferred stock has a priority (or senior status) relative to common stock in one or more areas, usually (1) dividends and (2) assets in case of liquidation. Preferred stock usually does not carry voting rights and can be convertible or callable. Convertibility permits the holder to convert preferred to common. Callability permits the issuer to buy back preferred stock under specified conditions.

C4 Explain the items reported in retained earnings. Many companies face statutory and contractual restrictions on retained earnings. Corporations can voluntarily appropriate retained earnings to inform others about their disposition. Prior period adjustments are corrections of errors in prior financial statements.

A1 Compute earnings per share and describe its use. A company with a simple capital structure computes basic EPS by dividing net income less any preferred dividends by the weighted-average number of outstanding common shares. A company with a complex capital structure must usually report both basic and diluted EPS.

A2 Compute price-earnings ratio and describe its use in analysis. A common stock's price-earnings (PE) ratio is computed by dividing the stock's market value (price) per share by its EPS. A stock's PE is based on expectations that can prove to be better or worse than eventual performance.

A3 Compute dividend yield and explain its use in analysis. Dividend yield is the ratio of a stock's annual cash dividends per share to its market value (price) per share. Dividend yield can be compared with the yield of other companies to determine whether the stock is expected to be an income or growth stock.

A4 Compute book value and explain its use in analysis. Book value per common share is equity applicable to common shares divided by the number of outstanding common shares. Book value per preferred share is equity applicable to preferred shares divided by the number of outstanding preferred shares.

P1 Record the issuance of corporate stock. When stock is issued, its par or stated value is credited to the stock account and any excess is credited to a separate contributed capital account. If a stock has neither par nor stated value, the entire proceeds are credited to the stock account. Stockholders must contribute assets equal to minimum legal capital or be potentially liable for the deficiency.

P2 Record transactions involving cash dividends. Cash dividends involve three events. On the date of declaration, the directors bind the company to pay the dividend. A dividend declaration reduces retained earnings and creates a current liability. On the date of record, recipients of the dividend are identified. On the date of payment, cash is paid to stockholders and the current liability is removed.

P3 Account for stock dividends and stock splits. Neither a stock dividend nor a stock split alters the value of the company. However, the value of each share is less due to the distribution of additional shares. The distribution of additional shares is according to individual stockholders' ownership percent. Small stock dividends ($\leq$25%) are recorded by capitalizing retained earnings equal to the market value of distributed shares. Large stock dividends (>25%) are recorded by capitalizing retained earnings equal to the par or stated value of distributed shares. Stock splits do not yield journal entries but do yield changes in the description of stock.

P4 Distribute dividends between common stock and preferred stock. Preferred stockholders usually hold the right to dividend distributions before common stockholders. When preferred stock is cumulative and in arrears, the amount in arrears must be distributed to preferred before any dividends are distributed to common.

P5 Record purchases and sales of treasury stock and the retirement of stock. When a corporation purchases its own previously issued stock, it debits the cost of these shares to Treasury Stock. Treasury stock is subtracted from equity in the balance sheet. If treasury stock is reissued, any proceeds in excess of cost are credited to Paid-In Capital, Treasury Stock. If the proceeds are less than cost, they are debited to Paid-In Capital, Treasury Stock to the extent a credit balance exists. Any remaining amount is debited to Retained Earnings. When stock is retired, all accounts related to the stock are removed.

Guidance Answers to **Decision Maker** and **Decision Ethics**

Entrepreneur The 50% stock dividend provides you no direct income. A stock dividend often reveals management's optimistic expectations about the future and can improve a stock's marketability by making it affordable to more investors. Accordingly, a stock dividend usually reveals "good news" and because of this, it likely increases (slightly) the market value for your stock. The same conclusions apply to the 3-for-2 stock split.

Concert Organizer You have two basic options: (1) different classes of common stock or (2) common and preferred stock. Your objective is to issue to yourself stock that has all or a majority of the voting power. The other class of stock would carry limited or no voting rights. In this way, you maintain control and are able to raise the necessary funds.

Money Manager Since one company requires a payment of $19 for each $1 of earnings, and the other requires $25, you would prefer the stock with the PE of 19; it is a better deal given identical prospects. You should make sure these companies' earnings computations are roughly the same, for example, no extraordinary items, unusual events, and so forth. Also, your PE estimates for these companies do matter. If you are willing to pay $29 for each $1 of earnings for these companies, you obviously expect both to exceed current market expectations.

Investor Book value reflects recorded values. BMX's book value is $4 per common share. Stock price reflects the market's expectation of net asset value (both tangible and intangible items). BMX's market value is $7 per common share. Comparing these figures suggests BMX's market value of net assets is higher than its recorded values (by an amount of $7 versus $4 per share).

Guidance Answers to **Quick Checks**

1. (*b*)
2. A corporation pays taxes on its income, and its stockholders normally pay personal income taxes (at the 15% rate or lower) on any cash dividends received from the corporation.
3. A proxy is a legal document used to transfer a stockholder's right to vote to another person.
4. (*a*)
5. A stock premium is an amount in excess of par (or stated) value paid by purchasers of newly issued stock.
6. Minimum legal capital intends to protect creditors of a corporation by obligating stockholders to some minimum level of equity financing and by constraining a corporation from excessive payments to stockholders.
7. Common Dividend Payable is a current liability account.
8. The date of declaration, date of record, and date of payment.
9. A dividend is a legal liability at the date of declaration, on which date it is recorded as a liability.
10. A stock dividend does not transfer assets to stockholders, but it does require an amount of retained earnings to be transferred to a contributed capital account(s).
11. A small stock dividend is 25% or less of the previous outstanding shares. A large stock dividend is more than 25%.
12. Retained earnings equal to the distributable shares' market value should be capitalized for a small stock dividend.
13. Typically, preferred stock has a preference in receipt of dividends and in distribution of assets.
14. (*a*)
15. (*b*)

Total cash dividend	$288,000
To preferred shareholders	135,000*
Remainder to common shareholders	$153,000

* 9,000 × $50 × 10% × 3 years = $135,000.

16. (*b*)
17. No. The shares are an investment for Southern Co. and are issued and outstanding shares for Northern Corp.
18. Treasury stock does not affect the number of authorized or issued shares, but it reduces the outstanding shares.
19. (*a*)

Key Terms

mhhe.com/wildFAP19e

Key Terms are available at the book's Website for learning and testing in an online Flashcard Format.

Appropriated retained earnings (p. 523)
Authorized stock (p. 509)
Basic earnings per share (p. 524)
Book value per common share (p. 526)
Book value per preferred share (p. 526)
Call price (p. 519)
Callable preferred stock (p. 519)
Capital stock (p. 509)
Changes in accounting estimates (p. 523)
Common stock (p. 508)
Complex capital structure (p. 524)
Convertible preferred stock (p. 518)
Corporation (p. 506)
Cumulative preferred stock (p. 517)
Date of declaration (p. 513)
Date of payment (p. 513)
Date of record (p. 513)
Diluted earnings per share (p. 524)
Dilutive securities (p. 524)
Discount on stock (p. 511)
Dividend in arrears (p. 517)
Dividend yield (p. 525)
Earnings per share (EPS) (p. 524)
Financial leverage (p. 519)
Large stock dividend (p. 514)
Liquidating cash dividend (p. 514)
Market value per share (p. 509)
Minimum legal capital (p. 509)
Noncumulative preferred stock (p. 517)
Nonparticipating preferred stock (p. 518)
No-par value stock (p. 509)
Organization expenses (pp. 507, 512)
Paid-in capital (p. 510)
Paid-in capital in excess of par value (p. 511)
Participating preferred stock (p. 518)
Par value (p. 509)
Par value stock (p. 509)
Preemptive right (p. 508)
Preferred stock (p. 516)
Premium on stock (p. 511)
Price-earnings (PE) ratio (p. 525)
Prior period adjustments (p. 523)
Proxy (p. 507)
Restricted retained earnings (p. 522)
Retained earnings (p. 510)
Retained earnings deficit (p. 513)
Reverse stock split (p. 516)
Simple capital structure (p. 524)
Small stock dividend (p. 514)
Stated value stock (p. 510)
Statement of stockholders' equity (p. 523)
Stock dividend (p. 514)
Stock options (p. 523)
Stock split (p. 516)
Stockholders' equity (p. 510)
Treasury stock (p. 520)

Multiple Choice Quiz

Answers on p. 547

mhhe.com/wildFAP19e

Additional Quiz Questions are available at the book's Website.

Quiz13

1. A corporation issues 6,000 shares of $5 par value common stock for $8 cash per share. The entry to record this transaction includes:
a. A debit to Paid-In Capital in Excess of Par Value for $18,000.
b. A credit to Common Stock for $48,000.
c. A credit to Paid-In Capital in Excess of Par Value for $30,000.
d. A credit to Cash for $48,000.
e. A credit to Common Stock for $30,000.

2. A company reports net income of $75,000. Its weighted-average common shares outstanding is 19,000. It has no other stock outstanding. Its earnings per share is:
a. $4.69
b. $3.95
c. $3.75
d. $2.08
e. $4.41

3. A company has 5,000 shares of $100 par preferred stock and 50,000 shares of $10 par common stock outstanding. Its total stockholders' equity is $2,000,000. Its book value per common share is:
a. $100.00
b. $ 10.00
c. $ 40.00
d. $ 30.00
e. $ 36.36

4. A company paid cash dividends of $0.81 per share. Its earnings per share is $6.95 and its market price per share is $45.00. Its dividend yield is:
a. 1.8%
b. 11.7%
c. 15.4%
d. 55.6%
e. 8.6%

5. A company's shares have a market value of $85 per share. Its net income is $3,500,000, and its weighted-average common shares outstanding is 700,000. Its price-earnings ratio is:
a. 5.9
b. 425.0
c. 17.0
d. 10.4
e. 41.2

Discussion Questions

1. What are organization expenses? Provide examples.
2. How are organization expenses reported?
3. Who is responsible for directing a corporation's affairs?
4. What is the preemptive right of common stockholders?
5. List the general rights of common stockholders.
6. What is the difference between authorized shares and outstanding shares?
7. Why would an investor find convertible preferred stock attractive?
8. What is the difference between the market value per share and the par value per share?
9. What is the difference between the par value and the call price of a share of preferred stock?
10. Identify and explain the importance of the three dates relevant to corporate dividends.
11. Why is the term *liquidating dividend* used to describe cash dividends debited against paid-in capital accounts?
12. How does declaring a stock dividend affect the corporation's assets, liabilities, and total equity? What are the effects of the eventual distribution of that stock?
13. What is the difference between a stock dividend and a stock split?
14. Courts have ruled that a stock dividend is not taxable income to stockholders. What justifies this decision?
15. How does the purchase of treasury stock affect the purchaser's assets and total equity?
16. Why do laws place limits on treasury stock purchases?
17. How are EPS results computed for a corporation with a simple capital structure?
18. What is a stock option?
19. How is book value per share computed for a corporation with no preferred stock? What is the main limitation of using book value per share to value a corporation?
20. Review the balance sheet for **Best Buy** in Appendix A and list the classes of stock that it has issued.
21. Refer to the balance sheet for **Circuit City** in Appendix A. What is the par value per share of its common stock? Suggest a rationale for the amount of par value it assigned.

22. Refer to **RadioShack**'s balance sheet in Appendix A. How many shares of common stock are authorized? How many shares of common stock are issued?
23. Refer to the financial statements for **Apple** in Appendix A. What are its cash proceeds from issuance of common stock and its cash repurchases of common stock for the year ended September 30, 2006? Explain.

Denotes Discussion Questions that involve decision making.

McGraw-Hill's HOMEWORK MANAGER® Available with McGraw-Hill's Homework Manager

QUICK STUDY

QS 13-1
Characteristics of corporations
C1

Of the following statements, which are true for the corporate form of organization?

1. Capital is more easily accumulated than with most other forms of organization.
2. Corporate income that is distributed to shareholders is usually taxed twice.
3. Owners have unlimited liability for corporate debts.
4. Ownership rights cannot be easily transferred.
5. Owners are not agents of the corporation.
6. It is a separate legal entity.
7. It has a limited life.

QS 13-2
Issuance of common stock
P1

Prepare the journal entry to record Miltone Company's issuance of 50,000 shares of $1 par value common stock assuming the shares sell for:

a. $1 cash per share.
b. $3 cash per share.

QS 13-3
Issuance of par and stated value common stock
P1

Prepare the journal entry to record Katrick Company's issuance of 75,000 shares of its common stock assuming the shares have a:

a. $5 par value and sell for $12 cash per share.
b. $5 stated value and sell for $12 cash per share.

QS 13-4
Issuance of no-par common stock
P1

Prepare the journal entry to record Gaylord Company's issuance of 52,000 shares of no-par value common stock assuming the shares:

a. Sell for $30 cash per share.

b. Are exchanged for land valued at $1,560,000.

QS 13-5
Issuance of common stock
P1

Prepare the issuer's journal entry for each separate transaction. (*a*) On March 1, Atlantic Co. issues 37,500 shares of $5 par value common stock for $300,000 cash. (*b*) On April 1, BP Co. issues no-par value common stock for $90,000 cash. (*c*) On April 6, MPG issues 3,500 shares of $10 par value common stock for $20,000 of inventory, $130,000 of machinery, and acceptance of a $75,000 note payable.

QS 13-6
Issuance of preferred stock
C3

a. Prepare the journal entry to record Tamar Company's issuance of 6,000 shares of $100 par value 6% cumulative preferred stock for $102 cash per share.

b. Assuming the facts in part 1, if Tamar declares a year-end cash dividend, what is the amount of dividend paid to preferred shareholders? (Assume no dividends in arrears.)

QS 13-7
Accounting for cash dividends
P2

Prepare journal entries to record the following transactions for Forrest Corporation.

May 15 Declared a $32,000 cash dividend payable to common stockholders.
June 30 Paid the dividend declared on May 15.

QS 13-8
Accounting for small stock dividend
C2 P3

The stockholders' equity section of Atari Company's balance sheet as of April 1 follows. On April 2, Atari declares and distributes a 10% stock dividend. The stock's per share market value on April 2 is $18 (prior to the dividend). Prepare the stockholders' equity section immediately after the stock dividend.

Common stock—$5 par value, 375,000 shares authorized, 200,000 shares issued and outstanding	$1,000,000
Paid-in capital in excess of par value, common stock	600,000
Retained earnings	833,000
Total stockholders' equity	$2,433,000

QS 13-9
Dividend allocation between classes of shareholders
P4  

Stockholders' equity of Marwick Company consists of 10,000 shares of $20 par value, 8% cumulative preferred stock and 400,000 shares of $1 par value common stock. Both classes of stock have been outstanding since the company's inception. Marwick did not declare any dividends in the prior year, but it now declares and pays a $92,000 cash dividend at the current year-end. Determine the amount distributed to each class of stockholders for this two-year-old company.

QS 13-10
Purchase and sale of treasury stock P5

On May 3, Winmac Corporation purchased 3,000 shares of its own stock for $45,000 cash. On November 4, Winmac reissued 850 shares of this treasury stock for $14,450. Prepare the May 3 and November 4 journal entries to record Winmac's purchase and reissuance of treasury stock.

QS 13-11
Accounting for changes in estimates; error adjustments
C4 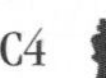

Answer the following questions related to a company's activities for the current year:

1. A review of the notes payable files discovers that three years ago the company reported the entire amount of a payment (principal and interest) on an installment note payable as interest expense. This mistake had a material effect on the amount of income in that year. How should the correction be reported in the current year financial statements?

2. After using an expected useful life of seven years and no salvage value to depreciate its office equipment over the preceding three years, the company decided early this year that the equipment will last only two more years. How should the effects of this decision be reported in the current year financial statements?

Campbell Company reports net income of $840,000 for the year. It has no preferred stock, and its weighted-average common shares outstanding is 300,000 shares. Compute its basic earnings per share.

QS 13-12
Basic earnings per share A1

Epic Company earned net income of $950,000 this year. The number of common shares outstanding during the entire year was 400,000, and preferred shareholders received a $40,000 cash dividend. Compute Epic Company's basic earnings per share.

QS 13-13
Basic earnings per share A1

Compute Tripp Company's price-earnings ratio if its common stock has a market value of $31.50 per share and its EPS is $3.75. Would an analyst likely consider this stock potentially over- or underpriced? Explain.

QS 13-14
Price-earnings ratio A2

Payne Company expects to pay a $1.62 per share cash dividend this year on its common stock. The current market value of Payne stock is $22.50 per share. Compute the expected dividend yield on the Payne stock. Would you classify the Payne stock as a growth or an income stock? Explain.

QS 13-15
Dividend yield
A3

The stockholders' equity section of Klaus Company's balance sheet follows. The preferred stock's call price is $25. Determine the book value per share of the common stock.

QS 13-16
Book value per common share
A4

Preferred stock—5% cumulative, $10 par value, 20,000 shares authorized, issued and outstanding	$ 200,000
Common stock—$5 par value, 200,000 shares authorized, 150,000 shares issued and outstanding	750,000
Retained earnings	889,500
Total stockholders' equity	$1,839,500

McGraw-Hill's HOMEWORK MANAGER® HM Available with McGraw-Hill's Homework Manager

EXERCISES

Describe how each of the following characteristics of organizations applies to corporations.

Exercise 13-1
Characteristics of corporations
C1

1. Owner authority and control	5. Duration of life
2. Ease of formation	6. Owner liability
3. Transferability of ownership	7. Legal status
4. Ability to raise large capital amounts	8. Tax status of income

Rodriguez Corporation issues 12,000 shares of its common stock for $182,700 cash on February 20. Prepare journal entries to record this event under each of the following separate situations.

1. The stock has neither par nor stated value.
2. The stock has a $12 par value.
3. The stock has a $6 stated value.

Exercise 13-2
Accounting for par, stated, and no-par stock issuances
P1

Prepare journal entries to record the following four separate issuances of stock.

1. A corporation issued 2,500 shares of no-par common stock to its promoters in exchange for their efforts, estimated to be worth $43,500. The stock has no stated value.
2. A corporation issued 2,500 shares of no-par common stock to its promoters in exchange for their efforts, estimated to be worth $43,500. The stock has a $2 per share stated value.
3. A corporation issued 5,000 shares of $30 par value common stock for $180,000 cash.
4. A corporation issued 1,250 shares of $100 par value preferred stock for $168,500 cash.

Exercise 13-3
Recording stock issuances
P1

Soku Company issues 12,000 shares of $9 par value common stock in exchange for land and a building. The land is valued at $75,000 and the building at $120,000. Prepare the journal entry to record issuance of the stock in exchange for the land and building.

Exercise 13-4
Stock issuance for noncash assets
P1

Exercise 13-5
Identifying characteristics of preferred stock
C2 C3

Match each description 1 through 6 with the characteristic of preferred stock that it best describes by writing the letter of that characteristic in the blank next to each description.

A. Convertible **B.** Cumulative **C.** Noncumulative
D. Nonparticipating **E.** Participating **F.** Callable

_______ **1.** Holders of the stock are entitled to receive current and all past dividends before common stockholders receive any dividends.
_______ **2.** The issuing corporation can retire the stock by paying a prespecified price.
_______ **3.** Holders of the stock can receive dividends exceeding the stated rate under certain conditions.
_______ **4.** Holders of the stock are not entitled to receive dividends in excess of the stated rate.
_______ **5.** Holders of this stock can exchange it for shares of common stock.
_______ **6.** Holders of the stock lose any dividends that are not declared in the current year.

Exercise 13-6
Stock dividends and splits
P3

On June 30, 2009, Samson Corporation's common stock is priced at $30.50 per share before any stock dividend or split, and the stockholders' equity section of its balance sheet appears as follows.

Common stock—$8 par value, 80,000 shares authorized, 32,000 shares issued and outstanding	$256,000
Paid-in capital in excess of par value, common stock	100,000
Retained earnings	356,000
Total stockholders' equity	$712,000

1. Assume that the company declares and immediately distributes a 100% stock dividend. This event is recorded by capitalizing retained earnings equal to the stock's par value. Answer these questions about stockholders' equity as it exists *after* issuing the new shares.
a. What is the retained earnings balance?
b. What is the amount of total stockholders' equity?
c. How many shares are outstanding?

Check (1b) $712,000

2. Assume that the company implements a 2-for-1 stock split instead of the stock dividend in part 1. Answer these questions about stockholders' equity as it exists *after* issuing the new shares.
a. What is the retained earnings balance?
b. What is the amount of total stockholders' equity?
c. How many shares are outstanding?

(2a) $356,000

3. Explain the difference, if any, to a stockholder from receiving new shares distributed under a large stock dividend versus a stock split.

Exercise 13-7
Stock dividends and per share book values
P3

The stockholders' equity of Tyron Company at the beginning of the day on February 5 follows.

Common stock—$25 par value, 150,000 shares authorized, 64,000 shares issued and outstanding	$1,600,000
Paid-in capital in excess of par value, common stock	525,000
Retained earnings	671,800
Total stockholders' equity	$2,796,800

On February 5, the directors declare a 15% stock dividend distributable on February 28 to the February 15 stockholders of record. The stock's market value is $50 per share on February 5 before the stock dividend. The stock's market value is $43.60 per share on February 28.

1. Prepare entries to record both the dividend declaration and its distribution.
2. One stockholder owned 900 shares on February 5 before the dividend. Compute the book value per share and total book value of this stockholder's shares immediately before *and* after the stock dividend of February 5.
3. Compute the total market value of the investor's shares in part 2 as of February 5 and February 28.

Check (2) Book value per share: before, $43.70; after, $38.00

Exercise 13-8
Dividends on common and noncumulative preferred stock
P4

Norton's outstanding stock consists of (*a*) 13,000 shares of noncumulative 8% preferred stock with a $10 par value and (*b*) 32,500 shares of common stock with a $1 par value. During its first four years of operation, the corporation declared and paid the following total cash dividends.

2009	$ 8,000
2010	24,000
2011	120,000
2012	197,000

Determine the amount of dividends paid each year to each of the two classes of stockholders. Also compute the total dividends paid to each class for the four years combined.

Check Total paid to preferred, $39,200

Exercise 13-9
Dividends on common and cumulative preferred stock P4

Use the data in Exercise 13-8 to determine the amount of dividends paid each year to each of the two classes of stockholders assuming that the preferred stock is cumulative. Also determine the total dividends paid to each class for the four years combined.

Exercise 13-10
Recording and reporting treasury stock transactions
P5

On October 10, the stockholders' equity of Syntax Systems appears as follows.

Common stock—$10 par value, 72,000 shares authorized, issued, and outstanding	$ 720,000
Paid-in capital in excess of par value, common stock	216,000
Retained earnings	864,000
Total stockholders' equity	$1,800,000

1. Prepare journal entries to record the following transactions for Syntax Systems.
  a. Purchased 5,000 shares of its own common stock at $22 per share on October 11.
  b. Sold 1,000 treasury shares on November 1 for $28 cash per share.
  c. Sold all remaining treasury shares on November 25 for $17 cash per share.
2. Explain how the company's equity section changes after the October 11 treasury stock purchase, and prepare the revised equity section of its balance sheet at that date.

Check (1c) Dr. Retained Earnings, $14,000

Exercise 13-11
Preparing a statement of retained earnings
C4

The following information is available for Arturo Company for the year ended December 31, 2009.
a. Balance of retained earnings, December 31, 2008, prior to discovery of error, $1,375,000.
b. Cash dividends declared and paid during 2009, $43,000.
c. It neglected to record 2007 depreciation expense of $55,500, which is net of $4,500 in income taxes.
d. The company earned $126,000 in 2009 net income.
Prepare a 2009 statement of retained earnings for Arturo Company.

Exercise 13-12
Earnings per share
A1

Grossmont Company reports $1,375,500 of net income for 2009 and declares $192,500 of cash dividends on its preferred stock for 2009. At the end of 2009, the company had 350,000 weighted-average shares of common stock.
1. What amount of net income is available to common stockholders for 2009?
2. What is the company's basic EPS for 2009?

Check (2) $3.38

Exercise 13-13
Earnings per share
A1

Franklin Company reports $1,875,000 of net income for 2009 and declares $262,500 of cash dividends on its preferred stock for 2009. At the end of 2009, the company had 250,000 weighted-average shares of common stock.
1. What amount of net income is available to common stockholders for 2009?
2. What is the company's basic EPS for 2009?

Check (2) $6.45

Exercise 13-14
Price-earnings ratio computation and interpretation
A2

Compute the price-earnings ratio for each of these four separate companies. Which stock might an analyst likely investigate as being potentially undervalued by the market? Explain.

Company	Earnings per Share	Market Value per Share
1	$12.00	$145.20
2	11.00	116.60
3	7.80	74.10
4	43.20	60.48

Exercise 13-15
Dividend yield computation and interpretation
A3

Compute the dividend yield for each of these four separate companies. Which company's stock would probably *not* be classified as an income stock? Explain.

Company	Annual Cash Dividend per Share	Market Value per Share
1	$14.00	$229.51
2	11.00	110.00
3	5.52	60.00
4	1.90	118.75

Exercise 13-16
Book value per share
A4

The equity section of Westchester Corporation's balance sheet shows the following.

Preferred stock—6% cumulative, $30 par value, $35 call price, 10,000 shares issued and outstanding	$300,000
Common stock—$10 par value, 35,000 shares issued and outstanding	350,000
Retained earnings	267,500
Total stockholders' equity	$917,500

Determine the book value per share of the preferred and common stock under two separate situations.

1. No preferred dividends are in arrears.
2. Three years of preferred dividends are in arrears.

Check (1) Book value of common, $16.21

Available with McGraw-Hill's Homework Manager McGraw-Hill's HOMEWORK MANAGER®

PROBLEM SET A

Problem 13-1A
Stockholders' equity transactions and analysis
C2 C3 P1 

Keshena Co. is incorporated at the beginning of this year and engages in a number of transactions. The following journal entries impacted its stockholders' equity during its first year of operations.

a.	Cash	320,000	
	Common Stock, $25 Par Value		250,000
	Paid-In Capital in Excess of Par Value, Common Stock		70,000
b.	Organization Expenses	160,000	
	Common Stock, $25 Par Value		125,000
	Paid-In Capital in Excess of Par Value, Common Stock		35,000

[continued on next page]

[continued from previous page]

c.	Cash	45,500	
	Accounts Receivable	16,000	
	Building	82,000	
	Notes Payable		59,500
	Common Stock, $25 Par Value		50,000
	Paid-In Capital in Excess of Par Value, Common Stock		34,000
d.	Cash	123,000	
	Common Stock, $25 Par Value		75,000
	Paid-In Capital in Excess of Par Value, Common Stock		48,000

Required

1. Explain the transaction(s) underlying each journal entry (*a*) through (*d*).

2. How many shares of common stock are outstanding at year-end?

3. What is the amount of minimum legal capital (based on par value) at year-end?

4. What is the total paid-in capital at year-end?

5. What is the book value per share of the common stock at year-end if total paid-in capital plus retained earnings equals $785,000?

Check (2) 20,000 shares
(3) $500,000
(4) $687,000

Problem 13-2A
Cash dividends, treasury stock, and statement of retained earnings

C2 C4 P2 P5

Rocklin Corporation reports the following components of stockholders' equity on December 31, 2009.

Common stock—$25 par value, 100,000 shares authorized, 45,000 shares issued and outstanding	$1,125,000
Paid-in capital in excess of par value, common stock	60,000
Retained earnings	460,000
Total stockholders' equity	$1,645,000

In year 2010, the following transactions affected its stockholders' equity accounts.

Jan. 1	Purchased 4,500 shares of its own stock at $25 cash per share.
Jan. 5	Directors declared a $3 per share cash dividend payable on Feb. 28 to the Feb. 5 stockholders of record.
Feb. 28	Paid the dividend declared on January 5.
July 6	Sold 1,688 of its treasury shares at $29 cash per share.
Aug. 22	Sold 2,812 of its treasury shares at $22 cash per share.
Sept. 5	Directors declared a $3 per share cash dividend payable on October 28 to the September 25 stockholders of record.
Oct. 28	Paid the dividend declared on September 5.
Dec. 31	Closed the $388,000 credit balance (from net income) in the Income Summary account to Retained Earnings.

Required

1. Prepare journal entries to record each of these transactions for 2010.

2. Prepare a statement of retained earnings for the year ended December 31, 2010.

3. Prepare the stockholders' equity section of the company's balance sheet as of December 31, 2010.

Check (2) Retained earnings, Dec. 31, 2010, $589,816.

Problem 13-3A
Equity analysis—journal entries and account balances

P2 P3

At September 30, the end of Chan Company's third quarter, the following stockholders' equity accounts are reported.

Common stock, $10 par value	$420,000
Paid-in capital in excess of par value, common stock	100,000
Retained earnings	400,000

In the fourth quarter, the following entries related to its equity are recorded.

Oct. 2	Retained Earnings	63,000	
	Common Dividend Payable		63,000
Oct. 25	Common Dividend Payable	63,000	
	Cash		63,000
Oct. 31	Retained Earnings	92,400	
	Common Stock Dividend Distributable		42,000
	Paid-In Capital in Excess of Par Value, Common Stock		50,400
Nov. 5	Common Stock Dividend Distributable	42,000	
	Common Stock, $10 Par Value		42,000
Dec. 1	Memo—Change the title of the common stock account to reflect the new par value of $5.		
Dec. 31	Income Summary	230,000	
	Retained Earnings		230,000

Required

1. Explain the transaction(s) underlying each journal entry.

2. Complete the following table showing the equity account balances at each indicated date (include the balances from September 30).

	Oct. 2	Oct. 25	Oct. 31	Nov. 5	Dec. 1	Dec. 31
Common stock	$____	$____	$____	$____	$____	$____
Common stock dividend distributable	____	____	____	____	____	____
Paid-in capital in excess of par, common stock	____	____	____	____	____	____
Retained earnings	____	____	____	____	____	____
Total equity	$____	$____	$____	$____	$____	$____

Check Total equity: Oct. 2, $857,000; Dec. 31, $1,087,000

Problem 13-4A
Analysis of changes in stockholders' equity accounts
C4 P2 P3 P5 

The equity sections from Sierra Group's 2009 and 2010 year-end balance sheets follow.

Stockholders' Equity (December 31, 2009)	
Common stock—$6 par value, 100,000 shares authorized, 45,000 shares issued and outstanding	$270,000
Paid-in capital in excess of par value, common stock	230,000
Retained earnings	340,000
Total stockholders' equity	$840,000

Stockholders' Equity (December 31, 2010)	
Common stock—$6 par value, 100,000 shares authorized, 53,200 shares issued, 4,000 shares in treasury	$319,200
Paid-in capital in excess of par value, common stock	262,800
Retained earnings ($60,000 restricted by treasury stock)	400,000
	982,000
Less cost of treasury stock	(60,000)
Total stockholders' equity	$922,000

The following transactions and events affected its equity during year 2010.

Jan. 5	Declared a $0.50 per share cash dividend, date of record January 10.
Mar. 20	Purchased treasury stock for cash.
Apr. 5	Declared a $0.50 per share cash dividend, date of record April 10.
July 5	Declared a $0.50 per share cash dividend, date of record July 10.
July 31	Declared a 20% stock dividend when the stock's market value is $10 per share.

Aug. 14 Issued the stock dividend that was declared on July 31.
Oct. 5 Declared a $0.50 per share cash dividend, date of record October 10.

Required

1. How many common shares are outstanding on each cash dividend date?
2. What is the total dollar amount for each of the four cash dividends?
3. What is the amount of the capitalization of retained earnings for the stock dividend?
4. What is the per share cost of the treasury stock purchased?
5. How much net income did the company earn during year 2010?

Check (3) $82,000
(4) $15
(5) $230,100

Problem 13-5A
Computation of book values and dividend allocations
C3 A4 P4 

Folsom Corporation's common stock is currently selling on a stock exchange at $183 per share, and its current balance sheet shows the following stockholders' equity section.

Preferred stock—5% cumulative, $___ par value, 1,000 shares authorized, issued, and outstanding	$ 85,000
Common stock—$___ par value, 4,000 shares authorized, issued, and outstanding	200,000
Retained earnings	350,000
Total stockholders' equity	$635,000

Required

1. What is the current market value (price) of this corporation's common stock?
2. What are the par values of the corporation's preferred stock and its common stock?
3. If no dividends are in arrears, what are the book values per share of the preferred stock and the common stock?
4. If two years' preferred dividends are in arrears, what are the book values per share of the preferred stock and the common stock?
5. If two years' preferred dividends are in arrears and the preferred stock is callable at $95 per share, what are the book values per share of the preferred stock and the common stock?
6. If two years' preferred dividends are in arrears and the board of directors declares cash dividends of $24,750, what total amount will be paid to the preferred and to the common shareholders? What is the amount of dividends per share for the common stock?

Check (4) Book value of common, $135.38
(5) Book value of common, $132.88
(6) Dividends per common share, $3.00

Analysis Component

7. What are some factors that can contribute to a difference between the book value of common stock and its market value (price)?

PROBLEM SET B

Problem 13-1B
Stockholders' equity transactions and analysis
C2 C3 P1 

Mayport Company is incorporated at the beginning of this year and engages in a number of transactions. The following journal entries impacted its stockholders' equity during its first year of operations.

a.	Cash	60,000	
	Common Stock, $1 Par Value		1,500
	Paid-In Capital in Excess of Par Value, Common Stock		58,500
b.	Organization Expenses	20,000	
	Common Stock, $1 Par Value		500
	Paid-In Capital in Excess of Par Value, Common Stock		19,500
c.	Cash	6,650	
	Accounts Receivable	4,000	
	Building	18,500	
	Notes Payable		9,150
	Common Stock, $1 Par Value		400
	Paid-In Capital in Excess of Par Value, Common Stock		19,600

[continued on next page]

[continued from previous page]

d.	Cash	30,000	
	Common Stock, $1 Par Value		600
	Paid-In Capital in Excess of Par Value, Common Stock		29,400

Required

1. Explain the transaction(s) underlying each journal entry (*a*) through (*d*).

2. How many shares of common stock are outstanding at year-end?

3. What is the amount of minimum legal capital (based on par value) at year-end?

4. What is the total paid-in capital at year-end?

5. What is the book value per share of the common stock at year-end if total paid-in capital plus retained earnings equals $141,500?

Check (2) 3,000 shares
(3) $3,000
(4) $130,000

Problem 13-2B
Cash dividends, treasury stock, and statement of retained earnings
C2 C4 P2 P5

San Marco Corp. reports the following components of stockholders' equity on December 31, 2009.

Common stock—$1 par value, 160,000 shares authorized, 100,000 shares issued and outstanding	$ 100,000
Paid-in capital in excess of par value, common stock	700,000
Retained earnings	1,080,000
Total stockholders' equity	$1,880,000

It completed the following transactions related to stockholders' equity in year 2010.

Jan. 10	Purchased 20,000 shares of its own stock at $12 cash per share.
Mar. 2	Directors declared a $1.50 per share cash dividend payable on March 31 to the March 15 stockholders of record.
Mar. 31	Paid the dividend declared on March 2.
Nov. 11	Sold 12,000 of its treasury shares at $13 cash per share.
Nov. 25	Sold 8,000 of its treasury shares at $9.50 cash per share.
Dec. 1	Directors declared a $2.50 per share cash dividend payable on January 2 to the December 10 stockholders of record.
Dec. 31	Closed the $536,000 credit balance (from net income) in the Income Summary account to Retained Earnings.

Required

1. Prepare journal entries to record each of these transactions for 2010.

2. Prepare a statement of retained earnings for the year ended December 31, 2010.

3. Prepare the stockholders' equity section of the company's balance sheet as of December 31, 2010.

Check (2) Retained earnings, Dec. 31, 2010, $1,238,000

Problem 13-3B
Equity analysis—journal entries and account balances
P2 P3

At December 31, the end of Santee Communication's third quarter, the following stockholders' equity accounts are reported.

Common stock, $10 par value	$480,000
Paid-in capital in excess of par value, common stock	192,000
Retained earnings	800,000

In the fourth quarter, the following entries related to its equity are recorded.

Jan. 17	Retained Earnings	48,000	
	Common Dividend Payable		48,000
Feb. 5	Common Dividend Payable	48,000	
	Cash		48,000

[continued on next page]

[continued from previous page]

Feb. 28	Retained Earnings	126,000	
	Common Stock Dividend Distributable		60,000
	Paid-In Capital in Excess of Par Value, Common Stock		66,000
Mar. 14	Common Stock Dividend Distributable	60,000	
	Common Stock, $10 Par Value		60,000
Mar. 25	Memo—Change the title of the common stock account to reflect the new par value of $5.		
Mar. 31	Income Summary	360,000	
	Retained Earnings		360,000

Required

1. Explain the transaction(s) underlying each journal entry.

2. Complete the following table showing the equity account balances at each indicated date (include the balances from December 31).

	Jan. 17	Feb. 5	Feb. 28	Mar. 14	Mar. 25	Mar. 31
Common stock	$____	$____	$____	$____	$____	$____
Common stock dividend distributable	____	____	____	____	____	____
Paid-in capital in excess of par, common stock	____	____	____	____	____	____
Retained earnings	____	____	____	____	____	____
Total equity	$____	$____	$____	$____	$____	$____

Check Total equity: Jan. 17, $1,424,000; Mar. 31, $1,784,000

Problem 13-4B
Analysis of changes in stockholders' equity accounts
C4 P2 P3 P5

The equity sections from Kiwa Corporation's 2009 and 2010 balance sheets follow.

Stockholders' Equity (December 31, 2009)	
Common stock—$20 par value, 15,000 shares authorized, 8,500 shares issued and outstanding	$170,000
Paid-in capital in excess of par value, common stock	30,000
Retained earnings	135,000
Total stockholders' equity	$335,000

Stockholders' Equity (December 31, 2010)	
Common stock—$20 par value, 15,000 shares authorized, 9,500 shares issued, 500 shares in treasury	$190,000
Paid-in capital in excess of par value, common stock	52,000
Retained earnings ($20,000 restricted by treasury stock)	147,600
	389,600
Less cost of treasury stock	(20,000)
Total stockholders' equity	$369,600

The following transactions and events affected its equity during year 2010.

Feb. 15	Declared a $0.40 per share cash dividend, date of record five days later.
Mar. 2	Purchased treasury stock for cash.
May 15	Declared a $0.40 per share cash dividend, date of record five days later.
Aug. 15	Declared a $0.40 per share cash dividend, date of record five days later.
Oct. 4	Declared a 12.5% stock dividend when the stock's market value is $42 per share.
Oct. 20	Issued the stock dividend that was declared on October 4.
Nov. 15	Declared a $0.40 per share cash dividend, date of record five days later.

Required

1. How many common shares are outstanding on each cash dividend date?
2. What is the total dollar amount for each of the four cash dividends?
3. What is the amount of the capitalization of retained earnings for the stock dividend?
4. What is the per share cost of the treasury stock purchased?
5. How much net income did the company earn during year 2010?

Check (3) $42,000
(4) $40
(5) $68,000

Problem 13-5B

Computation of book values and dividend allocations

C3 A4 P4 

Hansen Company's common stock is currently selling on a stock exchange at $90 per share, and its current balance sheet shows the following stockholders' equity section.

Preferred stock—8% cumulative, $___ par value, 1,500 shares authorized, issued, and outstanding . . .	$ 187,500
Common stock—$___ par value, 18,000 shares authorized, issued, and outstanding . . .	450,000
Retained earnings . . .	562,500
Total stockholders' equity . . .	$1,200,000

Required

1. What is the current market value (price) of this corporation's common stock?
2. What are the par values of the corporation's preferred stock and its common stock?
3. If no dividends are in arrears, what are the book values per share of the preferred stock and the common stock?
4. If two years' preferred dividends are in arrears, what are the book values per share of the preferred stock and the common stock?
5. If two years' preferred dividends are in arrears and the preferred stock is callable at $140 per share, what are the book values per share of the preferred stock and the common stock?
6. If two years' preferred dividends are in arrears and the board of directors declares cash dividends of $50,000, what total amount will be paid to the preferred and to the common shareholders? What is the amount of dividends per share for the common stock?

Check (4) Book value of common, $54.58
(5) Book value of common, $53.33
(6) Dividends per common share, $0.28

Analysis Component

7. Discuss why the book value of common stock is not always a good estimate of its market value.

SERIAL PROBLEM

Success Systems

(This serial problem began in Chapter 1 and continues through most of the book. If previous chapter segments were not completed, the serial problem can begin at this point. It is helpful, but not necessary, to use the Working Papers that accompany the book.)

SP 13 Adriana Lopez created Success Systems on October 1, 2009. The company has been successful, and Adriana plans to expand her business. She believes that an additional $86,000 is needed and is investigating three funding sources.

a. Adriana's sister Cicely is willing to invest $86,000 in the business as a common shareholder. Since Adriana currently has about $129,000 invested in the business, Cicely's investment will mean that Adriana will maintain about 60% ownership, and Cicely will have 40% ownership of Success Systems.
b. Adriana's uncle Marcello is willing to invest $86,000 in the business as a preferred shareholder. Marcello would purchase 860 shares of $100 par value, 7% preferred stock.
c. Adriana's banker is willing to lend her $86,000 on a 7%, 10-year note payable. Adriana would make monthly payments of $1,000 per month for 10 years.

Required

1. Prepare the journal entry to reflect the initial $86,000 investment under each of the options (a), (b), and (c).
2. Evaluate the three proposals for expansion, providing the pros and cons of each option.
3. Which option do you recommend Adriana adopt? Explain.

BEYOND THE NUMBERS

REPORTING IN ACTION

C2 C3 A1 A4

BTN 13-1 Refer to Best Buy's financial statements in Appendix A to answer the following.

1. How many shares of common stock are issued and outstanding at March 3, 2007, and February 25, 2006? How do these numbers compare with the basic weighted-average common shares outstanding at March 3, 2007, and February 25, 2006?
2. What is the book value of its entire common stock at March 3, 2007?
3. What is the total amount of cash dividends paid to common stockholders for the years ended March 3, 2007, and February 25, 2006?
4. Identify and compare basic EPS amounts across fiscal years 2007, 2006, and 2005. Identify and comment on any marked changes.
5. Does Best Buy hold any treasury stock as of March 3, 2007? As of February 25, 2006?

Fast Forward

6. Access Best Buy's financial statements for fiscal years ending after March 3, 2007, from its Website (BestBuy.com) or the SEC's EDGAR database (www.SEC.gov). Has the number of common shares outstanding increased since March 3, 2007? Has Best Buy increased the total amount of cash dividends paid compared to the total amount for fiscal year 2007?

COMPARATIVE ANALYSIS

A1 A2 A3 A4

BTN 13-2 Key comparative figures for Best Buy, Circuit City, and RadioShack follow.

Key Figures	Best Buy	Circuit City	RadioShack
Net income (in millions)	$1,377	$ (8)	$ 73
Cash dividends declared per common share	$ 0.36	$ 0.12	$ 0.25
Common shares outstanding (in millions)	481	171	136
Weighted-average common shares outstanding (in mil.)	482	170	136
Market value (price) per share	$46.35	$19.00	$16.78
Equity applicable to common shares (in millions)	$6,201	$1,791	$ 654

Required

1. Compute the book value per common share for each company using these data.
2. Compute the basic EPS for each company using these data.
3. Compute the dividend yield for each company using these data. Does the dividend yield of any of the companies characterize it as an income or growth stock? Explain.
4. Compute, compare, and interpret the price-earnings ratio for each company using these data.

ETHICS CHALLENGE

C4

BTN 13-3 Brianna Moore is an accountant for New World Pharmaceuticals. Her duties include tracking research and development spending in the new product development division. Over the course of the past six months, Brianna notices that a great deal of funds have been spent on a particular project for a new drug. She hears "through the grapevine" that the company is about to patent the drug and expects it to be a major advance in antibiotics. Brianna believes that this new drug will greatly improve company performance and will cause the company's stock to increase in value. Brianna decides to purchase shares of New World in order to benefit from this expected increase.

Required

What are Brianna's ethical responsibilities, if any, with respect to the information she has learned through her duties as an accountant for New World Pharmaceuticals? What are the implications to her planned purchase of New World shares?

COMMUNICATING IN PRACTICE

A1 A2

Hint: Make a transparency of each team's memo for a class discussion.

BTN 13-4 Teams are to select an industry, and each team member is to select a different company in that industry. Each team member then is to acquire the selected company's financial statements (or Form 10-K) from the SEC EDGAR site (www.SEC.gov). Use these data to identify basic EPS. Use the financial press (or finance.yahoo.com) to determine the market price of this stock, and then compute the price-earnings ratio. Communicate with teammates via a meeting, e-mail, or telephone to discuss the meaning of this ratio, how companies compare, and the industry norm. The team must prepare a single memorandum reporting the ratio for each company and identifying the team conclusions or consensus of opinion. The memorandum is to be duplicated and distributed to the instructor and teammates.

TAKING IT TO THE NET

C2

BTN 13-5 Access the February 26, 2007, filing of the 2006 calendar-year 10-K report of **McDonald's**, (ticker MCD) from www.SEC.gov.

Required

1. Review McDonald's balance sheet and identify how many classes of stock it has issued.
2. What are the par values, number of authorized shares, and issued shares of the classes of stock you identified in part 1?
3. Review its statement of cash flows and identify what total amount of cash it paid in 2006 to purchase treasury stock.
4. What amount did McDonald's pay out in common stock cash dividends for 2006?

TEAMWORK IN ACTION

P5

Hint: Instructor should be sure each team accurately completes part 1 before proceeding.

BTN 13-6 This activity requires teamwork to reinforce understanding of accounting for treasury stock.

1. Write a brief team statement (*a*) generalizing what happens to a corporation's financial position when it engages in a stock "buyback" and (*b*) identifying reasons that a corporation would engage in this activity.
2. Assume that an entity acquires 100 shares of its $100 par value common stock at a cost of $134 cash per share. Discuss the entry to record this acquisition. Next, assign *each* team member to prepare *one* of the following entries (assume each entry applies to all shares):
   a. Reissue treasury shares at cost.
   b. Reissue treasury shares at $150 per share.
   c. Reissue treasury shares at $120 per share; assume the paid-in capital account from treasury shares has a $1,500 balance.
   d. Reissue treasury shares at $120 per share; assume the paid-in capital account from treasury shares has a $1,000 balance.
   e. Reissue treasury shares at $120 per share; assume the paid-in capital account from treasury shares has a zero balance.
3. In sequence, each member is to present his/her entry to the team and explain the *similarities* and *differences* between that entry and the previous entry.

ENTREPRENEURIAL DECISION

C2 C3 P2 

BTN 13-7 Assume that Ali Perry, Byron Myers, and Brenton Taylor of **Inogen** decide to launch a new retail chain to market their portable oxygen systems. This chain, named O-to-Go, requires $500,000 of start-up capital. The three contribute $375,000 of personal assets in return for 15,000 shares of common stock, but they need to raise another $125,000 in cash. There are two alternative plans for raising the additional cash. Plan A is to sell 3,750 shares of common stock to one or more investors for $125,000 cash. Plan B is to sell 1,250 shares of cumulative preferred stock to one or more investors for $125,000 cash (this preferred stock would have a $100 par value, an annual 8% dividend rate, and be issued at par).

1. If the new business is expected to earn $72,000 of after-tax net income in the first year, what rate of return on beginning equity will the three (as a group) earn under each alternative? Which plan will provide the higher expected return to them?
2. If the new business is expected to earn $16,800 of after-tax net income in the first year, what rate of return on beginning equity will the three (as a group) earn under each alternative? Which plan will provide the higher expected return to them?
3. Analyze and interpret the differences between the results for parts 1 and 2.

HITTING THE ROAD

A1 A2 A3

BTN 13-8 Watch 30 to 60 minutes of financial news programming on television. Take notes on companies that are catching analysts' attention. You might hear reference to over- and undervaluation of firms and to reports about PE ratios, dividend yields, and earnings per share. Be prepared to give a brief description to the class of your observations.

GLOBAL DECISION

A1 C4

DSG

BTN 13-9 Financial information for **DSG international plc** (www.DSGiplc.com) follows.

Net income (in millions)	£ 207
Cash dividends declared per share	£ 0.07
Number of shares outstanding (in millions)*	1,843
Equity applicable to shares (in millions)	£1,304

* Assume that the year-end number of shares outstanding approximates the weighted-average shares outstanding.

Required

1. Compute book value per share for DSG.
2. Compute earnings per share (EPS) for DSG.
3. Compare DSG's dividends per share with its EPS. Is DSG paying out a large or small amount of its income as dividends? Explain.

ANSWERS TO MULTIPLE CHOICE QUIZ

1. e; Entry to record this stock issuance is:

Cash (6,000 × $8)	48,000	
Common Stock (6,000 × $5)		30,000
Paid-In Capital in Excess of Par Value, Common Stock		18,000

2. b; $75,000/19,000 shares = $3.95 per share
3. d; Preferred stock = 5,000 × $100 = $500,000
   Book value per share = ($2,000,000 − $500,000)/50,000 shares = $30 per common share
4. a; $0.81/$45.00 = 1.8%
5. c; Earnings per share = $3,500,000/700,000 shares = $5 per share
   PE ratio = $85/$5 = 17.0

A Look Back

Chapter 13 focused on corporate equity transactions, including stock issuances and dividends. We also explained how to report and analyze income, earnings per share, and retained earnings.

A Look at This Chapter

This chapter describes the accounting for and analysis of bonds and notes. We explain their characteristics, payment patterns, interest computations, retirement, and reporting requirements. An appendix to this chapter introduces leases and pensions.

A Look Ahead

Chapter 15 focuses on how to classify, account for, and report investments in both debt and equity securities. We also describe accounting for transactions listed in a foreign currency.

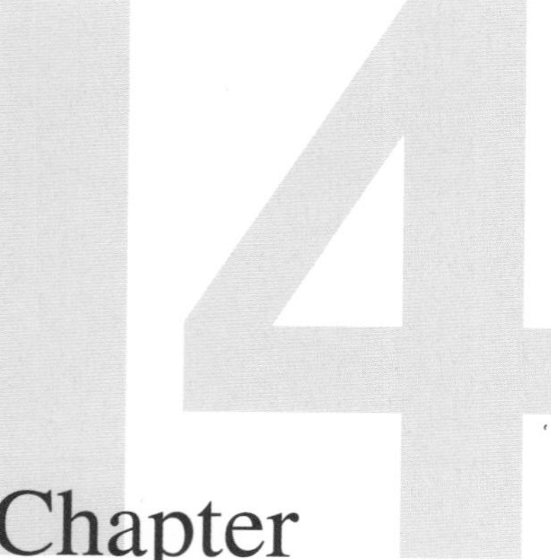

Chapter 14

Long-Term Liabilities

Learning Objectives

CAP

Conceptual

C1 Explain the types and payment patterns of notes. *(p. 560)*

C2 *Appendix 14A*—Explain and compute the present value of an amount(s) to be paid at a future date(s). *(p. 567)*

C3 *Appendix 14C*—Describe the accrual of bond interest when bond payments do not align with accounting periods. *(p. 571)*

C4 *Appendix 14D*—Describe accounting for leases and pensions. *(p. 573)*

Analytical

A1 Compare bond financing with stock financing. *(p. 550)*

A2 Assess debt features and their implications. *(p. 563)*

A3 Compute the debt-to-equity ratio and explain its use. *(p. 563)*

LP14

Procedural

P1 Prepare entries to record bond issuance and bond interest expense. *(p. 552)*

P2 Compute and record amortization of bond discount. *(p. 553)*

P3 Compute and record amortization of bond premium. *(p. 556)*

P4 Record the retirement of bonds. *(p. 559)*

P5 Prepare entries to account for notes. *(p. 562)*

Decision Feature

Hip-Hop Financing

"I wanted to create a product that inner-city kids would relate to" — James Lindsay

PHILADELPHIA—James "Fly" Lindsay and his three sisters were raised by their mother in North Philadelphia. "We were poor," explains Lindsay, "but we were rich in the values that she [mother] set forth." One of those values was education—Lindsay was the first in his family to attend college—and another was the commitment to community. "You have everyone taking away, but you have to give back," insists Lindsay.

Lindsay decided to give back by launching his own business. "I wanted to have a chip company that kids from the hood could relate to," he explains. But Lindsay knew that success depended on securing financing and then planning for long-term liabilities. "Between family and friends," explains Lindsay, "we put our dollars together and made it happen." With $40,000 in financing, he launched **Rap Snacks** **[RapSnacks.com]**, a maker of snack foods with a twist: Lindsay would sell his snacks with rappers on the wrappers. The snacks would be targeted to urban youth immersed in hip-hop culture. The snack wrappers, importantly, would include positive messages to the kids such as "Stay in School," "Respect Yourself," and "Money equals Education." Lindsay hopes they instill a craving for success, commitment, and entrepreneurship.

Lindsay insists that urban youth need to learn about business and how it can help a community. Basic accounting principles and financing concepts such as bonds and notes are like another language to urban youth, he explains. Lindsay struggles to change that and more. "You can make money," says Lindsay, "but also have a social responsibility to where you came from." Actor and rapper Lil' Romeo has recently taken a lead in funding Rap Snacks.

Making investments in urban youth is a priority, says Lindsay. He continually works to convey knowledge of business financing, with the belief that urban youth can be successful. He manages his own company's long-term liabilities, interest payments, and collateral agreements, and is confident that urban youth can do the same. Understanding liabilities is not easy, says Lindsay, but the costs of not understanding are more severe—and more of the same—for the kids. "It feels good to make a difference," says Lindsay, "and set an example for the kids."

[Sources: *Rap Snacks Website,* January 2009; *Entrepreneur,* July 2003; *Philadelphia Inquirer,* October 2002; *Source Magazine,* June 2001; *Maxim,* March 2002; *Business Review,* September 2007; *PR Newswire,* September 2007]

Chapter Preview

Individuals, companies, and governments issue bonds to finance their activities. In return for financing, bonds promise to repay the lender with interest. This chapter explains the basics of bonds and the accounting for their issuance and retirement. The chapter also describes long-term notes as another financing source. We explain how present value concepts impact both the accounting for and reporting of bonds and notes. Appendixes to this chapter discuss present value concepts applicable to liabilities, effective interest amortization, and the accounting for leases and pensions.

Long-Term Liabilities

Bond Basics
- Bond financing
- Bond trading
- Issuance procedures

Bond Issuances
- Issuance at par
- Issuance at a discount
- Issuance at a premium
- Bond pricing

Bond Retirement
- At maturity
- Before maturity
- By conversion

Long-Term Notes
- Installment notes
- Mortgage terms

Basics of Bonds

Video14.2

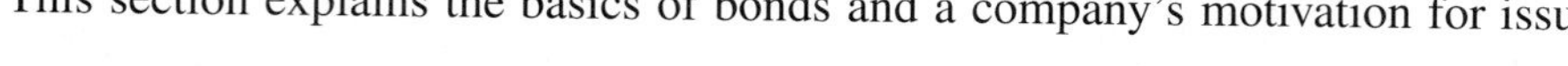

This section explains the basics of bonds and a company's motivation for issuing them.

A1 Compare bond financing with stock financing.

Bond Financing

Projects that demand large amounts of money often are funded from bond issuances. (Both for-profit and nonprofit companies, as well as governmental units, such as nations, states, cities, and school districts, issue bonds.) A **bond** is its issuer's written promise to pay an amount identified as the par value of the bond with interest. The **par value of a bond,** also called the *face amount* or *face value,* is paid at a specified future date known as the bond's *maturity date.* Most bonds also require the issuer to make semiannual interest payments. The amount of interest paid each period is determined by multiplying the par value of the bond by the bond's contract rate of interest. This section explains both advantages and disadvantages of bond financing.

Advantages of Bonds There are three main advantages of bond financing:

1. *Bonds do not affect owner control.* Equity financing reflects ownership in a company, whereas bond financing does not. A person who contributes $1,000 of a company's $10,000 equity financing typically controls one-tenth of all owner decisions. A person who owns a $1,000, 11%, 20-year bond has no ownership right. This person, or bondholder, is to receive from the bond issuer 11% interest, or $110, each year the bond is outstanding and $1,000 when it matures in 20 years.

2. *Interest on bonds is tax deductible.* Bond interest payments are tax deductible for the issuer, but equity payments (distributions) to owners are not. To illustrate, assume that a corporation with no bond financing earns $15,000 in income *before* paying taxes at a 40% tax rate, which amounts to $6,000 ($15,000 × 40%) in taxes. If a portion of its financing is in bonds, however, the resulting bond interest is deducted in computing taxable income. That is, if bond interest expense is $10,000, the taxes owed would be $2,000 ([$15,000 − $10,000] × 40%), which is less than the $6,000 owed with no bond financing.

3. *Bonds can increase return on equity.* A company that earns a higher return with borrowed funds than it pays in interest on those funds increases its return on equity. This process is called *financial leverage* or *trading on the equity.*

Point: Financial leverage reflects issuance of bonds, notes, or preferred stock.

To illustrate the third point, consider Magnum Co., which has $1 million in equity and is planning a $500,000 expansion to meet increasing demand for its product. Magnum predicts the

$500,000 expansion will yield $125,000 in additional income before paying any interest. It currently earns $100,000 per year and has no interest expense. Magnum is considering three plans. Plan A is to not expand. Plan B is to expand and raise $500,000 from equity financing. Plan C is to expand and issue $500,000 of bonds that pay 10% annual interest ($50,000). Exhibit 14.1 shows how these three plans affect Magnum's net income, equity, and return on equity (net income/equity). The owner(s) will earn a higher return on equity if expansion occurs. Moreover, the preferred expansion plan is to issue bonds. Projected net income under Plan C ($175,000) is smaller than under Plan B ($225,000), but the return on equity is larger because of less equity investment. Plan C has another advantage if income is taxable. This illustration reflects a general rule: *Return on equity increases when the expected rate of return from the new assets is higher than the rate of interest expense on the debt financing.*

Example: Compute return on equity for all three plans if Magnum currently earns $150,000 instead of $100,000.
Answer ($ 000s):
Plan A = 15% ($150/$1,000)
Plan B = 18.3% ($275/$1,500)
Plan C = 22.5% ($225/$1,000)

EXHIBIT 14.1

Financing with Bonds versus Equity

	Plan A: Do Not Expand	Plan B: Equity Financing	Plan C: Bond Financing
Income before interest expense	$ 100,000	$ 225,000	$ 225,000
Interest expense	—	—	(50,000)
Net income	**$ 100,000**	**$ 225,000**	**$ 175,000**
Equity	$1,000,000	$1,500,000	$1,000,000
Return on equity	**10.0%**	**15.0%**	**17.5%**

Disadvantages of Bonds The two main disadvantages of bond financing are these:

1. *Bonds can decrease return on equity.* When a company earns a lower return with the borrowed funds than it pays in interest, it decreases its return on equity. This downside risk of financial leverage is more likely to arise when a company has periods of low income or net losses.
2. *Bonds require payment of both periodic interest and the par value at maturity.* Bond payments can be especially burdensome when income and cash flow are low. Equity financing, in contrast, does not require any payments because cash withdrawals (dividends) are paid at the discretion of the owner (or board).

Point: Debt financing is desirable when interest is tax deductible, when owner control is preferred, and when return on equity exceeds the debt's interest rate.

A company must weigh the risks and returns of the disadvantages and advantages of bond financing when deciding whether to issue bonds to finance operations.

Bond Trading

Bonds are securities that can be readily bought and sold. A large number of bonds trade on both the New York Exchange and the American Exchange. A bond *issue* consists of a number of bonds, usually in denominations of $1,000 or $5,000, and is sold to many different lenders. After bonds are issued, they often are bought and sold by investors, meaning that any particular bond probably has a number of owners before it matures. Since bonds are exchanged (bought and sold) in the market, they have a market value (price). For convenience, bond market values are expressed as a percent of their par (face) value. For example, a company's bonds might be trading at 103½, meaning they can be bought or sold for 103.5% of their par value. Bonds can also trade below par value. For instance, if a company's bonds are trading at 95, they can be bought or sold at 95% of their par value.

Decision Insight

Bonds	Rate	Mat	Yld	Vol	Close	Chg
IBM	7	25	5.9	130	119¼	+1¼

Quotes The **IBM** bond quote here is interpreted (left to right) as **Bonds,** issuer name; **Rate,** contract interest rate (7%); **Mat,** matures in year 2025 when principal is paid; **Yld,** yield rate (5.9%) of bond at current price; **Vol,** daily dollar worth ($130,000) of trades (in 1,000s); **Close,** closing price (119.25) for the day as percentage of par value; **Chg,** change (+1.25) in closing price from prior day's close.

Bond-Issuing Procedures

State and federal laws govern bond issuances. Bond issuers also want to ensure that they do not violate any of their existing contractual agreements when issuing bonds. Authorization of bond issuances includes the number of bonds authorized, their par value, and the contract interest rate. The legal document identifying the rights and obligations of both the bondholders and the issuer is called the **bond indenture,** which is the legal contract between the issuer and the bondholders. A bondholder may also receive a bond certificate as evidence of the company's debt. A **bond certificate,** such as that shown in Exhibit 14.2, includes specifics such as the issuer's name, the par value, the contract interest rate, and the maturity date. Many companies reduce costs by not issuing paper certificates to bondholders.[1]

EXHIBIT 14.2

Bond Certificate

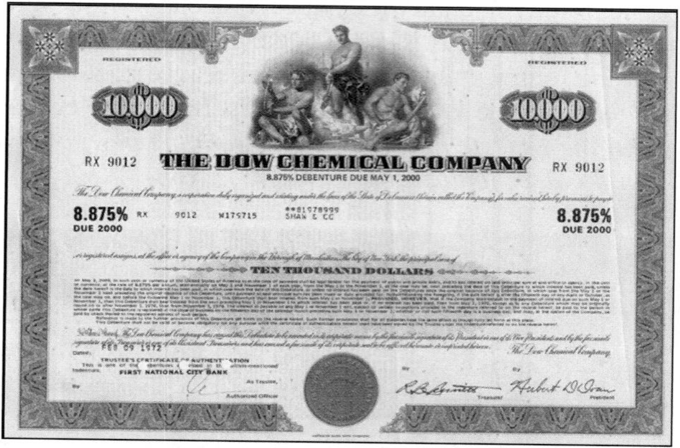

Point: *Indenture* refers to a bond's legal contract; *debenture* refers to an unsecured bond.

Bond Issuances

Video14.2

This section explains accounting for bond issuances at par, below par (discount), and above par (premium). It also describes how to amortize a discount or premium and record bonds issued between interest payment dates.

Issuing Bonds at Par

P1 Prepare entries to record bond issuance and bond interest expense.

To illustrate an issuance of bonds at par value, suppose a company receives authorization to issue $800,000 of 9%, 20-year bonds dated January 1, 2009, that mature on December 31, 2028, and pay interest semiannually on each June 30 and December 31. After accepting the bond indenture on behalf of the bondholders, the trustee can sell all or a portion of the bonds to an underwriter. If all bonds are sold at par value, the issuer records the sale as follows.

Assets	=	Liabilities	+	Equity
+800,000		+800,000		

2009			
Jan. 1	Cash	800,000	
	Bonds Payable.		800,000
	Sold bonds at par.		

This entry reflects increases in the issuer's cash *and* long-term liabilities.

The issuer records the first semiannual interest payment as follows.

Assets	=	Liabilities	+	Equity
−36,000				−36,000

2009			
June 30	Bond Interest Expense	36,000	
	Cash		36,000
	Paid semiannual interest (9% × $800,000 × ½ year).		

Point: The *spread* between the dealer's cost and what buyers pay can be huge. Dealers earn more than $25 billion in annual spread revenue.

Global: In the United Kingdom, government bonds are called *gilts*—short for gilt-edged investments.

[1] The issuing company normally sells its bonds to an investment firm called an *underwriter,* which resells them to the public. An issuing company can also sell bonds directly to investors. When an underwriter sells bonds to a large number of investors, a *trustee* represents and protects the bondholders' interests. The trustee monitors the issuer to ensure that it complies with the obligations in the bond indenture. Most trustees are large banks or trust companies. The trustee writes and accepts the terms of a bond indenture before it is issued. When bonds are offered to the public, called *floating an issue,* they must be registered with the Securities and Exchange Commission (SEC). SEC registration requires the issuer to file certain financial information. Most company bonds are issued in par value units of $1,000 or $5,000. *A baby bond* has a par value of less than $1,000, such as $100.

The issuer pays and records its semiannual interest obligation every six months until the bonds mature. When they mature, the issuer records its payment of principal as follows.

2028			
Dec. 31	Bonds Payable	800,000	
	Cash		800,000
	Paid bond principal at maturity.		

Assets = Liabilities + Equity
−800,000 −800,000

Bond Discount or Premium

The bond issuer pays the interest rate specified in the indenture, the **contract rate,** also referred to as the *coupon rate, stated rate,* or *nominal rate*. The annual interest paid is determined by multiplying the bond par value by the contract rate. The contract rate is usually stated on an annual basis, even if interest is paid semiannually. For example, if a company issues a $1,000, 8% bond paying interest semiannually, it pays annual interest of $80 (8% × $1,000) in two semiannual payments of $40 each.

The contract rate sets the amount of interest the issuer pays in *cash,* which is not necessarily the *bond interest expense* actually incurred by the issuer. Bond interest expense depends on the bond's market value at issuance, which is determined by market expectations of the risk of lending to the issuer. The bond's **market rate** of interest is the rate that borrowers are willing to pay and lenders are willing to accept for a particular bond and its risk level. As the risk level increases, the rate increases to compensate purchasers for the bonds' increased risk. Also, the market rate is generally higher when the time period until the bond matures is longer due to the risk of adverse events occurring over a longer time period.

Many bond issuers try to set a contract rate of interest equal to the market rate they expect as of the bond issuance date. When the contract rate and market rate are equal, a bond sells at par value, but when they are not equal, a bond does not sell at par value. Instead, it is sold at a *premium* above par value or at a *discount* below par value. Exhibit 14.3 shows the relation between the contract rate, market rate, and a bond's issue price.

EXHIBIT 14.3

Relation between Bond Issue Price, Contract Rate, and Market Rate

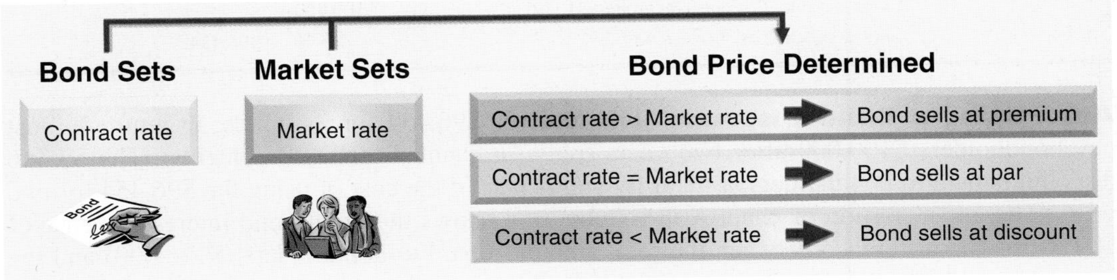

Quick Check

Answers—p. 576

1. Unsecured bonds backed only by the issuer's general credit standing are called (*a*) serial bonds, (*b*) debentures, (*c*) registered bonds, or (*d*) convertible bonds.
2. How do you compute the amount of interest a bond issuer pays in cash each year?
3. When the contract rate is above the market rate, do bonds sell at a premium or a discount? Do purchasers pay more or less than the par value of the bonds?

Issuing Bonds at a Discount

P2 Compute and record amortization of bond discount.

A **discount on bonds payable** occurs when a company issues bonds with a contract rate less than the market rate. This means that the issue price is less than par value. To illustrate, assume that **Fila** announces an offer to issue bonds with a $100,000 par value, an 8% annual contract rate (paid semiannually), and a two-year life. Also assume that the market rate for Fila

Point: The difference between the contract rate and the market rate of interest on a new bond issue is usually a fraction of a percent. We use a difference of 2% to emphasize the effects.

bonds is 10%. These bonds then will sell at a discount since the contract rate is less than the market rate. The exact issue price for these bonds is stated as 96.454 (implying 96.454% of par value, or $96,454); we show how to compute this issue price later in the chapter. These bonds obligate the issuer to pay two separate types of future cash flows:

1. Par value of $100,000 cash at the end of the bonds' two-year life.
2. Cash interest payments of $4,000 (4% × $100,000) at the end of each semiannual period during the bonds' two-year life.

The exact pattern of cash flows for the Fila bonds is shown in Exhibit 14.4.

EXHIBIT 14.4

Cash Flows for Fila Bonds

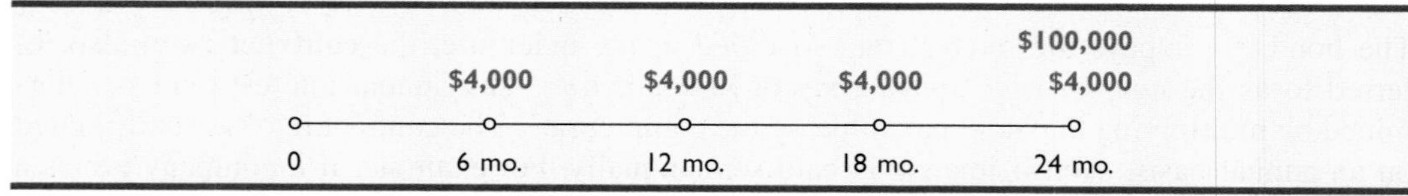

When Fila accepts $96,454 cash for its bonds on the issue date of December 31, 2009, it records the sale as follows.

Assets	=	Liabilities	+	Equity
+96,454		+100,000		
		−3,546		

Dec. 31	Cash	96,454	
	Discount on Bonds Payable	3,546	
	Bonds Payable		100,000
	Sold bonds at a discount on their issue date.		

Point: Book value at issuance always equals the issuer's cash borrowed.

These bonds are reported in the long-term liability section of the issuer's December 31, 2009, balance sheet as shown in Exhibit 14.5. A discount is deducted from the par value of bonds to yield the **carrying (book) value of bonds.** Discount on Bonds Payable is a contra liability account.

EXHIBIT 14.5

Balance Sheet Presentation of Bond Discount

Long-term liabilities		
Bonds payable, 8%, due December 31, 2011	$100,000	
Less discount on bonds payable	3,546	$96,454

Amortizing a Bond Discount Fila receives $96,454 for its bonds; in return it must pay bondholders $100,000 after two years (plus semiannual interest payments). The $3,546 discount is paid to bondholders at maturity and is part of the cost of using the $96,454 for two years. The upper portion of panel A in Exhibit 14.6 shows that total bond interest expense of $19,546 is the difference between the total amount repaid to bondholders ($116,000) and the amount borrowed from bondholders ($96,454). Alternatively, we can compute total bond interest expense as the sum of the four interest payments and the bond discount. This alternative computation is shown in the lower portion of panel A.

Point: *Zero-coupon bonds* do not pay periodic interest (contract rate is zero). These bonds always sell at a discount because their 0% contract rate is always below the market rate.

The total $19,546 bond interest expense must be allocated across the four semiannual periods in the bonds' life, and the bonds' carrying value must be updated at each balance sheet date. This is accomplished using the straight-line method (or the effective interest method in Appendix 14B). Both methods systematically reduce the bond discount to zero over the two-year life. This process is called *amortizing a bond discount.*

Video14.2

Straight-Line Method The **straight-line bond amortization** method allocates an equal portion of the total bond interest expense to each interest period. To apply the straight-line method to Fila's bonds, we divide the total bond interest expense of $19,546 by 4 (the number of semiannual periods in the bonds' life). This gives a bond interest expense of $4,887 per period, which is $4,886.5 rounded to the nearest dollar per period (all computations, including those for assignments, are rounded to the nearest whole dollar). Alternatively, we can find this number by first dividing the $3,546 discount by 4, which yields the $887 amount of discount to be amortized each interest period. When the $887 is added to the $4,000 cash payment, the

EXHIBIT 14.6

Interest Computation and Entry for Bonds Issued at a Discount

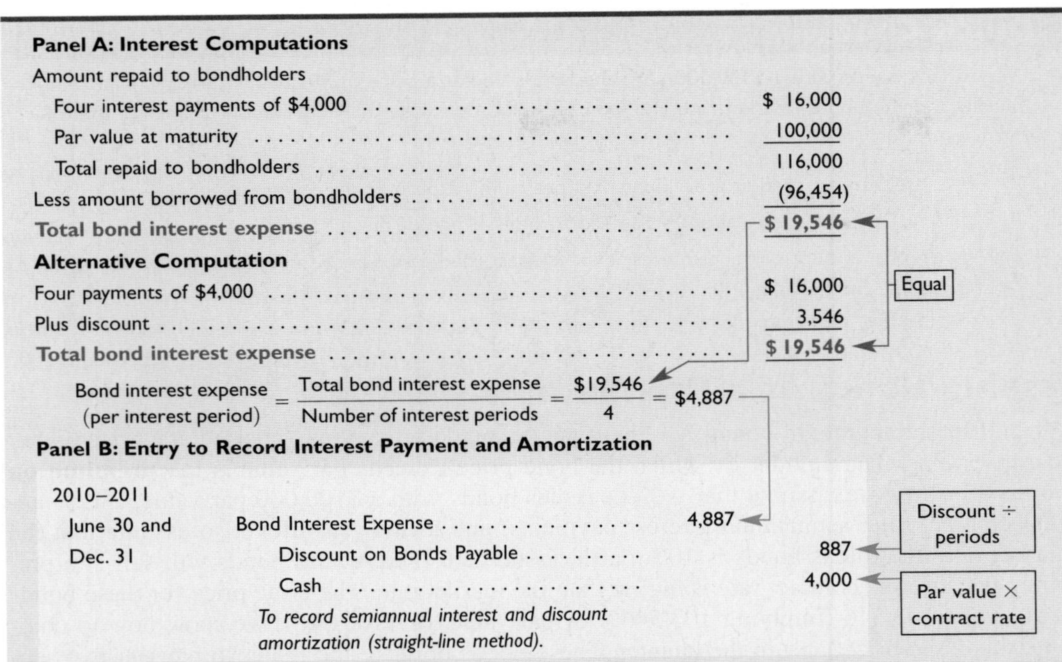

Panel A: Interest Computations

Amount repaid to bondholders	
Four interest payments of $4,000	$ 16,000
Par value at maturity	100,000
Total repaid to bondholders	116,000
Less amount borrowed from bondholders	(96,454)
Total bond interest expense	**$ 19,546**
Alternative Computation	
Four payments of $4,000	$ 16,000
Plus discount	3,546
Total bond interest expense	**$ 19,546**

Equal

$$\text{Bond interest expense (per interest period)} = \frac{\text{Total bond interest expense}}{\text{Number of interest periods}} = \frac{\$19{,}546}{4} = \$4{,}887$$

Panel B: Entry to Record Interest Payment and Amortization

2010–2011			
June 30 and Dec. 31	Bond Interest Expense	4,887	
	Discount on Bonds Payable		887
	Cash		4,000
	To record semiannual interest and discount amortization (straight-line method).		

Discount ÷ periods

Par value × contract rate

bond interest expense for each period is $4,887. Panel B of Exhibit 14.6 shows how the issuer records bond interest expense and updates the balance of the bond liability account at the end of *each* of the four semiannual interest periods (June 30, 2010, through December 31, 2011).

Exhibit 14.7 shows the pattern of decreases in the Discount on Bonds Payable account and the pattern of increases in the bonds' carrying value. The following points summarize the discount bonds' straight-line amortization:

EXHIBIT 14.7

Straight-Line Amortization of Bond Discount

Semiannual Period-End	Unamortized Discount*	Carrying Value†
(0) 12/31/2009	$3,546	$ 96,454
(1) 6/30/2010	2,659	97,341
(2) 12/31/2010	1,772	98,228
(3) 6/30/2011	885	99,115
(4) **12/31/2011**	**0‡**	**100,000**

* Total bond discount (of $3,546) less accumulated periodic amortization ($887 per semiannual interest period).

† Bond par value (of $100,000) less unamortized discount.

‡ Adjusted for rounding.

The two columns always sum to par value for a discount bond.

1. At issuance, the $100,000 par value consists of the $96,454 cash received by the issuer plus the $3,546 discount.
2. During the bonds' life, the (unamortized) discount decreases each period by the $887 amortization ($3,546/4), and the carrying value (par value less unamortized discount) increases each period by $887.
3. At maturity, the unamortized discount equals zero, and the carrying value equals the $100,000 par value that the issuer pays the holder.

We see that the issuer incurs a $4,887 bond interest expense each period but pays only $4,000 cash. The $887 unpaid portion of this expense is added to the bonds' carrying value. (The total $3,546 unamortized discount is "paid" when the bonds mature; $100,000 is paid at maturity but only $96,454 was received at issuance.)

Decision Insight

Ratings Game Many bond buyers rely on rating services to assess bond risk. The best known are **Standard & Poor's** and **Moody's**. These services focus on the issuer's financial statements and other factors in setting ratings. Standard & Poor's ratings, from best quality to default, are AAA, AA, A, BBB, BB, B, CCC, CC, C, and D. Ratings can include a plus (+) or minus (−) to show relative standing within a category.

Quick Check Answers—p. 576

Five-year, 6% bonds with a $100,000 par value are issued at a price of $91,893. Interest is paid semiannually, and the bonds' market rate is 8% on the issue date. Use this information to answer the following questions:

4. Are these bonds issued at a discount or a premium? Explain your answer.
5. What is the issuer's journal entry to record the issuance of these bonds?
6. What is the amount of bond interest expense recorded at the first semiannual period using the straight-line method?

Issuing Bonds at a Premium

P3 Compute and record amortization of bond premium.

When the contract rate of bonds is higher than the market rate, the bonds sell at a price higher than par value. The amount by which the bond price exceeds par value is the **premium on bonds.** To illustrate, assume that Adidas issues bonds with a $100,000 par value, a 12% annual contract rate, semiannual interest payments, and a two-year life. Also assume that the market rate for Adidas bonds is 10% on the issue date. The Adidas bonds will sell at a premium because the contract rate is higher than the market rate. The issue price for these bonds is stated as 103.546 (implying 103.546% of par value, or $103,546); we show how to compute this issue price later in the chapter. These bonds obligate the issuer to pay out two separate future cash flows:

1. Par value of $100,000 cash at the end of the bonds' two-year life.
2. Cash interest payments of $6,000 (6% × $100,000) at the end of each semiannual period during the bonds' two-year life.

The exact pattern of cash flows for the Adidas bonds is shown in Exhibit 14.8.

EXHIBIT 14.8

Cash Flows for Adidas Bonds

0	6 mo.	12 mo.	18 mo.	24 mo.
	$6,000	$6,000	$6,000	$100,000 $6,000

When Adidas accepts $103,546 cash for its bonds on the issue date of December 31, 2009, it records this transaction as follows.

Assets	=	Liabilities	+	Equity
+103,546		+100,000 +3,546		

Dec. 31	Cash	103,546	
	Premium on Bonds Payable		3,546
	Bonds Payable		100,000
	Sold bonds at a premium on their issue date.		

These bonds are reported in the long-term liability section of the issuer's December 31, 2009, balance sheet as shown in Exhibit 14.9. A premium is added to par value to yield the carrying (book) value of bonds. Premium on Bonds Payable is an adjunct (also called *accretion*) liability account.

EXHIBIT 14.9

Balance Sheet Presentation of Bond Premium

Long-term liabilities		
Bonds payable, 12%, due December 31, 2011	$100,000	
Plus premium on bonds payable	3,546	$103,546

Amortizing a Bond Premium Adidas receives $103,546 for its bonds; in return, it pays bondholders $100,000 after two years (plus semiannual interest payments). The $3,546 premium not repaid to issuer's bondholders at maturity goes to reduce the issuer's expense of using the $103,546 for two years. The upper portion of panel A of Exhibit 14.10 shows that total bond interest expense of $20,454 is the difference between the total amount repaid to bondholders ($124,000) and the amount borrowed from bondholders ($103,546). Alternatively, we can compute

EXHIBIT 14.10

Interest Computation and Entry for Bonds Issued at a Premium

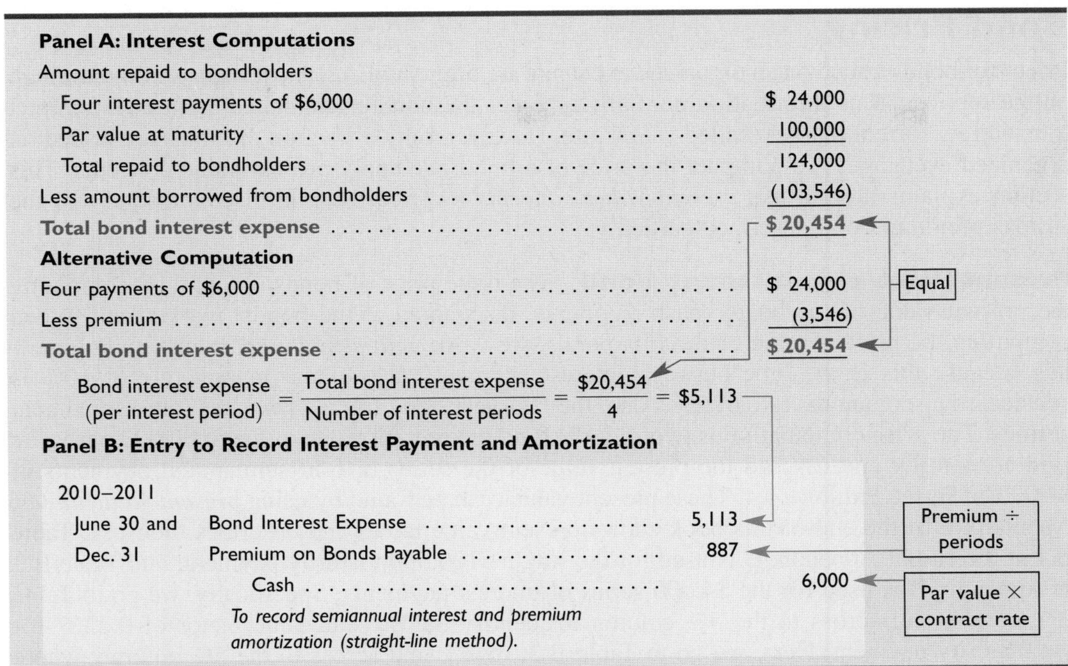

Panel A: Interest Computations

Amount repaid to bondholders	
Four interest payments of $6,000	$ 24,000
Par value at maturity	100,000
Total repaid to bondholders	124,000
Less amount borrowed from bondholders	(103,546)
Total bond interest expense	**$ 20,454**
Alternative Computation	
Four payments of $6,000	$ 24,000
Less premium	(3,546)
Total bond interest expense	**$ 20,454**

$$\text{Bond interest expense (per interest period)} = \frac{\text{Total bond interest expense}}{\text{Number of interest periods}} = \frac{\$20{,}454}{4} = \$5{,}113$$

Panel B: Entry to Record Interest Payment and Amortization

2010–2011			
June 30 and Dec. 31	Bond Interest Expense	5,113	
	Premium on Bonds Payable	887	
	Cash		6,000
	To record semiannual interest and premium amortization (straight-line method).		

total bond interest expense as the sum of the four interest payments less the bond premium. The premium is subtracted because it will not be paid to bondholders when the bonds mature; see the lower portion of panel A. Total bond interest expense must be allocated over the four semiannual periods using the straight-line method (or the effective interest method in Appendix 14B).

Point: A premium decreases Bond Interest Expense; a discount increases it.

Straight-Line Method The straight-line method allocates an equal portion of total bond interest expense to each of the bonds' semiannual interest periods. To apply this method to Adidas bonds, we divide the two years' total bond interest expense of $20,454 by 4 (the number of semiannual periods in the bonds' life). This gives a total bond interest expense of $5,113 per period, which is $5,113.5 rounded down so that the journal entry balances and for simplicity in presentation (alternatively, one could carry cents). Panel B of Exhibit 14.10 shows how the issuer records bond interest expense and updates the balance of the bond liability account for *each* semiannual period (June 30, 2010, through December 31, 2011).

EXHIBIT 14.11

Straight-Line Amortization of Bond Premium

Semiannual Period-End	Unamortized Premium*	Carrying Value†
(0) 12/31/2009	$3,546	$103,546
(1) 6/30/2010	2,659	102,659
(2) 12/31/2010	1,772	101,772
(3) 6/30/2011	885	100,885
(4) 12/31/2011	**0‡**	**100,000**

* Total bond premium (of $3,546) less accumulated periodic amortization ($887 per semiannual interest period).

† Bond par value (of $100,000) plus unamortized premium.

‡ Adjusted for rounding.

During the bond life, carrying value is adjusted to par and the amortized premium to zero.

Exhibit 14.11 shows the pattern of decreases in the unamortized Premium on Bonds Payable account and in the bonds' carrying value. The following points summarize straight-line amortization of the premium bonds:

1. At issuance, the $100,000 par value plus the $3,546 premium equals the $103,546 cash received by the issuer.
2. During the bonds' life, the (unamortized) premium decreases each period by the $887 amortization ($3,546/4), and the carrying value decreases each period by the same $887.
3. At maturity, the unamortized premium equals zero, and the carrying value equals the $100,000 par value that the issuer pays the holder.

The next section describes bond pricing. An instructor can choose to cover bond pricing or not. Assignments requiring the next section are Quick Study 14-4, Exercises 14-9 & 14-10, and Problems 14-1A & 14-1B and 14-4A & 14-4B.

Bond Pricing

Prices for bonds traded on an organized exchange are often published in newspapers and through online services. This information normally includes the bond price (called *quote*), its contract rate, and its current market (called *yield*) rate. However, only a fraction of bonds are traded on organized exchanges. To compute the price of a bond, we apply present value concepts. This section explains how to use *present value concepts* to price the Fila discount bond and the Adidas premium bond described earlier.

Point: InvestingInBonds.com is a bond research and learning source.

Present Value of a Discount Bond The issue price of bonds is found by computing the present value of the bonds' cash payments, discounted at the bonds' market rate. When computing the present value of the Fila bonds, we work with *semiannual* compounding periods because this is the time between interest payments; the annual market rate of 10% is considered a semiannual rate of 5%. Also, the two-year bond life is viewed as four semiannual periods. The price computation is twofold: (1) Find the present value of the $100,000 par value paid at maturity and (2) find the present value of the series of four semiannual payments of $4,000 each; see Exhibit 14.4. These present values can be found by using *present value tables.* Appendix B at the end of this book shows present value tables and describes their use. Table B.1 at the end of Appendix B is used for the single $100,000 maturity payment, and Table B.3 in Appendix B is used for the $4,000 series of interest payments. Specifically, we go to Table B.1, row 4, and across to the 5% column to identify the present value factor of 0.8227 for the maturity payment. Next, we go to Table B.3, row 4, and across to the 5% column, where the present value factor is 3.5460 for the series of interest payments. We compute bond price by multiplying the cash flow payments by their corresponding present value factors and adding them together; see Exhibit 14.12.

Point: A bond's market value (price) at issuance equals the present value of its future cash payments, where the interest (discount) rate used is the bond's market rate.

Point: Many calculators have present value functions for computing bond prices.

EXHIBIT 14.12

Computing Issue Price for the Fila Discount Bonds

Cash Flow	Table	Present Value Factor		Amount		Present Value
$100,000 par (maturity) value	B.1	0.8227	×	$100,000	=	$ 82,270
$4,000 interest payments	B.3	3.5460	×	4,000	=	14,184
Price of bond						$96,454

Present Value of a Premium Bond We find the issue price of the Adidas bonds by using the market rate to compute the present value of the bonds' future cash flows. When computing the present value of these bonds, we again work with *semiannual* compounding periods because this is the time between interest payments. The annual 10% market rate is applied as a semiannual rate of 5%, and the two-year bond life is viewed as four semiannual periods. The computation is twofold: (1) Find the present value of the $100,000 par value paid at maturity and (2) find the present value of the series of four payments of $6,000 each; see Exhibit 14.8. These present values can be found by using present value tables. First, go to Table B.1, row 4, and across to the 5% column where the present value factor is 0.8227 for the maturity payment. Second, go to Table B.3, row 4, and across to the 5% column, where the present value factor is 3.5460 for the series of interest payments. The bonds' price is computed by multiplying the cash flow payments by their corresponding present value factors and adding them together; see Exhibit 14.13.

Point: There are nearly 5 million individual U.S. bond issues, ranging from huge treasuries to tiny municipalities. This compares to about 12,000 individual U.S. stocks that are traded.

EXHIBIT 14.13

Computing Issue Price for the Adidas Premium Bonds

Cash Flow	Table	Present Value Factor		Amount		Present Value
$100,000 par (maturity) value	B.1	0.8227	×	$100,000	=	$ 82,270
$6,000 interest payments	B.3	3.5460	×	6,000	=	21,276
Price of bond						$103,546

Quick Check

Answers—p. 576

On December 31, 2008, a company issues 16%, 10-year bonds with a par value of $100,000. Interest is paid on June 30 and December 31. The bonds are sold to yield a 14% annual market rate at an issue price of $110,592. Use this information to answer questions 7 through 9:

7. Are these bonds issued at a discount or a premium? Explain your answer.
8. Using the straight-line method to allocate bond interest expense, the issuer records the second interest payment (on December 31, 2009) with a debit to Premium on Bonds Payable in the amount of (*a*) $7,470, (*b*) $530, (*c*) $8,000, or (*d*) $400.
9. How are these bonds reported in the long-term liability section of the issuer's balance sheet as of December 31, 2009?

Bond Retirement

This section describes the retirement of bonds (1) at maturity, (2) before maturity, and (3) by conversion to stock.

Bond Retirement at Maturity

P4 Record the retirement of bonds.

The carrying value of bonds at maturity always equals par value. For example, both Exhibits 14.7 (a discount) and 14.11 (a premium) show that the carrying value of bonds at the end of their lives equals par value ($100,000). The retirement of these bonds at maturity, assuming interest is already paid and entered, is recorded as follows:

2011			
Dec. 31	Bonds Payable	100,000	
	Cash		100,000
	To record retirement of bonds at maturity.		

Assets	= Liabilities	+ Equity
−100,000	−100,000	

Bond Retirement before Maturity

Issuers sometimes wish to retire some or all of their bonds prior to maturity. For instance, if interest rates decline greatly, an issuer may wish to replace high-interest-paying bonds with new low-interest bonds. Two common ways to retire bonds before maturity are to (1) exercise a call option or (2) purchase them on the open market. In the first instance, an issuer can reserve the right to retire bonds early by issuing callable bonds. The bond indenture can give the issuer an option to *call* the bonds before they mature by paying the par value plus a *call premium* to bondholders. In the second case, the issuer retires bonds by repurchasing them on the open market at their current price. Whether bonds are called or repurchased, the issuer is unlikely to pay a price that exactly equals their carrying value. When a difference exists between the bonds' carrying value and the amount paid, the issuer records a gain or loss equal to the difference.

Point: Bond retirement is also referred to as *bond redemption*.

Point: Gains and losses from retiring bonds were *previously* reported as extraordinary items. New standards require that they now be judged by the "unusual and infrequent" criteria for reporting purposes.

To illustrate the accounting for retiring callable bonds, assume that a company issued callable bonds with a par value of $100,000. The call option requires the issuer to pay a call premium of $3,000 to bondholders in addition to the par value. Next, assume that after the June 30, 2009, interest payment, the bonds have a carrying value of $104,500. Then on July 1, 2009, the issuer calls these bonds and pays $103,000 to bondholders. The issuer recognizes a $1,500 gain from the difference between the bonds' carrying value of $104,500 and the retirement price of $103,000. The issuer records this bond retirement as follows:

July 1	Bonds Payable	100,000	
	Premium on Bonds Payable	4,500	
	Gain on Bond Retirement		1,500
	Cash		103,000
	To record retirement of bonds before maturity.		

Assets	= Liabilities	+ Equity
−103,000	−100,000	+1,500
	−4,500	

An issuer usually must call all bonds when it exercises a call option. However, to retire as many or as few bonds as it desires, an issuer can purchase them on the open market. If it retires less than the entire class of bonds, it recognizes a gain or loss for the difference between the carrying value of those bonds retired and the amount paid to acquire them.

Bond Retirement by Conversion

Convertible Bond

Holders of convertible bonds have the right to convert their bonds to stock. When conversion occurs, the bonds' carrying value is transferred to equity accounts and no gain or loss is recorded. (We further describe convertible bonds in the Decision Analysis section of this chapter.)

To illustrate, assume that on January 1 the $100,000 par value bonds of **Converse**, with a carrying value of $100,000, are converted to 15,000 shares of $2 par value common stock. The entry to record this conversion follows (the market prices of the bonds and stock are *not* relevant to this entry; the material in Chapter 13 is helpful in understanding this transaction):

Assets	=	Liabilities	+	Equity
		−100,000		+30,000
				+70,000

Jan. 1	Bonds Payable	100,000	
	Common Stock		30,000
	Paid-In Capital in Excess of Par Value		70,000
	To record retirement of bonds by conversion.		

Decision Insight

Junk Bonds Junk bonds are company bonds with low credit ratings due to a higher than average likelihood of default. On the upside, the high risk of junk bonds can yield high returns if the issuer survives and repays its debt.

Quick Check

Answer—p. 576

10. Six years ago, a company issued $500,000 of 6%, eight-year bonds at a price of 95. The current carrying value is $493,750. The company decides to retire 50% of these bonds by buying them on the open market at a price of 102½. What is the amount of gain or loss on the retirement of these bonds?

Long-Term Notes Payable

Video14.1

Like bonds, notes are issued to obtain assets such as cash. Unlike bonds, notes are typically transacted with a *single* lender such as a bank. An issuer initially records a note at its selling price—that is, the note's face value minus any discount or plus any premium. Over the note's life, the amount of interest expense allocated to each period is computed by multiplying the market rate (at issuance of the note) by the beginning-of-period note balance. The note's carrying (book) value at any time equals its face value minus any unamortized discount or plus any unamortized premium; carrying value is also computed as the present value of all remaining payments, discounted using the market rate at issuance.

Installment Notes

C1 Explain the types and payment patterns of notes.

An **installment note** is an obligation requiring a series of payments to the lender. Installment notes are common for franchises and other businesses when lenders and borrowers agree to spread payments over several periods. To illustrate, assume that Foghog borrows $60,000 from a bank to purchase equipment. It signs an 8% installment note requiring six annual payments of principal plus interest and it records the note's issuance at January 1, 2009, as follows.

Assets	=	Liabilities	+	Equity
+60,000		+60,000		

Jan. 1	Cash	60,000	
	Notes Payable		60,000
	Borrowed $60,000 by signing an 8%, six-year installment note.		

Payments on an installment note normally include the accrued interest expense plus a portion of the amount borrowed (the *principal*). This section describes an installment note with equal payments.

The equal total payments pattern consists of changing amounts of both interest and principal. To illustrate, assume that Foghog borrows $60,000 by signing a $60,000 note that requires six *equal payments* of $12,979 at the end of each year. (The present value of an annuity of six annual payments of $12,979, discounted at 8%, equals $60,000; we show this computation in footnote 2 on the next page.) The $12,979 includes both interest and principal, the amounts of which change with each payment. Exhibit 14.14 shows the pattern of equal total payments and its two parts, interest and principal. Column A shows the note's beginning balance. Column B shows accrued interest for each year at 8% of the beginning note balance. Column C shows the impact on the note's principal, which equals the difference between the total payment in column D and the interest expense in column B. Column E shows the note's year-end balance.

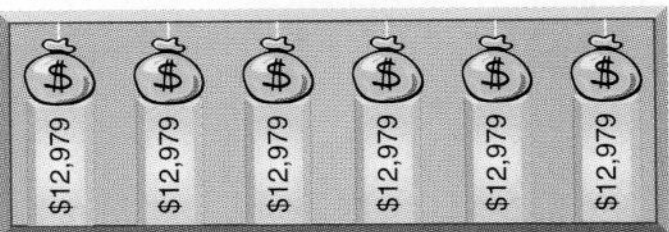

Point: Most consumer notes are installment notes that require equal total payments.

EXHIBIT 14.14

Installment Note: Equal Total Payments

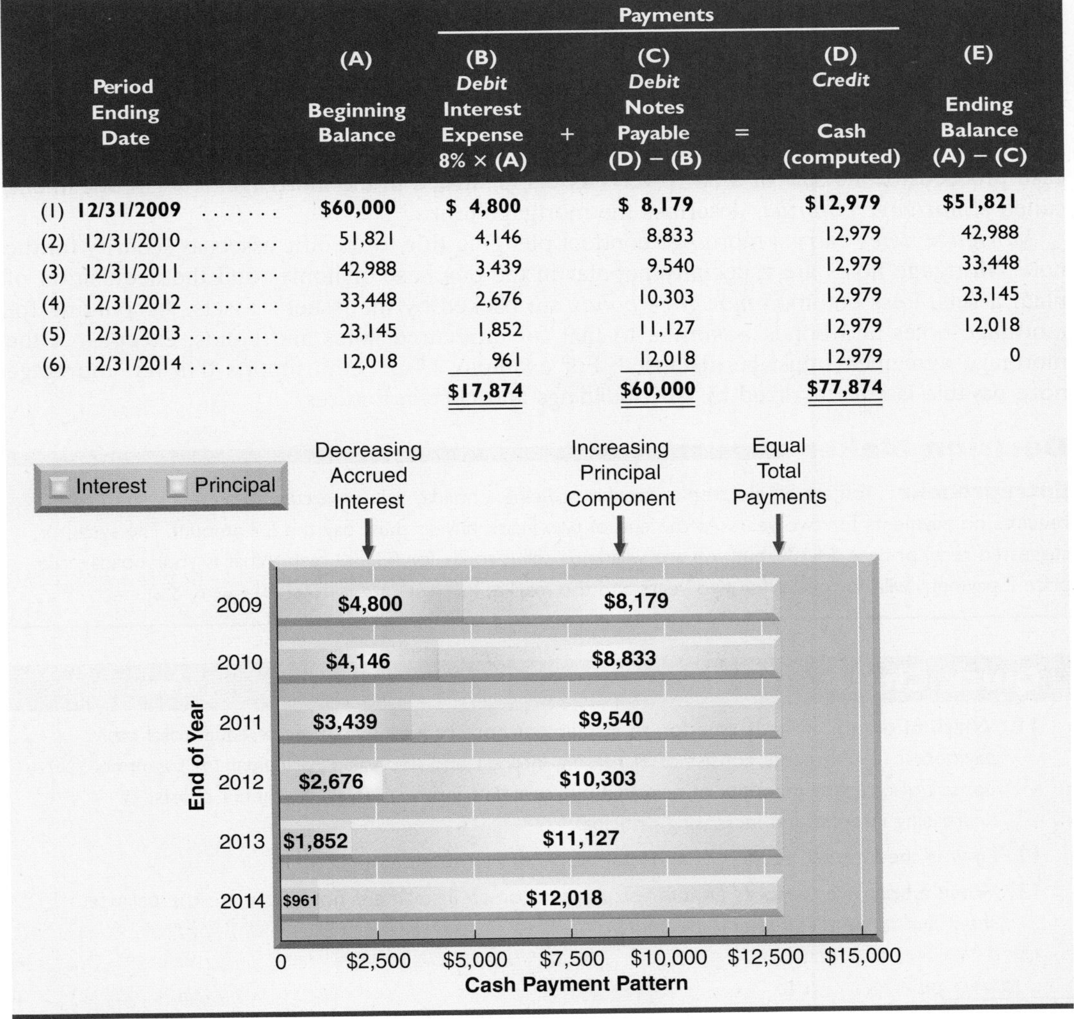

Period Ending Date	(A) Beginning Balance	Payments: (B) *Debit* Interest Expense 8% × (A)	+	(C) *Debit* Notes Payable (D) − (B)	=	(D) *Credit* Cash (computed)	(E) Ending Balance (A) − (C)
(1) **12/31/2009**	**$60,000**	**$ 4,800**		**$ 8,179**		**$12,979**	**$51,821**
(2) 12/31/2010	51,821	4,146		8,833		12,979	42,988
(3) 12/31/2011	42,988	3,439		9,540		12,979	33,448
(4) 12/31/2012	33,448	2,676		10,303		12,979	23,145
(5) 12/31/2013	23,145	1,852		11,127		12,979	12,018
(6) 12/31/2014	12,018	961		12,018		12,979	0
		$17,874		**$60,000**		**$77,874**	

Although the six cash payments are equal, accrued interest decreases each year because the principal balance of the note declines. As the amount of interest decreases each year, the portion of each payment applied to principal increases. This pattern is graphed in the lower part

Point: The Truth-in-Lending Act requires lenders to provide information about loan costs including finance charges and interest rate.

P5 Prepare entries to account for notes.

of Exhibit 14.14. Foghog uses the amounts in Exhibit 14.14 to record its first two payments (for years 2009 and 2010) as follows:

Assets	=	Liabilities	+	Equity
−12,979		−8,179		−4,800

Date	Account	Debit	Credit
2009			
Dec. 31	Interest Expense	4,800	
	Notes Payable	8,179	
	Cash		12,979
	To record first installment payment.		

Assets	=	Liabilities	+	Equity
−12,979		−8,833		−4,146

Date	Account	Debit	Credit
2010			
Dec. 31	Interest Expense	4,146	
	Notes Payable	8,833	
	Cash		12,979
	To record second installment payment.		

Foghog records similar entries but with different amounts for each of the remaining four payments. After six years, the Notes Payable account balance is zero.[2]

Mortgage Notes and Bonds

A **mortgage** is a legal agreement that helps protect a lender if a borrower fails to make required payments on notes or bonds. A mortgage gives the lender a right to be paid from the cash proceeds of the sale of a borrower's assets identified in the mortgage. A legal document, called a *mortgage contract,* describes the mortgage terms.

Mortgage notes carry a mortgage contract pledging title to specific assets as security for the note. Mortgage notes are especially popular in the purchase of homes and the acquisition of plant assets. Less common *mortgage bonds* are backed by the issuer's assets. Accounting for mortgage notes and bonds is similar to that for unsecured notes and bonds, except that the mortgage agreement must be disclosed. For example, **Musicland** reports that its "mortgage note payable is collateralized by land, buildings and certain fixtures."

Global: Countries vary in the preference given to debtholders vs. stockholders when a company is in financial distress. Some countries such as Germany, France, and Japan give preference to stockholders over debtholders.

Decision Maker

Entrepreneur You are an electronics retailer planning a holiday sale on a custom stereo system that requires no payments for two years. At the end of two years, buyers must pay the full amount. The system's suggested retail price is $4,100, but you are willing to sell it today for $3,000 cash. What is your holiday sale price if payment will not occur for two years and the market interest rate is 10%? [Answer—p. 576]

Quick Check

Answers—p. 576

11. Which of the following is true for an installment note requiring a series of equal total cash payments? (*a*) Payments consist of increasing interest and decreasing principal; (*b*) payments consist of changing amounts of principal but constant interest; or (*c*) payments consist of decreasing interest and increasing principal.

12. How is the interest portion of an installment note payment computed?

13. When a borrower records an interest payment on an installment note, how are the balance sheet and income statement affected?

---

[2] Table B.3 in Appendix B is used to compute the dollar amount of the six payments that equal the initial note balance of $60,000 at 8% interest. We go to Table B.3, row 6, and across to the 8% column, where the present value factor is 4.6229. The dollar amount is then computed by solving this relation:

Table	Present Value Factor		Dollar Amount		Present Value
B.3	4.6229	×	?	=	$60,000

The dollar amount is computed by dividing $60,000 by 4.6229, yielding $12,979.

Example: Suppose the $60,000 installment loan has an 8% interest rate with eight equal annual payments. What is the annual payment? *Answer* (using Table B.3): $60,000/5.7466 = $10,441

Debt Features and the Debt-to-Equity Ratio

Decision Analysis

Collateral agreements can reduce the risk of loss for both bonds and notes. Unsecured bonds and notes are riskier because the issuer's obligation to pay interest and principal has the same priority as all other unsecured liabilities in the event of bankruptcy. If a company is unable to pay its debts in full, the unsecured creditors (including the holders of debentures) lose all or a portion of their balances. These types of legal agreements and other characteristics of long-term liabilities are crucial for effective business decisions. The first part of this section describes the different types of features sometimes included with bonds and notes. The second part explains and applies the debt-to-equity ratio.

Features of Bonds and Notes

A2 Assess debt features and their implications.

This section describes common features of debt securities.

Secured or Unsecured **Secured bonds** (and notes) have specific assets of the issuer pledged (or *mortgaged*) as collateral. This arrangement gives holders added protection against the issuer's default. If the issuer fails to pay interest or par value, the secured holders can demand that the collateral be sold and the proceeds used to pay the obligation. **Unsecured bonds** (and notes), also called *debentures,* are backed by the issuer's general credit standing. Unsecured debt is riskier than secured debt. *Subordinated debentures* are liabilities that are not repaid until the claims of the more senior, unsecured (and secured) liabilities are settled.

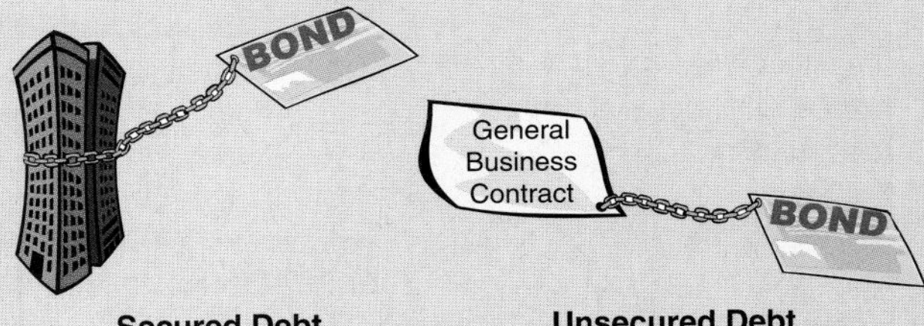

Secured Debt Unsecured Debt

Term or Serial **Term bonds** (and notes) are scheduled for maturity on one specified date. **Serial bonds** (and notes) mature at more than one date (often in series) and thus are usually repaid over a number of periods. For instance, $100,000 of serial bonds might mature at the rate of $10,000 each year from 6 to 15 years after they are issued. Many bonds are **sinking fund bonds,** which to reduce the holder's risk require the issuer to create a *sinking fund* of assets set aside at specified amounts and dates to repay the bonds.

Registered or Bearer Bonds issued in the names and addresses of their holders are **registered bonds.** The issuer makes bond payments by sending checks (or cash transfers) to registered holders. A registered holder must notify the issuer of any ownership change. Registered bonds offer the issuer the practical advantage of not having to actually issue bond certificates. Bonds payable to whoever holds them (the *bearer*) are called **bearer bonds** or *unregistered bonds.* Sales or exchanges might not be recorded, so the holder of a bearer bond is presumed to be its rightful owner. As a result, lost bearer bonds are difficult to replace. Many bearer bonds are also **coupon bonds.** This term reflects interest coupons that are attached to the bonds. When each coupon matures, the holder presents it to a bank or broker for collection. At maturity, the holder follows the same process and presents the bond certificate for collection. Issuers of coupon bonds cannot deduct the related interest expense for taxable income. This is to prevent abuse by taxpayers who own coupon bonds but fail to report interest income on their tax returns.

Convertible and/or Callable **Convertible bonds** (and notes) can be exchanged for a fixed number of shares of the issuing corporation's common stock. Convertible debt offers holders the potential to participate in future increases in stock price. Holders still receive periodic interest while the debt is held and the par value if they hold the debt to maturity. In most cases, the holders decide whether and when to convert debt to stock. **Callable bonds** (and notes) have an option exercisable by the issuer to retire them at a stated dollar amount before maturity.

Convertible Debt

Callable Debt

Decision Insight

Munis More than a million municipal bonds, or "munis," exist, and many are tax exempt. Munis are issued by state, city, town, and county governments to pay for public projects including schools, libraries, roads, bridges, and stadiums.

Debt-to-Equity Ratio

A3 Compute the debt-to-equity ratio and explain its use.

Beyond assessing different characteristics of debt as just described, we want to know the level of debt, especially in relation to total equity. Such knowledge helps us assess the risk of a company's financing structure. A company financed mainly with debt is more risky because liabilities must be repaid—usually with periodic interest—whereas equity financing does not. A measure to assess the risk of a company's financing structure is the **debt-to-equity ratio** (see Exhibit 14.15).

EXHIBIT 14.15
Debt-to-Equity Ratio

$$\textbf{Debt-to-equity} = \frac{\textbf{Total liabilities}}{\textbf{Total equity}}$$

The debt-to-equity ratio varies across companies and industries. Industries that are more variable tend to have lower ratios, while more stable industries are less risky and tend to have higher ratios. To apply the debt-to-equity ratio, let's look at this measure for Six Flags in Exhibit 14.16.

EXHIBIT 14.16
Six Flags' Debt-to-Equity Ratio

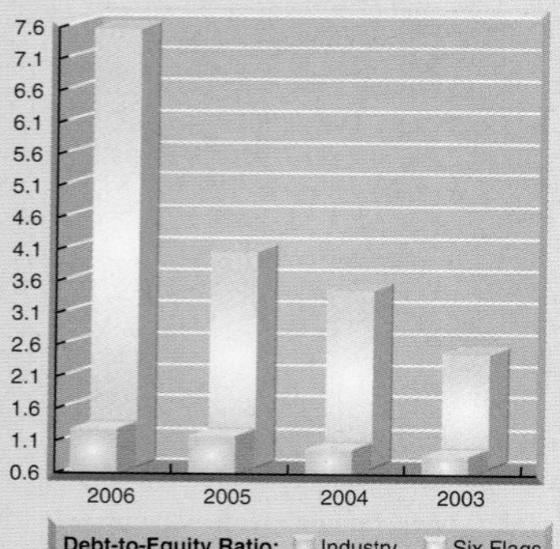

($ millions)	2006	2005	2004	2003
Total liabilities	$2,811	$2,799	$2,816	$3,321
Total equity	$ 376	$ 694	$ 826	$1,362
Debt-to-equity	7.5	4.0	3.4	2.4
Industry debt-to-equity	1.2	1.1	0.9	0.8

Six Flags' 2006 debt-to-equity ratio was 7.5, meaning that debtholders contributed $7.5 for each $1 contributed by equityholders. This implies a fairly risky financing structure for Six Flags. A similar concern is drawn from a comparison of Six Flags with its competitors, where the 2006 industry ratio was 1.2. Analysis across the years shows that Six Flags' financing structure has grown increasingly risky in recent years. Given its sluggish revenues and increasing operating expenses in recent years (see its annual report), Six Flags is increasingly at risk of financial distress.

Decision Maker

Bond Investor You plan to purchase debenture bonds from one of two companies in the same industry that are similar in size and performance. The first company has $350,000 in total liabilities, and $1,750,000 in equity. The second company has $1,200,000 in total liabilities, and $1,000,000 in equity. Which company's debenture bonds are less risky based on the debt-to-equity ratio? [Answer—p. 576]

Demonstration Problem

DP14

Water Sports Company (WSC) patented and successfully test-marketed a new product. To expand its ability to produce and market the new product, WSC needs to raise $800,000 of financing. On January 1, 2009, the company obtained the money in two ways:

a. WSC signed a $400,000, 10% installment note to be repaid with five equal annual installments to be made on December 31 of 2009 through 2013.

b. WSC issued five-year bonds with a par value of $400,000. The bonds have a 12% annual contract rate and pay interest on June 30 and December 31. The bonds' annual market rate is 10% as of January 1, 2009.

Required

1. For the installment note, (*a*) compute the size of each annual payment, (*b*) prepare an amortization table such as Exhibit 14.14, and (*c*) prepare the journal entry for the first payment.

2. For the bonds, (*a*) compute their issue price; (*b*) prepare the January 1, 2009, journal entry to record their issuance; (*c*) prepare an amortization table using the straight-line method; (*d*) prepare the June 30, 2009, journal entry to record the first interest payment; and (*e*) prepare a journal entry to record retiring the bonds at a $416,000 call price on January 1, 2011.

3.[B] Redo parts 2(*c*), 2(*d*), and 2(*e*) assuming the bonds are amortized using the effective interest method.

Planning the Solution

- For the installment note, divide the borrowed amount by the annuity factor (from Table B.3) using the 10% rate and five payments to compute the amount of each payment. Prepare a table similar to Exhibit 14.14 and use the numbers in the table's first line for the journal entry.
- Compute the bonds' issue price by using the market rate to find the present value of their cash flows (use tables found in Appendix B). Then use this result to record the bonds' issuance. Next, prepare an amortization table like Exhibit 14.11 (and Exhibit 14B.2) and use it to get the numbers needed for the journal entry. Also use the table to find the carrying value as of the date of the bonds' retirement that you need for the journal entry.

Solution to Demonstration Problem

Part 1: Installment Note

a. Annual payment = Note balance/Annuity factor = \$400,000/3.7908 = \$105,519 (The annuity factor is for five payments and a rate of 10%.)

b. An amortization table follows.

Annual Period Ending	(a) Beginning Balance	(b) Debit Interest Expense	+	(c) Debit Notes Payable	=	(d) Credit Cash	(e) Ending Balance
		Payments					
(1) 12/31/2009	\$400,000	\$ 40,000		\$ 65,519		\$105,519	\$334,481
(2) 12/31/2010	334,481	33,448		72,071		105,519	262,410
(3) 12/31/2011	262,410	26,241		79,278		105,519	183,132
(4) 12/31/2012	183,132	18,313		87,206		105,519	95,926
(5) 12/31/2013	95,926	9,593		95,926		105,519	0
		\$127,595		\$400,000		\$527,595	

c. Journal entry for December 31, 2009, payment.

		Debit	Credit
Dec. 31	Interest Expense	40,000	
	Notes Payable	65,519	
	Cash		105,519
	To record first installment payment.		

Part 2: Bonds (Straight-Line Amortization)

a. Compute the bonds' issue price.

Cash Flow	Table	Present Value Factor*	Amount	Present Value
Par (maturity) value	B.1 in App. B (PV of 1)	0.6139	× 400,000	= \$245,560
Interest payments	B.3 in App. B (PV of annuity)	7.7217	× 24,000	= 185,321
Price of bond				\$430,881

* Present value factors are for 10 payments using a semiannual market rate of 5%.

b. Journal entry for January 1, 2009, issuance.

		Debit	Credit
Jan. 1	Cash	430,881	
	Premium on Bonds Payable		30,881
	Bonds Payable		400,000
	Sold bonds at a premium.		

c. Straight-line amortization table for premium bonds.

Semiannual Period-End		Unamortized Premium	Carrying Value
(0)	1/1/2009	$30,881	$430,881
(1)	6/30/2009	27,793	427,793
(2)	12/31/2009	24,705	424,705
(3)	6/30/2010	21,617	421,617
(4)	12/31/2010	18,529	418,529
(5)	6/30/2011	15,441	415,441
(6)	12/31/2011	12,353	412,353
(7)	6/30/2012	9,265	409,265
(8)	12/31/2012	6,177	406,177
(9)	6/30/2013	3,089	403,089
(10)	12/31/2013	0*	400,000

* Adjusted for rounding.

d. Journal entry for June 30, 2009, bond payment.

June 30	Bond Interest Expense	20,912	
	Premium on Bonds Payable	3,088	
	Cash		24,000
	Paid semiannual interest on bonds.		

e. Journal entry for January 1, 2011, bond retirement.

Jan. 1	Bonds Payable	400,000	
	Premium on Bonds Payable	18,529	
	Cash		416,000
	Gain on Retirement of Bonds		2,529
	To record bond retirement (carrying value as of Dec. 31, 2010).		

Part 3: Bonds (Effective Interest Amortization)[B]

c. The effective interest amortization table for premium bonds.

Semiannual Interest Period		(A) Cash Interest Paid 6% × $400,000	(B) Interest Expense 5% × Prior (E)	(C) Premium Amortization (A) − (B)	(D) Unamortized Premium Prior (D) − (C)	(E) Carrying Value $400,000 + (D)
(0)	1/1/2009				$30,881	$430,881
(1)	6/30/2009	$ 24,000	$ 21,544	$ 2,456	28,425	428,425
(2)	12/31/2009	24,000	21,421	2,579	25,846	425,846
(3)	6/30/2010	24,000	21,292	2,708	23,138	423,138
(4)	12/31/2010	24,000	21,157	2,843	20,295	420,295
(5)	6/30/2011	24,000	21,015	2,985	17,310	417,310
(6)	12/31/2011	24,000	20,866	3,134	14,176	414,176
(7)	6/30/2012	24,000	20,709	3,291	10,885	410,885
(8)	12/31/2012	24,000	20,544	3,456	7,429	407,429
(9)	6/30/2013	24,000	20,371	3,629	3,800	403,800
(10)	12/31/2013	24,000	20,200*	3,800	0	400,000
		$240,000	$209,119	$30,881		

* Adjusted for rounding

d. Journal entry for June 30, 2009, bond payment.

June 30	Bond Interest Expense	21,544	
	Premium on Bonds Payable	2,456	
	Cash		24,000
	Paid semiannual interest on bonds.		

e. Journal entry for January 1, 2011, bond retirement.

Jan. 1	Bonds Payable	400,000	
	Premium on Bonds Payable	20,295	
	Cash		416,000
	Gain on Retirement of Bonds		4,295
	To record bond retirement (carrying value as of December 31, 2010).		

APPENDIX

Present Values of Bonds and Notes

14A

This appendix explains how to apply present value techniques to measure a long-term liability when it is created and to assign interest expense to the periods until it is settled. Appendix B at the end of the book provides additional discussion of present value concepts.

Present Value Concepts

C2 Explain and compute the present value of an amount(s) to be paid at a future date(s).

The basic present value concept is that cash paid (or received) in the future has less value now than the same amount of cash paid (or received) today. To illustrate, if we must pay $1 one year from now, its present value is less than $1. To see this, assume that we borrow $0.9259 today that must be paid back in one year with 8% interest. Our interest expense for this loan is computed as $0.9259 × 8%, or $0.0741. When the $0.0741 interest is added to the $0.9259 borrowed, we get the $1 payment necessary to repay our loan with interest. This is formally computed in Exhibit 14A.1. The $0.9259 borrowed is the present value of the $1 future payment. More generally, an amount borrowed equals the present value of the future payment. (This same interpretation applies to an investment. If $0.9259 is invested at 8%, it yields $0.0741 in revenue after one year. This amounts to $1, made up of principal and interest.)

EXHIBIT 14A.1

Components of a One-Year Loan

Amount borrowed	**$0.9259**
Interest for one year at 8%	0.0741
Amount owed after 1 year	$ 1.0000

To extend this example, assume that we owe $1 two years from now instead of one year, and the 8% interest is compounded annually. *Compounded* means that interest during the second period is based on the total of the amount borrowed plus the interest accrued from the first period. The second period's interest is then computed as 8% multiplied by the sum of the amount borrowed plus interest earned in the first period. Exhibit 14A.2 shows how we compute the present value of $1 to be paid in two years. This amount is $0.8573. The first year's interest of $0.0686 is added to the principal so that the second year's interest is based on $0.9259. Total interest for this two-year period is $0.1427, computed as $0.0686 plus $0.0741.

Point: Benjamin Franklin is said to have described compounding as "the money, money makes, makes more money."

EXHIBIT 14A.2

Components of a Two-Year Loan

Amount borrowed	**$0.8573**
Interest for first year ($0.8573 × 8%)	0.0686
Amount owed after 1 year	0.9259
Interest for second year ($0.9259 × 8%)	0.0741
Amount owed after 2 years	$ 1.0000

Present Value Tables

The present value of $1 that we must repay at some future date can be computed by using this formula: $1/(1 + i)^n$. The symbol i is the interest rate per period and n is the number of periods until the future payment must be made. Applying this formula to our two-year loan, we get $\$1/(1.08)^2$, or $0.8573. This is the same value shown in Exhibit 14A.2. We can use this formula to find any present value. However, a simpler method is to use a *present value table,* which lists present values computed with this formula for various interest rates and time periods. Many people find it helpful in learning present value concepts to first work with the table and then move to using a calculator.

EXHIBIT 14A.3

Present Value of 1

Periods	Rate 6%	8%	10%
1	0.9434	**0.9259**	0.9091
2	0.8900	**0.8573**	0.8264
3	0.8396	0.7938	0.7513
4	0.7921	0.7350	0.6830
5	0.7473	0.6806	0.6209
6	0.7050	0.6302	0.5645
7	0.6651	0.5835	0.5132
8	0.6274	0.5403	0.4665
9	0.5919	0.5002	0.4241
10	0.5584	0.4632	0.3855

Exhibit 14A.3 shows a present value table for a future payment of 1 for up to 10 periods at three different interest rates. Present values in this table are rounded to four decimal places. This table is drawn from the larger and more complete Table B.1 in Appendix B at the end of the book. Notice that the first value in the 8% column is 0.9259, the value we computed earlier for the present value of a $1 loan for one year at 8% (see Exhibit 14A.1). Go to the second row in the same 8% column and find the present value of 1 discounted at 8% for two years, or 0.8573. This $0.8573 is the present value of our obligation to repay $1 after two periods at 8% interest (see Exhibit 14A.2).

Example: Use Exhibit 14A.3 to find the present value of $1 discounted for 2 years at 6%. *Answer:* $0.8900

Applying a Present Value Table

To illustrate how to measure a liability using a present value table, assume that a company plans to borrow cash and repay it as follows: $2,000 after one year, $3,000 after two years, and $5,000 after three years. How much does this company receive today if the interest rate on this loan is 10%? To answer, we need to compute the present value of the three future payments, discounted at 10%. This computation is shown in Exhibit 14A.4 using present values from Exhibit 14A.3. The company can borrow $8,054 today at 10% interest in exchange for its promise to make these three payments at the scheduled dates.

EXHIBIT 14A.4

Present Value of a Series of Unequal Payments

Periods	Payments	Present Value of 1 at 10%	Present Value of Payments
1	$2,000	0.9091	$ 1,818
2	3,000	0.8264	2,479
3	5,000	0.7513	3,757
Present value of all payments			**$8,054**

Present Value of an Annuity

The $8,054 present value for the loan in Exhibit 14A.4 equals the sum of the present values of the three payments. When payments are not equal, their combined present value is best computed by adding the individual present values as shown in Exhibit 14A.4. Sometimes payments follow an **annuity,** which is a series of *equal* payments at equal time intervals. The present value of an annuity is readily computed.

To illustrate, assume that a company must repay a 6% loan with a $5,000 payment at each year-end for the next four years. This loan amount equals the present value of the four payments discounted at 6%. Exhibit 14A.5 shows how to compute this loan's present value of $17,326 by multiplying each payment by its matching present value factor taken from Exhibit 14A.3.

EXHIBIT 14A.5

Present Value of a Series of Equal Payments (Annuity) by Discounting Each Payment

Periods	Payments	Present Value of 1 at 6%	Present Value of Payments
1	$5,000	0.9434	$ 4,717
2	5,000	0.8900	4,450
3	5,000	0.8396	4,198
4	5,000	0.7921	3,961
Present value of all payments		**3.4651**	**$17,326**

However, the series of $5,000 payments is an annuity, so we can compute its present value with either of two shortcuts. First, the third column of Exhibit 14A.5 shows that the sum of the present values of 1 at 6% for periods 1 through 4 equals 3.4651. One shortcut is to multiply this total of 3.4651 by the $5,000 annual payment to get the combined present value of $17,326. It requires one multiplication instead of four.

The second shortcut uses an *annuity table* such as the one shown in Exhibit 14A.6, which is drawn from the more complete Table B.3 in Appendix B. We go directly to the annuity table to get the present value factor for a specific number of payments and interest rate. We then multiply this factor by the amount of the payment to find the present value of the annuity. Specifically, find the row for four periods and go across to the 6% column, where the factor is 3.4651. This factor equals the present value of an annuity with four payments of 1, discounted at 6%. We then multiply 3.4651 by $5,000 to get the $17,326 present value of the annuity.

EXHIBIT 14A.6

Present Value of an Annuity of 1

	Rate		
Periods	6%	8%	10%
1	0.9434	0.9259	0.9091
2	1.8334	1.7833	1.7355
3	2.6730	2.5771	2.4869
4	**3.4651**	3.3121	3.1699
5	4.2124	3.9927	3.7908
6	4.9173	4.6229	4.3553
7	5.5824	5.2064	4.8684
8	6.2098	5.7466	5.3349
9	6.8017	6.2469	5.7590
10	7.3601	6.7101	6.1446

Example: Use Exhibit 14A.6 to find the present value of an annuity of eight $15,000 payments with an 8% interest rate. *Answer:* $15,000 × 5.7466 = $86,199

Compounding Periods Shorter Than a Year

The present value examples all involved periods of one year. In many situations, however, interest is compounded over shorter periods. For example, the interest rate on bonds is usually stated as an annual rate but interest is often paid every six months (semiannually). This means that the present value of interest payments from such bonds must be computed using interest periods of six months.

Assume that a borrower wants to know the present value of a series of 10 *semiannual payments* of $4,000 made over five years at an *annual interest rate* of 12%. The interest rate is stated as an annual rate of 12%, but it is actually a rate of 6% per semiannual interest period. To compute the present value of this series of $4,000 payments, go to row 10 of Exhibit 14A.6 and across to the 6% column to find the factor 7.3601. The present value of this annuity is $29,440 (7.3601 × $4,000).

Appendix B further describes present value concepts and includes more complete present value tables and assignments.

Example: If this borrower makes five semiannual payments of $8,000, what is the present value of this annuity at a 12% rate? *Answer:* 4.2124 × $8,000 = $33,699

Quick Check Answers—p. 576

14. A company enters into an agreement to make four annual year-end payments of $1,000 each, starting one year from now. The annual interest rate is 8%. The present value of these four payments is (*a*) $2,923, (*b*) $2,940, or (*c*) $3,312.
15. Suppose a company has an option to pay either (*a*) $10,000 after one year or (*b*) $5,000 after six months and another $5,000 after one year. Which choice has the lower present value?

APPENDIX

Effective Interest Amortization 14B

Effective Interest Amortization of a Discount Bond

The straight-line method yields changes in the bonds' carrying value while the amount for bond interest expense remains constant. This gives the impression of a changing interest rate when users divide a constant bond interest expense over a changing carrying value. As a result, accounting standards allow use of the straight-line method only when its results do not differ materially from those obtained using the effective interest method. The **effective interest method,** or simply *interest method,* allocates total bond interest expense over the bonds' life in a way that yields a constant rate of interest. This constant rate of interest is the market rate at the issue date. Thus, bond interest expense for a period equals the carrying value of the bond at the beginning of that period multiplied by the market rate when issued.

Point: The effective interest method computes bond interest expense using the market rate at issuance. This rate is applied to a changing carrying value.

Exhibit 14B.1 shows an effective interest amortization table for the Fila bonds (as described in Exhibit 14.4). The key difference between the effective interest and straight-line methods lies in computing bond interest expense. Instead of assigning an equal amount of bond interest expense to each period, the effective interest method assigns a bond interest expense amount that increases over the life of a discount bond. **Both methods allocate the *same* $19,546 of total bond interest expense to the bonds' life, but in different patterns.** Specifically, the amortization table in Exhibit 14B.1 shows that the balance of the discount (column D) is amortized until it reaches zero. Also, the bonds' carrying value (column E) changes each period until it equals par value at maturity. Compare columns D and E to the corresponding columns in Exhibit 14.7 to see the amortization patterns. Total bond interest expense is $19,546, consisting of $16,000 of semiannual cash payments and $3,546 of the original bond discount, the same for both methods.

EXHIBIT 14B.1

Effective Interest Amortization of Bond Discount

Bonds: $100,000 Par Value, Semiannual Interest Payments, Two-Year Life, 4% Semiannual Contract Rate, 5% Semiannual Market Rate

Semiannual Interest Period-End	(A) Cash Interest Paid	(B) Bond Interest Expense	(C) Discount Amortization	(D) Unamortized Discount	(E) Carrying Value
(0) 12/31/2009				$3,546	$ 96,454
(1) 6/30/2010	$4,000	$4,823	$ 823	2,723	97,277
(2) 12/31/2010	4,000	4,864	864	1,859	98,141
(3) 6/30/2011	4,000	4,907	907	952	99,048
(4) 12/31/2011	4,000	4,952	952	0	100,000
	$16,000	$19,546	$3,546		

Column (**A**) is the par value ($100,000) multiplied by the semiannual contract rate (4%).
Column (**B**) is the prior period's carrying value multiplied by the semiannual market rate (5%).
Column (**C**) is the difference between interest paid and bond interest expense, or [(B) − (A)].
Column (**D**) is the prior period's unamortized discount less the current period's discount amortization.
Column (**E**) is the par value less unamortized discount, or [$100,000 − (D)].

Except for differences in amounts, journal entries recording the expense and updating the liability balance are the same under the effective interest method and the straight-line method. We can use the numbers in Exhibit 14B.1 to record each semiannual entry during the bonds' two-year life (June 30, 2010, through December 31, 2011). For instance, we record the interest payment at the end of the first semiannual period as follows:

Assets	=	Liabilities	+	Equity
−4,000		+823		−4,823

2010			
June 30	Bond Interest Expense	4,823	
	Discount on Bonds Payable		823
	Cash		4,000
	To record semiannual interest and discount amortization (effective interest method).		

Effective Interest Amortization of a Premium Bond

Exhibit 14B.2 shows the amortization table using the effective interest method for the Adidas bonds (as described in Exhibit 14.8). Column A lists the semiannual cash payments. Column B shows the amount of bond interest expense, computed as the 5% semiannual market rate at issuance multiplied by the beginning-of-period carrying value. The amount of cash paid in column A is larger than the bond interest expense because the cash payment is based on the higher 6% semiannual contract rate. The excess cash payment over the interest expense reduces the principal. These amounts are shown in column C. Column E

EXHIBIT 14B.2

Effective Interest Amortization of Bond Premium

Bonds: $100,000 Par Value, Semiannual Interest Payments, Two-Year Life, 6% Semiannual Contract Rate, 5% Semiannual Market Rate					
Semiannual Interest Period-End	(A) Cash Interest Paid	(B) Bond Interest Expense	(C) Premium Amortization	(D) Unamortized Premium	(E) Carrying Value
(0) **12/31/2009**				**$3,546**	**$103,546**
(1) **6/30/2010**	**$6,000**	**$5,177**	**$ 823**	**2,723**	**102,723**
(2) 12/31/2010	6,000	5,136	864	1,859	101,859
(3) 6/30/2011	6,000	5,093	907	952	100,952
(4) 12/31/2011	6,000	5,048	952	**0**	**100,000**
	$24,000	$20,454	$3,546		

Column (**A**) is the par value ($100,000) multiplied by the semiannual contract rate (6%).
Column (**B**) is the prior period's carrying value multiplied by the semiannual market rate (5%).
Column (**C**) is the difference between interest paid and bond interest expense, or [(A) − (B)].
Column (**D**) is the prior period's unamortized premium less the current period's premium amortization.
Column (**E**) is the par value plus unamortized premium, or [$100,000 + (D)].

shows the carrying value after deducting the amortized premium in column C from the prior period's carrying value. Column D shows the premium's reduction by periodic amortization. When the issuer makes the first semiannual interest payment, the effect of premium amortization on bond interest expense and bond liability is recorded as follows:

2010			
June 30	Bond Interest Expense	5,177	
	Premium on Bonds Payable.	823	
	Cash		6,000
	To record semiannual interest and premium amortization (effective interest method).		

Assets	=	Liabilities	+	Equity
−6,000		−823		−5,177

Similar entries with different amounts are recorded at each payment date until the bond matures at the end of 2011. The effective interest method yields decreasing amounts of bond interest expense and increasing amounts of premium amortization over the bonds' life.

Decision Insight

IFRSs Unlike U.S. GAAP, IFRSs require that interest expense be computed using the effective interest method with *no exemptions*.

APPENDIX

14C Issuing Bonds between Interest Dates

C3 Describe the accrual of bond interest when bond payments do not align with accounting periods.

An issuer can sell bonds at a date other than an interest payment date. When this occurs, the buyers normally pay the issuer the purchase price plus any interest accrued since the prior interest payment date. This accrued interest is then repaid to these buyers on the next interest payment date. To illustrate, suppose Avia sells $100,000 of its 9% bonds at par on March 1, 2009, 60 days after the stated issue date. The interest on Avia bonds is payable semiannually on each June 30 and December 31. Since 60 days have passed, the issuer collects accrued interest from the buyers at the time of issuance. This amount is $1,500 ($100,000 × 9% × 60/360 year). This case is reflected in Exhibit 14C.1.

EXHIBIT 14C.1

Accruing Interest between Interest Payment Dates

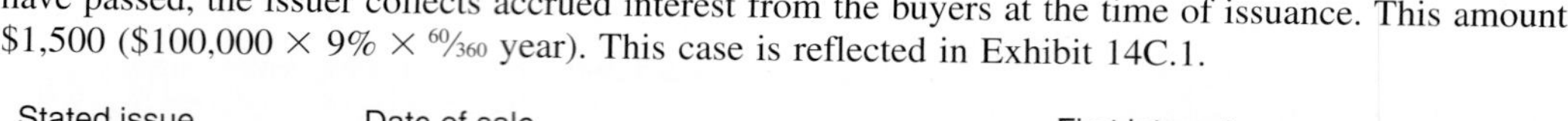

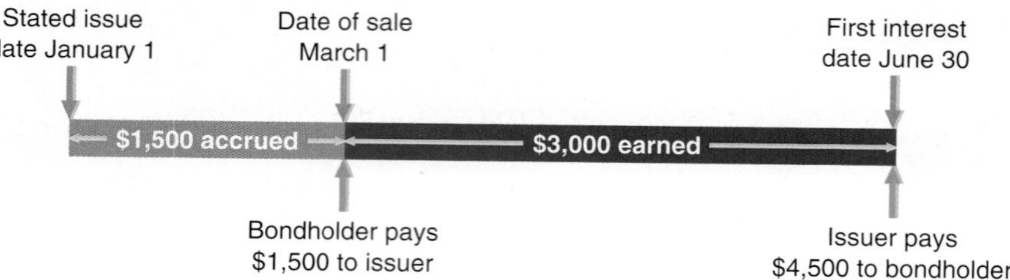

Avia records the issuance of these bonds on March 1, 2009, as follows:

Assets = Liabilities + Equity
+101,500 +100,000
+1,500

Mar. 1	Cash	101,500	
	Interest Payable		1,500
	Bonds Payable		100,000
	Sold bonds at par with accrued interest.		

Example: How much interest is collected from a buyer of $50,000 of Avia bonds sold at par 150 days after the contract issue date? *Answer:* $1,875 (computed as $50,000 × 9% × 150/360 year)

Liabilities for interest payable and bonds payable are recorded in separate accounts. When the June 30, 2009, semiannual interest date arrives, Avia pays the full semiannual interest of $4,500 ($100,000 × 9% × ½ year) to the bondholders. This payment includes the four months' interest of $3,000 earned by the bondholders from March 1 to June 30 *plus* the repayment of the 60 days' accrued interest collected by Avia when the bonds were sold. Avia records this first semiannual interest payment as follows:

Assets = Liabilities + Equity
−4,500 −1,500 −3,000

June 30	Interest Payable	1,500	
	Bond Interest Expense	3,000	
	Cash		4,500
	Paid semiannual interest on the bonds.		

The practice of collecting and then repaying accrued interest with the next interest payment is to simplify the issuer's administrative efforts. To explain, suppose an issuer sells bonds on 15 or 20 different dates between the stated issue date and the first interest payment date. If the issuer does not collect accrued interest from buyers, it needs to pay different amounts of cash to each of them according to the time that passed after purchasing the bonds. The issuer needs to keep detailed records of buyers and the dates they bought bonds. Issuers avoid this recordkeeping by having each buyer pay accrued interest at purchase. Issuers then pay the full semiannual interest to all buyers, regardless of when they bought bonds.

Accruing Bond Interest Expense

If a bond's interest period does not coincide with the issuer's accounting period, an adjusting entry is needed to recognize bond interest expense accrued since the most recent interest payment. To illustrate, assume that the stated issue date for Adidas bonds described in Exhibit 14.10 is September 1, 2009, instead of December 31, 2009, and that the bonds are sold on September 1, 2009. As a result, four months' interest (and premium amortization) accrue before the end of the 2009 calendar year. Interest for this period equals $3,409, or 4/6 of the first six months' interest of $5,113. Also, the premium amortization is $591, or 4/6 of the first six months' amortization of $887. The sum of the bond interest expense and the amortization is $4,000 ($3,409 + $591), which equals 4/6 of the $6,000 cash payment due on February 28, 2010. Adidas records these effects with an adjusting entry at December 31, 2009.

Point: Computation of accrued bond interest may use months instead of days for simplicity purposes. For example, the accrued interest computation for the Adidas bonds is based on months.

Assets = Liabilities + Equity
−591 −3,409
+4,000

Dec. 31	Bond Interest Expense	3,409	
	Premium on Bonds Payable	591	
	Interest Payable		4,000
	To record four months' accrued interest and premium amortization.		

Similar entries are made on each December 31 throughout the bonds' two-year life. When the $6,000 cash payment occurs on each February 28 interest payment date, Adidas must recognize bond interest expense and amortization for January and February. It must also eliminate the interest payable liability created by the December 31 adjusting entry. For example, Adidas records its payment on February 28, 2010, as follows:

Feb. 28	Interest Payable	4,000	
	Bond Interest Expense ($5,113 × 2/6)	1,705	
	Premium on Bonds Payable ($887 × 2/6)	295*	
	Cash		6,000
	To record 2 months' interest and amortization, and eliminate accrued interest liability.		

Assets	=	Liabilities	+	Equity
−6,000		−4,000		−1,705
		−295		

*Adjusted for rounding.

The interest payments made each August 31 are recorded as usual because the entire six-month interest period is included within this company's calendar-year reporting period.

Decision Maker

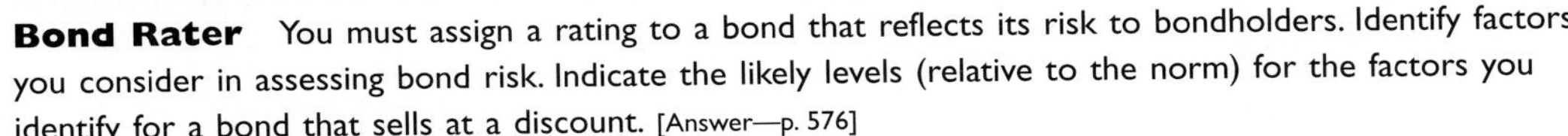

Bond Rater You must assign a rating to a bond that reflects its risk to bondholders. Identify factors you consider in assessing bond risk. Indicate the likely levels (relative to the norm) for the factors you identify for a bond that sells at a discount. [Answer—p. 576]

Quick Check

Answer—p. 576

16. On May 1, a company sells 9% bonds with a $500,000 par value that pay semiannual interest on each January 1 and July 1. The bonds are sold at par plus interest accrued since January 1. The issuer records the first semiannual interest payment on July 1 with (*a*) a debit to Interest Payable for $15,000, (*b*) a debit to Bond Interest Expense for $22,500, or (*c*) a credit to Interest Payable for $7,500.

APPENDIX

Leases and Pensions

This appendix briefly explains the accounting and analysis for both leases and pensions.

Lease Liabilities

C4 Describe accounting for leases and pensions.

A **lease** is a contractual agreement between a *lessor* (asset owner) and a *lessee* (asset renter or tenant) that grants the lessee the right to use the asset for a period of time in return for cash (rent) payments. Nearly one-fourth of all equipment purchases are financed with leases. The advantages of lease financing include the lack of an immediate large cash payment and the potential to deduct rental payments in computing taxable income. From an accounting perspective, leases can be classified as either operating or capital leases.

Point: Home Depot reports that its rental expenses from operating leases total more than $900 million.

Operating Leases **Operating leases** are short-term (or cancelable) leases in which the lessor retains the risks and rewards of ownership. Examples include most car and apartment rental agreements. The lessee records such lease payments as expenses; the lessor records them as revenue. The lessee does not report the leased item as an asset or a liability (it is the lessor's asset). To illustrate, if an employee of Amazon leases a car for $300 at an airport while on company business, Amazon (lessee) records this cost as follows:

July 4	Rental Expense	300	
	Cash		300
	To record lease rental payment.		

Assets	=	Liabilities	+	Equity
−300				−300

Capital Leases **Capital leases** are long-term (or noncancelable) leases by which the lessor transfers substantially all risks and rewards of ownership to the lessee.[3] Examples include most leases of airplanes and department store buildings. The lessee records the leased item as its own asset along with a lease liability at the start of the lease term; the amount recorded equals the present value of all lease payments. To illustrate, assume that K2 Co. enters into a six-year lease of a building in which it will sell sporting equipment. The lease transfers all building ownership risks and rewards to K2 (the present value of its $12,979 annual lease payments is $60,000). K2 records this transaction as follows:

Assets	= Liabilities	+ Equity
+60,000	+60,000	

2009			
Jan. 1	Leased Asset—Building	60,000	
	Lease Liability		60,000
	To record leased asset and lease liability.		

Point: Home Depot reports *"certain locations . . . are leased under capital leases."* The net present value of this Lease Liability is about $400 million.

K2 reports the leased asset as a plant asset and the lease liability as a long-term liability. The portion of the lease liability expected to be paid in the next year is reported as a current liability.[4] At each year-end, K2 records depreciation on the leased asset (assume straight-line depreciation, six-year lease term, and no salvage value) as follows:

Assets	= Liabilities	+ Equity
−10,000		−10,000

Dec. 31	Depreciation Expense—Building	10,000	
	Accumulated Depreciation—Building		10,000
	To record depreciation on leased asset.		

K2 also accrues interest on the lease liability at each year-end. Interest expense is computed by multiplying the remaining lease liability by the interest rate on the lease. Specifically, K2 records its annual interest expense as part of its annual lease payment ($12,979) as follows (for its first year):

Assets	= Liabilities	+ Equity
−12,979	−8,179	−4,800

2009			
Dec. 31	Interest Expense	4,800	
	Lease Liability	8,179	
	Cash		12,979
	*To record first annual lease payment.**		

* These numbers are computed from a *lease payment schedule.* For simplicity, we use the same numbers from Exhibit 14.14 for this lease payment schedule—with different headings as follows:

		Payments			
	(A)	(B) *Debit*	(C) *Debit*	(D) *Credit*	(E)
Period Ending Date	Beginning Balance of Lease Liability	Interest on Lease Liability 8% × (A) +	Lease Liability (D) − (B) =	Cash Lease Payment	Ending Balance of Lease Liability (A) − (C)
12/31/2009	**$60,000**	**$ 4,800**	**$ 8,179**	**$12,979**	**$51,821**
12/31/2010	51,821	4,146	8,833	12,979	42,988
12/31/2011	42,988	3,439	9,540	12,979	33,448
12/31/2012	33,448	2,676	10,303	12,979	23,145
12/31/2013	23,145	1,852	11,127	12,979	12,018
12/31/2014	12,018	961	12,018	12,979	0
		$17,874	**$60,000**	**$77,874**	

[3] A *capital lease* meets any one or more of four criteria: (1) transfers title of leased asset to lessee, (2) contains a bargain purchase option, (3) has a lease term that is 75% or more of the leased asset's useful life, or (4) has a present value of lease payments that is 90% or more of the leased asset's market value.

[4] Most lessees try to keep leased assets and lease liabilities off their balance sheets by failing to meet any one of the four criteria of a capital lease. This is because a lease liability increases a company's total liabilities, making it more difficult to obtain additional financing. The acquisition of assets without reporting any related liabilities (or other asset outflows) on the balance sheet is called **off-balance-sheet financing.**

Pension Liabilities

A **pension plan** is a contractual agreement between an employer and its employees for the employer to provide benefits (payments) to employees after they retire. Most employers pay the full cost of the pension, but sometimes employees pay part of the cost. An employer records its payment into a pension plan with a debit to Pension Expense and a credit to Cash. A *plan administrator* receives payments from the employer, invests them in pension assets, and makes benefit payments to *pension recipients* (retired employees). Insurance and trust companies often serve as pension plan administrators.

Many pensions are known as *defined benefit plans* that define future benefits; the employer's contributions vary, depending on assumptions about future pension assets and liabilities. Several disclosures are necessary in this case. Specifically, a pension liability is reported when the accumulated benefit obligation is *more than* the plan assets, a so-called *underfunded plan.* The accumulated benefit obligation is the present value of promised future pension payments to retirees. *Plan assets* refer to the market value of assets the plan administrator holds. A pension asset is reported when the accumulated benefit obligation is *less than* the plan assets, a so-called *overfunded plan.* An employer reports pension expense when it receives the benefits from the employees' services, which is sometimes decades before it pays pension benefits to employees. (*Other Postretirement Benefits* refer to nonpension benefits such as health care and life insurance benefits. Similar to a pension, costs of these benefits are estimated and liabilities accrued when the employees earn them.)

Point: Fringe benefits are often 40% or more of salaries and wages, and pension benefits make up nearly 15% of fringe benefits.

Point: Two types of pension plans are (1) *defined benefit plan*—the retirement benefit is defined and the employer estimates the contribution necessary to pay these benefits—and (2) *defined contribution plan*—the pension contribution is defined and the employer and/or employee contributes amounts specified in the pension agreement.

Summary

C1 Explain the types and payment patterns of notes. Notes repaid over a period of time are called *installment notes* and usually follow one of two payment patterns: (1) decreasing payments of interest plus equal amounts of principal or (2) equal total payments. Mortgage notes also are common.

C2[A] Explain and compute the present value of an amount(s) to be paid at a future date(s). The basic concept of present value is that an amount of cash to be paid or received in the future is worth less than the same amount of cash to be paid or received today. Another important present value concept is that interest is compounded, meaning interest is added to the balance and used to determine interest for succeeding periods. An annuity is a series of equal payments occurring at equal time intervals. An annuity's present value can be computed using the present value table for an annuity (or a calculator).

C3[C] Describe the accrual of bond interest when bond payments do not align with accounting periods. Issuers and buyers of debt record the interest accrued when issue dates or accounting periods do not coincide with debt payment dates.

C4[D] Describe accounting for leases and pensions. A lease is a rental agreement between the lessor and the lessee. When the lessor retains the risks and rewards of asset ownership (an *operating lease*), the lessee debits Rent Expense and credits Cash for its lease payments. When the lessor substantially transfers the risks and rewards of asset ownership to the lessee (a *capital lease*), the lessee capitalizes the leased asset and records a lease liability. Pension agreements can result in either pension assets or pension liabilities.

A1 Compare bond financing with stock financing. Bond financing is used to fund business activities. Advantages of bond financing versus stock include (1) no effect on owner control, (2) tax savings, and (3) increased earnings due to financial leverage. Disadvantages include (1) interest and principal payments and (2) amplification of poor performance.

A2 Assess debt features and their implications. Certain bonds are secured by the issuer's assets; other bonds, called *debentures,* are unsecured. Serial bonds mature at different points in time; term bonds mature at one time. Registered bonds have each bondholder's name recorded by the issuer; bearer bonds are payable to the holder. Convertible bonds are exchangeable for shares of the issuer's stock. Callable bonds can be retired by the issuer at a set price. Debt features alter the risk of loss for creditors.

A3 Compute the debt-to-equity ratio and explain its use. Both creditors and equity holders are concerned about the relation between the amount of liabilities and the amount of equity. A company's financing structure is at less risk when the debt-to-equity ratio is lower, as liabilities must be paid and usually with periodic interest.

P1 Prepare entries to record bond issuance and bond interest expense. When bonds are issued at par, Cash is debited and Bonds Payable is credited for the bonds' par value. At bond interest payment dates (usually semiannual), Bond Interest Expense is debited and Cash credited—the latter for an amount equal to the bond par value multiplied by the bond contract rate.

P2 Compute and record amortization of bond discount. Bonds are issued at a discount when the contract rate is less than the market rate, making the issue (selling) price less than par. When this occurs, the issuer records a credit to Bonds Payable (at par) and debits both Discount on Bonds Payable and Cash. The amount of bond interest expense assigned to each period is computed using either the straight-line or effective interest method.

P3 Compute and record amortization of bond premium. Bonds are issued at a premium when the contract rate is higher than the market rate, making the issue (selling) price greater than par. When this occurs, the issuer records a debit to Cash and credits both Premium on Bonds Payable and Bonds Payable (at par). The amount of bond interest expense assigned to each period is computed using either the straight-line or effective interest method. The Premium on Bonds Payable is allocated to reduce bond interest expense over the life of the bonds.

P4 Record the retirement of bonds. Bonds are retired at maturity with a debit to Bonds Payable and a credit to Cash at par value. The issuer can retire the bonds early by exercising a call option or purchasing them in the market. Bondholders can also retire bonds early by exercising a conversion feature on convertible bonds. The issuer recognizes a gain or loss for the difference between the amount paid and the bond carrying value.

P5 **Prepare entries to account for notes.** Interest is allocated to each period in a note's life by multiplying its beginning-period carrying value by its market rate at issuance. If a note is repaid with equal payments, the payment amount is computed by dividing the borrowed amount by the present value of an annuity factor (taken from a present value table) using the market rate and the number of payments.

Guidance Answers to **Decision Maker**

Entrepreneur This is a "present value" question. The market interest rate (10%) and present value ($3,000) are known, but the payment required two years later is unknown. This amount ($3,630) can be computed as $3,000 × 1.10 × 1.10. Thus, the sale price is $3,630 when no payments are received for two years. The $3,630 received two years from today is equivalent to $3,000 cash today.

Bond Investor The debt-to-equity ratio for the first company is 0.2 ($350,000/$1,750,000) and for the second company is 1.2 ($1,200,000/$1,000,000), suggesting that the financing structure of the second company is more risky than that of the first company. Consequently, as a buyer of unsecured debenture bonds, you prefer the first company (all else equal).

Bond Rater Bonds with longer repayment periods (life) have higher risk. Also, bonds issued by companies in financial difficulties or facing higher than normal uncertainties have higher risk. Moreover, companies with higher than normal debt and large fluctuations in earnings are considered of higher risk. Discount bonds are more risky on one or more of these factors.

Guidance Answers to **Quick Checks**

1. (*b*)

2. Multiply the bond's par value by its contract rate of interest.

3. Bonds sell at a premium when the contract rate exceeds the market rate and the purchasers pay more than their par value.

4. The bonds are issued at a discount, meaning that issue price is less than par value. A discount occurs because the bond contract rate (6%) is less than the market rate (8%).

5.

Cash	91,893	
Discount on Bonds Payable	8,107	
Bonds Payable		100,000

6. $3,811 (total bond interest expense of $38,107 divided by 10 periods; or the $3,000 semiannual cash payment plus the $8,107 discount divided by 10 periods).

7. The bonds are issued at a premium, meaning issue price is higher than par value. A premium occurs because the bonds' contract rate (16%) is higher than the market rate (14%).

8. (*b*) For each semiannual period: $10,592/20 periods = $530 premium amortization.

9.

Bonds payable, 16%, due 12/31/2018	$100,000	
Plus premium on bonds payable	9,532*	$109,532

* Original premium balance of $10,592 less $530 and $530 amortized on 6/30/2009 and 12/31/2009, respectively.

10. $9,375 loss, computed as the difference between the repurchase price of $256,250 [50% of ($500,000 × 102.5%)] and the carrying value of $246,875 (50% of $493,750).

11. (*c*)

12. The interest portion of an installment payment equals the period's beginning loan balance multiplied by the market interest rate at the time of the note's issuance.

13. On the balance sheet, the account balances of the related liability (note payable) and asset (cash) accounts are decreased. On the income statement, interest expense is recorded.

14. (*c*), computed as 3.3121 × $1,000 = $3,312.

15. The option of paying $10,000 after one year has a lower present value. It postpones paying the first $5,000 by six months. More generally, the present value of a further delayed payment is always lower than a less delayed payment.

16. (*a*) Reflects payment of accrued interest recorded back on May 1; $500,000 × 9% × 4/12 = $15,000.

Key Terms

mhhe.com/wildFAP19e

Key Terms are available at the book's Website for learning and testing in an online Flashcard Format.

Annuity (p. 568)
Bearer bonds (p. 563)
Bond (p. 550)
Bond certificate (p. 552)
Bond indenture (p. 552)
Callable bonds (p. 563)
Capital leases (p. 574)
Carrying (book) value of bonds (p. 554)
Contract rate (p. 553)
Convertible bonds (p. 563)
Coupon bonds (p. 563)
Debt-to-equity ratio (p. 564)
Discount on bonds payable (p. 553)
Effective interest method (p. 569)
Installment note (p. 560)
Lease (p. 573)
Market rate (p. 553)
Mortgage (p. 562)
Off-balance-sheet financing (p. 574)
Operating leases (p. 573)
Par value of a bond (p. 550)

Pension plan (p. 575)
Premium on bonds (p. 556)
Registered bonds (p. 563)
Secured bonds (p. 563)
Serial bonds (p. 563)
Sinking fund bonds (p. 563)
Straight-line bond amortization (p. 554)
Term bonds (p. 563)
Unsecured bonds (p. 563)

Multiple Choice Quiz

Answers on p. 589 mhhe.com/wildFAP19e

Additional Quiz Questions are available at the book's Website.

Quiz14

1. A bond traded at 97½ means that
   a. The bond pays 97½% interest.
   b. The bond trades at $975 per $1,000 bond.
   c. The market rate of interest is below the contract rate of interest for the bond.
   d. The bonds can be retired at $975 each.
   e. The bond's interest rate is 2½%.
2. A bondholder that owns a $1,000, 6%, 15-year bond has
   a. The right to receive $1,000 at maturity.
   b. Ownership rights in the bond issuing entity.
   c. The right to receive $60 per month until maturity.
   d. The right to receive $1,900 at maturity.
   e. The right to receive $600 per year until maturity.
3. A company issues 8%, 20-year bonds with a par value of $500,000. The current market rate for the bonds is 8%. The amount of interest owed to the bondholders for each semiannual interest payment is
   a. $40,000.
   b. $0.
   c. $20,000.
   d. $800,000.
   e. $400,000.
4. A company issued 5-year, 5% bonds with a par value of $100,000. The company received $95,735 for the bonds. Using the straight-line method, the company's interest expense for the first semiannual interest period is
   a. $2,926.50.
   b. $5,853.00.
   c. $2,500.00.
   d. $5,000.00.
   e. $9,573.50.
5. A company issued 8-year, 5% bonds with a par value of $350,000. The company received proceeds of $373,745. Interest is payable semiannually. The amount of premium amortized for the first semiannual interest period, assuming straight-line bond amortization, is
   a. $2,698.
   b. $23,745.
   c. $8,750.
   d. $9,344.
   e. $1,484.

Superscript letter B (C, D) denotes assignments based on Appendix 14B (14C, 14D).

Discussion Questions

1. What is the main difference between a bond and a share of stock?
2. What is the main difference between notes payable and bonds payable?
3. What is the advantage of issuing bonds instead of obtaining financing from the company's owners?
4. What are the duties of a trustee for bondholders?
5. What is a bond indenture? What provisions are usually included in it?
6. What are the *contract* rate and the *market* rate for bonds?
7. What factors affect the market rates for bonds?
8.[B] Does the straight-line or effective interest method produce an interest expense allocation that yields a constant rate of interest over a bond's life? Explain.
9.[C] Why does a company that issues bonds between interest dates collect accrued interest from the bonds' purchasers?
10. If you know the par value of bonds, the contract rate, and the market rate, how do you compute the bonds' price?
11. What is the issue price of a $2,000 bond sold at 98¼? What is the issue price of a $6,000 bond sold at 101½?
12. Describe the debt-to-equity ratio and explain how creditors and owners would use this ratio to evaluate a company's risk.
13. What obligation does an entrepreneur (owner) have to investors that purchase bonds to finance the business?
14. Refer to **Best Buy**'s annual report in Appendix A. Is there any indication that Best Buy has issued bonds?

15. Refer to the statement of cash flows for **Circuit City** in Appendix A. For the year ended February 28, 2007, what was the amount of principal payments on long-term debt?
16. Did **RadioShack**'s long-term debt increase or decrease during 2006?

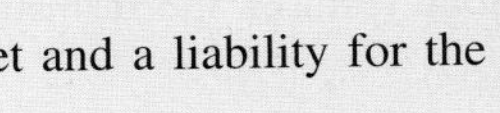

17. Refer to the annual report for **Apple** in Appendix A. For the year ended September 30, 2006, did it raise more cash by issuing stock or debt?
18.[D] When can a lease create both an asset and a liability for the lessee?
19.[D] Compare and contrast an operating lease with a capital lease.
20.[D] Describe the two basic types of pension plans.

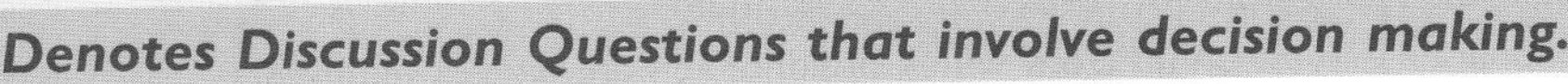

Available with McGraw-Hill's Homework Manager McGraw-Hill's HOMEWORK MANAGER®

QUICK STUDY

Round dollar amounts to the nearest whole dollar.

QS 14-1
Bond computations—straight-line P1 P2

Randell Company issues 7%, 10-year bonds with a par value of $150,000 and semiannual interest payments. On the issue date, the annual market rate for these bonds is 8%, which implies a selling price of 93¼. The straight-line method is used to allocate interest expense.

1. What are the issuer's cash proceeds from issuance of these bonds?
2. What total amount of bond interest expense will be recognized over the life of these bonds?
3. What is the amount of bond interest expense recorded on the first interest payment date?

QS 14-2[B]
Bond computations—effective interest
P1 P3

Elton Company issues 7%, 15-year bonds with a par value of $350,000 and semiannual interest payments. On the issue date, the annual market rate for these bonds is 6%, which implies a selling price of 109¾. The effective interest method is used to allocate interest expense.

1. What are the issuer's cash proceeds from issuance of these bonds?
2. What total amount of bond interest expense will be recognized over the life of these bonds?
3. What amount of bond interest expense is recorded on the first interest payment date?

QS 14-3
Journalize bond issuance P1

Prepare the journal entries for the issuance of the bonds in both QS 14-1 and QS 14-2. Assume that both bonds are issued for cash on January 1, 2009.

QS 14-4
Computing bond price P2 P3

Using the bond details in both QS 14-1 and QS 14-2, confirm that the bonds' selling prices given in each problem are approximately correct. Use the present value tables B.1 and B.3 in Appendix B.

QS 14-5
Recording bond issuance and discount amortization P1 P2

Boulware Company issues 8%, five-year bonds, on December 31, 2008, with a par value of $100,000 and semiannual interest payments. Use the following straight-line bond amortization table and prepare journal entries to record (*a*) the issuance of bonds on December 31, 2008; (*b*) the first interest payment on June 30, 2009; and (*c*) the second interest payment on December 31, 2009.

Semiannual Period-End	Unamortized Discount	Carrying Value
(0) 12/31/2008	$7,723	$92,277
(1) 6/30/2009	6,951	93,049
(2) 12/31/2009	6,179	93,821

QS 14-6
Bond retirement by call option
P4

On July 1, 2009, Teller Company exercises a $4,000 call option (plus par value) on its outstanding bonds that have a carrying value of $208,000 and par value of $200,000. The company exercises the call option after the semiannual interest is paid on June 30, 2009. Record the entry to retire the bonds.

QS 14-7
Bond retirement by stock conversion P4

On January 1, 2009, the $1,000,000 par value bonds of Staten Company with a carrying value of $1,000,000 are converted to 500,000 shares of $1.00 par value common stock. Record the entry for the conversion of the bonds.

QS 14-8
Computing payments for an installment note C1

Jordyn Company borrows $600,000 cash from a bank and in return signs an installment note for five annual payments of equal amount, with the first payment due one year after the note is signed. Use Table B.3 in Appendix B to compute the amount of the annual payment for each of the following annual market rates: (*a*) 4%, (*b*) 6%, and (*c*) 8%.

QS 14-9
Bond features and terminology
A2

Enter the letter of the description *A* through *H* that best fits each term or phrase 1 through 8.

A. Records and tracks the bondholders' names.
B. Is unsecured; backed only by the issuer's credit standing.
C. Has varying maturity dates for amounts owed.
D. Identifies rights and responsibilities of the issuer and the bondholders.
E. Can be exchanged for shares of the issuer's stock.
F. Is unregistered; interest is paid to whoever possesses them.
G. Maintains a separate asset account from which bondholders are paid at maturity.

H. Pledges specific assets of the issuer as collateral.

1. _______ Convertible bond
2. _______ Bond indenture
3. _______ Sinking fund bond
4. _______ Debenture
5. _______ Registered bond
6. _______ Serial bond
7. _______ Secured bond
8. _______ Bearer bond

QS 14-10
Debt-to-equity ratio
A2

Compute the debt-to-equity ratio for each of the following companies. Which company appears to have a riskier financing structure? Explain.

	NLF Company	ABL Company
Total liabilities	$615,000	$ 480,000
Total equity	820,000	1,500,000

QS 14-11[C]
Issuing bonds between interest dates
P1

Knapp Company plans to issue 6% bonds on January 1, 2009, with a par value of $2,000,000. The company sells $1,800,000 of the bonds on January 1, 2009. The remaining $200,000 sells at par on March 1, 2009. The bonds pay interest semiannually as of June 30 and December 31. Record the entry for the March 1 cash sale of bonds.

QS 14-12[D]
Recording operating leases C4

Lu Villena, an employee of ETrain.com, leases a car at O'Hare airport for a three-day business trip. The rental cost is $400. Prepare the entry by ETrain.com to record Lu Villena's short-term car lease cost.

QS 14-13[D]
Recording capital leases C4

Artel, Inc., signs a five-year lease for office equipment with Office Solutions. The present value of the lease payments is $13,500. Prepare the journal entry that Artel records at the inception of this capital lease.

McGraw-Hill's HOMEWORK MANAGER® Available with McGraw-Hill's Homework Manager

Round dollar amounts to the nearest whole dollar. Assume no reversing entries are used.

EXERCISES

Exercise 14-1
Recording bond issuance and interest
P1

On January 1, 2009, Bartel Enterprises issues bonds that have a $3,650,000 par value, mature in 20 years, and pay 10% interest semiannually on June 30 and December 31. The bonds are sold at par.

1. How much interest will Bartel pay (in cash) to the bondholders every six months?
2. Prepare journal entries to record (*a*) the issuance of bonds on January 1, 2009; (*b*) the first interest payment on June 30, 2009; and (*c*) the second interest payment on December 31, 2009.
3. Prepare the journal entry for issuance assuming the bonds are issued at (*a*) 98 and (*b*) 105.

Exercise 14-2
Straight-line amortization of bond discount
P2

Sears issues bonds with a par value of $175,000 on January 1, 2009. The bonds' annual contract rate is 4%, and interest is paid semiannually on June 30 and December 31. The bonds mature in three years. The annual market rate at the date of issuance is 6%, and the bonds are sold for $165,523.

1. What is the amount of the discount on these bonds at issuance?
2. How much total bond interest expense will be recognized over the life of these bonds?
3. Prepare an amortization table like the one in Exhibit 14.7 for these bonds; use the straight-line method to amortize the discount.

Exercise 14-3[B]
Effective interest amortization of bond discount
P2

Ritter issues bonds dated January 1, 2009, with a par value of $300,000. The bonds' annual contract rate is 9%, and interest is paid semiannually on June 30 and December 31. The bonds mature in three years. The annual market rate at the date of issuance is 12%, and the bonds are sold for $277,872.

1. What is the amount of the discount on these bonds at issuance?
2. How much total bond interest expense will be recognized over the life of these bonds?
3. Prepare an amortization table like the one in Exhibit 14B.1 for these bonds; use the effective interest method to amortize the discount.

Exercise 14-4
Straight-line amortization of bond premium P3

Dell Co. issues bonds dated January 1, 2009, with a par value of $450,000. The bonds' annual contract rate is 9%, and interest is paid semiannually on June 30 and December 31. The bonds mature in three years. The annual market rate at the date of issuance is 8%, and the bonds are sold for $461,795.

1. What is the amount of the premium on these bonds at issuance?
2. How much total bond interest expense will be recognized over the life of these bonds?
3. Prepare an amortization table like the one in Exhibit 14.11 for these bonds; use the straight-line method to amortize the premium.

Exercise 14-5[B]
Effective interest amortization of bond premium P3

Refer to the bond details in Exercise 14-4 and prepare an amortization table like the one in Exhibit 14B.2 for these bonds using the effective interest method to amortize the premium.

Exercise 14-6
Recording bond issuance and premium amortization
P1 P3

Anna Company issues 8%, five-year bonds, on December 31, 2008, with a par value of $100,000 and semiannual interest payments. Use the following straight-line bond amortization table and prepare journal entries to record (*a*) the issuance of bonds on December 31, 2008; (*b*) the first interest payment on June 30, 2009; and (*c*) the second interest payment on December 31, 2009.

Semiannual Period-End	Unamortized Premium	Carrying Value
(0) 12/31/2008	$7,720	$107,720
(1) 6/30/2009	6,948	106,948
(2) 12/31/2009	6,176	106,176

Exercise 14-7
Recording bond issuance and discount amortization
P1 P2

St. Charles Company issues 10%, four-year bonds, on December 31, 2009, with a par value of $100,000 and semiannual interest payments. Use the following straight-line bond amortization table and prepare journal entries to record (*a*) the issuance of bonds on December 31, 2009; (*b*) the first interest payment on June 30, 2010; and (*c*) the second interest payment on December 31, 2010.

Semiannual Period-End	Unamortized Discount	Carrying Value
(0) 12/31/2009	$8,000	$92,000
(1) 6/30/2010	7,000	93,000
(2) 12/31/2010	6,000	94,000

Exercise 14-8
Recording bond issuance and discount amortization
P1 P2

Zander Company issues 6%, two-year bonds, on December 31, 2009, with a par value of $100,000 and semiannual interest payments. Use the following straight-line bond amortization table and prepare journal entries to record (*a*) the issuance of bonds on December 31, 2009; (*b*) the first through fourth interest payments on each June 30 and December 31; and (*c*) the maturity of the bond on December 31, 2011.

Semiannual Period-End	Unamortized Discount	Carrying Value
(0) 12/31/2009	$4,000	$ 96,000
(1) 6/30/2010	3,000	97,000
(2) 12/31/2010	2,000	98,000
(3) 6/30/2011	1,000	99,000
(4) 12/31/2011	0	100,000

Exercise 14-9
Computing bond interest and price; recording bond issuance
P2

Target Company issues bonds with a par value of $950,000 on their stated issue date. The bonds mature in 15 years and pay 10% annual interest in semiannual payments. On the issue date, the annual market rate for the bonds is 12%.

1. What is the amount of each semiannual interest payment for these bonds?
2. How many semiannual interest payments will be made on these bonds over their life?
3. Use the interest rates given to determine whether the bonds are issued at par, at a discount, or at a premium.
4. Compute the price of the bonds as of their issue date.
5. Prepare the journal entry to record the bonds' issuance.

Check (4) $819,223

Exercise 14-10
Computing bond interest and price; recording bond issuance
P3 

Boston Company issues bonds with a par value of $160,000 on their stated issue date. The bonds mature in six years and pay 8% annual interest in semiannual payments. On the issue date, the annual market rate for the bonds is 6%.

1. What is the amount of each semiannual interest payment for these bonds?
2. How many semiannual interest payments will be made on these bonds over their life?
3. Use the interest rates given to determine whether the bonds are issued at par, at a discount, or at a premium.
4. Compute the price of the bonds as of their issue date.
5. Prepare the journal entry to record the bonds' issuance.

Check (4) $175,930

Exercise 14-11
Bond computations, straight-line amortization, and bond retirement
P2 P4

On January 1, 2009, Seldon issues $450,000 of 10%, 15-year bonds at a price of 93¼. Six years later, on January 1, 2015, Seldon retires 20% of these bonds by buying them on the open market at 109¾. All interest is accounted for and paid through December 31, 2014, the day before the purchase. The straight-line method is used to amortize any bond discount.

1. How much does the company receive when it issues the bonds on January 1, 2009?
2. What is the amount of the discount on the bonds at January 1, 2009?
3. How much amortization of the discount is recorded on the bonds for the entire period from January 1, 2009, through December 31, 2014?
4. What is the carrying (book) value of the bonds as of the close of business on December 31, 2014? What is the carrying value of the 20% soon-to-be-retired bonds on this same date?
5. How much did the company pay on January 1, 2015, to purchase the bonds that it retired?
6. What is the amount of the recorded gain or loss from retiring the bonds?
7. Prepare the journal entry to record the bond retirement at January 1, 2015.

Check (6) $12,420 loss

Exercise 14-12[C]
Recording bond issuance with accrued interest
C4 P1

On May 1, 2009, Bradley Enterprises issues bonds dated January 1, 2009, that have a $1,950,000 par value, mature in 20 years, and pay 8% interest semiannually on June 30 and December 31. The bonds are sold at par plus four months' accrued interest.

1. How much accrued interest do the bond purchasers pay Bradley on May 1, 2009?
2. Prepare Bradley's journal entries to record (*a*) the issuance of bonds on May 1, 2009; (*b*) the first interest payment on June 30, 2009; and (*c*) the second interest payment on December 31, 2009.

Check (1) $52,000

Exercise 14-13
Straight-line amortization and accrued bond interest expense
P1 P2

Stockton Co. issues four-year bonds with a $50,000 par value on June 1, 2009, at a price of $47,850. The annual contract rate is 8%, and interest is paid semiannually on November 30 and May 31.

1. Prepare an amortization table like the one in Exhibit 14.7 for these bonds. Use the straight-line method of interest amortization.
2. Prepare journal entries to record the first two interest payments and to accrue interest as of December 31, 2009.

Exercise 14-14
Installment note with equal total payments C1 P5

On January 1, 2009, American Eagle borrows $90,000 cash by signing a four-year, 5% installment note. The note requires four equal total payments of accrued interest and principal on December 31 of each year from 2009 through 2012.

1. Compute the amount of each of the four equal total payments.
2. Prepare an amortization table for this installment note like the one in Exhibit 14.14.

Check (1) $25,381

Exercise 14-15
Installment note entries P5

Use the information in Exercise 14-14 to prepare the journal entries for American Eagle to record the loan on January 1, 2009, and the four payments from December 31, 2009, through December 31, 2012.

Exercise 14-16
Applying debt-to-equity ratio
A3 

Motin Company is considering a project that will require a $250,000 loan. It presently has total liabilities of $110,000, and total assets of $310,000.

1. Compute Motin's (*a*) present debt-to-equity ratio and (*b*) the debt-to-equity ratio assuming it borrows $250,000 to fund the project.
2. Evaluate and discuss the level of risk involved if Motin borrows the funds to pursue the project.

Exercise 14-17[D]
Identifying capital and operating leases
C4

Indicate whether the company in each separate case 1 through 3 has entered into an operating lease or a capital lease.

1. The lessor retains title to the asset, and the lease term is three years on an asset that has a five-year useful life.
2. The title is transferred to the lessee, the lessee can purchase the asset for $1 at the end of the lease, and the lease term is five years. The leased asset has an expected useful life of six years.
3. The present value of the lease payments is 95% of the leased asset's market value, and the lease term is 70% of the leased asset's useful life.

Exercise 14-18[D]
Accounting for capital lease
C4

Hartel (lessee) signs a five-year capital lease for office equipment with a $21,000 annual lease payment. The present value of the five annual lease payments is $88,460, based on a 6% interest rate.

1. Prepare the journal entry Hartel will record at inception of the lease.
2. If the leased asset has a five-year useful life with no salvage value, prepare the journal entry Hartel will record each year to recognize depreciation expense related to the leased asset.

Exercise 14-19[D]
Analyzing lease options
C2 C3 C4

General Motors advertised three alternatives for a 25-month lease on a new Blazer: (1) zero dollars down and a lease payment of $2,522 per month for 25 months, (2) $5,000 down and $2,240 per month for 25 months, or (3) $55,000 down and no payments for 25 months. Use the present value Table B.3 in Appendix B to determine which is the best alternative (assume you have enough cash to accept any alternative and the annual interest rate is 12% compounded monthly).

Available with McGraw-Hill's Homework Manager — McGraw-Hill's HOMEWORK MANAGER®

PROBLEM SET A

Round dollar amounts to the nearest whole dollar. Assume no reversing entries are used.

Problem 14-1A
Computing bond price and recording issuance
P1 P2 P3

Harvard Research issues bonds dated January 1, 2009, that pay interest semiannually on June 30 and December 31. The bonds have a $45,000 par value and an annual contract rate of 6%, and they mature in six years.

Required

For each of the following three separate situations, (*a*) determine the bonds' issue price on January 1, 2009, and (*b*) prepare the journal entry to record their issuance.

1. The market rate at the date of issuance is 4%.
2. The market rate at the date of issuance is 6%.
3. The market rate at the date of issuance is 8%.

Check (1) Premium, $4,760

(3) Discount, $4,223

Problem 14-2A
Straight-line amortization of bond discount and bond premium
P1 P2 P3

mhhe.com/wildFAP19e

Braeburn issues $3,500,000 of 8%, 15-year bonds dated January 1, 2009, that pay interest semiannually on June 30 and December 31. The bonds are issued at a price of $3,024,000.

Required

1. Prepare the January 1, 2009, journal entry to record the bonds' issuance.
2. For each semiannual period, compute (*a*) the cash payment, (*b*) the straight-line discount amortization, and (*c*) the bond interest expense.
3. Determine the total bond interest expense to be recognized over the bonds' life.
4. Prepare the first two years of an amortization table like Exhibit 14.7 using the straight-line method.
5. Prepare the journal entries to record the first two interest payments.
6. Assume that the bonds are issued at a price of $4,284,000. Repeat parts 1 through 5.

Check (3) $4,676,000

(4) 12/31/2010 carrying value, $3,087,468

Problem 14-3A
Straight-line amortization of bond premium
P1 P3

mhhe.com/wildFAP19e

Jules issues 4.5%, five-year bonds dated January 1, 2009, with a $230,000 par value. The bonds pay interest on June 30 and December 31 and are issued at a price of $235,160. The annual market rate is 4% on the issue date.

Required

1. Calculate the total bond interest expense over the bonds' life.
2. Prepare a straight-line amortization table like Exhibit 14.11 for the bonds' life.
3. Prepare the journal entries to record the first two interest payments.

Check (2) 6/30/2009 carrying value, $234,644

Problem 14-4A[B]
Effective interest amortization of bond premium; computing bond price P1 P3

Refer to the bond details in Problem 14-3A.

Required

1. Compute the total bond interest expense over the bonds' life.
2. Prepare an effective interest amortization table like the one in Exhibit 14B.2 for the bonds' life.
3. Prepare the journal entries to record the first two interest payments.
4. Use the market rate at issuance to compute the present value of the remaining cash flows for these bonds as of December 31, 2011. Compare your answer with the amount shown on the amortization table as the balance for that date (from part 2) and explain your findings.

Check (2) 6/30/2011 carrying value, $232,704
(4) $232,179

Problem 14-5A
Straight-line amortization of bond discount
P1 P2

Legacy issues $345,000 of 5%, four-year bonds dated January 1, 2009, that pay interest semiannually on June 30 and December 31. They are issued at $332,888 and their market rate is 6% at the issue date.

Required

1. Prepare the January 1, 2009, journal entry to record the bonds' issuance.
2. Determine the total bond interest expense to be recognized over the bonds' life.
3. Prepare a straight-line amortization table like the one in Exhibit 14.7 for the bonds' first two years.
4. Prepare the journal entries to record the first two interest payments.

Check (2) $81,112
(3) 12/31/2010 carrying value, $338,944

Analysis Component

5. Assume the market rate on January 1, 2009, is 4% instead of 6%. Without providing numbers, describe how this change affects the amounts reported on Legacy's financial statements.

Problem 14-6A[B]
Effective interest amortization of bond discount P1 P2

Refer to the bond details in Problem 14-5A.

Required

1. Prepare the January 1, 2009, journal entry to record the bonds' issuance.
2. Determine the total bond interest expense to be recognized over the bonds' life.
3. Prepare an effective interest amortization table like the one in Exhibit 14B.1 for the bonds' first two years.
4. Prepare the journal entries to record the first two interest payments.

Check (2) $81,112
(3) 12/31/2010 carrying value, $338,586

mhhe.com/wildFAP19e

Problem 14-7A[B]
Effective interest amortization of bond premium; retiring bonds
P1 P3 P4

Shopko issues $185,000 of 12%, three-year bonds dated January 1, 2009, that pay interest semiannually on June 30 and December 31. They are issued at $189,620. Their market rate is 11% at the issue date.

Required

1. Prepare the January 1, 2009, journal entry to record the bonds' issuance.
2. Determine the total bond interest expense to be recognized over the bonds' life.
3. Prepare an effective interest amortization table like Exhibit 14B.2 for the bonds' first two years.
4. Prepare the journal entries to record the first two interest payments.
5. Prepare the journal entry to record the bonds' retirement on January 1, 2011, at 97.

Check (3) 6/30/2010 carrying value, $187,494
(5) $7,256 gain

Analysis Component

6. Assume that the market rate on January 1, 2009, is 13% instead of 11%. Without presenting numbers, describe how this change affects the amounts reported on Shopko's financial statements.

mhhe.com/wildFAP19e

Problem 14-8A
Installment notes
C1 P5

On November 1, 2009, Norwood borrows $700,000 cash from a bank by signing a five-year installment note bearing 7% interest. The note requires equal total payments each year on October 31.

Required

1. Compute the total amount of each installment payment.
2. Complete an amortization table for this installment note similar to the one in Exhibit 14.14.
3. Prepare the journal entries in which Norwood records (*a*) accrued interest as of December 31, 2009 (the end of its annual reporting period), and (*b*) the first annual payment on the note.

Check (2) 10/31/2013 ending balance, $159,556

Problem 14-9A
Applying the debt-to-equity ratio
A3

At the end of the current year, the following information is available for both the Pulaski Company and the Scott Company.

	Pulaski Company	Scott Company
Total assets	$1,800,000	$900,000
Total liabilities	723,600	478,800
Total equity	1,080,000	420,000

Required

1. Compute the debt-to-equity ratios for both companies.
2. Comment on your results and discuss the riskiness of each company's financing structure.

Problem 14-10A[D]
Capital lease accounting
C4

Thomas Company signs a five-year capital lease with Universal Company for office equipment. The annual lease payment is $14,000, and the interest rate is 8%.

Required

Check (1) $55,898

(3) Year 3 ending balance, $24,966

1. Compute the present value of Thomas's five-year lease payments.
2. Prepare the journal entry to record Thomas's capital lease at its inception.
3. Complete a lease payment schedule for the five years of the lease with the following headings. Assume that the beginning balance of the lease liability (present value of lease payments) is $55,898. (*Hint:* To find the amount allocated to interest in year 1, multiply the interest rate by the beginning-of-year lease liability. The amount of the annual lease payment not allocated to interest is allocated to principal. Reduce the lease liability by the amount allocated to principal to update the lease liability at each year-end.)

Period Ending Date	Beginning Balance of Lease Liability	Interest on Lease Liability	Reduction of Lease Liability	Cash Lease Payment	Ending Balance of Lease Liability

4. Use straight-line depreciation and prepare the journal entry to depreciate the leased asset at the end of year 1. Assume zero salvage value and a five-year life for the office equipment.

PROBLEM SET B

Round dollar amounts to the nearest whole dollar. Assume no reversing entries are used.

Problem 14-1B
Computing bond price and recording issuance
P1 P2 P3

Fortune Systems issues bonds dated January 1, 2009, that pay interest semiannually on June 30 and December 31. The bonds have a $35,000 par value and an annual contract rate of 4%, and they mature in four years.

Required

For each of the following three separate situations, (*a*) determine the bonds' issue price on January 1, 2009, and (*b*) prepare the journal entry to record their issuance.

Check (1) Premium, $2,679

(3) Discount, $2,457

1. The market rate at the date of issuance is 2%.
2. The market rate at the date of issuance is 4%.
3. The market rate at the date of issuance is 6%.

Problem 14-2B
Straight-line amortization of bond discount and bond premium
P1 P2 P3

Long Beach issues $4,500,000 of 8%, 15-year bonds dated January 1, 2009, that pay interest semiannually on June 30 and December 31. The bonds are issued at a price of $3,888,000.

Required

Check (3) $6,012,000

(4) 6/30/2010 carrying value, $3,949,200

1. Prepare the January 1, 2009, journal entry to record the bonds' issuance.
2. For each semiannual period, compute (*a*) the cash payment, (*b*) the straight-line discount amortization, and (*c*) the bond interest expense.
3. Determine the total bond interest expense to be recognized over the bonds' life.
4. Prepare the first two years of an amortization table like Exhibit 14.7 using the straight-line method.
5. Prepare the journal entries to record the first two interest payments.
6. Assume that the bonds are issued at a price of $5,508,000. Repeat parts 1 through 5.

Problem 14-3B
Straight-line amortization of bond premium
P1 P3

San Mateo Company issues 7%, five-year bonds dated January 1, 2009, with a $220,000 par value. The bonds pay interest on June 30 and December 31 and are issued at a price of $229,385. Their annual market rate is 6% on the issue date.

Required

1. Calculate the total bond interest expense over the bonds' life.
2. Prepare a straight-line amortization table like Exhibit 14.11 for the bonds' life.
3. Prepare the journal entries to record the first two interest payments.

Check (2) 12/31/2011 carrying value, $223,751

Problem 14-4B[B]
Effective interest amortization of bond premium; computing bond price P1 P3

Refer to the bond details in Problem 14-3B.

Required

1. Compute the total bond interest expense over the bonds' life.
2. Prepare an effective interest amortization table like the one in Exhibit 14B.2 for the bonds' life.
3. Prepare the journal entries to record the first two interest payments.
4. Use the market rate at issuance to compute the present value of the remaining cash flows for these bonds as of December 31, 2011. Compare your answer with the amount shown on the amortization table as the balance for that date (from part 2) and explain your findings.

Check (2) 6/30/2011 carrying value, $225,041
(4) $224,092

Problem 14-5B
Straight-line amortization of bond discount
P1 P2

Kelly issues $315,000 of 4%, 15-year bonds dated January 1, 2009, that pay interest semiannually on June 30 and December 31. They are issued at $253,263, and their market rate is 6% at the issue date.

Required

1. Prepare the January 1, 2009, journal entry to record the bonds' issuance.
2. Determine the total bond interest expense to be recognized over the life of the bonds.
3. Prepare a straight-line amortization table like the one in Exhibit 14.7 for the bonds' first two years.
4. Prepare the journal entries to record the first two interest payments.

Check (2) $250,737
(3) 6/30/2010 carrying value, $259,437

Problem 14-6B[B]
Effective interest amortization of bond discount
P1 P2 

Refer to the bond details in Problem 14-5B.

Required

1. Prepare the January 1, 2009, journal entry to record the bonds' issuance.
2. Determine the total bond interest expense to be recognized over the bonds' life.
3. Prepare an effective interest amortization table like the one in Exhibit 14B.1 for the bonds' first two years.
4. Prepare the journal entries to record the first two interest payments.

Check (2) $250,737;
(3) 6/30/2010 carrying value, $257,275

Problem 14-7B[B]
Effective interest amortization of bond premium; retiring bonds
P1 P3 P4

Kendall issues $175,000 of 11%, three-year bonds dated January 1, 2009, that pay interest semiannually on June 30 and December 31. They are issued at $179,439, and their market rate is 10% at the issue date.

Required

1. Prepare the January 1, 2009, journal entry to record the bonds' issuance.
2. Determine the total bond interest expense to be recognized over the bonds' life.
3. Prepare an effective interest amortization table like the one in Exhibit 14B.2 for the bonds' first two years.
4. Prepare the journal entries to record the first two interest payments.
5. Prepare the journal entry to record the bonds' retirement on January 1, 2011, at 105.

Check (3) 6/30/2010 carrying value, $177,380
(5) $7,126 loss

Analysis Component

6. Assume that the market rate on January 1, 2009, is 12% instead of 10%. Without presenting numbers, describe how this change affects the amounts reported on Kendall's financial statements.

Problem 14-8B
Installment notes
C1 P5

On October 1, 2009, Miami Enterprises borrows $200,000 cash from a bank by signing a three-year installment note bearing 7% interest. The note requires equal total payments each year on September 30.

Required

1. Compute the total amount of each installment payment.
2. Complete an amortization table for this installment note similar to the one in Exhibit 14.14.
3. Prepare the journal entries to record (*a*) accrued interest as of December 31, 2009 (the end of its annual reporting period) and (*b*) the first annual payment on the note.

Check (2) 9/30/2011 ending balance, $71,223

Problem 14-9B
Applying the debt-to-equity ratio
A3

At the end of the current year, the following information is available for both Caesar Company and Delta Company.

	Caesar Company	Delta Company
Total assets	$360,000	$1,500,000
Total liabilities	162,360	1,125,000
Total equity	198,000	375,000

Required

1. Compute the debt-to-equity ratios for both companies.
2. Comment on your results and discuss what they imply about the relative riskiness of these companies.

Problem 14-10B[D]
Capital lease accounting
C4

Allan Company signs a five-year capital lease with Vortal Company for office equipment. The annual lease payment is $13,000, and the interest rate is 8%.

Required

1. Compute the present value of Allan's lease payments.
2. Prepare the journal entry to record Allan's capital lease at its inception.
3. Complete a lease payment schedule for the five years of the lease with the following headings. Assume that the beginning balance of the lease liability (present value of lease payments) is $51,905. (*Hint:* To find the amount allocated to interest in year 1, multiply the interest rate by the beginning-of-year lease liability. The amount of the annual lease payment not allocated to interest is allocated to principal. Reduce the lease liability by the amount allocated to principal to update the lease liability at each year-end.)

Period Ending Date	Beginning Balance of Lease Liability	Interest on Lease Liability	Reduction of Lease Liability	Cash Lease Payment	Ending Balance of Lease Liability

4. Use straight-line depreciation and prepare the journal entry to depreciate the leased asset at the end of year 1. Assume zero salvage value and a five-year life for the office equipment.

Check (1) $51,905

(3) Year 3 ending balance, $23,182

SERIAL PROBLEM

Success Systems

(This serial problem began in Chapter 1 and continues through most of the book. If previous chapter segments were not completed, the serial problem can begin at this point. It is helpful, but not necessary, to use the Working Papers that accompany the book.)

SP 14 Adriana Lopez has consulted with her local banker and is considering financing an expansion of her business by obtaining a long-term bank loan. Selected account balances at March 31, 2010, for Success Systems follow.

Total assets	$129,909	Total liabilities	$875	Total equity	$129,034

Required

1. The bank has offered a long-term secured note to Success Systems. The bank's loan procedures require that a client's debt-to-equity ratio not exceed 0.8. As of March 31, 2010, what is the maximum amount that Success Systems could borrow from this bank (rounded to nearest dollar)?
2. If Success Systems borrows the maximum amount allowed from the bank, what percentage of assets would be financed (*a*) by debt and (*b*) by equity?
3. What are some factors Lopez should consider before borrowing the funds?

Check (1) $102,352

BEYOND THE NUMBERS

REPORTING IN ACTION

A1 A2

BTN 14-1 Refer to Best Buy's financial statements in Appendix A to answer the following.

1. Identify the items that make up Best Buy's long-term debt at March 3, 2007. (*Hint:* See note 5.)
2. How much annual cash interest must Best Buy pay on its 2.25% convertible subordinated debt?
3. Did it have any additions to long-term debt that provided cash for the year ending March 3, 2007?

Fast Forward

4. Access Best Buy's financial statements for the years ending after March 3, 2007, from its Website (BestBuy.com) or the SEC's EDGAR database (www.SEC.gov). Has it issued additional long-term debt since the year-end March 3, 2007? If yes, identify the amount(s).

COMPARATIVE ANALYSIS

A3

BTN 14-2 Key figures for Best Buy, Circuit City, and RadioShack follow.

($ millions)	Best Buy Current Year	Best Buy Prior Year	Circuit City Current Year	Circuit City Prior Year	RadioShack Current Year	RadioShack Prior Year
Total assets	$13,570	$11,864	$4,007	$4,069	$2,070	$2,205
Total liabilities	7,369	6,607	2,216	2,114	1,416	1,616
Total equity	6,201	5,257	1,791	1,955	654	589

Required

1. Compute the debt-to-equity ratios for Best Buy, Circuit City, and RadioShack for both the current year and the prior year.
2. Use the ratios you computed in part 1 to determine which company's financing structure is least risky. Assume an industry average of 0.24 for debt-to-equity.

ETHICS CHALLENGE

C4 A1

BTN 14-3 Brevard County needs a new county government building that would cost $24 million. The politicians feel that voters will not approve a municipal bond issue to fund the building since it would increase taxes. They opt to have a state bank issue $24 million of tax-exempt securities to pay for the building construction. The county then will make yearly lease payments (of principal and interest) to repay the obligation. Unlike conventional municipal bonds, the lease payments are not binding obligations on the county and, therefore, require no voter approval.

Required

1. Do you think the actions of the politicians and the bankers in this situation are ethical?
2. How do the tax-exempt securities used to pay for the building compare in risk to a conventional municipal bond issued by Brevard County?

COMMUNICATING IN PRACTICE

P3

BTN 14-4 Your business associate mentions that she is considering investing in corporate bonds currently selling at a premium. She says that since the bonds are selling at a premium, they are highly valued and her investment will yield more than the going rate of return for the risk involved. Reply with a memorandum to confirm or correct your associate's interpretation of premium bonds.

TAKING IT TO THE NET

A2

BTN 14-5 Access the March 29, 2007, filing of the 10-K report of **Home Depot** for the year ended January 28, 2007, from www.sec.gov (Ticker: HD). Refer to Home Depot's balance sheet, including its note 4 (on debt).

Required

1. Identify Home Depot's long-term liabilities and the amounts for those liabilities from Home Depot's balance sheet at January 28, 2007.
2. Review Home Depot's note 4. The note reports that it "issued \$3.0 billion of 5.875% senior notes due December 16, 2036, at a discount of \$42 million with interest payable semiannually on June 16 and December 16 each year."
   a. Why would Home Depot issue \$3.0 billion of its notes for only \$2.958 billion?
   b. How much cash interest must Home Depot pay each June 16 and December 16 on these notes?

TEAMWORK IN ACTION

P2 P3

BTN 14-6[B] Break into teams and complete the following requirements related to effective interest amortization for a premium bond.

1. Each team member is to independently prepare a blank table with proper headings for amortization of a bond premium. When all have finished, compare tables and ensure that all are in agreement.

Parts 2 and 3 require use of these facts: On January 1, 2008, BC issues \$100,000, 9%, five-year bonds at 104.1. The market rate at issuance is 8%. BC pays interest semiannually on June 30 and December 31.

2. In rotation, *each* team member must explain how to complete *one* line of the bond amortization table, including all computations for his or her line. (Round amounts to the nearest dollar.) All members are to fill in their tables during this process. You need not finish the table; stop after all members have explained a line.
3. In rotation, *each* team member is to identify a separate column of the table and indicate what the final number in that column will be and explain the reasoning.
4. Reach a team consensus as to what the total bond interest expense on this bond issue will be if the bond is not retired before maturity.
5. As a team, prepare a list of similarities and differences between the amortization table just prepared and the amortization table if the bond had been issued at a discount.

Hint: Rotate teams to report on parts 4 and 5. Consider requiring entries for issuance and interest payments.

ENTREPRENEURIAL DECISION

A1

BTN 14-7 James Lindsay is the founder of **Rap Snacks**. Assume that his company currently has \$250,000 in equity, and he is considering a \$100,000 expansion to meet increased demand. The \$100,000 expansion would yield \$16,000 in additional annual income before interest expense. Assume that the business currently earns \$40,000 annual income before interest expense of \$10,000, yielding a return on equity of 12% (\$30,000/\$250,000). To fund the expansion, he is considering the issuance of a 10-year, \$100,000 note with annual interest payments (the principal due at the end of 10 years).

Required

1. Using return on equity as the decision criterion, show computations to support or reject the expansion if interest on the \$100,000 note is (*a*) 10%, (*b*) 15%, (*c*) 16%, (*d*) 17%, and (*e*) 20%.
2. What general rule do the results in part 1 illustrate?

HITTING THE ROAD

A1

BTN 14-8 Visit your city or county library. Ask the librarian to help you locate the recent financial records of your city or county government. Examine those records.

Required

1. Determine the amount of long-term bonds and notes currently outstanding.
2. Read the supporting information to your municipality's financial statements and record
   a. The market interest rate(s) when the bonds and/or notes were issued.
   b. The date(s) when the bonds and/or notes will mature.
   c. Any rating(s) on the bonds and/or notes received from **Moody's**, **Standard & Poor's**, or another rating agency.

BTN 14-9 DSG international plc (www.DSGiplc.com), Best Buy, Circuit City, and RadioShack are competitors in the global marketplace. Selected results from these companies follow.

GLOBAL DECISION

A3 

DSG

	DSG (£ millions)		Best Buy ($ millions)		Circuit City ($ millions)		RadioShack ($ millions)	
Key Figures	Current Year	Prior Year	Current Year	Prior Year	Current Year	Prior Year	Current Year	Prior Year
Total assets	£3,977	£4,120	$13,570	$11,864	$4,007	$4,069	$2,070	$2,205
Total liabilities	2,673	2,696	7,369	6,607	2,216	2,114	1,416	1,616
Total equity	1,304	1,424	6,201	5,257	1,791	1,955	654	589
Debt-to-equity ratio	?	?	1.2	1.3	1.2	1.1	2.2	2.7

Required

1. Compute DSG's debt-to-equity ratios for the current year and the prior year.
2. Use the data provided and the ratios you computed in part 1 to determine which company's financing structure is least risky.

ANSWERS TO MULTIPLE CHOICE QUIZ

1. b
2. a
3. c; $500,000 × 0.08 × ½ year = $20,000
4. a; Cash interest paid = $100,000 × 5% × ½ year = $2,500
   Discount amortization = ($100,000 − $95,735)/10 periods = $426.50
   Interest expense = $2,500.00 + $426.50 = $2,926.50
5. e; ($373,745 − $350,000)/16 periods = $1,484

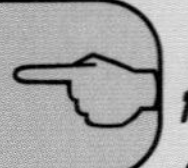

A Look Back

Chapter 14 focused on long-term liabilities—a main part of most companies' financing. We explained how to value, record, amortize, and report these liabilities in financial statements.

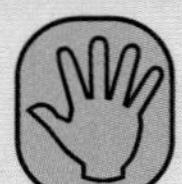

A Look at This Chapter

This chapter focuses on investments in securities. We explain how to identify, account for, and report investments in both debt and equity securities. We also explain accounting for transactions listed in a foreign currency.

A Look Ahead

Chapter 16 focuses on reporting and analyzing a company's cash flows. Special emphasis is directed at the statement of cash flows reported under the indirect method.

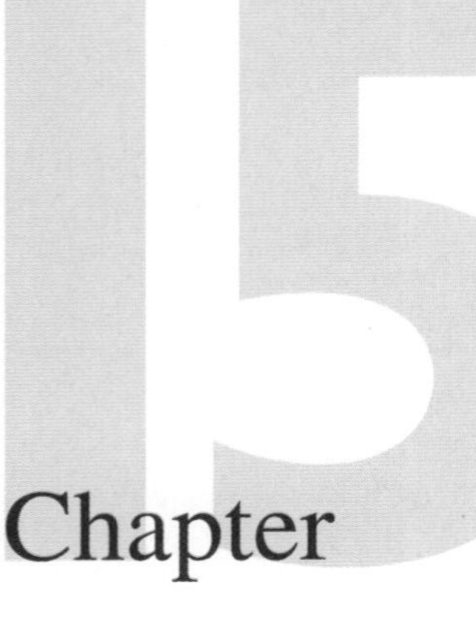

Chapter 15 Investments and International Operations

Learning Objectives

CAP

Conceptual

C1 Distinguish between debt and equity securities and between short-term and long-term investments. *(p. 592)*

C2 Identify and describe the different classes of investments in securities. *(p. 593)*

C3 Describe how to report equity securities with controlling influence. *(p. 599)*

C4 *Appendix 15A*—Explain foreign exchange rates between currencies. *(p. 605)*

Analytical

A1 Compute and analyze the components of return on total assets. *(p. 600)*

LP15

Procedural

P1 Account for trading securities. *(p. 595)*

P2 Account for held-to-maturity securities. *(p. 596)*

P3 Account for available-for-sale securities. *(p. 596)*

P4 Account for equity securities with significant influence. *(p. 598)*

P5 *Appendix 15A*—Record transactions listed in a foreign currency. *(p. 606)*

Decision Feature

Designing Business

"*It's a wonderful insanity*"—Amy Smilovic

NEW YORK—Amy Smilovic loves business. But after a job transfer by her husband to Hong Kong, she quit her job and looked for her next inspiration. Within weeks, and inspired by the colorful and comfortable clothing she saw, Amy jumped into the designing business of clothes. Her company, Tibi **(Tibi.com),** experienced quick success. "I really kind of just did it," explains Amy. "I tried not to over-think it."

What Amy did was design a clothing line that appeals to the masses. "Just design for yourself, someone will identify with it," insists Amy. "You'll find a group of people who want to dress like you." Amy has been so successful that this group of people extends to many countries. She now has showrooms in Milan and London, and locations in Australia, Hong Kong, Indonesia, Japan, Korea, Malaysia, New Zealand, and Singapore.

This broad reach became what she calls her biggest challenge: investments and international operations. "I didn't know what custom regulations were," explains Amy. "I didn't know how to set up a U.S. base and distribute . . . I was buying books on customs and international shipping laws." Her investments in international operations require Amy to translate their performance into U.S. dollars for financial reports. It also requires conducting international transactions and doing currency translations. Those tasks require knowledge of accounting and reporting requirements for investments, including investments in securities of other companies.

Amy's continued success is sure to include more investments and further international operations. She views her business background and focus as key to that success. "People always ask me, 'How are you a designer without a design background,'" laughs Amy. "I always wonder, 'How are you a designer without a business background?' It's really a business at the end of the day."

[Sources: *Tibi Website,* January 2009, *Entrepreneur,* October 2007; *Fashion Week,* 2008; *Ladies Who Launch,* 2008; *New York Social Diary,* April 2007]

Chapter Preview

This chapter's main focus is investments in securities. Many companies have investments, and many of these are in the form of debt and equity securities issued by other companies. We describe investments in these securities and how to account for them. An increasing number of companies also invest in international operations. We explain how to account for and report international transactions listed in foreign currencies.

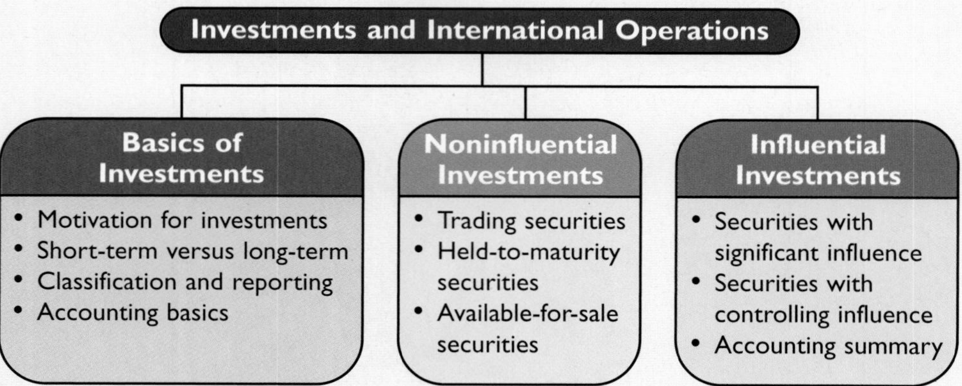

Basics of Investments

C1 Distinguish between debt and equity securities and between short-term and long-term investments.

This section describes the motivation for investments, the distinction between short- and long-term investments, and the different classes of investments.

Motivation for Investments

Companies make investments for at least three reasons. First, companies transfer *excess cash* into investments to produce higher income. Second, some entities, such as mutual funds and pension funds, are set up to produce income from investments. Third, companies make investments for strategic reasons. Examples are investments in competitors, suppliers, and even customers. Exhibit 15.1 shows short-term (S-T) and long-term (L-T) investments as a percent of total assets for several companies.

EXHIBIT 15.1

Investments of Selected Companies

Company	S-T	L-T
Pfizer	19%	4%
Gap	2%	1%
Starbucks	3%	5%
Coca-Cola	1%	18%

0% — 25%
Percent of total assets

Short-Term versus Long-Term

Cash equivalents are investments that are both readily converted to known amounts of cash and mature within three months. Many investments, however, mature between 3 and 12 months. These investments are **short-term investments,** also called *temporary investments* and *marketable securities*. Specifically, short-term investments are securities that (1) management intends to convert to cash within one year or the operating cycle, whichever is longer, and (2) are readily convertible to cash. Short-term investments are reported under current assets and serve a purpose similar to cash equivalents.

Long-term investments in securities are defined as those securities that are not readily convertible to cash or are not intended to be converted into cash in the short term. Long-term investments can also include funds earmarked for a special purpose, such as bond sinking funds and investments in land or other assets not used in the company's operations. Long-term investments are reported in the noncurrent section of the balance sheet, often in its own separate line titled *Long-Term Investments*.

Investments in securities can include both debt and equity securities. *Debt securities* reflect a creditor relationship such as investments in notes, bonds, and certificates of deposit; they are issued by governments, companies, and individuals. *Equity securities* reflect an owner relationship such as shares of stock issued by companies.

Classification and Reporting

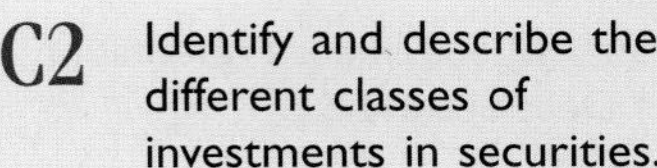
C2 Identify and describe the different classes of investments in securities.

Accounting for investments in securities depends on three factors: (1) security type, either debt or equity, (2) the company's intent to hold the security either short term or long term, and (3) the company's (investor's) percent ownership in the other company's (investee's) equity securities. Exhibit 15.2 identifies five classes of securities using these three factors. It describes each of these five classes of securities and the standard reporting required under each class.

EXHIBIT 15.2

Investments in Securities

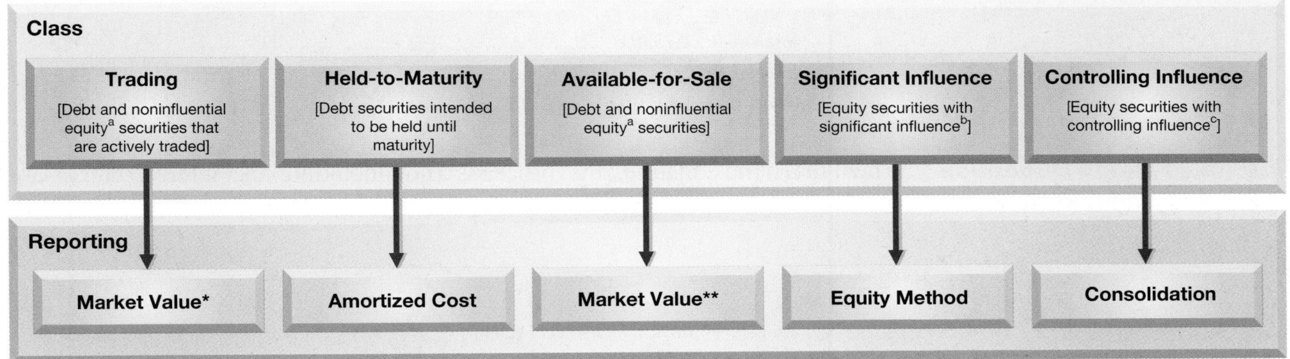

[a] Holding less than 20% of voting stock (equity securities only). [b] Holding 20% or more, but not more than 50%, of voting stock.
[c] Holding more than 50% of voting stock.
* Unrealized gains and losses reported on the income statement.
** Unrealized gains and losses reported in the equity section of the balance sheet and in comprehensive income.

Accounting Basics for Debt Securities

This section explains the accounting basics for *debt securities,* including that for acquisition, disposition, and any interest.

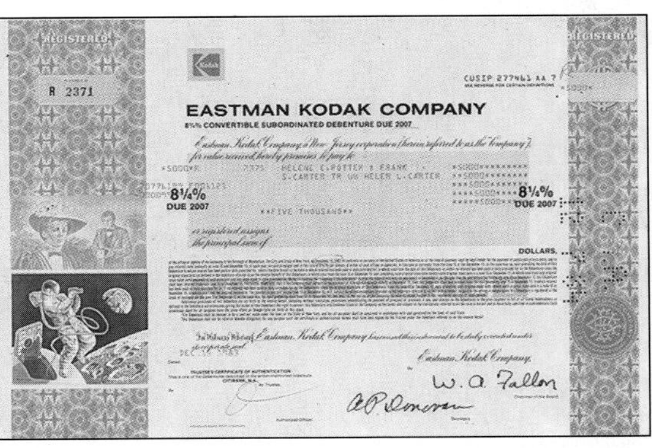

Acquisition. Debt securities are recorded at cost when purchased. To illustrate, assume that Music City paid $29,500 plus a $500 brokerage fee on September 1, 2008, to buy Dell's 7%, two-year bonds payable with a $30,000 par value. The bonds pay interest semiannually on August 31 and February 28. Music City intends to hold the bonds until they mature on August 31, 2010; consequently, they are classified as held-to-maturity (HTM) securities. The entry to record this purchase follows. (If the maturity of the securities was short term, and management's intent was to hold them until they mature, then they would be classified as Short-Term Investments—HTM.)

2008			
Sept. 1	Long-Term Investments—HTM (Dell)	30,000	
	Cash		30,000
	Purchased bonds to be held to maturity.		

Assets = Liabilities + Equity
+30,000
−30,000

Interest earned. Interest revenue for investments in debt securities is recorded when earned. To illustrate, on December 31, 2008, at the end of its accounting period, Music City accrues interest receivable as follows.

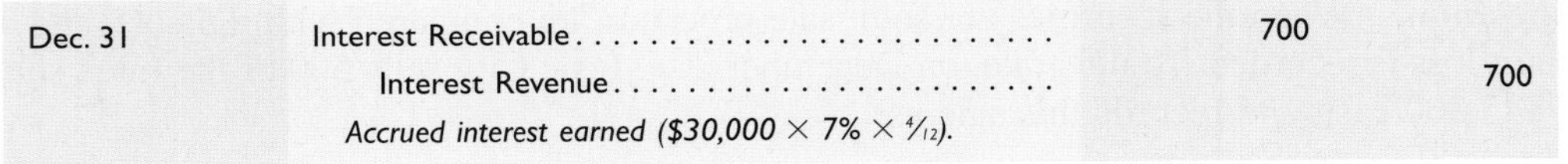

Dec. 31	Interest Receivable	700	
	Interest Revenue		700
	Accrued interest earned ($30,000 × 7% × 4/12).		

Assets = Liabilities + Equity
+700 +700

The $700 reflects 4/6 of the semiannual cash receipt of interest—the portion Music City earned as of December 31. Relevant sections of Music City's financial statements at December 31, 2008, are shown in Exhibit 15.3.

EXHIBIT 15.3

Financial Statement Presentation of Debt Securities

On the income statement for year 2008:	
Interest revenue	**$ 700**
On the December 31, 2008, balance sheet:	
Long-term investments—Held-to-maturity securities (at amortized cost)	**$30,000**

On February 28, 2009, Music City records receipt of semiannual interest.

Assets = Liabilities + Equity
+1,050 +350
−700

Feb. 28	Cash	1,050	
	Interest Receivable		700
	Interest Revenue		350
	Received six months' interest on Dell bonds.		

Disposition. When the bonds mature, the proceeds (not including the interest entry) are recorded as:

Assets = Liabilities + Equity
+30,000
−30,000

2010			
Aug. 31	Cash	30,000	
	Long-Term Investments—HTM (Dell)		30,000
	Received cash from matured bonds.		

Example: What is cost per share? *Answer:* Cost per share is the total cost of acquisition, including broker fees, divided by number of shares acquired.

The cost of a debt security can be either higher or lower than its maturity value. When the investment is long term, the difference between cost and maturity value is amortized over the remaining life of the security. We assume for ease of computations that the cost of a long-term debt security equals its maturity value.

Accounting Basics for Equity Securities

This section explains the accounting basics for *equity securities,* including that for acquisition, dividends, and disposition.

Acquisition. Equity securities are recorded at cost when acquired, including commissions or brokerage fees paid. To illustrate, assume that Music City purchases 1,000 shares of Intex common stock at par value for $86,000 on October 10, 2008. It records this purchase of available-for-sale (AFS) securities as follows.

Assets = Liabilities + Equity
+86,000
−86,000

Oct. 10	Long-Term Investments—AFS (Intex)	86,000	
	Cash		86,000
	Purchased 1,000 shares of Intex.		

Dividend earned. Any cash dividends received are credited to Dividend Revenue and reported in the income statement. To illustrate, on November 2, Music City receives a $1,720 quarterly cash dividend on the Intex shares, which it records as:

Assets = Liabilities + Equity
+1,720 +1,720

Nov. 2	Cash	1,720	
	Dividend Revenue		1,720
	Received dividend of $1.72 per share.		

Disposition. When the securities are sold, sale proceeds are compared with the cost, and any gain or loss is recorded. To illustrate, on December 20, Music City sells 500 of the Intex shares for $45,000 cash and records this sale as:

Assets = Liabilities + Equity
+45,000 +2,000
−43,000

Dec. 20	Cash	45,000	
	Long-Term Investments—AFS (Intex)		43,000
	Gain on Sale of Long-Term Investments		2,000
	Sold 500 Intex shares ($86,000 × 500/1,000).		

Reporting of *Non*influential Investments

Companies must value and report most noninfluential investments at *fair market value,* or simply *market value.* The exact reporting requirements depend on whether the investments are classified as (1) trading, (2) held-to-maturity, or (3) available-for-sale.

P1 Account for trading securities.

Trading Securities

Trading securities are *debt and equity securities* that the company intends to actively manage and trade for profit. Frequent purchases and sales are expected and are made to earn profits on short-term price changes. Trading securities are *always* reported as current assets.

Valuing and reporting trading securities. The entire portfolio of trading securities is reported at its market value; this requires a "market adjustment" from the cost of the portfolio. The term *portfolio* refers to a group of securities. Any **unrealized gain (or loss)** from a change in the market value of the portfolio of trading securities is reported on the income statement. Most users believe accounting reports are more useful when changes in market value for trading securities are reported in income.

To illustrate, TechCom's portfolio of trading securities had a total cost of $11,500 and a market value of $13,000 on December 31, 2008, the first year it held trading securities. The difference between the $11,500 cost and the $13,000 market value reflects a $1,500 gain. It is an unrealized gain because it is not yet confirmed by actual sales. The market adjustment for trading securities is recorded with an adjusting entry at the end of each period to equal the difference between the portfolio's cost and its market value. TechCom records this gain as follows.

Point: *'Unrealized gain (or loss)'* refers to a change in market value that is not yet realized through actual sale.

Point: 'Market Adjustment—Trading' is a *permanent account*, shown as a deduction or addition to 'Short-Term Investments—Trading.'

Dec. 31	Market Adjustment—Trading.	1,500	
	Unrealized Gain—Income		1,500
	To reflect an unrealized gain in market values of trading securities.		

Assets = Liabilities + Equity
+1,500 +1,500

The Unrealized Gain (or Loss) is reported in the Other Revenues and Gains (or Expenses and Losses) section on the income statement. Unrealized Gain (or Loss)—Income is a *temporary* account that is closed to Income Summary at the end of each period. Market Adjustment—Trading is a *permanent* account, which adjusts the reported value of the trading securities portfolio from its prior period market value to the current period market value. The total cost of the trading securities portfolio is maintained in one account, and the market adjustment is recorded in a separate account. For example, TechCom's investment in trading securities is reported in the current assets section of its balance sheet as follows.

Example: If TechCom's trading securities have a cost of $14,800 and a market of $16,100 at Dec. 31, 2009, its adjusting entry is
Unrealized Loss—Income 200
Market Adj.—Trading 200
This is computed as: $1,500 Beg. Dr. bal. + $200 Cr. = $1,300 End. Dr. bal.

Current Assets		
Short-term investments—Trading (at cost)	$11,500	
Market adjustment—Trading	1,500	
Short-term investments—Trading (at market)		$13,000
or simply		
Short-term investments—Trading (at market; cost is $11,500)		$13,000

Selling trading securities. When individual trading securities are sold, the difference between the net proceeds (sale price less fees) and the cost of the individual trading securities that are sold is recognized as a gain or a loss. Any prior period market adjustment to the portfolio is *not* used to compute the gain or loss from sale of individual trading securities. For example, if TechCom sold some of its trading securities that had cost $1,000 for $1,200 cash on January 9, 2009, it would record the following.

Point: Reporting securities at market value is referred to as *mark-to-market* accounting.

Assets = Liabilities + Equity
+1,200 +200
−1,000

Jan. 9	Cash	1,200	
	Short-Term Investments—Trading		1,000
	Gain on Sale of Short-Term Investments		200
	Sold trading securities costing $1,000 for $1,200 cash.		

A gain is reported in the Other Revenues and Gains section on the income statement, whereas a loss is shown in Other Expenses and Losses. When the period-end market adjustment for the portfolio of trading securities is computed, it excludes the cost and market value of any securities sold.

Held-to-Maturity Securities

P2 Account for held-to-maturity securities.

Held-to-maturity (HTM) securities are *debt* securities a company intends and is able to hold until maturity. They are reported in current assets if their maturity dates are within one year or the operating cycle, whichever is longer. HTM securities are reported in long-term assets when the maturity dates extend beyond one year or the operating cycle, whichever is longer. All HTM securities are recorded at cost when purchased, and interest revenue is recorded when earned.

The portfolio of HTM securities is usually reported at (amortized) cost, which is explained in advanced courses. There is no market adjustment to the portfolio of HTM securities—neither to the short-term nor long-term portfolios. The basics of accounting for HTM securities were described earlier in this chapter.

Point: Only debt securities can be classified as *held-to-maturity;* equity securities have no maturity date.

Decision Maker

Money Manager You expect interest rates to sharply fall within a few weeks and remain at this lower rate. What is your strategy for holding investments in fixed-rate bonds and notes? [Answer—p. 608]

Available-for-Sale Securities

P3 Account for available-for-sale securities.

Available-for-sale (AFS) securities are *debt and equity securities* not classified as trading or held-to-maturity securities. AFS securities are purchased to yield interest, dividends, or increases in market value. They are not actively managed like trading securities. If the intent is to sell AFS securities within the longer of one year or operating cycle, they are classified as short-term investments. Otherwise, they are classified as long-term.

Valuing and reporting available-for-sale securities. As with trading securities, companies adjust the cost of the portfolio of AFS securities to reflect changes in market value. This is done with a market adjustment to its total portfolio cost. However, any unrealized gain or loss for the portfolio of AFS securities is *not* reported on the income statement. Instead, it is reported in the equity section of the balance sheet (and is part of *comprehensive income,* explained later). To illustrate, assume that Music City had no prior period investments in available-for-sale securities other than those purchased in the current period. Exhibit 15.4 shows both the cost and market value of those investments on December 31, 2008, the end of its reporting period.

Example: If market value in Exhibit 15.4 is $70,000 (instead of $74,550), what entry is made? *Answer:*
Unreal. Loss—Equity . . . 3,000
Market Adj.—AFS 3,000

EXHIBIT 15.4

Cost and Market Value of Available-for-Sale Securities

	Cost	Market Value	Unrealized Gain (Loss)
Improv bonds	$30,000	$29,050	$ (950)
Intex common stock, 500 shares	43,000	45,500	2,500
Total	$73,000	$74,550	**$1,550**

The year-end adjusting entry to record the market value of these investments follows.

Assets = Liabilities + Equity
+1,550 +1,550

Dec. 31	Market Adjustment—Available-for-Sale (LT)	1,550	
	Unrealized Gain—Equity		1,550
	To record adjustment to market value of available-for-sale securities.		

Exhibit 15.5 shows the December 31, 2008, balance sheet presentation—it assumes these investments are long term, but they can also be short term. It is also common to combine the cost of investments with the balance in the Market Adjustment account and report the net as a single amount.

Point: 'Unrealized Loss—Equity' and 'Unrealized Gain—Equity' are *permanent* (balance sheet) equity *accounts*.

EXHIBIT 15.5

Balance Sheet Presentation of Available-for-Sale Securities

Assets		
Long-term investments—Available-for-sale (at cost)	$73,000	
Market adjustment— Available-for-sale	1,550	
Long-term investments—Available-for-sale (at market)		$74,550
or simply		
Long-term investments—Available-for-sale (at market; cost is $73,000)		$74,550
Equity		
... consists of usual equity accounts ...		
Add unrealized gain on available-for-sale securities*		$ 1,550

(Reconciled: Market adjustment ↔ Add unrealized gain)

* Often included under the caption Accumulated Other Comprehensive Income.

Let's extend this illustration and assume that at the end of its next calendar year (December 31, 2009), Music City's portfolio of long-term AFS securities has an $81,000 cost and an $82,000 market value. It records the adjustment to market value as follows.

Point: Income can be window-dressed upward by selling AFS securities with unrealized gains; income is reduced by selling those with unrealized losses.

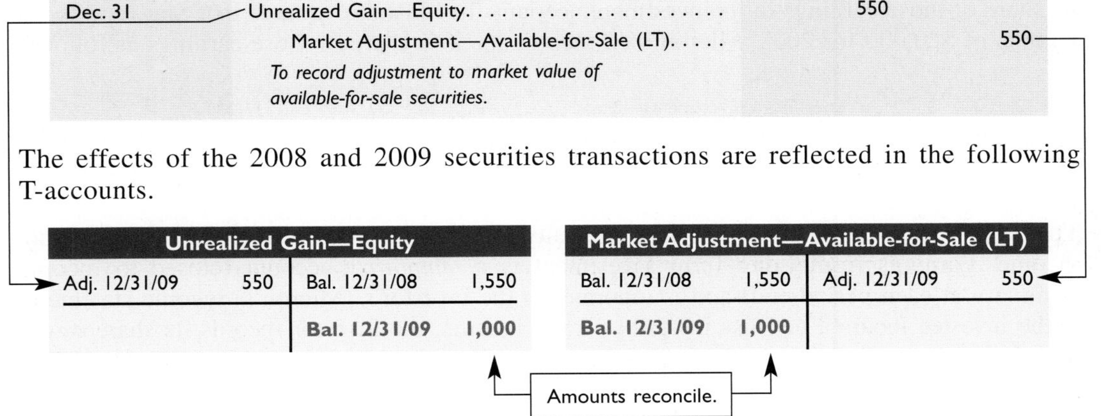

Dec. 31	Unrealized Gain—Equity	550	
	Market Adjustment—Available-for-Sale (LT)		550
	To record adjustment to market value of available-for-sale securities.		

Assets = Liabilities + Equity
−550 −550

The effects of the 2008 and 2009 securities transactions are reflected in the following T-accounts.

Unrealized Gain—Equity

Adj. 12/31/09	550	Bal. 12/31/08	1,550
		Bal. 12/31/09	**1,000**

Market Adjustment—Available-for-Sale (LT)

Bal. 12/31/08	1,550	Adj. 12/31/09	550
Bal. 12/31/09	**1,000**		

Amounts reconcile.

Example: If cost is $83,000 and market is $82,000 at Dec. 31, 2009, it records the following adjustment:

Unreal. Gain—Equity	1,550	
Unreal. Loss—Equity	1,000	
Mkt. Adj.—AFS		2,550

Selling available-for-sale securities. Accounting for the sale of individual AFS securities is identical to that described for the sale of trading securities. When individual AFS securities are sold, the difference between the cost of the individual securities sold and the net proceeds (sale price less fees) is recognized as a gain or loss.

Point: 'Market Adjustment—Available-for-Sale' is a permanent account, shown as a deduction or addition to the Investment account.

Quick Check

Answers—p. 608

1. How are short-term held-to-maturity securities reported (valued) on the balance sheet?
2. How are trading securities reported (valued) on the balance sheet?
3. Where are unrealized gains and losses on available-for-sale securities reported?
4. Where are unrealized gains and losses on trading securities reported?

Alert *The FASB released FAS 157 and FAS 159 that permit companies to use market value in reporting financial assets (referred to as the fair value option). This option allows companies to report any financial asset at fair market value and recognize value changes in income. This method was previously reserved only for trading securities, but would now be an option for available-for-sale and held-to-maturity securities (and other 'financial assets and liabilities' such as accounts and notes receivable, accounts and notes payable, and bonds). These standards also set a 3-level system to determine fair value:*
—Level 1: Use quoted market values
—Level 2: Use observable values from related assets or liabilities
—Level 3: Use unobservable values from estimates or assumptions
To date, a fairly small set of companies has chosen to broadly apply the fair value option—but, we continue to monitor its use...

Reporting of Influential Investments

Investment in Securities with Significant Influence

P4 Account for equity securities with significant influence.

A long-term investment classified as **equity securities with significant influence** implies that the investor can exert significant influence over the investee. An investor that owns 20% or more (but not more than 50%) of a company's voting stock is usually presumed to have a significant influence over the investee. In some cases, however, the 20% test of significant influence is overruled by other, more persuasive, evidence. This evidence can either lower the 20% requirement or increase it. The **equity method** of accounting and reporting is used for long-term investments in equity securities with significant influence, which is explained in this section.

Long-term investments in equity securities with significant influence are recorded at cost when acquired. To illustrate, Micron Co. records the purchase of 3,000 shares (30%) of Star Co. common stock at a total cost of $70,650 on January 1, 2008, as follows.

Assets	= Liabilities	+ Equity
+70,650		
−70,650		

Jan. 1	Long-Term Investments—Star	70,650	
	Cash		70,650
	To record purchase of 3,000 Star shares.		

The investee's (Star) earnings increase both its net assets and the claim of the investor (Micron) on the investee's net assets. Thus, when the investee reports its earnings, the investor records its share of those earnings in its investment account. To illustrate, assume that Star reports net income of $20,000 for 2008. Micron then records its 30% share of those earnings as follows.

Assets	= Liabilities	+ Equity
+6,000		+6,000

Dec. 31	Long-Term Investments—Star	6,000	
	Earnings from Long-Term Investment		6,000
	To record 30% equity in investee earnings.		

The debit reflects the increase in Micron's equity in Star. The credit reflects 30% of Star's net income. Earnings from Long-Term Investment is a *temporary* account (closed to Income Summary at each period-end) and is reported on the investor's (Micron's) income statement. If the investee incurs a net loss instead of a net income, the investor records its share of the loss and reduces (credits) its investment account. The investor closes this earnings or loss account to Income Summary.

The receipt of cash dividends is not revenue under the equity method because the investor has already recorded its share of the investee's earnings. Instead, cash dividends received by an investor from an investee are viewed as a conversion of one asset to another; that is, dividends reduce the balance of the investment account. To illustrate, Star declares and pays $10,000 in cash dividends on its common stock. Micron records its 30% share of these dividends received on January 9, 2009, as:

Assets	= Liabilities	+ Equity
+3,000		
−3,000		

Jan. 9	Cash	3,000	
	Long-Term Investments—Star		3,000
	To record share of dividend paid by Star.		

The book value of an investment under the equity method equals the cost of the investment plus (minus) the investor's equity in the *undistributed* (*distributed*) earnings of the investee. Once Micron records these transactions, its Long-Term Investments account appears as in Exhibit 15.6.

EXHIBIT 15.6

Investment in Star Common Stock (Ledger Account)

Date	Explanation	Debit	Credit	Balance
2008				
Jan. 1	Investment acquisition	70,650		70,650
Dec. 31	Share of earnings	6,000		76,650
2009				
Jan. 9	Share of dividend		3,000	73,650

Micron's account balance on January 9, 2009, for its investment in Star is $73,650. This is the investment's cost *plus* Micron's equity in Star's earnings since its purchase *less* Micron's equity in Star's cash dividends since its purchase. When an investment in equity securities is sold, the gain or loss is computed by comparing proceeds from the sale with the book value of the investment on the date of sale. If Micron sells its Star stock for $80,000 on January 10, 2009, it records the sale as:

Point: Security prices are sometimes listed in fractions. For example, a debt security with a price of $22\frac{1}{4}$, is the same as $22.25.

Jan. 10	Cash	80,000	
	Long-Term Investments—Star		73,650
	Gain on Sale of Investment		6,350
	Sold 3,000 shares of stock for $80,000.		

Assets	= Liabilities +	Equity
+80,000		+6,350
−73,650		

Investment in Securities with Controlling Influence

C3 Describe how to report equity securities with controlling influence.

A long-term investment classified as **equity securities with controlling influence** implies that the investor can exert a controlling influence over the investee. An investor who owns more than 50% of a company's voting stock has control over the investee. This investor can dominate all other shareholders in electing the corporation's board of directors and has control over the investee's management. In some cases, controlling influence can extend to situations of less than 50% ownership. Exhibit 15.7 summarizes the accounting for investments in equity securities based on an investor's ownership in the stock.

EXHIBIT 15.7

Accounting for Equity Investments by Percent of Ownership

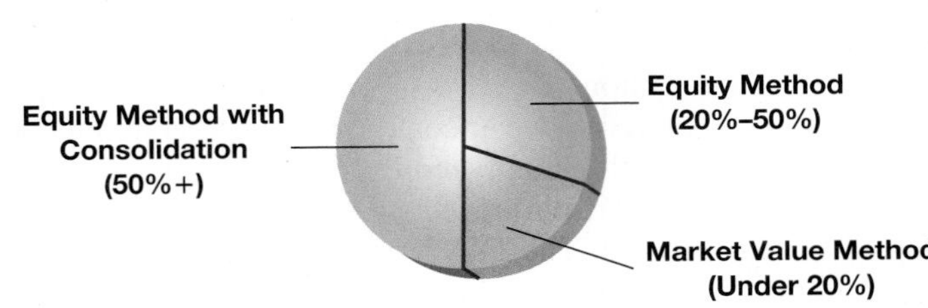

The *equity method with consolidation* is used to account for long-term investments in equity securities with controlling influence. The investor reports *consolidated financial statements* when owning such securities. The controlling investor is called the **parent,** and the investee is called the **subsidiary.** Many companies are parents with subsidiaries. Examples are (1) **McGraw-Hill**, the parent of *BusinessWeek,* Standard & Poor's, and Compustat; (2) **Gap, Inc.**, the parent of Gap, Old Navy, and Banana Republic; and (3) **Brunswick**, the parent of Mercury Marine, Sea Ray, and U.S. Marine. A company owning all the outstanding stock of a subsidiary can, if it desires, take over the subsidiary's assets, retire the subsidiary's stock, and merge the subsidiary into the parent. However, there often are financial, legal, and tax advantages if a business operates as a parent controlling one or more subsidiaries. When a company operates as a parent with subsidiaries, each entity maintains separate accounting records. From a legal viewpoint, the parent and each subsidiary are separate entities with all rights, duties, and responsibilities of individual companies.

Consolidated financial statements show the financial position, results of operations, and cash flows of all entities under the parent's control, including all subsidiaries. These statements are prepared as if the business were organized as one entity. The parent uses the equity method in its accounts, but the investment account is *not* reported on the parent's financial statements. Instead, the individual assets and liabilities of the parent and its subsidiaries are combined on one balance sheet. Their revenues and expenses also are combined on one income statement, and their cash flows are combined on one statement of cash flows. The procedures for preparing consolidated financial statements are in advanced courses.

Decision Insight

IFRSs Unlike U.S. GAAP, IFRSs require uniform accounting policies be used throughout the group of consolidated subsidiaries. Also, unlike U.S. GAAP, IFRSs offer no detailed guidance on valuation procedures.

Accounting Summary for Investments in Securities

Exhibit 15.8 summarizes the standard accounting for investments in securities. Recall that many investment securities are classified as either short term or long term depending on management's

EXHIBIT 15.8

Accounting for Investments in Securities

Classification	Accounting
Short-Term Investment in Securities	
Held-to-maturity (debt) securities	**Cost** (without any discount or premium amortization)
Trading (debt and equity) securities	**Market value** (with market adjustment to income)
Available-for-sale (debt and equity) securities	**Market value** (with market adjustment to equity)
Long-Term Investment in Securities	
Held-to-maturity (debt) securities	**Cost** (with any discount or premium amortization)
Available-for-sale (debt and equity) securities	**Market value** (with market adjustment to equity)
Equity securities with significant influence	Equity method
Equity securities with controlling influence	Equity method (with consolidation)

intent and ability to convert them in the future. Understanding the accounting for these investments enables us to draw better conclusions from financial statements in making business decisions.

Comprehensive Income The term **comprehensive income** refers to all changes in equity for a period except those due to investments and distributions to owners. This means that it includes (1) the revenues, gains, expenses, and losses reported in net income *and* (2) the gains and losses that bypass net income but affect equity. An example of an item that bypasses net income is unrealized gains and losses on available-for-sale securities. These items make up *other comprehensive income* and are usually reported as a part of the statement of stockholders' equity. (Two other options are as a second separate income statement or as a combined income statement of comprehensive income; these less common options are described in advanced courses.) Most often this simply requires one additional column for Other Comprehensive Income in the usual columnar form of the statement of stockholders' equity (the details of this are left for advanced courses). The FASB encourages, but does *not* require, other comprehensive income items to be grouped under the caption *Accumulated Other Comprehensive Income* in the equity section of the balance sheet, which would include unrealized gains and losses on available-for-sale securities. For instructional benefits, we use actual account titles for these items in the equity section instead of this general, less precise caption.

Point: Some users believe that since AFS securities are not actively traded, reporting market value changes in income would unnecessarily increase income variability and decrease usefulness.

Quick Check

Answers—p. 608

5. Give at least two examples of assets classified as long-term investments.
6. What are the requirements for an equity security to be listed as a long-term investment?
7. Identify similarities and differences in accounting for long-term investments in debt securities that are held-to-maturity versus those available-for-sale.
8. What are the three possible classifications of long-term equity investments? Describe the criteria for each class and the method used to account for each.

Decision Analysis — Components of Return on Total Assets

A1 Compute and analyze the components of return on total assets.

A company's **return on total assets** (or simply *return on assets*) is important in assessing financial performance. The return on total assets can be separated into two components, profit margin and total asset turnover, for additional analyses. Exhibit 15.9 shows how these two components determine return on total assets.

EXHIBIT 15.9

Components of Return on Total Assets

Return on total assets = Profit margin × Total asset turnover

$$\frac{\text{Net income}}{\text{Average total assets}} = \frac{\text{Net income}}{\text{Net sales}} \times \frac{\text{Net sales}}{\text{Average total assets}}$$

Profit margin reflects the percent of net income in each dollar of net sales. Total asset turnover reflects a company's ability to produce net sales from total assets. All companies desire a high return on total assets.

By considering these two components, we can often discover strengths and weaknesses not revealed by return on total assets alone. This improves our ability to assess future performance and company strategy.

To illustrate, consider return on total assets and its components for **Gap Inc.** in Exhibit 15.10.

Fiscal Year	Return on Total Assets	=	Profit Margin	×	Total Asset Turnover
2007	9.0%	=	4.9%	×	1.84
2006	11.8	=	6.9	×	1.70
2005	11.1	=	7.1	×	1.57

EXHIBIT 15.10

Gap's Components of Return on Total Assets

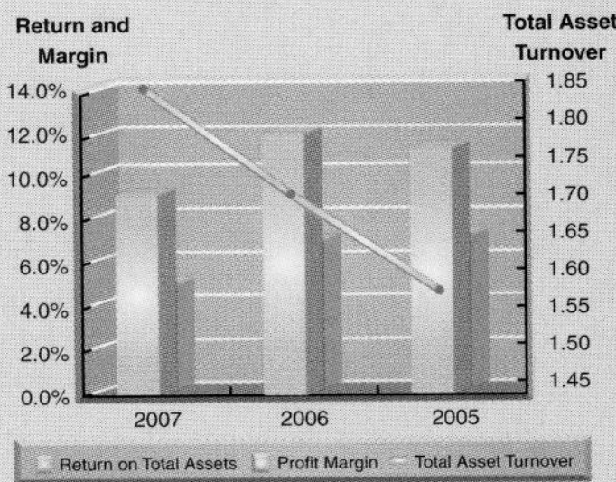

At least three findings emerge. First, Gap's return on total assets improved from 11.1% in 2005 to 11.8% in 2006, but then declined to 9.0% in 2007. Second, total asset turnover has markedly improved from 1.57 to 1.70 to 1.84 over this period. Third, Gap's profit margin steadily declined from 7.1% to 6.9%, and then to 4.9% over this period. These components reveal the dual role of profit margin and total asset turnover in determining return on total assets. They also reveal that the driver of Gap's recent decline is not total asset turnover but profit margin.

Generally, if a company is to maintain or improve its return on total assets, it must meet any decline in either profit margin or total asset turnover with an increase in the other. If not, return on assets will decline. Companies consider these components in planning strategies. A component analysis can also reveal where a company is weak and where changes are needed, especially in a competitor analysis. If asset turnover is lower than the industry norm, for instance, a company should focus on raising asset turnover at least to the norm. The same applies to profit margin.

Decision Maker

Retailer You are an entrepreneur and owner of a retail sporting goods store. The store's recent annual performance reveals (industry norms in parentheses): return on total assets = 11% (11.2%); profit margin = 4.4% (3.5%); and total asset turnover = 2.5 (3.2). What does your analysis of these figures reveal? [Answer—p. 608]

Demonstration Problem—1

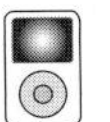

DP15

Garden Company completes the following selected transactions related to its short-term investments during 2008.

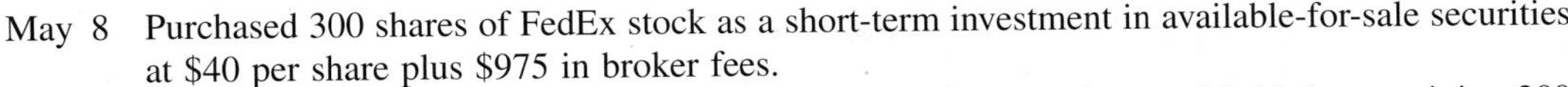

May 8 Purchased 300 shares of FedEx stock as a short-term investment in available-for-sale securities at $40 per share plus $975 in broker fees.

Sept. 2 Sold 100 shares of its investment in FedEx stock at $47 per share and held the remaining 200 shares; broker's commission was $225.

Oct. 2 Purchased 400 shares of Ajay stock for $60 per share plus $1,600 in commissions. The stock is held as a short-term investment in available-for-sale securities.

Required

1. Prepare journal entries for the above transactions of Garden Company for 2008.
2. Prepare an adjusting journal entry as of December 31, 2008, if the market prices of the equity securities held by Garden Company are $48 per share for FedEx and $55 per share for Ajay. (Year 2008 is the first year Garden Company acquired short-term investments.)

Solution to Demonstration Problem—1

1.

Date	Account	Debit	Credit
May 8	Short-Term Investments—AFS (FedEx)	12,975	
	Cash		12,975
	Purchased 300 shares of FedEx stock (300 × $40) + $975.		

[continued on next page]

[continued from previous page]

Sept. 2	Cash	4,475	
	Gain on Sale of Short-Term Investment		150
	Short-Term Investments—AFS (FedEx)		4,325
	Sold 100 shares of FedEx for $47 per share less a $225 commission. The original cost is ($12,975 × 100/300).		
Oct. 2	Short-Term Investments—AFS (Ajay)	25,600	
	Cash		25,600
	Purchased 400 shares of Ajay for $60 per share plus $1,600 in commissions.		

2. Computation of unrealized gain or loss follows.

Short-Term Investments in Available-for-Sale Securities	Shares	Cost per Share	Total Cost	Market Value per Share	Total Market Value	Unrealized Gain (Loss)
FedEx	200	$43.25	$ 8,650	$48.00	$ 9,600	
Ajay	400	64.00	25,600	55.00	22,000	
Totals			$34,250		$31,600	$(2,650)

The adjusting entry follows.

Dec. 31	Unrealized Loss—Equity	2,650	
	Market Adjustment—Available-for-Sale (ST)		2,650
	To reflect an unrealized loss in market values of available-for-sale securities.		

Demonstration Problem—2

The following transactions relate to Brown Company's long-term investments during 2008 and 2009. Brown did not own any long-term investments prior to 2008. Show (1) the appropriate journal entries and (2) the relevant portions of each year's balance sheet and income statement that reflect these transactions for both 2008 and 2009.

2008

Sept. 9 Purchased 1,000 shares of Packard, Inc., common stock for $80,000 cash. These shares represent 30% of Packard's outstanding shares.

Oct. 2 Purchased 2,000 shares of AT&T common stock for $60,000 cash. These shares represent less than a 1% ownership in AT&T.

17 Purchased as a long-term investment 1,000 shares of Apple Computer common stock for $40,000 cash. These shares are less than 1% of Apple's outstanding shares.

Nov. 1 Received $5,000 cash dividend from Packard.

30 Received $3,000 cash dividend from AT&T.

Dec. 15 Received $1,400 cash dividend from Apple.

31 Packard's net income for this year is $70,000.

31 Market values for the investments in equity securities are Packard, $84,000; AT&T, $48,000; and Apple Computer, $45,000.

31 For preparing financial statements, note the following post-closing account balances: Common Stock, $500,000, and Retained Earnings, $350,000.

2009

Jan. 1 Sold Packard, Inc., shares for $108,000 cash.

May 30 Received $3,100 cash dividend from AT&T.

June 15 Received $1,600 cash dividend from Apple.

Aug. 17 Sold the AT&T stock for $52,000 cash.
19 Purchased 2,000 shares of Coca-Cola common stock for $50,000 cash as a long-term investment. The stock represents less than a 5% ownership in Coca-Cola.
Dec. 15 Received $1,800 cash dividend from Apple.
31 Market values of the investments in equity securities are Apple, $39,000, and Coca-Cola, $48,000.
31 For preparing financial statements, note the following post-closing account balances: Common Stock, $500,000, and Retained Earnings, $410,000.

Planning the Solution

- Account for the investment in Packard under the equity method.
- Account for the investments in AT&T, Apple, and Coca-Cola as long-term investments in available-for-sale securities.
- Prepare the information for the two years' balance sheets by including the appropriate asset and equity accounts.

Solution to Demonstration Problem—2

1. Journal entries for 2008.

Date	Account and Explanation	Debit	Credit
Sept. 9	Long-Term Investments—Packard	80,000	
	Cash		80,000
	Acquired 1,000 shares, representing a 30% equity in Packard.		
Oct. 2	Long-Term Investments—AFS (AT&T)	60,000	
	Cash		60,000
	Acquired 2,000 shares as a long-term investment in available-for-sale securities.		
Oct. 17	Long-Term Investments—AFS (Apple)	40,000	
	Cash		40,000
	Acquired 1,000 shares as a long-term investment in available-for-sale securities.		
Nov. 1	Cash	5,000	
	Long-Term Investments—Packard		5,000
	Received dividend from Packard.		
Nov. 30	Cash	3,000	
	Dividend Revenue		3,000
	Received dividend from AT&T.		
Dec. 15	Cash	1,400	
	Dividend Revenue		1,400
	Received dividend from Apple.		
Dec. 31	Long-Term Investments—Packard	21,000	
	Earnings from Investment (Packard)		21,000
	To record 30% share of Packard's annual earnings of $70,000.		
Dec. 31	Unrealized Loss—Equity	7,000	
	Market Adjustment—Available-for-Sale (LT)*		7,000
	To record change in market value of long-term available-for-sale securities.		

* Market adjustment computations:

	Cost	Market Value	Unrealized Gain (Loss)
AT&T	$ 60,000	$48,000	$(12,000)
Apple	40,000	45,000	5,000
Total	$100,000	$93,000	$ (7,000)

Required balance of the Market Adjustment—Available-for-Sale (LT) account (credit)	$(7,000)
Existing balance	0
Necessary adjustment (credit)	$(7,000)

2. The December 31, 2008, selected balance sheet items appear as follows.

Assets	
Long-term investments	
Available-for-sale securities (at market; cost is $100,000)	$ 93,000
Investment in equity securities	96,000
Total long-term investments	189,000
Stockholders' Equity	
Common stock	500,000
Retained earnings	350,000
Unrealized loss—Equity	(7,000)

The relevant income statement items for the year ended December 31, 2008, follow.

Dividend revenue	$ 4,400
Earnings from investment	21,000

1. Journal entries for 2009.

Date	Account	Debit	Credit
Jan. 1	Cash	108,000	
	Long-Term Investments—Packard.		96,000
	Gain on Sale of Long-Term Investments		12,000
	Sold 1,000 shares for cash.		
May 30	Cash	3,100	
	Dividend Revenue		3,100
	Received dividend from AT&T.		
June 15	Cash	1,600	
	Dividend Revenue		1,600
	Received dividend from Apple.		
Aug. 17	Cash	52,000	
	Loss on Sale of Long-Term Investments	8,000	
	Long-Term Investments—AFS (AT&T).		60,000
	Sold 2,000 shares for cash.		
Aug. 19	Long-Term Investments—AFS (Coca-Cola)	50,000	
	Cash		50,000
	Acquired 2,000 shares as a long-term investment in available-for-sale securities.		
Dec. 15	Cash	1,800	
	Dividend Revenue		1,800
	Received dividend from Apple.		
Dec. 31	Market Adjustment—Available-for-Sale (LT)*	4,000	
	Unrealized Loss—Equity		4,000
	To record change in market value of long-term available-for-sale securities.		

* Market adjustment computations:

	Cost	Market Value	Unrealized Gain (Loss)
Apple	$40,000	$39,000	$(1,000)
Coca-Cola	50,000	48,000	(2,000)
Total	$90,000	$87,000	$(3,000)

Required balance of the Market Adjustment—Available-for-Sale (LT) account (credit)	$(3,000)
Existing balance (credit)	(7,000)
Necessary adjustment (debit)	$ 4,000

2. The December 31, 2009, balance sheet items appear as follows.

Assets	
Long-term investments	
Available-for-sale securities (at market; cost is $90,000)	$ 87,000
Stockholders' Equity	
Common stock	500,000
Retained earnings	410,000
Unrealized loss—Equity	(3,000)

The relevant income statement items for the year ended December 31, 2009, follow.

Dividend revenue	$ 6,500
Gain on sale of long-term investments	12,000
Loss on sale of long-term investments	(8,000)

APPENDIX

Investments in International Operations 15A

Many entities from small entrepreneurs to large corporations conduct business internationally. Some entities' operations occur in so many different countries that the companies are called **multinationals.** Many of us think of Coca-Cola and McDonald's, for example, as primarily U.S. companies, but most of their sales occur outside the United States. Exhibit 15A.1 shows the percent of international sales and income for selected U.S. companies. Managing and accounting for multinationals present challenges. This section describes some of these challenges and how to account for and report these activities.

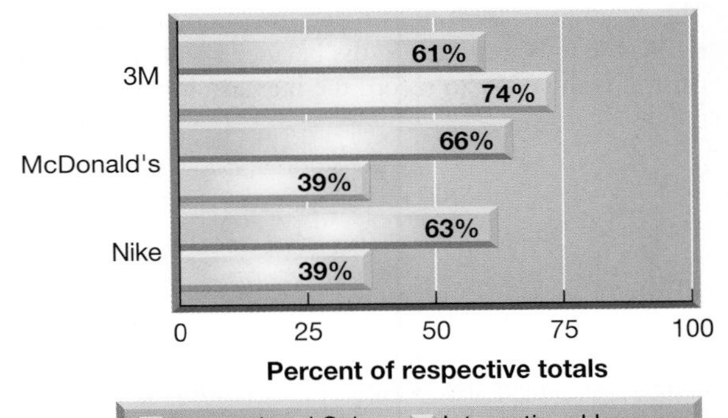

EXHIBIT 15A.1

International Sales and Income as a Percent of Their Totals

Two major accounting challenges that arise when companies have international operations relate to transactions that involve more than one currency. The first is to account for sales and purchases listed in a foreign currency. The second is to prepare consolidated financial statements with international subsidiaries. For ease in this discussion, we use companies with a U.S. base of operations and assume the need to prepare financial statements in U.S. dollars. This means the *reporting currency* of these companies is the U.S. dollar.

Point: Transactions *listed* or *stated* in a foreign currency are said to be *denominated* in that currency.

Exchange Rates between Currencies

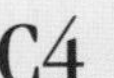

C4 Explain foreign exchange rates between currencies.

Markets for the purchase and sale of foreign currencies exist all over the world. In these markets, U.S. dollars can be exchanged for Canadian dollars, British pounds, Japanese yen, Euros, or any other legal currencies. The price of one currency stated in terms of another currency is called a **foreign exchange rate.** Exhibit 15A.2 lists recent exchange rates for selected currencies. The exchange rate for British pounds and U.S. dollars is $1.8054, meaning 1 British pound could be purchased for $1.8054. On that same day, the exchange rate between Mexican pesos and U.S. dollars is $0.0925, or 1 Mexican peso can be purchased for $0.0925. Exchange rates fluctuate due to changing economic and political conditions, including the supply and demand for currencies and expectations about future events.

Point: To convert currency, see XE.com

Decision Insight

Rush to Russia Investors are still eager to buy Russian equities even in the face of rampant crime, corruption, and slow economic growth. Why? Many argue Russia remains a bargain-priced, if risky, bet on future growth. Some analysts argue that natural-resource-rich Russia is one of the least expensive emerging markets.

EXHIBIT 15A.2

Foreign Exchange Rates for Selected Currencies*

Source (unit)	Price in $U.S.	Source (unit)	Price in $U.S.
Britain (pound)	$1.8980	Canada (dollar)	$0.9793
Mexico (peso)	0.0925	Japan (yen)	0.0090
Taiwan (dollar)	0.0305	Europe (Euro)	1.2920

* Rates will vary over time based on economic, political, and other changes.

Sales and Purchases Listed in a Foreign Currency

P5 Record transactions listed in a foreign currency.

When a U.S. company makes a credit sale to an international customer, accounting for the sale and the account receivable is straightforward if sales terms require the international customer's payment in U.S. dollars. If sale terms require (or allow) payment in a foreign currency, however, the U.S. company must account for the sale and the account receivable in a different manner.

Sales in a Foreign Currency To illustrate, consider the case of the U.S.-based manufacturer Boston Company, which makes credit sales to London Outfitters, a British retail company. A sale occurs on December 12, 2008, for a price of £10,000 with payment due on February 10, 2009. Boston Company keeps its accounting records in U.S. dollars. To record the sale, Boston Company must translate the sales price from pounds to dollars. This is done using the exchange rate on the date of the sale. Assuming the exchange rate on December 12, 2008, is $1.80, Boston records this sale as follows.

Assets	= Liabilities +	Equity
+18,000		+18,000

Dec. 12	Accounts Receivable—London Outfitters	18,000	
	Sales*		18,000
	To record a sale at £10,000, when the exchange rate equals $1.80. * (£10,000 × $1.80)		

When Boston Company prepares its annual financial statements on December 31, 2008, the current exchange rate is $1.84. Thus, the current dollar value of Boston Company's receivable is $18,400 (10,000 × $1.84). This amount is $400 higher than the amount recorded on December 12. Accounting principles require a receivable to be reported in the balance sheet at its current dollar value. Thus, Boston Company must make the following entry to record the increase in the dollar value of this receivable at year-end.

Assets	= Liabilities +	Equity
+400		+400

Dec. 31	Accounts Receivable—London Outfitters	400	
	Foreign Exchange Gain		400
	To record the increased value of the British pound for the receivable.		

Point: Foreign exchange gains are credits, and foreign exchange losses are debits.

On February 10, 2009, Boston Company receives London Outfitters' payment of £10,000. It immediately exchanges the pounds for U.S. dollars. On this date, the exchange rate for pounds is $1.78. Thus, Boston Company receives only $17,800 (£10,000 × $1.78). It records the cash receipt and the loss associated with the decline in the exchange rate as follows.

Assets	= Liabilities +	Equity
+17,800		−600
−18,400		

Feb. 10	Cash	17,800	
	Foreign Exchange Loss	600	
	Accounts Receivable—London Outfitters......		18,400
	Received foreign currency payment of an account and converted it into dollars.		

Gains and losses from foreign exchange transactions are accumulated in the Foreign Exchange Gain (or Loss) account. After year-end adjustments, the balance in the Foreign Exchange Gain (or Loss) account is reported on the income statement and closed to the Income Summary account.

Example: Assume that a U.S. company makes a credit purchase from a British company for £10,000 when the exchange rate is $1.62. At the balance sheet date, this rate is $1.72. Does this imply a gain or loss for the U.S. company? *Answer:* A loss.

Purchases in a Foreign Currency Accounting for credit purchases from an international seller is similar to the case of a credit sale to an international customer. In particular, if the U.S. company is required to make payment in a foreign currency, the account payable must be translated into dollars before the U.S. company can record it. If the exchange rate is different when preparing financial statements and when paying for the purchase, the U.S. company must recognize a foreign exchange gain or loss at those dates. To illustrate, assume NC Imports, a U.S. company, purchases products costing

€20,000 (euros) from Hamburg Brewing on January 15, when the exchange rate is $1.20 per euro. NC records this transaction as follows.

Jan. 15	Inventory	24,000	
	Accounts Payable—Hamburg Brewing.		24,000
	To record a €20,000 purchase when exchange rate is $1.20 (€20,000 × $1.20)		

Assets = Liabilities + Equity
+24,000 +24,000

NC Imports makes payment in full on February 14 when the exchange rate is $1.25 per euro, which is recorded as follows.

Feb. 14	Accounts Payable—Hamburg Brewing	24,000	
	Foreign Exchange Loss	1,000	
	Cash		25,000
	To record cash payment towards €20,000 account when exchange rate is $1.25 (€20.000 × $1.25).		

Assets = Liabilities + Equity
−25,000 −24,000 −1,000

Decision Insight

Global Greenback What do changes in foreign exchange rates mean? A decline in the price of the U.S. dollar against other currencies usually yields increased international sales for U.S. companies, without hiking prices or cutting costs, and puts them on a stronger competitive footing abroad. At home, they can raise prices without fear that foreign rivals will undercut them.

Consolidated Statements with International Subsidiaries

A second challenge in accounting for international operations involves preparing consolidated financial statements when the parent company has one or more international subsidiaries. Consider a U.S.-based company that owns a controlling interest in a French subsidiary. The reporting currency of the U.S. parent is the dollar. The French subsidiary maintains its financial records in euros. Before preparing consolidated statements, the parent must translate financial statements of the French company into U.S. dollars. After this translation is complete (including that for accounting differences), it prepares consolidated statements the same as for domestic subsidiaries. Procedures for translating an international subsidiary's account balances depend on the nature of the subsidiary's operations. The process requires the parent company to select appropriate foreign exchange rates and to apply those rates to the foreign subsidiary's account balances. This is described in advanced courses.

Global: A weaker U.S. dollar often increases global sales for U.S. companies.

Decision Maker

Entrepreneur You are a U.S. home builder that purchases lumber from mills in both the U.S. and Canada. The price of the Canadian dollar in terms of the U.S. dollar jumps from US$0.70 to US$0.80. Are you now more or less likely to buy lumber from Canadian or U.S. mills? [Answer—p. 608]

Summary

C1 Distinguish between debt and equity securities and between short-term and long-term investments. *Debt securities* reflect a creditor relationship and include investments in notes, bonds, and certificates of deposit. *Equity securities* reflect an owner relationship and include shares of stock issued by other companies. Short-term investments in securities are current assets that meet two criteria: (1) They are expected to be converted into cash within one year or the current operating cycle of the business, whichever is longer and (2) they are readily convertible to cash, or *marketable*. All other investments in securities are long-term. Long-term investments also include assets not used in operations and those held for special purposes, such as land for expansion.

C2 Identify and describe the different classes of investments in securities. Investments in securities are classified into one of five groups: (1) trading securities, which are always short-term, (2) debt securities held-to-maturity, (3) debt and equity securities available-for-sale, (4) equity securities in which an investor has a significant influence over the investee, and (5) equity securities in which an investor has a controlling influence over the investee.

C3 Describe how to report equity securities with controlling influence. If an investor owns more than 50% of another company's voting stock and controls the investee, the investor's financial reports are prepared on a consolidated basis. These reports are prepared as if the company were organized as one entity.

C4[A] Explain foreign exchange rates between currencies. A foreign exchange rate is the price of one currency stated in terms of another. An entity with transactions in a foreign currency when the exchange rate changes between the transaction dates and their settlement will experience exchange gains or losses.

A1 Compute and analyze the components of return on total assets. Return on total assets has two components: profit margin and total asset turnover. A decline in one component must be met with an increase in another if return on assets is to be maintained. Component analysis is helpful in assessing company performance compared to that of competitors and its own past.

P1 Account for trading securities. Investments are initially recorded at cost, and any dividend or interest from these investments is recorded in the income statement. Investments classified as trading securities are reported at market value. Unrealized gains and losses on trading securities are reported in income. When investments are sold, the difference between the net proceeds from the sale and the cost of the securities is recognized as a gain or loss.

P2 Account for held-to-maturity securities. Debt securities held-to-maturity are reported at cost when purchased. Interest revenue is recorded as it accrues. The cost of long-term held-to-maturity securities is adjusted for the amortization of any difference between cost and maturity value.

P3 Account for available-for-sale securities. Debt and equity securities available-for-sale are recorded at cost when purchased. Available-for-sale securities are reported at their market values on the balance sheet with unrealized gains or losses shown in the equity section. Gains and losses realized on the sale of these investments are reported in the income statement.

P4 Account for equity securities with significant influence. The equity method is used when an investor has a significant influence over an investee. This usually exists when an investor owns 20% or more of the investee's voting stock but not more than 50%. The equity method means an investor records its share of investee earnings with a debit to the investment account and a credit to a revenue account. Dividends received reduce the investment account balance.

P5[A] Record transactions listed in a foreign currency. When a company makes a credit sale to a foreign customer and sales terms call for payment in a foreign currency, the company must translate the foreign currency into dollars to record the receivable. If the exchange rate changes before payment is received, exchange gains or losses are recognized in the year they occur. The same treatment is used when a company makes a credit purchase from a foreign supplier and is required to make payment in a foreign currency.

Guidance Answers to **Decision Maker**

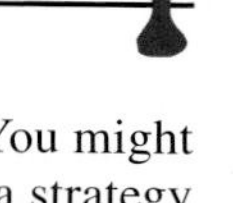

Money Manager If you have investments in fixed-rate bonds and notes when interest rates fall, the value of your investments increases. This is so because the bonds and notes you hold continue to pay the same (high) rate while the market is demanding a new lower interest rate. Your strategy is to continue holding your investments in bonds and notes, and, potentially, to increase these holdings through additional purchases.

Retailer Your store's return on assets is 11%, which is similar to the industry norm of 11.2%. However, disaggregation of return on assets reveals that your store's profit margin of 4.4% is much higher than the norm of 3.5%, but your total asset turnover of 2.5 is much lower than the norm of 3.2. These results suggest that, as compared with competitors, you are less efficient in using assets. You need to focus on increasing sales or reducing assets. You might consider reducing prices to increase sales, provided such a strategy does not reduce your return on assets. For instance, you could reduce your profit margin to 4% to increase sales. If total asset turnover increases to more than 2.75 when profit margin is lowered to 4%, your overall return on assets is improved.

Entrepreneur You are now less likely to buy Canadian lumber because it takes more U.S. money to buy a Canadian dollar (and lumber). For instance, the purchase of lumber from a Canadian mill with a $1,000 (Canadian dollars) price would have cost the U.S. builder $700 (U.S. dollars, computed as C$1,000 × US$0.70) before the rate change, and $800 (US dollars, computed as C$1,000 × US$0.80) after the rate change.

Guidance Answers to **Quick Checks**

1. Short-term held-to-maturity securities are reported at cost.
2. Trading securities are reported at market value.
3. The equity section of the balance sheet (and in comprehensive income).
4. The income statement.
5. Long-term investments include (1) long-term funds earmarked for a special purpose, (2) debt and equity securities that do not meet current asset requirements, and (3) long-term assets not used in the regular operations of the business.
6. An equity investment is classified as long term if it is not marketable or, if marketable, it is not held as an available source of cash to meet the needs of current operations.
7. Debt securities held-to-maturity and debt securities available-for-sale are both recorded at cost. Also, interest on both is accrued as earned. However, only long-term securities held-to-maturity require amortization of the difference between cost and maturity value. In addition, only securities available-for-sale require a period-end adjustment to market value.
8. Long-term equity investments are placed in one of three categories and accounted for as follows: (a) **available-for-sale** (noninfluential, less than 20% of outstanding stock)—market value; (b) **significant influence** (20% to 50% of outstanding stock)—equity method; and (c) **controlling influence** (holding more than 50% of outstanding stock)—equity method with consolidation.

Key Terms

mhhe.com/wildFAP19e

Key Terms are available at the book's Website for learning and testing in an online Flashcard Format.

Available-for-sale (AFS) securities (p. 596)
Comprehensive income (p. 600)
Consolidated financial statements (p. 599)
Equity method (p. 598)
Equity securities with controlling influence (p. 599)
Equity securities with significant influence (p. 598)
Foreign exchange rate (p. 605)
Held-to-maturity (HTM) securities (p. 596)
Long-term investments (p. 592)
Multinational (p. 605)
Parent (p. 599)
Return on total assets (p. 600)
Short-term investments (p. 592)
Subsidiary (p. 599)
Trading securities (p. 595)
Unrealized gain (loss) (p. 595)

Multiple Choice Quiz

Answers on p. 625 mhhe.com/wildFAP19e

Additional Quiz Questions are available at the book's Website.

1. A company purchased $30,000 of 5% bonds for investment purposes on May 1. The bonds pay interest on February 1 and August 1. The amount of interest revenue accrued at December 31 (the company's year-end) is:
a. $1,500
b. $1,375
c. $1,000
d. $625
e. $300

2. Earlier this period, Amadeus Co. purchased its only available-for-sale investment in the stock of Bach Co. for $83,000. The period-end market value of this stock is $84,500. Amadeus records a:
a. Credit to Unrealized Gain—Equity for $1,500
b. Debit to Unrealized Loss—Equity for $1,500
c. Debit to Investment Revenue for $1,500
d. Credit to Market Adjustment—Available-for-Sale for $3,500
e. Credit to Cash for $1,500

3. Mozart Co. owns 35% of Melody Inc. Melody pays $50,000 in cash dividends to its shareholders for the period. Mozart's entry to record the Melody dividend includes a:
a. Credit to Investment Revenue for $50,000.
b. Credit to Long-Term Investments for $17,500.
c. Credit to Cash for $17,500.
d. Debit to Long-Term Investments for $17,500.
e. Debit to Cash for $50,000.

4. A company has net income of $300,000, net sales of $2,500,000, and total assets of $2,000,000. Its return on total assets equals:
a. 6.7%
b. 12.0%
c. 8.3%
d. 80.0%
e. 15.0%

5. A company had net income of $80,000, net sales of $600,000, and total assets of $400,000. Its profit margin and total asset turnover are:

	Profit Margin	Total Asset Turnover
a.	1.5%	13.3
b.	13.3%	1.5
c.	13.3%	0.7
d.	7.0%	13.3
e.	10.0%	26.7

Superscript A denotes assignments based on Appendix 15A.

Discussion Questions

1. Under what two conditions should investments be classified as current assets?

2. On a balance sheet, what valuation must be reported for short-term investments in trading securities?

3. If a short-term investment in available-for-sale securities costs $6,780 and is sold for $7,500, how should the difference between these two amounts be recorded?

4. Identify the three classes of noninfluential and two classes of influential investments in securities.

5. Under what conditions should investments be classified as current assets? As long-term assets?

6. If a company purchases its only long-term investments in available-for-sale debt securities this period and their market value is below cost at the balance sheet date, what entry is required to recognize this unrealized loss?

7. On a balance sheet, what valuation must be reported for debt securities classified as available-for-sale?

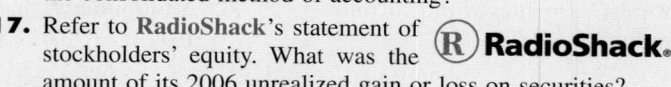

8. Under what circumstances are long-term investments in debt securities reported at cost and adjusted for amortization of any difference between cost and maturity value?
9. For investments in available-for-sale securities, how are unrealized (holding) gains and losses reported?
10. In accounting for investments in equity securities, when should the equity method be used?
11. Under what circumstances does a company prepare consolidated financial statements?
12.[A] What are two major challenges in accounting for international operations?
13.[A] Assume a U.S. company makes a credit sale to a foreign customer that is required to make payment in its foreign currency. In the current period, the exchange rate is $1.40 on the date of the sale and is $1.30 on the date the customer pays the receivable. Will the U.S. company record an exchange gain or loss?
14.[A] If a U.S. company makes a credit sale to a foreign customer required to make payment in U.S. dollars, can the U.S. company have an exchange gain or loss on this sale?
15. Refer to **Best Buy**'s statement of changes in shareholders' equity in Appendix A. What is the amount of foreign currency translation adjustment for the year ended March 3, 2007? Is this adjustment an unrealized gain or an unrealized loss?
16. Refer to the balance sheet of **Circuit City** in Appendix A. How can you tell that Circuit City uses the consolidated method of accounting?
17. Refer to **RadioShack**'s statement of stockholders' equity. What was the amount of its 2006 unrealized gain or loss on securities?
18. Refer to the financial statements of **Apple** in Appendix A. Compute its return on total assets for the year ended September 30, 2006.

Denotes Discussion Questions that involve decision making.

Available with McGraw-Hill's Homework Manager — McGraw-Hill's HOMEWORK MANAGER®

QUICK STUDY

QS 15-1
Short-term equity investments C2 P1

On April 18, Rollo Co. made a short-term investment in 600 common shares of TXT Co. The purchase price is $84 per share and the broker's fee is $500. The intent is to actively manage these shares for profit. On May 30, Rollo Co. receives $0.75 per share from TXT in dividends. Prepare the April 18 and May 30 journal entries to record these transactions.

QS 15-2
Available-for-sale securities
C2 P3

Malox Co. purchased short-term investments in available-for-sale securities at a cost of $100,000 on November 25, 2009. At December 31, 2009, these securities had a market value of $94,000. This is the first and only time the company has purchased such securities.

1. Prepare the December 31, 2009, year-end adjusting entry for the securities' portfolio.
2. For each account in the entry for part 1, explain how it is reported in financial statements.
3. Prepare the April 6, 2010, entry when Malox sells one-fourth of these securities for $27,000.

QS 15-3
Available-for-sale securities
C2 P3

Prepare Vikon Company's journal entries to reflect the following transactions for the current year.

May 7 Purchases 200 shares of Felton stock as a short-term investment in available-for-sale securities at a cost of $100 per share plus $400 in broker fees.
June 6 Sells 200 shares of its investment in Felton stock at $112 per share. The broker's commission on this sale is $250.

QS 15-4
Available-for-sale securities
C2 P3

Texar Company completes the following transactions during the current year.

May 9 Purchases 400 shares of Crayton stock as a short-term investment in available-for-sale securities at a cost of $30 per share plus $200 in broker fees.
June 2 Sells 200 shares of its investment in Crayton stock at $32 per share. The broker's commission on this sale is $120.
Dec. 31 The closing market price of the Crayton stock is $28 per share.

Prepare the May 9 and June 2 journal entries and the December 31 adjusting entry. This is the first and only time the company purchased such securities.

QS 15-5
Identifying long-term investments
C1

Which of the following statements are true of long-term investments?

a. They are held as an investment of cash available for current operations.
b. They can include funds earmarked for a special purpose, such as bond sinking funds.

c. They can include investments in trading securities.
d. They can include debt securities held-to-maturity.
e. They are always easily sold and therefore qualify as being marketable.
f. They can include debt and equity securities available-for-sale.
g. They can include bonds and stocks not intended to serve as a ready source of cash.

QS 15-6
Describing investments in securities
C1 C2 C3

Complete the following descriptions by filling in the blanks.

1. Equity securities giving an investor significant influence are accounted for using the ______ ______.
2. Trading securities are classified as ______ assets.
3. Accrual of interest on bonds held as long-term investments requires a credit to ______ ______.
4. The controlling investor (more than 50% ownership) is called the ______, and the investee company is called the ______.
5. Available-for-sale debt securities are reported on the balance sheet at ______ ______.

QS 15-7
Debt securities transactions
C2 P2

On February 1, 2009, Garzon purchased 6% bonds issued by Integal Utilities at a cost of $80,000, which is their par value. The bonds pay interest semiannually on July 31 and January 31. For 2009, prepare entries to record Garzon's July 31 receipt of interest and its December 31 year-end interest accrual.

QS 15-8
Recording equity securities
C2 P3

On May 20, 2009, Chiu Co. paid $1,500,000 to acquire 25,000 common shares (10%) of BBE Corp. as a long-term investment. On August 5, 2010, Chiu sold one-half of these shares for $937,500. What valuation method should be used to account for this stock investment? Prepare entries to record both the acquisition and the sale of these shares.

QS 15-9
Equity method transactions
C2 P4

Assume the same facts as in QS 15-8 except that the stock acquired represents 40% of BBE Corp.'s outstanding stock. Also assume that BBE Corp. paid a $150,000 dividend on November 1, 2009, and reported a net income of $1,050,000 for 2009. Prepare the entries to record (*a*) the receipt of the dividend and (*b*) the December 31, 2009, year-end adjustment required for the investment account.

QS 15-10
Recording market adjustment for securities
P3

During the current year, Marketplace Consulting Group acquired long-term available-for-sale securities at an $85,000 cost. At its December 31 year-end, these securities had a market value of $62,000. This is the first and only time the company purchased such securities.

1. Prepare the necessary year-end adjusting entry related to these securities.
2. Explain how each account used in part 1 is reported in the financial statements.

QS 15-11
Return on total assets A1

The return on total assets is the focus of analysts, creditors, and other users of financial statements.

1. How is the return on total assets computed?
2. What does this important ratio reflect?

QS 15-12
Component return on total assets A1

Return on total assets can be separated into two important components.

1. Write the formula to separate the return on total assets into its two basic components.
2. Explain how these components of the return on total assets are helpful to financial statement users for business decisions.

QS 15-13[A]
Foreign currency transactions
P5

A U.S. company sells a product to a British company with the transaction listed in British pounds. On the date of the sale, the transaction total of $40,600 is billed as £20,000, reflecting an exchange rate of 2.03 (that is, $2.03 per pound). Prepare the entry to record (1) the sale and (2) the receipt of payment in pounds when the exchange rate is 1.95.

QS 15-14[A]
Foreign currency transactions
P5

On March 1, 2009, a U.S. company made a credit sale requiring payment in 30 days from a Malaysian company, Hamac Sdn. Bhd., in 20,000 Malaysian ringgits. Assuming the exchange rate between Malaysian ringgits and U.S. dollars is $0.2963 on March 1 and $0.3005 on March 31, prepare the entries to record the sale on March 1 and the cash receipt on March 31.

EXERCISES

Exercise 15-1
Accounting for transactions in short-term securities
C2 P1 P2 P3

Check (c) Dr. Cash $173,400

(f) Dr. Cash $13,025

Prepare journal entries to record the following transactions involving the short-term securities investments of Bolton Co., all of which occurred during year 2009.

a. On February 15, paid $170,000 cash to purchase ACC's 90-day short-term debt securities ($170,000 principal), dated February 15, that pay 8% interest (categorized as held-to-maturity securities).

b. On March 22, purchased 850 shares of Ross Company stock at $21 per share plus a $100 brokerage fee. These shares are categorized as trading securities.

c. On May 16, received a check from ACC in payment of the principal and 90 days' interest on the debt securities purchased in transaction *a*.

d. On August 1, paid $70,000 cash to purchase Nita Co.'s 11% debt securities ($70,000 principal), dated July 30, 2009, and maturing January 30, 2010 (categorized as available-for-sale securities).

e. On September 1, received a $1.10 per share cash dividend on the Ross Company stock purchased in transaction *b*.

f. On October 8, sold 425 shares of Ross Co. stock for $31 per share, less a $150 brokerage fee.

g. On October 30, received a check from Nita Co. for 90 days' interest on the debt securities purchased in transaction *d*.

Exercise 15-2
Accounting for trading securities
C1 P1

Check (3) Gain, $2,250

Borchert Co. purchases various investments in trading securities at a cost of $76,000 on December 27, 2009. (This is its first and only purchase of such securities.) At December 31, 2009, these securities had a market value of $85,000.

1. Prepare the December 31, 2009, year-end adjusting entry for the trading securities' portfolio.

2. Explain how each account in the entry of part 1 is reported in financial statements.

3. Prepare the January 3, 2010, entry when Borchert sells a portion of its trading securities (that had originally cost $38,000) for $40,250.

Exercise 15-3
Adjusting available-for-sale securities to market
C2 P3

Check Unrealized loss, $392

On December 31, 2009, Tagert Company held the following short-term investments in its portfolio of available-for-sale securities. Tagert had no short-term investments in its prior accounting periods. Prepare the December 31, 2009, adjusting entry to report these investments at market value.

	Cost	Market Value
Verrizano Corporation bonds payable	$ 81,400	$92,000
Porter Corporation notes payable	54,900	47,928
Laverne Company common stock	100,500	96,480

Exercise 15-4
Transactions in short-term and long-term investments
C1 C2

LT 25,940
Cash 25,940

Prepare journal entries to record the following transactions involving both the short-term and long-term investments of Corveau Corp., all of which occurred during calendar year 2009. Use the account Short-Term Investments for any transactions that you determine are short term.

a. On February 15, paid $100,000 cash to purchase Anthem's 90-day short-term notes at par, which are dated February 15 and pay 6% interest (classified as held-to-maturity).

b. On March 22, bought 600 shares of Frain Industries common stock at $43 cash per share plus a $140 brokerage fee (classified as long-term available-for-sale securities).

c. On May 15, received a check from Anthem in payment of the principal and 90 days' interest on the notes purchased in transaction *a*.

d. On July 30, paid $30,000 cash to purchase Moto Electronics' 5% notes at par, dated July 30, 2009, and maturing on January 30, 2010 (classified as trading securities).

e. On September 1, received a $0.40 per share cash dividend on the Frain Industries common stock purchased in transaction *b*.

f. On October 8, sold 300 shares of Frain Industries common stock for $49 cash per share, less a $120 brokerage fee.

g. On October 30, received a check from Moto Electronics for three months' interest on the notes purchased in transaction *d*.

Exercise 15-5
Market adjustment to available-for-sale securities
P3

On December 31, 2009, Loren Co. held the following short-term available-for-sale securities.

	Cost	Market Value
Nintendo Co. common stock	$64,500	$70,305
Unilever bonds payable	25,800	23,994
Kellogg Co. notes payable	46,440	43,654
McDonald's Corp. common stock	87,075	82,721

Loren had no short-term investments prior to the current period. Prepare the December 31, 2009, year-end adjusting entry to record the market adjustment for these securities.

Exercise 15-6
Market adjustment to available-for-sale securities
P3

Patica Co. began operations in 2008. The cost and market values for its long-term investments portfolio in available-for-sale securities are shown below. Prepare Patica's December 31, 2009, adjusting entry to reflect any necessary market adjustment for these investments.

	Cost	Market Value
December 31, 2008	$67,842	$61,736
December 31, 2009	73,479	77,888

Exercise 15-7
Multi-year market adjustments to available-for-sale securities
P3

Basil Services began operations in 2007 and maintains long-term investments in available-for-sale securities. The year-end cost and market values for its portfolio of these investments follow. Prepare journal entries to record each year-end market adjustment for these securities.

	Cost	Market Value
December 31, 2007	$392,900	$381,113
December 31, 2008	447,906	474,780
December 31, 2009	609,152	720,627
December 31, 2010	919,820	818,640

Exercise 15-8
Classifying investments in securities; recording market values
C1 C2 P2 P3 P4 

Information regarding Seaton Company's individual investments in securities during its calendar-year 2009, along with the December 31, 2009, market values, follows.

a. Investment in Beeman Company bonds: $443,150 cost, $481,704 market value. Seaton intends to hold these bonds until they mature in 2014.

b. Investment in Baybridge common stock: 29,500 shares; $352,304 cost; $382,954 market value. Seaton owns 32% of Baybridge's voting stock and has a significant influence over Baybridge.

c. Investment in Carroll common stock: 12,000 shares; $181,692 cost; $195,864 market value. This investment amounts to 3% of Carroll's outstanding shares, and Seaton's goal with this investment is to earn dividends over the next few years.

d. Investment in Newtech common stock: 3,500 shares; $101,038 cost; $99,320 market value. Seaton's goal with this investment is to reap an increase in market value of the stock over the next three to five years. Newtech has 30,000 common shares outstanding.

e. Investment in Flock common stock: 16,300 shares; $110,788 cost; $117,657 market value. This stock is marketable and is held as an investment of cash available for operations.

Required

1. Identify whether each investment should be classified as a short-term or long-term investment. For each long-term investment, indicate in which of the long-term investment classifications it should be placed.

2. Prepare a journal entry dated December 31, 2009, to record the market value adjustment of the long-term investments in available-for-sale securities. Seaton had no long-term investments prior to year 2009.

Check (2) Unrealized gain, $12,454

Exercise 15-9
Securities transactions; equity method
P4 C2

Prepare journal entries to record the following transactions and events of Kareen Company.

2009

Jan. 2	Purchased 55,000 shares of Altus Co. common stock for $374,000 cash plus a broker's fee of $2,650 cash. Altus has 137,500 shares of common stock outstanding and its policies will be significantly influenced by Kareen.
Sept. 1	Altus declared and paid a cash dividend of $3.05 per share.
Dec. 31	Altus announced that net income for the year is $1,106,900.

2010

June 1	Altus declared and paid a cash dividend of $3.30 per share.
Dec. 31	Altus announced that net income for the year is $1,240,900.
Dec. 31	Kareen sold 11,000 shares of Altus for $294,250 cash.

Exercise 15-10
Return on total assets
A1

The following information is available from the financial statements of Interstate Industries. Compute Interstate's return on total assets for 2009 and 2010. (Round returns to one-tenth of a percent.) Comment on the company's efficiency in using its assets in 2009 and 2010.

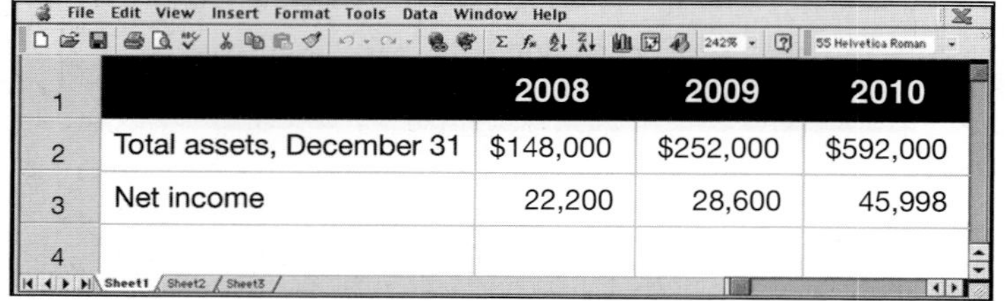

	2008	2009	2010
Total assets, December 31	$148,000	$252,000	$592,000
Net income	22,200	28,600	45,998

Exercise 15-11[A]
Foreign currency transactions
P5

Leigh of New York sells its products to customers in the United States and the United Kingdom. On December 16, 2009, Leigh sold merchandise on credit to Berton Ltd. of London at a price of 20,000 pounds. The exchange rate on that day for £1 was $2.0325. On December 31, 2009, when Leigh prepared its financial statements, the rate was £1 for $2.0292. Berton paid its bill in full on January 15, 2010, at which time the exchange rate was £1 for $2.0340. Leigh immediately exchanged the 20,000 pounds for U.S. dollars. Prepare Leigh's journal entries on December 16, December 31, and January 15 (round to the nearest dollar).

Exercise 15-12[A]
Computing foreign exchange gains and losses on receivables
C4 P5

On May 8, 2009, Jett Company (a U.S. company) made a credit sale to Munoz (a Mexican company). The terms of the sale required Munoz to pay 850,000 pesos on February 10, 2010. Jett prepares quarterly financial statements on March 31, June 30, September 30, and December 31. The exchange rates for pesos during the time the receivable is outstanding follow.

May 8, 2009	$0.0932
June 30, 2009	0.0941
September 30, 2009	0.0952
December 31, 2009	0.0935
February 10, 2010	0.0974

Compute the foreign exchange gain or loss that Jett should report on each of its quarterly income statements for the last three quarters of 2009 and the first quarter of 2010. Also compute the amount reported on Jett's balance sheets at the end of each of its last three quarters of 2009.

PROBLEM SET A

Problem 15-1A
Recording transactions and market adjustments for trading securities
C2 P1

Protom Company, which began operations in 2009, invests its idle cash in trading securities. The following transactions are from its short-term investments in its trading securities.

2009

Jan. 20	Purchased 800 shares of Ford Motor Co. at $26 per share plus a $120 commission.
Feb. 9	Purchased 2,600 shares of Lucent at $39 per share plus a $578 commission.
Oct. 12	Purchased 800 shares of Z-Seven at $7.50 per share plus a $200 commission.

2010

Apr. 15 Sold 800 shares of Ford Motor Co. at $30 per share less a $300 commission.
July 5 Sold 800 shares of Z-Seven at $11 per share less a $103 commission.
July 22 Purchased 2,000 shares of Hunt Corp. at $39 per share plus a $444 commission.
Aug. 19 Purchased 1,600 shares of Donna Karan at $19.50 per share plus a $290 commission.

2011

Feb. 27 Purchased 3,500 shares of HCA at $31 per share plus a $420 commission.
Mar. 3 Sold 2,000 shares of Hunt at $35 per share less a $250 commission.
June 21 Sold 2,600 shares of Lucent at $36.75 per share less a $420 commission.
June 30 Purchased 1,300 shares of Black & Decker at $47.50 per share plus a $595 commission.
Nov. 1 Sold 1,600 shares of Donna Karan at $19.50 per share less a $309 commission.

Required

1. Prepare journal entries to record these short-term investment activities for the years shown. (Ignore any year-end adjusting entries.)

2. On December 31, 2011, prepare the adjusting entry to record any necessary market adjustment for the portfolio of trading securities when HCA's share price is $33 and Black & Decker's share price is $43.50. (Assume the Market Adjustment—Trading account had an unadjusted balance of zero.)

Check (2) Dr. Market Adjustment—Trading $785

Problem 15-2A
Recording, adjusting, and reporting short-term available-for-sale securities
C2 P3

Freema Company had no short-term investments prior to year 2009. It had the following transactions involving short-term investments in available-for-sale securities during 2009.

Apr. 16 Purchased 8,000 shares of Gem Co. stock at $29.75 per share plus a $440 brokerage fee.
May 1 Paid $125,000 to buy 90-day U.S. Treasury bills (debt securities): $125,000 principal amount, 4% interest, securities dated May 1.
July 7 Purchased 4,000 shares of PepsiCo stock at $47.75 per share plus a $410 brokerage fee.
20 Purchased 2,000 shares of Xerox stock at $19.75 per share plus a $490 brokerage fee.
Aug. 3 Received a check for principal and accrued interest on the U.S. Treasury bills that matured on July 29.
15 Received a $0.90 per share cash dividend on the Gem Co. stock.
28 Sold 4,000 shares of Gem Co. stock at $36.50 per share less a $250 brokerage fee.
Oct. 1 Received a $1.75 per share cash dividend on the PepsiCo shares.
Dec. 15 Received a $1.05 per share cash dividend on the remaining Gem Co. shares.
31 Received a $1.30 per share cash dividend on the PepsiCo shares.

Required

1. Prepare journal entries to record the preceding transactions and events.

2. Prepare a table to compare the year-end cost and market values of Freema's short-term investments in available-for-sale securities. The year-end market values per share are: Gem Co., $32.00; PepsiCo, $45.00; and Xerox, $16.75.

Check (2) Cost = $350,620

3. Prepare an adjusting entry, if necessary, to record the year-end market adjustment for the portfolio of short-term investments in available-for-sale securities.

(3) Dr. Unrealized Loss—Equity $9,120

Analysis Component

4. Explain the balance sheet presentation of the market adjustment for Freema's short-term investments.

5. How do these short-term investments affect Freema's (*a*) income statement for year 2009 and (*b*) the equity section of its balance sheet at year-end 2009?

Problem 15-3A
Recording, adjusting, and reporting long-term available-for-sale securities
C2 P3

Tennant Security, which began operations in 2009, invests in long-term available-for-sale securities. Following is a series of transactions and events determining its long-term investment activity.

2009

Jan. 20 Purchased 1,200 shares of Johnson & Johnson at $20.50 per share plus a $240 commission.
Feb. 9 Purchased 1,000 shares of Sony at $46.20 per share plus a $220 commission.
June 12 Purchased 1,700 shares of Mattel at $28 per share plus a $195 commission.
Dec. 31 Per share market values for stocks in the portfolio are Johnson & Johnson, $21.50; Mattel, $26.50; Sony, $38.

2010

Apr. 15 Sold 1,200 shares of Johnson & Johnson at $23.50 per share less a $525 commission.
July 5 Sold 1,700 shares of Mattel at $26.50 per share less a $235 commission.
July 22 Purchased 500 shares of Sara Lee at $22.50 per share plus a $420 commission.
Aug. 19 Purchased 900 shares of Eastman Kodak at $14 per share plus a $198 commission.
Dec. 31 Per share market values for stocks in the portfolio are: Kodak, $16.25; Sara Lee, $20.00; Sony, $35.00.

2011

Feb. 27 Purchased 3,000 shares of Microsoft at $65 per share plus a $520 commission.
June 21 Sold 1,000 shares of Sony at $48.00 per share less an $880 commission.
June 30 Purchased 1,500 shares of Black & Decker at $38 per share plus a $435 commission.
Aug. 3 Sold 500 shares of Sara Lee at $16.25 per share less a $435 commission.
Nov. 1 Sold 900 shares of Eastman Kodak at $19.75 per share less a $625 commission.
Dec. 31 Per share market values for stocks in the portfolio are: Black & Decker, $41; Microsoft, $67.

Required

1. Prepare journal entries to record these transactions and events and any year-end market adjustments to the portfolio of long-term available-for-sale securities.

Check (2b) Market adjustment bal.: 12/31/09, $(10,205); 12/31/10; $(11,263)

2. Prepare a table that summarizes the (*a*) total cost, (*b*) total market adjustment, and (*c*) total market value of the portfolio of long-term available-for-sale securities at each year-end.

(3b) Unrealized Gain at 12/31/2011, $9,545

3. Prepare a table that summarizes (*a*) the realized gains and losses and (*b*) the unrealized gains or losses for the portfolio of long-term available-for-sale securities at each year-end.

Problem 15-4A
Long-term investment transactions; unrealized and realized gains and losses
C2 C3 P3 P4

Elevant Co.'s long-term available-for-sale portfolio at December 31, 2008, consists of the following.

Available-for-Sale Securities	Cost	Market Value
40,000 shares of Company A common stock	$535,300	$500,000
7,000 shares of Company B common stock	159,380	151,000
17,500 shares of Company C common stock	662,600	640,938

Elevant enters into the following long-term investment transactions during year 2009.

Jan. 29 Sold 3,500 shares of Company B common stock for $79,100 less a brokerage fee of $1,400.
Apr. 17 Purchased 9,900 shares of Company W common stock for $197,500 plus a brokerage fee of $2,300. The shares represent a 30% ownership in Company W.
July 6 Purchased 4,200 shares of Company X common stock for $118,125 plus a brokerage fee of $1,650. The shares represent a 10% ownership in Company X.
Aug. 22 Purchased 50,000 shares of Company Y common stock for $375,000 plus a brokerage fee of $1,100. The shares represent a 51% ownership in Company Y.
Nov. 13 Purchased 8,300 shares of Company Z common stock for $261,596 plus a brokerage fee of $2,350. The shares represent a 5% ownership in Company Z.
Dec. 9 Sold 40,000 shares of Company A common stock for $515,000 less a brokerage fee of $4,000.

The market values of its investments at December 31, 2009, are: B, $81,375; C, $610,312; W, $191,250; X, $110,250; Y, $531,250; and Z, $272,240.

Required

1. Determine the amount Elevant should report on its December 31, 2009, balance sheet for its long-term investments in available-for-sale securities.

Check (2) Cr. Unrealized Loss—Equity, $13,508

2. Prepare any necessary December 31, 2009, adjusting entry to record the market value adjustment for the long-term investments in available-for-sale securities.

3. What amount of gains or losses on transactions relating to long-term investments in available-for-sale securities should Elevant report on its December 31, 2009, income statement?

Problem 15-5A
Accounting for long-term investments in securities; with and without significant influence
C2 P3 P4

Selk Steel Co., which began operations on January 4, 2009, had the following subsequent transactions and events in its long-term investments.

2009

Jan. 5 Selk purchased 50,000 shares (20% of total) of Wulf's common stock for $1,567,000.
Oct. 23 Wulf declared and paid a cash dividend of $3.20 per share.
Dec. 31 Wulf's net income for 2009 is $1,164,000, and the market value of its stock at December 31 is $34 per share.

2010

Oct. 15 Wulf declared and paid a cash dividend of $2.50 per share.
Dec. 31 Wulf's net income for 2010 is $1,476,000, and the market value of its stock at December 31 is $36.00 per share.

2011

Jan. 2 Selk sold all of its investment in Wulf for $1,895,500 cash.

Part 1

Assume that Selk has a significant influence over Wulf with its 20% share of stock.

Required

1. Prepare journal entries to record these transactions and events for Selk.
2. Compute the carrying (book) value per share of Selk's investment in Wulf common stock as reflected in the investment account on January 1, 2011.
3. Compute the net increase or decrease in Selk's equity from January 5, 2009, through January 2, 2011, resulting from its investment in Wulf.

Check (2) Carrying value per share, $36.20

Part 2

Assume that although Selk owns 20% of Wulf's outstanding stock, circumstances indicate that it does not have a significant influence over the investee and that it is classified as an available-for-sale security investment.

Required

1. Prepare journal entries to record the preceding transactions and events for Selk. Also prepare an entry dated January 2, 2011, to remove any balance related to the market adjustment.
2. Compute the cost per share of Selk's investment in Wulf common stock as reflected in the investment account on January 1, 2011.
3. Compute the net increase or decrease in Selk's equity from January 5, 2009, through January 2, 2011, resulting from its investment in Wulf.

(1) 1/2/2011 Dr. Unrealized Gain—Equity $233,000

(3) Net increase, $613,500

Problem 15-6A[A]
Foreign currency transactions
C4 P5

Patriot Company, a U.S. corporation with customers in several foreign countries, had the following selected transactions for 2009 and 2010.

2009

Apr. 8 Sold merchandise to Salinas & Sons of Mexico for $27,456 cash. The exchange rate for pesos is $0.0932 on this day.
July 21 Sold merchandise on credit to Sumito Corp. in Japan. The price of 2.7 million yen is to be paid 120 days from the date of sale. The exchange rate for yen is $0.0082 on this day.
Oct. 14 Sold merchandise for 18,000 pounds to Smithers Ltd. of Great Britain, payment in full to be received in 90 days. The exchange rate for pounds is $2.0330 on this day.
Nov. 18 Received Sumito's payment in yen for its July 21 purchase and immediately exchanged the yen for dollars. The exchange rate for yen is $0.0079 on this day.
Dec. 20 Sold merchandise for 20,000 ringgits to Hamid Albar of Malaysia, payment in full to be received in 30 days. On this day, the exchange rate for ringgits is $0.2963.

Dec. 31 Recorded adjusting entries to recognize exchange gains or losses on Patriot's annual financial statements. Rates for exchanging foreign currencies on this day follow.

Pesos (Mexico)	$0.0937
Yen (Japan)	0.0075
Pounds (Britain)	2.0345
Ringgits (Malaysia)	0.2949

2010

Jan. 12 Received full payment in pounds from Smithers for the October 14 sale and immediately exchanged the pounds for dollars. The exchange rate for pounds is $2.0355 on this day.

Jan. 19 Received Hamid Albar's full payment in ringgits for the December 20 sale and immediately exchanged the ringgits for dollars. The exchange rate for ringgits is $0.2936 on this day.

Required

1. Prepare journal entries for the Patriot transactions and adjusting entries (round amounts to the nearest dollar).

Check (2) 2009 total foreign exchange loss, $811

2. Compute the foreign exchange gain or loss to be reported on Patriot's 2009 income statement.

Analysis Component

3. What actions might Patriot consider to reduce its risk of foreign exchange gains or losses?

PROBLEM SET B

Problem 15-1B
Recording transactions and market adjustments for trading securities

C2 P1

Harter Company, which began operations in 2009, invests its idle cash in trading securities. The following transactions relate to its short-term investments in its trading securities.

2009

Mar. 10 Purchased 900 shares of Timex at $28.00 per share plus a $125 commission.
May 7 Purchased 2,500 shares of MTV at $37.00 per share plus a $578 commission.
Sept. 1 Purchased 780 shares of UPS at $7.00 per share plus a $200 commission.

2010

Apr. 26 Sold 2,500 shares of MTV at $35.50 per share less a $295 commission.
Apr. 27 Sold 780 shares of UPS at $10.50 per share less a $103 commission.
June 2 Purchased 1,600 shares of SPW at $34.00 per share plus a $444 commission.
June 14 Purchased 1,600 shares of Wal-Mart at $20.00 per share plus a $290 commission.

2011

Jan. 28 Purchased 3,400 shares of PepsiCo at $38.00 per share plus a $400 commission.
Jan. 31 Sold 1,600 shares of SPW at $29.00 per share less a $250 commission.
Aug. 22 Sold 900 shares of Timex at $26.25 per share less a $420 commission.
Sept. 3 Purchased 1,500 shares of Vodaphone at $47.50 per share plus a $600 commission.
Oct. 9 Sold 1,600 shares of Wal-Mart at $22.50 per share less a $309 commission.

Required

1. Prepare journal entries to record these short-term investment activities for the years shown. (Ignore any year-end adjusting entries.)

Check (2) Cr. Market Adjustment—Trading $13,050

2. On December 31, 2011, prepare the adjusting entry to record any necessary market adjustment for the portfolio of trading securities when PepsiCo's share price is $36.00 and Vodaphone's share price is $44.00. (Assume the Market Adjustment—Trading account had an unadjusted balance of zero.)

Problem 15-2B
Recording, adjusting, and reporting short-term available-for-sale securities
C2 P3

SP Systems had no short-term investments prior to 2009. It had the following transactions involving short-term investments in available-for-sale securities during 2009.

Feb.	6	Purchased 6,000 shares of Nokia stock at $24.75 per share plus a $400 brokerage fee.
	15	Paid $250,000 to buy six-month U.S. Treasury bills (debt securities): $250,000 principal amount, 4% interest, securities dated February 15.
Apr.	7	Purchased 3,000 shares of Dell Co. stock at $49.25 per share plus a $370 brokerage fee.
June	2	Purchased 1,500 shares of Merck stock at $18.25 per share plus a $450 brokerage fee.
	30	Received a $0.19 per share cash dividend on the Nokia shares.
Aug.	11	Sold 1,500 shares of Nokia stock at $29.50 per share less a $490 brokerage fee.
	16	Received a check for principal and accrued interest on the U.S. Treasury bills purchased February 15.
	24	Received a $0.16 per share cash dividend on the Dell shares.
Nov.	9	Received a $0.20 per share cash dividend on the remaining Nokia shares.
Dec.	18	Received a $0.21 per share cash dividend on the Dell shares.

Required

1. Prepare journal entries to record the preceding transactions and events.

2. Prepare a table to compare the year-end cost and market values of the short-term investments in available-for-sale securities. The year-end market values per share are: Nokia, $23.50; Dell, $55.50; and Merck, $15.25.

Check (2) Cost = $287,620

3. Prepare an adjusting entry, if necessary, to record the year-end market adjustment for the portfolio of short-term investments in available-for-sale securities.

(3) Cr. Unrealized Gain—Equity, $7,505

Analysis Component

4. Explain the balance sheet presentation of the market adjustment to SP's short-term investments.

5. How do these short-term investments affect (*a*) its income statement for year 2009 and (*b*) the equity section of its balance sheet at the 2009 year-end?

Problem 15-3B
Recording, adjusting, and reporting long-term available-for-sale securities
C2 P3

Bleeker Enterprises, which began operations in 2009, invests in long-term available-for-sale securities. Following is a series of transactions and events involving its long-term investment activity.

2009

Mar. 10	Purchased 1,000 shares of Apple at $23.00 per share plus $240 commission.
Apr. 7	Purchased 1,300 shares of Ford at $46.20 per share plus $225 commission.
Sept. 1	Purchased 1,500 shares of Polaroid at $25.00 per share plus $195 commission.
Dec. 31	Per share market values for stocks in the portfolio are: Apple, $24.00; Ford, $42.00; Polaroid, $27.00.

2010

Apr. 26	Sold 1,300 shares of Ford at $44.00 per share less a $250 commission.
June 2	Purchased 1,800 shares of Duracell at $19.25 per share plus a $235 commission.
June 14	Purchased 600 shares of Sears at $22.50 per share plus a $470 commission.
Nov. 27	Sold 1,500 shares of Polaroid at $29 per share less a $198 commission.
Dec. 31	Per share market values for stocks in the portfolio are: Apple, $26.00; Duracell, $18.00; Sears, $25.00.

2011

Jan. 28	Purchased 2,000 shares of Coca-Cola Co. at $69 per share plus a $530 commission.
Aug. 22	Sold 1,000 shares of Apple at $20.00 per share less a $280 commission.
Sept. 3	Purchased 2,000 shares of Motorola at $36 per share plus a $435 commission.
Oct. 9	Sold 600 shares of Sears at $26.25 per share less a $435 commission.
Oct. 31	Sold 1,800 shares of Duracell at $15.00 per share less a $425 commission.
Dec. 31	Per share market values for stocks in the portfolio are: Coca-Cola, $75.00; Motorola, $32.00.

Required

1. Prepare journal entries to record these transactions and events and any year-end market adjustments to the portfolio of long-term available-for-sale securities.

2. Prepare a table that summarizes the (*a*) total cost, (*b*) total market adjustment, and (*c*) total market value for the portfolio of long-term available-for-sale securities at each year-end.

3. Prepare a table that summarizes (*a*) the realized gains and losses and (*b*) the unrealized gains or losses for the portfolio of long-term available-for-sale securities at each year-end.

Check (2b) Market adjustment bal.: 12/31/09, $(2,120); 12/31/10, $1,305

(3b) Unrealized Gain at 12/31/2011, $3,035

Problem 15-4B
Long-term investment transactions; unrealized and realized gains and losses
C2 C3 P3 P4

Chavez's long-term available-for-sale portfolio at December 31, 2008, consists of the following.

Available-for-Sale Securities	Cost	Market Value
55,000 shares of Company R common stock	$1,118,250	$1,198,000
17,000 shares of Company S common stock	600,600	586,500
22,000 shares of Company T common stock	294,590	303,600

Chavez enters into the following long-term investment transactions during year 2009.

Jan. 13 Sold 4,250 shares of Company S stock for $145,500 less a brokerage fee of $650.
Mar. 24 Purchased 31,000 shares of Company U common stock for $565,750 plus a brokerage fee of $1,980. The shares represent a 62% ownership interest in Company U.
Apr. 5 Purchased 85,000 shares of Company V common stock for $267,750 plus a brokerage fee of $1,125. The shares represent a 10% ownership in Company V.
Sept. 2 Sold 22,000 shares of Company T common stock for $313,500 less a brokerage fee of $2,700.
Sept. 27 Purchased 5,000 shares of Company W common stock for $101,000 plus a brokerage fee of $350. The shares represent a 25% ownership interest in Company W.
Oct. 30 Purchased 10,000 shares of Company X common stock for $97,500 plus a brokerage fee of $300. The shares represent a 13% ownership interest in Company X.

The market values of its investments at December 31, 2009, are: R, $1,136,250; S, $420,750; U, $545,600; V, $269,876; W, $109,378; and X, $91,250.

Required

1. Determine the amount Chavez should report on its December 31, 2009, balance sheet for its long-term investments in available-for-sale securities.

2. Prepare any necessary December 31, 2009, adjusting entry to record the market value adjustment of the long-term investments in available-for-sale securities.

3. What amount of gains or losses on transactions relating to long-term investments in available-for-sale securities should Chavez report on its December 31, 2009, income statement?

Check (2) Dr. Unrealized Gain—Equity, $74,660; Cr. Market Adjustment, $91,909

Problem 15-5B
Accounting for long-term investments in securities; with and without significant influence
C2 P3 P4

Devin Company, which began operations on January 3, 2009, had the following subsequent transactions and events in its long-term investments.

2009

Jan. 5 Devin purchased 25,000 shares (25% of total) of Bloch's common stock for $401,000.
Aug. 1 Bloch declared and paid a cash dividend of $1.10 per share.
Dec. 31 Bloch's net income for 2009 is $164,000, and the market value of its stock is $17.50 per share.

2010

Aug. 1 Bloch declared and paid a cash dividend of $1.30 per share.
Dec. 31 Bloch's net income for 2010 is $156,000, and the market value of its stock is $19.25 per share.

2011

Jan. 8 Devin sold all of its investment in Bloch for $550,000 cash.

Part 1

Assume that Devin has a significant influence over Bloch with its 25% share.

Required

1. Prepare journal entries to record these transactions and events for Devin.
2. Compute the carrying (book) value per share of Devin's investment in Bloch common stock as reflected in the investment account on January 7, 2011.
3. Compute the net increase or decrease in Devin's equity from January 5, 2009, through January 8, 2011, resulting from its investment in Bloch.

Check (2) Carrying value per share, $16.84

Part 2

Assume that although Devin owns 25% of Bloch's outstanding stock, circumstances indicate that it does not have a significant influence over the investee and that it is classified as an available-for-sale security investment.

Required

1. Prepare journal entries to record these transactions and events for Devin. Also prepare an entry dated January 8, 2011, to remove any balance related to the market adjustment.
2. Compute the cost per share of Devin's investment in Bloch common stock as reflected in the investment account on January 7, 2011.
3. Compute the net increase or decrease in Devin's equity from January 5, 2009, through January 8, 2011, resulting from its investment in Bloch.

(1) 1/8/2011 Dr. Unrealized Gain—Equity $80,250

(3) Net increase, $209,000

Problem 15-6B^A
Foreign currency transactions
C4 P5

Kitna, a U.S. corporation with customers in several foreign countries, had the following selected transactions for 2009 and 2010.

2009

May 26 Sold merchandise for 5.5 million yen to Fuji Company of Japan, payment in full to be received in 60 days. On this day, the exchange rate for yen is $0.0088.

June 1 Sold merchandise to Fordham Ltd. of Great Britain for $73,500 cash. The exchange rate for pounds is $2.0331 on this day.

July 25 Received Fuji's payment in yen for its May 26 purchase and immediately exchanged the yen for dollars. The exchange rate for yen is $0.0087 on this day.

Oct. 15 Sold merchandise on credit to Martinez Brothers of Mexico. The price of 425,000 pesos is to be paid 90 days from the date of sale. On this day, the exchange rate for pesos is $0.0932.

Dec. 6 Sold merchandise for 300,000 yuans to Chi-Ying Company of China, payment in full to be received in 30 days. The exchange rate for yuans is $0.1335 on this day.

Dec. 31 Recorded adjusting entries to recognize exchange gains or losses on Kitna's annual financial statements. Rates of exchanging foreign currencies on this day follow.

Yen (Japan)	$0.0089
Pounds (Britain)	2.0402
Pesos (Mexico)	0.0994
Yuans (China)	0.1351

2010

Jan. 5 Received Chi-Ying's full payment in yuans for the December 6 sale and immediately exchanged the yuans for dollars. The exchange rate for yuans is $0.1372 on this day.

Jan. 13 Received full payment in pesos from Martinez for the October 15 sale and immediately exchanged the pesos for dollars. The exchange rate for pesos is $0.0960 on this day.

Required

1. Prepare journal entries for the Kitna transactions and adjusting entries.
2. Compute the foreign exchange gain or loss to be reported on Kitna's 2009 income statement.

Check 2009 total foreign exchange gain, $2,565

Analysis Component

3. What actions might Kitna consider to reduce its risk of foreign exchange gains or losses?

SERIAL PROBLEM

Success Systems

(This serial problem began in Chapter 1 and continues through most of the book. If previous chapter segments were not completed, the serial problem can begin at this point. It is helpful, but not necessary, to use the Working Papers that accompany the book.)

SP 15 While reviewing the March 31, 2010, balance sheet of Success Systems, Adriana Lopez notes that the business has built a large cash balance of $77,845. Its most recent bank money market statement shows that the funds are earning an annualized return of 0.75%. Lopez decides to make several investments with the desire to earn a higher return on the idle cash balance. Accordingly, in April 2010, Success Systems makes the following investments in trading securities:

April 16 Purchases 400 shares of Johnson & Johnson stock at $50 per share plus $300 commission.
April 30 Purchases 200 shares of Starbucks Corporation at $22 per share plus $250 commission.

On June 30, 2010, the per share market price of the Johnson & Johnson shares is $55 and the Starbucks shares is $19.

Required

1. Prepare journal entries to record the April purchases of trading securities by Success Systems.
2. On June 30, 2010, prepare the adjusting entry to record any necessary market adjustment to its portfolio of trading securities.

BEYOND THE NUMBERS

REPORTING IN ACTION

C3 C4 A1 

BTN 15-1 Refer to **Best Buy**'s financial statements in Appendix A to answer the following.

1. Are Best Buy's financial statements consolidated? How can you tell?
2. What is Best Buy's *comprehensive income* for the year ended March 3, 2007?
3. Does Best Buy have any foreign operations? How can you tell?
4. Compute Best Buy's return on total assets for the year ended March 3, 2007.

Fast Forward

5. Access Best Buy's annual report for a fiscal year ending after March 3, 2007, from either its Website (BestBuy.com) or the SEC's EDGAR database (www.SEC.gov). Recompute Best Buy's return on total assets for the years subsequent to March 3, 2007.

COMPARATIVE ANALYSIS

A1 

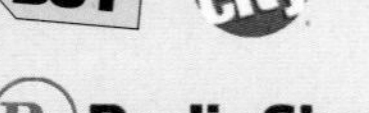

BTN 15-2 Key figures for **Best Buy**, **Circuit City**, and **RadioShack** follow.

	Best Buy			Circuit City			RadioShack		
($ millions)	Current Year	1 Year Prior	2 Years Prior	Current Year	1 Year Prior	2 Years Prior	Current Year	1 Year Prior	2 Years Prior
Net income	$ 1,377	$ 1,140	$ 984	$ (8)	$ 140	$ 62	$ 73	$ 267	$ 337
Net sales	35,934	30,848	27,433	12,430	11,514	10,414	4,778	5,082	4,841
Total assets	13,570	11,864	10,294	4,007	4,069	3,840	2,070	2,205	2,517

Required

1. Compute return on total assets for Best Buy, Circuit City, and RadioShack for the two most recent years.
2. Separate the return on total assets computed in part 1 into its components for all three companies and both years according to the formula in Exhibit 15.9.
3. Which company has the highest total return on assets? The highest profit margin? The highest total asset turnover? What does this comparative analysis reveal? (Assume an industry average of 8.2% for return on assets.)

BTN 15-3 Kaylee Wecker is the controller for Wildcat Company, which has numerous long-term investments in debt securities. Wildcat's investments are mainly in 10-year bonds. Wecker is preparing its year-end financial statements. In accounting for long-term debt securities, she knows that each long-term investment must be designated as a held-to-maturity or an available-for-sale security. Interest rates rose sharply this past year causing the portfolio's market value to substantially decline. The company does not intend to hold the bonds for the entire 10 years. Wecker also earns a bonus each year, which is computed as a percent of net income.

ETHICS CHALLENGE

C2 P2 P3

Required

1. Will Wecker's bonus depend in any way on the classification of the debt securities? Explain.
2. What criteria must Wecker use to classify the securities as held-to-maturity or available-for-sale?
3. Is there likely any company oversight of Wecker's classification of the securities? Explain.

BTN 15-4 Assume that you are Jackson Company's accountant. Company owner Abel Terrio has reviewed the 2009 financial statements you prepared and questions the $6,000 loss reported on the sale of its investment in Blackhawk Co. common stock. Jackson acquired 50,000 shares of Blackhawk's common stock on December 31, 2007, at a cost of $500,000. This stock purchase represented a 40% interest in Blackhawk. The 2008 income statement reported that earnings from all investments were $126,000. On January 3, 2009, Jackson Company sold the Blackhawk stock for $575,000. Blackhawk did not pay any dividends during 2008 but reported a net income of $202,500 for that year. Terrio believes that because the Blackhawk stock purchase price was $500,000 and was sold for $575,000, the 2009 income statement should report a $75,000 gain on the sale.

COMMUNICATING IN PRACTICE

C2 P4 

Required

Draft a one-half page memorandum to Terrio explaining why the $6,000 loss on sale of Blackhawk stock is correctly reported.

BTN 15-5 Access the August 3, 2007, 10-K filing (for year-end June 30, 2007) of **Microsoft** (MSFT) at www.SEC.gov. Review its note 3, "Investments."

TAKING IT TO THE NET

C1 C2 

Required

1. How does the cost-basis total for its investments as of June 30, 2007, compare to the prior year-end amount?
2. Identify at least eight types of short-term investments held by Microsoft as of June 30, 2007.
3. What were Microsoft's unrealized gains and its unrealized losses from its investments for 2007?
4. Was the cost or market ("recorded") value of the investments higher as of June 30, 2007?

BTN 15-6 Each team member is to become an expert on a specific classification of long-term investments. This expertise will be used to facilitate other teammates' understanding of the concepts and procedures relevent to the classification chosen.

TEAMWORK IN ACTION

C1 C2 C3 P1 P2 P3 P4

1. Each team member must select an area for expertise by choosing one of the following classifications of long-term investments.
   a. Held-to-maturity debt securities
   b. Available-for-sale debt and equity securities
   c. Equity securities with significant influence
   d. Equity securities with controlling influence
2. Learning teams are to disburse and expert teams are to be formed. Expert teams are made up of those who select the same area of expertise. The instructor will identify the location where each expert team will meet.
3. Expert teams will collaborate to develop a presentation based on the following requirements. Students must write the presentation in a format they can show to their learning teams in part (4).

Requirements for Expert Presentation

- **a.** Write a transaction for the acquisition of this type of investment security. The transaction description is to include all necessary data to reflect the chosen classification.
- **b.** Prepare the journal entry to record the acquisition.

  [*Note:* The expert team on equity securities with controlling influence will substitute requirements (*d*) and (*e*) with a discussion of the reporting of these investments.]
- **c.** Identify information necessary to complete the end-of-period adjustment for this investment.
- **d.** Assuming that this is the only investment owned, prepare any necessary year-end entries.
- **e.** Present the relevant balance sheet section(s).

4. Re-form learning teams. In rotation, experts are to present to their teams the presentations they developed in part 3. Experts are to encourage and respond to questions.

ENTREPRENEURIAL DECISION

C4 P5 

BTN 15-7[A] Refer to the opening feature in this chapter about Amy Smilovic and her company, **Tibi**. Assume that Amy must acquire the Japanese rights to certain clothing designs that will then be produced for sale to U.S. consumers. Amy acquires those rights on January 1, 2008, from a Japanese distributor and agrees to pay 12,000,000 yen per year for those rights. Quarterly payments are due March 31, June 30, September 30, and December 31 each year. On January 1, 2008, the yen is worth $0.00891.

Required

1. Prepare the journal entry to record the rights purchased on January 1, 2008.

2. Prepare the journal entries to record the payments on March 31, June 30, September 30, and December 31, 2008. The value of the yen on those dates follows.

March 31	$0.00893
June 30	0.00901
September 30	0.00902
December 31	0.00897

3. How can Amy protect herself from unanticipated gains and losses from currency translation if all of her payments are specified to be paid in yen?

HITTING THE ROAD C4

BTN 15-8[A] Assume that you are planning a spring break trip to Europe. Identify three locations where you can find exchange rates for the dollar relative to the Euro or other currencies.

GLOBAL DECISION

A1 

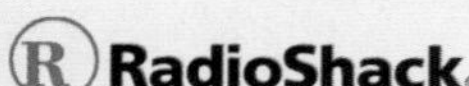

BTN 15-9 **DSG international**, **Best Buy**, **Circuit City**, and **RadioShack** are competitors in the global marketplace. Following are selected data from each company.

	DSG (£ millions)			Best Buy		Circuit City		RadioShack	
Key Figure	Current Year	One Year Prior	Two Years Prior	Current Year	Prior Year	Current Year	Prior Year	Current Year	Prior Year
Net income . . .	£ 18	£ 256	£ 212	—	—	—	—	—	—
Total assets . . .	3,977	4,120	4,104	—	—	—	—	—	—
Net sales	7,930	6,984	7,072	—	—	—	—	—	—
Profit margin . .	?	?	—	3.8%	3.7%	(0.1)%	1.2%	1.5%	5.3%
Total asset turnover	?	?	—	2.8	2.8	3.1	2.9	2.2	2.2

Required

1. Compute DSG's return on total assets for the most recent two years using the data provided.
2. Which of these four companies has the highest return on total assets? Highest profit margin? Highest total asset turnover?

ANSWERS TO MULTIPLE CHOICE QUIZ

1. d; $30,000 × 5% × 5/12 = $625
2. a; Unrealized gain = $84,500 − $83,000 = $1,500
3. b; $50,000 × 35% = $17,500
4. e; $300,000/$2,000,000 = 15%
5. b; Profit margin = $80,000/$600,000 = 13.3%
   Total asset turnover = $600,000/$400,000 = 1.5

A Look Back

Chapter 15 focused on how to identify, account for, and report investments in securities. We also accounted for transactions listed in a foreign currency.

A Look at This Chapter

This chapter focuses on reporting and analyzing cash inflows and cash outflows. We emphasize how to prepare and interpret the statement of cash flows.

A Look Ahead

Chapter 17 focuses on tools to help us analyze financial statements. We also describe comparative analysis and the application of ratios for financial analysis.

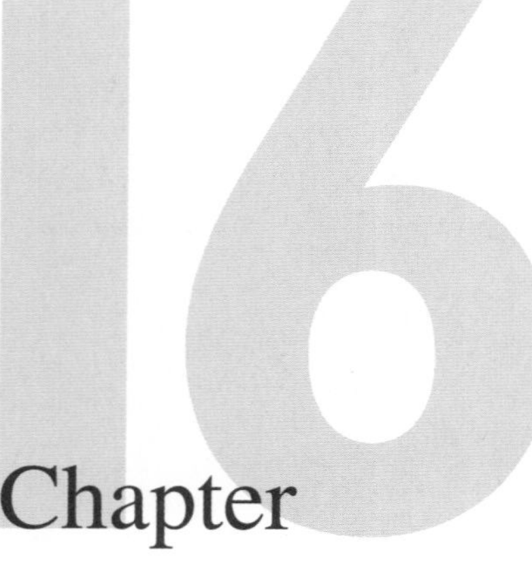

Chapter

Reporting the Statement of Cash Flows

Learning Objectives

CAP

Conceptual

C1 Explain the purpose and importance of cash flow information. *(p. 628)*

C2 Distinguish between operating, investing, and financing activities. *(p. 629)*

C3 Identify and disclose noncash investing and financing activities. *(p. 631)*

C4 Describe the format of the statement of cash flows. *(p. 631)*

Analytical

A1 Analyze the statement of cash flows. *(p. 645)*

A2 Compute and apply the cash flow on total assets ratio. *(p. 646)*

Procedural

P1 Prepare a statement of cash flows. *(p. 632)*

P2 Compute cash flows from operating activities using the indirect method. *(p. 635)*

P3 Determine cash flows from both investing and financing activities. *(p. 641)*

P4 *Appendix 16A*—Illustrate use of a spreadsheet to prepare a statement of cash flows. *(p. 649)*

P5 *Appendix 16B*—Compute cash flows from operating activities using the direct method. *(p. 652)*

LP16

Decision Feature

Wizard of Odd

"If you put enough energy into your dream, you can make anything happen" —Jim Bonaminio

FAIRFIELD, OH—Jim Bonaminio built his roadside produce stand while living in an abandoned gas station. "I would get up and leave at 4 in the morning to buy everything fresh [and] my wife opened the market at 8 a.m.," recalls Jim. "By 10 o'clock at night, we'd be sitting on the bed balancing the register receipts . . . we worked seven days a week." The fruit of those early efforts is **Jungle Jim's International Market (JungleJims.com).**

Jungle Jim's is no Wal-Mart wannabe, but it is arguably America's wackiest supermarket. Instead of trying to beat the big chains at the price-squeezing game, Jim's is a funhouse maze of a store. A seven-foot Elvis lion sings "Jailhouse Rock," an antique fire engine rests atop cases of hot sauce, port-a-potties lead to fancy restrooms, and Robin Hood greets customers with English food set within a 30-foot-tall Sherwood Forest. This is just a sampling.

"If you don't go out on a limb, then you're just like everybody else," insists Jim. "The stuff I've collected—all sorts of weird stuff—gets reused." Despite the wackiness, Jim is first and foremost a businessman. He learned firsthand about the importance of monitoring cash inflows and cash outflows. In the early days, recalls Jim, it was all about sales and profits. Then inventory and asset growth yielded negative cash flows, and Jim was in a pinch. That's when he realized that tracking cash flows was important, explains Jim.

Jim eventually learned how to monitor and control cash flows for each of his operating, investing, and financing activities. Today, says Jim, "I hire professional people to [help me monitor cash] . . . and to look for ways to make money." Yet Jim explains that he always reviews the statement of cash flows and the individual cash inflows and outflows.

Cash management has not curtailed Jim's fun-loving approach to business. "I'm trying to create something that has never been done," laughs Jim. "I just want to see if I can do it and have fun."

[Sources: *Jungle Jim's Website,* January 2009; *BusinessWeek,* April 2005; *Country Living,* November 2004; *Miamian,* Summer 2004; *Plain Dealer,* November 2004; *Supermarket News,* September 2006; *Cintas,* August 2007]

Chapter Preview

A company cannot achieve or maintain profits without carefully managing cash. Managers and other users of information pay attention to a company's cash position and the events and transactions affecting cash. This chapter explains how we prepare, analyze, and interpret a statement of cash flows. It also discusses the importance of cash flow information for predicting future performance and making managerial decisions. More generally, effectively using the statement of cash flows is crucial for managing and analyzing the operating, investing, and financing activities of businesses.

Reporting the Statement of Cash Flows

Basics of Cash Flow Reporting
- Purpose
- Importance
- Measurement
- Classification
- Noncash activities
- Format and preparation

Cash Flows from Operating
- Indirect and direct methods of reporting
- Application of indirect method of reporting
- Summary of indirect method adjustments

Cash Flows from Investing
- Three-stage process of analysis
- Analysis of noncurrent assets
- Analysis of other assets

Cash Flows from Financing
- Three-stage process of analysis
- Analysis of noncurrent liabilities
- Analysis of equity

Basics of Cash Flow Reporting

This section describes the basics of cash flow reporting, including its purpose, measurement, classification, format, and preparation.

Purpose of the Statement of Cash Flows

C1 Explain the purpose and importance of cash flow information.

The purpose of the **statement of cash flows** is to report cash receipts (inflows) and cash payments (outflows) during a period. This includes separately identifying the cash flows related to operating, investing, and financing activities. The statement of cash flows does more than simply report changes in cash. It is the detailed disclosure of individual cash flows that makes this statement useful to users. Information in this statement helps users answer questions such as these:

- How does a company obtain its cash?
- Where does a company spend its cash?
- What explains the change in the cash balance?

Point: Internal users rely on the statement of cash flows to make investing and financing decisions. External users rely on this statement to assess the amount and timing of a company's cash flows.

The statement of cash flows addresses important questions such as these by summarizing, classifying, and reporting a company's cash inflows and cash outflows for each period.

Importance of Cash Flows

Information about cash flows can influence decision makers in important ways. For instance, we look more favorably at a company that is financing its expenditures with cash from operations than one that does it by selling its assets. Information about cash flows helps users decide whether a company has enough cash to pay its existing debts as they mature. It is also relied upon to evaluate a company's ability to meet unexpected obligations and pursue unexpected opportunities. External information users especially want to assess a company's ability to take advantage of new business opportunities. Internal users such as managers use cash flow information to plan day-to-day operating activities and make long-term investment decisions.

Macy's striking turnaround is an example of how analysis and management of cash flows can lead to improved financial stability. Several years ago Macy's obtained temporary protection from bankruptcy, at which time it desperately needed to improve its cash flows. It did so by engaging in aggressive cost-cutting measures. As a result, Macy's annual cash flow rose to $210 million, up from a negative cash flow of $38.9 million in the prior year. Macy's eventually met its financial obligations and then successfully merged with **Federated Department Stores**.

The case of **W. T. Grant Co.** is a classic example of the importance of cash flow information in predicting a company's future performance and financial strength. Grant reported net income of more than $40 million per year for three consecutive years. At that same time, it was experiencing an alarming decrease in cash provided by operations. For instance, net cash outflow was more than $90 million by the end of that three-year period. Grant soon went bankrupt. Users who relied solely on Grant's income numbers were unpleasantly surprised. This reminds us that cash flows as well as income statement and balance sheet information are crucial in making business decisions.

Video16.1

Decision Insight

Cash Savvy "A lender must have a complete understanding of a borrower's cash flows to assess both the borrowing needs and repayment sources. This requires information about the major types of cash inflows and outflows. I have seen many companies, whose financial statements indicate good profitability, experience severe financial problems because the owners or managers lacked a good understanding of cash flows."—Mary E. Garza, **Bank of America**

Measurement of Cash Flows

Cash flows are defined to include both *cash* and *cash equivalents*. The statement of cash flows explains the difference between the beginning and ending balances of cash and cash equivalents. We continue to use the phrases *cash flows* and the *statement of cash flows,* but we must remember that both phrases refer to cash and cash equivalents. Recall that a cash equivalent must satisfy two criteria: (1) be readily convertible to a known amount of cash and (2) be sufficiently close to its maturity so its market value is unaffected by interest rate changes. In most cases, a debt security must be within three months of its maturity to satisfy these criteria. Companies must disclose and follow a clear policy for determining cash and cash equivalents and apply it consistently from period to period. **American Express**, for example, defines its cash equivalents as "time deposits and other highly liquid investments with original maturities of 90 days or less."

Classification of Cash Flows

C2 Distinguish between operating, investing, and financing activities.

Since cash and cash equivalents are combined, the statement of cash flows does not report transactions between cash and cash equivalents such as cash paid to purchase cash equivalents and cash received from selling cash equivalents. However, all other cash receipts and cash payments are classified and reported on the statement as operating, investing, or financing activities. Individual cash receipts and payments for each of these three categories are labeled to identify their originating transactions or events. A net cash inflow (source) occurs when the receipts in a category exceed the payments. A net cash outflow (use) occurs when the payments in a category exceed the receipts.

Operating Activities **Operating activities** include those transactions and events that determine net income. Examples are the production and purchase of merchandise, the sale of goods and services to customers, and the expenditures to administer the business. Not all items in income, such as unusual gains and losses, are operating activities (we discuss these exceptions later in the chapter). Exhibit 16.1 lists the more common cash inflows and outflows from operating activities. (Although cash receipts and cash payments from buying and selling trading

EXHIBIT 16.1

Cash Flows from Operating Activities

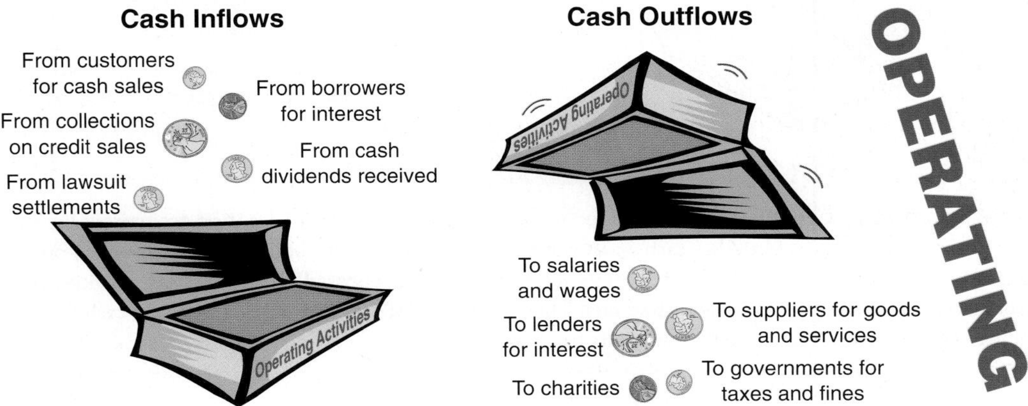

securities are often reported under operating activities, new standards require that these receipts and payments be classified based on the nature and purpose of those securities.)

Investing Activities **Investing activities** generally include those transactions and events that affect long-term assets—namely, the purchase and sale of long-term assets. They also include (1) the purchase and sale of short-term investments in the securities of other entities, other than cash equivalents and trading securities and (2) lending and collecting money for notes receivable. Exhibit 16.2 lists examples of cash flows from investing activities. Proceeds from collecting the principal amounts of notes deserve special mention. If the note results from sales to customers, its cash receipts are classed as operating activities whether short-term or long-term. If the note results from a loan to another party apart from sales, however, the cash receipts from collecting the note principal are classed as an investing activity. The FASB requires that the collection of interest on loans be reported as an operating activity.

Point: The FASB requires that *cash dividends received* and *cash interest received* be reported as operating activities.

EXHIBIT 16.2

Cash Flows from Investing Activities

Cash Inflows

From selling investments in securities

From collecting principal on loans

From selling long-term productive assets

From selling (discounting) of notes

Investing Activities

Cash Outflows

Investing Activities

To make loans to others

To purchase long-term productive assets

To purchase investments in securities

INVESTING

Financing Activities **Financing activities** include those transactions and events that affect long-term liabilities and equity. Examples are (1) obtaining cash from issuing debt and repaying the amounts borrowed and (2) receiving cash from or distributing cash to owners. These activities involve transactions with a company's owners and creditors. They also often involve borrowing and repaying principal amounts relating to both short- and long-term debt. GAAP requires that payments of interest expense be classified as operating activities. Also, cash payments to settle credit purchases of merchandise, whether on account or by note, are operating activities. Exhibit 16.3 lists examples of cash flows from financing activities.

EXHIBIT 16.3

Cash Flows from Financing Activities

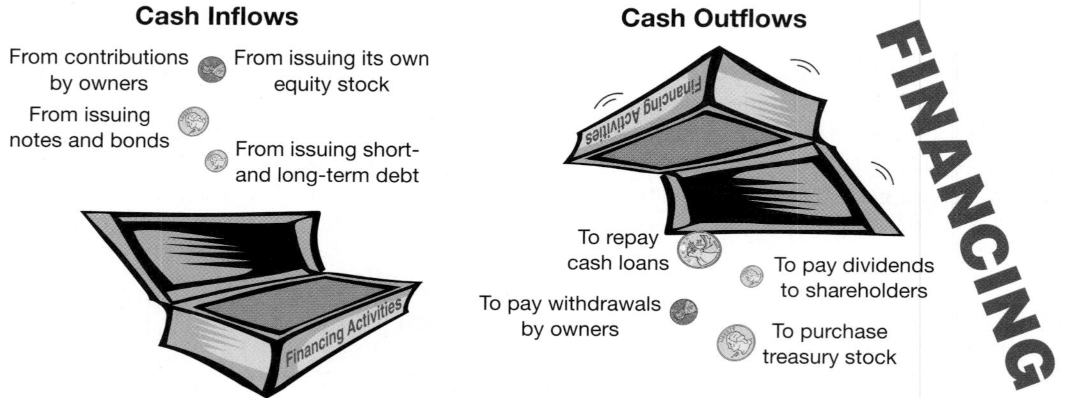

Point: Interest payments on a loan are classified as operating activities, but payments of loan principal are financing activities.

Decision Insight

Cash Reporting Cash flows can be delayed or accelerated at the end of a period to improve or reduce current period cash flows. Also, cash flows can be misclassified. Cash outflows reported under operations are interpreted as expense payments. However, cash outflows reported under investing activities are interpreted as a positive sign of growth potential. Thus, managers face incentives to misclassify cash flows. For these reasons, cash flow reporting warrants our scrutiny.

Noncash Investing and Financing

When important investing and financing activities do not affect cash receipts or payments, they are still disclosed at the bottom of the statement of cash flows or in a note to the statement because of their importance and the *full-disclosure principle*. One example of such a transaction is the purchase of long-term assets using a long-term note payable (loan). This transaction involves both investing and financing activities but does not affect any cash inflow or outflow and is not reported in any of the three sections of the statement of cash flows. This disclosure rule also extends to transactions with partial cash receipts or payments.

C3 Identify and disclose noncash investing and financing activities.

To illustrate, assume that Goorin purchases land for $12,000 by paying $5,000 cash and trading in used equipment worth $7,000. The investing section of the statement of cash flows reports only the $5,000 cash outflow for the land purchase. The $12,000 investing transaction is only partially described in the body of the statement of cash flows, yet this information is potentially important to users because it changes the makeup of assets. Goorin could either describe the transaction in a footnote or include information at the bottom of its statement that lists the $12,000 land purchase along with the cash financing of $5,000 and a $7,000 trade-in of equipment. As another example, Borg Co. acquired $900,000 of assets in exchange for $200,000 cash and a $700,000 long-term note, which should be reported as follows:

Point: A stock dividend transaction involving a transfer from retained earnings to common stock or a credit to contributed capital is *not* considered a noncash investing and financing activity because the company receives no consideration for shares issued.

Fair value of assets acquired	$900,000
Less cash paid	200,000
Liabilities incurred or assumed	$700,000

Exhibit 16.4 lists transactions commonly disclosed as noncash investing and financing activities.

EXHIBIT 16.4

Examples of Noncash Investing and Financing Activities

- Retirement of debt by issuing equity stock.
- Conversion of preferred stock to common stock.
- Lease of assets in a capital lease transaction.
- Purchase of long-term assets by issuing a note or bond.
- Exchange of noncash assets for other noncash assets.
- Purchase of noncash assets by issuing equity or debt.

Format of the Statement of Cash Flows

Accounting standards require companies to include a statement of cash flows in a complete set of financial statements. This statement must report information about a company's cash receipts and cash payments during the period. Exhibit 16.5 shows the usual format. A company must report cash flows from three activities: operating, investing, and financing. The statement explains how transactions and events impact the prior period-end cash (and cash equivalents) balance to produce its current period-end balance.

C4 Describe the format of the statement of cash flows.

EXHIBIT 16.5

Format of the Statement of Cash Flows

COMPANY NAME
Statement of Cash Flows
For *period* Ended *date*

Cash flows from operating activities	
[List of individual inflows and outflows]	
Net cash provided (used) by operating activities	$ #
Cash flows from investing activities	
[List of individual inflows and outflows]	
Net cash provided (used) by investing activities	#
Cash flows from financing activities	
[List of individual inflows and outflows]	
Net cash provided (used) by financing activities	#
Net increase (decrease) in cash	$ #
Cash (and equivalents) balance at prior period-end	#
Cash (and equivalents) balance at current period-end	$ #

Separate schedule or note disclosure of any "noncash investing and financing transactions" is required.

Decision Maker

Entrepreneur You are considering purchasing a start-up business that recently reported a $110,000 annual net loss and a $225,000 annual net cash inflow. How are these results possible? [Answer—p. 658]

Quick Check Answers—p. 658

1. Does a statement of cash flows report the cash payments to purchase cash equivalents? Does it report the cash receipts from selling cash equivalents?
2. Identify the three categories of cash flows reported separately on the statement of cash flows.
3. Identify the cash activity category for each transaction: (*a*) purchase equipment for cash, (*b*) cash payment of wages, (*c*) sale of common stock for cash, (*d*) receipt of cash dividends from stock investment, (*e*) cash collection from customers, (*f*) notes issued for cash.

Preparing the Statement of Cash Flows

P1 Prepare a statement of cash flows.

Preparing a statement of cash flows involves five steps: (1) compute the net increase or decrease in cash; (2) compute and report the net cash provided or used by operating activities (using either the direct or indirect method; both are explained); (3) compute and report the net cash provided or used by investing activities; (4) compute and report the net cash provided or used by financing activities; and (5) compute the net cash flow by combining net cash provided or used by operating, investing, and financing activities and then *prove it* by adding it to the beginning cash balance to show that it equals the ending cash balance.

Step 1: Compute net increase or decrease in cash

Step 2: Compute net cash from operating activities

Step 3: Compute net cash from investing activities

Step 4: Compute net cash from financing activities

Step 5: Prove and report beginning and ending cash balances

Computing the net increase or net decrease in cash is a simple but crucial computation. It equals the current period's cash balance minus the prior period's cash balance. This is the *bottom-line* figure for the statement of cash flows and is a check on accuracy. The information we need to prepare a statement of cash flows comes from various sources including comparative balance sheets at the beginning and end of the period, and an income statement for the period. There are two alternative approaches to preparing the statement: (1) analyzing the Cash account and (2) analyzing noncash accounts.

Analyzing the Cash Account A company's cash receipts and cash payments are recorded in the Cash account in its general ledger. The Cash account is therefore a natural place to look for information about cash flows from operating, investing, and financing activities. To illustrate, review the summarized Cash T-account of Genesis, Inc., in Exhibit 16.6. Individual cash transactions are summarized in this Cash account according to the major types of cash receipts and cash payments. For instance, only the total of cash receipts from all customers is listed. Individual cash transactions underlying these totals can number in the thousands. Accounting software is available to provide summarized cash accounts.

Preparing a statement of cash flows from Exhibit 16.6 requires determining whether an individual cash inflow or outflow is an operating, investing, or financing activity, and then listing each by

EXHIBIT 16.6

Summarized Cash Account

Accounting System:

File Edit Maintain Tasks Analysis Options Reports Window Help

Cash			
Balance, Dec. 31, 2008	12,000		
Receipts from customers	570,000	Payments for merchandise	319,000
Receipts from asset sales	12,000	Payments for wages and operating expenses	218,000
Receipts from stock issuance	15,000	Payments for interest	8,000
		Payments for taxes	5,000
		Payments for assets	10,000
		Payments for notes retirement	18,000
		Payments for dividends	14,000
Balance, Dec. 31, 2009	17,000		

Sales Purchases General Ledger Payroll Inventory Company Analysis

activity. This yields the statement shown in Exhibit 16.7. However, preparing the statement of cash flows from an analysis of the summarized Cash account has two limitations. First, most companies have many individual cash receipts and payments, making it difficult to review them all. Accounting software minimizes this burden, but it is still a task requiring professional judgment for many transactions. Second, the Cash account does not usually carry an adequate description of each cash transaction, making assignment of all cash transactions according to activity difficult.

Point: View the change in cash as a *target* number that we will fully explain and prove in the statement of cash flows.

EXHIBIT 16.7

Statement of Cash Flows—Direct Method

GENESIS
Statement of Cash Flows
For Year Ended December 31, 2009

Cash flows from operating activities		
Cash received from customers	$570,000	
Cash paid for merchandise	(319,000)	
Cash paid for wages and other operating expenses	(218,000)	
Cash paid for interest	(8,000)	
Cash paid for taxes	(5,000)	
Net cash provided by operating activities		$20,000
Cash flows from investing activities		
Cash received from sale of plant assets	12,000	
Cash paid for purchase of plant assets	(10,000)	
Net cash provided by investing activities		2,000
Cash flows from financing activities		
Cash received from issuing stock	15,000	
Cash paid to retire notes	(18,000)	
Cash paid for dividends	(14,000)	
Net cash used in financing activities		(17,000)
Net increase in cash		$ 5,000
Cash balance at prior year-end		12,000
Cash balance at current year-end		$17,000

Analyzing Noncash Accounts A second approach to preparing the statement of cash flows is analyzing noncash accounts. This approach uses the fact that when a company records cash inflows and outflows with debits and credits to the Cash account (see Exhibit 16.6), it also records credits and debits in noncash accounts (reflecting double-entry accounting). Many of these noncash accounts are balance sheet accounts—for instance, from the sale of land for cash. Others are revenue and expense accounts that are closed to equity. For instance, the sale of services for cash yields a credit to Services Revenue that is closed to Retained Earnings for a corporation. In sum, *all cash transactions eventually affect noncash balance sheet accounts*. Thus, we can determine cash inflows and outflows by analyzing changes in noncash balance sheet accounts.

Exhibit 16.8 uses the accounting equation to show the relation between the Cash account and the noncash balance sheet accounts. This exhibit starts with the accounting equation at the top. It is then expanded in line (2) to separate cash from noncash asset accounts. Line (3) moves noncash asset accounts to the right-hand side of the equality where they are subtracted. This shows that cash equals the sum of the liability and equity accounts *minus* the noncash asset accounts. Line (4) points

EXHIBIT 16.8

Relation between Cash and Noncash Accounts

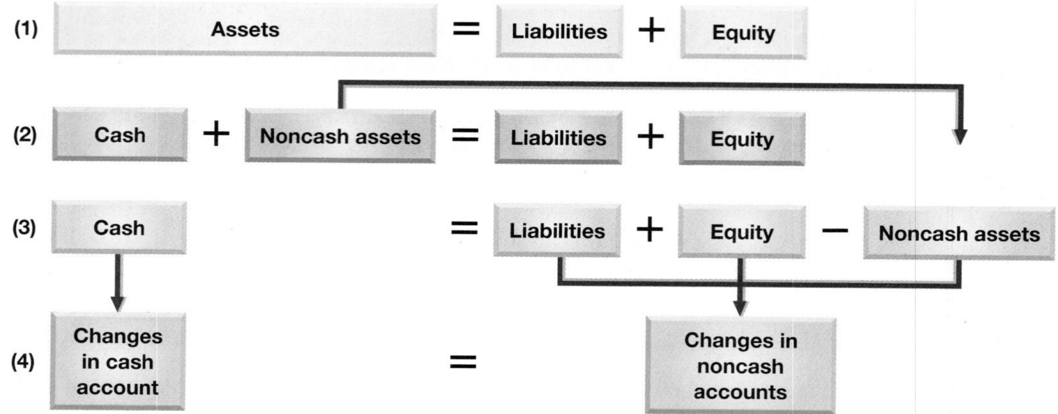

out that *changes* on one side of the accounting equation equal *changes* on the other side. It shows that we can explain changes in cash by analyzing changes in the noncash accounts consisting of liability accounts, equity accounts, and noncash asset accounts. By analyzing noncash balance sheet accounts and any related income statement accounts, we can prepare a statement of cash flows.

Information to Prepare the Statement Information to prepare the statement of cash flows usually comes from three sources: (1) comparative balance sheets, (2) the current income statement, and (3) additional information. Comparative balance sheets are used to compute changes in noncash accounts from the beginning to the end of the period. The current income statement is used to help compute cash flows from operating activities. Additional information often includes details on transactions and events that help explain both the cash flows and noncash investing and financing activities.

Decision Insight

e-Cash Every credit transaction on the Net leaves a trail that a hacker or a marketer can pick up. Enter e-cash—or digital money. The encryption of e-cash protects your money from snoops and thieves and cannot be traced, even by the issuing bank.

Cash Flows from Operating

Indirect and Direct Methods of Reporting

Cash flows provided (used) by operating activities are reported in one of two ways: the *direct method* or the *indirect method*. **These two different methods apply only to the operating activities section.**

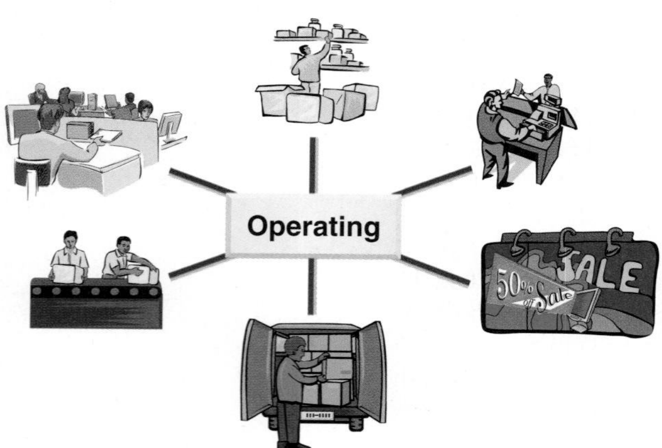

The **direct method** separately lists each major item of operating cash receipts (such as cash received from customers) and each major item of operating cash payments (such as cash paid for merchandise). The cash payments are subtracted from cash receipts to determine the net cash provided (used) by operating activities. The operating activities section of Exhibit 16.7 reflects the direct method of reporting operating cash flows.

The **indirect method** reports net income and then adjusts it for items necessary to obtain net cash provided or used by operating activities. It does *not* report individual items of cash inflows and cash outflows from operating activities. Instead, the indirect method reports the necessary adjustments to reconcile net income to net cash provided or used by operating activities. The operating activities section for Genesis prepared under the indirect method is shown in Exhibit 16.9.

Cash flows from operating activities		
Net income	$ 38,000	
Adjustments to reconcile net income to net cash provided by operating activities		
Increase in accounts receivable	(20,000)	
Increase in merchandise inventory	(14,000)	
Increase in prepaid expenses	(2,000)	
Decrease in accounts payable	(5,000)	
Decrease in interest payable	(1,000)	
Increase in income taxes payable	10,000	
Depreciation expense	24,000	
Loss on sale of plant assets	6,000	
Gain on retirement of notes	(16,000)	
Net cash provided by operating activities		**$20,000**

EXHIBIT 16.9

Operating Activities Section—Indirect Method

The net cash amount provided by operating activities is *identical* under both the direct and indirect methods. This equality always exists. The difference in these methods is with the computation and presentation of this amount. The FASB recommends the direct method, but because it is not required and the indirect method is arguably easier to compute, nearly all companies report operating cash flows using the indirect method.

Point: To better understand the direct and indirect methods of reporting operating cash flows, identify similarities and differences between Exhibits 16.7 and 16.11.

To illustrate, we prepare the operating activities section of the statement of cash flows for Genesis. Exhibit 16.10 shows the December 31, 2008 and 2009, balance sheets of Genesis along with its 2009 income statement. We use this information to prepare a statement of cash flows that explains the $5,000 increase in cash for 2009 as reflected in its balance sheets. This $5,000 is computed as Cash of $17,000 at the end of 2009 minus Cash of $12,000 at the end of 2008. Genesis discloses additional information on its 2009 transactions:

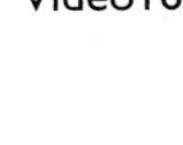

Video16.1

a. The accounts payable balances result from merchandise inventory purchases.

b. Purchased $70,000 in plant assets by paying $10,000 cash and issuing $60,000 of notes payable.

c. Sold plant assets with an original cost of $30,000 and accumulated depreciation of $12,000 for $12,000 cash, yielding a $6,000 loss.

d. Received $15,000 cash from issuing 3,000 shares of common stock.

e. Paid $18,000 cash to retire notes with a $34,000 book value, yielding a $16,000 gain.

f. Declared and paid cash dividends of $14,000.

The next section describes the indirect method. Appendix 16B describes the direct method. An instructor can choose to cover either one or both methods. Neither section depends on the other.

Application of the Indirect Method of Reporting

P2 Compute cash flows from operating activities using the indirect method.

Net income is computed using accrual accounting, which recognizes revenues when earned and expenses when incurred. Revenues and expenses do not necessarily reflect the receipt and payment of cash. The indirect method of computing and reporting net cash flows from operating activities involves adjusting the net income figure to obtain the net cash provided or used by operating activities. This includes subtracting noncash increases (credits) from net income and adding noncash charges (debits) back to net income.

To illustrate, the indirect method begins with Genesis's net income of $38,000 and adjusts it to obtain net cash provided by operating activities of $20,000. Exhibit 16.11 shows the results of the indirect method of reporting operating cash flows, which adjusts net income for three types of adjustments. There are adjustments ① to reflect changes in noncash current assets and current liabilities related to operating activities, ② to income statement items involving operating activities that do not affect cash inflows or outflows, and ③ to eliminate gains and losses resulting from investing and financing activities (not part of operating activities). This section describes each of these adjustments.

Point: *Noncash credits* refer to *revenue* amounts reported on the income statement that are *not collected in cash* this period. *Noncash charges* refer to *expense* amounts reported on the income statement that are *not paid* this period.

EXHIBIT 16.10

Financial Statements

GENESIS
Income Statement
For Year Ended December 31, 2009

Sales		$590,000
Cost of goods sold	$300,000	
Wages and other operating expenses	216,000	
Interest expense	7,000	
Depreciation expense	24,000	(547,000)
		43,000
Other gains (losses)		
Gain on retirement of notes	16,000	
Loss on sale of plant assets	(6,000)	10,000
Income before taxes		53,000
Income taxes expense		(15,000)
Net income		$ 38,000

GENESIS
Balance Sheets
December 31, 2009 and 2008

	2009	2008
Assets		
Current assets		
Cash	$ 17,000	$ 12,000
Accounts receivable	60,000	40,000
Merchandise inventory	84,000	70,000
Prepaid expenses	6,000	4,000
Total current assets	167,000	126,000
Long-term assets		
Plant assets	250,000	210,000
Accumulated depreciation	(60,000)	(48,000)
Total assets	$357,000	$288,000
Liabilities		
Current liabilities		
Accounts payable	$ 35,000	$ 40,000
Interest payable	3,000	4,000
Income taxes payable	22,000	12,000
Total current liabilities	60,000	56,000
Long-term notes payable	90,000	64,000
Total liabilities	150,000	120,000
Equity		
Common stock, $5 par	95,000	80,000
Retained earnings	112,000	88,000
Total equity	207,000	168,000
Total liabilities and equity	$357,000	$288,000

① Adjustments for Changes in Current Assets and Current Liabilities This section describes adjustments for changes in noncash current assets and current liabilities.

Point: Operating activities are typically those that determine income, which are often reflected in changes in current assets and current liabilities.

Adjustments for changes in noncash current assets. Changes in noncash current assets normally result from operating activities. Examples are sales affecting accounts receivable and building usage affecting prepaid rent. Decreases in noncash current assets yield the following adjustment:

Decreases in noncash current assets are added to net income.

To see the logic for this adjustment, consider that a decrease in a noncash current asset such as accounts receivable suggests more available cash at the end of the period compared to the beginning. This is so because a decrease in accounts receivable implies higher cash receipts than reflected in sales. We add these higher cash receipts (from decreases in noncash current assets) to net income when computing cash flow from operations.

In contrast, an increase in noncash current assets such as accounts receivable implies less cash receipts than reflected in sales. As another example, an increase in prepaid rent indicates that more cash is paid for rent than is deducted as rent expense. Increases in noncash current assets yield the following adjustment:

Increases in noncash current assets are subtracted from net income.

To illustrate, these adjustments are applied to the noncash current assets in Exhibit 16.10.

Accounts receivable. Accounts receivable *increase* $20,000, from a beginning balance of $40,000 to an ending balance of $60,000. This increase implies that Genesis collects less cash than is reported in sales. That is, some of these sales were in the form of accounts receivable and that amount increased during the period. To see this it is helpful to use *account analysis.* This usually involves setting up a T-account and reconstructing its major entries to compute cash receipts or payments. The following reconstructed Accounts Receivable T-account reveals that cash receipts are less than sales:

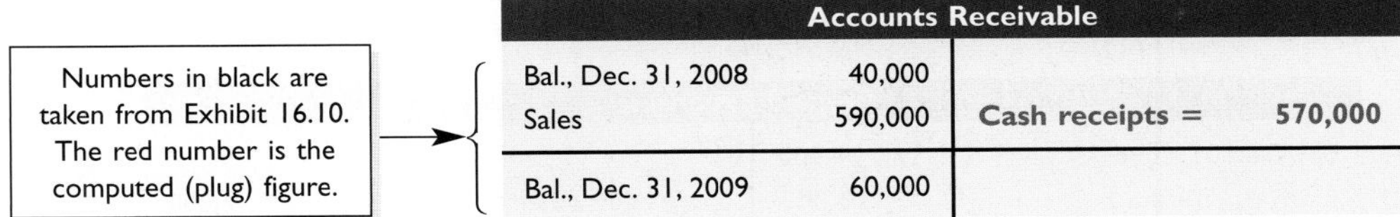

Numbers in black are taken from Exhibit 16.10. The red number is the computed (plug) figure.

Accounts Receivable			
Bal., Dec. 31, 2008	40,000		
Sales	590,000	**Cash receipts =**	**570,000**
Bal., Dec. 31, 2009	60,000		

We see that sales are $20,000 greater than cash receipts. This $20,000—as reflected in the $20,000 increase in Accounts Receivable—is subtracted from net income when computing cash provided by operating activities (see Exhibit 16.11).

Merchandise inventory. Merchandise inventory *increases* by $14,000, from a $70,000 beginning balance to an $84,000 ending balance. This increase implies that Genesis had greater cash purchases than cost of goods sold. This larger amount of cash purchases is in the form of inventory, as reflected in the following account analysis:

Merchandise Inventory			
Bal., Dec. 31, 2008	70,000		
Purchases =	**314,000**	Cost of goods sold	300,000
Bal., Dec. 31, 2009	84,000		

EXHIBIT 16.11

Statement of Cash Flows—Indirect Method

GENESIS Statement of Cash Flows For Year Ended December 31, 2009		
Cash flows from operating activities		
Net income	$ 38,000	
Adjustments to reconcile net income to net cash provided by operating activities		
① Increase in accounts receivable	(20,000)	
① Increase in merchandise inventory	(14,000)	
① Increase in prepaid expenses	(2,000)	
① Decrease in accounts payable	(5,000)	
① Decrease in interest payable	(1,000)	
① Increase in income taxes payable	10,000	
② Depreciation expense	24,000	
③ Loss on sale of plant assets	6,000	
③ Gain on retirement of notes	(16,000)	
Net cash provided by operating activities		$20,000
Cash flows from investing activities		
Cash received from sale of plant assets	12,000	
Cash paid for purchase of plant assets	(10,000)	
Net cash provided by investing activities		2,000
Cash flows from financing activities		
Cash received from issuing stock	15,000	
Cash paid to retire notes	(18,000)	
Cash paid for dividends	(14,000)	
Net cash used in financing activities		(17,000)
Net increase in cash		$ 5,000
Cash balance at prior year-end		12,000
Cash balance at current year-end		$17,000

Point: Refer to Exhibit 16.10 and identify the $5,000 change in cash. This change is what the statement of cash flows explains; it serves as a check.

The amount by which purchases exceed cost of goods sold—as reflected in the $14,000 increase in inventory—is subtracted from net income when computing cash provided by operating activities (see Exhibit 16.11).

Prepaid expenses. Prepaid expenses *increase* $2,000, from a $4,000 beginning balance to a $6,000 ending balance, implying that Genesis's cash payments exceed its recorded prepaid expenses. These higher cash payments increase the amount of Prepaid Expenses, as reflected in its reconstructed T-account:

Prepaid Expenses			
Bal., Dec. 31, 2008	4,000		
Cash payments =	**218,000**	Wages and other operating exp.	216,000
Bal., Dec. 31, 2009	6,000		

The amount by which cash payments exceed the recorded operating expenses—as reflected in the $2,000 increase in Prepaid Expenses—is subtracted from net income when computing cash provided by operating activities (see Exhibit 16.11).

Adjustments for changes in current liabilities. Changes in current liabilities normally result from operating activities. An example is a purchase that affects accounts payable. Increases in current liabilities yield the following adjustment to net income when computing operating cash flows:

Increases in current liabilities are added to net income.

To see the logic for this adjustment, consider that an increase in the Accounts Payable account suggests that cash payments are less than the related (cost of goods sold) expense. As another example, an increase in wages payable implies that cash paid for wages is less than the recorded wages expense. Since the recorded expense is greater than the cash paid, we add the increase in wages payable to net income to compute net cash flow from operations.

Conversely, when current liabilities decrease, the following adjustment is required:

Decreases in current liabilities are subtracted from net income.

To illustrate, these adjustments are applied to the current liabilities in Exhibit 16.10.

Accounts payable. Accounts payable *decrease* $5,000, from a beginning balance of $40,000 to an ending balance of $35,000. This decrease implies that cash payments to suppliers exceed purchases by $5,000 for the period, which is reflected in the reconstructed Accounts Payable T-account:

Accounts Payable			
		Bal., Dec. 31, 2008	40,000
Cash payments =	**319,000**	Purchases	314,000
		Bal., Dec. 31, 2009	35,000

The amount by which cash payments exceed purchases—as reflected in the $5,000 decrease in Accounts Payable—is subtracted from net income when computing cash provided by operating activities (see Exhibit 16.11).

Interest payable. Interest payable *decreases* $1,000, from a $4,000 beginning balance to a $3,000 ending balance. This decrease indicates that cash paid for interest exceeds interest expense by $1,000, which is reflected in the Interest Payable T-account:

Interest Payable			
		Bal., Dec. 31, 2008	4,000
Cash paid for interest =	**8,000**	Interest expense	7,000
		Bal., Dec. 31, 2009	3,000

The amount by which cash paid exceeds recorded expense—as reflected in the $1,000 decrease in Interest Payable—is subtracted from net income (see Exhibit 16.11).

Income taxes payable. Income taxes payable *increase* $10,000, from a $12,000 beginning balance to a $22,000 ending balance. This increase implies that reported income taxes exceed the cash paid for taxes, which is reflected in the Income Taxes Payable T-account:

Income Taxes Payable			
		Bal., Dec. 31, 2008	12,000
Cash paid for taxes =	5,000	Income taxes expense	15,000
		Bal., Dec. 31, 2009	22,000

Summary Adjustments for Changes in Current Assets and Current Liabilities		
Account	Increases	Decreases
Noncash current assets	Deduct from NI	Add to NI
Current liabilities	Add to NI	Deduct from NI

The amount by which cash paid falls short of the reported taxes expense—as reflected in the $10,000 increase in Income Taxes Payable—is added to net income when computing cash provided by operating activities (see Exhibit 16.11).

② Adjustments for Operating Items Not Providing or Using Cash The income statement usually includes some expenses that do not reflect cash outflows in the period. Examples are depreciation, amortization, depletion, and bad debts expense. The indirect method for reporting operating cash flows requires that

Expenses with no cash outflows are added back to net income.

To see the logic of this adjustment, recall that items such as depreciation, amortization, depletion, and bad debts originate from debits to expense accounts and credits to noncash accounts. These entries have *no* cash effect, and we add them back to net income when computing net cash flows from operations. Adding them back cancels their deductions.

Similarly, when net income includes revenues that do not reflect cash inflows in the period, the indirect method for reporting operating cash flows requires that

Revenues with no cash inflows are subtracted from net income.

We apply these adjustments to the Genesis operating items that do not provide or use cash.

Depreciation. Depreciation expense is the only Genesis operating item that has no effect on cash flows in the period. We must add back the $24,000 depreciation expense to net income when computing cash provided by operating activities. (We later explain that any cash outflow to acquire a plant asset is reported as an investing activity.)

③ Adjustments for Nonoperating Items Net income often includes losses that are not part of operating activities but are part of either investing or financing activities. Examples are a loss from the sale of a plant asset and a loss from retirement of notes payable. The indirect method for reporting operating cash flows requires that

Nonoperating losses are added back to net income.

To see the logic, consider that items such as a plant asset sale and a notes retirement are normally recorded by recognizing the cash, removing all plant asset or notes accounts, and recognizing any loss or gain. The cash received or paid is not part of operating activities but is part of either investing or financing activities. *No* operating cash flow effect occurs. However, because the nonoperating loss is a deduction in computing net income, we need to add it back to net income when computing cash flow from operations. Adding it back cancels the deduction.

Point: An income statement reports revenues, gains, expenses, and losses on an accrual basis. The statement of cash flows reports cash received and cash paid for operating, financing, and investing activities.

Similarly, when net income includes gains not part of operating activities, the indirect method for reporting operating cash flows requires that

Nonoperating gains are subtracted from net income.

To illustrate these adjustments, we consider the nonoperating items of Genesis.

Loss on sale of plant assets. Genesis reports a $6,000 loss on sale of plant assets as part of net income. This loss is a proper deduction in computing income, but it is *not part of operating activities*. Instead, a sale of plant assets is part of investing activities. Thus, the $6,000 nonoperating loss is added back to net income (see Exhibit 16.11). Adding it back cancels the loss. We later explain how to report the cash inflow from the asset sale in investing activities.

Gain on retirement of debt. A $16,000 gain on retirement of debt is properly included in net income, but it is *not part of operating activities*. This means the $16,000 nonoperating gain must be subtracted from net income to obtain net cash provided by operating activities (see Exhibit 16.11). Subtracting it cancels the recorded gain. We later describe how to report the cash outflow to retire debt.

Summary of Adjustments for Indirect Method

Exhibit 16.12 summarizes the most common adjustments to net income when computing net cash provided or used by operating activities under the indirect method.

EXHIBIT 16.12

Summary of Selected Adjustments for Indirect Method

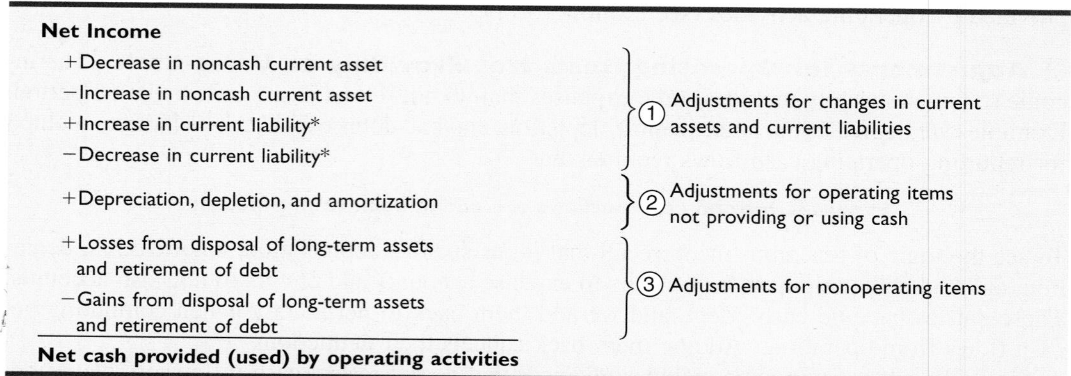

Net Income

+Decrease in noncash current asset	① Adjustments for changes in current assets and current liabilities
−Increase in noncash current asset	
+Increase in current liability*	
−Decrease in current liability*	
+Depreciation, depletion, and amortization	② Adjustments for operating items not providing or using cash
+Losses from disposal of long-term assets and retirement of debt	③ Adjustments for nonoperating items
−Gains from disposal of long-term assets and retirement of debt	

Net cash provided (used) by operating activities

* Excludes current portion of long-term debt and any (nonsales-related) short-term notes payable—both are financing activities.

The computations in determining cash provided or used by operating activities are different for the indirect and direct methods, but the result is identical. Both methods yield the same $20,000 figure for cash from operating activities for Genesis; see Exhibits 16.7 and 16.11.

Decision Insight

Cash or Income The difference between net income and operating cash flows can be large and sometimes reflects on the quality of earnings. This bar chart shows the net income and operating cash flows of three companies. Operating cash flows can be either higher or lower than net income.

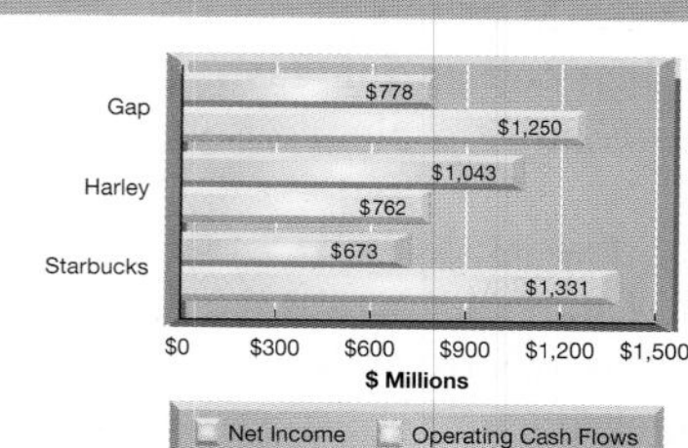

Quick Check

Answers—p. 658

4. Determine the net cash provided or used by operating activities using the following data: net income, $74,900; decrease in accounts receivable, $4,600; increase in inventory, $11,700; decrease in accounts payable, $1,000; loss on sale of equipment, $3,400; payment of cash dividends, $21,500.
5. Why are expenses such as depreciation and amortization added to net income when cash flow from operating activities is computed by the indirect method?
6. A company reports net income of $15,000 that includes a $3,000 gain on the sale of plant assets. Why is this gain subtracted from net income in computing cash flow from operating activities using the indirect method?

Cash Flows from Investing

The third major step in preparing the statement of cash flows is to compute and report cash flows from investing activities. We normally do this by identifying changes in (1) all noncurrent asset accounts and (2) the current accounts for both notes receivable and investments in securities (excluding trading securities). We then analyze changes in these accounts to determine their effect, if any, on cash and report the cash flow effects in the investing activities section of the statement of cash flows. **Reporting of investing activities is identical under the direct method and indirect method.**

Three-Stage Process of Analysis

Video16.1

Information to compute cash flows from investing activities is usually taken from beginning and ending balance sheets and the income statement. We use a three-stage process to determine cash provided or used by investing activities: (1) identify changes in investing-related accounts, (2) explain these changes using reconstruction analysis, and (3) report their cash flow effects.

Analysis of Noncurrent Assets

P3 Determine cash flows from both investing and financing activities.

Information about the Genesis transactions provided earlier reveals that the company both purchased and sold plant assets during the period. Both transactions are investing activities and are analyzed for their cash flow effects in this section.

Point: Investing activities include (1) purchasing and selling long-term assets, (2) lending and collecting on notes receivable, and (3) purchasing and selling short-term investments other than cash equivalents and trading securities.

Plant Asset Transactions The first stage in analyzing the Plant Assets account and its related Accumulated Depreciation is to identify any changes in these accounts from comparative balance sheets in Exhibit 16.10. This analysis reveals a $40,000 increase in plant assets from $210,000 to $250,000 and a $12,000 increase in accumulated depreciation from $48,000 to $60,000.

The second stage is to explain these changes. Items *b* and *c* of the additional information for Genesis (page 635) are relevant in this case. Recall that the Plant Assets account is affected by both asset purchases and sales, while its Accumulated Depreciation account is normally increased from depreciation and decreased from the removal of accumulated depreciation in asset sales. To explain changes in these accounts and to identify their cash flow effects, we prepare *reconstructed entries* from prior transactions; *they are not the actual entries by the preparer*.

Point: Financing and investing info is available in ledger accounts to help explain changes in comparative balance sheets. Post references lead to relevant entries and explanations.

To illustrate, item *b* reports that Genesis purchased plant assets of $70,000 by issuing $60,000 in notes payable to the seller and paying $10,000 in cash. The reconstructed entry for analysis of item *b* follows:

Reconstruction	Plant Assets	70,000	
	Notes Payable		60,000
	Cash		**10,000**

This entry reveals a $10,000 cash outflow for plant assets and a $60,000 noncash investing and financing transaction involving notes exchanged for plant assets.

Next, item *c* reports that Genesis sold plant assets costing $30,000 (with $12,000 of accumulated depreciation) for $12,000 cash, resulting in a $6,000 loss. The reconstructed entry for analysis of item *c* follows:

Reconstruction	**Cash**	**12,000**	
	Accumulated Depreciation	12,000	
	Loss on Sale of Plant Assets	6,000	
	Plant Assets		30,000

This entry reveals a $12,000 cash inflow from assets sold. The $6,000 loss is computed by comparing the asset book value to the cash received and does not reflect any cash inflow or outflow. We also reconstruct the entry for Depreciation Expense using information from the income statement.

Reconstruction	Depreciation Expense	24,000	
	Accumulated Depreciation		24,000

This entry shows that Depreciation Expense results in no cash flow effect. These three reconstructed entries are reflected in the following plant asset and related T-accounts.

Plant Assets			
Bal., Dec. 31, 2008	210,000		
Purchase	**70,000**	**Sale**	**30,000**
Bal., Dec. 31, 2009	250,000		

Accumulated Depreciation—Plant Assets			
		Bal., Dec. 31, 2008	48,000
Sale	**12,000**	**Depr. expense**	**24,000**
		Bal., Dec. 31, 2009	60,000

Example: If a plant asset costing $40,000 with $37,000 of accumulated depreciation is sold at a $1,000 loss, what is the cash flow? What is the cash flow if this asset is sold at a gain of $3,000? *Answers:* +$2,000; +$6,000.

This reconstruction analysis is complete in that the change in plant assets from $210,000 to $250,000 is fully explained by the $70,000 purchase and the $30,000 sale. Also, the change in accumulated depreciation from $48,000 to $60,000 is fully explained by depreciation expense of $24,000 and the removal of $12,000 in accumulated depreciation from an asset sale. (Preparers of the statement of cash flows have the entire ledger and additional information at their disposal, but for brevity reasons only the information needed for reconstructing accounts is given.)

The third stage looks at the reconstructed entries for identification of cash flows. The two identified cash flow effects are reported in the investing section of the statement as follows (also see Exhibit 16.7 or 16.11):

Cash flows from investing activities	
Cash received from sale of plant assets	$12,000
Cash paid for purchase of plant assets	(10,000)

The $60,000 portion of the purchase described in item *b* and financed by issuing notes is a noncash investing and financing activity. It is reported in a note or in a separate schedule to the statement as follows:

Noncash investing and financing activity	
Purchased plant assets with issuance of notes	$60,000

Analysis of Other Assets

Many other asset transactions (including those involving current notes receivable and investments in certain securities) are considered investing activities and can affect a company's cash flows. Since Genesis did not enter into other investing activities impacting assets, we do not need to extend our analysis to these other assets. If such transactions did exist, we would analyze them using the same three-stage process illustrated for plant assets.

Quick Check

Answer—p. 658

7. Equipment costing $80,000 with accumulated depreciation of $30,000 is sold at a loss of $10,000. What is the cash receipt from this sale? In what section of the statement of cash flows is this transaction reported?

Cash Flows from Financing

The fourth major step in preparing the statement of cash flows is to compute and report cash flows from financing activities. We normally do this by identifying changes in all noncurrent liability accounts (including the current portion of any notes and bonds) and the equity accounts. These accounts include long-term debt, notes payable, bonds payable, common stock, and retained earnings. Changes in these accounts are then analyzed using available information to determine their effect, if any, on cash. Results are reported in the financing activities section of the statement. **Reporting of financing activities is identical under the direct method and indirect method.**

Three-Stage Process of Analysis

Video16.1

We again use a three-stage process to determine cash provided or used by financing activities: (1) identify changes in financing-related accounts, (2) explain these changes using reconstruction analysis, and (3) report their cash flow effects.

Analysis of Noncurrent Liabilities

Information about Genesis provided earlier reveals two transactions involving noncurrent liabilities. We analyzed one of those, the $60,000 issuance of notes payable to purchase plant assets. This transaction is reported as a significant noncash investing and financing activity in a footnote or a separate schedule to the statement of cash flows. The other remaining transaction involving noncurrent liabilities is the cash retirement of notes payable.

Point: Financing activities generally refer to changes in the noncurrent liability and the equity accounts. Examples are (1) receiving cash from issuing debt or repaying amounts borrowed and (2) receiving cash from or distributing cash to owners.

Notes Payable Transactions The first stage in analysis of notes is to review the comparative balance sheets from Exhibit 16.10. This analysis reveals an increase in notes payable from $64,000 to $90,000.

The second stage explains this change. Item *e* of the additional information for Genesis (page 635) reports that notes with a carrying value of $34,000 are retired for $18,000 cash, resulting in a $16,000 gain. The reconstructed entry for analysis of item *e* follows:

Reconstruction	Notes Payable	34,000	
	Gain on retirement of debt		16,000
	Cash		**18,000**

This entry reveals an $18,000 cash outflow for retirement of notes and a $16,000 gain from comparing the notes payable carrying value to the cash received. This gain does not reflect any cash inflow or outflow. Also, item *b* of the additional information reports that Genesis purchased plant assets costing $70,000 by issuing $60,000 in notes payable to the seller and paying $10,000 in cash. We reconstructed this entry when analyzing investing activities: It showed a $60,000 increase to notes payable that is reported as a noncash investing and financing transaction. The Notes Payable account reflects (and is fully explained by) these reconstructed entries as follows:

Notes Payable			
		Bal., Dec. 31, 2008	64,000
Retired notes	**34,000**	**Issued notes**	**60,000**
		Bal., Dec. 31, 2009	90,000

The third stage is to report the cash flow effect of the notes retirement in the financing section of the statement as follows (also see Exhibit 16.7 or 16.11):

Cash flows from financing activities	
Cash paid to retire notes	$(18,000)

Analysis of Equity

The Genesis information reveals two transactions involving equity accounts. The first is the issuance of common stock for cash. The second is the declaration and payment of cash dividends. We analyze both.

Common Stock Transactions The first stage in analyzing common stock is to review the comparative balance sheets from Exhibit 16.10, which reveals an increase in common stock from $80,000 to $95,000.

The second stage explains this change. Item *d* of the additional information (page 635) reports that 3,000 shares of common stock are issued at par for $5 per share. The reconstructed entry for analysis of item *d* follows:

Reconstruction	**Cash**	**15,000**	
	Common Stock		15,000

This entry reveals a $15,000 cash inflow from stock issuance and is reflected in (and explains) the Common Stock account as follows:

Common Stock			
		Bal., Dec. 31, 2008	80,000
		Issued stock	**15,000**
		Bal., Dec. 31, 2009	95,000

The third stage discloses the cash flow effect from stock issuance in the financing section of the statement as follows (also see Exhibit 16.7 or 16.11):

Cash flows from financing activities	
Cash received from issuing stock	$15,000

Retained Earnings Transactions The first stage in analyzing the Retained Earnings account is to review the comparative balance sheets from Exhibit 16.10. This reveals an increase in retained earnings from $88,000 to $112,000.

The second stage explains this change. Item *f* of the additional information (page 635) reports that cash dividends of $14,000 are paid. The reconstructed entry follows:

Reconstruction	Retained Earnings..........................	14,000	
	Cash................................		**14,000**

This entry reveals a $14,000 cash outflow for cash dividends. Also see that the Retained Earnings account is impacted by net income of $38,000. (Net income was analyzed under the operating section of the statement of cash flows.) The reconstructed Retained Earnings account follows:

Retained Earnings			
		Bal., Dec. 31, 2008	88,000
Cash dividend	**14,000**	**Net income**	**38,000**
		Bal., Dec. 31, 2009	112,000

Point: Financing activities not affecting cash flow include *declaration* of a cash dividend, *declaration* of a stock dividend, payment of a stock dividend, and a stock split.

The third stage reports the cash flow effect from the cash dividend in the financing section of the statement as follows (also see Exhibit 16.7 or 16.11):

Cash flows from financing activities	
Cash paid for dividends	$(14,000)

Global: There are no requirements to separate domestic and international cash flows, leading some users to ask, "Where in the world is cash flow?"

We now have identified and explained all of the Genesis cash inflows and cash outflows and one noncash investing and financing transaction. Specifically, our analysis has reconciled changes in all noncash balance sheet accounts.

Proving Cash Balances

The fifth and final step in preparing the statement is to report the beginning and ending cash balances and prove that the *net change in cash* is explained by operating, investing, and financing cash flows. This step is shown here for Genesis.

Net cash provided by operating activities	$20,000
Net cash provided by investing activities	2,000
Net cash used in financing activities	(17,000)
Net increase in cash	**$ 5,000**
Cash balance at 2008 year-end	12,000
Cash balance at 2009 year-end	$ 17,000

The preceding table shows that the $5,000 net increase in cash, from $12,000 at the beginning of the period to $17,000 at the end, is reconciled by net cash flows from operating ($20,000 inflow), investing ($2,000 inflow), and financing ($17,000 outflow) activities. This is formally reported at the bottom of the statement of cash flows as shown in both Exhibits 16.7 and 16.11.

Decision Maker

Reporter Management is in labor contract negotiations and grants you an interview. It highlights a recent $600,000 net loss that involves a $930,000 extraordinary loss and a total net cash outflow of $550,000 (which includes net cash outflows of $850,000 for investing activities and $350,000 for financing activities). What is your assessment of this company? [Answer—p. 658]

Cash Flow Analysis

Decision Analysis

Analyzing Cash Sources and Uses

A1 Analyze the statement of cash flows.

Most managers stress the importance of understanding and predicting cash flows for business decisions. Creditors evaluate a company's ability to generate cash before deciding whether to lend money. Investors also assess cash inflows and outflows before buying and selling stock. Information in the statement of cash flows helps address these and other questions such as (1) How much cash is generated from or used in operations? (2) What expenditures are made with cash from operations? (3) What is the source of cash for debt payments? (4) What is the source of cash for distributions to owners? (5) How is the increase in investing activities financed? (6) What is the source of cash for new plant assets? (7) Why is cash flow from operations different from income? (8) How is cash from financing used?

To effectively answer these questions, it is important to separately analyze investing, financing, and operating activities. To illustrate, consider data from three different companies in Exhibit 16.13. These companies operate in the same industry and have been in business for several years.

EXHIBIT 16.13

Cash Flows of Competing Companies

($ thousands)	BMX	ATV	Trex
Cash provided (used) by operating activities	$90,000	$40,000	$(24,000)
Cash provided (used) by investing activities			
Proceeds from sale of plant assets			26,000
Purchase of plant assets	(48,000)	(25,000)	
Cash provided (used) by financing activities			
Proceeds from issuance of debt			13,000
Repayment of debt	(27,000)		
Net increase (decrease) in cash	$15,000	$15,000	$ 15,000

Each company generates an identical $15,000 net increase in cash, but its sources and uses of cash flows are very different. BMX's operating activities provide net cash flows of $90,000, allowing it to purchase plant assets of $48,000 and repay $27,000 of its debt. ATV's operating activities provide $40,000 of cash flows, limiting its purchase of plant assets to $25,000. Trex's $15,000 net cash increase is due to selling plant assets and incurring additional debt. Its operating activities yield a net cash outflow of $24,000.

Overall, analysis of these cash flows reveals that BMX is more capable of generating future cash flows than is ATV or Trex.

Decision Insight

Free Cash Flows Many investors use cash flows to value company stock. However, cash-based valuation models often yield different stock values due to differences in measurement of cash flows. Most models require cash flows that are "free" for distribution to shareholders. These *free cash flows* are defined as cash flows available to shareholders after operating asset reinvestments and debt payments. Knowledge of the statement of cash flows is key to proper computation of free cash flows. A company's growth and financial flexibility depend on adequate free cash flows.

Cash Flow on Total Assets

A2 Compute and apply the cash flow on total assets ratio.

Cash flow information has limitations, but it can help measure a company's ability to meet its obligations, pay dividends, expand operations, and obtain financing. Users often compute and analyze a cash-based ratio similar to return on total assets except that its numerator is net cash flows from operating activities. The **cash flow on total assets** ratio is in Exhibit 16.14.

EXHIBIT 16.14

Cash Flow on Total Assets

$$\textbf{Cash flow on total assets} = \frac{\textbf{Cash flow from operations}}{\textbf{Average total assets}}$$

This ratio reflects actual cash flows and is not affected by accounting income recognition and measurement. It can help business decision makers estimate the amount and timing of cash flows when planning and analyzing operating activities.

To illustrate, the 2007 cash flow on total assets ratio for Nike is 18.3%—see Exhibit 16.15. Is an 18.3% ratio good or bad? To answer this question, we compare this ratio with the ratios of prior years (we could also compare its ratio with those of its competitors and the market). Nike's cash flow on total assets ratio for several prior years is in the second column of Exhibit 16.15. Results show that its 18.3% return is the median of the prior years' returns.

EXHIBIT 16.15

Nike's Cash Flow on Total Assets

Year	Cash Flow on Total Assets	Return on Total Assets
2007	18.3%	14.5%
2006	17.9	14.9
2005	18.8	14.5
2004	20.6	12.8
2003	13.9	7.1

As an indicator of *earnings quality,* some analysts compare the cash flow on total assets ratio to the return on total assets ratio. Nike's return on total assets is provided in the third column of Exhibit 16.15. Nike's cash flow on total assets ratio exceeds its return on total assets in each of the five years, leading some analysts to infer that Nike's earnings quality is high for that period because more earnings are realized in the form of cash.

Decision Insight

Cash Flow Ratios Analysts use various other cash-based ratios, including the following two:

(1) $$\textbf{Cash coverage of growth} = \frac{\textbf{Operating cash flow}}{\textbf{Cash outflow for plant assets}}$$

where a low ratio (less than 1) implies cash inadequacy to meet asset growth, whereas a high ratio implies cash adequacy for asset growth.

(2) $$\textbf{Operating cash flow to sales} = \frac{\textbf{Operating cash flow}}{\textbf{Net sales}}$$

When this ratio substantially and consistently differs from the operating income to net sales ratio, the risk of accounting improprieties increases.

Point: The following ratio helps assess whether operating cash flow is adequate to meet long-term obligations:
Cash coverage of debt = Cash flow from operations ÷ Noncurrent liabilities.
A low ratio suggests a higher risk of insolvency; a high ratio suggests a greater ability to meet long-term obligations.

Demonstration Problem

Umlauf's comparative balance sheets, income statement, and additional information follow.

DP16

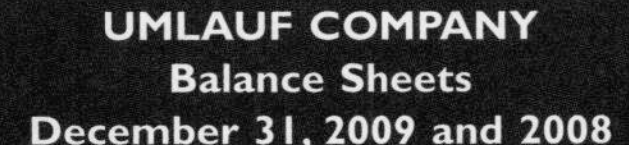

UMLAUF COMPANY
Balance Sheets
December 31, 2009 and 2008

	2009	2008
Assets		
Cash	$ 43,050	$ 23,925
Accounts receivable	34,125	39,825
Merchandise inventory	156,000	146,475
Prepaid expenses	3,600	1,650
Equipment	135,825	146,700
Accum. depreciation—Equipment	(61,950)	(47,550)
Total assets	$310,650	$311,025
Liabilities and Equity		
Accounts payable	$ 28,800	$ 33,750
Income taxes payable	5,100	4,425
Dividends payable	0	4,500
Bonds payable	0	37,500
Common stock, $10 par	168,750	168,750
Retained earnings	108,000	62,100
Total liabilities and equity	$310,650	$311,025

UMLAUF COMPANY
Income Statement
For Year Ended December 31, 2009

Sales		$446,100
Cost of goods sold	$222,300	
Other operating expenses	120,300	
Depreciation expense	25,500	(368,100)
		78,000
Other gains (losses)		
Loss on sale of equipment	3,300	
Loss on retirement of bonds	825	(4,125)
Income before taxes		73,875
Income taxes expense		(13,725)
Net income		$ 60,150

Additional Information

a. Equipment costing $21,375 with accumulated depreciation of $11,100 is sold for cash.
b. Equipment purchases are for cash.
c. Accumulated Depreciation is affected by depreciation expense and the sale of equipment.
d. The balance of Retained Earnings is affected by dividend declarations and net income.
e. All sales are made on credit.
f. All merchandise inventory purchases are on credit.
g. Accounts Payable balances result from merchandise inventory purchases.
h. Prepaid expenses relate to "other operating expenses."

Required

1. Prepare a statement of cash flows using the indirect method for year 2009.
2.[B] Prepare a statement of cash flows using the direct method for year 2009.

Planning the Solution

- Prepare two blank statements of cash flows with sections for operating, investing, and financing activities using the (1) indirect method format and (2) direct method format.
- Compute the cash paid for equipment and the cash received from the sale of equipment using the additional information provided along with the amount for depreciation expense and the change in the balances of equipment and accumulated depreciation. Use T-accounts to help chart the effects of the sale and purchase of equipment on the balances of the Equipment account and the Accumulated Depreciation account.
- Compute the effect of net income on the change in the Retained Earnings account balance. Assign the difference between the change in retained earnings and the amount of net income to dividends declared. Adjust the dividends declared amount for the change in the Dividends Payable balance.
- Compute cash received from customers, cash paid for merchandise, cash paid for other operating expenses, and cash paid for taxes as illustrated in the chapter.
- Enter the cash effects of reconstruction entries to the appropriate section(s) of the statement.
- Total each section of the statement, determine the total net change in cash, and add it to the beginning balance to get the ending balance of cash.

Solution to Demonstration Problem

Supporting computations for cash receipts and cash payments.

(1)	*Cost of equipment sold	$ 21,375
	Accumulated depreciation of equipment sold	(11,100)
	Book value of equipment sold	10,275
	Loss on sale of equipment	(3,300)
	Cash received from sale of equipment	$ 6,975
	Cost of equipment sold	$ 21,375
	Less decrease in the equipment account balance	(10,875)
	Cash paid for new equipment	$ 10,500
(2)	Loss on retirement of bonds	$ 825
	Carrying value of bonds retired	37,500
	Cash paid to retire bonds	$ 38,325
(3)	Net income	$ 60,150
	Less increase in retained earnings	45,900
	Dividends declared	14,250
	Plus decrease in dividends payable	4,500
	Cash paid for dividends	$ 18,750
(4)[B]	Sales	$ 446,100
	Add decrease in accounts receivable	5,700
	Cash received from customers	$451,800
(5)[B]	Cost of goods sold	$ 222,300
	Plus increase in merchandise inventory	9,525
	Purchases	231,825
	Plus decrease in accounts payable	4,950
	Cash paid for merchandise	$236,775
(6)[B]	Other operating expenses	$ 120,300
	Plus increase in prepaid expenses	1,950
	Cash paid for other operating expenses	$122,250
(7)[B]	Income taxes expense	$ 13,725
	Less increase in income taxes payable	(675)
	Cash paid for income taxes	$ 13,050

* Supporting T-account analysis for part 1 follows:

Equipment			
Bal., Dec. 31, 2008	146,700		
Cash purchase	10,500	Sale	21,375
Bal., Dec. 31, 2009	135,825		

Accumulated Depreciation—Equipment			
		Bal., Dec. 31, 2008	47,550
Sale	11,100	Depr. expense	25,500
		Bal., Dec. 31, 2009	61,950

UMLAUF COMPANY
Statement of Cash Flows (Indirect Method)
For Year Ended December 31, 2009

Cash flows from operating activities		
Net income	$60,150	
Adjustments to reconcile net income to net cash provided by operating activities		
Decrease in accounts receivable	5,700	
Increase in merchandise inventory	(9,525)	
Increase in prepaid expenses	(1,950)	
Decrease in accounts payable	(4,950)	
Increase in income taxes payable	675	
Depreciation expense	25,500	
Loss on sale of plant assets	3,300	
Loss on retirement of bonds	825	
Net cash provided by operating activities		$79,725

[continued on next page]

[continued from previous page]

Cash flows from investing activities		
Cash received from sale of equipment	6,975	
Cash paid for equipment	(10,500)	
Net cash used in investing activities		(3,525)
Cash flows from financing activities		
Cash paid to retire bonds payable	(38,325)	
Cash paid for dividends	(18,750)	
Net cash used in financing activities		(57,075)
Net increase in cash		$19,125
Cash balance at prior year-end		23,925
Cash balance at current year-end		$43,050

UMLAUF COMPANY
Statement of Cash Flows (Direct Method)
For Year Ended December 31, 2009

Cash flows from operating activities		
Cash received from customers	$451,800	
Cash paid for merchandise	(236,775)	
Cash paid for other operating expenses	(122,250)	
Cash paid for income taxes	(13,050)	
Net cash provided by operating activities		$79,725
Cash flows from investing activities		
Cash received from sale of equipment	6,975	
Cash paid for equipment	(10,500)	
Net cash used in investing activities		(3,525)
Cash flows from financing activities		
Cash paid to retire bonds payable	(38,325)	
Cash paid for dividends	(18,750)	
Net cash used in financing activities		(57,075)
Net increase in cash		$19,125
Cash balance at prior year-end		23,925
Cash balance at current year-end		$43,050

APPENDIX

Spreadsheet Preparation of the Statement of Cash Flows

16A

This appendix explains how to use a spreadsheet to prepare the statement of cash flows under the indirect method.

Preparing the Indirect Method Spreadsheet

P4 Illustrate use of a spreadsheet to prepare a statement of cash flows.

Analyzing noncash accounts can be challenging when a company has a large number of accounts and many operating, investing, and financing transactions. A *spreadsheet,* also called *work sheet* or *working paper,* can help us organize the information needed to prepare a statement of cash flows. A spreadsheet also makes it easier to check the accuracy of our work. To illustrate, we return to the comparative balance sheets and income statement shown in Exhibit 16.10. We use the following identifying letters *a* through *g* to code changes in accounts, and letters *h* through *m* for additional information, to prepare the statement of cash flows:

a. Net income is $38,000.
b. Accounts receivable increase by $20,000.
c. Merchandise inventory increases by $14,000.
d. Prepaid expenses increase by $2,000.
e. Accounts payable decrease by $5,000.
f. Interest payable decreases by $1,000.
g. Income taxes payable increase by $10,000.
h. Depreciation expense is $24,000.
i. Plant assets costing $30,000 with accumulated depreciation of $12,000 are sold for $12,000 cash. This yields a loss on sale of assets of $6,000.
j. Notes with a book value of $34,000 are retired with a cash payment of $18,000, yielding a $16,000 gain on retirement.
k. Plant assets costing $70,000 are purchased with a cash payment of $10,000 and an issuance of notes payable for $60,000.
l. Issued 3,000 shares of common stock for $15,000 cash.
m. Paid cash dividends of $14,000.

Exhibit 16A.1 shows the indirect method spreadsheet for Genesis. We enter both beginning and ending balance sheet amounts on the spreadsheet. We also enter information in the Analysis of Changes columns (keyed to the additional information items *a* through *m*) to explain changes in the accounts and determine the cash flows for operating, investing, and financing activities. Information about noncash investing and financing activities is reported near the bottom.

Entering the Analysis of Changes on the Spreadsheet

The following sequence of procedures is used to complete the spreadsheet after the beginning and ending balances of the balance sheet accounts are entered:

① Enter net income as the first item in the Statement of Cash Flows section for computing operating cash inflow (debit) and as a credit to Retained Earnings.

② In the Statement of Cash Flows section, adjustments to net income are entered as debits if they increase cash flows and as credits if they decrease cash flows. Applying this same rule, adjust net income for the change in each noncash current asset and current liability account related to operating activities. For each adjustment to net income, the offsetting debit or credit must help reconcile the beginning and ending balances of a current asset or current liability account.

③ Enter adjustments to net income for income statement items not providing or using cash in the period. For each adjustment, the offsetting debit or credit must help reconcile a noncash balance sheet account.

④ Adjust net income to eliminate any gains or losses from investing and financing activities. Because the cash from a gain must be excluded from operating activities, the gain is entered as a credit in the operating activities section. Losses are entered as debits. For each adjustment, the related debit and/or credit must help reconcile balance sheet accounts and involve reconstructed entries to show the cash flow from investing or financing activities.

⑤ After reviewing any unreconciled balance sheet accounts and related information, enter the remaining reconciling entries for investing and financing activities. Examples are purchases of plant assets, issuances of long-term debt, stock issuances, and dividend payments. Some of these may require entries in the noncash investing and financing section of the spreadsheet (reconciled).

⑥ Check accuracy by totaling the Analysis of Changes columns and by determining that the change in each balance sheet account has been explained (reconciled).

Point: Analysis of the changes on the spreadsheet are summarized here:

1. Cash flows from operating activities generally affect net income, current assets, and current liabilities.
2. Cash flows from investing activities generally affect noncurrent asset accounts.
3. Cash flows from financing activities generally affect noncurrent liability and equity accounts.

We illustrate these steps in Exhibit 16A.1 for Genesis:

Step	Entries
①	(*a*)
②	(*b*) through (*g*)
③	(*h*)
④	(*i*) through (*j*)
⑤	(*k*) through (*m*)

EXHIBIT 16A.1

Spreadsheet for Preparing Statement of Cash Flows—Indirect Method

GENESIS **Spreadsheet for Statement of Cash Flows—Indirect Method** **For Year Ended December 31, 2009**					
		Analysis of Changes			
	Dec. 31, 2008		**Debit**		**Credit**	**Dec. 31, 2009**

	Dec. 31, 2008		Debit		Credit	Dec. 31, 2009
Balance Sheet—Debits						
Cash	$ 12,000					$ 17,000
Accounts receivable	40,000	*(b)*	$ 20,000			60,000
Merchandise inventory	70,000	*(c)*	14,000			84,000
Prepaid expenses	4,000	*(d)*	2,000			6,000
Plant assets	210,000	*(k1)*	70,000	*(i)*	$ 30,000	250,000
	$336,000					$417,000
Balance Sheet—Credits						
Accumulated depreciation	$ 48,000	*(i)*	12,000	*(h)*	24,000	$ 60,000
Accounts payable	40,000	*(e)*	5,000			35,000
Interest payable	4,000	*(f)*	1,000			3,000
Income taxes payable	12,000			*(g)*	10,000	22,000
Notes payable	64,000	*(j)*	34,000	*(k2)*	60,000	90,000
Common stock, $5 par value	80,000			*(l)*	15,000	95,000
Retained earnings	88,000	*(m)*	14,000	*(a)*	38,000	112,000
	$336,000					$417,000
Statement of Cash Flows						
Operating activities						
Net income		*(a)*	38,000			
Increase in accounts receivable				*(b)*	20,000	
Increase in merchandise inventory				*(c)*	14,000	
Increase in prepaid expenses				*(d)*	2,000	
Decrease in accounts payable				*(e)*	5,000	
Decrease in interest payable				*(f)*	1,000	
Increase in income taxes payable		*(g)*	10,000			
Depreciation expense		*(h)*	24,000			
Loss on sale of plant assets		*(i)*	6,000			
Gain on retirement of notes				*(j)*	16,000	
Investing activities						
Receipts from sale of plant assets		*(i)*	12,000			
Payment for purchase of plant assets				*(k1)*	10,000	
Financing activities						
Payment to retire notes				*(j)*	18,000	
Receipts from issuing stock		*(l)*	15,000			
Payment of cash dividends				*(m)*	14,000	
Noncash Investing and Financing Activities						
Purchase of plant assets with notes		*(k2)*	60,000	*(k1)*	60,000	
			$337,000		$337,000	

Since adjustments *i, j,* and *k* are more challenging, we show them in the following debit and credit format. These entries are for purposes of our understanding; they are *not* the entries actually made in the journals. Changes in the Cash account are identified as sources or uses of cash.

i.	Loss from sale of plant assets	6,000	
	Accumulated depreciation	12,000	
	Receipt from sale of plant assets **(source of cash)**	12,000	
	Plant assets		30,000
	To describe sale of plant assets.		

[continued on next page]

[continued from previous page]

j.	Notes payable	34,000	
	Payments to retire notes **(use of cash)**		18,000
	Gain on retirement of notes		16,000
	To describe retirement of notes.		
k1.	Plant assets	70,000	
	Payment to purchase plant assets **(use of cash)**		10,000
	Purchase of plant assets financed by notes		60,000
	To describe purchase of plant assets.		
k2.	Purchase of plant assets financed by notes	60,000	
	Notes payable		60,000
	To issue notes for purchase of assets.		

APPENDIX

16B Direct Method of Reporting Operating Cash Flows

P5 Compute cash flows from operating activities using the direct method.

We compute cash flows from operating activities under the direct method by adjusting accrual-based income statement items to the cash basis. The usual approach is to adjust income statement accounts related to operating activities for changes in their related balance sheet accounts as follows:

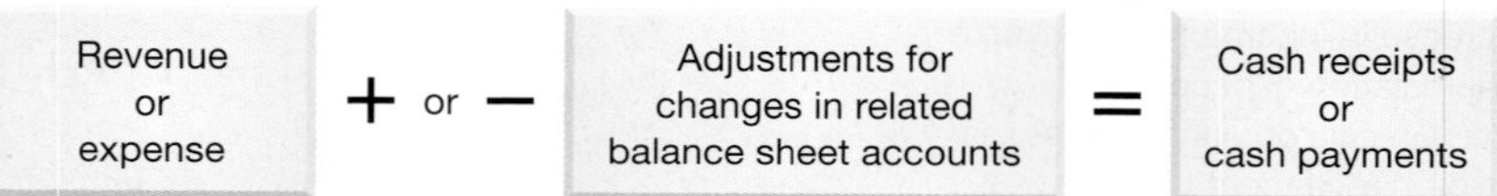

The framework for reporting cash receipts and cash payments for the operating section of the cash flow statement under the direct method is shown in Exhibit 16B.1. We consider cash receipts first and then cash payments.

EXHIBIT 16B.1

Major Classes of Operating Cash Flows

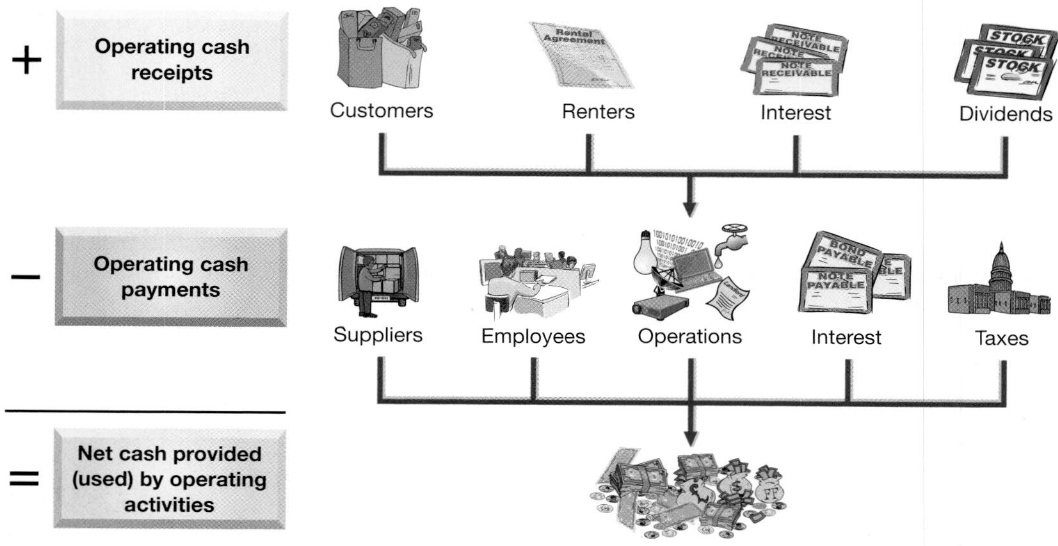

Operating Cash Receipts

A review of Exhibit 16.10 and the additional information reported by Genesis suggests only one potential cash receipt: sales to customers. This section, therefore, starts with sales to customers as reported on the income statement and then adjusts it as necessary to obtain cash received from customers to report on the statement of cash flows.

Cash Received from Customers If all sales are for cash, the amount received from customers equals the sales reported on the income statement. When some or all sales are on account, however, we must adjust the amount of sales for the change in Accounts Receivable. It is often helpful to use *account analysis* to do this. This usually involves setting up a T-account and reconstructing its major entries, with emphasis on cash receipts and payments. To illustrate, we use a T-account that includes accounts receivable balances for Genesis on December 31, 2008 and 2009. The beginning balance is $40,000 and the ending balance is $60,000. Next, the income statement shows sales of $590,000, which we enter on the debit side of this account. We now can reconstruct the Accounts Receivable account to determine the amount of cash received from customers as follows:

Point: An accounts receivable increase implies that cash received from customers is less than sales (the converse is also true).

Accounts Receivable			
Bal., Dec. 31, 2008	40,000		
Sales	590,000	**Cash receipts =**	**570,000**
Bal., Dec. 31, 2009	60,000		

This T-account shows that the Accounts Receivable balance begins at $40,000 and increases to $630,000 from sales of $590,000, yet its ending balance is only $60,000. This implies that cash receipts from customers are $570,000, computed as $40,000 + $590,000 − [?] = $60,000. This computation can be rearranged to express cash received as equal to sales of $590,000 minus a $20,000 increase in accounts receivable. This computation is summarized as a general rule in Exhibit 16B.2. The statement of cash flows in Exhibit 16.7 reports the $570,000 cash received from customers as a cash inflow from operating activities.

Example: If the ending balance of accounts receivable is $20,000 (instead of $60,000), what is cash received from customers? *Answer:* $610,000

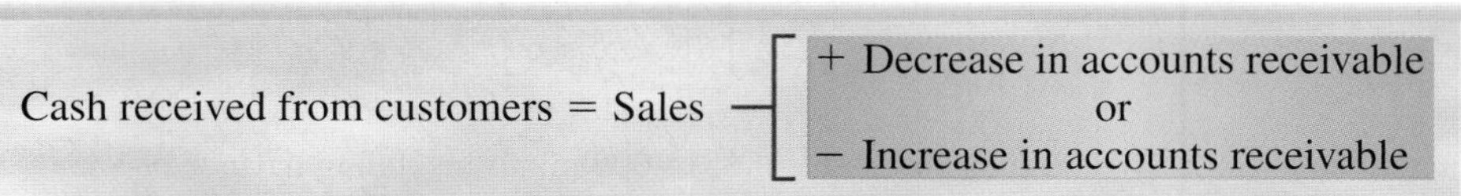

EXHIBIT 16B.2

Formula to Compute Cash Received from Customers—Direct Method

Other Cash Receipts While Genesis's cash receipts are limited to collections from customers, we often see other types of cash receipts, most commonly cash receipts involving rent, interest, and dividends. We compute cash received from these items by subtracting an increase in their respective receivable or adding a decrease. For instance, if rent receivable increases in the period, cash received from renters is less than rent revenue reported on the income statement. If rent receivable decreases, cash received is more than reported rent revenue. The same logic applies to interest and dividends. The formulas for these computations are summarized later in this appendix.

Point: Net income is measured using accrual accounting. Cash flows from operations are measured using cash basis accounting.

Operating Cash Payments

A review of Exhibit 16.10 and the additional Genesis information shows four operating expenses: cost of goods sold; wages and other operating expenses; interest expense; and taxes expense. We analyze each expense to compute its cash amounts for the statement of cash flows. (We then examine depreciation and the other losses and gains.)

Cash Paid for Merchandise We compute cash paid for merchandise by analyzing both cost of goods sold and merchandise inventory. If all merchandise purchases are for cash and the ending balance of Merchandise Inventory is unchanged from the beginning balance, the amount of cash paid for merchandise equals cost of goods sold—an uncommon situation. Instead, there normally is some change in the Merchandise Inventory balance. Also, some or all merchandise purchases are often made on credit, and this yields changes in the Accounts Payable balance. When the balances of both Merchandise Inventory and Accounts Payable change, we must adjust the cost of goods sold for changes in both accounts to compute cash paid for merchandise. This is a two-step adjustment.

First, we use the change in the account balance of Merchandise Inventory, along with the cost of goods sold amount, to compute cost of purchases for the period. An increase in merchandise inventory implies that we bought more than we sold, and we add this inventory increase to cost of goods sold to compute cost of purchases. A decrease in merchandise inventory implies that we bought less than we sold, and we subtract the inventory decrease from cost of goods sold to compute purchases. We illustrate the *first step* by reconstructing the Merchandise Inventory account of Genesis:

Merchandise Inventory			
Bal., Dec. 31, 2008	70,000		
Purchases =	**314,000**	Cost of goods sold	300,000
Bal., Dec. 31, 2009	84,000		

The beginning balance is $70,000, and the ending balance is $84,000. The income statement shows that cost of goods sold is $300,000, which we enter on the credit side of this account. With this information, we determine the amount for cost of purchases to be $314,000. This computation can be rearranged to express cost of purchases as equal to cost of goods sold of $300,000 plus the $14,000 increase in inventory.

The second step uses the change in the balance of Accounts Payable, and the amount of cost of purchases, to compute cash paid for merchandise. A decrease in accounts payable implies that we paid for more goods than we acquired this period, and we would then add the accounts payable decrease to cost of purchases to compute cash paid for merchandise. An increase in accounts payable implies that we paid for less than the amount of goods acquired, and we would subtract the accounts payable increase from purchases to compute cash paid for merchandise. The *second step* is applied to Genesis by reconstructing its Accounts Payable account:

Accounts Payable			
		Bal., Dec. 31, 2008	40,000
Cash payments =	**319,000**	Purchases	314,000
		Bal., Dec. 31, 2009	35,000

Example: If the ending balances of Inventory and Accounts Payable are $60,000 and $50,000, respectively (instead of $84,000 and $35,000), what is cash paid for merchandise? *Answer:* $280,000

Its beginning balance of $40,000 plus purchases of $314,000 minus an ending balance of $35,000 yields cash paid of $319,000 (or $40,000 + $314,000 − [?] = $35,000). Alternatively, we can express cash paid for merchandise as equal to purchases of $314,000 plus the $5,000 decrease in accounts payable. The $319,000 cash paid for merchandise is reported on the statement of cash flows in Exhibit 16.7 as a cash outflow under operating activities.

We summarize this two-step adjustment to cost of goods sold to compute cash paid for merchandise inventory in Exhibit 16B.3.

EXHIBIT 16B.3

Two Steps to Compute Cash Paid for Merchandise—Direct Method

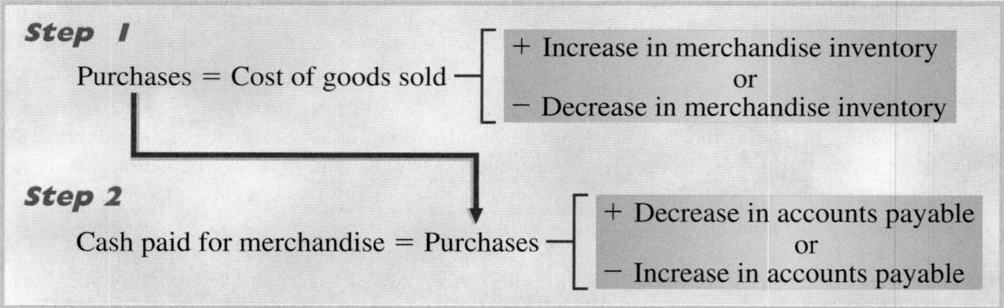

Cash Paid for Wages and Operating Expenses (Excluding Depreciation)

The income statement of Genesis shows wages and other operating expenses of $216,000 (see Exhibit 16.10). To compute cash paid for wages and other operating expenses, we adjust this amount for any changes in their related balance sheet accounts. We begin by looking for any prepaid expenses and accrued liabilities related to wages and other operating expenses in the balance sheets of Genesis in

Exhibit 16.10. The balance sheets show prepaid expenses but no accrued liabilities. Thus, the adjustment is limited to the change in prepaid expenses. The amount of adjustment is computed by assuming that all cash paid for wages and other operating expenses is initially debited to Prepaid Expenses. This assumption allows us to reconstruct the Prepaid Expenses account:

Prepaid Expenses			
Bal., Dec. 31, 2008	4,000		
Cash payments =	**218,000**	Wages and other operating exp.	216,000
Bal., Dec. 31, 2009	6,000		

Prepaid Expenses increase by $2,000 in the period, meaning that cash paid for wages and other operating expenses exceeds the reported expense by $2,000. Alternatively, we can express cash paid for wages and other operating expenses as equal to its reported expenses of $216,000 plus the $2,000 increase in prepaid expenses.[1]

Point: A decrease in prepaid expenses implies that reported expenses include an amount(s) that did not require a cash outflow in the period.

Exhibit 16B.4 summarizes the adjustments to wages (including salaries) and other operating expenses. The Genesis balance sheet did not report accrued liabilities, but we include them in the formula to explain the adjustment to cash when they do exist. A decrease in accrued liabilities implies that we paid cash for more goods or services than received this period, so we add the decrease in accrued liabilities to the expense amount to obtain cash paid for these goods or services. An increase in accrued liabilities implies that we paid cash for less than what was acquired, so we subtract this increase in accrued liabilities from the expense amount to get cash paid.

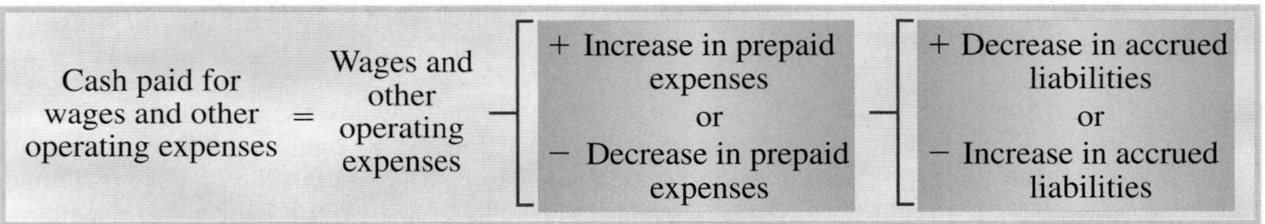

EXHIBIT 16B.4

Formula to Compute Cash Paid for Wages and Operating Expenses—Direct Method

Cash Paid for Interest and Income Taxes Computing operating cash flows for interest and taxes is similar to that for operating expenses. Both require adjustments to their amounts reported on the income statement for changes in their related balance sheet accounts. We begin with the Genesis income statement showing interest expense of $7,000 and income taxes expense of $15,000. To compute the cash paid, we adjust interest expense for the change in interest payable and then the income taxes expense for the change in income taxes payable. These computations involve reconstructing both liability accounts:

Interest Payable			
		Bal., Dec. 31, 2008	4,000
Cash paid for interest =	**8,000**	Interest expense	7,000
		Bal., Dec. 31, 2009	3,000

Income Taxes Payable			
		Bal., Dec. 31, 2008	12,000
Cash paid for taxes =	**5,000**	Income taxes expense	15,000
		Bal., Dec. 31, 2009	22,000

These accounts reveal cash paid for interest of $8,000 and cash paid for income taxes of $5,000. The formulas to compute these amounts are in Exhibit 16B.5. Both of these cash payments are reported as operating cash outflows on the statement of cash flows in Exhibit 16.7.

[1] The assumption that all cash payments for wages and operating expenses are initially debited to Prepaid Expenses is not necessary for our analysis to hold. If cash payments are debited directly to the expense account, the total amount of cash paid for wages and other operating expenses still equals the $216,000 expense plus the $2,000 increase in Prepaid Expenses (which arise from end-of-period adjusting entries).

EXHIBIT 16B.5

Formulas to Compute Cash Paid for Both Interest and Taxes—Direct Method

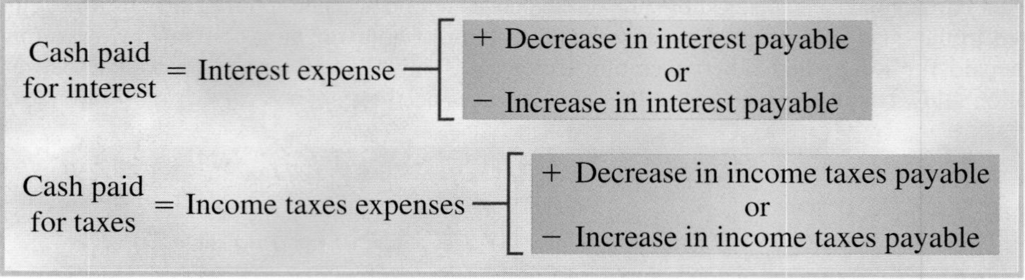

Cash paid for interest = Interest expense { + Decrease in interest payable or − Increase in interest payable }

Cash paid for taxes = Income taxes expenses { + Decrease in income taxes payable or − Increase in income taxes payable }

Analysis of Additional Expenses, Gains, and Losses Genesis has three additional items reported on its income statement: depreciation, loss on sale of assets, and gain on retirement of debt. We must consider each for its potential cash effects.

Depreciation Expense Depreciation expense is $24,000. It is often called a *noncash expense* because depreciation has no cash flows. Depreciation expense is an allocation of an asset's depreciable cost. The cash outflow with a plant asset is reported as part of investing activities when it is paid for. Thus, depreciation expense is *never* reported on a statement of cash flows using the direct method; nor is depletion or amortization expense.

Loss on Sale of Assets Sales of assets frequently result in gains and losses reported as part of net income, but the amount of recorded gain or loss does *not* reflect any cash flows in these transactions. Asset sales result in cash inflow equal to the cash amount received, regardless of whether the asset was sold at a gain or a loss. This cash inflow is reported under investing activities. Thus, the loss or gain on a sale of assets is *never* reported on a statement of cash flows using the direct method.

Point: The direct method is usually viewed as *user friendly* because less accounting knowledge is required to understand and use it.

Gain on Retirement of Debt Retirement of debt usually yields a gain or loss reported as part of net income, but that gain or loss does *not* reflect cash flow in this transaction. Debt retirement results in cash outflow equal to the cash paid to settle the debt, regardless of whether the debt is retired at a gain or loss. This cash outflow is reported under financing activities; the loss or gain from retirement of debt is *never* reported on a statement of cash flows using the direct method.

Summary of Adjustments for Direct Method

Exhibit 16B.6 summarizes common adjustments for net income to yield net cash provided (used) by operating activities under the direct method.

EXHIBIT 16B.6

Summary of Selected Adjustments for Direct Method

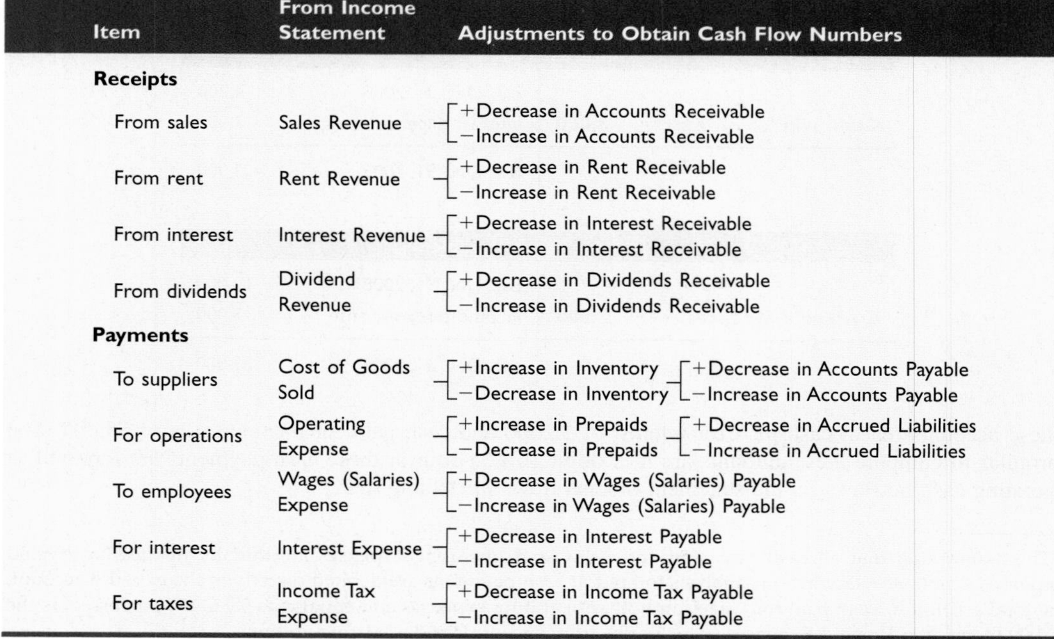

Item	From Income Statement	Adjustments to Obtain Cash Flow Numbers	
Receipts			
From sales	Sales Revenue	+Decrease in Accounts Receivable −Increase in Accounts Receivable	
From rent	Rent Revenue	+Decrease in Rent Receivable −Increase in Rent Receivable	
From interest	Interest Revenue	+Decrease in Interest Receivable −Increase in Interest Receivable	
From dividends	Dividend Revenue	+Decrease in Dividends Receivable −Increase in Dividends Receivable	
Payments			
To suppliers	Cost of Goods Sold	+Increase in Inventory −Decrease in Inventory	+Decrease in Accounts Payable −Increase in Accounts Payable
For operations	Operating Expense	+Increase in Prepaids −Decrease in Prepaids	+Decrease in Accrued Liabilities −Increase in Accrued Liabilities
To employees	Wages (Salaries) Expense	+Decrease in Wages (Salaries) Payable −Increase in Wages (Salaries) Payable	
For interest	Interest Expense	+Decrease in Interest Payable −Increase in Interest Payable	
For taxes	Income Tax Expense	+Decrease in Income Tax Payable −Increase in Income Tax Payable	

Direct Method Format of Operating Activities Section

Exhibit 16.7 shows the Genesis statement of cash flows using the direct method. Major items of cash inflows and cash outflows are listed separately in the operating activities section. The format requires that operating cash outflows be subtracted from operating cash inflows to get net cash provided (used) by operating activities. The FASB recommends that the operating activities section of the statement of cash flows be reported using the direct method, which is considered more useful to financial statement users. *However, the FASB requires a reconciliation of net income to net cash provided (used) by operating activities when the direct method is used* (which can be reported in the notes). This reconciliation is similar to preparation of the operating activities section of the statement of cash flows using the indirect method.

Point: Some preparers argue that it is easier to prepare a statement of cash flows using the indirect method. This likely explains its greater frequency in financial statements.

Decision Insight

IFRSs Like U.S. GAAP, IFRSs allow cash flows from operating activities to be reported using either the indirect method or the direct method.

Quick Check

Answers—p. 658

8. Net sales in a period are $590,000, beginning accounts receivable are $120,000, and ending accounts receivable are $90,000. What cash amount is collected from customers in the period?
9. The Merchandise Inventory account balance decreases in the period from a beginning balance of $32,000 to an ending balance of $28,000. Cost of goods sold for the period is $168,000. If the Accounts Payable balance increases $2,400 in the period, what is the cash amount paid for merchandise inventory?
10. This period's wages and other operating expenses total $112,000. Beginning-of-period prepaid expenses totaled $1,200, and its ending balance is $4,200. There were no beginning-of-period accrued liabilities, but end-of-period wages payable equal $5,600. How much cash is paid for wages and other operating expenses?

Summary

C1 **Explain the purpose and importance of cash flow information.** The main purpose of the statement of cash flows is to report the major cash receipts and cash payments for a period. This includes identifying cash flows as relating to operating, investing, or financing activities. Most business decisions involve evaluating activities that provide or use cash.

C2 **Distinguish between operating, investing, and financing activities.** Operating activities include transactions and events that determine net income. Investing activities include transactions and events that mainly affect long-term assets. Financing activities include transactions and events that mainly affect long-term liabilities and equity.

C3 **Identify and disclose noncash investing and financing activities.** Noncash investing and financing activities must be disclosed in either a note or a separate schedule to the statement of cash flows. Examples are the retirement of debt by issuing equity and the exchange of a note payable for plant assets.

C4 **Describe the format of the statement of cash flows.** The statement of cash flows separates cash receipts and cash payments into operating, investing, or financing activities.

A1 **Analyze the statement of cash flows.** To understand and predict cash flows, users stress identification of the sources and uses of cash flows by operating, investing, and financing activities. Emphasis is on operating cash flows since they derive from continuing operations.

A2 **Compute and apply the cash flow on total assets ratio.** The cash flow on total assets ratio is defined as operating cash flows divided by average total assets. Analysis of current and past values for this ratio can reflect a company's ability to yield regular and positive cash flows. It is also viewed as a measure of earnings quality.

P1 **Prepare a statement of cash flows.** Preparation of a statement of cash flows involves five steps: (1) Compute the net increase or decrease in cash; (2) compute net cash provided or used by operating activities (*using either the direct or indirect method*); (3) compute net cash provided or used by investing activities; (4) compute net cash provided or used by financing activities; and (5) report the beginning and ending cash balance and prove that it is explained by net cash flows. Noncash investing and financing activities are also disclosed.

P2 **Compute cash flows from operating activities using the indirect method.** The indirect method for reporting net cash provided or used by operating activities starts with net income and then adjusts it for three items: (1) changes in noncash current assets and current liabilities related to operating activities, (2) revenues and expenses not providing or using cash, and (3) gains and losses from investing and financing activities.

P3 **Determine cash flows from both investing and financing activities.** Cash flows from both investing and financing activities are determined by identifying the cash flow effects of transactions and events affecting each balance sheet account related to these activities. All cash flows from these activities are identified when we can explain changes in these accounts from the beginning to the end of the period.

P4[A] Illustrate use of a spreadsheet to prepare a statement of cash flows. A spreadsheet is a useful tool in preparing a statement of cash flows. Six key steps (see Appendix 16A) are applied when using the spreadsheet to prepare the statement.

P5[B] Compute cash flows from operating activities using the direct method. The direct method for reporting net cash provided or used by operating activities lists major operating cash inflows less cash outflows to yield net cash inflow or outflow from operations.

Guidance Answers to **Decision Maker**

Entrepreneur Several factors might explain an increase in net cash flows when a net loss is reported, including (1) early recognition of expenses relative to revenues generated (such as research and development), (2) cash advances on long-term sales contracts not yet recognized in income, (3) issuances of debt or equity for cash to finance expansion, (4) cash sale of assets, (5) delay of cash payments, and (6) cash prepayment on sales. Analysis needs to focus on the components of both the net loss and the net cash flows and their implications for future performance.

Reporter Your initial reaction based on the company's $600,000 loss with a $550,000 decrease in net cash flows is not positive. However, closer scrutiny reveals a more positive picture of this company's performance. Cash flow from operating activities is $650,000, computed as [?] − $850,000 − $350,000 = $(550,000). You also note that net income *before* the extraordinary loss is $330,000, computed as [?] − $930,000 = $(600,000).

Guidance Answers to **Quick Checks**

1. No to both. The statement of cash flows reports changes in the sum of cash plus cash equivalents. It does not report transfers between cash and cash equivalents.

2. The three categories of cash inflows and outflows are operating activities, investing activities, and financing activities.

3. **a.** Investing **c.** Financing **e.** Operating
b. Operating **d.** Operating **f.** Financing

4. $74,900 + $4,600 − $11,700 − $1,000 + $3,400 = $70,200

5. Expenses such as depreciation and amortization do not require current cash outflows. Therefore, adding these expenses back to net income eliminates these noncash items from the net income number, converting it to a cash basis.

6. A gain on the sale of plant assets is subtracted from net income because a sale of plant assets is not an operating activity; it is an investing activity for the amount of cash received from its sale. Also, such a gain yields no cash effects.

7. $80,000 − $30,000 − $10,000 = $40,000 cash receipt. The $40,000 cash receipt is reported as an investing activity.

8. $590,000 + ($120,000 − $90,000) = $620,000

9. $168,000 − ($32,000 − $28,000) − $2,400 = $161,600

10. $112,000 + ($4,200 − $1,200) − $5,600 = $109,400

Key Terms

mhhe.com/wildFAP19e

Key Terms are available at the book's Website for learning and testing in an online Flashcard Format.

Cash flow on total assets (p. 646)
Direct method (p. 634)
Financing activities (p. 630)
Indirect method (p. 634)
Investing activities (p. 630)
Operating activities (p. 629)
Statement of cash flows (p. 628)

Multiple Choice Quiz

Answers on p. 677

mhhe.com/wildFAP19e

Additional Quiz Questions are available at the book's Website.

Quiz16

1. A company uses the indirect method to determine its cash flows from operating activities. Use the following information to determine its net cash provided or used by operating activities.

Net income	$15,200
Depreciation expense	10,000
Cash payment on note payable	8,000
Gain on sale of land	3,000
Increase in inventory	1,500
Increase in accounts payable	2,850

a. $23,550 used by operating activities
b. $23,550 provided by operating activities
c. $15,550 provided by operating activities
d. $42,400 provided by operating activities
e. $20,850 provided by operating activities

2. A machine with a cost of $175,000 and accumulated depreciation of $94,000 is sold for $87,000 cash. The amount reported as a source of cash under cash flows from investing activities is
a. $81,000.
b. $6,000.
c. $87,000.
d. Zero; this is a financing activity.
e. Zero; this is an operating activity.

3. A company settles a long-term note payable plus interest by paying $68,000 cash toward the principal amount and $5,440 cash for interest. The amount reported as a use of cash under cash flows from financing activities is
a. Zero; this is an investing activity.
b. Zero; this is an operating activity.
c. $73,440.
d. $68,000.
e. $5,440.

4. The following information is available regarding a company's annual salaries and wages. What amount of cash is paid for salaries and wages?

Salaries and wages expense	$255,000
Salaries and wages payable, prior year-end	8,200
Salaries and wages payable, current year-end	10,900

a. $252,300
b. $257,700
c. $255,000
d. $274,100
e. $235,900

5. The following information is available for a company. What amount of cash is paid for merchandise for the current year?

Cost of goods sold	$545,000
Merchandise inventory, prior year-end	105,000
Merchandise inventory, current year-end	112,000
Accounts payable, prior year-end	98,500
Accounts payable, current year-end	101,300

a. $545,000
b. $554,800
c. $540,800
d. $535,200
e. $549,200

Superscript letter ^{A}(^{B}) denotes assignments based on Appendix 16A (16B).

Discussion Questions

1. What is the reporting purpose of the statement of cash flows? Identify at least two questions that this statement can answer.
2. Describe the direct method of reporting cash flows from operating activities.
3. When a statement of cash flows is prepared using the direct method, what are some of the operating cash flows?
4. Describe the indirect method of reporting cash flows from operating activities.
5. What are some investing activities reported on the statement of cash flows?
6. What are some financing activities reported on the statement of cash flows?
7. Where on the statement of cash flows is the payment of cash dividends reported?
8. Assume that a company purchases land for $100,000, paying $20,000 cash and borrowing the remainder with a long-term note payable. How should this transaction be reported on a statement of cash flows?
9. On June 3, a company borrows $50,000 cash by giving its bank a 160-day, interest-bearing note. On the statement of cash flows, where should this be reported?
10. If a company reports positive net income for the year, can it also show a net cash outflow from operating activities? Explain.
11. Is depreciation a source of cash flow?
12. Refer to **Best Buy**'s statement of cash flows in Appendix A. (*a*) Which method is used to compute its net cash provided by operating activities? (*b*) While its balance sheet shows an increase in receivables from fiscal years 2006 to 2007, why is this increase in receivables subtracted when computing net cash provided by operating activities for the year ended March 3, 2007?
13. Refer to **Circuit City**'s statement of cash flows in Appendix A. What are its cash flows from financing activities for the year ended February 28, 2007? List items and amounts.
14. Refer to **RadioShack**'s statement of cash flows in Appendix A. List its cash flows from operating activities, investing activities, and financing activities.
15. Refer to **Apple**'s statement of cash flows in Appendix A. What investing activities result in cash outflows for the year ended September 30, 2006? List items and amounts.

Denotes Discussion Questions that involve decision making.

Available with McGraw-Hill's Homework Manager McGraw-Hill's HOMEWORK MANAGER®

QUICK STUDY

QS 16-1
Statement of cash flows
C1 C2 C3

The statement of cash flows is one of the four primary financial statements.

1. Describe the content and layout of a statement of cash flows, including its three sections.
2. List at least three transactions classified as investing activities in a statement of cash flows.
3. List at least three transactions classified as financing activities in a statement of cash flows.
4. List at least three transactions classified as significant noncash financing and investing activities in the statement of cash flows.

QS 16-2
Transaction classification by activity
C2 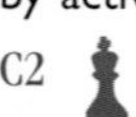

Classify the following cash flows as operating, investing, or financing activities.

1. Sold long-term investments for cash.
2. Received cash payments from customers.
3. Paid cash for wages and salaries.
4. Purchased inventories for cash.
5. Paid cash dividends.
6. Issued common stock for cash.
7. Received cash interest on a note.
8. Paid cash interest on outstanding notes.
9. Received cash from sale of land at a loss.
10. Paid cash for property taxes on building.

QS 16-3
Computing cash from operations (indirect)
P2

Use the following information to determine this company's cash flows from operating activities using the indirect method.

LOLLAND COMPANY
Selected Balance Sheet Information
December 31, 2009 and 2008

	2009	2008
Current assets		
Cash	$169,300	$ 53,600
Accounts receivable	50,000	64,000
Inventory	120,000	108,200
Current liabilities		
Accounts payable	60,800	51,400
Income taxes payable	4,100	4,400

LOLLAND COMPANY
Income Statement
For Year Ended December 31, 2009

Sales		$1,030,000
Cost of goods sold		663,200
Gross profit		366,800
Operating expenses		
Depreciation expense	$ 72,000	
Other expenses	243,000	315,000
Income before taxes		51,800
Income taxes expense		15,400
Net income		$ 36,400

QS 16-4
Computing cash from asset sales
P3

The following selected information is from Manning Company's comparative balance sheets.

At December 31	2009	2008
Furniture	$ 264,000	$ 369,000
Accumulated depreciation—Furniture	(174,400)	(221,400)

The income statement reports depreciation expense for the year of $36,000. Also, furniture costing $105,000 was sold for its book value. Compute the cash received from the sale of furniture.

QS 16-5
Computing financing cash flows
P3

The following selected information is from the Tanner Company's comparative balance sheets.

At December 31	2009	2008
Common stock, $10 par value	$ 210,000	$200,000
Paid-in capital in excess of par	1,134,000	684,000
Retained earnings	627,000	575,000

The company's net income for the year ended December 31, 2009, was $96,000.

1. Compute the cash received from the sale of its common stock during 2009.
2. Compute the cash paid for dividends during 2009.

Use the following balance sheets and income statement to answer QS 16-6 through QS 16-11.

QS 16-6
Computing cash from operations (indirect) P2 

Use the indirect method to prepare the cash provided or used from operating activities section only of the statement of cash flows for this company.

AMMONS, INC.
Comparative Balance Sheets
December 31, 2009

	2009	2008
Assets		
Cash	$189,600	$ 48,000
Accounts receivable, net	82,000	102,000
Inventory	171,600	191,600
Prepaid expenses	10,800	8,400
Furniture	218,000	238,000
Accum. depreciation—Furniture	(34,000)	(18,000)
Total assets	$638,000	$570,000
Liabilities and Equity		
Accounts payable	$ 30,000	$ 42,000
Wages payable	18,000	10,000
Income taxes payable	2,800	5,200
Notes payable (long-term)	58,000	138,000
Common stock, $5 par value	458,000	358,000
Retained earnings	71,200	16,800
Total liabilities and equity	$638,000	$570,000

AMMONS, INC.
Income Statement
For Year Ended December 31, 2009

Sales		$976,000
Cost of goods sold		628,000
Gross profit		348,000
Operating expenses		
Depreciation expense	$ 75,200	
Other expenses	178,200	253,400
Income before taxes		94,600
Income taxes expense		34,600
Net income		$ 60,000

QS 16-7
Computing cash from asset sales P3 

Refer to the data in QS 16-6.
Furniture costing $110,000 is sold at its book value in 2009. Acquisitions of furniture total $90,000 cash, on which no depreciation is necessary because it is acquired at year-end. What is the cash inflow related to the sale of furniture?

QS 16-8
Computing financing cash outflows P3

Refer to the data in QS 16-6.

1. Assume that all common stock is issued for cash. What amount of cash dividends is paid during 2009?
2. Assume that no additional notes payable are issued in 2009. What cash amount is paid to reduce the notes payable balance in 2009?

QS 16-9[B]
Computing cash received from customers P5

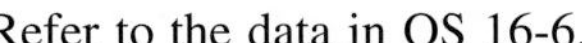
Refer to the data in QS 16-6.

1. How much cash is received from sales to customers for year 2009?
2. What is the net increase or decrease in cash for year 2009?

QS 16-10[B]
Computing operating cash outflows P5

Refer to the data in QS 16-6.

1. How much cash is paid to acquire merchandise inventory during year 2009?
2. How much cash is paid for operating expenses during year 2009?

QS 16-11[B]
Computing cash from operations (direct) P5 

Refer to the data in QS 16-6.
Use the direct method to prepare the cash provided or used from operating activities section only of the statement of cash flows for this company.

QS 16-12
Analyses of sources and uses of cash A1 A2 

Financial data from three competitors in the same industry follow.

1. Which of the three competitors is in the strongest position as shown by its statement of cash flows?
2. Analyze and compare the strength of Peña's cash flow on total assets ratio to that of Garcia.

($ thousands)	Peña	Garcia	Piniella
Cash provided (used) by operating activities	$ 140,000	$ 120,000	$ (48,000)
Cash provided (used) by investing activities			
Proceeds from sale of operating assets			52,000
Purchase of operating assets	(56,000)	(68,000)	
Cash provided (used) by financing activities			
Proceeds from issuance of debt			46,000
Repayment of debt	(12,000)		
Net increase (decrease) in cash	$ 72,000	$ 52,000	$ 50,000
Average total assets	$ 1,580,000	$ 1,250,000	$ 600,000

QS 16-13^A
Noncash accounts on a spreadsheet P4

When a spreadsheet for a statement of cash flows is prepared, all changes in noncash balance sheet accounts are fully explained on the spreadsheet. Explain how these noncash balance sheet accounts are used to fully account for cash flows on a spreadsheet.

QS 16-14
Computing cash flows from operations (indirect)
P2

For each of the following separate cases, compute cash flows from operations. The list includes all balance sheet accounts related to operating activities.

	Case A	Case B	Case C
Net income	$ 8,000	$200,000	$144,000
Depreciation expense	60,000	16,000	48,000
Accounts receivable increase (decrease)	80,000	40,000	(8,000)
Inventory increase (decrease)	(40,000)	(20,000)	21,000
Accounts payable increase (decrease)	48,000	(44,000)	28,000
Accrued liabilities increase (decrease)	(88,000)	24,000	(16,000)

QS 16-15
Computing cash flows from investing
P3

Compute cash flows from investing activities using the following company information.

Sale of short-term investments	$12,000
Cash collections from customers	32,000
Purchase of used equipment	10,000
Depreciation expense	4,000

QS 16-16
Computing cash flows from financing
P3

Compute cash flows from financing activities using the following company information.

Additional short-term borrowings	$40,000
Purchase of short-term investments	10,000
Cash dividends paid	32,000
Interest paid	16,000

Available with McGraw-Hill's Homework Manager McGraw-Hill's HOMEWORK MANAGER®

EXERCISES

Exercise 16-1
Cash flow from operations (indirect)
P2

Hehman Company reports net income of $530,000 for the year ended December 31, 2009. It also reports $95,400 depreciation expense and a $4,000 gain on the sale of machinery. Its comparative balance sheets reveal a $42,400 increase in accounts receivable, $21,730 increase in accounts payable, $11,660 decrease in prepaid expenses, and $16,430 decrease in wages payable.

Required

Prepare only the operating activities section of the statement of cash flows for 2009 using the *indirect method.*

Exercise 16-2
Cash flow classification (indirect) C2 C3 P2 

The following transactions and events occurred during the year. Assuming that this company uses the *indirect method* to report cash provided by operating activities, indicate where each item would appear on its statement of cash flows by placing an *x* in the appropriate column.

	Statement of Cash Flows			Noncash Investing and Financing Activities	Not Reported on Statement or in Notes
	Operating Activities	Investing Activities	Financing Activities		
a. Paid cash to purchase inventory.	____	____	____	____	____
b. Purchased land by issuing common stock.	____	____	____	____	____
c. Accounts receivable decreased in the year.	____	____	____	____	____
d. Sold equipment for cash, yielding a loss.	____	____	____	____	____
e. Recorded depreciation expense.	____	____	____	____	____
f. Income taxes payable increased in the year.	____	____	____	____	____
g. Declared and paid a cash dividend.	____	____	____	____	____
h. Accounts payable decreased in the year	____	____	____	____	____
i. Paid cash to settle notes payable	____	____	____	____	____
j. Prepaid expenses increased in the year	____	____	____	____	____

Exercise 16-3B
Cash flow classification (direct) C2 C3 P5

The following transactions and events occurred during the year. Assuming that this company uses the *direct method* to report cash provided by operating activities, indicate where each item would appear on the statement of cash flows by placing an *x* in the appropriate column.

	Statement of Cash Flows			Noncash Investing and Financing Activities	Not Reported on Statement or in Notes
	Operating Activities	Investing Activities	Financing Activities		
a. Retired long-term notes payable by issuing common stock	____	____	____	____	____
b. Recorded depreciation expense.	____	____	____	____	____
c. Paid cash dividend that was declared in a prior period.	____	____	____	____	____
d. Sold inventory for cash.	____	____	____	____	____
e. Borrowed cash from bank by signing a nine-month note payable.	____	____	____	____	____
f. Paid cash to purchase a patent.	____	____	____	____	____
g. Accepted six-month note receivable in exchange for plant assets.	____	____	____	____	____
h. Paid cash toward accounts payable.	____	____	____	____	____
i. Collected cash from sales.	____	____	____	____	____
j. Paid cash to acquire treasury stock.	____	____	____	____	____

Exercise 16-4
Cash flows from operating activities (indirect)
P2

Zander Company's calendar-year 2009 income statement shows the following: Net Income, $395,000; Depreciation Expense, $48,980; Amortization Expense, $9,875; Gain on Sale of Plant Assets, $4,900. An examination of the company's current assets and current liabilities reveals the following changes (all from operating activities): Accounts Receivable decrease, $7,600; Merchandise Inventory decrease, $22,040; Prepaid Expenses increase, $2,000; Accounts Payable decrease, $5,000; Other Payables increase, $760. Use the *indirect method* to compute cash flow from operating activities.

Exercise 16-5[B]
Computation of cash flows (direct)
P5

For each of the following three separate cases, use the information provided about the calendar-year 2010 operations of Kowa Company to compute the required cash flow information.

Case A: Compute cash received from customers:	
Sales	$590,000
Accounts receivable, December 31, 2009	38,000
Accounts receivable, December 31, 2010	52,440
Case B: Compute cash paid for rent:	
Rent expense	$117,400
Rent payable, December 31, 2009	6,700
Rent payable, December 31, 2010	5,561
Case C: Compute cash paid for merchandise:	
Cost of goods sold	$651,000
Merchandise inventory, December 31, 2009	201,810
Accounts payable, December 31, 2009	84,760
Merchandise inventory, December 31, 2010	165,484
Accounts payable, December 31, 2010	105,102

Exercise 16-6
Cash flows from operating activities (indirect)
P2

Use the following income statement and information about changes in noncash current assets and current liabilities to prepare only the cash flows from operating activities section of the statement of cash flows using the *indirect* method.

SEYMOUR COMPANY
Income Statement
For Year Ended December 31, 2009

Sales		$2,175,000
Cost of goods sold		1,065,750
Gross profit		1,109,250
Operating expenses		
Salaries expense	$297,975	
Depreciation expense	52,200	
Rent expense	58,725	
Amortization expenses—Patents	6,525	
Utilities expense	23,925	439,350
		669,900
Gain on sale of equipment		8,700
Net income		$ 678,600

Changes in current asset and current liability accounts for the year that relate to operations follow.

Accounts receivable	$45,300 increase	Accounts payable	$10,075 decrease
Merchandise inventory	35,150 increase	Salaries payable	4,750 decrease

Exercise 16-7[B]
Cash flows from operating activities (direct) P5

Refer to the information about Seymour Company in Exercise 16-6.
Use the *direct method* to prepare only the cash provided or used by operating activities section of the statement of cash flows for this company.

Exercise 16-8
Cash flows from investing activities
P3

Use the following information to determine this company's cash flows from investing activities.

a. Equipment with a book value of $72,500 and an original cost of $158,000 was sold at a loss of $22,000.
b. Paid $95,000 cash for a new truck.
c. Sold land costing $315,000 for $400,000 cash, yielding a gain of $15,000.
d. Long-term investments in stock were sold for $94,700 cash, yielding a gain of $5,750.

Exercise 16-9
Cash flows from financing activities
P3

Use the following information to determine this company's cash flows from financing activities.

a. Net income was $53,000.
b. Issued common stock for $75,000 cash.
c. Paid cash dividend of $13,000.
d. Paid $90,000 cash to settle a note payable at its $90,000 maturity value.
e. Paid $18,000 cash to acquire its treasury stock.
f. Purchased equipment for $67,000 cash.

Exercise 16-10
Preparation of statement of cash flows (indirect)
C2 A2 P1 P2 P3 

Use the following financial statements and additional information to (1) prepare a statement of cash flows for the year ended June 30, 2009, using the *indirect method,* and (2) compute the company's cash flow on total assets ratio for its fiscal year 2009.

BOULWARE INC.
Comparative Balance Sheets
June 30, 2009 and 2008

	2009	2008
Assets		
Cash	$ 84,663	$ 49,494
Accounts receivable, net	65,720	56,952
Inventory	62,620	106,107
Prepaid expenses	4,960	5,763
Equipment	118,387	131,532
Accum. depreciation—Equipment	(26,350)	(10,848)
Total assets	$310,000	$339,000
Liabilities and Equity		
Accounts payable	$ 24,490	$ 35,256
Wages payable	6,510	17,628
Income taxes payable	2,170	4,068
Notes payable (long term)	31,953	76,953
Common stock, $5 par value	208,000	158,000
Retained earnings	36,877	47,095
Total liabilities and equity	$310,000	$339,000

BOULWARE INC.
Income Statement
For Year Ended June 30, 2009

Sales		$976,600
Cost of goods sold		625,024
Gross profit		351,576
Operating expenses		
Depreciation expense	$ 88,753	
Other expenses	101,879	
Total operating expenses		190,632
		160,944
Other gains (losses)		
Gain on sale of equipment		3,125
Income before taxes		164,069
Income taxes expense		56,604
Net income		$107,465

Additional Information

a. A $45,000 note payable is retired at its carrying (book) value in exchange for cash.
b. The only changes affecting retained earnings are net income and cash dividends paid.
c. New equipment is acquired for $85,000 cash.
d. Received cash for the sale of equipment that had cost $98,145, yielding a $3,125 gain.
e. Prepaid Expenses and Wages Payable relate to Other Expenses on the income statement.
f. All purchases and sales of merchandise inventory are on credit.

Exercise 16-11[B]
Preparation of statement of cash flows (direct) C2 P1 P3 P5

Refer to the data in Exercise 16-10.
Using the *direct method,* prepare the statement of cash flows for the year ended June 30, 2009.

Exercise 16-12[B]
Preparation of statement of cash flows (direct) and supporting note
C2 C3 C4 P1

Use the following information about the cash flows of Valencia Company to prepare a complete statement of cash flows (*direct method*) for the year ended December 31, 2009. Use a note disclosure for any noncash investing and financing activities.

Cash and cash equivalents balance, December 31, 2008	$ 43,000
Cash and cash equivalents balance, December 31, 2009	120,916
Cash received as interest	4,300
Cash paid for salaries	124,700
Bonds payable retired by issuing common stock (no gain or loss on retirement)	180,000
Cash paid to retire long-term notes payable	215,000
Cash received from sale of equipment	105,350
Cash received in exchange for six-month note payable	43,000
Land purchased by issuing long-term note payable	104,400
Cash paid for store equipment	40,850
Cash dividends paid	25,800
Cash paid for other expenses	68,800
Cash received from customers	834,200
Cash paid for merchandise	433,784

Exercise 16-13[B]
Preparation of statement of cash flows (direct) from Cash T-account
C2 A1 P1 P3 P5 

The following summarized Cash T-account reflects the total debits and total credits to the Cash account of Clarett Corporation for calendar year 2009.

(1) Use this information to prepare a complete statement of cash flows for year 2009. The cash provided or used by operating activities should be reported using the *direct method.*

(2) Refer to the statement of cash flows prepared for part 1 to answer the following questions *a* through *d*: (*a*) Which section—operating, investing, or financing—shows the largest cash (i) inflow and (ii) outflow? (*b*) What is the largest individual item among the investing cash outflows? (*c*) Are the cash proceeds larger from issuing notes or issuing stock? (*d*) Does the company have a net cash inflow or outflow from borrowing activities?

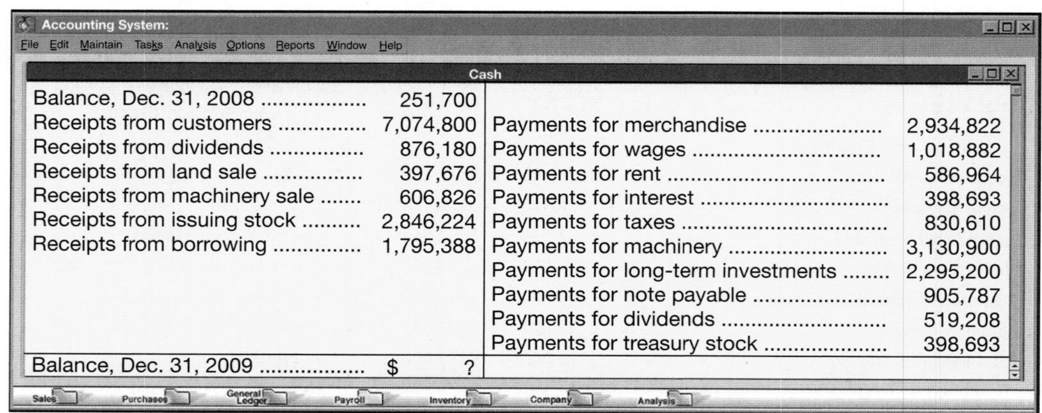

Cash			
Balance, Dec. 31, 2008	251,700		
Receipts from customers	7,074,800	Payments for merchandise	2,934,822
Receipts from dividends	876,180	Payments for wages	1,018,882
Receipts from land sale	397,676	Payments for rent	586,964
Receipts from machinery sale	606,826	Payments for interest	398,693
Receipts from issuing stock	2,846,224	Payments for taxes	830,610
Receipts from borrowing	1,795,388	Payments for machinery	3,130,900
		Payments for long-term investments	2,295,200
		Payments for note payable	905,787
		Payments for dividends	519,208
		Payments for treasury stock	398,693
Balance, Dec. 31, 2009	$?		

Exercise 16-14
Reporting cash flows from operations (indirect)
C4 P2

Woodlock Company reports the following information for its recent calendar year.

Sales	$80,000
Expenses	
Cost of goods sold	50,000
Salaries expense	12,000
Depreciation expense	6,000
Net income	$12,000
Accounts receivable increase	$ 5,000
Inventory decrease	8,000
Salaries payable increase	500

Required

Prepare the operating activities section of the statement of cash flows for Woodlock Company using the indirect method.

Exercise 16-15
Reporting and interpreting cash flows from operations (indirect)
C4 P2 

Portland Company disclosed the following information for its recent calendar year.

Revenues	$200,000
Expenses	
Salaries expense	168,000
Utilities expense	28,000
Depreciation expense	29,200
Other expenses	6,800
Net loss	$ (32,000)
Accounts receivable decrease	$ 48,000
Purchased a machine	20,000
Salaries payable increase	36,000
Other accrued liabilities decrease	16,000

Required

1. Prepare the operating activities section of the statement of cash flows using the indirect method.
2. What were the major reasons that this company was able to report a net loss but positive cash flow from operations?
3. Of the potential causes of differences between cash flow from operations and net income, which are the most important to investors?

McGraw-Hill's HOMEWORK MANAGER® Available with McGraw-Hill's Homework Manager

PROBLEM SET A

Problem 16-1A
Statement of cash flows (indirect method)
C2 C3 A1 P1 P2 P3

Georgia Company, a merchandiser, recently completed its calendar-year 2009 operations. For the year, (1) all sales are credit sales, (2) all credits to Accounts Receivable reflect cash receipts from customers, (3) all purchases of inventory are on credit, (4) all debits to Accounts Payable reflect cash payments for inventory, and (5) Other Expenses are paid in advance and are initially debited to Prepaid Expenses. The company's balance sheets and income statement follow.

GEORGIA COMPANY
Comparative Balance Sheets
December 31, 2009 and 2008

	2009	2008
Assets		
Cash	$ 49,800	$ 73,500
Accounts receivable	65,840	56,000
Merchandise inventory	277,000	252,000
Prepaid expenses	1,000	1,500
Equipment	158,500	107,500
Accum. depreciation—Equipment	(43,000)	(52,000)
Total assets	$509,140	$438,500
Liabilities and Equity		
Accounts payable	$ 42,965	$113,000
Short-term notes payable	10,000	7,000
Long-term notes payable	70,000	48,000
Common stock, $5 par value	162,750	151,000
Paid-in capital in excess of par, common stock	35,250	0
Retained earnings	188,175	119,500
Total liabilities and equity	$509,140	$438,500

GEORGIA COMPANY
Income Statement
For Year Ended December 31, 2009

Sales		$584,500
Cost of goods sold		281,000
Gross profit		303,500
Operating expenses		
Depreciation expense	$ 20,000	
Other expenses	132,800	152,800
Other gains (losses)		
Loss on sale of equipment		5,875
Income before taxes		144,825
Income taxes expense		24,250
Net income		$120,575

Additional Information on Year 2009 Transactions

a. The loss on the cash sale of equipment was $5,875 (details in *b*).

b. Sold equipment costing $46,500, with accumulated depreciation of $29,000, for $11,625 cash.

c. Purchased equipment costing $97,500 by paying $35,000 cash and signing a long-term note payable for the balance.

d. Borrowed $3,000 cash by signing a short-term note payable.

e. Paid $40,500 cash to reduce the long-term notes payable.

f. Issued 2,350 shares of common stock for $20 cash per share.

g. Declared and paid cash dividends of $51,900.

Required

Check Cash from operating activities, $42,075

1. Prepare a complete statement of cash flows; report its operating activities using the *indirect method.* Disclose any noncash investing and financing activities in a note.

Analysis Component

2. Analyze and discuss the statement of cash flows prepared in part 1, giving special attention to the wisdom of the cash dividend payment.

Problem 16-2A[A]
Cash flows spreadsheet (indirect method)
P1 P2 P3 P4

Refer to the information reported about Georgia Company in Problem 16-1A.

Required

Prepare a complete statement of cash flows using a spreadsheet as in Exhibit 16A.1; report its operating activities using the indirect method. Identify the debits and credits in the Analysis of Changes columns with letters that correspond to the following list of transactions and events.

a. Net income was $120,575.

b. Accounts receivable increased.

c. Merchandise inventory increased.

d. Prepaid expenses decreased.

e. Accounts payable decreased.

f. Depreciation expense was $20,000.

g. Sold equipment costing $46,500, with accumulated depreciation of $29,000, for $11,625 cash. This yielded a loss of $5,875.

h. Purchased equipment costing $97,500 by paying $35,000 cash and **(i.)** by signing a long-term note payable for the balance.

j. Borrowed $3,000 cash by signing a short-term note payable.

k. Paid $40,500 cash to reduce the long-term notes payable.

Check Analysis of Changes column totals, $594,850

l. Issued 2,350 shares of common stock for $20 cash per share.

m. Declared and paid cash dividends of $51,900.

Problem 16-3A[B]
Statement of cash flows (direct method) C3 P1 P3 P5

Check Cash used in financing activities, $(42,400)

Refer to Georgia Company's financial statements and related information in Problem 16-1A.

Required

Prepare a complete statement of cash flows; report its operating activities according to the *direct method.* Disclose any noncash investing and financing activities in a note.

Problem 16-4A
Statement of cash flows (indirect method) C3 P1 P2 P3

mhhe.com/wildFAP19e

Memphis Corp., a merchandiser, recently completed its 2009 operations. For the year, (1) all sales are credit sales, (2) all credits to Accounts Receivable reflect cash receipts from customers, (3) all purchases of inventory are on credit, (4) all debits to Accounts Payable reflect cash payments for inventory, (5) Other Expenses are all cash expenses, and (6) any change in Income Taxes Payable reflects the accrual and cash payment of taxes. The company's balance sheets and income statement follow.

MEMPHIS CORPORATION
Comparative Balance Sheets
December 31, 2009 and 2008

	2009	2008
Assets		
Cash	$ 165,000	$137,000
Accounts receivable	82,000	74,000
Merchandise inventory	620,000	525,000
Equipment	345,000	240,000
Accum. depreciation—Equipment	(159,000)	(102,000)
Total assets	$1,053,000	$874,000
Liabilities and Equity		
Accounts payable	$ 160,000	$ 96,000
Income taxes payable	22,000	19,000
Common stock, $2 par value	588,000	560,000
Paid-in capital in excess of par value, common stock	201,000	159,000
Retained earnings	82,000	40,000
Total liabilities and equity	$1,053,000	$874,000

MEMPHIS CORPORATION
Income Statement
For Year Ended December 31, 2009

Sales		$1,794,000
Cost of goods sold		1,088,000
Gross profit		706,000
Operating expenses		
Depreciation expense	$ 57,000	
Other expenses	500,000	557,000
Income before taxes		149,000
Income taxes expense		22,000
Net income		$ 127,000

Additional Information on Year 2009 Transactions

a. Purchased equipment for $105,000 cash.
b. Issued 14,000 shares of common stock for $5 cash per share.
c. Declared and paid $85,000 in cash dividends.

Required

Prepare a complete statement of cash flows; report its cash inflows and cash outflows from operating activities according to the *indirect method.*

Check Cash from operating activities, $148,000

Problem 16-5A[A]
Cash flows spreadsheet (indirect method)
P1 P2 P3 P4

mhhe.com/wildFAP19e

Refer to the information reported about Memphis Corporation in Problem 16-4A.

Required

Prepare a complete statement of cash flows using a spreadsheet as in Exhibit 16A.1; report operating activities under the indirect method. Identify the debits and credits in the Analysis of Changes columns with letters that correspond to the following list of transactions and events.

a. Net income was $127,000.
b. Accounts receivable increased.
c. Merchandise inventory increased.
d. Accounts payable increased.
e. Income taxes payable increased.
f. Depreciation expense was $57,000.
g. Purchased equipment for $105,000 cash.
h. Issued 14,000 shares at $5 cash per share.
i. Declared and paid $85,000 of cash dividends.

Check Analysis of Changes column totals, $614,000

Problem 16-6A[B]
Statement of cash flows (direct method) P1 P3 P5

mhhe.com/wildFAP19e

Refer to Memphis Corporation's financial statements and related information in Problem 16-4A.

Required

Prepare a complete statement of cash flows; report its cash flows from operating activities according to the *direct method.*

Check Cash used in financing activities, $(15,000)

Problem 16-7A
Computing cash flows from operations (indirect)
C4 P2

Rawling Company's 2009 income statement and selected balance sheet data at December 31, 2008 and 2009, follow ($ thousands).

RAWLING COMPANY
Selected Balance Sheet Accounts

At December 31	2009	2008
Accounts receivable	$280	$290
Inventory	99	77
Accounts payable	220	230
Salaries payable	44	35
Utilities payable	11	8
Prepaid insurance	13	14
Prepaid rent	11	9

RAWLING COMPANY
Income Statement

Sales revenue	$48,600
Expenses	
Cost of goods sold	21,000
Depreciation expense	6,000
Salaries expense	9,000
Rent expense	4,500
Insurance expense	1,900
Interest expense	1,800
Utilities expense	1,400
Net income	$ 3,000

Required

Prepare the cash flows from operating activities section only of the company's 2009 statement of cash flows using the indirect method.

Check Cash from operating activities, $8,989

Problem 16-8A[B]
Computing cash flows from operations (direct)
C4 P5

Refer to the information in Problem 16-7A.

Required

Prepare the cash flows from operating activities section only of the company's 2009 statement of cash flows using the direct method.

PROBLEM SET B

Problem 16-1B
Statement of cash flows (indirect method)
C2 C3 A1 P1 P2 P3

Wilson Corporation, a merchandiser, recently completed its calendar-year 2009 operations. For the year, (1) all sales are credit sales, (2) all credits to Accounts Receivable reflect cash receipts from customers, (3) all purchases of inventory are on credit, (4) all debits to Accounts Payable reflect cash payments for inventory, and (5) Other Expenses are paid in advance and are initially debited to Prepaid Expenses. The company's balance sheets and income statement follow.

WILSON CORPORATION
Comparative Balance Sheets
December 31, 2009 and 2008

	2009	2008
Assets		
Cash	$ 49,400	$ 74,000
Accounts receivable	65,830	55,000
Merchandise inventory	277,000	252,000
Prepaid expenses	1,250	1,600
Equipment	158,500	107,500
Accum. depreciation—Equipment	(36,625)	(46,000)
Total assets	$515,355	$444,100
Liabilities and Equity		
Accounts payable	$ 55,380	$112,000
Short-term notes payable	9,000	7,000
Long-term notes payable	70,000	48,250
Common stock, $5 par	162,500	150,750
Paid-in capital in excess of par, common stock	35,250	0
Retained earnings	183,225	126,100
Total liabilities and equity	$515,355	$444,100

WILSON CORPORATION
Income Statement
For Year Ended December 31, 2009

Sales		$585,000
Cost of goods sold		285,000
Gross profit		300,000
Operating expenses		
Depreciation expense	$ 20,000	
Other expenses	134,000	
Total operating expenses		154,000
		146,000
Other gains (losses)		
Loss on sale of equipment		5,625
Income before taxes		140,375
Income taxes expense		24,250
Net income		$116,125

Additional Information on Year 2009 Transactions

a. The loss on the cash sale of equipment was $5,625 (details in *b*).

b. Sold equipment costing $46,500, with accumulated depreciation of $29,375, for $11,500 cash.

c. Purchased equipment costing $97,500 by paying $25,000 cash and signing a long-term note payable for the balance.

d. Borrowed $2,000 cash by signing a short-term note payable.

e. Paid $50,750 cash to reduce the long-term notes payable.

f. Issued 2,350 shares of common stock for $20 cash per share.

g. Declared and paid cash dividends of $59,000.

Required

1. Prepare a complete statement of cash flows; report its operating activities using the *indirect method.* Disclose any noncash investing and financing activities in a note.

Check Cash from operating activities, $49,650

Analysis Component

2. Analyze and discuss the statement of cash flows prepared in part 1, giving special attention to the wisdom of the cash dividend payment.

Problem 16-2B[A]
Cash flows spreadsheet (indirect method)
P1 P2 P3 P4

Refer to the information reported about Wilson Corporation in Problem 16-1B.

Required

Prepare a complete statement of cash flows using a spreadsheet as in Exhibit 16A.1; report its operating activities using the *indirect method.* Identify the debits and credits in the Analysis of Changes columns with letters that correspond to the following list of transactions and events.

a. Net income was $116,125.

b. Accounts receivable increased.

c. Merchandise inventory increased.

d. Prepaid expenses decreased.

e. Accounts payable decreased.

f. Depreciation expense was $20,000.

g. Sold equipment costing $46,500, with accumulated depreciation of $29,375, for $11,500 cash. This yielded a loss of $5,625.

h. Purchased equipment costing $97,500 by paying $25,000 cash and **(i.)** by signing a long-term note payable for the balance.

j. Borrowed $2,000 cash by signing a short-term note payable.

k. Paid $50,750 cash to reduce the long-term notes payable.

l. Issued 2,350 shares of common stock for $20 cash per share.

m. Declared and paid cash dividends of $59,000.

Check Analysis of Changes column totals, $604,175

Problem 16-3B[B]
Statement of cash flows (direct method) C3 P1 P3 P5

Refer to Wilson Corporation's financial statements and related information in Problem 16-1B.

Required

Prepare a complete statement of cash flows; report its operating activities according to the *direct method.* Disclose any noncash investing and financing activities in a note.

Check Cash used in financing activities, $(60,750)

Problem 16-4B
Statement of cash flows (indirect method)
C3 P1 P2 P3

Prius Company, a merchandiser, recently completed its 2009 operations. For the year, (1) all sales are credit sales, (2) all credits to Accounts Receivable reflect cash receipts from customers, (3) all purchases of inventory are on credit, (4) all debits to Accounts Payable reflect cash payments for inventory, (5) Other Expenses are cash expenses, and (6) any change in Income Taxes Payable reflects the accrual and cash payment of taxes. The company's balance sheets and income statement follow.

PRIUS COMPANY
Comparative Balance Sheets
December 31, 2009 and 2008

	2009	2008
Assets		
Cash	$ 164,000	$ 131,000
Accounts receivable	82,000	70,000
Merchandise inventory	605,000	515,000
Equipment	350,000	276,000
Accum. depreciation—Equipment	(157,000)	(102,000)
Total assets	$1,044,000	$ 890,000
Liabilities and Equity		
Accounts payable	$ 173,000	$ 119,000
Income taxes payable	20,000	17,000
Common stock, $2 par value	580,000	560,000
Paid-in capital in excess of par, common stock	193,000	163,000
Retained earnings	78,000	31,000
Total liabilities and equity	$1,044,000	$ 890,000

PRIUS COMPANY
Income Statement
For Year Ended December 31, 2009

Sales		$1,792,000
Cost of goods sold		1,087,000
Gross profit		705,000
Operating expenses		
Depreciation expense	$ 55,000	
Other expenses	494,000	549,000
Income before taxes		156,000
Income taxes expense		24,000
Net income		$ 132,000

Additional Information on Year 2009 Transactions

a. Purchased equipment for $74,000 cash.
b. Issued 10,000 shares of common stock for $5 cash per share.
c. Declared and paid $85,000 of cash dividends.

Required

Prepare a complete statement of cash flows; report its cash inflows and cash outflows from operating activities according to the *indirect method*.

Check Cash from operating activities, $142,000

Problem 16-5B[A]
Cash flows spreadsheet (indirect method)
P1 P2 P3 P4

Refer to the information reported about Prius Company in Problem 16-4B.

Required

Prepare a complete statement of cash flows using a spreadsheet as in Exhibit 16A.1; report operating activities under the *indirect method.* Identify the debits and credits in the Analysis of Changes columns with letters that correspond to the following list of transactions and events.

a. Net income was $132,000.
b. Accounts receivable increased.
c. Merchandise inventory increased.
d. Accounts payable increased.
e. Income taxes payable increased.
f. Depreciation expense was $55,000.
g. Purchased equipment for $74,000 cash.
h. Issued 10,000 shares at $5 cash per share.
i. Declared and paid $85,000 of cash dividends.

Check Analysis of Changes column totals, $555,000

Problem 16-6B[B]
Statement of cash flows (direct method) P1 P3 P5

Refer to Prius Company's financial statements and related information in Problem 16-4B.

Required

Prepare a complete statement of cash flows; report its cash flows from operating activities according to the *direct method.*

Check Cash used by financing activities, $(35,000)

Problem 16-7B
Computing cash flows from operations (indirect)
C4 P2

Kodak Company's 2009 income statement and selected balance sheet data at December 31, 2008 and 2009, follow ($ thousands).

KODAK COMPANY
Income Statement

Sales revenue	$312,000
Expenses	
Cost of goods sold	144,000
Depreciation expense	64,000
Salaries expense	40,000
Rent expense	10,000
Insurance expense	5,200
Interest expense	4,800
Utilities expense	4,000
Net income	$ 40,000

KODAK COMPANY
Selected Balance Sheet Accounts

At December 31	2009	2008
Accounts receivable	$720	$600
Inventory	172	196
Accounts payable	480	520
Salaries payable	180	120
Utilities payable	40	0
Prepaid insurance	28	36
Prepaid rent	20	40

Required

Prepare the cash flows from operating activities section only of the company's 2009 statement of cash flows using the indirect method.

Check Cash from operating activities, $103,992

Problem 16-8B[B]
Computing cash flows from operations (direct)
C4 P5

Refer to the information in Problem 16-7B.

Required

Prepare the cash flows from operating activities section only of the company's 2009 statement of cash flows using the direct method.

SERIAL PROBLEM

Success Systems

(This serial problem began in Chapter 1 and continues through most of the book. If previous chapter segments were not completed, the serial problem can begin at this point. It is helpful, but not necessary, to use the Working Papers that accompany the book.)

SP 16 Adriana Lopez, owner of Success Systems, decides to prepare a statement of cash flows for her business. (Although the serial problem allowed for various ownership changes in earlier chapters, we will prepare the statement of cash flows using the following financial data.)

SUCCESS SYSTEMS
Income Statement
For Three Months Ended March 31, 2010

Computer services revenue		$25,160
Net sales		18,693
Total revenue		43,853
Cost of goods sold	$14,052	
Depreciation expense—Office equipment	400	
Depreciation expense—Computer equipment	1,250	
Wages expense	3,250	
Insurance expense	555	
Rent expense	2,475	
Computer supplies expense	1,305	
Advertising expense	600	
Mileage expense	320	
Repairs expense—Computer	960	
Total expenses		25,167
Net income		$18,686

SUCCESS SYSTEMS
Comparative Balance Sheets
December 31, 2009, and March 31, 2010

	2010	2009
Assets		
Cash	$ 77,845	$58,160
Accounts receivable	22,720	5,668
Merchandise inventory	704	0
Computer supplies	2,005	580
Prepaid insurance	1,110	1,665
Prepaid rent	825	825
Office equipment	8,000	8,000
Accumulated depreciation—Office equipment	(800)	(400)
Computer equipment	20,000	20,000
Accumulated depreciation—Computer equipment	(2,500)	(1,250)
Total assets	$129,909	$93,248
Liabilities and Equity		
Accounts payable	$ 0	$ 1,100
Wages payable	875	500
Unearned computer service revenue	0	1,500
Common stock	108,000	83,000
Retained earnings	21,034	7,148
Total liabilities and equity	$129,909	$93,248

Required

Prepare a statement of cash flows for Success Systems using the *indirect method* for the three months ended March 31, 2010. Recall that the owner Adriana Lopez contributed $25,000 to the business in exchange for additional stock in the first quarter of 2010 and has received $4,800 in cash dividends.

Check Cash flows used by operations: $(515)

BEYOND THE NUMBERS

REPORTING IN ACTION

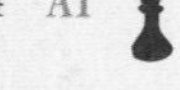

C4 A1

BTN 16-1 Refer to Best Buy's financial statements in Appendix A to answer the following.

1. Is Best Buy's statement of cash flows prepared under the direct method or the indirect method? How do you know?
2. For each fiscal year 2007, 2006, and 2005, is the amount of cash provided by operating activities more or less than the cash paid for dividends?
3. What is the largest amount in reconciling the difference between net income and cash flow from operating activities in 2007? In 2006? In 2005?
4. Identify the largest cash flows for investing and for financing activities in 2007 and in 2006.

Fast Forward

5. Obtain Best Buy's financial statements for a fiscal year ending after March 3, 2007, from either its Website (BestBuy.com) or the SEC's EDGAR database (www.sec.gov). Since March 3, 2007, what are Best Buy's largest cash outflows and cash inflows in the investing and in the financing sections of its statement of cash flow?

BTN 16-2 Key figures for Best Buy, Circuit City, and RadioShack follow.

COMPARATIVE ANALYSIS

A1 A2 

($ millions)	Best Buy			Circuit City			RadioShack		
	Current Year	1 Year Prior	2 Years Prior	Current Year	1 Year Prior	2 Years Prior	Current Year	1 Year Prior	2 Years Prior
Operating cash flows	$ 1,762	$ 1,740	$ 1,981	$ 316	$ 365	$ 389	$ 315	$ 363	$ 353
Total assets	13,570	11,864	10,294	4,007	4,069	3,840	2,070	2,205	2,517

Required

1. Compute the recent two years' cash flow on total assets ratios for Best Buy, Circuit City, and RadioShack.
2. What does the cash flow on total assets ratio measure?
3. Which company has the highest cash flow on total assets ratio for the periods shown?
4. Does the cash flow on total assets ratio reflect on the quality of earnings? Explain.

BTN 16-3 Kaelyn Gish is preparing for a meeting with her banker. Her business is finishing its fourth year of operations. In the first year, it had negative cash flows from operations. In the second and third years, cash flows from operations were positive. However, inventory costs rose significantly in year 4, and cash flows from operations will probably be down 25%. Gish wants to secure a line of credit from her banker as a financing buffer. From experience, she knows the banker will scrutinize operating cash flows for years 1 through 4 and will want a projected number for year 5. Gish knows that a steady progression upward in operating cash flows for years 1 through 4 will help her case. She decides to use her discretion as owner and considers several business actions that will turn her operating cash flow in year 4 from a decrease to an increase.

ETHICS CHALLENGE

C1 C2 A1 

Required

1. Identify two business actions Gish might take to improve cash flows from operations.
2. Comment on the ethics and possible consequences of Gish's decision to pursue these actions.

BTN 16-4 Your friend, Hanna Willard, recently completed the second year of her business and just received annual financial statements from her accountant. Willard finds the income statement and balance sheet informative but does not understand the statement of cash flows. She says the first section is especially confusing because it contains a lot of additions and subtractions that do not make sense to her. Willard adds, "The income statement tells me the business is more profitable than last year and that's most important. If I want to know how cash changes, I can look at comparative balance sheets."

COMMUNICATING IN PRACTICE

C1 C4 

Required

Write a half-page memorandum to your friend explaining the purpose of the statement of cash flows. Speculate as to why the first section is so confusing and how it might be rectified.

BTN 16-5 Access the April 19, 2007, filing of the 10-K report (for fiscal year ending February 3, 2007) of J. Crew Group, Inc., at www.SEC.gov.

TAKING IT TO THE NET

A1 

Required

1. Does J. Crew use the direct or indirect method to construct its consolidated statement of cash flows?
2. For the fiscal year ended February 3, 2007, what is the largest item in reconciling the net income to net cash provided by operating activities?
3. In the recent three years, has the company been more successful in generating operating cash flows or in generating net income? Identify the figures to support the answer.
4. In the year ended February 3, 2007, what was the largest cash outflow for investing activities and for financing activities?
5. What item(s) does J. Crew report as supplementary cash flow information?
6. Does J. Crew report any noncash financing activities for fiscal year 2007? Identify them, if any.

TEAMWORK IN ACTION

C1 C4 A1 P2 P5

BTN 16-6 Team members are to coordinate and independently answer one question within each of the following three sections. Team members should then report to the team and confirm or correct teammates' answers.

1. Answer *one* of the following questions about the statement of cash flows.
   a. What are this statement's reporting objectives?
   b. What two methods are used to prepare it? Identify similarities and differences between them.
   c. What steps are followed to prepare the statement?
   d. What types of analyses are often made from this statement's information?

2. Identify and explain the adjustment from net income to obtain cash flows from operating activities using the indirect method for *one* of the following items.
   a. Noncash operating revenues and expenses.
   b. Nonoperating gains and losses.
   c. Increases and decreases in noncash current assets.
   d. Increases and decreases in current liabilities.

3.[B] Identify and explain the formula for computing cash flows from operating activities using the direct method for *one* of the following items.
   a. Cash receipts from sales to customers.
   b. Cash paid for merchandise inventory.
   c. Cash paid for wages and operating expenses.
   d. Cash paid for interest and taxes.

Note: For teams of more than four, some pairing within teams is necessary. Use as an in-class activity or as an assignment. If used in class, specify a time limit on each part. Conclude with reports to the entire class, using team rotation. Each team can prepare responses on a transparency.

ENTREPRENEURIAL DECISION

C1 A1

BTN 16-7 Review the chapter's opener involving **Jungle Jim's International Market**.

Required

1. In a business such as Jungle Jim's, monitoring cash flow is always a priority. Even though Jungle Jim's now has thousands in annual sales and earns a positive net income, explain how cash flow can lag behind earnings.

2. Jungle Jim's is a closely held corporation. What are potential sources of financing for its future expansion?

C2 A1

BTN 16-8 Jenna and Matt Wilder are completing their second year operating Mountain High, a downhill ski area and resort. Mountain High reports a net loss of $(10,000) for its second year, which includes an $85,000 extraordinary loss from fire. This past year also involved major purchases of plant assets for renovation and expansion, yielding a year-end total asset amount of $800,000. Mountain High's net cash outflow for its second year is $(5,000); a summarized version of its statement of cash flows follows:

Net cash flow provided by operating activities	$295,000
Net cash flow used by investing activities	(310,000)
Net cash flow provided by financing activities	10,000

Required

Write a one-page memorandum to the Wilders evaluating Mountain High's current performance and assessing its future. Give special emphasis to cash flow data and their interpretation.

HITTING THE ROAD

C1

BTN 16-9 Visit **The Motley Fool**'s Website (**Fool.com**). Click on the sidebar link titled *Fool's School* (or *Fool.com/School*). Identify and select the link *How to Value Stocks*.

Required

1. Click on *Introduction to Valuation,* and then *Cash-Flow-Based Valuations.* How does the Fool's school define cash flow? What is the school's reasoning for this definition?

2. Per the school's instruction, why do analysts focus on earnings before interest and taxes (EBIT)?

3. Visit other links at this Website that interest you such as "How to Read a Balance Sheet," or find out what the "Fool's Ratio" is. Write a half-page report on what you find.

BTN 16-10 Key comparative information for **DSG international plc** (DSGiplc.com) follows.

(£ millions)	Current Year	1 Year Prior	2 Years Prior
Operating cash flows	£ 207	£ 338	£ 375
Total assets	3,977	4,120	4,104

GLOBAL DECISION

C1 C2 C4

DSG

Required

1. Compute the recent two years' cash flow on total assets ratio for DSG.
2. How does DSG's ratio compare to Best Buy's, Circuit City's, and RadioShack's ratios from BTN 16-2?

ANSWERS TO MULTIPLE CHOICE QUIZ

1. b;

Net income	$15,200
Depreciation expense	10,000
Gain on sale of land	(3,000)
Increase in inventory	(1,500)
Increase in accounts payable	2,850
Net cash provided by operations	$23,550

2. c; cash received from sale of machine is reported as an investing activity.
3. d; FASB requires cash interest paid to be reported under operating.
4. a; Cash paid for salaries and wages = $255,000 + $8,200 − $10,900 = $252,300
5. e; Increase in inventory = $112,000 − $105,000 = $7,000
   Increase in accounts payable = $101,300 − $98,500 = $2,800
   Cash paid for merchandise = $545,000 + $7,000 − $2,800 = $549,200

A Look Back

Chapter 16 focused on reporting and analyzing cash inflows and cash outflows. We explained how to prepare, analyze, and interpret the statement of cash flows.

A Look at This Chapter

This chapter emphasizes the analysis and interpretation of financial statement information. We learn to apply horizontal, vertical, and ratio analyses to better understand company performance and financial condition.

A Look Ahead

Chapter 18 introduces us to managerial accounting. We discuss its purposes, concepts, and roles in helping managers gather and organize information for decisions. We also explain basic management principles.

Chapter 17

Analysis of Financial Statements

Learning Objectives

CAP

Conceptual

C1 Explain the purpose of analysis. *(p. 680)*

C2 Identify the building blocks of analysis. *(p. 681)*

C3 Describe standards for comparisons in analysis. *(p. 682)*

C4 Identify the tools of analysis. *(p. 682)*

Analytical

A1 Summarize and report results of analysis. *(p. 700)*

A2 *Appendix 17A*—Explain the form and assess the content of a complete income statement. *(p. 703)*

LP17

Procedural

P1 Explain and apply methods of horizontal analysis. *(p. 682)*

P2 Describe and apply methods of vertical analysis. *(p. 687)*

P3 Define and apply ratio analysis. *(p. 690)*

Decision Feature

Motley Fool

"What goes on at The Motley Fool . . . is similar to what goes on in a library"
—Tom Gardner (David Gardner on left)

ALEXANDRIA, VA—In Shakespeare's Elizabethan comedy *As You Like It,* only the fool could speak truthfully to the King without getting his head lopped off. Inspired by Shakespeare's stage character, Tom and David Gardner vowed to become modern-day fools who tell it like it is. With under $10,000 in start-up money, the brothers launched **The Motley Fool (Fool.com).** And befitting of a Shakespearean play, the two say they are "dedicated to educating, amusing, and enriching individuals in search of the truth."

The Gardners do not fear the wrath of any King, real or fictional. They are intent on exposing the truth, as they see it, "that the financial world preys on ignorance and fear." As Tom explains, "There is such a great need in the general populace for financial information." Who can argue, given their brilliant success through practically every medium; including their Website, radio shows, newspaper columns, online store, investment newsletters, and global expansion.

Despite the brothers' best efforts, however, ordinary people still do not fully use information contained in financial statements. For instance, discussions keep appearing on The Motley Fool's online bulletin board that can be easily resolved using reliable and available accounting data. So, it would seem that the Fools must continue their work of "educating and enriching" individuals.

Resembling The Motley Fools' objectives, this chapter introduces horizontal and vertical analyses—tools used to reveal crucial trends and insights from financial information. It also expands on ratio analysis, which gives insight into a company's financial condition and performance. By arming ourselves with the information contained in this chapter and the investment advice of The Motley Fool, *we* can be sure to not play the fool in today's financial world.

[Sources: *Motley Fool Website,* January 2009; *Entrepreneur,* July 1997; *What to Do with Your Money Now,* June 2002; *USA Weekend,* July 2004; *Washington Post,* November 2007; *Money after 40,* April 2007]

Chapter Preview

This chapter shows how we use financial statements to evaluate a company's financial performance and condition. We explain financial statement analysis, its basic building blocks, the information available, standards for comparisons, and tools of analysis. Three major analysis tools are presented: horizontal analysis, vertical analysis, and ratio analysis. We apply each of these tools using **Best Buy**'s financial statements, and we introduce comparative analysis using **Circuit City** and **RadioShack**. This chapter expands and organizes the ratio analyses introduced at the end of each chapter.

Analysis of Financial Statements

Basics of Analysis	Horizontal Analysis	Vertical Analysis	Ratio Analysis
• Purpose • Building blocks • Information • Standards for comparisons • Tools	• Comparative balance sheets • Comparative income statements • Trend analysis	• Common-size balance sheet • Common-size income statement • Common-size graphics	• Liquidity and efficiency • Solvency • Profitability • Market prospects • Ratio summary

Basics of Analysis

Video17.1

Financial statement analysis applies analytical tools to general-purpose financial statements and related data for making business decisions. It involves transforming accounting data into more useful information. Financial statement analysis reduces our reliance on hunches, guesses, and intuition as well as our uncertainty in decision making. It does not lessen the need for expert judgment; instead, it provides us an effective and systematic basis for making business decisions. This section describes the purpose of financial statement analysis, its information sources, the use of comparisons, and some issues in computations.

Purpose of Analysis

C1 Explain the purpose of analysis.

Internal users of accounting information are those involved in strategically managing and operating the company. They include managers, officers, internal auditors, consultants, budget directors, and market researchers. The purpose of financial statement analysis for these users is to provide strategic information to improve company efficiency and effectiveness in providing products and services.

External users of accounting information are *not* directly involved in running the company. They include shareholders, lenders, directors, customers, suppliers, regulators, lawyers, brokers, and the press. External users rely on financial statement analysis to make better and more informed decisions in pursuing their own goals.

Point: Financial statement analysis tools are also used for personal financial investment decisions.

We can identify other uses of financial statement analysis. Shareholders and creditors assess company prospects to make investing and lending decisions. A board of directors analyzes financial statements in monitoring management's decisions. Employees and unions use financial statements in labor negotiations. Suppliers use financial statement information in establishing credit terms. Customers analyze financial statements in deciding whether to establish supply relationships. Public utilities set customer rates by analyzing financial statements. Auditors use financial statements in assessing the "fair presentation" of their clients' financial results. Analyst services such as **Dun & Bradstreet**, **Moody's**, and **Standard & Poor's** use financial statements in making buy-sell recommendations and in setting credit ratings. The common goal of these users is to evaluate company performance and financial condition. This includes evaluating (1) past and current performance, (2) current financial position, and (3) future performance and risk.

Point: Financial statement analysis is a topic on the CPA, CMA, CIA, and CFA exams.

Building Blocks of Analysis

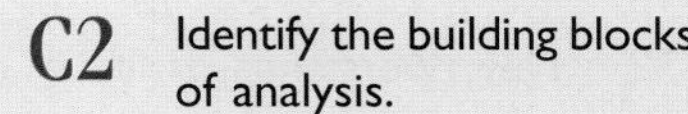

C2 Identify the building blocks of analysis.

Financial statement analysis focuses on one or more elements of a company's financial condition or performance. Our analysis emphasizes four areas of inquiry—with varying degrees of importance. These four areas are described and illustrated in this chapter and are considered the *building blocks* of financial statement analysis:

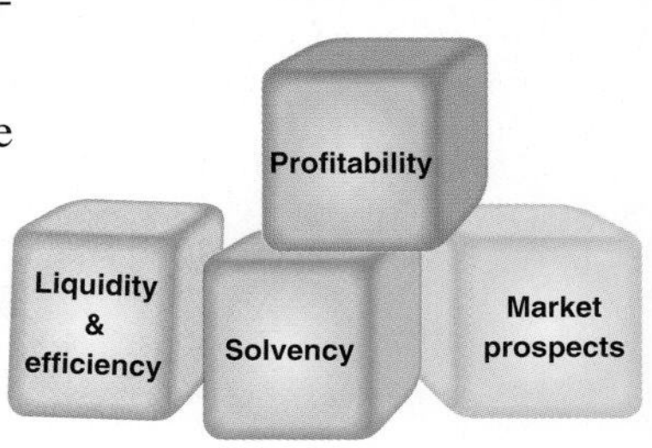

- **Liquidity** and **efficiency**—ability to meet short-term obligations and to efficiently generate revenues.
- **Solvency**—ability to generate future revenues and meet long-term obligations.
- **Profitability**—ability to provide financial rewards sufficient to attract and retain financing.
- **Market prospects**—ability to generate positive market expectations.

Applying the building blocks of financial statement analysis involves determining (1) the objectives of analysis and (2) the relative emphasis among the building blocks. We distinguish among these four building blocks to emphasize the different aspects of a company's financial condition or performance, yet we must remember that these areas of analysis are interrelated. For instance, a company's operating performance is affected by the availability of financing and short-term liquidity conditions. Similarly, a company's credit standing is not limited to satisfactory short-term liquidity but depends also on its profitability and efficiency in using assets. Early in our analysis, we need to determine the relative emphasis of each building block. Emphasis and analysis can later change as a result of evidence collected.

Decision Insight

Chips and Brokers The phrase *blue chips* refers to stock of big, profitable companies. The phrase comes from poker; where the most valuable chips are blue. The term *brokers* refers to those who execute orders to buy or sell stock. The term comes from wine retailers—individuals who broach (break) wine casks.

Information for Analysis

Some users, such as managers and regulatory authorities, are able to receive special financial reports prepared to meet their analysis needs. However, most users must rely on **general-purpose financial statements** that include the (1) income statement, (2) balance sheet, (3) statement of stockholders' equity (or statement of retained earnings), (4) statement of cash flows, and (5) notes to these statements.

Financial reporting refers to the communication of financial information useful for making investment, credit, and other business decisions. Financial reporting includes not only general-purpose financial statements but also information from SEC 10-K or other filings, press releases, shareholders' meetings, forecasts, management letters, auditors' reports, and Webcasts.

Management's Discussion and Analysis (MD&A) is one example of useful information outside traditional financial statements. **Best Buy**'s MD&A (available at BestBuy.com), for example, begins with an overview and strategic initiatives. It then discusses operating results followed by liquidity and capital resources—roughly equivalent to investing and financing. The final few parts discuss special financing arrangements, key accounting policies, interim results, and the next year's outlook. The MD&A is an excellent starting point in understanding a company's business activities.

Decision Insight

Analysis Online Many Websites offer free access and screening of companies by key numbers such as earnings, sales, and book value. For instance, **Standard & Poor's** has information for more than 10,000 stocks (StandardPoor.com).

Standards for Comparisons

C3 Describe standards for comparisons in analysis.

When interpreting measures from financial statement analysis, we need to decide whether the measures indicate good, bad, or average performance. To make such judgments, we need standards (benchmarks) for comparisons that include the following:

- *Intracompany*—The company under analysis can provide standards for comparisons based on its own prior performance and relations between its financial items. Best Buy's current net income, for instance, can be compared with its prior years' net income and in relation to its revenues or total assets.
- *Competitor*—One or more direct competitors of the company being analyzed can provide standards for comparisons. Coca-Cola's profit margin, for instance, can be compared with PepsiCo's profit margin.
- *Industry*—Industry statistics can provide standards of comparisons. Such statistics are available from services such as Dun & Bradstreet, Standard & Poor's, and Moody's.
- *Guidelines (rules of thumb)*—General standards of comparisons can develop from experience. Examples are the 2:1 level for the current ratio or 1:1 level for the acid-test ratio. Guidelines, or rules of thumb, must be carefully applied because context is crucial.

Point: Each chapter's *Reporting in Action* problems engage students in *intracompany* analysis, whereas *Comparative Analysis* problems require competitor analysis (Best Buy vs. Circuit City).

All of these comparison standards are useful when properly applied, yet measures taken from a selected competitor or group of competitors are often best. Intracompany and industry measures are also important. Guidelines or rules of thumb should be applied with care, and then only if they seem reasonable given past experience and industry norms.

Tools of Analysis

C4 Identify the tools of analysis.

Three of the most common tools of financial statement analysis are

1. **Horizontal analysis**—Comparison of a company's financial condition and performance across time.
2. **Vertical analysis**—Comparison of a company's financial condition and performance to a base amount.
3. **Ratio analysis**—Measurement of key relations between financial statement items.

The remainder of this chapter describes these analysis tools and how to apply them.

Quick Check

Answers—p. 706

1. Who are the intended users of general-purpose financial statements?
2. General-purpose financial statements consist of what information?
3. Which of the following is *least* useful as a basis for comparison when analyzing ratios? (*a*) Company results from a different economic setting. (*b*) Standards from past experience. (*c*) Rule-of-thumb standards. (*d*) Industry averages.
4. What is the preferred basis of comparison for ratio analysis?

Horizontal Analysis

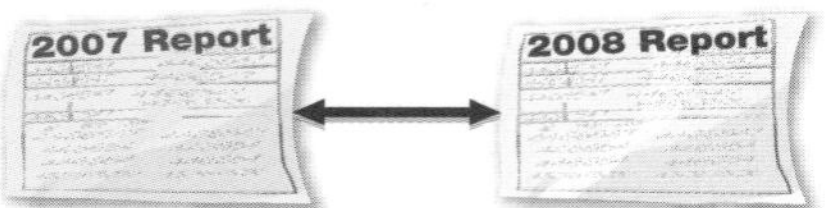

Analysis of any single financial number is of limited value. Instead, much of financial statement analysis involves identifying and describing relations between numbers, groups of numbers, and changes in those numbers. Horizontal analysis refers to examination of financial statement data *across time*. [The term *horizontal analysis* arises from the left-to-right (or right-to-left) movement of our eyes as we review comparative financial statements across time.]

Comparative Statements

P1 Explain and apply methods of horizontal analysis.

Comparing amounts for two or more successive periods often helps in analyzing financial statements. **Comparative financial statements** facilitate this comparison by showing financial amounts in side-by-side columns on a single statement, called a *comparative format*. Using figures from **Best Buy**'s financial statements, this section explains how to compute dollar changes and percent changes for comparative statements.

Video17.1

Computation of Dollar Changes and Percent Changes Comparing financial statements over relatively short time periods—two to three years—is often done by analyzing changes in line items. A change analysis usually includes analyzing absolute dollar amount changes and percent changes. Both analyses are relevant because dollar changes can yield large percent changes inconsistent with their importance. For instance, a 50% change from a base figure of $100 is less important than the same percent change from a base amount of $100,000 in the same statement. Reference to dollar amounts is necessary to retain a proper perspective and to assess the importance of changes. We compute the *dollar change* for a financial statement item as follows:

Example: What is a more significant change, a 70% increase on a $1,000 expense or a 30% increase on a $400,000 expense? *Answer:* The 30% increase.

Dollar change = Analysis period amount − Base period amount

Analysis period is the point or period of time for the financial statements under analysis, and *base period* is the point or period of time for the financial statements used for comparison purposes. The prior year is commonly used as a base period. We compute the *percent change* by dividing the dollar change by the base period amount and then multiplying this quantity by 100 as follows:

$$\textbf{Percent change (\%)} = \frac{\textbf{Analysis period amount} - \textbf{Base period amount}}{\textbf{Base period amount}} \times 100$$

We can always compute a dollar change, but we must be aware of a few rules in working with percent changes. To illustrate, look at four separate cases in this chart:

Case	Analysis Period	Base Period	Change Analysis: Dollar	Change Analysis: Percent
A	$ 1,500	$(4,500)	$ 6,000	—
B	(1,000)	2,000	(3,000)	—
C	8,000	—	8,000	—
D	0	10,000	(10,000)	(100%)

When a negative amount appears in the base period and a positive amount in the analysis period (or vice versa), we cannot compute a meaningful percent change; see cases A and B. Also, when no value is in the base period, no percent change is computable; see case C. Finally, when an item has a value in the base period and zero in the analysis period, the decrease is 100 percent; see case D.

Example: When there is a value in the base period and zero in the analysis period, the decrease is 100%. Why isn't the reverse situation an increase of 100%? *Answer:* A 100% increase of zero is still zero.

It is common when using horizontal analysis to compare amounts to either average or median values from prior periods (average and median values smooth out erratic or unusual fluctuations).[1] We also commonly round percents and ratios to one or two decimal places, but practice on this matter is not uniform. Computations are as detailed as necessary, which is judged by whether rounding potentially affects users' decisions. Computations should not be excessively detailed so that important relations are lost among a mountain of decimal points and digits.

Comparative Balance Sheets Comparative balance sheets consist of balance sheet amounts from two or more balance sheet dates arranged side by side. Its usefulness is often improved by showing each item's dollar change and percent change to highlight large changes.

[1] *Median* is the middle value in a group of numbers. For instance, if five prior years' incomes are (in 000s) $15, $19, $18, $20, and $22, the median value is $19. When there are two middle numbers, we can take their average. For instance, if four prior years' sales are (in 000s) $84, $91, $96, and $93, the median is $92 (computed as the average of $91 and $93).

Point: Spreadsheet programs can help with horizontal, vertical, and ratio analyses, including graphical depictions of financial relations.

Point: Business consultants use comparative statement analysis to provide management advice.

Analysis of comparative financial statements begins by focusing on items that show large dollar or percent changes. We then try to identify the reasons for these changes and, if possible, determine whether they are favorable or unfavorable. We also follow up on items with small changes when we expected the changes to be large.

Exhibit 17.1 shows comparative balance sheets for **Best Buy**. A few items stand out. Many asset categories substantially increase, which is probably not surprising because Best Buy is a growth company. Much of the increase in current assets is from the 20.7% increase in merchandise inventories. The long-term assets of property, equipment, and goodwill also increased. Of course, its sizeable total asset growth of 14.4% must be accompanied by future income to validate Best Buy's growth strategy.

We likewise see substantial increases on the financing side, the most notable ones being accounts payable and long-term debt totaling about $1,112 million. The increase in payables is related to the increase in cash levels, and the increase in debt is partly explained by the increase in long-term assets. Best Buy also reinvested much of its income as reflected in the $1,203 million increase in retained earnings. Again, we must monitor these increases in

EXHIBIT 17.1

Comparative Balance Sheets

BEST BUY
Comparative Balance Sheets
March 3, 2007, and February 25, 2006

(in millions)	2007	2006	Dollar Change	Percent Change
Assets				
Cash and cash equivalents	$ 1,205	$ 748	$ 457	61.1%
Short-term investments	2,588	3,041	(453)	(14.9)
Receivables, net	548	449	99	22.0
Merchandise inventories	4,028	3,338	690	20.7
Other current assets	712	409	303	74.1
Total current assets	9,081	7,985	1,096	13.7
Property and equipment	4,904	4,836	68	1.4
Less accumulated depreciation	1,966	2,124	(158)	(7.4)
Net property and equipment	2,938	2,712	226	8.3
Goodwill	919	557	362	65.0
Tradenames	81	44	37	84.1
Long-term investments	318	218	100	45.9
Other long-term assets	233	348	(115)	(33.0)
Total assets	$13,570	$11,864	$1,706	14.4
Liabilities				
Accounts payable	$ 3,934	$ 3,234	$ 700	21.6%
Unredeemed gift card liabilities	496	469	27	5.8
Accrued compensation and related expenses	332	354	(22)	(6.2)
Accrued liabilities	990	878	112	12.8
Accrued income taxes	489	703	(214)	(30.4)
Short-term debt	41	0	41	—
Current portion of long-term debt	19	418	(399)	(95.5)
Total current liabilities	6,301	6,056	245	4.0
Long-term liabilities	443	373	70	18.8
Long-term debt	590	178	412	231.5
Minority interests	35	0	35	—
Stockholders' Equity				
Common stock	48	49	(1)	(2.0)
Additional paid-in capital	430	643	(213)	(33.1)
Retained earnings	5,507	4,304	1,203	28.0
Accumulated other comprehensive income	216	261	(45)	(17.2)
Total stockholders' equity	6,201	5,257	944	18.0
Total liabilities and stockholders' equity	$13,570	$11,864	$1,706	14.4

investing and financing activities to be sure they are reflected in increased operating performance.

Comparative Income Statements Comparative income statements are prepared similarly to comparative balance sheets. Amounts for two or more periods are placed side by side, with additional columns for dollar and percent changes. Exhibit 17.2 shows Best Buy's comparative income statements.

EXHIBIT 17.2

Comparative Income Statements

BEST BUY
Comparative Income Statements
For Years Ended March 3, 2007, and February 25, 2006

(in millions, except per share data)	2007	2006	Dollar Change	Percent Change
Revenues	$35,934	$30,848	$5,086	16.5%
Cost of goods sold	27,165	23,122	4,043	17.5
Gross profit	8,769	7,726	1,043	13.5
Selling, general, and administrative expenses	6,770	6,082	688	11.3
Operating income	1,999	1,644	355	21.6
Net interest income (expense)	111	77	34	44.2
Gain on investments	20	0	20	—
Earnings from continuing operations before income taxes	2,130	1,721	409	23.8
Income tax expense	752	581	171	29.4
Minority interest in earnings	1	0	1	—
Net earnings	$ 1,377	$ 1,140	$ 237	20.8
Basic earnings per share	$ 2.86	$ 2.33	$ 0.53	22.7
Diluted earnings per share	$ 2.79	$ 2.27	$ 0.52	22.9

Best Buy has substantial revenue growth of 16.5% in 2007. This finding helps support management's growth strategy as reflected in the comparative balance sheets. Best Buy also reveals some ability to control cost of sales and general and administrative expenses, which increased 17.5% and 11.3%, respectively. Best Buy's net income growth of 20.8% on revenue growth of 16.5% is impressive.

Point: Percent change can also be computed by dividing the current period by the prior period and subtracting 1.0. For example, the 16.5% revenue increase of Exhibit 17.2 is computed as: ($35,934/$30,848) − 1.

Trend Analysis

Trend analysis, also called *trend percent analysis* or *index number trend analysis,* is a form of horizontal analysis that can reveal patterns in data across successive periods. It involves computing trend percents for a series of financial numbers and is a variation on the use of percent changes. The difference is that trend analysis does not subtract the base period amount in the numerator. To compute trend percents, we do the following:

1. Select a *base period* and assign each item in the base period a weight of 100%.
2. Express financial numbers as a percent of their base period number.

Specifically, a *trend percent,* also called an *index number,* is computed as follows:

$$\textbf{Trend percent (\%)} = \frac{\textbf{Analysis period amount}}{\textbf{Base period amount}} \times \textbf{100}$$

Point: *Index* refers to the comparison of the analysis period to the base period. Percents determined for each period are called *index numbers.*

To illustrate trend analysis, we use the Best Buy data shown in Exhibit 17.3.

EXHIBIT 17.3

Revenues and Expenses

(in millions)	2007	2006	2005	2004	2003
Revenues	$35,934	$30,848	$27,433	$24,548	$20,943
Cost of goods sold	27,165	23,122	20,938	18,677	15,998
Selling, general & administrative expenses	6,770	6,082	5,053	4,567	3,935

These data are from Best Buy's *Selected Financial Data* section. The base period is 2003 and the trend percent is computed in each subsequent year by dividing that year's amount by its 2003 amount. For instance, the revenue trend percent for 2007 is 171.6%, computed as \$35,934/\$20,943. The trend percents—using the data from Exhibit 17.3—are shown in Exhibit 17.4.

EXHIBIT 17.4

Trend Percents for Revenues and Expenses

	2007	2006	2005	2004	2003
Revenues	171.6%	147.3%	131.0%	117.2%	100.0%
Cost of goods sold	169.8	144.5	130.9	116.7	100.0
Selling, general & administrative expenses	172.0	154.6	128.4	116.1	100.0

Point: Trend analysis expresses a percent of base, not a percent of change.

Graphical depictions often aid analysis of trend percents. Exhibit 17.5 shows the trend percents from Exhibit 17.4 in a *line graph,* which can help us identify trends and detect changes in direction or magnitude. It reveals that the trend line for revenues consistently exceeds that for cost of goods sold. Moreover, the magnitude of that difference has slightly grown. This result bodes well for Best Buy because its cost of goods sold are by far its largest cost, and the company shows an ability to control these expenses as it expands. The line graph also reveals a consistent increase in each of these accounts, which is typical of growth companies. The trend line for selling, general and administrative expenses is less encouraging because it exceeds the revenue trend line in 2006–2007. The good news is that nearly all of that upward shift in costs occured in one year (2006). In other years, management appears to have limited those costs to not exceed revenue growth.

EXHIBIT 17.5

Trend Percent Lines for Revenues and Expenses of Best Buy

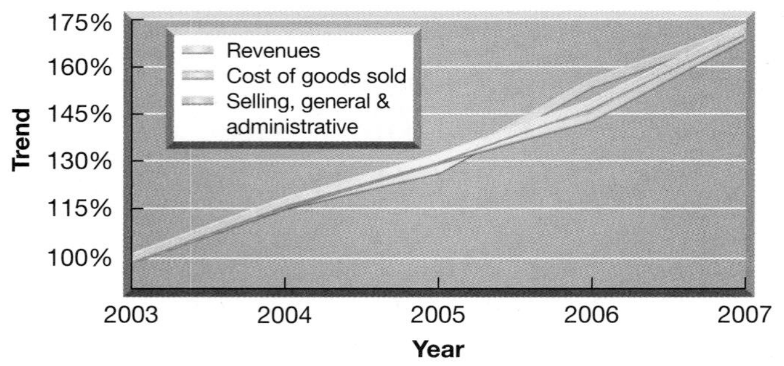

Exhibit 17.6 compares **Best Buy**'s revenue trend line to that of **Circuit City** and **RadioShack** for this same period. Best Buy's revenues sharply increased over this time period while those of Circuit City exhibited less growth, and those for RadioShack were flat. These data indicate that Best Buy's products and services have met with considerable consumer acceptance.

EXHIBIT 17.6

Trend Percent Lines—Best Buy, Circuit City, and RadioShack

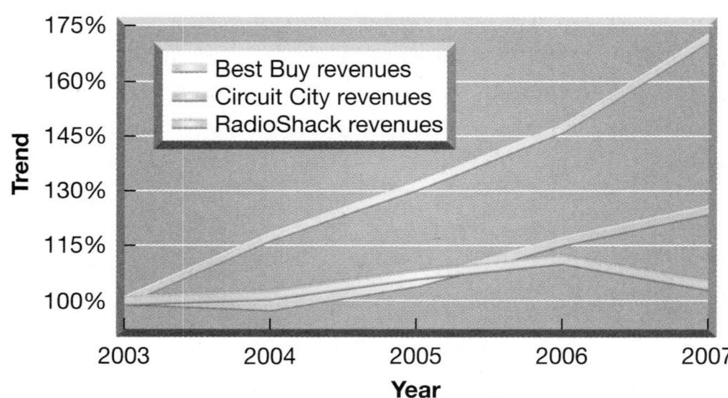

Trend analysis of financial statement items can include comparisons of relations between items on different financial statements. For instance, Exhibit 17.7 compares Best Buy's revenues and total assets. The rate of increase in total assets (176.4%) is greater than the increase in revenues (171.6%) since 2003. Is this result favorable or not? It suggests that Best Buy was slightly less efficient in using its assets in 2007. Management apparently is expecting future years' revenues to compensate for this asset growth.

Overall we must remember that an important role of financial statement analysis is identifying questions and areas of interest, which often direct us to important factors bearing

($ millions)	2007	2003	Trend Percent (2007 vs. 2003)
Revenues	$35,934	$20,943	171.6%
Total assets	13,570	7,694	176.4

EXHIBIT 17.7

Revenue and Asset Data for Best Buy

on a company's future. Accordingly, financial statement analysis should be seen as a continuous process of refining our understanding and expectations of company performance and financial condition.

Decision Maker

Auditor Your tests reveal a 3% increase in sales from $200,000 to $206,000 and a 4% decrease in expenses from $190,000 to $182,400. Both changes are within your "reasonableness" criterion of ±5%, and thus you don't pursue additional tests. The audit partner in charge questions your lack of follow-up and mentions the *joint relation* between sales and expenses. To what is the partner referring? [Answer—p. 706]

Vertical Analysis

Vertical analysis is a tool to evaluate individual financial statement items or a group of items in terms of a specific base amount. We usually define a key aggregate figure as the base, which for an income statement is usually revenue and for a balance sheet is usually total assets. This section explains vertical analysis and applies it to Best Buy. [The term *vertical analysis* arises from the up-down (or down-up) movement of our eyes as we review common-size financial statements. Vertical analysis is also called *common-size analysis*.]

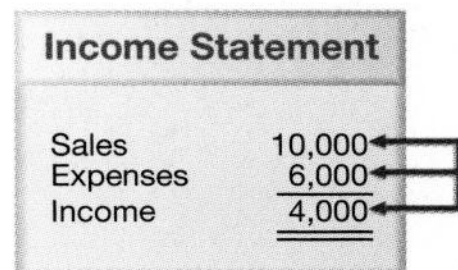

Common-Size Statements

P2 Describe and apply methods of vertical analysis.

The comparative statements in Exhibits 17.1 and 17.2 show the change in each item over time, but they do not emphasize the relative importance of each item. We use **common-size financial statements** to reveal changes in the relative importance of each financial statement item. All individual amounts in common-size statements are redefined in terms of common-size percents. A *common-size percent* is measured by dividing each individual financial statement amount under analysis by its base amount:

$$\textbf{Common-size percent (\%)} = \frac{\textbf{Analysis amount}}{\textbf{Base amount}} \times \textbf{100}$$

Common-Size Balance Sheets Common-size statements express each item as a percent of a *base amount,* which for a common-size balance sheet is usually total assets. The base amount is assigned a value of 100%. (This implies that the total amount of liabilities plus equity equals 100% since this amount equals total assets.) We then compute a common-size percent for each asset, liability, and equity item using total assets as the base amount. When we present a company's successive balance sheets in this way, changes in the mixture of assets, liabilities, and equity are apparent.

Point: The *base* amount in common-size analysis is an *aggregate* amount from that period's financial statement.

Exhibit 17.8 shows common-size comparative balance sheets for Best Buy. Some relations that stand out on both a magnitude and percentage basis include (1) a 41% increase in cash and equivalents, (2) a 6.5% decline in short-term investments as a percentage of assets, (3) a 1.2% decrease in net property and equipment as a percentage of assets, (4) a 1.7% increase in the percentage of accounts payable, (5) a 3.4% decline in the current portion of long-term debt, and (6) a marked increase in retained earnings. Most of these changes are characteristic of a successful growth/stable company. The concern, if any, is whether Best Buy can continue to generate sufficient revenues and income to support its asset buildup within a very competitive industry.

Point: Common-size statements often are used to compare two or more companies in the same industry.

Point: Common-size statements are also useful in comparing firms that report in different currencies.

EXHIBIT 17.8

Common-Size Comparative Balance Sheets

BEST BUY
Common-Size Comparative Balance Sheets
March 3, 2007, and February 25, 2006

(in millions)	2007	2006	Common-Size Percents* 2007	Common-Size Percents* 2006
Assets				
Cash and cash equivalents	$ 1,205	$ 748	8.9%	6.3%
Short-term investments	2,588	3,041	19.1	25.6
Receivables, net	548	449	4.0	3.8
Merchandise inventories	4,028	3,338	29.7	28.1
Other current assets	712	409	5.2	3.4
Total current assets	9,081	7,985	66.9	67.3
Property and equipment	4,904	4,836	36.1	40.8
Less accumulated depreciation	1,966	2,124	14.5	17.9
Net property and equipment	2,938	2,712	21.7	22.9
Goodwill	919	557	6.8	4.7
Tradenames	81	44	0.6	0.4
Long-term investments	318	218	2.3	1.8
Other long-term assets	233	348	1.7	2.9
Total assets	$13,570	$11,864	100.0%	100.0%
Liabilities				
Accounts payable	$ 3,934	$ 3,234	29.0%	27.3%
Unredeemed gift card liabilities	496	469	3.7	4.0
Accrued compensation and related expenses	332	354	2.4	3.0
Accrued liabilities	990	878	7.3	7.4
Accrued income taxes	489	703	3.6	5.9
Short-term debt	41	0	0.3	0.0
Current portion of long-term debt	19	418	0.1	3.5
Total current liabilities	6,301	6,056	46.4	51.0
Long-term liabilities	443	373	3.3	3.1
Long-term debt	590	178	4.3	1.5
Minority interests	35	0	0.3	0.0
Stockholders' Equity				
Common stock	48	49	0.4	0.4
Additional paid-in capital	430	643	3.2	5.4
Retained earnings	5,507	4,304	40.6	36.3
Accumulated other comprehensive income	216	261	1.6	2.2
Total stockholders' equity	6,201	5,257	45.7	44.3
Total liabilities and stockholders' equity	$13,570	$11,864	100.0%	100.0%

* Percents are rounded to tenths and thus may not exactly sum to totals and subtotals.

Global: International companies sometimes disclose "convenience" financial statements, which are statements translated in other languages and currencies. However, these statements rarely adjust for differences in accounting principles across countries.

Common-Size Income Statements Analysis also benefits from use of a common-size income statement. Revenues is usually the base amount, which is assigned a value of 100%. Each common-size income statement item appears as a percent of revenues. If we think of the 100% revenues amount as representing one sales dollar, the remaining items show how each revenue dollar is distributed among costs, expenses, and income.

Exhibit 17.9 shows common-size comparative income statements for each dollar of Best Buy's revenues. The past two years' common-size numbers are similar. The good news is that Best Buy has been able to squeeze an extra 0.1 cent in earnings per revenue dollar—evidenced by the 3.7% to 3.8% rise in earnings as a percentage of revenues. This implies that management is effectively controlling costs and/or the company is reaping growth benefits, so-called *economies of scale*. The bad news is that gross profit lost 0.6 cent per revenue dollar—evidenced by the 25.0% to 24.4% decline in gross profit as a percentage of revenues. This is a concern given the price-competitive

EXHIBIT 17.9

Common-Size Comparative Income Statements

BEST BUY Common-Size Comparative Income Statements For Years Ended March 3, 2007, and February 25, 2006			Common-Size Percents*	
($ millions)	2007	2006	2007	2006
Revenues	$35,934	$30,848	100.0%	100.0%
Cost of goods sold	27,165	23,122	75.6	75.0
Gross profit	8,769	7,726	24.4	25.0
Selling, general, and administrative expenses	6,770	6,082	18.8	19.7
Operating income	1,999	1,644	5.6	5.3
Net interest income (expense)	111	77	0.3	0.2
Gain on investments	20	0	0.1	0.0
Earnings from continuing operations before income taxes	2,130	1,721	5.9	5.6
Income tax expense	752	581	2.1	1.9
Minority interest in earnings	1	0	0.0	0.0
Net earnings	$ 1,377	$ 1,140	3.8%	3.7%

* Percents are rounded to tenths and thus may not exactly sum to totals and subtotals.

electronics market. Analysis here shows that common-size percents for successive income statements can uncover potentially important changes in a company's expenses. Evidence of no changes, especially when changes are expected, is also informative.

Common-Size Graphics

Two of the most common tools of common-size analysis are trend analysis of common-size statements and graphical analysis. The trend analysis of common-size statements is similar to that of comparative statements discussed under vertical analysis. It is not illustrated here because the only difference is the substitution of common-size percents for trend percents. Instead, this section discusses graphical analysis of common-size statements.

An income statement readily lends itself to common-size graphical analysis. This is so because revenues affect nearly every item in an income statement. Exhibit 17.10 shows Best Buy's 2007 common-size income statement in graphical form. This pie chart highlights the contribution of each component of revenues for net earnings.

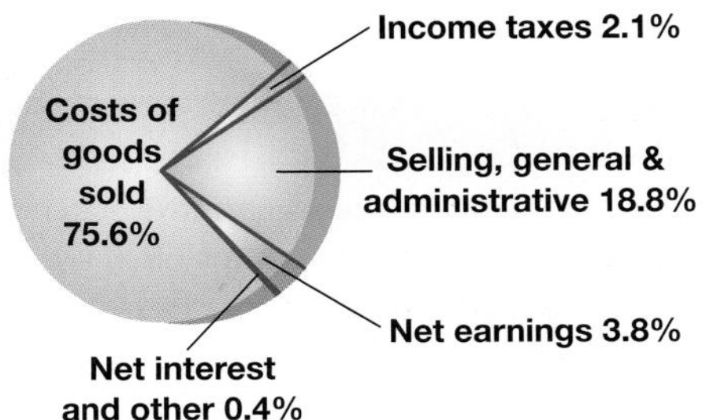

EXHIBIT 17.10

Common-Size Graphic of Income Statement

Exhibit 17.11 previews more complex graphical analyses available and the insights they provide. The data for this exhibit are taken from Best Buy's *Segments* footnote. Best Buy has two reportable segments: domestic and international.

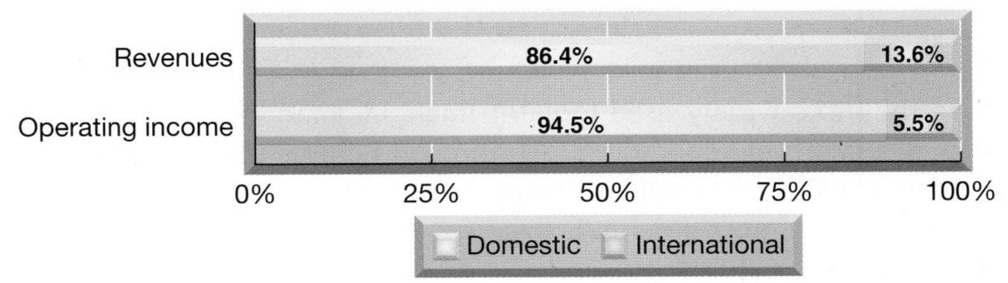

EXHIBIT 17.11

Revenue and Operating Income Breakdown by Segment

EXHIBIT 17.12

Common-Size Graphic of Asset Components

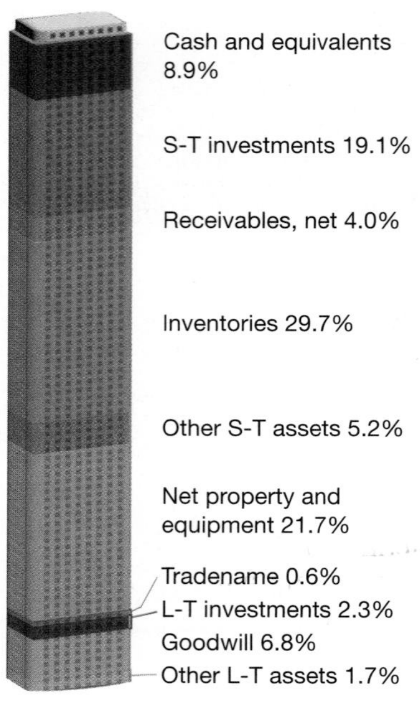

The upper bar in Exhibit 17.11 shows the percent of revenues from each segment. The major revenue source is Domestic (86.4%). The lower bar shows the percent of operating income from each segment. Although International provides 13.6% of revenues, it provides only 5.5% of operating income. This type of information can help users in determining strategic analyses and actions.

Graphical analysis is also useful in identifying (1) sources of financing including the distribution among current liabilities, noncurrent liabilities, and equity capital and (2) focuses of investing activities, including the distribution among current and noncurrent assets. As illustrative, Exhibit 17.12 shows a common-size graphical display of Best Buy's assets. Common-size balance sheet analysis can be extended to examine the composition of these subgroups. For instance, in assessing liquidity of current assets, knowing what proportion of current assets consists of inventories is usually important, and not simply what proportion inventories are of total assets.

Common-size financial statements are also useful in comparing different companies. Exhibit 17.13 shows common-size graphics of Best Buy, Circuit City, and RadioShack on financing sources. This graphic highlights the larger percent of equity financing for Best Buy and Circuit City than for RadioShack. It also highlights the much larger noncurrent (debt) financing of RadioShack. Comparison of a company's common-size statements with competitors' or industry common-size statistics alerts us to differences in the structure or distribution of its financial statements but not to their dollar magnitude.

EXHIBIT 17.13

Common-Size Graphic of Financing Sources—Competitor Analysis

	Best Buy	Circuit City	RadioShack
Current liabilities	46.4%	42.8%	47.5%
Noncurrent liabilities	7.9%	12.5%	20.9%
Equity	45.7%	44.7%	31.6%

Quick Check Answers—p. 706

5. Which of the following is true for common-size comparative statements? (*a*) Each item is expressed as a percent of a base amount. (*b*) Total assets often are assigned a value of 100%. (*c*) Amounts from successive periods are placed side by side. (*d*) All are true. (*e*) None is true.
6. What is the difference between the percents shown on a comparative income statement and those shown on a common-size comparative income statement?
7. Trend percents are (*a*) shown on comparative income statements and balance sheets, (*b*) shown on common-size comparative statements, or (*c*) also called *index numbers*.

Ratio Analysis

P3 Define and apply ratio analysis.

Ratios are among the more widely used tools of financial analysis because they provide clues to and symptoms of underlying conditions. A ratio can help us uncover conditions and trends difficult to detect by inspecting individual components making up the ratio. Ratios, like other analysis tools, are usually future oriented; that is, they are often adjusted for their probable future trend and magnitude, and their usefulness depends on skillful interpretation.

A ratio expresses a mathematical relation between two quantities. It can be expressed as a percent, rate, or proportion. For instance, a change in an account balance from \$100 to \$250 can be expressed as (1) 150%, (2) 2.5 times, or (3) 2.5 to 1 (or 2.5:1). Computation of a ratio is a simple arithmetic operation, but its interpretation is not. To be meaningful, a ratio must refer to an economically important relation. For example, a direct and crucial relation exists between an item's sales price and its cost. Accordingly, the ratio of cost of goods sold to sales is meaningful. In contrast, no obvious relation exists between freight costs and the balance of long-term investments.

This section describes an important set of financial ratios and its application. The selected ratios are organized into the four building blocks of financial statement analysis: (1) liquidity and efficiency, (2) solvency, (3) profitability, and (4) market prospects. All of these ratios were explained at relevant points in prior chapters. The purpose here is to organize and apply them under a summary framework. We use four common standards, in varying degrees, for comparisons: intracompany, competitor, industry, and guidelines.

Point: Some sources for industry norms are *Annual Statement Studies* by Robert Morris Associates, *Industry Norms & Key Business Ratios* by Dun & Bradstreet, *Standard & Poor's Industry Surveys,* and Reuters.com/finance.

Liquidity and Efficiency

Liquidity refers to the availability of resources to meet short-term cash requirements. It is affected by the timing of cash inflows and outflows along with prospects for future performance. Analysis of liquidity is aimed at a company's funding requirements. *Efficiency* refers to how productive a company is in using its assets. Efficiency is usually measured relative to how much revenue is generated from a certain level of assets.

Both liquidity and efficiency are important and complementary. If a company fails to meet its current obligations, its continued existence is doubtful. Viewed in this light, all other measures of analysis are of secondary importance. Although accounting measurements assume the company's continued existence, our analysis must always assess the validity of this assumption using liquidity measures. Moreover, inefficient use of assets can cause liquidity problems. A lack of liquidity often precedes lower profitability and fewer opportunities. It can foretell a loss of owner control. To a company's creditors, lack of liquidity can yield delays in collecting interest and principal payments or the loss of amounts due them. A company's customers and suppliers of goods and services also are affected by short-term liquidity problems. Implications include a company's inability to execute contracts and potential damage to important customer and supplier relationships. This section describes and illustrates key ratios relevant to assessing liquidity and efficiency.

Working Capital and Current Ratio The amount of current assets less current liabilities is called **working capital,** or *net working capital.* A company needs adequate working capital to meet current debts, to carry sufficient inventories, and to take advantage of cash discounts. A company that runs low on working capital is less likely to meet current obligations or to continue operating. When evaluating a company's working capital, we must not only look at the dollar amount of current assets less current liabilities, but also at their ratio. The *current ratio* is defined as follows (see Chapter 3 for additional explanation):

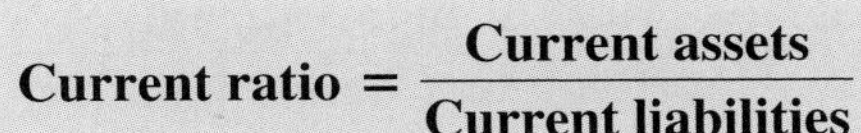

$$\textbf{Current ratio} = \frac{\textbf{Current assets}}{\textbf{Current liabilities}}$$

Drawing on information in Exhibit 17.1, Best Buy's working capital and current ratio for both 2007 and 2006 are shown in Exhibit 17.14. Circuit City (1.68), RadioShack (1.63), and the Industry's current ratio of 1.6 is shown in the margin. Best Buy's 2007 ratio (1.44) is lower than any of the comparison ratios, but it does not appear in danger of defaulting on loan payments. A high current ratio suggests a strong liquidity position and an ability to meet current obligations. A company can, however, have a current ratio that is too high. An excessively high current ratio means that the company has invested too much in current assets compared to its current obligations. An

EXHIBIT 17.14

Best Buy's Working Capital and Current Ratio

(\$ millions)	2007	2006
Current assets	\$ 9,081	\$ 7,985
Current liabilities	6,301	6,056
Working capital	**\$2,780**	**\$1,929**
Current ratio		
\$9,081/\$6,301	**1.44 to 1**	
\$7,985/\$6,056		**1.32 to 1**

Current ratio
Circuit City = 1.68
RadioShack = 1.63
Industry = 1.6

excessive investment in current assets is not an efficient use of funds because current assets normally generate a low return on investment (compared with long-term assets).

Many users apply a guideline of 2:1 (or 1.5:1) for the current ratio in helping evaluate a company's debt-paying ability. A company with a 2:1 or higher current ratio is generally thought to be a good credit risk in the short run. Such a guideline or any analysis of the current ratio must recognize at least three additional factors: (1) type of business, (2) composition of current assets, and (3) turnover rate of current asset components.

Type of business. A service company that grants little or no credit and carries few inventories can probably operate on a current ratio of less than 1:1 if its revenues generate enough cash to pay its current liabilities. On the other hand, a company selling high-priced clothing or furniture requires a higher ratio because of difficulties in judging customer demand and cash receipts. For instance, if demand falls, inventory may not generate as much cash as expected. Accordingly, analysis of the current ratio should include a comparison with ratios from successful companies in the same industry and from prior periods. We must also recognize that a company's accounting methods, especially choice of inventory method, affect the current ratio. For instance, when costs are rising, a company using LIFO tends to report a smaller amount of current assets than when using FIFO.

Point: When a firm uses LIFO in a period of rising costs, the standard for an adequate current ratio usually is lower than if it used FIFO.

Composition of current assets. The composition of a company's current assets is important to an evaluation of short-term liquidity. For instance, cash, cash equivalents, and short-term investments are more liquid than accounts and notes receivable. Also, short-term receivables normally are more liquid than inventory. Cash, of course, can be used to immediately pay current debts. Items such as accounts receivable and inventory, however, normally must be converted into cash before payment is made. An excessive amount of receivables and inventory weakens a company's ability to pay current liabilities. The acid-test ratio (see below) can help with this assessment.

Turnover rate of assets. Asset turnover measures a company's efficiency in using its assets. One relevant measure of asset efficiency is the revenue generated. A measure of total asset turnover is revenues divided by total assets, but evaluation of turnover for individual assets is also useful. We discuss both receivables turnover and inventory turnover on the next page.

Decision Maker

Banker A company requests a one-year, $200,000 loan for expansion. This company's current ratio is 4:1, with current assets of $160,000. Key competitors carry a current ratio of about 1.9:1. Using this information, do you approve the loan application? Does your decision change if the application is for a 10-year loan? [Answer—p. 706]

Acid-Test Ratio Quick assets are cash, short-term investments, and current receivables. These are the most liquid types of current assets. The *acid-test ratio,* also called *quick ratio,* and introduced in Chapter 5, reflects on a company's short-term liquidity.

$$\text{Acid-test ratio} = \frac{\text{Cash + Short-term investments + Current receivables}}{\text{Current liabilities}}$$

Best Buy's acid-test ratio is computed in Exhibit 17.15. Best Buy's 2007 acid-test ratio (0.69) is between that for Circuit City (0.65) and RadioShack (0.73), and less than the 1:1 common

EXHIBIT 17.15

Acid-Test Ratio

($ millions)	2007	2006
Cash and equivalents	$1,205	$ 748
Short-term investments	2,588	3,041
Current receivables	548	449
Total quick assets	$4,341	$4,238
Current liabilities	$6,301	$6,056
Acid-test ratio		
$4,341/$6,301	0.69 to 1	
$4,238/$6,056		0.70 to 1

Acid-test ratio
Circuit City = 0.65
RadioShack = 0.73
Industry = 0.7

guideline for an acceptable acid-test ratio; each of these ratios is similar to the 0.7 industry ratio. As with analysis of the current ratio, we need to consider other factors. For instance, the frequency with which a company converts its current assets into cash affects its working capital requirements. This implies that analysis of short-term liquidity should also include an analysis of receivables and inventories, which we consider next.

Global: Ratio analysis helps overcome currency translation problems, but it does *not* overcome differences in accounting principles.

Accounts Receivable Turnover We can measure how frequently a company converts its receivables into cash by computing the *accounts receivable turnover*. This ratio is defined as follows (see Chapter 9 for additional explanation):

$$\textbf{Accounts receivable turnover} = \frac{\textbf{Net sales}}{\textbf{Average accounts receivable, net}}$$

Short-term receivables from customers are often included in the denominator along with accounts receivable. Also, accounts receivable turnover is more precise if credit sales are used for the numerator, but external users generally use net sales (or net revenues) because information about credit sales is typically not reported. Best Buy's 2007 accounts receivable turnover is computed as follows ($ millions).

Point: Some users prefer using gross accounts receivable (before subtracting the allowance for doubtful accounts) to avoid the influence of a manager's bad debts estimate.

$$\frac{\$35{,}934}{(\$548 + \$449)/2} = 72.1 \text{ times}$$

Accounts receivable turnover
Circuit City = 41.2
RadioShack = 17.1

Best Buy's value of 72.1 is larger than Circuit City's 41.2 and RadioShack's 17.1. Accounts receivable turnover is high when accounts receivable are quickly collected. A high turnover is favorable because it means the company need not commit large amounts of funds to accounts receivable. However, an accounts receivable turnover can be too high; this can occur when credit terms are so restrictive that they negatively affect sales volume.

Point: Ending accounts receivable can be substituted for the average balance in computing accounts receivable turnover if the difference between ending and average receivables is small.

Inventory Turnover How long a company holds inventory before selling it will affect working capital requirements. One measure of this effect is *inventory turnover,* also called *merchandise turnover* or *merchandise inventory turnover,* which is defined as follows (see Chapter 6 for additional explanation):

$$\textbf{Inventory turnover} = \frac{\textbf{Cost of goods sold}}{\textbf{Average inventory}}$$

Using Best Buy's cost of goods sold and inventories information, we compute its inventory turnover for 2007 as follows (if the beginning and ending inventories for the year do not represent the usual inventory amount, an average of quarterly or monthly inventories can be used).

$$\frac{\$27{,}165}{(\$4{,}028 + \$3{,}338)/2} = 7.38 \text{ times}$$

Inventory turnover
Circuit City = 5.70
RadioShack = 2.96
Industry = 4.5

Best Buy's inventory turnover of 7.38 is higher than Circuit City's 5.70, RadioShack's 2.96, and the industry's 4.5. A company with a high turnover requires a smaller investment in inventory than one producing the same sales with a lower turnover. Inventory turnover can be too high, however, if the inventory a company keeps is so small that it restricts sales volume.

Days' Sales Uncollected Accounts receivable turnover provides insight into how frequently a company collects its accounts. Days' sales uncollected is one measure of this activity, which is defined as follows (Chapter 8 provides additional explanation):

$$\textbf{Days' sales uncollected} = \frac{\textbf{Accounts receivable, net}}{\textbf{Net sales}} \times \textbf{365}$$

Any short-term notes receivable from customers are normally included in the numerator.

Best Buy's 2007 days' sales uncollected follows.

Day's sales uncollected
Circuit City = 11.23
RadioShack = 18.94

$$\frac{\$548}{\$35{,}934} \times 365 = 5.57 \text{ days}$$

Both Circuit City's days' sales uncollected of 11.23 days and RadioShack's 18.94 days are longer than the 5.57 days for Best Buy. Days' sales uncollected is more meaningful if we know company credit terms. A rough guideline states that days' sales uncollected should not exceed $1\frac{1}{3}$ times the days in its (1) credit period, *if* discounts are not offered or (2) discount period, *if* favorable discounts are offered.

Days' Sales in Inventory *Days' sales in inventory* is a useful measure in evaluating inventory liquidity. Days' sales in inventory is linked to inventory in a way that days' sales uncollected is linked to receivables. We compute days' sales in inventory as follows (Chapter 6 provides additional explanation).

$$\textbf{Days' sales in inventory} = \frac{\textbf{Ending inventory}}{\textbf{Cost of goods sold}} \times \textbf{365}$$

Best Buy's days' sales in inventory for 2007 follows.

Days' sales in inventory
Circuit City = 62.9
RadioShack = 107.9

$$\frac{\$4{,}028}{\$27{,}165} \times 365 = 54.1 \text{ days}$$

Point: *Average collection period* is estimated by dividing 365 by the accounts receivable turnover ratio. For example, 365 divided by an accounts receivable turnover of 6.1 indicates a 60-day average collection period.

If the products in Best Buy's inventory are in demand by customers, this formula estimates that its inventory will be converted into receivables (or cash) in 54.1 days. If all of Best Buy's sales were credit sales, the conversion of inventory to receivables in 54.1 days *plus* the conversion of receivables to cash in 5.57 days implies that inventory will be converted to cash in about 59.67 days (54.1 + 5.57).

Total Asset Turnover *Total asset turnover* reflects a company's ability to use its assets to generate sales and is an important indication of operating efficiency. The definition of this ratio follows (Chapter 10 offers additional explanation).

$$\textbf{Total asset turnover} = \frac{\textbf{Net sales}}{\textbf{Average total assets}}$$

Best Buy's total asset turnover for 2007 follows and is less than Circuit City's, but greater than that for RadioShack.

Total asset turnover
Circuit City = 3.08
RadioShack = 2.24

$$\frac{\$35{,}934}{(\$13{,}570 + \$11{,}864)/2} = 2.83 \text{ times}$$

Quick Check

Answers—p. 706

8. Information from Paff Co. at Dec. 31, 2008, follows: cash, \$820,000; accounts receivable, \$240,000; inventories, \$470,000; plant assets, \$910,000; accounts payable, \$350,000; and income taxes payable, \$180,000. Compute its (*a*) current ratio and (*b*) acid-test ratio.
9. On Dec. 31, 2009, Paff Company (see question 8) had accounts receivable of \$290,000 and inventories of \$530,000. During 2009, net sales amounted to \$2,500,000 and cost of goods sold was \$750,000. Compute (*a*) accounts receivable turnover, (*b*) days' sales uncollected, (*c*) inventory turnover, and (*d*) days' sales in inventory.

Solvency

Solvency refers to a company's long-run financial viability and its ability to cover long-term obligations. All of a company's business activities—financing, investing, and operating—affect its solvency. Analysis of solvency is long term and uses less precise but more encompassing measures than liquidity. One of the most important components of solvency analysis is the composition of a company's capital structure. *Capital structure* refers to a company's financing sources. It ranges from relatively permanent equity financing to riskier or more temporary short-term financing. Assets represent security for financiers, ranging from loans secured by specific assets to the assets available as general security to unsecured creditors. This section describes the tools of solvency analysis. Our analysis focuses on a company's ability to both meet its obligations and provide security to its creditors *over the long run*. Indicators of this ability include *debt* and *equity* ratios, the relation between *pledged assets and secured liabilities*, and the company's capacity to earn sufficient income to *pay fixed interest charges*.

Debt and Equity Ratios One element of solvency analysis is to assess the portion of a company's assets contributed by its owners and the portion contributed by creditors. This relation is reflected in the debt ratio (also described in Chapter 2). The *debt ratio* expresses total liabilities as a percent of total assets. The **equity ratio** provides complementary information by expressing total equity as a percent of total assets. **Best Buy**'s debt and equity ratios follow.

Point: For analysis purposes, Minority Interest is usually included in equity.

($ millions)	2007	Ratios	
Total liabilities	$ 7,369	54.3%	[Debt ratio]
Total equity	6,201	45.7	[Equity ratio]
Total liabilities and equity	$13,570	100.0%	

Debt ratio :: Equity ratio
Circuit City = 55.3% :: 44.7%
RadioShack = 68.4% :: 31.6%

Best Buy's financial statements reveal more debt than equity. A company is considered less risky if its capital structure (equity and long-term debt) contains more equity. One risk factor is the required payment for interest and principal when debt is outstanding. Another factor is the greater the stockholder financing, the more losses a company can absorb through equity before the assets become inadequate to satisfy creditors' claims. From the stockholders' point of view, if a company earns a return on borrowed capital that is higher than the cost of borrowing, the difference represents increased income to stockholders. The inclusion of debt is described as *financial leverage* because debt can have the effect of increasing the return to stockholders. Companies are said to be highly leveraged if a large portion of their assets is financed by debt.

Point: Bank examiners from the FDIC and other regulatory agencies use debt and equity ratios to monitor compliance with regulatory capital requirements imposed on banks and S&Ls.

Debt-to-Equity Ratio The ratio of total liabilities to equity is another measure of solvency. We compute the ratio as follows (Chapter 14 offers additional explanation).

$$\textbf{Debt-to-equity ratio} = \frac{\textbf{Total liabilities}}{\textbf{Total equity}}$$

Best Buy's debt-to-equity ratio for 2007 is

$$\$7,369/\$6,201 = 1.19$$

Debt-to-equity
Circuit City = 1.24
RadioShack = 2.17
Industry = 0.99

Best Buy's 1.19 debt-to-equity ratio is less than the 1.24 ratio for Circuit City and the 2.17 for RadioShack, but greater than the industry ratio of 0.99. Consistent with our inferences from the debt ratio, Best Buy's capital structure has more debt than equity, which increases risk. Recall that debt must be repaid with interest, while equity does not. These debt requirements can be burdensome when the industry and/or the economy experience a downturn. A larger debt-to-equity ratio also implies less opportunity to expand through use of debt financing.

Times Interest Earned The amount of income before deductions for interest expense and income taxes is the amount available to pay interest expense. The following

Point: The times interest earned ratio and the debt and equity ratios are of special interest to bank lending officers.

times interest earned ratio reflects the creditors' risk of loan repayments with interest (see Chapter 11 for additional explanation).

$$\textbf{Times interest earned} = \frac{\textbf{Income before interest expense and income taxes}}{\textbf{Interest expense}}$$

The larger this ratio, the less risky is the company for creditors. One guideline says that creditors are reasonably safe if the company earns its fixed interest expense two or more times each year. Best Buy's times interest earned ratio follows; its value suggests that its creditors have little risk of nonrepayment.

Times interest earned
Circuit City = 12.5
RadioShack = 3.5

$$\frac{\$1{,}377 + \$31\text{ (see Best Buy note \#7)} + \$752}{\$31} = 69.7$$

Decision Insight

Bears and Bulls A *bear market* is a declining market. The phrase comes from bear-skin jobbers who often sold the skins before the bears were caught. The term *bear* was then used to describe investors who sold shares they did not own in anticipation of a price decline. A *bull market* is a rising market. This phrase comes from the once popular sport of bear and bull baiting. The term *bull* came to mean the opposite of *bear.*

Profitability

We are especially interested in a company's ability to use its assets efficiently to produce profits (and positive cash flows). *Profitability* refers to a company's ability to generate an adequate return on invested capital. Return is judged by assessing earnings relative to the level and sources of financing. Profitability is also relevant to solvency. This section describes key profitability measures and their importance to financial statement analysis.

Profit Margin A company's operating efficiency and profitability can be expressed by two components. The first is *profit margin,* which reflects a company's ability to earn net income from sales (Chapter 3 offers additional explanation). It is measured by expressing net income as a percent of sales (*sales* and *revenues* are similar terms). Best Buy's profit margin follows.

Profit margin
Circuit City = −0.1%
RadioShack = 1.5%

$$\textbf{Profit margin} = \frac{\textbf{Net income}}{\textbf{Net sales}} = \frac{\$1{,}377}{\$35{,}934} = 3.8\%$$

To evaluate profit margin, we must consider the industry. For instance, an appliance company might require a profit margin between 10% and 15%; whereas a retail supermarket might require a profit margin of 1% or 2%. Both profit margin and *total asset turnover* make up the two basic components of operating efficiency. These ratios reflect on management because managers are ultimately responsible for operating efficiency. The next section explains how we use both measures to analyze return on total assets.

Return on Total Assets *Return on total assets* is defined as follows.

$$\textbf{Return on total assets} = \frac{\textbf{Net income}}{\textbf{Average total assets}}$$

Best Buy's 2007 return on total assets is

Return on total assets
Circuit City = −0.2%
RadioShack = 3.4%
Industry = 3.0

$$\frac{\$1{,}377}{(\$13{,}570 + \$11{,}864)/2} = 10.8\%$$

Best Buy's 10.8% return on total assets is lower than that for many businesses but is higher than RadioShack's return of 3.4% and the industry's 3.0% return. We also should evaluate any trend in the rate of return.

Point: Many analysts add back *Interest expense* × *(1 − Tax rate)* to net income in computing return on total assets.

The following equation shows the important relation between profit margin, total asset turnover, and return on total assets.

$$\textbf{Profit margin} \times \textbf{Total asset turnover} = \textbf{Return on total assets}$$

or

$$\frac{\textbf{Net income}}{\textbf{Net sales}} \times \frac{\textbf{Net sales}}{\textbf{Average total assets}} = \frac{\textbf{Net income}}{\textbf{Average total assets}}$$

Both profit margin and total asset turnover contribute to overall operating efficiency, as measured by return on total assets. If we apply this formula to Best Buy, we get

$$3.8\% \times 2.83 = 10.8\%$$

Circuit City: −0.1% × 3.08 = −0.2%
RadioShack: 1.5% × 2.24 = 3.4%
(with rounding)

This analysis shows that Best Buy's superior return on assets versus that of Circuit City and RadioShack is driven mainly by its higher profit margin.

Return on Common Stockholders' Equity Perhaps the most important goal in operating a company is to earn net income for its owner(s). *Return on common stockholders' equity* measures a company's success in reaching this goal and is defined as follows.

$$\textbf{Return on common stockholders' equity} = \frac{\textbf{Net income} - \textbf{Preferred dividends}}{\textbf{Average common stockholders' equity}}$$

Best Buy's 2007 return on common stockholders' equity is computed as follows:

$$\frac{\$1{,}377 - \$0}{(\$6{,}236 + \$5{,}257)/2} = 24.0\%$$

Return on common equity
Circuit City = −0.4%
RadioShack = 11.8%

The denominator in this computation is the book value of common equity (including minority interest). In the numerator, the dividends on cumulative preferred stock are subtracted whether they are declared or are in arrears. If preferred stock is noncumulative, its dividends are subtracted only if declared.

Decision Insight

Wall Street *Wall Street* is synonymous with financial markets, but its name comes from the street location of the original New York Stock Exchange. The street's name derives from stockades built by early settlers to protect New York from pirate attacks.

Market Prospects

Market measures are useful for analyzing corporations with publicly traded stock. These market measures use stock price, which reflects the market's (public's) expectations for the company. This includes expectations of both company return and risk—as the market perceives it.

Price-Earnings Ratio Computation of the *price-earnings ratio* follows (Chapter 13 provides additional explanation).

$$\textbf{Price-earnings ratio} = \frac{\textbf{Market price per common share}}{\textbf{Earnings per share}}$$

Point: PE ratio can be viewed as an indicator of the market's expected growth and risk for a stock. High expected risk suggests a low PE ratio. High expected growth suggests a high PE ratio.

Predicted earnings per share for the next period is often used in the denominator of this computation. Reported earnings per share for the most recent period is also commonly used. In both cases, the ratio is used as an indicator of the future growth and risk of a company's earnings as perceived by the stock's buyers and sellers.

The market price of Best Buy's common stock at the start of fiscal year 2008 was $46.35. Using Best Buy's $2.86 basic earnings per share, we compute its price-earnings ratio as follows (some analysts compute this ratio using the median of the low and high stock price).

PE (year-end)
Circuit City = −380.0
RadioShack = 31.1

$$\frac{\$46.35}{\$2.86} = 16.2$$

Point: Some investors avoid stocks with high PE ratios under the belief they are "overpriced." Alternatively, some investors *sell these stocks short*—hoping for price declines.

Best Buy's price-earnings ratio is less than that for RadioShack, but is slightly higher than the norm. (Circuit City's ratio is negative due to its abnormally low earnings.) Best Buy's middle-of-the-pack ratio likely reflects investors' expectations of continued growth but normal earnings.

Dividend Yield *Dividend yield* is used to compare the dividend-paying performance of different investment alternatives. We compute dividend yield as follows (Chapter 13 offers additional explanation).

$$\textbf{Dividend yield} = \frac{\textbf{Annual cash dividends per share}}{\textbf{Market price per share}}$$

Best Buy's dividend yield, based on its fiscal year-end market price per share of $46.35 and its policy of $0.36 cash dividends per share, is computed as follows.

Dividend yield
Circuit City = 0.6%
RadioShack = 1.5%

$$\frac{\$0.36}{\$46.35} = 0.8\%$$

Some companies do not declare and pay dividends because they wish to reinvest the cash.

Summary of Ratios

Point: Corporate PE ratios and dividend yields are found in daily stock market quotations listed in *The Wall Street Journal, Investor's Business Daily,* or other publications and Web services.

Exhibit 17.16 summarizes the major financial statement analysis ratios illustrated in this chapter and throughout the book. This summary includes each ratio's title, its formula, and the purpose for which it is commonly used.

Decision Insight

Ticker Prices *Ticker prices* refer to a band of moving data on a monitor carrying up-to-the-minute stock prices. The phrase comes from *ticker tape,* a 1-inch-wide strip of paper spewing stock prices from a printer that ticked as it ran. Most of today's investors have never seen actual ticker tape, but the phrase survives.

Quick Check

Answers—p. 706

10. Which ratio best reflects a company's ability to meet immediate interest payments? (*a*) Debt ratio. (*b*) Equity ratio. (*c*) Times interest earned.

11. Which ratio best measures a company's success in earning net income for its owner(s)? (*a*) Profit margin. (*b*) Return on common stockholders' equity. (*c*) Price-earnings ratio. (*d*) Dividend yield.

12. If a company has net sales of $8,500,000, net income of $945,000, and total asset turnover of 1.8 times, what is its return on total assets?

EXHIBIT 17.16

Financial Statement Analysis Ratios*

Ratio	Formula	Measure of
Liquidity and Efficiency		
Current ratio	$= \dfrac{\text{Current assets}}{\text{Current liabilities}}$	Short-term debt-paying ability
Acid-test ratio	$= \dfrac{\text{Cash + Short-term investments + Current receivables}}{\text{Current liabilities}}$	Immediate short-term debt-paying ability
Accounts receivable turnover	$= \dfrac{\text{Net sales}}{\text{Average accounts receivable, net}}$	Efficiency of collection
Inventory turnover	$= \dfrac{\text{Cost of goods sold}}{\text{Average inventory}}$	Efficiency of inventory management
Days' sales uncollected	$= \dfrac{\text{Accounts receivable, net}}{\text{Net sales}} \times 365$	Liquidity of receivables
Days' sales in inventory	$= \dfrac{\text{Ending inventory}}{\text{Cost of goods sold}} \times 365$	Liquidity of inventory
Total asset turnover	$= \dfrac{\text{Net sales}}{\text{Average total assets}}$	Efficiency of assets in producing sales
Solvency		
Debt ratio	$= \dfrac{\text{Total liabilities}}{\text{Total assets}}$	Creditor financing and leverage
Equity ratio	$= \dfrac{\text{Total equity}}{\text{Total assets}}$	Owner financing
Debt-to-equity ratio	$= \dfrac{\text{Total liabilities}}{\text{Total equity}}$	Debt versus equity financing
Times interest earned	$= \dfrac{\text{Income before interest expense and income taxes}}{\text{Interest expense}}$	Protection in meeting interest payments
Profitability		
Profit margin ratio	$= \dfrac{\text{Net income}}{\text{Net sales}}$	Net income in each sales dollar
Gross margin ratio	$= \dfrac{\text{Net sales} - \text{Cost of goods sold}}{\text{Net sales}}$	Gross margin in each sales dollar
Return on total assets	$= \dfrac{\text{Net income}}{\text{Average total assets}}$	Overall profitability of assets
Return on common stockholders' equity	$= \dfrac{\text{Net income} - \text{Preferred dividends}}{\text{Average common stockholders' equity}}$	Profitability of owner investment
Book value per common share	$= \dfrac{\text{Shareholders' equity applicable to common shares}}{\text{Number of common shares outstanding}}$	Liquidation at reported amounts
Basic earnings per share	$= \dfrac{\text{Net income} - \text{Preferred dividends}}{\text{Weighted-average common shares outstanding}}$	Net income per common share
Market Prospects		
Price-earnings ratio	$= \dfrac{\text{Market price per common share}}{\text{Earnings per share}}$	Market value relative to earnings
Dividend yield	$= \dfrac{\text{Annual cash dividends per share}}{\text{Market price per share}}$	Cash return per common share

* Additional ratios also examined in previous chapters included credit risk ratio; plant asset useful life; plant asset age; days' cash expense coverage; cash coverage of growth; cash coverage of debt; free cash flow; cash flow on total assets; and payout ratio.

Decision Analysis — Analysis Reporting

A1 Summarize and report results of analysis.

Understanding the purpose of financial statement analysis is crucial to the usefulness of any analysis. This understanding leads to efficiency of effort, effectiveness in application, and relevance in focus. The purpose of most financial statement analyses is to reduce uncertainty in business decisions through a rigorous and sound evaluation. A *financial statement analysis report* helps by directly addressing the building blocks of analysis and by identifying weaknesses in inference by requiring explanation: It forces us to organize our reasoning and to verify its flow and logic. A report also serves as a communication link with readers, and the writing process reinforces our judgments and vice versa. Finally, the report helps us (re)evaluate evidence and refine conclusions on key building blocks. A good analysis report usually consists of six sections:

1. **Executive summary**—brief focus on important analysis results and conclusions.
2. **Analysis overview**—background on the company, its industry, and its economic setting.
3. **Evidential matter**—financial statements and information used in the analysis, including ratios, trends, comparisons, statistics, and all analytical measures assembled; often organized under the building blocks of analysis.
4. **Assumptions**—identification of important assumptions regarding a company's industry and economic environment, and other important assumptions for estimates.
5. **Key factors**—list of important favorable and unfavorable factors, both quantitative and qualitative, for company performance; usually organized by areas of analysis.
6. **Inferences**—forecasts, estimates, interpretations, and conclusions drawing on all sections of the report.

We must remember that the user dictates relevance, meaning that the analysis report should include a brief table of contents to help readers focus on those areas most relevant to their decisions. All irrelevant matter must be eliminated. For example, decades-old details of obscure transactions and detailed miscues of the analysis are irrelevant. Ambiguities and qualifications to avoid responsibility or hedging inferences must be eliminated. Finally, writing is important. Mistakes in grammar and errors of fact compromise the report's credibility.

Decision Insight

Short Selling *Short selling* refers to selling stock before you buy it. Here's an example: You borrow 100 shares of Nike stock, sell them at $40 each, and receive money from their sale. You then wait. You hope that Nike's stock price falls to, say, $35 each and you can replace the borrowed stock for less than you sold it for, reaping a profit of $5 each less any transaction costs.

Demonstration Problem

DP17

Use the following financial statements of Precision Co. to complete these requirements.

1. Prepare comparative income statements showing the percent increase or decrease for year 2009 in comparison to year 2008.
2. Prepare common-size comparative balance sheets for years 2009 and 2008.
3. Compute the following ratios as of December 31, 2009, or for the year ended December 31, 2009, and identify its building block category for financial statement analysis.
   - **a.** Current ratio
   - **b.** Acid-test ratio
   - **c.** Accounts receivable turnover
   - **d.** Days' sales uncollected
   - **e.** Inventory turnover
   - **f.** Debt ratio
   - **g.** Debt-to-equity ratio
   - **h.** Times interest earned
   - **i.** Profit margin ratio
   - **j.** Total asset turnover
   - **k.** Return on total assets
   - **l.** Return on common stockholders' equity

PRECISION COMPANY Comparative Balance Sheets December 31, 2009 and 2008	2009	2008
Assets		
Current assets		
Cash	$ 79,000	$ 42,000
Short-term investments	65,000	96,000
Accounts receivable, net	120,000	100,000
Merchandise inventory	250,000	265,000
Total current assets	514,000	503,000
Plant assets		
Store equipment, net	400,000	350,000
Office equipment, net	45,000	50,000
Buildings, net	625,000	675,000
Land	100,000	100,000
Total plant assets	1,170,000	1,175,000
Total assets	$1,684,000	$1,678,000
Liabilities		
Current liabilities		
Accounts payable	$ 164,000	$ 190,000
Short-term notes payable	75,000	90,000
Taxes payable	26,000	12,000
Total current liabilities	265,000	292,000
Long-term liabilities		
Notes payable (secured by mortgage on buildings)	400,000	420,000
Total liabilities	665,000	712,000
Stockholders' Equity		
Common stock, $5 par value	475,000	475,000
Retained earnings	544,000	491,000
Total stockholders' equity	1,019,000	966,000
Total liabilities and equity	$1,684,000	$1,678,000

PRECISION COMPANY Comparative Income Statements For Years Ended December 31, 2009 and 2008	2009	2008
Sales	$2,486,000	$2,075,000
Cost of goods sold	1,523,000	1,222,000
Gross profit	963,000	853,000
Operating expenses		
Advertising expense	145,000	100,000
Sales salaries expense	240,000	280,000
Office salaries expense	165,000	200,000
Insurance expense	100,000	45,000
Supplies expense	26,000	35,000
Depreciation expense	85,000	75,000
Miscellaneous expenses	17,000	15,000
Total operating expenses	778,000	750,000
Operating income	185,000	103,000
Interest expense	44,000	46,000
Income before taxes	141,000	57,000
Income taxes	47,000	19,000
Net income	$ 94,000	$ 38,000
Earnings per share	$ 0.99	$ 0.40

Planning the Solution

- Set up a four-column income statement; enter the 2009 and 2008 amounts in the first two columns and then enter the dollar change in the third column and the percent change from 2008 in the fourth column.
- Set up a four-column balance sheet; enter the 2009 and 2008 year-end amounts in the first two columns and then compute and enter the amount of each item as a percent of total assets.
- Compute the required ratios using the data provided. Use the average of beginning and ending amounts when appropriate (see Exhibit 17.16 for definitions).

Solution to Demonstration Problem

1.

PRECISION COMPANY Comparative Income Statements For Years Ended December 31, 2009 and 2008	2009	2008	Increase (Decrease) in 2009 Amount	Percent
Sales	$2,486,000	$2,075,000	$411,000	19.8%
Cost of goods sold	1,523,000	1,222,000	301,000	24.6
Gross profit	963,000	853,000	110,000	12.9
Operating expenses				
Advertising expense	145,000	100,000	45,000	45.0
Sales salaries expense	240,000	280,000	(40,000)	(14.3)
Office salaries expense	165,000	200,000	(35,000)	(17.5)

[continued on next page]

[continued from previous page]

Insurance expense	100,000	45,000	55,000	122.2
Supplies expense	26,000	35,000	(9,000)	(25.7)
Depreciation expense	85,000	75,000	10,000	13.3
Miscellaneous expenses	17,000	15,000	2,000	13.3
Total operating expenses	778,000	750,000	28,000	3.7
Operating income	185,000	103,000	82,000	79.6
Interest expense	44,000	46,000	(2,000)	(4.3)
Income before taxes	141,000	57,000	84,000	147.4
Income taxes	47,000	19,000	28,000	147.4
Net income	$ 94,000	$ 38,000	$ 56,000	147.4
Earnings per share	$ 0.99	$ 0.40	$ 0.59	147.5

2.

PRECISION COMPANY
Common-Size Comparative Balance Sheets
December 31, 2009 and 2008

	December 31		Common-Size Percents	
	2009	2008	2009*	2008*
Assets				
Current assets				
Cash	$ 79,000	$ 42,000	4.7%	2.5%
Short-term investments	65,000	96,000	3.9	5.7
Accounts receivable, net	120,000	100,000	7.1	6.0
Merchandise inventory	250,000	265,000	14.8	15.8
Total current assets	514,000	503,000	30.5	30.0
Plant Assets				
Store equipment, net	400,000	350,000	23.8	20.9
Office equipment, net	45,000	50,000	2.7	3.0
Buildings, net	625,000	675,000	37.1	40.2
Land	100,000	100,000	5.9	6.0
Total plant assets	1,170,000	1,175,000	69.5	70.0
Total assets	$1,684,000	$1,678,000	100.0	100.0
Liabilities				
Current liabilities				
Accounts payable	$ 164,000	$ 190,000	9.7%	11.3%
Short-term notes payable	75,000	90,000	4.5	5.4
Taxes payable	26,000	12,000	1.5	0.7
Total current liabilities	265,000	292,000	15.7	17.4
Long-term liabilities				
Notes payable (secured by mortgage on buildings)	400,000	420,000	23.8	25.0
Total liabilities	665,000	712,000	39.5	42.4
Stockholders' Equity				
Common stock, $5 par value	475,000	475,000	28.2	28.3
Retained earnings	544,000	491,000	32.3	29.3
Total stockholders' equity	1,019,000	966,000	60.5	57.6
Total liabilities and equity	$1,684,000	$1,678,000	100.0	100.0

* Columns do not always exactly add to 100 due to rounding.

3. Ratios for 2009:

a. Current ratio: $514,000/$265,000 = 1.9:1 (liquidity and efficiency)

b. Acid-test ratio: ($79,000 + $65,000 + $120,000)/$265,000 = 1.0:1 (liquidity and efficiency)

c. Average receivables: ($120,000 + $100,000)/2 = $110,000
Accounts receivable turnover: $2,486,000/$110,000 = 22.6 times (liquidity and efficiency)

d. Days' sales uncollected: ($120,000/$2,486,000) × 365 = 17.6 days (liquidity and efficiency)

e. Average inventory: ($250,000 + $265,000)/2 = $257,500
Inventory turnover: $1,523,000/$257,500 = 5.9 times (liquidity and efficiency)

f. Debt ratio: \$665,000/\$1,684,000 = 39.5% (solvency)

g. Debt-to-equity ratio: \$665,000/\$1,019,000 = 0.65 (solvency)

h. Times interest earned: \$185,000/\$44,000 = 4.2 times (solvency)

i. Profit margin ratio: \$94,000/\$2,486,000 = 3.8% (profitability)

j. Average total assets: (\$1,684,000 + \$1,678,000)/2 = \$1,681,000
Total asset turnover: \$2,486,000/\$1,681,000 = 1.48 times (liquidity and efficiency)

k. Return on total assets: \$94,000/\$1,681,000 = 5.6% or 3.8% × 1.48 = 5.6% (profitability)

l. Average total common equity: (\$1,019,000 + \$966,000)/2 = \$992,500
Return on common stockholders' equity: \$94,000/\$992,500 = 9.5% (profitability)

APPENDIX

17A Sustainable Income

A2 Explain the form and assess the content of a complete income statement.

When a company's revenue and expense transactions are from normal, continuing operations, a simple income statement is usually adequate. When a company's activities include income-related events not part of its normal, continuing operations, it must disclose information to help users understand these events and predict future performance. To meet these objectives, companies separate the income statement into continuing operations, discontinued segments, extraordinary items, comprehensive income, and earnings per share. For illustration, Exhibit 17A.1 shows such an income statement for ComUS. These separate distinctions help us measure *sustainable income,* which is the income level most likely to continue into the future. Sustainable income is commonly used in PE ratios and other market-based measures of performance.

Continuing Operations

The first major section (①) shows the revenues, expenses, and income from continuing operations. Users especially rely on this information to predict future operations. Many users view this section as the most important. Earlier chapters explained the items comprising income from continuing operations.

Discontinued Segments

A **business segment** is a part of a company's operations that serves a particular line of business or class of customers. A segment has assets, liabilities, and financial results of operations that can be distinguished from those of other parts of the company. A company's gain or loss from selling or closing down a segment is separately reported. Section ② of Exhibit 17A.1 reports both (1) income from operating the discontinued segment for the current period prior to its disposal and (2) the loss from disposing of the segment's net assets. The income tax effects of each are reported separately from the income taxes expense in section ①.

Extraordinary Items

Section ③ reports **extraordinary gains and losses,** which are those that are *both unusual* and *infrequent.* An **unusual gain or loss** is abnormal or otherwise unrelated to the company's regular activities and environment. An **infrequent gain or loss** is not expected to recur given the company's operating environment. Reporting extraordinary items in a separate category helps users predict future performance, absent the effects of extraordinary items. Items usually considered extraordinary include (1) expropriation (taking away) of property by a foreign government, (2) condemning of property by a domestic government body, (3) prohibition against using an asset by a newly enacted law, and (4) losses and gains from an unusual and infrequent calamity ("act of God"). Items *not* considered extraordinary include (1) write-downs

EXHIBIT 17A.1

Income Statement (all-inclusive) for a Corporation

ComUS Income Statement For Year Ended December 31, 2009			
①	Net sales		$8,478,000
	Operating expenses		
	Cost of goods sold	$5,950,000	
	Depreciation expense	35,000	
	Other selling, general, and administrative expenses	515,000	
	Interest expense	20,000	
	Total operating expenses		(6,520,000)
	Other gains (losses)		
	Loss on plant relocation		(45,000)
	Gain on sale of surplus land		72,000
	Income from continuing operations before taxes		1,985,000
	Income taxes expense		(595,500)
	Income from continuing operations		1,389,500
②	**Discontinued segment**		
	Income from operating Division A (net of $180,000 taxes)	420,000	
	Loss on disposal of Division A (net of $66,000 tax benefit)	(154,000)	266,000
	Income before extraordinary items		1,655,500
③	**Extraordinary items**		
	Gain on land expropriated by state (net of $85,200 taxes)	198,800	
	Loss from earthquake damage (net of $270,000 tax benefit)	(630,000)	(431,200)
	Net income		$1,224,300
④	**Earnings per common share (200,000 outstanding shares)**		
	Income from continuing operations		$ 6.95
	Discontinued operations		1.33
	Income before extraordinary items		8.28
	Extraordinary items		(2.16)
	Net income (basic earnings per share)		$ 6.12

of inventories and write-offs of receivables, (2) gains and losses from disposing of segments, and (3) financial effects of labor strikes.

Gains and losses that are neither unusual nor infrequent are reported as part of continuing operations. Gains and losses that are *either* unusual *or* infrequent, but *not* both, are reported as part of continuing operations *but* after the normal revenues and expenses.

Decision Maker

Small Business Owner You own an orange grove near Jacksonville, Florida. A bad frost destroys about one-half of your oranges. You are currently preparing an income statement for a bank loan. Can you claim the loss of oranges as extraordinary? [Answer—p. 706]

Earnings per Share

The final section ④ of the income statement in Exhibit 17A.1 reports earnings per share for each of the three subcategories of income (continuing operations, discontinued segments, and extraordinary items) when they exist. Earnings per share is discussed in Chapter 13.

Changes in Accounting Principles

The *consistency concept* directs a company to apply the same accounting principles across periods. Yet a company can change from one acceptable accounting principle (such as FIFO, LIFO, or weighted-average) to another as long as the change improves the usefulness of information in its financial statements. A footnote would describe the accounting change and why it is an improvement.

Point: Changes in principles are sometimes required when new accounting standards are issued.

Changes in accounting principles require retrospective application to prior periods' financial statements. *Retrospective application* involves applying a different accounting principle to prior periods as if that principle had always been used. Retrospective application enhances the consistency of financial information between periods, which improves the usefulness of information, especially with comparative analyses. (Prior to 2005, the cumulative effect of changes in accounting principles was recognized in net income in the period of the change.) Accounting standards also require that *a change in depreciation, amortization, or depletion method for long-term operating assets is accounted for as a change in accounting estimate*—that is, prospectively over current and future periods. This reflects the notion that an entity should change its depreciation, amortization, or depletion method only with changes in estimated asset benefits, the pattern of benefit usage, or information about those benefits.

Comprehensive Income

Comprehensive income is net income plus certain gains and losses that bypass the income statement. These items are recorded directly to equity. Specifically, comprehensive income equals the change in equity for the period, excluding investments from and distributions (dividends) to its stockholders. For Best Buy, it is computed as follows ($ millions):

Net income	$1,377
Accumulated other comprehensive income (loss)	(45)
Comprehensive income	$1,332

The most common items included in *accumulated other comprehensive income,* or *AOCI,* are unrealized gains and losses on available-for-sale securities and foreign currency translation adjustments. (Detailed computations for these items are in advanced courses.) Analysts disagree on how to treat these items. Some analysts believe that AOCI items should not be considered when predicting future performance, and some others believe AOCI items should be considered as they reflect on company and managerial performance. Whatever our position, we must be familiar with what AOCI items are as they are commonly reported in financial statements. Best Buy reports its comprehensive income in its statement of shareholders' equity (see Appendix A).

Quick Check

Answers—p. 706

13. Which of the following is an extraordinary item? (*a*) a settlement paid to a customer injured while using the company's product, (*b*) a loss to a plant from damages caused by a meteorite, or (*c*) a loss from selling old equipment.
14. Identify the four major sections of an income statement that are potentially reportable.
15. A company using FIFO for the past 15 years decides to switch to LIFO. The effect of this event on prior years' net income is (*a*) reported as if the new method had always been used; (*b*) ignored because it is a change in an accounting estimate; or (*c*) reported on the current year income statement.

Summary

C1 **Explain the purpose of analysis.** The purpose of financial statement analysis is to help users make better business decisions. Internal users want information to improve company efficiency and effectiveness in providing products and services. External users want information to make better and more informed decisions in pursuing their goals. The common goals of all users are to evaluate a company's (1) past and current performance, (2) current financial position, and (3) future performance and risk.

C2 **Identify the building blocks of analysis.** Financial statement analysis focuses on four "building blocks" of analysis: (1) liquidity and efficiency—ability to meet short-term obligations and efficiently generate revenues; (2) solvency—ability to generate future revenues and meet long-term obligations; (3) profitability—ability to provide financial rewards sufficient to attract and retain financing; and (4) market prospects—ability to generate positive market expectations.

C3 **Describe standards for comparisons in analysis.** Standards for comparisons include (1) intracompany—prior performance and relations between financial items for the company under analysis; (2) competitor—one or more direct competitors of the company; (3) industry—industry statistics; and (4) guidelines (rules of thumb)—general standards developed from past experiences and personal judgments.

C4 **Identify the tools of analysis.** The three most common tools of financial statement analysis are (1) horizontal analysis—comparing a company's financial condition and performance across time; (2) vertical analysis—comparing a company's financial condition and performance to a base amount such as revenues or total assets; and (3) ratio analysis—using and quantifying key relations among financial statement items.

A1 **Summarize and report results of analysis.** A financial statement analysis report is often organized around the building blocks of analysis. A good report separates interpretations and conclusions of analysis from the information underlying them. An analysis report often consists of six sections: (1) executive summary, (2) analysis overview, (3) evidential matter, (4) assumptions, (5) key factors, and (6) inferences.

A2[A] **Explain the form and assess the content of a complete income statement.** An income statement has four *potential* sections: (1) continuing operations, (2) discontinued segments, (3) extraordinary items, and (4) earnings per share.

P1 **Explain and apply methods of horizontal analysis.** Horizontal analysis is a tool to evaluate changes in data across time. Two important tools of horizontal analysis are comparative statements and trend analysis. Comparative statements show amounts for two or more successive periods, often with changes disclosed in both absolute and percent terms. Trend analysis is used to reveal important changes occurring from one period to the next.

P2 **Describe and apply methods of vertical analysis.** Vertical analysis is a tool to evaluate each financial statement item or group of items in terms of a base amount. Two tools of vertical analysis are common-size statements and graphical analyses. Each item in common-size statements is expressed as a percent of a base amount. For the balance sheet, the base amount is usually total assets, and for the income statement, it is usually sales.

P3 **Define and apply ratio analysis.** Ratio analysis provides clues to and symptoms of underlying conditions. Ratios, properly interpreted, identify areas requiring further investigation. A ratio expresses a mathematical relation between two quantities such as a percent, rate, or proportion. Ratios can be organized into the building blocks of analysis: (1) liquidity and efficiency, (2) solvency, (3) profitability, and (4) market prospects.

Guidance Answers to **Decision Maker**

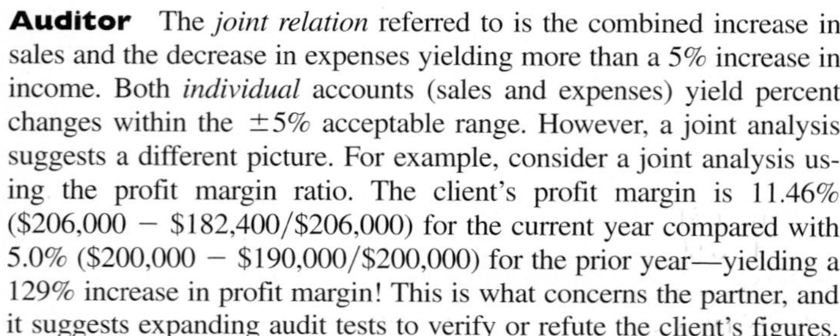

Auditor The *joint relation* referred to is the combined increase in sales and the decrease in expenses yielding more than a 5% increase in income. Both *individual* accounts (sales and expenses) yield percent changes within the ±5% acceptable range. However, a joint analysis suggests a different picture. For example, consider a joint analysis using the profit margin ratio. The client's profit margin is 11.46% (\$206,000 − \$182,400/\$206,000) for the current year compared with 5.0% (\$200,000 − \$190,000/\$200,000) for the prior year—yielding a 129% increase in profit margin! This is what concerns the partner, and it suggests expanding audit tests to verify or refute the client's figures.

Banker Your decision on the loan application is positive for at least two reasons. First, the current ratio suggests a strong ability to meet short-term obligations. Second, current assets of \$160,000 and a current ratio of 4:1 imply current liabilities of \$40,000 (one-fourth of current assets) and a working capital excess of \$120,000. This working capital excess is 60% of the loan amount. However, if the application is for a 10-year loan, our decision is less optimistic. The current ratio and working capital suggest a good safety margin, but indications of inefficiency in operations exist. In particular, a 4:1 current ratio is more than double its key competitors' ratio. This is characteristic of inefficient asset use.

Small Business Owner The frost loss is probably not extraordinary. Jacksonville experiences enough recurring frost damage to make it difficult to argue this event is both unusual and infrequent. Still, you want to highlight the frost loss and hope the bank views this uncommon event separately from continuing operations.

Guidance Answers to **Quick Checks**

1. General-purpose financial statements are intended for a variety of users interested in a company's financial condition and performance—users without the power to require specialized financial reports to meet their specific needs.
2. General-purpose financial statements include the income statement, balance sheet, statement of stockholders' (owner's) equity, and statement of cash flows plus the notes related to these statements.
3. *a*
4. Data from one or more direct competitors are usually preferred for comparative purposes.
5. *d*
6. Percents on comparative income statements show the increase or decrease in each item from one period to the next. On common-size comparative income statements, each item is shown as a percent of net sales for that period.
7. *c*
8. (*a*) (\$820,000 + \$240,000 + \$470,000)/(\$350,000 + \$180,000) = 2.9 to 1.
   (*b*) (\$820,000 + \$240,000)/(\$350,000 + \$180,000) = 2:1.
9. (*a*) \$2,500,000/[(\$290,000 + \$240,000)/2] = 9.43 times.
   (*b*) (\$290,000/\$2,500,000) × 365 = 42 days.
   (*c*) \$750,000/[(\$530,000 + \$470,000)/2] = 1.5 times.
   (*d*) (\$530,000/\$750,000) × 365 = 258 days.
10. *c*
11. *b*
12. $$\text{Profit margin} \times \frac{\text{Total asset}}{\text{turnover}} = \frac{\text{Return on}}{\text{total assets}}$$
    $$\frac{\$945,000}{\$8,500,000} \times 1.8 = 20\%$$
13. (*b*)
14. The four (potentially reportable) major sections are income from continuing operations, discontinued segments, extraordinary items, and earnings per share.
15. (*a*); known as retrospective application.

Key Terms

mhhe.com/wildFAP19e

Key Terms are available at the book's Website for learning and testing in an online Flashcard Format.

Business segment (p. 703)
Common-size financial statement (p. 687)
Comparative financial statements (p. 682)
Efficiency (p. 681)
Equity ratio (p. 695)
Extraordinary gains and losses (p. 703)
Financial reporting (p. 681)
Financial statement analysis (p. 680)
General-purpose financial statements (p. 681)
Horizontal analysis (p. 682)
Infrequent gain or loss (p. 703)
Liquidity (p. 681)
Market prospects (p. 681)
Profitability (p. 681)
Ratio analysis (p. 682)
Solvency (p. 681)
Unusual gain or loss (p. 703)
Vertical analysis (p. 682)
Working capital (p. 691)

Multiple Choice Quiz

Answers on p. 723

mhhe.com/wildFAP19e

Additional Quiz Questions are available at the book's Website.

Quiz17

1. A company's sales in 2008 were \$300,000 and in 2009 were \$351,000. Using 2008 as the base year, the sales trend percent for 2009 is:
a. 17%
b. 85%
c. 100%
d. 117%
e. 48%

Use the following information for questions 2 through 5.

GALLOWAY COMPANY Balance Sheet December 31, 2009	
Assets	
Cash	\$ 86,000
Accounts receivable	76,000
Merchandise inventory	122,000
Prepaid insurance	12,000
Long-term investments	98,000
Plant assets, net	436,000
Total assets	\$830,000
Liabilities and Equity	
Current liabilities	\$124,000
Long-term liabilities	90,000
Common stock	300,000
Retained earnings	316,000
Total liabilities and equity	\$830,000

2. What is Galloway Company's current ratio?
a. 0.69
b. 1.31
c. 3.88
d. 6.69
e. 2.39

3. What is Galloway Company's acid-test ratio?
a. 2.39
b. 0.69
c. 1.31
d. 6.69
e. 3.88

4. What is Galloway Company's debt ratio?
a. 25.78%
b. 100.00%
c. 74.22%
d. 137.78%
e. 34.74%

5. What is Galloway Company's equity ratio?
a. 25.78%
b. 100.00%
c. 34.74%
d. 74.22%
e. 137.78%

Superscript letter A denotes assignments based on Appendix 17A.

Discussion Questions

1. What is the difference between comparative financial statements and common-size comparative statements?

2. Which items are usually assigned a 100% value on (*a*) a common-size balance sheet and (*b*) a common-size income statement?

3. Explain the difference between financial reporting and financial statements.

4. What three factors would influence your evaluation as to whether a company's current ratio is good or bad?

5. Suggest several reasons why a 2:1 current ratio might not be adequate for a particular company.

6. Why is working capital given special attention in the process of analyzing balance sheets?

7. What does the number of days' sales uncollected indicate?

8. What does a relatively high accounts receivable turnover indicate about a company's short-term liquidity?

9. Why is a company's capital structure, as measured by debt and equity ratios, important to financial statement analysts?

10. How does inventory turnover provide information about a company's short-term liquidity?

11. What ratios would you compute to evaluate management performance?

12. Why would a company's return on total assets be different from its return on common stockholders' equity?

13. Where on the income statement does a company report an unusual gain not expected to occur more often than once every two years or so?

14. Use Best Buy's financial statements in Appendix A to compute its return on total assets for the years ended March 3, 2007, and February 25, 2006. Total assets at February 26, 2005, were $10,294 (in millions).

15. Refer to Circuit City's financial statements in Appendix A to compute its equity ratio as of February 28, 2007, and February 28, 2006.

16. Refer to RadioShack's financial statements in Appendix A. Compute its debt ratio as of December 31, 2006, and December 31, 2005.

17. Refer to Apple's financial statements in Appendix A. Compute its profit margin for the fiscal year ended September 30, 2006.

Denotes Discussion Questions that involve decision making.

Available with McGraw-Hill's Homework Manager

QUICK STUDY

QS 17-1
Financial reporting C1

Which of the following items (1) through (9) are part of financial reporting but are *not* included as part of general-purpose financial statements? (1) stock price information and analysis, (2) statement of cash flows, (3) management discussion and analysis of financial performance, (4) income statement, (5) company news releases, (6) balance sheet, (7) financial statement notes, (8) statement of shareholders' equity, (9) prospectus.

QS 17-2
Standard of comparison C3

What are four possible standards of comparison used to analyze financial statement ratios? Which of these is generally considered to be the most useful? Which one is least likely to provide a good basis for comparison?

QS 17-3
Common-size and trend percents P1 P2

Use the following information for Owens Corporation to determine (1) the 2008 and 2009 common-size percents for cost of goods sold using net sales as the base and (2) the 2008 and 2009 trend percents for net sales using 2008 as the base year.

($ thousands)	2009	2008
Net sales	$101,400	$58,100
Cost of goods sold	55,300	30,700

QS 17-4
Horizontal analysis P1

Compute the annual dollar changes and percent changes for each of the following accounts.

	2009	2008
Short-term investments	$110,000	$80,000
Accounts receivable	22,000	25,000
Notes payable	30,000	0

QS 17-5
Building blocks of analysis C2 C4 P3

Match the ratio to the building block of financial statement analysis to which it best relates.

A. Liquidity and efficiency
B. Solvency
C. Profitability
D. Market prospects

1. ______ Gross margin ratio
2. ______ Acid-test ratio
3. ______ Equity ratio
4. ______ Return on total assets
5. ______ Dividend yield
6. ______ Book value per common share
7. ______ Days' sales in inventory
8. ______ Accounts receivable turnover
9. ______ Debt-to-equity
10. ______ Times interest earned

QS 17-6
Identifying financial ratios
C4 P3

1. Which two short-term liquidity ratios measure how frequently a company collects its accounts?
2. What measure reflects the difference between current assets and current liabilities?
3. Which two ratios are key components in measuring a company's operating efficiency? Which ratio summarizes these two components?

QS 17-7
Ratio interpretation
P3

For each ratio listed, identify whether the change in ratio value from 2008 to 2009 is usually regarded as favorable or unfavorable.

Ratio	2009	2008	Ratio	2009	2008
1. Profit margin	10%	9%	5. Accounts receivable turnover	6.7	5.5
2. Debt ratio	43%	39%	6. Basic earnings per share	$1.25	$1.10
3. Gross margin	32%	44%	7. Inventory turnover	3.4	3.6
4. Acid-test ratio	1.20	1.05	8. Dividend yield	4%	3.2%

QS 17-8[A]
Error adjustments
A2

A review of the notes payable files discovers that three years ago the company reported the entire amount of a payment (principal and interest) on an installment note payable as interest expense. This mistake had a material effect on the amount of income in that year. How should the correction be reported in the current year financial statements?

McGraw-Hill's HOMEWORK MANAGER® Available with McGraw-Hill's Homework Manager

EXERCISES

Exercise 17-1
Computation and analysis of trend percents
P1

Compute trend percents for the following accounts, using 2007 as the base year. State whether the situation as revealed by the trends appears to be favorable or unfavorable for each account.

	2011	2010	2009	2008	2007
Sales	$282,700	$270,700	$252,500	$234,460	$150,000
Cost of goods sold	128,100	121,980	115,180	106,340	67,000
Accounts receivable	18,000	17,200	16,300	15,100	9,000

Exercise 17-2
Determination of income effects from common-size and trend percents
P1 P2

Common-size and trend percents for Danian Company's sales, cost of goods sold, and expenses follow. Determine whether net income increased, decreased, or remained unchanged in this three-year period.

	Common-Size Percents			Trend Percents		
	2010	2009	2008	2010	2009	2008
Sales	100.0%	100.0%	100.0%	104.9%	103.7%	100.0%
Cost of goods sold	67.7	61.2	58.4	102.5	108.6	100.0
Total expenses	14.4	13.9	14.2	106.5	101.5	100.0

Exercise 17-3
Common-size percent computation and interpretation
P2 

Express the following comparative income statements in common-size percents and assess whether or not this company's situation has improved in the most recent year.

MULAN CORPORATION Comparative Income Statements For Years Ended December 31, 2009 and 2008	2009	2008
Sales	$657,386	$488,400
Cost of goods sold	427,301	286,202
Gross profit	230,085	202,198
Operating expenses	138,051	94,750
Net income	$ 92,034	$107,448

Exercise 17-4
Analysis of short-term financial condition
A1 P3 

The following information is available for Orkay Company and Lowes Company, similar firms operating in the same industry. Write a half-page report comparing Orkay and Lowes using the available information. Your discussion should include their ability to meet current obligations and to use current assets efficiently.

Team Project: Assume that the two companies apply for a one-year loan from the team. Identify additional information the companies must provide before the team can make a loan decision.

	Orkay			Lowes		
	2010	2009	2008	2010	2009	2008
Current ratio	1.6	1.7	2.0	3.1	2.6	1.8
Acid-test ratio	0.9	1.0	1.1	2.7	2.4	1.5
Accounts receivable turnover	29.5	24.2	28.2	15.4	14.2	15.0
Merchandise inventory turnover	23.2	20.9	16.1	13.5	12.0	11.6
Working capital	$60,000	$48,000	$42,000	$121,000	$93,000	$68,000

Exercise 17-5
Analysis of efficiency and financial leverage
A1 P3 

Caren Company and Revlon Company are similar firms that operate in the same industry. Revlon began operations in 2009 and Caren in 2006. In 2011, both companies pay 7% interest on their debt to creditors. The following additional information is available.

	Caren Company			Revlon Company		
	2011	2010	2009	2011	2010	2009
Total asset turnover	3.0	2.7	2.9	1.6	1.4	1.1
Return on total assets	8.9%	9.5%	8.7%	5.8%	5.5%	5.2%
Profit margin ratio	2.3%	2.4%	2.2%	2.7%	2.9%	2.8%
Sales	$400,000	$370,000	$386,000	$200,000	$160,000	$100,000

Write a half-page report comparing Caren and Revlon using the available information. Your analysis should include their ability to use assets efficiently to produce profits. Also comment on their success in employing financial leverage in 2011.

Exercise 17-6
Common-size percents
P2 

Nabisco Company's year-end balance sheets follow. Express the balance sheets in common-size percents. Round amounts to the nearest one-tenth of a percent. Analyze and comment on the results.

At December 31	2010	2009	2008
Assets			
Cash	$ 36,229	$ 42,780	$ 44,562
Accounts receivable, net	106,073	76,377	57,087
Merchandise inventory	137,408	98,929	62,038
Prepaid expenses	11,548	11,003	4,903
Plant assets, net	335,317	311,062	272,710
Total assets	$626,575	$540,151	$441,300
Liabilities and Equity			
Accounts payable	$157,577	$ 94,024	$ 57,087
Long-term notes payable secured by mortgages on plant assets	116,618	127,962	99,478
Common stock, $10 par value	163,500	163,500	163,500
Retained earnings	188,880	154,665	121,235
Total liabilities and equity	$626,575	$540,151	$441,300

Refer to Nabisco Company's balance sheets in Exercise 17-6. Analyze its year-end short-term liquidity position at the end of 2010, 2009, and 2008 by computing (1) the current ratio and (2) the acid-test ratio. Comment on the ratio results. (Round ratio amounts to two decimals.)

Exercise 17-7
Liquidity analysis
P3

Refer to the Nabisco Company information in Exercise 17-6. The company's income statements for the years ended December 31, 2010 and 2009, follow. Assume that all sales are on credit and then compute: (1) days' sales uncollected, (2) accounts receivable turnover, (3) inventory turnover, and (4) days' sales in inventory. Comment on the changes in the ratios from 2009 to 2010. (Round amounts to one decimal.)

Exercise 17-8
Liquidity analysis and interpretation
P3

For Year Ended December 31	2010		2009	
Sales		$685,000		$557,000
Cost of goods sold	$417,850		$356,265	
Other operating expenses	207,282		141,971	
Interest expense	8,175		8,960	
Income taxes	12,900		12,450	
Total costs and expenses		646,207		519,646
Net income		$ 38,793		$ 37,354
Earnings per share		$ 2.37		$ 2.28

Refer to the Nabisco Company information in Exercises 17-6 and 17-8. Compare the company's long-term risk and capital structure positions at the end of 2010 and 2009 by computing these ratios: (1) debt and equity ratios, (2) debt-to-equity ratio, and (3) times interest earned. Comment on these ratio results.

Exercise 17-9
Risk and capital structure analysis
P3

Refer to Nabisco Company's financial information in Exercises 17-6 and 17-8. Evaluate the company's efficiency and profitability by computing the following for 2010 and 2009: (1) profit margin ratio, (2) total asset turnover, and (3) return on total assets. Comment on these ratio results.

Exercise 17-10
Efficiency and profitability analysis P3 

Refer to Nabisco Company's financial information in Exercises 17-6 and 17-8. Additional information about the company follows. To help evaluate the company's profitability, compute and interpret the following ratios for 2010 and 2009: (1) return on common stockholders' equity, (2) price-earnings ratio on December 31, and (3) dividend yield.

Exercise 17-11
Profitability analysis
P3

Common stock market price, December 31, 2010	$30.00
Common stock market price, December 31, 2009	28.00
Annual cash dividends per share in 2010	0.28
Annual cash dividends per share in 2009	0.24

In 2009, Simplon Merchandising, Inc., sold its interest in a chain of wholesale outlets, taking the company completely out of the wholesaling business. The company still operates its retail outlets. A listing of the major sections of an income statement follows:

A. Income (loss) from continuing operations
B. Income (loss) from operating, or gain (loss) from disposing, a discontinued segment
C. Extraordinary gain (loss)

Indicate where each of the following income-related items for this company appears on its 2009 income statement by writing the letter of the appropriate section in the blank beside each item.

Exercise 17-12[A]
Income statement categories
A2

Section	Item	Debit	Credit
______	1. Net sales		$3,000,000
______	2. Gain on state's condemnation of company property (net of tax)		330,000
______	3. Cost of goods sold	$1,580,000	
______	4. Income taxes expense	117,000	
______	5. Depreciation expense	332,500	
______	6. Gain on sale of wholesale business segment (net of tax)		875,000
______	7. Loss from operating wholesale business segment (net of tax)	544,000	
______	8. Salaries expense	740,000	

Exercise 17-13[A]
Income statement presentation
A2

Use the financial data for Simplon Merchandising, Inc., in Exercise 17-12 to prepare its income statement for calendar year 2009. (Ignore the earnings per share section.)

Available with McGraw-Hill's Homework Manager

McGraw-Hill's HOMEWORK MANAGER®

PROBLEM SET A

Problem 17-1A
Ratios, common-size statements, and trend percents
P1 P2 P3

mhhe.com/wildFAP19e

Selected comparative financial statements of Astalon Company follow.

ASTALON COMPANY
Comparative Income Statements
For Years Ended December 31, 2010, 2009, and 2008

	2010	2009	2008
Sales	$526,304	$403,192	$279,800
Cost of goods sold	316,835	255,624	179,072
Gross profit	209,469	147,568	100,728
Selling expenses	74,735	55,640	36,934
Administrative expenses	47,367	35,481	23,223
Total expenses	122,102	91,121	60,157
Income before taxes	87,367	56,447	40,571
Income taxes	16,250	11,572	8,236
Net income	$ 71,117	$ 44,875	$ 32,335

ASTALON COMPANY
Comparative Balance Sheets
December 31, 2010, 2009, and 2008

	2010	2009	2008
Assets			
Current assets	$ 48,242	$ 38,514	$ 51,484
Long-term investments	0	800	3,620
Plant assets, net	92,405	97,259	58,047
Total assets	$140,647	$136,573	$113,151
Liabilities and Equity			
Current liabilities	$ 20,534	$ 20,349	$ 19,801
Common stock	69,000	69,000	51,000
Other paid-in capital	8,625	8,625	5,667
Retained earnings	42,488	38,599	36,683
Total liabilities and equity	$140,647	$136,573	$113,151

Required

1. Compute each year's current ratio. (Round ratio amounts to one decimal.)

2. Express the income statement data in common-size percents. (Round percents to two decimals.)

3. Express the balance sheet data in trend percents with 2008 as the base year. (Round percents to two decimals.)

Check (3) 2010, Total assets trend, 124.30%

Analysis Component

4. Comment on any significant relations revealed by the ratios and percents computed.

Problem 17-2A
Calculation and analysis of trend percents

A1 P1

Selected comparative financial statements of Adobe Company follow.

ADOBE COMPANY
Comparative Income Statements
For Years Ended December 31, 2010–2004

($ thousands)	2010	2009	2008	2007	2006	2005	2004
Sales	$2,431	$2,129	$1,937	$1,776	$1,657	$1,541	$1,263
Cost of goods sold	1,747	1,421	1,223	1,070	994	930	741
Gross profit	684	708	714	706	663	611	522
Operating expenses	521	407	374	276	239	236	196
Net income	$ 163	$ 301	$ 340	$ 430	$ 424	$ 375	$ 326

ADOBE COMPANY
Comparative Balance Sheets
December 31, 2010–2004

($ thousands)	2010	2009	2008	2007	2006	2005	2004
Assets							
Cash	$ 163	$ 216	$ 224	$ 229	$ 238	$ 235	$ 242
Accounts receivable, net	1,173	1,232	1,115	855	753	714	503
Merchandise inventory	4,244	3,090	2,699	2,275	2,043	1,735	1,258
Other current assets	109	98	60	108	91	93	48
Long-term investments	0	0	0	334	334	334	334
Plant assets, net	5,192	5,172	4,526	2,553	2,639	2,345	2,015
Total assets	$10,881	$9,808	$8,624	$6,354	$6,098	$5,456	$4,400
Liabilities and Equity							
Current liabilities	$ 2,734	$2,299	$1,509	$1,255	$1,089	$1,030	$ 664
Long-term liabilities	2,924	2,547	2,478	1,151	1,176	1,273	955
Common stock	1,980	1,980	1,980	1,760	1,760	1,540	1,540
Other paid-in capital	495	495	495	440	440	385	385
Retained earnings	2,748	2,487	2,162	1,748	1,633	1,228	856
Total liabilities and equity	$10,881	$9,808	$8,624	$6,354	$6,098	$5,456	$4,400

Required

1. Compute trend percents for all components of both statements using 2004 as the base year. (Round percents to one decimal.)

Check (1) 2010, Total assets trend, 247.3%

Analysis Component

2. Analyze and comment on the financial statements and trend percents from part 1.

Problem 17-3A
Transactions, working capital, and liquidity ratios

P3

mhhe.com/wildFAP19e

Page Corporation began the month of May with $884,000 of current assets, a current ratio of 2.6:1, and an acid-test ratio of 1.5:1. During the month, it completed the following transactions (the company uses a perpetual inventory system).

May 2 Purchased $70,000 of merchandise inventory on credit.
8 Sold merchandise inventory that cost $60,000 for $130,000 cash.
10 Collected $30,000 cash on an account receivable.
15 Paid $31,000 cash to settle an account payable.

Check May 22: Current ratio, 2.23; Acid-test ratio, 1.37

May 29: Current ratio, 2.00; Working capital, $462,000

17 Wrote off a $5,000 bad debt against the Allowance for Doubtful Accounts account.
22 Declared a $1 per share cash dividend on its 67,000 shares of outstanding common stock.
26 Paid the dividend declared on May 22.
27 Borrowed $85,000 cash by giving the bank a 30-day, 10% note.
28 Borrowed $100,000 cash by signing a long-term secured note.
29 Used the $185,000 cash proceeds from the notes to buy new machinery.

Required

Prepare a table showing Page's (1) current ratio, (2) acid-test ratio, and (3) working capital, after each transaction. Round ratios to two decimals.

Problem 17-4A
Calculation of financial statement ratios
P3

mhhe.com/wildFAP19e

Selected year-end financial statements of Cadet Corporation follow. (All sales were on credit; selected balance sheet amounts at December 31, 2008, were inventory, $56,900; total assets, $219,400; common stock, $85,000; and retained earnings, $52,348.)

CADET CORPORATION
Income Statement
For Year Ended December 31, 2009

Sales	$456,600
Cost of goods sold	297,450
Gross profit	159,150
Operating expenses	99,400
Interest expense	3,900
Income before taxes	55,850
Income taxes	22,499
Net income	$ 33,351

CADET CORPORATION
Balance Sheet
December 31, 2009

Assets		Liabilities and Equity	
Cash	$ 20,000	Accounts payable	$ 21,500
Short-term investments	8,200	Accrued wages payable	4,400
Accounts receivable, net	29,400	Income taxes payable	3,700
Notes receivable (trade)*	7,000	Long-term note payable, secured by mortgage on plant assets	67,400
Merchandise inventory	34,150	Common stock	85,000
Prepaid expenses	2,700	Retained earnings	66,750
Plant assets, net	147,300		
Total assets	$248,750	Total liabilities and equity	$248,750

* These are short-term notes receivable arising from customer (trade) sales.

Required

Check Acid-test ratio, 2.2 to 1; Inventory turnover, 6.5

Compute the following: (1) current ratio, (2) acid-test ratio, (3) days' sales uncollected, (4) inventory turnover, (5) days' sales in inventory, (6) debt-to-equity ratio, (7) times interest earned, (8) profit margin ratio, (9) total asset turnover, (10) return on total assets, and (11) return on common stockholders' equity.

Problem 17-5A
Comparative ratio analysis A1 P3

Summary information from the financial statements of two companies competing in the same industry follows.

	Karto Company	Bryan Company		Karto Company	Bryan Company
Data from the current year-end balance sheets			**Data from the current year's income statement**		
Assets			Sales	$790,000	$897,200
Cash	$ 19,500	$ 36,000	Cost of goods sold	588,100	634,500
Accounts receivable, net	36,400	53,400	Interest expense	7,600	19,000
Current notes receivable (trade)	9,400	7,600	Income tax expense	15,185	24,769
Merchandise inventory	84,740	134,500	Net income	$179,115	$218,931
Prepaid expenses	6,200	7,250	Basic earnings per share	$ 4.71	$ 5.58
Plant assets, net	350,000	307,400			
Total assets	$506,240	$546,150			
			Beginning-of-year balance sheet data		
Liabilities and Equity			Accounts receivable, net	$ 26,800	$ 51,200
Current liabilities	$ 63,340	$ 73,819	Current notes receivable (trade)	0	0
Long-term notes payable	82,485	99,000	Merchandise inventory	55,600	107,400
Common stock, $5 par value	190,000	196,000	Total assets	408,000	422,500
Retained earnings	170,415	177,331	Common stock, $5 par value	190,000	196,000
Total liabilities and equity	$506,240	$546,150	Retained earnings	124,300	95,600

Required

1. For both companies compute the (*a*) current ratio, (*b*) acid-test ratio, (*c*) accounts (including notes) receivable turnover, (*d*) inventory turnover, (*e*) days' sales in inventory, and (*f*) days' sales uncollected. Identify the company you consider to be the better short-term credit risk and explain why.

Check (1) Bryan: Accounts receivable turnover, 16.0; Inventory turnover, 5.2

2. For both companies compute the (*a*) profit margin ratio, (*b*) total asset turnover, (*c*) return on total assets, and (*d*) return on common stockholders' equity. Assuming that each company paid cash dividends of $3.50 per share and each company's stock can be purchased at $85 per share, compute their (*e*) price-earnings ratios and (*f*) dividend yields. Identify which company's stock you would recommend as the better investment and explain why.

(2) Karto: Profit margin, 22.7%; PE, 18.0

Problem 17-6A[A]
Income statement computations and format
A2 

Selected account balances from the adjusted trial balance for Lindo Corporation as of its calendar year-end December 31, 2009, follow.

	Debit	Credit
a. Interest revenue		$ 15,000
b. Depreciation expense—Equipment	$ 35,000	
c. Loss on sale of equipment	26,850	
d. Accounts payable		45,000
e. Other operating expenses	107,400	
f. Accumulated depreciation—Equipment		72,600
g. Gain from settlement of lawsuit		45,000
h. Accumulated depreciation—Buildings		175,500
i. Loss from operating a discontinued segment (pretax)	19,250	
j. Gain on insurance recovery of tornado damage (pretax and extraordinary)		30,120
k. Net sales		999,500
l. Depreciation expense—Buildings	53,000	
m. Correction of overstatement of prior year's sales (pretax)	17,000	
n. Gain on sale of discontinued segment's assets (pretax)		35,000
o. Loss from settlement of lawsuit	24,750	
p. Income taxes expense	?	
q. Cost of goods sold	483,500	

Required

Answer each of the following questions by providing supporting computations.

1. Assume that the company's income tax rate is 30% for all items. Identify the tax effects and after-tax amounts of the four items labeled pretax.
2. What is the amount of income from continuing operations before income taxes? What is the amount of the income taxes expense? What is the amount of income from continuing operations?
3. What is the total amount of after-tax income (loss) associated with the discontinued segment?
4. What is the amount of income (loss) before the extraordinary items?
5. What is the amount of net income for the year?

Check (3) $11,025
(4) $241,325
(5) $262,409

PROBLEM SET B

Problem 17-1B

Ratios, common-size statements, and trend percents

P1 P2 P3 

Selected comparative financial statement information of Danno Corporation follows.

DANNO CORPORATION
Comparative Income Statements
For Years Ended December 31, 2010, 2009, and 2008

	2010	2009	2008
Sales	$392,000	$300,304	$208,400
Cost of goods sold	235,984	190,092	133,376
Gross profit	156,016	110,212	75,024
Selling expenses	55,664	41,442	27,509
Administrative expenses	35,280	26,427	17,297
Total expenses	90,944	67,869	44,806
Income before taxes	65,072	42,343	30,218
Income taxes	12,103	8,680	6,134
Net income	$ 52,969	$ 33,663	$ 24,084

DANNO CORPORATION
Comparative Balance Sheets
December 31, 2010, 2009, and 2008

	2010	2009	2008
Assets			
Current assets	$ 53,776	$ 42,494	$ 55,118
Long-term investments	0	400	4,110
Plant assets, net	99,871	106,303	64,382
Total assets	$153,647	$149,197	$123,610
Liabilities and Equity			
Current liabilities	$ 22,432	$ 22,230	$ 21,632
Common stock	70,000	70,000	52,000
Other paid-in capital	8,750	8,750	5,778
Retained earnings	52,465	48,217	44,200
Total liabilities and equity	$153,647	$149,197	$123,610

Required

1. Compute each year's current ratio. (Round ratio amounts to one decimal.)
2. Express the income statement data in common-size percents. (Round percents to two decimals.)
3. Express the balance sheet data in trend percents with 2008 as the base year. (Round percents to two decimals.)

Check (3) 2010, Total assets trend, 124.30%

Analysis Component

4. Comment on any significant relations revealed by the ratios and percents computed.

Problem 17-2B
Calculation and analysis of trend percents
A1 P1

Selected comparative financial statements of Park Company follow.

PARK COMPANY
Comparative Income Statements
For Years Ended December 31, 2010–2004

($ thousands)	2010	2009	2008	2007	2006	2005	2004
Sales	$570	$620	$640	$690	$750	$780	$870
Cost of goods sold	286	300	304	324	350	360	390
Gross profit	284	320	336	366	400	420	480
Operating expenses	94	114	122	136	150	154	160
Net income	$190	$206	$214	$230	$250	$266	$320

PARK COMPANY
Comparative Balance Sheets
December 31, 2010–2004

($ thousands)	2010	2009	2008	2007	2006	2005	2004
Assets							
Cash	$ 54	$ 56	$ 62	$ 64	$ 70	$ 72	$ 78
Accounts receivable, net	140	146	150	154	160	164	170
Merchandise inventory	176	182	188	190	196	200	218
Other current assets	44	44	46	48	48	50	50
Long-term investments	46	40	36	120	120	120	120
Plant assets, net	520	524	530	422	430	438	464
Total assets	$980	$992	$1,012	$998	$1,024	$1,044	$1,100
Liabilities and Equity							
Current liabilities	$158	$166	$ 196	$200	$ 220	$ 270	$ 290
Long-term liabilities	102	130	152	158	204	224	270
Common stock	180	180	180	180	180	180	180
Other paid-in capital	80	80	80	80	80	80	80
Retained earnings	460	436	404	380	340	290	280
Total liabilities and equity	$980	$992	$1,012	$998	$1,024	$1,044	$1,100

Required

1. Compute trend percents for all components of both statements using 2004 as the base year. (Round percents to one decimal.)

Check (1) 2010, Total assets trend, 89.1%

Analysis Component

2. Analyze and comment on the financial statements and trend percents from part 1.

Problem 17-3B
Transactions, working capital, and liquidity ratios
P3

Menardo Corporation began the month of June with $600,000 of current assets, a current ratio of 2.5:1, and an acid-test ratio of 1.4:1. During the month, it completed the following transactions (the company uses a perpetual inventory system).

June 1 Sold merchandise inventory that cost $150,000 for $240,000 cash.
3 Collected $176,000 cash on an account receivable.
5 Purchased $300,000 of merchandise inventory on credit.
7 Borrowed $200,000 cash by giving the bank a 60-day, 8% note.
10 Borrowed $240,000 cash by signing a long-term secured note.
12 Purchased machinery for $550,000 cash.
15 Declared a $1 per share cash dividend on its 160,000 shares of outstanding common stock.
19 Wrote off a $10,000 bad debt against the Allowance for Doubtful Accounts account.
22 Paid $24,000 cash to settle an account payable.
30 Paid the dividend declared on June 15.

Check June 3: Current ratio, 2.88; Acid-test ratio, 2.40

June 30: Working capital, $(20,000); Current ratio, 0.97

Required

Prepare a table showing the company's (1) current ratio, (2) acid-test ratio, and (3) working capital after each transaction. Round ratios to two decimals.

Problem 17-4B
Calculation of financial statement ratios
P3

Selected year-end financial statements of Steele Corporation follow. (All sales were on credit; selected balance sheet amounts at December 31, 2008, were inventory, $55,900; total assets, $249,400; common stock, $105,000; and retained earnings, $17,748.)

STEELE CORPORATION Income Statement For Year Ended December 31, 2009	
Sales	$447,600
Cost of goods sold	298,150
Gross profit	149,450
Operating expenses	98,500
Interest expense	4,600
Income before taxes	46,350
Income taxes	18,672
Net income	$ 27,678

STEELE CORPORATION Balance Sheet December 31, 2009			
Assets		**Liabilities and Equity**	
Cash	$ 8,000	Accounts payable	$ 25,500
Short-term investments	8,000	Accrued wages payable	3,000
Accounts receivable, net	28,800	Income taxes payable	4,000
Notes receivable (trade)*	8,000	Long-term note payable, secured by mortgage on plant assets	63,400
Merchandise inventory	34,150	Common stock, $5 par value	105,000
Prepaid expenses	2,750	Retained earnings	39,100
Plant assets, net	150,300		
Total assets	$240,000	Total liabilities and equity	$240,000

* These are short-term notes receivable arising from customer (trade) sales.

Required

Check Acid-test ratio, 1.6 to 1; Inventory turnover, 6.6

Compute the following: (1) current ratio, (2) acid-test ratio, (3) days' sales uncollected, (4) inventory turnover, (5) days' sales in inventory, (6) debt-to-equity ratio, (7) times interest earned, (8) profit margin ratio, (9) total asset turnover, (10) return on total assets, and (11) return on common stockholders' equity.

Problem 17-5B
Comparative ratio analysis A1 P3

Summary information from the financial statements of two companies competing in the same industry follows.

	Crisco Company	Silas Company		Crisco Company	Silas Company
Data from the current year-end balance sheets			**Data from the current year's income statement**		
Assets			Sales	$394,600	$668,500
Cash	$ 21,000	$ 37,500	Cost of goods sold	291,600	481,000
Accounts receivable, net	78,100	71,500	Interest expense	6,900	13,300
Current notes receivable (trade)	12,600	10,000	Income tax expense	6,700	14,300
Merchandise inventory	87,800	83,000	Net income	34,850	62,700
Prepaid expenses	10,700	11,100	Basic earnings per share	1.16	1.84
Plant assets, net	177,900	253,300			
Total assets	$388,100	$466,400			
			Beginning-of-year balance sheet data		
Liabilities and Equity			Accounts receivable, net	$ 73,200	$ 74,300
Current liabilities	$100,500	$ 98,000	Current notes receivable (trade)	0	0
Long-term notes payable	85,650	62,400	Merchandise inventory	106,100	81,500
Common stock, $5 par value	150,000	170,000	Total assets	384,400	444,000
Retained earnings	51,950	136,000	Common stock, $5 par value	150,000	170,000
Total liabilities and equity	$388,100	$466,400	Retained earnings	50,100	110,700

Required

1. For both companies compute the (*a*) current ratio, (*b*) acid-test ratio, (*c*) accounts (including notes) receivable turnover, (*d*) inventory turnover, (*e*) days' sales in inventory, and (*f*) days' sales uncollected. Identify the company you consider to be the better short-term credit risk and explain why.

2. For both companies compute the (*a*) profit margin ratio, (*b*) total asset turnover, (*c*) return on total assets, and (*d*) return on common stockholders' equity. Assuming that each company paid cash dividends of $1.10 per share and each company's stock can be purchased at $25 per share, compute their (*e*) price-earnings ratios and (*f*) dividend yields. Identify which company's stock you would recommend as the better investment and explain why.

Check (1) Crisco: Accounts receivable turnover, 4.8; Inventory turnover, 3.0

(2) Silas: Profit margin, 9.4%; PE, 13.6

Problem 17-6B[A]
Income statement computations and format

A2

Selected account balances from the adjusted trial balance for Harton Corp. as of its calendar year-end December 31, 2009, follow.

	Debit	Credit
a. Accumulated depreciation—Buildings		$ 410,000
b. Interest revenue		30,000
c. Net sales		2,650,000
d. Income taxes expense	$?	
e. Loss on hurricane damage (pretax and extraordinary)	74,000	
f. Accumulated depreciation—Equipment		230,000
g. Other operating expenses	338,000	
h. Depreciation expense—Equipment	110,000	
i. Loss from settlement of lawsuit	46,000	
j. Gain from settlement of lawsuit		78,000
k. Loss on sale of equipment	34,000	
l. Loss from operating a discontinued segment (pretax)	130,000	
m. Depreciation expense—Buildings	166,000	
n. Correction of overstatement of prior year's expense (pretax)		58,000
o. Cost of goods sold	1,050,000	
p. Loss on sale of discontinued segment's assets (pretax)	190,000	
q. Accounts payable		142,000

Required

Answer each of the following questions by providing supporting computations.

1. Assume that the company's income tax rate is 25% for all items. Identify the tax effects and after-tax amounts of the four items labeled pretax.

2. What is the amount of income from continuing operations before income taxes? What is the amount of income taxes expense? What is the amount of income from continuing operations?

3. What is the total amount of after-tax income (loss) associated with the discontinued segment?

4. What is the amount of income (loss) before the extraordinary items?

5. What is the amount of net income for the year?

Check (3) $(240,000)
(4) $520,500
(5) $465,000

SERIAL PROBLEM

Success Systems

(This serial problem began in Chapter 1 and continues through most of the book. If previous chapter segments were not completed, the serial problem can begin at this point. It is helpful, but not necessary, to use the Working Papers that accompany the book.)

SP 17 Use the following selected data from Success Systems' income statement for the three months ended March 31, 2010, and from its March 31, 2010, balance sheet to complete the requirements below: computer services revenue, $25,160; net sales (of goods), $18,693; total sales and revenue, $43,853; cost of goods sold, $14,052; net income, $18,686; quick assets, $100,205; current assets, $105,209; total assets, $129,909; current liabilities, $875; total liabilities, $875; and total equity, $129,034.

Required

1. Compute the gross margin ratio (both with and without services revenue) and net profit margin ratio.
2. Compute the current ratio and acid-test ratio.
3. Compute the debt ratio and equity ratio.
4. What percent of its assets are current? What percent are long term?

BEYOND THE NUMBERS

REPORTING IN ACTION

A1 P1 P2 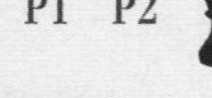

BTN 17-1 Refer to Best Buy's financial statements in Appendix A to answer the following.

1. Using fiscal 2005 as the base year, compute trend percents for fiscal years 2005, 2006, and 2007 for revenues, cost of sales, selling general and administrative expenses, income taxes, and net income. (Round to the nearest whole percent.)
2. Compute common-size percents for fiscal years 2007 and 2006 for the following categories of assets: (*a*) total current assets, (*b*) property and equipment, net, and (*c*) intangible assets. (Round to the nearest tenth of a percent.)
3. Comment on any significant changes across the years for the income statement trends computed in part 1 and the balance sheet percents computed in part 2.

Fast Forward

4. Access Best Buy's financial statements for fiscal years ending after March 3, 2007, from Best Buy's Website (BestBuy.com) or the SEC database (www.SEC.gov). Update your work for parts 1, 2, and 3 using the new information accessed.

COMPARATIVE ANALYSIS

C3 P2

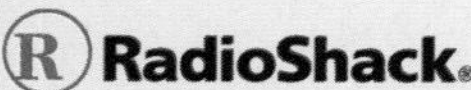

BTN 17-2 Key figures for Best Buy, Circuit City, and RadioShack follow.

($ millions)	Best Buy	Circuit City	RadioShack
Cash and equivalents	$ 1,205	$ 141	$ 472
Accounts receivable, net	548	383	248
Inventories	4,028	1,637	752
Retained earnings	5,507	1,336	1,781
Cost of sales	27,165	9,501	2,544
Revenues	35,934	12,430	4,778
Total assets	13,570	4,007	2,070

Required

1. Compute common-size percents for each of the companies using the data provided. (Round percents to one decimal.)
2. Which company retains a higher portion of cumulative net income in the company?
3. Which company has a higher gross margin ratio on sales?
4. Which company holds a higher percent of its total assets as inventory?

ETHICS CHALLENGE

A1

BTN 17-3 As Beacon Company controller, you are responsible for informing the board of directors about its financial activities. At the board meeting, you present the following information.

	2009	2008	2007
Sales trend percent	147.0%	135.0%	100.0%
Selling expenses to sales	10.1%	14.0%	15.6%
Sales to plant assets ratio	3.8 to 1	3.6 to 1	3.3 to 1
Current ratio	2.9 to 1	2.7 to 1	2.4 to 1
Acid-test ratio	1.1 to 1	1.4 to 1	1.5 to 1
Inventory turnover	7.8 times	9.0 times	10.2 times
Accounts receivable turnover	7.0 times	7.7 times	8.5 times
Total asset turnover	2.9 times	2.9 times	3.3 times
Return on total assets	10.4%	11.0%	13.2%
Return on stockholders' equity	10.7%	11.5%	14.1%
Profit margin ratio	3.6%	3.8%	4.0%

After the meeting, the company's CEO holds a press conference with analysts in which she mentions the following ratios.

	2009	2008	2007
Sales trend percent	147.0%	135.0%	100.0%
Selling expenses to sales	10.1%	14.0%	15.6%
Sales to plant assets ratio	3.8 to 1	3.6 to 1	3.3 to 1
Current ratio	2.9 to 1	2.7 to 1	2.4 to 1

Required

1. Why do you think the CEO decided to report 4 ratios instead of the 11 prepared?

2. Comment on the possible consequences of the CEO's reporting of the ratios selected.

BTN 17-4 Each team is to select a different industry, and each team member is to select a different company in that industry and acquire its financial statements. Use those statements to analyze the company, including at least one ratio from each of the four building blocks of analysis. When necessary, use the financial press to determine the market price of its stock. Communicate with teammates via a meeting, e-mail, or telephone to discuss how different companies compare to each other and to industry norms. The team is to prepare a single one-page memorandum reporting on its analysis and the conclusions reached.

COMMUNICATING IN PRACTICE

C2 A1 P3

BTN 17-5 Access the February 23, 2007, filing of the 2006 10-K report of the **Hershey Foods Corporation** (ticker HSY) at www.SEC.gov and complete the following requirements.

TAKING IT TO THE NET

C4 P3

Required

Compute or identify the following profitability ratios of Hershey for its years ending December 31, 2006, *and* December 31, 2005. Interpret its profitability using the results obtained for these two years.

1. Profit margin ratio.

2. Gross profit ratio.

3. Return on total assets. (Total assets in 2004 were $3,794,750,000.)

4. Return on common stockholders' equity. (Total shareholders' equity in 2004 was $1,137,103,000.)

5. Basic earnings per common share.

BTN 17-6 A team approach to learning financial statement analysis is often useful.

TEAMWORK IN ACTION

C2 P1 P2 P3

Required

1. Each team should write a description of horizontal and vertical analysis that all team members agree with and understand. Illustrate each description with an example.

2. *Each* member of the team is to select *one* of the following categories of ratio analysis. Explain what the ratios in that category measure. Choose one ratio from the category selected, present its formula, and explain what it measures.

a. Liquidity and efficiency **c.** Profitability

b. Solvency **d.** Market prospects

3. Each team member is to present his or her notes from part 2 to teammates. Team members are to confirm or correct other teammates' presentation.

Hint: Pairing within teams may be necessary for part 2. Use as an in-class activity or as an assignment. Consider presentations to the entire class using team rotation with transparencies.

ENTREPRENEURIAL DECISION

A1 P1 P2 P3 

BTN 17-7 Assume that David and Tom Gardner of **The Motley Fool** (Fool.com) have impressed you since you first heard of their rather improbable rise to prominence in financial circles. You learn of a staff opening at The Motley Fool and decide to apply for it. Your resume is successfully screened from the thousands received and you advance to the interview process. You learn that the interview consists of analyzing the following financial facts and answering analysis questions. (*Note:* The data are taken from a small merchandiser in outdoor recreational equipment.)

	2008	2007	2006
Sales trend percents	137.0%	125.0%	100.0%
Selling expenses to sales	9.8%	13.7%	15.3%
Sales to plant assets ratio	3.5 to 1	3.3 to 1	3.0 to 1
Current ratio	2.6 to 1	2.4 to 1	2.1 to 1
Acid-test ratio	0.8 to 1	1.1 to 1	1.2 to 1
Merchandise inventory turnover	7.5 times	8.7 times	9.9 times
Accounts receivable turnover	6.7 times	7.4 times	8.2 times
Total asset turnover	2.6 times	2.6 times	3.0 times
Return on total assets	8.8%	9.4%	11.1%
Return on equity	9.75%	11.50%	12.25%
Profit margin ratio	3.3%	3.5%	3.7%

Required

Use these data to answer each of the following questions with explanations.

1. Is it becoming easier for the company to meet its current liabilities on time and to take advantage of any available cash discounts? Explain.
2. Is the company collecting its accounts receivable more rapidly? Explain.
3. Is the company's investment in accounts receivable decreasing? Explain.
4. Is the company's investment in plant assets increasing? Explain.
5. Is the owner's investment becoming more profitable? Explain.
6. Did the dollar amount of selling expenses decrease during the three-year period? Explain.

HITTING THE ROAD

C1 P3

BTN 17-8 You are to devise an investment strategy to enable you to accumulate $1,000,000 by age 65. Start by making some assumptions about your salary. Next compute the percent of your salary that you will be able to save each year. If you will receive any lump-sum monies, include those amounts in your calculations. Historically, stocks have delivered average annual returns of 10–11%. Given this history, you should probably not assume that you will earn above 10% on the money you invest. It is not necessary to specify exactly what types of assets you will buy for your investments; just assume a rate you expect to earn. Use the future value tables in Appendix B to calculate how your savings will grow. Experiment a bit with your figures to see how much less you have to save if you start at, for example, age 25 versus age 35 or 40. (For this assignment, do not include inflation in your calculations.)

GLOBAL DECISION

A1

BTN 17-9 **DSG international plc** (www.DSGiplc.com), **Best Buy**, **Circuit City**, and **RadioShack** are competitors in the global marketplace. Key figures for DSG follow (in millions).

Cash and equivalents	£ 441
Accounts receivable, net	393
Inventories	1,031
Retained earnings	1,490
Cost of sales	7,285
Revenues	7,930
Total assets	3,977

Required

1. Compute common-sized percents for DSG using the data provided. (Round percents to one decimal.)
2. Compare the results with Best Buy, Circuit City, and RadioShack from BTN 17-2.

ANSWERS TO MULTIPLE CHOICE QUIZ

1. d; ($351,000/$300,000) × 100 = 117%
2. e; ($86,000 + $76,000 + $122,000 + $12,000)/$124,000 = 2.39
3. c; ($86,000 + $76,000)/$124,000 = 1.31
4. a; ($124,000 + $90,000)/$830,000 = 25.78%
5. d; ($300,000 + $316,000)/$830,000 = 74.22%

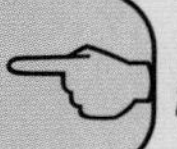

A Look Back

Chapter 17 described the analysis and interpretation of financial statement information. We applied horizontal, vertical, and ratio analyses to better understand company performance and financial condition.

A Look at This Chapter

We begin our study of managerial accounting by explaining its purpose and describing its major characteristics. We also discuss cost concepts and describe how they help managers gather and organize information for making decisions. The reporting of manufacturing activities is also discussed.

A Look Ahead

The remaining chapters discuss the types of decisions managers must make and how managerial accounting helps with those decisions. The first of these chapters, Chapter 19, considers how we measure costs assigned to certain types of projects.

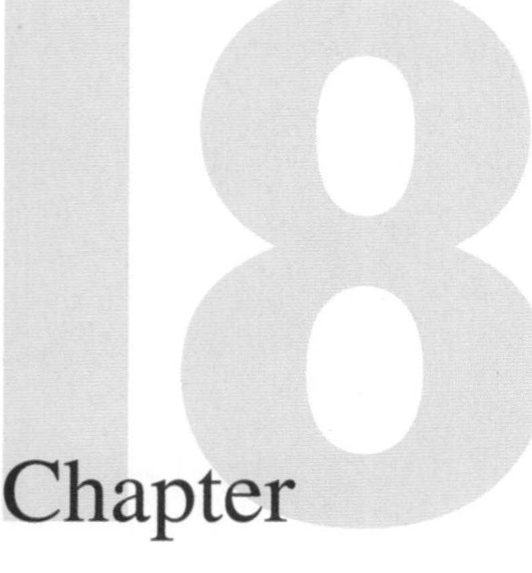

Chapter 18 Managerial Accounting Concepts and Principles

Learning Objectives

CAP

Conceptual

C1 Explain the purpose and nature of managerial accounting. *(p. 726)*

C2 Describe the lean business model. *(p. 729)*

C3 Describe fraud and the role of ethics in managerial accounting. *(p. 731)*

C4 Describe accounting concepts useful in classifying costs. *(p. 732)*

C5 Define product and period costs and explain how they impact financial statements. *(p. 733)*

C6 Explain how balance sheets and income statements for manufacturing and merchandising companies differ. *(p. 735)*

C7 Explain manufacturing activities and the flow of manufacturing costs. *(p. 739)*

Analytical

A1 Compute cycle time and cycle efficiency, and explain their importance to production management *(p. 742)*

Procedural

P1 Compute cost of goods sold for a manufacturer. *(p. 737)*

P2 Prepare a manufacturing statement and explain its purpose and links to financial statements. *(p. 740)*

LP18

Decision Feature

No Naked Popcorn

"Find a niche and stay focused"—Brian Taylor

ELK GROVE VILLAGE, IL—As a hungry college student, Brian Taylor liked to eat popcorn. Lots of it. Bored with "naked popcorn," Brian began experimenting with seasonings such as nacho cheese, cajun, jalapeño, and apple cinnamon. After he shared his concoctions with friends, dorm mates, and others, the demand for Brian's seasonings ballooned. In less than two years, Brian had the number one shake-on popcorn seasoning in the market, **Kernel Season's (KernelSeasons.com).**

Brian launched Kernel Season's with $7,000 he earned from giving tennis lessons and selling knives. In the beginning, he gave away his popcorn seasonings to local theaters to build awareness. Just like his college friends, moviegoers loved the all-natural, low-calorie seasonings. Soon theaters across the country were asking for his seasonings, and Brian worked hard to meet demand. "I was the only employee," explains Brian. "I made sales and shipped orders. I was figuring it out as I went along."

Well, business is now popping. Fourteen varieties of Kernel Season's are available in over 14,000 movie theaters and 15,000 grocery stores. Annual sales now exceed $5 million, and Brian is on Inc.com's "30 under 30," a list of America's coolest young entrepreneurs.

Brian believes college is the best time to start a new business. "Risk is low, and banks understand young entrepreneurs are trying to get things going," explains Brian. But Brian emphasizes that understanding basic managerial principles, product and period costs, manufacturing statements, and cost flows is equally crucial. "[I was] dedicated to business classes," says Brian, including my "accounting class." Brian uses managerial accounting information from his production process to monitor and control costs and to assess new business opportunities, including Kernel Season's apparel. Brian further stresses that company success and growth require him to develop budgets, monitor product performance, and make quick decisions.

Brian believes entrepreneurs fill a void by creating a niche. However, financial success depends on monitoring and controlling operations to best meet customer needs. Brian cautions would-be entrepreneurs to "stay focused" because in the absence of applying managerial accounting principles and concepts, it's just naked popcorn.

[Sources: *Kernel Season's Website,* January 2009; *Lake County News Sun,* October 2003; *Female Entrepreneur,* July/August 2003; *Chicago Tonight* interview, August 2007; *StartupNation.com,* May 2007; *Inc.com Website,* May 2008]

Chapter Preview

Managerial accounting, like financial accounting, provides information to help users make better decisions. However, managerial accounting and financial accounting differ in important ways, which this chapter explains. This chapter also compares the accounting and reporting practices used by manufacturing and merchandising companies. A merchandising company sells products without changing their condition. A manufacturing company buys raw materials and turns them into finished products for sale to customers. A third type of company earns revenues by providing services rather than products. The skills, tools, and techniques developed for measuring a manufacturing company's activities apply to service companies as well. The chapter concludes by explaining the flow of manufacturing activities and preparing the manufacturing statement.

Managerial Accounting Concepts and Principles

Managerial Accounting Basics
- Purpose of managerial accounting
- Nature of managerial accounting
- Managerial decisions
- Managerial accounting in business
- Fraud and ethics in managerial accounting

Managerial Cost Concepts
- Types of cost classifications
- Identification of cost classifications
- Cost concepts for service companies

Reporting Manufacturing Activities
- Balance sheet
- Income statement
- Flow of activities
- Manufacturing statement

Managerial Accounting Basics

Video18.1

Managerial accounting is an activity that provides financial and nonfinancial information to an organization's managers and other internal decision makers. This section explains the purpose of managerial accounting (also called *management accounting*) and compares it with financial accounting. The main purpose of the financial accounting system is to prepare general-purpose financial statements. That information is incomplete for internal decision makers who manage organizations.

Purpose of Managerial Accounting

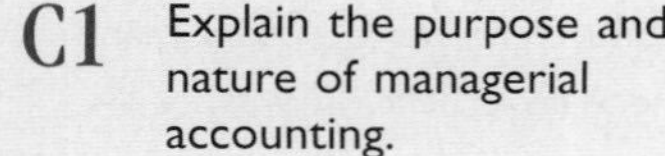

C1 Explain the purpose and nature of managerial accounting.

The purpose of both managerial accounting and financial accounting is providing useful information to decision makers. They do this by collecting, managing, and reporting information in demand by their users. Both areas of accounting also share the common practice of reporting monetary information, although managerial accounting includes the reporting of nonmonetary information. They even report some of the same information. For instance, a company's financial statements contain information useful for both its managers (insiders) and other persons interested in the company (outsiders).

Point: Nonfinancial information, also called nonmonetary information, includes customer and employee satisfaction data, the percentage of on-time deliveries, and product defect rates.

The remainder of this book looks carefully at managerial accounting information, how to gather it, and how managers use it. We consider the concepts and procedures used to determine the costs of products and services as well as topics such as budgeting, break-even analysis, product costing, profit planning, and cost analysis. Information about the costs of products and services is important for many decisions that managers make. These decisions include predicting the future costs of a product or service. Predicted costs are used in product pricing, profitability analysis, and in deciding whether to make or buy a product or component. More generally, much of managerial accounting involves gathering information about costs for planning and control decisions.

Point: Costs are important to managers because they impact both the financial position and profitability of a business. Managerial accounting assists in analysis, planning, and control of costs.

Planning is the process of setting goals and making plans to achieve them. Companies formulate long-term strategic plans that usually span a 5- to 10-year horizon and then refine them with medium-term and short-term plans. Strategic plans usually set a firm's long-term direction by developing a road map based on opportunities such as new products, new markets, and capital investments. A strategic plan's goals and objectives are broadly defined given its long-term orientation. Medium- and short-term plans are more operational in nature. They translate the strategic plan into actions. These plans are more concrete and consist of better defined objectives and goals. A short-term plan often covers a one-year period that, when translated in monetary terms, is known as a budget.

Control is the process of monitoring planning decisions and evaluating an organization's activities and employees. It includes the measurement and evaluation of actions, processes, and outcomes. Feedback provided by the control function allows managers to revise their plans. Measurement of actions and processes also allows managers to take corrective actions to avoid undesirable outcomes. For example, managers periodically compare actual results with planned results. Exhibit 18.1 portrays the important management functions of planning and control.

EXHIBIT 18.1
Planning and Control

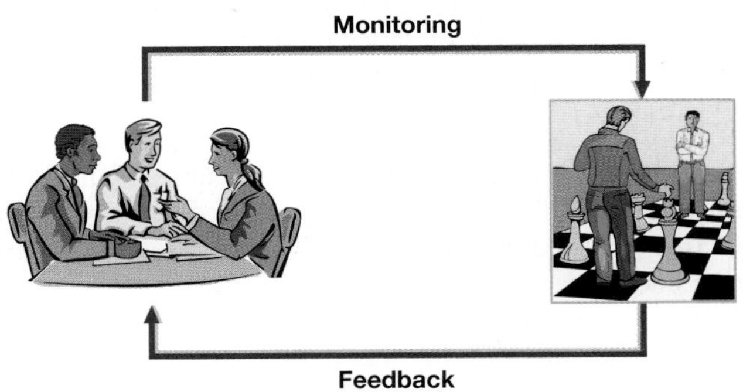
Monitoring
Feedback

Planning

- Strategic aims
- Long- & short-term
- Annual budgets

Control

- Measurement
- Evaluation
- Oversight

Managers use information to plan and control business activities. In later chapters, we explain how managers also use this information to direct and improve business operations.

Nature of Managerial Accounting

Managerial accounting has its own special characteristics. To understand these characteristics, we compare managerial accounting to financial accounting; they differ in at least seven important ways. These differences are summarized in Exhibit 18.2. This section discusses each of these characteristics.

EXHIBIT 18.2
Key Differences between Managerial Accounting and Financial Accounting

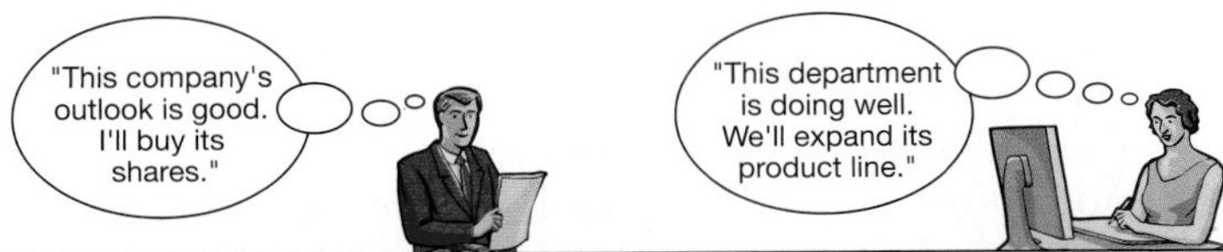
"This company's outlook is good. I'll buy its shares."
"This department is doing well. We'll expand its product line."

	Financial Accounting	Managerial Accounting
1. Users and decision makers	Investors, creditors, and other users external to the organization	Managers, employees, and decision makers internal to the organization
2. Purpose of information	Assist external users in making investment, credit, and other decisions	Assist managers in making planning and control decisions
3. Flexibility of practice	Structured and often controlled by GAAP	Relatively flexible (no GAAP constraints)
4. Timeliness of information	Often available only after an audit is complete	Available quickly without the need to wait for an audit
5. Time dimension	Focus on historical information with some predictions	Many projections and estimates; historical information also presented
6. Focus of information	Emphasis on whole organization	Emphasis on an organization's projects, processes, and subdivisions
7. Nature of information	Monetary information	Mostly monetary; but also nonmonetary information

Users and Decision Makers Companies accumulate, process, and report financial accounting and managerial accounting information for different groups of decision makers. Financial accounting information is provided primarily to external users including investors, creditors, analysts, and regulators. External users rarely have a major role in managing a company's daily activities. Managerial accounting information is provided primarily to internal users who are responsible for making and implementing decisions about a company's business activities.

Point: It is desirable to accumulate information for management reports in a database separate from financial accounting records.

Purpose of Information Investors, creditors, and other external users of financial accounting information must often decide whether to invest in or lend to a company. If they have already done so, they must decide whether to continue owning the company or carrying the loan. Internal decision makers must plan a company's future. They seek to take advantage of opportunities or to overcome obstacles. They also try to control activities and ensure their effective and efficient implementation. Managerial accounting information helps these internal users make both planning and control decisions.

Flexibility of Practice External users compare companies by using financial reports and need protection against false or misleading information. Accordingly, financial accounting relies on accepted principles that are enforced through an extensive set of rules and guidelines, or GAAP. Internal users need managerial accounting information for planning and controlling their company's activities rather than for external comparisons. They require different types of information depending on the activity. This makes standardizing managerial accounting systems across companies difficult. Instead, managerial accounting systems are flexible. The design of a company's managerial accounting system depends largely on the nature of the business and the arrangement of its internal operations. Managers can decide for themselves what information they want and how they want it reported. Even within a single company, different managers often design their own systems to meet their special needs. The important question a manager must ask is whether the information being collected and reported is useful for planning, decision making, and control purposes.

Point: The *Institute of Management Accountants* issues statements that govern the practice of managerial accounting. Accountants who pass a qualifying exam are awarded the CMA.

Timeliness of Information Formal financial statements reporting past transactions and events are not immediately available to outside parties. Independent certified public accountants often must *audit* a company's financial statements before it provides them to external users. Thus, because audits often take several weeks to complete, financial reports to outsiders usually are not available until well after the period-end. However, managers can quickly obtain managerial accounting information. External auditors need not review it. Estimates and projections are acceptable. To get information quickly, managers often accept less precision in reports. As an example, an early internal report to management prepared right after the year-end could report net income for the year between $4.2 and $4.8 million. An audited income statement could later show net income for the year at $4.6 million. The internal report is not precise, but its information can be more useful because it is available earlier.

Internal auditing plays an important role in managerial accounting. Internal auditors evaluate the flow of information not only inside but also outside the company. Managers are responsible for preventing and detecting fraudulent activities in their companies.

Point: Financial statements are usually issued several weeks after the period-end. GAAP requires the reporting of important events that occur while the statements are being prepared. These events are called *subsequent events.*

Point: Independent auditors test the integrity of managerial accounting records when they are used in preparing financial statements.

Time Dimension To protect external users from false expectations, financial reports deal primarily with results of both past activities and current conditions. While some predictions such as service lives and salvage values of plant assets are necessary, financial accounting avoids predictions whenever possible. Managerial accounting regularly includes predictions of conditions and events. As an example, one important managerial accounting report is a budget, which predicts revenues, expenses, and other items. If managerial accounting reports were restricted to the past and present, managers would be less able to plan activities and less effective in managing and evaluating current activities.

Focus of Information Companies often organize into divisions and departments, but investors rarely can buy shares in one division or department. Nor do creditors lend money to a company's single division or department. Instead, they own shares in or make loans to the entire company. Financial accounting focuses primarily on a company as a whole as depicted in Exhibit 18.3.

EXHIBIT 18.3

Focus of External Reports

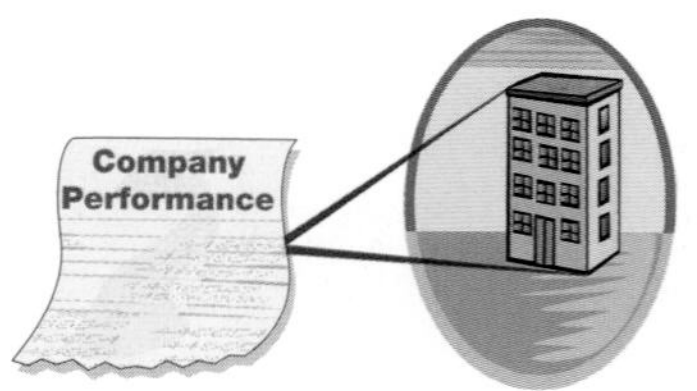

The focus of managerial accounting is different. While top-level managers are responsible for managing the whole company, most other managers are responsible for much smaller sets of activities. These middle-level and lower-level managers need managerial accounting reports dealing with specific activities, projects, and subdivisions for which they are responsible. For instance, division sales managers are directly responsible only for the results achieved in their divisions. Accordingly, division sales managers need information about results achieved in their own divisions to improve their performance. This information includes the level of success achieved by each individual, product, or department in each division as depicted in Exhibit 18.4.

EXHIBIT 18.4

Focus of Internal Reports

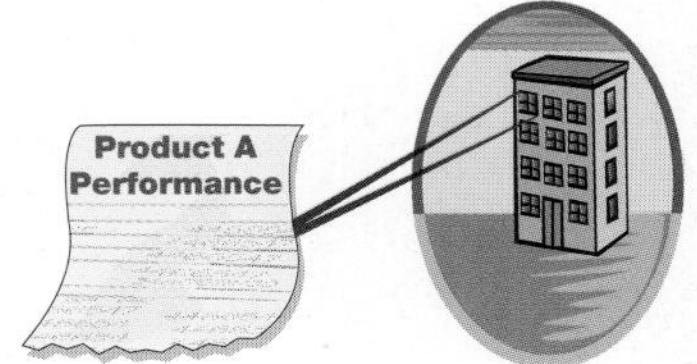

Nature of Information Both financial and managerial accounting systems report monetary information. Managerial accounting systems also report considerable nonmonetary information. Monetary information is an important part of managerial decisions, and nonmonetary information plays a crucial role, especially when monetary effects are difficult to measure. Common examples of nonmonetary information are the quality and delivery criteria of purchasing decisions.

Decision Ethics

Production Manager You invite three friends to a restaurant. When the dinner check arrives, David, a self-employed entrepreneur, picks it up saying, "Here, let me pay. I'll deduct it as a business expense on my tax return." Denise, a salesperson, takes the check from David's hand and says, "I'll put this on my company's credit card. It won't cost us anything." Derek, a factory manager for a company, laughs and says, "Neither of you understands. I'll put this on my company's credit card and call it overhead on a cost-plus contract my company has with a client." (*A cost-plus contract means the company receives its costs plus a percent of those costs.*) Adds Derek, "That way, my company pays for dinner *and* makes a profit." Who should pay the bill? Why? [Answer—p. 748]

Managerial Decision Making

The previous section emphasized differences between financial and managerial accounting, but they are not entirely separate. Similar information is useful to both external and internal users. For instance, information about costs of manufacturing products is useful to all users in making decisions. Also, both financial and managerial accounting affect peoples' actions. For example, Trek's design of a sales compensation plan affects the behavior of its salesforce. It also must estimate the dual effects of promotion and sales compensation plans on buying patterns of customers. These estimates impact the equipment purchase decisions for manufacturing and can affect the supplier selection criteria established by purchasing. Thus, financial and managerial accounting systems do more than measure; they also affect people's decisions and actions.

Managerial Accounting in Business

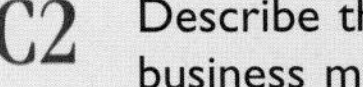

C2 Describe the lean business model.

We have explained the importance of managerial accounting for internal decision making. Although the analytical tools and techniques of managerial accounting have always been useful, their relevance and importance continue to increase. This is so because of changes in the business environment. This section describes some of these changes and their impact on managerial accounting.

Lean Business Model Two important factors have encouraged companies to be more effective and efficient in running their operations. First, there is an increased emphasis on *customers* as the most important constituent of a business. Customers expect to derive a certain value for the money they spend to buy products and services. Specifically, they expect that their suppliers will offer them the right service (or product) at the right time and the right price. This implies that companies accept the notion of **customer orientation,** which means that employees

understand the changing needs and wants of their customers and align their management and operating practices accordingly.

Second, our *global economy* expands competitive boundaries, thereby providing customers more choices. The global economy also produces changes in business activities. One notable case that reflects these changes in customer demand and global competition is auto manufacturing. The top three Japanese auto manufacturers (**Honda**, **Nissan**, and **Toyota**) once controlled more than 40% of the U.S. auto market. Customers perceived that Japanese auto manufacturers provided value not available from other manufacturers. Many European and North American auto manufacturers responded to this challenge and regained much of the lost market share.

Companies must be alert to these and other factors. Many companies have responded by adopting the **lean business model,** whose goal is to *eliminate waste* while "satisfying the customer" and "providing a positive return" to the company.

Lean Practices **Continuous improvement** rejects the notions of "good enough" or "acceptable" and challenges employees and managers to continuously experiment with new and improved business practices. This has led companies to adopt practices such as total quality management (TQM) and just-in-time (JIT) manufacturing. The philosophy underlying both practices is continuous improvement; the difference is in the focus.

Point: Goals of a TQM process include reduced waste, better inventory control, fewer defects, and continuous improvement. Just-in-time concepts have similar goals.

Total quality management focuses on quality improvement and applies this standard to all aspects of business activities. In doing so, managers and employees seek to uncover waste in business activities including accounting activities such as payroll and disbursements. To encourage an emphasis on quality, the U.S. Congress established the Malcolm Baldrige National Quality Award (MBQNA). Entrants must conduct a thorough analysis and evaluation of their business using guidelines from the Baldrige committee. **Ritz Carlton Hotel** is a recipient of the Baldrige award in the service category. The company applies a core set of values, collectively called *The Gold Standards,* to improve customer service.

Point: The time between buying raw materials and selling finished goods is called *throughput time.*

Just-in-time manufacturing is a system that acquires inventory and produces only when needed. An important aspect of JIT is that companies manufacture products only after they receive an order (a *demand-pull* system) and then deliver the customer's requirements on time. This means that processes must be aligned to eliminate any delays and inefficiencies including inferior inputs and outputs. Companies must also establish good relations and communications with their suppliers. On the downside, JIT is more susceptible to disruption than traditional systems. As one example, several **General Motors** plants were temporarily shut down due to a strike at an assembly division; the plants supplied components *just in time* to the assembly division.

Decision Insight

Global Lean **Toyota Motor Corporation** pioneered lean manufacturing, and it has since spread to other manufacturers throughout the world. The goals include improvements in quality, reliability, inventory turnover, productivity, exports, and—above all—sales and income.

Video18.3

Implications for Managerial Accounting Adopting the lean business model can be challenging because to foster its implementation, all systems and procedures that a company follows must be realigned. Managerial accounting has an important role to play by providing accurate cost and performance information. Companies must understand the nature and sources of cost and must develop systems that capture costs accurately. Developing such a system is important to measuring the "value" provided to customers. The price that customers pay for acquiring goods and services is an important determinant of value. In turn, the costs a company incurs are key determinants of price. All else being equal, the better a company is at controlling its costs, the better its performance.

Decision Insight

Balanced Scorecard The *balanced scorecard* aids continuous improvement by augmenting financial measures with information on the "drivers" (indicators) of future financial performance along four dimensions: (1) *financial*—profitability and risk, (2) *customer*—value creation and product and service differentiation, (3) *internal business processes*—business activities that create customer and owner satisfaction, and (4) *learning and growth*—organizational change, innovation, and growth.

Fraud and Ethics in Managerial Accounting

C3 Describe fraud and the role of ethics in managerial accounting.

Fraud, and the role of ethics in reducing fraud, are important factors in running business operations. Fraud involves the use of one's job for personal gain through the deliberate misuse of the employer's assets. Examples include theft of the employer's cash or other assets, overstating reimbursable expenses, payroll schemes, and financial statement fraud. Fraud affects all business and it is costly: A 2006 *Report to the Nation* from the Association of Certified Fraud Examiners estimates the average U.S. business loses 5% of its annual revenues to fraud.

The most common type of fraud, where employees steal or misuse the employer's resources, results in an average loss of $150,000 per occurrence. For example, in a billing fraud, an employee sets up a bogus supplier. The employee then prepares bills from the supplier and pays these bills from the employer's checking account. The employee cashes the checks sent to the bogus supplier and uses them for his or her own personal benefit.

Although there are many types of fraud schemes, all fraud

- Is done to provide direct or indirect benefit to the employee.
- Violates the employee's duties to his employer.
- Costs the employer money.
- Is secret.

Implications for Managerial Accounting Fraud increases a business's costs. Left undetected, these inflated costs can result in poor pricing decisions, an improper product mix, and faulty performance evaluations. Management can develop accounting systems to closely track costs and identify deviations from expected amounts. In addition, managers rely on an **internal control system** to monitor and control business activities. An internal control system is the policies and procedures managers use to

- Urge adherence to company policies.
- Promote efficient operations.
- Ensure reliable accounting.
- Protect assets.

Combating fraud and other dilemmas requires ethics in accounting. **Ethics** are beliefs that distinguish right from wrong. They are accepted standards of good and bad behavior. Identifying the ethical path can be difficult. The preferred path is a course of action that avoids casting doubt on one's decisions.

The **Institute of Management Accountants** (IMA), the professional association for management accountants, has issued a code of ethics to help accountants involved in solving ethical dilemmas. The IMA's Statement of Ethical Professional Practice requires that management accountants be competent, maintain confidentiality, act with integrity, and communicate information in a fair and credible manner.

The IMA provides a "road map" for resolving ethical conflicts. It suggests that an employee follow the company's policies on how to resolve such conflicts. If the conflict remains unresolved, an employee should contact the next level of management (such as the immediate supervisor) who is not involved in the ethical conflict.

Point: The IMA also issues the Certified Management Accountant (CMA) and the Certified Financial Manager (CFM) certifications. Employees with the CMA or CFM certifications typically earn higher salaries than those without.

Point: The **Sarbanes-Oxley Act** requires each issuer of securities to disclose whether it has adopted a code of ethics for its senior officers and the content of that code.

Quick Check Answers—p. 749

1. Managerial accounting produces information (*a*) to meet internal users' needs, (*b*) to meet a user's specific needs, (*c*) often focusing on the future, or (*d*) all of these.
2. What is the difference between the intended users of financial and managerial accounting?
3. Do generally accepted accounting principles (GAAP) control and dictate managerial accounting?
4. What is the basic objective for a company practicing total quality management?

Managerial Cost Concepts

C4 Describe accounting concepts useful in classifying costs.

An organization incurs many different types of costs that are classified differently, depending on management needs (different costs for different purposes). We can classify costs on the basis of their (1) behavior, (2) traceability, (3) controllability, (4) relevance, and (5) function. This section explains each concept for assigning costs to products and services.

Types of Cost Classifications

Video18.2

Classification by Behavior At a basic level, a cost can be classified as fixed or variable. A **fixed cost** does not change with changes in the volume of activity (within a range of activity known as an activity's *relevant range*). For example, straight-line depreciation on equipment is a fixed cost. A **variable cost** changes in proportion to changes in the volume of activity. Sales commissions computed as a percent of sales revenue are variable costs. Additional examples of fixed and variable costs for a bike manufacturer are provided in Exhibit 18.5. When cost items are combined, total cost can be fixed, variable, or mixed. *Mixed* refers to a combination of fixed and variable costs. Equipment rental often includes a fixed cost for some minimum amount and a variable cost based on amount of usage. Classification of costs by behavior is helpful in cost-volume-profit analyses and short-term decision making. We discuss these in Chapters 22 and 25.

EXHIBIT 18.5

Fixed and Variable Costs

Fixed Cost: Rent for Rocky Mountain Bikes' building is $22,000, and it doesn't change with the number of bikes produced.

Variable Cost: Cost of bicycle tires is variable with the number of bikes produced—this cost is $15 per pair.

Classification by Traceability A cost is often traced to a **cost object,** which is a product, process, department, or customer to which costs are assigned. **Direct costs** are those traceable to a single cost object. For example, if a product is a cost object, its material and labor costs are usually directly traceable. **Indirect costs** are those that cannot be easily and cost–beneficially traced to a single cost object. An example of an indirect cost is a maintenance plan that benefits two or more departments. Exhibit 18.6 identifies examples of both direct and indirect costs for the maintenance department in a manufacturing plant. Thus, salaries of Rocky Mountain Bikes' maintenance department employees are considered indirect if the cost object is bicycles and direct if the cost object is the maintenance department. Classification of costs by traceability is useful for cost allocation. This is discussed in Chapter 21.

Decision Maker

Entrepreneur You wish to trace as many of your assembly department's direct costs as possible. You can trace 90% of them in an economical manner. To trace the other 10%, you need sophisticated and costly accounting software. Do you purchase this software? [Answer—p. 748]

EXHIBIT 18.6

Direct and Indirect Costs of a Maintenance Department

Direct Costs		Indirect Costs	
• Salaries of maintenance department employees	• Materials purchased by maintenance department	• Factory accounting	• Factory light and heat
• Equipment purchased by maintenance department	• Maintenance department equipment depreciation	• Factory administration	• Factory internal audit
		• Factory rent	• Factory intranet
		• Factory managers' salary	• Insurance on factory

Classification by Controllability A cost can be defined as **controllable** or **not controllable.** Whether a cost is controllable or not depends on the employee's responsibilities, as shown in Exhibit 18.7. This is referred to as *hierarchical levels* in management, or *pecking order.* For example, investments in machinery are controllable by upper-level managers but not lower-level managers. Many daily operating expenses such as overtime often are controllable by lower-level managers. Classification of costs by controllability is especially useful for assigning responsibility to and evaluating managers.

EXHIBIT 18.7

Controllability of Costs

Senior Manager
Controls costs of investments in land, buildings, and equipment.

Supervisor
Controls daily expenses such as supplies, maintenance, and overtime.

Classification by Relevance A cost can be classified by relevance by identifying it as either a sunk cost or an out-of-pocket cost. A **sunk cost** has already been incurred and cannot be avoided or changed. It is irrelevant to future decisions. One example is the cost of a company's office equipment previously purchased. An **out-of-pocket cost** requires a future outlay of cash and is relevant for decision making. Future purchases of equipment involve out-of-pocket costs. A discussion of relevant costs must also consider opportunity costs. An **opportunity cost** is the potential benefit lost by choosing a specific action from two or more alternatives. One example is a student giving up wages from a job to attend evening classes. Consideration of opportunity cost is important when, for example, an insurance company must decide whether to outsource its payroll function or maintain it internally. This is discussed in Chapter 25.

Point: Opportunity costs are not recorded by the accounting system.

C5 Define product and period costs and explain how they impact financial statements.

Classification by Function Another cost classification (for manufacturers) is capitalization as inventory or to expense as incurred. Costs capitalized as inventory are called **product costs,** which refer to expenditures necessary and integral to finished products. They include direct materials, direct labor, and indirect manufacturing costs called *overhead costs*. Product costs pertain to activities carried out to manufacture the product. Costs expensed are called **period costs,** which refer to expenditures identified more with a time period than with finished products. They include selling and general administrative expenses. Period costs pertain to activities that are not part of the manufacturing process. A distinction between product and period costs is important because period costs are expensed in the income statement and product costs are assigned to inventory on the balance sheet until that inventory is sold. An ability to understand and identify product costs and period costs is crucial to using and interpreting a *manufacturing statement* described later in this chapter.

Exhibit 18.8 shows the different effects of product and period costs. Period costs flow directly to the current income statement as expenses. They are not reported as assets. Product costs are first assigned to inventory. Their final treatment depends on when inventory is sold or disposed of. Product costs assigned to finished goods that are sold in year 2009 are reported on the 2009 income statement as part of cost of goods sold. Product costs assigned to unsold inventory are carried forward on the balance sheet at the end of year 2009. If this inventory is sold in year 2010, product costs assigned to it are reported as part of cost of goods sold in that year's income statement.

Point: Only costs of production and purchases are classed as product costs.

EXHIBIT 18.8

Period and Product Costs in Financial Statements

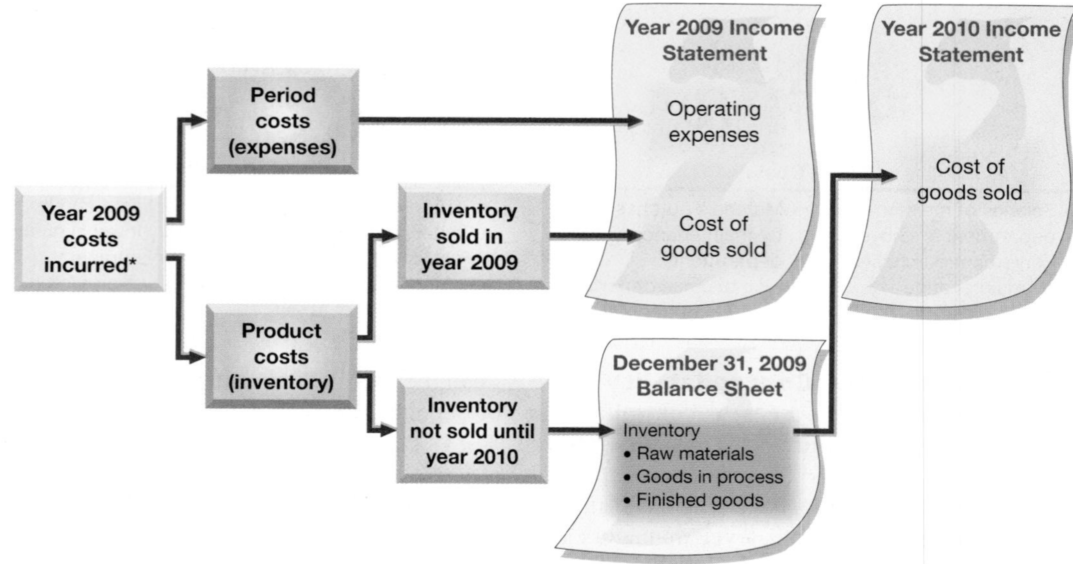

* This diagram excludes costs to acquire assets other than inventory.

Point: Product costs are either in the income statement as part of cost of goods sold or in the balance sheet as inventory. Period costs appear only on the income statement under operating expenses. See Exhibit 18.8.

Point: For a team approach to identifying period and product costs, see *Teamwork in Action* in the *Beyond the Numbers* section.

The difference between period and product costs explains why the year 2009 income statement does not report operating expenses related to either factory workers' wages or depreciation on factory buildings and equipment. Instead, both costs are combined with the cost of raw materials to compute the product cost of finished goods. A portion of these manufacturing costs (related to the goods sold) is reported in the year 2009 income statement as part of Cost of Goods Sold. The other portion is reported on the balance sheet at the end of that year as part of Inventory. The portion assigned to inventory could be included in any or all of raw materials, goods in process, or finished goods inventories.

Decision Maker

Purchase Manager You are evaluating two potential suppliers of seats for the manufacturing of motorcycles. One supplier (A) quotes a $145 price per seat and ensures 100% quality standards and on-time delivery. The second supplier (B) quotes a $115 price per seat but does not give any written assurances on quality or delivery. You decide to contract with the second supplier (B), saving $30 per seat. Does this decision have opportunity costs? [Answer—p. 749]

Identification of Cost Classifications

It is important to understand that a cost can be classified using any one (or combination) of the five different means described here. To do this we must understand costs and operations. Specifically, for the five classifications, we must be able to identify the *activity* for behavior, *cost object* for traceability, *management hierarchical level* for controllability, *opportunity cost* for relevance, and *benefit period* for function. Factory rent, for instance, can be classified as a product cost; it is fixed with respect to number of units produced, it is indirect with respect to the product, and it is not controllable by a production supervisor. Potential multiple classifications are shown in Exhibit 18.9 using different cost items incurred in manufacturing mountain bikes. The finished bike is the cost object. Proper allocation of these costs and the managerial decisions based on cost data depend on a correct cost classification.

Cost Concepts for Service Companies

Point: All expenses of service companies are period costs because these companies do not have inventory.

The cost concepts described are generally applicable to service organizations. For example, consider **Southwest Airlines**. Its cost of beverages for passengers is a variable cost based on number of passengers. The cost of leasing an aircraft is fixed with respect to number of passengers. We can also trace a flight crew's salary to a specific flight whereas we likely

Cost Item	By Behavior	By Traceability	By Function
Bicycle tires	Variable	Direct	Product
Wages of assembly worker*	Variable	Direct	Product
Advertising	Fixed	Indirect	Period
Production manager's salary	Fixed	Indirect	Product
Office depreciation	Fixed	Indirect	Period

* Although an assembly worker's wages are classified as variable costs, their actual behavior depends on how workers are paid and whether their wages are based on a union contract (such as piece rate or monthly wages).

EXHIBIT 18.9

Examples of Multiple Cost Classifications

cannot trace wages for the ground crew to a specific flight. Classification by function (such as product versus period costs) is not relevant to service companies because services are not inventoried. Instead, costs incurred by a service firm are expensed in the reporting period when incurred.

Managers in service companies must understand and apply cost concepts. They seek and rely on accurate cost estimates for many decisions. For example, an airline manager must often decide between canceling or rerouting flights. The manager must also be able to estimate costs saved by canceling a flight versus rerouting. Knowledge of fixed costs is equally important. We explain more about the cost requirements for these and other managerial decisions in Chapter 25.

Service Costs

- Beverages and snacks
- Cleaning fees
- Pilot and copilot salaries
- Attendant salaries
- Fuel and oil costs
- Travel agent fees
- Ground crew salaries

Quick Check Answers—p. 749

5. Which type of cost behavior increases total costs when volume of activity increases?
6. How could traceability of costs improve managerial decisions?

Reporting Manufacturing Activities

Companies with manufacturing activities differ from both merchandising and service companies. The main difference between merchandising and manufacturing companies is that merchandisers buy goods ready for sale while manufacturers produce goods from materials and labor. **Payless** is an example of a merchandising company. It buys and sells shoes without physically changing them. **Adidas** is primarily a manufacturer of shoes, apparel, and accessories. It purchases materials such as leather, cloth, dye, plastic, rubber, glue, and laces and then uses employees' labor to convert these materials to products. **Southwest Airlines** is a service company that transports people and items.

Manufacturing activities differ from both selling merchandise and providing services. Also, the financial statements for manufacturing companies differ slightly. This section considers some of these differences and compares them to accounting for a merchandising company.

Manufacturer's Balance Sheet

C6 Explain how balance sheets and income statements for manufacturing and merchandising companies differ.

Manufacturers carry several unique assets and usually have three inventories instead of the single inventory that merchandisers carry. Exhibit 18.10 shows three different inventories in the current asset section of the balance sheet for Rocky Mountain Bikes, a manufacturer. The three inventories are raw materials, goods in process, and finished goods.

Raw Materials Inventory **Raw materials inventory** refers to the goods a company acquires to use in making products. It uses raw materials in two ways: directly and indirectly. Most raw materials physically become part of a product and are identified with specific units or batches of a product. Raw materials used directly in a product are called *direct materials*. Other materials used to support production processes are sometimes not as clearly identified with specific units or batches of product. These materials are called **indirect materials** because they are not clearly identified with specific product units or batches. Items used as indirect materials often appear on a

Point: Reducing the size of inventories saves storage costs and frees money for other uses.

EXHIBIT 18.10

Balance Sheet for a Manufacturer

ROCKY MOUNTAIN BIKES
Balance Sheet
December 31, 2009

Assets		**Liabilities and Equity**	
Current assets		Current liabilities	
Cash	$ 11,000	Accounts payable	$ 14,000
Accounts receivable, net	30,150	Wages payable	540
Raw materials inventory	**9,000**	Interest payable	2,000
Goods in process inventory	**7,500**	Income taxes payable	32,600
Finished goods inventory	**10,300**	Total current liabilities	49,140
Factory supplies	350	Long-term liabilities	
Prepaid insurance	300	Long-term notes payable	50,000
Total current assets	68,600	Total liabilities	99,140
Plant assets			
Small tools, net	1,100	Stockholders' equity	
Delivery equipment, net	5,000	Common stock, $1.2 par	24,000
Office equipment, net	1,300	Paid-in capital	76,000
Factory machinery, net	65,500	Retained earnings	49,760
Factory building, net	86,700	Total stockholders' equity	149,760
Land	9,500	Total liabilities and equity	$248,900
Total plant assets, net	169,100		
Intangible assets (patents), net	11,200		
Total assets	$248,900		

balance sheet as factory supplies or are included in raw materials. Some direct materials are classified as indirect materials when their costs are low (insignificant). Examples include screws and nuts used in assembling mountain bikes and staples and glue used in manufacturing shoes. Using the *materiality principle,* individually tracing the costs of each of these materials and classifying them separately as direct materials does not make much economic sense. For instance, keeping detailed records of the amount of glue used to manufacture one shoe is not cost beneficial.

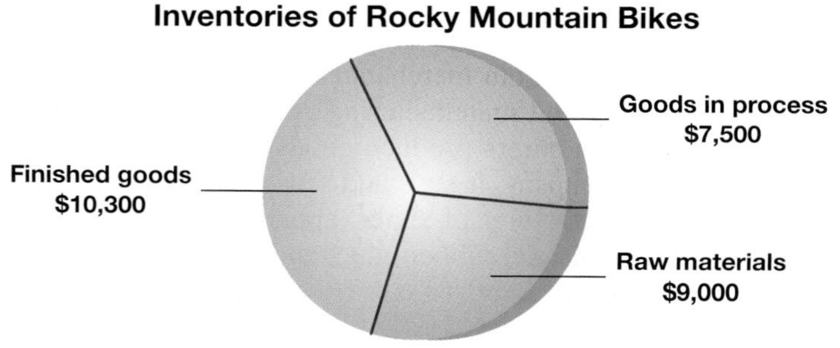

Goods in Process Inventory Another inventory held by manufacturers is **goods in process inventory,** also called *work in process inventory.* It consists of products in the process of being manufactured but not yet complete. The amount of goods in process inventory depends on the type of production process. If the time required to produce a unit of product is short, the goods in process inventory is likely small; but if weeks or months are needed to produce a unit, the goods in process inventory is usually larger.

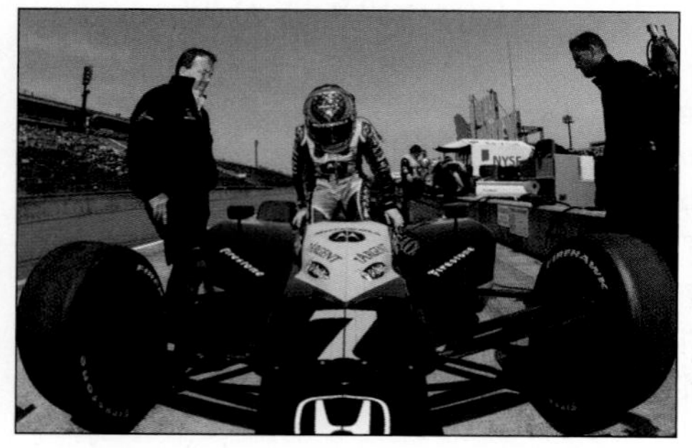

Finished Goods Inventory A third inventory owned by a manufacturer is **finished goods inventory,** which consists of completed products ready for sale. This inventory is similar to merchandise inventory owned by a merchandising company. Manufacturers also often own unique plant assets such as small tools, factory buildings, factory equipment, and patents to manufacture products. The balance sheet in Exhibit 18.10 shows that Rocky Mountain Bikes owns all of these assets. Some manufacturers invest millions or even billions of dollars in production facilities and patents. Briggs & Stratton's recent balance sheet shows about $1 billion net investment in land, buildings, machinery and equipment, much of which involves production facilities. It manufactures more racing engines than any other company in the world.

Manufacturer's Income Statement

P1 Compute cost of goods sold for a manufacturer.

The main difference between the income statement of a manufacturer and that of a merchandiser involves the items making up cost of goods sold. Exhibit 18.11 compares the components of cost of goods sold for a manufacturer and a merchandiser. A merchandiser adds cost of goods purchased to beginning merchandise inventory and then subtracts ending merchandise inventory to get cost of goods sold. A manufacturer adds cost of goods manufactured to beginning finished goods inventory and then subtracts ending finished goods inventory to get cost of goods sold.

EXHIBIT 18.11

Cost of Goods Sold Computation

Merchandiser	Manufacturer
Beginning merchandise inventory	Beginning finished goods inventory
+	+
Cost of goods purchased	Cost of goods manufactured
−	−
Ending merchandise inventory	Ending finished goods inventory
= Cost of goods sold	= Cost of goods sold

A merchandiser often uses the term *merchandise* inventory; a manufacturer often uses the term *finished goods* inventory. A manufacturer's inventories of raw materials and goods in process are not included in finished goods because they are not available for sale. A manufacturer also shows cost of goods *manufactured* instead of cost of goods *purchased*. This difference occurs because a manufacturer produces its goods instead of purchasing them ready for sale. We show later in this chapter how to derive cost of goods manufactured from the manufacturing statement.

The Cost of Goods Sold sections for both a merchandiser (Tele-Mart) and a manufacturer (Rocky Mountain Bikes) are shown in Exhibit 18.12 to highlight these differences. The remaining income statement sections are similar.

EXHIBIT 18.12

Cost of Goods Sold for a Merchandiser and Manufacturer

Merchandising (Tele-Mart) Company		Manufacturing (Rocky Mtn. Bikes) Company	
Cost of goods sold		Cost of goods sold	
Beginning *merchandise* inventory	$ 14,200	**Beginning *finished goods* inventory**	$ 11,200
Cost of merchandise *purchased*	234,150	**Cost of goods *manufactured****	170,500
Goods available for sale	248,350	Goods available for sale	181,700
Less ending *merchandise* inventory	12,100	**Less ending *finished goods* inventory**	10,300
Cost of goods sold	$236,250	Cost of goods sold	$171,400

* Cost of goods manufactured is reported in the income statement of Exhibit 18.14.

Although the cost of goods sold computations are similar, the numbers in these computations reflect different activities. A merchandiser's cost of goods purchased is the cost of buying products to be sold. A manufacturer's cost of goods manufactured is the sum of direct materials, direct labor, and factory overhead costs incurred in producing products. The remainder of this section further explains these three manufacturing costs and describes prime and conversion costs.

Direct Materials **Direct materials** are tangible components of a finished product. **Direct material costs** are the expenditures for direct materials that are separately and readily traced through the manufacturing process to finished goods. Examples of direct materials in manufacturing a mountain bike include its tires, seat, frame, pedals, brakes, cables, gears, and handlebars. The chart in the margin shows that direct materials generally make up about 45% of manufacturing costs in today's products, but this amount varies across industries and companies.

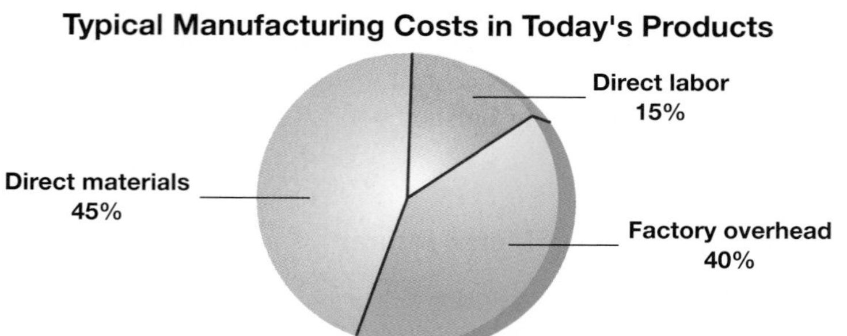

Direct Labor **Direct labor** refers to the efforts of employees who physically convert materials to finished product. **Direct labor costs** are the wages and salaries for direct labor that are separately and readily traced through the manufacturing process to finished goods. Examples of direct labor in manufacturing a mountain bike include operators directly involved in converting raw materials into finished products (welding, painting, forming) and assembly workers who attach materials such as tires, seats, pedals, and brakes to the bike frames. Costs of other workers on the assembly line who assist direct laborers are classified as **indirect labor costs. Indirect labor** refers to manufacturing workers' efforts not linked to specific units or batches of the product.

Point: Indirect labor costs are part of factory overhead.

Factory Overhead **Factory overhead** consists of all manufacturing costs that are not direct materials or direct labor. **Factory overhead costs** cannot be separately or readily traced to finished goods. These costs include indirect materials and indirect labor, costs not directly traceable to the product. Overtime paid to direct laborers is also included in overhead because overtime is due to delays, interruptions, or constraints not necessarily identifiable to a specific product or batches of product. Factory overhead costs also include maintenance of the mountain bike factory, supervision of its employees, repairing manufacturing equipment, factory utilities (water, gas, electricity), production manager's salary, factory rent, depreciation on factory buildings and equipment, factory insurance, property taxes on factory buildings and equipment, and factory accounting and legal services. Factory overhead does *not* include selling and administrative expenses because they are not incurred in manufacturing products. These expenses are called *period costs* and are recorded as expenses on the income statement when incurred.

Point: Factory overhead is also called *manufacturing overhead.*

EXHIBIT 18.13

Prime and Conversion Costs and Their Makeup

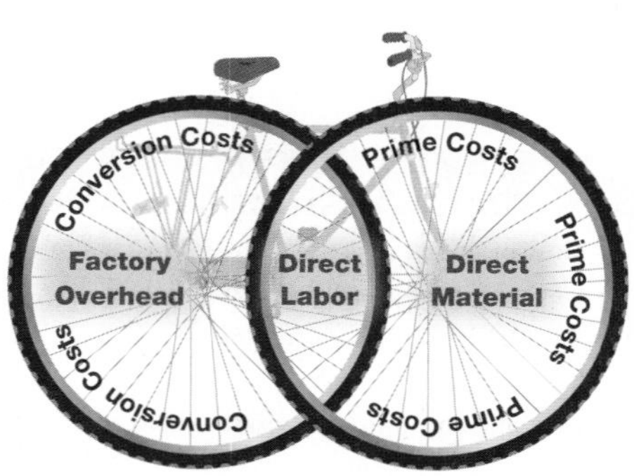

Prime and Conversion Costs Direct material costs and direct labor costs are also called **prime costs**—expenditures directly associated with the manufacture of finished goods. Direct labor costs and overhead costs are called **conversion costs**—expenditures incurred in the process of converting raw materials to finished goods. Direct labor costs are considered both prime costs and conversion costs. Exhibit 18.13 conveys the relation between prime and conversion costs and their components of direct material, direct labor, and factory overhead.

Point: Prime costs = Direct materials + Direct labor. Conversion costs = Direct labor + Factory overhead.

Reporting Performance Exhibit 18.14 shows the income statement for Rocky Mountain Bikes. Its operating expenses include sales salaries, office salaries, and depreciation of delivery and office equipment. Operating expenses do not include manufacturing costs such as factory workers' wages and depreciation of production equipment and the factory buildings. These manufacturing costs are reported as part of cost of goods manufactured and included in cost of goods sold. We explained why and how this is done in the section "Classification by Function."

Point: Manufacturers treat costs such as depreciation and rent as product costs if they are related to manufacturing.

EXHIBIT 18.14

Income Statement for a Manufacturer

ROCKY MOUNTAIN BIKES
Income Statement
For Year Ended December 31, 2009

Sales		$310,000
Cost of goods sold		
Finished goods inventory, Dec. 31, 2008	$ 11,200	
Cost of goods manufactured	**170,500**	
Goods available for sale	181,700	
Less finished goods inventory, Dec. 31, 2009	10,300	
Cost of goods sold		171,400
Gross profit		138,600
Operating expenses		
Selling expenses		
Sales salaries expense	18,000	
Advertising expense	5,500	
Delivery wages expense	12,000	
Shipping supplies expense	250	
Insurance expense—Delivery equipment	300	
Depreciation expense—Delivery equipment	2,100	
Total selling expenses		38,150
General and administrative expenses		
Office salaries expense	15,700	
Miscellaneous expense	200	
Bad debts expense	1,550	
Office supplies expense	100	
Depreciation expense—Office equipment	200	
Interest expense	4,000	
Total general and administrative expenses		21,750
Total operating expenses		59,900
Income before income taxes		78,700
Income taxes expense		32,600
Net income		$ 46,100
Net income per common share (20,000 shares)		$ 2.31

Quick Check

Answers—p. 749

7. What are the three types of inventory on a manufacturing company's balance sheet?
8. How does cost of goods sold differ for merchandising versus manufacturing companies?

Flow of Manufacturing Activities

C7 Explain manufacturing activities and the flow of manufacturing costs.

To understand manufacturing and its reports, we must first understand the flow of manufacturing activities and costs. Exhibit 18.15 shows the flow of manufacturing activities for a manufacturer. This exhibit has three important sections: *materials activity, production activity,* and *sales activity.* We explain each activity in this section.

Materials Activity The far left side of Exhibit 18.15 shows the flow of raw materials. Manufacturers usually start a period with some beginning raw materials inventory carried over from the previous period. The company then acquires additional raw materials in the current period. Adding these purchases to beginning inventory gives total raw materials available for use in production. These raw materials are then either used in production in the current period or remain in inventory at the end of the period for use in future periods.

Point: Knowledge of managerial accounting provides us a means of measuring manufacturing costs and is a sound foundation for studying advanced business topics.

Production Activity The middle section of Exhibit 18.15 describes production activity. Four factors come together in production: beginning goods in process inventory, direct materials,

EXHIBIT 18.15

Activities and Cost Flows in Manufacturing

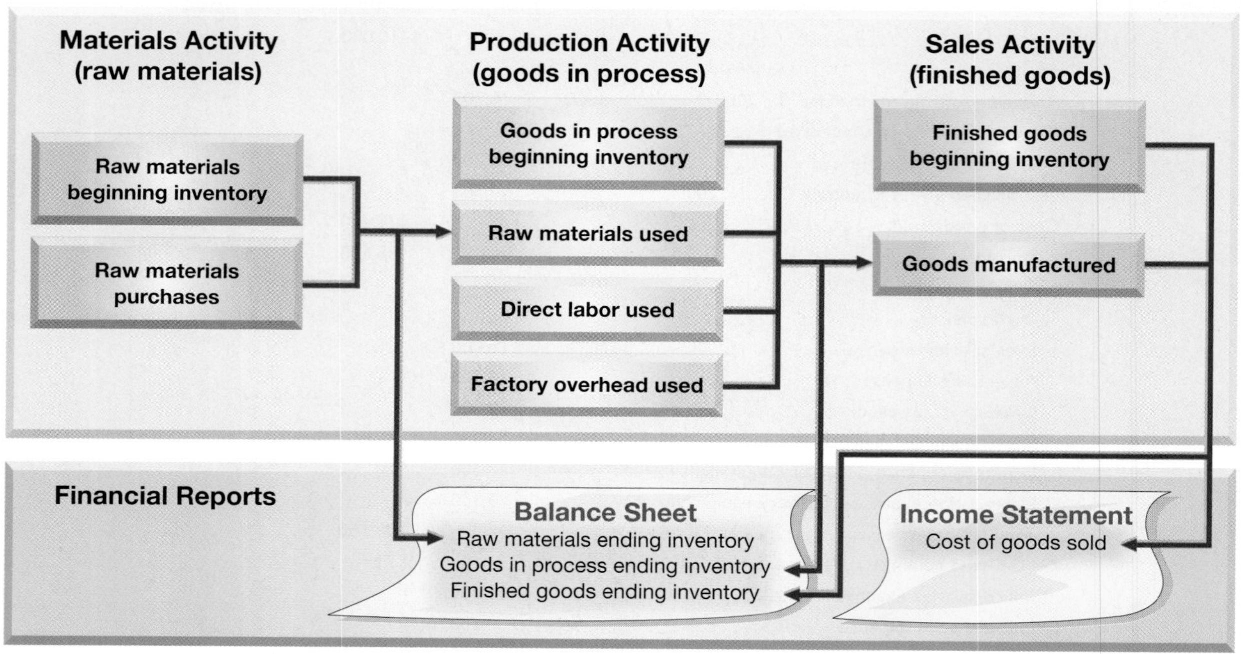

Point: The series of activities that add value to a company's products or services is called a **value chain.**

direct labor, and overhead. Beginning goods in process inventory consists of partly assembled products from the previous period. Production activity results in products that are either finished or remain unfinished. The cost of finished products makes up the cost of goods manufactured for the current period. Unfinished products are identified as ending goods in process inventory. The cost of unfinished products consists of direct materials, direct labor, and factory overhead, and is reported on the current period's balance sheet. The costs of both finished goods manufactured and goods in process are *product costs.*

Sales Activity The company's sales activity is portrayed in the far right side of Exhibit 18.15. Newly completed units are combined with beginning finished goods inventory to make up total finished goods available for sale in the current period. The cost of finished products sold is reported on the income statement as cost of goods sold. The cost of products not sold is reported on the current period's balance sheet as ending finished goods inventory.

Manufacturing Statement

P2 Prepare a manufacturing statement and explain its purpose and links to financial statements.

A company's manufacturing activities are described in a **manufacturing statement,** also called the *schedule of manufacturing activities* or the *schedule of cost of goods manufactured.* The manufacturing statement summarizes the types and amounts of costs incurred in a company's manufacturing process. Exhibit 18.16 shows the manufacturing statement for Rocky Mountain Bikes. The statement is divided into four parts: *direct materials, direct labor, overhead,* and *computation of cost of goods manufactured.* We describe each of these parts in this section.

① The manufacturing statement begins by computing direct materials used. We start by adding beginning raw materials inventory of $8,000 to the current period's purchases of $86,500. This yields $94,500 of total raw materials available for use. A physical count of inventory shows $9,000 of ending raw materials inventory. This implies a total cost of raw materials used during the period of $85,500 ($94,500 total raw materials available for use − $9,000 ending inventory). (*Note:* All raw materials are direct materials for Rocky Mountain Bikes.)

EXHIBIT 18.16
Manufacturing Statement

ROCKY MOUNTAIN BIKES Manufacturing Statement For Year Ended December 31, 2009			
①	**Direct materials**		
	Raw materials inventory, Dec. 31, 2008	$ 8,000	
	Raw materials purchases	86,500	
	Raw materials available for use	94,500	
	Less raw materials inventory, Dec. 31, 2009	9,000	
	Direct materials used		$ 85,500
②	**Direct labor**		60,000
③	**Factory overhead**		
	Indirect labor	9,000	
	Factory supervision	6,000	
	Factory utilities	2,600	
	Repairs—Factory equipment	2,500	
	Property taxes—Factory building	1,900	
	Factory supplies used	600	
	Factory insurance expired	1,100	
	Depreciation expense—Small tools	200	
	Depreciation expense—Factory equipment	3,500	
	Depreciation expense—Factory building	1,800	
	Amortization expense—Patents	800	
	Total factory overhead		30,000
④	Total manufacturing costs		175,500
	Add goods in process inventory, Dec. 31, 2008		2,500
	Total cost of goods in process		178,000
	Less goods in process inventory, Dec. 31, 2009		7,500
	Cost of goods manufactured		$170,500

② The second part of the manufacturing statement reports direct labor costs. Rocky Mountain Bikes had total direct labor costs of $60,000 for the period. This amount includes payroll taxes and fringe benefits.

③ The third part of the manufacturing statement reports overhead costs. The statement lists each important factory overhead item and its cost. Total factory overhead cost for the period is $30,000. Some companies report only *total* factory overhead on the manufacturing statement and attach a separate schedule listing individual overhead costs.

④ The final section of the manufacturing statement computes and reports the *cost of goods manufactured.* (Total manufacturing costs for the period are $175,500 [$85,500 + $60,000 + $30,000], the sum of direct materials used and direct labor and overhead costs incurred.) This amount is first added to beginning goods in process inventory. This gives the total goods in process inventory of $178,000 ($175,500 + $2,500). We then compute the current period's cost of goods manufactured of $170,500 by taking the $178,000 total goods in process and subtracting the $7,500 cost of ending goods in process inventory that consists of direct materials, direct labor, and factory overhead. The cost of goods manufactured amount is also called *net cost of goods manufactured* or *cost of goods completed*. Exhibit 18.14 shows that this item and amount are listed in the Cost of Goods Sold section of Rocky Mountain Bikes' income statement and the balance sheet.

Point: Direct material and direct labor costs increase with increases in production volume and are called *variable costs*. Overhead can be both variable and fixed. When overhead costs vary with production, they are called *variable overhead*. When overhead costs don't vary with production, they are called *fixed overhead*.

Point: Manufacturers sometimes report variable and fixed overhead separately in the manufacturing statement to provide more information to managers about cost behavior.

A managerial accounting system records costs and reports them in various reports that eventually determine financial statements. Exhibit 18.17 shows how overhead costs flow through the system: from an initial listing of specific costs, to a section of the manufacturing statement, to the reporting on the income statement and the balance sheet.

"My boss wants us to appeal to a younger and hipper crowd. So, I'd like to get a tattoo that says-- 'Accounting rules!'"

EXHIBIT 18.17

Overhead Cost Flows across Accounting Reports

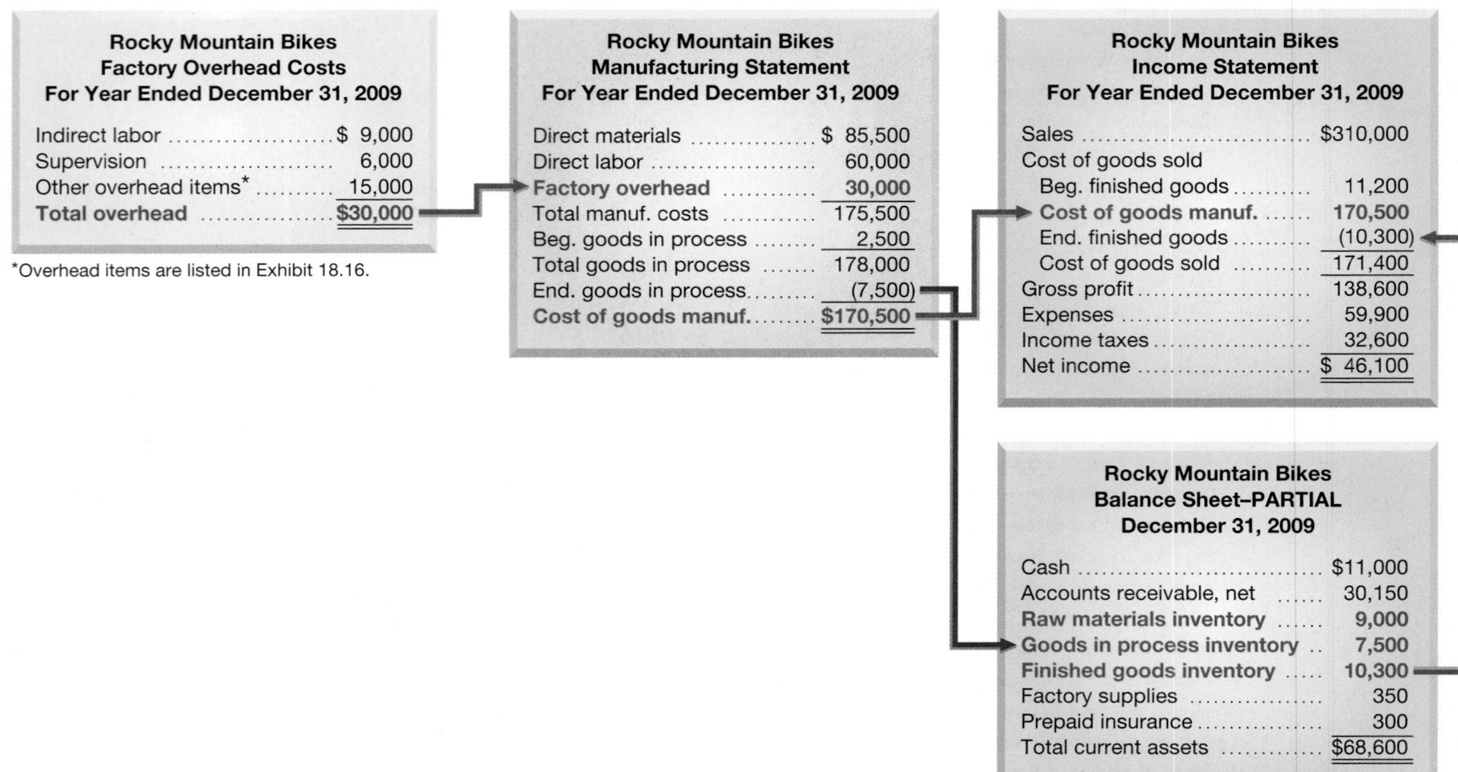

Rocky Mountain Bikes
Factory Overhead Costs
For Year Ended December 31, 2009

Indirect labor	$ 9,000
Supervision	6,000
Other overhead items*	15,000
Total overhead	**$30,000**

*Overhead items are listed in Exhibit 18.16.

Rocky Mountain Bikes
Manufacturing Statement
For Year Ended December 31, 2009

Direct materials	$ 85,500
Direct labor	60,000
Factory overhead	**30,000**
Total manuf. costs	175,500
Beg. goods in process	2,500
Total goods in process	178,000
End. goods in process	(7,500)
Cost of goods manuf.	**$170,500**

Rocky Mountain Bikes
Income Statement
For Year Ended December 31, 2009

Sales	$310,000
Cost of goods sold	
Beg. finished goods	11,200
Cost of goods manuf.	**170,500**
End. finished goods	(10,300)
Cost of goods sold	171,400
Gross profit	138,600
Expenses	59,900
Income taxes	32,600
Net income	$ 46,100

Rocky Mountain Bikes
Balance Sheet–PARTIAL
December 31, 2009

Cash	$11,000
Accounts receivable, net	30,150
Raw materials inventory	**9,000**
Goods in process inventory	**7,500**
Finished goods inventory	**10,300**
Factory supplies	350
Prepaid insurance	300
Total current assets	$68,600

Management uses information in the manufacturing statement to plan and control the company's manufacturing activities. To provide timely information for decision making, the statement is often prepared monthly, weekly, or even daily. In anticipation of release of its much-hyped iPhone, **Apple** grew its inventory of Flash-based memory chips, a critical component, and its finished goods inventory. The manufacturing statement contains information useful to external users but is not a general-purpose financial statement. Companies rarely publish the manufacturing statement because managers view this information as proprietary and potentially harmful to them if released to competitors.

Quick Check

Answers—p. 749

9. A manufacturing statement (*a*) computes cost of goods manufactured for the period, (*b*) computes cost of goods sold for the period, or (*c*) reports operating expenses incurred for the period.

10. Are companies required to report a manufacturing statement?

11. How are both beginning and ending goods in process inventories reported on a manufacturing statement?

Decision Analysis — Cycle Time and Cycle Efficiency

A1 Compute cycle time and cycle efficiency, and explain their importance to production management.

As lean manufacturing practices help companies move toward just-in-time manufacturing, it is important for these companies to reduce the time to manufacture their products and to improve manufacturing efficiency. One metric that measures that time element is **cycle time (CT).** A definition of cycle time is in Exhibit 18.18.

Cycle time = Process time + Inspection time + Move time + Wait time

EXHIBIT 18.18

Cycle Time

Process time is the time spent producing the product. *Inspection time* is the time spent inspecting (1) raw materials when received, (2) goods in process while in production, and (3) finished goods prior to shipment. *Move time* is the time spent moving (1) raw materials from storage to production and (2) goods in process from factory location to another factory location. *Wait time* is the time that an order or job sits with no production applied to it; this can be due to order delays, bottlenecks in production, and poor scheduling.

Process time is considered **value-added time** because it is the only activity in cycle time that adds value to the product from the customer's perspective. The other three time activities are considered **non-value-added time** because they add no value to the customer.

Companies strive to reduce non-value-added time to improve **cycle efficiency (CE).** Cycle efficiency is the ratio of value-added time to total cycle time—see Exhibit 18.19.

$$\textbf{Cycle efficiency} = \frac{\textbf{Value-added time}}{\textbf{Cycle time}}$$

EXHIBIT 18.19

Cycle Efficiency

To illustrate, assume that Rocky Mountain Bikes receives and produces an order for 500 Tracker® mountain bikes. Assume that the following times were measured during production of this order.

Process time... 1.8 days

Inspection time... 0.5 days

Move time... 0.7 days

Wait time... 3.0 days

In this case, cycle time is 6.0 days, computed as 1.8 days + 0.5 days + 0.7 days + 3.0 days. Also, cycle efficiency is 0.3, or 30%, computed as 1.8 days divided by 6.0 days. This means that Rocky Mountain Bikes spends 30% of its time working on the product (value-added time). The other 70% is spent on non-value-added activities.

If a company has a CE of 1, it means that its time is spent entirely on value-added activities. If the CE is low, the company should evaluate its production process to see if it can identify ways to reduce non-value-added activities. The 30% CE for Rocky Mountain Bikes is low and its management should look for ways to reduce non-value-added activities.

Demonstration Problem 1: Cost Behavior and Classification

Understanding the classification and assignment of costs is important. Consider a company that manufactures computer chips. It incurs the following costs in manufacturing chips and in operating the company.

DP18

1. Plastic board used to mount the chip, $3.50 each.
2. Assembly worker pay of $15 per hour to attach chips to plastic board.
3. Salary for factory maintenance workers who maintain factory equipment.
4. Factory supervisor pay of $55,000 per year to supervise employees.
5. Real estate taxes paid on the factory, $14,500.
6. Real estate taxes paid on the company office, $6,000.
7. Depreciation costs on machinery used by workers, $30,000.
8. Salary paid to the chief financial officer, $95,000.
9. Advertising costs of $7,800 paid to promote products.
10. Salespersons' commissions of $0.50 for each assembled chip sold.
11. Management has the option to rent the manufacturing plant to six local hospitals to store medical records instead of producing and assembling chips.

Classify each cost in the following table according to the categories listed in the table header. A cost can be classified under more than one category. For example, the plastic board used to mount chips is classified as a direct material product cost and as a direct unit cost.

	Period Costs	Product Costs			Unit Cost Classification		Sunk Cost	Opportunity Cost
Cost	Selling and Administrative	Direct Material (Prime Cost)	Direct Labor (Prime and Conversion)	Factory Overhead (Conversion Cost)	Direct	Indirect		
1. Plastic board used to mount the chip, $3.50 each		✔			✔			

Solution to Demonstration Problem 1

	Period Costs	Product Costs			Unit Cost Classification		Sunk Cost	Opportunity Cost
Cost*	Selling and Administrative	Direct Material (Prime Cost)	Direct Labor (Prime and Conversion)	Factory Overhead (Conversion Cost)	Direct	Indirect		
1.		✔			✔			
2.			✔		✔			
3.				✔		✔		
4.				✔		✔		
5.				✔		✔		
6.	✔							
7.				✔		✔	✔	
8.	✔							
9.	✔							
10.	✔							
11.								✔

* Costs 1 through 11 refer to the 11 cost items described at the beginning of the problem.

Demonstration Problem 2: Reporting for Manufacturers

A manufacturing company's balance sheet and income statement differ from those for a merchandising or service company.

Required

1. Fill in the [BLANK] descriptors on the partial balance sheets for both the manufacturing company and the merchandising company. Explain why a different presentation is required.

Manufacturing Company

ADIDAS GROUP
Partial Balance Sheet
December 31, 2009

Current assets	
Cash	$10,000
[BLANK]	8,000
[BLANK]	5,000
[BLANK]	7,000
Supplies	500
Prepaid insurance	500
Total current assets	$31,000

Merchandising Company

PAYLESS SHOE OUTLET
Partial Balance Sheet
December 31, 2009

Current assets	
Cash	$ 5,000
[BLANK]	12,000
Supplies	500
Prepaid insurance	500
Total current assets	$18,000

2. Fill in the [BLANK] descriptors on the income statements for the manufacturing company and the merchandising company. Explain why a different presentation is required.

Manufacturing Company

ADIDAS GROUP
Partial Income Statement
For Year Ended December 31, 2009

Sales	$200,000
Cost of goods sold	
Finished goods inventory, Dec. 31, 2008	10,000
[BLANK]	120,000
Goods available for sale	130,000
Finished goods inventory, Dec. 31, 2009	(7,000)
Cost of goods sold	123,000
Gross profit	$ 77,000

Merchandising Company

PAYLESS SHOE OUTLET
Partial Income Statement
For Year Ended December 31, 2009

Sales	$190,000
Cost of goods sold	
Merchandise inventory, Dec. 31, 2008	8,000
[BLANK]	108,000
Goods available for sale	116,000
Merchandise inventory, Dec. 31, 2009	(12,000)
Cost of goods sold	104,000
Gross profit	$ 86,000

3. The manufacturer's cost of goods manufactured is the sum of (a) ______, (b) ______, and (c) ______ costs incurred in producing the product.

Solution to Demonstration Problem 2

1. Inventories for a manufacturer and for a merchandiser.

Manufacturing Company

ADIDAS GROUP
Partial Balance Sheet
December 31, 2009

Current assets	
Cash	$10,000
Raw materials inventory	8,000
Goods in process inventory	5,000
Finished goods inventory	7,000
Supplies	500
Prepaid insurance	500
Total current assets	$31,000

Merchandising Company

PAYLESS SHOE OUTLET
Partial Balance Sheet
December 31, 2009

Current assets	
Cash	$ 5,000
Merchandise inventory	12,000
Supplies	500
Prepaid insurance	500
Total current assets	$18,000

Explanation: A manufacturing company must control and measure three types of inventories: raw materials, goods in process, and finished goods. In the sequence of making a product, the raw materials move into production—called *goods in process inventory*—and then to finished goods. All raw materials and goods in process inventory at the end of each accounting period are considered current assets. All unsold finished inventory is considered a current asset at the end of each accounting period. The merchandising company must control and measure only one type of inventory, purchased goods.

2. Cost of goods sold for a manufacturer and for a merchandiser.

Manufacturing Company

ADIDAS GROUP Partial Income Statement For Year Ended December 31, 2009	
Sales	$200,000
Cost of goods sold	
Finished goods inventory, Dec. 31, 2008	10,000
Cost of goods manufactured	120,000
Goods available for sale	130,000
Finished goods inventory, Dec. 31, 2009	(7,000)
Cost of goods sold	123,000
Gross profit	$ 77,000

Merchandising Company

PAYLESS SHOE OUTLET Partial Income Statement For Year Ended December 31, 2009	
Sales	$190,000
Cost of goods sold	
Merchandise inventory, Dec. 31, 2008	8,000
Cost of purchases	108,000
Goods available for sale	116,000
Merchandise inventory, Dec. 31, 2009	(12,000)
Cost of goods sold	104,000
Gross profit	$ 86,000

Explanation: Manufacturing and merchandising companies use different reporting terms. In particular, the terms *finished goods* and *cost of goods manufactured* are used to reflect the production of goods, yet the concepts and techniques of reporting cost of goods sold for a manufacturing company and merchandising company are similar.

3. A manufacturer's cost of goods manufactured is the sum of (a) *direct material,* (b) *direct labor,* and (c) *factory overhead* costs incurred in producing the product.

Demonstration Problem 3: Manufacturing Statement

The following account balances and other information are from SUNN Corporation's accounting records for year-end December 31, 2009. Use this information to prepare (1) a table listing factory overhead costs, (2) a manufacturing statement (show only the total factory overhead cost), and (3) an income statement.

Advertising expense	$ 85,000	Goods in process inventory, Dec. 31, 2008	$ 8,000
Amortization expense—Factory Patents	16,000	Goods in process inventory, Dec. 31, 2009	9,000
Bad debts expense	28,000	Income taxes	53,400
Depreciation expense—Office equipment	37,000	Indirect labor	26,000
Depreciation expense—Factory building	133,000	Interest expense	25,000
Depreciation expense—Factory equipment	78,000	Miscellaneous expense	55,000
Direct labor	250,000	Property taxes on factory equipment	14,000
Factory insurance expired	62,000	Raw materials inventory, Dec. 31, 2008	60,000
Factory supervision	74,000	Raw materials inventory, Dec. 31, 2009	78,000
Factory supplies used	21,000	Raw materials purchases	313,000
Factory utilities	115,000	Repairs expense—Factory equipment	31,000
Finished goods inventory, Dec. 31, 2008	15,000	Salaries expense	150,000
Finished goods inventory, Dec. 31, 2009	12,500	Sales	1,630,000

Planning the Solution

- Analyze the account balances and select those that are part of factory overhead costs.
- Arrange these costs in a table that lists factory overhead costs for the year.
- Analyze the remaining costs and select those related to production activity for the year; selected costs should include the materials and goods in process inventories and direct labor.

- Prepare a manufacturing statement for the year showing the calculation of the cost of materials used in production, the cost of direct labor, and the total factory overhead cost. When presenting overhead cost on this statement, report only total overhead cost from the table of overhead costs for the year. Show the costs of beginning and ending goods in process inventory to determine cost of goods manufactured.
- Organize the remaining revenue and expense items into the income statement for the year. Combine cost of goods manufactured from the manufacturing statement with the finished goods inventory amounts to compute cost of goods sold for the year.

Solution to Demonstration Problem 3

SUNN CORPORATION
Factory Overhead Costs
For Year Ended December 31, 2009

Amortization expense—Factory patents	$ 16,000
Depreciation expense—Factory building	133,000
Depreciation expense—Factory equipment	78,000
Factory insurance expired	62,000
Factory supervision	74,000
Factory supplies used	21,000
Factory utilities	115,000
Indirect labor	26,000
Property taxes on factory equipment	14,000
Repairs expense—Factory equipment	31,000
Total factory overhead	$570,000

SUNN CORPORATION
Manufacturing Statement
For Year Ended December 31, 2009

Direct materials	
Raw materials inventory, Dec. 31, 2008	$ 60,000
Raw materials purchase	313,000
Raw materials available for use	373,000
Less raw materials inventory, Dec. 31, 2009	78,000
Direct materials used	295,000
Direct labor	250,000
Factory overhead	570,000
Total manufacturing costs	1,115,000
Goods in process inventory, Dec. 31, 2008	8,000
Total cost of goods in process	1,123,000
Less goods in process inventory, Dec. 31, 2009	9,000
Cost of goods manufactured	$1,114,000

SUNN CORPORATION
Income Statement
For Year Ended December 31, 2009

Sales		$1,630,000
Cost of goods sold		
Finished goods inventory, Dec. 31, 2008	$ 15,000	
Cost of goods manufactured	1,114,000	
Goods available for sale	1,129,000	
Less finished goods inventory, Dec. 31, 2009	12,500	
Cost of goods sold		1,116,500
Gross profit		513,500
Operating expenses		
Advertising expense	85,000	
Bad debts expense	28,000	
Depreciation expense—Office equipment	37,000	
Interest expense	25,000	
Miscellaneous expense	55,000	
Salaries expense	150,000	
Total operating expenses		380,000
Income before income taxes		133,500
Income taxes		53,400
Net income		$ 80,100

Summary

C1 Explain the purpose and nature of managerial accounting. The purpose of managerial accounting is to provide useful information to management and other internal decision makers. It does this by collecting, managing, and reporting both monetary and nonmonetary information in a manner useful to internal users. Major characteristics of managerial accounting include (1) focus on internal decision makers, (2) emphasis on planning and control, (3) flexibility, (4) timeliness, (5) reliance on forecasts and estimates, (6) focus on segments and projects, and (7) reporting both monetary and nonmonetary information.

C2 Describe the lean business model. The main purpose of the lean business model is the elimination of waste. Concepts such as total quality management and just-in-time production often aid in effective application of the model.

C3 Describe fraud and the role of ethics in managerial accounting. Fraud involves the use of one's job for personal gain through deliberate misuse of the employer's assets. All fraud is secret, violates the employee's job duties, provides financial benefits to the employee, and costs the employer money. A code of ethical beliefs can be used to resolve ethical conflicts.

C4 Describe accounting concepts useful in classifying costs. We can classify costs on the basis of their (1) behavior—fixed vs. variable, (2) traceability—direct vs. indirect, (3) controllability—controllable vs. uncontrollable, (4) relevance—sunk vs. out of pocket, and (5) function—product vs. period. A cost can be classified in more than one way, depending on the purpose for which the cost is being determined. These classifications help us understand cost patterns, analyze performance, and plan operations.

C5 Define product and period costs and explain how they impact financial statements. Costs that are capitalized because they are expected to have future value are called *product costs;* costs that are expensed are called *period costs.* This classification is important because it affects the amount of costs expensed in the income statement and the amount of costs assigned to inventory on the balance sheet. Product costs are commonly made up of direct materials, direct labor, and overhead. Period costs include selling and administrative expenses.

C6 Explain how balance sheets and income statements for manufacturing and merchandising companies differ. The main difference is that manufacturers usually carry three inventories on their balance sheets—raw materials, goods in process, and finished goods—instead of one inventory that merchandisers carry. The main difference between income statements of manufacturers and merchandisers is the items making up cost of goods sold. A merchandiser adds beginning merchandise inventory to cost of goods purchased and then subtracts ending merchandise inventory to get cost of goods sold. A manufacturer adds beginning finished goods inventory to cost of goods manufactured and then subtracts ending finished goods inventory to get cost of goods sold.

C7 Explain manufacturing activities and the flow of manufacturing costs. Manufacturing activities consist of materials, production, and sales activities. The materials activity consists of the purchase and issuance of materials to production. The production activity consists of converting materials into finished goods. At this stage in the process, the materials, labor, and overhead costs have been incurred and the manufacturing statement is prepared. The sales activity consists of selling some or all of finished goods available for sale. At this stage, the cost of goods sold is determined.

A1 Compute cycle time and cycle efficiency, and explain their importance to production management. It is important for companies to reduce the time to produce their products and to improve manufacturing efficiency. One measure of that time is cycle time (CT), defined as Process time + Inspection time + Move time + Wait time. Process time is value-added time; the others are non-value-added time. Cycle efficiency (CE) is the ratio of value-added time to total cycle time. If CE is low, management should evaluate its production process to see if it can reduce non-value-added activities.

P1 Compute cost of goods sold for a manufacturer. A manufacturer adds beginning finished goods inventory to cost of goods manufactured and then subtracts ending finished goods inventory to get cost of goods sold.

P2 Prepare a manufacturing statement and explain its purpose and links to financial statements. The manufacturing statement reports computation of cost of goods manufactured for the period. It begins by showing the period's costs for direct materials, direct labor, and overhead and then adjusts these numbers for the beginning and ending inventories of the goods in process to yield cost of goods manufactured.

Guidance Answers to **Decision Maker** and **Decision Ethics**

Production Manager It appears that all three friends want to pay the bill with someone else's money. David is using money belonging to the tax authorities, Denise is taking money from her company, and Derek is defrauding the client. To prevent such practices, companies have internal audit mechanisms. Many companies also adopt ethical codes of conduct to help guide employees. We must recognize that some entertainment expenses are justifiable and even encouraged. For example, the tax law allows certain deductions for entertainment that have a business purpose. Corporate policies also sometimes allow and encourage reimbursable spending for social activities, and contracts can include entertainment as allowable costs. Nevertheless, without further details, payment for this bill should be made from personal accounts.

Entrepreneur Tracing all costs directly to cost objects is always desirable, but you need to be able to do so in an economically feasible manner. In this case, you are able to trace 90% of the assembly department's direct costs. It may not be economical to spend more money on a new software to trace the final 10% of costs. You need to make a cost–benefit trade-off. If the software offers benefits beyond tracing the remaining 10% of the assembly department's costs, your decision should consider this.

Purchase Manager Opportunity costs relate to the potential quality and delivery benefits given up by not choosing supplier (A). Selecting supplier (B) might involve future costs of poor-quality seats (inspection, repairs, and returns). Also, potential delivery delays could interrupt work and increase manufacturing costs. Your company could also incur sales losses if the product quality of supplier (B) is low. As purchase manager, you are responsible for these costs and must consider them in making your decision.

Guidance Answers to **Quick Checks**

1. *d*
2. Financial accounting information is intended for users external to an organization such as investors, creditors, and government authorities. Managerial accounting focuses on providing information to managers, officers, and other decision makers within the organization.
3. No, GAAP do not control the practice of managerial accounting. Unlike external users, the internal users need managerial accounting information for planning and controlling business activities rather than for external comparison. Different types of information are required, depending on the activity. Therefore it is difficult to standardize managerial accounting.
4. Under TQM, all managers and employees should strive toward higher standards in their work and in the products and services they offer to customers.
5. Variable costs increase when volume of activity increases.
6. By being able to trace costs to cost objects (say, to products and departments), managers better understand the total costs associated with a cost object. This is useful when managers consider making changes to the cost object (such as when dropping the product or expanding the department).
7. Raw materials inventory, goods in process inventory, and finished goods inventory.
8. The cost of goods sold for merchandising companies includes all costs of acquiring the merchandise; the cost of goods sold for manufacturing companies includes the three costs of manufacturing: direct materials, direct labor, and overhead.
9. *a*
10. No; companies rarely report a manufacturing statement.
11. Beginning goods in process inventory is added to total manufacturing costs to yield total goods in process. Ending goods in process inventory is subtracted from total goods in process to yield cost of goods manufactured for the period.

Key Terms mhhe.com/wildFAP19e

Key Terms are available at the book's Website for learning and testing in an online Flashcard Format.

Continuous improvement (p. 730)
Control (p. 727)
Controllable or not controllable cost (p. 733)
Conversion costs (p. 738)
Cost object (p. 732)
Customer orientation (p. 729)
Cycle efficiency (CE) (p. 743)
Cycle time (CT) (p. 742)
Direct costs (p. 732)
Direct labor (p. 738)
Direct labor costs (p. 738)
Direct material (p. 738)
Direct material costs (p. 738)
Ethics (p. 731)
Factory overhead (p. 738)
Factory overhead costs (p. 738)
Finished goods inventory (p. 736)
Fixed cost (p. 732)
Goods in process inventory (p. 736)
Indirect costs (p. 732)
Indirect labor (p. 738)
Indirect labor costs (p. 738)
Indirect material (p. 735)
Institute of Management Accountants (IMA) (p. 731)
Internal control system (p. 731)
Just-in-time (JIT) manufacturing (p. 730)
Lean business model (p. 730)
Managerial accounting (p. 726)
Manufacturing statement (p. 740)
Non-value-added time (p. 743)
Opportunity cost (p. 733)
Out-of-pocket cost (p. 733)
Period costs (p. 733)
Planning (p. 726)
Prime costs (p. 738)
Product costs (p. 733)
Raw materials inventory (p. 735)
Sunk cost (p. 733)
Total quality management (TQM) (p. 730)
Value-added time (p. 743)
Value chain (p. 740)
Variable cost (p. 732)

Multiple Choice Quiz Answers on p. 767 mhhe.com/wildFAP19e

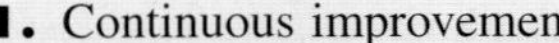

Additional Quiz Questions are available at the book's Website.

Quiz18

1. Continuous improvement
   a. Is used to reduce inventory levels.
   b. Is applicable only in service businesses.
   c. Rejects the notion of "good enough."
   d. Is used to reduce ordering costs.
   e. Is applicable only in manufacturing businesses.

2. A direct cost is one that is
   a. Variable with respect to the cost object.
   b. Traceable to the cost object.
   c. Fixed with respect to the cost object.
   d. Allocated to the cost object.
   e. A period cost.
3. Costs that are incurred as part of the manufacturing process, but are not clearly traceable to the specific unit of product or batches of product, are called
   a. Period costs.
   b. Factory overhead.
   c. Sunk costs.
   d. Opportunity costs.
   e. Fixed costs.
4. The three major cost components of manufacturing a product are
   a. Direct materials, direct labor, and factory overhead.
   b. Period costs, product costs, and sunk costs.
   c. Indirect labor, indirect materials, and fixed expenses.
   d. Variable costs, fixed costs, and period costs.
   e. Opportunity costs, sunk costs, and direct costs.
5. A company reports the following for the current year.

Finished goods inventory, beginning year	$6,000
Finished goods inventory, ending year	3,200
Cost of goods sold	7,500

   Its cost of goods manufactured for the current year is
   a. $1,500.
   b. $1,700.
   c. $7,500.
   d. $2,800.
   e. $4,700.

Discussion Questions

1. Describe the managerial accountant's role in business planning, control, and decision making.
2. Distinguish between managerial and financial accounting on
   a. Users and decision makers. b. Purpose of information.
   c. Flexibility of practice. d. Time dimension.
   e. Focus of information. f. Nature of information.
3. Identify the usual changes that a company must make when it adopts a customer orientation.
4. Distinguish between direct material and indirect material.
5. Distinguish between direct labor and indirect labor.
6. Distinguish between (*a*) factory overhead and (*b*) selling and administrative overhead.
7. What product cost is listed as both a prime cost and a conversion cost?
8. Assume that you tour **Apple**'s factory where it makes its products. List three direct costs and three indirect costs that you are likely to see.
9. Should we evaluate a manager's performance on the basis of controllable or noncontrollable costs? Why?
10. Explain why knowledge of cost behavior is useful in product performance evaluation.
11. Explain why product costs are capitalized but period costs are expensed in the current accounting period.
12. Explain how business activities and inventories for a manufacturing company, a merchandising company, and a service company differ.
13. Why does managerial accounting often involve working with numerous predictions and estimates?
14. How do an income statement and a balance sheet for a manufacturing company and a merchandising company differ?
15. Besides inventories, what other assets often appear on manufacturers' balance sheets but not on merchandisers' balance sheets?
16. Why does a manufacturing company require three different inventory categories?
17. Manufacturing activities of a company are described in the _______. This statement summarizes the types and amounts of costs incurred in its manufacturing _______.
18. What are the three categories of manufacturing costs?
19. List several examples of factory overhead.
20. List the four components of a manufacturing statement and provide specific examples of each for **Apple**.

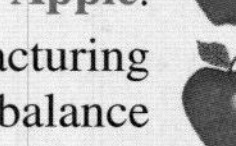

21. Prepare a proper title for the annual "manufacturing statement" of **Apple**. Does the date match the balance sheet or income statement? Why?
22. Describe the relations among the income statement, the manufacturing statement, and a detailed listing of factory overhead costs.
23. Define and describe *cycle time* and identify the components of cycle time.
24. Explain the difference between value-added time and non-value-added time.
25. Define and describe *cycle efficiency*.
26. Can management of a company such as **Best Buy** use cycle time and cycle efficiency as useful measures of performance? Explain.

27. Access **Anheuser-Busch**'s 2006 annual report (10-K) for the fiscal year ended December 31, 2006, at the SEC's EDGAR database (SEC.gov) or its Website (Anheuser-Busch.com). From its financial statement notes, identify the titles and amounts of its inventory components.

Denotes Discussion Questions that involve decision making.

QUICK STUDY

QS 18-1
Managerial accounting defined
C1

Managerial accounting (choose one)

1. Provides information that is widely available to all interested parties.
2. Is directed at reporting aggregate data on the company as a whole.
3. Must follow generally accepted accounting principles.
4. Provides information to aid management in planning and controlling business activities.

QS 18-2
Managerial accounting versus financial accounting
C1

Identify whether each description most likely applies to managerial or financial accounting.

1. _______ It is directed at external users in making investment, credit, and other decisions.
2. _______ Its information is often available only after an audit is complete.
3. _______ Its primary focus is on the organization as a whole.
4. _______ Its principles and practices are very flexible.
5. _______ Its primary users are company managers.

QS 18-3
Lean business concepts
C2

Match each lean business concept with its best description by entering its letter in the blank.

1. _______ Just-in-time manufacturing
2. _______ Continuous improvements
3. _______ Customer orientation
4. _______ Total quality management

A. Every manager and employee constantly looks for ways to improve company operations.
B. Focuses on quality throughout the production process.
C. Inventory is acquired or produced only as needed.
D. Flexible product designs can be modified to accommodate customer choices.

QS 18-4
Fixed and variable costs
C4

Which of these statements is true regarding fixed and variable costs?

1. Fixed costs increase and variable costs decrease in total as activity volume decreases.
2. Both fixed and variable costs stay the same in total as activity volume increases.
3. Both fixed and variable costs increase as activity volume increases.
4. Fixed costs stay the same and variable costs increase in total as activity volume increases.

QS 18-5
Direct and indirect costs
C4

Crosby Company produces sporting equipment, including footballs. Identify each of the following costs as direct or indirect if the cost object is a football produced by Crosby.

1. Depreciation on equipment used to produce footballs.
2. Salary of manager who supervises the entire plant.
3. Labor used on the football production line.
4. Electricity used in the production plant.
5. Materials used to produce footballs.

QS 18-6
Product and period costs
C5

Which of these statements is true regarding product and period costs?

1. Factory maintenance is a product cost and sales commission is a period cost.
2. Sales commission is a product cost and factory rent is a period cost.
3. Factory wages are a product cost and direct material is a period cost.
4. Sales commission is a product cost and depreciation on factory equipment is a product cost.

QS 18-7
Inventory reporting for manufacturers
C6

Three inventory categories are reported on a manufacturing company's balance sheet: (*a*) raw materials, (*b*) goods in process, and (*c*) finished goods. Identify the usual order in which these inventory items are reported on the balance sheet.

1. (*b*)(*c*)(*a*) **2.** (*c*)(*b*)(*a*) **3.** (*a*)(*b*)(*c*) **4.** (*b*)(*a*)(*c*)

QS 18-8
Cost of goods sold P1

A company has year-end cost of goods manufactured of $8,000, beginning finished goods inventory of $1,000, and ending finished goods inventory of $1,500. Its cost of goods sold is

1. $8,500 **2.** $8,000 **3.** $7,500 **4.** $7,800

QS 18-9
Manufacturing flows identified
C7

Identify the usual sequence of manufacturing activities by filling in the blank (with 1, 2, or 3) corresponding to its order: ______ Production activities; ______ sales activities; ______ materials activities.

QS 18-10
Cost of goods manufactured
P2

Prepare the 2009 manufacturing statement for Biron Company using the following information.

Direct materials	$381,000
Direct labor	126,300
Factory overhead costs	48,000
Goods in process, Dec. 31, 2008	315,200
Goods in process, Dec. 31, 2009	285,500

QS 18-11
Manufacturing cycle time and efficiency
A1

Compute and interpret (*a*) manufacturing cycle time and (*b*) manufacturing cycle efficiency using the following information from a manufacturing company.

Process time	7.5 hours
Inspection time	1.0 hours
Move time	3.2 hours
Wait time	18.3 hours

QS 18-12
Cost of goods sold
P1

Compute cost of goods sold for year 2009 using the following information.

Finished goods inventory, Dec. 31, 2008	$ 690,000
Goods in process inventory, Dec. 31, 2008	167,000
Goods in process inventory, Dec. 31, 2009	144,600
Cost of goods manufactured, year 2009	1,837,400
Finished goods inventory, Dec. 31, 2009	567,200

Available with McGraw-Hill's Homework Manager McGraw-Hill's HOMEWORK MANAGER®

EXERCISES

Exercise 18-1
Sources of accounting information
C1

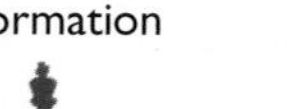

Both managerial accounting and financial accounting provide useful information to decision makers. Indicate in the following chart the most likely source of information for each business decision (a decision can require major input from both sources, in which case both can be marked).

Business Decision	Primary Information Source: Managerial	Primary Information Source: Financial
1. Determine amount of dividends to pay stockholders	____	____
2. Evaluate a purchasing department's performance	____	____
3. Report financial performance to board of directors	____	____
4. Estimate product cost for a new line of shoes	____	____
5. Plan the budget for next quarter	____	____
6. Measure profitability of all individual stores	____	____
7. Prepare financial reports according to GAAP	____	____
8. Determine location and size for a new plant	____	____

Exercise 18-2
Planning and control descriptions
C1

Complete the following statements by filling in the blanks.

1. ________ is the process of monitoring planning decisions and evaluating an organization's activities and employees.

2. ________ is the process of setting goals and making plans to achieve them.

3. ________ ________ usually covers a period of 5 to 10 years.

4. ________ ________ usually covers a period of one year.

Exercise 18-3
Characteristics of financial accounting and managerial accounting
C1

In the following chart, compare financial accounting and managerial accounting by describing how each differs for the items listed. Be specific in your responses.

	Financial Accounting	Managerial Accounting
1. Users and decision makers	________	________
2. Timeliness of information	________	________
3. Purpose of information	________	________
4. Nature of information	________	________
5. Flexibility of practice	________	________
6. Focus of information	________	________
7. Time dimension	________	________

Exercise 18-4
Customer orientation in practice
C2 

Customer orientation means that a company's managers and employees respond to customers' changing wants and needs. A manufacturer of plastic fasteners has created a customer satisfaction survey that it asks each of its customers to complete. The survey asks about the following factors: (A) lead time; (B) delivery; (C) price; (D) product performance. Each factor is to be rated as unsatisfactory, marginal, average, satisfactory, or very satisfied.

a. Match the competitive forces 1 through 4 to the factors on the survey. A factor can be matched to more than one competitive force.

Survey Factor	Competitive Force
A. Lead time	________ **1.** Cost
B. Delivery	________ **2.** Time
C. Price	________ **3.** Quality
D. Product performance	________ **4.** Flexibility of service

b. How can managers of this company use the information from this customer satisfaction survey to better meet competitive forces and satisfy their customers?

Exercise 18-5
Management concepts
C2 

Following are three separate events affecting the managerial accounting systems for different companies. Match the management concept(s) that the company is likely to adopt for the event identified. There is some overlap in the meaning of customer orientation and total quality management and, therefore, some responses can include more than one concept.

Event	Management Concept
________ 1. The company starts reporting measures on customer complaints and product returns from customers.	a. Total quality management (TQM)
	b. Continuous improvement (CI)
________ 2. The company starts reporting measures such as the percent of defective products and the number of units scrapped.	c. Customer orientation (CO)
________ 3. The company starts measuring inventory turnover and discontinues elaborate inventory records. Its new focus is to pull inventory through the system.	d. Just-in-time (JIT) system

Exercise 18-6
Cost analysis and identification
C4 C5

Georgia Pacific, a manufacturer, incurs the following costs. (1) Classify each cost as either a product or a period cost. If a product cost, identify it as a prime and/or conversion cost. (2) Classify each product cost as either a direct cost or an indirect cost using the product as the cost object.

Cost	Product Cost		Period Cost	Direct Cost	Indirect Cost
	Prime	Conversion			
1. Amortization of patents on factory machine	___	___	___	___	___
2. Payroll taxes for production supervisor	___	___	___	___	___
3. Accident insurance on factory workers	___	___	___	___	___
4. Depreciation—Factory building	___	___	___	___	___
5. State and federal income taxes	___	___	___	___	___
6. Wages to assembly workers	___	___	___	___	___
7. Direct materials used	___	___	___	___	___
8. Office supplies used	___	___	___	___	___
9. Bad debts expense	___	___	___	___	___
10. Small tools used	___	___	___	___	___
11. Factory utilities	___	___	___	___	___
12. Advertising	___	___	___	___	___

Exercise 18-7
Cost classifications C4

(1) Identify each of the five cost classifications discussed in the chapter. (2) List two purposes of identifying these separate cost classifications.

Exercise 18-8
Cost analysis and classification
C4

Listed here are product costs for the production of soccer balls. (1) Classify each cost (a) as either fixed or variable and (b) as either direct or indirect. (2) What pattern do you see regarding the relation between costs classified by behavior and costs classified by traceability?

Product Cost	Cost by Behavior		Cost by Traceability	
	Variable	Fixed	Direct	Indirect
1. Annual flat fee paid for office security	___	___	___	___
2. Leather covers for soccer balls	___	___	___	___
3. Lace to hold leather together	___	___	___	___
4. Wages of assembly workers	___	___	___	___
5. Coolants for machinery	___	___	___	___
6. Machinery depreciation	___	___	___	___
7. Taxes on factory	___	___	___	___

Exercise 18-9
Balance sheet identification and preparation
C6

Current assets for two different companies at calendar year-end 2009 are listed here. One is a manufacturer, Nordic Skis Mfg., and the other, Fresh Foods, is a grocery distribution company. (1) Identify which set of numbers relates to the manufacturer and which to the merchandiser. (2) Prepare the current asset section for each company from this information. Discuss why the current asset section for these two companies is different.

Account	Company 1	Company 2
Cash	$13,000	$11,000
Raw materials inventory	—	41,250
Merchandise inventory	44,250	—
Goods in process inventory	—	30,000
Finished goods inventory	—	50,000
Accounts receivable, net	62,000	81,000
Prepaid expenses	3,000	600

Exercise 18-10
Cost of goods sold computation
C6 P1

Compute cost of goods sold for each of these two companies for the year ended December 31, 2009.

		Computer Merchandising	Log Homes Manufacturing
1			
2			
3	Beginning inventory		
4	Merchandise	$301,000	
5	Finished goods		$602,000
6	Cost of purchases	580,000	
7	Cost of goods manufactured		790,000
8	Ending inventory		
9	Merchandise	201,000	
10	Finished goods		195,000

Check Computer Merchandising COGS, $680,000

Exercise 18-11
Cost of goods manufactured and cost of goods sold computation
P1 P2

Using the following data, compute (1) the cost of goods manufactured and (2) the cost of goods sold for both Jahmed Company and Kabiro Company.

	Jahmed Company	Kabiro Company
Beginning finished goods inventory	$15,000	$15,000
Beginning goods in process inventory	21,000	21,500
Beginning raw materials inventory	9,500	13,000
Rental cost on factory equipment	33,000	27,000
Direct labor	22,000	44,000
Ending finished goods inventory	19,500	12,000
Ending goods in process inventory	22,000	21,000
Ending raw materials inventory	10,500	9,400
Factory utilities	13,000	17,000
Factory supplies used	10,600	10,000
General and administrative expenses	22,000	54,000
Indirect labor	3,250	9,660
Repairs—Factory equipment	6,780	3,500
Raw materials purchases	24,000	47,000
Sales salaries	49,000	41,000

Check Jahmed COGS, $106,130

Exercise 18-12
Components of accounting reports
C7 P2

For each of the following account balances for a manufacturing company, place a ✔ in the appropriate column indicating that it appears on the balance sheet, the income statement, the manufacturing statement, and/or a detailed listing of factory overhead costs. Assume that the income statement shows the calculation of cost of goods sold and the manufacturing statement shows only the total amount of factory overhead. (An account balance can appear on more than one report.)

	Account	Balance Sheet	Income Statement	Manufacturing Statement	Overhead Report
3	Accounts receivable				
4	Computer supplies used in office				
5	Beginning finished goods inventory				
6	Beginning goods in process inventory				
7	Beginning raw materials inventory				
8	Cash				
9	Depreciation expense—Factory building				
10	Depreciation expense—Factory equipment				
11	Depreciation expense—Office building				
12	Depreciation expense—Office equipment				
13	Direct labor				
14	Ending finished goods inventory				
15	Ending goods in process inventory				
16	Ending raw materials inventory				
17	Factory maintenance wages				
18	Computer supplies used in factory				
19	Income taxes				
20	Insurance on factory building				
21	Rent cost on office building				
22	Office supplies used				
23	Property taxes on factory building				
24	Raw materials purchases				
25	Sales				

Exercise 18-13
Manufacturing statement preparation P2

Given the following selected account balances of Spalding Company, prepare its manufacturing statement for the year ended on December 31, 2009. Include a listing of the individual overhead account balances in this statement.

Sales	$1,363,000
Raw materials inventory, Dec. 31, 2008	40,000
Goods in process inventory, Dec. 31, 2008	53,600
Finished goods inventory, Dec. 31, 2008	60,400
Raw materials purchases	181,900
Direct labor	243,000
Factory computer supplies used	15,700
Indirect labor	54,000
Repairs—Factory equipment	7,250
Rent cost of factory building	56,000
Advertising expense	92,000
General and administrative expenses	140,000
Raw materials inventory, Dec. 31, 2009	44,000
Goods in process inventory, Dec. 31, 2009	41,200
Finished goods inventory, Dec. 31, 2009	66,200

Check Cost of goods manufactured, $566,250

Exercise 18-14
Income statement preparation P2

Use the information in Exercise 18-13 to prepare an income statement for Spalding Company (a manufacturer). Assume that its cost of goods manufactured is $566,250.

Exercise 18-15
Cost flows in manufacturing C7 P2

The following chart shows how costs flow through a business as a product is manufactured. Some boxes in the flowchart show cost amounts. Compute the cost amounts for the boxes that contain question marks.

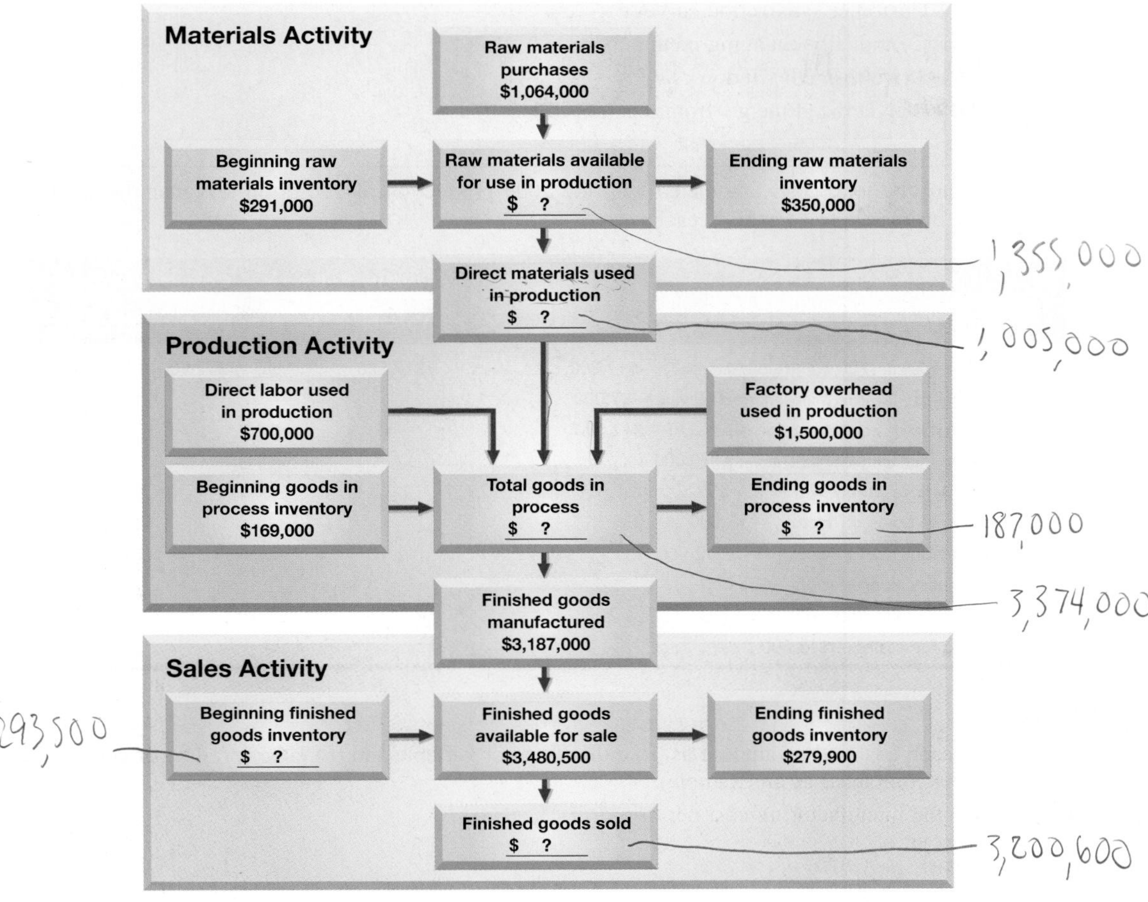

Exercise 18-16
C3

Fraud affects **Best Buy**. Refer to Best Buy's financial statements in Appendix A to answer the following:

1. Explain how inventory losses (such as theft) impact how Best Buy reports inventory on its balance sheet.

2. In what income statement account does Best Buy report inventory losses?

McGraw-Hill's HOMEWORK MANAGER® Available with McGraw-Hill's Homework Manager

PROBLEM SET A

Problem 18-1A
Managerial accounting role
C1 C2

This chapter explained the purpose of managerial accounting in the context of the current business environment. Review the *automobile* section of your local newspaper; the Sunday paper is often best. Review advertisements of sport-utility vehicles and identify the manufacturers that offer these products and the factors on which they compete.

Required

Discuss the potential contributions and responsibilities of the managerial accounting professional in helping an automobile manufacturer succeed. (*Hint:* Think about information and estimates that a managerial accountant might provide new entrants into the sport-utility market.)

Problem 18-2A
Lean business concepts
C2 

Many fast-food restaurants compete on lean business concepts. Match each of the following activities at a fast-food restaurant with the lean business concept it strives to achieve. Some activities might relate to more than one lean business concept.

_______ **1.** Courteous employees
_______ **2.** Food produced to order
_______ **3.** New product development
_______ **4.** Clean tables and floors
_______ **5.** Orders filled within three minutes
_______ **6.** Standardized food making processes

a. Just-in-time (JIT)
b. Continuous improvement (CI)
c. Total quality management (TQM)

[continued on next page]

_______ **7.** Customer satisfaction surveys
_______ **8.** Continually changing menus
_______ **9.** Drive-through windows
_______ **10.** Standardized menus from location to location

Problem 18-3A
Cost computation, classification, and analysis
C4 

Listed here are the total costs associated with the 2009 production of 700 drum sets manufactured by Roland. The drum sets sell for $600 each.

Costs	Cost by Behavior		Cost by Function	
	Variable	Fixed	Product	Period
1. Drum stands (700 stands outsourced)—$17,500	$17,500		$17,500	
2. Annual flat fee for maintenance service—$7,000				
3. Rent cost of equipment for sales staff—$12,000				
4. Upper management salaries—$170,000				
5. Wages of assembly workers—$59,500				
6. Property taxes on factory—$3,500				
7. Accounting staff salaries—$42,000				
8. Machinery depreciation—$28,000				
9. Sales commissions—$20 per unit				
10. Plastic for casing—$12,600				

Required

Check (1) Total variable manufacturing cost, $89,600

1. Classify each cost and its amount as (*a*) either fixed or variable and (*b*) either product or period (the last cost is completed as an example).

2. Compute the manufacturing cost per drum set.

Analysis Component

3. Assume that 1,000 drum sets are produced in the next month. What do you predict will be the total cost of plastic for the casings and the per unit cost of the plastic for the casings? Explain.

4. Assume that 1,000 drum sets are produced in the next month. What do you predict will be the total cost of property taxes and the per unit cost of the property taxes? Explain.

Problem 18-4A
Cost classification and explanation
C4 C5

Assume that you must make a presentation to the marketing staff explaining the difference between product and period costs. Your supervisor tells you the marketing staff would also like clarification regarding prime and conversion costs and an explanation of how these terms fit with product and period cost. You are told that many on the staff are unable to classify costs in their merchandising activities.

Required

Prepare a one-page memorandum to your supervisor outlining your presentation to the marketing staff.

Problem 18-5A
Opportunity cost estimation and application
C1 C4 

Refer to *Decision Maker,* **Purchase Manager,** in this chapter. Assume that you are the motorcycle manufacturer's managerial accountant. The purchasing manager asks you about preparing an estimate of the related costs for buying motorcycle seats from supplier (B). She tells you this estimate is needed because unless dollar estimates are attached to nonfinancial factors, such as lost production time, her supervisor will not give it full attention. The manager also shows you the following information.

- Production output is 1,000 motorcycles per year based on 250 production days a year.
- Production time per day is 8 hours at a cost of $2,000 per hour to run the production line.
- Lost production time due to poor quality is 1%.
- Satisfied customers purchase, on average, three motorcycles during a lifetime.
- Satisfied customers recommend the product, on average, to five other people.
- Marketing predicts that using seat (B) will result in five lost customers per year from repeat business and referrals.
- Average contribution margin per motorcycle is $3,000.

Required

Estimate the costs (including opportunity costs) of buying motorcycle seats from supplier (B). This problem requires that you think creatively and make reasonable estimates; thus there could be more than one correct answer. (*Hint:* Reread the answer to *Decision Maker* and compare the cost savings for buying from supplier [B] to the sum of lost customer revenue from repeat business and referrals and the cost of lost production time.)

Check Estimated cost of lost production time, $40,000

Problem 18-6A
Ending inventory computation and evaluation
C2 C6

Laredo Boot Company makes specialty boots for the rodeo circuit. On December 31, 2008, the company had (*a*) 300 pairs of boots in finished goods inventory and (*b*) 1,400 heels at a cost of $16 each in raw materials inventory. During 2009, the company purchased 46,000 additional heels at $16 each and manufactured 16,800 pairs of boots.

Required

1. Determine the unit and dollar amounts of raw materials inventory in heels at December 31, 2009.

Check (1) Ending (heel) inventory, 13,800 units; $220, 800

Analysis Component

2. Write a one-half page memorandum to the production manager explaining why a just-in-time inventory system for heels should be considered. Include the amount of working capital that can be reduced at December 31, 2009, if the ending heel raw material inventory is cut by 75%.

Problem 18-7A
Inventory computation and reporting
C4 C6 P1

excel
mhhe.com/wildFAP19e

Shown here are annual financial data at December 31, 2009, taken from two different companies.

	Active Sports Retail	Sno-Board Manufacturing
Beginning inventory		
Merchandise	$145,000	
Finished goods		$340,000
Cost of purchases	240,000	
Cost of goods manufactured		582,000
Ending inventory		
Merchandise	110,000	
Finished goods		150,000

Required

1. Compute the cost of goods sold section of the income statement at December 31, 2009, for each company. Include the proper title and format in the solution.

Check (1) Sno-Board's cost of goods sold, $772,000

2. Write a half-page memorandum to your instructor (*a*) identifying the inventory accounts and (*b*) describing where each is reported on the income statement and balance sheet for both companies.

Problem 18-8A
Manufacturing and income statements; inventory analysis P2

The following calendar year-end information is taken from the December 31, 2009, adjusted trial balance and other records of Gucci Company.

Advertising expense	$ 26,600	Direct labor	680,400
Depreciation expense—Office equipment	11,500	Income taxes expense	291,500
Depreciation expense—Selling equipment	10,800	Indirect labor	58,800
Depreciation expense—Factory equipment	38,200	Miscellaneous production costs	9,800
Factory supervision	105,700	Office salaries expense	74,000
Factory supplies used	7,800	Raw materials purchases	965,000
Factory utilities	34,000	Rent expense—Office space	23,000
Inventories		Rent expense—Selling space	25,200
Raw materials, December 31, 2008	165,900	Rent expense—Factory building	81,600
Raw materials, December 31, 2009	187,000	Maintenance expense—Factory equipment	37,100
Goods in process, December 31, 2008	18,100	Sales	4,630,000
Goods in process, December 31, 2009	24,600	Sales discounts	63,600
Finished goods, December 31, 2008	164,100	Sales salaries expense	398,400
Finished goods, December 31, 2009	135,900		

Required

Check (1) Cost of goods manufactured, $1,990,800

1. Prepare the company's 2009 manufacturing statement.
2. Prepare the company's 2009 income statement that reports separate categories for (*a*) selling expenses and (*b*) general and administrative expenses.

Analysis Component

3. Compute the (*a*) inventory turnover, defined as cost of goods sold divided by average inventory, and (*b*) days' sales in inventory, defined as 365 times ending inventory divided by cost of goods sold, for both its raw materials inventory and its finished goods inventory. (To compute turnover and days' sales in inventory for raw materials, use raw materials used rather than cost of goods sold.) Discuss some possible reasons for differences between these ratios for the two types of inventories.

Problem 18-9A
Manufacturing cycle time and efficiency
A1

Mission Oak Company produces oak bookcases to customer order. It received an order from a customer to produce 5,000 bookcases. The following information is available for the production of the bookcases.

Process time	18.0 days
Inspection time	2.0 days
Move time	4.4 days
Wait time	20.6 days

Required

1. Compute the company's manufacturing cycle time.

Check (2) Manufacturing cycle efficiency, 0.40

2. Compute the company's manufacturing cycle efficiency. Interpret your answer.

Analysis Component

3. Assume that Mission Oak wishes to increase its manufacturing cycle efficiency to 0.75. What are some ways that it can accomplish this?

PROBLEM SET B

Problem 18-1B
Managerial accounting role
C1 C2

This chapter described the purpose of managerial accounting in the context of the current business environment. Review the *home electronics* section of your local newspaper; the Sunday paper is often best. Review advertisements of home electronics and identify the manufacturers that offer these products and the factors on which they compete.

Required

Discuss the potential contributions and responsibilities of the managerial accounting professional in helping a home electronics manufacturer succeed. (*Hint:* Think about information and estimates that a managerial accountant might provide new entrants into the home electronics market.)

Problem 18-2B
Lean business concepts
C2

Eastman-Kodak manufactures digital cameras and must compete on lean manufacturing concepts. Match each of the following activities that it engages in with the lean manufacturing concept it strives to achieve. (Some activities might relate to more than one lean manufacturing concept.)

_______ 1. Lenses are received daily based on customer orders.

_______ 2. Customers receive a satisfaction survey with each camera purchased.

_______ 3. The manufacturing process is standardized and documented.

_______ 4. Cameras are produced in small lots, and only to customer order.

_______ 5. Manufacturing facilities are arranged to reduce move time and wait time.

_______ 6. Kodak conducts focus groups to determine new features that customers want in digital cameras.

a. Just-in-time (JIT)
b. Continuous improvement (CI)
c. Total quality management (TQM)

[continued on next page]

_______ **7.** Orders received are filled within two business days.

_______ **8.** Kodak works with suppliers to reduce inspection time of incoming materials.

_______ **9.** Kodak monitors the market to determine what features its competitors are offering on digital cameras.

_______ **10.** Kodak asks production workers for ideas to improve production.

Problem 18-3B
Cost computation, classification, and analysis
C4 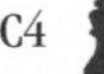

Listed here are the total costs associated with the production of 10,000 Blu-ray Discs (BDs) manufactured by New Age. The BDs sell for $15 each.

	Cost by Behavior		Cost by Function	
Costs	Variable	Fixed	Product	Period
1. Annual fixed fee for cleaning service—$3,000		$3,000		$3,000
2. Cost of office equipment rent—$700				
3. Upper management salaries—$100,000				
4. Labeling (10,000 outsourced)—$2,500				
5. Wages of assembly workers—$20,000				
6. Sales commissions—$0.50 per BD				
7. Machinery depreciation—$15,000				
8. Systems staff salaries—$10,000				
9. Cost of factory rent—$4,500				
10. Plastic for BDs—$1,000				

Required

1. Classify each cost and its amount as (*a*) either fixed or variable and (*b*) either product or period.

2. Compute the manufacturing cost per BD.

Check (2) Total variable manufacturing cost, $23,500

Analysis Component

3. Assume that 12,000 BDs are produced in the next month. What do you predict will be the total cost of plastic for the BDs and the per unit cost of the plastic for the BDs? Explain.

4. Assume that 12,000 BDs are produced in the next month. What do you predict will be the total cost of factory rent and the per unit cost of the factory rent? Explain.

Problem 18-4B
Cost classification and explanation
C4 C5

Assume that you must make a presentation to a client explaining the difference between prime and conversion costs. The client makes and sells 200,000 cookies per week. The client tells you that her sales staff also would like a clarification regarding product and period costs. She tells you that most of the staff lack training in managerial accounting.

Required

Prepare a one-page memorandum to your client outlining your planned presentation to her sales staff.

Problem 18-5B
Opportunity cost estimation and application
C1 C4 

Refer to *Decision Maker,* **Purchase Manager,** in this chapter. Assume that you are the motorcycle manufacturer's managerial accountant. The purchasing manager asks you about preparing an estimate of the related costs for buying motorcycle seats from supplier (B). She tells you this estimate is needed because unless dollar estimates are attached to nonfinancial factors such as lost production time, her supervisor will not give it full attention. The manager also shows you the following information.

- Production output is 1,000 motorcycles per year based on 250 production days a year.
- Production time per day is 8 hours at a cost of $500 per hour to run the production line.
- Lost production time due to poor quality is 1%.
- Satisfied customers purchase, on average, three motorcycles during a lifetime.
- Satisfied customers recommend the product, on average, to four other people.
- Marketing predicts that using seat (B) will result in four lost customers per year from repeat business and referrals.
- Average contribution margin per motorcycle is $4,000.

Required

Check Cost of lost customer revenue, $16,000

Estimate the costs (including opportunity costs) of buying motorcycle seats from supplier (B). This problem requires that you think creatively and make reasonable estimates; thus there could be more than one correct answer. (*Hint:* Reread the answer to *Decision Maker,* and compare the cost savings for buying from supplier [B] to the sum of lost customer revenue from repeat business and referrals and the cost of lost production time.)

Problem 18-6B
Ending inventory computation and evaluation
C2 C6 

CCMD Company makes specialty skates for the ice skating circuit. On December 31, 2008, the company had (*a*) 1,500 skates in finished goods inventory and (*b*) 2,500 blades at a cost of $15 each in raw materials inventory. During 2009, CCMD purchased 45,000 additional blades at $15 each and manufactured 20,000 pairs of skates.

Required

Check (1) Ending (blade) inventory, 7,500 units; $112,500

1. Determine the unit and dollar amounts of raw materials inventory in blades at December 31, 2009.

Analysis Component

2. Write a one-half page memorandum to the production manager explaining why a just-in-time inventory system for blades should be considered. Include the amount of working capital that can be reduced at December 31, 2009, if the ending blade raw material inventory is cut in half.

Problem 18-7B
Inventory computation and reporting
C4 C6 P1

Shown here are annual financial data at December 31, 2009, taken from two different companies.

	AAA Imports (Retail)	Marina Boats (Manufacturing)
Beginning inventory		
Merchandise	$ 50,000	
Finished goods		$200,000
Cost of purchases	350,000	
Cost of goods manufactured		686,000
Ending inventory		
Merchandise	25,000	
Finished goods		300,000

Required

Check (1) AAA Imports cost of goods sold, $375,000

1. Compute the cost of goods sold section of the income statement at December 31, 2009, for each company. Include the proper title and format in the solution.

2. Write a half-page memorandum to your instructor (*a*) identifying the inventory accounts and (*b*) identifying where each is reported on the income statement and balance sheet for both companies.

Problem 18-8B
Manufacturing and income statements; analysis of inventories
P2

The following calendar year-end information is taken from the December 31, 2009, adjusted trial balance and other records of Homestyle Furniture.

Advertising expense	$ 22,250	Direct labor	564,500
Depreciation expense—Office equipment	10,440	Income taxes expense	138,700
Depreciation expense—Selling equipment	12,125	Indirect labor	61,000
Depreciation expense—Factory equipment	37,400	Miscellaneous production costs	10,440
Factory supervision	123,500	Office salaries expense	72,875
Factory supplies used	8,060	Raw materials purchases	896,375
Factory utilities	39,500	Rent expense—Office space	25,625
Inventories		Rent expense—Selling space	29,000
Raw materials, December 31, 2008	42,375	Rent expense—Factory building	95,500
Raw materials, December 31, 2009	72,430	Maintenance expense—Factory equipment	32,375
Goods in process, December 31, 2008	14,500	Sales	5,002,000
Goods in process, December 31, 2009	16,100	Sales discounts	59,375
Finished goods, December 31, 2008	179,200	Sales salaries expense	297,300
Finished goods, December 31, 2009	143,750		

Required

1. Prepare the company's 2009 manufacturing statement.

2. Prepare the company's 2009 income statement that reports separate categories for (*a*) selling expenses and (*b*) general and administrative expenses.

Check (1) Cost of goods manufactured, $1,836,995

Analysis Component

3. Compute the (*a*) inventory turnover, defined as cost of goods sold divided by average inventory, and (*b*) days' sales in inventory, defined as 365 times ending inventory divided by cost of goods sold, for both its raw materials inventory and its finished goods inventory. (To compute turnover and days' sales in inventory for raw materials, use raw materials used rather than cost of goods sold.) Discuss some possible reasons for differences between these ratios for the two types of inventories.

Problem 18-9B
Manufacturing cycle time and efficiency

A1 

Fast Ink produces ink-jet printers for personal computers. It received an order for 400 printers from a customer. The following information is available for this order.

Process time	8.0 hours
Inspection time	1.7 hours
Move time	4.5 hours
Wait time	10.8 hours

Required

1. Compute the company's manufacturing cycle time.

2. Compute the company's manufacturing cycle efficiency. Interpret your answer.

Analysis Component

3. Assume that Fast Ink wishes to increase its manufacturing cycle efficiency to 0.80. What are some ways that it can accomplish this?

SERIAL PROBLEM

Success Systems

(This serial problem begins in Chapter 1 and continues through most of the book. If previous chapter segments were not completed, the serial problem can begin at this point. It is helpful, but not necessary, to use the Working Papers that accompany the book.)

SP 18 Adriana Lopez, owner of Success Systems, decides to diversify her business by also manufacturing computer workstation furniture.

Required

1. Classify the following manufacturing costs of Success Systems by behavior and traceability.

Product Costs	Cost by Behavior		Cost by Traceability	
	Variable	Fixed	Direct	Indirect
1. Monthly flat fee to clean workshop	___	___	___	___
2. Laminate coverings for desktops	___	___	___	___
3. Taxes on assembly workshop	___	___	___	___
4. Glue to assemble workstation component parts	___	___	___	___
5. Wages of desk assembler	___	___	___	___
6. Electricity for workshop	___	___	___	___
7. Depreciation on tools	___	___	___	___

2. Prepare a manufacturing statement for Success Systems for the month ended January 31, 2010. Assume the following manufacturing costs:

Direct materials: $2,200
Factory overhead: $490
Direct labor: $900
Beginning goods in process: none (December 31, 2009)
Ending goods in process: $540 (January 31, 2010)
Beginning finished goods inventory: none (December 31, 2009)
Ending finished goods inventory: $350 (January 31, 2010)

Check (3) COGS, $2,700

3. Prepare the cost of goods sold section of a partial income statement for Success Systems for the month ended January 31, 2010.

BEYOND THE NUMBERS

REPORTING IN ACTION

C1 C2 

BTN 18-1 Managerial accounting is more than recording, maintaining, and reporting financial results. Managerial accountants must provide managers with both financial and nonfinancial information including estimates, projections, and forecasts. There are many accounting estimates that management accountants must make, and **Best Buy** must notify shareholders of these estimates.

Required

1. Access and read Best Buy's "Critical Accounting Estimates" section (six pages), which is part of its *Management's Discussion and Analysis of Financial Condition and Results of Operations* section, from either its annual report or its 10-K for the year ended March 3, 2007 [BestBuy.com]. What are some of the accounting estimates that Best Buy made in preparing its financial statements? What are some of the effects if the actual results of Best Buy differ from its assumptions?

2. What is the management accountant's role in determining those estimates?

Fast Forward

3. Access **Best Buy**'s annual report for a fiscal year ending after March 3, 2007, from either its Website [BestBuy.com] or the SEC's EDGAR database [www.SEC.gov]. Answer the questions in parts (1) and (2) after reading the current MD&A section. Identify any major changes.

COMPARATIVE ANALYSIS

C1 C2 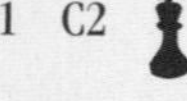

BTN 18-2 **Best Buy** and **RadioShack** are both merchandisers that rely on customer satisfaction. Access and read (1) Best Buy's "Business Strategy and Core Philosophies" section (one page) and (2) RadioShack's "Financial Impact of Turnaround Program" section (one page). Both sections are located in the respective company's *Management Discussion and Analysis of Financial Condition and Results of Operations* section from the annual report or 10-K. The Best Buy report is for the year ended March 3, 2007, and the RadioShack report is for the year ended December 31, 2006.

Required

1. Identify the strategic initiatives that each company put forward in its desire to better compete and succeed in the marketplace.

2. For each of these strategic initiatives for both companies, explain how it reflects (or does not reflect) a customer satisfaction focus.

ETHICS CHALLENGE

C3 C4 C5

BTN 18-3 Assume that you are the managerial accountant at Infostore, a manufacturer of hard drives, CDs, and diskettes. Its reporting year-end is December 31. The chief financial officer is concerned about having enough cash to pay the expected income tax bill because of poor cash flow management. On November 15, the purchasing department purchased excess inventory of CD raw materials in anticipation of rapid growth of this product beginning in January. To decrease the company's tax liability, the chief financial officer tells you to record the purchase of this inventory as part of supplies and expense it in the current year; this would decrease the company's tax liability by increasing expenses.

Required

1. In which account should the purchase of CD raw materials be recorded?

2. How should you respond to this request by the chief financial officer?

COMMUNICATING IN PRACTICE

BTN 18-4 Write a one-page memorandum to a prospective college student about salary expectations for graduates in business. Compare and contrast the expected salaries for accounting (including different subfields such as public, corporate, tax, audit, and so forth), marketing, management, and finance majors. Prepare a graph showing average starting salaries (and those for experienced professionals in those fields if available). To get this information, stop by your school's career services office; libraries also have this information. The Website JobStar.org (click on *Salary Info*) also can get you started.

TAKING IT TO THE NET

C1 C3 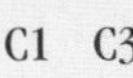 

BTN 18-5 Managerial accounting professionals follow a code of ethics. As a member of the Institute of Management Accountants, the managerial accountant must comply with Standards of Ethical Conduct.

Required

1. Identify, print, and read the *Statement of Ethical Professional Practice* posted at www.IMAnet.org. (Search using "ethical professional practice.")

2. What four overarching ethical principles underlie the IMA's statement?

3. Describe the courses of action the IMA recommends in resolving ethical conflicts.

TEAMWORK IN ACTION

C7 P2

BTN 18-6 The following calendar-year information is taken from the December 31, 2009, adjusted trial balance and other records of Dahlia Company.

Advertising expense	$ 19,125	Direct labor	650,750
Depreciation expense—Office equipment	8,750	Indirect labor	60,000
Depreciation expense—Selling equipment	10,000	Miscellaneous production costs	8,500
Depreciation expense—Factory equipment	32,500	Office salaries expense	100,875
Factory supervision	122,500	Raw materials purchases	872,500
Factory supplies used	15,750	Rent expense—Office space	21,125
Factory utilities	36,250	Rent expense—Selling space	25,750
Inventories		Rent expense—Factory building	79,750
Raw materials, December 31, 2008	177,500	Maintenance expense—Factory equipment	27,875
Raw materials, December 31, 2009	168,125	Sales	3,275,000
Goods in process, December 31, 2008	15,875	Sales discounts	57,500
Goods in process, December 31, 2009	14,000	Sales salaries expense	286,250
Finished goods, December 31, 2008	164,375		
Finished goods, December 31, 2009	129,000		

Required

1. *Each* team member is to be responsible for computing **one** of the following amounts. You are not to duplicate your teammates' work. Get any necessary amounts from teammates. Each member is to explain the computation to the team in preparation for reporting to class.

a. Materials used.
b. Factory overhead.
c. Total manufacturing costs.
d. Total cost of goods in process.
e. Cost of goods manufactured.

2. Check your cost of goods manufactured with the instructor. If it is correct, proceed to part (3).

3. *Each* team member is to be responsible for computing **one** of the following amounts. You are not to duplicate your teammates' work. Get any necessary amounts from teammates. Each member is to explain the computation to the team in preparation for reporting to class.

a. Net sales.
b. Cost of goods sold.
c. Gross profit.
d. Total operating expenses.
e. Net income or loss before taxes.

Point: Provide teams with transparencies and markers for presentation purposes.

ENTREPRENEURIAL DECISION

C1 C4 

BTN 18-7 Brian Taylor of **Kernel Season's** must understand his manufacturing costs to effectively operate and succeed as a profitable and efficient company.

Required

1. What are the three main categories of manufacturing costs that Brian must monitor and control? Provide examples of each.

2. How can Brian make the Kernel Season's manufacturing process more cost-effective? Provide examples of two useful managerial measures of time and efficiency.

3. What are four goals of a total quality management process? How can Kernel Season's use TQM to improve its business activities?

HITTING THE ROAD

C1 C5 

BTN 18-8 Visit your favorite fast-food restaurant. Observe its business operations.

Required

1. Describe all business activities from the time a customer arrives to the time that customer departs.

2. List all costs you can identify with the separate activities described in part 1.

3. Classify each cost from part 2 as fixed or variable, and explain your classification.

GLOBAL DECISION

DSG

BTN 18-9 Access DSG's annual report for the year ended April 28, 2007 (www.DSGiplc.com). Read the section "Corporate Governance" dealing with the responsibilities of the board of directors.

Required

1. Identify the responsibilities (see the "schedule of matters reserved for the board") of DSG's board of directors.
2. How would management accountants be involved in assisting the board of directors in carrying out their responsibilities? Explain.

ANSWERS TO MULTIPLE CHOICE QUIZ

1. c
2. b
3. b
4. a
5. Beginning finished goods + Cost of goods manufactured (COGM) − Ending finished goods = Cost of goods sold
   $6,000 + COGM − $3,200 = $7,500
   COGM = $4,700

A Look Back

Chapter 18 introduced managerial accounting and explained basic cost concepts. We also described the lean business model and the reporting of manufacturing activities, including the manufacturing statement.

A Look at This Chapter

We begin this chapter by describing a cost accounting system. We then explain the procedures used to determine costs using a job order costing system. We conclude with a discussion of over- and underapplied overhead.

A Look Ahead

Chapter 20 focuses on measuring costs in process production companies. We explain process production, describe how to assign costs to processes, and compute and analyze cost per equivalent unit.

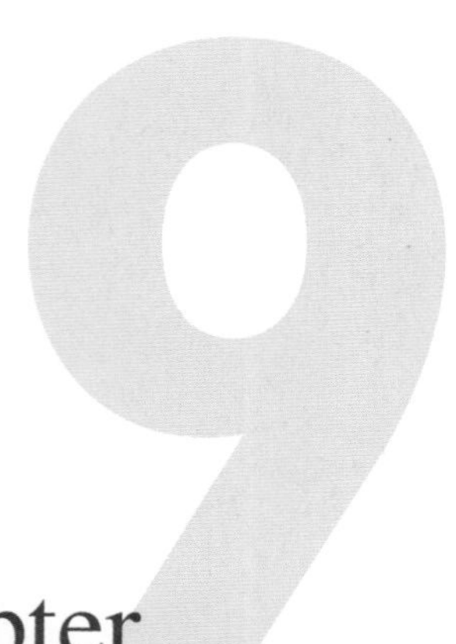

Chapter

Job Order Cost Accounting

Learning Objectives

CAP

Conceptual

C1 Explain the cost accounting system. *(p. 770)*

C2 Describe important features of job order production. *(p. 770)*

C3 Explain job cost sheets and how they are used in job order cost accounting. *(p. 772)*

Analytical

A1 Apply job order costing in pricing services. *(p. 782)*

Procedural

P1 Describe and record the flow of materials costs in job order cost accounting. *(p. 773)*

P2 Describe and record the flow of labor costs in job order cost accounting. *(p. 775)*

P3 Describe and record the flow of overhead costs in job order cost accounting. *(p. 776)*

P4 Determine adjustments for overapplied and underapplied factory overhead. *(p. 781)*

LP19

Decision Feature

Working the Field

"Being successful is having a vision which you are excited to follow without the fear of failure"
—Hank Julicher

PHILADELPHIA, PA—One size fits all? Not when it comes to synthetic turf for athletic fields—this according to Hank Julicher, founder of **Sprinturf (Sprinturf.com).** "Not all fields are exactly alike, because no two owners have the same exact needs," insists Hank. "Many variables must be considered, including playing requirements, climate, and financial considerations." Designing, installing, and servicing synthetic turf systems are Sprinturf's mission.

"There is much more to a playing field than just the surface," explains Hank. "Many would argue that the base is the most important—it needs the strength to support athletes and vehicles, while still being able to drain over 20″ of rainfall per hour." For this, Sprinturf relies on its all-rubber infill system for its installations. Still, understanding customer needs is key. In extremely hot, arid climates, Sprinturf uses light-colored rubber infill to reduce the temperature of playing surfaces. In cold areas, Sprinturf offers solutions to reduce snow and ice buildup. Hank has put in fields from Utah State University to University of Montana to Long Beach City College. While a touchdown is worth 6 points on every Sprinturf field, each field is otherwise unique.

Manufacturers of custom products, such as that from Sprinturf, use state-of-the-art job order cost accounting to track costs. This includes tracking the cost of materials, labor and overhead, and managing those expenses. To help control costs and ensure product quality, Sprinturf does not outsource any part of the design or installation process. Controlling all aspects of the process enables it to better isolate costs and avoid the run-away costs often experienced by startups that fail to use costing techniques. Recruiting top-notch personnel and experienced supervisors also helps control labor costs. Reflecting the unique nature of each field, each installation is videotaped to ensure it is done exactly according to customer specifications.

Hank Julicher stresses cost control as vital to Sprinturf's success. "To take on two 800 pound gorillas in our industry, we had to be more creative, efficient, and cost-effective to win," explains Hank. "We just hung in there until the public recognized our quality and value." This winning formula has led to product growth that any team would envy.

[Sources: *Sprinturf Website,* January 2009; *Entrepreneur,* 2007; *PanStadia,* February and November 2005]

Chapter Preview

This chapter introduces a system for assigning costs to the flow of goods through a production process. We then describe the details of a *job order cost accounting system*. Job order costing is frequently used by manufacturers of custom products or providers of custom services. Manufacturers that use job order costing typically base it on a perpetual inventory system, which provides a continuous record of materials, goods in process, and finished goods inventories.

Job Order Cost Accounting

Job Order Cost Accounting
- Cost accounting system
- Job order manufacturing
- Events in job order costing
- Job cost sheet

Job Order Cost Flows and Reports
- Materials cost flows and documents
- Labor cost flows and documents
- Overhead cost flows and documents
- Summary of cost flows

Adjustment of Overapplied or Underapplied Overhead
- Underapplied overhead
- Overapplied overhead

Job Order Cost Accounting

This section describes a cost accounting system and job order production and costing.

Cost Accounting System

C1 Explain the cost accounting system.

An ever-increasing number of companies use a cost accounting system to generate timely and accurate inventory information. A **cost accounting system** records manufacturing activities using a *perpetual* inventory system, which continuously updates records for costs of materials, goods in process, and finished goods inventories. A cost accounting system also provides timely information about inventories and manufacturing costs per unit of product. This is especially helpful for managers' efforts to control costs and determine selling prices. (A **general accounting system** records manufacturing activities using a *periodic* inventory system. Some companies still use a general accounting system, but its use is declining as competitive forces and customer demands have increased pressures on companies to better manage inventories.)

Point: Cost accounting systems accumulate costs and then assign them to products and services.

The two basic types of cost accounting systems are *job order cost accounting* and *process cost accounting*. We describe job order cost accounting in this chapter. Process cost accounting is explained in the next chapter.

Job Order Production

C2 Describe important features of job order production.

Many companies produce products individually designed to meet the needs of a specific customer. Each customized product is manufactured separately and its production is called **job order production,** or *job order manufacturing* (also called *customized production,* which is the production of products in response to special orders). Examples of such products include synthetic football fields, special-order machines, a factory building, custom jewelry, wedding invitations, and artwork.

The production activities for a customized product represent a **job.** The principle of customization is equally applicable to both manufacturing *and* service companies. Most service companies meet customers' needs by performing a custom service for a specific customer. Examples of such services include an accountant auditing a client's financial statements, an interior designer remodeling an office, a wedding consultant planning and supervising a reception, and a lawyer defending a client. Whether the setting is manufacturing or services, job order operations involve meeting the needs of customers by producing or performing custom jobs.

Boeing's aerospace division is one example of a job order production system. Its primary business is twofold: (1) design, develop, and integrate space carriers and (2) provide systems engineering and integration of Department of Defense (DoD) systems. Many of its orders are customized and produced through job order operations.

When a job involves producing more than one unit of a custom product, it is often called a **job lot.** Products produced as job lots could include benches for a church, imprinted T-shirts for a 10K race or company picnic, or advertising signs for a chain of stores. Although these orders involve more than one unit, the volume of production is typically low, such as 50 benches, 200 T-shirts, or 100 signs. Another feature of job order production is the diversity, often called *heterogeneity,* of the products produced. Namely, each customer order is likely to differ from another in some important respect. These variations can be minor or major.

Point: Many professional examinations including the CPA and CMA exams require knowledge of job order and process cost accounting.

Decision Insight

Custom Design Managers once saw companies as the center of a solar system orbited by suppliers and customers. Now the customer has become the center of the business universe. **Nike** allows custom orders over the Internet, enabling customers to select materials, colors, and to personalize their shoes with letters and numbers. Soon consumers may be able to personalize almost any product, from cellular phones to appliances to furniture.

Events in Job Order Costing

The initial event in a normal job order operation is the receipt of a customer order for a custom product. This causes the company to begin work on a job. A less common case occurs when management decides to begin work on a job before it has a signed contract. This is referred to as *jobs produced on speculation.*

Video19.1

The first step in both cases is to predict the cost to complete the job. This cost depends on the product design prepared by either the customer or the producer. The second step is to negotiate a sales price and decide whether to pursue the job. Other than for government or other cost-plus contracts, the selling price is determined by market factors. Producers evaluate the market price, compare it to cost, and determine whether the profit on the job is reasonable. If the profit is not reasonable, the producer would determine a desired **target cost.** The third step is for the producer to schedule production of the job to meet the customer's needs and to fit within its own production constraints. Preparation of this work schedule should consider workplace facilities including equipment, personnel, and supplies. Once this schedule is complete, the producer can place orders for raw materials. Production occurs as materials and labor are applied to the job.

Point: Some jobs are priced on a *cost-plus basis:* The customer pays the manufacturer for costs incurred on the job plus a negotiated amount or rate of profit.

An overview of job order production activity is shown in Exhibit 19.1. This exhibit shows the March production activity of Road Warriors, which manufactures security-equipped cars and trucks. The company converts any vehicle by giving it a diversity of security items such as alarms, reinforced exterior, bulletproof glass, and bomb detectors. The company began by catering to high-profile celebrities, but it now caters to anyone who desires added security in a vehicle.

Job order production for Road Warriors requires materials, labor, and overhead costs. Recall that direct materials are goods used in manufacturing that are clearly identified with a particular job. Similarly, direct labor is effort devoted to a particular job. Overhead costs support production of more than one job. Common overhead items are depreciation on factory buildings and equipment, factory supplies, supervision, maintenance, cleaning, and utilities.

Exhibit 19.1 shows that materials, labor, and overhead are added to Jobs B15, B16, B17, B18, and B19, which were started during March. Road Warriors completed Jobs B15, B16, and B17 in March and delivered Jobs B15 and B16 to customers. At the end of March, Jobs B18 and B19 remain in goods in process inventory and Job B17 is in finished goods inventory. Both labor and materials costs are also separated into their direct and indirect components. Their indirect amounts are added to overhead. Total overhead cost is then allocated to the various jobs.

Decision Insight

Target Costing Many producers determine a target cost for their jobs. Target cost is determined as follows: Expected selling price − Desired profit = Target cost. If the projected target cost of the job as determined by job costing is too high, the producer can apply *value engineering,* which is a method of determining ways to reduce job cost until the target cost is met.

EXHIBIT 19.1

Job Order Production Activities

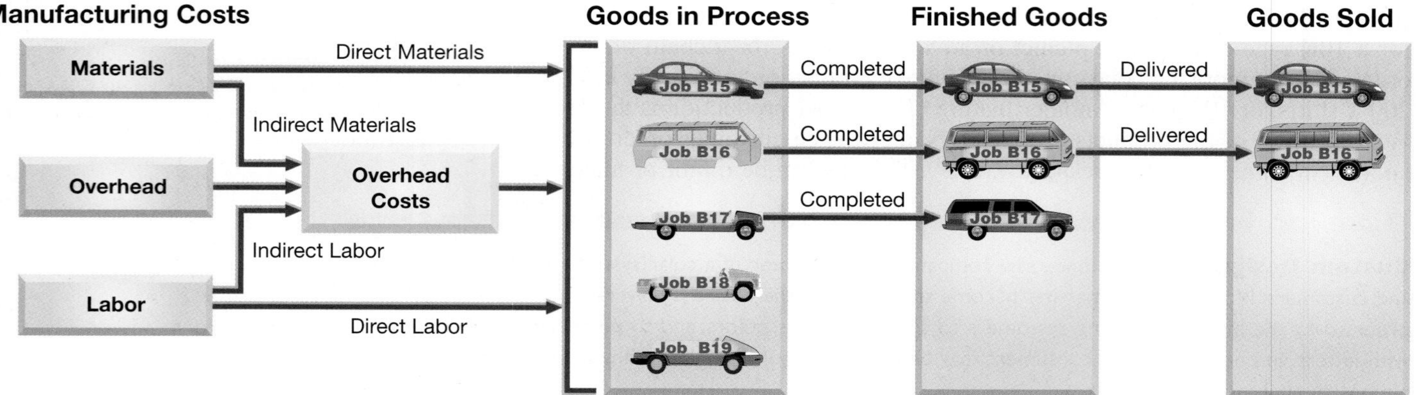

Job Cost Sheet

C3 Explain job cost sheets and how they are used in job order cost accounting.

General ledger accounts usually do not provide the accounting information that managers of job order cost operations need to plan and control production activities. This is so because the needed information often requires more detailed data. Such detailed data are usually stored in subsidiary records controlled by general ledger accounts. Subsidiary records store information about raw materials, overhead costs, jobs in process, finished goods, and other items. This section describes the use of these records.

A major aim of a **job order cost accounting system** is to determine the cost of producing each job or job lot. In the case of a job lot, the system also aims to compute the cost per unit. The accounting system must include separate records for each job to accomplish this, and it must capture information about costs incurred and charge these costs to each job.

A **job cost sheet** is a separate record maintained for each job. Exhibit 19.2 shows a job cost sheet for an alarm system that Road Warriors produced for a customer. This job cost sheet identifies the customer, the job number assigned, the product, and key dates. Costs incurred on the job are immediately recorded on this sheet. When each job is complete, the supervisor enters the date of completion, records any remarks, and signs the sheet. The job cost sheet in Exhibit 19.2 classifies costs as direct materials, direct labor, or overhead. It shows that a total of $600 in direct materials is added to Job B15 on four different dates. It also shows seven entries for direct labor costs that total $1,000. Road Warriors *allocates* (also termed *applies, assigns,* or *charges*) factory overhead costs of $1,600 to this job using an allocation rate of 160% of direct labor cost (160% × $1,000)—we discuss overhead allocation later in this chapter.

Point: Factory overhead consists of costs (other than direct materials and direct labor) that ensure the production activities are carried out.

While a job is being produced, its accumulated costs are kept in **Goods in Process Inventory.** The collection of job cost sheets for all jobs in process makes up a subsidiary ledger controlled by the Goods in Process Inventory account in the general ledger. Managers use job cost sheets to monitor costs incurred to date and to predict and control costs for each job.

Point: Documents (electronic and paper) are crucial in a job order system, and the job cost sheet is a cornerstone. Understanding it aids in grasping concepts of capitalizing product costs and product cost flow.

When a job is finished, its job cost sheet is completed and moved from the jobs in process file to the finished jobs file. This latter file acts as a subsidiary ledger controlled by the **Finished Goods Inventory** account. When a finished job is delivered to a customer, the job cost sheet is moved to a permanent file supporting the total cost of goods sold. This permanent file contains records from both current and prior periods.

Decision Maker

Management Consultant One of your tasks is to control and manage costs for a consulting company. At the end of a recent month, you find that three consulting jobs were completed and two are 60% complete. Each unfinished job is estimated to cost $10,000 and to earn a revenue of $12,000. You are unsure how to recognize goods in process inventory and record costs and revenues. Do you recognize any inventory? If so, how much? How much revenue is recorded for unfinished jobs this month? [Answer—p. 786]

EXHIBIT 19.2

Job Cost Sheet

Accounting System: Exhibit 19-2

File Edit Maintain Tasks Analysis Options Reports Window Help

Road Warriors, Los Angeles, California — **JOB COST SHEET**

Customer's Name Carroll Connor — **Job No.** B15

Address 1542 High Point Dr. — **City & State** Portland, Oregon

Job Description Level 1 Alarm System on Ford Expedition

Date promised March 15 — **Date started** March 3 — **Date completed** March 11

Direct Materials			Direct Labor			Overhead		
Date	**Requisition**	**Cost**	**Date**	**Time Ticket**	**Cost**	**Date**	**Rate**	**Cost**
3/3/2009	R-4698	100.00	3/3/2009	L-3393	120.00	3/11/2009	160% of Direct Labor Cost	1,600.00
3/7/2009	R-4705	225.00	3/4/2009	L-3422	150.00			
3/9/2009	R-4725	180.00	3/5/2009	L-3456	180.00			
3/10/2009	R-4777	95.00	3/8/2009	L-3479	60.00			
			3/9/2009	L-3501	90.00			
			3/10/2009	L-3535	240.00			
			3/11/2009	L-3559	160.00			
	Total	600.00		**Total**	1,000.00		**Total**	1,600.00

REMARKS: Completed job on March 11, and shipped to customer on March 15. Met all specifications and requirements.

SUMMARY:

Materials	600.00
Labor	1,000.00
Overhead	1,600.00
Total cost	3,200.00

Signed: *C. Luther, Supervisor*

Quick Check

Answers—p. 787

1. Which of these products is likely to involve job order production? (*a*) inexpensive watches, (*b*) racing bikes, (*c*) bottled soft drinks, or (*d*) athletic socks.
2. What is the difference between a job and a job lot?
3. Which of these statements is correct? (*a*) The collection of job cost sheets for unfinished jobs makes up a subsidiary ledger controlled by the Goods in Process Inventory account, (*b*) Job cost sheets are financial statements provided to investors, or (*c*) A separate job cost sheet is maintained in the general ledger for each job in process.
4. What three costs are normally accumulated on job cost sheets?

Job Order Cost Flows and Reports

Materials Cost Flows and Documents

P1 Describe and record the flow of materials costs in job order cost accounting.

This section focuses on the flow of materials costs and the related documents in a job order cost accounting system. We begin analysis of the flow of materials costs by examining Exhibit 19.3. When materials are first received from suppliers, the employees count and inspect them and record the items' quantity and cost on a receiving report. The receiving report serves as the *source document* for recording materials received in both a materials ledger card and in the general ledger. In nearly all job order cost systems, **materials ledger cards** (or files) are perpetual records that are updated each time units are purchased and each time units are issued for use in production.

Materials

Point: Some companies certify certain suppliers based on the quality of their materials. Goods received from these suppliers are not always inspected by the purchaser to save costs.

To illustrate the purchase of materials, Road Warriors acquired $450 of wiring and related materials on March 4, 2009. This purchase is recorded as follows.

Mar. 4	Raw Materials Inventory—M-347	450	
	Accounts Payable		450
	To record purchase of materials for production.		

Assets = Liabilities + Equity
+450 +450

EXHIBIT 19.3

Materials Cost Flows through Subsidiary Records

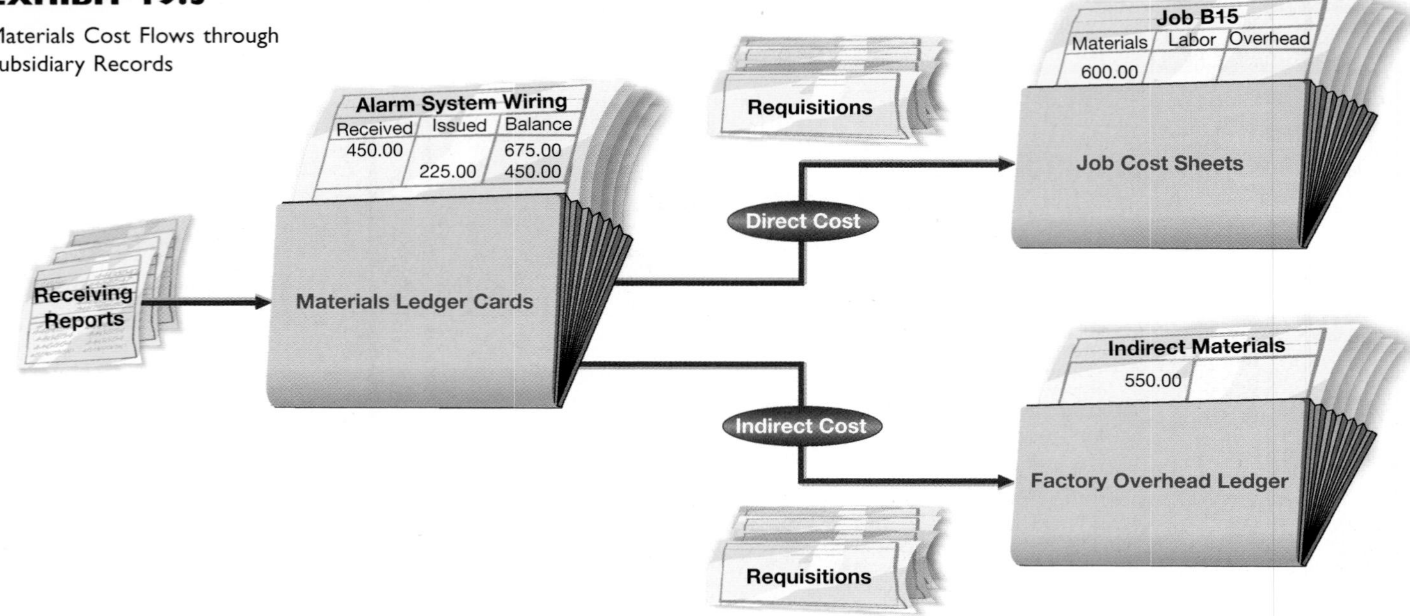

Exhibit 19.3 shows that materials can be requisitioned for use either on a specific job (direct materials) or as overhead (indirect materials). Cost of direct materials flows from the materials ledger card to the job cost sheet. The cost of indirect materials flows from the materials ledger card to the Indirect Materials account in the factory overhead ledger, which is a subsidiary ledger controlled by the Factory Overhead account in the general ledger.

Video19.1

Exhibit 19.4 shows a materials ledger card for material received and issued by Road Warriors. The card identifies the item as alarm system wiring and shows the item's stock number, its location in the storeroom, information about the maximum and minimum quantities that should be available, and the reorder quantity. For example, alarm system wiring is issued and recorded on March 7, 2009. The job cost sheet in Exhibit 19.2 showed that Job B15 used this wiring.

EXHIBIT 19.4

Materials Ledger Card

MATERIALS LEDGER CARD

Road Warriors
Los Angeles, California

Item	Alarm system wiring	Stock No.	M–347	Location in Storeroom	Bin 137
Maximum quantity	5 units	Minimum quantity	1 unit	Quantity to reorder	2 units

	Received				Issued				Balance		
Date	Receiving Report Number	Units	Unit Price	Total Price	Requisition Number	Units	Unit Price	Total Price	Units	Unit Price	Total Price
									1	225.00	225.00
3/ 4/2009	C-7117	2	225.00	450.00					3	225.00	675.00
3/ 7/2009					R–4705	1	225.00	225.00	2	225.00	450.00

When materials are needed in production, a production manager prepares a **materials requisition** and sends it to the materials manager. The requisition shows the job number, the type of material, the quantity needed, and the signature of the manager authorized to make the requisition. Exhibit 19.5 shows the materials requisition for alarm system wiring for Job B15. To see how this requisition ties to the flow of costs, compare the information on the requisition with the March 7, 2009, data in Exhibits 19.2 and 19.4.

Point: Requisitions are often accumulated and recorded in one entry. The frequency of entries depends on the job, the industry, and management procedures.

EXHIBIT 19.5

Materials Requisition

MATERIALS REQUISITION No. R–4705

Road Warriors
Los Angeles, California

Job No.	B15	**Date**	3/7/2009
Material Stock No.	M–347	**Material Description**	Alarm system wiring
Quantity Requested	1	**Requested By**	C. Luther
Quantity Provided	1	**Date Provided**	3/7/2009
Filled By	M. Bateman	**Material Received By**	C. Luther
Remarks			

The use of alarm system wiring on Job B15 yields the following entry (locate this cost item in the job cost sheet shown in Exhibit 19.2).

Mar. 7	Goods in Process Inventory—Job B15.	225	
	Raw Materials Inventory—M-347		225
	To record use of material on Job B15.		

Assets = Liabilities + Equity
+225
−225

This entry is posted both to its general ledger accounts and to subsidiary records. Posting to subsidiary records includes a debit to a job cost sheet and a credit to a materials ledger card. (*Note:* An entry to record use of indirect materials is the same as that for direct materials *except* the debit is to Factory Overhead. In the subsidiary factory overhead ledger, this entry is posted to Indirect Materials.)

Labor Cost Flows and Documents

P2 Describe and record the flow of labor costs in job order cost accounting.

Exhibit 19.6 shows the flow of labor costs from clock cards and the Factory Payroll account to subsidiary records of the job order cost accounting system. Recall that costs in subsidiary records give detailed information needed to manage and control operations.

Labor

EXHIBIT 19.6

Labor Cost Flows through Subsidiary Records

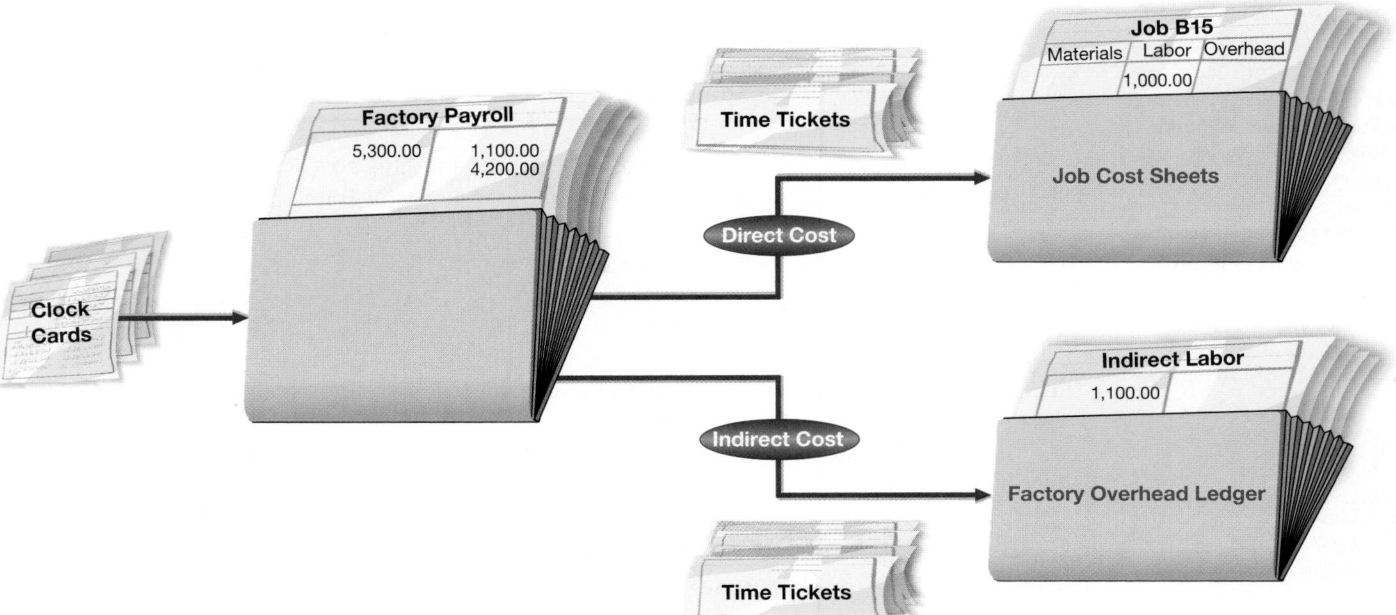

Point: Indirect materials are included in overhead on the job cost sheet. Assigning overhead costs to products is described in the next section.

Point: Many employee fraud schemes involve payroll, including overstated hours on clock cards.

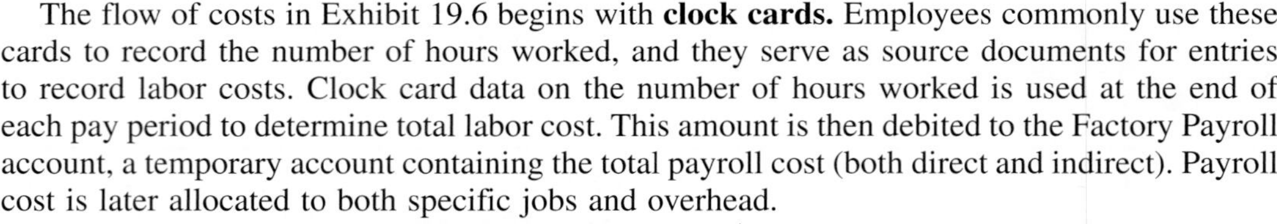

The flow of costs in Exhibit 19.6 begins with **clock cards.** Employees commonly use these cards to record the number of hours worked, and they serve as source documents for entries to record labor costs. Clock card data on the number of hours worked is used at the end of each pay period to determine total labor cost. This amount is then debited to the Factory Payroll account, a temporary account containing the total payroll cost (both direct and indirect). Payroll cost is later allocated to both specific jobs and overhead.

According to clock card data, workers earned $1,500 for the week ended March 5. Illustrating the flow of labor costs, the accrual and payment of these wages are recorded as follows.

Assets = Liabilities + Equity
−1,500 −1,500

Mar. 6	Factory payroll.	1,500	
	Cash		1,500
	To record the weekly payroll.		

"It's on Corporate Standard Time... It loses an hour of your pay every day."

To assign labor costs to specific jobs and to overhead, we must know how each employee's time is used and its costs. Source documents called **time tickets** usually capture these data. Employees regularly fill out time tickets to report how much time they spent on each job. An employee who works on several jobs during a day completes a separate time ticket for each job. Tickets are also prepared for time charged to overhead as indirect labor. A supervisor signs an employee's time ticket to confirm its accuracy.

Exhibit 19.7 shows a time ticket reporting the time a Road Warrior employee spent working on Job B15. The employee's supervisor signed the ticket to confirm its accuracy. The hourly rate and total labor cost are computed after the time ticket is turned in. To see the effect of this time ticket on the job cost sheet, look at the entry dated March 8, 2009, in Exhibit 19.2.

EXHIBIT 19.7

Time Ticket

ROAD WARRIORS
Road Warriors
Los Angeles, California

No. L–3479

TIME TICKET **Date** ..March 8.. **20** ..09..

Employee Name	Employee Number	Job No.
T. Zeller	3969	B15

TIME AND RATE INFORMATION:

Start Time	Finish Time	Elapsed Time	Hourly Rate
9:00	12:00	3.0	$20.00
Approved By ......C. Luther......		Total Cost	$60.00

Remarks
....

Point: In the accounting equation, we treat accounts such as Factory Overhead and Factory Payroll as temporary accounts, which hold various expenses until they are allocated to balance sheet or income statement accounts.

When time tickets report labor used on a specific job, this cost is recorded as direct labor. The following entry records the data from the time ticket in Exhibit 19.7.

Assets = Liabilities + Equity
+60 +60

Mar. 8	Goods in Process Inventory—Job B15	60	
	Factory Payroll		60
	To record direct labor used on Job B15.		

The debit in this entry is posted both to the general ledger account and to the appropriate job cost sheet. (*Note:* An entry to record indirect labor is the same as for direct labor *except* that it debits Factory Overhead and credits Factory Payroll. In the subsidiary factory overhead ledger, the debit in this entry is posted to the Indirect Labor account.)

P3 Describe and record the flow of overhead costs in job order cost accounting.

Overhead Cost Flows and Documents

Factory overhead (or simply overhead) cost flows are shown in Exhibit 19.8. Factory overhead includes all production costs other than direct materials and direct labor. Two sources of

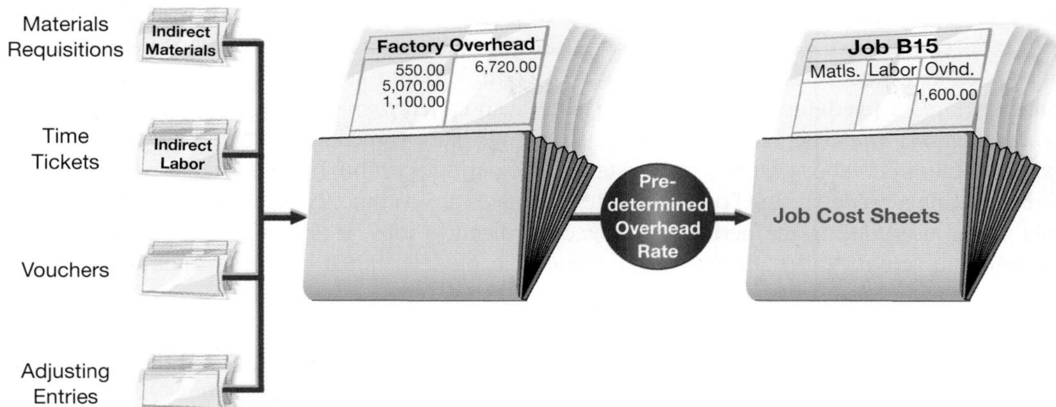

EXHIBIT 19.8

Overhead Cost Flows through Subsidiary Records

overhead costs are indirect materials and indirect labor. These costs are recorded from requisitions for indirect materials and time tickets for indirect labor. Two other sources of overhead are (1) vouchers authorizing payments for items such as supplies or utilities and (2) adjusting entries for costs such as depreciation on factory assets.

Overhead

Factory overhead usually includes many different costs and, thus, a separate account for each is often maintained in a subsidiary factory overhead ledger. This ledger is controlled by the Factory Overhead account in the general ledger. Factory Overhead is a temporary account that accumulates costs until they are allocated to jobs.

Recall that overhead costs are recorded with debits to the Factory Overhead account and with credits to other accounts such as Cash, Accounts Payable, and Accumulated Depreciation—Equipment. In the subsidiary factory overhead ledger, the debits are posted to their respective accounts such as Depreciation Expense—Equipment, Insurance Expense—Warehouse, or Amortization Expense—Patents.

To illustrate the recording of overhead, the following two entries reflect the depreciation of factory equipment and the accrual of utilities, respectively, for the week ended March 6.

Mar. 6	Factory Overhead	600	
	Accumulated Depreciation—Equipment		600
	To record depreciation on factory equipment.		
Mar. 6	Factory Overhead	250	
	Utilities Payable.		250
	To record the accrual of factory utilities.		

Assets	=	Liabilities	+	Equity
−600				−600

Assets	=	Liabilities	+	Equity
		+250		−250

Exhibit 19.8 shows that overhead costs flow from the Factory Overhead account to job cost sheets. Because overhead is made up of costs not directly associated with specific jobs or job lots, we cannot determine the dollar amount incurred on a specific job. We know, however, that overhead costs represent a necessary part of business activities. If a job cost is to include all costs needed to complete the job, some amount of overhead must be included. Given the difficulty in determining the overhead amount for a specific job, however, we allocate overhead to individual jobs in some reasonable manner.

We generally allocate overhead by linking it to another factor used in production, such as direct labor or machine hours. The factor to which overhead costs are linked is known as the *allocation base.* A manager must think carefully about how many and which allocation bases to use. This managerial decision influences the accuracy with which overhead costs are allocated to individual jobs. In turn, the cost of individual jobs might impact a manager's decisions for pricing or performance evaluation. In Exhibit 19.2, overhead is expressed as 160% of direct labor. We then allocate overhead by multiplying 160% by the estimated amount of direct labor on the jobs.

We cannot wait until the end of a period to allocate overhead to jobs because perpetual inventory records are part of the job order costing system (demanding up-to-date costs). Instead, we

Point: The predetermined overhead rate is computed at the start of the period and is used throughout the period to allocate overhead to jobs.

Point: Predetermined overhead rates can be estimated using mathematical equations, statistical analysis, or professional experience.

must predict overhead in advance and assign it to jobs so that a job's total costs can be estimated prior to its completion. This estimated cost is useful for managers in many decisions including setting prices and identifying costs that are out of control. Being able to estimate overhead in advance requires a **predetermined overhead rate,** also called *predetermined overhead allocation* (or *application*) *rate.* This rate requires an estimate of total overhead cost and an allocation factor such as total direct labor cost before the start of the period. Exhibit 19.9 shows the usual formula for computing a predetermined overhead rate (estimates are commonly based on annual amounts). This rate is used during the period to allocate overhead to jobs. It is common for companies to use multiple activity (allocation) bases and multiple predetermined overhead rates for different types of products and services.

EXHIBIT 19.9

Predetermined Overhead Allocation Rate Formula

$$\textbf{Predetermined overhead rate} = \textbf{Estimated overhead costs} \div \textbf{Estimated activity base}$$

Example: If management predicts total direct labor costs of $100,000 and total overhead costs of $200,000, what is its predetermined overhead rate? *Answer:* 200% of direct labor cost.

To illustrate, Road Warriors allocates overhead by linking it to direct labor. At the start of the current period, management predicts total direct labor costs of $125,000 and total overhead costs of $200,000. Using these estimates, management computes its predetermined overhead rate as 160% of direct labor cost ($200,000 ÷ $125,000). Specifically, reviewing the job order cost sheet in Exhibit 19.2, we see that $1,000 of direct labor went into Job B15. We then use the predetermined overhead rate of 160% to allocate $1,600 (equal to $1,000 × 1.60) of overhead to this job. The entry to record this allocation is

Assets	=	Liabilities	+	Equity
+1,600				+1,600

Mar. 11	Goods in Process Inventory—Job B15.	1,600	
	Factory Overhead.		1,600
	To assign overhead to Job B15.		

Since the allocation rate for overhead is estimated at the start of a period, the total amount assigned to jobs during a period rarely equals the amount actually incurred. We explain how this difference is treated later in this chapter.

Decision Ethics

Web Consultant You are working on seven client engagements. Two clients reimburse your firm for actual costs plus a 10% markup. The other five pay a fixed fee for services. Your firm's costs include overhead allocated at $47 per labor hour. The managing partner of your firm instructs you to record as many labor hours as possible to the two markup engagements by transferring labor hours from the other five. What do you do? [Answer—p. 786]

Summary of Cost Flows

Point: Study the flow of manufacturing costs through general ledger accounts and job cost sheets. Use Exhibit 19.11 as reinforcement.

We showed journal entries for charging Goods in Process Inventory (Job B15) with the cost of (1) direct materials requisitions, (2) direct labor time tickets, and (3) factory overhead. We made separate entries for each of these costs, but they are usually recorded in one entry. Specifically, materials requisitions are often collected for a day or a week and recorded with a single entry summarizing them. The same is done with labor time tickets. When summary entries are made, supporting schedules of the jobs charged and the types of materials used provide the basis for postings to subsidiary records.

To show all production cost flows for a period and their related entries, we again look at Road Warriors' activities. Exhibit 19.10 shows costs linked to all of Road Warriors' production activities for March. Road Warriors did not have any jobs in process at the beginning of March, but it did apply materials, labor, and overhead costs to five new jobs in March. Jobs B15 and B16 are completed and delivered to customers in March, Job B17 is completed but not delivered, and Jobs B18 and B19 are still in process. Exhibit 19.10 also shows purchases of raw materials for $2,750, labor costs incurred for $5,300, and overhead costs of $6,720.

The upper part of Exhibit 19.11 shows the flow of these costs through general ledger accounts and the end-of-month balances in key subsidiary records. Arrow lines are numbered

EXHIBIT 19.10

Job Order Costs of All Production Activities

ROAD WARRIORS
Job Order Manufacturing Costs
For Month Ended March 31, 2009

Explanation	Materials	Labor	Overhead Incurred	Overhead Allocated	Goods in Process	Finished Goods	Cost of Goods Sold
Job B15	$ 600	$1,000		$1,600			$3,200
Job B16	300	800		1,280			2,380
Job B17	500	1,100		1,760		$3,360	
Job B18	150	700		1,120	$1,970		
Job B19	250	600		960	1,810		
Total job costs	1,800	4,200		$6,720	$3,780	$3,360	$5,580
Indirect materials	550		$ 550				
Indirect labor		1,100	1,100				
Other overhead			5,070				
Total costs used in production	2,350	$5,300	$6,720				
Ending materials inventory	1,400						
Materials available	3,750						
Less beginning materials inventory	(1,000)						
Materials purchased	$2,750						

EXHIBIT 19.11

Job Order Cost Flows and Ending Job Cost Sheets

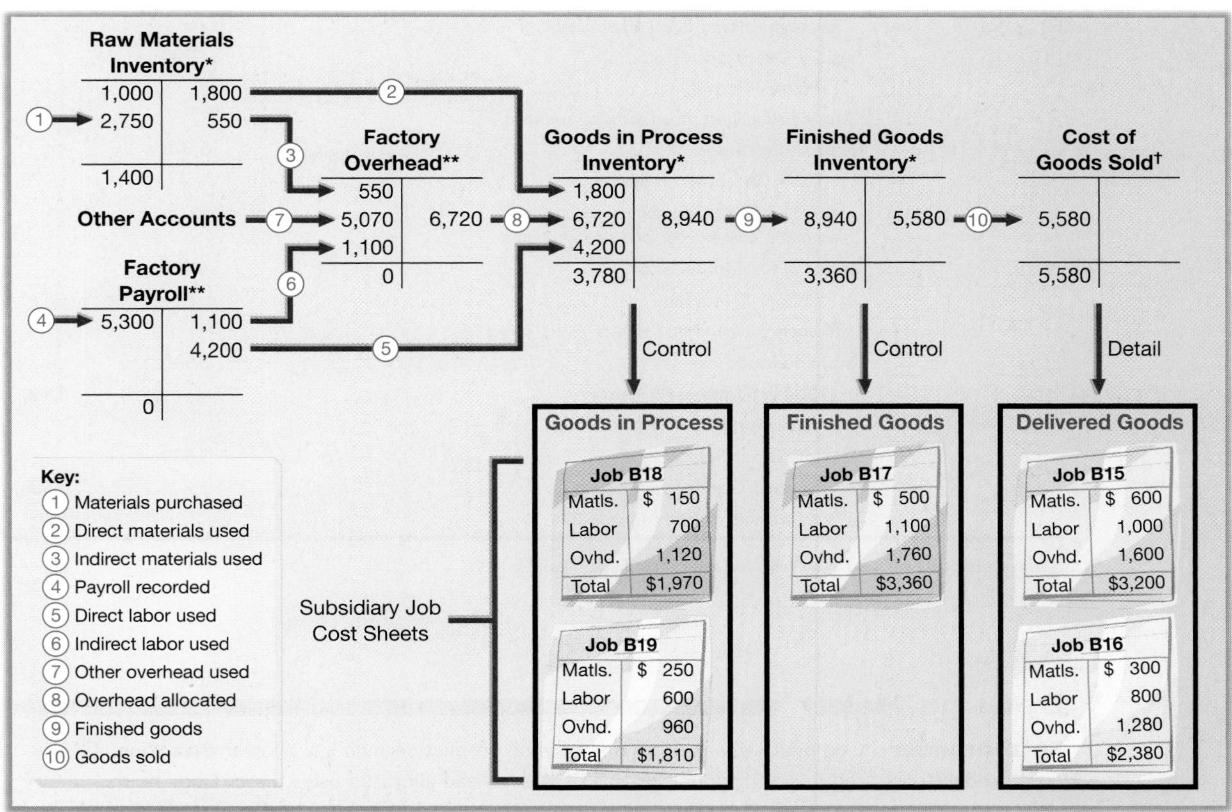

* The ending balances in the inventory accounts are carried to the balance sheet.

† The Cost of Goods Sold balance is carried to the income statement.

** Factory Payroll and Factory Overhead are considered temporary accounts; when these costs are allocated to jobs, the balances in these accounts are reduced.

to show the flows of costs for March. Each numbered cost flow reflects several entries made in March. The lower part of Exhibit 19.11 shows summarized job cost sheets and their status at the end of March. The sum of costs assigned to the jobs in process ($1,970 + $1,810) equals the $3,780 balance in Goods in Process Inventory shown in Exhibit 19.10. Also, costs assigned to Job B17 equal the $3,360 balance in Finished Goods Inventory. The sum of costs assigned to Jobs B15 and B16 ($3,200 + $2,380) equals the $5,580 balance in Cost of Goods Sold.

Exhibit 19.12 shows each cost flow with a single entry summarizing the actual individual entries made in March. Each entry is numbered to link with the arrow lines in Exhibit 19.11.

EXHIBIT 19.12

Entries for Job Order Production Costs*

①	Raw Materials Inventory	2,750	
	Accounts Payable		2,750
	Acquired materials on credit for factory use.		
②	Goods in Process Inventory	1,800	
	Raw Materials Inventory		1,800
	To assign costs of direct materials used.		
③	Factory Overhead	550	
	Raw Materials Inventory		550
	To record use of indirect materials.		
④	Factory Payroll	5,300	
	Cash (and other accounts)		5,300
	To record salaries and wages of factory workers (including various payroll liabilities).		
⑤	Goods in Process Inventory	4,200	
	Factory Payroll		4,200
	To assign costs of direct labor used.		
⑥	Factory Overhead	1,100	
	Factory Payroll		1,100
	To record indirect labor costs as overhead.		
⑦	Factory Overhead	5,070	
	Cash (and other accounts)		5,070
	To record factory overhead costs such as insurance, utilities, rent, and depreciation.		
⑧	Goods in Process Inventory	6,720	
	Factory Overhead		6,720
	To apply overhead at 160% of direct labor.		
⑨	Finished Goods Inventory	8,940	
	Goods in Process Inventory		8,940
	To record completion of Jobs B15, B16, and B17.		
⑩	Cost of Goods Sold	5,580	
	Finished Goods Inventory		5,580
	To record sale of Jobs B15 and B16.		

* Transactions are numbered to be consistent with arrow lines in Exhibit 19.11.

Point: *Actual* overhead is debited to Factory Overhead. *Allocated* overhead is credited to Factory Overhead.

Decision Maker

Entrepreneur Competitors' prices on one of your product segments are lower than yours. Of the total product cost used in setting your prices, 53% is overhead allocated using direct labor hours. You believe that product costs are distorted and wonder whether there is a better way to allocate overhead and to set product price. What do you suggest? [Answer—p. 787]

Quick Check Answers—p. 787

5. In job order cost accounting, which account is debited in recording a raw materials requisition? (*a*) Raw Materials Inventory, (*b*) Raw Materials Purchases, (*c*) Goods in Process Inventory if for a job, or (*d*) Goods in Process Inventory if they are indirect materials.
6. What are four sources of information for recording costs in the Factory Overhead account?
7. Why does job order cost accounting require a predetermined overhead rate?
8. What events result in a debit to Factory Payroll? What events result in a credit?

Adjustment of Overapplied or Underapplied Overhead

Refer to the debits in the Factory Overhead account in Exhibit 19.11 (or Exhibit 19.12). The total cost of factory overhead incurred during March is $6,720 ($550 + $5,070 + $1,100). The $6,720 exactly equals the amount assigned to goods in process inventory (see arrow line ⑧). Therefore, the overhead incurred equals the overhead applied in March. The amount of overhead incurred rarely equals the amount of overhead applied, however, because a job order cost accounting system uses a predetermined overhead rate in applying factory overhead costs to jobs. This rate is determined using estimated amounts before the period begins, and estimates rarely equal the exact amounts actually incurred. This section explains what we do when too much or too little overhead is applied to jobs.

Video19.1

Underapplied Overhead

P4 Determine adjustments for overapplied and underapplied factory overhead.

When less overhead is applied than is actually incurred, the remaining debit balance in the Factory Overhead account at the end of the period is called **underapplied overhead.** To illustrate, assume that Road Warriors actually incurred *other overhead costs* of $5,550 instead of the $5,070 shown in Exhibit 19.11. This yields an actual total overhead cost of $7,200 in March. Since the amount of overhead applied was only $6,720, the Factory Overhead account is left with a $480 debit balance as shown in the ledger account in Exhibit 19.13.

EXHIBIT 19.13

Underapplied Overhead in the Factory Overhead Ledger Account

Factory Overhead				Acct. No. 540
Date	**Explanation**	**Debit**	**Credit**	**Balance**
Mar. 31	Indirect materials cost	550		550 Dr.
31	Indirect labor cost	1,100		1,650 Dr.
31	Other overhead cost	5,550		7,200 Dr.
31	Overhead costs applied to jobs		6,720	480 Dr.

The $480 debit balance reflects manufacturing costs not assigned to jobs. This means that the balances in Goods in Process Inventory, Finished Goods Inventory, and Cost of Goods Sold do not include all production costs incurred. When the underapplied overhead amount is immaterial, it is allocated (closed) to the Cost of Goods Sold account with the following adjusting entry.

Example: If we do not adjust for underapplied overhead, will net income be overstated or understated? *Answer:* Overstated.

Mar. 31	Cost of Goods Sold	480	
	Factory Overhead		480
	To adjust for underapplied overhead costs.		

Assets = Liabilities + Equity
−480
+480

The $480 debit (increase) to Cost of Goods Sold reduces income by $480. (When the underapplied (or overapplied) overhead is material, the amount is normally allocated to the Cost of Goods Sold, Finished Goods Inventory, and Goods in Process Inventory accounts. This process is covered in advanced courses.)

Overapplied Overhead

When the overhead applied in a period exceeds the overhead incurred, the resulting credit balance in the Factory Overhead account is called **overapplied overhead.** We treat overapplied overhead at the end of the period in the same way we treat underapplied overhead, except that we debit Factory Overhead and credit Cost of Good Sold for the amount.

Decision Insight

Job Order Education Many companies invest in their employees, and the demand for executive education is strong. Annual spending on training and education exceeds $20 billion. Annual revenues for providers of executive education continue to rise, with about 40% of revenues coming from custom programs designed for one or a select group of companies.

Quick Check

Answers—p. 787

9. In a job order cost accounting system, why does the Factory Overhead account usually have an overapplied or underapplied balance at period-end?
10. When the Factory Overhead account has a debit balance at period-end, does this reflect overapplied or underapplied overhead?

Decision Analysis — Pricing for Services

A1 Apply job order costing in pricing services.

The chapter described job order costing mainly using a manufacturing setting. However, these concepts and procedures are applicable to a service setting. Consider AdWorld, an advertising agency that develops Web-based ads for small firms. Each of its customers has unique requirements, so costs for each individual job must be tracked separately.

AdWorld uses two types of labor: Web designers ($65 per hour) and computer staff ($50 per hour). It also incurs overhead costs that it assigns using two different predetermined overhead allocation rates: $125 per designer hour and $96 per staff hour. For each job, AdWorld must estimate the number of designer and staff hours needed. Then total costs pertaining to each job are determined using the procedures in the chapter. (*Note:* Most service firms have neither the category of materials cost nor inventory.)

To illustrate, a manufacturer of golf balls requested a quote from AdWorld for an advertising engagement. AdWorld estimates that the job will require 43 designer hours and 61 staff hours, with the following total estimated cost for this job.

Direct Labor		
Designers (43 hours × $65)	$ 2,795	
Staff (61 hours × $50)	3,050	
Total direct labor		$ 5,845
Overhead		
Designer related (43 hours × $125)	5,375	
Staff related (61 hours × $96)	5,856	
Total overhead		11,231
Total estimated job cost		$17,076

AdWorld can use this cost information to help determine the price quote for the job (see *Decision Maker,* **Sales Manager,** scenario in this chapter).

Another source of information that AdWorld must consider is the market, that is, how much competitors will quote for this job. Competitor information is often unavailable; therefore, AdWorld's managers must use estimates based on their assessment of the competitive environment.

Decision Maker

Sales Manager As AdWorld's sales manager, assume that you estimate costs pertaining to a proposed job as $17,076. Your normal pricing policy is to apply a markup of 18% from total costs. However, you learn that three other agencies are likely to bid for the same job, and that their quotes will range from $16,500 to $22,000. What price should you quote? What factors other than cost must you consider? [Answer—p. 787]

Demonstration Problem—Job Order Costing

DP19

The following information reflects Walczak Company's job order production activities for May.

Raw materials purchases	$16,000
Factory payroll cost	15,400
Overhead costs incurred	
Indirect materials	5,000
Indirect labor	3,500
Other factory overhead	9,500

Walczak's predetermined overhead rate is 150% of direct labor cost. Costs are allocated to the three jobs worked on during May as follows.

	Job 401	Job 402	Job 403
In-process balances on April 30			
Direct materials	$3,600		
Direct labor	1,700		
Applied overhead	2,550		
Costs during May			
Direct materials	3,550	$3,500	$1,400
Direct labor	5,100	6,000	800
Applied overhead	?	?	?
Status on May 31	**Finished (sold)**	**Finished (unsold)**	**In process**

Required

1. Determine the total cost of:
   a. The April 30 inventory of jobs in process.
   b. Materials used during May.
   c. Labor used during May.
   d. Factory overhead incurred and applied during May and the amount of any over- or underapplied overhead on May 31.
   e. Each job as of May 31, the May 31 inventories of both goods in process and finished goods, and the goods sold during May.
2. Prepare summarized journal entries for the month to record:
   a. Materials purchases (on credit), the factory payroll (paid with cash), indirect materials, indirect labor, and the other factory overhead (paid with cash).
   b. Assignment of direct materials, direct labor, and overhead costs to the Goods in Process Inventory account. (Use separate debit entries for each job.)
   c. Transfer of each completed job to the Finished Goods Inventory account.
   d. Cost of goods sold.
   e. Removal of any underapplied or overapplied overhead from the Factory Overhead account. (Assume the amount is not material.)
3. Prepare a manufacturing statement for May.

Planning the Solution

- Determine the cost of the April 30 goods in process inventory by totaling the materials, labor, and applied overhead costs for Job 401.

- Compute the cost of materials used and labor by totaling the amounts assigned to jobs and to overhead.
- Compute the total overhead incurred by summing the amounts for the three components. Compute the amount of applied overhead by multiplying the total direct labor cost by the predetermined overhead rate. Compute the underapplied or overapplied amount as the difference between the actual cost and the applied cost.
- Determine the total cost charged to each job by adding the costs incurred in April (if any) to the cost of materials, labor, and overhead applied during May.
- Group the costs of the jobs according to their completion status.
- Record the direct materials costs assigned to the three jobs, using a separate Goods in Process Inventory account for each job; do the same for the direct labor and the applied overhead.
- Transfer costs of Jobs 401 and 402 from Goods in Process Inventory to Finished Goods.
- Record the costs of Job 401 as cost of goods sold.
- Record the transfer of underapplied overhead from the Factory Overhead account to the Cost of Goods Sold account.
- On the manufacturing statement, remember to include the beginning and ending goods in process inventories and to deduct the underapplied overhead.

Solution to Demonstration Problem

1. Total cost of

a. April 30 inventory of jobs in process (Job 401).

Direct materials	$3,600
Direct labor	1,700
Applied overhead	2,550
Total cost	$7,850

b. Materials used during May.

Direct materials	
Job 401	$ 3,550
Job 402	3,500
Job 403	1,400
Total direct materials	8,450
Indirect materials	5,000
Total materials used	$13,450

c. Labor used during May.

Direct labor	
Job 401	$ 5,100
Job 402	6,000
Job 403	800
Total direct labor	11,900
Indirect labor	3,500
Total labor used	$15,400

d. Factory overhead incurred in May.

Actual overhead	
Indirect materials	$ 5,000
Indirect labor	3,500
Other factory overhead	9,500
Total actual overhead	18,000
Overhead applied (150% × $11,900)	17,850
Underapplied overhead	$ 150

e. Total cost of each job.

	401	402	403
In-process costs from April			
Direct materials	$ 3,600		
Direct labor	1,700		
Applied overhead*	2,550		
Cost incurred in May			
Direct materials	3,550	$ 3,500	$1,400
Direct labor	5,100	6,000	800
Applied overhead*	7,650	9,000	1,200
Total costs	$24,150	$18,500	$3,400

* Equals 150% of the direct labor cost.

Total cost of the May 31 inventory of goods in process (Job 403) = $3,400

Total cost of the May 31 inventory of finished goods (Job 402) = $18,500

Total cost of goods sold during May (Job 401) = $24,150

2. Journal entries.

a.

Account	Debit	Credit
Raw Materials Inventory	16,000	
Accounts Payable		16,000
To record materials purchases.		
Factory Payroll	15,400	
Cash		15,400
To record factory payroll.		
Factory Overhead	5,000	
Raw Materials Inventory		5,000
To record indirect materials.		
Factory Overhead	3,500	
Factory Payroll		3,500
To record indirect labor.		
Factory Overhead	9,500	
Cash		9,500
To record other factory overhead.		

b. Assignment of costs to Goods in Process Inventory.

Account	Debit	Credit
Goods in Process Inventory (Job 401)	3,550	
Goods in Process Inventory (Job 402)	3,500	
Goods in Process Inventory (Job 403)	1,400	
Raw Materials Inventory		8,450
To assign direct materials to jobs.		
Goods in Process Inventory (Job 401)	5,100	
Goods in Process Inventory (Job 402)	6,000	
Goods in Process Inventory (Job 403)	800	
Factory Payroll		11,900
To assign direct labor to jobs.		
Goods in Process Inventory (Job 401)	7,650	
Goods in Process Inventory (Job 402)	9,000	
Goods in Process Inventory (Job 403)	1,200	
Factory Overhead		17,850
To apply overhead to jobs.		

c. Transfer of completed jobs to Finished Goods Inventory.

Account	Debit	Credit
Finished Goods Inventory	42,650	
Goods in Process Inventory (Job 401)		24,150
Goods in Process Inventory (Job 402)		18,500
To record completion of jobs.		

d.

Account	Debit	Credit
Cost of Goods Sold	24,150	
Finished Goods Inventory		24,150
To record sale of Job 401.		

e.

Account	Debit	Credit
Cost of Goods Sold	150	
Factory Overhead		150
To assign underapplied overhead.		

3.

WALCZAK COMPANY Manufacturing Statement For Month Ended May 31		
Direct materials		$ 8,450
Direct labor		11,900
Factory overhead		
Indirect materials	$5,000	
Indirect labor	3,500	
Other factory overhead	9,500	18,000
Total production costs		38,350
Add goods in process, April 30		7,850
Total cost of goods in process		46,200
Less goods in process, May 31		3,400
Less underapplied overhead		150
Cost of goods manufactured		$42,650

Note how underapplied overhead is reported. Overapplied overhead is similarly reported, but is added.

Summary

C1 Explain the cost accounting system. A cost accounting system records production activities using a perpetual inventory system, which continuously updates records for transactions and events that affect inventory costs.

C2 Describe important features of job order production. Certain companies called *job order manufacturers* produce custom-made products for customers. These customized products are produced in response to a customer's orders. A job order manufacturer produces products that usually are different and, typically, produced in low volumes. The production systems of job order companies are flexible and are not highly standardized.

C3 Explain job cost sheets and how they are used in job order cost accounting. In a job order cost accounting system, the costs of producing each job are accumulated on a separate job cost sheet. Costs of direct materials, direct labor, and overhead are accumulated separately on the job cost sheet and then added to determine the total cost of a job. Job cost sheets for jobs in process, finished jobs, and jobs sold make up subsidiary records controlled by general ledger accounts.

A1 Apply job order costing in pricing services. Job order costing can usefully be applied to a service setting. The resulting job cost estimate can then be used to help determine a price for services.

P1 Describe and record the flow of materials costs in job order cost accounting. Costs of materials flow from receiving reports to materials ledger cards and then to either job cost sheets or the Indirect Materials account in the factory overhead ledger.

P2 Describe and record the flow of labor costs in job order cost accounting. Costs of labor flow from clock cards to the Factory Payroll account and then to either job cost sheets or the Indirect Labor account in the factory overhead ledger.

P3 Describe and record the flow of overhead costs in job order cost accounting. Overhead costs are accumulated in the Factory Overhead account that controls the subsidiary factory overhead ledger. Then, using a predetermined overhead rate, overhead costs are charged to jobs.

P4 Determine adjustments for overapplied and underapplied factory overhead. At the end of each period, the Factory Overhead account usually has a residual debit (underapplied overhead) or credit (overapplied overhead) balance. If the balance is not material, it is transferred to Cost of Goods Sold, but if it is material, it is allocated to Goods in Process Inventory, Finished Goods Inventory, and Cost of Goods Sold.

Guidance Answers to **Decision Maker** and **Decision Ethics**

Management Consultant Service companies (such as this consulting firm) do not recognize goods in process inventory or finished goods inventory—an important difference between service and manufacturing companies. For the two jobs that are 60% complete, you could recognize revenues and costs at 60% of the total expected amounts. This means you could recognize revenue of $7,200 (0.60 × $12,000) and costs of $6,000 (0.60 × $10,000), yielding net income of $1,200 from each job.

Web Consultant The partner has a monetary incentive to *manage* the numbers and assign more costs to the two cost-plus engagements. This also would reduce costs on the fixed-price engagements. To act in such a manner is unethical. As a professional and an honest person, it is your responsibility to engage in ethical behavior. You must not comply with the partner's instructions. If the partner insists you act in an unethical manner, you should report the matter to a higher authority in the organization.

Entrepreneur An inadequate cost system can distort product costs. You should review overhead costs in detail. Once you know the different cost elements in overhead, you can classify them into groups such as material related, labor related, or machine related. Other groups can also be formed (we discuss this in Chapter 21). Once you have classified overhead items into groups, you can better establish overhead allocation bases and use them to compute predetermined overhead rates. These multiple rates and bases can then be used to assign overhead costs to products. This will likely improve product pricing.

Sales Manager The price based on AdWorld's normal pricing policy is $20,150 ($17,076 × 1.18), which is within the price range offered by competitors. One option is to apply normal pricing policy and quote a price of $20,150. On the other hand, assessing the competition, particularly in terms of their service quality and other benefits they might offer, would be useful. Although price is an input customers use to select suppliers, factors such as quality and timeliness (responsiveness) of suppliers are important. Accordingly, your price can reflect such factors.

Guidance Answers to **Quick Checks**

1. *b*
2. A job is a special order for a custom product. A job lot consists of a quantity of identical, special-order items.
3. *a*
4. Three costs normally accumulated on a job cost sheet are direct materials, direct labor, and factory overhead.
5. *c*
6. Four sources of factory overhead are materials requisitions, time tickets, vouchers, and adjusting entries.
7. Since a job order cost accounting system uses perpetual inventory records, overhead costs must be assigned to jobs before the end of a period. This requires the use of a predetermined overhead rate.
8. Debits are recorded when wages and salaries of factory employees are paid or accrued. Credits are recorded when direct labor costs are assigned to jobs and when indirect labor costs are transferred to the Factory Overhead account.
9. Overapplied or underapplied overhead usually exists at the end of a period because application of overhead is based on estimates of overhead and another variable such as direct labor. Estimates rarely equal actual amounts incurred.
10. A debit balance reflects underapplied factory overhead.

Key Terms

mhhe.com/wildFAP19e

Key Terms are available at the book's Website for learning and testing in an online Flashcard Format.

Clock card (p. 776)
Cost accounting system (p. 770)
Finished Goods Inventory (p. 772)
General accounting system (p. 770)
Goods in Process Inventory (p. 772)
Job (p. 770)
Job cost sheet (p. 772)
Job lot (p. 771)
Job order cost accounting system (p. 772)
Job order production (p. 770)
Materials ledger card (p. 773)
Materials requisition (p. 774)
Overapplied overhead (p. 782)
Predetermined overhead rate (p. 778)
Target cost (p. 771)
Time ticket (p. 776)
Underapplied overhead (p. 781)

Multiple Choice Quiz

Answers on p. 805. mhhe.com/wildFAP19e

Additional Quiz Questions are available at the book's Website.

Quiz19

1. A company's predetermined overhead allocation rate is 150% of its direct labor costs. How much overhead is applied to a job that requires total direct labor costs of $30,000?
   a. $15,000
   b. $30,000
   c. $45,000
   d. $60,000
   e. $75,000
2. A company's cost accounting system uses direct labor costs to apply overhead to goods in process and finished goods inventories. Its production costs for the period are: direct materials, $45,000; direct labor, $35,000; and overhead applied, $38,500. What is its predetermined overhead allocation rate?
   a. 10%
   b. 110%
   c. 86%
   d. 91%
   e. 117%
3. A company's ending inventory of finished goods has a total cost of $10,000 and consists of 500 units. If the overhead applied to these goods is $4,000, and the predetermined

overhead rate is 80% of direct labor costs, how much direct materials cost was incurred in producing these 500 units?

a. $10,000
b. $ 6,000
c. $ 4,000
d. $ 5,000
e. $ 1,000

4. A company's Goods in Process Inventory T-account follows.

Goods in Process Inventory			
Beginning balance	9,000		
Direct materials	94,200		
Direct labor	59,200	?	Finished goods
Overhead applied	31,600		
Ending balance	17,800		

The cost of units transferred to Finished Goods inventory is

a. $193,000
b. $211,800
c. $185,000
d. $144,600
e. $176,200

5. At the end of its current year, a company learned that its overhead was underapplied by $1,500 and that this amount is not considered material. Based on this information, the company should

a. Close the $1,500 to Finished Goods Inventory.
b. Close the $1,500 to Cost of Goods Sold.
c. Carry the $1,500 to the next period.
d. Do nothing about the $1,500 because it is not material and it is likely that overhead will be overapplied by the same amount next year.
e. Carry the $1,500 to the Income Statement as "Other Expense."

Discussion Questions

1. Why must a company estimate the amount of factory overhead assigned to individual jobs or job lots?

2. ♝ The chapter used a percent of labor cost to assign factory overhead to jobs. Identify another factor (or base) a company might reasonably use to assign overhead costs.

3. ♝ What information is recorded on a job cost sheet? How do management and employees use job cost sheets?

4. In a job order cost accounting system, what records serve as a subsidiary ledger for Goods in Process Inventory? For Finished Goods Inventory?

5. What journal entry is recorded when a materials manager receives a materials requisition and then issues materials (both direct and indirect) for use in the factory?

6. ♝ How does the materials requisition help safeguard a company's assets?

7. What is the difference between a clock card and a time ticket?

8. What events cause debits to be recorded in the Factory Overhead account? What events cause credits to be recorded in the Factory Overhead account?

9. What account(s) is(are) used to eliminate overapplied or underapplied overhead from the Factory Overhead account, assuming the amount is not material?

10. ♝ Assume that **Apple** produces a batch of 1,000 iPods. Does it account for this as 1,000 individual jobs or as a job lot? Explain (consider costs and benefits).

11. Why must a company prepare a predetermined overhead rate when using job order cost accounting?

12. ♝ How would a hospital apply job order costing? Explain.

13. ♝ **Harley-Davidson** manufactures 30 custom-made, luxury-model motorcycles. Does it account for these motorcycles as 30 individual jobs or as a job lot? Explain.

Harley-Davidson

14. **Best Buy**'s GeekSquad performs computer and home theater installation and service, for an upfront flat price. How can Best Buy use a job order costing system?

♝ ***Denotes Discussion Questions that involve decision making.***

Available with McGraw-Hill's Homework Manager McGraw-Hill's HOMEWORK MANAGER®

QUICK STUDY

QS 19-1
Jobs and job lots
C2 ♝

Determine which products are most likely to be manufactured as a job and which as a job lot.

1. A custom-designed home.
2. Hats imprinted with company logo.
3. Little League trophies.
4. A hand-crafted table.
5. A 90-foot motor yacht.
6. Wedding dresses for a chain of stores.

QS 19-2
Job cost computation
C3

The following information is from the materials requisitions and time tickets for Job 9-1005 completed by Franklin Boats. The requisitions are identified by code numbers starting with the letter Q and the time tickets start with W. At the start of the year, management estimated that overhead cost would equal 110% of direct labor cost for each job. Determine the total cost on the job cost sheet for Job 9-1005.

Date	Document	Amount
7/1/2009	Q-4698	$2,500
7/1/2009	W-3393	1,200
7/5/2009	Q-4725	2,000
7/5/2009	W-3479	900
7/10/2009	W-3559	600

QS 19-3
Direct materials journal entries
P1

During the current month, a company that uses a job order cost accounting system purchases $25,000 in raw materials for cash. It then uses $6,000 of raw materials indirectly as factory supplies and uses $16,000 of raw materials as direct materials. Prepare entries to record these three transactions.

QS 19-4
Direct labor journal entries P2

During the current month, a company that uses a job order cost accounting system incurred a monthly factory payroll of $75,000, paid in cash. Of this amount, $29,000 is classified as indirect labor and the remainder as direct. Prepare entries to record these transactions.

QS 19-5
Factory overhead rates P3

A company incurred the following manufacturing costs this period: direct labor, $234,000; direct materials, $292,000; and factory overhead, $58,500. Compute its overhead cost as a percent of (1) direct labor and (2) direct materials.

QS 19-6
Factory overhead journal entries
P3

During the current month, a company that uses a job order cost accounting system incurred a monthly factory payroll of $350,000, paid in cash. Of this amount, $90,000 is classified as indirect labor and the remainder as direct for the production of Job 65A. Factory overhead is applied at 90% of direct labor. Prepare the entry to apply factory overhead to this job lot.

QS 19-7
Entry for over- or underapplied overhead P4

A company allocates overhead at a rate of 150% of direct labor cost. Actual overhead cost for the current period is $475,000, and direct labor cost is $300,000. Prepare the entry to close over- or underapplied overhead to cost of goods sold.

McGraw-Hill's HOMEWORK MANAGER® Available with McGraw-Hill's Homework Manager

EXERCISES

Exercise 19-1
Documents in job order cost accounting
C2 C3 P1 P2 P3

The left column lists the titles of documents and accounts used in job order cost accounting. The right column presents short descriptions of the purposes of the documents. Match each document in the left column to its numbered description in the right column.

A. Voucher
B. Materials requisition
C. Factory Overhead account
D. Clock card
E. Factory Payroll account
F. Materials ledger card
G. Time ticket

_____ **1.** Shows amount of time an employee works on a job.
_____ **2.** Temporarily accumulates incurred labor costs until they are assigned to specific jobs or to overhead.
_____ **3.** Shows only total time an employee works each day.
_____ **4.** Perpetual inventory record of raw materials received, used, and available for use.
_____ **5.** Shows amount approved for payment of an overhead or other cost.
_____ **6.** Temporarily accumulates the cost of incurred overhead until the cost is assigned to specific jobs.
_____ **7.** Communicates the need for materials to complete a job.

Exercise 19-2
Analysis of cost flows
C2 P1 P2 P3

As of the end of June, the job cost sheets at Tracer Wheels, Inc., show the following total costs accumulated on three custom jobs.

	Job 102	Job 103	Job 104
Direct materials	$25,000	$59,000	$56,000
Direct labor	14,000	26,700	40,000
Overhead	7,000	13,350	20,000

Job 102 was started in production in May and the following costs were assigned to it in May: direct materials, $13,000; direct labor, $3,600; and overhead, $1,600. Jobs 103 and 104 are started in June. Overhead cost is applied with a predetermined rate based on direct labor cost. Jobs 102 and 103 are finished in June, and Job 104 is expected to be finished in July. No raw materials are used indirectly in June. Using this information, answer the following questions. (Assume this company's predetermined overhead rate did not change across these months).

1. What is the cost of the raw materials requisitioned in June for each of the three jobs?
2. How much direct labor cost is incurred during June for each of the three jobs?
3. What predetermined overhead rate is used during June?
4. How much total cost is transferred to finished goods during June?

Check (4) $145,050

Exercise 19-3
Overhead rate; costs assigned to jobs
P3

In December 2008, Matsushi Electronics' management establishes the year 2009 predetermined overhead rate based on direct labor cost. The information used in setting this rate includes estimates that the company will incur $750,000 of overhead costs and $500,000 of direct labor cost in year 2009. During March 2009, Matsushi began and completed Job No. 13-56.

1. What is the predetermined overhead rate for year 2009?
2. Use the information on the following job cost sheet to determine the total cost of the job.

Check (2) $23,450

JOB COST SHEET

Customer's Name ESPN Co. **Job No.** 13-56

Job Description 5 plasma monitors—150 inch

	Direct Materials		Direct Labor		Overhead Costs Applied	
Date	**Requisition No.**	**Amount**	**Time-Ticket No.**	**Amount**	**Rate**	**Amount**
Mar. 8	4-129	$4,000	T-306	$ 680		
Mar. 11	4-142	7,450	T-432	1,280		
Mar. 18	4-167	3,800	T-456	1,320		
Totals						

Exercise 19-4
Analysis of costs assigned to goods in process
P3

Wilson Company uses a job order cost accounting system that charges overhead to jobs on the basis of direct material cost. At year-end, the Goods in Process Inventory account shows the following.

Accounting System
File Edit Maintain Tasks Analysis Options Reports Window Help

Goods in Process Inventory — Acct. No. 121

Date	Explanation	Debit	Credit	Balance
2009				
Dec. 31	Direct materials cost	1,200,000		1,200,000
31	Direct labor cost	270,000		1,470,000
31	Overhead costs	480,000		1,950,000
31	To finished goods		1,860,000	90,000

Sales Purchases General Ledger Payroll Inventory Company Analysis

1. Determine the overhead rate used (based on direct material cost).
2. Only one job remained in the goods in process inventory at December 31, 2009. Its direct materials cost is $40,000. How much direct labor cost and overhead cost are assigned to it?

Check (2) Direct labor cost, $34,000

Exercise 19-5
Cost flows in a job order cost system
C3 P3

The following information is available for SafeLife Company, which produces special-order security products and uses a job order cost accounting system.

	April 30	May 31
Inventories		
Raw materials	$27,000	$ 41,000
Goods in process	9,000	20,600
Finished goods	70,000	33,000
Activities and information for May		
Raw materials purchases (paid with cash)		183,000
Factory payroll (paid with cash)		500,000
Factory overhead		
Indirect materials		6,000
Indirect labor		74,000
Other overhead costs		95,500
Sales (received in cash)		1,500,000
Predetermined overhead rate based on direct labor cost		55%

Compute the following amounts for the month of May.

1. Cost of direct materials used.
2. Cost of direct labor used.
3. Cost of goods manufactured.
4. Cost of goods sold.*
5. Gross profit.
6. Overapplied or underapplied overhead.

*Do not consider any underapplied or overapplied overhead.

Check (3) $811,700

Exercise 19-6
Journal entries for a job order cost accounting system
P1 P2 P3 P4

Use information in Exercise 19-5 to prepare journal entries for the following events in May.

1. Raw materials purchases for cash.
2. Direct materials usage.
3. Indirect materials usage.
4. Factory payroll costs in cash.
5. Direct labor usage.
6. Indirect labor usage.
7. Factory overhead excluding indirect materials and indirect labor (record credit to Other Accounts).
8. Application of overhead to goods in process.
9. Transfer of finished jobs to the finished goods inventory.
10. Sale and delivery of finished goods to customers for cash (record unadjusted cost of sales).
11. Allocation (closing) of overapplied or underapplied overhead to Cost of Goods Sold.

Exercise 19-7
Factory overhead computed, applied, and adjusted
P3 P4

In December 2008, Dreamvision established its predetermined overhead rate for movies produced during year 2009 by using the following cost predictions: overhead costs, $1,700,000, and direct labor costs, $500,000. At year end 2009, the company's records show that actual overhead costs for the year are $1,710,000. Actual direct labor cost had been assigned to jobs as follows.

Movies completed and released	$400,000
Movies still in production	90,000
Total actual direct labor cost	$490,000

1. Determine the predetermined overhead rate for year 2009.
2. Set up a T-account for overhead and enter the overhead costs incurred and the amounts applied to movies during the year using the predetermined overhead rate.
3. Determine whether overhead is overapplied or underapplied (and the amount) during the year.
4. Prepare the adjusting entry to allocate any over- or underapplied overhead to Cost of Goods Sold.

Check (3) $44,000 underapplied

Exercise 19-8
Factory overhead computed, applied, and adjusted
P3 P4

In December 2008, Jens Company established its predetermined overhead rate for jobs produced during year 2009 by using the following cost predictions: overhead costs, $1,500,000, and direct labor costs, $1,250,000. At year end 2009, the company's records show that actual overhead costs for the year are $1,660,000. Actual direct labor cost had been assigned to jobs as follows.

Jobs completed and sold	$1,027,500
Jobs in finished goods inventory	205,500
Jobs in goods in process inventory	137,000
Total actual direct labor cost	$1,370,000

1. Determine the predetermined overhead rate for year 2009.
2. Set up a T-account for Factory Overhead and enter the overhead costs incurred and the amounts applied to jobs during the year using the predetermined overhead rate.
3. Determine whether overhead is overapplied or underapplied (and the amount) during the year.
4. Prepare the adjusting entry to allocate any over- or underapplied overhead to Cost of Goods Sold.

Check (3) $16,000 underapplied

Exercise 19-9
Overhead rate calculation, allocation, and analysis P3

Campton Company applies factory overhead based on direct labor costs. The company incurred the following costs during 2009: direct materials costs, $635,500; direct labor costs, $2,000,000; and factory overhead costs applied, $1,200,000.

1. Determine the company's predetermined overhead rate for year 2009.
2. Assuming that the company's $54,000 ending Goods in Process Inventory account for year 2009 had $13,000 of direct labor costs, determine the inventory's direct materials costs.
3. Assuming that the company's $337,435 ending Finished Goods Inventory account for year 2009 had $137,435 of direct materials costs, determine the inventory's direct labor costs and its overhead costs.

Check (3) $75,000 overhead costs

Exercise 19-10
Costs allocated to ending inventories
P3

Santana Company's ending Goods in Process Inventory account consists of 10,000 units of partially completed product, and its Finished Goods Inventory account consists of 12,000 units of product. The factory manager determines that Goods in Process Inventory includes direct materials cost of $20 per unit and direct labor cost of $14 per unit. Finished goods are estimated to have $24 of direct materials cost per unit and $18 of direct labor cost per unit. The company established the predetermined overhead rate using the following predictions: estimated direct labor cost, $600,000, and estimated factory overhead, $750,000. The company allocates factory overhead to its goods in process and finished goods inventories based on direct labor cost. During the period, the company incurred these costs: direct materials, $1,070,000; direct labor, $580,000; and factory overhead applied, $725,000.

1. Determine the predetermined overhead rate.
2. Compute the total cost of the two ending inventories.
3. Compute cost of goods sold for the year (assume no beginning inventories and no underapplied or overapplied overhead).

Check (3) Cost of goods sold, $1,086,000

Exercise 19-11
Cost-based pricing
A1

Clemente Corporation has requested bids from several architects to design its new corporate headquarters. Troy Architects is one of the firms bidding on the job. Troy estimates that the job will require the following direct labor.

	Labor	Estimated Hours	Hourly Rate
2	Architects	300	$400
3	Staff	300	65
4	Clerical	600	20

Troy applies overhead to jobs at 160% of direct labor cost. Troy would like to earn at least $90,000 profit on the architectural job. Based on past experience and market research, it estimates that the competition will bid between $450,000 and $550,000 for the job.

1. What is Troy's estimated cost of the architectural job?

2. What bid would you suggest that Troy submit?

Check (1) $393,900

McGraw-Hill's HOMEWORK MANAGER® Available with McGraw-Hill's Homework Manager

PROBLEM SET A

Problem 19-1A
Production costs computed and recorded; reports prepared

C3 P1 P2 P3 P4

Lemmon Co.'s March 31 inventory of raw materials is $170,000. Raw materials purchases in April are $310,000, and factory payroll cost in April is $224,000. Overhead costs incurred in April are: indirect materials, $25,000; indirect labor, $19,000; factory rent, $25,000; factory utilities, $13,000; and factory equipment depreciation, $41,000. The predetermined overhead rate is 65% of direct labor cost. Job 306 is sold for $400,000 cash in April. Costs of the three jobs worked on in April follow.

	Job 306	Job 307	Job 308
Balances on March 31			
Direct materials	$ 9,000	$ 17,000	
Direct labor	19,000	5,000	
Applied overhead	12,350	3,250	
Costs during April			
Direct materials	75,000	160,000	$ 65,000
Direct labor	31,000	74,000	100,000
Applied overhead	?	?	?
Status on April 30	Finished (sold)	Finished (unsold)	In process

Required

1. Determine the total of each production cost incurred for April (direct labor, direct materials, and applied overhead), and the total cost assigned to each job (including the balances from March 31).

2. Prepare journal entries for the month of April to record the following.

a. Materials purchases (on credit), factory payroll (paid in cash), and actual overhead costs including indirect materials and indirect labor. (Factory rent and utilities are paid in cash.)

b. Assignment of direct materials, direct labor, and applied overhead costs to the Goods in Process Inventory.

c. Transfer of Jobs 306 and 307 to the Finished Goods Inventory.

d. Cost of goods sold for Job 306.

e. Revenue from the sale of Job 306.

f. Assignment of any underapplied or overapplied overhead to the Cost of Goods Sold account. (The amount is not material.)

Check (2f) $10,250 overapplied

3. Prepare a manufacturing statement for April (use a single line presentation for direct materials and show the details of overhead cost).

(3) Cost of goods manufactured, $473,850

4. Compute gross profit for April. Show how to present the inventories on the April 30 balance sheet.

Analysis Component

5. The over- or underapplied overhead is closed to Cost of Goods Sold. Discuss how this adjustment impacts business decision making regarding individual jobs or batches of jobs.

Problem 19-2A
Source documents, journal entries, overhead, and financial reports

P1 P2 P3 P4

Mead Bay's computer system generated the following trial balance on December 31, 2009. The company's manager knows something is wrong with the trial balance because it does not show any balance for Goods in Process Inventory but does show balances for the Factory Payroll and Factory Overhead accounts.

	Debit	Credit
Cash	$ 40,000	
Accounts receivable	34,000	
Raw materials inventory	22,000	

[continued on next page]

[continued from previous page]

Goods in process inventory	0	
Finished goods inventory	12,000	
Prepaid rent	4,000	
Accounts payable		$ 8,500
Notes payable		11,500
Common stock		40,000
Retained earnings		84,000
Sales		178,000
Cost of goods sold	112,000	
Factory payroll	18,000	
Factory overhead	26,000	
Operating expenses	54,000	
Totals	$322,000	$322,000

After examining various files, the manager identifies the following six source documents that need to be processed to bring the accounting records up to date.

Materials requisition 21-3010:	$4,100 direct materials to Job 402
Materials requisition 21-3011:	$7,100 direct materials to Job 404
Materials requisition 21-3012:	$2,400 indirect materials
Labor time ticket 6052:	$2,000 direct labor to Job 402
Labor time ticket 6053:	$15,000 direct labor to Job 404
Labor time ticket 6054:	$1,000 indirect labor

Jobs 402 and 404 are the only units in process at year-end. The predetermined overhead rate is 150% of direct labor cost.

Required

1. Use information on the six source documents to prepare journal entries to assign the following costs.
   a. Direct materials costs to Goods in Process Inventory.
   b. Direct labor costs to Goods in Process Inventory.
   c. Overhead costs to Goods in Process Inventory.
   d. Indirect materials costs to the Factory Overhead account.
   e. Indirect labor costs to the Factory Overhead account.

Check (2) $3,900 underapplied overhead

2. Determine the revised balance of the Factory Overhead account after making the entries in part 1. Determine whether there is any under- or overapplied overhead for the year. Prepare the adjusting entry to allocate any over- or underapplied overhead to Cost of Goods Sold, assuming the amount is not material.

(3) T. B. totals, $322,000

3. Prepare a revised trial balance.

(4) Net income, $8,100

4. Prepare an income statement for year 2009 and a balance sheet as of December 31, 2009.

Analysis Component

5. Assume that the $2,400 on materials requisition 21-3012 should have been direct materials charged to Job 404. Without providing specific calculations, describe the impact of this error on the income statement for 2009 and the balance sheet at December 31, 2009.

Problem 19-3A
Source documents, journal entries, and accounts in job order cost accounting
P1 P2 P3

Challenger Watercraft's predetermined overhead rate for year 2009 is 200% of direct labor. Information on the company's production activities during May 2009 follows.

a. Purchased raw materials on credit, $200,000.
b. Paid $130,000 cash for factory wages.
c. Paid $16,000 cash to a computer consultant to reprogram factory equipment.
d. Materials requisitions record use of the following materials for the month.

Job 136	$ 50,000
Job 137	33,000
Job 138	19,800
Job 139	22,600
Job 140	6,800
Total direct materials	132,200
Indirect materials	20,000
Total materials used	$152,200

e. Time tickets record use of the following labor for the month.

Job 136	$ 12,100
Job 137	10,800
Job 138	37,500
Job 139	39,400
Job 140	3,200
Total direct labor	103,000
Indirect labor	27,000
Total	$130,000

f. Applied overhead to Jobs 136, 138, and 139.
g. Transferred Jobs 136, 138, and 139 to Finished Goods.
h. Sold Jobs 136 and 138 on credit at a total price of $550,000.
i. The company incurred the following overhead costs during the month (credit Prepaid Insurance for expired factory insurance).

Depreciation of factory building	$68,500
Depreciation of factory equipment	37,500
Expired factory insurance	11,000
Accrued property taxes payable	35,000

j. Applied overhead at month-end to the Goods in Process (Jobs 137 and 140) using the predetermined overhead rate of 200% of direct labor cost.

Required

1. Prepare a job cost sheet for each job worked on during the month. Use the following simplified form.

Job No. ________	
Materials	$ ________
Labor	________
Overhead	________
Total cost	$ ________

2. Prepare journal entries to record the events and transactions *a* through *j*.

Check (2f) Cr. Factory Overhead, $178,000

3. Set up T-accounts for each of the following general ledger accounts, each of which started the month with a zero balance: Raw Materials Inventory; Goods in Process Inventory; Finished Goods Inventory; Factory Payroll; Factory Overhead; Cost of Goods Sold. Then post the journal entries to these T-accounts and determine the balance of each account.

Check (3) Finished Goods Inventory, $140,800

4. Prepare a report showing the total cost of each job in process and prove that the sum of their costs equals the Goods in Process Inventory account balance. Prepare similar reports for Finished Goods Inventory and Cost of Goods Sold.

Problem 19-4A
Overhead allocation and adjustment using a predetermined overhead rate
C3 P3 P4

mhhe.com/wildFAP19e

In December 2008, Zander Company's manager estimated next year's total direct labor cost assuming 50 persons working an average of 2,000 hours each at an average wage rate of $30 per hour. The manager also estimated the following manufacturing overhead costs for year 2009.

Indirect labor	$ 339,200
Factory supervision	240,000
Rent on factory building	140,000
Factory utilities	318,000
Factory insurance expired	88,000
Depreciation—Factory equipment	480,000
Repairs expense—Factory equipment	60,000
Factory supplies used	88,800
Miscellaneous production costs	46,000
Total estimated overhead costs	$1,800,000

At the end of 2009, records show the company incurred $1,554,900 of actual overhead costs. It completed and sold five jobs with the following direct labor costs: Job 201, $604,000; Job 202, $573,000; Job 203, $318,000; Job 204, $726,000; and Job 205, $324,000. In addition, Job 206 is in process at the end of 2009 and had been charged $27,000 for direct labor. No jobs were in process at the end of 2008. The company's predetermined overhead rate is based on direct labor cost.

Required

1. Determine the following.
   a. Predetermined overhead rate for year 2009.
   b. Total overhead cost applied to each of the six jobs during year 2009.
   c. Over- or underapplied overhead at year-end 2009.
2. Assuming that any over- or underapplied overhead is not material, prepare the adjusting entry to allocate any over- or underapplied overhead to Cost of Goods Sold at the end of year 2009.

Check (1c) $11,700 underapplied
(2) Cr. Factory Overhead $11,700

Problem 19-5A
Production transactions; subsidiary records; and source documents
P1 P2 P3 P4

If the working papers that accompany this book are unavailable, do not attempt to solve this problem.
Morton Company manufactures variations of its product, a technopress, in response to custom orders from its customers. On May 1, the company had no inventories of goods in process or finished goods but held the following raw materials.

Material M	200 units @ $125 =	$25,000
Material R	95 units @ 90 =	8,550
Paint	55 units @ 40 =	2,200
Total cost		$35,750

On May 4, the company began working on two technopresses: Job 102 for Global Company and Job 103 for Kaddo Company.

Required

Follow the instructions in this list of activities and complete the sheets provided in the working papers.

a. Purchased raw materials on credit and recorded the following information from receiving reports and invoices.

Receiving Report No. 426, Material M, 250 units at $125 each.
Receiving Report No. 427, Material R, 90 units at $90 each.

Instructions: Record these purchases with a single journal entry and post it to general ledger T-accounts, using the transaction letter *a* to identify the entry. Enter the receiving report information on the materials ledger cards.

b. Requisitioned the following raw materials for production.

Requisition No. 35, for Job 102, 135 units of Material M.
Requisition No. 36, for Job 102, 72 units of Material R.
Requisition No. 37, for Job 103, 70 units of Material M.
Requisition No. 38, for Job 103, 38 units of Material R.
Requisition No. 39, for 15 units of paint.

Instructions: Enter amounts for direct materials requisitions on the materials ledger cards and the job cost sheets. Enter the indirect material amount on the materials ledger card and record a debit to the Indirect Materials account in the subsidiary factory overhead ledger. Do not record a journal entry at this time.

c. Received the following employee time tickets for work in May.

Time tickets Nos. 1 to 10 for direct labor on Job 102, $45,000.
Time tickets Nos. 11 to 30 for direct labor on Job 103, $32,500.
Time tickets Nos. 31 to 36 for equipment repairs, $9,625.

Instructions: Record direct labor from the time tickets on the job cost sheets and then debit indirect labor to the Indirect Labor account in the subsidiary factory overhead ledger. Do not record a journal entry at this time.

d. Paid cash for the following items during the month: factory payroll, $87,125, and miscellaneous overhead items, $51,000.

Instructions: Record these payments with journal entries and then post them to the general ledger accounts. Also record a debit in the Miscellaneous Overhead account in the subsidiary factory overhead ledger.

e. Finished Job 102 and transferred it to the warehouse. The company assigns overhead to each job with a predetermined overhead rate equal to 80% of direct labor cost.

Instructions: Enter the allocated overhead on the cost sheet for Job 102, fill in the cost summary section of the cost sheet, and then mark the cost sheet "Finished." Prepare a journal entry to record the job's completion and its transfer to Finished Goods and then post it to the general ledger accounts.

f. Delivered Job 102 and accepted the customer's promise to pay $200,000 within 30 days.

Instructions: Prepare journal entries to record the sale of Job 102 and the cost of goods sold. Post them to the general ledger accounts.

g. Applied overhead to Job 103 based on the job's direct labor to date.

Instructions: Enter overhead on the job cost sheet but do not make a journal entry at this time.

h. Recorded the total direct and indirect materials costs as reported on all the requisitions for the month.

Instructions: Prepare a journal entry to record these costs and post it to general ledger accounts.

Check (h) Dr. Goods in Process Inventory, $35,525

i. Recorded the total direct and indirect labor costs as reported on all time tickets for the month.

Instructions: Prepare a journal entry to record these costs and post it to general ledger accounts.

j. Recorded the total overhead costs applied to jobs.

Instructions: Prepare a journal entry to record the allocation of these overhead costs and post it to general ledger accounts.

Check Balance in Factory Overhead, $775 Cr., overapplied

PROBLEM SET B

Problem 19-1B
Production costs computed and recorded; reports prepared
C3 P1 P2 P3 P4 

Grant Co.'s August 31 inventory of raw materials is $75,000. Raw materials purchases in September are $200,000, and factory payroll cost in September is $110,000. Overhead costs incurred in September are: indirect materials, $15,000; indirect labor, $7,000; factory rent, $10,000; factory utilities, $6,000; and factory equipment depreciation, $15,000. The predetermined overhead rate is 50% of direct labor cost. Job 114 is sold for $190,000 cash in September. Costs for the three jobs worked on in September follow.

	Job 114	Job 115	Job 116
Balances on August 31			
Direct materials	$ 7,000	$ 9,000	
Direct labor	9,000	8,000	
Applied overhead	4,500	4,000	

[continued on next page]

[continued from previous page]

Costs during September			
Direct materials	50,000	85,000	$40,000
Direct labor	15,000	34,000	60,000
Applied overhead	?	?	?
Status on September 30	Finished (sold)	Finished (unsold)	In process

Required

1. Determine the total of each production cost incurred for September (direct labor, direct materials, and applied overhead), and the total cost assigned to each job (including the balances from August 31).

2. Prepare journal entries for the month of September to record the following.

a. Materials purchases (on credit), factory payroll (paid in cash), and actual overhead costs including indirect materials and indirect labor. (Factory rent and utilities are paid in cash.)

b. Assignment of direct materials, direct labor, and applied overhead costs to Goods in Process Inventory.

c. Transfer of Jobs 114 and 115 to the Finished Goods Inventory.

d. Cost of Job 114 in the Cost of Goods Sold account.

e. Revenue from the sale of Job 114.

Check (2f) $1,500 overapplied

f. Assignment of any underapplied or overapplied overhead to the Cost of Goods Sold account. (The amount is not material.)

(3) Cost of goods manufactured, $250,000

3. Prepare a manufacturing statement for September (use a single line presentation for direct materials and show the details of overhead cost).

4. Compute gross profit for September. Show how to present the inventories on the September 30 balance sheet.

Analysis Component

5. The over- or underapplied overhead adjustment is closed to Cost of Goods Sold. Discuss how this adjustment impacts business decision making regarding individual jobs or batches of jobs.

Problem 19-2B
Source documents, journal entries, overhead, and financial reports
P1 P2 P3 P4 

Coleman Company's computer system generated the following trial balance on December 31, 2009. The company's manager knows that the trial balance is wrong because it does not show any balance for Goods in Process Inventory but does show balances for the Factory Payroll and Factory Overhead accounts.

	Debit	Credit
Cash	$ 96,000	
Accounts receivable	84,000	
Raw materials inventory	52,000	
Goods in process inventory	0	
Finished goods inventory	18,000	
Prepaid rent	6,000	
Accounts payable		$ 21,000
Notes payable		27,000
Common stock		60,000
Retained earnings		174,000
Sales		360,000
Cost of goods sold	210,000	
Factory payroll	32,000	
Factory overhead	54,000	
Operating expenses	90,000	
Totals	$642,000	$642,000

After examining various files, the manager identifies the following six source documents that need to be processed to bring the accounting records up to date.

Materials requisition 94-231:	$9,200 direct materials to Job 603
Materials requisition 94-232:	$15,200 direct materials to Job 604
Materials requisition 94-233:	$4,200 indirect materials
Labor time ticket 765:	$10,000 direct labor to Job 603
Labor time ticket 766:	$16,000 direct labor to Job 604
Labor time ticket 777:	$6,000 indirect labor

Jobs 603 and 604 are the only units in process at year-end. The predetermined overhead rate is 200% of direct labor cost.

Required

1. Use information on the six source documents to prepare journal entries to assign the following costs.

a. Direct materials costs to Goods in Process Inventory.

b. Direct labor costs to Goods in Process Inventory.

c. Overhead costs to Goods in Process Inventory.

d. Indirect materials costs to the Factory Overhead account.

e. Indirect labor costs to the Factory Overhead account.

2. Determine the revised balance of the Factory Overhead account after making the entries in part 1. Determine whether there is under- or overapplied overhead for the year. Prepare the adjusting entry to allocate any over- or underapplied overhead to Cost of Goods Sold, assuming the amount is not material.

3. Prepare a revised trial balance.

4. Prepare an income statement for year 2009 and a balance sheet as of December 31, 2009.

Check (2) $12,200 underapplied overhead

(3) T. B. totals, $642,000

(4) Net income, $47,800

Analysis Component

5. Assume that the $4,200 indirect materials on materials requisition 94-233 should have been direct materials charged to Job 604. Without providing specific calculations, describe the impact of this error on the income statement for 2009 and the balance sheet at December 31, 2009.

Problem 19-3B
Source documents, journal entries, and accounts in job order cost accounting

P1 P2 P3

Bradley Company's predetermined overhead rate is 200% of direct labor. Information on the company's production activities during September 2009 follows.

a. Purchased raw materials on credit, $250,000.

b. Paid $168,000 cash for factory wages.

c. Paid $22,000 cash for miscellaneous factory overhead costs.

d. Materials requisitions record use of the following materials for the month.

Job 487	$60,000
Job 488	40,000
Job 489	24,000
Job 490	28,000
Job 491	8,000
Total direct materials	160,000
Indirect materials	24,000
Total materials used	$184,000

e. Time tickets record use of the following labor for the month.

Job 487	$ 16,000
Job 488	14,000
Job 489	50,000
Job 490	52,000
Job 491	4,000
Total direct labor	136,000
Indirect labor	32,000
Total	$168,000

f. Allocated overhead to Jobs 487, 489, and 490.
g. Transferred Jobs 487, 489, and 490 to Finished Goods.
h. Sold Jobs 487 and 489 on credit for a total price of $680,000.
i. The company incurred the following overhead costs during the month (credit Prepaid Insurance for expired factory insurance).

Depreciation of factory building	$74,000
Depreciation of factory equipment	42,000
Expired factory insurance	14,000
Accrued property taxes payable	62,000

j. Applied overhead at month-end to the Goods in Process (Jobs 488 and 491) using the predetermined overhead rate of 200% of direct labor cost.

Required

1. Prepare a job cost sheet for each job worked on in the month. Use the following simplified form.

Job No. ______	
Materials	$ ______
Labor	______
Overhead	______
Total cost	$ ______

Check (2f) Cr. Factory Overhead, $236,000

(3) Finished Goods Inventory, $184,000

2. Prepare journal entries to record the events and transactions *a* through *j*.
3. Set up T-accounts for each of the following general ledger accounts, each of which started the month with a zero balance: Raw Materials Inventory, Goods in Process Inventory, Finished Goods Inventory, Factory Payroll, Factory Overhead, Cost of Goods Sold. Then post the journal entries to these T-accounts and determine the balance of each account.
4. Prepare a report showing the total cost of each job in process and prove that the sum of their costs equals the Goods in Process Inventory account balance. Prepare similar reports for Finished Goods Inventory and Cost of Goods Sold.

Problem 19-4B
Overhead allocation and adjustment using a predetermined overhead rate
C3 P3 P4

In December 2008, Bigby Company's manager estimated next year's total direct labor cost assuming 100 persons working an average of 2,000 hours each at an average wage rate of $15 per hour. The manager also estimated the following manufacturing overhead costs for year 2009.

Indirect labor	$ 319,200
Factory supervision	240,000
Rent on factory building	140,000
Factory utilities	88,000
Factory insurance expired	68,000
Depreciation—Factory equipment	480,000
Repairs expense—Factory equipment	60,000
Factory supplies used	68,800
Miscellaneous production costs	36,000
Total estimated overhead costs	$1,500,000

At the end of 2009, records show the company incurred $1,450,000 of actual overhead costs. It completed and sold five jobs with the following direct labor costs: Job 625, $708,000; Job 626, $660,000; Job 627, $350,000; Job 628, $840,000; and Job 629, $368,000. In addition, Job 630 is in process at the end of 2009 and had been charged $20,000 for direct labor. No jobs were in process at the end of 2008. The company's predetermined overhead rate is based on direct labor cost.

Required

1. Determine the following.

a. Predetermined overhead rate for year 2009.

b. Total overhead cost applied to each of the six jobs during year 2009.

c. Over- or underapplied overhead at year-end 2009.

2. Assuming that any over- or underapplied overhead is not material, prepare the adjusting entry to allocate any over- or underapplied overhead to Cost of Goods Sold at the end of year 2009.

Check (1c) $23,000 overapplied

(2) Dr. Factory Overhead, $23,000

Problem 19-5B
Production transactions; subsidiary records; and source documents

P1 P2 P3 P4

If the working papers that accompany this book are unavailable, do not attempt to solve this problem. Parador Company produces variations of its product, a megatron, in response to custom orders from its customers. On June 1, the company had no inventories of goods in process or finished goods but held the following raw materials.

Material M	120 units @ $400 =	$48,000
Material R	80 units @ 320 =	25,600
Paint	44 units @ 144 =	6,336
Total cost		$79,936

On June 3, the company began working on two megatrons: Job 450 for Doso Company and Job 451 for Border, Inc.

Required

Follow instructions in this list of activities and complete the sheets provided in the working papers.

a. Purchased raw materials on credit and recorded the following information from receiving reports and invoices.

> Receiving Report No. 20, Material M, 150 units at $400 each.
> Receiving Report No. 21, Material R, 70 units at $320 each.

Instructions: Record these purchases with a single journal entry and post it to general ledger T-accounts, using the transaction letter *a* to identify the entry. Enter the receiving report information on the materials ledger cards.

b. Requisitioned the following raw materials for production.

> Requisition No. 223, for Job 450, 80 units of Material M.
> Requisition No. 224, for Job 450, 60 units of Material R.
> Requisition No. 225, for Job 451, 40 units of Material M.
> Requisition No. 226, for Job 451, 30 units of Material R.
> Requisition No. 227, for 12 units of paint.

Instructions: Enter amounts for direct materials requisitions on the materials ledger cards and the job cost sheets. Enter the indirect material amount on the materials ledger card and record a debit to the Indirect Materials account in the subsidiary factory overhead ledger. Do not record a journal entry at this time.

c. Received the following employee time tickets for work in June.

> Time tickets Nos. 1 to 10 for direct labor on Job 450, $80,000.
> Time tickets Nos. 11 to 20 for direct labor on Job 451, $64,000.
> Time tickets Nos. 21 to 24 for equipment repairs, $24,000.

Instructions: Record direct labor from the time tickets on the job cost sheets and then debit indirect labor to the Indirect Labor account in the subsidiary factory overhead ledger. Do not record a journal entry at this time.

d. Paid cash for the following items during the month: factory payroll, $168,000, and miscellaneous overhead items, $73,600.

Instructions: Record these payments with journal entries and post them to the general ledger accounts. Also record a debit in the Miscellaneous Overhead account in the subsidiary factory overhead ledger.

e. Finished Job 450 and transferred it to the warehouse. The company assigns overhead to each job with a predetermined overhead rate equal to 70% of direct labor cost.

Instructions: Enter the allocated overhead on the cost sheet for Job 450, fill in the cost summary section of the cost sheet, and then mark the cost sheet "Finished." Prepare a journal entry to record the job's completion and its transfer to Finished Goods and then post it to the general ledger accounts.

f. Delivered Job 450 and accepted the customer's promise to pay $580,000 within 30 days.

Instructions: Prepare journal entries to record the sale of Job 450 and the cost of goods sold. Post them to the general ledger accounts.

g. Applied overhead cost to Job 451 based on the job's direct labor used to date.

Instructions: Enter overhead on the job cost sheet but do not make a journal entry at this time.

Check (h) Dr. Goods in Process Inventory, $76,800

h. Recorded the total direct and indirect materials costs as reported on all the requisitions for the month.

Instructions: Prepare a journal entry to record these costs and post it to general ledger accounts.

i. Recorded the total direct and indirect labor costs as reported on all time tickets for the month.

Instructions: Prepare a journal entry to record these costs and post it to general ledger accounts.

j. Recorded the total overhead costs applied to jobs.

Check Balance in Factory Overhead, $1,472 Cr., overapplied

Instructions: Prepare a journal entry to record the allocation of these overhead costs and post it to general ledger accounts.

SERIAL PROBLEM

Success Systems

(This serial problem began in Chapter 1 and continues through most of the book. If previous chapter segments were not completed, the serial problem can begin at this point. It is helpful, but not necessary, to use the Working Papers that accompany the book.)

SP 19 The computer workstation furniture manufacturing that Adriana Lopez started in January is progressing well. As of the end of June, Success Systems' job cost sheets show the following total costs accumulated on three furniture jobs.

	Job 6.02	Job 6.03	Job 6.04
Direct materials	$1,500	$3,300	$2,700
Direct labor	800	1,420	2,100
Overhead	400	710	1,050

Job 6.02 was started in production in May, and these costs were assigned to it in May: direct materials, $600; direct labor, $180; and overhead, $90. Jobs 6.03 and 6.04 were started in June. Overhead cost is applied with a predetermined rate based on direct labor costs. Jobs 6.02 and 6.03 are finished in June, and Job 6.04 is expected to be finished in July. No raw materials are used indirectly in June. (Assume this company's predetermined overhead rate did not change over these months).

Required

Check (1) Total materials, $6,900

1. What is the cost of the raw materials used in June for each of the three jobs and in total?

2. How much total direct labor cost is incurred in June?

(3) 50%

3. What predetermined overhead rate is used in June?

4. How much cost is transferred to finished goods inventory in June?

BEYOND THE NUMBERS

REPORTING IN ACTION

C2 

BTN 19-1 Best Buy's financial statements and notes in Appendix A provide evidence of growth potential in its domestic sales.

Required

1. Identify at least two types of costs that will predictably increase as a percent of sales with growth in domestic sales.

2. Explain why you believe the types of costs identified for part 1 will increase, and describe how you might assess Best Buy's success with these costs. (*Hint:* You might consider the gross margin ratio.)

Fast Forward

3. Access Best Buy's annual report for a fiscal year ending after March 3, 2007, from its Website [BestBuy.com] or the SEC's EDGAR database [www.sec.gov]. Review and report its growth in sales along with its cost and income levels (including its gross margin ratio).

COMPARATIVE ANALYSIS

C1 

BTN 19-2 Retailers as well as manufacturers can apply just-in-time (JIT) to their inventory management. Both **Best Buy** and **Circuit City** want to know the impact of a JIT inventory system for their operating cash flows. Review each company's statement of cash flows in Appendix A to answer the following.

Required

1. Identify the impact on operating cash flows (increase or decrease) for changes in inventory levels (increase or decrease) for both companies for each of the three most recent years.

2. What impact would a JIT inventory system have on both Best Buy's and Circuit City's operating income? Link the answer to your response for part 1.

3. Would the move to a JIT system have a one-time or recurring impact on operating cash flow?

ETHICS CHALLENGE

P3

BTN 19-3 An accounting professional requires at least two skill sets. The first is to be technically competent. Knowing how to capture, manage, and report information is a necessary skill. Second, the ability to assess manager and employee actions and biases for accounting analysis is another skill. For instance, knowing how a person is compensated helps anticipate information biases. Draw on these skills and write a one-half page memo to the financial officer on the following practice of allocating overhead.

Background: Assume that your company sells portable housing to both general contractors and the government. It sells jobs to contractors on a bid basis. A contractor asks for three bids from different manufacturers. The combination of low bid and high quality wins the job. However, jobs sold to the government are bid on a cost-plus basis. This means price is determined by adding all costs plus a profit based on cost at a specified percent, such as 10%. You observe that the amount of overhead allocated to government jobs is higher than that allocated to contract jobs. These allocations concern you and motivate your memo.

Point: Students could compare responses and discuss differences in concerns with allocating overhead.

COMMUNICATING IN PRACTICE

C2 C3

BTN 19-4 Assume that you are preparing for a second interview with a manufacturing company. The company is impressed with your credentials but has indicated that it has several qualified applicants. You anticipate that in this second interview, you must show what you offer over other candidates. You learn the company currently uses a periodic inventory system and is not satisfied with the timeliness of its information and its inventory management. The company manufactures custom-order holiday decorations and display items. To show your abilities, you plan to recommend that it use a cost accounting system.

Required

In preparation for the interview, prepare notes outlining the following:

1. Your cost accounting system recommendation and why it is suitable for this company.

2. A general description of the documents that the proposed cost accounting system requires.

3. How the documents in part 2 facilitate the operation of the cost accounting system.

Point: Have students present a mock interview, one assuming the role of the president of the company and the other the applicant.

TAKING IT TO THE NET

C2 

BTN 19-5 Many contractors work on custom jobs that require a job order costing system.

Required

Access the Website AMSI.com and click on *Construction Management Software,* and then on STARBUILDER. Prepare a one-page memorandum for the CEO of a construction company providing information about the job order costing software this company offers. Would you recommend that the company purchase this software?

TEAMWORK IN ACTION

C2

BTN 19-6 Consider the activities undertaken by a medical clinic in your area.

Required

1. Do you consider a job order cost accounting system appropriate for the clinic?
2. Identify as many factors as possible to lead you to conclude that it uses a job order system.

ENTREPRENEURIAL DECISION

C2 

BTN 19-7 Refer to the chapter opener regarding Hank Julicher and his company, **Sprinturf**. All successful businesses track their costs, and it is especially important for startup businesses to monitor and control costs.

Required

1. Assume that Sprinturf uses a job order costing system. For the three basic cost categories of direct materials, direct labor, and overhead, identify at least two typical costs that would fall into each category for Sprinturf.
2. Assume a local high school expresses an interest in purchasing a synthetic field installation from Sprinturf. The high school's budget will allow them to pay no more than $600,000 for the field. How can Sprinturf use job cost information to assess whether to pursue this opportunity?

HITTING THE ROAD

C3 P2 P3 P4 

BTN 19-8 Job order cost accounting is frequently used by home builders.

Required

1. You (or your team) are to prepare a job cost sheet for a single-family home under construction. List four items of both direct materials and direct labor. Explain how you think overhead should be applied.
2. Contact a builder and compare your job cost sheet to this builder's job cost sheet. If possible, speak to that company's accountant. Write your findings in a short report.

GLOBAL DECISION

C1 

BTN 19-9 DSG, **Circuit City**, and **Best Buy** are competitors in the global marketplace. Access DSG's annual report (www.DSGiplc.com) for the year ended April 28, 2007. The following information is available for DSG.

(£ millions)	Current Year	One Year Prior	Two Years Prior
Inventories	£1,030	£873	£811

Required

1. Determine the change in DSG's inventories for the last two years. Then identify the impact on net resources generated by operating activities (increase or decrease) for changes in inventory levels (increase or decrease) for DSG for the last two years.

2. Would a move to a JIT system likely impact DSG more than it would Best Buy or Circuit City? Explain.

ANSWERS TO MULTIPLE CHOICE QUIZ

1. c; $30,000 × 150% = $45,000

2. b; $38,500/$35,000 = 110%

3. e; Direct materials + Direct labor + Overhead = Total cost;
Direct materials + ($4,000/.80) + $4,000 = $10,000
Direct materials = $1,000

4. e; $9,000 + $94,200 + $59,200 + $31,600 − Finished goods = $17,800
Thus, finished goods = $176,200

5. b

A Look Back

Chapter 18 introduced managerial accounting and described cost concepts and the reporting of manufacturing activities. Chapter 19 explained job order costing—an important cost accounting system for customized products and services.

A Look at This Chapter

This chapter focuses on how to measure and account for costs in process operations. We explain process production, describe how to assign costs to processes, and compute cost per equivalent unit for a process.

A Look Ahead

Chapter 21 explains how to allocate factory overhead costs to different products and introduces the activity-based costing method of overhead allocation. It also explains responsibility accounting and measures of departmental performance.

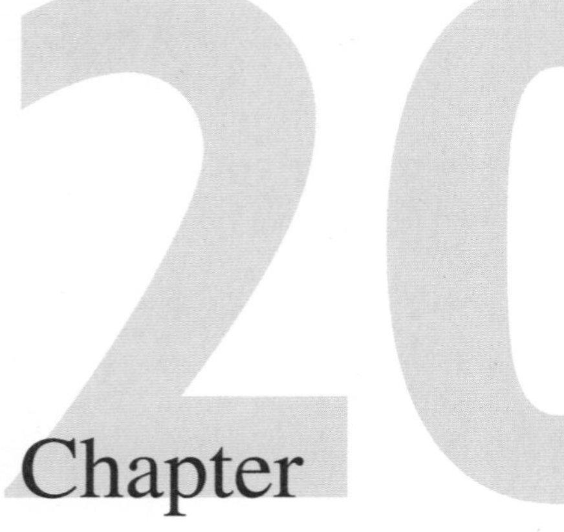

Chapter 20 Process Cost Accounting

Learning Objectives

CAP

Conceptual

C1 Explain process operations and the way they differ from job order operations. *(p. 808)*

C2 Define equivalent units and explain their use in process cost accounting. *(p. 815)*

C3 Explain the four steps in accounting for production activity in a period. *(p. 816)*

C4 Define a process cost summary and describe its purposes. *(p. 820)*

C5 *Appendix 20A*—Explain and illustrate the four steps in accounting for production activity using FIFO. *(p. 827)*

Analytical

A1 Compare process cost accounting and job order cost accounting. *(p. 809)*

A2 Explain and illustrate a hybrid costing system. *(p. 823)*

Procedural

P1 Record the flow of direct materials costs in process cost accounting. *(p. 812)*

P2 Record the flow of direct labor costs in process cost accounting. *(p. 813)*

P3 Record the flow of factory overhead costs in process cost accounting. *(p. 813)*

P4 Compute equivalent units produced in a period. *(p. 815)*

P5 Prepare a process cost summary. *(p. 820)*

P6 Record the transfer of completed goods to Finished Goods Inventory and Cost of Goods Sold. *(p. 821)*

LP20

Decision Feature

The Big Apple

"If we are willing to eat it, we're willing to squeeze it"
—David Ryan

HOOD RIVER, OR—After a few years of working in the family business of growing apples and making cider, David Ryan launched his own company, **Hood River Juice Company** [HRJCO.com], to focus on the processing stage of apple juice and cider. Like many entrepreneurs, David sought guidance from experienced mentors, in his case the Small Business Development Center located in the local community college. These mentors explained managerial accounting and the financial aspects of successful manufacturing.

Today, before an apple enters David's production process, it is inspected by his drivers when the apples are loaded from the field. A foreman then inspects the apples again when unloading them at his factory. David's factory employees then wash and hand select the best apples from those that survive the previous two inspections.

Apple quality is paramount. Explains David, "If we are willing to eat it, we're willing to squeeze it." From cutting apples into small pieces and squeezing those pieces into juice, through filtering the juice and packaging the finished product, David's production process is monitored and accounting reports are produced.

Entrepreneurs such as David are aided by process cost summaries that help them monitor and control the costs of material, labor, and overhead applied to production processes. For example, David tries to maintain regular full-time employees to better manage costs. Thus, he purchases and processes apples year-round as opposed to only seasonal production. David estimates this year-round process reduces his overhead costs by 40%. "Needless to say, every company has their own overhead they have to deal with," explains David. "If your total throughput is down by 35%, you must look elsewhere to get the margin to be sustainable. The only way to do that is to cut your overhead." Managerial accounting information aids in his decisions.

David's focus on cost management minimizes the risk of bad decisions, and his passion for quality control enables him to improve process operations. His overriding goal is customer satisfaction. That focus has led him to produce bulk apple juice for use in protein shakes and smoothies, and it has allowed his customers to select from over 50 varieties of apples for a custom-blended juice. Juice drinkers seem happy: From an initial investment of $36,000 in 2000, David's annual sales now exceed $14 million. Those are juicy numbers.

[Sources: *Hood River Juice Company Website*, January 2009; *Yakima-Herald.com*, March 2008; *Hood River News*, February 2006; *Entrepreneur*, April 2008]

Chapter Preview

The type of product or service a company offers determines its cost accounting system. Job order costing is used to account for custom products and services that meet the demands of a particular customer. Not all products are manufactured in this way; many carry standard designs so that one unit is no different than any other unit. Such a system often produces large numbers of units on a continuous basis, all of which pass through similar processes. This chapter describes how to use a process cost accounting system to account for these types of products. It also explains how costs are accumulated for each process and then assigned to units passing through those processes. This information helps us understand and estimate the cost of each process as well as find ways to reduce costs and improve processes.

Process Cost Accounting

Process Operations
- Comparing job order and process operations
- Organization of process operations
- GenX Company—an illustration

Process Cost Accounting
- Direct and indirect costs
- Accounting for materials costs
- Accounting for labor costs
- Accounting for factory overhead

Equivalent Units of Production (EUP)
- Accounting for goods in process
- Differences between EUP for materials, labor, and overhead

Process Costing Illustration
- Physical flow of units
- EUP
- Cost per EUP
- Cost reconciliation
- Process cost summary
- Transfers to finished goods and to cost of goods sold

Process Operations

C1 Explain process operations and the way they differ from job order operations.

Process operations, also called *process manufacturing* or *process production,* is the mass production of products in a continuous flow of steps. This means that products pass through a series of sequential processes. Petroleum refining is a common example of process operations. Crude oil passes through a series of steps before it is processed into different grades of petroleum. **Exxon Mobil**'s oil activities reflect a process operation. An important characteristic of process operations is the high level of standardization necessary if the system is to produce large volumes of products. Process operations also extend to services. Examples include mail sorting in large post offices and order processing in large mail-order firms such as **L.L. Bean**. The common feature in these service organizations is that operations are performed in a sequential manner using a series of standardized processes. Other companies using process operations include **Kellogg** (cereals), **Pfizer** (drugs), **Procter & Gamble** (household products), **Xerox** (copiers), **Coca-Cola** (soft drinks), **Heinz** (ketchup), **Penn** (tennis balls), and **Hershey** (chocolate). For a virtual tour of tennis ball manufacturing, see pennracquet.com/factory.html.

Each of these examples of products and services involves operations having a series of *processes,* or steps. Each process involves a different set of activities. A production operation that processes chemicals, for instance, might include the four steps shown in Exhibit 20.1. Understanding such processes for companies with process operations is crucial for measuring their costs. Increasingly, process operations use machines and automation to control product quality and reduce manufacturing costs.

EXHIBIT 20.1
Process Operations: Chemicals

Comparing Job Order and Process Operations

A1 Compare process cost accounting and job order cost accounting.

Job order and process operations can be considered as two ends of a continuum. Important features of both systems are shown in Exhibit 20.2. We often describe job order and process operations with manufacturing examples, but both also apply to service companies. In a job order costing system, the measurement focus is on the individual job or batch. In a process costing system, the measurement focus is on the process itself and the standardized units produced.

EXHIBIT 20.2
Comparing Job Order and Process Operations

Job Order Operations	Process Operations
• Custom orders	• Repetitive procedures
• Heterogeneous products and services	• Homogeneous products and services
• Low production volume	• High production volume
• High product flexibility	• Low product flexibility
• Low to medium standardization	• High standardization

Organization of Process Operations

In a process operation, each process is identified as a separate *production department, workstation,* or *work center.* With the exception of the first process or department, each receives the output from the prior department as a partially processed product. Depending on the nature of the process, a company applies direct labor, overhead, and, perhaps, additional direct materials to move the product toward completion. Only the final process or department in the series produces finished goods ready for sale to customers.

Tracking costs for several related departments can seem complex. Yet because process costing procedures are applied to the activity of each department or process separately, we need to consider only one process at a time. This simplifies the procedures.

When the output of one department becomes an input to another department, as is the case in sequential processing, we simply transfer the costs associated with those units from the first department into the next. We repeat these steps from department to department until the final process is complete. At that point the accumulated costs are transferred with the product from Goods in Process Inventory to Finished Goods Inventory. The next section illustrates a company with a single process, but the methods illustrated apply to a multiprocess scenario as each department's costs are handled separately for each department.

Decision Insight

Accounting for Health Many service companies use process departments to perform specific tasks for consumers. Hospitals, for instance, have radiology and physical therapy facilities with special equipment and trained employees. When patients need services, they are processed through departments to receive prescribed care. Service companies need process cost accounting information as much as manufacturers to estimate costs of services, to plan future operations, to control costs, and to determine customer charges.

GenX Company—An Illustration

The GenX Company illustrates process operations. It produces Profen®, an over-the-counter pain reliever for athletes. GenX sells Profen to wholesale distributors, who in turn sell it to retailers. Profen is produced by mixing its active ingredient, Profelene, with flavorings and preservatives, molding it into Profen tablets, and packaging the tablets. Exhibit 20.3 shows a summary floor plan of the GenX factory, which has five areas.

[1] *Storeroom*—materials are received and then distributed when requisitioned.
[2] *Production support offices*—used by administrative and maintenance employees who support manufacturing operations.
[3] *Locker rooms*—workers change from street clothes into sanitized uniforms before working in the factory.
[4] *Production floor*—area where the powder is processed into tablets.
[5] *Warehouse*—finished products are stored before being shipped to wholesalers.

Point: Electronic monitoring of operations is common in factories.

EXHIBIT 20.3
Floor Plan of GenX's Factory

The first step in process manufacturing is to decide when to produce a product. Management determines the types and quantities of materials and labor needed and then schedules the work. Unlike a job order process, where production often begins only after receipt of a custom order, managers of companies with process operations often forecast the demand expected for their products. Based on these plans, production begins. The flowchart in Exhibit 20.4 shows the production steps for GenX. The following sections explain how GenX uses a process cost accounting system to compute these costs. Many of the explanations refer to this exhibit and its numbered cost flows ① through ⑩. (*Hint:* The amounts for the numbered cost flows in Exhibit 20.4 are summarized in Exhibit 20.21. Those amounts are explained in the following pages, but it can help to refer to Exhibit 20.21 as we proceed through the explanations.)

EXHIBIT 20.4
Process Operations and Costs: GenX

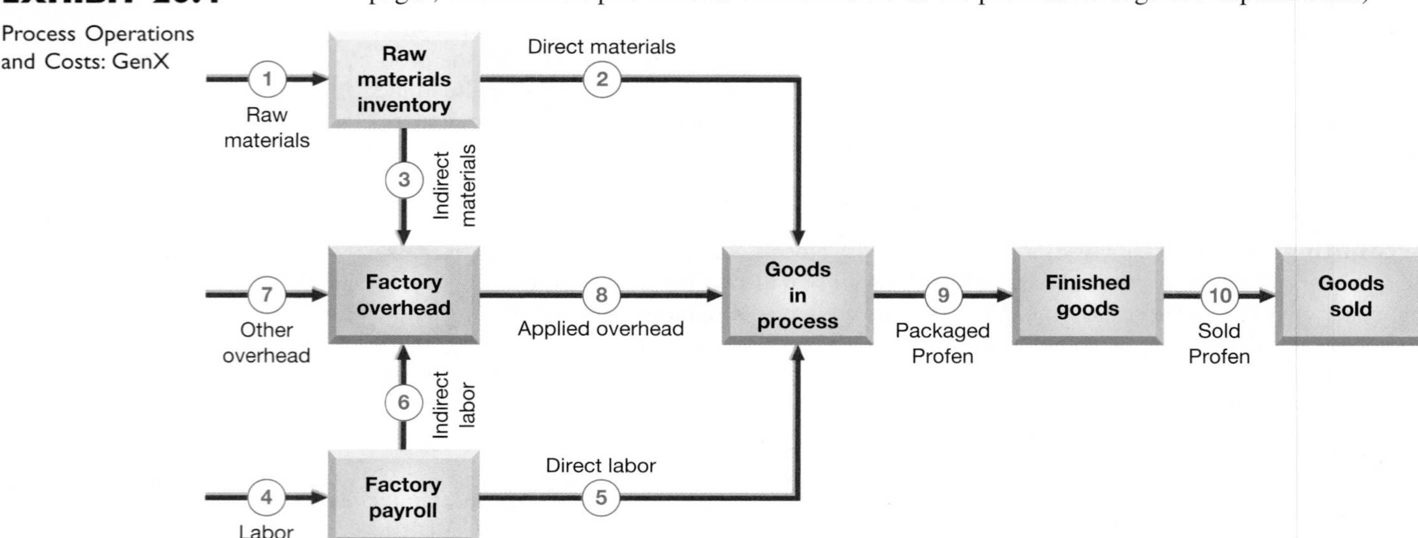

Process Cost Accounting

Process and job order operations are similar in that both combine materials, labor, and overhead in the process of producing products. They differ in how they are organized and managed. The measurement focus in a job order costing system is on the individual job or batch, whereas in a process costing system, it is on the individual process. Regardless of the measurement focus, we are ultimately interested in determining the cost per unit of product (or service) resulting from either system.

Video20.1

Specifically, the **job order cost accounting system** assigns direct materials, direct labor, and overhead to jobs. The total job cost is then divided by the number of units to compute a cost per unit for that job. The **process cost accounting system** assigns direct materials, direct labor, and overhead to specific processes (or departments). The total costs associated with each process are then divided by the number of units passing through that process to determine the cost per equivalent unit (defined later in the chapter) for that process. Differences in the way these two systems apply materials, labor, and overhead costs are highlighted in Exhibit 20.5.

Point: The cost object in a job order system is the specific job; the cost object in a process costing system is the process.

EXHIBIT 20.5

Comparing Job Order and Process Cost Accounting Systems

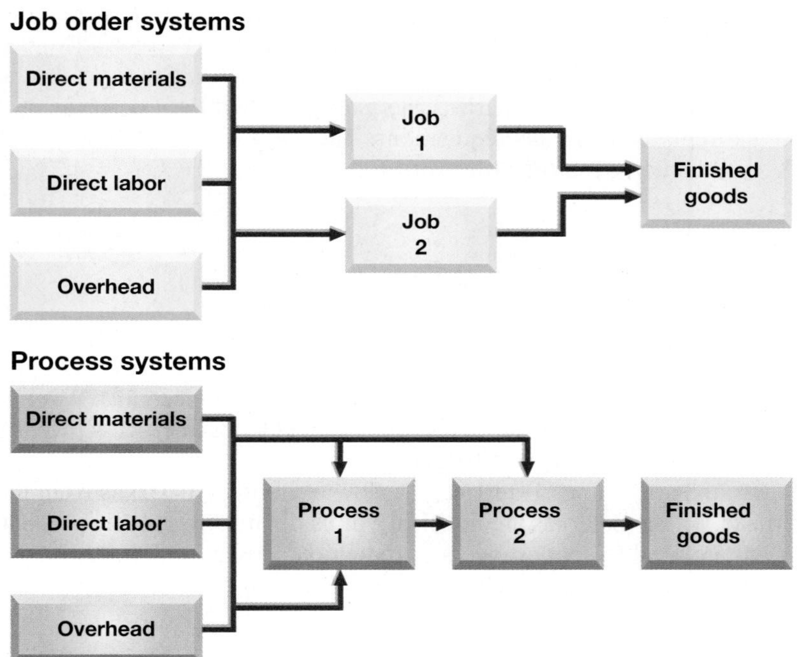

Direct and Indirect Costs

Like job order operations, process cost accounting systems use the concepts of direct and indirect costs. Materials and labor that can be traced to specific processes are assigned to those processes as direct costs. Materials and labor that cannot be traced to a specific process are indirect costs and are assigned to overhead. Some costs classified as overhead in a job order system may be classified as direct costs in process cost accounting. For example, depreciation of a machine used entirely by one process is a direct cost of that process.

Point: If a cost can be traced to the cost object, it is direct; if it cannot, it is indirect.

Decision Insight

JIT Boon to Process Operations Companies that adopt JIT manufacturing often organize their production system as a series of sequential processes. One survey found 60% of companies that converted to JIT used process operations; this compares to only 20% before converting to JIT.

P1 Record the flow of direct materials costs in process cost accounting.

Accounting for Materials Costs

In Exhibit 20.4, arrow line ① reflects the arrival of materials at GenX's factory. These materials include Profelene, flavorings, preservatives, and packaging. They also include supplies for the production support office. GenX uses a perpetual inventory system and makes all purchases on credit. The summary entry for receipts of raw materials in April follows (dates in journal entries numbered ① through ⑩ are omitted because they are summary entries, often reflecting two or more transactions or events).

Assets = Liabilities + Equity
+11,095 +11,095

①	Raw Materials Inventory	11,095	
	Accounts Payable		11,095
	Acquired materials on credit for factory use.		

Arrow line ② in Exhibit 20.4 reflects the flow of direct materials to production, where they are used to produce Profen. Most direct materials are physically combined into the finished product; the remaining direct materials include those used and clearly linked with a specific process. The manager of a process usually obtains materials by submitting a *materials requisition* to the materials storeroom manager. In some situations, materials move continuously from raw materials inventory through the manufacturing process. **Pepsi Bottling**, for instance, uses a process in which inventory moves continuously through the system. In these cases, a **materials consumption report** summarizes the materials used by a department during a reporting period and replaces materials requisitions. The entry to record the use of direct materials by GenX's production department in April follows.

Assets = Liabilities + Equity
+9,900
−9,900

②	Goods in Process Inventory	9,900	
	Raw Materials Inventory		9,900
	To assign costs of direct materials used in production.		

This entry transfers costs from one asset account to another asset account. (When two or more production departments exist, a company uses two or more Goods in Process Inventory accounts to separately accumulate costs incurred by each.)

Example: What types of materials might the flow of arrow line ③ in Exhibit 20.4 reflect? *Answer:* Goggles, gloves, protective clothing, recordkeeping supplies, and cleaning supplies.

In Exhibit 20.4, the arrow line ③ reflects the flow of indirect materials from the storeroom to factory overhead. These materials are not clearly linked with any specific production process or department but are used to support overall production activity. The following entry records the cost of indirect materials used by GenX in April.

Assets = Liabilities + Equity
−1,195 −1,195

③	Factory Overhead	1,195	
	Raw Materials Inventory		1,195
	To record indirect materials used in April.		

After the entries for both direct and indirect materials are posted, the Raw Materials Inventory account appears as shown in Exhibit 20.6. The April 30 balance sheet reports the $4,000 Raw Materials Inventory account as a current asset.

EXHIBIT 20.6

Raw Materials Inventory

Raw Materials Inventory					Acct. No. 132
Date		Explanation	Debit	Credit	Balance
Mar.	31	Balance			4,000
Apr.	30	Materials purchases	11,095		15,095
	30	Direct materials usage		9,900	5,195
	30	Indirect materials usage		1,195	4,000

Accounting for Labor Costs

P2 Record the flow of direct labor costs in process cost accounting.

Exhibit 20.4 shows GenX factory payroll costs as reflected in arrow line ④. Total labor costs of $8,920 are paid in cash and are recorded in the Factory Payroll account.

		Debit	Credit
④	Factory Payroll	8,920	
	Cash		8,920
	To record factory wages for April.		

Assets = Liabilities + Equity
−8,920 −8,920

Time reports from the production department and the production support office triggered this entry. (For simplicity, we do not separately identify withholdings and additional payroll taxes for employees.) In a process operation, the direct labor of a production department includes all labor used exclusively by that department. This is the case even if the labor is not applied to the product itself. If a production department in a process operation, for instance, has a full-time manager and a full-time maintenance worker, their salaries are direct labor costs of that process and are not factory overhead.

Arrow line ⑤ in Exhibit 20.4 shows GenX's use of direct labor in the production department. The following entry transfers April's direct labor costs from the Factory Payroll account to the Goods in Process Inventory account.

		Debit	Credit
⑤	Goods in Process Inventory	5,700	
	Factory Payroll		5,700
	To assign costs of direct labor used in production.		

Assets = Liabilities + Equity
+5,700 +5,700

Arrow line ⑥ in Exhibit 20.4 reflects GenX's indirect labor costs. These employees provide clerical, maintenance, and other services that help produce Profen efficiently. For example, they order materials, deliver them to the factory floor, repair equipment, operate and program computers used in production, keep payroll and other production records, clean up, and move the finished goods to the warehouse. The following entry charges these indirect labor costs to factory overhead.

Point: A department's indirect labor cost might include an allocated portion of the salary of a manager who supervises two or more departments. Allocation of costs between departments is discussed in a later chapter.

		Debit	Credit
⑥	Factory Overhead	3,220	
	Factory Payroll		3,220
	To record indirect labor as overhead.		

Assets = Liabilities + Equity
−3,220
+3,220

After these entries for both direct and indirect labor are posted, the Factory Payroll account appears as shown in Exhibit 20.7. The temporary Factory Payroll account is now closed to another temporary account, Factory Overhead, and is ready to receive entries for May. Next we show how to apply overhead to production and close the temporary Factory Overhead account.

EXHIBIT 20.7
Factory Payroll

Factory Payroll					Acct. No. 530
Date		**Explanation**	**Debit**	**Credit**	**Balance**
Mar.	31	Balance			0
Apr.	30	Total payroll for April	8,920		8,920
	30	Direct labor costs		5,700	3,220
	30	Indirect labor costs		3,220	0

Accounting for Factory Overhead

P3 Record the flow of factory overhead costs in process cost accounting.

Overhead costs other than indirect materials and indirect labor are reflected by arrow line ⑦ in Exhibit 20.4. These overhead items include the costs of insuring production assets, renting the factory building, using factory utilities, and depreciating equipment not directly related to a specific process. The following entry records overhead costs for April.

Assets = Liabilities + Equity
−180 +645 −2,425
−750
−850

		Debit	Credit
⑦	Factory Overhead	2,425	
	Prepaid Insurance		180
	Utilities Payable		645
	Cash		750
	Accumulated Depreciation—Factory Equipment . .		850
	To record overhead items incurred in April.		

After this entry is posted, the Factory Overhead account balance is $6,840, comprising indirect materials of $1,195, indirect labor of $3,220, and $2,425 of other overhead.

Arrow line ⑧ in Exhibit 20.4 reflects the application of factory overhead to production. Factory overhead is applied to processes by relating overhead cost to another variable such as direct labor hours or machine hours used. With increasing automation, companies with process operations are more likely to use machine hours to allocate overhead. In some situations, a single allocation basis such as direct labor hours (or a single rate for the entire plant) fails to provide useful allocations. As a result, management can use different rates for different production departments. Based on an analysis of its operations, GenX applies its April overhead at a rate of 120% of direct labor cost, as shown in Exhibit 20.8.

Point: The time it takes to process (cycle) products through a process is sometimes used to allocate costs.

EXHIBIT 20.8

Applying Factory Overhead

	Direct Labor Cost	Predetermined Rate	Overhead Applied
Production Department	$5,700	120%	$6,840

GenX records its applied overhead with the following entry.

Assets = Liabilities + Equity
+6,840 +6,840

		Debit	Credit
⑧	Goods in Process Inventory	6,840	
	Factory Overhead.		6,840
	Allocated overhead costs to production at 120% of direct labor cost.		

After posting this entry, the Factory Overhead account appears as shown in Exhibit 20.9. For GenX, the amount of overhead applied equals the actual overhead incurred during April. In most cases, using a predetermined overhead rate leaves an overapplied or underapplied balance in the Factory Overhead account. At the end of the period, this overapplied or underapplied balance should be closed to the Cost of Goods Sold account, as described in the job order costing chapter.

EXHIBIT 20.9

Factory Overhead

Factory Overhead					Acct. No. 540
Date		Explanation	Debit	Credit	Balance
Mar.	31	Balance			0
Apr.	30	Indirect materials usage	1,195		1,195
	30	Indirect labor costs	3,220		4,415
	30	Other overhead costs	2,425		6,840
	30	Applied to production departments		6,840	0

Example: If applied overhead results in a $6,940 credit to the factory overhead account, does it yield an over- or underapplied overhead amount? *Answer:* $100 overapplied overhead

Decision Ethics

Budget Officer You are working to identify the direct and indirect costs of a new processing department that has several machines. This department's manager instructs you to classify a majority of the costs as indirect to take advantage of the direct labor-based overhead allocation method so it will be charged a lower amount of overhead (because of its small direct labor cost). This would penalize other departments with higher allocations. It also will cause the performance ratings of managers in these other departments to suffer. What action do you take? [Answer—p. 832]

Quick Check Answers—p. 833

1. A process operation (*a*) is another name for a job order operation, (*b*) does not use the concepts of direct materials or direct labor, or (*c*) typically produces large quantities of homogeneous products or services.
2. Under what conditions is a process cost accounting system more suitable for measuring production costs than a job order cost accounting system?
3. When direct materials are assigned and used in production, the entry to record their use includes (*a*) a credit to Goods in Process Inventory, (*b*) a debit to Goods in Process Inventory, or (*c*) a debit to Raw Materials Inventory.
4. What are the three cost categories incurred by both job order and process operations?
5. How many Goods in Process Inventory accounts are needed in a process cost system?

Equivalent Units of Production

C2 Define equivalent units and explain their use in process cost accounting.

We explained how materials, labor, and overhead costs for a period are accumulated in the Goods in Process Inventory account, but we have not explained the arrow lines labeled ⑨ and ⑩ in Exhibit 20.4. These lines reflect the transfer of products from the production department to finished goods inventory, and from finished goods inventory to cost of goods sold. To determine the costs recorded for these flows, we must first determine the cost per unit of product and then apply this result to the number of units transferred.

Accounting for Goods in Process

If a process has *no beginning and no ending goods in process inventory,* the unit cost of goods transferred out of a process is computed as follows.

Video20.1

$$\frac{\textbf{Total cost assigned to the process (direct materials, direct labor, and overhead)}}{\textbf{Total number of units started and finished in the period}}$$

If a process has a beginning or ending inventory of partially processed units (or both), then the total cost assigned to the process must be allocated to all completed and incomplete units worked on during the period. Therefore, the denominator must measure the entire production activity of the process for the period, called **equivalent units of production** (or **EUP**), a phrase that refers to the number of units that could have been started *and* completed given the cost incurred during a period. This measure is then used to compute the cost per equivalent unit and to assign costs to finished goods and goods in process inventory.

To illustrate, assume that GenX adds (or introduces) 100 units into its process during a period. Suppose at the end of that period, the production supervisor determines that those 100 units are 60% of the way through the process. Therefore, equivalent units of production for that period total 60 EUP (100 units × 60%). This means that with the resources used to put 100 units 60% of the way through the process, GenX could have started and completed 60 whole units.

Point: For GenX, "units" might refer to individual Profen tablets. For a juice maker, units might refer to gallons.

Differences in Equivalent Units for Materials, Labor, and Overhead

P4 Compute equivalent units produced in a period.

In many processes, the equivalent units of production for direct materials are not the same with respect to direct labor and overhead. To illustrate, consider a five-step process operation shown in Exhibit 20.10.

EXHIBIT 20.10

An Illustrative Five-Step Process Operation

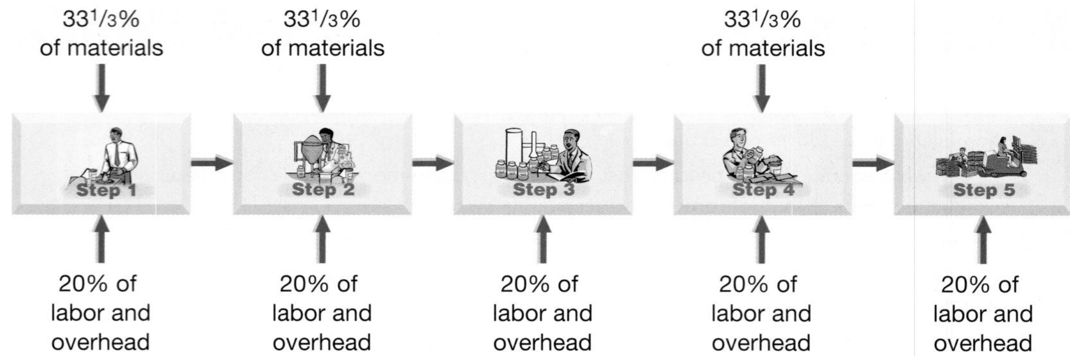

This exhibit shows that one-third of the direct material cost is added at each of three steps: 1, 2, and 4. One-fifth of the direct labor cost is added at each of the five steps. One-fifth of the overhead also is added at each step because overhead is applied as a percent of direct labor for this company.

When units finish step 1, they are one-third complete with respect to direct materials but only one-fifth complete with respect to direct labor and overhead. When they finish step 2, they are two-thirds complete with respect to direct materials but only two-fifths complete with respect to direct labor and overhead. When they finish step 3, they remain two-thirds complete with respect to materials but are now three-fifths complete with respect to labor and overhead. When they finish step 4, they are 100% complete with respect to materials (all direct materials have been added) but only four-fifths complete with respect to labor and overhead.

For example, if 300 units of product are started and processed through step 1 of Exhibit 20.10, they are said to be one-third complete *with respect to materials*. Expressed in terms of equivalent finished units, the processing of these 300 units is equal to finishing 100 EUP with respect to materials (300 units × 33⅓%). However, only one-fifth of direct labor and overhead has been applied to the 300 units at the end of step 1. This means that the equivalent units of production *with respect to labor and overhead* total 60 EUP (300 units × 20%).

Decision Insight

Process Services Customer interaction software is a hot item in customer service processes. Whether in insurance, delivery, or technology services, companies are finding that this software can turn their customer service process into an asset. How does it work? For starters, it cuts time spent on service calls because a customer describes a problem only once. It also yields a database of customer questions and complaints that gives insights into needed improvements. It recognizes incoming phone numbers and accesses previous dealings.

Process Costing Illustration

C3 Explain the four steps in accounting for production activity in a period.

This section applies process costing concepts and procedures to GenX. **This illustration uses the weighted-average method for inventory costs. The FIFO method is illustrated in Appendix 20A.** (Assume a weighted-average cost flow for all computations and assignments in this chapter unless explicitly stated differently. When using a just-in-time inventory system, different inventory methods yield similar results because inventories are immaterial.)

Exhibit 20.11 shows selected information from the production department for the month of April. Accounting for a department's activity for a period includes four steps involving analysis of (1) physical flow, (2) equivalent units, (3) cost per equivalent unit, and (4) cost assignment and reconciliation. The next sections describe each step.

EXHIBIT 20.11
Production Data

Beginning goods in process inventory (March 31)	
Units of product	30,000
Percentage of completion—Direct materials	100%
Percentage of completion—Direct labor	65%
Direct materials costs	$ 3,300
Direct labor costs	$ 600
Factory overhead costs applied (120% of direct labor)	$ 720
Activities during the current period (April)	
Units started this period	90,000
Units transferred out (completed)	100,000
Direct materials costs	$ 9,900
Direct labor costs	$ 5,700
Factory overhead costs applied (120% of direct labor)	$ 6,840
Ending goods in process inventory (April 30)	
Units of product	20,000
Percentage of completion—Direct materials	100%
Percentage of completion—Direct labor	25%

Step 1: Determine the Physical Flow of Units

Video20.1

A *physical flow reconciliation* is a report that reconciles (1) the physical units started in a period with (2) the physical units completed in that period. A physical flow reconciliation for GenX is shown in Exhibit 20.12 for April.

EXHIBIT 20.12
Physical Flow Reconciliation

Units to Account For		Units Accounted For	
Beginning goods in process inventory	30,000 units	Units completed and transferred out	100,000 units
Units started this period	90,000 units	Ending goods in process inventory	20,000 units
Total units to account for	**120,000 units**	Total units accounted for	**120,000 units**

reconciled

The weighted-average method does not require us to separately track the units in beginning work in process from those units started this period. Instead, the units are treated as part of a large pool with an average cost per unit.

Step 2: Compute Equivalent Units of Production

The second step is to compute *equivalent units of production* for direct materials, direct labor, and factory overhead for April. Overhead is applied using direct labor as the allocation base for GenX. This also implies that equivalent units are the same for both labor and overhead.

GenX used its direct materials, direct labor, and overhead to make finished units of Profen and to begin processing some units that are not yet complete. We must convert the physical units measure to equivalent units based on how each input has been used. Equivalent units are computed by multiplying the number of physical units by the percentage of completion for each input—see Exhibit 20.13.

EXHIBIT 20.13
Equivalent Units of Production—Weighted Average

Equivalent Units of Production	Direct Materials	Direct Labor	Factory Overhead
Equivalent units completed and transferred out (100,000 × 100%)	100,000 EUP	100,000 EUP	100,000 EUP
Equivalent units for ending goods in process			
Direct materials (20,000 × 100%)	20,000		
Direct labor (20,000 × 25%)		5,000	
Factory overhead (20,000 × 25%)			5,000
Equivalent units of production	120,000 EUP	105,000 EUP	105,000 EUP

The first row of Exhibit 20.13 reflects units transferred out in April. The production department entirely completed its work on the 100,000 units transferred out. These units have 100% of the materials, labor, and overhead required, or 100,000 equivalent units of each input (100,000 × 100%).

The second row references the ending goods in process, and rows three, four, and five break it down by materials, labor, and overhead. For direct materials, the units in ending goods in process inventory (20,000 physical units) include all materials required, so there are 20,000 equivalent units (20,000 × 100%) of materials in the unfinished physical units. Regarding labor, the units in ending goods in process inventory include 25% of the labor required, which implies 5,000 equivalent units of labor (20,000 × 25%). These units are only 25% complete and labor is used uniformly through the process. Overhead is applied on the basis of direct labor for GenX, so equivalent units for overhead are computed identically to labor (20,000 × 25%).

The final row reflects the whole units of product that could have been manufactured with the amount of inputs used to create some complete and some incomplete units. For GenX, the amount of inputs used to produce 100,000 complete units and to start 20,000 additional units is equivalent to the amount of direct materials in 120,000 whole units, the amount of direct labor in 105,000 whole units, and the amount of overhead in 105,000 whole units.

Step 3: Compute the Cost per Equivalent Unit

Equivalent units of production for each product (from step 2) is used to compute the average cost per equivalent unit. Under the **weighted-average method,** the computation of EUP does not separate the units in beginning inventory from those started this period; similarly, this method combines the costs of beginning goods in process inventory with the costs incurred in the current period. This process is illustrated in Exhibit 20.14.

EXHIBIT 20.14

Cost per Equivalent Unit of Production—Weighted Average

Cost per Equivalent Unit of Production	Direct Materials	Direct Labor	Factory Overhead
Costs of beginning goods in process inventory	$ 3,300	$ 600	$ 720
Costs incurred this period	9,900	5,700	6,840
Total costs	$13,200	$6,300	$7,560
÷ Equivalent units of production (from Step 2)	120,000 EUP	105,000 EUP	105,000 EUP
= Cost per equivalent unit of production	$0.11 per EUP*	$0.06 per EUP†	$0.072 per EUP‡

*$13,200 ÷ 120,000 EUP †$6,300 ÷ 105,000 EUP ‡$7,560 ÷ 105,000 EUP

For direct materials, the cost averages $0.11 per EUP, computed as the sum of direct materials cost from beginning goods in process inventory ($3,300) and the direct materials cost incurred in April ($9,900), and this sum ($13,200) is then divided by the 120,000 EUP for materials (from step 2). The costs per equivalent unit for labor and overhead are similarly computed. Specifically, direct labor cost averages $0.06 per EUP, computed as the sum of labor cost in beginning goods in process inventory ($600) and the labor costs incurred in April ($5,700), and this sum ($6,300) divided by 105,000 EUP for labor. Overhead costs averages $0.072 per EUP, computed as the sum of overhead cost in the beginning goods in process inventory ($720) and the overhead costs applied in April ($6,840), and this sum ($7,560) divided by 105,000 EUP for overhead.

Step 4: Assign and Reconcile Costs

The EUP from step 2 and the cost per EUP from step 3 are used in step 4 to assign costs to (a) units that production completed and transferred to finished goods and (b) units that remain in process. This is illustrated in Exhibit 20.15.

EXHIBIT 20.15

Report of Costs Accounted For—Weighted Average

Cost of units completed and transferred out		
Direct materials (100,000 EUP × $0.11 per EUP)	$11,000	
Direct labor (100,000 EUP × $0.06 per EUP)	6,000	
Factory overhead (100,000 EUP × $0.072 per EUP)	7,200	
Cost of units completed this period		$ 24,200
Cost of ending goods in process inventory		
Direct materials (20,000 EUP × $0.11 per EUP)	2,200	
Direct labor (5,000 EUP × $0.06 per EUP)	300	
Factory overhead (5,000 EUP × $0.072 per EUP)	360	
Cost of ending goods in process inventory		2,860
Total costs accounted for		$27,060

Cost of Units Completed and Transferred The 100,000 units completed and transferred to finished goods inventory required 100,000 EUP of direct materials. Thus, we assign $11,000 (100,000 EUP × $0.11 per EUP) of direct materials cost to those units. Similarly, those units had received 100,000 EUP of direct labor and 100,000 EUP of factory overhead (recall Exhibit 20.13). Thus, we assign $6,000 (100,000 EUP × $0.06 per EUP) of direct labor and $7,200 (100,000 EUP × $0.072 per EUP) of overhead to those units. The total cost of the 100,000 completed and transferred units is $24,200 ($11,000 + $6,000 + $7,200) and their average cost per unit is $0.242 ($24,200 ÷ 100,000 units).

Cost of Units for Ending Goods in Process There are 20,000 incomplete units in goods in process inventory at period-end. For direct materials, those units have 20,000 EUP of material (from step 2) at a cost of $0.11 per EUP (from step 3), which yields the materials cost of goods in process inventory of $2,200 (20,000 EUP × $0.11 per EUP). For direct labor, the in-process units have 25% of the required labor, or 5,000 EUP (from step 2). Using the $0.06 labor cost per EUP (from step 3) we obtain the labor cost of goods in process inventory of $300 (5,000 EUP × $0.06 per EUP). For overhead, the in-process units reflect 5,000 EUP (from step 2). Using the $0.072 overhead cost per EUP (from step 3) we obtain overhead costs with in-process inventory of $360 (5,000 EUP × $0.072 per EUP). Total cost of goods in process inventory at period-end is $2,860 ($2,200 + $300 + $360).

As a check, management verifies that total costs assigned to those units completed and transferred plus the costs of those in process (from Exhibit 20.15) equal the costs incurred by production. Exhibit 20.16 shows the costs incurred by production this period. We then reconcile the *costs accounted for* in Exhibit 20.15 with the *costs to account for* in Exhibit 20.16.

EXHIBIT 20.16

Report of Costs to Account For—Weighted Average

Cost of beginning goods in process inventory		
Direct materials	$3,300	
Direct labor	600	
Factory overhead	720	$ 4,620
Cost incurred this period		
Direct materials	9,900	
Direct labor	5,700	
Factory overhead	6,840	22,440
Total costs to account for		$27,060

At GenX, the production department manager is responsible for $27,060 in costs: $4,620 that is assigned to the goods in process at the start of the period plus $22,440 of materials, labor, and overhead incurred in the period. At period-end, that manager must show where these costs are assigned. The manager for GenX reports that $2,860 are assigned to units in process and $24,200 are assigned to units completed (per Exhibit 20.15). The sum of these amounts equals $27,060. Thus, the total *costs to account for* equal the total *costs accounted for* (minor differences can sometimes occur from rounding).

C4 Define a process cost summary and describe its purposes.

Point: Managers can examine changes in monthly costs per equivalent unit to help control the production process. When prices are set in a competitive market, managers can use process cost summary information to determine which costs should be cut to achieve a profit.

Process Cost Summary An important managerial accounting report for a process cost accounting system is the **process cost summary** (also called *production report*), which is prepared separately for each process or production department. Three reasons for the summary are to (1) help department managers control and monitor their departments, (2) help factory managers evaluate department managers' performances, and (3) provide cost information for financial statements. A process cost summary achieves these purposes by describing the costs charged to each department, reporting the equivalent units of production achieved by each department, and determining the costs assigned to each department's output. For our purposes, it is prepared using a combination of Exhibits 20.13, 20.14, 20.15, and 20.16.

P5 Prepare a process cost summary.

The process cost summary for GenX is shown in Exhibit 20.17. The report is divided into three sections. Section ① lists the total costs charged to the department, including direct materials, direct labor, and overhead costs incurred, as well as the cost of the beginning goods in process inventory. Section ② describes the equivalent units of production for the department. Equivalent units for materials, labor, and overhead are in separate columns. It also reports direct

EXHIBIT 20.17

Process Cost Summary

GenX COMPANY
Process Cost Summary
For Month Ended April 30, 2009

① **Costs Charged to Production**

Costs of beginning goods in process		
Direct materials	$3,300	
Direct labor	600	
Factory overhead	720	$ 4,620
Costs incurred this period		
Direct materials	9,900	
Direct labor	5,700	
Factory overhead	6,840	22,440
Total costs to account for		$27,060

② **Unit Cost Information**

Units to account for:		Units accounted for:	
Beginning goods in process	30,000	Completed and transferred out	100,000
Units started this period	90,000	Ending goods in process	20,000
Total units to account for	120,000	Total units accounted for	120,000

Equivalent Units of Production (EUP)	Direct Materials	Direct Labor	Factory Overhead
Units completed and transferred out	100,000 EUP	100,000 EUP	100,000 EUP
Units of ending goods in process			
Direct materials (20,000 × 100%)	20,000		
Direct labor (20,000 × 25%)		5,000	
Factory overhead (20,000 × 25%)			5,000
Equivalent units of production	120,000 EUP	105,000 EUP	105,000 EUP

Cost per EUP	Direct Materials	Direct Labor	Factory Overhead
Costs of beginning goods in process	$ 3,300	$ 600	$ 720
Costs incurred this period	9,900	5,700	6,840
Total costs	$13,200	$6,300	$7,560
÷EUP	120,000 EUP	105,000 EUP	105,000 EUP
Cost per EUP	$0.11 per EUP	$0.06 per EUP	$0.072 per EUP

③ **Cost Assignment and Reconciliation**

Costs transferred out (cost of goods manufactured)		
Direct materials (100,000 EUP × $0.11 per EUP)	$11,000	
Direct labor (100,000 EUP × $0.06 per EUP)	6,000	
Factory overhead (100,000 EUP × $0.072 per EUP)	7,200	$ 24,200
Costs of ending goods in process		
Direct materials (20,000 EUP × $0.11 per EUP)	2,200	
Direct labor (5,000 EUP × $0.06 per EUP)	300	
Factory overhead (5,000 EUP × $0.072 per EUP)	360	2,860
Total costs accounted for		$27,060

reconciled

materials, direct labor, and overhead costs per equivalent unit. Section ③ allocates total costs among units worked on in the period. The $24,200 is the total cost of goods transferred out of the department, and the $2,860 is the cost of partially processed ending inventory units. The assigned costs are then added to show that the total $27,060 cost charged to the department in section ① is now assigned to the units in section ③.

Quick Check Answers—p. 833

6. Equivalent units are (*a*) a measure of a production department's productivity in using direct materials, direct labor, or overhead; (*b*) units of a product produced by a foreign competitor that are similar to units produced by a domestic company; or (*c*) generic units of a product similar to brand name units of a product.
7. Interpret the meaning of a department's equivalent units with respect to direct labor.
8. A department began the period with 8,000 units that were one-fourth complete with respect to direct labor. It completed 58,000 units, and ended with 6,000 units that were one-third complete with respect to direct labor. What were its direct labor equivalent units for the period using the weighted-average method?
9. A process cost summary for a department has three sections. What information is presented in each of them?

Transfers to Finished Goods Inventory and Cost of Goods Sold

P6 Record the transfer of completed goods to Finished Goods Inventory and Cost of Goods Sold.

Arrow line ⑨ in Exhibit 20.4 reflects the transfer of completed products from production to finished goods inventory. The process cost summary shows that the 100,000 units of finished Profen are assigned a cost of $24,200. The entry to record this transfer follows.

		Debit	Credit
⑨	Finished Goods Inventory	24,200	
	Goods in Process Inventory		24,200
	To record transfer of completed units.		

Assets = Liabilities + Equity
+24,200
−24,200

The credit to Goods in Process Inventory reduces that asset balance to reflect that 100,000 units are no longer in production. The cost of these units has been transferred to Finished Goods Inventory, which is recognized as a $24,200 increase in this asset. After this entry is posted, there remains a balance of $2,860 in the Goods in Process Inventory account, which is the amount computed in Step 4 previously. The cost of units transferred from Goods in Process Inventory to Finished Goods Inventory is called the **cost of goods manufactured.** Exhibit 20.18 reveals the activities in the Goods in Process Inventory account for this period. The ending balance of this account equals the cost assigned to the partially completed units in section ③ of Exhibit 20.17.

EXHIBIT 20.18

Goods in Process Inventory

Goods in Process Inventory					Acct. No. 134
Date		Explanation	Debit	Credit	Balance
Mar.	31	Balance			4,620
Apr.	30	Direct materials usage	9,900		14,520
	30	Direct labor costs incurred	5,700		20,220
	30	Factory overhead applied	6,840		27,060
	30	Transfer completed product to warehouse		24,200	2,860

Arrow line ⑩ in Exhibit 20.4 reflects the sale of finished goods. Assume that GenX sold 106,000 units of Profen this period, and that its beginning inventory of finished goods consisted of 26,000 units with a cost of $6,292. Also assume that its ending finished goods inventory consists of 20,000 units at a cost of $4,840. Using this information, we can compute its cost of goods sold for April as shown in Exhibit 20.19.

Point: We omit the journal entry for sales, but it totals the number of units sold times price per unit.

EXHIBIT 20.19
Cost of Goods Sold

Beginning finished goods inventory	$ 6,292
+ Cost of goods manufactured this period	24,200
= Cost of goods available for sale	$30,492
− Ending finished goods inventory	4,840
= Cost of goods sold	$25,652

The summary entry to record cost of goods sold for this period follows.

Assets	= Liabilities +	Equity
−25,652		−25,652

		Debit	Credit
⑩	Cost of Goods Sold	25,652	
	Finished Goods Inventory		25,652
	To record cost of goods sold for April.		

The Finished Goods Inventory account now appears as shown in Exhibit 20.20.

EXHIBIT 20.20
Finished Goods Inventory

Finished Goods Inventory					Acct. No. 135
Date		Explanation	Debit	Credit	Balance
Mar.	31	Balance			6,292
Apr.	30	Transfer in cost of goods manufactured	24,200		30,492
	30	Cost of goods sold		25,652	4,840

Summary of Cost Flows Exhibit 20.21 shows GenX's manufacturing cost flows for April. Each of these cost flows and the entries to record them have been explained. The flow of costs through the accounts reflects the flow of production activities and products.

EXHIBIT 20.21*
Cost Flows through GenX

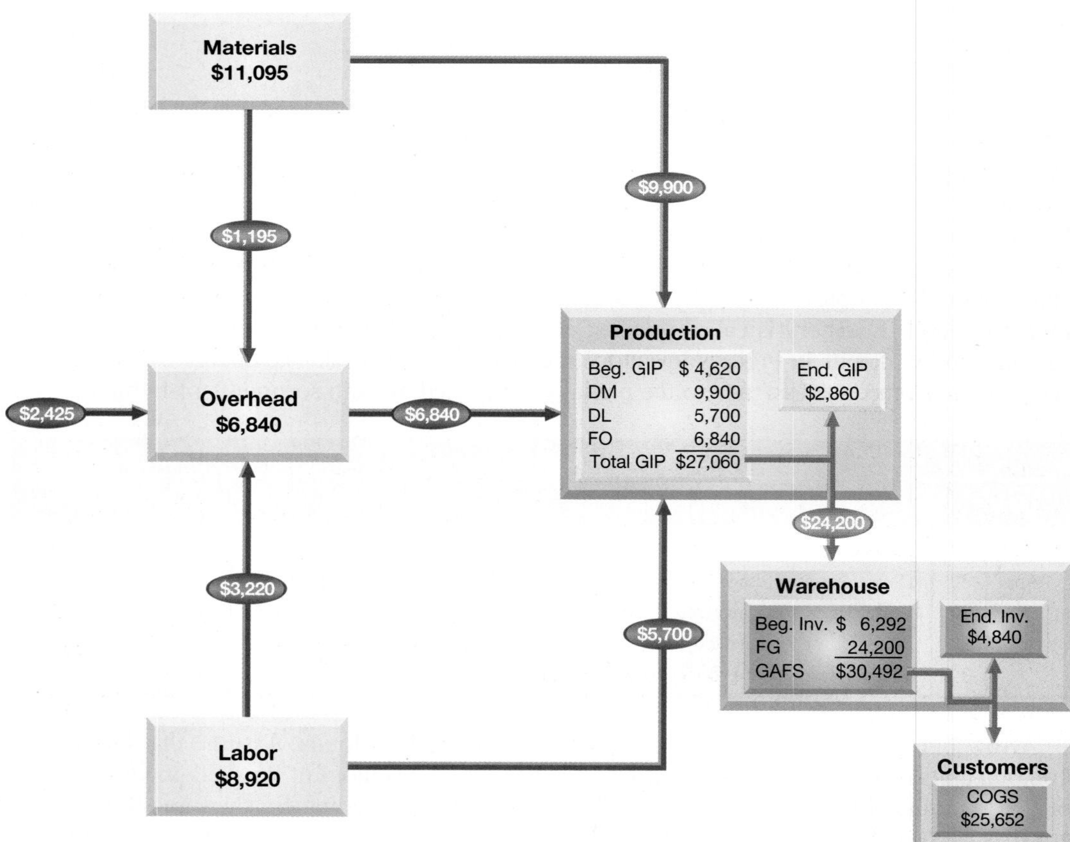

*Abbreviations: GIP (goods in process); DM (direct materials); DL (direct labor); FO (factory overhead); FG (finished goods); GAFS (goods available for sale); COGS (cost of goods sold).

Decision Insight

Best of Both Customer orientation demands both flexibility and standardization. Flexibility allows companies to supply products or services to a customer's specifications as in a job order setting, and standardization helps achieve efficiencies and lower costs as in a process operation.

Effect of the Lean Business Model on Process Operations

Adopting lean business practices often yields changes in process operations. Management concerns with throughput and just-in-time manufacturing, for instance, cause boundary lines between departments to blur. In some cases, higher quality and better efficiency are obtained by entirely reorganizing production processes. For example, instead of producing different types of computers in a series of departments, a separate work center for each computer can be established in one department. When such a rearrangement occurs, the process cost accounting system is changed to account for each work center's costs.

To illustrate, when a company adopts a just-in-time inventory system, its inventories can be minimal. If raw materials are not ordered or received until needed, a Raw Materials Inventory account may be unnecessary. Instead, materials cost is immediately debited to the Goods in Process Inventory account. Similarly, a Finished Goods Inventory account may not be needed. Instead, cost of finished goods may be immediately debited to the Cost of Goods Sold account.

Decision Insight

Lean Machine Attention to customer orientation has led to improved processes for companies. A manufacturer of control devices improved quality and reduced production time by forming teams to study processes and suggest improvements. Another company set up project groups to evaluate its production processes.

Hybrid Costing System

Decision Analysis

A2 Explain and illustrate a hybrid costing system.

This chapter explained the process costing system and contrasted it with the job order costing system. Many organizations use a *hybrid system* that contains features of both process and job order operations. A recent survey of manufacturers revealed that a majority use hybrid systems.

To illustrate, consider a car manufacturer's assembly line. On one hand, the line resembles a process operation in that the assembly steps for each car are nearly identical. On the other hand, the specifications of most cars have several important differences. At the Ford Mustang plant, each car assembled on a given day can be different from the previous car and the next car. This means that the costs of materials (subassemblies or components) for each car can differ. Accordingly, while the conversion costs (direct labor and overhead) can be accounted for using a process costing system, the component costs (direct materials) are accounted for using a job order system (separately for each car or type of car).

A hybrid system of processes requires a *hybrid costing system* to properly cost products or services. In the Ford plant, the assembly costs per car are readily determined using process costing. The costs of additional components can then be added to the assembly costs to determine each car's total cost (as in job order costing). To illustrate, consider the following information for a daily assembly process at Ford.

Assembly process costs	
Direct materials	$10,600,000
Direct labor	$5,800,000
Factory overhead	$6,200,000
Number of cars assembled	1,000
Costs of three different types of steering wheels	$240, $330, $480
Costs of three different types of seats	$620, $840, $1,360

The assembly process costs $22,600 per car. Depending on the type of steering wheel and seats the customer requests, the cost of a car can range from $23,460 to $24,440 (a $980 difference).

Today companies are increasingly trying to standardize processes while attempting to meet individual customer needs. To the extent that differences among individual customers' requests are large, understanding the costs to satisfy those requests is important. Thus, monitoring and controlling both process and job order costs are important.

Decision Ethics

Entrepreneur You operate a process production company making similar products for three different customers. One customer demands 100% quality inspection of products at your location before shipping. The added costs of that inspection are spread across all customers, not just the one demanding it. If you charge the added costs to that customer, you could lose that customer and experience a loss. Moreover, your other two customers have agreed to pay 110% of full costs. What actions (if any) do you take?

[Answer—pp. 832–833]

Demonstration Problem

Pennsylvania Company produces a product that passes through a single production process. Then completed products are transferred to finished goods in its warehouse. Information related to its manufacturing activities for July follows.

Raw Materials		Production Department	
Beginning inventory	$100,000	Beginning goods in process inventory (units)	5,000
Raw materials purchased on credit	211,400	Percentage completed—Materials	100%
Direct materials used	(190,000)	Percentage completed—Labor and overhead	60%
Indirect materials used	(51,400)	Beginning goods in process inventory (costs)	
Ending inventory	$ 70,000	Direct materials used	$ 20,000
		Direct labor incurred	9,600
Factory Payroll		Overhead applied (200% of direct labor)	19,200
Direct labor incurred	$ 55,500	Total costs of beginning goods in process	$ 48,800
Indirect labor incurred	50,625		
Total payroll (paid in cash)	$106,125	Units started this period	20,000
		Units completed this period	17,000
Factory Overhead			
Indirect materials used	$ 51,400	Ending goods in process inventory (units)	8,000
Indirect labor used	50,625	Percentage completed—Materials	100%
Other overhead costs	71,725	Percentage completed—Labor and overhead	20%
Total factory overhead incurred	$173,750		
		Finished Goods Inventory	
Factory Overhead Applied		Beginning finished goods inventory	$ 96,400
Overhead applied (200% of direct labor)	$111,000	Cost transferred in from production	321,300
		Cost of goods sold	(345,050)
		Ending finished goods inventory	$ 72,650

DP20

Required

1. Prepare a physical flow reconciliation for July as illustrated in Exhibit 20.12.
2. Compute the equivalent units of production in July for direct materials, direct labor, and factory overhead.
3. Compute the costs per equivalent units of production in July for direct materials, direct labor, and factory overhead.
4. Prepare a report of costs accounted for and a report of costs to account for.

5. Prepare summary journal entries to record the transactions and events of July for (a) raw materials purchases, (b) direct materials usage, (c) indirect materials usage, (d) factory payroll costs, (e) direct labor usage, (f) indirect labor usage, (g) other overhead costs (credit Other Accounts), (h) application of overhead to production, (i) transfer of finished goods from production, and (j) the cost of goods sold.

Planning the Solution

- Track the physical flow to determine the number of units completed in July.
- Compute the equivalent unit of production for direct materials, direct labor, and factory overhead.
- Compute the costs per equivalent unit of production with respect to direct materials, direct labor, and overhead; and determine the cost per unit for each.
- Compute the total cost of the goods transferred to production by using the equivalent units and unit costs. Determine (a) the cost of the beginning in-process inventory, (b) the materials, labor, and overhead costs added to the beginning in-process inventory, and (c) the materials, labor, and overhead costs added to the units started and completed in the month.
- Determine the cost of goods sold using balances in finished goods and cost of units completed this period.
- Use the information to record the summary journal entries for July.

Solution to Demonstration Problem

1. Physical flow reconciliation.

Units to Account For		Units Accounted For	
Beginning goods in process inventory	5,000 units	Units completed and transferred out	17,000 units
Units started this period	20,000 units	Ending goods in process inventory	8,000 units
Total units to account for	**25,000 units**	Total units accounted for	**25,000 units**

reconciled

2. Equivalent units of production.

Equivalent Units of Production	Direct Materials	Direct Labor	Factory Overhead
Equivalent units completed and transferred out	17,000 EUP	17,000 EUP	17,000 EUP
Equivalent units in ending goods in process			
Direct materials (8,000 × 100%)	8,000		
Direct labor (8,000 × 20%)		1,600	
Factory overhead (8,000 × 20%)			1,600
Equivalent units of production	25,000 EUP	18,600 EUP	18,600 EUP

3. Costs per equivalent unit of production.

Costs per Equivalent Unit of Production	Direct Materials	Direct Labor	Factory Overhead
Costs of beginning goods in process	$ 20,000	$ 9,600	$ 19,200
Costs incurred this period	190,000	55,500	111,000**
Total costs	$210,000	$65,100	$130,200
÷ Equivalent units of production (from part 2)	25,000 EUP	18,600 EUP	18,600 EUP
= Costs per equivalent unit of production	$8.40 per EUP	$3.50 per EUP	$7.00 per EUP

**Factory overhead applied

4. Reports of costs accounted for and of costs to account for

Report of Costs Accounted For		
Cost of units transferred out (cost of goods manufactured)		
Direct materials ($8.40 per EUP × 17,000 EUP)	$142,800	
Direct labor ($3.50 per EUP × 17,000 EUP)	59,500	
Factory overhead ($7.00 per EUP × 17,000 EUP)	119,000	
Cost of units completed this period		$ 321,300
Cost of ending goods in process inventory		
Direct materials ($8.40 per EUP × 8,000 EUP)	67,200	
Direct labor ($3.50 per EUP × 1,600 EUP)	5,600	
Factory overhead ($7.00 per EUP × 1,600 EUP)	11,200	
Cost of ending goods in process inventory		84,000
Total costs accounted for		**$405,300**

Report of Costs to Account For		
Cost of beginning goods in process inventory		
Direct materials	$ 20,000	
Direct labor	9,600	
Factory overhead	19,200	$ 48,800
Cost incurred this period		
Direct materials	190,000	
Direct labor	55,500	
Factory overhead	111,000	356,500
Total costs to account for		**$405,300**

reconciled

5. Summary journal entries for the transactions and events in July.

a.	Raw Materials Inventory	211,400	
	Accounts Payable		211,400
	To record raw materials purchases.		
b.	Goods in Process Inventory	190,000	
	Raw Materials Inventory		190,000
	To record direct materials usage.		
c.	Factory Overhead	51,400	
	Raw Materials Inventory		51,400
	To record indirect materials usage.		
d.	Factory Payroll	106,125	
	Cash		106,125
	To record factory payroll costs.		
e.	Goods in Process Inventory	55,500	
	Factory Payroll		55,500
	To record direct labor usage.		
f.	Factory Overhead	50,625	
	Factory Payroll		50,625
	To record indirect labor usage.		
g.	Factory Overhead	71,725	
	Other Accounts		71,725
	To record other overhead costs.		

[continued on next page]

[continued from previous page]

h.	Goods in Process Inventory	111,000	
	Factory Overhead		111,000
	To record application of overhead.		
i.	Finished Goods Inventory	321,300	
	Goods in Process Inventory		321,300
	To record transfer of finished goods from production.		
j.	Cost of Goods Sold	345,050	
	Finished Goods Inventory		345,050
	To record cost of goods sold.		

APPENDIX

FIFO Method of Process Costing

20A

C5 Explain and illustrate the four steps in accounting for production activity using FIFO.

The **FIFO method** of process costing assigns costs to units assuming a first-in, first-out flow of product. The objectives, concepts, and journal entries (not amounts) are the same as for the weighted-average method, but computation of equivalent units of production and cost assignment are slightly different.

Exhibit 20A.1 shows selected information from GenX's production department for the month of April. Accounting for a department's activity for a period includes four steps: (1) determine physical flow, (2) compute equivalent units, (3) compute cost per equivalent unit, and (4) determine cost assignment and reconciliation. This appendix describes each of these steps using the FIFO method for process costing.

EXHIBIT 20A.1

Production Data

Beginning goods in process inventory (March 31)	
Units of product	30,000
Percentage of completion—Direct materials	100%
Percentage of completion—Direct labor	65%
Direct materials costs	$ 3,300
Direct labor costs	$ 600
Factory overhead costs applied (120% of direct labor)	$ 720
Activities during the current period (April)	
Units started this period	90,000
Units transferred out (completed)	100,000
Direct materials costs	$ 9,900
Direct labor costs	$ 5,700
Factory overhead costs applied (120% of direct labor)	$ 6,840
Ending goods in process inventory (April 30)	
Units of product	20,000
Percentage of completion—Direct materials	100%
Percentage of completion—Direct labor	25%

Step 1: Determine Physical Flow of Units

A *physical flow reconciliation* is a report that reconciles (1) the physical units started in a period with (2) the physical units completed in that period. The physical flow reconciliation for GenX is shown in Exhibit 20A.2 for April.

EXHIBIT 20A.2

Physical Flow Reconciliation

Units to Account For		Units Accounted For	
Beginning goods in process inventory	30,000 units	Units completed and transferred out	100,000 units
Units started this period	90,000 units	Ending goods in process inventory	20,000 units
Total units to account for	**120,000 units**	Total units accounted for	**120,000 units**

reconciled

FIFO assumes that the 100,000 units transferred to finished goods during April include the 30,000 units from the beginning goods in process inventory. The remaining 70,000 units transferred out are from units started in April. Of the total 90,000 units started in April, 70,000 were completed, leaving 20,000 units unfinished at period-end.

Step 2: Compute Equivalent Units of Production—FIFO

GenX used its direct materials, direct labor, and overhead both to make complete units of Profen and to start some units that are not yet complete. We need to convert the physical measure of units to equivalent units based on how much of each input has been used. We do this by multiplying the number of physical units by the percentage of processing applied to those units in the current period; this is done for each input (materials, labor, and overhead). The FIFO method accounts for cost flow in a sequential manner—earliest costs are the first to flow out. (This is different from the weighted-average method, which combines prior period costs—those in beginning Goods in Process Inventory—with costs incurred in the current period.)

Three distinct groups of units must be considered in determining the equivalent units of production under the FIFO method: (a) units in beginning Goods in Process Inventory that were completed this period, (b) units started *and* completed this period, and (c) units in ending Goods in Process Inventory. We must determine how much material, labor, and overhead are used for each of these unit groups. These computations are shown in Exhibit 20A.3. The remainder of this section explains these computations.

EXHIBIT 20A.3

Equivalent Units of Production—FIFO

Equivalent Units of Production	Direct Materials	Direct Labor	Factory Overhead
(a) Equivalent units to complete beginning goods in process			
Direct materials (30,000 × 0%)	0 EUP		
Direct labor (30,000 × 35%)		10,500 EUP	
Factory overhead (30,000 × 35%)			10,500 EUP
(b) Equivalent units started and completed*	70,000	70,000	70,000
(c) Equivalent units in ending goods in process			
Direct materials (20,000 × 100%)	20,000		
Direct labor (20,000 × 25%)		5,000	
Factory overhead (20,000 × 25%)			5,000
Equivalent units of production	90,000 EUP	85,500 EUP	85,500 EUP

*Units completed this period	100,000 units
Less units in beginning goods in process	30,000
Units started and completed this period	70,000 units

(a) Beginning Goods in Process Under FIFO, we assume that production first completes any units started in the prior period. There were 30,000 physical units in beginning goods in process inventory. Those units were 100% complete with respect to direct materials as of the end of the prior period. This means that no materials (0%) are needed in April to complete those 30,000 units. So the equivalent units of *materials* to complete beginning goods in process are zero (30,000 × 0%)—see first row under row "(a)" in Exhibit 20A.3. The units in process as of April 1 had already been through 65% of production prior to this period and need only go through the remaining 35% of production. The equivalent units of *labor* to complete the beginning goods in process are 10,500 (30,000 × 35%)—

see the second row under row "(a)." This implies that the amount of labor required this period to complete the 30,000 units started in the prior period is the amount of labor needed to make 10,500 units, start-to-finish. Finally, overhead is applied based on direct labor costs, so GenX computes equivalent units for overhead as it would for direct labor.

(b) Units Started and Completed This Period After completing any beginning goods in process, FIFO assumes that production begins on newly started units. GenX began work on 90,000 new units this period. Of those units, 20,000 remain incomplete at period-end. This means that 70,000 of the units started in April were completed in April. These complete units have received 100% of materials, labor, and overhead. Exhibit 20A.3 reflects this by including 70,000 equivalent units (70,000 × 100%) of materials, labor, and overhead in its equivalent units of production—see row "(b)."

(c) Ending Goods in Process The 20,000 units started in April that GenX was not able to complete by period-end consumed materials, labor, and overhead. Specifically, those 20,000 units received 100% of materials and, therefore, the equivalent units of materials in ending goods in process inventory are 20,000 (20,000 × 100%)—see the first row under row "(c)." For labor and overhead, the units in ending goods in process were 25% complete in production. This means the equivalent units of labor and overhead for those units are 5,000 (20,000 × 25%) as GenX incurs labor and overhead costs uniformly throughout its production process. Finally, for each input (direct materials, direct labor, and factory overhead), the equivalent units for each of the unit groups (a), (b), and (c) are added to determine the total equivalent units of production with respect to each—see the final row in Exhibit 20A.3.

Step 3: Compute Cost per Equivalent Unit—FIFO

To compute cost per equivalent unit, we take the product costs (for each of direct materials, direct labor, and factory overhead from Exhibit 20A.1) added in April and divide by the equivalent units of production from step 2. Exhibit 20A.4 illustrates these computations.

EXHIBIT 20A.4

Cost per Equivalent Unit of Production—FIFO

Cost per Equivalent Unit of Production	Direct Materials	Direct Labor	Factory Overhead
Costs incurred this period	$9,900	$5,700	$6,840
÷ Equivalent units of production (from Step 2)	90,000 EUP	85,500 EUP	85,500 EUP
Cost per equivalent unit of production	$0.11 per EUP	$0.067 per EUP	$0.08 per EUP

It is essential to compute costs per equivalent unit for *each* input because production inputs are added at different times in the process. The FIFO method computes the cost per equivalent unit based solely on this period's EUP and costs (unlike the weighted-average method, which adds in the costs of the beginning goods in process inventory).

Step 4: Assign and Reconcile Costs

The equivalent units determined in step 2 and the cost per equivalent unit computed in step 3 are both used to assign costs (1) to units that the production department completed and transferred to finished goods and (2) to units that remain in process at period-end.

In Exhibit 20A.5, under the section for cost of units transferred out, we see that the cost of units completed in April includes the $4,620 cost carried over from March for work already applied to the 30,000 units that make up beginning Goods in Process Inventory, plus the $1,544 incurred in April to complete those units. This section also includes the $17,990 of cost assigned to the 70,000 units started and completed this period. Thus, the total cost of goods manufactured in April is $24,154 ($4,620 + $1,544 + $17,990). The average cost per unit for goods completed in April is $0.242 ($24,154 ÷ 100,000 completed units).

The computation for cost of ending goods in process inventory is in the lower part of Exhibit 20A.5. The cost of units in process includes materials, labor, and overhead costs corresponding to the percentage of these resources applied to those incomplete units in April. That cost of $2,935 ($2,200 + $335 + $400) also is the ending balance for the Goods in Process Inventory account.

EXHIBIT 20A.5

Report of Costs Accounted For—FIFO

Cost of units transferred out (cost of goods manufactured)		
Cost of beginning goods in process inventory		$ 4,620
Cost to complete beginning goods in process		
Direct materials ($0.11 per EUP × 0 EUP)	$ 0	
Direct labor ($0.067 per EUP × 10,500 EUP)	704	
Factory overhead ($0.08 per EUP × 10,500 EUP)	840	1,544
Cost of units started and completed this period		
Direct materials ($0.11 per EUP × 70,000 EUP)	7,700	
Direct labor ($0.067 per EUP × 70,000 EUP)	4,690	
Factory overhead ($0.08 per EUP × 70,000 EUP)	5,600	17,990
Total cost of units finished this period		24,154
Cost of ending goods in process inventory		
Direct materials ($0.11 per EUP × 20,000 EUP)	2,200	
Direct labor ($0.067 per EUP × 5,000 EUP)	335	
Factory overhead ($0.08 per EUP × 5,000 EUP)	400	
Total cost of ending goods in process inventory		2,935
Total costs accounted for		**$27,089**

Management verifies that the total costs assigned to units transferred out and units still in process equal the total costs incurred by production. We reconcile the costs accounted for (in Exhibit 20A.5) to the costs that production was charged for as shown in Exhibit 20A.6.

EXHIBIT 20A.6

Report of Costs to Account For—FIFO

Cost of beginning goods in process inventory		
Direct materials	$3,300	
Direct labor	600	
Factory overhead	720	$ 4,620
Costs incurred this period		
Direct materials	9,900	
Direct labor	5,700	
Factory overhead	6,840	22,440
Total costs to account for		**$27,060**

The production manager is responsible for $27,060 in costs: $4,620 that had been assigned to the department's Goods in Process Inventory as of April 1 plus $22,440 of materials, labor, and overhead costs the department incurred in April. At period-end, the manager must identify where those costs were assigned. The production manager can report that $24,154 of cost was assigned to units completed in April and $2,935 was assigned to units still in process at period-end. The sum of these amounts is $29 different from the $27,060 total costs incurred by production due to rounding in step 3—rounding errors are common and not a concern.

The final report is the process cost summary, which summarizes key information from Exhibits 20A.3, 20A.4, 20A.5, and 20A.6. Reasons for the summary are to (1) help managers control and monitor costs, (2) help upper management assess department manager performance, and (3) provide cost information for financial reporting. The process cost summary, using FIFO, for GenX is in Exhibit 20A.7. Section ① lists the total costs charged to the department, including direct materials, direct labor, and overhead costs incurred, as well as the cost of the beginning goods in process inventory. Section ② describes the equivalent units of production for the department. Equivalent units for materials, labor, and overhead are in separate columns. It also reports direct materials, direct labor, and overhead costs per equivalent unit. Section ③ allocates total costs among units worked on in the period.

Decision Maker

Cost Manager As cost manager for an electronics manufacturer, you apply a process costing system using FIFO. Your company plans to adopt a just-in-time system and eliminate inventories. What is the impact of the use of FIFO (versus the weighted-average method) given these plans? [Answer—p. 833]

EXHIBIT 20A.7

Process Cost Summary

GenX COMPANY
Process Cost Summary
For Month Ended April 30, 2009

①	**Costs charged to production**		
	Costs of beginning goods in process inventory		
	Direct materials	$3,300	
	Direct labor	600	
	Factory overhead	720	$ 4,620
	Costs incurred this period		
	Direct materials	9,900	
	Direct labor	5,700	
	Factory overhead	6,840	22,440
	Total costs to account for		$27,060

②	**Unit cost information**			
	Units to account for		Units accounted for	
	Beginning goods in process	30,000	Transferred out	100,000
	Units started this period	90,000	Ending goods in process	20,000
	Total units to account for	120,000	Total units accounted for	120,000

Equivalent units of production	**Direct Materials**	**Direct Labor**	**Factory Overhead**
Equivalent units to complete beginning goods in process			
Direct materials (30,000 × 0%)	0 EUP		
Direct labor (30,000 × 35%)		10,500 EUP	
Factory overhead (30,000 × 35%)			10,500 EUP
Equivalent units started and completed	70,000	70,000	70,000
Equivalent units in ending goods in process			
Direct materials (20,000 × 100%)	20,000		
Direct labor (20,000 × 25%)		5,000	
Factory overhead (20,000 × 25%)			5000
Equivalent units of production	90,000 EUP	85,500 EUP	85,500 EUP

Cost per equivalent unit of production	**Direct Materials**	**Direct Labor**	**Factory Overhead**
Costs incurred this period	$9,900	$5,700	$6,840
÷ Equivalent units of production	90,000 EUP	85,500 EUP	85,500 EUP
Cost per equivalent unit of production	$0.11 per EUP	$0.067 per EUP	$0.08 per EUP

③	**Cost assignment and reconciliation (cost of units completed and transferred out)**		
	Cost of beginning goods in process		$ 4,620
	Cost to complete beginning goods in process		
	Direct materials ($0.11 per EUP × 0 EUP)	$ 0	
	Direct labor ($0.067 per EUP × 10,500 EUP)	704	
	Factory overhead ($0.08 per EUP × 10,500 EUP)	840	1,544
	Cost of units started and completed this period		
	Direct materials ($0.11 per EUP × 70,000 EUP)	7,700	
	Direct labor ($0.067 per EUP × 70,000 EUP)	4,690	
	Factory overhead ($0.08 per EUP × 70,000 EUP)	5,600	17,990
	Total cost of units finished this period		24,154
	Cost of ending goods in process		
	Direct materials ($0.11 per EUP × 20,000 EUP)	2,200	
	Direct labor ($0.067 per EUP × 5,000 EUP)	335	
	Factory overhead ($0.08 per EUP × 5,000 EUP)	400	
	Total cost of ending goods in process		2,935
	Total costs accounted for		$27,089*

(Total costs to account for $27,060 and total costs accounted for $27,089: reconciled)

*$29 difference due to rounding

Summary

C1 Explain process operations and the way they differ from job order operations. Process operations produce large quantities of similar products or services by passing them through a series of processes, or steps, in production. Like job order operations, they combine direct materials, direct labor, and overhead in the operations. Unlike job order operations that assign the responsibility for each job to a manager, process operations assign the responsibility for each *process* to a manager.

C2 Define equivalent units and explain their use in process cost accounting. Equivalent units of production measure the activity of a process as the number of units that would be completed in a period if all effort had been applied to units that were started and finished. This measure of production activity is used to compute the cost per equivalent unit and to assign costs to finished goods and goods in process inventory.

C3 Explain the four steps in accounting for production activity in a period. The four steps involved in accounting for production activity in a period are (1) recording the physical flow of units, (2) computing the equivalent units of production, (3) computing the cost per equivalent unit of production, and (4) reconciling costs. The last step involves assigning costs to finished goods and goods in process inventory for the period.

C4 Define a process cost summary and describe its purposes. A process cost summary reports on the activities of a production process or department for a period. It describes the costs charged to the department, the equivalent units of production for the department, and the costs assigned to the output. The report aims to (1) help managers control their departments, (2) help factory managers evaluate department managers' performances, and (3) provide cost information for financial statements.

C5 Explain and illustrate the four steps in accounting for production activity using FIFO. The FIFO method for process costing is applied and illustrated to (1) report the physical flow of units, (2) compute the equivalent units of production, (3) compute the cost per equivalent unit of production, and (4) assign and reconcile costs.

A1 Compare process cost accounting and job order cost accounting. Process and job order manufacturing operations are similar in that both combine materials, labor, and factory overhead to produce products or services. They differ in the way they are organized and managed. In job order operations, the job order cost accounting system assigns materials, labor, and overhead to specific jobs. In process operations, the process cost accounting system assigns materials, labor, and overhead to specific processes. The total costs associated with each process are then divided by the number of units passing through that process to get cost per equivalent unit. The costs per equivalent unit for all processes are added to determine the total cost per unit of a product or service.

A2 Explain and illustrate a hybrid costing system. A hybrid costing system contains features of both job order and process costing systems. Generally, certain direct materials are accounted for by individual products as in job order costing, but direct labor and overhead costs are accounted for similar to process costing.

P1 Record the flow of direct materials costs in process cost accounting. Materials purchased are debited to a Raw Materials Inventory account. As direct materials are issued to processes, they are separately accumulated in a Goods in Process Inventory account for that process.

P2 Record the flow of direct labor costs in process cost accounting. Direct labor costs are initially debited to the Factory Payroll account. The total amount in it is then assigned to the Goods in Process Inventory account pertaining to each process.

P3 Record the flow of factory overhead costs in process cost accounting. The different factory overhead items are first accumulated in the Factory Overhead account and are then allocated, using a predetermined overhead rate, to the different processes. The allocated amount is debited to the Goods in Process Inventory account pertaining to each process.

P4 Compute equivalent units produced in a period. To compute equivalent units, determine the number of units that would have been finished if all materials (or labor or overhead) had been used to produce units that were started and completed during the period. The costs incurred by a process are divided by its equivalent units to yield cost per unit.

P5 Prepare a process cost summary. A process cost summary includes the physical flow of units, equivalent units of production, costs per equivalent unit, and a cost reconciliation. It reports the units and costs to account for during the period and how they were accounted for during the period. In terms of units, the summary includes the beginning goods in process inventory and the units started during the month. These units are accounted for in terms of the goods completed and transferred out, and the ending goods in process inventory. With respect to costs, the summary includes materials, labor, and overhead costs assigned to the process during the period. It shows how these costs are assigned to goods completed and transferred out, and to ending goods in process inventory.

P6 Record the transfer of completed goods to Finished Goods Inventory and Cost of Goods Sold. As units complete the final process and are eventually sold, their accumulated cost is transferred to Finished Goods Inventory and finally to Cost of Goods Sold.

Guidance Answers to **Decision Maker** and **Decision Ethics**

Budget Officer By instructing you to classify a majority of costs as indirect, the manager is passing some of his department's costs to a common overhead pool that other departments will partially absorb. Since overhead costs are allocated on the basis of direct labor for this company and the new department has a relatively low direct labor cost, the new department will be assigned less overhead. Such action suggests unethical behavior by this manager. You must object to such reclassification. If this manager refuses to comply, you must inform someone in a more senior position.

Entrepreneur By spreading the added quality-related costs across three customers, the entrepreneur is probably trying to remain competitive with respect to the customer that demands the 100% quality inspection. Moreover, the entrepreneur is partly covering the added costs by recovering two-thirds of them from the other two customers who are paying 110% of total costs. This act likely breaches the trust placed by the two customers in this entrepreneur's application of its costing system. The costing system should be changed, and the entrepreneur should consider renegotiating the pricing and/or quality test agreement with this one customer (at the risk of losing this currently loss-producing customer).

Cost Manager Differences between the FIFO and weighted-average methods are greatest when large work in process inventories exist and when costs fluctuate. The method used if inventories are eliminated does not matter; both produce identical costs.

Guidance Answers to **Quick Checks**

1. *c*
2. When a company produces large quantities of similar products/services, a process cost system is often more suitable.
3. *b*
4. The costs are direct materials, direct labor, and overhead.
5. A goods in process inventory account is needed for *each* production department.
6. *a*
7. Equivalent units with respect to direct labor are the number of units that would have been produced if all labor had been used on units that were started and finished during the period.
8.

Units completed and transferred out	58,000 EUP
Units of ending goods in process	
Direct labor (6,000 × 1/3)	2,000 EUP
Units of production	60,000 EUP

9. The first section shows the costs charged to the department. The second section describes the equivalent units produced by the department. The third section shows the assignment of total costs to units worked on during the period.

Key Terms

mhhe.com/wildFAP19e

Key Terms are available at the book's Website for learning and testing in an online Flashcard Format.

Cost of goods manufactured (p. 821)
Equivalent units of production (EUP) (p. 815)
FIFO method (p. 827)
Job order cost accounting system (p. 811)
Materials consumption report (p. 812)
Process cost accounting system (p. 811)
Process cost summary (p. 820)
Process operations (p. 808)
Weighted-average method (p. 818)

Multiple Choice Quiz

Answers on p. 849

mhhe.com/wildFAP19e

Additional Quiz Questions are available at the book's Website.

Quiz20

1. Equivalent units of production are equal to
   a. Physical units that were completed this period from all effort being applied to them.
   b. The number of units introduced into the process this period.
   c. The number of finished units actually completed this period.
   d. The number of units that could have been started and completed given the cost incurred.
   e. The number of units in the process at the end of the period.
2. Recording the cost of raw materials purchased for use in a process costing system includes a
   a. Credit to Raw Materials Inventory.
   b. Debit to Goods in Process Inventory.
   c. Debit to Factory Overhead.
   d. Credit to Factory Overhead.
   e. Debit to Raw Materials Inventory.
3. The production department started the month with a beginning goods in process inventory of $20,000. During the month, it was assigned the following costs: direct materials, $152,000; direct labor, $45,000; overhead applied at the rate of 40% of direct labor cost. Inventory with a cost of $218,000 was transferred to finished goods. The ending balance of goods in process inventory is
   a. $330,000.
   b. $ 17,000.
   c. $220,000.
   d. $112,000.
   e. $118,000.
4. A company's beginning work in process inventory consists of 10,000 units that are 20% complete with respect to direct labor costs. A total of 40,000 units are completed this period. There

are 15,000 units in goods in process, one-third complete for direct labor, at period-end. The equivalent units of production (EUP) with respect to direct labor at period-end, assuming the weighted average method, are

a. 45,000 EUP.
b. 40,000 EUP.
c. 5,000 EUP.
d. 37,000 EUP.
e. 43,000 EUP.

5. Assume the same information as in question 4. Also assume that beginning work in process had $6,000 in direct labor cost and that $84,000 in direct labor is added during this period. What is the cost per EUP for labor?

a. $0.50 per EUP
b. $1.87 per EUP
c. $2.00 per EUP
d. $2.10 per EUP
e. $2.25 per EUP

Assume the weighted-average inventory method is used for all assignments unless stated differently. Superscript letter [A] denotes assignments based on Appendix 20A.

Discussion Questions

1. Can services be delivered by means of process operations? Support your answer with an example.

2. What is the main factor for a company in choosing between the job order costing and process costing accounting systems? Give two likely applications of each system.

3. Identify the control document for materials flow when a materials requisition slip is not used.

4. The focus in a job order costing system is the job or batch. Identify the main focus in process costing.

5. Are the journal entries that match cost flows to product flows in process costing primarily the same or much different than those in job order costing? Explain.

6. Explain in simple terms the notion of equivalent units of production (EUP). Why is it necessary to use EUP in process costing?

7. What are the two main inventory methods used in process costing? What are the differences between these methods?

8. Why is it possible for direct labor in process operations to include the labor of employees who do not work directly on products or services?

9. Assume that a company produces a single product by processing it first through a single production department. Direct labor costs flow through what accounts in this company's process cost system?

10. After all labor costs for a period are allocated, what balance should remain in the Factory Payroll account?

11. Is it possible to have under- or overapplied overhead costs in a process cost accounting system? Explain.

12. Explain why equivalent units of production for both direct labor and overhead can be the same as, and why they can be different from, equivalent units for direct materials.

13. List the four steps in accounting for production activity in a reporting period (for process operations).

14. What purposes does a process cost summary serve?

15. Are there situations where **Best Buy** can use process costing? Identify at least one and explain it.

16. **Apple** produces iMacs with a multiple production line. Identify and list some of its production processing steps and departments.

Denotes Discussion Questions that involve decision making.

Available with McGraw-Hill's Homework Manager — McGraw-Hill's HOMEWORK MANAGER®

QUICK STUDY

QS 20-1
Matching of product to cost accounting system
C1

For each of the following products and services, indicate whether it is most likely produced in a process operation or in a job order operation.

1. Door hinges
2. Cut flower arrangements
3. House paints
4. Concrete swimming pools
5. Custom tailored suits
6. Grand pianos
7. Wall clocks
8. Sport shirts
9. Bolts and nuts
10. Folding chairs
11. Headphones
12. Designed boathouse

QS 20-2
Recording costs of direct materials
P1

Industrial Boxes makes cardboard shipping cartons in a single operation. This period, Industrial purchased $124,000 in raw materials. Its production department requisitioned $100,000 of those materials for use in producing cartons. Prepare journal entries to record its (1) purchase of raw materials and (2) requisition of direct materials.

QS 20-3
Recording costs of direct labor
P2

Refer to the information in QS 20-2. Industrial Boxes incurred $270,000 in factory payroll costs, of which $250,000 was direct labor. Prepare journal entries to record its (1) total factory payroll incurred and (2) direct labor used in production.

QS 20-4
Recording costs of factory overhead
P3

Refer to the information in QS 20-2 and QS 20-3. Industrial Boxes requisitioned $18,000 of indirect materials from its raw materials and used $20,000 of indirect labor in its production of boxes. Also, it incurred $312,000 of other factory overhead costs. It applies factory overhead at the rate of 135% of direct labor costs. Prepare journal entries to record its (1) indirect materials requisitioned, (2) indirect labor used in production, (3) other factory overhead costs incurred, and (4) application of overhead to production.

QS 20-5
Recording transfer of costs to finished goods P6

Refer to the information in QS 20-2, QS 20-3, and QS 20-4. Industrial Boxes completed 40,000 boxes costing $550,000 and transferred them to finished goods. Prepare its journal entry to record the transfer of the boxes from production to finished goods inventory.

QS 20-6
Computing equivalent units of production
P4

The following refers to units processed in Sunflower Printing's binding department in March. Compute the total equivalent units of production with respect to labor for March using the weighted-average inventory method.

	Units of Product	Percent of Labor Added
Beginning goods in process	75,000	85%
Goods started	155,000	100
Goods completed	170,000	100
Ending goods in process	60,000	25

QS 20-7
Computing EUP cost C4 P5

The cost of beginning inventory plus the costs added during the period should equal the cost of units ____________ plus the cost of ____________.

QS 20-8
Hybrid costing system A2

Explain a hybrid costing system. Identify a product or service operation that might well fit a hybrid costing system.

QS 20-9A
Computing equivalent units—FIFO C2 C5 P4

Refer to QS 20-6 and compute the total equivalent units of production with respect to labor for March using the FIFO inventory method.

McGraw-Hill's HOMEWORK MANAGER® Available with McGraw-Hill's Homework Manager

EXERCISES

Exercise 20-1
Terminology in process cost accounting
C1 A1 P1 P2 P3

Match each of the following items A through G with the best numbered description of its purpose.

A. Raw Materials Inventory account
B. Materials requisition
C. Finished Goods Inventory account
D. Factory Overhead account
E. Process cost summary
F. Equivalent units of production
G. Goods in Process Inventory

_______ **1.** Notifies the materials manager to send materials to a production department.
_______ **2.** Holds costs of indirect materials, indirect labor, and similar costs until assigned to production.
_______ **3.** Holds costs of direct materials, direct labor, and applied overhead until products are transferred from production to finished goods (or another department).
_______ **4.** Standardizes partially completed units into equivalent completed units.
_______ **5.** Holds costs of finished products until sold to customers.
_______ **6.** Describes the activity and output of a production department for a period.
_______ **7.** Holds costs of materials until they are used in production or as factory overhead.

Exercise 20-2
Journal entries in process cost accounting
P1 P2 P3

Festive Toy Company manufactures toy trucks. Prepare journal entries to record its following production activities for January.

1. Purchased $40,000 of raw materials on credit.
2. Used $17,000 of direct materials in production.
3. Used $20,500 of indirect materials.

4. Incurred total labor cost of $77,000, which is paid in cash.
5. Used $58,000 of direct labor in production.
6. Used $19,000 of indirect labor.
7. Incurred overhead costs of $22,000 (paid in cash).
8. Applied overhead at 90% of direct labor costs.
9. Transferred completed products with a cost of $137,000 to finished goods inventory.
10. Sold $450,000 of products on credit. Their cost is $150,000.

Check (8) Cr. Factory Overhead, $52,200

Exercise 20-3
Recording cost flows in a process cost system
P1 P2 P3 P6

Seattle Lumber produces bagged bark for use in landscaping. Production involves packaging bark chips in plastic bags in a bagging department. The following information describes production operations for October.

		Bagging Department
3	Direct materials used	$ 460,000
4	Direct labor used	$ 76,000
5	Predetermined overhead rate (based on direct labor)	180%
6	Goods transferred from bagging to finished goods	$(407,000)

The company's revenue for the month totaled $900,000 from credit sales, and its cost of goods sold for the month is $500,000. Prepare summary journal entries dated October 31 to record its October production activities for (1) direct material usage, (2) direct labor usage, (3) overhead allocation, (4) goods transfer from production to finished goods, and (5) sales.

Check (3) Cr. Factory Overhead, $136,800

Exercise 20-4
Interpretation of journal entries in process cost accounting
P1 P2 P3 P6

The following journal entries are recorded in Lewis Co.'s process cost accounting system. Lewis produces apparel and accessories. Overhead is applied to production based on direct labor cost for the period. Prepare a brief explanation (including any overhead rates applied) for each journal entry *a* through *j*.

	Account	Debit	Credit
a.	Raw Materials Inventory	52,000	
	Accounts Payable		52,000
b.	Goods in Process Inventory	42,000	
	Raw Materials Inventory		42,000
c.	Goods in Process Inventory	26,000	
	Factory Payroll		26,000
d.	Factory Payroll	32,000	
	Cash		32,000
e.	Factory Overhead	10,000	
	Cash		10,000
f.	Factory Overhead	10,000	
	Raw Materials Inventory		10,000
g.	Factory Overhead	6,000	
	Factory Payroll		6,000
h.	Goods in Process Inventory	32,500	
	Factory Overhead		32,500
i.	Finished Goods Inventory	88,000	
	Goods in Process Inventory		88,000
j.	Accounts Receivable	250,000	
	Sales		250,000
	Cost of Goods Sold	100,000	
	Finished Goods Inventory		100,000

Exercise 20-5
Computing equivalent units of production—weighted average
C2 P4

During April, the production department of a process manufacturing system completed a number of units of a product and transferred them to finished goods. Of these transferred units, 30,000 were in process in the production department at the beginning of April and 120,000 were started and completed in April. April's beginning inventory units were 60% complete with respect to materials and 40% complete with respect to labor. At the end of April, 41,000 additional units were in process in the production department and were 80% complete with respect to materials and 30% complete with respect to labor.

1. Compute the number of units transferred to finished goods.

2. Compute the number of equivalent units with respect to both materials used and labor used in the production department for April using the weighted-average method.

Check (2) EUP for materials, 182,800

Exercise 20-6
Costs assigned to output and inventories—weighted average
C3 P4 P5 

The production department described in Exercise 20-5 had \$425,184 of direct materials and \$326,151 of direct labor cost charged to it during April. Also, its beginning inventory included \$59,236 of direct materials cost and \$22,794 of direct labor.

1. Compute the direct materials cost and the direct labor cost per equivalent unit for the department.

2. Using the weighted-average method, assign April's costs to the department's output—specifically, its units transferred to finished goods and its ending goods in process inventory.

Check (2) Costs accounted for, \$833,365

Exercise 20-7[A]
Computing equivalent units of production—FIFO
C5 P4

Refer to the information in Exercise 20-5 to compute the number of equivalent units with respect to both materials used and labor used in the production department for April using the FIFO method.

Exercise 20-8[A]
Costs assigned to output—FIFO
C5 P4 P5

Refer to the information in Exercise 20-6 and complete its parts (1) and (2) using the FIFO method.

Exercise 20-9
Equivalent units computed—weighted average
C2 P4 P5 

The production department in a process manufacturing system completed 383,000 units of product and transferred them to finished goods during a recent period. Of these units, 63,000 were in process at the beginning of the period. The other 320,000 units were started and completed during the period. At period-end, 59,000 units were in process. Compute the department's equivalent units of production with respect to direct materials under each of three separate assumptions:

1. All direct materials are added to products when processing begins.

2. Direct materials are added to products evenly throughout the process. Beginning goods in process inventory was 40% complete, and ending goods in process inventory was 75% complete.

3. One-half of direct materials is added to products when the process begins and the other half is added when the process is 75% complete as to direct labor. Beginning goods in process inventory is 40% complete as to direct labor, and ending goods in process inventory is 60% complete as to direct labor.

Check (3) EUP for materials, 412,500

Exercise 20-10[A]
Equivalent units computed—FIFO
C5 P4

Refer to the information in Exercise 20-9 and complete it for each of the three separate assumptions using the FIFO method for process costing.

Check (3) EUP for materials, 381,000

Exercise 20-11
Flowchart of costs for a process operation P1 P2 P3 P6

The following flowchart shows the August production activity of the Jez Company. Use the amounts shown on the flowchart to compute the missing four numbers identified by blanks.

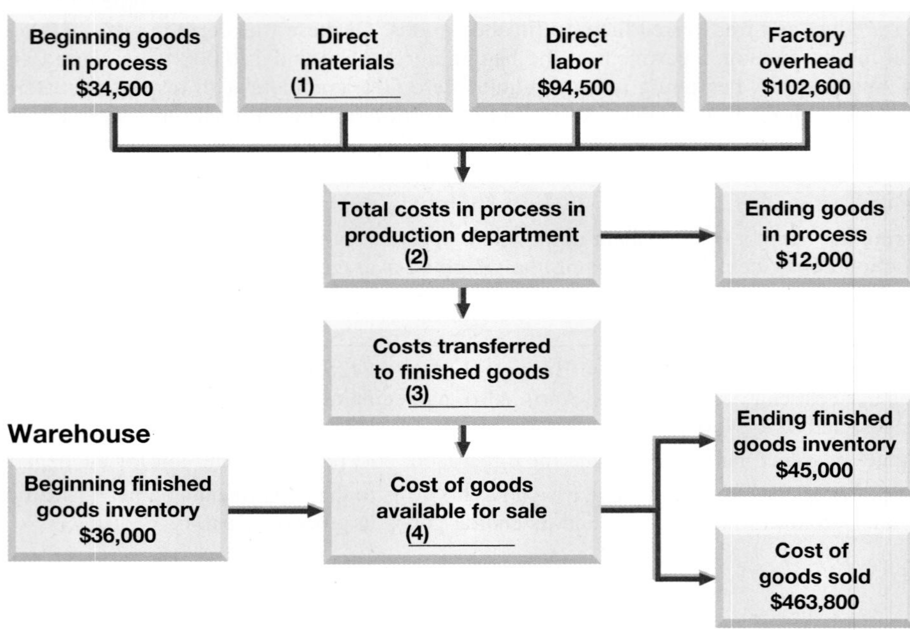

Exercise 20-12
Completing a process cost summary
P5

The following partially completed process cost summary describes the July production activities of Anton Company. Its production output is sent to its warehouse for shipping. Prepare its process cost summary using the weighted-average method.

Equivalent Units of Production	Direct Materials	Direct Labor	Factory Overhead
Units transferred out	64,000	64,000	64,000
Units of ending goods in process	5,000	3,000	3,000
Equivalent units of production	69,000	67,000	67,000
Costs per EUP	**Direct Materials**	**Direct Labor**	**Factory Overhead**
Costs of beginning goods in process	$ 37,100	$ 1,520	$ 3,040
Costs incurred this period	715,000	125,780	251,560
Total costs	$752,100	$127,300	$254,600
Units in beginning goods in process			4,000
Units started this period			65,000
Units completed and transferred out			64,000
Units in ending goods in process			5,000

Exercise 20-13
Process costing—weighted average
P1 P2 P6

Nu-Test Company uses the weighted-average method of process costing to assign production costs to its products. Information for September follows. Assume that all materials are added at the beginning of its production process, and that direct labor and factory overhead are added uniformly throughout the process.

Goods in process inventory, September 1 (4,000 units, 100% complete with respect to direct materials, 80% complete with respect to direct labor and overhead; includes $90,000 of direct material cost, $51,200 in direct labor cost, $61,440 overhead cost)	$202,640
Units started in September	56,000
Units completed and transferred to finished goods inventory	46,000
Goods in process inventory, September 30 (? units, 100% complete with respect to direct materials, 40% complete with respect to direct labor and overhead)	?

[continued on next page]

[continued from previous page]

Costs incurred in September	
Direct materials	$750,000
Direct labor	$310,000
Overhead applied at 120% of direct labor cost	?

Required

Fill in the blanks labeled *a* through *uu* in the following process cost summary.

NU-TEST COMPANY
Process Cost Summary
For Month Ended September 30

Costs Charged to Production		
Costs of beginning goods in process		
Direct materials	$ 90,000	
Direct labor	51,200	
Factory overhead	61,440	$202,640
Costs incurred this period		
Direct materials	$750,000	
Direct labor	310,000	
Factory overhead	(a)____	(b)____
Total costs to account for		(c)____

Check (c) $1,634,640

Unit Cost Information			
Units to account for		Units accounted for	
Beginning goods in process	4,000	Completed and transferred out	46,000
Units started this period	56,000	Ending goods in process	(d)____
Total units to account for	(e)____	Total units accounted for	(f)____

Equivalent Units of Production (EUP)				**Direct Materials**	**Direct Labor**	**Factory Overhead**
Units completed and transferred out				(g)____EUP	(h)____EUP	(i)____EUP
Units of ending goods in process						
Materials	(j)____	×	100%	(k)____EUP		
Direct labor	(l)____	×	40%		(m)____EUP	
Factory overhead	(n)____	×	40%	____	____	(o)____EUP
Equivalent units of production (EUP)				(p)____EUP	(q)____EUP	(r)____EUP

Cost per EUP	**Direct Materials**	**Direct Labor**	**Factory Overhead**
Costs of beginning goods in process	$ 90,000	$ 51,200	$61,440
Costs incurred this period	750,000	310,000	(s)____
Total costs	$840,000	$361,200	(t)____
÷ EUP	(u)____	(v)____	(w)____
Cost per EUP	(x)____	(y)____	(z)____

(z) $8.40 per EUP

Cost Assignment and Reconciliation					
Costs transferred out	Cost/EUP	×	EUP		
Direct materials	(aa)____	×	(bb)____	(cc)____	
Direct labor	(dd)____	×	(ee)____	(ff)____	
Factory overhead	(gg)____	×	(hh)____	(ii)____	
Costs of goods completed and transferred out					(jj)____
Costs of ending goods in process					
Direct materials	(kk)____	×	(ll)____	(mm)____	
Direct labor	(nn)____	×	(oo)____	(pp)____	
Factory overhead	(qq)____	×	(rr)____	(ss)____	
Costs of ending goods in process					(tt)____
Total costs accounted for					(uu)____

Available with McGraw-Hill's Homework Manager — McGraw-Hill's HOMEWORK MANAGER®

PROBLEM SET A

Problem 20-1A
Production cost flow and measurement; journal entries

P1 P2 P3 P6

Harvey Company manufactures woven blankets and accounts for product costs using process costing. The following information is available regarding its May inventories.

	Beginning Inventory	Ending Inventory
Raw materials inventory	$ 56,000	$ 51,000
Goods in process inventory	441,500	504,000
Finished goods inventory	638,000	554,000

The following additional information describes the company's production activities for May.

Raw materials purchases (on credit)	$ 270,000
Factory payroll cost (paid in cash)	1,583,000
Other overhead cost (Other Accounts credited)	86,000
Materials used	
Direct	$ 187,000
Indirect	62,000
Labor used	
Direct	$ 704,000
Indirect	879,000
Overhead rate as a percent of direct labor	110%
Sales (on credit)	$3,000,000

Required

Check (1b) Cost of goods sold $1,686,900

1. Compute the cost of (a) products transferred from production to finished goods, and (b) goods sold.

2. Prepare summary journal entries dated May 31 to record the following production activities during May: (a) raw materials purchases, (b) direct materials usage, (c) indirect materials usage, (d) payroll costs, (e) direct labor costs, (f) indirect labor costs, (g) other overhead costs, (h) overhead applied, (i) goods transferred from production to finished goods, and (j) sale of finished goods.

Problem 20-2A
Cost per equivalent unit; costs assigned to products

P4 P5

mhhe.com/wildFAP19e

Carmen Company uses weighted-average process costing to account for its production costs. Direct labor is added evenly throughout the process. Direct materials are added at the beginning of the process. During November, the company transferred 735,000 units of product to finished goods. At the end of November, the goods in process inventory consists of 207,000 units that are 90% complete with respect to labor. Beginning inventory had $244,920 of direct materials and $69,098 of direct labor cost. The direct labor cost added in November is $1,312,852, and the direct materials cost added is $1,639,080.

Required

Check (2) Direct labor cost per equivalent unit, $1.50

(3b) $693,450

1. Determine the equivalent units of production with respect to (a) direct labor and (b) direct materials.

2. Compute both the direct labor cost and the direct materials cost per equivalent unit.

3. Compute both direct labor cost and direct materials cost assigned to (a) units completed and transferred out, and (b) ending goods in process inventory.

Analysis Component

4. The company sells and ships all units to customers as soon as they are completed. Assume that an error is made in determining the percentage of completion for units in ending inventory. Instead of being 90% complete with respect to labor, they are actually 75% complete. Write a one-page memo to the plant manager describing how this error affects its November financial statements.

Problem 20-3A
Journalizing in process costing; equivalent units and costs
P1 P2 P3 P4 P6

Crystal Company produces large quantities of a standardized product. The following information is available for its production activities for March.

Raw materials		Factory overhead incurred	
Beginning inventory	$ 26,000	Indirect materials used	$ 81,500
Raw materials purchased (on credit)	255,000	Indirect labor used	50,000
Direct materials used	(172,000)	Other overhead costs	159,308
Indirect materials used	(81,500)	Total factory overhead incurred	$290,808
Ending inventory	$ 27,500		
Factory payroll		**Factory overhead applied (140% of direct labor cost)**	
Direct labor used	$207,720	Total factory overhead applied	$290,808
Indirect labor used	50,000		
Total payroll cost (paid in cash)	$257,720		

Additional information about units and costs of production activities follows.

Units		Costs		
Beginning goods in process inventory	2,200	Beginning goods in process inventory		
Started	30,000	Direct materials	$3,500	
Ending goods in process inventory	5,900	Direct labor	3,225	
		Factory overhead	4,515	$ 11,240
Status of ending goods in process inventory		Direct materials added		172,000
Materials—Percent complete	50%	Direct labor added		207,720
Labor and overhead—Percent complete	65%	Overhead applied (140% of direct labor)		290,808
		Total costs		$681,768
		Ending goods in process inventory		$ 82,128

During March, 25,000 units of finished goods are sold for $85 cash each. Cost information regarding finished goods follows.

Beginning finished goods inventory	$155,000
Cost transferred in	599,640
Cost of goods sold	(612,500)
Ending finished goods inventory	$142,140

Required

1. Prepare journal entries dated March 31 to record the following March activities: (a) purchase of raw materials, (b) direct materials usage, (c) indirect materials usage, (d) factory payroll costs, (e) direct labor costs used in production, (f) indirect labor costs, (g) other overhead costs—credit Other Accounts, (h) overhead applied, (i) goods transferred to finished goods, and (j) sale of finished goods.
2. Prepare a process cost summary report for this company, showing costs charged to production, units cost information, equivalent units of production, cost per EUP, and its cost assignment and reconciliation.

Check (2) Cost per equivalent unit: materials, $6.00; labor, $7.00; overhead, $9.80

Analysis Component

3. The company provides incentives to its department managers by paying monthly bonuses based on their success in controlling costs per equivalent unit of production. Assume that the production department underestimates the percentage of completion for units in ending inventory with the result that its equivalent units of production in ending inventory for March are understated. What impact does this error have on the March bonuses paid to the production managers? What impact, if any, does this error have on April bonuses?

Problem 20-4A
Process cost summary; equivalent units
P4 P5 P6

mhhe.com/wildFAP19e

King Co. produces its product through a single processing department. Direct materials are added at the start of production, and direct labor and overhead are added evenly throughout the process. The company uses monthly reporting periods for its weighted-average process cost accounting system. Its Goods in Process Inventory account follows after entries for direct materials, direct labor, and overhead costs for October.

Goods in Process Inventory					Acct. No. 133
Date		Explanation	Debit	Credit	Balance
Oct.	1	Balance			348,638
	31	Direct materials	104,090		452,728
	31	Direct labor	416,360		869,088
	31	Applied overhead	244,920		1,114,008

Its beginning goods in process consisted of $60,830 of direct materials, $176,820 of direct labor, and $110,988 of factory overhead. During October, the company started 140,000 units and transferred 153,000 units to finished goods. At the end of the month, the goods in process inventory consisted of 20,600 units that were 80% complete with respect to direct labor and factory overhead.

Required

1. Prepare the company's process cost summary for October using the weighted-average method.
2. Prepare the journal entry dated October 31 to transfer the cost of the completed units to finished goods inventory.

Check (1) Costs transferred to finished goods, $1,002,150

Problem 20-5A
Process cost summary; equivalent units; cost estimates
P4 P5 

Cisneros Co. manufactures a single product in one department. All direct materials are added at the beginning of the manufacturing process. Direct labor and overhead are added evenly throughout the process. The company uses monthly reporting periods for its weighted-average process cost accounting. During May, the company completed and transferred 11,100 units of product to finished goods inventory. Its 1,500 units of beginning goods in process consisted of $9,900 of direct materials, $61,650 of direct labor, and $49,320 of factory overhead. It has 1,200 units (100% complete with respect to direct materials and 80% complete with respect to direct labor and overhead) in process at month-end. After entries to record direct materials, direct labor, and overhead for May, the company's Goods in Process Inventory account follows.

Goods in Process Inventory					Acct. No. 133
Date		Explanation	Debit	Credit	Balance
May	1	Balance			120,870
	31	Direct materials	248,400		369,270
	31	Direct labor	601,650		970,920
	31	Applied overhead	481,320		1,452,240

Required

1. Prepare the company's process cost summary for May.
2. Prepare the journal entry dated May 31 to transfer the cost of completed units to finished goods inventory.

Check (1) EUP for labor and overhead, 12,060 EUP

(2) Cost transferred to finished goods, $1,332,000

Analysis Components

3. The cost accounting process depends on numerous estimates.
   a. Identify two major estimates that determine the cost per equivalent unit.
   b. In what direction might you anticipate a bias from management for each estimate in part 3a (assume that management compensation is based on maintaining low inventory amounts)? Explain your answer.

Problem 20-6A[A]
Process cost summary; equivalent units; cost estimates—FIFO
C5 P5 P6

Refer to the data in Problem 20-5A. Assume that Cisneros uses the FIFO method to account for its process costing system. The following additional information is available:

- Beginning goods in process consisted of 1,500 units that were 100% complete with respect to direct materials and 40% complete with respect to direct labor and overhead.
- Of the 11,100 units completed, 1,500 were from beginning goods in process. The remaining 9,600 were units started and completed during May.

Required

1. Prepare the company's process cost summary for May using FIFO.
2. Prepare the journal entry dated May 31 to transfer the cost of completed units to finished goods inventory.

Check (1) EUP for labor and overhead, 11,460 EUP

(2) Cost transferred to finished goods, $1,333,920

PROBLEM SET B

Problem 20-1B
Production cost flow and measurement; journal entries

P1 P2 P3 P6

Select Toys Company manufactures video game consoles and accounts for product costs using process costing. The following information is available regarding its June inventories.

	Beginning Inventory	Ending Inventory
Raw materials inventory	$36,000	$ 55,000
Goods in process inventory	78,000	125,000
Finished goods inventory	80,000	99,000

The following additional information describes the company's production activities for June.

Raw materials purchases (on credit)	$100,000
Factory payroll cost (paid in cash)	200,000
Other overhead cost (Other Accounts credited)	85,250
Materials used	
Direct	$ 60,000
Indirect	21,000
Labor used	
Direct	$175,000
Indirect	25,000
Overhead rate as a percent of direct labor	75%
Sales (on credit)	$500,000

Required

1. Compute the cost of (a) products transferred from production to finished goods, and (b) goods sold.
2. Prepare journal entries dated June 30 to record the following production activities during June: (a) raw materials purchases, (b) direct materials usage, (c) indirect materials usage, (d) payroll costs, (e) direct labor costs, (f) indirect labor costs, (g) other overhead costs, (h) overhead applied, (i) goods transferred from production to finished goods, and (j) sale of finished goods.

Check (1b) Cost of goods sold, $300,250

Problem 20-2B
Cost per equivalent unit; costs assigned to products

P4 P5

Maximus Company uses process costing to account for its production costs. Direct labor is added evenly throughout the process. Direct materials are added at the beginning of the process. During September, the production department transferred 40,000 units of product to finished goods. Beginning goods in process had $116,000 of direct materials and $172,800 of direct labor cost. At the end of September, the goods in process inventory consists of 4,000 units that are 25% complete with respect to labor. The direct materials cost added in September is $1,424,000, and direct labor cost added is $3,960,000.

Required

1. Determine the equivalent units of production with respect to (a) direct labor and (b) direct materials.
2. Compute both the direct labor cost and the direct materials cost per equivalent unit.
3. Compute both direct labor cost and direct materials cost assigned to (a) units completed and transferred out, and (b) ending goods in process inventory.

Check (2) Direct labor cost per equivalent unit, $100.80

(3b) $240,800

Analysis Component

4. The company sells and ships all units to customers as soon as they are completed. Assume that an error is made in determining the percentage of completion for units in ending inventory. Instead of being 25% complete with respect to labor, they are actually 75% complete. Write a one-page memo to the plant manager describing how this error affects its September financial statements.

Problem 20-3B
Journalizing in process costing; equivalent units and costs
P1 P2 P3 P4 P6

Fantasia Company produces large quantities of a standardized product. The following information is available for its production activities for May.

Raw materials		**Factory overhead incurred**	
Beginning inventory	$ 16,000	Indirect materials used	$20,280
Raw materials purchased (on credit)	110,560	Indirect labor used	18,160
Direct materials used	(98,560)	Other overhead costs	17,216
Indirect materials used	(20,280)	Total factory overhead incurred	$55,656
Ending inventory	$ 7,720	**Factory overhead applied (90% of direct labor cost)**	
Factory payroll		Total factory overhead applied	$55,656
Direct labor used	$ 61,840		
Indirect labor used	18,160		
Total payroll cost (paid in cash)	$ 80,000		

Additional information about units and costs of production activities follows.

Units		**Costs**		
Beginning goods in process inventory	8,000	Beginning goods in process inventory		
Started	24,000	Direct materials	$2,240	
Ending goods in process inventory	6,000	Direct labor	1,410	
		Factory overhead	1,269	$ 4,919
Status of ending goods in process inventory		Direct materials added		98,560
Materials—Percent complete	100%	Direct labor added		61,840
Labor and overhead—Percent complete	25%	Overhead applied (90% of direct labor)		55,656
		Total costs		$220,975
		Ending goods in process inventory		$ 25,455

During May, 30,000 units of finished goods are sold for $30 cash each. Cost information regarding finished goods follows.

Beginning finished goods inventory	$ 74,200
Cost transferred in from production	195,520
Cost of goods sold	(225,000)
Ending finished goods inventory	$ 44,720

Required

1. Prepare journal entries dated May 31 to record the following May activities: (a) purchase of raw materials, (b) direct materials usage, (c) indirect materials usage, (d) factory payroll costs, (e) direct labor costs used in production, (f) indirect labor costs, (g) other overhead costs—credit Other Accounts, (h) overhead applied, (i) goods transferred to finished goods, and (j) sale of finished goods.
2. Prepare a process cost summary report for this company, showing costs charged to production, unit cost information, equivalent units of production, cost per EUP, and its cost assignment and reconciliation.

Check (2) Cost per equivalent unit: materials, $3.15; labor, $2.30; overhead, $2.07

Analysis Component

3. This company provides incentives to its department managers by paying monthly bonuses based on their success in controlling costs per equivalent unit of production. Assume that production overestimates the percentage of completion for units in ending inventory with the result that its equivalent units of production in ending inventory for May are overstated. What impact does this error have on bonuses paid to the managers of the production department? What impact, if any, does this error have on these managers' June bonuses?

Problem 20-4B
Process cost summary; equivalent units
P4 P5 P6

Paloma Company produces its product through a single processing department. Direct materials are added at the beginning of the process. Direct labor and overhead are added to the product evenly throughout the process. The company uses monthly reporting periods for its weighted-average process cost accounting. Its Goods in Process Inventory account follows after entries for direct materials, direct labor, and overhead costs for November.

Goods in Process Inventory					Acct. No. 133
Date		Explanation	Debit	Credit	Balance
Nov.	1	Balance			10,650
	30	Direct materials	58,200		68,850
	30	Direct labor	213,400		282,250
	30	Applied overhead	320,100		602,350

The 3,750 units of beginning goods in process consisted of $3,400 of direct materials, $2,900 of direct labor, and $4,350 of factory overhead. During November, the company finished and transferred 50,000 units of its product to finished goods. At the end of the month, the goods in process inventory consisted of 6,000 units that were 100% complete with respect to direct materials and 25% complete with respect to direct labor and factory overhead.

Required

1. Prepare the company's process cost summary for November using the weighted-average method.

2. Prepare the journal entry dated November 30 to transfer the cost of the completed units to finished goods inventory.

Check (1) Cost transferred to finished goods, $580,000

Problem 20-5B
Process cost summary; equivalent units; cost estimates
P4 P5

Foster Co. manufactures a single product in one department. Direct labor and overhead are added evenly throughout the process. Direct materials are added as needed. The company uses monthly reporting periods for its weighted-average process cost accounting. During January, Foster completed and transferred 220,000 units of product to finished goods inventory. Its 10,000 units of beginning goods in process consisted of $8,400 of direct materials, $13,960 of direct labor, and $34,900 of factory overhead. 40,000 units (50% complete with respect to direct materials and 30% complete with respect to direct labor and overhead) are in process at month-end. After entries for direct materials, direct labor, and overhead for January, the company's Goods in Process Inventory account follows.

Goods in Process Inventory					Acct. No. 133
Date		Explanation	Debit	Credit	Balance
Jan.	1	Balance			57,260
	31	Direct materials	111,600		168,860
	31	Direct labor	176,280		345,140
	31	Applied overhead	440,700		785,840

Required

1. Prepare the company's process cost summary for January.

2. Prepare the journal entry dated January 31 to transfer the cost of completed units to finished goods inventory.

Check (1) EUP for labor and overhead, 232,000
(2) Cost transferred to finished goods, $741,400

Analysis Components

3. The cost accounting process depends on several estimates.

a. Identify two major estimates that affect the cost per equivalent unit.

b. In what direction might you anticipate a bias from management for each estimate in part 3a (assume that management compensation is based on maintaining low inventory amounts)? Explain your answer.

Problem 20-6B^A
Process cost summary; equivalent units; cost estimates—FIFO
C5 P5 P6

Refer to the information in Problem 20-5B. Assume that Foster uses the FIFO method to account for its process costing system. The following additional information is available.

- Beginning goods in process consists of 10,000 units that were 75% complete with respect to direct materials and 60% complete with respect to direct labor and overhead.
- Of the 220,000 units completed, 10,000 were from beginning goods in process; the remaining 210,000 were units started and completed during January.

Check (1) Labor and overhead EUP, 226,000
(2) Cost transferred, $743,480

Required

1. Prepare the company's process cost summary for January using FIFO. Round cost per EUP to one-tenth of a cent.
2. Prepare the journal entry dated January 31 to transfer the cost of completed units to finished goods inventory.

SERIAL PROBLEM

Success Systems
C1 A1

(This serial problem began in Chapter 1 and continues through most of the book. If previous chapter segments were not completed, the serial problem can begin at this point.)

SP 20 The computer workstation furniture manufacturing that Adriana Lopez started is progressing well. At this point, Adriana is using a job order costing system to account for the production costs of this product line. Adriana has heard about process costing and is wondering whether process costing might be a better method for her to keep track of and monitor her production costs.

Required

1. What are the features that distinguish job order costing from process costing?
2. Do you believe that Adriana should continue to use job order costing or switch to process costing for her workstation furniture manufacturing? Explain.

COMPREHENSIVE PROBLEM

Major League Bat Company
(Review of Chapters 2, 5, 18, 20)

CP 20 Major League Bat Company manufactures baseball bats. In addition to its goods in process inventories, the company maintains inventories of raw materials and finished goods. It uses raw materials as direct materials in production and as indirect materials. Its factory payroll costs include direct labor for production and indirect labor. All materials are added at the beginning of the process, and direct labor and factory overhead are applied uniformly throughout the production process.

Required

You are to maintain records and produce measures of inventories to reflect the July events of this company. Set up the following general ledger accounts and enter the June 30 balances: Raw Materials Inventory, $25,000; Goods in Process Inventory, $8,135 ($2,660 of direct materials, $3,650 of direct labor, and $1,825 of overhead); Finished Goods Inventory, $110,000; Sales, $0; Cost of Goods Sold, $0; Factory Payroll, $0; and Factory Overhead, $0.

1. Prepare journal entries to record the following July transactions and events.
   a. Purchased raw materials for $125,000 cash (the company uses a perpetual inventory system).
   b. Used raw materials as follows: direct materials, $52,440; and indirect materials, $10,000.
   c. Incurred factory payroll cost of $227,250 paid in cash (ignore taxes).
   d. Assigned factory payroll costs as follows: direct labor, $202,250; and indirect labor, $25,000.
   e. Incurred additional factory overhead costs of $80,000 paid in cash.
   f. Allocated factory overhead to production at 50% of direct labor costs.
2. Information about the July inventories follows. Use this information with that from part 1 to prepare a process cost summary, assuming the weighted-average method is used.

Check (1f) Cr. Factory Overhead, $101,125

Check (2) EUP for overhead, 14,200

Units	
Beginning inventory	5,000 units
Started	14,000 units
Ending inventory	8,000 units
Beginning inventory	
Materials—Percent complete	100%
Labor and overhead—Percent complete	75%
Ending inventory	
Materials—Percent complete	100%
Labor and overhead—Percent complete	40%

3. Using the results from part 2 and the available information, make computations and prepare journal entries to record the following:
   a. Total costs transferred to finished goods for July (label this entry g). (3a) $271,150
   b. Sale of finished goods costing $265,700 for $625,000 in cash (label this entry h).
4. Post entries from parts 1 and 3 to the ledger accounts set up at the beginning of the problem.
5. Compute the amount of gross profit from the sales in July. (*Note:* Add any underapplied overhead to, or deduct any overapplied overhead from, the cost of goods sold. Ignore the corresponding journal entry.)

BEYOND THE NUMBERS

REPORTING IN ACTION

C2 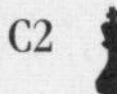

BTN 20-1 Best Buy reports in notes to its financial statements that, in addition to its merchandise sold, it includes the following costs (among others) in cost of goods sold: freight expenses associated with moving inventories from vendors to distribution centers, costs of services provided, customer shipping and handling expenses, costs associated with operating its distribution network, and freight expenses associated with moving merchandise from distribution centers to retail stores.

Required

1. Why do you believe Best Buy includes these costs in its cost of goods sold?
2. What effect does this cost accounting policy for its cost of goods sold have on Best Buy's financial statements and any analysis of these statements? Explain.

Fast Forward

3. Access Best Buy's financial statements for the fiscal years after March 3, 2007, from its Website (BestBuy.com) or the SEC's EDGAR Website (sec.gov). Review its footnote relating to Cost of Goods Sold and Selling, General, and Administrative Expense. Has Best Buy changed its policy with respect to what costs are included in the cost of goods sold? Explain.

COMPARATIVE ANALYSIS

C1

BTN 20-2 Retailers such as Best Buy, Circuit City, and RadioShack usually work to maintain a high-quality and low-cost operation. One ratio routinely computed for this assessment is the cost of goods sold divided by total expenses. A decline in this ratio can mean that the company is spending too much on selling and administrative activities. An increase in this ratio beyond a reasonable level can mean that the company is not spending enough on selling activities. (Assume for this analysis that total expenses equal the cost of goods sold plus selling, general, and administrative expenses.)

Required

1. For Best Buy, Circuit City, and RadioShack refer to Appendix A and compute the ratios of cost of goods sold to total expenses for their two most recent fiscal years.
2. Comment on the similarities or differences in the ratio results across both years among the companies.

ETHICS CHALLENGE

C1 C3

BTN 20-3 Many accounting and accounting-related professionals are skilled in financial analysis, but most are not skilled in manufacturing. This is especially the case for process manufacturing environments (for example, a bottling plant or chemical factory). To provide professional accounting and financial services, one must understand the industry, product, and processes. We have an ethical responsibility to develop this understanding before offering services to clients in these areas.

Required

Write a one-page action plan, in memorandum format, discussing how you would obtain an understanding of key business processes of a company that hires you to provide financial services. The memorandum should specify an industry, a product, and one selected process and should draw on at least one reference, such as a professional journal or industry magazine.

COMMUNICATING IN PRACTICE

A1 C1 P1 P2

BTN 20-4 You hire a new assistant production manager whose prior experience is with a company that produced goods to order. Your company engages in continuous production of homogeneous products that go through various production processes. Your new assistant e-mails you questioning some cost classifications on an internal report—specifically why the costs of some materials that do not actually become part of the finished product, including some labor costs not directly associated with producing the product, are classified as direct costs. Respond to this concern via memorandum.

TAKING IT TO THE NET

C1 C3

BTN 20-5 Many companies acquire software to help them monitor and control their costs and as an aid to their accounting systems. One company that supplies such software is **proDacapo** (prodacapo.com). There are many other such vendors. Access proDacapo's Website, click on "Business Process Management," and review the information displayed.

Required

How is process management software helpful to businesses? Explain with reference to costs, efficiency, and examples, if possible.

TEAMWORK IN ACTION

C1 P1 P2 P3 P6

BTN 20-6 The purpose of this team activity is to ensure that each team member understands process operations and the related accounting entries. Find the activities and flows identified in Exhibit 20.4 with numbers ①–⑩. Pick a member of the team to start by describing activity number ① in this exhibit, then verbalizing the related journal entry, and describing how the amounts in the entry are computed. The other members of the team are to agree or disagree; discussion is to continue until all members express understanding. Rotate to the next numbered activity and next team member until all activities and entries have been discussed. If at any point a team member is uncertain about an answer, the team member may pass and get back in the rotation when he or she can contribute to the team's discussion.

ENTREPRENEURIAL DECISION

C4 A2

BTN 20-7 Read the chapter opener about **Hood River Juice Company**. David Ryan explained that purchasing apples year-round and processing them immediately reduces costs, and that his company blends juices to fit customer needs.

Required

1. How does not holding raw materials inventories (apples) reduce costs? If the items are not used in production, how can they affect profits? Explain.
2. Explain why Hood River Juice Company might use a hybrid costing system.

HITTING THE ROAD

C3 

BTN 20-8 In process costing, the process is analyzed first and then a unit measure is computed in the form of equivalent units for direct materials, direct labor, overhead, and all three combined. The same analysis applies to both manufacturing and service processes.

Required

Visit your local **U.S. Mail** center. Look into the back room, and you will see several ongoing processes. Select one process, such as sorting, and list the costs associated with this process. Your list should include materials, labor, and overhead; be specific. Classify each cost as fixed or variable. At the bottom of your list, outline how overhead should be assigned to your identified process. The following format (with an example) is suggested.

Point: The class can compare and discuss the different processes studied and the answers provided.

Cost Description	Direct Material	Direct Labor	Overhead	Variable Cost	Fixed Cost
Manual sorting		X		X	
⋮					
Overhead allocation suggestions:					

GLOBAL DECISION

C1 

BTN 20-9 **DSG international plc**, **Best Buy**, **Circuit City**, and **RadioShack** are competitors in the global marketplace. Selected data for DSG follow.

(millions of pounds)	Current Year	Prior Year
Cost of goods sold	£7,285	£6,369
General, selling, and administrative expenses	381	339
Total expenses	£7,666	£6,708

Required

1. Review the discussion of the importance of the cost of goods sold divided by total expenses ratio in BTN 20-2. Compute the cost of goods sold to total expenses ratio for DSG for the two years of data provided.
2. Comment on the similarities or differences in the ratio results calculated in part 1 and in BTN 20-2 across years and companies.

ANSWERS TO MULTIPLE CHOICE QUIZ

1. d
2. e
3. b; $20,000 + $152,000 + $45,000 + $18,000 − $218,000 = $17,000
4. a; 40,000 + (15,000 × 1/3) = 45,000 EUP
5. c; ($6,000 + $84,000) ÷ 45,000 EUP = $2 per EUP

A Look Back

Chapter 20 focused on how to measure and account for costs in process operations. It explained process production, described how to assign costs to processes, and computed cost per equivalent unit.

A Look at This Chapter

This chapter describes cost allocation and activity-based costing. It identifies managerial reports useful in directing a company's activities. It also describes responsibility accounting, measuring departmental performance, transfer pricing, and allocating common costs across departments.

A Look Ahead

Chapter 22 looks at cost behavior and explains how its identification is useful to managers in performing cost-volume-profit analyses. It also shows how to apply cost-volume-profit analysis for managerial decisions.

Chapter 21

Cost Allocation and Performance Measurement

Learning Objectives

CAP

Conceptual

C1 Explain departmentalization and the role of departmental accounting. *(p. 858)*

C2 Distinguish between direct and indirect expenses. *(p. 859)*

C3 Identify bases for allocating indirect expenses to departments. *(p. 860)*

C4 Explain controllable costs and responsibility accounting. *(p. 869)*

C5 *Appendix* 21A—Explain transfer pricing and methods to set transfer prices. *(p. 875)*

C6 *Appendix 21B*—Describe allocation of joint costs across products. *(p. 877)*

Analytical

A1 Analyze investment centers using return on assets, residual income, and balanced scorecard. *(p. 867)*

A2 Analyze investment centers using profit margin and investment turnover. *(p. 872)*

Procedural

P1 Assign overhead costs using two-stage cost allocation. *(p. 852)*

P2 Assign overhead costs using activity-based costing. *(p. 854)*

P3 Prepare departmental income statements. *(p. 861)*

P4 Prepare departmental contribution reports. *(p. 866)*

LP21

Decision Feature

On The Green

"The more clicks we can get, the better our future"
—Todd Rath

ROCHESTER, NY—Brothers Tom and Todd Rath paid their college tuition by diving for lost golf balls and then reselling them. Today, their company **RockBottomGolf.com** applies a similar strategy of buying leftover products and reselling them. "Some of our critics refer to us as the 'graveyard of golf,'" explains Tom. "Oftentimes, we may be selling the last 3,000 drivers a manufacturer has ever made. If anyone can find a home for it, we can." The company boasts over 500,000 customers, affectionately referred to as "Rock Heads."

RockBottom's warehouse sports signs with "Scratch," the company's cartoonish, red-bearded caveman mascot. Scratch is surrounded with slogans such as: "A Clean Cave Is a Happy Cave" and "A Happy Rock Head Stays a Rock Head." Though Scratch is goofy, the company is all business. Offering a wide inventory of well-known brands of golf clubs, bags, balls, apparel, and accessories, the company buys in large lots and strives to keep overhead low. For example, they located their distribution center in Virginia—enabling them to ship to over 60% of the U.S. population within two days. Also, they pack items in small, uniformly sized boxes to lower costs and offer free shipping on certain orders.

Many other cost management procedures are applied. For example, they analyze "checkout flow," providing details on the point at which potential customers drop out of the checkout process and how many drop out. "If I had a 50% checkout success rate one day and 23% the next day, this lets me see that," explains Todd. This mix of financial and nonfinancial information helps Todd steer more customers through the checkout process. He also tracks customer approval ratings, currently above 99%, as a performance measure.

The diversity of its product offerings requires additional cost management. Company managers monitor direct, indirect, and controllable costs, and allocate them to departments and products. Understanding how it's product lines—such as clubs, bags, apparel—are performing and their contribution margins helps them plan for expansion. As Todd emphasizes, "We use tools to measure our ROI (return on investment). We will only expand as long as there are customers to win."

Their expansion plans do not stop with golf. RockBottomGolf wants to become RockBottomSports, with many other sporting goods products available. This increased departmentalization will require them to monitor contribution margins, return on investment, checkout flow, and customer approval. With its fast-paced growth and position as the top golf retailer on the Internet, RockBottomGolf is "on the green."

[Sources: *RockBottomGolf.com Website,* January 2009; *Internet Retailer,* July 2007; *Inside Business-Hampton Roads,* October 2006.]

Chapter Preview

This chapter describes how to allocate costs shared by more than one product across those different products and how to allocate indirect costs of shared items such as utilities, advertising, and rent. The chapter also describes activity-based costing and how it traces the costs of individual activities. This knowledge helps managers better understand how to assign costs and assess company performance. The chapter also introduces additional managerial accounting reports useful in managing a company's activities and explains how and why management divides companies into departments.

Cost Allocation and Performance Measurement

Overhead Cost Allocation Methods
- Two-stage cost allocation
- Activity-based cost allocation
- Comparison of allocation methods

Departmental Accounting
- Motivation for departmentalization
- Departmental evaluation
- Departmental reporting and analysis

Departmental Expense Allocation
- Direct and indirect expenses
- Allocation of indirect expenses
- Departmental income statements
- Departmental contribution to overhead

Investment Centers
- Financial measures of performance
- Nonfinancial measures of performance
- Balanced scorecard

Responsibility Accounting
- Controllable versus direct costs
- Responsibility accounting system
- Transfer pricing

Section 1—Allocating Costs for Product Costing

Video21.1

This first of two sections in this chapter focuses on alternatives for allocation of costs to products and services. We explain and illustrate two basic methods: (1) traditional two-state cost allocation and (2) activity-based cost allocation. The second section describes and illustrates the allocation of costs for performance evaluation.

Overhead Cost Allocation Methods

P1 Assign overhead costs using two-stage cost allocation.

Point: Use of a single overhead allocation rate is known as using a *plantwide rate.*

We previously explained how to assign overhead costs to jobs (and processes) by using a predetermined overhead rate per unit of an allocation base such as direct labor cost. When a single overhead rate is used on a companywide basis, all overhead is lumped together, and a predetermined overhead rate per unit of an allocation base is computed and used to assign overhead to jobs (and processes). The use of a single predetermined overhead rate suggests that this allocation process is simple. In reality, it can be complicated. This chapter explains the traditional two-stage cost allocation procedure and then introduces the activity-based cost allocation procedure.

Two-Stage Cost Allocation

An organization incurs overhead costs in many activities. These activities can be identified with various departments, which can be broadly classified as either operating or service departments. *Operating departments* perform an organization's main functions. For example, an accounting firm's main functions usually include auditing, tax, and advisory services. Similarly, the production and selling departments of a manufacturing firm perform its main functions and serve as operating departments. *Service departments* provide support to an organization's operating departments. Examples of service departments are payroll, human resource management, accounting, and executive management. Service departments do not engage in activities that generate revenues, yet their support is crucial for the operating departments' success. In this section, we apply a two-stage cost allocation procedure to assign (1) service department costs to operating departments and (2) operating department costs, including those assigned from service departments, to the organization's output.

Illustration of Two-Stage Cost Allocation Exhibit 21.1 shows the two-allocation procedure. This exhibit uses data from **AutoGrand**, a custom automobile turer. AutoGrand has five manufacturing-related departments: janitorial, maintenance, accounting, machining, and assembly. Expenses incurred by each of these departments a sidered product costs. There are three service departments—janitorial, maintenance, ar tory accounting; each incurs expenses of $10,000, $15,000 and $8,000, respectively. As s in Exhibit 21.1, the first stage of the two-stage procedure involves allocating the costs o three service departments to the two operating departments (machining and assembly). two operating departments use the resources of these service departments.

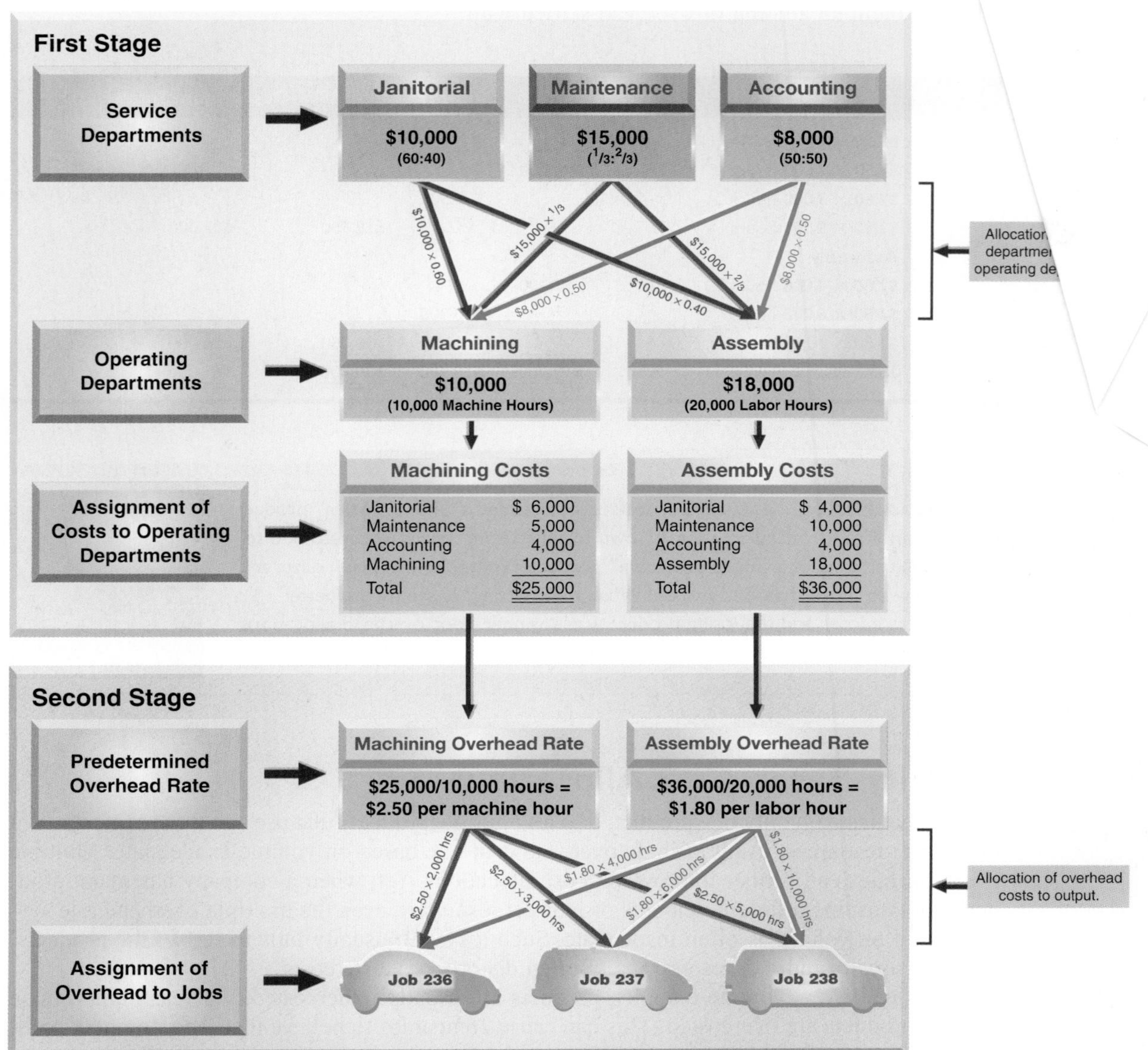

To illustrate the first stage of cost allocation, we use the janitorial department. Its costs are allocated to machining and assembly in the ratio 60 : 40. This means that 60%, or $6,000, of janitorial costs are assigned to the machining department and 40%, or $4,000, to the assembly department. The expenses incurred by the maintenance and factory accounting departments are similarly assigned to machining and assembly. We then add the expenses directly incurred by each operating department to these assigned costs to determine the total expenses for each operating department. This yields total costs of $25,000 for machining and $36,000 for assembly.

Point: Use of a separate overhead allocation rate for each department is known as using *departmental rates.*

In the second stage, predetermined overhead rates are computed for each operating department. The allocation base is machine hours for machining and labor hours for assembly. The

Char

ermined overhead rate is $2.50 per machine hour for the machining department and $1.80 abor hour for the assembly department. These predetermined overhead rates are then used assign overhead to output.

To illustrate this second stage, assume that three jobs were started and finished in a recent month. These jobs consumed resources as follows: Job 236—2,000 machine hours in machining and 4,000 labor hours in assembly; Job 237—3,000 machine hours and 6,000 labor hours; Job 238—5,000 machine hours and 10,000 labor hours. The overhead assigned to these three jobs is shown with the arrow lines in the bottom row of Exhibit 21.1.

Exhibit 21.2 summarizes these allocations. Total overhead allocated to Jobs 236, 237, and 238, is $12,200, $18,300, and $30,500, respectively. These allocated costs sum to $61,000, which is the total amount of overhead started with.

nead

	Job 236	Job 237	Job 238	Department Totals
Machining				
$2.50 × 2,000 hours	$ 5,000			
$2.50 × 3,000 hours		$ 7,500		
$2.50 × 5,000 hours			$12,500	$25,000
Assembly				
$1.80 × 4,000 hours	7,200			
$1.80 × 6,000 hours		10,800		
$1.80 × 10,000 hours			18,000	36,000
Total overhead assigned	$12,200	$18,300	$30,500	$61,000

Decision Insight

Overhead Misled Futura Computer outsourced a "money-losing" product to a Korean firm for manufacturing. Its own manufacturing facility was retooled to produce extra units of a "more profitable" product. Profits did not materialize, and losses grew to more than $20 million! What went wrong? It seems the better product was a loser and the losing product was a winner. Poor overhead allocations misled Futura's management.

Video21.2

P2 Assign overhead costs using activity-based costing.

Activity-Based Cost Allocation

For companies with only one product, or with multiple products that use about the same amount of indirect resources, using a single overhead cost rate based on volume is adequate. Multiple overhead rates can further improve on cost allocations. Yet, when a company has many products that consume different amounts of indirect resources, even the multiple overhead rate system based on volume is often inadequate. Such a system usually fails to reflect the products' different uses of indirect resources and often distorts products costs.

Specifically, low-volume complex products are usually undercosted, whereas high-volume simpler products are overcosted. This can cause companies to believe that their complex products are more profitable than they really are, which can lead those companies to focus on them to the detriment of high-volume simpler products. This creates a demand for a better cost allocation system for these indirect (overhead) costs.

Point: A survey found that most respondents believe that activity-based costing is worth the investment because of improved management decisions.

Activity-based costing (ABC) attempts to better allocate costs to the proper users of overhead by focusing on *activities.* Costs are traced to individual activities and then allocated to cost objects. Exhibit 21.3 shows the (two-stage) activity-based cost allocation method. The first stage identifies the activities involved in processing Jobs 236, 237, and 238 and forms activity cost *pools* by combining those activities. The second stage involves computing predetermined overhead cost rates for each cost pool and then assigning costs to jobs.

We begin our explanation at the top of Exhibit 21.3. The first stage identifies individual activities, which are pooled in a logical manner into homogenous groups, or *cost pools.* A

EXHIBIT 21.3

Activity-Based Cost Allocation

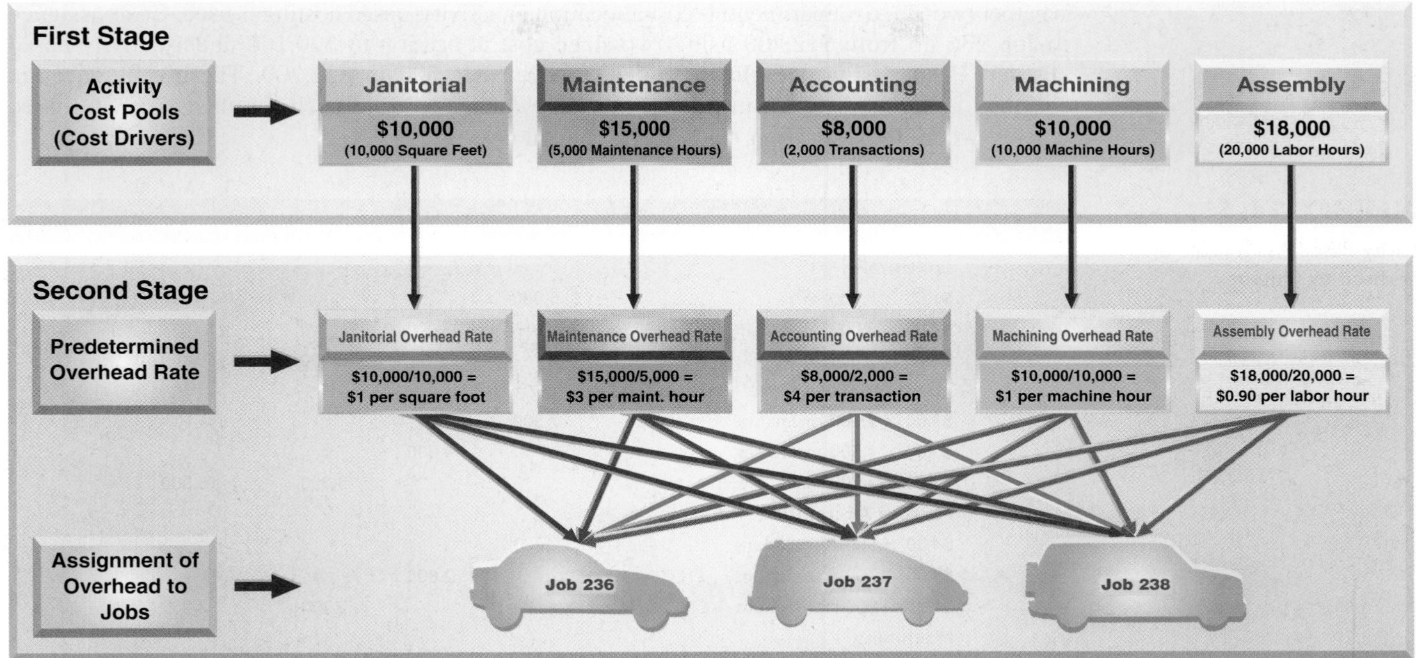

homogenous cost pool consists of activities that belong to the same process and/or are caused by the same cost driver. An **activity cost driver,** or simply *cost driver,* is a factor that causes the cost of an activity to go up or down. For example, preparing an invoice, checking it, and dispatching it are activities of the "invoicing" process and can therefore be grouped in a single cost pool. Moreover, the number of invoices processed likely drives the costs of these activities.

Point: A cost driver is different from an allocation base. An allocation base is used as a basis for assigning overhead but need not have a cause-effect relation with the costs assigned. However, a cost driver has a cause-effect relation with the cost assigned.

An **activity cost pool** is a temporary account accumulating the costs a company incurs to support an identified set of activities. Costs accumulated in an activity cost pool include the variable and fixed costs of the activities in the pool. Variable costs pertain to resources acquired as needed (such as materials); fixed costs pertain to resources acquired in advance (such as equipment). An activity cost pool account is handled like a factory overhead account.

In the second stage, after all activity costs are accumulated in an activity cost pool account, overhead rates are computed. Then, costs are allocated to cost objects (users) based on cost drivers (allocation bases).

Decision Insight

Measuring Health Activity-based costing is used in many settings. A study found that activity-based costing improves health care costing accuracy, enabling improved profitability analysis and decision making. However, identifying cost drivers in a health care setting is challenging.

Illustration of Activity-Based Costing To illustrate, let's return to AutoGrand's three jobs. Assume that resources used to complete Jobs 236, 237, and 238 are shown in Exhibit 21.4.

EXHIBIT 21.4

Activity Resource Use

Resources Used	Job 236	Job 237	Job 238
Square feet of space	5,000	3,000	2,000
Maintenance hours	2,500	1,500	1,000
Number of transactions	500	700	800
Machine hours	2,000	3,000	5,000
Direct labor hours	4,000	6,000	10,000

The $61,000 of total costs are assigned to these three jobs using activity-based costing as shown in Exhibit 21.5 (rates are taken from the second stage of Exhibit 21.3). Comparing Exhibits 21.2 and 21.5, we see that the costs assigned to the three jobs vary markedly depending on whether two-stage (departmental) cost allocation or activity-based costing is used. Costs assigned to Job 236 go from $12,200 using two-stage cost allocation to $20,100 under activity-based costing. Costs assigned to Job 238 decline from $30,500 to $22,200. These differences in assigned amounts result from more accurately tracing costs to each job using activity-based costing where the allocation bases reflect actual cost drivers.

EXHIBIT 21.5

Activity-Based Assignment of Overhead to Output

	Job 236	Job 237	Job 238	Activity Totals
Janitorial				
$1.00 × 5,000 sq. ft.	$ 5,000			
$1.00 × 3,000 sq. ft.		$ 3,000		
$1.00 × 2,000 sq. ft.			$ 2,000	$10,000
Maintenance				
$3.00 × 2,500 maint. hrs.	7,500			
$3.00 × 1,500 maint. hrs.		4,500		
$3.00 × 1,000 maint. hrs.			3,000	15,000
Factory Accounting				
$4.00 × 500 transactions	2,000			
$4.00 × 700 transactions		2,800		
$4.00 × 800 transactions			3,200	8,000
Machining				
$1.00 × 2,000 machine hrs.	2,000			
$1.00 × 3,000 machine hrs.		3,000		
$1.00 × 5,000 machine hrs.			5,000	10,000
Assembly				
$0.90 × 4,000 labor hrs.	3,600			
$0.90 × 6,000 labor hrs.		5,400		
$0.90 × 10,000 labor hrs.			9,000	18,000
Total overhead assigned	$20,100	$18,700	$22,200	$61,000

Decision Maker

Director of Operations Two department managers at your ad agency complain to you that overhead costs assigned to them are too high. Overhead is assigned on the basis of labor hours for designers. These managers argue that overhead depends not only on designers' hours but on many activities unrelated to these hours. What is your response? [Answer—p. 879]

Comparison of Two-Stage and Activity-Based Cost Allocation

Traditional cost systems capture overhead costs by individual department (or function) and accumulate these costs in one or more overhead accounts. Companies then assign these overhead costs using a single allocation base such as direct labor or multiple volume-based allocation bases. Unfortunately, traditional cost systems have tended to use allocation bases that are often not closely related to the way these costs are actually incurred.

In contrast, activity-based cost systems capture costs by individual activity. These activities and their costs are then accumulated into activity cost pools. A company selects a cost driver (allocation base) for each activity pool. It uses this cost driver to assign the accumulated activity costs to cost objects (such as jobs or products) benefiting from the activity.

An activity-based costing system commonly consists of more allocation bases as compared to a traditional cost system. For example, a Chicago-based manufacturer currently uses nearly

EXHIBIT 21.6

Cost Pools and Cost Drivers in Activity-Based Costing

Activity Cost Pool	Cost Driver
Materials purchasing	Number of purchase orders
Materials handling	Number of materials requisitions
Personnel processing	Number of employees hired or laid off
Equipment depreciation	Number of products produced or hours of use
Quality inspection	Number of units inspected
Indirect labor in setting up equipment	Number of setups required
Engineering costs for product modifications	Number of modifications (engineering change orders)

20 different activity cost drivers to assign overhead costs to its products. Exhibit 21.6 lists common examples of overhead cost pools and their usual cost drivers.

Activity-based costing is especially effective when the same department or departments produce many different types of products. For instance, more complex products often require more help from service departments such as engineering, maintenance, and materials handling. If the same amount of direct labor is applied to the complex and simple products, a traditional overhead allocation system assigns the same overhead cost to both. With activity-based costing, however, the complex products are assigned a larger portion of overhead. The difference in overhead assigned can affect product pricing, make-or-buy, and other managerial decisions.

Activity-based costing encourages managers to focus on *activities* as well as the use of those activities. For instance, assume AutoGrand can reduce the number of transactions processed in Factory Accounting to 1,500 (375 transactions for Job 236, 525 transactions for Job 237, and 600 transactions for Job 238) and that through continuous improvement it can reduce costs of processing those transactions to $4,500. The resulting rate to process a transaction is $3 per transaction ($4,500/1,500 transactions—down from $4 per Exhibit 21.3). The cost of transaction processing is reduced for all jobs (Job 236, $1,125; Job 237, $1,575; Job 238, $1,800). However, if those accounting costs were grouped in a single overhead cost pool, it is more difficult to identify cost savings and understand their effects on product costs.

Activity-based costing requires managers to look at each item and encourages them to manage each cost to increase the benefit from each dollar spent. It also encourages managers to cooperate because it shows how their efforts are interrelated. This results in *activity-based management.*

Decision Ethics

Accounting Officer Your company produces expensive garments, whose production involves many complex and specialized activities. Your general manager recently learned about activity-based costing (ABC) and asks your advice. However, your supervisor does not want to disturb the existing cost system and instructs you to prepare a report stating that "implementation of ABC is a complicated process involving too many steps and not worth the effort." You believe ABC will actually help the company identify sources of costs and control them. What action do you take? [Answer—p. 879]

Quick Check

Answers—p. 880

1. What is a cost driver?
2. When activity-based costing is used rather than traditional allocation methods, (*a*) managers must identify cost drivers for various items of overhead cost, (*b*) individual cost items in service departments are allocated directly to products or services, (*c*) managers can direct their attention to the activities that drive overhead cost, or (*d*) all of the above.

Section 2—Allocating Costs for Performance Evaluation

This second section of the chapter describes and illustrates allocation of costs for performance evaluation. We begin with departmental accounting and expense allocations and conclude with responsibility accounting.

Departmental Accounting

Video21.3

Companies are divided into *departments,* also called *subunits,* when they are too large to be managed effectively as a single unit. Managerial accounting for departments has two main goals. The first is to set up a **departmental accounting system** to provide information for managers to evaluate the profitability or cost effectiveness of each department's activities. The second goal is to set up a **responsibility accounting system** to control costs and expenses and evaluate managers' performances by assigning costs and expenses to the managers responsible for controlling them. Departmental and responsibility accounting systems are related and share much information.

Motivation for Departmentalization

C1 Explain departmentalization and the role of departmental accounting.

Many companies are so large and complex that they are broken into separate divisions for efficiency and/or effectiveness purposes. Divisions then are usually organized into separate departments. When a company is departmentalized, each department is often placed under the direction of a manager. As a company grows, management often divides departments into new departments so that responsibilities for a department's activities do not overwhelm the manager's ability to oversee and control them. A company also creates departments to take advantage of the skills of individual managers. Departments are broadly classified as either operating or service departments.

Departmental Evaluation

Point: To improve profitability, **Sears, Roebuck & Co.** eliminated several departments, including its catalog division.

When a company is divided into departments, managers need to know how each department is performing. The accounting system must supply information about resources used and outputs achieved by each department. This requires a system to measure and accumulate revenue and expense information for each department whenever possible.

Departmental information is rarely distributed publicly because of its potential usefulness to competitors. Information about departments is prepared for internal managers to help control operations, appraise performance, allocate resources, and plan strategy. If a department is highly profitable, management may decide to expand its operations, or if a department is performing poorly, information about revenues or expenses can suggest useful changes.

More companies are emphasizing customer satisfaction as a main responsibility of many departments. This has led to changes in the measures reported. Increasingly, financial measurements are being supplemented with quality and customer satisfaction indexes. **Motorola**, for instance, uses two key measures: the number of defective parts per million parts produced and the percent of orders delivered on time to customers. (Note that some departments have only "internal customers.")

Financial information used to evaluate a department depends on whether it is evaluated as a profit center, cost center, or investment center. A **profit center** incurs costs and generates revenues; selling departments are often evaluated as profit centers. A **cost center** incurs costs without directly generating revenues. An **investment center** incurs costs and generates revenues, and is responsible for effectively using center assets. The manufacturing departments of a manufacturer and its service departments such as accounting, advertising, and purchasing, are all cost centers.

Point: Selling departments are often treated as *revenue centers;* their managers are responsible for maximizing sales revenues.

Evaluating managers' performance depends on whether they are responsible for profit centers, cost centers, or investment centers. Profit center managers are judged on their abilities to generate revenues in excess of the department's costs. They are assumed to influence both revenue generation and cost incurrence. Cost center managers are judged on their abilities to control costs by keeping them within a satisfactory range under an assumption that only they influence costs. Investment center managers are evaluated on their use of center assets to generate income.

Decision Insight

Nonfinancial Measures A majority of companies now report nonfinancial performance measures to management. Common measures are cycle time, defect rate, on-time deliveries, inventory turnover, customer satisfaction, and safety. When nonfinancial measures are used with financial measures, the performance measurement system resembles a ***balanced scorecard.*** Many of these companies also use activity-based management as part of their performance measurement system.

Departmental Reporting and Analysis

Companies use various measures (financial and nonfinancial) and reporting formats to evaluate their departments. The type and form of information depend on management's focus and philosophy. **Hewlett-Packard**'s statement of corporate objectives, for instance, indicates that its goal is to satisfy customer needs. Its challenge is to set up managerial accounting systems to provide relevant feedback for evaluating performance in terms of its stated objectives. Also, the means used to obtain information about departments depend on how extensively a company uses computer and information technology.

When accounts are not maintained separately in the general ledger by department, a company can create departmental information by using a *departmental spreadsheet analysis.* For example, after recording sales in its usual manner, a company can compute daily total sales by department and enter these totals on a sales spreadsheet. At period-end, column totals of the spreadsheet show sales by department. The combined total of all columns equals the balance of the Sales account. A merchandiser that uses a spreadsheet analysis of department sales often uses separate spreadsheets to accumulate sales, sales returns, purchases, and purchases returns by department. If each department keeps a count of its inventory, it can also compute its gross profit (assuming it's a profit center).

Point: Many retailers use a point-of-sales system capturing sales data and creating requests to release inventory from the warehouse and order more merchandise. **Wal-Mart**'s sales system not only collects data for internal use but also is used by **Procter & Gamble** to plan its production and product deliveries to **Wal-Mart.**

Point: **Link Wood Products,** a manufacturer of lawn and garden products, records each sale by department on a spreadsheet. Daily totals are accumulated in another spreadsheet to obtain monthly totals by department.

Quick Check

Answers—p. 880

3. What is the difference between a departmental accounting system and a responsibility accounting system?
4. Service departments (*a*) manufacture products, (*b*) make sales directly to customers, (*c*) produce revenues, (*d*) assist operating departments.
5. Explain the difference between a cost center and a profit center. Cite an example of each.

Departmental Expense Allocation

When a company computes departmental profits, it confronts some accounting challenges that involve allocating its expenses across its operating departments.

Direct and Indirect Expenses

C2 Distinguish between direct and indirect expenses.

Direct expenses are costs readily traced to a department because they are incurred for that department's sole benefit. They require no allocation across departments. For example, the salary of an employee who works in only one department is a direct expense of that one department.

Indirect expenses are costs that are incurred for the joint benefit of more than one department and cannot be readily traced to only one department. For example, if two or more departments share a single building, all enjoy the benefits of the expenses for rent, heat, and light. Indirect expenses are allocated across departments benefiting from them when we need information about departmental profits. Ideally, we allocate indirect expenses by using a cause-effect

Point: Utility expense has elements of both direct and indirect expenses.

relation. When we cannot identify cause-effect relations, we allocate each indirect expense on a basis approximating the relative benefit each department receives. Measuring the benefit for each department from an indirect expense can be difficult.

Illustration of Indirect Expense Allocation To illustrate how to allocate an indirect expense, we consider a retail store that purchases janitorial services from an outside company. Management allocates this cost across the store's three departments according to the floor space each occupies. Costs of janitorial services for a recent month are $300. Exhibit 21.7 shows the square feet of floor space each department occupies. The store computes the percent of total square feet allotted to each department and uses it to allocate the $300 cost.

EXHIBIT 21.7

Indirect Expense Allocation

Department	Square Feet	Percent of Total	Allocated Cost
Jewelry	2,400	60%	$180
Watch repair	600	15	45
China and silver	1,000	25	75
Totals	4,000	100%	$300

Specifically, because the jewelry department occupies 60% of the floor space, 60% of the total $300 cost is assigned to it. The same procedure is applied to the other departments. When the allocation process is complete, these and other allocated costs are deducted from the gross profit for each department to determine net income for each. One consideration in allocating costs is to motivate managers and employees to behave as desired. As a result, a cost incurred in one department might be best allocated to other departments when one of the other departments caused the cost.

Allocation of Indirect Expenses

C3 Identify bases for allocating indirect expenses to departments.

This section describes how to identify the bases used to allocate indirect expenses across departments. No standard rule identifies the best basis because expense allocation involves several factors, and the relative importance of these factors varies across departments and organizations. Judgment is required, and people do not always agree. In our discussion, note the parallels between activity-based costing and the departmental expense allocation procedures described here.

Point: Expense allocations cannot always avoid some arbitrariness.

Wages and Salaries Employee wages and salaries can be either direct or indirect expenses. If their time is spent entirely in one department, their wages are direct expenses of that department. However, if employees work for the benefit of more than one department, their wages are indirect expenses and must be allocated across the departments benefited. An employee's contribution to a department usually depends on the number of hours worked in contributing to that department. Thus, a reasonable basis for allocating employee wages and salaries is the *relative amount of time spent in each department.* In the case of a supervisor who manages more than one department, recording the time spent in each department may not always be practical. Instead, a company can allocate the supervisor's salary to departments on the basis of the number of employees in each department—a reasonable basis if a supervisor's main task is managing people. Another basis of allocation is on sales across departments, also a reasonable basis if a supervisor's job reflects on departmental sales.

Point: Some companies ask supervisors to estimate time spent supervising specific departments for purposes of expense allocation.

Rent and Related Expenses Rent expense for a building is reasonably allocated to a department on the basis of floor space it occupies. Location can often make some floor space more valuable than other space. Thus, the allocation method can charge departments that occupy more valuable space a higher expense per square foot. Ground floor retail space, for instance, is often more valuable than basement or upper-floor space because all customers pass departments near the entrance but fewer go beyond the first floor. When no precise measures of floor space values exist, basing allocations on data such as customer traffic and real estate assessments is helpful. When a company owns its building, its expenses for depreciation, taxes, insurance, and other related building expenses are allocated like rent expense.

Advertising Expenses Effective advertising of a department's products increases its sales and customer traffic. Moreover, advertising products for some departments usually helps other departments' sales because customers also often buy unadvertised products. Thus, many stores treat advertising as an indirect expense allocated on the basis of each department's proportion of total sales. For example, a department with 10% of a store's total sales is assigned 10% of advertising expense. Another method is to analyze each advertisement to compute the Web/newspaper space or TV/radio time devoted to the products of a department and charge that department for the proportional costs of advertisements. Management must consider whether this more detailed and costly method is justified.

Equipment and Machinery Depreciation Depreciation on equipment and machinery used only in one department is a direct expense of that department. Depreciation on equipment and machinery used by more than one department is an indirect expense to be allocated across departments. Accounting for each department's depreciation expense requires a company to keep records showing which departments use specific assets. The number of hours that a department uses equipment and machinery is a reasonable basis for allocating depreciation.

Point: Employee morale suffers when allocations are perceived as unfair. Thus, it is important to carefully design and explain the allocation of service department costs.

Utilities Expenses Utilities expenses such as heating and lighting are usually allocated on the basis of floor space occupied by departments. This practice assumes their use is uniform across departments. When this is not so, a more involved allocation can be necessary, although there is often a trade-off between the usefulness of more precise allocations and the effort to compute them.

Point: Manufacturers often allocate electricity cost to departments on the basis of the horsepower of equipment located in each department.

Service Department Expenses To generate revenues, operating departments require support services provided by departments such as personnel, payroll, advertising, and purchasing. Such service departments are typically evaluated as cost centers because they do not produce revenues. (Evaluating them as profit centers requires the use of a system that "charges" user departments a price that then serves as the "revenue" generated by service departments.) A departmental accounting system can accumulate and report costs incurred directly by each service department for this purpose. The system then allocates a service department's expenses to operating departments benefiting from them. This is often done, for example, using traditional two-stage cost allocation (see Exhibit 21.1). Exhibit 21.8 shows some commonly used bases for allocating service department expenses to operating departments.

Point: When a service department "charges" its user departments within a company, a *transfer pricing system* must be set up to determine the "revenue" from its services provided.

EXHIBIT 21.8

Bases for Allocating Service Department Expenses

Service Department	Common Allocation Bases
Office expenses	Number of employees or sales in each department
Personnel expenses	Number of employees in each department
Payroll expenses	Number of employees in each department
Advertising expenses	Sales or amount of advertising charged directly to each department
Purchasing costs	Dollar amounts of purchases or number of purchase orders processed
Cleaning expenses	Square feet of floor space occupied
Maintenance expenses	Square feet of floor space occupied

Departmental Income Statements

P3 Prepare departmental income statements.

An income statement can be prepared for each operating department once expenses have been assigned to it. Its expenses include both direct expenses and its share of indirect expenses. For this purpose, compiling all expenses incurred in service departments before assigning them to operating departments is useful. We illustrate the steps to prepare departmental income statements using A-1 Hardware and its five departments. Two of them (office and purchasing) are

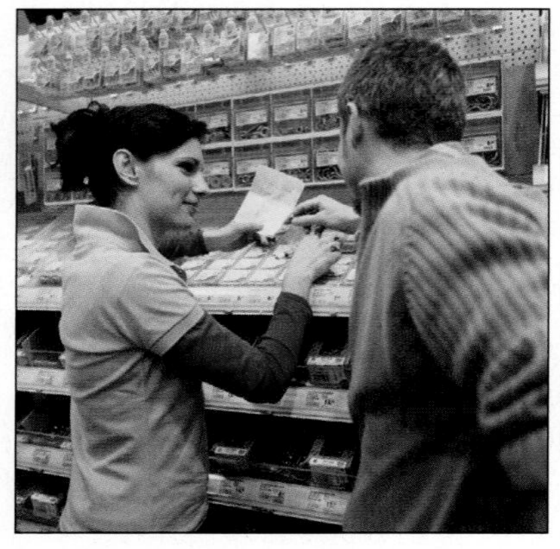

service departments and the other three (hardware, housewares, and appliances) are operating (selling) departments. Allocating costs to operating departments and preparing departmental income statements involves four steps. (1) Accumulating direct expenses by department. (2) Allocating indirect expenses across departments. (3) Allocating service department expenses to operating department. (4) Preparing departmental income statements.

Step 1 Step 1 accumulates direct expenses for each service and operating department as shown in Exhibit 21.9. Direct expenses include salaries, wages, and other expenses that each department incurs but does not share with any other department. This information is accumulated in departmental expense accounts.

EXHIBIT 21.9

Step 1: Direct Expense Accumulation

Direct Expenses → General office
Direct Expenses → Purchasing
Direct Expenses → Hardware
Direct Expenses → Housewares
Direct Expenses → Appliances

Point: We sometimes allocate service department costs across other service departments before allocating them to operating departments. This "step-wise" process is in advanced courses.

Step 2 Step 2 allocates indirect expenses across all departments as shown in Exhibit 21.10. Indirect expenses can include items such as depreciation, rent, advertising, and any other expenses that cannot be directly assigned to a department. Indirect expenses are recorded in company expense accounts, an allocation base is identified for each expense, and costs are allocated using a *departmental expense allocation spreadsheet* described in step 3.

EXHIBIT 21.10

Step 2: Indirect Expense Allocation

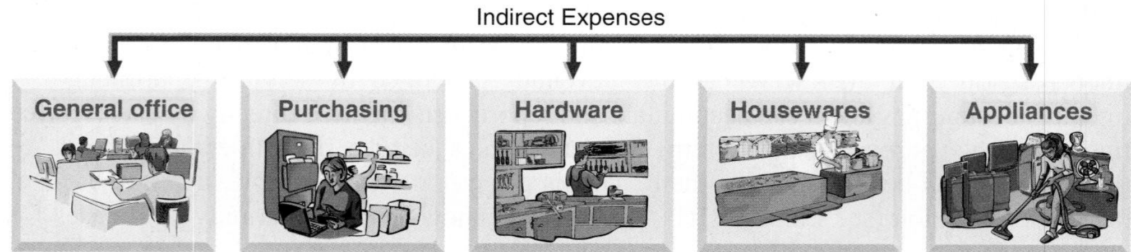

Step 3 Step 3 allocates expenses of the service departments (office and purchasing) to the operating departments. Service department costs are not allocated to other service departments. Exhibit 21.11 reflects the allocation of service department expenses using the allocation base(s). All of the direct and indirect expenses of service departments are allocated to operating departments.[1]

EXHIBIT 21.11

Step 3: Service Department Expense Allocation to Operating Departments

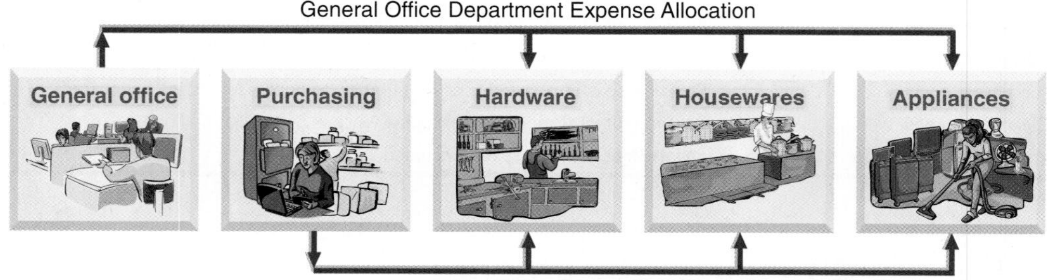

[1] In some cases we allocate a service department's expenses to other service departments when they use its services. For example, expenses of a payroll office benefit all service and operating departments and can be assigned to all departments. Nearly all examples and assignment materials in this book allocate service expenses only to operating departments for simplicity.

Computations for both steps 2 and 3 are commonly made using a departmental expense allocation spreadsheet as shown in Exhibit 21.12. The first two sections of this spreadsheet list direct expenses and indirect expenses by department. The third section lists the service department expenses and their allocations to operating departments. The allocation bases are identified in the second column, and total expense amounts are reported in the third column.

EXHIBIT 21.12

Departmental Expense Allocation Spreadsheet

A-1 HARDWARE
Departmental Expense Allocations
For Year Ended December 31, 2009

	Allocation Base	Expense Account Balance	General Office Dept.	Purchasing Dept.	Hardware Dept.	Housewares Dept.	Appliances Dept.
			Allocation of Expenses to Departments				
Direct expenses							
Salaries expense	Payroll records	$51,900	$13,300	$8,200	$15,600	$ 7,000	$ 7,800
Depreciation—Equipment	Depreciation records	1,500	500	300	400	100	200
Supplies expense	Requisitions	900	200	100	300	200	100
Indirect expenses							
Rent expense	Amount and value of space	12,000	600	600	4,860	3,240	2,700
Utilities expense	Floor space	2,400	300	300	810	540	450
Advertising expense	Sales	1,000			500	300	200
Insurance expense	Value of insured assets	2,500	400	200	900	600	400
Total department expenses		72,200	15,300	9,700	23,370	11,980	11,850
Service department expenses							
General office department	Sales		(15,300)		7,650	4,590	3,060
Purchasing department	Purchase orders			(9,700)	3,880	2,630	3,190
Total expenses allocated to operating departments		$72,200	$ 0	$ 0	$34,900	$19,200	$18,100

The departmental expense allocation spreadsheet is useful in implementing the first three steps. To illustrate, first (step 1) the three direct expenses of salaries, depreciation, and supplies are accumulated in each of the five departments.

Second (step 2), the four indirect expenses of rent, utilities, advertising, and insurance are allocated to all departments using the allocation bases identified. For example, consider rent allocation. Exhibit 21.13 lists the five departments' square footage of space occupied.

EXHIBIT 21.13

Departments' Allocation Bases

Department	Floor Space (Square Feet)	Value of Insured Assets ($)	Sales ($)	Number of Purchase Orders
General office	1,500	$ 38,000		—
Purchasing	1,500	19,000		—*
Hardware	4,050	85,500	$119,500	394
Housewares	2,700	57,000	71,700	267
Appliances	2,250	38,000	47,800	324
Total	12,000	$237,500	$239,000	985

* Purchasing department tracks purchase orders by department.

The two service departments (office and purchasing) occupy 25% of the total space (3,000 sq. feet/12,000 sq. feet). However, they are located near the back of the building, which is of lower value than space near the front that is occupied by operating departments. Management estimates that space near the back accounts for $1,200 of the total rent expense of $12,000. Exhibit 21.14 shows how we allocate the $1,200 rent expense between

EXHIBIT 21.14

Allocating Indirect (Rent) Expense to Service Departments

Department	Square Feet	Percent of Total	Allocated Cost
General office	1,500	50.0%	$ 600
Purchasing	1,500	50.0	600
Totals	3,000	100.0%	$1,200

these two service departments in proportion to their square footage. Exhibit 21.14 shows a simple rule for cost allocations: Allocated cost = Percentage of allocation base × Total cost. We then allocate the remaining $10,800 of rent expense to the three operating departments as shown in Exhibit 21.15. We continue step 2 by allocating the $2,400 of utilities expense to all departments based on the square footage occupied as shown in Exhibit 21.16.

EXHIBIT 21.15

Allocating Indirect (Rent) Expense to Operating Departments

Department	Square Feet	Percent of Total	Allocated Cost
Hardware	4,050	45.0%	$ 4,860
Housewares	2,700	30.0	3,240
Appliances	2,250	25.0	2,700
Totals	9,000	100.0%	$10,800

EXHIBIT 21.16

Allocating Indirect (Utilities) Expense to All Departments

Department	Square Feet	Percent of Total	Allocated Cost
General office	1,500	12.50%	$ 300
Purchasing	1,500	12.50	300
Hardware	4,050	33.75	810
Housewares	2,700	22.50	540
Appliances	2,250	18.75	450
Totals	12,000	100.00%	$2,400

Exhibit 21.17 shows the allocation of $1,000 of advertising expense to the three operating departments on the basis of sales dollars. We exclude service departments from this allocation because they do not generate sales.

EXHIBIT 21.17

Allocating Indirect (Advertising) Expense to Operating Departments

Department	Sales	Percent of Total	Allocated Cost
Hardware	$119,500	50.0%	$ 500
Housewares	71,700	30.0	300
Appliances	47,800	20.0	200
Totals	$239,000	100.0%	$1,000

To complete step 2 we allocate insurance expense to each service and operating department as shown in Exhibit 21.18.

EXHIBIT 21.18

Allocating Indirect (Insurance) Expense to All Departments

Department	Value of Insured Assets	Percent of Total	Allocated Cost
General Office	$ 38,000	16.0%	$ 400
Purchasing	19,000	8.0	200
Hardware	85,500	36.0	900
Housewares	57,000	24.0	600
Appliances	38,000	16.0	400
Total	$237,500	100.0%	$2,500

Third (step 3), total expenses of the two service departments are allocated to the three operating departments as shown in Exhibits 21.19 and 21.20.

Department	Sales	Percent of Total	Allocated Cost
Hardware	$119,500	50.0%	$ 7,650
Housewares	71,700	30.0	4,590
Appliances	47,800	20.0	3,060
Total	$239,000	100.0%	$15,300

EXHIBIT 21.19

Allocating Service Department (General Office) Expenses to Operating Departments

Department	Number of Purchase Orders	Percent of Total	Allocated Cost
Hardware	394	40.00%	$3,880
Housewares	267	27.11	2,630
Appliances	324	32.89	3,190
Total	985	100.00%	$9,700

EXHIBIT 21.20

Allocating Service Department (Purchasing) Expenses to Operating Departments

Step 4 The departmental expense allocation spreadsheet can now be used to prepare performance reports for the company's service and operating departments. The general office and purchasing departments are cost centers, and their managers will be evaluated on their control of costs. Actual amounts of service department expenses can be compared to budgeted amounts to help assess cost center manager performance.

Amounts in the operating department columns are used to prepare departmental income statements as shown in Exhibit 21.21. This exhibit uses the spreadsheet for its operating expenses; information on sales and cost of goods sold comes from departmental records.

Example: If the $15,300 general office expenses in Exhibit 21.12 are allocated equally across departments, what is net income for the hardware department and for the combined company? *Answer:* Hardware income, $13,350; combined income, $19,000.

A-1 HARDWARE
Departmental Income Statements
For Year Ended December 31, 2009

	Hardware Department	Housewares Department	Appliances Department	Combined
Sales	$119,500	$71,700	$47,800	$239,000
Cost of goods sold	73,800	43,800	30,200	147,800
Gross profit	45,700	27,900	17,600	91,200
Operating expenses				
Salaries expense	15,600	7,000	7,800	30,400
Depreciation expense—Equipment	400	100	200	700
Supplies expense	300	200	100	600
Rent expense	4,860	3,240	2,700	10,800
Utilities expense	810	540	450	1,800
Advertising expense	500	300	200	1,000
Insurance expense	900	600	400	1,900
Share of general office expenses	7,650	4,590	3,060	15,300
Share of purchasing expenses	3,880	2,630	3,190	9,700
Total operating expenses	34,900	19,200	18,100	72,200
Net income (loss)	**$ 10,800**	**$ 8,700**	**$ (500)**	**$ 19,000**

EXHIBIT 21.21

Departmental Income Statements

Departmental Contribution to Overhead

P4 Prepare departmental contribution reports.

Point: Net income is the same in Exhibits 21.21 and 21.22. The method of reporting indirect expenses in Exhibit 21.22 does not change total net income but does identify each department's contribution to overhead and net income.

Data from departmental income statements are not always best for evaluating each profit center's performance, especially when indirect expenses are a large portion of total expenses and when weaknesses in assumptions and decisions in allocating indirect expenses can markedly affect net income. In these and other cases, we might better evaluate profit center performance using the **departmental contribution to overhead,** which is a report of the amount of sales less *direct* expenses.[2] We can also examine cost center performance by focusing on control of direct expenses.

The upper half of Exhibit 21.22 shows a departmental (profit center) contribution to overhead as part of an expanded income statement. This format is common when reporting departmental contributions to overhead. Using the information in Exhibits 21.21 and 21.22, we can evaluate the profitability of the three profit centers. For instance, let's compare the performance of the appliances department as described in these two exhibits. Exhibit 21.21 shows a $500 net loss resulting from this department's operations, but Exhibit 21.22 shows a $9,500 positive contribution to overhead, which is 19.9% of the appliance department's sales. The contribution of the appliances department is not as large as that of the other selling departments, but a $9,500 contribution to overhead is better than a $500 loss. This tells us that the appliances department is not a money loser. On the contrary, it is contributing $9,500 toward defraying total indirect expenses of $40,500.

EXHIBIT 21.22

Departmental Contribution to Overhead

A-1 HARDWARE
Income Statement Showing Departmental Contribution to Overhead
For Year Ended December 31, 2009

	Hardware Department	Housewares Department	Appliances Department	Combined
Sales	$119,500	$ 71,700	$47,800	$239,000
Cost of goods sold	73,800	43,800	30,200	147,800
Gross profit	45,700	27,900	17,600	91,200
Direct expenses				
Salaries expense	15,600	7,000	7,800	30,400
Depreciation expense—Equipment	400	100	200	700
Supplies expense	300	200	100	600
Total direct expenses	16,300	7,300	8,100	31,700
Departmental contributions to overhead	**$ 29,400**	**$20,600**	**$ 9,500**	**$ 59,500**
Indirect expenses				
Rent expense				10,800
Utilities expense				1,800
Advertising expense				1,000
Insurance expense				1,900
General office department expense				15,300
Purchasing department expense				9,700
Total indirect expenses				40,500
Net income				**$ 19,000**
Contribution as percent of sales	24.6%	28.7%	19.9%	24.9%

[2] A department's contribution is said to be "to overhead" because of the practice of considering all indirect expenses as overhead. Thus, the excess of a department's sales over direct expenses is a contribution toward at least a portion of its total overhead.

Quick Check Answers—p. 880

6. If a company has two operating (selling) departments (shoes and hats) and two service departments (payroll and advertising), which of the following statements is correct? (*a*) Wages incurred in the payroll department are direct expenses of the shoe department, (*b*) Wages incurred in the payroll department are indirect expenses of the operating departments, or (*c*) Advertising department expenses are allocated to the other three departments.
7. Which of the following bases can be used to allocate supervisors' salaries across operating departments? (*a*) Hours spent in each department, (*b*) number of employees in each department, (*c*) sales achieved in each department, or (*d*) any of the above, depending on which information is most relevant and accessible.
8. What three steps are used to allocate expenses to operating departments?
9. An income statement showing departmental contribution to overhead, (*a*) subtracts indirect expenses from each department's revenues, (*b*) subtracts only direct expenses from each department's revenues, or (*c*) shows net income for each department.

Evaluating Investment Center Performance

This section introduces both financial and nonfinancial measures of investment center performance.

Financial Performance Evaluation Measures

A1 Analyze investment centers using return on assets, residual income, and balanced scorecard.

Investment center managers are typically evaluated using performance measures that combine income and assets. Consider the following data for ZTel, a company which operates two divisions: LCD and S-Phone. The LCD division manufactures liquid crystal display (LCD) touch-screen monitors and sells them for use in computers, cellular phones, and other products. The S-Phone division sells smartphones, mobile phones that also function as personal computers, MP3 players, cameras, and global positioning satellite (GPS) systems. Exhibit 21.23 shows current year income and assets for those divisions.

EXHIBIT 21.23

Investment Center Income and Assets

	LCD	S-Phone
Net income	$ 526,500	$ 417,600
Average invested assets	2,500,000	1,850,000

Investment Center Return on Total Assets One measure to evaluate division performance is the **investment center return on total assets,** also called *return on investment* (ROI). This measure is computed as follows

$$\textbf{Return on investment} = \frac{\textbf{Investment center net income}}{\textbf{Investment center average invested assets}}$$

The return on investment for the LCD division is 21% (rounded), computed as $526,500/$2,500,000. The S-Phone division's return on investment is 23% (rounded), computed as $417,600/$1,850,000. Though the LCD division earned more dollars of net income, it was less efficient in using its assets to generate income compared to the S-Phone division.

Investment Center Residual Income Another way to evaluate division performance is to compute **investment center residual income,** which is computed as follows

$$\textbf{Residual income} = \begin{array}{c}\textbf{Investment center}\\ \textbf{net income}\end{array} - \begin{array}{c}\textbf{Target investment center}\\ \textbf{net income}\end{array}$$

Assume ZTel's top management sets target net income at 8% of divisional assets. For an investment center, this **hurdle rate** is typically the cost of obtaining financing. Applying this hurdle rate using the data from Exhibit 21.23 yields the residual income for ZTel's divisions in Exhibit 21.24:

EXHIBIT 21.24

Investment Center Residual Income

	LCD	S-Phone
Net income	$526,500	$417,600
Less: Target net income		
$2,500,000 × 8%	200,000	
$1,850,000 × 8%		148,000
Investment center residual income	$326,500	$269,600

Unlike return on assets, residual income is expressed in dollars. The LCD division outperformed the S-Phone division on the basis of residual income. However, this result is due in part to the LCD division having a larger asset base than the S-Phone division.

Using residual income to evaluate division performance encourages division managers to accept all opportunities that return more than the target net income, thus increasing company value. For example, the S-Phone division might not want to accept a new customer that will provide a 15% return on investment, since that will reduce the S-Phone division's overall return on investment (23% as shown above). However, the S-Phone division should accept this opportunity because the new customer would increase residual income by providing net income above the target net income.

Point: Residual income is also called *economic value added* (EVA).

Nonfinancial Performance Evaluation Measures

Evaluating performance solely on financial measures such as return on investment or residual income has limitations. For example, some investment center managers might forgo profitable opportunities to keep their return on investment high. Also, residual income is less useful when comparing investment centers of different size. And, both return on investment and residual income can encourage managers to focus too heavily on short-term financial goals.

In response to these limitations, companies consider nonfinancial measures. For example, a delivery company such as FedEx might track the percentage of on-time deliveries. The percentage of defective tennis balls manufactured can be used to assess performance of Penn's production managers. Walmart's credit card screens commonly ask customers at check-out whether the cashier was friendly or the store was clean. This kind of information can help division managers run their divisions and help top management evaluate division manager performance.

Balanced Scorecard The **balanced scorecard** is a system of performance measures, including nonfinancial measures, used to assess company and division manager performance. The balanced scorecard requires managers to think of their company from four perspectives:

1. *Customer:* What do customers think of us?
2. *Internal processes:* Which of our operations are critical to meeting customer needs?
3. *Innovation and learning:* How can we improve?
4. *Financial:* What do our owners think of us?

Point: One survey indicates that nearly 60% of global companies use some form of balanced scorecard.

The balanced scorecard collects information on several key performance indicators within each of the four perspectives. These key indicators vary across companies. Exhibit 21.25 lists common performance measures.

After selecting key performance indicators, companies collect data on each indicator and compare actual amounts to expected amounts to assess performance. For example, a company might have a goal of filling 98% of customer orders within two hours. Balanced scorecard reports are often presented in graphs or tables that can be updated frequently. Such timely information aids division managers in their decisions, and can be used by top management to evaluate division manager performance.

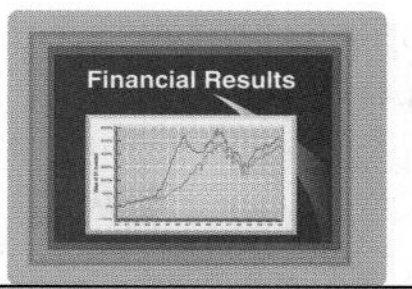

Customer	Internal Process	Innovation/Learning	Financial
• Customer satisfaction rating • # of new customers acquired • % of on-time deliveries • % of sales from new products • Time to fill orders % of sales returned	• Defect rates • Cycle time • Product costs • Labor hours per order • Production days without an accident	• Employee satisfaction • Employee turnover • $ spent on training • # of new products • # of patents • $ spent on research	• Net income • ROI • Sales growth • Cash flow • Residual income • Stock price

EXHIBIT 21.25

Balanced Scorecard Performance Indicators

Exhibit 21.26 is an example of balanced scorecard reporting on the customer perspective for an Internet retailer. This scorecard reports for example that the retailer is getting 62% of its potential customers successfully through the checkout process, and that 2.2% of all orders are returned. The *color* of the arrows in the right-most column reveals whether the company is exceeding its goal (green), barely meeting the goal (yellow), or not meeting the goal (red). The *direction* of the arrows reveals any trend in performance: an upward arrow indicates improvement, a downward arrow indicates declining performance, and an arrow pointing sideways indicates no change. A review of these arrows' color and direction suggests the retailer is meeting or exceeding its goals on checkout success, orders returned, and customer satisfaction. Further, checkout success and customer satisfaction are improving. The red arrow shows the company has received more customer complaints than was hoped for; however, the number of customer complaints is declining. A manager would combine this information with similar information on the internal process, innovation and learning, and financial perspectives to get an overall view of division performance.

EXHIBIT 21.26

Balanced Scorecard Reporting: Internet Retailer

Customer Perspective	Actual	Goal
Checkout success	62%	↑
Orders returned	2.20%	↔
Customer satisfaction rating	9.5	↑
Number of customer complaints	142	↓

Decision Maker

Center Manager Your center's usual return on total assets is 19%. You are considering two new investments for your center. The first requires a $250,000 average investment and is expected to yield annual net income of $50,000. The second requires a $1 million average investment with an expected annual net income of $175,000. Do you pursue either? [Answer—pp. 879–880]

Responsibility Accounting

C4 Explain controllable costs and responsibility accounting.

Departmental accounting reports often provide data used to evaluate a department's performance, but are they useful in assessing how well a department *manager* performs? Neither departmental income nor its contribution to overhead may be useful because many expenses can be outside a manager's control. Instead, we often evaluate a manager's performance using responsibility accounting reports that describe a department's activities in terms of **controllable costs.**[3] A cost is controllable if a manager has the power to determine or at least significantly affect the amount incurred. **Uncontrollable costs** are not within the manager's control or influence.

Controllable versus Direct Costs

Controllable costs are not always the same as direct costs. Direct costs are readily traced to a department, but the department manager might or might not control their amounts. For example, department managers often have little or no control over depreciation expense because they cannot affect the amount of equipment assigned to their departments. Also, department managers rarely control their own salaries. However, they can control or influence items such as the cost of supplies used in their department. When evaluating managers' performances, we should use data reflecting their departments' outputs along with their controllable costs and expenses.

Distinguishing between controllable and uncontrollable costs depends on the particular manager and time period under analysis. For example, the cost of property insurance is usually not controllable at the department manager's level but by the executive responsible for obtaining the company's insurance coverage. Likewise, this executive might not control costs resulting from insurance policies already in force. However, when a policy expires, this executive can renegotiate a replacement policy and then controls these costs. Therefore, all costs are controllable at some management level if the time period is sufficiently long. We must use good judgment in identifying controllable costs.

Responsibility Accounting System

A *responsibility accounting system* uses the concept of controllable costs to assign managers the responsibility for costs and expenses under their control. Prior to each reporting period, a company prepares plans that identify costs and expenses under each manager's control. These plans are called **responsibility accounting budgets.** To ensure the cooperation of managers and the reasonableness of budgets, managers should be involved in preparing their budgets.

A responsibility accounting system also involves performance reports. A **responsibility accounting performance report** accumulates and reports costs and expenses that a manager is responsible for and their budgeted amounts. Management's analysis of differences between budgeted amounts and actual costs and expenses often results in corrective or strategic managerial actions. Upper-level management uses performance reports to evaluate the effectiveness of lower-level managers in controlling costs and expenses and keeping them within budgeted amounts.

A responsibility accounting system recognizes that control over costs and expenses belongs to several levels of management. We illustrate this by considering the organization chart in Exhibit 21.27. The lines in this chart connecting the managerial positions reflect channels of authority. For example, the four department managers of this consulting firm (benchmarking, cost management, outsourcing, and service) are responsible for controllable costs and expenses incurred in their

EXHIBIT 21.27

Organizational Responsibility Chart

Board of Directors
President
Executive Vice President Marketing
Executive Vice President Operations
Executive Vice President Finance
Vice President Operational Consulting
Vice President Strategic Consulting
Manager Benchmarking Department
Manager Cost Management Department
Manager Outsourcing Department
Manager Service Department

[3] The terms *cost* and *expense* are often used interchangeably in managerial accounting, but they are not necessarily the same. *Cost* often refers to the monetary outlay to acquire some resource that can have present and future benefit. *Expense* usually refers to an expired cost. That is, as the benefit of a resource expires, a portion of its cost is written off as an expense.

departments, but these same costs are subject to the overall control of the vice president (VP) for operational consulting. Similarly, this VP's costs are subject to the control of the executive vice president (EVP) for operations, the president, and, ultimately, the board of directors.

At lower levels, managers have limited responsibility and relatively little control over costs and expenses. Performance reports for low-level management typically cover few controllable costs. Responsibility and control broaden for higher-level managers; therefore, their reports span a wider range of costs. However, reports to higher-level managers seldom contain the details reported to their subordinates but are summarized for two reasons: (1) lower-level managers are often responsible for these detailed costs and (2) detailed reports can obscure broader, more important issues facing a company.

Point: Responsibility accounting does not place blame. Instead, responsibility accounting is used to identify opportunities for improving performance.

Exhibit 21.28 shows summarized performance reports for the three management levels identified in Exhibit 21.27. Exhibit 21.28 shows that costs under the control of the benchmarking department manager are totaled and included among controllable costs of the VP for operational consulting. Also, costs under the control of the VP are totaled and included among controllable costs of the EVP for operations. In this way, a responsibility accounting system provides relevant information for each management level.

EXHIBIT 21.28

Responsibility Accounting Performance Reports

Executive Vice President, Operations

Controllable Costs	For July Budgeted Amount	 Actual Amount	 Over (Under) Budget
Salaries, VPs	$ 80,000	$ 80,000	$ 0
Quality control costs	21,000	22,400	1,400
Office costs	29,500	28,800	(700)
Operational consulting	**276,700**	**279,500**	**2,800**
Strategic consulting	390,000	380,600	(9,400)
Totals	$ 797,200	$ 791,300	$ (5,900)

Vice President, Operational Consulting

Controllable Costs	For July Budgeted Amount	 Actual Amount	 Over (Under) Budget
Salaries, department managers	$ 75,000	$ 78,000	$ 3,000
Depreciation	10,600	10,600	0
Insurance	6,800	6,300	(500)
Benchmarking department	**79,600**	**79,900**	**300**
Cost management department	61,500	60,200	(1,300)
Outsourcing department	24,300	24,700	400
Service department	18,900	19,800	900
Totals	**$276,700**	**$279,500**	**$2,800**

Manager, Benchmarking Department

Controllable Costs	For July Budgeted Amount	 Actual Amount	 Over (Under) Budget
Salaries	$ 51,600	$ 52,500	$ 900
Supplies	8,000	7,800	(200)
Other controllable costs	20,000	19,600	(400)
Totals	**$ 79,600**	**$ 79,900**	**$ 300**

Technological advances increase our ability to produce vast amounts of information that often exceed our ability to use it. Good managers select relevant data for planning and controlling the areas under their responsibility. A good responsibility accounting system makes every effort to provide relevant information to the right person (the one who controls the cost) at the right time (before a cost is out of control).

Point: Responsibility accounting usually divides a company into subunits, or *responsibility centers.* A center manager is evaluated on how well the center performs, as reported in responsibility accounting reports.

Quick Check

Answers—p. 880

10. Are the reports of departmental net income and the departmental contribution to overhead useful in assessing a department manager's performance? Explain.

11. Performance reports to evaluate managers should (*a*) include data about controllable expenses, (*b*) compare actual results with budgeted levels, or (*c*) both (*a*) and (*b*).

Decision Analysis — Investment Center Profit Margin and Investment Turnover

A2 Analyze investment centers using profit margin and investment turnover.

We can further examine investment center (division) performance by splitting return on investment into **profit margin** and **investment turnover** as follows

$$\textbf{Return on investment} = \textbf{Profit margin} \times \textbf{Investment turnover}$$

$$\frac{\text{Investment center net income}}{\text{Investment center average assets}} = \frac{\text{Investment center net income}}{\text{Investment center sales}} \times \frac{\text{Investment center sales}}{\text{Investment center average assets}}$$

Profit margin measures the income earned per dollar of sales. **Investment turnover** measures how efficiently an investment center generates sales from its invested assets. Higher profit margin and higher investment turnover indicate better performance. To illustrate, consider Best Buy which reports in Exhibit 21.29 results for two divisions (segments): Domestic and International.

EXHIBIT 21.29

Best Buy Division Sales, Income, and Assets

($ millions)	Domestic	International
Sales	$24,616	$2,817
Net income	1,393	49
Average invested assets	8,372	1,922

Profit margin and investment turnover for its Domestic and International divisions are computed and shown in Exhibit 21.30:

EXHIBIT 21.30

Best Buy Division Profit Margin and Investment Turnover

($ millions)	Domestic	International
Profit Margin		
$1,393/$24,616	5.66%	
$49/$2,817		1.74%
Investment Turnover		
$24,616/$8,372	2.94	
$2,817/$1,922		1.47

Best Buy's Domestic division generates 5.66 cents of profit per $1 of sales, while its International division generates only 1.74 cents of profit per dollar of sales. Its Domestic division also uses its assets more efficiently; its investment turnover of 2.94 is twice that of its International division's 1.47. Top management can use profit margin and investment turnover to evaluate the performance of division managers. The measures can also aid management when considering further investment in its divisions.

Decision Maker

Division Manager You manage a division in a highly competitive industry. You will receive a cash bonus if your division achieves an ROI above 12%. Your division's profit margin is 7%, equal to the industry average, and your division's investment turnover is 1.5. What actions can you take to increase your chance of receiving the bonus? [Answer—p. 880]

Demonstration Problem

DP21

Management requests departmental income statements for Hacker's Haven, a computer store that has five departments. Three are operating departments (hardware, software, and repairs) and two are service departments (general office and purchasing).

	General Office	Purchasing	Hardware	Software	Repairs
Sales	—	—	$960,000	$600,000	$840,000
Cost of goods sold	—	—	500,000	300,000	200,000
Direct expenses					
Payroll	$60,000	$45,000	80,000	25,000	325,000
Depreciation	6,000	7,200	33,000	4,200	9,600
Supplies	15,000	10,000	10,000	2,000	25,000

The departments incur several indirect expenses. To prepare departmental income statements, the indirect expenses must be allocated across the five departments. Then the expenses of the two service departments must be allocated to the three operating departments. Total cost amounts and the allocation bases for each indirect expense follow.

Indirect Expense	Total Cost	Allocation Basis
Rent	$150,000	Square footage occupied
Utilities	50,000	Square footage occupied
Advertising	125,000	Dollars of sales
Insurance	30,000	Value of assets insured
Service departments		
General office	?	Number of employees
Purchasing	?	Dollars of cost of goods sold

The following additional information is needed for indirect expense allocations.

Department	Square Feet	Sales	Insured Assets	Employees	Cost of Goods Sold
General office	500		$ 60,000		
Purchasing	500		72,000		
Hardware	4,000	$ 960,000	330,000	5	$ 500,000
Software	3,000	600,000	42,000	5	300,000
Repairs	2,000	840,000	96,000	10	200,000
Totals	10,000	$2,400,000	$600,000	20	$1,000,000

Required

1. Prepare a departmental expense allocation spreadsheet for Hacker's Haven.
2. Prepare a departmental income statement reporting net income for each operating department and for all operating departments combined.

Planning the Solution

- Set up and complete four tables to allocate the indirect expenses—one each for rent, utilities, advertising, and insurance.
- Allocate the departments' indirect expenses using a spreadsheet like the one in Exhibit 21.12. Enter the given amounts of the direct expenses for each department. Then enter the allocated amounts of the indirect expenses that you computed.
- Complete two tables for allocating the general office and purchasing department costs to the three operating departments. Enter these amounts on the spreadsheet and determine the total expenses allocated to the three operating departments.
- Prepare departmental income statements like the one in Exhibit 21.17. Show sales, cost of goods sold, gross profit, individual expenses, and net income for each of the three operating departments and for the combined company.

Solution to Demonstration Problem

Allocations of the four indirect expenses across the five departments.

Rent	Square Feet	Percent of Total	Allocated Cost
General office	500	5.0%	$ 7,500
Purchasing	500	5.0	7,500
Hardware	4,000	40.0	60,000
Software	3,000	30.0	45,000
Repairs	2,000	20.0	30,000
Totals	10,000	100.0%	$150,000

Utilities	Square Feet	Percent of Total	Allocated Cost
General office	500	5.0%	$ 2,500
Purchasing	500	5.0	2,500
Hardware	4,000	40.0	20,000
Software	3,000	30.0	15,000
Repairs	2,000	20.0	10,000
Totals	10,000	100.0%	$50,000

Advertising	Sales Dollars	Percent of Total	Allocated Cost
Hardware	$ 960,000	40.0%	$ 50,000
Software	600,000	25.0	31,250
Repairs	840,000	35.0	43,750
Totals	$2,400,000	100.0%	$125,000

Insurance	Assets Insured	Percent of Total	Allocated Cost
General office	$ 60,000	10.0%	$ 3,000
Purchasing	72,000	12.0	3,600
Hardware	330,000	55.0	16,500
Software	42,000	7.0	2,100
Repairs	96,000	16.0	4,800
Totals	$600,000	100.0%	$30,000

1. Allocations of service department expenses to the three operating departments.

General Office Allocations to	Employees	Percent of Total	Allocated Cost
Hardware	5	25.0%	$23,500
Software	5	25.0	23,500
Repairs	10	50.0	47,000
Totals	20	100.0%	$94,000

Purchasing Allocations to	Cost of Goods Sold	Percent of Total	Allocated Cost
Hardware	$ 500,000	50.0%	$37,900
Software	300,000	30.0	22,740
Repairs	200,000	20.0	15,160
Totals	$1,000,000	100.0%	$75,800

HACKER'S HAVEN
Departmental Expense Allocations
For Year Ended December 31, 2009

	Allocation Base	Expense Account Balance	General Office Dept.	Purchasing Dept.	Hardware Dept.	Software Dept.	Repairs Dept.
Direct Expenses							
Payroll		$ 535,000	$ 60,000	$ 45,000	$ 80,000	$ 25,000	$ 325,000
Depreciation		60,000	6,000	7,200	33,000	4,200	9,600
Supplies		62,000	15,000	10,000	10,000	2,000	25,000
Indirect Expenses							
Rent	Square ft.	150,000	7,500	7,500	60,000	45,000	30,000
Utilities	Square ft.	50,000	2,500	2,500	20,000	15,000	10,000
Advertising	Sales	125,000	—	—	50,000	31,250	43,750
Insurance	Assets	30,000	3,000	3,600	16,500	2,100	4,800
Total expenses		**1,012,000**	**94,000**	**75,800**	269,500	124,550	448,150
Service Department Expenses							
General office	Employees		**(94,000)**		23,500	23,500	47,000
Purchasing	Goods sold			**(75,800)**	37,900	22,740	15,160
Total expenses allocated to operating departments		**$1,012,000**	**$ 0**	**$ 0**	**$330,900**	**$170,790**	**$510,310**

2. Departmental income statements for Hacker's Haven.

HACKER'S HAVEN Departmental Income Statements For Year Ended December 31, 2009	Hardware	Software	Repairs	Combined
Sales	$ 960,000	$ 600,000	$ 840,000	$2,400,000
Cost of goods sold	500,000	300,000	200,000	1,000,000
Gross profit	460,000	300,000	640,000	1,400,000
Expenses				
Payroll	80,000	25,000	325,000	430,000
Depreciation	33,000	4,200	9,600	46,800
Supplies	10,000	2,000	25,000	37,000
Rent	60,000	45,000	30,000	135,000
Utilities	20,000	15,000	10,000	45,000
Advertising	50,000	31,250	43,750	125,000
Insurance	16,500	2,100	4,800	23,400
Share of general office	23,500	23,500	47,000	94,000
Share of purchasing	37,900	22,740	15,160	75,800
Total expenses	330,900	170,790	510,310	1,012,000
Net income	**$129,100**	**$129,210**	**$129,690**	**$ 388,000**

APPENDIX

Transfer Pricing 21A

C5 Explain transfer pricing and methods to set transfer prices.

Divisions in decentralized companies sometimes do business with one another. For example, a separate division of **Harley-Davidson** manufactures its plastic and fiberglass parts used in the company's motorcycles. **Anheuser-Busch**'s metal container division makes cans and lids used in its brewing operations, and also sells cans and lids to soft-drink companies. A division of **Prince** produces strings used in tennis rackets made by **Prince** and other manufacturers.

Determining the price that should be used to record transfers between divisions in the same company is the focus of this appendix. Because these transactions are transfers within the same company, the price to record them is called the **transfer price.** In decentralized organizations, division managers have input on or decide those prices. Transfer prices can be used in cost, profit, and investment centers. Since these transfers are not with customers outside the company, the transfer price has no direct impact on the company's overall profits. However, transfer prices can impact performance evaluations and, if set incorrectly, lead to bad decisions.

Point: Transfer pricing can impact company profits when divisions are located in countries with different tax rates; this is covered in advanced courses.

Alternative Transfer Prices

Exhibit 21A.1 reports data on the LCD division of ZTel. LCD manufactures liquid crystal display (LCD) touch-screen monitors for use in ZTel's S-Phone division's smartphones, which sell for $400 each. The monitors can also be used in other products. So, LCD can sell its monitors to buyers other than S-Phone. Likewise, the S-Phone division can purchase monitors from suppliers other than LCD.

Exhibit 21A.1 reveals the range of transfer prices for transfers of monitors from LCD to S-Phone. The manager of LCD wants to report a division profit; thus, this manager will not accept a transfer price less than $40 (variable manufacturing cost per unit) because doing so would cause the division to lose

EXHIBIT 21A.1

LCD Division Manufacturing Information—Monitors

Production capacity	100,000 units
Selling price per unit to outside customers	$80
Variable manufacturing costs per unit	$40
Fixed manufacturing costs	$2,000,000

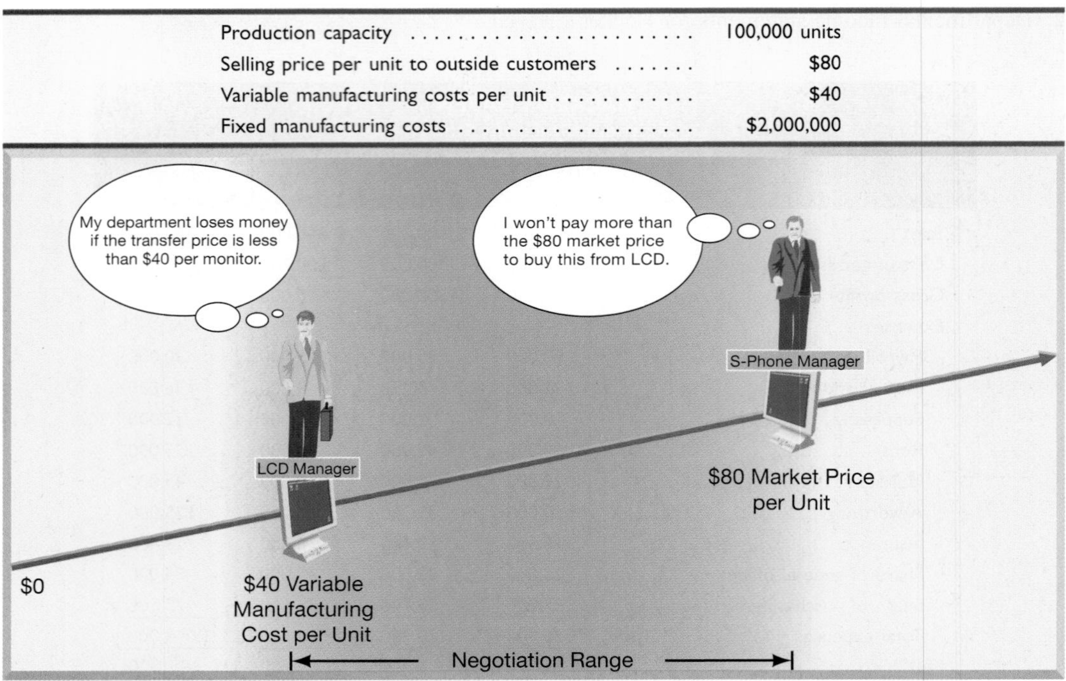

money on each monitor transferred. The LCD manager will only consider transfer prices of $40 or more. On the other hand, the S-Phone division manager also wants to report a division profit. Thus, this manager will not pay more than $80 per monitor because similar monitors can be bought from outside suppliers at that price. The S-Phone manager will only consider transfer prices of $80 or less. As any transfer price between $40 and $80 per monitor is possible, how does ZTel determine the transfer price? The answer depends in part on whether the LCD division has excess capacity to manufacture monitors.

No Excess Capacity Assume the LCD division can sell every monitor it produces, and thus is producing 100,000 units. In that case, a **market-based transfer price** of $80 per monitor is preferred. At that price, the LCD division manager is willing to either transfer monitors to S-Phone or sell to outside customers. The S-Phone manager cannot buy monitors for less than $80 from outside suppliers, so the $80 price is acceptable. Further, with a transfer price of $80 per monitor, top management of ZTel is indifferent to S-Phone buying from LCD or buying similar-quality monitors from outside suppliers.

With no excess capacity, the LCD manager will not accept a transfer price less than $80 per monitor. For example, suppose the S-Phone manager suggests a transfer price of $70 per monitor. At that price the LCD manager incurs an unnecessary *opportunity cost* of $10 per monitor (computed as $80 market price minus $70 transfer price). This would lower the LCD division's income and hurt its performance evaluation.

Excess Capacity Assume that the LCD division has excess capacity. For example, the LCD division might currently be producing only 80,000 units. Because LCD has $2,000,000 of fixed manufacturing costs, both LCD and the top management of ZTel prefer that S-Phone purchases its monitors from LCD. For example, if S-Phone purchases its monitors from an outside supplier at the market price of $80 each, LCD manufactures no units. Then, LCD reports a division loss equal to its fixed costs, and ZTel overall reports a lower net income as its costs are higher. Consequently, with excess capacity, LCD should accept any transfer price of $40 per unit or greater and S-Phone should purchase monitors from LCD. This will allow LCD to recover some (or all) of its fixed costs and increase ZTel's overall profits. For example, if a transfer price of $50 per monitor is used, the S-Phone manager is pleased to buy from LCD, since that price is below the market price of $80. For each monitor transferred from LCD to S-Phone at $50, the LCD division receives a *contribution margin* of $10 (computed as $50 transfer price less $40 variable cost) to contribute towards recovering its fixed costs. This form of transfer pricing is called **cost-based transfer pricing.** Under this approach the transfer price might be based on variable costs, total costs, or variable costs plus a markup. Determining the transfer price under excess capacity is complex and is covered in advanced courses.

Additional Issues in Transfer Pricing Several additional issues arise in determining transfer prices which include the following:

- **No market price exists.** Sometimes there is no market price for the product being transferred. The product might be a key component that requires additional conversion costs at the next stage and is not easily replicated by an outside company. For example, there is no market for a console for a **Nissan** Maxima and there is no substitute console **Nissan** can use in assembling a Maxima. In this case a market-based transfer price cannot be used.
- **Cost control.** To provide incentives for cost control, transfer prices might be based on standard, rather than actual costs. For example, if a transfer price of actual variable costs plus a markup of $20 per unit is used in the case above, LCD has no incentive to control its costs.
- **Division managers' negotiation.** With excess capacity, division managers will often negotiate a transfer price that lies between the variable cost per unit and the market price per unit. In this case, the **negotiated transfer price** and resulting departmental performance reports reflect, in part, the negotiating skills of the respective division managers. This might not be best for overall company performance.
- **Nonfinancial factors.** Factors such as quality control, reduced lead times, and impact on employee morale can be important factors in determining transfer prices.

Transfer Pricing Approaches Used by Companies

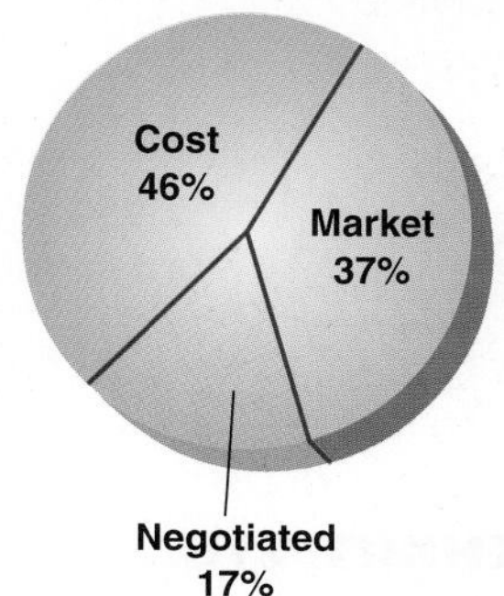

APPENDIX

Joint Costs and Their Allocation 21B

C6 Describe allocation of joint costs across products.

Most manufacturing processes involve **joint costs,** which refer to costs incurred to produce or purchase two or more products at the same time. A joint cost is like an indirect expense in the sense that more than one cost object share it. For example, a sawmill company incurs a joint cost when it buys logs that it cuts into lumber as shown in Exhibit 21B.1. The joint cost includes the logs (raw material) and its cutting (conversion) into boards classified as Clear, Select, No. 1 Common, No. 2 Common, No. 3 Common, and other types of lumber and by-products.

EXHIBIT 21B.1

Joint Products from Logs

Joint Cost: Cutting of Logs → Split-off Point → Joint Products: Clear, Select, No. 1 Common, No. 2 Common, No. 3 Common

When a joint cost is incurred, a question arises as to whether to allocate it to different products resulting from it. The answer is that when management wishes to estimate the costs of individual products, joint costs are included and must be allocated to these joint products. However, when management needs information to help decide whether to sell a product at a certain point in the production process or to process it further, the joint costs are ignored.

Financial statements prepared according to GAAP must assign joint costs to products. To do this, management must decide how to allocate joint costs across products benefiting from these costs. If some products are sold and others remain in inventory, allocating joint costs involves assigning costs to both cost of goods sold and ending inventory.

The two usual methods to allocate joint costs are the (1) *physical basis* and (2) the *value basis.* The physical basis typically involves allocating joint cost using physical characteristics such as the ratio of pounds, cubic feet, or gallons of each joint product to the total pounds, cubic feet, or gallons of all joint products flowing from the cost. This method is not preferred because the resulting cost allocations do not reflect the relative market values the joint cost generates. The preferred approach is the value basis, which allocates joint cost in proportion to the sales value of the output produced by the process at the "split-off point"; see Exhibit 21B.1.

Physical Basis Allocation of Joint Cost To illustrate the physical basis of allocating a joint cost, we consider a sawmill that bought logs for $30,000. When cut, these logs produce 100,000 board feet of lumber in the grades and amounts shown in Exhibit 21B.2. The logs produce 20,000 board feet of No. 3 Common lumber, which is 20% of the total. With physical allocation, the No. 3 Common lumber is assigned 20% of the $30,000 cost of the logs, or $6,000 ($30,000 × 20%). Because this low-grade lumber sells for $4,000, this allocation gives a $2,000 loss from its production and sale. The physical basis for allocating joint costs does not reflect the extra value flowing into some products or the inferior value flowing into others. That is, the portion of a log that produces Clear and Select grade lumber is worth more than the portion used to produce the three grades of common lumber, but the physical basis fails to reflect this.

EXHIBIT 21B.2

Allocating Joint Costs on a Physical Basis

Grade of Lumber	Board Feet Produced	Percent of Total	Allocated Cost	Sales Value	Gross Profit
Clear and Select	10,000	10.0%	$ 3,000	$12,000	$ 9,000
No. 1 Common	30,000	30.0	9,000	18,000	9,000
No. 2 Common	40,000	40.0	12,000	16,000	4,000
No. 3 Common	20,000	20.0	6,000	4,000	(2,000)
Totals	100,000	100.0%	$30,000	$50,000	$20,000

Value Basis Allocation of Joint Cost Exhibit 21B.3 illustrates the value basis method of allocation. It determines the percents of the total costs allocated to each grade by the ratio of each grade's sales value to the total sales value of $50,000 (sales value is the unit selling price multiplied by the number of units produced). The Clear and Select lumber grades receive 24% of the total cost ($12,000/$50,000) instead of the 10% portion using a physical basis. The No. 3 Common lumber receives only 8% of the total cost, or $2,400, which is much less than the $6,000 assigned to it using the physical basis.

EXHIBIT 21B.3

Allocating Joint Costs on a Value Basis

Grade of Lumber	Sales Value	Percent of Total	Allocated Cost	Gross Profit
Clear and Select	$12,000	24.0%	$ 7,200	$ 4,800
No. 1 Common	18,000	36.0	10,800	7,200
No. 2 Common	16,000	32.0	9,600	6,400
No. 3 Common	4,000	8.0	2,400	1,600
Totals	$50,000	100.0%	$30,000	$20,000

Example: Refer to Exhibit 21B.3. If the sales value of Clear and Select lumber is changed to $10,000, what is the revised ratio of the market value of No. 1 Common to the total?
Answer: $18,000/$48,000 = 37.5%

An outcome of value basis allocation is that *each* grade produces exactly the same 40% gross profit at the split-off point. This 40% rate equals the gross profit rate from selling all the lumber made from the $30,000 logs for a combined price of $50,000.

Quick Check

Answers—p. 880

12. A company produces three products, B1, B2, and B3. The joint cost incurred for the current month for these products is $180,000. The following data relate to this month's production:

Product	Units Produced	Unit Sales Value
B1	96,000	$3.00
B2	64,000	6.00
B3	32,000	9.00

The amount of joint cost allocated to product B3 using the value basis allocation is (*a*) $30,000, (*b*) $54,000, or (*c*) $90,000.

Summary

C1 Explain departmentalization and the role of departmental accounting. Companies are divided into departments when they are too large to be effectively managed as a single unit. Operating departments carry out an organization's main functions. Service departments support the activities of operating departments. Departmental accounting systems provide information for evaluating departmental performance.

C2 Distinguish between direct and indirect expenses. Direct expenses are traced to a specific department and are incurred for the sole benefit of that department. Indirect expenses benefit more than one department. Indirect expenses are allocated to departments when computing departmental net income.

C3 Identify bases for allocating indirect expenses to departments. Ideally, we allocate indirect expenses by using a cause-effect relation for the allocation base. When a cause-effect relation is not identifiable, each indirect expense is allocated on a basis reflecting the relative benefit received by each department.

C4 Explain controllable costs and responsibility accounting. A controllable cost is one that is influenced by a specific management level. The total expenses of operating a department often include some items a department manager does not control. Responsibility accounting systems provide information for evaluating the performance of department managers. A responsibility accounting system's performance reports for evaluating department managers should include only the expenses (and revenues) that each manager controls.

C5 Explain transfer pricing and methods to set transfer prices. Transfer prices are used to record transfers of items between divisions of the same company. Transfer prices can be based on costs or market prices, or can be negotiated by division managers.

C6 Describe allocation of joint costs across products. A joint cost refers to costs incurred to produce or purchase two or more products at the same time. When income statements are prepared, joint costs are usually allocated to the resulting joint products using either a physical or value basis.

A1 Analyze investment centers using return on assets, residual income, and balanced scorecard. A financial measure often used to evaluate an investment center manager is the *investment center return on total assets,* also called *return on investment.* This measure is computed as the center's net income divided by the center's average total assets. Residual income, computed as investment center net income minus a target net income is an alternative financial measure of investment center performance. A balanced scorecard uses a combination of financial and non-financial measures to evaluate performance.

A2 Analyze investment centers using profit margin and investment turnover. Return on investment can also be computed as profit margin times investment turnover. Profit margin (equal to net income/sales) measures the income earned per dollar of sales and investment turnover (equal to sales/assets) measures how efficiently a division uses its assets.

P1 Assign overhead costs using two-stage cost allocation. In the traditional two-stage cost allocation procedure, service department costs are first assigned to operating departments. Then, in the second stage, a predetermined overhead allocation rate is computed for each operating department and is used to assign overhead to output.

P2 Assign overhead costs using activity-based costing. In activity-based costing, the costs of related activities are collected and then pooled in some logical manner into activity cost pools. After all activity costs have been accumulated in an activity cost pool account, users of the activity, termed *cost objects,* are assigned a portion of the total activity cost using a cost driver (allocation base).

P3 Prepare departmental income statements. Each profit center (department) is assigned its expenses to yield its own income statement. These costs include its direct expenses and its share of indirect expenses. The departmental income statement lists its revenues and costs of goods sold to determine gross profit. Its operating expenses (direct expenses and its indirect expenses allocated to the department) are deducted from gross profit to yield departmental net income.

P4 Prepare departmental contribution reports. The departmental contribution report is similar to the departmental income statement in terms of computing the gross profit for each department. Then the direct operating expenses for each department are deducted from gross profit to determine the contribution generated by each department. Indirect operating expenses are deducted *in total* from the company's combined contribution.

Guidance Answers to **Decision Maker** and **Decision Ethics**

Director of Operations You should collect details on overhead items and review them to see whether direct labor drives these costs. If it does not, overhead might be improperly assigned to departments. The situation also provides an opportunity to consider other overhead allocation bases, including use of activity-based costing.

Accounting Officer You should not author a report that you disagree with. You are responsible for ascertaining all the facts of ABC (implementation procedures, advantages and disadvantages, and costs). You should then approach your supervisor with these facts and suggest that you would like to modify the report to request, for example, a pilot test. The pilot test will allow you to further assess the suitability of ABC. Your suggestion might be rejected, at which time you may wish to speak with a more senior-level manager.

Center Manager We must first realize that the two investment opportunities are not comparable on the basis of absolute dollars of income or on assets. For instance, the second investment provides a higher income in absolute dollars but requires a higher investment. Accordingly, we need to compute return on total assets for each alternative: (1) $50,000 ÷ $250,000 = 20%, and (2) $175,000 ÷ $1 million = 17.5%. Alternative 1 has the higher return and is preferred over alternative 2. Do you pursue one, both, or neither? Because alternative 1's return is higher than the center's usual return of 19%, it should be pursued, assuming its risks are acceptable. Also, since alternative 1 requires a small investment, top management is likely to be more agreeable to pursuing it. Alternative 2's return is lower than the usual 19% and is not likely to be acceptable.

Division Manager Your division's ROI without further action is 10.5% (equal to 7% × 1.5). In a highly competitive industry, it is difficult to increase profit margins by raising prices. Your division might be better able to control its costs to increase its profit margin. In addition, you might engage in a marketing program to increase sales without increasing your division's invested assets. Investment turnover and thus ROI will increase if the marketing campaign attracts customers.

Guidance Answers to **Quick Checks**

1. Cost drivers are the factors that have a cause-effect relation with costs (or activities that pertain to costs).
2. *d*
3. A departmental accounting system provides information used to evaluate the performance of *departments.* A responsibility accounting system provides information used to evaluate the performance of *department managers.*
4. *d*
5. A cost center, such as a service department, incurs costs without directly generating revenues. A profit center, such as a product division, incurs costs but also generates revenues.
6. *b*
7. *d*
8. (1) Assign the direct expenses to each department. (2) Allocate indirect expenses to all departments. (3) Allocate the service department expenses to the operating departments.
9. *b*
10. No, because many expenses that enter into these calculations are beyond the manager's control, and managers should not be evaluated using costs they do not control.
11. *c*
12. *b*; $180,000 × ([32,000 × $9]/[96,000 × $3 + 64,000 × $6 + 32,000 × $9]) = $54,000.

Key Terms

mhhe.com/wildFAP19e

Key Terms are available at the book's Website for learning and testing in an online Flashcard Format.

Activity-based costing (ABC) (p. 854)
Activity cost driver (p. 855)
Activity cost pool (p. 855)
Balanced scorecard (p. 868)
Controllable costs (p. 870)
Cost-based transfer pricing (p. 876)
Cost center (p. 858)
Departmental accounting system (p. 858)
Departmental contribution to overhead (p. 866)
Direct expenses (p. 859)
Hurdle rate (p. 868)
Indirect expenses (p. 859)
Investment center (p. 858)
Investment center residual income (p. 867)
Investment center return on total assets (p. 867)
Investment turnover (p. 872)
Joint cost (p. 877)
Market-based transfer price (p. 876)
Negotiated transfer price (p. 877)
Profit center (p. 858)
Profit margin (p. 872)
Responsibility accounting budget (p. 870)
Responsibility accounting performance report (p. 870)
Responsibility accounting system (p. 858)
Transfer price (p. 875)
Uncontrollable costs (p. 870)

Multiple Choice Quiz

Answers on p. 899

mhhe.com/wildFAP19e

Additional Quiz Questions are available at the book's Website.

Quiz21

1. A retailer has three departments—housewares, appliances, and clothing—and buys advertising that benefits all departments. Advertising expense is $150,000 for the year, and departmental sales for the year follow: housewares, $356,250; appliances, $641,250; clothing, $427,500. How much advertising expense is allocated to appliances if allocation is based on departmental sales?
   a. $37,500
   b. $67,500
   c. $45,000
   d. $150,000
   e. $641,250
2. An activity-based costing system
   a. Does not require the level of detail that a traditional costing system requires.
   b. Does not enable the calculation of unit cost data.
   c. Allocates costs to products on the basis of activities performed on them.

d. Cannot be used by a service company.
e. Allocates costs to products based on the number of direct labor hours used.

3. A company produces two products, Grey and Red. The following information is available relating to those two products. Assume that the company's total setup cost is $162,000. Using activity-based costing, how much setup cost is allocated to each unit of Grey?

	Grey	Red
Units produced	500	40,000
Number of setups	50	100
Direct labor hours per unit	15	15

a. $1,080
b. $ 72
c. $ 162
d. $2,000
e. $ 108

4. A company operates three retail departments as profit centers, and the following information is available for each. Which department has the largest dollar amount of departmental contribution to overhead and what is the dollar amount contributed?

Department	Sales	Cost of Goods Sold	Direct Expenses	Allocated Indirect Expenses
X	$500,000	$350,000	$50,000	$40,000
Y	200,000	75,000	20,000	50,000
Z	350,000	150,000	75,000	10,000

a. Department Y, $ 55,000
b. Department Z, $125,000
c. Department X, $500,000
d. Department Z, $200,000
e. Department X, $ 60,000

5. Using the data in question 4, Department X's contribution to overhead as a percentage of sales is
a. 20%
b. 30%
c. 12%
d. 48%
e. 32%

Superscript letter A(B) denotes assignments based on Appendix 21A (21B).

Discussion Questions

1. Why are many companies divided into departments?

2. Complete the following for a traditional two-stage allocation system: In the first stage, service department costs are assigned to _____ departments. In the second stage, a predetermined overhead rate is computed for each operating department and used to assign overhead to _____.

3. What is the difference between operating departments and service departments?

4. What is activity-based costing? What is its goal?

5. Identify at least four typical cost pools for activity-based costing in most organizations.

6. In activity-based costing, costs in a cost pool are allocated to _____ using predetermined overhead rates.

7. What company circumstances especially encourage use of activity-based costing?

8. What are two main goals in managerial accounting for reporting on and analyzing departments?

9. Is it possible to evaluate a cost center's profitability? Explain.

10. What is the difference between direct and indirect expenses?

11. Suggest a reasonable basis for allocating each of the following indirect expenses to departments: (a) salary of a supervisor who manages several departments, (b) rent, (c) heat, (d) electricity for lighting, (e) janitorial services, (f) advertising, (g) expired insurance on equipment, and (h) property taxes on equipment.

12. How is a department's contribution to overhead measured?

13. What are controllable costs?

14. Controllable and uncontrollable costs must be identified with a particular _____ and a definite _____ period.

15. Why should managers be closely involved in preparing their responsibility accounting budgets?

16. In responsibility accounting, who receives timely cost reports and specific cost information? Explain.

17.[A] What is a transfer price? Under what conditions is a market-based transfer price most likely to be used?

18.[B] What is a joint cost? How are joint costs usually allocated among the products produced from them?

19.[B] Give two examples of products with joint costs.

20. Each retail store of **Best Buy** has several departments. Why is it useful for its management to (a) collect accounting information about each department and (b) treat each department as a profit center?

21. **Apple** delivers its products to locations around the world. List three controllable and three uncontrollable costs for its delivery department.

Denotes Discussion Questions that involve decision making.

Available with McGraw-Hill's Homework Manager

McGraw-Hill's HOMEWORK MANAGER®

QUICK STUDY

QS 21-1

Allocation and measurement terms

C1 C2 C3 C4 A1

In each blank next to the following terms, place the identifying letter of its best description.

1. ______ Cost center
2. ______ Investment center
3. ______ Departmental accounting system
4. ______ Operating department
5. ______ Profit center
6. ______ Responsibility accounting system
7. ______ Service department

A. Engages directly in manufacturing or in making sales directly to customers.
B. Does not directly manufacture products but contributes to profitability of the entire company.
C. Incurs costs and also generates revenues.
D. Provides information used to evaluate the performance of a department.
E. Incurs costs without directly yielding revenues.
F. Provides information used to evaluate the performance of a department manager.
G. Holds manager responsible for revenues, costs, and investments.

QS 21-2

Basis for cost allocation

C3 

For each of the following types of indirect expenses and service department expenses, identify one allocation basis that could be used to distribute it to the departments indicated.

1. Computer service expenses of production scheduling for operating departments.
2. General office department expenses of the operating departments.
3. Maintenance department expenses of the operating departments.
4. Electric utility expenses of all departments.

QS 21-3

Activity-based costing and overhead cost allocation

P2

The following is taken from Mortan Co.'s internal records of its factory with two operating departments. The cost driver for indirect labor and supplies is direct labor costs, and the cost driver for the remaining overhead items is number of hours of machine use. Compute the total amount of overhead cost allocated to Operating Department 1 using activity-based costing.

	Direct Labor	Machine Use Hours
Operating department 1	$18,800	2,000
Operating department 2	13,200	1,200
Totals	$32,000	3,200
Factory overhead costs		
Rent and utilities		$12,200
Indirect labor		5,400
General office expense		4,000
Depreciation—Equipment		3,000
Supplies		2,600
Total factory overhead		$27,200

Check Dept. 1 allocation, $16,700

QS 21-4

Departmental contribution to overhead

P4

Use the information in the following table to compute each department's contribution to overhead (both in dollars and as a percent). Which department contributes the largest dollar amount to total overhead? Which contributes the highest percent (as a percent of sales)?

	Dept. A	Dept. B	Dept. C
Sales	$106,000	$360,000	$168,000
Cost of goods sold	68,370	207,400	99,120
Gross profit	37,630	152,600	68,880
Total direct expenses	6,890	74,120	15,120
Contribution to overhead	$ ______	$ ______	$ ______
Contribution percent	______ %	______ %	______ %

QS 21-5
Investment center analysis
A1 

Compute return on assets for each of these **Best Buy** divisions (each is an investment center). Comment on the relative performance of each investment center.

Investment Center	Net Income	Average Assets	Return on Assets
Cameras and camcorders	$4,500,000	$20,000,000	______
Phones and communications	1,500,000	12,500,000	______
Computers and accessories	800,000	10,000,000	______

QS 21-6
Computing residual income
A1

Refer to information in QS 21-5. Assume a target income of 12% of average invested assets. Compute residual income for each of Best Buy's divisions.

QS 21-7
Computing performance measures
A1 A2

A company's shipping division (an investment center) has sales of $2,700,000, net income of $216,000, and average invested assets of $2,000,000. Compute the division's return on invested assets, profit margin, and investment turnover.

QS 21-8
Performance measures
A1 A2

Fill in the blanks in the schedule below for two separate investment centers A and B.

	Investment Center	
	A	B
Sales	$______	$3,200,000
Net income	$126,000	$______
Average invested assets	$700,000	______
Profit margin	6%	______%
Investment turnover	______	1.6
Return on assets	______%	10%

QS 21-9
Performance measures—balanced scorecard
A1

Classify each of the performance measures below into the most likely balanced scorecard perspective it relates to. Label your answers using C (customer), P (internal process), I (innovation and growth), or F (financial).

1. Change in market share ______
2. Employee training sessions attended ______
3. Number of days of employee absences ______
4. Customer wait time ______
5. Number of new products introduced ______
6. Length of time raw materials are in inventory ______
7. Profit margin ______
8. Customer satisfaction index ______

QS 21-10
Performance measures—balanced scorecard
A1

Walt Disney reports the following information for its two Parks and Resorts divisions.

	East Coast		West Coast	
	Current year	Prior year	Current year	Prior year
Hotel occupancy rates	89%	86%	92%	93%

Assume **Walt Disney** uses a balanced scorecard and sets a target of 90% occupancy in its resorts. Using Exhibit 21.26 as a guide, show how the company's performance on hotel occupancy would appear on a balanced scorecard report.

QS 21-11[A]
Determining transfer prices without excess capacity
C5

The Windshield division of Chee Cycles makes windshields for use in Chee's Assembly division. The Windshield division incurs variable costs of $175 per windshield and has capacity to make 50,000 windshields per year. The market price is $300 per windshield. The Windshield division incurs total fixed costs of $1,500,000 per year. If the Windshield division is operating at full capacity, what transfer price should be used on transfers between the Windshield and Assembly divisions? Explain.

QS 21-12[A]
Determining transfer prices with excess capacity
C5

Refer to information in QS 21-11. If the Windshield division has excess capacity, what is the range of possible transfer prices that could be used on transfers between the Windshield and Assembly divisions? Explain.

QS 21-13[B]
Joint cost allocation
C6

A company purchases a 10,020 square foot commercial building for $500,000 and spends an additional $50,000 to divide the space into two separate rental units and prepare it for rent. Unit A, which has the desirable location on the corner and contains 3,340 square feet, will be rented for $2.00 per square foot. Unit B contains 6,680 square feet and will be rented for $1.50 per square foot. How much of the joint cost should be assigned to Unit B using the value basis of allocation?

EXERCISES

Exercise 21-1
Departmental expense allocations
P1 C3

Firefly Co. has four departments: materials, personnel, manufacturing, and packaging. In a recent month, the four departments incurred three shared indirect expenses. The amounts of these indirect expenses and the bases used to allocate them follow.

Indirect Expense	Cost	Allocation Base
Supervision	$ 80,000	Number of employees
Utilities	61,000	Square feet occupied
Insurance	16,700	Value of assets in use
Total	$157,700	

Departmental data for the company's recent reporting period follow.

Department	Employees	Square Feet	Asset Values
Materials	40	27,000	$ 60,000
Personnel	22	5,000	1,200
Manufacturing	104	45,000	42,000
Packaging	34	23,000	16,800
Total	200	100,000	$120,000

(1) Use this information to allocate each of the three indirect expenses across the four departments. (2) Prepare a summary table that reports the indirect expenses assigned to each of the four departments.

Check (2) Total of $40,820 assigned to Materials Dept.

Exercise 21-2
Activity-based costing of overhead
P2 

Northwest Company produces two types of glass shelving, rounded edge and squared edge, on the same production line. For the current period, the company reports the following data.

	Rounded Edge	Squared Edge	Total
Direct materials	$19,000	$ 43,200	$ 62,200
Direct labor	12,200	23,800	36,000
Overhead (300% of direct labor cost)	36,600	71,400	108,000
Total cost	$67,800	$138,400	$206,200
Quantity produced	10,500 ft.	14,100 ft.	
Average cost per ft.	$ 6.46	$ 9.82	

Northwest's controller wishes to apply activity-based costing (ABC) to allocate the $108,000 of overhead costs incurred by the two product lines to see whether cost per foot would change markedly from that reported above. She has collected the following information.

Overhead Cost Category (Activity Cost Pool)	Cost
Supervision	$ 5,400
Depreciation of machinery	56,600
Assembly line preparation	46,000
Total overhead	$108,000

She has also collected the following information about the cost drivers for each category (cost pool) and the amount of each driver used by the two product lines.

Overhead Cost Category (Activity Cost Pool)	Driver	Usage		
		Rounded Edge	Squared Edge	Total
Supervision	Direct labor cost($)	$12,200	$23,800	$36,000
Depreciation of machinery	Machine hours	500 hours	1,500 hours	2,000 hours
Assembly line preparation	Setups (number)	40 times	210 times	250 times

Use this information to (1) assign these three overhead cost pools to each of the two products using ABC, (2) determine average cost per foot for each of the two products using ABC, and (3) compare the average cost per foot under ABC with the average cost per foot under the current method for each product. For part 3, explain why a difference between the two cost allocation methods exists.

Check (2) Rounded edge, $5.19; Squared edge, $10.76

Exercise 21-3
Rent expense allocated to departments
P1 C3

Expert Garage pays $128,000 rent each year for its two-story building. The space in this building is occupied by five departments as specified here.

Paint department	1,200 square feet of first-floor space
Engine department	3,600 square feet of first-floor space
Window department	1,920 square feet of second-floor space
Electrical department	1,056 square feet of second-floor space
Accessory department	1,824 square feet of second-floor space

The company allocates 65% of total rent expense to the first floor and 35% to the second floor, and then allocates rent expense for each floor to the departments occupying that floor on the basis of space occupied. Determine the rent expense to be allocated to each department. (Round percents to the nearest one-tenth and dollar amounts to the nearest whole dollar.)

Check Allocated to Paint Dept., $20,800

Exercise 21-4
Departmental expense allocation spreadsheet
C3 P1

Off-Road Cycle Shop has two service departments (advertising and administration) and two operating departments (cycles and clothing). During 2009, the departments had the following direct expenses and occupied the following amount of floor space.

Department	Direct Expenses	Square Feet
Advertising	$ 21,000	1,820
Administrative	15,000	1,540
Cycles	102,000	6,440
Clothing	12,000	4,200

The advertising department developed and distributed 100 advertisements during the year. Of these, 72 promoted cycles and 28 promoted clothing. The store sold $300,000 of merchandise during the year. Of this amount, $228,000 is from the cycles department, and $72,000 is from the clothing department. The utilities expense of $65,000 is an indirect expense to all departments. Prepare a departmental expense allocation spreadsheet for Off-Road Cycle Shop. The spreadsheet should assign (1) direct expenses to each of the four departments, (2) the $65,000 of utilities expense to the four departments on the basis of floor space occupied, (3) the advertising department's expenses to the two operating departments on the basis of the number of ads placed that promoted a department's products, and (4) the administrative department's expenses to the two operating departments based on the amount of sales. Provide supporting computations for the expense allocations.

Check Total expenses allocated to Cycles Dept., $169,938

Exercise 21-5
Service department expenses allocated to operating departments
P3

The following is a partially completed lower section of a departmental expense allocation spreadsheet for Haston Bookstore. It reports the total amounts of direct and indirect expenses allocated to its five departments. Complete the spreadsheet by allocating the expenses of the two service departments (advertising and purchasing) to the three operating departments.

	Allocation Base	Expense Account Balance	Advertising Dept.	Purchasing Dept.	Books Dept.	Magazines Dept.	Newspapers Dept.
			Allocation of Expenses to Departments				
Total department expenses..........		$653,000	$23,000	$30,000	$426,000	$85,000	$89,000
Service department expenses							
Advertising department.............	Sales		?		?	?	?
Purchasing department.............	Purch. orders			?	?	?	?
Total expenses allocated to operating departments..............		?	$ 0	$ 0	?	?	?

Advertising and purchasing department expenses are allocated to operating departments on the basis of dollar sales and purchase orders, respectively. Information about the allocation bases for the three operating departments follows.

Department	Sales	Purchase Orders
Books	$440,000	400
Magazines	160,000	250
Newspapers	200,000	350
Total	$800,000	1,000

Check Total expenses allocated to Books Dept., $450,650

Exercise 21-6
Indirect payroll expense allocated to departments
C3

Jaria Stevens works in both the jewelry department and the hosiery department of a retail store. Stevens assists customers in both departments and arranges and stocks merchandise in both departments. The store allocates Stevens' $35,000 annual wages between the two departments based on a sample of the time worked in the two departments. The sample is obtained from a diary of hours worked that Stevens kept in a randomly chosen two-week period. The diary showed the following hours and activities spent in the two departments. Allocate Stevens' annual wages between the two departments.

Selling in jewelry department	41 hours
Arranging and stocking merchandise in jewelry department	4 hours
Selling in hosiery department	24 hours
Arranging and stocking merchandise in hosiery department	6 hours
Idle time spent waiting for a customer to enter one of the selling departments	5 hours

Check Assign $14,000 to Hosiery

Exercise 21-7
Managerial performance evaluation
C4

Rex Stanton manages an auto dealership's service department. The recent month's income statement for his department follows. (1) Analyze the items on the income statement and identify those that definitely should be included on a performance report used to evaluate Stanton's performance. List them and explain why you chose them. (2) List and explain the items that should definitely be excluded. (3) List the items that are not definitely included or excluded and explain why they fall into that category.

Revenues		
Sales of parts	$ 72,000	
Sales of services	105,000	$177,000
Costs and expenses		
Cost of parts sold	30,000	
Building depreciation	9,300	
Income taxes allocated to department	8,700	
Interest on long-term debt	7,500	
Manager's salary	12,000	
Payroll taxes	8,100	
Supplies	15,900	
Utilities	4,400	
Wages (hourly)	16,000	
Total costs and expenses		111,900
Departmental net income		$ 65,100

Exercise 21-8
Investment center analysis
A1

You must prepare a return on investment analysis for the regional manager of Veggie Burgers. This growing chain is trying to decide which outlet of two alternatives to open. The first location (A) requires a $500,000 investment and is expected to yield annual net income of $85,000. The second location (B) requires a $200,000 investment and is expected to yield annual net income of $42,000. Compute the return on investment for each Veggie Burgers alternative and then make your recommendation in a one-half page memorandum to the regional manager. (The chain currently generates an 18% return on total assets.)

Exercise 21-9
Computing performance measures
A1

ZMart, a retailer of consumer goods, provides the following information on two of its departments (each considered an investment center).

Investment Center	Sales	Net Income	Average Invested Assets
Electronics	$10,000,000	$750,000	$3,750,000
Sporting goods	8,000,000	800,000	5,000,000

(1) Compute return on investment for each department. Using return on investment, which department is most efficient at using assets to generate returns for the company? (2) Assume a target income level of 12% of average invested assets. Compute residual income for each department. Which department generated the most residual income for the company? (3) Assume the Electronics department is presented with a new investment opportunity that will yield a 15% return on assets. Should the new investment opportunity be accepted? Explain.

Exercise 21-10
Computing performance measures A2

Refer to information in Exercise 21-9. Compute profit margin and investment turnover for each department. Which department generates the most net income per dollar of sales? Which department is most efficient at generating sales from average invested assets?

Exercise 21-11
Performance measures—balanced scorecard
A1

MidCoast Airlines uses the following performance measures. Classify each of the performance measures below into the most likely balanced scorecard perspective it relates to. Label your answers using C (customer), P (internal process), I (innovation and growth), or F (financial).

1. Percentage of ground crew trained _______
2. On-time flight percentage _______
3. Percentage of on-time departures _______
4. Market value _______
5. Flight attendant training sessions attended _______
6. Revenue per seat _______
7. Customer complaints _______
8. Time airplane is on ground between flights _______
9. Number of reports of mishandled or lost baggage _______
10. Cash flow from operations _______
11. Accidents or safety incidents per mile flown _______
12. Airplane miles per gallon of fuel _______
13. Return on investment _______
14. Cost of leasing airplanes _______

Exercise 21-12[A]
Determining transfer prices
C5

The Trailer department of Sprint Bicycles makes bike trailers that attach to bicycles and can carry children or cargo. The trailers have a retail price of $100 each. Each trailer incurs $40 of variable manufacturing costs. The Trailer department has capacity for 20,000 trailers per year, and incurs fixed costs of $500,000 per year.

Required

1. Assume the Assembly division of Sprint Bicycles wants to buy 5,000 trailers per year from the Trailer division. If the Trailer division can sell all of the trailers it manufactures to outside customers, what price should be used on transfers between Sprint Bicycle's divisions? Explain.
2. Assume the Trailer division currently only sells 10,000 trailers to outside customers, and the Assembly division wants to buy 5,000 trailers per year from the Trailer division. What is the range of acceptable prices that could be used on transfers between Sprint Bicycle's divisions? Explain.
3. Assume transfer prices of either $40 per trailer or $70 per trailer are being considered. Comment on the preferred transfer prices from the perspectives of the Trailer division manager, the Assembly division manager, and the top management of Sprint Bicycles.

Exercise 21-13[B]
Joint real estate costs assigned
C6

Check Total Canyon cost, $2,700,000

Mountain Home Properties is developing a subdivision that includes 300 home lots. The 225 lots in the Canyon section are below a ridge and do not have views of the neighboring canyons and hills; the 75 lots in the Hilltop section offer unobstructed views. The expected selling price for each Canyon lot is $50,000 and for each Hilltop lot is $100,000. The developer acquired the land for $2,500,000 and spent another $2,000,000 on street and utilities improvements. Assign the joint land and improvement costs to the lots using the value basis of allocation and determine the average cost per lot.

Exercise 21-14[B]
Joint product costs assigned
C6

Ocean Seafood Company purchases lobsters and processes them into tails and flakes. It sells the lobster tails for $20 per pound and the flakes for $15 per pound. On average, 100 pounds of lobster are processed into 57 pounds of tails and 24 pounds of flakes, with 19 pounds of waste. Assume that the company purchased 3,000 pounds of lobster for $6.00 per pound and processed the lobsters with an additional labor cost of $1,800. No materials or labor costs are assigned to the waste. If 1,570 pounds of tails and 640 pounds of flakes are sold, what is (1) the allocated cost of the sold items and (2) the allocated cost of the ending inventory? The company allocates joint costs on a value basis. (Round the dollar cost per pound to the nearest thousandth.)

Check (2) Inventory cost, $1,760

McGraw-Hill's HOMEWORK MANAGER® Available with McGraw-Hill's Homework Manager

PROBLEM SET A

Problem 21-1A
Allocation of building occupancy costs to departments
C3 P1 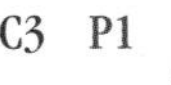

mhhe.com/wildFAP19e

Citizens Bank has several departments that occupy both floors of a two-story building. The departmental accounting system has a single account, Building Occupancy Cost, in its ledger. The types and amounts of occupancy costs recorded in this account for the current period follow.

Depreciation—Building	$ 31,500
Interest—Building mortgage	47,000
Taxes—Building and land	14,000
Gas (heating) expense	4,425
Lighting expense	5,250
Maintenance expense	9,625
Total occupancy cost	$111,800

The building has 5,000 square feet on each floor. In prior periods, the accounting manager merely divided the $111,800 occupancy cost by 10,000 square feet to find an average cost of $11.18 per square foot and then charged each department a building occupancy cost equal to this rate times the number of square feet that it occupied.

Helen Lanya manages a first-floor department that occupies 1,000 square feet, and Jose Jimez manages a second-floor department that occupies 1,700 square feet of floor space. In discussing the departmental reports, the second-floor manager questions whether using the same rate per square foot for all departments makes sense because the first-floor space is more valuable. This manager also references a recent real estate study of average local rental costs for similar space that shows first-floor space worth $40 per square foot and second-floor space worth $10 per square foot (excluding costs for heating, lighting, and maintenance).

Required

1. Allocate occupancy costs to the Lanya and Jimez departments using the current allocation method.

2. Allocate the depreciation, interest, and taxes occupancy costs to the Lanya and Jimez departments in proportion to the relative market values of the floor space. Allocate the heating, lighting, and maintenance costs to the Lanya and Jimez departments in proportion to the square feet occupied (ignoring floor space market values).

Check (1) Total allocated to Lanya and Jimez, $30,186 (2) Total occupancy cost to Lanya, $16,730

Analysis Component

3. Which allocation method would you prefer if you were a manager of a second-floor department? Explain.

Problem 21-2A
Activity-based costing
P2

Health Co-op is an outpatient surgical clinic that was profitable for many years, but Medicare has cut its reimbursements by as much as 40%. As a result, the clinic wants to better understand its costs. It decides to prepare an activity-based cost analysis, including an estimate of the average cost of both general surgery and orthopedic surgery. The clinic's three cost centers and their cost drivers follow.

Cost Center	Cost	Cost Driver	Driver Quantity
Professional salaries	$1,600,000	Professional hours	10,000
Patient services and supplies	27,000	Number of patients	600
Building cost	150,000	Square feet	1,500

The two main surgical units and their related data follow.

Service	Hours	Square Feet*	Patients
General surgery	2,500	600	400
Orthopedic surgery	7,500	900	200

* Orthopedic surgery requires more space for patients, supplies, and equipment.

Required

1. Compute the cost per cost driver for each of the three cost centers.

2. Use the results from part 1 to allocate costs from each of the three cost centers to both the general surgery and the orthopedic surgery units. Compute total cost and average cost per patient for both the general surgery and the orthopedic surgery units.

Check (2) Average cost of general (orthopedic) surgery, $1,195 ($6,495) per patient

Analysis Component

3. Without providing computations, would the average cost of general surgery be higher or lower if all center costs were allocated based on the number of patients? Explain.

Problem 21-3A
Departmental income statements; forecasts
P3 

mhhe.com/wildFAP19e

Warton Company began operations in January 2009 with two operating (selling) departments and one service (office) department. Its departmental income statements follow.

WARTON COMPANY Departmental Income Statements For Year Ended December 31, 2009	Clock	Mirror	Combined
Sales	$170,000	$95,000	$265,000
Cost of goods sold	83,300	58,900	142,200
Gross profit	86,700	36,100	122,800
Direct expenses			
Sales salaries	21,000	7,100	28,100
Advertising	2,100	700	2,800
Store supplies used	550	350	900
Depreciation—Equipment	2,300	900	3,200
Total direct expenses	25,950	9,050	35,000
Allocated expenses			
Rent expense	7,040	3,780	10,820
Utilities expense	2,800	1,600	4,400
Share of office department expenses	13,500	6,500	20,000
Total allocated expenses	23,340	11,880	35,220
Total expenses	49,290	20,930	70,220
Net income	$ 37,410	$15,170	$ 52,580

Warton plans to open a third department in January 2010 that will sell paintings. Management predicts that the new department will generate $50,000 in sales with a 45% gross profit margin and will require the following direct expenses: sales salaries, $8,500; advertising, $1,100; store supplies, $400; and equipment depreciation, $1,000. It will fit the new department into the current rented space by taking some square footage from the other two departments. When opened the new painting department will fill one-fifth of the space presently used by the clock department and one-fourth used by the mirror department. Management does not predict any increase in utilities costs, which are allocated to the departments in proportion to occupied space (or rent expense). The company allocates office department expenses to the operating departments in proportion to their sales. It expects the painting department to increase total office department expenses by $8,000. Since the painting department will bring new customers into the store, management expects sales in both the clock and mirror departments to increase by 8%. No changes for those departments' gross profit percents or their direct expenses are expected except for store supplies used, which will increase in proportion to sales.

Required

Prepare departmental income statements that show the company's predicted results of operations for calendar year 2010 for the three operating (selling) departments and their combined totals. (Round percents to the nearest one-tenth and dollar amounts to the nearest whole dollar.)

Check 2010 forecasted combined net income (sales), $65,832 ($336,200)

Problem 21-4A
Responsibility accounting performance reports; controllable and budgeted costs
C4 P4

Billie Whitehorse, the plant manager of Travel Free's Ohio plant, is responsible for all of that plant's costs other than her own salary. The plant has two operating departments and one service department. The camper and trailer operating departments manufacture different products and have their own managers. The office department, which Whitehorse also manages, provides services equally to the two operating departments. A budget is prepared for each operating department and the office department. The company's responsibility accounting system must assemble information to present budgeted and actual costs in performance reports for each operating department manager and the plant manager. Each performance report includes only those costs that a particular operating department manager can control: raw materials, wages, supplies used, and equipment depreciation. The plant manager is responsible for the department managers' salaries, utilities, building rent, office salaries other than her own, and other office costs plus all costs controlled by the two operating department managers. The annual departmental budgets and actual costs for the two operating departments follow.

	Budget			Actual		
	Campers	Trailers	Combined	Campers	Trailers	Combined
Raw materials	$195,900	$276,200	$ 472,100	$194,800	$273,600	$ 468,400
Employee wages	104,200	205,200	309,400	107,200	208,000	315,200
Dept. manager salary	44,000	53,000	97,000	44,800	53,900	98,700
Supplies used	34,000	92,200	126,200	32,900	91,300	124,200
Depreciation—Equip.	63,000	127,000	190,000	63,000	127,000	190,000
Utilities	3,600	5,200	8,800	4,500	4,700	9,200
Building rent	5,700	10,000	15,700	6,200	9,300	15,500
Office department costs	67,750	67,750	135,500	68,550	68,550	137,100
Totals	$518,150	$836,550	$1,354,700	$521,950	$836,350	$1,358,300

The office department's annual budget and its actual costs follow.

	Budget	Actual
Plant manager salary	$100,000	$ 84,000
Other office salaries	46,500	30,100
Other office costs	22,000	21,000
Totals	$168,500	$135,100

Required

1. Prepare responsibility accounting performance reports like those in Exhibit 21.28 that list costs controlled by the following:

a. Manager of the camper department.

b. Manager of the trailer department.

c. Manager of the Ohio plant.

In each report, include the budgeted and actual costs and show the amount that each actual cost is over or under the budgeted amount.

Check (1a) $800 total over budget

(1c) Ohio plant controllable costs, $15,400 total under budget

Analysis Component

2. Did the plant manager or the operating department managers better manage costs? Explain.

Problem 21-5A[B]
Allocation of joint costs
C6

Florida Orchards produced a good crop of peaches this year. After preparing the following income statement, the company believes it should have given its No. 3 peaches to charity and saved its efforts.

FLORIDA ORCHARDS Income Statement For Year Ended December 31, 2009	No. 1	No. 2	No. 3	Combined
Sales (by grade)				
No. 1: 300,000 lbs. @ $1.50/lb	$450,000			
No. 2: 250,000 lbs. @ $0.75/lb		$187,500		
No. 3: 600,000 lbs. @ $0.50/lb			$300,000	
Total sales				$937,500
Costs				
Tree pruning and care @ $0.40/lb	120,000	100,000	240,000	460,000
Picking, sorting, and grading @ $0.10/lb	30,000	25,000	60,000	115,000
Delivery costs	15,000	15,000	37,500	67,500
Total costs	165,000	140,000	337,500	642,500
Net income (loss)	$285,000	$ 47,500	$ (37,500)	$295,000

In preparing this statement, the company allocated joint costs among the grades on a physical basis as an equal amount per pound. The company's delivery cost records show that $30,000 of the $67,500 relates to crating the No. 1 and No. 2 peaches and hauling them to the buyer. The remaining $37,500 of delivery costs is for crating the No. 3 peaches and hauling them to the cannery.

Required

Check (1) $147,200 tree pruning and care costs allocated to No. 3

1. Prepare reports showing cost allocations on a sales value basis to the three grades of peaches. Separate the delivery costs into the amounts directly identifiable with each grade. Then allocate any shared delivery costs on the basis of the relative sales value of each grade.

(2) Net income from No. 1 & No. 2 peaches, $152,820 & $63,680

2. Using your answers to part 1, prepare an income statement using the joint costs allocated on a sales value basis.

Analysis Component

3. Do you think delivery costs fit the definition of a joint cost? Explain.

PROBLEM SET B

Problem 21-1B
Allocation of building occupancy costs to departments
C3 P1 

Marshall's has several departments that occupy all floors of a two-story building that includes a basement floor. Marshall rented this building under a long-term lease negotiated when rental rates were low. The departmental accounting system has a single account, Building Occupancy Cost, in its ledger. The types and amounts of occupancy costs recorded in this account for the current period follow.

Building rent	$320,000
Lighting expense	20,000
Cleaning expense	32,000
Total occupancy cost	$372,000

The building has 7,500 square feet on each of the upper two floors but only 5,000 square feet in the basement. In prior periods, the accounting manager merely divided the $372,000 occupancy cost by 20,000 square feet to find an average cost of $18.60 per square foot and then charged each department a building occupancy cost equal to this rate times the number of square feet that it occupies.

Riley Miller manages a department that occupies 2,000 square feet of basement floor space. In discussing the departmental reports with other managers, she questions whether using the same rate per square foot for all departments makes sense because different floor space has different values. Miller checked a recent real estate report of average local rental costs for similar space that shows first-floor space worth $48 per square foot, second-floor space worth $24 per square foot, and basement space worth $12 per square foot (excluding costs for lighting and cleaning).

Required

1. Allocate occupancy costs to Miller's department using the current allocation method.

2. Allocate the building rent cost to Miller's department in proportion to the relative market value of the floor space. Allocate to Miller's department the lighting and heating costs in proportion to the square feet occupied (ignoring floor space market values). Then, compute the total occupancy cost allocated to Miller's department.

Check Total costs allocated to Miller's Dept., (1) $37,200; (2) $18,000

Analysis Component

3. Which allocation method would you prefer if you were a manager of a basement department?

Problem 21-2B
Activity-based costing
P2

Verdant Landscape Architects has enjoyed profits for many years, but new competition has cut service revenue by as much as 30%. As a result, the company wants to better understand its costs. It decides to prepare an activity-based cost analysis, including an estimate of the average cost of both general landscaping services and custom design landscaping services. The company's three cost centers and their cost drivers follow.

Cost Center	Cost	Cost Driver	Driver Quantity
Professional salaries	$900,000	Professional hours	15,000
Customer supplies	225,000	Number of customers	1,200
Building cost	360,000	Square feet	3,750

The two main landscaping units and their related data follow.

Service	Hours	Square Feet*	Customers
General landscaping	3,750	1,500	900
Custom design landscaping	11,250	2,250	300

* Custom design landscaping requires more space for equipment, supplies, and planning.

Required

1. Compute the cost per cost driver for each of the three cost centers.

2. Use the results from part 1 to allocate costs from each of the three cost centers to both the general landscaping and the custom design landscaping units. Compute total cost and average cost per customer for both the general landscaping and the custom design landscaping units.

Check (2) Average cost of general (custom) landscaping, $597.50 ($3,157.50) per customer

Analysis Component

3. Without providing computations, would the average cost of general landscaping be higher or lower if all center costs were allocated based on the number of customers? Explain.

Problem 21-3B
Departmental income statements; forecasts
P3

Collosal Entertainment began operations in January 2009 with two operating (selling) departments and one service (office) department. Its departmental income statements follow.

COLLOSAL ENTERTAINMENT
Departmental Income Statements
For Year Ended December 31, 2009

	Movies	Video Games	Combined
Sales	$900,000	$300,000	$1,200,000
Cost of goods sold	630,000	231,000	861,000
Gross profit	270,000	69,000	339,000
Direct expenses			
Sales salaries	55,500	22,500	78,000
Advertising	18,750	9,000	27,750
Store supplies used	6,000	1,500	7,500
Depreciation—Equipment	6,750	4,500	11,250
Total direct expenses	87,000	37,500	124,500

[continued on next page]

[continued from previous page]

Allocated expenses			
Rent expense	61,500	13,500	75,000
Utilities expense	11,070	2,430	13,500
Share of office department expenses	84,375	28,125	112,500
Total allocated expenses	156,945	44,055	201,000
Total expenses	243,945	81,555	325,500
Net income (loss)	$ 26,055	$ (12,555)	$ 13,500

The company plans to open a third department in January 2010 that will sell compact discs. Management predicts that the new department will generate $450,000 in sales with a 35% gross profit margin and will require the following direct expenses: sales salaries, $27,000; advertising, $15,000; store supplies, $3,000; and equipment depreciation, $1,800. The company will fit the new department into the current rented space by taking some square footage from the other two departments. When opened, the new compact disc department will fill one-fourth of the space presently used by the movie department and one-third of the space used by the video game department. Management does not predict any increase in utilities costs, which are allocated to the departments in proportion to occupied space (or rent expense). The company allocates office department expenses to the operating departments in proportion to their sales. It expects the compact disc department to increase total office department expenses by $15,000. Since the compact disc department will bring new customers into the store, management expects sales in both the movie and video game departments to increase by 8%. No changes for those departments' gross profit percents or for their direct expenses are expected, except for store supplies used, which will increase in proportion to sales.

Required

Check 2010 forecasted movies net income (sales), $78,674 ($972,000)

Prepare departmental income statements that show the company's predicted results of operations for calendar year 2010 for the three operating (selling) departments and their combined totals. (Round percents to the nearest one-tenth and dollar amounts to the nearest whole dollar.)

Problem 21-4B
Responsibility accounting performance reports; controllable and budgeted costs
C4 P4

Warren Brown, the plant manager of LMN Co.'s San Diego plant, is responsible for all of that plant's costs other than his own salary. The plant has two operating departments and one service department. The refrigerator and dishwasher operating departments manufacture different products and have their own managers. The office department, which Brown also manages, provides services equally to the two operating departments. A monthly budget is prepared for each operating department and the office department. The company's responsibility accounting system must assemble information to present budgeted and actual costs in performance reports for each operating department manager and the plant manager. Each performance report includes only those costs that a particular operating department manager can control: raw materials, wages, supplies used, and equipment depreciation. The plant manager is responsible for the department managers' salaries, utilities, building rent, office salaries other than his own, and other office costs plus all costs controlled by the two operating department managers. The April departmental budgets and actual costs for the two operating departments follow.

	Budget			Actual		
	Refrigerators	Dishwashers	Combined	Refrigerators	Dishwashers	Combined
Raw materials	$ 480,000	$240,000	$ 720,000	$ 462,000	$242,400	$ 704,400
Employee wages	204,000	96,000	300,000	209,640	97,800	307,440
Dept. manager salary	66,000	58,800	124,800	66,000	55,800	121,800
Supplies used	18,000	10,800	28,800	16,800	11,640	28,440
Depreciation—Equip.	63,600	44,400	108,000	63,600	44,400	108,000
Utilities	36,000	21,600	57,600	41,400	24,840	66,240
Building rent	75,600	20,400	96,000	78,960	19,800	98,760
Office department costs	84,600	84,600	169,200	90,000	90,000	180,000
Totals	$1,027,800	$576,600	$1,604,400	$1,028,400	$586,680	$1,615,080

The office department's budget and its actual costs for April follow.

	Budget	Actual
Plant manager salary	$ 96,000	$102,000
Other office salaries	48,000	42,240
Other office costs	25,200	35,760
Totals	$169,200	$180,000

Required

1. Prepare responsibility accounting performance reports like those in Exhibit 21.28 that list costs controlled by the following:

a. Manager of the refrigerator department.

b. Manager of the dishwasher department.

c. Manager of the San Diego plant.

In each report, include the budgeted and actual costs for the month and show the amount by which each actual cost is over or under the budgeted amount.

Check (1a) $13,560 total under budget

(1c) San Diego plant controllable costs, $4,680 total over budget

Analysis Component

2. Did the plant manager or the operating department managers better manage costs? Explain.

Problem 21-5B[B]
Allocation of joint costs
C6 

Rita and Rick Redding own and operate a tomato grove. After preparing the following income statement, Rita believes they should have offered the No. 3 tomatoes to the public for free and saved themselves time and money.

RITA AND RICK REDDING
Income Statement
For Year Ended December 31, 2009

	No. 1	No. 2	No. 3	Combined
Sales (by grade)				
No. 1: 600,000 lbs. @ $1.80/lb	$1,080,000			
No. 2: 480,000 lbs. @ $1.25/lb		$600,000		
No. 3: 120,000 lbs. @ $0.40/lb			$ 48,000	
Total sales				$1,728,000
Costs				
Land preparation, seeding, and cultivating @ $0.70/lb	420,000	336,000	84,000	840,000
Harvesting, sorting, and grading @ $0.04/lb	24,000	19,200	4,800	48,000
Delivery costs	20,000	14,000	6,000	40,000
Total costs	464,000	369,200	94,800	928,000
Net income (loss)	$ 616,000	$230,800	$(46,800)	$ 800,000

In preparing this statement, Rita and Rick allocated joint costs among the grades on a physical basis as an equal amount per pound. Also, their delivery cost records show that $34,000 of the $40,000 relates to crating the No. 1 and No. 2 tomatoes and hauling them to the buyer. The remaining $6,000 of delivery costs is for crating the No. 3 tomatoes and hauling them to the cannery.

Required

1. Prepare reports showing cost allocations on a sales value basis to the three grades of tomatoes. Separate the delivery costs into the amounts directly identifiable with each grade. Then allocate any shared delivery costs on the basis of the relative sales value of each grade. (Round percents to the nearest one-tenth and dollar amounts to the nearest whole dollar.)

Check (1) $1,344 harvesting, sorting and grading costs allocated to No. 3

(2) Net income from No. 1 & No. 2 tomatoes, $503,138 & $279,726

2. Using your answers to part 1, prepare an income statement using the joint costs allocated on a sales value basis.

Analysis Component

3. Do you think delivery costs fit the definition of a joint cost? Explain.

BEYOND THE NUMBERS

REPORTING IN ACTION

C4

BTN 21-1 Review **Best Buy**'s income statement in Appendix A and identify its revenues for the years ended March 3, 2007, February 25, 2006, and February 26, 2005. For the year ended March 3, 2007, Best Buy reports the following product revenue mix. (Assume that its product revenue mix is the same for each of the three years reported when answering the requirements.)

Home Office	Entertainment Software	Consumer Electronics	Appliances
33%	12%	45%	10%

Required

1. Compute the amount of revenue from each of its product lines for the years ended March 3, 2007, February 25, 2006, and February 26, 2005.

2. If Best Buy wishes to evaluate each of its product lines, how can it allocate its operating expenses to each of them to determine each product line's profitability?

Fast Forward

3. Access Best Buy's annual report for a fiscal year ending after March 3, 2007, from its Website (BestBuy.com) or the SEC's EDGAR database (sec.gov). Compute its revenues for its product lines for the most recent year(s). Compare those results to those from part 1. How has its product mix changed?

COMPARATIVE ANALYSIS

P3

BTN 21-2 **Best Buy**, **Circuit City**, and **RadioShack** compete across the country in several markets. The most common competitive markets for these companies are by location.

Required

1. Design a three-tier responsibility accounting organizational chart assuming that you have available internal information for all three companies. Use Exhibit 21.19 as an example. The goal of this assignment is to design a reporting framework for the companies; numbers are not required. Limit your reporting framework to sales activity only.

2. Explain why it is important to have similar performance reports when comparing performance within a company (and across different companies). Be specific in your response.

ETHICS CHALLENGE

P3

BTN 21-3 Senior Security Co. offers a range of security services for senior citizens. Each type of service is considered within a separate department. Mary Pincus, the overall manager, is compensated partly on the basis of departmental performance by staying within the quarterly cost budget. She often revises operations to make sure departments stay within budget. Says Pincus, "I will not go over budget even if it means slightly compromising the level and quality of service. These are minor compromises that don't significantly affect my clients, at least in the short term."

Required

1. Is there an ethical concern in this situation? If so, which parties are affected? Explain.

2. Can Mary Pincus take action to eliminate or reduce any ethical concerns? Explain.

3. What is Senior Security's ethical responsibility in offering professional services?

COMMUNICATING IN PRACTICE

C4 C5 P3

BTN 21-4 Home Station is a national home improvement chain with more than 100 stores throughout the country. The manager of each store receives a salary plus a bonus equal to a percent of the store's net income for the reporting period. The following net income calculation is on the Denver store manager's performance report for the recent monthly period.

Sales	$2,500,000
Cost of goods sold	800,000
Wages expense	500,000
Utilities expense	200,000
Home office expense	75,000
Net income	$925,000
Manager's bonus (0.5%)	$ 4,625

In previous periods, the bonus had also been 0.5%, but the performance report had not included any charges for the home office expense, which is now assigned to each store as a percent of its sales.

Required

Assume that you are the national office manager. Write a one-half page memorandum to your store managers explaining why home office expense is in the new performance report.

TAKING IT TO THE NET

A1

BTN 21-5 This chapter described and used spreadsheets to prepare various managerial reports (see Exhibit 21-12). You can download from Websites various tutorials showing how spreadsheets are used in managerial accounting and other business applications.

Required

1. Link to the Website Lacher.com. Scroll down past "Microsoft Excel Examples" and select "Business Solutions." Identify and list three tutorials for review.

2. Describe in a one-half page memorandum to your instructor how the applications described in each tutorial are helpful in business and managerial decision making.

TEAMWORK IN ACTION

C1 C2

BTN 21-6 Activity-based costing (ABC) is increasingly popular as a useful managerial tool to (1) measure the cost of resources consumed and (2) assign cost to products and services. This managerial tool has been available to accounting and business decision makers for more than 25 years.

Required

Break into teams and identify at least three likely reasons that activity-based costing has gained popularity in recent years. Be prepared to present your answers in a class discussion. (*Hint:* What changes have occurred in products and services over the past 25 years?)

ENTREPRENEURIAL DECISION

P3

BTN 21-7 **RockBottomGolf** is an Internet retailer and the focus of this chapter's opener. It sells discounted golf merchandise through departments such as clubs, bags, apparel, and accessories. The company plans to expand to include many other types of sporting goods.

Required

1. How can RockBottomGolf use departmental income statements to assist in understanding and controlling operations?
2. Are departmental income statements always the best measure of a department's performance? Explain.
3. Provide examples of nonfinancial performace indicators RockBottomGolf might use as part of a balanced scorecard system of performance evaluation.

HITTING THE ROAD

P3

BTN 21-8 Visit a local movie theater and check out both its concession area and its showing areas. The manager of a theater must confront questions such as:

- How much return do we earn on concessions?
- What types of movies generate the greatest sales?
- What types of movies generate the greatest net income?

Required

Assume that you are the new accounting manager for a 16-screen movie theater. You are to set up a responsibility accounting reporting framework for the theater.

1. Recommend how to segment the different departments of a movie theater for responsibility reporting.
2. Propose an expense allocation system for heat, rent, insurance, and maintenance costs of the theater.

GLOBAL DECISION

DSG

BTN 21-9 Selected product data from **DSG international plc** (www.DSGiplc.com) follow.

Product Segment for Year Ended (£ millions)	Net Sales		Operating Income	
	April 28, 2007	April 29, 2006	April 28, 2007	April 29, 2006
Computing	£2,198	£2,040	£97	£107
Electrical	5,281	4,912	193	198
e-commerce	451	26	1	0

Required

1. Compute the percentage growth in net sales for each product line from fiscal year 2006 to 2007.
2. Which product line's net sales grew the fastest?
3. Which segment was the most profitable?
4. How can DSG's managers use this information?

ANSWERS TO MULTIPLE CHOICE QUIZ

1. b; [$641,250/($356,250 + $641,250 + $427,500)] × $150,000 = $67,500
2. c;
3. e; $162,000 × 50/150 setups = $54,000; $54,000/500 units = $108 per Grey unit. (Red is $2.70 per unit.)
4. b;

	Department X	Department Y	Department Z
Sales	$500,000	$200,000	$350,000
Cost of goods sold	350,000	75,000	150,000
Gross profit	150,000	125,000	200,000
Direct expenses	50,000	20,000	75,000
Departmental contribution	$100,000	$105,000	$125,000

5. a; $100,000/$500,000 = 20%

A Look Back

Chapter 21 focused on cost allocation, activity-based costing, and performance measurement. We identified ways to measure and analyze company activities, its departments, and its managers.

A Look at This Chapter

This chapter shows how information on both costs and sales behavior is useful to managers in performing cost-volume-profit analysis. This analysis is an important part of successful management and sound business decisions.

A Look Ahead

Chapter 23 introduces and describes the budgeting process and its importance to management. It also explains the master budget and its usefulness to the planning of future company activities.

Chapter 22 Cost-Volume-Profit Analysis

Learning Objectives

CAP

Conceptual

C1 Describe different types of cost behavior in relation to production and sales volume. *(p. 902)*

C2 Identify assumptions in cost-volume-profit analysis and explain their impact. *(p. 911)*

C3 Describe several applications of cost-volume-profit analysis. *(p. 913)*

Analytical

A1 Compare the scatter diagram, high-low, and regression methods of estimating costs. *(p. 907)*

A2 Compute the contribution margin and describe what it reveals about a company's cost structure. *(p. 908)*

A3 Analyze changes in sales using the degree of operating leverage. *(p. 918)*

Procedural

P1 Determine cost estimates using three different methods. *(p. 905)*

P2 Compute the break-even point for a single product company. *(p. 909)*

P3 Graph costs and sales for a single product company. *(p. 910)*

P4 Compute the break-even point for a multiproduct company. *(p. 915)*

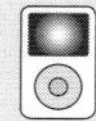

LP22

Decision Feature

Recipe for Growth

"Don't sit on the sidelines talking about your dream . . . get out and make it happen"—Martin Sprock

"Welcome to Moe's"! A chorus of welcomes greets each customer at **Moe's Southwest Grill (Moes.com),** a chain of quirky Tex-Mex restaurants, which is part of Raving Brands. The zaniness continues with menu items such as Art Vandalay, Joey Bag of Donuts, the Close Talker, and the Billy Barou. They play music from "dead rock stars" like the Beatles, Elvis Presley, and Jimi Hendrix because "Moe wanted to pay tribute to his heroes who have passed on and would never have a chance to taste his food."

Moe's founder Martin Sprock explains, "We make a point of having the happiest associates. You feel good visiting our stores, and that means something to me. I'd go so far as to say I'd actually be willing to take a date to them."

But there is more to Moe's than fun. Moe's features burritos, tacos, quesadillas, and salads. To appeal to health-conscious diners, Moe's does not use frozen ingredients or microwaves or cook with fat. This recipe has resulted in Moe's being one of the fastest-growing "fast casual" restaurants.

With such rapid growth, an understanding of cost behavior is critical. Identifying fixed and variable costs is key to understanding break-even points and maintaining the right mix of menu choices. Each Moe's manager earns a degree from "Moe's Training School," where the finer points of cost management are taught. Moe's online ordering and payment system is linked with its cash registers to enable managers to better determine which menu items are in demand. An understanding of how costs relate to sales volume and profits helps drive the menu options.

Martin Sprock's vision is to run a chain of restaurants that treats employees as well as they treat owners. This family-first mentality and service-oriented approach have spurred Moe's growth. Sprock, a former ski bum, encourages potential entrepreneurs to get out and make it happen. "I had no money when I started trying to fulfill my ambitions . . . I just did it."

[Sources: *Moe's Southwest Grill Website*, January 2009; *Go AirTran Airways* Magazine, 2005; *Atlanta Business Chronicle,* May 2008; *Pittsburgh Business Times,* March 2008.

Chapter Preview

This chapter describes different types of costs and shows how changes in a company's operating volume affect these costs. The chapter also analyzes a company's costs and sales to explain how different operating strategies affect profit or loss. Managers use this type of analysis to forecast what will happen if changes are made to costs, sales volume, selling prices, or product mix. They then use these forecasts to select the best business strategy for the company.

Cost-Volume-Profit Analysis

Identifying Cost Behavior
- Fixed costs
- Variable costs
- Mixed costs
- Step-wise costs
- Curvilinear costs

Measuring Cost Behavior
- Scatter diagrams
- High-low method
- Least-squares regression
- Comparison of cost estimation methods

Using Break-Even Analysis
- Computing contribution margin
- Computing break-even
- Preparing a cost-volume-profit chart
- Making assumptions in cost-volume-profit analysis

Applying Cost-Volume-Profit Analysis
- Computing income from sales and costs
- Computing sales for target income
- Computing margin of safety
- Using sensitivity analysis
- Computing multiproduct break-even

Identifying Cost Behavior

Video22.1

Point: *Profit* is another term for *income.*

Planning a company's future activities and events is a crucial phase in successful management. One of the first steps in planning is to predict the volume of activity, the costs to be incurred, sales to be made, and profit to be received. An important tool to help managers carry out this step is **cost-volume-profit (CVP) analysis,** which helps them predict how changes in costs and sales levels affect income. In its basic form, CVP analysis involves computing the sales level at which a company neither earns an income nor incurs a loss, called the *break-even point.* For this reason, this basic form of cost-volume-profit analysis is often called *break-even analysis.* Managers use variations of CVP analysis to answer questions such as these:

- What sales volume is needed to earn a target income?
- What is the change in income if selling prices decline and sales volume increases?
- How much does income increase if we install a new machine to reduce labor costs?
- What is the income effect if we change the sales mix of our products or services?

Consequently, cost-volume-profit analysis is useful in a wide range of business decisions.

Conventional cost-volume-profit analysis requires management to classify all costs as either *fixed* or *variable* with respect to production or sales volume. The remainder of this section discusses the concepts of fixed and variable cost behavior as they relate to CVP analysis.

Decision Insight

No Free Lunch Hardly a week goes by without a company advertising a free product with the purchase of another. Examples are a free printer with a digital camera purchase or a free monitor with a computer purchase. Can these companies break even, let alone earn profits? We are reminded of the *no-free-lunch* adage, meaning that companies expect profits from the companion or add-on purchase to make up for the free product.

C1 Describe different types of cost behavior in relation to production and sales volume.

Fixed Costs

A *fixed cost* remains unchanged in amount when the volume of activity varies from period to period within a relevant range. For example, $5,000 in monthly rent paid for a factory building remains the same whether the factory operates with a single eight-hour shift or around the clock with three shifts. This means that rent cost is the same each month at any level of output from zero to the plant's full productive capacity. Notice that while *total* fixed cost does not change as the level of production changes, the fixed cost *per unit* of output decreases as volume increases. For instance, if 20 units are produced when monthly rent is $5,000, the average rent cost per unit is $250 (computed as $5,000/20 units). When production increases to 100 units per month, the average cost per unit decreases to $50 (computed as $5,000/100 units). The average cost decreases to $10 per unit if production increases to 500 units per month. Common examples of fixed costs include depreciation, property taxes, office salaries, and many service department costs.

Point: Fixed costs do not change when volume changes, but the per unit cost declines as volume increases.

When production volume and costs are graphed, units of product are usually plotted on the *horizontal axis* and dollars of cost are plotted on the *vertical axis.* Fixed costs then are represented as a horizontal line because they remain constant at all levels of production. To illustrate, the graph in Exhibit 22.1 shows that fixed costs remain at $32,000 at all production levels up to the company's monthly capacity of 2,000 units of output. The *relevant range* for fixed costs in Exhibit 22.1 is 0 to 2,000 units. If the relevant range changes (that is, production capacity extends beyond this range), the amount of fixed costs will likely change.

Example: If the fixed cost line in Exhibit 22.1 is shifted upward, does the total cost line shift up, down, or remain in the same place? *Answer:* It shifts up by the same amount.

EXHIBIT 22.1

Relations of Fixed and Variable Costs to Volume

Monthly Capacity

Cost: $80,000; 70,000; 60,000; 50,000; 40,000; 30,000; 20,000; 10,000; 0

Volume (Units): 0; 200; 400; 600; 800; 1,000; 1,200; 1,400; 1,600; 1,800; 2,000

Fixed Costs, $32,000 — Variable Costs, $20 per unit — Total (Mixed) Costs

Example: If the level of fixed costs in Exhibit 22.1 changes, does the slope of the total cost line change? *Answer:* No, the slope doesn't change. The total cost line is simply shifted upward or downward.

Variable Costs

A *variable cost* changes in proportion to changes in volume of activity. The direct materials cost of a product is one example of a variable cost. If one unit of product requires materials costing $20, total materials costs are $200 when 10 units of product are manufactured, $400 for 20 units, $600 for 30 units, and so on. Notice that variable cost *per unit* remains constant but the *total* amount of variable cost changes with the level of production. In addition to direct materials, common variable costs include direct labor (if employees are paid per unit), sales commissions, shipping costs, and some overhead costs.

Point: Fixed costs are constant in total but vary (decline) per unit as more units are produced. Variable costs vary in total but are fixed per unit.

When variable costs are plotted on a graph of cost and volume, they appear as a straight line starting at the zero cost level. This straight line is upward (positive) sloping. The line rises as volume of activity increases. A variable cost line using a $20 per unit cost is graphed in Exhibit 22.1.

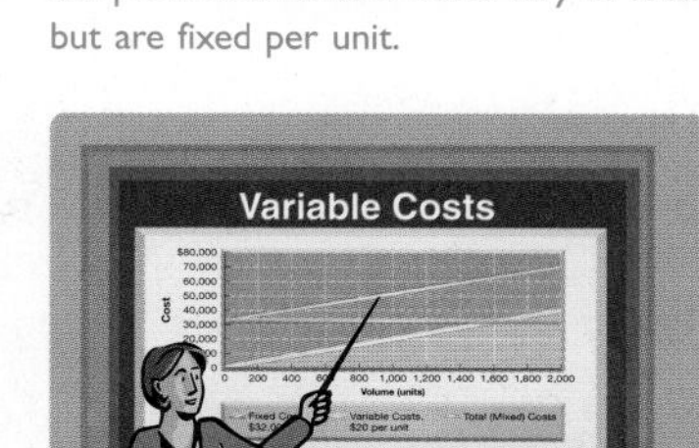

Mixed Costs

A **mixed cost** includes both fixed and variable cost components. For example, compensation for sales representatives often includes a fixed monthly salary and a variable commission based on sales. The total cost line in Exhibit 22.1 is a mixed cost. Like a fixed cost, it is greater than zero when volume is zero; but unlike a fixed cost, it increases steadily in proportion to increases in volume. The mixed cost line in Exhibit 22.1 starts on the vertical axis at the $32,000 fixed cost point. Thus, at the zero volume level, total cost equals the fixed costs. As the activity level increases, the mixed cost line increases at an amount equal to the variable cost per unit. This line is highest when volume of activity is at 2,000 units (the end point of the relevant range). In CVP analysis, mixed costs are often separated into fixed and variable components. The fixed component is added to other fixed costs, and the variable component is added to other variable costs.

Step-Wise Costs

A **step-wise cost** reflects a step pattern in costs. Salaries of production supervisors often behave in a step-wise manner in that their salaries are fixed within a *relevant range* of the current production volume. However, if production volume expands significantly (for example, with the addition of another shift), additional supervisors must be hired. This means that the total cost for supervisory salaries goes up by a lump-sum amount. Similarly, if volume takes another significant step up, supervisory salaries will increase by another lump sum. This behavior reflects a step-wise cost, also known as a *stair-step cost,* which is graphed in Exhibit 22.2. See how the step-wise cost line is flat within ranges (steps). Then, when volume significantly changes, it shifts to another level for that range (step).

EXHIBIT 22.2

Step-Wise and Curvilinear Costs

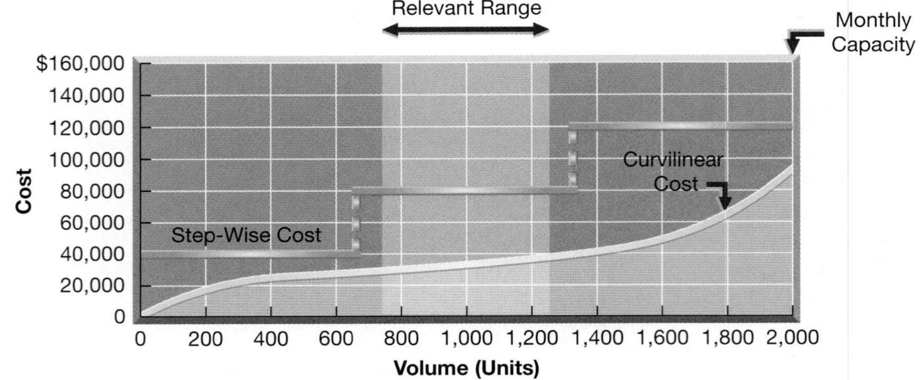

In a conventional CVP analysis, a step-wise cost is usually treated as either a fixed cost or a variable cost. This treatment involves manager judgment and depends on the width of the range and the expected volume. To illustrate, suppose after the production of every 25 snowboards, an operator lubricates the finishing machine. The cost of this lubricant reflects a step-wise pattern. Also, suppose that after the production of every 1,000 units, the snowboard cutting tool is replaced. Again, this is a step-wise cost. Note that the range of 25 snowboards is much narrower than the range of 1,000 snowboards. Some managers might treat the lubricant cost as a variable cost and the cutting tool cost as a fixed cost.

Point: Computer spreadsheets are important and effective tools for CVP analysis and for analyzing alternative "what-if" strategies.

Curvilinear Costs

A variable cost, as explained, is a *linear* cost; that is, it increases at a constant rate as volume of activity increases. A **curvilinear cost,** also called a *nonlinear cost,* increases at a nonconstant rate as volume increases. When graphed, curvilinear costs appear as a curved line. Exhibit 22.2 shows a curvilinear cost beginning at zero when production is zero and then increasing at different rates.

An example of a curvilinear cost is total direct labor cost when workers are paid by the hour. At low to medium levels of production, adding more employees allows each of them to specialize by doing certain tasks repeatedly instead of doing several different tasks. This often yields additional units of output at lower costs. A point is eventually reached at which adding more employees creates inefficiencies. For instance, a large crew demands more time and effort in communicating and coordinating their efforts. While adding employees in this case increases output, the labor cost per unit increases, and the total labor cost goes up at a steeper slope. This pattern is seen in Exhibit 22.2 where the curvilinear cost curve starts at zero, rises, flattens out, and then increases at a faster rate as output nears the maximum.

Point: Cost-volume-profit analysis helped Rod Canion, Jim Harris, and Bill Murto raise start-up capital of $20 million to launch **Compaq Computer.** They showed that break-even volumes were attainable within the first year.

Quick Check

Answers—p. 922

1. Which of the following statements is typically true? (*a*) Variable cost per unit increases as volume increases, (*b*) fixed cost per unit decreases as volume increases, or (*c*) a curvilinear cost includes both fixed and variable elements.
2. Describe the behavior of a fixed cost.
3. If cost per unit of activity remains constant (fixed), why is it called a variable cost?

Measuring Cost Behavior

P1 Determine cost estimates using three different methods.

Identifying and measuring cost behavior requires careful analysis and judgment. An important part of this process is to identify costs that can be classified as either fixed or variable, which often requires analysis of past cost behavior. Three methods are commonly used to analyze past costs: scatter diagrams, high-low method, and least-squares regression. Each method is discussed in this section using the unit and cost data shown in Exhibit 22.3, which are taken from a start-up company that uses units produced as the activity base in estimating cost behavior.

EXHIBIT 22.3

Data for Estimating Cost Behavior

Month	Units Produced	Total Cost
January	17,500	$20,500
February	27,500	21,500
March	25,000	25,000
April	35,000	21,500
May	47,500	25,500
June	22,500	18,500
July	30,000	23,500
August	52,500	28,500
September	37,500	26,000
October	57,500	26,000
November	62,500	31,000
December	67,500	29,000

Scatter Diagrams

Scatter diagrams display past cost and unit data in graphical form. In preparing a scatter diagram, units are plotted on the horizontal axis and cost is plotted on the vertical axis. Each individual point on a scatter diagram reflects the cost and number of units for a prior period. In Exhibit 22.4, the prior 12 months' costs and numbers of units are graphed. Each point reflects total costs incurred and units produced for one of those months. For instance, the point labeled March had units produced of 25,000 and costs of $25,000.

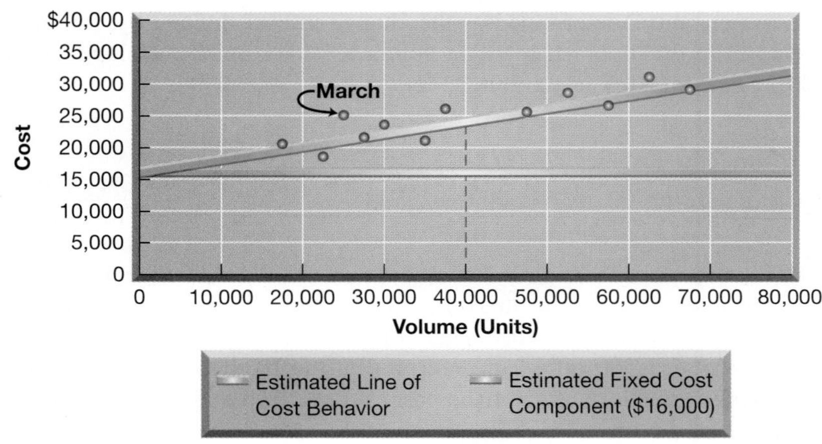

EXHIBIT 22.4

Scatter Diagram

The **estimated line of cost behavior** is drawn on a scatter diagram to reflect the relation between cost and unit volume. This line best visually "fits" the points in a scatter diagram. Fitting this line demands judgment. The line drawn in Exhibit 22.4 intersects the vertical axis at approximately $16,000, which reflects fixed cost. To compute variable cost per unit, or the slope, we perform three steps. First, we select any two points on the horizontal axis (units), say 0 and 40,000. Second, we draw a vertical line from each of these points to intersect the estimated line of cost behavior. The point on the vertical axis (cost) corresponding to the 40,000 units point that intersects the estimated line is roughly $24,000. Similarly, the cost corresponding to zero units is $16,000 (the fixed cost point). Third, we compute the slope of the line, or variable cost, as the change in cost divided by the change in units. Exhibit 22.5 shows this computation.

EXHIBIT 22.5

Variable Cost per Unit (Scatter Diagram)

$$\frac{\textbf{Change in cost}}{\textbf{Change in units}} = \frac{\$24{,}000 - \$16{,}000}{40{,}000 - 0} = \frac{\$8{,}000}{40{,}000} = \$0.20 \text{ per unit}$$

Example: In Exhibits 22.4 and 22.5, if units are projected at 30,000, what is the predicted cost? *Answer:* Approximately $22,000.

Variable cost is $0.20 per unit. Thus, the cost equation that management will use to estimate costs for different unit levels is **$16,000 plus $0.20 per unit**.

High-Low Method

The **high-low method** is a way to estimate the cost equation by graphically connecting the two cost amounts at the highest and lowest unit volumes. In our case, the lowest number of units is 17,500, and the highest is 67,500. The costs corresponding to these unit volumes are $20,500 and $29,000, respectively (see the data in Exhibit 22.3). The estimated line of cost behavior for the high-low method is then drawn by connecting these two points on the scatter diagram corresponding to the lowest and highest unit volumes as follows.

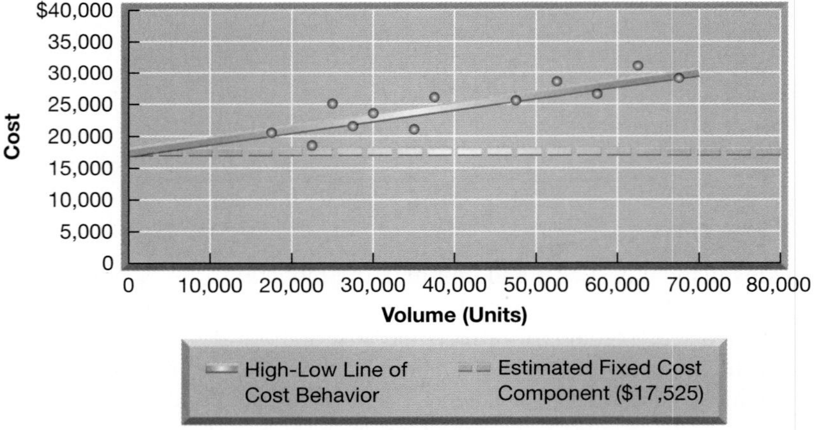

Point: Note that the high-low method identifies the high and low points of the volume (activity) base, and the costs linked with those extremes—which may not be the highest and lowest costs.

The variable cost per unit is determined as the change in cost divided by the change in units and uses the data from the high and low unit volumes. This results in a slope, or variable cost per unit, of $0.17 as computed in Exhibit 22.6.

EXHIBIT 22.6

Variable Cost per Unit (High-Low Method)

$$\frac{\textbf{Change in cost}}{\textbf{Change in units}} = \frac{\$29{,}000 - \$20{,}500}{67{,}500 - 17{,}500} = \frac{\$8{,}500}{50{,}000} = \$0.17 \text{ per unit}$$

To estimate the fixed cost for the high-low method, we use the knowledge that total cost equals fixed cost plus variable cost per unit times the number of units. Then we pick either the high or low point to determine the fixed cost. This computation is shown in Exhibit 22.7—where we use the high point (67,500 units) in determining the fixed cost of $17,525. Use of the low point (17,500 units) yields the same fixed cost estimate: $20,500 = Fixed cost + ($0.17 per unit × 17,500), or Fixed cost = $17,525.

Total cost	=	Fixed cost + (Variable cost × Units)
$29,000	=	Fixed cost + ($0.17 per unit × 67,500 units)
Then,		Fixed cost = $17,525

EXHIBIT 22.7

Fixed Cost (High-Low Method)

Thus, the cost equation used to estimate costs at different units is **$17,525 plus $0.17 per unit**. This cost equation differs slightly from that determined from the scatter diagram method. A deficiency of the high-low method is that it ignores all cost points except the highest and lowest. The result is less precision because the high-low method uses the most extreme points rather than the more usual conditions likely to recur.

Least-Squares Regression

Least-squares regression is a statistical method for identifying cost behavior. For our purposes, we use the cost equation estimated from this method but leave the computational details for more advanced courses. Such computations for least-squares regression are readily done using most spreadsheet programs or calculators. We illustrate this using Excel® in Appendix 22A.

The regression cost equation for the data presented in Exhibit 22.3 is **$16,947 plus $0.19 per unit**; that is, the fixed cost is estimated as $16,947 and the variable cost at $0.19 per unit. Both costs are reflected in the following graph.

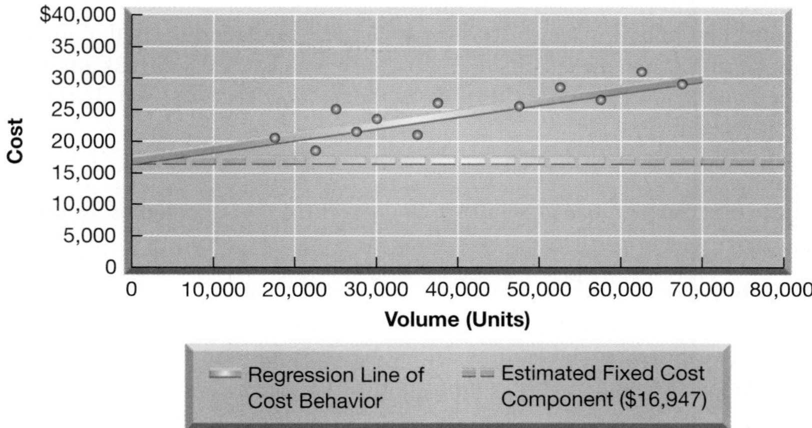

Comparison of Cost Estimation Methods

A1 Compare the scatter diagram, high-low, and regression methods of estimating costs.

The three cost estimation methods result in slightly different estimates of fixed and variable costs as summarized in Exhibit 22.8. Estimates from the scatter diagram are based on a visual fit of the cost line and are subject to interpretation. Estimates from the high-low method use only two sets of values corresponding to the lowest and highest unit volumes. Estimates from least-squares regression use a statistical technique and all available data points.

EXHIBIT 22.8

Comparison of Cost Estimation Methods

Estimation Method	Fixed Cost	Variable Cost
Scatter diagram	$16,000	$0.20 per unit
High-low method	17,525	0.17 per unit
Least-squares regression	16,947	0.19 per unit

We must remember that all three methods use *past data*. Thus, cost estimates resulting from these methods are only as good as the data used for estimation. Managers must establish that the data are reliable in deriving cost estimates for the future.

Quick Check Answers—p. 922

4. Which of the following methods is likely to yield the most precise estimated line of cost behavior? (*a*) High-low, (*b*) least-squares regression, or (*c*) scatter diagram.
5. What is the primary weakness of the high-low method?
6. Using conventional CVP analysis, a mixed cost should be (*a*) disregarded, (*b*) treated as a fixed cost, or (*c*) separated into fixed and variable components.

Using Break-Even Analysis

Video22.2

Break-even analysis is a special case of cost-volume-profit analysis. This section describes break-even analysis by computing the break-even point and preparing a CVP (or break-even) chart.

Contribution Margin and Its Measures

A2 Compute the contribution margin and describe what it reveals about a company's cost structure.

We explained how managers classify costs by behavior. This often refers to classifying costs as being fixed or variable with respect to volume of activity. In manufacturing companies, volume of activity usually refers to the number of units produced. We then classify a cost as either fixed or variable, depending on whether total cost changes as the number of units produced changes. Once we separate costs by behavior, we can then compute a product's contribution margin. **Contribution margin per unit,** or *unit contribution margin,* is the amount by which a product's unit selling price exceeds its total unit variable cost. This excess amount contributes to covering fixed costs and generating profits on a per unit basis. Exhibit 22.9 shows the contribution margin per unit formula.

EXHIBIT 22.9
Contribution Margin per Unit

Contribution margin per unit = Sales price per unit − Total variable cost per unit

The **contribution margin ratio,** which is the percent of a unit's selling price that exceeds total unit variable cost, is also useful for business decisions. It can be interpreted as the percent of each sales dollar that remains after deducting the total unit variable cost. Exhibit 22.10 shows the formula for the contribution margin ratio.

EXHIBIT 22.10
Contribution Margin Ratio

$$\text{Contribution margin ratio} = \frac{\text{Contribution margin per unit}}{\text{Sales price per unit}}$$

To illustrate the use of contribution margin, let's consider **Rydell,** which sells footballs for $100 per unit and incurs variable costs of $70 per unit sold. Its fixed costs are $24,000 per month with monthly capacity of 1,800 units (footballs). Rydell's contribution margin per unit is $30, which is computed as follows.

Selling price per unit	$100
Variable cost per unit	70
Contribution margin per unit	$ 30

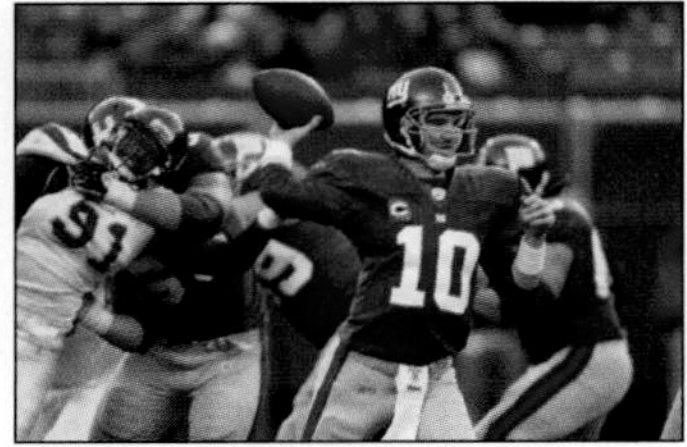

Its contribution margin ratio is 30%, computed as $30/$100. This reveals that for each unit sold, Rydell has $30 that contributes to covering fixed cost and profit. If we consider sales in dollars, a contribution margin of 30% implies that for each $1 in sales, Rydell has $0.30 that contributes to fixed cost and profit.

Decision Maker

Sales Manager You are evaluating orders from two customers but can accept only one of the orders because of your company's limited capacity. The first order is for 100 units of a product with a contribution margin ratio of 60% and a selling price of $1,000. The second order is for 500 units of a product with a contribution margin ratio of 20% and a selling price of $800. The incremental fixed costs are the same for both orders. Which order do you accept? [Answer—p. 921]

Computing the Break-Even Point

P2 Compute the break-even point for a single product company.

The **break-even point** is the sales level at which a company neither earns a profit nor incurs a loss. The concept of break-even is applicable to nearly all organizations, activities, and events. One of the most important items of information when launching a project is whether it will break even—that is, whether sales will at least cover total costs. The break-even point can be expressed in either units or dollars of sales.

To illustrate the computation of break-even analysis, let's again look at Rydell, which sells footballs for $100 per unit and incurs $70 of variable costs per unit sold. Its fixed costs are $24,000 per month. Rydell breaks even for the month when it sells 800 footballs (sales volume of $80,000). We compute this break-even point using the formula in Exhibit 22.11. This formula uses the contribution margin per unit, which for Rydell is $30 ($100 − $70). From this we can compute the break-even sales volume as $24,000/$30, or 800 units per month.

EXHIBIT 22.11

Formula for Computing Break-Even Sales (in Units)

$$\textbf{Break-even point in units} = \frac{\textbf{Fixed costs}}{\textbf{Contribution margin per unit}}$$

At a price of $100 per unit, monthly sales of 800 units yield sales dollars of $80,000 (called *break-even sales dollars*). This $80,000 break-even sales can be computed directly using the formula in Exhibit 22.12.

Point: The break-even point is where total expenses equal total sales and the profit is zero.

EXHIBIT 22.12

Formula for Computing Break-Even Sales (in Dollars)

$$\textbf{Break-even point in dollars} = \frac{\textbf{Fixed costs}}{\textbf{Contribution margin ratio}}$$

Rydell's break-even point in dollars is computed as $24,000/0.30, or $80,000 of monthly sales. To verify that Rydell's break-even point equals $80,000 (or 800 units), we prepare a simplified income statement in Exhibit 22.13. It shows that the $80,000 revenue from sales of 800 units exactly equals the sum of variable and fixed costs.

Point: Even if a company operates at a level in excess of its break-even point, management may decide to stop operating because it is not earning a reasonable return on investment.

EXHIBIT 22.13

Contribution Margin Income Statement for Break-Even Sales

RYDELL COMPANY
Contribution Margin Income Statement (at Break-Even)
For Month Ended January 31, 2009

Sales (800 units at $100 each)	$80,000
Variable costs (800 units at $70 each)	56,000
Contribution margin	24,000
Fixed costs	24,000
Net income	$ 0

Point: A contribution margin income statement is also referred to as a ***variable costing income statement.*** This differs from the traditional ***absorption costing*** approach where all product costs are assigned to units sold and to units in ending inventory. Recall that variable costing expenses all fixed product costs. Thus, income for the two approaches differs depending on the level of finished goods inventory; the lower inventory is, the more similar the two approaches are.

The statement in Exhibit 22.13 is called a *contribution margin income statement.* It differs in format from a conventional income statement in two ways. First, it separately classifies costs and expenses as variable or fixed. Second, it reports contribution margin (Sales − Variable costs). The contribution margin income statement format is used in this chapter's assignment materials because of its usefulness in CVP analysis.

Preparing a Cost-Volume-Profit Chart

P3 Graph costs and sales for a single product company.

Exhibit 22.14 is a graph of Rydell's cost-volume-profit relations. This graph is called a **cost-volume-profit (CVP) chart,** or a *break-even chart* or *break-even graph.* The horizontal axis is the number of units produced and sold and the vertical axis is dollars of sales and costs. The lines in the chart depict both sales and costs at different output levels.

EXHIBIT 22.14

Cost-Volume-Profit Chart

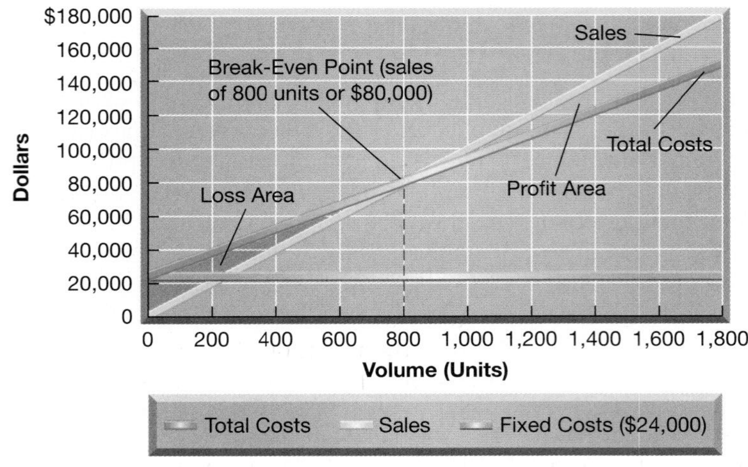

We follow three steps to prepare a CVP chart, which can also be drawn with computer programs that convert numeric data to graphs:

1. Plot fixed costs on the vertical axis ($24,000 for Rydell). Draw a horizontal line at this level to show that fixed costs remain unchanged regardless of output volume (drawing this fixed cost line is not essential to the chart).
2. Draw the total (variable plus fixed) costs line for a relevant range of volume levels. This line starts at the fixed costs level on the vertical axis because total costs equal fixed costs at zero volume. The slope of the total cost line equals the variable cost per unit ($70). To draw the line, compute the total costs for any volume level, and connect this point with the vertical axis intercept ($24,000). Do not draw this line beyond the productive capacity for the planning period (1,800 units for Rydell).
3. Draw the sales line. Start at the origin (zero units and zero dollars of sales) and make the slope of this line equal to the selling price per unit ($100). To sketch the line, compute dollar sales for any volume level and connect this point with the origin. Do not extend this line beyond the productive capacity. Total sales will be at the highest level at maximum capacity.

Example: In Exhibit 22.14, the sales line intersects the total cost line at 800 units. At what point would the two lines intersect if selling price is increased by 20% to $120 per unit? *Answer:* $24,000/($120 − $70) = 480 units

The total costs line and the sales line intersect at 800 units in Exhibit 22.14, which is the break-even point—the point where total dollar sales of $80,000 equals the sum of both fixed and variable costs ($80,000).

On either side of the break-even point, the vertical distance between the sales line and the total costs line at any specific volume reflects the profit or loss expected at that point. At volume levels to the left of the break-even point, this vertical distance is the amount of the expected loss because the total costs line is above the total sales line. At volume levels to the right of the break-even point, the vertical distance represents the expected profit because the total sales line is above the total costs line.

Decision Maker

Operations Manager As a start-up manufacturer, you wish to identify the behavior of manufacturing costs to develop a production cost budget. You know three methods can be used to identify cost behavior from past data, but past data are unavailable because this is a start-up. What do you do? [Answer—p. 921]

Making Assumptions in Cost-Volume-Profit Analysis

Cost-volume-profit analysis assumes that relations can normally be expressed as simple lines similar to those in Exhibits 22.4 and 22.14. Such assumptions allow users to answer several important questions, but the usefulness of the answers depends on the validity of three assumptions: (1) constant selling price per unit, (2) constant variable costs per unit, and (3) constant total fixed costs. These assumptions are not always realistic, but they do not necessarily limit the usefulness of CVP analysis as a way to better understand costs and sales. This section discusses these assumptions and other issues for CVP analysis.

C2 Identify assumptions in cost-volume-profit analysis and explain their impact.

Working with Assumptions The behavior of individual costs and sales often is not perfectly consistent with CVP assumptions. If the expected costs and sales behavior differ from the assumptions, the results of CVP analysis can be limited. Still, we can perform useful analyses in spite of limitations with these assumptions for several reasons.

Summing costs can offset individual deviations. Deviations from assumptions with individual costs are often minor when these costs are summed. That is, individual variable cost items may not be perfectly variable, but when we sum these variable costs, their individual deviations can offset each other. This means the assumption of variable cost behavior can be proper for total variable costs. Similarly, an assumption that total fixed costs are constant can be proper even when individual fixed cost items are not exactly constant.

Point: CVP analysis can be very useful for business decision making even when its assumptions are not strictly met.

CVP is applied to a relevant range of operations. Sales, variable costs, and fixed costs often are reasonably reflected in straight lines on a graph when the assumptions are applied over a relevant range. The **relevant range of operations** is the normal operating range for a business. Except for unusually difficult or prosperous times, management typically plans for operations within a range of volume neither close to zero nor at maximum capacity. The relevant range excludes extremely high and low operating levels that are unlikely to occur. The validity of assuming that a specific cost is fixed or variable is more acceptable when operations are within the relevant range. As shown in Exhibit 22.2, a curvilinear cost can be treated as variable and linear if the relevant range covers volumes where it has a nearly constant slope. If the normal range of activity changes, some costs might need reclassification.

Video22.2

CVP analysis yields estimates. CVP analysis yields approximate answers to questions about costs, volumes, and profits. These answers do not have to be precise because the analysis makes rough estimates about the future. As long as managers understand that CVP analysis gives estimates, it can be a useful tool for starting the planning process. Other qualitative factors also must be considered.

Working with Output Measures CVP analysis usually describes the level of activity in terms of *sales volume,* which can be expressed in terms of either units sold or dollar sales. However, other measures of output exist. For instance, a manufacturer can use the number of units produced as a measure of output. Also, to simplify analysis, we sometimes assume that the production level is the same as the sales level. That is, inventory levels do not change. This often is justified by arguing that CVP analysis provides only approximations.

Example: If the selling price declines, what happens to the break-even point? *Answer:* It increases.

Quick Check

Answers—p. 922

7. Fixed cost divided by the contribution margin ratio yields the (*a*) break-even point in dollars, (*b*) contribution margin per unit, or (*c*) break-even point in units.
8. A company sells a product for $90 per unit with variable costs of $54 per unit. What is the contribution margin ratio?
9. Refer to Quick Check (8). If fixed costs for the period are $90,000, what is the break-even point in dollars?
10. What three basic assumptions are used in CVP analysis?

Working with Changes in Estimates Because CVP analysis uses estimates, knowing how changes in those estimates impact breakeven is useful. For example, a manager might form three estimates for each of the components of breakeven: optimistic, most likely, and pessimistic. Then ranges of break-even points in units can be computed using the formula in Exhibit 22.11.

To illustrate, assume Rydell's managers provide the set of estimates in Exhibit 22.15.

EXHIBIT 22.15

Alternative Estimates for Break-Even Analysis

	Selling Price per Unit	Variable Cost per Unit	Total Fixed Costs
Optimistic	$105	$68	$21,000
Most likely	100	70	24,000
Pessimistic	95	72	27,000

If, for example, Rydell's managers believe they can raise the selling price of a football to $105, without any change in variable or fixed costs, then the revised contribution margin per football is $35, and the revised break-even in units follows in Exhibit 22.16.

EXHIBIT 22.16

Revised Break-Even in Units

$$\text{Revised break-even point in units} = \frac{\$24{,}000}{\$35} = 686 \text{ units}$$

EXHIBIT 22.17

Scatter Diagrams—Break-Even Points for Alternative Estimates

Repeating this calculation using each of the other eight separate estimates above, and graphing the results, yields the three scatter diagrams in Exhibit 22.17.

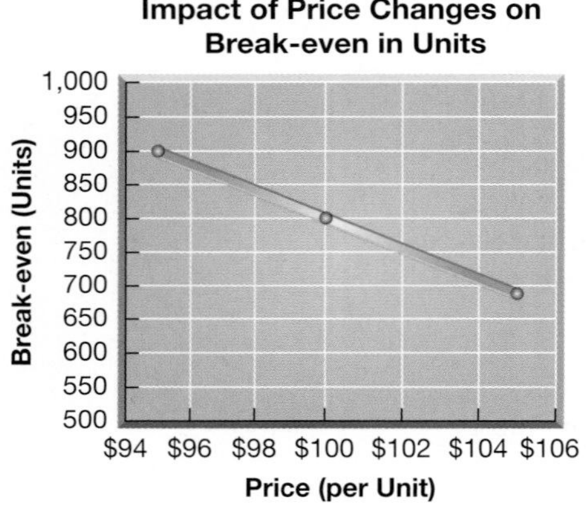

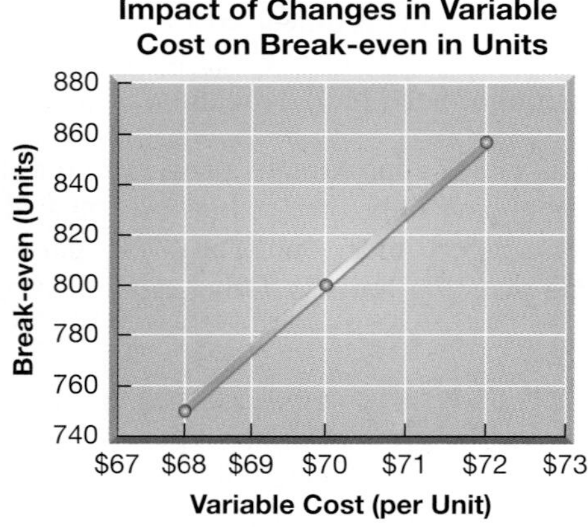

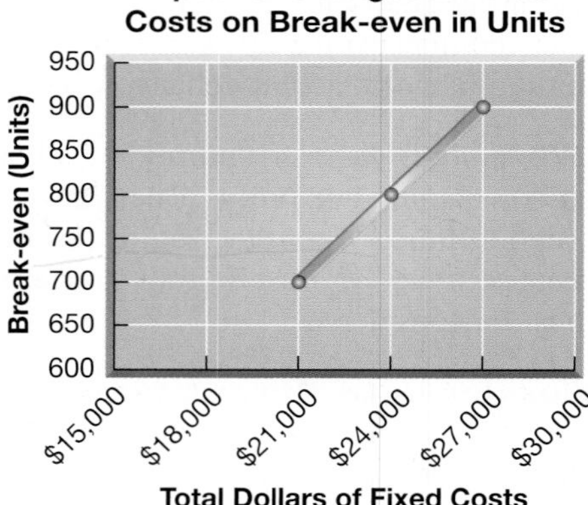

These scatter diagrams show how changes in selling prices, variable costs, and fixed costs impact break-even. When selling prices can be increased without impacting costs, break-even decreases. When competition drives selling prices down, and the company cannot reduce costs, break-even increases. Increases in either variable or fixed costs, if they cannot be passed on to customers via higher selling prices, will increase break-even. If costs can be reduced and selling prices held constant, the break-even decreases.

Point: This analysis changed only one estimate at a time; managers can examine how combinations of changes in estimates will impact break-even.

Applying Cost-Volume-Profit Analysis

Managers consider a variety of strategies in planning business operations. Cost-volume-profit analysis is useful in helping managers evaluate the likely effects of these strategies, which is the focus of this section.

Computing Income from Sales and Costs

C3 Describe several applications of cost-volume-profit analysis.

An important question managers often need an answer to is "What is the predicted income from a predicted level of sales?" To answer this, we look at four variables in CVP analysis. These variables and their relations to income (pretax) are shown in Exhibit 22.18. We use these relations to compute expected income from predicted sales and cost levels.

EXHIBIT 22.18

Income Relations in CVP Analysis

Sales
− Variable costs
Contribution margin
− Fixed costs
Income (pretax)

To illustrate, let's assume that Rydell's management expects to sell 1,500 units in January 2009. What is the amount of income if this sales level is achieved? Following Exhibit 22.18, we compute Rydell's expected income in Exhibit 22.19.

EXHIBIT 22.19

Computing Expected Pretax Income from Expected Sales

RYDELL COMPANY
Contribution Margin Income Statement
For Month Ended January 31, 2009

Sales (1,500 units at \$100 each)	\$150,000
Variable costs (1,500 units at \$70 each)	105,000
Contribution margin	45,000
Fixed costs	24,000
Income (pretax)	\$ 21,000

The \$21,000 income is pretax. To find the amount of *after-tax* income from selling 1,500 units, management must apply the proper tax rate. Assume that the tax rate is 25%. Then we can prepare the after-tax income statement shown in Exhibit 22.20. We can also compute pretax income as after-tax income divided by (1 − tax rate); for Rydell, this is \$15,750/(1 − 0.25), or \$21,000.

EXHIBIT 22.20

Computing Expected After-Tax Income from Expected Sales

RYDELL COMPANY
Contribution Margin Income Statement
For Month Ended January 31, 2009

Sales (1,500 units at \$100 each)	\$150,000
Variable costs (1,500 units at \$70 each)	105,000
Contribution margin	45,000
Fixed costs	24,000
Pretax income	21,000
Income taxes (25%)	5,250
Net income (after tax)	\$ 15,750

Management then assesses whether this income is an adequate return on assets invested. Management should also consider whether sales and income can be increased by raising or lowering prices. CVP analysis is a good tool for addressing these kinds of "what-if" questions.

Computing Sales for a Target Income

"How many
units must I sell
to earn \$50,000?"

Many companies' annual plans are based on certain income targets (sometimes called *budgets*). Rydell's income target for this year is to increase income by 10% over the prior year. When prior year income is known, Rydell easily computes its target income. CVP analysis helps to determine the sales level needed to achieve the target income. Computing this sales level is important because planning for the year is then based on this level. We use the formula shown in Exhibit 22.21 to compute sales for a target *after-tax* income.

EXHIBIT 22.21

Computing Sales (Dollars) for a Target After-Tax Income

$$\textbf{Dollar sales at target after-tax income} = \frac{\textbf{Fixed costs} + \textbf{Target pretax income}}{\textbf{Contribution margin ratio}}$$

To illustrate, Rydell has monthly fixed costs of \$24,000 and a 30% contribution margin ratio. Assume that it sets a target monthly after-tax income of \$9,000 when the tax rate is 25%. This means the pretax income is targeted at \$12,000 [\$9,000/(1 − 0.25)] with a tax expense of \$3,000. Using the formula in Exhibit 22.21, we find that \$120,000 of sales are needed to produce a \$9,000 after-tax income as shown in Exhibit 22.22.

EXHIBIT 22.22

Rydell's Dollar Sales for a Target Income

$$\text{Dollar sales at target after-tax income} = \frac{\$24{,}000 + \$12{,}000}{30\%} = \$120{,}000$$

Point: Break-even is a special case of the formulas in Exhibits 22.21 and 22.23; simply set target pretax income to \$0 and the formulas reduce to those in Exhibits 22.11 and 22.12.

We can alternatively compute *unit sales* instead of dollar sales. To do this, we substitute *contribution margin per unit* for the contribution margin ratio in the denominator. This gives the number of units to sell to reach the target after-tax income. Exhibit 22.23 illustrates this for Rydell. The two computations in Exhibits 22.22 and 22.23 are equivalent because sales of 1,200 units at \$100 per unit equal \$120,000 of sales.

EXHIBIT 22.23

Computing Sales (Units) for a Target After-Tax Income

$$\textbf{Unit sales at target after-tax income} = \frac{\textbf{Fixed costs} + \textbf{Target pretax income}}{\textbf{Contribution margin per unit}}$$

$$= \frac{\$24{,}000 + \$12{,}000}{\$30} = 1{,}200 \text{ units}$$

Computing the Margin of Safety

All companies wish to sell more than the break-even number of units. The excess of expected sales over the break-even sales level is called a company's **margin of safety,** the amount that sales can drop before the company incurs a loss. It can be expressed in units, dollars, or even as a percent of the predicted level of sales. To illustrate, if Rydell's expected sales are \$100,000, the margin of safety is \$20,000 above break-even sales of \$80,000. As a percent, the margin of safety is 20% of expected sales as shown in Exhibit 22.24.

EXHIBIT 22.24

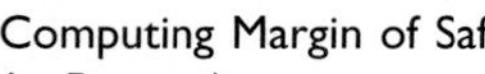

Computing Margin of Safety (in Percent)

$$\textbf{Margin of safety (in percent)} = \frac{\textbf{Expected sales} - \textbf{Break-even sales}}{\textbf{Expected sales}}$$

$$= \frac{\$100{,}000 - \$80{,}000}{\$100{,}000} = 20\%$$

Management must assess whether the margin of safety is adequate in light of factors such as sales variability, competition, consumer tastes, and economic conditions.

Decision Ethics

Supervisor Your team is conducting a cost-volume-profit analysis for a new product. Different sales projections have different incomes. One member suggests picking numbers yielding favorable income because any estimate is "as good as any other." Another member points to a scatter diagram of 20 months' production on a comparable product and suggests dropping unfavorable data points for cost estimation. What do you do? [Answer—p. 921]

Using Sensitivity Analysis

Earlier we showed how changing one of the estimates in a CVP analysis impacts breakeven. We can also examine strategies that impact several estimates in the CVP analysis. For instance, we might want to know what happens to income if we automate a currently manual process. We can use CVP analysis to predict income if we can describe how these changes affect a company's fixed costs, variable costs, selling price, and volume.

To illustrate, assume that Rydell Company is looking into buying a new machine that would increase monthly fixed costs from $24,000 to $30,000 but decrease variable costs from $70 per unit to $60 per unit. The machine is used to produce output whose selling price will remain unchanged at $100. This results in increases in both the unit contribution margin and the contribution margin ratio. The revised contribution margin per unit is $40 ($100 − $60), and the revised contribution margin ratio is 40% of selling price ($40/$100). Using CVP analysis, Rydell's revised break-even point in dollars would be $75,000 as computed in Exhibit 22.25.

Example: If fixed costs decline, what happens to the break-even point? *Answer:* It decreases.

EXHIBIT 22.25

Revising Break-even When Changes Occur

$$\textbf{Revised break-even point in dollars} = \frac{\textbf{Revised fixed costs}}{\textbf{Revised contribution margin ratio}} = \frac{\$30{,}000}{40\%} = \$75{,}000$$

The revised fixed costs and the revised contribution margin ratio can be used to address other issues including computation of (1) expected income for a given sales level and (2) the sales level needed to earn a target income. Once again, we can use sensitivity analysis to generate different sets of revenue and cost estimates that are *optimistic, pessimistic,* and *most likely.* Different CVP analyses based on these estimates provide different scenarios that management can analyze and use in planning business strategy.

Point: Price competition led paging companies to give business to resellers—companies that lease services at a discount and then resell to subscribers. **Paging Network** charged some resellers under $1 per month, less than a third of what was needed to break even. Its CEO now admits the low-price strategy was flawed.

Decision Insight

Eco-CVP Ford Escape, Toyota Prius, and Honda Insight are hybrids. Many promise to save owners $1,000 or more a year in fuel costs relative to comparables, and they generate fewer greenhouse gases. Are these models economically feasible? Analysts estimate that **Ford** can break even with its Escape when a $3,000 premium is paid over comparable gas-based models.

Quick Check

Answers—p. 922

11. A company has fixed costs of $50,000 and a 25% contribution margin ratio. What dollar sales are necessary to achieve an after-tax net income of $120,000 if the tax rate is 20%? (*a*) $800,000, (*b*) $680,000, or (*c*) $600,000.

12. If a company's contribution margin ratio decreases from 50% to 25%, what can be said about the unit sales needed to achieve the same target income level?

13. What is a company's margin of safety?

Computing a Multiproduct Break-Even Point

P4 Compute the break-even point for a multiproduct company.

To this point, we have looked only at cases where the company sells a single product or service. This was to keep the basic CVP analysis simple. However, many companies sell multiple products or services, and we can modify the CVP analysis for use in these cases. An important assumption in a multiproduct setting is that the sales mix of different products is known and remains constant during the planning period. **Sales mix** is the ratio (proportion) of the sales volumes for the various products. For instance, if a company normally sells 10,000 footballs, 5,000 softballs, and 4,000 basketballs per month, its sales mix can be expressed as 10:5:4 for footballs, softballs, and basketballs.

To apply multiproduct CVP analysis, we can estimate the break-even point by using a **composite unit,** which consists of a specific number of units of each product in proportion to their expected sales mix. Multiproduct CVP analysis treats this composite unit as a single product. To illustrate, let's look at **Hair-Today,** a styling salon that offers three cuts: basic, ultra, and budget in the ratio of 4 basic units to 2 ultra units to 1 budget unit (expressed as 4:2:1). Management wants to estimate its break-even point for next year. Unit selling prices for these three cuts are basic, \$20; ultra, \$32; and budget, \$16. Using the 4:2:1 sales mix, the selling price of a composite unit of the three products is computed as follows.

4 units of basic @ \$20 per unit	\$ 80
2 units of ultra @ \$32 per unit	64
1 unit of budget @ \$16 per unit	16
Selling price of a composite unit	**\$160**

Point: Selling prices and variable costs are usually expressed in per unit amounts. Fixed costs are usually expressed in total amounts.

Hair-Today's fixed costs are \$192,000 per year, and its variable costs of the three products are basic, \$13; ultra, \$18.00; and budget, \$8.00. Variable costs for a composite unit of these products follow.

4 units of basic @ \$13 per unit	\$52
2 units of ultra @ \$18 per unit	36
1 unit of budget @ \$8 per unit	8
Variable costs of a composite unit	**\$96**

Hair-Today's \$64 contribution margin for a composite unit is computed by subtracting the variable costs of a composite unit (\$96) from its selling price (\$160). We then use the contribution margin to determine Hair-Today's break-even point in composite units in Exhibit 22.26.

EXHIBIT 22.26

Break-Even Point in Composite Units

$$\textbf{Break-even point in composite units} = \frac{\textbf{Fixed costs}}{\textbf{Contribution margin per composite unit}}$$

$$= \frac{\$192{,}000}{\$64} = 3{,}000 \text{ composite units}$$

Point: The break-even point in dollars for Exhibit 22.26 is \$192,000/(\$64/\$160) = \$480,000.

This computation implies that Hair-Today breaks even when it sells 3,000 composite units. To determine how many units of each product it must sell to break even, we multiply the number of units of each product in the composite by 3,000 as follows.

Basic:	4 × 3,000	12,000 units
Ultra:	2 × 3,000	6,000 units
Budget:	1 × 3,000	3,000 units

Instead of computing contribution margin per composite unit, a company can compute a **weighted-average contribution margin.** Given the 4:2:1 product mix, basic cuts comprise 57.14% (computed as 4/7) of the company's haircuts, ultra makes up 14.29% of its business, and budget cuts comprise 28.57%. The weighted-average contribution margin follows in Exhibit 22.27.

EXHIBIT 22.27

Weighted-Average Contribution Margin

	Unit contribution margin	×	Percentage of sales mix	=	Weighted unit contribution margin
Basic	\$ 7		57.14%		\$4.000
Ultra	14		28.57		4.000
Budget	8		14.29		1.143
Weighted-average contribution margin					**\$9.143**

The company's break-even point in units is computed as follows:

$$\textbf{Break-even point in units} = \frac{\textbf{Fixed costs}}{\textbf{Weighted-average contribution margin}}$$

$$= \frac{\$192{,}000}{\$9.143} = 21{,}000 \text{ units}$$

EXHIBIT 22.28

Break-Even in Units using Weighted-Average Contribution Margin

We see that the weighted-average contribution margin method yields 21,000 whole units as the break-even amount, the same total as the composite unit approach.

Exhibit 22.29 verifies the results for composite units by showing Hair-Today's sales and costs at this break-even point using a contribution margin income statement.

EXHIBIT 22.29

Multiproduct Break-Even Income Statement

HAIR-TODAY
Forecasted Contribution Margin Income Statement (at Breakeven)

	Basic	Ultra	Budget	Totals
Sales				
Basic (12,000 @ $20)	$240,000			
Ultra (6,000 @ $32)		$192,000		
Budget (3,000 @ $16)			$48,000	
Total sales				$480,000
Variable costs				
Basic (12,000 @ $13)	156,000			
Ultra (6,000 @ $18)		108,000		
Budget (3,000 @ $8)			24,000	
Total variable costs				288,000
Contribution margin	$ 84,000	$ 84,000	$24,000	192,000
Fixed costs				192,000
Net income				$ 0

A CVP analysis using composite units can be used to answer a variety of planning questions. Once a product mix is set, all answers are based on the assumption that the mix remains constant at all relevant sales levels as other factors in the analysis do. We also can vary the sales mix to see what happens under alternative strategies.

Decision Maker

Entrepreneur A CVP analysis indicates that your start-up, which markets electronic products, will break even with the current sales mix and price levels. You have a target income in mind. What analysis might you perform to assess the likelihood of achieving this income? [Answer—p. 921]

Quick Check

Answers—p. 922

14. The sales mix of a company's two products, X and Y, is 2:1. Unit variable costs for both products are $2, and unit sales prices are $5 for X and $4 for Y. What is the contribution margin per composite unit? (*a*) $5, (*b*) $10, or (*c*) $8.

15. What additional assumption about sales mix must be made in doing a conventional CVP analysis for a company that produces and sells more than one product?

Decision Analysis — Degree of Operating Leverage

A3 Analyze changes in sales using the degree of operating leverage.

CVP analysis is especially useful when management begins the planning process and wishes to predict outcomes of alternative strategies. These strategies can involve changes in selling prices, fixed costs, variable costs, sales volume, and product mix. Managers are interested in seeing the effects of changes in some or all of these factors.

One goal of all managers is to get maximum benefits from their fixed costs. Managers would like to use 100% of their output capacity so that fixed costs are spread over the largest number of units. This would decrease fixed cost per unit and increase income. The extent, or relative size, of fixed costs in the total cost structure is known as **operating leverage.** Companies having a higher proportion of fixed costs in their total cost structure are said to have higher operating leverage. An example of this is a company that chooses to automate its processes instead of using direct labor, increasing its fixed costs and lowering its variable costs. A useful managerial measure to help assess the effect of changes in the level of sales on income is the **degree of operating leverage (DOL)** defined in Exhibit 22.30.

EXHIBIT 22.30

Degree of Operating Leverage

DOL = Total contribution margin (in dollars)/Pretax income

To illustrate, let's return to Rydell Company. At a sales level of 1,200 units, Rydell's total contribution margin is $36,000 (1,200 units × $30 contribution margin per unit). Its pretax income, after subtracting fixed costs of $24,000, is $12,000 ($36,000 − $24,000). Rydell's degree of operating leverage at this sales level is 3.0, computed as contribution margin divided by pretax income ($36,000/$12,000). We then use DOL to measure the effect of changes in the level of sales on pretax income. For instance, suppose Rydell expects sales to increase by 10%. If this increase is within the relevant range of operations, we can expect this 10% increase in sales to result in a 30% increase in pretax income computed as DOL multiplied by the increase in sales (3.0 × 10%). Similar analyses can be done for expected decreases in sales.

Demonstration Problem

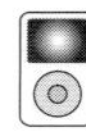

DP22

Sport Caps Co. manufactures and sells caps for different sporting events. The fixed costs of operating the company are $150,000 per month, and the variable costs for caps are $5 per unit. The caps are sold for $8 per unit. The fixed costs provide a production capacity of up to 100,000 caps per month.

Required

1. Use the formulas in the chapter to compute the following:
   a. Contribution margin per cap.
   b. Break-even point in terms of the number of caps produced and sold.
   c. Amount of net income at 30,000 caps sold per month (ignore taxes).
   d. Amount of net income at 85,000 caps sold per month (ignore taxes).
   e. Number of caps to be produced and sold to provide $45,000 of after-tax income, assuming an income tax rate of 25%.
2. Draw a CVP chart for the company, showing cap output on the horizontal axis. Identify (*a*) the break-even point and (*b*) the amount of pretax income when the level of cap production is 70,000. (Omit the fixed cost line.)
3. Use the formulas in the chapter to compute the
   a. Contribution margin ratio.
   b. Break-even point in terms of sales dollars.
   c. Amount of net income at $250,000 of sales per month (ignore taxes).
   d. Amount of net income at $600,000 of sales per month (ignore taxes).
   e. Dollars of sales needed to provide $45,000 of after-tax income, assuming an income tax rate of 25%.

Planning the Solution

- Identify the formulas in the chapter for the required items expressed in units and solve them using the data given in the problem.

- Draw a CVP chart that reflects the facts in the problem. The horizontal axis should plot the volume in units up to 100,000, and the vertical axis should plot the total dollars up to $800,000. Plot the total cost line as upward sloping, starting at the fixed cost level ($150,000) on the vertical axis and increasing until it reaches $650,000 at the maximum volume of 100,000 units. Verify that the break-even point (where the two lines cross) equals the amount you computed in part 1.
- Identify the formulas in the chapter for the required items expressed in dollars and solve them using the data given in the problem.

Solution to Demonstration Problem

1. a. Contribution margin per cap = Selling price per unit − Variable cost per unit = $8 − $5 = $3

b. Break-even point in caps $= \dfrac{\text{Fixed costs}}{\text{Contribution margin per cap}} = \dfrac{\$150{,}000}{\$3} = 50{,}000 \text{ caps}$

c. Net income at 30,000 caps sold = (Units × Contribution margin per unit) − Fixed costs = (30,000 × $3) − $150,000 = $(60,000) loss

d. Net income at 85,000 caps sold = (Units × Contribution margin per unit) − Fixed costs = (85,000 × $3) − $150,000 = $105,000 profit

e. Pretax income = $45,000/(1 − 0.25) = $60,000
Income taxes = $60,000 × 25% = $15,000

Units needed for $45,000 income $= \dfrac{\text{Fixed costs} + \text{Target pretax income}}{\text{Contribution margin per cap}} = \dfrac{\$150{,}000 + \$60{,}000}{\$3} = 70{,}000 \text{ caps}$

2. CVP chart.

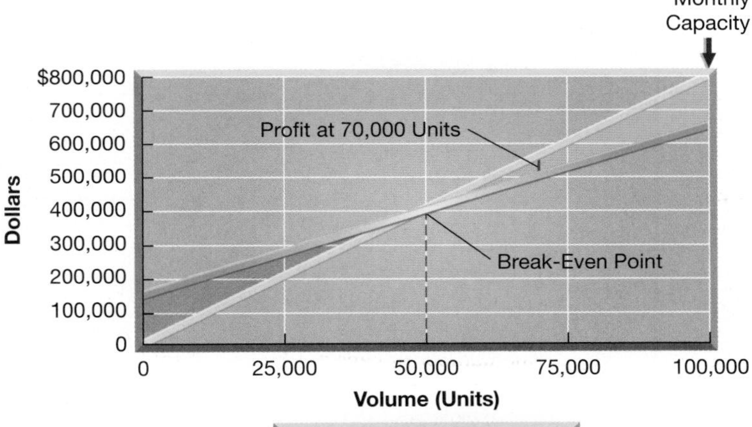

3. a. Contribution margin ratio $= \dfrac{\text{Contribution margin per unit}}{\text{Selling price per unit}} = \dfrac{\$3}{\$8} = 0.375, \text{ or } 37.5\%$

b. Break-even point in dollars $= \dfrac{\text{Fixed costs}}{\text{Contribution margin ratio}} = \dfrac{\$150{,}000}{37.5\%} = \$400{,}000$

c. Net income at sales of $250,000 = (Sales × Contribution margin ratio) − Fixed costs = ($250,000 × 37.5%) − $150,000 = $(56,250) loss

d. Net income at sales of $600,000 = (Sales × Contribution margin ratio) − Fixed costs = ($600,000 × 37.5%) − $150,000 = $75,000 income

e. Dollars of sales to yield $45,000 after-tax income $= \dfrac{\text{Fixed costs} + \text{Target pretax income}}{\text{Contribution margin ratio}} = \dfrac{\$150{,}000 + \$60{,}000}{37.5\%} = \$560{,}000$

APPENDIX

22A Using Excel to Estimate Least-Squares Regression

Microsoft Excel® 2007 and other spreadsheet software can be used to perform least-squares regressions to identify cost behavior. In Excel®, the INTERCEPT and SLOPE functions are used. The following screen shot reports the data from Exhibit 22.3 in cells Al through C13 and shows the cell contents to find the intercept (cell B16) and slope (cell B17). Cell B16 uses Excel® to find the intercept from a least-squares regression of total cost (shown as C2:C13 in cell B16) on units produced (shown as B2:B13 in cell B16). Spreadsheet software is useful in understanding cost behavior when many data points (such as monthly total costs and units produced) are available.

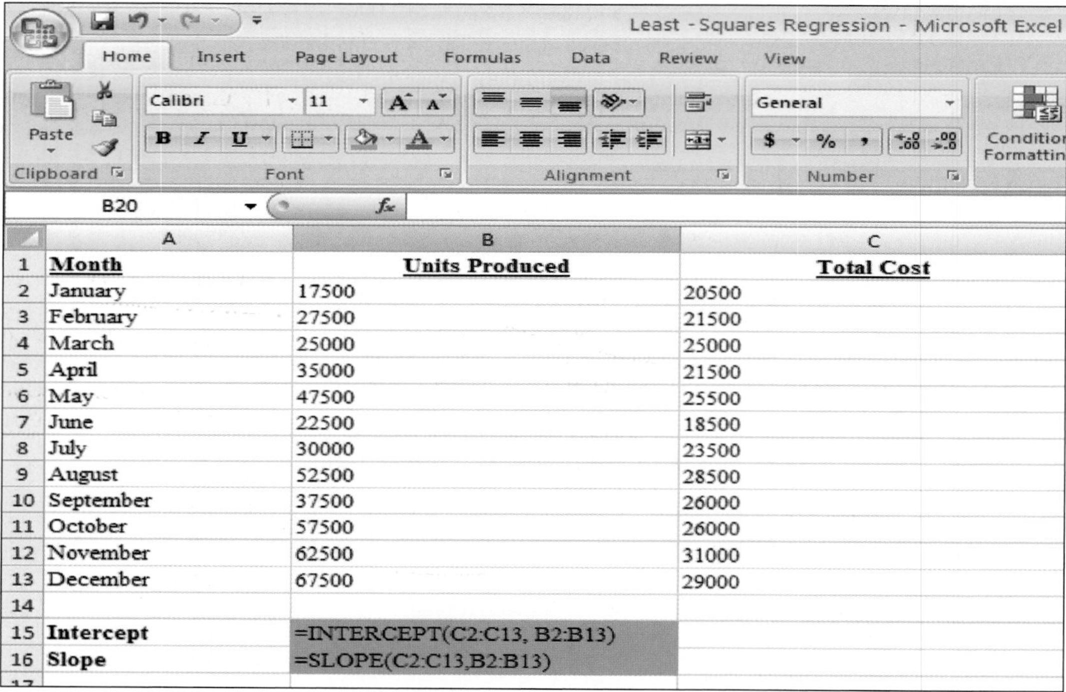

	A	B	C
1	**Month**	**Units Produced**	**Total Cost**
2	January	17500	20500
3	February	27500	21500
4	March	25000	25000
5	April	35000	21500
6	May	47500	25500
7	June	22500	18500
8	July	30000	23500
9	August	52500	28500
10	September	37500	26000
11	October	57500	26000
12	November	62500	31000
13	December	67500	29000
14			
15	**Intercept**	=INTERCEPT(C2:C13, B2:B13)	
16	**Slope**	=SLOPE(C2:C13,B2:B13)	

Excel® can also be used to create scatter diagrams such as that in Exhibit 22.4. In contrast to visually drawing a line that "fits" the data, Excel® more precisely fits the regression line. To draw a scatter diagram with a line of fit, follow these steps:

1. Highlight the data cells you wish to diagram; in this example, start from cell C13 and highlight through cell B2.
2. Then select "Insert" and "Scatter" from the drop-down menus. Selecting the chart type in the upper left corner of the choices under Scatter will produce a diagram that looks like that in Exhibit 22.4, without a line of fit.
3. To add a line of fit (also called trend line), select "Layout" and "Trendline" from the drop-down menus. Selecting "Linear Trendline" will produce a diagram that looks like that in Exhibit 22.4, including the line of fit.

Summary

C1 Describe different types of cost behavior in relation to production and sales volume. Cost behavior is described in terms of how its amount changes in relation to changes in volume of activity within a relevant range. Fixed costs remain constant to changes in volume. Total variable costs change in direct proportion to volume changes. Mixed costs display the effects of both fixed and variable components. Step-wise costs remain constant over a small volume range, then change by a lump sum and remain constant over another volume range, and so on. Curvilinear costs change in a nonlinear relation to volume changes.

C2 Identify assumptions in cost-volume-profit analysis and explain their impact. Conventional cost-volume-profit analysis is based on assumptions that the product's selling price remains constant and that variable and fixed costs behave in a manner consistent with their variable and fixed classifications.

C3 Describe several applications of cost-volume-profit analysis. Cost-volume-profit analysis can be used to predict what can happen under alternative strategies concerning sales volume, selling prices, variable costs, or fixed costs. Applications include "what-if" analysis, computing sales for a target income, and break-even analysis.

A1 Compare the scatter diagram, high-low, and regression methods of estimating costs. Cost estimates from a scatter diagram are based on a visual fit of the cost line. Estimates from the high-low method are based only on costs corresponding to the lowest and highest sales. The least-squares regression method is a statistical technique and uses all data points.

A2 Compute the contribution margin and describe what it reveals about a company's cost structure. Contribution margin per unit is a product's sales price less its total variable costs. Contribution margin ratio is a product's contribution margin per unit divided by its sales price. Unit contribution margin is the amount received from each sale that contributes to fixed costs and income. The contribution margin ratio reveals what portion of each sales dollar is available as contribution to fixed costs and income.

A3 Analyze changes in sales using the degree of operating leverage. The extent, or relative size, of fixed costs in a company's total cost structure is known as *operating leverage.* One tool useful in assessing the effect of changes in sales on income is the degree of operating leverage, or DOL. DOL is the ratio of the contribution margin divided by pretax income. This ratio can be used to determine the expected percent change in income given a percent change in sales.

P1 Determine cost estimates using three different methods. Three different methods used to estimate costs are the scatter diagram, the high-low method, and least-squares regression. All three methods use past data to estimate costs.

P2 Compute the break-even point for a single product company. A company's break-even point for a period is the sales volume at which total revenues equal total costs. To compute a break-even point in terms of sales units, we divide total fixed costs by the contribution margin per unit. To compute a break-even point in terms of sales dollars, divide total fixed costs by the contribution margin ratio.

P3 Graph costs and sales for a single product company. The costs and sales for a company can be graphically illustrated using a CVP chart. In this chart, the horizontal axis represents the number of units sold and the vertical axis represents dollars of sales or costs. Straight lines are used to depict both costs and sales on the CVP chart.

P4 Compute the break-even point for a multiproduct company. CVP analysis can be applied to a multiproduct company by expressing sales volume in terms of composite units. A composite unit consists of a specific number of units of each product in proportion to their expected sales mix. Multiproduct CVP analysis treats this composite unit as a single product.

Guidance Answers to **Decision Maker** and **Decision Ethics**

Sales Manager The contribution margin per unit for the first order is \$600 (60% of \$1,000); the contribution margin per unit for the second order is \$160 (20% of \$800). You are likely tempted to accept the first order based on its high contribution margin per unit, but you must compute the total contribution margin based on the number of units sold for each order. Total contribution margin is \$60,000 (\$600 per unit × 100 units) and \$80,000 (\$160 per unit × 500 units) for the two orders, respectively. The second order provides the largest return in absolute dollars and is the order you would accept. Another factor to consider in your selection is the potential for a long-term relationship with these customers including repeat sales and growth.

Operations Manager Without the availability of past data, none of the three methods described in the chapter can be used to measure cost behavior. Instead, the manager must investigate whether data from similar manufacturers can be accessed. This is likely difficult due to the sensitive nature of such data. In the absence of data, the manager should develop a list of the different production inputs and identify input-output relations. This provides guidance to the manager in measuring cost behavior. After several months, actual cost data will be available for analysis.

Supervisor Your dilemma is whether to go along with the suggestions to "manage" the numbers to make the project look like it will achieve sufficient profits. You should not succumb to these suggestions. Many people will likely be affected negatively if you manage the predicted numbers and the project eventually is unprofitable. Moreover, if it does fail, an investigation would likely reveal that data in the proposal were "fixed" to make it look good. Probably the only benefit from managing the numbers is the short-term payoff of pleasing those who proposed the product. One way to deal with this dilemma is to prepare several analyses showing results under different assumptions and then let senior management make the decision.

Entrepreneur You must first compute the level of sales required to achieve the desired net income. Then you must conduct sensitivity analysis by varying the price, sales mix, and cost estimates. Results from the sensitivity analysis provide information you can use to assess the possibility of reaching the target sales level. For instance, you might have to pursue aggressive marketing strategies to push the high-margin products, or you might have to cut prices to increase sales and profits, or another strategy might emerge.

Guidance Answers to **Quick Checks**

1. *b*
2. A fixed cost remains unchanged in total amount regardless of output levels. However, fixed *cost per unit* declines with increased output.
3. Such a cost is considered variable because the *total* cost changes in proportion to volume changes.
4. *b*
5. The high-low method ignores all costs and sales (activity base) volume data points except the costs corresponding to the highest and lowest (most extreme) sales (activity base) volume.
6. *c*
7. *a*
8. ($90 − $54)/$90 = 40%
9. $90,000/40% = $225,000
10. Three basic CVP assumptions are that (1) selling price per unit is constant, (2) variable costs per unit are constant, and (3) total fixed costs are constant.
11. a; Two steps are required for explanation:
    (1) Pretax income = $120,000/(1 − 0.20) = $150,000
    (2) $\frac{\$50{,}000 + \$150{,}000}{25\%} = \$800{,}000$
12. If the contribution margin ratio decreases from 50% to 25%, unit sales would have to double.
13. A company's margin of safety is the excess of the predicted sales level over its break-even sales level.
14. *c*; Selling price of a composite unit:

2 units of X @ $5 per unit	$10
1 unit of Y @ $4 per unit	4
Selling price of a composite unit	$14
Variable costs of a composite unit:	
2 units of X @ $2 per unit	$4
1 unit of Y @ $2 per unit	2
Variable costs of a composite unit	$6

    Therefore, the contribution margin per composite unit is $8.
15. It must be assumed that the sales mix remains unchanged at all sales levels in the relevant range.

Key Terms

mhhe.com/wildFAP19e

Key Terms are available at the book's Website for learning and testing in an online Flashcard Format.

Break-even point (p. 909)
Composite unit (p. 916)
Contribution margin per unit (p. 908)
Contribution margin ratio (p. 908)
Cost-volume-profit (CVP) analysis (p. 902)
Cost-volume-profit (CVP) chart (p. 910)
Curvilinear cost (p. 904)
Degree of operating leverage (DOL) (p. 918)
Estimated line of cost behavior (p. 906)
High-low method (p. 906)
Least-squares regression (p. 907)
Margin of safety (p. 914)
Mixed cost (p. 903)
Operating leverage (p. 918)
Relevant range of operations (p. 911)
Sales mix (p. 915)
Scatter diagram (p. 905)
Step-wise cost (p. 904)
Weighted-average contribution margin (p. 916)

Multiple Choice Quiz

Answers on p. 937

mhhe.com/wildFAP19e

Additional Quiz Questions are available at the book's Website.

Quiz22

1. A company's only product sells for $150 per unit. Its variable costs per unit are $100, and its fixed costs total $75,000. What is its contribution margin per unit?
   a. $50
   b. $250
   c. $100
   d. $150
   e. $25
2. Using information from question 1, what is the company's contribution margin ratio?
   a. 66⅔%
   b. 100%
   c. 50%
   d. 0%
   e. 33⅓%
3. Using information from question 1, what is the company's break-even point in units?
   a. 500 units
   b. 750 units
   c. 1,500 units
   d. 3,000 units
   e. 1,000 units
4. A company's forecasted sales are $300,000 and its sales at break-even are $180,000. Its margin of safety in dollars is
   a. $180,000.
   b. $120,000.

c. \$480,000.
d. \$60,000.
e. \$300,000.

5. A product sells for \$400 per unit and its variable costs per unit are \$260. The company's fixed costs are \$840,000. If the company desires \$70,000 pretax income, what is the required dollar sales?

a. \$2,400,000
b. \$200,000
c. \$2,600,000
d. \$2,275,000
e. \$1,400,000

Superscript letter A denotes assignments based on Appendix 22A.

Discussion Questions

1. How is cost-volume-profit analysis useful?
2. What is a variable cost? Identify two variable costs.
3. When output volume increases, do variable costs per unit increase, decrease, or stay the same within the relevant range of activity? Explain.
4. When output volume increases, do fixed costs per unit increase, decrease, or stay the same within the relevant range of activity? Explain.
5. How do step-wise costs and curvilinear costs differ?
6. Define and describe *contribution margin* per unit.
7. Define and explain the *contribution margin ratio.*
8. Describe the contribution margin ratio in layperson's terms.
9. In performing CVP analysis for a manufacturing company, what simplifying assumption is usually made about the volume of production and the volume of sales?
10. What two arguments tend to justify classifying all costs as either fixed or variable even though individual costs might not behave exactly as classified?
11. How does assuming that operating activity occurs within a relevant range affect cost-volume-profit analysis?
12. List three methods to measure cost behavior.
13. How is a scatter diagram used to identify and measure the behavior of a company's costs?
14. In cost-volume-profit analysis, what is the estimated profit at the break-even point?
15. Assume that a straight line on a CVP chart intersects the vertical axis at the level of fixed costs and has a positive slope that rises with each additional unit of volume by the amount of the variable costs per unit. What does this line represent?
16. Why are fixed costs depicted as a horizontal line on a CVP chart?
17. Each of two similar companies has sales of \$20,000 and total costs of \$15,000 for a month. Company A's total costs include \$10,000 of variable costs and \$5,000 of fixed costs. If Company B's total costs include \$4,000 of variable costs and \$11,000 of fixed costs, which company will enjoy more profit if sales double?
18. ______ of ______ reflects expected sales in excess of the level of break-even sales.
19. **Apple** produces iPods for sale. Identify some of the variable and fixed product costs associated with that production. [*Hint:* Limit costs to product costs.]
20. Should **Best Buy** use single product or multi-product break-even analysis? Explain.

21. **Apple** is thinking of expanding sales of its most popular Macintosh model by 65%. Do you expect its variable and fixed costs for this model to stay within the relevant range? Explain.

Denotes Discussion Questions that involve decision making.

McGraw-Hill's HOMEWORK MANAGER® Available with McGraw-Hill's Homework Manager

QUICK STUDY

QS 22-1
Cost behavior identification
C1

Determine whether each of the following is best described as a fixed, variable, or mixed cost with respect to product units.

1. Packaging expense.
2. Factory supervisor's salary.
3. Taxes on factory building.
4. Depreciation expense of warehouse.
5. Rubber used to manufacture athletic shoes.
6. Maintenance of factory machinery.
7. Wages of an assembly-line worker paid on the basis of acceptable units produced.

QS 22-2
Cost behavior identification
C1

Listed here are four series of separate costs measured at various volume levels. Examine each series and identify whether it is best described as a fixed, variable, step-wise, or curvilinear cost. (It can help to graph the cost series.)

Volume (Units)	Series 1	Series 2	Series 3	Series 4
0	$450	$ 0	$ 800	$100
100	450	800	800	105
200	450	1,600	800	120
300	450	2,400	1,600	145
400	450	3,200	1,600	190
500	450	4,000	2,400	250
600	450	4,800	2,400	320

QS 22-3
Cost behavior estimation
C1 P1

This scatter diagram reflects past maintenance hours and their corresponding maintenance costs.

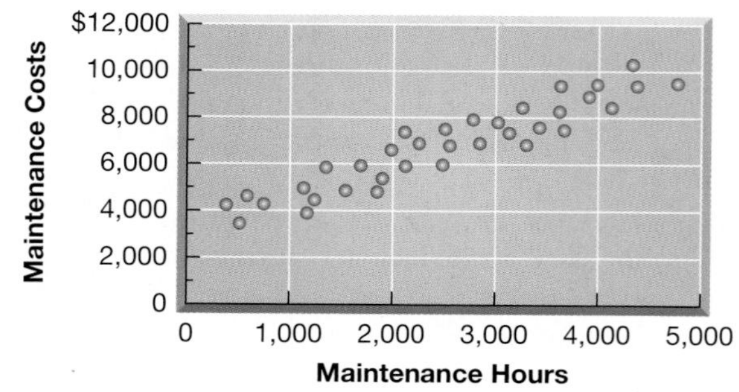

1. Draw an estimated line of cost behavior.
2. Estimate the fixed and variable components of maintenance costs.

QS 22-4
Cost behavior estimation—high-low method
C1 P1

The following information is available for a company's maintenance cost over the last seven months. Using the high-low method, estimate both the fixed and variable components of its maintenance cost.

Month	Maintenance Hours	Maintenance Cost
June	18	$5,450
July	36	6,900
August	24	5,100
September	30	6,000
October	42	6,900
November	48	8,100
December	12	3,600

QS 22-5
Contribution margin ratio
A2 

Compute and interpret the contribution margin ratio using the following data: sales, $100,000; total variable cost, $60,000.

QS 22-6
Contribution margin per unit and break-even units
A2 P2

BSD Phone Company sells its cordless phone for $150 per unit. Fixed costs total $270,000, and variable costs are $60 per unit. Determine the (1) contribution margin per unit and (2) break-even point in units.

QS 22-7
Assumptions in CVP analysis
C2

Refer to the information from QS 22-6. How will the break-even point in units change in response to each of the following independent changes in selling price per unit, variable cost per unit, or total fixed costs? Use I for increase and D for decrease. (It is not necessary to compute new break-even points.)

Change	Breakeven in Units Will
1. Variable cost to $50 per unit	________
2. Total fixed cost to $272,000	________
3. Selling price per unit to $145	________
4. Total fixed cost to $260,000	________
5. Variable cost to $67 per unit	________
6. Selling price per unit to $160	________

QS 22-8
Contribution margin ratio and break-even dollars
P2

Refer to QS 22-6. Determine the (1) contribution margin ratio and (2) break-even point in dollars.

QS 22-9
CVP analysis and target income
C3 P2

Refer to QS 22-6. Assume that BSD Phone Co. is subject to a 30% income tax rate. Compute the units of product that must be sold to earn after-tax income of $252,000.

QS 22-10
CVP assumptions
C2

Which one of the following is an assumption that underlies cost-volume-profit analysis?

1. For costs classified as variable, the costs per unit of output must change constantly.
2. For costs classified as fixed, the costs per unit of output must remain constant.
3. All costs have approximately the same relevant range.
4. The selling price per unit must change in proportion to the number of units sold.

QS 22-11
Operating leverage analysis
A3

A high proportion of Company A's total costs are variable with respect to units sold; a high proportion of Company B's total costs are fixed with respect to units sold. Which company is likely to have a higher degree of operating leverage (DOL)? Explain.

QS 22-12
Multiproduct break-even
P4

Call Me Company manufactures and sells two products, green beepers and gold beepers, in the ratio of 5:3. Fixed costs are $66,500, and the contribution margin per composite unit is $95. What number of both green and gold beepers is sold at the break-even point?

McGraw-Hill's HOMEWORK MANAGER® Available with McGraw-Hill's Homework Manager

EXERCISES

Exercise 22-1
Measurement of cost behavior using a scatter diagram
P1

A company reports the following information about its sales and its cost of sales. Each unit of its product sells for $1,000. Use these data to prepare a scatter diagram. Draw an estimated line of cost behavior and determine whether the cost appears to be variable, fixed, or mixed.

Period	Sales	Cost of Sales
1	$45,000	$30,300
2	34,500	22,500
3	31,500	21,000
4	22,500	16,500
5	27,000	18,000
6	37,500	28,500

Exercise 22-2
Cost behavior in graphs
C1

Following are five graphs representing various cost behaviors. (1) Identify whether the cost behavior in each graph is mixed, step-wise, fixed, variable, or curvilinear. (2) Identify the graph (by number) that best illustrates each cost behavior: (a) Factory policy requires one supervisor for every 30 factory workers; (b) real estate taxes on factory; (c) electricity charge that includes the standard monthly charge plus a charge for each kilowatt hour; (d) commissions to salespersons; and (e) costs of hourly paid workers

[continued on next page]

that provide substantial gains in efficiency when a few workers are added but gradually smaller gains in efficiency when more workers are added.

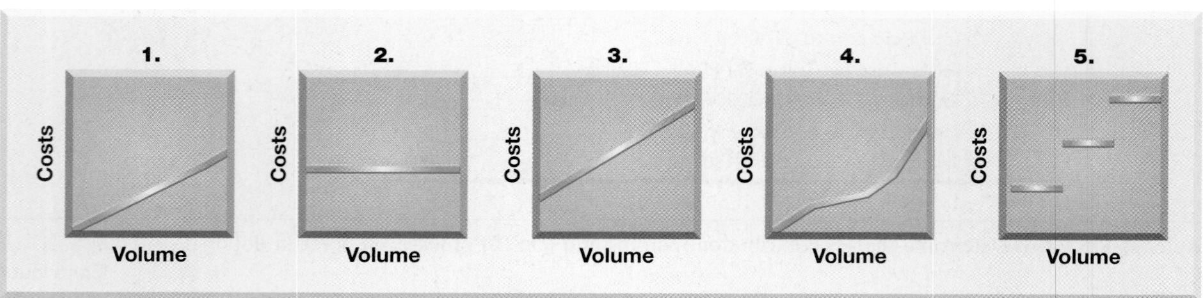

Exercise 22-3
Cost behavior defined
C1

The left column lists several cost classifications. The right column presents short definitions of those costs. In the blank space beside each of the numbers in the right column, write the letter of the cost best described by the definition.

A. Total cost
B. Variable cost
C. Fixed cost
D. Mixed cost
E. Curvilinear cost
F. Step-wise cost

_____ **1.** This cost is the combined amount of all the other costs.

_____ **2.** This cost remains constant over a limited range of volume; when it reaches the end of its limited range, it changes by a lump sum and remains at that level until it exceeds another limited range.

_____ **3.** This cost has a component that remains the same over all volume levels and another component that increases in direct proportion to increases in volume.

_____ **4.** This cost increases when volume increases, but the increase is not constant for each unit produced.

_____ **5.** This cost remains constant over all volume levels within the productive capacity for the planning period.

_____ **6.** This cost increases in direct proportion to increases in volume; its amount is constant for each unit produced.

Exercise 22-4
Cost behavior identification
C1

Following are five series of costs *A* through *E* measured at various volume levels. Examine each series and identify which is fixed, variable, mixed, step-wise, or curvilinear.

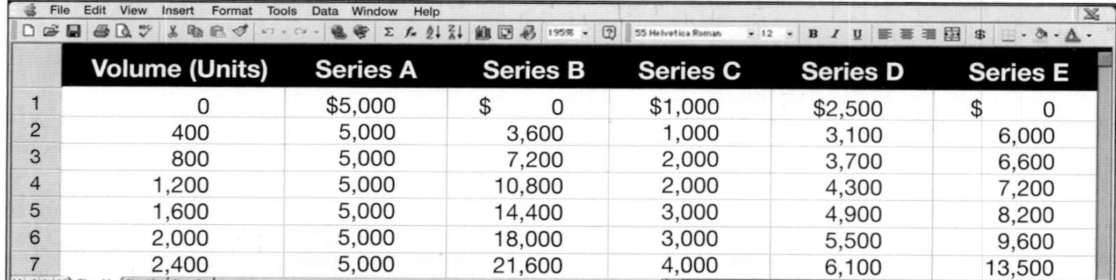

	Volume (Units)	Series A	Series B	Series C	Series D	Series E
1	0	$5,000	$ 0	$1,000	$2,500	$ 0
2	400	5,000	3,600	1,000	3,100	6,000
3	800	5,000	7,200	2,000	3,700	6,600
4	1,200	5,000	10,800	2,000	4,300	7,200
5	1,600	5,000	14,400	3,000	4,900	8,200
6	2,000	5,000	18,000	3,000	5,500	9,600
7	2,400	5,000	21,600	4,000	6,100	13,500

Exercise 22-5
Predicting sales and variable costs using contribution margin
C3

Stewart Company management predicts that it will incur fixed costs of $230,000 and earn pretax income of $350,000 in the next period. Its expected contribution margin ratio is 25%. Use this information to compute the amounts of (1) total dollar sales and (2) total variable costs.

Exercise 22-6
Scatter diagram and measurement of cost behavior
P1

Use the following information about sales and costs to prepare a scatter diagram. Draw a cost line that reflects the behavior displayed by this cost. Determine whether the cost is variable, step-wise, fixed, mixed, or curvilinear.

Period	Sales	Costs	Period	Sales	Costs
1	$1,520	$1,180	9	$1,160	$ 780
2	1,600	1,120	10	640	480
3	400	460	11	480	460
4	800	800	12	1,440	1,100
5	960	780	13	560	520
6	1,240	1,100	14	880	820
7	1,360	1,180	15	760	520
8	1,080	860			

Exercise 22-7
Cost behavior estimation—scatter diagram and high-low
P1

A company reports the following information about its sales and cost of sales. Draw an estimated line of cost behavior using a scatter diagram, and compute fixed costs and variable costs per unit sold. Then use the high-low method to estimate the fixed and variable components of the cost of sales.

Period	Units Sold	Cost of Sales	Period	Units Sold	Cost of Sales
1	0	$2,500	6	2,000	5,500
2	400	3,100	7	2,400	6,100
3	800	3,700	8	2,800	6,700
4	1,200	4,300	9	3,200	7,300
5	1,600	4,900	10	3,600	7,900

Exercise 22-8[A]
Measurement of cost behavior using regression
P1

Refer to the information from Exercise 22-7. Use spreadsheet software to use ordinary least-squares regression to estimate the cost equation, including fixed and variable cost amounts.

Exercise 22-9
Contribution margin, break-even, and CVP chart
P2 P3 A2

Seton Company manufactures a single product that sells for $360 per unit and whose total variable costs are $270 per unit. The company's annual fixed costs are $1,125,000. (1) Use this information to compute the company's (a) contribution margin, (b) contribution margin ratio, (c) break-even point in units, and (d) break-even point in dollars of sales. (2) Draw a CVP chart for the company.

Exercise 22-10
Income reporting and break-even analysis
C3

Refer to Exercise 22-9. (1) Prepare a contribution margin income statement for Seton Company showing sales, variable costs, and fixed costs at the break-even point. (2) If the company's fixed costs increase by $270,000, what amount of sales (in dollars) is needed to break even? Explain.

Exercise 22-11
Computing sales to achieve target income
C3

Seton Company management (in Exercise 22-9) targets an annual after-tax income of $1,620,000. The company is subject to a 20% income tax rate. Assume that fixed costs remain at $1,125,000. Compute the (1) unit sales to earn the target after-tax net income and (2) dollar sales to earn the target after-tax net income.

Exercise 22-12
Forecasted income statement
C3

Seton Company sales manager (in Exercise 22-9) predicts that annual sales of the company's product will soon reach 80,000 units and its price will increase to $400 per unit. According to the production manager, the variable costs are expected to increase to $280 per unit but fixed costs will remain at $1,125,000. The income tax rate is 20%. What amounts of pretax and after-tax income can the company expect to earn from these predicted changes? (*Hint:* Prepare a forecasted contribution margin income statement as in Exhibit 22.20.)

Check Forecasted income, $6,780,000

Exercise 22-13
Predicting unit and dollar sales
C3

Maya Company management predicts $600,000 of variable costs, $700,000 of fixed costs, and a pretax income of $110,000 in the next period. Management also predicts that the contribution margin per unit will be $9. Use this information to compute the (1) total expected dollar sales for next period and (2) number of units expected to be sold next period.

Exercise 22-14
Computation of variable and fixed costs; CVP chart
P3

Corveau Company expects to sell 400,000 units of its product next year, which would generate total sales of $34 million. Management predicts that pretax net income for next year will be $2,500,000 and that the contribution margin per unit will be $50. (1) Use this information to compute next year's total expected (a) variable costs and (b) fixed costs. (2) Prepare a CVP chart from this information.

Exercise 22-15
CVP analysis using composite units P4

Check (3) 1,500 units

Modern Home sells windows and doors in the ratio of 9:1 (windows:doors). The selling price of each window is $90 and of each door is $250. The variable cost of a window is $60 and of a door is $220. Fixed costs are $450,000. Use this information to determine the (1) selling price per composite unit, (2) variable costs per composite unit, (3) break-even point in composite units, and (4) number of units of each product that will be sold at the break-even point.

Exercise 22-16
CVP analysis using weighted-average contribution margin
P4

Refer to the information from Exercise 22-15. Use the information to determine the (1) weighted-average contribution margin, (2) break-even point in units, and (3) number of units of each product that will be sold at the break-even point.

Exercise 22-17
CVP analysis using composite units
P4

Precision Tax Service offers tax and consulting services to individuals and small businesses. Data for fees and costs of three types of tax returns follow. Precision provides services in the ratio of 5:3:2 (easy, moderate, business). Fixed costs total $18,000 for the tax season. Use this information to determine the (1) selling price per composite unit, (2) variable costs per composite unit, (3) break-even point in composite units, and (4) number of units of each product that will be sold at the break-even point.

Type of Return	Fee Charged	Variable Cost per Return
Easy (form 1040EZ)	$ 50	$ 30
Moderate (form 1040)	125	75
Business	275	100

Exercise 22-18
CVP analysis using weighted-average contribution margin
P4

Refer to the information from Exercise 22-17. Use the information to determine the (1) weighted-average contribution margin, (2) break-even point in units, and (3) number of units of each product that will be sold at the break-even point.

Exercise 22-19
Operating leverage computed and applied
A3 

Company A is a manufacturer with current sales of $1,500,000 and a 60% contribution margin. Its fixed costs equal $650,000. Company B is a consulting firm with current service revenues of $1,500,000 and a 25% contribution margin. Its fixed costs equal $125,000. Compute the degree of operating leverage (DOL) for each company. Identify which company benefits more from a 20% increase in sales and explain why.

McGraw-Hill's HOMEWORK MANAGER® Available with McGraw-Hill's Homework Manager

PROBLEM SET A

Problem 22-1A
Contribution margin income statement and contribution margin ratio
A2

The following costs result from the production and sale of 2,000 drum sets manufactured by Harris Drum Company for the year ended December 31, 2009. The drum sets sell for $500 each. The company has a 25% income tax rate.

Variable production costs	
Plastic for casing	$ 34,000
Wages of assembly workers	164,000
Drum stands	52,000
Variable selling costs	
Sales commissions	30,000
Fixed manufacturing costs	
Taxes on factory	10,000
Factory maintenance	20,000
Factory machinery depreciation	80,000
Fixed selling and administrative costs	
Lease of equipment for sales staff	20,000
Accounting staff salaries	70,000
Administrative management salaries	250,000

Check (1) Net income, $202,500

Required

1. Prepare a contribution margin income statement for the company.

2. Compute its contribution margin per unit and its contribution margin ratio.

Analysis Component

3. Interpret the contribution margin and contribution margin ratio from part 2.

Problem 22-2A
CVP analysis and charting
P2 P3

mhhe.com/wildFAP19e

Extreme Equipment Co. manufactures and markets a number of rope products. Management is considering the future of Product HG, a special rope for hang gliding, that has not been as profitable as planned. Since Product HG is manufactured and marketed independently of the other products, its total costs can be precisely measured. Next year's plans call for a $200 selling price per 100 yards of HG rope. Its fixed costs for the year are expected to be $330,000, up to a maximum capacity of 20,000,000 yards of rope. Forecasted variable costs are $170 per 100 yards of HG rope.

Required

1. Estimate Product HG's break-even point in terms of (a) sales units and (b) sales dollars.

Check (1) Break-even sales, 11,000 units or $2,200,000

2. Prepare a CVP chart for Product HG like that in Exhibit 22.14. Use 20,000,000 yards as the maximum number of sales units on the horizontal axis of the graph, and $4,000,000 as the maximum dollar amount on the vertical axis.

3. Prepare a contribution margin income statement showing sales, variable costs, and fixed costs for Product HG at the break-even point.

Problem 22-3A
Scatter diagram and cost behavior estimation
P1

Alden Co.'s monthly sales and cost data for its operating activities of the past year follow. Management wants to use these data to predict future fixed and variable costs.

Period	Sales	Total Cost	Period	Sales	Total Cost
1	$325,000	$162,500	7	$355,000	$242,000
2	170,000	106,250	8	275,000	156,750
3	270,000	210,600	9	75,000	60,000
4	210,000	105,000	10	155,000	135,625
5	295,000	206,500	11	99,000	99,000
6	195,000	117,000	12	105,000	76,650

Required

1. Prepare a scatter diagram for these data with sales volume (in $) plotted on the horizontal axis and total cost plotted on the vertical axis.

Check (2) Variable costs, $0.65 per sales dollar; fixed costs, $11,250

2. Estimate both the variable costs per sales dollar and the total monthly fixed costs using the high-low method. Draw the total costs line on the scatter diagram in part 1.

3. Use the estimated line of cost behavior and results from part 2 to predict future total costs when sales volume is (a) $210,000 and (b) $300,000.

Problem 22-4A
Break-even analysis; income targeting and forecasting
C3 P2

Teller Co. sold 20,000 units of its only product and incurred a $70,000 loss (ignoring taxes) for the current year as shown here. During a planning session for year 2010's activities, the production manager notes that variable costs can be reduced 50% by installing a machine that automates several operations. To obtain these savings, the company must increase its annual fixed costs by $210,000. The maximum output capacity of the company is 40,000 units per year.

TELLER COMPANY Contribution Margin Income Statement For Year Ended December 31, 2009	
Sales	$1,000,000
Variable costs	800,000
Contribution margin	200,000
Fixed costs	270,000
Net loss	$ (70,000)

Required

1. Compute the break-even point in dollar sales for year 2009.

2. Compute the predicted break-even point in dollar sales for year 2010 assuming the machine is installed and there is no change in the unit sales price.

Check (3) Net income, $120,000

3. Prepare a forecasted contribution margin income statement for 2010 that shows the expected results with the machine installed. Assume that the unit sales price and the number of units sold will not change, and no income taxes will be due.

(4) Required sales, $1,300,000 or 26,000 units

4. Compute the sales level required in both dollars and units to earn $210,000 of after-tax income in 2010 with the machine installed and no change in the unit sales price. Assume that the income tax rate is 30%. (*Hint:* Use the procedures in Exhibits 22.21 and 22.23.)

(5) Net income, $210,000

5. Prepare a forecasted contribution margin income statement that shows the results at the sales level computed in part 4. Assume an income tax rate of 30%.

Problem 22-5A
Break-even analysis, different cost structures, and income calculations
C3

Shol Co. produces and sells two products, T and O. It manufactures these products in separate factories and markets them through different channels. They have no shared costs. This year, the company sold 51,000 units of each product. Sales and costs for each product follow.

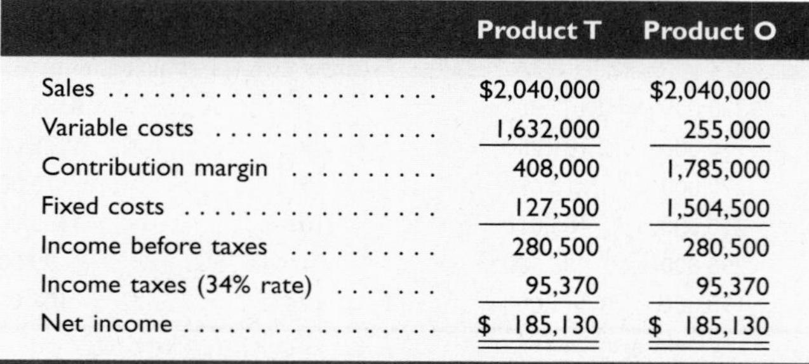

	Product T	Product O
Sales	$2,040,000	$2,040,000
Variable costs	1,632,000	255,000
Contribution margin	408,000	1,785,000
Fixed costs	127,500	1,504,500
Income before taxes	280,500	280,500
Income taxes (34% rate)	95,370	95,370
Net income	$ 185,130	$ 185,130

Required

1. Compute the break-even point in dollar sales for each product.
2. Assume that the company expects sales of each product to decline to 40,000 units next year with no change in unit sales price. Prepare forecasted financial results for next year following the format of the contribution margin income statement as just shown with columns for each of the two products (assume a 34% tax rate). Also, assume that any loss before taxes yields a 34% tax savings.
3. Assume that the company expects sales of each product to increase to 65,000 units next year with no change in unit sales price. Prepare forecasted financial results for next year following the format of the contribution margin income statement shown with columns for each of the two products (assume a 34% tax rate).

Check (2) After-tax income: T, $127,050; O, $(68,970)

(3) After-tax income: T, $259,050; O, $508,530

Analysis Component

4. If sales greatly decrease, which product would experience a greater loss? Explain.
5. Describe some factors that might have created the different cost structures for these two products.

Problem 22-6A
Analysis of price, cost, and volume changes for contribution margin and net income
C3 P2

mhhe.com/wildFAP19e

This year Calypso Company sold 60,000 units of its only product for $20 per unit. Manufacturing and selling the product required $97,500 of fixed manufacturing costs and $157,500 of fixed selling and administrative costs. Its per unit variable costs follow.

Material	$8.00
Direct labor (paid on the basis of completed units)	5.00
Variable overhead costs	1.60
Variable selling and administrative costs	0.40

Next year the company will use new material, which will reduce material costs by 50% and direct labor costs by 60% and will not affect product quality or marketability. Management is considering an increase in the unit sales price to reduce the number of units sold because the factory's output is nearing its annual output capacity of 65,000 units. Two plans are being considered. Under plan 1, the company will keep the price at the current level and sell the same volume as last year. This plan will increase income because of the reduced costs from using the new material. Under plan 2, the company will increase price by 25%. This plan will decrease unit sales volume by 15%. Under both plans 1 and 2, the total fixed costs and the variable costs per unit for overhead and for selling and administrative costs will remain the same.

Required

1. Compute the break-even point in dollar sales for both (a) plan 1 and (b) plan 2.
2. Prepare a forecasted contribution margin income statement with two columns showing the expected results of plan 1 and plan 2. The statements should report sales, total variable costs, contribution margin, total fixed costs, income before taxes, income taxes (30% rate), and net income.

Check (1) Breakeven: Plan 1, $425,000; Plan 2, $375,000

(2) Net income: Plan 1, $325,500; Plan 2, $428,400

Problem 22-7A
Break-even analysis with composite units
P4 C3 

Patriot Co. manufactures and sells three products: red, white, and blue. Their unit sales prices are red, $74; white, $108; and blue, $99. The per unit variable costs to manufacture and sell these products are red, $48; white, $75; and blue, $90. Their sales mix is reflected in a ratio of 5:4:2 (red:white:blue). Annual fixed costs shared by all three products are $179,200. One type of raw material has been used to manufacture all three products. The company has developed a new material of equal quality for less cost. The new material would reduce variable costs per unit as follows: red, by $10; white, by $16; and blue, by $13. However, the new material requires new equipment, which will increase annual fixed costs by $22,400. (Round answers to whole composite units.)

Required

1. If the company continues to use the old material, determine its break-even point in both sales units and sales dollars of each individual product.
2. If the company uses the new material, determine its new break-even point in both sales units and sales dollars of each individual product.

Check (1) Old plan breakeven, 640 composite units

(2) New plan breakeven, 480 composite units

Analysis Component

3. What insight does this analysis offer management for long-term planning?

PROBLEM SET B

Problem 22-1B
Contribution margin income statement and contribution margin ratio

A2

The following costs result from the production and sale of 240,000 CD sets manufactured by Jawan Company for the year ended December 31, 2009. The CD sets sell for $9 each. The company has a 25% income tax rate.

Variable manufacturing costs	
Plastic for CD sets	$ 21,600
Wages of assembly workers	300,000
Labeling	43,200
Variable selling costs	
Sales commissions	24,000
Fixed manufacturing costs	
Rent on factory	100,000
Factory cleaning service	75,000
Factory machinery depreciation	125,000
Fixed selling and administrative costs	
Lease of office equipment	120,000
Systems staff salaries	600,000
Administrative management salaries	300,000

Required

Check (1) Net income, $338,400

1. Prepare a contribution margin income statement for the company.

2. Compute its contribution margin per unit and its contribution margin ratio.

Analysis Component

3. Interpret the contribution margin and contribution margin ratio from part 2.

Problem 22-2B
CVP analysis and charting

P2 P3

Tip-Top Co. manufactures and markets several products. Management is considering the future of one product, electronic keyboards, that has not been as profitable as planned. Since this product is manufactured and marketed independently of the other products, its total costs can be precisely measured. Next year's plans call for a $175 selling price per unit. The fixed costs for the year are expected to be $420,000, up to a maximum capacity of 10,000 units. Forecasted variable costs are $105 per unit.

Required

Check (1) Break-even sales, 6,000 units or $1,050,000

1. Estimate the keyboards' break-even point in terms of (a) sales units and (b) sales dollars.

2. Prepare a CVP chart for keyboards like that in Exhibit 22.14. Use 10,000 keyboards as the maximum number of sales units on the horizontal axis of the graph, and $1,600,000 as the maximum dollar amount on the vertical axis.

3. Prepare a contribution margin income statement showing sales, variable costs, and fixed costs for keyboards at the break-even point.

Problem 22-3B
Scatter diagram and cost behavior estimation

P1

Merdam Co.'s monthly sales and costs data for its operating activities of the past year follow. Management wants to use these data to predict future fixed and variable costs.

Period	Sales	Total Cost	Period	Sales	Total Cost
1	$390	$194	7	$290	$186
2	250	174	8	370	210
3	210	146	9	270	170
4	310	178	10	170	116
5	190	162	11	350	190
6	430	220	12	230	158

Required

1. Prepare a scatter diagram for these data with sales volume (in $) plotted on the horizontal axis and total costs plotted on the vertical axis.

2. Estimate both the variable costs per sales dollar and the total monthly fixed costs using the high-low method. Draw the total costs line on the scatter diagram in part 1.

3. Use the estimated line of cost behavior and results from part 2 to predict future total costs when sales volume is (a) $200 and (b) $340.

Check (2) Variable costs, $0.40 per sales dollar; fixed costs, $48

Problem 22-4B
Break-even analysis; income targeting and forecasting
C3 P2

Noru Co. sold 30,000 units of its only product and incurred a $75,000 loss (ignoring taxes) for the current year as shown here. During a planning session for year 2010's activities, the production manager notes that variable costs can be reduced 40% by installing a machine that automates several operations. To obtain these savings, the company must increase its annual fixed costs by $220,000. The maximum output capacity of the company is 50,000 units per year.

NORU COMPANY Contribution Margin Income Statement For Year Ended December 31, 2009	
Sales	$1,125,000
Variable costs	900,000
Contribution margin	225,000
Fixed costs	300,000
Net loss	$ (75,000)

Required

1. Compute the break-even point in dollar sales for year 2009.
2. Compute the predicted break-even point in dollar sales for year 2010 assuming the machine is installed and no change occurs in the unit sales price. (Round the change in variable costs to a whole number.)
3. Prepare a forecasted contribution margin income statement for 2010 that shows the expected results with the machine installed. Assume that the unit sales price and the number of units sold will not change, and no income taxes will be due.
4. Compute the sales level required in both dollars and units to earn $104,000 of after-tax income in 2010 with the machine installed and no change in the unit sales price. Assume that the income tax rate is 20%. (*Hint:* Use the procedures in Exhibits 22.21 and 22.23.)
5. Prepare a forecasted contribution margin income statement that shows the results at the sales level computed in part 4. Assume an income tax rate of 20%.

Check (3) Net income, $65,000

(4) Required sales, $1,250,000 or 33,334 units

(5) Net income, $104,000 (rounded)

Problem 22-5B
Break-even analysis, different cost structures, and income calculations
C3 

Best Co. produces and sells two products, BB and TT. It manufactures these products in separate factories and markets them through different channels. They have no shared costs. This year, the company sold 100,000 units of each product. Sales and costs for each product follow.

	Product BB	Product TT
Sales	$1,600,000	$1,600,000
Variable costs	1,120,000	200,000
Contribution margin	480,000	1,400,000
Fixed costs	200,000	1,120,000
Income before taxes	280,000	280,000
Income taxes (32% rate)	89,600	89,600
Net income	$ 190,400	$ 190,400

Required

1. Compute the break-even point in dollar sales for each product.
2. Assume that the company expects sales of each product to decline to 67,000 units next year with no change in the unit sales price. Prepare forecasted financial results for next year following the format of the contribution margin income statement as shown here with columns for each of the two products (assume a 32% tax rate, and that any loss before taxes yields a 32% tax savings).
3. Assume that the company expects sales of each product to increase to 125,000 units next year with no change in the unit sales prices. Prepare forecasted financial results for next year following the format of the contribution margin income statement as shown here with columns for each of the two products (assume a 32% tax rate).

Check (2) After-tax income: BB, $82,688; TT, $(123,760)

(3) After-tax income: BB, $272,000; TT, $428,400

Analysis Component

4. If sales greatly increase, which product would experience a greater increase in profit? Explain.

5. Describe some factors that might have created the different cost structures for these two products.

Problem 22-6B
Analysis of price, cost, and volume changes for contribution margin and net income
C3 P2

This year Blanko Company earned a disappointing 3.85% after-tax return on sales (Net income/Sales) from marketing 50,000 units of its only product. The company buys its product in bulk and repackages it for resale at the price of $20 per unit. Blanko incurred the following costs this year.

Total variable unit costs	$400,000
Total variable packaging costs	50,000
Fixed costs	$495,000
Income tax rate	30%

The marketing manager claims that next year's results will be the same as this year's unless some changes are made. The manager predicts the company can increase the number of units sold by 60% if it reduces the selling price by 20% and upgrades the packaging. This change would increase variable packaging costs by 20%. Increased sales would allow the company to take advantage of a 25% quantity purchase discount on the cost of the bulk product. Neither the packaging change nor the volume discount would affect fixed costs, which provide an annual output capacity of 100,000 units.

Required

Check (1) Breakeven for new strategy, $900,000

(2) Net income: Existing strategy, $38,500; new strategy, $146,300

1. Compute the break-even point in dollar sales under the (a) existing business strategy and (b) new strategy that alters both unit sales price and variable costs.

2. Prepare a forecasted contribution margin income statement with two columns showing the expected results of (a) the existing strategy and (b) changing to the new strategy. The statements should report sales, total variable costs (unit and packaging), contribution margin, fixed costs, income before taxes, income taxes, and net income. Also determine the after-tax return on sales for these two strategies.

Problem 22-7B
Break-even analysis with composite units
P4 C3 

Milagro Co. manufactures and sells three products: product 1, product 2, and product 3. Their unit sales prices are product 1, $200; product 2, $150; and product 3, $100. The per unit variable costs to manufacture and sell these products are product 1, $150; product 2, $75; and product 3, $40. Their sales mix is reflected in a ratio of 6:4:2. Annual fixed costs shared by all three products are $5,400,000. One type of raw material has been used to manufacture products 1 and 2. The company has developed a new material of equal quality for less cost. The new material would reduce variable costs per unit as follows: product 1 by $50, and product 2, by $25. However, the new material requires new equipment, which will increase annual fixed costs by $200,000.

Required

Check (1) Old plan breakeven, 7,500 composite units

(2) New plan breakeven, 5,000 composite units

1. If the company continues to use the old material, determine its break-even point in both sales units and sales dollars of each individual product.

2. If the company uses the new material, determine its new break-even point in both sales units and sales dollars of each individual product.

Analysis Component

3. What insight does this analysis offer management for long-term planning?

SERIAL PROBLEM

Success Systems

(This serial problem began in Chapter 1 and continues through most of the book. If previous chapter segments were not completed, the serial problem can begin at this point. It is helpful, but not necessary, to use the working papers that accompany the book.)

SP 22 Success Systems sells upscale modular desk units and office chairs in the ratio of 3:2 (desk unit:chair). The selling prices are $1,250 per desk unit and $500 per chair. The variable costs are $750 per desk unit and $250 per chair. Fixed costs are $120,000.

Required

1. Compute the selling price per composite unit.
2. Compute the variable costs per composite unit.
3. Compute the break-even point in composite units.
4. Compute the number of units of each product that would be sold at the break-even point.

Check (3) 60 composite units

BEYOND THE NUMBERS

REPORTING IN ACTION

C1 

BTN 22-1 Best Buy offers services to customers that help them use products they purchase from Best Buy. One of these services is its Geek Squad, which is Best Buy's 24-hour computer support task force. As you complete the following requirements, assume that the Geek Squad uses many of Best Buy's existing resources such as its purchasing department and its buildings and equipment.

Required

1. Identify several of the variable, mixed, and fixed costs that the Geek Squad is likely to incur in carrying out its services.
2. Assume that Geek Squad revenues are expected to grow by 25% in the next year. How do you expect the costs identified in part 1 to change, if at all?
3. How is your answer to part 2 different from many of the examples discussed in the chapter? (*Hint:* Consider how the contribution margin ratio changes as volume—sales or customers served—increases.)

COMPARATIVE ANALYSIS

P2 C3 A2 

BTN 22-2 Both Best Buy and Circuit City sell numerous consumer products, and each of these companies has a different product mix.

Required

1. Assume the following data are available for both companies. Compute each company's break-even point in unit sales. (Each company sells many products at many different selling prices, and each has its own variable costs. This assignment assumes an *average* selling price per unit and an *average* cost per item.)

	Best Buy	Circuit City
Average selling price per item sold	$90	$40
Average variable cost per item sold	$64	$30
Total fixed costs	$5,980 million	$2,570 million

2. If unit sales were to decline, which company would experience the larger decline in operating profit? Explain.

ETHICS CHALLENGE

C1

BTN 22-3 Labor costs of an auto repair mechanic are seldom based on actual hours worked. Instead, the amount paid a mechanic is based on an industry average of time estimated to complete a repair job. The repair shop bills the customer for the industry average amount of time at the repair center's billable cost per hour. This means a customer can pay, for example, $120 for two hours of work on a car when the actual time worked was only one hour. Many experienced mechanics can complete repair jobs faster than the industry average. The average data are compiled by engineering studies and surveys conducted in the auto repair business. Assume that you are asked to complete such a survey for a repair center. The survey calls for objective input, and many questions require detailed cost data and analysis. The mechanics and owners know you have the survey and encourage you to complete it in a way that increases the average billable hours for repair work.

Required

Write a one-page memorandum to the mechanics and owners that describes the direct labor analysis you will undertake in completing this survey.

COMMUNICATING IN PRACTICE

C2

BTN 22-4 Several important assumptions underlie CVP analysis. Assumptions often help simplify and focus our analysis of sales and costs. A common application of CVP analysis is as a tool to forecast sales, costs, and income.

Required

Assume that you are actively searching for a job. Prepare a one-half page report identifying (1) three assumptions relating to your expected revenue (salary) and (2) three assumptions relating to your expected costs for the first year of your new job. Be prepared to discuss your assumptions in class.

TAKING IT TO THE NET

C1 C3 

BTN 22-5 Access and review the entrepreneurial information at **Business Owner's Toolkit** [Toolkit.cch.com]. Access and review its *New Business Cash Needs Estimate* under the Business Tools/Business Finance menu bar or similar worksheets related to controls of cash and costs.

Required

Write a one-half page report that describes the information and resources available at the Business Owner's Toolkit to help the owner of a start-up business to control and monitor its costs.

TEAMWORK IN ACTION

C2

BTN 22-6 A local movie theater owner explains to you that ticket sales on weekends and evenings are strong, but attendance during the weekdays, Monday through Thursday, is poor. The owner proposes to offer a contract to the local grade school to show educational materials at the theater for a set charge per student during school hours. The owner asks your help to prepare a CVP analysis listing the cost and sales projections for the proposal. The owner must propose to the school's administration a charge per child. At a minimum, the charge per child needs to be sufficient for the theater to break even.

Required

Your team is to prepare two separate lists of questions that enable you to complete a reliable CVP analysis of this situation. One list is to be answered by the school's administration, the other by the owner of the movie theater.

ENTREPRENEURIAL DECISION

C1 

BTN 22-7 Martin Sprock is a diligent businessman. He continually searches for new menu items to further increase the profitability of **Moe's Southwest Grill**.

Required

1. What information should Sprock search for to help him decide whether to add new menu items or other products to existing Moe's product lines?
2. What managerial tools are available to Sprock to help make the decisions in part 1?

HITTING THE ROAD

P4 

BTN 22-8 Multiproduct break-even analysis is often viewed differently when actually applied in practice. You are to visit a local fast-food restaurant and count the number of items on the menu. To apply multiproduct break-even analysis to the restaurant, similar menu items must often be fit into groups. A reasonable approach is to classify menu items into approximately five groups. We then estimate average selling price and average variable cost to compute average contribution margin. (*Hint:* For fast-food restaurants, the highest contribution margin is with its beverages, at about 90%.)

Required

1. Prepare a one-year multiproduct break-even analysis for the restaurant you visit. Begin by establishing groups. Next, estimate each group's volume and contribution margin. These estimates are necessary to compute each group's contribution margin. Assume that annual fixed costs in total are $500,000 per year. (*Hint:* You must develop your own estimates on volume and contribution margin for each group to obtain the break-even point and sales.)
2. Prepare a one-page report on the results of your analysis. Comment on the volume of sales necessary to break even at a fast-food restaurant.

BTN 22-9 Access and review DSG's Website (www.DSGiplc.com) to answer the following questions.

GLOBAL DECISION

C3 

DSG

1. Do you believe that DSG's managers use single product CVP analysis or multiproduct break-even point analysis? Explain.
2. How does the addition of a new product line affect DSG's CVP analysis?
3. How does the addition of a new store affect DSG's CVP analysis?

ANSWERS TO MULTIPLE CHOICE QUIZ

1. a; $150 − $100 = $50
2. e; ($150 − $100)/$150 = 33⅓%
3. c; $75,000/$50 CM per unit = 1,500 units
4. b; $300,000 − $180,000 = $120,000
5. c; Contribution margin ratio = ($400 − $260)/$400 = 0.35
   Targeted sales = ($840,000 + $70,000)/0.35 = $2,600,000

A Look Back

Chapter 22 looked at cost behavior and its use by managers in performing cost-volume-profit analysis. It also illustrated the application of cost-volume-profit analysis.

A Look at This Chapter

This chapter explains the importance of budgeting and describes the master budget and its preparation. It also discusses the value of the master budget to the planning of future business activities.

A Look Ahead

Chapter 24 focuses on flexible budgets, standard costs, and variance reporting. It explains the usefulness of these procedures and reports for business decisions.

Chapter 23 Master Budgets and Planning

Learning Objectives

CAP

Conceptual

C1 Describe the importance and benefits of budgeting. *(p. 940)*

C2 Explain the process of budget administration. *(p. 942)*

C3 Describe a master budget and the process of preparing it. *(p. 944)*

Analytical

A1 Analyze expense planning using activity-based budgeting. *(p. 953)*

LP23

Procedural

P1 Prepare each component of a master budget and link each to the budgeting process. *(p. 946)*

P2 Link both operating and capital expenditures budgets to budgeted financial statements. *(p. 950)*

P3 *Appendix 23A*—Prepare production and manufacturing budgets. *(p. 959)*

Decision Feature

Lucky Charms

"The Number One thing is you have got to take the chance"—Rich Schmelzer

BOULDER, CO—Each pair of **Crocs (Crocs.com)** shoes includes ventilation holes for breathability and to filter water out. Sheri Schmelzer and her kids thought it more fun to use clay and rhinestones to decorate the holes with fun charms. Sheri's husband Rich, an entrepreneur, immediately saw the profit potential—within 48 hours the Schmelzer's had filed patents for the design of **Jibbitz (Jibbitz.com),** which are small accessories made to fit in the holes of Crocs. Today, Jibbitz accessories come in various shapes and sizes, and include more than 1100 designs such as peace signs, flowers, musical notes, sports gear, and letters to spell out words.

Jibbitz started small, with an assembly line in the family's basement and a Website to process orders. Like many new businesses, Jibbitz began with few formal budgets or plans. "We didn't write a business plan" admits Sheri. Rich explains "We recalibrated our business every week depending on what we sold. We were very nimble." Soon, Jibbitz was processing hundreds of orders per day. "It turned from a very simple business to a very complex business," says Rich.

As business grew, master budgets and the budgeting process became more important. Budgets helped formalize business plans and goals, and helped direct employees—a team of staff designers and warehouse personnel in Boulder, and a manufacturing group in Asia. Realizing that a too-rapid sales growth could strain its capacity to meet customer expectations, Jibbitz avoids advertising and has turned down some large retailers' bids to carry its products. An understanding of sales budgets and their link to expense budgets was vital in making these decisions. Likewise, production and manufacturing budgets helped plan for use of materials, labor, and overhead.

Eventually, Rich and Sheri teamed up with Crocs. Now operating as a division within Crocs, budgeting remains important. If Jibbitz meets certain sales and income targets, Rich and Sheri will receive an additional payment from Crocs. Linking their budgeted data to budgeted income statements, and using that information to control costs, is key to that future payment. Still, both Sheri and Rich stress the importance of having fun and a passion for what they do as keys to success. "I'm having a blast," explains Sheri. "I don't want it to stop."

[Sources: *Jibbitz Website,* January 2009; *Crocs Website,* January 2009; *Crocs 2007 10-K report; Rocky Mountain News,* September 2007; *Ladies Who Launch Magazine,* March 2008; *Business 2.0,* November 2006; *Boulder Daily Camera,* August 2006; *Denverpost.com,* October 2006]

Chapter Preview

Management seeks to turn its strategies into action plans. These action plans include financial details that are compiled in a master budget. The budgeting process serves several purposes, including motivating employees and communicating with them. The budget process also helps coordinate a company's activities toward common goals and is useful in evaluating results and management performance. This chapter explains how to prepare a master budget and use it as a formal plan of a company's future activities. The ability to prepare this type of plan is of enormous help in starting and operating a company. Such planning gives managers a glimpse into the future, and it can help translate ideas into actions.

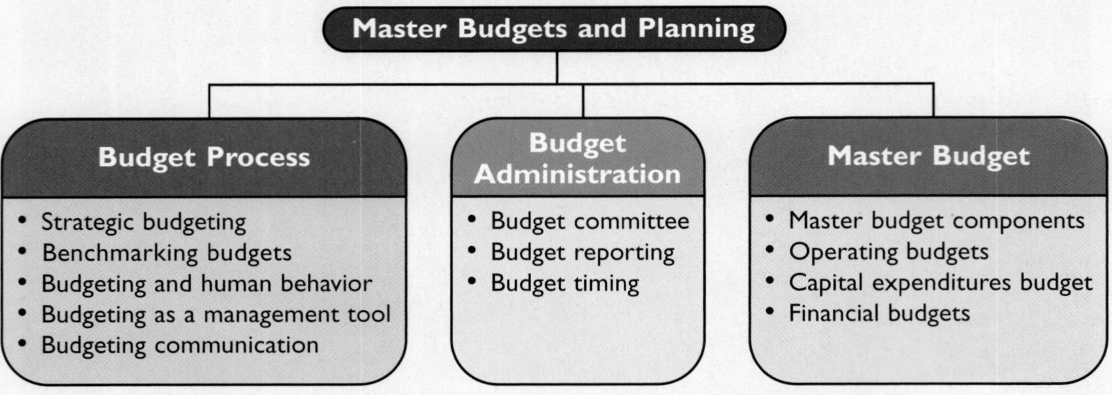

Budget Process

Strategic Budgeting

C1 Describe the importance and benefits of budgeting.

Most companies prepare long-term strategic plans spanning 5 to 10 years. They then fine-tune them in preparing medium-term and short-term plans. Strategic plans usually set a company's long-term direction. They provide a road map for the future about potential opportunities such as new products, markets, and investments. The strategic plan can be inexact, given its long-term focus. Medium- and short-term plans are more operational and translate strategic plans into actions. These action plans are fairly concrete and consist of defined objectives and goals.

Short-term financial plans are called *budgets* and typically cover a one-year period. A **budget** is a formal statement of a company's future plans. It is usually expressed in monetary terms because the economic or financial aspects of the business are the primary factors driving management's decisions. All managers should be involved in **budgeting,** the process of planning future business actions and expressing them as formal plans. Managers who plan carefully and formalize plans in a budgeting process increase the likelihood of both personal and company success. (Although most firms prepare annual budgets, it is not unusual for organizations to prepare three-year and five-year budgets that are revised at least annually.)

Companies Performing Annual Budgeting

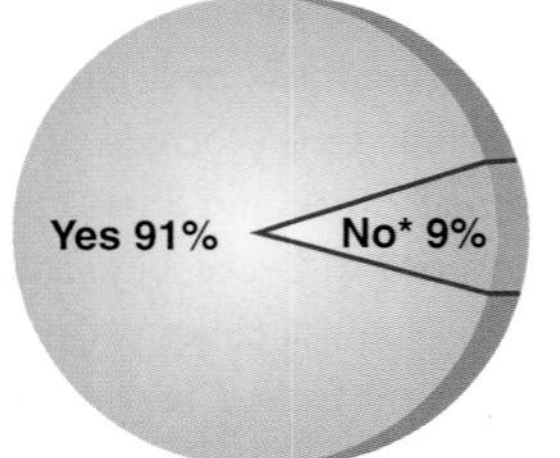

*Most of the 9% have eliminated annual budgeting in favor of rolling or continual budgeting.

The relevant focus of a budgetary analysis is the future. Management must focus on future transactions and events and the opportunities available. A focus on the future is important because the pressures of daily operating problems often divert management's attention and take precedence over planning. A good budgeting system counteracts this tendency by formalizing the planning process and demanding relevant input. Budgeting makes planning an explicit management responsibility.

Benchmarking Budgets

The control function requires management to evaluate (benchmark) business operations against some norm. Evaluation involves comparing actual results against one of two usual alternatives: (1) past performance or (2) expected performance.

An evaluation assists management in identifying problems and taking corrective actions if necessary. Evaluation using expected, or budgeted, performance is potentially superior to using past performance to decide whether actual results trigger a need for corrective actions. This is so because past performance fails to consider several changes that can affect current and future activities. Changes in economic conditions, shifts in competitive advantages within the industry, new product developments, increased or decreased advertising, and other factors reduce the usefulness of comparisons with past results. In hi-tech industries, for instance, increasing competition, technological advances, and other innovations often reduce the usefulness of performance comparisons across years.

Video23.1

Budgeted performance is computed after careful analysis and research that attempts to anticipate and adjust for changes in important company, industry, and economic factors. Therefore, budgets usually provide management an effective control and monitoring system.

Point: Managers can evaluate performance by preparing reports that compare actual results to budgeted plans.

Budgeting and Human Behavior

Budgeting provides standards for evaluating performance and can affect the attitudes of employees evaluated by them. It can be used to create a positive effect on employees' attitudes, but it can also create negative effects if not properly applied. Budgeted levels of performance, for instance, must be realistic to avoid discouraging employees. Personnel who will be evaluated should be consulted and involved in preparing the budget to increase their commitment to meeting it. Performance evaluations must allow the affected employees to explain the reasons for apparent performance deficiencies.

The budgeting process has three important guidelines: (1) Employees affected by a budget should be consulted when it is prepared (*participatory budgeting*), (2) goals reflected in a budget should be attainable, and (3) evaluations should be made carefully with opportunities to explain any failures. Budgeting can be a positive motivating force when these guidelines are followed. Budgeted performance levels can provide goals for employees to attain or even exceed as they carry out their responsibilities. This is especially important in organizations that consider the annual budget a "sacred" document.

Point: The practice of involving employees in the budgeting process is known as *participatory budgeting.*

Decision Insight

Budgets Exposed When companies go public and trade their securities on an organized exchange, management usually develops specific future plans and budgets. For this purpose, companies often develop detailed six- to twelve-month budgets and less-detailed budgets spanning 2 to 5 years.

Budgeting as a Management Tool

An important management objective in large companies is to ensure that activities of all departments contribute to meeting the company's overall goals. This requires coordination. Budgeting provides a way to achieve this coordination.

We describe later in this chapter that a company's budget, or operating plan, is based on its objectives. This operating plan starts with the sales budget, which drives all other budgets including production, materials, labor, and overhead. The budgeting process coordinates the activities of these various departments to meet the company's overall goals.

Budgeting Communication

Managers of small companies can adequately explain business plans directly to employees through conversations and other informal communications. However, conversations can create uncertainty and confusion if not supported by clear documentation of the plans. A written budget is preferred and can inform employees in all types of organizations about management's plans. The budget can also communicate management's specific action plans for the employees in the budget period.

Decision Ethics

Budget Staffer Your company's earnings for the current period will be far below the budgeted amount reported in the press. One of your superiors, who is aware of the upcoming earnings shortfall, has accepted a management position with a competitor. This superior is selling her shares of the company. What are your ethical concerns, if any? [Answer—p. 961]

Budget Administration

Budget Committee

C2 Explain the process of budget administration.

The task of preparing a budget should not be the sole responsibility of any one department. Similarly, the budget should not be simply handed down as top management's final word. Instead, budget figures and budget estimates developed through a *bottom-up* process usually are more useful. This includes, for instance, involving the sales department in preparing sales estimates. Likewise, the production department should have initial responsibility for preparing its own expense budget. Without active employee involvement in preparing budget figures, there is a risk these employees will feel that the numbers fail to reflect their special problems and needs.

Most budgets should be developed by a bottom-up process, but the budgeting system requires central guidance. This guidance is supplied by a budget committee of department heads and other executives responsible for seeing that budgeted amounts are realistic and coordinated. If a department submits initial budget figures not reflecting efficient performance, the budget committee should return them with explanatory comments on how to improve them. Then the originating department must either adjust its proposals or explain why they are acceptable. Communication between the originating department and the budget committee should continue as needed to ensure that both parties accept the budget as reasonable, attainable, and desirable.

Point: In a large company, developing a budget through a bottom-up process can involve hundreds of employees and take several weeks to finalize.

The concept of continuous improvement applies to budgeting as well as production. **BP**, one of the world's largest energy companies, streamlined its monthly budget report from a one-inch-thick stack of monthly control reports to a tidy, two-page flash report on monthly earnings and key production statistics. The key to this efficiency gain was the integration of new budgeting and cost allocation processes with its strategic planning process. BP's controller explained the new role of the finance department with respect to the budgetary control process as follows: "there's less of an attitude that finance's job is to control. People really have come to see that our job is to help attain business objectives."

Budget Reporting

The budget period usually coincides with the accounting period. Most companies prepare at least an annual budget, which reflects the objectives for the next year. To provide specific guidance, the annual budget usually is separated into quarterly or monthly budgets. These short-term budgets allow management to periodically evaluate performance and take needed corrective action.

Managers can compare actual results to budgeted amounts in a report such as that shown in Exhibit 23.1. This report shows actual amounts, budgeted amounts, and their differences. A difference is called a *variance*. Management examines variances to identify areas for improvement and corrective action.

Budget Timing

The time period required for the annual budgeting process can vary considerably. For example, budgeting for 2010 can begin as early as January 2009 or as late as December 2009. Large, complex organizations usually require a longer time to prepare their budgets than do smaller organizations. This is so because considerable effort is required to coordinate the different units (departments) within large organizations.

Companies Using Rolling Budgets

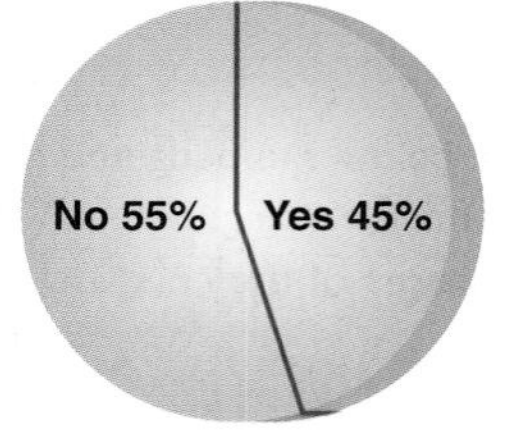

Many companies apply **continuous budgeting** by preparing **rolling budgets.** As each monthly or quarterly budget period goes by, these companies revise their entire set of budgets for the months or quarters remaining and add new monthly or quarterly budgets to replace the ones that have lapsed. At any point in time, monthly or quarterly budgets are available for the next

EXHIBIT 23.1

Comparing Actual Performance with Budgeted Performance

ECCENTRIC MUSIC Income Statement with Variations from Budget For Month Ended April 30, 2009	Actual	Budget	Variance
Net sales	$60,500	$57,150	$+3,350
Cost of goods sold	41,350	39,100	+2,250
Gross profit	19,150	18,050	+1,100
Operating expenses			
Selling expenses			
Sales salaries	6,250	6,000	+250
Advertising	900	800	+100
Store supplies	550	500	+50
Depreciation—Store equipment	1,600	1,600	
Total selling expenses	9,300	8,900	+400
General and administrative expenses			
Office salaries	2,000	2,000	
Office supplies used	165	150	+15
Rent	1,100	1,100	
Insurance	200	200	
Depreciation—Office equipment	100	100	
Total general and administrative expenses	3,565	3,550	+15
Total operating expenses	12,865	12,450	+415
Net income	$ 6,285	$ 5,600	$ +685

Example: Assume that you must explain variances to top management. Which variances in Exhibit 23.1 would you research and why? *Answer:* Sales and cost of goods sold—due to their large variances.

12 months or four quarters. Exhibit 23.2 shows rolling budgets prepared at the end of five consecutive periods. The first set (at top) is prepared in December 2008 and covers the four calendar quarters of 2009. In March 2009, the company prepares another rolling budget for the next four quarters through March 2010. This same process is repeated every three months. As a result, management is continuously planning ahead.

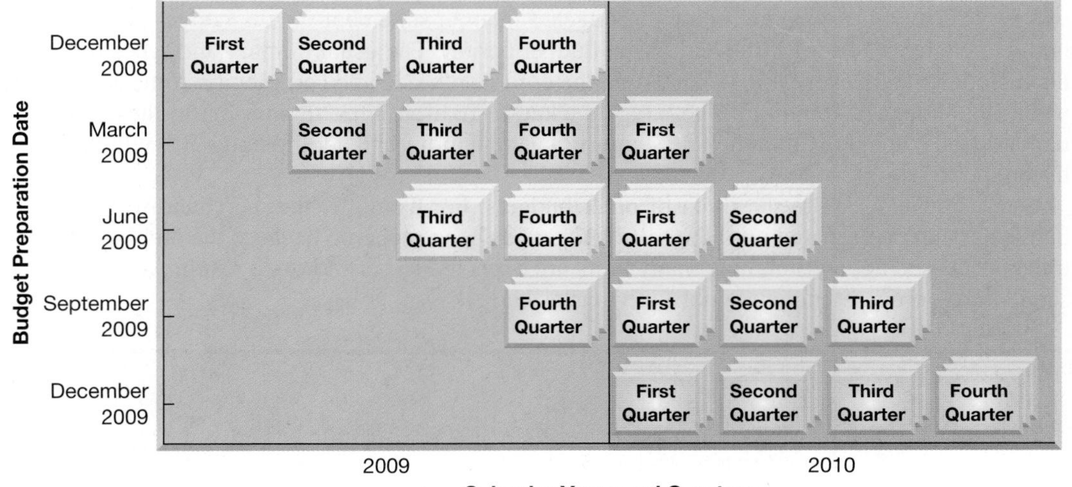

EXHIBIT 23.2

Rolling Budgets

Exhibit 23.2 reflects an annual budget composed of four quarters prepared four times per year using the most recent information available. For example, the budget for the fourth quarter of 2009 is prepared in December 2008 and revised in March, June, and September of 2009. When continuous budgeting is not used, the fourth-quarter budget is nine months old and perhaps out of date when applied.

Decision Insight

Budget Calendar Many companies use long-range operating budgets. For large companies, three groups usually determine or influence the budgets: creditors, directors, and management. All three are interested in the companies' future cash flows and earnings. The annual budget process often begins six months or more before the budget is due to the board of directors. A typical budget calendar, shown here, provides insight into the budget process during a typical calendar year.

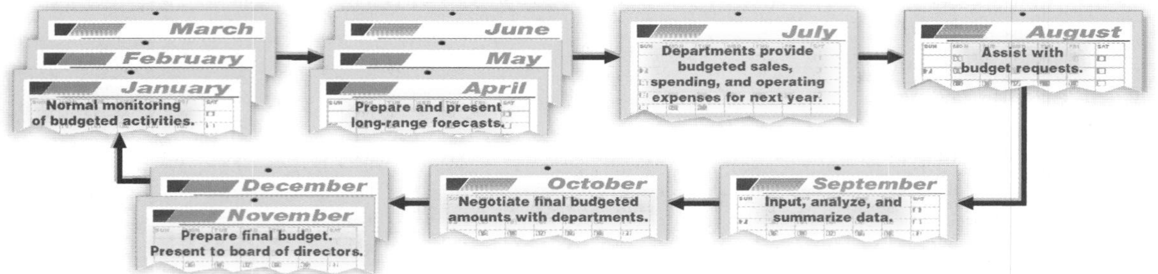

Quick Check

Answers—p. 961

1. What are the major benefits of budgeting?
2. What is the main responsibility of the budget committee?
3. What is the usual time period covered by a budget?
4. What are rolling budgets?

Master Budget

C3 Describe a master budget and the process of preparing it.

A **master budget** is a formal, comprehensive plan for a company's future. It contains several individual budgets that are linked with each other to form a coordinated plan.

Master Budget Components

The master budget typically includes individual budgets for sales, purchases, production, various expenses, capital expenditures, and cash. Managers often express the expected financial results of these planned activities with both a budgeted income statement for the budget period and a budgeted balance sheet for the end of the budget period. The usual number and types of budgets included in a master budget depend on the company's size and complexity. A master budget should include, at a minimum, the budgets listed in Exhibit 23.3. In addition to these individual budgets, managers often include supporting calculations and additional tables with the master budget.

Some budgets require the input of other budgets. For example, the merchandise purchases budget cannot be prepared until the sales budget has been prepared because the number of units to be purchased depends on how many units are expected to be sold. As a result, we often must sequentially prepare budgets within the master budget.

EXHIBIT 23.3

Basic Components of a Master Budget

Operating budgets
- *Sales budget*
- For merchandisers add: *Merchandise purchases budget* (units to be purchased)
- For manufacturers add: *Production budget* (units to be produced)
  Manufacturing budget (manufacturing costs)
- *Selling expense budget*
- *General and administrative expense budget*

Capital expenditures budget (expenditures for plant assets)

Financial budgets
- *Cash budget* (cash receipts and disbursements)
- *Budgeted income statement*
- *Budgeted balance sheet*

Decision Insight

Budgeting Targets Budgeting is a crucial part of any acquisition. Analysis begins by projecting annual sales volume and prices. It then estimates cost of sales, expenses, and income for the next several years. Using the present value of this projected income stream, buyers determine an offer price.

Video23.1

A typical sequence for a quarterly budget consists of the five steps in Exhibit 23.4. Any stage in this budgeting process might reveal undesirable outcomes, so changes often must be made to prior budgets by repeating the previous steps. For instance, an early version of the cash budget could show an insufficient amount of cash unless cash outlays are reduced. This could yield a reduction in planned equipment purchases. A preliminary budgeted balance sheet could also reveal too much debt from an ambitious capital expenditures budget. Findings such as these often result in revised plans and budgets.

EXHIBIT 23.4

Master Budget Sequence

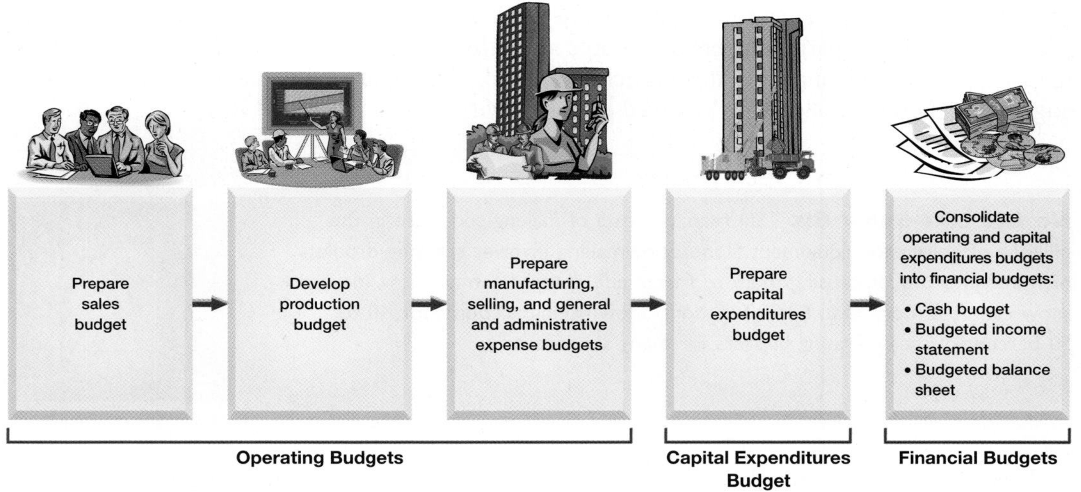

The remainder of this section explains how Hockey Den (HD), a retailer of youth hockey sticks, prepares its master budget. Its master budget includes operating, capital expenditures, and cash budgets for each month in each quarter. It also includes a budgeted income statement for each quarter and a budgeted balance sheet as of the last day of each quarter. We show how HD prepares budgets for October, November, and December 2009. Exhibit 23.5 presents HD's balance sheet at the start of this budgeting period, which we often refer to as we prepare the component budgets.

EXHIBIT 23.5

Balance Sheet Prior to the Budgeting Periods

HOCKEY DEN
Balance Sheet
September 30, 2009

Assets		
Cash		$ 20,000
Accounts receivable		42,000
Inventory (900 units @ $60)		54,000
Equipment*	$200,000	
Less accumulated depreciation	36,000	164,000
Total assets		$280,000
Liabilities and Equity		
Liabilities		
Accounts payable	$ 58,200	
Income taxes payable (due 10/31/2009)	20,000	
Note payable to bank	10,000	$ 88,200
Stockholders' equity		
Common stock	150,000	
Retained earnings	41,800	191,800
Total liabilities and equity		$280,000

* Equipment is depreciated on a straight-line basis over 10 years (salvage value is $20,000).

Operating Budgets

P1 Prepare each component of a master budget and link each to the budgeting process.

This section explains HD's preparation of operating budgets. Its operating budgets consist of the sales budget, merchandise purchases budget, selling expense budget, and general and administrative expense budget. HD does not prepare production and manufacturing budgets because it is a merchandiser. (The preparation of production budgets and manufacturing budgets is described in Appendix 23A.)

Sales Budget The first step in preparing the master budget is planning the **sales budget,** which shows the planned sales units and the expected dollars from these sales. The sales budget is the starting point in the budgeting process because plans for most departments are linked to sales.

The sales budget should emerge from a careful analysis of forecasted economic and market conditions, business capacity, proposed selling expenses (such as advertising), and predictions of unit sales. A company's sales personnel are usually asked to develop predictions of sales for each territory and department because people normally feel a greater commitment to goals they help set. Another advantage to this participatory budgeting approach is that it draws on knowledge and experience of people involved in the activity.

Decision Insight

No Biz Like Snow Biz Ski resorts' costs of making snow are in the millions of dollars for equipment alone. Snowmaking involves spraying droplets of water into the air, causing them to freeze and come down as snow. Making snow can cost more than $2,000 an hour. Snowmaking accounts for 40 to 50 percent of the operating budgets for many ski resorts.

Example: Assume a company's sales force receives a bonus when sales exceed the budgeted amount. How would this arrangement affect the bottom-up process of sales forecasts? *Answer:* Sales reps may understate their budgeted sales.

To illustrate, in September 2009, HD sold 700 hockey sticks at $100 per unit. After considering sales predictions and market conditions, HD prepares its sales budget for the next quarter (three months) plus one extra month (see Exhibit 23.6). The sales budget includes January 2010 because the purchasing department relies on estimated January sales to decide on December 2009 inventory purchases. The sales budget in Exhibit 23.6 includes forecasts of both unit sales and unit prices. Some sales budgets are expressed only in total sales dollars, but most are more detailed. Management finds it useful to know budgeted units and unit prices for many different products, regions, departments, and sales representatives.

EXHIBIT 23.6

Sales Budget for Planned Unit and Dollar Sales

HOCKEY DEN
Monthly Sales Budget
October 2009–January 2010

	Budgeted Unit Sales	Budgeted Unit Price	Budgeted Total Sales
September 2009 (actual)	700	$100	$ 70,000
October 2009	1,000	$100	$100,000
November 2009	800	100	80,000
December 2009	1,400	100	140,000
Totals for the quarter	3,200	100	$320,000
January 2010	900	100	$ 90,000

Decision Maker

Entrepreneur You run a start-up that manufactures designer clothes. Business is seasonal, and fashions and designs quickly change. How do you prepare reliable annual sales budgets? [Answer—p. 961]

Merchandise Purchases Budget Companies use various methods to help managers make inventory purchasing decisions. These methods recognize that the number of units added to inventory depends on budgeted sales volume. Whether a company manufactures or purchases the product it sells, budgeted future sales volume is the primary factor in most inventory management decisions. A company must also consider its inventory system and other factors that we discuss next.

Just-in-time inventory systems. Managers of *just-in-time* (JIT) inventory systems use sales budgets for short periods (often as few as one or two days) to order just enough merchandise or materials to satisfy the immediate sales demand. This keeps the amount of inventory to a minimum (or zero in an ideal situation). A JIT system minimizes the costs of maintaining inventory, but it is practical only if customers are content to order in advance or if managers can accurately determine short-term sales demand. Suppliers also must be able and willing to ship small quantities regularly and promptly.

Point: Accurate estimates of future sales are crucial in a JIT system.

Safety stock inventory systems. Market conditions and manufacturing processes for some products do not allow use of a just-in-time system. Companies in these cases maintain sufficient inventory to reduce the risk and cost of running short. This practice requires enough purchases to satisfy the budgeted sales amounts and to maintain a **safety stock,** a quantity of inventory that provides protection against lost sales caused by unfulfilled demands from customers or delays in shipments from suppliers.

Merchandise purchases budget preparation. A merchandiser usually expresses a **merchandise purchases budget** in both units and dollars. Exhibit 23.7 shows the general layout for this budget in equation form. If this formula is expressed in units and only one product is involved, we can compute the number of dollars of inventory to be purchased for the budget by multiplying the units to be purchased by the cost per unit.

Inventory to be purchased	=	Budgeted ending inventory	+	Budgeted cost of sales for the period	−	Budgeted beginning inventory

EXHIBIT 23.7

General Formula for a Merchandise Purchases Budget

To illustrate, after assessing the cost of keeping inventory along with the risk and cost of inventory shortages, HD decided that the number of units in its inventory at each month-end should equal 90% of next month's predicted sales. For example, inventory at the end of October should equal 90% of budgeted November sales, and the November ending inventory should equal 90% of budgeted December sales, and so on. Also, HD's suppliers expect the September 2009 per unit cost of $60 to remain unchanged through January 2010. This information along with knowledge of 900 units in inventory at September 30 (see Exhibit 23.5) allows the company to prepare the merchandise purchases budget shown in Exhibit 23.8.

Example: Assume Hockey Den adopts a JIT system in purchasing merchandise. How will its sales budget differ from its merchandise purchases budget? *Answer:* The two budgets will be similar because future inventory should be near zero.

EXHIBIT 23.8

Merchandise Purchases Budget

HOCKEY DEN
Merchandise Purchases Budget
October 2009–December 2009

	October	November	December
Next month's budgeted sales (units)	800	1,400	900
Ratio of inventory to future sales	× 90%	× 90%	× 90%
Budgeted ending inventory (units)	720	1,260	810
Add budgeted sales (units)	1,000	800	1,400
Required units of available merchandise	1,720	2,060	2,210
Deduct beginning inventory (units)	900	720	1,260
Units to be purchased	820	1,340	950
Budgeted cost per unit	$ 60	$ 60	$ 60
Budgeted cost of merchandise purchases	$49,200	$80,400	$57,000

Example: If ending inventory in Exhibit 23.8 is required to equal 80% of next month's predicted sales, how many units must be purchased each month? *Answer:* Budgeted ending inventory: Oct. = 640 units; Nov. = 1,120 units; Dec. = 720 units. Required purchases: Oct. = 740 units; Nov. = 1,280 units; Dec. = 1,000 units.

The first three lines of HD's merchandise purchases budget determine the required ending inventories (in units). Budgeted unit sales are then added to the desired ending inventory to give the required units of available merchandise. We then subtract beginning inventory to determine the budgeted number of units to be purchased. The last line is the budgeted cost of the purchases, computed by multiplying the number of units to be purchased by the predicted cost per unit.

We already indicated that some budgeting systems describe only the total dollars of budgeted sales. Likewise, a system can express a merchandise purchases budget only in terms of the total cost of merchandise to be purchased, omitting the number of units to be purchased. This method assumes a constant relation between sales and cost of goods sold. HD, for instance, might assume the expected cost of goods sold to be 60% of sales, computed from the budgeted unit cost of $60 and the budgeted sales price of $100. However, it still must consider the effects of changes in beginning and ending inventories in determining the amounts to be purchased.

Selling Expense Budget The **selling expense budget** is a plan listing the types and amounts of selling expenses expected during the budget period. Its initial responsibility usually rests with the vice president of marketing or an equivalent sales manager. The selling expense budget is normally created to provide sufficient selling expenses to meet sales goals reflected in the sales budget. Predicted selling expenses are based on both the sales budget and the experience of previous periods. After some or all of the master budget is prepared, management might decide that projected sales volume is inadequate. If so, subsequent adjustments in the sales budget can require corresponding adjustments in the selling expense budget.

Example: If sales commissions in Exhibit 23.9 are increased, which budgets are affected? *Answer:* Selling expenses budget, cash budget, and budgeted income statement.

To illustrate, HD's selling expense budget is in Exhibit 23.9. The firm's selling expenses consist of commissions paid to sales personnel and a $2,000 monthly salary paid to the sales manager. Sales commissions equal 10% of total sales and are paid in the month sales occur. Sales commissions are variable with respect to sales volume, but the sales manager's salary is fixed. No advertising expenses are budgeted for this particular quarter.

EXHIBIT 23.9

Selling Expense Budget

HOCKEY DEN
Selling Expense Budget
October 2009–December 2009

	October	November	December	Totals
Budgeted sales	$100,000	$80,000	$140,000	$320,000
Sales commission percent	× 10%	× 10%	× 10%	× 10%
Sales commissions	10,000	8,000	14,000	32,000
Salary for sales manager	2,000	2,000	2,000	6,000
Total selling expenses	$ 12,000	$10,000	$ 16,000	$ 38,000

General and Administrative Expense Budget The **general and administrative expense budget** plans the predicted operating expenses not included in the selling expenses budget. General and administrative expenses can be either variable or fixed with respect to sales volume. The office manager responsible for general administration often is responsible for preparing the initial general and administrative expense budget.

Interest expense and income tax expense are often classified as general and administrative expenses in published income statements, but normally cannot be planned at this stage of the budgeting process. The prediction of interest expense follows the preparation of the cash budget and the decisions regarding debt. The predicted income tax expense depends on the budgeted amount of pretax income. Both interest and income taxes are usually beyond the control of the office manager. As a result, they are not used in comparison to the budget to evaluate that person's performance.

Exhibit 23.10 shows HD's general and administrative expense budget. It includes salaries of $54,000 per year, or $4,500 per month (paid each month when they are earned). Using information in Exhibit 23.5, the depreciation on equipment is computed as $18,000 per year [($200,000 − $20,000)/10 years], or $1,500 per month ($18,000/12 months).

EXHIBIT 23.10

General and Administrative Expense Budget

HOCKEY DEN General and Administrative Expense Budget October 2009–December 2009	October	November	December	Totals
Administrative salaries	$4,500	$4,500	$4,500	$13,500
Depreciation of equipment	1,500	1,500	1,500	4,500
Total general and administrative expenses	$6,000	$6,000	$6,000	$18,000

Example: In Exhibit 23.10, how would a rental agreement of $5,000 per month plus 1% of sales affect the general and administrative expense budget? (Budgeted sales are in Exhibit 23.6.)
Answer: Rent expense: Oct. = $6,000; Nov. = $5,800; Dec. = $6,400; Total = $18,200; *Revised total general and administrative expenses:* Oct. = $12,000; Nov. = $11,800; Dec. = $12,400; Total = $36,200.

Quick Check Answers—p. 961

5. What is a master budget?
6. A master budget (*a*) always includes a manufacturing budget specifying the units to be produced; (*b*) is prepared with a process starting with the operating budgets and continues with the capital expenditures budget and then financial budgets; or (*c*) is prepared with a process ending with the sales budget.
7. What are the three primary categories of budgets in the master budget?
8. In preparing monthly budgets for the third quarter, a company budgeted sales of 120 units for July and 140 units for August. Management wants each month's ending inventory to be 60% of next month's sales. The June 30 inventory consists of 50 units. How many units of product for July acquisition should the merchandise purchases budget specify for the third quarter? (*a*) 84, (*b*) 120, (*c*) 154, or (*d*) 204.
9. How do the operating budgets for merchandisers and manufacturers differ?
10. How does a just-in-time inventory system differ from a safety stock system?

Capital Expenditures Budget

The **capital expenditures budget** lists dollar amounts to be both received from plant asset disposals and spent to purchase additional plant assets to carry out the budgeted business activities. It is usually prepared after the operating budgets. Since a company's plant assets determine its productive capacity, this budget is usually affected by long-range plans for the business. Yet the process of preparing a sales or purchases budget can reveal that the company requires more (or less) capacity, which implies more (or less) plant assets.

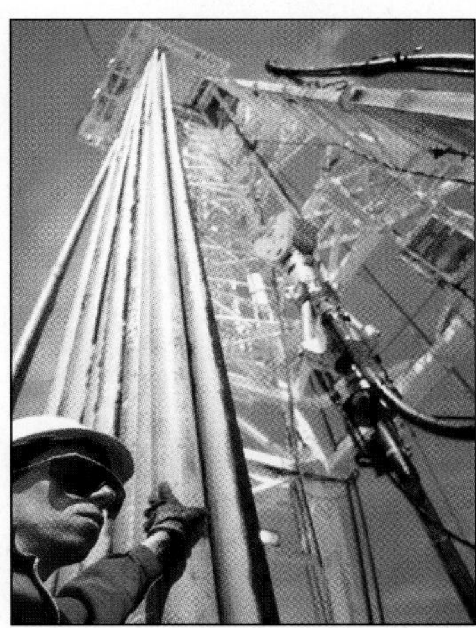

Capital budgeting is the process of evaluating and planning for capital (plant asset) expenditures. This is an important management task because these expenditures often involve long-run commitments of large amounts, affect predicted cash flows, and impact future debt and equity financing. This means that the capital expenditures budget is often linked with management's evaluation of the company's ability to take on more debt. We describe capital budgeting in Chapter 25.

Hockey Den does not anticipate disposal of any plant assets through December 2009, but it does plan to acquire additional equipment for $25,000 cash near the end of December 2009. This is the only budgeted capital expenditure from October 2009 through January 2010. Thus, no separate budget is shown. The cash budget in Exhibit 23.11 reflects this $25,000 planned expenditure.

Financial Budgets

After preparing its operating and capital expenditures budgets, a company uses information from these budgets to prepare at least three financial budgets: the cash budget, budgeted income statement, and budgeted balance sheet.

EXHIBIT 23.11

Cash Budget

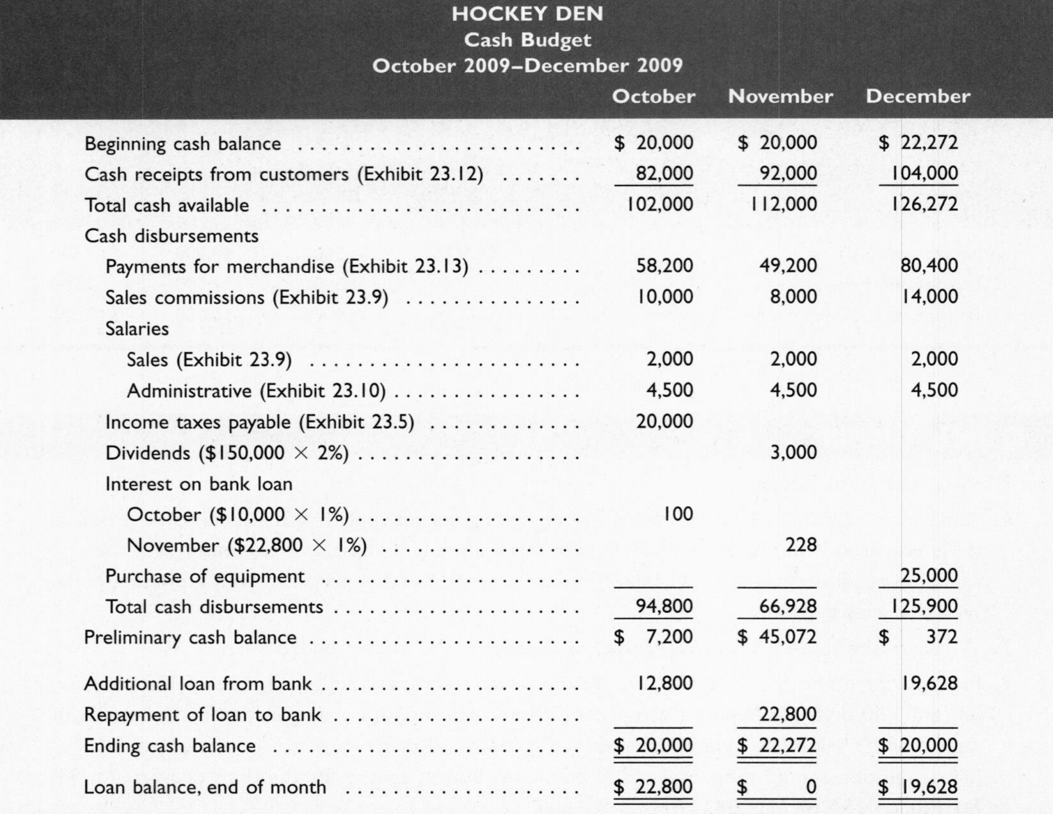

HOCKEY DEN Cash Budget October 2009–December 2009	October	November	December
Beginning cash balance	$ 20,000	$ 20,000	$ 22,272
Cash receipts from customers (Exhibit 23.12)	82,000	92,000	104,000
Total cash available	102,000	112,000	126,272
Cash disbursements			
Payments for merchandise (Exhibit 23.13)	58,200	49,200	80,400
Sales commissions (Exhibit 23.9)	10,000	8,000	14,000
Salaries			
Sales (Exhibit 23.9)	2,000	2,000	2,000
Administrative (Exhibit 23.10)	4,500	4,500	4,500
Income taxes payable (Exhibit 23.5)	20,000		
Dividends ($150,000 × 2%)		3,000	
Interest on bank loan			
October ($10,000 × 1%)	100		
November ($22,800 × 1%)		228	
Purchase of equipment			25,000
Total cash disbursements	94,800	66,928	125,900
Preliminary cash balance	$ 7,200	$ 45,072	$ 372
Additional loan from bank	12,800		19,628
Repayment of loan to bank		22,800	
Ending cash balance	$ 20,000	$ 22,272	$ 20,000
Loan balance, end of month	$ 22,800	$ 0	$ 19,628

Example: If the minimum ending cash balance in Exhibit 23.11 is changed to $25,000 for each month, what is the projected loan balance at Dec. 31, 2009?

Answer:

Loan balance, Oct. 31	$27,800
November interest	278
November payment	25,022
Loan balance, Nov. 30	2,778
December interest	28
Additional loan in Dec.	21,928
Loan balance, Dec. 31	$24,706

P2 Link both operating and capital expenditures budgets to budgeted financial statements.

Cash Budget After developing budgets for sales, merchandise purchases, expenses, and capital expenditures, the next step is to prepare the **cash budget,** which shows expected cash inflows and outflows during the budget period. It is especially important to maintain a cash balance necessary to meet ongoing obligations. By preparing a cash budget, management can prearrange loans to cover anticipated cash shortages before they are needed. A cash budget also helps management avoid a cash balance that is too large. Too much cash is undesirable because it earns a relatively low (if any) return.

When preparing a cash budget, we add expected cash receipts to the beginning cash balance and deduct expected cash disbursements. If the expected ending cash balance is inadequate, additional cash requirements appear in the budget as planned increases from short-term loans. If the expected ending cash balance exceeds the desired balance, the excess is used to repay loans or to acquire short-term investments. Information for preparing the cash budget is mainly taken from the operating and capital expenditures budgets.

To illustrate, Exhibit 23.11 presents HD's cash budget. The beginning cash balance for October is taken from the September 30, 2009, balance sheet in Exhibit 23.5. The remainder of this section describes the computations in the cash budget.

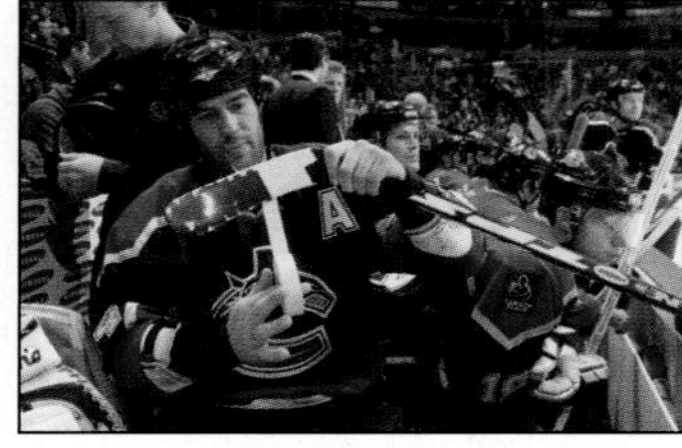

We begin with reference to HD's budgeted sales (Exhibit 23.6). Analysis of past sales indicates that 40% of the firm's sales are for cash. The remaining 60% are credit sales; these customers are expected to pay in full in the month following the sales. We now can compute the budgeted cash receipts from customers as shown in Exhibit 23.12. October's budgeted cash receipts consist of $40,000 from expected cash sales ($100,000 × 40%) plus the anticipated collection of $42,000 of accounts receivable from the end of September. Each month's cash receipts from customers are transferred to the second line of Exhibit 23.11.

Next, we see that HD's merchandise purchases are entirely on account. It makes full payment during the month following its purchases. Therefore, cash disbursements for

EXHIBIT 23.12

Computing Budgeted Cash Receipts

	September	October	November	December
Sales	$70,000	$100,000	$80,000	$140,000
Less ending accounts receivable (60%)	42,000	60,000	48,000	84,000
Cash receipts from				
Cash sales (40% of sales)		40,000	32,000	56,000
Collections of prior month's receivables		42,000	60,000	48,000
Total cash receipts		$ 82,000	$92,000	$104,000

purchases can be computed from the September 30, 2009, balance sheet (Exhibit 23.5) and the merchandise purchases budget (Exhibit 23.8). This computation is shown in Exhibit 23.13.

EXHIBIT 23.13

Computing Cash Disbursements for Purchases

October payments (September 30 balance)	$58,200
November payments (October purchases)	49,200
December payments (November purchases)	80,400

The monthly budgeted cash disbursements for sales commissions and salaries are taken from the selling expense budget (Exhibit 23.9) and the general and administrative expense budget (Exhibit 23.10). The cash budget is unaffected by depreciation as reported in the general and administrative expenses budget.

Income taxes are due and payable in October as shown in the September 30, 2009, balance sheet (Exhibit 23.5). The cash budget in Exhibit 23.11 shows this $20,000 expected payment in October. Predicted income tax expense for the quarter ending December 31 is 40% of net income and is due in January 2010. It is therefore not reported in the October–December 2009 cash budget but in the budgeted income statement as income tax expense and on the budgeted balance sheet as income tax liability.

Hockey Den also pays a cash dividend equal to 2% of the par value of common stock in the second month of each quarter. The cash budget in Exhibit 23.11 shows a November payment of $3,000 for this purpose (2% of $150,000; see Exhibit 23.5).

Hockey Den has an agreement with its bank that promises additional loans at each month-end, if necessary, to keep a minimum cash balance of $20,000. If the cash balance exceeds $20,000 at a month-end, HD uses the excess to repay loans. Interest is paid at each month-end at the rate of 1% of the beginning balance of these loans. For October, this payment is 1% of the $10,000 amount reported in the balance sheet of Exhibit 23.5. For November, HD expects to pay interest of $228, computed as 1% of the $22,800 expected loan balance at October 31. No interest is budgeted for December because the company expects to repay the loans in full at the end of November. Exhibit 23.11 shows that the October 31 cash balance declines to $7,200 (before any loan-related activity). This amount is less than the $20,000 minimum. Hockey Den will bring this balance up to the minimum by borrowing $12,800 with a short-term note. At the end of November, the budget shows an expected cash balance of $45,072 before any loan activity. This means that HD expects to repay $22,800 of debt. The equipment purchase budgeted for December reduces the expected cash balance to $372, far below the $20,000 minimum. The company expects to borrow $19,628 in that month to reach the minimum desired ending balance.

Example: Give one reason for maintaining a minimum cash balance when the budget shows extra cash is not needed. *Answer:* For unexpected events.

Decision Insight

Netting Cash The Hockey Company—whose brands include CCM, JOFA, and KOHO—reported net cash outflows for investing activities of $32 million. Much of this amount was a prepayment to the NHL for a 10-year license agreement.

Budgeted Income Statement One of the final steps in preparing the master budget is to summarize the income effects. The **budgeted income statement** is a managerial accounting report showing predicted amounts of sales and expenses for the budget period. Information needed for preparing a budgeted income statement is primarily taken from already prepared budgets. The volume of information summarized in the budgeted income statement is so large for some companies that they often use spreadsheets to accumulate the budgeted transactions and classify them by their effects on income. We condense HD's budgeted income statement and show it in Exhibit 23.14. All information in this exhibit is taken from earlier budgets. Also, we now can predict the amount of income tax expense for the quarter, computed as 40% of the budgeted pretax income. This amount is included in the cash budget and/or the budgeted balance sheet as necessary.

EXHIBIT 23.14
Budgeted Income Statement

HOCKEY DEN
Budgeted Income Statement
For Three Months Ended December 31, 2009

Sales (Exhibit 23.6, 3,200 units @ $100)		$320,000
Cost of goods sold (3,200 units @ $60)		192,000
Gross profit		128,000
Operating expenses		
Sales commissions (Exhibit 23.9)	$32,000	
Sales salaries (Exhibit 23.9)	6,000	
Administrative salaries (Exhibit 23.10)	13,500	
Depreciation on equipment (Exhibit 23.10)	4,500	
Interest expense (Exhibit 23.11)	328	56,328
Income before income taxes		71,672
Income tax expense ($71,672 × 40%)		28,669
Net income		$ 43,003

Point: Lending institutions often require potential borrowers to provide cash budgets, budgeted income statements, and budgeted balance sheets, as well as data on past performance.

Budgeted Balance Sheet The final step in preparing the master budget is summarizing the company's financial position. The **budgeted balance sheet** shows predicted amounts for the company's assets, liabilities, and equity as of the end of the budget period. HD's budgeted balance sheet in Exhibit 23.15 is prepared using information from the other budgets. The sources of amounts are reported in the notes to the budgeted balance sheet.[1]

Decision Insight

Plan Ahead Most companies allocate dollars based on budgets submitted by department managers. These managers verify the numbers and monitor the budget. Managers must remember, however, that a budget is judged by its success in helping achieve the company's mission. One analogy is that a hiker must know the route to properly plan a hike and monitor hiking progress.

[1] An eight-column spreadsheet, or work sheet, can be used to prepare a budgeted balance sheet (and income statement). The first two columns show the ending balance sheet amounts from the period prior to the budget period. The budgeted transactions and adjustments are entered in the third and fourth columns in the same manner as adjustments are entered on an ordinary work sheet. After all budgeted transactions and adjustments have been entered, the amounts in the first two columns are combined with the budget amounts in the third and fourth columns and sorted to the proper Income Statement (fifth and sixth columns) and Balance Sheet columns (seventh and eighth columns). Amounts in these columns are used to prepare the budgeted income statement and balance sheet.

EXHIBIT 23.15

Budgeted Balance Sheet

HOCKEY DEN Budgeted Balance Sheet December 31, 2009		
Assets		
Cash[a]		$ 20,000
Accounts receivable[b]		84,000
Inventory[c]		48,600
Equipment[d]	$225,000	
Less accumulated depreciation[e]	40,500	184,500
Total assets		$337,100
Liabilities and Equity		
Liabilities		
Accounts payable[f]	$ 57,000	
Income taxes payable[g]	28,669	
Bank loan payable[h]	19,628	$105,297
Stockholders' equity		
Common stock[i]	150,000	
Retained earnings[j]	81,803	231,803
Total liabilities and equity		$337,100

[a] Ending balance for December from the cash budget in Exhibit 23.11.
[b] 60% of $140,000 sales budgeted for December from the sales budget in Exhibit 23.6.
[c] 810 units in budgeted December ending inventory at the budgeted cost of $60 per unit (from the purchases budget in Exhibit 23.8).
[d] September 30 balance of $200,000 from the beginning balance sheet in Exhibit 23.5 plus $25,000 cost of new equipment from the cash budget in Exhibit 23.11.
[e] September 30 balance of $36,000 from the beginning balance sheet in Exhibit 23.5 plus $4,500 expense from the general and administrative expense budget in Exhibit 23.10.
[f] Budgeted cost of purchases for December from the purchases budget in Exhibit 23.8.
[g] Income tax expense from the budgeted income statement for the fourth quarter in Exhibit 23.14.
[h] Budgeted December 31 balance from the cash budget in Exhibit 23.11.
[i] Unchanged from the beginning balance sheet in Exhibit 23.5.
[j] September 30 balance of $41,800 from the beginning balance sheet in Exhibit 23.5 plus budgeted net income of $43,003 from the budgeted income statement in Exhibit 23.14 minus budgeted cash dividends of $3,000 from the cash budget in Exhibit 23.11.

Quick Check

Answers—p. 961

11. In preparing a budgeted balance sheet, (*a*) plant assets are determined by analyzing the capital expenditures budget and the balance sheet from the beginning of the budget period, (*b*) liabilities are determined by analyzing the general and administrative expense budget, or (*c*) retained earnings are determined from information contained in the cash budget and the balance sheet from the beginning of the budget period.
12. What sequence is followed in preparing the budgets that constitute the master budget?

Activity-Based Budgeting

Decision Analysis

A1 Analyze expense planning using activity-based budgeting.

Activity-based budgeting (ABB) is a budget system based on expected activities. Knowledge of expected activities and their levels for the budget period enables management to plan for resources required to perform the activities. To illustrate, we consider the budget of a company's accounting department. Traditional budgeting systems list items such as salaries, supplies, equipment, and utilities. Such an itemized budget informs management of the use of the funds budgeted (for example, salaries), but management cannot assess the basis for increases or decreases in budgeted amounts as compared to prior periods. Accordingly, management often makes across-the-board cuts or increases. In contrast, ABB requires management to list activities performed by, say, the accounting department such as auditing, tax reporting, financial reporting, and cost accounting. Exhibit 23.16 contrasts a traditional budget with an activity-based budget for a company's accounting department. An understanding of the resources required to perform the activities, the costs associated with these resources,

EXHIBIT 23.16

Activity-Based Budgeting versus Traditional Budgeting (for an accounting department)

Activity-Based Budget		Traditional Budget	
Auditing	$ 58,000	Salaries	$152,000
Tax reporting	71,000	Supplies	22,000
Financial reporting	63,000	Depreciation	36,000
Cost accounting	32,000	Utilities	14,000
Total	$224,000	Total	$224,000

and the way resource use changes with changes in activity levels allows management to better assess how expenses will change to accommodate changes in activity levels. Moreover, by knowing the relation between activities and costs, management can attempt to reduce costs by eliminating nonvalue-added activities.

Decision Maker

Environmental Manager You hold the new position of environmental control manager for a chemical company. You are asked to develop a budget for your job and identify job responsibilities. How do you proceed? [Answer—p. 961]

Demonstration Problem

DP23

Wild Wood Company's management asks you to prepare its master budget using the following information. The budget is to cover the months of April, May, and June of 2009.

WILD WOOD COMPANY
Balance Sheet
March 31, 2009

Assets		Liabilities and Equity	
Cash	$ 50,000	Accounts payable	$156,000
Accounts receivable	175,000	Short-term notes payable	12,000
Inventory	126,000	Total current liabilities	168,000
Total current assets	351,000	Long-term note payable	200,000
Equipment, gross	480,000	Total liabilities	368,000
Accumulated depreciation	(90,000)	Common stock	235,000
Equipment, net	390,000	Retained earnings	138,000
		Total stockholders' equity	373,000
Total assets	$741,000	Total liabilities and equity	$741,000

Additional Information

a. Sales for March total 10,000 units. Each month's sales are expected to exceed the prior month's results by 5%. The product's selling price is $25 per unit.

b. Company policy calls for a given month's ending inventory to equal 80% of the next month's expected unit sales. The March 31 inventory is 8,400 units, which complies with the policy. The purchase price is $15 per unit.

c. Sales representatives' commissions are 12.5% of sales and are paid in the month of the sales. The sales manager's monthly salary will be $3,500 in April and $4,000 per month thereafter.

d. Monthly general and administrative expenses include $8,000 administrative salaries, $5,000 depreciation, and 0.9% monthly interest on the long-term note payable.

e. The company expects 30% of sales to be for cash and the remaining 70% on credit. Receivables are collected in full in the month following the sale (none is collected in the month of the sale).

f. All merchandise purchases are on credit, and no payables arise from any other transactions. One month's purchases are fully paid in the next month.

g. The minimum ending cash balance for all months is $50,000. If necessary, the company borrows enough cash using a short-term note to reach the minimum. Short-term notes require an interest payment of 1% at each month-end (before any repayment). If the ending cash balance exceeds the minimum, the excess will be applied to repaying the short-term notes payable balance.

h. Dividends of $100,000 are to be declared and paid in May.

i. No cash payments for income taxes are to be made during the second calendar quarter. Income taxes will be assessed at 35% in the quarter.

j. Equipment purchases of $55,000 are scheduled for June.

Required

Prepare the following budgets and other financial information as required:

1. Sales budget, including budgeted sales for July.
2. Purchases budget, the budgeted cost of goods sold for each month and quarter, and the cost of the June 30 budgeted inventory.
3. Selling expense budget.
4. General and administrative expense budget.
5. Expected cash receipts from customers and the expected June 30 balance of accounts receivable.
6. Expected cash payments for purchases and the expected June 30 balance of accounts payable.
7. Cash budget.
8. Budgeted income statement.
9. Budgeted statement of retained earnings.
10. Budgeted balance sheet.

Planning the Solution

- The sales budget shows expected sales for each month in the quarter. Start by multiplying March sales by 105% and then do the same for the remaining months. July's sales are needed for the purchases budget. To complete the budget, multiply the expected unit sales by the selling price of $25 per unit.
- Use these results and the 80% inventory policy to budget the size of ending inventory for April, May, and June. Add the budgeted sales to these numbers and subtract the actual or expected beginning inventory for each month. The result is the number of units to be purchased each month. Multiply these numbers by the per unit cost of $15. Find the budgeted cost of goods sold by multiplying the unit sales in each month by the $15 cost per unit. Compute the cost of the June 30 ending inventory by multiplying the expected units available at that date by the $15 cost per unit.
- The selling expense budget has only two items. Find the amount of the sales representatives' commissions by multiplying the expected dollar sales in each month by the 12.5% commission rate. Then include the sales manager's salary of $3,500 in April and $4,000 in May and June.
- The general and administrative expense budget should show three items. Administrative salaries are fixed at $8,000 per month, and depreciation is $5,000 per month. Budget the monthly interest expense on the long-term note by multiplying its $200,000 balance by the 0.9% monthly interest rate.
- Determine the amounts of cash sales in each month by multiplying the budgeted sales by 30%. Add to this amount the credit sales of the prior month (computed as 70% of prior month's sales). April's cash receipts from collecting receivables equals the March 31 balance of $175,000. The expected June 30 accounts receivable balance equals 70% of June's total budgeted sales.
- Determine expected cash payments on accounts payable for each month by making them equal to the merchandise purchases in the prior month. The payments for April equal the March 31 balance of accounts payable shown on the beginning balance sheet. The June 30 balance of accounts payable equals merchandise purchases for June.
- Prepare the cash budget by combining the given information and the amounts of cash receipts and cash payments on account that you computed. Complete the cash budget for each month by either borrowing enough to raise the preliminary balance to the minimum or paying off short-term debt as much as the balance allows without falling below the minimum. Show the ending balance of the short-term note in the budget.
- Prepare the budgeted income statement by combining the budgeted items for all three months. Determine the income before income taxes and multiply it by the 35% rate to find the quarter's income tax expense.
- The budgeted statement of retained earnings should show the March 31 balance plus the quarter's net income minus the quarter's dividends.

- The budgeted balance sheet includes updated balances for all items that appear in the beginning balance sheet and an additional liability for unpaid income taxes. Amounts for all asset, liability, and equity accounts can be found either in the budgets, other calculations, or by adding amounts found there to the beginning balances.

Solution to Demonstration Problem

1. Sales budget

	April	May	June	July
Prior period's unit sales	10,000	10,500	11,025	11,576
Plus 5% growth	500	525	551	579
Projected unit sales	10,500	11,025	11,576	12,155

	April	May	June	Quarter
Projected unit sales	10,500	11,025	11,576	
Selling price per unit	× $25	× $25	× $25	
Projected sales	$262,500	$275,625	$289,400	$827,525

2. Purchases budget

	April	May	June	Quarter
Next period's unit sales (part 1)	11,025	11,576	12,155	
Ending inventory percent	× 80%	× 80%	× 80%	
Desired ending inventory	8,820	9,261	9,724	
Current period's unit sales (part 1)	10,500	11,025	11,576	
Units to be available	19,320	20,286	21,300	
Less beginning inventory	8,400	8,820	9,261	
Units to be purchased	10,920	11,466	12,039	
Budgeted cost per unit	× $15	× $15	× $15	
Projected purchases	$163,800	$171,990	$180,585	$516,375

Budgeted cost of goods sold

	April	May	June	Quarter
This period's unit sales (part 1)	10,500	11,025	11,576	
Budgeted cost per unit	× $15	× $15	× $15	
Projected cost of goods sold	$157,500	$165,375	$173,640	$496,515

Budgeted inventory for June 30

Units (part 2)	9,724
Cost per unit	× $15
Total	$145,860

3. Selling expense budget

	April	May	June	Quarter
Budgeted sales (part 1)	$262,500	$275,625	$289,400	$827,525
Commission percent	× 12.5%	× 12.5%	× 12.5%	× 12.5%
Sales commissions	32,813	34,453	36,175	103,441
Manager's salary	3,500	4,000	4,000	11,500
Projected selling expenses	$ 36,313	$ 38,453	$ 40,175	$114,941

4. General and administrative expense budget

	April	May	June	Quarter
Administrative salaries	$ 8,000	$ 8,000	$ 8,000	$24,000
Depreciation	5,000	5,000	5,000	15,000
Interest on long-term note payable (0.9% × $200,000)	1,800	1,800	1,800	5,400
Projected expenses	$14,800	$14,800	$14,800	$44,400

5. Expected cash receipts from customers

	April	May	June	Quarter
Budgeted sales (part 1)	$262,500	$275,625	$289,400	
Ending accounts receivable (70%)	$183,750	$192,938	$202,580	
Cash receipts				
Cash sales (30%)	$ 78,750	$ 82,687	$ 86,820	$248,257
Collections of prior period's receivables	175,000	183,750	192,938	551,688
Total cash to be collected	$253,750	$266,437	$279,758	$799,945

6. Expected cash payments to suppliers

	April	May	June	Quarter
Cash payments (equal to prior period's purchases)	$156,000	$163,800	$171,990	$491,790
Expected June 30 balance of accounts payable (June purchases)			$180,585	

7. Cash budget

	April	May	June
Beginning cash balance	$ 50,000	$ 89,517	$ 50,000
Cash receipts (part 5)	253,750	266,437	279,758
Total cash available	303,750	355,954	329,758
Cash payments			
Payments for merchandise (part 6)	156,000	163,800	171,990
Sales commissions (part 3)	32,813	34,453	36,175
Salaries			
Sales (part 3)	3,500	4,000	4,000
Administrative (part 4)	8,000	8,000	8,000
Interest on long-term note (part 4)	1,800	1,800	1,800
Dividends		100,000	
Equipment purchase			55,000
Interest on short-term notes			
April ($12,000 × 1.0%)	120		
June ($6,099 × 1.0%)			61
Total cash payments	202,233	312,053	277,026
Preliminary balance	101,517	43,901	52,732
Additional loan		6,099	
Loan repayment	(12,000)		(2,732)
Ending cash balance	$ 89,517	$ 50,000	$ 50,000
Ending short-term notes	$ 0	$ 6,099	$ 3,367

8.

WILD WOOD COMPANY
Budgeted Income Statement
For Quarter Ended June 30, 2009

Sales (part 1)		$ 827,525
Cost of goods sold (part 2)		496,515
Gross profit		331,010
Operating expenses		
Sales commissions (part 3)	$103,441	
Sales salaries (part 3)	11,500	
Administrative salaries (part 4)	24,000	
Depreciation (part 4)	15,000	
Interest on long-term note (part 4)	5,400	
Interest on short-term notes (part 7)	181	
Total operating expenses		159,522
Income before income taxes		171,488
Income taxes (35%)		60,021
Net income		$ 111,467

9.

WILD WOOD COMPANY
Budgeted Statement of Retained Earnings
For Quarter Ended June 30, 2009

Beginning retained earnings (given)	$138,000
Net income (part 8)	111,467
	249,467
Less cash dividends (given)	100,000
Ending retained earnings	$149,467

10.

WILD WOOD COMPANY
Budgeted Balance Sheet
June 30, 2009

Assets		
Cash (part 7)		$ 50,000
Accounts receivable (part 5)		202,580
Inventory (part 2)		145,860
Total current assets		398,440
Equipment (given plus purchase)	$535,000	
Less accumulated depreciation (given plus expense)	105,000	430,000
Total assets		$828,440
Liabilities and Equity		
Accounts payable (part 6)		$180,585
Short-term notes payable (part 7)		3,367
Income taxes payable (part 8)		60,021
Total current liabilities		243,973
Long-term note payable (given)		200,000
Total liabilities		443,973
Common stock (given)		235,000
Retained earnings (part 9)		149,467
Total stockholders' equity		384,467
Total liabilities and equity		$828,440

APPENDIX

23A Production and Manufacturing Budgets

P3 Prepare production and manufacturing budgets.

Unlike a merchandising company, a manufacturer must prepare a **production budget** instead of a merchandise purchases budget. A production budget, which shows the number of units to be produced each month, is similar to merchandise purchases budgets except that the number of units to be purchased each month (as shown in Exhibit 23.8) is replaced by the number of units to be manufactured each month. A production budget does not show costs; it is *always expressed in units of product*. Exhibit 23A.1 shows the production budget for **Toronto Sticks Company (TSC),** a manufacturer of hockey sticks. TSC is an exclusive supplier of hockey sticks to Hockey Den, meaning that TSC uses HD's budgeted sales figures (Exhibit 23.6) to determine its production and manufacturing budgets.

EXHIBIT 23A.1
Production Budget

TSC
Production Budget
October 2009–December 2009

	October	November	December
Next period's budgeted sales (units)	800	1,400	900
Ratio of inventory to future sales	× 90%	× 90%	× 90%
Budgeted ending inventory (units)	720	1,260	810
Add budgeted sales for the period (units)	1,000	800	1,400
Required units of available production	1,720	2,060	2,210
Deduct beginning inventory (units)	(900)	(720)	(1,260)
Units to be produced	820	1,340	950

A **manufacturing budget** shows the budgeted costs for direct materials, direct labor, and overhead. It is based on the budgeted production volume from the production budget. The manufacturing budget for most companies consists of three individual budgets: direct materials budget, direct labor budget, and overhead budget. Exhibits 23A.2–23A.4 show these three manufacturing budgets for TSC. These budgets yield the total expected cost of goods to be manufactured in the budget period.

The *direct materials budget* is driven by the budgeted materials needed to satisfy each month's production requirement. To this we must add the desired ending inventory requirements. The desired ending inventory of direct materials as shown in Exhibit 23A.2 is 50% of next month's budgeted materials requirements of wood. For instance, in October 2009, an ending inventory of 335 units of material is desired (50% of November's 670 units). The desired ending inventory for December 2009 is 225 units,

EXHIBIT 23A.2
Direct Materials Budget

TSC
Direct Materials Budget
October 2009–December 2009

	October	November	December
Budget production (units)	820	1,340	950
Materials requirements per unit	× 0.5	× 0.5	× 0.5
Materials needed for production (units)	410	670	475
Add budgeted ending inventory (units)	335	237.5	225
Total materials requirements (units)	745	907.5	700
Deduct beginning inventory (units)	(205)	(335)	(237.5)
Materials to be purchased (units)	540	572.5	462.5
Material price per unit	$ 20	$ 20	$ 20
Total cost of direct materials purchases	$10,800	$11,450	$9,250

computed from the direct material requirement of 450 units for a production level of 900 units in January 2010. The total materials requirements are computed by adding the desired ending inventory figures to that month's budgeted production material requirements. For October 2009, the total materials requirement is 745 units (335 + 410). From the total materials requirement, we then subtract the units of materials available in beginning inventory. For October 2009, the materials available from September 2009 are computed as 50% of October's materials requirements to satisfy production, or 205 units (50% of 410). Therefore, direct materials purchases in October 2009 are budgeted at 540 units (745 − 205). See Exhibit 23A.2.

TSC's *direct labor budget* is shown in Exhibit 23A.3. About 15 minutes of labor time is required to produce one unit. Labor is paid at the rate of $12 per hour. Budgeted labor hours are computed by multiplying the budgeted production level for each month by one-quarter (0.25) of an hour. Direct labor cost is then computed by multiplying budgeted labor hours by the labor rate of $12 per hour.

EXHIBIT 23A.3

Direct Labor Budget

TSC
Direct Labor Budget
October 2009–December 2009

	October	November	December
Budgeted production (units)	820	1,340	950
Labor requirements per unit (hours)	× 0.25	× 0.25	× 0.25
Total labor hours needed	205	335	237.5
Labor rate (per hour)	$ 12	$ 12	$ 12
Labor dollars	$2,460	$4,020	$2,850

TSC's *factory overhead budget* is shown in Exhibit 23A.4. The variable portion of overhead is assigned at the rate of $2.50 per unit of production. The fixed portion stays constant at $1,500 per month. The budget in Exhibit 23A.4 is in condensed form; most overhead budgets are more detailed, listing each overhead cost item.

EXHIBIT 23A.4

Factory Overhead Budget

TSC
Factory Overhead Budget
October 2009–December 2009

	October	November	December
Budgeted production (units)	820	1,340	950
Variable factory overhead rate	× $2.50	× $2.50	× $2.50
Budgeted variable overhead	2,050	3,350	2,375
Budgeted fixed overhead	1,500	1,500	1,500
Budgeted total overhead	$3,550	$4,850	$3,875

Summary

C1 **Describe the importance and benefits of budgeting.** Planning is a management responsibility of critical importance to business success. Budgeting is the process management uses to formalize its plans. Budgeting promotes management analysis and focuses its attention on the future. Budgeting also provides a basis for evaluating performance, serves as a source of motivation, is a means of coordinating activities, and communicates management's plans and instructions to employees.

C2 **Explain the process of budget administration.** Budgeting is a detailed activity that requires administration. At least three aspects are important: budget committee, budget reporting, and budget timing. A budget committee oversees the budget preparation. The budget period pertains to the time period for which the budget is prepared such as a year or month.

C3 **Describe a master budget and the process of preparing it.** A master budget is a formal overall plan for a company. It consists of plans for business operations and capital expenditures, plus the financial results of those activities. The budgeting process begins with a sales budget. Based on expected sales volume, companies can budget purchases, selling expenses, and administrative expenses. Next, the capital expenditures budget is prepared, followed by the cash budget and budgeted financial statements. Manufacturers also must budget production quantities, materials purchases, labor costs, and overhead.

A1 **Analyze expense planning using activity-based budgeting.** Activity-based budgeting requires management to identify activities performed by departments, plan necessary activity levels, identify resources required to perform these activities, and budget the resources.

P1 Prepare each component of a master budget and link each to the budgeting process. The term *master budget* refers to a collection of individual component budgets. Each component budget is designed to guide persons responsible for activities covered by that component. A master budget must reflect the components of a company and their interaction in pursuit of company goals.

P2 Link both operating and capital expenditures budgets to budgeted financial statements. The operating budgets, capital expenditures budget, and cash budget contain much of the information to prepare a budgeted income statement for the budget period and a budgeted balance sheet at the end of the budget period. Budgeted financial statements show the expected financial consequences of the planned activities described in the budgets.

P3 Prepare production and manufacturing budgets. A manufacturer must prepare a *production budget* instead of a purchases budget. A *manufacturing budget* shows the budgeted production costs for direct materials, direct labor, and overhead.

Guidance Answers to **Decision Maker** and **Decision Ethics**

Budget Staffer Your superior's actions appear unethical because she is using private information for personal gain. As a budget staffer, you are low in the company's hierarchical structure and probably unable to confront this superior directly. You should inform an individual with a position of authority within the organization about your concerns.

Entrepreneur You must deal with two issues. First, because fashions and designs frequently change, you cannot heavily rely on previous budgets. As a result, you must carefully analyze the market to understand what designs are in vogue. This will help you plan the product mix and estimate demand. The second issue is the budgeting period. An annual sales budget may be unreliable because tastes can quickly change. Your best bet might be to prepare monthly and quarterly sales budgets that you continuously monitor and revise.

Environmental Manager You are unlikely to have data on this new position to use in preparing your budget. In this situation, you can use activity-based budgeting. This requires developing a list of activities to conduct, the resources required to perform these activities, and the expenses associated with these resources. You should challenge yourself to be absolutely certain that the listed activities are necessary and that the listed resources are required.

Guidance Answers to **Quick Checks**

1. Major benefits include promoting a focus on the future; providing a basis for evaluating performance; providing a source of motivation; coordinating the departments of a business; and communicating plans and instructions.
2. The budget committee's responsibility is to provide guidance to ensure that budget figures are realistic and coordinated.
3. Budget periods usually coincide with accounting periods and therefore cover a month, quarter, or a year. Budgets can also be prepared for longer time periods, such as five years.
4. Rolling budgets are budgets that are periodically revised in the ongoing process of continuous budgeting.
5. A master budget is a comprehensive or overall plan for the company that is generally expressed in monetary terms.
6. *b*
7. The master budget includes operating budgets, the capital expenditures budget, and financial budgets.
8. *c*; Computed as $(60\% \times 140) + 120 - 50 = 154$.
9. Merchandisers prepare merchandise purchases budgets; manufacturers prepare production and manufacturing budgets.
10. A just-in-time system keeps the level of inventory to a minimum and orders merchandise or materials to meet immediate sales demand. A safety stock system maintains an inventory that is large enough to meet sales demands plus an amount to satisfy unexpected sales demands and an amount to cover delayed shipments from suppliers.
11. *a*
12. (a) Operating budgets (such as sales, selling expense, and administrative budgets), (b) capital expenditures budget, (c) financial budgets: cash budget, budgeted income statement, and budgeted balance sheet.

Key Terms

mhhe.com/wildFAP19e

Key Terms are available at the book's Website for learning and testing in an online Flashcard Format.

Activity-based budgeting (ABB) (p. 953)
Budget (p. 940)
Budgeted balance sheet (p. 952)
Budgeted income statement (p. 952)
Budgeting (p. 940)
Capital expenditures budget (p. 949)
Cash budget (p. 950)
Continuous budgeting (p. 942)
General and administrative expense budget (p. 948)
Manufacturing budget (p. 959)
Master budget (p. 944)
Merchandise purchases budget (p. 947)
Production budget (p. 959)
Rolling budgets (p. 942)
Safety stock (p. 947)
Sales budget (p. 946)
Selling expense budget (p. 948)

Multiple Choice Quiz

Answers on p. 977 mhhe.com/wildFAP19e

Additional Quiz Questions are available at the book's Website.

Quiz23

1. A plan that reports the units or costs of merchandise to be purchased by a merchandising company during the budget period is called a
   - **a.** Capital expenditures budget.
   - **b.** Cash budget.
   - **c.** Merchandise purchases budget.
   - **d.** Selling expenses budget.
   - **e.** Sales budget.
2. A hardware store has budgeted sales of $36,000 for its power tool department in July. Management wants to have $7,000 in power tool inventory at the end of July. Its beginning inventory of power tools is expected to be $6,000. What is the budgeted dollar amount of merchandise purchases?
   - **a.** $36,000
   - **b.** $43,000
   - **c.** $42,000
   - **d.** $35,000
   - **e.** $37,000
3. A store has the following budgeted sales for the next five months.

May	$210,000
June	186,000
July	180,000
August	220,000
September	240,000

Cash sales are 25% of total sales and all credit sales are expected to be collected in the month following the sale. The total amount of cash expected to be received from customers in September is
   - **a.** $240,000
   - **b.** $225,000
   - **c.** $ 60,000
   - **d.** $165,000
   - **e.** $220,000
4. A plan that shows the expected cash inflows and cash outflows during the budget period, including receipts from loans needed to maintain a minimum cash balance and repayments of such loans, is called
   - **a.** A rolling budget.
   - **b.** An income statement.
   - **c.** A balance sheet.
   - **d.** A cash budget.
   - **e.** An operating budget.
5. [A] The following sales are predicted for a company's next four months.

	September	October	November	December
Unit sales	480	560	600	480

Each month's ending inventory of finished goods should be 30% of the next month's sales. At September 1, the finished goods inventory is 140 units. The budgeted production of units for October is
   - **a.** 572 units.
   - **b.** 560 units.
   - **c.** 548 units.
   - **d.** 600 units.
   - **e.** 180 units.

Superscript letter [A] denotes assignments based on Appendix 23A.

Discussion Questions

1. Identify at least three roles that budgeting plays in helping managers control and monitor a business.
2. What two common benchmarks can be used to evaluate actual performance? Which of the two is generally more useful?
3. What is the benefit of continuous budgeting?
4. Identify three usual time horizons for short-term planning and budgets.
5. Why should each department participate in preparing its own budget?
6. How does budgeting help management coordinate and plan business activities?
7. Why is the sales budget so important to the budgeting process?
8. What is a selling expense budget? What is a capital expenditures budget?
9. Budgeting promotes good decision making by requiring managers to conduct ______ and by focusing their attention on the ______.
10. What is a cash budget? Why must operating budgets and the capital expenditures budget be prepared before the cash budget?
11. [A] What is the difference between a production budget and a manufacturing budget?
12. Would a manager of a **Best Buy** retail store participate more in budgeting than a manager at the corporate offices? Explain.

13. Does the manager of a local **Circuit City** retail store participate in long-term budgeting? Explain.
14. Assume that **Apple**'s iMac division is charged with preparing a master budget. Identify the participants—for example, the sales manager for the sales budget—and describe the information each person provides in preparing the master budget.

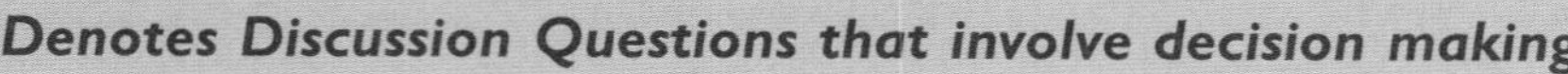

Denotes Discussion Questions that involve decision making.

McGraw-Hill's HOMEWORK MANAGER® Available with McGraw-Hill's Homework Manager

QUICK STUDY

QS 23-1
Components of a master budget C3

Which one of the following sets of items are all necessary components of the master budget?

1. Prior sales reports, capital expenditures budget, and financial budgets.
2. Sales budget, operating budgets, and historical financial budgets.
3. Operating budgets, financial budgets, and capital expenditures budget.
4. Operating budgets, historical income statement, and budgeted balance sheet.

QS 23-2
Budget motivation C1

The motivation of employees is one goal of budgeting. Identify three guidelines that organizations should follow if budgeting is to serve effectively as a source of motivation for employees.

QS 23-3
Purchases budget P1

Brill Company's July sales budget calls for sales of $800,000. The store expects to begin July with $30,000 of inventory and to end the month with $35,000 of inventory. Gross margin is typically 40% of sales. Determine the budgeted cost of merchandise purchases for July.

QS 23-4
Budgeting process C2

Good management includes good budgeting. (1) Explain why the bottom-up approach to budgeting is considered a more successful management technique than a top-down approach. (2) Provide an example of implementation of the bottom-up approach to budgeting.

QS 23-5
Computing budgeted accounts receivable P2

RedTop Company anticipates total sales for June and July of $540,000 and $472,000, respectively. Cash sales are normally 30% of total sales. Of the credit sales, 25% are collected in the same month as the sale, 70% are collected during the first month after the sale, and the remaining 5% are collected in the second month. Determine the amount of accounts receivable reported on the company's budgeted balance sheet as of July 31.

QS 23-6
Cash budget P1 P2

Use the following information to prepare a cash budget for the month ended on March 31 for Grant Company. The budget should show expected cash receipts and cash disbursements for the month of March and the balance expected on March 31.

a. Beginning cash balance on March 1, $75,000.
b. Cash receipts from sales, $315,000.
c. Budgeted cash disbursements for purchases, $204,000.
d. Budgeted cash disbursements for salaries, $90,000.
e. Other budgeted cash expenses, $30,000.
f. Cash repayment of bank loan, $25,000.

QS 23-7
Activity-based budgeting A1

Activity-based budgeting is a budget system based on *expected activities.* (1) Describe activity-based budgeting, and explain its preparation of budgets. (2) How does activity-based budgeting differ from traditional budgeting?

QS 23-8A
Production budget P3

Luna Company manufactures watches and has a JIT policy that ending inventory must equal 8% of the next month's sales. It estimates that October's actual ending inventory will consist of 24,000 watches. November and December sales are estimated to be 300,000 and 250,000 watches, respectively. Compute the number of watches to be produced that would appear on the company's production budget for the month of November.

QS 23-9A
Factory overhead budget P3

Refer to information from QS 23-8A. Luna Company assigns variable overhead at the rate of $1.75 per unit of production. Fixed overhead equals $5,000,000 per month. Prepare a factory overhead budget for November.

QS 23-10
Sales budget P1

Tech-Cam sells miniature digital cameras for $800 each. 450 units were sold in May, and it forecasts 2% growth in unit sales each month. Determine (a) the number of camera sales and (b) the dollar amount of camera sales for the month of June.

QS 23-11
Selling expense budget P1

Refer to information from QS 23-10. Tech-Cam pays a sales manager a monthly salary of $3,000 and a commission of 7.5% of camera sales (in dollars). Prepare a selling expense budget for the month of June.

QS 23-12
Cash budget P1

Refer to information from QS 23-10. Assume 30% of Tech-Cam's sales are for cash. The remaining 70% are credit sales; these customers pay in the month following the sale. Compute the budgeted cash receipts for June.

QS 23-13
Budgeted financial statements
P2

Following are selected accounts for a company. For each account, indicate whether it will appear on a budgeted income statement (BIS) or a budgeted balance sheet (BBS). If an item will not appear on either budgeted financial statement, label it NA.

Sales	______
Administrative salaries paid	______
Accumulated depreciation	______
Depreciation expense	______
Interest paid on bank loan	______
Cash dividends paid	______
Bank loan owed	______

Available with McGraw-Hill's Homework Manager

McGraw-Hill's HOMEWORK MANAGER®

EXERCISES

Exercise 23-1
Preparation of merchandise purchases budgets (for three periods)
C3 P1

Troy Company prepares monthly budgets. The current budget plans for a September ending inventory of 38,000 units. Company policy is to end each month with merchandise inventory equal to a specified percent of budgeted sales for the following month. Budgeted sales and merchandise purchases for the three most recent months follow. (1) Prepare the merchandise purchases budget for the months of July, August, and September. (2) Compute the ratio of ending inventory to the next month's sales for each budget prepared in part 1. (3) How many units are budgeted for sale in October?

	Sales (Units)	Purchases (Units)
July	170,000	200,000
August	320,000	312,000
September	280,000	262,000

Check July budgeted ending inventory, 64,000

Exercise 23-2
Preparation of cash budgets (for three periods)
C3 P2

Franke Co. budgeted the following cash receipts and cash disbursements for the first three months of next year.

	Cash Receipts	Cash Disbursements
January	$525,000	$484,000
February	411,000	350,000
March	456,000	520,000

According to a credit agreement with the company's bank, Franke promises to have a minimum cash balance of $20,000 at each month-end. In return, the bank has agreed that the company can borrow up to $160,000 at an annual interest rate of 12%, paid on the last day of each month. The interest is computed based on the beginning balance of the loan for the month. The company has a cash balance of $20,000 and a loan balance of $40,000 at January 1. Prepare monthly cash budgets for each of the first three months of next year.

Check January ending cash balance, $20,600

Exercise 23-3
Preparation of a cash budget
C3 P2

Use the following information to prepare the July cash budget for Anker Co. It should show expected cash receipts and cash disbursements for the month and the cash balance expected on July 31.

a. Beginning cash balance on July 1: $63,000.

b. Cash receipts from sales: 30% is collected in the month of sale, 50% in the next month, and 20% in the second month after sale (uncollectible accounts are negligible and can be ignored). Sales amounts are: May (actual), $1,700,000; June (actual), $1,200,000; and July (budgeted), $1,400,000.

c. Payments on merchandise purchases: 90% in the month of purchase and 10% in the month following purchase. Purchases amounts are: June (actual), $620,000; and July (budgeted), $790,000.
d. Budgeted cash disbursements for salaries in July: $220,000.
e. Budgeted depreciation expense for July: $11,000.
f. Other cash expenses budgeted for July: $230,000.
g. Accrued income taxes due in July: $50,000.
h. Bank loan interest due in July: $7,000.

Check Ending cash balance, $143,000

Exercise 23-4
Preparing a budgeted income statement and balance sheet
C3 P2

Use the information in Exercise 23-3 and the following additional information to prepare a budgeted income statement for the month of July and a budgeted balance sheet for July 31.
a. Cost of goods sold is 60% of sales.
b. Inventory at the end of June is $80,000 and at the end of July is $30,000.
c. Salaries payable on June 30 are $50,000 and are expected to be $60,000 on July 31.
d. The equipment account balance is $1,600,000 on July 31. On June 30, the accumulated depreciation on equipment is $280,000.
e. The $7,000 cash payment of interest represents the 1% monthly expense on a bank loan of $700,000.
f. Income taxes payable on July 31 are $24,600, and the income tax rate applicable to the company is 30%.
g. The only other balance sheet accounts are: Common Stock, with a balance of $850,000 on June 30; and Retained Earnings, with a balance of $931,000 on June 30.

Check Net income, $57,400; Total assets, $2,702,000

Exercise 23-5
Computing budgeted cash payments for purchases
C3 P2

DeVon Company's cost of goods sold is consistently $30 per unit. The company plans to carry ending merchandise inventory for each month equal to 20% of the next month's budgeted unit sales; August beginning inventory is 2,000 units. All merchandise is purchased on credit, and 40% of the purchases made during a month is paid for in that month. Another 25% is paid for during the first month after purchase, and the remaining 35% is paid for during the second month after purchase. Expected unit sales are: August (actual), 10,000; September (actual), 9,500; October (estimated), 8,750; November (estimated), 8,250. Use this information to determine October's expected cash payments for purchases. (*Hint:* Use the layout of Exhibit 23.8, but revised for the facts given here.)

Check Budgeted purchases: August, $297,000; October, $259,500

Exercise 23-6
Computing budgeted purchases and costs of goods sold
C3 P1 P2

Dollar Value Company purchases all merchandise on credit. It recently budgeted the following month-end accounts payable balances and merchandise inventory balances. Cash payments on accounts payable during each month are expected to be: May, $1,500,000; June, $1,530,000; July, $1,350,000; and August, $1,495,000. Use the available information to compute the budgeted amounts of (1) merchandise purchases for June, July, and August, and (2) cost of goods sold for June, July, and August.

	Accounts Payable	Merchandise Inventory
May 31	$120,000	$250,000
June 30	170,000	200,000
July 31	300,000	250,000
August 31	150,000	350,000

Check June purchases, $1,580,000; June cost of goods sold, $1,630,000

Exercise 23-7
Computing budgeted accounts payable and purchases—sales forecast in dollars
P1 P2

E-Sound, a merchandising company specializing in home computer speakers, budgets its monthly cost of goods sold to equal 50% of sales. Its inventory policy calls for ending inventory in each month to equal 40% of the next month's budgeted cost of goods sold. All purchases are on credit, and 40% of the purchases in a month is paid for in the same month. Another 40% is paid for during the first month after purchase, and the remaining 20% is paid for in the second month after purchase. The following sales budgets are set: July, $200,000; August, $140,000; September, $170,000; October, $125,000; and

Check July purchases, $88,000; Sept. payments on accts. pay., $78,400

November, $115,000. Compute the following: (1) budgeted merchandise purchases for July, August, September, and October; (2) budgeted payments on accounts payable for September and October; and (3) budgeted ending balances of accounts payable for September and October. (*Hint:* For part 1, refer to Exhibits 23.7 and 23.8 for guidance, but note that budgeted sales are in dollars for this assignment.)

Exercise 23-8A
Preparing production budgets (for two periods) P3

Check Second quarter production, 381,300 units

Electro Company manufactures an innovative automobile transmission for electric cars. Management predicts that ending inventory for the first quarter will be 38,500 units. The following unit sales of the transmissions are expected during the rest of the year: second quarter, 221,000 units; third quarter, 497,000 units; and fourth quarter, 243,500 units. Company policy calls for the ending inventory of a quarter to equal 40% of the next quarter's budgeted sales. Prepare a production budget for both the second and third quarters that shows the number of transmissions to manufacture.

Exercise 23-9A
Direct materials budget P3

Refer to information from Exercise 23-8A. Electro Company reports direct materials requirements of 0.60 per unit. It also aims to end each quarter with an ending inventory of direct materials equal to 40% of next quarter's budgeted materials requirements. Direct materials cost $175 per unit. Prepare a direct materials budget for the second quarter.

Exercise 23-10A
Direct labor budget P3

Refer to information from Exercise 23-8A. Each transmission requires 2 direct labor hours, at a cost of $18 per hour. Prepare a direct labor budget for the second quarter.

Available with McGraw-Hill's Homework Manager — McGraw-Hill's HOMEWORK MANAGER®

PROBLEM SET A

Problem 23-1A
Preparation and analysis of merchandise purchases budgets
C3 P1

mhhe.com/wildFAP19e

Herron Supply is a merchandiser of three different products. The company's February 28 inventories are footwear, 18,500 units; sports equipment, 80,000 units; and apparel, 50,000 units. Management believes that excessive inventories have accumulated for all three products. As a result, a new policy dictates that ending inventory in any month should equal 29% of the expected unit sales for the following month. Expected sales in units for March, April, May, and June follow.

	Budgeted Sales in Units			
	March	April	May	June
Footwear	15,000	26,500	31,500	35,000
Sports equipment	70,500	89,000	96,000	89,500
Apparel	40,000	38,000	34,000	23,000

Required

Check (1) March budgeted purchases Footwear, 4,185; Sports equip., 16,310; Apparel, 1,020

1. Prepare a merchandise purchases budget (in units) for each product for each of the months of March, April, and May.

Analysis Component

2. The purchases budgets in part 1 should reflect fewer purchases of all three products in March compared to those in April and May. What factor caused fewer purchases to be planned? Suggest business conditions that would cause this factor to both occur and impact the company in this way.

Problem 23-2A
Preparation of cash budgets (for three periods) C3 P2

mhhe.com/wildFAP19e

During the last week of August, Muir Company's owner approaches the bank for a $100,000 loan to be made on September 2 and repaid on November 30 with annual interest of 12%, for an interest cost of $3,000. The owner plans to increase the store's inventory by $80,000 during September and needs the loan to pay for inventory acquisitions. The bank's loan officer needs more information about Muir's ability to repay the loan and asks the owner to forecast the store's November 30 cash position. On September 1, Muir is expected to have a $4,000 cash balance, $152,000 of accounts receivable, and $115,000 of accounts payable. Its budgeted sales, merchandise purchases, and various cash disbursements for the next three months follow.

Budgeted Figures*	September	October	November
Sales	$350,000	$400,000	$425,000
Merchandise purchases	275,000	185,000	180,000
Cash disbursements			
Payroll	25,000	30,000	35,000
Rent	12,000	12,000	12,000
Other cash expenses	38,000	29,000	24,500
Repayment of bank loan			100,000
Interest on the bank loan			3,000

* Operations began in August; August sales were $200,000 and purchases were $115,000.

The budgeted September merchandise purchases include the inventory increase. All sales are on account. The company predicts that 24% of credit sales is collected in the month of the sale, 44% in the month following the sale, 21% in the second month, 8% in the third, and the remainder is uncollectible. Applying these percents to the August credit sales, for example, shows that $88,000 of the $200,000 will be collected in September, $42,000 in October, and $16,000 in November. All merchandise is purchased on credit; 85% of the balance is paid in the month following a purchase, and the remaining 15% is paid in the second month. For example, of the $115,000 August purchases, $97,750 will be paid in September and $17,250 in October.

Required

Prepare a cash budget for September, October, and November for Muir Company. Show supporting calculations as needed.

Check Budgeted cash balance: September, $103,250; October, $73,250; November, $67,750

Problem 23-3A
Preparation and analysis of cash budgets with supporting inventory and purchases budgets
C3 P2

Culver Company sells its product for $165 per unit. Its actual and projected sales follow.

	Units	Dollars
April (actual)	4,000	$ 660,000
May (actual)	2,200	363,000
June (budgeted)	5,000	825,000
July (budgeted)	6,500	1,072,500
August (budgeted)	3,700	610,500

All sales are on credit. Recent experience shows that 28% of credit sales is collected in the month of the sale, 42% in the month after the sale, 25% in the second month after the sale, and 5% proves to be uncollectible. The product's purchase price is $110 per unit. All purchases are payable within 10 days. Thus, 60% of purchases made in a month is paid in that month and the other 40% is paid in the next month. The company has a policy to maintain an ending monthly inventory of 19% of the next month's unit sales plus a safety stock of 135 units. The April 30 and May 31 actual inventory levels are consistent with this policy. Selling and administrative expenses for the year are $1,140,000 and are paid evenly throughout the year in cash. The company's minimum cash balance at month-end is $60,000. This minimum is maintained, if necessary, by borrowing cash from the bank. If the balance exceeds $60,000, the company repays as much of the loan as it can without going below the minimum. This type of loan carries an annual 12% interest rate. On May 31, the loan balance is $39,000, and the company's cash balance is $60,000.

Required

1. Prepare a table that shows the computation of cash collections of its credit sales (accounts receivable) in each of the months of June and July.
2. Prepare a table that shows the computation of budgeted ending inventories (in units) for April, May, June, and July.
3. Prepare the merchandise purchases budget for May, June, and July. Report calculations in units and then show the dollar amount of purchases for each month.

Check (1) Cash collections: June, $548,460; July, $737,550

(3) Budgeted purchases: May, $300,520; June, $581,350

4. Prepare a table showing the computation of cash payments on product purchases for June and July.

(5) Budgeted ending loan balance: June, $54,948; July, $39,375

5. Prepare a cash budget for June and July, including any loan activity and interest expense. Compute the loan balance at the end of each month.

Analysis Component

6. Refer to your answer to part 5. Culver's cash budget indicates the company will need to borrow more than $15,000 in June and will be able to pay most of it back in July. Suggest some reasons that knowing this information in May would be helpful to management.

Problem 23-4A
Preparation and analysis of budgeted income statements
C3 P2 

Poole, a one-product mail-order firm, buys its product for $75 per unit and sells it for $140 per unit. The sales staff receives a 10% commission on the sale of each unit. Its December income statement follows.

POOLE COMPANY Income Statement For Month Ended December 31, 2009	
Sales	$1,400,000
Cost of goods sold	750,000
Gross profit	650,000
Expenses	
Sales commissions (10%)	140,000
Advertising	215,000
Store rent	26,000
Administrative salaries	42,000
Depreciation	52,000
Other expenses	13,000
Total expenses	488,000
Net income	$ 162,000

Management expects December's results to be repeated in January, February, and March of 2010 without any changes in strategy. Management, however, has an alternative plan. It believes that unit sales will increase at a rate of 10% *each* month for the next three months (beginning with January) if the item's selling price is reduced to $125 per unit and advertising expenses are increased by 15% and remain at that level for all three months. The cost of its product will remain at $75 per unit, the sales staff will continue to earn a 10% commission, and the remaining expenses will stay the same.

Required

Check (1) Budgeted net income: January, $32,250; February, $73,500; March, $118,875

1. Prepare budgeted income statements for each of the months of January, February, and March that show the expected results from implementing the proposed changes. Use a three-column format, with one column for each month.

Analysis Component

2. Use the budgeted income statements from part 1 to recommend whether management should implement the proposed changes. Explain.

Problem 23-5A
Preparation of a complete master budget
C2 C3 P1 P2

Near the end of 2009, the management of Nygaard Sports Co., a merchandising company, prepared the following estimated balance sheet for December 31, 2009.

NYGAARD SPORTS COMPANY Estimated Balance Sheet December 31, 2009		
Assets		
Cash		$ 35,000
Accounts receivable		520,000
Inventory		142,500
Total current assets		697,500
Equipment	$540,000	
Less accumulated depreciation	67,500	472,500
Total assets		$1,170,000

[continued on next page]

[continued from previous page]

Liabilities and Equity		
Accounts payable	$345,000	
Bank loan payable	14,000	
Taxes payable (due 3/15/2010)	91,000	
Total liabilities		$ 450,000
Common stock	473,000	
Retained earnings	247,000	
Total stockholders' equity		720,000
Total liabilities and equity		$1,170,000

To prepare a master budget for January, February, and March of 2010, management gathers the following information.

a. Nygaard Sports' single product is purchased for $30 per unit and resold for $53 per unit. The expected inventory level of 4,750 units on December 31, 2009, is more than management's desired level for 2010, which is 20% of the next month's expected sales (in units). Expected sales are: January, 7,500 units; February, 9,250 units; March, 10,750 units; and April, 10,500 units.

b. Cash sales and credit sales represent 20% and 80%, respectively, of total sales. Of the credit sales, 57% is collected in the first month after the month of sale and 43% in the second month after the month of sale. For the December 31, 2009, accounts receivable balance, $130,000 is collected in January and the remaining $390,000 is collected in February.

c. Merchandise purchases are paid for as follows: 20% in the first month after the month of purchase and 80% in the second month after the month of purchase. For the December 31, 2009, accounts payable balance, $70,000 is paid in January and the remaining $275,000 is paid in February.

d. Sales commissions equal to 20% of sales are paid each month. Sales salaries (excluding commissions) are $72,000 per year.

e. General and administrative salaries are $156,000 per year. Maintenance expense equals $2,100 per month and is paid in cash.

f. Equipment reported in the December 31, 2009, balance sheet was purchased in January 2009. It is being depreciated over eight years under the straight-line method with no salvage value. The following amounts for new equipment purchases are planned in the coming quarter: January, $36,000; February, $96,000; and March, $28,800. This equipment will be depreciated under the straight-line method over eight years with no salvage value. A full month's depreciation is taken for the month in which equipment is purchased.

g. The company plans to acquire land at the end of March at a cost of $155,000, which will be paid with cash on the last day of the month.

h. Nygaard Sports has a working arrangement with its bank to obtain additional loans as needed. The interest rate is 12% per year, and interest is paid at each month-end based on the beginning balance. Partial or full payments on these loans can be made on the last day of the month. The company has agreed to maintain a minimum ending cash balance of $25,000 in each month.

i. The income tax rate for the company is 43%. Income taxes on the first quarter's income will not be paid until April 15.

Required

Prepare a master budget for each of the first three months of 2010; include the following component budgets (show supporting calculations as needed, and round amounts to the nearest dollar):

1. Monthly sales budgets (showing both budgeted unit sales and dollar sales).

2. Monthly merchandise purchases budgets.

3. Monthly selling expense budgets.

4. Monthly general and administrative expense budgets.

5. Monthly capital expenditures budgets.

6. Monthly cash budgets.

7. Budgeted income statement for the entire first quarter (not for each month).

8. Budgeted balance sheet as of March 31, 2010.

Check (2) Budgeted purchases: January, $138,000; February, $286,500
(3) Budgeted selling expenses: January, $85,500; February, $104,050

(6) Ending cash bal.: January, $25,000; February, $175,308
(8) Budgeted total assets at March 31, $1,527,448

Problem 23-6A[A]
Preparing production and direct materials budgets
C3 P3

Black Diamond Company produces snow skis. Each ski requires 2 pounds of carbon fiber. The company's management predicts that 4,800 skis and 6,100 pounds of carbon fiber will be in inventory on June 30 of the current year and that 152,000 skis will be sold during the next (third) quarter. Management wants to end the third quarter with 3,700 skis and 4,200 pounds of carbon fiber in inventory. Carbon fiber can be purchased for $15 per pound.

Required

Check (1) Units manuf., 150,900; (2) Cost of carbon fiber purchases, $4,498,500

1. Prepare the third-quarter production budget for skis.
2. Prepare the third-quarter direct materials (carbon fiber) budget; include the dollar cost of purchases.

PROBLEM SET B

Problem 23-1B
Preparation and analysis of merchandise purchases budgets
C3 P1

Water Sports Corp. is a merchandiser of three different products. The company's March 31 inventories are water skis, 60,000 units; tow ropes, 45,000 units; and life jackets, 75,000 units. Management believes that excessive inventories have accumulated for all three products. As a result, a new policy dictates that ending inventory in any month should equal 10% of the expected unit sales for the following month. Expected sales in units for April, May, June, and July follow.

	Budgeted Sales in Units			
	April	May	June	July
Water skis	105,000	135,000	195,000	150,000
Tow ropes	50,000	45,000	55,000	50,000
Life jackets	80,000	95,000	100,000	60,000

Required

Check (1) April budgeted purchases: Water skis, 58,500; Tow ropes, 9,500; Life jackets, 14,500

1. Prepare a merchandise purchases budget (in units) for each product for each of the months of April, May, and June.

Analysis Component

2. The purchases budgets in part 1 should reflect fewer purchases of all three products in April compared to those in May and June. What factor caused fewer purchases to be planned? Suggest business conditions that would cause this factor to both occur and affect the company as it has.

Problem 23-2B
Preparation of cash budgets (for three periods)
C3 P2

During the last week of March, Harlan Stereo's owner approaches the bank for an $80,000 loan to be made on April 1 and repaid on June 30 with annual interest of 12%, for an interest cost of $2,400. The owner plans to increase the store's inventory by $120,000 in April and needs the loan to pay for inventory acquisitions. The bank's loan officer needs more information about Harlan Stereo's ability to repay the loan and asks the owner to forecast the store's June 30 cash position. On April 1, Harlan Stereo is expected to have a $6,000 cash balance, $270,000 of accounts receivable, and $200,000 of accounts payable. Its budgeted sales, merchandise purchases, and various cash disbursements for the next three months follow.

	Budgeted Figures*	April	May	June
2	Sales	$440,000	$600,000	$760,000
3	Merchandise purchases	420,000	360,000	440,000
4	Cash disbursements			
5	Payroll	32,000	34,000	36,000
6	Rent	12,000	12,000	12,000
7	Other cash expenses	128,000	16,000	14,000
8	Repayment of bank loan			80,000
9	Interest on the bank loan.........			2,400

* Operations began in March; March sales were $360,000 and purchases were $200,000.

The budgeted April merchandise purchases include the inventory increase. All sales are on account. The company predicts that 25% of credit sales is collected in the month of the sale, 45% in the month following the sale, 20% in the second month, 9% in the third, and the remainder is uncollectible. Applying these percents to the March credit sales, for example, shows that $162,000 of the $360,000 will be collected in April, $72,000 in May, and $32,400 in June. All merchandise is purchased on credit; 80% of the balance is paid in the month following a purchase and the remaining 20% is paid in the second month. For example, of the $200,000 March purchases, $160,000 will be paid in April and $40,000 in May.

Required

Prepare a cash budget for April, May, and June for Harlan Stereo. Show supporting calculations as needed.

Check Budgeted cash balance: April, $26,000; May, $8,000; June, $72,000

Problem 23-3B
Preparation and analysis of cash budgets with supporting inventory and purchases budgets
C3 P2 

Parador Company sells its product for $22 per unit. Its actual and projected sales follow.

	Units	Dollars
January (actual)	9,000	$198,000
February (actual)	11,250	247,500
March (budgeted)	9,500	209,000
April (budgeted)	9,375	206,250
May (budgeted)	10,500	231,000

All sales are on credit. Recent experience shows that 40% of credit sales is collected in the month of the sale, 35% in the month after the sale, 23% in the second month after the sale, and 2% proves to be uncollectible. The product's purchase price is $12 per unit. All purchases are payable within 21 days. Thus, 30% of purchases made in a month is paid in that month and the other 70% is paid in the next month. The company has a policy to maintain an ending monthly inventory of 20% of the next month's unit sales plus a safety stock of 100 units. The January 31 and February 28 actual inventory levels are consistent with this policy. Selling and administrative expenses for the year are $960,000 and are paid evenly throughout the year in cash. The company's minimum cash balance for month-end is $25,000. This minimum is maintained, if necessary, by borrowing cash from the bank. If the balance exceeds $25,000, the company repays as much of the loan as it can without going below the minimum. This type of loan carries an annual 12% interest rate. At February 28, the loan balance is $20,000, and the company's cash balance is $25,000.

Required

1. Prepare a table that shows the computation of cash collections of its credit sales (accounts receivable) in each of the months of March and April.
2. Prepare a table showing the computations of budgeted ending inventories (units) for January, February, March, and April.
3. Prepare the merchandise purchases budget for February, March, and April. Report calculations in units and then show the dollar amount of purchases for each month.
4. Prepare a table showing the computation of cash payments on product purchases for March and April.
5. Prepare a cash budget for March and April, including any loan activity and interest expense. Compute the loan balance at the end of each month.

Check (1) Cash collections: March, $215,765; April, $212,575

(3) Budgeted purchases: February, $130,800; March, $113,700

(5) Ending cash balance: March, $25,000, April, $33,219

Analysis Component

6. Refer to your answer to part 5. Parador's cash budget indicates whether the company must borrow additional funds at the end of March. Suggest some reasons that knowing the loan needs in advance would be helpful to management.

Problem 23-4B
Preparation and analysis of budgeted income statements
C3 P2

Tech-Media buys its product for $90 and sells it for $200 per unit. The sales staff receives a 12% commission on the sale of each unit. Its June income statement follows.

TECH-MEDIA COMPANY Income Statement For Month Ended June 30, 2009	
Sales	$2,000,000
Cost of goods sold	900,000
Gross profit	1,100,000
Expenses	
Sales commissions (12%)	240,000
Advertising	225,000
Store rent	32,000
Administrative salaries	75,000
Depreciation	80,000
Other expenses	25,000
Total expenses	677,000
Net income	$ 423,000

Management expects June's results to be repeated in July, August, and September without any changes in strategy. Management, however, has another plan. It believes that unit sales will increase at a rate of 10% *each* month for the next three months (beginning with July) if the item's selling price is reduced to $180 per unit and advertising expenses are increased by 20% and remain at that level for all three months. The cost of its product will remain at $90 per unit, the sales staff will continue to earn a 12% commission, and the remaining expenses will stay the same.

Required

Check Budgeted net income: July, $270,400; August, $345,640; September, $428,404

1. Prepare budgeted income statements for each of the months of July, August, and September that show the expected results from implementing the proposed changes. Use a three-column format, with one column for each month.

Analysis Component

2. Use the budgeted income statements from part 1 to recommend whether management should implement the proposed plan. Explain.

Problem 23-5B
Preparation of a complete master budget
C2 C3 P1 P2

Near the end of 2009, the management of Pak Corp., a merchandising company, prepared the following estimated balance sheet for December 31, 2009.

PAK CORPORATION Estimated Balance Sheet December 31, 2009		
Assets		
Cash		$ 36,000
Accounts receivable		470,000
Inventory		300,000
Total current assets		806,000
Equipment	$1,080,000	
Less accumulated depreciation	135,000	945,000
Total assets		$1,751,000
Liabilities and Equity		
Accounts payable	$395,000	
Bank loan payable	25,000	
Taxes payable (due 3/15/2010)	20,000	
Total liabilities		$ 440,000
Common stock	550,000	
Retained earnings	761,000	
Total stockholders' equity		1,311,000
Total liabilities and equity		$1,751,000

To prepare a master budget for January, February, and March of 2010, management gathers the following information.

a. Pak Corp.'s single product is purchased for $30 per unit and resold for $45 per unit. The expected inventory level of 10,000 units on December 31, 2009, is more than management's desired level for 2010, which is 25% of the next month's expected sales (in units). Expected sales are: January, 12,000 units; February, 16,000 units; March, 20,000 units; and April, 18,000 units.

b. Cash sales and credit sales represent 25% and 75%, respectively, of total sales. Of the credit sales, 60% is collected in the first month after the month of sale and 40% in the second month after the month of sale. For the $470,000 accounts receivable balance at December 31, 2009, $330,000 is collected in January 2010 and the remaining $140,000 is collected in February 2010.

c. Merchandise purchases are paid for as follows: 20% in the first month after the month of purchase and 80% in the second month after the month of purchase. For the $395,000 accounts payable balance at December 31, 2009, $207,000 is paid in January 2010 and the remaining $188,000 is paid in February 2010.

d. Sales commissions equal to 20% of sales are paid each month. Sales salaries (excluding commissions) are $180,000 per year.

e. General and administrative salaries are $540,000 per year. Maintenance expense equals $6,000 per month and is paid in cash.

f. Equipment reported in the December 31, 2009, balance sheet was purchased in January 2009. It is being depreciated over 8 years under the straight-line method with no salvage value. The following amounts for new equipment purchases are planned in the coming quarter: January, $72,000; February, $96,000; and March, $28,800. This equipment will be depreciated using the straight-line method over 8 years with no salvage value. A full month's depreciation is taken for the month in which equipment is purchased.

g. The company plans to acquire land at the end of March at a cost of $150,000, which will be paid with cash on the last day of the month.

h. Pak Corp. has a working arrangement with its bank to obtain additional loans as needed. The interest rate is 12% per year, and interest is paid at each month-end based on the beginning balance. Partial or full payments on these loans can be made on the last day of the month. Pak has agreed to maintain a minimum ending cash balance of $36,000 in each month.

i. The income tax rate for the company is 30%. Income taxes on the first quarter's income will not be paid until April 15.

Required

Prepare a master budget for each of the first three months of 2010; include the following component budgets (show supporting calculations as needed, and round amounts to the nearest dollar):

1. Monthly sales budgets (showing both budgeted unit sales and dollar sales).

2. Monthly merchandise purchases budgets.

3. Monthly selling expense budgets.

4. Monthly general and administrative expense budgets.

5. Monthly capital expenditures budgets.

6. Monthly cash budgets.

7. Budgeted income statement for the entire first quarter (not for each month).

8. Budgeted balance sheet as of March 31, 2010.

Check (2) Budgeted purchases: January, $180,000; February, $510,000;
(3) Budgeted selling expenses: January, $123,000; February, $159,000

(6) Ending cash bal.: January, $36,000; February, $55,617
(8) Budgeted total assets at March 31, $2,355,317

Problem 23-6B[A]

Preparing production and direct materials budgets

C3 P3

Thorpe Company produces baseball bats. Each bat requires 3 pounds of aluminum alloy. Management predicts that 4,000 bats and 7,500 pounds of aluminum alloy will be in inventory on March 31 of the current year and that 125,000 bats will be sold during this year's second quarter. Management wants to end the second quarter with 3,000 finished bats and 6,000 pounds of aluminum alloy in inventory. Aluminum alloy can be purchased for $4 per pound.

Required

1. Prepare the second-quarter production budget for bats.

2. Prepare the second-quarter direct materials (aluminum alloy) budget; include the dollar cost of purchases.

Check (1) Units manuf., 124,000;
(2) Cost of aluminum purchases, $1,482,000

SERIAL PROBLEM

Success Systems

(This serial problem began in Chapter 1 and continues through most of the book. If previous chapter segments were not completed, the serial problem can begin at this point. It is helpful, but not necessary, to use the Working Papers that accompany the book.)

SP 23 Adriana Lopez expects second quarter 2010 sales of her new line of computer furniture to be the same as the first quarter's sales (reported below) without any changes in strategy. Monthly sales averaged 40 desk units (sales price of $1,250) and 20 chairs (sales price of $500).

SUCCESS SYSTEMS Segment Income Statement* For Quarter Ended March 31, 2010	
Sales†	$180,000
Cost of goods sold‡	115,000
Gross profit	65,000
Expenses	
Sales commissions (10%)	18,000
Advertising expenses	9,000
Other fixed expenses	18,000
Total expenses	45,000
Net income	$ 20,000

* Reflects revenue and expense activity only related to the computer furniture segment.

† Revenue: (120 desks × $1,250) + (60 chairs × $500) = $150,000 + $30,000 = $180,000

‡ Cost of goods sold: (120 desks × $750) + (60 chairs × $250) + $10,000 = $115,000

Lopez believes that sales will increase each month for the next three months (April, 48 desks, 32 chairs; May, 52 desks, 35 chairs; June, 56 desks, 38 chairs) *if* selling prices are reduced to $1,150 for desks and $450 for chairs, and advertising expenses are increased by 10% and remain at that level for all three months. The products' variable cost will remain at $750 for desks and $250 for chairs. The sales staff will continue to earn a 10% commission, the fixed manufacturing costs per month will remain at $10,000 and other fixed expenses will remain at $6,000 per month.

Required

Check (1) Budgeted income (loss): April, $(660); May, $945

1. Prepare budgeted income statements for each of the months of April, May, and June that show the expected results from implementing the proposed changes. Use a three-column format, with one column for each month.
2. Use the budgeted income statements from part 1 to recommend whether Lopez should implement the proposed changes. Explain.

BEYOND THE NUMBERS

REPORTING IN ACTION

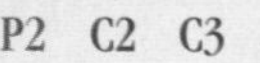

P2 C2 C3

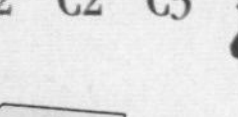

BTN 23-1 Financial statements often serve as a starting point in formulating budgets. You are assigned to review Best Buy's financial statements to determine its cash paid for dividends in the current year and the budgeted cash needed to pay its next year's dividend.

Required

1. Which financial statement(s) reports the amount of (a) cash dividends paid and (b) annual cash dividends declared? Explain where on the statement(s) this information is reported.
2. Indicate the amount of cash dividends (a) paid in the year ended March 3, 2007, and (b) to be paid (budgeted for) next year under the assumption that annual cash dividends equal 20% of the prior year's net income.

Fast Forward

3. Access Best Buy's financial statements for a fiscal year ending after March 3, 2007, from either its Website [BestBuy.com] or the SEC's EDGAR database [www.SEC.gov]. Compare your answer for part 2 with actual cash dividends paid for that fiscal year. Compute the error, if any, in your estimate. Speculate as to why dividends were higher or lower than budgeted.

COMPARATIVE ANALYSIS

P2

BTN 23-2 One source of cash savings for a company is improved management of inventory. To illustrate, assume that **Best Buy** and **Circuit City** both have $300,000 per month in sales in the Virginia area, and both forecast this level of sales per month for the next 24 months. Also assume that both Best Buy and Circuit City have a 20% contribution margin and equal fixed costs, and that cost of goods sold is the only variable cost. Assume that the main difference between Best Buy and Circuit City is the distribution system. Best Buy uses a just-in-time system and requires ending inventory of only 10% of next month's sales in inventory at each month-end. However, Circuit City is building an improved distribution system and currently requires 40% of next month's sales in inventory at each month-end.

Required

1. Compute the amount by which Circuit City can reduce its inventory level if it can match Best Buy's system of maintaining an inventory equal to 10% of next month's sales. (*Hint:* Focus on the facts given and only on the Virginia area.)

2. Explain how the analysis in part 1 that shows ending inventory levels for both the 40% and 10% required inventory policies can help justify a just-in-time inventory system. You can assume a 15% interest cost for resources that are tied up in ending inventory.

ETHICS CHALLENGE

C1 C2

BTN 23-3 Both the budget process and budgets themselves can impact management actions, both positively and negatively. For instance, a common practice among not-for-profit organizations and government agencies is for management to spend any amounts remaining in a budget at the end of the budget period, a practice often called "use it or lose it." The view is that if a department manager does not spend the budgeted amount, top management will reduce next year's budget by the amount not spent. To avoid losing budget dollars, department managers often spend all budgeted amounts regardless of the value added to products or services. All of us pay for the costs associated with this budget system.

Required

Write a one-half page report to a local not-for-profit organization or government agency offering a solution to the "use it or lose it" budgeting problem.

COMMUNICATING IN PRACTICE

P1

BTN 23-4 The sales budget is usually the first and most crucial of the component budgets in a master budget because all other budgets usually rely on it for planning purposes.

Required

Assume that your company's sales staff provides information on expected sales and selling prices for items making up the sales budget. Prepare a one-page memorandum to your supervisor outlining concerns with the sales staff's input in the sales budget when its compensation is at least partly tied to these budgets. More generally, explain the importance of assessing any potential bias in information provided to the budget process.

TAKING IT TO THE NET

C2 P1 P2

BTN 23-5 Access information on e-budgets through The Manage Mentor:
http://www.themanagementor.com/kuniverse/kmailers_universe/finance_kmailers/cfa/budgeting2.htm
Read the information provided.

Required

1. Assume the role of a senior manager in a large, multidivision company. What are the benefits of using e-budgets?
2. As a senior manager, what concerns do you have with the concept and application of e-budgets?

TEAMWORK IN ACTION

A1

BTN 23-6 Your team is to prepare a budget report outlining the costs of attending college (full-time) for the next two semesters (30 hours) or three quarters (45 hours). This budget's focus is solely on attending college; do not include personal items in the team's budget. Your budget must include tuition, books, supplies, club fees, food, housing, and all costs associated with travel to and from college. This budgeting exercise is similar to the initial phase in activity-based budgeting. Include a list of any assumptions you use in completing the budget. Be prepared to present your budget in class.

ENTREPRENEURIAL DECISION

C1

BTN 23-7 Jibbitz produces charms to fit in the holes of Crocs shoes. Assume Jibbitz is considering expanding its product line to include necklaces that hold the charms. They plan on meeting with a financial institution for potential funding and have asked by its loan officers for their business plan.

Required

1. What should Jibbitz's business plan include?
2. How can budgeting help the owners efficiently develop and operate their business?

HITTING THE ROAD

C3 P1

BTN 23-8 To help understand the factors impacting a sales budget, you are to visit three businesses with the same ownership or franchise membership. Record the selling prices of two identical products at each location, such as regular and premium gas sold at Chevron stations. You are likely to find a difference in prices for at least one of the three locations you visit.

Required

1. Identify at least three external factors that must be considered when setting the sales budget. (*Note:* There is a difference between internal and external factors that impact the sales budget.)
2. What factors might explain any differences identified in the prices of the businesses you visited?

GLOBAL DECISION

BTN 23-9 Access DSG's income statement (www.DSGiplc.com) for the year ended April 28, 2007.

Required

1. Is DSG's administrative expense budget likely to be an important budget in its master budgeting process? Explain. (*Hint:* Review its Note 3.)
2. Identify three types of expenses that would be reported as administrative expenses on DSG's income statement.
3. Who likely has the initial responsibility for DSG's administrative expense budget? Explain.

ANSWERS TO MULTIPLE CHOICE QUIZ

1. c

2. e; Budgeted purchases = $36,000 + $7,000 − $6,000 = $37,000

3. b; Cash collected = 25% of September sales + 75% of August sales = (0.25 × $240,000) + (0.75 × $220,000) = $225,000

4. d

5. a; 560 units + (0.30 × 600 units) − (0.30 × 560 units) = 572 units

A Look Back

Chapter 23 explained the master budget and its component budgets as well as their usefulness for planning and monitoring company activities.

A Look at This Chapter

This chapter describes flexible budgets, variance analysis, and standard costs. It explains how each is used for purposes of better controlling and monitoring business activities.

A Look Ahead

Chapter 25 focuses on capital budgeting decisions. It also explains and illustrates several procedures used in evaluating short-term managerial decisions.

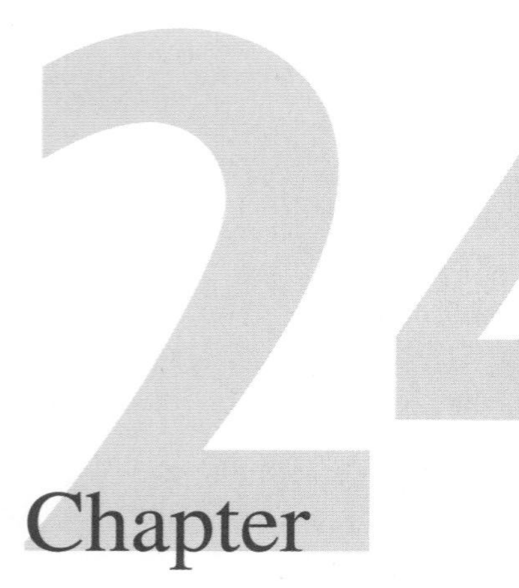

Chapter 24 Flexible Budgets and Standard Costs

Learning Objectives

CAP

Conceptual

C1 Define *standard costs* and explain their computation and uses. *(p. 985)*

C2 Describe variances and what they reveal about performance. *(p. 986)*

C3 Explain how standard cost information is useful for management by exception. *(p. 996)*

Analytical

A1 Compare fixed and flexible budgets. *(p. 982)*

A2 Analyze changes in sales from expected amounts. *(p. 998)*

LP24

Procedural

P1 Prepare a flexible budget and interpret a flexible budget performance report. *(p. 982)*

P2 Compute materials and labor variances. *(p. 988)*

P3 Compute overhead variances. *(p. 992)*

P4 Prepare journal entries for standard costs and account for price and quantity variances. *(p. 996)*

Decision Feature

Good Vibrations

"Look at each part of the process and improve it"
—Chris Martin

NAZARETH, PA—Eric Clapton. Paul McCartney. Johnny Cash. Jimi Hendrix. What do these musical legends have in common? All played guitars manufactured by the **Martin Guitar Company** **(MartinGuitar.com).** Martin manufactures high-quality guitars and recently sold its millionth. This family-owned company, headed by Christian (Chris) F. Martin, has prospered by hurdling challenges facing all manufacturers—materials quality, product design, quality control, manufacturing methods, and new investment.

Chris' entrepreneurial spirit stimulated innovative product design and growth while adhering closely to product quality. Understanding cost analysis and variances, flexible and fixed budgets, and standard costs helps his company control its production process. Martin's "X" bracing system is a key part of the distinctive Martin guitar tone. The company also embraces continuous improvement. Recently it began a lean manufacturing project to improve production efficiency, work flow, and cycle time in one of its plants.

Martin Guitar adheres to tight standards variances. Vince Gentilcore, Martin's director of quality, classifies production problems into three types: materials, process, and employee. Developing managerial accounting systems to evaluate its performance on each of these dimensions is key. "[Defects] in wood affect yield, productivity, and costs of quality," explains Vince. "We have exacting specifications and controls in place to detect problems; we don't allow material to go into a guitar that doesn't satisfy our requirements." As for process, he closely monitors the company's computer-controlled machines to ensure excessive tool wear does not impair product quality. Another key to process control, explains Vince, is "the moisture content of the wood, which we track on a regular basis." Regarding employee costs, Chris Martin explains that "we have work quotas; we know how much labor costs and how long it takes."

Achieving high standards is the goal at Martin Guitar. "We're trying to make the best," proclaims Chris. "We are doing so much more volume today, even with all those competitors. [Our workers] hold the company to an extraordinarily high standard." With standards like these, Chris' company produces a pretty tune.

[Sources: *Martin Guitar Website,* January 2009; *Quality Digest,* November 2007; *Modern Guitars Magazine,* December and March 2005; For a virtual tour of Martin Guitars see MartinGuitar.com/visit/vtour.php]

Chapter Preview

Budgeting helps organize and formalize management's planning activities. This chapter extends the study of budgeting to look more closely at the use of budgets to evaluate performance. Evaluations are important for controlling and monitoring business activities. This chapter also describes and illustrates the use of standard costs and variance analyses. These managerial tools are useful for both evaluating and controlling organizations and for the planning of future activities.

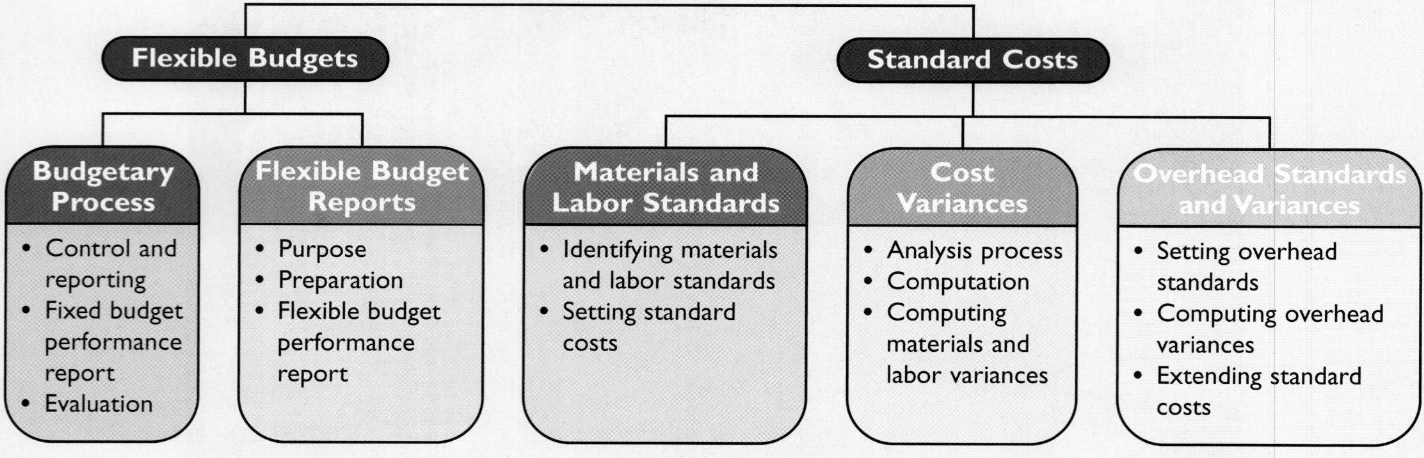

Section 1—Flexible Budgets

This section introduces fixed budgets and fixed budget performance reports. It then introduces flexible budgets and flexible budget performance reports and illustrates their advantages.

Budgetary Process

Video24.2

A master budget reflects management's planned objectives for a future period. The preparation of a master budget is based on a predicted level of activity such as sales volume for the budget period. This section discusses the effects on the usefulness of budget reports when the actual level of activity differs from the predicted level.

Budgetary Control and Reporting

Budgetary control refers to management's use of budgets to monitor and control a company's operations. This includes using budgets to see that planned objectives are met. **Budget reports** contain relevant information that compares actual results to planned activities. This comparison is motivated by a need to both monitor performance and control activities. Budget reports are sometimes viewed as progress reports, or *report cards,* on management's performance in achieving planned objectives. These reports can be prepared at any time and for any period. Three common periods for a budget report are a month, quarter, and year.

Point: Budget reports are often used as a base to determine bonuses of managers.

The budgetary control process involves at least four steps: (1) develop the budget from planned objectives, (2) compare actual results to budgeted amounts and analyze any differences, (3) take corrective and strategic actions, and (4) establish new planned objectives and prepare a new budget. Exhibit 24.1 shows this continual process of budgetary control. Budget

EXHIBIT 24.1

Process of Budgetary Control

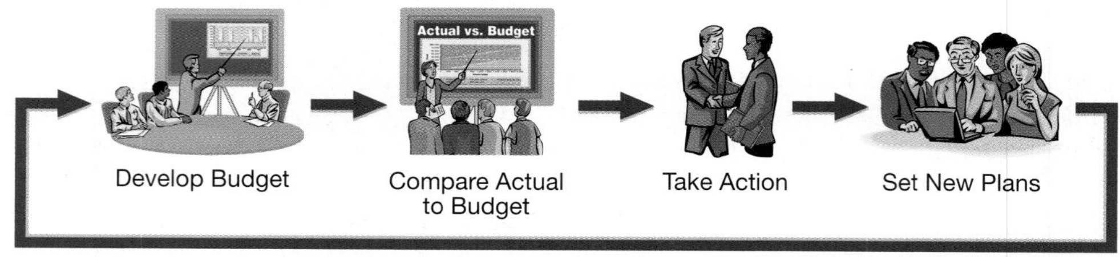

reports and related documents are effective tools for managers to obtain the greatest benefits from this budgetary process.

Fixed Budget Performance Report

In a fixed budgetary control system, the master budget is based on a single prediction for sales volume or other activity level. The budgeted amount for each cost essentially assumes that a specific (or *fixed*) amount of sales will occur. A **fixed budget,** also called a *static budget,* is based on a single predicted amount of sales or other measure of activity.

One benefit of a budget is its usefulness in comparing actual results with planned activities. Information useful for analysis is often presented for comparison in a performance report. As shown in Exhibit 24.2, a **fixed budget performance report** for **Optel** compares actual results for January 2009 with the results expected under its fixed budget that predicted 10,000 (composite) units of sales. Optel manufactures inexpensive eyeglasses, frames, contact lens, and related supplies. For this report, its production volume equals sales volume (its inventory level did not change).

EXHIBIT 24.2
Fixed Budget Performance Report

OPTEL
Fixed Budget Performance Report
For Month Ended January 31, 2009

	Fixed Budget	Actual Results	Variances*
Sales (in units)	10,000	12,000	
Sales (in dollars)	$100,000	$125,000	$25,000 F
Cost of goods sold			
Direct materials	10,000	13,000	3,000 U
Direct labor	15,000	20,000	5,000 U
Overhead			
Factory supplies	2,000	2,100	100 U
Utilities	3,000	4,000	1,000 U
Depreciation—machinery	8,000	8,000	0
Supervisory salaries	11,000	11,000	0
Selling expenses			
Sales commissions	9,000	10,800	1,800 U
Shipping expenses	4,000	4,300	300 U
General and administrative expenses			
Office supplies	5,000	5,200	200 U
Insurance expenses	1,000	1,200	200 U
Depreciation—office equipment	7,000	7,000	0
Administrative salaries	13,000	13,000	0
Total expenses	88,000	99,600	11,600 U
Income from operations	$ 12,000	$ 25,400	$13,400 F

* F = Favorable variance; U = Unfavorable variance.

This type of performance report designates differences between budgeted and actual results as variances. We see the letters *F* and *U* located beside the numbers in the third number column of this report. Their meanings are as follows:

F = **Favorable variance** When compared to budget, the actual cost or revenue contributes to a *higher* income. That is, actual revenue is higher than budgeted revenue, or actual cost is lower than budgeted cost.

U = **Unfavorable variance** When compared to budget, the actual cost or revenue contributes to a *lower* income; actual revenue is lower than budgeted revenue, or actual cost is higher than budgeted cost.

This convention is common in practice and is used throughout this chapter.

Example: How is it that the favorable sales variance in Exhibit 24.2 is linked with so many unfavorable cost and expense variances? *Answer:* Costs have increased with the increase in sales.

Budget Reports for Evaluation

A primary use of budget reports is as a tool for management to monitor and control operations. Evaluation by Optel management is likely to focus on a variety of questions that might include these:

- Why is actual income from operations $13,400 higher than budgeted?
- Are amounts paid for each expense item too high?
- Is manufacturing using too much direct material?
- Is manufacturing using too much direct labor?

The performance report in Exhibit 24.2 provides little help in answering these questions because actual sales volume is 2,000 units higher than budgeted. A manager does not know if this higher level of sales activity is the cause of variations in total dollar sales and expenses or if other factors have influenced these amounts. This inability of fixed budget reports to adjust for changes in activity levels is a major limitation of a fixed budget performance report. That is, it fails to show whether actual costs are out of line due to a change in actual sales volume or some other factor.

Decision Insight

Green Budget Budget reporting and evaluation are used at the **Environmental Protection Agency (EPA)**. It regularly prepares performance plans and budget requests that describe performance goals, measure outcomes, and analyze variances.

Flexible Budget Reports

Purpose of Flexible Budgets

A1 Compare fixed and flexible budgets.

To help address limitations with the fixed budget performance report, particularly from the effects of changes in sales volume, management can use a flexible budget. A **flexible budget,** also called a *variable budget,* is a report based on predicted amounts of revenues and expenses corresponding to the actual level of output. Flexible budgets are useful both before and after the period's activities are complete.

Video24.2

A flexible budget prepared before the period is often based on several levels of activity. Budgets for those different levels can provide a "what-if" look at operations. The different levels often include both a best case and worst case scenario. This allows management to make adjustments to avoid or lessen the effects of the worst case scenario.

A flexible budget prepared after the period helps management evaluate past performance. It is especially useful for such an evaluation because it reflects budgeted revenues and costs based on the actual level of activity. Thus, comparisons of actual results with budgeted performance are more likely to identify the causes of any differences. This can help managers focus attention on real problem areas and implement corrective actions. This is in contrast to a fixed budget, whose primary purpose is to assist managers in planning future activities and whose numbers are based on a single predicted amount of budgeted sales or production.

Preparation of Flexible Budgets

P1 Prepare a flexible budget and interpret a flexible budget performance report.

A flexible budget is designed to reveal the effects of volume of activity on revenues and costs. To prepare a flexible budget, management relies on the distinctions between fixed and variable costs. Recall that the cost per unit of activity remains constant for variable costs so that the total amount of a variable cost changes in direct proportion to a change in activity level. The total amount of fixed cost remains unchanged regardless of changes in the level of activity within a relevant (normal) operating range. (Assume that costs can be reasonably classified as variable or fixed within a relevant range.)

When we create the numbers constituting a flexible budget, we express each variable cost as either a constant amount per unit of sales or as a percent of a sales dollar. In the case of a fixed cost, we express its budgeted amount as the total amount expected to occur at any sales volume within the relevant range.

Exhibit 24.3 shows a set of flexible budgets for Optel in January 2009. Seven of its expenses are classified as variable costs. Its remaining five expenses are fixed costs. These classifications result from management's investigation of each expense. Variable and fixed expense categories are *not* the same for every company, and we must avoid drawing conclusions from specific cases. For example, depending on the nature of a company's operations, office supplies expense can be either fixed or variable with respect to sales.

Point: The usefulness of a flexible budget depends on valid classification of variable and fixed costs. Some costs are mixed and must be analyzed to determine their variable and fixed portions.

EXHIBIT 24.3

Flexible Budgets

OPTEL
Flexible Budgets
For Month Ended January 31, 2009

	Flexible Budget				
	Variable Amount per Unit	Total Fixed Cost	Flexible Budget for Unit Sales of 10,000	Flexible Budget for Unit Sales of 12,000	Flexible Budget for Unit Sales of 14,000
Sales	$10.00		$100,000	$120,000	$140,000
Variable costs					
Direct materials	1.00		10,000	12,000	14,000
Direct labor	1.50		15,000	18,000	21,000
Factory supplies	0.20		2,000	2,400	2,800
Utilities	0.30		3,000	3,600	4,200
Sales commissions	0.90		9,000	10,800	12,600
Shipping expenses	0.40		4,000	4,800	5,600
Office supplies	0.50		5,000	6,000	7,000
Total variable costs	4.80		48,000	57,600	67,200
Contribution margin	$ 5.20		$ 52,000	$ 62,400	$ 72,800
Fixed costs					
Depreciation—machinery		$ 8,000	8,000	8,000	8,000
Supervisory salaries		11,000	11,000	11,000	11,000
Insurance expense		1,000	1,000	1,000	1,000
Depreciation—office equipment		7,000	7,000	7,000	7,000
Administrative salaries		13,000	13,000	13,000	13,000
Total fixed costs		$40,000	40,000	40,000	40,000
Income from operations			$ 12,000	$ 22,400	$ 32,800

The layout for the flexible budgets in Exhibit 24.3 follows a *contribution margin format*—beginning with sales followed by variable costs and then fixed costs. Both the expected individual and total variable costs are reported and then subtracted from sales. The difference between sales and variable costs equals contribution margin. The expected amounts of fixed costs are listed next, followed by the expected income from operations before taxes.

Example: Using Exhibit 24.3, what is the budgeted income from operations for unit sales of (a) 11,000 and (b) 13,000? *Answers:* $17,200 for unit sales of 11,000; $27,600 for unit sales of 13,000.

The first and second number columns of Exhibit 24.3 show the flexible budget amounts for variable costs per unit and each fixed cost for any volume of sales in the relevant range. The third, fourth, and fifth columns show the flexible budget amounts computed for three different sales volumes. For instance, the third column's flexible budget is based on 10,000 units. These numbers are the same as those in the fixed budget of Exhibit 24.2 because the expected volumes are the same for these two budgets.

Recall that Optel's actual sales volume for January is 12,000 units. This sales volume is 2,000 units more than the 10,000 units originally predicted in the master budget. When differences between actual and predicted volume arise, the usefulness of a flexible budget is apparent. For instance, compare the flexible budget for 10,000 units in the third column (which is the same as the fixed budget in Exhibit 24.2) with the flexible budget for 12,000 units in the

Point: Flexible budgeting allows a budget to be prepared at the *actual* output level. Performance reports are then prepared comparing the flexible budget to actual revenues and costs.

Point: A flexible budget yields an "apples to apples" comparison because budgeted activity levels are the same as the actual.

fourth column. The higher levels for both sales and variable costs reflect nothing more than the increase in sales activity. Any budget analysis comparing actual with planned results that ignores this information is less useful to management.

To illustrate, when we evaluate Optel's performance, we need to prepare a flexible budget showing actual and budgeted values at 12,000 units. As part of a complete profitability analysis, managers could compare the actual income of $25,400 (from Exhibit 24.2) with the $22,400 income expected at the actual sales volume of 12,000 units (from Exhibit 24.3). This results in a total favorable income variance of $3,000 to be explained and interpreted. This variance is markedly lower from the $13,400 favorable variance identified in Exhibit 24.2 using a fixed budget, but still suggests good performance. After receiving the flexible budget based on January's actual volume, management must determine what caused this $3,000 difference. The next section describes a flexible budget performance report that provides guidance in this analysis.

Decision Maker

Entrepreneur The heads of both the strategic consulting and tax consulting divisions of your financial services firm complain to you about the unfavorable variances on their performance reports. "We worked on more consulting assignments than planned. It's not surprising our costs are higher than expected. To top it off, this report characterizes our work as *poor!*" How do you respond? [Answer—p. 1004]

Flexible Budget Performance Report

A **flexible budget performance report** lists differences between actual performance and budgeted performance based on actual sales volume or other activity level. This report helps direct management's attention to those costs or revenues that differ substantially from budgeted amounts. Exhibit 24.4 shows Optel's flexible budget performance report for January. We prepare this report after the actual volume is known to be 12,000 units. This report shows a $5,000 favorable variance in total dollar sales. Because actual and budgeted volumes are both 12,000 units, the $5,000 sales variance must have resulted from a higher than expected selling price. Further analysis of the facts surrounding this $5,000 sales variance reveals a favorable sales variance per unit of nearly $0.42 as shown here:

Actual average price per unit (rounded to cents)	$125,000/12,000 =	$10.42
Budgeted price per unit	$120,000/12,000 =	10.00
Favorable sales variance per unit	$5,000/12,000 =	$ 0.42

The other variances in Exhibit 24.4 also direct management's attention to areas where corrective actions can help control Optel's operations. Each expense variance is analyzed as the sales variance was. We can think of each expense as the joint result of using a given number of units of input and paying a specific price per unit of input. Optel's expense variances total $2,000 unfavorable, suggesting poor control of some costs, particularly direct materials and direct labor.

Each variance in Exhibit 24.4 is due in part to a difference between *actual price* per unit of input and *budgeted price* per unit of input. This is a **price variance.** Each variance also can be due in part to a difference between *actual quantity* of input used and *budgeted quantity* of input. This is a **quantity variance.** We explain more about this breakdown, known as **variance analysis,** later in the standard costs section.

Quick Check

Answers—p. 1004

1. A flexible budget (*a*) shows fixed costs as constant amounts of cost per unit of activity, (*b*) shows variable costs as constant amounts of cost per unit of activity, or (*c*) is prepared based on one expected amount of budgeted sales or production.
2. What is the initial step in preparing a flexible budget?
3. What is the main difference between a fixed and a flexible budget?
4. What is the contribution margin?

EXHIBIT 24.4

Flexible Budget Performance Report

OPTEL Flexible Budget Performance Report For Month Ended January 31, 2009	Flexible Budget	Actual Results	Variances*
Sales (12,000 units)	$120,000	$125,000	$5,000 F
Variable costs			
Direct materials	12,000	13,000	1,000 U
Direct labor	18,000	20,000	2,000 U
Factory supplies	2,400	2,100	300 F
Utilities	3,600	4,000	400 U
Sales commissions	10,800	10,800	0
Shipping expenses	4,800	4,300	500 F
Office supplies	6,000	5,200	800 F
Total variable costs	57,600	59,400	1,800 U
Contribution margin	62,400	65,600	3,200 F
Fixed costs			
Depreciation—machinery	8,000	8,000	0
Supervisory salaries	11,000	11,000	0
Insurance expense	1,000	1,200	200 U
Depreciation—office equipment	7,000	7,000	0
Administrative salaries	13,000	13,000	0
Total fixed costs	40,000	40,200	200 U
Income from operations	$ 22,400	$ 25,400	$3,000 F

* F = Favorable variance; U = Unfavorable variance.

Section 2—Standard Costs

C1 Define *standard costs* and explain their computation and uses.

Standard costs are preset costs for delivering a product or service under normal conditions. These costs are established by personnel, engineering, and accounting studies using past experiences and data. Management uses these costs to assess the reasonableness of actual costs incurred for producing the product or service. When actual costs vary from standard costs, management follows up to identify potential problems and take corrective actions.

Standard costs are often used in preparing budgets because they are the anticipated costs incurred under normal conditions. Terms such as *standard materials cost, standard labor cost,* and *standard overhead cost* are often used to refer to amounts budgeted for direct materials, direct labor, and overhead.

Point: Since standard costs are often budgeted costs, they can be used to prepare both fixed budgets and flexible budgets.

Materials and Labor Standards

This section explains how to set materials and labor standards and how to prepare a standard cost card.

Video24.1

Identifying Standard Costs

Managerial accountants, engineers, personnel administrators, and other managers combine their efforts to set standard costs. To identify standards for direct labor costs, we can conduct time and motion studies for each labor operation in the process of providing a product or service. From these studies, management can learn the best way to perform the operation and then set the standard labor time required for the operation under normal conditions. Similarly, standards for materials are set by studying the quantity, grade, and cost of each material used. Standards for overhead costs are explained later in the chapter.

Point: Business practice often uses the word *budget* when speaking of total amounts and *standard* when discussing per unit amounts.

Regardless of the care used in setting standard costs and in revising them as conditions change, actual costs frequently differ from standard costs, often as a result of one or more factors. For instance, the actual quantity of material used can differ from the standard, or the price paid per unit of material can differ from the standard. Quantity and price differences from

Example: What factors might be considered when deciding whether to revise standard costs? *Answer:* Changes in the processes and/or resources needed to carry out the processes.

standard amounts can also occur for labor. That is, the actual labor time and actual labor rate can vary from what was expected. The same analysis applies to overhead costs.

Decision Insight

Cruis'n Standards The **Corvette** consists of hundreds of parts for which engineers set standards. Various types of labor are also involved in its production, including machining, assembly, painting, and welding, and standards are set for each. Actual results are periodically compared with standards to assess performance.

Setting Standard Costs

To illustrate the setting of a standard cost, we consider a professional league baseball bat manufactured by **ProBat.** Its engineers have determined that manufacturing one bat requires 0.90 kg. of high-grade wood. They also expect some loss of material as part of the process because of inefficiencies and waste. This results in adding an *allowance* of 0.10 kg., making the standard requirement 1.0 kg. of wood for each bat.

The 0.90 kg. portion is called an *ideal standard;* it is the quantity of material required if the process is 100% efficient without any loss or waste. Reality suggests that some loss of material usually occurs with any process. The standard of 1.0 kg. is known as the *practical standard,* the quantity of material required under normal application of the process.

Point: Companies promoting continuous improvement strive to achieve ideal standards by eliminating inefficiencies and waste.

High-grade wood can be purchased at a standard price of $25 per kg. The purchasing department sets this price as the expected price for the budget period. To determine this price, the purchasing department considers factors such as the quality of materials, future economic conditions, supply factors (shortages and excesses), and any available discounts. The engineers also decide that two hours of labor time (after including allowances) are required to manufacture a bat. The wage rate is $20 per hour (better than average skilled labor is required). ProBat assigns all overhead at the rate of $10 per labor hour. The standard costs of direct materials, direct labor, and overhead for one bat are shown in Exhibit 24.5 in what is called a *standard cost card.* These cost amounts are then used to prepare manufacturing budgets for a budgeted level of production.

EXHIBIT 24.5
Standard Cost Card

STANDARD COST CARD

Production factor	Cost factor	Total
Direct materials (wood)	1 kg. @ $25 per kg.	$25
Direct labor	2 hours @ $20 per hour	40
Overhead	2 labor hours @ $10 per hour	20
	Total	$85

REMARKS:
Based on standard costs of direct materials, direct labor, and overhead for a single ProBat

SUMMARY:	
Materials	$25
Labor	40
Overhead	20
Total cost	$85

Cost Variances

C2 Describe variances and what they reveal about performance.

A **cost variance,** also simply called a *variance,* is the difference between actual and standard costs. A cost variance can be favorable or unfavorable. A variance from standard cost is considered favorable if actual cost is less than standard cost. It is considered unfavorable if actual cost is more than standard cost.[1] This section discusses variance analysis.

[1] Short-term favorable variances can sometimes lead to long-term unfavorable variances. For instance, if management spends less than the budgeted amount on maintenance or insurance, the performance report would show a favorable variance. Cutting these expenses can lead to major losses in the long run if machinery wears out prematurely or insurance coverage proves inadequate.

Cost Variance Analysis

Video24.1

Variances are usually identified in performance reports. When a variance occurs, management wants to determine the factors causing it. This often involves analysis, evaluation, and explanation. The results of these efforts should enable management to assign responsibility for the variance and then to take actions to correct the situation.

To illustrate, ProBat's standard materials cost for producing 500 bats is $12,500. Assume, that its actual materials cost for those 500 bats proved to be $13,000. The $500 unfavorable variance raises questions that call for answers that, in turn, can lead to changes to correct the situation and eliminate this variance in the next period. A performance report often identifies the existence of a problem, but we must follow up with further investigation to see what can be done to improve future performance.

Exhibit 24.6 shows the flow of events in the effective management of variance analysis. It shows four steps: (1) preparing a standard cost performance report, (2) computing and analyzing variances, (3) identifying questions and their explanations, and (4) taking corrective and strategic actions. These variance analysis steps are interrelated and are frequently applied in good organizations.

EXHIBIT 24.6

Variance Analysis

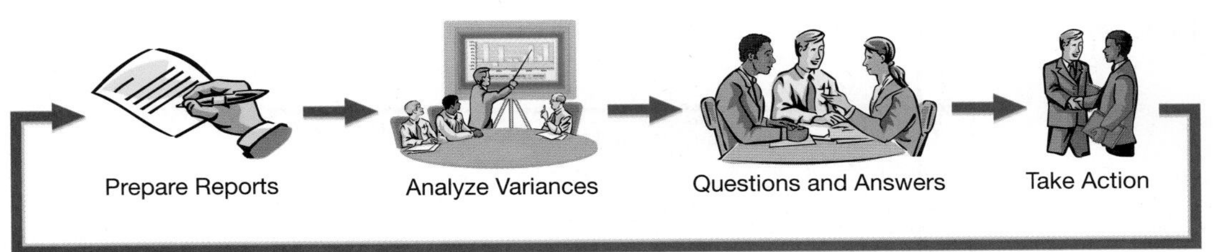

Cost Variance Computation

Management needs information about the factors causing a cost variance, but first it must properly compute the variance. In its most simple form, a cost variance (CV) is computed as the difference between actual cost (AC) and standard cost (SC) as shown in Exhibit 24.7.

EXHIBIT 24.7

Cost Variance Formulas

Cost Variance (CV) = **Actual Cost** (AC) − **Standard Cost** (SC)

where:

Actual Cost (AC) = **Actual Quantity** (AQ) × **Actual Price** (AP)

Standard Cost (SC) = **Standard Quantity** (SQ) × **Standard Price** (SP)

A cost variance is further defined by its components. Actual quantity (AQ) is the input (material or labor) used to manufacture the quantity of output. Standard quantity (SQ) is the expected input for the quantity of output. Actual price (AP) is the amount paid to acquire the input (material or labor), and standard price (SP) is the expected price.

Two main factors cause a cost variance: (1) the difference between actual price and standard price results in a *price* (or rate) *variance* and (2) the difference between actual quantity and standard quantity results in a *quantity* (or usage or efficiency) *variance*. To assess the impacts of these two factors in a cost variance, we use the formulas in Exhibit 24.8.

Point: Price and quantity variances for direct labor are nearly always referred to as *rate* and *efficiency variances*, respectively.

EXHIBIT 24.8

Price Variance and Quantity Variance Formulas

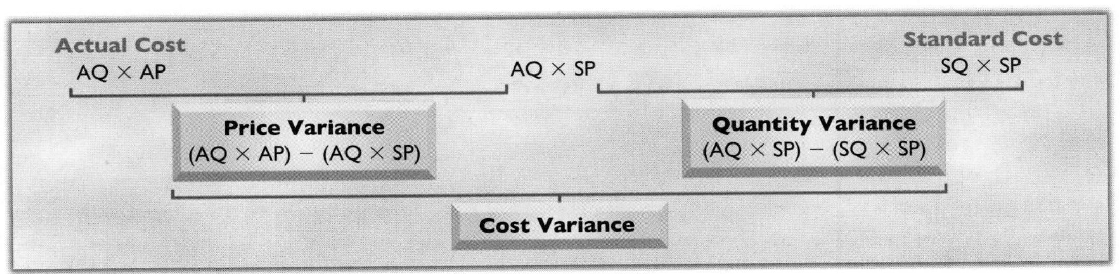

In computing a price variance, the quantity (actual) is held constant. In computing a quantity variance, the price (standard) is held constant. The cost variance, or total variance, is the sum of the price and quantity variances. These formulas identify the sources of the cost variance. Managers sometimes find it useful to apply an alternative (but equivalent) computation for the price and quantity variances as shown in Exhibit 24.9.

EXHIBIT 24.9

Alternative Price Variance and Quantity Variance Formulas

Price Variance (PV) = [**Actual Price** (AP) − **Standard Price** (SP)] × **Actual Quantity** (AQ)

Quantity Variance (QV) = [**Actual Quantity** (AQ) − **Standard Quantity** (SQ)] × **Standard Price** (SP)

The results from applying the formulas in Exhibits 24.8 and 24.9 are identical.

Computing Materials and Labor Variances

P2 Compute materials and labor variances.

We illustrate the computation of the materials and labor cost variances using data from **G-Max,** a company that makes specialty golf equipment and accessories for individual customers. This company has set the following standard quantities and costs for materials and labor per unit for one of its hand-crafted golf clubheads:

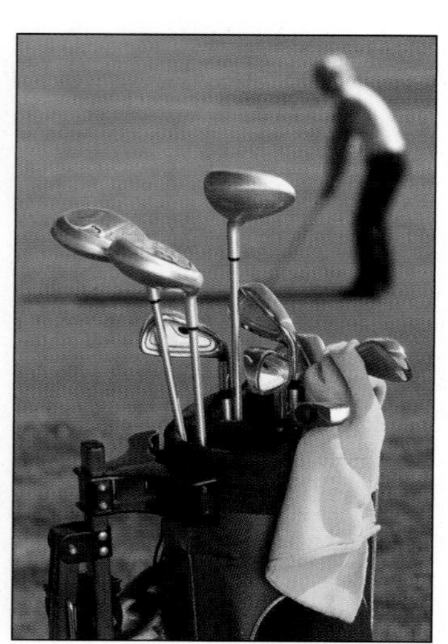

Direct materials (1 lb. per unit at $1 per lb.)	$1.00
Direct labor (1 hr. per unit at $8 per hr.)	8.00
Total standard direct cost per unit	$9.00

Materials Cost Variances During May 2009, G-Max budgeted to produce 4,000 clubheads (units). It actually produced only 3,500 units. It used 3,600 pounds of direct materials (titanium) costing $1.05 per pound, meaning its total materials cost was $3,780. This information allows us to compute both actual and standard direct materials costs for G-Max's 3,500 units and its direct materials cost variance as follows:

Actual cost	3,600 lbs. @ $1.05 per lb.	= $3,780
Standard cost	3,500 lbs. @ $1.00 per lb.	= 3,500
Direct materials cost variance (unfavorable)		= $ 280

To better isolate the causes of this $280 unfavorable total direct materials cost variance, the materials price and quantity variances for these G-Max clubheads are computed and shown in Exhibit 24.10.

EXHIBIT 24.10

Materials Price and Quantity Variances

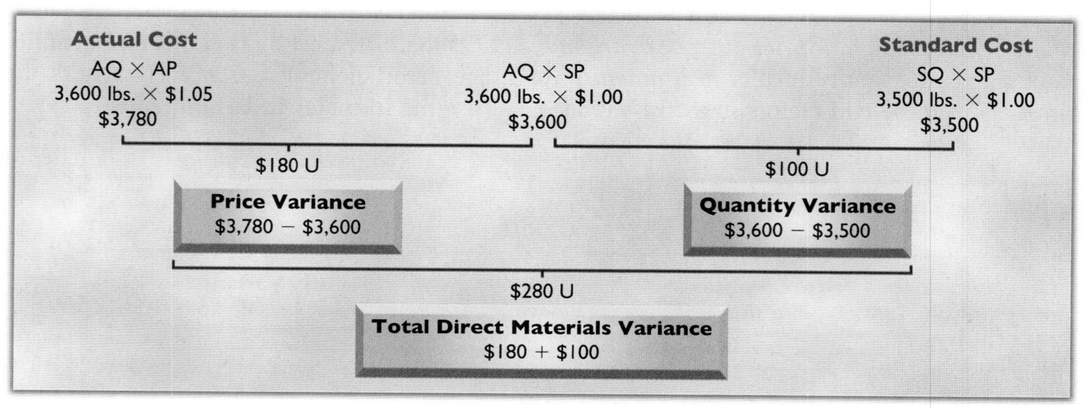

The $180 unfavorable price variance results from paying 5 cents more per unit than the standard price, computed as 3,600 lbs. × $0.05. The $100 unfavorable quantity variance is due to using 100 lbs. more materials than the standard quantity, computed as 100 lbs. × $1. The total direct materials variance is $280 and it is unfavorable. This information allows management to ask the responsible individuals for explanations and corrective actions.

Example: Identify at least two factors that might have caused the $100 unfavorable quantity variance and the $180 unfavorable price variance in Exhibit 24.10. *Answer:* Poor quality materials or untrained workers for the former; poor price negotiation or higher-quality materials for the latter.

The purchasing department is usually responsible for the price paid for materials. Responsibility for explaining the price variance in this case rests with the purchasing manager if a price higher than standard caused the variance. The production department is usually responsible for the amount of material used and in this case is responsible for explaining why the process used more than the standard amount of materials.

Variance analysis presents challenges. For instance, the production department could have used more than the standard amount of material because its quality did not meet specifications and led to excessive waste. In this case, the purchasing manager is responsible for explaining why inferior materials were acquired. However, the production manager is responsible for explaining what happened if analysis shows that waste was due to inefficiencies, not poor quality material.

In evaluating price variances, managers must recognize that a favorable price variance can indicate a problem with poor product quality. **Redhook Ale**, a micro brewery in the Pacific Northwest, can probably save 10% to 15% in material prices by buying six-row barley malt instead of the better two-row from Washington's Yakima valley. Attention to quality, however, has helped Redhook Ale become the first craft brewer to be kosher certified. Redhook's purchasing activities are judged on both the quality of the materials and the purchase price variance.

Labor Cost Variances Labor cost for a specific product or service depends on the number of hours worked (quantity) and the wage rate paid to employees (price). When actual amounts for a task differ from standard, the labor cost variance can be divided into a rate (price) variance and an efficiency (quantity) variance.

To illustrate, G-Max's direct labor standard for 3,500 units of its hand-crafted clubheads is one hour per unit, or 3,500 hours at $8 per hour. Since only 3,400 hours at $8.30 per hour were actually used to complete the units, the actual and standard labor costs are

Actual cost	3,400 hrs. @ $8.30 per hr.	= $28,220
Standard cost	3,500 hrs. @ $8.00 per hr.	= 28,000
Direct labor cost variance (unfavorable)		= $ 220

This analysis shows that actual cost is merely $220 over the standard and suggests no immediate concern. Computing both the labor rate and efficiency variances reveals a different picture, however, as shown in Exhibit 24.11.

EXHIBIT 24.11

Labor Rate and Efficiency Variances*

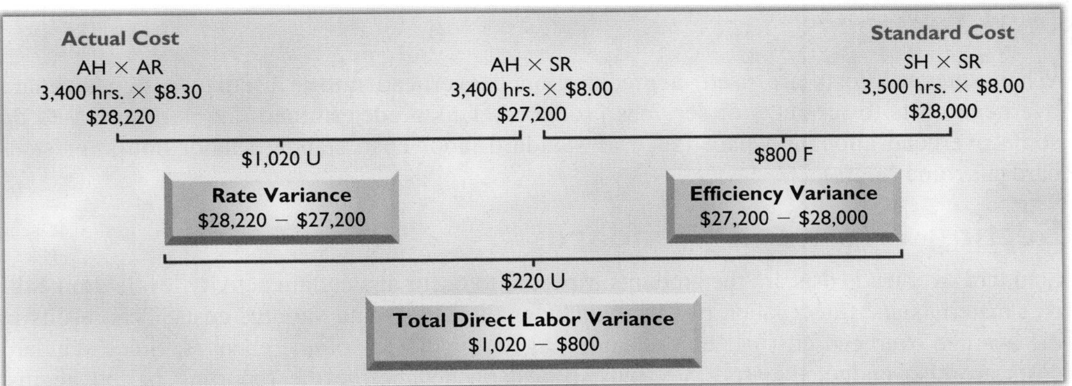

* AH is actual direct labor hours: AR is actual wage rate; SH is standard direct labor hours allowed for actual output; SR is standard wage rate.

Example: Compute the rate variance and the efficiency variance for Exhibit 24.11 if 3,700 actual hours are used at an actual price of $7.50 per hour. *Answer:* $1,850 favorable labor rate variance and $1,600 unfavorable labor efficiency variance.

The analysis in Exhibit 24.11 shows that an $800 favorable efficiency variance results from using 100 fewer direct labor hours than standard for the units produced, but this favorable variance is more than offset by a wage rate that is $0.30 per hour higher than standard. The personnel administrator or the production manager needs to explain why the wage rate is higher than expected. The production manager should also explain how the labor hours were reduced. If this experience can be repeated and transferred to other departments, more savings are possible.

One possible explanation of these labor rate and efficiency variances is the use of workers with different skill levels. If this is the reason, senior management must discuss the implications with the production manager who has the responsibility to assign workers to tasks with the appropriate skill level. In this case, an investigation might show that higher-skilled workers were used to produce 3,500 units of hand-crafted clubheads. As a result, fewer labor hours might be required for the work, but the wage rate paid these workers is higher than standard because of their greater skills. The effect of this strategy is a higher than standard total cost, which would require actions to remedy the situation or adjust the standard.

Decision Maker

Human Resource Manager You receive the manufacturing variance report for June and discover a large unfavorable labor efficiency (quantity) variance. What factors do you investigate to identify its possible causes? [Answer—p. 1004]

Quick Check

Answers—pp. 1004–1005

5. A standard cost (*a*) changes in direct proportion to changes in the level of activity, (*b*) is an amount incurred at the actual level of production for the period, or (*c*) is an amount incurred under normal conditions to provide a product or service.
6. What is a cost variance?
7. The following information is available for York Company.

Actual direct labor hours per unit	2.5 hours
Standard direct labor hours per unit	2.0 hours
Actual production (units)	2,500 units
Budgeted production (units)	3,000 units
Actual rate per hour	$3.10
Standard rate per hour	$3.00

The labor efficiency variance is (*a*) $3,750 U, (*b*) $3,750 F, or (*c*) $3,875 U.

8. Refer to Quick Check 7; the labor rate variance is (*a*) $625 F or (*b*) $625 U.
9. If a materials quantity variance is favorable and a materials price variance is unfavorable, can the total materials cost variance be favorable?

Overhead Standards and Variances

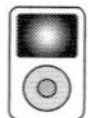

Video24.1&24.3

When standard costs are used, a predetermined overhead rate is used to assign standard overhead costs to products or services produced. This predetermined rate is often based on some overhead allocation base (such as standard labor cost, standard labor hours, or standard machine hours).

Setting Overhead Standards

Standard overhead costs are the amounts expected to occur at a certain activity level. Unlike direct materials and direct labor, overhead includes fixed costs and variable costs. This results in the average overhead cost per unit changing as the predicted volume changes. Since standard costs are also budgeted costs, they must be established before the reporting period begins. Standard overhead costs are therefore average per unit costs based on the predicted activity level.

To establish the standard overhead cost rate, management uses the same cost structure it used to construct a flexible budget at the end of a period. This cost structure identifies the different overhead cost components and classifies them as variable or fixed. To get the standard overhead rate, management selects a level of activity (volume) and predicts total overhead cost. It then divides this total by the allocation base to get the standard rate. Standard direct labor hours expected to be used to produce the predicted volume is a common allocation base and is used in this section.

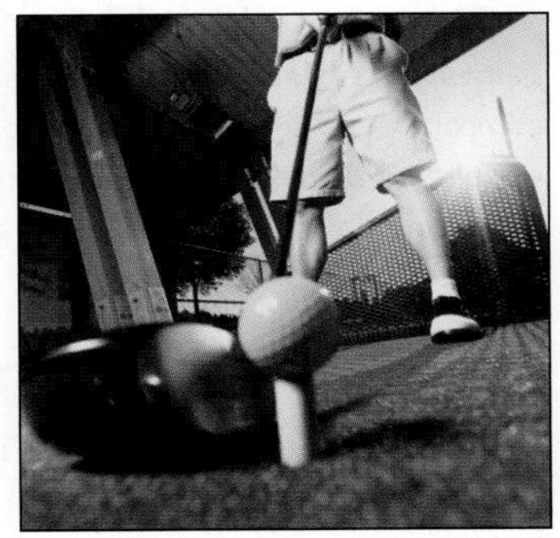

To illustrate, Exhibit 24.12 shows the overhead cost structure used to develop G-Max's flexible overhead budgets for May 2009. The predetermined standard overhead rate for May is set before the month begins. The first two number columns list the per unit amounts of variable costs and the monthly amounts of fixed costs. The four right-most columns show the costs expected to occur at four different levels of production activity. The predetermined overhead rate per labor hour is smaller as volume of activity increases because total fixed costs remain constant.

G-Max managers predicted an 80% activity level for May, or a production volume of 4,000 clubheads. At this volume, they budget \$8,000 as the May total overhead. This choice implies a \$2 per unit (labor hour) average overhead cost (\$8,000/4,000 units). Since G-Max has a standard of one direct labor hour per unit, the predetermined standard overhead rate for May is \$2 per standard direct labor hour. The variable overhead rate remains constant at \$1 per direct labor hour regardless of the budgeted production level. The fixed overhead rate changes according to the budgeted production volume. For instance, for the predicted level of 4,000 units of production, the fixed rate is \$1 per hour (\$4,000 fixed costs/4,000 units). For a production level of 5,000 units, however, the fixed rate is \$0.80 per hour (\$4,000 fixed costs/5,000 units).

When choosing the predicted activity level, management considers many factors. The level can be set as high as 100% of capacity, but this is rare. Factors causing the activity level to

Point: Managers consider the types of overhead costs when choosing the basis for assigning overhead costs to products.

Point: With increased automation, machine hours are frequently used in applying overhead instead of labor hours.

Point: Variable costs per unit remain constant, but fixed costs per unit decline with increases in volume. This means the average total overhead cost per unit declines with increases in volume.

EXHIBIT 24.12

Flexible Overhead Budgets

G-MAX
Flexible Overhead Budgets
For Month Ended May 31, 2009

	Flexible Budget					
	Variable Amount per Unit	Total Fixed Cost	Flexible Budget at 70% Capacity	Flexible Budget at 80% Capacity	Flexible Budget at 90% Capacity	Flexible Budget at 100% Capacity
Production (in units)	1 unit		3,500	4,000	4,500	5,000
Factory overhead						
Variable costs						
Indirect labor	\$0.40/unit		\$1,400	\$1,600	\$1,800	\$2,000
Indirect materials	0.30/unit		1,050	1,200	1,350	1,500
Power and lights	0.20/unit		700	800	900	1,000
Maintenance	0.10/unit		350	400	450	500
Total variable overhead costs	\$1.00/unit		3,500	4,000	4,500	5,000
Fixed costs (per month)						
Building rent		\$1,000	1,000	1,000	1,000	1,000
Depreciation—machinery		1,200	1,200	1,200	1,200	1,200
Supervisory salaries		1,800	1,800	1,800	1,800	1,800
Total fixed overhead costs		\$4,000	4,000	4,000	4,000	4,000
Total factory overhead			\$7,500	\$8,000	\$8,500	\$9,000
Standard direct labor hours 1 hr./unit			3,500 hrs.	4,000 hrs.	4,500 hrs.	5,000 hrs.
Predetermined overhead rate per standard direct labor hour			\$ 2.14	\$ 2.00	\$ 1.89	\$ 1.80

be less than full capacity include difficulties in scheduling work, equipment under repair or maintenance, and insufficient product demand. Good long-run management practices often call for some plant capacity in excess of current operating needs to allow for special opportunities and demand changes.

Decision Insight

Measuring Up In the spirit of continuous improvement, competitors compare their processes and performance standards against benchmarks established by industry leaders. Those that use **benchmarking** include Precision Lube, Jiffy Lube, All Tune and Lube, and Speedee Oil Change and Tune-Up.

Computing Overhead Cost Variances

When standard costs are used, the cost accounting system applies overhead to the good units produced using the predetermined standard overhead rate. At period-end, the difference between the total overhead cost applied to products and the total overhead cost actually incurred is called an **overhead cost variance** (total overhead variance), which is defined in Exhibit 24.13.

EXHIBIT 24.13
Overhead Cost Variance

Overhead cost variance (OCV) = Actual overhead incurred (AOI) − Standard overhead applied (SOA)

EXHIBIT 24.14
Framework for Understanding Total Overhead Variance

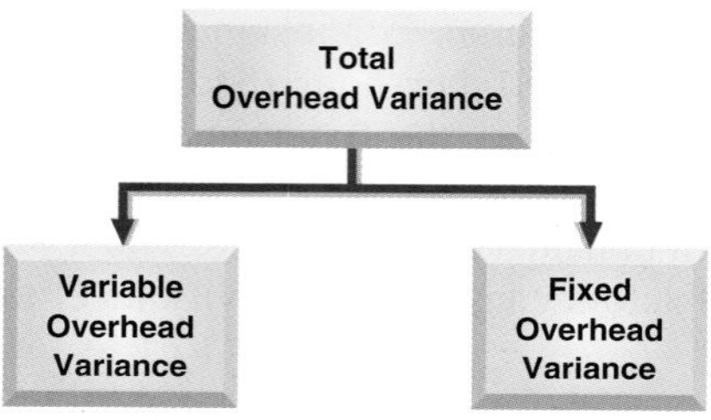

To help identify factors causing the overhead cost variance, managers analyze this variance separately for variable and fixed overhead, as illustrated in Exhibit 24.14. The results provide information useful for taking strategic actions to improve company performance.

P3 Compute overhead variances.

Computing Variable and Fixed Overhead Cost Variances To illustrate the computation of overhead cost variances, we return to the G-Max data. We know that G-Max produced 3,500 units when 4,000 units were budgeted. Additional data from cost reports show that the actual overhead cost incurred is $7,650 (the variable portion of $3,650 and the fixed portion of $4,000). Recall from Exhibit 24.12 that each unit requires 1 hour of direct labor, that variable overhead is applied at a rate of $1.00 per direct labor hour, and that the predetermined fixed overhead rate is $1.00 per direct labor hour. Using this information, we can compute overhead variances for both variable and fixed overhead as follows:

Actual variable overhead (given)	$3,650
Applied variable overhead (3,500 × $1.00)	3,500
Unfavorable variable overhead variance	$ 150

Actual fixed overhead (given)	$4,000
Applied fixed overhead (3,500 × $1.00)	3,500
Unfavorable fixed overhead variance	$ 500

"Well, according to the books, you've got too much overhead."

Management should seek to determine the causes of these unfavorable variances and take corrective action. To help better isolate the causes of these variances, more detailed overhead variances can be used, as shown in the next section.

Computing Controllable Overhead Variances and Volume Variances The total overhead variance for G-Max is $650 unfavorable, consisting of $150 unfavorable variable overhead variance and $500 unfavorable fixed overhead variance.

EXHIBIT 24.15

Expanded Framework for Total Overhead Variance

Total Overhead Variance

Variable Overhead Variance | Fixed Overhead Variance

Spending Variance | Efficiency Variance | Spending Variance | Volume Variance

Controllable Variance

Similar to analysis of direct materials and direct labor, both the variable and fixed overhead variances can be separately analyzed. Exhibit 24.15 shows an expanded framework for understanding these component overhead variances. A **spending variance** occurs when management pays an amount different than the standard price to acquire an item. For instance, the actual wage rate paid to indirect labor might be higher than the standard rate. Similarly, actual supervisory salaries might be different than expected. Spending variances such as these cause management to investigate the reasons that the amount paid differs from the standard. Both variable and fixed overhead costs can yield their own spending variances. Analyzing variable overhead includes computing an **efficiency variance,** which occurs when standard direct labor hours (the allocation base) expected for actual production differ from the actual direct labor hours used. This efficiency variance reflects on the cost-effectiveness in using the overhead allocation base (such as direct labor).

A **volume variance** occurs when a difference occurs between the actual volume of production and the standard volume of production. The budgeted fixed overhead amount is the same regardless of the volume of production (within the relevant range). This budgeted amount is computed based on the standard direct labor hours that the budgeted production volume allows. The applied overhead is based, however, on the standard direct labor hours allowed for the actual volume of production. A difference between budgeted and actual production volumes results in a difference in the standard direct labor hours allowed for these two production levels. This situation yields a volume variance different from zero.

Example: Does an unfavorable volume variance indicate poor management performance? *Answer.* No, it only indicates production volume was less than expected. This can be due to many factors, such as falling demand for company products, that are usually viewed outside a manager's control.

We can combine the variable overhead spending variance, the fixed overhead spending variance, and the variable overhead efficiency variance to get **controllable variance.** The controllable variance is so named because it refers to activities usually under management control. Exhibit 24.16

EXHIBIT 24.16

Variable and Fixed Overhead Variances

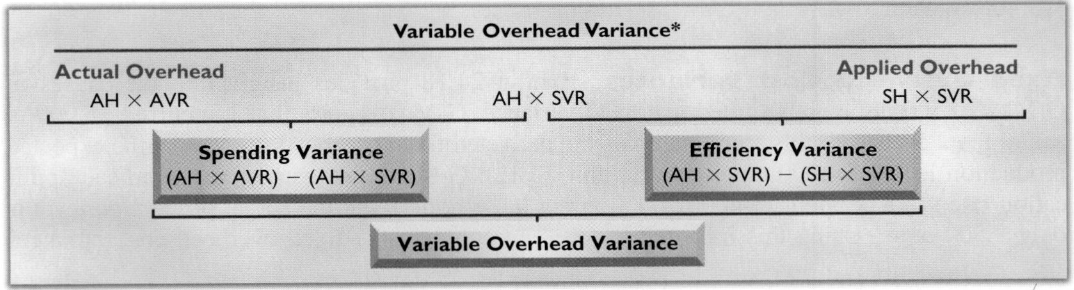

* AH = actual direct labor hours; AVR = actual variable overhead rate; SH = standard direct labor hours; SVR = standard variable overhead rate.

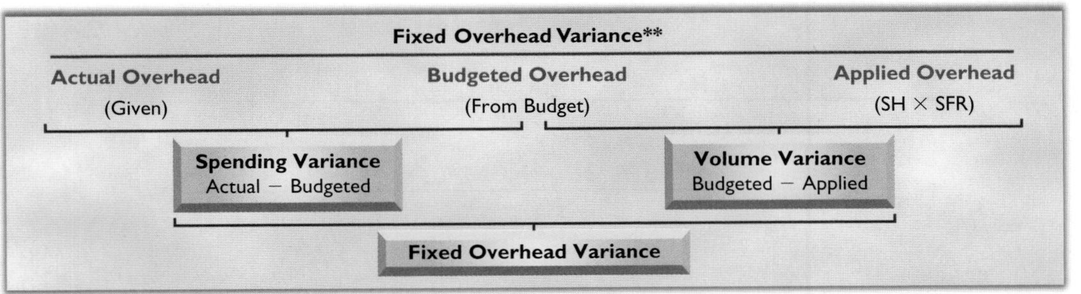

** SH = standard direct labor hours; SFR = standard fixed overhead rate.

shows formulas to use in computing detailed overhead variances that can better identify reasons for variances.

Variable Overhead Cost Variances Exhibit 24.17 offers insight into the causes of G-Max's $150 unfavorable variable overhead cost variance. Recall that G-Max applies overhead based on direct labor hours as the allocation base. We know that it used 3,400 direct labor hours to produce 3,500 units. This compares favorably to the standard requirement of 3,500 direct labor hours at one labor hour per unit. At a standard variable overhead rate of $1.00 per direct labor hour, this should have resulted in variable overhead costs of $3,400 (middle column of Exhibit 24.17).

EXHIBIT 24.17

Computing Variable Overhead Cost Variances

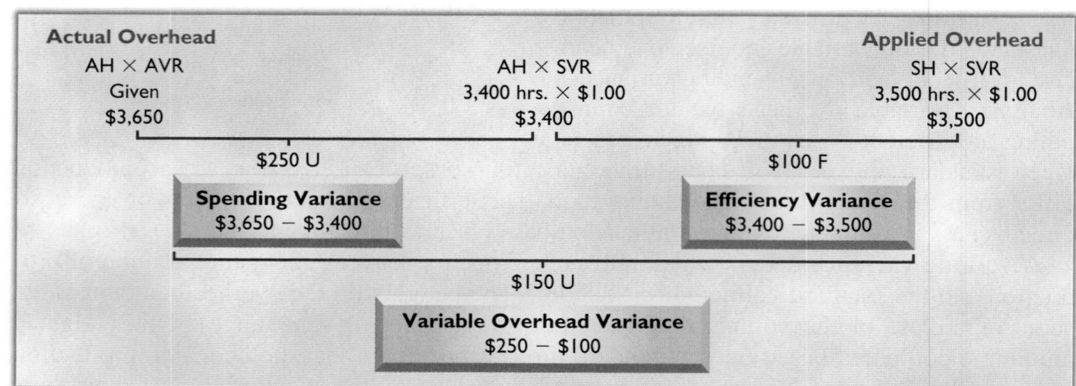

G-Max's cost records, however, report actual variable overhead of $3,650, or $250 higher than expected. This means G-Max has an unfavorable variable overhead spending variance of $250 ($3,650 − $3,400). On the other hand, G-Max used 100 fewer labor hours than expected to make 3,500 units, and its actual variable overhead is lower than its applied variable overhead. Thus, G-Max has a favorable variable overhead efficiency variance of $100 ($3,400 − $3,500).

Fixed Overhead Cost Variances Exhibit 24.18 provides insight into the causes of G-Max's $500 unfavorable fixed overhead variance. G-Max reports that it incurred $4,000 in actual fixed overhead; this amount equals the budgeted fixed overhead for May at the expected production level of 4,000 units (see Exhibit 24.12). G-Max's budgeted fixed overhead application rate is $1 per hour ($4,000/4,000 direct labor hours), but the actual production level is only 3,500 units. Using this information, we can compute the fixed overhead cost variances

EXHIBIT 24.18

Computing Fixed Overhead Cost Variances

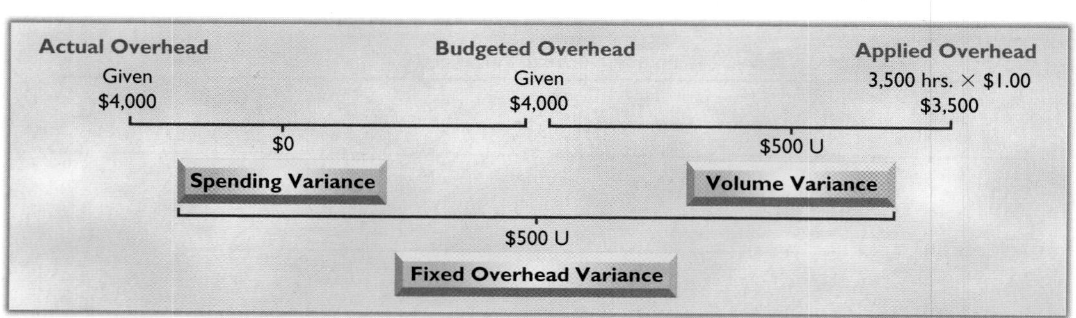

shown in Exhibit 24.18. The applied fixed overhead is computed by multiplying 3,500 standard hours allowed for the actual production by the $1 fixed overhead allocation rate. Exhibit 24.18 reveals that the fixed overhead spending variance is zero, suggesting good control of fixed overhead costs. The volume variance of $500 occurs because 500 fewer units are produced than budgeted; namely, 80% of the manufacturing capacity is budgeted but only 70% is used.

An unfavorable volume variance implies that the company did not reach its predicted operating level. Management needs to know why the actual level of performance differs from the expected level. The main purpose of the volume variance is to identify what portion of the total variance is caused by failing to meet the expected volume level. This information permits management to focus on the controllable variance.

Overhead Variance Reports Using the information from Exhibits 24.17 and 24.18, we compute the total controllable overhead variance as $150 unfavorable ($250 U + $100 F + $0). To help management isolate the reasons for this controllable variance, an *overhead variance report* can be prepared.

A complete overhead variance report provides managers information about specific overhead costs and how they differ from budgeted amounts. Exhibit 24.19 shows G-Max's overhead variance report for May. It reveals that (1) fixed costs and maintenance cost were incurred as expected, (2) costs for indirect labor and power and lights were higher than expected, and (3) indirect materials cost was less than expected.

The total controllable variance amount is also readily available from Exhibit 24.19. The overhead variance report shows the total volume variance as $500 unfavorable (shown at the top) and the $150 unfavorable controllable variance (reported at the bottom right). The sum of the controllable variance and the volume variance equals the total (fixed and variable) overhead variance of $650 unfavorable.

EXHIBIT 24.19

Overhead Variance Report

G-MAX
Overhead Variance Report
For Month Ended May 31, 2009

Volume Variance			
Expected production level	80% of capacity		
Production level achieved	70% of capacity		
Volume variance	**$500 (unfavorable)**		
Controllable Variance	**Flexible Budget**	**Actual Results**	**Variances***
Variable overhead costs			
Indirect labor	$1,400	$1,525	$125 U
Indirect materials	1,050	1,025	25 F
Power and lights	700	750	50 U
Maintenance	350	350	0
Total variable overhead costs	3,500	3,650	150 U†
Fixed overhead costs			
Building rent	1,000	1,000	0
Depreciation—machinery	1,200	1,200	0
Supervisory salaries	1,800	1,800	0
Total fixed overhead costs	4,000	4,000	0‡
Total overhead costs	$7,500	$7,650	$150 U

* F = Favorable variance; U = Unfavorable variance.

† Total variable overhead (spending and efficiency) variance.

‡ Fixed overhead spending variance.

Extensions of Standard Costs

This section extends the application of standard costs for control purposes, for service companies, and for accounting systems.

Standard Costs for Control

C3 Explain how standard cost information is useful for management by exception.

To control business activities, top management must be able to affect the actions of lower-level managers responsible for the company's revenues and costs. After preparing a budget and establishing standard costs, management should take actions to gain control when actual costs differ from standard or budgeted amounts.

Reports such as the ones illustrated in this chapter call management's attention to variances from business plans and other standards. When managers use these reports to focus on problem areas, the budgeting process contributes to the control function. In using budgeted performance reports, practice of management by exception is often useful. **Management by exception** means that managers focus attention on the most significant variances and give less attention to areas where performance is reasonably close to the standard. This practice leads management to concentrate on the exceptional or irregular situations. Management by exception is especially useful when directed at controllable items.

Decision Ethics

Internal Auditor You discover a manager who always spends exactly what is budgeted. About 30% of her budget is spent just before the period-end. She admits to spending what is budgeted, whether or not it is needed. She offers three reasons: (1) she doesn't want her budget cut, (2) "management by exception" focuses on budget deviations; and (3) she believes the money is budgeted to be spent. What action do you take? [Answer—p. 1004]

Standard Costs for Services

Many managers use standard costs and variance analysis to investigate manufacturing costs. Many managers also recognize that standard costs and variances can help them control *nonmanufacturing* costs. Companies providing services instead of products can benefit from the use of standard costs. Application of standard costs and variances can be readily adapted to nonmanufacturing situations. To illustrate, many service providers use standard costs to help control expenses. First, they use standard costs as a basis for budgeting all services. Second, they use periodic performance reports to compare actual results to standards. Third, they use these reports to identify significant variances within specific areas of responsibility. Fourth, they implement the appropriate control procedures.

Decision Insight

Health Budget Medical professionals continue to struggle with business realities. Quality medical service is paramount, but efficiency in providing that service also is important. The use of budgeting and standard costing is touted as an effective means to control and monitor medical costs, especially overhead.

Standard Cost Accounting System

P4 Prepare journal entries for standard costs and account for price and quantity variances.

We have shown how companies use standard costs in management reports. Most standard cost systems also record these costs and variances in accounts. This practice simplifies recordkeeping and helps in preparing reports. Although we do not need knowledge of standard cost accounting practices to understand standard costs and their use, we must know how to interpret the accounts in which standard costs and variances are recorded. The entries in this section briefly illustrate the important aspects of this process for G-Max's standard costs and variances for May.

The first of these entries records standard materials cost incurred in May in the Goods in Process Inventory account. This part of the entry is similar to the usual accounting entry, but the amount of the debit equals the standard cost ($3,500) instead of the actual cost ($3,780).

This entry credits Raw Materials Inventory for actual cost. The difference between standard and actual direct materials costs is recorded with debits to two separate materials variance accounts (recall Exhibit 21.10). Both the materials price and quantity variances are recorded as debits because they reflect additional costs higher than the standard cost (if actual costs were less than the standard, they are recorded as credits). This treatment (debit) reflects their unfavorable effect because they represent higher costs and lower income.

May 31	Goods in Process Inventory	3,500	
	Direct Materials Price Variance*	**180**	
	Direct Materials Quantity Variance	**100**	
	Raw Materials Inventory		3,780
	To charge production for standard quantity of materials used (3,500 lbs.) at the standard price ($1 per lb.), and to record material price and material quantity variances.		

Assets	= Liabilities	+ Equity
+3,500		−100
−3,780		−180

* Many companies record the materials price variance when materials are purchased. For simplicity, we record both the materials price and quantity variances when materials are issued to production.

The second entry debits Goods in Process Inventory for the standard labor cost of the goods manufactured during May ($28,000). Actual labor cost ($28,220) is recorded with a credit to the Factory Payroll account. The difference between standard and actual labor costs is explained by two variances (see Exhibit 21.11). The direct labor rate variance is unfavorable and is debited to that account. The direct labor efficiency variance is favorable and that account is credited. The direct labor efficiency variance is favorable because it represents a lower cost and a higher net income.

May 31	Goods in Process Inventory	28,000	
	Direct Labor Rate Variance	**1,020**	
	Direct Labor Efficiency Variance		**800**
	Factory Payroll		28,220
	To charge production with 3,500 standard hours of direct labor at the standard $8 per hour rate, and to record the labor rate and efficiency variances.		

Assets	= Liabilities	+ Equity
+28,000		+28,220
		− 1,020
		+ 800

The entry to assign standard predetermined overhead to the cost of goods manufactured must debit the $7,000 predetermined amount to the Goods in Process Inventory account. Actual overhead costs of $7,650 were debited to Factory Overhead during the period (entries not shown here). Thus, when Factory Overhead is applied to Goods in Process Inventory, the actual amount is credited to the Factory Overhead account. To account for the difference between actual and standard overhead costs, the entry includes a $250 debit to the Variable Overhead Spending Variance, a $100 credit to the Variable Overhead Efficiency Variance, and a $500 debit to the Volume Variance (recall Exhibits 21.17 and 21.18). An alternative (simpler) approach is to record the difference with a $150 debit to the Controllable Variance account and a $500 debit to the Volume Variance account (recall from Exhibit 21.15 that controllable variance is the sum of both variable overhead variances and the fixed overhead spending variance).

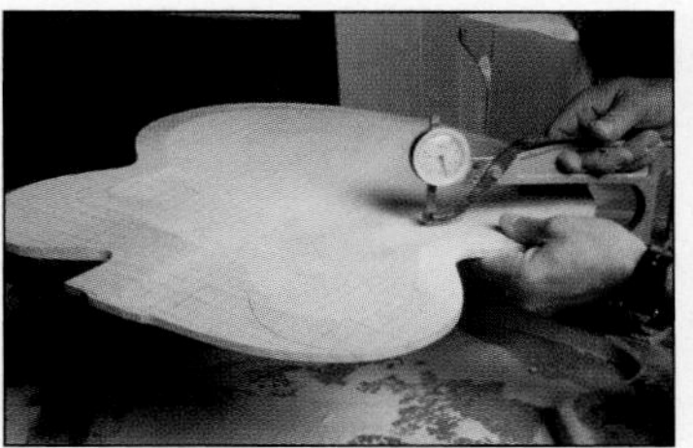

May 31	Goods in Process Inventory	7,000	
	Volume Variance	**500**	
	Variable Overhead Spending Variance	**250**	
	Variable Overhead Efficiency Variance		**100**
	Factory Overhead		7,650
	To apply overhead at the standard rate of $2 per standard direct labor hour (3,500 hours), and to record overhead variances.		

Assets	= Liabilities	+ Equity
+7,000		+7,650
		− 250
		− 500
		+ 100

The balances of these different variance accounts accumulate until the end of the accounting period. As a result, the unfavorable variances of some months can offset the favorable variances of other months.

These ending variance account balances, which reflect results of the period's various transactions and events, are closed at period-end. If the amounts are *immaterial,* they are added to or subtracted from the balance of the Cost of Goods Sold account. This process is similar to that shown in the job order costing chapter for eliminating an underapplied or overapplied balance in the Factory Overhead account. (*Note:* These variance balances, which represent differences between actual and standard costs, must be added to or subtracted from the materials, labor, and overhead costs recorded. In this way, the recorded costs equal the actual costs incurred in the period; a company must use actual costs in external financial statements prepared in accordance with generally accepted accounting principles.)

Point: If variances are material they can be allocated between Goods in Process Inventory, Finished Goods Inventory, and Cost of Goods Sold. This closing process is explained in advanced courses.

Quick Check

Answers—p. 1005

10. Under what conditions is an overhead volume variance considered favorable?
11. To use management by exception with standard costs, a company (*a*) must record standard costs in its accounting, (*b*) should compute variances from flexible budget amounts to allow management to focus its attention on significant differences between actual and budgeted results, or (*c*) should analyze only variances for direct materials and direct labor.
12. A company uses a standard cost accounting system. Prepare the journal entry to record these direct materials variances:

Direct materials cost actually incurred	$73,200
Direct materials quantity variance (favorable)	3,800
Direct materials price variance (unfavorable)	1,300

13. If standard costs are recorded in the manufacturing accounts, how are recorded variances treated at the end of an accounting period?

Decision Analysis — Sales Variances

A2 Analyze changes in sales from expected amounts.

This chapter explained the computation and analysis of cost variances. A similar variance analysis can be applied to sales. To illustrate, consider the following sales data from G-Max for two of its golf products, Excel golf balls and Big Bert® drivers.

	Budgeted	Actual
Sales of Excel golf balls (units)	1,000 units	1,100 units
Sales price per Excel golf ball	$10	$10.50
Sales of Big Bert® drivers (units)	150 units	140 units
Sales price per Big Bert® driver	$200	$190

Using this information, we compute both the *sales price variance* and the *sales volume variance* as shown in Exhibit 24.20. The total sales price variance is $850 unfavorable, and the total sales volume variance is $1,000 unfavorable. Neither variance implies anything positive about these two products. However, further analysis of these total sales variances reveals that both the sales price and sales volume variances for Excel golf balls are favorable, meaning that both the unfavorable total sales price variance and the unfavorable total sales volume variance are due to the Big Bert driver.

EXHIBIT 24.20

Computing Sales Variances*

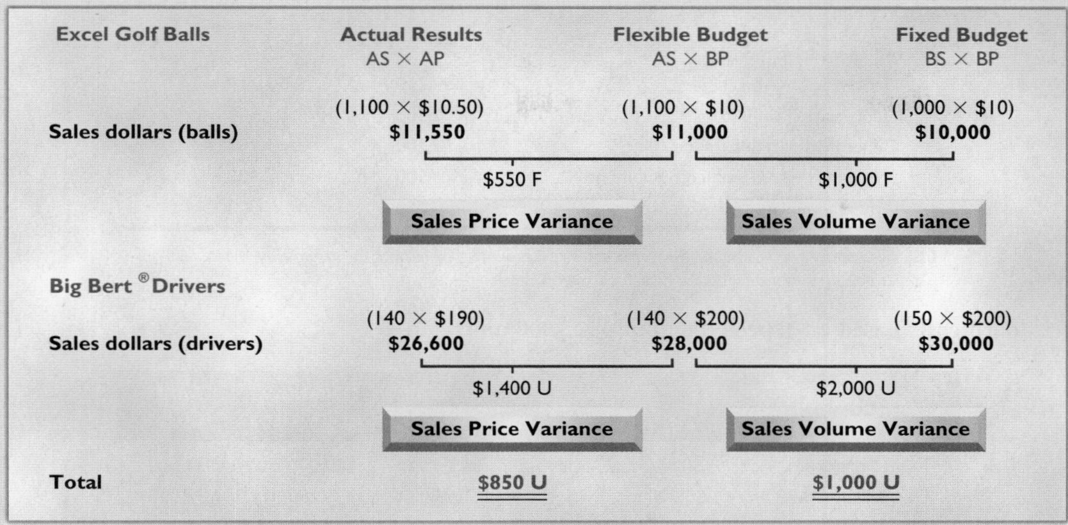

* AS = actual sales units; AP = actual sales price; BP = budgeted sales price; BS = budgeted sales units (fixed budget).

Managers use sales variances for planning and control purposes. The sales variance information is used to plan future actions to avoid unfavorable variances. G-Max sold 90 total combined units (both balls and drivers) more than planned, but these 90 units were not sold in the proportion budgeted. G-Max sold fewer than the budgeted quantity of the higher-priced driver, which contributed to the unfavorable total sales variances. Managers use such detail to question what caused the company to sell more golf balls and fewer drivers. Managers also use this information to evaluate and even reward their salespeople. Extra compensation is paid to salespeople who contribute to a higher profit margin. Finally, with multiple products, the sales volume variance can be separated into a *sales mix variance* and a *sales quantity variance*. The sales mix variance is the difference between the actual and budgeted sales mix of the products. The sales quantity variance is the difference between the total actual and total budgeted quantity of units sold.

Decision Maker

Sales Manager The current performance report reveals a large favorable sales volume variance but an unfavorable sales price variance. You did not expect to see a large increase in sales volume. What steps do you take to analyze this situation? [Answer—p. 1004]

Demonstration Problem

DP24

Pacific Company provides the following information about its budgeted and actual results for June 2009. Although the expected June volume was 25,000 units produced and sold, the company actually produced and sold 27,000 units as detailed here:

	Budget (25,000 units)	Actual (27,000 units)
Selling price	$5.00 per unit	$5.23 per unit
Variable costs (per unit)		
Direct materials	1.24 per unit	1.12 per unit
Direct labor	1.50 per unit	1.40 per unit
Factory supplies*	0.25 per unit	0.37 per unit
Utilities*	0.50 per unit	0.60 per unit
Selling costs	0.40 per unit	0.34 per unit

[continued on next page]

[continued from previous page]

Fixed costs (per month)		
Depreciation—machinery*	$3,750	$3,710
Depreciation—building*	2,500	2,500
General liability insurance	1,200	1,250
Property taxes on office equipment	500	485
Other administrative expense	750	900

* Indicates factory overhead item; $0.75 per unit or $3 per direct labor hour for variable overhead, and $0.25 per unit or $1 per direct labor hour for fixed overhead.

Standard costs based on expected output of 25,000 units

	Per Unit of Output	Quantity to Be Used	Total Cost
Direct materials, 4 oz. @ $0.31/oz.	$1.24/unit	100,000 oz.	$31,000
Direct labor, 0.25 hrs. @ $6.00/hr.	1.50/unit	6,250 hrs.	37,500
Overhead	1.00/unit		25,000

Actual costs incurred to produce 27,000 units

	Per Unit of Output	Quantity Used	Total Cost
Direct materials, 4 oz. @ $0.28/oz.	$1.12/unit	108,000 oz.	$30,240
Direct labor, 0.20 hrs. @ $7.00/hr.	1.40/unit	5,400 hrs.	37,800
Overhead	1.20/unit		32,400

Standard costs based on expected output of 27,000 units

	Per Unit of Output	Quantity to Be Used	Total Cost
Direct materials, 4 oz. @ $0.31/oz.	$1.24/unit	108,000 oz.	$33,480
Direct labor, 0.25 hrs. @ $6.00/hr.	1.50/unit	6,750 hrs.	40,500
Overhead			26,500

Required

1. Prepare June flexible budgets showing expected sales, costs, and net income assuming 20,000, 25,000, and 30,000 units of output produced and sold.
2. Prepare a flexible budget performance report that compares actual results with the amounts budgeted if the actual volume had been expected.
3. Apply variance analysis for direct materials, for direct labor, and for overhead.
4. Prepare journal entries to record standard costs, and price and quantity variances, for direct materials, direct labor, and factory overhead.

Planning the Solution

- Prepare a table showing the expected results at the three specified levels of output. Compute the variable costs by multiplying the per unit variable costs by the expected volumes. Include fixed costs at the given amounts. Combine the amounts in the table to show total variable costs, contribution margin, total fixed costs, and income from operations.
- Prepare a table showing the actual results and the amounts that should be incurred at 27,000 units. Show any differences in the third column and label them with an *F* for favorable if they increase income or a *U* for unfavorable if they decrease income.
- Using the chapter's format, compute these total variances and the individual variances requested:
  - Total materials variance (including the direct materials quantity variance and the direct materials price variance).

- Total direct labor variance (including the direct labor efficiency variance and rate variance).
- Total overhead variance (including both variable and fixed overhead variances and their component variances).

Solution to Demonstration Problem

1.

PACIFIC COMPANY
Flexible Budgets
For Month Ended June 30, 2009

	Flexible Budget		Flexible Budget for Unit Sales of 20,000	Flexible Budget for Unit Sales of 25,000	Flexible Budget for Unit Sales of 30,000
	Variable Amount per Unit	Total Fixed Cost			
Sales	$5.00		$100,000	$125,000	$150,000
Variable costs					
Direct materials	1.24		24,800	31,000	37,200
Direct labor	1.50		30,000	37,500	45,000
Factory supplies	0.25		5,000	6,250	7,500
Utilities	0.50		10,000	12,500	15,000
Selling costs	0.40		8,000	10,000	12,000
Total variable costs	3.89		77,800	97,250	116,700
Contribution margin	$1.11		22,200	27,750	33,300
Fixed costs					
Depreciation—machinery		$3,750	3,750	3,750	3,750
Depreciation—building		2,500	2,500	2,500	2,500
General liability insurance		1,200	1,200	1,200	1,200
Property taxes on office equipment		500	500	500	500
Other administrative expense		750	750	750	750
Total fixed costs		$8,700	8,700	8,700	8,700
Income from operations			$ 13,500	$ 19,050	$ 24,600

2.

PACIFIC COMPANY
Flexible Budget Performance Report
For Month Ended June 30, 2009

	Flexible Budget	Actual Results	Variance*
Sales (27,000 units)	$135,000	$141,210	$6,210 F
Variable costs			
Direct materials	33,480	30,240	3,240 F
Direct labor	40,500	37,800	2,700 F
Factory supplies	6,750	9,990	3,240 U
Utilities	13,500	16,200	2,700 U
Selling costs	10,800	9,180	1,620 F
Total variable costs	105,030	103,410	1,620 F
Contribution margin	29,970	37,800	7,830 F
Fixed costs			
Depreciation—machinery	3,750	3,710	40 F
Depreciation—building	2,500	2,500	0
General liability insurance	1,200	1,250	50 U
Property taxes on office equipment	500	485	15 F
Other administrative expense	750	900	150 U
Total fixed costs	8,700	8,845	145 U
Income from operations	$ 21,270	$ 28,955	$7,685 F

* F = Favorable variance; U = Unfavorable variance.

3. Variance analysis of materials, labor, and overhead costs.

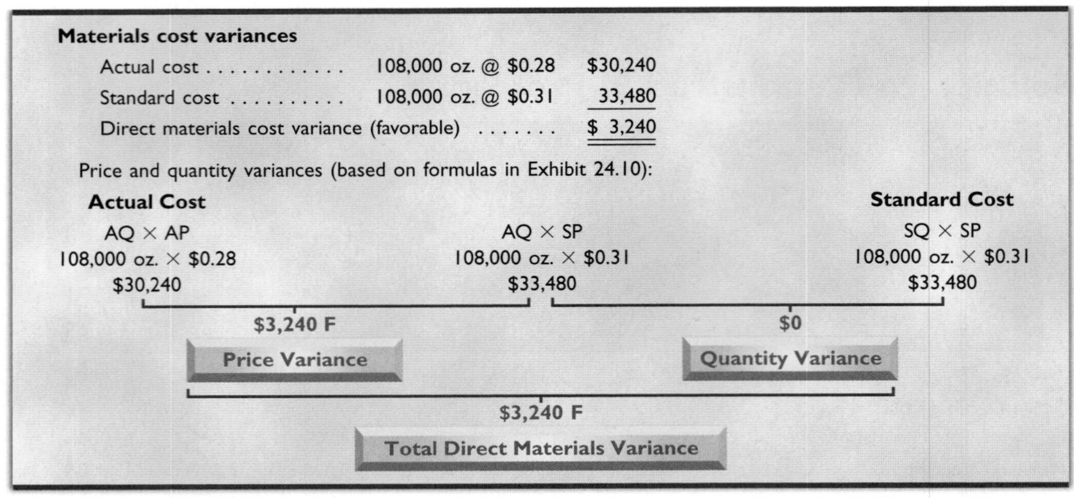

Materials cost variances

Actual cost	108,000 oz. @ $0.28	$30,240
Standard cost	108,000 oz. @ $0.31	33,480
Direct materials cost variance (favorable)		$ 3,240

Price and quantity variances (based on formulas in Exhibit 24.10):

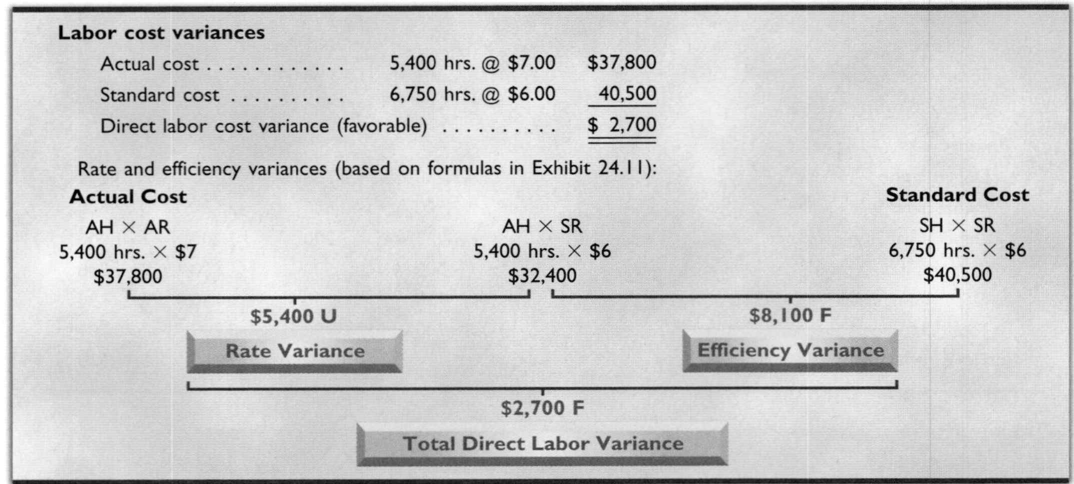

Labor cost variances

Actual cost	5,400 hrs. @ $7.00	$37,800
Standard cost	6,750 hrs. @ $6.00	40,500
Direct labor cost variance (favorable)		$ 2,700

Rate and efficiency variances (based on formulas in Exhibit 24.11):

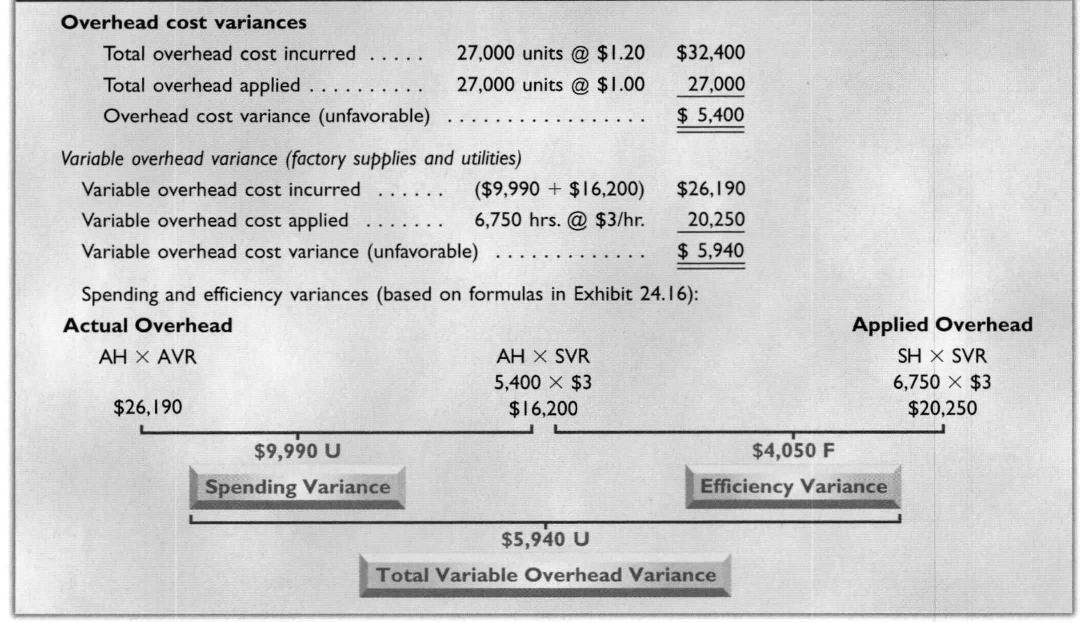

Overhead cost variances

Total overhead cost incurred	27,000 units @ $1.20	$32,400
Total overhead applied	27,000 units @ $1.00	27,000
Overhead cost variance (unfavorable)		$ 5,400

Variable overhead variance (factory supplies and utilities)

Variable overhead cost incurred	($9,990 + $16,200)	$26,190
Variable overhead cost applied	6,750 hrs. @ $3/hr.	20,250
Variable overhead cost variance (unfavorable)		$ 5,940

Spending and efficiency variances (based on formulas in Exhibit 24.16):

[continued on next page]

[continued from previous page]

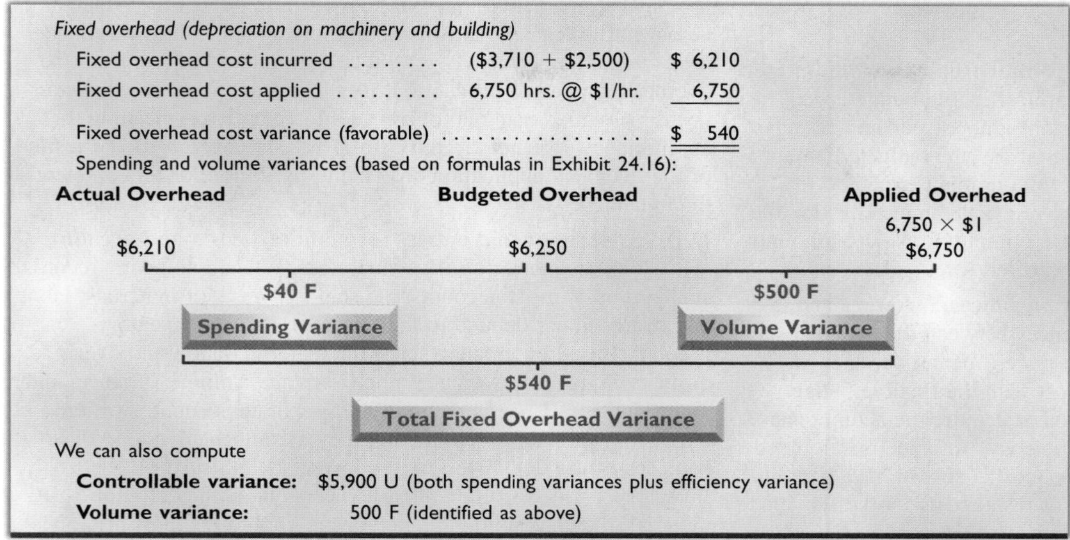

Fixed overhead (depreciation on machinery and building)
Fixed overhead cost incurred .... ($3,710 + $2,500) $ 6,210
Fixed overhead cost applied .... 6,750 hrs. @ $1/hr. 6,750
Fixed overhead cost variance (favorable) .... $ 540
Spending and volume variances (based on formulas in Exhibit 24.16):
Actual Overhead
Budgeted Overhead
Applied Overhead
6,750 × $1
$6,210
$6,250
$6,750
$40 F
Spending Variance
$500 F
Volume Variance
$540 F
Total Fixed Overhead Variance
We can also compute
Controllable variance: $5,900 U (both spending variances plus efficiency variance)
Volume variance: 500 F (identified as above)

4.

Account	Debit	Credit
Goods in Process Inventory	33,480	
Direct Materials Price Variance		3,240
Raw Materials Inventory		30,240
Goods in Process Inventory	40,500	
Direct Labor Rate Variance	5,400	
Direct Labor Efficiency Variance		8,100
Factory Payroll		37,800
Goods in Process Inventory*	27,000	
Variable Overhead Spending Variance	9,990	
Variable Overhead Efficiency Variance		4,050
Fixed Overhead Spending Variance		40
Fixed Overhead Volume Variance		500
Factory Overhead**		32,400

* $20,250 + $6,750 **$26,190 + $6,210

Summary

C1 Define *standard costs* and explain their computation and uses. Standard costs are the normal costs that should be incurred to produce a product or perform a service. They should be based on a careful examination of the processes used to produce a product or perform a service as well as the quantities and prices that should be incurred in carrying out those processes. On a performance report, standard costs (which are flexible budget amounts) are compared to actual costs, and the differences are presented as variances.

C2 Describe variances and what they reveal about performance. Management can use variances to monitor and control activities. Total cost variances can be broken into price and quantity variances to direct management's attention to those responsible for quantities used and prices paid.

C3 Explain how standard cost information is useful for management by exception. Standard cost accounting provides management information about costs that differ from budgeted (expected) amounts. Performance reports disclose the costs or areas of operations that have significant variances from budgeted amounts. This allows managers to focus attention on the exceptions and less attention on areas proceeding normally.

A1 Compare fixed and flexible budgets. A fixed budget shows the revenues and costs expected to occur at a specified volume level. If actual volume is at some other level, the amounts in the fixed budget do not provide a reasonable basis for evaluating actual performance. A flexible budget expresses variable costs in per unit terms so that it can be used to develop budgeted amounts for any volume level within the relevant range. Thus, managers compute budgeted amounts for evaluation after a period for the volume that actually occurred.

A2 Analyze changes in sales from expected amounts. Actual sales can differ from budgeted sales, and managers can investigate this difference by computing both the sales price and sales volume variances. The *sales price variance* refers to that portion of total variance resulting from a difference between actual and budgeted selling prices. The *sales volume variance* refers to that portion of total variance resulting from a difference between actual and budgeted sales quantities.

P1 **Prepare a flexible budget and interpret a flexible budget performance report.** To prepare a flexible budget, we express each variable cost as a constant amount per unit of sales (or as a percent of sales dollars). In contrast, the budgeted amount of each fixed cost is expressed as a total amount expected to occur at any sales volume within the relevant range. The flexible budget is then determined using these computations and amounts for fixed and variable costs at the expected sales volume.

P2 **Compute materials and labor variances.** Materials and labor variances are due to differences between the actual costs incurred and the budgeted costs. The price (or rate) variance is computed by comparing the actual cost with the flexible budget amount that should have been incurred to acquire the actual quantity of resources. The quantity (or efficiency) variance is computed by comparing the flexible budget amount that should have been incurred to acquire the actual quantity of resources with the flexible budget amount that should have been incurred to acquire the standard quantity of resources.

P3 **Compute overhead variances.** Overhead variances are due to differences between the actual overhead costs incurred and the overhead applied to production. An overhead spending variance arises when the actual amount incurred differs from the budgeted amount of overhead. An overhead efficiency (or volume) variance arises when the flexible overhead budget amount differs from the overhead applied to production. It is important to realize that overhead is assigned using an overhead allocation base, meaning that an efficiency variance (in the case of variable overhead) is a result of the overhead application base being used more or less efficiently than planned.

P4 **Prepare journal entries for standard costs and account for price and quantity variances.** When a company records standard costs in its accounts, the standard costs of materials, labor, and overhead are debited to the Goods in Process Inventory account. Based on an analysis of the material, labor, and overhead costs, each quantity variance, price variance, volume variance, and controllable variance is recorded in a separate account. At period-end, if the variances are material, they are allocated among the balances of the Goods in Process Inventory, Finished Goods Inventory, and Cost of Goods Sold accounts. If they are not material, they are simply debited or credited to the Cost of Goods Sold account.

Guidance Answers to **Decision Maker** and **Decision Ethics**

Entrepreneur From the complaints, this performance report appears to compare actual results with a fixed budget. This comparison is useful in determining whether the amount of work actually performed was more or less than planned, but it is not useful in determining whether the divisions were more or less efficient than planned. If the two consulting divisions worked on more assignments than expected, some costs will certainly increase. Therefore, you should prepare a flexible budget using the actual number of consulting assignments and then compare actual performance to the flexible budget.

Human Resource Manager As HR manager, you should investigate the causes for any labor-related variances although you may not be responsible for them. An unfavorable labor efficiency variance occurs because more labor hours than standard were used during the period. There are at least three possible reasons for this: (1) materials quality could be poor, resulting in more labor consumption due to rework; (2) unplanned interruptions (strike, breakdowns, accidents) could have occurred during the period; and (3) the production manager could have used a different labor mix to expedite orders. This new labor mix could have consisted of a larger proportion of untrained labor, which resulted in more labor hours.

Internal Auditor Although the manager's actions might not be unethical, this action is undesirable. The internal auditor should report this behavior, possibly recommending that for the purchase of such discretionary items, the manager must provide budgetary requests using an activity-based budgeting process. The internal auditor would then be given full authority to verify this budget request.

Sales Manager The unfavorable sales price variance suggests that actual prices were lower than budgeted prices. As the sales manager, you want to know the reasons for a lower than expected price. Perhaps your salespeople lowered the price of certain products by offering quantity discounts. You then might want to know what prompted them to offer the quantity discounts (perhaps competitors were offering discounts). You want to break the sales volume variance into both the sales mix and sales quantity variances. You could find that although the sales quantity variance is favorable, the sales mix variance is not. Then you need to investigate why the actual sales mix differs from the budgeted sales mix.

Guidance Answers to **Quick Checks**

1. *b*
2. The first step is classifying each cost as variable or fixed.
3. A fixed budget is prepared using an expected volume of sales or production. A flexible budget is prepared using the actual volume of activity.
4. The contribution margin equals sales less variable costs.
5. *c*
6. It is the difference between actual cost and standard cost.

7. *a*; Total actual hours: 2,500 × 2.5 = 6,250
Total standard hours: 2,500 × 2.0 = 5,000
Efficiency variance = (6,250 − 5,000) × $3.00
= $3,750 U

8. *b*; Rate variance = ($3.10 − $3.00) × 6,250 = $625 U

9. Yes, this will occur when the materials quantity variance is more than the materials price variance.

10. The overhead volume variance is favorable when the actual operating level is higher than the expected level.

11. *b*

12.

Goods in Process Inventory	75,700	
Direct Materials Price Variance	1,300	
Direct Materials Quantity Variance		3,800
Raw Materials Inventory		73,200

13. If the variances are material, they should be prorated among the Goods in Process Inventory, Finished Goods Inventory, and Cost of Goods Sold accounts. If they are not material, they can be closed to Cost of Goods Sold.

Key Terms

mhhe.com/wildFAP19e

Key Terms are available at the book's Website for learning and testing in an online Flashcard Format.

Budgetary control (p. 980)
Budget report (p. 980)
Controllable variance (p. 993)
Cost variance (p. 986)
Efficiency variance (p. 993)
Favorable variance (p. 981)
Fixed budget (p. 981)
Fixed budget performance report (p. 981)
Flexible budget (p. 982)
Flexible budget performance report (p. 984)
Management by exception (p. 996)
Overhead cost variance (p. 992)
Price variance (p. 984)
Quantity variance (p. 984)
Spending variance (p. 993)
Standard costs (p. 985)
Unfavorable variance (p. 981)
Variance analysis (p. 984)
Volume variance (p. 993)

Multiple Choice Quiz

Answers on p. 1021

mhhe.com/wildFAP19e

Additional Quiz Questions are available at the book's Website.

Quiz24

1. A company predicts its production and sales will be 24,000 units. At that level of activity, its fixed costs are budgeted at $300,000, and its variable costs are budgeted at $246,000. If its activity level declines to 20,000 units, what will be its fixed costs and its variable costs?
a. Fixed, $300,000; variable, $246,000
b. Fixed, $250,000; variable, $205,000
c. Fixed, $300,000; variable, $205,000
d. Fixed, $250,000; variable, $246,000
e. Fixed, $300,000; variable, $300,000

2. Using the following information about a single product company, compute its total actual cost of direct materials used.
- Direct materials standard cost: 5 lbs. × $2 per lb. = $10.
- Total direct materials cost variance: $15,000 unfavorable.
- Actual direct materials used: 300,000 lbs.
- Actual units produced: 60,000 units.

a. $585,000
b. $600,000
c. $300,000
d. $315,000
e. $615,000

3. A company uses four hours of direct labor to produce a product unit. The standard direct labor cost is $20 per hour. This period the company produced 20,000 units and used 84,160 hours of direct labor at a total cost of $1,599,040. What is its labor rate variance for the period?
a. $83,200 F
b. $84,160 U
c. $84,160 F
d. $83,200 U
e. $ 960 F

4. A company's standard for a unit of its single product is $6 per unit in variable overhead (4 hours × $1.50 per hour). Actual data for the period show variable overhead costs of $150,000 and production of 24,000 units. Its total variable overhead cost variance is
a. $ 6,000 F.
b. $ 6,000 U.
c. $114,000 U.
d. $114,000 F.
e. $ 0.

5. A company's standard for a unit of its single product is $4 per unit in fixed overhead ($24,000 total/6,000 units budgeted). Actual data for the period show total actual fixed overhead of $24,100 and production of 4,800 units. Its volume variance is
a. $4,800 U.
b. $4,800 F.
c. $ 100 U.
d. $ 100 F.
e. $4,900 U.

Discussion Questions

1. What limits the usefulness to managers of fixed budget performance reports?
2. Identify the main purpose of a flexible budget for managers.
3. Prepare a flexible budget performance report title (in proper form) for Spalding Company for the calendar year 2009. Why is a proper title important for this or any report?
4. What type of analysis does a flexible budget performance report help management perform?
5. In what sense can a variable cost be considered constant?
6. What department is usually responsible for a direct labor rate variance? What department is usually responsible for a direct labor efficiency variance? Explain.
7. What is a price variance? What is a quantity variance?
8. What is the purpose of using standard costs?
9. In an analysis of fixed overhead cost variances, what is the volume variance?
10. What is the predetermined standard overhead rate? How is it computed?
11. In general, variance analysis is said to provide information about ________ and ________ variances.
12. In an analysis of overhead cost variances, what is the controllable variance and what causes it?
13. What are the relations among standard costs, flexible budgets, variance analysis, and management by exception?
14. How can the manager of a music department of a **Best Buy** retail store use flexible budgets to enhance performance?

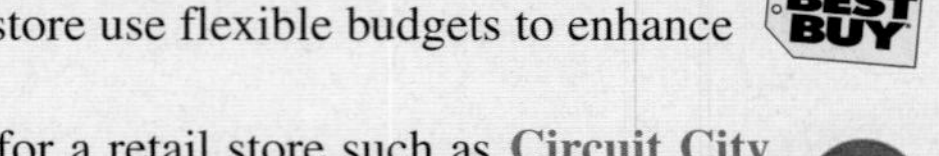

15. Is it possible for a retail store such as **Circuit City** to use variances in analyzing its operating performance? Explain.

16. Assume that **Apple** is budgeted to operate at 80% of capacity but actually operates at 75% of capacity. What effect will the 5% deviation have on its controllable variance? Its volume variance?

Denotes Discussion Questions that involve decision making.

Available with McGraw-Hill's Homework Manager

QUICK STUDY

QS 24-1
Flexible budget performance report
P1

Quail Company reports the following selected financial results for May. For the level of production achieved in May, the budgeted amounts would be sales, $650,000; variable costs, $375,000; and fixed costs, $150,000. Prepare a flexible budget performance report for May.

Sales (100,000 units)	$637,500
Variable costs	356,250
Fixed costs	150,000

QS 24-2
Labor cost variances
C2 P2

Martin Company's output for the current period results in a $10,000 unfavorable direct labor rate variance and a $5,000 unfavorable direct labor efficiency variance. Production for the current period was assigned a $200,000 standard direct labor cost. What is the actual total direct labor cost for the current period?

QS 24-3
Materials cost variances
C2 P2

Blanda Company's output for the current period was assigned a $300,000 standard direct materials cost. The direct materials variances included a $24,000 favorable price variance and a $4,000 favorable quantity variance. What is the actual total direct materials cost for the current period?

QS 24-4
Materials cost variances
C2 P2

For the current period, Roja Company's manufacturing operations yield an $8,000 unfavorable price variance on its direct materials usage. The actual price per pound of material is $156; the standard price is $154. How many pounds of material are used in the current period?

QS 24-5
Management by exception
C3

Managers use *management by exception* for control purposes. (1) Describe the concept of management by exception. (2) Explain how standard costs help managers apply this concept to monitor and control costs.

QS 24-6
Overhead cost variances P3

Gohan Company's output for the current period yields a $12,000 favorable overhead volume variance and a $21,500 unfavorable overhead controllable variance. Standard overhead charged to production for the period is $410,000. What is the actual total overhead cost incurred for the period?

Refer to the information in QS 24-6. Gohan records standard costs in its accounts. Prepare the journal entry to charge overhead costs to the Goods in Process Inventory account and to record any variances.

QS 24-7
Preparing overhead entries P4

Wills Company specializes in selling used trucks. During the first six months of 2009, the dealership sold 50 trucks at an average price of $18,000 each. The budget for the first six months of 2009 was to sell 45 trucks at an average price of $19,000 each. Compute the dealership's sales price variance and sales volume variance for the first six months of 2009.

QS 24-8
Computing sales price and volume variances
A2

Harp Company applies overhead using machine hours and reports the following information. Compute the total variable overhead cost variance.

QS 24-9
Overhead cost variances
P3

Actual machine hours used	4,950 hours
Standard machine hours	5,000 hours
Actual variable overhead rate per hour	$2.10
Standard variable overhead rate per hour	$2.00

Refer to the information from QS 24-9. Compute the variable overhead spending variance and the variable overhead efficiency variance.

QS 24-10
Overhead spending and efficiency variances P3

McGraw-Hill's HOMEWORK MANAGER® Available with McGraw-Hill's Homework Manager

EXERCISES

Exercise 24-1
Preparation of flexible budgets
P1

Tryon Company's fixed budget for the first quarter of calendar year 2009 reveals the following. Prepare flexible budgets following the format of Exhibit 24.3 that show variable costs per unit, fixed costs, and three different flexible budgets for sales volumes of 14,500, 15,000, and 15,500 units.

Sales (15,000 units)		$3,030,000
Cost of goods sold		
Direct materials	$345,000	
Direct labor	705,000	
Production supplies	405,000	
Plant manager salary	90,000	1,545,000
Gross profit		1,485,000
Selling expenses		
Sales commissions	150,000	
Packaging	240,000	
Advertising	100,000	490,000
Administrative expenses		
Administrative salaries	110,000	
Depreciation—office equip.	60,000	
Insurance	48,000	
Office rent	54,000	272,000
Income from operations		$ 723,000

Check Income (at 14,500 units), $683,500

RTEX Company manufactures and sells mountain bikes. It normally operates eight hours a day, five days a week. Using this information, classify each of the following costs as fixed or variable. If additional information would affect your decision, describe the information.

Exercise 24-2
Classification of costs as fixed or variable
P1 

a. Management salaries
b. Incoming shipping expenses
c. Office supplies
d. Taxes on property
e. Gas used for heating
f. Direct labor
g. Repair expense for tools
h. Depreciation on tools
i. Pension cost
j. Bike frames
k. Screws for assembly

Exercise 24-3
Preparation of a flexible budget performance report
A1

Hall Company's fixed budget performance report for June follows. The $660,000 budgeted expenses include $450,000 variable expenses and $210,000 fixed expenses. Actual expenses include $200,000 fixed expenses. Prepare a flexible budget performance report that shows any variances between budgeted results and actual results. List fixed and variable expenses separately.

	Fixed Budget	Actual Results	Variances
Sales (in units)	9,000	7,900	
Sales (in dollars)	$720,000	$647,800	$72,200 U
Total expenses	660,000	606,850	53,150 F
Income from operations	$ 60,000	$ 40,950	$19,050 U

Check Income variance, $13,950 F

Exercise 24-4
Preparation of a flexible budget performance report
A1

Burton Company's fixed budget performance report for July follows. The $675,000 budgeted expenses include $634,500 variable expenses and $40,500 fixed expenses. Actual expenses include $52,500 fixed expenses. Prepare a flexible budget performance report showing any variances between budgeted and actual results. List fixed and variable expenses separately.

	Fixed Budget	Actual Results	Variances
Sales (in units)	9,000	11,400	
Sales (in dollars)	$900,000	$1,140,000	$240,000 F
Total expenses	675,000	810,000	135,000 U
Income from operations	$225,000	$ 330,000	$105,000 F

Check Income variance, $34,200 F

Exercise 24-5
Computation and interpretation of labor variances C2 P2

Check October rate variance, $14,880 F

After evaluating Pima Company's manufacturing process, management decides to establish standards of 1.4 hours of direct labor per unit of product and $15 per hour for the labor rate. During October, the company uses 3,720 hours of direct labor at a $40,920 total cost to produce 4,000 units of product. In November, the company uses 4,560 hours of direct labor at a $54,720 total cost to produce 3,500 units of product. (1) Compute the rate variance, the efficiency variance, and the total direct labor cost variance for each of these two months. (2) Interpret the October direct labor variances.

Exercise 24-6
Computation and interpretation of total variable and fixed overhead variances
C2 P3 

Venture Company set the following standard costs for one unit of its product for 2009.

Direct material (20 lbs. @ $5.00 per lb.)	$100.00
Direct labor (10 hrs. @ $16.00 per hr.)	160.00
Factory variable overhead (10 hrs. @ $8.00 per hr.)	80.00
Factory fixed overhead (10 hrs. @ $3.20 per hr.)	32.00
Standard cost	$372.00

The $11.20 ($8.00 + $3.20) total overhead rate per direct labor hour is based on an expected operating level equal to 75% of the factory's capacity of 50,000 units per month. The following monthly flexible budget information is also available.

	Operating Levels (% of capacity)		
	70%	75%	80%
Budgeted output (units)	35,000	37,500	40,000
Budgeted labor (standard hours)	350,000	375,000	400,000
Budgeted overhead (dollars)			
Variable overhead	$2,800,000	$3,000,000	$3,200,000
Fixed overhead	1,200,000	1,200,000	1,200,000
Total overhead	$4,000,000	$4,200,000	$4,400,000

During the current month, the company operated at 70% of capacity, employees worked 340,000 hours, and the following actual overhead costs were incurred.

Variable overhead costs	$2,750,000
Fixed overhead costs	1,257,200
Total overhead costs	$4,007,200

(1) Show how the company computed its predetermined overhead application rate per hour for total overhead, variable overhead, and fixed overhead. (2) Compute the variable and fixed overhead variances.

Check (2) Variable overhead cost variance, $50,000 F

Exercise 24-7
Computation and interpretation of overhead spending, efficiency, and volume variances P3

Refer to the information from Exercise 24-6. Compute and interpret the following.

1. Variable overhead spending and efficiency variances.
2. Fixed overhead spending and volume variances.
3. Controllable variance.

Check (2) Variable overhead: Spending, $30,000 U; efficiency, $80,000 F

Exercise 24-8
Computation and interpretation of materials variances
C2 P2 

Listor Company made 3,800 bookshelves using 23,200 board feet of wood costing $290,000. The company's direct materials standards for one bookshelf are 8 board feet of wood at $12 per board foot. (1) Compute the direct materials variances incurred in manufacturing these bookshelves. (2) Interpret the direct materials variances.

Check Price variance, $11,600 U

Exercise 24-9
Materials variances recorded and closed
C3 P4

Refer to Exercise 24-8. Listor Company records standard costs in its accounts and its material variances in separate accounts when it assigns materials costs to the Goods in Process Inventory account. (1) Show the journal entry that both charges the direct materials costs to the Goods in Process Inventory account and records the materials variances in their proper accounts. (2) Assume that Listor's material variances are the only variances accumulated in the accounting period and that they are immaterial. Prepare the adjusting journal entry to close the variance accounts at period-end. (3) Identify the variance that should be investigated according to the management by exception concept. Explain.

Check (2) Cr. to cost of goods sold, $74,800

Exercise 24-10
Computation of total variable and fixed overhead variances
P3

Integra Company expects to operate at 80% of its productive capacity of 52,000 units per month. At this planned level, the company expects to use 26,000 standard hours of direct labor. Overhead is allocated to products using a predetermined standard rate based on direct labor hours. At the 80% capacity level, the total budgeted cost includes $57,200 fixed overhead cost and $280,800 variable overhead cost. In the current month, the company incurred $320,000 actual overhead and 23,000 actual labor hours while producing 37,000 units. (1) Compute its overhead application rate for total overhead, variable overhead, and fixed overhead. (2) Compute its total overhead variance.

Check (1) Variable overhead rate, $10.80 per hour

Exercise 24-11
Computation of volume and controllable overhead variances P3

Refer to the information from Exercise 24-10. Compute the (1) overhead volume variance and (2) overhead controllable variance.

Check (2) $13,050 U

Exercise 24-12
Computing and interpreting sales variances A2

Wiz Electronics sells computers. During May 2009, it sold 500 computers at a $1,000 average price each. The May 2009 fixed budget included sales of 550 computers at an average price of $950 each. (1) Compute the sales price variance and the sales volume variance for May 2009. (2) Interpret the findings.

McGraw-Hill's HOMEWORK MANAGER® Available with McGraw-Hill's Homework Manager

PROBLEM SET A

Problem 24-1A
Computation of materials, labor, and overhead variances
C2 P2 P3

Beck Company set the following standard unit costs for its single product.

Direct materials (26 lbs. @ $4 per lb.)	$104.00
Direct labor (8 hrs. @ $8 per hr.)	64.00
Factory overhead—variable (8 hrs. @ $5 per hr.)	40.00
Factory overhead—fixed (8 hrs. @ $7 per hr.)	56.00
Total standard cost	$264.00

mhhe.com/wildFAP19e

The predetermined overhead rate is based on a planned operating volume of 70% of the productive capacity of 50,000 units per quarter. The following flexible budget information is available.

	Operating Levels		
	60%	70%	80%
Production in units	30,000	35,000	40,000
Standard direct labor hours	240,000	280,000	320,000
Budgeted overhead			
Fixed factory overhead	$1,960,000	$1,960,000	$1,960,000
Variable factory overhead	$1,200,000	$1,400,000	$1,600,000

During the current quarter, the company operated at 80% of capacity and produced 40,000 units of product; actual direct labor totaled 178,600 hours. Units produced were assigned the following standard costs:

Direct materials (1,040,000 lbs. @ $4 per lb.)	$ 4,160,000
Direct labor (320,000 hrs. @ $8 per hr.)	2,560,000
Factory overhead (320,000 hrs. @ $12 per hr.)	3,840,000
Total standard cost	$10,560,000

Actual costs incurred during the current quarter follow:

Direct materials (1,035,000 lbs. @ $4.10)	$ 4,243,500
Direct labor (327,000 hrs. @ $7.75)	2,534,250
Fixed factory overhead costs	1,875,000
Variable factory overhead costs	1,482,717
Total actual costs	$10,135,467

Required

Check (1) Materials variances: Price, $103,500 U; quantity, $20,000 F. (2) Labor variances: Rate, $81,750 F; efficiency, $56,000 U

1. Compute the direct materials cost variance, including its price and quantity variances.
2. Compute the direct labor variance, including its rate and efficiency variances.
3. Compute the total variable overhead and total fixed overhead variances.
4. Compute these variances: (a) variable overhead spending and efficiency, (b) fixed overhead spending and volume, and (c) total overhead controllable.

Problem 24-2A
Preparation and analysis of a flexible budget
P1 A1

Major Company's 2009 master budget included the following fixed budget report. It is based on an expected production and sales volume of 15,000 units.

MAJOR COMPANY Fixed Budget Report For Year Ended December 31, 2009		
Sales		$3,300,000
Cost of goods sold		
Direct materials	$960,000	
Direct labor	240,000	
Machinery repairs (variable cost)	60,000	
Depreciation—plant equipment	300,000	
Utilities ($60,000 is variable)	180,000	
Plant management salaries	210,000	1,950,000
Gross profit		1,350,000
Selling expenses		
Packaging	75,000	
Shipping	105,000	
Sales salary (fixed annual amount)	235,000	415,000
General and administrative expenses		
Advertising expense	100,000	
Salaries	241,000	
Entertainment expense	85,000	426,000
Income from operations		$ 509,000

Required

1. Classify all items listed in the fixed budget as variable or fixed. Also determine their amounts per unit or their amounts for the year, as appropriate.

2. Prepare flexible budgets (see Exhibit 24.3) for the company at sales volumes of 14,000 and 16,000 units.

3. The company's business conditions are improving. One possible result is a sales volume of approximately 18,000 units. The company president is confident that this volume is within the relevant range of existing capacity. How much would operating income increase over the 2009 budgeted amount of $509,000 if this level is reached without increasing capacity?

4. An unfavorable change in business is remotely possible; in this case, production and sales volume for 2009 could fall to 12,000 units. How much income (or loss) from operations would occur if sales volume falls to this level?

Check (2) Budgeted income at 16,000 units, $629,000

(4) Potential operating income, $149,000

Problem 24-3A
Preparation and analysis of a flexible budget performance report

P1 A2

excel

mhhe.com/wildFAP19e

Refer to the information in Problem 24-2A. Major Company's actual income statement for 2009 follows.

MAJOR COMPANY Statement of Income from Operations For Year Ended December 31, 2009		
Sales (18,000 units)		$3,948,000
Cost of goods sold		
Direct materials	$1,160,000	
Direct labor	293,000	
Machinery repairs (variable cost)	63,000	
Depreciation—plant equipment	300,000	
Utilities (fixed cost is $147,500)	215,500	
Plant management salaries	220,000	2,251,500
Gross profit		1,696,500
Selling expenses		
Packaging	87,500	
Shipping	118,500	
Sales salary (annual)	253,000	459,000
General and administrative expenses		
Advertising expense	107,000	
Salaries	241,000	
Entertainment expense	88,500	436,500
Income from operations		$ 801,000

Required

1. Prepare a flexible budget performance report for 2009.

Check (1) Variances: Fixed costs, $66,000 U; income, $68,000 U

Analysis Component

2. Analyze and interpret both the (a) sales variance and (b) direct materials variance.

Problem 24-4A
Flexible budget preparation; computation of materials, labor, and overhead variances; and overhead variance report

P1 P2 P3 C2

Silver Company set the following standard costs for one unit of its product.

Direct materials (5 lbs. @ $6 per lb.)	$30.00
Direct labor (2 hrs. @ $12 per hr.)	24.00
Overhead (2 hrs. @ $16.65 per hr.)	33.30
Total standard cost	$87.30

The predetermined overhead rate ($16.65 per direct labor hour) is based on an expected volume of 75% of the factory's capacity of 20,000 units per month. Following are the company's budgeted overhead costs per month at the 75% level.

Overhead Budget (75% Capacity)		
Variable overhead costs		
Indirect materials	$ 21,000	
Indirect labor	96,000	
Power	22,500	
Repairs and maintenance	57,000	
Total variable overhead costs		$196,500
Fixed overhead costs		
Depreciation—building	23,000	
Depreciation—machinery	71,000	
Taxes and insurance	18,000	
Supervision	191,000	
Total fixed overhead costs		303,000
Total overhead costs		$499,500

The company incurred the following actual costs when it operated at 75% of capacity in October.

Direct materials (75,500 lbs. @ $6.10 per lb.)		$ 460,550
Direct labor (29,000 hrs. @ $12.20 per hr.)		353,800
Overhead costs		
Indirect materials	$ 22,500	
Indirect labor	88,800	
Power	21,500	
Repairs and maintenance	60,250	
Depreciation—building	23,000	
Depreciation—machinery	65,000	
Taxes and insurance	18,100	
Supervision	185,000	484,150
Total costs		$1,298,500

Required

1. Examine the monthly overhead budget to (a) determine the costs per unit for each variable overhead item and its total per unit costs, and (b) identify the total fixed costs per month.

2. Prepare flexible overhead budgets (as in Exhibit 24.12) for October showing the amounts of each variable and fixed cost at the 65%, 75%, and 85% capacity levels.

3. Compute the direct materials cost variance, including its price and quantity variances.

4. Compute the direct labor cost variance, including its rate and efficiency variances.

5. Compute the (a) variable overhead spending and efficiency variances, (b) fixed overhead spending and volume variances, and (c) total overhead controllable variance.

6. Prepare a detailed overhead variance report (as in Exhibit 24.19) that shows the variances for individual items of overhead.

Check (2) Budgeted total overhead at 13,000 units, $473,300.

(3) Materials variances: Price, $7,550 U; quantity, $3,000 U

(4) Labor variances: Rate, $5,800 U; efficiency, $12,000 F

Problem 24-5A
Materials, labor, and overhead variances; and overhead variance report
C2 P2 P3

Green Company has set the following standard costs per unit for the product it manufactures.

Direct materials (15 lbs. @ $3.90 per lb.)	$ 58.50
Direct labor (4 hrs. @ $18 per hr.)	72.00
Overhead (4 hrs. @ $4.20 per hr.)	16.80
Total standard cost	$147.30

The predetermined overhead rate is based on a planned operating volume of 80% of the productive capacity of 10,000 units per month. The following flexible budget information is available.

	Operating Levels		
	70%	80%	90%
Production in units	7,000	8,000	9,000
Standard direct labor hours	28,000	32,000	36,000
Budgeted overhead			
Variable overhead costs			
Indirect materials	$ 14,000	$ 16,000	$ 18,000
Indirect labor	20,300	23,200	26,100
Power	5,600	6,400	7,200
Maintenance	38,500	44,000	49,500
Total variable costs	78,400	89,600	100,800
Fixed overhead costs			
Rent of factory building	15,000	15,000	15,000
Depreciation—machinery	10,000	10,000	10,000
Supervisory salaries	19,800	19,800	19,800
Total fixed costs	44,800	44,800	44,800
Total overhead costs	$123,200	$134,400	$145,600

During May, the company operated at 90% of capacity and produced 9,000 units, incurring the following actual costs.

Direct materials (139,000 lbs. @ $3.80 per lb.)		$ 528,200
Direct labor (33,000 hrs. @ $18.50 per hr.)		610,500
Overhead costs		
Indirect materials	$16,000	
Indirect labor	27,500	
Power	7,200	
Maintenance	42,000	
Rent of factory building	15,000	
Depreciation—machinery	10,000	
Supervisory salaries	24,000	141,700
Total costs		$1,280,400

Required

1. Compute the direct materials variance, including its price and quantity variances.

2. Compute the direct labor variance, including its rate and efficiency variances.

3. Compute these variances: (a) variable overhead spending and efficiency, (b) fixed overhead spending and volume, and (c) total overhead controllable.

4. Prepare a detailed overhead variance report (as in Exhibit 24.19) that shows the variances for individual items of overhead.

Check (1) Materials variances: Price, $13,900 F; quantity, $15,600 U (2) Labor variances: Rate, $16,500 U; efficiency, $54,000 F

Problem 24-6A
Materials, labor, and overhead variances recorded and analyzed
C3 P4 

Brose Company's standard cost accounting system recorded this information from its December operations.

Standard direct materials cost	$104,000
Direct materials quantity variance (unfavorable)	3,000
Direct materials price variance (favorable)	550
Actual direct labor cost	90,000
Direct labor efficiency variance (favorable)	6,850
Direct labor rate variance (unfavorable)	1,200
Actual overhead cost	375,000
Volume variance (unfavorable)	13,000
Controllable variance (unfavorable)	9,000

Check (1) Dr. Goods in Process Inventory (for overhead), $353,000

Required

1. Prepare December 31 journal entries to record the company's costs and variances for the month. (Do not prepare the journal entry to close the variances.)

Analysis Component

2. Identify the areas that would attract the attention of a manager who uses management by exception. Explain what action(s) the manager should consider.

PROBLEM SET B

Problem 24-1B
Computation of materials, labor, and overhead variances
C2 P2 P3

Krug Company set the following standard unit costs for its single product.

Direct materials (5 lbs. @ $2 per lb.)	$10.00
Direct labor (0.3 hrs. @ $15 per hr.)	4.50
Factory overhead—variable (0.3 hrs. @ $10 per hr.)	3.00
Factory overhead—fixed (0.3 hrs. @ $14 per hr.)	4.20
Total standard cost	$21.70

The predetermined overhead rate is based on a planned operating volume of 80% of the productive capacity of 600,000 units per quarter. The following flexible budget information is available.

	Operating Levels		
	70%	80%	90%
Production in units	420,000	480,000	540,000
Standard direct labor hours	126,000	144,000	162,000
Budgeted overhead			
Fixed factory overhead	$2,016,000	$2,016,000	$2,016,000
Variable factory overhead	1,260,000	1,440,000	1,620,000

During the current quarter, the company operated at 70% of capacity and produced 420,000 units of product; direct labor hours worked were 125,000. Units produced were assigned the following standard costs:

Direct materials (2,100,000 lbs. @ $2 per lb.)	$4,200,000
Direct labor (126,000 hrs. @ $15 per hr.)	1,890,000
Factory overhead (126,000 hrs. @ $24 per hr.)	3,024,000
Total standard cost	$9,114,000

Actual costs incurred during the current quarter follow:

Direct materials (2,000,000 lbs. @ $2.15)	$4,300,000
Direct labor (125,000 hrs. @ $15.50)	1,937,500
Fixed factory overhead costs	1,960,000
Variable factory overhead costs	1,200,000
Total actual costs	$9,397,500

Required

Check (1) Materials variances: Price, $300,000 U; quantity, $200,000 F (2) Labor variances: Rate, $62,500 U; efficiency, $15,000 F

1. Compute the direct materials cost variance, including its price and quantity variances.
2. Compute the direct labor variance, including its rate and efficiency variances.
3. Compute the total variable overhead and total fixed overhead variances.
4. Compute these variances: (a) variable overhead spending and efficiency, (b) fixed overhead spending and volume, and (c) total overhead controllable.

Problem 24-2B
Preparation and analysis of a flexible budget P1 A1

Toronto Company's 2009 master budget included the following fixed budget report. It is based on an expected production and sales volume of 10,000 units.

TORONTO COMPANY Fixed Budget Report For Year Ended December 31, 2009		
Sales		$1,500,000
Cost of goods sold		
Direct materials	$600,000	
Direct labor	130,000	
Machinery repairs (variable cost)	28,500	
Depreciation—machinery	125,000	
Utilities (25% is variable cost)	100,000	
Plant manager salaries	70,000	1,053,500
Gross profit		446,500
Selling expenses		
Packaging	40,000	
Shipping	58,000	
Sales salary (fixed annual amount)	80,000	178,000
General and administrative expenses		
Advertising	40,500	
Salaries	120,500	
Entertainment expense	45,000	206,000
Income from operations		$ 62,500

Required

1. Classify all items listed in the fixed budget as variable or fixed. Also determine their amounts per unit or their amounts for the year, as appropriate.

2. Prepare flexible budgets (see Exhibit 24.3) for the company at sales volumes of 9,500 and 10,500 units.

Check (2) Budgeted income at 10,500 units, $93,425

3. The company's business conditions are improving. One possible result is a sales volume of approximately 12,000 units. The company president is confident that this volume is within the relevant range of existing capacity. How much would operating income increase over the 2009 budgeted amount of $62,500 if this level is reached without increasing capacity?

4. An unfavorable change in business is remotely possible; in this case, production and sales volume for 2009 could fall to 8,000 units. How much income (or loss) from operations would occur if sales volume falls to this level?

(4) Potential operating loss, $(61,200)

Problem 24-3B
Preparation and analysis of a flexible budget performance report
P1 A2

Refer to the information in Problem 24-2B. Toronto Company's actual income statement for 2009 follows.

TORONTO COMPANY Statement of Income from Operations For Year Ended December 31, 2009		
Sales (10,500 units)		$1,596,000
Cost of goods sold		
Direct materials	$612,500	
Direct labor	157,500	
Machinery repairs (variable cost)	26,250	
Depreciation—machinery	125,000	
Utilities (variable cost, $28,000)	105,000	
Plant manager salaries	77,500	1,103,750
Gross profit		492,250
Selling expenses		
Packaging	39,375	
Shipping	54,250	
Sales salary (annual)	81,000	174,625
General and administrative expenses		
Advertising expense	52,000	
Salaries	116,000	
Entertainment expense	50,000	218,000
Income from operations		$ 99,625

Required

Check (1) Variances: Fixed costs, $22,500 U; income, $6,200 F

1. Prepare a flexible budget performance report for 2009.

Analysis Component

2. Analyze and interpret both the (a) sales variance and (b) direct materials variance.

Problem 24-4B
Flexible budget preparation; computation of materials, labor, and overhead variances; and overhead variance report
P1 P2 P3 C2

Stevens Company set the following standard costs for one unit of its product.

Direct materials (9 lb. @ $6 per lb.)	$ 54.00
Direct labor (3 hrs. @ $16 per hr.)	48.00
Overhead (3 hrs. @ $11.75 per hr.)	35.25
Total standard cost	137.25

The predetermined overhead rate ($11.75 per direct labor hour) is based on an expected volume of 75% of the factory's capacity of 20,000 units per month. Following are the company's budgeted overhead costs per month at the 75% level.

Overhead Budget (75% Capacity)		
Variable overhead costs		
Indirect materials	$ 33,750	
Indirect labor	135,000	
Power	22,500	
Repairs and maintenance	67,500	
Total variable overhead costs		$258,750
Fixed overhead costs		
Depreciation—building	36,000	
Depreciation—machinery	108,000	
Taxes and insurance	27,000	
Supervision	99,000	
Total fixed overhead costs		270,000
Total overhead costs		$528,750

The company incurred the following actual costs when it operated at 75% of capacity in December.

Direct materials (139,000 lbs. @ $6.10)		$ 847,900
Direct labor (43,500 hrs. @ $16.30)		709,050
Overhead costs		
Indirect materials	$ 31,600	
Indirect labor	133,400	
Power	23,500	
Repairs and maintenance	69,700	
Depreciation—building	36,000	
Depreciation—machinery	110,000	
Taxes and insurance	24,500	
Supervision	99,000	527,700
Total costs		$2,084,650

Required

1. Examine the monthly overhead budget to (a) determine the costs per unit for each variable overhead item and its total per unit costs, and (b) identify the total fixed costs per month.

Check (2) Budgeted total overhead at 17,000 units, $563,250

2. Prepare flexible overhead budgets (as in Exhibit 24.12) for December showing the amounts of each variable and fixed cost at the 65%, 75%, and 85% capacity levels.

(3) Materials variances: Price, $13,900 U; quantity, $24,000 U

3. Compute the direct materials cost variance, including its price and quantity variances.

4. Compute the direct labor cost variance, including its rate and efficiency variances.

(4) Labor variances: Rate, $13,050 U; efficiency, $24,000 F

5. Compute the (a) variable overhead spending and efficiency variances, (b) fixed overhead spending and volume variances, and (c) total overhead controllable variance.

6. Prepare a detailed overhead variance report (as in Exhibit 24.19) that shows the variances for individual items of overhead.

Problem 24-5B
Materials, labor, and overhead variances; and overhead variance report
C2 P2 P3

Harris Company has set the following standard costs per unit for the product it manufactures.

Direct materials (5 lbs. @ $3.00 per lb.)	$15
Direct labor (2 hr. @ $20 per hr.)	40
Overhead (2 hr. @ $10 per hr.)	20
Total standard cost	$75

The predetermined overhead rate is based on a planned operating volume of 80% of the productive capacity of 10,000 units per month. The following flexible budget information is available.

	Operating Levels		
	70%	80%	90%
Production in units	7,000	8,000	9,000
Standard direct labor hours	14,000	16,000	18,000
Budgeted overhead			
Variable overhead costs			
Indirect materials	$ 17,500	$ 20,000	$22,500
Indirect labor	28,000	32,000	36,000
Power	7,000	8,000	9,000
Maintenance	3,500	4,000	4,500
Total variable costs	56,000	64,000	72,000
Fixed overhead costs			
Rent of factory building	24,000	24,000	24,000
Depreciation—machinery	40,000	40,000	40,000
Taxes and insurance	4,800	4,800	4,800
Supervisory salaries	27,200	27,200	27,200
Total fixed costs	96,000	96,000	96,000
Total overhead costs	$152,000	$160,000	$168,000

During March, the company operated at 90% of capacity and produced 9,000 units, incurring the following actual costs.

Direct materials (46,000 lbs. @ $2.95 per lb.)		$ 135,700
Direct labor (18,800 hrs. @ $20.10 per hr.)		377,880
Overhead costs		
Indirect materials	$22,000	
Indirect labor	32,000	
Power	9,600	
Maintenance	4,750	
Rent of factory building	24,000	
Depreciation—machinery	39,400	
Taxes and insurance	5,200	
Supervisory salaries	28,000	164,950
Total costs		$678,530

Required

Check (1) Materials variances: Price, $2,300 F; quantity, $3,000 U
(2) Labor variances: Rate, $1,880 U; efficiency, $16,000 U

1. Compute the direct materials cost variance, including its price and quantity variances.
2. Compute the direct labor variance, including its rate and efficiency variances.
3. Compute these variances: (a) variable overhead spending and efficiency, (b) fixed overhead spending and volume, and (c) total overhead controllable.
4. Prepare a detailed overhead variance report (as in Exhibit 24.19) that shows the variances for individual items of overhead.

Problem 24-6B
Materials, labor, and overhead variances recorded and analyzed
C3 P4 

Del Company's standard cost accounting system recorded this information from its June operations.

Standard direct materials cost	$260,000
Direct materials quantity variance (favorable)	10,000
Direct materials price variance (favorable)	3,000
Actual direct labor cost	130,000
Direct labor efficiency variance (favorable)	6,000
Direct labor rate variance (unfavorable)	1,000
Actual overhead cost	500,000
Volume variance (unfavorable)	24,000
Controllable variance (unfavorable)	16,000

Required

Check (1) Dr. Goods in Process Inventory (for overhead), $460,000

1. Prepare journal entries dated June 30 to record the company's costs and variances for the month. (Do not prepare the journal entry to close the variances.)

Analysis Component

2. Identify the areas that would attract the attention of a manager who uses management by exception. Describe what action(s) the manager should consider.

SERIAL PROBLEM

Success Systems

(This serial problem began in Chapter 1 and continues through most of the book. If previous chapter segments were not completed, the serial problem can begin at this point. It is helpful, but not necessary, to use the working papers that accompany the book.)

SP 24 Success Systems' second quarter 2010 fixed budget performance report for its computer furniture operations follows. The $156,000 budgeted expenses include $108,000 in variable expenses for desks and $18,000 in variable expenses for chairs, as well as $30,000 fixed expenses. The actual expenses include $31,000 fixed expenses. Prepare a flexible budget performance report that shows any variances between budgeted results and actual results. List fixed and variable expenses separately.

	Fixed Budget	Actual Results	Variances
Desk sales (in units)	144	150	
Chair sales (in units)	72	80	
Desk sales (in dollars)	$180,000	$186,000	$6,000 F
Chair sales (in dollars)	$ 36,000	$ 41,200	$5,200 F
Total expenses	$156,000	$163,880	$7,880 U
Income from operations	$ 60,000	$ 63,320	$3,320 F

Check Variances: Fixed expenses, $1,000 U

BEYOND THE NUMBERS

REPORTING IN ACTION

C1 

BTN 24-1 Analysis of flexible budgets and standard costs emphasizes the importance of a similar unit of measure for meaningful comparisons and evaluations. When **Best Buy** compiles its financial reports in compliance with GAAP, it applies the same unit of measurement, U.S. dollars, for most measures of business operations. One issue for Best Buy is how best to adjust account values for its subsidiaries that compile financial reports in currencies other than the U.S. dollar.

Required

1. Read Best Buy's Note 1 in Appendix A and identify the financial statement where it reports the annual adjustment for foreign currency translation.
2. Record the annual amount of its foreign currency translation adjustment for the fiscal years 2005 through 2007.

Fast Forward

3. Access Best Buy's financial statements for a fiscal year ending after March 3, 2007, from either its Website [BestBuy.com] or the SEC's EDGAR database [www.SEC.gov]. (a) Identify its foreign currency translation adjustment. (b) Does this adjustment increase or decrease net income? Explain.

COMPARATIVE ANALYSIS

A2

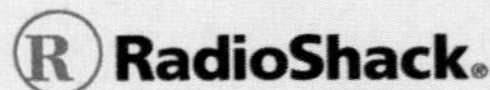

BTN 24-2 The usefulness of budgets, variances, and related analyses often depends on the accuracy of management's estimates of future sales activity.

Required

1. Identify and record the prior three years' sales (in dollars) for both **Best Buy**, **Circuit City**, and **RadioShack** using their financial statements in Appendix A.
2. Using the data in part 1, predict all three companies' sales activity for the next two to three years. (If possible, compare your predictions to actual sales figures for these years.)

ETHICS CHALLENGE

C1

BTN 24-3 Setting materials, labor, and overhead standards is challenging. If standards are set too low, companies might purchase inferior products and employees might not work to their full potential. If standards are set too high, companies could be unable to offer a quality product at a profitable rate and employees could be overworked. The ethical challenge is to set a high but reasonable standard. Assume that as a manager, you are asked to set the standard materials price and quantity for the new 1,000 CKB Mega-Max chip, a technically advanced product. To properly set the price and quantity standards, you assemble a team of specialists to provide input.

Required

Identify four types of specialists that you would assemble to provide information to help set the materials price and quantity standards. Briefly explain why you chose each individual.

COMMUNICATING IN PRACTICE

P4 C2 

BTN 24-4 The reason we use the words *favorable* and *unfavorable* when evaluating variances is made clear when we look at the closing of accounts. To see this, consider that (1) all variance accounts are closed at the end of each period (temporary accounts), (2) a favorable variance is always a credit balance, and (3) an unfavorable variance is always a debit balance. Write a one-half page memorandum to your instructor with three parts that answer the three following requirements. (Assume that variance accounts are closed to Cost of Goods Sold.)

Required

1. Does Cost of Goods Sold increase or decrease when closing a favorable variance? Does gross margin increase or decrease when a favorable variance is closed to Cost of Goods Sold? Explain.
2. Does Cost of Goods Sold increase or decrease when closing an unfavorable variance? Does gross margin increase or decrease when an unfavorable variance is closed to Cost of Goods Sold? Explain.
3. Explain the meaning of a favorable variance and an unfavorable variance.

TAKING IT TO THE NET

C1  

BTN 24-5 Access **iSixSigma**'s Website (**iSixSigma.com**) to search for and read information about *benchmarking* to complete the following requirements.

Required

1. Write a one-paragraph explanation (in layperson's terms) of benchmarking.
2. How does standard costing relate to benchmarking?

TEAMWORK IN ACTION

C2 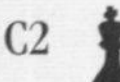

BTN 24-6 Many service industries link labor rate and time (quantity) standards with their processes. One example is the standard time to board an aircraft. The reason time plays such an important role in the service industry is that it is viewed as a competitive advantage: best service in the shortest amount of time. Although the labor rate component is difficult to observe, the time component of a service delivery standard is often readily apparent—for example, "Lunch will be served in less than five minutes, or it is free."

Required

Break into teams and select two service industries for your analysis. Identify and describe all the time elements each industry uses to create a competitive advantage.

ENTREPRENEURIAL DECISION

C1 C2 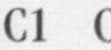 

BTN 24-7 Entrepreneur Chris Martin of **Martin Guitar Company** (see Chapter opener) uses a costing system with standard costs for direct materials, direct labor, and overhead costs. Two comments frequently are mentioned in relation to standard costing and variance analysis: "Variances are not explanations" and "Management's goal is not to minimize variances."

Required

Write Chris Martin a short memo (no more than 1 page) interpreting these two comments.

HITTING THE ROAD

C1 

BTN 24-8 Training employees to use standard amounts of materials in production is common. Typically large companies invest in this training but small organizations do not. One can observe these different practices in a trip to two different pizza businesses. Visit both a local pizza business and a national pizza chain business and then complete the following.

Required

1. Observe and record the number of raw material items used to make a typical cheese pizza. Also observe how the person making the pizza applies each item when preparing the pizza.
2. Record any differences in how items are applied between the two businesses.
3. Estimate which business is more profitable from your observations. Explain.

GLOBAL DECISION

BTN 24-9 Access the annual report of **DSG** (at **www.DSGiplc.com**) for the year ended April 28, 2007. The usefulness of its budgets, variances, and related analyses depends on the accuracy of management's estimates of future sales activity.

Required

1. Identify and record the prior two years' sales (in pounds) for DSG from its income statement.
2. Using the data in part 1, predict sales activity for DSG for the next two years. Explain your prediction process.

ANSWERS TO MULTIPLE CHOICE QUIZ

1. c; Fixed costs remain at $300,000; Variable costs = ($246,000/24,000 units) × 20,000 units = $205,000.
2. e; Budgeted direct materials + Unfavorable variance = Actual cost of direct materials used; or, 60,000 units × $10 per unit = $600,000 + $15,000 U = $615,000.
3. c; (AH × AR) − (AH × SR) = $1,599,040 – (84,160 hours × $20 per hour) = $84,160 F.
4. b; Actual variable overhead − Variable overhead applied to production = Variable overhead cost variance; or $150,000 − (96,000 hours × $1.50 per hour) = $6,000 U.
5. a; Budgeted fixed overhead − Fixed overhead applied to production = Volume variance; or $24,000 − (4,800 units × $4 per unit) = $4,800 U.

A Look Back

Chapter 24 discussed flexible budgets, variance analysis, and standard costs. It explained how management uses each to control and monitor business activities.

A Look at This Chapter

This chapter focuses on evaluating capital budgeting decisions. It also explains several tools and procedures used in making and evaluating short-term managerial decisions.

Chapter 25 Capital Budgeting and Managerial Decisions

Learning Objectives

CAP

Conceptual

C1 Explain the importance of capital budgeting. *(p. 1024)*

C2 Describe the selection of a hurdle rate for an investment. *(p. 1033)*

C3 Describe the importance of relevant costs for short-term decisions. *(p. 1035)*

Analytical

A1 Evaluate short-term managerial decisions using relevant costs. *(p. 1035)*

A2 Analyze a capital investment project using break-even time. *(p. 1043)*

LP25

Procedural

P1 Compute payback period and describe its use. *(p. 1025)*

P2 Compute accounting rate of return and explain its use. *(p. 1027)*

P3 Compute net present value and describe its use. *(p. 1029)*

P4 Compute internal rate of return and explain its use. *(p. 1031)*

Decision Feature

Batter Up

"Now batting, a 34-ounce Prairie Sticks double-dipped black maple bat!"—PA Announcer

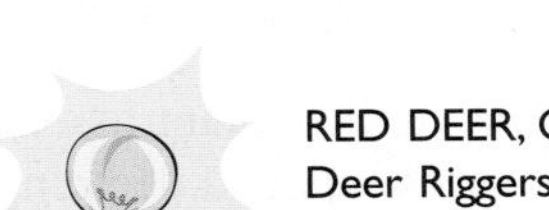

RED DEER, CANADA—Jared Greenberg, of the Red Deer Riggers, and Dan Zinger of the Red Deer Stags, dream to make it to the major leagues . . . not as players, but as makers of baseball bats. Their start-up company, **Prairie Sticks Bat Company (PrairieSticks.com),** started in Jared's workshop with a hand lathe and a piece of wood when local amateur players had trouble getting maple bats from manufacturers. Jared says he began producing bats for his teammates and friends "just like you would do in your middle school shop class."

Prairie Sticks' bats are made from four different types of wood, each with different prices (the company also makes fungo bats and training bats). Jared and Dan use product contribution margins in determining their best sales mix. This is especially important given their constraints on machine hours and labor—they have only one hydraulic tracing lathe and no other employees that make bats.

This past year they sold 1,500 bats. With production growth comes new business questions. Do we take a one-time deal with a buyer? Do we scrap or rework unacceptable inventory? Do we make or buy certain raw materials? These questions need answers. Jared and Dan focus on relevant costs and incremental revenues for insight into answering those questions. If a customer wants a bat in a color Prairie Sticks does not stock, the company charges a higher price to cover the incremental cost of the new color. The company makes novelty bats, unusable for play but fine for gifts and awards, out of inferior wood. These novelty bats sell at reduced prices, but enable the company to avoid costly rework and processing costs. They also sell apparel and hats, made by outside manufacturers.

Prairie Sticks now makes bats for big leaguers. It uses the same wood as the major batmakers; and $100,000 worth of equipment, including the hydraulic lathe, can turn out an unfinished bat in less than two minutes. Soon, they hope to step to the plate to accept additional business. Jared and Dan apply capital budgeting methods to help assess payback periods, rates of return, present values, and break-even points to their equipment and other expenditures. This aids them in decisions on what, and when, to buy.

A recent news release reported that a minor league player had been traded for "10 Prairie Sticks double-dipped maple bats, black," which led to major publicity and a surge in orders. "It's been crazy," says Jared. "[Since] this story has broken . . . we're on the verge of picking up our Major League vendor's license," explains Dan. That would be a tape-measure home run.

[Sources: *Prairie Sticks Bat Company Website,* January 2009; *AlbertaLocalNews.com,* May 2008; *Fox Sports on MSN.com,* May 2008; *Edmonton CityTV.com* interview, May 2008]

Chapter Preview

Making business decisions involves choosing between alternative courses of action. Many factors affect business decisions, yet analysis typically focuses on finding the alternative that offers the highest return on investment or the greatest reduction in costs. Some decisions are based on little more than an intuitive understanding of the situation because available information is too limited to allow a more systematic analysis. In other cases, intangible factors such as convenience, prestige, and environmental considerations are more important than strictly quantitative factors. In all situations, managers can reach a sounder decision if they identify the consequences of alternative choices in financial terms. This chapter explains several methods of analysis that can help managers make those business decisions.

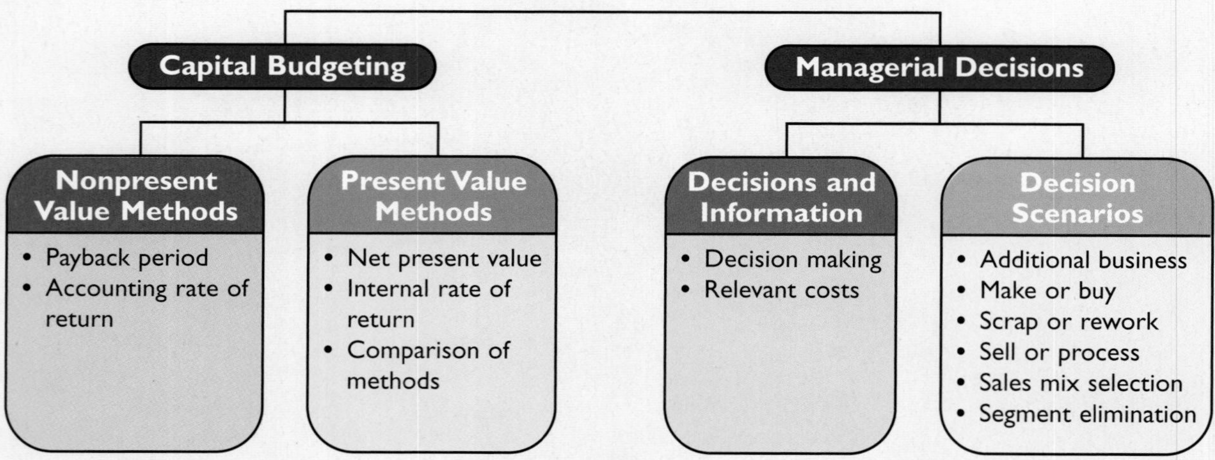

Section 1—Capital Budgeting

C1 Explain the importance of capital budgeting.

The capital expenditures budget is management's plan for acquiring and selling plant assets. **Capital budgeting** is the process of analyzing alternative long-term investments and deciding which assets to acquire or sell. These decisions can involve developing a new product or process, buying a new machine or a new building, or acquiring an entire company. An objective for these decisions is to earn a satisfactory return on investment.

Video25.2

Capital budgeting decisions require careful analysis because they are usually the most difficult and risky decisions that managers make. These decisions are difficult because they require predicting events that will not occur until well into the future. Many of these predictions are tentative and potentially unreliable. Specifically, a capital budgeting decision is risky because (1) the outcome is uncertain, (2) large amounts of money are usually involved, (3) the investment involves a long-term commitment, and (4) the decision could be difficult or impossible to reverse, no matter how poor it turns out to be. Risk is especially high for investments in technology due to innovations and uncertainty.

Managers use several methods to evaluate capital budgeting decisions. Nearly all of these methods involve predicting cash inflows and cash outflows of proposed investments, assessing the risk of and returns on those flows, and then choosing the investments to make. Management often restates future cash flows in terms of their present value. This approach applies the time value of money: A dollar today is worth more than a dollar tomorrow. Similarly, a dollar tomorrow is worth less than a dollar today. The process of restating future cash flows in terms of their present value is called *discounting*. The time value of money is important when evaluating capital investments, but managers sometimes apply evaluation methods that ignore present value. This section describes four methods for comparing alternative investments.

Point: The nature of capital spending has changed with the business environment. Budgets for information technology have increased from about 25% of corporate capital spending 20 years ago to an estimated 35% today.

Methods Not Using Time Value of Money

All investments, whether they involve the purchase of a machine or another long-term asset, are expected to produce net cash flows. *Net cash flow* is cash inflows minus cash outflows. Sometimes managers perform simple analyses of the financial feasibility of an investment's net cash flow without using the time value of money. This section explains two of the most common methods in this category: (1) payback period and (2) accounting rate of return.

Payback Period

P1 Compute payback period and describe its use.

An investment's **payback period (PBP)** is the expected time period to recover the initial investment amount. Managers prefer investing in assets with shorter payback periods to reduce the risk of an unprofitable investment over the long run. Acquiring assets with short payback periods reduces a company's risk from potentially inaccurate long-term predictions of future cash flows.

Computing Payback Period with Even Cash Flows To illustrate use of the payback period for an investment with even cash flows, we look at data from FasTrac, a manufacturer of exercise equipment and supplies. (*Even cash flows* are cash flows that are the same each and every year; *uneven cash flows* are cash flows that are not all equal in amount.) FasTrac is considering several different capital investments, one of which is to purchase a machine to use in manufacturing a new product. This machine costs $16,000 and is expected to have an eight-year life with no salvage value. Management predicts this machine will produce 1,000 units of product each year and that the new product will be sold for $30 per unit. Exhibit 25.1 shows the expected annual net cash flows for this asset over its life as well as the expected annual revenues and expenses (including depreciation and income taxes) from investing in the machine.

EXHIBIT 25.1
Cash Flow Analysis

FASTRAC
Cash Flow Analysis—Machinery Investment
January 15, 2009

	Expected Accrual Figures	Expected Net Cash Flows
Annual sales of new product	$30,000	$30,000
Deduct annual expenses		
Cost of materials, labor, and overhead (except depreciation)	15,500	15,500
Depreciation—Machinery	2,000	
Additional selling and administrative expenses	9,500	9,500
Annual pretax accrual income	3,000	
Income taxes (30%)	900	900
Annual net income	$ 2,100	
Annual net cash flow		**$ 4,100**

The amount of net cash flow from the machinery is computed by subtracting expected cash outflows from expected cash inflows. The cash flows column of Exhibit 25.1 excludes all noncash revenues and expenses. Depreciation is FasTrac's only noncash item. Alternatively, managers can adjust the projected net income for revenue and expense items that do not affect cash flows. For FasTrac, this means taking the $2,100 net income and adding back the $2,000 depreciation.

Point: Annual net cash flow in Exhibit 25.1 equals net income plus depreciation (a noncash expense).

The formula for computing the payback period of an investment that yields even net cash flows is in Exhibit 25.2.

EXHIBIT 25.2
Payback Period Formula with Even Cash Flows

$$\textbf{Payback period} = \frac{\textbf{Cost of investment}}{\textbf{Annual net cash flow}}$$

The payback period reflects the amount of time for the investment to generate enough net cash flow to return (or pay back) the cash initially invested to purchase it. FasTrac's payback period for this machine is just under four years:

$$\text{Payback period} = \frac{\$16{,}000}{\$4{,}100} = 3.9 \text{ years}$$

Example: If an alternative machine (with different technology) yields a payback period of 3.5 years, which one does a manager choose? Answer: The alternative (3.5 is less than 3.9).

The initial investment is fully recovered in 3.9 years, or just before reaching the halfway point of this machine's useful life of eight years.

Decision Insight

Payback Phones Profits of telecoms have declined as too much capital investment chased too little revenue. Telecom success depends on new technology, and communications gear is evolving at a dizzying rate. Consequently, managers of telecoms often demand short payback periods and large expected net cash flows to compensate for the investment risk.

Computing Payback Period with Uneven Cash Flows Computing the payback period in the prior section assumed even net cash flows. What happens if the net cash flows are uneven? In this case, the payback period is computed using the *cumulative total of net cash flows.* The word *cumulative* refers to the addition of each period's net cash flows as we progress through time. To illustrate, consider data for another investment that FasTrac is considering. This machine is predicted to generate uneven net cash flows over the next eight years. The relevant data and payback period computation are shown in Exhibit 25.3.

EXHIBIT 25.3

Payback Period Calculation with Uneven Cash Flows

Period*	Expected Net Cash Flows	Cumulative Net Cash Flows
Year 0	$(16,000)	$(16,000)
Year 1	3,000	(13,000)
Year 2	4,000	(9,000)
Year 3	4,000	(5,000)
Year 4	4,000	(1,000)
Year 5	5,000	4,000
Year 6	3,000	7,000
Year 7	2,000	9,000
Year 8	2,000	11,000
		Payback period = 4.2 years

* All cash inflows and outflows occur uniformly during the year.

Example: Find the payback period in Exhibit 25.3 if net cash flows for the first 4 years are:
Year 1 = $6,000; Year 2 = $5,000;
Year 3 = $4,000; Year 4 = $3,000.
Answer: 3.33 years

Year 0 refers to the period of initial investment in which the $16,000 cash outflow occurs at the end of year 0 to acquire the machinery. By the end of year 1, the cumulative net cash flow is reduced to $(13,000), computed as the $(16,000) initial cash outflow plus year 1's $3,000 cash inflow. This process continues throughout the asset's life. The cumulative net cash flow amount changes from negative to positive in year 5. Specifically, at the end of year 4, the cumulative net cash flow is $(1,000). As soon as FasTrac receives net cash inflow of $1,000 during the fifth year, it has fully recovered the investment. If we assume that cash flows are received uniformly *within* each year, receipt of the $1,000 occurs about one-fifth of the way through the year. This is computed as $1,000 divided by year 5's total net cash flow of $5,000, or 0.20. This yields a payback period of 4.2 years, computed as 4 years plus 0.20 of year 5.

Using the Payback Period Companies desire a short payback period to increase return and reduce risk. The more quickly a company receives cash, the sooner it is available for other uses and the less time it is at risk of loss. A shorter payback period also improves the company's ability to respond to unanticipated changes and lowers its risk of having to keep an unprofitable investment.

Payback period should never be the only consideration in evaluating investments. This is so because it ignores at least two important factors. First, it fails to reflect differences in the timing of net cash flows within the payback period. In Exhibit 25.3, FasTrac's net cash flows in the first five years were \$3,000, \$4,000, \$4,000, \$4,000, and \$5,000. If another investment had predicted cash flows of \$9,000, \$3,000, \$2,000, \$1,800, and \$1,000 in these five years, its payback period would also be 4.2 years, but this second alternative could be more desirable because it provides cash more quickly. The second important factor is that the payback period ignores *all* cash flows after the point where its costs are fully recovered. For example, one investment might pay back its cost in 3 years but stop producing cash after 4 years. A second investment might require 5 years to pay back its cost yet continue to produce net cash flows for another 15 years. A focus on only the payback period would mistakenly lead management to choose the first investment over the second.

"So what if I underestimated costs and overestimated revenues? It all averages out in the end."

Quick Check

Answers—p. 1049

1. Capital budgeting is (*a*) concerned with analyzing alternative sources of capital, including debt and equity, (*b*) an important activity for companies when considering what assets to acquire or sell, or (*c*) best done by intuitive assessments of the value of assets and their usefulness.
2. Why are capital budgeting decisions often difficult?
3. A company is considering purchasing equipment costing \$75,000. Future annual net cash flows from this equipment are \$30,000, \$25,000, \$15,000, \$10,000, and \$5,000. The payback period is (*a*) 4 years, (*b*) 3.5 years, or (*c*) 3 years.
4. If depreciation is an expense, why is it added back to an investment's net income to compute the net cash flow from that investment?
5. If two investments have the same payback period, are they equally desirable? Explain.

Accounting Rate of Return

P2 Compute accounting rate of return and explain its use.

The **accounting rate of return,** also called *return on average investment,* is computed by dividing a project's after-tax net income by the average amount invested in it. To illustrate, we return to FasTrac's \$16,000 machinery investment described in Exhibit 25.1. We first compute (1) the after-tax net income and (2) the average amount invested. The \$2,100 after-tax net income is already available from Exhibit 25.1. To compute the average amount invested, we assume that net cash flows are received evenly throughout each year. Thus, the average investment for each year is computed as the average of its beginning and ending book values. If FasTrac's \$16,000 machine is depreciated \$2,000 each year, the average amount invested in the machine for each year is computed as shown in Exhibit 25.4. The average for any year is the average of the beginning and ending book values.

EXHIBIT 25.4

Computing Average Amount Invested

	Beginning Book Value	Annual Depreciation	Ending Book Value	Average Book Value
Year 1	\$16,000	\$2,000	\$14,000	\$15,000
Year 2	14,000	2,000	12,000	13,000
Year 3	12,000	2,000	10,000	11,000
Year 4	10,000	2,000	8,000	9,000
Year 5	8,000	2,000	6,000	7,000
Year 6	6,000	2,000	4,000	5,000
Year 7	4,000	2,000	2,000	3,000
Year 8	2,000	2,000	0	1,000
All years				\$ 8,000

Next we need the average book value for the asset's entire life. This amount is computed by taking the average of the individual yearly averages. This average equals \$8,000, computed as \$64,000 (the sum of the individual years' averages) divided by eight years (see last column of Exhibit 25.4).

Point: General formula for *annual average investment* is the sum of individual years' average book values divided by the number of years of the planned investment.

If a company uses straight-line depreciation, we can find the average amount invested by using the formula in Exhibit 25.5. Because FasTrac uses straight-line depreciation, its average amount invested for the eight years equals the sum of the book value at the beginning of the asset's investment period and the book value at the end of its investment period, divided by 2, as shown in Exhibit 25.5.

EXHIBIT 25.5

Computing Average Amount Invested under Straight-Line Depreciation

$$\textbf{Annual average investment} = \frac{\textbf{Beginning book value + Ending book value}}{\textbf{2}}$$

(straight-line case only)

$$= \frac{\$16{,}000 + \$0}{2} = \$8{,}000$$

If an investment has a salvage value, the average amount invested when using straight-line depreciation is computed as (Beginning book value + Salvage value)/2.

Once we determine the after-tax net income and the average amount invested, the accounting rate of return on the investment can be computed from the annual after-tax net income divided by the average amount invested, as shown in Exhibit 25.6.

EXHIBIT 25.6

Accounting Rate of Return Formula

$$\textbf{Accounting rate of return} = \frac{\textbf{Annual after-tax net income}}{\textbf{Annual average investment}}$$

This yields an accounting rate of return of 26.25% ($2,100/$8,000). FasTrac management must decide whether a 26.25% accounting rate of return is satisfactory. To make this decision, we must factor in the investment's risk. For instance, we cannot say an investment with a 26.25% return is preferred over one with a lower return unless we recognize any differences in risk. Thus, an investment's return is satisfactory or unsatisfactory only when it is related to returns from other investments with similar lives and risk.

When accounting rate of return is used to choose among capital investments, the one with the least risk, the shortest payback period, and the highest return for the longest time period is often identified as the best. However, use of accounting rate of return to evaluate investment opportunities is limited because it bases the amount invested on book values (not predicted market values) in future periods. Accounting rate of return is also limited when an asset's net incomes are expected to vary from year to year. This requires computing the rate using *average* annual net incomes, yet this accounting rate of return fails to distinguish between two investments with the same average annual net income but different amounts of income in early years versus later years or different levels of income variability.

Quick Check Answers—p. 1049

6. The following data relate to a company's decision on whether to purchase a machine:

Cost	$180,000
Salvage value	15,000
Annual after-tax net income	40,000

The machine's accounting rate of return, assuming the even receipt of its net cash flows during the year and use of straight-line depreciation, is (*a*) 22%, (*b*) 41%, or (*c*) 21%.

7. Is a 15% accounting rate of return for a machine a good rate?

Methods Using Time Value of Money

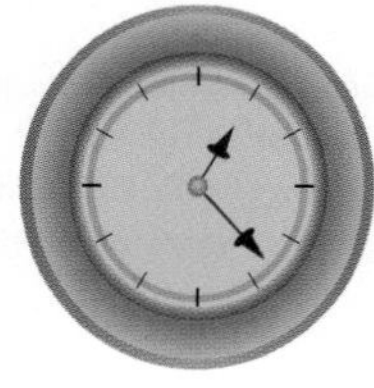

This section describes two methods that help managers with capital budgeting decisions and that use the time value of money: (1) net present value and (2) internal rate of return. *(To apply these methods, you need a basic understanding of the concept of present value. An expanded explanation of present value concepts is in Appendix B near the end of the book. You can use the present value tables at the end of Appendix B to solve many of this chapter's assignments that use the time value of money.)*

Net Present Value

P3 Compute net present value and describe its use.

Net present value analysis applies the time value of money to future cash inflows and cash outflows so management can evaluate a project's benefits and costs at one point in time. Specifically, **net present value (NPV)** is computed by discounting the future net cash flows from the investment at the project's required rate of return and then subtracting the initial amount invested. A company's required return, often called its hurdle rate, is typically its **cost of capital,** which is the rate the company must pay to its long-term creditors and shareholders.

To illustrate, let's return to FasTrac's proposed machinery purchase described in Exhibit 25.1. Does this machine provide a satisfactory return while recovering the amount invested? Recall that the machine requires a $16,000 investment and is expected to provide $4,100 annual net cash inflows for the next eight years. If we assume that net cash flows from this machine are received at each year-end and that FasTrac requires a 12% annual return, net present value can be computed as in Exhibit 25.7.

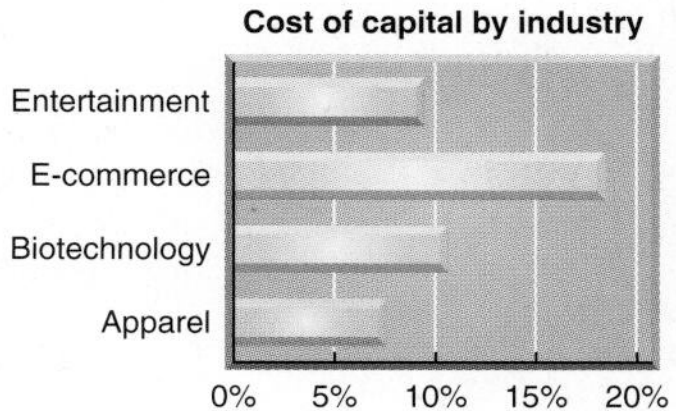

EXHIBIT 25.7

Net Present Value Calculation with Equal Cash Flows

	Net Cash Flows*	Present Value of 1 at 12%†	Present Value of Net Cash Flows
Year 1	$ 4,100	0.8929	$ 3,661
Year 2	4,100	0.7972	3,269
Year 3	4,100	0.7118	2,918
Year 4	4,100	0.6355	2,606
Year 5	4,100	0.5674	2,326
Year 6	4,100	0.5066	2,077
Year 7	4,100	0.4523	1,854
Year 8	4,100	0.4039	1,656
Totals	$32,800		$20,367
Amount invested			(16,000)
Net present value			$ 4,367

* Cash flows occur at the end of each year.

† Present value of 1 factors are taken from Table B.1 in Appendix B.

The first number column of Exhibit 25.7 shows the annual net cash flows. Present value of 1 factors, also called *discount factors,* are shown in the second column. Taken from Table B.1 in Appendix B, they assume that net cash flows are received at each year-end. *(To simplify present value computations and for assignment material at the end of this chapter, we assume that net cash flows are received at each year-end.)* Annual net cash flows from the first column of Exhibit 25.7 are multiplied by the discount factors in the second column to give present values shown in the third column. The last three lines of this exhibit show the final NPV computations. The asset's $16,000 initial cost is deducted from the $20,367 total present value of all future net cash flows to give this asset's NPV of $4,367. The machine is thus expected to (1) recover its cost, (2) provide a 12% compounded return, and (3) generate $4,367 above cost. We summarize this analysis by saying the present value of this machine's future net cash flows to FasTrac exceeds the $16,000 investment by $4,367.

Point: The assumption of end-of-year cash flows simplifies computations and is common in practice.

Point: The amount invested includes all costs that must be incurred to get the asset in its proper location and ready for use.

Example: What is the net present value in Exhibit 25.7 if a 10% return is applied? *Answer:* $5,873

Net Present Value Decision Rule The decision rule in applying NPV is as follows: When an asset's expected cash flows are discounted at the required rate and yield a *positive* net present value, the asset should be acquired. This decision rule is reflected in the graphic below. When comparing several investment opportunities of about the same cost and same risk, we prefer the one with the highest positive net present value.

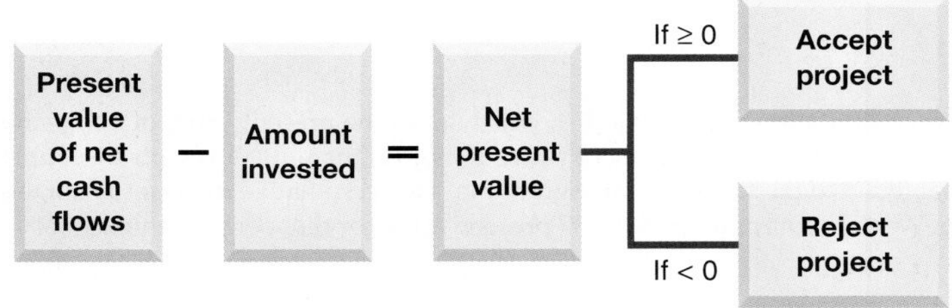

Example: Why does the net present value of an investment increase when a lower discount rate is used? *Answer:* The present value of net cash flows increases.

Simplifying Computations The computations in Exhibit 25.7 use separate present value of 1 factors for each of the eight years. Each year's net cash flow is multiplied by its present value of 1 factor to determine its present value. The individual present values for each of the eight net cash flows are added to give the asset's total present value. This computation can be simplified in two ways if annual net cash flows are equal in amount. One way is to add the eight annual present value of 1 factors for a total of 4.9676 and multiply this amount by the annual \$4,100 net cash flow to get the \$20,367 total present value of net cash flows.[1] A second simplification is to use a calculator with compound interest functions or a spreadsheet program. We show how to use Excel functions to compute net present value in this chapter's Appendix. Whatever procedure you use, it is important to understand the concepts behind these computations.

Decision Ethics

Systems Manager Top management adopts a policy requiring purchases in excess of \$5,000 to be submitted with cash flow projections to the cost analyst for capital budget approval. As systems manager, you want to upgrade your computers at a \$25,000 cost. You consider submitting several orders all under \$5,000 to avoid the approval process. You believe the computers will increase profits and wish to avoid a delay. What do you do? [Answer—p. 1048]

Uneven Cash Flows Net present value analysis can also be applied when net cash flows are uneven (unequal). To illustrate, assume that FasTrac can choose only one capital investment from among projects A, B, and C. Each project requires the same \$12,000 initial investment. Future net cash flows for each project are shown in the first three number columns of Exhibit 25.8.

EXHIBIT 25.8

Net Present Value Calculation with Uneven Cash Flows

	Net Cash Flows			Present Value of 1 at 10%	Present Value of Net Cash Flows		
	A	B	C		A	B	C
Year 1	\$ 5,000	\$ 8,000	\$ 1,000	0.9091	\$ 4,546	\$ 7,273	\$ 909
Year 2	5,000	5,000	5,000	0.8264	4,132	4,132	4,132
Year 3	5,000	2,000	9,000	0.7513	3,757	1,503	6,762
Totals	\$15,000	\$15,000	\$15,000		12,435	12,908	11,803
Amount invested					(12,000)	(12,000)	(12,000)
Net present value					\$ 435	\$ 908	\$ (197)

Example: If 12% is the required return in Exhibit 25.8, which project is preferred? *Answer:* Project B. Net present values are: A = \$10; B = \$553; C = \$(715).

Example: Will the rankings of Projects A, B, and C change with the use of different discount rates, assuming the same rate is used for all projects? *Answer:* No; only the NPV amounts will change.

The three projects in Exhibit 25.8 have the same expected total net cash flows of \$15,000. Project A is expected to produce equal amounts of \$5,000 each year. Project B is expected to produce a larger amount in the first year. Project C is expected to produce a larger amount in the third year. The fourth column of Exhibit 25.8 shows the present value of 1 factors from Table B.1 assuming 10% required return.

Computations in the right-most columns show that Project A has a \$435 positive NPV. Project B has the largest NPV of \$908 because it brings in cash more quickly. Project C has a \$(197) *negative* NPV because its larger cash inflows are delayed. If FasTrac requires a 10% return, it should reject Project C because its NPV implies a return *under* 10%. If only one project can be accepted, project B appears best because it yields the highest NPV.

[1] We can simplify this computation using Table B.3, which gives the present value of 1 to be received periodically for a number of periods. To determine the present value of these eight annual receipts discounted at 12%, go down the 12% column of Table B.3 to the factor on the eighth line. This cumulative discount factor, also known as an *annuity* factor, is 4.9676. We then compute the \$20,367 present value for these eight annual \$4,100 receipts, computed as 4.9676 × \$4,100.

Salvage Value and Accelerated Depreciation FasTrac predicted the $16,000 machine to have zero salvage value at the end of its useful life (recall Exhibit 25.1). In many cases, assets are expected to have salvage values. If so, this amount is an additional net cash inflow received at the end of the final year of the asset's life. All other computations remain the same.

Point: Projects with higher cash flows in earlier years generally yield higher net present values.

Depreciation computations also affect net present value analysis. FasTrac computes depreciation using the straight-line method. Accelerated depreciation is also commonly used, especially for income tax reports. Accelerated depreciation produces larger depreciation deductions in the early years of an asset's life and smaller deductions in later years. This pattern results in smaller income tax payments in early years and larger payments in later years. Accelerated depreciation does not change the basics of a present value analysis, but it can change the result. Using accelerated depreciation for tax reporting affects the NPV of an asset's cash flows because it produces larger net cash inflows in the early years of the asset's life and smaller ones in later years. Being able to use accelerated depreciation for tax reporting always makes an investment more desirable because early cash flows are more valuable than later ones.

Example: When is it appropriate to use different discount rates for different projects? *Answer:* When risk levels are different.

Point: Tax savings from depreciation is called: **depreciation tax shield.**

Use of Net Present Value In deciding whether to proceed with a capital investment project, we approve the proposal if the NPV is positive but reject it if the NPV is negative. When considering several projects of similar investment amounts and risk levels, we can compare the different projects' NPVs and rank them on the basis of their NPVs. However, if the amount invested differs substantially across projects, the NPV is of limited value for comparison purposes. One means to compare projects, especially when a company cannot fund all positive net present value projects, is to use the **profitability index,** which is computed as:

$$\textbf{Profitability index} = \frac{\textbf{Net present value of cash flows}}{\textbf{Investment}}$$

A higher profitability index suggests a more desirable project. To illustrate, suppose that Project X requires a $1 million investment and provides a $100,000 NPV. Project Y requires an investment of only $100,000 and returns a $75,000 NPV. Ranking on the basis of NPV puts Project X ahead of Y, yet X's profitability index is only 0.10 ($100,000/$1,000,000) whereas Y's profitability index is 0.75. We must also remember that when reviewing projects with different risks, we computed the NPV of individual projects using different discount rates. The higher the risk, the higher the discount rate.

Inflation Large price-level increases should be considered in NPV analyses. Hurdle rates already include investor's inflation forecasts. Net cash flows can be adjusted for inflation by using *future value* computations. For example, if the expected net cash inflow in year 1 is $4,100 and 5% inflation is expected, then the expected net cash inflow in year 2 is $4,305, computed as $4,100 × 1.05 (1.05 is the future value of $1 (Table B.2) for 1 period with a 5% rate).

Internal Rate of Return

P4 Compute internal rate of return and explain its use.

Another means to evaluate capital investments is to use the **internal rate of return (IRR),** which equals the rate that yields an NPV of zero for an investment. This means that if we compute the total present value of a project's net cash flows using the IRR as the discount rate and then subtract the initial investment from this total present value, we get a zero NPV.

To illustrate, we use the data for FasTrac's Project A from Exhibit 25.8 to compute its IRR. Exhibit 25.9 shows the two-step process in computing IRR.

EXHIBIT 25.9

Computing Internal Rate of Return (with even cash flows)

Step 1: Compute the present value factor for the investment project.

$$\textbf{Present value factor} = \frac{\textbf{Amount invested}}{\textbf{Net cash flows}} = \frac{\$12{,}000}{\$5{,}000} = 2.4000$$

Step 2: Identify the discount rate (IRR) yielding the present value factor

Search Table B.3 for a present value factor of 2.4000 in the three-year row (equaling the 3-year project duration). The 12% discount rate yields a present value factor of 2.4018. This implies that the IRR is approximately 12%.*

* Since the present value factor of 2.4000 is not exactly equal to the 12% factor of 2.4018, we can more precisely estimate the IRR as follows:

Discount rate	Present Value Factor from Table B.3
12%	2.4018
15%	2.2832
	0.1186 = difference

$$\text{Then, IRR} = 12\% + \left[(15\% - 12\%) \times \frac{2.4018 - 2.4000}{0.1186}\right] = \underline{\underline{12.05\%}}$$

When cash flows are equal, as with Project A, we compute the present value factor (as shown in Exhibit 25.9) by dividing the initial investment by its annual net cash flows. We then use an annuity table to determine the discount rate equal to this present value factor. For FasTrac's Project A, we look across the three-period row of Table B.3 and find that the discount rate corresponding to the present value factor of 2.4000 roughly equals the 2.4018 value for the 12% rate. This row is reproduced here:

Present Value of an Annuity of 1 for Three Periods

	Discount Rate				
Periods	1%	5%	10%	12%	15%
3	2.9410	2.7232	2.4869	2.4018	2.2832

The 12% rate is the Project's IRR. A more precise IRR estimate can be computed following the procedure shown in the note to Exhibit 25.9. Spreadsheet software and calculators can also compute this IRR. We show how to use an Excel function to compute IRR in this chapter's Appendix.

Uneven Cash Flows If net cash flows are uneven, we must use trial and error to compute the IRR. We do this by selecting any reasonable discount rate and computing the NPV. If the amount is positive (negative), we recompute the NPV using a higher (lower) discount rate. We continue these steps until we reach a point where two consecutive computations result in NPVs having different signs (positive and negative). Because the NPV is zero using IRR, we know that the IRR lies between these two discount rates. We can then estimate its value. Spreadsheet programs and calculators can do these computations for us.

Decision Insight

Fun-IRR Many theme parks use both financial and nonfinancial criteria to evaluate their investments in new rides and activities. The use of IRR is a major part of this evaluation. This requires good estimates of future cash inflows and outflows. It also requires risk assessments of the uncertainty of the future cash flows.

Use of Internal Rate of Return When we use the IRR to evaluate a project, we compare it to a predetermined **hurdle rate,** which is a minimum acceptable rate of return and is applied as follows.

C2 Describe the selection of a hurdle rate for an investment.

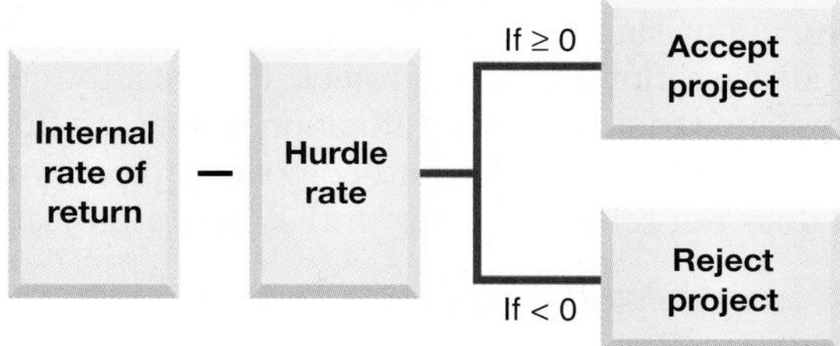

Top management selects the hurdle rate to use in evaluating capital investments. Financial formulas aid in this selection, but the choice of a minimum rate is subjective and left to management. For projects financed from borrowed funds, the hurdle rate must exceed the interest rate paid on these funds. The return on an investment must cover its interest and provide an additional profit to reward the company for its risk. For instance, if money is borrowed at 10%, an average risk investment often requires an after-tax return of 15% (or 5% above the borrowing rate). Remember that lower-risk investments require a lower rate of return compared with higher-risk investments.

If the project is internally financed, the hurdle rate is often based on actual returns from comparable projects. If the IRR is higher than the hurdle rate, the project is accepted. Multiple projects are often ranked by the extent to which their IRR exceeds the hurdle rate. The hurdle rate for individual projects is often different, depending on the risk involved. IRR is not subject to the limitations of NPV when comparing projects with different amounts invested because the IRR is expressed as a percent rather than as an absolute dollar value in NPV.

Example: How can management evaluate the risk of an investment? *Answer:* It must assess the uncertainty of future cash flows.

Point: A survey reports that 41% of top managers would reject a project with an internal rate of return *above* the cost of capital, *if* the project would cause the firm to miss its earnings forecast. The roles of benchmarks and manager compensation plans must be considered in capital budgeting decisions.

Decision Maker

Entrepreneur You are developing a new product and you use a 12% discount rate to compute its NPV. Your banker, from whom you hope to obtain a loan, expresses concern that your discount rate is too low. How do you respond? [Answer—p. 1048]

Comparison of Capital Budgeting Methods

We explained four methods that managers use to evaluate capital investment projects. How do these methods compare with each other? Exhibit 25.10 addresses that question. Neither the payback period nor the accounting rate of return considers the time value of money. On the other hand, both the net present value and the internal rate of return do.

EXHIBIT 25.10

Comparing Capital Budgeting Methods

	Payback Period	Accounting Rate of Return	Net Present Value	Internal Rate of Return
Measurement basis	■ Cash flows	■ Accrual income	■ Cash flows ■ Profitability	■ Cash flows ■ Profitability
Measurement unit	■ Years	■ Percent	■ Dollars	■ Percent
Strengths	■ Easy to understand ■ Allows comparison of projects	■ Easy to understand ■ Allows comparison of projects	■ Reflects time value of money ■ Reflects varying risks over project's life	■ Reflects time value of money ■ Allows comparisons of dissimilar projects
Limitations	■ Ignores time value of money ■ Ignores cash flows after payback period	■ Ignores time value of money ■ Ignores annual rates over life of project	■ Difficult to compare dissimilar projects	■ Ignores varying risks over life of project

The payback period is probably the simplest method. It gives managers an estimate of how soon they will recover their initial investment. Managers sometimes use this method when they have limited cash to invest and a number of projects to choose from. The accounting rate of return yields a percent measure computed using accrual income instead of cash flows. The accounting rate of return is an average rate for the entire investment period. Net present value considers all estimated net cash flows for the project's expected life. It can be applied to even and uneven cash flows and can reflect changes in the level of risk over a project's life. Since it yields a dollar measure, comparing projects of unequal sizes is more difficult. The internal rate of return considers all cash flows from a project. It is readily computed when the cash flows are even but requires some trial and error estimation when cash flows are uneven. Because the IRR is a percent measure, it is readily used to compare projects with different investment amounts. However, IRR does not reflect changes in risk over a project's life.

Decision Insight

And the Winner Is . . . How do we choose among the methods for evaluating capital investments? Management surveys consistently show the internal rate of return (IRR) as the most popular method followed by the payback period and net present value (NPV). Few companies use the accounting rate of return (ARR), but nearly all use more than one method.

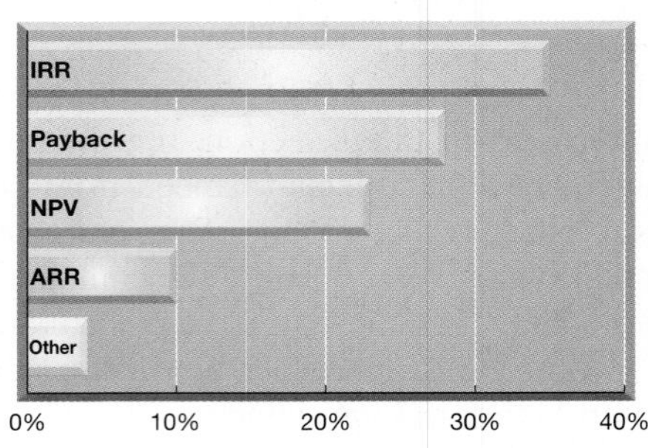

Company Usage for Capital Budgeting Methods

Quick Check

Answers—p. 1049

8. A company can invest in only one of two projects, A or B. Each project requires a $20,000 investment and is expected to generate end-of-period, annual cash flows as follows:

	Year 1	Year 2	Year 3	Total
Project A	$12,000	$8,500	$4,000	$24,500
Project B	4,500	8,500	13,000	26,000

Assuming a discount rate of 10%, which project has the higher net present value?

9. Two investment alternatives are expected to generate annual cash flows with the same net present value (assuming the same discount rate applied to each). Using this information, can you conclude that the two alternatives are equally desirable?
10. When two investment alternatives have the same total expected cash flows but differ in the timing of those flows, which method of evaluating those investments is superior, (*a*) accounting rate of return or (*b*) net present value?

Section 2—Managerial Decisions

This section focuses on methods that use accounting information to make several important managerial decisions. Most of these involve short-term decisions. This differs from methods used for longer-term managerial decisions that are described in the first section of this chapter and in several other chapters of this book.

Decisions and Information

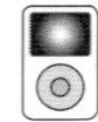
Video25.1

This section explains how managers make decisions and the information relevant to those decisions.

Decision Making

Managerial decision making involves five steps: (1) define the decision task, (2) identify alternative courses of action, (3) collect relevant information and evaluate each alternative,

(4) select the preferred course of action, and (5) analyze and assess decisions made. These five steps are illustrated in Exhibit 25.11.

EXHIBIT 25.11
Managerial Decision Making

Both managerial and financial accounting information play an important role in most management decisions. The accounting system is expected to provide primarily *financial* information such as performance reports and budget analyses for decision making. *Nonfinancial* information is also relevant, however; it includes information on environmental effects, political sensitivities, and social responsibility.

Relevant Costs

C3 Describe the importance of relevant costs for short-term decisions.

Most financial measures of revenues and costs from accounting systems are based on historical costs. Although historical costs are important and useful for many tasks such as product pricing and the control and monitoring of business activities, we sometimes find that an analysis of *relevant costs,* or *avoidable costs,* is especially useful. Three types of costs are pertinent to our discussion of relevant costs: sunk costs, out-of-pocket costs, and opportunity costs.

A *sunk cost* arises from a past decision and cannot be avoided or changed; it is irrelevant to future decisions. An example is the cost of computer equipment previously purchased by a company. Most of a company's allocated costs, including fixed overhead items such as depreciation and administrative expenses, are sunk costs.

An *out-of-pocket cost* requires a future outlay of cash and is relevant for current and future decision making. These costs are usually the direct result of management's decisions. For instance, future purchases of computer equipment involve out-of-pocket costs.

Example: Depreciation and amortization are allocations of the original cost of plant and intangible assets. Are they out-of-pocket costs? *Answer:* No; they are sunk costs.

An *opportunity cost* is the potential benefit lost by taking a specific action when two or more alternative choices are available. An example is a student giving up wages from a job to attend summer school. Companies continually must choose from alternative courses of action. For instance, a company making standardized products might be approached by a customer to supply a special (nonstandard) product. A decision to accept or reject the special order must consider not only the profit to be made from the special order but also the profit given up by devoting time and resources to this order instead of pursuing an alternative project. The profit given up is an opportunity cost. Consideration of opportunity costs is important. The implications extend to internal resource allocation decisions. For instance, a computer manufacturer must decide between internally manufacturing a chip versus buying it externally. In another case, management of a multidivisional company must decide whether to continue operating or close a particular division.

Point: Opportunity costs are not entered in accounting records. This does not reduce their relevance for managerial decisions.

Besides relevant costs, management must also consider the relevant benefits associated with a decision. **Relevant benefits** refer to the additional or *incremental* revenue generated by selecting a particular course of action over another. For instance, a student must decide the relevant benefits of taking one course over another. In sum, both relevant costs and relevant benefits are crucial to managerial decision making.

Managerial Decision Scenarios

A1 Evaluate short-term managerial decisions using relevant costs.

Managers experience many different scenarios that require analyzing alternative actions and making a decision. We describe several different types of decision scenarios in this section. We set these tasks in the context of FasTrac, an exercise supplies and equipment manufacturer introduced earlier. *We treat each of these decision tasks as separate from each other.*

Video25.1

Additional Business

FasTrac is operating at its normal level of 80% of full capacity. At this level, it produces and sells approximately 100,000 units of product annually. Its per unit and annual total costs are shown in Exhibit 25.12.

EXHIBIT 25.12

Selected Operating Income Data

	Per Unit	Annual Total
Sales (100,000 units)	$10.00	$1,000,000
Direct materials	(3.50)	(350,000)
Direct labor	(2.20)	(220,000)
Overhead	(1.10)	(110,000)
Selling expenses	(1.40)	(140,000)
Administrative expenses	(0.80)	(80,000)
Total costs and expenses	(9.00)	(900,000)
Operating income	$ 1.00	$ 100,000

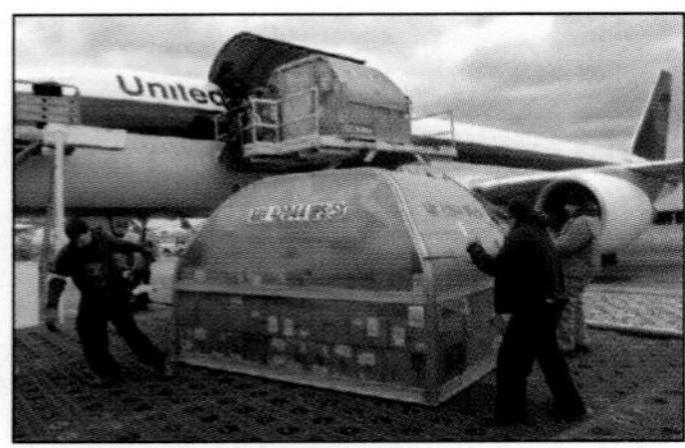

A current buyer of FasTrac's products wants to purchase additional units of its product and export them to another country. This buyer offers to buy 10,000 units of the product at $8.50 per unit, or $1.50 less than the current price. The offer price is low, but FasTrac is considering the proposal because this sale would be several times larger than any single previous sale and it would use idle capacity. Also, the units will be exported, so this new business will not affect current sales.

To determine whether to accept or reject this order, management needs to know whether accepting the offer will increase net income. The analysis in Exhibit 25.13 shows that if management relies on per unit historical costs, it would reject the sale because it yields a loss. However, historical costs are *not* relevant to this decision. Instead, the relevant costs are the additional costs, called **incremental costs.** These costs, also called *differential costs,* are the additional costs incurred if a company pursues a certain course of action. FasTrac's incremental costs are those related to the added volume that this new order would bring.

EXHIBIT 25.13

Analysis of Additional Business Using Historical Costs

	Per Unit	Total
Sales (10,000 additional units)	$ 8.50	$ 85,000
Direct materials	(3.50)	(35,000)
Direct labor	(2.20)	(22,000)
Overhead	(1.10)	(11,000)
Selling expenses	(1.40)	(14,000)
Administrative expenses	(0.80)	(8,000)
Total costs and expenses	(9.00)	(90,000)
Operating loss	$(0.50)	$ (5,000)

To make its decision, FasTrac must analyze the costs of this new business in a different manner. The following information regarding the order is available:

- Manufacturing 10,000 additional units requires direct materials of $3.50 per unit and direct labor of $2.20 per unit (same as for all other units).
- Manufacturing 10,000 additional units adds $5,000 of incremental overhead costs for power, packaging, and indirect labor (all variable costs).
- Incremental commissions and selling expenses from this sale of 10,000 additional units would be $2,000 (all variable costs).
- Incremental administrative expenses of $1,000 for clerical efforts are needed (all fixed costs) with the sale of 10,000 additional units.

We use this information, as shown in Exhibit 25.14, to assess how accepting this new business will affect FasTrac's income.

EXHIBIT 25.14

Analysis of Additional Business Using Relevant Costs

	Current Business	Additional Business	Combined
Sales	$1,000,000	$ 85,000	$1,085,000
Direct materials	(350,000)	(35,000)	(385,000)
Direct labor	(220,000)	(22,000)	(242,000)
Overhead	(110,000)	(5,000)	(115,000)
Selling expenses	(140,000)	(2,000)	(142,000)
Administrative expense	(80,000)	(1,000)	(81,000)
Total costs and expenses	(900,000)	(65,000)	(965,000)
Operating income	$ 100,000	$ 20,000	$ 120,000

The analysis of relevant costs in Exhibit 25.14 suggests that the additional business be accepted. It would provide $85,000 of added revenue while incurring only $65,000 of added costs. This would yield $20,000 of additional pretax income, or a pretax profit margin of 23.5%. More generally, FasTrac would increase its income with any price that exceeded $6.50 per unit ($65,000 incremental cost/10,000 additional units).

An analysis of the incremental costs pertaining to the additional volume is always relevant for this type of decision. We must proceed cautiously, however, when the additional volume approaches or exceeds the factory's existing available capacity. If the additional volume requires the company to expand its capacity by obtaining more equipment, more space, or more personnel, the incremental costs could quickly exceed the incremental revenue. Another cautionary note is the effect on existing sales. All new units of the extra business will be sold outside FasTrac's normal domestic sales channels. If accepting additional business would cause existing sales to decline, this information must be included in our analysis. The contribution margin lost from a decline in sales is an opportunity cost. If future cash flows over several time periods are affected, their net present value also must be computed and used in this analysis.

Example: Exhibit 25.14 uses quantitative information. Suggest some qualitative factors to be considered when deciding whether to accept this project. *Answer:* (1) Impact on relationships with other customers and (2) Improved relationship with customer buying additional units.

The key point is that *management must not blindly use historical costs, especially allocated overhead costs.* Instead, the accounting system needs to provide information about the incremental costs to be incurred if the additional business is accepted.

Decision Maker

Partner You are a partner in a small accounting firm that specializes in keeping the books and preparing taxes for clients. A local restaurant is interested in obtaining these services from your firm. Identify factors that are relevant in deciding whether to accept the engagement. [Answer—p. 1048]

Make or Buy

The managerial decision to make or buy a component for one of its current products is commonplace and depends on incremental costs. To illustrate, FasTrac has excess productive capacity it can use to manufacture Part 417, a component of the main product it sells. The part is currently purchased and delivered to the plant at a cost of $1.20 per unit. FasTrac estimates that making Part 417 would cost $0.45 for direct materials, $0.50 for direct labor, and an undetermined amount for overhead. The task is to determine how much overhead to add to these costs so we can decide whether to make or buy Part 417. If FasTrac's normal predetermined overhead application rate is 100% of direct labor cost, we might be tempted to conclude that overhead cost is $0.50 per unit, computed as 100% of the $0.50 direct labor cost. We would then mistakenly conclude that total cost is $1.45 ($0.45 of materials + $0.50 of labor + $0.50 of overhead). A wrong decision in this case would be to conclude that the company is better off buying the part at $1.20 each than making it for $1.45 each.

Instead, as we explained earlier, only incremental overhead costs are relevant in this situation. Thus, we must compute an *incremental overhead rate.* Incremental overhead costs might include, for example, additional power for operating machines, extra supplies, added cleanup costs, materials handling, and quality control. We can prepare a per unit analysis in this case as shown in Exhibit 25.15.

EXHIBIT 25.15

Make or Buy Analysis

	Make	Buy
Direct materials	$0.45	—
Direct labor	0.50	—
Overhead costs	[?]	—
Purchase price	—	$ 1.20
Total incremental costs	**$0.95 + [?]**	**$1.20**

We can see that if incremental overhead costs are less than $0.25 per unit, the total cost of making the component is less than the purchase price of $1.20 and FasTrac should make the part. FasTrac's decision rule in this case is that any amount of overhead less than $0.25 per unit yields a total cost for Part 417 that is less than the $1.20 purchase price. FasTrac must consider several nonfinancial factors in the make or buy decision, including product quality, timeliness of delivery (especially in a just-in-time setting), reactions of customers and suppliers, and other intangibles such as employee morale and workload. It must also consider whether making the part requires incremental fixed costs to expand plant capacity. When these added factors are considered, small cost differences may not matter.

Point: Managers must consider nonfinancial factors when making decisions.

Decision Insight

Make or Buy Services Companies apply make or buy decisions to their services. Many now outsource their payroll activities to a payroll service provider. It is argued that the prices paid for such services are close to what it costs them to do it, and without the headaches.

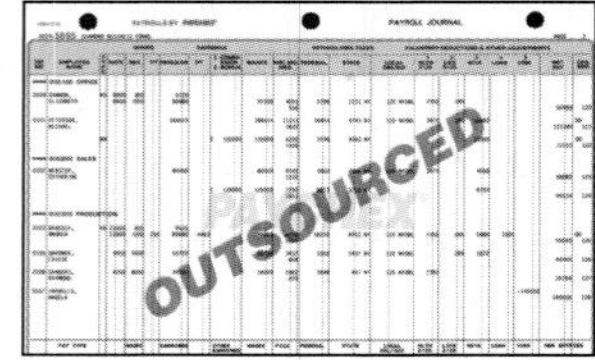

Scrap or Rework

Managers often must make a decision on whether to scrap or rework products in process. Remember that costs already incurred in manufacturing the units of a product that do not meet quality standards are sunk costs that have been incurred and cannot be changed. Sunk costs are irrelevant in any decision on whether to sell the substandard units as scrap or to rework them to meet quality standards.

To illustrate, assume that FasTrac has 10,000 defective units of a product that have already cost $1 per unit to manufacture. These units can be sold as is (as scrap) for $0.40 each, or they can be reworked for $0.80 per unit and then sold for their full price of $1.50 each. Should FasTrac sell the units as scrap or rework them?

To make this decision, management must recognize that the already incurred manufacturing costs of $1 per unit are sunk (unavoidable). These costs are *entirely irrelevant* to the decision. In addition, we must be certain that all costs of reworking defects, including interfering with normal operations, are accounted for in our analysis. For instance, reworking the defects means that FasTrac is unable to manufacture 10,000 *new* units with an incremental cost of $1 per unit and a selling price of $1.50 per unit, meaning it incurs an opportunity cost equal to the lost $5,000 net return from making and selling 10,000 new units. This opportunity cost is the difference between the $15,000 revenue (10,000 units × $1.50) from selling these new units and their $10,000 manufacturing costs (10,000 units × $1). Our analysis is reflected in Exhibit 25.16.

EXHIBIT 25.16

Scrap or Rework Analysis

	Scrap	Rework
Sale of scrapped/reworked units	$ 4,000	$ 15,000
Less costs to rework defects		(8,000)
Less opportunity cost of not making new units		**(5,000)**
Incremental net income	**$4,000**	**$ 2,000**

The analysis yields a $2,000 difference in favor of scrapping the defects, yielding a total incremental net income of $4,000. If we had failed to include the opportunity costs of $5,000, the rework option would have shown an income of $7,000 instead of $2,000, mistakenly making the reworking appear more favorable than scrapping.

Quick Check Answers—p. 1049

11. A company receives a special order for 200 units that requires stamping the buyer's name on each unit, yielding an additional fixed cost of $400 to its normal costs. Without the order, the company is operating at 75% of capacity and produces 7,500 units of product at the following costs:

Direct materials	$37,500
Direct labor	60,000
Overhead (30% variable)	20,000
Selling expenses (60% variable)	25,000

The special order will not affect normal unit sales and will not increase fixed overhead and selling expenses. Variable selling expenses on the special order are reduced to one-half the normal amount. The price per unit necessary to earn $1,000 on this order is (*a*) $14.80, (*b*) $15.80, (*c*) $19.80, (*d*) $20.80, or (*e*) $21.80.

12. What are the incremental costs of accepting additional business?

Sell or Process

The managerial decision to sell partially completed products as is or to process them further for sale depends significantly on relevant costs. To illustrate, suppose that FasTrac has 40,000 units of partially finished Product Q. It has already spent $0.75 per unit to manufacture these 40,000 units at a $30,000 total cost. FasTrac can sell the 40,000 units to another manufacturer as raw material for $50,000. Alternatively, it can process them further and produce finished products X, Y, and Z at an incremental cost of $2 per unit. The added processing yields the products and revenues shown in Exhibit 25.17. FasTrac must decide whether the added revenues from selling finished products X, Y, and Z exceed the costs of finishing them.

Product	Price	Units	Revenues
Product X	$4.00	10,000	$ 40,000
Product Y	6.00	22,000	132,000
Product Z	8.00	6,000	48,000
Spoilage	—	2,000	0
Totals		40,000	$220,000

EXHIBIT 25.17

Revenues from Processing Further

Exhibit 25.18 shows the two-step analysis for this decision. First, FasTrac computes its incremental revenue from further processing Q into products X, Y, and Z. This amount is the difference between the $220,000 revenue from the further processed products and the $50,000 FasTrac will give up by not selling Q as is (a $50,000 opportunity cost). Second, FasTrac computes its incremental costs from further processing Q into X, Y, and Z. This amount is $80,000 (40,000 units × $2 incremental cost). The analysis shows that FasTrac can earn incremental net income of $90,000 from a decision to further process Q. (Notice that the earlier incurred $30,000 manufacturing cost for the 40,000 units of Product Q does not appear in Exhibit 25.18 because it is a sunk cost and as such is irrelevant to the decision.)

Example: Does the decision change if incremental costs in Exhibit 25.18 increase to $4 per unit and the opportunity cost increases to $95,000? *Answer:* Yes. There is now an incremental net loss of $35,000.

EXHIBIT 25.18
Sell or Process Analysis

Revenue if processed	$220,000
Revenue if sold as is	(50,000)
Incremental revenue	170,000
Cost to process	(80,000)
Incremental net income	**$ 90,000**

Quick Check

Answers—p. 1049

13. A company has already incurred a $1,000 cost in partially producing its four products. Their selling prices when partially and fully processed follow with additional costs necessary to finish these partially processed units:

Product	Unfinished Selling Price	Finished Selling Price	Further Processing Costs
Alpha	$300	$600	$150
Beta	450	900	300
Gamma	275	425	125
Delta	150	210	75

Which product(s) should *not* be processed further, (*a*) Alpha, (*b*) Beta, (*c*) Gamma, or (*d*) Delta?

14. Under what conditions is a sunk cost relevant to decision making?

Sales Mix Selection

Point: A method called *linear programming* is useful for finding the optimal sales mix for several products subject to many market and production constraints. This method is described in advanced courses.

When a company sells a mix of products, some are likely to be more profitable than others. Management is often wise to concentrate sales efforts on more profitable products. If production facilities or other factors are limited, an increase in the production and sale of one product usually requires reducing the production and sale of others. In this case, management must identify the most profitable combination, or *sales mix* of products. To identify the best sales mix, management must know the contribution margin of each product, the facilities required to produce each product, any constraints on these facilities, and its markets.

To illustrate, assume that FasTrac makes and sells two products, A and B. The same machines are used to produce both products. A and B have the following selling prices and variable costs per unit:

	Product A	Product B
Selling price per unit	$5.00	$7.50
Variable costs per unit	3.50	5.50
Contribution margin per unit	$1.50	$2.00

The variable costs are included in the analysis because they are the incremental costs of producing these products within the existing capacity of 100,000 machine hours per month. We consider three separate cases.

Case 1: Assume that (1) each product requires 1 machine hour per unit for production and (2) the markets for these products are unlimited. Under these conditions, FasTrac should produce as much of Product B as it can because of its larger contribution margin of $2 per unit. At full capacity, FasTrac would produce $200,000 of total contribution margin per month, computed as $2 per unit times 100,000 machine hours.

Case 2: Assume that (1) Product A requires 1 machine hour per unit, (2) Product B requires 2 machine hours per unit, and (3) the markets for these products are unlimited. Under these conditions, FasTrac should produce as much of Product A as it can because it has a contribution margin of $1.50 per machine hour compared with only $1 per machine hour for Product B. Exhibit 25.19 shows the relevant analysis.

EXHIBIT 25.19

Sales Mix Analysis

	Product A	Product B
Selling price per unit	$ 5.00	$ 7.50
Variable costs per unit	3.50	5.50
Contribution margin per unit	$ 1.50	$ 2.00
Machine hours per unit	1.0	2.0
Contribution margin per machine hour	**$1.50**	**$1.00**

At its full capacity of 100,000 machine hours, FasTrac would produce 100,000 units of Product A, yielding $150,000 of total contribution margin per month. In contrast, if it uses all 100,000 hours to produce Product B, only 50,000 units would be produced yielding a contribution margin of $100,000. These results suggest that when a company faces excess demand and limited capacity, only the most profitable product per input should be manufactured.

Case 3: The need for a mix of different products arises when market demand is not sufficient to allow a company to sell all that it produces. For instance, assume that (1) Product A requires 1 machine hour per unit, (2) Product B requires 2 machine hours per unit, and (3) the market for Product A is limited to 80,000 units. Under these conditions, FasTrac should produce no more than 80,000 units of Product A. This would leave another 20,000 machine hours of capacity for making Product B. FasTrac should use this spare capacity to produce 10,000 units of Product B. This sales mix would maximize FasTrac's total contribution margin per month at an amount of $140,000.

Example: For Case 2, if Product B's variable costs per unit increase to $6, Product A's variable costs per unit decrease to $3, and the same machine hours per unit are used, which product should FasTrac produce? *Answer:* Product A. Its contribution margin of $2 per machine hour is higher than B's $.75 per machine hour.

Decision Insight

Companies such as **Gap, Abercrombie & Fitch,** and **American Eagle** must continuously monitor and manage the sales mix of their product lists. Selling their products in hundreds of countries and territories further complicates their decision process. The contribution margin of each product is crucial to their product mix strategies.

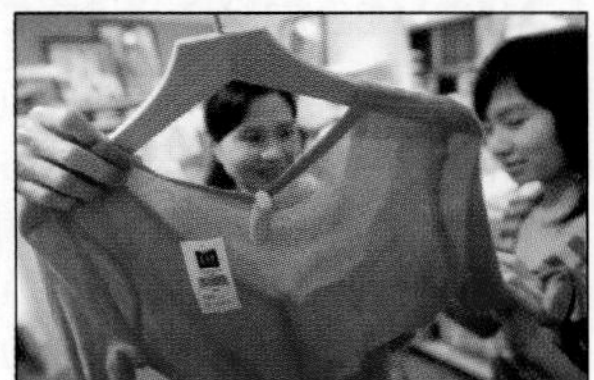

Segment Elimination

When a segment such as a department or division is performing poorly, management must consider eliminating it. Segment information on either net income (loss) or its contribution to overhead is not sufficient for this decision. Instead, we must look at the segment's avoidable expenses and unavoidable expenses. **Avoidable expenses,** also called *escapable expenses,* are amounts the company would not incur if it eliminated the segment. **Unavoidable expenses,** also called *inescapable expenses,* are amounts that would continue even if the segment is eliminated.

To illustrate, FasTrac considers eliminating its treadmill division because its $48,300 total expenses are higher than its $47,800 sales. Classification of this division's operating expenses into avoidable or unavoidable expenses is shown in Exhibit 25.20.

Point: FasTrac might consider buying another machine to reduce the constraint on production. A strategy designed to reduce the impact of constraints or bottlenecks, on production, is called the *theory of constraints.*

EXHIBIT 25.20

Classification of Segment Operating Expenses for Analysis

	Total	Avoidable Expenses	Unavoidable Expenses
Cost of goods sold	$ 30,000	$ 30,000	—
Direct expenses			
Salaries expense	7,900	7,900	—
Depreciation expense—Equipment	200	—	$ 200
Indirect expenses			
Rent and utilities expense	3,150	—	3,150
Advertising expense	400	400	—
Insurance expense	400	300	100
Service department costs			
Share of office department expenses	3,060	2,200	860
Share of purchasing expenses	3,190	1,000	2,190
Total	**$48,300**	**$41,800**	**$6,500**

Example: How can insurance be classified as either avoidable or unavoidable? *Answer:* Depends on whether the assets insured can be removed and the premiums canceled.

FasTrac's analysis shows that it can avoid $41,800 expenses if it eliminates the treadmill division. Because this division's sales are $47,800, eliminating it will cause FasTrac to lose $6,000 of income. *Our decision rule is that a segment is a candidate for elimination if its revenues are less than its avoidable expenses.* Avoidable expenses can be viewed as the costs to generate this segment's revenues.

Example: Give an example of a segment that a company might profitably use to attract customers even though it might incur a loss. *Answer:* Warranty and post-sales services.

When considering elimination of a segment, we must assess its impact on other segments. A segment could be unprofitable on its own, but it might still contribute to other segments' revenues and profits. It is possible then to continue a segment even when its revenues are less than its avoidable expenses. Similarly, a profitable segment might be discontinued if its space, assets, or staff can be more profitably used by expanding existing segments or by creating new ones. Our decision to keep or eliminate a segment requires a more complex analysis than simply looking at a segment's performance report. Such reports provide useful information, but they do not provide all the information necessary for this decision.

Qualitative Decision Factors

Managers must consider qualitative factors in making managerial decisions. Consider a decision on whether to buy a component from an outside supplier or continue to make it. Several qualitative decision factors must be considered. For example, the quality, delivery, and reputation of the proposed supplier are important. The effects from deciding not to make the component can include potential layoffs and impaired worker morale. Consider another situation in which a company is considering a one-time sale to a new customer at a special low price. Qualitative factors to consider in this situation include the effects of a low price on the company's image and the threat that regular customers might demand a similar price. The company must also consider whether this customer is really a one-time customer. If not, can it continue to offer this low price in the long run? Clearly, management cannot rely solely on financial data to make such decisions.

Quick Check Answers—p. 1049

15. What is the difference between avoidable and unavoidable expenses?

16. A segment is a candidate for elimination if (*a*) its revenues are less than its avoidable expenses, (*b*) it has a net loss, (*c*) its unavoidable expenses are higher than its revenues.

Break-Even Time

Decision Analysis

A2 Analyze a capital investment project using break-even time.

The first section of this chapter explained several methods to evaluate capital investments. Break-even time of an investment project is a variation of the payback period method that overcomes the limitation of not using the time value of money. **Break-even time (BET)** is a time-based measure used to evaluate a capital investment's acceptability. Its computation yields a measure of expected time, reflecting the time period until the *present value* of the net cash flows from an investment equals the initial cost of the investment. In basic terms, break-even time is computed by restating future cash flows in terms of present values and then determining the payback period using these present values.

To illustrate, we return to the FasTrac case described in Exhibit 25.1 involving a $16,000 investment in machinery. The annual net cash flows from this investment are projected at $4,100 for eight years. Exhibit 25.21 shows the computation of break-even time for this investment decision.

EXHIBIT 25.21

Break-Even Time Analysis*

Year	Cash Flows	Present Value of 1 at 10%	Present Value of Cash Flows	Cumulative Present Value of Cash Flows
0	$(16,000)	1.0000	$(16,000)	$(16,000)
1	4,100	0.9091	3,727	(12,273)
2	4,100	0.8264	3,388	(8,885)
3	4,100	0.7513	3,080	(5,805)
4	4,100	0.6830	2,800	(3,005)
5	4,100	0.6209	2,546	**(459)**
6	4,100	0.5645	2,314	**1,855**
7	4,100	0.5132	2,104	3,959
8	4,100	0.4665	1,913	5,872

* The time of analysis is the start of year 1 (same as end of year 0). All cash flows occur at the end of each year.

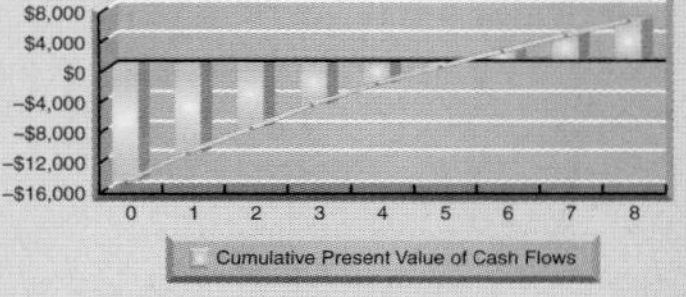

The right-most column of this exhibit shows that break-even time is between 5 and 6 years, or about 5.2 years—also see margin graph (where the line crosses the zero point). This is the time the project takes to break even after considering the time value of money (recall that the payback period computed without considering the time value of money was 3.9 years). We interpret this as cash flows earned after 5.2 years contribute to a positive net present value that, in this case, eventually amounts to $5,872.

Break-even time is a useful measure for managers because it identifies the point in time when they can expect the cash flows to begin to yield net positive returns. Managers expect a positive net present value from an investment if break-even time is less than the investment's estimated life. The method allows managers to compare and rank alternative investments, giving the project with the shortest break-even time the highest rank.

Decision Maker

Investment Manager Management asks you, the investment manager, to evaluate three alternative investments. Investment recovery time is crucial because cash is scarce. The time value of money is also important. Which capital budgeting method(s) do you use to assess the investments? [Answer—p. 1048]

Demonstration Problem

DP25

Determine the appropriate action in each of the following managerial decision situations.

1. Packer Company is operating at 80% of its manufacturing capacity of 100,000 product units per year. A chain store has offered to buy an additional 10,000 units at $22 each and sell them to customers so as not to compete with Packer Company. The following data are available.

Costs at 80% Capacity	Per Unit	Total
Direct materials	$ 8.00	$ 640,000
Direct labor	7.00	560,000
Overhead (fixed and variable)	12.50	1,000,000
Totals	$27.50	$2,200,000

In producing 10,000 additional units, fixed overhead costs would remain at their current level but incremental variable overhead costs of $3 per unit would be incurred. Should the company accept or reject this order?

2. Green Company uses Part JR3 in manufacturing its products. It has always purchased this part from a supplier for $40 each. It recently upgraded its own manufacturing capabilities and has enough excess capacity (including trained workers) to begin manufacturing Part JR3 instead of buying it. The company prepares the following cost projections of making the part, assuming that overhead is allocated to the part at the normal predetermined rate of 200% of direct labor cost.

Direct materials	$11
Direct labor	15
Overhead (fixed and variable) (200% of direct labor)	30
Total	$56

The required volume of output to produce the part will not require any incremental fixed overhead. Incremental variable overhead cost will be $17 per unit. Should the company make or buy this part?

3. Gold Company's manufacturing process causes a relatively large number of defective parts to be produced. The defective parts can be (a) sold for scrap, (b) melted to recover the recycled metal for reuse, or (c) reworked to be good units. Reworking defective parts reduces the output of other good units because no excess capacity exists. Each unit reworked means that one new unit cannot be produced. The following information reflects 500 defective parts currently available.

Proceeds of selling as scrap	$2,500
Additional cost of melting down defective parts	400
Cost of purchases avoided by using recycled metal from defects	4,800
Cost to rework 500 defective parts	
Direct materials	0
Direct labor	1,500
Incremental overhead	1,750
Cost to produce 500 new parts	
Direct materials	6,000
Direct labor	5,000
Incremental overhead	3,200
Selling price per good unit	40

Should the company melt the parts, sell them as scrap, or rework them?

4. White Company can invest in one of two projects, TD1 or TD2. Each project requires an initial investment of $100,000 and produces the year-end cash inflows shown in the following table. Use net present values to determine which project, if any, should be chosen. Assume that the company requires a 10% return from its investments.

	Net Cash Flows	
	TD1	TD2
Year 1	$ 20,000	$ 40,000
Year 2	30,000	40,000
Year 3	70,000	40,000
Totals	$120,000	$120,000

Planning the Solution

- Determine whether Packer Company should accept the additional business by finding the incremental costs of materials, labor, and overhead that will be incurred if the order is accepted. Omit fixed costs that the order will not increase. If the incremental revenue exceeds the incremental cost, accept the order.
- Determine whether Green Company should make or buy the component by finding the incremental cost of making each unit. If the incremental cost exceeds the purchase price, the component should be purchased. If the incremental cost is less than the purchase price, make the component.
- Determine whether Gold Company should sell the defective parts, melt them down and recycle the metal, or rework them. To compare the three choices, examine all costs incurred and benefits received from the alternatives in working with the 500 defective units versus the production of 500 new units. For the scrapping alternative, include the costs of producing 500 new units and subtract the $2,500 proceeds from selling the old ones. For the melting alternative, include the costs of melting the defective units, add the net cost of new materials in excess over those obtained from recycling, and add the direct labor and overhead costs. For the reworking alternative, add the costs of direct labor and incremental overhead. Select the alternative that has the lowest cost. The cost assigned to the 500 defective units is sunk and not relevant in choosing among the three alternatives.
- Compute White Company's net present value of each investment using a 10% discount rate.

Solution to Demonstration Problem

1. This decision involves accepting additional business. Since current unit costs are $27.50, it appears initially as if the offer to sell for $22 should be rejected, but the $27.50 cost includes fixed costs. When the analysis includes only *incremental* costs, the per unit cost is as shown in the following table. The offer should be accepted because it will produce $4 of additional profit per unit (computed as $22 price less $18 incremental cost), which yields a total profit of $40,000 for the 10,000 additional units.

Direct materials	$ 8.00
Direct labor	7.00
Variable overhead (given)	3.00
Total incremental cost	$18.00

2. For this make or buy decision, the analysis must not include the $13 nonincremental overhead per unit ($30 − $17). When only the $17 incremental overhead is included, the relevant unit cost of manufacturing the part is shown in the following table. It would be better to continue buying the part for $40 instead of making it for $43.

Direct materials	$11.00
Direct labor	15.00
Variable overhead	17.00
Total incremental cost	$43.00

3. The goal of this scrap or rework decision is to identify the alternative that produces the greatest net benefit to the company. To compare the alternatives, we determine the net cost of obtaining 500 marketable units as follows:

Incremental Cost to Produce 500 Marketable Units	Sell as Is	Melt and Recycle	Rework Units
Direct materials			
New materials	$ 6,000	$6,000	
Recycled metal materials		(4,800)	
Net materials cost		1,200	
Melting costs		400	
Total direct materials cost	6,000	1,600	
Direct labor	5,000	5,000	$1,500
Incremental overhead	3,200	3,200	1,750
Cost to produce 500 marketable units	14,200	9,800	3,250
Less proceeds of selling defects as scrap	(2,500)		
Opportunity costs*			5,800
Net cost	$11,700	$9,800	$9,050

* The $5,800 opportunity cost is the lost contribution margin from not being able to produce and sell 500 units because of reworking, computed as ($40 − [$14,200/500 units]) × 500 units.

The incremental cost of 500 marketable parts is smallest if the defects are reworked.

4. TD1:

	Net Cash Flows	Present Value of 1 at 10%	Present Value of Net Cash Flows
Year 1	$ 20,000	0.9091	$ 18,182
Year 2	30,000	0.8264	24,792
Year 3	70,000	0.7513	52,591
Totals	$120,000		95,565
Amount invested			(100,000)
Net present value			**$ (4,435)**

TD2:

	Net Cash Flows	Present Value of 1 at 10%	Present Value of Net Cash Flows
Year 1	$ 40,000	0.9091	$ 36,364
Year 2	40,000	0.8264	33,056
Year 3	40,000	0.7513	30,052
Totals	$120,000		99,472
Amount invested			(100,000)
Net present value			**$ (528)**

White Company should not invest in either project. Both are expected to yield a negative net present value, and it should invest only in positive net present value projects.

APPENDIX

Using Excel to Compute Net Present Value and Internal Rate of Return

25A

Computing present values and internal rates of return for projects with uneven cash flows is tedious and error prone. These calculations can be performed simply and accurately by using functions built into Excel. Many calculators and other types of spreadsheet software can perform them too. To illustrate, consider Fastrac, a company that is considering investing in a new machine with the expected cash flows shown in the following spreadsheet. Cash outflows are entered as negative numbers, and cash inflows are entered as positive numbers. Assume Fastrac requires a 12% annual return, entered as 0.12 in cell C1.

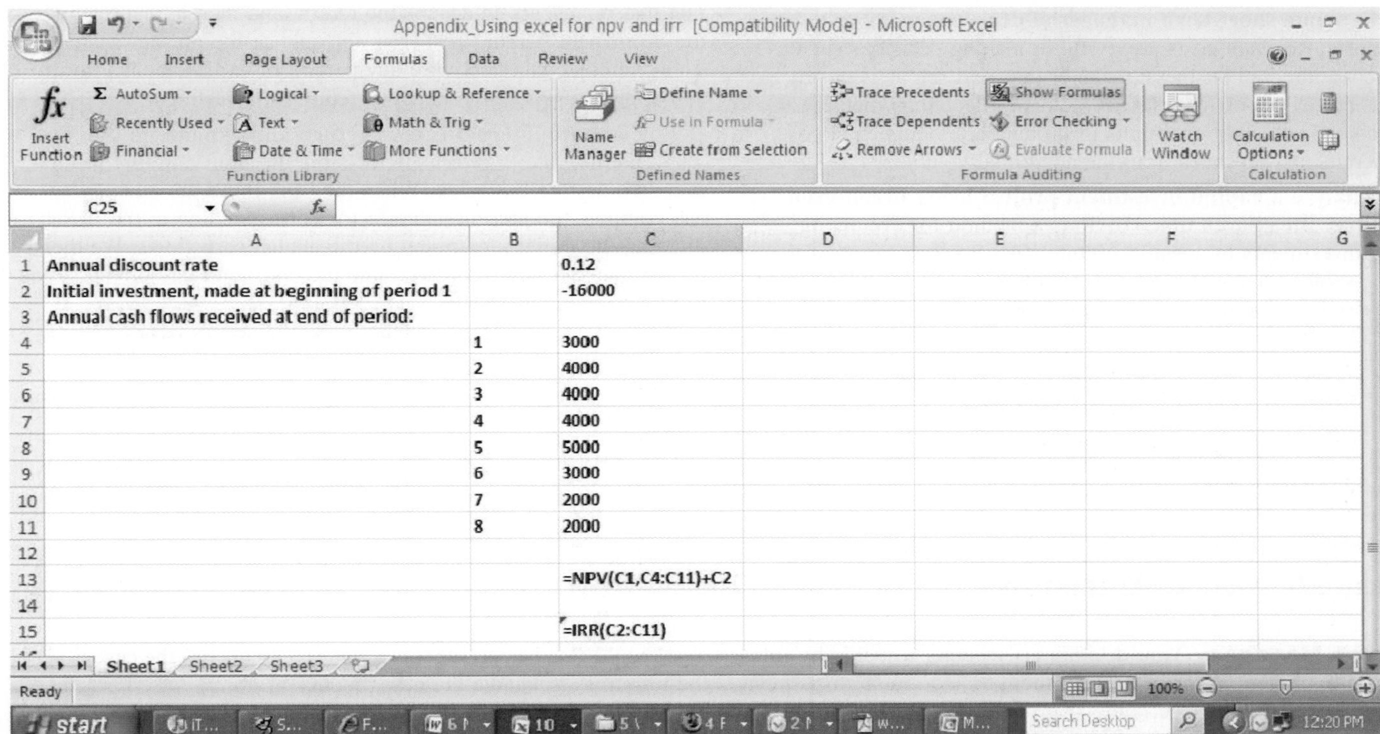

	A	B	C
1	Annual discount rate		0.12
2	Initial investment, made at beginning of period 1		-16000
3	Annual cash flows received at end of period:		
4		1	3000
5		2	4000
6		3	4000
7		4	4000
8		5	5000
9		6	3000
10		7	2000
11		8	2000
12			
13			=NPV(C1,C4:C11)+C2
14			
15			=IRR(C2:C11)

To compute the net present value of this project, the following is entered into cell C13:

=NPV(C1,C4:C11)+C2.

This instructs Excel to use its NPV function to compute the present value of the cash flows in cells C4 through C11, using the discount rate in cell C1, and then add the amount of the (negative) initial investment. For this stream of cash flows and a discount rate of 12%, the net present value is $1,326.03.

To compute the internal rate of return for this project, the following is entered into cell C15:

=IRR(C2:C11).

This instructs Excel to use its IRR function to compute the internal rate of return of the cash flows in cells C2 through C11. By default, Excel starts with a guess of 10%, and then uses trial and error to find the IRR. The IRR equals 14% for this project.

Summary

C1 Explain the importance of capital budgeting. Capital budgeting is the process of analyzing alternative investments and deciding which assets to acquire or sell. It involves predicting the cash flows to be received from the alternatives, evaluating their merits, and then choosing which ones to pursue.

C2 Describe the selection of a hurdle rate for an investment. Top management should select the hurdle (discount) rate to use in evaluating capital investments. The required hurdle rate should be at least higher than the interest rate on money borrowed because the return on an investment must cover the interest and provide an additional profit to reward the company for risk.

C3 Describe the importance of relevant costs for short-term decisions. A company must rely on relevant costs pertaining to alternative courses of action rather than historical costs. Out-of-pocket expenses and opportunity costs are relevant because these are avoidable; sunk costs are irrelevant because they result from past decisions and are therefore unavoidable. Managers must also consider the relevant benefits associated with alternative decisions.

A1 Evaluate short-term managerial decisions using relevant costs. Relevant costs are useful in making decisions such as to accept additional business, make or buy, and sell as is or process further. For example, the relevant factors in deciding whether to produce and sell additional units of product are incremental costs and incremental revenues from the additional volume.

A2 Analyze a capital investment project using break-even time. Break-even time (BET) is a method for evaluating capital investments by restating future cash flows in terms of their present values (discounting the cash flows) and then calculating the payback period using these present values of cash flows.

P1 Compute payback period and describe its use. One way to compare potential investments is to compute and compare their payback periods. The payback period is an estimate of the expected time before the cumulative net cash inflow from the investment equals its initial cost. A payback period analysis fails to reflect risk of the cash flows, differences in the timing of cash flows within the payback period, and cash flows that occur after the payback period.

P2 Compute accounting rate of return and explain its use. A project's accounting rate of return is computed by dividing the expected annual after-tax net income by the average amount of investment in the project. When the net cash flows are received evenly throughout each period and straight-line depreciation is used, the average investment is computed as the average of the investment's initial book value and its salvage value.

P3 Compute net present value and describe its use. An investment's net present value is determined by predicting the future cash flows it is expected to generate, discounting them at a rate that represents an acceptable return, and then by subtracting the investment's initial cost from the sum of the present values. This technique can deal with any pattern of expected cash flows and applies a superior concept of return on investment.

P4 Compute internal rate of return and explain its use. The internal rate of return (IRR) is the discount rate that results in a zero net present value. When the cash flows are equal, we can compute the present value factor corresponding to the IRR by dividing the initial investment by the annual cash flows. We then use the annuity tables to determine the discount rate corresponding to this present value factor.

Guidance Answers to **Decision Maker** and **Decision Ethics**

Systems Manager Your dilemma is whether to abide by rules designed to prevent abuse or to bend them to acquire an investment that you believe will benefit the firm. You should not pursue the latter action because breaking up the order into small components is dishonest and there are consequences of being caught at a later stage. Develop a proposal for the entire package and then do all you can to expedite its processing, particularly by pointing out its benefits. When faced with controls that are not working, there is rarely a reason to overcome its shortcomings by dishonesty. A direct assault on those limitations is more sensible and ethical.

Entrepreneur The banker is probably concerned because new products are risky and should therefore be evaluated using a higher rate of return. You should conduct a thorough technical analysis and obtain detailed market data and information about any similar products available in the market. These factors might provide sufficient information to support the use of a lower return. You must convince yourself that the risk level is consistent with the discount rate used. You should also be confident that your company has the capacity and the resources to handle the new product.

Partner You should identify the differences between existing clients and this potential client. A key difference is that the restaurant business has additional inventory components (groceries, vegetables, meats, etc.) and is likely to have a higher proportion of depreciable assets. These differences imply that the partner must spend more hours auditing the records and understanding the business, regulations, and standards that pertain to the restaurant business. Such differences suggest that the partner must use a different "formula" for quoting a price to this potential client vis-à-vis current clients.

Investment Manager You should probably focus on either the payback period or break-even time because both the time value of money and recovery time are important. Break-even time method is superior because it accounts for the time value of money, which is an important consideration in this decision.

Guidance Answers to **Quick Checks**

1. *b*

2. A capital budgeting decision is difficult because (1) the outcome is uncertain, (2) large amounts of money are usually involved, (3) a long-term commitment is required, and (4) the decision could be difficult or impossible to reverse.

3. *b*

4. Depreciation expense is subtracted from revenues in computing net income but does not use cash and should be added back to net income to compute net cash flows.

5. Not necessarily. One investment can continue to generate cash flows beyond the payback period for a longer time period than the other. The timing of their cash flows within the payback period also can differ.

6. *b*; Annual average investment = ($180,000 + $15,000)/2
= $97,500

Accounting rate of return = $40,000/$97,500 = 41%

7. For this determination, we need to compare it to the returns expected from alternative investments with similar risk.

8. Project A has the higher net present value as follows:

		Project A		Project B	
Year	Present Value of 1 at 10%	Net Cash Flows	Present Value of Net Cash Flows	Net Cash Flows	Present Value of Net Cash Flows
1	0.9091	$12,000	$10,909	$ 4,500	$ 4,091
2	0.8264	8,500	7,024	8,500	7,024
3	0.7513	4,000	3,005	13,000	9,767
Totals		$24,500	$20,938	$26,000	$20,882
Amount invested			(20,000)		(20,000)
Net present value			**$ 938**		**$ 882**

9. No, the information is too limited to draw that conclusion. For example, one investment could be riskier than the other, or one could require a substantially larger initial investment.

10. *b*

11. *e*; Variable costs per unit for this order of 200 units follow:

Direct materials ($37,500/7,500)	$ 5.00
Direct labor ($60,000/7,500)	8.00
Variable overhead [(0.30 × $20,000)/7,500]	0.80
Variable selling expenses [(0.60 × $25,000 × 0.5)/7,500]	1.00
Total variable costs per unit	$14.80

Cost to produce special order: (200 × $14.80) + $400
= $3,360.

Price per unit to earn $1,000: ($3,360 + $1,000)/200 = 21.80.

12. They are the additional (new) costs of accepting new business.

13. *d*;

	Incremental benefits		Incremental costs
Alpha	$300 ($600 − $300)	>	$150 (given)
Beta	$450 ($900 − $450)	>	$300 (given)
Gamma	$150 ($425 − $275)	>	$125 (given)
Delta	$ 60 ($210 − $150)	<	$ 75 (given)

14. A sunk cost is *never* relevant because it results from a past decision and is already incurred.

15. Avoidable expenses are ones a company will not incur by eliminating a segment; unavoidable expenses will continue even after a segment is eliminated.

16. *a*

Key Terms

mhhe.com/wildFAP19e

Key Terms are available at the book's Website for learning and testing in an online Flashcard Format.

Accounting rate of return (p. 1027)
Avoidable expense (p. 1041)
Break-even time (BET) (p. 1043)
Capital budgeting (p. 1024)
Cost of capital (p. 1029)
Hurdle rate (p. 1033)
Incremental cost (p. 1036)
Internal rate of return (IRR) (p. 1031)
Net present value (NPV) (p. 1029)
Payback period (PBP) (p. 1025)
Profitability index (p. 1031)
Relevant benefits (p. 1035)
Unavoidable expense (p. 1041)

Multiple Choice Quiz

Answers on p. 1064

mhhe.com/wildFAP19e

Additional Quiz Questions are available at the book's Website.

Quiz25

1. A company inadvertently produced 3,000 defective MP3 players. The players cost $12 each to produce. A recycler offers to purchase the defective players as they are for $8 each. The production manager reports that the defects can be corrected for $10 each, enabling them to be sold at their regular market price of $19 each. The company should:

a. Correct the defect and sell them at the regular price.
b. Sell the players to the recycler for $8 each.
c. Sell 2,000 to the recycler and repair the rest.
d. Sell 1,000 to the recycler and repair the rest.
e. Throw the players away.

2. A company's productive capacity is limited to 480,000 machine hours. Product X requires 10 machine hours to produce; and Product Y requires 2 machine hours to produce. Product X sells for $32 per unit and has variable costs of $12 per unit; Product Y sells for $24 per unit and has variable costs of $10 per unit. Assuming that the company can sell as many of either product as it produces, it should:
   a. Produce X and Y in the ratio of 57% and 43%.
   b. Produce X and Y in the ratio of 83% X and 17% Y.
   c. Produce equal amounts of Product X and Product Y.
   d. Produce only Product X.
   e. Produce only Product Y.
3. A company receives a special one-time order for 3,000 units of its product at $15 per unit. The company has excess capacity and it currently produces and sells the units at $20 each to its regular customers. Production costs are $13.50 per unit, which includes $9 of variable costs. To produce the special order, the company must incur additional fixed costs of $5,000. Should the company accept the special order?
   a. Yes, because incremental revenue exceeds incremental costs.
   b. No, because incremental costs exceed incremental revenue.
   c. No, because the units are being sold for $5 less than the regular price.
   d. Yes, because incremental costs exceed incremental revenue.
   e. No, because incremental cost exceeds $15 per unit when total costs are considered.
4. A company is considering the purchase of equipment for $270,000. Projected annual cash inflow from this equipment is $61,200 per year. The payback period is:
   a. 0.2 years
   b. 5.0 years
   c. 4.4 years
   d. 2.3 years
   e. 3.9 years
5. A company buys a machine for $180,000 that has an expected life of nine years and no salvage value. The company expects an annual net income (after taxes of 30%) of $8,550. What is the accounting rate of return?
   a. 4.75%
   b. 42.75%
   c. 2.85%
   d. 9.50%
   e. 6.65%

Discussion Questions

1. What is capital budgeting?
2. Identify four reasons that capital budgeting decisions by managers are risky.
3. Capital budgeting decisions require careful analysis because they are generally the ________ ________ and ________ decisions that management faces.
4. Identify two disadvantages of using the payback period for comparing investments.
5. Why is an investment more attractive to management if it has a shorter payback period?
6. What is the average amount invested in a machine during its predicted five-year life if it costs $200,000 and has a $20,000 salvage value? Assume that net income is received evenly throughout each year and straight-line depreciation is used.
7. If the present value of the expected net cash flows from a machine, discounted at 10%, exceeds the amount to be invested, what can you say about the investment's expected rate of return? What can you say about the expected rate of return if the present value of the net cash flows, discounted at 10%, is less than the investment amount?
8. Why is the present value of $100 that you expect to receive one year from today worth less than $100 received today? What is the present value of $100 that you expect to receive one year from today, discounted at 12%?
9. Why should managers set the required rate of return higher than the rate at which money can be borrowed when making a typical capital budgeting decision?
10. Why does the use of the accelerated depreciation method (instead of straight line) for income tax reporting increase an investment's value?
11. What is an out-of-pocket cost? What is an opportunity cost? Are opportunity costs recorded in the accounting records?
12. Why are sunk costs irrelevant in deciding whether to sell a product in its present condition or to make it into a new product through additional processing?
13. Identify the incremental costs incurred by **Best Buy** for shipping one additional iPod from a warehouse to a retail store along with the store's normal order of 75 iPods.

14. **Circuit City** is considering expanding a store. Identify three methods management can use to evaluate whether to expand.
15. Assume that **Apple** manufactures and sells 500,000 units of a product at $30 per unit in domestic markets. It costs $20 per unit to manufacture ($13 variable cost per unit, $7 fixed cost per unit). Can you describe a situation under which the company is willing to sell an additional 25,000 units of the product in an international market at $15 per unit?

Denotes Discussion Questions that involve decision making.

McGraw-Hill's HOMEWORK MANAGER® Available with McGraw-Hill's Homework Manager

QUICK STUDY

QS 25-1
Analyzing payback periods P1

Trek Company is considering two alternative investments. The payback period is 2.5 years for Investment A and 3 years for Investment B. (1) If management relies on the payback period, which investment is preferred? (2) Why might Trek's analysis of these two alternatives lead to the selection of B over A?

QS 25-2
Payback period P1

Foster Company is considering an investment that requires immediate payment of $360,000 and provides expected cash inflows of $120,000 annually for four years. What is the investment's payback period?

QS 25-3
Computation of net present value P3

If Kimball Company invests $100,000 today, it can expect to receive $20,000 at the end of each year for the next seven years plus an extra $12,000 at the end of the seventh year. What is the net present value of this investment assuming a required 8% return on investments?

QS 25-4
Computation of accounting rate of return P2

Camino Company is considering an investment expected to generate an average net income after taxes of $3,825 for three years. The investment costs $90,000 and has an estimated $12,000 salvage value. Compute the accounting rate of return for this investment; assume the company uses straight-line depreciation. Hint: Use the formula in Exhibit 25.5 when computing the average annual investment.

QS 25-5
Selection of sales mix C3 A1

Flash Memory Company can sell all units of computer memory X and Y that it can produce, but it has limited production capacity. It can produce four units of X per hour *or* six units of Y per hour, and it has 16,000 production hours available. Contribution margin is $10 for Product X and $8 for Product Y. What is the most profitable sales mix for this company?

QS 25-6
Analysis of incremental costs C3 A1

Falcon Company incurs a $18 per unit cost for Product A, which it currently manufactures and sells for $27 per unit. Instead of manufacturing and selling this product, the company can purchase Product B for $10 per unit and sell it for $24 per unit. If it does so, unit sales would remain unchanged and $10 of the $18 per unit costs assigned to Product A would be eliminated. Should the company continue to manufacture Product A or purchase Product B for resale?

QS 25-7
Computation of break-even time A2

Fast Feet, a shoe manufacturer, is evaluating the costs and benefits of new equipment that would custom fit each pair of athletic shoes. The customer would have his or her foot scanned by digital computer equipment; this information would be used to cut the raw materials to provide the customer a perfect fit. The new equipment costs $300,000 and is expected to generate an additional $105,000 in cash flows for five years. A bank will make a $300,000 loan to the company at a 8% interest rate for this equipment's purchase. Use the following table to determine the break-even time for this equipment. (Round the present value of cash flows to the nearest dollar.)

Year	Cash Flows*	Present Value of 1 at 8%	Present Value of Cash Flows	Cumulative Present Value of Cash Flows
0	$(300,000)	1.0000		
1	105,000	0.9259		
2	105,000	0.8573		
3	105,000	0.7938		
4	105,000	0.7350		
5	105,000	0.6806		

* All cash flows occur at year-end.

QS 25-8
Profitability index P3

Jemak Company is considering two alternative projects. Project 1 requires an initial investment of $800,000 and has a net present value of cash flows of $1,600,000. Project 2 requires an initial investment of $4,000,000 and has a net present value of cash flows of $2,000,000. Compute the profitability index for each project. Based on the profitability index, which project should the company prefer? Explain.

Available with McGraw-Hill's Homework Manager McGraw-Hill's HOMEWORK MANAGER®

EXERCISES

Exercise 25-1
Payback period computation; even cash flows
P1

Compute the payback period for each of these two separate investments (round the payback period to two decimals):

a. A new operating system for an existing machine is expected to cost $250,000 and have a useful life of four years. The system yields an incremental after-tax income of $72,000 each year after deducting its straight-line depreciation. The predicted salvage value of the system is $10,000.

b. A machine costs $180,000, has a $12,000 salvage value, is expected to last eight years, and will generate an after-tax income of $39,000 per year after straight-line depreciation.

Exercise 25-2
Payback period computation; uneven cash flows
P1

Walker Company is considering the purchase of an asset for $90,000. It is expected to produce the following net cash flows. The cash flows occur evenly throughout each year. Compute the payback period for this investment.

	Year 1	Year 2	Year 3	Year 4	Year 5	Total
Net cash flows	$40,000	$30,000	$40,000	$70,000	$29,000	$209,000

Check 2.5 years

Exercise 25-3
Payback period computation; declining-balance depreciation
P1

A machine can be purchased for $600,000 and used for 5 years, yielding the following net incomes. In projecting net incomes, double-declining balance depreciation is applied, using a 5-year life and a zero salvage value. Compute the machine's payback period (ignore taxes). (Round the payback period to two decimals.)

	Year 1	Year 2	Year 3	Year 4	Year 5
Net incomes	$40,000	$100,000	$200,000	$150,000	$400,000

Check 2.27 years

Exercise 25-4
Accounting rate of return P2

A machine costs $200,000 and is expected to yield an after-tax net income of $5,040 each year. Management predicts this machine has a 12-year service life and a $40,000 salvage value, and it uses straight-line depreciation. Compute this machine's accounting rate of return.

Exercise 25-5
Payback period and accounting rate of return on investment
P1 P2

MLM Co. is considering the purchase of equipment that would allow the company to add a new product to its line. The equipment is expected to cost $324,000 with a 12-year life and no salvage value. It will be depreciated on a straight-line basis. The company expects to sell 128,000 units of the equipment's product each year. The expected annual income related to this equipment follows. Compute the (1) payback period and (2) accounting rate of return for this equipment.

Sales	$200,000
Costs	
Materials, labor, and overhead (except depreciation)	107,000
Depreciation on new equipment	27,000
Selling and administrative expenses	20,000
Total costs and expenses	154,000
Pretax income	46,000
Income taxes (30%)	13,800
Net income	$ 32,200

Check (1) 5.47 years (2) 19.88%

Exercise 25-6
Computing net present value P3

After evaluating the risk of the investment described in Exercise 25-5, MLM Co. concludes that it must earn at least a 10% return on this investment. Compute the net present value of this investment. (Round the net present value to the nearest dollar.)

Exercise 25-7
Computation and interpretation of net present value and internal rate of return
P3 P4 

Cerritos Company can invest in each of three cheese-making projects: C1, C2, and C3. Each project requires an initial investment of $438,374 and would yield the following annual cash flows.

	C1	C2	C3
Year 1	$ 24,000	$192,000	$360,000
Year 2	216,000	192,000	120,000
Year 3	336,000	192,000	96,000
Totals	$576,000	$576,000	$576,000

(1) Assuming that the company requires a 12% return from its investments, use net present value to determine which projects, if any, should be acquired. (2) Using the answer from part 1, explain whether the internal rate of return is higher or lower than 12% for project C2. (3) Compute the internal rate of return for project C2.

Check (3) IRR = 15%

Exercise 25-8
NPV and profitability index
P3

Following is information on two alternative investments being considered by Jakem Company. The company requires a 10% return from its investments.

	Project A	Project B
Initial investment	$(180,325)	$(150,960)
Expected net cash flows in year:		
1	45,000	35,000
2	50,000	52,000
3	82,295	58,000
4	86,400	75,000
5	64,000	29,000

For each alternative project compute the (a) net present value, and (b) profitability index. If the company can only select one project, which should it choose? Explain.

Exercise 25-9A
Using Excel to compute IRR
P4

Refer to the information in Exercise 25-8. Create an Excel spreadsheet to compute the internal rate of return for each of the projects. Round the percentage return to two decimals.

Exercise 25-10
Decision to accept additional business or not
C3 A1 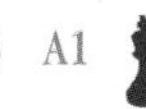

Harlan Co. expects to sell 300,000 units of its product in the next period with the following results.

Sales (300,000 units)	$4,500,000
Costs and expenses	
Direct materials	600,000
Direct labor	1,200,000
Overhead	300,000
Selling expenses	450,000
Administrative expenses	771,000
Total costs and expenses	3,321,000
Net income	$1,179,000

The company has an opportunity to sell 30,000 additional units at $13 per unit. The additional sales would not affect its current expected sales. Direct materials and labor costs per unit would be the same for the additional units as they are for the regular units. However, the additional volume would create the following incremental costs: (1) total overhead would increase by 16% and (2) administrative expenses would increase by $129,000. Prepare an analysis to determine whether the company should accept or reject the offer to sell additional units at the reduced price of $13 per unit.

Check Income increase, $33,000

Exercise 25-11
Make or buy decision
C3 A1 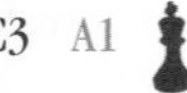

Check $1,600 increased costs to make

Simons Company currently manufactures one of its crucial parts at a cost of $2.72 per unit. This cost is based on a normal production rate of 40,000 units per year. Variable costs are $1.20 per unit, fixed costs related to making this part are $40,000 per year, and allocated fixed costs are $50,000 per year. Allocated fixed costs are unavoidable whether the company makes or buys the part. Simons is considering buying the part from a supplier for a quoted price of $2.16 per unit guaranteed for a three-year period. Should the company continue to manufacture the part, or should it buy the part from the outside supplier? Support your answer with analyses.

Exercise 25-12
Sell or process decision
C3 A1

Starr Company has already manufactured 50,000 units of Product A at a cost of $50 per unit. The 50,000 units can be sold at this stage for $1,250,000. Alternatively, it can be further processed at a $750,000 total additional cost and be converted into 10,000 units of Product B and 20,000 units of Product C. Per unit selling price for Product B is $75 and for Product C is $50. Prepare an analysis that shows whether the 50,000 units of Product A should be processed further or not.

Exercise 25-13
Analysis of income effects from eliminating departments
C3 A1 

Johns Co. expects its five departments to yield the following income for next year.

		Dept. M	Dept. N	Dept. O	Dept. P	Dept. T
2	Sales	$34,000	$23,500	$33,000	$27,500	$ 10,500
3	Expenses					
4	Avoidable	4,700	18,900	15,800	8,000	14,900
5	Unavoidable	20,000	5,100	2,900	15,000	5,900
6	Total expenses	24,700	24,000	18,700	23,000	20,800
7	Net income (loss)	$ 9,300	$ (500)	$14,300	$ 4,500	$(10,300)

Check Total income (2) $17,100, (3) $21,700

Recompute and prepare the departmental income statements (including a combined total column) for the company under each of the following separate scenarios: Management (1) does not eliminate any department, (2) eliminates departments with expected net losses, and (3) eliminates departments with sales dollars that are less than avoidable expenses. Explain your answers to parts 2 and 3.

Exercise 25-14
Sales mix determination and analysis
C3 A1

Jersey Company owns a machine that can produce two specialized products. Production time for Product TLX is two units per hour and for Product MTV is five units per hour. The machine's capacity is 2,200 hours per year. Both products are sold to a single customer who has agreed to buy all of the company's output up to a maximum of 3,740 units of Product TLX and 2,090 units of Product MTV. Selling prices and variable costs per unit to produce the products follow. Determine (1) the company's most profitable sales mix and (2) the contribution margin that results from that sales mix.

	Product TLX	Product MTV
Selling price per unit	$11.50	$6.90
Variable costs per unit	3.45	4.14

Check (2) $34,661

Exercise 25-15
Comparison of payback and BET
P1 A2 

This chapter explained two methods to evaluate investments using recovery time, the payback period and break-even time (BET). Refer to QS 25-7 and (1) compute the recovery time for both the payback period and break-even time, (2) discuss the advantage(s) of break-even time over the payback period, and (3) list two conditions under which payback period and break-even time are similar.

McGraw-Hill's HOMEWORK MANAGER® Available with McGraw-Hill's Homework Manager

PROBLEM SET A

Problem 25-1A
Computation of payback period, accounting rate of return, and net present value

P1 P2 P3

mhhe.com/wildFAP19e

Burtle Company is planning to add a new product to its line. To manufacture this product, the company needs to buy a new machine at a $488,000 cost with an expected four-year life and a $15,200 salvage value. All sales are for cash, and all costs are out of pocket except for depreciation on the new machine. Additional information includes the following.

Expected annual sales of new product	$1,870,000
Expected annual costs of new product	
Direct materials	465,000
Direct labor	680,000
Overhead excluding straight-line depreciation on new machine	335,000
Selling and administrative expenses	158,000
Income taxes	40%

Required

1. Compute straight-line depreciation for each year of this new machine's life. (Round depreciation amounts to the nearest dollar.)

2. Determine expected net income and net cash flow for each year of this machine's life. (Round answers to the nearest dollar.)

3. Compute this machine's payback period, assuming that cash flows occur evenly throughout each year. (Round the payback period to two decimals.)

4. Compute this machine's accounting rate of return, assuming that income is earned evenly throughout each year. (Round the percentage return to two decimals.)

5. Compute the net present value for this machine using a discount rate of 8% and assuming that cash flows occur at each year-end. (*Hint:* Salvage value is a cash inflow at the end of the asset's life. Round the net present value to the nearest dollar.)

Check (4) 27.14%

(5) $140,794

Problem 25-2A
Analysis and computation of payback period, accounting rate of return, and net present value

P1 P2 P3 

Jackson Company has an opportunity to invest in one of two new projects. Project Y requires a $360,000 investment for new machinery with a four-year life and no salvage value. Project Z requires a $360,000 investment for new machinery with a three-year life and no salvage value. The two projects yield the following predicted annual results. The company uses straight-line depreciation, and cash flows occur evenly throughout each year.

	Project Y	Project Z
Sales	$355,000	$265,000
Expenses		
Direct materials	49,700	30,125
Direct labor	71,000	36,750
Overhead including depreciation	127,800	129,250
Selling and administrative expenses	25,000	20,000
Total expenses	273,500	216,125
Pretax income	81,500	48,875
Income taxes (30%)	24,450	14,663
Net income	$ 57,050	$ 34,212

Required

1. Compute each project's annual expected net cash flows. (Round the net cash flows to the nearest dollar.)

2. Determine each project's payback period. (Round the payback period to two decimals.)

3. Compute each project's accounting rate of return. (Round the percentage return to one decimal.)

4. Determine each project's net present value using 6% as the discount rate. For part 4 only, assume that cash flows occur at each year-end. (Round the net present value to the nearest dollar.)

Check For Project Y: (2) 2.45 years, (3) 31.7%, (4) $149,543

Analysis Component

5. Identify the project you would recommend to management and explain your choice.

Problem 25-3A
Computation of cash flows and net present values with alternative depreciation methods

P3 

Deandra Corporation is considering a new project requiring a $97,500 investment in test equipment with no salvage value. The project would produce $71,000 of pretax income before depreciation at the end of each of the next six years. The company's income tax rate is 32%. In compiling its tax return and computing its income tax payments, the company can choose between the two alternative depreciation schedules shown in the table.

	Straight-Line Depreciation	MACRS Depreciation*
Year 1	$ 9,750	$19,500
Year 2	19,500	31,200
Year 3	19,500	18,720
Year 4	19,500	11,232
Year 5	19,500	11,232
Year 6	9,750	5,616
Totals	$97,500	$97,500

* The modified accelerated cost recovery system (MACRS) for depreciation is discussed in Chapter 10.

Required

1. Prepare a five-column table that reports amounts (assuming use of straight-line depreciation) for each of the following for each of the six years: (a) pretax income before depreciation, (b) straight-line depreciation expense, (c) taxable income, (d) income taxes, and (e) net cash flow. Net cash flow equals the amount of income before depreciation minus the income taxes. (Round answers to the nearest dollar.)

2. Prepare a five-column table that reports amounts (assuming use of MACRS depreciation) for each of the following for each of the six years: (a) pretax income before depreciation, (b) MACRS depreciation expense, (c) taxable income, (d) income taxes, and (e) net cash flow. Net cash flow equals the income amount before depreciation minus the income taxes. (Round answers to the nearest dollar.)

Check Net present value: (3) $135,347, (4) $136,893

3. Compute the net present value of the investment if straight-line depreciation is used. Use 10% as the discount rate. (Round the net present value to the nearest dollar.)

4. Compute the net present value of the investment if MACRS depreciation is used. Use 10% as the discount rate. (Round the net present value to the nearest dollar.)

Analysis Component

5. Explain why the MACRS depreciation method increases this project's net present value.

Problem 25-4A
Analysis of income effects of additional business

C3 A1

mhhe.com/wildFAP19e

Ingraham Products manufactures and sells to wholesalers approximately 200,000 packages per year of underwater markers at $4 per package. Annual costs for the production and sale of this quantity are shown in the table.

Direct materials	$256,000
Direct labor	64,000
Overhead	192,000
Selling expenses	80,000
Administrative expenses	53,000
Total costs and expenses	$645,000

A new wholesaler has offered to buy 33,000 packages for $3.44 each. These markers would be marketed under the wholesaler's name and would not affect Ingraham Products' sales through its normal channels. A study of the costs of this additional business reveals the following:

- Direct materials costs are 100% variable.
- Per unit direct labor costs for the additional units would be 50% higher than normal because their production would require overtime pay at one-and-one-half times the usual labor rate.
- 35% of the normal annual overhead costs are fixed at any production level from 150,000 to 300,000 units. The remaining 65% of the annual overhead cost is variable with volume.

- Accepting the new business would involve no additional selling expenses.
- Accepting the new business would increase administrative expenses by a $5,000 fixed amount.

Required

Prepare a three-column comparative income statement that shows the following:

1. Annual operating income without the special order (column 1).
2. Annual operating income received from the new business only (column 2).
3. Combined annual operating income from normal business and the new business (column 3).

Check Operating income: (1) $155,000, (2) $29,848

Problem 25-5A
Analysis of sales mix strategies
C3 A1

Virginia Company is able to produce two products, G and B, with the same machine in its factory. The following information is available.

	Product G	Product B
Selling price per unit	$280	$240
Variable costs per unit	130	60
Contribution margin per unit	$150	$180
Machine hours to produce 1 unit	0.2 hours	2.0 hours
Maximum unit sales per month	1,200 units	200 units

The company presently operates the machine for a single eight-hour shift for 22 working days each month. Management is thinking about operating the machine for two shifts, which will increase its productivity by another eight hours per day for 22 days per month. This change would require $63,000 additional fixed costs per month.

Required

1. Determine the contribution margin per machine hour that each product generates.
2. How many units of Product G and Product B should the company produce if it continues to operate with only one shift? How much total contribution margin does this mix produce each month?
3. If the company adds another shift, how many units of Product G and Product B should it produce? How much total contribution margin would this mix produce each month? Should the company add the new shift? Explain.
4. Suppose that the company determines that it can increase Product G's maximum sales to 1,400 units per month by spending $24,000 per month in marketing efforts. Should the company pursue this strategy and the double shift? Explain.

Check Units of Product G: (2) 880, (3) 1,200, (4) 1,400

Problem 25-6A
Analysis of possible elimination of a department
C3 A1

Eclectic Decor Company's management is trying to decide whether to eliminate Department 200, which has produced losses or low profits for several years. The company's 2009 departmental income statement shows the following.

ECLECTIC DECOR COMPANY
Departmental Income Statements
For Year Ended December 31, 2009

	Dept. 100	Dept. 200	Combined
Sales	$437,000	$280,000	$717,000
Cost of goods sold	263,000	207,000	470,000
Gross profit	174,000	73,000	247,000
Operating expenses			
Direct expenses			
Advertising	17,500	13,500	31,000
Store supplies used	5,000	4,600	9,600
Depreciation—Store equipment	4,200	3,000	7,200
Total direct expenses	26,700	21,100	47,800

[continued on next page]

[continued from previous page]

Allocated expenses			
Sales salaries	52,000	31,200	83,200
Rent expense	9,500	4,750	14,250
Bad debts expense	9,500	7,400	16,900
Office salary	15,600	10,400	26,000
Insurance expense	1,900	1,000	2,900
Miscellaneous office expenses	2,500	1,700	4,200
Total allocated expenses	91,000	56,450	147,450
Total expenses	117,700	77,550	195,250
Net income (loss)	$ 56,300	$ (4,550)	$ 51,750

In analyzing whether to eliminate Department 200, management considers the following:

a. The company has one office worker who earns $500 per week, or $26,000 per year, and four salesclerks who each earn $400 per week, or $20,800 per year.

b. The full salaries of two salesclerks are charged to Department 100. The full salary of one sales clerk is charged to Department 200. The salary of the fourth clerk, who works half-time in both departments, is divided evenly between the two departments.

c. Eliminating Department 200 would avoid the sales salaries and the office salary currently allocated to it. However, management prefers another plan. Two salesclerks have indicated that they will be quitting soon. Management believes that their work can be done by the other two clerks if the one office worker works in sales half-time. Eliminating Department 200 will allow this shift of duties. If this change is implemented, half the office worker's salary would be reported as sales salaries and half would be reported as office salary.

d. The store building is rented under a long-term lease that cannot be changed. Therefore, Department 100 will use the space and equipment currently used by Department 200.

e. Closing Department 200 will eliminate its expenses for advertising, bad debts, and store supplies; 70% of the insurance expense allocated to it to cover its merchandise inventory; and 25% of the miscellaneous office expenses presently allocated to it.

Required

Check (1) Total expenses: (a) $665,250, (b) $275,225

1. Prepare a three-column report that lists items and amounts for (a) the company's total expenses (including cost of goods sold)—in column 1, (b) the expenses that would be eliminated by closing Department 200—in column 2, and (c) the expenses that will continue—in column 3.

(2) Forecasted net income without Department 200, $46,975

2. Prepare a forecasted annual income statement for the company reflecting the elimination of Department 200 assuming that it will not affect Department 100's sales and gross profit. The statement should reflect the reassignment of the office worker to one-half time as a salesclerk.

Analysis Component

3. Reconcile the company's combined net income with the forecasted net income assuming that Department 200 is eliminated (list both items and amounts). Analyze the reconciliation and explain why you think the department should or should not be eliminated.

PROBLEM SET B

Problem 25-1B
Computation of payback period, accounting rate of return, and net present value

P1 P2 P3

Sorbo Company is planning to add a new product to its line. To manufacture this product, the company needs to buy a new machine at a $600,000 cost with an expected four-year life and a $20,000 salvage value. All sales are for cash and all costs are out of pocket, except for depreciation on the new machine. Additional information includes the following.

Expected annual sales of new product	$2,300,000
Expected annual costs of new product	
Direct materials	600,000
Direct labor	840,000
Overhead excluding straight-line depreciation on new machine	420,000
Selling and administrative expenses	200,000
Income taxes	30%

Required

1. Compute straight-line depreciation for each year of this new machine's life. (Round depreciation amounts to the nearest dollar.)

2. Determine expected net income and net cash flow for each year of this machine's life. (Round answers to the nearest dollar.)

3. Compute this machine's payback period, assuming that cash flows occur evenly throughout each year. (Round the payback period to two decimals.)

4. Compute this machine's accounting rate of return, assuming that income is earned evenly throughout each year. (Round the percentage return to two decimals.)

5. Compute the net present value for this machine using a discount rate of 7% and assuming that cash flows occur at each year-end. (*Hint:* Salvage value is a cash inflow at the end of the asset's life.)

Check (4) 21.45%

(5) $131,650

Problem 25-2B
Analysis and computation of payback period, accounting rate of return, and net present value

P1 P2 P3

Morris Company has an opportunity to invest in one of two projects. Project A requires a $480,000 investment for new machinery with a four-year life and no salvage value. Project B also requires a $480,000 investment for new machinery with a three-year life and no salvage value. The two projects yield the following predicted annual results. The company uses straight-line depreciation, and cash flows occur evenly throughout each year.

	Project A	Project B
Sales	$500,000	$400,000
Expenses		
Direct materials	70,000	50,000
Direct labor	100,000	60,000
Overhead including depreciation	180,000	180,000
Selling and administrative expenses	36,000	36,000
Total expenses	386,000	326,000
Pretax income	114,000	74,000
Income taxes (30%)	34,200	22,200
Net income	$ 79,800	$ 51,800

Required

1. Compute each project's annual expected net cash flows. (Round net cash flows to the nearest dollar.)

2. Determine each project's payback period. (Round the payback period to two decimals.)

3. Compute each project's accounting rate of return. (Round the percentage return to one decimal.)

4. Determine each project's net present value using 8% as the discount rate. For part 4 only, assume that cash flows occur at each year-end. (Round net present values to the nearest dollar.)

Check For Project A: (2) 2.4 years, (3) 33.3%, (4) $181,758

Analysis Component

5. Identify the project you would recommend to management and explain your choice.

Problem 25-3B
Computation of cash flows and net present values with alternative depreciation methods

P3 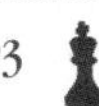 

Lee Corporation is considering a new project requiring a $300,000 investment in an asset having no salvage value. The project would produce $125,000 of pretax income before depreciation at the end of each of the next six years. The company's income tax rate is 35%. In compiling its tax return and computing its income tax payments, the company can choose between two alternative depreciation schedules as shown in the table.

	Straight-Line Depreciation	MACRS Depreciation*
Year 1	$ 30,000	$ 60,000
Year 2	60,000	96,000
Year 3	60,000	57,600
Year 4	60,000	34,560
Year 5	60,000	34,560
Year 6	30,000	17,280
Totals	$300,000	$300,000

* The modified accelerated cost recovery system (MACRS) for depreciation is discussed in Chapter 10.

Required

1. Prepare a five-column table that reports amounts (assuming use of straight-line depreciation) for each of the following items for each of the six years: (a) pretax income before depreciation, (b) straight-line depreciation expense, (c) taxable income, (d) income taxes, and (e) net cash flow. Net cash flow equals the amount of income before depreciation minus the income taxes. (Round answers to the nearest dollar.)

2. Prepare a five-column table that reports amounts (assuming use of MACRS depreciation) for each of the following items for each of the six years: (a) income before depreciation, (b) MACRS depreciation expense, (c) taxable income, (d) income taxes, and (e) net cash flow. Net cash flow equals the amount of income before depreciation minus the income taxes. (Round answers to the nearest dollar.)

Check Net present value: (3) $129,846, (4) $135,050

3. Compute the net present value of the investment if straight-line depreciation is used. Use 10% as the discount rate. (Round the net present value to the nearest dollar.)

4. Compute the net present value of the investment if MACRS depreciation is used. Use 10% as the discount rate. (Round the net present value to the nearest dollar.)

Analysis Component

5. Explain why the MACRS depreciation method increases the net present value of this project.

Problem 25-4B
Analysis of income effects of additional business
C3 A1

Wyn Company manufactures and sells to local wholesalers approximately 150,000 units per month at a sales price of $4 per unit. Monthly costs for the production and sale of this quantity follow.

Direct materials	$192,000
Direct labor	48,000
Overhead	144,000
Selling expenses	60,000
Administrative expenses	40,000
Total costs and expenses	$484,000

A new out-of-state distributor has offered to buy 25,000 units next month for $3.44 each. These units would be marketed in other states and would not affect Wyn's sales through its normal channels. A study of the costs of this new business reveals the following:

- Direct materials costs are 100% variable.
- Per unit direct labor costs for the additional units would be 50% higher than normal because their production would require time-and-a-half overtime pay to meet the distributor's deadline.
- Twenty-five percent of the normal annual overhead costs are fixed at any production level from 125,000 to 200,000 units. The remaining 75% is variable with volume.
- Accepting the new business would involve no additional selling expenses.
- Accepting the new business would increase administrative expenses by a $2,000 fixed amount.

Required

Prepare a three-column comparative income statement that shows the following:

Check Operating income: (1) $116,000, (2) $22,000

1. Monthly operating income without the special order (column 1).

2. Monthly operating income received from the new business only (column 2).

3. Combined monthly operating income from normal business and the new business (column 3).

Problem 25-5B
Analysis of sales mix strategies
C3 A1

Verto Company is able to produce two products, R and T, with the same machine in its factory. The following information is available.

	Product R	Product T
Selling price per unit	$ 120	$160
Variable costs per unit	65	90
Contribution margin per unit	$ 55	$ 70
Machine hours to produce 1 unit	0.2 hours	0.5 hours
Maximum unit sales per month	1,100 units	350 units

The company presently operates the machine for a single eight-hour shift for 22 working days each month. Management is thinking about operating the machine for two shifts, which will increase its productivity by another eight hours per day for 22 days per month. This change would require $30,000 additional fixed costs per month.

Required

1. Determine the contribution margin per machine hour that each product generates.
2. How many units of Product R and Product T should the company produce if it continues to operate with only one shift? How much total contribution margin does this mix produce each month?
3. If the company adds another shift, how many units of Product R and Product T should it produce? How much total contribution margin would this mix produce each month? Should the company add the new shift? Explain.
4. Suppose that the company determines that it can increase Product R's maximum sales to 1,350 units per month by spending $9,000 per month in marketing efforts. Should the company pursue this strategy and the double shift? Explain.

Check Units of Product R: (2) 880, (3) 1,100, (4) 1,350

Problem 25-6B
Analysis of possible elimination of a department

C3 A1 

Kumar Company's management is trying to decide whether to eliminate Department Z, which has produced low profits or losses for several years. The company's 2009 departmental income statement shows the following.

KUMAR COMPANY
Departmental Income Statements
For Year Ended December 31, 2009

	Dept. A	Dept. Z	Combined
Sales	$1,050,000	$262,500	$1,312,500
Cost of goods sold	691,950	187,650	879,600
Gross profit	358,050	74,850	432,900
Operating expenses			
Direct expenses			
Advertising	40,500	4,500	45,000
Store supplies used	8,400	2,100	10,500
Depreciation—Store equipment	21,000	10,500	31,500
Total direct expenses	69,900	17,100	87,000
Allocated expenses			
Sales salaries	105,300	35,100	140,400
Rent expense	33,120	8,280	41,400
Bad debts expense	31,500	6,000	37,500
Office salary	31,200	7,800	39,000
Insurance expense	6,300	2,100	8,400
Miscellaneous office expenses	2,550	3,750	6,300
Total allocated expenses	209,970	63,030	273,000
Total expenses	279,870	80,130	360,000
Net income (loss)	$ 78,180	$ (5,280)	$ 72,900

In analyzing whether to eliminate Department Z, management considers the following items:

a. The company has one office worker who earns $750 per week or $39,000 per year and four salesclerks who each earn $675 per week or $35,100 per year.

b. The full salaries of three salesclerks are charged to Department A. The full salary of one salesclerk is charged to Department Z.

c. Eliminating Department Z would avoid the sales salaries and the office salary currently allocated to it. However, management prefers another plan. Two salesclerks have indicated that they will be quitting soon. Management believes that their work can be done by the two remaining clerks if the one office worker works in sales half time. Eliminating Department Z will allow this shift of duties. If this change is implemented, half the office worker's salary would be reported as sales salaries and half would be reported as office salary.

d. The store building is rented under a long-term lease that cannot be changed. Therefore, Department A will use the space and equipment currently used by Department Z.

e. Closing Department Z will eliminate its expenses for advertising, bad debts, and store supplies; 65% of the insurance expense allocated to it to cover its merchandise inventory; and 30% of the miscellaneous office expenses presently allocated to it.

Required

Check (1) Total expenses: (a) $1,239,600, (b) $272,940

(2) Forecasted net income without Department Z, $83,340

1. Prepare a three-column report that lists items and amounts for (a) the company's total expenses (including cost of goods sold)—in column 1, (b) the expenses that would be eliminated by closing Department Z—in column 2, and (c) the expenses that will continue—in column 3.

2. Prepare a forecasted annual income statement for the company reflecting the elimination of Department Z assuming that it will not affect Department A's sales and gross profit. The statement should reflect the reassignment of the office worker to one-half time as a salesclerk.

Analysis Component

3. Reconcile the company's combined net income with the forecasted net income assuming that Department Z is eliminated (list both items and amounts). Analyze the reconciliation and explain why you think the department should or should not be eliminated.

SERIAL PROBLEM

Success Systems

(This serial problem began in Chapter 1 and continues through most of the book. If previous chapter segments were not completed, the serial problem can begin at this point. It is helpful, but not necessary, to use the Working Papers that accompany the book.)

SP 25 Adriana Lopez is considering the purchase of equipment for Success Systems that would allow the company to add a new product to its computer furniture line. The equipment is expected to cost $300,000 and to have a six-year life and no salvage value. It will be depreciated on a straight-line basis. Success Systems expects to sell 100 units of the equipment's product each year. The expected annual income related to this equipment follows.

Sales	$375,000
Costs	
Materials, labor, and overhead (except depreciation)	200,000
Depreciation on new equipment	50,000
Selling and administrative expenses	37,500
Total costs and expenses	287,500
Pretax income	87,500
Income taxes (30%)	26,250
Net income	$ 61,250

Required

Compute the (1) payback period and (2) accounting rate of return for this equipment.

BEYOND THE NUMBERS

REPORTING IN ACTION

C1 A1 P3 

BTN 25-1 In fiscal 2007, **Best Buy** invested $251 million in store-related projects that included store remodels, relocations, expansions, and various merchandising projects. Assume that these projects have a seven-year life, and that Best Buy requires a 12% internal rate of return on these projects.

Required

1. What is the amount of annual cash flows that Best Buy must earn from these projects to have a 12% internal rate of return? (*Hint:* Identify the seven-period, 12% factor from the present value of an annuity table, and then divide $251 million by this factor to get the annual cash flows necessary.)

Fast Forward

2. Access Best Buy's financial statements for fiscal years ended after March 3, 2007, from its Website (BestBuy.com) or the SEC's Website (SEC.gov).
   a. Determine the amount that Best Buy invested in similar store-related projects for the most recent year.
   b. Assume a seven-year life and a 12% internal rate of return. What is the amount of cash flows that Best Buy must earn on these new projects?

COMPARATIVE ANALYSIS

C3 

BTN 25-2 Best Buy, Circuit City, and RadioShack sell several different products; most are profitable but some are not. Teams of employees in each company make advertising, investment, and product mix decisions. A certain portion of advertising for both companies is on a local basis to a target audience.

Required

1. Find one major advertisement of a product or group of products for each company in your local newspaper. Contact the newspaper and ask the approximate cost of this ad space (for example, cost of one page or one-half page of advertising).
2. Estimate how many products this advertisement must sell to justify its cost. Begin by taking the product's sales price advertised for each company and assume a 20% contribution margin.
3. Prepare a one-half page memorandum explaining the importance of effective advertising when making a product mix decision. Be prepared to present your ideas in class.

ETHICS CHALLENGE

P3

BTN 25-3 A consultant commented that "too often the numbers look good but feel bad." This comment often stems from estimation error common to capital budgeting proposals that relate to future cash flows. Three reasons for this error often exist. First, reliably predicting cash flows several years into the future is very difficult. Second, the present value of cash flows many years into the future (say, beyond 10 years) is often very small. Third, it is difficult for personal biases and expectations not to unduly influence present value computations.

Required

1. Compute the present value of $100 to be received in 10 years assuming a 12% discount rate.
2. Why is understanding the three reasons mentioned for estimation errors important when evaluating investment projects? Link this response to your answer for part 1.

COMMUNICATING IN PRACTICE

P1 P2 P3 P4

BTN 25-4 Payback period, accounting rate of return, net present value, and internal rate of return are common methods to evaluate capital investment opportunities. Assume that your manager asks you to identify the type of measurement basis and unit that each method offers and to list the advantages and disadvantages of each. Present your response in memorandum format of less than one page.

TAKING IT TO THE NET

A1 

BTN 25-5 Many companies must determine whether to internally produce their component parts or to outsource them. Further, some companies now outsource key components or business processes to international providers. Access the Website BizBrim.com and review the available information on outsourcing—especially as it relates to both the advantages and the negative effects of outsourcing.

Required

1. What does Bizbrim identify as the major advantages and the major disadvantages of outsourcing?
2. Does it seem that Bizbrim is generally in favor of or opposed to outsourcing? Explain.

TEAMWORK IN ACTION

P1 P3

BTN 25-6 Break into teams and identify four reasons that an international airline such as Southwest, Northwest, or American would invest in a project when its direct analysis using both payback period and net present value indicate it to be a poor investment. (*Hint:* Think about qualitative factors.) Provide an example of an investment project supporting your answer.

ENTREPRENEURIAL DECISION

A1  

BTN 25-7 Jared Greenberg and Dan Zinger of **Prairie Sticks Bat Company** make baseball bats. They must decide on the best sales mix. Assume their company has a capacity of 80 hours of lathe/processing time available each month and it makes two types of bats, Deluxe and Premium. Information on these bats follows.

	Deluxe	Premium
Selling price per bat	$70	$90
Variable costs per bat	$40	$50
Lathe/processing minutes per bat	6 minutes	12 minutes

Required

1. Assume the markets for both models of bats are unlimited. How many Deluxe bats and how many Premium bats should the company make each month? Explain. How much total contribution margin does this mix produce each month?
2. Assume the market for Deluxe bats is limited to 600 bats per month, with no market limit for Premium bats. How many Deluxe bats and how many Premium bats should the company make each month? Explain. How much total contribution margin does this mix produce each month?

HITTING THE ROAD

C1 P3 

BTN 25-8 Visit or call a local auto dealership and inquire about leasing a car. Ask about the down payment and the required monthly payments. You will likely find the salesperson does not discuss the cost to purchase this car but focuses on the affordability of the monthly payments. This chapter gives you the tools to compute the cost of this car using the lease payment schedule in present dollars and to estimate the profit from leasing for an auto dealership.

Required

1. Compare the cost of leasing the car to buying it in present dollars using the information from the dealership you contact. (Assume you will make a final payment at the end of the lease and then own the car.)
2. Is it more costly to lease or buy the car? Support your answer with computations.

GLOBAL DECISION

C1

BTN 25-9 Access **DSG**'s 2006 annual report dated April 29, 2006, from its Website www.DSGiplc.com. Identify its report on corporate responsibility.

Required

Dixons reports that it recycled 25,607 tons of waste. Efforts such as these can be costly to a company. Why would a company like DSG pursue these costly efforts?

ANSWERS TO MULTIPLE CHOICE QUIZ

1. a; Reworking provides incremental revenue of $11 per unit ($19 − $8); and, it costs $10 to rework them. The company is better off by $1 per unit when it reworks these products and sells them at the regular price.
2. e; Product X has a $2 contribution margin per machine hour [($32 − $12)/10 MH]; Product Y has a $7 contribution margin per machine hour [($24 − $10)/2 MH]. It should produce as much of Product Y as possible.
3. a; Total revenue from the special order = 3,000 units × $15 per unit = $45,000; and, Total costs for the special order = (3,000 units × $9 per unit) + $5,000 = $32,000. Net income from the special order = $45,000 − $32,000 = $13,000. Thus, yes, it should accept the order.
4. c; Payback = $270,000/$61,200 per year = 4.4 years.
5. d; Accounting rate of return = $8,550/[($180,000 + $0)/2] = 9.5%.

A Financial Statement Information

This appendix includes financial information for (1) **Best Buy**, (2) **Circuit City**, (3) **RadioShack**, and (4) **Apple**. This information is taken from their annual 10-K reports filed with the SEC. An **annual report** is a summary of a company's financial results for the year along with its current financial condition and future plans. This report is directed to external users of financial information, but it also affects the actions and decisions of internal users.

A company uses an annual report to showcase itself and its products. Many annual reports include attractive photos, diagrams, and illustrations related to the company. The primary objective of annual reports, however, is the *financial section,* which communicates much information about a company, with most data drawn from the accounting information system. The layout of an annual report's financial section is fairly established and typically includes the following:

- Letter to Shareholders
- Financial History and Highlights
- Management Discussion and Analysis
- Management's Report on Financial Statements and on Internal Controls
- Report of Independent Accountants (Auditor's Report) and on Internal Controls
- Financial Statements
- Notes to Financial Statements
- List of Directors and Officers

This appendix provides the financial statements for Best Buy (plus selected notes), Circuit City, RadioShack, and Apple. The appendix is organized as follows:

- **Best Buy** **A-2** through **A-18**
- **Circuit City** **A-19** through **A-23**
- **RadioShack** **A-24** through **A-28**
- **Apple Computer** **A-29** through **A-33**

Many assignments at the end of each chapter refer to information in this appendix. We encourage readers to spend time with these assignments; they are especially useful in showing the relevance and diversity of financial accounting and reporting.

Special note: The SEC maintains the EDGAR (**E**lectronic **D**ata **G**athering, **A**nalysis, and **R**etrieval) database at www.sec.gov. The **Form 10-K** is the annual report form for most companies. It provides electronically accessible information. The **Form 10-KSB** is the annual report form filed by small businesses. It requires slightly less information than the Form 10-K. One of these forms must be filed within 90 days after the company's fiscal year-end. (Forms 10-K405, 10-KT, 10-KT405, and 10-KSB405 are slight variations of the usual form due to certain regulations or rules.)

Financial Report

Selected Financial Data

The following table presents our selected financial data. In fiscal 2004, we sold our interest in Musicland. All fiscal years presented reflect the classification of Musicland's financial results as discontinued operations.

Five-Year Financial Highlights

$ in millions, except per share amounts

Fiscal Year	2007[(1)]	2006	2005	2004	2003
Consolidated Statements of Earnings Data					
Revenue	$35,934	$30,848	$27,433	$24,548	$20,943
Operating income	1,999	1,644	1,442	1,304	1,010
Earnings from continuing operations	1,377	1,140	934	800	622
Loss from discontinued operations, net of tax	—	—	—	(29)	(441)
Gain (loss) on disposal of discontinued operations, net of tax	—	—	50	(66)	—
Cumulative effect of change in accounting principles, net of tax	—	—	—	—	(82)
Net earnings	1,377	1,140	984	705	99
Per Share Data					
Continuing operations	$ 2.79	$ 2.27	$ 1.86	$ 1.61	$ 1.27
Discontinued operations	—	—	—	(0.06)	(0.89)
Gain (loss) on disposal of discontinued operations	—	—	0.10	(0.13)	—
Cumulative effect of accounting changes	—	—	—	—	(0.16)
Net earnings	2.79	2.27	1.96	1.42	0.20
Cash dividends declared and paid	0.36	0.31	0.28	0.27	—
Common stock price:					
High	59.50	56.00	41.47	41.80	35.83
Low	43.51	31.93	29.25	17.03	11.33
Operating Statistics					
Comparable store sales gain	5.0%	4.9%	4.3%	7.1%	2.4%
Gross profit rate	24.4%	25.0%	23.7%	23.9%	23.6%
Selling, general and administrative expenses rate	18.8%	19.7%	18.4%	18.6%	18.8%
Operating income rate	5.6%	5.3%	5.3%	5.3%	4.8%
Year-End Data					
Current ratio	1.4	1.3	1.4	1.3	1.3
Total assets	$13,570	$11,864	$10,294	$ 8,652	$ 7,694
Debt, including current portion	650	596	600	850	834
Total shareholders' equity	6,201	5,257	4,449	3,422	2,730
Number of stores					
Domestic	868	774	694	631	567
International	304	167	144	127	112
Total	1,172	941	838	758	679
Retail square footage (000s)					
Domestic	33,959	30,826	28,465	26,640	24,432
International	7,926	3,564	3,139	2,800	2,375
Total	41,885	34,390	31,604	29,440	26,807

(1) Fiscal 2007 included 53 weeks. All other periods presented included 52 weeks.

Consolidated Balance Sheets
$ in millions, except per share amounts

	March 3, 2007	February 25, 2006
Assets		
Current Assets		
Cash and cash equivalents	$ 1,205	$ 748
Short-term investments	2,588	3,041
Receivables	548	449
Merchandise inventories	4,028	3,338
Other current assets	712	409
Total current assets	9,081	7,985
Property and Equipment		
Land and buildings	705	580
Leasehold improvements	1,540	1,325
Fixtures and equipment	2,627	2,898
Property under capital lease	32	33
	4,904	4,836
Less accumulated depreciation	1,966	2,124
Net property and equipment	2,938	2,712
Goodwill	919	557
Tradenames	81	44
Long-Term Investments	318	218
Other Assets	233	348
Total Assets	$13,570	$11,864
Liabilities and Shareholders' Equity		
Current Liabilities		
Accounts payable	$ 3,934	$ 3,234
Unredeemed gift card liabilities	496	469
Accrued compensation and related expenses	332	354
Accrued liabilities	990	878
Accrued income taxes	489	703
Short-term debt	41	—
Current portion of long-term debt	19	418
Total current liabilities	6,301	6,056
Long-Term Liabilities	443	373
Long-Term Debt	590	178
Minority Interests	35	—
Shareholders' Equity		
Preferred stock, $1.00 par value: Authorized — 400,000 shares; Issued and outstanding — none	—	—
Common stock, $.10 par value: Authorized — 1 billion shares; Issued and outstanding — 480,655,000 and 485,098,000 shares, respectively	48	49
Additional paid-in capital	430	643
Retained earnings	5,507	4,304
Accumulated other comprehensive income	216	261
Total shareholders' equity	6,201	5,257
Total Liabilities and Shareholders' Equity	$13,570	$11,864

Consolidated Statements of Earnings
$ in millions, except per share amounts

Fiscal Years Ended	March 3, 2007	February 25, 2006	February 26, 2005
Revenue	$35,934	$30,848	$27,433
Cost of goods sold	27,165	23,122	20,938
Gross profit	8,769	7,726	6,495
Selling, general and administrative expenses	6,770	6,082	5,053
Operating income	1,999	1,644	1,442
Net interest income	111	77	1
Gain on investments	20	—	—
Earnings from continuing operations before income tax expense	2,130	1,721	1,443
Income tax expense	752	581	509
Minority interest in earnings	1	—	—
Earnings from continuing operations	1,377	1,140	934
Gain on disposal of discontinued operations, net of tax	—	—	50
Net earnings	$ 1,377	$ 1,140	$ 984
Basic earnings per share:			
Continuing operations	$ 2.86	$ 2.33	$ 1.91
Gain on disposal of discontinued operations	—	—	0.10
Basic earnings per share	$ 2.86	$ 2.33	$ 2.01
Diluted earnings per share:			
Continuing operations	$ 2.79	$ 2.27	$ 1.86
Gain on disposal of discontinued operations	—	—	0.10
Diluted earnings per share	$ 2.79	$ 2.27	$ 1.96
Basic weighted-average common shares outstanding (in millions)	482.1	490.3	488.9
Diluted weighted-average common shares outstanding (in millions)	496.2	504.8	505.0

Consolidated Statements of Changes in Shareholders' Equity
$ and shares in millions

	Common Shares	Common Stock	Additional Paid-In Capital	Retained Earnings	Accumulated Other Comprehensive Income	Total
Balances at February 28, 2004	**487**	**$49**	**$819**	**$ 2,468**	**$ 86**	**$3,422**
Net earnings	—	—	—	984	—	984
Other comprehensive income, net of tax:						
Foreign currency translation adjustments	—	—	—	—	59	59
Other	—	—	—	—	4	4
Total comprehensive income						1,047
Stock options exercised	10	1	219	—	—	220
Tax benefit from stock options exercised and employee stock purchase plan	—	—	60	—	—	60
Issuance of common stock under employee stock purchase plan	2	—	36	—	—	36
Vesting of restricted stock awards	—	—	1	—	—	1
Common stock dividends, $0.28 per share	—	—	—	(137)	—	(137)
Repurchase of common stock	(6)	(1)	(199)	—	—	(200)
Balances at February 26, 2005	**493**	**49**	**936**	**3,315**	**149**	**4,449**
Net earnings	—	—	—	1,140	—	1,140
Other comprehensive income, net of tax:						
Foreign currency translation adjustments	—	—	—	—	101	101
Other	—	—	—	—	11	11
Total comprehensive income						1,252
Stock options exercised	9	1	256	—	—	257
Tax benefit from stock options exercised and employee stock purchase plan	—	—	55	—	—	55
Issuance of common stock under employee stock purchase plan	1	—	35	—	—	35
Stock-based compensation	—	—	132	—	—	132
Common stock dividends, $0.31 per share	—	—	—	(151)	—	(151)
Repurchase of common stock	(18)	(1)	(771)	—	—	(772)
Balances at February 25, 2006	**485**	**49**	**643**	**4,304**	**261**	**5,257**
Net earnings	—	—	—	1,377	—	1,377
Other comprehensive loss, net of tax:						
Foreign currency translation adjustments	—	—	—	—	(33)	(33)
Other	—	—	—	—	(12)	(12)
Total comprehensive income						1,332
Stock options exercised	7	1	167	—	—	168
Tax benefit from stock options exercised and employee stock purchase plan	—	—	47	—	—	47
Issuance of common stock under employee stock purchase plan	1	—	49	—	—	49
Stock-based compensation	—	—	121	—	—	121
Common stock dividends, $0.36 per share	—	—	—	(174)	—	(174)
Repurchase of common stock	(12)	(2)	(597)	—	—	(599)
Balances at March 3, 2007	**481**	**$48**	**$430**	**$5,507**	**$216**	**$6,201**

Consolidated Statements of Cash Flows
$ in millions

Fiscal Years Ended	March 3, 2007	February 25, 2006	February 26, 2005
Operating Activities			
Net earnings	$1,377	$1,140	$ 984
Gain from disposal of discontinued operations, net of tax	—	—	(50)
Earnings from continuing operations	1,377	1,140	934
Adjustments to reconcile earnings from continuing operations to total cash provided by operating activities from continuing operations:			
Depreciation	509	456	459
Asset impairment charges	32	4	22
Stock-based compensation	121	132	(1)
Deferred income taxes	82	(151)	(28)
Excess tax benefits from stock-based compensation	(50)	(55)	—
Other, net	(11)	(3)	24
Changes in operating assets and liabilities, net of acquired assets and liabilities:			
Receivables	(70)	(43)	(30)
Merchandise inventories	(550)	(457)	(240)
Other assets	(47)	(11)	(50)
Accounts payable	320	385	347
Other liabilities	185	165	243
Accrued income taxes	(136)	178	301
Total cash provided by operating activities from continuing operations	1,762	1,740	1,981
Investing Activities			
Additions to property and equipment, net of $75 and $117 noncash capital expenditures in fiscal 2006 and 2005, respectively	(733)	(648)	(502)
Purchases of available-for-sale securities	(4,541)	(4,319)	(8,517)
Sales of available-for-sale securities	4,886	4,187	7,730
Acquisitions of businesses, net of cash acquired	(421)	—	—
Proceeds from disposition of investments	24	—	—
Change in restricted assets	—	(20)	(140)
Other, net	5	46	7
Total cash used in investing activities from continuing operations	(780)	(754)	(1,422)
Financing Activities			
Repurchase of common stock	(599)	(772)	(200)
Issuance of common stock under employee stock purchase plan and for the exercise of stock options	217	292	256
Dividends paid	(174)	(151)	(137)
Repayments of debt	(84)	(69)	(371)
Proceeds from issuance of debt	96	36	—
Excess tax benefits from stock-based compensation	50	55	—
Other, net	(19)	(10)	(7)
Total cash used in financing activities from continuing operations	(513)	(619)	(459)
Effect of Exchange Rate Changes on Cash	(12)	27	9
Increase in Cash and Cash Equivalents	457	394	109
Cash and Cash Equivalents at Beginning of Year	748	354	245
Cash and Cash Equivalents at End of Year	$1,205	$ 748	$ 354
Supplemental Disclosure of Cash Flow Information			
Income taxes paid	$ 804	$ 547	$ 241
Interest paid	14	16	35

SELECTED Notes to Consolidated Financial Statements

$ in millions, except per share amounts

1. Summary of Significant Accounting Policies

Description of Business

Best Buy Co., Inc. is a specialty retailer of consumer electronics, home-office products, entertainment software, appliances and related services, with fiscal 2007 revenue from continuing operations of $35.9 billion.

We operate two reportable segments: Domestic and International. The Domestic segment is comprised of all U.S. store and online operations of Best Buy, Geek Squad, Magnolia Audio Video and Pacific Sales Kitchen and Bath Centers, Inc. ("Pacific Sales"). We acquired Pacific Sales on March 7, 2006. U.S. Best Buy stores offer a wide variety of consumer electronics, home-office products, entertainment software, appliances and related services through 822 stores at the end of fiscal 2007. Geek Squad provides residential and commercial computer repair, support and installation services in all U.S. Best Buy stores and at 12 stand-alone stores at the end of fiscal 2007. Magnolia Audio Video stores offer high-end audio and video products and related services through 20 stores at the end of fiscal 2007. Pacific Sales stores offer high-end home-improvement products, appliances and related services through 14 stores at the end of fiscal 2007.

Fiscal Year

Our fiscal year ends on the Saturday nearest the end of February. Fiscal 2007 included 53 weeks and fiscal 2006 and 2005 each included 52 weeks.

Cash and Cash Equivalents

Cash primarily consists of cash on hand and bank deposits. Cash equivalents primarily consist of money market accounts and other highly liquid investments with an original maturity of three months or less when purchased. We carry these investments at cost, which approximates market value. The amounts of cash equivalents at March 3, 2007, and February 25, 2006, were $695 and $350, respectively, and the weighted-average interest rates were 4.8% and 3.3%, respectively.

Outstanding checks in excess of funds on deposit ("book overdrafts") totaled $183 and $230 at March 3, 2007, and February 25, 2006, respectively, and are reflected as current liabilities in our consolidated balance sheets.

Merchandise Inventories

Merchandise inventories are recorded at the lower of average cost or market. In-bound freight-related costs from our vendors are included as part of the net cost of merchandise inventories. Also included in the cost of inventory are certain vendor allowances that are not a reimbursement of specific, incremental and identifiable costs to promote a vendor's products. Other costs associated with acquiring, storing and transporting merchandise inventories to our retail stores are expensed as incurred and included in cost of goods sold.

Our inventory loss reserve represents anticipated physical inventory losses (e.g., theft) that have occurred since the last physical inventory date. Independent physical inventory counts are taken on a regular basis to ensure that the inventory reported in our consolidated financial statements is properly stated. During the interim period between physical inventory counts, we reserve for anticipated physical inventory losses on a location-by-location basis.

Property and Equipment

Property and equipment are recorded at cost. We compute depreciation using the straight-line method over the estimated useful lives of the assets. Leasehold improvements are depreciated over the shorter of their estimated useful lives or the period from the date the assets are placed in service to the end of the initial lease term. Leasehold improvements made significantly after the initial lease term are depreciated over the shorter of their estimated useful lives or the remaining lease term, including renewal periods, if reasonably assured.

$ in millions, except per share amounts

Accelerated depreciation methods are generally used for income tax purposes.

When property is fully depreciated, retired or otherwise disposed of, the cost and accumulated depreciation are removed from the accounts and any resulting gain or loss is reflected in the consolidated statement of earnings.

Repairs and maintenance costs are charged directly to expense as incurred. Major renewals or replacements that substantially extend the useful life of an asset are capitalized and depreciated.

Estimated useful lives by major asset category are as follows:

Asset	Life (in years)
Buildings	30–40
Leasehold improvements	3–25
Fixtures and equipment	3–20
Property under capital lease	3–20

Impairment of Long-Lived Assets

We account for the impairment or disposal of long-lived assets in accordance with SFAS No. 144, *Accounting for the Impairment* or *Disposal of Long-Lived Assets,* which requires long-lived assets, such as property and equipment, to be evaluated for impairment whenever events or changes in circumstances indicate the carrying value of an asset may not be recoverable. Factors considered important that could result in an impairment review include, but are not limited to, significant underperformance relative to historical or planned operating results, significant changes in the manner of use of the assets or significant changes in our business strategies. An impairment loss is recognized when the estimated undiscounted cash flows expected to result from the use of the asset plus net proceeds expected from disposition of the asset (if any) are less than the carrying value of the asset. When an impairment loss is recognized, the carrying amount of the asset is reduced to its estimated fair value based on quoted market prices or other valuation techniques.

Leases

We conduct the majority of our retail and distribution operations from leased locations. The leases require payment of real estate taxes, insurance and common area maintenance, in addition to rent. The terms of our lease agreements generally range from 10 to 20 years. Most of the leases contain renewal options and escalation clauses, and certain store leases require contingent rents based on factors such as specified percentages of revenue or the consumer price index. Other leases contain covenants related to the maintenance of financial ratios.

Goodwill and Intangible Assets

Goodwill

Goodwill is the excess of the purchase price over the fair value of identifiable net assets acquired in business combinations accounted for under the purchase method. We do not amortize goodwill but test it for impairment annually, or when indications of potential impairment exist, utilizing a fair value approach at the reporting unit level. A reporting unit is the operating segment, or a business unit one level below that operating segment, for which discrete financial information is prepared and regularly reviewed by segment management.

Tradenames

We have an indefinite-lived intangible asset related to our Pacific Sales tradename which is included in the Domestic segment. We also have indefinite-lived intangible assets related to our Future Shop and Five Star tradenames which are included in the International segment.

We determine fair values utilizing widely accepted valuation techniques, including discounted cash flows and market multiple analyses. During the fourth quarter of fiscal 2007, we completed our annual impairment testing of our goodwill and tradenames, using the valuation techniques as described above, and determined there was no impairment.

Lease Rights

Lease rights represent costs incurred to acquire the lease of a specific commercial property. Lease rights are recorded at cost and are amortized to rent expense over the remaining lease term, including renewal periods, if reasonably assured. Amortization periods range up to 16 years, beginning with the date we take possession of the property.

$ in millions, except per share amounts

Investments

Short-term and long-term investments are comprised of municipal and United States government debt securities as well as auction-rate securities and variable-rate demand notes. In accordance with SFAS No. 115, *Accounting for Certain Investments in Debt and Equity Securities,* and based on our ability to market and sell these instruments, we classify auction-rate securities, variable-rate demand notes and other investments in debt securities as available-for-sale and carry them at amortized cost, which approximates fair value. Auction-rate securities and variable-rate demand notes are similar to short-term debt instruments because their interest rates are reset periodically. Investments in these securities can be sold for cash on the auction date. We classify auction-rate securities and variable-rate demand notes as short-term or long-term investments based on the reset dates.

We also hold investments in marketable equity securities and classify them as available-for-sale. Investments in marketable equity securities are included in other assets in our consolidated balance sheets. Investments in marketable equity securities are reported at fair value, based on quoted market prices when available. All unrealized holding gains or losses are reflected net of tax in accumulated other comprehensive income in shareholders' equity.

We review the key characteristics of our debt and marketable equity securities portfolio and their classification in accordance with GAAP on an annual basis, or when indications of potential impairment exist. If a decline in the fair value of a security is deemed by management to be other than temporary, the cost basis of the investment is written down to fair value, and the amount of the write-down is included in the determination of net earnings.

Income Taxes

We account for income taxes under the liability method. Under this method, deferred tax assets and liabilities are recognized for the estimated future tax consequences attributable to differences between the financial statement carrying amounts of existing assets and liabilities and their respective tax bases, and operating loss and tax credit carryforwards. Deferred tax assets and liabilities are measured using enacted income tax rates in effect for the year in which those temporary differences are expected to be recovered or settled. The effect on deferred tax assets and liabilities of a change in income tax rates is recognized in our consolidated statement of earnings in the period that includes the enactment date. A valuation allowance is recorded to reduce the carrying amounts of deferred tax assets if it is more likely than not that such assets will not be realized.

Long-Term Liabilities

The major components of long-term liabilities at March 3, 2007, and February 25, 2006, included long-term rent-related liabilities, deferred compensation plan liabilities, self-insurance reserves and advances received under vendor alliance programs.

Foreign Currency

Foreign currency denominated assets and liabilities are translated into U.S. dollars using the exchange rates in effect at our consolidated balance sheet date. Results of operations and cash flows are translated using the average exchange rates throughout the period. The effect of exchange rate fluctuations on translation of assets and liabilities is included as a component of shareholders' equity in accumulated other comprehensive income. Gains and losses from foreign currency transactions, which are included in SG&A, have not been significant.

Revenue Recognition

We recognize revenue when the sales price is fixed or determinable, collectibility is reasonably assured and the customer takes possession of the merchandise, or in the case of services, at the time the service is provided.

$ in millions, except per share amounts

Amounts billed to customers for shipping and handling are included in revenue. Revenue is reported net of estimated sales returns and excludes sales taxes.

We estimate our sales returns reserve based on historical return rates. We initially established our sales returns reserve in the fourth quarter of fiscal 2005. Our sales returns reserve was $104 and $78, at March 3, 2007, and February 25, 2006, respectively.

We sell extended service contracts on behalf of an unrelated third party. In jurisdictions where we are not deemed to be the obligor on the contract, commissions are recognized in revenue at the time of sale. In jurisdictions where we are deemed to be the obligor on the contract, commissions are recognized in revenue ratably over the term of the service contract. Commissions represented 2.2%, 2.5% and 2.6% of revenues in fiscal 2007, 2006 and 2005, respectively.

For revenue transactions that involve multiple deliverables, we defer the revenue associated with any undelivered elements. The amount of revenue deferred in connection with the undelivered elements is determined using the relative fair value of each element, which is generally based on each element's relative retail price. See additional information regarding our customer loyalty program in *Sales Incentives* below.

Gift Cards

We sell gift cards to our customers in our retail stores, through our Web sites, and through selected third parties. We do not charge administrative fees on unused gift cards and our gift cards do not have an expiration date. We recognize income from gift cards when: (i) the gift card is redeemed by the customer; or (ii) the likelihood of the gift card being redeemed by the customer is remote ("gift card breakage") and we determine that we do not have a legal obligation to remit the value of unredeemed gift cards to the relevant jurisdictions. We determine our gift card breakage rate based upon historical redemption patterns. Based on our historical information, the likelihood of a gift card remaining unredeemed can be determined 24 months after the gift card is issued. At that time, we recognize breakage income for those cards for which the likelihood of redemption is deemed remote and we do not have a legal obligation to remit the value of such unredeemed gift cards to the relevant jurisdictions. Gift card breakage income is included in revenue in our consolidated statements of earnings.

We began recognizing gift card breakage income during the third quarter of fiscal 2006. Gift card breakage income was as follows in fiscal 2007, 2006 and 2005:

	2007[1]	2006[1]	2005
Gift card breakage income	$46	$43	$ —

(1) Due to the resolution of certain legal matters associated with gift card liabilities, we recognized $19 and $27 of gift card breakage income in fiscal 2007 and 2006, respectively, that related to prior fiscal years.

Sales Incentives

We frequently offer sales incentives that entitle our customers to receive a reduction in the price of a product or service. Sales incentives include discounts, coupons and other offers that entitle a customer to receive a reduction in the price of a product or service by submitting a claim for a refund or rebate. For sales incentives issued to a customer in conjunction with a sale of merchandise or services, for which we are the obligor, the reduction in revenue is recognized at the time of sale, based on the retail value of the incentive expected to be redeemed.

Customer Loyalty Program

We have a customer loyalty program which allows members to earn points for each qualifying purchase. Points earned enable members to receive a certificate that may be redeemed on future purchases at U.S. Best Buy stores.

$ in millions, except per share amounts

Cost of Goods Sold and Selling, General and Administrative Expenses

The following table illustrates the primary costs classified in each major expense category:

Cost of Goods Sold

- Total cost of products sold including:
  - —Freight expenses associated with moving merchandise inventories from our vendors to our distribution centers;
  - —Vendor allowances that are not a reimbursement of specific, incremental and identifiable costs to promote a vendor's products; and
  - —Cash discounts on payments to vendors;
- Cost of services provided including:
  - —Payroll and benefits costs for services employees; and
  - —Cost of replacement parts and related freight expenses;
- Physical inventory losses;
- Markdowns;
- Customer shipping and handling expenses;
- Costs associated with operating our distribution network, including payroll and benefit costs, occupancy costs, and depreciation;
- Freight expenses associated with moving merchandise inventories from our distribution centers to our retail stores; and
- Promotional financing costs.

SG&A

- Payroll and benefit costs for retail and corporate employees;
- Occupancy costs of retail, services and corporate facilities;
- Depreciation related to retail, services and corporate assets;
- Advertising;
- Vendor allowances that are a reimbursement of specific, incremental and identifiable costs to promote a vendor's products;
- Charitable contributions;
- Outside service fees;
- Long-lived asset impairment charges; and
- Other administrative costs, such as credit card service fees, supplies, and travel and lodging.

Advertising Costs

Advertising costs, which are included in SG&A, are expensed the first time the advertisement runs. Advertising costs consist primarily of print and television advertisements as well as promotional events. Net advertising expenses were $692, $644 and $597 in fiscal 2007, 2006 and 2005, respectively. Allowances received from vendors for advertising of $140, $123 and $115, in fiscal 2007, 2006 and 2005, respectively, were classified as reductions of advertising expenses.

$ in millions, except per share amounts

4. Investments

Short-Term and Long-Term Investments

The following table presents the amortized principal amounts, related weighted-average interest rates, maturities and major security types for our investments:

	March 3, 2007		Feb. 25, 2006	
	Amortized Principal Amount	**Weighted-Average Interest Rate**	**Amortized Principal Amount**	**Weighted-Average Interest Rate**
Short-term investments (less than one year)	$2,588	5.68%	$3,041	4.76%
Long-term investments (one to three years)	318	5.68%	218	4.95%
Total	$2,906		$3,259	
Municipal debt securities	$2,840		$3,155	
Auction-rate and asset-backed securities	66		97	
Debt securities issued by U.S. Treasury and other U.S. government entities	—		7	
Total	$2,906		$3,259	

The carrying value of our investments approximated fair value at March 3, 2007, and February 25, 2006, due to the rapid turnover of our portfolio and the highly liquid nature of these investments. Therefore, there were no significant realized or unrealized gains or losses.

Marketable Equity Securities

The carrying values of our investments in marketable equity securities at March 3, 2007, and February 25, 2006, were $4 and $28, respectively. Net unrealized (loss)/gain, net of tax, included in accumulated other comprehensive income was ($1) and $12 at March 3, 2007, and February 25, 2006, respectively.

$ in millions, except per share amounts

5. Debt

Short-term debt consisted of the following:

	March 3, 2007	Feb. 25, 2006
Notes payable to banks, secured, interest rates ranging from 3.5% to 6.7%	$ 21	$ —
Revolving credit facility, secured, variable interest rate of 5.6% at March 3, 2007	20	—
Total short-term debt	$ 41	$ —
Weighted-average interest rate	5.3%	—

Long-term debt consisted of the following:

	March 3, 2007	Feb. 25, 2006
Convertible subordinated debentures, unsecured, due 2022, interest rate 2.25%	$402	$ 402
Financing lease obligations, due 2009 to 2023, interest rates ranging from 3.0% to 6.5%	171	157
Capital lease obligations, due 2008 to 2026, interest rates ranging from 1.8% to 8.0%	24	27
Other debt, due 2010, interest rate 8.8%	12	10
Total debt	609	596
Less: current portion[1]	(19)	(418)
Total long-term debt	$590	$ 178

[1] Since holders of our debentures due in 2022 could have required us to purchase all or a portion of their debentures on January 15, 2007, we classified our debentures in the current portion of long-term debt at February 25, 2006. However, no holders of our debentures exercised this put option on January 15, 2007. The next time the holders of our debentures could require us to purchase all or a portion of their debentures is January 15, 2012. Therefore, we classified our debentures as long-term debt at March 3, 2007.

Certain debt is secured by property and equipment with a net book value of $80 and $41 at March 3, 2007, and February 25, 2006, respectively.

Convertible Debentures

In January 2002, we sold convertible subordinated debentures having an aggregate principal amount of $402. The proceeds from the offering, net of $6 in offering expenses, were $396. The debentures mature in 2022 and are callable at par, at our option, for cash on or after January 15, 2007.

Holders may require us to purchase all or a portion of their debentures on January 15, 2012, and January 15, 2017, at a purchase price equal to 100% of the principal amount of the debentures plus accrued and unpaid interest up to but not including the date of purchase. We have the option to settle the purchase price in cash, stock, or a combination of cash and stock.

$ in millions, except per share amounts

Other

The fair value of debt approximated $683 and $693 at March 3, 2007, and February 25, 2006, respectively, based on the ask prices quoted from external sources, compared with carrying values of $650 and $596, respectively.

At March 3, 2007, the future maturities of long-term debt, including capitalized leases, consisted of the following:

Fiscal Year	
2008	$ 19
2009	18
2010	27
2011	18
2012	420
Thereafter	107
	$609

Earnings per Share

Our basic earnings per share calculation is computed based on the weighted-average number of common shares outstanding. Our diluted earnings per share calculation is computed based on the weighted-average number of common shares outstanding adjusted by the number of additional shares that would have been outstanding had the potentially dilutive common shares been issued. Potentially dilutive shares of common stock include stock options, nonvested share awards and shares issuable under our ESPP, as well as common shares that would have resulted from the assumed conversion of our convertible debentures. Since the potentially dilutive shares related to the convertible debentures are included in the calculation, the related interest expense, net of tax, is added back to earnings from continuing operations, as the interest would not have been paid if the convertible debentures had been converted to common stock. Nonvested market-based share awards and nonvested performance-based share awards are included in the average diluted shares outstanding each period if established market or performance criteria have been met at the end of the respective periods.

The following table presents a reconciliation of the numerators and denominators of basic and diluted earnings per share from continuing operations in fiscal 2007, 2006 and 2005:

	2007	2006	2005
Numerator:			
Earnings from continuing operations, basic	$1,377	$1,140	$ 934
Adjustment for assumed dilution:			
Interest on convertible debentures due in 2022, net of tax	7	7	7
Earnings from continuing operations, diluted	$1,384	$1,147	$ 941
Denominator (in millions):			
Weighted-average common shares outstanding	482.1	490.3	488.9
Effect of potentially dilutive securities:			
Shares from assumed conversion of convertible debentures	8.8	8.8	8.8
Stock options and other	5.3	5.7	7.3
Weighted-average common shares outstanding, assuming dilution	496.2	504.8	505.0
Basic earnings per share — continuing operations	$ 2.86	$ 2.33	$ 1.91
Diluted earnings per share — continuing operations	$ 2.79	$ 2.27	$ 1.86

$ in millions, except per share amounts

Comprehensive Income

Comprehensive income is computed as net earnings plus certain other items that are recorded directly to shareholders' equity. In addition to net earnings, the significant components of comprehensive income include foreign currency translation adjustments and unrealized gains and losses, net of tax, on available-for-sale marketable equity securities. Foreign currency translation adjustments do not include a provision for income tax expense when earnings from foreign operations are considered to be indefinitely reinvested outside the United States. Comprehensive income was $1,332, $1,252 and $1,047 in fiscal 2007, 2006 and 2005, respectively.

7. Net Interest Income

Net interest income was comprised of the following in fiscal 2007, 2006 and 2005:

	2007	2006	2005
Interest income	$142	$103	$ 45
Interest expense	(31)	(30)	(44)
Dividend income	—	4	—
Net interest income	$111	$ 77	$ 1

8. Leases

The composition of net rent expense for all operating leases, including leases of property and equipment, was as follows in fiscal 2007, 2006 and 2005:

	2007	2006	2005
Minimum rentals	$679	$569	$516
Contingent rentals	1	1	1
Total rent expense	680	570	517
Less: sublease income	(20)	(18)	(16)
Net rent expense	$660	$552	$501

$ in millions, except per share amounts

The future minimum lease payments under our capital, financing and operating leases by fiscal year (not including contingent rentals) at March 3, 2007, were as follows:

Fiscal Year	Capital Leases	Financing Leases	Operating Leases
2008	$ 6	$ 23	$ 741
2009	4	23	715
2010	4	23	672
2011	3	23	632
2012	1	23	592
Thereafter	17	112	3,316
Subtotal	35	227	$6,668
Less: imputed interest	(11)	(56)	
Present value of lease obligations	$24	$171	

Total minimum lease payments have not been reduced by minimum sublease rent income of approximately $119 due under future noncancelable subleases.

10. Income Taxes

The following is a reconciliation of the federal statutory income tax rate to income tax expense from continuing operations in fiscal 2007, 2006 and 2005:

	2007	2006	2005
Federal income tax at the statutory rate	$ 747	$ 603	$ 505
State income taxes, net of federal benefit	38	34	29
Benefit from foreign operations	(36)	(37)	(7)
Non-taxable interest income	(34)	(28)	(22)
Other	37	9	4
Income tax expense	$ 752	$ 581	$ 509
Effective income tax rate	35.3%	33.7%	35.3%

Income tax expense was comprised of the following in fiscal 2007, 2006 and 2005:

	2007	2006	2005
Current:			
Federal	$609	$640	$502
State	45	78	36
Foreign	16	14	(1)
	670	732	537
Deferred:			
Federal	51	(131)	(4)
State	19	(14)	(20)
Foreign	12	(6)	(4)
	82	(151)	(28)
Income tax expense	$752	$581	$509

Deferred taxes are the result of differences between the bases of assets and liabilities for financial reporting and income tax purposes.

$ in millions, except per share amounts

11. Segment and Geographic Information

Segment Information

We operate two reportable segments: Domestic and International. The Domestic segment is comprised of U.S. store and online operations, including Best Buy, Geek Squad, Magnolia Audio Video and Pacific Sales. The International segment is comprised of all Canada store and online operations, including Best Buy, Future Shop and Geek Squad, as well as our Five Star and Best Buy retail and online operations in China.

The following tables present our business segment information for continuing operations in fiscal 2007, 2006 and 2005:

	2007	2006	2005
Revenue			
Domestic	$31,031	$27,380	$24,616
International	4,903	3,468	2,817
Total revenue	$35,934	$30,848	$27,433
Operating Income			
Domestic	$ 1,889	$ 1,588	$ 1,393
International	110	56	49
Total operating income	1,999	1,644	1,442
Net interest income	111	77	1
Gain on investments	20	—	—
Earnings from continuing operations before income tax expense	$ 2,130	$ 1,721	$ 1,443
Assets			
Domestic	$10,614	$ 9,722	$ 8,372
International	2,956	2,142	1,922
Total assets	$13,570	$11,864	$10,294

12. Contingencies and Commitments

Contingencies

On December 8, 2005, a purported class action lawsuit captioned, *Jasmen Holloway, et cl. v. Best Buy Co., Inc.*, was filed in the U.S. District Court for the Northern District of California alleging we discriminate against women and minority individuals on the basis of gender, race, color and/or national origin with respect to our employment policies and practices. The action seeks an end to discriminatory policies and practices, an award of back and front pay, punitive damages and injunctive relief, including rightful place relief for all class members. As of March 3, 2007, no accrual had been established as it was not possible to estimate the possible loss or range of loss because this matter had not advanced to a stage where we could make any such estimate. We believe the allegations are without merit and intend to defend this action vigorously.

We are involved in various other legal proceedings arising in the normal course of conducting business. We believe the amounts provided in our consolidated financial statements, as prescribed by GAAP, are adequate in light of the probable and estimable liabilities. The resolution of those other proceedings is not expected to have a material impact on our results of operations or financial condition.

Commitments

We engage Accenture LLP ("Accenture") to assist us with improving our operational capabilities and reducing our costs in the information systems, procurement and human resources areas. Our future contractual obligations to Accenture are expected to range from $76 to $334 per year through 2012, the end of the contract period. Prior to our engagement of Accenture, a significant portion of these costs were incurred as part of normal operations.

We had outstanding letters of credit for purchase obligations with a fair value of $85 at March 3, 2007.

At March 3, 2007, we had commitments for the purchase and construction of facilities valued at approximately $69. Also, at March 3, 2007, we had entered into lease commitments for land and buildings for 115 future locations. These lease commitments with real estate developers provide for minimum rentals ranging from seven to 20 years, which if consummated based on current cost estimates, will approximate $84 annually over the initial lease terms. These minimum rentals have been included in the future minimum lease payments included in Note 8, Leases.

Financial Reports

CIRCUIT CITY STORES, INC.

Circuit City Stores, Inc.
CONSOLIDATED BALANCE SHEETS

(Amounts in thousands except share data)	At February 28 2007	2006
ASSETS		
CURRENT ASSETS:		
Cash and cash equivalents	**$ 141,141**	$ 315,970
Short-term investments	**598,341**	521,992
Accounts receivable, net of allowance for doubtful accounts	**382,555**	220,869
Merchandise inventory	**1,636,507**	1,698,026
Deferred income taxes	**34,868**	29,598
Income tax receivable	**42,722**	5,571
Prepaid expenses and other current assets	**47,378**	41,315
TOTAL CURRENT ASSETS	**2,883,512**	2,833,341
Property and equipment, net of accumulated depreciation	**921,027**	839,356
Deferred income taxes	**31,910**	97,889
Goodwill	**121,774**	223,999
Other intangible assets, net of accumulated amortization	**19,285**	30,372
Other assets	**29,775**	44,087
TOTAL ASSETS	**$4,007,283**	$4,069,044
LIABILITIES AND STOCKHOLDERS' EQUITY		
CURRENT LIABILITIES:		
Merchandise payable	**$ 922,205**	$ 850,359
Expenses payable	**281,709**	202,300
Accrued expenses and other current liabilities	**404,444**	379,768
Accrued compensation	**98,509**	84,743
Accrued income taxes	**–**	75,909
Short-term debt	**–**	22,003
Current installments of long-term debt	**7,162**	7,248
TOTAL CURRENT LIABILITIES	**1,714,029**	1,622,330
Long-term debt, excluding current installments	**50,487**	51,985
Accrued straight-line rent and deferred rent credits	**277,636**	256,120
Accrued lease termination costs	**76,326**	79,091
Other liabilities	**97,561**	104,885
TOTAL LIABILITIES	**2,216,039**	2,114,411
Commitments and contingent liabilities		
STOCKHOLDERS' EQUITY:		
Common stock, $0.50 par value; 525,000,000 shares authorized; 170,689,406 shares issued and outstanding (174,789,390 in 2006)	**85,345**	87,395
Additional paid-in capital	**344,144**	458,211
Retained earnings	**1,336,317**	1,364,740
Accumulated other comprehensive income	**25,438**	44,287
TOTAL STOCKHOLDERS' EQUITY	**1,791,244**	1,954,633
TOTAL LIABILITIES AND STOCKHOLDERS' EQUITY	**$4,007,283**	$4,069,044

Circuit City Stores, Inc.
CONSOLIDATED STATEMENTS OF OPERATIONS

	Years Ended February 28					
(Amounts in thousands except per share data)	**2007**	%	2006	%	2005	%
NET SALES	**$12,429,754**	100.0	$11,514,151	100.0	$10,413,524	100.0
Cost of sales, buying and warehousing	**9,501,438**	76.4	8,703,683	75.6	7,861,364	75.5
GROSS PROFIT	**2,928,316**	23.6	2,810,468	24.4	2,552,160	24.5
Selling, general and administrative expenses	**2,841,619**	22.9	2,595,706	22.5	2,470,712	23.7
Impairment of goodwill	**92,000**	0.7	–	–	–	–
Finance income	**–**	**–**	–	–	5,564	0.1
OPERATING (LOSS) INCOME	**(5,303)**	–	214,762	1.9	87,012	0.8
Interest income	**27,150**	0.2	21,826	0.2	14,404	0.1
Interest expense	**1,519**	–	3,143	–	4,451	–
Earnings from continuing operations before income taxes	**20,328**	0.2	233,445	2.0	96,965	0.9
Income tax expense	**30,510**	0.2	85,996	0.7	36,396	0.3
NET (LOSS) EARNINGS FROM CONTINUING OPERATIONS	**(10,182)**	(0.1)	147,449	1.3	60,569	0.6
EARNINGS (LOSS) FROM DISCONTINUED OPERATIONS, NET OF TAX	**128**	**–**	(5,350)	–	1,089	–
CUMULATIVE EFFECT OF CHANGE IN ACCOUNTING PRINCIPLES, NET OF TAX	**1,773**	–	(2,353)	–	–	–
NET (LOSS) EARNINGS	**$ (8,281)**	(0.1)	$ 139,746	1.2	$ 61,658	0.6

	2007	2006	2005
Weighted average common shares:			
Basic	**170,448**	177,456	193,466
Diluted	**170,448**	180,653	196,227
(LOSS) EARNINGS PER SHARE:			
Basic:			
Continuing operations	**$ (0.06)**	$ 0.83	$ 0.31
Discontinued operations	**$ –**	$ (0.03)	$ 0.01
Cumulative effect of change in accounting principles	**$ 0.01**	$ (0.01)	$ –
Basic (loss) earnings per share	**$ (0.05)**	$ 0.79	$ 0.32
Diluted:			
Continuing operations	**$ (0.06)**	$ 0.82	$ 0.31
Discontinued operations	**$ –**	$ (0.03)	$ 0.01
Cumulative effect of change in accounting principles	**$ 0.01**	$ (0.01)	$ –
Diluted (loss) earnings per share	**$ (0.05)**	$ 0.77	$ 0.31

CIRCUIT CITY

Circuit City Stores, Inc.

CONSOLIDATED STATEMENTS OF STOCKHOLDERS' EQUITY AND COMPREHENSIVE INCOME

(Amounts in thousands except per share data)	Common Stock Shares	Common Stock Amount	Additional Paid-in Capital	Retained Earnings	Accumulated Other Comprehensive Income	Total
BALANCE AT FEBRUARY 29, 2004	203,899	$101,950	$922,600	$1,191,904	$ –	$2,216,454
Comprehensive income:						
Net earnings	–	–	–	61,658	–	61,658
Other comprehensive income, net of taxes:						
Foreign currency translation adjustment (net of deferred taxes of $13,707)	–	–	–	–	25,100	25,100
Comprehensive income						86,758
Repurchases of common stock	(19,163)	(9,582)	(250,250)	–	–	(259,832)
Compensation for stock awards	–	–	18,305	–	–	18,305
Exercise of common stock options	3,489	1,745	26,761	–	–	28,506
Shares issued under stock-based incentive plans, net of cancellations, and other	(75)	(38)	(1,312)	–	–	(1,350)
Tax effect from stock issued	–	–	(1,564)	–	–	(1,564)
Shares issued in acquisition of InterTAN, Inc.	–	–	6,498	–	–	6,498
Dividends – common stock ($0.07 per share)	–	–	–	(13,848)	–	(13,848)
BALANCE AT FEBRUARY 28, 2005	188,150	94,075	721,038	1,239,714	25,100	2,079,927
Comprehensive income:						
Net earnings	–	–	–	139,746	–	139,746
Other comprehensive income (loss), net of taxes:						
Foreign currency translation adjustment (net of deferred taxes of $11,316)	–	–	–	–	19,500	19,500
Minimum pension liability adjustment (net of deferred taxes of $182)	–	–	–	–	(313)	(313)
Comprehensive income						158,933
Repurchases of common stock	(19,396)	(9,698)	(328,778)	–	–	(338,476)
Compensation for stock awards	–	–	24,386	–	–	24,386
Exercise of common stock options	3,830	1,915	36,752	–	–	38,667
Shares issued under stock-based incentive plans, net of cancellations, and other	2,205	1,103	(2,160)	–	–	(1,057)
Tax effect from stock issued	–	–	6,973	–	–	6,973
Redemption of preferred share purchase rights	–	–	–	(1,876)	–	(1,876)
Dividends – common stock ($0.07 per share)	–	–	–	(12,844)	–	(12,844)
BALANCE AT FEBRUARY 28, 2006	174,789	87,395	458,211	1,364,740	44,287	1,954,633
Comprehensive loss:						
Net loss	–	–	–	(8,281)	–	(8,281)
Other comprehensive (loss) income, net of taxes:						
Foreign currency translation adjustment (net of deferred taxes of $3,630)	–	–	–	–	(7,793)	(7,793)
Unrealized gain on available-for-sale securities (net of deferred taxes of $219)	–	–	–	–	377	377
Minimum pension liability adjustment (net of deferred taxes of $136)	–	–	–	–	(229)	(229)
Comprehensive loss						(15,926)
Adjustment to initially apply SFAS No. 158 (net of deferred taxes of $6,628)	–	–	–	–	(11,204)	(11,204)
Repurchases of common stock	(10,032)	(5,016)	(232,187)	–	–	(237,203)
Compensation for stock awards	–	–	26,727	–	–	26,727
Adjustment to initially apply SFAS No. 123(R)	–	–	(2,370)	–	–	(2,370)
Exercise of common stock options, net	5,767	2,883	86,228	–	–	89,111
Shares issued under stock-based incentive plans, net of cancellations, and other	165	83	(1,027)	–	–	(944)
Tax effect from stock issued	–	–	8,562	–	–	8,562
Dividends – common stock ($0.115 per share)	–	–	–	(20,142)	–	(20,142)
BALANCE AT FEBRUARY 28, 2007	170,689	$85,345	$344,144	$1,336,317	$25,438	$1,791,244

Circuit City Stores, Inc.

CONSOLIDATED STATEMENTS OF CASH FLOWS

(Amounts in thousands)	Years Ended February 28 2007	2006	2005[a]
OPERATING ACTIVITIES:			
Net (loss) earnings	**$ (8,281)**	$ 139,746	$ 61,658
Adjustments to reconcile net (loss) earnings to net cash provided by operating activities of continuing operations:			
Net (earnings) loss from discontinued operations	**(128)**	5,350	(1,089)
Depreciation expense	**177,828**	160,608	151,597
Amortization expense	**3,645**	2,618	1,851
Impairment of goodwill	**92,000**	–	–
Stock-based compensation expense	**26,727**	24,386	18,305
(Gain) loss on dispositions of property and equipment	**(1,439)**	2,370	(206)
Provision for deferred income taxes	**72,717**	(14,252)	(116,300)
Cumulative effect of change in accounting principles	**(1,773)**	2,353	–
Other	**1,689**	(1,726)	–
Changes in operating assets and liabilities:			
Accounts receivable, net	**(133,152)**	16,552	(58,738)
Retained interests in securitized receivables	**–**	–	32,867
Merchandise inventory	**49,352**	(231,114)	160,037
Prepaid expenses and other current assets	**(9,580)**	(17,341)	7,207
Other assets	**535**	(3,061)	3,816
Merchandise payable	**73,317**	211,362	28,199
Expenses payable	**55,722**	40,921	(17,372)
Accrued expenses and other current liabilities, and accrued income taxes	**(81,364)**	43,202	54,021
Other long-term liabilities	**(1,474)**	(17,032)	63,494
NET CASH PROVIDED BY OPERATING ACTIVITIES OF CONTINUING OPERATIONS	**316,341**	364,942	389,347
INVESTING ACTIVITIES:			
Purchases of property and equipment	**(285,725)**	(254,451)	(261,461)
Proceeds from sales of property and equipment	**38,620**	55,421	106,369
Purchases of investment securities	**(2,002,123)**	(1,409,760)	(125,325)
Sales and maturities of investment securities	**1,926,086**	1,014,910	–
Other investing activities	**(11,567)**	–	–
Proceeds from the sale of the private-label finance operation	**–**	–	475,857
Acquisitions, net of cash acquired of $30,615	**–**	–	(262,320)
NET CASH USED IN INVESTING ACTIVITIES OF CONTINUING OPERATIONS	**(334,709)**	(593,880)	(66,880)
FINANCING ACTIVITIES:			
Proceeds from short-term borrowings	**35,657**	73,954	12,329
Principal payments on short-term borrowings	**(56,912)**	(53,893)	(13,458)
Proceeds from long-term debt	**1,216**	1,032	–
Principal payments on long-term debt	**(6,724)**	(1,829)	(28,008)
Changes in overdraft balances	**19,347**	(22,540)	36,329
Repurchases of common stock	**(237,203)**	(338,476)	(259,832)
Issuances of common stock	**89,662**	38,038	27,156
Dividends paid	**(20,126)**	(12,844)	(13,848)
Excess tax benefit from stock-based payments	**15,729**	–	–
Redemption of preferred share purchase rights	**–**	(1,876)	–
Other financing activities	**(1,424)**	–	–
NET CASH USED IN FINANCING ACTIVITIES OF CONTINUING OPERATIONS	**(160,778)**	(318,434)	(239,332)
DISCONTINUED OPERATIONS:			
Operating cash flows	**3,310**	(9,884)	(7,193)
Investing cash flows	**2,958**	(8,089)	(6,615)
Financing cash flows	**(592)**	–	(724)
NET CASH PROVIDED BY (USED IN) DISCONTINUED OPERATIONS	**5,676**	(17,973)	(14,532)
EFFECT OF EXCHANGE RATE CHANGES ON CASH	**(1,359)**	1,655	2,016
(Decrease) increase in cash and cash equivalents	**(174,829)**	(563,690)	70,619
Cash and cash equivalents at beginning of year	**315,970**	879,660	809,041
CASH AND CASH EQUIVALENTS AT END OF YEAR	**$ 141,141**	$ 315,970	$879,660

CIRCUIT CITY

RADIOSHACK

Alabama Albertville, Alexander City, Andalusia, Arab, Ardmore, Athens, Atmore, Attalla, Bay Minette, Bayou La Batre, Bessemer, Birmingham, Butler, Calera, Camden, Center Point, Centre, Childersburg, Clanton, C
Daphne, Decatur, Demopolis, Dothan, Enterprise, Fairfield, Fairhope, Florence, Foley, Fort Payne, Gadsden, Gardendale, Gulf Shores, Guntersville, Haleyville, Hamilton, Hartselle, Hoover, Huntsville, Jackson, Jasper,
Linden, Luverne, Madison, Marion, Mobile, Montgomery, Moulton, Northport, Opelika, Opp, Oxford, Pelham, Pell City, Phenix City, Piedmont, Prattville, Robertsdale, Rogersville, Russellville, Saraland, Scottsboro,
Sumiton, Sylacauga, Tallassee, Thomasville, Troy, Tuscaloosa *Alaska* Anchorage, Bethel, Cordova, Craig, Eagle River, Fairbanks, Glennallen, Haines, Homer, Juneau, Kenai, Ketchikan, Kodiak, Petersburg, Seward
Skagway, Soldotna, Valdez, Wasilla *Arizona* Ajo, Apache Junction, Avondale, Benson, Bullhead City, Camp Verde, Casa Grande, Chandler, Chino Valley, Colorado City, Coolidge, Cottonwood, Douglas, Flagstaff, Fl
Fort Mohave, Fountain Hills, Gilbert, Glendale, Greenvalley, Heber, Holbrook, Kayenta, Kingman, Lake Havasu, Lakeside, Maricopa, Mesa, Miami, Morenci, New River, Nogales, Oro Valley, Parker, Payson, Peoria, P
Prescott, Prescott Valley, Quartzsite, Safford, San Manuel, Scottsdale, Sedona, Show Low, Sierra Vista, Springerville, St. Johns, Sun City, Surprise, Taylor, Tempe, Thatcher, Tuba City, Tucson, Wickenburg, Willcox
Arkansas Arkadelphia, Ash Flat, Batesville, Beebe, Benton, Bentonville, Berryville, Brinkley, Bryant, Cabot, Camden, Cave City, Clarksville, Clinton, Conway, Danville, De Queen, De Witt, Dumas, El Dorado, Fayetteville,
Forrest City, Fort Smith, Glenwood, Harrison, Heber Springs, Hope, Hot Springs, Jacksonville, Jasper, Jonesboro, Little Rock, Magnolia, Malvern, Mammoth Springs, Marshall, Melbourne, Mena, Mountain Home, M
View, North Little Rock, Nashville, Newport, Paragould, Paris, Pine Bluff, Prescott, Rogers, Russellville, Salem, Searcy, Sheridan, Siloam Springs, Springdale, Star City, Stuttgart, Van Buren, West Helena, West Me
Wynne *California* Agoura, Alameda, Albany, Alhambra, Alta Loma, Alturas, American Canyon, Anaheim, Anaheim Hills, Anderson, Angels Camp, Antioch, Apple Valley, Arcadia, Arcata, Arnold, Arroyo Grande, Atas
Atwater, Auburn, Avalon, Azusa, Bakersfield, Baldwin Park, Barstow, Beaumont, Bell, Belmont, Benicia, Berkeley, Beverly Hills, Big Bear Lake, Bishop, Blue Jay, Blythe, Brawley, Brea, Buellton, Buena Park, B
Burlingame, Calexico, California City, Camarillo, Canoga Park, Canyon Country, Capitola, Carlsbad, Carmichael, Carpinteria, Carson, Castro Valley, Cathedral City, Cerritos, Chatsworth, Chico, Chino, Chino Hills, Chul
Citrus Heights, City of Industry, Clearlake, Cloverdale, Clovis, Coachella, Coalinga, Colton, Colusa, Compton, Concord, Corcoran, Corning, Corona, Corte Madera, Costa Mesa, Covina, Crescent City, Crestline, Culv
Cupertino, Cypress, Daly City, Dana Point, Danville, Davis, Del Mar, Delano, Desert Hot Springs, Diamond Bar, Dinuba, Downey, Duarte, Dublin, E Los Angeles, El Cajon, El Centro, El Cerrito, El Monte, Elk Grove, Eme
Encinitas, Encino, Escondido, Eureka, Fairfield, Fall River Mills, Fallbrook, Folsom, Fontana, Foothill Ranch, Fortuna, Foster City, Fountain Valley, Freedom, Fremont, Fresno, Fort Bragg, Fullerton, Garberville, Garden
Gardena, Gilroy, Glendale, Glendora, Goleta, Gonzales, Granada Hills, Grass Valley, Greenfield, Grover Beach, Hanford, Harbor City, Hawthorne, Hayward, Hemet, Hercules, Hesperia, Highland, Hollister, Holl
Huntington Beach, Huntington Park, Indio, Inglewood, Irvine, Jackson, King City, La Habra, La Jolla, La Mesa, La Mirada, La Puente, La Quinta, La Verne, Lafayette, Laguna Hills, Laguna Niguel, Lake Elsinore, Lake I
Lakeport, Lakewood, Lancaster, Lawndale, Lemoore, Lincoln Heights, Livermore, Lodi, Lompoc, Long Beach, Los Alamitos, Los Angeles, Los Banos, Los Gatos, Los Osos, Lynwood, Madera, Malibu, Mammoth
Manhattan Beach, Manteca, Marina Del Rey, Martinez, Marysville, Maywood, Merced, Milpitas, Mission Hills, Modesto, Mojave, Monrovia, Montclair, Montebello, Monterey, Monterey Park, Montrose, Moorpark, I
Valley, Morgan Hill, Morro Bay, Mountain View, Mount Shasta, Murrieta, Napa, National City, Newbury Park, Newhall, Newport Beach, North Highlands, North Hollywood, Northridge, Norwalk, Novato, Oakdale, Oa
Oakland, Oakley, Oceanside, Ojai, Ontario, Orange, Orangevale, Orland, Oroville, Oxnard, Pacifica, Palm Desert, Palm Springs, Palmdale, Palo Alto, Panorama City, Paradise, Paramount, Pasadena, Paso Robles, Pat
Perris, Petaluma, Phelan, Pico Rivera, Pinole, Pittsburg, Placentia, Placerville, Pleasant Hill, Pleasanton, Pollock Pines, Pomona, Porterville, Poway, Quincy, Ramona, Rancho Cordova, Rancho Cucamonga, Rancho
Margarita, Red Bluff, Redding, Redlands, Redondo Beach, Redwood City, Reedley, Rialto, Ridgecrest, Rio Vista, Riverbank, Riverside, Rocklin, Rohnert Park, Rolling Hills, Rosamond, Rosemead, Roseville, Rowland H
Sacramento, Salinas, San Bernardino, San Bruno, San Clemente, San Diego, San Dimas, San Francisco, San Jose, San Leandro, San Luis Obispo, San Marcos, San Mateo, San Pablo, San Pedro, San Rafael, San I
Sanger, Santa Ana, Santa Barbara, Santa Clara, Santa Cruz, Santa Maria, Santa Monica, Santa Paula, Santa Rosa, Santee, Saugus, Scotts Valley, Seal Beach, Seaside, Sebastopol, Selma, Sherman Oaks, Signal Hil
San Juan Capistrano, Soledad, Sonoma, Sonora, South Gate, South Lake Tahoe, South Pasadena, South San Francisco, Spring Valley, Stockton, Studio City, Sun Valley, Sunnyvale, Susanville, Sylmar, Taft, Teh
Temecula, Temple City, Thousand Oaks, Torrance, Tracy, Truckee, Tujunga, Tulare, Turlock, Tustin, Twentynine Palms, Ukiah, Union City, Upland, Vacaville, Valencia, Vallejo, Valley Springs, Van Nuys, Venice, V
Victorville, Visalia, Vista, Walnut Creek, Wasco, Watsonville, Weaverville, West Covina, West Hollywood, West Los Angeles, West Sacramento, Westchester, Westminster, Whittier, Willits, Willows, Wilmington, W
Woodland, Woodland Hills, Yorba Linda, Yreka, Yuba City, Yucaipa, Yucca Valley *Colorado* Alamosa, Arvada, Aspen, Aurora, Avon, Bayfield, Bennett, Boulder, Brighton, Broomfield, Buena Vista, Burlington, Canon City,
Rock, Castle Rock, Centennial, Center, Colorado Springs, Conifer, Cortez, Craig, Crested Butte, Denver, Durango, Elizabeth, Englewood, Estes Park, Evergreen, Flagler, Fort Collins, Fountain, Fraser, Frisco, Glenwood
Golden, Grand Junction, Greeley, Greenwood Village, Gunnison, Highlands Ranch, Holyoke, Idaho Springs, La Junta, Lafayette, Lakewood, Lamar, Limon, Littleton, Longmont, Loveland, Meeker, Monte Vista, Mo
Monument, Northglenn, Pagosa Springs, Paonia, Parachute, Parker, Pueblo, Rifle, Salida, Springfield, Steamboat Springs, Sterling, Thornton, Westminster, Woodland Park, Wray, Yuma *Connecticut* Avon, Barkha
Bloomfield, Branford, Bridgeport, Bristol, Canaan, Cheshire, Clinton, Cos Cob, Cromwell, Danbury, Derby, East Haven, Enfield, Fairfield, Farmington, Glastonbury, Groton, Guilford, Hamden, Hartford, Manchester, M
Middletown, Milford, Naugatuck, New Britain, New Canaan, New Haven, New London, New Milford, Newington, Newtown, North Haven, Norwalk, Norwich, Old Saybrook, Orange, Plainfield, Putnam, Ridgefield, Riv
Southbury, Southington, Stamford, Torrington, Trumbull, Vernon, Wallingford, Waterbury, Waterford, Watertown, West Hartford, Westport, Wethersfield, Willimantic, Wilton, Windsor, *D.C.* Washington *Delawar*
Claymont, Dover, Georgetown, Middletown, Milford, New Castle, Newark, Rehoboth Beach, Seaford, Smyrna, Wilmington *Florida* Alachua, Altamonte, Altamonte Springs, Apopka, Arcadia, Atlantic Beach, Auburndale
Park, Bartow, Bayonet Point, Belle Glade, Belleview, Big Pine Key, Boca Raton, Bonita Springs, Boynton Beach, Bradenton, Brandon, Branford, Brooksville, Callaway, Cape Coral, Casselberry, Century, Chiefland, C
Clearwater, Clermont, Clewiston, Cocoa, Cocoa Beach, Cooper City, Coral Gables, Coral Springs, Crawfordville, Crestview, Crystal River, Davie, Daytona Beach, Deerfield, Deerfield Beach, Defuniak Springs, Deland,
Beach, Deltona, Destin, Dunnellon, Englewood, Fernandina Beach, Florida City, Fort Lauderdale, Fort Myers, Fort Pierce, Fort Walton Beach, Gainesville, Greenacres, Gulf Breeze, Haines City, Hialeah, Hilliard, H
Hollywood, Homestead, Homosassa, Immokalee, Indiantown, Inverness, Jacksonville, Jensen Beach, Jupiter, Key Largo, Key West, Keystone Heights, Kissimmee, Lady Lake, Lake City, Lake Mary, Lake Placid, Lake
Lake Worth, Lakeland, Lantana, Largo, Lauderdale Lakes, Lauderhill, Leesburg, Lehigh Acres, Live Oak, Longwood, Lutz, Macclenny, Madison, Marathon, Marco Island, Margate, Marianna, Mary Esther, Melbourne,
Island, Miami, Miami Beach, Milton, Miramar, Monticello, Mount Dora, North Fort Myers, North Miami Beach, Naples, Navarre, New Port Richey, New Smyrna Beach, Niceville, Oakland Park, Ocala, Ocoee, Okeec
Orange City, Orange Park, Orlando, Ormond Beach, Oviedo, Palatka, Palm Bay, Palm Beach Garden, Palm Coast, Palm Harbor, Panama City, Pembroke Pines, Pensacola, Perry, Plant City, Plantation, Pompano Beac
Charlotte, Port Orange, Port Richey, Port St. Joe, Port St. Lucie, Punta Gorda, Riverview, Royal Palm Beach, Ruskin, Sanford, Santa Rosa Beach, Sarasota, Satellite Beach, Sebastian, Sebring, Seffner, Seminole,
Daytona, Spring Hill, St. Augustine, St. Cloud, St. Petersburg, Starke, Stuart, Sunrise, Tallahassee, Tampa, Tarpon Springs, Temple Terrace, Tequesta, Titusville, Venice, Vero Beach, Wauchula, Wellington, West Palm
Weston, Wildwood, Wilton Manors, Winter Haven, Winter Park, Winter Springs, Zephyrhills *Georgia* Adel, Albany, Alpharetta, Americus, Athens, Atlanta, Augusta, Austell, Bainbridge, Barnesville, Baxley, Blairsville, E
Blue Ridge, Brunswick, Buford, Cairo, Calhoun, Canton, Carrollton, Cartersville, Cedartown, Centerville, Chamblee, Chatsworth, Clayton, Cleveland, Columbus, Commerce, Conyers, Cordele, Cornelia, Covington, Cur
Cuthbert, Dahlonega, Dalton, Dawson, Dawsonville, Decatur, Donalsonville, Douglas, Douglasville, Dublin, Duluth, East Ellijay, Elberton, Fayetteville, Fitzgerald, Folkston, Forest Park, Forsyth, Fort Gaines, Fort Ogle
Fort Valley, Gainesville, Griffin, Hampton, Hartwell, Hazlehurst, Hiawassee, Hinesville, Hiram, Homerville, Jackson, Jasper, Jesup, Kennesaw, Lafayette, Lagrange, Lawrenceville, Lilburn, Lincolnton, Lithonia, I
Madison, Marietta, Martinez, Mc Rae, McDonough, Metter, Milledgeville, Monroe, Monticello, Morrow, Moultrie, Nashville, Newnan, Norcross, Oakwood, Peachtree City, Perry, Quitman, Richmond Hill, Riverdale, Roc
Rome, Roswell, Royston, Savannah, Smyrna, Snellville, St. Marys, St. Simons Island, Statesboro, Stockbridge, Stone Mountain, Summerville, Suwanee, Sylvania, Sylvester, Thomaston, Thomasville, Thomson, Tifton, T
Trenton, Union City, Valdosta, Vidalia, Villa Rica, Warner Robins, Washington, Waycross, Winder, Woodbury, Woodstock *Hawaii* Aiea, Ewa Beach, Haleiwa, Hilo, Honolulu, Kahului, Kailua, Kailua-Kona, Kamuela, Ka
Kapolei, Kihei, Lahaina, Lihue, Mililani, Wahiawa, Waianae, Waipahu *Idaho* American Falls, Blackfoot, Boise, Bonners Ferry, Buhl, Burley, Caldwell, Chubbuck, Coeur d'Alene, Cottonwood, Driggs, Emmett, Grangeville,
Idaho Falls, Lewiston, McCall, Meridian, Montpelier, Moscow, Mountain Home, Nampa, Orofino, Pocatello, Post Falls, Rexburg, Rigby, Salmon, Sandpoint, Twin Falls, Wendell *Illinois* Aledo, Alton, Anna, Antioch,
Arlington Heights, Arthur, Aurora, Bartlett, Batavia, Belleville, Belvidere, Bensenville, Benton, Berwyn, Bloomingdale, Bloomington, Blue Island, Bolingbrook, Bourbonnais, Burbank, Calumet City, Canton, Carbo
Carlinville, Carmi, Centralia, Champaign, Channahon, Chester, Chicago, Chicago Heights, Cicero, Collinsville, Crystal Lake, Danville, Decatur, Des Plaines, Dixon, Dolton, Downers Grove, Du Quoin, Dwight, East Peori
St Louis, Effingham, El Paso, Elgin, Elk Grove Village, Eureka, Evanston, Fairbury, Fairfield, Fairview Heights, Flora, Fox Lake, Frankfort, Freeport, Galesburg, Geneseo, Gibson City, Glen Carbon, Glen Ellyn, Glencoe, Gle
Granite City, Greenville, Gurnee, Harrisburg, Havana, Highland, Highland Park, Hoffman Estates, Homer Glen, Homewood, Hoopeston, Jacksonville, Jerseyville, Joliet, Kankakee, Kewanee, La Grange, Lake Zurich, La
Lemont, Lincoln, Litchfield, Lake in the Hills, Lombard, Machesney Park, Macomb, Marengo, Marion, Markham, Mascoutah, Matteson, Mattoon, McHenry, Melrose Park, Mendota, Midlothian, Moline, Montgomery, I
Mount Vernon, Mundelein, Naperville, Nashville, Niles, Norridge, North Riverside, Oak Lawn, Oak Park, Olney, Ottawa, Palatine, Palos Heights, Paris, Pekin, Peoria, Peru, Petersburg, Pontiac, Princeton, Quincy, Rob
Rochelle, Rockford, Round Lake Beach, Salem, Sandwich, Savanna, Savoy, Schaumburg, Seneca, Shiloh, Skokie, South Elgin, South Holland, Sparta, Springfield, St. Charles, Staunton, Sterling, Streator, Sullivan, Syca
Tinley Park, Tuscola, Urbana, Vernon Hills, Villa Park, Virden, Waterloo, Watseka, Waukegan, West Dundee, Wheaton, Wheeling, Willowbrook, Wilmington, Wood River, Yorkville, Zion *Indiana* Anderson, Angola,
Auburn, Aurora, Avon, Batesville, Bedford, Berne, Bicknell, Bloomington, Bluffton, Brazil, Bremen, Brook, Brookville, Brownsburg, Brownstown, Cannelton, Carmel, Clarksville, Columbia City, Columbus, Corydon, Covi
Crawfordsville, Crown Point, Decatur, Demotte, Elkhart, Elwood, Evansville, Fishers, Fort Wayne, Fowler, Frankfort, Franklin, Gary, Goshen, Greencastle, Greenfield, Greensburg, Greenwood, Griffith, Hammond, H
Huntington, Indianapolis, Jasper, Kendallville, Knox, Kokomo, La Porte, Lafayette, Lagrange, Lebanon, Ligonier, Linton, Madison, Marion, Martinsville, Merrillville, Michigan City, Mishawaka, Monticello, Mooresville, M
Munster, Nappanee, New Albany, New Carlisle, New Castle, New Haven, Noblesville, North Manchester, North Vernon, Paoli, Peru, Petersburg, Plainfield, Plymouth, Portage, Portland, Princeton, Rensselaer, Rich
Rising Sun, Rochester, Rockport, Rockville, Rushville, Schererville, Seymour, Shelbyville, South Bend, Syracuse, Terre Haute, Tipton, Valparaiso, Vincennes, W Lafayette, Wabash, Warsaw, Washington, Winamac, Winc
Iowa Adel, Altoona, Ames, Ankeny, Atlantic, Belle Plaine, Boone, Carroll, Cedar Falls, Cedar Rapids, Chariton, Charles City, Cherokee, Clarinda, Clinton, Coralville, Council Bluffs, Cresco, Creston, Davenport, De
Denison, Des Moines, Dubuque, Dyersville, Eagle Grove, Estherville, Fairfield, Fort Dodge, Fort Madison, Garner, Glenwood, Greenfield, Grinnell, Hampton, Harlan, Humboldt, Independence, Iowa City, Iowa Falls, Jef
Keokuk, Knoxville, Le Mars, Logan, Manchester, Maquoketa, Marengo, Marshalltown, Mason City, Mount Pleasant, Muscatine, New Hampton, Newton, Orange City, Osage, Osceola, Ottumwa, Pella, Perry, Pocahonta
Oak, Rock Valley, Sac City, Sheldon, Sioux Center, Sioux City, Spencer, Spirit Lake, Stuart, Vinton, Washington, Waterloo, Webster City, West Burlington, West Des Moines, West Union, Winterset *Kansas* Abilene, Ar
Arkansas City, Atchison, Atwood, Bonner Springs, Burlington, Chanute, Clay Center, Colby, Columbus, Concordia, Derby, Dodge City, El Dorado, Ellsworth, Emporia, Fort Scott, Garden City, Garnett, Girard, Goodland
Bend, Hays, Hillsboro, Horton, Hutchinson, Independence, Iola, Junction City, Kansas City, Lawrence ... Lenexa, Liberal, Manhattan, McPherson, Mission, Newton, Oakley, Olathe, Osage City, Osawa
Ottawa, Overland Park, Parsons, Pittsburg, Pratt, Salina, Scott City, Seneca, Shawnee Mission ... Wellington, Wichita, Winfield *Kentucky* Alexandria, Ashland, Barbourville, Bardstown, Ba
Beaver Dam, Berea, Bowling Green, Brandenburg, Cadiz, Campbellsville, Campton, Carrollt ... al City, Columbia, ... a, Danville, Dry Ridge, Elizabethtown, Erlanger, Falmouth, Flemingsburg, Florence, Fra
Franklin, Georgetown, Glasgow, Grayson, Hazard, Henderson, Hopkinsville, Jackson, ... er, La Grange, Latonia, Leba ... exington, London, Louisville, Madisonville, Mayfield, Maysville, Middlesboro, Mon
Morehead, Morgantown, Mount Sterling, Mount Vernon, Murray, Newport, Nichol ... e, Owensboro, Paducah, Paris, Pike ... Pineville, Prestonsburg, Princeton, Radcliff, Richmond, Russell Springs, Russe
Salyersville, Scottsville, Shelbyville, Somerset, South Williamson, Stanton, Taylorsvil ... arsaw, West Liberty, Whitley City, Williams ... Winchester *Louisiana* Abbeville, Alexandria, Bastrop, Baton Rouge, Bogalusa, E
City, Boutte, Crowley, Cut Off, Denham Springs, Deridder, Eunice, Franklinton, G ... ies, Hammond, Harahan, Harvey, Houma, Jea ... te, Jena, Jennings, Kenner, Kentwood, La Place, Lafayette, Lake Charles, Le
Mandeville, Mansfield, Many, Metairie, Minden, Monroe, Morgan City, Natchit ... s, New Iberia, New Orleans, New Roads, Oakdale ... elousas, Pineville, Plaquemine, Rayne, Ruston, Shreveport, Slidell, Springh
Francisville, Sulphur, Thibodaux, Ville Platte, West Monroe, Westwego, Winfiel ... innsboro, Zachary *Maine* Auburn, Augusta, Bangor, B ... arbor, Belfast, Biddeford, Boothbay Harbor, Brunswick, Bucksport, Damaris
Dover-Foxcroft, Ellsworth, Falmouth, Farmington, Fort Kent, Lewiston, Mac ... s, Madawaska, Mexico, Millinocket, Oxford, Portland, P ... ue Isle, Rockland, Sanford, Skowhegan, South Portland, Standish, Top
Waterville, Wells, Windham *Maryland* Aberdeen, Annapolis, Baltimore, Bel ... Beltsville. Bethesda. Bowie. Burtonsville, Cambridge, Cator ... le, Charlotte Hall, Chestertown, Clinton, Cockeysville, College Park, Col
Denton, Derwood, Dunkirk, Easton, Edgewood, Eldersburg, Elkton, Ellicott ... y, F ... rederick, Gaithersburg, Gan ... ls, Germantown, Glen Burnie, Greenbelt, Hagerstown, Hampstead, Ha
Hyattsville, Kensington, La Plata, La Vale, Largo, Laurel, Leonardtown, Lexi ... on Park, M ... ights, M ... Mount Airy, Oakland, Ocea ... ity, Odenton, Olney, Owings Mills, Oxon Hill, Oxon Hill, Pasadena, Poc
City, Potomac, Prince Fredrick, Randallstown, Reisterstown, Rockville, S ... bury, Seat ... Severna ... er Spring, Stevensville, ... oma Park, Towson, Waldorf, Westminster, Wheaton *Massachusetts*
Andover, Ashland, Athol, Auburn, Bedford, Beverly, Billerica, Boston, Bra ... ee, Bridgev ... ckton, Br ... rlington, Cambridge, Che ... sford, Chicopee, Danvers, Dedham, Dorchester, East Boston, East W
East Wareham, Fairhaven, Fall River, Falmouth, Fitchburg, Foxboro, Frar ... , Gardner, ... r, Greenfi ... Barrington, Hadley, Hano ... Haverhill, Holyoke, Hyannis, Kingston, Lanesborough, Lenox, Leom
Lowell, Lynn, Malden, Marlborough, Marshfield, Medford, Milford, Nan ... ket, Natick, ... ford, New ... Newton, North Adams, N ... Attleboro, North Dartmouth, Northampton, Orleans, Peabody, Pit
Plymouth, Quincy, Raynham, Revere, Roslindale, Saugus, South Attle ... o, South Bo ... uth Denni ... Easton, South Lawrence, ... th Yarmouth, Southbridge, Springfield, Stoneham, Stoughton, Su
Swampscott, Swansea, Taunton, Vineyard Haven, Waltham, Watertow ... Webster, We ... field, We ... n, Westfield, Westford, Whi ... sville, Wilmington, Woburn, Worcester *Michigan* Adrian, Albion, Al
Alpena, Ann Arbor, Auburn Hills, Bad Axe, Battle Creek, Bay City, Bellaire ... lleville, Ben ... ig Rapids, Birmingham, Bloomfie ... Boyne City, Brighton, Brooklyn, Brown City, Burton, Byron Center, Ca
Canton, Caro, Carson City, Cass City, Center Line, Charlevoix, Cheboygan ... esaning, Cl ... n Township, Clio, Coldwater, Co ... erce, Davison, Dearborn, Dearborn Heights, Detroit, Dowagiac, Eastp
Eaton Rapids, Escanaba, Evart, Farmington Hills, Farmington, Fenton, Fe ... dale, Flint, F ... t, Fran ... Fremont, Gaylord, Grand Bla ... Grand Haven, Grand Rapids, Grayling, Greenville, Grosse Pointe, Ha
Hemlock, Highland Park, Holland, Houghton Lake, Howell, Imlay City, Ion ... ron Mount ... ood, Isl ... ackson, Jenison, Jonesville ... lamazoo, Kalkaska, Kentwood, Lake Orion, L'Anse, Lansing, Lapeer, L
Park, Livonia, Ludington, Madison Heights, Manistee, Manistique, Marin ... ty, Marqu ... hall, Mi ... ord, Monroe, Mount Pleas ... , Munising, Muskegon, Newberry, Niles, Novi, Oak Park, Okemos, O
Owosso, Petoskey, Pinconning, Plainwell, Pontiac, Port Huron, Portage, Re ... d, Reed C ... ond, Roc ... s, Rogers City, Roseville, ... al Oak, Saginaw, Sandusky, Sault St. Marie, Shelby, South Haven, Sout
Southgate, St. Ignace, St. Johns, Standish, Stanton, Sturgis, Suttons Bay, ... as City, ... ee Oaks, ... rs, Traverse City, Troy, ... a, Vassar, Washington Township, Waterford, Wayne, Westland, White
White Pigeon, Whitehall, Woodhaven, Wyoming, Ypsilanti *Minnesota* Ada, ... n, Alba ... Lea, Alex ... ole Valley, Austin, Bau ... e, Baxter, Bemidji, Benson, Blaine, Bloomington, Brooklyn Center, Burr
Cambridge, Chanhassen, Coon Rapids, Cottage Grove, Crystal, Detroit Lakes, ... uth, Du ... gan, Eden ... na, Elk River, Erskine ... irmont, Faribault, Fergus Falls, Forest Lake, Golden Valley, Grand Marais,
Rapids, Hibbing, Hilltop, Hutchinson, International Falls, Jackson, Lake City, ... ueur ... Little Fall ... airie, Mankato, M ... Grove, Marshall, Minneapolis, Minnetonka, Montevideo, Monticello, Moo
Moose Lake, Mora, Morris, New Ulm, North Branch, Ortonville, Owatonna, Pa ... pids, Pi ... eton, Red W ... Redwood Falls, Richfield, Rochester, Roseau, Roseville, Saint Cloud, Sain
Savage, Shakopee, Sleepy Eye, St. Cloud, St. James, St. Louis Park, St. Paul, S ... er, Stillwater, Thief River Falls, Vadnais Height ... rginia, Walker, Warroad, Waseca, Wayzata, Willmar, Windom, Winona, Woo
Worthington, Young America *Mississippi* Amory, Batesville, Biloxi, Booneville, Bro ... ven, Canton, Carthage, Clarksdale, Clevel ... Clinton, Columbia, Columbus, Corinth, Crystal Springs, D'Iberville, Flora, Gree
Greenwood, Grenada, Gulfport, Hattiesburg, Houston, Jackson, Laurel, Lucedale, Mag ... McComb, Mendenhall, Meridian, Mont ... , Morton, Natchez, New Albany, Ocean Springs, Olive Branch, Oxford, Pascagoula,
Philadelphia, Picayune, Pontotoc, Poplarville, Prentiss, Purvis, Quitman, Ridgeland, Se ... , Southaven, Starkville, Tupel ... rtown, Vicksburg, Waynesboro, West Point, Wiggins, Yazoo City *Missouri* Alton, A

RadioShack Corporation | Financial Reports

CONSOLIDATED BALANCE SHEETS
RadioShack Corporation and Subsidiaries

	December 31,	
(In millions, except for share amounts)	**2006**	**2005**
Assets		
Current assets:		
Cash and cash equivalents	$ 472.0	$ 224.0
Accounts and notes receivable, net	247.9	309.4
Inventories	752.1	964.9
Other current assets	127.6	129.0
Total current assets	1,599.6	1,627.3
Property, plant and equipment, net	386.3	476.2
Other assets, net	84.1	101.6
Total assets	$ 2,070.0	$ 2,205.1
Liabilities and Stockholders' Equity		
Current liabilities:		
Short-term debt, including current maturities of long-term debt	$ 194.9	$ 40.9
Accounts payable	254.5	490.9
Accrued expenses and other current liabilities	442.2	379.5
Income taxes payable	92.6	75.0
Total current liabilities	984.2	986.3
Long-term debt, excluding current maturities	345.8	494.9
Other non-current liabilities	86.2	135.1
Total liabilities	1,416.2	1,616.3
Commitments and contingent liabilities		
Stockholders' equity:		
Preferred stock, no par value, 1,000,000 shares authorized:		
Series A junior participating, 300,000 shares designated and none issued	—	—
Common stock, $1 par value, 650,000,000 shares authorized; 191,033,000 shares issued	191.0	191.0
Additional paid-in capital	92.6	87.7
Retained earnings	1,780.9	1,741.4
Treasury stock, at cost; 55,196,000 and 56,071,000 shares, respectively	(1,409.1)	(1,431.6)
Accumulated other comprehensive (loss) income	(1.6)	0.3
Total stockholders' equity	653.8	588.8
Total liabilities and stockholders' equity	$ 2,070.0	$ 2,205.1

RADIOSHACK

CONSOLIDATED STATEMENTS OF INCOME
RadioShack Corporation and Subsidiaries

	Year Ended December 31,					
	2006		2005		2004	
(In millions, except per share amounts)	Dollars	% of Revenues	Dollars	% of Revenues	Dollars	% of Revenues
Net sales and operating revenues	$ 4,777.5	100.0%	$ 5,081.7	100.0%	$ 4,841.2	100.0%
Cost of products sold	2,544.4	53.3	2,706.3	53.3	2,406.7	49.7
Gross profit	2,233.1	46.7	2,375.4	46.7	2,434.5	50.3
Operating expenses:						
Selling, general and administrative	1,903.7	39.8	1,901.7	37.4	1,774.8	36.7
Depreciation and amortization	128.2	2.7	123.8	2.4	101.4	2.1
Impairment of long-lived assets and other charges	44.3	0.9	—	—	—	—
Total operating expenses	2,076.2	43.4	2,025.5	39.8	1,876.2	38.8
Operating income	156.9	3.3	349.9	6.9	558.3	11.5
Interest income	7.4	0.1	5.9	0.1	11.4	0.2
Interest expense	(44.3)	(0.9)	(44.5)	(0.8)	(29.6)	(0.5)
Other (loss) income	(8.6)	(0.2)	10.2	0.2	2.0	—
Income before income taxes	111.4	2.3	321.5	6.4	542.1	11.2
Income tax provision	38.0	0.8	51.6	1.0	204.9	4.2
Income before cumulative effect of change in accounting principle	73.4	1.5	269.9	5.4	337.2	7.0
Cumulative effect of change in accounting principle, net of $1.8 million tax benefit in 2005	—	—	(2.9)	(0.1)	—	—
Net income	$ 73.4	1.5%	$ 267.0	5.3%	$ 337.2	7.0%
Net income per share						
Basic:						
Income before cumulative effect of change in accounting principle	$ 0.54		$ 1.82		$ 2.09	
Cumulative effect of change in accounting principle, net of taxes	—		(0.02)		—	
Basic income per share	$ 0.54		$ 1.80		$ 2.09	
Assuming dilution:						
Income before cumulative effect of change in accounting principle	$ 0.54		$ 1.81		$ 2.08	
Cumulative effect of change in accounting principle, net of taxes	—		(0.02)		—	
Diluted income per share	$ 0.54		$ 1.79		$ 2.08	
Shares used in computing income per share:						
Basic	136.2		148.1		161.0	
Diluted	136.2		148.8		162.5	

RADIOSHACK

CONSOLIDATED STATEMENTS OF STOCKHOLDERS' EQUITY AND COMPREHENSIVE INCOME
RadioShack Corporation and Subsidiaries

	Shares at December 31,			Dollars at December 31,		
(In millions)	2006	2005	2004	2006	2005	2004
Common stock						
Beginning and end of year	191.0	191.0	191.0	$ 191.0	$ 191.0	$ 191.0
Treasury stock						
Beginning of year	(56.0)	(32.8)	(28.5)	$ (1,431.6)	$ (859.4)	$ (707.2)
Purchase of treasury stock	—	(25.3)	(8.0)	—	(625.8)	(246.9)
Issuance of common stock	0.6	1.2	1.3	18.6	31.8	33.8
Exercise of stock options and grant of stock awards	0.2	0.9	2.4	3.9	21.8	60.9
End of year	(55.2)	(56.0)	(32.8)	$ (1,409.1)	$ (1,431.6)	$ (859.4)
Additional paid-in capital						
Beginning of year				$ 87.7	$ 82.7	$ 75.2
Issuance of common stock				(5.7)	3.5	5.7
Exercise of stock options and grant of stock awards				(1.7)	(5.0)	(9.5)
Stock option compensation				12.0	—	—
Stock option income tax benefits				0.3	6.5	11.3
End of year				$ 92.6	$ 87.7	$ 82.7
Retained earnings						
Beginning of year				$ 1,741.4	$ 1,508.1	$ 1,210.6
Net income				73.4	267.0	337.2
Common stock cash dividends declared				(33.9)	(33.7)	(39.7)
End of year				$ 1,780.9	$ 1,741.4	$ 1,508.1
Accumulated other comprehensive (loss) income						
Beginning of year				$ 0.3	$ (0.3)	$ (0.3)
Pension adjustments, net of tax				(1.0)	—	—
Other comprehensive (loss) income				(0.9)	0.6	—
End of year				$ (1.6)	$ 0.3	$ (0.3)
Total stockholders' equity				$ 653.8	$ 588.8	$ 922.1
Comprehensive income						
Net income				$ 73.4	$ 267.0	$ 337.2
Other comprehensive income, net of tax:						
Foreign currency translation adjustments				0.3	(0.4)	0.1
Amortization of gain on cash flow hedge				(0.1)	(0.1)	(0.1)
Unrealized (loss) gain on securities				(1.1)	1.1	—
Other comprehensive (loss) income				(0.9)	0.6	—
Comprehensive income				$ 72.5	$ 267.6	$ 337.2

CONSOLIDATED STATEMENTS OF CASH FLOWS
RadioShack Corporation and Subsidiaries

(In millions)	Year Ended December 31, 2006	2005	2004
Cash flows from operating activities:			
Net income	$ 73.4	$ 267.0	$ 337.2
Adjustments to reconcile net income to net cash provided by operating activities:			
Depreciation and amortization	128.2	123.8	101.4
Cumulative effect of change in accounting principle	—	4.7	—
Impairment of long-lived assets and other charges	44.3	—	—
Stock option compensation	12.0	—	—
Deferred income taxes and other items	(27.6)	(76.9)	50.2
Provision for credit losses and bad debts	0.4	0.1	(0.3)
Changes in operating assets and liabilities, excluding acquisitions:			
Accounts and notes receivable	61.8	(68.2)	(53.0)
Inventories	212.8	38.8	(234.2)
Other current assets	2.5	28.5	(7.5)
Accounts payable, accrued expenses, income taxes payable and other	(193.0)	45.1	158.7
Net cash provided by operating activities	314.8	362.9	352.5
Cash flows from investing activities:			
Additions to property, plant and equipment	(91.0)	(170.7)	(229.4)
Proceeds from sale of property, plant and equipment	11.1	226.0	2.5
Purchase of kiosk business	—	—	(59.1)
Other investing activities	0.6	(16.0)	(4.2)
Net cash (used in) provided by investing activities	(79.3)	39.3	(290.2)
Cash flows from financing activities:			
Purchases of treasury stock	—	(625.8)	(251.1)
Sale of treasury stock to employee benefit plans	10.5	30.1	35.4
Proceeds from exercise of stock options	1.7	17.4	50.4
Payments of dividends	(33.9)	(33.7)	(39.7)
Changes in short-term borrowings and outstanding checks in excess of cash balances, net	42.2	(4.0)	(14.0)
Reductions of long-term borrowings	(8.0)	(0.1)	(40.1)
Net cash provided by (used in) financing activities	12.5	(616.1)	(259.1)
Net increase (decrease) in cash and cash equivalents	248.0	(213.9)	(196.8)
Cash and cash equivalents, beginning of period	224.0	437.9	634.7
Cash and cash equivalents, end of period	$ 472.0	$ 224.0	$ 437.9
Supplemental cash flow information:			
Interest paid	$ 44.0	$ 43.4	$ 29.3
Income taxes paid	52.9	158.5	182.7

RADIOSHACK

Apple Financial Report

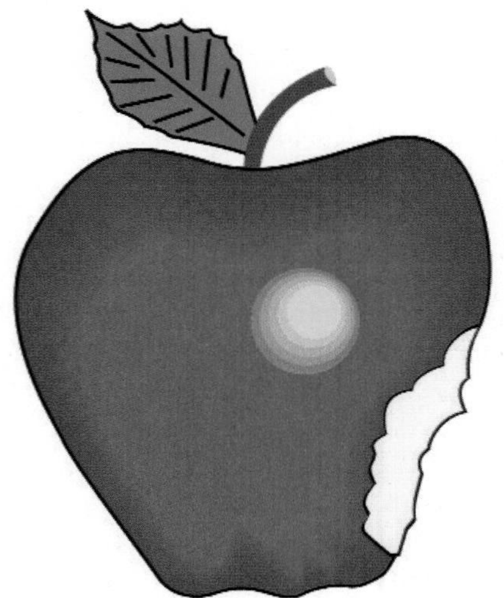

CONSOLIDATED BALANCE SHEETS
(In millions, except share amounts)

	September 30, 2006	September 24, 2005
ASSETS		
Current assets:		
Cash and cash equivalents	$ 6,392	$ 3,491
Short-term investments	3,718	4,770
Accounts receivable, less allowances of $52 and $46, respectively	1,252	895
Inventories	270	165
Deferred tax assets	607	331
Other current assets	2,270	648
Total current assets	14,509	10,300
Property, plant, and equipment, net	1,281	817
Goodwill	38	69
Acquired intangible assets, net	139	27
Other assets	1,238	303
Total assets	$17,205	$11,516
LIABILITIES AND SHAREHOLDERS' EQUITY		
Current liabilities:		
Accounts payable	$ 3,390	$ 1,779
Accrued expenses	3,081	1,708
Total current liabilities	6,471	3,487
Noncurrent liabilities	750	601
Total liabilities	7,221	4,088
Commitments and contingencies		
Shareholders' equity:		
Common stock, no par value; 1,800,000,000 shares authorized; 855,262,568 and 835,019,364 shares issued and outstanding, respectively	4,355	3,564
Deferred stock compensation	—	(61)
Retained earnings	5,607	3,925
Accumulated other comprehensive income	22	—
Total shareholders' equity	9,984	7,428
Total liabilities and shareholders' equity	$17,205	$11,516

APPLE

CONSOLIDATED STATEMENTS OF OPERATIONS
(In millions, except share and per share amounts)

Three fiscal years ended September 30, 2006	2006	2005	2004
Net sales	$ 19,315	$ 13,931	$ 8,279
Cost of sales	13,717	9,889	6,022
Gross margin	5,598	4,042	2,257
Operating expenses:			
Research and development	712	535	491
Selling, general, and administrative	2,433	1,864	1,430
Restructuring costs	—	—	23
Total operating expenses	3,145	2,399	1,944
Operating income	2,453	1,643	313
Other income and expense	365	165	57
Income before provision for income taxes	2,818	1,808	370
Provision for income taxes	829	480	104
Net income	$ 1,989	$ 1,328	$ 266
Earnings per common share:			
Basic	$ 2.36	$ 1.64	$ 0.36
Diluted	$ 2.27	$ 1.55	$ 0.34
Shares used in computing earnings per share (in thousands):			
Basic	844,058	808,439	743,180
Diluted	877,526	856,878	774,776

CONSOLIDATED STATEMENTS OF SHAREHOLDERS' EQUITY
(In millions, except share amounts which are in thousands)

	Common Stock Shares	Common Stock Amount	Deferred Stock Compensation	Retained Earnings	Accumulated Other Comprehensive Income (Loss)	Total Shareholders' Equity
Balances as of September 27, 2003 as previously reported	733,454	$1,926	$ (62)	$2,394	$(35)	$4,223
Adjustments to opening shareholders' equity	—	85	(22)	(63)	—	—
Balance as of September 27, 2003 as restated	733,454	$2,011	$ (84)	$2,331	$(35)	$4,223
Components of comprehensive income:						
Net income	—	—	—	266	—	266
Change in foreign currency translation	—	—	—	—	13	13
Change in unrealized gain on available-for-sale securities, net of tax	—	—	—	—	(5)	(5)
Change in unrealized loss on derivative investments, net of tax	—	—	—	—	12	12
Total comprehensive income						286
Issuance of stock-based compensation awards	—	63	(63)	—	—	—
Adjustment to common stock related to a prior year acquisition	(159)	(2)	—	—	—	(2)
Stock-based compensation	—	—	46	—	—	46
Common stock issued under stock plans	49,592	427	—	—	—	427
Tax benefit related to stock options	—	83	—	—	—	83
Balances as of September 25, 2004	782,887	$2,582	$(101)	$2,597	$(15)	$5,063
Components of comprehensive income:						
Net income	—	—	—	1,328	—	1,328
Change in foreign currency translation	—	—	—	—	7	7
Change in unrealized gain on derivative investments, net of tax	—	—	—	—	8	8
Total comprehensive income						1,343
Issuance of stock-based compensation awards	—	7	(7)	—	—	—
Stock-based compensation	—	—	47	—	—	47
Common stock issued under stock plans	52,132	547	—	—	—	547
Tax benefit related to stock options	—	428	—	—	—	428
Balances as of September 24, 2005	835,019	$3,564	$ (61)	$3,925	$ —	$7,428
Components of comprehensive income:						
Net income	—	—	—	1,989	—	1,989
Change in foreign currency translation	—	—	—	—	19	19
Change in unrealized gain on available-for-sale securities, net of tax	—	—	—	—	4	4
Change in unrealized loss on derivative investments, net of tax	—	—	—	—	(1)	(1)
Total comprehensive income						2,011
Common stock repurchased	(4,574)	(48)	—	(307)	—	(355)
Stock-based compensation	—	163	—	—	—	163
Deferred compensation	—	(61)	61	—	—	—
Common stock issued under stock plans	24,818	318	—	—	—	318
Tax benefit related to stock-based compensation	—	419	—	—	—	419
Balances as of September 30, 2006	855,263	$4,355	$ —	$5,607	$ 22	$9,984

CONSOLIDATED STATEMENTS OF CASH FLOWS
(In millions)

Three fiscal years ended September 30, 2006	2006	2005	2004
Cash and cash equivalents, beginning of the year	$ 3,491	$ 2,969	$ 3,396
Operating Activities:			
Net income	1,989	1,328	266
Adjustments to reconcile net income to cash generated by operating activities:			
Depreciation, amortization and accretion	225	179	150
Stock-based compensation expense	163	49	46
Provision for deferred income taxes	53	50	19
Excess tax benefits from stock options	—	428	83
Gain on sale of PowerSchool net assets	(4)	—	—
Loss on disposition of property, plant, and equipment	15	9	7
Gains on sales of investments, net	—	—	(5)
Changes in operating assets and liabilities:			
Accounts receivable	(357)	(121)	(8)
Inventories	(105)	(64)	(45)
Other current assets	(1,626)	(150)	(176)
Other assets	(1,040)	(35)	(25)
Accounts payable	1,611	328	297
Other liabilities	1,296	534	325
Cash generated by operating activities	2,220	2,535	934
Investing Activities:			
Purchases of short-term investments	(7,255)	(11,470)	(3,270)
Proceeds from maturities of short-term investments	7,226	8,609	1,141
Proceeds from sales of investments	1,086	586	806
Purchases of long-term investments	(25)	—	—
Proceeds from sale of PowerSchool net assets	40	—	—
Purchases of property, plant, and equipment	(657)	(260)	(176)
Other	(58)	(21)	11
Cash generated by (used for) investing activities	357	(2,556)	(1,488)
Financing Activities:			
Payment of long-term debt	—	—	(300)
Proceeds from issuance of common stock	318	543	427
Excess tax benefits from stock-based compensation	361	—	—
Repurchases of common stock	(355)	—	—
Cash generated by financing activities	324	543	127
Increase (decrease) in cash and cash equivalents	2,901	522	(427)
Cash and cash equivalents, end of the year	$ 6,392	$ 3,491	$ 2,969
Supplemental cash flow disclosures:			
Cash paid during the year for interest	$ —	$ —	$ 10
Cash paid (received) for income taxes, net	$ 194	$ 17	$ (7)

APPLE

B Time Value of Money

Learning Objectives

CAP

Conceptual

C1 Describe the earning of interest and the concepts of present and future values. *(p. B-1)*

Procedural

P1 Apply present value concepts to a single amount by using interest tables. *(p. B-3)*

P2 Apply future value concepts to a single amount by using interest tables. *(p. B-4)*

P3 Apply present value concepts to an annuity by using interest tables. *(p. B-5)*

P4 Apply future value concepts to an annuity by using interest tables. *(p. B-6)*

Appendix Preview

The concepts of present and future values are important to modern business, including the preparation and analysis of financial statements. The purpose of this appendix is to explain, illustrate, and compute present and future values. This appendix applies these concepts with reference to both business and everyday activities.

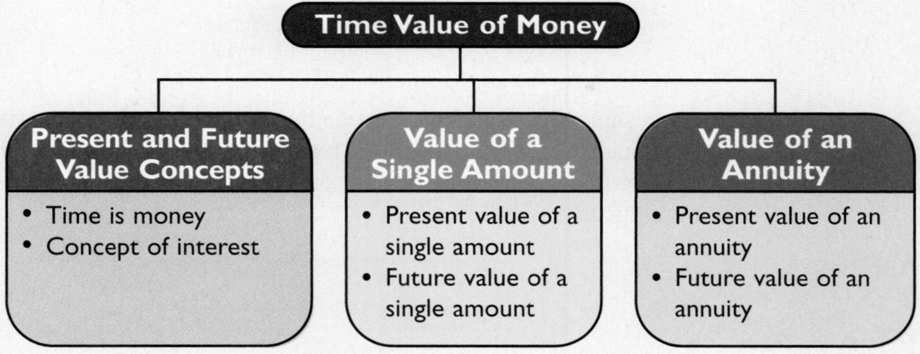

Present and Future Value Concepts

C1 Describe the earning of interest and the concepts of present and future values.

The old saying "Time is money" reflects the notion that as time passes, the values of our assets and liabilities change. This change is due to *interest,* which is a borrower's payment to the owner of an asset for its use. The most common example of interest is a savings account asset. As we keep a balance of cash in the account, it earns interest that the financial institution pays us. An example of a liability is a car loan. As we carry the balance of the loan, we accumulate interest costs on it. We must ultimately repay this loan with interest.

Present and future value computations enable us to measure or estimate the interest component of holding assets or liabilities over time. The present value computation is important when we want to know the value of future-day assets *today.* The future value computation is important when we want to know the value of present-day assets *at a future date.* The first section focuses on the present value of a single amount. The second section focuses on the future value of a single amount. Then both the present and future values of a series of amounts (called an *annuity*) are defined and explained.

Decision Insight

Keep That Job Lottery winners often never work again. Kenny Dukes, a recent Georgia lottery winner, doesn't have that option. He is serving parole for burglary charges, and Georgia requires its parolees to be employed (or in school). For his lottery winnings, Dukes had to choose between $31 million in 30 annual payments or $16 million in one lump sum ($10.6 million after-tax); he chose the latter.

Present Value of a Single Amount

We graphically express the present value, called *p*, of a single future amount, called *f*, that is received or paid at a future date in Exhibit B.1.

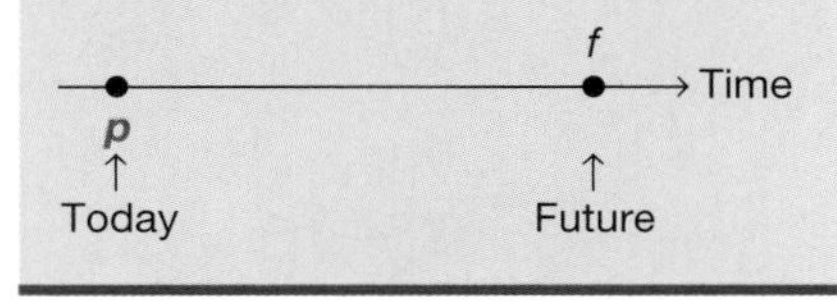

EXHIBIT B.1

Present Value of a Single Amount Diagram

The formula to compute the present value of a single amount is shown in Exhibit B.2, where p = present value; f = future value; i = rate of interest per period; and n = number of periods. (Interest is also called the *discount,* and an interest rate is also called the *discount rate.*)

EXHIBIT B.2

Present Value of a Single Amount Formula

$$p = \frac{f}{(1 + i)^n}$$

To illustrate present value concepts, assume that we need $220 one period from today. We want to know how much we must invest now, for one period, at an interest rate of 10% to provide for this $220. For this illustration, the p, or present value, is the unknown amount—the specifics are shown graphically as follows:

(i = 0.10)
f = $220
p = ?

Conceptually, we know p must be less than $220. This is obvious from the answer to this question: Would we rather have $220 today or $220 at some future date? If we had $220 today, we could invest it and see it grow to something more than $220 in the future. Therefore, we would prefer the $220 today. This means that if we were promised $220 in the future, we would take less than $220 today. But how much less? To answer that question, we compute an estimate of the present value of the $220 to be received one period from now using the formula in Exhibit B.2 as follows:

$$p = \frac{f}{(1 + i)^n} = \frac{\$220}{(1 + 0.10)^1} = \$200$$

We interpret this result to say that given an interest rate of 10%, we are indifferent between $200 today or $220 at the end of one period.

We can also use this formula to compute the present value for *any number of periods.* To illustrate, consider a payment of $242 at the end of two periods at 10% interest. The present value of this $242 to be received two periods from now is computed as follows:

$$p = \frac{f}{(1 + i)^n} = \frac{\$242}{(1 + 0.10)^2} = \$200$$

I will pay your allowance at the end of the month. Do you want to wait or receive its present value today?

Together, these results tell us we are indifferent between $200 today, or $220 one period from today, or $242 two periods from today given a 10% interest rate per period.

The number of periods (n) in the present value formula does not have to be expressed in years. Any period of time such as a day, a month, a quarter, or a year can be used. Whatever period is used, the interest rate (i) must be compounded for the same period. This means that if a situation expresses n in months and i equals 12% per year, then i is transformed into interest earned per month (or 1%). In this case, interest is said to be *compounded monthly.*

A present value table helps us with present value computations. It gives us present values (factors) for a variety of both interest rates (i) and periods (n). Each present value in a present value table assumes that the future value (f) equals 1. When the future value (f) is different from 1, we simply multiply the present value (p) from the table by that future value to give us the estimate. The formula used to construct a table of present values for a single future amount of 1 is shown in Exhibit B.3.

EXHIBIT B.3

Present Value of 1 Formula

$$p = \frac{1}{(1 + i)^n}$$

This formula is identical to that in Exhibit B.2 except that f equals 1. Table B.1 at the end of this appendix is such a present value table. It is often called a **present value of 1 table.** A present value table involves three factors: p, i, and n. Knowing two of these three factors allows us to compute the third. (A fourth is f, but as already explained, we need only multiply the 1 used in the formula by f.) To illustrate the use of a present value table, consider three cases.

P1 Apply present value concepts to a single amount by using interest tables.

Case 1 (solve for p when knowing i and n). To show how we use a present value table, let's look again at how we estimate the present value of \$220 (the f value) at the end of one period ($n = 1$) where the interest rate (i) is 10%. To solve this case, we go to the present value table (Table B.1) and look in the row for 1 period and in the column for 10% interest. Here we find a present value (p) of 0.9091 based on a future value of 1. This means, for instance, that \$1 to be received one period from today at 10% interest is worth \$0.9091 today. Since the future value in this case is not \$1 but \$220, we multiply the 0.9091 by \$220 to get an answer of \$200.

Case 2 (solve for n when knowing p and i). To illustrate, assume a \$100,000 future value (f) that is worth \$13,000 today (p) using an interest rate of 12% (i) but where n is unknown. In particular, we want to know how many periods (n) there are between the present value and the future value. To put this in context, it would fit a situation in which we want to retire with \$100,000 but currently have only \$13,000 that is earning a 12% return and we will be unable to save any additional money. How long will it be before we can retire? To answer this, we go to Table B.1 and look in the 12% interest column. Here we find a column of present values (p) based on a future value of 1. To use the present value table for this solution, we must divide \$13,000 (p) by \$100,000 (f), which equals 0.1300. This is necessary because *a present value table defines* f *equal to 1, and* p *as a fraction of 1.* We look for a value nearest to 0.1300 (p), which we find in the row for 18 periods (n). This means that the present value of \$100,000 at the end of 18 periods at 12% interest is \$13,000; alternatively stated, we must work 18 more years.

Case 3 (solve for i when knowing p and n). In this case, we have, say, a \$120,000 future value (f) worth \$60,000 today (p) when there are nine periods (n) between the present and future values, but the interest rate is unknown. As an example, suppose we want to retire with \$120,000, but we have only \$60,000 and we will be unable to save any additional money, yet we hope to retire in nine years. What interest rate must we earn to retire with \$120,000 in nine years? To answer this, we go to the present value table (Table B.1) and look in the row for nine periods. To use the present value table, we must divide \$60,000 (p) by \$120,000 (f), which equals 0.5000. Recall that this step is necessary because a present value table defines f equal to 1 and p as a fraction of 1. We look for a value in the row for nine periods that is nearest to 0.5000 (p), which we find in the column for 8% interest (i). This means that the present value of \$120,000 at the end of nine periods at 8% interest is \$60,000 or, in our example, we must earn 8% annual interest to retire in nine years.

Quick Check Answer—p. B-7

1. A company is considering an investment expected to yield \$70,000 after six years. If this company demands an 8% return, how much is it willing to pay for this investment?

Future Value of a Single Amount

We must modify the formula for the present value of a single amount to obtain the formula for the future value of a single amount. In particular, we multiply both sides of the equation in Exhibit B.2 by $(1 + i)^n$ to get the result shown in Exhibit B.4.

EXHIBIT B.4

Future Value of a Single Amount Formula

$$f = p \times (1 + i)^n$$

The future value (f) is defined in terms of p, i, and n. We can use this formula to determine that \$200 (p) invested for 1 (n) period at an interest rate of 10% (i) yields a future value of

\$220 as follows:

$$f = p \times (1 + i)^n$$
$$= \$200 \times (1 + 0.10)^1$$
$$= \$220$$

P2 Apply future value concepts to a single amount by using interest tables.

This formula can also be used to compute the future value of an amount for *any number of periods* into the future. To illustrate, assume that \$200 is invested for three periods at 10%. The future value of this \$200 is \$266.20, computed as follows:

$$f = p \times (1 + i)^n$$
$$= \$200 \times (1 + 0.10)^3$$
$$= \$266.20$$

A future value table makes it easier for us to compute future values (f) for many different combinations of interest rates (i) and time periods (n). Each future value in a future value table assumes the present value (p) is 1. As with a present value table, if the future amount is something other than 1, we simply multiply our answer by that amount. The formula used to construct a table of future values (factors) for a single amount of 1 is in Exhibit B.5.

EXHIBIT B.5

Future Value of 1 Formula

$$f = (1 + i)^n$$

Table B.2 at the end of this appendix shows a table of future values for a current amount of 1. This type of table is called a **future value of 1 table**.

There are some important relations between Tables B.1 and B.2. In Table B.2, for the row where $n = 0$, the future value is 1 for each interest rate. This is so because no interest is earned when time does not pass. We also see that Tables B.1 and B.2 report the same information but in a different manner. In particular, one table is simply the *inverse* of the other. To illustrate this inverse relation, let's say we invest \$100 for a period of five years at 12% per year. How much do we expect to have after five years? We can answer this question using Table B.2 by finding the future value (f) of 1, for five periods from now, compounded at 12%. From that table we find $f = 1.7623$. If we start with \$100, the amount it accumulates to after five years is \$176.23 (\$100 × 1.7623). We can alternatively use Table B.1. Here we find that the present value (p) of 1, discounted five periods at 12%, is 0.5674. Recall the inverse relation between present value and future value. This means that $p = 1/f$ (or equivalently, $f = 1/p$). We can compute the future value of \$100 invested for five periods at 12% as follows: $f = \$100 \times (1/0.5674) = \$176.24$ (which equals the \$176.23 just computed, except for a 1 cent rounding difference).

A future value table involves three factors: f, i, and n. Knowing two of these three factors allows us to compute the third. To illustrate, consider these three possible cases.

Case 1 (solve for f when knowing i and n). Our preceding example fits this case. We found that \$100 invested for five periods at 12% interest accumulates to \$176.24.

Case 2 (solve for n when knowing f and i). In this case, we have, say, \$2,000 (p) and we want to know how many periods (n) it will take to accumulate to \$3,000 (f) at 7% (i) interest. To answer this, we go to the future value table (Table B.2) and look in the 7% interest column. Here we find a column of future values (f) based on a present value of 1. To use a future value table, we must divide \$3,000 (f) by \$2,000 (p), which equals 1.500. This is necessary because *a future value table defines* p *equal to 1, and* f *as a multiple of 1.* We look for a value nearest to 1.50 (f), which we find in the row for six periods (n). This means that \$2,000 invested for six periods at 7% interest accumulates to \$3,000.

Case 3 (solve for i when knowing f and n). In this case, we have, say, \$2,001 (p), and in nine years (n) we want to have \$4,000 (f). What rate of interest must we earn to accomplish this? To answer that, we go to Table B.2 and search in the row for nine periods. To use a future value table, we must divide \$4,000 (f) by \$2,001 (p), which equals 1.9990. Recall that this is necessary because a future value table defines p equal to 1 and f as a multiple of 1. We look for a value nearest to 1.9990 (f), which we find in the column for 8% interest (i). This means that \$2,001 invested for nine periods at 8% interest accumulates to \$4,000.

Quick Check Answer—p. B-7

2. Assume that you win a $150,000 cash sweepstakes. You decide to deposit this cash in an account earning 8% annual interest, and you plan to quit your job when the account equals $555,000. How many years will it be before you can quit working?

Present Value of an Annuity

An *annuity* is a series of equal payments occurring at equal intervals. One example is a series of three annual payments of $100 each. An *ordinary annuity* is defined as equal end-of-period payments at equal intervals. An ordinary annuity of $100 for three periods and its present value (p) are illustrated in Exhibit B.6.

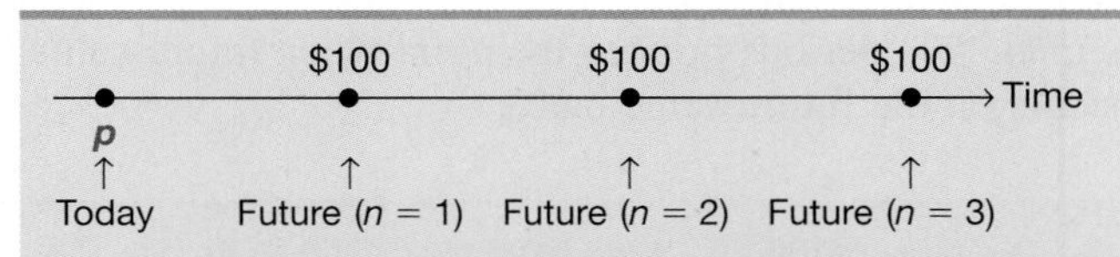

EXHIBIT B.6

Present Value of an Ordinary Annuity Diagram

P3 Apply present value concepts to an annuity by using interest tables.

One way to compute the present value of an ordinary annuity is to find the present value of each payment using our present value formula from Exhibit B.3. We then add each of the three present values. To illustrate, let's look at three $100 payments at the end of each of the next three periods with an interest rate of 15%. Our present value computations are

$$p = \frac{\$100}{(1 + 0.15)^1} + \frac{\$100}{(1 + 0.15)^2} + \frac{\$100}{(1 + 0.15)^3} = \$228.32$$

This computation is identical to computing the present value of each payment (from Table B.1) and taking their sum or, alternatively, adding the values from Table B.1 for each of the three payments and multiplying their sum by the $100 annuity payment.

A more direct way is to use a present value of annuity table. Table B.3 at the end of this appendix is one such table. This table is called a **present value of an annuity of 1 table**. If we look at Table B.3 where $n = 3$ and $i = 15\%$, we see the present value is 2.2832. This means that the present value of an annuity of 1 for three periods, with a 15% interest rate, equals 2.2832.

A present value of an annuity formula is used to construct Table B.3. It can also be constructed by adding the amounts in a present value of 1 table. To illustrate, we use Tables B.1 and B.3 to confirm this relation for the prior example:

From Table B.1		From Table B.3	
$i = 15\%, n = 1$	0.8696		
$i = 15\%, n = 2$	0.7561		
$i = 15\%, n = 3$	0.6575		
Total	2.2832	$i = 15\%, n = 3$	2.2832

We can also use business calculators or spreadsheet programs to find the present value of an annuity.

Decision Insight

Better Lucky Than Good "I don't have good luck—I'm blessed," proclaimed Andrew "Jack" Whittaker, 55, a sewage treatment contractor, after winning the largest ever undivided jackpot in a U.S. lottery. Whittaker had to choose between $315 million in 30 annual installments or $170 million in one lump sum ($112 million after-tax).

Quick Check Answer—p. B-7

3. A company is considering an investment paying $10,000 every six months for three years. The first payment would be received in six months. If this company requires an 8% annual return, what is the maximum amount it is willing to pay for this investment?

Future Value of an Annuity

The future value of an *ordinary annuity* is the accumulated value of each annuity payment with interest as of the date of the final payment. To illustrate, let's consider the earlier annuity of three annual payments of $100. Exhibit B.7 shows the point in time for the future value (f). The first payment is made two periods prior to the point when future value is determined, and the final payment occurs on the future value date.

EXHIBIT B.7

Future Value of an Ordinary Annuity Diagram

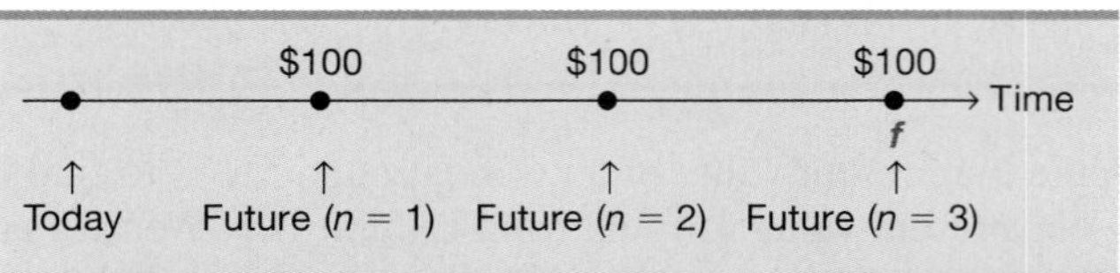

One way to compute the future value of an annuity is to use the formula to find the future value of *each* payment and add them. If we assume an interest rate of 15%, our calculation is

$$f = \$100 \times (1 + 0.15)^2 + \$100 \times (1 + 0.15)^1 + \$100 \times (1 + 0.15)^0 = \$347.25$$

This is identical to using Table B.2 and summing the future values of each payment, or adding the future values of the three payments of 1 and multiplying the sum by $100.

P4 Apply future value concepts to an annuity by using interest tables.

A more direct way is to use a table showing future values of annuities. Such a table is called a **future value of an annuity of 1 table**. Table B.4 at the end of this appendix is one such table. Note that in Table B.4 when $n = 1$, the future values equal 1 ($f = 1$) for all rates of interest. This is so because such an annuity consists of only one payment and the future value is determined on the date of that payment—no time passes between the payment and its future value. The future value of an annuity formula is used to construct Table B.4. We can also construct it by adding the amounts from a future value of 1 table. To illustrate, we use Tables B.2 and B.4 to confirm this relation for the prior example:

From Table B.2		From Table B.4	
$i = 15\%, n = 0$	1.0000		
$i = 15\%, n = 1$	1.1500		
$i = 15\%, n = 2$	1.3225		
Total	3.4725	$i = 15\%, n = 3$	3.4725

Note that the future value in Table B.2 is 1.0000 when $n = 0$, but the future value in Table B.4 is 1.0000 when $n = 1$. Is this a contradiction? No. When $n = 0$ in Table B.2, the future value is determined on the date when a single payment occurs. This means that no interest is earned because no time has passed, and the future value equals the payment. Table B.4 describes annuities with equal payments occurring at the end of each period. When $n = 1$, the annuity has one payment, and its future value equals 1 on the date of its final and only payment. Again, no time passes between the payment and its future value date.

Quick Check

Answer—p. B-7

4. A company invests $45,000 per year for five years at 12% annual interest. Compute the value of this annuity investment at the end of five years.

Summary

C1 **Describe the earning of interest and the concepts of present and future values.** Interest is payment by a borrower to the owner of an asset for its use. Present and future value computations are a way for us to estimate the interest component of holding assets or liabilities over a period of time.

P1 **Apply present value concepts to a single amount by using interest tables.** The present value of a single amount received at a future date is the amount that can be invested now at the specified interest rate to yield that future value.

P2 **Apply future value concepts to a single amount by using interest tables.** The future value of a single amount invested at a specified rate of interest is the amount that would accumulate by the future date.

P3 **Apply present value concepts to an annuity by using interest tables.** The present value of an annuity is the amount that can be invested now at the specified interest rate to yield that series of equal periodic payments.

P4 **Apply future value concepts to an annuity by using interest tables.** The future value of an annuity invested at a specific rate of interest is the amount that would accumulate by the date of the final payment.

Guidance Answers to Quick Checks

1. $70,000 × 0.6302 = $44,114 (use Table B.1, $i = 8\%$, $n = 6$).
2. $555,000/$150,000 = 3.7000; Table B.2 shows this value is not achieved until after 17 years at 8% interest.
3. $10,000 × 5.2421 = $52,421 (use Table B.3, $i = 4\%$, $n = 6$).
4. $45,000 × 6.3528 = $285,876 (use Table B.4, $i = 12\%$, $n = 5$).

McGraw-Hill's HOMEWORK MANAGER® HM Available with McGraw-Hill's Homework Manager

QUICK STUDY

QS B-1 Identifying interest rates in tables C1

Assume that you must make future value estimates using the *future value of 1 table* (Table B.2). Which interest rate column do you use when working with the following rates?

1. 8% compounded quarterly
2. 12% compounded annually
3. 6% compounded semiannually
4. 12% compounded monthly

QS B-2 Interest rate on an investment P1

Ken Francis is offered the possibility of investing $2,745 today and in return to receive $10,000 after 15 years. What is the annual rate of interest for this investment? (Use Table B.1.)

QS B-3 Number of periods of an investment P1

Megan Brink is offered the possibility of investing $6,651 today at 6% interest per year in a desire to accumulate $10,000. How many years must Brink wait to accumulate $10,000? (Use Table B.1.)

QS B-4 Present value of an amount P1

Flaherty is considering an investment that, if paid for immediately, is expected to return $140,000 five years from now. If Flaherty demands a 9% return, how much is she willing to pay for this investment?

QS B-5 Future value of an amount P2

CII, Inc., invests $630,000 in a project expected to earn a 12% annual rate of return. The earnings will be reinvested in the project each year until the entire investment is liquidated 10 years later. What will the cash proceeds be when the project is liquidated?

QS B-6
Present value of an annuity P3

Beene Distributing is considering a project that will return $150,000 annually at the end of each year for six years. If Beene demands an annual return of 7% and pays for the project immediately, how much is it willing to pay for the project?

QS B-7
Future value of an annuity P4

Claire Fitch is planning to begin an individual retirement program in which she will invest $1,500 at the end of each year. Fitch plans to retire after making 30 annual investments in the program earning a return of 10%. What is the value of the program on the date of the last payment?

EXERCISES

Exercise B-1
Number of periods of an investment P2

Bill Thompson expects to invest $10,000 at 12% and, at the end of a certain period, receive $96,463. How many years will it be before Thompson receives the payment? (Use Table B.2.)

Exercise B-2
Interest rate on an investment P2

Ed Summers expects to invest $10,000 for 25 years, after which he wants to receive $108,347. What rate of interest must Summers earn? (Use Table B.2.)

Exercise B-3
Interest rate on an investment P3

Jones expects an immediate investment of $57,466 to return $10,000 annually for eight years, with the first payment to be received one year from now. What rate of interest must Jones earn? (Use Table B.3.)

Exercise B-4
Number of periods of an investment P3

Keith Riggins expects an investment of $82,014 to return $10,000 annually for several years. If Riggins earns a return of 10%, how many annual payments will he receive? (Use Table B.3.)

Exercise B-5
Interest rate on an investment P4

Algoe expects to invest $1,000 annually for 40 years to yield an accumulated value of $154,762 on the date of the last investment. For this to occur, what rate of interest must Algoe earn? (Use Table B.4.)

Exercise B-6
Number of periods of an investment P4

Kate Beckwith expects to invest $10,000 annually that will earn 8%. How many annual investments must Beckwith make to accumulate $303,243 on the date of the last investment? (Use Table B.4.)

Exercise B-7
Present value of an annuity P3

Sam Weber finances a new automobile by paying $6,500 cash and agreeing to make 40 monthly payments of $500 each, the first payment to be made one month after the purchase. The loan bears interest at an annual rate of 12%. What is the cost of the automobile?

Exercise B-8
Present value of bonds P1 P3

Spiller Corp. plans to issue 10%, 15-year, $500,000 par value bonds payable that pay interest semiannually on June 30 and December 31. The bonds are dated December 31, 2008, and are issued on that date. If the market rate of interest for the bonds is 8% on the date of issue, what will be the total cash proceeds from the bond issue?

Exercise B-9
Present value of an amount P1

McAdams Company expects to earn 10% per year on an investment that will pay $606,773 six years from now. Use Table B.1 to compute the present value of this investment. (Round the amount to the nearest dollar.)

Exercise B-10
Present value of an amount and of an annuity P1 P3

Compute the amount that can be borrowed under each of the following circumstances:

1. A promise to repay $90,000 seven years from now at an interest rate of 6%.
2. An agreement made on February 1, 2008, to make three separate payments of $20,000 on February 1 of 2009, 2010, and 2011. The annual interest rate is 10%.

Exercise B-11
Present value of an amount P1

On January 1, 2008, a company agrees to pay $20,000 in three years. If the annual interest rate is 10%, determine how much cash the company can borrow with this agreement.

Exercise B-12
Present value of an amount P1

Find the amount of money that can be borrowed today with each of the following separate debt agreements *a* through *f*. (Round amounts to the nearest dollar.)

Case	Single Future Payment	Number of Periods	Interest Rate
a.	$40,000	3	4%
b.	75,000	7	8
c.	52,000	9	10
d.	18,000	2	4
e.	63,000	8	6
f.	89,000	5	2

Exercise B-13
Present values of annuities P3

C&H Ski Club recently borrowed money and agrees to pay it back with a series of six annual payments of $5,000 each. C&H subsequently borrows more money and agrees to pay it back with a series of four annual payments of $7,500 each. The annual interest rate for both loans is 6%.

1. Use Table B.1 to find the present value of these two separate annuities. (Round amounts to the nearest dollar.)

2. Use Table B.3 to find the present value of these two separate annuities. (Round amounts to the nearest dollar.)

Exercise B-14
Present value with semiannual compounding C1 P3

Otto Co. borrows money on April 30, 2008, by promising to make four payments of $13,000 each on November 1, 2008; May 1, 2009; November 1, 2009; and May 1, 2010.

1. How much money is Otto able to borrow if the interest rate is 8%, compounded semiannually?

2. How much money is Otto able to borrow if the interest rate is 12%, compounded semiannually?

3. How much money is Otto able to borrow if the interest rate is 16%, compounded semiannually?

Exercise B-15
Future value of an amount P2

Mark Welsch deposits $7,200 in an account that earns interest at an annual rate of 8%, compounded quarterly. The $7,200 plus earned interest must remain in the account 10 years before it can be withdrawn. How much money will be in the account at the end of 10 years?

Exercise B-16
Future value of an annuity P4

Kelly Malone plans to have $50 withheld from her monthly paycheck and deposited in a savings account that earns 12% annually, compounded monthly. If Malone continues with her plan for two and one-half years, how much will be accumulated in the account on the date of the last deposit?

Exercise B-17
Future value of an amount plus an annuity P2 P4

Starr Company decides to establish a fund that it will use 10 years from now to replace an aging production facility. The company will make a $100,000 initial contribution to the fund and plans to make quarterly contributions of $50,000 beginning in three months. The fund earns 12%, compounded quarterly. What will be the value of the fund 10 years from now?

Exercise B-18
Future value of an amount P2

Catten, Inc., invests $163,170 today earning 7% per year for nine years. Use Table B.2 to compute the future value of the investment nine years from now. (Round the amount to the nearest dollar.)

Exercise B-19
Using present and future value tables C1 P1 P2 P3 P4

For each of the following situations, identify (1) the case as either (*a*) a present or a future value and (*b*) a single amount or an annuity, (2) the table you would use in your computations (but do not solve the problem), and (3) the interest rate and time periods you would use.

a. You need to accumulate $10,000 for a trip you wish to take in four years. You are able to earn 8% compounded semiannually on your savings. You plan to make only one deposit and let the money accumulate for four years. How would you determine the amount of the one-time deposit?

b. Assume the same facts as in part (*a*) except that you will make semiannual deposits to your savings account.

c. You want to retire after working 40 years with savings in excess of $1,000,000. You expect to save $4,000 a year for 40 years and earn an annual rate of interest of 8%. Will you be able to retire with more than $1,000,000 in 40 years? Explain.

d. A sweepstakes agency names you a grand prize winner. You can take $225,000 immediately or elect to receive annual installments of $30,000 for 20 years. You can earn 10% annually on any investments you make. Which prize do you choose to receive?

TABLE B.1

Present Value of 1

$$p = 1/(1 + i)^n$$

Periods	Rate 1%	2%	3%	4%	5%	6%	7%	8%	9%	10%	12%	15%
1	0.9901	0.9804	0.9709	0.9615	0.9524	0.9434	0.9346	0.9259	0.9174	0.9091	0.8929	0.8696
2	0.9803	0.9612	0.9426	0.9246	0.9070	0.8900	0.8734	0.8573	0.8417	0.8264	0.7972	0.7561
3	0.9706	0.9423	0.9151	0.8890	0.8638	0.8396	0.8163	0.7938	0.7722	0.7513	0.7118	0.6575
4	0.9610	0.9238	0.8885	0.8548	0.8227	0.7921	0.7629	0.7350	0.7084	0.6830	0.6355	0.5718
5	0.9515	0.9057	0.8626	0.8219	0.7835	0.7473	0.7130	0.6806	0.6499	0.6209	0.5674	0.4972
6	0.9420	0.8880	0.8375	0.7903	0.7462	0.7050	0.6663	0.6302	0.5963	0.5645	0.5066	0.4323
7	0.9327	0.8706	0.8131	0.7599	0.7107	0.6651	0.6227	0.5835	0.5470	0.5132	0.4523	0.3759
8	0.9235	0.8535	0.7894	0.7307	0.6768	0.6274	0.5820	0.5403	0.5019	0.4665	0.4039	0.3269
9	0.9143	0.8368	0.7664	0.7026	0.6446	0.5919	0.5439	0.5002	0.4604	0.4241	0.3606	0.2843
10	0.9053	0.8203	0.7441	0.6756	0.6139	0.5584	0.5083	0.4632	0.4224	0.3855	0.3220	0.2472
11	0.8963	0.8043	0.7224	0.6496	0.5847	0.5268	0.4751	0.4289	0.3875	0.3505	0.2875	0.2149
12	0.8874	0.7885	0.7014	0.6246	0.5568	0.4970	0.4440	0.3971	0.3555	0.3186	0.2567	0.1869
13	0.8787	0.7730	0.6810	0.6006	0.5303	0.4688	0.4150	0.3677	0.3262	0.2897	0.2292	0.1625
14	0.8700	0.7579	0.6611	0.5775	0.5051	0.4423	0.3878	0.3405	0.2992	0.2633	0.2046	0.1413
15	0.8613	0.7430	0.6419	0.5553	0.4810	0.4173	0.3624	0.3152	0.2745	0.2394	0.1827	0.1229
16	0.8528	0.7284	0.6232	0.5339	0.4581	0.3936	0.3387	0.2919	0.2519	0.2176	0.1631	0.1069
17	0.8444	0.7142	0.6050	0.5134	0.4363	0.3714	0.3166	0.2703	0.2311	0.1978	0.1456	0.0929
18	0.8360	0.7002	0.5874	0.4936	0.4155	0.3503	0.2959	0.2502	0.2120	0.1799	0.1300	0.0808
19	0.8277	0.6864	0.5703	0.4746	0.3957	0.3305	0.2765	0.2317	0.1945	0.1635	0.1161	0.0703
20	0.8195	0.6730	0.5537	0.4564	0.3769	0.3118	0.2584	0.2145	0.1784	0.1486	0.1037	0.0611
25	0.7798	0.6095	0.4776	0.3751	0.2953	0.2330	0.1842	0.1460	0.1160	0.0923	0.0588	0.0304
30	0.7419	0.5521	0.4120	0.3083	0.2314	0.1741	0.1314	0.0994	0.0754	0.0573	0.0334	0.0151
35	0.7059	0.5000	0.3554	0.2534	0.1813	0.1301	0.0937	0.0676	0.0490	0.0356	0.0189	0.0075
40	0.6717	0.4529	0.3066	0.2083	0.1420	0.0972	0.0668	0.0460	0.0318	0.0221	0.0107	0.0037

TABLE B.2

Future Value of 1

$$f = (1 + i)^n$$

Periods	Rate 1%	2%	3%	4%	5%	6%	7%	8%	9%	10%	12%	15%
0	1.0000	1.0000	1.0000	1.0000	1.0000	1.0000	1.0000	1.0000	1.0000	1.0000	1.0000	1.0000
1	1.0100	1.0200	1.0300	1.0400	1.0500	1.0600	1.0700	1.0800	1.0900	1.1000	1.1200	1.1500
2	1.0201	1.0404	1.0609	1.0816	1.1025	1.1236	1.1449	1.1664	1.1881	1.2100	1.2544	1.3225
3	1.0303	1.0612	1.0927	1.1249	1.1576	1.1910	1.2250	1.2597	1.2950	1.3310	1.4049	1.5209
4	1.0406	1.0824	1.1255	1.1699	1.2155	1.2625	1.3108	1.3605	1.4116	1.4641	1.5735	1.7490
5	1.0510	1.1041	1.1593	1.2167	1.2763	1.3382	1.4026	1.4693	1.5386	1.6105	1.7623	2.0114
6	1.0615	1.1262	1.1941	1.2653	1.3401	1.4185	1.5007	1.5869	1.6771	1.7716	1.9738	2.3131
7	1.0721	1.1487	1.2299	1.3159	1.4071	1.5036	1.6058	1.7138	1.8280	1.9487	2.2107	2.6600
8	1.0829	1.1717	1.2668	1.3686	1.4775	1.5938	1.7182	1.8509	1.9926	2.1436	2.4760	3.0590
9	1.0937	1.1951	1.3048	1.4233	1.5513	1.6895	1.8385	1.9990	2.1719	2.3579	2.7731	3.5179
10	1.1046	1.2190	1.3439	1.4802	1.6289	1.7908	1.9672	2.1589	2.3674	2.5937	3.1058	4.0456
11	1.1157	1.2434	1.3842	1.5395	1.7103	1.8983	2.1049	2.3316	2.5804	2.8531	3.4785	4.6524
12	1.1268	1.2682	1.4258	1.6010	1.7959	2.0122	2.2522	2.5182	2.8127	3.1384	3.8960	5.3503
13	1.1381	1.2936	1.4685	1.6651	1.8856	2.1329	2.4098	2.7196	3.0658	3.4523	4.3635	6.1528
14	1.1495	1.3195	1.5126	1.7317	1.9799	2.2609	2.5785	2.9372	3.3417	3.7975	4.8871	7.0757
15	1.1610	1.3459	1.5580	1.8009	2.0789	2.3966	2.7590	3.1722	3.6425	4.1772	5.4736	8.1371
16	1.1726	1.3728	1.6047	1.8730	2.1829	2.5404	2.9522	3.4259	3.9703	4.5950	6.1304	9.3576
17	1.1843	1.4002	1.6528	1.9479	2.2920	2.6928	3.1588	3.7000	4.3276	5.0545	6.8660	10.7613
18	1.1961	1.4282	1.7024	2.0258	2.4066	2.8543	3.3799	3.9960	4.7171	5.5599	7.6900	12.3755
19	1.2081	1.4568	1.7535	2.1068	2.5270	3.0256	3.6165	4.3157	5.1417	6.1159	8.6128	14.2318
20	1.2202	1.4859	1.8061	2.1911	2.6533	3.2071	3.8697	4.6610	5.6044	6.7275	9.6463	16.3665
25	1.2824	1.6406	2.0938	2.6658	3.3864	4.2919	5.4274	6.8485	8.6231	10.8347	17.0001	32.9190
30	1.3478	1.8114	2.4273	3.2434	4.3219	5.7435	7.6123	10.0627	13.2677	17.4494	29.9599	66.2118
35	1.4166	1.9999	2.8139	3.9461	5.5160	7.6861	10.6766	14.7853	20.4140	28.1024	52.7996	133.1755
40	1.4889	2.2080	3.2620	4.8010	7.0400	10.2857	14.9745	21.7245	31.4094	45.2593	93.0510	267.8635

$$p = \left[1 - \frac{1}{(1 + i)^n}\right] / i$$

TABLE B.3

Present Value of an Annuity of 1

Periods	1%	2%	3%	4%	5%	6%	7%	8%	9%	10%	12%	15%
	Rate											
1	0.9901	0.9804	0.9709	0.9615	0.9524	0.9434	0.9346	0.9259	0.9174	0.9091	0.8929	0.8696
2	1.9704	1.9416	1.9135	1.8861	1.8594	1.8334	1.8080	1.7833	1.7591	1.7355	1.6901	1.6257
3	2.9410	2.8839	2.8286	2.7751	2.7232	2.6730	2.6243	2.5771	2.5313	2.4869	2.4018	2.2832
4	3.9020	3.8077	3.7171	3.6299	3.5460	3.4651	3.3872	3.3121	3.2397	3.1699	3.0373	2.8550
5	4.8534	4.7135	4.5797	4.4518	4.3295	4.2124	4.1002	3.9927	3.8897	3.7908	3.6048	3.3522
6	5.7955	5.6014	5.4172	5.2421	5.0757	4.9173	4.7665	4.6229	4.4859	4.3553	4.1114	3.7845
7	6.7282	6.4720	6.2303	6.0021	5.7864	5.5824	5.3893	5.2064	5.0330	4.8684	4.5638	4.1604
8	7.6517	7.3255	7.0197	6.7327	6.4632	6.2098	5.9713	5.7466	5.5348	5.3349	4.9676	4.4873
9	8.5660	8.1622	7.7861	7.4353	7.1078	6.8017	6.5152	6.2469	5.9952	5.7590	5.3282	4.7716
10	9.4713	8.9826	8.5302	8.1109	7.7217	7.3601	7.0236	6.7101	6.4177	6.1446	5.6502	5.0188
11	10.3676	9.7868	9.2526	8.7605	8.3064	7.8869	7.4987	7.1390	6.8052	6.4951	5.9377	5.2337
12	11.2551	10.5753	9.9540	9.3851	8.8633	8.3838	7.9427	7.5361	7.1607	6.8137	6.1944	5.4206
13	12.1337	11.3484	10.6350	9.9856	9.3936	8.8527	8.3577	7.9038	7.4869	7.1034	6.4235	5.5831
14	13.0037	12.1062	11.2961	10.5631	9.8986	9.2950	8.7455	8.2442	7.7862	7.3667	6.6282	5.7245
15	13.8651	12.8493	11.9379	11.1184	10.3797	9.7122	9.1079	8.5595	8.0607	7.6061	6.8109	5.8474
16	14.7179	13.5777	12.5611	11.6523	10.8378	10.1059	9.4466	8.8514	8.3126	7.8237	6.9740	5.9542
17	15.5623	14.2919	13.1661	12.1657	11.2741	10.4773	9.7632	9.1216	8.5436	8.0216	7.1196	6.0472
18	16.3983	14.9920	13.7535	12.6593	11.6896	10.8276	10.0591	9.3719	8.7556	8.2014	7.2497	6.1280
19	17.2260	15.6785	14.3238	13.1339	12.0853	11.1581	10.3356	9.6036	8.9501	8.3649	7.3658	6.1982
20	18.0456	16.3514	14.8775	13.5903	12.4622	11.4699	10.5940	9.8181	9.1285	8.5136	7.4694	6.2593
25	22.0232	19.5235	17.4131	15.6221	14.0939	12.7834	11.6536	10.6748	9.8226	9.0770	7.8431	6.4641
30	25.8077	22.3965	19.6004	17.2920	15.3725	13.7648	12.4090	11.2578	10.2737	9.4269	8.0552	6.5660
35	29.4086	24.9986	21.4872	18.6646	16.3742	14.4982	12.9477	11.6546	10.5668	9.6442	8.1755	6.6166
40	32.8347	27.3555	23.1148	19.7928	17.1591	15.0463	13.3317	11.9246	10.7574	9.7791	8.2438	6.6418

$$f = [(1 + i)^n - 1] / i$$

TABLE B.4

Future Value of an Annuity of 1

Periods	1%	2%	3%	4%	5%	6%	7%	8%	9%	10%	12%	15%
	Rate											
1	1.0000	1.0000	1.0000	1.0000	1.0000	1.0000	1.0000	1.0000	1.0000	1.0000	1.0000	1.0000
2	2.0100	2.0200	2.0300	2.0400	2.0500	2.0600	2.0700	2.0800	2.0900	2.1000	2.1200	2.1500
3	3.0301	3.0604	3.0909	3.1216	3.1525	3.1836	3.2149	3.2464	3.2781	3.3100	3.3744	3.4725
4	4.0604	4.1216	4.1836	4.2465	4.3101	4.3746	4.4399	4.5061	4.5731	4.6410	4.7793	4.9934
5	5.1010	5.2040	5.3091	5.4163	5.5256	5.6371	5.7507	5.8666	5.9847	6.1051	6.3528	6.7424
6	6.1520	6.3081	6.4684	6.6330	6.8019	6.9753	7.1533	7.3359	7.5233	7.7156	8.1152	8.7537
7	7.2135	7.4343	7.6625	7.8983	8.1420	8.3938	8.6540	8.9228	9.2004	9.4872	10.0890	11.0668
8	8.2857	8.5830	8.8923	9.2142	9.5491	9.8975	10.2598	10.6366	11.0285	11.4359	12.2997	13.7268
9	9.3685	9.7546	10.1591	10.5828	11.0266	11.4913	11.9780	12.4876	13.0210	13.5795	14.7757	16.7858
10	10.4622	10.9497	11.4639	12.0061	12.5779	13.1808	13.8164	14.4866	15.1929	15.9374	17.5487	20.3037
11	11.5668	12.1687	12.8078	13.4864	14.2068	14.9716	15.7836	16.6455	17.5603	18.5312	20.6546	24.3493
12	12.6825	13.4121	14.1920	15.0258	15.9171	16.8699	17.8885	18.9771	20.1407	21.3843	24.1331	29.0017
13	13.8093	14.6803	15.6178	16.6268	17.7130	18.8821	20.1406	21.4953	22.9534	24.5227	28.0291	34.3519
14	14.9474	15.9739	17.0863	18.2919	19.5986	21.0151	22.5505	24.2149	26.0192	27.9750	32.3926	40.5047
15	16.0969	17.2934	18.5989	20.0236	21.5786	23.2760	25.1290	27.1521	29.3609	31.7725	37.2797	47.5804
16	17.2579	18.6393	20.1569	21.8245	23.6575	25.6725	27.8881	30.3243	33.0034	35.9497	42.7533	55.7175
17	18.4304	20.0121	21.7616	23.6975	25.8404	28.2129	30.8402	33.7502	36.9737	40.5447	48.8837	65.0751
18	19.6147	21.4123	23.4144	25.6454	28.1324	30.9057	33.9990	37.4502	41.3013	45.5992	55.7497	75.8364
19	20.8109	22.8406	25.1169	27.6712	30.5390	33.7600	37.3790	41.4463	46.0185	51.1591	63.4397	88.2118
20	22.0190	24.2974	26.8704	29.7781	33.0660	36.7856	40.9955	45.7620	51.1601	57.2750	72.0524	102.4436
25	28.2432	32.0303	36.4593	41.6459	47.7271	54.8645	63.2490	73.1059	84.7009	98.3471	133.3339	212.7930
30	34.7849	40.5681	47.5754	56.0849	66.4388	79.0582	94.4608	113.2832	136.3075	164.4940	241.3327	434.7451
35	41.6603	49.9945	60.4621	73.6522	90.3203	111.4348	138.2369	172.3168	215.7108	271.0244	431.6635	881.1702
40	48.8864	60.4020	75.4013	95.0255	120.7998	154.7620	199.6351	259.0565	337.8824	442.5926	767.0914	1,779.0903

Glossary

Absorption costing Costing method that assigns both variable and fixed costs to products.

Accelerated depreciation method Method that produces larger depreciation charges in the early years of an asset's life and smaller charges in its later years. *(p. 398)*

Account Record within an accounting system in which increases and decreases are entered and stored in a specific asset, liability, equity, revenue, or expense. *(p. 49)*

Account balance Difference between total debits and total credits (including the beginning balance) for an account. *(p. 53)*

Account form balance sheet Balance sheet that lists assets on the left side and liabilities and equity on the right. *(p. 18)*

Account payable Liability created by buying goods or services on credit; backed by the buyer's general credit standing. *(p. 50)*

Accounting Information and measurement system that identifies, records, and communicates relevant information about a company's business activities. *(p. 4)*

Accounting cycle Recurring steps performed each accounting period, starting with analyzing transactions and continuing through the post-closing trial balance (or reversing entries). *(p. 144)*

Accounting equation Equality involving a company's assets, liabilities, and equity; Assets = Liabilities + Equity; also called *balance sheet equation*. *(p. 12)*

Accounting information system People, records, and methods that collect and process data from transactions and events, organize them in useful forms, and communicate results to decision makers. *(p. 268)*

Accounting period Length of time covered by financial statements; also called *reporting period*. *(p. 92)*

Accounting rate of return Rate used to evaluate the acceptability of an investment; equals the after-tax periodic income from a project divided by the average investment in the asset; also called *rate of return on average investment*. *(p. 1027)*

Accounts payable ledger Subsidiary ledger listing individual creditor (supplier) accounts. *(p. 273)*

Accounts receivable Amounts due from customers for credit sales; backed by the customer's general credit standing. *(p. 358)*

Accounts receivable ledger Subsidiary ledger listing individual customer accounts. *(p. 273)*

Accounts receivable turnover Measure of both the quality and liquidity of accounts receivable; indicates how often receivables are received and collected during the period; computed by dividing net sales by average accounts receivable. *(p. 372)*

Accrual basis accounting Accounting system that recognizes revenues when earned and expenses when incurred; the basis for GAAP. *(p. 93)*

Accrued expenses Costs incurred in a period that are both unpaid and unrecorded; adjusting entries for recording accrued expenses involve increasing expenses and increasing liabilities. *(p. 99)*

Accrued revenues Revenues earned in a period that are both unrecorded and not yet received in cash (or other assets); adjusting entries for recording accrued revenues involve increasing assets and increasing revenues. *(p. 101)*

Accumulated depreciation Cumulative sum of all depreciation expense recorded for an asset. *(p. 97)*

Acid-test ratio Ratio used to assess a company's ability to settle its current debts with its most liquid assets; defined as quick assets (cash, short-term investments, and current receivables) divided by current liabilities. *(p. 193)*

Activity-based budgeting (ABB) Budget system based on expected activities. *(p. 953)*

Activity-based costing (ABC) Cost allocation method that focuses on activities performed; traces costs to activities and then assigns them to cost objects. *(p. 854)*

Activity cost driver Variable that causes an activity's cost to go up or down; a causal factor. *(p. 855)*

Activity cost pool Temporary account that accumulates costs a company incurs to support an activity. *(p. 855)*

Adjusted trial balance List of accounts and balances prepared after period-end adjustments are recorded and posted. *(p. 104)*

Adjusting entry Journal entry at the end of an accounting period to bring an asset or liability account to its proper amount and update the related expense or revenue account. *(p. 94)*

Aging of accounts receivable Process of classifying accounts receivable by how long they are past due for purposes of estimating uncollectible accounts. *(p. 366)*

Allowance for Doubtful Accounts Contra asset account with a balance approximating uncollectible accounts receivable; also called *Allowance for Uncollectible Accounts*. *(p. 363)*

Allowance method Procedure that (a) estimates and matches bad debts expense with its sales for the period and/or (b) reports accounts receivable at estimated realizable value. *(p. 362)*

Amortization Process of allocating the cost of an intangible asset to expense over its estimated useful life. *(p. 407)*

Annual financial statements Financial statements covering a one-year period; often based on a calendar year, but any consecutive 12-month (or 52-week) period is acceptable. *(p. 92)*

Annual report Summary of a company's financial results for the year with its current financial condition and future plans; directed to external users of financial information. *(p. A-1)*

Annuity Series of equal payments at equal intervals. *(p. 568)*

Appropriated retained earnings Retained earnings separately reported to inform stockholders of funding needs. *(p. 523)*

Asset book value (See *book value.*)

Assets Resources a business owns or controls that are expected to provide current and future benefits to the business. *(p. 13)*

Audit Analysis and report of an organization's accounting system, its records, and its reports using various tests. *(p. 11)*

Auditors Individuals hired to review financial reports and information systems. *Internal auditors* of a company are employed to assess and evaluate its system of internal controls, including the resulting reports. *External auditors* are independent of a company and are hired to assess and evaluate the "fairness" of financial statements (or to perform other contracted financial services) *(p. 11).*

Authorized stock Total amount of stock that a corporation's charter authorizes it to issue. *(p. 509)*

Available-for-sale (AFS) securities Investments in debt and equity securities that are not classified as trading securities or held-to-maturity securities. *(p. 596)*

Average cost See *weighted average. (p. 230)*

Avoidable expense Expense (or cost) that is relevant for decision making; expense that is not incurred if a department, product, or service is eliminated. *(p. 1041)*

Bad debts Accounts of customers who do not pay what they have promised to pay; an expense of selling on credit; also called *uncollectible accounts. (p. 361)*

Balance column account Account with debit and credit columns for recording entries and another column for showing the balance of the account after each entry. *(p. 56)*

Balance sheet Financial statement that lists types and dollar amounts of assets, liabilities, and equity at a specific date. *(p. 18)*

Balance sheet equation (See *accounting equation.*)

Bank reconciliation Report that explains the difference between the book (company) balance of cash and the cash balance reported on the bank statement. *(p. 329)*

Bank statement Bank report on the depositor's beginning and ending cash balances, and a listing of its changes, for a period. *(p. 327)*

Basic earnings per share Net income less any preferred dividends and then divided by weighted-average common shares outstanding. *(p. 524)*

Batch processing Accumulating source documents for a period of time and then processing them all at once such as once a day, week, or month. *(p. 282)*

Bearer bonds Bonds made payable to whoever holds them (the *bearer*); also called *unregistered bonds. (p. 563)*

Benchmarking Practice of comparing and analyzing company financial performance or position with other companies or standards.

Betterments Expenditures to make a plant asset more efficient or productive; also called *improvements. (p. 403)*

Bond Written promise to pay the bond's par (or face) value and interest at a stated contract rate; often issued in denominations of $1,000. *(p. 550)*

Bond certificate Document containing bond specifics such as issuer's name, bond par value, contract interest rate, and maturity date. *(p. 552)*

Bond indenture Contract between the bond issuer and the bondholders; identifies the parties' rights and obligations. *(p. 552)*

Book value Asset's acquisition costs less its accumulated depreciation (or depletion, or amortization); also sometimes used synonymously as the *carrying value* of an account. *(p. 98)*

Book value per common share Recorded amount of equity applicable to common shares divided by the number of common shares outstanding. *(p. 526)*

Book value per preferred share Equity applicable to preferred shares (equals its call price [or par value if it is not callable] plus any cumulative dividends in arrears) divided by the number of preferred shares outstanding. *(p. 526)*

Bookkeeping (See *recordkeeping.*)

Break-even point Output level at which sales equals fixed plus variable costs; where income equals zero. *(p. 909)*

Break-even time (BET) Time-based measurement used to evaluate the acceptability of an investment; equals the time expected to pass before the present value of the net cash flows from an investment equals its initial cost. *(p. 1043)*

Budget Formal statement of future plans, usually expressed in monetary terms. *(p. 940)*

Budget report Report comparing actual results to planned objectives; sometimes used as a progress report. *(p. 980)*

Budgetary control Management use of budgets to monitor and control company operations. *(p. 980)*

Budgeted balance sheet Accounting report that presents predicted amounts of the company's assets, liabilities, and equity balances as of the end of the budget period. *(p. 952)*

Budgeted income statement Accounting report that presents predicted amounts of the company's revenues and expenses for the budget period. *(p. 952)*

Budgeting Process of planning future business actions and expressing them as formal plans. *(p. 940)*

Business An organization of one or more individuals selling products and/or services for profit. *(p. 10)*

Business entity assumption Principle that requires a business to be accounted for separately from its owner(s) and from any other entity. *(p. 10)*

Business segment Part of a company that can be separately identified by the products or services that it provides or by the geographic markets that it serves; also called *segment. (p. 703)*

C corporation Corporation that does not qualify for nor elect to be treated as a proprietorship or partnership for income tax purposes and therefore is subject to income taxes; also called *C corp. (p. 480)*

Call price Amount that must be paid to call and retire a callable preferred stock or a callable bond. *(p. 519)*

Callable bonds Bonds that give the issuer the option to retire them at a stated amount prior to maturity. *(p. 563)*

Callable preferred stock Preferred stock that the issuing corporation, at its option, may retire by paying the call price plus any dividends in arrears. *(p. 519)*

Canceled checks Checks that the bank has paid and deducted from the depositor's account. *(p. 328)*

Capital budgeting Process of analyzing alternative investments and deciding which assets to acquire or sell. *(p. 1024)*

Capital expenditures Additional costs of plant assets that provide material benefits extending beyond the current period; also called *balance sheet expenditures. (p. 402)*

Capital expenditures budget Plan that lists dollar amounts to be both received from disposal of plant assets and spent to purchase plant assets. *(p. 949)*

Capital leases Long-term leases in which the lessor transfers substantially all risk and rewards of ownership to the lessee. *(p. 574)*

Capital stock General term referring to a corporation's stock used in obtaining capital (owner financing). *(p. 509)*

Capitalize Record the cost as part of a permanent account and allocate it over later periods. *(p. 402)*

Carrying (book) value of bonds Net amount at which bonds are reported on the balance sheet; equals the par value of the bonds less any unamortized discount or plus any unamortized premium; also called *carrying amount* or *book value. (p. 554)*

Cash Includes currency, coins, and amounts on deposit in bank checking or savings accounts. *(p. 319)*

Cash basis accounting Accounting system that recognizes revenues when cash is received and records expenses when cash is paid. *(p. 93)*

Cash budget Plan that shows expected cash inflows and outflows during the budget period, including receipts from loans needed to maintain a minimum cash balance and repayments of such loans. *(p. 950)*

Cash disbursements journal Special journal normally used to record all payments of cash; also called *cash payments journal. (p. 280)*

Cash discount Reduction in the price of merchandise granted by a seller to a buyer when payment is made within the discount period. *(p. 181)*

Cash equivalents Short-term, investment assets that are readily convertible to a known cash amount or sufficiently close to their maturity date (usually within 90 days) so that market value is not sensitive to interest rate changes. *(p. 319)*

Cash flow on total assets Ratio of operating cash flows to average total assets; not sensitive to income recognition and measurement; partly reflects earnings quality. *(p. 646)*

Cash Over and Short Income statement account used to record cash overages and cash shortages arising from errors in cash receipts or payments. *(p. 321)*

Cash receipts journal Special journal normally used to record all receipts of cash. *(p. 277)*

Change in an accounting estimate Change in an accounting estimate that results from new information, subsequent developments, or improved judgment that impacts current and future periods. *(pp. 401 & 523)*

Chart of accounts List of accounts used by a company; includes an identification number for each account. *(p. 52)*

Check Document signed by a depositor instructing the bank to pay a specified amount to a designated recipient. *(p. 326)*

Check register Another name for a cash disbursements journal when the journal has a column for check numbers. *(pp. 280 & 337)*

Classified balance sheet Balance sheet that presents assets and liabilities in relevant subgroups, including current and noncurrent classifications. *(p. 145)*

Clock card Source document used to record the number of hours an employee works and to determine the total labor cost for each pay period. *(p. 776)*

Closing entries Entries recorded at the end of each accounting period to transfer end-of-period balances in revenue, gain, expense, loss, and withdrawal (dividend for a corporation) accounts to the capital account (to retained earnings for a corporation). *(p. 141)*

Closing process Necessary end-of-period steps to prepare the accounts for recording the transactions of the next period. *(p. 140)*

Columnar journal Journal with more than one column. *(p. 274)*

Common stock Corporation's basic ownership share; also generically called *capital stock. (pp. 11 & 508)*

Common-size financial statement Statement that expresses each amount as a percent of a base amount. In the balance sheet, total assets is usually the base and is expressed as 100%. In the income statement, net sales is usually the base. *(p. 687)*

Comparative financial statement Statement with data for two or more successive periods placed in side-by-side columns, often with changes shown in dollar amounts and percents. *(p. 682)*

Compatibility principle Information system principle that prescribes an accounting system to conform with a company's activities, personnel, and structure. *(p. 269)*

Complex capital structure Capital structure that includes outstanding rights or options to purchase common stock, or securities that are convertible into common stock. *(p. 524)*

Components of accounting systems Five basic components of accounting systems are source documents, input devices, information processors, information storage, and output devices. *(p. 269)*

Composite unit Generic unit consisting of a specific number of units of each product; unit comprised in proportion to the expected sales mix of its products. *(p. 916)*

Compound journal entry Journal entry that affects at least three accounts. *(p. 61)*

Comprehensive income Net change in equity for a period, excluding owner investments and distributions. *(p. 600)*

Computer hardware Physical equipment in a computerized accounting information system.

Computer network Linkage giving different users and different computers access to common databases and programs. *(p. 283)*

Computer software Programs that direct operations of computer hardware.

Conservatism constraint Principle that prescribes the less optimistic estimate when two estimates are about equally likely. *(p. 234)*

Consignee Receiver of goods owned by another who holds them for purposes of selling them for the owner. *(p. 224)*

Consignor Owner of goods who ships them to another party who will sell them for the owner. *(p. 224)*

Consistency concept Principle that prescribes use of the same accounting method(s) over time so that financial statements are comparable across periods. *(p. 233)*

Consolidated financial statements Financial statements that show all (combined) activities under the parent's control, including those of any subsidiaries. *(p. 599)*

Contingent liability Obligation to make a future payment if, and only if, an uncertain future event occurs. *(p. 445)*

Continuous budgeting Practice of preparing budgets for a selected number of future periods and revising those budgets as each period is completed. *(p. 942)*

Continuous improvement Concept requiring every manager and employee continually to look to improve operations. *(p. 730)*

Contra account Account linked with another account and having an opposite normal balance; reported as a subtraction from the other account's balance. *(p. 97)*

Contract rate Interest rate specified in a bond indenture (or note); multiplied by the par value to determine the interest paid each period; also called *coupon rate, stated rate,* or *nominal rate. (p. 553)*

Contributed capital Total amount of cash and other assets received from stockholders in exchange for stock; also called *paid-in capital. (p. 13)*

Contributed capital in excess of par value Difference between the par value of stock and its issue price when issued at a price above par.

Contribution margin Sales revenue less total variable costs.

Contribution margin income statement Income statement that separates variable and fixed costs; highlights the contribution margin, which is sales less variable expenses.

Contribution margin per unit Amount that the sale of one unit contributes toward recovering fixed costs and earning profit; defined as sales price per unit minus variable expense per unit. *(p. 908)*

Contribution margin ratio Product's contribution margin divided by its sale price. *(p. 908)*

Control Process of monitoring planning decisions and evaluating the organization's activities and employees. *(p. 727)*

Control principle Information system principle that prescribes an accounting system to aid managers in controlling and monitoring business activities. *(p. 268)*

Controllable costs Costs that a manager has the power to control or at least strongly influence. *(pp. 733 & 870)*

Controllable variance Combination of both overhead spending variances (variable and fixed) and the variable overhead efficiency variance. *(p. 993)*

Controlling account General ledger account, the balance of which (after posting) equals the sum of the balances in its related subsidiary ledger. *(p. 273)*

Conversion costs Expenditures incurred in converting raw materials to finished goods; includes direct labor costs and overhead costs. *(p. 738)*

Convertible bonds Bonds that bondholders can exchange for a set number of the issuer's shares. *(p. 563)*

Convertible preferred stock Preferred stock with an option to exchange it for common stock at a specified rate. *(p. 518)*

Copyright Right giving the owner the exclusive privilege to publish and sell musical, literary, or artistic work during the creator's life plus 70 years. *(p. 408)*

Corporation Business that is a separate legal entity under state or federal laws with owners called *shareholders* or *stockholders. (pp. 11 & 506)*

Cost All normal and reasonable expenditures necessary to get an asset in place and ready for its intended use. *(p. 393)*

Cost accounting system Accounting system for manufacturing activities based on the perpetual inventory system. *(p. 770)*

Cost-benefit principle Information system principle that prescribes the benefits from an activity in an accounting system to outweigh the costs of that activity. *(p. 269)*

Cost center Department that incurs costs but generates no revenues; common example is the accounting or legal department. *(p. 858)*

Cost object Product, process, department, or customer to which costs are assigned. *(p. 732)*

Cost of goods available for sale Consists of beginning inventory plus net purchases of a period.

Cost of goods manufactured Total manufacturing costs (direct materials, direct labor, and factory overhead) for the period plus beginning goods in process less ending goods in process; also called *net cost of goods manufactured* and *cost of goods completed. (p. 821)*

Cost of goods sold Cost of inventory sold to customers during a period; also called *cost of sales. (p. 178)*

Cost principle Accounting principle that prescribes financial statement information to be based on actual costs incurred in business transactions. *(p. 9)*

Cost variance Difference between the actual incurred cost and the standard cost. *(p. 986)*

Cost-volume-profit (CVP) analysis Planning method that includes predicting the volume of activity, the costs incurred, sales earned, and profits received. *(p. 902)*

Cost-volume-profit (CVP) chart Graphic representation of cost-volume-profit relations. *(p. 910)*

Coupon bonds Bonds with interest coupons attached to their certificates; bondholders detach coupons when they mature and present them to a bank or broker for collection. *(p. 563)*

Credit Recorded on the right side; an entry that decreases asset and expense accounts, and increases liability, revenue, and most equity accounts; abbreviated Cr. *(p. 53)*

Credit memorandum Notification that the sender has credited the recipient's account in the sender's records. *(p. 187)*

Credit period Time period that can pass before a customer's payment is due. *(p. 181)*

Credit terms Description of the amounts and timing of payments that a buyer (debtor) agrees to make in the future. *(p. 181)*

Creditors Individuals or organizations entitled to receive payments. *(p. 50)*

Cumulative preferred stock Preferred stock on which undeclared dividends accumulate until paid; common stockholders cannot receive dividends until cumulative dividends are paid. *(p. 517)*

Current assets Cash and other assets expected to be sold, collected, or used within one year or the company's operating cycle, whichever is longer. *(p. 146)*

Current liabilities Obligations due to be paid or settled within one year or the company's operating cycle, whichever is longer. *(pp. 147 & 435)*

Current portion of long-term debt Portion of long-term debt due within one year or the operating cycle, whichever is longer; reported under current liabilities. *(p. 442)*

Current ratio Ratio used to evaluate a company's ability to pay its short-term obligations, calculated by dividing current assets by current liabilities. *(p. 147)*

Curvilinear cost Cost that changes with volume but not at a constant rate. *(p. 904)*

Customer orientation Company position that its managers and employees be in tune with the changing wants and needs of consumers. *(p. 729)*

Cycle efficiency (CE) A measure of production efficiency, which is defined as value-added (process) time divided by total cycle time. *(p. 743)*

Cycle time (CT) A measure of the time to produce a product or service, which is the sum of process time, inspection time, move time, and wait time; also called *throughput time. (p. 742)*

Date of declaration Date the directors vote to pay a dividend. *(p. 513)*

Date of payment Date the corporation makes the dividend payment. *(p. 513)*

Date of record Date directors specify for identifying stockholders to receive dividends. *(p. 513)*

Days' sales in inventory Estimate of number of days needed to convert inventory into receivables or cash; equals ending inventory divided by cost of goods sold and then multiplied by 365; also called *days' stock on hand. (p. 237)*

Days' sales uncollected Measure of the liquidity of receivables computed by dividing the current balance of receivables by the annual credit (or net) sales and then multiplying by 365; also called *days' sales in receivables. (p. 332)*

Debit Recorded on the left side; an entry that increases asset and expense accounts, and decreases liability, revenue, and most equity accounts; abbreviated Dr. *(p. 53)*

Debit memorandum Notification that the sender has debited the recipient's account in the sender's records. *(p. 182)*

Debt ratio Ratio of total liabilities to total assets; used to reflect risk associated with a company's debts. *(p. 67)*

Debt-to-equity ratio Defined as total liabilities divided by total equity; shows the proportion of a company financed by non-owners (creditors) in comparison with that financed by owners. *(p. 564)*

Debtors Individuals or organizations that owe money. *(p. 49)*

Declining-balance method Method that determines depreciation charge for the period by multiplying a depreciation rate (often twice the straight-line rate) by the asset's beginning-period book value. *(p. 398)*

Deferred income tax liability Corporation income taxes that are deferred until future years because of temporary differences between GAAP and tax rules. *(p. 457)*

Degree of operating leverage (DOL) Ratio of contribution margin divided by pretax income; used to assess the effect on income of changes in sales. *(p. 918)*

Departmental accounting system Accounting system that provides information useful in evaluating the profitability or cost effectiveness of a department. *(p. 858)*

Departmental contribution to overhead Amount by which a department's revenues exceed its direct expenses. *(p. 866)*

Depletion Process of allocating the cost of natural resources to periods when they are consumed and sold. *(p. 406)*

Deposit ticket Lists items such as currency, coins, and checks deposited and their corresponding dollar amounts. *(p. 326)*

Deposits in transit Deposits recorded by the company but not yet recorded by its bank. *(p. 329)*

Depreciable cost Cost of a plant asset less its salvage value.

Depreciation Expense created by allocating the cost of plant and equipment to periods in which they are used; represents the expense of using the asset. *(pp. 97 & 395)*

Diluted earnings per share Earnings per share calculation that requires dilutive securities be added to the denominator of the basic EPS calculation. *(p. 524)*

Dilutive securities Securities having the potential to increase common shares outstanding; examples are options, rights, convertible bonds, and convertible preferred stock. *(p. 524)*

Direct costs Costs incurred for the benefit of one specific cost object. *(p. 732)*

Direct expenses Expenses traced to a specific department (object) that are incurred for the sole benefit of that department. *(p. 859)*

Direct labor Efforts of employees who physically convert materials to finished product. *(p. 738)*

Direct labor costs Wages and salaries for direct labor that are separately and readily traced through the production process to finished goods. *(p. 738)*

Direct material Raw material that physically becomes part of the product and is clearly identified with specific products or batches of product. *(p. 738)*

Direct material costs Expenditures for direct material that are separately and readily traced through the production process to finished goods. *(p. 738)*

Direct method Presentation of net cash from operating activities for the statement of cash flows that lists major operating cash receipts less major operating cash payments. *(p. 634)*

Direct write-off method Method that records the loss from an uncollectible account receivable at the time it is determined to be uncollectible; no attempt is made to estimate bad debts. *(p. 361)*

Discount on bonds payable Difference between a bond's par value and its lower issue price or carrying value; occurs when the contract rate is less than the market rate. *(p. 553)*

Discount on note payable Difference between the face value of a note payable and the (lesser) amount borrowed; reflects the added interest to be paid on the note over its life.

Discount on stock Difference between the par value of stock and its issue price when issued at a price below par value. *(p. 511)*

Discount period Time period in which a cash discount is available and the buyer can make a reduced payment. *(p. 181)*

Discount rate Expected rate of return on investments; also called *cost of capital, hurdle rate,* or *required rate of return. (p. B-2)*

Discounts lost Expenses resulting from not taking advantage of cash discounts on purchases. *(p. 338)*

Dividend in arrears Unpaid dividend on cumulative preferred stock; must be paid before any regular dividends on preferred stock and before any dividends on common stock. *(p. 517)*

Dividend yield Ratio of the annual amount of cash dividends distributed to common shareholders relative to the common stock's market value (price). *(p. 525)*

Dividends Corporation's distributions of assets to its owners.

Double-declining-balance (DDB) depreciation Depreciation equals beginning book value multiplied by 2 times the straight-line rate.

Double-entry accounting Accounting system in which each transaction affects at least two accounts and has at least one debit and one credit. *(p. 53)*

Double taxation Corporate income is taxed and then its later distribution through dividends is normally taxed again for shareholders. *(p. 11)*

Earnings (See *net income.*)

Earnings per share (EPS) Amount of income earned by each share of a company's outstanding common stock; also called *net income per share. (p. 524)*

Effective interest method Allocates interest expense over the bond life to yield a constant rate of interest; interest expense for a period is found by multiplying the balance of the liability at the beginning of the period by the bond market rate at issuance; also called *interest method. (p. 569)*

Efficiency Company's productivity in using its assets; usually measured relative to how much revenue a certain level of assets generates. *(p. 681)*

Efficiency variance Difference between the actual quantity of an input and the standard quantity of that input. *(p. 993)*

Electronic funds transfer (EFT) Use of electronic communication to transfer cash from one party to another. *(p. 327)*

Employee benefits Additional compensation paid to or on behalf of employees, such as premiums for medical, dental, life, and disability insurance, and contributions to pension plans. *(p. 443)*

Employee earnings report Record of an employee's net pay, gross pay, deductions, and year-to-date payroll information. *(p. 454)*

Enterprise resource planning (ERP) software Programs that manage a company's vital operations, which range from order taking to production to accounting. *(p. 283)*

Entity Organization that, for accounting purposes, is separate from other organizations and individuals. *(p. 10)*

EOM Abbreviation for *end of month;* used to describe credit terms for credit transactions. *(p. 181)*

Equity Owner's claim on the assets of a business; equals the residual interest in an entity's assets after deducting liabilities; also called *net assets. (p. 13)*

Equity method Accounting method used for long-term investments when the investor has "significant influence" over the investee. *(p. 598)*

Equity ratio Portion of total assets provided by equity, computed as total equity divided by total assets. *(p. 695)*

Equity securities with controlling influence Long-term investment when the investor is able to exert controlling influence over the investee; investors owning 50% or more of voting stock are presumed to exert controlling influence. *(p. 599)*

Equity securities with significant influence Long-term investment when the investor is able to exert significant influence over the investee; investors owning 20 percent or more (but less than 50 percent) of voting stock are presumed to exert significant influence. *(p. 598)*

Equivalent units of production (EUP) Number of units that would be completed if all effort during a period had been applied to units that were started and finished. *(p. 815)*

Estimated liability Obligation of an uncertain amount that can be reasonably estimated. *(p. 443)*

Estimated line of cost behavior Line drawn on a graph to visually fit the relation between cost and sales. *(p. 906)*

Ethics Codes of conduct by which actions are judged as right or wrong, fair or unfair, honest or dishonest. *(pp. 8, 731)*

Events Happenings that both affect an organization's financial position and can be reliably measured. *(p. 14)*

Expanded accounting equation Assets = Liabilities + Equity; Equity equals [Owner capital − Owner withdrawals + Revenues − Expenses] for a noncorporation; Equity equals [Contributed capital + Retained earnings + Revenues − Expenses] for a corporation where dividends are subtracted from retained earnings. *(p. 13)*

Expenses Outflows or using up of assets as part of operations of a business to generate sales. *(p. 13)*

External transactions Exchanges of economic value between one entity and another entity. *(p. 13)*

External users Persons using accounting information who are not directly involved in running the organization. *(p. 5)*

Extraordinary gains or losses Gains or losses reported separately from continuing operations because they are both unusual and infrequent. *(p. 703)*

Extraordinary repairs Major repairs that extend the useful life of a plant asset beyond prior expectations; treated as a capital expenditure. *(p. 403)*

Factory overhead Factory activities supporting the production process that are not direct material or direct labor; also called *overhead* and *manufacturing overhead. (p. 738)*

Factory overhead costs Expenditures for factory overhead that cannot be separately or readily traced to finished goods; also called *overhead costs.* *(p. 738)*

Favorable variance Difference in actual revenues or expenses from the budgeted amount that contributes to a higher income. *(p. 981)*

Federal depository bank Bank authorized to accept deposits of amounts payable to the federal government. *(p. 452)*

Federal Insurance Contributions Act (FICA) Taxes Taxes assessed on both employers and employees; for Social Security and Medicare programs. *(p. 440)*

Federal Unemployment Taxes (FUTA) Payroll taxes on employers assessed by the federal government to support its unemployment insurance program. *(p. 442)*

FIFO method (See *first-in, first-out.*)

Financial accounting Area of accounting mainly aimed at serving external users. *(p. 5)*

Financial Accounting Standards Board (FASB) Independent group of full-time members responsible for setting accounting rules. *(p. 9)*

Financial leverage Earning a higher return on equity by paying dividends on preferred stock or interest on debt at a rate lower than the return earned with the assets from issuing preferred stock or debt; also called *trading on the equity.* *(p. 519)*

Financial reporting Process of communicating information relevant to investors, creditors, and others in making investment, credit, and business decisions. *(p. 681)*

Financial statement analysis Application of analytical tools to general-purpose financial statements and related data for making business decisions. *(p. 680)*

Financial statements Includes the balance sheet, income statement, statement of owner's (or stockholders') equity, and statement of cash flows. *(p. 17)*

Financing activities Transactions with owners and creditors that include obtaining cash from issuing debt, repaying amounts borrowed, and obtaining cash from or distributing cash to owners. *(p. 630)*

Finished goods inventory Account that controls the finished goods files, which acts as a subsidiary ledger (of the Inventory account) in which the costs of finished goods that are ready for sale are recorded. *(pp. 736 & 772)*

First-in, first-out (FIFO) Method to assign cost to inventory that assumes items are sold in the order acquired; earliest items purchased are the first sold. *(pp. 229, 827)*

Fiscal year Consecutive 12-month (or 52-week) period chosen as the organization's annual accounting period. *(p. 93)*

Fixed budget Planning budget based on a single predicted amount of volume; unsuitable for evaluations if the actual volume differs from predicted volume. *(p. 981)*

Fixed budget performance report Report that compares actual revenues and costs with fixed budgeted amounts and identifies the differences as favorable or unfavorable variances. *(p. 981)*

Fixed cost Cost that does not change with changes in the volume of activity. *(p. 732)*

Flexibility principle Information system principle that prescribes an accounting system be able to adapt to changes in the company, its operations, and needs of decision makers. *(p. 269)*

Flexible budget Budget prepared (using actual volume) once a period is complete that helps managers evaluate past performance; uses fixed and variable costs in determining total costs. *(p. 982)*

Flexible budget performance report Report that compares actual revenues and costs with their variable budgeted amounts based on actual sales volume (or other level of activity) and identifies the differences as variances. *(p. 958)*

FOB Abbreviation for *free on board;* the point when ownership of goods passes to the buyer; *FOB shipping point* (or *factory*) means the buyer pays shipping costs and accepts ownership of goods when the seller transfers goods to carrier; *FOB destination* means the seller pays shipping costs and buyer accepts ownership of goods at the buyer's place of business. *(p. 183)*

Foreign exchange rate Price of one currency stated in terms of another currency. *(p. 605)*

Form 940 IRS form used to report an employer's federal unemployment taxes (FUTA) on an annual filing basis. *(p. 452)*

Form 941 IRS form filed to report FICA taxes owed and remitted. *(p. 450)*

Form 10-K (or 10-KSB) Annual report form filed with SEC by businesses (small businesses) with publicly traded securities. *(p. A-1)*

Form W-2 Annual report by an employer to each employee showing the employee's wages subject to FICA and federal income taxes along with amounts withheld. *(p. 452)*

Form W-4 Withholding allowance certificate, filed with the employer, identifying the number of withholding allowances claimed. *(p. 454)*

Franchises Privileges granted by a company or government to sell a product or service under specified conditions. *(p. 408)*

Full disclosure principle Principle that prescribes financial statements (including notes) to report all relevant information about an entity's operations and financial condition. *(p. 10)*

GAAP (See *generally accepted accounting principles.*)

General accounting system Accounting system for manufacturing activities based on the *periodic* inventory system. *(p. 770)*

General and administrative expenses Expenses that support the operating activities of a business. *(p. 191)*

General and administrative expense budget Plan that shows predicted operating expenses not included in the selling expenses budget. *(p. 948)*

General journal All-purpose journal for recording the debits and credits of transactions and events. *(pp. 54 & 272)*

General ledger (See *ledger.*)

General partner Partner who assumes unlimited liability for the debts of the partnership; responsible for partnership management. *(p. 479)*

General partnership Partnership in which all partners have mutual agency and unlimited liability for partnership debts. *(p. 479)*

Generally accepted accounting principles (GAAP) Rules that specify acceptable accounting practices. *(p. 8)*

Generally accepted auditing standards (GAAS) Rules that specify acceptable auditing practices.

General-purpose financial statements Statements published periodically for use by a variety of interested parties; includes the income statement, balance sheet, statement of owner's equity (or statement of retained earnings for a corporation), statement of cash flows, and notes to these statements. *(p. 681)*

Going-concern assumption Principle that prescribes financial statements to reflect the assumption that the business will continue operating. *(p. 10)*

Goods in process inventory Account in which costs are accumulated for products that are in the process of being produced but are not yet complete; also called *work in process inventory. (pp. 736 & 772)*

Goodwill Amount by which a company's (or a segment's) value exceeds the value of its individual assets less its liabilities. *(p. 409)*

Gross margin (See *gross profit.*)

Gross margin ratio Gross margin (net sales minus cost of goods sold) divided by net sales; also called *gross profit ratio. (p. 193)*

Gross method Method of recording purchases at the full invoice price without deducting any cash discounts. *(p. 338)*

Gross pay Total compensation earned by an employee. *(p. 440)*

Gross profit Net sales minus cost of goods sold; also called *gross margin. (p. 178)*

Gross profit method Procedure to estimate inventory when the past gross profit rate is used to estimate cost of goods sold, which is then subtracted from the cost of goods available for sale. *(p. 248)*

Held-to-maturity (HTM) securities Debt securities that a company has the intent and ability to hold until they mature. *(p. 596)*

High-low method Procedure that yields an estimated line of cost behavior by graphically connecting costs associated with the highest and lowest sales volume. *(p. 906)*

Horizontal analysis Comparison of a company's financial condition and performance across time. *(p. 682)*

Hurdle rate Minimum acceptable rate of return (set by management) for an investment. *(pp. 868 & 1033)*

Impairment Diminishment of an asset value. *(pp. 402 & 407)*

Imprest system Method to account for petty cash; maintains a constant balance in the fund, which equals cash plus petty cash receipts.

Inadequacy Condition in which the capacity of plant assets is too small to meet the company's production demands. *(p. 395)*

Income (See *net income.*)

Income statement Financial statement that subtracts expenses from revenues to yield a net income or loss over a specified period of time; also includes any gains or losses. *(p. 17)*

Income Summary Temporary account used only in the closing process to which the balances of revenue and expense accounts (including any gains or losses) are transferred; its balance is transferred to the capital account (or retained earnings for a corporation). *(p. 141)*

Incremental cost Additional cost incurred only if a company pursues a specific course of action. *(p. 1036)*

Indefinite life Asset life that is not limited by legal, regulatory, contractual, competitive, economic, or other factors. *(p. 407)*

Indirect costs Costs incurred for the benefit of more than one cost object. *(p. 732)*

Indirect expenses Expenses incurred for the joint benefit of more than one department (or cost object). *(p. 859)*

Indirect labor Efforts of production employees who do not work specifically on converting direct materials into finished products and who are not clearly identified with specific units or batches of product. *(p. 738)*

Indirect labor costs Labor costs that cannot be physically traced to production of a product or service; included as part of overhead. *(p. 738)*

Indirect material Material used to support the production process but not clearly identified with products or batches of product. *(p. 735)*

Indirect method Presentation that reports net income and then adjusts it by adding and subtracting items to yield net cash from operating activities on the statement of cash flows. *(p. 634)*

Information processor Component of an accounting system that interprets, transforms, and summarizes information for use in analysis and reporting. *(p. 270)*

Information storage Component of an accounting system that keeps data in a form accessible to information processors. *(p. 270)*

Infrequent gain or loss Gain or loss not expected to recur given the operating environment of the business. *(p. 703)*

Input device Means of capturing information from source documents that enables its transfer to information processors. *(p. 270)*

Installment note Liability requiring a series of periodic payments to the lender. *(p. 560)*

Institute of Management Accountants (IMA) A professional association of management accountants. *(p. 731)*

Intangible assets Long-term assets (resources) used to produce or sell products or services; usually lack physical form and have uncertain benefits. *(pp. 147 & 407)*

Interest Charge for using money (or other assets) loaned from one entity to another. *(p. 368)*

Interim financial statements Financial statements covering periods of less than one year; usually based on one-, three-, or six-month periods. *(pp. 92 & 247)*

Internal controls or **Internal control system** All policies and procedures used to protect assets, ensure reliable accounting, promote efficient operations, and urge adherence to company policies. *(pp. 268 & 314)*

Internal rate of return (IRR) Rate used to evaluate the acceptability of an investment; equals the rate that yields a net present value of zero for an investment. *(p. 1031)*

Internal transactions Activities within an organization that can affect the accounting equation. *(p. 13)*

Internal users Persons using accounting information who are directly involved in managing the organization. *(p. 6)*

International Accounting Standards Board (IASB) Group that identifies preferred accounting practices and encourages global acceptance; issues International Financial Reporting Standards (IFRS). *(p. 9)*

Inventory Goods a company owns and expects to sell in its normal operations. *(p. 179)*

Inventory turnover Number of times a company's average inventory is sold during a period; computed by dividing cost of goods sold by average inventory; also called *merchandise turnover*. *(p. 236)*

Investing activities Transactions that involve purchasing and selling of long-term assets, includes making and collecting notes receivable and investments in other than cash equivalents. *(p. 630)*

Investment center Center of which a manager is responsible for revenues, costs, and asset investments. *(p. 858)*

Investment center residual income The net income an investment center earns above a target return on average invested assets. *(p. 867)*

Investment center return on total assets Center net income divided by average total assets for the center. *(p. 867)*

Investment turnover The efficiency with which a company generates sales from its available assets; computed as sales divided by average invested assets. *(p. 872)*

Invoice Itemized record of goods prepared by the vendor that lists the customer's name, items sold, sales prices, and terms of sale. *(p. 335)*

Invoice approval Document containing a checklist of steps necessary for approving the recording and payment of an invoice; also called *check authorization*. *(p. 336)*

Job Production of a customized product or service. *(p. 770)*

Job cost sheet Separate record maintained for each job. *(p. 772)*

Job lot Production of more than one unit of a customized product or service. *(p. 771)*

Job order cost accounting system Cost accounting system to determine the cost of producing each job or job lot. *(pp. 772 & 811)*

Job order production Production of special-order products; also called *customized production*. *(p. 770)*

Joint cost Cost incurred to produce or purchase two or more products at the same time. *(p. 877)*

Journal Record in which transactions are entered before they are posted to ledger accounts; also called *book of original entry*. *(p. 54)*

Journalizing Process of recording transactions in a journal. *(p. 54)*

Just-in-time (JIT) manufacturing Process of acquiring or producing inventory only when needed. *(p. 730)*

Known liabilities Obligations of a company with little uncertainty; set by agreements, contracts, or laws; also called *definitely determinable liabilities*. *(p. 436)*

Land improvements Assets that increase the benefits of land, have a limited useful life, and are depreciated. *(p. 394)*

Large stock dividend Stock dividend that is more than 25% of the previously outstanding shares. *(p. 514)*

Last-in, first-out (LIFO) Method to assign cost to inventory that assumes costs for the most recent items purchased are sold first and charged to cost of goods sold. *(p. 229)*

Lean business model Practice of eliminating waste while meeting customer needs and yielding positive company returns. *(p. 730)*

Lease Contract specifying the rental of property. *(pp. 409 & 573)*

Leasehold Rights the lessor grants to the lessee under the terms of a lease. *(p. 409)*

Leasehold improvements Alterations or improvements to leased property such as partitions and storefronts. *(p. 409)*

Least-squares regression Statistical method for deriving an estimated line of cost behavior that is more precise than the high-low method and the scatter diagram. *(p. 907)*

Ledger Record containing all accounts (with amounts) for a business; also called *general ledger*. *(p. 49)*

Lessee Party to a lease who secures the right to possess and use the property from another party (the lessor). *(p. 409)*

Lessor Party to a lease who grants another party (the lessee) the right to possess and use its property. *(p. 409)*

Liabilities Creditors' claims on an organization's assets; involves a probable future payment of assets, products, or services that a company is obligated to make due to past transactions or events. *(p. 12)*

Licenses (See *franchises*.) *(p. 408)*

Limited liability Owner can lose no more than the amount invested. *(p. 11)*

Limited liability company Organization form that combines select features of a corporation and a limited partnership; provides limited liability to its members (owners), is free of business tax, and allows members to actively participate in management. *(p. 480)*

Limited liability partnership Partnership in which a partner is not personally liable for malpractice or negligence unless that partner is responsible for providing the service that resulted in the claim. *(p. 479)*

Limited life (See *useful life*.)

Limited partners Partners who have no personal liability for partnership debts beyond the amounts they invested in the partnership. *(p. 479)*

Limited partnership Partnership that has two classes of partners, limited partners and general partners. *(p. 479)*

Liquid assets Resources such as cash that are easily converted into other assets or used to pay for goods, services, or liabilities. *(p. 319)*

Liquidating cash dividend Distribution of assets that returns part of the original investment to stockholders; deducted from contributed capital accounts. *(p. 514)*

Liquidation Process of going out of business; involves selling assets, paying liabilities, and distributing remainder to owners.

Liquidity Availability of resources to meet short-term cash requirements. *(pp. 319 & 681)*

List price Catalog (full) price of an item before any trade discount is deducted. *(p. 180)*

Long-term investments Long-term assets not used in operating activities such as notes receivable and investments in stocks and bonds. *(pp. 147 & 592)*

Long-term liabilities Obligations not due to be paid within one year or the operating cycle, whichever is longer. *(pp. 147 & 435)*

Lower of cost or market (LCM) Required method to report inventory at market replacement cost when that market cost is lower than recorded cost. *(p. 233)*

Maker of the note Entity who signs a note and promises to pay it at maturity. *(p. 368)*

Management by exception Management process to focus on significant variances and give less attention to areas where performance is close to the standard. *(p. 996)*

Managerial accounting Area of accounting mainly aimed at serving the decision-making needs of internal users; also called *management accounting. (pp. 6 & 726)*

Manufacturer Company that uses labor and operating assets to convert raw materials to finished goods. *(p. 13)*

Manufacturing budget Plan that shows the predicted costs for direct materials, direct labor, and overhead to be incurred in manufacturing units in the production budget. *(p. 959)*

Manufacturing statement Report that summarizes the types and amounts of costs incurred in a company's production process for a period; also called *cost of goods manufacturing statement. (p. 740)*

Margin of safety Excess of expected sales over the level of break-even sales. *(p. 914)*

Market prospects Expectations (both good and bad) about a company's future performance as assessed by users and other interested parties. *(p. 681)*

Market rate Interest rate that borrowers are willing to pay and lenders are willing to accept for a specific lending agreement given the borrowers' risk level. *(p. 553)*

Market value per share Price at which stock is bought or sold. *(p. 509)*

Master budget Comprehensive business plan that includes specific plans for expected sales, product units to be produced, merchandise (or materials) to be purchased, expenses to be incurred, plant assets to be purchased, and amounts of cash to be borrowed or loans to be repaid, as well as a budgeted income statement and balance sheet. *(p. 944)*

Matching principle Prescribes expenses to be reported in the same period as the revenues that were earned as a result of the expenses. *(pp. 10 & 362)*

Materiality constraint Prescribes that accounting for items that significantly impact financial statement and any inferences from them adhere strictly to GAAP. *(p. 362)*

Materials consumption report Document that summarizes the materials a department uses during a reporting period; replaces materials requisitions. *(p. 812)*

Materials ledger card Perpetual record updated each time units are purchased or issued for production use. *(p. 773)*

Materials requisition Source document production managers use to request materials for production; used to assign materials costs to specific jobs or overhead. *(p. 774)*

Maturity date of a note Date when a note's principal and interest are due. *(p. 368)*

Merchandise (See *merchandise inventory.*)

Merchandise inventory Goods that a company owns and expects to sell to customers; also called *merchandise* or *inventory. (p. 179)*

Merchandise purchases budget Plan that shows the units or costs of merchandise to be purchased by a merchandising company during the budget period. *(p. 947)*

Merchandiser Entity that earns net income by buying and selling merchandise. *(p. 178)*

Merit rating Rating assigned to an employer by a state based on the employer's record of employment. *(p. 442)*

Minimum legal capital Amount of assets defined by law that stockholders must (potentially) invest in a corporation; usually defined as par value of the stock; intended to protect creditors. *(p. 509)*

Mixed cost Cost that behaves like a combination of fixed and variable costs. *(p. 903)*

Modified Accelerated Cost Recovery System (MACRS) Depreciation system required by federal income tax law. *(p. 400)*

Monetary unit assumption Principle that assumes transactions and events can be expressed in money units. *(p. 10)*

Mortgage Legal loan agreement that protects a lender by giving the lender the right to be paid from the cash proceeds from the sale of a borrower's assets identified in the mortgage. *(p. 562)*

Multinational Company that operates in several countries. *(p. 605)*

Multiple-step income statement Income statement format that shows subtotals between sales and net income, categorizes expenses, and often reports the details of net sales and expenses. *(p. 191)*

Mutual agency Legal relationship among partners whereby each partner is an agent of the partnership and is able to bind the partnership to contracts within the scope of the partnership's business. *(p. 478)*

Natural business year Twelve-month period that ends when a company's sales activities are at their lowest point. *(p. 93)*

Natural resources Assets physically consumed when used; examples are timber, mineral deposits, and oil and gas fields; also called *wasting assets. (p. 406)*

Net assets (See *equity.*)

Net income Amount earned after subtracting all expenses necessary for and matched with sales for a period; also called *income, profit,* or *earnings. (p. 13)*

Net loss Excess of expenses over revenues for a period. *(p. 18)*

Net method Method of recording purchases at the full invoice price less any cash discounts. *(p. 338)*

Net pay Gross pay less all deductions; also called *take-home pay. (p. 440)*

Net present value (NPV) Dollar estimate of an asset's value that is used to evaluate the acceptability of an investment; computed by discounting future cash flows from the investment at a satisfactory rate and then subtracting the initial cost of the investment. *(p. 1029)*

Net realizable value Expected selling price (value) of an item minus the cost of making the sale. *(p. 224)*

Noncumulative preferred stock Preferred stock on which the right to receive dividends is lost for any period when dividends are not declared. *(p. 517)*

Noninterest-bearing note Note with no stated (contract) rate of interest; interest is implicitly included in the note's face value.

Nonparticipating preferred stock Preferred stock on which dividends are limited to a maximum amount each year. *(p. 518)*

No-par value stock Stock class that has not been assigned a par (or stated) value by the corporate charter. *(p. 509)*

Nonsufficient funds (NSF) check Maker's bank account has insufficient money to pay the check; also called *hot check.*

Non-value-added time The portion of cycle time that is not directed at producing a product or service; equals the sum of inspection time, move time, and wait time. *(p. 743)*

Not controllable costs Costs that a manager does not have the power to control or strongly influence. *(p. 733)*

Note (See promissory note.)

Note payable Liability expressed by a written promise to pay a definite sum of money on demand or on a specific future date(s).

Note receivable Asset consisting of a written promise to receive a definite sum of money on demand or on a specific future date(s).

Objectivity principle Principle that prescribes independent, unbiased evidence to support financial statement information. *(p. 9)*

Obsolescence Condition in which, because of new inventions and improvements, a plant asset can no longer be used to produce goods or services with a competitive advantage. *(p. 395)*

Off-balance-sheet financing Acquisition of assets by agreeing to liabilities not reported on the balance sheet. *(p. 574)*

Online processing Approach to inputting data from source documents as soon as the information is available. *(p. 282)*

Operating activities Activities that involve the production or purchase of merchandise and the sale of goods or services to customers, including expenditures related to administering the business. *(p. 629)*

Operating cycle Normal time between paying cash for merchandise or employee services and receiving cash from customers. *(p. 145)*

Operating leases Short-term (or cancelable) leases in which the lessor retains risks and rewards of ownership. *(p. 573)*

Operating leverage Extent, or relative size, of fixed costs in the total cost structure. *(p. 918)*

Opportunity cost Potential benefit lost by choosing a specific action from two or more alternatives. *(p. 733)*

Ordinary repairs Repairs to keep a plant asset in normal, good operating condition; treated as a revenue expenditure and immediately expensed. *(p. 402)*

Organization expenses (costs) Costs such as legal fees and promoter fees to bring an entity into existence. *(pp. 507 & 512)*

Out-of-pocket cost Cost incurred or avoided as a result of management's decisions. *(p. 733)*

Output devices Means by which information is taken out of the accounting system and made available for use. *(p. 271)*

Outsourcing Manager decision to buy a product or service from another entity; part of a *make-or-buy* decision; also called *make or buy.*

Outstanding checks Checks written and recorded by the depositor but not yet paid by the bank at the bank statement date. *(p. 329)*

Outstanding stock Corporation's stock held by its shareholders.

Overapplied overhead Amount by which the overhead applied to production in a period using the predetermined overhead rate exceeds the actual overhead incurred in a period. *(p. 782)*

Overhead cost variance Difference between the total overhead cost applied to products and the total overhead cost actually incurred. *(p. 992)*

Owner, Capital Account showing the owner's claim on company assets; equals owner investments plus net income (or less net losses) minus owner withdrawals since the company's inception; also referred to as *equity. (p. 13)*

Owner investment Assets put into the business by the owner. *(p. 13)*

Owner's equity (See *equity.*)

Owner, Withdrawals Account used to record asset distributions to the owner. (See also *withdrawals.*) *(pp. 13 & 51)*

Paid-in capital (See *contributed capital.*) *(p. 510)*

Paid-in capital in excess of par value Amount received from issuance of stock that is in excess of the stock's par value. *(p. 511)*

Par value Value assigned a share of stock by the corporate charter when the stock is authorized. *(p. 509)*

Par value of a bond Amount the bond issuer agrees to pay at maturity and the amount on which cash interest payments are based; also called *face amount* or *face value* of a bond. *(p. 550)*

Par value stock Class of stock assigned a par value by the corporate charter. *(p. 509)*

Parent Company that owns a controlling interest in a corporation (requires more than 50% of voting stock). *(p. 599)*

Participating preferred stock Preferred stock that shares with common stockholders any dividends paid in excess of the percent stated on preferred stock. *(p. 518)*

Partner return on equity Partner net income divided by average partner equity for the period. *(p. 490)*

Partnership Unincorporated association of two or more persons to pursue a business for profit as co-owners. *(pp. 10 & 478)*

Partnership contract Agreement among partners that sets terms under which the affairs of the partnership are conducted; also called *articles of partnership. (p. 478)*

Partnership liquidation Dissolution of a partnership by (1) selling noncash assets and allocating any gain or loss according to partners' income-and-loss ratio, (2) paying liabilities, and (3) distributing any remaining cash according to partners' capital balances. *(p. 488)*

Patent Exclusive right granted to its owner to produce and sell an item or to use a process for 20 years. *(p. 408)*

Payback period (PBP) Time-based measurement used to evaluate the acceptability of an investment; equals the time expected to pass before an investment's net cash flows equal its initial cost. *(p. 1025)*

Payee of the note Entity to whom a note is made payable. *(p. 368)*

Payroll bank account Bank account used solely for paying employees; each pay period an amount equal to the total employees' net pay is deposited in it and the payroll checks are drawn on it. *(p. 454)*

Payroll deductions Amounts withheld from an employee's gross pay; also called *withholdings. (p. 440)*

Payroll register Record for a pay period that shows the pay period dates, regular and overtime hours worked, gross pay, net pay, and deductions. *(p. 452)*

Pension plan Contractual agreement between an employer and its employees for the employer to provide benefits to employees after they retire; expensed when incurred. *(p. 575)*

Period costs Expenditures identified more with a time period than with finished products costs; includes selling and general administrative expenses. *(p. 733)*

Periodic inventory system Method that records the cost of inventory purchased but does not continuously track the quantity available or sold to customers; records are updated at the end of each period to reflect the physical count and costs of goods available. *(p. 180)*

Permanent accounts Accounts that reflect activities related to one or more future periods; balance sheet accounts whose balances are not closed; also called *real accounts. (p. 140)*

Perpetual inventory system Method that maintains continuous records of the cost of inventory available and the cost of goods sold. *(p. 180)*

Petty cash Small amount of cash in a fund to pay minor expenses; accounted for using an imprest system. *(p. 323)*

Planning Process of setting goals and preparing to achieve them. *(p. 726)*

Plant assets Tangible long-lived assets used to produce or sell products and services; also called *property, plant and equipment (PP&E)* or *fixed assets. (p. 403)*

Pledged assets to secured liabilities Ratio of the book value of a company's pledged assets to the book value of its secured liabilities.

Post-closing trial balance List of permanent accounts and their balances from the ledger after all closing entries are journalized and posted. *(p. 144)*

Posting Process of transferring journal entry information to the ledger; computerized systems automate this process. *(p. 54)*

Posting reference (PR) column A column in journals in which individual ledger account numbers are entered when entries are posted to those ledger accounts. *(p. 56)*

Predetermined overhead rate Rate established prior to the beginning of a period that relates estimated overhead to another variable, such as estimated direct labor, and is used to assign overhead cost to production. *(p. 778)*

Preemptive right Stockholders' right to maintain their proportionate interest in a corporation with any additional shares issued. *(p. 508)*

Preferred stock Stock with a priority status over common stockholders in one or more ways, such as paying dividends or distributing assets. *(p. 516)*

Premium on bonds Difference between a bond's par value and its higher carrying value; occurs when the contract rate is higher than the market rate; also called *bond premium. (p. 556)*

Premium on stock (See *contributed capital in excess of par value.*) *(p. 511)*

Prepaid expenses Items paid for in advance of receiving their benefits; classified as assets. *(p. 95)*

Price-earnings (PE) ratio Ratio of a company's current market value per share to its earnings per share; also called *price-to-earnings. (p. 525)*

Price variance Difference between actual and budgeted revenue or cost caused by the difference between the actual price per unit and the budgeted price per unit. *(p. 984)*

Prime costs Expenditures directly identified with the production of finished goods; include direct materials costs and direct labor costs. *(p. 738)*

Principal of a note Amount that the signer of a note agrees to pay back when it matures, not including interest. *(p. 368)*

Principles of internal control Principles prescribing management to establish responsibility, maintain records, insure assets, separate recordkeeping from custody of assets, divide responsibility for related transactions, apply technological controls, and perform reviews. *(p. 315)*

Prior period adjustment Correction of an error in a prior year that is reported in the statement of retained earnings (or statement of stockholders' equity) net of any income tax effects. *(p. 523)*

Pro forma financial statements Statements that show the effects of proposed transactions and events as if they had occurred. *(p. 140)*

Process cost accounting system System of assigning direct materials, direct labor, and overhead to specific processes; total costs associated with each process are then divided by the number of units passing through that process to determine the cost per equivalent unit. *(p. 811)*

Process cost summary Report of costs charged to a department, its equivalent units of production achieved, and the costs assigned to its output. *(p. 820)*

Process operations Processing of products in a continuous (sequential) flow of steps; also called *process manufacturing* or *process production. (p. 808)*

Product costs Costs that are capitalized as inventory because they produce benefits expected to have future value; include direct materials, direct labor, and overhead. *(p. 733)*

Production budget Plan that shows the units to be produced each period. *(p. 959)*

Profit (See *net income.*)

Profit center Business unit that incurs costs and generates revenues. *(p. 858)*

Profit margin Ratio of a company's net income to its net sales; the percent of income in each dollar of revenue; also called *net profit margin. (pp. 106 & 872)*

Profitability Company's ability to generate an adequate return on invested capital. *(p. 681)*

Profitability index A measure of the relation between the expected benefits of a project and its investment, computed as the present value of expected future cash flows from the investment divided by the cost of the investment; a higher value indicates a more desirable investment, and a value below 1 indicates an unacceptable project. *(p. 1031)*

Promissory note (or **note**) Written promise to pay a specified amount either on demand or at a definite future date; is a *note receivable* for the lender but a *note payable* for the lendee. *(p. 368)*

Proprietorship (See *sole proprietorship.*)

Proxy Legal document giving a stockholder's agent the power to exercise the stockholder's voting rights. *(p. 507)*

Purchase discount Term used by a purchaser to describe a cash discount granted to the purchaser for paying within the discount period. *(p. 181)*

Purchase order Document used by the purchasing department to place an order with a seller (vendor). *(p. 335)*

Purchase requisition Document listing merchandise needed by a department and requesting it be purchased. *(p. 335)*

Purchases journal Journal normally used to record all purchases on credit. *(p. 279)*

Quantity variance Difference between actual and budgeted revenue or cost caused by the difference between the actual number of units and the budgeted number of units. *(p. 984)*

Ratio analysis Determination of key relations between financial statement items as reflected in numerical measures. *(p. 682)*

Raw materials inventory Goods a company acquires to use in making products. *(p. 735)*

Realizable value Expected proceeds from converting an asset into cash. *(p. 363)*

Receiving report Form used to report that ordered goods are received and to describe their quantity and condition. *(p. 336)*

Recordkeeping Part of accounting that involves recording transactions and events, either manually or electronically; also called *bookkeeping*. *(p. 4)*

Registered bonds Bonds owned by investors whose names and addresses are recorded by the issuer; interest payments are made to the registered owners. *(p. 563)*

Relevance principle Information system principle prescribing that its reports be useful, understandable, timely, and pertinent for decision making. *(p. 268)*

Relevant benefits Additional or incremental revenue generated by selecting a particular course of action over another. *(p. 1035)*

Relevant range of operations Company's normal operating range; excludes extremely high and low volumes not likely to occur. *(p. 911)*

Report form balance sheet Balance sheet that lists accounts vertically in the order of assets, liabilities, and equity. *(p. 18)*

Responsibility accounting budget Report of expected costs and expenses under a manager's control. *(p. 870)*

Responsibility accounting performance report Responsibility report that compares actual costs and expenses for a department with budgeted amounts. *(p. 870)*

Responsibility accounting system System that provides information that management can use to evaluate the performance of a department's manager. *(p. 858)*

Restricted retained earnings Retained earnings not available for dividends because of legal or contractual limitations. *(p. 510)*

Retail inventory method Method to estimate ending inventory based on the ratio of the amount of goods for sale at cost to the amount of goods for sale at retail. *(p. 247)*

Retailer Intermediary that buys products from manufacturers or wholesalers and sells them to consumers. *(p. 178)*

Retained earnings Cumulative income less cumulative losses and dividends. *(p. 510)*

Retained earnings deficit Debit (abnormal) balance in Retained Earnings; occurs when cumulative losses and dividends exceed cumulative income; also called *accumulated deficit*. *(p. 513)*

Return Monies received from an investment; often in percent form. *(p. 23)*

Return on assets (See *return on total assets*) *(p. 20)*

Return on equity Ratio of net income to average equity for the period.

Return on total assets Ratio reflecting operating efficiency; defined as net income divided by average total assets for the period; also called *return on assets* or *return on investment*. *(p. 600)*

Revenue expenditures Expenditures reported on the current income statement as an expense because they do not provide benefits in future periods. *(p. 402)*

Revenue recognition principle The principle prescribing that revenue is recognized when earned. *(p. 10)*

Revenues Gross increase in equity from a company's business activities that earn income; also called *sales*. *(p. 13)*

Reverse stock split Occurs when a corporation calls in its stock and replaces each share with less than one new share; increases both market value per share and any par or stated value per share. *(p. 516)*

Reversing entries Optional entries recorded at the beginning of a period that prepare the accounts for the usual journal entries as if adjusting entries had not occurred in the prior period. *(p. 151)*

Risk Uncertainty about an expected return. *(p. 24)*

Rolling budget New set of budgets a firm adds for the next period (with revisions) to replace the ones that have lapsed. *(p. 942)*

S corporation Corporation that meets special tax qualifications so as to be treated like a partnership for income tax purposes. *(p. 480)*

Safety stock Quantity of inventory or materials over the minimum needed to satisfy budgeted demand. *(p. 947)*

Sales (See *revenues*.)

Sales budget Plan showing the units of goods to be sold or services to be provided; the starting point in the budgeting process for most departments. *(p. 946)*

Sales discount Term used by a seller to describe a cash discount granted to buyers who pay within the discount period. *(p. 181)*

Sales journal Journal normally used to record sales of goods on credit. *(p. 274)*

Sales mix Ratio of sales volumes for the various products sold by a company. *(p. 915)*

Salvage value Estimate of amount to be recovered at the end of an asset's useful life; also called *residual value* or *scrap value*. *(p. 395)*

Sarbanes-Oxley Act (SOX) Created the Public *Company Accounting Oversight Board*, regulates analyst conflicts, imposes corporate governance requirements, enhances accounting and control disclosures, impacts insider transactions and executive loans, establishes new types of criminal conduct, and expands penalties for violations of federal securities laws. *(pp. 11 & 314)*

Scatter diagram Graph used to display data about past cost behavior and sales as points on a diagram. *(p. 905)*

Schedule of accounts payable List of the balances of all accounts in the accounts payable ledger and their total. *(p. 280)*

Schedule of accounts receivable List of the balances for all accounts in the accounts receivable ledger and their total. *(p. 275)*

Secured bonds Bonds that have specific assets of the issuer pledged as collateral. *(p. 563)*

Securities and Exchange Commission (SEC) Federal agency Congress has charged to set reporting rules for organizations that sell ownership shares to the public. *(p. 9)*

Segment return on assets Segment operating income divided by segment average (identifiable) assets for the period. *(p. 284)*

Selling expense budget Plan that lists the types and amounts of selling expenses expected in the budget period. *(p. 948)*

Selling expenses Expenses of promoting sales, such as displaying and advertising merchandise, making sales, and delivering goods to customers. *(p. 191)*

Serial bonds Bonds consisting of separate amounts that mature at different dates. *(p. 563)*

Service company Organization that provides services instead of tangible products.

Shareholders Owners of a corporation; also called *stockholders. (p. 11)*

Shares Equity of a corporation divided into ownership units; also called *stock. (p. 11)*

Short-term investments Debt and equity securities that management expects to convert to cash within the next 3 to 12 months (or the operating cycle if longer); also called *temporary investments* or *marketable securities. (p. 592)*

Short-term note payable Current obligation in the form of a written promissory note. *(p. 437)*

Shrinkage Inventory losses that occur as a result of theft or deterioration. *(p. 188)*

Signature card Includes the signatures of each person authorized to sign checks on the bank account. *(p. 326)*

Simple capital structure Capital structure that consists of only common stock and nonconvertible preferred stock; consists of no dilutive securities. *(p. 524)*

Single-step income statement Income statement format that includes cost of goods sold as an expense and shows only one subtotal for total expenses. *(p. 192)*

Sinking fund bonds Bonds that require the issuer to make deposits to a separate account; bondholders are repaid at maturity from that account. *(p. 563)*

Small stock dividend Stock dividend that is 25% or less of a corporation's previously outstanding shares. *(p. 514)*

Social responsibility Being accountable for the impact that one's actions might have on society. *(p. 8)*

Sole proprietorship Business owned by one person that is not organized as a corporation; also called *proprietorship. (p. 10)*

Solvency Company's long-run financial viability and its ability to cover long-term obligations. *(p. 681)*

Source documents Source of information for accounting entries that can be in either paper or electronic form; also called *business papers. (p. 49)*

Special journal Any journal used for recording and posting transactions of a similar type. *(p. 272)*

Specific identification Method to assign cost to inventory when the purchase cost of each item in inventory is identified and used to compute cost of inventory. *(p. 227)*

Spending variance Difference between the actual price of an item and its standard price. *(p. 993)*

Spreadsheet Computer program that organizes data by means of formulas and format; also called *electronic work sheet.*

Standard costs Costs that should be incurred under normal conditions to produce a product or component or to perform a service. *(p. 985)*

State Unemployment Taxes (SUTA) State payroll taxes on employers to support its unemployment programs. *(p. 442)*

Stated value stock No-par stock assigned a stated value per share; this amount is recorded in the stock account when the stock is issued. *(p. 510)*

Statement of cash flows A financial statement that lists cash inflows (receipts) and cash outflows (payments) during a period; arranged by operating, investing, and financing. *(p. 628)*

Statement of owner's equity Report of changes in equity over a period; adjusted for increases (owner investment and net income) and for decreases (withdrawals and net loss). *(p. 17)*

Statement of partners' equity Financial statement that shows total capital balances at the beginning of the period, any additional investment by partners, the income or loss of the period, the partners' withdrawals, and the partners' ending capital balances; also called *statement of partners' capital. (p. 483)*

Statement of retained earnings Report of changes in retained earnings over a period; adjusted for increases (net income), for decreases (dividends and net loss), and for any prior period adjustment. *(p. 7)*

Statement of stockholders' equity Financial statement that lists the beginning and ending balances of each major equity account and describes all changes in those accounts. *(p. 523)*

Statements of Financial Accounting Standards (SFAS) FASB publications that establish U.S. GAAP. *(p. 9)*

Step-wise cost Cost that remains fixed over limited ranges of volumes but changes by a lump sum when volume changes occur outside these limited ranges. *(p. 904).*

Stock (See *shares.*)

Stock dividend Corporation's distribution of its own stock to its stockholders without the receipt of any payment. *(p. 514)*

Stock options Rights to purchase common stock at a fixed price over a specified period of time. *(p. 523)*

Stock split Occurs when a corporation calls in its stock and replaces each share with more than one new share; decreases both the market value per share and any par or stated value per share. *(p. 516)*

Stock subscription Investor's contractual commitment to purchase unissued shares at future dates and prices.

Stockholders (See *shareholders.*) *(p. 11)*

Stockholders' equity A corporation's equity; also called *shareholders' equity* or *corporate capital*. *(p. 510)*

Straight-line depreciation Method that allocates an equal portion of the depreciable cost of plant asset (cost minus salvage) to each accounting period in its useful life. *(pp. 97 & 396)*

Straight-line bond amortization Method allocating an equal amount of bond interest expense to each period of the bond life. *(p. 554)*

Subsidiary Entity controlled by another entity (parent) in which the parent owns more than 50% of the subsidiary's voting stock. *(p. 599)*

Subsidiary ledger List of individual sub-accounts and amounts with a common characteristic; linked to a controlling account in the general ledger. *(p. 272)*

Sunk cost Cost already incurred and cannot be avoided or changed. *(p. 733)*

Supplementary records Information outside the usual accounting records; also called *supplemental records*. *(p. 184)*

Supply chain Linkages of services or goods extending from suppliers, to the company itself, and on to customers.

T-account Tool used to show the effects of transactions and events on individual accounts. *(p. 53)*

Target cost Maximum allowable cost for a product or service; defined as expected selling price less the desired profit. *(p. 771)*

Temporary accounts Accounts used to record revenues, expenses, and withdrawals (dividends for a corporation); they are closed at the end of each period; also called *nominal accounts*. *(p. 140)*

Term bonds Bonds scheduled for payment (maturity) at a single specified date. *(p. 563)*

Throughput time (See *cycle time.*)

Time period assumption Assumption that an organization's activities can be divided into specific time periods such as months, quarters, or years. *(p. 92)*

Time ticket Source document used to report the time an employee spent working on a job or on overhead activities and then to determine the amount of direct labor to charge to the job or the amount of indirect labor to charge to overhead. *(p. 776)*

Times interest earned Ratio of income before interest expense (and any income taxes) divided by interest expense; reflects risk of covering interest commitments when income varies. *(p. 447)*

Total asset turnover Measure of a company's ability to use its assets to generate sales; computed by dividing net sales by average total assets. *(p. 410)*

Total quality management (TQM) Concept calling for all managers and employees at all stages of operations to strive toward higher standards and reduce number of defects. *(p. 730)*

Trade discount Reduction from a list or catalog price that can vary for wholesalers, retailers, and consumers. *(p. 180)*

Trademark or **Trade (Brand) name** Symbol, name, phrase, or jingle identified with a company, product, or service. *(p. 408)*

Trading on the equity (See *financial leverage.*)

Trading securities Investments in debt and equity securities that the company intends to actively trade for profit. *(p. 595)*

Transaction Exchange of economic consideration affecting an entity's financial position that can be reliably measured. *(p. 13)*

Treasury stock Corporation's own stock that it reacquired and still holds. *(p. 520)*

Trial balance List of accounts and their balances at a point in time; total debit balances equal total credit balances. *(p. 63)*

Unadjusted trial balance List of accounts and balances prepared before accounting adjustments are recorded and posted. *(p. 104)*

Unavoidable expense Expense (or cost) that is not relevant for business decisions; an expense that would continue even if a department, product, or service is eliminated. *(p. 1041)*

Unclassified balance sheet Balance sheet that broadly groups assets, liabilities, and equity accounts. *(p. 145)*

Uncontrollable costs Costs that a manager does not have the power to determine or strongly influence. *(p. 870)*

Underapplied overhead Amount by which overhead incurred in a period exceeds the overhead applied to that period's production using the predetermined overhead rate. *(p. 781)*

Unearned revenue Liability created when customers pay in advance for products or services; earned when the products or services are later delivered. *(pp. 50 & 98)*

Unfavorable variance Difference in revenues or costs, when the actual amount is compared to the budgeted amount, that contributes to a lower income. *(p. 981)*

Unit contribution margin Amount a product's unit selling price exceeds its total unit variable cost.

Units-of-production depreciation Method that charges a varying amount to depreciation expense for each period of an asset's useful life depending on its usage. *(p. 397)*

Unlimited liability Legal relationship among general partners that makes each of them responsible for partnership debts if the other partners are unable to pay their shares. *(p. 479)*

Unrealized gain (loss) Gain (loss) not yet realized by an actual transaction or event such as a sale. *(p. 595)*

Unsecured bonds Bonds backed only by the issuer's credit standing; almost always riskier than secured bonds; also called *debentures*. *(p. 563)*

Unusual gain or loss Gain or loss that is abnormal or unrelated to the company's ordinary activities and environment. *(p. 703)*

Useful life Length of time an asset will be productively used in the operations of a business; also called *service life* or *limited life*. *(p. 395)*

Value-added time The portion of cycle time that is directed at producing a product or service; equals process time. *(p. 743)*

Value chain Sequential activities that add value to an entity's products or services; includes design, production, marketing, distribution, and service. *(p. 740)*

Variable cost Cost that changes in proportion to changes in the activity output volume. *(p. 732)*

Variance analysis Process of examining differences between actual and budgeted revenues or costs and describing them in terms of price and quantity differences. *(p. 984)*

Vendee Buyer of goods or services. *(p. 335)*

Vendor Seller of goods or services. *(p. 335)*

Vertical analysis Evaluation of each financial statement item or group of items in terms of a specific base amount. *(p. 682)*

Volume variance Difference between two dollar amounts of fixed overhead cost; one amount is the total budgeted overhead cost, and the other is the overhead cost allocated to products using the predetermined fixed overhead rate. *(p. 993)*

Voucher Internal file used to store documents and information to control cash disbursements and to ensure that a transaction is properly authorized and recorded. *(p. 322)*

Voucher register Journal (referred to as *book of original entry*) in which all vouchers are recorded after they have been approved. *(p. 337)*

Voucher system Procedures and approvals designed to control cash disbursements and acceptance of obligations. *(p. 322)*

Wage bracket withholding table Table of the amounts of income tax withheld from employees' wages. *(p. 454)*

Warranty Agreement that obligates the seller to correct or replace a product or service when it fails to perform properly within a specified period. *(p. 444)*

Weighted average Method to assign inventory cost to sales; the cost of available-for-sale units is divided by the number of units available to determine per unit cost prior to each sale that is then multiplied by the units sold to yield the cost of that sale. *(pp. 230, 245 & 818)*

Weighted-average method (See *weighted average.*)

Wholesaler Intermediary that buys products from manufacturers or other wholesalers and sells them to retailers or other wholesalers. *(p. 178)*

Withdrawals Payment of cash or other assets from a proprietorship or partnership to its owner or owners. *(p. 13)*

Work sheet Spreadsheet used to draft an unadjusted trial balance, adjusting entries, adjusted trial balance, and financial statements. *(p. 136)*

Working capital Current assets minus current liabilities at a point in time. *(p. 691)*

Working papers Analyses and other informal reports prepared by accountants and managers when organizing information for formal reports and financial statements. *(p. 136)*

Credits

Chapter 12
Page 477 Courtesy of Samantashoes.com
Page 479 © David R. Frazier Photolibrary, Inc./Alamy
Page 481 © Erik Freeland/Corbis
Page 484 © Bill Bachmann/Alamy
Page 487 © James Doberman/Getty Images/Iconica

Chapter 13
Page 505 Courtesy of Inogen, Inc.
Page 508 Courtesy of Green Bay Packers, Inc.

Chapter 14
Page 549 © James Wasserman
Page 552 Courtesy of Dow Chemicals

Chapter 15
Page 591 Courtesy of Tibi.com
Page 593 Courtesy of Scripophily.com
Page 595 AP Images/Richard Drew

Chapter 16
Page 627 Courtesy of Jungle Jim's International Market
Page 628 AP Images/Brian Kersey
Page 634 © Corbis Super RF/Alamy

Chapter 17
Page 679 © Dirck Halstead/Liaison/Getty Images
Page 693 AP Images/Ben Margot
Page 698 AP Images/Chuck Burton

Chapter 18
Page 725 Courtesy of Kernel Seasons
Page 729 AP Images/KEYSTONE/Alessandro Della Bella
Page 735 © Jeff Greenberg/The Image Works
Page 736 © Darrell Ingham/Getty Images
Page 742 © Gabriel Bouys/AFP/Getty Images

Chapter 19
Page 769 Courtesy of Specialty Surfaces International, Inc. d/b/a Sprinturf
Page 770 Peter Newcomb/Bloomberg News/Landov
Page 777 © Photodisc/Alamy
Page 782 image100/Punchstock

Chapter 20
Page 807 Courtesy of Hood River Juice Company
Page 807 © Grant V. Faint/Getty Images/Iconica
Page 808 © INGO WAGNER/dpa/Landov
Page 812 © Tom Tracy Photography/Alamy
Page 818 © Lester Lefkowitz/Getty Images/Stone
Page 823 © Frederic Pitchal/Sygma/Corbis

Chapter 21
Page 851 Courtesy of RockBottomGolf.com
Page 852 © ACE STOCK LIMITED/Alamy
Page 854 © Royalty-Free/Corbis
Page 862 © Christian Hoehn/Getty Images/Stone

Chapter 22
Page 901 Courtesy of Moe's Southwest Grill
Page 903 © Royalty Free/Corbis
Page 909 © Rob Tringali/Sportschrome/Getty Images
Page 915 © Jonathan Fickies/Getty Images
Page 916 © Nick Onken/Getty Images/UpperCut Images

Chapter 23
Page 939 AP Images/David Zalubowski
Page 942 Courtesy © B.P., pl.l.c
Page 946 © Royalty Free/CORBIS
Page 950 © Jeff Vinnick/Getty Images
Page 951 © Jim McIsaac/Getty Images
Page 952 © Teo Lannie/Getty Images/PhotoAlto Agency RF Collections

Chapter 24
Page 979 Courtesy of Martin Guitar Co.
Page 982 © Ken Samuelsen/PhotoDisc/Getty Images
Page 984 © Royalty-Free/Corbis
Page 986 AP Images/David Vincent
Page 988 © Kristjan Maack/Getty Images/Nordic Photos
Page 989 © ColorBlind Images/Corbis
Page 991 © Patrik Giardino/Photolibrary
Page 992 © Nick Daly/Getty Images/Digital Vision
Page 997 © Photodisc/Alamy

Chapter 25
Page 1023 Courtesy of Danielle Greenberg/Prairie Sticks Bat Company
Page 1026 AP Images/Katsumi Kasahara
Page 1032 © Philip Lee Harvey/Getty Images/Taxi
Page 1036 © JOHN SOMMERS/Reuters/Corbis
Page 1041 © Dimas Ardian/Getty Images
Page 1042 © Louie Psihoyos/Corbis

Index

Note: Page numbers followed by *n* indicate material in footnotes.

ABB (activity-based budgeting), 953–954, 960, 961
ABC. *See* **Activity-based costing**
Abercrombie & Fitch, 1041
Absorption costing approach, 910
Accelerated depreciation methods, 1031
Account analysis, 637, 653
Accounting
  changes in accounting principles, 704–705
  for corporations. *See* Corporations
Accounting errors, 523
Accounting firms, 479
Accounting information
  in cost estimation, 907, 910, 921
  for financial statement analysis, 681
  purpose of, 728
  segment and geographic information, A-18
Accounting periods, B-2
Accounting rate of return, 1025, **1027–**1028, 1033–1034, 1048
Accounting software, 633
Accounts payable, 637, 638
Accounts receivable, 636–637
Accounts receivable turnover ratio, 693, 702
Accretion liability accounts, 556
Accrual basis accounting, 635
Accrued interest
  on bonds, 557, 572–573, 575
  on installment notes payable, 561
Accumulated deficit, 513
Accumulated other comprehensive income (AOCI), 600, 705
ACFE (Association of Certified Fraud Examiners), 731
Acid-test ratio, 692–693, 702
Activity-based budgeting (ABB), 953–954, 960, 961
Activity-based costing (ABC), 854–856, 879
  comparison with two-stage costing, 855, 856–857
  cost drivers in, 856, 879
  cost pools in, 854–855, 856, 857
  ethical issues, 857, 879
  illustration of, 854, 855–856
Activity-based management, 857, 859
Activity cost driver, 855
Activity cost pools, 854–855, 856, 857
Activity levels, 991–992
"Acts of God," 703
Additional business decision, 1036–1037
  customer relationships and, 1037, 1048
  demonstration of, 1043–1044, 1045
  qualitative factors in, 1042
Adidas, 556, 558, 735
Adjunct liability accounts, 556
Adjusting entries
  for factory overhead, 777
  short-term investments, 601–602
  for trading securities, 595
  for unrealized gain (loss) on AFS securities, 596
Administrative expenses, incremental, 1036
Admission of partner, 484–486, 492–493
  investment of assets, 485–486
    bonus to new partner, 485–486
    bonus to old partners, 484, 485
  purchase of partnership interest, 484–485
Advertising costs, 861, 863, 864, A-12
AFS (available-for-sale) securities, 593, **596–**597, 600, 608
After-tax income, computing sales for, 913–914, 919
Allocation base
  in allocating indirect expenses, 863–864
  in computing predetermined overhead rates, 853–854
  cost driver contrasted, 855
  in establishing standard costs, 991
  for factory overhead, 777
  (*See also* Cost allocation; Overhead cost allocation)
Allocation rate formula, 778
Allowance(s), in standard setting, 986
All Tune and Lube, 992
Altria Group, 525
Amazon.com, 511, 514, 525
American Eagle, 1041
American Express, 629
American Greetings, 509
American Stock Exchange, 551
Amortization
  of bond discount, 554–555, 575
    effective interest amortization on bonds, 554, 569–570
    straight-line amortization, 554–555
  of bond premium, 556–557, 575
    effective interest amortization on bonds, 556, 570–571
    straight-line amortization, 557
  change in method, 705
  journal entries for bond amortization, 570
  reporting securities at amortized cost, 593
Amortization tables, 565, 566
Analysis overview, 700
Analysis period, 683
Analyst services, 680
Andreessen, Marc, 507
"Angel" investors, 508
Anheuser-Busch, 875
Annual average investment, 1027–1028
Annual financial statements
  in annual report, A-1
  examples of, A-1–A-33
    Apple Computer, A-29–A-33
    Best Buy, A-2–A-18
    Circuit City, A-19–A-23
    RadioShack, A-24–A-28
  footnotes to, A-1, A-8–A-18
Annual interest rate, 569
Annual report, A-1
Annual Statement Studies (Robert Morris Associates), 691
Annuity, 568, B-5
  ordinary annuity
    future value of, B-6–B-7, B-10, B-11
    present value of, B-2, B-5, B-7, B-10, B-11
  present value of, 568–569, B-11
Annuity factor, 1030*n*, B-11
AOCI (accumulated other comprehensive income), 705
Apple Computer, 523, 524, 742
  consolidated financial statements of, A-29–A-33
    balance sheets, A-30
    statements of cash flows, A-33
    statements of operations, A-31
    statements of shareholders' equity, A-32
Appropriated retained earnings, 523
Articles of copartnership, 478, 482, 487
Asset(s)
  acquired with off-balance-sheet financing, 574*n*
  analysis of transactions, 642
  book value of
    in computing accounting rate of return, 1027–1028
    dissolution of partnership and, 487, 493
  cash flow on total assets ratio, 646, 657, 699*n*
  cash flows from investing activities
    noncurrent asset analysis, 641–642
    other asset transactions, 642
    short-term assets, 630
  current. *See* Current assets
  distribution of. *See* Dividend(s)
  investment in partnership, 485–486
  legal prohibition against use of, 703
  long-lived assets. *See* Plant assets
  loss on sale of, 637, 640, 656
  noncash, issuing common stock for, 512, 517
  pledged assets, 563, 695
Asset turnover rate, 692
Association of Certified Fraud Examiners (ACFE), 731
Assumptions
  in analysis reports, 700
  in CVP analysis, 905, 911–912, 921
    changes in estimates, 912
    limitations of, 911
    output measures, 911
Auditor(s), 680, 728
Auditor's report, A-1
Authorized stock, 509
AutoGrand, 853–854

Automobile industry, 986
Available-for-sale (AFS) securities, 596, 608
classification and reporting, 593
selling, 597
valuing and reporting, 596–597, 600
Average collection period, 694
Average values, comparison of, 683
Avia, 571–572
Avoidable costs, 1035
Avoidable expenses, 1041

Baby bond, 552*n*
Balanced scorecard, 731, **859, 868–**869
Balance sheet
AFS securities shown in, 597
budgeted, 952, 952*n*, 953, 958
common-size, 687, 688, 702
comparative. *See* Comparative balance sheets
consolidated, A-4, A-20, A-25, A-30
cost flows to, 742
long-term investments shown in, 604, 605
long-term liability section, 554
for manufacturers
different from merchandiser, 745, 748
finished goods inventory, 736
goods in process inventory, 736
raw materials inventory, 735–736
off-balance-sheet financing, 574*n*
for partnerships, 484
stockholders' equity shown in, 528
use in preparation of budgets, 945
Balance sheet accounts, 633
Baldrige Award (MBQNA), 730
Banana Republic, 599
Banking activities
brokerage houses, 509
notes payable. *See* Notes payable
Bank of America, 629
Bankruptcy, 628, 629
Bargain purchase option, 574*n*
Base amount, 687
Base period, 683, 685
Basic earnings per share, 524
Bearer, 563
Bearer bonds, 563
Bear market, 696
Behavior, classification of costs by, 732, 734, 743–744
Benchmarking, 682, 992
Benchmarking budgets, 940–941
Berkshire Hathaway, 516
Best Buy, 509, 510, 872
analysis of financial statements, 680, 682
common-size graphics, 689–690
comparative format, 682–685
ratio analysis, 691–699
trend analysis, 685–687
vertical analysis, 687–689
consolidated financial statements of, A-2–A-18
balance sheets, A-4
selected financial data, A-3
selected notes to, A-8–A-18
statements of cash flows, A-7
statements of changes in shareholders' equity, A-6
statements of earnings, A-5
Management's Discussion and Analysis (MD&A), 681
BestBuy.com, 681
Best case scenario, 982
BET (break-even time), 1043, 1048
Big River Productions, 481
Blue chips, 681
BMX, 526, 531
Board of directors
corporate, 507
declaration of dividends by, 518
influence on budgets, 944
list of, in annual report, A-1
use of financial statement analysis, 680
Boeing, 770–771
Bonaminio, Jim, 627
Bond(s), 550
bond financing, 575
advantages of bonds, 550–551
disadvantages of bonds, 551
convertible and/or callable, 559, 560, 563
demonstration problem, 565–566
discount on. *See* **Discount on bonds payable**
effective interest amortization, 569–571
of discount bond, 554, 569–570
of premium bond, 556, 570–571
interest on, 575
accrued interest, 557, 572–573
issuing bonds between interest dates, 571–573, 575
market rate, 553
payment of, 551
recording semiannual obligation, 553
tax deductible, 550
issuances, 552–559, 575
accruing interest expense on, 557, 572–573, 575
at discount, 553–555
between interest dates, 571–573
at par, 552–553
at premium, 556–557
pricing bonds, 558
procedures for, 552, 552*n*
junk bonds, 560
premium on. *See* **Premium on bonds**
present values of, 567–569
applying present value table, 568
compounding periods shorter than one year, 569
present value concepts, 567
present value of annuity, 568–569, B-11
present value tables, 567, 568, B-10
rating services, 555
reading bond quotes, 551
registered or bearer bonds, 563
retirement of, 575–576
by conversion, 560
demonstration of, 566, 567
at maturity, 559
before maturity, 559–560
secured or unsecured, 563
term or serial, 563
trading, 551
(*See also* Long-term notes payable)
Bond certificate, 552, 552*n*
Bond indenture, 552
Bond quotations, 551
Bond redemption, 559–560
Bonuses (bonus plans)
effect on sales forecasts, 946
partnerships
investment in partnership, 485–486
withdrawal of partner, 486–487
Book (carrying) value of bonds, 554, 557
Book value of assets
in computing accounting rate of return, 1027–1028
dissolution of partnership and, 487, 493
Book value per common share ratio, 511, **526,** 530, 531
Book value per preferred share ratio, **526,** 530
Boston Celtics LP, 490
Boston Celtics LP I, 490
Bottom-line figure, 632
Bottom-up process, budgeting as, 942
BP, 942
Break-even analysis, 908–912
break-even point computations, 909
contribution margin, 908
cost-volume-profit chart, 910, 919
making assumptions in, 905, 911–912, 921
multiproduct break-even point, 915–917, 921
Break-even chart (graph), 910
Break-even point, 909, 919, 921
fixed costs and, 915
multiproduct, 915–917, 921
revised, 905, 909, 912, 915
selling price and, 911, 912
Break-even sales dollars, 909
Break-even time (BET), 1043, 1048
Briggs & Stratton, 736
Brokerage houses (investment bankers), 509
Brokers, 681
Brunswick, 599
Budget(s), 726–727, **940**
Budget administration, 942–944, 960
budget committee, 942
reporting, 942, 943
timing, 942–943
Budgetary control, 980–981
Budget calendar, 944
Budget committee, 942
Budgeted balance sheet, 952, 952*n*, 953, 958
Budgeted costs. *See* **Standard cost(s)**
Budgeted income statement, 952, 958
Budgeted performance, 940, 941
Budgeting, 938–961
activity-based, 953–954, 960–961
budget administration, 942–944, 960
budget committee, 942
reporting, 942, 943
timing, 942–943
capital budgets. *See* **Capital budgeting**
continuous budgeting, 940, 942–943
defined, **940**
demonstration problem, 954–958
flexible budgets. *See* **Flexible budget(s)**
for manufacturers, 961
manufacturing budgets, 946, 959–960
production budgets, 946, 947, 949
master budgets, 942–953
capital expenditures budget, 944, 945, 949, 950, 961
components of, 942–945
financial budgets, 949–953
operating budgets, 946–949
for merchandisers, 946
merchandise purchases budget, 947–948
operating budgets, 944
as process. *See* Budget process
production and manufacturing budgets, 946, 947, 959–960, 961
standard costs. *See* **Standard cost(s); Variance analysis**

Budget process
  benchmarking budgets, 940–941
  communication of, 941
  flexible budgets, 980–982
    budgetary control and reporting, 980–981
    budget reports for evaluation, 982
    fixed budget performance report, 981
  human behavior and, 941
  as management tool, 941
  strategic budgeting, 940
Budget reports, 980, 982–985
  fixed budget performance report, 981
  flexible budget performance report, 984, 985, 1001, 1005
  preparation of, 981, 982–984
  purpose of, 982
  as tool for evaluation, 982
Building blocks of financial statement analysis, 681
Building expenses, 861
Bull market, 696
Business consultants, 684
Business description, A-8
Business entity assumption, 506
Business environment, 729–731
  implications of, 730–731
  lean business model, 729–730, 748
  lean practices, 730
Business segments, 703
  discontinued segments, 703, 704
  information regarding, A-18
  reportable, graphical analysis of, 689
  segment elimination decision, 1041–1042
BusinessWeek magazine, 599
Bylaws, corporate, 507

Callable bonds (and notes), 559, **563**
Callable preferred stock, 519
Call option on bonds, 559
Call premium on bonds, 559
Call price, 519
Cancelable (short-term) leases, 573
Canion, Rod, 904
Capacity to pay fixed interest charges, 695
Capital
  corporate capital. *See* **Stockholders' equity**
  ease of accumulation for corporations, 506
  minimum legal capital, 509
  net working capital, 691
  paid-in capital, 510
  paid-in capital in excess of par value, 511
  working capital, 691
Capital accounts, 481
Capital balances method, 482
Capital budgeting, 949, **1024–**1034, 1048
  break-even time and, 1043
  comparison of methods, 1033–1034
  ethical issues, 1030, 1048
  master budgets, 949, 950
  not using time value of money, 1025–1028
    accounting rate of return, 1025, 1027–1028, 1033–1034, 1048
    payback period, 1025–1027
  using time value of money, 1028–1034
    internal rate of return, 1030, 1031–1033, 1047, 1048, B-11
    net present value. *See* **Net present value**
Capital deficiency (partnerships)
  no deficiency, 488, 488*n*
  partner cannot pay deficiency, 489
  partner pays deficiency, 489
Capital expenditures budget, 944, 945, **949,** 950, 961
Capitalization of retained earnings. *See* **Stock dividends**
Capital leases, 574, 574*n*
Capital stock, 509–510
Capital structure of company, 695
Carrying (book) value of bonds, 554, 557
Cash, 629
  cash flows. *See* Cash flow(s)
  distributed to or received from owners, 630
  e-cash, 634
  excess, investing, 592
  note regarding, A-8
Cash, Johnny, 979
Cash accounts
  analyzing, 632–633
  relation to noncash accounts, 633, 634
Cash balances, proving, 632, 637, 645
Cash budget, 945, 946, 948, 949, **950–**951, 957
Cash controls, ratio analysis in, 693–694, 702
Cash coverage of debt ratio, **646,** 699*n*
Cash coverage of growth ratio, **646,** 699*n*
Cash disbursements, 950–951
Cash dividends, 513–514, 530
  accounting for, 513, 513*n*
  budgeting for, 951
  declaration of, 644
  deficits and, 513–514
  on equity securities, 594
  payout ratio on, 525
  receipt of, under equity method, 598
Cash equivalents, 629, A-8
Cash flow(s)
  bankruptcy and, 628, 629
  classification of
    financing activities, 630
    investing activities, 630
    operating activities, 629–630
  declaration of dividend and, 644
  delaying or accelerating, 630
  for discount on bonds payable, 554
  even
    computing IRR with, 1032, B-11
    computing payback period with, 1025–1026
  expected future cash flows, 525
  free cash flows, 646, 699*n*
  inflows, 629, 632–633
  measurement of, 629
  misclassification of, 630
  net. *See* Net cash flows
  outflows, 629, 632–633
  predicting, 1024
  sources and uses of cash, 645–646
  uneven
    computing net present value, 1030–1031
    computing payback period with, 1026
Cash flow analysis, 645–646, 1025
Cash flow on total assets ratio, **646,** 657, 699*n*
Cash receipts, reporting
  operating cash payments, 636, 652, 653–656
  operating cash receipts, 633, 636, 652, 653
Cause-effect relation, 860, 879
C Corporations, 480
CE (cycle efficiency), 743, 748
Celtics LP, 490
CEO (chief executive officer), 507
Certified Financial Manager (CFM), 731
Certified Management Accountant (CMA), 728, 731
CFM (Certified Financial Manager), 731
Change in accounting estimate, 523, 705
Change in accounting principle
  consistency concept and, 704
  and retrospective application of, 705
  sustainable income and, 704–705
Chief executive officer (CEO), 507
Chief operating officer (COO), 507–508
Circuit City, 522
  analysis of financial statements, 680, 682
    graphical analysis, 686, 690
    ratio analysis, 691–699
  consolidated financial statements of, A-19–A-23
    balance sheets, A-20
    statements of cash flows, A-23
    statements of operations, A-21
    statements of stockholders' equity and comprehensive income, A-22
Clapton, Eric, 979
Clock cards, 775, **776**
Closely held corporations, 506
CMA (Certified Management Accountant), 728, 731
CMA exams, 771
Coca-Cola Company, 605, 682, 808
Collateral agreements, 563
Commissions
  budgeting for, 951
  incremental, 1036
  in selling expense budget, 948
  as variable costs, 903
Common-size analysis. *See* **Vertical analysis**
Common-size financial statements, 687
  balance sheets, 687, 688, 702
  in vertical analysis, 684, 685, 687–689
    balance sheets, 687, 688
    income statements, 688–689
Common-size graphics, 689–690
Common stock, 508, 530
  analysis of transactions, 644
  classes of, 509
  dividends on. *See* Dividend(s)
  interpreting stock quotes, 510
  issuance of, 510–512, 530
    for noncash assets, 512, 517
    no-par value stock, 511
    at par, 510–511
    stated value stock, 512
Company performance, 686–687
Compaq Computer, 904
Comparative balance sheets, 634, 636, 683–685, 701
  analysis of equity, 644
  demonstration problem, 647–649
  investing activities, 641–642
  notes payable transactions, 643
Comparative financial statements, 682
  in horizontal analysis, 682–685
    comparative balance sheets. *See* Comparative balance sheets
    comparative income statements, 685, 701–702
    computation of dollar and percent changes, 683, 683*n*
Comparative income statements, 685, 701–702
Competitor benchmarks, 682
Complex capital structure, 524*n*
Component costs, 823
Composite unit, 916
Compounding, 567

Compounding periods, 553, 558, B-2
Comprehensive income, 596, **600,** 705, A-16
Compustat, 599
Condemnation of property, 703
Consistency concept, 704
Consolidated financial statements, 599
examples of
balance sheets, A-4, A-20, A-25, A-30
income statements, A-5, A-21, A-26, A-31
statements of cash flows, A-7, A-23, A-28, A-33
statements of shareholders' equity, A-6, A-22, A-27, A-32
foreign currencies and
foreign exchange rate, 605, 606
international subsidiaries, 607
sales and purchases, 606–607, 608
Consolidation, 593
Continuing operations, 703
Continuous budgeting, 940, **942–**943
Continuous improvement, 730, 942
Continuous life of corporation, 506
Contract(s), 729
Contract rate on bonds, **553**
Contractual restrictions on retained earnings, 522
Contribution margin, 909, 921
in sales mix selection decision, 1040, 1041
in transfer pricing, 876
weighted-average, 916–917
Contribution margin format, 983
Contribution margin income statement, 909, 910
Contribution margin per unit, 908, 914, 919
Contribution margin ratio, 908, 909, 914, 919, 920
Control (control function), **727,** 940–941, 996
Controllability, classification of costs by, 733, 734
Controllable costs, 733, 870, 870*n,* 879
Controllable variances, 993, 995, 997
Controlling influence, 593
Control of corporation, 519, 531
"Convenience" financial statements, 688
Converse, 560
Conversion, bond retirement by, 560
Conversion costs, 738, 823
Convertible bonds (and notes), 559, 560, **563**
Convertible debentures, A-14
Convertible preferred stock, 518–519
COO (chief operating officer), 507–508
Corporate capital. *See* **Stockholders' equity**
Corporations, 504–531
C corporations, 480
common stock issued by. *See* **Common stock**
defined, **506**
demonstration problems
dividends, 529–530
stockholders' equity, 527–529
equity reporting, 522–524
statement of retained earnings, 522–523
statement of stockholders' equity, 523, 524, A-6, A-22, A-27, A-32
stock option reporting, 523–524
form of organization, 506–510, 530
basics of capital stock, 509–510
characteristics, 506–507
organization and management, 507–508
stockholders, 506, 508–509
preferred stock issued by. *See* **Preferred stock**
ratio analysis, 524–526
book value per common share, 511, 526, 530, 531
book value per preferred share, 526, 530
dividend yield, 525, 530, 698
earnings per share, 524, 524*n*
price-earnings ratio, 525, 530, 531, 697–698
S corporations, 480
simple or complex structure of, 524*n*
treasury stock, 520–522
purchase of, 520–521
reissuing, 521–522
retiring, 522
Corvette, 986
Cost(s)
of advertising, 861, 863, 864, A-12
allocation of. *See* Cost allocation
analysis of. *See* **Cost-volume-profit (CVP) analysis**
assigning and reconciling
by FIFO method, 828, 829–830, 831
by weighted average method, 818–821
behavior of. *See* Cost behavior
classifications of, 732–734, 748
by behavior, 732, 734, 743–744
by controllability, 733, 734
demonstration problem, 743–744
by function, 733–734
identifying, 734, 735
indirect costs (ethical issue), 814, 832
materiality principal and, 736
by relevance, 733, 734
by traceability, 732, 733, 734, 748
computing income from, 913
controllable, 733, 870, 870*n,* 879
conversion costs, 738, 823
costs accounted for, 819, 826, 830
costs to account for, 819, 826, 830
curvilinear costs, 904, 911
expenses contrasted, 870*n*
for factory overhead, 733
incremental, 1036, 1037
shown in manufacturing statement, 741
fixed. *See* **Fixed cost(s)**
joint costs, 877–878, 879
managerial cost concepts, 732–735
of materials. *See* Materials costs
mixed costs, 732, 903–904, 983
opportunity costs, 733, 734, 749, 1035
of additional business decision, 1037
of scrap or rework decision, 1038
in transfer pricing, 876
predicted, in job order costing, 771
for service companies, 734–735
step-wise costs, 904
uncontrollable, 733, 870
variable. *See* **Variable cost(s)**
Cost accounting system, 770, 786
Cost allocation
classification of costs by traceability, 732, 733, 734, 748
joint costs, 877–878, 879
for operating departments, 852, 853–854
overhead cost allocation. *See* Overhead cost allocation
for performance evaluation, 858–872
demonstration problem, 872–875
departmental accounting, 858–859
departmental expense allocation, 859–869
profit margin and investment turnover, 872
responsibility accounting, 869–872, 870*n*
process cost allocation, 814, 830, 831
transfer pricing, 875–877
two-stage, 879
activity-based costing compared, 855, 856–857
illustration of, 853–854
indirect expenses, 853, 861
in product costing, 852–854
value basis of allocating joint costs, 877, 878
Cost-based transfer pricing, 876
Cost behavior, 920
classification of costs by, 732, 734, 743–744
demonstration of, 743–744
estimated line of cost behavior, 906
identification of, 902–905
curvilinear costs, 904, 911
fixed costs, 902–903
mixed costs, 903–904
step-wise costs, 904
variable costs, 903
measurement of past behavior, 905–908, 921
comparison of methods, 907
high-low method, 905, 906–907
least-squares regression, 905, 907, 920
scatter diagrams, 905–906, 907
Cost centers, 858, 859, 861
Cost classification, identification of, 734, 735
Cost control, 877
Cost drivers, 855, 856, 857
Cost estimates, 915
Cost flow(s)
in job order costing
labor costs, 773, 775–776
materials costs, 773–775, 784, 786
overhead costs, 773, 776–778
summary of, 778–780
for manufacturing activities, 739–740, 748
materials activity, 739
production activity, 739–740
sales activity, 740
for process operations, 810, 822
Cost method, 520–521
Cost objects, 732, 748, 811
Cost of capital, 1029
Cost of goods manufactured, 737, 745, 746, **821**
Cost of goods sold
budgeted, 956
constant relation with sales, 948
job order costing, 785
note regarding, A-12
shown in manufacturing statement, 741, 748
summary entries for, 822
transfers to, 810, 821–822, 832
Cost per equivalent unit of production, 815
changes to, 820
computing, 818, 827, 829
demonstration of, 825
Cost per unit of output, 903
Cost-plus basis, 771
Cost-plus contract, 729
Costs accounted for, 819, 826, 830
Costs to account for, 819, 826, 830
Cost variances (CV), 986–990, 986*n*
computation of, 987–988
demonstration of, 1002–1003
labor cost variances, 1004
demonstration of, 1002
investigating causes of, 990, 1004
rate and efficiency variance, 989–990
materials cost variances, 988–989, 1002, 1004

overhead cost variances, 992, 1004
demonstration of, 1002–1003
fixed, 991, 992–995
graphical analysis, 993, 994, 995
variable. *See* Variable overhead cost variance
variance analysis, 987
Cost-volume-profit (CVP) analysis, 900–921
applications of, 912–917
income computed from sales and costs, 913
margin of safety, 914
multiproduct break-even point, 915–917, 921
sales computed for target income, 913–914
sensitivity analysis, 915
break-even analysis used in, 908–912
break-even point computations, 909
contribution margin, 908
cost-volume-profit chart, 910, 919
making assumptions in, 905, 911–912, 921
multiproduct break-even point, 915–917, 921
defined, **902**
degree of operating leverage, 918
demonstration problem, 918–919
identifying cost behavior, 902–905
curvilinear costs, 904, 911
fixed costs, 902–903
mixed costs, 903–904
step-wise costs, 904
variable costs, 903
measuring past cost behavior, 905–908
comparison of methods, 907
high-low method, 905, 906–907
least-squares regression, 905, 907, 920
scatter diagrams, 905–906, 907
Cost-volume-profit (CVP) chart, 910, 919
Coupon bonds, 563
Coupon rate on bonds, 553
CPA exams, 771
Creditors
influence on budgets, 944
use of financial statement analysis, 680
Credit risk ratio, 699*n*
Crocs, 939
Crocs.com, 939
CT (cycle time), 742–743, 748, 814, 859
Cumulative discount factor, 1030*n,* B-11
Cumulative preferred stock, 517–518
Cumulative total of net cash flows, 1026
Currency. *See* Foreign currencies
Current assets
adjustments for changes in, 636–639
accounts receivable, 636–637
merchandise inventory, 637–638
prepaid expenses, 637, 638
composition of, 692
Current liabilities
adjustments for changes in, 636, 638–639
accounts payable, 637, 638
income taxes payable, 637, 639
interest payable, 637, 638–639
ratio analysis of, 695–696, 703
Current ratio, 684, 691–692, 699, 702
Curvilinear costs, 904, 911
Customer(s)
balanced scorecard and, 731, 868, 869
use of financial statement analysis, 680
Customer-interaction software, 816
Customer loyalty programs, A-11
Customer orientation, 729–730
flexibility and standardization, 823, 824
lean business model and, 823
Customer relationships, 1037, 1048
Customer satisfaction, 858, 859
Customer service, 816
Customized production. *See* Job order cost accounting
CV. *See* **Cost variances**
CVP analysis. *See* **Cost-volume-profit (CVP) analysis**
CVP chart. *See* **Cost-volume-profit (CVP) chart**
Cycle efficiency (CE), 743, 748
Cycle time (CT), 742–743, 748
as nonfinancial performance measure, 859
in process operations, 814

Date of declaration, 513
Date of payment, 513, 515
Date of record, 513
Days' cash expense coverage ratio, 699*n*
Days' sales in inventory ratio, 694, 702
Days' sales uncollected ratio, 693–694, 702
Death of partner, 487, 493
Debenture bonds, 564, 576
Debentures, 563, A-14
Debt
fair value of, A-15
gain on retirement of, 656
issuing and repaying, 630
note regarding, A-14
Debt ratio, 695, 703
Debt securities, 592, 607
accounting for
acquisition of, 593
disposition of, 594
interest earned, 593–594
available-for-sale securities, 593, 596–597
bonds. *See* **Bond(s)**
common features of, 563, 575
disposition of, 594
held-to-maturity securities, 593, 596, 608
notes. *See* Notes payable
Debt-to-equity ratio, 563–**564,** 575, 695, 703
Decentralized companies, 875–877
Decision making. *See* Managerial decisions
Decision rule for applying NPV, 1029–1030
Declaration of cash dividend, 644
Defect rates, 859
Defined benefit pension plans, 575
Defined contribution pension plans, 575
Degree of operating leverage (DOL), 918, 921
Demand-pull system, 730
Department (subunit), 858
operating departments, 852, 853–854
production departments, 809
purchasing department, 986, 989
service departments
allocating expenses of, 862–865, 862*n*
cost allocation for, 852, 853
fixed costs of, 903
Departmental accounting system, 858–859
departmental evaluation, 858–859
motivation for departmentalization, 858, 879
reporting and analysis, 859
Departmental contribution reports, 866, 866*n,* 879
Departmental contribution to overhead, 866, 866*n,* 876
Departmental evaluation, 858–859
Departmental expense allocation, 859–869
contribution to overhead, 866, 866*n*
direct expenses, 859, 862, 879
indirect expenses, 859–860, 879
advertising expenses, 861, 863, 864
demonstration problem, 873, 874
depreciation expense, 861
illustration of, 860
rent and related expenses, 860–861, 863–864
service department expenses, 853, 861
utilities expense, 861, 863, 864
wages and salaries, 860
investment center performance, 867–869
Departmental income statements, 861–865, 879
allocation of direct and indirect expenses, 862
demonstration of, 875
departmental expense allocation sheet, 862–865, 862*n*
Departmentalization, 858, 879
Departmental rates, 853
Departmental spreadsheet analysis, 859
Department managers, 813, 820, 830, 858–859
Depletion method, change in, 705
Depreciation
adjustments for, 639
changes in depreciation methods, 705
as direct cost, 811
effects on NPV analysis, 1025, 1031
as factory overhead item, 813
as fixed cost, 903
as product cost for manufacturers, 738
Depreciation expense, 656, 861, 949
Depreciation tax shield, 1031
Differential costs, 1036
Diluted earnings per share, 524*n*
Dilutive securities, 524*n*
Direct costs, 732, 733, 811, 870
Direct expenses, 859, 862, 879
Direct labor, 738
computing cost per equivalent unit for, 818
computing EUP for, 818
in cost of goods manufactured, 746
flow of, 810, 813, 832
Direct labor budget, 960
Direct labor costs, 738, 740–741
Direct labor hours, 780, 787
Direct labor rate variance, 997
Direct materials, 735, 736, **738**
computing cost per equivalent unit for, 818
computing EUP for, 818
in cost of goods manufactured, 746
flow of, 810, 812, 832
incremental costs and, 1036
shown in manufacturing statement, 740–741
Direct materials budget, 959–960
Direct materials costs, 738
Direct method of reporting operating cash flows, **634,** 657
demonstration of, 648, 649
financing activities, 643–645
format of operating activities section, 657
indirect method compared, 633, 634–635, 637
investing activities, 641–642
operating cash payments, 636, 653–656
for interest and income taxes, 632, 655–656
for merchandise, 653–654
other expenses, gains, and losses, 656
for wages and operating expenses, 636, 654–655, 655*n*
operating cash receipts, 636, 653
from customers, 633, 653
other receipts, 653
summary of adjustments, 656

Disclosures
of noncash investing or financing activities, 631, 642
in notes to financial statements
in annual report, A-1
examples of, A-8–A-18
Discontinued segments, 703, 704
Discount(s), 999, 1004, B-2
Discount factors, 1029, 1030*n,* B-11
Discounting, 1024
Discount on bonds payable, 553–555
amortizing bond discount, 554–555, 575
effective interest amortization on bonds, 554, 569–570
straight-line amortization, 554–555
present value concepts, 558, B-10, B-11
recording, 553–554
Discount on stock, 511
Discount rate, B-2
in computing NPV, 1033, 1048
yielding present value factor, 1032
Dissolution of partnership, 478, 487
Distributable, 515
Distributed earnings, 598
Dividend(s), 513–516
in cash
accounting for, 513, 513*n*
declaration of, 644
deficits and, 513–514
on equity securities, 594
payout ratio on, 525
receipt of, under equity method, 598
reported as operating activities, 630
demonstration problem, 529–530
liability for, 518
stock dividends
accounting for, 514–516
cash flow not affected by, 644
large stock dividend, 514, 515–516
not considered noncash investing or financing, 631
reasons for, 514
small stock dividend, 514, 515
value of investment and, 516, 531
stock splits, 516
Dividend allocation table, 528
Dividend in arrears, 517–518
Dividend preference, 531
cumulative or noncumulative, 517–518
participating or nonparticipating, 518, 518*n*
Dividend rights of stockholders, 508
Dividend yield ratio, **525,** 530, 698
Division managers, 877
DOL (degree of operating leverage), 918, 921
Dollar changes, 683
Dollar sales, 913–914
Double taxation, 507
Dow Chemical Company, 552
Dukes, Kenny, B-2
Dun & Bradstreet, 680, 682, 691

Earnings per share (EPS), 524, 524*n,* 530
basic, 524
diluted, 524*n*
note regarding, A-15
sustainable income and, 704
Earnings quality, 646, 657, 699*n*
Earnings statement. *See* **Income statement**
Eastman Kodak Company, 593
E-cash, 634
Economic value added (EVA), 868
Economies of scale, 688
EDGAR (Electronic Data Gathering, Analysis, and Retrieval) database (SEC), A-1
Effective interest method of amortization, **569–**571
amortization table, 566
discount bond, 554, 569–570
premium bond, 556, 570–571
Efficiency, 681, 691, 858, 879
Efficiency variance, 989–990, **993**
Electronic Data Gathering, Analysis, and Retrieval (EDGAR) database (SEC), A-1
Employee(s)
active involvement in budgeting, 941, 942
employer investment in training, 782
fraud committed by, 731, 776
morale of, perceived unfairness and, 861
use of financial statement analysis, 680
Entrepreneur magazine, 851
Entrepreneurship
budgeting process, 939
cost management, 807, 851
CVP analysis, 901
equity financing, 505
financial statements and, 627, 679
international operations, 591
job order cost accounting, 769
liabilities management, 549
managerial accounting, 725, 979
managerial decisions, 1023
partnerships, 477
Environmental Protection Agency (EPA), 982
EPA (Environmental Protection Agency), 982
EPS. *See* **Earnings per share**
Equity
applicable to preferred shares, 526
effect of treasury stock on, 521
minority interest, 697
reporting. *See* Equity reporting
return on equity, 550, 551
"trading on the equity," 519, 550
transactions affecting, 630
unaffected by stock splits, 516
(*See also* **Stockholders' equity**)
Equity analysis, 644
Equity method, 593, **598–**599
Equity method with consolidation, 599
Equity ratio, 695
Equity reporting, 522–524
statement of retained earnings
closing process, 523
prior period adjustments, 523
restrictions and appropriations, 522–523
statement of stockholders' equity, 523, 524, A-6, A-22, A-27, A-32
stock option reporting, 523–524
Equity securities, 592, 607
acquisition of, 594
available-for-sale securities, 593, 596–597
disposition of, 594
dividend earned, 594
note regarding, A-13
Equity securities with controlling influence, 593, **599,** 607
Equity securities with significant influence, 593, **598–**599, 608
Equivalent units of production (EUP), 810, **815–**816, 832
cost per equivalent unit, 815
changes to, 820
computing, 818, 827, 829
demonstration of, 825
differences in, 815–816
demonstration of, 825
FIFO method of process costing
computing cost per unit, 827, 829
computing EUP, 828–829, 830
units started and completed, 828, 829
goods in process, 815
shown in process cost summary, 830, 831
weighted average method
computing cost per unit, 818
computing EUP, 817–818
Ernst & Young LLP, 477
Estimate(s), 921
changes in, 905, 909, 912
in job order costing, 777–778
yielded by CVP analysis, 911
Estimated line of cost behavior, 906
Ethics, 731
activity-based costing, 857, 879
added costs, 824, 832–833
in budgeting, 941, 961
capital budgeting, 1030, 1048
dissolution of partnership, 487, 493
in job order costing, 778, 786
in managerial accounting, 729, 748
"managing" the numbers, 914, 921
misclassification of cash flows, 630
process costing, 814, 832
spending budgeted amounts, 996, 1004
in treasury stock purchases, 520
EUP. *See* **Equivalent units of production**
EVA (economic value added), 868
Even cash flows
computing IRR with, 1032, B-11
computing payback period with, 1025–1026
Evidential matter, 700
EVP (executive vice president), 871
Excel® (Microsoft), 1030
computing IRR and NPV, 1047
in least squares regression, 907, 920
Excess cash, 592, 950
Executive summary, 700
Executive vice president (EVP), 871
Expected activities, 953
Expected performance, 940, 941
Expense(s), 870*n,* A-12
Expense accounts, 655*n*
Expropriation of property, 703
External users of accounting information, 628, 680, 728
Extraordinary gains and losses, 703–704
Extraordinary items, 704, 706
Exxon Mobil, 808

Face amount (face value) of bond, 550
Factory floor plan, 810
Factory overhead, 738
Factory overhead budget, 960
Factory overhead costs, 738
allocation of, 772, 778
components of, 813
in cost of goods manufactured, 746
demonstration of, 747
predetermined overhead rate for, 777, 778
in process cost accounting system, 810, 813–814
Factory overhead ledger, 775
Fair market value, 595
Falcon Cable Communications LLC, 487

Favorable variances, 981, 984, 986*n*
FDIC (Federal Deposit Insurance Corporation), 695
Federal Deposit Insurance Corporation (FDIC), 695
Federal Express (FedEx), 868
Federal laws on bond issuances, 552
Federated Department Stores, 628
FedEx (Federal Express), 868
Feedback from control function, 727
FIFO method of process costing, **827–**831, 832
  assigning and reconciling costs, 828, 829–830, 831
  computing cost per equivalent unit, 827, 829
  computing equivalent units of production, 828
    beginning goods in process, 828–829
    ending goods in process, 828, 829, 830
    units started and completed, 828, 829
  determining physical flow of units, 827–828
Fila, 553, 554, 558, 570
Financial Accounting Standards Board (FASB)
  on comprehensive income reporting, 600
  on reporting operating activities, 630
  Statements of Financial Accounting Standards (SFAS)
    No. 157: "Fair Value Measurements," 597
    No. 159: "The Fair Value Option for Financial Assets and Financial Liabilities—Including an amendment of FASB Statement No. 115," 597
Financial budgets, 949–953
  budgeted balance sheet, 952, 952*n*, 953, 958
  budgeted income statement, 952, 958
  cash budget, 945, 946, 948, 949, 950–951, 957
  components of, 944
  preparation of, 945
Financial calculators, 558, 1030, B-5
Financial condition of company, 686–687
Financial dimension of balanced scorecard, 731
Financial history and highlights, A-1, A-3
Financial information for managerial decisions, 1035
Financial leverage, 519, 550, 695
Financial reporting, 681
Financial results (balanced scorecard), 868, 869
Financial statement(s)
  budgeted, 961
    balance sheet, 952, 952*n*, 953, 958
    income statement, 952, 958
    statement of retained earnings, 958
  managerial accounting system and, 741, 742
  partnerships, 483–484
  preparation of, 820
Financial statement analysis, 678–706
  analysis reporting, 700
  building blocks of, 681, 705
  defined, **680**
  demonstration problem, 700–703
  horizontal analysis, 682–687
    comparative statements, 682–685
    dollar and percent changes, 683, 683*n*
    trend analysis, 685–687
  information for analysis, 681
  purpose of, 680, 705
  ratio analysis. *See* Ratio analysis
  standards for comparison, 682, 705
  sustainable income and. *See* Sustainable income
  tools of, 682, 705–706
  vertical analysis, 687–690
    common-size graphics, 689–690
    common-size statements, 684, 685, 687–689
Financial statement analysis reports, 700, 706
Financing, obtaining, hurdle rate and, 868
Financing activities, 630, 657
  cash flows from, 643–645, 657–658
    analysis of changes in, 650
    equity analysis, 644
    noncurrent liabilities, 643
    proving cash balances, 632, 637, 645
    sources and uses of cash, 645–646
    three-stage analysis of, 643
  graphical analysis of, 690
  monitoring, 684–685
Finished goods inventory, 736, 746
  assigning and reconciling costs, 818
  sale of, 810, 821
  term used by manufacturers, 737
  transfers to, 810, 821–822, 832
Finished Goods Inventory account, **772,** 785
First-in, first-out (FIFO) method. *See* **FIFO method** of process costing
Fiscal year, A-8
Fixed budget, 981, 1003
Fixed budget performance report, 981
Fixed cost(s), 732, 735
  break-even point and, 915
  in computing multiproduct break-even point, 916
  constant, 911
  cost behavior, 902–903
  estimating for high-low method, 906–907
  plotting for CVP chart, 910
  relevant range for, 903, 904
  variable costs distinguished, 982–983
Fixed overhead, 741
Fixed overhead cost variance, 991, 992–995
Flexibility, 823, 824
Flexibility of practice, 728
Flexible budget(s), 980–985, 1003, 1004
  budgetary process, 980–982
    budgetary control and reporting, 980–981
    budget reports for evaluation, 982
    fixed budget performance report, 981
  budget reports, 982–985
    flexible budget performance report, 984, 985, 1001, 1005
    preparation of, 981, 982–984
    purpose of, 982
  defined, **982**
Flexible budget performance report, 984, 985, 1001, 1004
Flexible overhead budget, 991
Floating an issue, 552*n*
Flow of manufacturing activities and costs, 739–740, 748
  materials activity, 739
  production activity, 739–740
  sales activity, 740
Focus of information, 727, 728–729
Fool.com, 679
Ford Motor Company, 823, 915
Foreign currencies, 608
  common-size financial statements and, 687
  foreign exchange rate and, 605, 606, 608
  international subsidiaries, 607
  note regarding, A-10
  ratio analysis and, 693
  sales and purchases in, 606–607, 608
  value of U.S. dollar against, 607, 608
Foreign exchange rate, 605, 606, 608
Formats of financial statements, 631, 657, 983
Form 10-K, A-1
Form 10-KSB, A-1
Forms of business organization
  corporate. *See* **Corporations**
  partnerships, 492
    characteristics, 478–479
    limited liability company, 480
    limited liability partnerships, 479
    limited partnerships, 479
    S corporations, 480
Fractional shares in partnerships, 482
Franklin, Benjamin, 567
Fraud, 731, 748, 776
Free cash flows, 646, 699*n*
Fringe benefits, 575
Full-disclosure principle
  amount of preferred dividends in arrears, 518
  noncash investing and financing, 631
Function, classification of costs by, 733–734
Futura Computer, 854
Future value(s)
  concepts, B-1, B-7
  of ordinary annuity, B-6–B-7, B-10, B-11
  of single amount, B-3–B-4, B-7, B-10
Future value calculations, 1031, B-10
Future value of annuity of 1 table, B-6, B-7, B-11
Future value of 1 table, B-4, B-6, B-7, B-10

Gains, 656
  from bond retirement, 559
  from disposition of segments, not considered extraordinary items, 704
  extraordinary, 703–704
  infrequent, 703
  from liquidation of partnership, 488, 488*n*
  nonoperating items, 639–640
  on retirement of debt, 637, 640, 656
  on sale of trading securities, 595–596
  unrealized, 595, 596, 602
  unusual, 703
  (*See also* Losses)
Gap, Inc., 592, 599, 601, 640, 1041
Gardner, David, 679
Gardner, Tom, 679
Garza, Mary E., 629
General accounting system, 770
General and administrative expense budget, 945, **948–**949, 957
General and administrative expenses, A-12
General ledger, cash account, 632–633
Generally accepted accounting principles (GAAP), managerial accounting and, 728
General Motors, 730
General partner, 479
General partnership, 479
General-purpose financial statements, 681
General rights of stockholders, 508
Gentilcore, Vince, 979
Geographic information, A-18
Gift cards, A-11
Gilts, 552
Global economy, 730
Global issues
  balanced scorecard, 868
  corporate labels, 508
  domestic v. international cash flows, 644
  gilts, 552
  preference to debtholders v. stockholders, 562
  reporting standards. *See* International Financial Reporting Standards
  Russian market, 605

Global issues—*Cont.*
in transfer pricing, 875
value of U.S. dollar, 607
(*See also* **Multinationals**)
Goals in budgets, 941
The Gold Standards, 730
Goods in process inventory, 736, 745–746
assigning and reconciling costs, 818
equivalent units of production, 815
FIFO method of process costing, 828
beginning goods in process, 828–829
ending goods in process, 828, 829, 830
Goods in Process Inventory account, **772**
recognition of, 772, 785, 786
unfavorable variances in, 996–997
Goodwill, A-9
Government bonds, 552
Government regulation, 507
Grades of material, 985
Graphical analysis, 921
balanced scorecard reporting, 868, 869
of break-even time, 1043
common-size graphics, 689–690
in CVP analysis
cost behavior, 903
cost-volume-profit (CVP) chart, 910, 919
high-low method, 906
scatter diagrams, 905–906, 907, 912, 921
step-wise and curvilinear costs, 904
overhead variances, 993, 994, 995
trend percent analysis, 686
Green Bay Packers, 508
Greenberg, Jared, 1023
Gross profit, decline in, 688–689
Growth, balanced scorecard and, 731
Growth companies, 684
Growth stocks, 525
Guidelines (rules of thumb), 682

H. J. Heinz Co., 808
Harley-Davidson, 516, 640, 875
Harris, Jim, 904
Health care industry
activity-based costing in, 855
process operations in, 809
standard costing in, 996
Held-to-maturity (HTM) securities, 593, **596,** 608
Hendrix, Jimi, 979
Hershey, 808
Heterogeneity of products, 771
Hewlett-Packard, 859
Hierarchical levels of management, 733, 734
High-low method, 905, **906–**907, 921
Historical costs, 1037
The Hockey Company, 951
Home Depot, 523, 573
Honda, 730, 915
Hood River Juice Company, 807
Horizontal analysis, 682–687, 706
comparative financial statements, 682–685
comparative balance sheets. *See* Comparative balance sheets
comparative income statements, 685
computation of dollar and percent changes, 683, 683*n*
trend analysis, 685–687
Hostile takeover, 518*n*
HRJCO.com, 807
HTM (held-to-maturity) securities, 593, **596,** 608
Human behavior, 941
Hurdle rate, 868, 1033, 1048
Hybrid costing system, 823–824, 832

IBM, 551
Ideal standard, 986
IMA (Institute of Management Accountants), 728, 731
Immaterial variances, 998
Impairment of asset value, A-9
Income
comprehensive, A-16
computed from sales and costs, 913–914, 919
effect of automation on income, 915, 918
investment center residual income, 867–868
partnerships
allocation on capital balances, 482
allocation on services, capital, and stated ratios, 482–483
allocation on stated ratios, 482
target income, 913–914, 917, 919, 921
Income-and-loss-sharing ratio method, 482, 488*n,* 489
Income statement
budgeted income statement, 952, 958
common-size, 688–689
comparative, 685, 701–702
consolidated, A-5, A-21, A-26, A-31
contribution margin income statement, 909, 910
cost flows to, 742
departmental contribution to overhead shown in, 866, 866*n*
departmental income statements, 861–865, 879
allocation of direct and indirect expenses, 862
demonstration of, 875
departmental expense allocation sheet, 862–865, 862*n*
earnings from long-term investments shown in, 604, 605
form and content of, 607
manufacturer's income statement, 737–739, 747
different from merchandiser's statement, 745, 746, 748
direct labor, 738
direct materials, 738
factory overhead, 738
prime and conversion costs, 738
reporting performance in, 738, 739
in preparing statement of cash flows, 634
variable costing income statement, 909, 910
Income stocks, 525
Income taxes
corporate, 507
interest on bonds deductible, 550
for limited liability companies, 480
note regarding, A-10, A-17
not paid by S corporations, 480
partnerships, 478
payable, budgeting for, 951
reporting operating cash payments for, 632, 655–656
taxes payable, adjustments for changes in current liabilities, 637, 639
tax exemption for municipal bonds, 563
Income tax expense, 948
Income variance, 984
Incorporation, 507
Incorporators (promoters), 507, 512
Incremental costs, 1036, 1037, 1046
Incremental overhead rate, 1037
Incremental revenue, 1035, 1039
Index numbers, 685, 686
Index number trend analysis, 685–687
Indirect costs, 732, 733
classification of, 814, 832
in process cost accounting system, 811
Indirect expenses, 859, 879
allocation of. *See* Departmental expense allocation
direct expenses compared, 859–860
Indirect labor, 738, 810, 813
Indirect labor costs, 738
Indirect materials, 735–736
flow of, 810, 812
included in overhead, 775
recording use of, 775
Indirect method of reporting operating cash flows, **634,** 635–640, 657
adjustments to net income, 635
analysis of changes, 650
for changes in current assets and liabilities, 636–639
for nonoperating items, 639–640
for operating items not providing or using cash, 639
demonstration of, 648–649
direct method compared, 633, 634–635, 637
spreadsheet preparation of statement, 636, 649–650, 651
summary of adjustments, 633, 637, 640
Industry benchmarks, 682
Industry comparisons, 687
Industry Norms & Key Business Ratios (Dun & Bradstreet), 691
Inescapable expenses, 1041
Inferences, in analysis reports, 700
Inflation, NPV analysis and, 1031, B-10
Influential investments, 598–600
with controlling influence, 593, 599, 607
with significant influence, 593, 598–599, 608
Informal communications, 941
Information storage, 728
Information technology budgets, 1024
Infrequent gains and losses, 703
Initial public offering (IPO), 509
Innovation/learning, 868, 869
Inogen, 505
Inogen.net, 505
Inspection time, 743
Installment notes, 560–562, 562*n,* 565, B-11
Institute of Management Accountants (IMA), 728, 731
Insurance, allocation of indirect expense, 863, 864
Intangible assets, A-9
Intercompany comparisons, 690
Interest
accrued
on bonds, 557, 572–573
on installment notes payable, 561
on bonds, 575
accrued interest, 557, 572–573
issuing bonds between interest dates, 571–573, 575
market rate, 553
payment of, 551
recording semiannual obligation, 553
tax deductible, 550
capacity to pay fixed interest charges, 695
times interest earned ratio, 695–696, 703
"Interest allowances," 482–483
Interest dates, issuing bonds between, 571–573, 575

Interest expense, 948
Interest in business, 484–485, 697
Interest income, A-16
Interest method. *See* **Effective interest method** of amortization
Interest payable, 572, 637, 638–639
Interest payments, 630, 632, 655–656
Interest rate, 553, 569
Interest revenue, 593–594, 630
Internal auditing, 728
Internal control systems, 731
Internal processes, 731, 868, 869
Internal rate of return (IRR), 1030, **1031–**1033, 1048, B-11
  compared with other methods, 1033, 1034
  computing with Excel®, 1047
  with uneven cash flows, 1032, B-11
  use of, 1033
Internal users of accounting information
  financial statement analysis and, 680
  managerial accounting information, 727, 728
  statement of cash flows, 628
International Financial Reporting Standards (IFRS) of IASB
  on consolidated subsidiaries, 599
  on effective interest method, 571
  preferred stock requirements, 519
  on reporting cash flows, 657
International operations, 605–607
  consolidated financial statements, 607
  by entrepreneurs, 591
  foreign exchange rates and, 605, 606
  sales and purchases, 606–607
Intracompany benchmarks, 682
Inventory(ies)
  adjustments for changes in, 637–638
  budgeted, 956
  choice of inventory method, 692
  for manufacturer v. merchandiser, 745, 746, 748
  note regarding, A-8
  ratio analysis of
    days' sales in inventory ratio, 694, 702
    inventory turnover ratio, 693, 702
  term used by merchandisers, 737
Inventory turnover ratio, 693, 699, 702, 859
Investing activities, 630
  cash flows from, 630, 657–658
    analysis of changes in, 650
    noncurrent asset analysis, 641–642
    other asset transactions, 642
    short-term assets, 630
    sources and uses of cash, 645–646
    three-stage analysis of, 641
  graphical analysis of, 690
  monitoring, 684–685
InvestingInBonds.com, 558
Investment(s), 590–608
  accounting summary for, 599–600
  basics of, 592–594
    classification and reporting of, 593, 607
    for debt securities, 593–594
    for equity securities, 594
    motivation for, 592
    short-term v. long-term, 592
  comprehensive income from, 600
  demonstration problems
    long-term investments, 602–605
    short-term investments, 601–602
  evaluating using return on total assets, 869, 879
  influential, reporting, 598–600, 607
    securities with controlling influence, 593, 599, 607
    securities with significant influence, 593, 598–599, 608
  in international operations, 605–607
    consolidated financial statements, 607
    foreign exchange rates and, 605, 606
    sales and purchases, 606–607
  noninfluential, reporting, 595–597, 607
    available-for-sale securities, 596–597, 600
    held-to-maturity securities, 596
    trading securities, 595–596
  notes regarding, A-10, A-13
  in partnership
    accounting for, 481
    bonus to new partner, 485–486
    bonus to old partners, 484, 485
  strategies for holding, 596, 608
Investment center(s), 858, 859
  evaluating performance of
    investment center residual income, 867–868
    investment center return on total assets, 867
    with nonfinancial measures, 868–869
Investment center residual income, 867–868
Investment center return on total assets, 867, 872
Investment turnover, 872
Investors, 508, 518*n*
IPO (initial public offering), 509
IRR. *See* **Internal rate of return**

Jibbitz, 939
Jibbitz.com, 939
Jiffy Lube, 992
JIT (just-in-time) inventory systems, 947
JIT (just-in-time) manufacturing, 730, 811, 830, 833
Job, 770
Job cost sheets, 772, 773, 775, 786
  factory overhead recorded, 777
  labor costs, 776
  materials requisitioned, 775
  summary, 779
Job lot, 771
Job order cost accounting, 768–786
  allocation rates (ethical issue), 778, 786
  cost accounting system, 770
  demonstration problem, 783–786
  events in job costing, 771, 772, 784
  job cost sheet, 772, 773
  job order cost flows and reports, 773–781
    labor cost flows and documents, 773, 775–776, 784, 786
    materials cost flows and documents, 773–775
    overhead cost flows and documents, 773, 776–778
    summary of, 778–780
  job order production, 770–771
  overapplied overhead, 782
  pricing for services, 782, 783, 786, 787
  underapplied overhead, 779, 781
Job order cost accounting system, 770, **772, 811,** 832
Job order operations, 809
Job order production, 770–771, 786
Jobs produced on speculation, 771
Joint costs, 877–878, 879
Joint relation between sales and expenses, 687, 706
Joseph, Kelvin, 477
Joseph, Samanta, 477
Journal entries
  for bond amortization, 570
  in job order costing, 778, 780, 785
  for long-term investments, 603, 604
  relating to stockholders' equity, 527–528, 529
  summary entries, 812
Journalizing transactions, 492
Julicher, Hank, 769
JungleJim.com, 627
Jungle Jim's International Market, 627
Junk bonds, 560
Just-in-time (JIT) inventory systems, 947
Just-in-time (JIT) manufacturing, 730, 811, 830, 833

Kellogg, 808
Kernel Season's, 725
KernelSeasons.com, 725
Key factors in analysis reports, 700
Key performance indicators, 868, 869

Labor contract negotiations, 645, 658
Labor costs
  cost flows and documents, 773, 776–778, 784, 786
  differences in equivalent units, 815–816
  direct labor
    computing cost per equivalent unit for, 818
    computing EUP for, 818
    in cost of goods manufactured, 746
    flow of, 810, 813, 832
  flow of indirect labor, 810, 813
  indirect labor costs, 738
  in process cost accounting system, 810, 813
  quantity and price variations in, 985–986
Labor cost variances, 1004
  demonstration of, 1002
  investigating causes of, 990, 1004
  rate and efficiency variance, 989–990
Labor hours, 993
Labor operations, 985
Labor strikes, 704
Labor unions, 680
Lack of mutual agency, 506
Land, 631
Large stock dividend, 514, 515–516
Lean business model, 729–730, 748, 823
Lean practices, 729–730
Learning, 731
Lease(s), 573
  lease liabilities, 573–574
    capital leases, 574, 574*n*
    operating leases, 573
  note regarding, A-9, A-16–A-17
Lease rights, A-9
Lease term, 574*n*
Least-squares regression, 907
  measurement of past behavior, 905, 907, 920, 921
  prepared using Excel®, 907, 920
Legal issues
  prohibition against use of asset, 703
  statutory restrictions on retained earnings, 522
Lenders (creditors), 629
Lending, 692, 696, 706
Lessee, 573
Lessor, 573
Letter to shareholders, A-1
Liabilities
  long-term. *See* **Long-term liabilities**
  note regarding, A-10
  partnerships, 487, 493
  unlimited, 479

Lil' Romeo, 549
Limited liability company (LLC), 480
Limited liability of stockholders, 506
Limited liability partnership (LLP), 479
Limited life of partnership, 478
Limited partners, 479
Limited partnership (LP), 479
Lindsay, James "Fly," 549
Linear costs. *See* Variable cost(s)
Linear programming, 1041
Line graph, 686
Link Wood Products, 859
Liquidating cash dividend, 514
Liquidity, 681, 691
Liquidity and efficiency ratios, 691–694, 699
  accounts receivable turnover, 693, 702
  days' sales in inventory, 694, 702
  days' sales uncollected, 693–694, 702
  inventory turnover, 693, 699, 702
  total asset turnover, 601, 608, 694, 696, 703
List of directors and officers, A-1
L.L. Bean, 808
LLC (limited liability company), 480
LLP (limited liability partnership), 479
Loans, repayment of, 951
Long-range operating budgets, 944
Long-term (noncancelable) leases. *See* **Capital leases**
Long-term assets, 630
Long-term investments, 592, 600, 607
  demonstration problem, 602–605
  note regarding, A-13
  short-term investments compared, 592
  shown in balance sheet, 604, 605
Long-term liabilities, 548–576
  bonds. *See* **Bond(s)**
  demonstration problem, 564–567
  discount on bonds payable, 554
  lease liabilities, 573–574
    capital leases, 574, 574*n*
    operating leases, 573
  note regarding, A-10
  pension liabilities, 575
  ratio analysis
    debt-to-equity ratio, 563–564, 575, 695, 703
    features of bonds and notes, 563
  transactions affecting, 630
Long-term notes payable, 560–562
  installment notes, 560–562, 562*n*, B-11
  mortgage notes and bonds, 562
  (*See also* **Bond(s);** Notes payable)
Long-term planning, 726
Long-term strategic plans, 940
Losses, 656
  from bond retirement, 559
  from disposition of segments, 704
  extraordinary, 703–704
  infrequent, 703
  from liquidation of partnership, 488, 488*n*
  nonoperating items, 639–640
  on sale of assets, 637, 640, 656
  on sale of trading securities, 595–596
  sharing agreement loss (deficit), 482
  unrealized, 595, 596, 602
  unusual, 703
  (*See also* Gains)
Lottery winners, B-1, B-5
LP (limited partnership), 479

Machine hours, 814
Macy's, 628
Make or buy decision, 1037–1038, 1042, 1044, 1045
Malcolm Baldrige National Quality Award (MBQNA), 730
Management
  budgeting as tool of, 941
  hierarchical levels in, 733, 734
  influence on budgets, 944
  middle- and lower-level managers, 729
    department managers, 813, 820, 830
    division managers, 868, 872, 877
    evaluating performance of, 820, 830, 858–859
    in service companies, 735
  summary reports for top-level managers, 871
  (*See also* Responsibility accounting)
Management accounting. *See* Managerial accounting
Management by exception, 996, 1003
Management Discussion and Analysis (MD&A), 681, A-1
Management of corporation, 507–508
Management report on controls, A-1
Managerial accounting
  demonstration problems, 743–747
    cost behavior and classification, 743–744
    manufacturing statement, 746–747
    reporting, 744–746
  for entrepreneurs, 725, 979
  ethical issues in, 729, 748
  job order cost accounting. *See* Job order cost accounting
  managerial cost concepts, 732–735
    cost classifications, 732–734, 748
    identification of classifications, 734, 735
    for service companies, 734–735
  performance measurement. *See* Performance evaluation
  process cost accounting. *See* Process cost accounting
  reporting manufacturing activities, 735–742, 748
    demonstration problem, 744–746
    flow of activities, 739–740
    manufacturer's balance sheet, 735–736
    manufacturer's income statement, 737–739
    manufacturing statement, 740–742
Managerial accounting, 726
  allocation of costs. *See* Cost allocation
  basic concepts, 726–732, 748
    business environment and, 729–731
    fraud and ethics, 731, 748
    managerial decision making. *See* Managerial decisions
    nature of, 727–729
    purposes, 726–727
  budgeting. *See* **Budgeting; Capital budgeting; Flexible budget(s)**
  cost-volume-profit analysis. *See* **Cost-volume-profit (CVP) analysis**
  cycle efficiency, 743, 748
  cycle time, 742–743, 748, 814, 859
Managerial cost concepts, 732–735
Managerial decisions, 729, 1034–1042, 1048
  about capital budgeting. *See* **Capital budgeting**
  demonstration problem, 1043–1046
  by entrepreneurs, 1023
  relevant costs and, 1035, 1036–1037
  role of managerial accounting in, 726
  scenarios for, 1035–1042
    additional business, 1036–1037, 1042, 1043–1044, 1045, 1048
    make or buy, 1037–1038, 1042, 1044, 1045
    qualitative decision factors, 1042
    sales mix selection, 1040–1041
    scrap or rework, 1038–1039, 1044, 1046
    segment elimination, 1041–1042
    sell or process, 1039, 1040
  steps in decision making, 1034–1035
  use of CVP analysis in. *See* **Cost-volume-profit (CVP) analysis**
Managers. *See* Management
Manufacturers
  budgeting for, 961
    manufacturing budgets, 946, 959–960
    operating budgets, 944
    production budgets, 946, 947, 949
  depreciation as product cost for, 738
  inventories of
    finished goods inventory, 737
    merchandisers compared, 745, 746, 748
Manufacturer's balance sheet
  different from merchandiser, 745, 748
  finished goods inventory, 736
  goods in process inventory, 736
  raw materials inventory, 735–736
Manufacturer's income statement, 737–739, 747
  different from merchandiser's statement, 745, 746, 748
  direct labor, 738
  direct materials, 738
  factory overhead, 738
  prime and conversion costs, 738
  reporting performance in, 738, 739
Manufacturing activities
  job order manufacturing. *See* Job order cost accounting
  just-in-time (JIT) manufacturing, 730, 811, 830, 833
  reporting. *See* **Managerial accounting**
Manufacturing budgets, 946, **959–**960, 961
Manufacturing statement, 740–742, 748
  demonstration problem, 746–747
  direct labor costs, 740, 741
  direct materials computations, 740–741
  job order costing, 786
  product costs in, 733
Margin of safety, 914
Marketable securities. *See* **Short-term investments**
Market-based transfer price, 876
Market prices, 877
Market prospects, 681
Market prospects ratios, 699
  dividend yield ratio, 525, 530, 698
  price-earnings ratio, 525, 530, 531, 697–698
Market rate of interest on bonds, **553**
Market value
  adjustment of AFS securities to, 596, 597
  of bonds traded, 551, 558
  reporting noninfluential investments, 595
  reporting securities at, 593
  of stock or noncash asset, 512
Market value method, 599
Market value per share, 509
Martin, Christian F., 979
MartinGuitar.com, 979
Martin Guitar Company, 979
Master budgets, 944–953, 960, 961
  capital expenditures budget, 949, 950
  components of, 942–945
  financial budgets, 949–953
    budgeted balance sheet, 952, 952*n*, 953, 958
    budgeted income statement, 952

cash budget, 945, 946, 948, 949, 950–951, 957
components of, 944
preparation of, 945
operating budgets, 946–949
components of, 944
general and administrative expense budget, 945, 948–949
merchandise purchases budget, 947–948
sales budget, 941, 945, 946, 956
selling expense budget, 948
Materiality constraint
classification of costs and, 736
variances and, 998
Materials activity, flow of, 739, 740
Materials and labor standards, 985–986
Materials consumption report, 812
Materials costs
cost flows and documents, 773–775, 784, 786
differences in equivalent units, 815–816
direct materials
computing cost per equivalent unit for, 818
computing EUP for, 818
in cost of goods manufactured, 746
flow of, 810, 812, 832
incremental costs and, 1036
shown in manufacturing statement, 740–741
indirect materials, 735–736, 775, 810, 812
in process cost accounting system, 810, 812
role of purchasing department, 986, 989
standards for, 985
Materials cost variances, 988–989, 1002, 1004
Materials ledger cards, 773, 774
Materials requisitions, 774, 775, 777, 778, 812
Maturity date of bond, 550, 551, 559–560
MBQNA. *See* Malcolm Baldrige National Quality Award
McCartney, Paul, 979
McDonald's, 605
McGraw-Hill, 599
MD&A (Management Discussion and Analysis), 681, A-1
Median values, 683, 683*n*
Medium-term strategic plans, 940
Members in LLCs, 480
Merchandise, cash paid for, 653–654
Merchandise purchases, 950–951
Merchandise purchases budget, 947
demonstration of, 956
just-in-time inventory systems and, 947
preparation of, 945, 947–948
safety stock and, 947
Merchandisers, budgeting for, 946
merchandise purchases budget, 947–948
operating budgets, 944
Mercury Marine, 599
Mergers and acquisitions, 945
Microsoft Corporation, 592
Microsoft Excel®
computing IRR and NPV, 1047
in least squares regression, 907, 920
Minimum cash balance, 951
Minimum legal capital, 509
Minority interest, 697
Misclassification of cash flows, 630
Mixed costs, 732, **903–**904, 983
Moes.com, 901
Moe's Southwest Grill, 901
Monthly budgets, 942, 943
Moody's, 555, 680, 682
Mortgage, 562
Mortgage bonds, 562
Mortgage contract, 562
Mortgaged assets, 563
Mortgage notes, 562
Most likely cost estimates, 915
Motivation for investments, 592
The Motley Fool, 679
Motorola, 858
Mt. Macadamia Orchards LP, 479
Move time, 743
Multinationals, 605, 605–607
consolidated financial statements, 607
foreign exchange rates and, 605, 606
sales and purchases, 606–607
Multiproduct break-even point, 915–917, 921
Municipal bonds ("munis"), 563
Murto, Bill, 904
Musicland, 562
Mutual agency, 478–479, 506
Myers, Byron, 505

National Hockey League (NHL), 951
National Venture Capital Association, 508
Negative values, 683
Negotiated transfer price, 877
Net cash flows
cumulative total of, 1026
investments expected to produce, 1025
net present value method, 1029
present value of, 1043
Net cash inflow, 629
Net cash outflow, 629
Net cost of goods manufactured, 741
Net income
computed using accrual accounting, 635
from long-term investments, 604, 605
operating cash flows contrasted, 640
undistributed, book value of investment and, 598
Net income per share. *See* **Earnings per share**
Net interest income, A-16
Net present value (NPV), 1025, **1029–**1031, 1048
compared with other methods, 1033, 1034
computing with Excel®, 1047
decision rule for, 1029–1030
demonstration of, 1044–1045, 1046
inflation and, 1031, B-10
salvage value and accelerated depression, 1025, 1031
simplified computations, 1030, 1030*n,* B-11
with uneven cash flows, 1030–1031
using, 1031
Netscape, 507
Net working capital, 691
New York Stock Exchange (NYSE), 551, 697
NHL (National Hockey League), 951
Nike, 605, 646, 700, 771
Nissan, 730, 877
No-free-lunch adage, 902
Nominal rate on bonds, 553
Noncancelable (long-term) leases. *See* **Capital leases**
Noncash accounts, 633–634
Noncash assets, 512, 517
Noncash charges, 635
Noncash credits, 635
Noncash expenses, 639
Noncash investing and financing activities, 631, 642, 657
Noncash revenues, 639
Non-controllable costs, 733
Noncumulative preferred stock, 517–518
Noncurrent assets, 641–642
Noncurrent liabilities, 632, 636, 637, 643
Nonfinancial criteria, 877, 1032, 1038
Nonfinancial information, 727, 1035
Nonfinancial performance measures, 859
Noninfluential investments, 595–597
available-for-sale securities, 596–597, 600
held-to-maturity securities, 596
trading securities, 595–596
Nonlinear costs, 904
Nonoperating items, 639–640
Nonparticipating preferred stock, 518
Non-value-added time, 743
No-par value stock, 509, 511, 517
Notes payable
analysis of, 632, 636, 637, 643
convertible and/or callable, 563
long-term, 560–562, 575
installment notes, 560–562, 562*n,* B-11
mortgage notes and bonds, 562
(*See also* **Bond(s)**)
present values of, 567–569
applying present value table, 568
compounding periods shorter than one year, 569
present value concepts, 567
present value of annuity, 568–569, B-11
present value tables, 567, 568, B-10
secured or unsecured, 563
term or serial, 563
Notes receivable, 630
Notes to financial statements
in annual report, A-1
examples of, A-8–A-18
advertising costs, A-12
business description, A-8
cash and cash equivalents, A-8
comprehensive income, A-16
convertible debentures, A-14
cost of goods sold, A-12
customer loyalty programs, A-11
debt, A-14, A-15
earnings per share (EPS), A-15
equity securities, A-13
expenses, A-12
fiscal year, A-8
foreign currency translation, A-10
geographic information, A-18
gift cards, A-11
goodwill, A-9
impairment of asset value, A-9
income taxes, A-10, A-17
intangible assets, A-9
inventories, A-8
investments, A-10, A-13
lease rights, A-9
leases, A-9, A-16–A-17
long-term liabilities, A-10
net interest income, A-16
plant assets, A-8–A-9
revenue recognition, A-10–A-11
sales incentives, A-11
selling, general, and administrative expenses, A-12
short-term investments, A-13
noncash investing and financing, 642
NPV. *See* **Net present value**
NVCA.org, 508
NYSE (New York Stock Exchange), 551, 697

Off-balance-sheet financing, 574*n*
Old Navy, 599
100% quality inspection, 824, 832–833
On-time deliveries, 859
Operating activities, 629–630, 657
cash flows from. *See* Operating cash flows
indirect method of reporting
adjustments to net income, 635, 636–640, 650
demonstration of, 648–649
direct method compared, 633, 634–635, 637
summary of adjustments, 633, 637, 640
Operating budgets, 946–949, 961
components of, 944
general and administrative expense budget, 945, 948–949
merchandise purchases budget, 947–948
sales budget, 941, 945, 946, 956
selling expense budget, 948
Operating cash flows, 634–640
direct method of reporting, 652–657
demonstration of, 649
financing activities, 643–645
format of operating activities section, 657
indirect method compared, 633, 634–635
investing activities, 641–642
operating cash payments, 636, 653–656
operating cash receipts, 636, 653
summary of adjustments for, 656
indirect method of reporting, 634, 635–640, 657
adjustments to net income, 635, 636–640, 650
demonstration of, 648–649
direct method compared, 633, 634–635, 637
spreadsheet preparation of, 636, 649–650, 651
summary of adjustments, 633, 637, 640
net income contrasted, 640
sources and uses of cash, 645–646
Operating cash flow to sales ratio, **646**
Operating cash payments, 636, 652, 653–656
for interest and income taxes, 632, 655–656
for merchandise, 653–654
other expenses, gains, and losses, 656
for wages and operating expenses, 636, 654–655, 655*n*
Operating cash receipts, 633, 636, 652, 653
Operating departments, 852, 853–854
Operating efficiency, 696, 697
Operating expenses, 636, 654–655, 655*n*
Operating leases, 573
Operating leverage, 918
Opportunity costs, 733, 734, 749, 1035
of additional business decision, 1037
of scrap or rework decision, 1038
in transfer pricing, 876
Optimistic cost estimates, 915
Ordinary annuity
future value of, B-6–B-7, B-10, B-11
present value of, B-2, B-5, B-7, B-10, B-11
Organizational responsibility chart, 870
Organization expenses, 507, 512
Other comprehensive income, 600
Out-of-pocket costs, 733, 1035
Output measures, 911
Outsourcing, 854, 1038
Outstanding stock, 509, 511, 526
Overapplied overhead, 781, **782,** 786, 814
Overfunded pension plan, 575
Overhead
actual v. allocated, 780
allocation of overhead costs. *See* Overhead cost allocation
computing cost per equivalent unit for, 818
computing EUP for, 818
cost flows and documents, 773, 776–778, 784, 786
departmental contribution to, 866, 866*n,* 876
depreciation, rent, and utilities as, 813
differences in equivalent units, 815–816
fixed overhead, 741
flow of costs, 810, 813–814, 832
indirect materials included in, 775
overapplied, 781, 782, 786, 814
in process cost accounting system, 810, 813–814
underapplied, 779, 781, 786, 814
variable, 741
Overhead allocation rate, 852
Overhead cost(s), 733
incremental, 1036, 1037
shown in manufacturing statement, 741
Overhead cost allocation, 852–857
activity-based method, 854–856
actual overhead and, 780
allocation base for, 777
allocation rate formula, 778
allocation rates, 778, 786
comparison of methods, 855, 856–857
factory overhead costs, 772, 778
two-stage method. *See* Two-stage cost allocation
using direct labor hours, 780, 787
Overhead cost rates, 854
Overhead cost variances, 992, 1004
demonstration of, 1002–1003
fixed, 991, 992–995
graphical analysis of, 993, 994, 995
reporting, 994–995
variable. *See* Variable overhead cost variance
Overhead standards and variances, 990–995
overhead cost variance analysis, 992
setting standards, 990–992
variable and fixed variances, 992–995
Overhead variance reports, 994–995
Overpriced stocks, 698
Owner control, 550
Ownership, 506, 574*n*

Paging Network, 915
Paid-in capital, 510
Paid-in capital in excess of par value, 511
Parent, 599
Participating preferred stock, 518, 518*n*
Participatory budgeting, 941, 942, 945, 946, 956
Partner return on equity ratio, **490,** 492
Partnership(s), 476–493
admission of partner, 484–486, 492–493
investing assets in partnership, 485–486
purchase of interest, 484–485
basic accounting for, 481–484, 492
dividing income or loss, 481–483
financial statements, 483–484
organizing partnership, 481
death of partner, 487
defined, **478**
demonstration problem, 490–492
form of organization, 492
characteristics, 478–479
limited liability company, 480
limited liability partnerships, 479
limited partnerships, 479
S corporations, 480
liquidation of, 487–489, 493
capital deficiency, 489
no capital deficiency, 488, 488*n*
partner return on equity ratio, 490, 492
statement of partners' equity, 483–484, 491–492
Uniform Partnership Act, 479
withdrawal of partner, 485, 493
bonus to remaining partners, 486–487
bonus to withdrawing partner, 487
no bonus, 486
Partnership contract, 478, 482, 487
Partnership liquidation, 487–489, 493
capital deficiency, 489
defined, **488**
no capital deficiency, 488, 488*n*
Par value, 509
Par value method, 520
Par value of bond, 550, 551
Par value stock, 509, 510, 511
Past performance, 940, 941, 982
Payback period (PBP), 1025–1027, 1048
compared with other methods, 1033, 1034
computing with even cash flows, 1025–1026
computing with uneven cash flows, 1026
using, 1026–1027
Payless, 735
Payments
of interest on bonds, 551
operating cash payments, 636, 652, 653–656
for interest and income taxes, 632, 655–656
for merchandise, 653–654
other expenses, gains, and losses, 656
for wages and operating expenses, 636, 654–655, 655*n*
payment pattern on installment notes, 561
repayment of loans, 951
from trading securities, 629–630
Payout ratio, 525, 699*n*
Payroll activities, 1038
employee fraud, 776
payroll costs, 810, 813
PBP. *See* **Payback period**
PE (price-earnings) ratio, 525, 530, 531, 697–698
Pecking order, 733, 734
Penn Inc., 808, 868
Pension liabilities, 575
Pension plans, 575
defined benefit/defined contribution, 575
plan assets, 599, 699*n*
Pension recipients, 575
Pepsi Bottling, 812
PepsiCo, 682
Percent changes, computing, 683, 683*n*
Performance evaluation
cost allocation for, 858–872
demonstration problem, 872–875
departmental accounting, 858–859
departmental expense allocation, 859–869
profit margin and investment turnover, 872
responsibility accounting, 869–872, 870*n*
manager evaluation
department managers, 820, 830, 858–859
division managers, 868, 872
Performance reporting, 738, 739
fixed budget performance report, 981
in manufacturer's income statement, 738, 739
in responsibility accounting, 870–871
for service companies, 996
variances identified in, 987

Period costs, 733, 734, 738, 748
Periodic inventory system, 770
Permanent accounts, 595
Perpetual inventory system, 770, 777–778
Perry, Ali, 505
Personalized products, 771
Personal transactions, 484–485
Pessimistic cost estimates, 915
Petroleum refining, 808
Pfizer, 592, 808
Physical basis of allocating joint costs, 877, 878
Physical flow reconciliation, 817, 825, 827–828
Plan administrator (pension plans), 575
Plan assets (pension plans), 575, 699*n*
Planning, 726–727
  long-term strategic plans, 940
  use of CVP analysis in. *See* **Cost-volume-profit (CVP) analysis**
Plant assets
  impairment of asset value, A-9
  loss on sale of, 637, 640
  note regarding, A-8–A-9
  purchase or disposal of, 949
  reconstruction analysis of, 632, 636, 637, 641–642
  total asset turnover ratio, 601, 608, 694, 696, 703
Plantwide overhead allocation rate, 852
Pledged assets, 563, 695
P&L ratio method, 482
Point-of-sale systems, 859
Poison pill, 518*n*
Portfolio, 595, 596
Practical standard, 986
Prairie Sticks Bat Company, 1023
PrairieSticks.com, 1023
Precision Lube, 992
Predetermined overhead rate, 778
  departmental rates, 853
  for factory overhead, 777, 778
  used with standard costs, 990, 991
Predicted costs, 726
Preemptive right, 508
Preferred stock, 516–520, 530
  callable preferred stock, 519
  convertible preferred stock, 518–519
  dividend preference, 531
    cumulative or noncumulative, 517–518
    participating or nonparticipating, 518, 518*n*
  issuance of, 517
  no-par value stock, 517
  reasons for issuing, 519
  special rights of, 516–517
Premium, 511
Premium on bonds, 553, **556**
  amortizing, 556–557, 575
    effective interest method, 556, 570–571
    straight-line method, 557
  present value concepts, 556, 558, B-11
Premium on stock, 511
Prepaid expenses (deferrals), 637, 638
Present value(s)
  of annuity of 1, 1032
  of bonds and notes, 567–569, 576
    applying present value table, 568
    compounding periods shorter than one year, 569
    present value concepts, 567
    present value of annuity, 568–569, B-11
    present value tables, 567, 568, B-10
  of net cash flows, 1043
  of ordinary annuity, B-2, B-5, B-7, B-10, B-11
  of single amount, B-1–B-3, B-7, B-10
Present value concepts, 575, 576, 1028–1029, B-1, B-7
  bond pricing
    discount bond, 558, B-10, B-11
    premium bond, 556, 558, B-11
  holiday sales promotions, 562, 576
Present value function in calculators, 558
Present value of annuity of 1 table, 568, 569, B-5, B-7, B-11
Present value of 1 factors, 1029
Present value of 1 table, 568, B-3, B-5, B-7, B-10
Present value tables, 1029, B-10, B-11
Price-earnings (PE) ratio, 525, 530, 531, 697–698
Price variance, 984
  computing, 987–988
  materials variances, 988–989, 1002, 1004
Prime costs, 738
Prince Sports Inc., 875
Principal of note, 561
Prior period adjustments, 523
Privately held corporations, 506
Process(es), 808
Process cost accounting, 770, 806–832
  accounting system. *See* **Process cost accounting system**
  demonstration problem, 824–827
  equivalent units of production, 810, 815–816
    differences for materials, labor, and overhead, 815–816
    goods in process, 815
  FIFO method, 827–831
    assigning and reconciling costs, 828, 829–830, 831
    computing cost per equivalent unit, 827, 829
    computing equivalent units of production, 828–829
    determining physical flow of units, 827–828
  hybrid costing system, 823–824
  illustration of. *See* Process costing illustration
  process cost summary, 817, 818, 819, 820–821, 832
    equivalent units of production shown in, 830, 831
    use in preparing financial statements, 820
  process operations, 808–810
    effect of lean business model on, 823
    illustration of, 809–810, 822
    job order operations compared, 809
    organization of, 809
Process cost accounting system, 811–815
  direct and indirect costs, 811
  factory overhead costs, 810, 813–814
  labor costs, 810, 813
  materials costs, 810, 812
Process cost allocation, 814, 830, 831
Process costing illustration, 816–823, 832
  assigning and reconciling costs, 818–821
    cost of units completed and transferred, 819
    cost of units for ending goods in process, 819
    process cost summary, 817, 818, 819, 820–821, 830, 831
  computing cost per equivalent unit, 818
  computing equivalent units of production, 817–818
  determining physical flow of units, 817
  summary of cost flows, 822
  transfers to finished goods and cost of goods sold, 810, 821–822
Process cost summary, 817, 818, 819, **820–**821, 832
  equivalent units of production shown in, 830, 831
  use in preparing financial statements, 820
Process operations, 808–810, 832
  effects of lean business model on, 823
  illustration of, 809–810, 822
  job order operations compared, 809
  organization of, 809
Process time, 743
Procter & Gamble, 808, 859
Product cost(s), 733, 734, 740, 748
Product costing, 852–857
  overhead cost allocation for. *See* Overhead cost allocation
  two-stage cost allocation, 852–854
Production activity, 739–740
Production budgets, 945, 946, 947, **949,** 961
Production capacity, 1041
Production departments, 809
Production report, 817, 818, 819, 820–821
Production scheduling, 771
Production volume, 903
Product mix, 916–917
Profitability, 681, 696
Profitability index, 1031
Profitability ratios, 696–697, 699
  profit margin, 696, 703
  return on common stockholders' equity, 697, 703
  return on total assets, 600–601, 696–697, 703
  return on total assets ratio, 869, 879
Profit and loss ratio method, 482
Profit centers, 858–859, 861
Profit margin, 872
Profit margin ratio, 696, 703
Progress reports, 980
Promoters (incorporators), 507, 512
Property taxes, 903
Prospectus, 509
Proxy, 507
Publicly held corporations, 506
Publicly traded companies, 941
Public sale, 506
Public utilities, 680
Purchases, in foreign currency, 606–607, 608
Purchasing department, 986, 989

Qualitative decision factors, 1042
Quality
  as factor in variances, 989
  Malcolm Baldrige National Quality Award (MBQNA), 730
  100% quality inspection, 824, 832–833
  total quality management, 730
Quantity discounts, 999, 1004
Quantity of input, 984
Quantity of material, 985
Quantity variance, 984
  computing, 987–988
  materials variances, 988–989
Quarterly budgets, 942, 943
Quick ratio. *See* Acid-test ratio

RadioShack
  analysis of financial statements, 680
    graphical analysis, 686, 690
    ratio analysis, 691–699

RadioShack—*Cont.*
consolidated financial statements of, A-24–A-28
balance sheets, A-25
income statements, A-26
statements of cash flows, A-28
statements of stockholders' equity and comprehensive income, A-27
Rap Snacks, 549
RapSnacks.com, 549
Rate variance (labor), 989–990
Rath, Todd, 851
Rath, Tom, 851
Rating services, bonds, 555
Ratio(s), 690–691
Ratio analysis, 682, 706
accounts receivable turnover ratio, 693, 702
book value per common share ratio, 511, 526, 530, 531
book value per preferred share ratio, 526, 530
cash coverage of debt ratio, 646, 699*n*
cash coverage of growth ratio, 646, 699*n*
cash flow on total assets ratio, 646, 657, 699*n*
contribution margin ratio, 908, 909, 914, 919, 920
for corporations, 524–526
credit risk ratio, 699*n*
days' sales in inventory ratio, 694, 702
days' sales uncollected, 693–694, 702
debt-to-equity ratio, 563–564, 575, 695, 703
dividend yield, 525, 530, 698
equity ratio, 695
of financial statements, 690–699
foreign currency translation and, 693
inventory ratios
days' sales in inventory ratio, 694, 702
inventory turnover ratio, 693, 702
inventory turnover ratio, 693, 702
liquidity and efficiency ratios, 691–694, 699
accounts receivable turnover, 693, 702
days' sales in inventory, 694, 702
days' sales uncollected, 693–694, 702
inventory turnover, 693, 702
total asset turnover, 601, 608, 694, 696, 703
market prospects ratios, 699
dividend yield ratio, 525, 530, 698
price-earnings ratio, 525, 530, 531, 697–698
operating cash flow to sales ratio, 646
partner return on equity ratio, 490, 492
price-earnings ratio, 525, 530, 531, 697–698
profitability ratios, 696–697, 699
profit margin, 696, 703
return on common stockholders' equity, 697, 703
return on total assets, 600–601, 608, 696–697, 703
return on total assets ratio, 869, 879
profit margin, 696, 703
return on common stockholders' equity, 697, 703
return on total assets ratio, 600–601, 608, 696–697, 703, 869, 879
solvency ratios, 695–696, 699
debt-to-equity ratio, 563–564, 575, 695, 703
equity ratio, 695
times interest earned ratio, 695–696, 703
summary of ratios, 698, 699
times interest earned ratio, 695–696, 703
total asset turnover ratio, 601, 608, 694, 696, 703
Raw materials, ordered, 771
Raw materials inventory, 735–736
Raw Materials Inventory account, 812
"Reasonableness" criterion, 687, 706
Receiving report, 773, 774
Reconstructed entries, 641–642, 643
Reconstruction analysis, 632, 636, 637
common stock transactions, 644
merchandise purchases, 653–654
notes payable transactions, 643
plant asset transactions, 632, 636, 637, 641–642
retained earnings transactions, 644
Recording transactions
capitalizing retained earnings, 514
cash dividends, 513, 513*n*
discount on bonds payable, 553–554
equity securities
with controlling influence, 593, 599, 607
with significant influence, 593, 598–599, 608
interest payable, 572
in job order costing
factory overhead, 777
use of indirect materials, 775
semiannual interest obligation on bonds, 553
stock retirement, 522
Redemption value of preferred stock, 519
Redhook Ale, 989
Registered bonds, 563
Registrar, 508–509
Regulatory agencies, 695
Relevance, classification of costs by, 733, 734
Relevant benefits, 1035
Relevant costs, 1048
managerial decisions and, 1035, 1036–1037
in sell or process decision, 1039
Relevant range for fixed costs, 903, 904
Relevant range of operations, 911
Rent expense
allocation of indirect expenses, 860–861, 863–864
as factory overhead item, 813
as product cost for manufacturers, 738
Reportable segments, 689
Report cards, 980
Reporting
balanced scorecard reporting, 868, 869
in budgetary process, 980–981
equity reporting. *See* Equity reporting; **Stockholders' equity**
graphical analysis of reportable segments, 689
manufacturing activities, 735–742, 748
demonstration problem, 744–746
flow of activities, 739–740
manufacturer's balance sheet, 735–736
manufacturer's income statement, 737–739
manufacturing statement, 740–742
of operating cash flows
by direct method. *See* **Direct method**
by indirect method. *See* **Indirect method**
statement of cash flows. *See* **Statement of cash flows**
overhead cost variances, 994–995
performance reporting, 738, 739
fixed budget performance report, 981
in manufacturer's income statement, 738, 739
in responsibility accounting, 870–871
for service companies, 996
variances identified in, 987
quarterly or monthly budgets, 942, 943
stock options, 523–524
Reporting currency, 605
Reports
departmental contribution reports, 866, 866*n*, 879
fixed budget performance report, 981
flexible budget performance report, 984, 985, 1001, 1004
in job order cost accounting, 773–781
materials consumption report, 812
overhead variance reports, 994–995
physical flow reconciliation, 817
production reports, 817, 818, 819, 820–821
receiving reports, 773, 774
responsibility accounting performance report, 870–871
summary reports for top-level managers, 871
2006 Report to the Nation (ACFE), 731
Residual income, 867–868
Responsibility accounting, 869–872, 870*n*, 879
controllable v. direct costs, 870
responsibility accounting system, 870–871
Responsibility accounting budgets, 870
Responsibility accounting performance report, 870–871
Responsibility accounting system, 858
Responsibility centers, 871
Restricted retained earnings, 522–523
Retained earnings, 510, 530
appropriated, 523
capitalizing, 514
equity analysis of transactions, 644
negative, 511
restricted, 522–523
statutory or contractual restrictions on, 522–523
Retained earnings deficit, 513
Retirement of bonds, 575–576
by conversion, 560
demonstration problem, 566, 567
at maturity, 559
before maturity, 559–560
Retirement of debt, 637, 640, 656
Retirement of stock, 522
Retrospective application of change in accounting principle, 705
Return on assets (ROA), 600–601, 608, 867, 872
Return on average investment, 1025, 1027–1028
Return on common stockholders' equity ratio, 697, 703
Return on equity, 550, 551
Return on total assets ratio, **600–**601, 608, 696–697, 703, 869, 879
Reuters.com/finance, 691
Revenue recognition, A-10–A-11
Reverse stock split, 516
Revised break-even point, 905, 909, 912, 915
Risk(s)
accounting rate of return and, 1028
bond ratings and, 573, 576
in capital budgeting decisions, 1024
of mutual agency in partnerships, 479
reducing using payback period, 1026–1027
Ritz-Carlton Hotel, 730
Robert Morris Associates, 691
RockBottomGolf.com, 851
Rolling budgets, 940, **942,** 943
Ryan, David, 807

Safety, 859
Safety stock, 947
Salaries
allocation of indirect expenses, 860
budgeting for, 951

of department managers, 813
as fixed cost, 903
in general and administrative budget, 948–949
"Salary allowances" in partnerships, 481–483
Sales
break-even sales dollars, 909
computed for target income, 913–914, 919
computing income from, 913
constant relation with cost of goods sold, 948
in foreign currency, 606–607, 608
free product with purchase, 902
graphing in CVP chart, 910
Sales activity, flow of, 740
Sales budget, 941, 945, **946,** 956
Sales forecasts, 946
Sales incentives, A-11
Sales mix, 915, 1040, 1041
Sales mix selection decision, 1040–1041
Sales price, negotiated, 771
Sales promotions, 562, 576
Sales representatives, 903–904
Sales variances
price variance, 998–999, 1003–1004
volume variance, 998–999, 1004
Sales volume, 911, 983–984
Salvage value of asset, 1025, 1031
Samanta Shoes, 477
SamantaShoes.com, 477
Sarbanes-Oxley Act of 2002 (SOX), 731
SBA.gov, 480
Scatter diagrams, 905, 914, 921
changes in estimates, 912
created with Excel®, 907, 920
measurement of past behavior, 905–906, 907
Schedule of cost of goods manufactured. *See* **Manufacturing statement**
Schedule of manufacturing activities. *See* **Manufacturing statement**
Schmelzer, Rich, 939
Schmelzer, Sheri, 939
S Corporations, 480
Scrap or rework decision, 1038–1039, 1044, 1046
Sea Ray, 599
Sears, Roebuck & Co., 858
Seasonal businesses, 946, 961
SEC (Securities and Exchange Commission), 552*n,* A-1
Secured bonds (and notes), **563**
Secured liabilities, 695
Securities. *See* **Bond(s);** Debt securities
Securities and Exchange Commission (SEC), 552*n,* A-1
Segment elimination decision, 1041–1042
Selling expense(s), 1036
Selling expense budget, 945, **948,** 956
Selling expenses, A-12
Selling price per unit, 911
Selling stock directly/indirectly, 509
Sell or process decision, 1039, 1040
Semiannual compounding periods, 553, 558, 569
Sensitivity analysis, 915
Separate legal entity, 506
Sequential processing, 809, 811
Serial bonds (and notes), **563**
Service companies
cost concepts for, 734–735
current ratio of, 692
executive education, 782
expenses of, 734
job order cost accounting for, 770
process operations in, 808, 809
use of standard costs, 996
Service departments
allocating expenses of, 862–865, 862*n*
allocation of costs to operating departments, 852, 853
evaluating as profit centers, 861
fixed costs of, 903
Services
capital, and stated ratio method
when allowances exceed income, 482–483
when income exceeds allowance, 482, 483
pricing using job order costing, 782, 783, 786, 787
use of standard costs for, 996
SFAS. See Statements of Financial Accounting Standards (SFAS)
Shareholders. *See* Stockholders
Shareholders' equity. *See* **Stockholders' equity**
Share rights of stockholders, 508
Sharing agreement loss (deficit), 482
Short selling, 698, 700
Short-term (cancelable) leases, 573
Short-term financial plans. *See* **Budget(s)**
Short-term investments, 592, 600, 607
demonstration problem, 601–602
note regarding, A-13
in securities, 630
Short-term loans, 950
Short-term notes, 693
Short-term planning, 726–727
Short-term strategic plans, 940
Significant influence, 593
Simple capital structure, 524*n*
Sinking fund, 563
Sinking fund bonds, 563
Six Flags, 564
Ski resorts, 946
Small Business Administration, 480
Small businesses, 478
Small stock dividend, 514, 515
Smilovic, Amy, 591
Solvency, 681
Solvency ratios, 695–696, 699
debt-to-equity ratio, 563–564, 575, 695, 703
equity ratio, 695
times interest earned ratio, 695–696, 703
Source documents
job cost sheets, 772, 773
receiving reports, 773
Sources of cash, 645–646
Sources of information, for preparing statement of cash flows, 634, 636
analysis of equity, 644
financing activities, 630, 643
investing activities, 641–642
notes payable transactions, 643
Southwest Airlines, 734–735
SOX (Sarbanes-Oxley Act) of 2002, 731
Specific rights of stockholders, 508
Speedee Oil Change and Tune-Up, 992
Spending variance, 993
"Split-off point," 877
Spread, in bond trading, 552
Spreadsheet(s)
in computing NPV, 1030
departmental expense allocation sheet, 862–865, 862*n,* 874
departmental spreadsheet analysis, 859
in financial statement analysis, 684
in finding present value of annuity, B-5
in preparing budgeted balance sheet, 952*n*
preparing statement of cash flows, 649–652, 658
analysis of changes, 650–652
indirect method spreadsheet, 636, 649–650, 651
use in CVP analysis, 904
use preparing least squares regression, 907, 920
Sprinturf, 769
Sprinturf.com, 769
Sprock, Martin, 901
Stair-step costs, 904
Standard cost(s), 985–998, 1003
cost variances, 986–990, 986*n*
computation of, 987–988
materials and labor variances, 988–990
variance analysis, 987
extensions of, 996–998
for control, 996
for services, 996
standard cost accounting system, 898, 988, 993, 994, 996–998
materials and labor standards
identifying, 985–986
setting, 986
overhead standards and variances, 990–995
overhead cost variance analysis, 992
setting standards, 990–992
variable and fixed variances, 992–995
Standard cost accounting system, 988, 989, 993, 994, 996–998, 1004
Standard cost card, 986
Standardization, 823, 824
Standard overhead cost rate, 991
StandardPoor.com, 681
Standard & Poor's, 555, 599, 680, 681, 682
Standard & Poor's Industry Surveys, 691
Standard setting, 986
Standards for comparison (benchmarks), 682
Starbucks, 523, 592, 640
Start-up businesses, 632, 658
Start-up money, 508
Stated rate on bonds, 553
Stated ratio method, 482
Stated value stock, 510, 512
State laws, 514, 552
Statement of cash flows, 626–658
basics of, 657
classification of cash flows, 629–630
format of, 631, 657
importance of, 628–629
measuring cash flows, 629
noncash investing and financing, 631
purpose of, 628
cash flows from financing activities, 643–645
analysis of changes in, 650
equity analysis, 644
noncurrent liabilities, 643
proving cash balances, 632, 637, 645
sources and uses of cash, 645–646
three-stage analysis of, 643
cash flows from investing activities
analysis of changes in, 650
noncurrent asset analysis, 641–642
other asset transactions, 630, 642
sources and uses of cash, 645–646
three-stage analysis of, 641
cash flows from operating activities. *See* Operating cash flows
consolidated, A-7, A-23, A-28, A-33

Statement of cash flows—*Cont.*
defined, **628**
demonstration problem, 647–649
direct method of reporting, 634, 657
demonstration of, 648, 649
financing activities, 643–645
format of operating activities section, 657
indirect method compared, 633, 634–635, 637
investing activities, 641–642
operating cash payments, 636, 653–656
operating cash receipts, 636, 653
summary of adjustments, 656
indirect method of reporting, 634, 635–640, 657
adjustments to net income, 635
demonstration of, 648–649
direct method compared, 633, 634–635, 637
summary of adjustments, 633, 637, 640
preparation of, 632–634, 657
analyzing cash account, 632–633
analyzing noncash accounts, 633–634
sources of information for, 634
using spreadsheet, 636, 649–652, 658
spreadsheet preparation of, 649–652, 658
analysis of changes, 650–652
indirect method spreadsheet, 636, 649–650, 651
Statement of Ethical Professional Practice (of IMA), 731
Statement of partners' capital, 483–484, 491–492
Statement of partners' equity, 483–484, 491–492
Statement of retained earnings
budgeted, 958
closing process, 523
prior period adjustments to, 523
restrictions and appropriations, 522–523
Statement of stockholders' equity, 523, 524, A-6, A-22, A-27, A-32
Statements of Financial Accounting Standards (SFAS) of FASB
No. 157: "Fair Value Measurements," 597
No. 159: "The Fair Value Option for Financial Assets and Financial Liabilities—Including an amendment of FASB Statement No. 115," 597
Static (fixed) budget, 981
Statutory (legal) restrictions on retained earnings, 522
Step-wise costs, 904
"Step-wise" expense allocation, 862
Stock
authorized stock, 509
capital stock, 509–510
common stock. *See* **Common stock**
discount on stock, 511
dividends. *See* **Stock dividends**
growth stocks, 525
income stocks, 525
market value of, 512
no-par value stock, 509, 511, 517
outstanding stock, 509, 511, 526
overpriced stocks, 698
par value stock, 509, 510, 511
preferred stock. *See* **Preferred stock**
retirement of, recording, 522
stock options, 523–524
stock splits, 516, 530–531, 644
treasury stock. *See* **Treasury stock**
unissued, 520
Stock buybacks. *See* **Treasury stock**
Stock certificate, 508
Stock dividends, 514–516, 530
cash flow not affected by, 644
large stock dividend, 514, 515–516
not considered noncash investing or financing, 631
reasons for, 514
small stock dividend, 514, 515
value of investment and, 516, 531
Stockholders, 506, 508–509
advantages of corporation to, 506
distribution of assets to. *See* Dividend(s)
global issues, 562
lack of mutual agency for, 506
limited liability of, 506
registrar and transfer agents, 508–509
rights of, 508
stock certificates and transfer, 508
use of financial statement analysis, 680
Stockholders' equity, 510, 530
balance sheet presentation of, 528
demonstration problem, 527–529
journal entries relating to, 527–528, 529
return on common stockholders' equity, 697, 703
Stock market, 605, 696
Stock options, 523–524
Stock splits, 516, **516,** 530–531, 644
Straight-line bond amortization, 554–555, 557, 566
Straight-line depreciation
average book value of asset and, 1028
effects on NPV analysis, 1031
Strategic budgeting, 940
Sub-Chapter C corporations, 480
Sub-Chapter S corporations, 480
Subordinated debentures, 563
Subsequent events, 728
Subsidiaries, 599, 607
Subsidiary ledgers, job cost sheets as, 772, 773
Subunit. *See* Department
Summaries of reports, 871
Summary entries, 822, 826–827
Sunk costs, 733, 1035, 1038, 1039
Suppliers, 680, 773
Sustainable income, 703–705
changes in accounting principles, 704–705
comprehensive income, 704–705
continuing operations, 703
discontinued segments, 703, 704
earnings per share, 704
extraordinary items, 703–704

T-accounts, 637, 653
Target cost, 771
Target income, 913–914, 917, 919, 921
Target numbers, 633
Taxation
depreciation tax shield, 1031
double taxation, 507
income taxes. *See* Income taxes
property taxes, 903
tax rate, 913
Taylor, Brenton, 505
Taylor, Brian, 725
Technology
automation, 808, 814, 915, 918
information produced due to, 871
Telecom companies, 1026
Temporary accounts
investments, 598
in job order costing, 776
trading securities, 595
variance accounts, 997–998
Temporary investments. *See* **Short-term investments**
Term bonds (and notes), **563**
Theme parks, 1032
3M, 605
Three-stage analysis, 641–642, 643
Tibi, 591
Tibi.com, 591
Ticker prices, 698
Ticker tape, 698
Time and motion studies, 985
Time dimension of managerial accounting, 728
Timeliness of information, 728
Time period, 942–943
Times interest earned ratio, 695–696, 699, 703
Time tickets, 775, **776,** 777, 778
Time value of money, B–B-7, B-10, B-11
capital budgeting with, 1024, 1028–1034
internal rate of return, 1030, 1031–1033, 1047, 1048, B-11
net present value. *See* **Net present value**
capital budgeting without, 1025–1028
accounting rate of return, 1025, 1027–1028, 1033–1034, 1048
payback period, 1025–1027
in evaluating alternative investments, 1043, 1048
future value of ordinary annuity, B-6–B-7, B-10, B-11
future value of single amount, B-3–B-4, B-7, B-10
present and future value concepts, B-1, B-7
present value of ordinary annuity, B-2, B-5, B-7, B-10, B-11
present value of single amount, B-1–B-3, B-7, B-10
Timing of budgets, 942–943
Tools of financial statement analysis, 682
Total asset turnover ratio, 601, 608, 694, 696, 703
Total costs, graphing in CVP chart, 910
Total overhead variance, 992
Total quality management (TQM), 730
Total variable costs, 911
Toyota Motor Corporation, 730, 915
TQM (total quality management), 730
Traceability
classification of costs by, 732, 733, 734, 748
of factory overhead costs, 738
Trade names, A-9
Trading bonds, 551
Trading on the equity, 519, 550
Trading securities, 595, 608
cash receipts and payments from, 629–630
classification and reporting, 593
selling, 595–596
valuing and reporting, 595
Training and education, 782
Transaction analysis, 634
financing activities, 630, 643
investing activities, 630, 642
notes payable transactions, 632, 636, 637, 643
plant asset transactions, 632, 636, 637, 641–642
Transfer agent, 509
Transfer prices, 875–876, 877
Transfer pricing system, 861, 875–877, 879
additional issues in, 877
alternative transfer prices, 875–876

Treasury stock, 520–522, 531
purchase of, 520–521
reissuing, 521–522
retiring, 522
Trek, 729
Trend analysis, 685–687
Trend percents, 685, 686
Trump Entertainment Resorts LP, 484
Trustee, 552*n*
Truth-in-Lending Act, 561
TSC (Toronto Sticks Company), 959
Two-stage cost allocation, 879
activity-based costing compared, 855, 856–857
illustration of, 853–854
indirect expenses, 853, 861
in product costing, 852–854

Unavoidable expenses, 1041
Uncontrollable costs, 733, 870
Underapplied overhead, 779, **781,** 786, 814
Underfunded pension plan, 575
Underwriters, 509, 552, 552*n*
Undistributed earnings, 598
Uneven cash flows, 1025
computing IRR with, 1032
computing NPV with, 1030–1031
computing payback period with, 1026
Unfavorable variances, 981
in Goods in Process Inventory account, 996–997
long-term, 986*n*
preparation of flexible budgets and, 984, 1004
"Unfriendly" investors, 518*n*
Uniform Partnership Act, 479
Unissued stock, 520
Unit contribution margin, 908, 914
Unit cost, 815
U.S. Marine Co., 599
Unit of output, 984
Unit sales, 914
Units of product, 959
Unlimited liability, 479
Unrealized gain (loss), 595, 596, 602
Unregistered bonds, 563
Unsecured bonds (and notes), **563**
"Unusual and infrequent" criteria, 559
Unusual gains and losses, 703
Useful life of asset, 523
Users of accounting information
external, 628, 680, 728
financial statement analysis, 680, 700
internal, 628, 680, 727, 728
managerial accounting, 727, 728
statement of cash flows, 628
Utilities expense, 813, 859, 861, 863, 864

Value, determinants of, 730–731
Value-added time, 743
Value basis of allocating joint costs, 877, 878
Value chain, 740
Value engineering, 771
Variable budgets. *See* **Flexible budget(s)**
Variable cost(s), 732
in computing multiproduct break-even point, 916
cost behavior, 903
cost per unit, 903, 906, 911
fixed costs distinguished, 982–983
in sales mix selection decision, 1040
total variable costs, 911
Variable costing income statement, 909, 910
Variable overhead, 741
Variable overhead cost variance, 992–995
computing
controllable overhead and volume variances, 993
fixed overhead cost variances, 991, 994
overhead variance reports, 994–995
variable overhead cost variances, 993–994
controllable variances, 993, 995, 997
volume variances, 993, 994, 997
Variance(s), 942, 943, 1003
controllable variances, 993, 995, 997
cost variances. *See* **Cost variances (CV)**
efficiency variance, 989–990, 993
favorable, 981, 984, 986*n*
immaterial, 998
labor variances
direct labor rate variance, 997
labor cost variance, 989–990, 1002, 1004
materiality principle and, 998
materials cost variances, 988–989, 1002, 1004
overhead cost variances. *See* **Overhead cost variances**
price variance, 984, 987–989, 1002, 1004
quality as factor in, 989
quantity variance, 984, 987–989
sales variances
price variance, 998–999, 1003–1004
volume variance, 998–999, 1004
spending variance, 993
unfavorable, 981, 984, 986*n,* 996–997, 1004
Variance accounts, 997–998
Variance analysis, 984, 986–990, 986*n*
Venture capitalists, 508
Vertical analysis, 682, 687–690, 706
common-size graphics, 689–690
common-size statements, 684, 685, 687–689
balance sheets, 687, 688, 702
income statements, 688–689
Vice-president (VP), 507–508, 871
Volume variances, 993, 994, 997
Voluntary association, 478
Voting rights of stockholders, 508
Vouchers, 777
VP (vice president), 507–508, 871

W. T. Grant Co., 629
Wages
allocation of indirect expenses, 860
cash paid for, 636, 654–655, 655*n*
spending variances, 993
Wait time, 743
Wall Street, 697
The Wall Street Journal, 509, 698
Wal-Mart Stores, Inc., 627, 859, 868
Waste, elimination of, 730
Weighted-average contribution margin, 916–917
Weighted average method, 816–823, **818**
"What-if" situations
CVP analysis and, 913
flexible budgets for, 982
spreadsheets used in analyzing, 904
Whittaker, Andrew "Jack," B-5
Withdrawal of partner, 485, 493
bonus to remaining partners, 486–487
bonus to withdrawing partners, 487
no bonus, 486
Work centers, 809
Working capital, 691
Working paper. *See* Spreadsheet(s)
Work in process inventory, 736
Work sheet. *See* Spreadsheet(s)
Workstations, 809
Worst case scenario, 982
Write-downs of inventory, 703–704
Write-offs of receivables, 703–704
Written budgets, 941
www.SEC.gov, A-1

XE.com, 605
Xerox, 808

Zinger, Dan, 1023